THE LIGHT AND SMITH MANUAL

The Light and Smith Manual

INTERTIDAL INVERTEBRATES
FROM CENTRAL CALIFORNIA TO OREGON

FOURTH EDITION, COMPLETELY REVISED AND EXPANDED

Edited by

JAMES T. CARLTON

UNIVERSITY OF CALIFORNIA PRESS
Berkeley   Los Angeles   London

University of California Press, one of the most distinguished university presses in the United States, enriches lives around the world by advancing scholarship in the humanities, social sciences, and natural sciences. Its activities are supported by the UC Press Foundation and by philanthropic contributions from individuals and institutions. For more information, visit www.ucpress.edu.

University of California Press
Berkeley and Los Angeles, California

University of California Press, Ltd.
London, England

Library of Congress Cataloging-in-Publication Data

Light, Sol Felty, 1886–1947.
  The Light & Smith manual : intertidal invertebrates from central California to Oregon / edited by James T. Carlton. — 4th ed.
    p. cm.
  Includes bibliographical references.
  ISBN-13: 978-0-520-23939-5 (case : alk. paper)
  ISBN-10: 0-520-23939-3 (case : alk. paper)
  1. Marine invertebrates—California.   2. Marine invertebrates—Pacific Coast (North America) I. Carlton, James T.   II. Light, Sol Felty, 1886–1947. Light's manual.   III. Title.

QL164.L53 2007
592'.177—dc22   2007005842

Manufactured in Canada
10 09 08 07
10 9 8 7 6 5 4 3 2 1

The paper used in this publication meets the minimum requirements of ANSI/NISO Z39.48-1992 (R 1997) (*Permanence of Paper*). ∞

Cover photograph: The rocky shore at Bastendorff Beach, Coos Bay, Oregon, July 2005. Photograph by Deniz Haydar.

Title page artwork: The "Big Three" (*Pisaster ochraceus, Mytilus californianus, Pollicipes polymerus*) of the outer wave-swept coast. Drawing by Joel W. Hedgpeth.

CONTENTS

LIST OF CONTRIBUTORS

DONALD P. ABBOTT (DECEASED) Hopkins Marine Station, Stanford University, Pacific Grove, California

ILSE BARTSCH Forschungsinstitut Senckenberg, Hamburg, Germany

DENTON BELK (DECEASED) Our Lady of the Lake University, San Antonio, Texas

PETER BELLINGER (DECEASED) California State University, Northridge

JAMES A. BLAKE ENSR Marine and Coastal Center, Woods Hole, Massachusetts

EDWARD L. BOUSFIELD Victoria, British Columbia, Canada

DARL E. BOWERS Oakland, California

RALPH O. BRINKHURST Lebanon, Tennessee

RICHARD C. BRUSCA Arizona-Sonora Desert Museum, Tucson, Arizona

EUGENE M. BURRESON Department of Environmental and Aquatic Animal Health, Virginia Institute of Marine Science, College of William and Mary, Gloucester Point

DALE R. CALDER Department of Natural History, Royal Ontario Museum, Toronto, Ontario, Canada

ROY L. CALDWELL Department of Integrative Biology, University of California, Berkeley

CHRISTOPHER B. CAMERON Sciences Biologiques, Université de Montréal, Montréal, Québec, Canada

ERNESTO CAMPOS Laboratorio de Sistemática de Invertebrados, Facultad de Ciencias, Universidad Autonoma de Baja California, Baja California, Mexico

JAMES T. CARLTON Maritime Studies Program, Williams College–Mystic Seaport, Mystic, Connecticut

ALBERT CARRANZA Bodega Marine Laboratory, University of California, Bodega Bay

HENRY W. CHANEY Santa Barbara Museum of Natural History, California

JOHN W. CHAPMAN Department of Fisheries and Wildlife, Mark O. Hatfield Marine Science Center, Oregon State University, Newport

C. ALLAN CHILD Bethesda, Maryland

KENNETH CHRISTIANSEN Department of Biology, Grinnell College, Iowa

ROGER N. CLARK Eagle Mountain, Utah

EUGENE V. COAN Palo Alto, California

VÂNIA R. COELHO Department of Natural Sciences and Mathematics, Dominican University of California, San Rafael

ANDREW N. COHEN San Francisco Estuary Institute, Oakland, California

ANNE C. COHEN Bodega Bay, California

DAVID G. COOK Greely, Ontario, Canada

JEFFERY R. CORDELL School of Aquatic and Fishery Sciences, University of Washington, Seattle

HOWELL V. DALY Kensington, California

M. CRISTINA DÍAZ Museo Marino de Margarita, Boca del Rio, Peninsula de Macanao, Nueva Esparta, Venezuela

ANTHONY DRAEGER Kensington, California

CASEY W. DUNN Kewalo Marine Laboratory, University of Hawaii, Honolulu

DOUGLAS J. EERNISSE Department of Biological Science, California State University, Fullerton

CHRISTER ERSÉUS Department of Zoology, Göteborg University, Sweden

RICHARD FARRIS Biology Department, Linfield College, McMinnville, Oregon

DAPHNE G. FAUTIN Department of Ecology and Evolutionary Biology and Natural History Museum, University of Kansas, Lawrence

STEVEN C. FRADKIN Lake Crescent Laboratory, Olympic National Park, Port Angeles, Washington

JONATHAN B. GELLER Moss Landing Marine Laboratories, California

JEFFREY H. R. GODDARD Marine Science Institute, University of California, Santa Barbara

TERRENCE M. GOSLINER Department of Invertebrate Zoology and Geology, California Academy of Sciences, San Francisco

STEVEN H. D. HADDOCK Monterey Bay Aquarium Research Institute, Moss Landing, California

CADET HAND (DECEASED) Bodega Marine Laboratory, University of California, Bodega Bay

TODD A. HANEY Natural History Museum of Los Angeles County, California

WILLARD D. HARTMAN Peabody Museum of Natural History, Yale University, New Haven, Connecticut

DENIZ HAYDAR Department of Marine Biology, CEES, University of Groningen, Haren, Netherlands

JOEL W. HEDGPETH (DECEASED) Santa Rosa, California

GORDON HENDLER Natural History Museum of Los Angeles County, California

ROBERT HESSLER Scripps Institution of Oceanography, University of California, San Diego, La Jolla

ROBERT P. HIGGINS Asheville, North Carolina

F. G. HOCHBERG Santa Barbara Museum of Natural History, California

RICHARD HOFFMAN Virginia Museum of Natural History, Martinsville

JOHN J. HOLLEMAN San Andreas, California

MATTHEW D. HOOGE Department of Biological Sciences, University of Maine, Orono

W. DUANE HOPE Department of Invertebrate Zoology, National Museum of Natural History, Smithsonian Institution, Washington, D.C.

WILLIAM D. HUMMON Department of Biological Sciences, Athens, Ohio

EUGENE N. KOZLOFF Friday Harbor Laboratories, University of Washington, Washington

ARMAND M. KURIS Department of Ecology, Evolution and Marine Biology, University of California, Santa Barbara

CAROL M. LALLI Institute of Ocean Sciences, Department of Fisheries and Oceans, Sidney, British Columbia, Canada

CHARLES C. LAMBERT Seattle, Washington, and Friday Harbor Laboratories, University of Washington, Washington

GRETCHEN LAMBERT Seattle, Washington and Friday Harbor Laboratories, University of Washington, Washington

PHILIP LAMBERT Royal British Columbia Museum, Victoria, British Columbia, Canada

STEPHEN C. LANDERS Department of Biological and Environmental Sciences, Troy University, Alabama

RONALD J. LARSON U.S. Fish and Wildlife Service, Klamath Falls, Oregon

BERTHA E. LAVANIEGOS Departamento de Oceanografía Biológica, Centro de Investigación Científica y Educación Superior de Ensenada, Baja California, Mexico

VINCENT F. LEE Department of Entomology, California Academy of Sciences, San Francisco

WELTON L. LEE Oakland, California

DAVID R. LINDBERG Department of Integrative Biology, University of California, Berkeley

ROSALIE F. MADDOCKS Department of Geosciences, University of Houston, Texas

LAURENCE P. MADIN Department of Biology, Woods Hole Oceanographic Institution, Massachusetts

CHRISTOPHER MAH Department of Invertebrate Zoology, National Museum of Natural History, Smithsonian Institution, Washington, D.C.

RICHARD N. MARISCAL Department of Biological Science, Florida State University, Tallahassee

ANTONIO C. MARQUES Departmento de Zoologia, Instituto de Biociencias, Universidade de Sao Paulo, Brasil

JOEL W. MARTIN Natural History Museum of Los Angeles County, California

SVETLANA MASLAKOVA Friday Harbor Laboratories, University of Washington, Washington

GARY R. MCDONALD Joseph M. Long Marine Laboratory, Institute of Marine Sciences, University of California, Santa Cruz

MARY MCGANN U.S. Geological Survey, Coastal and Marine Geology Team, Menlo Park, California

JAMES H. MCLEAN Natural History Museum of Los Angeles County, California

CHARLES G. MESSING Nova Southeastern University, Oceanographic Center, Dania Beach, Florida

ALVARO E. MIGOTTO Centro de Biologia Marinha, Universidade de São Paulo, São Sebastião, SP, Brasil

CLAUDIA E. MILLS Friday Harbor Laboratories, University of Washington, Washington

RICHARD F. MODLIN Department of Biological Sciences, University of Alabama in Huntsville, Alabama

RICH MOOI Department of Invertebrate Zoology and Geology, California Academy of Sciences, San Francisco

PENNY A. MORRIS Department of Natural Science, University of Houston–Downtown, Texas

AMANDA NELSON Hancock Biological Station, Murray State University, Kentucky

A. TODD NEWBERRY Institute of Marine Sciences, University of California, Santa Cruz

IRWIN M. NEWELL (DECEASED) University of California, Riverside

WILLIAM A. NEWMAN Scripps Institution of Oceanography, University of California, San Diego, La Jolla

THOMAS M. NIESEN Department of Biology, San Francisco State University, California

JON L. NORENBURG Department of Invertebrate Zoology, National Museum of Natural History, Smithsonian Institution, Washington, D.C.

JAMES W. NYBAKKEN Moss Landing Marine Laboratories, California

JOHN S. PEARSE University of California, Santa Cruz

VICKI BUCHSBAUM PEARSE University of California, Santa Cruz

DAWN E. PETERSON Museum of Paleontology, University of California, Berkeley

LELAND W. POLLOCK Department of Biology, Drew University, Madison, New Jersey

PHILIP R. PUGH National Oceanography Centre, Southampton, United Kingdom

LANGDON QUETIN Marine Science Institute, University of California, Santa Barbara

JOHN T. REES Department of Biological Sciences, California State University, East Bay, Hayward, California

DAVID REID Department of Zoology, Natural History Museum, London, United Kingdom

MARY E. RICE Smithsonian Marine Station at Fort Pierce, Smithsonian Institution, Florida

PAMELA ROE Department of Biological Sciences, California State University, Stanislaus Turlock

ROBIN ROSS Marine Science Institute, University of California, Santa Barbara

R. EUGENE RUFF Ruff Systematics, Puyallup, Washington

PATRICIA S. SADEGHIAN Santa Barbara Museum of Natural History, California

AMELIE H. SCHELTEMA Department of Biology, Woods Hole Oceanographic Institution, Massachusetts

EVERT I. SCHLINGER The World Spider-Endoparasitoid Laboratory, Santa Ynez, California

ROGER R. SEAPY Department of Biological Science, California State University, Fullerton

RONALD L. SHIMEK Wilsall, Montana

DANNA JOY SHULMAN Stanford University, Hopkins Marine Station, Pacific Grove, California

DOROTHY F. SOULE (DECEASED) University of Southern California, Los Angeles

JOHN D. SOULE (DECEASED) University of Southern California, Los Angeles

HELMUT STURM University of Hildesheim, Germany

STEFANO TAITI Istituto per lo Studio degli Ecosistemi, CNR, Sesto Fiorentino (Firenze), Italy

ERIK V. THUESEN Laboratory One, The Evergreen State College, Olympia, Washington

PAUL VALENTICH-SCOTT Santa Barbara Museum of Natural History, California

ROBERT VAN SYOC Department of Invertebrate Zoology and Geology, California Academy of Sciences, San Francisco

ERIC W. VETTER Hawaii Pacific University, Kaneohe

KERSTIN WASSON Elkhorn Slough National Estuarine Research Reserve, Watsonville, California

LES WATLING Department of Zoology, University of Hawaii at Manoa, Honolulu

DAVID WHITE Hancock Biological Station, Murray State University, Kentucky

GARY C. WILLIAMS Department of Invertebrate Zoology and Geology, California Academy of Sciences, San Francisco

KEITH H. WOODWICK Fresno, California

RUSSEL ZIMMER Department of Biological Sciences, University of Southern California, Los Angeles

PREFACE

This manual represents a progress report on the state of our knowledge of the intertidal and selected planktonic and shallow-water invertebrates of a portion of the Pacific coast of North America. Since the third edition, 30 years have passed, and one of many results is that less than 10 percent of the previous book has been carried forward to this new edition. The book continues to grow, roughly doubling in size with every edition; the much-increased length of the present book over the last edition is in part disguised by the larger format you hold in your hands. More phyla are treated now, and groups once passed over in a paragraph, such as nematodes, now properly constitute an entire chapter. Formally founded in 1941 as a class syllabus for undergraduates in invertebrate zoology, this manual has become a guidebook for many users including graduate students, professional zoologists, and ecologists. As noted in the previous edition, we cannot go backward, we cannot make it simpler or easier to use, but we hope we have made it better.

The team of authors has likewise grown with each edition: From 22 (second edition, 1954) to 43 (third edition, 1975) to 120 (in this fourth edition). Five authors from the second edition are still represented, four posthumously (Donald Abbott, Cadet Hand, Joel Hedgpeth, and Irwin Newell), and 25 authors from the third edition rejoin this new effort; of these, in addition to Abbott, Hand, Hedgpeth, and Newell, John and Dorothy Soule, the bryozoologists, have also passed away. We are saddened to note that none of the previous editors are still with us; S. F. Light died in 1947, Don Abbott in 1986, Ralph Smith in 1993, Frances Weesner in 2002, and Frank Pitelka in 2003. We present a memorial to Light and Smith following this preface. Joel W. Hedgpeth, who died at 94 in July 2006, contributed to Light's unpublished class syllabus in 1937, and authored or coauthored the sea-spider chapter in 1941, 1954, 1975, and 2007. We also note the passing of contributors Denton Belk and Peter Bellinger while this edition was in preparation.

The previous editions focused on the central California region encompassed by the radius of teachings of Dr. Light based in Berkeley and on the regions most useful for those working at and between Hopkins Marine Station on the Monterey Peninsula and Bodega Marine Laboratory on Bodega Head. For the present work, we have sought to expand the coverage south to Point Conception and north to the Oregon-Washington border. North of Oregon, the monographic works of Eugene Kozloff and colleagues document the Pacific Northwest fauna. South of Point Conception, a warmer-water biota adds many hundreds, if not thousands, of species that could not be treated here. Thus, the utility of this manual will lessen with distance from these boundaries. Some chapters, however, treat species north and south of the book's limits, especially if the addition of several more species essentially completed the treatment of the genus, family, or other group in question. In all, over 3,700 species are keyed or discussed in this fourth edition. Distributions are shown for species whose ranges are limited or unusual and for groups that are poorly known. We have generally (but without complete consistency) not shown the distribution for the many species ranging from Alaska or British Columbia to Mexico.

The treatment of planktonic organisms has expanded since the previous edition for the benefit of those working with a dipnet around piers or floats or in a small boat in places such as Monterey Bay, San Francisco Bay, or Coos Bay. However, the treatment has not been exhaustive across the phyla, nor intended to cover all nearshore or deeper waters, nor enough to change the title of the book. Also included are more species found just below the tides—additions Ralph Smith referred to as the contributions of long-armed and water-resistant intertidal collectors. Gone are chapters on intertidal seaweeds and fishes, represented in previous editions, but now replaced by large and readily available monographs and books.

We have permitted the authors a good deal of nomenclatural and systematic flexibility, although no new species are described. We are given to understand that most of the taxonomic changes shown here will, in due course, be published in the journal literature, and our judgment was to proceed with expert opinion such that this work would not go out of date too quickly.

As is true for all works of this kind, and by nature of the animals themselves and those who study them, the user of this book will find the treatment of groups uneven and, very likely, will wish for more. The keys will be found to vary in their completeness of coverage, geographical range of usefulness, ease of use, and accuracy, an inevitable result of the differences in the numbers of species in various groups, the ease of separating species from each other, the completeness of knowledge of the group, the professional background of each author, and his or her success in constructing the key. Only use by students and investigators will tell us how suitable a given key may be: The editor and contributors welcome comments and information leading to improvements, corrections, and revisions.

Species lists follow most keys. These lists are of species reliably reported from central California to Oregon, although we have no doubt that some species have been inadvertently omitted. Taxa not included in the keys are noted in the lists by asterisks. Some lists and species have been annotated with general information about habitat, ecology, or references that may be interesting or helpful; no collection of references can now be complete, and most users will turn to electronic resources for additional information and journal literature. Users of this book will doubtless add to these notations, and such marginalia will be of great value in future revisions.

Finally, because of the much more extensive treatment of most invertebrate groups in this edition, because the systematics and classification of the invertebrates of coastal California and Oregon are better documented for some groups than others, and because of the very wide variety of taxonomic hierarchies employed in some chapters and not others, we have not formatted the same taxonomic levels (such as classes, orders, or families) in the same manner across all phyla. In the previous edition, family names were presented in all capital letters in all chapters. Family names—and most other hierarchical levels—across phyla are not biologically co-equal. In this edition, taxonomic levels are formatted within the context of the surrounding hierarchies.

The second and third editions of this book carried on a tradition, not previously mentioned in these pages and born at Hopkins Marine Station in Pacific Grove, that we would be remiss not to mention, and which readers who have a hand lens and a copy of either edition can discover for themselves. In the late 1940s, an intertidal rock near Hopkins, with a peculiarly prominent proboscis-like portion, became popularly known among students as "Snadrock." The rock was the subject of a study on vertical zonation of barnacles by Don Abbott, John Davis, and Cadet Hand. "Snadrock," whose name (as "Professor Snadrock") would appear on the blackboard of the invertebrate zoology teaching lab in the Agassiz Building at the station, became a mythical figure.

If one takes a hand lens to the drawing of the tusk shell (which has not been reproduced in this fourth edition), on page 213 of the second edition or on page 498 of the third edition, and looks carefully on the line representing the sediment surface, the word "Snadrock" can be found written as part of the mud ripple. The drawing was made by Ralph Smith. A small sketch (drawn by Abbott) itself forms the chapter head block on page xi, and the word "Snadrock" appears on a book spine in the chapter block on page 685 of the third edition. While Ralph permitted us to index the word "mermaid" in the third edition (although no one has inquired about it in the past 30 years)—amazing us by actually reading the draft index for the last edition—he removed the entry for "Snadrock."

While this book was in press, we learned of several dozen additional changes that would alter treatments here, only a few of which we were able to capture. This is a cause for celebration: While larval recruitment to systematics and taxonomy has steadily declined (only 20 percent of our authors are under the age of 50), such changes indicate that work continues. As Ralph Smith remarked near the completion of the previous edition, "the job will never really be ended." Much remains to be learned about the thousands of species of invertebrates that live along the shores of the northeastern Pacific Ocean.

James T. Carlton
Stonington, Connecticut
January 2007

ACKNOWLEDGMENTS

For many years, almost 120 zoologists donated their time to this book, which generates no royalties, and I am indebted to them all. Many bore considerable personal costs, freely given to this labor of cooperation and devotion with no material reward, supported by many hundreds of assistants, spouses, and significant others, who have sacrificed, along with the authors, and with endless patience and understanding, uncountable days, nights, and weekends.

I am grateful to Bonnie Bain, Patrick Baker, Kitty Brown, Barbara Butler, John Chapman, Victor Chow, Andrew Cohen, Peter Connors, Howell Daly, Stanley Dodson, Richard Everett, Jonathan Geller, Jeff Goddard, the late Cadet Hand, Chad Hewitt, Janet Hodder, Armand Kuris, Jody Martin, Claudia Mills, Chris Patton, John Pearse, Vicki Pearse, Gregory Ruiz, Isabel Stirling, Les Watling, and Joseph Wible, all of whom contributed importantly along the way with information or assistance. Deniz Haydar logged many hours assisting with artwork for a number of chapters. My good staff of the Williams College–Mystic Seaport Maritime Studies Program in Mystic, Connecticut, aided in all aspects of production over the years, and legions of Williams-Mystic students, exploring with me the tidepools of the California and Oregon coasts, helped keep me from switching Atlantic and Pacific names too often.

Debby Carlton and Bridget Holohan spent hundreds of hours reading the page proofs of this Fourth Edition, and the result is, thus, immeasurably improved. The staff of the University of California Press, including Charles Crumly, Danette Davis, and Scott Norton, provided constant support and encouragement. It was also a pleasure to work with Joanne Bowser of Aptara during the production phase.

You would not be holding this book were it not for Bridget Holohan. Bridget has been my left and right hands throughout the book's creation, from working at the beginning with individual authors to, toward the end, producing the final artwork inventories for the entire book. Bridget's memory of events over the past decade has been critical in solving hundreds of challenges—virtually bridging the unbridgeable. After a long workday at her "regular" job, Bridget would start work all over again at 5 or 6 p.m. and would urge, cajole, and encourage me and everyone else to push on regardless. And so it is that you now hold the *Light and Smith Manual*—because Bridget would call and ask me a thousand times over a thousand days, "How's it going?"

My wife Debby has been my anchor since 1975; it impossible to imagine how far I would have drifted without her. I cannot imagine anyone else who would have had the lifetime of tolerance and patience that Debby has shown and allowed me, oblivious, to keep my head in the tidepools and in vast piles of paper while the real world flowed by.

James T. Carlton
Stonington, Connecticut
January 2007

S. F. LIGHT AND R. I. SMITH

The namesakes of this manual are two zoology professors both of whom enjoyed productive and inspiring careers at the University of California at Berkeley: Sol Felty Light (1886–1947) and Ralph Ingram Smith (1916–1993).

S. F. Light was born in Elm Mills, Kansas, was an undergraduate at Park College, Missouri, and took his Ph.D. at Berkeley in 1926 under Charles A. Kofoid. Light stayed at Berkeley until his death in 1947 when he drowned in Clear Lake while on a fishing holiday. Light's research career focused on termite systematics and biology (particularly symbiotic flagellates) and freshwater copepods. The 1937 anniversary "Golden Book of California" (University of California Alumni Association) featured a photo of Light standing before large racks of what the late Ted Bullock described as "staggering numbers of laboratory colonies" of termites.

Most students who worked with Light referred to him as "Dr. Light"; his wife Mary referred to him in correspondence

Sol Felty Light

after his death as "S. F." Light taught "Zoology 112," Berkeley's core course in invertebrate zoology. In the 1930s, he began taking students to the seashore for 5-week summer courses, based in various locales, including the dining room of a tavern-restaurant in Moss Beach (south of San Francisco) and a former dance hall in the, then, beach-resort of Dillon Beach (north of the city). Many of the undergraduate student papers produced in Light's classes, beginning in the 1920s, on marine and freshwater organisms of the San Francisco Bay area still remain and are housed in the Cadet Hand Library at the Bodega Marine Laboratory (having been moved there from the Berkeley campus).

By the late 1930s, Light had developed an extensive "Laboratory Syllabus" for his course, with species lists, keys, and laboratory and field exercises. A 1937 copy of this syllabus is 122 pages long; recommended for the reading list for Moss Beach that summer is "Between Pacific Tides" by E. F. Ricketts and J. Calvin, although it was not to be published until 1939. Among the contributors to the 1937 syllabus was the 25-year-old Joel W. Hedgpeth, who wrote a one-page key on pycnogonids (and is late co-author of the same chapter in the present edition, 70 years later).

Light's laboratory syllabi formed the foundation for the "Laboratory and Field Text in Invertebrate Zoology" published in 1941 by the "Associated Students Store" of the University. This constituted the first edition of what was eventually to be known as *Light's Manual*; a rich resource reflecting the central California biota of the 1920s and 1930s, this green paperback is surely one of the rarest marine biology books of the 20th century.

"In those days," wrote Joel Hedgpeth in 1985, "more was being done for the cause of what we now so glibly call marine biology by a somewhat old-fashioned professor at Berkeley, S. F. Light, who conducted field trips to such places as Moss Beach attired in his gray business suit, complete with vest and starched collar." Ted Bullock and colleagues, in their 1947 *Memoriam*, wrote that Light "took great pains in planning and executing his courses, which were outstanding in their appeal and challenge to the serious student. . . . His unique courses in marine zoölogy given at the seashore under difficult conditions . . . maintained standards of excellence unsurpassed by any center of instruction in marine biology in the country. . . . Those who were taken behind an outer reserve found a

Ralph I. Smith (photo by Donald L. Mykles)

warm and sensitive personality with a discerning appreciation of the good in others and the values to be found in even the most trying situations. In all of his friends and students, whether intimates or not, he inspired something more than respect—a personal confidence and an attachment to his ideals that could hardly be separated from an attachment to the man. He will be remembered for his modesty, extending to an underestimation of self, for exacting criticism in the use of words and ideas, which drove him now to caution and again to very forward positions, for a sincere interest in human relations, and for a strong appreciation of natural beauty."

Remarkably, this description would also fit Light's successor, Ralph I. Smith, who arrived in Berkeley in the fall of 1946. Smith was born in Cambridge, Massachusetts and took all of his degrees from Harvard University, including the Ph.D. in 1942 under John H. Welsh. Smith rose to full professor at Berkeley where he stayed until his retirement in 1987. In his years at Berkeley, he taught invertebrate zoology and invertebrate physiology. For 30 years, Smith was also a Sierra Club ski tour leader, and for over a decade, led Boy Scout backpacking trips in the mountains. In this love of the mountains, he mirrored a long tradition at Berkeley: Joseph LeConte, one of the founders of science at the University of California, similarly spent as much time as he could in the Sierra Nevada. LeConte's interest had been seeded by an excursion with John Muir in the summer of 1870.

Smith's research career focused on physiology (often on adaptations to variable salinity regimes) and reproduction.

Much of his work was on nereid polychaetes; in the late 1980s, Smith's attention was drawn to terebellid biology as well. Smith's remarkable academic tree—a "Ph.D.logenetic tree" as he called it—of his first, second, third, and fourth generation students was published in 1988.

Smith, like Light, lost no time upon his arrival in Berkeley heading for the seashore with his students. He was scheduled to co-teach invertebrate zoology with Light at Hopkins Marine Station in Pacific Grove in the summer of 1947, but Light passed away in June. Smith was joined that summer by Ted Bullock and Frank Pitelka. In the course were graduate students Cadet Hand (late co-author of the Hydrozoan and anthozoan chapters in the present edition) and Donald Abbott (late co-author of the ascidian chapter in this edition).

Smith took over revisions of Light's 1941 manual and became lead editor of both the second (1954) and third (1975) editions of this book. Remarkably, Smith also found time to produce the *vade mecum* guide to the marine invertebrates of Cape Cod, on the eastern shore of the United States in 1964.

I did not know Light; he died the year before I was born. But I first met Ralph in the summer of 1965 and had the immense pleasure of knowing him for 28 years. I was in high school when John Holleman (author of the flatworm chapter of the present edition) directed me to Ralph on the Berkeley campus, so that I could examine some of Light's students' papers of the 1920s and 1930s on Lake Merritt in Oakland. Ralph allowed me access to the famous invertebrate reading room (now gone) on the fourth floor of Berkeley's Life Sciences Building (now renovated). After several hours of locating and scanning (with my eyes) the papers I needed, I returned to Ralph's office to ask if I could photocopy some of the documents. Ralph looked at me and said, "No—you should read them instead."

By 1973, I was once again spending many hours in that room, this time helping Ralph with the third edition of *Light's Manual*. Ralph's dry humor was remarked upon by many. At one point, a graduate student had written on the old black board in the invertebrate library, "Study Nature, Not Books—Louis Agassiz." Underneath was written, "Then Return the Books—Ralph Smith."

Colin Hermans, in his 1997 obituary of Ralph, noted Ralph's prodigious field stamina. One of my most vivid memories of him is during an early morning low tide in late July 1973 at Bodega Bay. A year before, in the summer of 1972, Ralph became interested in the identity of a large tube-dwelling cerianthid anemone, living in the goopy, sloppy mudflats in the back corner of Bodega Harbor where, as Ralph wrote to Mary Needler Arai, "No one in his right mind goes collecting." By the end of that summer Ralph had described this elusive animal as "too slippery to grab, too deep to dig," but suggested that he might eventually "be able to outwit one." In the summer of 1973, convinced it was important to include this animal in the next (third) edition of *Light's Manual*, Ralph laid out plans to tackle this creature living one meter straight down into soft black mud. Unhappily, in early June 1973, Ralph failed to negotiate a turn around a steel post near the Berkeley campus while riding his bicycle and ended up with his left arm in a cast.

And so off we marched at 5:30 a.m. on July 30. Every step out was a foot down into the oozing sediment as Ralph proceeded steadily, *the injured arm in a sling,* far out to the water's edge. Shoveling with one hand, Ralph soon had himself prostrate on the flats, his good arm up to his shoulder in the mud, doing battle with a massive gelatinous contortion that eventually produced most of the anemone.

Ralph wrote to Ted Bullock two weeks later, "I have had a good summer with the seashore course, though somewhat slowed down with my left arm in a full cast. . . . Still, I am right-handed, and am now perhaps the first person to have dug out a cerianthid with one arm from three feet of mud. That sentence rather sounds as if one-armed cerianthus were rare, but it is the sort of writing I get from my contributors. It was a messy job—I'm not sure Dr. Light would have approved."

In 1974, several years into nearly continuous work on the next edition of the manual, Eugene Kozloff (author of the chapter on orthonectids in this edition), knowing of Ralph's long struggle with endless name changes perpetrated upon well-known species, wrote to Ralph referring to him as "one of you systematists." Ralph wrote back:

"I can hardly take that lying down! I am not a systematist by any stretch of the imagination, which is one reason why I work on things like the *Light Manual*. If I were a systematist, I would be getting everyone confused instead of trying to clarify matters, but just because I end up getting everyone confused does not of itself make me a systematist."

Reminiscent of his predecessor's predilection "for exacting criticism in the use of words and ideas," Ralph spent hundreds of hours copy-editing the third edition, taking on matters ranging from the proper rendition of women's hyphenated maiden-married names, to whether "intertidal" was a noun, to the elimination of the umlaut over the second o of zoöid and zoölogy. In responding to an inquiry from Donald Abbott on the latter question, Ralph wrote, "In respect to umlauts: I realize that modern zoölogists do not spell zoölogy with an ö, and I will try to coöperate with you in deleting them." In the final days of production, Ralph patiently removed scores of commas inserted by UC Press editors. Responding to the Press's meticulous touching up of the artwork for the book, Ralph sent a note back that "one very small animal was removed entirely—apparently mistaken for a dirt spot."

Ralph Smith died in May 1993 at the age of 76, following a heart attack after finishing a 3-kilometer foot race in Santa Rosa, California—the "Human Race," a charity event. At the time, Ralph had begun writing contributors about a revision of the third edition of this book. A rich recounting of Ralph's life and character is found in Colin Hermans' 1997 reminiscence. Ralph, at one and the same time, offered an often firm and demanding exterior over a generous and kind personality of great depth; in the California mountains his concern for others and his great generosity would emerge, as it did when he guided undergraduates, graduates, and colleagues in the laboratory and field. Smith was, above all, as Eugene Kozloff once remarked, a "Practical Zoologist. First Class."

This book is the living tribute to the careers that S. F. Light and Ralph Smith invested in students and the ocean.

James T. Carlton

Bullock, Theodore H. 1947. S. F. Light 1886–1947. Science 106: 483–484.
Carlton, James T., Daphne G. Fautin, Michael G. Kellogg, Barbara E. Weitbrecht, and Armand M. Kuris. 1988. Professor Ralph I. Smith: A tribute to his manuals of marine invertebrates and to his academic progeny. Veliger 31: 135–138 (from which some of the above material was directly drawn).
Hermans, Colin O. 1997. Ralph Ingram Smith, July 3, 1916–May 12, 1993. Bulletin of Marine Science 60: 224–234.

INTRODUCTION

Intertidal Habitats and Marine Biogeography
of the Oregonian Province

THOMAS M. NIESEN

We enter the 21st century when accurate identification of marine organisms is essential in the face of declining biodiversity (Carlton 1993, Tegner et al. 1996, Carlton et al. 1999), documented faunal shifts linked to coastal warming (Barry et al. 1995, Larson et al. 1997, Sagarin et al. 1999), and, in some habitats, the wholesale reshuffling of the faunal deck by the unchecked barrage of introduced species (Cohen and Carlton 1995). I have approached this introductory chapter with the history of this manual's utilization in mind, knowing it will continue to be used by students in upper division and graduate courses with limited familiarity of our varied coastal environments. Therefore, I hope to introduce this variety briefly and pass on some suggestions for successful field observation of marine invertebrates. Borrowing from the previous edition, I will first characterize the main biological and physical considerations at play in the intertidal zone, briefly characterize the main intertidal habitats of the region, and finally suggest the biogeographical underpinnings of our varied marine invertebrate fauna.

Field Studies

The Physical Setting

As you round Point Conception heading north and encounter the spectacle of Morro Rock, you realize you aren't in southern California any more. Gone is the west to east tending sea coast sheltered by the offshore Channel Islands. In its place, an exposed north to south coastline prevails, unprotected from the onslaught of the Pacific swells. Also gone is the dry, semi-desert weather of southern California. As you proceed north into more rainy climes, you are confronted with the reality of large watersheds draining into rivers, which harbor extensive estuaries at their drowned mouths. The rugged coastal mountain chains reflect the tectonically active nature of this coastline with protected bays and estuaries, like San Francisco and Tomales Bays, nestled in their folds. Soaring coastal headlands and rocky points have extensive exposed rocky intertidal habitat at their bases and protected embayments in their lee. It is truly a wondrous place to study marine invertebrates.

No picture of organisms that ignores their physical and organic environment can be even approximately complete. Studies of dead animals or their parts, or even of living animals in the laboratory, give but a partial picture. For fuller understanding, we must seek firsthand knowledge of living organisms in their natural settings. Field trips are of prime importance in gaining such knowledge and understanding. They make possible a study of the environment itself. This includes the study of the distribution of organisms within specific habitats, the behavior and interrelationships of species, and the influence of physical and biotic factors on the distribution of organisms. Only through information gained in field studies is it possible to establish correlations between the structure and behavior of an organism and its habitat and ecological niche. One of the original purposes of this manual was to aid in the inclusion of extensive field studies of marine intertidal invertebrates in courses for advanced undergraduates and graduate students. We hope this tradition continues and offer these suggestions for successful field trips.

Where to Look?

Virtually any site that is accessible along central and northern California's and Oregon's coastal counties can be a rewarding place to investigate marine invertebrates. Be careful when descending coastal bluffs because many are unstable and a serious fall can result. Two of the most useful books available to the student are the *California Coastal Access Guide* and its companion text, the *California Coastal Resource Guide*. These books provide maps and information about access for all the coastal counties and include details about parking, camping facilities, and special points of interest. They are widely available at local libraries, as well as at nature and sporting goods stores, museums, and state parks.

When searching soft sediment habitats, most discoveries will be made by digging. Sandy beaches have virtually no organisms that live exposed on the surface. Protected sand flats and mud flats may have a few hardy species obvious on the surface, but the majority of the organisms will be burrowed beneath the substrate. Most will be in the upper 15–20 cm of the substrate to maintain contact with the water to breathe.

Some of the larger clam species in protected embayments are found considerably deeper, thus only the most dedicated student will see them on any given day. Remember as you dig that many of the most interesting organisms will be quite small, a few centimeters or less, so adjust your search image accordingly.

In rocky habitats, it is easy to be drawn to the large and colorful organisms first. However, after you've run out of big, obvious organisms, once again adjust your search pattern and explore the diversity of the smaller, more cryptic organisms. Many small animals seek the shelter of crevices or seaweed, and others live under loose rocks and boulders. Careful searching of any of these microhabitats will be rewarded with new treasures.

A word of caution about turning rocks: animals live attached to the rock's bottom as well as underneath it, so be careful when turning the rock not to crush any of the inhabitants. When you've finished exploring, turn the rock back over the way you found it—but again, make sure not to crush the animals you just observed! Place the free-living animals under other rocks or among seaweed so they won't dry out. Remember that organisms in the rocky intertidal zone are usually found in fairly specific locations relative to the tide and exposure. Don't move them out of their preferred locations.

When to Look

The serious student soon realizes that more can be seen during low tide than high, therefore it becomes necessary to learn about tides and tide tables to guarantee a successful field trip. Tidal height on the West Coast is measured from an arbitrary zero point called "mean lower low water" (MLLW).

This zero point is the average of all the lower low tides that occur in a year. A high tide listed in the tide table as +6.0 ft means the tide at its peak will be 6 ft above 0 MLLW, again the average level of lower low water. A low tide listed as −1.5 ft means the tide will be a foot and a half below this average level. The latter tide is referred to as a "minus tide" and represents the best time to explore, as more of the intertidal zone will be exposed.

Tides are caused by the gravitational pull of the sun and the moon on the earth's surface. Tide tables can be made up years in advance because the position and effect of the sun and the moon are highly predictable. However, tide tables cannot anticipate local weather conditions, which can significantly alter the actual tide that occurs on a given day. Large storm waves pushed by strong onshore winds can completely wash out a scheduled low tide and cause a high tide to be several feet above the tide table prediction. Likewise, a strong high-pressure area over the coast can push down on the water causing both high and low tides to be lower than predicted.

Intertidal observation at low tide is usually best approximately two hours before to two hours after the scheduled low tide. Obviously, the lower the tide, the more area that will be exposed. This is important for rocky habitats, as the organisms are distributed in somewhat distinct tidal zones. The lower intertidal zones are the more diverse because they experience less exposure to drying. The recommended method is to explore the lowest exposed tidal zone first and move up into the intertidal as the tide rises.

Biological and Physical Factors Influencing Marine Invertebrates of the Intertidal Zone

To be of maximum value, studies of marine invertebrates should be cumulative and comparative. Each new situation should be compared to others already studied, noting similarities and differences among faunas and the environmental conditions they experience.

In pursuing field studies, it is well to keep in mind certain fundamentals. All animals have similar basic needs or requirements. Ultimately these can be reduced to food, oxygen, protection, and the proper conditions for reproduction. In situations where an animal occurs regularly and in high numbers, we may be sure that these needs are met, though the manner in which they are met is not always obvious.

FOOD as used here, includes all substances (except oxygen and water) from the environment necessary to provide animals with energy and body-building materials. The intertidal is a region of abundant light and food. In areas of hard substrate, there is often much plant life. This includes not only the larger algae and occasional flowering plants, such as the surfgrass *Phyllospadix*, but also the film of microscopic plant life growing on exposed surfaces of rocks and larger plants. A few organisms (e.g., the kelp crab *Pugettia producta,* the red sea urchin *Strongylocentrotus franciscanus,* and the limpet *Lottia instabilis*) graze directly on the attached larger algae. Seaweeds broken loose and washed into crevices and pools or ashore on beaches provide a rich source of food for other forms (e.g., the purple sea urchin *Strongylocentrotus purpuratus* and the beach amphipods *Traskorchestia* and *Megalorchestia*).

Plant material, ground into a fine organic detritus by the action of turbulent waters against the substrate, provides food for a host of creatures that feed in many different ways. Detritus suspended in moving waters is taken, together with living plankton, by a variety of particle feeders. Some use mucous

FIGURE 1 The moonlike landscape of holes and burrows created by the intertidal ghost shrimp *Neotrypaea californiensis* (photo by Thomas Niesen).

nets or webs to trap this food (e.g., the echiuran *Urechis,* the polychaete *Chaetopterus* and the attached gastropod *Petaloconchus*), whereas others use cilia, often in conjunction with sheets or strands of mucus (e.g., most bivalves, brachiopods, bryozoans, tunicates, serpulid and sabellid polychaetes, sand dollars, sponges). Still other feeders on suspended detritus and plankton use combs of fine bristles or setae to catch particles (e.g., the decapods *Petrolisthes* and *Emerita,* barnacles, and many other lower crustaceans; yet others use tentacles (e.g., the sea cucumber *Pseudocnus curatus,* terebellid and spionid polychaetes, and ctenophores).

Organic detritus of plant and animal origin also accumulates on and in soft substrates, and here it forms food for another complex of organisms. Some of these organisms "vacuum" up the surface layer of detritus (e.g., some species of the clam *Macoma*); others, such as many annelids, swallow the muddy substrate more or less unselectively. Still others burrow and sift the bottom material for edible particles (a good example is the shrimplike *Neotrypaea,* fig. 1).

Finally, organic detritus tends to cling to exposed surfaces of rocks and plants. Here, macroscopic encrusting algae, together with the microscopic plants, bacteria, and protozoa, serve as food for animals with rasping or scraping organs, particularly the gastropods and chitons.

The variety of organisms living on plant and animal detritus and plankton provides food in plenty for intertidal predators. Micropredators, like hydroids and the sea anemone *Metridium,* feed upon animal plankton brought in by the tides. Predators on larger organisms (e.g., most sea stars [but not *Patiria*]; the gastropods *Nucella, Acanthinucella,* and *Ceratostoma;* crabs, nemerteans) find a rich diet of clams, snails, worms, barnacles, and other small crustaceans. Fishes as well as invertebrates are significant predators when the tides are high; whereas shore birds and humans, and a surprising variety of other mammals (Carlton and Hodder 2003) have their impact when the waters recede.

When studying an organism keep in mind these questions: What does it eat? How does it get its food? What is its role in the food web of the area?

OXYGEN presents a lesser problem for most intertidal organisms. In most areas, continual movement of the shallow waters ensures full oxygenation at all times. The oxygen tension in air is the same as in saturated water, and as long as animals remain damp, atmospheric oxygen can readily diffuse across respiratory membranes. The oxygen requirements of sessile and sedentary animals are not large, and some species are capable of withstanding temporary anaerobic conditions.

Some intertidal organisms, however, do have adaptations to meet specialized needs with respect to oxygen. Species living high in the intertidal zone may show structural modifications, enabling them to carry on aerial respiration. Some species of the periwinkle *Littorina* have considerable vascularization of the wall of the mantle cavity, which serves in some degree as a lung. Crabs, such as *Pachygrapsus,* have gills reduced in size and stiff enough for self-support, in addition to arrangements for retaining water in the gill cavity and vascularization of the wall of the cavity itself. Mudflat forms, stranded in the substrate with a limited water and oxygen supply, face a severe problem. Yet even in this harsh environment, there are physiological mechanisms to cope.

REPRODUCTION can be more difficult to study in the field. Many intertidal benthic forms shed eggs and sperm into the sea, where developing embryos and feeding larvae lead a pelagic existence for a time. Pelagic stages may be collected in plankton tows but otherwise are seldom seen. However, a large number and variety of marine and brackish-water organisms retain, carry, or brood their eggs (e.g., the sea cucumbers *Pseudocnus curatus* and *Lissothuria nutriens,* the sea star *Leptasterias,* many crustaceans, colonial ascidians, and the viviparous polychaete *Neanthes limnicola*).

Other invertebrates produce characteristic egg cases (e.g., those of the snails *Nucella* and *Acanthinucella,* most nudibranchs and cephalopods, certain polychaete worms). These early stages should be observed and, if possible, connected with the animals that produced them. Still other animals reproduce asexually and may form extensive colonies or aggregations of individuals (the aggregating anemone *Anthopleura elegantissima,* colonial ascidians, hydroids, bryozoans, the hydrocoral *Stylantheca,* the alcyonarian *Clavularia,* sponges). Observations of such reproductive features in the field or in individuals collected and brought to the laboratory are important in natural history studies and may well result in previously unrecorded discoveries.

Physical Factors

The questions of protection and of the environmental conditions from which organisms require shelter are crucial to field studies. Animals persistently present at any particular spot must be adapted to survive the most unfavorable extremes of environmental conditions that occur there. All mechanisms—morphological, physiological, and behavioral—through which an animal internally preserves the conditions for cellular life are implied here in the term protection. Therefore, protection includes the sheltered places sought by motile forms like crabs, the production of anchored fortresses of lime by barnacles, the capacity of brackish-water (and freshwater) organisms to osmoregulate, the ability of the brine shrimp to tolerate warm and hypersaline medium, and a myriad other adaptations that enable organisms to withstand conditions externally that they could not tolerate internally.

Owing to the regular rise and fall of the sea, the intertidal zone is a region of great periodic fluctuations in environmental conditions. It is a region of transition between sea and land, but the boundary between aquatic and terrestrial conditions is less sharp and far more complex than that at the shorelines of lakes and ponds. In the intertidal region, animals are subjected to conditions that are alternately marine and semi-terrestrial.

Even with their various protective adaptations, animals at the seashore are more or less limited to particular habitats. Physical factors of the environment, coupled with specific requirements and adaptations of animals (for food, oxygen, reproduction, and protection, as discussed below), definitely limit the distribution of each species. The most important of the physical environmental factors with which we will be concerned are outlined here.

NATURE OF THE SUBSTRATE The nature of the substrate is of cardinal importance in limiting animal and plant distribution. The substrate may vary from nearly unbroken cliffs and rocky ledges, through a series of such intergrades as broken rocky reefs, boulders, and pebbles, to the finer substrates of sand, mud, and clay. Rocky substrate habitats provide many places for organisms to live. Depending on the nature of the rock and how it weathers, rocky areas can have considerable topographic relief. Broad rocky reefs with ledges, crevices, outcroppings, boulder fields, and tide pools provide myriad microhabitats for organisms to occupy. Even the hardness of the rock (e.g., hard granite versus relatively soft sandstone or shale) is of great importance, particularly for attached and rock-boring forms. Soft substrates, such as the sand on a sandy beach or the mud of an estuarine mud flat, provide many fewer sites for organisms. Because the surface of a soft substrate provides no fixed sites for attachment and is usually quite flat and exposed, the only place to live is beneath the substrate surface. As a general rule the marine community living in a soft substrate habitat will have a relatively low diversity of organisms but can have high numbers of individual species. Hard substrate communities tend to be highly diverse with smaller numbers of individual organisms.

DEGREE OF WAVE SHOCK AND CURRENT ACTION The pounding and abrasive action of waves has many effects on both environment and biota (Koehl and Rosenfeld 2006). Where severe, wave action prevents the accumulation of substrates of mud and clay, it restricts the distribution of animals that are not tough, flexible, firmly rooted, or capable of clinging tightly to (or burrowing or boring into) the substrate. Wave action also makes possible the presence of aquatic or semi-aquatic forms in a splash zone above the highest level reached by the surface of the sea.

Currents may erode soft substrates in one area and redeposit them elsewhere. The shifting bottom materials and the scouring action of particle-laden water may have important effects on the bottom-dwelling organisms present. Ricketts et al. (1985), in *Between Pacific Tides,* clearly recognized the importance of the degree of exposure to wave shock as a factor in limiting animal distribution; their primary ecologic divisions of the Pacific intertidal zone, based on this factor, are "open coast," "protected outer coast," and "bay and estuary," with appropriate subdivisions of each according to the nature of the substrate.

ALTERED TEMPERATURES At low tides on sunny summer days and windy winter nights, the temperature extremes for exposed animals markedly exceed those found in the sea. Ability to withstand great changes in temperature, particularly the higher temperatures, is one important factor in determining the intertidal distribution of organisms.

DESICCATION Most dwellers in the intertidal zone have come up from the sea; few are invaders from the land. It follows that withstanding desiccation is a great problem for many intertidal species, and surviving prolonged submersion is far less so. Marine organisms must remain moist in order to breathe, and exposure to air dries them out, particularly during low tides on hot, sunny days, and the problem is closely related to that of withstanding higher temperatures. The higher in the intertidal zone an organism is located relative to the low tide mark, the longer it will be exposed to air. Not surprisingly then, the only organisms found in the highest tidal zone will be those that have special adaptations or behaviors to avoid drying out. Motile forms may show a protective behavior, shifting to positions under rocks or ledges or seeking the slighter protection of depressions and shallow crevices. Sedentary and sessile forms may have protective shells, adherent layers of sand or gravel, tough and often mucus-covered integuments, or large internal stores of water. As you move lower into the intertidal zone, the degree of exposure an organism encounters is reduced, and more diversity will be discovered, with the most diverse areas occurring at tidal levels exposed by only the lowest tides. This effect is especially noticeable in rocky substrate habitats where organisms are attached to the rocks and directly exposed to air. The effect of tidal position can be somewhat reduced for organisms living in soft substrate habitats if the substrate remains saturated with water at low tide.

ALTERED SALINITY On hot days at low tide, the salinity of high tide pools may rise owing to evaporation. During rains, direct precipitation and runoff from streams and beaches may greatly reduce the salinity in some areas. Organisms on high rocks, in high pools, and near stream mouths must be able to avoid, regulate against, or tolerate at least temporary hypersaline and/or brackish conditions.

LOWERED OXYGEN AVAILABILITY Only in certain instances is this an important factor. It becomes important where intertidal forms are desiccated to a degree that exposed surfaces no longer serve as respiratory membranes. It may also be important for dwellers in mudflats exposed by tides.

As a result of all these factors and of many others that usually are less accessible to study, intertidal environments, floras, and faunas may vary greatly, even within relatively restricted regions of the coast. Therefore, along with this review of the main biological and physical environmental factors influencing marine invertebrates, a brief overview of how these factors interact to shape our marine intertidal habitats available for study is appropriate.

Intertidal Habitats of Central and Northern California and Oregon

Sandy Beaches

Sandy beaches dominate much of the open coastline of central and northern California and Oregon, stretching uninterrupted for miles in many regions. These sandy beaches represent the most physically controlled of all the intertidal marine habitats and, as such, one of the most difficult to live in. Sandy beaches have one thing in common: all are composed of sediment particles that overlay a rocky beach platform. Beyond that, they can vary considerably. The sediments that make up these beaches are small pieces of rock, mainly the minerals quartz and feldspar. These minerals are either weathered from the continental rocks and washed down to the ocean in rivers or are the products of coastal erosion. Once these sediment particles reach the sea, they are carried along the coast until they find their way onto a sandy beach or offshore and perhaps down a submarine canyon. The transport of sediment along the coast requires water movement in the form of waves and currents.

Waves come from all directions along the open coast, and the direction is related to prevailing wind patterns. When the waves come onshore and break, they often break at an angle to the shoreline, generating currents that run along the beach. These are called longshore currents, and as they move along the coast, they carry sand particles with them. This movement of sediment along the beach is called longshore transport. In the summer, the prevailing longshore current flows toward the south, and the net longshore transport of sediment is likewise southerly. However, the net longshore current direction and transport of sediment are to the north in the winter. It should be pointed out that along the coast local conditions can vary considerably, and the deposition of sediment in the form of sand bars and sand spits can likewise vary considerably from place to place.

Because sandy beaches are subject to relatively continuous wave action, only the larger sediment particles, known as sands, accumulate and remain on the beach. The size of the sand particles on the beach is directly related to the size of the waves that hit the beach. On beaches that experience vigorous wave action, the finer sand grains will be resuspended and removed, leaving only large sand grains behind that feel coarse to the touch. Beaches with fine sands develop when only gentle wave action occurs, and thus only finer particles are transported to and deposited on them. Beaches will also undergo drastic changes in the type of sediment present. Many beaches go through an annual cycle of sand accumulation during the spring and summer months as small waves return fine grains to the beach, creating a gentle, sloping beach. This is followed by a drastic loss of sand with the first strong winter storm that leaves the beach in a winter condition. The storm waves strip the beach of sand or leave only the largest particles behind.

Sandy beaches are truly dynamic habitats. What's it like to live there? First there is the sediment to contend with. It doesn't stay put; living on the surface or trying to excavate and live in a permanent burrow are not options. Because the sediment can shift so rapidly and unpredictably, any organism living on the beach must be able to swim and/or burrow rapidly to keep from being swept away (good examples are the sand crab *Emerita* and haustoriid amphipods). Then there is a problem with drainage. As the tide recedes, water drains from between the sand grains, leaving organisms burrowed in the sand in danger of drying out. The spaces between the grains on coarse beaches are too large to hold water by capillary action. These beaches can become quite dry at high tide, especially if the sun is shining or the wind is blowing. This is less of a problem on beaches made up of fine sand grains, as water is held in the small spaces between the grains by capillary action. Finally, there are predators. Fishes, crabs, shrimps, worms, and an occasional predatory snail forage on the beaches at high tide. Birds, small mammals, and insects take over at low tide.

Why live on a sandy beach? Because there is abundant food. The same wave and current action that accounts for the beach's dynamic physical nature also delivers food to the sandy beach. Food can vary from the carcasses of large animals, such as fish and marine mammals, to large seaweeds that have been torn form their attachment offshore. However, the majority of food available to beach dwellers is detritus. Detrital particles are derived from large and small organisms produced elsewhere in the ocean that have been broken up and carried along by the moving water. Not surprisingly the marine invertebrates found on sandy beaches are typically rapidly burrowing filter feeders (e.g., the sand crab *Emerita analoga*, and the razor clam *Siliqua patula*) or animals that live high up on the beach and feed on deposited plant material known as wrack (e.g., beach hoppers, *Megalorchestia* spp.).

Quiet Water, Soft Bottom Habitats

The coastline of the northeastern Pacific is constantly changing. Sandy beaches can be transformed in a matter of hours during winter storms. Coastal bluffs are continuously weathered by wind and waves. Sea level changes and plate tectonic activity over geologic time have alternately inundated and uncovered vast areas of coastal land. Against this backdrop of change, a number of habitats become shielded from the onslaught of waves and develop into quiet water habitats.

For example, estuaries are typically formed when a large river is flooded by rising sea level. The sediment carried from the land accumulates as a delta in the drowned river mouth. Sand flats, mudflats, and finally salt marshes develop as the delta grows and stabilizes. A long, fingerlike sand spit may form along the interface between the estuary and the ocean, as is seen all along the coast of northern California and central Oregon. Another example would be a strong longshore current transporting sediment along the coastline and depositing it behind a headland. The deposited sediment builds up a sand bar, and finally a sand spit gradually cuts off a cove behind the headland, forming a protected coastal embayment. As tidal action moves water in and out of the embayment, the fine sediments are deposited and sand flats and mudflats are formed.

Quiet water, soft bottom habitats provide an environment that is very different from that of the sandy beach. Here organisms can create burrows that will be somewhat permanent in the soft substrate. An organism can move across the substrate at high tide and not be washed away by wave action. Tidal action continues to bring suspended food into these habitats, which combines with local plankton production to provide ample resources for a host of filter feeders. In estuarine areas, organic detritus is brought downstream by fresh water flows. Finally, as the water becomes quiet in the most sheltered regions of these habitats, the fine organic material carried in suspension is deposited, providing food for deposit feeders.

As the types of substrates and available food are similar in quiet water habitats, they tend to host similar types of animals. However, there are some differences. Coastal embayments typically contain normal, undiluted seawater. In comparison, estuaries have a gradient of salinities, from fresh water at the head of the estuary, where the river or other freshwater source enters, to pure seawater at the mouth of the estuary where it enters the sea. Salt water lagoons vary in salinity depending on the source(s) of water supplying them. I will briefly describe the unique features of estuaries, and then treat the common protected soft bottom habitats collectively.

ESTUARIES

Estuaries are an aquatic interface between the riverine (freshwater) and marine environments. They serve a vital role for many organisms whose life cycles require both these environments. Most obvious of these are the salmonid fishes that spawn in upstream freshwater habitats and then migrate to the sea to feed and mature. These fishes typically pass through an estuary to reach the sea. The graded range of salinities found in the estuary allows the young salmonids to acclimate gradually from fresh to salt water. Another good example is the Chinese mitten crab *Eriocheir sinensis,* which lives in large populations in the freshwater rivers of central California and migrates to the estuaries to reproduce. Species that go to fresh water to reproduce are *anadromous*; species that return to the sea from fresh water to reproduce are *catadromous*.

Estuaries also play important roles as nurseries for early stages in the life cycles of many marine organisms. A key commercial species that uses the estuary as juveniles is the Dungeness crab, *Cancer magister*. Advanced larvae enter the estuary from offshore and soon metamorphose and settle to the bottom as young crabs. The crabs may spend up to three years feeding and growing in the estuary before they move toward the mouth and out to sea. In large estuaries such as Humboldt Bay, adult populations of Dungeness crabs can be found year-round in the more saline waters of the estuary mouth.

The internal body fluids of marine invertebrates are essentially isotonic to pure seawater. When the salinity of the water varies, these animals become osmotically stressed. The more saline water of a lagoon, a coastal embayment, and at the mouth of an estuary can all potentially support similar marine fauna. However the upper, less saline reaches of the estuary will harbor a unique fauna adapted to the lower salinities that occur there. Very few marine invertebrate groups have successfully evolved the ability to withstand the low and fluctuating salinities found in estuaries. On the other hand, those groups that have adapted to estuarine conditions tend to be hardy. Unfortunately, such creatures are easily transported by human conveyances, such as in the ballast water tanks of ships, or among estuarine food species imported for human consumption, such as oysters. The estuarine invertebrate fauna along the coast of the northeastern Pacific is poorly developed, and the native fauna found in a given estuary contains comparatively few species. Because the estuaries of the northeastern Pacific have such limited faunas, and because estuaries, such as San Francisco and Humboldt Bays tend to be such intense sites of international human commerce, large numbers of estuarine animals from around the world have been successfully introduced there.

THE MUDFLAT

Although mudflats and sand flats are treated as separate habitats, they are part of a continuum of sediment types that grade from the finest clay mud to cobblestone beaches. In the quiet water habitats discussed here, the soft substrate is often a mixture of several sediment sizes reflecting the different types of water movement that occur over a particular area. Marine biologists might describe one such sediment as a sandy mud and another, with slightly more sand, as a muddy sand—that is to say, arbitrary terms suggest the general composition of the substrate. Therefore, although this section is titled mudflats, many of the invertebrates mentioned may be found over a range of soft sediment habitats.

Because of the gradual way mudflats are formed, they tend to be quite flat with little physical relief. However, the surface signs of the burrowing and feeding activities of the resident organisms can sometimes be quite extensive (fig. 1). Mudflats might seem to be the last place to find seaweeds, but sometimes extensive growths of a variety of species of the enteromorphine green algae *Ulva* can be found. Red algae will also sometimes grow attached to an exposed shell or other debris. These algae occasionally accumulate in dense layers and smother fragile organisms in the sediment below.

The surface of the mudflat is often covered with a slick coating of benthic diatoms, adding a further source of local primary production. Finally, mud flats may often be fringed by salt marsh vegetation, which provides a rich source of particulate plant detritus for mudflat dwellers.

Particulate organic matter settles out of the water at about the same rate as clay particles do. Therefore, mudflats are rich in deposited organic matter, both on the surface and buried within the substrate. Mud snails, tube-dwelling amphipods, and vacuum cleaner clams feed on the surface deposit, while lugworms, sea cucumbers, sipunculans, ghost shrimp, and others feed on organic matter buried by sedimentation. Thus the mudflat can support a variety of species on this rich, abundant source of food.

The very small silt and clay particles that make up mudflats develop a very strong capillary action. As a result, mudflats typically do not dry out during low tide. However, with the exception of a shallow surface sediment layer, there is little circulation with the water above the mudflat and the water in the sediment. Beneath this thin surface layer, oxygen is depleted by decomposing bacteria, and the substrate is colored dark brown to black by hydrogen sulfide and has the odor of rotten eggs. Obviously this is not a hospitable place for an aerobic (oxygen-consuming) organism, and many infaunal invertebrates are excluded. However it is perfect for invertebrates like clams that can burrow safely into this anaerobic mud and reach the surface with their siphons. Often the shells of clams such as *Tresus nuttallii* and *Saxidomus* spp. taken from such a substrate will be stained black.

THE SAND FLAT

Many mudflat invertebrates overlap into sand flats, especially the burrowers. Sand flats occur in protected environments where there is an appreciable flow of water and a source of sediment. For example, in an estuary, sand flats typically form near the estuarine mouth adjacent to the main channel where relatively constant tidal currents keep the finer sediment particles in suspension and only the larger sand grains settle out. Sand flats are also common in coastal embayments for similar reasons. The combination of tidal currents and locally generated waves transport sand into the embayment where it is deposited when the water movement slows. The most attractive feature of sand flats to marine invertebrate enthusiasts is that they are firm and easily traversed. With a little patience, a good shovel and a decent low tide, sand flats are very rewarding places to explore.

The sand flat is a more physically active habitat than the mudflat. It is subjected to stronger water currents and sometimes, especially in coastal embayments, wave action. For the animals that live here, this means that a burrow may occasionally be disrupted or buried. For example, sand flat clams live more shallowly in the substrate than those on mudflats and have correspondingly shorter siphons. To compensate for the occasional disruption that can sometimes wash them out of the sand, these clams, such as *Clinocardium nuttalli* and *Leukoma staminea,* have a large digging foot and are excellent burrowers.

Sandy sediments pack more loosely than mud, which creates more space between the particles. This allows for an effective circulation of oxygenated water through the sediment, and an anaerobic layer does not readily form. Because water movement over sand flats is more rapid than over mudflats, particulate organic matter tends to remain in suspension. Thus sand flats are typically dominated by beds of filter feeders like clams and echiuran and phoronid worms. These beds can be quite extensive. On some sand flats in central California, such as those in Elkhorn Slough in Monterey County, or Drakes Estero on Point Reyes in Marin County, patches of substrate in the low intertidal zone seem suddenly firm under foot. Careful excavation will reveal the closely packed, flexible chitinous tubes of the phoronid worm *Phoronopsis harmeri.* The phoronid bed may consist of many thousands of these slender, 10-cm-long worms.

More obvious, motile invertebrates of the sand flats include the scavenging olive snail, *Callianax biplicata,* and the predatory moon snail, *Euspira lewisii,* whose telltale calling card—a clam shell with a circular, counter-sunk bore hole (fig. 2)—is a familiar sight on these flats. Other large invertebrate predators on sand flats include sea stars (such as *Pisaster brevispinus* and *Pycnopodia helianthoides*) and cancroid crabs (*Cancer gracilis,* juvenile *Cancer magister, Cancer productus* and *Cancer antennarius*).

SALT MARSHES

Protected coastal environments are typically created by the flow of sediment into a bay, river mouth, or other shallow area, forming a delta. As the delta builds up, it can be colonized by special plants adapted to seawater-inundated soils, and a salt marsh is born. Once the salt marsh plants take hold, their spreading root and underground stem systems trap more sediment. The level of the marsh is raised, and the margin of the marsh increases outward until it reaches a dynamic equilibrium with the local pattern of water movement that controls sedimentation. As the marsh becomes more extensive, drainage channels, called tidal creeks, develop that direct the flow of water out of the marsh as the tide recedes. These tidal creeks can become deeply eroded, and their bottoms and banks create a habitat for marine invertebrates.

The West Coast does not contain the vast acreage of salt marshes as is seen along our Atlantic and Gulf coasts. In addi-

FIGURE 2 A moon snail bore hole in the clam *Leukoma staminea* (photo by Thomas Niesen).

tion, much of the salt marsh habitat has been diked, drained, and turned into pasture or commercial real estate in California and Oregon. The value of salt marshes and other coastal wetlands has become apparent to policymakers, and movements to preserve and even restore wetlands are in place, including coastal salt marshes. What salt marsh habitat we do have is now generally well protected by regulations and will hopefully be preserved.

The average student of marine invertebrates will not venture too far into a salt marsh. The vegetation is thick and the underfooting often unsure. Likewise, the bottom of tidal creeks can be very muddy and hard to traverse. Many salt marshes in California, and some in Oregon, do provide viewing platforms, and a few, such as the Palo Alto marsh on San Francisco Bay, have catwalks that extend over the marsh or have trails built along the marsh uplands.

The portion of the marsh affected by the tides is dominated by two plants. Along the lower intertidal, a fringe of the tall (up to 1 m) cord grass *Spartina* spp. can be found. In San Francisco Bay, the native species *Spartina foliosa* is being replaced by an East Coast species, *Spartina alterniflora,* and in some places they have hybridized. The *Spartina* fringe may be fairly extensive but eventually gives way to species of the shorter pickleweed or glasswart, *Salicornia* spp., that grow at higher tidal elevations and typically dominate the salt marsh. Above the pickleweed is a community of upland marsh plants. The tidal marsh is incised by meandering tidal creeks that usually drain out across a low intertidal flat.

The marsh itself is home to a mixture of terrestrial and marine organisms. The terrestrial component includes many insects, small mammals, and of course birds. Living among the plants are many small marine invertebrates, including crustaceans and mollusks. The tidal creeks are invaded by many of the same animals that occur on mud- and sand flats.

EELGRASS BEDS

A final protected, soft substrate habitat that may be visited is the eelgrass bed. Eelgrass, *Zostera marina,* grows along the lower intertidal edge of mud- and sand flats and in shallow, subtidal protected habitats. In Oregon, this native seagrass is joined by extensive mid- and upper-intertidal meadows of the introduced Japanese eelgrass *Zostera japonica.* If the water is clear, subtidal beds of *Z. marina* can be quite extensive, such as those of Elkhorn Slough and Humboldt Bay. Eelgrass beds create important habitat for fishes, shrimps, juvenile crabs, and water fowl.

The eelgrass beds accessible intertidally will yield a number of familiar marine invertebrates, including many clam and worm species. The matted root mass of the grass is a particularly good place to observe polychaetes. There are also a number of unique marine invertebrates that can be found here (e.g., the attached jellyfish *Haliclystus* sp., and Taylor's sea hare, *Phyllaplysia taylori*). The California sea hare, *Aplysia californica,* comes into eelgrass beds to mate, sometimes in large numbers. After exchanging sperm, the sea hares extrude long, sticky strings of jelly-encased fertilized eggs that they attach to the eel grass.

Hard Substrates

PIER PILINGS

Pier pilings represent a hard substrate that receives normal tidal exposure. As a result, it harbors a variety of organisms that prefer and/or tolerate these conditions. The upper part of a piling is usually the domain of hardy barnacles and a smattering of limpets and shore crabs (*Pachygrapsus crassipes*) typical of the upper intertidal zone of the rocky coast. Below the barnacles, at approximately middle tide level, mussels (*Mytilus* spp.) will be common, often in sizable clumps. The mussels provide attachment sites for a variety of organisms, including barnacles, limpets, sea anemones, and encrusting forms such as sponges and bryozoans. If the clump is especially well developed, it can harbor sea star predators and a myriad of small invertebrates.

FLOATING DOCKS

The sides and bottoms of floating boat docks, such as in marinas, are the other common hard substrates that may be found in quiet water habitats. These are typically made up of panels of high-density plastic foams that have high flotation properties and are imperious to seawater. However, they are not imperious to marine invertebrates. Unlike the pier piling discussed above, these substrates are never exposed to tidal action. Consequently, they harbor a suite of fragile, space-grabbing invertebrates, many of which grow in flat, encrusting colonies. Here are the bright reds and yellows of marine sponges, the purple and orange colonies of compound tunicates, and the delicate basket-weave patterns of encrusting bryozoan colonies. Erect or branching bryozoan colonies are also common, along with a variety of hydroid cnidarians. Barnacles and mussels will also grow on these structures, providing yet more substrate for the encrusting species. Marina floats are richly encrusted with many nonnative species in our estuaries.

Rocky Intertidal Zone

The interaction of physical and biological factors outlined earlier in this chapter reach their zenith in the rocky intertidal zone. The main physical factors affecting the distribution of marine intertidal organisms—tidal exposure, degree of exposure to wave action, and the type of substrate—merit particular consideration for the rocky intertidal zone.

The degree of wave exposure a given habitat receives is due to a combination of the proximity of offshore protection, the direction of the prevailing waves, and the geographic orientation of the habitat. For example, along the coast of central and northern California and Oregon, there are no large islands and few offshore submarine banks to intercept waves. Waves approach the shoreline primarily from the northwest in the summer and from the west and southwest in the winter. Therefore, a rocky headland facing the west will experience the full brunt of the heavy Pacific waves. The only organisms capable of living here will have to be able to attach and grow under the onslaught of the pounding waves. Behind the headland, a small sheltered cove that faces southeast will experience decidedly less wave action and many, more fragile organisms may survive and flourish here.

Rocky intertidal substrates vary in their composition. Remember that the basic materials making up the rocky coastline you see today probably had their origin in a previous geological epoch. The rocks were most likely formed under vastly different circumstances compared to their present situation (McPhee 1993). Some habitats are hewn from sturdy igneous rocks that originated from fiery volcanic activities. Other rocky habitats consist of sedimentary rocks, formed by pressure on the compacted layers of sands, silts, and clays that once formed the bottoms of coastal embayments or estuaries, or the conglomerate rocks composed of compressed glacial till. Sedimentary rocks like limestone, sandstone, siltstone, and mudstone are softer and provide less secure attachment sites for animals like barnacles and mussels. These substrates are eroded by wave action in different patterns than are the igneous rocks. Sedimentary rocks are also more easily penetrated by boring organisms, which in turn weaken the rocky substrates and influence the way they will erode.

Very hard, dense, igneous rocks like basalts will resist erosion and tend to weather evenly. Sedimentary rocks are more easily broken up by wave action, and often intertidal boulder fields will be found, consisting of broken pieces of the reef surface. Likewise, the weathering of coastal cliffs can contribute loose rocky material to the intertidal zone at their base. In places where different types of rocks intergrade, an uneven pattern of erosion can result. Wave-cut surge channels, tunnels and caves, shallow and deep tide pools, upraised outcroppings, fringing tidal reefs, and even towering headlands and sea-stacks, so familiar to Pacific seascapes, all can be formed from differential erosion patterns.

The extent of the rocky intertidal zone depends on the slope of the wave-cut bench. This slope is again related to the type of rock and the immediate geography of the area. Rocky intertidal areas are often found at the base of steep, rocky cliffs. These areas tend to be likewise steep with vary narrow, vertical intertidal habitats. Other rocky intertidal areas occur on broad, wave-cut terraces. These tend to have extensive rocky intertidal areas with a very gradual slope.

The result of the interaction of wave and substrate is a rocky intertidal habitat that can vary considerably from place to place.

The reef may be simple bedrock with little diversity of habitat. In contrast, it may consist of a mix of flat areas strewn with algae and boulder fields, upraised substrate, tide pools, and surge channels and contain a myriad of microhabitats for organisms to inhabit. The extensive rocky intertidal of the Monterey Peninsula is a striking example of such habitat diversity.

TIDAL ZONATION SCHEME

The rocky intertidal habitat is really several subhabitats stacked vertically on one another along a gradient of tidal exposure. The rocky intertidal zones, as suggested by the pioneer marine biologist Ed Ricketts, are the standard utilized to delineate the intertidal habitats found along the coast of the northeastern Pacific (Ricketts et al. 1985). He provided a general scheme of intertidal zonation for rocky shorelines in which each zone is described separately. These zones and the tidal elevations relative to MLLW (mean lower low water, the zero point of the tide tables) that they typically encompass in California and Oregon are:

1. the high intertidal zone, which includes the uppermost area wetted by the sea down to 5 ft above MLLW
2. the upper intertidal zone, which includes the tidal elevations from 5 ft to 2.5 ft above MLLW
3. the middle intertidal zone, which extends from 2.5 ft down to 0 ft MLLW
4. the low intertidal zone, which extends from zero ft MLLW down to the lowest level the tides reach

Exposure to wave action, tidal level, presence or absence of standing water, slope, rocky substrate type, and erosion pattern, and many other factors contribute to the make-up of these zones. Therefore, although these zones are delineated separately, you will observe that they are seldom discrete units.

The pattern and causation of rocky intertidal zonation are thoroughly described in Ricketts et al. (1985), and an attempt to recreate that description here would be redundant. However a brief review of the main subhabitats of the rocky intertidal zone is appropriate. These are the supralittoral splash zone, high tide pool, exposed-rock surface, and mussel clump subhabitats of the high, upper, and middle intertidal zones, and the open reef flat, under-rock, low tide pool, and surge channel subhabitats of the middle and low intertidal zones. Although these descriptions are presented separately, you will observe that they often intergrade with and overlap one another. Similarly, many invertebrates will be present in several different subhabitats, while others will be unique to only one.

SUPRALITTORAL

An easily accessed, extraordinarily interesting, and yet rarely studied marine habitat is the uppermost fringe of the intertidal shore, submerged by the highest tides or by storm action. This facies has been called the supralittoral zone, the wrack line, the drift line, the maritime zone, and the strand line. The supralittoral, an area hardly more than 1 or 2 m in width and bounded by an inhospitable world above and below, is an ecotonal habitat that suffers from low subscription by either terrestrial or marine biologists. This habitat has been subjected to extensive obliteration, having been replaced in many regions by sea-

walls, marinas, riprap, parking lots, beach tourism development, housing developments, and so forth.

Marine invertebrates living in the supralittoral occur under rocks, stones, and drift materials; in recent centuries, human-generated debris has added to the beach habitat. Most animals living here are adapted to the vagaries of both terrestrial conditions (e.g., desiccation and exposure to fresh water) and marine conditions (e.g., immersion and exposure to salt). A variety of terrestrial invaders, such as spiders, may co-occur with truly halophilic species as well.

A window into this world can be had by placing "pit traps" on the beach at dusk and retrieving them at dawn. A simple trap can be made from a large coffee can with a funnel fitted at the top (into which the animals may fall but generally cannot exit) and a finer-gauge mesh secured at the bottom for drainage. The can is placed flush with the surface. Quantitative and experimental work can be conducted by anchoring "litter bags" (filled with different drift materials and of different mesh sizes) on the upper shore.

The supralittoral biota differs on high and low energy shores. On more exposed coasts, talitrid amphipods (the sand hoppers or beach hoppers), oniscid isopods (especially the sand-burrowing *Alloniscus*), the pseudoscorpion *Garypus californicus*, mites, geophilid centipedes, and a host of other arthropods (e.g., the silverfish *Neomachilis halophila*, the seaside earwig *Anisolabis maritima*, and a superb assortment of staphylinid beetles and coelopid and helcoymizid kelp and sea-beach flies) are characteristic elements along the Californian and Oregon coasts.

On the lower-energy shores of bays and estuaries, the semi-terrestrial nemertean *"Pantinonemertes" californiensis* occurs under logs and stones that turn over infrequently. Look for small white worms beneath damp wood; these are likely enchytraeid "oligochaetes." An often overlooked polychaete worm in this habitat is *Namanereis pontica*. Oniscid isopods of the genera *Armadilloniscus* and *Detonella*, talitrid amphipods (a mixture of open-coast and quiet water species), the pseudoscorpion *Halobisium occidentale* (also found under cobblestones on the open coast), geophilid centipedes, mites, and a variety of insects co-occur in these quieter water high-shore communities, as do the snails *Myosotella myosotis*, *Assiminea californica*, and *Littorina subrotundata*, which are also common in salt marshes.

SPLASH ZONE

The splash zone subhabitat, called the uppermost horizon or Zone 1 by Ricketts, is the region covered only by the highest tides and the narrow strip above the high-water mark that is still influenced by the sea, primarily by the splash from waves and windborne sea spray. This is a zone of transition between the land and the sea, and many of the organisms dwelling here take advantage of aspects of both environments. Here you will often find lichens at the top of the uppermost tidal horizon. Prominent here also are algae, such as the enteromorphine *Ulva* spp., that depend on the sea spray and fresh water seeping out of cliffs that often back rocky intertidal zones.

A few hardy rockweeds (e.g., *Fucus* sp.) hang on in the lower part of this subhabitat. Other algae grow here, such as small blue-green algae and diatoms that grow close to the rock surface and are not readily visible. These microscopic plants are most important to the few invertebrates that live here—primarily marine snails that feed on the thin algal film growing on the rocks (e.g., the periwinkle *Littorina keenae* and the limpets *Lottia digitalis* and *Lottia scabra*).

A semi-terrestrial isopod, *Ligia pallasii*, known as the rock louse, treads the fine line between air and water in this zone. A more obvious crustacean, the green-lined shore crab *Pachygrapsus crassipes*, is sometimes found here tucked into crevices.

HIGH TIDE POOLS

Tide pools are considered by many to be the most interesting of all the intertidal subhabitats. Tide pools are formed by depressions in the rocky substrate that trap water at low tide and thus provide a refuge from desiccation. Therefore tide pools may contain a more diverse and often different association of organisms than found on adjacent, exposed substrate. Tide pools are not without physical stresses, however—especially during low-tide periods. A small volume of seawater trapped in a tide pool can experience a critical elevation in temperature on a warm, sunny day. Similarly, evaporation caused by wind and sun can raise the salinity to a sometimes dangerous level. Conversely, a very cold day can cause a decrease in water temperature during low tide, and a rainy low-tide period may cause a reduction in salinity.

The change in physical factors experienced within a tide pool is related to the tidal level of the pool, its shape, and the volume of water it contains. A small, shallow pool located in the high intertidal zone would experience the greatest fluctuations, while a large, deep pool in the low intertidal zone would experience the least change. Therefore, instead of visualizing tide pools as a single subhabitat, it becomes obvious that they constitute a continuum of subhabitats, depending on the particular situation of the individual pool. With this in mind, two tide pool subhabitats are described here.

The first, called the high tide pool, encompasses the tide pools of the upper, high, and middle tidal zones, from approximately +9 to 0.0 ft above MLLW. In all but the highest pools, some form of coralline algae will appear. Because of their crusty texture, coralline algae are tough fodder for most intertidal herbivores.

The industrious hermit crabs (*Pagurus* spp.) can be seen moving around the pool during the day and night. Other rapidly moving, but less-often seen, animals are small shrimp, *Heptacarpus* spp. These small shrimp (2.5 cm long or smaller) can be quite numerous in lower pools. A variety of carnivores occur in tide pools, including the stationary hunters, the anemones (*Anthopleura elegantissima, Anthopleura xanthogrammica,* and *Epiactis prolifera*) and sea stars (*Leptasterias* spp., *Dermasterias imbricata, Patiria miniata,* and *Pisaster ochraceus*).

If a mid-level pool is fairly deep and contains small boulders and a little sediment, another group of invertebrates may occur. In this under-rock habitat, look for crabs (e.g., *Cancer antennarius* and *Lophopanopeus bellus*) and small, delicate brittle stars (*Amphipholis squamata* and *Amphiodia occidentalis*). Look on the undersides of the rocks for motile invertebrates (e.g., flatworms and the isopods *Idotea* spp. and *Cirolana* spp.) and attached (spirorbid worms and the rock scallop *Crassadoma gigantea*) and sedentary species (e.g., chitons and juvenile *Strongylocentrotus purpuratus*).

If there is a substantial amount of fleshy algae in a pool, searching through it will usually produce a swarm of small amphipod crustaceans and a number of small snails and perhaps the spidery-looking kelp crab, *Pugettia producta*.

EXPOSED ROCK/OUTCROPPINGS

Often, portions of the rocky intertidal reef rise up above the flat bedrock of the reef face. These upraised areas are called out-

croppings, and they provide rocky surfaces exposed to desiccation more than adjacent flat reef areas. Similarly, large boulders that remain relatively stationary also provide this elevated subhabitat. In unprotected intertidal areas exposed to considerable wave action, this upper exposed rock surface subhabitat grades into the mussel clump subhabitat as subtle changes in tidal height and the degree of wave exposure occur. Thus, many species occur in both subhabitats. In general, the exposed rock subhabitat is higher in tidal elevation and thus more subject to drying than is the mussel clump. The main line of demarcation between these two subhabitats represents the upper physiological limit of the California sea mussel, *Mytilus californianus*, which provides the superstructure of the mussel clump subhabitat.

The most abundant invertebrate found on these exposed rocks often is the volcano-shaped, acorn barnacle. The small barnacles found highest on the outcrop are *Balanus glandula* and *Chthamalus dalli*, the most desiccation-resistant of the common barnacles. A larger species, *Semibalanus cariosus*, is found lower on the sides of the outcrop mixed with *Balanus glandula* and *Chthamalus*. A fourth barnacle species, the stalked barnacle *Pollicipes polymerus*, will also be found in sheltered cracks on the outcrop but reaches its peak abundance in mussel clumps. The ribbed and digit limpets, *Lottia scabra* and *Lottia digitalis*, share the highest elevations of the outcropping with the barnacles.

In the lower, more shaded portion of the exposed rock subhabitat, the abundant, stationary acorn barnacles provide ready prey items for several carnivorous snails: the emarginate dogwinkle or whelk, *Nucella emarginata*; the angular unicorn, *Acanthinucella spirata*; and circled dogwhelk, *Ocinebrina circumtexta*. These barnacle-eating snails share the lower outcropping with several species of herbivorous limpets (the shield limpet, *Lottia pelta*; the file limpet, *Lottia limatula*; the plate limpet, *Lottia scutum*; and the owl limpet *Lottia gigantea*). The shore crabs *Hemigrapsus nudus* and *Pachygrapsus crassipes* can also be found in this exposed rock subhabitat, holed up in cracks and crevices.

MUSSEL CLUMP

Large beds or clumps of the California sea mussel *Mytilus californianus* are common all along the open coast of the northeastern Pacific. The mussel thrives in areas of high wave energy, and indeed, its distribution corresponds to the most exposed, wave-tossed rocky intertidal habitat. The mussel's upper distribution is limited by its physiological tolerance to exposure. In areas of consistent wave action, the mussels can inhabit high intertidal zones successfully because of the wave splash, but in areas of periodic calm, their upper limit is the middle intertidal zone. Another control of the mussel is the Pacific sea star, *Pisaster ochraceus*. *Pisaster's* predation on the mussel is sufficient to preclude it from moving into and monopolizing the lower intertidal zone in the way it can dominate the middle intertidal zone. Thus *Pisaster* keeps the lower substrate open for other species to colonize and inhabit, allowing for a more diverse assemblage of organisms attached to primary space than in the mussel clump subhabitat (Paine 1974).

In addition to *Pisaster* and *Mytilus*, several other prominent organisms inhabit the mussel clump. The stalked barnacle, *Pollicipes polymerus*, occurs in round aggregations, sometimes surrounded by mussels. The solitary giant owl limpet, *Lottia gigantea*, is a conspicuous loner compared to the "togetherness" of *Mytilus* and *Pollicipes*, with individual *L. gigantea*

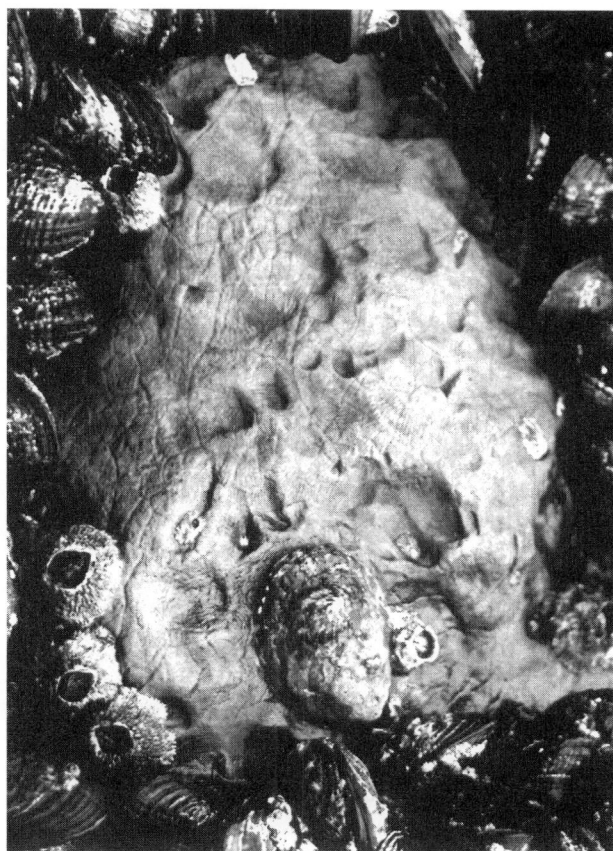

FIGURE 3 A grazed-out territory created by the limpet *Lottia gigantea* (photo by Thomas Niesen).

actively maintaining territories in the middle of a dense mussel clump (fig. 3).

The surface of the mussels' shells also serves as available living space for acorn barnacles, numerous small limpets of several species, and anemones. With this abundance of small invertebrate prey, it is not surprising to find that the small, predatory six-rayed sea star, *Leptasterias* spp., and the barnacle-eating whelk, *Nucella emarginata,* occur here. These are only the larger, obvious animals of this subhabitat.

Close scrutiny of the clump will reveal myriad smaller, motile animals and encrusting forms that combine to form one of the most diverse intertidal assemblages. In addition to the shells of the mussels, the webs created by their collective attachment (byssal) fibers provide a maze of nooks and crannies for marine invertebrates to inhabit. Young sea urchins find shelter at the base of the clump. Worms are especially able to maneuver here, and a wide variety of species occurs. Prominent are the polychaetes (e.g., *Nereis vexillosa* and *Halosynda brevisetosa*), nemerteans, and flat worms. Crabs abound in the interstices of the mussel clump. Young of the shore crabs *Pachygrapsus crassipes* and *Hemigrapsus nudus* are common. Other common crabs are the porcelain crabs (*Petrolisthes* spp., *Pachycheles* spp.).

REEF FLAT

In many rocky intertidal habitats with a gradual slope, the middle and low intertidal zones can be quite extensive. In addition to the upraised rocks and outcrops, the middle and low intertidal zones can contain extensive stretches of flat, open reef and areas where small boulders have accumulated, called boulder fields. Both of these subhabitats grade relatively seamlessly between the middle and low intertidal zones (below 0.0 MLLW).

The flat, open areas of the middle intertidal zone are usually characterized by lawns of short-growing (5 cm–10 cm) red algae and are sometimes referred to as red algal "turfs." The flat reef has little relief to break up the sweeping surf, and consequently, the diversity of large invertebrates is relatively low here. The lush growth of red algae does provide ample fodder for herbivorous snails such as the turbans *Chlorostoma funebralis* and *Chlorostoma brunnea*. The trapped moisture makes a hospitable setting for clones of the aggregating sea anemone, *Anthopleura elegantissima,* which can be quite extensive. A layer of sand and small pebbles accumulates at the base of these plants, harboring many small invertebrates, including numerous worms, small snails, and a variety of crustaceans (amphipods and juveniles of the crabs *Pugettia producta* and *Cancer antennarius*). Interspersed among the algal turf are occasional large invertebrate predators (e.g., the sea stars *Pisaster ochraceus* and *Dermasterias imbricata*).

SURFGRASS

The red of the algal turf is occasionally broken up by the bright green of the surfgrass *Phyllospadix*, of which three species occur in our region (*P. scouleri, P. torreyi,* and *P. serrulatus,* the latter as far south as Cape Arago in southern Oregon). Unlike the algae, which attach to bare rock with their holdfasts, surfgrass is a flowering plant that sinks its roots into soft sediment. Surfgrass takes up nutrients necessary for continued growth principally through the root system, while algae absorb nutrients directly from the water across their entire surface. The root mass of the surfgrass traps sediments and provides a living space for many small invertebrates, such as sipunculans, polychaetes, and small crustaceans. As sediments accumulate, the surfgrass sends out rhizomes trapping still more sediment and allowing it to increase its coverage.

The lower intertidal zone remains submerged much longer than the middle intertidal, and surfgrass can grow quite lush here as a result. Extensive surfgrass beds that cover not only the solid rock substrate, but also small boulders and sandy areas as well, are common sights in many rocky intertidal areas of the California and Oregon coasts. Surfgrass is a favorite low-tide resting spot of large rock crabs, *Cancer* spp., and mated pairs can often be discovered in the spring. Hermit crabs, *Pagurus* spp., and kelp crabs, *Pugettia producta,* are also common. Although they probably spend low tide under or among adjacent rocks, octopuses are occasionally seen during low tide seeking the refuge of the surfgrass bed, especially if it is still partially covered by water. The surfgrass beds and the standing pools of kelp like oar-weed, *Laminaria* spp., left by the lowest of the low tides often harbor nudibranchs among the vegetation (common are *Hermissenda crassicornis, Aeolidia papillosa, Peltodoris nobilis, Doris montereyensis, Diaulula sandiegensis, Triopha catalinae,* and *Okenia rosacea*).

BOULDER FIELD

In low-lying regions of the middle and low rocky intertidal, small boulders will accumulate, forming boulder fields. In open, exposed rocky intertidal habitats, small boulders are frequently moved around and broken up, or swept off the reef platform entirely by wave action. Therefore, boulder fields are more characteristic of rocky intertidal areas that are semi-protected from direct wave action. Here rocks of dinner plate size and larger stand a good chance of staying in place for sustained period of time. Thus the boulder field subhabitat provides a relatively protected, stable, under-rock environment that supports a rich and varied assemblage of invertebrates. Look for rocks that have well-developed algal growth. This is an indication that they have remained in place for some time and will constitute fruitful searching. The common seaweed species include the flat-bladed *Mastocarpus* spp., *Chondracanthus* spp., and *Mazzaella* spp. growing in profusion.

The under-rock habitat is prime crab-viewing territory for the student. Among the most prominent crabs are the purple shore crab, *Hemigrapsus nudus*; the large, red rock crabs, *Cancer antennarius* and *Cancer productus*; and the small pygmy cancer crab, *Cancer oregonensis*. Several spider crabs are also found here (*Pugettia* spp., *Mimulus foliatus*, and *Scyra acutifrons*). Also, hermit crabs and hordes of the flattened porcelain crab *Petrolisthes* spp. are common under rocks.

If there is any sediment beneath the boulder, look for sipunculan worms, shallow-burrowed polychaete worms, brittle stars (*Ophiopteris papillosa*), and small sea stars (*Leptasterias* spp. and *Henricia* spp.). Larger sea stars can be occasionally seen on the tops and sides of rocks (*Patiria miniata, Pisaster ochraceus, Pisaster giganteus,* and *Pycnopodia helianthoides*). Sea cucumbers are very successful dwellers of the tight under-rock quarters (*Eupentacta quinquesemita* and *Cucumaria miniata*); small (<2 cm in diameter) purple sea urchins, *Strongylocentrotus purpuratus*, also occur here. Among the rocks of the middle and low intertidal zones, a number of shelled gastropods can be found (*Haliotis rufescens, Diodora aspera, Fissurellidea bimaculata, Acmaea mitra, Ceratostoma foliatum, Lirabuccinum dirum, Chlorostoma* spp., *Calliostoma ligatum,* and *Amphissa versicolor*).

Attached to the bottom of the rocks will be sea anemones (*Epiactis prolifera*), the calcium carbonate tubes of filter-feeding polychaete worms (spirorbid worms and *Serpula columbiana*), and the intertwined shells of vermetid gastropods (*Serpulorbis squamigerus*). Also seen attached to the under-rock surface are attached bivalve mollusks (*Crassadoma gigantea, Pododesmus macrochisma*), and a variety of chitons (*Mopalia* spp., *Tonicella* spp., *Placiphorella velata,* and *Lepidozona mertensii*).

The rocks and boulders found in intertidal boulder fields are derived from a variety of sources. Some material may come from the weathering of cliffs that abut the intertidal zone. Much rocky debris may be thrown onto the reef from the subtidal zone by wave action. However, the majority of rocks are often composed of pieces broken from the solid reef substrate itself, especially if the reef is made up of soft, sedimentary rock like shale, sandstone, or siltstone.

If the rocks of the boulder field are soft, the student might notice round holes in them. These are the burrows of the rock-boring bivalve mollusks known as pholads or rock piddocks (e.g., *Penitella penita, Zirfaea pilsbryi*). The burrows excavated by these unique clams appear round from the surface and conical when viewed from the side. These clams riddle the soft, rocky reef in the mid- and lower-intertidal zones until it becomes unstable and pieces break away under the pounding surf. When the boring clams die, the now vacant burrows become the home of a variety of other animals (sipunculans, polychaetes, snails, small sea stars, nestling clams, sponges, bryozoans, and the lumpy porcelain crab, *Pachycheles rudis*).

LOW TIDE POOL

Low intertidal pools are unique subhabitats. The nature of the rocky substrate has to be such that dished-out areas occur in the low intertidal region, and the exposure to wave action plays an important role, as well. Therefore students won't find tide pools at every rocky area they visit, and of course, viewing them requires that the tide for the day be below the zero tidal level (MLLW). However, they are easily recognized for the unique group of organisms they support.

The first and foremost inhabitant is the purple sea urchin, *Strongylocentrotus purpuratus,* one of the most common marine animals along the open coast. At very low tides (−1.0' MLLW or lower), large numbers of urchins can be found exposed on flat, seaward portions of the lower reef. In both this exposed, lowest intertidal region and in the low tide pools, the urchins occur in rounded depressions or pits, thought to be excavated by the urchins using their spines and their special, five-jawed chewing apparatus called Aristotle's lantern. Look for small purple-colored invertebrates, including crustaceans and flatworms, in these urchin holes.

In the low tide pool, the dark purple of the urchins is contrasted against the lighter pastel purple and pinks of coralline algae, abundant here because their calcium carbonate–impregnated tissues are too hard for the urchins to graze. Other bright spots of color on the walls and floor of the low tide pool are provided by sponges and brightly colored cnidarians (*Balanophyllia elegans, Corynactis californica, Epiactis prolifera, Anthopleura xanthogrammica,* and *Urticina* spp.). Tube-dwelling polychaetes are also common here (*Dodecaceria fewkesi, Serpula columbiana*). Several mollusks contribute to these colorful tide pools: the lined chiton, *Tonicella* spp.; the dunce cap limpet, *Acmaea mitra*; the beaded top snail, *Calliostoma annulatum*; and the rock scallop, *Crassadoma gigantea*, distinguished by its bright orange mantle and blue eyespots that are visible when the shell gapes open.

During the quiet morning, minus low tides of spring and summer, the low tide pools feature the showiest of all the mollusks, the nudibranchs, or sea slugs. It is in the low tide pool, with its smaller area and beautiful background colors, that nudibranchs can be most appreciated. Because the pools harbor sponges, cnidarians, and bryozoans that serve as food for sea slugs, it is not unusual to find several species in a single pool. Look among the coralline algae and on the surface of the pool itself for nudibranchs crawling along upside down, using the surface tension created at the air-water interface as a foothold.

Sea stars are seen in the low tide pools. The Pacific sea star *Pisaster ochraceus*; sea bat *Patiria miniata*; leather star, *Dermasterias imbricata*; and the sunflower star *Pycnopodia helianthoides* are all occasionally discovered here. Smaller species of sea stars more commonly seen in these pools include small blood stars, *Henricia* sp., and six-rayed sea stars, *Leptasterias* spp. These two small stars often sport mottled colors that blend in with the coralline algae.

SURGE CHANNEL

Surge channels are formed by the differential weathering of the reef platform by the ocean. They are sometimes cut below the tidal level and thus never completely drain, even during the lowest tides. These submerged channels are typically at the very edge of the reef and support large stands of the oar-weed kelp, *Laminaria* spp., and other kelps on the bottom, as well as feather-boa kelp, *Egregia menziesii*, on the sides. Other surge channels extend well into the reef, in some cases reaching up into the mid-tidal level and above.

Besides tidal level, another variable in the surge channel subhabitat is orientation. Channels that extend directly into the reef, essentially perpendicular to the reef's edge, receive the direct force of the waves. Surge channels that turn to parallel the reef's edge are quieter and receive a somewhat less forceful flow of water. Finally, the shape of the surge channels must be considered. Channels with straight sides tend to harbor a more meager cast of organisms than do channels that have substantial undercutting and overhangs. Channels with dished out bottoms tend to trap small boulders that bounce around and scour the walls, while channels with bottoms that slope continuously seaward are swept free of such material by wave action.

The surge channel subhabitat is thus a varied one. What you discover in a given channel depends on all the variables listed above and a number of others, including the time of the year. One thing the student should remember about surge channels: they are high-energy environments. They require that organisms be able to attach and hold on against the movement of strong water currents. They are also food-rich environments, in that the surging water contains many small organisms and organic detritus swept from the reef and brought in from offshore. It is not surprising then that many surge channel animals are attached filter-feeders that take advantage of this waveborne bounty. A final note of caution: the same high energy that characterizes the surge channel can catch you off guard! Watch out for waves. These are fascinating areas to explore, and you are more often than not bent over or on your hands and knees. Have a lookout watch for incoming swells and unannounced surges of chilling seawater. Remember the surge channels are the avenues through which the tide floods into the intertidal zone. Be alert.

The organisms occurring along the top of the surge channel walls reflect the general organismal association for the particular tidal height and exposure. Thus, some walls are relatively barren, others are cloaked in thick algal growth, and still others support the spill-over from a well-developed mussel clump. However, it is in the shade of deeply undercut or overhanging walls in the lowest intertidal that the surge channel subhabitat achieves its glory. These strongly undercut habitats are also found along the low intertidal, seaward edge of some semi-protected reefs. When you first observe one of these well-developed, low intertidal surge channel overhangs, you will be taken by the variety of colors, shapes, and textures.

Because the overhang is in deep shade, only the hardiest encrusting coralline algae will occur, and these are in competition with a variety of space-monopolizing, encrusting marine invertebrates. The roof of the overhang often has sea anemones (*Epiactis prolifera* and *Anthopleura elegantissima*) and hydroids (*Aglaophenia latirostris*, *Pinauay* sp., *Abietinaria* spp.) hanging down and, occasionally, large barnacles (*Balanus* spp. and *Tetraclita rubescens*). The back wall of an overhang typically harbors several species of sponge growing in red, yellow, purple, brown, gray, and off-white sheets; compound and solitary tunicates, such as *Aplidium californicum* and *Styela montereyensis*; and clones of the asexually produced light-bulb tunicate, *Clavelina huntsmani*. Scattered among the other encrusting invertebrates are colonies of bryozoans such as *Eurystomella bilabiata*; solitary orange cup corals, *Balanophyllia elegans*; and the tubes of feather-duster worms (*Serpula columbiana* and *Eudistylia polymorpha*).

Invertebrates on the bottom of the surge channel overhang include more compound tunicates, sponges, and patches of the colonial polychaete, *Dodecaceria fewkesi*. The giant green sea anemone, *Anthopleura xanthogrammica*, is frequently found on the bottom of surge channels, waiting patiently for dislodged prey to be swept into its grasp.

Motile animals are also found in surge channels. Large cancer crabs, especially female *Cancer antennarius* brooding embryos, seek out cracks and crevices along the overhang walls. Several sea stars appear to use surge channels as avenues in and out of the intertidal zone and may be encountered here on occasion (*Pisaster* spp., *Pycnopodia helianthoides*, and *Dermasterias imbricata*). Six-rayed sea stars, *Leptasterias* spp.; blood stars, *Henricia* spp.; and sea bats, *Patiria miniata* appear to be more permanent residents.

Gastropods with tenacious grips are found in surge channels. *Chlorostoma* spp., *Calliostoma annulatum*, *Ceratostoma foliatum*, *Haliotis rufescens*, and *Diodora aspera* are among the common species usually seen. Nudibranchs also occur occasionally. The sea lemon *Peltodoris nobilis*; the ring-spotted dorid *Diaulula sandiegensis*; and the red sponge nudibranch, *Rostanga pulchra* are common species. Chitons are sometimes seen in surge channels, with the vividly colored lined chiton, *Tonicella* spp., being the most common.

Biogeography of the Oregonian Province

To a new student of marine invertebrate zoology, the diversity of the shallow water invertebrate fauna of the northeastern Pacific is initially staggering. To master the variety of types, let alone the names of the many species, seems impossible. However it can be done, and after becoming acquainted with the modern species, the student's curiosity often questions where all these invertebrates came from. The answer lies in the province of biogeography; as an example, here we discuss zoogeography, the study of the origin and distribution of animals.

Marine biogeographers look at the geographic distributions of all the members of a group, extinct and extant. Based on the patterns of co-occurrence they see, they determine the degree of relatedness among faunas. Biogeographers use the taxonomic hierarchy as their yardstick. For example, two areas harboring many of the same species probably were connected more recently than two areas that share few species but have many genera in common. For the northeastern Pacific biogeographers recognize a distinct cold-water province, the Arctic Province, that includes much of Alaska, and a warm-water or tropical province, the Panamic Province, which includes the Gulf of California and the tip of Baja California. In between these two is an area referred to as a cold-temperate region, including California, Oregon, Washington, and British Columbia. The cold temperate region is divided into several provinces—the Canadian, Oregonian, and Californian. This manual covers a good portion of the Oregonian Province, with Point Conception generally recognized as its southern boundary. There is less than unanimous agreement on the location of the northern boundary of the Oregonian Province, which has been designated as occurring as far south as central Oregon, to as far north as Dixon Landing, Alaska (Ekman 1953; Briggs 1974).

Before considering the modern distribution and relationship of these faunal provinces of the northeastern Pacific, some background is necessary. All shallow water marine faunas are related to the Tethyan fauna, which originated 190 million years ago when there was a single landmass known as Pangea.

When Pangea divided into northern and southern continental land masses, Laurasia and Gondwana, the Tethys Sea was formed in between as a shallow, trans-equatorial sea. The Tethys Sea is considered the evolutionary point of origin for all modern marine invertebrate groups. Remnants of this ancient tropical fauna and flora can be seen today in the circumtropical distribution of many families and genera. Anyone who has observed coral reefs in both the Caribbean and equatorial west Pacific cannot help but notice the striking similarity in reef faunas, especially among the conspicuous reef-forming corals and brightly colored reef fishes.

The Laurasia and Gondwana landmasses divided into smaller continents, which gradually aligned in their current north-south orientation separated by the Pacific and newly formed Atlantic Ocean basins. As this alignment developed, the Tethys Sea became divided, and the two isolated portions of Tethyan fauna evolved into two distinct shallow water tropical faunas, the Indo-West Pacific and the Atlanto-East Pacific. These names reflect the global consequences of continental drift and sea floor spreading. As the continents were moving about and realigning, the shallow water tropical marine habitats of the East and West East Pacific remained continuously separated by a large expanse of deep oceanic water. The Atlantic and tropical East Pacific were joined across the Isthmus of Panama and shared a similar fauna, the Atlanto-East Pacific. The Indian Ocean and the vast island arcs of the tropical West Pacific shared the Indo-West Pacific warm water fauna.

During the Tertiary (70 mya) global climate became colder, more water was tied up as ice at the poles, and as the ocean level dropped, land emerged. The Isthmus of Panama was closed by the formation of a land bridge dividing the Atlanto-East Pacific faunal region. The ocean temperature cooled with an equator-to-polar decrease in temperature resulting in distinct zones or "provinces" of water temperature that were to some degree physically separated. This physical separation set the stage for the formation of the modern shallow water marine faunal provinces along the continental coasts that we recognize today. But how were these modern shallow water provinces derived and how are they related to each other? Since the Tertiary, the global climate has been dominated by periods of glaciation and warmer interglacial periods. In the recent past, the coast of the northeastern Pacific has been influenced by the last ice age. About 12,000–15,000 years ago, much of the northern hemisphere was covered by a thick ice sheet. This ice sheet spread as far south as Puget Sound. The water sequestered in glaciers worldwide caused sea level to drop about 400 ft. This exposed much of the shallow continental shelf. As the ice sheet melted and sea level rose, many changes occurred along the coast. The advancing water weathered the soft sedimentary rocks of the Coastal Range, cutting flat marine terraces that typify much of our coast and providing much of the sediment that formed many of our sandy beaches and sand dune systems. Harder basaltic rocks that had been formed by volcanic processes and intruded into the sedimentary rock of the Coastal Range resisted erosion and became sea stacks and coastal headlands, such as Morro Rock and Point Sur. Coastal river valleys were flooded with the rising seawater. As sediment washed down from the mountains, it became trapped in these drowned river mouths to form estuaries such as Humboldt Bay.

This last Pleistocene glaciation was only one in a series of such events that occurred periodically over the past 100,000 years. These were periods of vigorous physical and biological disturbance and change that collectively have contributed most directly to the fauna we see today. For example, as sea-water temperatures have fluctuated along our coast over geological time, dynamic faunal changes have likewise occurred. During periods of warmer water, elements of the tropical Panamic fauna to the south have migrated northward. Conversely, when the water cooled, elements of the Arctic fauna moved southward. When the seawater temperature began to change again, these animals either adjusted to the new water temperature, retreated back to their respective provinces, or evolved into new species capable of existing with the new seawater temperature.

The effect of these dynamic changes in species distribution with seawater temperature can be seen by looking at the extant fauna in our Oregonian Province. There is an element of the warm water province present, albeit faint. The spiny lobster, *Panulirus interruptus*, which is seen occasionally as far north as the Monterey Bay subtidal, is a reminder of the subtropical influence. The relationship with the cold water faunal provinces is much more apparent. For example the echinoderm genera represented in our fauna, *Strongylocentrotus, Leptasterias*, and *Henricia*, all have their origin in the Arctic.

The list of cold water genera and even families of marine invertebrates represented in our Oregonian fauna is quite extensive, and it is easily understood why there is a stronger representation of cold water elements than warm. First, the cold waters of the southerly flowing California Current bathes the Oregonian Province. Second, the Oregonian Province experiences the strongest and most persistent spring and summer upwelling of the entire northeastern Pacific coast (Parrish et al. 1981, Bakun 1996). The cold, upwelled water inundates the nearshore, negating any summer warming of seawater temperature. Obviously these cold water influences would more strongly favor the success of cold water faunal invaders than warm.

Finally, we come to the fact that on average across all faunal groups, approximately half of the species of the Oregonian Province are endemic (i.e., evolved in our province). The explanation for this unusually high degree of species endemism (provincial endemism usually ranges from 10%–25%) is rooted in the dynamic scenario of glaciations and interglacials outlined above, along with strong seasonal upwelling. This glaciation/interglaciations account for waves of cold water invaders followed by periods of relative isolation and biological accommodation, while seasonal upwelling potentates the food chain with a high level of predictable primary productivity. This combination of highly transitory faunal distributions and high levels of available food energy fostered spectacular adaptive radiation and speciation across the range of invertebrate phyla (e.g., the decapod genera *Cancer* and *Crangon* and the gastropod genera *Haliotis* and *Lottia*).

In summary, the Oregonian faunal province that we deal with in this manual thus has a high degree of species endemism, with little affinity to the warm water region to the south. Although the ancient tropical faunas were the seed beds of evolution of modern marine invertebrates, the tropical influence on the Oregonian fauna is seen only at a distance via limited overlap at the generic level. This suggests that for some time the fauna of the Oregonian Province developed independently of the present day warm water fauna. What we can see in present day species and particularly genera found in the Oregonian Province is the high degree of overlap with the Arctic cold water fauna, which demonstrates the importance of the migration of northern fauna down the coast as temperatures cooled since the late Tertiary.

What does all this mean to today's student exploring the modern shallow water invertebrate fauna of the Oregonian Province?

Basically, it means when you explore a habitat in the southern end of the area covered by this manual, you will encounter some marine invertebrates that have the majority of their close relatives in the warm water provinces to the south. These animals usually also have a distinct northern limit to their distribution along the open coast. An example of this would be Kellet's whelk, *Kelletia kelletii,* a southern species seen in the shallow subtidal of Monterey Bay, but rare north of Monterey Bay.

Conversely, exploring habitats at the northern end of region covered by this manual will reveal a certain portion of the organisms with affinities to cold water relatives and distinct southern limits to their distribution. An example of this is the sun star, *Solaster dawsoni*. This sea star occurs in the rocky intertidal zone in Humboldt County but is only found subtidally south to Monterey Bay (Morris et al. 1980).

Remember, however, along with the warm and cold water species, you will find a much larger group of invertebrates that sweep along the entire coast from Alaska to Baja California, including such wonderful and charismatic species as the ochre seastar *Pisaster ochraceus* and the sea mussel *Mytilus californianus*—two of thousands of species that form some of the most spectacular marine provinces anywhere in the world.

References

Bakun, A. 1996. Patterns in the ocean: ocean processes and marine population dynamics. California Sea Grant College System, NOAA, 323 pp.

Barry, J. P., C. H. Baxter, R. D. Sagarin, and S. E. Gilman. 1995. Climate-related, long-term faunal changes in a California rocky intertidal community. Science 267: 872–875.

Briggs, J. C. 1974. Marine zoogeography. McGraw-Hill.

California Coastal Commission. 2003. California coastal access guide. 6th ed. Berkeley: University of California Press, 304 pp.

California Coastal Commission. 1987. California coastal resource guide. Berkeley: University of California Press, 384 pp.

Carlton, J. T. 1993. Neoextinctions of marine invertebrates. Amer. Zool. 33: 499–509.

Carlton, J. T. and J. Hodder. 2003. Maritime mammals: terrestrial mammals as consumers in marine intertidal communities. Mar. Ecol. Prog. Ser. 256: 271–286.

Carlton, J. T., J. B. Geller, M. L. Reaka-Kudla, and E. A. Norse. 1999. Historical extinction in the sea. Annu. Rev. Ecol. Syst. 30: 515–538.

Cohen, A. and J. T. Carlton. 1995. Nonindigenous aquatic species in a United States estuary: a case study of the biological invasions of San Francisco Bay and Delta. Report for the U.S. Fish and Wildlife Service and National Sea Grant Program, Connecticut Sea Grant. No. PB96-166525.

Ekman, S. 1953. Zoogeography of the sea. Sidgwick and Jackson, London 417 pp.

Koehl, M. and A. W. Rosenfeld. 2006. Wave-swept shore. The rigors of life on a rocky coast. University of California Press, Berkeley, 179 pp.

Larson, R. J., W. S. Alevison, T. M. Niesen, and S. L. Clark. 1997. Changes in fish communities off Santa Cruz Island, California, over the last 25 years: Effects of ocean warming and habitat change. Abstracts Amer. Soc. Ichthyologists and Herpetologists; 77th annual meeting; Seattle.

McPhee, J. A. 1993. Assembling California. Farrar, Straus and Giroux, New York, 224 pp.

Morris, R. H., D. P. Abbott, and E. C. Haderlie. 1980. Intertidal invertebrates of California. Stanford University Press, Stanford, California, 690 pp.

Parrish, R. H., C. S. Nelson, and A. Bakun. 1981. Transport mechanisms and reproductive success in fishes of the California Current. Biol. Oceanog. 1: 175–203.

Paine, R. T. 1974. Intertidal community structure. Experimental studies on the relationship between a dominant competitor and its principal predator. Oecologia 15: 93–120.

Ricketts, E., F. J. Calvin, J. W. Hedgpeth, and D. Phillips. 1985. Between Pacific Tides. 5th ed. Stanford University Press, Stanford, California, 652 pp.

Sagarin, R. D., J. P. Barry, S. E. Gilman, and C. H. Baxter. 1999. Climate related changes in an intertidal community over short and long time scales. Ecol. Monogr. 69: 465–490.

Tegner, M. J., L. V. Basch, and P. K. Dayton. 1996. Near extinction of an exploited marine invertebrate. Trends Ecol. Evol. 11: 278–280.

Intertidal Meiobenthos

JAMES W. NYBAKKEN AND ROBERT P. HIGGINS

In addition to the larger intertidal benthic invertebrates associated with sand and mud and visible to the naked eye that inhabit the intertidal areas of the Pacific coast, there are a large number of microscopic benthic invertebrates hidden in sediment. These organisms, which when sieved pass through a 1-mm-mesh sieve and are retained on a 62-μm-mesh sieve, are termed *meiobenthos* or *meiofauna*.

Meiobenthic invertebrates can be further divided according to their specific habitat: those living on the surface of the sediment at the water-sediment interface are referred to as *epibenthic*. Those living in sediment where they displace sediment particles in their movement (burrowing) are called *endobenthic*. Those living in the interstices of particulate matter are referred to as *mesobenthic* or *interstitial*. In such habitats as coarse shell-gravel, all three types may exist because fine particles may be present in all or part of the interstitial spaces. Note that the term *interstitial* applies to any organism living in interstices. Meiobenthic invertebrates, which are exclusively sand dwellers, may be referred to as *psammon* from the Greek term *psammos* (= sand). In such cases, the terms *epipsammon, mesopsammon,* and *endopsammon* are applicable.

In the marine environment, six animal phyla are exclusively meiobenthic: Placozoa, Kinorhyncha, Tardigrada, Loricifera, Gastrotricha, and Gnathostomulida. The meiobenthos includes the smallest members of 16 other phyla, such as Cnidaria, Mollusca, and Annelida, that otherwise are dominated by relatively large organisms. Some of these phyla, for example Brachiopoda and Sipuncula, may have only one or two species that can be considered meiobenthic, in contrast with the Arthropoda, Annelida, Platyhelminthes, and Nematoda, which have numerous meiobenthic representatives.

The meiobenthos is relatively well-studied in Europe and on the Atlantic coast of North America, but these animals are virtually unknown on the Pacific coast. We take this opportunity to strongly encourage students to enter this open field of research. That much remains to be known is revealed by our sampling of the shores of Monterey Bay, in central California, alone. Here we found many remarkable species, including at least three taxa previously unknown from the Eastern Pacific Ocean: the hydroid *Halammohydra*, the nudibranch *Pseudovermis*, and the tiny mystacocarid crustaceans.

Two essential texts for meiobenthic studies are Higgins and Thiel (1988) and Giere (1993).

Habitat

Meiobenthic diversity usually is highest in subtidal and, to a lesser extent, intertidal coarse sand, such as that found on high-energy marine beaches (fig. 4A), and lowest in the finer sediments, especially in fine sand. In terms of abundance, intertidal mudflats may have as many as 8 million meiobenthic organisms per square meter, which occupy only the upper few centimeters depending on the amount of available oxygen. The depth at which meiofauna are found in sediment is primarily dependent upon the oxygen gradient. In fine sediments as silt or mud, meiobenthos rarely penetrate more than a few centimeters of the substrate. In coarse sediments (e.g., medium sand, coarse sand, or shell gravel), especially where these sediments are part of a high-energy beach ecosystem, meiobenthos may be found as deep as 1 m or more (Kristensen and Higgins 1984). In general, they are not found in fine sand, but occasionally they may be present in the narrow stratum of silt or flocculent material at the fine sand-water interface.

Most meiobenthos will exist only in the oxygenated sediment. The depth to which meiobenthos penetrate, therefore, is also dependent on the particle size. In fine sediment, both intertidal and subtidal, only the upper few centimeters of substrate should be sampled; in coarse sediments, one should concentrate on sampling the upper strata (between 0–10 cm in subtidal sediment, 5–50 cm in intertidal high-energy beach habitat, and between 0–20 cm in other intertidal coarse sediments). The most likely habitat in which to find meiobenthos in both numbers and diversity is a stable mud or muddy-sand bottom where salinity is at or slightly below 34%.

Beaches and mud or sand flats exposed during low tide also require the meiobenthos to be euryhaline and eurythermal. Exposure to intense solar radiation as well as torrential rainfall at low tide require the meiobenthic organisms to be tolerant of wide variations of physical factors. Horn (1978), for example, found the kinorhynch *Echinoderes coulli* in intertidal mud in North Carolina with salinities ranging from 1 ppt to 42 ppt during a tidal cycle.

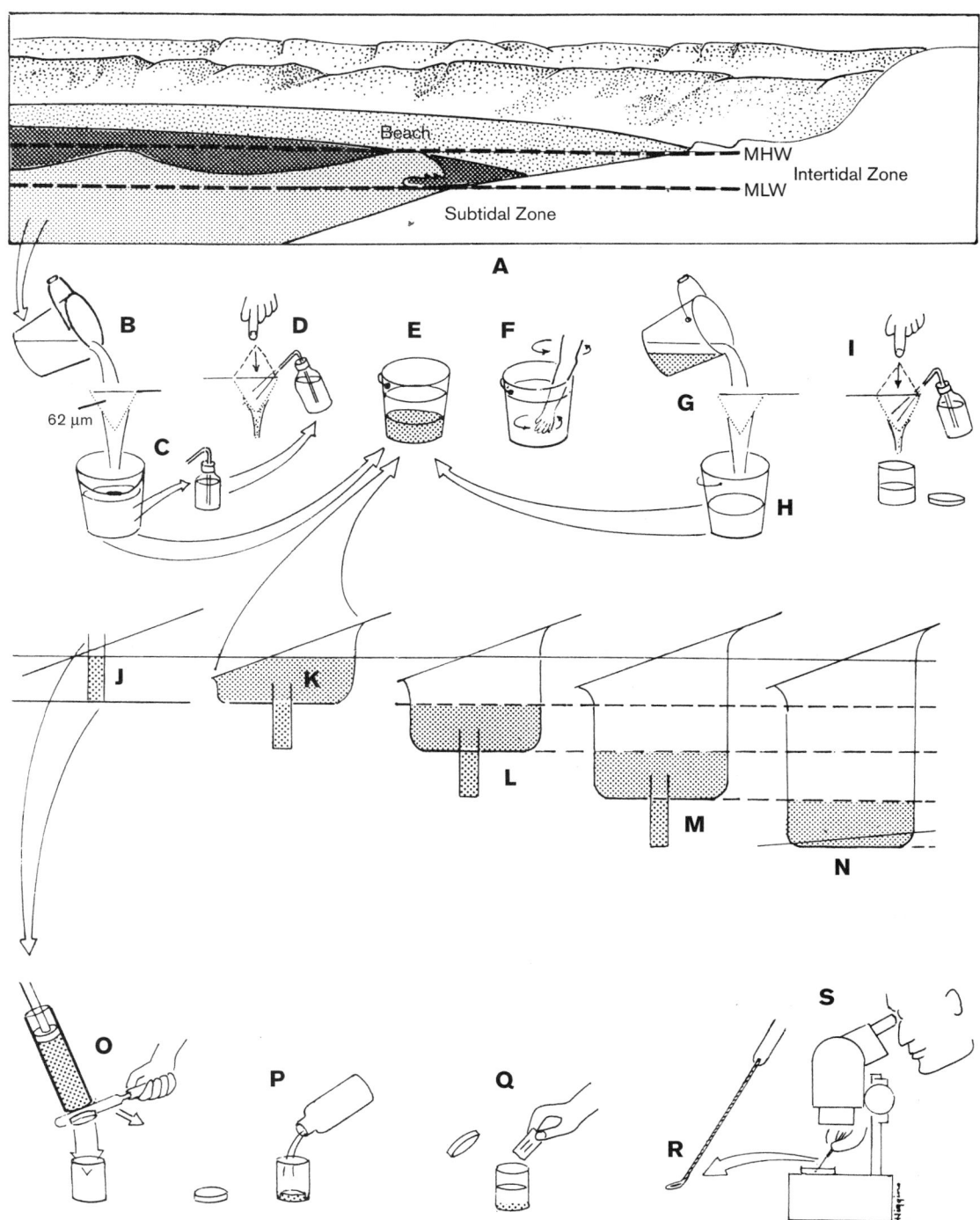

FIGURE 4 Collecting and processing techniques for sampling intertidal sand. A, Profile of typical beach showing intertidal zone marked by mean high water (MHW) and mean low water (MLW); B, obtain seawater and pour through 62-μm mesh sieve into, C, clean bucket; D, transfer some filtered seawater to wash bottle and "back-wash" (clean) the plankton from the sieve so as to avoid contamination in later steps; from a determined depth on the beach (e.g., the first 10-cm layer) remove sand (K) and place in bucket of filtered seawater (E); F, with hand or some item suitable for stirring, mix the sand with filtered seawater to suspend the meiobenthos; G, decant the water with suspended meiobenthos into a clean (D) 62-μm mesh sieve and catch filtered seawater for additional use; L–N, repeat this step, each time sampling the next 10 cm horizon until ground water is encountered (N); figures K–M also indicate coring procedure to qualitatively sample these same horizons; O, quantitative sample core may be subdivided into 1-cm layers if desirable; I, P–R, show how to preserve samples (either qualitative or quantitative) by (P) adding sufficient formalin to achieve a 5%–10% formalin fixation; Q, both material that is fixed or unfixed for live-sorting should have a label placed in the container; R, preserved or live material should be sorted with the aid of a stereomicroscope with 25×–50× magnification; S, specimens, both live and preserved, should be removed from the sample with the finest of Irwin loops, usually indicated by blue color of handle (see Higgins and Thiel 1988 for further information).

Meiobenthos are also often associated with plant material or even other organisms. It is not uncommon to find interesting meiobenthos associated with the holdfasts of macroalgae (Moore 1973), or even among the filaments of some of the smaller, attached algae. And meiobenthos have been found in colonies of bryozoans (Higgins 1977a), in the cavities of sponges (Higgins 1977b), and occasionally in the alimentary cavities of larger invertebrates (Martorelli and Higgins 2004) or fishes (Millward 1982).

Collecting and Processing Meiobenthos

A complete description of sampling meiofauna is found in *Introduction to the Study of Meiofauna* (Higgins and Thiel 1988). The first consideration is whether one is studying the quantitative (number of given taxon per unit volume, usually per 10 sq cm) or qualitative (taxa only) aspects of meiofauna.

INTERTIDAL QUANTITATIVE SAMPLING—MUD Intertidal quantitative sampling generally requires the use of a coring device. Several replicate cores are required to have sufficient statistical confidence. Cores are best taken using a plastic cylinder that has a cross-sectional area of 10 sq cm (see Higgins and Thiel 1988). The generally accepted procedure is to take a series of cores, slice off 1-cm-thick sections (fig. 4O), preserve the section using 5% formalin (fig. 4P), and add Rose Bengal if staining is preferred; if material is to be processed alive, pass the sections through a 62-µm mesh sieve to eliminate the finest fraction, and then place it in a dish and sort the organisms using a stereomicroscope with at least 50× magnification (fig. 4R). Some investigators pass each section through a series of graded sieves, but each sieving runs the risk of losing specimens, and the small amount of material from a core usually make this process unnecessary. Because meiobenthos rarely exist deeper than 4 cm in mud, make a series of test cores to determine the extinction level and then slice off only those strata known to contain meiobenthos.

INTERTIDAL QUANTITATIVE SAMPLING—SAND (figs. 4J–M, 4O–R). Because depth of oxygen penetration increases considerably in a sandy habitat, coring down to as much as 1 m may be necessary. It is nearly impossible to take a core in sand that will properly penetrate more than 10 cm without significant compaction. Thus, some investigators find it more useful to take a series of progressively deeper 5-cm or 10-cm cores (figs. 4K–M) and treat each of these as a unit for counting and identifying the meiobenthic components rather than slicing off portions (fig. 4O). After the first 5-cm or 10-cm sample has been extracted, we have found it best to carefully remove all surrounding sand from that zone and then proceed to sample the next stratum. This allows for sufficient replicate samples and, at the same time, facilitates deeper penetration until the extinction zone is reached. Each fraction is treated the same as the mud fractions described previously (figs. 4P–R).

INTERTIDAL QUALITATIVE SAMPLING—MUD Intertidal qualitative sampling can be done in muddy substrates; the meiobenthic zone rarely exceeds 3 cm deep. Because this procedure is intended to determine what taxa are present and to obtain some idea as to their relative abundance in the sample, most any device can be used to collect a 3-cm-deep sample. A spatula is an excellent tool for this. Depending on the purpose of the sampling, one can use a series of sieves or simply sieve whatever material necessary through the smallest sieve, usually one with a 62-µm mesh size. If material is to be examined alive, small amounts can be placed in a dish, which is then placed under a

stereomicroscope and sorted. Otherwise, the sample may be fixed and stained.

INTERTIDAL QUALITATIVE SAMPLING—SAND Sampling the sand from a high-energy beach is usually very productive. Equipment includes at least three clean plastic pails, a 62-µm mesh sieve, 0.5 liter or 1.0 liter wash bottle, and formalin if material is to be fixed in the field and sorted in the laboratory.

First, obtain a bucket of seawater (figs. 4A, B). Pour the seawater through the 62-µm mesh net into a clean bucket (fig. 4C). Use some of this filtered seawater to back-flush the sieve to remove any organisms (fig. 4D). Fill the wash bottle with filtered seawater. This procedure is necessary to ensure that there is no plankton contamination in the sample. In the area to be sampled, carefully remove the upper few centimeters of sand; this removes plankton that has been deposited in the uppermost sediment.

With a shovel or trowel, remove whatever amount of sand is desired (figs. 4K–N)—we prefer a volume of about 3 liters—and place it into a bucket half-filled with filtered seawater (fig. 4E). With a stirring device or simply by using one's arm, stir this mixture vigorously until all of the sediment is suspended (fig. 4F). Immediately pour off the water and suspended meiobenthos through the sieve (fig. 4G) and catch the decanted water in a third clean bucket (fig. 4H). Repeat this process at least five times, each time retaining the seawater that passes through the sieve. Depending on the amount of silt in the sand, it may be necessary to use new seawater source after a few sievings. If the filtered seawater becomes unusable because of an excess of silt in the sand, simply refilter a new bucket of seawater and continue the process.

After several such sample treatments, the material in the sieve should be carefully washed either into a container (fig. 4I) that can be used for live examination, or into a container that can be fixed (figs. 4P–R) in the manner noted previously. Live sorting of this material is often rewarding. If the sample(s) are to be transported back to the laboratory, keep the sample cool. Note that preserved samples of coarse sediment must be transported with great care. Samples transported by car over a bumpy road for any length of time literally "sandpapers" the meiobenthos, destroying most of the soft-bodied taxa and variously damaging the hard-bodied taxa.

Some meiobenthos cling tenaciously to the sand particles. A method developed by Higgins and Kristensen (1986) involves a technique called "freshwater shock." This method simply causes osmotic imbalance in the organisms when the sand sample is placed the first bucket, which now contains fresh water. The sample is treated identically with the first step outlined above, but it must be done rapidly, and the material collected in the sieve must be immersed carefully in a second bucket of filtered seawater after each treatment (an alternative is to rinse the material in the sieve with seawater from a wash bottle). When sieving the material, catch the fresh water in a labeled bucket so the procedure can be repeated successfully. When the sampling is completed, the material in sieve is washed into containers for live sorting, or fixed in 5% formalin.

MEIOBENTHOS ASSOCIATED WITH PLANTS OR OTHER ANIMALS Plant material, especially, can provide some interesting meiobenthos. Place the plant material in a plastic bag of filtered seawater and agitate. Then decant the seawater and (hopefully) detached meiobenthos through a series of sieves, or at least a 62-µm mesh sieve, and treat as above. A small amount of ethanol added to the seawater in the plastic bag may assist in narcotizing any attached meiofauna. An interesting assemblage of meiobenthos exists in macroalgae holdfasts. Just the holdfast needs be rinsed in the plastic bag. Similarly, meiobenthos also associate with bryozoan colonies, sponges, and other organisms and can be treated in a manner similar to that used for plant material. Certain meiobenthos, however, remain attached by specialized devices and will not be shaken free.

FLOTATION TECHNIQUE The so-called "flotation-technique" developed by Higgins (1964) effectively removes many hard-bodied meiobenthos (meiobenthos with cuticle) from subtidal mud (Higgins and Thiel 1988). *Subtidal* is the focal point: this technique is often ineffective when used to remove meiobenthos from intertidal mud. Such organisms are often adapted to resist being trapped in the surface tension of seawater. Without such adaptation, they would be removed by each incoming tide. The "flotation technique" is not quantitative, nor is it effective with all taxa in subtidal mud, but for organisms such as crustaceans, nematodes, kinorhynchs, and many others, bubbles passing upward through a mixture of mud and seawater cause the organisms to be trapped on the surface film.

Originally, this technique involved taking several liters of mud, preferably the uppermost layer, adding seawater (at least twice the volume of the mud), stirring the mud to make a soup-like slurry, and then pump air into the mixture using any pump device attached to an aquarium air-stone. We have determined that a quicker way to process this material is merely to lift the slurry mixture about a meter and pour the contents rapidly into a second, clean bucket. After either bubbling or pouring methods have brought the organisms to the surface film, gently place a piece of copy paper on the surface, quickly remove it, and, using a wash-bottle filled with filtered seawater, wash the material adhering to the paper into a fine-mesh net. Repeating this process—pouring or bubbling, blotting, and washing—one can often obtain an interesting assemblage of meiobenthos. This method has been used by Higgins for more than 50 years to sample Kinorhyncha, but it is excellent for ostracodes, amphipods, and many other crustaceans as well.

CENTRIFUGATION Investigators have found that the use of an artificial medium in conjunction with centrifugation may successfully remove many if not most of the intertidal mud-inhabiting meiobenthos. A review of the literature about meiobenthos extraction techniques will provide many suggestions, but we have found that the use of a sugar solution is very effective, inexpensive, and nontoxic. To prepare the solution, place enough sugar in a flask to reach the 1-liter mark. Then pour in boiling water so that after going into solution, the water level is at the 1-liter mark. After cooling, mix equal volume of the sugar solution with an equal volume of (drained) intertidal mud sample (untreated or fixed in formalin). Gently shake the mixture to suspend the particulate material, pour equal amounts into 50 ml centrifuge tubes and centrifuge the samples at relatively slow rate. The centrifuge speed and duration must be adjusted to the sample. The decanted material can be sieved through a 62-μm mesh using filtered seawater to re-establish osmotic equilibrium and then can be placed in a sorting dish or in container where a proper formalin concentration and Rose Bengal fixes and preserves the meiobenthos for later study.

THE IRWIN LOOP (fig. 4S). A final note in the processing of a sample. Sorting—that is, the removal of organisms from whatever medium is involved—should be done with Irwin loops (see Higgins and Thiel 1988). Using pipettes, especially with living material, is hazardous. Fixed or unfixed, these tiny organisms have a tendency to adhere to the inside of glass pipettes. The careful use of an Irwin loop can make a difference.

THE "MERMAID BRA" (figs. 4B, 4I). This useful and well-known sieve, made by the junior author's wife, is a sieve with a base diameter of about 28 cm made with 62-μm mesh nylon cloth. Seams are sealed with silicon caulking. The supporting loop is made from plastic coated "clothesline" wire. The circumference is obtained by bending the wire around the lid-groove of a U.S. 1 gallon paint can. The handle is made from plastic tubing of suitable diameter and length.

Taxonomic Survey of Meiobenthos

Placozoa

There are no reports of placozoans occurring on the Pacific Coast.

Cnidaria

Only a few cnidarians are known to be meiofaunal in size. Although all the classes are represented in the meiofauna, only the hydrozoans are represented by more than one species. The chapter by Mills and colleagues on hydroids notes that polyps of *Euphysa* are interstitial and that the tiny *Protohydra leuckarti* occurs in sandy mud in Puget Sound. The hydrozoan subclass Actinulidae includes meiofaunal species that are solitary, ciliated individuals presumably moving in the interstices by ciliary motion. We have found the actinulid *Halammohydra* sp. on a sandy beach near Moss Landing in Monterey Bay.

Platyhelminthes

"Turbellarian" flatworms are abundant in marine sediments of all kinds. They are often conspicuous members of the meiofauna. Turbellarians can usually be extracted by the seawater ice method or the magnesium chloride method. Holmquist and Karling (1972) described two interstitial species from Oregon and California (see the chapter on "Turbellaria" by Holleman).

Nemertea

Nemerteans in the genus *Ototyphlonemertes* may occur on intertidal sand beaches. See the nemertean chapter by Roe, Norenburg, and Maslakova.

Nematoda

Nematodes are exceedingly abundant in meiofauna samples and easily placed into the phylum because of their shape and characteristic writhing movement. Consult the chapter by Hope.

Gastrotricha

Gastrotrichs are not uncommon in sand beaches of Oregon and California. See the gastrotrich chapter by Hummon.

Loricifera

This phylum was described by Kristensen (1983) from specimens found along the Normandy coast. Eight other species representing two additional genera were found off the North Carolina coast, a tenth species was found at hadal depths in the North Pacific, and an additional species was described from shallow water off the coast of Italy. Additional specimens are known from many other places. One specimen was collected in the shallow water sand of a South Pacific reef, a larval loriciferan was found off the west coast of Panama, and others have been collected from varying depths at other localities around the world. Loriciferans thus appear to occur in a wide variety of sediment from shallow to hadal depths. The freshwater shock method is a preferred extraction technique but is not a requirement. Where general meiobenthic collections have been looked at more carefully, loriciferans have been found. Their abundance and distribution may depend more on the diligence and expertise of sorting than on their biology. Illustrations of an adult and Higgins-larva of an example of the two most abundant known genera are found in the Kinorhyncha, Loricifera, and Priapulida chapter.

Kinorhyncha

Most kinorhynch taxa are subtidal; few have been found intertidally, and only two species are known from the Pacific coast of the United States: *Cephalorhyncha nybakkeni* from medium to coarse sand at Carmel Beach, California, and *Echinoderes kozloffi* from a tidally exposed stony habitat at Friday Harbor, Washington. Although both of these represent the order Cyclorhagida, it is likely that some of the homalorhagid taxa, especially *Kinorhynchus ilyocryptus,* may be found on intertidal or shallow subtidal mudflats. Careful sampling of the region's tidally-exposed muddy creek sediments has a highly probable chance of yielding species of *Echinoderes* similar to the group typified by *E. coulli* Higgins, 1977a, found in these habitats along the southeastern coast of the United States and other similar habitats throughout the world. Some cyclorhagids are associated with algal holdfasts. Occasionally kinorhynchs, especially members of the genus *Echinoderes,* are washed up on beaches by heavy wave action. For additional details, including illustrations, see the Kinorhyncha, Loricifera, and Priapulida chapter.

Priapulida

Of the 19 described species of Priapulida, half are of meiobenthic size but only the Indo-Pacific *Meiopriapulus fijiensis* Morse, 1981, has been found intertidally in coarse shell-gravel beach sediment. The remaining meiobenthic representatives are restricted to subtidal habitats. Among the temporary meiobenthos (young stages), priapulids are represented in California and Oregon intertidal mudflats by the larval stages of *Priapulus caudatus.* See the Kinorhyncha, Loricifera and Priapulida chapter for additional details, including illustrations.

Gnathostomulida

On the Pacific coast gnathostomulids have been found in certain anaerobic sediments but are often not seen because they are difficult to extract from the sediments and are extremely fragile, often disintegrating before one can get them under a microscope. Another habitat to look for these animals is in the anaerobic sand around the roots of the surfgrass *Phyllospadix.* For further information, see the chapter on Gnathostomulida by Farris.

Rotifera

Rotifers are reported to be abundant in the meiofauna of fairly coarse marine intertidal sands generally in the top 2–4 cm. In our experience, however, they do not appear to be abundant on Monterey Bay beaches. See the Rotifera chapter by Fradkin, who notes that species in the genera *Philodina* and *Rotatoria* are to be expected.

Tardigrada

A number of tardigrade species have been reported along the Oregon and California coasts. These are treated in the chapter on Tardigrada by Pollock and Carranza.

Annelida

Tiny polychaetes (grouped, in the past, in a taxon called Archiannelida) are common inhabitants of the meiobenthos. Blake's chapter on Polychaeta covers the nine families of meiofaunal polychaetes found in our region.

Although oligochaetes are relatively abundant in marine and estuarine sediments, little attention has been paid to them on the Pacific coast. The chapter by Cook et al. on tubificid, enchytraeid, and randiellid oligochaetes should be consulted. Among the few regional interstitial species described are *Aktedrilus locyi, A. oregonensis,* and *Randiella litoralis.*

Crustacea

COPEPODA

Most crustaceans are large enough that they do not qualify as meiofaunal organisms as adults. However, there are a number of crustacean groups that have abundant meiofaunal representatives. Perhaps the most abundant meiofaunal crustaceans are the harpacticoid copepods and may only be second in abundance to the ubiquitous nematodes. The chapter by Cordell on free-living copepods provides an introduction.

OSTRACODA

Another abundant meiofaunal group is the Ostracoda. Some of these bivalved crustaceans are meiofaunal in size and may be abundant in sediments. See the chapter by Cohen et al., which provides further information on interstitial taxa.

ISOPODA

Interstitial isopods include *Caecianiropsis psammophila* from Tomales Point and Asilomar and *Coxicerberus abbotti* from the sandy beach at Hopkins Marine Station. See the chapter by Brusca et al. on Isopoda.

MYSTACOCARIDA

A final crustacean group is the Mystacocarida. There are only a few species in this class, and they have not been previously reported from Pacific coast beaches. However, the senior author has extracted specimens from a sandy beach at Moss Landing in Monterey Bay.

Pycnogonida

The interstitial *Rhynchothorax philopsammum* is recorded from central California. Joel Hedgpeth, in the second edition (1954) of this manual, noted that on the inner side of Tomales Point "it occurs several inches beneath the surface of the sand in association with several other forms, including harpacticoids, small holothurians and isopods."

Acari

Interstitial species in the genera *Scaptognathus, Anomalohalacarus,* and *Actacarus* in coarse sand are to be expected in our region. See the chapter by Newell and Bartsch.

Collembola

Interstitial collembolans may occur; see the chapter by Christiansen and Bellinger.

Mollusca

Although many mollusks are tiny and may in early life history stages be small enough to be considered meiofaunal, as adults they usually exceed the size range of meiofaunal organisms. Exceptions include the tiny tubular snails in the Caecidae and the 800-µm-tall scissurellid snail *Sinezona rimuloides,* both treated in the gastropod chapter by McLean.

Also found interstitially are microscopic opisthobranch slugs. Acochlidiacea are found interstitially in both marine and freshwater habitats. Robilliard and Kozloff (1996) report an undescribed acochlidiacean slug in sand in Puget Sound and around the San Juan Archipelago. We have found a species of the nudibranch *Pseudovermis* in coarse sand intertidally in Monterey Bay.

Other Phyla

Although meiobenthic representatives are known from other phyla including Echinodermata, Bryozoa, Brachiopoda, Sipuncula, and Tunicata (Urochordata), virtually all are found in subtidal habitats and even then, uncommonly.

References

Giere, O. 1993. Meiobenthology: the microscopic fauna in aquatic sediments. Berlin, Springer-Verlag, 328 pp.

Higgins, R. P. 1964. A method for meiobenthic invertebrate collection. Am. Zool. 4: 291.

Higgins, R. P. 1977a. Two new species of *Echinoderes* (Kinorhyncha) from South Carolina. Trans. Amer. Micros. Soc. 96: 340–354.

Higgins, R. P. 1977b. Redescription of *Echinoderes dujardinii* (Kinorhyncha) with descriptions of closely related species. Smithson. Contr. Zool. 248: 1–26.

Higgins, R. P. and R. M. Kristensen. 1986. Kinorhyncha from Disko Island, West Greenland. Smithson. Contr. Zool. 458: 1–56.

Higgins, R. P. and J. Thiel, eds. 1988. Introduction to the study of meiofauna. Smithsonian Institution Press, Washington.

Holmquist, C. and T. G. Karling 1972. Two new species of interstitial marine triclads from the North American Pacific coast, with comments on evolutionary trends and systematics in Tricladida (Turbellaria). Zool. Scripta 1: 175–184.

Horn, T. D. 1978. The distribution of *Echinoderes coulli* (Kinorhynchs) along an interstitial salinity gradient. Trans. Amer. Micros. Soc. 97: 586–589.

Kristensen, R. M. 1983. Loricifera, a new phylum with Aschelminthes characters from the meiobenthos. Zeitschrift für zoologische Systematik und Evolutionsforschung 21: 163–180.

Kristensen, R. M. and R. P. Higgins. 1984. A new family of Arthrotardigrada (Tardigrada: Heterotardigrada) from the Atlantic coast of Florida, U.S.A. Trans. Amer. Micros. Soc. 103: 295–311.

Martorelli, S. and R. P. Higgins. 2004. Kinorhyncha from the stomach of the shrimp *Pleoticus muelleri* (Bate 1888) from Comodoro Rivadavia, Argentina. Zool. Anz. 243: 85–98.

Millward, G. E. 1982. Mangrove-dependent biota, pp. 121–139. In Mangrove ecosystems in Australia. Structure, function and management. B. F. Clough, ed. Australian Institute of Marine Science, Townsville.

Moore, P. G. (1973). *Campyloderes macquariae* Johnston, 1938 (Kinorhyncha: Cyclorhagida) from the northern hemisphere. Journal of Natural History 7: 341–354.

Morse, P. 1981. *Meiopriapulus fijiensis* n. sp.: an interstitial priapulid from coarse sand in Fiji. Trans. Am. Micros. Soc. 100: 239–252.

Robilliard, G. A. and E. N. Kozloff. 1996. Subclass Opisthobranchia, pp. 232–258. In Marine invertebrates of the Pacific Northwest. E. N. Kozloff, ed. University of Washington Press, Seattle, 539 pp.

Intertidal Parasites and Commensals

ARMAND M. KURIS

$E = mc^2$. Without mass (m) there can be no energy (E). Although parasites, being often small and hidden, are usually out of our sight lines, they are sometimes present in impressive numbers, in critical locations, influence food webs, and sometimes have considerable biomass. Some intriguing parasites and commensals that demonstrate these aspects can be readily observed in our region in the marine intertidal zone.

Beyond their importance—and their medical and economic significance is well described in parasitology books—parasites are also interesting. Although the "degenerate" epithet is often tossed their way, some features of parasites are fascinating. They have highly evolved, distinctive adaptations, complex life cycles, and sometimes alter host behavior, physiology, and morphology.

Parasites are certainly small compared to the size of their hosts. Most parasites are less than 1% of the weight of their hosts (Lafferty and Kuris 2002). However, compared to their free-living relatives, parasites are quite large. The largest dinoflagellates are parasitic. So are the largest flatworms (compare even the biggest polyclad with a beef tapeworm) and nematodes (compare any free-living nematode with the human roundworm, *Ascaris lumbricoides*). Parasitic entoniscid isopods are large compared to most free-living isopods, rhizocephalans and whale barnacles are large compared to most filter-feeding free-living barnacles, and almost all parasitic copepods are much larger than free-living copepods.

Parasites are also long-lived. Although a free-living flatworm usually lives for weeks or months, at most a few years, human schistosomes may live for decades, larval trematodes in snails for more than a decade and some filarial worms may also live for many years.

A remarkable adaptation for many parasitic groups is hypermetamorphosis. Many marine invertebrates undergo remarkable morphological transformations (metamorphoses) to complete their life cycles. Adding a metamorphosis to a life cycle makes it hypermetamorphic. An intertidal crustacean parasite offers a striking example (Høeg 1995). A free-living barnacle hatches from an egg and completes two metamorphoses: nauplius to cyprid, cyprid to pinhead postlarval barnacle. This grows as a juvenile until it gradually reaches adulthood. For a rhizocephalan barnacle, such as *Heterosaccus californicus,* which parasitizes various California spider crabs, after the nauplius,

the cyprid, having found a crab, metamorphoses into the kentrogon, adding a metamorphosis to the typical barnacle life cycle. The kentrogon (a living hypodermic needle) then metamorphoses into a vermigon (a wormlike form) that is injected into the crab. The vermigon then metamorphoses into the rootlike interna, which finally metamorphoses to the adult when the virgin externa emerges from the cuticle of the crab to await fertilization from a male. This totals three additional metamorphoses compared to free-living barnacles. Meanwhile, the male rhizocephalan also adds some radical changes to complete its life cycle. Its cyprid attaches to the virgin externa and metamorphoses to a trichogon, a bristly stage that makes its way to a chamber in the female externa, molts again, and becomes little more than a clump of testicular cells (one additional metamorphosis compared to a free-living barnacle).

Behavior modification by parasites is increasingly being studied, although few associations with invertebrates have been investigated on the Pacific coast. To return to the rhizocephalans, infected males develop female secondary sexual characteristics (become feminized) and exhibit brood care behavior (grooming the parasites' externa instead of its own egg mass). Infected male crabs also make a breeding migration with normal females (Rasmussen 1959). Many other examples of parasite-induced behavior modification are described in Zimmer (2000).

Pacific coast invertebrates harbor some interesting and often species-rich parasite groups. These include larval trematodes in snails and clams, rhombozoans in cephalopods, pea crabs in bivalves, ciliates and turbellarians in echinoderms, and nemerteans, parasitic barnacles, parasitic isopods in crustaceans and copepods on many kinds of hosts. Beyond the scope of this guide, many marine fishes have rich parasite faunas including Myxozoa, Monogenea, Digenea, Cestoda, and Copepoda of diverse forms and habits.

Some Easily Observed and Abundant Parasites of Invertebrates

Certain snails are hosts for a diverse assemblage of larval trematodes. These block reproduction of their snail hosts and alter their growth and abundance. These larval trematodes vigorously compete for resources in the snail to the extent that dominant species eliminate subordinate species. These interactions have been extensively studied for the 18+ species in the native horn snail *Cerithidea californica,* abundant in the salt marshes of central and southern California. (Sousa and Gleason 1989, Kuris 1990, Lafferty 1993, Kuris and Lafferty 1994). Other snails that are often heavily parasitized include species of *Littorina, Callianax biplicata,* the introduced *Batillaria attramentaria* and *Ilyanassa obsoleta,* and the small clams *Nutricola* spp. All of these are easily observed. Potential hosts can be kept alive in seawater and the container checked the next day for swimming cercariae released from the snail. Snail shells can then be cracked, releasing hundreds of wormlike sporocysts, rediae, and swimming cercariae.

Ching (1991) provides a list of larval helminthes from Pacific coast invertebrates. The list is remarkably short, and many more interesting species remain to be discovered. It includes 73 trematodes, seven cestodes, five acanthocephalans, and five nematodes. Almost half come from only four genera of hosts (*Cerithidea, Littorina, Leukoma,* and *Macoma*). Some readily available and virtually unexamined groups of hosts include polychaetes, sea anemones, small species of bivalves and gastropods, nudibranchs, small crustaceans (peracarids, juvenile crabs), shrimps, seastars and brittle stars. Some of these are hard to dissect (echinoderms, polychaetes, sea anemones), and others are just too pretty to cut up (nudibranchs). However, there will be some remarkable findings as a reward for those who do investigate this vast, unknown fauna.

Intertidal octopuses and nearshore squids and sepiolids will almost always have their kidneys filled with Rhombozoa. Young hosts may harbor the nematogen phase that asexually produces more of the same. When the host ceases to grow, most of the nematogens become rhombogens and sexually produce the infusoriform larvae that exit the host.

Various clams in mud flats sometimes contain a large white nemertean, *Malacobdella grossa.* Its relations with the clam are still not well understood. Mussels on the outer coast fairly often harbor large soft-shelled females of the pea crab, *Fabia subquadrata,* which causes damage to the gills (Pearce 1966). Male pea crabs are small and hard-shelled, moving between mussels that contain the female crabs. Some limpets are specialized feeders and are essentially parasitic on their host plants. These include *Lottia paleacea* on surfgrasses, *L. instabilis* on *Laminaria dentigera,* and *L. insessa* on the kelp *Egregia menziesii.*

Echinoderms harbor very interesting parasites. The common intertidal sea urchins, *Strongylocentrotus* spp., have a diverse array of ciliates in the gut. They are also often parasitized by two species of turbellarians—the red *Syndisyrinx franciscanus* and the tan *Syndesmis* sp. Both are often spotted when they move across the yellow gonads during an urchin dissection. *Syndisyrinx franciscanus* sometimes feeds on the rich gut ciliate fauna of the host (Shinn 1981). Related dallyellioid turbellarians can be found in sand dollars and *Stichopus* spp. sea cucumbers. Sea cucumbers also have interesting large ciliates in their respiratory trees. Further examination may turn up little-studied parasitic castrators. These include wormlike gastropods in the body cavities of sea cucumbers and starfishes, *Orchitophrya* sp. ciliates in the testes of starfishes, orthonectids in the bursas of brittle stars, and ascothoracican barnacles in starfishes.

Rhizocephalan barnacles parasitize several decapods including majids, xanthids, hermit crabs, and porcellanids. All are parasitic castrators with feminizing effects. Most remain to be investigated for prevalences and behavior modification. A variety of caridean shrimps, thalassinids, hermit crabs, and porce-

lain crabs are sometimes parasitized by bopyrid isopods. Their relations with host reproduction, behavior, and molting are all topics worth pursuing. Female bopyrids are readily detected. Most species are in the gill chamber. They cause the host to expand the carapace over the gill chamber, forming an obvious blister to cover the isopod. Other species are under the abdomen. Dwarf males cling to the female isopod. In contrast, the even more unusual entoniscid isopods are fully internal parasites enclosed in a sheath formed by host blood cells (Kuris et al. 1980). They are common in *Hemigrapsus oregonensis* and are sometimes found in *H. nudus.* To see them, one must lift the carapace. To extract an intact adult, the posterior of the midgut of the host must be carefully severed; the large female isopod can then be gently lifted and flopped into the carapace. The anterior and posterior ends of the isopod are freed by carefully teasing away host tissue. The males are dwarf and are not highly modified, looking like little isopods. They crawl about the female. Other hosts that may harbor undiscovered entoniscids are the spider crabs, snapping shrimp, and anomurans.

To study behavior modification leading to trophic transmission of helminths, one could investigate a variety of mollusks such as razor clams, shore crabs, or mole crabs, as these often harbor a variety of larval digenean trematodes, tapeworms, acanthocephalans, and nematodes. Final hosts can be fishes, shore birds or sea otters.

Symbiotic Egg Predators

Ovigerous female brachyuran crabs, anomurans, and lobsters often harbor symbiotic nemertean egg predators, mostly in the genus *Carcinonemertes.* Carcinonemerteans are usually pink and readily seen with the naked eye under the abdominal flap of their crab hosts or in the limb axillae. They are not parasites because they do not attack the host crab itself. Instead, they wait for the female crab to oviposit her eggs, migrate to the egg mass, and begin to feed on the embryos. Hence, they are egg predators. They live in a durable, intimate association with their host; hence they are considered "symbiotic." *Carcinonemertes errans* on the Dungeness crab, *Cancer magister,* may have delayed or prevented the recovery of that important fishery (Hobbs and Botsford 1989). Locally, other nemerteans have been recovered from several other *Cancer* species, grapsid crabs, pea crabs, spider crabs, the spiny lobster *Panulirus interruptus,* and the pebble crab *Randallia ornata* (Wickham and Kuris 1988, Sadeghian and Kuris 2001). Their life cycles are often complex, and they have remarkable adaptations for surviving on nonovigerous hosts (Kuris 1993). Some amphipods associated with spider crabs are egg predators. Examination of other decapod crustaceans will likely reveal further new and interesting species of symbiotic egg predators.

Some Abundant and Interesting Commensals

Tube dwellers and burrowing invertebrates offer an array of commensals whose interactions with their hosts merit further study. For example, the rich fauna of symbionts associated with the innkeeper echiuran, *Urechis caupo,* includes the current-stealing clam *Cryptomya californica,* shrimp, pea crabs, and scaleworms. Some of these are also associated with the large burrowing mud shrimp, *Upogebia pugettensis,* and other larger infaunal species. A notable association is the frequent occurrence of the highly modified pea crab, *Pinnixa longipes* (several times wider than it is long), with the maldanid bamboo worm, *Axiothella rubrocincta.*

Several large, slow-moving invertebrates harbor scale worms, other polychaetes, and symbiotic shrimps. Some of these are host specific. Some also may be mutualistic associations because the large invertebrate provides a home while the crustacean or worm bites the potential predator. Some species of amphipods are associated with sea urchins, others with compound tunicates or large crabs. In most cases the nature of the relationships is not known. The snails *Chlorostoma* spp. and *Norrisia norrisi,* often serve as hosts for filter-feeding species of *Crepidula,* the grazing specialized limpet *Lottia asmi* and as a nursery for juvenile limpets, particularly *L. strigatella* (Jessica Bean, personal communication).

The clam, *Mytilimeria nuttallii,* is highly evolved for a specialized habitat, embedded in compound tunicates, most often in sea pork, *Cystodytes lobatus.* The shell of this clam is paper-thin. How this interaction is initiated is not known, nor is how the clam avoids decalcification by the highly acidic fluids in its host.

Some small snail species are micropredators, the marine equivalents of mosquitoes or ticks. They take a small bite from a host and may move on to another for their next meal. These include several species of pyramidellid and eulimid snails. Some can be seen on *Mytilus californianus,* the boring clam, *Netastoma rostratum,* sea urchins, and sea stars. The beautiful *Epitonium tinctum* is a micropredator of sea anemones, *Anthopleura* spp. Many species of nudibranchs also feed as micropredators on solitary hosts or by killing zooids of clonal hosts. Good examples are *Aeolidia papillosa* feeding on sea anemones and *Rostanga pulchra* feeding on sponges. Most pycnogonids are also micropredators of hydroids or bryozoans, although some are truly parasitic embedded in the tissues of hosts such as abalone.

A number of encrusting and bioeroding invertebrates form a facultative association with snail shells occupied by hermit crabs. The living snail prevents fouling within the shell aperture and umbilicus. The association of organisms with hermit crab shells is called *pagurization* because the shell-dwelling organisms are at those locations that the living snail's mantle keeps completely free of fouling organisms. Most of these species can be found elsewhere, but some are predominantly hermit crab–associated. Along the coast of California these associates are most often seen on hermit crab–occupied shells of *Chlorostoma funebralis* and *C. brunnea.* They include several species of spirorbid polychaetes (usually the first organisms to pagurize the shell), serpulid polychaetes, acorn barnacles, and hydroids. *Hydractinia* hydroids are almost always seen on shells inhabited by *Isocheles pilosus* on sandy beaches. Studies elsewhere indicate that these are actually mutualistic associates, reducing the risk of the occupant hermit crab to octopus predation and to loss of the shell to competitor hermit crabs (Wright 1973).

More invasive, sometimes even eroding shells of living snails, are endolithic fungi, the boring sponge, *Cliona californiana,* a boring bryozoan, *Immergentia californica* (recognizable as a row of little pores in the shell aperture), and *Polydora* spp. (spionid polychaetes, recognizable by paired openings often at the base of the columella; see Walker 1988). The distinctive acrothoracican barnacles, *Trypetesa* spp., bore into the upper whorls of hermit crab shells and can only be seen by cracking open the spire of the shell and examining the inner surface of the upper whorls.

Pagurization can be detected on 50% of the *Chlorostoma funebralis* shells within six weeks of hermit crab occupation.

A *C. funebralis* shell completely breaks down in 12–15 months, mostly from the activities of the boring organisms (A. Kuris, J. T. Carlton, and M. Brody, unpublished observations). Information about hermit crab associates has been well summarized by Walker (1992) and Williams and McDermott (2004). The dynamics of the pagurization process are a fruitful topic for observation and experimental study.

Large mollusks such as rock scallops, jingle shells, abalone, and Kellet's whelks are often eroded by several interesting species. These can include large colonies of the boring sponge, *Cliona californiana;* several spionids, mostly in the genus *Polydora;* and a small species of pholad clam *Penitella conradi.* These often greatly weaken the shell and cause the living mollusk to secrete additional laminar shell that may protrude as a blister on the inner surface of the shell. Once the mollusk has died, these organisms quickly degrade the rest of the shell until it breaks apart. Because these shell borers may affect the fitness of their hosts, these relationships are equivalent to parasitism.

Although it appears to be extinct in nature in California, it is worth mentioning the introduced species *Terebrasabella heterouncinata.* This small sabellid polychaete was accidentally introduced to California abalone mariculture facilities from its native South Africa. It escaped into the rocky intertidal zone near Cayucos, California, but was fortunately apparently eradicated (Culver and Kuris 2000). This was the first successful eradication of an established introduced marine pest. The biology of *T. heterouncinata* is quite interesting because it has a unique ability to foil the defenses against fouling organisms provided by the mantle of gastropods. The newly settled worm cannot be killed or dislodged by the mantle. It is then able to pervert the next line of defense of the host, stimulating the mantle to secrete additional layers of nacre, guided by the worm to form a tube for itself (Kuris and Culver 1999). As the worms grow they can subsequently enlarge these tubes.

To advance our knowledge of these and the many other parasitic and commensal relationships in the marine environment, perhaps the most efficient starting point is to carefully sample the hosts and detail the information on the distribution and abundance of the symbiotic relationship. A quantitative analysis of host and site specificity can also be very informative. The extent to which the symbiont is aggregated among hosts, changes in abundance with host size, sex, and location will generally lead to interesting and testable hypotheses about the nature of the relationship between the host and symbiont and perhaps the effect of the symbiont on host physiology, host population dynamics, and their role in community structure.

References

Ching, H. L. 1991. Lists of larval worms from marine invertebrates of the Pacific Coast of North America. J. Helminthol. Soc. Wash. 58: 57–68.

Culver, C. S. and A. M. Kuris. 2000. The apparent eradication of a locally established introduced marine pest. Biol. Invasions 2: 245–253.

Culver, C. S., A. M. Kuris, and B. Beede. 1997. Identification and Management of the Exotic Sabellid Pest in California Cultured Abalone. La Jolla, California Sea Grant College Program, Publ. T-041, 29 pp.

Hobbs, R. C. and L. W. Botsford. 1989. Dynamics of an age-structured prey with density-, and predator-dependent recruitment: the Dungeness crab and a nemertean egg predator worm. Theor. Pop. Biol. 36: 1–22.

Hoeg, J. T. 1995. The biology and life cycle of the Rhizocephala (Cirripedia). J. Mar. Biol. Assoc. U.K. 75: 517–550.

Kuris, A. M., G. O. Poinar, and R. T. Hess. 1980. Post-larval mortality of the endoparasitic isopod castrator *Portunion conformis* (Epicaridea: Entoniscidae) in the shore crab, *Hemigrapsus oregonensis,* with a description of the host response. Parasitology 80: 211–232.

Kuris, A. K. 1990. Guild structure of larval trematodes in molluscan hosts: prevalence, dominance and significance of competition, In Parasite Communities: Patterns and Processes. G. W. Esch, A. O. Bush, and J. M. Aho, eds. Chapman and Hall, pp. 69–100.

Kuris, A. M. 1993. Life cycles of nemerteans that are symbiotic egg predators of decapod Crustacea: adaptations to host life histories. Hydrobiologia 266: 1–14.

Kuris, A. M. and K. D. Lafferty. 1994. Community structure: Larval trematodes in snail hosts. Ann. Rev. Ecol. Syst. 25: 189–217.

Lafferty, K. D. 1993. Effects of parasitic castration on growth, reproduction and population dynamics of the marine snail *Cerithidea californica*. Mar. Ecol. Prog. Ser. 96: 229–237.

Lafferty, K. D. and A. M. Kuris. 2002. Trophic strategies, animal diversity and body size. Trends Ecol. Evol. 17: 507–513.

Pearce, J. B. 1966. The biology of the mussel crab, *Fabia subquadrata,* from the waters of the San Juan Archipelago, Washington. Pac. Sci. 20: 3–35.

Rasmussen, E. 1959. Behaviour of sacculinized shore crabs (*Carcinus maenas* Pennant). Nature 183: 479–480.

Sadeghian, P. S. and A. M. Kuris, 2001. Distribution and abundance of a nemertean egg predator (*Carcinonemertes* sp.) on a leucosiid crab, *Randallia ornata*. Hydrobiologia 456: 59–63.

Shinn, G. L. 1981. The diet of three species of umagillid neorhabdocoel turbellarians inhabiting the intestine of echinoids. Hydrobiologia 84: 155–162.

Sousa, W. P. and M. Gleason. 1989. Does parasitic infection compromise host survival under extreme environmental conditions?: the case for *Cerithidea californica* (Gastropoda: Prosobranchia). Oecologia 80: 456–464.

Walker, S. E. 1988. Taphonomic significance of hermit crabs (Anomura: Paguridea): epifaunal hermit crab—infaunal gastropod example. Palaeogeogr. Palaeoclim. Palaeoecol. 63: 45–71.

Walker, S. E. 1992. Criteria for recognizing marine hermit crabs in the fossil record using gastropod shells. J. Paleontol. 66: 535–558.

Wickham, D. E. and A. M. Kuris. 1988. Diversity among nemertean egg predators of decapod crustaceans. Hydrobiologia 156: 23–30.

Williams, J. D. and J. J. McDermott. 2004. Hermit crab biocoenoses: a worldwide review of the diversity and natural history of hermit crab associates. J. Exp. Mar. Biol. Ecol. 305: 1–128.

Wright, H. O. 1973. Effect of commensal hydroids on hermit crab competition in the littoral zone of Texas. Nature 241: 139–140.

Zimmer, C. 2000. Parasite Rex. New York: Free Press, 298 pp.

Introduced Marine and Estuarine Invertebrates

JAMES T. CARLTON AND ANDREW N. COHEN

The arrival of numerous species from other parts of the world has complicated the task of identifying Pacific coast invertebrates, especially in estuaries, bays, ports, and harbors. Although this manual contains many of the introduced species documented on this coast, there are doubtless many nonnative species that remain misidentified or mistaken as native species, and new invaders continue to arrive on a steady basis. Such further additions to our fauna will of course not "key out" in this manual.

An introduced species is one that has been transported by human activities to a region where it is not native and that has become established there by maintaining a reproducing population. More than 150 years of transporting marine organisms to the Pacific coast, either intentionally or accidentally, has resulted in the establishment of a minimum of 300 introduced species of nonnative protists, invertebrates, fish, algae, and seagrasses, although the actual number may be many times this. All larger and most smaller invertebrate phyla are represented among the nonnative biota, with the exception of the Echinodermata, indicating that few groups are immune to potential taxonomic challenges. Most of these species are restricted to bay and estuarine environments, although the open rocky coast, open sandy beaches, and the continental shelf itself have documented invasions as well. This largely estuarine restriction may be the result, in part, of the paucity of transport mechanisms that would serve to introduce species to nonestuarine areas from other regions of the world.

Transport Vectors of Introduced Species

Ship Fouling

Beginning as long as 500 years ago, the earliest wooden sailing vessels that entered (and occasionally sank in) Pacific coast bays may have introduced hull fouling and boring organisms from the Atlantic Ocean or from the southern or western Pacific Ocean. A signature event in the history of marine invasions on the Pacific coast was the California Gold Rush commencing in 1849: a great influx of ships from around the world arrived in a matter of a few years, and many of these lay at anchor or pierside for many weeks or months, or were entirely abandoned.

Thus the North Atlantic barnacle *Balanus improvisus* was first collected in 1853 in San Francisco Bay, and many of our nonnative hydroids may have arrived in these early years, as well. In later years, ships from Asia brought the Japanese isopod *Synidotea laevidorsalis,* now one of the most abundant fouling organisms on hydroid substrates in San Francisco Bay; ships from New Zealand brought the burrowing isopod *Sphaeroma quoianum,* and its tiny commensal isopod *Iais californica* (the latter being one of a number of introductions to the Pacific coast that were mistakenly described as native species and given local geographic names, the southern hemisphere hydroid *Garveia franciscana* being another). Wood-boring invertebrates, including gribbles (small isopods in the genus *Limnoria*) and shipworms (highly modified bivalve mollusks) in the genera *Lyrodus* and *Teredo,* arrived by the 19th and early 20th centuries, causing catastrophic damage to unprotected wharves and piers. Near the close of World War I, the Australian tubeworm *Ficopomatus enigmaticus* appeared in San Francisco Bay, where its large aggregations were described in local newspapers as "coral reefs."

The density of ship-fouling was reduced over the last century by the development of ever more effective antifouling paints, faster ships, and faster port turnarounds by cargo vessels (with less time for organisms to attach to hulls). Nevertheless, the introduction of species as fouling organisms continues: today there are more vessels, with much larger hull surfaces, making more voyages than in previous centuries; often there are some areas on a ship's hull where antifouling paint is not applied or where it wears away on poorly maintained vessels; certain fouling organisms, such as the encrusting bryozoan *Watersipora subtorquata,* are unaffected by some of the toxic materials used in antifouling paints, and when such organisms colonize painted hull surfaces they provide a nontoxic substrate for other, more susceptible, species. A vessel's seachest (the compartment where water is drawn into the ship to supply the ballast, engine-cooling, and fire-fighting systems) may also harbor a well-developed fouling community. Further, the characteristic modern transporter of significant fouling communities may not be a cargo ship in normal operation, but rather a vessel that remains in one place for a long period without hull cleaning or maintenance and then sails or is towed at relatively low speed to a new location, such as an ocean-going barge or a semi-submersible drilling platform.

Recent examples of probable hull-fouling introductions include a large array of seasquirts (ascidians) that arrived in California harbors over the past few decades. A noninvertebrate example is the Asian seaweed *Undaria pinnatifida,* discovered in Los Angeles Harbor in 2000. It may have initially arrived as fouling on a cargo ship, while its rapid spread up the coast to Monterey Bay was probably accomplished as fouling on the hulls of pleasure craft, which mechanism must also account for the spread of many ballast water introductions to boat harbors that are distant from cargo ports.

Ship Ballast

Ships transport many species in ballast water, which is taken aboard to reduce buoyancy and thus increase stability when a cargo ship is empty or lightly loaded, to adjust trim, and for other purposes related to navigation and cargo management. Water taken aboard from a bay or estuary or from coastal waters on shores as distant as Australia or the Atlantic and released in a bay along the Pacific coast may lead to the successful introduction of nonnative species. Examples include the arrival of the Asian shrimp *Palaemon macrodactylus* in San Francisco Bay in the early 1950s, coincident with increased vessel traffic during the Korean War, and, commencing in the 1970s and 1980s, the arrival in San Francisco Bay of various Asian copepods and mysid shrimp, as well as the Asian clam *Corbula amurensis,* coincident with increased vessel traffic from mainland China , Japan, and other Pacific Rim countries linked to the liberalization and expansion of international trade. Many other invasions along the Pacific coast are also believed to be ballast-mediated.

Prior to the 1900s, ballast was often carried in the form of rocks, sand, and other heavy and cheap materials from coastal environments. Many maritime plants and insects were distributed globally by this "dry" or "solid" ballast. Lumber ships returning in beach ballast from Chile or New Zealand thus served to bring the beach hopper *Transorchestia enigmatica* to California. In this interesting case, this species remains known only from San Francisco Bay, although recognized as a member of the *Transorchestia chiliensis* species complex native to the southern hemisphere.

Although the dumping of solid ballast was regulated in California waters by the first half of the 19th century (largely due to concerns about filling navigable channels), the discharge of water ballast was erroneously believed to be entirely benign and thus remained unregulated over most of its history. In 2000, California became the first U.S. state to require mid-ocean exchange of ballast water originating from overseas to reduce the release of nonnative coastal organisms.

Oyster Industry and Other Mariculture

Extensive importations of adult and seed oysters of *Crassostrea virginica* from the Atlantic coast (from the 1860s to the 1920s, and continuing in small numbers thereafter) and *Crassostrea gigas* from Japan (commencing in commercial quantities in the 1920s) led to the translocation of a large number of estuarine organisms from the western Atlantic and western Pacific. These oysters were imported and planted in northeastern Pacific estuaries with the hopes of establishing a "natural" fishery, for growing to market size, temporarily relaid for freshening purposes,

or simply held for the market. Early shipments in particular were apparently rich with epizoic organisms and sediments bearing infaunal organisms, including sponges, cnidarians, polychaetes, mollusks, crustaceans, bryozoans, and other invertebrates. Ironically, many of these species became established and proliferated on the Pacific coast, although the oysters themselves failed to become established at most sites (the Pacific or Japanese oyster *C. gigas* reproduces and maintains natural beds in Washington and British Columbia; it may also be established in southern California). Atlantic oyster-mediated contributions to our molluscan fauna include the slipper limpets *Crepidula fornicata, Crepidula convexa,* and *Crepidula plana;* the mudsnail *Ilyanassa obsoleta;* the oyster drill *Urosalpinx cinerea;* and the bivalves *Gemma gemma, Macoma petalum, Geukensia demissa, Mya arenaria,* and *Petricolaria pholadiformis.* Contributions from Japan include the snail *Batillaria attramentaria* and the bivalves *Musculista senhousia* and *Venerupis philippinarum.*

Although the importation of ship- or plane-loads of living oysters from distant lands for laying out in Pacific coast estuaries has ceased and cleaner processing has greatly reduced the inadvertent transport of associated organisms in oyster shipments compared to the "old days" when oysters, mud, and water were roughly packed in barrels or boxes, substantial quantities of oysters continue to be shipped between bays and countries on the Pacific coast (between Mexico, the United States, and Canada), and these are sometimes heavily fouled. Shipments between certain locations are banned, and inspections of shipments are required in other cases, but these inspections are focused on a small number of known oyster pests and, even for those species, are not likely to detect small infestations. Meanwhile, some shipments no doubt move under the regulatory radar screen, and small quantities of oysters from overseas may continue to make their way to our shores, sanctioned or not, for "experimental" mariculture operations. More recently developed forms of mariculture may also introduce new and harmful species, as demonstrated by the importation, statewide distribution, and accidental release of a South African shell parasite, the sabellid worm *Terebrasabella heterouncinata,* by California abalone farms.

Fishing Bait

Starting in the early 1960s, live marine polychaete worms (glycerids and nereids) harvested in New England have been packed in seaweed, primarily the intertidal rockweed *Ascophyllum nodosum,* and shipped by air express to California bait shops. The worms, still packed in seaweed, are sold to anglers, who frequently discard the seaweed and any New England invertebrates contained therein into local waters. Occasional specimens of the Atlantic periwinkle snail *Littorina littorea* turn up on rocks in San Francisco and Newport Bays, and another Atlantic periwinkle, *Littorina saxatilis,* has established populations throughout San Francisco Bay; both of these snails are abundant in the worm shipments, which undoubtedly transported them to the Pacific. Juveniles of the Atlantic shore crab *Carcinus maenas* also occur in the seaweed packing; *Carcinus* became established in San Francisco Bay by the early 1990s and has spread along the coast. Other species are to be expected from the same pathway. Meanwhile the global trade in live marine bait, primarily consisting of polychaetes but sometimes including other types of organisms, is expanding, and new species, packed in a variety of materials, are sometimes imported and sold in California. This opens up opportunities for introductions of other bait species or associated organisms from other regions of the world.

Aquariums, Seafood, Research, and Education

A web-based trade in marine aquarium organisms and live seafood, the use of live nonnative marine specimens in research and education, and the legal importation of these species into Pacific states provides a source for additional invasions. Unwanted aquarium organisms and live specimens used in research or education may be discarded directly into coastal waters or down the storm drains that lead to them by individuals unaware of the potential consequences, and live seafood species may be released in deliberate attempts to establish them in California. To be expected are occasional specimens of the Atlantic quahog *Mercenaria mercenaria*, oysters of a variety of species, the Atlantic crab *Callinectes sapidus*, Atlantic lobsters *Homarus americanus*, the Atlantic horseshoe crab *Limulus polyphemus*, and other popular aquarium, research, or seafood species. The Chinese mitten crab *Eriocheir sinensis*, which is considered a delicacy in parts of Asia, may also have been released intentionally into San Francisco Bay in the early 1990s; although it has been illegal to import this species into California since 1987, persons arriving from Asia and carrying live mitten crabs were regularly intercepted at California airports in the late 1980s and early 1990s.

Introduced species are not restricted to sites where they are initially released; secondary distribution can and does occur. This distribution may be mediated by either natural processes (e.g., by currents moving planktonic larvae, or by rafting of adults) or human activities (e.g., by pleasure craft, fishing boats, or cargo ships moving along the coast). Species initially introduced with oyster shipments or by cargo vessels may thus occur in lagoons or estuaries that have had neither oyster culture nor ship traffic.

Identification of Introduced Species

The recognition of a species as introduced and its correct identification are complicated by several factors. First, the systematics of the endemic species of a group on the Pacific coast may not be sufficiently advanced to distinguish natives from nonnatives. Second, the source or donor area of an introduced species could potentially be in any climatically appropriate coastal region of the world, requiring a command of the knowledge of the world fauna of a given taxon. Third, in some cases an introduced species may not yet be recognized or described in its native region.

Secondary evidence may suggest whether a species is introduced, including highly localized occurrence (disjunct distributions), recent spread to new areas on the coast, direct association with a suitable transport vector, and absence from the recent fossil record and from Native American shellmounds of species that would be expected to occur there (for the latter, edible shelled species, as well as certain epizoics such as barnacles and bryozoans). Disjunct distributions are often obscured in the published literature by over-generalizations of a species' range: we may thus find a certain species reported as occurring from "Puget Sound to San Francisco," when in fact it may be known only from Puget Sound and San Francisco Bay. Recent arrivals of introduced species must be distinguished from seasonal fluctuations in abundance of endemic species and from temporary range extensions of southern species during El Niño years.

It is sometimes mistakenly assumed that a taxon must be endemic if it cannot be referred to a species described elsewhere. However, many species remain undescribed or unknown in their native regions, and this especially holds true for smaller or taxonomically more difficult groups. The tubeworm *Ficopomatus enigmaticus*, native to western Australia, was first described from France, the Japanese oyster flatworm *Pseudostylochus ostreophagus* was first described after its accidental introduction to Puget Sound, and the South African shell parasite worm *Terebrasabella heterouncinata* was recognized and described in the 1990s only after it had been accidentally introduced into and infested California abalone farms.

In addition, not a few introduced species have been redescribed as new species through failure to match the species to a known entity elsewhere. Such erroneous "new" species names given to introduced species may last for several months to hundreds of years, masking the species' true status, before they are corrected. The Atlantic soft-shelled clam *Mya arenaria* was surprisingly redescribed as *Mya hemphilli* from San Francisco Bay shortly after its introduction; the Japanese clam *Venerupis philippinarum* was described as a new species *Paphia bifurcata*, from British Columbia; the Japanese oyster copepod *Mytilicola orientalis* as *M. ostrea*; the Atlantic ostracode *Eusarsiella zostericola* as *Sarsiella tricostata*; and many similar cases are known among polychaetes, seasquirts, mollusks, and crustaceans on the Pacific coast.

Establishment and Ecology of Introduced Species

The reasons for the successful introduction of so many nonnative species on the Pacific coast are varied and not fully known. There are, as described above, numerous vectors that have transported marine and estuarine species to the Pacific coast from other regions of the world. The extensive modification of Pacific coast estuaries through filling, dredging, damming, pollution, and the construction of wharves and marinas greatly modified aboriginal habitats and in some cases may have eliminated native populations, while at the same time providing a great deal of new hard-substrate habitat for fouling species to colonize.

Certain areas on the Pacific coast are now dominated by an introduced fauna that would not be unfamiliar to biologists from Massachusetts or Japan. Consider the south end of San Francisco Bay, where the majority of macroscopic invertebrates (and several of the fish and algae) are introduced species. On the pilings and floats are rich assemblages of organisms hailing from other coasts, including the beautiful orange-pink colonies of the tubularian hydroid *Pinauay crocea* (and its tiny nudibranch predator *Tenellia adspersa*), the white barnacle *Balanus improvisus*, the bryozoan *Conopeum tenuissimum*, the kamptozoan *Barentsia benedeni*, the seasquirt *Molgula manhattensis*, and the sea anemone *Diadumene lineata*. The Asian isopod *Synidotea laevidorsalis* is abundant, as are several species of introduced amphipods in the genus *Corophium*. Boring into the pilings is the introduced gribble *Limnoria tripunctata* (whose native region is unknown) and into the styrofoam of the floats the New Zealand isopod *Sphaeroma quoianum* (with its introduced commensal *Iais californica* riding on its belly), which also burrows in mud banks.

On the mud bottom below the marina floats are great hordes of the mudsnail *Ilyanassa obsoleta* from New England and dense nests of the little green mussel *Musculista senhousia* from Japan. Burrowing in the mud are the Japanese cockle *Venerupis philippinarum,* the Atlantic softshell clam *Mya arenaria,* and the Japanese amphipod *Grandidierella japonica.* The Korean shrimp *Palaemon macrodactylus* swims rapidly over the bottom and rests in fouling communities. In nearby salt marshes are dense stands of the New England mussel *Geukensia demissa* and abundant populations of the Atlantic marsh snail *Myosotella myosotis.* The striking faunal changes caused by the introductions of these estuarine wanderers warrants detailed investigation of their distributions and ecological relationships.

Control of Invasions

There have been relatively few efforts to control or eradicate populations of introduced marine or estuarine invertebrates. Several general considerations—including the high fecundity of many marine invertebrates relative to freshwater or terrestrial species, the rapid and wide dispersal of planktonic life stages, and the relative difficulty of locating populations, effectively applying biocides, finding acceptable species-specific biocontrol agents, or otherwise working cheaply and efficiently in marine environments—suggest that such efforts will be difficult or impossible in most individual cases and, as an overall approach to the aggregate problem of marine invasions, unworkable. There have been only two successful eradications of introduced marine invertebrates to date: the elimination of an infestation in northern Australia of the black-striped mussel *Mytilopsis sallei* by pouring large quantities of biocides (chlorine and copper sulphate) into an enclosed boat basin (connected to the ocean by a lock), and the removal of a very localized infestation of the South African fanworm *Terebrasabella heterouncinata* from a small rocky cove in southern California by reducing (through manual collection) the density of its potential host snails.

It is possible that advances in molecular genetics may someday allow the construction of safe and effective "killer genes" that would remove introduced marine species from invaded regions. Until that time, it appears that the greatest gains in managing marine introductions are to be had by reducing the flood of organisms being transported around the world by shipping, mariculture, the bait trade, and other vectors. It thus behooves students of Pacific coast marine biodiversity to remain on the watch for, to document, and to report on the arrival of new species and the extent and impacts of their introduction to provide support for and track the success of vector management efforts.

References

Byers, J. E. 1999. The distribution of an introduced mollusc and its role in the long-term demise of a native confamilial species. Biol. Invasions 1: 339–353. (*Batillaria* and the native *Cerithidea*)

Byers, J. E. 2000. Mechanisms of competition between two estuarine snails: implications for exotic species invasion. Ecology 81: 1225–1239. (*Batillaria* and the native *Cerithidea*)

Carlton, J. T. 1996. Biological invasions and cryptogenic species. Ecology 77: 1653–1655.

Carlton, J. T. 2001. Introduced species in U.S. coastal waters: environmental impacts and management priorities. Pew Oceans Commission, Arlington, Virginia, 28 pp.

Carlton, J. T. and A. N. Cohen. 2003. Episodic global dispersal in shallow water marine organisms: the case history of the European shore crabs *Carcinus maenas* and *Carcinus aestuarii.* J. Biogeogr. 30: 1809–1820.

Carlton, J. T. and J. B. Geller. 1993. Ecological roulette: the global transport of nonindigenous marine organisms. Science 261: 78–82.

Carlton, J. T. and J. Hodder. 1995. Biogeography and dispersal of coastal marine organisms: experimental studies on a replica of a 16th-century sailing vessel. Mar. Biol. 121: 721–730.

Carlton, J. T. and G. M. Ruiz. 2005. The magnitude and consequences of bioinvasions in marine ecosystems: implications for conservation biology, pp. 123–148, in: Elliott A. Norse and Larry B. Crowder, eds. Marine Conservation Biology: The Science of Maintaining the Sea's Biodiversity. Island Press, Washington, D.C., 470 pp.

Carlton, J. T., J. K. Thompson, L. E. Schemel, and F. H. Nichols. 1990. Remarkable invasion of San Francisco Bay (California, U.S.A.) by the Asian clam *Potamocorbula amurensis.* I. Introduction and dispersal. Marine Ecology Progress Series 66: 81–94.

Cohen , A. N. and J. T. Carlton. 1995. Biological Study. Nonindigenous Aquatic Species in a United States Estuary: A Case Study of the Biological Invasions of the San Francisco Bay and Delta. A Report for the United States Fish and Wildlife Service, Washington, D.C., and The National Sea Grant College Program, Connecticut Sea Grant, NTIS Report Number PB96-166525, 246 pp.

Cohen, A. N. and J. T. Carlton. 1997. Transoceanic transport mechanisms: The introduction of the Chinese mitten crab, *Eriocheir sinensis,* to California. Pac. Sci. 51: 1–11.

Cohen, A. N. and J. T. Carlton. 1998. Accelerating invasion rate in a highly invaded estuary. Science 279: 555–558.

Culver, C. S. and A. M. Kuris. 2000. The apparent eradication of a locally established introduced marine pest. Biol. Invasions 2: 245–253 (the abalone worm *Terebrasabella heterouncinata*)

Lambert, C. C. and G. Lambert. 2003. Persistence and differential distribution of nonindigenous ascidians in harbors of the Southern California Bight. Marine Ecology Progress Series 259: 145–161.

Nichols, F. H. and J. K. Thompson. 1985. Persistence of an introduced mudflat community in South San Francisco Bay, California. Mar. Ecol. Prog. Ser. 24: 83–97.

Race, M. S. 1982. Competitive displacement and predation between introduced and native mud snails. Oecologia 54: 337–347. (*Cerithidea* and *Ilyanassa*).

Ruiz, G. M., P. W. Fofonoff, J. T. Carlton, M. J. Wonham, and A. H. Hines. 2000. Invasion of coastal marine communities in North America: apparent patterns, processes, and biases. Ann. Rev. Ecol. Syst. 31: 481–531.

Wasson, K., C. J. Zabin, L. Bedinger, C. M. Diaz, and J. S. Pearse. 2001. Biological invasions of estuaries without international shipping: the importance of intraregional transport. Biol. Conserv. 102: 143–153 (invasions of Elkhorn Slough in Monterey Bay).

Molecular Identification

JONATHAN B. GELLER

The previous edition (1975) of this work was published as genetic methods began to be widely applied to a study of the systematics and phylogenetics of Pacific coast marine invertebrates. Such investigations revealed several taxa that were in fact groups of two or more species that are difficult to distinguish with morphological characters but demonstrate genetic distinctiveness (appendix 2). Work prompted by these earlier discoveries and enthusiasm for phylogeography will uncover many additional cases, and identification of such species will be an increasing challenge. Because molecular techniques will be an increasingly applied identification tool, this chapter will discuss molecular identification of sibling species and other morphologically challenging specimens.

The existence of large numbers of sibling or cryptic species in the marine environment is an important revelation (Knowlton 1993), and understanding why genetic differentiation is often not reflected in apparent morphological change will be equally important. However, there is a difference between real and apparent absence of morphological divergence among species. Size differences, if not standard morphological characters, that distinguish solitary and clonal varieties of the sea anemone *Anthopleura elegantissima* were long known before allozyme analyses indicated genetic differences that revealed two distinct species were involved (Hand 1955, McFadden et al. 1997, Pearse and Francis 2000). In other cases, multiple morphological differences were noticed after genetic tools have confirmed that multiple species are present; the periwinkle snails *Littorina scutulata* and *L. plena* are examples (Mastro et al. 1982, Murray 1982, Chow 1987).

For other taxa, it will be difficult or impossible to identify species without genetic techniques. For example, mussels of the *Mytilus edulis* complex cannot reliably be identified by morphology due to overlapping shell variation. In California, both the native *M. trossulus* and the introduced *M. galloprovincialis* may be encountered (McDonald and Koehn 1988, McDonald et al. 1991, Sarver 1991, Geller et al. 1994, Suchanek et al. 1997). Without a genetic test, one can only guess to which species an individual belongs using habitat (which is not reliable, because both species can occur together) and perhaps geography (because the introduced species is most common in southern California). Critical determination requires genetics, especially in

central California where the two species geographically overlap, and genetics will provide the tools necessary to detect a new invasion should the Atlantic *Mytilus edulis* appear in our area. In other cases, extensive morphological variation has been noticed, but whether this variation might sort individuals into groups conforming to species concepts is not clear. For example, the dogwhelk *Nucella lamellosa* exhibits a wide array of shell thicknesses and sculpture (Kincaid 1957, Spight 1973), yet no discrete morphological boundaries appear to separate them.

At times, species level uncertainty may not be a severe handicap. It may be sufficient to know the genus of an organism for environmental monitoring, or the functional or trophic group of an organism in an ecological study. However, it is clearly important to have species-level information when diversity or biogeographic patterns are at issue. Also, sibling species are probably never entirely ecologically equivalent, so misidentification or lumping of two or more species together will raise the noise relative to signal in community-level manipulative experiments or correlative studies. Harger (1968, 1972) performed elegant experiments to demonstrate the competitive superiority of "*Mytilus edulis*" over *M. californianus* in a protected southern Californian embayment. We now know that Harger was almost certainly studying *Mytilus galloprovincialis*, although the extent to which the native *M. trossulus* may have been involved in his studies we do not know.

Studies of population dynamics will also suffer when different species are ignored or inadvertently grouped together: if similar-looking species are not adequately resolved, both short- and long-term changes in the community may go undetected. For example, the remarkable replacement of the native *Mytilus trossulus* by the introduced *Mytilus galloprovincialis* in southern California in the mid-20th century was undetected as it transpired (Geller 1999).

Even where morphological differences between species are profound and easily noted, this may not be true for all life stages. The morphology of larval and juvenile life stages are described for a minority of marine invertebrates, and these stages are often notoriously difficult to identify, even when descriptions exist. In contrast, genetic markers are not dependent on life stage and can therefore be more generally applied. For example, genetic markers for *Mytilus* have been used to detect newly settled juveniles in recruitment studies on the Pacific coast (Martel et al. 1999; Johnson and Geller 2006).

For all these reasons, there is a good reason to consider the use of genetic markers for the purpose of identification. It has been proposed that standardized genetic markers be developed for all living organisms (Tautz et al. 2002; Hebert et al. 2003a, b; Tautz et al. 2003). Should such a database be accomplished for the marine invertebrates of the Pacific coast, it will be a great boon to systematic, phylogenetic, and ecological research in our region.

Molecular Identification vs. Molecular Classification

A common misunderstanding is the difference between molecular identification and molecular classification. Molecular identification is the use of genetic markers to assign an individual specimen to an already recognized and properly described species. This is useful when gross morphological differences are absent or slight, or when detection of diagnostic characters requires a high level of expertise or extensive sample preparation. Molecular identification is also useful when diagnostic characteristics are known only for a particular life stage, typically the adult, but the specimen belongs to a different life stage. Molecular identification does not replace the traditional systematist, who remains called upon to investigate and adjudicate species boundaries. Instead, it provides a potentially applicable and available set of tools for students and professional biologists who are not taxonomic experts. Indeed, molecular tools may relieve taxonomic experts from the burden of identification services and allow greater attention to actual taxonomic problems.

At the species level, molecular classification is the use of genetic markers to discover discontinuities among genotypes of a collection of organisms previously thought to represent one species, either within one region, or over a larger geographic distance. When discrete groups based on genetic markers are detected, the existence of species boundaries may be suggested and new species described. Similar procedures could apply to higher taxonomic levels, as well. A group of organisms diagnosed by shared molecular markers, or monophyly in gene phylogenetic trees (the phylogenetic species concept), does not conform to the biological species concept of Mayr (1963) that requires evidence for reproductive isolation from other groups. Of course, this is not different from classical systematics, which diagnoses groups of organisms by shared morphological characters, and thus also does not conform to the biological species concept.

In practice, the systematist combines molecular and morphological methods. For example, species differences in the snails *Littorina* and *Nucella* and the sea anemone *Anthopleura* were suspected from morphological or life history differences, and molecular classification supported the hypotheses (Mastro et al. 1982, Palmer et al. 1990, McFadden et al. 1997, Marko et al. 2003).

Future Approaches to Molecular Identification

Methods for molecular identification are likely to change considerably, and any detailed protocols given here would be shortly outdated. Hillis et al. (1996) and Avise (2004) have provided comprehensive reviews of methods of genetic analysis,

and numerous practical guides exist. Although the methods by which DNA-based identification are done will change over time, in all cases DNA from specimens will be needed. The basic steps for preparing specimens for DNA analysis are shown in appendix 1.

Exhaustive databases of DNA sequences for Californian and Oregonian intertidal invertebrates do not yet exist. Molecular markers that are useful for identification exist where researchers have conducted population genetic, phylogeographic, or phylogenetic studies of the marine animals of California and Oregon, but few of these concern taxa where morphology is not a more convenient indicator of identity, at least for adults. However, several researchers have proposed a system of "DNA barcodes" (Tautz et al. 2002; Hebert et al. 2003a, b; Tautz et al. 2003). This would entail the development of a database of unique, identifying DNA sequences (the "barcodes") for all described species. In principle, one could isolate DNA from an animal of unknown identity, use the polymerase chain reaction (PCR) to amplify a gene present in all animals using universal primers, and sequence the resulting PCR products (or use other methods to detect specific DNA sequences, such as hybridization with DNA probes). By comparing the newly determined sequence to the DNA database, one would determine the identity of the unknown organism. Hebert et al. (2003a, b) has proposed that the gene encoding mitochondrial cytochrome oxidase c subunit I (COI) is an appropriate choice for the DNA barcode database because interspecific variation in COI is said to be sufficient to find unique sequences for all taxa and because universal primers (Folmer et al. 1994) have been used on a wide variety of organisms with success. Criticism of DNA barcoding has focused on its potential use for molecular classification (Lipscomb et al. 2003, Seberg et al. 2003, Will and Rubinoff 2004), although its greater value will be for molecular identification.

Despite some encouraging results (Hebert et al. 2004), there are practical problems with DNA barcoding. First, taxa such as anthozoan cnidarians show very slow rates of COI evolution, to the extent that no or little variation exists within genera (Shearer et al. 2002). Second, it will be difficult to establish that a given COI sequence is truly unique to one species, or conversely universal within that species, at least until the database is complete. Third, although so called "universal" primers may indeed be used to amplify PCR products from a large array of taxa, it will nonetheless be necessary to develop specific primers for many others. Last, a database of a single genetic locus may not provide the information needed for reliable molecular identification (Belfiore et al. 2003). However, to the extent that genetic polymorphisms do in fact delineate species, multilocus databases can provide for the needed reliability, and sequence detection technologies will be used to quickly determine DNA sequences of PCR products from unidentified samples.

Regardless of whether there is a compelling practical need for genetic identification in any particular case, it is likely that the needed DNA sequence information will become available for a great many of the marine invertebrates of the Pacific coast, and genetic identification will become an increasing option. Nonetheless, it is hoped that this book will remain a primary tool for identification, for it is both the organism's phenotype and genotype that is of interest to the naturalist, biogeographer, ecologist, and evolutionary biologist.

Appendix 1: How to Preserve Tissue Samples for Later Molecular Identification; DNA Preparation

In all cases, it is important to retain whole animals as morphological vouchers; label specimens with as much taxonomic, geographic, and habitat information as possible, and to cross index morphological, tissue, and DNA samples (e.g., with corresponding sample numbers) such that they can be confidently matched for subsequent work, perhaps decades or centuries later. Work must be done in as clean a manner as possible, as cross contamination of samples can easily occur and lead to false results. Thus, always clean dissection surfaces and tools with dilute bleach between samples. Alternatively, use sheets of plastic wrap or aluminum foil as disposable dissection surfaces, and use single-use razor blades to excise tissue samples.

DNA Preparation

Purified DNA is a stable molecule that can be stored frozen in aqueous solution or dry at room temperature and remain viable for PCR for years (Gerstein 2001). Thus, if one has the time and resources to isolate DNA from fresh tissue, this is the best option for long-term storage. General procedures for DNA isolation can be found in molecular biology handbooks (Gerstein 2001, Sambrook and Russell 2001, Ausubel 2002) or guides specifically written for molecular systematists (Hillis et al. 1996). In general, tissues are homogenized in one of many possible lysis buffers, and DNA is separated from other constituents by organic extraction and precipitation. Some commercially available kits for DNA isolation work well and avoid the need for toxic organic compounds such as phenol or chloroform by binding DNA from the lysis buffer to resins or solid substrates, from which they can be eluted into an aqueous solution.

As a rule, no single method works optimally for every organism or tissue, so some experimentation is usually needed. When possible, work with soft, easily homogenized tissues, and avoid tissues that produce large amounts of mucous compounds. Little tissue is needed to produce sufficient DNA for genetic analysis, so limit tissue slices to 100 mg when working with lysis buffer in volumes <1 ml. For all methods, essential equipment includes pipetters, heated water bath or dry heat block, and a microcentrifuge.

Simply boiling a tissue sample in the chromatography resin Chelex-100™ (Bio-Rad) may be sufficient to produce low-quality DNA that is nevertheless sufficient for PCR. This method has worked well for small crustaceans (Sotka et al. 2004). Digestion of tissue homogenates with proteinase-K, followed by isolation of DNA by organic extraction (Sambrook and Russell 2001) or capture on commercially available affinity columns (e.g., Qiagen™ or Promega™) often produces high-quality DNA. Some invertebrate tissues become viscous from mucus, which interferes with purification. Incubation in hexadecyltriethylamine bromide (CTAB) (Gerstein 2001) can assist in reducing sample viscosity.

In general, consult recent publications reporting results of molecular studies on taxa closely related to a novel study organism.

Tissue Storage for Later DNA Isolation

The best method is to freeze a small (5 mm^3–10 mm^3) section of soft, nucleated cellular tissue in liquid nitrogen and store at −70°C until needed. This is often not practical for lack of liquid nitrogen and ultra low freezers. Freezing a sample in a consumer quality manual defrost freezer (−20°C) will usually allow later isolation of DNA usable for PCR. Repeated defrost cycling in most conventional automatic defrost household freezers will result in greatly degraded DNA, which may prove unreliable for genetic analysis.

Preservation of a small (5 mm^3–10 mm^3) section of soft, nucleated cellular tissue in 70%–100% ethanol is often satisfactory. Isopropanol, available widely in drugstores, can be used when ethanol is unavailable. DNA from alcohol-preserved tissues is usually highly degraded, yet PCR amplification of targets <1000 bp is often successful. The outer soft tissues of whole animals preserved in ethanol are useful for DNA extraction, but animals entirely enclosed by exoskeletons may need to be cracked before immersion in ethanol. For general purposes, ethanol is a good single choice that preserves DNA and morphology reasonably well. Individual chapters in this book provide specific guidelines for how to preserve different types of taxa, as does the introductory chapter on methods of preservation and anesthetization.

Other preservation solutions yield higher quality DNA than will ethanol, in my experience. For example, DNE (20% DMSO, 500 mM EDTA, and NaCl at saturation) can be used successfully on a wide variety of invertebrate tissues. However, the high EDTA content makes DNE inappropriate for morphological preservation of specimens with calcareous parts: these will be dissolved.

If tissue needs to be stored for only a short time (a few weeks) between collecting in the field and returning to the laboratory, I have found it useful to homogenize the sample in DNAzol™ (Molecular Research Center, Inc.) or a Proteinase-K lysis buffer with Proteinase-K at 0.1 mg/ml and store at ambient temperature. The DNA extraction can then be conventionally completed in the laboratory with good yields and quality.

There may be a need to preserve RNA as well as DNA. For example, it might become important to forensically determine past patterns of expression of genes that are markers of stress due to, say, an oil spill episode. For this, freezing tissues in liquid nitrogen and storing at −70°C is the only reliable method for the long term. However, proprietary products such as RNAlater™ (Ambion) can be used to store tissues at +4°C or −20°C for weeks to months.

Appendix 2: Examples of California and Oregon Intertidal Invertebrates in Which Sibling Species Have Been Discovered or Verified by Molecular Methods

Genus	References
MOLLUSCA	
Littorina (Gastropoda)	Mastro et al. 1982, Chow 1987
Nucella (Gastropoda)	Palmer et al. 1990, Marko 1998, Marko et al. 2003
Lottia austrodigitalis (Gastropoda)	Murphy 1978
Mytilus (Bivalvia)	McDonald and Koehn 1988, McDonald et al. 1991, Sarver 1991, Sarver and Foltz 1993, Geller and Powers 1994, Suchanek et al. 1997
Macoma (Bivalvia)	Meehan et al. 1989
CNIDARIA	
Anthopleura (Anthozoa)	McFadden et al. 1997, Pearse and Francis 2000, Geller and Walton 2001
Epiactis (Anthozoa)	Edmands 1996
Metridium (Anthozoa)	Bucklin and Hedgecock 1982
Aurelia (Scyphozoa)	Dawson and Martin 2001
CRUSTACEA	
Carcinus (Decapoda)	Geller et al. 1997
Emerita (Decapoda)	Tam et al. 1996
Cletocamptus (Harpacticoida)	Castro-Longoria et al. 2003
Tigriopus (Harpacticoida)	Ganz and Burton 1995
Chthamalus (Cirripedia)	Hedgecock 1979
ECHINODERMATA	
Leptasterias (Asteroidea)	Kwast et al. 1990, Foltz et al. 1996a, Foltz et al. 1996b, Hrincevich and Foltz 1996, Flowers and Foltz 2001
BRYOZOA	
Bugula	Davidson and Haygood 1999
Membranipora	Schwaninger 1999

References

Ausubel, F. M. 2002. Short protocols in molecular biology: a compendium of methods from Current protocols in molecular biology. New York: Wiley.

Avise, J. C. 2004. Molecular markers, natural history, and evolution. Sunderland, MA: Sinauer Associates, Inc.

Belfiore, N. M., F. G. Hoffman, R. J. Baker, and J. A. DeWoody. 2003. The use of nuclear and mitochondrial single nucleotide polymorphisms to identify cryptic species. Mol. Ecol. 12: 2011–2017.

Bucklin, A. and D. Hedgecock. 1982. Biochemical evidence for a third species of *Metridium* (Coelenterata, Actiniaria). Mar. Biol. 66: 1–7.

Castro-Longoria, E., J. Alvarez-Borrego, A. Rocha-Olivares, S. Gomez, and V. Kober. 2003. Power of a multidisciplinary approach: Use of morphological, molecular and digital methods in the study of harpacticoid cryptic species. Mar. Ecol. Prog. Ser. 249: 297–303.

Chow, V. 1987. Morphological classification of sibling species of *Littorina* (Gastropoda: Prosobranchia); discretionary use of discriminant analysis. Veliger 29: 359–366.

Davidson, S. K. and M. G. Haygood. 1999. Identification of sibling species of the bryozoan *Bugula neritina* that produce different anticancer bryostatins and harbor distinct strains of the bacterial symbiont "*Candidatus Endobugula sertula.*" Biol. Bull. 196: 273–280.

Dawson, M. N. and L. E. Martin. 2001. Geographic variation and ecological adaptation in *Aurelia* (Scyphozoa, Semaeostomeae): Some implications from molecular phylogenetics. Hydrobiologia 451: 259–273.

Edmands S. 1996. The evolution of mating systems in a group of brooding sea anemones (*Epiactis*). Invertebr. Reprod. Dev. 30: 227–237.

Flowers, J. M. and D. W. Foltz. 2001. Reconciling molecular systematics and traditional taxonomy in a species-rich clade of sea stars (*Leptasterias* subgenus *Hexasterias*). Mar. Biol. 139: 475–483.

Folmer, O., M. Black, W. Hoeh, R. A. Lutz, and R. Vrijenhoek. 1994. DNA primers for amplification of mitochondrial cytochrome c oxidase subunit I from diverse metazoan invertebrates. Mol. Mar. Biol. Biotechnol. 3: 294–299.

Foltz, D., W., J. P. Breaux, E. L. Campagnaro, S. W. Herke, A. E. Himel, A. W. Hrincevich, J. W. Tamplin, and W. B. Stickle. 1996a. Limited morphological differences between genetically identified cryptic species within the *Leptasterias* species complex (Echinodermata: Asteroidea). Can. J. Zool. 74: 1275–1283.

Foltz, D., W., W. B. Stickle, E. L. Campagnaro, and A. E. Himel. 1996b. Mitochondrial DNA polymorphisms reveal additional genetic heterogeneity within the *Leptasterias hexactis* (Echinodermata: Asteroidea) species complex. Mar. Biol. 125: 569–578.

Ganz, H. H. and R. S. Burton. 1995. Genetic differentiation and reproductive incompatibility among Baja California populations of the copepod *Tigriopus californicus*. Mar. Biol. 123: 821–827.

Geller, J. B. 1999. Decline of a native mussel masked by sibling species invasion. Conserv. Biol. 13: 661–664.

Geller, J. B. J. T. Carlton, and D. A. Powers. 1994. PCR-based detection of mtDNA haplotypes of native and invading mussels on the northeastern Pacific coast: Latitudinal pattern of invasion. Mar. Biol. 119: 243–249.

Geller, J. B. and D. A. Powers. 1994. Site-directed mutagenesis with the polymerase chain reaction for identification of sibling species of *Mytilus*. Nautilus 108 (Supplement 2): 141–144.

Geller, J. B. and E. D. Walton. 2001. Breaking up and getting together: Evolution of symbiosis and cloning by fission in sea anemones (Genus *Anthopleura*). Evolution 55: 1781–1794.

Geller, J. B., E. D. Walton, E. D. Grosholz, and G. M. Ruiz. 1997. Cryptic invasions of the crab *Carcinus* detected by molecular phylogeography. Mol. Ecol. 6: 901–906.

Gerstein, A. S. 2001. Molecular biology problem solver: a laboratory guide. New York: Wiley-Liss.

Hand, C. 1955. The sea anemones of central California part II. The endomyarian and mesomyarian anemones. Wasmann J. Biol. 13: 37–99.

Harger, J. R. E. 1968. The role of behavioral traits in influencing the distribution of two species of sea mussel, *Mytilus edulis* and *Mytilus californianus*. Veliger 11: 45–49.

Harger, J. R. E. 1972 Competitive co-existence: maintenance of inter-acting associations of the sea mussels *Mytilus edulis* and *Mytilus californianus*. Veliger 14: 387–410.

Hebert, P. D. N., A. Cywinska, S. L. Ball, and J. R. deWaard. 2003a. Biological identifications through DNA barcodes. Proceedings of the Royal Society of London, Series B. 270: 313–321.

Hebert, P. D. N., S. Ratnasingham, and J. R. deWaard. 2003b. Barcoding animal life: cytochrome c oxidase subunit 1 divergences among closely related species. Proceedings of the Royal Society of London Series B. Suppl. S270: S96–S99.

Hebert, P. D. N., E. H. Penton, J. M. Burns, D. H. Janzen, and W. Hallwachs. 2004. Ten species in one: DNA barcoding reveals cryptic species in the neotropical skipper butterfly *Astraptes fulgerator*. Proc. Nat. Acad. Sci. U.S.A. 101: 14812–14817.

Hedgecock, D. 1979. Biochemical genetic variation and evidence of speciation in *Chthamalus* barnacles of the tropical eastern Pacific. Mar. Biol. 54: 207–214.

Hillis, D. M., C. Moritz, and B. K. Mable. 1996. Molecular systematics. Sunderland, MA: Sinauer Associates, Inc.

Hrincevich, A. W. and D. W. Foltz. 1996. Mitochondrial DNA sequence variation in a sea star (*Leptasterias* spp.) species complex. Mol. Phylogenet. Evol. 6: 408–415.

Johnson, S. B. and J. B. Geller. 2006. Larval settlement can explain the adult distribution of *Mytilus californianus* Conrad but not of *M. galloprovincialis* Lamarck or *M. trossulus* Gould in Moss Landing, central California: evidence from genetic identification of spat. J. Exp. Mar. Biol. Ecol. 328: 136–145.

Kincaid, T. 1957. Local races and clines in the marine gastropod *Thais lamellosa* Gmelin. A population study. Seattle: Calliostoma Press.

Knowlton, N. 1993. Sibling species in the sea. Annu. Rev. Ecol. Syst. 24: 189–216.

Kwast, K. E., D. W. Foltz, and W. B. Stickle. 1990. Population genetics and systematics of *Leptasterias hexactis* (Echinodermata: Asteroidea) species complex. Mar. Biol. 105: 477–489.

Lipscomb, D. A., N. Platnick, and Q. Wheeler. 2003. The intellectual content of taxonomy: a comment on DNA taxonomy. Trends Ecol. Evol. 18: 64–66.

Marko, P. B. 1998. Historical allopatry and the biogeography of speciation in the prosobranch snail genus *Nucella*. Evolution 52: 757–774.

Marko, P. B., A. R. Palmer, and G. J. Vermeij. 2003. Resurrection of *Nucella ostrina* (Gould, 1852), lectotype designation for *N. emarginata* (Deshayes 1839), and molecular genetic evidence of Pleistocene speciation. Veliger 46: 77–85.

Martel, A. L., B. S. Baldwin, R. M. Dermott, and R. A. Lutz. 1999. Distinguishing early juveniles of Eastern Pacific mussels (*Mytilus* spp.) using morphology and genomic DNA. Invertebr. Biol. 118: 149–164.

Mastro, E., V. Chow, and D. Hedgecock. 1982. *Littorina scutulata* and *Littorina plena*: sibling species status of two prosobranch gastropod species confirmed by electrophoresis. Veliger 24: 239–246.

Mayr, E. 1963. Animal species and evolution. Cambridge, MA: Belknap Press.

McDonald, J. H. and R. K. Koehn. 1988. The mussels *Mytilus galloprovincialis* and *M. trossulus* on the Pacific coast of North America. Mar. Biol. 99: 111–118.

McDonald, J. H., R. Seed, and R. H. Koehn. 1991. Allozymes and morphometric characters of three species of *Mytilus* in the northern and southern hemispheres. Mar. Biol. 111: 323–333.

McFadden, C. S., R. K. Grosberg, B. B. Cameron, D. P. Karlton, and D. Secord. 1997. Genetic relationships within and between clonal and solitary forms of the sea anemone *Anthopleura elegantissima* revisited: evidence for the existence of two species. Mar. Biol. 128: 127–139.

Meehan, B. W., J. T. Carlton, and R. Wenne. 1989. Genetic affinities of the bivalve *Macoma balthica* from the Pacific coast of North America: evidence for recent introduction and historical distribution. Mar. Biol. 102: 235–241.

Murphy, P. G. 1978. *Collisella austrodigitalis* sp. nov.: a sibling species of limpet (Acmaeidae) discovered by electrophoresis. Biol. Bull. 155: 193–206.

Murray, T. E. 1982. Morphological characterization of the *Littorina scutulata* species complex. Veliger 24: 233–238.

Palmer, A. R., S. D. Gayron, and D. S. Woodruff. 1990. Reproductive, morphological and genetic evidence for two cryptic species of northeastern Pacific *Nucella*. Veliger 33: 325–338.

Pearse, V. and L. Francis. 2000. *Anthopleura sola*, a new species, solitary sibling species to the aggregating sea anemone, *A. elegantissima* (Cnidaria: Anthozoa: Actiniaria: Actiniidae). Proc. Biol. Soc. Wash. 113: 596–608.

Sambrook, J. and D. W. Russell. 2001. *Molecular cloning: a laboratory manual*. Cold Spring Harbor, N.Y.: Cold Spring Harbor Laboratory Press.

Sarver, S. K. 1991. Genetic and biochemical studies of the octopine dehydrogenase polymorphism in *Mytilus trossulus*: The effect of genotype on specific activity. Am. Zool. 31: 115A.

Sarver, S. K. and D. W. Foltz. 1993. Genetic population structure of a species' complex of blue mussels (*Mytilus* spp.). Mar. Biol. 117: 105–112.

Schwaninger, H. R. 1999. Population structure of the widely dispersing marine bryozoan *Membranipora membranacea* (Cheilostomata): implications for population history, biogeography, and taxonomy. Mar. Biol. 135: 411–423.

Seberg, O., C. J. Humphries, S. Knapp, D. W. Stevenson, G. Petersen, N. Scharff, and N. M. Andersen. 2003. Shortcuts in systematics? A commentary on DNA-based taxonomy. Trends in Ecology and Evolution 18(2): 63–64.

Shearer, T. L., M. J. H. van Oppen, S. L. Romano, and G. Woerheide. 2002. Slow mitochondrial DNA sequence evolution in the Anthozoa (Cnidaria). Mol. Ecol. 11: 2475–2487.

Sotka, E. E., J. P. Wares, J. A. Barth, R. K. Grosberg, and S. R. Palumbi. 2004. Strong genetic clines and geographical variation in gene flow in the rocky intertidal barnacle *Balanus glandula*. Mol. Ecol. 13: 2143–2156.

Spight, T. M. 1973. Ontogeny, environment, and shape of a marine snail, *Thais lamellosa* Gmelin. J. Exp. Mar. Biol. Ecol. 13: 215–228.

Suchanek, T. H., J. B. Geller, B. R. Kreiser, and J. B. Mitton. 1997. Zoogeographic distributions of the sibling species *Mytilus galloprovincialis* and *M. trossulus* (Bivalvia: Mytilidae) and their hybrids in the North Pacific. Biol. Bull. 193: 187–194.

Tam, Y. K., I. Kornfield, and F. P. Ojeda. 1996. Divergence and zoogeography of mole crabs, *Emerita* spp. (Decapoda: Hippidae), in the Americas. Mar. Biol. 125: 489–497.

Tautz, D., P. Arctander, A. Minellli, R. H. Thomas, and A. P. Vogler. 2002. DNA points the way ahead in taxonomy. Nature 418: 479.

Tautz, D., P. Arctander, A. Minellli, R. H. Thomas, and A. P. Vogler. 2003. A plea for DNA taxonomy. Trends Ecol. Evol. 18: 71–74.

Will, K. E. and D. Rubinoff. 2004. Myth of the molecule: DNA barcodes for species cannot replace morphology for identification and classification. Cladistics 20: 47–55.

Methods of Preservation and Anesthetization of Marine Invertebrates

GARY C. WILLIAMS AND ROBERT VAN SYOC

It is important to properly prepare and preserve biological samples for future examination. Improperly preserved material may be rendered virtually useless for character analysis, dissection, histological, or genetic study. The accompanying tables provides a brief synopsis of techniques for the anesthetization, fixation, and preservation that are widely used for the most commonly encountered groups of organisms that inhabit the littoral zone of temperate west coast North America.

Histological Study

For histological study, many organisms, including sponges, tubicolous polychaetes, opisthobranchs, and hemichordates, are best fixed in "Bouin's solution" for at least 12–24 hours. This fluid tends to harden soft cuticles and tissues and dissolve calcareous structures. It should therefore not be used when one wishes to preserve for study those structures that are composed of calcium carbonate. Bouin's solution can be prepared by mixing 15 parts saturated aqueous picric acid, with five parts 100% formalin (= 37% formaldehyde), and one part glacial acetic acid. Solid picric acid can be explosive, and because of this, only saturated aqueous solutions should be used. It has been suggested that glutaraldehyde in seawater is a better fixative for some sponges, in which a detailed comparison of cytological and anatomical analyses is necessary for taxonomic identification.

Anesthetization

A variety of anesthetizing chemical agents has proven to be useful for certain animal groups. These include menthol, hydrated magnesium sulfate (Epsom salts), chloral hydrate, ethyl m-aminobenzoate (also known as MS 222), propylene phenoxetol (stock solution of 1.5% is made using warm tap water; a working solution of 0.15% is made from one part stock solution and nine parts seawater), ethyl alcohol, benzamine hydrochloride/cellosolve mixture, eucaine (b-eucaine hydrochloride), and stovaine (amyl chlorohydrin). See Lincoln and Sheals (1979) for detailed information. One of the most commonly used agents is isotonic "magnesium chloride"

(73 g $MgCl_2 \cdot 6 H_2O$ per liter of tap water; or 80 g of crystals per liter of seawater) used full strength or added in various proportions to containers of seawater. As a last resort, a few shreds of pipe tobacco added to a small bowl of seawater can be used when improvising is necessary!

Preservation

For preservation formal and alcohol are commonly used. "Formalin" is an aqueous solution made from the gas "formaldehyde" (CH_2O), usually with a small amount of "methanol" added. Formalin becomes acidic with age and can readily dissolve or degrade the fine sculpture of structures composed of calcium carbonate. Because of this, "buffered formalin" (formalin in which sodium carbonate [Na_2CO_3], hexamine, or a phosphate [$NaH_2PO_4 \cdot H_2O$] has been added to buffer or neutralize the acidity) should be used.

Of the three common types of "alcohol" used as preservatives, "ethanol" (ethyl alcohol) is the most effective and widely used, the others being "isopropyl alcohol" (rubbing alcohol) and "methanol" (methyl alcohol). Ninety-five percent ethanol can be used as an initial preservative, while 75% ethanol should be used for the long-term storage of most invertebrates.

Comb jellies, polyclad flatworms, and nemerteans are particularly difficult to relax and preserve. A summary of the special techniques required for these groups is as follows.

The delicate tissues of comb jellies (Ctenophora) make preservation difficult, particularly regarding larger animals (>50 mm), which require special treatment. Anesthetizing can be accomplished using chloral hydrate crystals added to a bowl of seawater containing comb jellies. Specimens are best fixed in a mixture of chromic/osmic acid (50 parts 1% chromic acid to one part 1% osmic acid—*a highly toxic mixture!*) for 15 minutes for small animals, and up to 60 minutes for larger ones. The specimen needs to then be slowly transferred through various concentrations of alcohol—30%, 40%, 50%, 60%, and finally 70%. Permanent storage should be in 70% ethanol. Several authorities agree that formalin should not be used with comb jellies.

Polyclad flatworms are especially difficult because of the thin and fragile nature of their bodies. A variety of methods have been employed for fixing and preserving, including the use of Bouin's solution, formalin/acetic acid mixture, or freezing. Perhaps one of the best (but most difficult) techniques is described by Hyman (1953). The procedure is summarized as follows: (1) saturate seawater with mercuric chloride ($HgCl_2$)—bring the solution to boil; (2) place the worm in a clean petri dish with only a small amount of seawater—allow the worm to spread out; (3) pour the boiling solution over the worm and allow the solution to cool for at least 30 minutes; (4) transfer the worm to another petri dish containing fresh water, decant, then rinse with three to five changes of fresh water; (5) to flatten a curled specimen, place it between two microscope slides held together with rubber bands—the bands should not be too tight as the specimen should only be flattened, not squeezed; (6) the specimen is then passed through a series of alcohol baths—20%, 30%, 50%, and finally in 70% with a little iodine added—if the brown dye of the iodine fades, there is still $HgCl_2$ remaining, so add small amounts of iodine until fading no longer takes place; (7) transfer the specimen to fresh 70% ethanol as a final preservative.

Alternatively, a much easier method has successfully been used for some specimens. Robust, less delicate flatworms may be gently placed between two glass microscope slides fastened together with a rubber band. This assemblage is then submerged in solutions of magnesium chloride to gradually relax the animal, then fixed in 10% formalin, and later transferred to 95% ethanol, and finally to 75% ethanol for long-term storage.

Preserving Material for DNA Analysis

DNA strands are broken into small pieces if tissues are fixed in formalin or preserved in low concentrations of alcohol. Ideally, fresh tissues or those that have been frozen or rapidly dried, should be used for DNA extraction. However, if tissues are initially preserved and maintained in 95% ethyl or isopropyl alcohol, one may possibly extract high-molecular weight DNA from them, years or even decades later. An alternate preservative is a solution of 20% DMSO (dimethyl sulfoxide) saturated with sodium chloride (NaCl). Tissues preserved in plain table salt (NaCl) can also be used for DNA analysis.

Appendix 1: Summary of Preservation Techniques

Taxon	Anesthetization	Wet Preserved		Dry Preserved
		Initial Fixative	Final Preservative	
Protozoa	—	3%–5% buffered formalin or 50% ethanol	70%–90% ethanol	Shells of radiolarians and foraminiferans can be air dried
Porifera	Not required	Complete immersion in 95% ethanol for 12 hours, then repeat using fresh alcohol	95% ethanol	Soak in fresh water, to eliminate salts, then air dry (not recommended for research specimens)
Hydrozoa (hydroids and hydromedusae)	In bowl of seawater with a few crystals of menthol added, or 8% magnesium chloride gradually added to seawater	20% buffered formalin	10% formalin or 70% ethanol	—
Scyphozoa	Usually not required	20% buffered formalin	10% formalin	—
Anthozoa (octocorals)	Bowl of seawater with 7.5% magnesium chloride gradually added	70% ethanol gradually added after relaxation	70% ethanol	Gorgonians can be sun dried after washing in fresh water
Anthozoa (anemones)	In a bowl of seawater with a few crystals of menthol added	20% buffered formalin	10% formalin	—
Ctenophora	See procedures described in text			
Platyhelminthes (polyclads)	See procedures described in text			
Nemertea	See Nemertea chapter			
Gnathostomulida, Rotifera, Gastrotricha, Kinorhyncha, Nematoda, Nematomorpha, Acanthocephala	—	For most taxa, 3%–10% formalin	70%–90% ethanol or 2%–5% formalin	—

Taxon	Anesthetization	Wet Preserved		Dry Preserved
		Initial Fixative	Final Preservative	
Annelida (clitellates)	0.15% propylene phenoxetol for 10–15 minutes	10% formalin for 24–48 hours	70%–90% ethanol	—
Annelida (polychaetes)	7.5% magnesium chloride solution in bowl of seawater	Bouin's solution for tubicolous worms to remove calcareous tubes; 5%–10% formalin for free-living taxa	70% ethanol	—
Priapulida	7.5% solution of magnesium chloride in bowl of seawater	5% formalin; or Bouin's solution for 24 hours; then 50% ethanol for one hour	70% ethanol	—
Sipuncula	Small amounts of ethanol or magnesium chloride added to a bowl of seawater	70%–90% ethanol or 4% formalin for 12 hours	70% ethanol	—
Echiura	Bowl of seawater with a few menthol crystals; or small amounts of ethanol or 7% magnesium chloride	5% buffered formalin	70% ethanol	—
Mollusca (chitons)	Magnesium chloride (see chiton chapter)	"Chiton stick" with specimen bound flat on a board, then immersed in ethanol 70% or stronger	70%–75% ethanol	—
Mollusca (bivalves)	0.15% propylene phenoxetol; see also bivalve chapter	5% buffered formalin	70% ethanol	Shells can be cleaned in boiling water then dried
Mollusca (shelled gastropods)	Menthol, magnesium sulfate, or magnesium chloride added to bowl of seawater; also freezing	5% buffered formalin	70% ethanol	Shells can be cleaned in boiling water then dried
Mollusca (opisthobranchs)	A bowl of seawater with 8% magnesium chloride gradually added; also freezing	Bouin's solution for 24 hours	70% ethanol	—
Mollusca (cephalopods)	Freezing or anesthesia (see cephalopod chapter)	4% buffered formalin; larger animals should have full strength formalin injected into viscera	70% ethanol	—
Tardigrada	—	5% formalin or 70%–80% ethanol	5% formalin or 70%–80% ethanol	—
Crustacea	Choral hydrate for large animals; propylene phenoxetol for smaller ones	5% buffered formalin for 24–94 hours	70%–90% ethanol after washing in fresh water	—
Pycnogonida	—	70%–90% ethanol	70%–90% ethanol	—
Arachnids and insects	—	70%–90% ethanol	70%–90% ethanol	—
Bryozoa	Bowl of seawater to which 1% stovaine or eucaine is added drop by drop; or menthol or magnesium sulfate crystals added slowly	Bouin's solution or 5% buffered formalin	Washed in fresh water then 70%–90% ethanol	—

(continued)

| Taxon | Anesthetization | Wet Preserved | | Dry Preserved |
		Initial Fixative	Final Preservative	
Kamptozoa	Bowl of seawater to which 1% stovaine or eucaine is added drop by drop; or menthol or magnesium sulfate crystals added slowly; or slow addition of 7.5% magnesium chloride in seawater	3%–5% formalin	70%–90% ethanol	—
Phoronida	7.5% magnesium chloride in a bowl of seawater	Bouin's solution 12–24 hours or 5% buffered formalin	70% ethanol	—
Brachiopoda	Gradually add alcohol to a bowl of seawater not to exceed 10% of the volume	70–90% ethanol	70%–90% ethanol	—
Echinodermata (echinoids)	—	10%–12% buffered formalin or 95%–100% ethanol	70%–90% ethanol	Air dried
Echinodermata (asteroids)	—	10%–12% buffered formalin or 95%–100% ethanol	70%–90% ethanol	Air dried
Echinodermata (ophiuroids)	Magnesium chloride; Epsom salts; see also ophiuroid chapter	10%–12% buffered formalin or 95%–100% ethanol	70%–90% ethanol	Air dried
Echinodermata (holothuroids)	In bowl of seawater to which a few crystals of magnesium sulfate or menthol are added	10%–12% buffered formalin or 95%–100% ethanol	70%–90% ethanol	—
Hemichordata	7% magnesium chloride in bowl of seawater 24 hours	Bouin's solution for	70% ethanol	—
Urochordata	In bowl of seawater containing a few crystals of magnesium sulfate or menthol	Buffered formalin gradually added after relaxation	70%–90% ethanol or buffered 5% formalin	—
Fishes	—	10% buffered formalin	45%–50% isopropyl alcohol or 75 % ethanol	—
Algae and eelgrasses	—	—	—	Submerge in 5% formalin, then place on herbarium sheet in pan of seawater, then affix to herbarium sheet and dry

Appendix 2: The California Academy of Sciences Procedural Protocol for Invertebrate Preservation

Taxon	Field	Lab	Final Preservative
Sponges	95% EtOH	95% EtOH	95% EtOH
Hydroids	7.5% $MgCl_2$ in FW	75–95% EtOH	75% EtOH
Jellyfish	Photograph live	10% formalin in SW	10% formalin
Anemones	7.5% $MgCl_2$ in FW	10% formalin in SW	10% formalin
Octocorals	7.5% $MgCl_2$ in FW	75% EtOH	75% EtOH
Ctenophores	7.5% $MgCl_2$ in FW	Trichloroacetic acid (1 g) in SW (99 ml) for 1/2 hour	70% EtOH
Flatworms	Photograph live	Allow to extend, then 10% formalin for 1/2 hour	75% EtOH
Nemerteans	7.5% $MgCl_2$ in FW	50–75% EtOH	75% EtOH
Polychaetes	7.5% $MgCl_2$ in FW	75–95% EtOH	75% EtOH
Chitons	7.5% $MgCl_2$ in FW	75–95% EtOH	75% EtOH
Shelled gastropods	7.5% $MgCl_2$ in FW	75–95% EtOH	75% EtOH
Shelled opisthobranchs	Photograph live	7.5% $MgCl_2$ in FW; then Bouin's (for histological purposes if shell is not to be retained), or buffered formalin (if shell is to preserved)	75% EtOH
Nudibranchs	Photograph live	7.5% $MgCl_2$ in FW; then Bouin's (for species without calcareous structures in the notum); or formalin (for species with calcareous structures)	75% EtOH
Bivalves	7.5% $MgCl_2$ in FW	75–95% EtOH	75–95% EtOH
Scaphopods	7.5% $MgCl_2$ in FW	75–95% EtOH	75–95% EtOH
Bryozoans	7.5% $MgCl_2$ in FW	75–95% EtOH	75–95% EtOH
Pycnogonids	95% EtOH	75–95% EtOH	75–95% EtOH
Ostracodes	95% EtOH	75–95% EtOH	75–95% EtOH
Barnacles	95% EtOH	75–95% EtOH	75–95% EtOH
Shrimp and crabs	95% EtOH	75–95% EtOH	75–95% EtOH
Isopods and amphipods	95% EtOH	75–95% EtOH	75–95% EtOH
Sea stars	95% EtOH	75–95% EtOH	75–95% EtOH
Ophiuroids	95% EtOH	75–95% EtOH	75–95% EtOH
Holothurians	7.5% $MgCl_2$ in FW	75–95% EtOH	75–95% EtOH
Sand dollars	95% EtOH	75% EtOH	75–95% EtOH
Urchins	95% EtOH (injection)	75% EtOH	75–95% EtOH
Tunicates	7.5% $MgCl_2$ in FW	10% SW formalin	10% formalin

References

Eschmeyer, W. N. and E. S. Herald. 1983. A field guide to Pacific Coast Fishes North America—Peterson Field Guide Series. Boston: Houghton Mifflin Company, 336 pp.

Hyman, L. H. 1953. The polyclad flatworms of the Pacific coast of North America. Bull. Am. Mus. Nat. Hist. 100: 265–392.

Lincoln, R.J. and J. G. Sheals. 1979. Invertebrate Animals—collection and preservation. Cambridge: Cambridge University Press and the British Museum (Natural History), 150 pp.

Russell, H. D. 1963. Notes on methods for the narcotization, killing, fixation, and preservation of marine organisms. Woods Hole, Massachusetts: Systematics-Ecology Program—Marine Biological Laboratory, 70 pp.

Smaldon, G. and E. W. Lee. 1979. A synopsis of methods for the narcotisation of marine invertebrates. Natural History 6, Royal Scottish Museum, Information Series, 96 pp.

Smithsonian Institution web site [http://www.nmnh.si.edu/iz/usap/usapspec.html]: Museum collection management terms and invertebrate processing procedures: methods of fixation and preservation. (This is a detailed and highly useful summary of techniques including the use of various relaxing agents, fixatives, washes, and final preservatives for many animal taxa.)

Steedman, H. F. 1976. Zooplankton fixation and preservation. Monographs on oceanographic methodology 4. Paris: The Unesco Press, 350 pp.

TAXONOMIC ACCOUNTS

Protista

(Plates 1–15)

Introduction

(Plates 1–2)

Protists include a vast group of often unrelated and (for our purposes here) largely microscopic eukaryotes. For decades, many of these magnificent yet understudied organisms were referred to as "protozoans," a term that captured a variety of kingdoms and phyla. Although protozoans are beyond the general scope of this manual, certain groups are so commonly encountered (e.g., foraminifers, members of the Rhizaria) that we include treatment of them here. In addition, in part to draw more attention to these underappreciated groups, we include chapters introducing symbiotic, commensal, and parasitic protozoa, especially those associated with some of the marine invertebrates covered here.

The student will encounter, as examples of free-living marine protests, a dizzying array of free-living ciliates in plankton samples, in benthic sediments, and on hydroids, bryozoans, seaweeds, and many other substrates. Particularly noticeable as epibiotic sessile organisms and often abundant on fouling panels are stalked suctorians (e.g., *Acineta*) and stalked "peritrichous" ciliates in genera such as *Zoothamnium, Vorticella,* and *Cothurnia. Zoothamnium* is sometimes sufficiently abundant as to form dense clusters of white "clouds" only 1 mm or 2 mm tall on bryozoans such as *Bugula* or on fouling panels. They are notable for the ability of a single colony of many zooids to contract in unison suddenly when disturbed, and students may spend hours in careful study of these colonies under the microscope.

Also at times abundant on a variety of substrates are several species of the well-known "bottle animalcules," or folliculinids (plate 1). They may be common on hydroid stems and may literally dot the surface of fouling panels if one examines the microscopic layers. Often with a vivid blue or green tinge, folliculinids typically occupy a small attached flasklike case, out of which emerge prominent ciliated, somewhat flattened "wings." Folliculinids are sometimes mistaken for tiny egg capsules.

Free-living marine protists (other than foraminifers) have received little modern attention on the California and Oregon coasts, and those that have been identified often bear older European names that may not be applicable to our region. However, a large number of estuarine and marine protists that have been brought to Pacific coast estuaries in ship fouling, with oyster introductions from Japan or the American Atlantic coast, and with ballast water and sediment are to be expected.

Works ranging from pictorial classics such as Jahn et al. (1978) to monographic treatments such as those of Kudo (1966), Carey (1992), Hausmann et al. (2003), and Lee et al. (2002) provide an introduction to the variety of taxa that may be encountered. Although largely covering the New England coast, Borror (1973) also provides a useful start for a variety of marine ciliate genera.

We would be remiss not to mention the famous marine rhizopod *Gromia* "*oviformis* Dujardin, 1835" (the name is placed in quotation marks, as it likely represents a species complex around the world). Often seen but unrecognized by both marine ecologists and invertebrate zoologists, *Gromia* is a small (up to perhaps 5 mm, but more typically 1–2 mm) seemingly round (but in reality slightly more pyriform) brown "ball" (plate 2), which is, as Professor Zach Arnold noted in the previous edition of this manual, "mistaken for eggs or fecal pellets unless left undisturbed to develop its characteristic web of nongranular pseudopodia." Arnold (1972) is a monographic review of the species. Arnold (1980, plate 174, fig. 25.9b) provides an excellent color photograph of *Gromia* on the small decorator crab *Scyra acutifrons* at Pacific Grove.

Although typically brown in color, *Gromia* may turn yellow during the spring on the Monterey Peninsula, when their test becomes filled with Monterey pine pollen. *Gromia* are milky white when filled with gametes but may become wine-red when living on kelp holdfasts. Arnold (1980) briefly discusses these various colors and colorful states of *Gromia*.

Gromia is sometimes common in algal and surfgrass (*Phyllospadix*) holdfasts, among tufts of seaweed, in eelgrass roots, or on sponges, hydroids, bryozoans, oyster shells, and so forth. It can also be found in fouling communities on marina floats and pontoons. Molecular studies indicate that *Gromia* is closely related to the foraminifers (Burki et al. 2002; Longet et al. 2004).

Also commonly overlooked or misidentified by many students are the agglutinating (sometimes called arenaceous) foraminifers that form what appear to be nondescript tiny accumulations of sand grains: these little cemented piles can be common on a variety of substrates, including rocks (e.g., the shale of Monterey

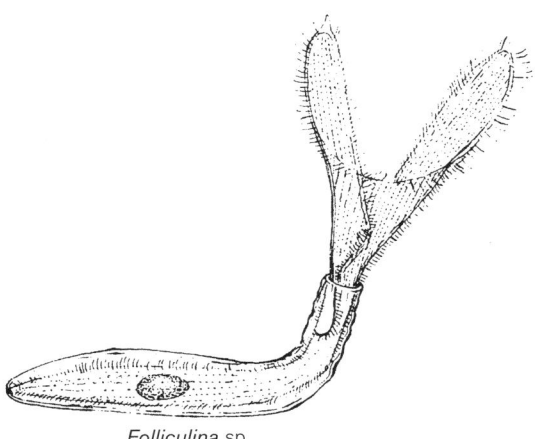

Folliculina sp.

PLATE 1 *Folliculina* sp., 200–500 μm (from Andrews 1921, Hadzi 1947).

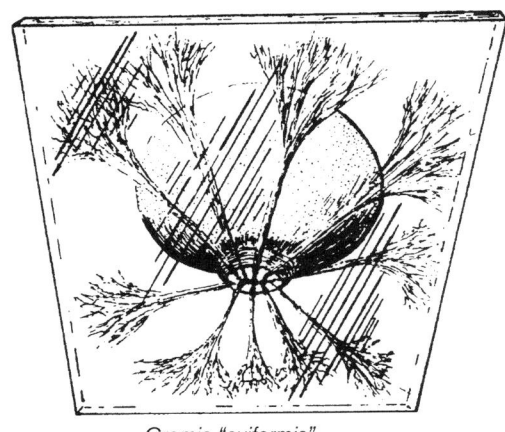

Gromia "oviformis"

PLATE 2 *Gromia "oviformis,"* to 5 mm (Zach Arnold, original).

Bay), eelgrass blades, and hydroids. Common species are in the genus *Iridia,* discussed in the following foram chapter.

References

Arnold, Z. M. 1972. Observations on the biology of the protozoan *Gromia oviformis* Dujardin. Univ. Calif. Publ. Zool. 100: 1–168.

Arnold, Z. M. 1980. Foraminifera: shelled protozoans, pp. 9–20. In Intertidal invertebrates of California. R. H. Morris, D. P. Abbott, and E. C. Haderlie, eds. Stanford University Press, 690 pp.

Borror, A. C. 1973. Marine flora and fauna of the northeastern United States: Protozoa: Ciliophora. NOAA Technical Report, NMFS CIRC (378). Seattle: United States Department of Commerce, National Oceanic and Atmospheric Administration, 62 pp.

Burki, F., C. Berney, and J. Pawlowski. 2002. Phylogenetic position of *Gromia oviformis* Dujardin inferred from nuclear-encoded small subunit ribosomal DNA. *Protist* 153: 251–260.

Capriulo, G. M., ed., 1990. Ecology of marine protozoa. Oxford: Oxford University Press, 384 pp.

Carey, P. C. 1992. Marine interstitial ciliates—an illustrated key. New York: Chapman and Hall, 351 pp.

Hausmann, K., N. Hulsmann, and R. Radek. 2003. Protistology. 3rd ed. Schweizerbart' sche. Stuttgart: Verlagsbuchhandlung.

Jahn, T. L., E. C. Bovee, F. F. Jahn, J. Bamrick, E. T. Cawley, and W. G. Jaques. 1978. How to know the Protozoa. Second ed. New York: McGraw-Hill, 304 pp.

Kudo, R. R. 1966. Protozoology. 5th ed. Springfield: C. C. Thomas, 1188 pp.

Lee, J. J., G. F. Leedale, and P. Bradbury, eds., 2002. An illustrated guide to the Protozoa. 2nd ed. Boston: Blackwell Publishers, 1432 pp.

Longet, D. F. Burki, J. Flakowski, C. Berney, S. Polet, J. Fahrni, and J. Pawlowski. 2004. Multigene evidence for close evolutionary relations between *Gromia* and Foraminifera. Acta Protozool. 43: 303–311.

Foraminiferida

MARY McGANN

(Plates 3–11)

The foraminifera are largely microscopic protists with often beautifully sculptured tests of calcium carbonate laid out in either a spiral arrangement or a row. The name "foraminifera" comes from "foramina," which refers to the presence of a large opening between chambers of the organism's shell, or test.

These single-celled animals are amoeboid in form with granular, netlike pseudopodia (plate 3A). The pseudopodia serve a vital role for the organism in locomotion; the capture, transport, and digestion of prey; expulsion of debris; construction of the test; and in certain species, attachment of the organism to hard substrates. Most foraminifera are microscopic; however, some living South Pacific and fossil species may be more than 1 cm in diameter. All intertidal foraminifera of the eastern Pacific are microscopic.

Foraminifera live in brackish and marine waters. They occur in both floating (planktonic) and bottom-dwelling (benthic) forms. Because planktonic forms live primarily in the upper water column over the deep ocean and shelf break and only very rarely over the continental shelf to the intertidal zone, they are not discussed here. In contrast, benthic foraminifera, which are the object of this key, are abundant, ubiquitous, and easily recoverable and live in the intertidal to abyssal zone. Most live freely, moving on or within the substrate, although a few are temporarily or permanently attached to mollusk shells, seaweed, coral, or other hard surfaces.

Many classification systems have been applied to the order Foraminiferida. A summary of some of the early attempts is provided in Loeblich and Tappan (1964). The classification used by most workers was first proposed by Loeblich and Tappan in 1964 and later revised in 1988. No attempt is made to classify the taxa in this key according to their systematics. Instead, they are separated by recognizable characters for ease of identification.

Terminology

The test of a foraminifera is readily preserved and is used to differentiate species. The general features are illustrated in plate 3A and 3B. Tests are composed of one or more **CHAMBERS**, which are the divisions of the test composed of the cell **WALL** surrounding a cavity or void where the animals live. A **SEPTUM** is a wall between adjacent chambers, which is expressed as a **SUTURE** on the outside surface of the test. The first chamber of a foraminiferal test is referred to as the **PROLOCULUS**. A single revolution (360 degrees) of a test that is wound around in a

LIVING ANIMAL

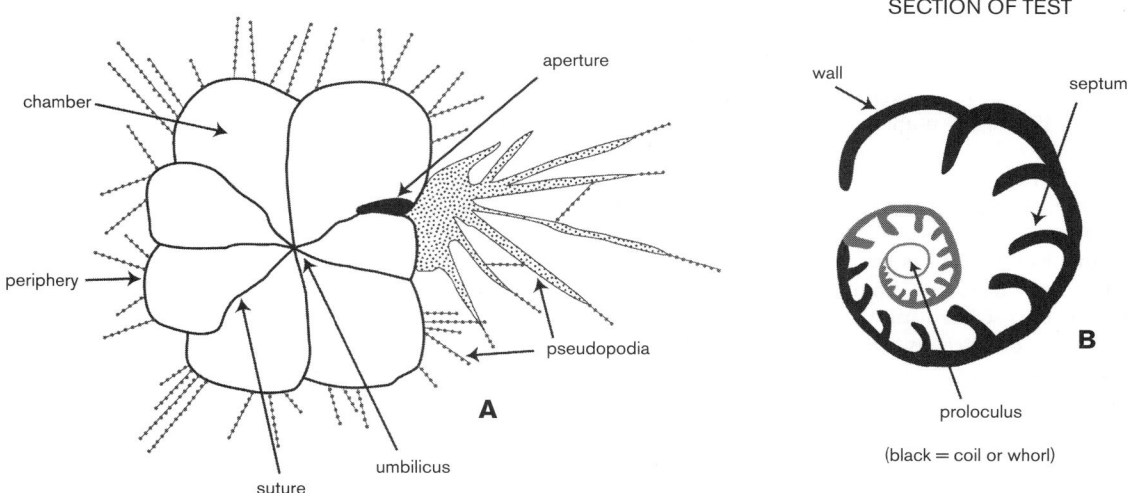

SECTION OF TEST

WALL TYPES

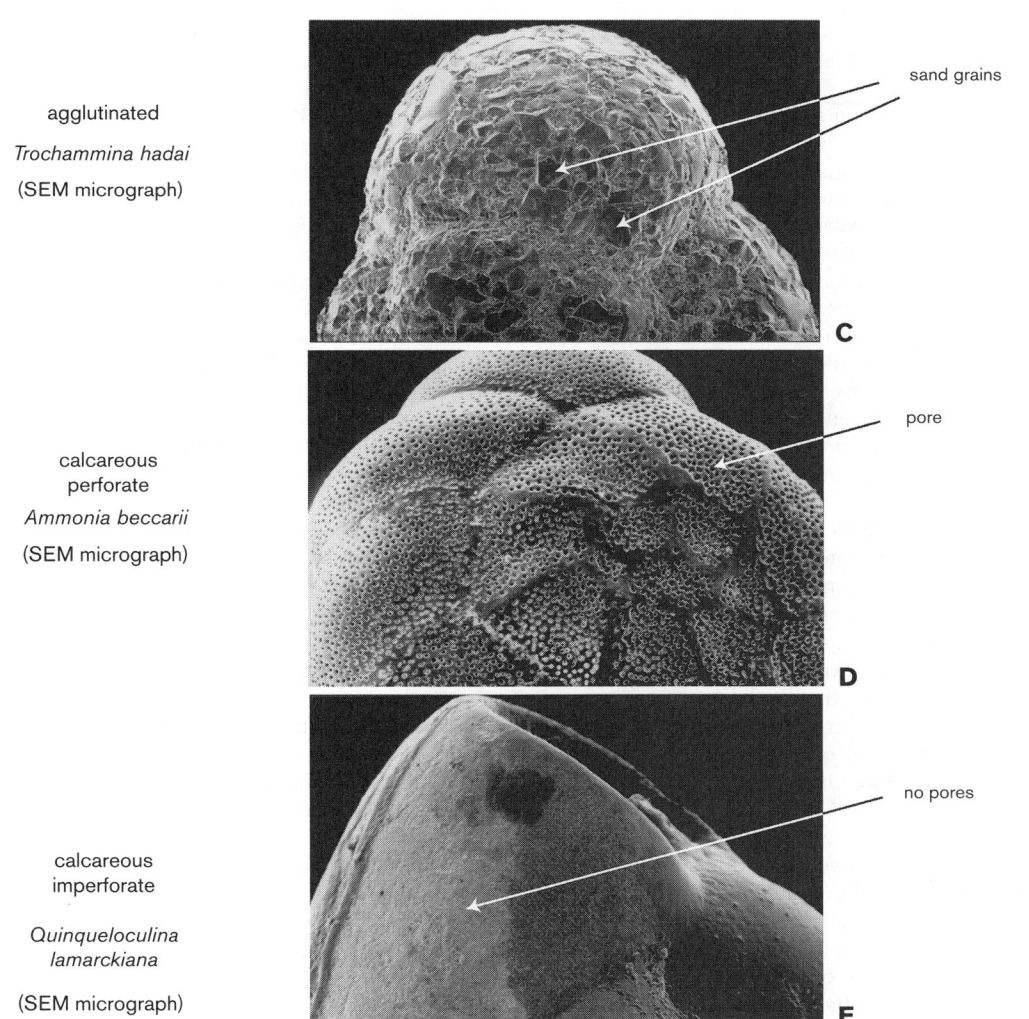

agglutinated

Trochammina hadai

(SEM micrograph)

calcareous
perforate

Ammonia beccarii

(SEM micrograph)

calcareous
imperforate

*Quinqueloculina
lamarckiana*

(SEM micrograph)

PLATE 3 Living animal, section of test, and wall types.

circle or spiral composed of one or, more commonly, many chambers is referred to as a **COIL** or **WHORL**. The perimeter or outside edge of the chambers is the **PERIPHERY**. An **APERTURE** is one or more main openings from the interior to the exterior of the test. **PSEUDOPODIA**, slender threads of protoplasm, extend out into the surrounding environment through both the aperture and, if present, numerous microscopic openings in the test wall, known as **PORES** (plate 3D).

Most foraminifera have rigid tests with walls built of various materials (plate 3C–3E). Some have a rough, granular appearance because they construct their tests by collecting foreign objects such as sand grains, sponge spicules, echinoid spines, fragments of molluscan shells, and other foraminiferal tests, cementing them together with organic, iron, calcareous, or siliceous cement. These tests are referred to as **AGGLUTINATED** or **ARENACEOUS**. A far greater percentage of foraminifera, however, have smooth tests, which they form by extracting calcium carbonate or, in rare cases, aragonite or silica from seawater and then secreting them in one of two forms: with or without pores. The first is referred to as a **CALCAREOUS PERFORATE** or **HYALINE** test and is characterized by a glassy, translucent appearance; the latter is a **CALCAREOUS IMPERFORATE** test and is dense, milky white, and porcelainlike. There are also a very small number of foraminiferal taxa whose tests are composed of **PSEUDOCHITINOUS** material, which is naturally flexible and easily deformed, especially when wet (plate 9P).

Reproduction in foraminifera may occur sexually or asexually, with most species alternating between the two in their life cycles. As a result, species occasionally exhibit **SEXUAL DIMORPHISM**, or differences in test appearance, based on whether they represent the sexual or asexual generation (plate 4A–4B). These differences may include size of the proloculus, overall test size and proportions, and arrangement of the chambers, particularly in the initial chambers. Sexual reproduction is the result of gametes from two individuals meeting and fusing. The gametes may be released haphazardly into the water by the parent or may be released between two fused parental tests (**PLASTOGAMY**; plates 5L, 11E), thus ensuring fertilization. A test resulting from sexual reproduction has a small proloculus, is generally larger in overall size, and is referred to as **MICROSPHERIC** (plate 4A1, 4B1).

In contrast, asexual reproduction results from budding or multiple fission of the parent protoplasm such that each embryo receives a part of the parent nucleus and the parental test is left empty. A test resulting from asexual reproduction is characterized by the presence of a larger proloculus and a smaller overall size, which is called **MEGALOSPHERIC** (plate 4A2, 4B2). Differences in the appearance of the microspheric and megalospheric tests of the same species may complicate systematics. Understanding the concept of sexual dimorphism and consulting more specialized foraminiferal literature may aid in identifying foraminifera.

The general shape of the test is highly variable in foraminifera (plate 4C–4I). Some are rather indistinct, such as an irregular, low dome attached to a shell fragment. More often, though, tests are clearly recognizable shapes, such as a subspherical pouch (**SAC**) or an egg or cucumber seed (**OVATE**). Coiled tests are generally of one of three types as viewed in edge view: (1) **PLANO-CONVEX**—having one side flat and the other curved outward, (2) **BICONVEX**—having both sides inflated or curved outward, and (3) **CONCAVO-CONVEX**—having one side curved inward or hollowed and the other curved outward. In addition, the periphery of the test may be referred to

as **LOBULATE** if the outline of the chambers in side view appears scalloped or like a series of lobes.

The outward appearance of the test also depends on the manner in which the chambers are arranged or modified (plate 5). The simplest form is a test composed of only one chamber, which is referred to as being **SINGLE-CHAMBERED**. Multichambered tests, which are uncoiled, are described by the number of rows in which the chambers are arranged: (1) **UNISERIAL**—in a single row, usually stacked end to end; (2) **BISERIAL**—two rows, which may be symmetrical or offset; (3) **TRISERIAL**—three rows or columns, which also may be symmetrical or offset, with three chambers per whorl; and (4) **MULTISERIAL**—numerous chambers per whorl. These chambers may be inflated or twisted along the long axis of the test. When tests with chambers form a circle or ring, the chamber arrangement is referred to as **ANNULAR**. The chambers of calcareous imperforate tests are usually arranged such that three chambers are visible on one side of the test and two on the other (**TRILOCULINE**), or four chambers are visible on one side and three on the other (**QUINQUELOCULINE**). Occasionally, species are encountered that are characterized by modified chambers covering the central portion of the test. Included are **LOBES**, which are swollen, distended portions of the chamber sometimes with fingerlike projections, and **SUPPLEMENTARY CHAMBERS**, which are smaller chambers either covering or in addition to the larger chambers.

If a foraminiferal test is coiled (plate 6), it may be either **EVOLUTE**, in which all of the earlier whorls and chambers are visible, or **INVOLUTE**, where the later whorls strongly overlap the previous ones, leaving only the final whorl and chambers visible. If the coiling of the test is in one plane and the test looks the same on both sides, it is referred to as **PLANISPIRAL COILING**. These characteristics can be combined to form a **PLANISPIRAL EVOLUTE** test in which all chambers are visible and the test looks the same on both sides (plate 6A), or a **PLANISPIRAL INVOLUTE** test with both sides appearing similar, but only the last-formed chambers are visible (plate 6B).

When the test is coiled in a spire instead of a plane and the coiling is evolute on one side of the test and involute on the other, this is known as a **TROCHOSPIRAL** or **TROCHOIDAL** test (plate 6C). In a trochospiral test, the evolute side is referred to as the **SPIRAL**, **COILED**, or **DORSAL** side, and the involute side is the **UMBILICAL** or **VENTRAL** side. Often, the aperture is located on the umbilical side of the test. The **UMBILICUS** or **UMBILICAL REGION** is the center of the involute side where the sutures of the chambers meet. It is usually slightly depressed and may be filled with secretions of calcite, such as knobs, thickenings, or a plug (plates 4F, 8I). Tests belonging to the family Cassidulinidae appear to be coiled but in fact are biserially arranged chambers that are rolled up (**ENROLLED**), such that portions of some chambers are seen on both sides of the test (plate 6D). In contrast, most calcareous imperforate tests are not enrolled, but display **MILIOLINE** coiling in which only two chambers, usually thin and elongate, make up each whorl (plate 6E).

Foraminifera come in contact with the outside world primarily through one or more apertures. Most tests have a **PRIMARY APERTURE**, which is the main opening of the test and usually located somewhere on the final chamber. These openings come in a variety of forms (plate 7A–7I). The simplest of these is an opening at the end of a tube. Others occur at the end of the final chamber and are referred to as **TERMINAL**. The terminal aperture may be a simple circular opening, a slit, or a round opening with a projection of calcite, referred to as a

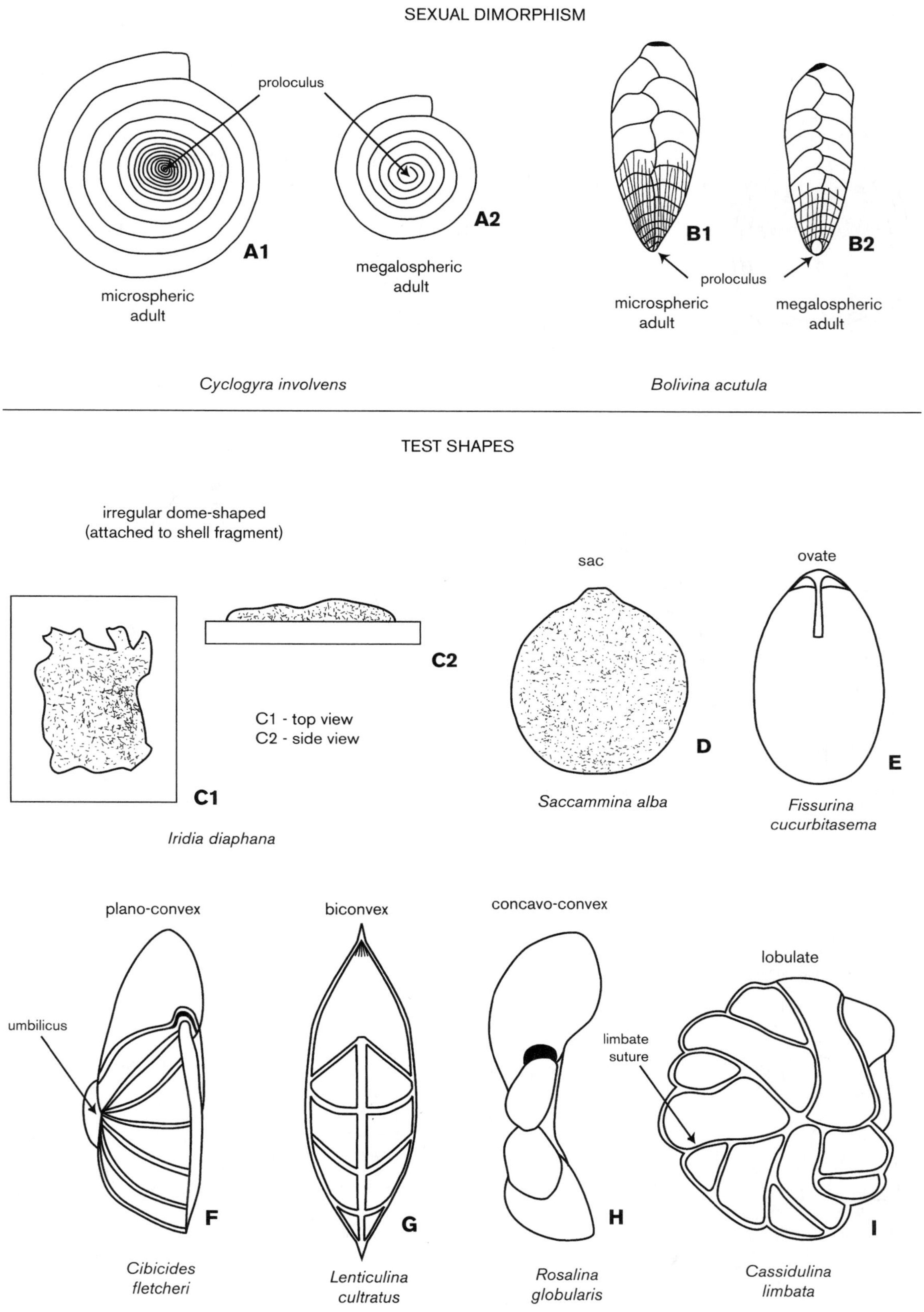

SEXUAL DIMORPHISM

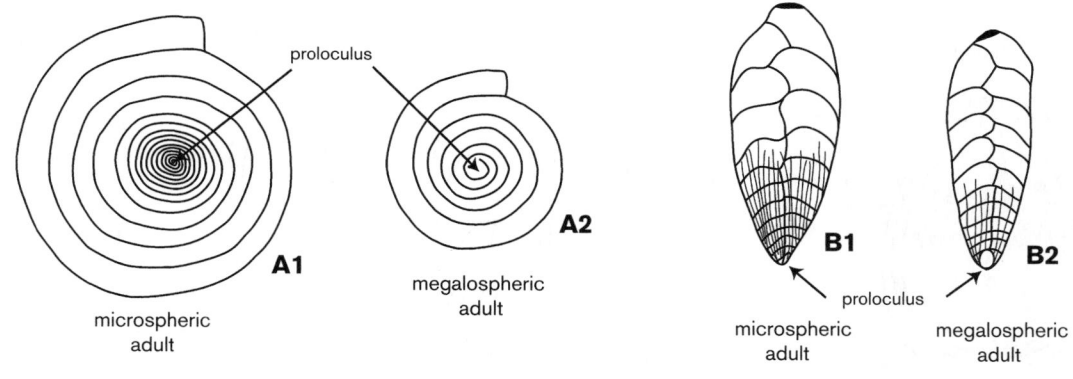

proloculus

A1 microspheric adult

A2 megalospheric adult

Cyclogyra involvens

B1 microspheric adult

B2 megalospheric adult

proloculus

Bolivina acutula

TEST SHAPES

irregular dome-shaped
(attached to shell fragment)

C2

C1 - top view
C2 - side view

C1

Iridia diaphana

sac

D

Saccammina alba

ovate

E

*Fissurina
cucurbitasema*

plano-convex

umbilicus

F

*Cibicides
fletcheri*

biconvex

G

*Lenticulina
cultratus*

concavo-convex

H

*Rosalina
globularis*

lobulate

limbate
suture

I

*Cassidulina
limbata*

PLATE 4 Sexual dimorphism and test shapes.

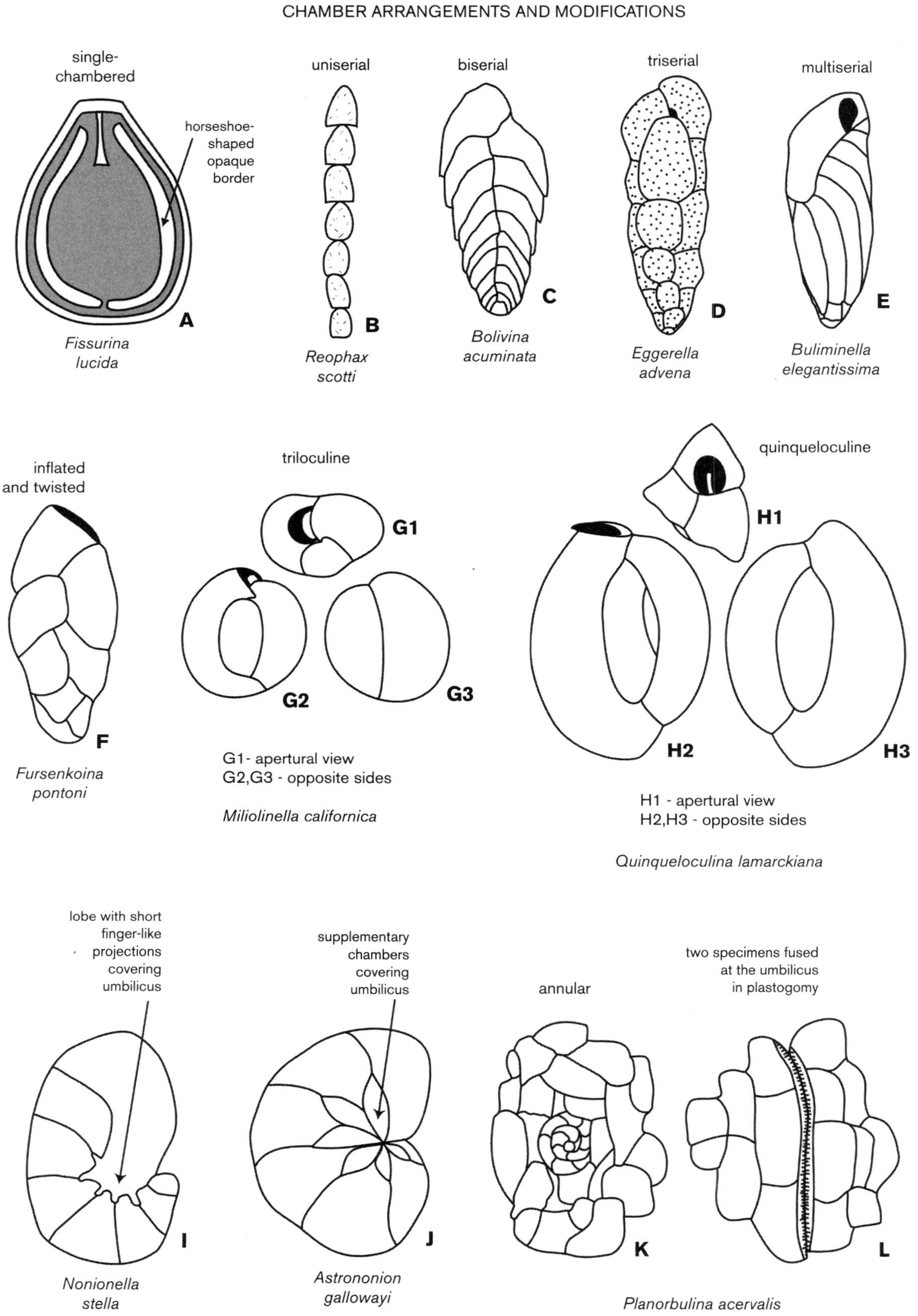

CHAMBER ARRANGEMENTS AND MODIFICATIONS

single-chambered

horseshoe-shaped opaque border

A

Fissurina lucida

uniserial

B

Reophax scotti

biserial

C

Bolivina acuminata

triserial

D

Eggerella advena

multiserial

E

Buliminella elegantissima

inflated and twisted

F

Fursenkoina pontoni

triloculine

G1

G2

G3

G1 - apertural view
G2,G3 - opposite sides

Miliolinella californica

quinqueloculine

H1

H2

H3

H1 - apertural view
H2,H3 - opposite sides

Quinqueloculina lamarckiana

lobe with short finger-like projections covering umbilicus

I

Nonionella stella

supplementary chambers covering umbilicus

J

Astrononion gallowayi

annular

K

two specimens fused at the umbilicus in plastogomy

L

Planorbulina acervalis

PLATE 5 Chamber arrangements and modifications.

TEST COILING

planispiral evolute

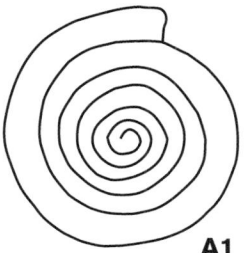  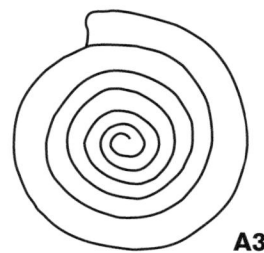

Cyclogyra involvens
A1, A3 - opposite sides
A2 - apertural view

A1   **A2**   **A3**

planispiral involute

*Haplophragmoides
columbiense*

B1, B3 - opposite sides
B2 - apertural view

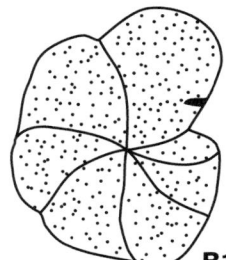 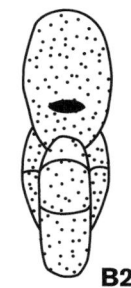 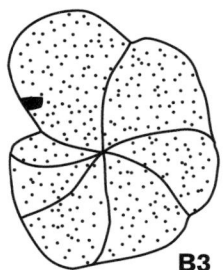

B1   **B2**   **B3**

trochospiral

umbilicus

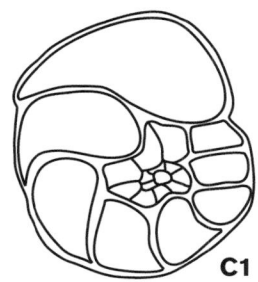 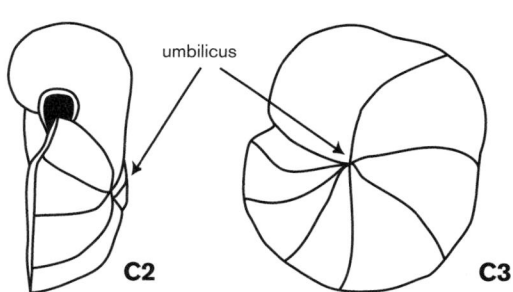

Cibicides lobatulus
C1 - spiral view
C2 - apertural view
C3 - umbilical view

C1   **C2**   **C3**

biserial, enrolled   milioline

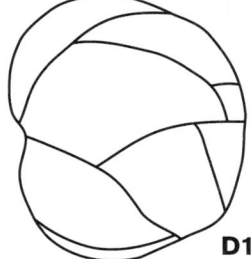 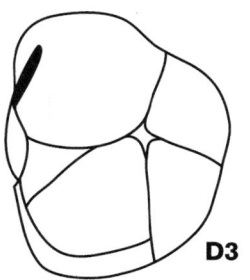 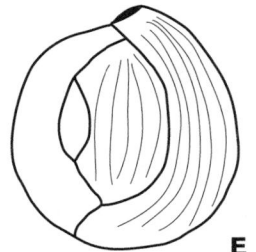

D1   **D2**   **D3**   **E**

Cassidulina subglobosa

D1,D3 - opposite sides
D2 - apertural view

*Quinqueloculina
vulgaris*

E - side view

PLATE 6 Test coiling.

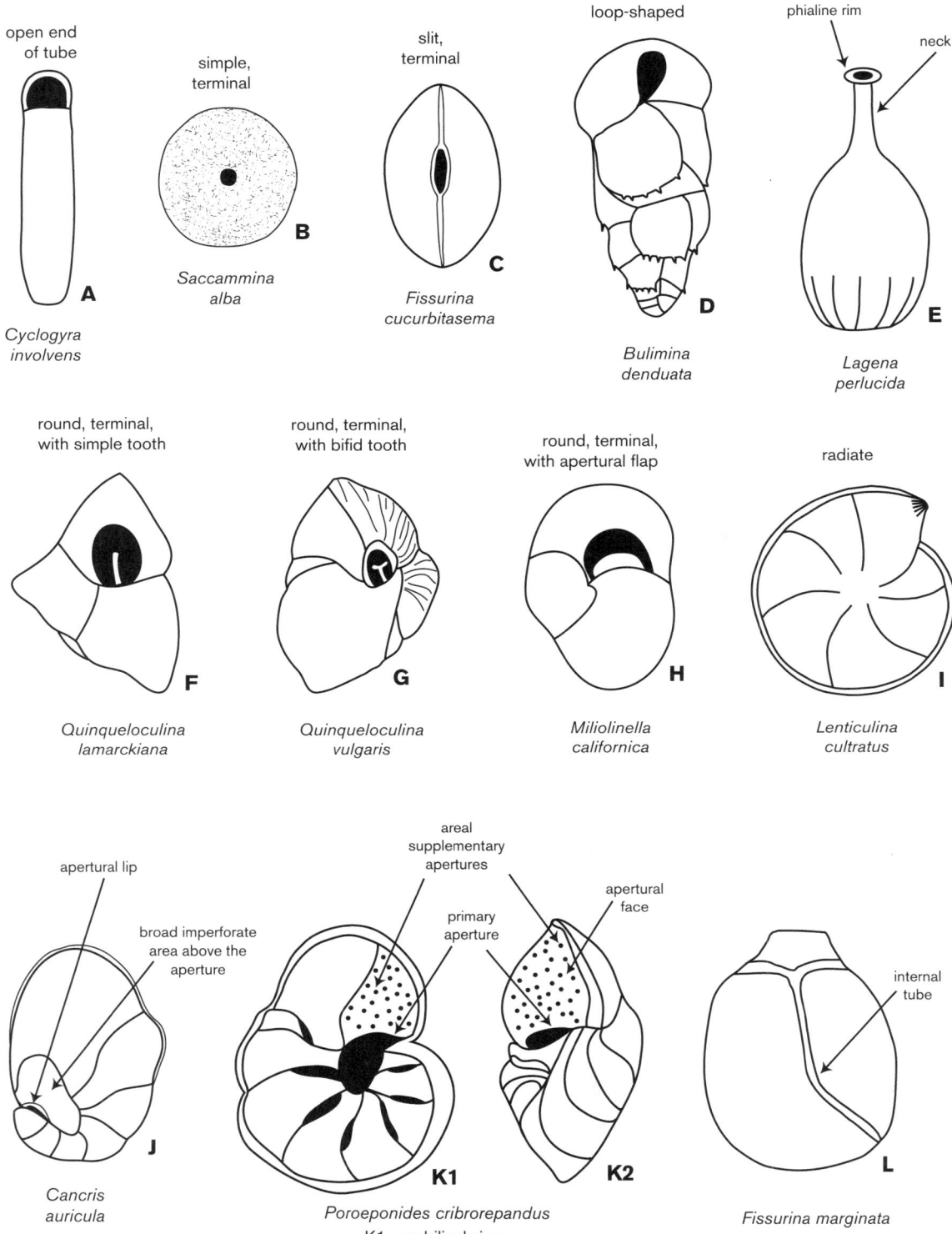

open end
of tube

A

*Cyclogyra
involvens*

simple,
terminal

B

*Saccammina
alba*

slit,
terminal

C

*Fissurina
cucurbitasema*

loop-shaped

D

*Bulimina
denduata*

phialine rim

neck

E

*Lagena
perlucida*

round, terminal,
with simple tooth

F

*Quinqueloculina
lamarckiana*

round, terminal,
with bifid tooth

G

*Quinqueloculina
vulgaris*

round, terminal,
with apertural flap

H

*Miliolinella
californica*

radiate

I

*Lenticulina
cultratus*

apertural lip

broad imperforate
area above the
aperture

J

*Cancris
auricula*

areal
supplementary
apertures

primary
aperture

apertural
face

K1

K2

Poroeponides cribrorepandus
K1 - umbilical view
K2 - apertural view

internal
tube

L

Fissurina marginata

PLATE 7 Apertures and internal apertural modifications.

TOOTH. Teeth commonly come in two forms in foraminifera from central California through Oregon: a single projection (**SIMPLE TOOTH**) or one that splits into two branches (**BIFID**). Occasionally there may be a thickening or flap of calcite covering part of the aperture, referred to as an **APERTURAL LIP** (plate 7J) or **APERTURAL FLAP** (plate 7H), respectively. A terminal aperture may also come at the end of a thin, elongate portion of the final chamber, or **NECK**, which is sometimes surrounded by an outward-turned rim, known as a **PHIALINE LIP**. Although rare, sometimes species exhibit a **RADIATE** aperture, in which numerous short, diverging slits surround the terminal opening.

Apertures may also be located on the face (**AREAL**; plate 7K) or along the final suture at the bottom of the last chamber (**INTERIOMARGINAL**; plate 6C2) as round, slit, **LOOP-SHAPED** (plate 7D), or multiple pores. Interiomarginal apertures in coiled tests may be present in the umbilicus (**UMBILICAL**), along the periphery (**PERIPHERAL**), or halfway between the two (**EXTRAUMBILICAL**) and are best viewed with the test in edge view (**APERTURAL VIEW**), which shows the **APERTURAL FACE** (plate 7K2), or final chamber and aperture on edge. An **EQUATORIAL** aperture (plate 6B2) is a symmetrical opening on the edge of a planispiral test, perpendicular to the axis of coiling, which can be either areal or interiomarginal. **SUPPLEMENTARY APERTURES** are openings larger than pores that occur in addition to, or in place of, the primary aperture and are often areal in location (plate 7K). Occasionally, internal modifications are also visible, such as a **HORSESHOE-SHAPED OPAQUE BORDER** or an **INTERNAL TUBE** inside the tests of *Fissurina* (plates 5A, 7L, 10I, 10K).

Ornamentation on foraminiferal tests occurs in many forms (plate 8). Tests may be **SPINOSE**, in which short, sharp projections of calcite (**SPINES**) cover the entire surface or hang off of individual chambers, or **COSTATE**, as the calcite forms raised ridges (**COSTAE**) on the outside of the test wall. The surface ornamentation may be irregular and rough, often forming ridges (**RUGOSE**), or the ridges may join into a network, meshwork, or box, referred to as **RETICULATE**. If the test has parallel lines, bands, grooves or furrows, it is **STRIATE**. **PUSTULOSE** or **PAPILLATE** refers to tiny raised knobs (**PUSTULES**) or nipplelike projections that are sometimes seen covering the apertural face or umbilicus. The presence of a keel, flange, or thickening—most often found along the periphery—is called **CARINATE**. A thickening at the edge of a chamber, usually referring to sutures, is termed **LIMBATE** (plates 4I, 8M).

More complex types of ornamentation include **RETRAL PROCESSES**, which are fingerlike projections of the chamber that extend backward, usually over a depressed suture. Large deposits of calcite may occur in the umbilicus, known as an **UMBILICAL PLUG**, or on the spiral side, which is referred to as a **DORSAL PLUG**. Imperforate normally refers to entire tests with no pores, but the term also applies to test ornamentation, as tests are occasionally seen with distinct imperforate areas on otherwise perforate tests (plate 7J).

Ecology

The intertidal zone from central California through Oregon consists of rocky shorelines, beaches, and estuaries. Rocky shorelines are adverse environments for foraminifera because the fauna are affected by the pounding of waves, exposure to drying, and a great range in temperature. In this turbulent nearshore environment, most foraminifera live close to the low-water line. They encrust rocks, bryozoan and hydrozoan colonies, other invertebrate shells, or are mixed in with sediment and organic debris in tide pools and rock crevices. These tiny organisms may also attach themselves to plant surfaces such as blades, holdfasts, and floats of algae or surfgrasses.

Rocky shorelines, such as Moss Beach, may have a diverse fauna with as many as 60 species, but are dominated by only a few: *Rosalina globularis* and *Cibicides lobatulus* attached to hydroids and bryozoa, *Glabratella ornatissima* on rock surfaces and surfgrass holdfasts, and *Elphidium* sp. and *Saccammina* aff. *alba* in tide pools (Steinker 1976). *Patellina corrugata* and *Spirillina vivipara* may occur in the folds of red and green algae (Arnold 1980). Other rocky tide pools along California and Oregon also contain common *Rosalina globularis* and *Glabratella ornatissima*, as well as *Cibicides fletcheri, Elphidiella hannai, Elphidium* spp., *Glabratella pyramidalis, Haplophragmoides columbiense, Neoconorbina opercularis*, and various miliolids (Detling 1958, Cooper 1961). Upon death, most tests are either destroyed by abrasion or transported to a depositional site offshore. As a result, foraminiferal assemblages collected along rocky shorelines bear little resemblance to the original fauna, often preferentially concentrating the more robust tests.

Beach environments are also characterized by pounding surf, desiccation, and extreme temperature variations. Again, faunas here may be diverse but are dominated by a few taxa, including *Trochammina kellettae, Buccella tenerrima, Glabratella ornatissima, Cassidulina limbata, Elphidiella hannai, Discorbis monicana, Quinqueloculina ackneriana,* and *Miliolinella californica* (Cooper 1961). Erskian and Lipps (1987, J. Foram. Res. 17: 240–256) found *Glabratella ornatissima* particularly well adapted to the shallow turbulent zone in Horseshoe Cove near Bodega Bay. Juvenile asexual (**AGAMONT**) individuals are released and settle in the red alga *Corallina chilensis*. In spring through summer, as upwelling increases the food supply, the adult agamonts reproduce, forming a sexual (**GAMONT**) generation. In the fall, the gamonts associate in plastogamy, and their abundance in the algae decreases. In the winter, the plastogamous pairs are weakly attached to the algae. Because wave action is strongest during this season, these pairs and other adult agamonts present are transported by waves, surge, and currents and dispersed with intertidal and subtidal sediments. This cycle ensures that the abundance of new recruits is highest when food is readily available and takes advantage of intense winter wave action to disperse them.

Foraminiferal assemblages obtained on exposed sandy beaches, as with the rocky shorelines faunas, are highly modified. Abrasion of tests by wave action leads to loss of fragile tests and loss of diversity, leaving behind the larger, more resilient tests. Occasionally, tests of deeper-dwelling foraminifera or older, fossilized individuals from nearby weathered exposures may be included in the beach sediments.

Estuaries off central California and Oregon are typically low-energy environments characterized by extreme ranges in temperature and salinity (brackish to fully marine), depending on tidal range and the amount of freshwater entering the system, with sediments ranging from sand to clay. They are further subdivided into tidal marshes, tidal mudflats, and subtidal regions, with each habitat having a distinct flora and foraminiferal fauna (Hunger 1966, M.S. thesis, Oregon State University, Corvallis, 112 p.; Phleger 1967; Scott 1974, Northwest Science 48: 211–218; Jennings and Nelson 1992).

Tidal marshes form the upper half of the intertidal zone and can be divided into high and low marsh. High marsh faunas are dominated by *Trochammina macrescens, T. inflata, Miliammina fusca,* and *Haplophragmoides subinvolutum*. Low marsh faunas include *T. macrescens, T. inflata, M. fusca, Ammotium salsum,* and *Ammobaculites exiguus*. Tidal mudflat assemblages

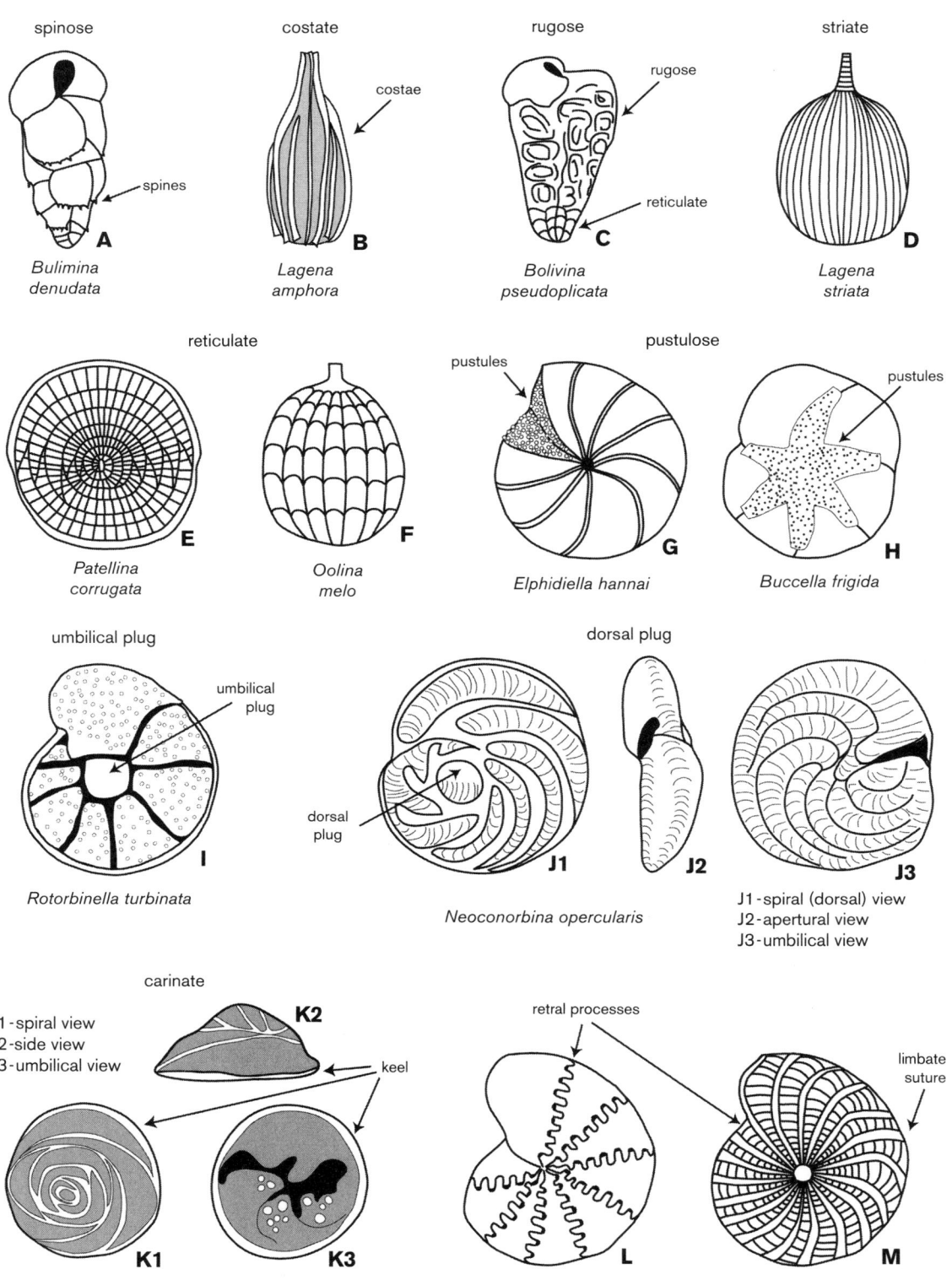

spinose

costae

costate

rugose

rugose

reticulate

striate

spines

Bulimina denudata

Lagena amphora

Bolivina pseudoplicata

Lagena striata

reticulate

pustulose

pustules

pustules

Patellina corrugata

Oolina melo

Elphidiella hannai

Buccella frigida

umbilical plug

dorsal plug

umbilical plug

dorsal plug

Rotorbinella turbinata

Neoconorbina opercularis

J1 - spiral (dorsal) view
J2 - apertural view
J3 - umbilical view

carinate

K1 - spiral view
K2 - side view
K3 - umbilical view

keel

retral processes

limbate suture

Neoconorbina terquemi

Elphidium excavatum

Elphidium crispum

PLATE 8 Test ornamentation.

contain *M. fusca, A. salsum, A. exiguus, Reophax nana,* and rare calcareous species such as *Ammonia beccarii,* and *Elphidium* spp. (*E. excavatum* in San Francisco Bay and *E. frigidum* off Oregon). The shallow subtidal region, subject to less extreme environmental conditions, is characterized by a more diverse fauna with fewer agglutinated species and more abundant calcareous perforate forms. Included are abundant *Ammonia beccarii* and *Elphidium* spp., as well as less common *Elphidiella hannai, Haynesina germanica, Eggerella advena,* and *Bolivina acutula.* The introduced species *Trochammina hadai* dominates the subtidal assemblage in San Francisco Bay, comprises a small percentage of the faunas in Bodega Harbor and Tomales Bay, and has not yet been found in Coos Bay. Unlike the rocky shorelines and beach environments, tests in estuaries are transported little, if any, and are not subject to mechanical abrasion. However, some marshes have highly acidic sediment, causing postmortem dissolution of calcareous tests. As a result, some intertidal estuarine foraminiferal faunas may appear exclusively agglutinated even though they originally had a calcareous component.

Collection and Preservation

Along rocky shorelines, foraminifera are best recovered attached to algae or surfgrass from the tide pools. If possible, attempt to recover 1 quart of algae from each site. Foraminifera can also be recovered in sediment and organic debris collected from the tide pools and rock crevices, or encrusting rocks, shells, or other invertebrate colonies. On beaches, their tests may be collected by scooping up the top one-quarter inch of sand along the strandline of the last wave and continuing this process until about a cup or more is gathered. In estuarine marshes or tidal mudflats, use a small scoop or a small open-ended tube pushed an inch down into the substrate to recover a cup of sediment. Shallow subtidal estuarine deposits are most easily obtained off piers, docks, or small boats by using a mini-grab sampler. Again, a cup of sediment should provide a sufficient number of foraminifera in most cases.

In the laboratory, samples to be studied for living foraminifera may be preserved in isopropyl alcohol or a 10–12% solution of formaldehyde and seawater buffered with 1 tablespoon borax/sample and allowed to stand until they are washed. One gram of rose Bengal stain is often added per 1 liter preservative to aid in the identification of living material, as the protoplasm readily takes up this stain and appears dark red (Walton 1952; Bernhard 1988). In this way, it is possible to determine which specimens were collected alive and which were only empty shells as the result of reproduction or death. The living organism occupies only the last few chambers, so if the entire test stains, it is likely due to an organic coating of algae or bacteria.

After the samples have sat in the preservative and stain for at least a day, they may be washed with tap water over nested screens (1 mm [16 mesh] on top, 0.150 mm [100 mesh] in the middle, and 0.63 mm [230 mesh] on the bottom) to separate the different size fractions and remove the finest sediment (<0.063 mm) where few foraminifera are present. If algae, surfgrass, or other plant material is used, it must be agitated or even scrubbed over the nested screens to remove the attached foraminifera. The specimens and sediment are then air- or oven-dried at a low (<40°C) temperature. Foraminifera may be recovered from sandy sediment more easily by immersing the samples in nontoxic sodium polytungstate at a specific gravity of about 2.2–2.4, which causes the tests to float to the surface. Additional sampling and processing techniques are presented by Murray (1973, 1991).

Once the samples are dry, the foraminifera are isolated from the matrix. This is accomplished by spreading out each fraction on a tray and scanning it under a microscope at magnifications of about ×10–×90. Specimens of interest are picked out using a very fine (000-sized) brush, moistened with water, and placed on a small cardboard slide precoated with a water-soluble glue, gum tragacanth. Often, these slides are covered with a glass slide and held together with an aluminum or plastic clip to protect the specimens.

Key to Intertidal Marine and Brackish-Estuarine Benthic Foraminifers

Foraminifera display a wide range in morphology, making them amenable to classification in a key. Classification is based on the wall type (e.g., agglutinated, calcareous perforate, calcareous imperforate, and pseudochitinous), the number and arrangement of chambers, the shape and number of apertures, the character of test ornamentation, and whether the organism lives freely or is attached. The following dichotomous key should permit identification of most central California through Oregon intertidal and estuarine benthic foraminifera and is patterned after one presented by Todd and Low (1981) for benthic foraminifera of the northeastern United States. Among the most useful publications describing and illustrating assemblages from California and Oregon are Arnold (1980), Cushman and McCulloch (1939, 1940, 1942, 1948, 1950), Lalicker and McCulloch (1940), and Lankford and Phleger (1973).

1. Test agglutinated 2
— Test secreted. 21
2. Test single-chambered 3
— Test more than one chamber 4
3. Test attached; irregular dome-shaped chamber with short projections; finely or coarsely arenaceous; no aperture present; length variable, 1.0–5.0 mm (plates 4C, 9A) . . . . . .
. . . . *Iridia diaphana*
— Test not attached; subspherical; wall consists of large grains; aperture a simple opening at the end of a neck; diameter 0.70–0.80 mm (plates 4D, 7B, 9B) . . . .
. . . . *Saccammina alba*
4. Chambers arranged in a row or rows 5
— Chambers arranged in a coil. 11
5. Test uniserial 6
— Test multiserial 9
6. Test coiled at the beginning 7
— Test not coiled at the beginning. 8
7. Test has an initial coil and an uncoiled part; the uncoiled part straight, with chambers stacked on top of one another; width of coiled part, 0.2 mm; overall length 0.6 mm (plate 9C) . . . . *Ammobaculites exiguus*
— Test only partially uncoiled; the later chambers reach back toward the initial coil (plate 9D) . . . . . . *Ammotium salsum*
8. Test flexible; thin and elongate; largely composed of chitin; chambers subcylindrical near the base, later ones overhanging previous chambers; aperture tiny and terminal; length ≥1.0 mm (plates 5B, 9E) . . . . *Reophax scotti*

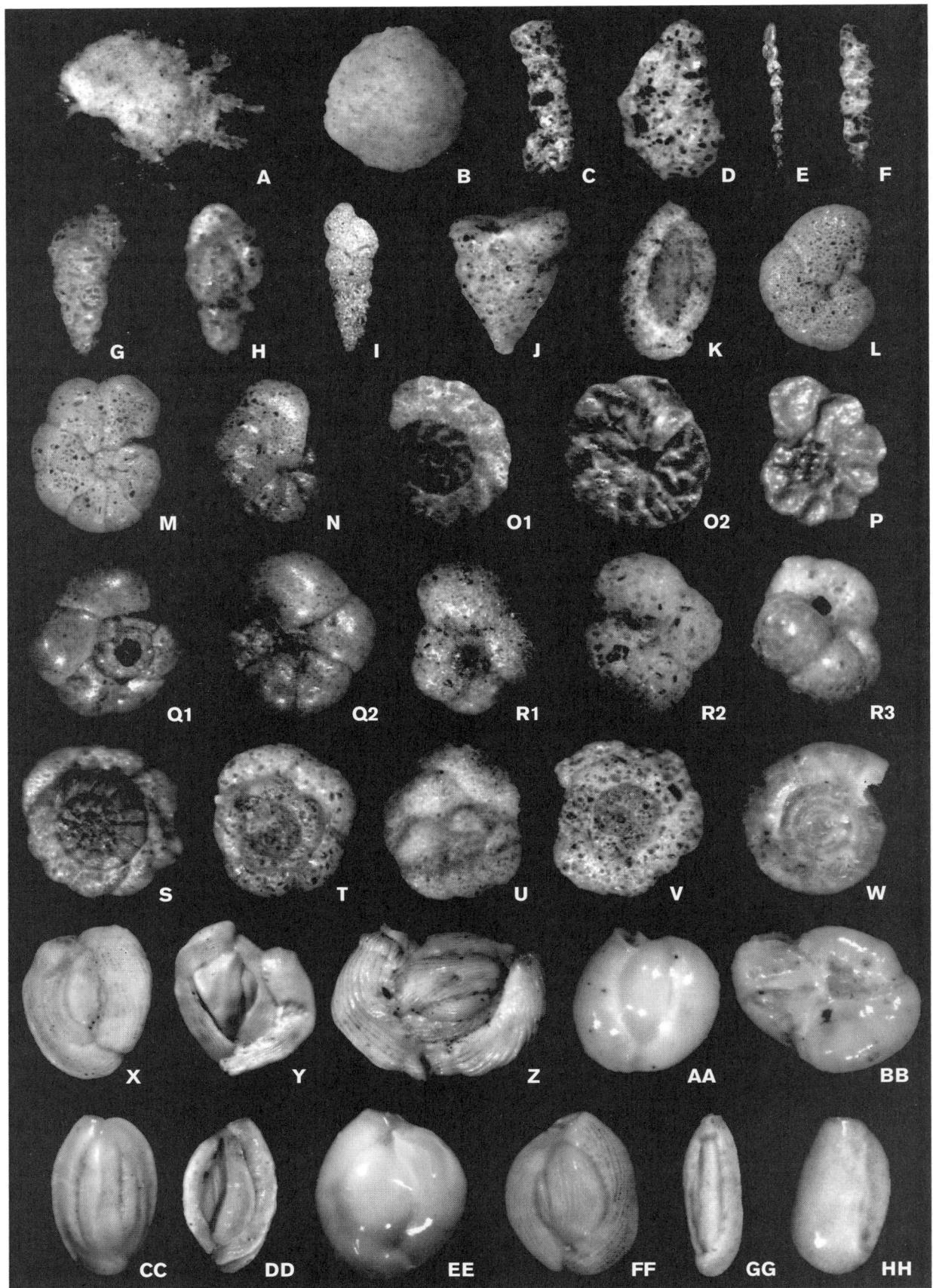

PLATE 9 A, *Iridia diaphana;* B, *Saccammina alba;* C, *Ammobaculites exiguus;* D, *Ammotium salsum;* E, *Reophax scotti;* F, *Reophax nana;* G, H, *Eggerella advena;* I, *Textularia earlandi;* J, *Textularia schencki;* K, *Miliammina fusca;* L, *Haplophragmoides columbiense;* M, *Haplophragmoides columbiense evolutum;* N, *Haplophragmoides subinvolutum;* O1–O2, *Trochammina kellettae;* P, *Trochammina macrescens;* Q1–Q2, *Trochammina inflata;* R1–R3, *Trochammina hadai;* S, *Trochammina squamiformis;* T, *Trochammina pacifica;* U, *Trochammina charlottensis;* V, *Trochammina rotaliformis;* W, *Cyclogyra involvens;* X, *Massilina pulchra;* Y, Z, *Massilina* sp.; AA, *Miliolinella californica;* BB, *Pateoris hauerinoides;* CC, *Quinqueloculina lamarckiana;* DD, *Quinqueloculina angulostriata;* EE, *Quinqueloculina vulgaris;* FF, *Quinqueloculina flexuosa;* GG, *Quinqueloculina laevigata;* HH, *Quinqueloculina bellatula.*

— Test rigid; small and compressed; composed of a large number of chambers; chambers roughly rectangular, more broad than high; aperture terminal; length 0.80 mm (plate 9F) . . . . *Reophax nana*

9. Test biserial 10
— Test triserial; wall finely agglutinated, smoothly finished; aperture a low slit near the base of the apertural face; length up to 0.65 mm, width 0.20 mm (plates 5D, 9G, 9H) . . . . *Eggerella advena*

10. Test very elongate, minute; straight or slightly curved, 12 or more chambers; aperture a slit at the base of the apertural face; length 0.4–0.5 mm, width up to 0.1 mm (plate 9I) . . . . *Textularia earlandi*
— Test broadly triangular in front view; short; about six pairs of chambers, rapidly increasing in size; aperture a slit at the base of the apertural face; length up to 0.90 mm, width 0.50–0.60 mm (plate 9J) . . . . *Textularia schencki*

11. Coiling milioline; chambers in quinqueloculine arrangement; very finely agglutinated and smoothly finished; test siliceous on an organic base and insoluble in hydrochloric acid; length 0.30–0.80 mm, width 0.15–0.50 mm (plate 9K) . . . . *Miliammina fusca*
— Coiling spiral 12

12. Coiling planispiral and involute; the same on both sides 13
— Coiling trochoidal; different on the two sides 14

13. Test flattened in side view, sometimes becoming slightly evolute; six to seven chambers, triangular-shaped; chambers slightly inflated; finely agglutinated; sides somewhat flattened; umbilicus well developed on both sides; aperture equatorial, a slit perpendicular to the coiling axis up in the apertural face; length 0.35–0.70 mm, width 0.25–0.50 mm, width 0.15–0.20 mm (plates 6B, 9L, 9M) . . . . *Haplophragmoides columbiense* and *H. c. evolutum*
— Test more inflated in side view and strongly involute; seven to eight chambers, triangular-shaped; wall very smoothly finished; aperture a low arched opening at the base of the apertural face with an overhanging lip; diameter 0.40–0.65 mm, thickness 0.15–0.30 mm (plate 9N) . . . . *Haplophragmoides subinvolutum*

14. Test flexible when wet; largely composed of chitin 15
— Test rigid, not flexible when wet 16

15. Many chambers (10 or more) in last whorl; test extremely flat, scale-like; three to four whorls; ventral side slightly concave with gently curving sutures or lobed near edge of test; dorsal side slightly convex; diameter 0.25–0.40 mm (plate 9O) . . . . *Trochammina kelletae*
— Fewer chambers (five to eight) in last whorl; test flattened; three whorls; sutures nearly at right angles to edge of test; concavity of individual chambers due to the contraction of the test upon drying; 0.25–0.35 mm (plate 9P) . . . . *Trochammina macrescens*

16. Fine-grained, smooth surface with a dull shine; chambers inflated; five to six chambers in last whorl; sutures distinct and nearly at right angles to edge of test; umbilicus open and deep; early chambers usually accentuated by being darker brown than the remaining chambers; aperture at base of final chamber; diameter 0.50–1.0 mm (plate 9Q) . . . . *Trochammina inflata*
— Medium- to coarse-grained 17

17. Test high in side view; occasionally four but usually five inflated, subglobose chambers in the last whorl; umbilicus deep; sutures slightly curved on spiral side, nearly radial ventrally; wall grainy, coarsely arenaceous; color white, reddish brown, or yellowish brown; diameter 0.40–0.52 mm (plates 3C, 9R) . . . . *Trochammina hadai*
— Test low in side view 18

18. Numerous chambers (eight to 15) in initial whorls on spiral side, higher than broad; chambers reduced to only six in final whorl; six chambers visible on ventral side, triangular, often with slightly raised lip-like portion over the aperture; spiral side convex, ventral side concave; diameter 0.25–0.30 mm (plate 9S) . . . . *Trochammina squamiformis*
— Few chambers (four to seven) in initial whorls on spiral side 19

19. Chambers longer than broad on spiral side, somewhat rectangular; those on ventral side roughly triangular; sutures radial; four to five chambers in last coil; somewhat flat on ventral side, but not as flat as *T. charlottensis*; aperture a narrow slit at the base of the ventral face of the last-formed chamber; diameter 0.30–0.50 mm (plate 9T) . . . . *Trochammina pacifica*
— Chambers half moon–shaped on spiral side 20

20. Generally six chambers on ventral side, roughly triangular to bell-shaped; sutures on spiral side long and curved; test flat on ventral side; umbilicus slightly depressed; aperture narrow, ventral near the umbilicus; diameter 0.30–0.40 mm (plate 9U) . . . . *Trochammina charlottensis*
— Generally four chambers on ventral side; sutures on spiral side long and curved; test flat on ventral side; umbilicus slightly depressed; aperture narrow, ventral near the umbilicus; diameter 0.30–0.50 mm (plate 9V) . . . . *Trochammina rotaliformis*

21. Wall imperforate 22
— Wall perforate 31

22. Coiling planispiral; initial chamber globular but may be obscured by subsequent chamber; second chamber undivided, tubular, enrolled; test large with many overlapping whorls; surface smooth or with transverse growth lines; aperture at open end of tube; diameter 0.30–0.40 mm (plates 4A, 6A, 7A, 9W) . . . . *Cyclogyra involvens*
— Coiling milioline 23

23. Chamber arrangement quinqueloculine in early stages, but not later 24
— Chamber arrangement quinqueloculine throughout 25

24. Chamber arrangement quinqueloculine in early stages, later ones added in a single plane; test compressed with an acute periphery; numerous costae, curved and oblique; length 1.0–2.5 mm (plate 9X–9Z) . . . . *Massilina pulchra*
— Chamber arrangement quinqueloculine in early stages, triloculine later; test smooth; aperture with flaplike tooth; diameter 0.40 mm (plates 5G, 7H, 9AA) . . . . *Miliolinella californica*

25. Aperture lacks a tooth; test is flattened; nearly circular in outline; rounded on the periphery; diameter up to 0.85 mm (plate 9BB) . . . . *Pateoris hauerinoides*
— Aperture has a tooth 26

26. Test angled on the periphery 27
— Test rounded on the periphery 28

27. One keel on each chamber; test smooth; triangular in apertural view; aperture terminal, round, with simple tooth; length 0.80 mm, width 0.60 mm (plates 3E, 5H, 7F, 9CC) . . . . *Quinqueloculina lamarckiana*

— Two keels on each chamber; chambers quadrate in apertural view; fine longitudinal ridges; length 1.5 mm, width 0.80 mm (plate 9DD) . . . . . . *Quinqueloculina angulostriata*

28. Test elongate 29

— Test nearly as wide as long; lenticular; smooth or with low costae on later chambers; aperture terminal, round, with bifid tooth; length ≥1.0 mm, width 0.60 mm (plates 6E, 7G, 9EE) . . . . *Quinqueloculina vulgaris*

29. Wall with deeply incised longitudinal grooves, sometimes anastomosing; length 0.90 mm, width 0.70 mm (plate 9FF) . . . . *Quinqueloculina flexuosa*

— Wall smooth 30

30. Last-formed chambers thinner than earlier ones; chambers generally parallel to long axis of test; test length 2.5 times width; sutures depressed; length 0.80 mm, width 0.40 mm (plate 9GG) . . . . *Quinqueloculina laevigata*

— Chambers curved and of fairly uniform diameter; test length two times width; sutures little depressed; length 0.50 mm, width 0.25 mm (plate 9HH) . . . . *Quinqueloculina bellatula*

31. Test single-chambered 32

— Test more than one chamber 42

32. Test a simple sac with an elongate neck 33

— Test a simple sac; neck absent or very short 36

33. Test generally globular 34

— Test elongate, flask-shaped 35

34. Test subspherical; striate; 15–30 thin longitudinal costae seen in side view; neck ornamented with concentric rings; aperture terminal, at end of neck; length 0.35 mm, width 0.25 mm (plates 8D, 10A) . . . . *Lagena striata*

— Test flattened on bottom; greatest width near base; in side view, five to 10 raised costae radiating out from the bottom and only extending at most halfway up the test; tapering neck ornamented with longitudinal costae, occasionally becoming slightly spiral; length 0.30–0.40 mm, diameter 0.25–0.28 mm (plate 10B) . . . . *Lagena pliocenica*

35. Wall is smooth except for a few short longitudinal costae near the base; test thin; base slightly truncated; short cylindrical neck with slightly developed lip; aperture terminal, at end of neck; length 0.40 mm, width 0.15 mm (plates 7E, 10C) . . . . *Lagena perlucida*

— Wall with numerous high, elongate costae, some restricted to body, most from the aperture to the base of the test; basal end slightly rounded; aperture terminal, at end of neck; length 0.40 mm, width 0.15 mm (plates 8B, 10D) . . . . *Lagena amphora*

36. Test circular in section 37

— Test flattened in section 40

37. Ornamentation consists of a reticulated pattern, with cells lined up in vertical rows 38

— Ornamentation consists of longitudinal ridges 39

38. Cells are few in number, about six per vertical row with about 14 vertical rows present; aperture on a very short neck; length and diameter about 0.40 mm (plates 8F, 10E) . . . . *Oolina melo*

— Cells are numerous in number, about 12 per vertical row with about 32 vertical rows present; aperture on a very short neck; length and diameter about 0.40 mm (plate 10F). . . . . *Oolina catenulata*

39. Longitudinal ridges few in number (about 10), thin, sometimes bladelike, and coalescing at the upper end to form a solid collar; test subglobular to pyriform; small; apertural end tapering to a very short, slender neck; length 0.30 mm,

width 0.20 mm (plate 10G) . . . . *Oolina acuticosta*

— Longitudinal ridges numerous (about 24), blunt and thickened, and coalescing at the upper end to form a solid collar; test subglobular; large; round aperture on a very short neck; length up to 1 mm, width up to 0.8 mm (plate 10H) . . . . *Oolina costata*

40. Test translucent, no opaque borders 41

— Horseshoe-shaped opaque border near the outer margin, clear central portion; test outline circular to ovate; aperture at end of test, oval to elongate; internal tube short and seen more easily when test is wet; length 0.30–0.40 mm, width 0.20–0.30 mm (plates 5A, 10I) . . . . *Fissurina lucida*

41. Test ovate and elongate, flattened like a melon seed; smooth, finely perforate wall; thin, marginal keel; aperture terminal, slit, with a slight lip; internal tube extending about one-half the length of the test; length 0.15–0.45, width 0.10–0.20 mm (plates 4E, 7C, 10J) . . . . *Fissurina cucurbitasema*

— Test rounded to ovate; surface smooth; narrow marginal keel; aperture terminal, elongate slit to ovate, with a clear collarlike portion surrounding it; internal tube short to elongate, free for upper one-third to one-half of test, then attached to the wall at the lower end; length 0.15–0.45 mm, width 0.10–0.25 mm (plates 7L, 10K) . . . . *Fissurina marginata*

42. Test two-chambered; coiling primarily planispiral; initial chamber globular; second chamber undivided, tubular, enrolled; test small with many overlapping whorls; surface smooth, hyaline; aperture at open end of tube; diameter 0.2–0.30 mm (plate 10L) . . . . *Spirillina vivipara*

— Test more than two chambers 43

43. Chamber arrangement generally in straight rows 44

— Chamber arrangement coiled 54

44. Chamber arrangement biserial 45

— Chambers arranged in more than two rows 51

45. Chambers inflated and twisted around long axis; wall smooth; aperture elongate, narrow; length up to 0.90 mm (plate 5F) . . . . *Fursenkoina pontoni*

— Plane of test flat 46

46. Periphery serrated; spines formed by the pointed ends of the chambers extending downward; length up to 0.60 mm (plates 5C, 10M) . . . . *Bolivina acuminata*

— Periphery nonserrated 47

47. Wall smooth, no ornamentation; length up to 0.75 mm (plate 10N) . . . . *Bolivina compacta*

— Wall ornamented 48

48. Longitudinal costae on early chambers; length up to 0.75 mm (plates 4B, 10O) . . . . *Bolivina acutula*

— No longitudinal costae 49

49. Chambers reticulated on the bottom portion of the test, becoming rugose on the upper portion; length up to 0.60 mm (plates 8C, 10P) . . . . *Bolivina pseudoplicata*

— Chambers not reticulated 50

50. Chambers with raised lobes on each side of the median line; test small; length up to 0.40 mm (plate 10Q) . . . . *Bolivina vaughani*

— Chambers excavated on each side of two raised median lines; test small; length 0.30 mm, width 0.20 mm (plate 10R) . . . . *Bolivina subexcavata*

51. Chamber arrangement triserial 52

— Chamber arrangement multiserial; test small, slender, elongate; only two to three whorls with as many as 10 chambers in each whorl; aperture a loop in the face of the final chamber, broadest at upper end; length 0.20 mm,

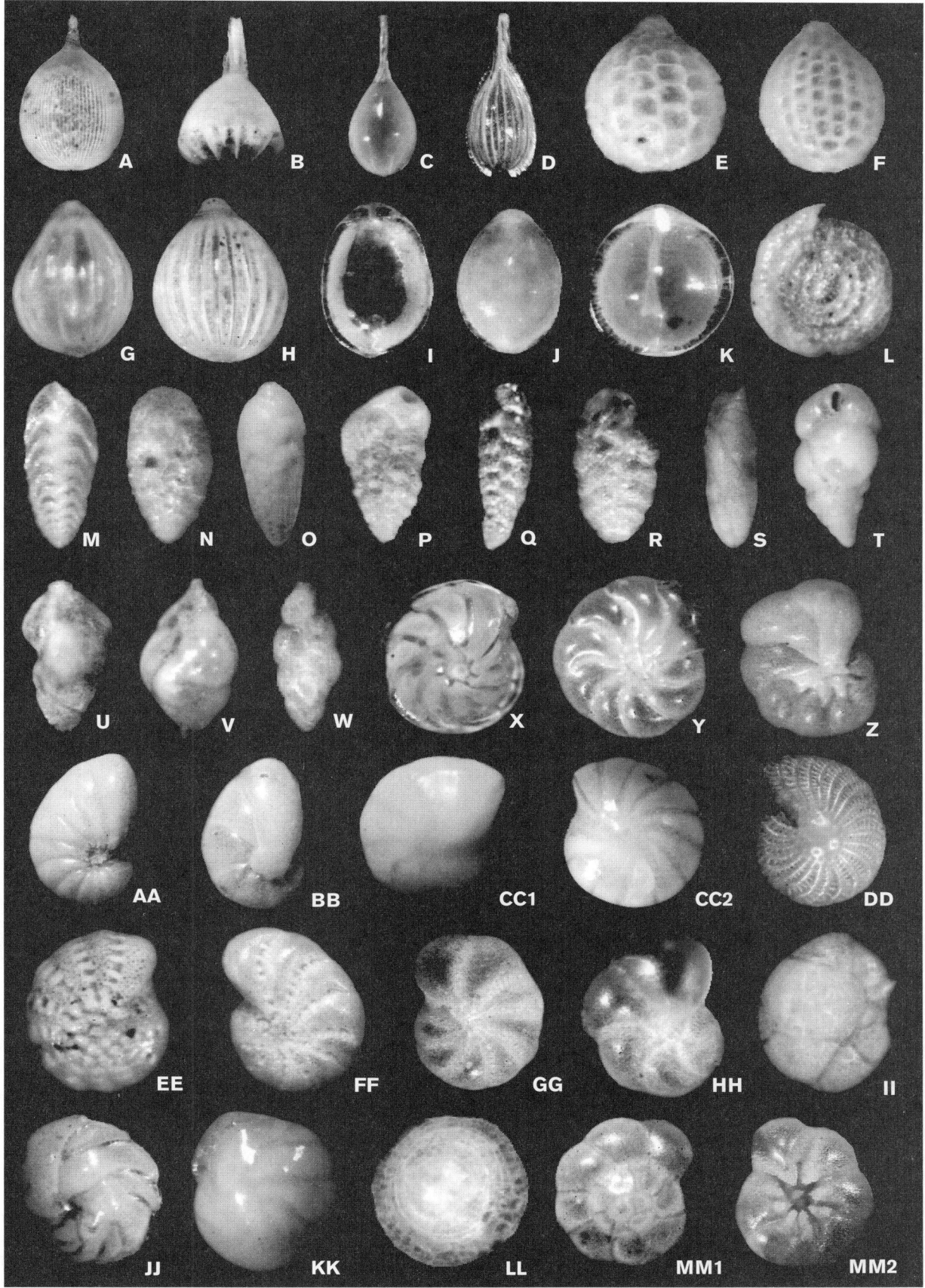

PLATE 10 A, *Lagena striata*; B, *Lagena pliocenica*; C, *Lagena perlucida*; D, *Lagena amphora*; E, *Oolina melo*; F, *Oolina catenulata*; G, *Oolina acuticosta*; H, *Oolina costata*; I, *Fissurina lucida*; J, *Fissurina cucurbitasema*; K, *Fissurina marginata*; L, *Spirillina vivipara*; M, *Bolivina acuminata*; N, *Bolivina compacta*; O, *Bolivina acutula*; P, *Bolivina pseudoplicata*; Q, *Bolivina vaughani*; R, *Bolivina subexcavata*; S, *Buliminella elegantissima*; T, *Bulimina denudata*; U, *Angulogerina angulosa*; V, *Angulogerina baggi*; W, *Hopkinsina pacifica*; X, *Lenticulina cultratus*; Y, *Haynesina germanica*; Z, *Astrononion gallowayi*; AA, *Nonionella basispinata*; BB, *Nonionella stella*; CC1–CC2, *Elphidiella hannai*; DD, *Elphidium crispum*; EE, *Elphidium gunteri*; FF, *Elphidium excavatum*; GG, *Elphidium frigidum*; HH, *Elphidium magellanicum*; II, *Cassidulina subglobosa*; JJ, *Cassidulina limbata*; KK, *Cassidulina tortuosa*; LL, *Patellina corrugata*; MM1–MM2, *Ammonia beccarii*.

width 0.10 mm (plates 5E, 10S) . . . .
. . . . *Buliminella elegantissima*

52. Aperture loop-shaped; test triserial in early stage, sometimes becoming almost uniserial later; chambers not inflated, strongly overhanging previous chamber, lower margin of each with short, downward-angling spines; length 0.35–0.55 mm (plates 7D, 8A, 10T) . . . .
. . . .*Bulimina denudata*
— Aperture terminal, on short neck 53

53. Test roughly triangular in cross section; triserial throughout; considerable variation in chamber appearance (angular to inflated) and surface ornamentation (smooth or with numerous low costae, which are not continuous across the sutures); aperture at the end of a very short, slightly flaring neck; length 0.35–0.50 mm (plate 10U, 10V). . . . . *Angulogerina angulosa*
— Test roughly circular in cross section; initial chambers triserial, later biserial; wall smooth, finely perforate; aperture at the end of a short neck, bordered by a slightly thickened phialine lip; length 0.20–0.30 mm (plate 10W) . . . . *Hopkinsina pacifica*

54. Test coiled, one chamber per row 55
— Test coiled, two chambers per row 66
55. Coiling planispiral; the same on both sides. 56
— Coiling trochoidal; different on the two sides. 68
56. Aperture radiate, protruding and located at the periphery; six to seven chambers, not inflated; sutures flush with the test wall; keel present; length 0.30 mm, width 0.25 mm (plates 4G, 7I, 10X) . . . . *Lenticulina cultratus*
— Aperture not radiate 57
57. Sutures simple 58
— Sutures complex 61
58. Coiling tight 59
— Coiling expanding. 60
59. Wall uncomplicated, no supplementary chambers; fine papillae covering slightly depressed umbilicus; chambers somewhat inflated and translucent; no retral processes; diameter 0.30–0.55 mm (plate 10Y) . . . .
. . . . *Haynesina germanica*
— Supplementary chambers covering umbilicus, forming star pattern; diameter 0.30–0.35 mm (plates 5J, 10Z) . . . .
. . . . *Astrononion gallowayi*
60. Chambers narrow and elongate, 10–15 in last whorl, slowly increasing in size; fringe of small spinose to papillate processes along the outer margin and toward the base of the last-formed chambers; length 0.40–0.80 mm (plate 10AA) . . . . *Nonionella basispinata*
— Chambers roughly wedge-shaped, eight to 10 in the last whorl; last-formed chamber on ventral side developing a starlike lobe covering the umbilicus; length 0.30–0.45 mm (plates 5I, 10BB). . . . . *Nonionella stella*
61. Sutures crossed by retral processes 62
— No retral processes, but two rows of pores adjacent to sutures; pustules present on chamber below aperture; 12–15 chambers in the last whorl; aperture a series of very fine pores along the base of the apertural face and numerous others scattered irregularly over the apertural face; diameter 1.0 mm (plates 8G, 10CC) . . . . *Elphidiella hannai*
62. Large number (>20) of chambers in outer whorl; retral processes pronounced; sutures limbate, so test appears radially ribbed; one row of pores adjacent to sutures; central portion with thickening of clear shell material; aperture a row of small pores at the base of the apertural face; diameter 1.0 mm (plates 8M, 10DD) . . . . *Elphidium crispum*

— Fewer number of chambers (<15) in outer whorl 63
63. Wall coarsely porous and rugose; about 14 chambers in the last whorl; rectangular-shaped retral processes fuse together at the base of the chambers; umbilical region with a group of irregular, slightly raised areas of clear shell material; aperture of several rounded interiomarginal openings; diameter 0.40–0.50 mm (plate 10EE) . . . .
. . . . *Elphidium gunteri*
— Wall finely perforate and smooth. 64
64. Sutures and retral processes distinct; nine to 13 chambers in the final whorl; single row of sutural pores; umbilical region may have an elevated thickening of clear shell material, papillae, or collar; aperture composed of multiple small pores in a single row at the base of the apertural face; diameter 0.23–0.70 mm (plates 8L, 10FF) . . . .
. . . . *Elphidium excavatum*
— Sutures and retral processes inconspicuous 65
65. Sutures and umbilicus depressed and covered by micropapillae; nine to 10 chambers in the last whorl; chambers generally wedge-shaped; final chambers enlarged and projecting beyond the general contour of the test; last-formed chambers often having distinct elongate markings; aperture numerous fine pores at the base of the apertural face; length 0.60–1.0 mm (plate 10GG) . . . .
. . . . *Elphidium frigidum*
— Sutures and umbilicus depressed and filled with very fine granular material, giving them a snowlike appearance and hiding small retral processes; five to six chambers in the last whorl; chambers inflated and sometimes irregular in size; length up to 0.35 mm (plate 10HH) . . . .
. . . . *Elphidium magellanicum*
66. Test nearly circular in side view 67
— Test subglobular and thick in side view; irregular contour; biserial enrolled; three to four pairs of chambers in the last-formed coil; aperture an oblique loop on the face of the last chamber, not in line with the axis of coiling; length 0.70 mm (plates 6D, 10II) . . . . *Cassidulina subglobosa*
67. Sutures distinct, very broadly limbate and raised, with central portion of each chamber "pinched-in"; test biconvex; periphery slightly lobulate, carinate; six pairs of chambers very distinct in the last-formed coil; central portion with thickening of clear shell material; aperture narrow, elongate, parallel to the general axis of coiling; length 0.75 mm, thickness 0.45 mm (plates 4I, 10JJ) . . . .
. . . . *Cassidulina limbata*
— Sutures not raised, barely if at all limbate, distinct when wet, with each chamber much curved, almost becoming spiral; test biconvex and broadly ovate in side view; central portion very thick, often half the length; periphery subacute; six to seven pairs of chambers in last whorl; central area with thickening of clear material; aperture generally parallel to axis of coiling; length up to 0.50 mm, thickness 0.25 mm (plate 10KK) . . . . . *Cassidulina tortuosa*
68. Coiled side convex 69
— Coiled side flat. 83
69. Outline of the test circular 70
— Outline of the test not circular, but notched at the final chamber. 71
70. Complex fluting on umbilical side; test conical in side view, concave on the underside; periphery sharp; chambers crescentic or semicircular, obscured by incomplete transverse septa resulting in a reticulated pattern on the spiral side; diameter 0.18–0.25 mm (plates 8E, 10LL) . . . .
. . . . *Patellina corrugata*

— Umbilical side simple, with extension of chambers triangular to platelike; test low conical in side view; umbilical side concave; periphery acute to carinate; chambers subglobular early, becoming very low and crescentic later; four chambers in final whorl; aperture beneath umbilical flap; diameter 0.30–0.50 mm (plate 8K) . . . .
. . . . *Neoconorbina terquemi*

71. Test has plug of clear shell material 72
— Test lacks plug of clear shell material 75
72. Plug of clear shell material on spiral side; spiral side nearly flat; ventral side flat to slightly concave; nine chambers in last whorl; chambers on ventral side have pinwheel appearance; last few chambers extending down toward the umbilicus; aperture on the umbilical side, beneath the chamber extensions; diameter 0.20–0.40 mm (plate 8J)
. . . . *Neoconorbina opercularis*
— Plug of clear shell material on ventral side, over umbilicus
. . . . 73
73. Test biconvex, round on periphery; presence or absence of umbilical plug may depend on the environment; five to 10 chambers in the last whorl; chambers separate toward the umbilicus, sometimes extending to points, forming angular open spaces; aperture a narrow slit beneath the inner angle of the last chamber, often supplemented by small openings under the umbilical extensions; diameter 0.40–0.90 mm (plates 3D, 10MM) . . . . . *Ammonia beccarii*
— Test plano-convex, angled on periphery 74
74. Dorsal side convex, in a low spire; umbilical plug small; ventral side flat or slightly concave except for the last-formed chamber, which projects; periphery limbate; about six chambers in the last-formed whorl; primary aperture an elongate slit from periphery to umbilicus; secondary aperture umbilical, under the chamber extensions, showing along the sutures and central plug; diameter 0.25–0.40 mm (plate 11A) . . . . *Rotorbinella campanulata*
— Dorsal side very strongly convex in a high, rounded spire; umbilical plug large; ventral side flat to slightly convex; periphery bluntly keeled; usually seven chambers in the last whorl; primary aperture an elongate slit extending from the periphery to the umbilicus; secondary aperture umbilical, under the chamber extensions, showing along the sutures and central plug; diameter 0.27–0.30 mm (plates 8I, 11B) . . . . *Rotorbinella turbinata*
75. Test has umbilical flaps; dorsal side convex, ventral side flat to moderately concave; about five chambers in the last whorl; flap extending from the basal portion of the last four to five chambers toward the umbilical region; arched primary aperture from near periphery to umbilicus; secondary sutural openings at posterior margin of umbilical flap remaining open as later chambers are formed; diameter 0.40 mm (plate 11C) . . . . *Discorbis monicana*
— Test lacks umbilical flaps. 76
76. Umbilicus covered with pustules 77
— Umbilicus open 79
77. Pustules and highly sculptured shell material obscure all but last chamber; five to nine chambers in last whorl; dorsal sutures flush with surface, gently curved, broad, and of darker shell material than chamber walls; aperture ventral, at base of last chamber, usually hidden under the irregular supplemental shell material; diameter 0.60–0.80 mm (plate 11D, 11E). *Glabratella ornatissima*
— Pustules do not obscure chambers 78
78. Periphery slightly lobulate and broadly rounded; five to seven chambers in last whorl, two to three times longer than high; all apertures and sutures on ventral side, umbilicus, and bottom of the last-formed chamber concealed by a thick coat of opaque pustules; arched primary aperture on final chamber midway between umbilicus and periphery; supplementary sutural apertures near the periphery; diameter 0.30–0.50 mm (plates 8H, 11F) . . . . .
. . . . *Buccella frigida*
— Periphery acute and limbate; seven to nine chambers in last whorl, long and thin, four to seven times longer than high; umbilicus, sutures, and bottom of the last-formed chamber with a thick coating of pustulose material; primary and supplementary apertures present, but the only visible ones located on the ventral side at the outer margin of each suture; diameter up to 0.60 mm (plate 11G)
. . . . *Buccella tenerrima*
79. Chambers surrounding open umbilicus ornamented with numerous fine papillae arranged in closely spaced, radiating rows. 80
— Chambers surrounding open umbilicus smooth, not ornamented. 81
80. Test small with dorsal side a low, rounded dome of two whorls, each with five chambers; ventrally flat to concave; aperture ventral, umbilical; sutures on umbilical side radial, spiral side oblique; fused pairs commonly seen (plate 11H, 11I). *Glabratella californiana*
— Dorsal side very high, shaped like a four-sided pyramid of three to five whorls, each with four chambers; widest portion at the top of the last-formed chambers; fused pairs commonly seen; diameter 0.08–0.28 mm, height 0.05–0.35 mm (plate 11J) . . . . *Glabratella pyramidalis*
81. Numerous round areal supplementary apertures scattered over apertural face; primary aperture from umbilicus to periphery; keeled; six to eight chambers per whorl; diameter up to 1.2 mm (plates 7K, 11K) . . . .
. . . . *Poroeponides cribrorepandus*
— No supplementary apertures present 82
82. Test not elongate, sometimes irregular due to attachment to substrate; concavo-convex; chambers increasing gradually in size; about six chambers per whorl; early spiral sutures limbate and flush, in adult whorl nonlimbate and depressed; aperture irregular, low, from periphery to umbilicus, frequently connecting with the aperture of the previous chamber; diameter 0.30–0.50 mm (plates 4H, 11L).
. . . . *Rosalina globularis*
— Test elongate; chambers in last whorl (six to seven) increasing rapidly in size; may have peripheral keel; apertural lip extending over umbilicus; broad imperforate area above the aperture; length 0.80 mm, height 0.60 mm (plates 7J, 11M) . . . . *Cancris auricula*
83. Coiled chambers on the flat side surrounded by annular chambers; chambers numerous; multiple peripheral apertures, commonly one to two oval to semilunar openings on each chamber of the final whorl, each with a narrow bordering lip; smaller supplementary openings occur on both sides; diameter ≥0.80 mm (plates 5K, 5L, 11N) . . . .
. . . . *Planorbulina acervalis*
— Coiled chambers on the flat side make up the entire test
. . . . 84
84. Spiral side strongly evolute, ventral side strongly involute, even in adults 85
— Spiral side strongly evolute, ventral side partially evolute in adults; test flattened; walls coarsely perforate; about 10 chambers in last whorl; sutures distinct, limbate, raised, and confluent with the blunt keel; aperture with a lip;

diameter 0.55–1.0 mm, thickness 0.20–0.25 mm (plate 11O). . . . . *Planulina ornata*

85. Clear shell material present on central portion of involute side. 86

— Clear shell material absent from central portion of involute side 87

86. Convex side low to moderate height; test plano-convex to slightly concavo-convex; about 12 chambers in last whorl; sutures on flat side limbate; sometimes a circle of clear material on spiral side in addition to that on involute side; wall coarsely perforate; arched peripheral aperture, extending along the suture between the last two coils for four to five chambers on the spiral side, with a distinct lip; diameter 0.65 mm, height 0.15 mm (plates 4F, 11P). . . . . *Cibicides fletcheri*

— Convex side very high; test concavo-convex; about 12 chambers in the last whorl; periphery narrowly rounded, not carinate; top view lobulate; sutures slightly limbate and gently curved on both sides; aperture a small opening on the periphery and extending along the suture line of the last two to three chambers on the spiral side; diameter 0.48 mm, height 0.32 mm (plate 11Q) . . . . *Cibicides conoideus*

87. Sutural slits present on involute (convex) side, deep near center; flattened spiral side may be distorted due to attachment to substrate; six to seven chambers in final whorl; aperture on the periphery with a lip, then extending along the suture between the whorls to form an open channel on flat side; coarsely perforate on dorsal convex side, not on flat side; periphery keeled; diameter 0.7 mm, thickness 0.27 mm (plate 11R). . . . . *Montfortella bramlettei*

— Sutural slits absent on involute side; test plano-convex; about eight to nine chambers in the last whorl; test may be deformed due to attachment to substrate; spiral (flat) side more coarsely perforate; sutures curved on both sides, limbate on spiral side only; arched peripheral aperture with a lip; diameter 0.70–1.0 mm (plates 6C, 11S) . . . . *Cibicides lobatulus*

List of Species

The foraminiferal families listed below are based on the classification of Loeblich and Tappan (1988). Geographic ranges of the species are based primarily on the publications of Cooper (1961), Culver and Buzas (1985, 1986, 1987), Cushman and McCulloch (1939, 1940, 1942, 1948, 1950), Lalicker and McCulloch (1940), and Lankford and Phleger (1973), as well as those of Arnal et al. (1980), Arnold (1980), Connor (1975), Detling (1958), Jennings and Nelson (1992), Locke (1971), Martin (1932), McCormick et al. (1994), Means (1965), Quinterno (1968), Riechers (1943), Slater (1965), Steinker (1976), Stinemeyer and Reiter (1958), and Wagner (1978). Mode of life (e.g., infaunal, epifaunal, clinging, or attached) and feeding strategy (e.g., herbivore or detritivore) applies to genera only if Murray (1991) is cited.

Most of the species keyed here occur along much of the West Coast; more limited ranges are noted.

HEMISPHAERAMMINIDAE

Iridia diaphana Heron-Allen and Earland, 1914. Coastal, middle to low intertidal; attached to rocks, invertebrates, or the bases of marine plants; abundant; nondescript test that could be easily overlooked; attached, irregular dome-shaped chamber with agglutinated particles; commonly associated with *I. serialis* Le Calvez, 1935, a one- to two- or more chambered agglutinated form, and *I. lucida* Le Calvez, 1936, which produces a single-chambered, transparent membranous test with no agglutinated material; Monterey Bay to about Point Arena; see Arnold 1980 (morphology, distribution, reproduction); Cushman 1920, Proc. U.S. Nat. Mus. 57: 153–158 (biology); Le Calvez 1936, Arch. Zool. Expér. Gén. 78: 115–131 (morphology); Marszalek 1969, J. Protozool. 16: 599–611 (growth, reproduction).

SACCAMMINIDAE

Saccammina alba Hedley, 1962. Coastal; abundant at Moss Beach, particularly on bushy bryozoans; see Steinker 1976 (distribution).

LITUOLIDAE

Ammobaculites exiguus Cushman and Bronnimann, 1948. Rare to common on mudflats and in marsh channels in Coos Bay, Siuslaw River estuary, Netarts Bay, Gray's Harbor, and Frasier River; infaunal, detritivore (Murray 1991); see Hunger 1966, master's thesis, Oregon State University, Corvallis (distribution); Jennings and Nelson 1992 (distribution); Phleger 1967 (distribution).

Ammotium salsum (Cushman and Bronnimann, 1948) (=*Ammobaculites salsum*). Rare to common in marshes of Coos Bay, Siuslaw River estuary, Netarts Bay, Gray's Harbor, and Frasier River; infaunal, detritivore (Murray 1991); see Hunger 1966, master's thesis, Oregon State University, Corvallis (distribution); Jennings and Nelson 1992 (distribution); Phleger 1967 (distribution).

HORMOSINIDAE

Reophax nana Rhumbler, 1911 (=*Protoschista findens* in Phleger 1967). Rare in coastal intertidal; few in brackish, estuarine; particularly abundant on mudflats; infaunal, detritivore (Murray 1991); see Jennings and Nelson 1992 (distribution); Phleger 1967 (distribution).

Reophax scotti Chaster, 1892. Coastal, in strandline or tide pool deposits; uncommon; infaunal, detritivore (Murray 1991).

EGGERELLIDAE

Eggerella advena (Cushman, 1921) (=*Veneuilina advena*). Minute size; coastal; few to common; locally abundant, particularly in polluted or organically enriched environments; pioneer colonizer of impacted regions; infaunal, detritivore (Murray 1991); see Bandy, Ingle, and Resig 1965, Limnol. Oceanogr 10: 314–332 (pollution); McGann et al. 2003, Mar. Environ. Res. 56: 299–342 (pollution).

TEXTULARIIDAE

Textularia earlandi Parker, 1952 (=*Textularia tenuissima* Earland, 1933). Coastal and estuarine; coastal Mexico to Crescent City; epifaunal, ?clinging, ?detritivore (Murray 1991).

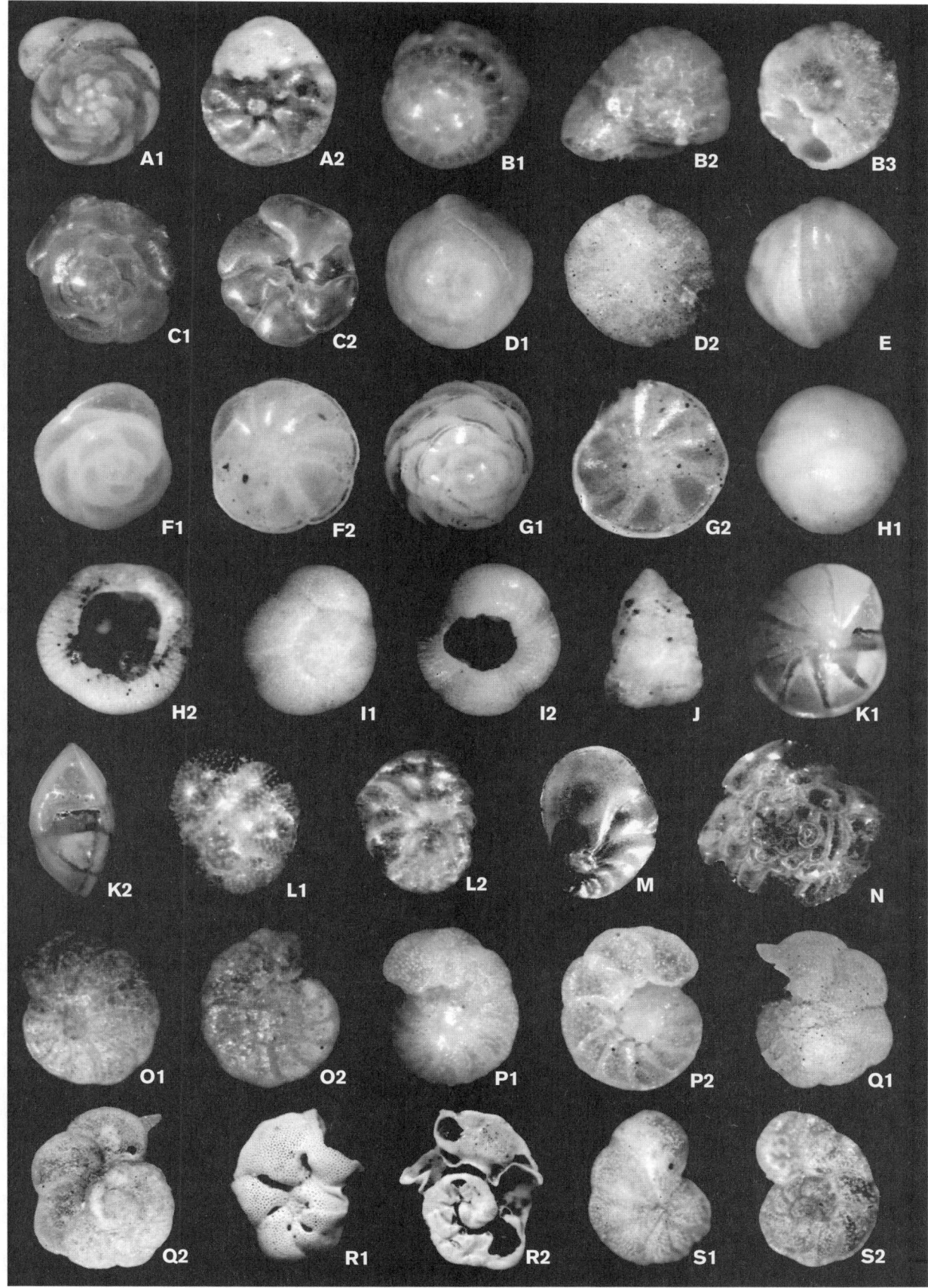

PLATE 11 A1–A2, *Rotorbinella campanulata*; B1–B3, *Rotorbinella turbinata*; C1–C2, *Discorbis monicana*; D1–D2, *Glabratella ornatissima*; E, Plastogamous pair of *Glabratella ornatissima*; F1–F2, *Buccella frigida*; G1–G2, *Buccella tenerrima*; H1–H2, *Glabratella californiana*, H1—dorsal side, H2—ventral side showing dissolved wall and presence of zygotes within test; I1–I2, *Glabratella californiana*; J, *Glabratella pyramidalis*; K1–K2, *Poroeponides cribrorepandus*; L1–L2, *Rosalina globularis*; M, *Cancris auricula*; N, *Planorbulina acervalis*; O1–O2, *Planulina ornata*; P1–P2, *Cibicides fletcheri*; Q1–Q2, *Cibicides conoideus*; R1–R2, *Montfortella bramlettei*; S1–S2, *Cibicides lobatulus*.

Textularia schencki Cushman and Valentine, 1930. Coastal, in strandline or tide pool deposits; uncommon; epifaunal, ?clinging, ?detritivore (Murray 1991).

RZEHAKINIDAE

Miliammina fusca (Brady, 1870) (=*Quinqueloculina fusca*). Brackish, estuarine; locally abundant, particularly in marshes and on mudflats; infaunal, detritivore (Murray 1991); see Jennings and Nelson 1992 (distribution); Phleger 1967 (distribution).

HAPLOPHRAGMOIDIDAE

Haplophragmoides columbiense Cushman, 1925, including the variety *Haplophragmoides columbiense evolutum* Cushman and McCulloch, 1939. Coastal, in strandline or tide pool deposits; occasionally estuarine; rare to common; infaunal, ?detritivore (Murray 1991).

Haplophragmoides subinvolutum Cushman and McCulloch, 1939. Brackish, estuarine; coastal, in strandline or tide pool deposits; few to common; infaunal, ?detritivore (Murray 1991); see Phleger 1967 (distribution).

TROCHAMMINIDAE

Trochammina charlottensis Cushman, 1925. Rare to few in coastal intertidal, and rare in brackish, estuarine; epifaunal or infaunal, herbivore or detritivore (Murray 1991).

Trochammina hadai Uchio, 1962. Introduced from Japan; mudflats, brackish, and estuarine in bays and harbors; locally abundant, particularly in polluted environments; San Diego to Alaska; epifaunal or infaunal, herbivore or detritivore (Murray 1991); see McGann and Sloan 1996, 1999, and McGann et al. 2000 (introduction).

Trochammina inflata (Montagu, 1808) (=*Nautilus inflatus*). Brackish, estuarine; locally abundant, particularly in marshes and on mudflats; epifaunal or infaunal, herbivore or detritivore (Murray 1991); see Jennings and Nelson 1992 (distribution); Phleger 1967 (distribution).

Trochammina kellettae Thalmann, 1932 (=*Trochammina peruviana* Cushman and Kellett, 1929). Coastal, brackish, estuarine; generally few, but locally abundant in coastal beach and tide pool faunas of Monterey Bay and San Francisco; rare in estuaries; Peru to Oregon; epifaunal or infaunal, herbivore or detritivore (Murray 1991).

Trochammina macrescens Brady, 1870 (=*Trochammina inflata* var. *macrescens*). Synonymized with *Jadammina polystoma* Bartenstein and Brand, 1938, in Loeblich and Tappan (1988), but retained here because of its common usage along the Pacific coast and because specimens lack the cribrate openings and equatorial apertures characteristic of *Jadammina*; brackish, estuarine; locally abundant, particularly in marshes and on mudflats; epifaunal or infaunal, herbivore or detritivore (Murray 1991); see Jennings and Nelson 1992 (distribution); Phleger 1967 (distribution).

Trochammina pacifica Cushman, 1925. Coastal; common species of the open continental shelf, where it is locally abundant, particularly in polluted or organically enriched environments; few intertidal, in strandline or tide pool deposits; epifaunal or infaunal, herbivore or detritivore (Murray 1991); see Bandy, Ingle, and Resig 1965, Limnol. Ocean. 10: 314–332 (pollution); Hedley, Hurdle, and Burdett 1964, New Zeal. J. Sci. 7: 417–426 (taxonomy); Watkins 1961, Micropaleo. 7: 199–206 (pollution); McGann et al. 2003, Mar. Environ. Res. 56: 299–342 (pollution).

Trochammina rotaliformis J. Wright, in Heron-Allen and Earland, 1911. Coastal, in strandline or tide pool deposits; few; epifaunal or infaunal, herbivore or detritivore (Murray 1991).

Trochammina squamiformis Cushman and McCulloch, 1939. Coastal, in strandline or tide pool deposits; uncommon; epifaunal or infaunal, herbivore or detritivore (Murray 1991).

CORNUSPIRIDAE

Cyclogyra involvens (Reuss, 1850) (=*Operculina involvens* and *Cornuspira involvens*). Possibly synonymous with *Cyclogyra lajollaensis* (Uchio, 1960) and *C. planorbis* (Schultze, 1854). Coastal, low intertidal to subtidal; on algal fronds (*Ulva*), invertebrates, rocks and debris; rare to common; specimens on test panels in Monterey Bay greatest in August through October, with highest abundances on floating test panels than those in intertidal or subtidal zones (Arnold 1980); see Gougé 1971, master's thesis, Paleontology, University of California, Berkeley (biology); Haderlie 1968, Veliger 10: 327–341 and 1969, Veliger 12: 182–192 (fouling).

HAUERINIDAE

Massilina pulchra Cushman and Gray, 1946. Middle and low intertidal; in washings from algal holdfasts; few; lives in protected areas; epiphytic, herbivore (Murray 1991).

Miliolinella californica Rhumbler, 1936. In some literature as *M. circularis*; uncommon; in coastal and marine-like environments of estuaries; Panama to Oregon; epifaunal, clinging, plants and hard substrate, herbivore (Murray 1991).

Pateoris hauerinoides Rhumbler, 1936 (=*Quinqueloculina subrotunda* forma *hauerinoides*). Distinct, gleaming-white test, with live individuals pinkish in color due to colored protoplasm; great variety in chamber arrangement; cold water; Moss Beach to Sunset Bay; one of the most abundant species in the tide pools of Sunset Bay; see Detling 1958 (distribution).

Quinqueloculina angulostriata Cushman and Valentine, 1930. Coastal; most common in subtidal regions with coarse sediments, but also present in low intertidal, often in washings from bryozoans, coralline algae, or algal holdfasts; few; Baja California to Moss Beach; epifaunal, free or clinging to plants or sediment, herbivore (Murray 1991).

Quinqueloculina bellatula Bandy, 1950 (=*Quinqueloculina akneriana* var. *bellatula*). Coastal, in strandline or tide pool deposits, and brackish to estuarine environments; uncommon; epifaunal, free or clinging to plants or sediment, herbivore (Murray 1991).

Quinqueloculina flexuosa d'Orbigny, 1839. Coastal, in strandline or tide pool deposits; uncommon; Baja California to Moss Beach; epifaunal, free or clinging to plants or sediment, herbivore (Murray 1991).

Quinqueloculina laevigata d'Orbigny, 1826. Coastal, in strandline or tide pool deposits; uncommon; Baja California to Monterey Bay; epifaunal, free or clinging to plants or sediment, herbivore (Murray 1991).

Quinqueloculina lamarckiana d'Orbigny, 1839. Coastal, in strandline or tide pool deposits; uncommon; epifaunal, free or clinging to plants or sediment, herbivore (Murray 1991).

Quinqueloculina vulgaris d'Orbigny, 1826. Found in low intertidal zone, in strandline or tide pool deposits, or among holdfasts of surfgrass and algae, bushy bryozoans, and coralline algae; few to common; epifaunal, free or clinging to plants or sediment, herbivore (Murray 1991).

LAGENIDAE

All coastal, in strandline, or tide pool deposits, uncommon.

Lagena amphora Reuss, 1862. Possibly includes *Lagena sesquistriata* Bagg, 1912, in Lankford and Phleger 1973.

Lagena perlucida (Montagu, 1803; =*Vermiculum perlucidum*).

Lagena pliocenica Cushman and Gray, 1946.

Lagena striata (d'Orbigny, 1839) (=*Oolina striata*).

ELLIPSOLAGENIDAE

All coastal, in strandline, or tide pool deposits, uncommon.

Fissurina cucurbitasema Loeblich and Tappan, 1953. Monterey Bay to Alaska.

Fissurina lucida (Williamson, 1848) (=*Entosolenia marginata* var. *lucida*). Central America to Alaska.

Fissurina marginata Sequenza, 1862.

Oolina acuticosta (Reuss, 1862) (=*Lagena acuticosta*).

Oolina catenulata (Jeffreys, 1848) [=*Entosolenia squamosa* (Montagu) var. α *catenulata* Jeffreys, in Williamson 1848]. Rare; approximately Pismo Beach to Half Moon Bay.

Oolina costata (Williamson, 1858; =*Entosolenia costata*). Baja California to Tomales Bay.

Oolina melo d'Orbigny, 1839. Probably includes *Lagena scalariformis* (Williamson, 1848) [=*Entosolenia squamosa* (Montagu) var. β, *scalariformis*].

SPIRILLINIDAE

Spirillina vivipara Ehrenberg, 1843. Coastal, in strandline or tide pool deposits; prefers rock and coarse sand deposits; found in folds of green and red algae (Arnold 1980); uncommon; epifaunal, clinging to hard substrates (Murray 1991).

FURSENKOINIDAE

Fursenkoina pontoni (Cushman, 1932; =*Virgulina pontoni*). Strandline, tide pool, or estuarine deposits; uncommon; Mexico to Tomales Bay; infaunal, detritivore (Murray 1991).

BOLIVINIDAE

Bolivina acuminata (Natland, 1946) (=*Bolivina subadvena* var. *acuminata*); synonymies: *Bolivina subadvena* Cushman var. *serrata* Natland, 1938, and *Brizalina acuminata*. Strandline, tide pool, or estuarine deposits; few; Panama to Tomales Bay; infaunal to epifaunal, ?detritivore (Murray 1991).

Bolivina acutula (Bandy, 1953) (=*Bolivina advena* var. *acutula*); synonymies: *Brizalina acutula* and *Bolivina advena* Cushman var. *striatella* Cushman, 1925. Possibly includes *Bolivina striatula* Cushman, 1922 identified in San Francisco Bay (Means 1965; Quinterno 1968; Wagner 1978; Arnal et al. 1980; Sloan 1980–1981 unpublished data; Lesen 2003). Strandline, tide pool, or estuarine deposits; rare to common; Panama to Tomales Bay; infaunal to epifaunal, ?detritivore (Murray 1991).

Bolivina compacta Sidebottom, 1905. Strandline, tide pool, or estuarine deposits; rare to common; infaunal to epifaunal, ?detritivore (Murray 1991).

Bolivina pseudoplicata Heron-Allen and Earland, 1930. Coastal, in strandline or tide pool deposits; few; infaunal to epifaunal, ?detritivore (Murray 1991).

Bolivina subexcavata Cushman and Wickenden, 1929. Strandline, tide pool, or estuarine deposits; rare to common; Ecuador to Bodega Bay; infaunal to epifaunal, ?detritivore (Murray 1991).

Bolivina vaughani Natland, 1938. Strandline, tide pool, or estuarine deposits; generally few, but locally abundant, particularly in polluted environments; Panama to Tomales Bay; infaunal to epifaunal, ?detritivore (Murray 1991); see Bandy, Ingle, and Resig 1964, Limnol. Oceanogr. 9: 112–123 (pollution).

BULIMINELLIDAE

Buliminella elegantissima (d'Orbigny, 1839; =*Bulimina elegantissima*). Rare in strandline or tide pool deposits; locally common in shallow estuaries near the open ocean, such as the Berkeley harbor or Tomales Bay; locally abundant in polluted environments; infaunal, ?detritivore (Murray 1991); see Bandy, Ingle, and Resig 1964, Limnol. Oceanogr. 9: 112–123, and 1965, Limnol. Oceanogr 10: 314–332 (pollution).

BULIMINIDAE

Bulimina denudata Cushman and Parker, 1938 (=*Bulimina pagoda* var. *denudata*). Coastal, in strandline or tide pool deposits; generally few, but locally abundant, particularly in polluted or organically enriched environments; Ecuador to Tomales Bay; ?infaunal, ?detritivore (Murray 1991); see Bandy, Ingle and Resig 1965, Limnol. Oceanogr. 10: 314–332 (pollution); McGann et al. 2003, Mar. Environ. Res. 56: 299–342 (pollution).

UVIGERINIDAE

Angulogerina angulosa (Williamson, 1858) [=*Uvigerina angulosa*, *Trifarina angulosa* (Williamson)]. Specimens of *A. baggi* (Galloway and Wissler), *A. hughesi* (Galloway and Wissler), *A. fluens* Todd; and *A. hughesi* var. *picta* Todd, included here as they appear intergradational in morphological characteristics (see discussion in Lankford and Phleger 1973); coastal, in strandline or tide pool deposits; uncommon; infaunal, ?detritivore (Murray 1991).

STAINFORTHIIDAE

Hopkinsina pacifica Cushman, 1933. Strandline, tide pool, or estuarine deposits; rare to common; Santa Catalina Island to Tomales Bay.

VAGINULINIDAE

Lenticulina cultratus (Montfort, 1808) (=*Robulus cultratus*, Montfort). Coastal, in strandline or tide pool deposits; uncommon; Panama to Monterey Bay; epifaunal, ?detritivore (Murray 1991).

NONIONIDAE

Haynesina germanica (Ehrenberg, 1840) [=*Nonionina germanica, Elphidium incertum obscurum* (Williamson, 1858)], and *Nonion pauciloculum* Cushman, 1944, as identified in San Francisco Bay (Arnal et al. 1980 and Locke 1971, respectively). Common at extreme ends of San Francisco Bay in very brackish settings; common in estuaries worldwide; infaunal, ?detritivore (Murray 1991); see Murray 1991 (distribution).

Astrononion gallowayi Loeblich and Tappan, 1953 (=*Astrononion stellatum* Cushman and Edwards, 1937). Coastal, in strandline or tide pool deposits; uncommon; Mexico to Alaska; more abundant in colder waters, British Columbia and northward; epifaunal to infaunal, free or clinging, ?detritivore (Murray 1991).

Nonionella basispinata (Cushman and Moyer, 1930) (=*Nonion pizarrense* var. *basispinata*). Uncommon in coastal, strandline or tide pool deposits, or among algal holdfasts; rare on continental shelves in sewage effluent areas; infaunal, ?detritivore (Murray 1991); see Bandy, Ingle and Resig 1965, Limnol. Ocean. 10: 314–332 (pollution).

Nonionella stella Cushman and Moyer, 1930. Primarily coastal, in strandline or tide pool deposits; uncommon; occasionally found in ocean outlet regions of estuaries (San Francisco, Tomales Bay, Russian River); infaunal, ?detritivore (Murray 1991).

ELPHIDIIDAE

Elphidiella hannai (Cushman and Grant, 1927) [=*Elphidium hannai, E. nitida* Cushman, 1941 (in Detling 1958)]. Coastal and estuarine; common to abundant in strandline and tide pool deposits, washings of algal holdfasts, bryozoans, and coralline algae, and marine-like estuarine settings; prefers cold water; Mexico to Alaska, but primarily north of Point Conception; ?infaunal, ?detritivore (Murray 1991).

Elphidium crispum (Linnaeus, 1758) (=*Nautilus crispus*). Coastal; common in strandline deposits and among algal holdfasts, bryozoans, and coralline algae in rock and coarse sand deposits; may survive and reproduce intertidally; epifaunal, herbivore (Murray 1991); see Arnold 1980 (reproduction); Rosset-Moulinier 1961, Rev. Micropaléontol. 14: 76–81 (taxonomy, ecology).

Elphidium excavatum (Terquem, 1876) (=*Polystomella excavata*); includes several subspecies (*E. excavatum clavatum* Cushman, 1930; *E. excavatum lidoensis* Cushman, 1936; and *E. excavatum selseyensis* Heron-Allen and Earland, 1911, among others), which are identified by morphological variations in the umbilical region, such as the presence of an elevated thickening of clear shell material, papillae, or collar; synonymies: *Cribroelphidium, Protoelphidium, Elphidium translucens* Natland, 1938; *Elphidium tumidum* Natland, 1938; and subspecies of *Elphidium incertum* (Williamson, 1858); Monterey Bay to Alaska; one of the most abundant species in estuaries, particularly San Francisco Bay; infaunal, herbivore or detritivore (Murray 1991); see Miller 1982, J. Foram. Res. 12: 116–144 (taxonomy); Wilkinson 1979, J. Paleontology 53: 628–641 (taxonomy).

Elphidium frigidum Cushman, 1933 (=*Cribrononion frigidum*). Typically an Arctic species but extends southward in cold water; abundant; Mexico to Alaska, but primarily Bodega Bay northward; infaunal, herbivore or detritivore (Murray 1991).

Elphidium gunteri Cole, 1931 (=*Cellanthus*). Estuarine; few at extreme ends of San Francisco Bay in very brackish settings; Panama to San Francisco Bay; infaunal, herbivore or detritivore (Murray 1991).

Elphidium magellanicum Heron-Allen and Earland, 1932. Rare to few in estuaries, particularly San Francisco Bay, and rare in coastal, strandline and tide pool deposits; Monterey Bay to Alaska; infaunal, herbivore or detritivore (Murray 1991).

CASSIDULINIDAE

Cassidulina limbata Cushman and Hughes, 1925. Coastal, in strandline or tide pool deposits, and washings of plant holdfasts, bryozoans, and coralline algae; sometimes in estuarine regions bordering the open ocean, such as Tomales Bay; rare to common; Gulf of California to Alaska, with greatest abundance north of Point Conception; infaunal, detritivore (Murray 1991).

Cassidulina subglobosa H. B. Brady, 1881 (=*Globocassidulina subglobosa*). Coastal, in strandline or tide pool deposits; uncommon; infaunal, detritivore (Murray 1991).

Cassidulina tortuosa Cushman and Hughes, 1925 (=*Cassidulina reflexa* Galloway and Wissler, 1927). Coastal, in strandline or tide pool deposits; uncommon; infaunal, detritivore (Murray 1991).

PATELLINIDAE

Patellina corrugata Williamson, 1858. Coastal, in strandline or tide pool deposits; prefers rock and coarse sand; found in folds of green and red algae (Arnold 1980); uncommon; epifaunal, clinging, ?herbivore (Murray 1991).

ROSALINIDAE

Neoconorbina opercularis (d'Orbigny, 1839) (=*Rosalina opercularis, Discorbis opercularis*). Coastal, in strandline or tide pool deposits; rare to abundant; Baja California to Oregon; particularly abundant in tide pool faunas from San Francisco to Point Arena; epifaunal, clinging, ?herbivore (Murray 1991).

Neoconorbina terquemi (Rzehak, 1888) [=*Discorbina terquemi*; =*D. orbicularis* (Terquem, 1876), homonym of *Rosalina orbicularis* d'Orbigny, 1850]. Coastal, in strandline or tide pool deposits; uncommon; epifaunal, clinging, ?herbivore (Murray 1991).

Rosalina globularis d'Orbigny, 1826. Commonly referred to as *Rosalina columbiensis* (Cushman, 1925), but synonymized by Douglas and Sliter (1965, Tulane Stud. Geol. 3: 149–164); includes *Discorbis isabelleanus* (d'Orbigny, 1839) described in Monterey Bay by Martin (1932) and Stinemeyer and Reiter (1958). Coastal; test highly variable because it is usually securely attached to marine plants, invertebrates, rocks, wood, and pier pilings; one of the most abundant intertidal species, which can live and reproduce in turbulent rocky intertidal environments; early chambers wine-red color; at Moss Beach, very abundant on the hydroid *Abietinaria inconstans* and the bushy bryozoans *Crisia occidentalis* and *Scrupocellaria californica*. Rarely, a planktonic reproductive stage is recovered in sediments, originally described as *Tretomphalus bulloides* (d'Orbigny, 1839) (=*Rosalina bulloides*); characterized by a regularly coiled form lying on top of a final bulbous float chamber; epifaunal, clinging or attached, ?herbivore, omnivore (Murray 1991); see Arnold 1980 (ecology); Lankford and Phleger 1973 (reproduction, morphology); Steinker 1976 (ecology).

ROTALIIDAE

Ammonia beccarii (Linnaeus, 1758) (=*Nautilus beccarii*); general name assigned to a number of species of *Ammonia* (including *A. beccarii, A. parkinsonia* [d'Orbigny], 1839, and *A. tepida* Cushman, 1926, among others), which were previously distinguished by morphological variations in the number, size, and thickness of chambers, and presence or absence of limbate sutures, sutural ornamentation, and umbilical fillings of various sizes, but now recognized to include possibly >25–30 molecular types worldwide (Hayward et al. 2004, Mar. Micropaleontol. 50: 237–271); one of the most abundant species in brackish waters of bays, estuaries and lagoons around the world; the most northern occurrence of the species on the west coast of North America may be Samish Bay, Washington (Scott 1974, Northwest Sci. 48: 211–218; Jones and Ross 1979, J. Paleontol. 53: 245–257); species highly tolerant of variable temperatures and salinity, reproducing in waters as low as 20°C and salinities of 15 psu, whereas optimum ranges for the species are 24–30°C and 20–40 psu, respectively (Bradshaw 1957, J. Paleontol. 31: 1138–1147); may be a facultative anaerobe at least for a very short time (Moodley and Hess 1992, Biol. Bull. 183: 94–98; Kitazato 1994, Mar. Micropaleontol. 24: 29–41); living specimens often very abundant in polluted and/or oxygen-depleted sediments (Sen Gupta et al. 1996, Geol. 24: 227–230); infaunal, ?herbivore (Murray 1991); see Banner and Williams 1973, J. Foram. Res. 3: 49–69 (morphology, biology); Bradshaw 1961, Contrib. Cushman Found. Foram. Res. 12: 87–106 (ecology); Brooks 1967, Limnol. Oceanogr. 12: 667–684 (standing crop, vertical distribution, morphometrics); Schnitker 1974, J. Foram. Res. 4: 216–223 (ecotypic variation); Walton and Sloan 1990, J. Foram. Res. 20: 128–156 (distribution and morphologic variability).

DISCORBIDAE

Discorbis monicana Zalesny, 1959 (=*Discorbis rosacea, Rotalia rosacea* d'Orbigny, 1826, a homonym). Cosmopolitan in beach and tide pool faunas from San Diego to Oregon; rare to abundant; epifaunal, clinging or attached, herbivore (Murray 1991).

Rotorbinella campanulata (Galloway and Wissler, 1930) (=*Globorotalia campanulata*); synonymies: *Discorbis, Gavelinopsis,* and *Trochulina,* but *Rotorbinella* is retained here because of its common usage along the Pacific coast. Coastal, in strandline or tide pool deposits, rock scrapings, and washings of bushy bryozoans and surfgrass; Mexico to the Russian River; sometimes in estuarine regions bordering the open ocean, such as Tomales Bay and the Russian River estuary; abundant; epifaunal, clinging, ?passive suspension feeder (as *Galvelinopsis,* Murray 1991).

Rotorbinella turbinata (Cushman and Valentine, 1930) (=*Rotalia turbinata*); synonymies: *Gavelinopsis* and *Trochulina,* but *Rotorbinella* is retained here because of its common usage along the Pacific coast. Coastal, in strandline or tide pool deposits; few; Baja California to Half Moon Bay; epifaunal-infaunal, ?detritivore (as *Trochulina,* Murray 1991).

TRICHOHYALIDAE

Buccella frigida (Cushman, 1922) (=*Pulvinulina frigida*); includes *Buccella parkerae* Anderson, 1952, as used by Lankford and Phleger (1973); *Eponides frigida* (Stinemeyer and Reiter, 1958). Coastal, in strandline or tide pool deposits; sometimes in estuarine regions bordering the open ocean, such as Tomales Bay, the Russian River, and San Francisco Bay; cold water; few to common; ?infaunal, ?detritivore (Murray 1991).

Buccella tenerrima (Bandy, 1950) (=*Rotalia tenerrima*); *Buccella inusitata* Anderson, 1952. Primarily coastal, in strandline or tide pool deposits; cold water; few to abundant; abundant in Tomales Bay; ?infaunal, ?detritivore (Murray 1991).

GLABRATELLIDAE

Glabratella californiana Lankford, in Lankford and Phleger 1973. Identified as *Glabratella lauriei* (Heron-Allen and Earland, 1924) [=*Discorbina lauriei,* synonymy: *Glabratellina lauriei* in Seiglie and Bermudez 1965, Geos. 12: 41] by earlier workers. Coastal, in strandline or tide pool deposits; rare to few; geographic range at least San Diego to about Cape Blanco; two tests fused for sexual reproduction often seen; epifaunal, clinging or attached to hard substrate, ?herbivore (Murray 1991); see Lankford and Phleger 1973 (taxonomy), Seiglie and Bermudez 1965, Geos. 12: 15–65 (taxonomy); Cooper 1961 (distribution).

Glabratella ornatissima (Cushman, 1925) [=*Discorbis ornatissima; Trichohyalus ornatissima* (Lankford and Phleger, 1973) and *Eponides columbiensis* (Cushman, 1925)]. One of most abundant species in strandline or tide pool deposits, rock scrapings, and washings of bushy bryozoans and surfgrass; sometimes in estuarine regions bordering the open ocean, such as Tomales Bay and the Russian River estuary; two tests fused for sexual reproduction often seen, more commonly in late winter and spring; when separated, the ventral sides and inner walls are dissolved away, leaving a large cavity with numerous two-chambered progeny; Morro Bay to Alaska; epifaunal, clinging or attached to hard substrate, ?herbivore (Murray 1991); see Erskian and Lipps 1987, J. Foram. Res. 17: 240–256 (population dynamics); Lankford and Phleger 1973 (biology); Lipps and Erskian 1969, J. Protozool. 16: 422–425 (reproduction); Steinker 1973, Compass of Sigma Gamma Epsilon 50: 10–21 (morphology).

Glabratella pyramidalis (Heron-Allen and Earland, 1924) (=*Discorbina pyramidalis*); *Glabratellina pyramidalis* in Seiglie and Bermudez 1965, Geos. 12: 41–42. Coastal, in strandline or tide pool deposits; rare to few; geographic range at least southern California to Point Arena; two tests fused for sexual reproduction often seen; epifaunal, clinging or attached to hard substrate, ?herbivore (Murray 1991); see Seiglie and Bermudez 1965, Geos. 12: 15–65 (taxonomy); Cooper 1961 (distribution).

EPONIDIDAE

Poroeponides cribrorepandus Asano and Uchio, 1951 [=*Eponides repandus* (Fichtel and Moll, 1798) (=*Nautilus repandus*) as used by many early workers off California]. Coastal, in strandline or tide pool deposits; most typically on rocky and coarse-detritus substrates (Lankford and Phleger 1973); rare; epifaunal, clinging to hard substrate, ?herbivore or passive suspension feeder (Murray 1991).

BAGGINIDAE

Cancris auricula (Fichtel and Moll, 1798) (=*Nautilus auriculus*). Coastal, in strandline or tide pool deposits; rare; coastal Mexico to Monterey Bay; epifaunal, ?detritivore (Murray 1991).

PLANORBULINIDAE

Planorbulina acervalis Brady, 1884. Coastal, in strandline or tide pool deposits; uncommon; Panama to Sunset Bay; epifaunal, attached to hard substrate, ?passive suspension feeder (Murray 1991).

PLANULINIDAE

Planulina ornata (d'Orbigny, 1839) (=*Truncatulina ornata*). Coastal, in strandline or tide pool deposits; rare; Central America to Oregon; epifaunal, clinging to hard substrate, ?passive suspension feeder (Murray 1991).

CIBICIDIDAE

Cibicides conoideus Galloway and Wissler, 1927. Coastal, in strandline or tide pool deposits, few; Baja California to Monterey Bay; epifaunal, attached to hard substrate, ?passive suspension feeder (Murray 1991).

Cibicides fletcheri Galloway and Wissler, 1927. Coastal, in strandline or tide pool deposits; sometimes in estuarine regions bordering the open ocean, such as Tomales Bay and the Russian River estuary; test attached to plants in life and sometimes slightly distorted; few to abundant; Panama to Alaska; epifaunal, ?passive suspension feeder (Murray 1991).

Cibicides lobatulus (Walker and Jacob, 1798) (=*Nautilus lobatulus*); often mistakenly identified as *Cibicides lobatus* (e.g., Cooper 1961). One of the most common species in strandline or tide pool deposits, or attached to rocks, shells, plant holdfasts, eelgrass, algae fronds, bryozoans or hydroids; Panama to Alaska; sometimes in estuarine regions bordering the open ocean, such as Tomales Bay and the Russian River estuary; test often distorted due to the configuration of the surface to which it is attached; common; epifaunal, ?passive suspension feeder (Murray 1991).

Montfortella bramlettei Loeblich and Tappan, 1963 (=*Cibicides tenuimargo* Galloway and Wissler, 1927, *C. gallowayi* Cushman and Valentine, 1930). Coastal, in strandline or tide pool deposits; uncommon.

References

Arnal, R. E., P. J. Quinterno, T. J. Conomos, and R. Gram. 1980. Trends in the distribution of recent foraminifera in San Francisco Bay. In Studies in marine micropaleontology and paleoecology. A memorial volume to Orville L. Bandy. W. V. Sliter (ed.). Cushman Found. Foram. Res. Spec. Publ. 19: 17–39.

Arnold, Z. M. 1980. Foraminifera: shelled protozoans. In Intertidal invertebrates of California. R. H. Morris, D. P. Abbott, and E. C. Haderlie (eds.), pp. 9–20. Stanford University Press.

Bernhard, J. M. 1988. Postmortem vital staining in benthic foraminifera: duration and importance in population and distributional studies. J. Foram. Res. 18: 143–146.

Bush, J. 1930. Foraminifera of Tomales Bay, California. Micropaleontology Bull. 2: 38–42.

Connor, C. L. 1975. Holocene sedimentation history of Richardson Bay, California. M.S. thesis, Stanford University, Stanford, CA: 1–112.

Cooper, W. C. 1961. Intertidal foraminifera of the California and Oregon coast. Contrib. Cushman Found. Foram. Res. 12: 47–63.

Culver, S. J. and M. A. Buzas. 1985. Distribution of recent benthic foraminifera off the North American Pacific Coast from Oregon to Alaska. Smithsonian Contrib. Marine Sci. 26: 1–234.

Culver, S. J. and M. Buzas. 1986. Distribution of recent benthic foraminifera off the North American Pacific Coast from California to Baja Smithsonian Contrib. Marine Sci. 28: 1–634.

Culver, S. J. and M. A. Buzas. 1987. Distribution of recent benthic foraminifera off the Pacific Coast of Mexico and Central America. Smithsonian Contrib. Marine Sci. 30: 1–184.

Cushman, J. A. and I. McCulloch. 1939. A report on some arenaceous foraminifer. Allan Hancock Pac. Exped. 6: 1–113.

Cushman, J. A. and I. McCulloch. 1940. Some Nonionidae in the collections of the Allan Hancock Foundation. Allan Hancock Pac. Exped. 6: 145–178.

Cushman, J. A. and I. McCulloch. 1942. Some Virgulininae in the collections of the Allan Hancock Foundation. Allan Hancock Pac. Exped. 6: 179–230.

Cushman, J. A. and I. McCulloch. 1948. The species of *Bulimina* and related genera in the collections of the Allan Hancock Foundation. Allan Hancock Pac. Exped. 6: 231–294.

Cushman, J. A. and I. McCulloch. 1950. Some Lagenidae in the collections of the Allan Hancock Foundation. Allan Hancock Pac. Exped. 6: 295–376.

Detling, M. R. 1958. Some littoral foraminifera from Sunset Bay, Coos County, Oregon. Contrib. Cushman Found. Foram. Res. 9: 25–31.

Erskian, M. G. and J. H. Lipps. 1977. Distribution of foraminifera in the Russian River estuary, northern California. Micropaleontology 23: 453–469.

Hanna, G D. and C. C. Church. 1927. A collection of Recent foraminifera taken off San Francisco Bay, California. J. Paleontology 1: 195–202.

Jennings, E. and R. Nelson. 1992. Foraminiferal assemblage zones in Oregon tidal marshes-relation to marsh floral zones and sea level. J. Foram. Res. 22: 13–29.

Lalicker, C. G. and I. McCulloch. 1940. Some Textulariidae of the Pacific Ocean. Allan Hancock Pac. Exped. 6: 115–143.

Lankford, R. R. and F. B. Phleger. 1973. Foraminifera from the nearshore turbulent zone. J. Foram. Res. 3: 101–132.

Lesen, E. 2003. Benthic foraminifera in San Francisco Bay: environment, ecology and paleoecology. Ph.D. Dissertation, University of California, Berkeley, CA: 1–181.

Locke, J. L. 1971. Sedimentation and foraminiferal aspects of the recent sediments of San Pablo Bay. M.S. Thesis, San Jose State College, San Jose, CA: 1–100.

Loeblich, R., Jr. and H. Tappan. 1964. Sarcodina chiefly "Thecamoebians" and Foraminiferida. In Treatise on Invertebrate Paleontology, Part C, Protista 2, R. C. Moore (ed.), 900 pp. Geol. Soc. America and University of Kansas Press.

Loeblich, R., Jr. and H. Tappan. 1988. Foraminiferal genera and their classification. Van Nostrand Reinhold: 1–970.

Martin, L. T. 1930. Foraminifera from the intertidal zone of Monterey Bay, California. Micropaleontol. Bull. 2, Stanford University: 50–54.

Martin, L. T. 1931. Additional notes on the foraminifera from the intertidal zone of Monterey Bay, California. Micropaleontol. Bull. 3, Stanford University: 13–14.

Martin, L. T. 1932. Observations on living foraminifera from the intertidal zone of Monterey Bay, California. M.A. Thesis, Stanford University, Stanford, CA: 1–66.

Maurer, D. 1968. Preliminary report of some littoral foraminifera from Tomales Bay, California. Contrib. Cushman Found. Foram. Res. 19: 163–164.

McCormick, J. M., K. P. Severin, and J. H. Lipps. 1994. Summer and winter distribution of foraminifera in Tomales Bay, Northern California. Cushman Found. Foram. Res., Spec. Pub. 32: 69–101.

McDonald, J. A. and P. L. Diediker. 1930. A preliminary report on the foraminifera of San Francisco Bay, California. Micropaleontology Bulletin 2: 33–37.

McGann, M. 2002. Historical and modern distributions of benthic foraminifers on the continental shelf of Monterey Bay, California. Mar. Geol. 181: 115–156.

McGann, M. and D. Sloan. 1996. Recent introduction of the foraminifer *Trochammina hadai* Uchio into San Francisco Bay, California USA. Mar. Micropaleontol. 28: 1–3.

McGann, M. and D. Sloan. 1999. Benthic foraminifers in the Regional Monitoring Program's San Francisco Estuary samples. In 1997 Annual Report for the Regional Monitoring Program for Trace Substances in the San Francisco Estuary. San Francisco Estuary Institute, Richmond, CA: 249–258.

McGann, M., D. Sloan, and A. N. Cohen. 2000. Invasion by a Japanese marine microorganism in western North America. Hydrobiologia 421: 25–30.

Means, K. E. 1965. Sediments and foraminifera of Richardson Bay, California. M.A. Thesis, University of Southern California, Los Angeles, CA: 1–80.

Murray, J. W. 1973. Distribution and ecology of living benthic foraminiferids. New York, Crane, Russak & Company: 1–274.

Murray, J. W. 1991. Ecology and palaeoecology of benthic foraminifera. New York, Wiley: 1–397.

Phleger, F. B. 1967. Marsh foraminiferal patterns, Pacific coast of North America. An. Inst. Bio. Univ. Natl. Auton. Mexico 38, Ser. Cienc. Del Mar y Limnol. 1: 11–38.

Quinterno, P. J. 1968. Distribution of Recent foraminifera in central and south San Francisco Bay. M.S. Thesis, San Jose State College, San Jose, CA: 1–83.

Riechers, M. 1943. A survey of the genera of the foraminifera of the littoral zone in the Coos Bay area. M.A. Thesis, University of Oregon, Eugene, OR: 1–53.

Slater, R. A. 1965. Sedimentary environments in Suisun Bay, California. M.A. Thesis, University of Southern California, Los Angeles, CA: 1–104.

Steinker, D. C. 1976. Foraminifera of the rocky tidal zone, Moss Beach, California. Maritime Sediments Spec. Pub.1, pt. A: 181–193.

Stinemeyer, E. H. and M. Reiter. 1958. Ecology of the foraminifera of Monterey Bay, California. Shell Oil Company, Exploration Misc. Rept. 1621, Appendix, Part II: 1–157.

Todd, R. and D. Low. 1981. Marine flora and fauna of the Northeastern United States. Protozoa: Sarcodina: Benthic foraminifera. NOAA Technical Report NMFS Circular 439: 1–51.

Wagner, D. B. 1978. Environmental history of central San Francisco Bay with emphasis on foraminiferal paleontology and clay mineralogy. Ph.D. Dissertation, University of California, Berkeley, CA: 1–274.

Walton, W. R. 1952. Techniques for recognition of living foraminifera. Contrib. Cushman Found. Foram. Res. 3: 56–60.

Parasitic and Commensal Marine Protozoa

ARMAND M. KURIS

Many fascinating species of protozoans live as parasites or commensals with other organisms. We highlight here, by way of introduction to this under-explored and understudied group of organisms, some associations with our marine invertebrates. Modern protozoan systematics is based on molecular genetic comparisons or on complex ultrastructural characters that are not readily observed. A comprehensive atlas is now available (Lee et al. 2002). Here, higher level groups will be referred to by widely accepted, common names.

Although many species of flagellates and amoebae are parasites and commensals of vertebrates, few are prominently associated with marine invertebrates. This may merely be due to lack of investigation. The dinoflagellates, however, include some highly evolved species that are remarkable parasites of marine invertebrates. These include species of *Hematodinium* that cause systemic and ultimately fatal infections of commercial crab species, including the blue crab, *Callinectes sapidus*, on the Atlantic coast of North America (Messick and Shields 2000), and bitter crab disease of tanner crab, *Chionoecetes bairdi*, in southeastern Alaska (Meyers et al. 1987). In California, the gut of a maldanid polychaete, the bamboo worm, *Axiothella rubrocincta*, is commonly parasitized by the dinoflagellate *Haplozoon axiothellae* (Siebert 1973). This dinoflagellate is multicellular with cell differentiation, a representative of the little studied radiation of multicellular forms among the dinoflagellates. The enigmatic Ellobiopsidae may represent the culmination of these multicellular dinoflagellates (Galt and Whisler 1970). These are

often parasitic castrators of crustaceans. Some exert a feminizing effect on males (Shields 1994). Chytriodinid dinoflagellates are common parasites of crustacean eggs, and an infection is fatal to the egg. Shields (1994) provides an excellent review of these parasites with many Pacific coast examples.

Some marine invertebrates are host to an array of commensal ciliates (see the accompanying chapter by Stephen Landers). Most of these are external on feeding structures or in the mantle cavity, and many use specialized sites on the cuticle of crustaceans. However, some are endocommensals in the gut or in other cavities with openings to the exterior of the host. Particularly interesting are the ciliates of the respiratory trees of sea cucumbers. These include the genera *Boveria* and *Licnophora*. The latter are readily visible to the naked eye and are among the most morphologically complex cells known (Balamuth 1941). Their ecology has not been investigated. It would be interesting to follow their populations through the annual cycles of evisceration and regeneration experienced by its sea cucumber host, *Parastichopus californicus*.

The sea urchins, *Strongylocentrotus purpuratus* and *Syndisyrinx franciscanus*, have a species-rich and diverse (several families) array of commensal ciliates in their guts (Lynch 1929). These ingest breakdown products of algae eaten by the urchins. The herbivorous species are fed on by predatory forms, such as *Plagiopyliella pacifica*. All of these may, in turn, be consumed by the dalyellioid turbellarian *Syndesmis franciscanus*, which is also commonly found in the sea urchin digestive tract (Shinn 1981). These complex interactions merit further study. Other species of sea urchins and sea cucumbers in California have been scantily investigated, if at all, for commensal ciliates. Commensal ciliates may also be found in the mantle cavities of bivalves and snails. Many species use crustaceans as hosts and can be found on the cuticle, in the gills, and in the egg masses.

Interesting and important parasites of invertebrates include the Microspora and Haplosporidia. They often present overwhelming infections and can be recognized by an array of small (usually <20 μm), regular, highly refractile spores that spill out from a dissected host or are abundant in tissue squashes. Infected tissues are often opaque white and may have a mushy or jellied texture. These diseases are increasingly important in mariculture facilities because high host densities permit these inefficiently transmitted pathogens to infect a high proportion of the cultured animals.

The Phylum Apicomplexa are all parasites having at least one intracellular phase in their life cycle. They include many important human and animal parasites (e.g., malaria, toxoplasmosis, cryptosporidiosis). In marine invertebrates, however, the most common apicomplexan parasites are the gregarines, which are often avirulent or only moderately pathogenic. There are many gregarine species. Most have a trophozoite stage ("troph") found in the lumen of the gut or in other host organs with a passage to the exterior environment. Trophs are usually large enough to be visible to the naked eye. Gregarines are often quite host- and site-specific and are commonly seen in the midguts or digestive diverticulae of marine crustaceans and polychaetes. The gut of the echiuran *Urechis caupo* often contains many white, shmoo-like trophozoites of the gregarine *Zygosoma globosum* (Noble 1938). Only a few of these interesting gregarines have been described, and none have received much ecological study although they are common and easily quantified.

Aggregata spp. are coccidians with a two-host life cycle. White spots are commonly seen under the skin of intertidal octopuses. These are cysts of *Aggregata* that are filled with spores. When these spores are released they are inadvertently ingested by crabs. The life cycle will be completed after the crab has been eaten by the octopus. This parasite has also received little study.

This short survey notes some easily seen and studied protozoan parasites and commensals. Surveys of almost any invertebrate host group will reveal new and potentially interesting and ecologically important species and are well worth additional attention.

References

Balamuth, W. 1941. Studies on the organization of ciliated Protozoa. I. J. Morph. 68: 241–261.

Galt, J. H. and H. C. Whisler. 1970. Differentiation of flagellated spores in *Thalassomyces* ellobiopsid parasites of marine Crustacea. Arch. Mikrobiol. 71: 295–303.

Lee, J. J., G. F. Leedale, and P. C. Bradbury. (Eds.) 2000. An illustrated guide to the protozoa, volumes 1 & 2, 2nd ed. Lawrence, KS: Society of Protozoologists, 1432 pp.

Lynch, J. E. 1929. Studies on the ciliates from the intestine of *Strongylocentrotus*. I. *Entorhipidium* gen. nov. Univ. Calif. Publ. Zool. 33: 27–56.

Lynn, D. H. and J. Berger. 1972. Morphology, systematics and demic variation of *Plagiopyliella pacifica* Poljansky, 1951 (Ciliatea: Philasterina), an entocommensal of strongylocentrotid echinoids. Trans. Am. Micr. Soc. 91: 310–336.

Messick, G. A. and J. D. Shields. 2000. Epizootiology of the parasitic dinoflagellate *Hematodinium* sp. in the American blue crab *Callinectes sapidus*. Dis. Aquat. Org. 43: 139–152.

Meyers, T., C. Botelho, T. M. Koeneman, S. Short, and K. Imamura. 1990. Distribution of bitter crab dinoflagellate syndrome in southeast Alaskan Tanner Crabs *Chionoecetes bairdi*. Dis. Aquat. Org. 9: 37–43.

Noble, E. R. 1938. The life cycle of *Zygosoma globosum* sp. nov., a gregarine parasite of *Urechis caupo*. Univ. Calif. Publ. Zool. 43: 41–66.

Shields, J. D. 1994. The parasitic dinoflagellates of marine crustaceans. Ann. Rev. Fish Dis. 4: 241–271.

Shinn, G. L. 1981. The diet of three species of umagillid neorhabdocoel turbellarians inhabiting the intestine of echinoids. Hydrobiologia 84: 155–162.

Siebert, A. E. 1973. A description of *Haplozoon axiothellae* n. sp., an endosymbiont of the polychaete *Axiothella rubrocincta*. J. Phycol. 9: 185–190.

Symbiotic and Attached Ciliated Protozoans

STEPHEN C. LANDERS

(Plates 12–15)

Ciliated protozoans are common episymbionts and endosymbionts of marine invertebrates. In addition, many ciliates attach to or crawl about the surface of marine organisms as a substrate but are not true symbionts with the species on which they occur. Ciliates can be found crawling on their host's outer surface, swimming within the intestine, encysted on the surface, attached with a stalk, or attached to gills with their mouth embedded in their host. There are many host-symbiont records of marine invertebrates and ciliates known from the Pacific coast, but these records barely scratch the surface of the incidence of these symbionts. The student who makes a critical study of marine invertebrate symbionts will regularly find new species of ciliates and new host-symbiont records!

This chapter is not as much an identification guide to ciliates as it is an introduction to the subject, with sample illustrations, references for further study, and a listing of records from the Pacific coast. Although vertebrate-ciliate symbioses are beyond the scope of this review, general references are provided for fish symbionts (Bradbury 1994a, Lom and Dykova 1992, Woo 1995).

Observation of Ciliates

The simplest method for observing ciliated protozoans is in the living condition, using a dissecting microscope and a compound microscope. Small crustaceans, mollusks, hydroids, and other invertebrates can be observed whole or after simple dissection to reveal their ciliated guests. Identification of ciliates to the family or genus level is possible with living material, using a good reference such as *The Illustrated Guide to the Protozoa,* second edition (Lee et al. 2002), which has useful keys to ciliates based on shape, color, and ciliature (see Lynn and Small 2002, in the previous reference). Another excellent, although older, resource for ciliate identification is *The Ciliated Protozoa* (Corliss 1979). If precise identification of the ciliates is necessary, the cells will need to be fixed and stained. A number of resources are available with methods for fixing and staining protozoans (Foissner 1991, Lee et al. 1985, Lee and Soldo 1992).

Symbiotic ciliates are found in many locations and in many forms on marine invertebrates, including the exoskeletons of crustacea, the shells of mollusks, and hydroid perisarcs. These epibionts may need to be scraped from their substrate and pipetted to a slide for observation. Alternatively, sessile attached ciliates can be collected using slide traps suspended in the ocean (see next section). Other habitats of ciliates are the soft tissues of the host: gills, intestines, gonads, respiratory trees, mantle epithelia, and molluscan foot surfaces. To examine these ciliates, the smear technique (Galigher and Kozloff 1971) is recommended: a piece of tissue is rubbed on a glass slide to dissociate the tissue and dislodge the protozoans. The protozoan smears and tissue preparations should be viewed using at least a 10× objective lens with a compound microscope.

The following section suggests starting points for observing symbiotic ciliates. These techniques can be adapted to suit the host you are studying. Plates 12–15 illustrate many ciliates associated with marine invertebrates and hard substrates. The Appendix lists marine invertebrate-ciliate records from the Pacific coast of the United States. References to the symbiotic ciliate examples given below and techniques are found in the Appendix or are indicated in the text.

SYMBIONTS OF MOLLUSKS AND SEA SQUIRTS

(Plates 12, 13)

Gastropods, chitons, and bivalves have symbiotic ciliates crawling on or attached to their mantle, foot, and gills. Ciliates may also be sought on the branchial basket of sea squirts. To obtain ciliates from the mollusk foot, rub the foot of the animal onto a glass slide to create a thin mucous layer, then add a drop of seawater and a coverslip for observation. To observe ciliates from the mantle cavity, gill, or branchial basket, the host or tissue will have to be removed from the shell or ascidian tunic and rubbed onto a slide with forceps. Ciliate examples include *Ancistrocoma, Boveria, Cochliophilus, Eupoterion, Raabella,* and *Scyphidia.* In addition to the above hosts, various cephalopods, including cuttlefish, squid, and octopus, are known to harbor the ciliates *Chromidina* and *Opalinopsis* in their kidneys (Hochberg 1990). Intertidal cephalopods from the Pacific coast may yield similar infections.

SYMBIONTS OF ECHINODERMS

(Plate 13)

SEA CUCUMBERS have ciliates in their respiratory tree. Smear and, with forceps, mince a piece of respiratory tree onto a slide

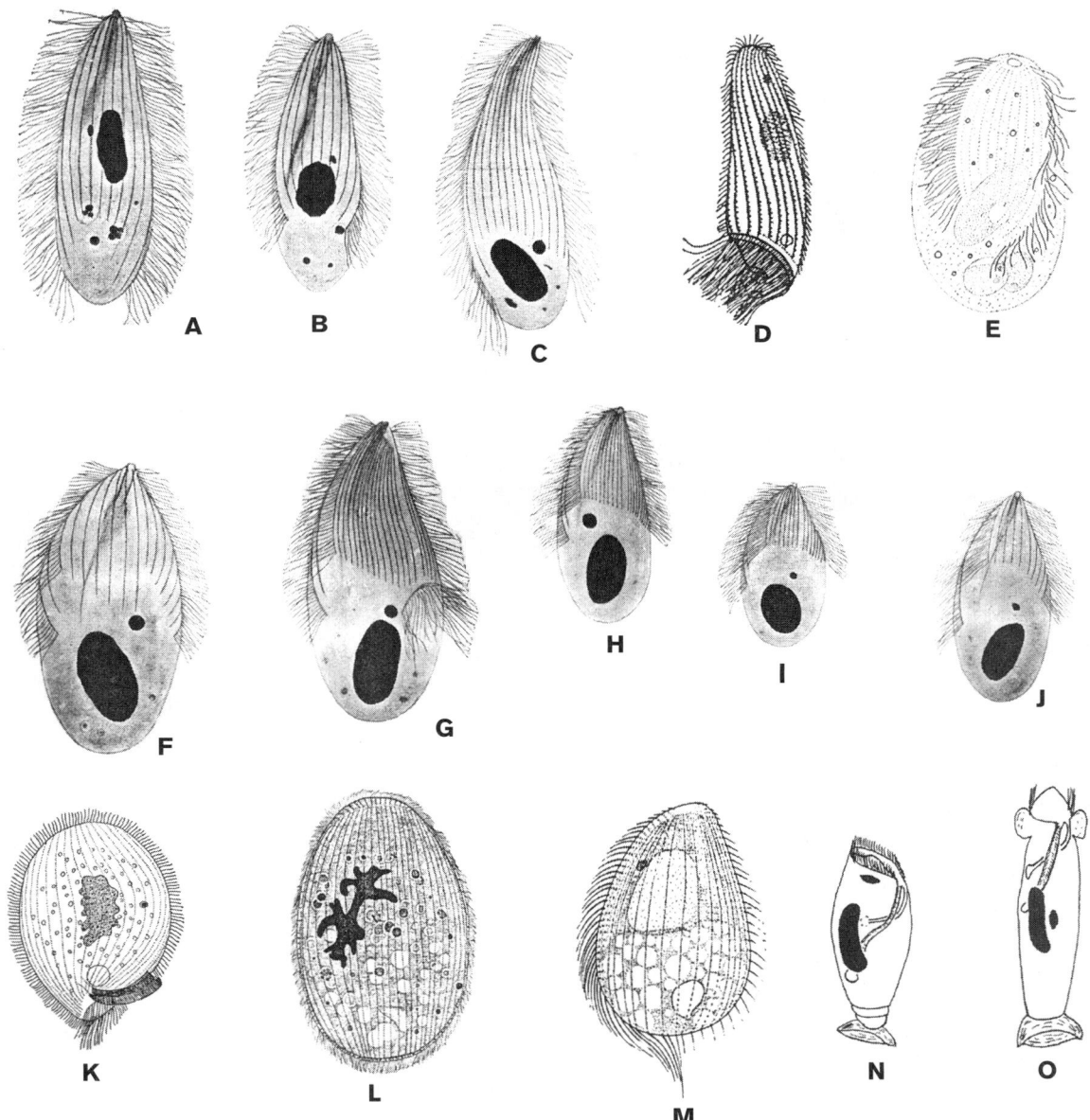

PLATE 12 Ciliates from mollusks. Sizes are cell lengths. A, *Ancistrocoma pelseneeri* 50–83 μm; B, *Ancistrocoma dissimilis* 33–51 μm; C, *Hypocomagalma pholadidis* 63–89 μm; D, *Boveria teredinidi* ~ 76 μm; E, *Hypocomella katharinae* ~ 25–38 μm; F, *Raabella helensis* 34–48 μm; G, *Insignicoma venusta* 42–52 μm; H, *Raabella botulae* 31–39 μm; I, *Raabella parva* 21–29 μm; J, *Raabella kelliae* 31–37 μm; K, *Cochliophilus minor* ~ 63 μm; L, *Conchopthirius caryoclada;* M, *Eupoterion pernix* ~ 58 μm; N–O, *Scyphidia ubiquita* ~ 30–100 μm (A–C, Kozloff 1946c; F–J, Kozloff 1946b; D, Pickard 1927; E, Kozloff 1961; K, Kozloff 1945a; L, Kidder 1933; M, MacLennan and Connell 1931; N–O, Hirshfield 1949 [D, M, with permission of the Regents of the University of California; N, O with permission of Wiley-Liss, Inc., a subsidiary of John Wiley & Sons, Inc.])

and add seawater and a coverslip. Ciliate examples include *Boveria* and *Licnophora*.

SEA URCHINS have ciliates in their intestine. Remove the intestine from an urchin and place it in a small dish filled with seawater. Then open the entire length of the intestine with fine scissors to release the ciliated protozoans, which can be pipetted to a slide. Ciliate examples include *Entorhipidium, Lechriopyla,* and *Schizocaryum.*

BRITTLE STARS have apostome ciliates in their gut cavity. Open the brittle star with forceps and pipette the gut con-

tents to a slide for observation. Ciliate examples include *Ophiuraespira.*

SYMBIONTS OF POLYCHAETES

(Plate 13)

Polychaetes have ciliated protozoans on their surface and within their intestine. To observe surface dwelling symbionts, the worm should be rubbed onto a slide to create a thin mucous layer. Additionally, the radioles or tentacles

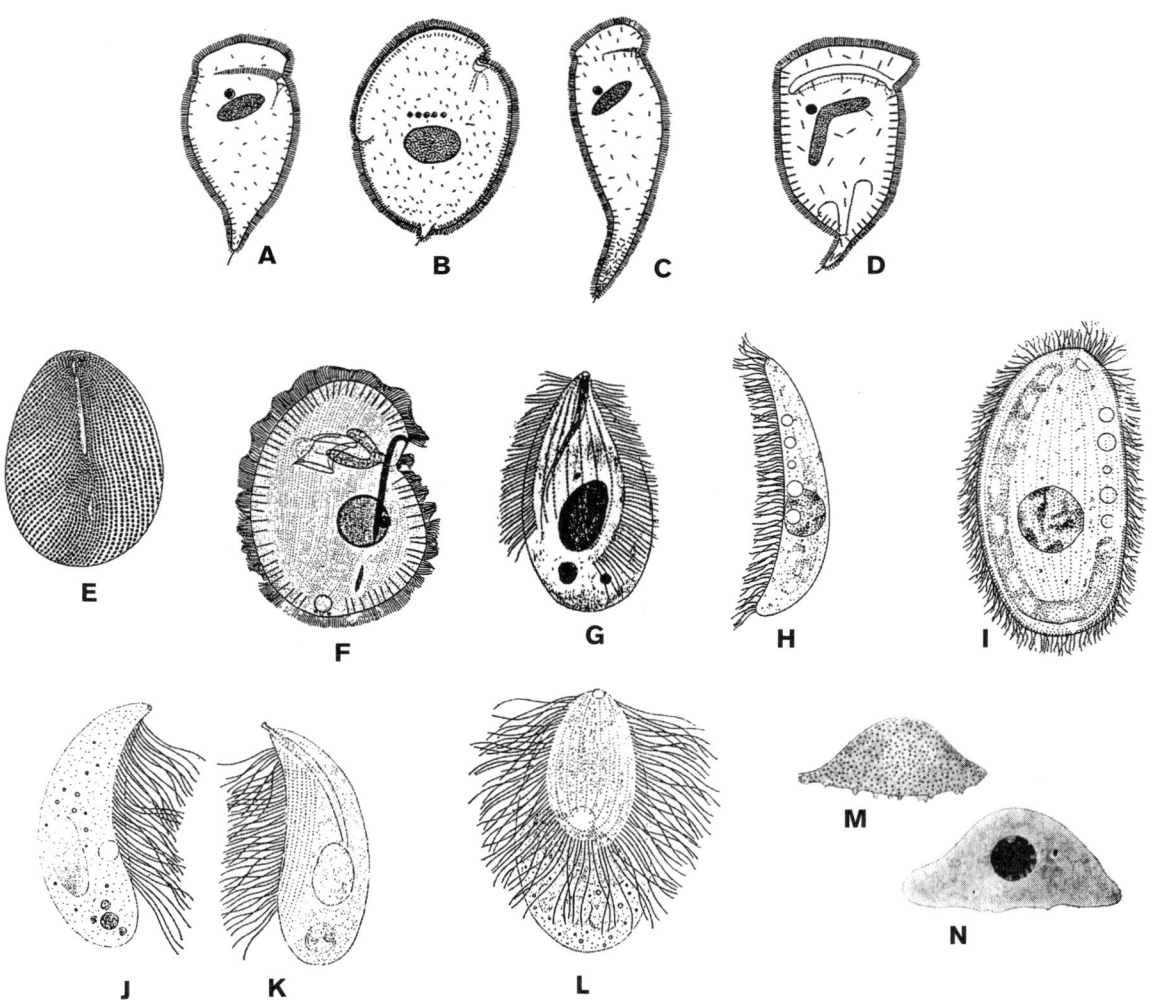

PLATE 13 Ciliates from echinoderms (A–F), phoronids (G), tunicates (H–I), and annelids (J–N). Sizes are cell lengths. A, *Entorhipidium echini* ~ 253 μm; B, *Entorhipidium multimicronucleatum* ~ 252 μm; C, *Entorhipidium tenue* ~ 314 μm; D, *Entorhipidium pilatum* ~ 267 μm; E, *Schizocaryum dogieli* ~ 162 μm; F, *Lechriopyla mystax* ~ 142 μm; G, *Kozloffiella phoronopsidis* 26–37 μm; H–I, *Parahypocoma rhamphisokarya* ~ 76 μm; J–K, *Ignotocoma sabellarum* ~ 23–33 μm; L, *Colligocineta furax* ~ 27–38 μm; M–N, *Phalacrocleptes verruciformis* ~ 27–50 μm. (A–D, Lynch 1929; F, Lynch 1930; E, Berger 1961; G, Kozloff 1945b; H–I, Burreson 1973; J–K, Kozloff 1961; L, Kozloff 1965; M–N, Kozloff 1966. Lynch figures with permission of the Regents of the University of California).

can be mounted on a slide with seawater for direct observation. To observe intestinal symbionts, the worm should be minced onto a slide with forceps. The entire length of the worm should be dissected, section by section, as some symbionts are found at different locations along the alimentary tract. Minced and smeared polychaete tissue will also likely contain gregarine parasites (Phylum Apicomplexa). Ciliate examples include *Colligocineta* and *Ignotocoma*.

SYMBIONTS WITHIN CRUSTACEAN EXOSKELETONS

(Plate 14)

Freshly molted exoskeletons of marine crustacea (including barnacles) contain apostome ciliates, which are very common. The feeding stages will appear as bloated bubbles (>50 μm in diameter) swimming within the exoskeleton. The

apostomes may be clear or may be a variety of colors from yellow to orange to blue, as they concentrate pigments from their host's cuticle during feeding. Pipette the apostomes from the inside of the exoskeleton and observe on a slide for the best view. Ciliate examples include *Gymnodinioides* and *Hyalophysa*.

SYMBIONTS OF CRUSTACEANS

(Plates 14, 15)

COPEPODS, **AMPHIPODS**, and **DECAPODS** will have stalked, encysted, or otherwise attached protozoans on their exoskeleton. Using a depression slide, small crustacea can be examined directly for apostome cysts (*Gymnodinioides*, *Hyalophysa*, *Mycodinium*), or chonotrich ciliates (*Lobochona*). Larger crustaceans must be dissected to observe the gills or appendages in de-

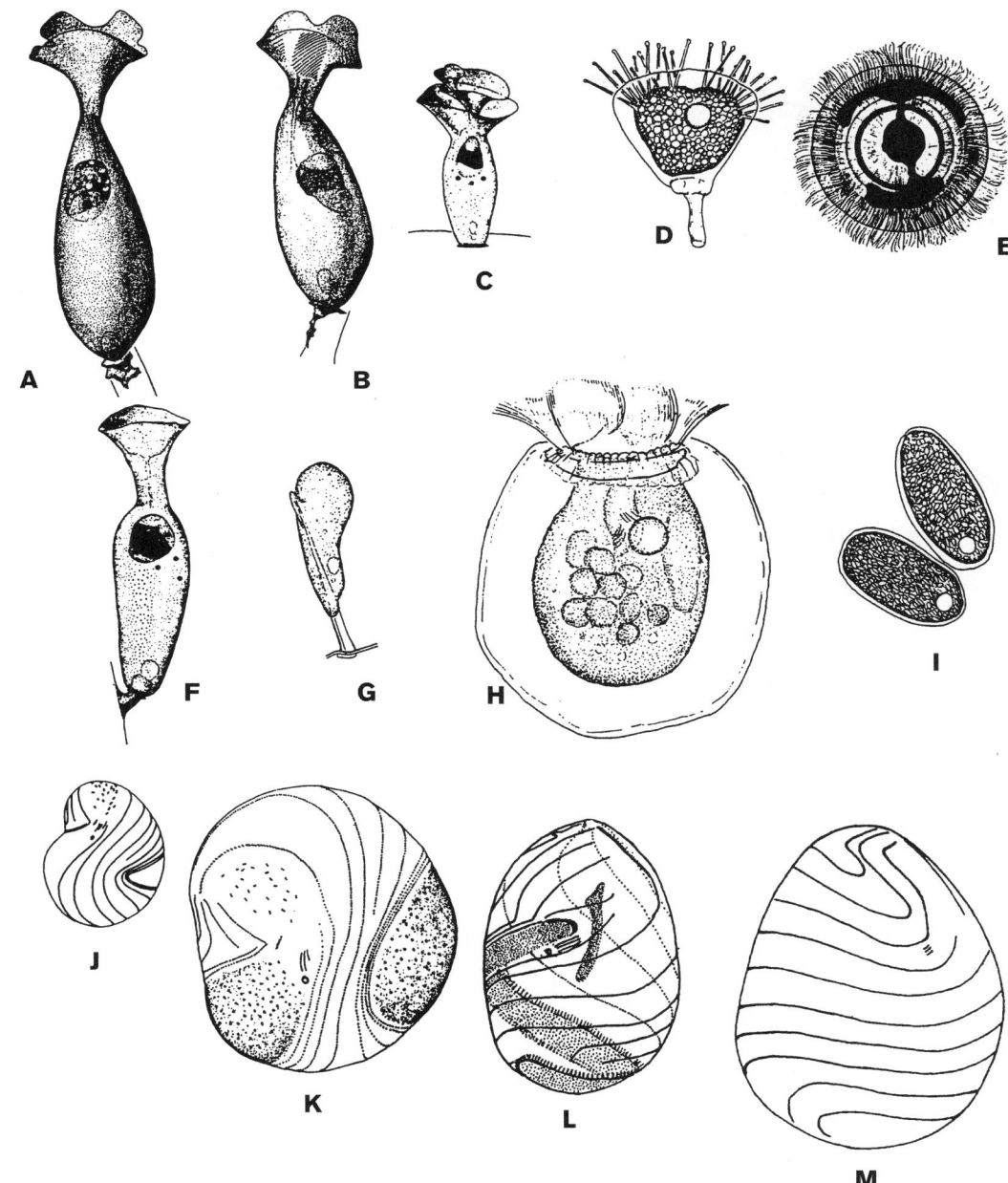

PLATE 14 Ciliates of arthropods (A–D, F–L), echiurans (E), and cnidarians (M). Sizes are cell lengths unless otherwise indicated. A, *Lobochona limnoriae* ~ 50 μm; B, *Lobochona prorates* ~ 43 μm; C, *Spirochona halophila* ~ 36–40 μm; D, *Acineta foetida,* tentacles project from the top surface, ~ 38 μm; E, *Trichodina urechi* ~ 58 μm; F, *Oenophorachona ectenolaemus* ~ 42–84 μm; G, *Mycodinium pilisuctor* ~ 40 μm; H, *Lagenophrys* sp., lorica length ~ 54 μm; I, *Gymnodinioides inkystans,* phoront cyst ~ 33 μm; J–K, *Hyalophysa chattoni,* lines represent ciliary rows: J, young trophont, 46 μm long, K, engorged trophont, 88 μm long; L, *Gymnodinioides caridinae;* lines represent ciliary rows ~ 56–87 mm; M, *Foettingeria actiniarum,* trophont, lines represent ciliary rows, 300 μm (A, B, Mohr et al. 1963; E, Noble 1940; C, F, Matsudo and Mohr 1968; D, G–I, Fenchel 1965b; J–K, Bradbury et al. 1997; L, Bradbury et al. 1996; M, original).

pression slides. Uropods make particularly good preparations because they are flat and often have loricate peritrichs (*Lagenophrys*) attached to the surface and apostome cysts (*Hyalophysa*) between the setae. Setae projecting from the appendages of caridean shrimp and amphipods have the apostome *Mycodinium* (*Conidophrys*) impaled on the tip, strobilating daughter cells. Additionally, the large peritrich colony

Zoothamnium can be found on crustacean exoskeletons and numerous other substrates (Carlton 1979).

Interesting hosts for epizoic protists are isopods in the genus *Limnoria,* which are known to have a variety of symbionts, particularly folliculinids, peritrichs, chonotrichs, and apostomes. These can be easily observed by placing a whole animal in a depression slide, or by dissecting off the telson.

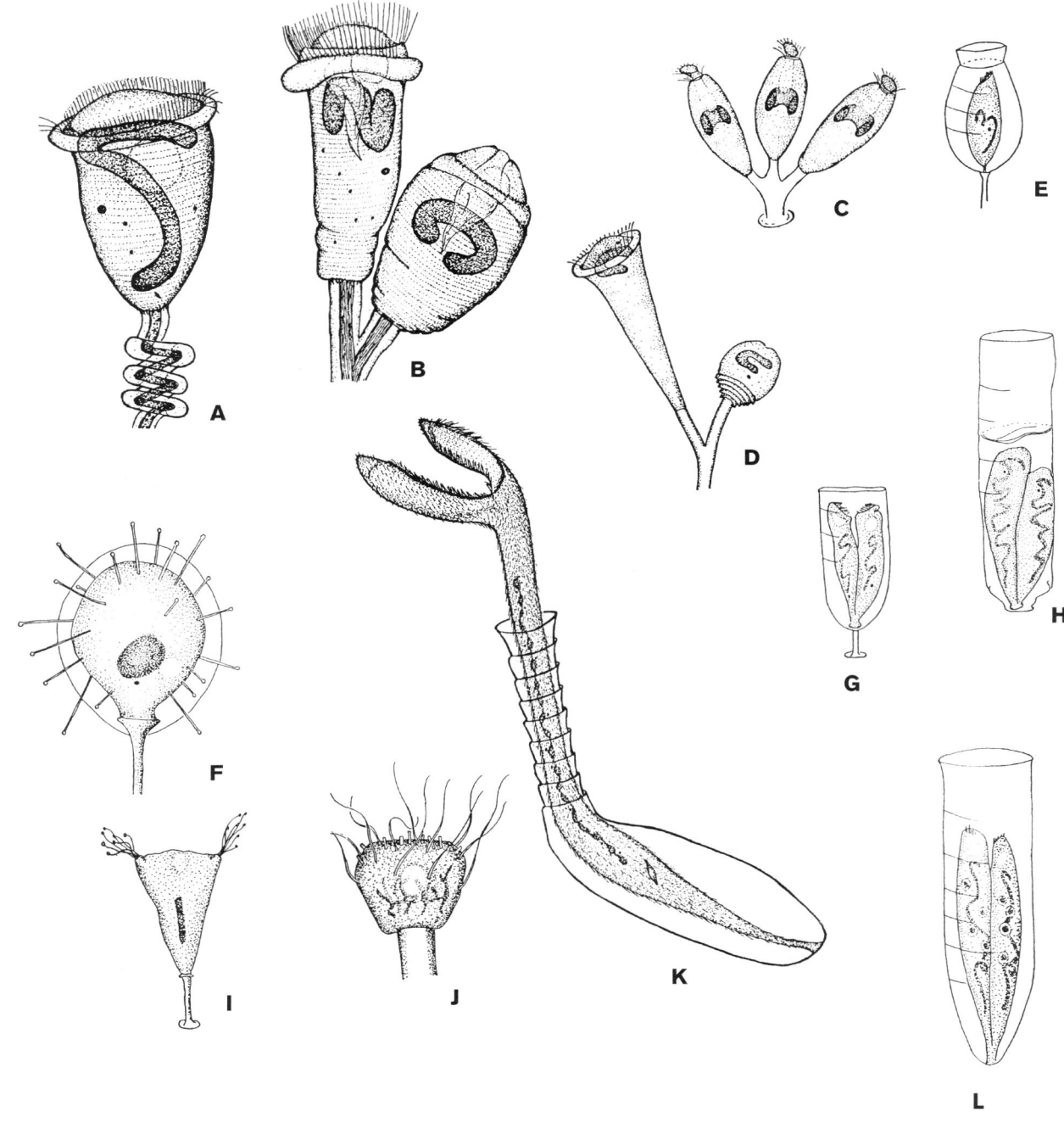

PLATE 15 Ciliates that colonize hard surfaces. Sizes are cell lengths. A, *Vorticella nebulifera* ~ 65–70 μm; B, *Zoothamnium mucedo,* large colonies of cells may form, ~ 30–60 μm; C, *Opercularia longigula,* large colonies of cells may form, ~ 80 μm; D, *Epistylis hentscheli,* large colonies of cells may form, ~ 120 μm; E, *Cothurnia limnoriae* ~ 37 μm; F, *Paracineta limbata,* tentacles project from the cell soma, ~ 50 μm; G, *Cothurnia fecunda* ~ 82 μm; H, *Thuricola valvata* ~ 190–240 μm; I, *Acineta foetida,* tentacles project from the cell soma. ~ 90 μm; J, *Ephelota gemmipara* ~ 40–110 μm; K, *Metafolliculina andrewsi* ~ 400–875 μm; L, *Vaginicola crystallina* ~ 130 μm (all from Jones 1974).

SYMBIONTS OF CNIDARIANS

(Plates 14, 15)

SEA ANEMONES have apostome trophonts (*Foettingeria*) in their gastrovascular cavity. The gut contents can be pipetted out of the mouth of the anemone or the host can be dissected to obtain the protozoans. Be cautious about protozoans that

may be associated with the anemone's food, as opposed to resident anemone symbionts.

HYDROIDS have sessile protozoans on their outer surface. Whole mount observation of the hydroid may reveal "suctorian" ciliates, which are not ciliated at all, but rather have numerous tentacles projected from the main cell body. Suctorians can be either directly attached to a substrate or projecting from a stalk. Also frequently abundant are stalked peritrichs. Ciliate examples include *Ephelota* and *Paracineta*.

SESSILE OR FOULING PROTOZOA THAT ATTACH TO HARD SURFACES

(Plate 15)

An easy way to collect some of the sessile protozoans also found on mollusk shells, crustaceans, and hydroids is to have the ciliates colonize glass slides in "slide traps" (Jones 1974). For this collection technique, a plastic slide box (25-slide capacity) is modified by having large holes cut in the sides. The box is then filled with cleaned slides, wired shut, and lowered into the water or tied to a dock piling. After one to two weeks, the box is collected and transferred to the lab submerged in seawater. Then a slide is removed, one side is wiped clean, and a drop of seawater and a large rectangular coverslip are added to the other side. Numerous invertebrates (e.g., hydroids, bryozoans, nematodes, barnacles) and protists of all types will have colonized the slides. This technique will not collect protozoans that are obligate epibionts, but rather those that attach to hard substrates in general. This preparation will also demonstrate protozoans directly attached to small invertebrates such as barnacles and hydroids. A useful exercise is to experiment with different colonization variables, such as deploying your slide trap at different depths or salinities, during different seasons, or for different lengths of time. Ciliate examples include *Acineta, Cothurnia, Ephelota, Vorticella, Zoothamnium*, and many others (Jones 1974, Landers and Phipps 2003).

Acknowledgments

I thank Eugene Kozloff and Denis Lynn for their review of this chapter and B. J. Bateman for help scanning drawings from the literature. I thank my past instructors and mentors over the years, especially Phyllis Bradbury and Eugene Kozloff, who fostered my enthusiasm for protozoology and first introduced me to many of the ciliates in this chapter.

Appendix: Symbiotic Ciliates Reported from Intertidal Marine Invertebrates on the Pacific Coast of the United States

Use this Appendix as a guide, recognizing that almost all invertebrates have symbiotic ciliates, with potentially hundreds of species remaining to be discovered. Modern scientific names of hosts are provided. Some hosts listed here from southern California and Washington are not keyed in this book. Additional papers review symbiotic ciliates from various localities and hosts (Arvy et al. 1969, Bradbury 1994b, Carlton 1979, Chatton and Lwoff 1935, Couch 1977, Fenchel 1965a, 1965b, Fernandez-Leborans 2001, Fernandez-Leborans and Tato-Porto 2000a, 2000b, Landers 2004, Morado and Small 1995).

Host	Ciliated Protozoan	Reference
	Cnidaria	
ANTHOZOA		
Anthopleura xanthogrammica	*Foettingeria* sp.	Ball & Moebius 1955
Aulactinia incubans	*Foettingeria actiniarum* (Claparède, 1863)	S. Landers (pers. obsv.)
HYDROZOA		
Obelia longissima	*Acineta* sp.	S. Landers (pers. obsv.)
	Ephelota sp.	S. Landers (pers. obsv.)
	Paracineta sp.	S. Landers (pers. obsv.)
Plumularia sp.	*Paranophrys marina* Thompson & Berger, 1965	Thompson & Berger 1965
	Echiura	
Urechis caupo	*Trichodina urechi* Noble, 1940	Noble 1940
	Annelida	
Eudistylia vancouveri	*Ignotocoma sabellarum* Kozloff, 1961	Kozloff 1961
Laonome kröyeri	*Colligocineta furax* Kozloff, 1965	Kozloff 1965
Schizobranchia insignis	*Ignotocoma sabellarum* Kozloff, 1961	Kozloff 1961
	Phalacrocleptes verruciformis Kozloff, 1966	Kozloff 1966
	Arthropoda	
AMPHIPODA		
Ampithoe lacertosa	*Trichochona lecythoides* Mohr, 1948	Mohr 1948, Mohr et al. 1970
	Cothurnia sp.	Mohr 1948
Ampithoe valida	*Trichochona* sp.	Mohr et al. 1970

(continued)

Host	Ciliated Protozoan	Reference

Arthropoda

Host	Ciliated Protozoan	Reference
Anisogammarus sp.	*Oenophorachona ectenolaemus* Matsudo & Mohr, 1968	Matsudo & Mohr 1968
	Spirochona halophila Matsudo & Mohr, 1968	Matsudo & Mohr 1968
Chelura terebrans	*Lobochona prorates* Mohr, LeVeque & Matsudo, 1963	Mohr, et al. 1963
Monocorophium acherusicum	*Mycodinium pilisuctor* (Chatton & Lwoff, 1934)	
	(*Conidophrys pilisuctor*)	Mohr & LeVeque 1948a

ISOPODA

Host	Ciliated Protozoan	Reference
Limnoria sp.	*Folliculina gunneri* Dons	Mohr & LeVeque 1948b
Limnoria algarum	*Lobochona* sp.	Mohr et al. 1970
Limnoria lignorum	*Mycodinium pilisuctor* (Chatton & Lwoff, 1934)	
	(*Conidophrys pilisuctor*)	Mohr & LeVeque 1948a
	Lobochona limnoriae Dons, 1940	Mohr et al. 1963
	Mirofolliculina limnoriae (Girard, 1883) Dons, 1927	Mohr & LeVeque 1948b
Limnoria quadripunctata	*Epistylis* sp.	Mohr 1951, Kofoid & Miller 1927
	Lobochona sp.	Mohr et al. 1963
	Mirofolliculina limnoriae (Girard, 1883) Dons, 1927	Mohr 1959
	Opercularia sp.	Kofoid & Miller 1927
	Vorticella sp.	Mohr 1951
	Vorticella microstoma Ehrenberg, 1830	Mohr 1959
Limnoria tripunctata	*Acineta* sp.	Mohr 1951, 1959
	Mycodinium pilisuctor (Chatton & Lwoff, 1934)	
	(*Conidophrys pilisuctor*)	Mohr, et al. 1970
	Cothurnia limnoriae Dons, 1928	Mohr 1951
	Epistylis sp.	Mohr 1951, Kofoid & Miller 1927
	Lagenophrys sp.	Mohr 1951, 1959
	Lobochona prorates Mohr, LeVeque & Matsudo, 1963	Mohr, et al. 1963
	Mirofolliculina limnoriae (Girard, 1883) Dons, 1927	Mohr 1959, Turner, et al. 1969
	Opercularia sp.	Kofoid & Miller 1927
	Vaginicola sp.	Mohr 1951, 1959
	Vorticella sp.	Mohr 1951
	Vorticella microstoma Ehrenberg, 1830	Mohr 1959

LEPTOSTRACA

Host	Ciliated Protozoan	Reference
"*Epinebalia*" sp.	*Stylochona* sp.	Mohr et al. 1970
Nebalia "*pugettensis*"	*Kentrochona* sp.	Matsudo & Mohr 1968

DECAPODA, BRACHYURA (CRABS)

Host	Ciliated Protozoan	Reference
Cancer magister	*Mesanophrys pugettensis* Morado & Small, 1994	Morado & Small 1994
Cancer oregonensis	*Hyalophysa chattoni* Bradbury, 1966	Bradbury 1966
Cancer productus	*Mesanophrys pugettensis* Morado & Small, 1994	Morado & Small 1994
Lophopanopeus bellus	*Hyalophysa chattoni* Bradbury, 1966	Bradbury 1966
Pugettia producta	*Mesanophrys pugettensis* Morado & Small, 1994	Morado & Small 1994

DECOPODA, ANOMURA (HERMIT CRABS)

Host	Ciliated Protozoan	Reference
Pagurus alaskensis	*Gymnodinioides inkystans* Minkiewicz, 1913	Bradbury 1966, 1994b
	Hyalophysa chattoni Bradbury, 1966	Bradbury 1966
Pagurus hemphilli	*Folliculina simplex* (Dons, 1917)(*Lagotia simplex*)	Andrews 1944
		Andrews & Reinhard 1943
Pagurus granosimanus	*Hyalophysa chattoni* Bradbury, 1966	Bradbury 1966
Pagurus hirsutiusculus	*Gymnodinioides* sp.	Bradbury 1966
	Hyalophysa chattoni Bradbury, 1966	Bradbury 1966
Pagurus samuelis	*Hyalophysa chattoni* Bradbury, 1966	Bradbury 1966

Mollusca

GASTROPODA (SNAILS)

Host	Ciliated Protozoan	Reference
Chlorostoma aureotincta	*Scyphidia* sp.	Hirshfield 1950
Chlorostoma brunnea	*Enerthecoma tegularum* (Kozloff, 1946)	Kozloff 1946a
Chlorostoma funebralis	*Scyphidia ubiquita* Hirshfield, 1949	Hirshfield 1949, 1950
	Trichodina tegula Hirshfield, 1949	Hirshfield 1949, 1950

Host	Ciliated Protozoan	Reference
	Mollusca	
Chlorostoma ligulata	*Scyphidia ubiquita* Hirshfield, 1949	Hirshfield 1949, 1950
	Trichodina tegula Hirshfield, 1949	Hirshfield 1949, 1950
Fissurella volcano	*Licnophora conklini* Stevens, 1904	Hirshfield 1950
	Scyphidia ubiquita Hirshfield, 1949	Hirshfield 1949, 1950
Lottia digitalis	*Eupoterion pernix* MacLennan & Connell, 1931	Hirshfield 1950
	Scyphidia ubiquita Hirshfield, 1949	Hirshfield 1949,1950
	Urceolaria karyolobia Hirshfield, 1949	Hirshfield 1949,1950
	Urceolaria korschelti Zick, 1928	Hirshfield 1950
Lottia fenestrata	*Eupoterion pernix* MacLennan & Connell, 1931	Hirshfield 1950
	Scyphidia ubiquita Hirshfield, 1949	Hirshfield 1949, 1950
	Urceolaria karyolobia Hirshfield, 1949	Hirshfield 1949, 1950
	Urceolaria korschelti Zick, 1928	Hirshfield 1950
Lottia gigantea	*Eupoterion pernix* MacLennan & Connell, 1931	Hirshfield 1950
	Scyphidia ubiquita Hirshfield, 1949	Hirshfield 1949, 1950
	Urceolaria karyolobia Hirshfield, 1949	Hirshfield 1949, 1950
	Urceolaria korschelti Zick, 1928	Hirshfield 1950
Lottia insessa	*Licnophora conklini* Stevens, 1904	Hirshfield 1950
	Scyphidia ubiquita Hirshfield, 1949	Hirshfield 1949, 1950
	Urceolaria karyolobia Hirshfield, 1949	Hirshfield 1949, 1950
	Urceolaria korschelti Zick, 1928	Hirshfield 1950
Lottia limatula	*Eupoterion pernix* MacLennan & Connell, 1931	Hirshfield 1950
	Scyphidia ubiquita Hirshfield, 1949	Hirshfield 1949, 1950
	Urceolaria karyolobia Hirshfield, 1949	Hirshfield 1949, 1950
	Urceolaria korschelti Zick, 1928	Hirshfield 1950
Lottia pelta	*Eupoterion pernix* MacLennan & Connell, 1931	Hirshfield 1950
	Scyphidia ubiquita Hirshfield, 1949	Hirshfield 1949, 1950
	Urceolaria karyolobia Hirshfield, 1949	Hirshfield 1949, 1950
	Urceolaria korschelti Zick, 1928	Hirshfield 1950
Lottia persona	*Eupoterion pernix* MacLennan & Connell, 1931	MacLennan & Connell 1931
Lottia scabra	*Eupoterion pernix* MacLennan & Connell, 1931	Hirshfield 1950
	Scyphidia ubiquita Hirshfield, 1949	Hirshfield 1949, 1950
	Urceolaria karyolobia Hirshfield, 1949	Hirshfield 1949, 1950
	Urceolaria korschelti Zick, 1928	Hirshfield 1950
Myosotella myosotis	*Cochliophilus depressus* Kozloff, 1945	Kozloff 1945a
	Cochliophilus minor Kozloff, 1945	Kozloff 1945a

BIVALVIA (CLAMS, MUSSELS, SHIPWORMS)

Host	Ciliated Protozoan	Reference
Adula californiensis	*Raabella botulae* (Kozloff, 1946)(*Hypocomides botulae*)	Kozloff 1946b
	Raabella parva (Kozloff, 1946)(*Hypocomides parva*)	Kozloff 1946b
	Insignicoma venusta Kozloff, 1946	Kozloff 1946b
Bankia setacea	*Boveria teredinidi* (Nelson, 1923)	Pickard 1927
Cryptomya californica	*Ancistrocoma pelseneeri* Chatton & Lwoff, 1926(*A. myae?*)	Kozloff 1946c
	Sphenophrya dosiniae Chatton & Lwoff, 1950, 1921	Kozloff 1946c
Kellia suborbicularis	*Raabella kelliae* (Kozloff, 1946)(*Hypocomides kelliae*)	Kozloff 1946b
Lyrodus pedicellatus	*Boveria teredinidi* (Nelson, 1923)	Horvath 1951
Macoma balthica	*Ancistrocoma pelseneeri* Chatton & Lwoff, 1926(*A. myae?*)	Kozloff 1946c
Macoma inquinata	*Ancistrocoma pelseneeri* Chatton & Lwoff, 1926(*A. myae?*)	Kozloff 1946c
Macoma nasuta	*Ancistrocoma pelseneeri* Chatton & Lwoff, 1926(*A. myae?*)	Kozloff 1946c
Macoma secta	*Ancistrocoma pelseneeri* Chatton & Lwoff, 1926(*A. myae?*)	Kozloff 1946c
Mya arenaria	*Ancistrocoma pelseneeri* Chatton & Lwoff, 1926(*A. myae?*)	Kofoid & Bush 1936, Kozloff 1946c
	Ancistrum cyclidioides (Issel)	Kozloff 1946c
	Sphenophrya dosiniae Chatton & Lwoff, 1950, 1921	Kozloff 1946c
Mytilus trossulus	*Ancistrum mytili* (Quennerstedt, 1967) Maupas,1883?	Pauley et al. 1966, Kozloff 1946b
	Ancistrum caudatum (Fenchel, 1964)?	Pauley et al. 1966
	Crebricoma carinata (Raabe, 1934) Kozloff, 1946	Kozloff 1946a,b
	Raabella helensis (Chatton & Lwoff, 1950)(*Hypocomides mytili*)	Kozloff 1946a,b
Penitella penita	*Ancistrocoma dissimilis* Kozloff, 1946	Kozloff 1946c
	Boveria sp.	Kozloff 1946c
	Hypocomagalma pholadidis Kozloff, 1946	Kozloff 1946c
	Sphenophrya sp.	Kozloff 1946c

(continued)

Host	Ciliated Protozoan	Reference

Mollusca

Petricolaria pholadiformis	*Ancistrumina kofoidi* (Bush, 1937)	Bush 1937
Siliqua patula	*Conchophthirus caryoclada* Kidder, 1933	Kidder 1933
Teredo navalis	*Boveria teredinidi* (Nelson, 1923)	Pickard 1927

POLYPLACOPHORA (CHITONS)

Katharina tunicata	*Hypocomella katharinae* Kozloff, 1961	Kozloff 1961
	Scyphidia sp.	Kozloff 1961
	Trichodina sp.	Kozloff 1961

Phoronida

Phoronopsis viridis	*Kozloffiella phoronopsidis* (Kozloff, 1945) Raabe, 1970	Kozloff 1945b

Brachiopoda

Hemithiris psittacea	*Urceolaria kozloffi* Bradbury, 1970	Bradbury 1970
Terebratalia transversa	*Urceolaria kozloffi* Bradbury, 1970	Bradbury 1970

Echinodermata

OPHIUROIDEA (BRITTLE STARS)

Amphiodia occidentalis	*Ophiuraespira weilli* Chatton & Lwoff, 1930	E. Kozloff (pers. comm)

HOLOTHUROIDEA (SEA CUCUMBERS)

Parastichopus californicus	*Boveria subcylindrica* Stevens, 1901	Balamuth 1941
	Licnophora macfarlandi Stevens, 1901	Balamuth 1941, Stevens 1901
		Lynn & Strüder-Kypke 2002
	Uronychia sp.	Balamuth 1941

ASTEROIDEA (SEA STARS)

Pisaster ochraceus	*Orchitophrya stellarum* Cépède, 1907	Leighton, et al. 1991

ECHINOIDEA (SEA URCHINS)

Strongylocentrotus droebachiensis	*Anophrys dogieli* Poljansky & Golikova, 1959	Berger 1960
	Cryptochilidium sigmoides Yagiu, 1934	Berger 1960
	Cyclidium stercoris Powers, 1935	Berger 1960
	Entodiscus borealis (Hentschel, 1924) Madsen, 1931	Berger 1960
	Euplotes balteatus (Dujardin, 1841) Kahl, 1932	Berger 1960
	Madsenia indomita (Madsen, 1931) Kahl, 1934	Berger 1960
	Plagiopyliella sp.	Berger 1960
	Schizocaryum dogieli Poljansky & Golikova, 1957	Berger 1961
	Thyrophylax vorax (Strelkow, 1959) Lynn & Berger, 1973	Lynn & Berger 1973
Strongylocentrotus franciscanus	*Entodiscus borealis* (Hentschel, 1924) Madsen, 1931	Berger 1960
	Entorhipidium echini Lynch, 1929	Berger 1960
	Plagiopyliella sp.	Berger 1960
	Lechriopyla mystax Lynch, 1930	Lynch 1930
	Schizocaryum dogieli Poljansky & Golikova, 1957	Berger 1961
	Thyrophylax vorax (Strelkov, 1959) Lynn & Berger, 1973	Lynn & Berger 1973
Strongylocentrotus purpuratus	*Cryptochilidium caudatum* Poljanksy, 1951	Berger 1960
	Cryptochilidium sigmoides Yagiu, 1934	Berger 1960
	Entorhipidium echini Lynch, 1929	Lynch 1929
	Entorhipidium tenue Lynch, 1929	Lynch 1929
	Entorhipidium pilatum Lynch, 1929	Lynch 1929
	E. multimicronucleatum Lynch, 1929	Lynch 1929
	Lechriopyla mystax Lynch, 1930	Lynch 1930
	Madsenia indomita (Madsen, 1931) Kahl, 1934	Berger 1960
	Schizocaryum dogieli Poljansky & Golikova, 1957	Berger 1961
	Thyrophylax vorax (Strelkov, 1959) Lynn & Berger, 1973	Lynn & Berger 1973

Host	Ciliated Protozoan	Reference

Chordata

Host	Ciliated Protozoan	Reference
Ascidia callosa	*Euplotaspis cionaecola* Chatton & Seguela, 1936	Burreson 1973
	Parahypocoma rhamphisokarya Burreson, 1973	Burreson 1973
	Trichophrya salparum Entz, 1884	Burreson 1973
Ascidia ceratodes	*Cryptolembus gongi* Gunderson, 1985	Gunderson 1985
Ascidia paratropa	*Euplotaspis cionaecola* Chatton & Seguela, 1936	Burreson 1973
	Trichophrya salparum Entz, 1884	Burreson 1973
Boltenia villosa	*Parahypocoma rhamphisokarya* Burreson, 1973	Burreson 1973
	Trichophrya salparum Entz, 1884	Burreson 1973
Pyura haustor	*Euplotaspis cionaecola* Chatton & Seguela, 1936	Burreson 1973
	Trichophrya salparum Entz, 1884	Burreson 1973
	Parahypocoma rhamphisokarya Burreson, 1973	Burreson 1973

References

GENERAL IDENTIFICATION OF CILIATES, REFERENCES, AND METHODS

Arvy, L., H. Batisse, and D. Lacombe. 1969. Péritriches épizoiques dans la chambre branchiale des Balanidae (Crustacea: Cirripedia) *Epistylis nigrelli* n. sp. *E. horizontalis* (Chatton 1930). Ann. Parasit. Hum. Comp. 44: 351–374. (Review of peritrichs on crustaceans worldwide.)

Bradbury, P. C. 1994a. Ciliates of fish, pp. 81–138. In: Parasitic Protozoa, 2nd ed., Vol. 8. J. P. Kreier, ed. New York: Academic Press.

Bradbury, P. C. 1994b. Parasitic protozoa of mollusks and crustaceans, pp. 139–264. In: Parasitic Protozoa, 2nd ed., Vol. 8. J. P. Kreier, ed. New York: Academic Press. (Review of parasitic protozoa, with invertebrate hosts listed.)

Bradbury, P. C., W. Song, and L. Zhang. 1997. Stomatogenesis during the formation of the tomite of *Hyalophysa chattoni* (Hymenostomatida: Ciliophora). Europ. J. Protistol. 33: 409–419.

Bradbury, P. C., L. Zhan, and X. Shi. 1996. A redescription of *Gymnodinioides caridinae* (Miyashita 1933) from *Palaemonetes sinesis* (Sollaud 1911) in the Songhua river. J. Euk. Microbiol. 43: 404–408.

Chatton, É. and A. Lwoff. 1935. Les Ciliés apostomes. 1. Aperçu historique et général; étude mongraphique des genres et des espèces. Arch. Zool. Exp. Gén. 77: 1–453. (Review of apostome ciliates.)

Corliss, J. O. 1979. The ciliated Protozoa. Characterization, classification and guide to the literature. 2nd ed. Elmsford, NY: Pergamon Press. 455 pp. (Review of all ciliate groups; well illustrated.)

Couch, J. A. 1977. Diseases, parasites, and toxic responses of commercial penaeid shrimps of the Gulf of Mexico and South Atlantic coasts of North America. Fishery Bull. 76: 1–44. (Protozoa of decapods.)

Fenchel, T. 1965a. Ciliates from Scandinavian molluscs. Ophelia 2: 71–174. (Well illustrated.)

Fenchel, T. 1965b. On the ciliate fauna associated with the marine species of the amphipod genus *Gammarus* J. G. Fabricius. Ophelia 2: 281–303. (The amphipod exoskeleton as an ecosystem of ciliates.)

Fernandez-Leborans, G. 2001. A review of the species of protozoan epibionts on crustaceans. III. Chonotrich Ciliates. Crustaceana 74: 581–607.

Fernandez-Leborans, G. and M. L. Tato-Porto. 2000a. A review of the species of protozoan epibionts on crustaceans. I. Peritrich Ciliates. Crustaceana 73: 643–683.

Fernandez-Leborans, G. and M. L. Tato-Porto. 2000b. A review of the species of protozoan epibionts on crustaceans. II. Suctorian Ciliates. Crustaceana 73: 1205–1237.

Foissner, W. 1991. Basic light and scanning electron microscopic methods for taxonomic studies of ciliated protozoa. Europ. J. Protistol. 27: 313–330.

Galigher, A. E. and E. N. Kozloff. 1971. Essentials of practical microtechnique, 2nd ed. Lea and Febiger. Philadelphia, 531 pp.

Hochberg, F. G. 1990. Diseases of Mollusca: Cephalopoda. Chp. 1.2. Diseases caused by protistans and metazoans, pp. 47–227. In Diseases of Marine Animals. Vol. 3. O. Kinne, ed. Hamburg: Biologische Anstalt Helgoland. (Ciliate infections of cephalopods are reviewed.)

Jones, E. E. 1974. The Protozoa of Mobile Bay, Alabama. University of South Alabama Monograph Series. 113 pp. (Well-illustrated monograph on benthic, sessile, and planktonic estuarine protozoa.)

Landers, S. C. and S. W. Phipps. 2003. Ciliated protozoan colonization of substrates from Weeks Bay, Alabama. Gulf of Mexico Science 2003(1): 79–85.

Landers, S. C. 2004. Exuviotrophic apostome ciliates from crustaceans of St. Andrew Bay, Florida, and a description of *Gymnodinioides kozloffi* n. sp. Journal of Eukaryotic Microbiology 51: 644–650. (Apostomes from crustacean molts.)

Lee, J. J., E. B. Small, D. H. Lynn, and E. C. Bovee. 1985. Some techniques for collecting, cultivating, and observing Protozoa, pp. 1–7. In: Lee, J. J., S. H. Hutner, and E. C. Bovee, eds. An illustrated guide to the Protozoa. Lawrence, KS: Society of Protozoologists.

Lee, J. J., G. F. Leedale, and P. Bradbury, eds. 2002. An illustrated guide to the Protozoa, 2nd ed., Vols. I, II. Lawrence, KS: Society of Protozoologists. (Copyrighted 2000, but not released until 2002).

Lee, J. J. and A. T. Soldo, eds. 1992. Protocols in Protozoology. Lawrence, KS: Society of Protozoolgists.

Lom, J. and I. Dykova. 1992. Protozoan parasites of fishes. Developments in Aquaculture and Fisheries Science, 26. Amsterdam: Elsevier.

Lynn, D. H. and E. B. Small. 2002. Phylum Ciliophora Doflein, 1901, pp. 371–656. In An illustrated guide to the Protozoa, 2nd ed., Vol. I. Lee, J. J., G. F. Leedale, and P. Bradbury, eds. Lawrence, KS: Society of Protozoologists.

Morado, J. F. and E. B. Small. 1995. Ciliate parasites and related diseases of Crustacea: A Review. Reviews in Fisheries Science 3: 275–354.

Woo, P. T. K. 1995. Fish diseases and disorders Vol. 1. Protozoan and Metazoan Infections. Wallingford, U.K.: CAB International.

REFERENCES FROM THE PACIFIC COAST OF THE UNITED STATES (CALIFORNIA–WASHINGTON)

Andrews, E. A. and E. G. Reinhard. 1943. A folliculinid associated with a hermit crab. J. Wash. Acad. Sci. 33: 216–223.

Andrews, E. A. 1944. Folliculinid Protozoa on North American coasts. Am. Midl. Nat. 31: 592–599.

Balamuth, W. 1941. Studies on the organization of ciliate protozoa. I. Microscopic anatomy of *Licnophora macfarlandi*. J. Morph. 68: 241–277.

Ball, G. H. and R. E. Moebius. 1955. A foettingeriid (Apostomea) from the sea anemone *Anthopleura xanthogrammica*. J. Protozool. 2 (Suppl.), 2.

Berger, J. 1960. The entocommensal ciliate fauna of *Strongylocentrotus* spp. from the northeast Pacific. J. Protozool. 7 (suppl.): 17.

Berger, J. 1961. Morphology and systematic position of *Schizocaryum dogieli*, a ciliate entocommensal in strongylocentrotid echinoids (Ciliata: Trichostomatida). J. Protozool. 8: 363–369.

Bradbury, P. C. 1966. The life cycle and morphology of the apostomatous ciliate, *Hyalophysa chattoni* n.g., n.sp. J. Protozool. 13: 209–225.

Bradbury, P. C. 1970. *Urceolaria kozloffi* sp. n., a symbiont of brachiopods. Acta Protozool. 7: 465–475.

Burreson, E. M. 1973. Symbiotic ciliates from solitary ascidians in the Pacific Northwest, with a description of *Parahypocoma rhamphisokarya* n. sp. Trans. Am. Miroscop. Soc. 92: 517–522.

Bush, M. 1937. *Ancistrina kofoidi* sp. nov., a ciliate in *Petricola pholadiformis* Lamarck from San Francisco Bay, California. Arch. Protistenk. 89: 100–103.

Carlton, J. T. 1979. History, biogeopgraphy, and ecology of the introduced marine and estuarine invertebrates of the pacific coast of North America. Ph.D. dissertation, Univ. California, Davis.

Gunderson, J. 1985. *Cryptolembus gongi* n. gen., n. sp., a philasterine scuticociliate from the intestine of the tunicate *Ascidia ceratodes* (Huntsman, 1912). J. Protozool. 32: 181–183.

Hirshfield, H. 1949. The morphology of *Urceolaria karyolobia,* sp. nov., *Trichodina tegula,* sp. nov., and *Scyphidia ubiquita,* sp. nov., three new ciliates from southern California limpets and turbans. J. Morph. 85: 1–33.

Hirshfield, H. I. 1950. The protozoan fauna of some species of intertidal invertebrates in southern California. J. Parasit. 36: 107–112.

Horvath, C. 1951. A study of the Teredinidae of Los Angeles–Long Beach Harbors. University of Southern California. M.S. thesis. 145 pp.

Kidder, G. W. 1933. *Conchophthirius caryoclada* sp. nov. (Protozoa, Ciliata). Biol. Bull. 65: 175–178.

Kofoid, C. A. and M. Bush. 1936. The life cycle of *Parachaenia myae* gen. nov., sp. nov., a ciliate parasitic in *Mya arenaria* Linn. from San Francisco Bay, California. Bull. Mus. R. Hist. Nat. Belg. 12: 1–15.

Kofoid, C. A. and R.C. Miller. 1927. Biological section, pp. 188–343. In Marine borers and their relation to marine construction on the Pacific coast. Hill, C. L. and Kofoid, C. A., eds.. San Francisco, 357 pp.

Kozloff, E. N. 1945a. *Cochiliophilus depressus* gen. nov, sp. nov. and *Cochliophilus minor* sp. nov., Holotrichous ciliates from the mantle cavity of *Phytia setifer* (Cooper). Biol. Bull. 89: 95–102.

Kozloff, E. N. 1945b. *Heterocineta phoronopsidis* sp. nov., a ciliate from the tentacles of *Phoronopsis viridis* Hilton. Biol. Bull. 89: 180–183.

Kozloff, E. N. 1946a. Studies on the ciliates of the family Ancistrocomidae Chatton and Lwoff (order Holotricha, suborder Thigmotricha). I. *Hypocomina tegularum* sp. nov. and *Crebricoma* gen. nov. Biol. Bull. 90: 1–7.

Kozloff, E. N. 1946b. Studies on the ciliates of the family Ancistrocomidae Chatton and Lwoff (order Holotricha, suborder Thigmotricha). II. *Hypocomides mytili* Chatton and Lwoff, *Hypocomides botulae* sp. nov., *Hypocomides parva* sp.nov., *Hypocomides kelliae,* sp. nov., and *Insignicoma venusta* gen. nov., sp. nov. Biol. Bull. 90: 200–212.

Kozloff, E. N. 1946c. Studies on the ciliates of the family Ancistrocomidae Chatton and Lwoff (order Holotricha, suborder Thigmotricha). III. *Ancistrocoma pelseneeri* Chatton and Lwoff, *Ancistrocoma dissimilis* sp. nov., and *Hypocomagalma pholadidis* sp. nov. Biol. Bull. 91: 189–199.

Kozloff, E. N. 1961. A new genus and two new species of ancistrocomid ciliates (Holotricha: Thigmotricha) from sabellid polychaetes and from a chiton. J. Protozool. 8: 60–63.

Kozloff, E. N. 1965. *Colligocineta furax* gen. nov., sp. nov., an ancistrocomid ciliate (Holotricha: Thigmotricha) from the sabellid polychaete *Laonome kroyeri* Malmgren. J. Protozool. 12: 333–334.

Kozloff, E. N. 1966. *Phalacrocleptes verruciformis* gen. nov., sp. nov., an unciliated ciliate from the sabellid polychaete *Schizobranchia insignis* Bush. Biol. Bull. 130: 202–210.

Leighton, B. J., J. D. G. Boom, C. Bouland, E. B. Hartwick, and M. J. Smith. 1991. Castration and mortality in *Pisaster ochraceus* parasitized by *Orchitophrya stellarum* (Ciliophora). Diseases Aquat. Org. 10: 71–73.

Lynch, J. E. 1929. Studies on the ciliates from the intestine of *Strongylocentrotus* I. *Entorhipidium* gen. nov. Univ. Calif. Pub. Zool. 33: 27–56.

Lynch, J. E. 1930. Studies on the ciliates from the intestine of *Strongylocentrotus* II. *Lechriopyla mystax,* gen. nov., sp. nov. Univ. Calif. Pub. Zool. 33: 307–350.

Lynn, D. H. and J. Berger. 1973. The Thryophylacidae, a family of carnivorous philasterine ciliates entocommensal in strongylocentrotid echinoids. Trans. Am. Microsc. Soc. 92: 533–557.

Lynn, D. H. and M. Strüder-Kypke. 2002. Phylogenetic position of *Licnophora, Lechriopyla,* and *Schizocaryum,* three unusual ciliates (Phylum Ciliophora) endosymbiotic in echinoderms (Phylum Echinodermata). J. Euk. Microbiol. 49: 460–468.

MacLennan, R. F. and F. H. Connell. 1931. The morphology of *Eupoterion pernix* gen. nov., sp. nov., a holotrichous ciliate from the intestine of *Acmaea persona* Eschscholtz. Univ. Cal. Publ. Zool. 36: 141–156.

Matsudo, H. and J. L. Mohr. 1968. *Oenophorachona ectenolaemus* n.g., n.sp. and *Spirochona halophila* n.sp., two new marine chonotrichous ciliates. J. Protozool. 15: 280–284.

Mohr, J. L. 1948. *Trichochona lecythoides,* a new genus and species of marine chonotrichous ciliate from California, with a consideration of the composition of the order Chonotricha Wallengren, 1895. Occ. Pap. Allan Hancock Fdn. No 5: 1–21.

Mohr, J. L. 1951. Common commensals of *Limnoria,* In Rept. Marine Borer Conf. Port Hueneme, CA: U.S. Naval Civil Eng. pp. R1–R4.

Mohr, J. L. 1959. On the protozoan associates of *Limnoria,* In Marine boring and fouling organisms. D. L. Ray, ed. Seattle: U. Washington Press. pp. 84–95.

Mohr, J. L. and J. A. LeVeque. 1948a. Occurrence of *Conidophrys pilisuctor* on *Corophium acherusicum* in Californian waters. J. Parasit. 34: 253.

Mohr, J. L. and J. A. LeVeque. 1948b. Folliculinids associated with *Limnoria* in California and Washington. J. Parsit. 34(2)Suppl., 26.

Mohr, J. L., J. A. LeVeque, and H. Matsudo. 1963. On a new collar ciliate of a gribble: *Lobochona prorates* n. sp. on *Limnoria tripunctata.* J. Protozool. 10: 226–233.

Mohr, J. L., H. Matsudo, and Y. M. Leung. 1970. The ciliate taxon Chonotricha. Oceanogr. Mar. Biol. Ann. Rev. 8: 415–456.

Morado, J. F. and E. B. Small. 1994. Morphology and stomatogenesis of *Mesanophrys pugettensis* n. sp. (Scuticociliatida: Orchitophryidae), a facultative parasitic ciliate of the dungeness crab, *Cancer magister* (Crustacea: Decapoda). Trans. Am. Microsc. Soc. 113: 343–364.

Noble, G. A. 1940. *Trichodina urechi* n.sp., an entozoic ciliate from the echiuroid worm, *Urechis caupo.* J. Parasit. 26: 387–405.

Pauley, G. B., A. K. Sparks, K. K. Chew, and E. J. Robbins. 1966. Infection of pacific coast mollusks by thigmotrichid ciliates. Proc. Natl. Shellfish Assoc. 58: 8.

Pickard, E. A. 1927. The neuromotor apparatus of *Boveria teredinidi* Nelson, a ciliate from the gills of *Teredo navalis.* Univ. Calif. Publ. Zool. 29: 405–428.

Stevens, N. 1901. Studies on ciliate infusoria. Proc. Calif. Acad. Sci., Ser. 3 (Zool.), Vol. 3: 1–42.

Thompson, J. C. and J. Berger. 1965. *Paranophrys marina* n.g., n.sp., a new ciliate associated with a hydroid from the northeast Pacific (Ciliata: Hymenostomatida). J. Protozool. 12: 527–531.

Turner, C. H., E. E. Ebert, and R. R. Given. 1969. Man-made reef ecology. Calif. Dept. Fish Game, Fish Bull. 146: 221 pp.

Placozoa

(Plate 16)

VICKI BUCHSBAUM PEARSE

While placozoans are not yet known from the central California or Oregon coasts, some evidence points to their presence off southern California; the nearest confirmed records are from Hawaii and the Pacific coast of Panama. The animals are abundant and globally distributed in shallow tropical and subtropical marine waters, but they could prove responsive to increases in sea temperature and should thus be watched for along the California coast as temperatures increase or during El Niño years. They swim free in the water column and glide slowly over substrates, digesting flagellates, algal films, and detritus. They settle preferentially on bacterial films and probably feed on them as well.

The simplest metazoans known, and with one of the smallest animal genomes, placozoans look like thin microscopic pancakes and are commonly <1 mm across, only about 20 μm thick, and nearly transparent. They are easily overlooked and are most often spotted on aquarium glass. The body consists of upper and lower epithelia, both flagellated but lacking a basal lamina; they enclose between them a fluid-filled space containing a network of contractile cells. There are none of the specialized cells, tissues, or organ systems typical of most animals (digestive, circulatory, excretory, neurosensory, or muscular), and there is no fixed symmetry. Despite this extreme morphological simplicity, placozoans are capable of simple behaviors, such as withdrawal and reversal of direction if poked with a needle on one edge, or righting if dislodged from the substrate and overturned. However, they are not easily dislodged: the specific name *Trichoplax adhaerens* von Schultze, 1883 reflects their ability to stick to a surface.

Other aspects of the biology remain murky, including the full life-history. Placozoans proliferate by pulling into two or more pieces or by budding, and can increase rapidly in number. But sex in placozoans remains elusive: although cells that look like oocytes are occasionally seen, meiosis, fertilization, and embryonic development have never been documented. Yet the findings of a molecular genetic analysis imply that sexual recombination does occur (Signorovitch et al. 2005). Likewise, the phylum appears to lack morphological diversity and currently includes only a single named species. In contrast, evidence of substantial genetic diversity strongly suggests multiple species or even higher level taxa and the need to construct a taxonomy for placozoans (Voigt et al. 2004). The position of placozoans near the base of the animal tree has increased interest in genomic and developmental research on these paradoxical animals whose very simplicity and transparency obscures our path toward understanding them.

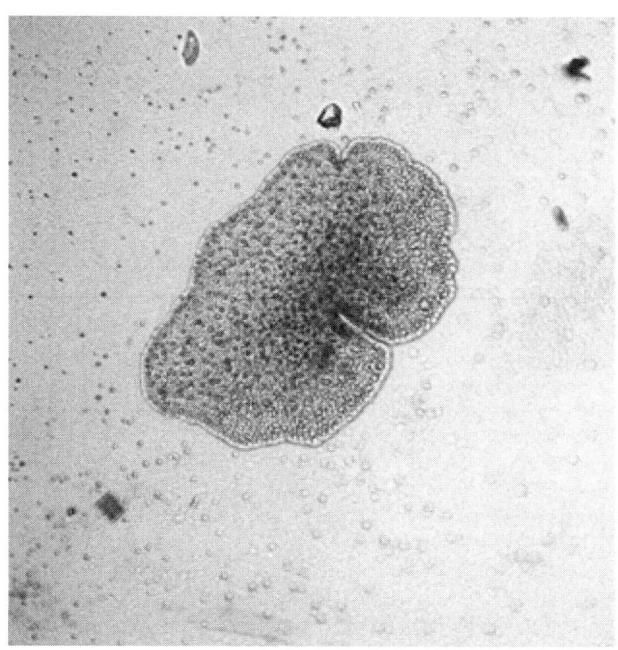

PLATE 16 Placozoan, *Trichoplax adhaerens,* about 300 microns diameter. Hong Kong (photo S. Dobretsov and V. B. Pearse).

References

Collins, A. G. "Introduction to Placozoa." http://www.ucmp.berkeley.edu/phyla/placozoa/placozoa.html

Grell, K. G., and A. Ruthmann. 1991. Placozoa, pp. 13–27. In Microscopic anatomy of invertebrates. Vol. 2. Placozoa, porifera, cnidaria, and ctenophora, F. W. Harrison and J. A. Westfall, eds. New York: Wiley-Liss.

Nielsen, C. 2001. Phylum Placozoa, pp. 48–50. In Animal evolution/interrelationships of the living phyla, 2nd ed. Oxford: Oxford University Press.

Pearse, V. B. 2003. Placozoa, p. 55. In Interdisciplinary encyclopedia of marine sciences. Vol. 3. J. W. Nybakken, W. W. Broenkow, and T. L. Vallier, eds. New York: Grolier Academic Reference.

Pearse, V., J. Pearse, M. Buchsbaum, and R. Buchsbaum. 1987. Placozoans, pp. 91–94. In Living invertebrates. Pacific Grove, CA: Boxwood Press.

Pearse, V. B., and O. Voigt. 2007. Field biology of placozoans (*Trichoplax*): distribution, diversity, biotic interactions. Integrative & Comparative Biology 47.

Signorovitch, A. Y., S. L. Dellaporta, and L. W. Buss. 2005. Molecular signatures for sex in the Placozoa. Proceedings of the National Academy of Sciences 102: 15518–15522.

Syed, T. and B. Schierwater. 2002. *Trichoplax adhaerens:* discovered as a missing link, forgotten as a hydrozoan, re-discovered as a key to metazoan evolution. Vie et Milieu 52: 177–188.

Voigt, O., A. G. Collins, V. B. Pearse, J. S. Pearse, A. Ender, H. Hadrys, and B. Schierwater. 2004. Placozoa—no longer a phylum of one. Current Biology 14: R944–R945.

Porifera

(Plates 17–37)

WELTON L. LEE, WILLARD D. HARTMAN, AND M. CRISTINA DÍAZ

Sponges are an important component of the intertidal biota of California and Oregon, but our knowledge of their biology and systematics is still rudimentary. Since the third edition of this key in 1975, several taxonomic studies, mostly involving Southern and Central California, have been conducted (see annotated references). There are well over 250 species of intertidal and subtidal sponges in California alone, many of which require revision and new names. Undescribed species will turn up in unexplored areas and as old collections are studied and revised. In the present key, 13 species have been added, and we present standardized synoptic descriptions of all 78 species included.

Important comprehensive systematic revisions are those of Bergquist (1980), De Weerdt (1985, 1986, 2000), Hooper (1996), Van Soest et al. (1990), and Diaz and Van Soest (1994). The *Systema Porifera* (John N.A. Hooper and Rob W. M. Van Soest, eds. 2002) is a synoptic comprehensive suprageneric classification of extant and extinct Porifera. The classification followed here is that given in the *Systema,* with occasional exceptions.

The sponge body plan is unique among the members of the animal kingdom. A sponge consists of loose aggregations of cells differentiated into ill-defined tissues, such as epithelia and mesenchyma, but organs are not formed. Despite the simplicity of their organizational plan, their mode of life requires complex cellular organization (see De Vos et al., 1991, for excellent scanning electron micrographs of sponge morphology). A mouth and a digestive tract are lacking; instead, the surface of the sponge is perforated by small incurrent pores (**OSTIA**) through which ambient water enters. The excurrent apertures, called **OSCULES**, are larger and relatively few in number. Water entering through the ostia passes through a system of canals and chambers lined by flagellated **CHOANOCYTES** (collar cells) and leaves the sponge through the oscules. This **AQUIFEROUS SYSTEM** presents various grades of complexity, ranging from those with a single central cavity lined by choanocytes (e.g., *Leucosolenia eleanor*) to those with numerous subdivisions of the internal channels and with choanocytes restricted to spherical, tubular, or ellipsoidal chambers located along the lengths of the channels (e.g., *Leucandra heathi*).

Most sponges are active filter feeders, efficiently capturing ultraplankton (<10 μm) and picoplankton (<2 μm), such as heterotrophic bacteria, autotrophic bacteria (prochlorophytes containing chlorophyll a and b), coccoid cyanobacteria, and autotrophic eukaryotes (Reiswig 1971; Pile et al. 1996; Pile et al. 1997). However, one family of deep-sea sponges, the Cladorhizidae, lacks choanocyte chambers and an aquiferous system. One genus of this family, *Asbestopluma*, in the case of a species living in a subtidal Mediterranean cave, has been shown to be carnivorous, capturing minute crustaceans by means of a surface layer of anisochelate microscleres on filaments that protrude from the distal end of the stalk (Vacelet and Boury-Esnault 1995, 1996). A species of *Cladorhiza*, living at nearly 5000 m in the Barbados Trench, harbors methanotrophic bacterial symbionts that it phagocytizes and digests, as well as capturing swimming prey by means of a surface layer of microscleres as in *Asbestopluma* (Vacelet et al. 1995, Vacelet et al. 1996). Several species of *Cladorhiza* and a species of *Asbestopluma* occur in deep water off California.

Many species of sponges possess symbiotic bacteria, cyanobacteria, zooxanthellae, and even branching green and red algae; these symbionts are known to provide a supplementary source of nutrients to their sponge hosts (see Wilkinson 1983, 1992).

Sponges also possess an extraordinary diversity of chemical substances that function in their defense against predators and potential fouling by other organisms (Faulkner 1984 and later publications). Nevertheless, predators are known. The yellow nudibranch *Doris montereyensis* is commonly seen feeding on the yellow sponge *Halichondria panicea* in and around mussel beds, and the red nudibranch *Rostanga pulchra* lives on red sponges, which both it and its eggs match closely in color. Some keyhole limpets, other gastropods, chitons, and fish feed on sponges, as well.

Many animals find sponges favorable sites in which to live. For example, amphipods, polychaetes, and shrimps are common in certain sponges, and masking crabs, such as *Loxorhynchus crispatus*, often place pieces of sponges on their backs. Other sessile organisms such as hydroids, kamptozoans, barnacles, bryozoans, and ascidians often grow on sponges.

The morphological diversity of sponges is quite large. Sponges may be massive, tubular, vaselike, lamellar, branched, or encrusting. Some sponges have a definite, radially symmetrical form, such as *Leucandra heathi* and *Tethya californiana,* among the local species. In the intertidal environment, encrusting sponges are most common, with shapes that follow the profile of their substrate. Encrusting sponges can have a thickness ranging from less than 1 mm to several centimeters. Deep-sea sponges tend to have more regular and definitive shapes than intertidal and shallow-water sponges.

The consistency of sponges varies exceedingly—from hard and stony to friable, rubbery, or gelatinous—depending upon the nature and arrangement of the skeletal elements. In all sponges, the morphology depends on organic skeletal support, most often reinforced by mineral material. The organic skeleton is composed of collagen organized in fibril fascicles or in spongin fibers. The mineral skeleton is formed of spicules (calcareous or siliceous) that are free or fused. A few sponges, such as the species of *Halisarca* (e.g., Halisarcidae, Dendroceratida) and *Oscarella* (e.g., Plakinidae, Homosclerophorida) lack a mineral skeleton, but possess fibrillar collagen. The classification of the main subdivisions of the phylum is based largely on the nature of the organic and mineral skeletal elements. The skeleton may consist of calcareous or siliceous spicules alone, of spongin fibers alone, or of a combination of siliceous spicules and spongin fibers. All sponges contain fibrillar collagen.

Some intertidal sponges are easily recognized and live in distinctive habitats. For example, *Leucilla nuttingi* is an easily identified, rocky intertidal sponge that occurs in clusters of individuals, each having a slender basal stalk and a brownish white tube up to an inch tall. *Clathria prolifera,* introduced with oysters from the Atlantic coast, forms striking, red-orange, branched growths in the brackish waters of San Francisco Bay. Quite resistant to exposure to air, the green *Halichondria panicea* and the purple *Haliclona* sp. A are both regularly found in beds of the mussel *Mytilus californianus.* The thin, encrusting, red *Clathria pennata* occurs commonly in midtidal areas on the sides and lower surfaces of rocks. Other species, such as the orange *Tethya californiana* and the yellow *Polymastia pachymastia,* are common in deeper water and can be collected intertidally on exposed coasts only at the lowest tides.

An interesting family of sponges that has received very little study on the West Coast is the Clionaidae, species of which excavate into calcareous materials such as barnacles, molluscan shells, coral skeletons, and limestone. One local *Cliona* is commonly associated with encrusting calcareous algae, living in a layer under the algal colony and excavating up through the alga to the surface in places to allow the exit of contractile tubules on which are borne the ostia and oscules. Other common species (of which no adequate taxonomic study has been made) bore into abalone or bivalve shells and, in these the openings, through which the ostia and oscular tubules extend, often form regular circular patterns on the surface of the shells. If an abalone shell inhabited by *Cliona* is broken, the yellow sponge can be seen filling extensive galleries that it has bored in the interior of the shell; shells riddled in this way are greatly weakened. *Cliona* can be a nuisance on oyster beds.

Classification of sponges depends largely upon skeletal characteristics—although the skeleton, sometimes absent or highly homogeneous (e.g., *Haliclona* spp.), is not always adequate for their identification. Embryological and life history studies, biochemical characteristics, and cytological details have helped in understanding the relationships, especially at generic and specific levels. Details of the morphology of choanocyte chambers and canal systems aid in the differentiation of some orders or families. Nuclear and mitochondrial DNA studies are now important to determine phylogenetic patterns.

There are three commonly recognized classes of sponges:

CLASS CALCAREA: spicules of calcium carbonate laid down as calcite; spongin absent.

CLASS DEMOSPONGIAE: skeleton consists of spicules of silicon dioxide (laid down in a hydrated form related to opal), **SPONGIN** (composed of a bundle of collagen fibrils each <10 nm in diameter), or both siliceous spicules and spongin.

CLASS HEXACTINELLIDA: skeleton consists of siliceous spicules with three axes (triaxons); not found intertidally, although well represented on the continental shelf and slope of California.

Skeletal characters are the most practical to employ in identification, but great care must be exercised in their use. Certain spicules, especially microscleres, may be absent in some members of a population; and failure to examine a number of specimens of the same species may lead one astray. For example, toxons are rare or absent in some specimens of the common red, encrusting sponge *Clathria* (*Thalysias*) *originalis*. *Penares saccharis* differs from *Penares cortius* in the absence of asters among its microscleres. It is possible that this character is a function of age and that the two are conspecific. Color is useful in separating some species, but it can be variable. Thus the common *Clathria* (*Microciona*) *pennata* is usually some shade of red, but some specimens are orange, yellow, or buff to light tan. Careful attention to habitat may also be useful in separating similar species, although relatively little is known about this aspect of sponge biology. For example, *Clathria* (*Microciona*) *pennata* usually occurs higher in the intertidal zone than its relative *Clathria* (*Thalysias*) *originalis*.

This key provides a synoptic representation of our current knowledge of sponge diversity in our local intertidal habitats. We hope it will facilitate and inspire further studies on both the taxonomy and biology of this fascinating group of organisms.

Collection, Study, and Preparation Techniques

We encourage the most careful and detailed preparation and observation possible. To this end, we include a detailed account of the processes and procedures used by the professional. This section is followed by a section about how to use the key. Here we not only explain how the key operates, but we have also included some shortcuts that allow the beginning investigator to proceed with reasonable success.

It is important to keep field notes on every specimen collected before fixation. Each specimen should be accompanied by a label with at least the date and location where it was collected. Color in life (best by reference to a color dictionary), oscular sizes in life, oscular morphology and localization, notable grouping of pores, surface features, consistency, distinctive odor, tendency to give off mucus, and the presence of buds, gemmules, or larvae are characteristics to look for. Notes on habitat, including tidal level, are essential. Note that part of the substrate should be collected along with encrusting sponges because certain spicules might be localized at the base, embedded in spongin affixed to the substrate. Collect an adequate sample size from which to make sections and spicule slides and to note surface and basal features of the sponge. Perpendicular cuts through the sponge will also allow observation of features of the water canal system.

Fixation may be done routinely in neutralized 10% formalin (4% formaldehyde) solution in seawater (Baker's calcium formaldehyde is a good fixative), followed by transfer to 75% ethyl alcohol within a few days. In certain sponges, such as *Haliclona* sp. B and *Isodictya quatsinoensis,* cells tend to become disassociated from the skeleton in formalin, but most sponges can be successfully fixed in formalin, which is better for histological work than if preserved directly in alcohol. Special fixatives, such as those of Bouin, Zenker, or Carnoy or glutaraldehyde, may be necessary for specific histological or ultrastructural studies. If it is impossible or impractical to fix recently collected sponges, the material should be placed directly in 95% or 75% ethyl alcohol (the former preferred). *This can be successful only if several changes of alcohol are made (two or three over several weeks, depending on the size of the sponge) because sponges hold large amounts of water that will dilute the preservative and prevent a satisfactory level of preservation.*

Because skeletal structures are most useful in sponge identification, two types of preparations are needed: (1) sections perpendicular and tangential to the surface, and (2) dissociated spicule mounts. Sections of fixed material may be made freehand with a sharp single-edged razor blade and stained in 1% basic fuchsin in 95% ethyl alcohol, followed by dehydration, clearing, and mounting on a slide in Canada balsam, Permount, or other synthetic resin. Note that basic fuchsin stains human skin as well. Sections might be blotted briefly on a paper towel after each step of the staining and dehydration process. It is easier to cut thin, freehand sections of many sponges after embedding a small piece of the sponge in paraffin, but this takes time and is not absolutely necessary for preliminary taxonomic studies. Such stained sections provide information about the arrangement of spicules in the skeletal framework and about the localization of spicule types. Thin microtome sections are necessary for cellular detail. The distribution of spongin is also clearer in thinner sections prepared with a collagen stain. A short cut involves cutting sections from a sponge preserved in 95% or 75% ethyl alcohol, not treated with stain, and, after placing on a microscope slide, dried on a hot plate or slide-warmer. Once dried, a small drop of xylene or toluene is deposited on the preparation along with a small amount of mounting media. A coverslip is placed on top of the now cleared section. Although such a shortened method does not last well, it does allow one to examine the skeletal structure of the sponge with spicules in place.

On the California and Oregon Coast, calcareous sponges may be distinguished from siliceous ones by the presence of triradiate spicules in the former (plate 17D). If doubts exist, place a small piece of the sponge on a microscope slide and add a small drop of acetic or hydrochloric acid. If the sample fizzes, calcareous matter is present. However, bits of included shell and other calcareous materials may cause fizzing in a siliceous sponge. To make certain of a correct determination, add a coverslip and examine the preparation under the microscope after the fizzing ceases to note whether the spicules are partially dissolved away.

Spicules not intrinsic to the sponge in question might adhere to the surface of the sponge or become incorporated into the ectosome or choanosome and hence will turn up on spicule slides. This can cause problems for both the beginners and the experienced investigator. Examination of a second sample from another part of the sponge or a second specimen and examination of sections of the skeletal framework will usually enable one to decide whether the spicule is foreign or proper to the sponge. Sponges with spongin fibers only (e.g., Orders Dictyoceratida, Dendroceratida, and Verongida) often incorporate the spicules of other sponges into their fibers. Spicules are usually recognizable as foreign when most are broken or when they represent a random miscellany of spicule types.

Temporary spicule preparations may be made by placing a small piece of the sponge, including both surface and interior tissues, on a slide and adding a few drops of sodium hypochlorite solution, such as Clorox. After bubbling has subsided, a cover glass should be added to the slide and the preparation should be examined under a compound microscope provided with an oil immersion objective and ocular micrometer. The use of too much Clorox will lead to diversionary activities of cleaning the microscope.

Permanent spicule mounts may be made by boiling a small piece (<1 cm^3) of sponge in concentrated nitric acid (or 10% KOH if the sponge has calcareous spicules) under a hood until the organic matter has disappeared. Be careful in handling hot nitric acid or KOH; use a test tube holder and keep the flask in motion to avoid local hot spots, which can lead to splattering. This must be followed by thorough washing away of the acid or alkali, which is best accomplished by transferring the preparation to a test tube and spinning down the spicules three times with a change of water each time. Before the final centrifugation, add 95% ethyl alcohol to speed up drying. When decanting the last fluid from the tube, leave a small amount to facilitate removal of the spicules, which, after being brought into suspension in the remaining fluid, are poured out on a slide on a warming plate. Check the tube to make certain that all spicules have been transferred to the slide.

Allow the slide to dry thoroughly, and before adding the mounting medium, add a drop of solvent for that medium to help prevent excessive bubble formation in the mounting medium. Next add mounting medium and a coverslip. Damar, Piccolyte, Canada balsam or Permount are satisfactory media. As little mounting medium should be added as feasible because too thick a slide cannot be examined with high power or oil immersion objectives. Such permanent spicule preparations are useful for measuring spicules with an ocular micrometer and for preparing drawings. Temporary mounts are sufficient for most routine identification.

Measurements should be made of the lengths and widths of all spicule types (microscleres do not need width measurements). It is advised that *no fewer* than 20 measurements be made for any spicule type. What is lost in time in this process is gained in accuracy. Spicules in ethyl alcohol solution (95%) may be transferred to absolute ethyl alcohol or acetone, critically point dried (using liquid carbon dioxide), mounted, and sputter-coated with gold for scanning electron micrograph (SEM) studies. In the key, measurements are given as overall ranges of several specimens examined. In some cases, a range of means is given as well.

After having made notes on the sponge as a living animal in its environment and having collected a portion of the sponge from which slides of sections and spicules have been prepared, one is ready to use the key. Note that temporary spicule preparations made by separating the spicules from the sponge tissue using Clorox and temporary unstained sections, dehydrated and cleared in xylol and mounted on a slide with an available mounting medium, will serve for preliminary identifications in many cases.

How to Use the Key

This key is composed of a series of simple dichotomies that present one, two, or more parallel sets of choices; additional information of value in determining the species is included immediately below the dichotomy. The data are organized into the following categories:

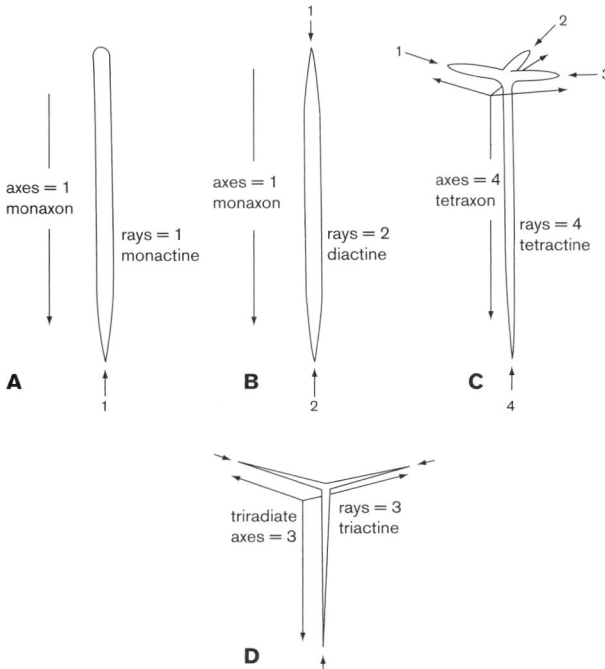

PLATE 17 Monactine, diactine, tetractine, and triactine spicule morphology and terminology.

SPONGE: this covers information about sponge shape, size, surface features, etc.

COLOR: this presents all known color forms reported in the literature.

SPICULES: all spicule types are enumerated, and for each, the ranges in length and width are given in micrometers. Spicules are represented by megascleres first, followed by microscleres. Each spicule type is numbered, so a quick glance will give the user the total number of spicule types present. In many cases, one or more spicule types may occur with distinct and nonoverlapping ranges in size. The number of these size classes is given when they occur. However, because the measurements used to determine the existence and range of size classes may be based on information from several specimens, the ranges may appear to overlap. A spicule type with two size classes given as 30–50 μm and 47–70 μm might be based on two specimens where the ranges measured were: size class 1, 30–40 μm and 37–50 μm and size class 2, 47–60 μm and 59–70 μm.

SKELETON: information describing the skeletal structure is included. It may be a description of spicule arrangements and alignments; configuration of fibers, if present; or location in the skeleton of specific spicule types. Other information, such as habitat notes, may be appended following description of the skeleton.

Remember that there are a large number of undescribed species on the West Coast. Anyone collecting in California and Oregon waters can be assured that encountering new species is not a rare event. This means that if the characters given in the key *do not fit* your specimen, it is distinctly possible that you *may* have a species not described in the key.

Illustrations

Illustrations of spicules are presented as SEMs that reveal extraordinary detail. Once such detail is clearly seen through SEM, observation of these same details through the light microscope is greatly enhanced. Illustrations of entire specimens of sponges are invariably of preserved specimens. Illustrations of fibrous skeletons and those showing spicules held together by spongin are made by photography through a light microscope. No attempt has been made to depict spicules to scale. This can best be determined by referring to the size ranges given for all spicules in the text.

The Sponge Skeleton

The identification of sponges is most often dependent upon knowledge of skeletal characteristics. Sponge skeletons are usually composed of mineralized elements called spicules secreted by specialized cells. These structures are formed of calcium carbonate or silicon dioxide. Sponges with calcareous spicules belong to the Class Calcarea; those with siliceous spicules belong to either the Class Demospongiae or Hexactinellida. In all three classes, there are species that construct rigid skeletons through the interlocking or fusion of spicular elements. Among the Demospongiae, there are forms that lay down basal skeletons of calcium carbonate, and one genus of the Calcarea lays down a basal skeleton of calcite. Such species are not found in our intertidal zone, nor, as noted earlier, are hexactinellids.

Organic Components of the Skeletons of Sponges

The intercellular matrix of sponges includes fibrillar collagen made up of fibrils that average 20 nm in diameter. The fibrils have a distinct transverse banding pattern consisting of light rings that occur at regular intervals of about 66 nm. These fibrils are elaborated by collenocytes, exopinacocytes, and some varieties of cells with inclusions. Such fibrils occur in all three classes of sponges.

Spongin is found only in the Class Demospongiae and is made up of intertwined or massed microfibrils, each varying from 5 to 10 nm in diameter; banding patterns, when present, occur at intervals of 60–65 nm. Spongin fibers, secreted by specialized cells called spongocytes, may bind together the tips of siliceous spicules to form a reticulate skeleton, or they may be secreted over tracts of spicules and joined with other like fibers to form a reinforced, strong skeleton of such a size that it is generally visible to the naked eye. Sponges belonging to three orders of the Class Demospongiae lack siliceous spicules and have only spongin fibers joined together to form networks of varying configurations (e.g., Dictyoceratida, Verongida) or systems of branching fibers that do not anastomose (e.g., Dendroceratida). Spongin fibers may include foreign mineralized elements such as sand grains, fragments of calcareous skeletons of plants or animals, and foreign sponge spicules, broken or not, all picked up from the environment.

The Kinds of Spicules

Spicules of demosponges may be separated according to size, with large spicules being known as megascleres and small ones as microscleres. These two terms are relative, however, depending upon the species of sponge. In species with megascleres ranging from 2,000 to 4,000 μm, the microscleres may be 200–400 μm, while species with megascleres ranging from 100 to 200 μm may have microscleres of 10 to 25 μm. Megascleres form the supporting framework of a sponge, usually in association with spongin fibers in demosponges, whereas microscleres may be scattered in the matrix (mesohyl) of the sponge, serving as strengthening agents. They may also be localized at the surface of the sponge or in the lining of the exhalant canals.

Another way of differentiating spicules is based on the number of axes or rays (points). Names for spicules are coined by adding the appropriate numerical prefix to the ending **-AXON** (when referring to the number of axes) or **-ACTINE** (when referring to the number of rays or points). For example, an important category of spicules consists of **MONAXONS**, formed by growth along a single axis. If growth occurs in a single direction (indicated by the presence of one point), the spicule is a **MONACTINAL MONAXON** (plates 17A, 20A–20D); if growth occurs in both directions, it is a **DIACTINAL MONAXON** (plates 17B, 20E–20J). Both monactines and diactines are usually megascleres, but in some instances, if larger spicules are also present and the **DIACTINES** are relatively small, they are considered microscleres (plate 23C–23D).

Another important category of spicules comprises **TETRAXONS** (plate 17C) with four axes and rays (points), each pointed in a different direction. The rays may be equal in length and separated by equal angles, giving rise to a three-dimensional spicule called a **CALTHROP** (plate 21A), or one ray (called the **RHABDOME**) may be longer than the others (called **CLADS**), in which case the spicules are referred to as **TRIAENES** (plate 21B–21F) because three of the rays (clads) are identical and separated from each other by equal angles. Clads may be lost from triaenes; if one is lost, leaving two, the spicule is called a **DIAENE** (plate 21G–21J), and if two clads are lost, leaving one, the spicule is known as a **MONAENE** (plate 21K–21L).

The spicules of sponges of the Class Calcarea have one to four rays and are named, as in demosponges, by adding a numerical prefix to the ending **-ACTINE** (meaning ray or point). Thus, there are monactines, diactines, triactines, and tetractines. The last two spicule types mentioned are also called **TRIRADIATES** (plates 17D, 20K, 20L) and **QUADRIRADIATES** (plate 17C). The three rays of a triradiate may lie in the same plane, but more often the spicules take on the gentle or more marked curvature of the surface of the sponge where they are laid down. **REGULAR TRIRADIATES** (plate 20K) consist of three similar rays of equal size with equal angles between the rays. In **SAGITTAL TRIRADIATES** (plate 20L), a single plane of symmetry exists with two paired (when the lateral rays are parallel, you have a tuning fork spicule) and an unpaired posterior ray. Two types of sagittal spicules can be recognized. In one, the angles between the rays are equal, and a condition of bilateral symmetry exists that is determined by one ray that is either longer or shorter than the other two. In another form, the angles may vary as well as the rays, with two lateral paired angles occurring with an anterior unpaired angle. Any of the variants of the triradiate spicules may be transformed into a **QUADRIRADIATE** (plates 17C, 20M) by the addition of a fourth ray arising from the center of the point of junction of the other three rays. This fourth ray may be straight or curved, is usually shorter than the other three rays, and may bear spines in some cases.

The following is a summary of the kinds of spicules that occur in the intertidal sponges of central and northern California and Oregon.

Glossary of Spicule Types

Sponges indicated are examples only.

Megascleres

MONACTINAL MONAXONS

STYLES rounded at one end, pointed at the other (*Hymeniacidon ungodon*); the adjectival form is "stylote." Plate 20A.

TYLOSTYLES with a distinct knob at one end, pointed at the other, resembling a marlinspike (*Cliona californiana*). Plate 20C.

SUBTYLOSTYLES tylostylelike with an indistinct knob on one end and a point at the other (*Clathria* [*Microciona*] *pennata*). Plate 20D.

ACANTHO—a prefix denoting that the spicule is covered with thorny processes (e.g., acanthostyles [plate 20B] in *Clathria* [*Microciona*] *microjoanna*); the adjectival form is "acanthose."

DIACTINAL MONAXONS

OXEAS pointed at both ends (*Neopetrosia vanilla*); the adjectival form is "oxeote." Plate 20E.

STRONGYLES rounded at both ends (*Tethya californiana*); the adjectival form is "strongylote." Plate 20F.

TORNOTES lance-headed at both ends (*Acanthancora cyanocrypta*). Plate 20G.

TYLOTES knobbed at both ends (*Antho* [*Acarnia*] *karykina*). Plate 20H.

SUBTYLOTE a diactinal spicule with indistinct knobs on both ends.

CLADOTYLOTES tylotes with more or less recurved clads at each end, the shaft spined or not (*Acarnus erithacus*). Plate 20I.

CENTROTYLOTES having a knob near the center of the shaft of a spicule such as a centrotylote oxea. Plate 20J.

TETRAXONS

CALTHROPS the four rays equal or nearly so (*Poecillastra tenuilaminaris*, an offshore species). Plate 21A.

TRIAENES tetraxons with one long ray (the rhabdome) and three short, equal rays (clads). Plate 21B–21F.

ORTHOTRIAENES with clads making an angle of about 90 degrees with the axis of the rhabdome (*Stelletta clarella*). Plate 21C.

PLAGIOTRIAENES with clads directed forward and making an angle of ca. 45 degrees with the produced axis of the rhabdome (*Stelletta clarella*). Plate 21D.

PROTRIAENES with clads directed forward and making an angle of <45 degrees with the produced axis of the rhabdome (*Tetilla arb*). Plate 21E.

ANATRIAENES with the clads directed backwards (*Tetilla arb*). Plate 21F.

DICHOTRIAENES with forked clads (*Stelletta clarella*). Plate 21B.

DIAENES triaenes modified through the loss of one clad (*Tetilla arb*). Plate 21G–21J.

MONAENES highly modified triaenes with only one clad (?*Tetilla* sp. B). Plate 21K, 21L.

ANAMONAENES with a single clad directed backward like a hook (?*Tetilla* sp. B). Plate 21K.

PROMONAENES with a single clad directed forward, at an angle of ca. 45 degrees. It is possible to confuse promonaenes with

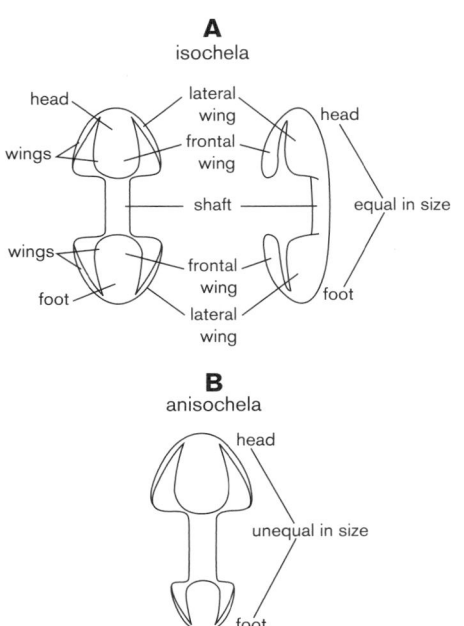

A
isochela

head
lateral wing
frontal wing
head
wings
shaft
equal in size
wings
frontal wing
foot
lateral wing
foot

B
anisochela

head

unequal in size

foot

PLATE 18 Isochela and anisochela spicule morphology and terminology.

oxeas. Remember that promonaenes are long, very thin spicules, the summits of which bend at an angle of ca. 45 degrees with the produced main axis. They are restricted in occurrence to the orders Spirophorida and Astrophorida. Oxeas are thicker and shorter than promonaenes. Unfortunately, there are oxeas with angulated summits. Plate 21L.

TRIRADIATES spicules with three rays more or less in the same plane or sometimes bent strongly out of the same plane; common in calcareous sponges such as *Leucosolenia*. Plates 17D, 20K, 20L.

REGULAR TRIRADIATES equirayed and equiangular. Plate 20K.

SAGITTAL TRIRADIATES have the unpaired ray longer or shorter than the paired rays or the paired lateral angles differ from the anterior, unpaired angle. Plate 20L.

QUADRIRADIATES derived from triradiates by the addition of a short ray, straight or curved and occasionally spined, at the junction of the other three rays. Common in calcareous sponges such as *Leucosolenia*. Plate 20M.

Microscleres

CHELAS short, straight, or curved shaft with ends reflexed and variously elaborated. Plate 18.

ISOCHELAS chela with ends equal in size, of several sorts as follows. Plates 18A, 22A–22D.

PALMATE ISOCHELAS the entire length of the lateral wings fused to the shaft; the single frontal wing totally formed and free (*Clathria* [*Clathria*] *prolifera*). Plates 19A, 22A.

ARCUATE ISOCHELAS half the length of the lateral wings fused to the shaft and half the length free. Frontal wing totally formed and free. Often but not always with shaft bent or arched (*Clathria* [*Microciona*] *spongigartina*). Plates 19B, 22B.

ANCHORATE ISOCHELAS lateral wings and frontal wings free; in addition, a ridge extends along each side of the shaft (*Myxilla* [*Myxilla*] *incrustans*). The wings may subdivide to form a variable number of teeth, producing an "unguiferate isochela," (plates 19E, 22D) (*Plocamiancora igzo*). Plates 19C, 22C, 22D.

ANISOCHELAS chelas with ends of unequal size. Plates 18B, 22E–22G.

PALMATE (plate 19D) and "anchorate" (plate 22G) forms of anisochelas may occur (e.g., palmate anisochelas, of *Mycale* [*Mycale*] *macginitiei*).

SIGMAS c- (plate 22H) or s-shaped (plate 22I) microscleres; at some times both occur, at others only c-shaped forms occur (*Lissodendoryx* [*Lissodendoryx*] *firma*, both c- and s-shaped forms occur). In some sponges large, c-shaped sigmas occur with "serrated ends" (plate 22J; *Mycale* [*Paresperella*] *psila*).

DIANCISTRAS microscleres with hooked, penknife-shaped ends, notched where the ends join the shaft and in the middle of the shaft (*Hamacantha* [*Vomerula*] *hyaloderma*). Plate 23E.

TOXAS bow-shaped microscleres (*Clathria* [*Microciona*] *pennata*). In some species, the tips of toxas may be "microspined" (plate 22M). Plate 22K, 22L.

FORCEPS spined u-shaped microscleres (*Forcepia* [*Forcepia*] *hartmani*). The spines may be exceedingly small and visible only in SEM photographs. Plate 23A, 23B.

RAPHIDES or "microxeas" very thin, hairlike microscleres, often occurring in bundles called "trichodragmas" (plate 23F). Plate 23C.

ONYCHAETES or "microspined microxeas" a long, thin, finely spined microsclere (*Tedania* [*Tedania*] *obscurata*). Plate 23D.

MICRO—a prefix used to refer to a microsclere that is similar in form to a megasclere (e.g., microxea).

ASTERS microsclere with short rays, radiating from a point, sphere or rod. Plate 24.

EUASTERS a collective term for asters in which the rays radiate from a central point. Plate 24A–24E.

OXYASTERS a euaster with pointed free rays coming from a small central body less than one third the diameter of the whole spicule (*Stelletta clarella*). Plate 24A.

OXYSPHERASTER or "spheroxyasters" a euaster with a central body that is more than one third the total diameter of the spicule (*Stelletta clarella*). Plate 24E.

STRONGYLASTERS a euaster with rounded free rays (*Stelletta clarella*). Plate 24B.

SPHERASTERS a euaster with short rays and a thick central body; the diameter of the central body is more than half the total diameter and exceeds the length of the rays (*Tethya californiana*). Plate 24D.

TYLASTER a euaster with free rays that have a knobbed end (*Tethya californiana*). Plate 24C.

STREPTASTERS a collective term for asters in which the rays come out from an axis that is usually a spiral. Plate 24F–24H.

SPIRASTERS streptasters with a spiral, rod-shaped axis and peripherally arranged spines (*Cliona* sp.). Plate 24H.

AMPHIASTERS a microsclere with microspined rays extending from both ends of a shaft, with the rays shorter than the shaft (*Acanthancora cyanocrypta*). Plate 24F.

SIGMASPIRES c- or s-shaped microspined spicules, often contorted; characteristic of the family Tetillidae (*Tetilla arb*). Plate 24I.

Glossary of Other Terms Used in the Key

ANISO a prefix denoting that the spicules are not equal or not alike, as in anisostrongyles, anisotornotes, and anisosubtylosles.

ANISOTROPIC RETICULATION reticulate skeleton with distinct primary and secondary tracts, lines, or fibers.

ATRIUM preoscular cavity. An exhalant cavity receiving water from one or more exhalant canals or syconoid chambers (in

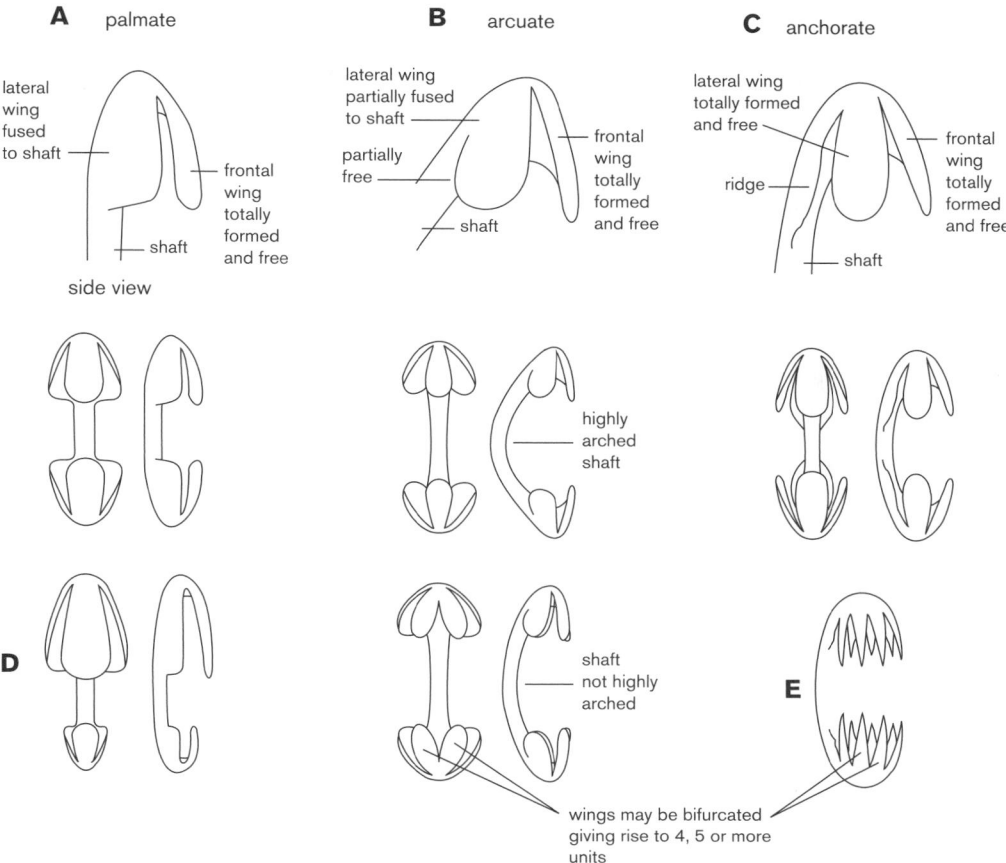

A palmate

lateral wing fused to shaft

side view

frontal wing totally formed and free

shaft

B arcuate

lateral wing partially fused to shaft

partially free

shaft

frontal wing totally formed and free

C anchorate

lateral wing totally formed and free

ridge

frontal wing totally formed and free

shaft

highly arched shaft

D

shaft not highly arched

E

wings may be bifurcated giving rise to 4, 5 or more units

PLATE 19 Palmate, arcuate, and anchorate spicule morphology and terminology.

calcareous sponges) and conducting it to one or more, usually terminal, oscules.

CLOACA see "atrium."

CONULOSE surface with numerous cone-shaped projections (conules) raised up by the underlying skeleton.

CORTEX the ectosome, when thick and gelatinous, fibrous (as in *Tethya californiana*), or packed with spicules of a special type (as in *Geodia*).

CHOANOCYTE cell having a flagellum that is surrounded by a collar of cytoplasmic microvilli.

CHOANOSOME interior region of the sponge lying below the ectosome, containing choanocyte chambers.

ECHINATE spicules that project from a spongin fiber either at right or acute angles are said to "echinate" the fiber.

ECTOSOME superficial peripheral region of the sponge that lacks choanocyte chambers.

ENDOSOME see "choanosome."

GASTRAL CAVITY see "atrium."

GEMMULES a resistant asexual reproductive body, composed of cells charged with reserves and enclosed in a noncellular protective envelope or coat often strengthened by spicules.

HISPID cellular surface or surface of spongin fibers covered by projecting spicules.

ISODICTYAL SKELETON isotropic reticulation in which the meshes are triangular and have sides one spicule long.

ISOTROPIC RETICULATION reticulation without differentiation into primary and secondary fibers, tracts, or lines.

LEUCONOID aquiferous system in which the choanocytes are restricted to discrete choanocyte chambers usually grouped in the interior of the sponge.

MUCRONATE adjective referring to the somewhat nipplelike point of a megasclere where the rounded end is produced axially to a point.

MULTISPICULAR refers to tracts or fibers consisting of six or more spicules adjacent to one another.

PAUCISPICULAR refers to tracts or fibers consisting of two to five spicules adjacent to one another.

PINACOCYTES cells delimiting the sponge from the external milieu and always only in a layer one-cell deep.

PINACODERM surface lined by pinnacocytes.

PLUMOSE SKELETON (TRACTS) a type of skeletal construction made up of primary fibers or spicule tracts from which the skeletal elements radiate obliquely.

PRIMARY FIBER an ascending fiber or tract ending at or near a right angle to the surface.

PORE SIEVES a specialized area of the ectosome with a cluster of ostia and an underlying inhalant cavity called the vestibule.

SECONDARY FIBER in a reticulate skeleton, a fiber or tract that links the primary fibers.

SPONGOCOELE see "atrium."

SUBDERMAL below the ectosome.

SYCONOID aquiferous system with elongate choanocyte chambers containing free distal cones or extending from cortex to atrium.

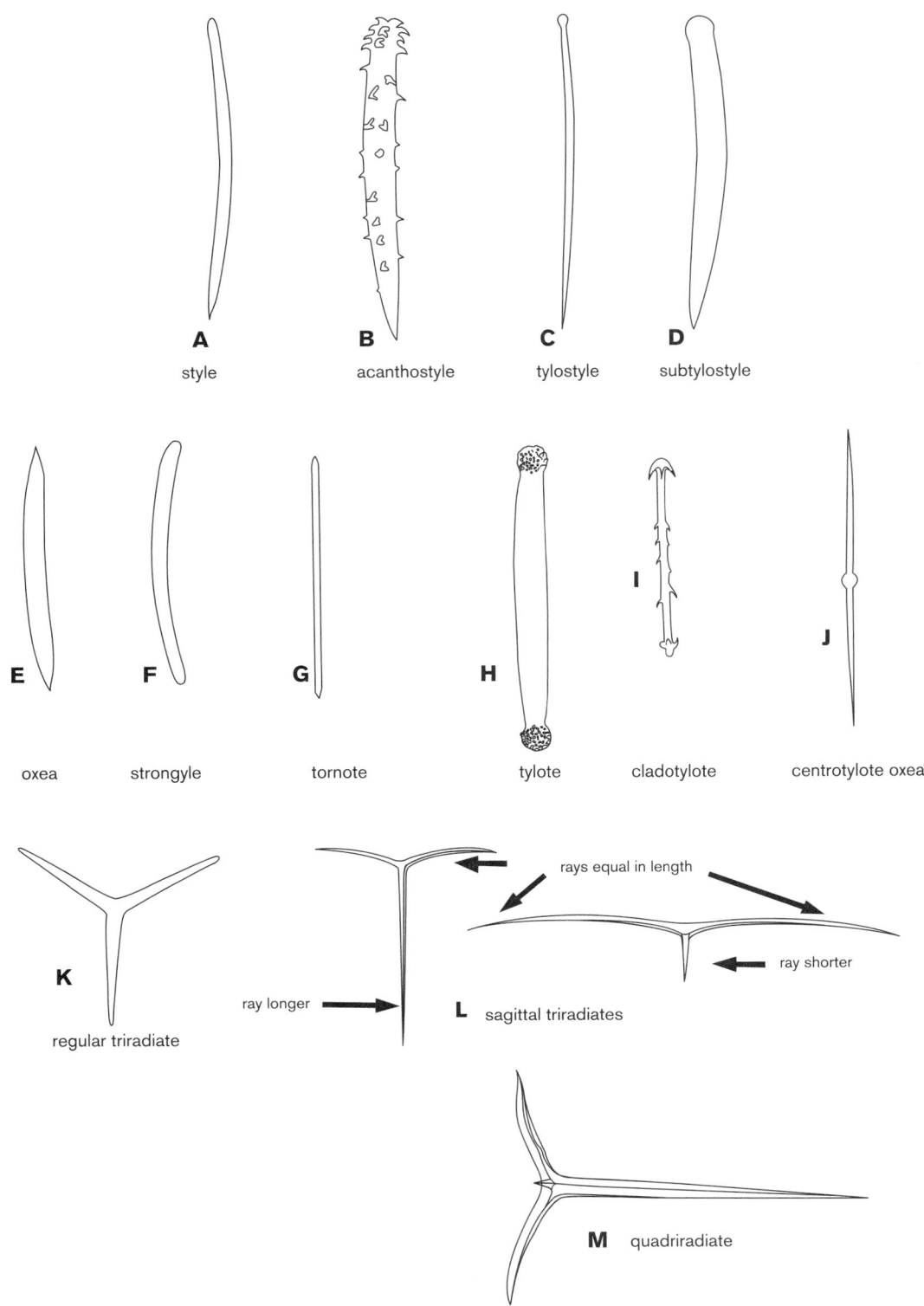

A style
B acanthostyle
C tylostyle
D subtylostyle

E oxea
F strongyle
G tornote
H tylote
I cladotylote
J centrotylote oxea

K regular triradiate

L sagittal triradiates

ray longer

rays equal in length

ray shorter

M quadriradiate

PLATE 20 Types of spicules.

TANGENTIAL SKELETON ectosomal skeleton arranged parallel to the surface.

Acknowledgments

As is always the case, contributions such as this frequently rest upon the shoulders of others whose help and efforts contribute heavily to the final product. One of us (Welton Lee) gratefully thanks the David and Lucile Packard Foundation for their support of the California Sponges Project, which provided funds for SEM use as well as basic research on the California sponge fauna. Our deepest gratitude to Henry Reiswig, John Hooper, Rob Van Soest, Walentina De Weerdt, Eduardo Hajdu, Jerry Bakus, and Klaus Reutzler who all contributed immensely with information,

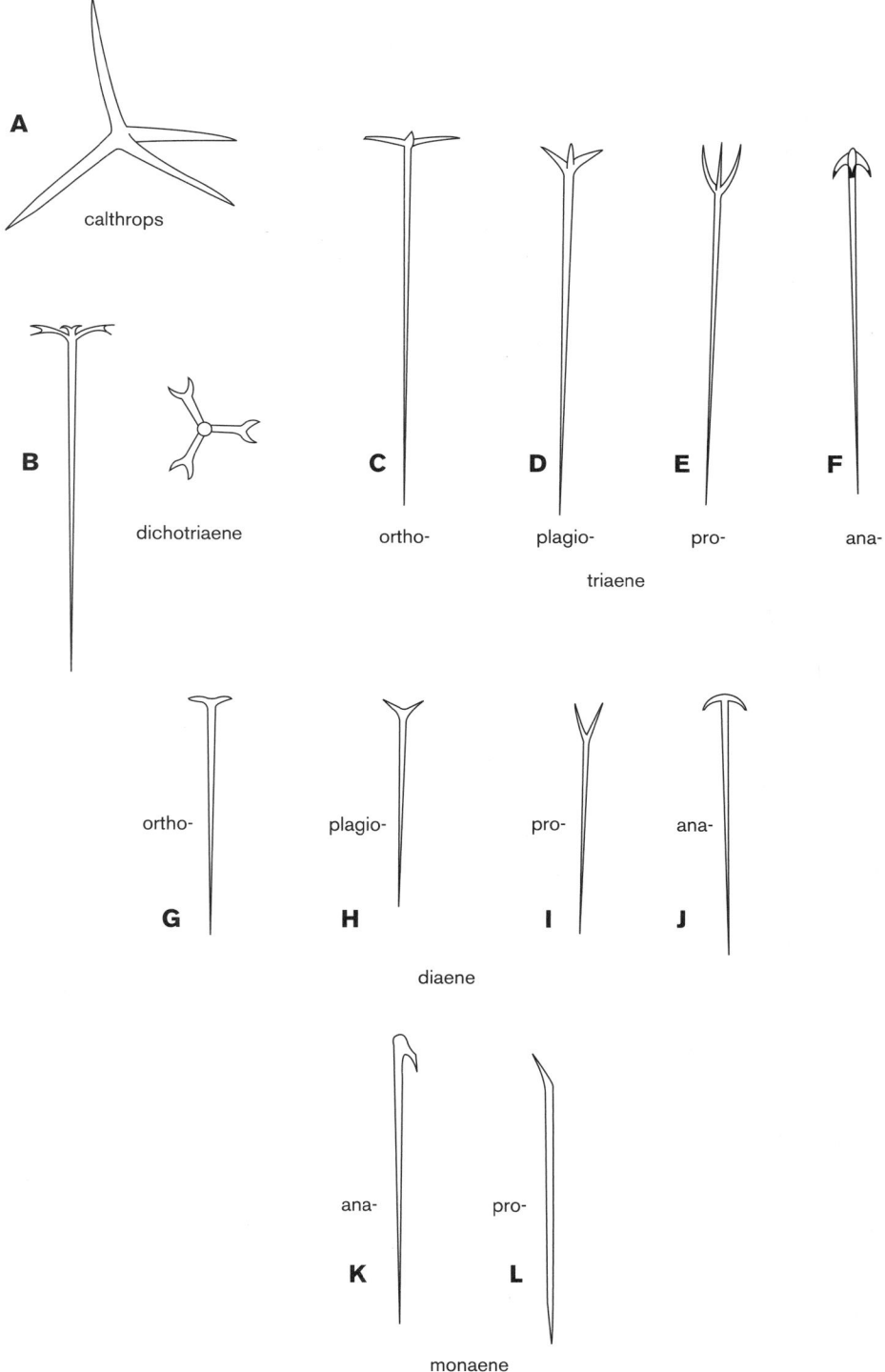

A calthrops

B dichotriaene

C ortho-

D plagio-

E pro-

F ana-

triaene

G ortho-

H plagio-

I pro-

J ana-

diaene

K ana-

L pro-

monaene

PLATE 21 Types of spicules.

advice, specimens, and critical reviews. Thanks go to the staffs of the California Academy of Sciences, the University of California Scripps Institution Benthic Invertebrate collection, and the Peabody Museum of Natural History—most notably Bob Van Syoc, Elizabeth Kools, Daryll Ubick, Chris Mah (California Academy of Sciences), Larry Lovell (Scripps), and Eric Lazo Wasem (Peabody Museum). Very special thanks go to Bill Austin who graciously gave permission to use and/or modify many of his fine drawings of sponge spicules, many of which were used to illustrate the Austin and Ott 1987 work on northern Pacific coast sponges, and to John Pearse who supplied photographs and valuable information on *Oscarella carmela*.

Key to Intertidal Sponges

1. Skeleton present; may consist of spicules, spicules and fibers, or fibers alone. 3

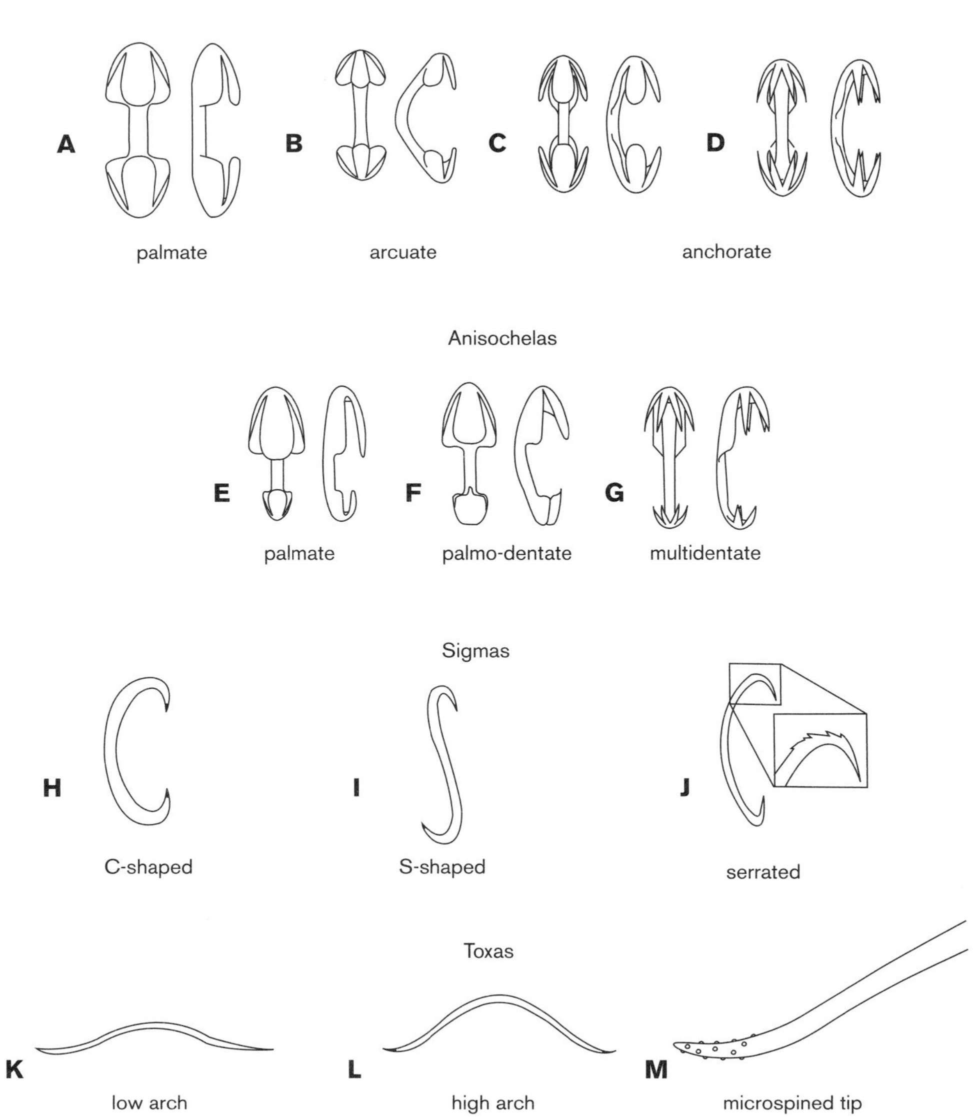

Anisochelas

A
palmate

B
arcuate

C
anchorate

D

E
palmate

F
palmo-dentate

G
multidentate

Sigmas

H
C-shaped

I
S-shaped

J
serrated

Toxas

K
low arch

L
high arch

M
microspined tip

PLATE 22 Types of spicules.

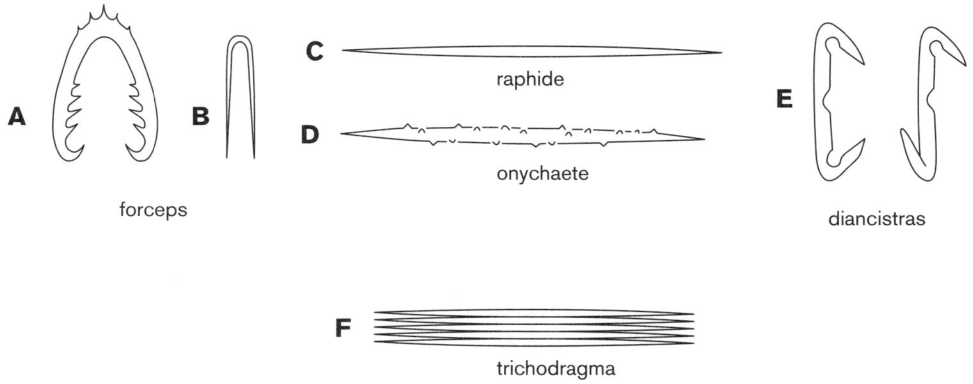

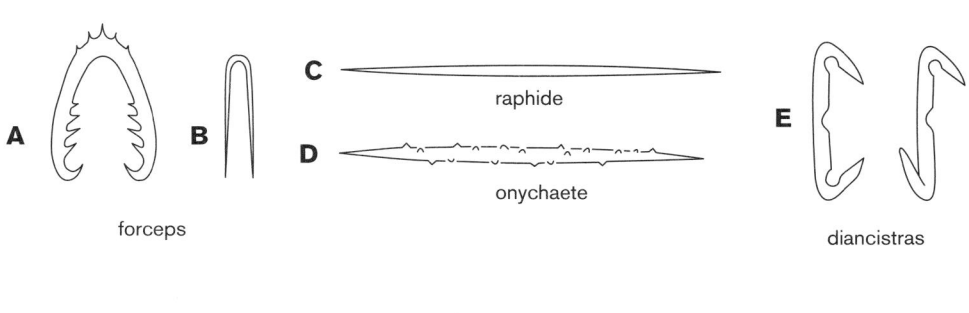

A
forceps

B

C
raphide

D
onychaete

E
diancistras

F
trichodragma

PLATE 23 Types of spicules.

EUASTERS

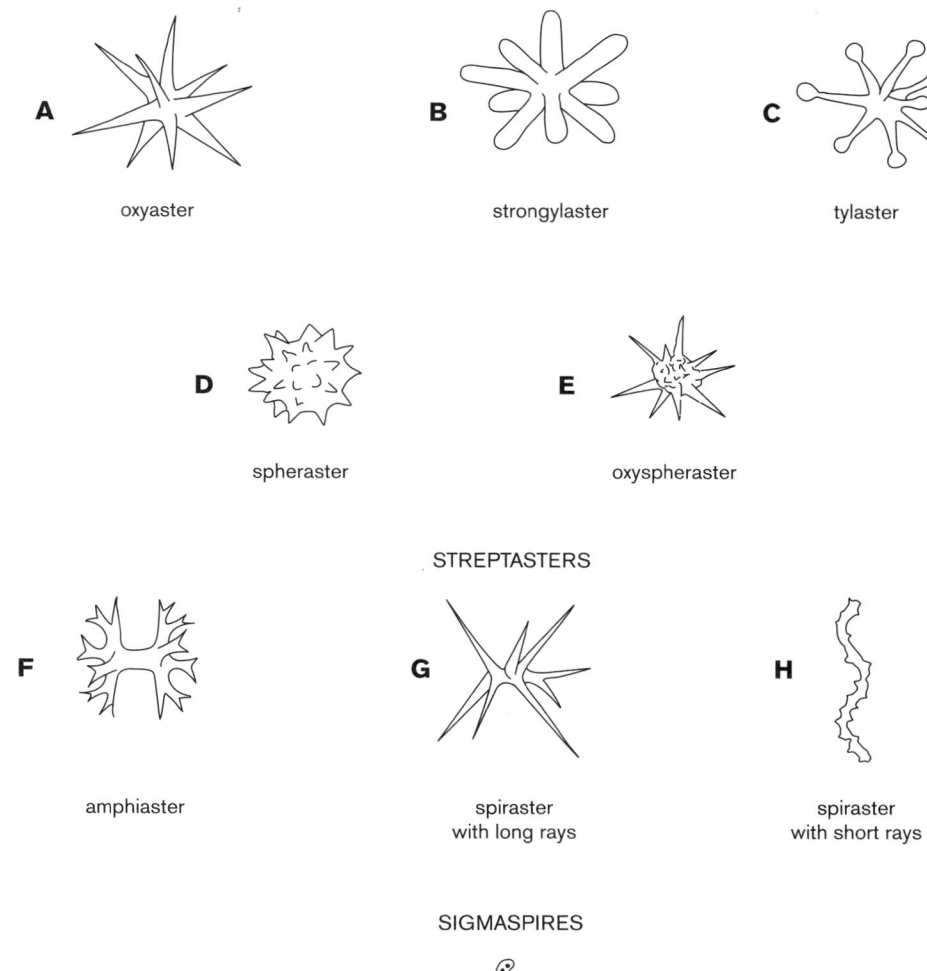

A oxyaster

B strongylaster

C tylaster

D spheraster

E oxyspheraster

STREPTASTERS

F amphiaster

G spiraster
with long rays

H spiraster
with short rays

SIGMASPIRES

I

PLATE 24 Types of spicules.

— Skeleton absent (neither spicules nor spongin present) 2
2. Choanocyte chambers tubular, elongated; pinacocytes without flagella; thin, encrusting; pale tan, yellow-tan, or light brown in color; may be mistaken for a colonial ascidian or the bryozoan *Alcyonidium* (plate 25A) . . . . *Halisarca* sp.
— Choanocyte chambers ovoid or round; pinacocytes flagellated; thin encrusting; pale tan to yellow tan in color; commonly observed in enclosed seawater tanks, aquaria (plate 25B) . . . . *Oscarella carmela*
3. Skeleton of spicules present; spicules may be calcareous or siliceous 6
— Skeleton of spicules absent; fiber skeleton present; fibers may contain foreign spicules, sand grains, etc.; foreign spicules usually broken, of improbable combinations such as both calcareous and siliceous spicules together, and/or of low numbers 4

4. Fibers arise from a basal plate of spongin (plate 25C); primary choanosomal fibers may branch but *do not* anastomose to form a reticulation; ectosome may have a fibrous reticulation in one plane; each fiber has a cortex of concentric layers of spongin surrounding a central pith; choanocyte chambers large, up to 75 μm or greater in diameter 5
— Fibers *do not* arise from a basal plate of spongin; primary and secondary choanosomal fibers anastomose to form a reticulation; ectosome opaque and melanistic, with no skeletal reticulation; sponge is massive, surface conulose; slate-gray; fibers may be covered with foreign spicules or sand grains; choanocyte chambers small, round, 25–30 μm in diameter (plate 25D). . . . . *Spongia* (*Spongia*) *idia*
5. Color deep purple; ectosome bears a reticulation of fibers packed with foreign spicules; sponge surface with scattered conules, 1 mm high; skeleton has primary fibers 80–180 μm

with no inclusions; cells packed with purple granules; soft, thickly encrusting to 5 mm; gives off purple pigment when collected (plate 25E) . . . . *Aplysilla polyraphus*

— Color olive-tan, salmon pink, or coral; ectosome, a dermis with *no* inclusions; sponge is encrusting to 1–3 mm; surface conulose, with conules 1–2 mm high and 2–3 mm apart; soft, slippery; skeleton has primary fibers 80–255 μm with no inclusions (plate 25F) . . . . . *Aplysilla glacialis*

6. Spicules calcareous; sponges often tubular or subspherical, with few local species; tan or whitish 74
— Spicules siliceous; sponges of various shapes with many local species; colors variable 7

7. Spicules are diactines with or without microscleres 8
— Spicules include monactines, monactines and diactines, or monactines, diactines, and tetractines; microscleres present or absent 21

8. Diactines are oxeas (have sharp points at each end, plate 25G) 9
— Diactines range from strongyles (rounded at each end) to subtylotes (with inconspicuously knobbed ends that are microspined); sponge is encrusting to 12 mm thick, exudes mucus; bright yellow-tan; spicules: (1) strongyles to subtylotes with microspined ends; of two size classes, 182–245 × 5.5–7.0 μm and 249–339 × 10.0–13.0 μm; (2) onychaetes 78–110 μm; skeleton has vague tracts of strongyles that run to the surface (plate 25H) . . . .
. . . . *Tedania (Tedania) obscurata*

9. Range of spicule length usually greater than 150 μm but never less than 100 μm in any one specimen; skeleton confused, with or without loose spicule tracts; surface with or without a detachable ectosomal skeleton bearing a more or less reticulate pattern of spicule tracts arranged tangentially 10
— Range of spicule length usually less than 35 μm but never greater than 60 μm in any one specimen; skeleton more or less reticulate, with or without defined spicule tracts, or so densely packed as to obscure the basic reticulate pattern; surface with or without detachable ectosomal skeleton
. . . . 11

10. Surface opaque; ectosomal skeleton with regular, close reticulation of multispicular tracts, pores closely spaced in areas between tracts; tangential skeleton not easily removed; sponge typically encrusting with oscules on ridges but may also have tubelike oscules or anastomosing hollow branches with oscules at the tip; oscules obvious; sponge typically encrusting, sometimes more massive, often with rows of oscules borne on ridges; chrome yellow, yellow-tan, and, especially in well-lighted situations, olive or green at surface spicules: oxeas, 160–420 × 1.0–11 μm; on open ocean coast (plate 26A) . . . .
. . . . *Halichondria (Halichondria) panicea*

Note: This species is most often found on the open coast. It may intrude into bays where salinity and current regimes approximate those of outer coast regimes. In bay localities this species tends to form branches similar to those of *H. bowerbanki*. No single character can unequivocally separate this and the next species. All characteristics must be reviewed.

— Surface translucent or nearly so; ectosomal skeleton of widely spaced, more or less parallel multispicular tracts, the areas between these tracts further subdivided by a pattern of overlapping individual spicules, pores more widely spaced; tangential skeleton easily removed; sponge typically encrusting or, more often, branching or in flattened lobes

with somewhat serrated edges or round, raised tubes; oscules frequently not obvious, scattered; yellow- tan, gold, or olive-brown; spicules: oxeas 120–480 × 5.0–12 μm; in bays and harbors (plate 26B) . . . .
. . . . *Halichondria (Halichondria) bowerbanki*

Note: This species is found in bay localities. However, under certain conditions this species takes on some of the features of *H. panicea*. These include a more strictly encrusting form with little or no branching. These two species are often difficult to distinguish. This may be exacerbated by the possibility that we may have a few additional "sister" species that could be confused with these common halichondriids.

11. Sponge with anastomosing spongin fibers (50–200 μm in diameter) filled with many rows of oxeas; also a secondary isodictyal (triangular) reticulation of oxeas of similar size with spongin joining the spicules where they meet to form a network; sponge is encrusting, 5–30 mm thick; oscules to 4 mm, flush with surface or with rims raised up to 2 mm above the surface; lavender or pale rose-gray; spicules: oxeas, 70–160 μm × 3.3–10.6 μm (plate 26C) . . . . *Niphates lunisimilis*
— Sponge *without* anastomosing spongin fibers bearing many rows of oxeas; no secondary isodictyal (triangular) reticulation of oxeas present 12

12. Sponge with hard, rocklike consistency; spicules densely packed 13
— Sponge with soft consistency; spicules not densely packed
. . . . 15
— Sponge crisp and brittle; spicules moderately densely packed; sponge encrusting, 2–4 mm thick; oscules few, irregular in shape, 1 mm in diameter, often with raised collars; surface superficially smooth; color when alive is pale lavender, color when preserved is cream; spicules: oxeas 105–112–122 × 3.0–6.2–6.9 μm; skeleton ectosomal crust of tangential oxeas, easily removed in flakes; choanosome of rather dense, confused reticulation (plate 37) . . . .
. . . . *Haliclona (Halichoclona) gellindra*

13. Sponge rough to touch (like fine sandpaper) resulting from ends of spicule tracts (up to 360 μm high and spaced 400–900 μm apart) protruding through surface of sponge; sponge encrusting (4–12 mm thick) or erectly lobate and laterally compressed; oscules 0.6–5.0 mm in diameter, flush with surface, arranged in lines on top edge of upright specimens or scattered over the surface of encrustations; dark brown, with internal color lighter brown; spicules: oxeas ranging from 115–260 × 2–20 μm; many immature stages may be present (plate 26D) . . . . *Amphimedon trindanea*
— Sponge smooth to touch. 14

14. Ectosomal skeleton indistinct; (2) choanosomal skeleton consisting of rounded meshes of spicules with a superimposed anisotropic reticulation of irregular tracts with short connecting tracts and with little spongin, especially toward the surface; (3) color buff, tan or dull orange brown; sponge encrusting, usually thin (5 mm) but up to 15 mm thick; oscules up to 2 mm across, flush with surface or with slightly raised rims; spicules: oxeas, 150–160 × 11–12 μm, some with a bend at each end (plate 26E) . . . .
. . . . *Neopetrosia vanilla*
— Ectosomal skeleton consisting of tangential oxeas arranged in tracts, two to five spicules wide, and forming an irregular polygonal network; choanosomal skeleton a reticulation of multispicular tracts, six to 10 spicules wide, varying to a confused arrangement of spicules; color white; sponge encrusting to 4–8 mm thick; oscules 0.8–1.3 mm in diameter,

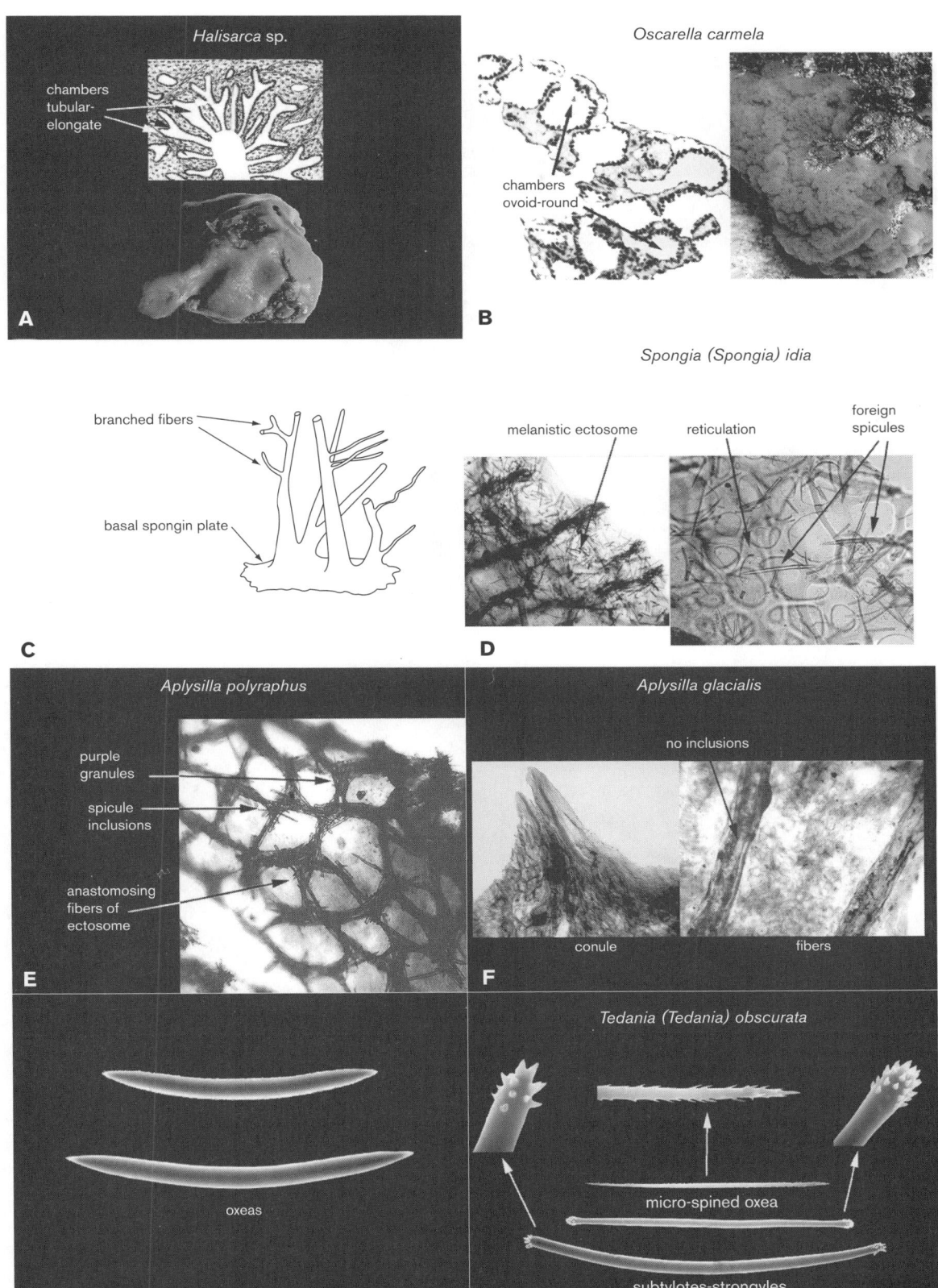

PLATE 25 A, *Halisarca* sp. because spicules are lacking, species identification is dependent upon histology, details of germ cells and larvae; B, *Oscarella carmela* identification is dependent upon histological features (e.g., choanocyte chamber shape and size, cell types), and reproductive biology (e.g., details of germ cells and larvae, time of reproduction); C, fibers arise from a basal plate of spongin affixed to the substrate and branch but do not anastomose to form a reticulation; D, *Spongia (Spongia) idia*; E, *Aplysilla polyraphus*; F, *Aplysilla glacialis*; G, diactines with sharp points at each end are called oxeas; H, *Tedania (Tedania) obscurata*.

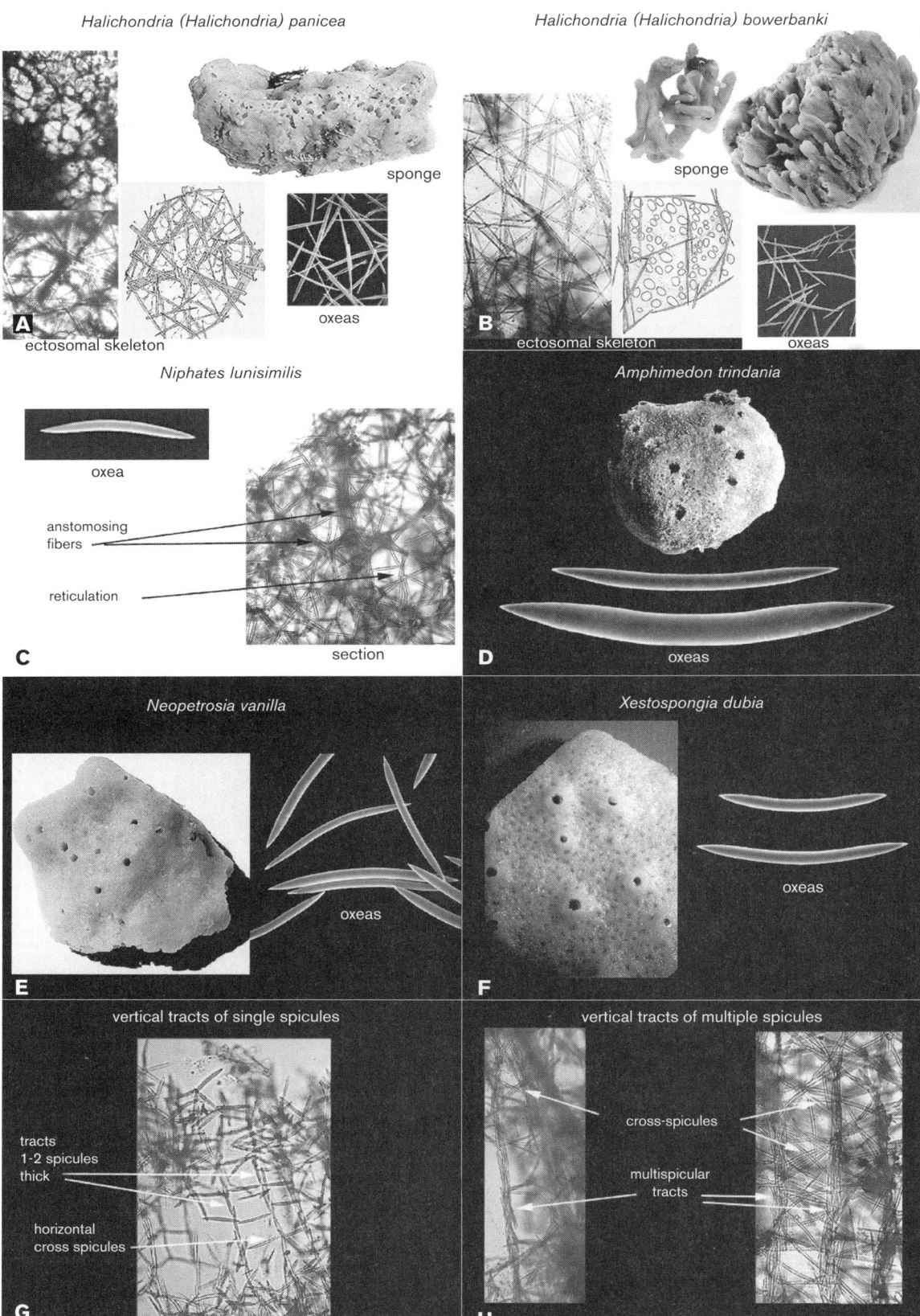

PLATE 26 A, *Halichondria (Halichondria) panicea*; B, *Halichondria (Halichondria) bowerbanki*; C, *Niphates lunisimilis*; D, *Amphimedon trindania*; E, *Neopetrosia vanilla*; F, *Xestospongia dubia*; G, vertical tracts of single or only a few spicules (paucispicular), cross spicules are chiefly horizontal, forming square or polygonal meshes. H, vertical tracts of multiple spicules (multispicular), cross spicules horizontal, diagonal, or irregular in arrangement.

3–5 mm apart, either with volcanolike rims 0.5–1.5 mm high, or flush with surface; spicules: oxeas 140–180 × 10–16 μm, straight to slightly curved; skeleton: subectosomal spaces supported by columns, four to 10 oxeas wide, may be present; some sponges show growth lines representing former ectosomal layers (plate 26F) ..... *Xestospongia dubia*

15. Spicule complement composed of megascleres only 16
— Spicule complement composed of megascleres and microscleres 19
16. Sponge encrusting with more or less flat surface; oscules flush or with raised rims; on open coast 17
— Sponge lobate to tubular, may branch or anastomose, the tubes up to 6.5 cm high; oscules commonly terminal on tips of tubes; in bays, commonly on floats; light tan to orange-brown or cinnamon; spicules: oxeas, 66–85–156–185 × 2.0–10.0 μm; skeleton vertical tracts of two to six or more spicules, most commonly multispicular; cross-spicules chiefly single and horizontal forming loose reticulation; produces gemmules; highly variable in both shape and spicule size (plate 27B) .... *Chalinula loosanoffi*
17. Vertical tracts (plate 26G) composed most often of a single row of spicules or of only a few spicules (=paucispicular); cross-spicules chiefly horizontal, forming square to polygonal meshes; oscules 1–3 mm across, closely spaced, almost flush with surface, or with raised rims, 1–5 mm high; oxeas, range of *mean* length 75–102 μm; on open coast; sponge thinly encrusting to 2–3 mm thick; mauve, buff, or purple; spicules: oxeas, 67–121 × 3.0–9.0 μm (plate 27A) .... *Haliclona (Haliclona)* sp. A
— Vertical tracts (plate 26H) multispicular; cross-spicules horizontal, diagonal, or irregular in arrangement; spicules: oxeas, larger, 32–188 μm, *mean* lengths 140–165 μm, *mean* widths 6.0–7.5 μm 18
18. Encrusting, up to 2 cm thick, surface irregular; oscules 2–7 mm across, sometimes arranged in rows on ridges flush with surface or with raised rims up to 10 mm high; color rose-lavender, lavender, or light gray-brown; spicules: oxeas, 130–188 × 3.0–11.0 μm (plate 27C) ....
.... *?Haliclona (Rhizoniera)* sp. A
— Encrusting up to 5 mm thick, surface smooth; oscules <1–2 mm across, barely raised above surface of sponge; color gray-blue; spicules: oxeas, 120–174 × 4.0–10.0 μm (plate 27D) .... *?Haliclona (Rhizoniera)* sp. B
19. Microscleres toxas; sponge encrusting, up to 15 mm thick; oscules, 2–3 mm across, sometimes on low mounds; olivebeige, gray-brown, or deep chrome yellow; spicules: (1) oxeas, 88–134 × 3.4–9.5 μm; (2) toxas, 35–96 × 2.0 μm; skeleton with paucispicular tracts and horizontal cross-spicules (plate 27E) (see note p. 115) ....
.... *?Haliclona (Reniera)* sp.
— Microscleres sigmas 20
20. Sigmas of two size classes; (2) oxeas >225 μm in length; sponge encrusting to 2 cm thick; buff-white in color; spicules: (1) oxeas, 260–275–300 × 12–13–15 μm; (2) sigmas, c-shaped with equal arms, 30–52–118 μm; (3) sigmas with unequal arms, greatest chord length, 75–87–118 μm; skeleton: vertical paucispicular tracts with cross-spicules oriented horizontally and diagonally; ectosome with a reticulation near the surface provided with much spongin; choanosomal skeleton with a greater abundance of confused spicules, but a basic isodictyal pattern remains (plate 27F) .... *Xestospongia edapha*
— Sigmas of a single size class; oxeas <225 μm in length; sponge encrusting to 12 mm thick, often overgrowing the sponge

Stelletta clarella. Color: rose-lavender; spicules: (1) oxeas, mean length 170–190 × 7.0–9.0 μm; (2) sigmas, 24–33–36 μm; skeleton: some oxeas at surface of dermal layer, others, standing on end, pierce the surface; sigmas and spongin abundant in surface layer; subdermal cavities present; vertical unispicular and paucispicular tracts descend into the choanosome with horizontal and diagonal cross-spicules; deeper in the choanosome a greater number of confused spicules occurs but a basic isodictyal pattern remains (plate 27G) (see note on p. 115) .... *?Haliclona (Gellius)* sp.

21. Spicules monactines (plate 27H), with or without microscleres 22
— Spicules monactines and diactines or may include tetractines or their derivatives, monaenes and diaenes; with or without microscleres (plate 28A) 48
22. Spicules monactines only 23
— Spicules monactines with microscleres 31
23. Spicules styles only 24
— Spicules tylostyles only, or styles to tylostyles 27
24. Spicule size range *in any one specimen* no >80 μm ... 25
— Spicule size range *in any one specimen* >200 μm 26
25. Surface more or less flat and superficially smooth; ectosome thick with layers of styles parallel to the surface and penetrated at relatively even intervals by emerging tracts; easily detachable; tracts appear only near the surface or may be absent; choanosome dominated by scattered spicules with no orientation; sponge encrusting to 2.7 cm; consistency firm; oscules, 0.3–1.3 mm across, are scattered over the surface; bright yellow in color; spicules: styles, 165–220 × 4.0–8.0 μm; skeleton has tangential ectosomal skeleton (80–120 μm thick) made up of one or two layers of horizontal styles, the ectosomal styles form multispicular tracts two to five spicules wide that unite into a polygonal network or are matted together to form a smooth, solid surface; the polygonal areas surround groups of pores, 10–30 μm across (plate 28B) .... *Hymeniacidon actites*
— Surface with long, often branched ridges and coarsely rugose; ectosome thick with spicules packed in every orientation, not readily detachable; tracts occur in both the ectosome and choanosome; choanosome with tracts of multiple spicules and frequently with spicules strewn in confusion; sponge encrusting to massive; oscules infrequent, oval, the long axis about 1 mm; found on open coast; spicules: styles, 130–252 × 3.4–10.2 μm; mahogany brown ectosome over yellow-drab choanosome; bright yellow, yellow-orange, cinnamon brown, or purple with yellow-ochre choanosome (plate 28C) .... *Hymeniacidon ungodon*
26. Surface smooth with scattered low conules <1mm high; color bright yellow, often with orange tints; on open coast and in bays; sponge encrusting, up to 10 cm high, 20 cm in diameter; oscules often raised on processes, diameter about 2 mm; spicules: styles, 130–460 μm × 3.4–10.2 μm; skeleton: ectosome developed into a thin, transparent, fleshy dermis containing few spicules; choanosome with spicules in confusion but occasionally organized into tracts directed vertically with points of styles up and expanding near the surface into subdermal brushes (plate 28D) .... *Hymeniacidon sinapium*
— Surface smooth with pointed processes 2.0 cm or greater in length; orange-brown to orange-tan in color; sponge thickly encrusting; spicules: styles, 127–363 × 3.6–12.0 μm; skeleton: spicules arranged in vague plumose tracts with numerous irregularly arranged spicules between; occurs in bays (plate 28E) .... *Hymeniacidon* sp. A

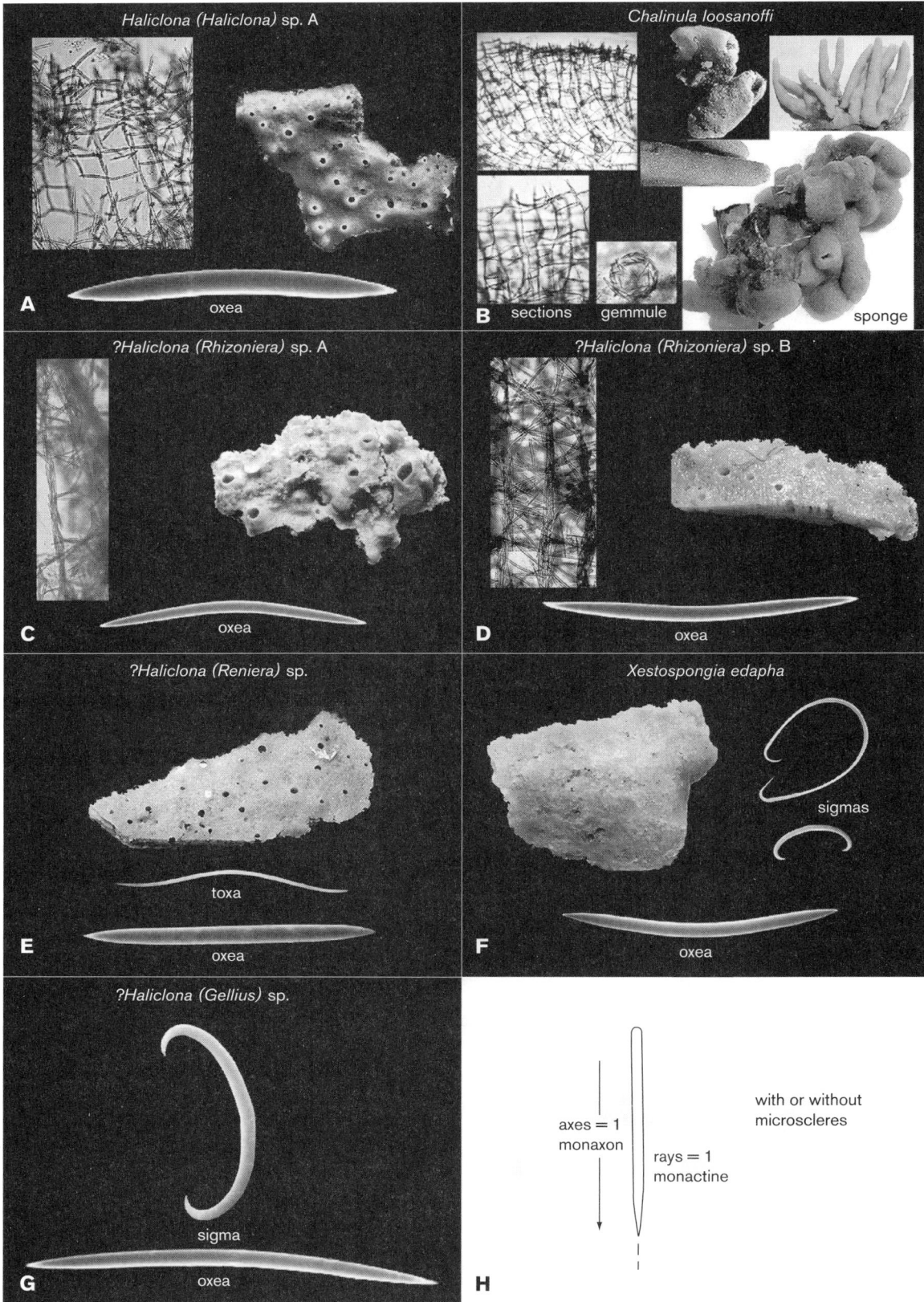

PLATE 27 A, *Haliclona (Haliclona)* sp. A; B, *Chalinula loosanoffi*; C, *?Haliclona (Rhizoniera)* sp. A; D, *?Haliclona (Rhizoniera)* sp. B; E, *?Haliclona (Reniera)* sp.; F, *Xestospongia edapha*; G, *?Haliclona (Gellius)* sp.; H, Megascleres are monactines only and these are megascleres with one axis and a single point.

27. An encrustation (up to 2 cm or more thick) from which arise numerous rounded to tapering papillae, 1–2 cm high; spicules are tylostyles and styles to subtylostyles; surface smooth to furry; sponge encrusting to 2 cm or more thick; lemon yellow to deep yellow in color; spicules: (1) mostly tylostyles of several size categories, up to 2,000 μm or greater, the heads small, the shaft narrowed at the head end, widest near the middle; heads terminal to subterminal; (2) thin subtylostyles to styles, long (up to 2,500 μm), usually project from the ectosome, giving rise to the fur-like surface (plate 28F) . . . . *Polymastia pachymastia*
— Surface of sponge without papillae; spicules are tylostyles only; surface variable 28
28. Tylostyles range in length from 190 to 340 μm; spicule range in length (from smallest to largest) *in any given sponge* is 80–150 μm; pore sieves present 29
— Tylostyles range in length from 200 to 925 μm; spicule range (from smallest to largest) *in any given sponge* is 140–750 μm; pore sieves absent 30
29. Sponge large, massive, up to 14 cm thick, 70 cm in diameter; ectosome packed with spicules to a depth of about 2 mm, the outer ones oriented with the points directed outward; choanosome with tracts of spicules of same size as ectosomal ones; these tracts form a vague reticulation; spicule range (from smallest to largest) *in any given sponge*, to 120 μm; sponge: massive; oscules up to 1 cm in diameter; pore fields present with pores localized in areas several square centimeters in size; pore canals, 1.5–3.0 mm in diameter, closed at the surface by a sieve, 300 μm in diameter and with three to six openings; lavender gray, purplish gray on outer surface, choanosome tan; spicules: tylostyles, 200–320 μm × 6.0–14.0 μm (plate 28G) . . . . . . *Spheciospongia confoederata*
— Sponge, during the first part of its life living in cavities it excavates in calcareous substrata such as shells of mollusks and barnacles; oscules and pores protrude through holes in the calcareous substratum; older sponges overgrow the substratum and may become free-living sponges up to 9 cm long, 6.5 cm wide, and 6.4 cm high; ectosome with scattered tylostyles; choanosome with scattered tylostyles; spicule range (from smallest to largest), *in any given sponge*, to 80 μm; bright yellow or pale to dark brown; spicules: tylostyles, some with subterminal heads, 190–340 × 2.0–13.6 μm (plate 28H) . . . . *Cliona californiana*
30. Spicules tylostyles ranging in length from 200–460 μm; sponge encrusting to massive, up to 6 cm thick and 10 cm in diameter; ectosome with small tylostyles; densely packed, with points directed outward; choanosome with larger scattered tylostyles; color light to dark chrome yellow, orange, pinkish-orange, cinnamon-brown, tan, orange-brown; on open coast (plate 29A) . . . . . . *Suberites* sp.
— Spicules tylostyles, ranging in length from 175–925 μm; sponge thinly encrusting, up to 2.0 mm thick; ectosome with densely packed tylostyles, points directed outward; choanosome with more or less well formed tracts of spicules running to the surface; color gold, yellow-brown, olive-brown, or hazel; in bays (plate 29B) . . . . . . *Prosuberites* sp.
31. Acanthose megascleres present 32
— Acanthose megascleres absent 40
32. Microscleres palmate isochelas only (plate 29C) 33
— Microscleres otherwise 34
33. Acanthose megascleres with heads embedded in a basal layer of spongin; encrusting to <1.0 mm thick; color yellow; spicules: (1) acanthosubtylostyles, sparsely spined, of two indistinct size classes, 92–100 × 6.0–11.0 μm and 135–470

× 6.5–14.5 μm; (2) styles to tylostyles with minutely microspined heads, the pointed ends sometimes truncate and microspined, 180–260 μm × 3.5–5.0 μm; (3) palmate isochelas, 3.0–14.0 μm (plate 29D) . . . .
. . . . *Clathria (Clathria) asodes*
— Acanthose megascleres echinating vertical branching spongin fibers; encrusting to 4 mm thick; color bright to deep scarlet; spicules: (1) acanthostyles of two size classes, often markedly curved, with spined heads and sparsely spined shafts; larger spicules often with less spination, 80–182 × 5.0–13.0 μm and 303–509 × 15.0–25.0 μm; (2) styles to tylostyles or subtylostyles with heads microspined 125–255 × 3.5–7.5 μm, (3) palmate isochelas 10.0–17.0 μm; skeleton: acanthose megascleres form echinated vertical tracts; styles, tylostyles to subtylostyles arranged vertically or at random at surface and associated with the main tracts (plate 29E) . . . . *Clathria (Microciona)* sp.
34. Microscleres toxas only; sponge: encrusting to 1 cm thick; color yellow; spicules: (1) acanthostyles of two size classes, the larger often less heavily spined, 60–85 × 5.0–8.0 μm, 125–290 × 7.0–13.0 μm; (2) tylostyles, usually with microspined heads but sometimes smooth, 230–410 × 5.0–13.5 μm; (3) toxas, 40–72 μm (plate 29F) . . . .
. . . . *Clathria (Wilsonella) pseudonapya*
— Microscleres otherwise 35
35. Microscleres arcuate isochelas only (plate 29G) 36
— Microscleres palmate isochelas and toxas (plate 29G)
. . . . 37
36. Acanthose spicules arranged vertically with heads embedded in basal spongin; color salmon red; arcuate isochelas of one size class, 17–25 μm; sponge thinly encrusting to <1 mm; spicules: (1) acanthosubtylostyles of two size classes, the larger often curved, heads heavily spined, 95–155 × 8.0–12.5 μm and 250–470 × 11.0–19.5 μm; (2) styles, 165–230 × 2.5–4.0 μm; (3) arcuate isochelas, 17–25 μm; skeleton: acanthose spicules with heads embedded in basal spongin; styles in tracts running to surface where they end in tufts (plate 29H) . . . . *Clathria (Microciona) brepha*
— Acanthose spicules forming branching vertical tracts; color: brick red or deep brown; arcuate isochelas of two size classes, 15–32 μm, 30–59 μm; sponge encrusting to 5 mm thick; spicules: (1) acanthostyles of two size classes, the larger with spines often confined to head, the smaller entirely spined, 164–497 × 10.5–12.6 μm and 73–145 × 7.0–8.5 μm; (2) subtylostyles to tylostyles with heads microspined, 152–390 × 4.0–4.5 μm; (3) arcuate isochelas, 15–32 μm and 30–59 μm; skeleton: long acanthostyles arranged in branching vertical tracts with spongin, the pointed ends up and protruding from the tracts at an angle; short acanthostyles standing erect in basal spongin and echinating the vertical tracts; subtylostyles to tylostyles in vague tracts near surface, arranged vertically or at random at the surface and accompanying the tracts of acanthostyles (plate 30A) . . . .
. . . . *Clathria (Microciona) spongigartina*
37. Skeleton of acanthostyles to acanthostrongyles in a reticulate pattern with styles to subtylostyles arising at nodes or near the surface; acanthostyles to acanthostrongyles of two size classes; sponge encrusting to 3 cm thick; bright scarlet, reddish orange, or brick red; spicules: (1) acanthostyles to acanthostrongyles of two size classes, 120–160 × 11.0–12.5 μm and 180–210 × 11.5–13.0 μm; (2) styles, 200–320 × 12.5–14.5 μm; (3) styles to subtylostyles often with microspined heads, 180–305 × 3.5–4.5 μm; (4)

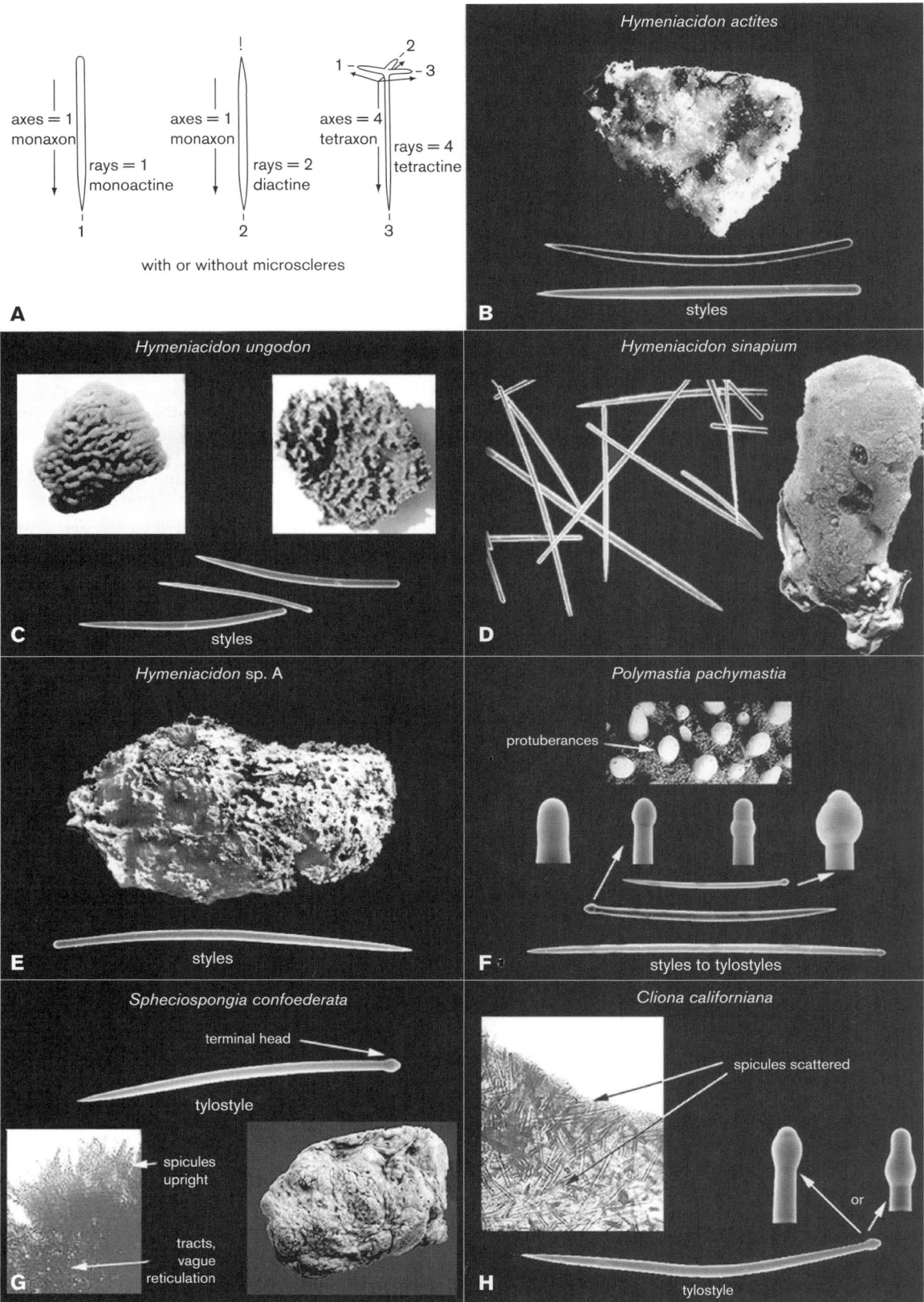

PLATE 28 A, Megascleres are monactines (one axis, one point), diactines (one axis, two points) or tetractines (four axes, four points) or their derivatives, including monaenes and diaenes; B, *Hymeniacidon actites*; C, *Hymeniacidon ungodon*; D, *Hymeniacidon sinapium*; E, *Hymeniacidon* sp. A; F, *Polymastia pachymastia*; G, *Spheciospongia confoederata*; H, *Cliona californiana*.

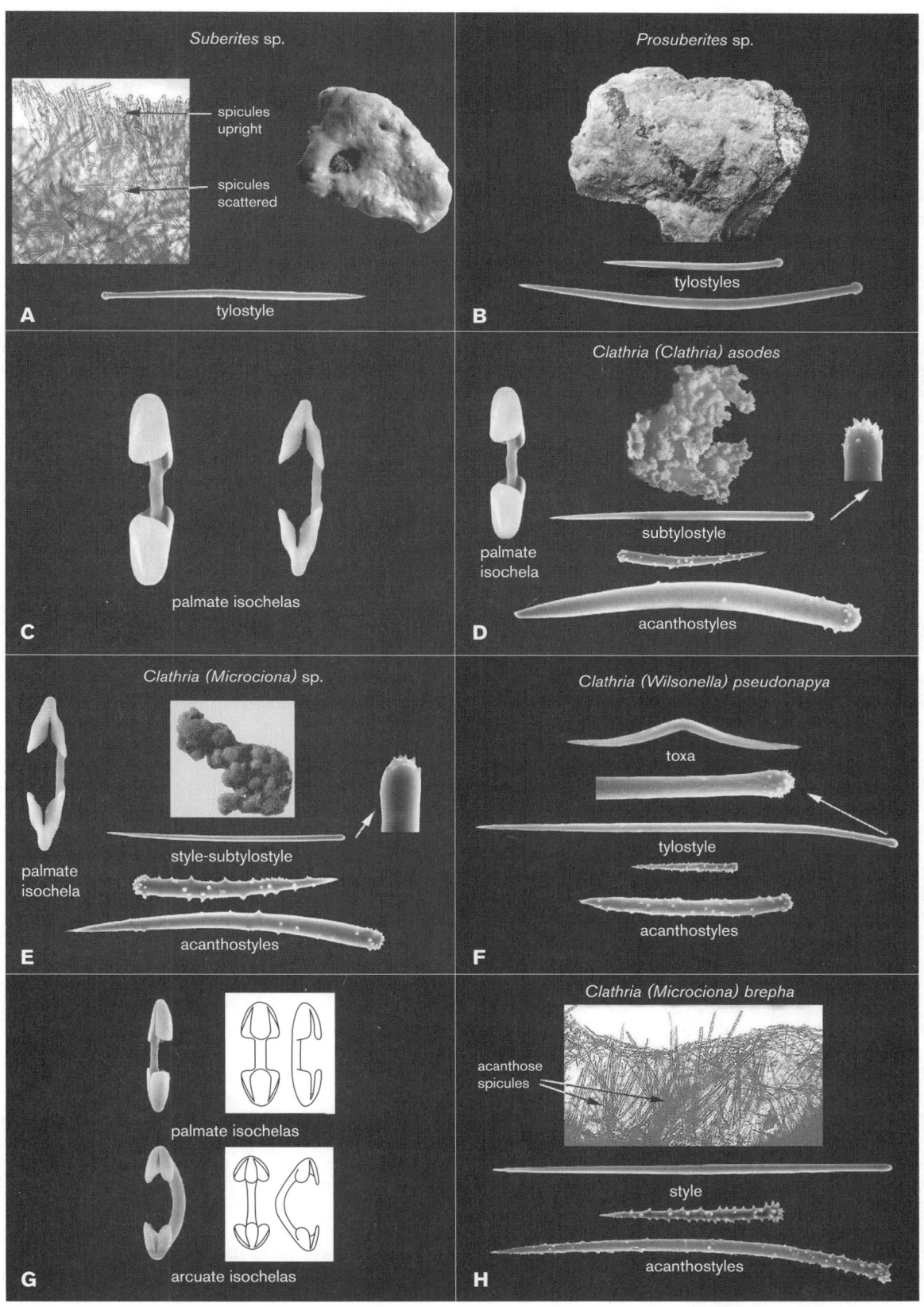

PLATE 29 A, *Suberites* sp.; B, *Prosuberites* sp.; C, palmate isochelas have the lateral wings fused to the shaft, which tends to be straight and not arched, and the frontal wing fully formed and free; D, *Clathria (Clathria) asodes*; E, *Clathria (Microciona)* sp.; F, *Clathria (Wilsonella) pseudonapya*; G, palmate isochelas have the lateral wing fully fused to the shaft and a frontal wing totally formed and free; arcuate isochelas have a lateral wing partially fused to the shaft, which is usually arched, and a frontal wing fully formed and free; H, *Clathria (Microciona) brepha*.

palmate isochelas, 19–24 μm; (5) toxas, 23–260 μm; skeleton: reticulate skeleton of acanthostyles to acanthostrongyles, with the longer acanthostyles and styles arising at the nodes, these often penetrating the surface making it hispid; thin styles to subtylostyles standing erect at or near the surface (plate 30B) . . . .
. . . . *Antho* (*Antho*) *lithophoenix*
— Skeleton of styles to subtylostyles in ascending tracts (plate 30C); acanthostyles of one size class 38
38. Acanthostyles <91 μm; sponge encrusting when young; formed of anastomosing branches when mature; habitat in bays; bright-red to orange-brown; spicules: (1) acanthostyles, 80–90 × 7.0–8.0 μm; (2) styles to subtylostyles of two size classes, the larger often with microspined heads, 155–175 × 3.0–3.5 μm and 180–225 × 10.0–12.5 μm; (3) palmate isochelas, 16–20 μm; (4) toxas, 15–100 μm; skeleton: ascending plumose tracts of large subtylostyles to styles with terminal tufts that extend beyond the surface, making it hispid; smaller subtylostyles to styles grouped at the surface; acanthostyles sparsely echinating the tracts (plate 30D) . . . . *Clathria* (*Clathria*) *prolifera*
— Acanthostyles >100 μm; sponge encrusting; habitat open coast. 39
39. Toxas, 15–325 μm, with microspined ends; sponge encrusting, up to 2 cm thick; salmon to salmon-red in color; spicules: (1) acanthostyles with heavily spined heads, 85–115 μm; (2) styles, often with minutely spined heads, 275–325 × 15.0–17.5 μm; (3) subtylostyles with microspined heads, 180–260 μm; (4) palmate isochelas, 12–16 μm; (5) toxas, 15–325 μm; skeleton: ascending plumose tracts of smooth styles with terminal tufts that penetrate the surface making it hispid; subtylostyles mostly grouped at the surface; acanthostyles sparsely echinating the tracts of styles (plate 30E) . . . . *Clathria* (*Microciona*) *microjoanna*
— Toxas, 25–100 μm with smooth ends; sponge encrusting to 3–20 mm thick; red to orange-brown; spicules: (1) acanthostyles, fully spined, 100–170 μm; (2) styles, 130–475 μm; (3) subtylostyles often with microspined heads, 150–300 μm; (4) palmate isochelas, 20–30 μm; (5) toxas, 14–100 μm; skeleton: ascending plumose tracts of smooth styles, the tracts penetrating the surface; subtylostyles mostly grouped at the surface; acanthostyles sparsely echinating the tracts of styles (plate 30F) . . . . *Clathria* (*Microciona*) *parthena*
40. Anisochelas present (plate 30G) 41
— Anisochelas absent 45
41. Microscleres include giant serrated sigmas of two size classes; sponge encrusting to massive; gold, beige, yellow, golden-brown, brown-black; spicules: (1) subtylostyles, 224–485 × 3.0–12.0 μm; (2) palmate anisochelas of two size; classes; 11–25 μm and 25–43 μm; (3) sigmas of two size classes, 29–125 μm and 70–209 μm; (4) toxas (rare), 12–159 μm; skeleton: branching and anastomosing tracts forming an irregular reticulation (plate 30H) . . . .
. . . . *Mycale* (*Paresperella*) *psila*
— Microscleres may include sigmas but these are never serrated or giant. 42
42. Sigmalike microscleres with low arch, straight back, and sharply bent tips; most often the bent tips differ in size, one smaller than the other and the shaft appears thickened; it is difficult to determine whether these are indeed true sigmas, or just various immature stages of anisochelae; sponge encrusting to 1 cm thick, often spherical; on rocks or attached to hydroids and bryozoans; light yellow, gold-beige, bronze, buff tan; spicules: (1) styles, 170–412 × 5.0–13.0 μm; (2)

palmate anisochelas, 26–68 μm; (3) "sigmas," 23–59 μm; skeleton: tracts of two to six styles radiating to surface with lateral connections forming a loose ladderlike skeleton (plate 31A) . . . . *Mycale* (*Mycale*) *hispida*
— C-shaped Sigmas always present. 43
43. Megascleres are sinuous subtylostyles; anisochelas of two or three size classes; sponge encrusting to 1.5 cm thick; chrome yellow, olive-brown, tan; spicules: (1) subtylostyles, sinuous, greatest width near center of spicules, 158–321 × 4.0–12.0 μm; (2) sigmas of two size classes, the smaller often missing, 14–33 μm and 43–86 μm; (3) palmate anisochelas of two or three size classes, the smaller two of which may appear to be a single size class, 11–17 μm, 17–30 μm and 31–49 μm; (4) toxas, may be rare, 5–118 μm; skeleton: spicule tracts forming a vague reticulation (plate 31B) . . . . *Mycale* (*Mycale*) *macginitiei*
— Megascleres are straight or slightly curved styles; anisochelas one or two to three size classes 44
44. Anisochelas of two or three size classes; megascleres fusiform styles to subtylostyles, the latter often with a distinct and extended narrowing before the slightly expanded head end; sponge: thin, encrusting, reported on *Chlamys* or free-living, amorphous, massive; color: light brown, yellow-brown, violet; spicules: (1) styles to subtylostyles, 242–461 × 6.0–15.0 μm; (2) anisochelas of three size classes, but sometimes either the smallest or largest is missing, 15–29 μm, 22–48 μm, and 51–92 μm; (3) sigmas, may be rare, 19–89 μm; (4) raphides, 29–111 μm, may be missing; skeleton: thin dermal membrane with reticulation of megascleres parallel to surface; choanosome with parallel tracts running to surface, often dendritic (plate 31C) . . . . . . *Mycale* (*Aegographila*) *adhaerens*
— Anisochelas of one size class; megascleres are styles, not fusiform in shape; sponge: encrusting to 4 mm thick, often wrapped around algae, hydroids, etc; oscules 1 mm across on low tubules, 1.0–1.5 mm high; color: yellow-rose, gold-beige; spicules: (1) styles, 125–212 × 5.0–10.0 μm; (2) sigmas, sometimes rare, 14–28 μm; (3) anisochelas, 25–39 μm, often aggregated in rosettes; skeleton: parallel tracts joined by horizontal connections to form ladderlike skeleton; often spreading tufts at surface (plate 31D) . . . .
. . . . *Mycale* (*Carmia*) *richardsoni*
45. Microscleres are sigmas and diancistras; sponge: thinly encrusting, to 1–3 mm thick; color: deep violet-blue, gray-lavender, or yellow-brown; spicules: (1) tylostyles, 142–227 × 5.0–7.0 μm, (2) sigmas, 12–65 μm, (3) diancistras, 23–41 μm (plate 31E) . . . . . . *Hamacantha* (*Vomerula*) *hyaloderma*
— Microscleres otherwise 46
46. Toxas absent, microscleres palmate isochelas only; sponge: lobate, flabellate, or ramose, and often washed up on shore; color: burnt orange; spicules: (1) styles, 135–200 × 11.0–24.0 μm, (2) palmate isochelas, 20–27 μm (plate 31F)
. . . . *Isodictya quatsinoensis*
— Toxas present, may or may not have palmate isochelas
. . . . 47
47. Microscleres are toxas only; sponge: encrusting to 8 mm thick; color; scarlet, burnt sienna, salmon-red, terra cotta, orange-brown, yellow-tan, or mustard; spicules: (1) thin styles to subtylostyles, with microspined heads, 165–225 × 2.0–8.0 μm, (2) thick styles to subtylostyles, rarely with microspined heads, 200–240 × 14–36 μm, (3) toxas with low arch, 26–165 μm; skeleton: thick styles to subtylostyles in ascending echinated tracts joined by crossspicules, thin styles to subtylostyles localized at surface and scattered within (plate 31G) . . . .

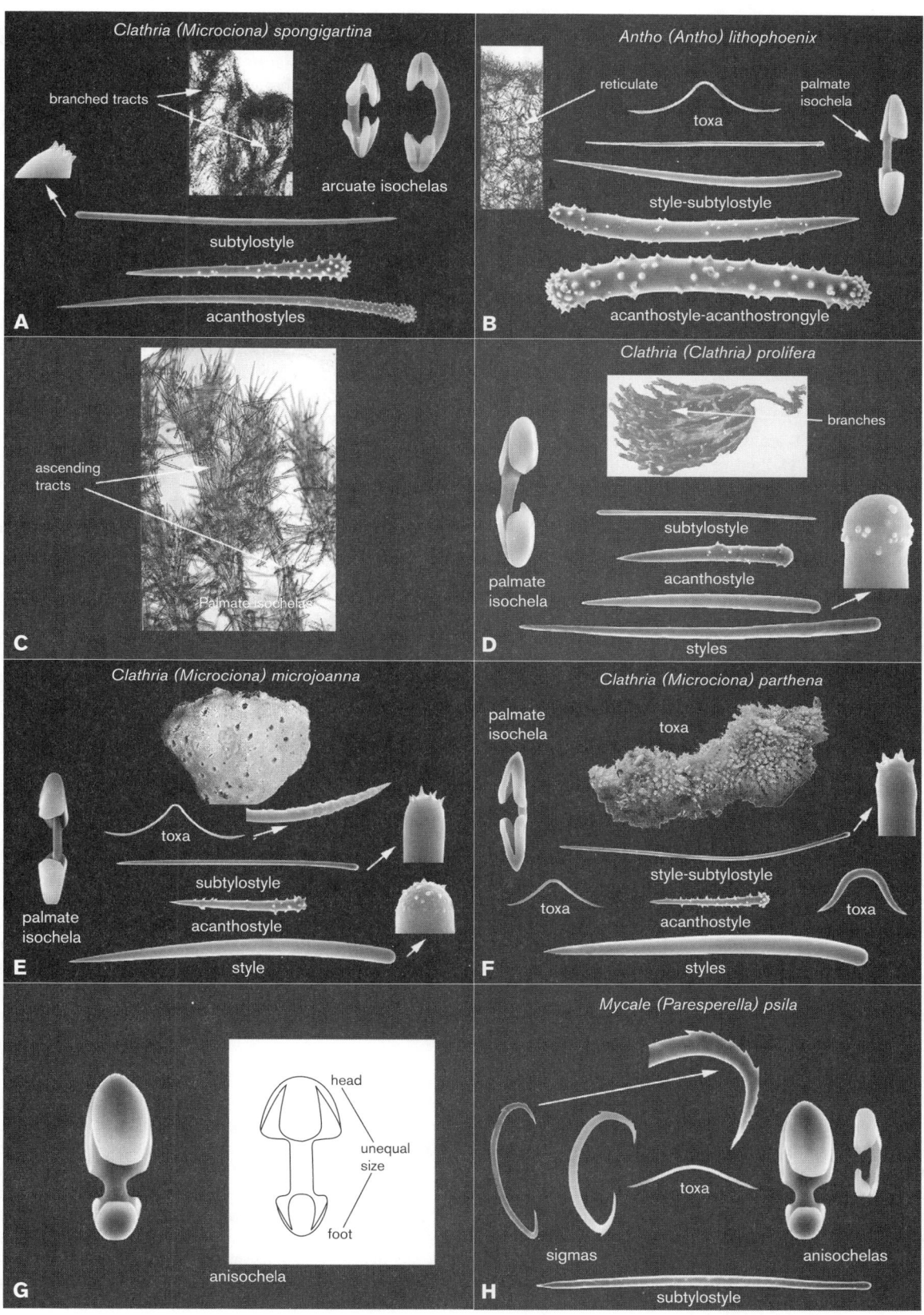

PLATE 30 A, *Clathria (Microciona) spongigartina*; B, *Antho (Antho) lithophoenix*; C, ascending tracts of styles to subtylostyles; D, *Clathria (Clathria) prolifera*; E, *Clathria (Microciona) microjoanna*; F, *Clathria (Microciona) parthena*; G, anisochelas are like isochelas except that one end (the foot) is smaller than the other (the head); H, *Mycale (Paresperella) psila*.

.... *Clathria (Microciona) pennata*

— Microscleres are toxas and palmate isochelas; sponge: encrusting to 3–16 mm thick; color: scarlet to vermilion, coral, salmon-pink, pale brick red, orange-brown, or pale yellow ochre; spicules: (1) thin styles to subtylostyles, 90–200 × 2.0–5.0 μm, (2) thick styles to subtylostyles, 130–165 × 9.0–12.0 μm, (3) toxas, with high arch, 46–112 μm, may be absent, (4) palmate isochelas, 12.0–19.0 μm; skeleton: thick styles to subtylostyles in ascending echinated tracts joined by cross-spicules, thin styles to subtylostyles localized at surface and scattered within (plate 31H) ....
.... *Clathria (Thalysias) originalis*

48. Megascleres are monactines and diactines only (plate 32A); microscleres absent 49

— Megascleres are monactines and diactines only or may include tetractines as well (plate 32B); microscleres present 50

49. Acanthostyles present; sponge: thinly encrusting, 1–2 mm thick; color: deep violet-blue, gray-lavender, or yellow brown; spicules: (1) acanthostyles, 72–157 × 5.3–6.7 μm, (2) tylotes to subtylotes, 96–169 × 2.3–3.4 μm; skeleton: monactines oriented vertically with heads against the substrate, diactines with two ends often subequal, one end being more strongylote (plate 32C)...... *Hymedesmia (Stylopus) arndti*

— Acanthostyles absent; sponge: thickly encrusting to massive, up to 7.5 × 5.0 × 5.0 cm; color: salmon pink, dull orange-red, dull orange-brown, burnt sienna, or pale terra cotta; spicules: (1) styles to subtylostyles, sometimes sparsely spined at head end, 222–398 × 6.5–7.0 μm, (2) subtylotes to tylotes, sometimes gently sinuous, 195–333 × 4.5–7.0 μm; skeleton: a thin dermal membrane contains tangentially arranged tylotes; the dermis is underlain by upright or randomly oriented tylotes to subtylotes, which also occur interstitially in the choanosome; multispicular tracts of subtylostyles to styles run a sinuous course into the choanosome (plate 32D)
.... *Lissodendoryx (Lissodendoryx) topsenti*

50. Megascleres are monactines and diactines with microscleres; no tetractines or their derivatives present........ 51

— Megascleres include tetractines (or their derivatives) with microscleres.................................. 69

51. Microscleres are onychaetes 52

— Microscleres otherwise 53

52. Megascleres include tornotes; sponge: encrusting, up to 2 cm high, cushion-shaped to amorphous; color: light brown, yellow, orange, tan; spicules: (1) styles to subtylostyles, 218–345 × 6.0–11.0 μm, (2) smooth tornotes, 171–241 × 4.0–6.0 μm, (3) onychaetes of two size classes, 51–99 μm and 159–260 μm (plate 32E)..............
.... *Tedania (Trachytedania) gurjanovae*

— Megascleres include tylotes to subtylotes, no tornotes; sponge: encrusting to massive; color: brownish orange or brownish red; spicules: (1) styles, mean 300 × 1.8 μm, (2) tylotes–subtylotes, mean 220 × 4.6 μm, (3) onychaetes, two size classes, means 60 × 2.3 μm and 180 × 1.8 μm ....
.... *Tedania (Trachytedania) toxicalis*

53. Acanthostyles with multiple large, obvious spines present (plate 32F) 62

— Acanthostyles absent; may have styles with only a few small spines on head or tip (plate 32F) 54

54. Microscleres include palmate isochelas with toxas (plate 32G).. 60

— Microscleres include asters, or anchorate or arcuate isochelas (plate 32H)............................. 55

55. Microscleres include asters; sponge: globular or subglobu-

lar in shape, up to 6 cm in diameter; surface warty or tuberculate, the tubercles flattened, 2–4 mm wide and 1–2 mm high; a two-layered cortex present, 0.5–1.0 mm thick, the inner layer fibrous; color: deep orange, orange-red, yellow ochre, or gold; spicules: (1) fusiform strongyles (anisostrongyles), 1,250–2,300 μm, (2) fusiform styles to subtylostyles, 500–1,800 μm, (3) oxyspherasters, with rays often bent or mucronate, or rarely bifid, 30–90 μm in diameter, (4) strongylasters, usually with somewhat rounded rays, occasionally knobbed, 8–25 μm across (plate 33A)
.... *Tethya californiana*

— Microscleres include anchorate or arcuate isochelas 56

56. Microscleres include anchorate isochelas........... 59

— Microscleres include arcuate isochelas.............. 57

57. Forceps present, *but exceedingly small and difficult to see*; sponge: encrusting to massive, up to 1.5 cm thick, surface nodular and ridged, ridges somewhat hispid, delineating shallow strands of grooves; color: yellow, light orange, buffy pale yellow; spicules: (1) styles to acanthostyles with few spines, 169–281 × 7.0–10.0 μm, (2) tylotes to subtylotes, with tips tending to be elongated, 137–205 × 4.0–6.0 μm, (3) arcuate isochelas, 18–38 μm, (4) sigmas, 30–55 μm, (5) forceps, heavily spined, 5.0–11.0 μm; skeleton: ectosome dense, made up of tangential tylotes to subtylotes; choanosome an irregular reticulation of wide tracts of styles with an overlying, looser, less-structured reticulation of random styles and some tylotes (plate 33B) ....
.... *Forcepia (Forcepia) hartmani*

— Forceps absent.................................. 58

58. Sigmas present; sponge: encrusting to massive, up to 4.5 cm thick, surface smooth to rugose; color: buff, light tan, gold, yellow, cinnamon brown; spicules: (1) styles, smooth or with a few spines on head, 190–311 × 6.0-11.0 μm, (2) tylotes to subtylotes, 193–260 × 3.0–7.0 μm, (3) arcuate isochelas, 21–35 μm, (3) sigmas, 30–64 μm; skeleton: vague to irregular reticulation giving rise to ill-defined tracts of styles which run to the surface (plate 33C) ....
.... *Lissodendoryx (Lissodendoryx) firma*

— Sigmas absent; sponge: encrusting to massive, up to 17 mm thick; color: brown-orange; spicules: (1) styles of two size classes, may have a very few spines, 176–243 × 10.0–16.0 μm, 362–472 × 19.0–26.0 μm, (2) subtylotes, 251–370 × 5.0–9.0 μm, (3) arcuate isochelas, 30–43 μm; skeleton: dermal membrane with numerous isochelas, ectosome with scattered subtylotes parallel to surface; in the endosome, tracts of styles form a loose reticulation (plate 33D) .... *Lissodendoryx (Lissodendoryx) kyma*

59. Diactines include tylotes with microspined heads, 125–130 × 9.5 μm, (2) monactines are tylostyles to subtylostyles with microspined heads, 130–310 × 10.0–17.0 μm; sponge: encrusting, up to 1 cm thick; color: carmine; spicules: (1) tylotes, heads coarsely to finely microspined, (2) tylostyles to subtylostyles, (3) anisotornotes to anisosubtylostyles, straight, thin (one end indistinctly knobbed, the other truncate, rounded, pointed, or with smaller knob) with ends microspined, 115–156 × 4.0 μm, (4) anchorate isochelas, 13–14 μm; skeleton: ascending, branching, plumose tracts of tylostyles to subtylostyles; secondary spicules are tylotes occurring interstitially or in the plumose tracts; in addition there are anisotornotes or anisosubtylostyles occurring interstitially, often with the variable end pointing toward the surface; anchorate isochelas are scattered through the choanosome and often concentrated near or at the surface (plate 33E)........

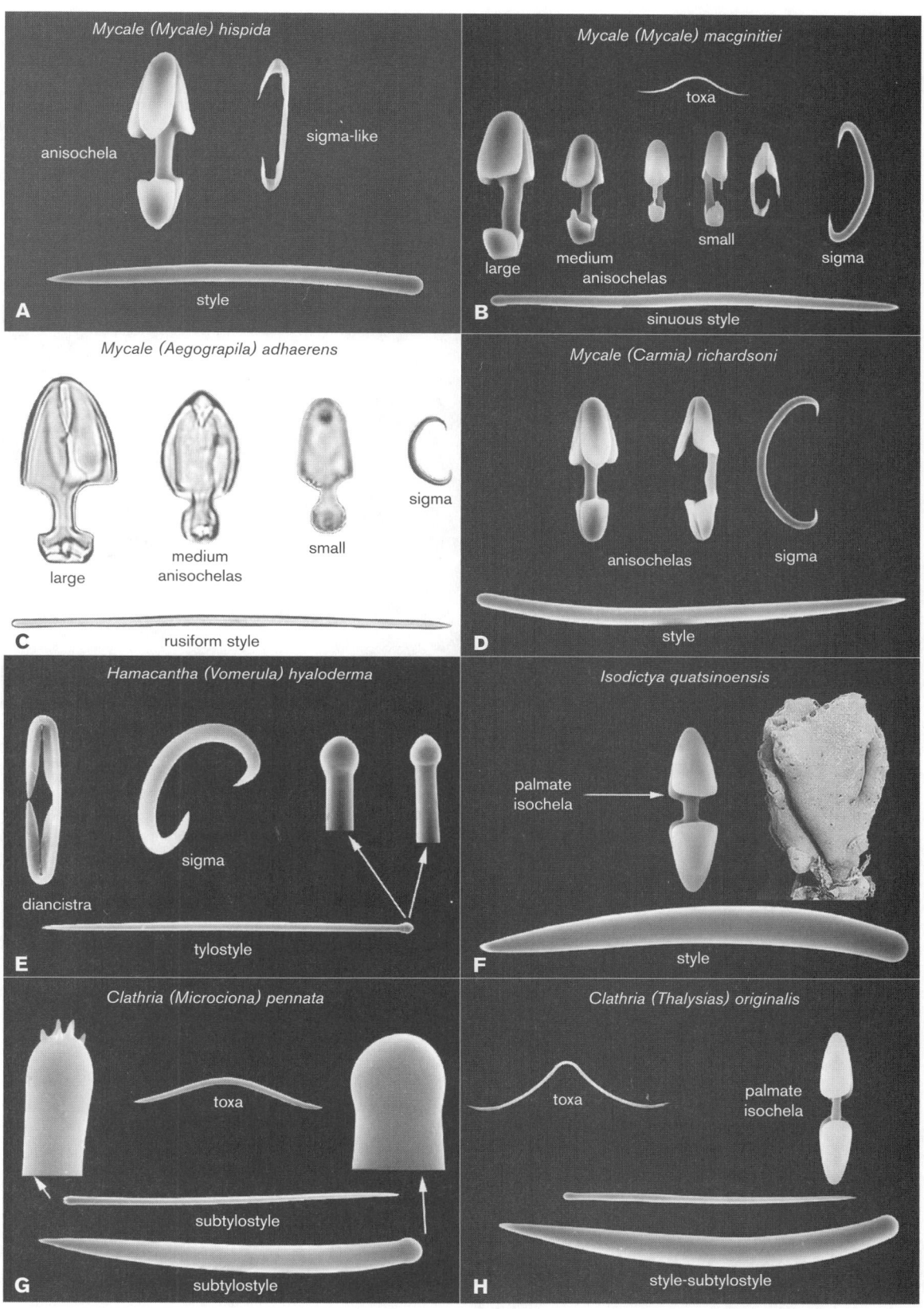

PLATE 31 A, *Mycale (Mycale) hispida*; B, *Mycale (Mycale) macginitiei*; C, *Mycale (Aegograpila) adhaerens*; D, *Mycale (Carmia) richardsoni*; E, *Hamacantha (Vomerula) hyaloderma*; F, *Isodictya quatsinoensis*; G, *Clathria (Microciona) pennata*; H, *Clathria (Thalysias) originalis*.

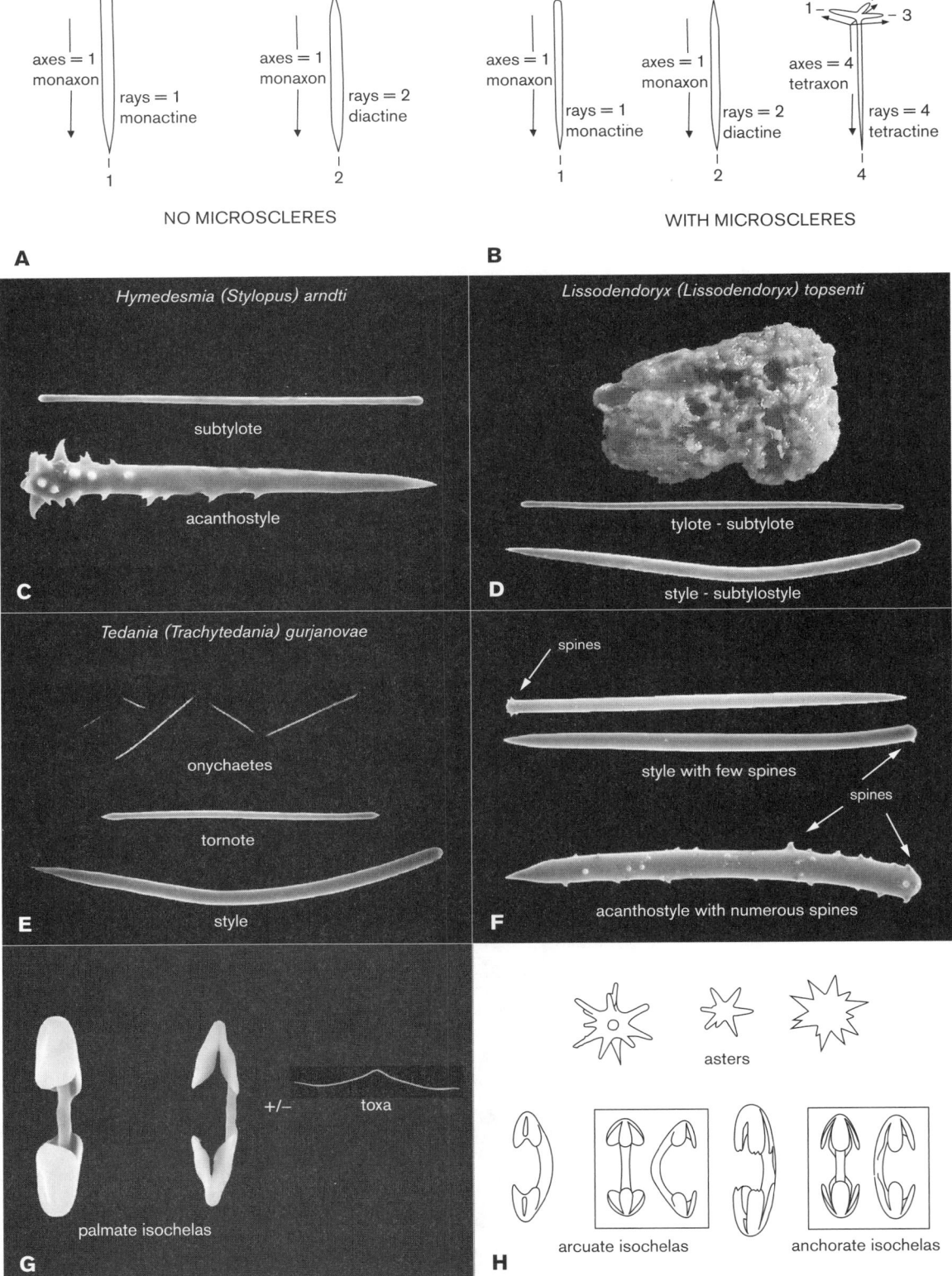

NO MICROSCLERES

WITH MICROSCLERES

A

B

axes = 1
monaxon

rays = 1
monactine

1

axes = 1
monaxon

rays = 2
diactine

2

axes = 1
monaxon

rays = 1
monactine

1

axes = 1
monaxon

rays = 2
diactine

2

axes = 4
tetraxon

rays = 4
tetractine

4

Hymedesmia (Stylopus) arndti

C

subtylote

acanthostyle

Lissodendoryx (Lissodendoryx) topsenti

D

tylote - subtylote

style - subtylostyle

Tedania (Trachytedania) gurjanovae

E

onychaetes

tornote

style

F

spines

style with few spines

spines

acanthostyle with numerous spines

G

palmate isochelas

+/−   toxa

H

asters

arcuate isochelas   anchorate isochelas

PLATE 32 A, monactines, megascleres with one axis and one point; diactines, megascleres with one axis and two points; without microscleres. B, monactines and diactines only, or, may include tetractines (megascleres with four axes and four points), with microscleres. C, *Hymedesmia (Stylopus) arndti*; D, *Lissodendoryx (Lissodendoryx) topsenti*; E, *Tedania (Trachytedania) gurjanovae*; F, technically, both of these spicule types are acanthostyles but frequently they can be separated as illustrated. G, palmate isochelas with or without toxas. H, arcuate and anchorate isochelas (see plate 19 for comparison of arcuate and anchorate isochelas). Asters are microscleres with short, radiating or starlike rays, radiating from a point, sphere or rod. These may vary greatly in size.

.... *Plocamiancora igzo*

— Diactines include tornotes to strongyles with blunt, multispined tips, 99–186 × 3.0–7.0 μm; monactines are styles to subtylostyles, sometimes with a very few spines on head, 145–227 × 5.0–11.0 μm; sponge: encrusting to 2.5 cm thick; color: unknown; spicules: (1) styles to subtylostyles, (2) tornotes to strongyles, most with ends microspined, (3) sigmas of two size classes, 9–28 and 25–42 μm, (4) anchorate isochelas, 14–30 μm (plate 33F) ....
.... *Myxilla (Myxilla) agennes*

60. Cladotylotes present; sponge: encrusting, up to 5 cm thick, often with oscular chimneys; color: bright scarlet to terra cotta; spicules: (1) styles to subtylostyles, 290–425 × 15.0–18.0 μm, (2) strongyles to subtylotes with microspined heads, 185–220 × 4.4–4.7 μm, (3) cladotylotes of two size classes, the larger (sometimes missing) with a smooth shaft and the smaller with the shaft spined, 95–120 × 4.0–6.0 μm, 220–230 × 10.0–11.0 μm, (4) palmate isochelas, 11–20 μm, (5) toxas of two size classes, 15–135 μm, 190–440 μm; skeleton: styles to subtylostyles forming ascending tracts echinated by cladotylotes; strongyles to subtylotes in ectosome (plate 33G) ....
.... *Acarnus erithacus*

— Cladotylotes absent 61

61. Palmate isochelas with central wing-plate on shaft (may be difficult to see under microscope); styles, 168–600 μm; sponge: thinly encrusting, up to 1 mm; color: orange; spicules: (1) styles to subtylostyles (with microspined heads), 168–600 × 8.0–24.0 μm, (2) subtylostyles to tylostyles, 94–270 × 3.0–6.0 μm, (3) tylostyles, 98–196 × 6.0–14.0 μm, (4) tylotes with microspined heads, 66–137 × 5.0–12.0 μm, (5) palmate isochelas, 10–16 μm, (6) toxas (may be absent), 31–63 μm; skeleton: regular reticulation of tylotes supporting a dermal arrangement of styles and tylostyles pointing outward (plate 33H) ....
.... *Antho (Acarnia) lambei*

— Palmate isochelas lack central wing-plates on shaft; styles, 180–215 μm; sponge: encrusting, up to 2.5 cm thick; color: scarlet to salmon-red or dull salmon-orange; spicules: (1) thick styles to subtylostyles, heads smooth or microspined, 180–215 × 15.0–17.0 μm, (2) thin tylostyles to subtylostyles, 145–200 × 3.3–3.7 μm, (3) tylotes with heads microspined, 175–240 × 17.0–24.0 μm, (4) palmate isochelas, 10–20 μm, (5) toxas, 15–100 μm; skeleton: ascending tracts of styles that penetrate the surface; tracts interconnected by tylotes; tylostyles to subtylostyles occur near the surface (plate 34A) ....
.... *Antho (Acarnia) karykina*

62. Microscleres, anchorate isochelas; sponge: thin and encrusting to massive, reported on *Chlamys* spp., or free-living, up to 8 cm thick; color: gold to light gold-brown, yellow; spicules: (1) acanthostyles, 104–248 × 3.0–14.0 μm, (2) tornotes with ends sharply tapered to small spine, 116–205 × 2.0–6.0 μm, (3) anchorate isochelas of two size classes, 11–26 μm, 27–59 μm, (4) sigmas of two size classes, 12–34 μm, 31–54 μm; skeleton: heavy dermal membrane with numerous irregularly distributed tornotes parallel to surface; regular tracts perpendicular to and near the surface, with choanosomal reticulation (plate 34B) ....
.... *Myxilla (Myxilla) incrustans*

— Microscleres otherwise 63

63. Acanthostrongyles present (plate 34C) 64

— Acanthostrongyles absent 65

64. Acanthostyles and acanthostrongyles, each of one size class;

thin subtylostyles with microspined heads; sponge: encrusting, up to 7 mm thick; surface frequently roughened with tubercles and ridges; color: tan, orange, red; spicules: (1) styles of two size classes, 140–240 × 2.0–4.0 μm (with microspined heads), 222–680 × 6.0–11.0 μm, (2) acanthostyles, 83–219 × 5.0–14.0 μm, (3) acanthostrongyles, 87–160 × 9.0–11.0 μm, (4) palmate isochelas, 16–35 μm, (5) toxas, 18–229 μm, (6) raphides (may be absent), 61–100 μm; skeleton: reticulation of acanthostrongyles with echinating acanthostyles; styles attached to reticulation at and penetrating the surface (plate 34D) .... *Antho (Acarnia) illgi*

— Acanthostyles and acanthostrongyles of two size classes; thin styles without microspined heads present; sponge: encrusting, up to 3 cm thick; color: bright scarlet, reddish orange, or brick red; spicules: (1) acanthostyles to acanthostrongyles of two size classes, 120–160 × 11.0–12.5 μm and 180–210 × 11.5–13.0 μm, (2) styles, 200–320 × 12.5–14.5 μm, (3) styles to subtylostyles often with microspined heads, 180–305 × 3.5–4.5 μm, (4) palmate isochelas, 19–24 μm, (5) toxas, 23–260 μm; skeleton: reticulate skeleton of acanthostyles to acanthostrongyles, with the longer acanthostyles and styles arising at the nodes, these often penetrating the surface, making it hispid; thin styles to subtylostyles standing erect at or near the surface (plate 34E) .... *Antho (Antho) lithophoenix*

65. Microscleres are palmate anisochelas; sponge: encrusting, to 5 mm thick; color: light orange to light cinnamon; spicules: (1) acanthostyles of from one to three size classes, most frequently of two size classes, 84–192 × 6.0–14.0 μm and 149–265 × 6.0–14.0 μm, (2) subtylotes, strongylotes, or tylostrongyles (one end usually knobbed and the other end thinner and rounded), 109–186 × 3.0–4.0 μm, (3) subtylostyles, 174–206 μm, (4) palmate anisochelas with short "tail" on the smaller end, 5.0–21.0 μm; skeleton: two size classes of erect acanthostyles are attached to the substratum by means of spongin; tracts of subtylostyles and/or subtylotes ascend from near the sponge base to or beneath the dermal membrane; abundant palmate anisochelas and some individual subtylostyles or subtylotes are scattered through the choanosome (plate 34F) .... *Iophon rayae*

— Microscleres otherwise 66

66. Microscleres include amphiasters; sponge: encrusting, to 4 mm thick; color: deep blue or light orange; spicules: (1) acanthostyles, 75–280 × 8.0–8.5 μm, (2) tornotes, very thin, 125–170 × 3.2 μm, (3) amphiasters, 10–17 μm (plate 34G) .... *Acanthancora cyanocrypta*

— Microscleres arcuate isochelas 67

67. Acanthostyles standing erect with heads embedded in basal spongin or forming short tracts to the surface 68

— Acanthostyles forming a basal reticulation from which arise plumose tracts; sponge: encrusting, up to 2 cm thick; color: orange, yellow-ochre, olive-tan, burnt sienna, pale terra cotta; spicules: (1) acanthostyles of two size classes, 63–159 × 7.0–7.5 μm, 116–301 × 8.5–9.5 μm, (2) tornotes, 92–152 × 3.4–4.5 μm, (3) arcuate isochelas, 12–31 μm; skeleton: basal reticulation of small acanthostyles, large acanthostyles forming plumose tracts that protrude from the surface; tornotes grouped at surface (plate 34H) .... *Plocamionida lyoni*

68. Acanthostyles of two size classes; diactines are tornotes (plate 20G); sponge: encrusting, to 1.5 mm thick; color: salmon, salmon-orange, pale terra cotta, brick red, or burnt sienna; spicules: (1) acanthostyles of two size classes, 72–112 × 6.5–7.0 μm and 200–290 × 8.0–8.5 μm, (2) tornotes to subtylotes, 121–180 × 3.5 μm, (3) arcuate

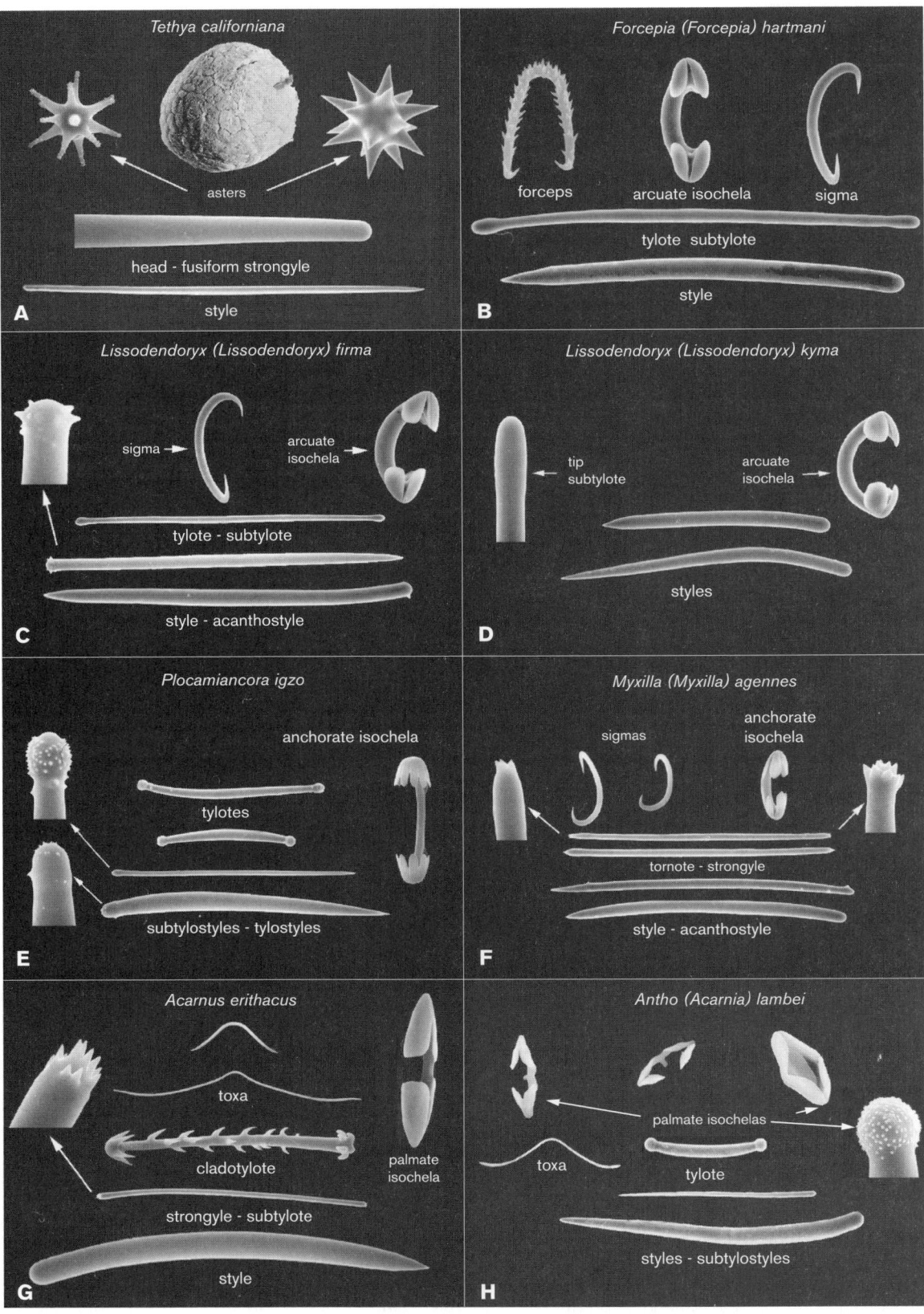

PLATE 33 A, *Tethya californiana*; B, *Forcepia (Forcepia) hartmani*; C, *Lissodendoryx (Lissodendoryx) firma*; D, *Lissodendoryx (Lissodendoryx) kyma*; E, *Plocamiancora igzo*; F, *Myxilla (Myxilla) agennes*; G, *Acarnus erithacus*; H, *Antho (Acarnia) lambei*.

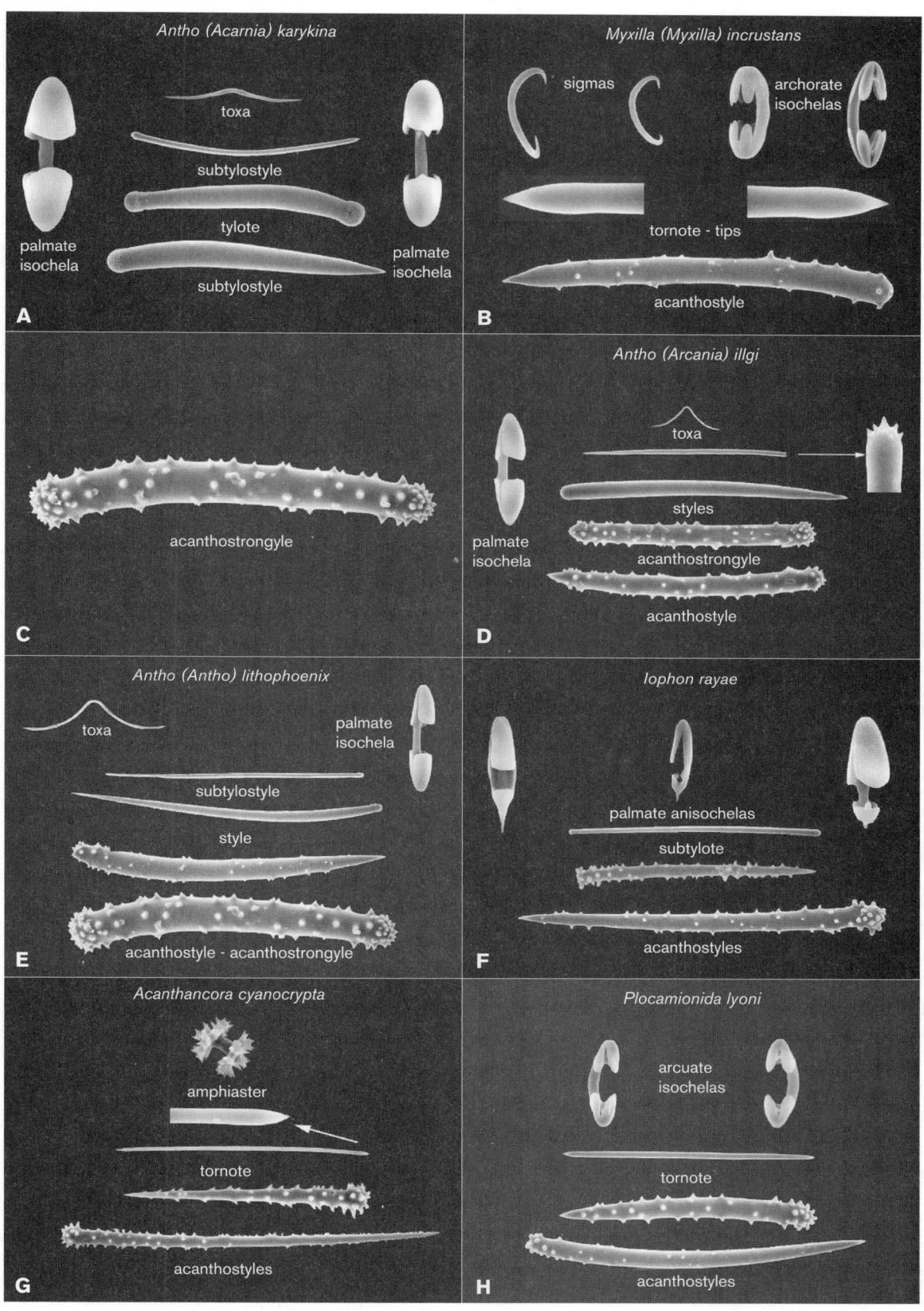

PLATE 34 A, *Antho (Acarnia) karykina*; B, *Myxilla (Myxilla) incrustans*; C, Acanthostrongyle; a spined diactinal spicule with equal rounded ends; D, *Antho illgi*; E, *Antho (Antho) lithophoenix*; F, *Iophon rayae*; G, *Acanthancora cyanocrypta*; H, *Plocamionida lyoni*.

isochelas, 15–30 μm (plate 35A).....................
.... *Hymedesmia (Hymedesmia)* sp. A
— Acanthostyles of a single size class; diactines are tylotes (plate 20H); sponge: encrusting, to <1 mm thick; color: salmon; spicules: (1) acanthostyles, 96–248 × 8.0 μm, (2) tylotes, 141–204 × 5.5 μm, (3) arcuate isochelas, 21–43 μm (plate 35B) .... *Hymedesmia (Hymedesmia)* sp. B
69. Microscleres sigmaspires (plate 35C) 70
— Microscleres otherwise 72
70. Sponge club-shaped, up to 22 mm high by 18 mm wide, with conspicuous anchoring root tufts of bundles of thin anatriaenes; on mud flats in San Francisco Bay; spicules: (1) oxeas, 381–1699 μm, (2) protriaenes and prodiaenes, 939–1900 μm or greater, (3) styles, 436–1333 μm, (4) anatriaenes, apparently only in anchor root, (5) sigmaspires (plate 35D) .... *Tetilla* sp. A
— Sponge spherical or subspherical, (2) attached to rocks on open coast 71
71. With a conspicuous crown of spicules up to 5 mm high around the oscules; color: chrome yellow to yellow-tan; spicules: (1) oxeas, 1200 mm to several cm, (2) anamonaenes, (3) protriaenes, (4) prodiaenes, (5) anatriaenes, (6) anadiaenes, (7) sigmaspires; skeleton: among the radiating spicule bundles are tracts of anamonaenes, usually provided with a basal mat of long, thin spicules, mostly anamonaenes (plate 35E) .... ?*Tetilla* sp. B
— Without a conspicuous crown of spicules around the oscules; color: pale tan when young, gray with buff choanosome when large; spicules: (1) oxeas, 2–3 cm, (2) anatriaenes, clads 50–90 μm, rhabdome >10 mm, (3) protriaenes, clads, 8–30 μm, rhabdome to 32 mm, (4) sigmaspires, 7.0–9.0 μm; skeleton: among the radiating spicule bundles are tracts of anatriaenes; young specimens usually without a well-developed basal mat of spicules while large specimens may have such a mat (plate 35F) .... *Craniella arb*
72. Microscleres are microxeas; sponge: encrusting, up to 1.5 mm thick; color: dark lavender-brown or white; spicules: (1) dichotriaenes, clads, 120–210 × 10.0–30.0 μm, rhabdomes, 320–435 × 20.0–30.0 μm, (2) oxeas, 400–890 × 10.0–25.0 μm, (3) microxeas, often bent once or twice, sometimes indistinctly centrotylote, 35–180 × 4.0–10.0 μm; skeleton: dichotriaenes with clads spreading out in the cortex; oxeas forming vague tracts in the choanosome; microxeas packed densely in cortex and occurring more sparsely in the choanosome (plate 35G) .... *Penares saccharis*
— Microscleres include asters 73
73. Microscleres are microstrongyles and oxyasters; sponge: encrusting, to 4 cm thick; color: gray to dark brown; spicules: (1) oxeas, 400–1000 × 10.0–25.0 μm, (2) dichotriaenes, rhabdomes to 400 × 50 μm, clads, 310 × 50 μm, (3) microstrongyles with two bends and often faintly centrotylote, 50–160 × 3.0–8.0 μm, (4) oxyspherasters, 7–25 μm; skeleton: dichotriaenes with clads spreading out in the cortex; oxeas forming vague tracts in the choanosome; oxyspherasters in the choanosome; microstrongyles packed densely in the cortex and occurring more sparsely in the choanosome (plate 35H) .... *Penares cortius*
— Microscleres are asters only; sponge: encrusting, up to 7 cm thick; color: white with buff interior, the surface sometimes tinged with mauve or pink; spicules and skeleton: (1) oxeas of two size classes, 1,400 × 15 μm and 3,500 × 50 μm, (2) interstitial anatriaenes: rhabdomes 1,100–2,000 × 9–15 μm; intercladal chords, 45–90 μm, (3) triaenes with clads directed forward, ranging from orthotriaenes to

plagiotriaenes to dichotriaenes, the last with their clads spread out at or near the surface; dichotriaenes with intercladal chords, 120–180 μm; microscleres, minutely microspined euasters of two kinds, abundant at surface and gradually decreasing in number in interior; (a) oxyasters, 7–15 μm, less common, (b) irregular oxyspherasters, 3–5 μm, very common (plate 36A) .... *Stelletta clarella*
74. Sponge subspherical to pear-shaped; conspicuous fringe of thin, monaxonid spicules, up to 10 mm high, around terminal oscule; sponge size, up to 9 × 11 cm in height × diameter; whitish in color; spicules: (1) coronal oxeas 5,000–10,000 × 4–12 μm; (2) ectosomal oxeas, of two size classes 700–2,000 × 50 μm, and 3,400–5,000 × 30–150 μm; (3) sagittal triradiates with rays 140–200 × 10 μm; (4) endosomal triradiates with rays 130–250 × 10 μm; (5) microxeas 50–140 × 4 μm; skeleton: surface hispid due to thick, ectosomal oxeas that project from the surface, dense layer of vertical microxeas at surface underlain by a tangential layer of triradiates; choanosome with triradiates without order; atrium lined by tangentially placed triradiates; choanocyte chambers leuconoid (plate 36B) ....
.... *Leucandra heathi*
— Sponge shape otherwise; oscular fringe may or may not be present but, if present, never as high as 7 mm 75
75. Sponge vase-shaped or tubular.................... 76
— Sponge sac-shaped or consisting of thin branching or anastomosing tubes, 4 mm or less in diameter........... 77
76. Sponge vase-shaped or tubular; smooth; borne on a narrow stalk up to 2 cm high; sponge: up to 1cm in diameter, narrowing distally; total height, up to 5 cm; commonly occurring in groups; white to tan in color; spicules: (1) oxeas 50–160 μm; (2) coronal oxeas to 1,250 μm (may be missing); (3) hypodermal triradiates and quadriradiates with rays 200–900 μm; (4) choanosomal triradiates and quadriradiates with rays 230 μm; (5) cloacal triradiates and quadriradiates rays 20–150 μm; (6) vertical microxeas 45 μm long, projecting from the dermis (plate 36C) ....
.... *Leucilla nuttingi*
— Sponge tubular, hispid; stalk absent; sponge: with or without oscular fringe; chambers syconoid and radially arranged; tan to white in color; skeleton consists of three layers; cortical, tubar (triradiates arranged in a series of circles around the chamber wall), and gastral quadriradiates with one ray extending into the spongocoele (plate 36D) ...... *Sycon* spp.
77. Sponge sac-shaped, laterally compressed; sponge: oscular fringe absent; whitish in color; skeleton: (a) a dermal cortical skeleton of tangentially arranged triradiates with (b) tufts of oxeas extending beyond the surface and (c) a tangential layer of endosomal triradiates and quadriradiates .... *Grantia* sp.
— Sponge consisting of thin tubes, 4 mm or less in diameter, which branch and may anastomose to form a reticulate mass ending in a cloacalike oscule................. 78
78. Triradiate spicules equiangular with rays equal or subequal; sponge: a tight network of tubes forming masses up to 4 cm across and 1.5 cm high; white in color; spicules: (1) rays of equiangular triradiates 20–65 μm; (2) to the equirayed, equiangular triradiates may be added sagittal triradiates in certain parts of the sponge as well as (3) oxeas (plate 36E)....................... *Clathrina* sp.
— Triradiate spicules with one ray longer than the other two predominate 79
79. Oxeas to <700 μm and of a single size class 80
— Oxeas to 1,000 μm and of two size classes; sponge: a basal

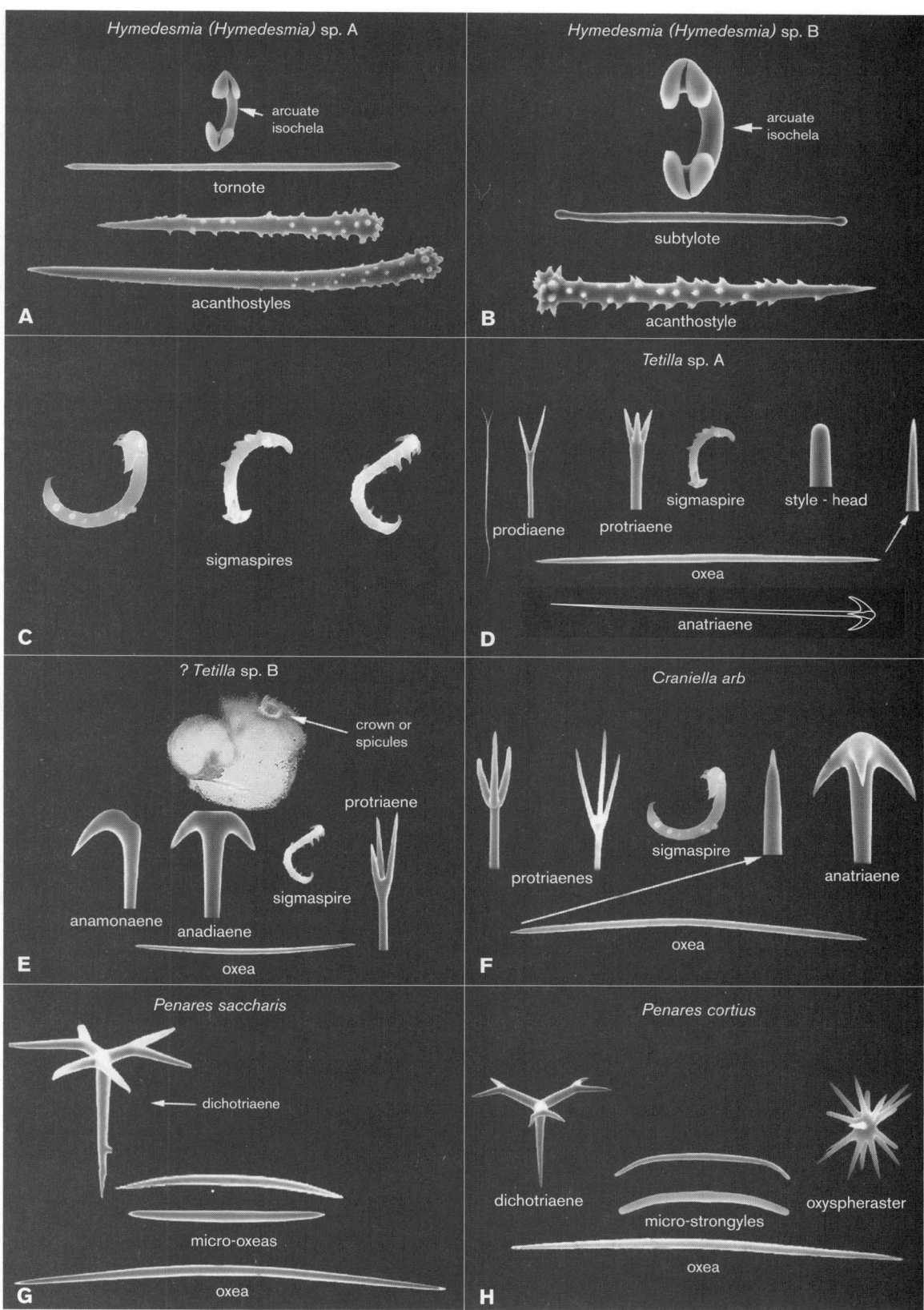

PLATE 35 A, *Hymedesmia* (*Hymedesmia*) sp. A; B, *Hymedesmia* (*Hymedesmia*) sp. B; C, sigmaspires are spiral C- or S-shaped and usually microspined microscleres; D, *Tetilla* sp. A; E, ?*Tetilla* sp. B; F, *Craniella arb*; G, *Penares saccharis*; H, *Penares cortius*.

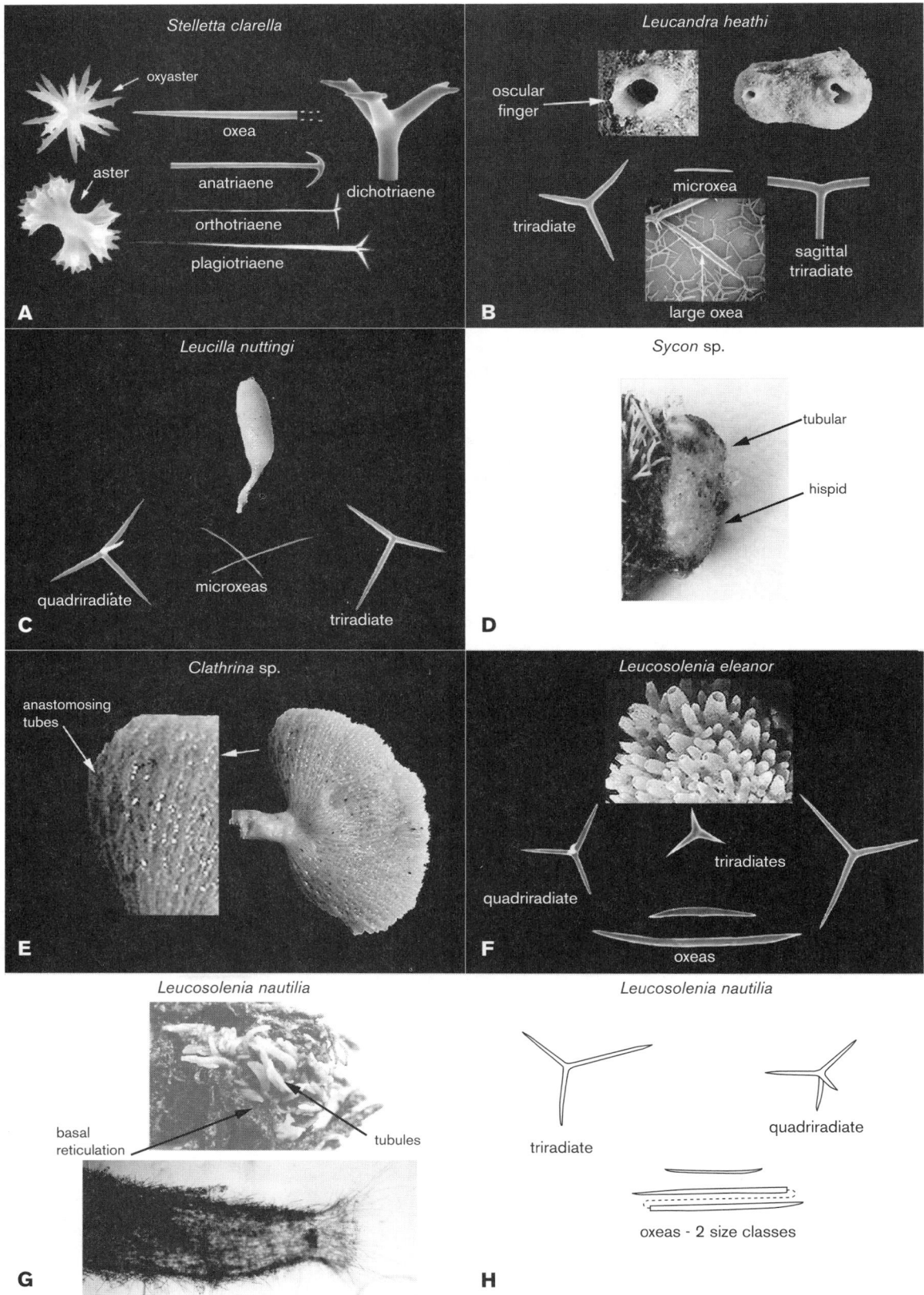

A *Stelletta clarella*

oxyaster
aster
oxea
anatriaene
orthotriaene
plagiotriaene
dichotriaene

B *Leucandra heathi*

oscular finger
triradiate
microxea
sagittal triradiate
large oxea

C *Leucilla nuttingi*

quadriradiate
microxeas
triradiate

D *Sycon* sp.

tubular
hispid

E *Clathrina* sp.

anastomosing tubes

F *Leucosolenia eleanor*

quadriradiate
triradiates
oxeas

G *Leucosolenia nautilia*

basal reticulation
tubules

H *Leucosolenia nautilia*

triradiate
quadriradiate
oxeas - 2 size classes

PLATE 36 A, *Stelletta clarella*; B, *Leucandra heathi*; C, *Leucilla nuttingi*; D, *Sycon* sp.; E, *Clathrina* sp.; F, *Leucosolenia eleanor*; G, *Leucosolenia nautilia*; H, *Leucosolenia nautilia*.

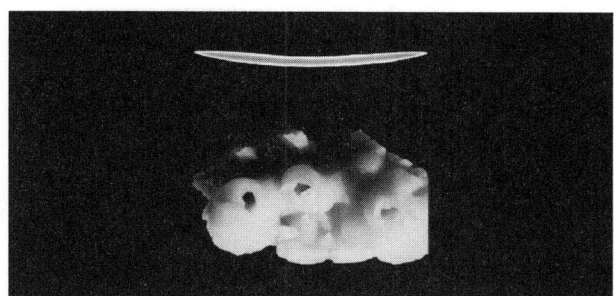

PLATE 37 *Haliclona (Halichoclona) gellindra.*

reticulation from which arise numerous long individual, hispid, tubes; tubes 0.2–2.0 mm × 20 mm; white in color; oscules apical; spicules: (1) oxeas (small) 80–160 μm; (2) oxeas (large) 320–1,000 μm; (3) quadriradiates (apical ray) 20–50 μm, (paired rays) 80–120 μm, (basal ray) 120–150 μm; (4) triradiates (paired rays) 50–100 μm, (basal ray) 50–140 μm; habitat: San Francisco Bay, on floats (plate 36G, 36H) . . . . *Leucosolenia nautilia*

80. Diameter of oscule-bearing tubules usually <1 mm but occasionally up to 1.7 mm; sponge a loose network of branching and anastomosing tubules, forming masses up to 10 cm across and 3 cm high; sponge: tubules 0.3–1.7 mm in diameter; oscules simple, few, 1 mm in diameter; spicules: (1) oxeas 70–435 × 4.0–9.0 μm; (2) quadriradiates with rays up to 140 × 9.0 μm; (3) sagittal alate (wing-like) triradiates with rays to 80 × 7.0 μm, the posterior ray being straight and the lateral two curved; (4) triradiates with rays subequal in length and up to 140 μm long × 7 μm wide; habitat: open coast, under boulders and undersides of rock overhangs (plate 36F) . . . . *Leucosolenia eleanor*

— Diameter of oscule-bearing tubules up to 3.8 mm in greatest width; sponge consists of basal anastomosis from which arise numerous branching oscule-bearing tubules; spicules: (1) oxeas 175–725 × 3.0–9.0 μm; (2) triradiates, sagittal, mostly inequiangular or sagittal, almost equiangular; habitat: harbors . . . . *Leucosolenia* sp.

List of Species

See also: W. Lee, D. Elvin, H. Reiswig. 2007. The Sponges of California. Vermont Information Systems, Shelburne, 395 pp.

Demospongiae

HOMOSCLEROPHORIDA

PLAKINIDAE

Oscarella carmela Muricy and Pearse, 2004. Known from aquaria, running water tanks, etc., as well as along the shores of Monterey and Carmel Bays, at Cape Arago (Oregon), and doubtless along much of the coast. See Muricy and Pearse 2004, Proc. Calif. Acad. Sci. (4) 55: 598–612 for detailed description and distinctions from *Halisarca.*

SPIROPHORIDA

TETILLIDAE

Craniella arb (de Laubenfels, 1930). Recently moved from *Tetilla* to *Craniella.* Very low intertidal; uncommon.

Tetilla sp. A Hartman, 1975. Formerly found locally on mudflats in San Francisco Bay; present status unknown. Megascleres in this genus are often very large and break easily, making measurement difficult.

?*Tetilla* sp. B Hartman, 1975. Low to very low intertidal; uncommon. Generic status uncertain.

ASTROPHORIDA

ANCORINIDAE

Penares cortius de Laubenfels, 1930. Intertidal; rare.

Penares saccharis (de Laubenfels, 1930). Originally placed in the genus *Papyrula.* Very low intertidal; rare. Possibly a juvenile of *P. cortius,* lacking asters.

Stelletta clarella de Laubenfels, 1930. Very low intertidal; moderately common. Megascleres are often large and break easily, making measurement difficult.

HADROMERIDA

CLIONAIDAE

Cliona californiana (de Laubenfels, 1932) (=*Cliona celata californiana*). Low to very low intertidal; common, excavating mollusc shells. Several other microsclere-bearing species of this genus occur on the Pacific coast. In the early part of its life history, *Cliona* burrows into calcareous substrata; oscules and pores protrude through holes in the calcareous substratum; the former with one contractile papillate opening per shell perforation, the latter with several pores opening on the upper surface of a small mushroom-shaped contractile papilla. When the papillae are contracted, two sizes of openings appear distributed over the calcareous substratum. As the sponges grow older they overgrow the calcareous substrata and become thick encrustations and eventual free-living sponges up to 9 cm long, 6.5 cm wide, and 6.4 cm high.

Spheciospongia confoederata de Laubenfels, 1930. Very low intertidal; rare.

POLYMASTIIDAE

Polymastia pachymastia de Laubenfels, 1932 Very low intertidal and subtidal; rare intertidally.

SUBERITIDAE

Prosuberites sp. Hartman, 1975. Low intertidal; locally common in San Francisco Bay, probably introduced from the Atlantic coast of the United States with oysters. This may represent a complex of more than one species.

Suberites sp. Hartman, 1975. Low to very low intertidal; common; two species may be present.

TETHYIDAE

Tethya californiana (de Laubenfels, 1932) (=*Tethya aurantia* var. *californiana*). Low to very low intertidal; moderately common. See Sarà and Corriero (1993).

POECILOSCLERIDA

ACARNIDAE

Acarnus erithacus de Laubenfels, 1927. Low to very low intertidal; moderately common.

Iophon rayae (Bakus, 1966) (=*Hymedesanisochela rayae*) Very low intertidal and subtidal; uncommon.

MICROCIONIDAE (See HOOPER, 1996)

Antho (*Acarnia*) *illgi* (Bakus, 1966) (=*Plocamilla illgi*) Low intertidal; moderately common.

Antho (*Acarnia*) *karykina* (de Laubenfels, 1927) (=*Plocamia karykina*). Mid to very low intertidal; common.

Antho (*Acarnia*) *lambei* (Burton, 1935). Mid to very low intertidal, in caves where there is standing water; rare.

Antho (*Antho*) *lithophoenix* (de Laubenfels, 1927) (=*Antho lithophoenix*). Mid to low intertidal; moderately common.

Clathria (*Clathria*) *asodes* (de Laubenfels, 1930) (=*Leptoclathria asodes*). Intertidal; rare.

Clathria (*Clathria*) *prolifera* (Ellis and Solander, 1786) (=*Microciona prolifera*). On pilings and floats in San Francisco Bay; introduced from the Atlantic coast; locally common and conspicuous; the only branching red sponge of the central California intertidal fauna.

Clathria (*Microciona*) *brepha* (de Laubenfels, 1930) (=*Hymedesmia brepha*). Intertidal; rare.

Clathria (*Microciona*) *microjoanna* (de Laubenfels, 1930) (=*Microciona microjoanna*). Very low intertidal; uncommon.

Clathria (*Microciona*) *parthena* (de Laubenfels, 1930) (=*Microciona parthena*). Very low intertidal; rare.

Clathria (*Microciona*) *pennata* (Lambe, 1895) (=*Ophlitaspongia pennata*). Mid to very low intertidal; common.

Clathria (*Microciona*) sp. Hartman, 1975. Very low intertidal; uncommon.

Clathria (*Microciona*) *spongigartina* (de Laubenfels, 1930) (=*Anaata spongigartina*). Low to very low intertidal; uncommon.

Clathria (*Thalysias*) *originalis* (de Laubenfels, 1930) (=*Axocielita originalis*). Very low intertidal; common.

Clathria (*Wilsonella*) *pseudonapya* (de Laubenfels, 1930) (=*Clathriopsamma pseudonapya*). Intertidal; rare.

COELOSPHAERIDAE

Forcepia (*Forcepia*) *hartmani*. Lee, 2001. Low to very low intertidal; common. This species is easily confused with *Lissodendoryx firma*. Its forceps are extremely small and easily overlooked.

Lissodendoryx (*Lissodendoryx*) *firma* (Lambe, 1895). Low to very low intertidal; common.

Lissodendoryx (*Lissodendoryx*) *kyma* de Laubenfels, 1930. Mostly subtidal; sometimes in very low intertidal or washed

up. The description given by Green and Bakus (1994) is more representative of the species than that of de Laubenfels' 1930 original description and figures.

Lissodendoryx (*Lissodendoryx*) *topsenti* (de Laubenfels, 1930). Low to very low intertidal; common. The "microscleres" reported by de Laubenfels 1932 for *Tedania topsenti* are immature styles.

HYMEDESMIIDAE

Acanthancora cyanocrypta (de Laubenfels, 1932) (=*Hymenamphiastra cyanocrypta*).Very low intertidal; uncommon. Two color forms exist, one deep blue and the other light orange. The deep blue form is most obvious in the field.

Hymedesmia (*Stylopus*) *arndti* (de Laubenfels, 1930) (=*Astylinifer arndti*). Very low intertidal; uncommon.

Hymedesmia (*Hymedesmia*) sp. A Hartman, 1975. Very low intertidal; moderately common. Numerous species within this genus have been reported from California but not adequately described.

Hymedesmia (*Hymedesmia*) sp. B Hartman, 1975. Very low intertidal; rare.

Plocamionida lyoni (Bakus, 1966) (=*Hymendectyon lyoni*). Mid to very low intertidal; uncommon.

MYXILLIDAE

Myxilla (*Myxilla*) *agennes* de Laubenfels, 1930. Very low intertidal; rare.

Myxilla (*Myxilla*) *incrustans* (Esper 1805–1814) of Bakus, 1966). Low intertidal to subtidal. Specimens from deeper water tend to be found on shells of the bivalve *Chlamys* as thin encrustations. Intertidal specimens tend to be encrusting and massive. This species has long been confused with *Myxilla parasitica,* which has not yet been documented from California. Three variations of this species occur in California. The details given in the key apply to species from San Francisco north. From Monterey to southern California, the most common form is one in which the acanthostyles are similar to those found in the northern form but are extremely thin and with fewer spines. The variant also has tornotes with dissimilar ends, 147–205 × 2.0–6.0 mm. The third variation which is found in southern California is similar to that from Monterey south but has generally larger tornotes, 149–265 mm.

Plocamiancora igzo (de Laubenfels, 1932) (=*Plocamia igzo*). Very low intertidal; uncommon.

TEDANIIDAE

Tedania (*Tedania*) *obscurata* (de Laubenfels, 1930). Very low intertidal; rare (was *Tedanione obscurata*).

Tedania (*Trachytedania*) *gurjanovae* Koltun, 1958. Low intertidal to subtidal; rare.

Tedania (*Trachytedania*) *toxicalis* de Laubenfels, 1930.

HAMACANTHIDAE

Hamacantha (*Vomerula*) *hyaloderma* (de Laubenfels, 1932) (=*Zygherpe hyaloderma*). Low to very low intertidal; rare.

ISODICTYIDAE

Isodictya quatsinoensis (Lambe, 1892). A subtidal species found occasionally in the intertidal waters below mean lower low water, or washed up on shore; systematic placement uncertain.

MYCALIDAE

Mycale (Aegograpila) adhaerens (Lambe, 1894). Very low intertidal to subtidal. Found on shells of *Chlamys* or free-living and massive; rare in intertidal.

Mycale (Mycale) hispida (Lambe, 1894). Very low intertidal to subtidal; rare in intertidal. May be confused with *Mycale richardsoni*.

Mycale (Mycale) macginitiei de Laubenfels, 1930. Low to very low intertidal; moderately common.

Mycale (Carmia) richardsoni Bakus, 1966. Very low intertidal; rare. There is a local variant of this species with two size classes of anisochelas.

Mycale (Paresperella) psila (de Laubenfels, 1930). Low intertidal to subtidal; rare (was *Paresperella psila*).

HALICHONDRIIDA

HALICHONDRIIDAE

Halichondria (Halichondria) bowerbanki Burton, 1930. Mid to low intertidal; common on pilings and floats, especially in San Francisco Bay where it was introduced from the Atlantic coast of the United States.

Halichondria (Halichondria) panicea (Pallas, 1766). Mid to very low intertidal; common on open coast, often in well-lighted situations. This may represent a group of closely related species.

Hymeniacidon sp. A Hartman, 1975. Low intertidal; locally common in Tomales Bay; possibly conspecific with *H. sinapium*.

Hymeniacidon actites (Ristau, 1978) (=*Leucophloeus actites*). Low intertidal and subtidal; uncommon.

Hymeniacidon sinapium de Laubenfels, 1930 (=?*Hymeniacidon* sp. A Hartman, 1975, of previous edition). Low intertidal. First described from southern California where it is abundant, but since the 1990s it is found on mudflats in Elkhorn Slough and elsewhere in central California.

Hymeniacidon ungodon de Laubenfels, 1932. Low to very low intertidal; moderately common. A number of species within this genus have been reported but are not adequately described.

HAPLOSCLERIDA

CHALINIDAE

Haliclona (Haliclona) sp. A Hartman, 1975. Mid to low intertidal; common, often in well-lighted situations. The purple, supposedly cosmopolitan species, *H. permolis*, is not cosmopolitan and use of the name for this species was abandoned in the previous edition.

Chalinula loosanoffi (Hartman, 1958) (=*Haliclona loosanoffi*). An interesting species that produces gemmules and was first described from the East Coast. Occurs on floats in San Francisco Bay and Bodega Harbor (and extensively in southern California in similar habitats). Formerly known here as *Haliclona* sp. B Hartman, 1975. Externally similar to *H. ecbasis* de Laubenfels from southern California (see Fell 1970). It is possible that some specimens that have been identified as *Haliclona* sp. B may well be something other than *C. loosanoffi*. The same is true for specimens of *C. loosanoffi*. Greater investigation of the species in this difficult group must be made before unequivocal identifications can be made.

Haliclona (Halichoclona) gellindra de Laubenfels, 1932. Intertidal. Not common.

?*Haliclona (Rhizoniera)* sp. A Hartman, 1975. Low to very low intertidal; uncommon. *Reniera* is now considered a subgenus of *Haliclona,* but this species does not match the characteristics of the subgenus *Reniera*. A thorough review of this and the next species is necessary (was *Reniera* sp. A).

?*Haliclona (Rhizoniera)* sp. B Hartman, 1975. Very low intertidal; rare (was *Reniera* sp. B).

?*Haliclona (Gellius)* sp. (Hartman, 1975) (was *Sigmadocia* sp.). Low to very low intertidal; uncommon. *Sigmadocia* is now considered to be within *Haliclona (Gellius)*. As such, this species needs review.

?*Haliclona (Reniera)* sp. Hartman, 1975 (was *Toxadocia*). Low to very low intertidal; uncommon. *Toxadocia* is no longer a valid category; the genus has been placed in *Haliclona (Reniera)* in the Chalinidae. Its placement here is in question, as it may be in the genus *Xestospongia*. A thorough review of this species needs to be made.

NIPHATIDAE

Niphates lunisimilis (de Laubenfels, 1930) (=*Haliclona lunisimilis*; as ?*Pachychalina lunisimilis* in the previous edition). Low to very low intertidal; rare.

PETROSIIDAE

Neopetrosia vanilla (de Laubenfels, 1930) (=*Xestospongia vanilla*) Low to very low intertidal; common.

Xestospongia dubia (Ristau, 1978) (=*Adocia dubia*). Very low intertidal; occasionally found in the subtidal and rarely in the intertidal.

Xestospongia edapha (de Laubenfels, 1930) (=*Sigmadocia edaphus*). Very low intertidal; rare.

Amphimedon trindanea (Ristau, 1978) (was *Xestospongia*). Low intertidal to subtidal and washed on shore; rare.

DICTYOCERATIDA

SPONGIIDAE

Spongia (Spongia) idia de Laubenfels, 1932. Very low intertidal; rare.

DENDROCERATIDA

DARWINELLIDAE

Aplysilla glacialis (Merejkowsky, 1878). Mid to very low intertidal; moderately common.

Aplysilla polyraphus de Laubenfels, 1930. Intertidal; rare.

HALISARCIDA

HALISARCIDAE

Halisarca sp. Hartman, 1975. Mid to very low intertidal; uncommon. See Muricy and Pearse 2004, Proc. Calif. Acad. Sci. (4) 55: 598–612 for discussion of this species and distinction from *Oscarella*.

Calcarea

CLATHRINIDA

CLATHRINIDAE

Clathrina sp. Hartman, 1975. Very low intertidal; fairly common. These sponges are made up of a tight network of tubes best seen under a dissecting microscope.

LEUCOSOLENIDA

LEUCOSOLENIIDAE

Leucosolenia eleanor Urban, 1905. Very low intertidal zone of open coast; moderately common.
Leucosolenia sp. Hartman, 1975.
Leucosolenia nautilia de Laubenfels, 1930. In bays and harbors; often on *Mytilus*.

SYCETTIDAE

Sycon spp. Hartman, 1975. Occur on floats in harbors and on rocks intertidally. Several species are present in California but are not yet described or include introduced species.

GRANTIIDAE

Grantia sp. Hartman, 1975.
Leucandra heathi (Urban, 1905). Very low intertidal zone; locally moderately common.

AMPHORISCIDAE

Leucilla nuttingi (Urban, 1902) (=*Rhabdodermella nuttingi*). Low to very low intertidal zone; common.

Annotated References

Austin, W. C. 1985. An annotated checklist of marine invertebrates of the cold temperate northeast Pacific. Cowichan Bay, British Columbia: Khoyatan Marine Laboratory. 682 pp. Contains an extensive list of Pacific Northwest sponges, many of which occur in California.

Austin, W. C. and B. Ott. 1987. *Phylum Porifera* in: Seashore life of the northern Pacific coast. Kozloff, E. N. ed. Seattle: University of Washington Press. 370 pp. *A work covering the Pacific Northwest. Includes approximately 115 species, some of which may be new to science and many that also occur in California and Oregon.*

Bakus, G. J. 1966. Marine poeciliscleridan sponges of the San Juan Archipelago, Washington. J. Zool. 149: 415–531. *Covers the Poecilosclerida of the Northeast Pacific but many may be found in California and Oregon. Mostly intertidal.*

Bakus, G. J. and K. D. Green. 1987. The distribution of marine sponges collected from the 1976–1978 Bureau of Land Management Southern California Bight Program. Bull. Southern California Acad. Sci. 86: 57–88. *Southern California Bight Program material. Includes a total of 58 species, with eight to nine possible new species that have not been published.*

Burton, M. 1963. A revision of the Classification of the Calcareous sponges. William Clowes & Sons Ltd. London Ltd.: 1–663. *See comments under Urban 1902.*

Duplessis, K. and H. M. Reiswig. 2000. Description of a new deep-water calcareous sponge (Porifera: Calcarea) from Northern California. Pacific Science 54: 10–14. *Most recent description of a California calcareous sponge.*

Green, K. D. and G. J. Bakus. 1994. Taxonomic atlas of the benthic fauna of the Santa Maria Basin and western Santa Barbara Channel. Volume 2. The Porifera. Santa Barbara Museum of Natural History, Santa Barbara California, 82 pp. *Covers the Santa Barbara Channel. Includes 43 species of which 21 may be new to science but have not been published.*

Hartman, W. D. 1958. Natural history of the marine sponges of southern New England. Bull. Peabody Mus. Nat. Hist. 12: 1–155. *Excellent discussions of species and genera now found in California and Oregon.*

Hartman, W. D. 1975. Phylum Porifera. Pp. 32–54. In: Light's manual. 3rd ed. R. I. Smith and J. T. Carlton (eds). *The predecessor to the present volume. Next to de Laubenfels, 1932, the largest published work on California sponges restricted to intertidal forms. Sixty seven species, including 19 possible new species and two from publications after de Laubenfels, 1932.*

Hooper, J. N. A. 1997. "Sponguide." Guide to Sponge Collection and Identification. Queensland Museum: Brisbane. Electronic version. [http://www.qmuseum.qld.gov.au/organisation/sections/Sessile MarineInvertebrates (Guide to sponge collection etc.)]. *Dr. Hooper and the Queensland Museum began the development of this guide to aid with the requests that the museum had received to identify sponges. The museum later became a prime mover in the Systema Porifera project, which was developed to bring about a major revision of the higher classification of sponges. The guide includes well-organized material on collection, preservation, and identification of sponges.*

Hozawa, S. 1929. Studies on the calcareous sponges of Japan. J. Fac. Sci. Imp. Univ. Tokyo, Sect. 4, Zoology, 1: 277–389. *See notes Urban 1902, below.*

Klontz, S. W. 1989. Ecology and systematics of the intertidal sponges of Southeast Farallon Island. MA thesis. 144 pp. San Francisco State University, San Francisco. *Covers the distribution of Farallon Island sponges. All are intertidal, including cave habitats. Forty five species, of which four may be new to science.*

Koltun, V. M. 1959. Corneosiliceous sponges of the northern and far eastern seas of the U.S.S.R. Opredeliteli po faune SSSR, izdavaemye zoologischeskim Institutom Akademija Nauk SSSR, 67 Izdatjelstvo Nauko. Moskva-Leningrad 67: 1–235 (in Russian). (Translated into English by the Fisheries Research Board of Canada Translation series, Number 1842, 1971).

Koltun, V. M. 1966. Four-rayed sponges of the northern and far eastern seas of the U.S.S.R. Opredeliteli po faune SSSR, izdavaemye zoologischeskim Institutom Akademija Nauk SSSR, 90 Izdatjelstvo Nauko. Moskva-Leningrad, 90: 1–107 (in Russian). (Translated into English by the Fisheries Research Board of Canada Translation series, Number 1785, 1971). *These extensive works by Koltun cover the sponges of the northern and far eastern seas of the USSR. Some of these species may be found in California and Oregon.*

Lambe, L. M. [1892] 1893. On some sponges from the Pacific coast of Canada and Behring Sea. Trans. Royal Soc. Canada 10: 67–78. *Sponges of the Northeast Pacific but includes many species that occur in California and Oregon.*

Lambe, L. M. 1893 [1894]. Sponges from the Pacific coast of Canada. Trans. Royal Soc. Canada 11: 113–148.

Lambe, L. M. 1894 [1895]. Sponges from the western coast of North America. Trans. Royal Soc. Canada 12: 113–138. *The only paper of this series that actually relates to material collected in California.*

Laubenfels, M. W. De. 1926. New sponges from California. Ann. Mag. Nat. Hist. (9) 17: 567–573. *Describes three California sponges.*

Laubenfels, M. W. De. 1927. The red sponges of Monterey Peninsula, California. Ann. Mag. Nat. Hist. (9) 19: 258–266. *Describes four prominent, red, California sponges.*

Laubenfels, M. W. De. 1930. The sponges of California. Stanford Univ. Bulletin ser. 5, 5(98): 24–29. *A synopsis of the doctoral dissertation that preceded his major (1932) work on California sponges.*

Laubenfels, M. W. De. 1932. The marine and fresh-water sponges of California. Proc. U.S. Nat. Mus. 81: 1–140. *Still the major published work on California sponges. In many instances, generic assignments have changed; includes mostly intertidal forms and a few deep water species. SEM was not available at this time. Minimal detail given. No keys. One hundred species.*

Lee, W. L. 2001. Four new species of *Forcepia* (Porifera, Demospongiae, Poecilosclerida, Coelosphaeridae) from California, and synonymy of *Wilsa* de Laubenfels, 1930, with *Forcepia*, Carter, 1874. Proc. Cal. Acad. Sci., 52(18): 227–244.

Ristau, D. A. 1978. Six new species of shallow-water marine demosponges from California. Proc. Biol. Soc. Wash. 91: 569–589. *The result of a graduate thesis. Covers six new species from central California.*

Sarà, M. and G. Corriero. 1993. Redescription of *Tethya californiana* de Laubenfels as a valid species for *Tethya aurantia* var. *californiana*

(Porifera, Demospongiae). Ophelia 37: 203–211. *A Restudy of Tethya aurantia var. californiana showed the California species to be distinct from T. aurantia.*

Sim, C. J. and J. G. Bakus 1986. Marine sponges of Santa Catalina Island, California. Allan Hancock Foundation Occasional Paper, New series 5: 1–23. *Includes 45 species, of which three may be new.*

Tanita, S. 1943a. Key to all described species of the genus *Leucosolenia* and their distribution. Sci. Rep. Tohoku Imp. Univ. (4) 17: 71–93.

Tanita, S. 1943b. Studies on the Calcarea of Japan. Sci. Rep. Tohoku Imp. Univ. (4) 17: 353–490.

Urban, F. 1902. *Rhabdodermella nuttingi*, nov. gen. et nov. spec. Zeitsch. wiss. Zool., 71: 268–275. *The papers by Urban (1902) and Tanita (1943b) represent the first and last major works that involve California calcareous sponges. Between these two papers, Hozawa (1929) and Tanita (1943a) described and revised the group on the basis of Japanese Calcarea. Burton (1963) revised the Calcarea relative to the holdings of the British Museum.*

References

Bergquist, P. 1980. A revision of the supraspecific classification of the orders Dictyoceratida, Dendroceratida, and Verongida (Class Demospongiae). N. Z. J. Zool. 7: 443–503.

Bakus, G. J. 1966. Marine poeciloscleridan sponges of the San Juan Archipelago, Wash. J. Zool. 149: 415–531.

De Vos, L., K. Rützler, N. Boury-Esnault, and C. Donadey, and Jean Vacelet. 1991. Atlas of Sponge Morphology. Washington and London: Smithsonian Institution Press. 117 pp.

De Weerdt, W. H. 1985. A systematic revision of the northeastern Atlantic shallow-water Haplosclerida (Porifera, Demospongiae), part 1: Introduction, Oceanapiidae and Petrosiidae. Beaufortia 35: 61–91.

De Weerdt, W. H. 1986. A systematic revision of the northeastern Atlantic shallow-water Haplosclerida (Porifera, Demospongiae), part 2: Chalinidae. Beaufortia 36: 81–165.

De Weerdt, W. H. 2000. A Monograph of the shallow-water Chalinidae (Porifera, Haplosclerida) of the Caribbean. Beaufortia 50: 1–67.

Diaz, M. C., and R. W. M. van Soest. 1994. The Plakinidae: a systematic review. In Sponges in time and space. Van Soest, R. W. M., T. M. G. van Kempen, and J. C. Braekman, eds. pp. 93–109. Balkema, Rotterdam.

Faulkner, D. J. 1984. Marine Natural Products. Natural Products Reports, Journal of Current Developments in Bio-organic Chemistry. Roy. Soc. Chem. G. B.

Fell, P. E. 1970. The natural history of *Haliclona ecbasis* de Laubenfels, a siliceous sponge of California. Pac. Sci. 24: 380–386.

Green, K. D. and G. J. Bakus. 1994. Taxonomic atlas of the benthic fauna of the Santa Maria Basin and western Santa Barbara Channel. *The Porifera.* Vol. 2. Santa Barbara, CA: Santa Barbara Museum of Natural History. 82 pp.

Hartman, W. D. 1975. Phylum Porifera. pp. 32–54. In: Light's manual: intertidal invertebrates of the central California coast. 3rd edition.

Smith, R. I. and J. T. Carlton, eds. Berkeley: University of California Press.

Hooper, J. N. A. 1996. Revision of the Microcionidae (Porifera: Poecilosclerida: Demospongiae), with description of Australian species. Mem. Queensland Museum 40: 1–626.

Hooper, J. N. A. and R. W. M. van Soest, eds. 2002. *Systema Porifera*, a guide to the classification of sponges. Vols. 1 and 2. Dordrecht: Kluwer Academic Publishers.

Laubenfels, M. W. De. 1932. The marine and fresh-water sponges of California. Proc. U.S. Natl. Mus. 81: 1–140.

Lee, W. L. 2001. Four new species of *Forcepia* (Porifera, Demospongiae, Poecilosclerida, Coelosphaeridae) from California, and synonymy of *Wilsa* de Laubenfels, 1930, with *Forcepia*, Carter, 1874. Proc. Cal. Acad. Sci. (4) 52: 227–244.

Pile, A. J., M. R. Patterson, and J. D. Witman. 1996. In situ grazing on plankton <10 μm by the boreal sponge *Mycale lingua*. Mar. Ecol. Prog. Ser. 141: 95–102.

Pile, A. J., M. R. Patterson, Michael Savarese, V. I. Chernykh, and V. A. Fialkov. 1997. Trophic effects of sponge feeding within Lake Baikal's littoral zone. 2. Sponge abundance, diet, feeding efficiency, and carbon flux. Limnol. Ocean. 42: 178–184.

Preston, C. M., K. Y Wu, T. F. Molinski, and E. F. De Long. 1996. A psychrophilic crenarchaeon inhabits a marine sponge: *Cenarchaeum symbiosum* gen. nov., sp. nov. Proc. Natl. Acad. Sci. USA. 93: 6241–6246.

Reiswig, H. M. 1971. Particle feeding in natural populations of three marine demosponges. Biol. Bull. 141: 568–591.

Reiswig, H. M. and G. O. Mackie. 1983. Studies on hexactinellid sponges. III. The taxonomic status of Hexactinellida within the Porifera. Phil. Trans. Roy. Soc. London. B. Biol. Sci. 301: (1107): 419–428.

Sarà, M. and G. Corriero 1993. Redescription of *Tethya californiana* de Laubenfels as a valid species for *Tethya aurantia* var. *californiana* (Porifera, Demospongiae). Ophelia 37: 203–211.

Soest, R. W., M. Van. 2002. Family Hymedesmiidae Topsent, 1928. In: *Systema Porifera*, a guide to the classification of sponges. Hooper, J. N. A. and R.W.M. Van Soest, eds. 2002. *Systema Porifera*, a guide to the classification of sponges. Vol. 1. Dordrecht: Kluwer Academic Publishers.

Soest, R., W. M. Van, M.C. Diaz, and S. A. Pomponi. 1990. Phylogenetic classification of the halichondrids (Porifera, Demospongiae). Beaufortia 40: 15–62.

Vacelet, J. and N. Boury-Esnault. 1995. Carnivorous sponges. Nature 373: 333–335.

Vacelet, J. and N. Boury-Esnault. 1996. A new species of carnivorous sponge (Demospongiae: Cladorhizidae) from a Mediterranean cave. Bull. Inst. Roy. Sci. Nat. Belg. Biol. 66 suppl: 109–115.

Vacelet, J., A. Fiala-Médioni, C. R. Fisher, and N. Boury-Esnault. 1996. Symbiosis between methane-oxidizing bacteria and a deep-sea carnivorous cladorhizid sponge. Mar. Ecol. Prog. Ser. 145: 77–85.

Wilkinson, C. 1983. Net primary productivity in coral reef sponges. Science 219: 410–412.

Wilkinson, C. 1993. Symbiotic interactions between marine sponges and algae. In Algae and symbiosis: plants, animals, fungi, viruses, interactions explored. W. Reisser. pp. 111–151. Bristol: Biopress Limited.

Cnidaria

(Plates 38–71)

Cnidarians, like sponges, are an ancient group, relatively simple in structural organization, wholly aquatic, and most greatly developed in the sea, where they occur from the shore to abyssal depths, both in the plankton and the benthos. At the seashore, cnidarians are confined with few exceptions to lower tidal levels or below because, like the sponges, bryozoans, and ascidians, they are not adapted to withstand exposure. But in contrast to these latter groups, which are mostly all filter feeders, cnidarians are primarily predators. Their success seems to be explained by two devices for food-getting and defense—tentacles and cnidocysts, by an effective means of distribution, the ciliated planula larva, and in many cases (most Scyphozoa, many Hydrozoa) by a free-swimming sexual medusa.

Hydrozoa: Polyps, Hydromedusae, and Siphonophora

CLAUDIA E. MILLS, ANTONIO C. MARQUES, ALVARO E. MIGOTTO,
DALE R. CALDER, AND CADET HAND

(Plates 38–60)

Hydrozoa, of which there are roughly 3,000 species (Schuchert 1998), are abundantly represented in the intertidal zone by "hydroids," the sessile polypoid stages of these cnidarians. Hydroids vary tremendously in form, from tiny individuals to large and showy colonies. The life cycles of hydrozoans (plate 38) often include a sexual **MEDUSA** stage, which exists free in the plankton, but even some hydroid colonies exist free-living in the plankton. In the intertidal, the medusa stage is more commonly retained upon the polypoid generation as an attached **MEDUSOID** or as an even more reduced **SPOROSAC**.

The existence of free-living medusa stages in some life cycles has led to difficult problems in taxonomy. In many cases, the polyp and the medusa of a single species have been described under different genus and species names, some of which have persisted in common usage even after the two forms have been recognized as stages in the life cycle of one species. We provide separate keys for the attached polypoid, or "hydroid" forms, for the hydromedusae, and for the siphonophores. The polyp phase is chiefly encountered in intertidal collecting, while medusae and siphonophores are generally taken in plankton tows or by dip-netting in pools, in harbors around floats, or among *Zostera* or macroalgae, and all might be collected by snorkeling or scuba diving.

The hydroids, apparently a nonmonophyletic group, include representatives of the subclasses **ANTHOATHECATA** (also known as **ANTHOMEDUSAE**), **LEPTOTHECATA** (also known as **LEP-TOMEDUSAE**), and **LIMNOMEDUSAE**, which account for almost all local species of hydroids and their medusae. Also included among the hydroids are the calcareous "hydrocorals," which are represented intertidally on this coast by one species of the family Stylasteridae, namely the lavender, encrusting *Stylantheca porphyra*. Subtidally, species of the stylasterid genus *Stylaster* occur as pink, encrusting and branching growths.

The anthoathecate family Porpitidae is often abundantly represented on our beaches by the blue "by-the-wind sailor" *Velella* (plate 43A), which may be blown ashore in vast numbers. Its floating hydroid colonies are composed of a series of gastrozooids, gonozooids bearing medusa buds, and dactylozooids surrounding a central mouth.

The pelagic colonial **SIPHONOPHORA** are now placed in the pelagic hydromedusae (Bouillon and Boero 2000; Collins 2000, 2002; Marques 2001; Marques and Collins 2004), but they receive their own key here for ease of identification. Hydromedusan affinities were earlier suggested by Petersen (1979, 1990) and Schuchert (1996). Siphonophores have a much more complex organization than other hydromedusae (Totton 1965, Kirkpatrick and Pugh 1984, Pugh 1999, Bouillon et al. 2004). Each colony has a distinct form, size, and arrangement of its members. The colony, which may be supported by a float, forms a complex array of polyps and medusoids specialized for feeding, swimming, reproduction, or other functions. When collected in plankton tows and preserved in formalin, these colonies fragment into bits, on which much of the identification has traditionally been based. Onshore currents and winds carry the floating hydroid *Velella*, as well as some siphonophores, to our coast, but both are more characteristic of oceanic waters.

The subclasses **NARCOMEDUSAE** and **TRACHYMEDUSAE** consist of hydromedusae that lack a polyp stage in the life cycle; these medusae are occasionally taken in our plankton, but most are essentially oceanic forms. Many of the holoplanktonic siphonophores, narcomedusae, and trachymedusae are considered to be cosmopolitan species. Molecular studies will eventually reveal the amounts of gene flow between animals

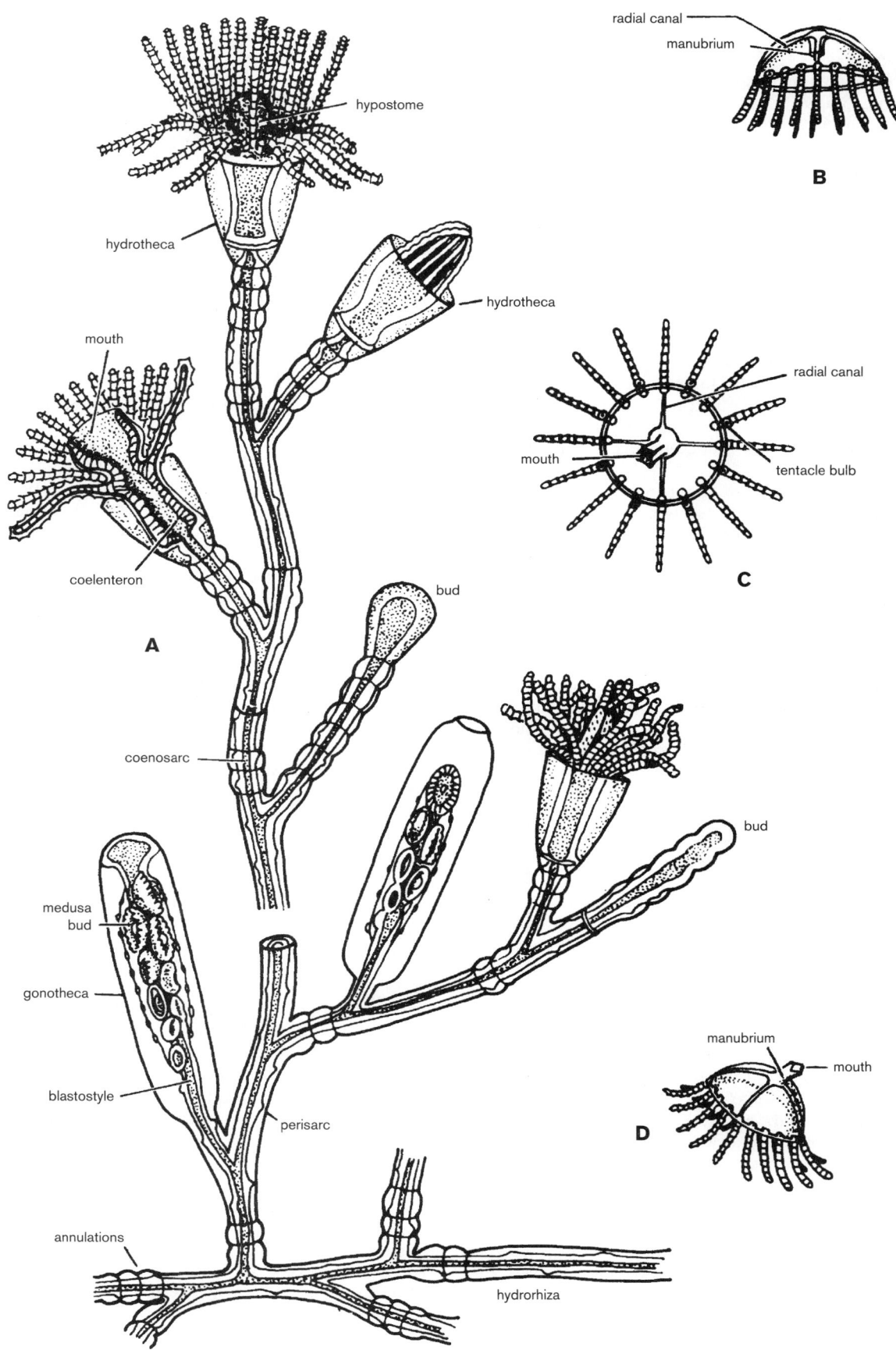

PLATE 38 Hydrozoan polymorphism (*Obelia*). A, the polypoid colony; B and C, immature medusae; D, medusa swimming inside out; note that *Obelia* lacks the typical hydromedusan velum (modified from Hand in Marshall and Williams 1972; used with permission of The Macmillan Company).

presently bearing the same names in widely separated oceans. The subclass **ACTINULIDAE**, considered a near-relative of the **NARCOMEDUSAE**, is composed of highly reduced interstitial medusae; one actinulid species of *Halammohydra* has been found on California beaches.

Hydroids include many local intertidal species of such varied form and structure that a detailed account is desirable. Certain terms are widely used for the parts of the hydroid colony. Unfortunately, they by no means always have the same meaning nor are they always consistently applied. The only monographic treatment for the West Coast is that of Fraser (1937), but the distinctions he makes are often obscure, his terminology complex, his illustrations sketchy at best, and in his work little or no attention was paid to the medusoid stages.

The hydroid colony (plate 38) is a continuous, often branching, cellular tube—the **COENOSARC**. The coenosarc consists of a layer of ectoderm separated by a thin layer of noncellular **MESOGLEA** from an inner layer of endoderm that surrounds a continuous central cavity, the **COELENTERON**, or gastrovascular cavity. Partly or completely surrounding the coenosarc is a thin, chitinous, noncellular, nonliving layer, the **PERISARC**.

Hydroid polyps show considerable polymorphism, and the zooids may be of several different types, named according to their specialized function: nutritive **GASTROZOOIDS**, generative **GONOZOOIDS**, or defensive **DACTYLOZOOIDS**. The term **HYDRANTH** is used to designate the terminal part of a nutritive zooid but does not include associated perisarcal structures such as the **HYDROTHECA**. The hydranth is therefore entirely coenosarcal, consisting of the body, hypostome, mouth, and tentacles.

The term **GONOSOME** is used to include all the specialized generative zooids of the colony and the perisarcal structures associated with them; the term **TROPHOSOME** refers to the rest of the colony. In thecate hydroids (such as *Obelia*), the gonosome includes the asexual generative zooids or **BLASTOSTYLES**, which produce sexual zooids—**MEDUSAE, MEDUSOIDS**, or **SPOROSACS**—by budding, together with the **GONOTHECAE** or cases enclosing the whole set.

The sexual zooids are termed **GONOPHORES** by some, or are referred to in general as the "medusoid" stage or generation, in contrast to the "hydroid" or polyp(oid) stage. However, Fraser (1937) uses gonophore as a synonym of blastostyle, but often includes the budding sexual zooids and the protective theca as well. We and others call this assemblage a **GONANGIUM**, a term Fraser uses to mean gonotheca. Such a reproductive element of a colony (blastostyle with buds and protective covering, if any) is occasionally still spoken of as a "fruiting body." The term "gonophore" is used, therefore, with the most diverse meanings, and we need to know with which of these meanings it is used in each case. Dr. Light remarked that, "This necessity of using terms whose meanings differ with the author, while annoying for the moment, affords very excellent intellectual experience."

The sexual zooids produce gametes from which, by fertilization, arise zygotes that develop into **PLANULA LARVAE**. Each larva can give rise to a new individual polyp or hydroid colony (or, in the case of species that do not produce polyps, will develop directly into a new medusa). All the zooids of a given colony are derived from a single zygote; hence the sexual zooids of a colony are clones, all of the same sex, and we speak of the colony as being male or female (in a few cases, colonies are hermaphroditic).

The generalized hydromedusa (plate 39) is a free-swimming animal consisting of a gelatinous **BELL** that can range from bell- to saucer-shaped with all gradations between. The outer sur-

face of the umbrella is known as the **EXUMBRELLAR SURFACE**, the inner as the **SUBUMBRELLAR SURFACE**. From the center of the subumbrellar surface hangs the **MANUBRIUM**, which can be of various lengths and, in some species, is mounted upon a gelatinous **PEDUNCLE**. The oral opening is terminal on the manubrium. It frequently carries lobes (often spoken of as "lips"), frills, or tentacles, all of which are liberally provided with cnidocysts. Where the manubrium joins the bell or peduncle there is usually a gastric cavity. **RADIAL CANALS** arise from the gastric cavity and course along the bell to the margin, where they join the ring canal. There are usually four radial canals, but other numbers commonly occur (e.g., six, eight, numerous). In a few species the radial canals are branched, while in others **CENTRIPETAL CANALS** rise upward from the **RING CANAL** but may not reach the stomach.

Hydromedusae and siphonophores are typically "craspedote"—that is, they possess a **VELUM** or membrane that partly closes off the subumbrellar space at the level of the bell margin (plate 39). The velum is occasionally lacking, as in *Obelia* (plate 38B–38D).

The bell margin is usually simple and unscalloped. Tentacles usually arise from the bell margin and may be simple, few, or many in number, occurring singly or in groups, or they may be branched (e.g., *Cladonema,* plate 54F–54I) or rudimentary. The margin may also be provided with specialized sense organs. Chief among these are **OCELLI** and **STATOCYSTS**. Ocelli occur as dark pigmented spots, usually one on each tentacle bulb, if present. Statocysts are **MARGINAL VESICLES**, or open pits, or dangling marginal clubs containing one or more concretions known as **STATOLITHS**.

Medusae are almost always of separate sexes, although most siphonophores are hermaphrodites. The gonads are epidermal structures on the radial canals, peduncle, or manubrium.

Many of the characteristics customarily used in the classification of hydrozoans are now recognized as varying markedly with environmental conditions and developmental stage, the result of which means that many species should be reexamined and their validity established (e.g., Boero 1987, Widmer 2004). In addition, for many species the complete life cycle is still not known, resulting, as noted above, in a curious double taxonomy in which polyp and medusa of the same animal have been described under separate names. Thus, the polyp originally named *Lar* is now known to give rise to the hydromedusa *Proboscidactyla*, and when such life cycles are established, the older name takes precedence. For many species, the polyp is known, but not the medusa, or vice versa.

Positive identification of many hydrozoan polyps cannot be made unless the fixed gonophores or sexually mature medusae associated with them are known. Specimens such as these are presently best keyed out only to genus. Some hydroids, including *Proboscidactyla* spp. and *Cladonema* spp., can be identified to species only if their medusae are raised to maturity because the species-distinguishing features lie only in the medusan portion of the life cycle (see both keys in this case for positive identification). Russell (1953), Naumov (1969), Kramp (1961), Millard (1975), Calder (1988, 1991, 1997), Cornelius (1995a, b), and Vervoort and Watson (2003) are of particular help for those who wish to pursue the taxonomy of the group.

For hydromedusae that cannot be identified by the following key, the most useful single reference is Russell (1953). The serious student will also find Kramp (1961) of great assistance because that monograph defines all families and genera of medusae known through 1960 and gives a brief diagnosis of each species. Kramp's *Dana Reports for the Pacific Ocean* (1965,

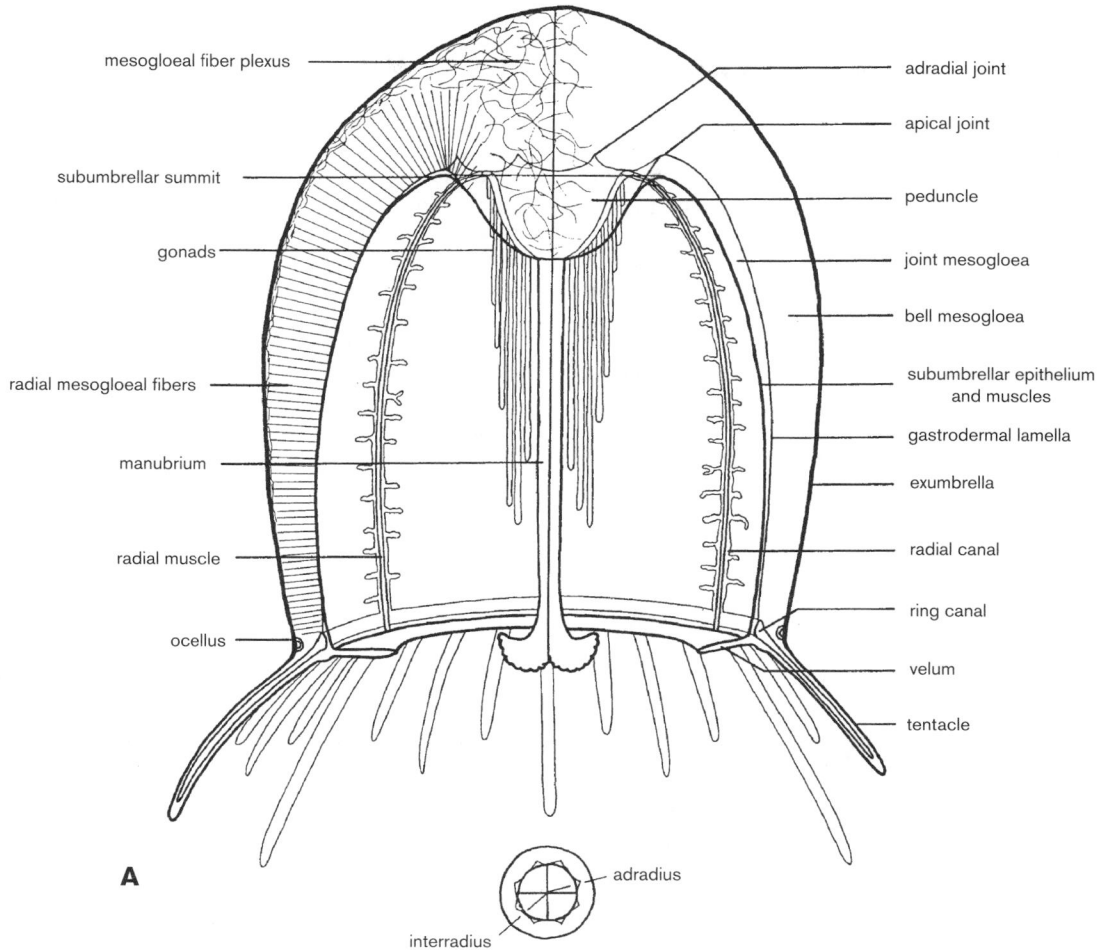

mesogloeal fiber plexus

subumbrellar summit

gonads

radial mesogloeal fibers

manubrium

radial muscle

ocellus

adradial joint

apical joint

peduncle

joint mesogloea

bell mesogloea

subumbrellar epithelium
and muscles

gastrodermal lamella

exumbrella

radial canal

ring canal

velum

tentacle

adradius

interradius

A

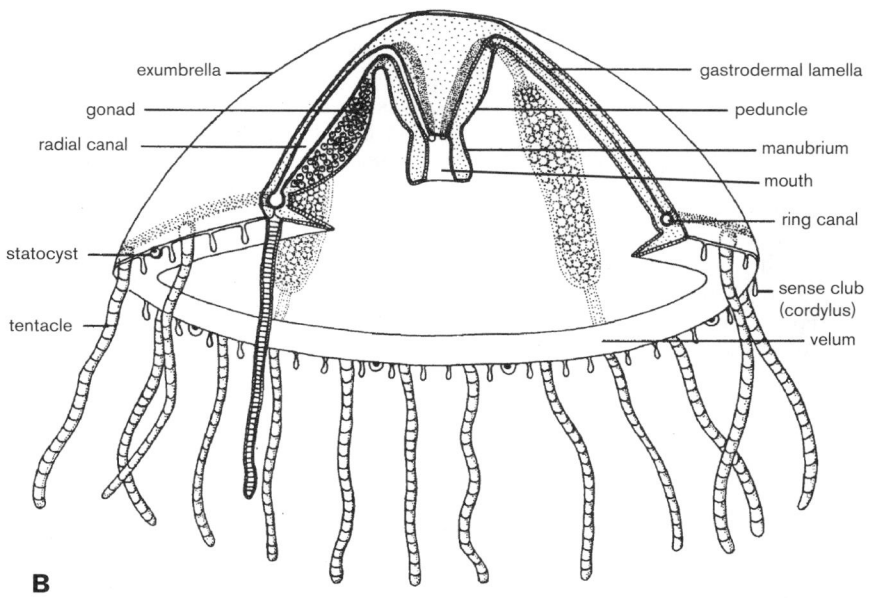

exumbrella

gonad

radial canal

statocyst

tentacle

gastrodermal lamella

peduncle

manubrium

mouth

ring canal

sense club
(cordylus)

velum

B

PLATE 39 Hydromedusa structure diagrammatic, with sections of bells removed. A, Anthomedusa *Polyorchis penicillatus* (from Gladfelter 1975, Helgoländer wiss. Meeresunters. 23: 41, fig. 1); B, generalized Leptomedusa.

1968) are a little less inclusive, but they provide illustrations and keys not found in the 1961 synopsis. Totton (1965), Kirkpatrick and Pugh (1984), and Pugh (1999) are indispensable for detailed study of siphonophores. Although it covers a different geographic region, a monograph on hydrozoa of the Mediterranean by Bouillon et al. (2004) has extensive descriptions of hydroid, hydromedusa, and siphonophore morphology (including a few species present or introduced on the West Coast) and good illustrations; this monograph includes descriptions of many families described in the past 50 years and not in Kramp (1961). Some of these important monographs are now available online at no cost (see "References" at the end of this section).

Hydrozoans are best examined alive, but it is frequently necessary to anesthetize them. To do so, a solution of magnesium chloride (73.2 g of $MgCl_2 \cdot 6 H_2O$ per liter of fresh water) is recommended in the proportion of 10–40% added to the water containing the animals. Relaxation will take several minutes. Preservation for general morphological studies should be in 5–10% formalin, which tends to dissolve the statoliths of medusae but leaves the structure of the statocyst intact. For histological work, Bouin's fixative is recommended. Specimens for molecular studies should be preserved in 95% ethanol.

Following is a combined glossary containing terms used for hydroid polyps, hydromedusae, and siphonophores. The hydroid polyps selected for inclusion in this key are those species found in the intertidal or the shallow subtidal in our area, or those raised from their medusae in the lab, whose field distribution is unknown. Hydromedusae and siphonophores selected for inclusion in the keys are those found near shore over the same geographic range—note that a few of these species are oceanic but may occasionally drift in from the high seas and have previously been collected in the study area.

In the unified annotated species list that follows the three keys, we have attempted to integrate the separate taxonomies of hydroid polyps and their medusae. Families used in the species list are a composite, found mostly in Kramp (1961), Totton (1965), Bouillon (1994), Bouillon et al. (2004), and adopted by Cairns et al. (2002).

ACKNOWLEDGMENTS

For information and assistance, we thank Charles M. D. Santos, Jeff Goddard, Stephen Cairns, Cathy McFadden, David Wrobel, John Moore, and Garry McCarthy. ACM received financial support from Fundação de Amparo à Pesquisa do Estado de São Paulo. DRC received financial support from the Natural Sciences and Engineering Research Council of Canada and the National Science Foundation Partnerships for Enhancing Expertise in Taxonomy.

Glossary of Hydrozoa

See text above for additional definitions. For additional discussion of terms, see Millard (1975), Cornelius (1995a), and Bouillon et al. (2004). Schuchert (2004–present) provides an extensive online, illustrated hydrozoan glossary at http://www.ville-ge.ch/musinfo/mhng/hydrozoa/glossary/glossary.htm.

P: refers to polypoid terminology.
M: refers to medusoid terminology.
S: refers to siphonophore terminology.

ABCAULINE (P) facing away from the stem or branch.

ABCAULINE CAECUM (P) a digit-shaped "blind sac" appearing on the abcauline wall of the contracted hydranth of some sertulariids; also known as the "abcauline diverticulum."

ABORAL (P/M) opposite to the location of the mouth.

ACTINULA (P/M) a larva resembling a polyp and typically having two whorls of tentacles.

ADCAULINE (P) facing towards the stem or branch.

ADNATE (P/M) in contact with (i.e., having one side of a structure adjoining that of another).

ANNULATION (P) ringed constriction of the perisarc, frequently in series.

APICAL PROJECTION (M) a glob or particularly thick portion of the umbrellar mesogloea at the top of the bell and often pinched off to some extent with a constriction.

BELL MARGIN (M) the broad or open edge of the umbrella, or bell-shaped jellyfish body.

BIMUCRONATE (P) with two sharp points; hydrothecal cusps each having two lateral points.

BRACT (S) a transparent protective, gelatinous structure covering other parts attached to the stem of siphonophores.

CAMPANULATE (P) bell-shaped.

CAPITATE (P/M) bulbous; having an enlarged tip (e.g., a tentacle or nematophore).

CENTRIPETAL CANAL (M) an outpocketing of the ring canal extending upward toward the manubrium.

CIRRI (M) small, solid tentaclelike organs situated on the umbrella margin between true tentacles and always without a basal bulb.

CLASPERS (P) tentaclelike structures supporting the gonophores in some candelabrid hydroids.

CNIDOCYST (=**NEMATOCYST**) (P/M/S) stinging organelle characteristic of the Cnidaria that consists of a double-walled capsule containing a fluid and a long tubule that everts and straightens when the capsule discharges upon stimulation; used for prey capture, defense, and attachment (also called "stinging cell"). Cnidocyst morphology, particularly of the everted tubule and its spines, is often used for taxonomic determinations.

CNIDOCYST RING (M) cnidocysts organized in such a way to form a distinct ringlike or annular structure on the tentacles of some hydromedusae.

COENOSARC (P) a living cellular tube of ectoderm and endoderm connecting polyps of a hydroid colony; usually covered by chitinous perisarc.

CORBULA (P) a protective basketlike structure that encloses several gonothecae and is composed of modified hydrocladia.

CORDYLI (M) minute, marginal club-shaped structures present instead of marginal vesicles in some Leptomedusae (Laodiceidae).

CRENULATE (P) scalloped or weakly notched.

CUSP (P) a toothlike projection; usually occurring as a series of prominences on the margin of a hydrotheca or gonotheca.

DACTYLOZOOIDS (P) elongate, slender, atentaculate polyps in some polymorphic hydroids; believed to be either defensive or chemosensory.

DESMONEME (P/M) a type of cnidocyst having a tightly coiled thread when discharged.

DISTO-LATERAL SPINE (P) a spine located to one side near the end of a structure, such as a gonotheca.

DIVERTICULA (singular diverticulum) (M) outpocketings (e.g., the blind sidebranches of some radial canals).

ERECT COLONY (P) a colony in which upright stems bearing multiple hydranths arise from the stolons (see "stolonal colony" for contrast).

EXUMBRELLA (M) the upper, or outer (aboral), surface of a jellyfish "bell."

FILIFORM (P/M) threadlike (e.g., a filiform tentacle is uniform in diameter throughout, or gradually tapering from end to end, without knoblike or beadlike concentrations of cnidocysts).

GASTROZOOID (P/S) a feeding polyp, with mouth and usually with oral tentacles (see "hydranth") in a hydroid; with basal tentacle in a siphonophore.

GONOPHORE (P/S) a reproductive structure bearing the gonads; gonophores may remain fixed or may be liberated as a medusa.

GONOTHECA (P) a capsule of perisarc enclosing and protecting a gonophore.

GONOZOOID (P) a reproductive polyp, capable of forming gonophores; sometimes derived from a gastrozooid.

HOLOPLANKTONIC (M/S) describes species whose entire life cycle takes place in the water column (without any benthic component of the life cycle).

HYDRANTH (P) a feeding polyp, usually with a mouth and tentacles (see "gastrozooid").

HYDROCAULUS (P) main stem of a hydroid.

HYDROCLADIUM (P) an ultimate branchlet, arising from a stem or a branch and usually bearing one or more hydranths.

HYDROECIUM (S) gutterlike furrow on a swimming bell; the stem is attached within the hydroecium and may retract partially or wholly into it.

HYDRORHIZA (P) a structure anchoring a hydroid to its substrate, varying from a system of stolons to an encrusting mat.

HYDROTHECA (P) a cuplike capsule of perisarc, usually capable of enclosing and protecting a hydranth.

HYPOSTOME (P) a part of the hydranth surrounding the mouth.

INTERNODE (P) a segment of a stem or branch, delimited at either end by a constriction or node, more appropriately called a segment.

LATERAL NEMATOTHECA (P) a nematotheca occurring lateral to the hydrothecal aperture; usually occurs in one or more pairs (see "nematotheca").

MANTLE CANALS (S) superficial canals present on the apex of sexual medusoid (gonophore).

MANUBRIUM (M) saclike feeding structure that hangs down from the subumbrellar apex or sometimes from a gastric peduncle, containing the stomach, the mouth, and often distinct lips of the mouth that may be ornamented in varying ways.

MARGINAL SENSORY CLUB (M) pendant, microscopic, statocyst structure located at the bell margin, containing one or more spherical statoliths, which are usually stacked vertically within it.

MARGINAL VESICLE (M) a microscopic statocyst structure located at the bell margin, composed of a cavity and one or more statoliths inside the cavity.

MESENTERIES (M) colorless tissue connections in a few species of hydromedusae connecting the wall of the manubrium to the upper portions of the radial canals.

MESOGLOEA (P/M/S) noncellular substance (e.g., the "jelly") lying between the ectoderm and endoderm of a hydrozoan. This forms the gelatinous bulk of the umbrella of a hydromedusa and a lamellalike layer in hydroids.

MONILIFORM (P/M) beadlike (e.g., a moniliform tentacle bears a series of annular swellings, armed with cnidocysts, along its length).

MONOSIPHONIC (P) single hydroid stem, not bundled with others (see "polysiphonic" for contrast).

MOUTHPLATE (S) process extending below the subumbrella opening of a swimming bell on the side where the stem is attached.

NEMATOCYST (P/M/S) a stinging capsule (see "cnidocyst").

NEMATOPHORE (P) a highly modified defensive zooid armed with cnidocysts.

NEMATOTHECA (P) a perisarcal sheath protecting a nematophore.

OCELLUS (plural ocelli) (M) a small dark spot at the bell margin, usually located on a tentacle bulb, or occasionally associated with a marginal vesicle.

ORAL (P/M) referring to the mouth.

ORAL TENTACLES (P/M) short, sometimes dichotomously branched tentacles located around or just above the mouth rim in some hydromedusae; also refers to tentacles immediately below the hypostome of hydroids.

OPERCULUM (P) a chitinous lid that closes the aperture of a hydrotheca or gonotheca; usually composed of one or more valves.

OTOPORPAE (M) elongate vertical tracks with bristles and cnidocysts, running upward from each marginal sensory club, only found in some Narcomedusae.

PALPON (S) a polyp, with or without a basal tentacle, that serves a defensive, excretory, or food-handling function; appears to be a reduced gastrozooid.

PEDICEL (P) a stalk that supports a hydranth (or a hydrotheca) or a gonophore (or a gonotheca).

PEDICELLATE (P) stalked; having a stalk or pedicel.

PEDUNCLE (M) a round to cone-shaped extension of the mesogloea down from the apex of the subumbrella, bearing the manubrium terminally; the radial canals run down the peduncle to reach the manubrium.

PERISARC (P) the chitinous exoskeleton enclosing and protecting the living tissues of a hydroid.

PLANKTONIC (P/M/S) free-living within the water column and being a sufficiently weak swimmer to be at the general mercy of the ocean currents.

PLANULA (P/M/S) a usually ciliated, embryonic, dispersal stage.

PLEUSTONIC (P/M/S) floating at the sea surface.

PODOCYST (P) a resting body, covered by perisarc.

POLYSIPHONIC (P) composed of two or more united tubes; a composite stem or branch (synonymous with "fascicled").

PSEUDOHYDROTHECA (P) a filmy covering of perisarc at the base of a hydranth resembling a hydrotheca, varying in shape and often with transverse wrinkles.

RADIAL CANALS (M/S) circulatory tubes running from the manubrium or apex of the bell to (usually) a ring canal circling the bell margin. In hydromedusae, the radial canals attach at the manubrium and usually are straight and narrow, but they may be broad, have wavy or even diverticulated walls; in a few species the radial canals branch one or more times before reaching the bell margin, particularly in siphonophores.

RUDIMENTARY TENTACLE OR BULB (M) a permanently minute or undeveloped marginal tentacle and its bulb; in some species there are a number of rudimentary tentacles or tentacle bulbs that never develop into the normal size and length of the other tentacles and bulbs.

SOMATOCYST (S) tubular or bulblike structure running up through the jelly in a calycophoran siphonophore from its origin at the base of the hydroecium; probably a caecal extension of the original, larval gastrovascular cavity.

STATOCYST (M) microscopic structure apparently used for orientation; always in symmetrical multiple arrangement and located around the bell margin, containing one or more hard concretions, or statoliths. Hydromedusan statocysts take the form of either marginal vesicles or marginal sensory clubs.

STATOLITH (M) spheroidal or polygonal concretion in a statocyst structure; the marginal vesicles or marginal sensory clubs.

STEM (P) see "hydrocaulus."

STOLON (P) a tube of coenosarc, covered with perisarc; basal stolons usually grow over the substrate and anchor the hydroid to its substrate.

STOLONAL COLONY (P) a colony in which hydranths arise singly from stolons, either with or without a pedicel, and without an upright stem.

STOMACH POUCH (M) well-defined and symmetrical outpocketing of the stomach surface, in some Narcomedusae.

SUBUMBRELLA (M/S) the concave underside of a jellyfish, which in siphonophores and most hydromedusae forms a distinct cavity with a small opening.

SWIMMING BELL (S) asexual medusoid present at or close to the apex of a siphonophore colony that propels the colony through the water; often arranged in a series with species- or genus-specific arrangement (also called nectophore).

TENTACLE BULB (M) swelling at the bell margin that forms the base of the hollow marginal tentacles of many species of hydromedusae. The bulb usually forms before the tentacle, and thus in a developing medusa there may be more bulbs than tentacles.

TENTACULA (singular tentaculum) (M) small solid marginal tentacles, usually without marginal bulbs, located between normal hollow tentacles.

TENTACULAR RUDIMENT (M) see "rudimentary tentacle."

VALVE (P) see "operculum."

ZOOXANTHELLAE symbiotic, photosynthetic single-celled organisms that live within the tissues of a number of cnidarians, including corals, some sea anemones, and some medusae.

Key to the Polypoid Stages of Hydrozoa

ANTONIO C. MARQUES, ALVARO E. MIGOTTO, DALE R. CALDER, AND CLAUDIA E. MILLS

Plates refer to polyp plates 40–49.

1. Hydranths enclosed by a distinct hydrotheca of definite shape; gonophores protected by gonothecae or similar structures . . . . Leptothecata 42
— Hydranths not enclosed by a hydrotheca; gonophores not protected by a gonotheca or thin perisarc sheath . . . . Anthoathecata, Limnomedusae 2
2. Exoskeleton a calcareous encrustation, vivid purple in color . . . . Stylantheca porphyra
— Exoskeleton not a calcareous encrustation 3
3. Polyps planktonic or pleustonic, with no vestige of stem 4
— Polyps benthic, sedentary, with or without stems and/or pedicels 5
4. Polyps pleustonic (floating at the sea surface), deep blue or purple, with oval float and upright triangular sail; colonies polymorphic, with central gastrozooid (plate 43A) . . . . Velella velella
— Polyps planktonic and solitary, with no float or sail; tentacles arranged more or less in definite circlets (plate 42I) . . . . Climacocodon ikarii
5. Polyps without tentacles and with mouth surrounded by cnidocysts; minute, fairly transparent and very cryptic 6
— Polyps with tentacles; of various sizes 8

6. Solitary; elongate and wormlike, with nematocysts sprinkled over the body surface as well as concentrated near the mouth; interstitial in fine sediment or in intertidal algal mats (plate 49R, 49S) . . . . Protohydra leuckarti
— Solitary or in small clusters; not wormlike in character 7
7. In fresh water (plate 49O) . . . . Craspedacusta sowerbii
— In estuarine to full salinity water (plate 49P, 49N) . . . . Maeotias marginata and Aglauropsis aeora
8. Hydranths with one to two tentacles 9
— Hydranths with three or more tentacles 10
9. Gastrozooids with two filiform tentacles; gonozooids without mouth or tentacles; colonies commensal on tubes of sabellid polychaetes; budding off medusae (see key to hydromedusae, couplets 34–36) (plate 41A) . . . . Proboscidactyla spp.
— Gastrozooids with one or rarely two capitate tentacles; dactylozooids lacking tentacles but with distal cnidocyst cluster; hydrorhizae mostly covered by bryozoan skeleton; polyps arising at intersection of bryozoan zooecia, or directly in front of zooecial openings (plate 43I, 43J) . . . . Zanclella bryozoophila
10. Mature hydranths with at least some capitate tentacles 11
— Mature hydranths with all tentacles filiform or appearing essentially filiform, or in a few cases moniliform 21
11. Hydroids solitary or with two to three interconnected polyps only 12
— Hydroids colonial 13
12. Hydranths large (ca. 10 mm or more), with numerous (up to 500) irregularly arranged tentacles; gonophores fixed; perisarcal spines absent; with clasper tentacles attaching to developing embryos in the blastostyle-bearing region, at least during the reproductive season (plate 43O) . . . . Candelabrum fritchmanii
— Hydranths small (0.15–0.2 mm), commonly with four tentacles (exceptionally with three to eight tentacles), fingerlike perisarcal projection (spine), protecting zooid, present on a circular thecal base (plate 42J) . . . . Halimedusa typus
13. Hydranths with both capitate and filiform tentacles (the filiform tentacles may be difficult to see) 14
— Hydranths with capitate tentacles only 16
14. Hydranths with an oral whorl of (usually) four capitate tentacles and an aboral whorl of four filiform tentacles (sometimes rudimentary) (plate 41H, 41I) . . . . Cladonema californicum and Cladonema radiatum
— Hydranths with about 10–20 capitate tentacles, arranged in two whorls or scattered along hydranth body, and an aboral whorl of four filiform tentacles (sometimes rudimentary) 15
15. Capitate tentacles eight to 10, in two whorls (plate 42E) . . . . Dipurena bicircella
— Capitate tentacles four to five in an oral whorl, plus three to four additional whorls of ca. four tentacles each (plate 42A) . . . . Coryne japonica
16. Capitate tentacles (usually) four, in a single oral whorl (plate 41E–41G) . . . . Cladonema myersi and Cladonema pacificum
— Capitate tentacles 15 to as many as 70, on hydranth body (although some hydranths in a colony may have only four capitate tentacles, around the mouth) 17
17. Hydranths large (generally more than 1 cm when extended); tentacles 30–70, arranged in five to six tight distal circlets (plate 42H) . . . . Hydrocoryne bodegensis

PLATE 40 Athecata. A, *Bougainvillia muscus*; B, *Garveia annulata*; C–D, *Garveia franciscana*; E, *Rhizorhagium formosum*; F, *Clava multicornis*; G, *Cordylophora caspia*; H, *Turritopsis* sp. I, *Eudendrium californicum*; J, K, *Hydractinia armata*; L–M, N (spine), *Hydractinia milleri*; O–Q, R (spine), *Hydractinia laevispina* (A, after Calder 1988; D, after Torrey 1902; G, after Schuchert 1996; H, after Mayer 1910; I, after Marques et al. 2000; J–K, after Fraser 1940; rest, after Fraser 1937).

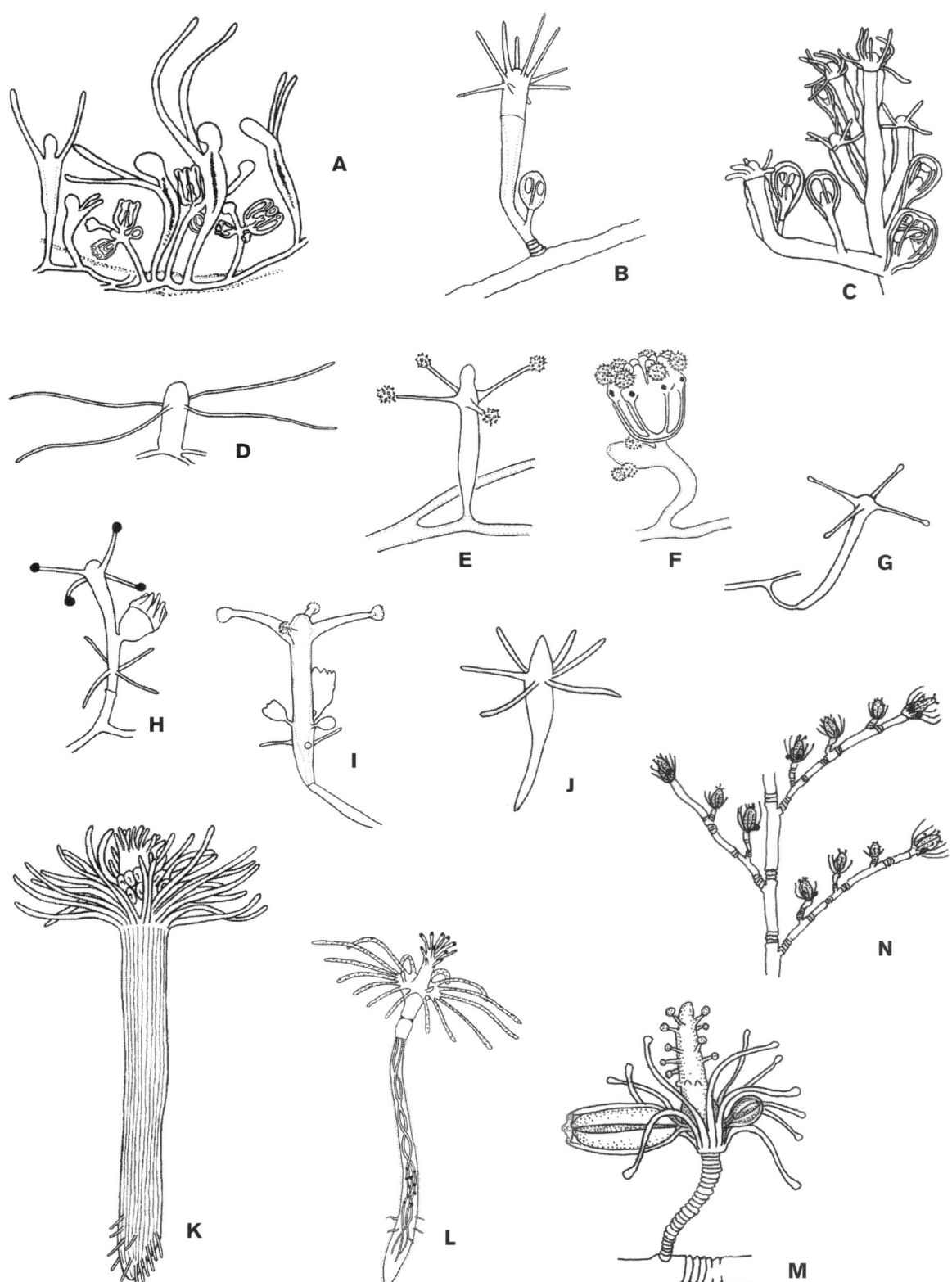

PLATE 41 Athecata (continued). A, *Proboscidactyla flavicirrata*; B, *Amphinema* sp.; C, *Leuckartiara octona*; D, *Rathkea octopunctata*; E, F, *Cladonema myersi*; G, *Cladonema pacificum*; H, *Cladonema radiatum*; I, *Cladonema californicum*; J, unidentified corymorphid from Lake Merritt, Oakland, in San Francisco Bay; K, *Corymorpha palma*; L, *Euphysora bigelowi*; M, N, *Pennaria disticha* (A, after Naumov 1969; B, from unpublished photographs by J. T. Rees; C, H, after Migotto 1996; D, after Schuchert 1996; E, F, after Rees 1949; G, after Hirohito 1988; I, after Rees 1979; J, from 1967 drawing by Jim Carlton and Dustin Chivers; K, after Fraser 1937; L, after Sassaman and Rees 1978; M, after Mayer 1910; N, after Millard 1975).

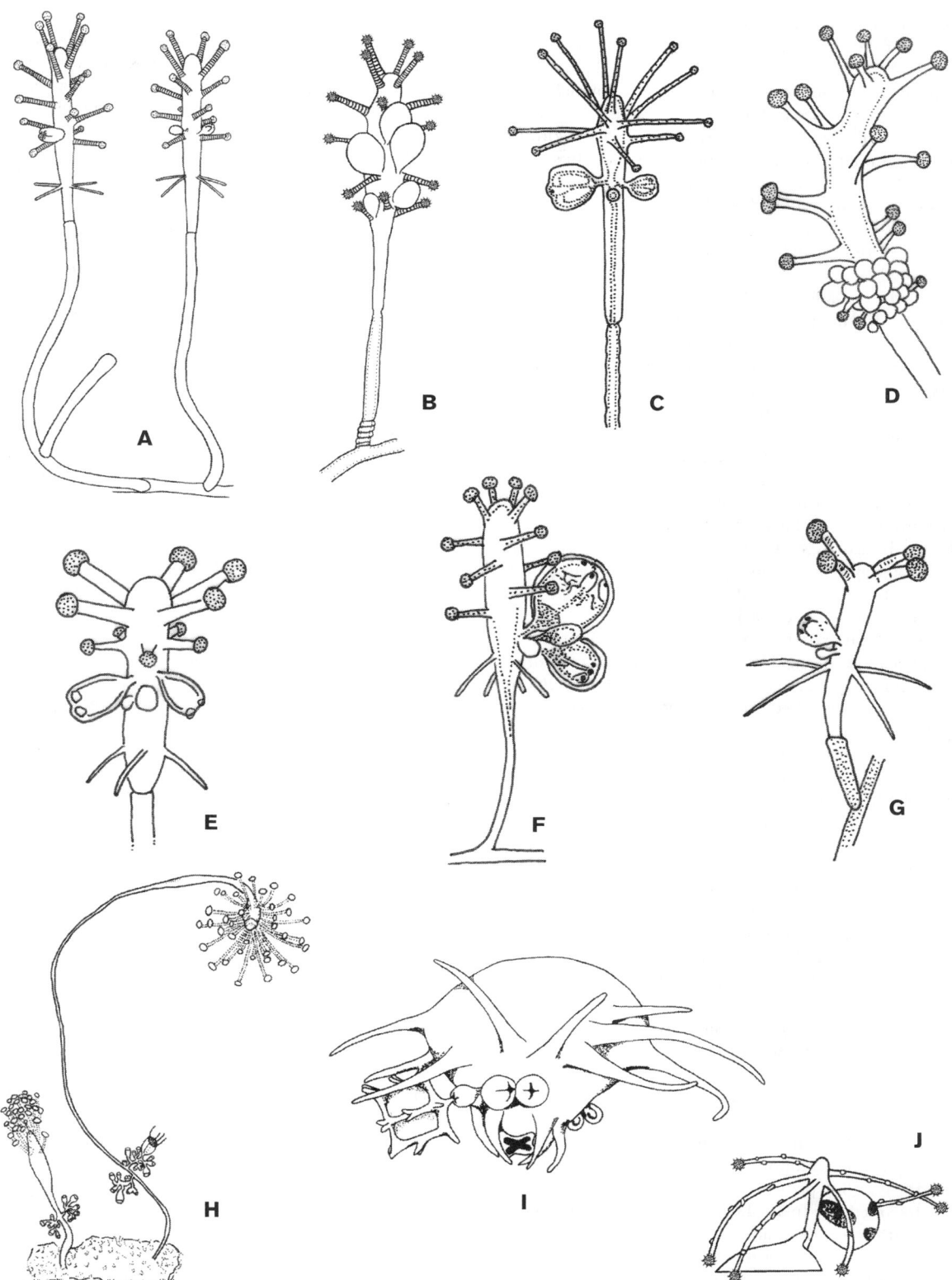

PLATE 42 Athecata (continued). A, *Coryne japonica*; B, *Coryne eximia*; C, *Sarsia tubulosa*; D, *Coryne* sp.; E, *Dipurena bicircella*; F, *Dipurena ophiogaster*; G, *Dipurena reesi*; H, *Hydrocoryne bodegensis*; I, *Climacocodon ikarii*; J, *Halimedusa typus* (A, after Schuchert 1996; B, after Brinckmann-Voss 1989; C, E–G, after Schuchert 2001; D, from unpublished photographs by J. T. Rees; H, after Rees et al. 1976; I, after Uchida 1924; J, after Mills 2000).

— Hydranths small (generally <5 mm long); tentacles 15–40 or less, scattered along hydranth body or in whorls 18

18. Tentacles up to 40; colony branched or unbranched, not growing on a bryozoan host 19
— Tentacles up to about 15, colony stolonal with hydrorhiza growing under the skeleton of a bryozoan host (plate 43K–43N) . . . . Zanclea bomala

19. Colonies much branched; perisarc irregularly annulated (plate 42B) . . . . Coryne eximia
— Colonies unbranched or slightly branched; perisarc not annulated 20

20. Hydranths 0.8–2.6 mm tall with 12–20 capitate tentacles scattered or arranged in lose whorls and confined to distal one-half of the hydranth; with one to eight medusa buds in an irregular whorl mid-hydranth, below the tentacles (plate 42C) . . . . Sarsia tubulosa
— Hydranths 2.0–4.5 mm tall with four or five loosely arranged whorls of five capitate tentacles; with up to 80 medusa buds on a single hydranth (plate 42D) . . . . . .
. . . . Coryne sp. (undescribed)

21. Hydranths with scattered tentacles 22
— Hydranths with tentacles in defined whorls 25

22. Hydranths usually solitary; when colonial, limited to a few budding polyps arising from hydrorhiza; tentacles moniliform (estuarine) (plate 43B) . . . . Moerisia sp.
— Hydranths colonial; tentacles filiform 23

23. Colony stolonal; hydranths naked; perisarc limited to hydrorhiza and forming a shallow collar at base of hydranth (plate 40F) . . . . Clava multicornis
— Colony erect, branched; stems, branches and pedicels covered by perisarc 24

24. Branches adnate to stem for some distance; gonophores forming free medusae; marine (plate 40H) . . . . . . Turritopsis sp.
— Branches free from stem except at base; gonophores fixed; brackish or fresh water (plate 40G) . . . .
. . . . Cordylophora caspia

25. Hydranths with tentacles in two whorls, one oral and one aboral; gonophores borne between tentacle whorls 26
— Hydranths with tentacles concentrated at oral end; gonophores borne below tentacles 31

26. Perisarc of stem well-developed, usually stiff, reaching to base of hydranth body; anchoring filaments absent . . . . .
. . . . 27
— Perisarc restricted to base of stem, weakly developed; anchoring filaments present 30

27. Hydranths solitary; gonophores fixed or releasing free medusae 28
— Hydranths colonial; gonophores fixed 29

28. Stem with distinct, deeply annulated expansion just below the hydranth; gonophores attached directly to the hydranth, without pedicels, asymmetrical and developing into free medusae (plate 43G) . . . . Hybocodon prolifer
— Stem with a few deep annulations, but not as a series of several rings just below the hydranth; gonophores fixed, each with three to four tentaclelike projections, borne in long, densely crowded pedicellate bunches, like grapes (plate 43C, 43D) . . . . Tubularia harrimani

29. Female gonophores with short distal crests; colony branched, forming a large tangled tuft to 15 cm high (plate 43H) . . . . Pinauay crocea
— Female gonophores with fingerlike, long processes distally; colony unbranched, polyps solitary or in small groups, to 5 cm long (plate 43E, 43F) . . . . Pinauay marina

30. Aboral tentacles 50–70; gonophores fixed (plate 41K)
. . . . Corymorpha palma
— Aboral tentacles 15–20; gonophores free medusae (plate 41L). . . . . Euphysora bigelowi

31. Hydranths with trumpet- or urn-shaped hypostome; body vasiform. 32
— Hydranths with conical, dome-, or nipple-shaped hypostome; body elongated to cylindrical (vasiform only when hypostome nipple-shaped) 33

32. Stems and branches heavily annulated throughout; hydranth somewhat squarish (plate 40I) . . . .
. . . . Eudendrium californicum
— Stems and branches generally annulated at origins but not throughout entire length; hydranth not squarish . . . .
. . . . Eudendrium spp.

33. Hydranth body covered to some extent by a variously developed pseudohydrotheca, usually encrusted with detritus. 34
— Hydranth without a pseudohydrotheca. 38

34. Hypostomes nipple-shaped; colonies with unbranched stems having a single terminal hydranth (rarely another lateral one) (plate 40E) . . . . Rhizorhagium formosum
— Hypostomes dome-shaped or conical; colonies erect, with branched stems 35

35. Colonies stolonal; on shells (plate 41C). . . . .
. . . . Leuckartiara octona
— Colonies erect, branched; on many kinds of substrates
. . . . 36

36. Stems polysiphonic; gonophores free medusae (plate 40A)
. . . . Bougainvillia muscus
— Stems monosiphonic or polysiphonic, gonophores fixed
. . . . 37

37. Stems strongly polysiphonic, annulated throughout; marine (bright yellow or orange in color) (plate 40B) . . . .
. . . . Garveia annulata
— Stems monosiphonic, slightly or not at all annulated; estuarine (plate 40C, 40D) . . . . Garveia franciscana

38. Hydranths partially covered by perisarc; bright orange-red (plate 41B). . . . . Amphinema sp.
— Hydranths naked. 39

39. Hydranths polymorphic, with gonophores borne on gonozooids. 40
— Hydranths not polymorphic, with highly extensible threadlike tentacles; medusa-buds borne on hydrorhiza (plate 41D) . . . . Rathkea octopunctata

40. Gastrozooids up to 5 mm; tentacles usually 12–20; four or fewer gonophores borne about midway between tentacles and base of hydranth, in a single whorl (plate 40L–40N)
. . . . Hydractinia milleri
— Gastrozooids up to 2.5 mm; tentacles usually 12 or fewer; four or more gonophores borne about midway between tentacles and base of hydranth, in a double or single whorl 41

41. Gastrozooids with nine to 12 tentacles; gonophores usually several, in a double whorl; gonozooids with a distinctive dark cap of closely packed cnidocysts that forms the distal half of the proboscis; spines numerous, smooth and long (about 1.2 mm), often with broken tips (plate 40J, 40K) . . . . Hydractinia armata
— Gastrozooids usually with eight tentacles; gonophores usually four, in a single whorl; gonozooids without distinctive dark cap of cnidocysts; spines few, smooth, short (about 0.5 mm) and blunt, slightly curved (plate 40O–40R). . . . . Hydractinia laevispina

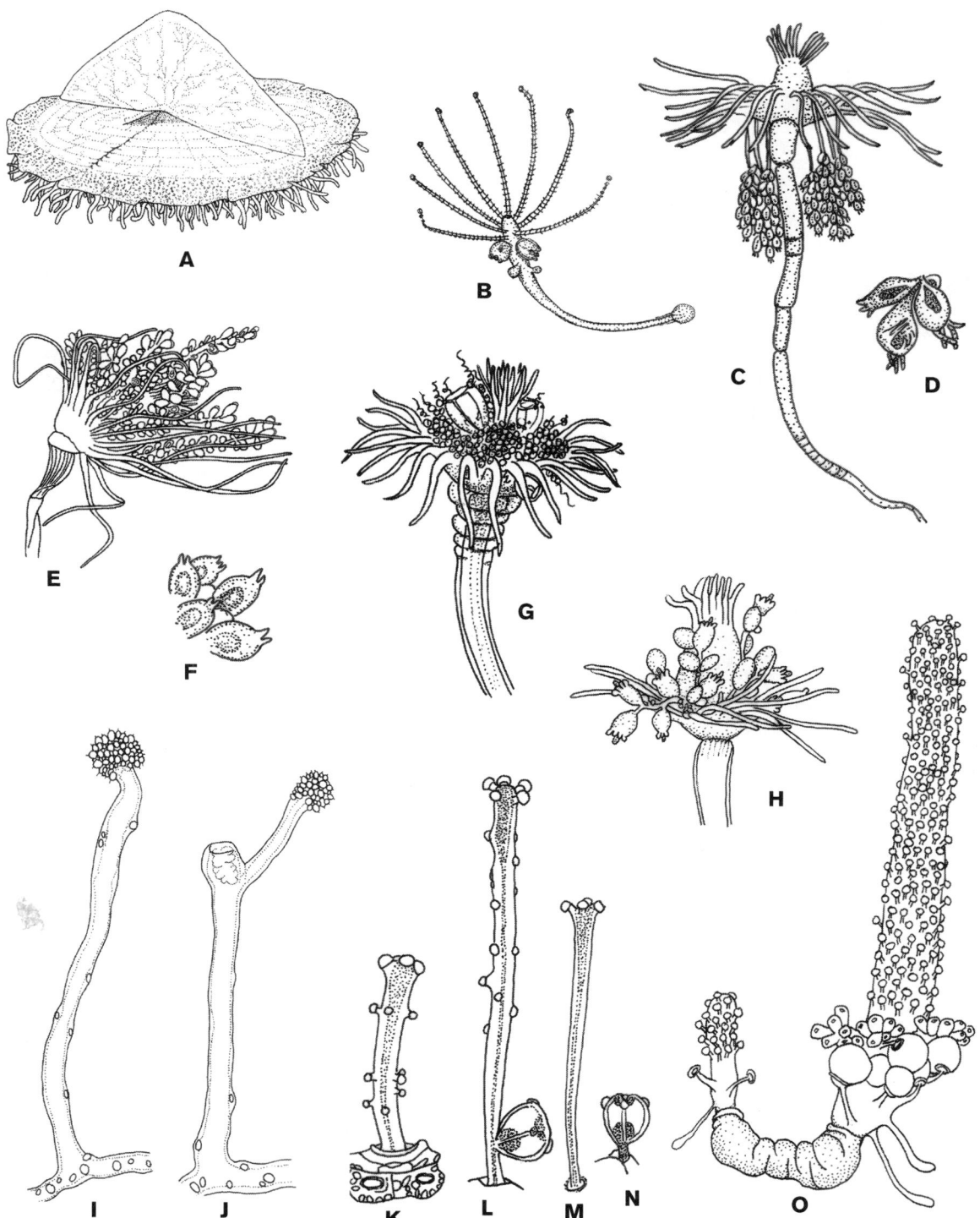

PLATE 43 Athecata (continued). A, *Velella velella*; B, *Moerisia* sp.; C, D, *Tubularia harrimani*; E, F, *Pinauay marina*; G, *Hybocodon prolifer*; H, *Pinauay crocea*; I, J, *Zanclella bryozoophila*; K–N, *Zanclea bomala*; O, *Candelabrum fritchmanii* (A, after Schuchert 1996; B, from unpublished photographs by J. T. Rees; C, D, F, G, after Fraser 1937 [G after Agassiz]; E, after Petersen 1990; H, after Brinckmann-Voss 1970; I, J, after Boero and Hewitt 1992; K–N, after Boero et al. 2002; O, after Hewitt and Goddard 2001, and unpublished photographs by Jeff Goddard).

42. Hydrothecae saucer- to basin-shaped, usually wider than deep, too small to contain contracted hydranth, margin entire (plate 45A, 45B) . . . . *Halecium* spp.
— Hydrothecae as deep or deeper than wide, able to contain contracted hydranth 43
43. Stem with segments becoming progressively longer distally,

in the form of sausage-shaped links; hydroids epizoic on mollusk shells on sandy beaches (plate 48A–48C) . . . .
. . . . *Eucheilota bakeri*
— Stem smooth or annulated, segments not sausage-shaped; hydroids not living on sandy beaches 44
44. Hydrothecae with an operculum 45

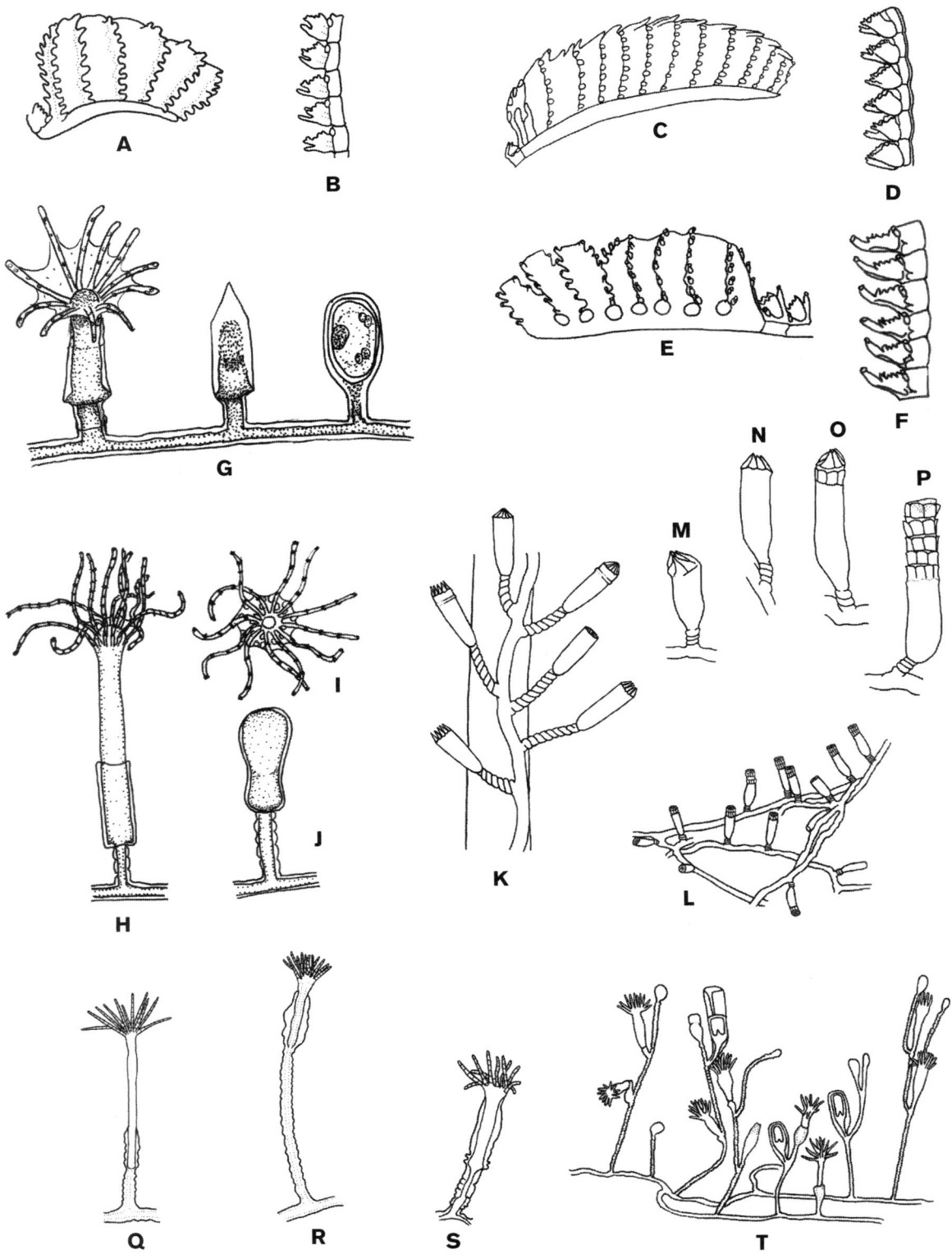

PLATE 44 Thecata. A, B, *Aglaophenia inconspicua*; C, D, *Aglaophenia struthionides*; E, F, *Aglaophenia latirostris*; G, *Blackfordia virginica*; H–J, *Aequorea victoria*; K–P, *Calycella syringa*; Q–T, *Eutonina indicans* (A–F, after Fraser 1937; G, after Mills and Rees 2000, and C. E. Mills unpublished observations; H–J, after Strong 1925; K, T, after Cornelius 1995; L–P, after Hirohito 1995; Q–S, after Naumov 1969).

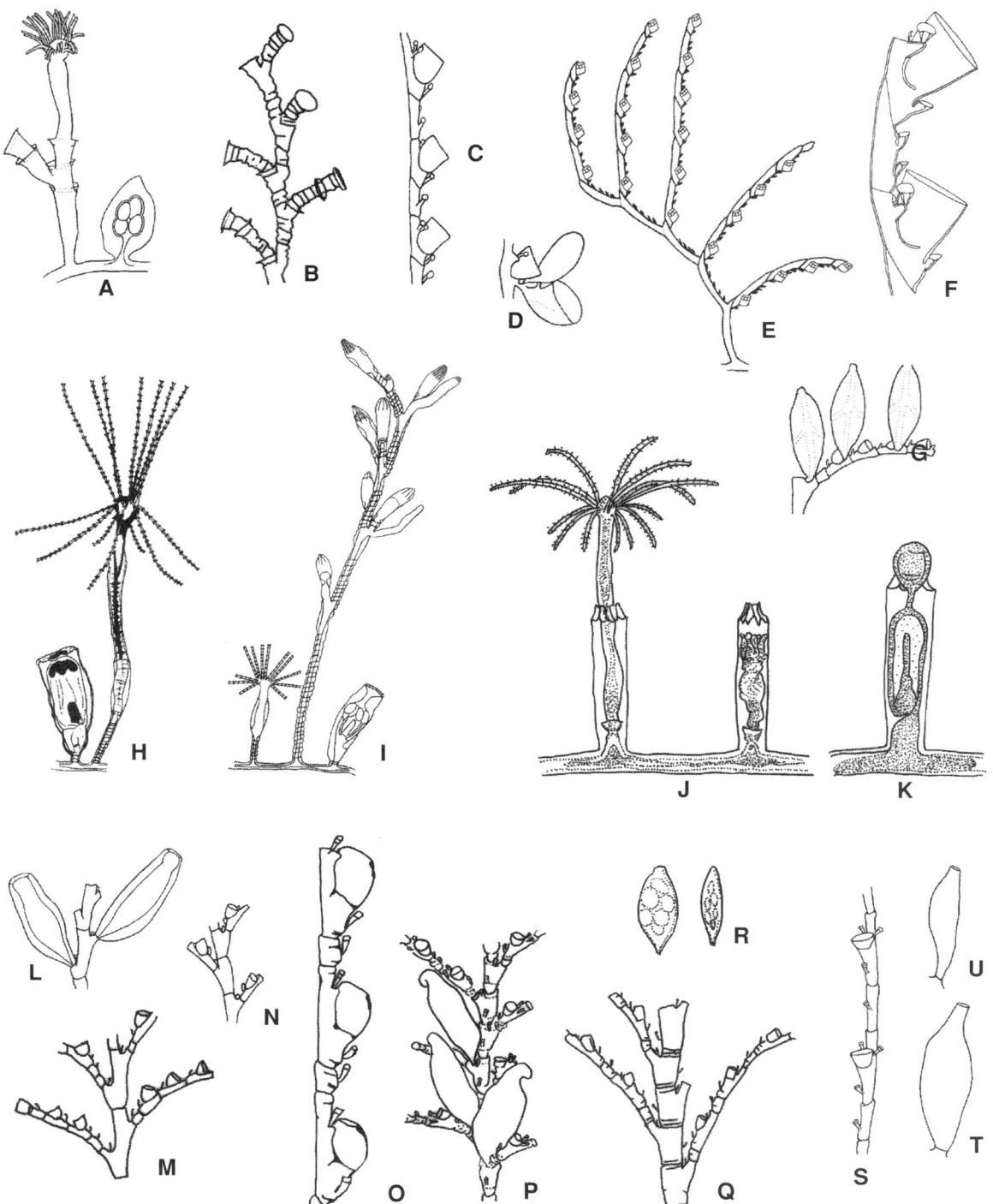

PLATE 45 Thecata (continued). A, *Halecium tenellum*; B, *Halecium annulatum*; C, D, *Antennella avalonia*; E, F, *Monostaechas quadridens*; G, *Kirchenpaueria plumularioides*; H, *Phialella zappai*; I, *Phialella fragilis*; J, K, *Mitrocoma cellularia*; L–N, *Plumularia goodei*; O–R, *Plumularia lagenifera*; S–U, *Plumularia setacea* (A, after Migotto 1996; B, after Torrey 1902; E, F, after Schuchert 1997; H, I, after Boero 1987; J, K, after Widmer 2004; L, M, after Millard 1975; S–U, after Cornelius 1995b; rest, after Fraser 1937).

71. Colonies with erect stems, each supporting several to many hydrothecae. 72
— Colonies stolonal, each pedicel (stalk) supporting a single hydrotheca (occasionally with a few branched upright stems). 79
72. Hydrothecal rims even, sinuous, or crenulate, sometimes abrading to smooth (note: search for an undamaged specimen) 73
— Hydrothecal rims cusped (note: stain with chlorazol black, if necessary). 77
73. Hydrothecae bell-shaped and distinctly concave beneath rim; female gonothecae folded over distally (plate 48S–48U)

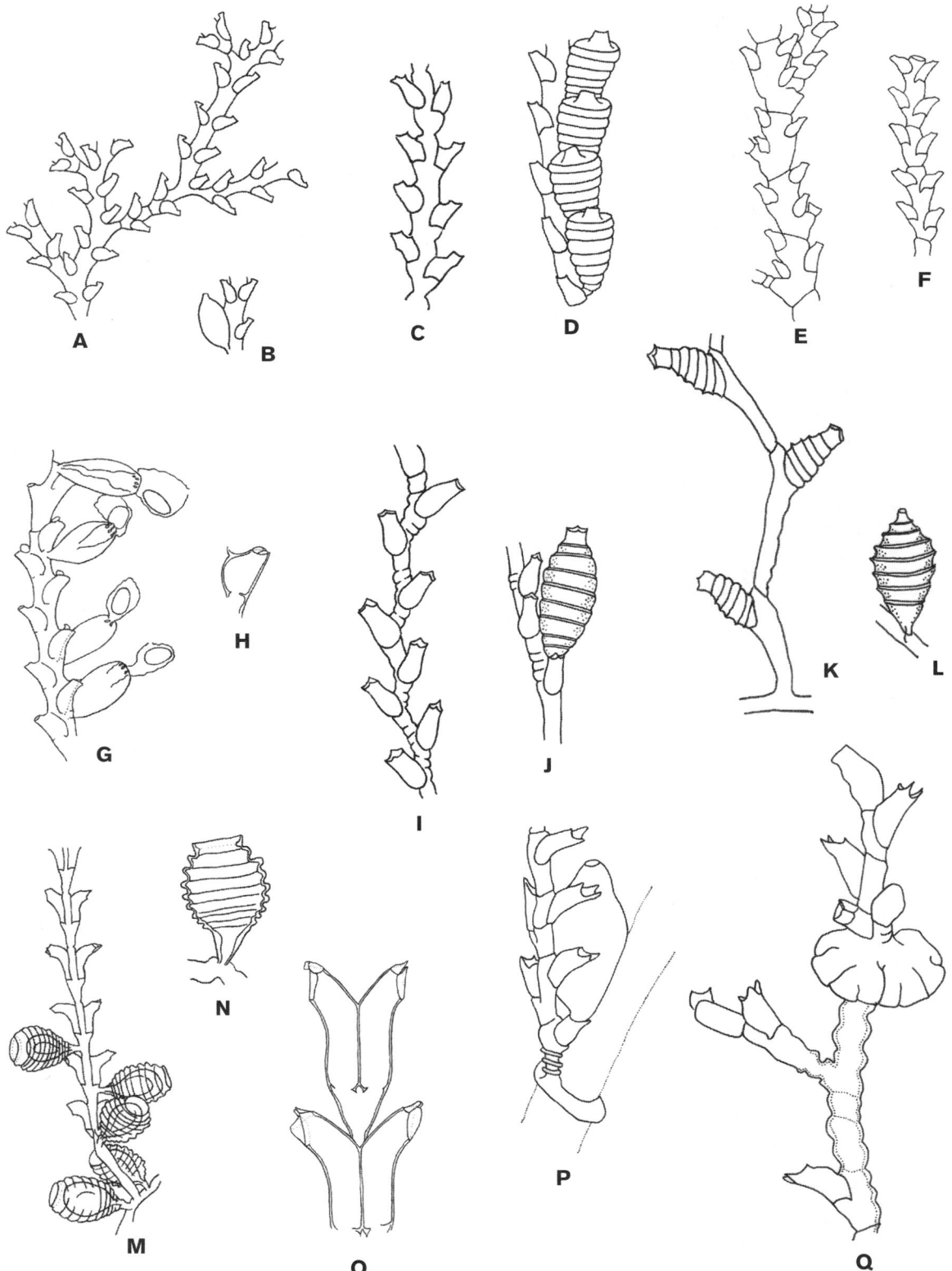

PLATE 46 Thecata (continued). A, B, *Abietinaria filicula*; C, D, *Abietinaria greenei*; E, F, *Abietinaria inconstans*; G, H, *Abietinaria traski*; I, J, *Sertularella fusiformis*; K, L, *Sertularella tenella*; M–O, *Dynamena disticha*; P, Q, *Fraseroscyphus sinuosus* (G, H, M–O, after Hirohito 1995; P, Q, after Boero and Bouillon 1993; rest, after Fraser 1937).

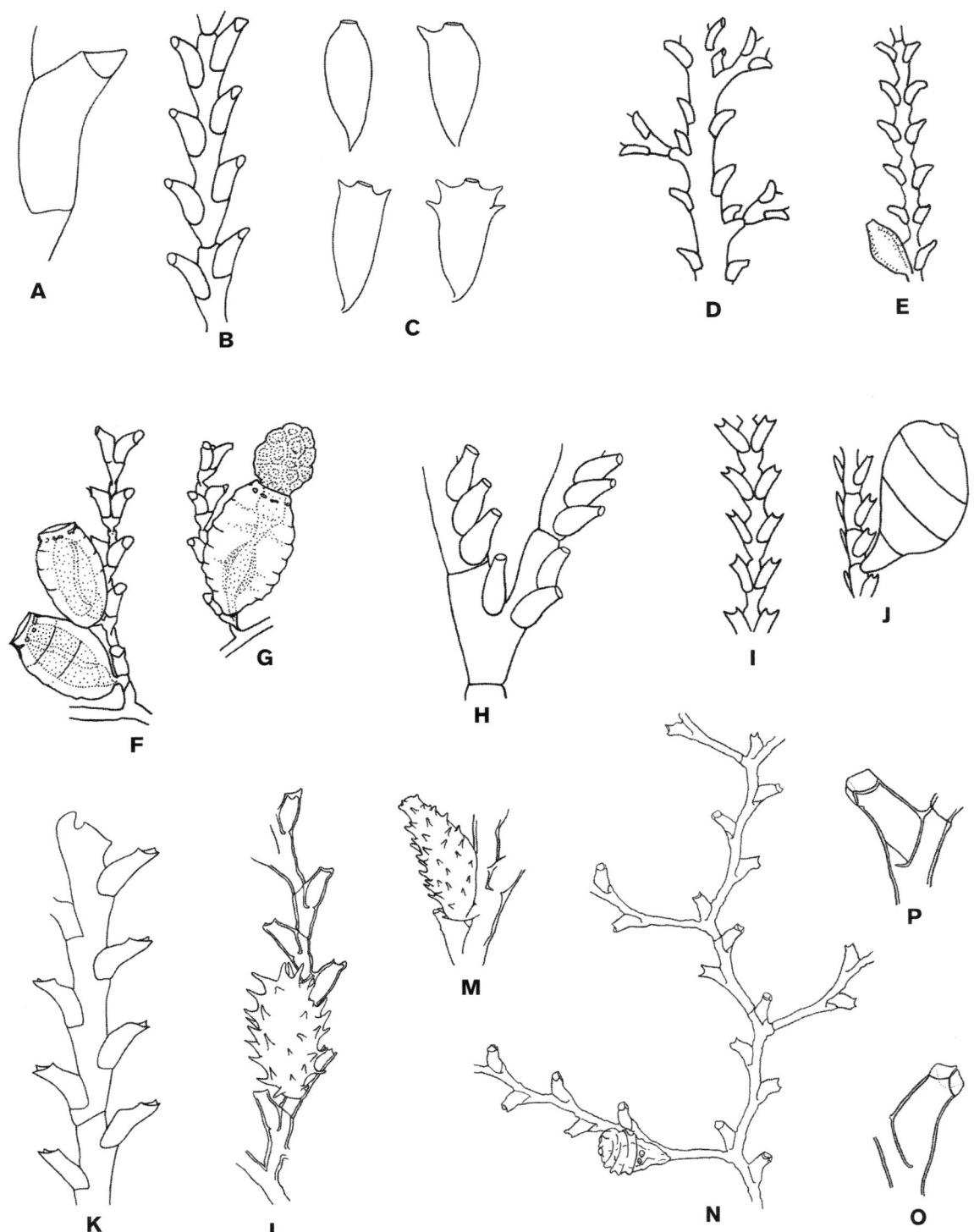

PLATE 47 Thecata (continued). A–C, *Sertularia argentea*; D, E, *Sertularia similis*; F, G, *Salacia desmoides*; H, *Hydrallmania franciscana*; I, J, *Amphisbetia furcata*; K, *Symplectoscyphus erectus*; L, M, *Symplectoscyphus turgidus*; N–P, *Symplectoscyphus tricuspidatus* (A–C, after Cornelius 1995b; F, G, after Millard 1975; L–P, after Hirohito 1995; rest, after Fraser 1937).

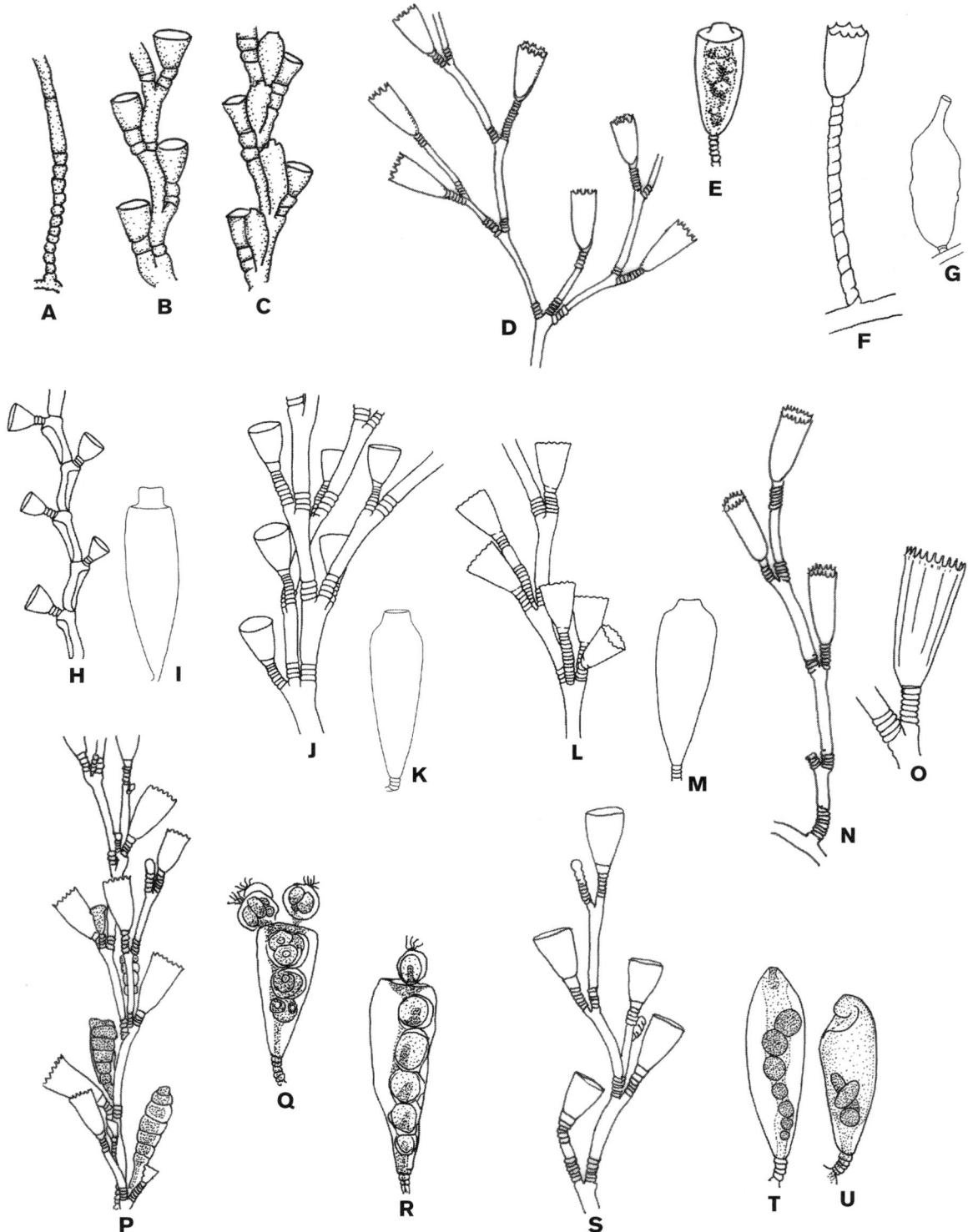

PLATE 48 Thecata (continued). A–C, *Eucheilota bakeri*; D, E, *Hartlaubella gelatinosa*; F, G, *Campanularia volubilis*; H, I, *Obelia geniculata*; J, K, *Obelia dichotoma*; L, M, *Obelia longissima*; N, O, *Obelia bidentata*; P–R, *Gonothyraea loveni*; S–U, *Laomedea calceolifera* (A–E, after Fraser 1937; F–M, after Cornelius 1995b; N, O, after Calder 1991; P–U, after Millard 1975).

older colonies ceasing growth at roughly uniform length, but gradually shorter towards growing tip (plate 48L, 48M) . . . . *Obelia longissima*
— Colonies loosely fan-shaped; stem monosiphonic or polysiphonic basally, pale to medium brown, never black; side branches typically irregular in length (plate 48J, 48K)

. . . . *Obelia dichotoma*
77. Cusps on hydrothecal rim flattened to slightly notched distally; stem monosiphonic (plate 48P–48R). . . . .
. . . . *Gonothyraea loveni*
— Cusps on hydrothecal rim distinctly bimucronate; stem monosiphonic or polysiphonic 78

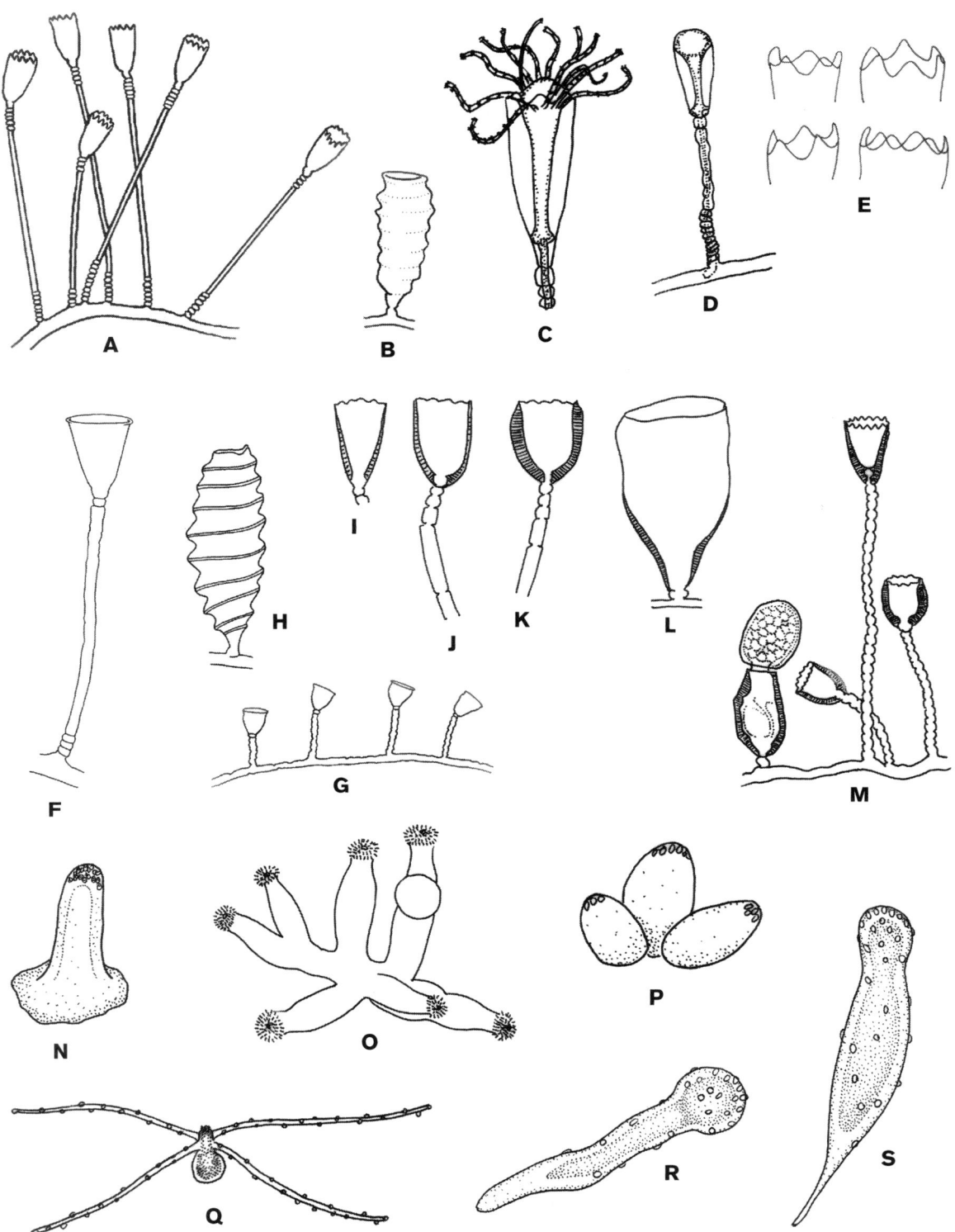

PLATE 49 Thecata (continued), polyps of Limnomedusae, and *Protohydra*. A, B, *Clytia hemisphaerica*; C–E, *Clytia gregaria*; F–H, *Orthopyxis integra*; I–L, *Orthopyxis compressa*; M, *Orthopyxis everta*; N, *Aglauropsis aeora*; O, *Craspedacusta sowerbii*; P, *Maeotias marginata*; Q, *Gonionemus vertens*; R, S, *Protohydra leuckarti* (A, B, F–H, after Cornelius 1995b; C–E, after Strong 1925; I–K, after Torrey 1902 and Naumov 1969; L, M, after Naumov 1969; N, after Mills et al. 1976; O, after Russell 1953; P, after Rees and Gershwin 2000; Q–S, original drawings by Claudia E. Mills).

78. Two points of each bimucronate cusp separated by a U-shaped notch; stem monosiphonic or polysiphonic; gonophores if present release free medusae (plate 48N, 48O) . . . . *Obelia bidentata*
— Two points of each bimucronate cusp separated by a V-shaped notch; stem strongly polysiphonic; gonophores if present fixed sporosacs (plate 48D, 48E) . . . .
. . . . *Hartlaubella gelatinosa*
79. Hydrothecal rims with long, triangular cusps; gonothecae ovate with strong transverse annulations, produce swimming medusae (plate 49A, 49B) . . . .
. . . . *Clytia hemisphaerica*
— Hydrothecal rims even or undulating; gonothecae may or may not produce swimming medusae 80
80. Hydrothecal rims undulating or with low cusps 81
— Hydrothecal rims even 82
81. Hydrothecae fairly cylindrical, with about 10 or more very low, rounded cusps or undulations; gonothecae flask-shaped with narrow neck, smooth, do not produce swimming medusae (plate 48F, 48G) . . . .
. . . . *Campanularia volubilis*
— Hydrothecae bell-shaped, rim variable with as few as five to six low cusps, or up to 12 more-triangular (but fragile and easily broken) cusps; gonothecae ovate and smooth (without transverse annulations), produce swimming medusae (plate 49C–49E) . . . . *Clytia gregaria*
82. Pedicels of hydrothecae wavy or annulated (plate 49M) . . . . *Orthopyxis everta*
— Pedicels of hydrothecae smooth 83
83. Gonothecae decidedly flattened laterally (plate 49I–49L) . . . . *Orthopyxis compressa*
— Gonothecae round to oval in cross section (plate 49F–49H) . . . . *Orthopyxis integra*

Key to the Hydromedusae

CLAUDIA E. MILLS AND JOHN T. REES

Plates refer to hydromedusa plates 50–57.
1. With one or more tentacles and a manubrium (unless the specimen is damaged); size variable, but mature specimens generally >2 mm high 3
— Without tentacles and with or without a manubrium; bell <2 mm high 2
2. With fully developed gonads along each of the four radial canals; without a manubrium (plate 52N–52P) . . . . *Orthopyxis* spp.
— With fully developed gonads covering the lower two-thirds of the manubrium (plate 52I) . . . . *Zanclella bryozoophila*
3. All or most tentacles originating at the margin of the bell (although they may then run through grooves in the jelly in a few cases) 4
— All or most tentacles originating decidedly above the margin of the bell 84
4. With four or fewer tentacles, with or without additional marginal tentacula or rudimentary tentacles 5
— With more than four tentacles, which may be of two types 27
5. With two tentacles, or two pairs of tentacles, situated on opposite sides of the bell (there may also be rudimentary tentacles, cirri, or tentacula elsewhere on the bell margin, but not at the bases of these tentacles) 6
— Arrangement of tentacles not as above 11
6. Bell rounded, <2 mm in diameter 7

— Bell usually pointed, may be conical or sharply pointed (occasionally rounded) and usually >2 mm in diameter 8
7. Bell with numerous zooxanthellae and a row of nematocysts running up the exumbrella above each tentacle bulb; with two opposite pairs of tentacles in mature specimens, or with only 2 tentacles in immature specimens; tentacles each with prominent terminal cnidocyst cluster (rare in the field) (plate 52E, 52F) . . . . *Velella velella*
— Bell without zooxanthellae; with two tentacles; tentacles with many stalked cnidocyst clusters, each attached to the main tentacle by a fine contractile filament (plate 52K) . . . . young medusa of *Zanclea bomala*
8. With a large manubrium on a broad gelatinous peduncle that hangs well below the bell margin; with two long tentacles and numerous (about 80) rudimentary tentacles around the bell margin (plate 52D) . . . . *Stomotoca atra*
— With manubrium hanging directly from the roof of the subumbrella (not on a gelatinous peduncle); with two long tentacles and numerous tentacula or marginal swellings around the bell margin (rare) 9
9. With rudimentary marginal tentacle bulbs at the bases of short tentacula 10
— With rudimentary marginal tentacle bulbs that do not produce tentacles or tentacula; without ocelli (this is the same species as in couplet 38 of the key to polypoid stages) (plate 52G) . . . . *Amphinema* sp.
10. Gonads highly folded, extending from sides of manubrium out along most of the radial canals; with conical tentacle bulbs on the two large tentacles, and with about 14 small pendent solid tentacula, each with a red ocellus (plate 52C) . . . . *Amphinema turrida*
— Gonads covering manubrium smooth or lumpy, but not folded; with broad marginal bulbs on the two large tentacles and 26 small pendent solid tentacula, without ocelli (plate 52H) . . . . *Amphinema platyhedos*
11. With three or four tentacles, each originating from separate tentacle bulbs, one opposite pair may be very reduced, the other opposite pair may each have a pair of associated cirri 12
— With one to four tentacles originating from a single tentacle bulb (medusae are also produced asexually from this bulb); with three other rudimentary tentacle bulbs and five exumbrellar cnidocyst tracks running from the tentacle bulbs toward the apex (plate 52A, 52B) . . . . *Hybocodon prolifer*
12. With three tentacles when mature (but medusae <2 mm high may have only one tentacle and a rounded bell profile; young *E. flammea* medusae have one to three tentacles, but later develop four) (plate 53K) . . . .
. . . . *Euphysa tentaculata*
— With four tentacles (many small medusae with four tentacles are young Anthomedusae or Leptomedusae that will develop more tentacles as they mature; immature stages are usually difficult to identify and most will not be dealt with in this key) 13
13. With marginal vesicles (four or eight) 14
— Without marginal vesicles, but there may be ocelli on the tentacle bulbs 15
14. With four marginal vesicles, without ocelli; with two well-developed opposite tentacles, each with a pair of cirri at the base, with two opposite very small reduced tentacles without cirri, and with a total of only two opposite gonads on the four radial canals (plate 52M) . . . .

PLATE 50 Limnomedusae and Anthomedusae. A, *Aglauropsis aeora*; B, *Eperetmus typus*; C, *Gonionemus vertens*; D, *Maeotias marginata*; E, *Vallentinia adherens*; F, *Halimedusa typus*; G, *Proboscidactyla flavicirrata*; H, *Climacocodon ikarii*; I, *Scrippsia pacifica*; J, *Polyorchis haplus*; K, *Polyorchis penicillatus* (E, David Wrobel; H, Richard Emlet; I, Garry McCarthy; J, Paul Foretic; all with permission; rest, Claudia E. Mills).

. . . . young medusa of *Eucheilota bakeri*
— With eight marginal vesicles, each with a black ocellus, spaced evenly around the bell margin . . . .
. . . . young medusa of *Tiaropsidium kelseyi*
15. Tentacles with many stalked cnidocyst clusters, each attached to the main tentacle by a fine contractile filament; bell with four prominent gelatinous bumps bearing cnidocysts just above the four tentacle bulbs (plate 52L) . . . . *Zanclea bomala*
— Tentacles without stalked cnidocyst clusters and bell with-

out gelatinous cnidocyst-bearing bumps just above tentacle bulbs 16
16. With four tentacles, all alike; bell apex rounded 17
— With one long tentacle armed with round cnidocyst clusters on the subumbrellar side and with three short tentacles; bell with pointed apical projection (plate 53I) . . . . *Euphysora bigelowi*
17. Broad, vase-shaped manubrium covered by four gonads; with uniformly scattered cnidocysts covering the exumbrella even on mature specimens 18

PLATE 51 Leptomedusae, Trachymedusae and Narcomedusae. A, *Clytia gregaria*; B, *Blackfordia virginica*; C, *Melicertum octocostatum*; D, *Eutonina indicans*; E, *Mitrocoma cellularia*; F, *Aglantha digitale*; G, *Aglaura hemistoma*; H, *Aequorea victoria*; I, *Solmaris* sp.; J, *Solmissus marshalli* (D, I, David Wrobel; G, Kevin Raskoff, MBARI; all with permission; rest, Claudia E. Mills).

— Tubular manubrium covered by cylindrical gonad or gonads in mature individuals (or without gonad on young medusae); exumbrella with or without scattered or clusters of cnidocysts on young medusae, but most cnidocysts not remaining on the exumbrellas of mature specimens 19

18. Manubrium attached directly to the subumbrella without a gelatinous peduncle; tentacles with a prominent, elongate, terminal swelling in which cnidocysts are concen-

trated (scattered cnidocysts may also occur in the tentacles, but they are not in swollen clusters); basal tentacle bulbs small to nonexistent and without ocelli, although they have more diffuse reddish-brown pigment; bell to about 5 mm tall (rare) (plate 54D) . . . . *Bythotiara stilbosa*

— Manubium attached to a broad, rounded gelatinous peduncle; tentacles with a round terminal knob of cnidocysts and distinct swollen clusters of cnidocysts along the entire

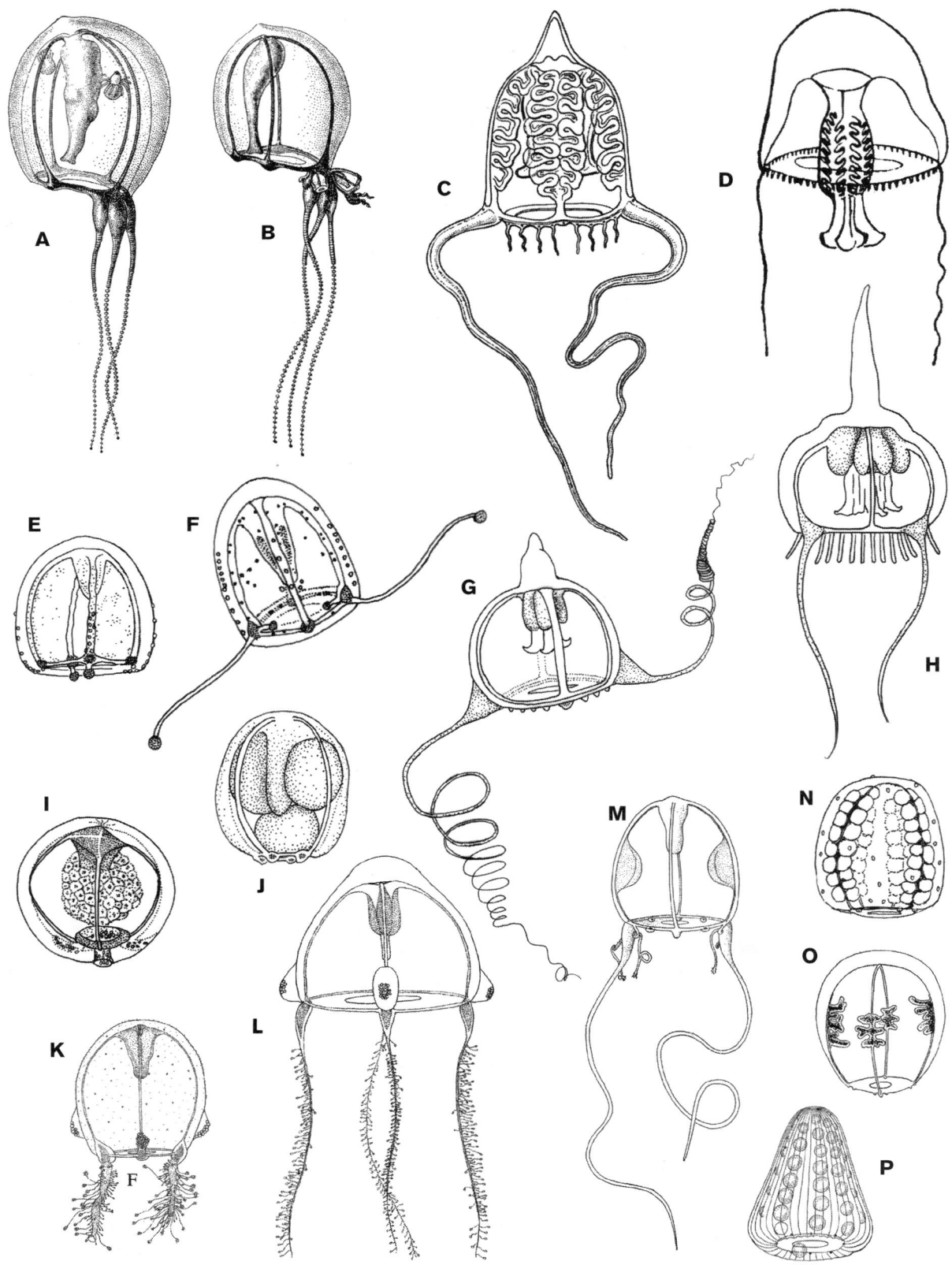

PLATE 52 Anthomedusae with four or fewer tentacles. A, B, *Hybocodon prolifer*; C, *Amphinema turrida*; D, *Stomotoca atra*; E, F, *Velella velella*; G, *Amphinema* sp. from Bodega Harbor; H, *Amphinema platyhedos*; I, *Zanclella bryozoophila*; J, *Pennaria disticha*; K, L, *Zanclea bomala*; M, *Eucheilota bakeri*; N, *Orthopyxis compressa*; O, P, *Orthopyxis integra* (A, B, from Naumov 1960; C, from Kramp 1968, after Mayer; D, from Kramp 1968, after Hartlaub; E, F, after Brinckmann-Voss 1970; G, H, original drawings by C. E. Mills; I, after Boero and Hewitt, 1992; J, after Hirohito, 1988; K, L, from Boero et al. 2000 with permission; M, from Torrey 1909; N, after Miller 1978; O, from Mayer 1910, after Browne; P, from Mayer 1910, after von Lendenfeld).

length; tentacle bulbs with red ocelli; bell to about 2.5 mm tall (rare) (plate 54A, 54B) . . . . *Hydrocoryne bodegensis*

19. With a red or black ocellus on each tentacle bulb; manubrium of variable length 20
— Without ocelli, although tentacle bulbs may be red; manubrium not reaching the level of the margin of the bell and often with bright red pigment (plate 53K, 53L) . . . . *Euphysa* spp.

20. With 24 small, well-defined clusters of cnidocysts on the exumbrella (with six clusters per quadrant, in three rows of two, spaced approximately evenly over bell surface) (plate 53F) . . . . . . young medusa of *Polyorchis penicillatus*
— With scattered cnidocysts on the exumbrella, or other than six distinct clusters per quadrant, or without cnidocysts on the exumbrella 21

21. Manubrium not longer than the bell cavity in mature medusae 22
— Manubrium hanging well below the base of the bell in mature medusae 24

22. Exumbrella without bubblelike vesicles 23
— Exumbrella with numerous, closely set bubblelike vesicles (about 10 μm diameter, and 20–70 μm apart); bell to 3 mm high and slightly less wide with jelly substantially thicker at the apex; manubrium with short, rounded apical chamber protruding into the mesogloea above the subumbrella; egg size 70 μm (plate 53E) . . . . . *Coryne* sp. (undescribed)

23. Bell 2–3 mm high and slightly less wide with jelly of nearly uniform thickness; manubrium without pointed apical chamber protruding into the mesogloea above the subumbrella; egg size 180–200 μm (plate 53H) . . . . *Coryne eximia*
— Bell 3–6 mm high and slightly less wide with jelly substantially thicker at the apex; manubrium with or without pointed apical chamber protruding into the mesogloea above the subumbrella; egg size 90–120 μm (plate 53A) . . . . *Coryne japonica*

24. Gonad a single continuous cylinder, covering nearly all of the manubrium 25
— Gonad divided into two or more separate cylinders around the manubrium 26

25. Bell 6–10 mm high, rounded, higher than wide, with jelly of nearly uniform thickness and with four interradial exumbrellar furrows beginning at the bell margin; apex of subumbrella rounded, not pointed; manubrium usually with short, rounded apical chamber protruding into the mesogloea above the subumbrella; manubrium long, two to three times the bell height, with a short gonad-free portion at the apex and the stomach also free of gonad; stomach swollen and spindle-shaped and covered at upper end with warts with cnidocysts (plate 53D) . . . . *Sarsia tubulosa*
— Not as above (includes plate 53B, 53C) . . . . *Sarsia* spp. or *Coryne* spp.

26. Bell 1.6–2.1 mm high and wide; manubrium about twice as long as bell height with two gonad rings and a bullet-shaped apical chamber protruding into the mesogloea above the subumbrella; tentacles short, only about as long as the bell is tall (plate 53G) . . . . *Dipurena bicircella*
— Bell 2–6 mm high and wide; manubrium about twice as long as bell height with two to four gonad rings and a bullet-shaped apical chamber protruding into the mesogloea above the subumbrella; tentacles much longer than the bell height (plate 53J) . . . . *Dipurena reesi* and *Dipurena ophiogaster*

27. Tentacles branched (medusae small, <4 mm high). 28
— Tentacles not branched 31

28. Tentacles (usually nine) branched once or twice, one branch ending in a sucker, the other one or two branches having swollen clusters of cnidocysts at the tip and along the length; medusa with nine unbranched radial canals (polyps with two whorls of four tentacles each, the upper ones capitate and the lower ones filiform, but may be difficult to see) (plate 54H, 54I) . . . . *Cladonema californicum*
— Tentacles (five to eleven) branched several times, the lower one to four branches ending in a sucker, the upper four to seven branches bearing swollen clusters of cnidocysts at the tip and along the length (polyps with either two whorls of tentacles or with only a single whorl of capitate tentacles) 29

29. With usually seven (rarely, five or six) unbranched radial canals and the same number of branched tentacles (polyps with only a single whorl of [usually four] capitate tentacles and without filiform tentacles) (plate 54G) . . . . *Cladonema myersi*
— With at least some dichotomously branched radial canals, resulting in five to 11 radial canals reaching the bell margin and the same number of branched tentacles 30

30. With usually five (sometimes four to seven) radial canals emanating from the manubrium, most of which branch dichotomously so that five to 11 reach the bell margin; usually with 10 branched tentacles, corresponding to the number of radial canals (polyps with two whorls of [usually four] tentacles each, the upper ones capitate and the lower ones filiform [but may be difficult to see]) (plate 54F) . . . . *Cladonema radiatum*
— With usually six radial canals emanating from the manubrium, with every other one branching dichotomously so that nine radial canals reach the bell margin, and with nine branched tentacles (polyps with only a single whorl of [usually four] capitate tentacles and without filiform tentacles) . . . . *Cladonema pacificum*

31. Manubrium suspended far below the bell margin from a long, slender, gelatinous peduncle 32
— Manubrium not suspended far below the bell margin, from a long, slender, gelatinous peduncle 33

32. With four broad radial canals and four gonads on the radial canals; with four long tentacles with cnidocyst rings, each at the end of a radial canal and four small solid tentacles between these; with eight marginal marginal vesicles (plate 57N) . . . . *Liriope tetraphylla*
— With six broad radial canals and six gonads on the radial canals; with six long tentacles with cnidocyst rings, each at the end of a radial canal and six small solid tentacles between these; with 12 marginal marginal vesicles (plate 57O) . . . . *Geryonia proboscidalis*

33. Radial canals branching, with all branches reaching the bell margin 34
— Radial canals not branching (they may, however, have numerous lateral diverticula) 38

34. Bell to 10 mm tall and as wide or a bit wider, with jelly thickest at apex, bell diameter usually broadening toward the margin; manubrium with four paired gonads, with or without irregular folds; with four radial canals that give rise to lateral branches at several levels so that about 20–70 canals reach the ring canal; with about 20–70 tentacles with swollen marginal basal bulbs, but without large terminal knob of cnidocysts; exumbrella with small clusters

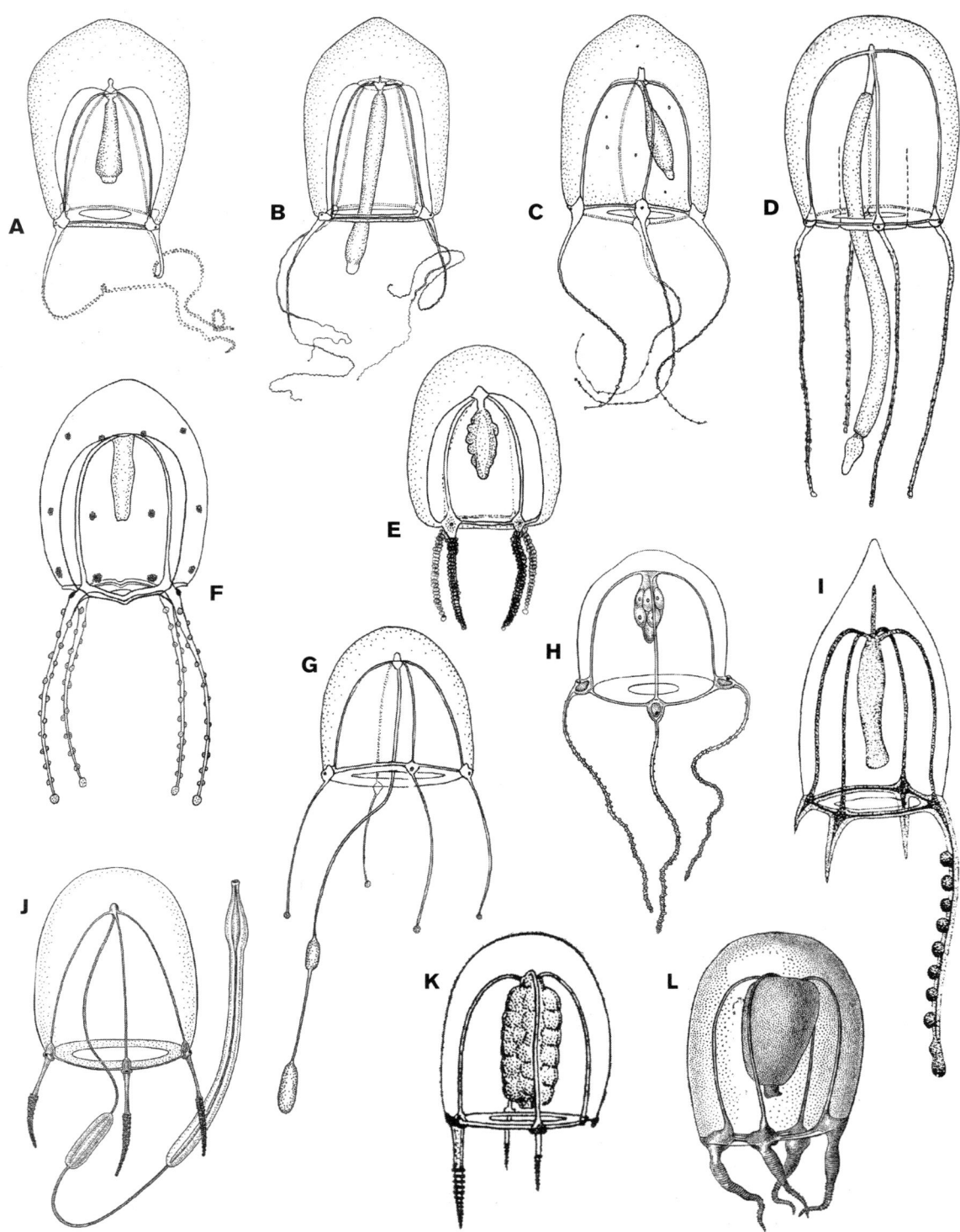

PLATE 53 Anthomedusae with three or four tentacles. A, *Coryne japonica*; B, *Sarsia* or *Coryne* sp. A from Bodega Harbor; C, *Coryne* sp. B from Bodega Harbor; D, *Sarsia tubulosa*; E, *Coryne* sp. (undescribed) from San Francisco Bay; F, *Polyorchis penicillatus*, young juvenile; G, *Dipurena bicircella*; H, *Coryne eximia*; I, *Euphysora bigelowi*; J, *Dipurena ophiogaster*; K, *Euphysa tentaculata*; L, *Euphysa flammea* (indistinguishable from mature *E. japonica* medusa) (A–C, after Rees 1975; D, after Schuchert 2001; E, F, original drawings by Claudia E. Mills; G, after Rees 1977; H, J, from Mayer 1910; I, from Sassaman and Rees 1978, Biol. Bull. fig 1c, p. 488, reprinted with permission from the Marine Biological Laboratory, Woods Hole, MA; K, from Naumov 1960, after Kramp; L, from Naumov, 1960).

of cnidocysts called cnidothalacies between the tentacles, on tracks, varying distances above the bell margin 35

— Bell to 30–40 mm tall and a little less wide, broadly rounded with jelly of fairly uniform thickness; manubrium with eight laterally folded gonads; with four radial canals that give rise to lateral branches at several levels so that about 40 canals reach the ring canal; with about 20–50 tentacles without marginal basal bulbs, but each terminating distally in a large knob of cnidocysts; exumbrella without clusters of cnidocysts above the bell margin. 37

35. Manubrium with barely folded, recurved lips; gonads comprise four swollen, but not folded, paired masses covering the manubrium; bell to 2.5 mm high and a little wider, apical jelly to about one-fourth bell height; with up to 32 tentacles when mature (central California) (plate 55B) . . . . *Proboscidactyla circumsabella*

— Manubrium with highly folded lips; gonads comprise four irregularly folded paired masses covering the manubrium; bell to 10 mm high and about as wide, but may be much less, with apical jelly accounting for about one-half bell height; with 40–72 tentacles when mature. 36

36. Bell to 10 mm high and a little wider, usually considerably wider at the base than at the apex, but sometimes nearly hemispherical; with 50–70 radial canals reaching the margin and about the same number or more of tentacles in fully grown medusae (Japan to central Oregon) (plates 50G, 55A) . . . . *Proboscidactyla flavicirrata*

— Bell to 3.5 mm high and wide, often tending to appear taller than a hemisphere and uniformly rounded; with up to 40 radial canals reaching the margin and about the same number of tentacles (southern California) (plate 55C) . . . . *Proboscidactyla occidentalis*

37. With 20–40 tentacles all of one type; manubrium with lightly folded lips that are not edged with cnidocyst clusters (rare) (plate 55J) . . . . *Sibogita geometrica*

— With four or more large tentacles and up to 45 small tentacles; manubrium with much-folded lips edged with a row of prominent cnidocyst clusters (rare) (plate 55M) . . . . *Calycopsis nematophora*

38. With four or eight radial canals (may have additional centripetal canals originating from the ring canal, see couplet 56) 40

— With numerous radial canals 39

39. Bell to about 10 cm wide, with up to 100 (or more) symmetrical radial canals (in mature specimens, all radial canals reach the bell margin, and gonads extend along nearly the entire length of the radial canals); with approximately as many tentacles as radial canals (plate 51H) . . . . *Aequorea victoria*

— Bell to 25 cm wide, with up to 100 (or more) symmetrical radial canals (in mature specimens, all radial canals reach the bell margin, and gonads extend along nearly the entire length of the radial canals); with three to six times as many tentacles as radial canals (uncommon) . . . . *Aequorea ?coerulescens*

40. With four radial canals (may have additional centripetal canals originating from the ring canal, see couplet 55) 43

— With 8 radial canals 41

41. Tubular or globular gonads attached to the radial canals only at the apex of the subumbrella or on the peduncle 42

— Sinuous gonads attached to most of the length of the

radial canals beginning some distance below the apex of the subumbrella; with up to 90 large tentacles alternating with as many small ones; without any marginal sense organs; bell transparent, gonads orangish (plate 51C) . . . . *Melicertum octocostatum*

42. Bell to about 10–40 mm tall and about half as wide; manubrium suspended on a long peduncle; tubular gonads attached to the eight radial canals at the apex of the subumbrella near the base of the peduncle (plate 51F) . . . . *Aglantha digitale*

— Bell to about 6 mm tall and about half as wide; manubrium suspended on a long peduncle; sausage-shaped gonads attached to the eight radial canals near the midpoint of the peduncle (plate 51G) . . . . *Aglaura hemistoma*

43. Tentacles evenly distributed around the bell margin, although some may exit the jelly some distance above the base of the bell 47

— Most tentacles arranged in clusters, although some single tentacles may also be present 44

44. With a cluster of several tentacles in line with each radial canal and with or without additional clusters of tentacles between these; manubrium with four oral tentacles, which are either simple or much branched; gonad without dark horizonal line 45

— With a single marginal tentacle in line with each radial canal, and with an interradial cluster of tentacles (these arising from separate bulbs) in each quadrant; manubrium with simple lips and without oral tentacles of any kind; tentacle bulbs with red or black ocelli; gonad with distinctive dark horizontal line near the midpoint (plates 50F, 54C) . . . . *Halimedusa typus*

45. With eight clusters of tentacles on the bell margin originating from eight tentacle bulbs (clusters with three and five tentacles, or one and three tentacles, usually alternate); with four short oral tentacles (these may be branched) at the corners of the mouth; may be budding small medusae from the manubrium walls (plate 55F) . . . . *Rathkea octopunctata*

— With four clusters of tentacles on the bell margin originating from four broad tentacle bulbs; with a simple tubular mouth and four dichotomously branched oral tentacles inserted above the mouth opening; never budding medusae from the manubrium walls. 46

46. Oral tentacles short, branching one or two times; with up to nine marginal tentacles originating from each of four bulbs; black ocelli on the tentacle bulbs; <4 mm high at maturity (plate 55D, 55G) . . . . *Bougainvillia muscus*

— Not as above, but with several times branched oral tentacles, with numerous marginal tentacles originating from each of four bulbs; with ocelli on the tentacle bulbs or on the bases of tentacles (includes plate 55E) . . . . *Bougainvillia* spp.

47. Gonads either covering the manubrium, or as four clusters of numerous pendant tubes associated with the four radial canals, but not running along the lengths of the radial canals 48

— Gonads linear or globular, attached to and running along the four radial canals (and may *also* cover the manubrium) 62

48. Gonads as four clusters of numerous pendant tubes associated with the radial canals. 49

— Gonads covering the manubrium 51

49. Manubrium suspended from a short, rounded, or conical gelatinous peduncle extending less than one-fourth the length of the subumbrella; numerous tubular, fingerlike

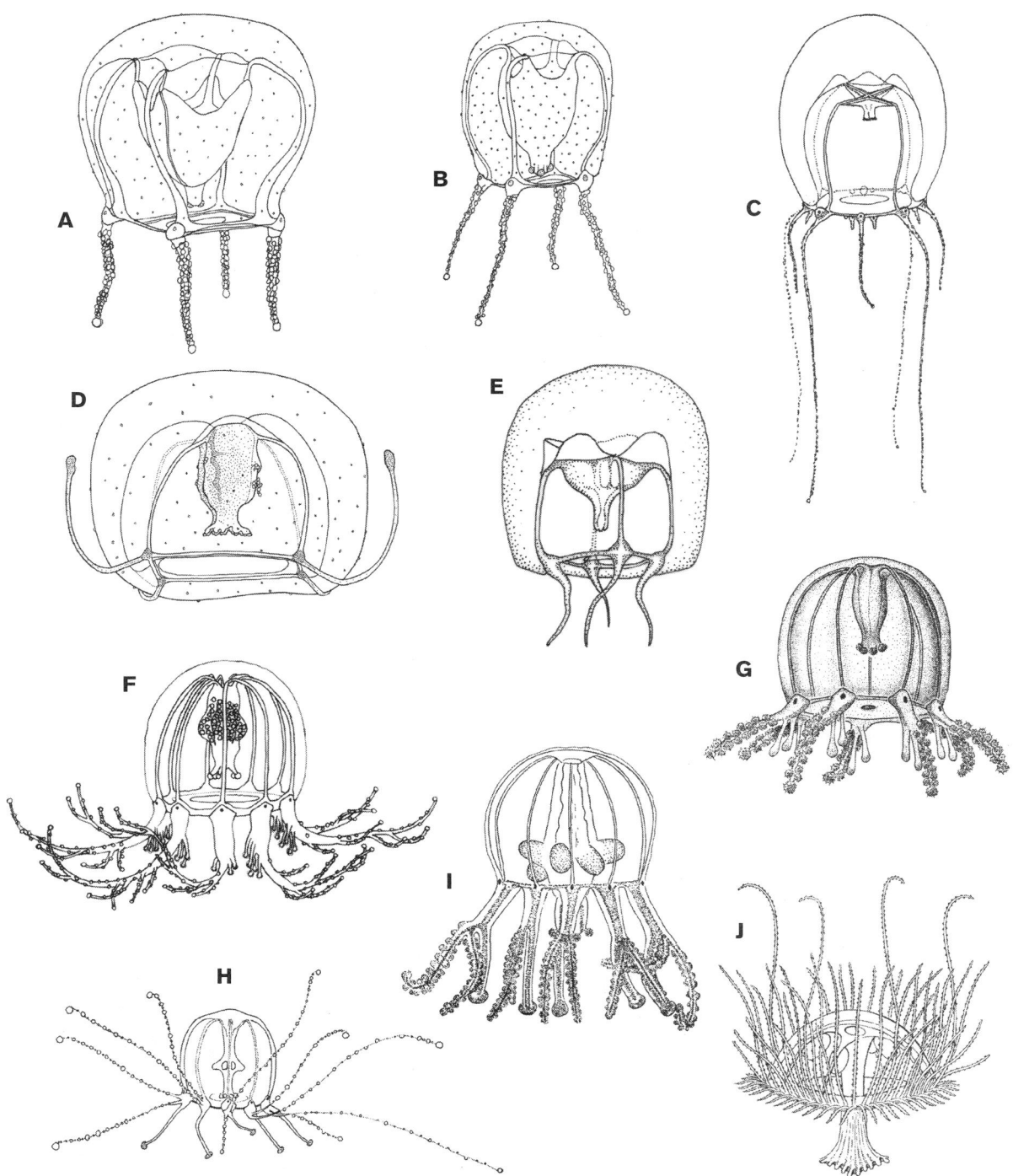

PLATE 54 Anthomedusae with four or more tentacles and one Limnomedusa. A, B (juvenile), *Hydrocoryne bodegensis*; C, *Halimedusa typus*, juvenile; D, *Bythotiara stilbosa*; E, *Protiara* sp.; F, *Cladonema radiatum*; G, *Cladonema myersi*; H, I, *Cladonema californicum*; J, *Craspedacusta sowerbii* (A, B, from Rees, Hand, and Mills 1976; C, from Mills 2000; D, from Mills and Rees 1979; E, modified from unpublished drawing by Ronald Larson, with permission; F, after Hirohito 1985; G, from Rees 1949 with permission; H, from Rees 1979; I, from Hyman 1947 with permission; J, from Mayer 1910, after Allman; A–E, drawings all by Claudia E. Mills; H; drawing by John T. Rees).

gonads in four groups, attached to the radial canals running up the peduncle to the bell apex, hanging down into the subumbrellar cavity; tentacles nearly all in a single row around the bell margin 50
— Manubrium suspended from a large, conical gelatinous peduncle extending about half the length of the subumbrella; numerous tubular, fingerlike gonads in four groups, attached to the radial canals along the surface of the gelatinous pe-

duncle below the apex of the subumbrella; some tentacles exiting from the jelly well above the bell margin (plate 50I) . . . . *Scrippsia pacifica*
50. With up to 160 tentacles and 60 mm bell height; radial canals with 15–25 pairs of short lateral diverticula that are longer than twice the width of the radial canal (plate 50K) . . . . *Polyorchis penicillatus*
— With up to 30 tentacles and 20 mm bell height; radial

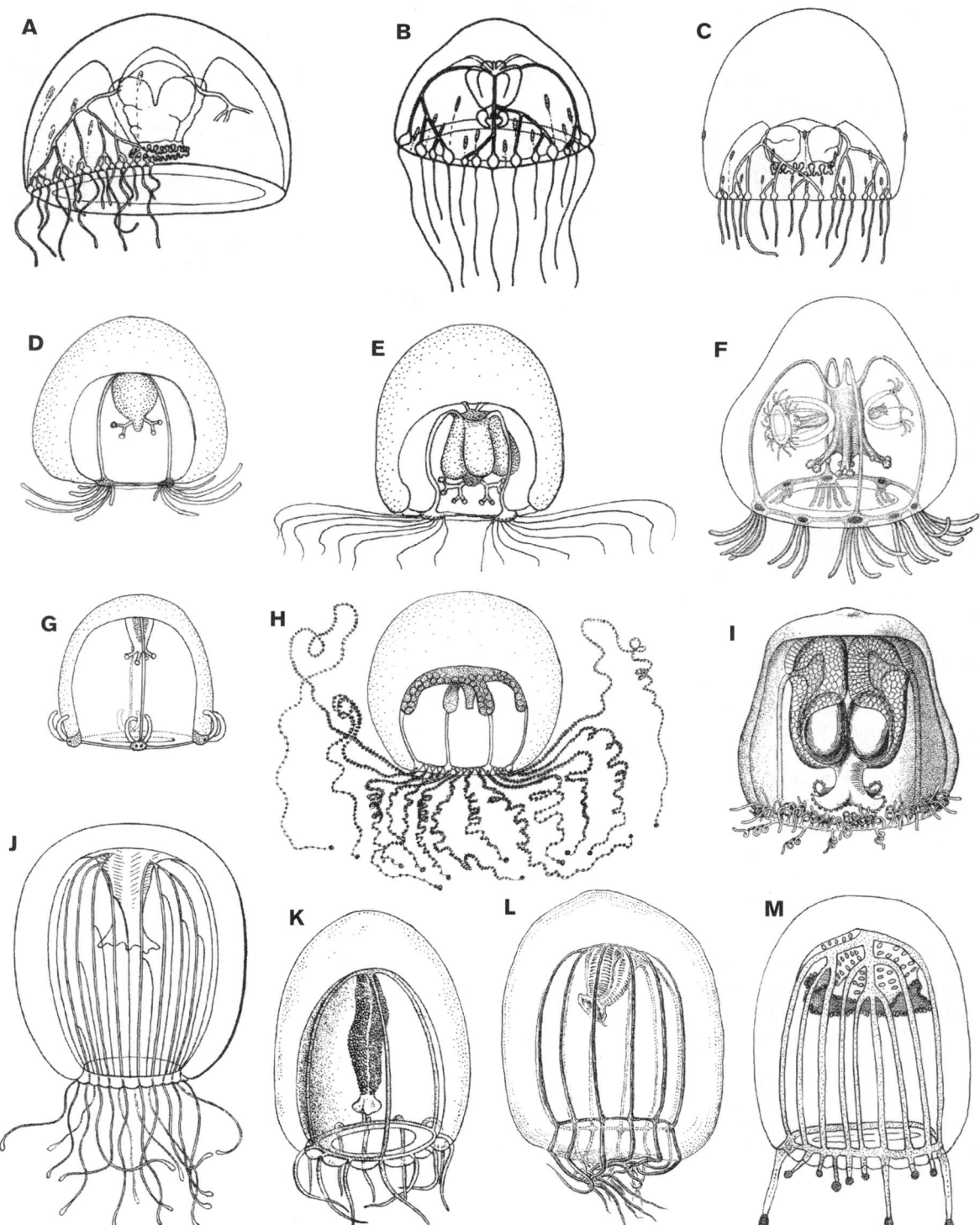

PLATE 55 Anthomedusae with more than four tentacles. A, *Proboscidactyla flavicirrata*; B, *Proboscidactyla circumsabella*; C, *Proboscidactyla occidentalis*; D, *Bougainvillia muscus*; E, *Bougainvillia* sp. from Bodega Harbor; F, *Rathkea octopunctata*; G, *Bougainvillia muscus* juvenile; H, *Moerisia* sp. from San Francisco Bay; I, *Turritopsis* sp.; J, *Sibogita geometrica*; K, *Heterotiara anonyma*; L, *Calycopsis simulans*; M, *Calycopsis nematophora* (A–C, from Hand 1954 with permission; F, from Naumov 1960; I, from Mayer 1910, after Brooks; J, from Mayer 1910; K, L, from Kramp 1968, after Bigelow; M, after Arai and Brinckmann-Voss 1980; D, E, G, H, original drawings by Claudia E. Mills).

canals largely without diverticula, although large specimens may have closely set knoblike branches on the radial canals (rare) (plate 50J) . . . . *Polyorchis haplus*

51. With more than 60 marginal tentacles, densely packed
. . . . 52

— With <60 marginal tentacles 54

52. Manubrium with mass of vacuolated cells above the digestive part, the lips lined with a single row of cnidocyst knobs; gonads on manubrium, without folds or pits, red in life, without "mesentery" connections to radial canals (see

below); bell to 7 mm high and less wide; jelly uniformly thin; with about 80–120 marginal tentacles, each with an ocellus on the inner surface of the basal bulbs (plate 55I)........ .... *Turritopsis* sp.

— Manubrium without masses of vacuolated cells above the digestive part and lips not edged with cnidocyst knobs; gonads on manubrium with peripheral folds and also a central pitted region, attached to radial canals by "mesentery" tissue connection on upper third to half of the manubrium; without ocelli 53

53. Bell to 45 mm high and less wide; jelly thick at the apex; with more than 100 tentacles, on laterally compressed marginal bulbs; radial canals broad, with jagged edges; gonads, manubrium and lips orange to rosy red in life (plate 56E) .... *Neoturris breviconis*

— Bell to 80 mm high and 65 mm wide, jelly slightly thicker at apex, with or without a small pointed apical process; with about 80 tentacles on laterally compressed marginal bulbs; radial canals broad with about 10 broad, tablike diverticula on each edge; colorless in life (plate 56F)...... .... *Neoturris* sp. (undescribed)

54. Bell with mesogloea of nearly uniform thickness, or somewhat thicker at the apex, but the apex is not pinched off from the lower portion of the bell or distinctly pointed 55

— Bell with an apical projection of mesogloea that is noticeably pinched off from the lower portion of the bell, or conically pointed................................... 57

55. With up to 44, highly contractile, tapering, filamentous tentacles, these with conical marginal bulbs, each with an ocellus, but without any specialized structures at the tips; manubrium broad and extending out onto radial canals, with vertically or obliquely folded gonads; bell to 23 mm high and wide (rare) (plate 56D)............. .... *Annatiara affinis*

— With up to about 12–15 thick, not very contractile tentacles, these without marginal bulbs, but with distinctive terminal bulbs or tapering tip structures; without ocelli; with cylindrical manubrium........................ 56

56. With four radial canals and 12 or more centripetal canals arising at the ring canal, most of which meet the manubrium, but not at the corners; with 12 or more large tentacles and a few small ones, each with a distinctive tapering tip; bell to 40 mm high and a little less wide, with deep brick red gonads on the manubrium (rare) (plate 55L) .... *Calycopsis simulans*

— With four radial canals and no centripetal canals; with six to 12 tentacles, each with a swollen knob of cnidocysts at the tip; bell to 22 mm high and a little less wide, with bright orange-red tentacle tips and yellowish manubrium in life (rare) (plate 55K) .... *Heterotiara anonyma*

57. Upper one-third or more of the manubrium attached to the four radial canals by thin tissue "mesenteries"; radial canals broad and often with jagged edges; with or without rudimentary tentacles along with the normal tentacles at the bell margin 58

— Manubrium not attached to the radial canals by mesenteries, although the top of manubrium may be somewhat elongated along the proximal portions of the radial canals; radial canals not broad, with mostly smooth edges; with four to 16 tentacles of varying sizes (these with ocelli), and with rudimentary marginal bulbs (with ocelli) between tentacles; with bulbous or conical apical projection (includes plate 56C).................... *Halitholus* spp.

58. Gonads (four) horseshoe-shaped with folds directed towards the radial canals 59

— Gonads (four) with folds and/or papillae and also a central pitted region, or only with pits or reticulations throughout 60

59. Bell to 20 mm high with conical or spherical apical projection, manubrium broad with gonads (red) on the whole surface; with 12–24 (usually 16) tentacles, each with a pronounced elongate spur directed upwards a short distance on the exumbrella from its tentacle bulb, and with 16 or more rudimentary marginal bulbs or tentacles; tentacle bulbs and rudiments with red ocelli; radial canals with smooth or slightly jagged edges (plate 56B)........ .... *Leuckartiara octona*

— Not as above (includes plate 56A) ...... *Leuckartiara* spp.

60. With ocelli; manubrium filling less than one-half of the subumbrellar space, covered with gonads without folds, but forming a complex, reticulated pitted surface; bell to 21 mm high, tall with a conical apex and with longitudinal ribs and ridges; radial canals fairly narrow and smooth-walled, with mesenteries connected to most of the manubrium; with 12–24 tentacles with laterally compressed marginal bulbs (plate 56J) .... *Pandea conica*

— Without ocelli; manubrium large, nearly filling the subumbrellar space and covered by gonads with folds facing the radial canals; bell to about 20–25 mm high, but otherwise not as above.......................... 61

61. Bell barrel-shaped, with bulbous, pointed apical projection; gonads irregularly transversely folded with numerous papillae and with a central pitted area; with 30–32 tentacles, with large, laterally compressed bulbs, these without exumbrellar spurs or pores; radial canals broad, with smooth or somewhat jagged margins (plate 56I) .... *Neoturris pelagica*

— Bell-shaped, with small apical projection over thickened apical jelly; gonads on manubrium with many transverse folds; with 16–20 large tentacles of varying sizes with laterally compressed bulbs, each with a distinct pore and spur pointed upward on exumbrella, and with about 40 rudimentary tentacles or bulbs; radial canals broad, with glandular diverticula throughout their lengths (plate 56H) .... *Neoturris fontata*

62. Tentacles with prominent rings of cnidocysts along their entire length 63

— Tentacles without prominent rings of cnidocysts along their entire lengths 71

63. Medusa bell-shaped, nearly hemispherical or wider; bell of mature specimens usually >6 mm in diameter....... 64

— Medusa nearly disk-shaped, not deeply convex; bell of mature specimens not more than 6 mm in diameter 71

64. Tentacles with adhesive discs, allowing the medusae to adhere to algae; medusae orange, brownish orange, or reddish 65

— Tentacles without adhesive discs; medusae pale pink, blue, greenish, tan, or colorless...................... 66

65. With about 60–80 tentacles, each with an adhesive pad located about midway, and characteristically angled where the pad is located; bell to about 25 mm in diameter (plate 50C)........................... *Gonionemus vertens*

— With up to about 40 tentacles, most with a terminal adhesive disk; bell to about 8 mm in diameter (plate 50E) .... *Vallentinia adherens*

66. With (microscopic) marginal vesicles at the bell margin, manubrium with four distinct lips................. 67

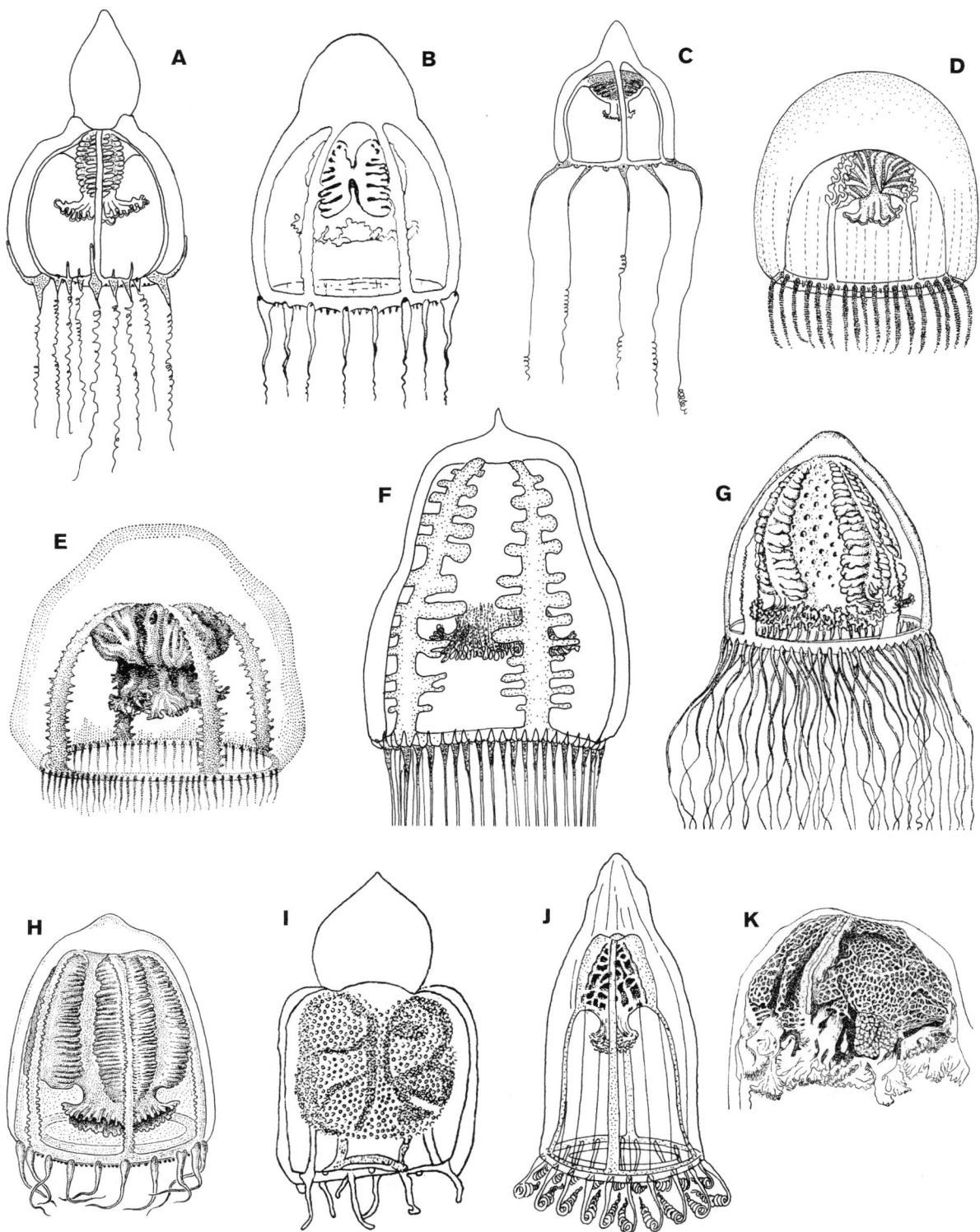

PLATE 56 Anthomedusae with many tentacles (all family Pandeidae). A, *Leuckartiara* sp. from Monterey Bay; B, *Leuckartiara octona*;
C, *Halitholus sp.* from Friday Harbor; D, *Annatiara affinis*; E, *Neoturris breviconis*; F, *Neoturris* sp. (undescribed); G, *Neoturris pileata*;
H, *Neoturris fontata*; I, *Neoturris pelagica*; J, *Pandea conica*; K, *Pandea rubra* manubrium (A, C, D, F, original drawings by Claudia. E. Mills;
B, G, H, I, from Kramp 1968, each after Russell, Hartlaub, Bigelow, and Foerster, respectively, with permission; E, from Naumov 1960;
J, from Kramp 1968 with permission; K, from Naumov 1969 after Bigelow).

— Without marginal vesicles, manubrium tubular without distinct lips; with cruciform gonad covering the manubrium and extending out about one-half the length of the radial canals, with very thick jelly, bell to about 8 mm in height; in low salinity (plate 55H) . . . . *Moerisia* sp.

67. With eight marginal vesicles; with up to 16 or 36 marginal tentacles. 76

— With numerous marginal vesicles, usually one or two between every pair of tentacles; with hundreds of tentacles, some of which may emerge through grooves in the jelly some distance above the bell margin 68

68. Marine, although may be up rivers, in very low salinity 69

— In fresh water; with up to 400 tentacles in several series; mouth with four slightly folded short lips; bell to about 20 mm in diameter (plate 54J) . . . . *Craspedacusta sowerbii*

69. With up to 200 tentacles, ring canal with or without centripetal canals extending upward in each quadrant; gonads hang as a wavy curtain from, and restricted to, the four radial canals; mouth with four frilly, short lips; in full salinity on outer coasts. 70

— With up to 600 tentacles, ring canal with several centripetal canals extending upward in each quadrant; gonads hang curtainlike from the four radial canals, extending onto "arms" of the manubrium out along each radial canal; mouth with four very elongate, frilly lips; bell to about 55 mm diameter; in low salinity, often up coastal rivers (plate 50D) . . . . *Maeotias marginata*

70. With up to 200 tentacles, ring canal without centripetal canals extending upward in each quadrant; bell to about 20 mm diameter (plate 50A) . . . . *Aglauropsis aeora*

— With up to 100 tentacles, ring canal with up to six broad centripetal canals extending upward in each quadrant; bell to about 45 mm diameter (rare) (plate 50B) . . . . *Eperetmus typus*

71. Gonads ovoid or nearly spherical, occupying only a short section of each radial canal; tentacles more or less of fixed length (not highly extensile); bell not >6 mm in diameter in mature specimens and very flat (plate 38) . . . . *Obelia* spp.

— Gonads elongated, associated with the radial canals for a substantial portion of their length; tentacles highly extensile; bell usually more than 6 mm in diameter in mature specimens and bell-shaped, not flat. 72

72. With ocelli on the tentacle bulbs or on the marginal vesicles. 73

— Without ocelli 74

73. With ocelli on some or all of the tentacle bulbs; without marginal vesicles, but with short marginal clubs; <40 mm bell diameter (rare). . . . . *Laodicea* sp.

— Without ocelli on the tentacle bulbs, but with a black ocellus on each of the eight marginal vesicles; with eight to 16 large tentacles and up to 128 additional short or rudimentary tentacles; lips of mouth slightly frilled; large—to 80 mm bell diameter (rare) (plate 57C) . . . . *Tiaropsidium kelseyi*

74. With manubrium directly attached to the subumbrella 75

— With manubrium suspended on a gelatinous peduncle 82

75. With only eight marginal vesicles around the bell margin; tentacles moniliform, ending in a distinct terminal cnidocyst cluster. 76

— With 16 or more marginal vesicles or marginal clubs

(cordyli) around the bell margin; tentacles filiform, without a distinct terminal cnidocyst cluster 77

76. With 16 tentacles, gonads linear, attached to lower one-third length of radial canals; mouth lips upturned and smooth, not frilly (plate 57H) . . . . *Phialella fragilis*

— With 36 tentacles, gonads short and rather hemispherical, in the middle of the radial canals; mouth lips upturned and frilly (plate 57G) . . . . *Phialella zappai*

77. With marginal vesicles (containing statoliths) at the bell margin 78

— With free-hanging marginal clubs (cordyli) at the bell margin 81

78. With up to 80 tentacles; with one to two marginal vesicles between each two tentacles, each vesicle with one to three statoliths; lips of mouth relatively short; bell diameter <30 mm 79

— With up to 350 tentacles; with a total of 16–24 marginal vesicles, each with numerous statoliths; lips of mouth long and extended; bell diameter to 90 mm (plate 51E) . . . . *Mitrocoma cellularia*

79. Tentacle bulbs without fingerlike extensions into the bell margin, marginal vesicles each with a single statolith 80

— Tentacle bulbs each with a distinctive fingerlike extension pointing inward toward the subumbrella, marginal vesicles each with two to three statoliths (estuarine) (plate 51B) . . . . *Blackfordia virginica*

80. Gonads not usually mature until diameter of bell is >1.5 cm; gonads, when sectioned transversely, with an elliptical outline; gonads usually light-colored, but may have a stripe of dark pigment running lengthwise; bell margin may have a ring of dark pigment; up to 20 tentacles per quadrant at maturity (plates 51A, 57A) . . . . *Clytia gregaria*

— Gonads mature by the time the bell reaches a diameter of 1 cm; gonads, when sectioned transversely, with a circular outline; gonads usually fairly dark in color (brown, gray, or yellowish); bell margin without a ring of dark pigment; <10 tentacles per quadrant at maturity (plate 57B) . . . . *Clytia lomae*

81. Bell to 10 mm wide and nearly as high; with about 48 tentacles and one to five cordyli between tentacles; gonads with 12–14 folds in proximal half of radial canals (rare) (plate 57I) . . . . *Ptychogena californica*

— Bell to 90 mm wide and 30 mm tall; with 300–500 tentacles and as many cordyli; gonads on 20–30 lamelliform diverticula of radial canals along their entire length (rare) (plate 57F). . . . . *Ptychogena lactea*

82. With eight marginal vesicles; with up to 200 marginal tentacles; (colorless or with white or sepia pigment in the manubrium, gonads, and tentacle bases); conical peduncle bearing manubrium may extend nearly to the bell margin (plate 51D, 57D) . . . . *Eutonina indicans*

— With many more than eight marginal vesicles; usually with <200 marginal tentacles; peduncle low, with manubrium not reaching to bell margin 83

83. With up to 180 marginal tentacles and with as many or more marginal vesicles; manubrium, gonads, and radial canals faintly yellow, tentacle bulbs brick red in life (rare) (plate 57E) . . . . *Eirene mollis*

— With up to 150 marginal tentacles and about half as many marginal vesicles; manubrium and radial canals (and sometimes gonads) purple (rare) (plate 57J) . . . . *Foersteria purpurea*

84. Bell about 1 mm tall and a little less wide, with very thin jelly; with four radial canals and with several pairs of

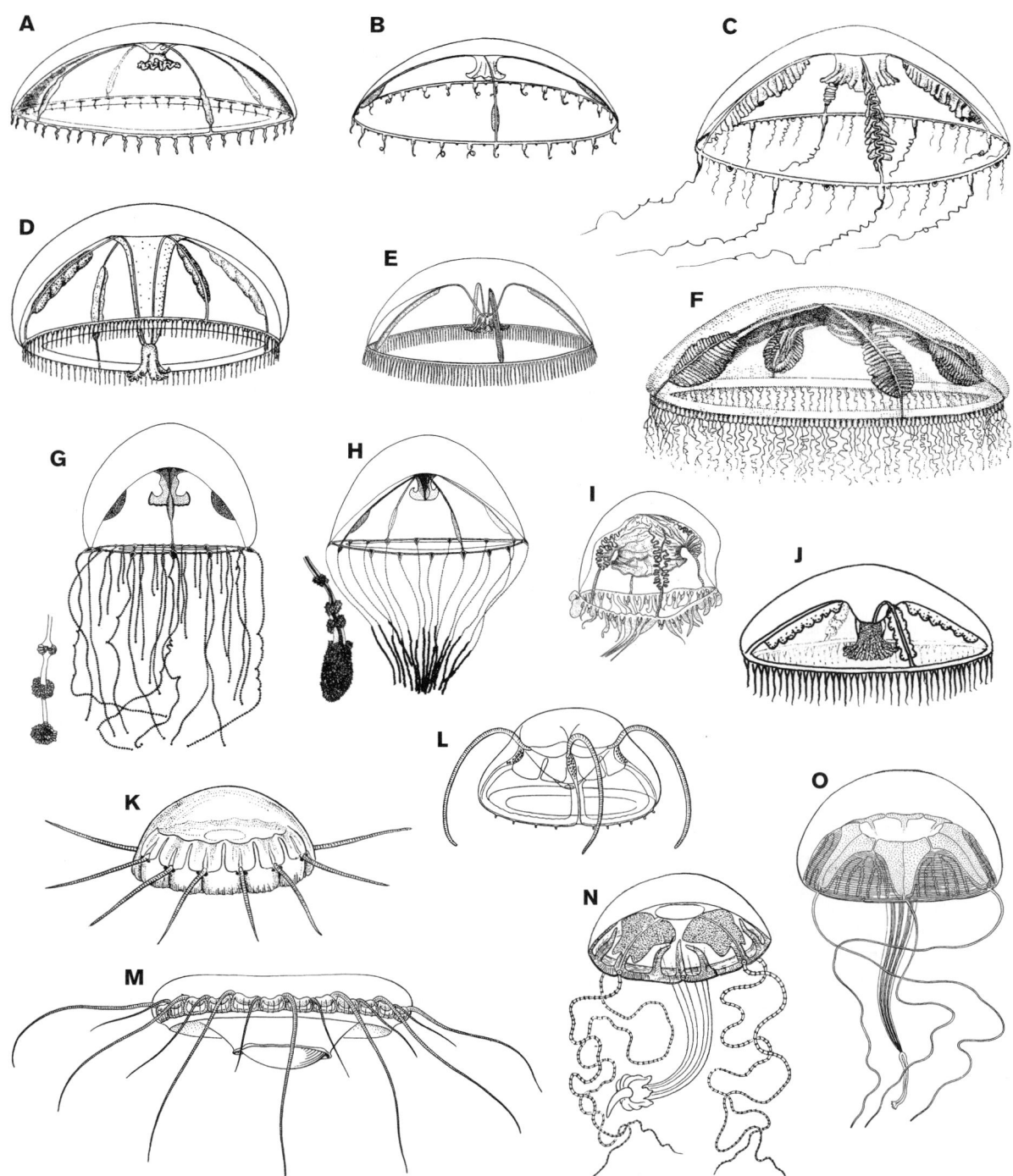

PLATE 57 Leptomedusae, Narcomedusae, and Trachymedusae. A, *Clytia gregaria*; B, *Clytia lomae*; C, *Tiaropsidium kelseyi*; D, *Eutonina indicans*; E, *Eirene mollis*; F, *Ptychogena lactea*; G, *Phialella zappai*; H, *Phialella fragilis*; I, *Ptychogena californica*; J, *Foersteria purpurea*; K, *Cunina peregrina*; L, *Aegina citrea*; M, *Pegantha clara*; N, *Liriope tetraphylla*; O, *Geryonia proboscidalis* (A, from Kramp 1968, after Murbach and Shearer; B, E, I, from Torrey 1909; D, K, from Kramp 1968; C, after Torrey 1909, and from life; F, from Naumov 1960; G, H, from Boero 1987, reprinted by permission of Taylor & Francis Ltd, http://www.tandf.co.uk/journals; J, original drawing by Claudia E. Mills; L, from Kramp 1968, after Mayer; M, O, from Mayer 1910; N, from Mayer 1910, after Haeckel).

tentacles arising at several levels on the exumbrella in line with each radial canal (plate 50H). . . . . *Climacocodon ikarii*
— Bell broad with thick jelly; without radial canals and with tentacles arising all at the same level, some distance above the bell margin 85
85. With stomach pouches 86

— Without stomach pouches 89
86. With four tentacles (rarely, five or six), and usually eight (rarely, to 12) well-defined stomach pouches containing the gonads (plate 57L). . . . . *Aegina citrea*
— With more than six tentacles, and an equal number of tentacles and stomach pouches 87

87. With marginal sensory clubs and otoporpae (bristly tracks of ectodermal cells running up the exumbrella from the bell margin) (rare) (plate 57K) . . . . *Cunina* spp.
— With marginal sensory clubs, but without otoporpae 88
88. With eight to 20 (usually 16) tentacles and stomach pouches; each marginal lappet with up to 20 marginal sensory clubs (plate 51J) . . . . *Solmissus marshalli*
— With 20–40 tentacles and stomach pouches; each marginal lappet with two to five marginal sensory clubs . . . . *Solmissus incisa*
89. Without peripheral canal system, without otoporpae (bristly tracks of ectodermal cells running up the exumbrella from the bell margin); with simple annular gonad (plate 51I) . . . . *Solmaris* spp.
— With peripheral canal system, with otoporpae (bristly tracks of ectodermal cells running up the exumbrella from the bell margin); gonads forming diverticula of the margin of the oral wall of the stomach (rare) (plate 57M) . . . . . . *Pegantha* spp.

Key to the Siphonophora

CLAUDIA E. MILLS, STEVEN H. D. HADDOCK, CASEY W. DUNN, AND PHILIP R. PUGH

Plates refer to siphonophore plates 58–60.

This key covers the life stages of some siphonophore species most likely to be encountered near shore, where they might be dipped from harbors and marinas or observed by snorkellers or scuba divers.

1. With a gas-filled float, and with numerous swimming bells arranged below this, followed by a stem region bearing groups of feeding, reproductive, and buoyant zooids (see supplementary key for stem pieces of *Apolemia* species) 2
— Without a gas-filled float, and usually with only one or two swimming bells (see supplementary key for stem pieces of *Rosacea* and *Praya* species) 6
2. Stem elongate with feeding and reproductive zooids along the entire length 3
— Stem reduced and laterally expanded into a bulbous structure, with zooids arranged spirally around it (plate 58C) . . . . *Physophora hydrostatica*
3. Swimming bells arranged in two rows along opposite sides of stem 4
— Swimming bells numerous and whorled (not in two rows), packed into a characteristic cone-shaped or cylindrical arrangement (plate 58F, 58G) . . . . *Forskalia* spp.
4. Elongate polyps present between swimming bells; stem with "woolly" appearance; tentacles without side branches (plates 58E, 59A) . . . . *Apolemia* spp.
— Polyps not present between swimming bells; tentacles with side branches 5
5. Stem densely covered with bracts (most noticeable when removed from the water); gastrozooids few and far between; tentacles all hanging down along one side of stem; stem cannot contract (plates 58A, 59B) . . . . *Agalma elegans*
— Stem with small inconspicuous bracts; gastrozooids relatively numerous; tentacles emerge from anywhere along the stem (not all from one side); stem can contract when the colony is disturbed (plates 58B, 59D) . . . . *Nanomia bijuga*

6. Swimming bells rounded, without ridges, without coming to a point; if two bells, then attached side by side 7
— Swimming bells elongate, pointed or faceted; if two bells, then not attached side by side, but slightly offset along the stem 10
7. Single, soft, spherical, colorless, and very transparent swimming bell, up to 8 mm in length; gastrozooids minute (plate 60A) . . . . *Sphaeronectes gracilis*
— Usually with two robust swimming bells; gastrozooids and tentacles yellow 8
8. Swimming bells with a simple, slender, tubular somatocyst; pair of deeply sinusoidal radial canals on subumbrella (plate 59G, 59H) . . . . *Rosacea* spp.
— Swimming bells with a complexly branched somatocyst; radial canals on subumbrella branch many times 9
9. No cross-links between branches of radial canals (plates 58D, 58H, 59E) . . . . *Praya dubia*
— Cross-links between branches of radial canals, forming a meshlike pattern (plate 59F) . . . . *Praya reticulata*
10. Anterior (upper) swimming bell roughly conical and larger than or approximately equal in size to posterior (lower) one 11
— Anterior (upper) swimming bell polyhedral; posterior (lower) bell considerably larger, with two prominent basal teeth (plate 60B) . . . . *Abylopsis tetragona*
11. Anterior swimming bell without ridges, with rounded apex 12
— Anterior swimming bell with ridges, with pointed apex 13
12. Anterior swimming bell with divided mouthplate and with four "teeth" on opening of subumbrella; with slender tubular somatocyst about one-third the length of the nectosac; posterior bell with characteristic constriction in middle of subumbrella (plate 60D) . . . . *Sulculeolaria quadrivalvis*
— Anterior swimming bell with undivided mouthplate and without "teeth" on opening of subumbrella, with carrot-shaped somatocyst about two-thirds the length of the subumbrella; posterior bell reduced, but rarely present (plate 60E) . . . . *Dimophyes arctica*
13. Ridges on swimming bells spirally twisted and prominently serrated (plate 60F) . . . . *Eudoxoides spiralis*
— Ridges on swimming bells straight or slightly curved, may or may not show serrations 14
14. Anterior swimming bell very stiff, with five or six ridges at base and only three or four ridges at apex. 15
— Anterior swimming bells not as above. 16
15. Anterior swimming bell to 20 mm long; somatocyst swollen and fusiform; small claw-shaped hydroecium only open at base (plate 60G) . . . . *Chelophyes appendiculata*
— Anterior swimming bell to 35 mm long; somatocyst with two prominent lateral swellings, forming a T shape; rounded hydroecium to half bell length, open at base and along one side (plate 60P) . . . . *Chuniphyes multidentata*
16. Somatocyst of anterior swimming bell extending to at least one-half of its length 17
— Somatocyst of the anterior swimming bell substantially less than one-quarter its length. 20
17. Hydroecium one-third to one-half the length of the anterior swimming bell 18
— Hydroecium very shallow; anterior swimming bell to 20 mm long with fusiform somatocyst extending to over one-half its length (plate 60L) . . . . *Lensia conoidea*

18. None of the ridges of the anterior swimming bell serrated, without teeth on the opening of the subumbrella; mouth-plate divided; hydroecium one-third the length of the bell; somatocyst long and slender, reaching to the apex of the subumbrella (plate 60I) . . . . *Muggiaea atlantica*
— At least some of the ridges of the anterior swimming bell serrated; with three conspicuous teeth around the opening of the subumbrella; mouthplate undivided 19
19. Anterior swimming bell to 35 mm in length, hydroecium extends one-half the length of the bell, somatocyst long and slender, subumbrella with distinct fingerlike extension at its apex (plate 60J) . . . . *Diphyes dispar*
— Anterior swimming bell to 14 mm in length, hydroecium extends nearly one-third the length of the bell, somatocyst fusiform, subumbrella tapering apically, without distinct extension (plate 60K) . . . . *Diphyes bojani*
20. Anterior swimming bell with five ridges and no obvious hydroecium 21
— Anterior swimming bell with numerous ridges; hydroecium relatively deep; inverted heart-shaped somatocyst (plate 60M) . . . . *Lensia hostile*
21. Anterior swimming bell with spherical, egg-shaped, or flattened somatocyst; shallow hydroecial cavity (plate 60O) . . . . *Lensia challengeri*
— Anterior swimming bell with oblique, ovate somatocyst; hydroecial cavity reduced to only a slight depression (plate 60N) . . . . *Lensia hotspur*

Supplementary Key to Some Stem Fragments of Siphonophores

1. Stem with overall "woolly" appearance, with numerous gastrozooids, palpons and bracts; gastrozooids red or white (plates 58E, 59A) . . . . *Apolemia* spp.
— Stem with distinct and repetitive groups of zooids (called "cormidia"), each with a single bract and gastrozooid; gastrozooids and tentacles yellow 2
2. Bracts hemispherical, gonophores with two mantle canals (plate 59G, 59H) . . . . *Rosacea* spp.
— Bracts somewhat flattened, gonophores with three mantle canals (plates 58D, 58H, 59E, 59F) . . . . *Praya* spp.

Note: The characters distinguishing the bracts of *Praya* species are often difficult to make out and are omitted here.

Combined Species List of Hydroids, Hydromedusae, and Siphonophores

CLAUDIA E. MILLS, DALE R. CALDER, ANTONIO C. MARQUES, ALVARO E. MIGOTTO, STEVEN H. D. HADDOCK, CASEY W. DUNN, AND PHILIP R. PUGH

HYDROZOA SUBCLASS ANTHOATHECATA (also known as ANTHOMEDUSAE and ATHECATA)

ORDER FILIFERA

BOUGAINVILLIIDAE

Bougainvillia muscus (Allman, 1863). Hydroid and medusa. Synonyms in Calder, (1988, pp. 24–25). The name replaces *B. ramosa* (van Beneden, 1844), which is an invalid junior homonym. Probably introduced, present in bays and harbors. Remarkable color illustration of hydroid and medusa from Naples in Brinckmann-Voss 1970, plate 9.

Bougainvillia spp. Hydroid and medusa. Unidentified hydroids of *Bougainvillia* occur in San Francisco Bay, and may be introduced species. Other *Bougainvillia,* of the same or different species in Bodega Harbor, have been collected and raised by J. T. Rees and C. E. Mills.

Garveia annulata Nutting, 1901. Hydroid. Hydroids conspicuous, with bright orange to yellow colonies and deeper orange gonophores (Torrey 1902; Fraser 1937; Haderlie et al. 1980—color photograph 3.6, plate 15). Rocky intertidal zones of the open coast, especially in late winter and spring, frequent on sponges and coralline algae; also reported subtidally to 117 m; Alaska to the Channel Islands (Fraser 1937, 1946).

Garveia franciscana (Torrey, 1902) (=*Bimeria franciscana*). Hydroid. A robust and conspicuous fouling species, abundant on floats and pilings in areas of low salinity in the San Francisco Bay area. Female gonophores a distinctive blue-purple, with a red-orange spadix. Lower intertidal and shallow subtidal. Reported in harbors from San Francisco Bay to San Diego (Torrey 1902; Fraser 1937, 1948). Introduced, but original provenance unknown.

Rhizorhagium formosum (Fewkes, 1889). Hydroid. Synonyms in Hochberg and Ljubenkov (1998, p. 9). A small and poorly known species, growing on gastropod shells and other hard substrates. Intertidal to 550 m; San Francisco Bay to Baja California (Fraser 1937, 1946; Hochberg and Ljubenkov 1998).

*Unidentified bougainvillioid(?). Hydroid and possibly medusa. Hand and Jones (see below) described and illustrated a light, flesh-pink, translucent hydroid collected from 10 m off Point Richmond in San Francisco Bay that underwent curious asexual reproduction involving changes in polarity. The tiny polyps (1–1.5 mm in length) supported four to 12 filiform tentacles inserted in a single cycle at the base of the proboscis. Bavestrello et al. (see below) observed a similar hydroid in the Genoa Aquarium that underwent both asexual reproduction and sexual reproduction with medusae and which they placed in the superfamily Bougainvillioidea; see Hand and Jones 1957, Biol. Bull. 112: 349–357; Bavestrello et al. 2000, Sci. Mar. 64 (Suppl. 1): 147–150.

BYTHOTIARIDAE (=CALYCOPSIDAE, a junior synonym)

Bythotiara stilbosa Mills and Rees, 1979. Medusa (hydroid unknown). Known only from newly released medusae collected off docks in Mason's Marina, Bodega Harbor, and raised in the laboratory. Some bythotiarid polyps are symbiotic in tunicates, including *Bythotiara huntsmani* (Fraser, 1911) in Washington and British Columbia. See Mills and Rees 1979, J. Nat. Hist. 13; 285–293. Color photograph in Wrobel and Mills 1998 and 2003, p. 25.

Calycopsis nematophora Bigelow, 1913. Medusa (hydroid unknown). Apparently a Pacific oceanic species that is occasionally found near shore (illustration in Arai and Brinckmann-Voss 1980, p. 68).

Calycopsis simulans (Bigelow, 1909). Medusa (hydroid unknown). Apparently a Pacific oceanic species that is occasionally found near shore. Color photograph in Wrobel and Mills 1998 and 2003, p. 25.

Heterotiara anonyma Maas, 1905. Medusa (hydroid unknown). An oceanic species of the Pacific, Atlantic, and Indian Oceans that is occasionally found near shore (illustration in Arai and Brinckmann-Voss 1980, p. 70).

Sibogita geometrica Maas, 1905. Medusa (hydroid unknown). An oceanic species of the Pacific, Atlantic, and Indian Oceans that is occasionally found near shore.

* = Not in key.

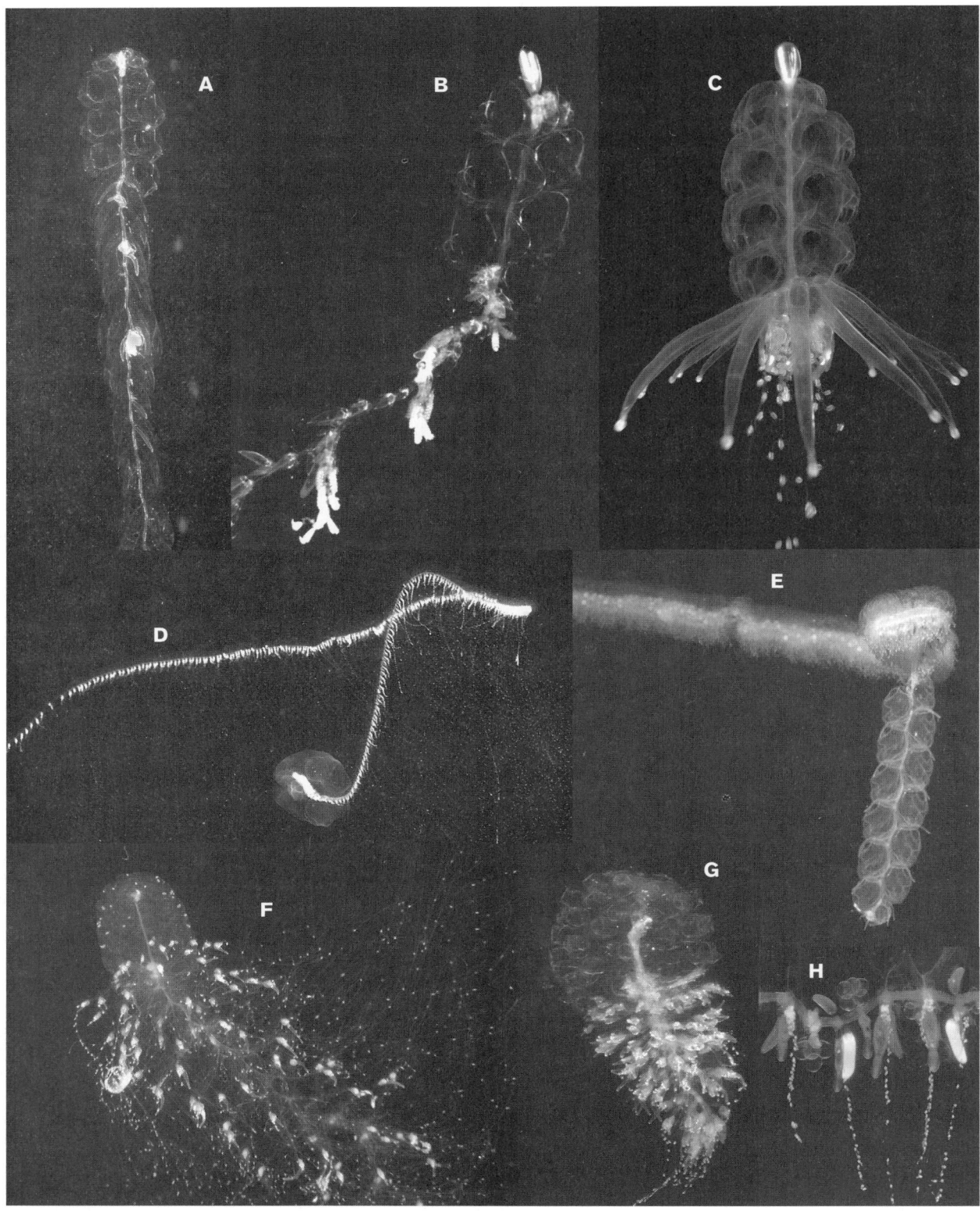

PLATE 58 Whole siphonophores, live. A, *Agalma elegans*; B, *Nanomia bijuga*; C, *Physophora hydrostatica*; D, *Praya dubia*; E, *Apolemia* sp.; F, *Forskalia* sp. 1; G, *Forskalia* sp. 2; H, *Praya dubia*, close-up of portion of the stem (A, F, G, photographs by Casey W. Dunn; B, H, photographs by Claudia E. Mills; C, photograph by Steven H. D. Haddock; D–E, in situ photograph from Monterey Bay Aquarium Research Institute [MBARI]).

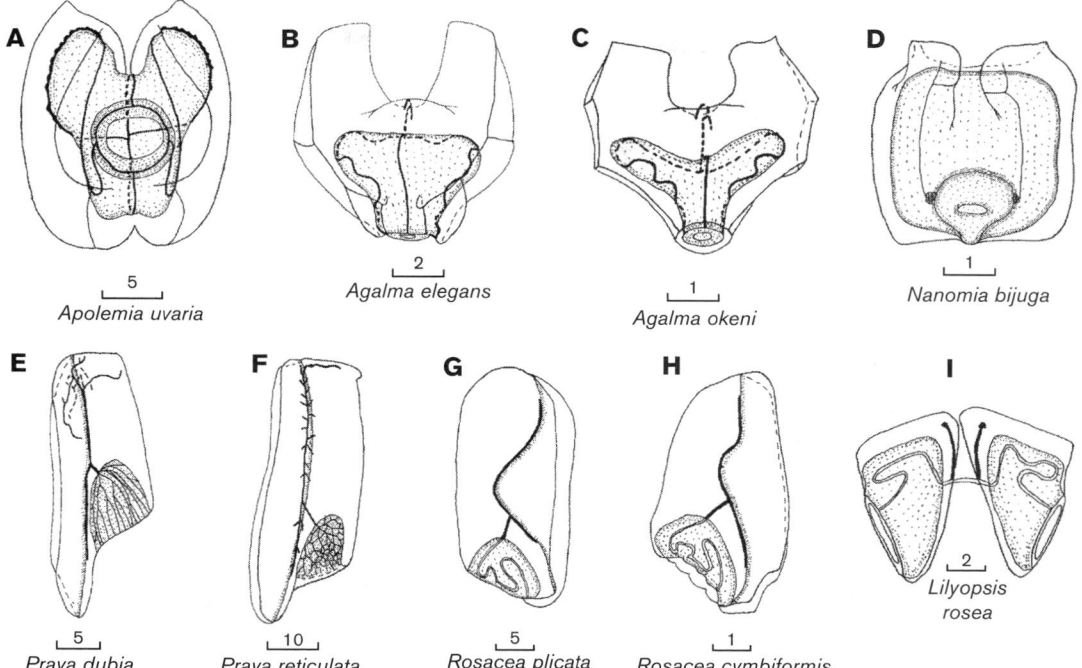

PLATE 59 Nectophores. Views from above (looking down the stem from the float–B, C) and distal views (stem running down the plane of the page–A, D) of nectophores of physonect siphonophores (A–D) and lateral views of nectophores of calycophoran siphonophores (E–I); scale bars in mm. For additional drawings of posterior nectophores, bracts, and gonophores, see Pugh (1999). A, *Apolemia uvaria*; B, *Agalma elegans*; C, *Agalma okeni*; D, *Nanomia bijuga*; E, *Praya dubia*; F, *Praya reticulata*; G, *Rosacea plicata sensu* Bigelow, 1911; H, *Rosacea cymbiformis*; I, *Lilyopsis rosea* (all from Pugh 1999).

EUDENDRIIDAE

Eudendrium californicum Torrey, 1902. Hydroid. Once a common intertidal and subtidal species in areas of open rocky coast, but apparently less frequent in recent decades. British Columbia to southern California, 4–115 m (Fraser 1937, 1946; Haderlie et al. 1980—see color photograph 3.7, plate 15).

Eudendrium spp. Hydroid. Species of this genus are characterized by typically styloid gonophores, trumpet-shaped hypostomes, and absence of desmonemes in the cnidome. Identification based on gross morphology alone is questionable and should include examination of cnidocysts. Various species of *Eudendrium* have been reported from the region, but the absence of information on cnidocyst complement makes these identifications uncertain (see Marques et al. 2000a, Zool. Meded., Leiden 74: 75–118; Marques et al. 2000b, J. Zoology, London 252: 197–213).

HYDRACTINIIDAE

Clava multicornis (Forsskål, 1775) (=*C. leptostyla* L. Agassiz, 1862). Hydroid. Additional synonyms in Edwards and Harvey 1975, J. Mar. Biol. Assoc. U.K. 55: 879–886. This species is included in the family Hydractiniidae here following Schuchert (2001a); molecular studies show that *Clava* and hydractiniids should be assigned to the same family. A cold-water species largely inhabiting the intertidal zone of bays and estuaries, sometimes forming large colonies due to stolonal growth. Hydranths and male gonophores pink, those of the female purple. Whether this well-known Atlantic hydroid still occurs in San Francisco Bay (its only known West Coast location) is uncertain.

First reported in 1895 from San Francisco Bay (to which it was introduced in ship fouling), it is unclear when it was last seen in the Bay. Light et al. (1954) noted that it was "abundant on Fruitvale and Bay Farm Island bridges, Oakland, in spring," but we find no further actual records in the past 50 years. We did not find it in surveys of San Francisco Bay between 1993 and 2004.

Hydractinia armata Fraser, 1940. Hydroid. Female gonophores bearing only a single egg. Found in association with *H. milleri* (Fraser 1946), on coralline algae in tide pools at Moss Beach, a rocky intertidal site on the open coast just south of San Francisco Bay first visited by S. F. Light and his students in the 1920s and 1930s, and in later decades the most popular tide pool site for hundreds of thousands of school children from central California schools. It would be interesting to determine if these millions of little feet have obliterated this hydroid from Moss Beach.

Hydractinia laevispina Fraser, 1922. Hydroid. Female gonophores bearing only a single egg. British Columbia to central California in the low subtidal to at least 20 m deep; on kelp off Coast Guard breakwater in Monterey Harbor (Light et al. 1954).

Hydractinia milleri Torrey, 1902. Hydroid. Female gonophores bearing only a single egg. Colonies often growing in patches, sometimes covering several square centimeters on rocks exposed to breakers of the open sea (Torrey 1902). British Columbia to central California in the lower intertidal (Fraser 1937, 1946; Haderlie et al. 1980).

Hydractinia spp. Hydroid. Additional species of *Hydractinia* occur in the region (R. Grosberg pers. comm.).

* = Not in key.

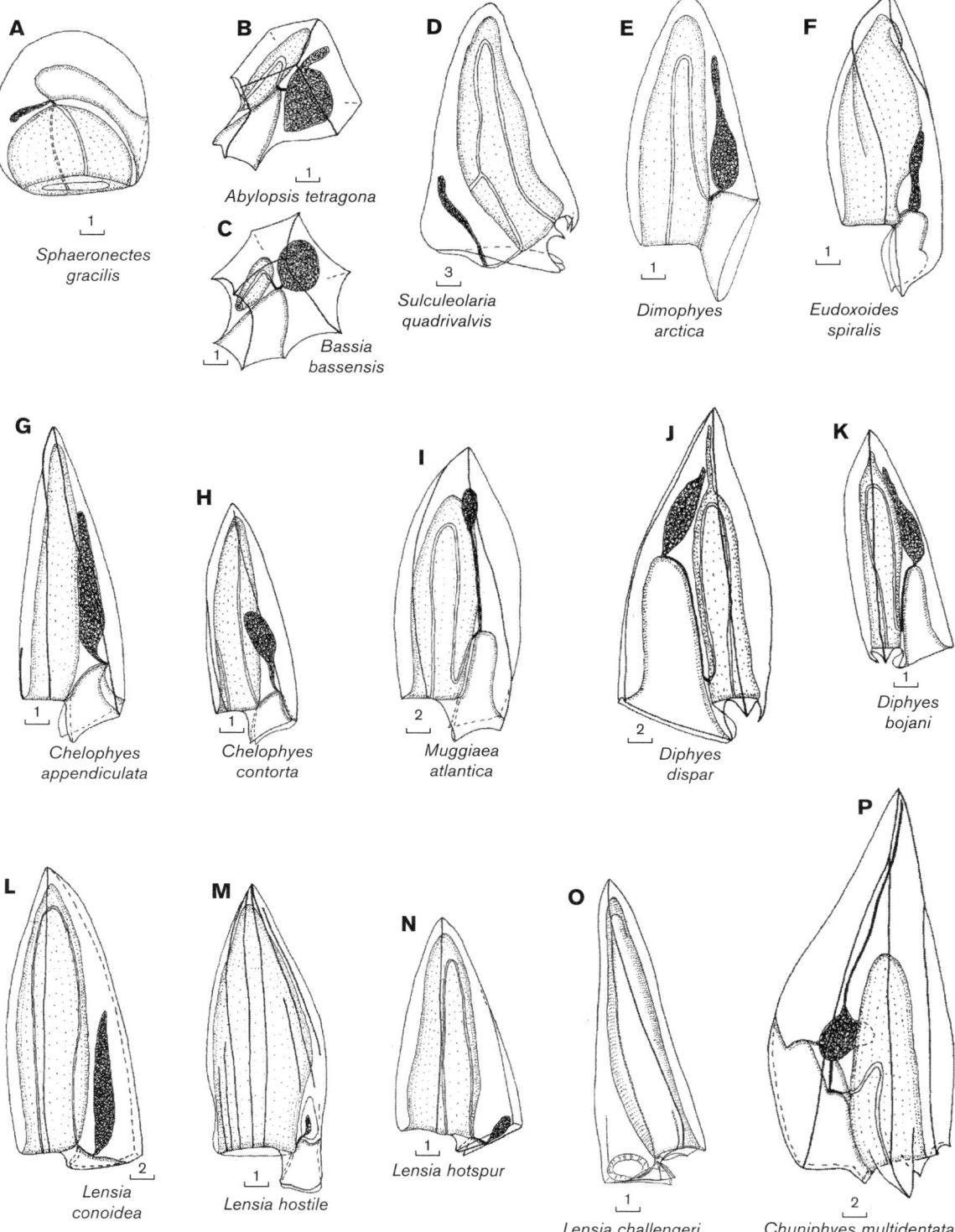

A, *Sphaeronectes gracilis*

B, *Abylopsis tetragona*

C, *Bassia bassensis*

D, *Sulculeolaria quadrivalvis*

E, *Dimophyes arctica*

F, *Eudoxoides spiralis*

G, *Chelophyes appendiculata*

H, *Chelophyes contorta*

I, *Muggiaea atlantica*

J, *Diphyes dispar*

K, *Diphyes bojani*

L, *Lensia conoidea*

M, *Lensia hostile*

N, *Lensia hotspur*

O, *Lensia challengeri*

P, *Chuniphyes multidentata*

PLATE 60 Lateral views of anterior nectophores of calycophoran siphonophores; scale bars in mm. For additional drawings of posterior nectophores, bracts, and gonophores, see Pugh (1999). A, *Sphaeronectes gracilis*; B, *Abylopsis tetragona*; C, *Bassia bassensis*; D, *Sulculeolaria quadrivalvis*; E, *Dimophyes arctica*; F, *Eudoxoides spiralis*; G, *Chelophyes appendiculata*; H, *Chelophyes contorta*; I, *Muggiaea atlantica*; J, *Diphyes dispar*; K, *Diphyes bojani*; L, *Lensia conoidea*; M, *Lensia hostile*; N, *Lensia hotspur*; O, *Lensia challengeri*; P, *Chuniphyes multidentata* (all from Pugh 1999, except O, from Totton 1954, with permission, Discovery Rep. 27, 1–162, text figures 24B, 43C, 54A, 54D, 65A, 66B).

OCEANIDAE (=CORDYLOPHORIDAE, a junior synonym)

Cordylophora caspia (Pallas, 1771) (=*C. lacustris* Allman, 1844). Hydroid. Additional synonyms in Schuchert (1996, p. 15; 2004, p. 346). Hydroids phenotypically variable, in part related to salinity (Roch 1924, Z. Morph. ökol. Tiere 2, 350–426). Eggs develop into planulae within the gonangia (see Schuchert 1996). Restricted to areas of brackish or fresh water. Colonies occurring on a wide variety of substrates including pilings, barnacles, floats, and tubes of the polychaete *Ficopomatus*; abundant in the San Francisco Bay Delta and recorded in Lake Merced on the San Francisco peninsula. Thought to be Ponto-Caspian in origin, it appears to have a remarkable temperature breadth (if only one species is involved), being introduced in ship fouling from British Columbia to Panama in the shallow subtidal.

Turritopsis sp. Hydroid and medusa. Previously identified from San Francisco Bay as *Turritopsis nutricula* McCrady, 1857; but from Schuchert's (2004) revision of the genus *Turritopsis*, it is now evident that there are several similar-looking species, so we do not attempt here a specific identification of this seemingly non-native species. *Turritopsis* medusae still found occasionally in San Francisco Bay in the late 1990s (these to 8 mm tall, with about 25 tentacles and pale red manubrium), but the hydroid has not been recorded there since 1940; San Francisco Bay, 0–9 m (Fraser 1937, 1946). What was previously seen as a widely introduced single species, usually in estuaries and bays, with affinities to warmer waters in many locations worldwide, has now been identified as at least four species, some or one of which may be quite widely introduced.

PANDEIDAE

Several species of Pandeidae have been assigned in the past to *Perigonimus*, a junior synonym of *Bougainvillia* (see Calder 1988). These are now referred to genera such as *Amphinema*, *Leuckartiara*, *Neoturris* and *Catablema* (Bouillon 1985; Calder 1988).

Amphinema platyhedos Arai and Brinckmann-Voss, 1983. Medusa (hydroid unknown). Rare; known from deep water in British Columbia and southern California. Compare with other *Amphinema* medusae (below) in the area.

Amphinema turrida (Mayer, 1900). Medusa (hydroid unknown). Rare; known from surface waters from Monterey Bay south into Mexico; Pacific and Atlantic. The two stout yellow tentacles and folded gonads that extend most of the way down the radial canals are distinctive. Color photograph in Wrobel and Mills 1998 and 2003, p. 26.

Amphinema sp. Hydroid and medusa. Hydranths bright orange-red; when disturbed, they bend over nearly 180° towards the substrate. This hydroid resembles *Amphinema rugosum* (Mayer 1900) in having medusa buds on the hydrocaulus (Schuchert 1996), but the medusae showed significant differences from *A. rugosum*. Colony occurred on non-native encrusting bryozoan *Watersipora* sp. on floating docks in Bodega Harbor. Medusa known only from rearing in the laboratory from field-collected hydroid. See Rees 2000, Sci. Mar. 64 (Suppl. 1): 165–172. Probably introduced.

Annatiara affinis (Hartlaub, 1913). Medusa (hydroid unknown). Genus name was created by reversal of *Tiaranna* in which this species was originally placed. A species that has been infrequently collected in the Pacific, Atlantic, and Indian Oceans; considered to live in deep water, but several medusae have been found near the surface in Monterey Bay. Color photograph in Wrobel and Mills 1998 and 2003, p. 26.

Halitholus spp. Medusa (hydroid unknown). Similar to and usually smaller than *Leuckartiara*, there are at least two undescribed *Halitholus* species on our coast. Color photograph in Mills and Wrobel 1998 and 2003, p. 27.

Leuckartiara octona (Fleming, 1823) (=*Perigonimus repens* [Wright, 1857]). Hydroid and medusa. Synonyms in Schuchert (1996, p. 68). Stolonal colonies on shells of gastropods including *Hima mendica* and *Callianax biplicata*. Colony form varies according to the host shell and the amount of friction to which it is subjected. The best-developed colonies, sometimes with branches and producing gonophores, are usually found on the upper surface of the shell where there is no abrasion (Torrey 1902; Millard 1975). Alaska to tropics, including San Francisco Bay (Fraser 1937, 1946, as *P. repens*). *L. octona* medusae collected in Bodega Bay and Bodega Harbor by Rees (1975). Medusae that meet this description except are two times larger (to 40 mm tall) and with up to 72 tentacular rudiments are sometimes very abundant offshore in central California near Monterey. This species, said to occur all over the world, likely represents a species complex, especially considering its specialized shallow-water habitat association with gastropods.

*Leuckartiara spp. Hydroid and medusa. Other species of *Leuckartiara* with *Perigonimus*-like polyps that are not *L. octona* are occasionally collected in our area. The shape of the apical projection on *Leuckartiara* and other pandeid species is rather plastic and, tempting though it may be, is not the best diagnostic character to use when trying to make identifications. Color photograph in Wrobel and Mills 1998 and 2003, p. 27.

Neoturris breviconis (Murbach and Shearer, 1902). Medusa (hydroid unknown). Known from both the North Pacific and North Atlantic, occurs on the West Coast from the Bering Sea to Monterey Bay. Color photograph in Wrobel and Mills 1998 and 2003, p. 28.

Neoturris fontata (Bigelow, 1909). Medusa (hydroid unknown). Medusa collected only once, at the surface near the coast, Baja California (Bigelow 1909, Mem. Mus. Comp. Zool. Harvard 37: 1–243, pls. 1–48; Kramp 1968, p. 49).

Neoturris pelagica (A. Agassiz and Mayer, 1902). Medusa (hydroid unknown). North and South Pacific, including coast of California; rare.

Neoturris sp. (undescribed). Medusa (hydroid unknown) that corresponds more or less to *N. pileata* (Forsskål, 1775) except for its enormous size (to 80 mm tall, whereas *N. pileata* [see hydromedusa plate 56G] is described as 40 mm tall) and the very broad, distinctive, tablike diverticula off its radial canals has been collected infrequently offshore in southern California (C. E. Mills, unpublished observations); a record of "*N. pileata*" off Vancouver Island (see Kramp 1968, p. 50) may be the same species.

Pandea conica (Quoy and Gaimard, 1827). Medusa (hydroid unknown). With characteristic, usually pointed bell, to 21 mm high, a shelf species perhaps, but occasionally encountered near shore in California; found in the Atlantic, Pacific, and Mediterranean (Kramp 1961, 1968). Remarkable color illustration of hydroid and medusa from Naples in Brinckmann-Voss 1970, plate 11.

*Pandea rubra Bigelow, 1913. Medusa (hydroid unknown). A larger (to 75 mm high and wide) flat-topped deep sea species that occurs off California (as well as in the Atlantic and Antarctic) with characteristic deep red gut, gonads, tentacles and subumbrella; unlikely to be found near shore (compare with *P. conica*) (Kramp 1961, 1968).

* = Not in key.

Stomotoca atra L. Agassiz, 1862. Medusa (hydroid unknown). A common near-shore medusa on the entire West Coast that swims in a sinusoidal pattern trailing its two long tentacles, feeding primarily on other small hydromedusae (Mills 1981). Many have attempted to raise its hydroid in the lab: the stolons elongate, but hydranths do not develop (Volker Schmid pers. comm.); the hydroid may be symbiotic on another organism. It is unlikely that our temperate West Coast *S. atra* is conspecific with the Papua New Guinea medusa of the same name. Color photograph in Wrobel and Mills 1998 and 2003, p. 28.

PROBOSCIDACTYLIDAE

Hydroids of the various species of *Proboscidactyla* strongly resemble each other and are differentiated largely on the basis of medusoid characters (Millard 1975). The family has been classified among the Limnomedusae by several authors (e.g., Kramp 1961; Naumov 1969; Millard 1975; Bouillon 1985, 1995a). Following Petersen (1990), Schuchert (1996), and others, we assign it to the Anthoathecata because of the presence of a stolon system, desmoneme cnidocysts, and the absence of statocysts in the medusae. Gastrozooids in all species of *Proboscidactyla* are arranged in a circle around the rim of the tube of a sabellid polychaete host. The two tentacles of each hydranth are directed toward the opening of the tube, and the asymmetrically placed pad of cnidocysts on the hypostome is oriented away from it. Polyps are active, securing food from the tentacles of the sabellid or consuming spawned eggs of the host. In bending over periodically towards the central axis of the polychaete tube, with the two tentacles extended and the hypostome resembling a human head, a hydranth resembles a person praying. Nonfeeding polyps sometimes occur some distance from the tube rim. The association of *Proboscidactyla* spp. with its polychaete substrate is obligatory: planulae settle and develop only on the worm tube (Hand 1954—see below; Millard 1975; Rees 1979—see below).

Proboscidactyla circumsabella Hand, 1954. Hydroid and medusa. Hydroid found at Pacific Grove, on tubes of the sabellid *Pseudopotamilla ocellata* (and rarely on *P. intermedia*), on rocks at low tide; medusae known from Monterey Bay (see Hand and Hendrickson 1950, Biol. Bull. 99: 74–87 and Hand, 1954, Pacific Sci. 8: 51–67).

Proboscidactyla flavicirrata Brandt, 1835. Hydroid and medusa. A northern species occurring from China and Japan to central Oregon, where it is rare. Hydroid known from Friday Harbor commensal on the sabellids *Schizobranchia insignis* and *Pseudopotamilla ocellata*. For life cycle, see Hand 1954, Pac. Sci. 8: 51–67 and Rees 1979, Can. J. Zool. 57: 551–557; for host specificity and description of larvae settling on sabellid worm tubes, see Campbell 1968, Pac. Sci. 22: 336–339 and Donaldson 1974, Biol. Bull. 147: 573–585; for medusa feeding behavior and diet, see Mills (1981) and Costello and Colin (2002). Photograph in Wrobel and Mills 1998 and 2003, p. 38.

Proboscidactyla occidentalis (Fewkes, 1889). Hydroid and medusa. Hydroids on tubes of sabellids (*Potamilla neglecta* and *Pseudopotamilla intermedia*) from a kelp holdfast, La Jolla, 12–15 m; medusae known from Santa Cruz Island, La Jolla, and San Diego Bay (see Hand 1954, Pac. Sci. 8: 51–67).

PROTIARIDAE

Protiara sp. Medusa (hydroid unknown). Two medusae collected in July in Bodega Bay (Rees 1975, p. 252a) and several others collected May through July in Yaquina Bay (1969–1974, R. J. Larson, personal communication). The Oregon specimens reached 5 mm in bell height and appear to be an undescribed species (see hydromedusa plate 54E) (R. J. Larson, personal communication).

RATHKEIDAE

Rathkea octopunctata (M. Sars, 1835). Hydroid and medusa. Synonyms in Schuchert (1996, p. 59). Medusa known from Alaska to central California; the miniscule hydroid, seldom seen, is described by Schuchert (1996). The medusae go through a period of asexually producing more medusae from the manubrium walls before sexual gonads develop in the same location. Reported from around the world. Color photograph in Wrobel and Mills 1998 and 2003, p. 28.

STYLASTERIDAE

Stylantheca porphyra Fisher, 1931 (=*Allopora porphyra*). Hydroid. Colonies forming vivid purple calcareous encrustations varying in size from tiny patches to large sheets (Haderlie et al. 1980). On exposed, rocky coasts, Monterey Bay and vicinity, low intertidal (Cairns 1983, Bull. Mar. Sci. 33: 427–508 (taxonomy); Alberto Lindner pers. comm.). The commensal polychaete *Polydora alloporis* forms distinctive paired holes in the colony.

*Stylaster californicus (Verrill, 1866) (=*Allopora californica*). Hydroid. The common sublittoral, branched representative of the alloporas, frequently seen by snorkelers and divers at Carmel, color ranging from orange to pink into purple. See Ostarello 1973, Biol. Bull. 145: 548–564 (natural history); color photograph in Haderlie et al. 1980, 3.17, plate 18. Unpublished molecular work by Alberto Lindner and Steve Blair indicates that this branched form is conspecific with *S. porphyra*, above, and possibly with other species further north; the *S. californicus* name has precedence (pers. comm.).

ORDER CAPITATA

CANDELABRIDAE

Candelabrum fritchmanii Hewitt and Goddard, 2001. Hydroid. Hydranths large (total body length to 100 mm when relaxed, to <20 mm when disturbed or contracted), with clasper tentacles attached to developing embryos in the blastostyle-bearing region, at least during the reproductive season. Colony attached to hard substrate by laterally flattened, basal hydrorhiza (this often with short, tentacular processes); naked attachment tentacles with chitinous pads also arise from the hydranth base. Under intertidal boulders in southern Oregon. Feeds on idoteid isopods and small gammarid amphipods; life cycle includes an actinula. See Hewitt and Goddard 2001, Can. J. Zool. 79: 2280–2288. *Note:* Hand and Gwilliam (1951, J. Wash. Acad. Sci. 41: 206–209) described solitary *Candelabrum* sp. from Pigeon Point, San Mateo Co., California. These specimens are similar to solitary *C. fritchmanii*, but owing to their lack of reproductive structures, cannot be specifically identified based on morphological characters alone.

CLADONEMATIDAE

The medusae of *Cladonema* are benthic, adhering to substrates by specialized, adhesive tentacles. The hydroids are frequently introduced accidentally to aquaria (Brinckmann-Voss 1970) and

* = Not in key.

can be kept for long periods (as many as 30 years, Russell 1953, p. 108). The medusae are exceptionally long-lived (several months) for such small species. The taxonomy of the genus is confused (Rees 1979—see below). In coastal California, two groups of species are distinguishable based on the hydroids: *Cladonema californicum* and *C. radiatum* (polyps with two whorls of tentacles, one capitate and the other filiform), and *Cladonema myersi* and *C. pacificum* (polyps with capitate tentacles only). Two different groups are recognized from the medusae: *C. californicum* in one, and *C. myersi*, *C. radiatum* and *C. pacificum* in another (Rees 1949—see below; Kramp 1961; Rees 1979, 1982—see below; Hirohito 1988). A combination of characters thus distinguishes the three species.

Cladonema californicum Hyman, 1947 (*californica* is a misspelling). Hydroid and medusa. For life cycle, see Rees 1979, J. Nat. Hist. 13: 295–302. Medusae present all year clinging to shallow eelgrass or *Ulva* in bays and harbors (see Hyman 1947, Trans. Am. Microsc. Soc. 66: 262–268). The medusae feed on benthic fauna associated with eelgrass and seaweed, capturing prey passively while remaining attached to the plants, but they swim actively when disturbed. Medusa found from British Columbia to Long Beach; hydroid known from Bodega Harbor. For lab culture, feeding, carbon, and nitrogen budgets, see Costello 1988, J. Exp. Mar. Biol. Ecol. 123: 177–188; Costello 1991, Mar. Biol. 108: 119–128; Costello 1998, J. Exp. Mar. Biol. Ecol. 225: 13–28. Color photograph in Wrobel and Mills 1998 and 2003, p. 29.

Cladonema myersi Rees, 1949. Hydroid and medusa. Hydroids on coralline algae; can be kept in aquaria for long periods (see Rees 1949, Proc. Zool. Soc. London 119: 861–865). Described from La Jolla, where the hydroid, releasing medusae, was found in 1934.

Cladonema pacificum Naumov, 1955 (=*C. uchidai* Hirai, 1958; see Hirohito 1988 for its complex taxonomic history). Hydroid and medusa. Found in display tank at the University of California, Berkeley, with material in the tank from San Francisco Bay and presumed to occur there (Rees 1982, Pac. Sci. 36: 439–444). Originally described from the Sea of Japan. An introduction from east Asia (Rees 1982 above).

Cladonema radiatum Dujardin, 1843. Hydroid and medusa. Synonyms in Schuchert (1996, p. 131). European species common amongst the eelgrass *Zostera* in Padilla Bay, Washington, adjacent to oil refineries on March Point, and likely to occur elsewhere on the West Coast because of frequent tanker traffic to refineries up and down the West Coast. Spectacular color illustration of hydroid and medusa in Brinckmann-Voss 1970, plate 5. Introduced.

CORYMORPHIDAE

Corymorpha palma Torrey, 1902. Hydroid. Synonyms in Hochberg and Ljubenkov (1998, p. 13). We follow Petersen (1990) and Schuchert (1996) in assigning this species to *Corymorpha*. In being an easy species to cultivate in the laboratory, aspects of its behavior, developmental biology, associations with algae, and statocysts have been investigated (see Haderlie et al. 1980 and Hochberg and Ljubenkov 1998, for references; color photograph in Haderlie et al., 3.4, P15). A southern species ranging at least north to Santa Barbara Channel, lower intertidal to 60 m; occurring where currents are moderate in turbid bays and estuaries, and on muddy bottoms offshore (Fraser 1937; Hochberg and Ljubenkov 1998).

Euphysa spp. Hydroid and medusa. A group of boreal (North Pacific, North Atlantic, and Arctic) species that are commonly found as far south as Washington and occasionally are col-

lected at least to central Oregon (R. J. Larson unpublished observations). Discussion of synonomies and northeast Pacific species in Arai and Brinckmann-Voss (1980: 6–9). *Euphysa tentaculata* medusae have only three tentacles at maturity (4–6 mm bell height); *E. flammea* and *E. japonica* both have four tentacles at maturity (about 1 cm tall) and are indistinguishable at that point. *E. flammea* has only one tentacle in the young stages, whereas *E. japonica* has four already at very small size. Some species of *Euphysa,* including *E. ruthae* near Friday Harbor, have interstitial polyps (see Norenburg and Morse 1983, Trans. Am. Microsc. Soc. 102: 1–17). Color photograph in Wrobel and Mills 1998 and 2003, p. 31.

Euphysora bigelowi Maas, 1905. Hydroid and medusa. Synonyms in Hochberg and Ljubenkov (1998, p. 16). Hydroid raised in lab from mature medusae collected in Monterey Bay (see Sassaman and Rees 1978, Biol. Bull. 154: 485–496). Typical of species of this family, the oral tentacles not distinctly capitate but may be thickened at their tips, especially in young polyps (Sassaman and Rees 1978; Petersen 1990). Hydroid generally found on soft bottoms, anchoring rootlets with inflated tips being used for attachment. Reproducing asexually with part of the base of the parental polyp detaching, developing tentacles, and breaking free (Hochberg and Ljubenkov 1998). Period of medusa liberation prolonged due to asynchronous development of the gonophores (Sassaman and Rees 1978). Assigned to *Corymorpha* by Sassaman and Rees (1978), Petersen (1990), and Cairns et al. (2002), but we refer it to *Euphysora* (Kramp 1961, 1968; Hochberg and Ljubenkov 1998; Bouillon et al. 2004). Indo-Pacific and Mediterranean.

*Unidentified corymorphid. A tiny (2–3 mm), orange-tinted polyp (see hydroid plate 44J), similar to the European *Corymorpha nutans* M. Sars 1835 collected from soft mud at Point Richmond in 1955 by Meredith Jones and again (if the same species) by James Carlton in Lake Merritt, Oakland, in 1967 (Cohen and Carlton 1995, p. 33).

CORYNIDAE

Coryne eximia Allman, 1859 (=*Syncoryne eximia*). Hydroid and medusa. Additional synonyms in Schuchert (2001b, p. 773). Hydroid easily kept in laboratory. See Schuchert (2001b) for differences between hydroid and medusa stages of *C. eximia* and *C. japonica*. Rocky intertidal and subtidal zones, Alaska to Monterey Bay, 0–18 m (Fraser 1937, 1946). Some reports of *C. eximia* could be *C. cliffordi* (Brinckmann-Voss 1989), a similar species known only from British Columbia (Schuchert 2001b).

Coryne japonica (Nagao, 1962) (=*Stauridiosarsia japonica*). Hydroid and medusa. Additional synonyms in Schuchert (2001b, p. 757). Known from Bodega Harbor and adjacent outer coast (Rees 1975); North and South Pacific, intertidal and shallow subtidal, on rocks and shells. See Arai and Brinckmann-Voss 1980; Schuchert 2001b.

Coryne sp. (undescribed). Hydroid and medusa. Large showy unbranched colonies of polyps (to 6 cm tall) growing on floating wood nearly always within 2–3 cm of the surface at Alameda Point, eastern shore of San Francisco Bay; large hydranths bear as many as 80 medusa buds at one time. Medusae grown in culture to maturity in 2 to 3 weeks, distinguished by bubblelike vesicles on the exumbrella (J. T. Rees unpublished observations). Probably an introduced species that is either undescribed or yet to be matched to a known species elsewhere in the world.

* = Not in key.

Dipurena bicircella Rees, 1977. Hydroid and medusa. Larger, mature polyps occurring at center of colony; smaller polyps with fewer or no medusa buds at periphery. Adult medusae sluggish swimmers, spending most of their time on the bottom in culture. On granite rock and shell, 3–10 m in Horseshoe Cove, Bodega Head (Rees 1977, Mar. Biol. 39: 197–202). See Schuchert (2001b) for comparison with other species of *Dipurena*.

Dipurena ophiogaster Haeckel, 1879. Hydroid and medusa. Hydroid (see hydroid plate 42F) grows on algae, barnacles and rock in Europe (see Schuchert 2001b, p. 803). Medusae with three or four gonads on the manubrium (McCormick 1969a, b; R. J. Larson unpublished observations) identified as *D. ophiogaster* in Yaquina Bay; and medusae without description identified as *D. ophiogaster*? in Bodega Harbor (Rees 1975, p. 252a), but this medusa is easily confused with *D. reesi*. *D. ophiogaster* is known from a variety of locations in both the Pacific and Atlantic Oceans (see Schuchert 2001b).

Dipurena reesi Vannucci, 1956. Hydroid and medusa. Barely detectable hydroid (see hydroid plate 42G) grows on rock and shell in Europe (see Schuchert 2001b, p. 805). Medusa reported from southern California and Mexico (known also from Brazil, Mediterranean and Bay of Biscay). Difficult to distinguish from *D. ophiogaster* (see Schuchert 2001b).

Sarsia tubulosa (M. Sars, 1835). Hydroid and medusa. As *Syncoryne mirabilis* (Agassiz, 1862) in Torrey (1902) and Fraser (1911, 1937, 1946), but that name, originally used for a medusa from New England, appears to have been applied to a complex of corynid species and is best not used for West Coast material pending further studies. In bays and harbors, Alaska to southern California, 0–22 m (Fraser 1937, 1946). Identifications questionable due to many similar species; see Miller 1982, J. Exp. Mar. Biol. Ecol. 62: 153–172 (species of Friday Harbor, including *S. tubulosa*), but see also Schuchert (2001b) for critical discussion and synonomies. Medusa feeding behavior and diet reported by Costello and Colin (2002) may represent more than one species, which may or may not be *S. tubulosa*. Present in San Francisco Bay in 2000 (J. T. Rees unpublished collection); likely introduced.

Sarsia spp. and *Coryne* spp. Species of *Sarsia* and *Coryne* are difficult to identify because of a lack of morphological characters (Brinckmann-Voss 1985, Can. J. Zool. 63: 673–681; Schuchert 2001b). Two unnamed species of the *"Sarsia/Coryne eximia"* group occur in our area, *Coryne* sp. A and B (Rees 1975). The *"eximia"* group, now referred to *Coryne*, comprise medusae with a short manubrium; the *"tubulosa"* group, now referred to *Sarsia*, include medusae with a long manubrium. Rees (1975) reports at least five unidentified species of *"Sarsia"* (includes *Coryne* by present generic definitions) present in the plankton in Bodega Bay. Color photograph in Wrobel and Mills 1998 and 2003, p. 29.

HALIMEDUSIDAE

This family was assigned to the Order Filifera (Bouillon 1985) prior to life-cycle studies. The hydranths have capitate tentacles (Mills and Miller 1987; Wrobel and Mills 1998; Mills 2000), indicating that the Halimedusidae should be assigned to the Capitata (Mills 2000).

Halimedusa typus Bigelow, 1916. Hydroid and medusa. Hydroid reared in the laboratory. Hydranths feed on nematodes and rotifers; their solitary condition and minuteness make them inconspicuous in nature (see Mills 2000, Sci. Mar. 64 [Suppl. 1]: 97–106). Medusae occur at least in Yaquina Bay, Coos Bay, Humboldt Bay, and Bodega Bay and Bodega Harbor. Color photograph in Wrobel and Mills 1998 and 2003, p. 26.

HYDROCORYNIDAE

Hydrocoryne bodegensis Rees, Hand and Mills, 1976. Hydroid and medusa. Polyps with large hydranths (0.5–6.0 cm high, depending on state of contraction); these move with the current, swinging like maces through the water and along the substrate, quickly contracting when disturbed. They feed on relatively large (2–4 mm) gammarid amphipods and other macrobenthic crustaceans (see Rees et al. 1976, Wasmann J. Biol. 34: 108–118). Intertidal and subtidal on jetty at entrance to Bodega Harbor, on rock faces exposed to moderately strong water movements; occasionally on kelps; medusa in Bodega Harbor and Bodega Bay.

MARGELOPSIDAE

Climacocodon ikarii Uchida, 1924. Hydroid and medusa. An oceanic species, with hydroids free-living in the plankton; occasionally collected near shore in central California and central Oregon. Polyps highly differentiated, budding medusae, which are released but do not grow much beyond 2 mm, making them among the smallest of the hydromedusae (see Uchida 1924, Japanese J. Zool. 1: 59–65).

MOERISIIDAE

Moerisia sp. Hydroid and medusa. Although this species was assigned to *Moerisia lyonsi* Boulenger, 1908 by Cairns et al. (2002), both the medusa and polyp in California have certain substantial differences from that species, as noted by Rees and Gershwin (2000), which prevent a positive species identification. It may also be conspecific with material from the Chesapeake Bay region identified as *M. gangetica* Kramp, 1958 by Petersen (1990). A study of variability in *Moerisia* is needed: if a single species, it is highly morphologically variable. Hydroid found on floats in low-salinity rivers and sloughs feeding into San Francisco Bay, reproducing asexually several ways, including frustulation and formation of podocysts, from which hydranths arise. Hydranth tentacles appear filiform but are moniliform, sometimes with a slightly capitate tip. Hydroids of this species, from the Petaluma River in San Francisco Bay, were incorrectly assigned to *Maeotias* by Mills and Sommer (1995); found in same location during 2004 on tubes of the polychaete *Ficopomatus enigmaticus*.

PENNARIIDAE

Pennaria disticha Goldfuss, 1820 (=*Halocordyle disticha*, =*Pennaria tiarella* [Ayres, 1854]). Hydroid and ephemeral medusa (see hydromedusa plate 52J), which sometimes release their gametes without ever being released from the hydroid colony, other times swim freely for a few hours (Mayer 1910). Reported from San Francisco Bay as *Pennaria tiarella* by Fraser (1937) without a collection date; then reported as *Pennaria* sp. on fouling panels at Mare Island, SF Bay, in 1944–1947; no further records (Cohen and Carlton 1995, Appendix 2, p. 1). See hydroid plate 52J, 52K.

POLYORCHIDAE

Polyorchis haplus Skogsberg, 1948. Medusa (hydroid unknown). Medusae found from Bodega Harbor to Scripps Pier,

* = Not in key.

La Jolla; rare. See Rees and Larson 1980, Can. J. Zool. 58: 2089–2095 (morphological variation). Color photograph in Wrobel and Mills 1998 and 2003, p. 30.

Polyorchis penicillatus (Eschscholtz, 1829). Medusa (hydroid unknown; see Brinckmann-Voss 2000, Sci. Mar. 64 [Suppl. 1]: 189–195 for retraction of discovery of "polyp of *P. penicillatus*"). Medusae found from Aleutian Islands to Sea of Cortez, primarily in protected bays. Once locally abundant throughout their range, they have disappeared or substantially decreased in numbers in the past several decades from highly urbanized or otherwise developed areas, especially in California. Medusae feed both in the water column and on the bottom, spending much of their time perched on their tentacles on the sea floor (Mills 1981). See Arai and Brinckmann-Voss (1980) and Rees and Larson 1980, Can. J. Zool. 58: 2089–2095 for morphological variability and discussion of the synonymy between *P. montereyensis* Skogsberg, 1948 and *P. penicillatus*. A number of scientists have devoted many hours to trying to culture the polyp of this charismatic species, with no success, leading us to imagine that it is likely symbiotic on or in some other organism. Newly released medusae, 1–2 mm in diameter, occur for example around marina floats heavily coated with fouling organisms in San Francisco Bay and Bodega Harbor and above a mud-bottom with abundant eelgrass and common infaunal bivalves, suggesting that a cryptic polyp is frustratingly close by. See Gladfelter 1972, Helgol. wiss. Meeresunters. 23: 38–79 (locomotor system); Haderlie et al. 1980 (general review, literature, see color photograph 3.8, plate 16, as *P. montereyensis*). Color photograph in Wrobel and Mills 1998 and 2003, p. 30.

Scrippsia pacifica Torrey, 1909. Medusa (hydroid unknown). These magnificent medusae are usually born in mid-winter, reaching adulthood in the summer months, when they are occasionally seen in large numbers in central California bays, at the surface or washed up on ocean beaches. They have been seen offshore on the bottom from about 20 m in La Jolla (by scuba divers) to as deep as 367 m in Monterey Bay (by ROV); their range is from northern California to Baja California. Color photograph in Wrobel and Mills 1998 and 2003, p. 30.

PORPITIDAE

Velella velella (Linnaeus, 1758). Hydroid and medusa. "By-the-wind sailor." Colonies polymorphic, deep blue, floating on the sea surface worldwide. Occasionally washed ashore across the region (massive strandings can be from British Columbia to central California) in large numbers, spring and summer (Alvariño 1971; Arai and Brinckmann-Voss 1980, and earlier papers by Edwards, Mackie, and others, cited therein). Color photograph in Wrobel and Mills 1998 and 2003, p. 31. Although released from the floating hydroid in copious numbers, the medusae are rarely collected in the field.

PROTOHYDRIDAE

Protohydra leuckarti Greeff, 1868. Hydroid only. Microscopic solitary polyp known mostly from Europe, with occasional sightings elsewhere including Puget Sound (1950s—see Wieser 1958, Pac. Sci. 12: 106–108; 1970s and 2005, C. E. Mills and E. N. Kozloff unpublished observations). This tentacleless hydroid (see plate 49R, 49S) is less than 1 mm in length and occurs in organically rich, sandy mud or mats of algae in brackish water and feeds on nematodes, harpacticoid copepods, and ostracodes. It is likely to be found elsewhere on the

Pacific coast as interstitial meiofauna in the low intertidal, a community that has rarely been closely investigated in our region.

TUBULARIIDAE

Hybocodon prolifer L. Agassiz, 1862. Hydroid and medusa. Synonyms in Schuchert (1996, p. 113). Hydroid with stolons often buried in sponges. Hydranths pink. Although solitary, several polyps may occur together; branched hydrocauli observed under cultivation (Petersen 1990; Schuchert 1996). Medusae with asymmetrical umbrella produce additional medusae from the one well-developed tentacle bulb; the same individuals later reproduce sexually, at which time eggs develop into actinulae, still attached to the manubrium. Medusa incorrectly recorded as "*Sarsia prolifer*" in Yaquina Bay (McCormick 1969a, b).

Pinauay crocea (L. Agassiz, 1862) (=*Ectopleura crocea*, =*Tubularia crocea*). Hydroid. Additional synonyms in Schuchert (1996, p. 107), genus *Pinauay* proposed by Marques and Migotto 2001, Pap. Avulsos Zool. (São Paulo) 41(25): 465–488, leading to the sequential (as it passed through *Ectopleura*) demise of one of the most well-known names (*Tubularia crocea*) in the hydroid and fouling literature. *Pinauay*, which means "water palm tree" in the pre-Columbian Tupi language of Brazil (because this hydroid, like the other "Tubularias," looks like a palm tree underwater), is pronounced "pin-áw-wa-i." A primary fouling species with a wide distribution, mainly in estuaries (Torrey 1902; Petersen 1990). Colonies forming conspicuous clusters on pilings and floats in bays and harbors. Gulf of Alaska to southern California, low tide to 40 m (Fraser 1937; Haderlie et al. 1980—see color photograph 3.3, p. 14). Introduced on ship bottoms from the North Atlantic.

Pinauay marina (Torrey, 1902) (=*Ectopleura marina*, =*Tubularia marina*). Hydroid. Colonies on the lee side of rocks exposed to breakers on the open seacoast; active during winter (Haderlie *et al.* 1980—see color photograph 3.2, plate 14). More delicate and sparse than *P. crocea*. British Columbia to California, lower intertidal to 37 m (Torrey 1902; Fraser 1937; Petersen 1990).

Tubularia harrimani Nutting, 1901. Hydroid. Characterized by having well-developed tentacles on the gonophores. Alaska to Monterey Bay (Fraser 1937). Sometimes found at Pescadero Point, on Monterey Peninsula before 1954 (Light et al. 1954, p. 32); present status in our area unknown.

ZANCLEIDAE

Zanclea bomala Boero, Bouillon and Gravili, 2000. Hydroid and medusa. Hydroid on unidentified bryozoan, with reticulate hydrorhiza growing under the bryozoan skeleton. Medusae released with two tentacles, but mature with four tentacles. Bodega Harbor, shallow water. See Boero et al. 2000, Italian J. Zool. 67: 93–124. There are scattered records of *Z. costata* Gegenbauer, 1856 in central California, including Rees (1975), who reports collecting a "*Z. costata*" medusa in a plankton tow in Bodega Harbor and also a zancleid polyp on floating docks in Bodega Harbor on a *Membranipora* colony, and a note from Ralph Smith that *Z. costata* is found "on pink bryozoa under boulders, Schuster's Rock (Doran Beach). In their zancleid review, Boero et al. caution that *Z. costata* is known with certainty only from the Mediterranean and that all other records must be confirmed by a study of the cnidocysts. Since two zancleids have now been described from Bodega Harbor, both associated with bryozoans, we remove *Z. costata* from our species list until such time as its presence on the West Coast is verified. Young *Z. bomala* medusae are nearly identical to those of *Z. costata*.

Zanclella bryozoophila Boero and Hewitt, 1992. Hydroid and medusa. Originally in *Zanclella*, synonymized with *Zanclea* (Schuchert 1996, p. 93), separated out again by Boero et al. (2000—see entry above). Hydroid epizoic on the bryozoan *Schizoporella* in shallow subtidal and protected by the bryozoan surface. Believed to feed on material gathered by the bryozoan lophophore. Diet microphagous, with hypostome and tentacles causing no retraction of the host's lophophores. Medusae, released with mature gonads, live only a few hours, sometimes spawning without actually being released from parent hydroid. Known only from the north jetty of Bodega Harbor. See Boero and Hewitt 1992, Can. J. Zool. 70: 1645–1651.

HYDROZOA SUBCLASS LEPTOTHECATA
(also known as LEPTOMEDUSAE and THECATA)

ORDER CONICA

AEQUOREIDAE

There seem to be at least two species of *Aequorea* on the West Coast of North America; a variety of names are available, and final identification awaits molecular genetic studies. The name *Aequorea victoria* (below) was assigned to the smaller, common, nearshore representative of the two obvious species by Arai and Brinckmann-Voss (1980), but this species, the source of both green-fluorescent protein (GFP) and aequorin, a bioluminescent protein, has also been called *A. aequorea* and *A. forskalea* in the literature. A much larger species of *Aequorea* is more oceanic off Washington, Oregon, and California, but nearshore in Alaska. These large offshore individuals in California, which occasionally make it to the coastline, seem to agree with *A. coerulescens* (below). They may or may not be the same as those in Alaska designated *A. aequorea* var. *albida* by Bigelow (1913, Proc. U.S. Natl. Mus. 44: 1–119).

Aequorea ?coerulescens (Brandt, 1835). Medusa (hydroid unknown). Very large *Aequorea* with large numbers of radial canals occasionally occur along the West Coast; these seem to be referable to this species, which was originally described from the open ocean 35°N, 144°W. This name is also used for some *Aequorea* medusae in Japan.

Aequorea victoria (Murbach and Shearer, 1902). Hydroid and medusa. Hydroid known mostly from lab culture (Strong 1925, Publ. Puget Sound Biol. Station 3: 383–399), where it grows easily. Medusae from Alaska to California, less abundant off Oregon and California than in Alaska, British Columbia, and Washington. This medusa was the source of the now laboratory-produced luminescent and fluorescent proteins aequorin and GFP. The taxonomy of this species remains confusing: *A victoria* may or may not be conspecific with *A. aequorea* (Forsskål 1775) and/or *A. forskalea* Péron and Lesueur, 1809 (see http://faculty.washington.edu/cemills/Aequorea.html). For medusa feeding behavior and diet, see Costello and Colin (2002). Color photograph in Wrobel and Mills 1998 and 2003, p. 32.

AGLAOPHENIIDAE

Aglaophenia inconspicua Torrey, 1904. Hydroid. Colonies small (35–40 mm), delicate, on algae. A southern species occuring as far north as Oregon, 0–154 m (Fraser 1937, 1946).

Aglaophenia latirostris Nutting, 1900. Hydroid (see hydroid plate 44E, 44F). Common on rocks and large red and brown algae from the low intertidal to 35 m, this species shares the common name "ostrich plume hydroid" with *A. stuthionides*. These robust hydroids are frequently seen cast ashore on beaches after storms. *A. latirostris* may have prominent reproductive structures called corbulae, containing the reduced meduoids (see color photograph in Haderlie et al. 1980, 3.15, plate 18); if ripe, these will produce planulae readily when left in the sun for approximately an hour (Light et al. 1954, p. 33). Fraser (1937, 1948) recorded it from San Pedro, Santa Barbara, and San Francisco Bay, with an overall range from British Columbia to Central America,

Aglaophenia struthionides (Murray, 1860). Hydroid. Synonyms in Fraser (1937, p. 180). This is a large (about 6.5 cm), conspicuous species in the lower intertidal and shallow subtidal, sometimes known as the "ostrich plume hydroid," a common name that it shares with *A. latirostris*. Exposed rocky shores. Alaska to southern California, low tide to 160 m (Fraser 1937, 1946). Haderlie et al. 1980 note that the nudibranch *Dendronotus subramosus* "often occurs on and closely resembles this hydroid in appearance"; see also color photograph 3.14, plate 18 (Haderlie et al. 1980).

Aglaophenia epizoica Fraser, 1948. Hydroid. A subtidal species, with large colonies (20 cm or more) and long corbulae. Abundant in the Channel Islands region (see Fraser 1948). A wealth of material of this species exists in collections from Allan Hancock Pacific Expeditions at the Santa Barbara Museum of Natural History. Santa Barbara to Baja California, 15–150 m.

Aglaophenia spp. Hydroid. Several other species of this genus have been reported from our region. See Fraser (1937, 1948) for records, and Cairns et al. (2002) for names.

BLACKFORDIIDAE

Blackfordia virginica Mayer, 1910. Hydroid and medusa. Hydroid colonies tiny (0.5 mm tall), in brackish waters; on introduced Atlantic barnacle *Amphibalanus improvisus*, attached to floats in rivers and sloughs feeding into San Francisco Bay. Medusae collected in tributaries to San Francisco Bay and Coos Bay (Mills and Rees 2000). Introduced, original provenance perhaps Ponto-Caspian; earliest collections on the West Coast are of medusae in the Napa and Petaluma Rivers in 1970 and 1974 (as "*Phialidium*," which they closely resemble).

CALYCELLIDAE

Calycella syringa (Linnaeus, 1767). Hydroid. Synonyms in Cornelius (1995a, p. 186). A stolonal epibiont on stems of other hydroids (e.g., *Tubularia*, *Hydrallmania*, *Lafoea*), as well as algae and similar substrates (Hochberg and Ljubenkov 1998). Tolerant of brackish water (Cornelius 1995a). Alaska to Baja California, intertidal to 300 m (Fraser 1937, 1946; Hochberg and Ljubenkov 1998). [*Calicella* Hincks, 1861 is not the same as *Calycella* Hincks, 1864, but it is instead a junior synonym of *Lafoea* Lamouroux, 1821; the specific name *syringa* has occasionally been combined with the genus *Calicella* in error.]

* = Not in key.

EIRENIDAE

Eirene mollis Torrey, 1909. Medusa (hydroid unknown). This West Coast species is rarely reported, possibly because it is easily overlooked among the similar, but much more abundant *Eutonina indicans* (as noted by Arai and Brinckmann-Voss, 1980). Its range is at least British Columbia to San Diego.

Eutonina indicans (Romanes, 1876). Hydroid and medusa. North Pacific and North Atlantic; Bodega Harbor, on *Zostera*, crabs, and rocks, shallow water (see Rees 1978, Wasmann J. Biol. 36: 201–209). Operculum sometimes shed in older colonies (Cornelius 1995a). Medusae abundant in summer plankton and may litter ocean beaches where the gelatinous peduncle shows up as a colorless, nipplelike piece of jelly. Color photograph in Wrobel and Mills 1998 and 2003, p. 33. It seems possible that the rarely reported *Eutimalphes brownei* medusa of Torrey (1909) (collected at that time in substantial numbers in San Diego Bay) is a junior synonym of *E. indicans*.

HALECIIDAE

Halecium spp. Hydroid. Several species of this diverse and difficult genus, including *Halecium annulatum* Torrey, 1902, *H. corrugatum* Nutting, 1899, and *H. kofoidi* Torrey, 1902 have been reported from our area. Identification is difficult or impossible unless gonophores are present. See Fraser (1937, 1946) for accounts, and Cairns et al. (2002) for nomenclature.

HALOPTERIDIDAE

Antennella avalonia Torrey, 1902. Hydroid. Synonyms in Schuchert (1997 [see below], p. 18). This species is considered to be of doubtful validity; possibly conspecific with the more cosmopolitan *A. secundaria* (Gmelin, 1791) by Schuchert (1997, Zool. Verh. Leiden 309: 1–162), but could also be a California regional species. Santa Catalina Island and southward, 4–64 m (Fraser 1937, 1946).

Monostaechas quadridens (McCrady, 1859). Hydroid. Synonyms in Hirohito (1995, pp. 249, 251) and Schuchert (1997, p. 130). A warm-water species, sometimes epizoic on other hydroids. San Francisco Bay and southward, 5–84 m (Fraser 1946, 1948).

KIRCHENPAUERIIDAE

Kirchenpaueria plumularioides (Clark, 1877) (=*Plumularia plumularioides*). Hydroid. Specific name misspelled *plumularoides* in Fraser (1937, 1946). Small colonies (30 mm), growing in clusters, lacking lateral nematothecae. Bering Sea to San Diego, intertidal to 46 m (Fraser 1946).

LAFOEIDAE

*Filellum serpens (Hassall, 1848). Hydroid. Alaska to San Diego, from low tide to 160 m (Fraser 1937); also reported from the Atlantic and Arctic Oceans (Cairns *et al.* 2002).

*Lafoea dumosa (Fleming, 1820). Hydroid. Alaska to San Pedro, from low tide to 110 m (Fraser 1937); also reported from the Atlantic and Arctic Oceans (Cairns et al. 2002).

LAODICEIDAE

Laodicea sp. Hydroid and medusa. Occasional medusae referable to the genus *Laodicea* have been collected in our area (Rees

* = Not in key.

1975, p. 2523b). Their specific identification is not known; the polyp of *L. undulata* is shown in Russell (1953, p. 238).

Ptychogena californica Torrey, 1909. Medusa (hydroid unknown). San Diego to the Bering Sea; possibly a young stage of *P. lactea* (see Arai and Brinckmann-Voss 1980, p. 83).

Ptychogena lactea A. Agassiz, 1865. Hydroid and medusa. San Diego to the Bering Sea, Arctic. Naumov (1969, p. 322) ascribes a deep water hydroid (250–520 m) to this species because they have the same general distribution. Color photograph in Wrobel and Mills, 1998 and 2003, p. 33.

LOVENELLIDAE

Eucheilota bakeri (Torrey, 1904) (=*Clytia bakeri*). Hydroid and medusa. Hydroid with stalks to 12 cm tall, epizoic on live gastropod and bivalve shells on sandy beaches, lower intertidal to subtidal; the small medusae sometimes abundant in near-shore plankton, San Francisco to Baja California, surf zone to more than 70 m (Fraser 1937; Haderlie et al. 1980, see color photograph of hydroid 3.9, p16, as *Clytia bakeri*).

MELICERTIDAE

Melicertum octocostatum (M. Sars, 1835). Hydroid and medusa. North boreal circumpolar nearshore species occasionally collected as medusae as far south as central Oregon. Further studies may restore the similar West Coast species *M. georgicum* A. Agassiz, 1862, which has been given junior synonym status by recent authors (Brinckmann-Voss and Arai 1980). Color photograph in Wrobel and Mills 1998 and 2003, p. 33. Hydroid described in Russell (1953, p. 250), but not yet collected on West Coast.

MITROCOMIDAE

Foersteria purpurea (Foerster, 1923). Medusa (hydroid unknown). Rare, possibly deep-water species, known only from British Columbia, where it occurs just off the bottom, and central California, where it has occasionally been found at the surface in Monterey Bay. Color photograph in Wrobel and Mills 1998 and 2003, p. 34.

Mitrocoma cellularia (A. Agassiz, 1865). Hydroid and medusa. The hydroid has been raised in the laboratory (Widmer 2004), but is not known from the field; it resembles those assigned to the catch-all genus *Cuspidella*. The medusae are found from the Bering Sea to southern California, from near shore to deep water (see Raskoff 2000, Scientia Marina, 64 [Suppl. 1]: 151–155); for feeding behavior and diet see Costello and Colin (2002). Color photograph in Wrobel and Mills 1998 and 2003, p. 34.

PHIALELLIDAE

Phialella fragilis (Uchida, 1938). Hydroid and medusa. Hydroid on mussels (*Mytilus*) on floating docks, with colonies best developed near edge of the shell beside the inhalant and exhalant currents. Occasionally on the same shells as *P. zappai* and having similar ecological requirements. Medusa to 10 mm. Rarely collected; known only from Japan and Bodega Harbor (Boero 1987). Possibly introduced.

Phialella zappai Boero, 1987. Hydroid and medusa; *Phialella* sp. in Boero (1987) and *Phialella* n. sp. in Rees (1975) are *P. zappai* (F. Boero pers. comm.). Hydroid on mussels (*Mytilus*) on floating docks and difficult to distinguish from *P. fragilis* without raising the medusae. Medusae to about 7 mm. Known only from Bodega Harbor. Named after Nando Boero's favorite musician, Frank Zappa, which afforded an entrée to a friendship between the two men (http://homepage.ntlworld.com/andy murkin/Resources/MusicRes/ZapRes/jellyfish.html).

PLUMULARIIDAE

Plumularia goodei Nutting, 1900. Hydroid. Small hydroids (25 mm), on shore and near-shore rocks; British Columbia to southern California (Fraser 1946).

Plumularia lagenifera Allman, 1885. Hydroid. Synonyms in Millard (1975, p. 392). Colonies to 5–10 cm high. Widely distributed in San Francisco Bay (Fraser 1937); found at Pier 39, San Francisco in 2004, and occurs all along the West Coast from the intertidal to offshore waters (Alaska to the tropics, 0–146 m) (Fraser 1937, 1946).

Plumularia setacea (Linnaeus, 1758). Hydroid. Synonyms in Cornelius (1995, p. 158). Life cycle information in Hughes 1986, Proc. R. Soc. (B) 228 (1251): 113–125. An abundant species in our region, substrate generalist; colony to 5 cm high locally; form varied, especially in epizoic colonies (Millard 1975). Recorded from boreal to tropical waters in the Eastern Pacific (Fraser 1937, 1946), the name *setacea* is applied worldwide and probably involves a multispecies complex.

SERTULARIIDAE

Abietinaria filicula (Ellis and Solander, 1786). Hydroid. Synonyms in Vervoort (1993, p. 99) and Cornelius (1995b, p. 27). Species believed to have retreated northwards in Europe over the past 150 years (Cornelius 1995). Different forms have been described (Naumov, 1969). Colonies usually less than 10 cm high. Known south to San Francisco Bay, on intertidal rocks to 66 m (Torrey, 1902; Fraser, 1937, 1946); genetic comparisons to Atlantic populations would be of interest to determine if this is indeed the same species.

Abietinaria greenei (Murray, 1860). Hydroid. Synonyms in Vervoort (1993, p. 99). Colonies in clusters, to 3 cm high. A northern species occurring south to Monterey Bay, intertidal to 37 m (Fraser 1937, 1946).

Abietinaria inconstans (Clark, 1876). Hydroid. Synonyms in Vervoort (1993, p. 99). Synonyms in Fraser (1937) include *A. amphora* Nutting, 1904 and *A. costata* Nutting, 1901. Colonies to 4 cm high, with thick stem. Alaska to Mexico, including San Francisco Bay, intertidal to 313 m (Fraser 1937, 1946, as *A. amphora*).

Abietinaria traski (Torrey, 1902). Hydroid. Synonyms in Vervoort (1993, p. 99) and Hirohito (1995, p. 156). Abundant in parts of its range and conspicuous because of the symmetry and whiteness of its colonies (Fraser 1946). Colonies pinnate, to 6 cm high. Alaska to Baja California, 10–400 m (Fraser 1937, 1946; Hochberg and Ljubenkov 1998).

Amphisbetia furcata (Trask, 1857) (=*Sertularia furcata*). Hydroid. Synonyms in Fraser (1937, p. 162). Colonies small (8 mm), common on algae, surfgrass, and shore rocks, in large patches (Fraser 1946). British Columbia to southern California and southward to the tropics, intertidal to 82 m (Fraser 1937, 1946). See color photograph in Haderlie et al., 3.11, plate 17 as *Sertularia furcata*.

Dynamena disticha (Bosc, 1802) (=*Dynamena cornicina* auct., =*Sertularia cornicina* auct.). Hydroid. Synonyms in Vervoort (1993, p. 108) and Hirohito (1995, pp. 167, 170). Colonies small (15 mm), unbranched; a warm-water species occurring north to San Francisco Bay in shallow waters (Fraser 1937,

1946, as *Sertularia cornicina*). This is another "global" species name, with numerous populations around the world requiring molecular genetic analysis.

Fraseroscyphus sinuosus (Fraser, 1948). Hydroid. New genus described by Boero and Bouillon 1993, Can J. Zool. 71: 1061-1064. Colonies small (to 20 mm), abundant on coralline algae in the shallow subtidal (6 m) on the exposed outer coast at Horseshoe Cove, Bodega Bay, in front of the Bodega Marine Laboratory.

Hydrallmania franciscana (Trask, 1857). Hydroid. Synonyms in Vervoort (1993, p. 187). According to Fraser (1946), *H. franciscana* and *H. distans* Nutting, 1899 (reported from British Columbia to San Francisco Bay) are virtually indistinguishable. Colonies 15–20 cm high. Known only from San Francisco Bay and not recorded since its original description (Vervoort 1993).

Salacia desmoides (Torrey, 1902) (=*Sertularia desmoides*). Hydroid (see hydroid plate 47F, 47G). Additional synonyms in Millard (1975, p. 274). Creeping stolon with stems to 4–24 mm high, San Francisco Bay to southern California, 2–150 m (Fraaser 1937). Also reported in South Africa and southern Indian Ocean (Millard 1975).

Sertularella fusiformis Hincks, 1861. Hydroid. Synonyms in Vervoort (1993, p. 190). Colonies small (20 mm), with few or no branches. Oregon and San Francisco Bay to the Galápagos, 11–366 m (Fraser 1937, 1946).

Sertularella spp. Hydroid. Additional species of this genus including *S. tenella* (Alder, 1856) (see hydroid plate 46K, 46L) occur in our region, as noted by Fraser (1937, 1946, 1948).

Sertularia argentea Linnaeus, 1758. Hydroid. Synonyms in Cornelius (1995b, p. 84). Similar to *Sertularia cupressina* Linnaeus, 1758. Colonies large (2–30 cm or more), growing in clusters on shore rocks. A cold-water species occurring south to San Francisco Bay, 9–119 m (Torrey 1902; Fraser 1937, 1946). As with many other Linnean (and other older Atlantic-based) taxa in our list, these species could represent either valid amphiboreal distributions or undescribed Pacific species with misapplied Atlantic names.

Sertularia spp. Hydroid. Other species of this genus, such as *Sertularia similis* Clark, 1876 (see hydroid plate 57D, 57E) occur in the region.

Symplectoscyphus erectus (Fraser, 1938) (=*Sertularella erecta* Fraser, 1938, sometimes placed in the genus *Amphisbetia*). Hydroid. Additional synonyms in Vervoort (1993, p. 239). Although still considered valid, this species (see hydroid plate 47K) is not included in the key because the description by Fraser is inconclusive. Colonies small (10 mm), erect, unbranched; southern California to the tropics, intertidal to 13 m (Fraser 1946).

Symplectoscyphus tricuspidatus (Alder, 1856) (=*Sertularella tricuspidata*). Hydroid. Additional synonyms in Vervoort (1993, p. 241), Cornelius (1995b, p. 94), and Hirohito (1995, p. 225). An epibiont on other hydroids and on mussels (Cornelius 1995b). Colonies small (to 4 cm), irregular in shape. Alaska to Baja California, 1–500 m (Fraser 1937, 1946; Hochberg and Ljubenkov 1998).

Symplectoscyphus turgidus (Trask, 1857) (=*Sertularia turgida*, *Sertularella turgida*). Hydroid. Additional synonyms in Vervoort (1993, p. 241) and Hirohito (1995, p. 225). A common species, growing on rocky bottoms and serving as a substrate for various epibionts (Hochberg and Ljubenkov 1998). Colonies small (3 cm), stiff, little branched. Alaska to Baja California, intertidal to 200 m (Fraser 1937, 1946; Haderlie et al. 1980;

* = Not in key.

Hochberg and Ljubenkov 1998). See color photograph in Haderlie et al. 1980, 3.12, plate 17 as *Sertularella turgida*.

TIAROPSIDAE

Tiaropsidium kelseyi Torrey, 1909. Medusa (hydroid unknown). Occasionally collected in San Diego, Monterey Bay, Friday Harbor, British Columbia (Arai and Brinckmann-Voss 1980; Wrobel and Mills 1998 and 2003). Color photograph in Wrobel and Mills 1998 and 2003, p. 35.

ORDER PROBOSCOIDA

CAMPANULARIIDAE

Campanularia volubilis (Linnaeus, 1758). Hydroid. Synonyms in Cornelius (1995, p. 232). Dispersive stage a planula brooded inside the female gonotheca (Cornelius 1995). Colonies stolonal. Reported from the Bering Sea to the tropics (Fraser 1937, 1946), but likely actually limited to Arctic and boreal waters.

*Clytia attenuata (Calkins, 1899). Hydroid and medusa. A little-known species of the shallow subtidal; the hydroid is known from Vancouver Island to southern California and is also reported from the Panama Canal and Brazil (Fraser 1946, 1948; West and Renshaw [see below]). Laboratory life cycle including both hydroid and medusa from material collected at Santa Catalina Island in West and Renshaw 1970, Mar. Biol. 7: 332–339. The synonymy between *"Phialidium" lomae* (see below, as *Clytia lomae*) and *Clytia attenuata* remains open to question according to Arai and Brinckmann-Voss (1980, p. 108). They found that medusae raised by West and Renshaw from the hydroid of *C. attenuata* were smaller in size, had gonads of a different shape, and fewer tentacles than the medusa *C. lomae*, and concluded "until *Clytia attenuata* can be reared from typical *Phialidium lomae* medusae the synonymy must be considered tentative."

Clytia gregaria (A. Agassiz, 1862) (=*Phialidium gregarium*, former name of medusa). Hydroid and medusa. *Phialidium* Leuckart, 1856, long used as a generic name in medusa literature, has been shown through life cycle studies to be a junior synonym of *Clytia* Lamouroux, 1812, a name originally applied to hydroids. The medusa, which swims in bursts followed by slow, upside-down sinking (Mills 1981), is often present in large numbers, thus the species name; for feeding behavior and diet, see also Costello and Colin (2002). Color photograph in Wrobel and Mills 1998 and 2003, p. 36. Hydroid described from the laboratory, raised from gametes from medusae (Strong 1925, Publ. Puget Sound Biol. Station 3: 383–399; Roosen-Runge 1970, Biol. Bull. 139: 203–221) and not corresponding well to any known species of *Clytia* from the field (Fraser 1937, 1946; Arai and Brinckmann-Voss 1980). Alaska to central Oregon (Wrobel and Mills 1998).

Clytia hemisphaerica (Linnaeus, 1767) (=*Phialidium hemisphaericum*, =*Clytia johnstoni*). Hydroid (medusa unknown on the West Coast of North America). Synonyms in Calder (1991, pp. 57–58) and Cornelius (1995, p. 252). More than one species may exist under this name, which seems not to be applicable to any of the *"Phialidium"* medusae on the West Coast, so not included in the key to hydromedusae. Hydroid colonies at least partly stolonal. On many substrates, and apparently tolerating lower salinities (Cornelius 1995; Fraser 1937 as *C. johnstoni* (Alder), central California to Alaska).

Clytia lomae (Torrey, 1909) (=*Phialidium lomae*). Medusa (hydroid unknown). Medusa described from San Diego and perhaps present along the entire West Coast. Arai and Brinckmann-Voss (1980) suspect that there are two species of *"Phialidium"* medusae along our coast, and assign the name *C. lomae* to the slightly smaller species with fewer tentacles, acknowledging that extensive life cycle studies could show only one, variable species to be here (see note under *Clytia attenuata*, above).

Gonothyraea loveni (Allman, 1859) (=*Gonothyraea clarki* [Marktanner-Turnerestcher, 1895]). Hydroid. The synonymy of *G. loveni* and *G. clarki* is generally accepted but needs confirmation; additional synonymy in Cornelius (1982, p. 92). This is the only species of *Gonothyraea* on the West Coast. The two other species assigned to that genus by Fraser are now referred to other genera (*G. gracilis* to *Clytia*; *G. inornata* to *Laomedea*); both occur north of our area. Predominantly estuarine, in colder waters. Colonies small (to 2.5 cm). Widespread in San Francisco Bay shallow waters in 2004, and occurring from there north to Alaska; intertidal to 124 m (Fraser 1937, 1946, as *G. clarki*). An abundant fouling organism.

Hartlaubella gelatinosa (Pallas, 1766) (=*Campanularia gelatinosa*). Hydroid. A boreal species found in both Atlantic and Pacific Oceans; Queen Charlotte Islands to central California, from lower intertidal to 150 m (Fraser 1937), often in harbors or estuaries. Recorded from 1859 to 1912 in San Francisco Bay including records of Agassiz 1865 and Torrey 1902, with no subsequent records (Cohen and Carlton 1995, Appendix 2, p. 1).

Laomedea calceolifera (Hincks, 1871) (=*Campanularia calceolifera*). Hydroid. Synonymy in Cornelius (1982, p. 102). A fouling species, frequent in harbors and estuaries. Colonies small (to 2.5 cm high), with sexually dimorphic gonothecae. Introduced from the Atlantic; on floating docks at Richmond Marina and Coyote Point Marina, San Francisco Bay, 2004.

Obelia bidentata Clark, 1875 (=*Obelia bicuspidata* Clark, 1875). Hydroid and medusa. While the names *O. bidentata* and *O. bicuspidata* were introduced in the same paper, the former was assigned priority under the First Reviser Principle in nomenclature by Jäderholm [1903].) Additional synonymy in Calder (1991, pp. 70–71). Medusae infrequently observed anywhere in the world. On floating dock at Pier 39, San Francisco, 2004. Hydroid throughout San Francisco Bay, 13–22 m (Fraser 1937, as *O. bicuspidata*).

Obelia dichotoma (Linnaeus, 1758) (=*Obelia commisssuralis* McCrady, 1859). Hydroid and medusa. Additional synonyms in Calder (1991, pp. 72–73), Cornelius (1995, p. 296), and Hirohito (1995, pp. 74–75). Species highly varied in form, occurring on many different substrates including swimming animals (ranging from sharks to copepods). Alaska to the tropics, frequent in our region (Fraser 1937, 1946). Standing (1976, pp. 155–164 in *Coelenterate Ecology and Behavior* [G. O. Mackie, ed.], Plenum) reports on the role of *O. dichotoma* in fouling community structure in Bodega Harbor.

Obelia geniculata (Linnaeus, 1758). Hydroid and medusa. Synonyms in Cornelius (1995, p. 301) and Hirohito (1995, p. 76). Colonies small (25 mm), immediately distinguished by asymmetrically thickened internodes and hydrothecae (Cornelius 1995), although occasional specimens occur with unthickened perisarc. On various substrates, especially algae; frequent in brackish water. British Columbia to the tropics, including our region (Fraser 1937, 1946).

Obelia longissima (Pallas, 1766). Hydroid and medusa. A large (up to 60 cm) fouling hydroid, common in harbors. Easy to confuse with *O. dichotoma*. Alaska to southern California, low

* = Not in key.

tide to 128 m (Fraser 1937, 1946). Our harbor-dwelling *Obelia* hydroids are probable ship fouling introductions. They are commonly fed upon by several nudibranch species.

Orthopyxis compressa (Clark, 1877) (=*Eucopella compressa*). Hydroid and ephemeral medusa. Discussion of synonymies and taxonomic confusion in Arai and Brinckmann-Voss (1980: 101–104). Colonies stolonal, with perisarc of varied thickness, sometimes thin but often very thick; pedicels smooth. Alaska to San Diego, 5–37 m (Fraser 1937). Common on larger hydroids or on red algae. Medusae are shed sequentially at dusk, with females released about 15–20 minutes before males from nearby colonies; the medusae live free for less than one hour, only long enough to shed gametes (see Miller 1978. J. Exp. Zool. 205: 385–392, misidentified as *O. caliculata*).

Orthopyxis everta (Clark, 1876) (=*Eucopella everta*). Hydroid. Retains gametes. Colonies stolonal, with perisarc of varied thickness, pedicels wavy or annulated. British Columbia to San Diego, 2–77 m (Fraser 1937). Sometimes abundant on kelp.

Orthopyxis integra (Macgillivray, 1842). (=*Eucopella caliculata*, =*Agastra mira*). Hydroid and ephemeral medusa. Synonyms in Cornelius (1982, p. 61; 1995, p. 235). Colonies largely stolonal. Alaska to southern California, low tide to 439 m (Fraser, 1937, 1946); cosmopolitan species.

SUBCLASS LIMNOMEDUSAE

OLINDIIDAE (formerly as OLINDIASIDAE)

Aglauropsis aeora Mills, Rees and Hand, 1976. Hydroid and medusa. Medusae collected primarily washed up on open beaches from Bodega Bay to Monterey Bay; minute polyp without tentacles known only from the laboratory. See Mills et al. 1976, Wasmann J. Biol. 34: 23–42. Color photograph in Wrobel and Mills 1998 and 2003, p. 36.

Craspedacusta sowerbii Lankester, 1880 (=*C. sowerbyi*, a misspelling). Hydroid and medusa. Introduced; now worldwide in fresh water, including in the upper Sacramento River near Redding, and in quarry lakes and reservoirs in many other areas in California. See Russell (1953) for detailed discussion. The simple, well-known hydroid, without tentacles, looks very much like that raised in the laboratory only through the primary polyp stage of both *Aglauropsis aeora* and *Maeotias marginata*.

Eperetmus typus Bigelow, 1915. Medusa (hydroid unknown). Distinguished from *Aglauropsis aeora* by the smaller number of thicker tentacles and presence of centripetal canals. From Alaska to Washington, where it becomes uncommon; rare sightings in Coos and Yaquina Bays. Usually pale pink. Color photograph in Wrobel and Mills 1998 and 2003, p. 37. Records from Japan are an undescribed *Aglauropsis*; see Mills et al. 1976, Wasmann J. Biol. 34: 23–42.

Gonionemus vertens A. Agassiz, 1862. Hydroid and medusa. In the shallow subtidal, usually seen clinging to algae or eelgrass, but may also be free-swimming near the surface in protected bays. Indigenous from Alaska to Washington, but known from a variety of locations worldwide and might be expected south of Washington. See plate 49Q for tiny, cryptic solitary polyp; color photograph of medusa in Wrobel and Mills 1998 and 2003, p. 37. A virulent-stinging variety or separate species occurs in the Russian Far East. See Edwards 1976, Adv. Mar. Biol. 14: 251–284 for a global review.

Maeotias marginata (Modeer, 1791). (=*Maeotias inexspectata* Ostroumoff, 1896 [misspelled occasionally as *inexpectata*]; see Mills and Rees 2000). Hydroid and medusa. Most medusae found in the San Francisco Bay system are males, but a few

females discovered in 1998 allowed for the culture of embryos. Hydroids of *Maeotias* are known only from juvenile polyps raised under laboratory conditions (Rees and Gershwin 2000); these are miniscule and morphologically simple, with a cluster of cnidocysts around the mouth and without tentacles. A brackish to freshwater species, introduced to the San Francisco Bay area by the 1980s or 1990s, with an unconfirmed observation in 1959 (see Mills and Rees 2000). Color photograph in Wrobel and Mills 1998 and 2003, p. 29. A hydroid identified as this species by Mills and Sommer (1995) from the San Francisco area is *Moerisia* sp. instead (see Mills and Rees 2000).

Vallentinia adherens Hyman, 1947. Medusa (hydroid unknown). Occurs near shore, clinging to algae (see Hyman 1947, Trans. Am. Microsc. Soc. 66: 262–268); known only from the Pacific Grove area (where it is found on the kelp *Macrocystis* off Hopkins Marine Station [between the breakwater and Point Piños], Freya Sommer personal communication), and Santa Barbara (Wrobel and Mills 1998 and 2003); rare. Color photograph in Wrobel and Mills 1998 and 2003, p. 37.

SUBCLASS SIPHONOPHORA

ORDER PHYSONECTAE

AGALMATIDAE

Agalma elegans (*pro parte* M. Sars, 1846)—Sars' original description included more than one species, and authorship is thus noted as *pro parte*. A cosmopolitan species, which can be found anywhere from Alaska to Mexico. Easily distinguished from *A. okeni* as it has a long stem with leaflike bracts, while in *A. okeni* the stem is short so the bracts, with two (young) or four (mature) distal facets, interlock with each other. Pacific, Indian, and Atlantic Oceans and the Mediterranean.

*Agalma okeni Eschscholtz, 1825. This second cosmopolitan species of *Agalma* (siphonophore plate 59C) is also present on our coast, but more likely to be encountered off southern California and Baja California. Further distinguished from *A. elegans* by the distinctive ridges on the swimming bells, and Y-shape of the subumbrella when viewed from above. Pacific, Indian, and Atlantic Oceans and the Mediterranean.

Nanomia bijuga (delle Chiaje, 1841). Probably the most common physonect off the West Coast, thought to be responsible in some regions for the deep scattering layer (Barham 1963, Science 140: 826–828; Barham 1966, Science 151: 1399–1403), but occurs to the surface; Pacific, Indian, and Atlantic Oceans and the Mediterranean. Color photograph in Wrobel and Mills, 1998 and 2003, p. 46.

APOLEMIIDAE

Apolemia spp. This genus is quite diverse on our coast, with several undescribed species (the name "*Apolemia uvaria*" has been applied rather indiscriminately in past West Coast literature). The colonies are often tens of meters long and in deep water, but many-centimeter-long fragments can be encountered near shore at the surface. They have an overall "fuzzy" appearance, with red or white gastrozooids, and pack a substantial sting. The flimsy, jelly-filled bracts also contain patches of stinging cells on their upper surfaces. Species occur in the Pacific, Indian, and Atlantic Oceans and Mediterranean. Color photograph in Wrobel and Mills 1998 and 2003, p. 45.

* = Not in key.

FORSKALIIDAE

Forskalia spp. Several species of *Forskalia,* which are difficult to distinguish, might be encountered along our coast. Divers sometimes liken the overall aspect of *Forskalia* to a Christmas tree: conical, widening at the base, with fine tentacles coming out from within the overall shape. They are active, strong swimmers, often spiraling around as they move. When disturbed they may release clouds of pigmented, bioluminescent material. Species occur in the Pacific, Indian, and Atlantic Oceans and the Mediterranean. Color photograph in Wrobel and Mills 1998 and 2003, p. 46.

PHYSOPHORIDAE

Physophora hydrostatica Forsskål, 1775. Colonies are typically several centimeters high and the compact complexity, symmetry, and pastel blue and pink colors of this worldwide species are sure to engender wonder in anyone who sees it; Pacific, Indian, and Atlantic Oceans and the Mediterranean. Color photograph in Wrobel and Mills 1998 and 2003, p. 46.

ORDER CALYCOPHORAE

ABYLIDAE

Abylopsis tetragona (Otto, 1823). This distinctive species is likely to be encountered only at the southern end of the range of this book; Pacific, Indian, and Atlantic Oceans and the Mediterranean.

Bassia bassensis (Quoy and Gaimard, 1833). Another polyhedral species (siphonophore plate 60C) similar to *Abylopsis tetragona,* which is also likely to be encountered only at the southern end of the range of this book; the ridges of the swimming bells have a bluish tinge; Pacific, Indian, and Atlantic Oceans and the Mediterranean.

CLAUSOPHYIDAE

Chuniphyes multidentata Lens and van Riemsdijk, 1908. An abundant midwater species that is occasionally encountered at the surface in central California; Pacific, Indian, and Atlantic Oceans.

DIPHYIDAE

Chelophyes appendiculata (Eschscholtz, 1829). One of the most common epipelagic temperate and tropical oceanic siphonophore species and likely to be seen anywhere along the Pacific West Coast including Baja California; Pacific, Indian, and Atlantic Oceans and the Mediterranean. When present in substantial numbers, this species has enough sting to be quite bothersome to divers and snorkellers.

Chelophyes contorta (Lens and van Riemsdijk, 1908). Perhaps a near-shore, rather than oceanic species (siphonophore plate 60H) (Totton 1965), with a somewhat more southern distribution that *C. appendiculata,* so to be expected only in the southern range of this book, continuing down into Mexico; appears to have an Indo-Pacific distribution (Bouillon et al 2004).

Dimophyes arctica (Chun, 1897). In spite of its name, this is a cosmopolitan species found in all oceans including the Arctic and Antarctic and can be encountered anywhere along the Pacific coast of North America.

Diphyes bojani (Eschscholtz, 1829). Might be encountered anywhere along the California and Baja California coasts; Pacific, Indian, and Atlantic Oceans and the Mediterranean.

Diphyes dispar Chamisso and Eysenhardt, 1821. More likely to be encountered in the southern range of this book, continuing down Baja California, but has worldwide distribution in warmer waters. Color photograph in Wrobel and Mills 1998 and 2003, p. 47.

Eudoxoides spiralis (Bigelow, 1911). Epipelagic species found usually south of about 40°N on our coast; Pacific, Indian, and Atlantic Oceans and the Mediterranean.

Lensia challengeri Totton, 1954. Can be encountered anywhere along the Californian and Baja Californian coast; found throughout the Pacific, usually south of about 40°N, usually near shore.

Lensia conoidea (Keferstein and Ehlers, 1860). Cosmopolitan species and likely to be seen anywhere along the Pacific West Coast including Baja California.

Lensia hostile Totton, 1941. A typically deep-water species found off California; Pacific, Indian, and Atlantic Oceans.

Lensia hotspur Totton, 1941. Can be encountered anywhere from Oregon to Baja California; Pacific, Indian, and Atlantic Oceans and the Mediterranean.

Lensia spp. Several other little *Lensia*s are found off the West Coast; only the most common species have been included in the key.

Muggiaea atlantica Cunningham, 1892. A coastal species of the temperate Pacific, Indian, and Atlantic Oceans and the Mediterranean that can be found throughout the study area. In some localities, *M. atlantica* can be replaced by *M. kochi* at different times of year, which might be related to water temperature, but the two species appear to be mutually exclusive. *M. atlantica* has been collected in Bodega Harbor. Photograph in Wrobel and Mills 1998 and 2003, p. 47.

Sulculeolaria quadrivalvis Blainville, 1834. The looping radial canals and lack of ridges are distinctive of this genus among the Diphyidae; Pacific, Indian, and Atlantic Oceans and the Mediterranean. Color photograph in Wrobel and Mills 1998 and 2003, p. 47.

PRAYIDAE

Desmophyes annectens Haeckel, 1888. Shaped like *Praya* and *Rosacea,* but with minute red pigment flecks around the opening of the subumbrella in the swimming bells in life, and with four straight radial canals; large, spherical, white somatocyst. Uncommon; Pacific, Indian, and Atlantic Oceans and the Mediterranean.

Lilyopsis rosea Chun, 1885. An uncommon prayid species (siphonophore plate 59I) with a large subumbrellar cavity, which has been seen in central California; Pacific, and Atlantic Oceans and the Mediterranean.

Praya dubia (Quoy and Gaimard, 1827). These siphonophores are often tens of meters long and in deep water, but several cm-long pieces of the colonies may be encountered near shore at the surface. The bright yellow color of the gastrozooids is striking; they have a substantial sting. Pacific, Indian, and Atlantic Oceans.

Praya reticulata (H. B. Bigelow, 1911). Similar to above species, but with reticulate pattern of canals on the subumbrella. The branching pattern of the somatocyst also distinguishes it. Pacific, Indian, and Atlantic Oceans.

* = Not in key.

Rosacea spp. Several species of *Rosacea* occur off the West Coast of North America and are difficult to identify and often confused with *Praya* spp. Typically, the bracts are hemispherical while those of *Praya* spp. are flattened. A near-shore observer is most likely to run into fragments of one of these colonies, which can reach many meters in length when undamaged and will sting.

SPHAERONECTIDAE

Sphaeronectes gracilis (Claus, 1873, 1874). Sometimes occurs near shore from Monterey Bay south, but can be very difficult to see; Pacific, Indian, and Atlantic Oceans and the Mediterranean.

SUBCLASS NARCOMEDUSAE

AEGINIDAE

Aegina citrea Eschscholtz, 1829. Medusa only. A variable worldwide, oceanic species (that may turn out with molecular study to be a species complex) that may occasionally be seen near shore; sometimes infused with yellow pigment. Color photograph in Wrobel and Mills 1998 and 2003, p. 38. Undescribed aeginids are also present in deep water.

CUNINIDAE

Cunina spp. Medusa only. Worldwide, oceanic species that occasionally come near shore. These are typically 10–60 mm in bell diameter and may be transparent and colorless or have some color. See Kramp (1961, 1968) for specific characters. Color photograph in Wrobel and Mills 1998 and 2003, p. 39.

Solmissus incisa (Fewkes, 1886). Medusa only. A worldwide, oceanic species, this is the larger (to 100 mm bell diameter) and less common *Solmissus*; it can be colorless or sometimes infused with transparent purple color (see Kramp 1961, 1968).

Solmissus marshalli A. Agassiz and Mayer, 1902. Medusa only. A worldwide, oceanic species, this is the smaller and more common of the two *Solmissus* that might be encountered near shore. It is usually colorless and <60 mm in bell diameter (see Kramp 1961, 1968). Color photograph in Wrobel and Mills 1998 and 2003, p. 39.

SOLMARISIDAE

Pegantha spp. Medusa only. Worldwide, oceanic species that occasionally come near shore. These are typically 25–50 mm in bell diameter and may be transparent and colorless or have some color. See Kramp (1961, 1968) for specific characters. Color photograph in Wrobel and Mills 1998 and 2003, p. 38.

Solmaris spp. Medusa only. Worldwide, oceanic and coastal species that are sometimes encountered near shore, sometimes in great numbers. These are small transparent medusae with a rapid pulsation rate. See Kramp (1961, 1968) for specific characters. Color photograph in Wrobel and Mills 1998 and 2003, p. 39.

TETRAPLATIDAE

Tetraplatia volitans Busch, 1851. Highly reduced narcomedusa up to about 1 cm long that looks more like a flying worm or pteropod than a jellyfish, with a ringlike constriction fairly near the midpoint dividing the oral and aboral ends, which are connected by four flying buttress–like structures. Oceanic in the upper 900 m, but occasionally found near shore; feeds on zooplankton (see Hand 1955, Pac. Sci. 9: 332–348; color photograph in Wrobel and Mills 1998 and 2003, p. 52, and at http://jellieszone.com/tetraplatia.htm). The two species of *Tetraplatia* have been proposed as both coronate scyphomedusae and as narcomedusae, but a genetic study by Collins et al. (2006) has placed these unusual medusae in the hydrozoan Narcomedusae.

SUBCLASS ACTINULIDAE (=HALAMMOHYDROIDA)

HALAMMOHYDRIDAE

Halammohydra sp. Medusa only. Minute (0.5–2 mm), highly reduced medusa (without a polyp phase, although it looks like a polyp), living interstitially in sand; solitary not colonial form. Found on a beach near Moss Landing (as reported by Robert Higgins and James Nybbaken). The entirely ciliated animal consists mostly of a manubrium with two whorls of long, contractile tentacles; there is a statocyst between each pair of tentacles and an aboral adhesive organ. A number of species have been described.

SUBCLASS TRACHYMEDUSAE

GERYONIDAE

Geryonia proboscidalis (Forsskål, 1775). Medusa only. Oceanic warm waters in the Pacific, Atlantic, and Mediterranean; occasionally seen near shore in the southern range of this book; six-part symmetry distinguishes this less-common species from *Liriope tetraphylla*. Color photograph in Wrobel and Mills 1998 and 2003, p. 40.

Liriope tetraphylla (Chamisso and Eysenhardt, 1821). Medusa only. Oceanic, Pacific, Atlantic, and the Mediterranean; occasionally seen near shore throughout the range of this book, sometimes in great numbers in warm water masses; four-part symmetry distinguishes this species from the less-common *Geryonia proboscidalis*. Color photograph in Wrobel and Mills 1998 and 2003, p. 40.

RHOPALONEMATIDAE

Aglantha digitale (O. F. Müller, 1776). Medusa only. A common species in the North Pacific, North Atlantic, and Arctic, typical of the upper 200 m and sometimes found nearshore. This species has two modes of swimming: a general slow swim and a strong escape swim separately mediated by giant axons. Usually colorless, but may have red, pink or orange color on the tentacles. For feeding behavior and diet see Costello and Colin (2002). Color photograph in Wrobel and Mills 1998 and 2003, p. 42.

Aglaura hemistoma Péron and Lesueur, 1809. Medusa only. Oceanic, Pacific, Atlantic, and the Mediterranean; occasionally nearshore, between about 40°N and 40°S, replacing *Aglantha* as the most abundant epipelagic species in warmer waters. It is smaller and more fragile than *Aglantha* and has two modes of swimming (slow feeding mode and fast escape swim). Color photograph in Wrobel and Mills 1998 and 2003, p. 42.

* = Not in key.

References

Alvariño, A. 1971. Siphonophores of the Pacific, with a review of the world distribution. Bulletin of the Scripps Institute of Oceanography, University of California Technical Series 16: 1–432.

Arai, M. N. and A. Brinckmann-Voss 1980. Hydromedusae of British Columbia and Puget Sound. Can. Bull. Fish. Aquatic Sci. 204: 192 pp.

Boero, F. 1987. Life cycles of *Phialella zappai* n. sp., *Phialella fragilis* and *Phialella* sp. (Cnidaria, Leptomedusae, Phialellidae) from Central California. J. Nat. Hist. 21: 465–480.

Boero, F., J. Bouillon and C. Gravili 2000. A survey of *Zanclea, Halocoryne* and *Zanclella* (Cnidaria, Hydrozoa, Anthomedusae, Zancleidae) with description of new species. Italian Journal of Zoology 67: 93–124.

Bouillon, J. 1985. Essai de classification des Hydropolypes—Hydroméduses (Hydrozoa-Cnidaria). Indo-Malayan Zool. 2: 29-243, tabs 1–32.

Bouillon, J. 1995a. Cnidaires: Généralités. In Traité de Zoologie, Vol. 3. P. P. Grassé, ed. Fascicule 2. Masson, Paris.

Bouillon, J. 1995b. Classe des Hydrozoaires. In Traité de Zoologie, Vol. 3. P. P. Grassé, ed. Fascicule 2. Masson, Paris.

Bouillon, J. and F. Boero 2000. The Hydrozoa: a new classification in the light of old knowledge. Thal. Salent. 24: 1–45.

Bouillon, J., M. D. Medel, F. Pagès, J. M. Gili, F. Boero, and C. Gravili 2004. Fauna of the Mediterranean Hydrozoa. Sci. Mar. 68 (Suppl. 2): 1–449. (Full text and illustrations available online at http://www. icm.csic.es/scimar/vol68s2.html.)

Brinckmann-Voss, A. 1970. Anthomedusae/Athecatae (Hydrozoa, Cnidaria) of the Mediterranean. Part I. Capitata (with 11 colour plates including Filifera). Fauna e Flora del Golfo di Napoli 39, 96 pp plus 11 plates.

Cairns, S. D., D. R. Calder, A. Brinckmann-Voss, C. B. Castro, D. G. Fautin, P. R. Pugh, C. E. Mills, W. C. Jaap, M. N. Arai, S. H. D. Haddock and D. M. Opresko 2002. Common and Scientific Names of Aquatic Invertebrates from the United States and Canada: Cnidaria and Ctenophora – Second Edition. American Fisheries Society Special Publication No. 28, Bethesda, Maryland, 115 pp.

Calder, D. R. 1988. Shallow-water hydroids of Bermuda: The Athecatae. Royal Ontario Museum, Life Sci. Contrib. 148: 1–107.

Calder, D. R. 1991. Shallow-water hydroids of Bermuda: The Thecatae, exclusive of Plumularioidea. Royal Ontario Museum, Life Sci. Contrib. 154: 1–140.

Calder, D. R. 1997. Shallow-water hydroids of Bermuda: superfamily Plumularioidea. Royal Ontario Museum, Life Sci. Contrib. 161: 1–107.

Cohen, A. N. and J. T. Carlton 1995. Biological Study. Nonindigenous Aquatic Species in a United States Estuary: A Case Study of the Biological Invasions of the San Francisco Bay and Delta. A Report for the United States Fish and Wildlife Service, Washington, D.C., and The National Sea Grant College Program, Connecticut Sea Grant, NTIS Report Number PB96-166525, 246 pp. + Appendices.

Collins, A. G. 2000. Towards understanding the phylogenetic history of Hydrozoa: hypothesis testing with 18S gene sequence data. Sci. Mar. 64 (Suppl. 1): 5–22.

Collins, A. G. 2002. Phylogeny of Medusozoa and the evolution of cnidarian life cycles. J. Evol. Biol. 15: 418–432.

Collins, A. G., B. Bentlage, G. I. Matsumoto, S. H. D. Haddock, K. J. Osborn and B. Schierwater 2006. Solution to the phylogenetic enigma of *Tetraplatia*, a worm-shaped cnidarian. Biology Letters 2: 120–124.

Cornelius, P. F. S. 1982. Hydroids and medusae of the family Campanulariidae recorded from the eastern North Atlantic, with a world synopsis of genera. Bull. British Museum (Natural History), Zool. 42: 37–148.

Cornelius, P. F. S. 1995a. North-west European thecate hydroids and their medusae. Part I. Introduction, Laodiceidae to Haleciidae. Synopses of the British Fauna (n.s.) 50: 347 pp.

Cornelius, P. F. S. 1995b. North-west European thecate hydroids and their medusae. Part 2. Sertulariidae to Campanulariidae. Synopses of the British Fauna (n.s.) 50: 386 pp.

Costello, J. H. and S. P. Colin 2002. Prey resource use by coexistent hydromedusae from Friday Harbor, Washington. Limnol. Oceanogr. 47: 934–942.

Fraser, C. M. 1937. Hydroids of the pacific coast of Canada and the United States. Toronto: University of Toronto Press, 207 pp.

Fraser, C. M. 1946. Distribution and relationship in American hydroids. Toronto: University of Toronto Press, 451 pp.

Fraser, C. M. 1948. Hydroids of the Allan Hancock Pacific Expeditions since March 1938. Allan Hancock Pacific Expedition, 4(5): 179-343, pls 22-42.

Haderlie, E. C., C. Hand and W. B. Gladfelter 1980. Cnidaria (Coelenterata): the sea anemones and allies. In Intertidal invertebrates of California, pp. 40–75. R. H. Morris, D. P. Abbott, and E. C. Haderlie. Stanford,CA: Stanford University Press.

Hirohito, Emperor of Japan 1988. The hydroids of Sagami Bay. (Part 1. Athecata). Publs. Biol. Lab., Imp. Household, Tokyo, 1988. i–x, 1–179 (English text), 1–110 (Japanese text), figs. 1–54, pls 1–4, 2 maps.

Hirohito, Emperor of Japan 1995. The hydroids of Sagami Bay. II. Thecata. Publs. Biol. Lab., Imp. Household, Tokyo, 1995. i–x, 1–355 (English text), 1–245 (Japanese text), figs. 1–106, pls. 1–13, 2 maps.

Hochberg, F. G. and J. C. Ljubenkov 1998. Class Hydrozoa. In Taxonomic atlas of the benthic fauna of the Santa Maria Basin and the western Santa Barbara Channel. 3. The Cnidaria, pp. 1–54. P. V. Scott and J. A. Blake, eds. Santa Barbara, CA: Santa Barbara Museum of Natural History, 150 pp.

Kirkpatrick, P. A. and P. R. Pugh 1984. Siphonophores and Velellids. E. J. Brill, London.

Kramp, P. L. 1961. Synopsis of the Medusae of the World. J. Mar. Biol. Assoc. U. K. 40: 1–469. (Available online at http://www.mba.ac.uk/ nmbl/publications/jmba_40/jmba_40.htm.)

Kramp, P. L. 1965. The Hydromedusae of the Pacific and Indian Oceans. Dana Report 63: 1–162.

Kramp, P. L. 1968. The Hydromedusae of the Pacific and Indian Oceans. Sections II and III. Dana Report 72: 1–200.

Marques, A. C. 2001. Simplifying hydrozoan classification: inappropriateness of the group Hydroidomedusae in a phylogenetic context. Contr. Zool. 70: 175–179.

Marques, A. C. and A. G. Collins 2004. Cladistic analysis of Medusozoa and cnidarian evolution. Invertebrate Biology, 123: 23–42.

McCormick, J. M. 1969a. Hydrographic and trophic relationships of hydromedusae in Yaquina Bay, Oregon. PhD dissertation, Oregon State University, Corvallis, Oregon, 125 pp.

McCormick, J. M. 1969b. Trophic relationships of hydromedusae in Yaquina Bay, Oregon. Northwest Science 43: 207–214.

Millard, N. A. H. 1975. Monograph on the Hydroida of southern Africa. Annals South African Museum 68: 1–513.

Mills, C. E. 1981. Diversity of swimming behaviors in hydromedusae as related to feeding and utilization of space. Mar. Biol. 64: 185–189.

Mills, C. E. 1987. Key to the Hydromedusae. In Marine invertebrates of the Pacific Northwest, pp. 32–44. E. N. Kozloff, ed. Seattle: University of Washington Press.

Mills, C. E. 1987. Key to the Order Siphonophora. In Marine Invertebrates of the Pacific Northwest, pp. 62–65. E. N. Kozloff, ed. Seattle: University of Washington Press.

Mills, C. E. 1996. Additions and corrections to the keys to Hydromedusae, Hydroid polyps, Siphonophora, Stauromedusan Scyphozoa, Actiniaria, and Ctenophora. In Marine invertebrates of the Pacific Northwest, with revisions and corrections, pp. 487–491. E. N. Kozloff, ed. Seattle: University of Washington Press.

Mills, C. E. and R. L. Miller 1987. Key to the Hydroid polyps. In Marine Invertebrates of the Pacific Northwest, pp. 44–61. E. N. Kozloff, ed. Seattle: University of Washington Press.

Mills, C. E. and J. T. Rees 2000. New observations and corrections concerning the trio of invasive hydromedusae *Maeotias marginata* (=*M. inexpectata*), *Blackfordia virginica*, and *Moerisia* sp. in the San Francisco Estuary. Scientia Marina 64 (Suppl. 1): 151–155.

Mills, C. E. and F. Sommer 1995. Invertebrate introductions in marine habitats: two species of hydromedusae (Cnidaria) native to the Black Sea, *Maeotias inexpectata* and *Blackfordia virginica*, invade San Francisco Bay. Mar. Biol. 122: 279–288.

Naumov, D. V. 1969. Hydroids and Hydromedusae of the USSR. Translated from Russian by the Israel Program for Scientific Translations, Jerusalem, 660 pp. (Naumov, D. V., 1960. Gidroidi i gidromedusy morskikh, solonovatovodnykh i presnovodnykh basseinov SSSR—Opredeleteli po faune SSSR, Izdavaemye Zoologicheskim Institutom Akademii Nauk SSSR 70, 626 pp.)

Petersen, K. W. 1979. Development of coloniality in Hydrozoa. In Biology and systematics of colonial organisms. G. Larwood and B. R. Rosen, eds. pp. 105–139. London: Academic Press, London.

Petersen, K. W. 1990. Evolution and taxonomy in capitate hydroids and medusae. Zool. J. Linnean Soc. 100: 101–231.

Pugh, P. R. 1999. Siphonophorae. In South Atlantic Zooplankton I. D. Boltovskoy, ed. Backhuys, Leiden, pp. 467–511.

Rees, J. T. 1975. Studies on Hydrozoa of the central California coast: aspects of systematics and ecology. PhD dissertation, University of California at Berkeley, 267 pp.

Rees, J. T. and L. Gershwin 2000. Non-indigenous hydromedusae in California's upper San Francisco Estuary: life cycles, distribution, and potential environmental impacts. Sci. Mar. 64 (Suppl. 1): 73–86.

Russell, F. S. 1953. The Medusae of the British Isles. Cambridge: Cambridge University Press,Cambridge, xiii + 530 pp. (Full text and illustrations available online at http://www.mba.ac.uk/nmbl/publications/medusae_1/medusae_1.htm.)

Russell, F. S. 1970. The Medusae of the British Isles. Vol. II. Pelagic Scyphozoa, with a supplement to the first volume on Hydromedusae. Cambridge: Cambridge University Press, 284 pp. (Full text and illustrations available online at http://www.mba.ac.uk/nmbl/publications/medusae_2/medusae_2.htm.)

Schuchert, P. 1996. The marine fauna of New Zealand: athecate hydroids and their medusae. New Zealand Oceanographic Institute Memoir 106: 1–159.

Schuchert, P. 1998. How many hydrozoan species are there? Zool. Verh. Leiden 323: 209–219.

Schuchert, P. 2001a. Hydroids of Greenland and Iceland (Cnidaria, Hydrozoa). Meddelelser om Grønland, Bioscience 53: 1–184.

Schuchert, P. 2001b. Survey of the family Corynidae (Cnidaria, Hydrozoa). Rev. Suisse Zool. 108: 739–878.

Schuchert, P. 2004. Revision of the European athecate hydroids and their medusae (Hydrozoa, Cnidaria): Families Oceanidae and Pachycordylidae. Rev. Suisse Zool. 111: 315–369.

Schuchert, P. 2004–present. The Hydrozoa Directory. Online web site at http://www.ville-ge.ch/musinfo/mhng/hydrozoa/hydrozoa-directory.htm.

Strong, L. H. 1925. Development of certain Puget Sound hydroids and medusae. Publications of the Puget Sound Biological Station 3: 383–399.

Torrey, H. B. 1902. The Hydroida of the Pacific coast of North America, with especial reference to the species in the collection of the University of California. Univ. Cal. Publ. Zool. 1: 1–104.

Torrey, H. B. 1909. The Leptomedusae of the San Diego region. Univ. Cal. Publ. Zool. 6: 11–31.

Totton, A. K. 1965. A Synopsis of the Siphonophora. London: British Museum (Natural History).

Vervoort, W. 1993. Cnidaria, Hydrozoa, Hydroida: Hydroids from the Western Pacific (Philippines, Indonesia and New Caledonia) I: Sertulariidae (Part 1). In: Résultats des Campagnes MUSORSTOM, 11. Mém. Mus. natn. Hist. nat. Paris, 158, Zool.: 89–298.

Vervoort, W. and J. E. Watson 2003. The marine fauna of New Zealand: Leptothecata (Cnidaria: Hydrozoa) (Thecate Hydroids). NIWA Biodiversity Memoir 119, 538 pp.

Widmer, C. L. 2004. The hydroid and early medusa stages of Mitrocoma cellularia (Hydrozoa, Mitrocomidae). Mar. Biol. 145: 315–321.

Wrobel, D. and C. Mills 1998, reprinted with corrections 2003. Pacific coast pelagic invertebrates: a guide to the common gelatinous animals. Monterey, California: Sea Challengers and the Monterey Bay Aquarium, 108 pp.

Scyphozoa: Scyphomedusae, Stauromedusae, and Cubomedusae

CLAUDIA E. MILLS AND RONALD J. LARSON

(Plates 61–63)

The relatedness of the 200 worldwide species of scyphozoan jellyfish (Mianzan and Cornelius 1999) known as semaeostome medusae, rhizostome medusae, coronate medusae, stauromedusae, and cubomedusae remains unclear (Dawson 2004b, Marques and Collins 2004). Here we use the names for the three jellyfish groups that we treat in this section without making phylogenetic judgments.

The jellyfish we cover are those that are likely to be found in the intertidal, shallow subtidal, or in bays and harbors along the coast. The semaeostome scyphomedusae are usually large and often colorful, pelagic medusae that are not encountered in the intertidal zone as adults except when cast ashore, but may often be seen in harbors. Nearshore species of semaeostome scyphomedusae are considered here. The stauromedusae, in contrast, are small, inconspicuous, stalked medusae that are found either in protected bays or in high-current or wave-swept areas in the lower intertidal and subtidal,

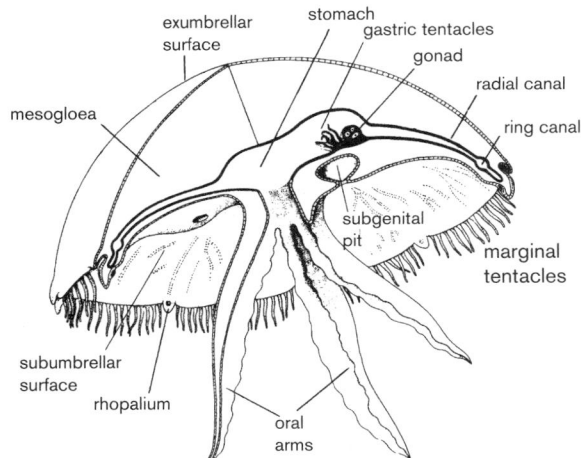

PLATE 61 Scyphozoa. Scyphomedusan structure, diagrammatic (modified after Naumov from Bayer and Owre, *The Free-Living Lower Invertebrates*, 1968; used with permission of The MacMillan Company).

attached by an aboral stalk to eelgrass, seaweed, or rock. One cubomedusa, *Carybdea* sp., may be found from Santa Barbara south in the shallow subtidal along the open coast. Two species of rhizostome scyphomedusae, *Phyllorhiza punctata* and *Stomolophus meleagris*, may also occasionally be found in Southern California.

Scyphomedusae (sometimes rather objectionably called the "true" jellyfish) can be distinguished from hydromedusae by their usually larger size, frilly mouth lobes, scalloped margins bearing lappets, absence of a velum, presence of marginal rhopalia, and often-complex pattern of radial canals (plates 61 and 62). In contrast, hydromedusae (see plate 39) usually are small, often glassy-clear, and possess a velum; most have four simple radial canals. For an account of the more fundamental morphological differences, the student is referred to Hyman (1940a), Russell (1970), or any detailed invertebrate zoology textbook.

Semaeostome scyphomedusae that live near our coast have an attached polypoid part of their life cycle: the soft, white "scyphistoma" stage, which can be encountered in great numbers under shaded parts of floats in harbors or marinas or on boat hulls. Most (probably all) of these scyphistomae (plate 62A, 62B) encountered in harbors on the West Coast of America belong to *Aurelia*. The polyp phases of other West Coast scyphomedusae have not been located in the field, although most have now been cultured in the laboratory by curators at the Monterey Bay Aquarium (some are described in Gershwin and Collins 2002).

Semaeostome scyphomedusae, when they are plentiful, play a significant role in coastal food webs by consuming a variety of zooplankton prey, ranging from small copepods to ctenophores and other large medusae. A number of other species depend on them for food or protection. Jellyfish are eaten by sunfish (*Mola mola*) and leatherback turtles. Gotshall et al. (1965) note that in California the blue rockfish *Sebastes mystinus* "stalks" *Chrysaora*, biting off pieces of the oral arms while avoiding contact with the marginal tentacles, and show a photograph of about a dozen fish feeding on a single *C. fuscescens*. Other jellyfish found in guts of the blue rockfish included *Aurelia* and "*Pelagia*" (= *C. colorata*?). A variety of species "hitchhikes" on the larger medusae, especially crustaceans such as amphipods and crab larvae.

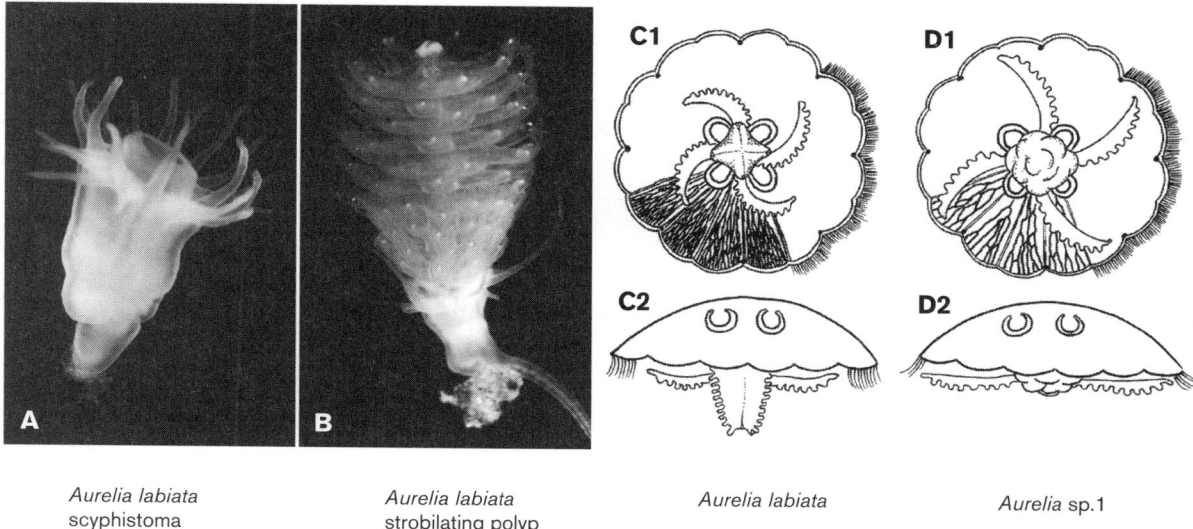

C1   D1

C2   D2

Aurelia labiata   Aurelia labiata   Aurelia labiata   Aurelia sp.1
scyphistoma   strobilating polyp

PLATE 62 *Aurelia* spp. A, *Aurelia labiata*. scyphistoma; B, *Aurelia labiata*. strobilating polyp; C1, C2, *Aurelia labiata* (oral and side views); D1, D2, *Aurelia* sp. 1 (oral and side views) (all images by Claudia E. Mills).

Many of the medusae discussed in this section are large and awkward to study whole under a microscope. Preservation for morphological purposes should be done in 5% to 10% percent buffered formalin; specimens or portions of specimens for molecular studies should be preserved in >70% ethanol and stored in a refrigerator or freezer (Dawson 2004b).

Beautiful color images of most of the scyphomedusae in this key can be found on the Internet. See especially http:// jel-lieszone.com/scyphomedusae.htm and http://divebums.com/main.html for particularly nice photographs taken mostly in California. Caution must be exercised in the accuracy of web identifications of some of the species; use keys such as this one for the authoritative name for a species.

Families used in the species list are from Cairns et al. (2002).

Glossary of Scyphozoa

ABORAL the side of the jellyfish opposite to that on which the mouth opens (the upper surface of the bell in pelagic species).

CALYX the flaring bell-like portion of a stauromedusa, containing the gonads, central manubrium and marginal tentacle clusters.

CNIDOCYST (=**NEMATOCYST**) stinging organelle characteristic of the Cnidaria, consisting of a double-walled capsule containing fluid and a long tubule that everts and straightens when the capsule discharges upon stimulation; used for prey capture, defense and attachment (also inaccurately called "stinging cell").

CNIDOCYST VESICLES round, white, cnidocyst-filled structures visible through the calyx wall of some species of stauromedusae.

EXUMBRELLA the outer, aboral, surface of a jellyfish "bell."

GONADIAL SACS the many round structures that fill the gonads of stauromedusae (also called "genital sacs," "vesicles," or "follicles").

LAPPET one of the lobes separated by clefts at the margin of a pelagic scyphomedusa.

MANUBRIUM central stomach structure that attaches to the center of the subumbrella.

MARGINAL ANCHOR marginal structure between tentacle clusters in some species of stauromedusae; lacking any special sensory capabilities, these are morphologically merely reduced tentacles (sometimes called "rhopalioids").

ORAL the side of the jellyfish to which the mouth opens.

ORAL ARM feeding and prey capture structure attached to the manubrium and surrounding the mouth. On a scyphomedusa, there are nearly always several (usually four in semaeostome and eight in rhizostome medusae) oral arms surrounding the mouth; they may be simple or frilly and membranaceous, or very complicated and fleshy; digestion often begins while prey are still within the oral arms.

PLANULA (plural planulae) microscopic, egg-shaped or subcylindrical larva that develops from a cnidarian egg.

RADIAL CANALS circulatory canals on the subumbrellar surface of a jellyfish emanating from the mouth toward the bell margin; these may be simple or may branch one or more times.

RHOPALIUM (plural rhopalia) one of the numerous marginal sense organs in clefts between lappets at the margin of a scyphozoan bell. Rhopalia usually have a gravity receptor (statocyst) and one or more "eyes" (ocelli). Cubomedusae are very sensitive to light and have multiple eyes, some with large lenses, on each rhopalium.

SCYPHISTOMA the asexual benthic (bottom-living) polyp stage of semaeostome and rhizostome scyphomedusae. These are often capable of reproduction by fission or pedal laceration and also metamorphose so as to produce stacks of tiny jellyfish (ephyrae) that break loose one by one (a process known as strobilation) and mature in the plankton, usually over several months.

STALK elongate portion at the aboral end of a stauromedusa, with which it attaches to rock, eelgrass, or seaweed.

STROBILATION asexual process in which a semaeostome scyphistoma absorbs its tentacles and produces instead a stack of developing jellyfish by transverse fission (the parent polyp and the resulting jellyfish are all the same sex, either male or female, and thus are clones). The base of the polyp remains attached to the substratum and grows new tentacles when strobilation is complete.

SUBUMBRELLA the inner, or underneath, surface of a jellyfish "bell".

VERMIFORM wormlike.

Key to the Scyphomedusae, including the attached Stauromedusae

1. Free-swimming bell-like forms, usually greater than 1 cm . . . . Semaeostomeae 2
— Small, inconspicuous, attached medusoid forms having a stalk and calyx with eight marginal clusters of short, knobbed tentacles. . . . . Stauromedusae 8
2. With 24 or fewer marginal tentacles 3
— With more than 24 tentacles, too numerous to easily count 5
3. With eight marginal tentacles alternating with eight rhopalia; color magenta, brown, and blue, combining to form a striking deep purple pattern on the pale whitish exumbrella composed of an apical ring and 16 radiating streaks . . . . Chrysaora colorata
— With 24 marginal tentacles, alternating in groups of three with eight rhopalia 4
4. Basic color of the exumbrella dark amber, darkest near the margin, sometimes with a fainter radiating starlike pattern; oral arms strongly spiraled in a screwlike manner; subumbrella without 16 radiating dark streaks. . . . . Chrysaora fuscescens
— Basic color of the exumbrella whitish and pale, with 16 dark brownish streaks radiating outward from a broad central ring; between the 16 steaks are usually found 32 dark crescents of pigment near the margin; oral arms twisted, but not in a tight spiral; subumbrella pale with 16 radiating dark streaks . . . . Chrysaora melanaster
5. Bell opaque whitish, although sometimes infused with pink, lavender, tan, or yellow; four horseshoe-shaped gonads visible from the exumbrellar side; tentacles originating at the bell margin and very small and numerous; four moderately frilly, mostly transparent, oral arms 6
— Bell mostly pigmented in tones of brown, red, or yellow; tentacles originating from the subumbrellar surface; oral arms extensively frilly, complicated, and prominent. 7
6. Bell margin scalloped into 16 lobes, separated into pairs by eight rhopalia; with a large conical manubrium that is one fifth to nearly one half as long as the bell is wide, at the center of the subumbrella, with oral arms attached to this structure; the oral arms each about two thirds to three quarters as long as the bell radius, frequently bent in a counterclockwise direction, or straight; radial canals numerous and many times branched in larger specimens; lavender or white planula larvae brooded on the manubrium in females (plate 62C) . . . . Aurelia labiata
— Bell margin scalloped into 16 lobes, separated into pairs by eight rhopalia; with a short rounded manubrium that is less than one sixth the bell diameter, at the center of the subumbrella with oral arms attached to this structure; the oral arms each about four fifths to nine tenths as long as the bell radius, straight; radial canals moderately numerous and branched; white, ochre, or orange planula larvae brooded on the manubrium in females (plate 62D) . . . . .

. . . . Aurelia sp. 1
7. With eight rhopalia; with eight clusters of up to 150 tentacles each, arranged in up to four rows originating from deep under the bell in U-shaped bases that are evident from the top of the bell; color usually yellowish-brown to reddish-brown to brick red; bell to about 1 m diameter, but usually much less . . . . Cyanea capillata
— With 16 rhopalia; with 16 linear groups of up to 25 tentacles each in a single row that originate on the subumbrella just inside of the bell margin; the umbrella is usually transparent to milky white, with a bright egg-yolk-yellow center; bell to about 60 cm diameter. . . . . Phacellophora camtschatica
8. Animal goblet-shaped, like glass stemware, with flared calyx of approximately equal length to stem; up to 4 cm long 9
— Animal mostly tubular; calyx flaring near the apex, stalk flaring near the base; up to 2 cm long (rare and inconspicuous) 13
9. Calyx broader than it is tall, widely open like a martini glass, with eight well-developed, equally spaced, marginal arms or lobes, each with a terminal cluster of up to 250 knobbed tentacles; tentacles all alike and without cushionlike swellings at their bases 10
— Calyx longer than it is broad, more or less wine glass-shaped, with eight poorly developed arms or lobes, each with a terminal cluster of up to 30 tentacles; the outermost tentacles in each cluster with cushionlike swellings at their bases. 11
10. Eight prominent coffee bean–shaped marginal anchors alternate with eight arms terminating in clusters of up to 130 tentacles; gonads extend the full length of each arm, each with 200–300 gonadial sacs, irregularly, but tightly packed, with as many as 10–22 abreast in the broadest part of each gonad; with a few white cnidocyst vesicles lined up adjacent to each gonad out near the tip of each arm; color variable in shades of green, brown, olive, yellow, orange, pink, purple; found in protected or semi-exposed waters (plate 63A) . . . . Haliclystus sp. (undescribed species)
— Eight elongate, trumpet-shaped marginal anchors alternate with eight arms terminating in clusters of up to 250 tentacles; gonads extend the full length of each arm, each with 80–120 gonadial sacs, irregularly packed, with two to eight (usually four to six) abreast in the broadest part of each gonad; with a few white cnidocyst vesicles irregularly arranged along the outer edges of each gonad, especially on the distal half; color usually golden brown or tan, occasionally greenish; found in protected bays (plate 63B) . . . . Haliclystus salpinx
11. Stalk gradually flaring into the calyx as in a trumpet or distinctly demarcated like a goblet; with eight marginal clusters of 15–25 tentacles and with single row of white cnidocyst vesicles along the margin between the tentacle clusters; overall color usually green to yellow-green or red, with four distinctive, linear paler "windows" on the exumbrella; found in protected waters (plate 63C). . . . . Manania handi
— Stalk distinctly demarcated from the calyx as in a goblet; with eight marginal clusters of 15–30 tentacles; color usually cream to tan or some shade of red, but without the distinctive four linear stripes on the exumbrella; found on the open coast in areas of considerable wave action. 12
12. Calyx and stalk usually dark red, but varying from tan to

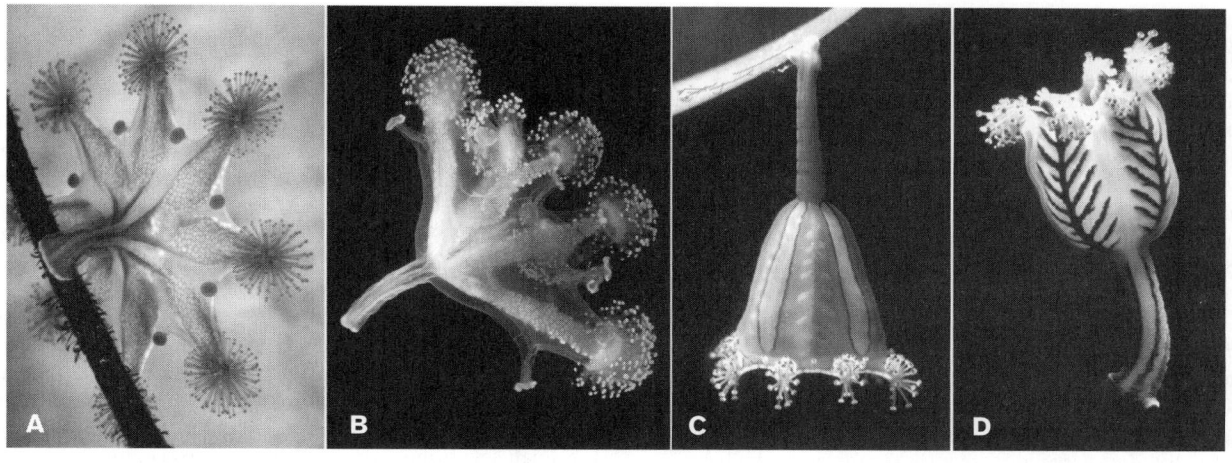

Haliclystus sp.   Haliclystus salpinx   Manania handi   Manania distincta
("sanjuanensis")

PLATE 63 Stauromedusae. A, *Haliclystus* sp. *"sanjuanensis"*; B, *Haliclystus salpinx*; C, *Manania handi*; D, *Manania distincta* (photographs A–C by Claudia E. Mills; photograph D by Ronald J. Larson).

magenta, with eight marginal clusters of 15–30 tentacles and with distinctive, contrasting patches of white subumbrellar cnidocyst vesicles between the pairs of tentacle clusters . . . . *Manania gwilliami*
— Calyx and stalk usually light tan to cream in color, with up to 26 tentacles in each of eight marginal clusters; with a distinctive dark brown herringbone pattern on the calyx that extends down the stalk as four dark lines (plate 63D) . . . . *Manania distincta*
13. Vermiform, with small flared calyx and flared base; with eight groups of about 25 tentacles alternating with eight small single tentacles; color purple with small white spots . . . . *Kyopoda lamberti*
— Vermiform, with a long tubular calyx demarcated by a constriction from a very short flared stalk; with eight groups of eight to 12 tentacles alternating with four rudimentary solid structures at the calyx margin; color greenish-brown with minute white flecks . . . . Undescribed genus and species ("*Stenoscyphopsis vermiformis*" of Gwilliam 1956)

List of Species

CUBOMEDUSAE

CARYBDEIDAE

*Carybdea sp. The only cubomedusa (box jelly) known from the West Coast of North America, can be common during autumn in the surf zone just inside the kelp beds at Santa Barbara. This medusa, bell to 4 cm tall, has also been collected at Redondo Beach, Malibu, and La Jolla, all in Southern California. Previously known in the West Coast literature as the Western Pacific *C. rastoni* Haacke, 1886, and the Atlantic *C. marsupialis* (Linnaeus 1778), ours is an undescribed species (L. Gershwin personal communication 2005, based on specimens collected at Redondo Beach). See Larson and Arneson, 1990; color photograph in Wrobel and Mills 1998 and 2003, p. 48.

* = Not in key.

STAUROMEDUSAE

DEPASTRIDAE

Manania distincta (Kishinouye, 1910). Northern Japan to Oregon, along the open coast in areas of wave action; to about 4 cm long. Very few specimens have been collected in the eastern Pacific (see Larson and Fautin 1989). Color photograph in Wrobel and Mills 1998 and 2003, p. 49.

Manania gwilliami Larson and Fautin, 1989. Vancouver Island and from northern California to Baja, California, this species probably occurs along the entire West Coast; to about 4 cm long. Attached to rocks and algae in surf-swept areas; very cryptic near coralline algae (so much so that the animal is essentially invisible when a color photograph is converted to black and white).

Manania handi Larson and Fautin, 1989. Known from Vancouver Island, Washington, and central California, attached to *Zostera* or algae in semiprotected subtidal habitats; to about 4 cm long. Often with *Haliclystus*; uncommon. Color photograph in Wrobel and Mills 1998 and 2003, p. 50.

KYOPODIIDAE

Kyopoda lamberti Larson, 1988. British Columbia and Southern California in the subtidal on cobble with, or nearby, crustose coralline algae in areas of wave surge. Extremely cryptic (to about 2 cm long), and thus probably occurs along much of the North American coast.

LUCERNARIIDAE

Haliclystus salpinx Clark, 1863. Known mostly from the North Atlantic, and on the Pacific coast from only from a small number of isolated bays in the San Juan Islands (Washington) and southern Vancouver Island, this boreal species, to about 3 cm, might be expected to be found elsewhere on the West Coast (unless it was introduced locally in early coastal exploration and shipping); its trumpet-shaped marginal anchors are distinctive.

Haliclystus sp. (=*H.* "sanjuanensis," *nomen nudum*). Hirano (1997), having studied *Haliclystus* from all over the world, concluded that West Coast material from at least British Columbia to California is an undescribed species (it had been previously called *H. auricula, H. octoradiatus, H. stejnegeri,* and *H. sanjuanensis*). It is the same as *H.* "sanjuanensis," which is, however, an unpublished manuscript name (Gellerman 1926) and thus not available despite its occasional appearance in the literature (Guberlet 1936, 1949; Hyman 1940b). Look closely to distinguish the superficially similar *H. salpinx* (see above). Color photograph in Wrobel and Mills 1998 and 2003, p. 49.

*Haliclystus sp. (=*H.* "californiensis" of Gwilliam 1956). This unpublished manuscript name (Gwilliam 1956) applies to a single specimen of about 2 cm, found by Paul Silva at 29 m, off Christy Cove on Santa Cruz Island. The species is distinct in having a very short stalk topped by a calyx that is wide open at the top, but very elongate, nearly tubular, at the base.

Undescribed genus and species ("*Stenoscyphopsis vermiformis*" of Gwilliam 1956). This unpublished manuscript name (Gwilliam 1956) applies to eight specimens, found only once in 1952, on *Macrocystis* off the Hopkins Marine Station in Pacific Grove. Gwilliam stated that this species (to about 1.5 cm long) was extremely cryptic, matching the color of the seaweed; he was unable to find it again.

SEMAEOSTOMEAE SCYPHOMEDUSAE

CYANEIDAE

Cyanea capillata (Linnaeus, 1758). The "lion's mane," especially abundant from Washington northward to the Bering Sea (and in the Atlantic) in the summer and fall, this cold-water species is less numerous off Oregon and only occasionally seen in northern California. Usually <50 cm in diameter. Color photograph in Wrobel and Mills 1998 and 2003, p. 54.

PELAGIIDAE

*Chrysaora achlyos Martin, Gershwin, Burnett, Cargo, and Bloom, 1997. This species can be enormous, with oral arms extending as much as 6 m beyond the 1 m diameter bell; it has a dark purple, nearly black, umbrella and was seen along the coast of Southern California and Baja, California, in 1989, 1999, and 2005. Where it occurs most years (when it is not found off California) is not known.

Chrysaora colorata (Russell, 1964). Formerly known as *Pelagia colorata* (see Gershwin and Collins 2002) and often misidentified as *P. noctiluca* (see below a separate species of worldwide distribution, found usually south of California in the eastern Pacific). The apparently oceanic *C. colorata* reaches about 70 cm in diameter and regularly washes ashore on the beaches of Southern California and is occasionally found as far north as San Francisco Bay and Bodega Bay; most sightings are in the late spring. Color photograph in Wrobel and Mills 1998 and 2003, p. 54.

Chrysaora fuscescens Brandt, 1835. This is the common central California and Oregon species of *Chrysaora*, which may occur in shoals just offshore and stranded on beaches. It can be found from the Gulf of Alaska to Mexico and usually has up to about 30 cm bell diameter. There is much confusion in the literature about the name of this animal, which has frequently been referred to as *C. melanaster*, a separate species (see below). Other names used in the literature for this species include *C. helvola* Brandt, 1838 and *C. gilberti* Kishinouye,

1899, both junior synonyms of *C. fuscescens* (see Larson 1990; color photograph in Wrobel and Mills 1998 and 2003, p. 53).

Chrysaora melanaster Brandt, 1835. Although this name is frequently seen in the older California literature, it properly belongs to a more strongly patterned species (to 60 cm diameter) found commonly in Alaska and perhaps British Columbia and occasionally drifting to the south at least as far as Oregon, perhaps with rare sightings in California (see Larson 1990; color photograph of underside in Wrobel and Mills 1998 and 2003, p. 53).

*Pelagia noctiluca (Forsskål, 1775). An oceanic warm water species of all seas most likely to be encountered south of the range of this book, distinguished by its 16 marginal lappets, with eight tentacles alternating with eight rhopalia around the margin, and conspicuous exumbrellar "warts." *P. noctiluca* is a small scyphomedusa, generally not exceeding 9 cm in diameter; color photograph in Wrobel and Mills 1998 and 2003, p. 54.

ULMARIDAE

*Aurelia aurita (Linnaeus, 1758; occasionally as *Aurellia*). This species, which may reach 50 cm in diameter, is apparently endemic to the North Atlantic and Baltic (see Gershwin 2001 and Dawson 2003, 2004a). It is characterized by eight, rather than 16, marginal lobes and unbranched, rather than anastomosing, adradial canals. Despite the common use of this name in West Coast literature, *A. aurita* is not known from the American Pacific coast.

Aurelia labiata Chamisso and Eysenhardt, 1821. Originally described from the San Francisco area, this once largely forgotten name is the correct one for a Northeast Pacific endemic species (to about 40 cm diameter) that includes most of the *Aurelia* on the West Coast, from central California to Oregon, with a northern variant extending to Alaska which appears to be an evolutionarily distinct lineage (Gershwin 2001; Dawson and Jacobs 2001; Dawson 2003, 2004a). Some specimens in Alaska with brown marginal pigment correspond to another apparently valid species, *Aurelia limbata* Brandt, 1835. *Aurelia* typically occurs in large aggregations, rather than singly, although the mechanisms for keeping such aggregations together are unknown. Color photograph in Wrobel and Mills 1998 and 2003, p. 55.

Aurelia sp. 1 (of Dawson 2003, 2004a, b). *Aurelia* medusae in Southern California are morphologically different but of similar size to those further north (Gershwin 2001) and have been shown using molecular techniques (Dawson and Jacobs 2001; Dawson 2003, 2004a, b) to be a separate species of worldwide distribution, including Los Angeles to San Diego, Tokyo Bay, and northern Japan, Australia, Atlantic and Mediterranean coasts of France, and probably also in south San Francisco Bay. This species may have been widely distributed by ships. Color photograph in Wrobel and Mills 1998 and 2003, p. 55 (top).

Phacellophora camtschatica Brandt, 1835. Occurs in the Pacific from Kamchatka to Alaska to Chile, but is more common at higher latitudes; also known from the North and South Atlantic and Mediterranean. Occasionally seen in bays in California and Oregon; to about 60 cm bell diameter. Color photograph in Wrobel and Mills 1998 and 2003, p. 55.

* = Not in key.

RHIZOSTOMEAE SCYPHOMEDUSAE

MASTIGIIDAE

Phyllorhiza punctata von Lendenfeld, 1884. An introduced Indo-Pacific species known also from Hawaii, the western tropical Atlantic, and the Mediterranean, and occasionally seen in Southern California in San Diego Bay and Mission Bay (Larson and Arneson 1990); bell to about 50 cm diameter. Color photograph in Wrobel and Mills 1998 and 2003, p. 57.

RHIZOSTOMATIDAE

Stomolophus meleagris L. Agassiz, 1862. Coastal species of the warm Atlantic, Caribbean, and Gulf of Mexico; occasionally collected in the Pacific from Southern California and the Gulf of California to Ecuador; bell to about 20 cm diameter. Color photograph in Wrobel and Mills 1998 and 2003, p. 57.

References

Cairns, S. D., D. R. Calder, A. Brinckmann-Voss, C. B. Castro, D. G. Fautin, P. R. Pugh, C. E. Mills, W. C. Jaap, M. N. Arai, S. H. D. Haddock and D. M. Opresko 2002. Common and scientific names of aquatic invertebrates from the United States and Canada: Cnidaria and Ctenophora. 2nd edition. Bethesda, MD: American Fisheries Society Special Publication #28, 115 pp.

Dawson, M. N. 2003. Macro-morphological variation among cryptic species of the moon jellyfish, *Aurelia* (Cnidaria: Scyphozoa). Marine Biology 143: 369–379.

Dawson, M. N. 2004a. Erratum: Macro-morphological variation among cryptic species of the moon jellyfish, *Aurelia* (Cnidaria: Scyphozoa). Marine Biology 144: 203.

Dawson, M. N. 2004b. Some implications of molecular phylogenetics for understanding biodiversity in jellyfishes, with emphasis on Scyphozoa. In Coelenterate biology 2003: trends in research on Cnidaria and Ctenophora, D. G. Fautin, J. A. Westfall, P. Cartwright, M. Daly, and C. R. Wyttenbach, eds. Hydrobiologia 530/531: 249–260.

Dawson, M. N. and D. K. Jacobs 2001. Molecular evidence for cryptic species of *Aurelia aurita* (Cnidaria, Scyphozoa). Biological Bulletin 200: 92–96.

Gellerman, M. P. 1926. Medusae of the San Juan Archipelago. M.S. thesis, University of Washington, 100 pp. plus 37 plates.

Gershwin, L. 2001. Systematics and biogeography of the jellyfish *Aurelia labiata* (Cnidaria: Scyphozoa). Biological Bulletin 201: 104–119.

Gershwin, L. and A. G. Collins 2002. A preliminary phylogeny of Pelagiidae (Cnidaria, Scyphozoa), with new observations of *Chrysaora colorata* comb. nov. Journal of Natural History 36: 127–148.

Gotshall, D. W., J. G. Smith, and A. Holbart 1965. Food of the blue rockfish *Sebastodes mystinus*. Calif. Fish Game 51: 147–162.

Guberlet, M. L. 1936 and 1949. Animals of the seashore. Portland: Binsfords and Mort, 410 pp.

Gwilliam, G. F. 1956. Studies on West Coast Stauromedusae. Ph.D. dissertation, University of California, Berkeley, 191 pp.

Hirano, Y. M. 1986. Species of stauromedusae from Hokkaido, with notes on their metamorphosis. Journal of the Faculty of Science, Hokkaido University, Series VI, Zoology 24: 182–201.

Hirano, Y. M. 1997. A review of a supposedly circumboreal species of stauromedusa, *Haliclystus auricula* (Rathke, 1806). In Proceedings of the 6th International Conference on Coelenterate Biology, *1995*, pp. 247–252.

Hyman, L. H. 1940a. The invertebrates: Protozoa through Ctenophora. Vol. I. McGraw-Hill (Scyphozoa, pp. 497–538; see also Vol. V, 1959, Retrospect, pp. 718–729).

Hyman, L. H. 1940b. Observations and experiments on the physiology of medusae. Biological Bulletin 79: 282–296.

Kramp, P. L. 1961. Synopsis of the medusae of the world. Journal of the Marine Biological Association of the United Kingdom 40: 1–469. (Full text available online at http://www.mba.ac.uk/nmbl/publications/jmba_40/jmba_40.htm).

Larson, R. J. 1988. *Kyopoda lamberti* gen. nov., sp. nov., an atypical stauromedusa (Scyphozoa, Cnidaria) from the eastern Pacific, representing a new family. Canadian Journal of Zoology 66: 2301–2303.

Larson, R. J. 1990. Scyphomedusae and Cubomedusae from the Eastern Pacific. Bulletin of Marine Science 47: 546–556.

Larson, R. J. and A. C. Arneson 1990. Two medusae new to the coast of California: *Carybdea marsupialis* (Linnaeus, 1758), a cubomedusa and *Phyllorhiza punctata* von Lendenfeld, 1884, a rhizostome scyphomedusa. Bulletin of the Southern California Academy of Science 89: 130–136.

Larson, R. J. and D. G. Fautin 1989. Stauromedusae of the genus *Manania* (=*Thaumatoscyphus*) (Cnidaria, Scyphozoa) in the northeast Pacific, including descriptions of new species *Manania gwilliami* and *Manania handi*. Canadian Journal of Zoology 67: 1543–1549.

Marques, A. C. and A. G. Collins 2004. Cladistic analysis of Medusozoa and cnidarian evolution. Invertebrate Biology 123: 23–42.

Martin, J. W., L. Gershwin, J. W. Burnett, D. G. Cargo, and D. A. Bloom 1997. *Chrysaora achlyos*, a remarkable new species of scyphozoan from the Eastern Pacific. Biological Bulletin 193: 8–13.

Mayer, A. G. 1910. Medusae of the World. Vol. III. The Scyphomedusae. Carnegie Institution of Washington Publication No. 109: 499–735. (Full text available online at http://www2.eve.ucdavis.edu/mndawson/tS/tsPDF/Mayer1910/Mayer1910_0Cover.html).

Mianzan, H. W. and P. F. S. Cornelius 1999. Cubomedusae and scyphomedusae. In *South Atlantic Zooplankton I*. D. Boltovskoy, ed. Backhuys, Leiden, pp. 513–559.

Mills, C. E. 1987. Class Scyphozoa: Order Semaeostomae. In Marine invertebrates of the Pacific Northwest. E. N. Kozloff, pp. 65–67. Seattle: University of Washington Press.

Mills, C. E. 1996. Additions and corrections to the keys to Hydromedusae, Hydroid polyps, Siphonophora, Stauromedusan Scyphozoa, Actiniaria, and Ctenophora. In Marine invertebrates of the Pacific Northwest, with revisions and corrections. E. N. Kozloff, pp. 487–491. Seattle: University of Washington Press.

Mills, C. E. Internet since 1999. Stauromedusae: list of all valid species names. Electronic Internet document available at http://faculty.washington.edu/cemills/Staurolist.html. Published by the author, web page first established October 1999, frequently updated (see date at end of page).

Russell, F. S. 1970. The Medusae of the British Isles. Vol. II. Pelagic Scyphozoa, with a supplement to the first volume on Hydromedusae. Cambridge: Cambridge University Press, 284 pp., 15 plates. (Full text and illustrations available online at http://www.mba.ac.uk/nmbl/publications/medusae_2/medusae_2.htm).

Wrobel, D. and C. Mills 1998, reprinted with corrections 2003. Pacific coast pelagic invertebrates: a guide to the common gelatinous animals. Monterey, CA: Sea Challengers and the Monterey Bay Aquarium, iv + 108 pp.

Anthozoa

DAPHNE G. FAUTIN AND CADET HAND

(Plates 64–68)

Anthozoa, the largest class of cnidarians, including the familiar sea anemones and corals, contains more than 6,000 species. The most diverse anthozoan fauna is in the coral-reef areas of tropical and subtropical seas, where representatives occur of some orders not found from central California to Oregon. Some tropical anthozoans are also much more massive than those on the shores of California and Oregon, but on some shores of the northeast Pacific, anthozoans dominate space, so animals of this class play a prominent role in the ecology of both tropical and temperate seas.

Unlike cnidarians of the other classes, anthozoans do not possess a medusa phase. Thus the anthozoan polyp reproduces sexually and, in most species, asexually as well. Gametes are derived from the endoderm and develop in the mesenteries. In most species, gametes are spawned through the mouth into the sea, where fertilization takes place and a planktonic planula larva develops. Ultimately, it metamorphoses and settles onto or into the substratum. Several species of local anemones, by contrast, brood their young. In *Epiactis ritteri*, for example,

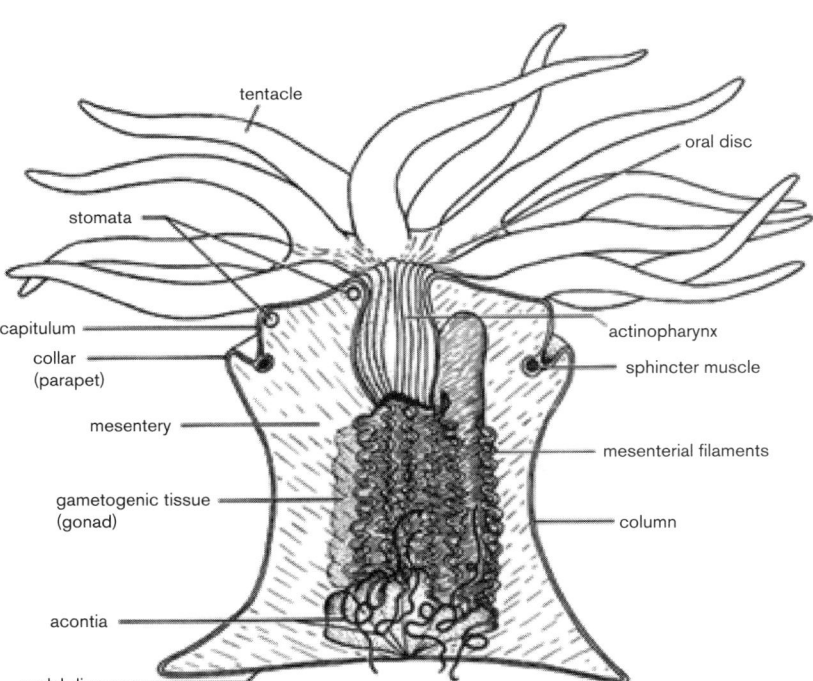

PLATE 64 Diagrammatic longitudinal section of a sea anemone (drawing by Emily Reid).

Labels on figure: tentacle, oral disc, stomata, capitulum, collar (parapet), mesentery, gametogenic tissue (gonad), acontia, pedal disc, actinopharynx, sphincter muscle, mesenterial filaments, column

freely spawned sperm seem to find their way into the coelenteron of a female, where the eggs are fertilized and develop. The small anemones that are released settle down directly, without having passed through a dispersal stage. Likewise, in *E. prolifera,* development is direct, but the young develop attached to the outside of the parent's column and simply crawl off when mature. It appears that these sexually produced young are being budded, but *Epiactis,* like most anemones, does not reproduce asexually, to the best of our knowledge.

Among the local sea anemones that undergo asexual reproduction are *Diadumene lineata* and *Metridium senile.* In the process known as "pedal laceration," little bits rip from the edge of the pedal disc of an individual of *M. senile* as it remains in one spot or as it glides along the substratum, a daughter anemone developing from each bit of tissue. Thus dense clones of the animals develop, with the original individual encircled or trailed by its genetically identical progeny. Clones are also characteristic of *Anthopleura elegantissima,* which propagates asexually by longitudinal binary fission (as does *D. lineata*), and of *Nematostella vectensis,* which propagates asexually by transverse binary fission.

Most local anthozoans belong to the subclass Hexacorallia (or Zoantharia). Hexacorallians are characterized by simple or branched hollow tentacles that are typically numerous, often a multiple of six in number. Sea anemones, members of the hexacorallian order Actiniaria, are the most diverse intertidal anthozoans in Central California. All sea anemones are solitary (although they may be clonal), whereas at least some members of most other orders are colonial (that is, clonemates are physically attached to one another). Another difference between sea anemones and hexacorallians of most other orders is that anemones lack skeletons.

Plate 64 illustrates the anatomy of a typical sea anemone. Its form is that of an animal attached by its base; *Flosmaris grandis* (plate 66) illustrates a typical burrowing anemone. At the basal end of the body column is the pedal disc, the site of attachment of the polyp to the substratum, or, in most burrowing anemones, the bulbous physa. In species in which the column is divisible into regions that differ morphologically, the scapus is just distal to (above) the physa or the pedal disc; the scapus is typically the longest region of the column. Distal to it is the scapulus and/or the capitulum. The column may have a collar (parapet) in its distal part and bear protuberances of various sorts; verrucae (Latin for warts) are present on many anemones of California and Oregon (plate 65). Hollow tentacles used in food capture and ingestion are arrayed in circlets and/or radial rows on the oral disc.

In anemones of some taxa, below the edge of the oral disc, at the top of the column, is a circle of small tentaclelike structures termed "marginal spherules"; when inflated, these are used in aggression (see, for example, Francis 1973b). The top of the column can be drawn over the tentacles through contraction of the circularly arrayed sphincter muscle. The slitlike form of the mouth, which is in the center of the oral disc, reflects the biradial symmetry of an anthozoan polyp. The long axis of the mouth is termed the "directive axis." The mouth leads into the actinopharynx (gullet), a tube of tissue that opens into the body space, the coelenteron.

The coelenteron is divided by numerous radially arrayed longitudinal sheets of tissue (mesenteries) that extend from the column wall and some of which attach at their opposite edge to the actinopharynx. (Mesenteries have also been called septa, a term that should be reserved for the calcareous partitions of scleractinian corals that are flanked and secreted by the mesenteries.) The edges of the mesenteries are thickened into mesenterial filaments, which bear gland cells that serve in digestion, and cilia that keep the fluids within the coelenteron moving. Gametes develop in some or all of the mesenteries. Extending from the lower edge of some mesenteries of certain taxa are threadlike structures, acontia. Acontia can be emitted though the mouth or through pores in the column wall (cinclides), presumably as a defense.

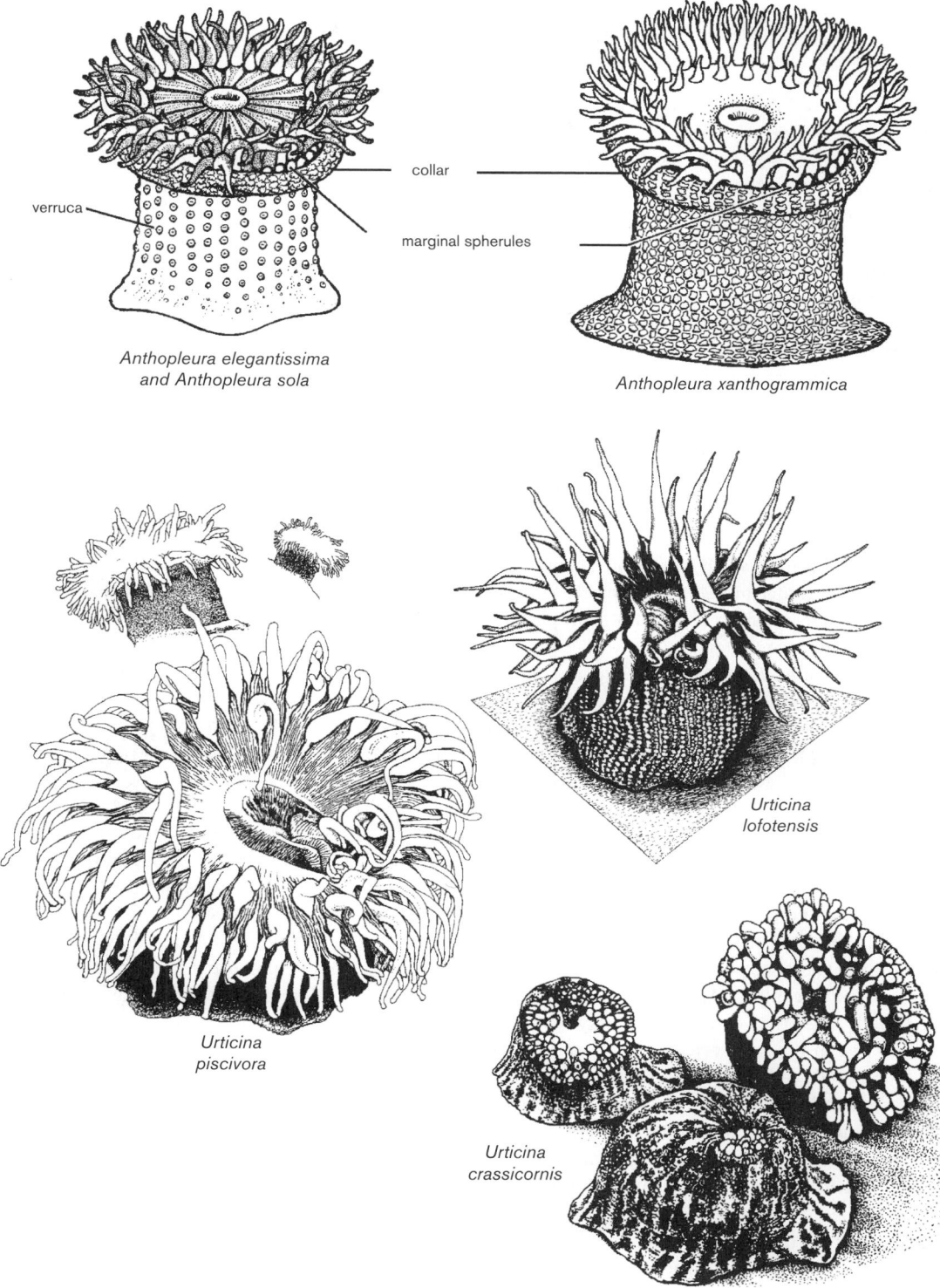

verruca

collar

marginal spherules

*Anthopleura elegantissima
and Anthopleura sola*

Anthopleura xanthogrammica

*Urticina
piscivora*

*Urticina
lofotensis*

*Urticina
crassicornis*

PLATE 65 Large sea anemones. Individuals of some species may exceed 500 mm in diameter, and most individuals of these species (except for *A. elegantissima*) reach 100 mm. All are members of family Actiniidae. Appearance of the species *Anthopleura elegantissima* and *A. sola* can be so similar they were long considered to belong to a single species. However, *A. elegantissima* is clonal whereas the larger *A. sola* is solitary. The two also differ in habitat. *Anthopleura xanthogrammica*, which can grow to a diameter of 500 mm, is typically somewhat smaller—about the same size as *A. sola,* with which it can be confused. These drawings are semidiagrammatic to emphasize the differences between them (Original drawings by Vicki B. Pearse, redrawn by Mildred Waltrip). Members of the genus *Urticina* are typically red in color. One of the largest anemones on this coast, *Urticina piscivora* has a column that is solid velvety-red and tentacles that are all either pink or white (Drawing by Steven Sechovec). The white verrucae and pigment spots of *U. lofotensis* may run together when the animal contracts, giving it a red-and-white-striped appearance (Drawing by Steven Sechovec). Some individuals of *Urticina crassicornis* are mottled with green, and the rare one may lack red altogether (drawing by Steven Sechovec).

SYMBIOTIC ALGAE
Daphne G. Fautin

Individuals of the sea anemones *Anthopleura elegantissima, A. sola,* and *A. xanthogrammica* typically posses zooxanthellae. These dinoflagellates live within cells of the endoderm, the inner cell layer of anemones. Several plant cells may live within one animal cell (Shick 1991). In *Anthopleura,* they are densest at the distal end and diminish in density proximally (that is, toward the base). Nearly all species of reef-forming corals are also zooxanthellate, as are giant clams, some foraminiferans, many species of tropical sea anemones, and some other tropical cnidarians. Outside the tropics and subtropics, some New Zealand anemones are also zooxanthellate (Buddemeier and Fautin 1996).

Zooxanthellae were once considered to belong to a single species, *Symbiodinium microadriaticum,* which was described from a tropical jellyfish. It is now known that there are many taxa of zooxanthellae and that there is not a simple 1:1 relationship between taxa of hosts and algal symbionts (e.g., Kinzie 1999). According to LaJeunesse and Trench (2000), some individuals of *A. elegantissima* in central and southern California possess simultaneously zooxanthellae of two species, *S. muscatinei* and *S. californium.*

A large proportion of the carbon fixed by the zooxanthellae through photosynthesis may be translocated to the animal, a process that was first demonstrated in *A. elegantissima* (see Muscatine and Hand 1958). Thus, although the host cnidarian cannot digest its symbiotic algae, it does benefit nutritionally from them. The symbiosis is mutualistic, the algae taking up products of the host's metabolism (Shick 1991). In a manner that is not understood, possession of zooxanthellae allows precipitation of massive calcium carbonate skeletons by reef-forming corals and giant clams (Gattuso et al. 1999). Calcification is also enhanced in zooxanthellate foraminiferans.

Individuals of *A. elegantissima, A. sola,* and *A. xanthogrammica* living in the dark, e.g., in caves (as in Oregon and Washington), under piers (as on the old Cannery Row of Monterey), or even under overhangs or deep in tide channels, lack symbiotic algae (Secord and Muller-Parker 2005). Such animals may be white rather than the typical green color; animals that live in dim light have low densities of zooxanthellae and most are intermediate in color. The green color, however, is not that of the algae, but is produced by the animals (Pearse 1974). The pigment functions to block potentially harmful radiation (Shick 1991).

Symbiotic algae affect not only the biochemistry but also the behavior of their host anemones. Individuals of *A. elegantissima* with zooxanthellae move toward light whereas those lacking them do not (Pearse 1974). Some tropical anemones possess outgrowths of their bodies in which zooxanthellae are concentrated and that extend during daylight and retract at night (Gladfelter 1975).

In addition to or instead of zooxanthellae, some individuals of *A. elegantissima* and *A. xanthogrammica* north of San Francisco possess unicellular symbiotic green algae, called zoochlorellae (Muscatine 1971, Lewis and Muller-Parker 2004). Anemones that contain mainly or only zoochlorellae tend to occur in places with lower light levels than those with mainly or only zooxanthellae (Secord and Augustine 2000). Zoochlorellae may live in the same cells as the zooxanthellae (Muller-Parker, personal communication). An individual zoochlorella cell photosynthesizes at a lower rate than an individual zooxanthella cell, and much more carbon is translocated to the anemones from zooxanthellae than from zoochlorellae (Verde and McCloskey 1996, Engebretson and Muller-Parker 1999).

Stress can disrupt the animal-alga symbiosis. Atypically high or low temperature, salinity, or sunlight, high UV insolation, or noxious chemicals can result in the break-down of the symbiosis. This phenomenon is termed "bleaching" because in corals it results in the normally brownish animal appearing white. The transparent tissue of the coral animal allows sunlight to reach the zooxanthellae, the color of which is usually perceived; when algae are absent, the white calcium carbonate skeleton is visible through the animal tissue. Bleaching is often explained as the host expelling the algae, but it is unclear whether that is happening or whether the dinoflagellates initiate the break-down in the symbiosis. Initiation probably varies with the stress and the species involved. Animals may reacquire zooxanthellae, so bleaching is not invariably lethal. In fact, a low level of algal loss appears to be normal (e.g., Fitt et al. 1997, 2000). Such "background bleaching" may be a mechanism of changing symbiotic partners (Buddemeier and Fautin 1994). *Anthopleura elegantissima, A. sola,* and *A. xanthogrammica* appear to be far less susceptible to bleaching than is typical of tropical animals. Moreover, individuals living in dark places appear not to suffer from their lack of symbiotic algae.

The flagellated, motile phase of the zooxanthella life cycle is poorly understood. Some species of cnidarians transmit the algae in their eggs (Fadlallah 1983), but others, such as *A. elegantissima* and *A. xanthogrammica*, do not (Siebert 1974), which obliges larval or juvenile animals to acquire their symbiotic algae from the environment (Schwarz et al. 2002), just as bleached animals presumably do. It is thought that algae may be concentrated in the feces of predators of anemones, such as nudibranchs (Muller Parker 1984), which thereby serve as vectors.

Members of most species of order Zoanthidea, also called Zoanthiniaria, propagate asexually, and the progeny remain tightly or loosely connected to one another. The tentacles of each polyp arise only at the margin. Grains of sand and other hard objects of similar size may be incorporated into the body wall. Zoanthids are mostly tropical; the lone local species, *Epizoanthus scotinus*, is subtidal in California and Oregon but may be intertidal in Washington.

Animals of order Corallimorpharia are commonly termed sea anemones but are actually more like scleractinian corals, except they lack a skeleton. Among the distinguishing features that can be seen in the lone local intertidal representative of this taxon, *Corynactis californica*, are the very large nematocysts (to more than 100 μm in length).

The stony corals, order Scleractinia, are represented in the local intertidal fauna by the solitary orange cup-coral, *Balanophyllia elegans*.

The tube anemones constitute order Ceriantharia. Almost exclusively subtidal, these animals have a ring of short tentacles immediately around the mouth and a ring of long tentacles at the margin of the oral disc. There are few cerianthid species in the world, and the only one locally is rarely found intertidally.

Members of the subclass Octocorallia (or Alcyonaria) are predominantly colonial, and each polyp has eight hollow tentacles that are pinnately branched.

Key to Subclasses of Anthozoa

1. Each polyp with eight, pinnately branched tentacles; polyps typically connected with one another in a colony ..... Subclass Octocorallia (Alcyonaria) (end of this section)
— Each polyp with simple (nonbranched) tentacles, which typically number more than eight .... Subclass Hexacorallia (Zoantharia), below

Key to Orders of Subclass Hexacorallia (Zoantharia)

1. Polyps solitary—not connected with one another, although they may be very near one another 2
— Polyps connected with one another at base; incorporated

into body wall are grains of sand and similar material so animal feels rough; tentacles only at margin of polyp .... Order Zoanthidea (=Zoanthiniaria)

Note: Only species that may occur locally and intertidally is *Epizoanthus scotinus* (plate 67).

2. Polyp lives within hard, calcareous exoskeleton or soft, slimy tube 3
— Polyp naked, usually without external covering of any sort, although those of some species may have a cuticular sheath or adherent sand grains, etc. 4
3. Each polyp with two circlets of tentacles—long ones around the margin and short ones around the mouth; column burrowed into mud or sand, ensheathed in a tube .... Order Ceriantharia ("tube anemones")

Note: Only local species that may occur intertidally is *Pachycerianthus fimbriatus*.

— Each polyp with a calcareous exoskeleton .... Order Scleractinia ("hard corals")

Note: Only local intertidal species is the bright orange *Balanophyllia elegans*.

4. Tentacles capitate (knobbed at tip); polyps typically in dense groups, each group of similar color (e.g., orange, pink, lavender, red, brown, or white)................ .... Order Corallimorpharia

Note: Only local intertidal species is *Corynactis californica* (plate 66).

— Tentacles not capitate; polyps in groups or solitary ...... Order Actiniaria (sea anemones, strictly defined) (below).

Key to Actiniaria

1. Base not attached to substratum or to buried shells or stones; slender, elongate column buried in soft sediments; only oral disc and tentacles protrude, so are the parts typically visible; tentacles 24 or fewer.................. 2
— Base attached to substratum or solid object; 24 or more tentacles except in very young individuals.............. 5

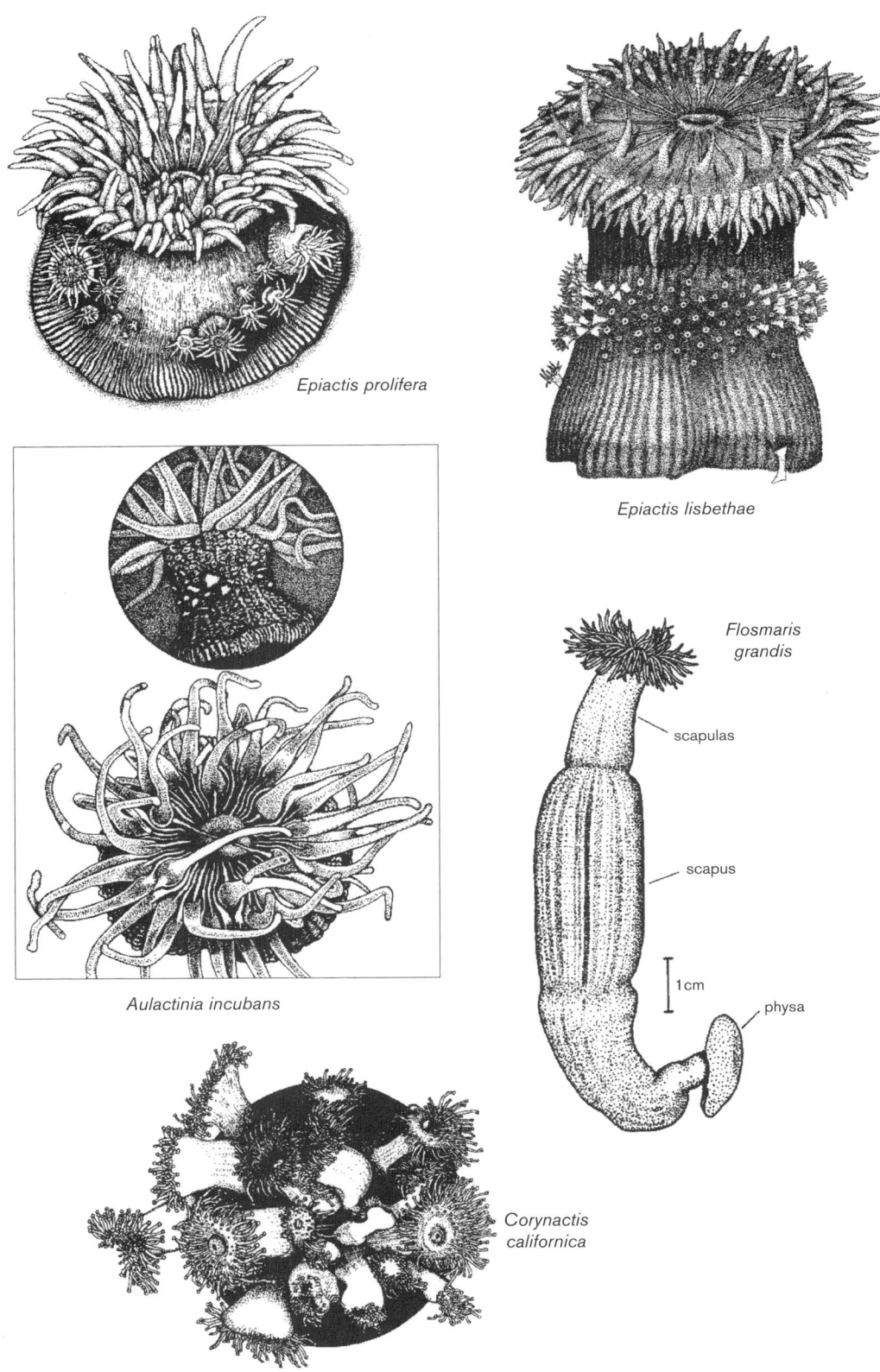

Epiactis prolifera

Epiactis lisbethae

Aulactinia incubans

Flosmaris
grandis

scapulas

scapus

1cm

physa

Corynactis
californica

PLATE 66 Some smaller sea anemones. *Epiactis prolifera* rarely exceeds a basal diameter of 30 mm. Its externally brooded young are typically of various sizes (drawing by Steven Sechovec). *Epiactis lisbethae* grows to a maximum basal diameter of 80 mm. The tentacles of this specimen are contracted, but typically the tentacles droop down over the column. The young are all of similar size. The oral disc of *Aulactinia incubans* resembles that of *E. prolifera* and *E. lisbethae*, but the lines on its directive axis run into the mouth. The young of *A. incubans* are brooded internally. Individuals of this species are similar in size to those of *Epiactis* (drawing by Steven Sechovec). *Flosmaris grandis* is a large burrowing species. This drawing illustrates the divisions of the body typical of only some anemones (drawing from Hand and Bushnell 1967). *Corynactis californica* is not a true sea anemone—it is actually a coral without a skeleton. Individual polyps are about 10 mm in diameter (drawing by Steven Sechovec).

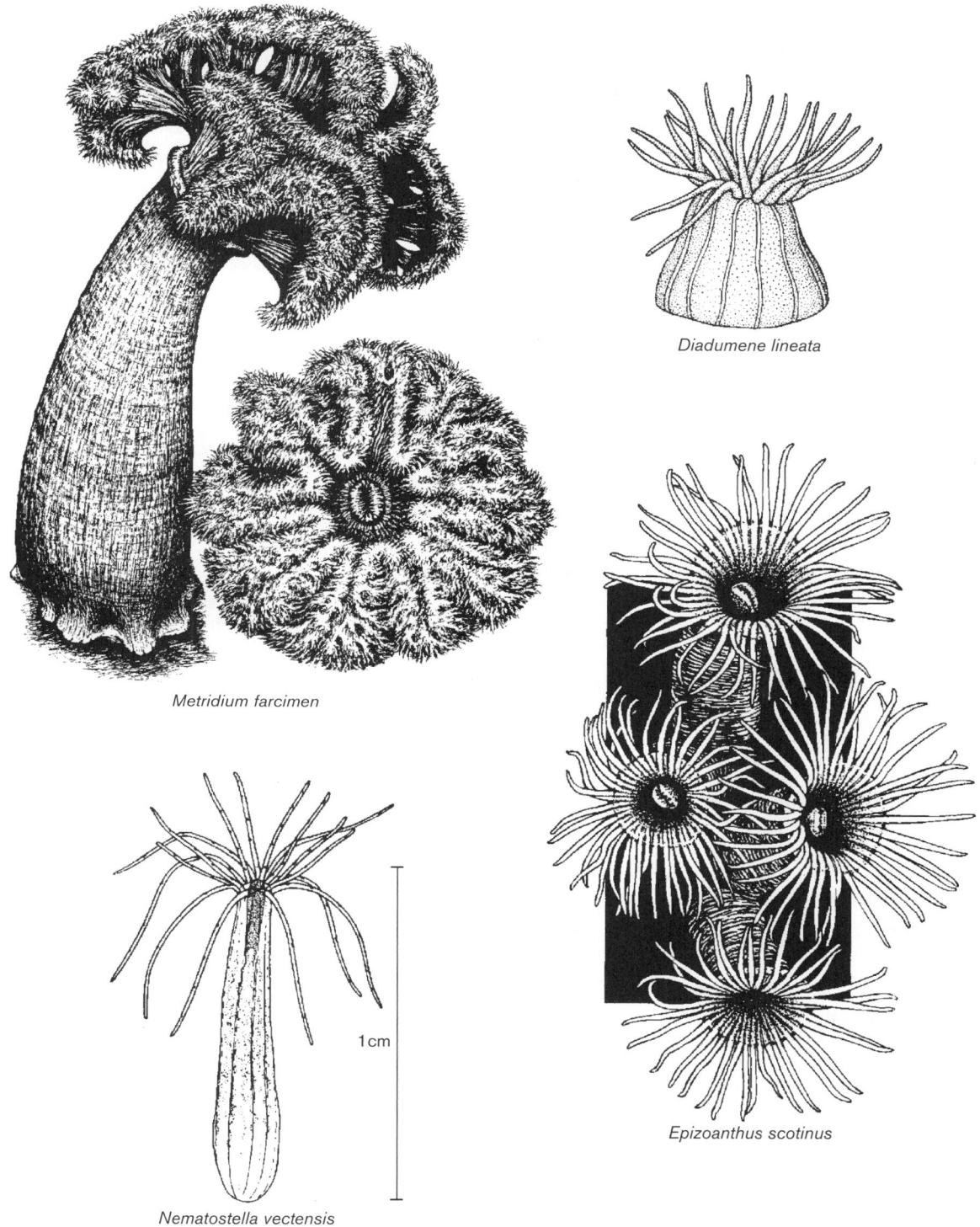

Metridium farcimen

Diadumene lineata

Epizoanthus scotinus

Nematostella vectensis

1 cm

PLATE 67 Sea anemones. The specific name of *Metridium farcimen* refers to its sausagelike appearance. Subtidal individuals may attain nearly a meter in length. Its size, habitat, and highly lobed oral disc are features that distinguish *M. farcimen* from the smaller, intertidal *M. senile* (drawing by Steven Sechovec). *Diadumene lineata,* which rarely is as large as 10 mm across, is probably the most widely distributed species of sea anemone in the world (after Hargitt 1914, redrawn by Emily Reid). Although typically no longer than 10 mm in the field, *Nematostella vectensis* can grow much larger under laboratory conditions (after Crowell 1946, redrawn by Emily Reid). The tentacles of zoanthid *Epizoanthus scotinus* arise only at the margin (drawing by Steven Sechovec).

2. Tentacles 10 in number; column to 60 mm long and 6 mm diameter (plate 68). . . . . *Halcampa decementaculata*
— Tentacles 12–18 in number. .3
3. Tentacles usually 16 in number but vary from 12 to 18; nematosomes (tiny ciliated spheres) circulate in coelenteron; column transparent in expansion, typically 10–15 mm long but can grow to several times that in laboratory (plate 67). *Nematostella vectensis*
— Tentacles usually 12 in number; column to 50 mm long; no nematosomes. 4
4. Column 5–6 mm diameter, opaque, creamy white, with crinkled, cuticular appearance; sand grains attached to column . . . . *Halcampa crypta*
— Column 1–3 mm diameter; divisible into physa, scapus, scapulus, and capitulum; scapus with brownish, well-developed cuticle . . . . *Edwardsia* sp.
5. Tentacles 24 in number, orange (with brown markings in some individuals); column with cuticular sheath (plate 68) . . . . *Cactosoma arenaria*
— Tentacles >24 in number 6
6. With acontia that are extruded through pores or breaks in the body wall, or through the mouth when the animal is disturbed, detached, or handled roughly. 7
— Without acontia 15
7. Column translucent or pale white, vermiform (to 500 mm long by 15 mm diameter); burrows in sandy mud with base attached to shells or pebbles (plate 66) . . . . *Flosmaris grandis*
— Neither vermiform nor burrowing 8
8. Margin of oral disc frilled; tentacles short and numerous so there is little tentacle-free area around mouth 9
— Margin of oral disc not frilled; tentacles fewer than 100, so a fairly large area around mouth is tentacle-free; animal typically no more than 20 mm tall, 10 mm diameter; commonly occur in aggregations due to asexual reproduction 10
9. Typically large (to 500 mm long), solitary, and subtidal; oral disc lobed and covered with short tentacles that may number in the hundreds; commonly white, uncommonly salmon, brown, or speckled (plate 67) . . . . *Metridium farcimen*
— Typically small (no more than 50 mm long), living in clusters, and intertidal; oral disc covered with many short tentacles but not lobed or only shallowly so; commonly white, orange, or brown (plate 68) . . . . . . *Metridium senile*
10. Directive tentacles (those nearest each end of mouth slit) with yellow bases; column transparent in extension, but cream, gray, or light green when contracted, commonly with longitudinal white stripes (plate 68) . . . . *Diadumene franciscana*
— Directive tentacles not marked differently from others 11
11. Column some shade of green, commonly with longitudinal single or double stripes of white, yellow, or orange; lives high in intertidal zone (plate 67) . . . . *Diadumene lineata*
— Column (and usually tentacles) yellow, orange, reddish, or pink 12
12. Column and tentacles pink- to salmon-colored, although upper parts of column may show tints of green; long and slender when extended (plate 68) . . . . *Diadumene leucolena*
— Column and tentacles yellow, orange, or reddish; to 20 mm diameter but commonly half that or less. 13
13. Extended column about as tall as wide or slightly taller

than wide; sometimes reddish; to 10 mm diameter. . . . . *Metridium exilis*
— Extended column at least twice as tall as wide; transparent light orange or yellow; commonly no more than 5 mm diameter 14
14. Lives on open coast; column commonly with bulges or kinks, and contracts asymmetrically (plate 68). . . . . *Diadumene lighti*
— Lives in bays and estuaries; column without bulges or kinks; rare individuals 20 mm diameter. . . . . *Diadumene* sp.
15. Column many times longer than broad (to 200 mm long by 15 mm diameter); burrows in sand or mud, attached to stones, worm tubes, or shells (plate 68) . . . . *Zaolutus actius*
— Column no more than about five times taller than broad 16
16. Column with verrucae to which debris (bits of gravel, shell, etc.) may adhere 17
— Column smooth, but may nonetheless be capable of holding debris. 24
17. Without marginal spherules at top of column just outside and below tentacles. 18
— With white to yellow marginal spherules; column green, gray, yellow, or white 21
18. Column cream, dull green, orangish, or brick red; diameter to 35 mm; oral disc with striking white radial lines, of which those on the directive axis reach the mouth (plate 66). . . . . *Aulactinia incubans*
— Column bright red, in some animals with green or brown patches; commonly more than 100 mm diameter. 19
19. Column densely covered with verrucae to which gravel, shells, etc. strongly adhere; anemone usually buried in gravel or shell debris . . . . *Urticina coriacea*
— Column verrucae sparse, weakly adhesive 20
20. Column scarlet; white verrucae in longitudinal rows make contracted animal appear striped (plate 65) . . . . *Urticina lofotensis*
— Red column may have patches of green or brownish green, or column may be predominantly green; verrucae that are color of column weak, sparse, or even absent (plate 65) . . . . *Urticina crassicornis*
21. Column capable of great elongation—animal lives buried in sand or gravel, with base attached to rock, and extends so tentacles are at surface; verrucae on upper two-thirds of column only; upper one-third of column black or gray, lower two-thirds white or pink; tentacles or oral disc commonly bright pink or orange . . . . . . *Anthopleura artemisia*
— Column generally green, entirely covered with verrucae 22
22. Column to 200 mm diameter, typically deep green; verrucae irregular, compound, not in clearly longitudinal rows; tentacles uniform in color, disc uniform or with faint radial stripes (plate 65) . . . . *Anthopleura xanthogrammica*
— Column green to yellow or white; verrucae circular, in longitudinal rows; tentacles may be tipped with pink or variously marked; oral disc commonly with conspicuous dark radial lines. 23
23. Column typically no more than 50 mm diameter; densely massed on rocks of the medium to high intertidal (plate 65) . . . . *Anthopleura elegantissima*
— Column typically >50 mm diameter; solitary; lives in medium to low intertidal (plate 65). . . . . *Anthopleura sola*

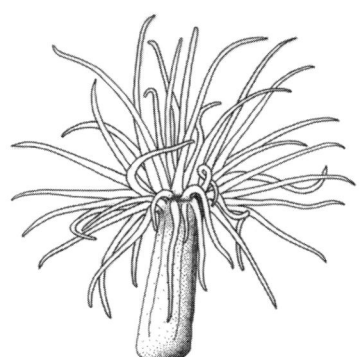

Diadumene spp.

Metridium senile

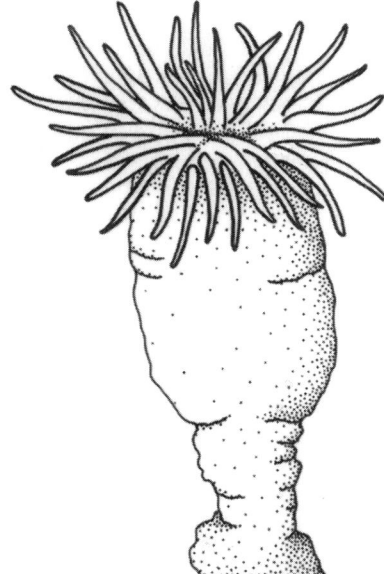

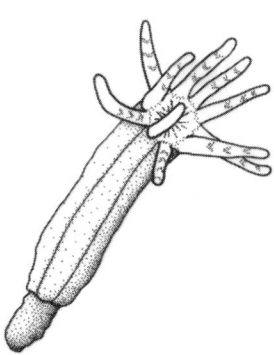

Halcampa decemtentaculata

Zaolutus actius

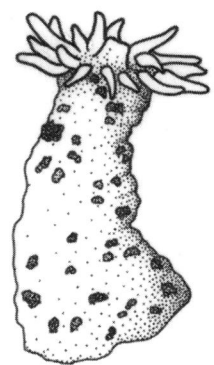

Cactosoma arenaria

PLATE 68 More small sea anemones. Looking very similar, an individual of *Diadumene* sp., *D. leucolena,* and *D. lighti* can resemble the anemone in this drawing. In California and Oregon, a specimen of *Metridium senile* is typically no longer than 50 mm. The name was also previously used also for specimens of *M. farcimen,* which can grow to more than 10 times that length. *Halcampa decemtentaculata* has, as its name indicates, 10 tentacles; it has been erroneously identified as the 12-tentacled *H. duodecimcirrata,* which occurs in Europe. *Zaolutus actius* burrows in soft sediments, its pedal disc attached to buried objects. Unlike most anemones lacking a true pedal disc, *Cactosoma arenaria* does not burrow; rather, it adheres to a firm substratum as in an anemone with a true pedal disc (drawings by Tracy R. White).

24. Column 100 mm or more diameter, rich red with velvety appearance; tentacles white or pink (plate 65) . . . . *Urticina piscivora*
— Column no more than 60 mm diameter, commonly reddish, brown, or greenish in color; white radial stripes around pedal disc and on oral disc 25
25. Column dull red to brown, very flat when contracted; sand may adhere to lower column; young brooded internally . . . . *Epiactis ritteri*
— Column dull red, brown, or greenish, domed when contracted, without adherent sand; young anemones may be attached to parent column 26
26. Young anemones on one parent differ in size; parent no more than 25 mm diameter (plate 66) . . . . *Epiactis prolifera*
— Young anemones on one parent of similar size; parent typically >50 mm diameter; bold striping around pedal disc may extend up column (plate 66) . . . . . . *Epiactis lisbethae*

List of Species

ORDER ACTINIARIA

ACTINIIDAE

Anthopleura artemisia (Pickering in Dana, 1846). On open coasts, solitary individuals live in pholad holes; in estuaries, groups of individuals attach to stones below surface of muddy sand, giving the appearance of being burrowers. See Smith and Potts 1987, Mar. Biol. 94: 537–546 (population genetics).

Anthopleura elegantissima (Brandt, 1835). Produces clonal aggregations by longitudinal fission; in bays or on rocks of sandy shores, may be buried by substratum, especially during winter. See Ford 1964, Pac. Sci. 18: 138-145 (reproduction); Francis 1973, Biol. Bull. 144: 64–72 (clone-specific segregation), 1973, Biol. Bull. 144: 73–92 (aggression), 1976, Biol. Bull. 150: 361–376 (social organization); Pearse 1974, Biol. Bull. 147: 641–651, and 1974, Biol. Bull. 147: 630–640 (influence of symbiotic algae on behavior); Hart and Crowe 1977, Trans. Amer. Micros. Soc. 96: 28–41 (role of attached gravel in reducing desiccation); Jennison 1979, Can. J. Zool. 57: 403–411 (reproduction); Smith and Potts 1987, Mar. Biol. 94: 537–546 (population genetics), Tsuchida and Potts 1994, J. Exp. Mar. Biol. Ecol. 183: 227–242 (growth); Yoshiyama et al. 1996, J. Exp. Mar. Biol. Ecol. 204: 23–42 (predation by the sculpin *Clinocottus*); Augustine and Muller-Parker 1998, Limnol. Oceanogr. 43: 711–715 (predation by *Clinocottus*); Seavy and Muller-Parker 2002, Invert. Biol. 121: 115–125 (predation by the nudibranch *Aeolidia*); Geller et al. 2005, Integr. Comp. Biol. 45: 615–622 (evolution of fission).

Anthopleura sola Pearse and Francis, 2000. Common on open coasts south of San Francisco, reported as far north as Coos Bay; formerly known as the solitary ecotype of *A. elegantissima;* can grow nearly as large as *A. xanthogrammica.*

Anthopleura xanthogrammica (Brandt, 1835). The "giant green anemone" of tide pools; differs from *A. sola* in having solid green tentacles and either a solid green oral disc or one on which radial stripes are faint. See Smith and Potts 1987, Mar. Biol. 94: 537–546 (population genetics); Hand 1996, Wasmann J. Biol. 51: 9–23 (predation by the sculpin *Clinocottus* and the nudibranch *Aeolidia*).

Aulactinia incubans Dunn, Chia, and Levine, 1980. Described from San Juan Island, Washington; occasionally found on the coast north of Santa Cruz in protected low-intertidal areas such as under overhangs; broods its young internally.

Epiactis lisbethae Fautin and Chia, 1986. Described from the San Juan Islands, Washington, where it can be confused with the more common *E. prolifera,* but it attains greater size and its externally brooded young are all the same size; rare in California and Oregon. See Edmands 1995, Mar. Biol. 123: 723–733, 1996, Invert. Repro. Dev. 123: 227–237 (mating system).

Epiactis prolifera Verrill, 1869. Ubiquitous in bays and on open coast on rocks, algae, and seagrass; externally broods young of various sizes simultaneously. See Dunn 1975, Nature 253: 528–529 (sexuality), 1975, Biol. Bull. 148: 199–218 (reproduction), 1977, Mar. Biol. 39: 41–49 (brooding), 1977, Mar. Biol. 39: 67–70 (locomotion), 1977, J. Nat. Hist. 11: 457–463 (variability); Edmands, 1995, Mar. Biol. 123: 723–733, 1996, Invert. Repro. Dev. 123: 227–237 (mating system).

Epiactis ritteri Torrey, 1902 (=*Cnidopus ritteri;* see Fautin and Chia, 1986, Can. J. Zool. 64: 1665–1674). Under and on rocks in protected places of outer coast; resembles *E. prolifera,* but young are brooded internally. See Hand and Dunn 1974, Wasmann J. Biol. 32: 187–194 (redescription); Edmands 1995, Mar. Biol. 123: 723–733, 1996, Invert. Repro. Dev. 123: 227–237 (mating system).

Urticina coriacea (Cuvier, 1798) (=*Tealia coriacea*). Middle to low intertidal and deeper, in bays and on outer coast; buried in sand or gravel, which adheres to column. See Hand 1955, Wasmann J. Biol. 12: 345–375 (redescription).

Urticina crassicornis (Müller, 1776) (=*Tealia crassicornis*). Middle to low intertidal and deeper, on outer coast; on undersides of large rocks and in protected pools. See Hand 1955, Wasmann J. Biol. 12: 345–375 (redescription).

Urticina lofotensis (Danielssen, 1890) (=*Tealia lofotensis*). Low intertidal and deeper, on outer coast; on rocks and in protected pools. See Hand 1955, Wasmann J. Biol. 12: 345–375 (redescription); Wedi and Dunn 1983, Biol. Bull. 165: 458–472 (reproduction); Sebens and Laakso 1978, Wasmann J. Biol. 35: 152–168 (redescription).

Urticina piscivora (Sebens and Laakso, 1978) (=*Tealia piscivora*). Strictly subtidal; on rock promontories.

DIADUMENIDAE

Diadumene sp. In embayments such as the Oakland Estuary.

Diadumene franciscana Hand, 1956. Southern and central California; in San Francisco Bay, sporadically abundant in Aquatic Park, Berkeley, on pilings in Lake Merritt in summer, etc. Considered an introduced species by Cohen and Carlton 1895 (see references in "Introduced Species," at the beginning of this book).

Diadumene leucolena (Verrill, 1866). On floats, pilings, stone, oyster shells, etc. in bays; introduced from the Atlantic Ocean.

Diadumene lighti Hand, 1956. In sand among algal holdfasts along edges of rocky shore tide channels.

Diadumene lineata (Verrill, 1871) (=*Haliplanella luciae* Verrill, 1898, and the combinations of the two generic and specific names). High intertidal of bays and estuaries, commonly in barnacle tests and crevices of rotting wood, and common in fouling communities. Transported by humans, it is the most widely distributed sea anemone in the world; it is probably a native of northeast Asia. See Hand 1956 (for 1955), Wasmann

J. Biol. 13: 189–251 (redescription). There is an extensive literature on this species.

EDWARDSIIDAE

Edwardsia sp. A small undescribed or unidentified species occurs in Bodega Harbor among *Phoronopsis* on muddy sand flats. May be the same species as has been called *Edwardsiella* from Tomales Bay.

Nematostella vectensis Stephenson, 1935. Widely distributed in salt marshes of both Atlantic and Pacific coasts of the United States (including San Francisco, Tomales, and Coos Bays), and the Atlantic coast of Europe (it is considered an endangered species in Britain, where its habitat is diminishing); reproduces asexually by transverse fission. See Hand 1955 (for 1954), Wasmann J. Biol. 12: 345–375 (redescription); Hand and Uhlinger 1994, Estuaries 17: 501–508 (biology and ecology); Hand and Uhlinger 1995, Invert. Biol. 114: 9–18 (asexual reproduction).

HALCAMPIDAE

Cactosoma arenaria Carlgren, 1931. On open coasts; frequents kelp holdfasts.

Halcampa crypta Siebert and Hand, 1974. In muddy shale gravel of inshore pools of Duxbury Reef, Bolinas; also found at Friday Harbor.

Halcampa decemtentaculata Hand, 1955. Found among roots of surfgrass *Phyllospadix*, holdfasts of laminarians, and gravelly pools of the low rocky intertidal; also found at Friday Harbor.

ISANTHIDAE

Zaolutus actius Hand, 1955. Specimens from Elkhorn Slough referred to as *Harenactis attenuata* by MacGinitie (1935); rare in central California, more common subtidally to the south; burrows in muddy sand.

ISOPHELLIIDAE

Flosmaris grandis Hand and Bushnell, 1967. Burrows in sand and sandy mud; known only from San Francisco Bay.

METRIDIIDAE

Metridium exilis Hand, 1956. Under rocks and ledges on open coast. See Bucklin 1987, J. Exp. Mar. Biol. Ecol. 110: 41–52 (growth and asexual reproduction).

Metridium farcimen (Brandt, 1835) (=*M. giganteum* Fautin, Bucklin, and Hand 1990, Wasmann Jour. Biol. (for 1989) 47: 77–85). This distinctive large, solitary, subtidal species that lives on pilings in bays and on rocks and shells off the coast was once considered an ecotype of *M. senile*. See Bucklin 1987, J. Exp. Mar. Biol. Ecol. 110: 41–52 (growth and asexual reproduction); Kramer and Francis 2004, Biol. Bull. 207: 130–140 (predation resistance and nematocyst scaling).

Metridium senile (Linnaeus, 1767). Common on pilings, rock jetties, and floats of bays and harbors; reproduces asexually by pedal laceration to form dense clonal groups. See Purcell and Kitting 1982, Biol. Bull. 162: 345–359 (population biology and intraspecific aggression); Bucklin 1987, J. Exp. Mar. Biol. Ecol. 110: 41–52 (growth and asexual reproduction); Kramer and Francis 2004, Biol. Bull. 207: 130–140 (predation resistance and nematocyst scaling). There is an extensive literature on this species.

ORDER CERIANTHARIA

CERIANTHIDAE

Pachycerianthus fimbriatus McMurrich, 1910 (=*P. torreyi* Arai, 1965, and *P. plicatulus* Carlgren, 1924). Rare intertidally in very soft mud; the thick, tough, soft, black, slimy tube extends to a depth of one meter or more. See Arai 1971, J. Fish. Res. Bd. Canada 28: 1677–1680 (taxonomy).

ORDER CORALLIMORPHARIA

CORALLIMORPHIDAE

Corynactis californica Carlgren, 1936. Low intertidal on rocky open coasts; reproduces asexually by longitudinal fission to form dense clusters. See Hand 1955 (for 1954), Wasmann J. Biol. 12: 345–375 (redescription); Chadwick 1987, Biol. Bull. 173: 110–125 (interspecific aggression); Chadwick and Adams 1991, Hydrobiologia 216–217: 263–269 (locomotion, reproduction, ecology); Holts and Beauchamp 1993, Mar. Biol. 116: 129–136 (sexual reproduction).

ORDER SCLERACTINIA

DENDROPHYLLIIDAE

Balanophyllia elegans Verrill, 1864. Low intertidal on rocky open coasts, typically in surge channels, often under overhangs; easily distinguished from *C. californica*, which it may resemble, by its hard skeleton. See Fadlallah and Pearse 1982, Mar. Biol. 71: 223–231 (sexual reproduction); Fadlallah 1983, Oecologia 58: 200–207 (population dynamics, life history, central California); Hellberg 1995, Mar. Biol. 123: 573–581 (gene flow); Hellberg and Taylor 2002, Mar. Biol. 141: 629–637 (sexual reproduction, genetics).

ORDER ZOANTHIDEA (=ZOANTHINIARIA)

EPIZOANTHIDAE

Epizoanthus scotinus Wood, 1958. Strictly subtidal in California, but intertidal in the San Juan Islands.

REFERENCES

Buddemeier, R. W. and D. G. Fautin 1993. Coral bleaching as an adaptive mechanism: a testable hypothesis. BioScience 43: 320–326.

Buddemeier, R. W. and D. G. Fautin 1996. Saturation state and the evolution and biogeography of symbiotic calcification. Bull. Inst. Océanogr. Monaco, Special No. 14: 23–32.

Carlgren, O. 1949. A survey of the Ptychodactiaria, Corallimorpharia and Actiniaria. K. Svenska Vetenskapsakad. Handl., ser. 4, 1(1): 1–121.

Carlgren, O. 1952. Actiniaria from North America. Ark. Zool., ser. 2, 3: 373–390.

Dunn, D. F., F.-S. Chia, and R. Levine 1980. Nomenclature of *Aulactinia* (=*Bunodactis*), with description of *Aulactinia incubans* n. sp. (Coelenterata: Actiniaria), an internally brooding sea anemone from Puget Sound. Can. J. Zool. 58: 2071–2080.

Engebretson, H. E. and G. Muller-Parker 1999. Translocation of photosynthetic carbon from two algal symbionts to the sea anemone *Anthopleura elegantissima*. Biol. Bull. 197: 72–81.

Fadlallah, Y. H. 1983. Sexual reproduction, development and larval biology in scleractinian corals. Coral Reefs 2: 129–150.

Fautin, D. G. 2006. Hexacorallians of the world—sea anemones, corals, and their allies: catalogue of species, bibliography of literature in which they were described, inventory of type specimens, distribution maps, and images (a component of Biogeoinformatics of hexacorals, compiled by Daphne G. Fautin and Robert W. Buddemeier). http://hercules.kgs.ku.edu/Hexacoral/Anemone2.

Fautin, D. G. and S. Romano 1997. Cnidaria (Coelenterata) [for the Tree of Life project]. http://tolweb.org/tree?group=Cnidaria&contgroup=Animals

Fitt, W. K., F. K. McFarland, and M. E. Warner 1997. Seasonal cycles of tissue biomass and zooxanthellae densities in Caribbean reef corals: a new definition [of bleaching] (abstract). Amer. Zool. 37(5): 72A.

Fitt, W. K., F. K. McFarland, M. E. Warner, and G. C. Chilcoat 2000. Seasonal patterns of tissue biomass and densities of symbiotic dinoflagellates in reef corals and relation to coral bleaching. Limnol. Oceanogr. 45: 677–685.

Gattuso, J.-P., D. Allemand, and M. Frankignoulle 1999. Photosynthesis and calcification at cellular, organismal and community levels in coral reefs: a review on interactions and control by carbonate chemistry. Amer. Zool. 39: 160–183.

Gladfelter, W. B. 1975. Sea anemone with zooxanthellae: simultaneous contraction and expansion in response to changing light intensity. Science 189: 570–571.

Hand, C. and R. Bushnell 1967. A new species of burrowing acontiate anemone from California (Isophelliidae: *Flosmaris*). Proc. U. S. Nat. Mus. 120: 1–8.

Kinzie, R. A., III 1999. Sex, symbiosis and coral reef communities. Amer. Zool. 39: 80–91.

LaJeunesse, T. C. and R. K. Trench 2000. Biogeography of two species of *Symbiodinium* (Freudenthal) inhabiting the intertidal sea anemone *Anthopleura elegantissima* (Brandt). Biol. Bull. 199: 126–134.

Lewis, L. A. and G. Muller-Parker 2004. Phylogenetic placement of "zoochlorellae" (Chlorophyta), algal symbiont of the temperate sea anemone *Anthopleura elegantissima*. Biol. Bull. 207: 87–92.

MacGinitie, G. E. 1935. Ecological aspects of a California marine estuary. Amer. Midl. Nat. 16: 629–765.

Muller Parker, G. 1984. Dispersal of zooxanthellae on coral reefs by predators on cnidarians. Biol. Bull. 167: 159–167.

Muscatine, L. 1971. Experiments on green algae coexistent with zooxanthellae in sea anemones. Pac. Sci. 25: 13–21.

Muscatine, L. and C. Hand 1958. Direct evidence for the transfer of materials from symbiotic algae to the tissues of a coelenterate. Proc. Nat. Acad. Sci. 44: 1259–1263.

Pearse, V. B. 1974. Modification of sea anemone behavior by symbiotic zooxanthellae: expansion and contraction. Biol. Bull. 147: 641–651.

Pearse, V. B. and L. Francis 2000. *Anthopleura sola*, a new species, solitary sibling species to the aggregating sea anemone, *A. elegantissima* (Cnidaria: Anthozoa: Actiniaria: Actiniidae). Proc. Biol. Soc. Wash. 113: 596–608.

Schwarz, J. A., V. M. Weis, and D. C. Potts 2002. Feeding behavior and acquisition of zooxanthellae by planula larvae of the sea anemone *Anthopleura elegantissima*. Mar. Biol. 140: 471–478.

Secord, D. and L. Augustine 2000. Biogeography and microhabitat variation in temperate algal-invertebrate symbioses: zooxanthellae and zoochlorellae in two Pacific intertidal sea anemones, *Anthopleura elegantissima* and *A. xanthogrammica*. Invert. Biol. 119: 139–146.

Secord, D. and G. Muller-Parker 2005. Symbiont distribution along a light gradient within an intertidal cave. Limnol. Oceanogr. 50: 272–278.

Shick, J. M. 1991. A functional biology of sea anemones. London and other cities: Chapman and Hall. 395 pp.

Siebert, A. E., Jr. 1974. A description of the embryology, larval development, and feeding of the sea anemones *Anthopleura elegantissima* and *A. xanthogrammica*. Can. J. Zool. 52: 1383–1388.

Siebert A. E., Jr. and C. Hand 1974. A description of the sea anemone *Halcampa crypta*, new species. Wasmann J. Biol. 32: 327–336.

Stephenson, T. A. 1928. The British sea anemones, Vol. I. London: Ray Society. 148 pp.

Stephenson, T. A. 1935. The British sea anemones, Vol. II. London: Ray Society. 426 pp.

Verde, E. A. and L. R. McCloskey 1996. Photosynthesis and respiration of two species of algal symbionts in the anemone *Anthopleura elegantissima* (Brandt) (Cnidaria; Anthozoa). J. Exp. Mar. Biol. Ecol. 195: 187–202.

Anthozoa: Octocorallia

GARY C. WILLIAMS

(Plates 69–71)

The octocorals are a morphologically diverse group easily distinguished from other cnidarians by having eight pinnate tentacles surrounding the mouth of each polyp. Most West Coast species are strictly subtidal, and several shallow water species of gorgonians (sea fans) occur from Monterey Bay to Baja, California. However, at least seven species of inconspicuous soft corals are occasionally encountered under rocky overhangs in the low intertidal from Southern California to Oregon. The sea pens are subtidal for the most part, but they may be encountered in protected bays at extreme low tide or washed ashore following storms. See the websites "Octocoral Research Center" or "Octocoral Home Page" for details about the group. The "Research Techniques" portion of this website describes a method to isolate sclerites (also known as spicules, the free calcitic skeletal elements of octocorals) from the surrounding tissue for examination.

KEY TO OCTOCORALLIA

1. Colonies attached by a common basal stolon or holdfast to hard substrata in the low rocky intertidal 2
— Colonies partly imbedded in soft sediments of bays or sloughs by a basal peduncle, in extreme low intertidal to shallow subtidal, or occasionally washed ashore 8
2. Polyps arise from ribbonlike or spreading basal stolons 3
— Colonies arise from a basal holdfast; colony has a membranous shape or is disc-shaped or lobate; polyps contained in globular bodies of tissue (>5 mm high) to form coherent colonies; color cream or white to rose or red 5
3. Polyps are cylindrical and tall (5–12 mm) 4
— Individual polyps appear as densely set rounded mounds (<5 mm high) that arise from a common and broad basal stolon; color cream or tan to gray (plate 69A) . . . . *Cryptophyton goddardi*
4. Whitish polyps are coated with rust-orange sponge; some polyps with one or two daughter polyps emanating from the lateral walls of the parent polyp (plate 70C, 70D) . . . . *Telesto* sp.
— Daughter polyps absent, not coated with sponge (plate 70E) . . . . *Clavularia* spp.
5. Colonies globular, with projecting lobes, grey or rose to red in color (plate 69B) . . . . *Alcyonium* sp.
— Colonies not lobate 6
6. Colonies membranous, usually somewhat flattened and disc-shaped or globular without lobes, round to oval, polyps fully retractile, pale salmon pink to white in color (plate 69C) . . . . *Discophyton rudyi*
— Colonies membranous, irregularly shaped; polyps retractile and clustered atop moundlike or calyxlike protuberances

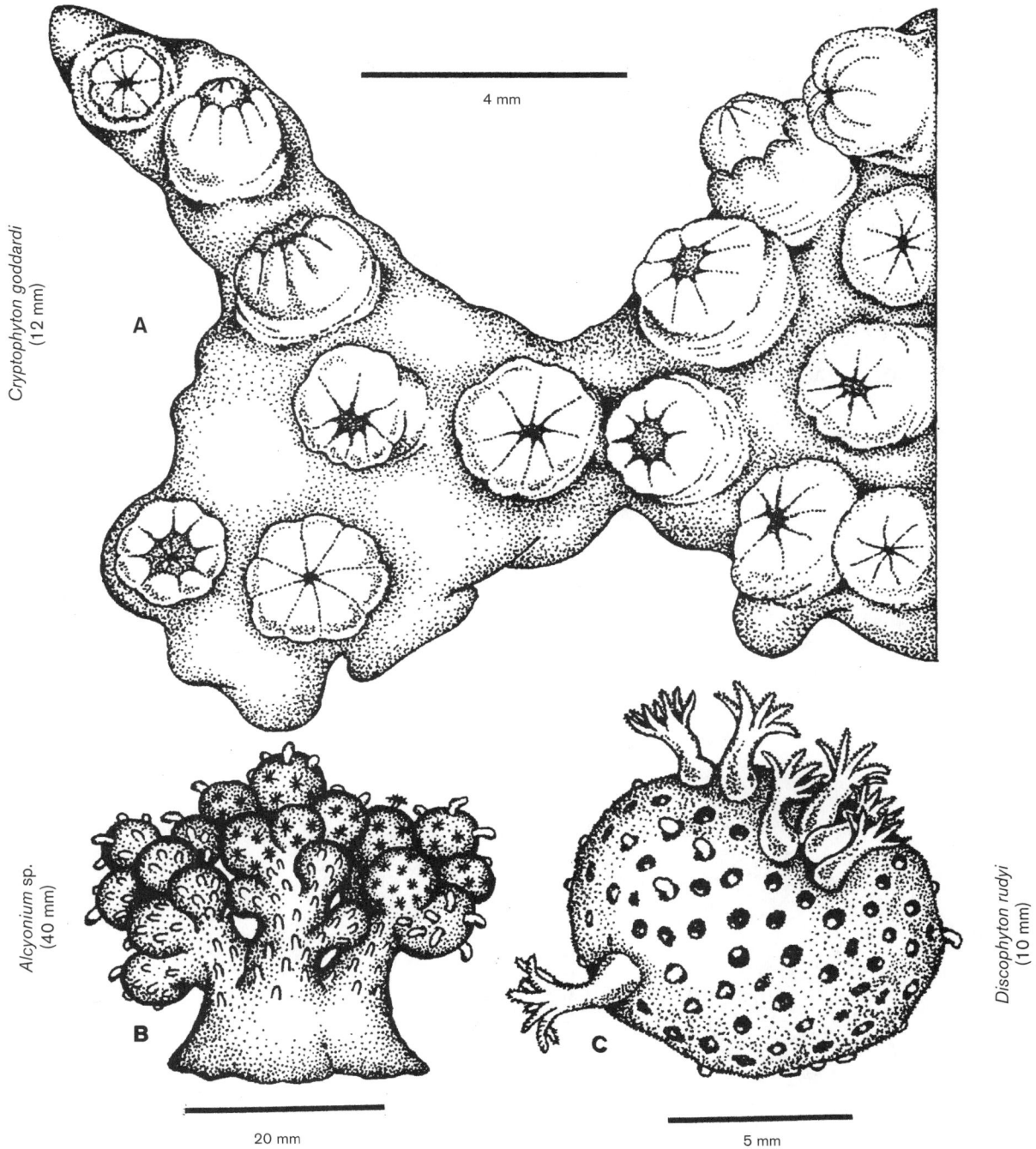

Cryptophyton goddardi
(12 mm)

A

Alcyonium sp.
(40 mm)

B

4 mm

C

Discophyton rudyi
(10 mm)

20 mm

5 mm

PLATE 69 A, *Cryptophyton goddardi*, scale bar = 4 mm; B, *Alcyonium* sp., scale bar = 20 mm; C, *Discophyton rudyi*, scale bar = 5 mm (all by G. C. Williams).

that are separated from one another by thin membranous regions devoid of polyps. 7

7. Sclerites are mostly coarsely tuberculated spindles; polyps heavily armored (plate 70F, 70G) . . . .
. . . . *Thrombophyton coronatum*

— Sclerites are coarsely tuberculated and robust spindles to ovoid or elliptical forms that give the surface of the colony a rough texture; sparsely tuberculated rods in the colony interior; polyps weakly armored (plate 70A, 70B) . . . .
. . . . *Thrombophyton trachydermum*

8. Polyp-bearing rachis is heart-shaped and flattened without lateral polyp leaves, lying upon the surface of sand with a wormlike peduncle buried in the sand below; color usually violet to brownish-purple (plate 71A) . . . . *Renilla amethystina*

— Polyp-bearing rachis is erect and stem- or featherlike with lateral polyp leaves; color usually orange, cream, or white and red-purple. 9

9. Stout and fleshy with dense, overlapping, large, and thickened polyp leaves; color often bright orange (plate 71E)
. . . . *Ptilosarcus gurneyi*

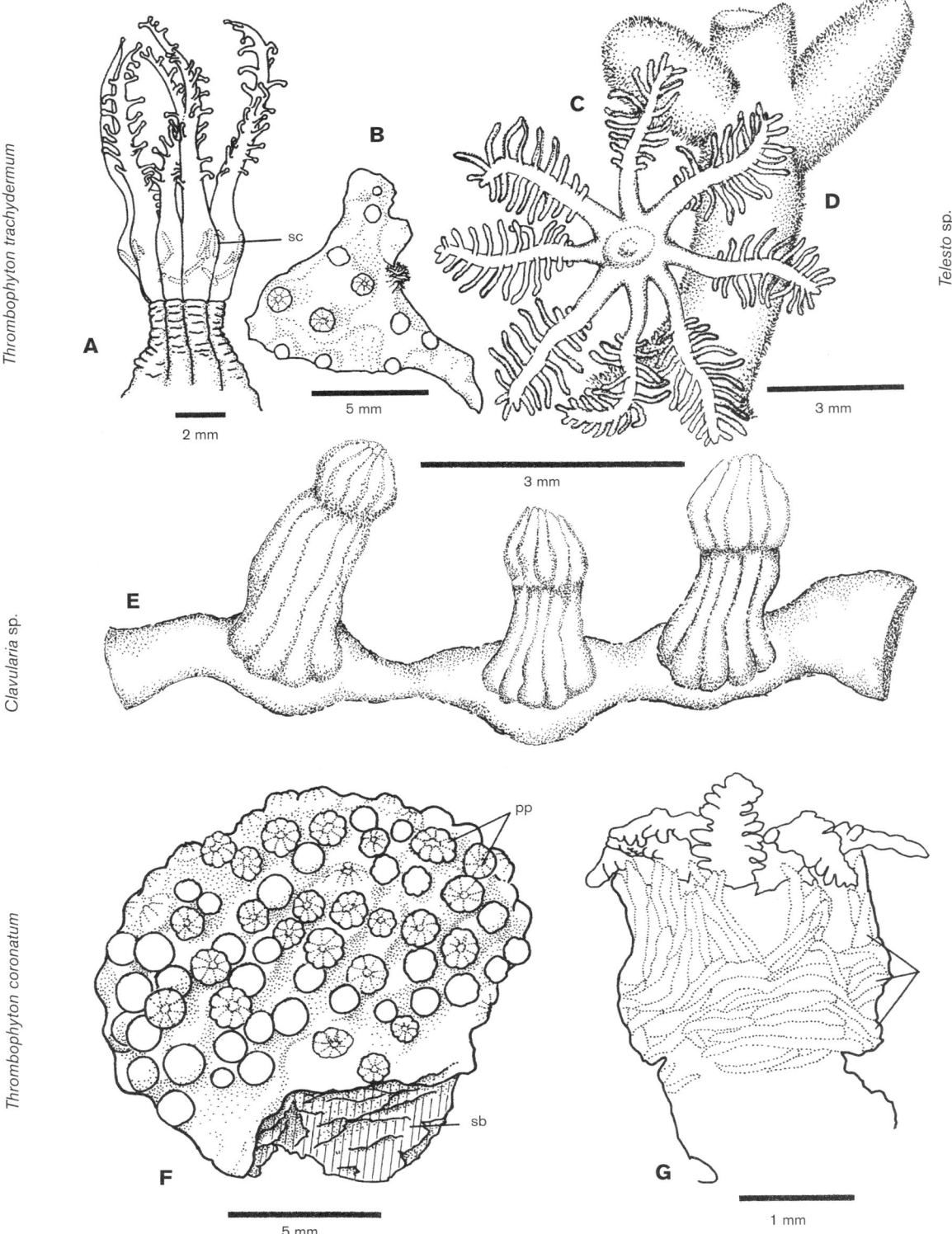

Thrombophyton trachydermum

Clavularia sp.

Thrombophyton coronatum

Telesto sp.

sc

B

A

2 mm

5 mm

C

D

3 mm

E

3 mm

F

pp

sb

5 mm

G

1 mm

PLATE 70 A, *Thrombophyton trachydermum*, single polyp, scale bar = 2 mm; B, *Thrombophyton trachydermum*, whole colony, scale bar = 5 mm; C, *Telesto* sp., expanded tentacles from a single polyp, scale bar = 3 mm; D, *Telesto* sp., single polyp with two daughter polyps, scale bar = 3 mm; E, *Clavularia* sp., part of a colony, scale bar = 3 mm; F, *Thrombophyton coronatum*, whole colony, scale bar = 5 mm; G, *Thrombophyton coronatum*, single polyp, scale bar = 1 mm (A, B, F, G after McFadden and Hochberg 2003, drawn by Linda D. Nelson; C, D, E by G. C. Williams).

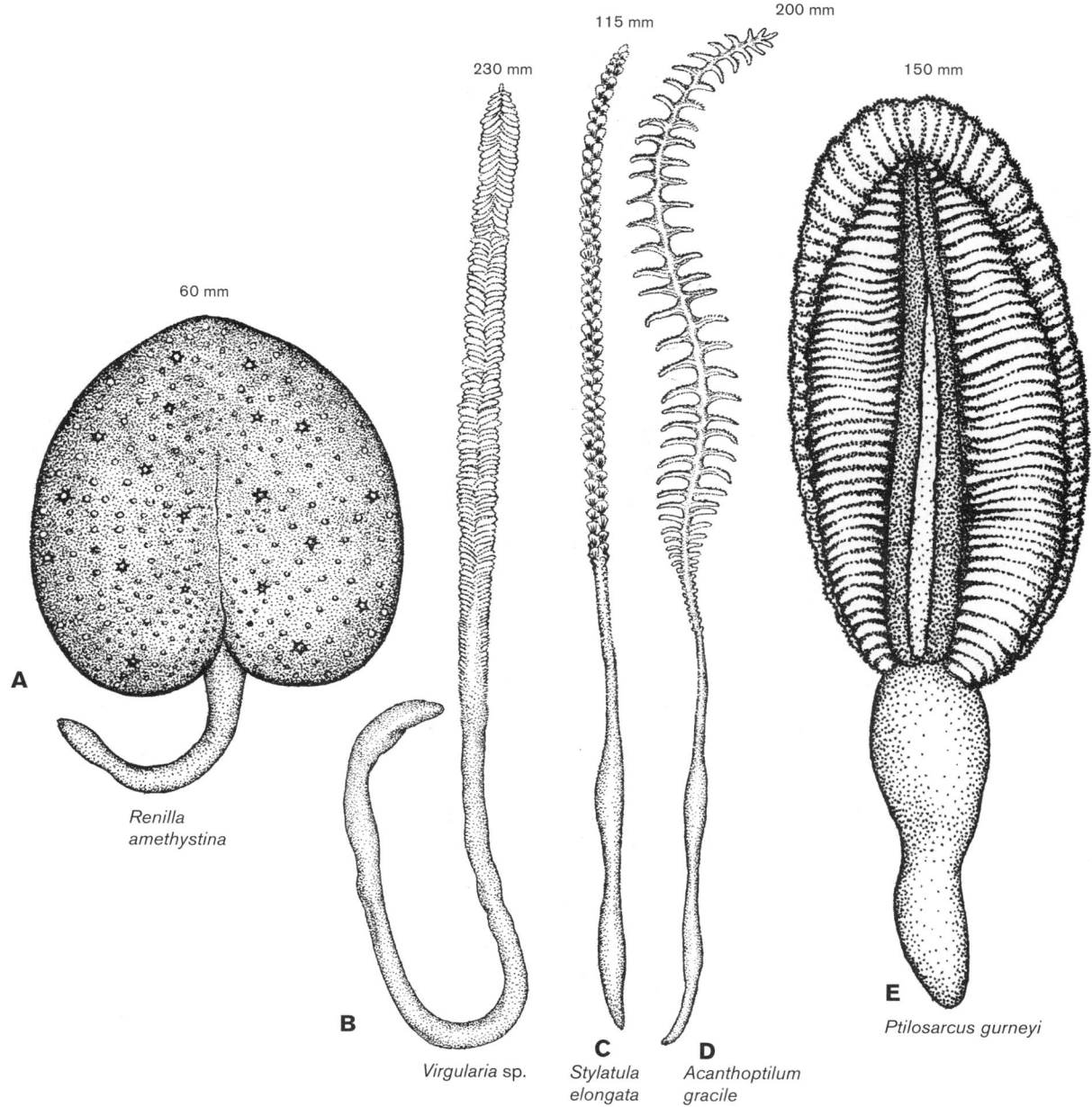

A
60 mm

*Renilla
amethystina*

B

230 mm

Virgularia sp.

C

115 mm

*Stylatula
elongata*

D

200 mm

*Acanthoptilum
gracile*

E

150 mm

Ptilosarcus gurneyi

PLATE 71 Measurements are lengths. A, *Renilla amethystina*, 60 mm; B, *Virgularia* sp., 230 mm; C, *Stylatula elongata*, 115 mm; D, *Acanthoptilum gracile*, 200 mm; E, *Ptilosarcus gurneyi*, 150 mm (all by G. C. Williams).

— Slender and elongate with short delicate polyp leaves; often colorless, milky white, or translucent gray to rose or light pink 10
10. Spindlelike sclerites present in polyps or at the bases of polyp leaves 11
— Sclerites absent except for minute oval bodies present mostly in the peduncle (plate 71B) . . . . *Virgularia* spp.
11. Needlelike sclerites form a fan-shaped armature at the base of each polyp leaf; colonies firm and scabrous (plate 71C) . . . . *Stylatula elongata*
— Short, three-flanged sclerites may form a weak cluster at the base of polyp leaf or at least are present elsewhere in the colony; colonies soft and flexible (plate 71D) . . . . *Acanthoptilum gracile*

LIST OF SPECIES

ORDER ALCYONACEA

CLAVULARIIDAE

Cryptophyton goddardi Williams, 1999. This stoloniferous taxon inhabits the underside of rocks and overhangs in low intertidal areas; known from the Oregon coast, but to be expected further south into California; possibly referred to as *"Clavularia"* sp. in some publications. More than one species of stoloniferous octocoral may inhabit the West Coast intertidal (see *Clavularia* spp., below). The genus is characterized by having polyps without sclerites and irregularly shaped sclerites in other parts of the colony. Preyed upon by the nudibranch *Tritonia festiva*.

Clavularia spp. Williams (2000), McFadden and Hochberg (2003) and J. Goddard (pers. comm.) report that members of this genus may inhabit intertidal and nearshore habitats along the California coast. This genus differs from *Cryptophyton* by the presence cylindrical polyps armed with crown and points and sclerites that are primarily spindles.

Telesto sp. An unidentified species has been found intertidally in exposed open coast situations of Santa Barbara County (J. Goddard, pers. comm.). The polyps form relatively dense clusters. Each polyp is usually 8–10 mm tall, and the polyp walls are thinly coated with a pale orange sponge. A few of the polyps have daughter polyps.

ALCYONIIDAE

Alcyonium sp. Sea strawberry. An undescribed lobate soft coral; Alaska to Sonoma County, California. The name "*Gersemia rubiformis*" (Ehrenberg 1834) has been applied by many authors to this taxon, but this application is erroneous because the genus *Gersemia* belongs to the related soft coral family Nephtheidae, and the species *G. rubiformis* (which also may belong to the genus *Alcyonium*) is native to the northern Atlantic. More than one species of *Alcyonium* may inhabit the Pacific coast intertidal area. This soft coral is preyed upon by the nudibranch *Tritonia festiva* (J. Goddard, pers. comm.).

*Alcyonium pacificum Yamada, 1950. McFadden and Hochberg (2003) report this northern Pacific species to the north and west of the California coast from the Alaskan Aleutian Islands to Japan; rocky substrata of kelp beds, 5–20 m.

Discophyton rudyi (Verseveldt and Ofwegen, 1992). An inconspicuous, disc-shaped soft coral found in the lower intertidal area, usually 8–15 mm in diameter; Vancouver Island to Point Lobos, Central California. The nudibranch *Tritonia festiva* is a reported predator.

Thrombophyton coronatum McFadden and Hochberg, 2003. Southern California (Palos Verdes to San Diego) and Catalina Island.

Thrombophyton trachydermum McFadden and Hochberg, 2003. Vancouver Island to Ano Nuevo Point, California.

ORDER PENNATULACEA

See Williams (1995) for a key and synopses of the sea pen genera. All sea pens in our area are fed upon by the nudibranch *Tritonia diomedea* (Behrens 2004. Proc. Calif. Acad. Sci. 55[2]: 43).

* = Not in key.

RENILLIDAE

Renilla amethystina Verrill, 1864 (=*Renilla koellikeri* Pfeffer, 1886). Sea pansy; common on low intertidal or shallow subtidal sandflats; Santa Barbara County to Baja California. The nudibranch *Armina californica* feeds on this species as well as some other sea pens.

VIRGULARIIDAE

Acanthoptilum gracile (Gabb, 1864). Occasionally encountered subtidally in bays; common in Tomales Bay; known from Central California (Sonoma to Monterey Counties). The genus is in need of revision, and other species are likely present, particularly in Southern California.

Stylatula elongata (Gabb, 1862). Often common in low intertidal areas of sandy mud; more than one species may inhabit the West Coast. The related genus *Virgularia* and this genus are easily confused: *Stylatula* differs from *Virgularia* by having each polyp leaf subtended by a fanlike armature of spindle-shaped or needlelike sclerites (Williams 1995).

Virgularia spp. Several species whose identification has not been verified occur from Monterey Bay and south, forming subtidal beds in the northern part of their range on gentle slopes or flats of sand or silt, but may be encountered on mud flats at low tide in the south.

PENNATULIDAE

Ptilosarcus gurneyi (Gray, 1860). Shallow subtidal; sometimes washed ashore. See Batie 1972, Northwest Sci. 46: 290–300 (taxonomy); Birkeland 1974, Eco. Mono. 44: 211–232 (predator/prey interactions). Preyed upon by the nudibranchs *Tritonia festiva, T. diomedea,* and *Toquina tetraquetra.* The more southern species of *Ptilosarcus, P. undulatus,* is found from Baja California to Peru.

REFERENCES

McFadden, C. S. and F. G. Hochberg 2003. Biology and taxonomy of encrusting alcyoniid soft corals in the northeastern Pacific Ocean with descriptions of two new genera (Cnidaria, Anthozoa, Octocorallia). Invertebrate Biology 122: 93–113.

Williams, G. C. 1995. Living genera of sea pens (Coelenterata: Octocorallia: Pennatulacea): illustrated key and synopses. Zoological Journal of the Linnean Society 113: 93–140.

Williams, G. C. 2000. A new genus of stoloniferous octocoral (Anthozoa: Clavulariidae) from the Pacific coast of North America. Zoologische Mededelingen Leiden 73: 333–343.

Verseveldt, J. and L. P. van Ofwegen 1992. New and redescribed species of *Alcyonium* Linnaeus, 1758 (Anthozoa: Alcyonacea). Zoologische Mededelingen 66: 155–181.

Ctenophora

(Plates 72–76)

CLAUDIA E. MILLS AND STEVEN H. D. HADDOCK

The ctenophores, or comb jellies, are transparent animals belonging to a small and entirely marine phylum of about 100–150 species, many (perhaps 50) of which are still undescribed. Although ctenophores are like medusae in being more or less transparent carnivores that often use tentacles to capture their prey, these similarities reflect a convergence in life styles rather than close evolutionary ties. Ctenophores are biradially symmetrical (discussion in Hyman 1940: 665), and nearly all West Coast species except *Pleurobrachia bachei* and some look-alikes are bioluminescent (Haddock and Case 1995). All ctenophores have a distinctive, complex, ciliated statocyst organ at the aboral pole that is used for orientation.

Ctenophores are so-named in recognition of the eight rows of ciliary comb plates, or ctenes, used for locomotion. All ctenophores bear these comb plates at some point in their life cycle, although some species can also move by flapping or waving other structures (the lobes and/or auricles). Most ctenophores use paired tentacles to capture prey, and these tentacles usually carry colloblasts—special, very sticky adhesive cells. Some ctenophore species capture prey on large mucous-laden, muscular surfaces known as lobes with the assistance of greatly reduced tentacles within the lobes. *Beroe,* lacking both lobes and tentacles, swallow their gelatinous prey whole.

The most commonly seen West Coast ctenophore is *Pleurobrachia bachei,* which is often cast up on beaches looking like a clear, watery marble and called a sea gooseberry or cat's eye. Most other species of ctenophores are not so robust and have little to no recognizable form out of water. Because no ctenophores are strictly intertidal invertebrates (although some may be found, on occasion, stranded in pools at low tide), we have taken some liberties in selecting the species to be keyed out here. The common coastal species and those species often encountered by blue-water divers offshore are included, as well as a small number of distinctive and robust deeper species likely to be recovered in midwater trawls not far from the coast.

Benthic ctenophores, known as platyctenes, account for nearly one-third of the known species. The body is highly flattened and modified, so they creep on a "foot" and can look much like flatworms, often living epizoically on other organisms such as sponges or seaweed (although their planktonic larvae are ciliated like other ctenophores). Although found in many locations worldwide, platyctenid ctenophores occur in warm water and are unlikely to be found within the geographic range of this book. The platyctene *Vallicula multiformis* was present briefly in shallow water in San Diego in 1997, being perhaps a short-lived introduction from the Caribbean.

Preservation of ctenophores is a difficult and frustration-filled endeavor. Some species, including *Pleurobrachia bachei* and some *Beroe,* preserve well, at least some of the time, in a weak (≤5%) formalin solution. Some species that leave only bits of debris in the bottle when preserved this way can be better preserved using glutaraldehyde instead of formaldehyde, and storage in a refrigerator after preservation seems to help maintain some specimens. Many fragile ctenophores, however, are simply unpreservable, in which case video, a good set of photographs, and drawings will have to suffice. For further techniques, see the introductory section here on preservation by Williams and Van Syoc.

Families used in the species list are from Cairns et al. (2002). However, a molecular phylogeny (Podar et al. 2001) supports the assertion that the phylum needs to be reorganized (Harbison 1985).

Glossary

(Plate 72)

ABORAL the end opposite the mouth, bearing a ciliated sense organ known as the "statocyst."

AURICLES four slender gelatinous appendages or processes at the base of the lobes and near the mouth (of a lobate ctenophore), with a ciliated edge; these may be short or long, straight or coiled, and may beat in such a way as to help move water past the lobes.

COMB PLATES transverse rows of thousands of fused macrocilia, shaped roughly like a flat paint brush and comprising subunits that are stacked together to form comb rows (also known as "ctenes"); the effector organs of swimming in a ctenophore (see Tamm 1973).

COMB ROWS eight longitudinal rows or costae of comb plates (ctenes), tightly spaced and overlying the meridional canals, which beat in metachronal waves that pass from the aboral end

189

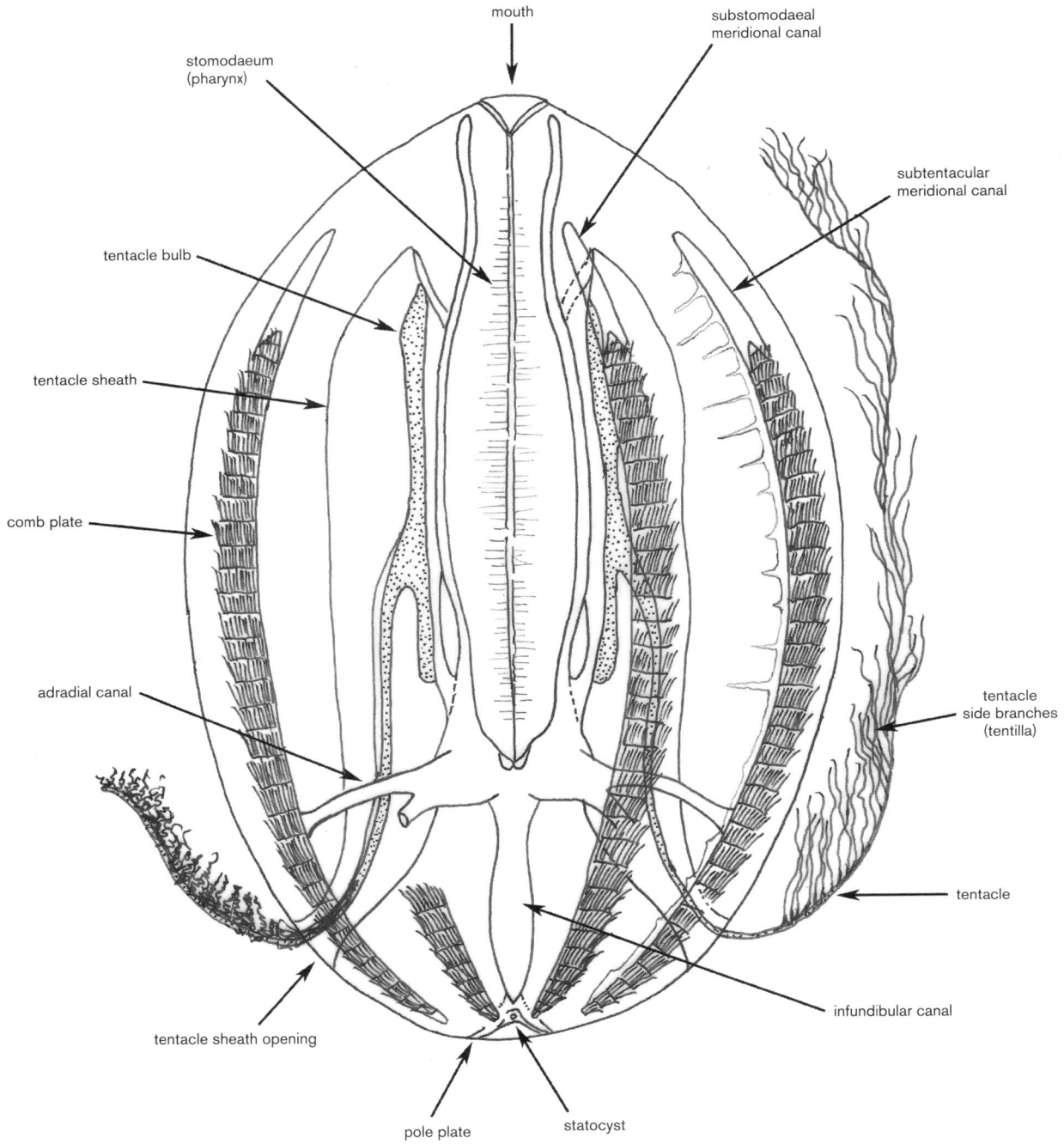

mouth

stomodaeum
(pharynx)

substomodaeal
meridional canal

subtentacular
meridional canal

tentacle bulb

tentacle sheath

comb plate

tentacle
side branches
(tentilla)

adradial canal

tentacle

infundibular canal

tentacle sheath opening

pole plate   statocyst

PLATE 72 *Hormiphora* sp., example of a cydippid ctenophore with labeled features (undescribed species, redrawn from Stanford, 1931, pp. 32 and 33, figs. 3 and 4).

toward the mouth, and thereby move the animal through the water by ciliary action.

CTENES see "comb plates."

DIVERTICULA (singular diverticulum) side branches or outpocketings emerging from any of the canals; if these meet and join, it is said that they "anastomose."

GASTRIC CAVITY see "stomodaeum."

INFUNDIBULAR CANAL broad, often funnel-shaped, canal running between the base of the stomodaeum and the aboral pole (also known as the "aboral canal").

LOBES a pair of thin, muscular, cuplike oral extensions of the body; they may be as large or larger than the body itself, on both sides of the mouth.

MEDUSIFORM shaped like a jellyfish (a medusa).

MERIDIONAL CANALS usually eight, but in a few species only four, circulatory canals that run under the comb rows, which are connected by smaller internal canals running through the jelly to the stomodaeum.

MOUTH opening into the stomodaeum at the oral end.

ORAL the end of a ctenophore bearing the mouth.

PARAGASTRIC CANALS pair of canals originating at the base of the stomodaeum, running upward along each flattened surface of the stomodaeum toward the mouth (also known as the pharyngeal canals).

STATOCYST ciliated orientation organ at the aboral pole, which may be flush with the surface or sunken to some extent, and which is connected to the eight comb rows by ciliated grooves (also known as the aboral sense organ).

STOMODAEUM the gastric cavity that opens from the mouth; it is highly expandable (also known as the pharynx).

SUBSTOMODAEAL CANALS the four meridional canals closest to the plane of the stomodaeum.

SUBTENTACULAR CANALS the four meridional canals that lie on either side of the openings of the tentacle sheaths, closest to the plane that runs longitudinally through the two tentacle bulbs.

TENTACLE elongate, highly extensile, (paired) filamentous, feeding appendage, emerging from a swollen tentacle bulb at its base, often with side branches (also known as "tentilla") that are finer than the main, lateral tentacular filament. Tentacles are equipped with thousands of microscopic adhesive structures called colloblasts, with which they capture prey. The tentacles are often fragile and may be partially or entirely broken off during collection.

TENTACLE SIDE BRANCHES fine side filaments on the tentacles of many (but not all) tentaculate species, which may be closely spaced or set rather far apart (also known as "tentilla").

TENTACLE BULBS the (two) swellings at the base of the tentacles, often very elongate, up to half of the body length in some species, from which the tentacular filaments emerge; attached to the inner wall of the tentacle sheaths.

TENTACLE SHEATHS the (two) tubelike, or funnel-shaped, openings from the base of the tentacle filaments at their juncture with the tentacle bulbs, through which the tentacles emerge to the outside.

TENTILLA (singular tentillum) see "tentacle side branches."

VESTIBULE large preoral cavity inside the "mouth" of *Dryodora*, which may hold prey but is separate from the much smaller stomodaeum beneath it.

Key to Ctenophora

1. Body like a sac, mouth opens wide, gastric cavity occupies nearly entire interior; without tentacles or lobes (see plate 76) . . . . Order Beroida 23
— Body otherwise and usually with some sort of tentacles 2
2. Body long and flat, like an airplane wing, with fine tentacles along the entire leading edge that sweep back over the body surface . . . . Order Cestida 22
— Body not like an airplane wing 3
3. Body medusiform, open like a shower cap, with eight short comb rows on the upper surface and two small branched tentacles hanging into the open cavity (Order Thalassocalycida) (plate 75G, 75H) . . . . *Thalassocalyce* sp.
— Body not like a shower cap. 4
4. Body solid, basically rounded at both ends, with a pair of tentacles arising in sheaths on opposite sides of the body; without oral lobes (see plate 72, 73, 76) . . . . Order Cydippida 5
— Body with a pair of muscular, bowl-like lobes at the oral end; usually with small, inconspicuous tentacles between the lobes (see plates 74, 75) . . . . Order Lobata 16

5. Tentacles with side branches (tentilla), exiting body closer to the aboral pole or near the midpoint of the body; mouth typically small, often slightly pointed 6
— Tentacles without side branches, exiting body closer to the oral pole or near the midpoint of the body; mouth very plastic and able to open very wide like a *Beroe* 13
6. Tentacle bulbs short; tentacle sheaths exit near the midpoint of the body 12
— Tentacle bulbs short or long; tentacle sheaths exit toward the aboral pole 7
7. Body transparent, firm, and unpigmented; rounded like a clear gooseberry or grape 8
— Body not as above 9
8. Tentacle bulbs short, with tentacle sheaths angling out at nearly 45°; stomodaeum less than half the body length; to about 15 mm diameter (plate 73B). . . . . *Pleurobrachia bachei*
— Tentacle bulbs one-quarter to one half body length; tentacle sheaths parallel the stomodaeum for some distance; stomodaeum greater than one half body length 9
9. Body with tentacle bulbs very close to walls of stomodaeum; transparent and mostly unpigmented; with side branches that extend out when the tentacle is relaxed, forming an ordered, comblike pattern rather than a disorganized mass of fine filaments 10
— Body with tentacle bulbs parallel to, but somewhat removed from, the walls of the stomodaeum; tentacles either nearly colorless and with a small number of side branches that are kept tightly coiled most of the time, or tentacles deeply colored with many disorganized, fine side branches 11
10. Body to 30 mm in length, moderately compressed, somewhat tapered at the aboral end; narrowing toward the mouth on small specimens, but becoming distinctly wider at the oral end on large ones; comb rows extend from very near the aboral pole to four-fifths body length or more; meridional canals about as long as comb rows and without diverticula; not luminescent (plate 73A). . . . . *Hormiphora californensis*
— Body to 35 mm in length, nearly circular in cross-section, somewhat flattened at the aboral end and narrowing toward the mouth; comb rows extend from very near the aboral pole to at least three-quarters body length; meridional canals extend beyond comb rows and branch inward toward the center of the body; luminescent characteristics not known (plate 72) . . . . *Hormiphora* sp. (undescribed)
11. Body to 20 mm in length, tentacles unpigmented or with a little red, with few side branches that are held coiled up except when actively capturing prey; body transparent, sometimes with gelatinous extensions protruding below the statocyst; fast and maneuverable swimmer (plate 73C) . . . . *Euplokamis dunlapae*
— Body to 20 mm in length, tentacles noticeably pink, red, or purplish, with fine and very numerous side branches that extend out to form a diaphanous cloud of filaments; body somewhat opaque, without gelatinous extensions at the aboral pole; not particularly fast swimming (plate 73D) . . . . undescribed mertensiid
12. Body to several centimeters long, circular in cross-section; with highly mobile and extensible mouth and with voluminous stomodaeum; young stages can flatten out onto salp body surface to look like a parasite (plate 73E) . . . . *Lampea* spp.

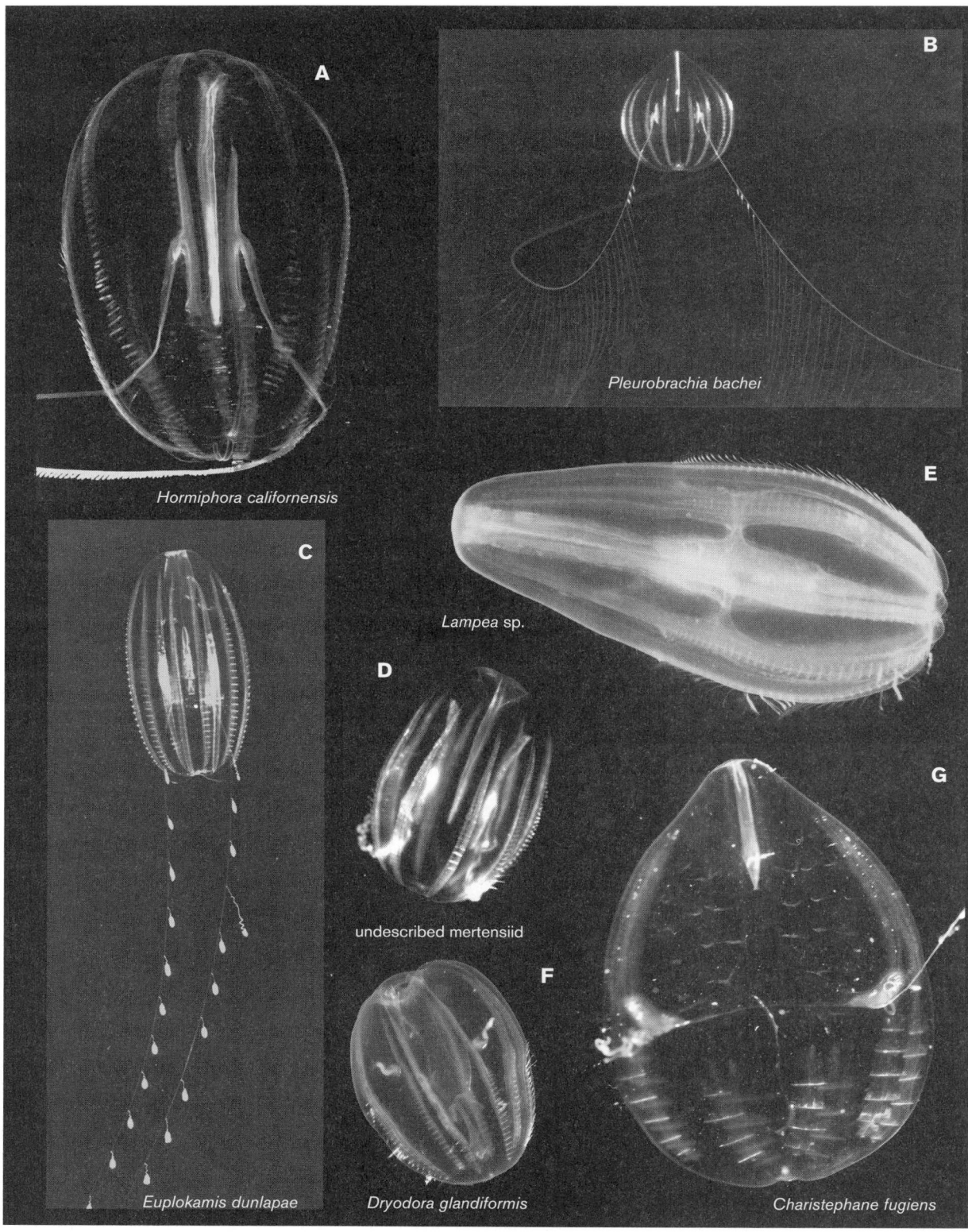

PLATE 73 A, *Hormiphora californensis*; B, *Pleurobrachia bachei*; C, *Euplokamis dunlapae*; D, undescribed mertensiid; E, *Lampea* sp.; F, *Dryodora glandiformis*; G, *Charistephane fugiens* (photographs A–B, D–G by Steven H. D. Haddock; photograph C by Claudia E. Mills).

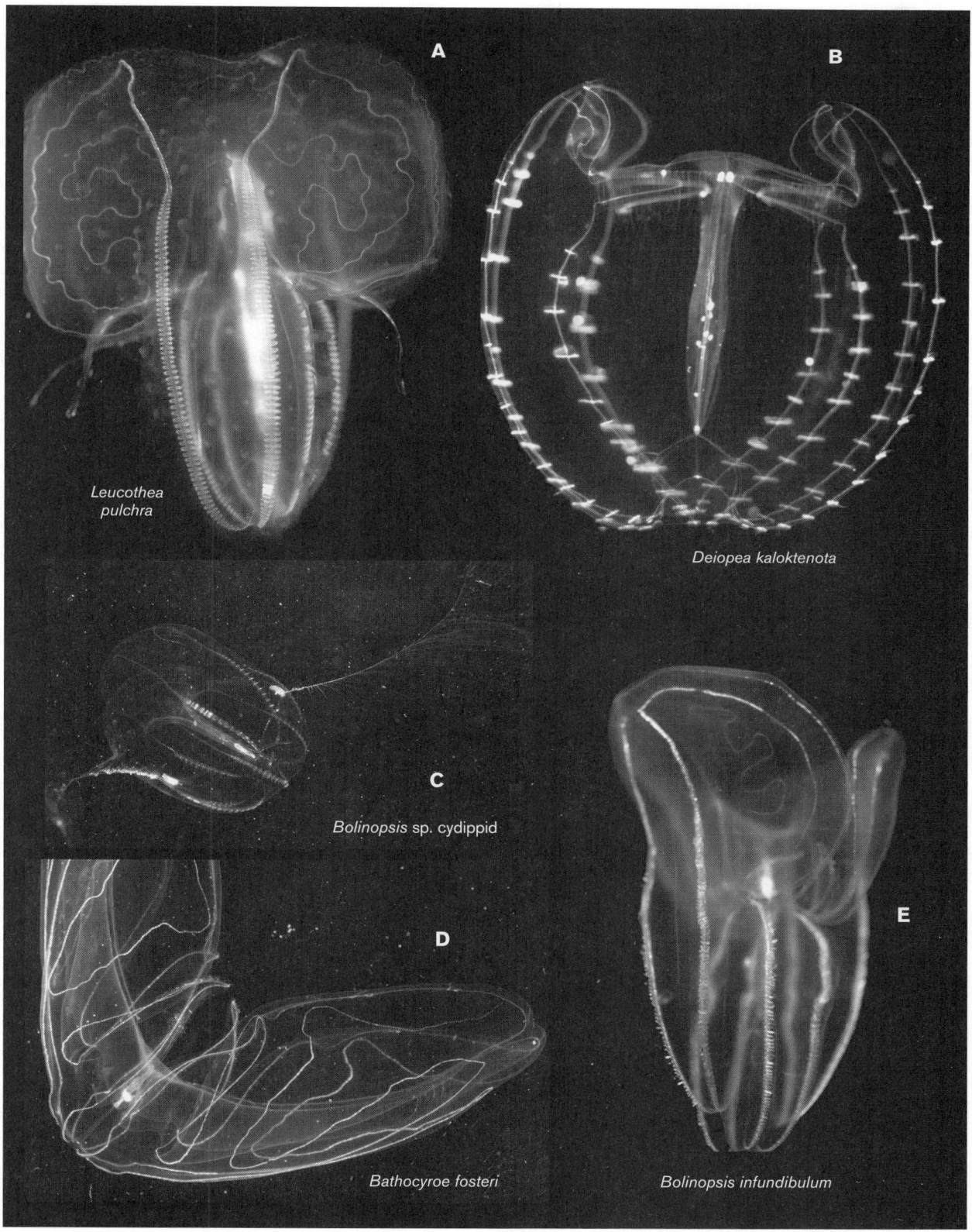

A

B

Leucothea pulchra

Deiopea kaloktenota

C

Bolinopsis sp. cydippid

D

Bathocyroe fosteri

E

Bolinopsis infundibulum

PLATE 74 A, *Leucothea pulchra*; B, *Deiopea kaloktenota*; C, cydippid larva of *Bolinopsis infundibulum*; D, *Bathocyroe fosteri*; E, *Bolinopsis infundibulum*. (all photographs A–E by Steven H. D. Haddock).

— Body to 20 mm tall, highly flattened in cross-section; tentacle sheaths short, angled, and near outer body wall; comb rows limited to aboral end of body, but subtentacular meridional canals extending three-quarters of the way to the oral end; highly transparent and unpigmented, or with light tan tentacles; deep water (plate 73G) . . . . *Charistephane fugiens*

13. Tentacle bulbs small, round, and very near the surface, with sheaths so short that the tentacles are unable to withdraw completely inside; stomodaeum deep inside a large "vestibule" at the oral end of the animal; body transparent, without a greenish tint; a boreal species seen only rarely as far south as California (plate 73F) . . . . *Dryodora glandiformis*

— Tentacle bulbs very long, typically more than one half the body length and with long, orally exiting tentacle sheaths; stomodaeum extends all the way to oral end of animal (this feature is true of nearly all ctenophores except *Dryodora*, above); body usually with a slight greenish tint 14

14. Meridional canals unequal in length, maximum are three-quarters of body length; comb rows one half to three-fifths of body length; without stomodeal canals; body ellipsoid in cross-section, usually transparent, though may have a greenish cast 15

— Meridional canals all equal in length and nearly as long as the body; comb rows less than one half body length; with stomodeal canals; body nearly circular in cross-section; usually <10 mm long, with long tentacle sheaths exiting the body near the mouth; without distinctive orange pigment spots, but usually slightly opaque body may have red pigment diffusely distributed along canals (plate 76G) . . . . *Haeckelia beehleri*

15. Very small, usually <3 mm, with rounded oral end; with long tentacle sheaths exiting about one-quarter of the body length below the mouth; with large orange-red pigment spots on the canals around the base of the stomodaeum and tentacles, and with small red spots along the comb rows (plate 76J) . . . . *Haeckelia bimaculata*

— Small, usually under 7 mm long, with pointed oral end; with long tentacle sheaths exiting the body near the mouth; with two pairs of orange-red pigment spots—on the tentacle bases and about two-thirds to three-quarters down the tentacle sheaths (plate 76H, 76I) . . . . *Haeckelia rubra*

16. Body with two rounded, lobe-like extensions used to catch prey and with four elongate, ciliated auricles that help move water across the lobes 17

— Body with two rounded, lobe-like extensions used to catch prey, but without auricles; body to 20 cm, rounded with very large lobes, colorless but not very transparent, comb rows end on stubby protrusions near mouth, in the place of auricles (usually in deep water) (plate 75A) . . . . undescribed "giant lobate"

17. Body surface smooth, mostly colorless; lobes either as large as the rest of the body or considerably smaller; with static auricles 18

— Body surface covered with brownish orange papillae; with very large lobes that can extend perpendicularly as much as a full body length; with long slender auricles that beat rhythmically (plate 74A) . . . . . *Leucothea pulchra*

18. Lobes usually shorter than stomodaeum, not capable of flap/swimming (see choice below) 19

— With large, muscular lobes that are used for feeding but can also be clapped together for escape swimming in a movement reminiscent of the swimmer's "frog kick" 21

19. Body oblong and rounded at both ends, not highly flattened; may have rows of dark spots on the lobes in line with the canals, which in larger specimens may coalesce into dark pigmented lines; total length to 15 cm (this is the common Pacific coast lobate) (plate 74E) . . . . *Bolinopsis infundibulum*

— Body highly flattened (rare coastal visitors from deep water) 20

20. Body rounded to oval in outline, highly flattened; with a pair of relatively small oral lobes; with only a few ctenes, spaced far apart on each comb row; total length not exceeding 5 cm; rare (plate 74B) . . . . *Deiopea kaloktenota*

— Body semi-circular to V-shaped, highly flattened; unpigmented, with a pair of large oral lobes; total length to 30 cm; rare (plate 75F) . . . . *Kiyohimea* spp.

21. Body flattened and reduced; may have a pair of diffuse dark spots on the inner surface of each lobe; gut colorless; near-surface species (plate 75B) . . . . . . *Ocyropsis maculata*

— Body rounded, much shorter than lobes; with small, dark-red stomodaeum; deep water species (plate 74D) . . . . *Bathocyroe fosteri*

22. Length to 1.5 m, but usually <80 cm, the set of canals that run out the midline of the body originate near the base of the stomodaeum, rapidly curving up to the midline; gonads continuous; escapes by graceful rapid undulating of the body (plate 75D) . . . . *Cestum veneris*

— Usually <20 cm long, the set of canals that run out the midline of the body originate near the center of the stomodaeum, without curves; gonads visible as dashed white line; escape swimming very rapid wriggling (plate 75C, 75E) . . . . *Velamen parallelum*

23. Body highly flattened 24

— Body cylindrical to moderately compressed 25

24. Body to 15 cm, shaped like a compressed cone, with pointed aboral end and broad oral end with very large mouth; comb rows all the same length and extend three-quarters to five-sixths of body length; with highly branched meridional canals forming a network whose branches anastomose (merge) with one another; usually tinged rose pink, with darker red along the comb rows (plate 76A) . . . . *Beroe forskalii*

— Body to about 6 cm, soft and flaccid, with aboral end round to slightly tapered and oral end very broad with nearly semicircular mouth; comb rows of two lengths, extending about one half and two-thirds to three-quarters body length; meridional canals with many branches that all bend toward the mouth and do not merge with one another; overall color white or pale pink, with a diffuse orange or red spot or pair of stripes on either side of the body overlying the stomodaeum (plate 76E, 76F) . . . . *Beroe mitrata*

25. Body slender and elongate, rarely longer than 3 cm and usually much less; meridional canals with very few branches, and those present extending inward toward the stomodaeum rather than in the plane of the body surface (plate 76D) . . . . *Beroe gracilis*

— Body substantial and thick-walled, typically reaching several centimeters in length, meridional canals branched and running in the plane of the body surface, but few if any of the branches merge 26

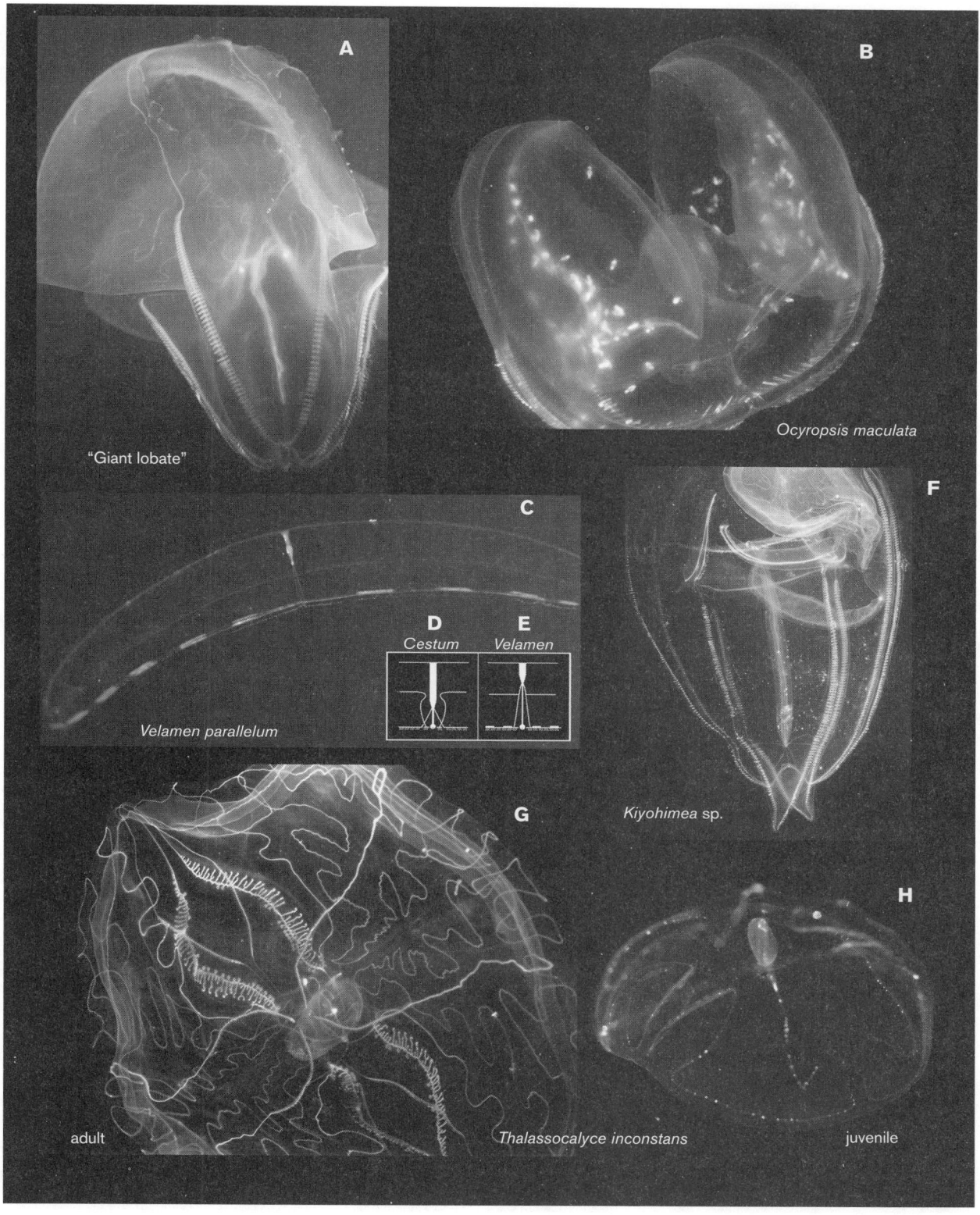

PLATE 75 A, Undescribed "giant lobate"; B, *Ocyropsis maculata* (white spots are symbiotic amphipods); C, *Velamen parallelum*; D, *Cestum veneris*; E, *Velamen parallelum*; F, *Kiyohimea* sp.; G, adult *Thalassocalyce inconstans*; H, juvenile *Thalassocalyce inconstans* (photographs A, B, C, F, G, H by Steven H. D. Haddock).

PLATE 76 A, *Beroe forskalii*; B, *Beroe ?cucumis*; C, *Beroe abyssicola*; D, *Beroe gracilis*; E and F, *Beroe mitrata*; G, *Haeckelia beehleri*, H and I, *Haeckelia rubra* (arrows indicate distinctive spots of red-orange pigment); J, *Haeckelia bimaculata* (photographs A, B, D, G, H, I, J by Steven H. D. Haddock; photographs C, E, F by Claudia E. Mills).

26. Body to 10 cm, usually much less, comb rows of two lengths, extending about one half and two-thirds the distance from aboral pole towards the mouth; transparent and colorless or with golden tan and/or orange-pink pigment on outer surface or along stomodaeum; paragastric canals unbranched; found near the surface, as well as to about 500 m (common) (plate 76B). . . . . *Beroe ?cucumis*

— Body to 7 cm, comb rows of equal lengths, extending one half to two-thirds the distance from aboral pole towards the mouth; body somewhat opaque and usually with intense red, purple or nearly black color lining the stomodaeum; paragastric canals branched—with diverticula; deep water (plate 76D) . . . . *Beroe abyssicola*

List of Species

Order Cydippida

DRYODORIDAE

Dryodora glandiformis (Mertens, 1833). A boreal species known from both northern and southern oceans; found near Friday Harbor most years in the late winter and spring and rarely seen off central California; usually <1 cm long in our waters, but two or three times that size in the Arctic. Feeds, perhaps exclusively, on appendicularians. Color photograph in Wrobel and Mills 1998 and 2003, p. 63.

EUPLOKAMIDIDAE

Euplokamis dunlapae Mills, 1987. This species (to 2 cm long) is well-known from fjords in Washington and British Columbia. An unidentified *Euplokamis*, which may prove to be *E. dunlapae*, has been seen occasionally on video in Monterey Bay. See Mills 1987, Can. J. Zool. 65: 2661–2668 (description, behavior, larvae, depth distribution in British Columbian fjords); Mackie, Mills, and Singla 1988, Zoomorphology 107: 319–337 (prey capture and morphology of prehensile side branches on tentacles); Mackie, Mills, and Singla 1992, Biol. Bull. 182: 248–256 (giant axons and escape swimming). Color photograph in Wrobel and Mills 1998 and 2003, p. 62.

HAECKELIIDAE

Haeckelia beehleri (Mayer, 1912). Usually <1 cm, occasionally seen near the surface by divers, sometimes together with *H. bimaculata* and/or *H. rubra*. Body wall has a distinct green tint. Color photograph in Wrobel and Mills 1998 and 2003, p. 59.

Haeckelia bimaculata Carré and Carré, 1989. A tiny (<3 mm long) species, sometimes seen near the surface by very observant divers. See Carré and Carré 1989, Comptes Rendus de l'Académie des Sciences, Paris, 308 (série III): 321–327. Like *H. rubra*, this species uses cnidocysts from narcomedusan prey in its tentacles instead of colloblasts. Color photograph in Wrobel and Mills 1998 and 2003, p. 59.

Haeckelia rubra (Kölliker, 1853) (=*Euchlora rubra*). Usually <1 cm and has cnidocysts instead of colloblasts on its tentacles; the cnidocysts are obtained from narcomedusan prey, much in the way some nudibranchs obtain them from hydroids or sea anemones. See Mills and Miller 1984, Mar. Biol. 78: 215–221 (feeding behavior); Carré and Carré 1980, Cah. Biol. Mar. 21: 221–226 (cnidocysts and feeding); Carré, Carré, and Mills 1989, Tissue and Cell 21: 723–734 (cnidocysts in *H. rubra* and narcomedusan prey). Color photograph in Wrobel and Mills 1998 and 2003, p. 59.

LAMPEIDAE

Lampea pancerina (Chun, 1879). A rare visitor to central and southern California. Several species of *Lampea* could be found in our region, all reaching several centimeters in length. All feed, perhaps exclusively, on salps, to which they may attach and flatten out, looking like parasites, as they feed; large *Lampea* can ingest whole salp chains. Color photograph in Wrobel and Mills 1998 and 2003, p. 60.

MERTENSIIDAE

Charistephane fugiens Chun, 1879. The refractive index of this highly flattened species is so similar to that of water that it is easily missed among other animals in midwater trawl samples; up to about 2 cm in length. Front and side-view photographs in Wrobel and Mills 1998 and 2003, p. 62.

Undescribed mertensiid. A cydippid to about 2 cm in length, with pink tentacles having high numbers of side branches (tentilla) that look like a fuzzy cloud when extended underwater, is collected occasionally at least from central California and the Puget Sound/Strait of Georgia region. This may be the species erroneously called "*Mertensia ovum*" in the previous edition; *M. ovum* is an Atlantic boreal and arctic species not known from the Pacific (alternately, Torrey [1904] mistakenly labeled his drawing of *Hormiphora californensis* (see below) as *M. ovum*, which may be the origin of this name in the California fauna in previous editions of this book). Color photograph in Wrobel and Mills 1998 and 2003, p. 63.

PLEUROBRACHIIDAE

Hormiphora californensis (Torrey, 1904). The original description, as *Euplokamis californensis*, is incomplete, and the figure accompanying it is unfortunately mislabeled as "*Mertensia ovum*"; occurs at least off San Diego and Santa Barbara and in Friday Harbor with *Pleurobrachia bachei* and may be substantially larger—to about 3 cm long—but most casual observers will not notice this second species as different. Color photograph in Wrobel and Mills 1998 and 2003, p. 60.

Hormiphora sp. (undescribed). This approximately 3-cm-long Monterey Bay species was treated in the unpublished master's thesis of Stanford (1931) under the *nomen nudum Hormiphora coeca*. It is possible that it is the same as *H. californensis*, but the branched or diverticulate meridional canals described by Stanford appear to differentiate the two species. *Hormiphora* spp. on our coast are rarely recognized by the casual observer to be different from *Pleurobrachia bachei*.

Pleurobrachia bachei A. Agassiz, 1860. Probably the most common ctenophore in northeast Pacific coastal waters, to about 1.5 cm long, frequently occurring in great numbers and sometimes left by outgoing tides on the beaches, looking like glassy marbles. See Hirota 1974, Fish. Bull. 72: 295–335 (natural history); Tamm and Moss 1985, J. Exp. Biol. 114: 443–461 (spin feeding in the related *P. pileus*). Color photograph in Wrobel and Mills 1998 and 2003, p. 62.

Order Platyctenida

COELOPLANIDAE

Vallicula multiformis Rankin, 1956. This small warm-water benthic ctenophore, apparently native to the Caribbean, was seen briefly in shallow water in San Diego late in the El Niño summer of 1997, but disappeared when the water cooled down to normal temperatures (Constance Gramlich and George Matsumoto, personal communication). It can occur on a variety of substrates including algae, eelgrass, rock, and floats. It has also been collected in Hawaii on floats.

Order Thalassocalycida

THALASSOCALYCIDAE

Thalassocalyce sp. Probably *T. inconstans* Madin and Harbison, 1978, described from the Sargasso Sea and Northwest Atlantic slope water, which can reach at least 15 cm in diameter, but specimens seen near shore in our waters are usually much smaller juveniles (plate 75H). As it has become evident that this fragile species is not so uncommon, one wonders if more than one species of *Thalassocalyce* exist under a single name; specific differences have not yet been established. See Madin and Harbison 1978, Bull. Mar. Sci. 28: 680–687. Color photograph in Wrobel and Mills 1998 and 2003, p. 63.

Order Lobata

BATHOCYROIDAE

Bathocyroe fosteri Madin and Harbison, 1978. This not-un-common species, reaching 4 cm across the lobes, was described from the northwest Atlantic, and it is possible that more than one species of *Bathocyroe* exists under this single name; it is common in deep water off California and easily distinguished by its flap-swimming (but see also *Ocyropsis*). See Madin and Harbison 1978, Bull. Mar. Sci. 58: 559–564. Color photograph in Wrobel and Mills 1998 and 2003, p. 64.

BOLINOPSIDAE

Bolinopsis infundibulum (O. F. Müller, 1776). This common arctic and boreal lobate species occurs from the Bering Sea to California, from the surface to about 400 m, reaching lengths of up to 15 cm, but near-shore specimens along our coast are usually much smaller. (*Bolinopsis* off central Mexico are apparently *B. vitrea* [L. Agassiz, 1860], a subtropical species). Feeding behavior described by Matsumoto and Harbison (1993). Color photograph in Wrobel and Mills 1998 and 2003, p. 64.

EURHAMPHAEIDAE

Deiopea kaloktenota Chun, 1879. Known from several sites worldwide, this distinctive smaller lobate (to 4.5 cm) may be a younger (and usually shallower) stage of *Kiyohimea*, the only other known ctenophore with widely spaced ctenes, but the morphology of young stages of *Kiyohimea* is not yet known. Color photograph in Wrobel and Mills 1998 and 2003, p. 65.

Eurhamphaea vexilligera Gegenbauer, 1856. A warm water species to several centimeters in length that, in some years, is encountered in the southern end of our area. This distinctive lobate has a pair of long filamentous "tails" trailing from the aboral end and releases luminescent ink from sacs near the comb rows when disturbed. Feeding behavior described by Matsumoto and Harbison (1993).

Kiyohimea aurita Komai and Tokioka, 1940. Described from a specimen found near the Seto Marine Biological Laboratory in Japan that had found its way inshore and was possibly damaged, having tentacle bulbs but no tentacles; body semicircular at the aboral end. Entire, undamaged specimens of this extremely fragile species are rarely collected, but may reach 30 cm or more in length.

Kiyohimea usagi Matsumoto and Robison, 1992. A large, very fragile, deep water species described from Monterey Bay, differentiated from *K. aurita* by having tentacles and being more V-shaped. See Matsumoto and Robison 1992, Bull. Mar. Sci. 51: 19–29; color photograph in Wrobel and Mills 1998 and 2003, p. 65. The two species of *Kiyohimea* could very likely be the same, in which case the name *K. aurita* has precedence.

LEUCOTHEIDAE

Leucothea pulchra Matsumoto, 1988. Described from Monterey Bay and near Catalina Island, this species, which may be 25 cm or more in length, is similar to the more cosmopolitan *L. multicornis* (Quoy and Gaimard, 1824). See Hamner, Strand, Matsumoto, and Hamner 1987, Limnol. Oceanogr. 32: 645–652 (foraging behavior); Matsumoto 1988, J. Plankton Res. 10: 301–311 (description and feeding); Matsumoto and Hamner 1988, Mar. Biol. 97: 551–558 (swimming and feeding behavior). Color photograph in Wrobel and Mills 1998 and 2003, p. 65.

OCYROPSIDAE

Ocyropsis maculata (Rang, 1828). This flap-swimming, surface-dwelling species to several centimeters long is occasionally seen off Catalina Island and other southern California sites, where its occurrence probably best corresponds with strong El Niños, including that of 1997. The species name recalls the pair of large, dark, diffuse spots on the lobes—other species of *Ocyropsis* do not have these spots. Feeding behavior described by Matsumoto and Harbison (1993). Although most ctenophores are hermaphrodites, this genus usually has separate sexes (see Harbison and Miller 1986, Mar. Biol. 90: 413–424). Color photograph in Wrobel and Mills 1998 and 2003, p. 66.

FAMILY UNKNOWN

Undescribed "giant lobate." Distinctive species only seen as an adult in deep water. Its younger stages have not been described, but should be readily distinguished by the lack of auricles.

Order Cestida

CESTIDAE

Cestum veneris Lesueur, 1813. Venus' girdle. Uncommon from central California south; cosmopolitan in warmer waters where

* = Not in key.

it may reach over 1 m in length. Feeding behavior described by Matsumoto and Harbison (1993). Color photograph in Wrobel and Mills 1998 and 2003, p. 66.

Velamen parallelum (Fol, 1869). This very fast (often described as "darting"), small (to 20 cm) "Venus' girdle" occurs fairly often from central California south. Feeding behavior described by Matsumoto and Harbison (1993). Color photograph in Wrobel and Mills 1998 and 2003, p. 67.

Order Beroida

BEROIDAE

Feeding behavior described by Matsumoto and Harbison (1993). Diversity of oral macrocilia and possible concomitant diet differences are described by Tamm and Tamm (1993).

Beroe abyssicola Mortensen, 1927. A usually deep water species reaching several centimeters in length, but sometimes also found at the surface, distinguished by its darkly pigmented gut; Vancouver Island to central California. See Arai 1988, Contrib. Nat. Sci. Roy. British Columbia Mus. 9: 1–7 (redescription); Mills and McLean 1991, Dis. Aquat. Org. 10: 211–216 (parasitic dinoflagellates on comb rows). Color photograph in Wrobel and Mills 1998 and 2003, p. 67.

Beroe ?cucumis Fabricius, 1780. A several-centimeter-long surface species in central California that may also be taken as deep as several hundred meters off southern California; corresponds fairly well to the description of *B. cucumis,* a bipolar arctic species, but it has not been demonstrated that the California specimens are really the same as those known from the Arctic. Color photograph in Wrobel and Mills 1998 and 2003, p. 68.

Beroe forskalii Milne Edwards, 1841. Probably an oceanic species, but seen fairly often along the California coast, where it reaches several cm in length. See Tamm and Tamm 1991, Biol. Bull. 181: 463–473 (how *Beroe* keeps its mouth shut using epithelial adhesion while swimming in a forward direction). Color photograph in Wrobel and Mills 1998 and 2003, p. 68.

Beroe gracilis Künne, 1939. A small species, often <1 cm long, described from near Helgoland in the northeast Atlantic, seen occasionally in swarms off central California and also collected individually in Friday Harbor. Probably more widely distributed, but not often specifically identified. Color photograph in Wrobel and Mills 1998 and 2003, p. 68.

Beroe mitrata (Moser, 1907). An infrequent visitor, to several centimeters cm in length. This highly mobile ctenophore (which can turn itself inside out), known also from Japan and the Mediterranean, may be an oceanic resident that sometimes is transported inshore. Color photograph in Wrobel and Mills 1998 and 2003, p. 69.

References

Bigelow, H. B. 1912. Reports on the scientific results of the expedition to the eastern tropical Pacific 1904–05. XXVI. The ctenophores. Bull. Mus. Comp. Zool. Harvard 54: 369–404.

Cairns, S. D., D. R. Calder, A. Brinckmann-Voss, C. B. Castro, D. G. Fautin, P. R. Pugh, C. E. Mills, W. C. Jaap, M. N. Arai, S. H. D. Haddock, and D. M. Opresko 2002. Common and scientific names of aquatic invertebrates from the United States and Canada: Cnidaria and Ctenophora, 2nd edition. American Fisheries Society Special Publication No. 28, Bethesda, MD, 115 pp.

Chun, C. 1880. Die Ctenophoren des Golfes von Neapel. Fauna und Flora des Golfes von Neapel 1: 313 pp.

Esterly, C. O. 1914. A study of the occurrence and manner of distribution of the Ctenophora of the San Diego region. Univ. Cal. Publ. Zool. 13: 21–38.

Haddock, S. H. D. and J. F. Case 1995. Not all ctenophores are bioluminescent: *Pleurobrachia*. Biol. Bull. 189: 356–362.

Harbison, G.R. 1985. On the classification and evolution of the Ctenophora. In The origins and relationships of lower invertebrates. S. Conway Morris, J. D. George, R. Gibson, H. M. Platt, eds. London: Systematics Assoc. Spec. Vol. 28, London, pp. 78–100.

Hyman, L. H. 1940. The Invertebrates: Protozoa through Ctenophora, Vol. I. McGraw-Hill (Ctenophora, pp. 662–696; also Vol. V, 1959, Retrospect, pp. 730–731).

Komai, T. 1918. On ctenophores of the neighborhood of Misaki. Annot. Zool. Jap. 9: 452–474.

Matsumoto, G. I. and G. R. Harbison 1993. In situ observations of foraging, feeding, and escape behavior in three orders of oceanic ctenophores; Lobata, Cestida and Beroida. Mar. Biol. 117: 279–287.

Mayer, A. G. 1911. Ctenophores of the Atlantic coast of North America. Carnegie Institution of Washington Publication 162, 58 pp.

Mills, C. E. 1987. Phylum Ctenophora. In Marine invertebrates of the Pacific Northwest, pp. 79–81. E. N. Kozloff, ed. Seattle: University of Washington Press.

Mills, C. E. 1996. Additions and corrections to the keys to Hydromedusae, Hydroid polyps, Siphonophora, Stauromedusan Scyphozoa, Actiniaria, and Ctenophora, pp. 487–491. In Marine invertebrates of the Pacific Northwest, with Revisions and Corrections. E. N. Kozloff, ed. Seattle: University of Washington Press.

Mills, C. E. Internet since 1998. Phylum Ctenophora: list of all valid species names. Electronic internet document available at http://faculty.washington.edu/cemills/Ctenolist.html. Published by the author, Web page established March 1998, frequently updated (see date at end of Web page).

Podar, M., S. H. D. Haddock, M. L. Sogin, and G. R. Harbison 2001. A molecular phylogenetic framework for the phylum Ctenophora using 18S rRNA genes. Molecular Phylogenetics and Evolution 21: 218–230.

Stanford, L. O. 1931. Studies on the ctenophores of Monterey Bay. Master's of arts thesis, Department of Biology, Stanford University, 67 pp.

Tamm, S. 1973. Mechanisms of ciliary co-ordination in ctenophores. J. Exp. Biol. 59: 231–245.

Tamm, S. L. and S. Tamm 1993. Diversity of macrociliary size, tooth patterns, and distribution in *Beroe* (Ctenophora). Zoomorphology 113: 79–89.

Torrey, H. B. 1904. The ctenophores of the San Diego Region. Univ. Cal. Publ. Zool. 2: 45–51.

Wrobel, D. and C. Mills 1998 and 2003. Pacific Coast Pelagic Invertebrates: a Guide to the Common Gelatinous Animals. Sea Challengers and the Monterey Bay Aquarium, Monterey, California, 108 pp.

Dicyemida (Rhombozoa)

(Plate 77)

DANNA JOY SHULMAN AND F. G. HOCHBERG

Dicyemids are tiny marine parasites that only occur in the kidneys of cephalopods, a distinctive habitat in which they flourish. In an individual adult host, thousands of dicyemids can be found. In host populations, prevalence of infection often approaches 100%. As new cephalopod species are described, they are usually found to harbor new dicyemid species.

The phylogenetic position of the Dicyemida is uncertain. Traditional classifications grouped them with the Orthonectida as Mesozoa, "middle animals" somewhere between truly multicellular Eumetazoa and the unicellular Protozoa. However, molecular phylogenies, as well as their life cycle within a molluscan host, place them close to the Platyhelminthes (flatworms).

On the Pacific coast of North America, 18 species of dicyemids are recognized, infecting eight host species. Although not strictly host specific, most dicyemid species are recovered from only one or two host species. When a species infects multiple hosts, the hosts are usually closely related. Some otherwise cryptic host species (e.g., *Octopus bimaculatus* and *O. bimaculoides*) are more readily identified by their dicyemid fauna than by their own morphology.

The anatomy of dicyemids has been well studied. Adults range in length from 0.1–9.0 mm, and they can be easily viewed through a light microscope. Dicyemids display eutely, a condition in which each adult individual of a given species has the same number of cells, making cell number a useful identifying character. Their structure is simple (plate 77): a single axial cell is surrounded by a jacket of ciliated cells. The anterior region of the organism is termed a "calotte" and attaches the dicyemid to folds on the surface of its host's renal appendages.

Dicyemids have two asexual adult forms (nematogens and rhombogens) that are anatomically similar, and referred to as the vermiform stages. Nematogens, which predominate in juvenile and immature hosts, produce vermiform larvae, which initially mature through direct development to form more nematogens. In this way, nematogens proliferate in young cephalopods, filling the kidneys.

As the infection ages, perhaps as the nematogens reach a certain density, vermiform larvae develop into rhombogens, rather than more nematogens. This sort of density-responsive reproductive cycle is reminiscent of the asexual reproduction of sporocysts or rediae in larval trematode infections of snails. As the snail tissue fills with these asexual stages, the larval parthenitae convert to production of cercariae, which exit the snail. As with the trematode asexual stages, a few nematogens can usually be found in older hosts. Their function may be to increase the population of the parasite to keep up with the growth of the host.

However, rhombogens are the predominant form in mature hosts. They contain sexual (hermaphroditic) infusorigens which produce infusoriform larvae. These larvae possess a very distinctive morphology. They typically have refringent bodies that resemble headlights and they swim about with long cilia on the posterior part of the body. It has long been assumed that this sexually produced infusoriform, which is released when the host eliminates urine from the kidneys, is both the dispersal and the infectious stage. The mechanism of infection, however, remains unknown, as are the effects, if any, of dicyemids on their hosts.

Some part of the dicyemid life cycle may be tied to temperate benthic environments, where they occur in greatest abundance. Although dicyemids have occasionally been found in the tropics, the infection rates are typically quite low, and many potential host species are not infected. Dicyemids have never been reported from truly oceanic cephalopods, which instead host a parasitic ciliate fauna.

Dicyemids are divided into two families, Conocyemidae and Dicyemidae. The four genera in Dicyemidae all have four propolar cells, and three of these four genera are found off the temperate West Coast.

Key to the Genera

— Four metapolar cells...........................*Dicyema*
— Five metapolar cells ....*Dicyemennea*
— Six metapolar cells......................*Dicyemodeca*

List of Species by Host

The range of dicyemids is host dependent (see the cephalopod chapter for host ranges).

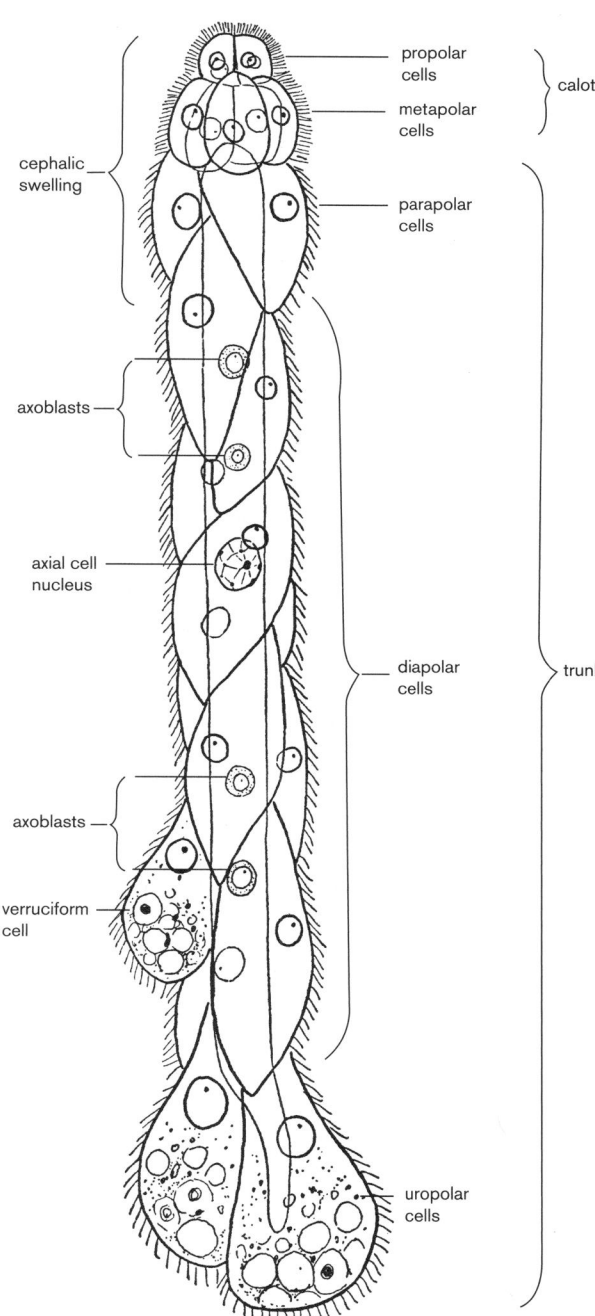

PLATE 77 Dicyemida. Generalized drawing of young nematogen stage with terms applied to various parts of the body (from McConnaughey 1951).

Labels on figure:
- propolar cells
- metapolar cells
- calotte
- cephalic swelling
- parapolar cells
- axoblasts
- axial cell nucleus
- diapolar cells
- trunk
- axoblasts
- verruciform cell
- uropolar cells

Dicyemodeca deca (McConnaughey, 1957) (=*Conocyema deca*). Vermiform stages: lengths <1,000 µm; with 23–24 cells; calotte flattened; verruciform cells absent. Infusoriform embryos: lengths to 35 µm; refringent bodies present.

Octopus bimaculatus

Dicyemennea abelis McConnaughey, 1949. Vermiform stages: lengths to 2,500 µm; with 24–27 cells; calotte rounded; verruciform cells present (diapolars + uropolars). Infusoriform embryos: lengths to 45 µm; refringent bodies present.

Dicyemennea californica McConnaughey, 1941 (=*D. granularis* McConnaughey, 1949). Vermiform stages: lengths to 4,000 µm; with 36 cells; calotte rounded; verruciform cells present (diapolars + uropolars). Infusoriform embryos: lengths to 45 µm; refringent bodies present.

Octopus bimaculoides

Dicyema sullivani McConnaughey, 1949. Vermiform stages: lengths to 1,500 µm; with 32 cells; calotte rounded; verruciform cells absent. Infusoriform embryos: lengths to 48 µm; refringent bodies present.

Dicyemennea californica. See *O. bimaculatus.*

Octopus bimaculatus or *O. bimaculoides*

Host identifications uncertain for these species.

Dicyema acheroni McConnaughey, 1949. Vermiform stages: lengths to 1,500 µm; with 23–28 cells; calotte pointed; verruciform cells absent. Infusoriform embryos: lengths unknown; refringent bodies present.

Dicyema acciaccatum McConnaughey, 1949. Vermiform stages: lengths <1,000 µm; with 22 cells; calotte rounded; verruciform cells absent. Infusoriform embryos: lengths to 37 µm; refringent bodies present.

Dicyemennea abasi McConnaughey, 1949. Vermiform stages: lengths <1,000 µm; with 22–26 cells; calotte rounded; verruciform cells absent. Infusoriform embryos; unknown.

Dicyemennea abbreviata McConnaughey, 1949. Vermiform stages: lengths >1,000 µm; with 25 cells; calotte flattened; verruciform cells present (uropolars). Infusoriform embryos: lengths to 41 µm; refringent bodies present.

"*Octopus*" *microprysus*

Dicyemennea sp. nov. A.
Dicyemennea sp. nov. B.

"*Octopus*" *rubescens* (=*O. apollyon*)

Dicyema apollyoni Nouvel, 1947 (=*D. balamuthi* McConnaughey, 1949). Vermiform stages: lengths less than 3,000 µm; with 22 cells; calotte rounded; verruciform cells absent. Infusoriform embryos: lengths to 35 µm; refringent bodies present.

Dicyemennea adminicula (McConnaughey, 1949) (=*Conocyema adminicula*). Vermiform stages: lengths <1,000 µm; with 17 cells; calotte flattened; verruciform cells present (uropolars). Infusoriform embryos: lengths to 32 µm; refringent bodies present.

Enteroctopus dofleini (=*Octopus dofleini*)

Dicyemennea abreida McConnaughey, 1957. Vermiform stages: lengths greater than 1,000 µm; with 28–30 cells; calotte rounded; verruciform cells absent. Infusoriform embryos: lengths to 38 µm; refringent bodies present.

Dicyemennea nouveli McConnaughey, 1959. Vermiform stages: lengths <500 µm; with 33–35 cells; calotte rounded; verruciform cells absent. Infusoriform embryos: lengths to 38 µm; refringent bodies present.

Dicyemennea adscita McConnaughey, 1949. Vermiform stages: lengths to 3,000 μm; with 23 cells; calotte rounded; verruciform cells present (diapolars + uropolars). Infusoriform embryos: lengths to 40 μm; refringent bodies present.

Dicyemennea brevicephalum McConnaughey, 1941. Vermiform stages: lengths to 2,000 μm; with 27 cells; calotte rounded; verruciform cells present (uropolars). Infusoriform embryos: lengths to 40 μm; refringent bodies present.

Rossia pacifica

Dicyemennea brevicephaloides Bogolepova-Dobrokhotova, 1962. Vermiform stages: lengths to 3,000 μm; with 23 cells; calotte flattened; verruciform cells absent. Infusoriform embryos: lengths to 31 μm; refringent bodies absent.

Dicyemennea filiformis Bogolepova-Dobrokhotova, 1965 (=*D. parva* Hoffman, 1965). Vermiform stages: lengths to 1,500 μm; with 25 cells; calotte pointed; verruciform cells absent. Infusoriform embryos: lengths to 32 μm; refringent bodies absent.

Doryteuthis opalescens (=*Loligo opalescens*)

Dicyemennea nouveli (not confirmed in this host). See *O. dofleini.*

References

Bogolepova-Dobrokhotova, I. I. 1962. Dicyemidae of the far-eastern seas. II. New species of the genus *Dicyemennea*. Zoologicheskii Zhurnal 41: 503–518. (In Russian.)

Furuya, H. 2002. Chapter 6. Phyla Dicyemida and Orthonectida. pp. 149–161. In Atlas of marine invertebrate larvae. C. M. Young, M. A. Sewell, and M. E. Rice, eds. San Diego: Academic Press.

Furuya, H. and K. Tsuneki 2003. Biology of Dicyemid Mesozoans. Zoological Science, 20: 519–532.

Furuya, H., F. G. Hochberg, and K. Tsuneki 2003. Reproductive traits in dicyemids. Marine Biology 143: 693–706.

Furuya, H., F. G. Hochberg, and K. Tsuneki 2003. Calotte morphology in the phylum Dicyemida: niche separation and *convergence*. Journal of Zoology 259: 361–373.

Furuya, H., F. G. Hochberg, and K. Tsuneki 2004. Cell number and cellular composition in infusoriform larvae of dicyemid mesozoans (Phylum Dicyemida). Zoological Science 21: 877–889.

Furuya, H., M. Ota, R. Kimura, and K. Tsuneki 2004. The renal organs of cephalopods: a habitat for dicyemids and chromidinids. Journal of Morphology 262: 629–643.

Hochberg, F. G. 1982. The "kidneys" of cephalopods: a unique habitat for parasites. Malacologia 23: 121–134.

Hochberg, F. G. 1990. Diseases of Cephalopoda: diseases caused by protistans and metazoans. pp. 47–227. In Diseases of marine animals. Vol. III. O. Kinne, ed. Hamburg, Germany: Biologische Anstalt Helgoland.

Hoffman, E. G. 1965. Mesozoa of the sepiolid, *Rossia pacifica* (Berry). Journal of Parasitology 51: 313–320.

McConnaughey, B. H. 1941. Two new Mesozoa from California, *Dicyemennea californica* and *Dicyemennea brevicephala* (Dicyemidae). Journal of Parasitology 27: 63–69.

McConnaughey, B. H. 1949a. Mesozoa of the Family Dicyemidae from California. University of California Publications in Zoology 55: 1–34.

McConnaughey, B. H. 1949b. *Dicyema sullivani*, a new mesozoan from Lower California. Journal of Parasitology 35: 122–124.

McConnaughey, B. H. 1951. The life cycle of the dicyemid Mesozoa. University of California Publications in Zoology 55: 295–336.

McConnaughey, B. H. 1957. Two new Mesozoa from the Pacific Northwest. Journal of Parasitology 43: 358–361.

McConnaughey, B. H. 1959. *Dicyema nouveli*, a new mesozoan from central California. Journal of Parasitology 45: 533–537.

McConnaughey, B. H. 1960. The rhombogen phase of *Dicyema sullivani* McConnaughey. University of California Publications in Zoology 46: 608–610.

Orthonectida

(Plate 78)

EUGENE N. KOZLOFF

Although there are fewer than 25 named species of orthonectids worldwide, these microscopic marine organisms parasitize a wide variety of hosts: turbellarians, nemerteans, polychaete annelids, gastropod and bivalve mollusks, brittle stars, and an ascidian. The species whose structure is at least reasonably well known fit into four genera: *Rhopalura, Intoshia, Ciliocincta,* and *Stoecharthrum.* Four species belonging to three of these genera are known to occur on the Pacific coast of North America.

In any marine province, including the Pacific coast, there are likely to be many species of orthonectids, but finding them may not be easy because the incidence of parasitism of a particular host species is often low. For instance, in the polychaete *Sabellaria cementarium,* collected by dredging at one locality in the San Juan Islands, fewer than 2% of more than 100 worms of one collection were parasitized. We would perhaps not even know that this species is host to an orthonectid if the first specimen examined had not been the lucky one. In the case of the ascidian *Ascidia callosa,* hundreds of which have been dissected for anatomical study, only a few specimens from one collection were parasitized, and this orthonectid has not been seen again, even in the same ascidian from the same locality.

What one sees in tissue of a parasitized host are sexual individuals in various stages of development. When released into seawater, those that are mature or close to mature swim in a nearly straight line, which gave rise to the name Orthonectida (straight-swimming). When adults escape normally from the host, however, they typically swim with a spiral motion.

The epidermis of a male or female orthonectid consists of a single layer of cells arranged in successive rings. The cells of some rings are ciliated, in whole or in part; those of other rings are not ciliated. In one ring, there are typically one or two genital pores, each surrounded by four or more small cells. Beneath the epidermis there are contractile cells and gametes or precursors of gametes, along with other cells, which may or may not be conspicuously specialized. The general appearance and structure of males and females are distinctly different. Species of *Stoecharthrum,* so far as known, are hermaphroditic.

The free-living stage is short-lived. During mating, males and females that have emerged from a parasitized host come together, with their genital pores touching. Sperm, with short tails, enter the inner mass of the female and fertilize the eggs. The males die soon after mating, but females live long enough to incubate their eggs until they have developed into swimming larvae. These consist of just a few cells and represent the infective stage.

The penetration of a susceptible host by larvae has been seen in only one species, *Rhopalura ophiocomae* (plate 78A), which parasitizes the brittle star *Amphipholis squamata.* About a hundred years ago, French zoologists extensively studied this orthonectid. They observed larvae entering the bursal pockets of the host and concluded that one or more cells of a larva multiply within the host as germinal cells. The rather homogeneous mass (plasmodium) within which the germinal cells are scattered was also thought to belong to the parasite. It is now known that the multiplication and differentiation of germinal cells takes place in muscle cells located beneath the epithelium lining the bursal pockets and gut of *Amphipholis.* The plasmodium is in fact the greatly enlarged mass of cytoplasm of an affected muscle cell. After the males or females that have developed in a hypertrophied muscle cell are ready to mate, they somehow escape from the mass and leave the host by way of the bursal pockets, and it seems likely that they also escape through the mouth.

In the case of *Ciliocincta sabellariae* (plate 78B), found in the polychaete *Sabellaria cementarium,* the hypertrophied mass within which germinal cells multiply and develop into adults is part of the epidermis. Presumably the still-functional layer of epidermis at the surface ruptures to allow males and females to escape. In the case of a species of *Intoshia* parasitizing a turbellarian in northern Russia, it was claimed that the mass containing germinal cells and developing males and females had nuclei of its own, suggesting that it may be a plasmodium of the sort claimed by early students of these parasites. The matter needs further study.

List of Species

There are four species of orthonectids known from the Pacific coast.

Rhopalura ophiocomae Giard, 1877. In the brittle star *Amphipholis squamata*; San Juan Island, Washington, and Point Pinos (Pacific Grove) California. Originally described from and widely distributed in Europe. Abbott (1987, p. 108) provides a sketch and notes on specimens from Point Pinos.

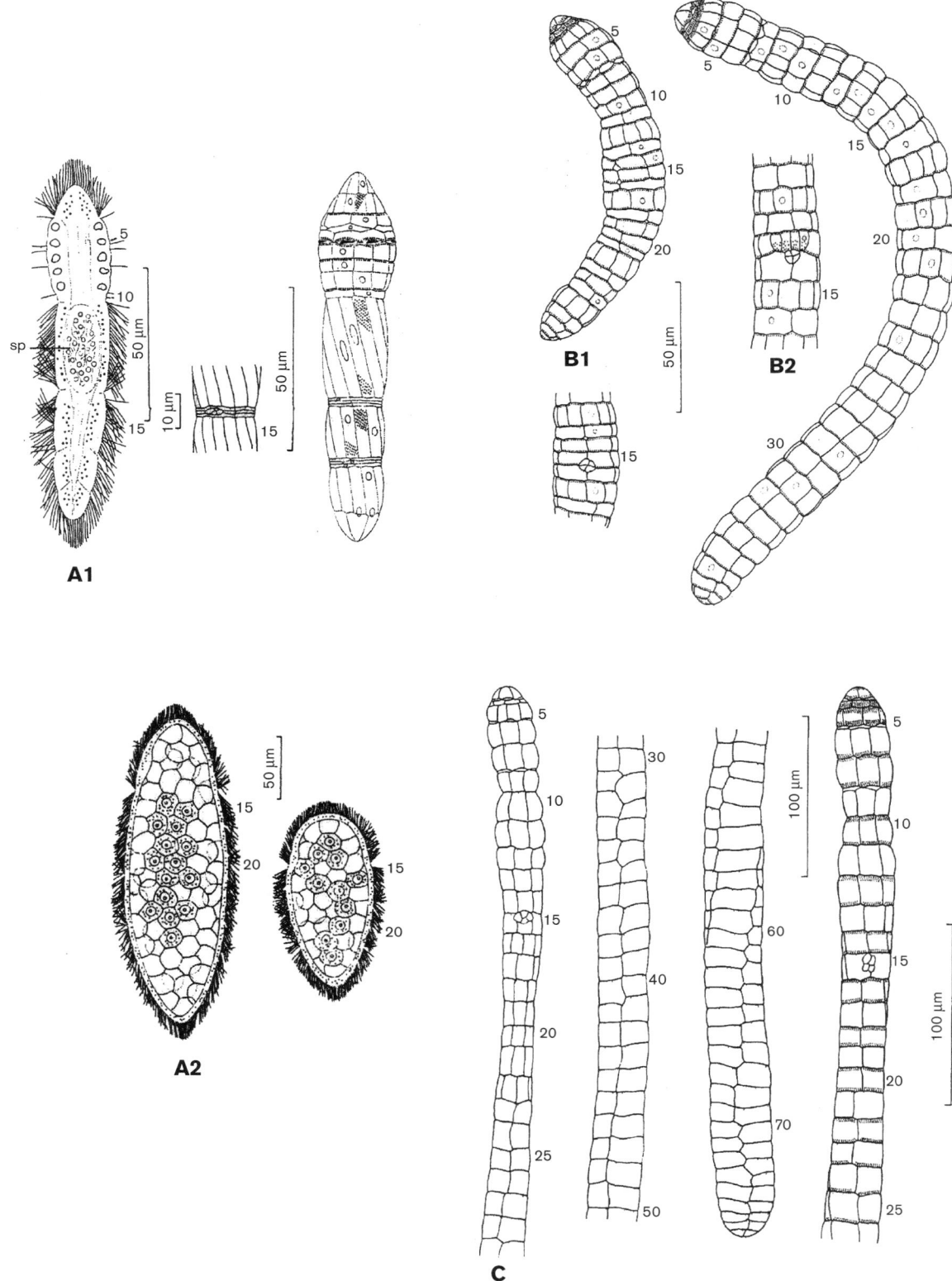

PLATE 78 A, *Rhopalura ophiocomae*, the number refers to the rings of epidermal cells. A1, male living specimen in optical section showing crystalline inclusions in five rings of epidermal cells between three and nine, specimen impregnated with silver to show cell boundaries of the epidermis and the genital pore, and a composite drawing showing arrangement of kinetosomes of some of the cilia; sp, sperm; A2, females of two types, these, so far as known, always in separate hypertrophied masses of affected muscle cells; B, *Ciliocincta sabellariae*, B1, male, composite drawings based on silver impregnations, showing cell boundaries, distribution of kinetosomes of cilia, and genital pore; B2, female; C, *Stoecharthrum fosterae*, a complete specimen showing boundaries of epidermal cells and the anterior portion of one showing arrangement of kinetosomes of cilia—this and other known species of *Stoecharthrum* are hermaphroditic, sperm production being limited to one or a few relatively small areas within the body (A, from Kozloff, 1969; B, from Kozloff, 1992; C, from Kozloff, 1993).

Ciliocincta sabellariae Kozloff, 1965. In the polychaete *Sabellaria cementarium*; San Juan Islands region.

Stoecharthrum fosterae Kozloff, 1993 (plate 78C). In the mussel *Mytilus trossulus*; Totten Inlet, southern Puget Sound.

Stoecharthrum burresoni Kozloff, 1993. In the seasquirt *Ascidia callosa*; San Juan Island.

References

Abbott, D. P. 1987. Observing marine invertebrates. Drawings from the laboratory. G. H. Hilgard, ed. Stanford, CA: Stanford University Press, 380 pp.

Kozloff, E. N. 1969. Morphology of the orthonectid *Rhopalura ophiocomae*. Journal of Parasitology 55: 171–195.

Kozloff, E. N. 1971. Morphology of the orthonectid *Ciliocincta sabellariae*. Journal of Parasitology 57: 585–597.

Kozloff, E. N. 1992. The genera of the phylum Orthonectida. Cahiers de Biologie Marine 33: 377–406.

Kozloff, E. N. 1993. Three new species of *Stoecharthrum* (phylum Orthonectida). Cahiers de Biologie Marine 34: 523–534.

Kozloff, E. N. 1994. The structure and origin of the plasmodium of *Rhopalura ophiocomae* (phylum Orthonectida). Acta Zoologica 75: 191–199.

Kozloff, E. N. 1997. Studies on the so-called plasmodium of *Ciliocincta sabellariae* (phylum Orthonectida), with notes on an associated microsporan parasite. Cahiers de Biologie Marine 38: 151–159.

Platyhelminthes

(Plates 79–85)

The Platyhelminthes is the lowest of the truly bilateral phyla and include a wide variety of both parasitic and free-living species. The parasitic groups include the Trematoda (flukes), the Cestoda (tapeworms), the Fecampiida, and the Monogenea; these are treated in this Manual in brief separate chapters that follow. The largely free-living "Turbellaria" encompass many thousands of species (a vast number of which remain undescribed, even in the world's shallowest waters). The "Turbellaria" are now recognized as a polyphyletic group; the tiny acoel flatworms (the Acoelomorpha), for example, are not closely related to other "turbellarians" (Hooge and Tyler 2006.)

"Turbellaria"

JOHN J. HOLLEMAN

(Plates 79–81)

The turbellarians are composed of herbivorous or predaceous free-living forms with a few commensal or parasitic members. Historically these have included the following groups that occur on the California and Oregon coasts. These are treated here together for convenience of discussion and identification, rather than as an indication of their phylogenetic relationships:

ACOELA are small marine forms without an intestine; these are treated separately by Matthew Hooge at the end of this section.

RHABDOCOELA are small, typically narrow forms in which the intestine is a single, straight tube; common in fresh water and represented in the local intertidal by what are likely a large number of free-living species. There are at least three symbiotic species: *Syndisyrinx franciscanus* in the intestine of sea urchins (plate 81A), *Syndesmis dendrastrorum* in sand dollars, and *Collastoma pacifica* in the sipunculan *Themiste pyroides*.

PROSERIATA are mostly marine forms, differing from rhabdocoels in having a more complex intestine and other features; they are small and commonly overlooked. Proseriata are common interstitially in mud, sand, and gravel, but they can also be found on seaweed and in tide pools.

TRICLADIDA, which include the familiar freshwater and terrestrial planarians, have a three-branched gut and a protrusible pharynx. Although common in fresh water and certain

damp situations on land, triclads are uncommon in the sea. One local intertidal triclad, *Nexilis epichitonius* (plate 81C), is commensal in the mantle cavity of the chiton *Mopalia hindsii*, and Holmquist and Karling (1972) have described two interstitial species, *Pacificides psammophilus* (plate 81D) and *Oregoniplana opisthopora*.

PROLECITHOPHORA are small turbellarians with a saclike gut and pharynx bulbous or with an annular fold. Commonly found in tide pools with seaweed and interstitially in sand and gravel.

POLYCLADIDA are a large marine group, characterized by an intestine with many branches (plates 79, 80). Abundant locally, some polyclads attain large size, and many are active and strikingly colored. There are many species of polyclads on the Pacific coast, often abundant under boulders and on pilings. Polyclads are difficult to identify except for highly colored or distinctively marked species. Exact determination often requires studying serial sagittal sections of the copulatory apparatus. Eye arrangement is sometimes clear in living worms, but in darkly pigmented species, eyes cannot be seen except in fixed, dehydrated, and cleared specimens. Characteristic and often diagnostic features, such as the nature of the penis stylet, Lang's vesicle, and the spermiducal bulbs, may sometimes be seen by carefully compressing living animals between slides and examining by transmitted light. Color and color patterns are often distinctive, but are often variable and may depend upon ingested food. The sizes given in the key are averages; many specimens collected will be larger or smaller.

For more details of Pacific coast polyclads and for species not treated in the following key, students should consult the valuable papers of Hyman (1953, 1955, 1959). Pictures of Pacific coast polyclads can be viewed at http://www.rzuser. uni-heidelberg.de/~bu6/.

Glossary

CEREBRAL EYES eyespots, usually of dark color and paired clusters over the brain area.

COMMON GENITAL PORE single pore to outside from both male and female reproductive systems.

FRONTAL EYES eyes spread over the region between the brain and the anterior margin

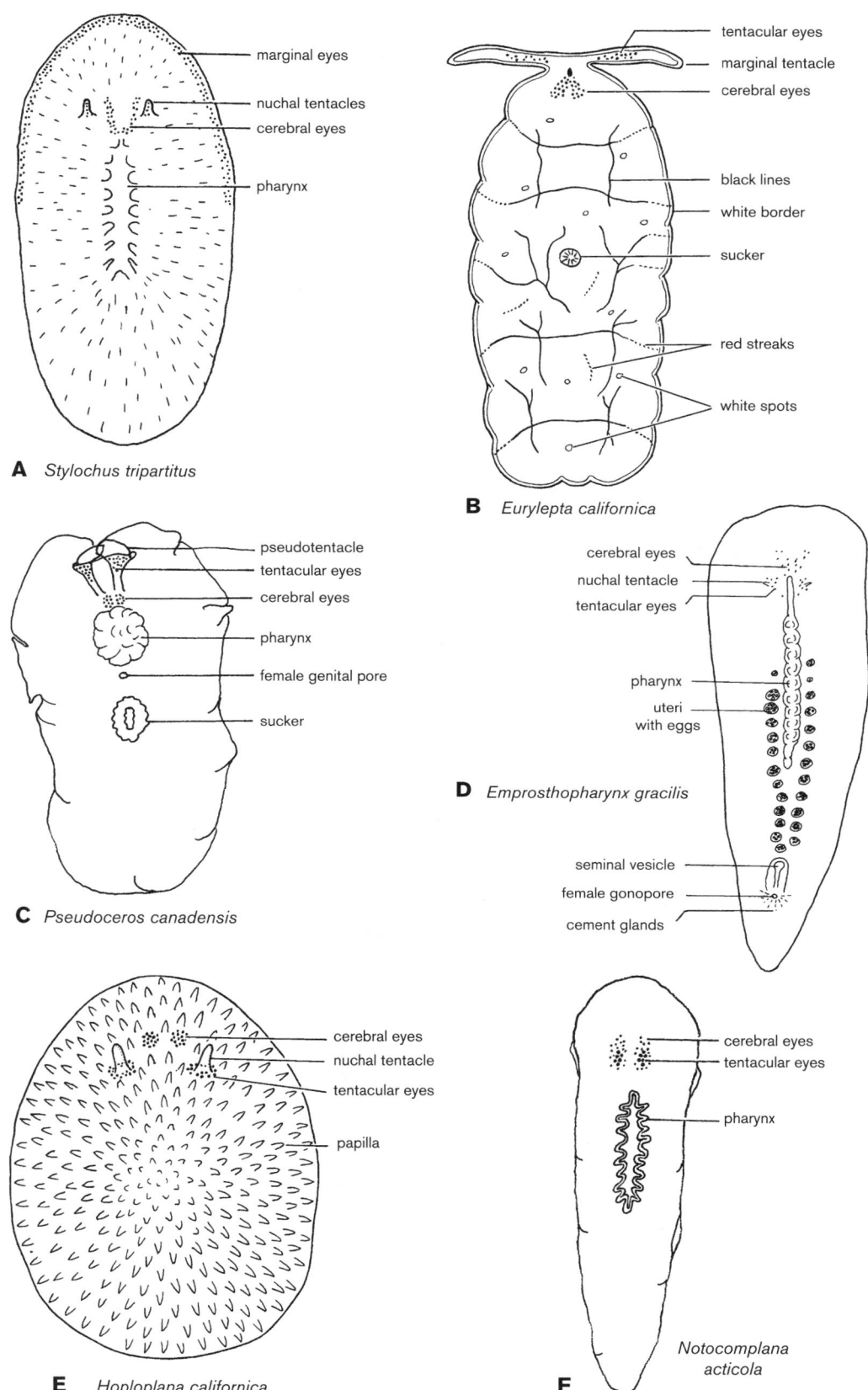

A *Stylochus tripartitus*

marginal eyes
nuchal tentacles
cerebral eyes
pharynx

B *Eurylepta californica*

tentacular eyes
marginal tentacle
cerebral eyes
black lines
white border
sucker
red streaks
white spots

C *Pseudoceros canadensis*

pseudotentacle
tentacular eyes
cerebral eyes
pharynx
female genital pore
sucker

D *Emprosthopharynx gracilis*

cerebral eyes
nuchal tentacle
tentacular eyes
pharynx
uteri with eggs
seminal vesicle
female gonopore
cement glands

E *Hoploplana californica*

cerebral eyes
nuchal tentacle
tentacular eyes
papilla

F *Notocomplana acticola*

cerebral eyes
tentacular eyes
pharynx

PLATE 79 Flatworms, polyclads, scales various (A, C, D, F, redrawn from Hyman; B, redrawn from a sketch by David H. Montgomery; E, Eugene Haderlie, from life).

anterior

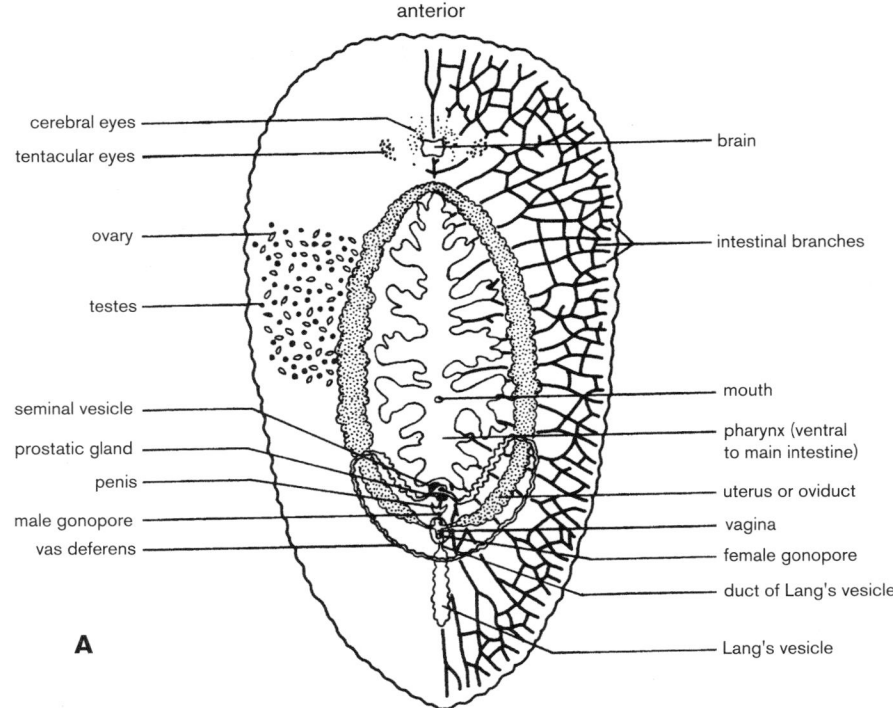

cerebral eyes

tentacular eyes

ovary

testes

seminal vesicle

prostatic gland

penis

male gonopore

vas deferens

brain

intestinal branches

mouth

pharynx (ventral
to main intestine)

uterus or oviduct

vagina

female gonopore

duct of Lang's vesicle

Lang's vesicle

A

anterior

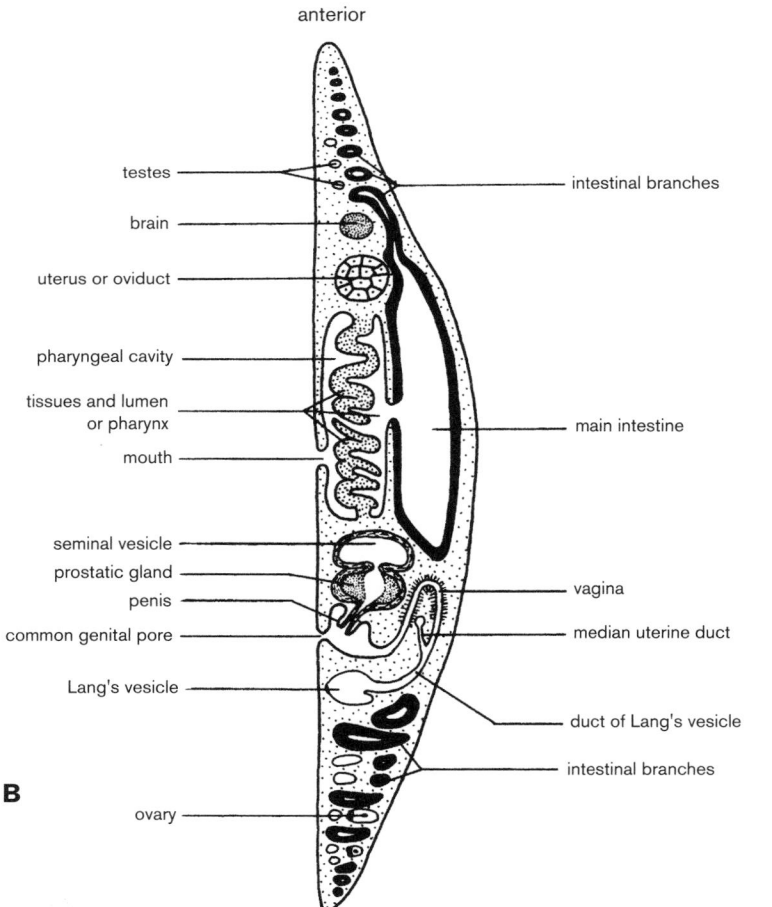

testes

brain

uterus or oviduct

pharyngeal cavity

tissues and lumen
or pharynx

mouth

seminal vesicle

prostatic gland

penis

common genital pore

Lang's vesicle

ovary

intestinal branches

main intestine

vagina

median uterine duct

duct of Lang's vesicle

intestinal branches

B

PLATE 80 Flatworms, polyclad anatomy; A, diagrammatic ventral view of *Notocomplana timida;* B, diagrammatic median longitudinal section of *Emprosthopharynx gracilis* (A, redrawn from Bresslau, after
Heath and McGregor; B, redrawn from Bresslau, after Lang).

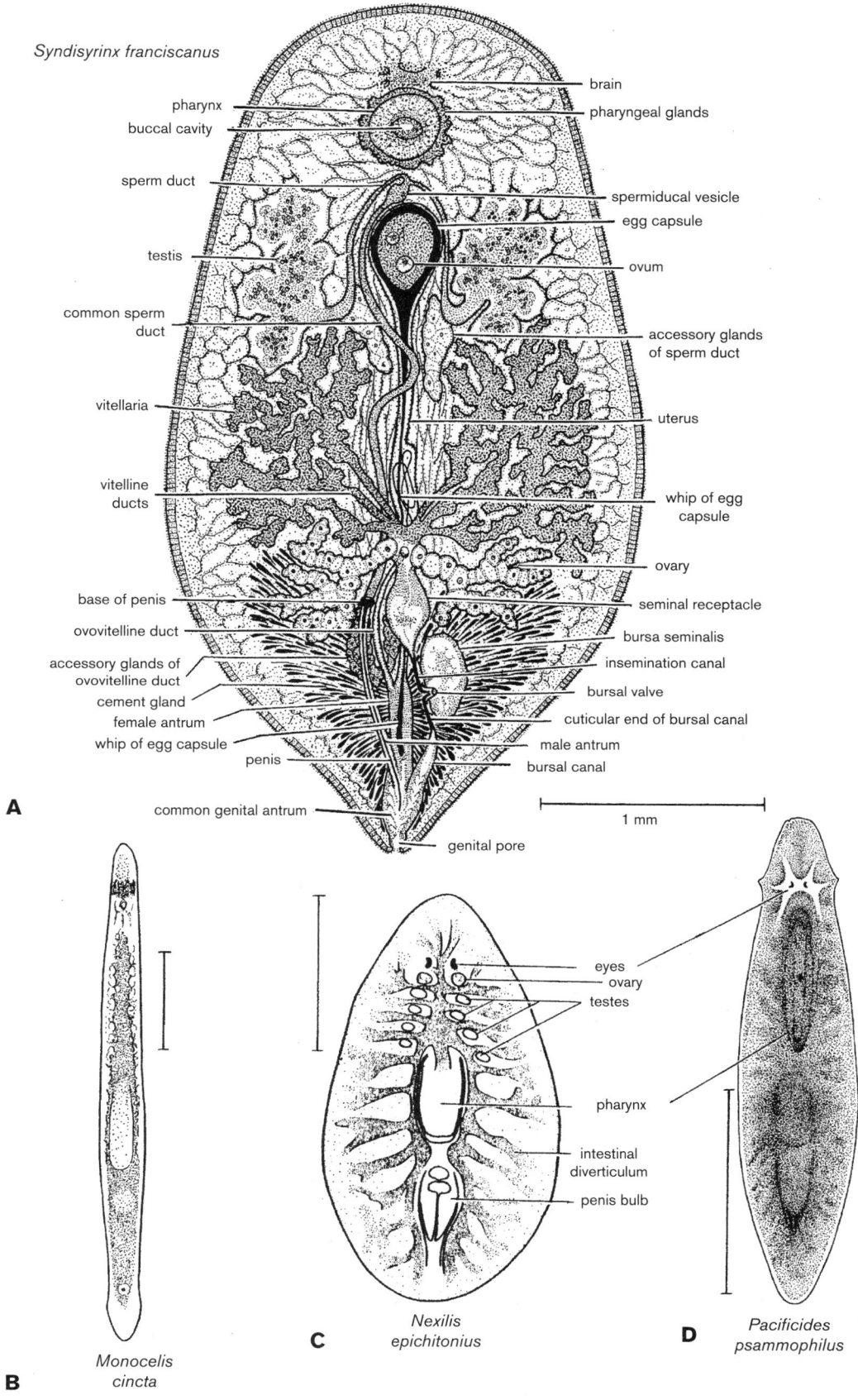

Syndisyrinx franciscanus

pharynx
buccal cavity
sperm duct
testis
common sperm duct
vitellaria
vitelline ducts
base of penis
ovovitelline duct
accessory glands of ovovitelline duct
cement gland
female antrum
whip of egg capsule
penis
common genital antrum
genital pore

brain
pharyngeal glands
spermiducal vesicle
egg capsule
ovum
accessory glands of sperm duct
uterus
whip of egg capsule
ovary
seminal receptacle
bursa seminalis
insemination canal
bursal valve
cuticular end of bursal canal
male antrum
bursal canal

1 mm

A

B

Monocelis cincta

C

Nexilis epichitonius

eyes
ovary
testes
pharynx
intestinal diverticulum
penis bulb

D

Pacificides psammophilus

PLATE 81 Flatworms, smaller Turbellaria, scale marks are 1 mm (A, after Lehman 1946; B, Karling 1966; C, Holleman and Hand 1962; D, Holmquist and Karling 1972).

LANG'S VESICLE a receptacle for sperm storage at upper end of the vagina.

MARGINAL EYES eyespots in a band along the whole or anterior part of body margin.

MARGINAL TENTACLES tentacles on anterior margin.

NUCHAL TENTACLES tentacles located over the lateral brain area, well back from anterior margin.

PENIS muscular projection that terminates the male copulatory system and is employed by simple protrusion to the outside.

PENIS STYLET a hard hollow tube—short, straight, curved or coiled—that is attached to the penis.

PSEUDOTENTACLES uplifted folds of the anterior margin which may be simple folds or earlike and pointed or square in the Cotylea.

SPERMIDUCAL BULBS heavily muscularized terminations of sperm ducts.

SUCKER a glandulo-muscular adhesive organ forming a noticeable elevation or depression. Its presence or absence determines the primary division of the polyclads into the Cotylea and Acotylea.

TENTACULAR EYES eyespots located in or around the nuchal tentacles or at sites where nuchal tentacles would be located if present.

Key to Polycladida

1. Ventral surface without a sucker; tentacles when present of nuchal type 2
— With sucker on ventral surface; tentacles, when present, situated at anterior margin 25
2. Dorsal surface covered with pointed papillae; pair of conspicuous, pointed, conical nuchal tentacles; tentacular eyes in rings at base of tentacles, cerebral eyes in two clusters medial and anterior to the tentacles; body shape oval to round, fairly thick; color orange-red to tan with dark spots on dorsum; living on encrusting ectoprocts; to 10 mm long (plate 79E) . . . . *Hoploplana californica*
— Dorsal surface not papillate 3
3. Eyes in tentacular, cerebral and frontal clusters and in submarginal and marginal areas of the body; eyes rarely absent. 4
— Without marginal eyes; eyes limited to clusters over the brain region or in and around tentacles 11
4. Marginal eyes present 5
— Eyes few or absent. 6
5. Marginal band of eyes limited to anterior one-quarter to one half of body margin; with nuchal tentacles 7
— Marginal band of eyes completely encircling body; with or without nuchal tentacles 8
6. Lacking eyes; small elliptical or fusiform body, bluntly pointed at both ends, thick and opaque; color grayish brown to brown; 10–14 m long. . . . . *Plehnia caeca*
— Eyes few in number in two loose cerebral groups that are composed of very small eyes; shape and size of the body and color same as *Plehnia caeca*. . . . . *Plehnia caeca* var. *oculifera*
7. Small, oval, to 10 mm long; pair of small inconspicuous nuchal tentacles; marginal eyes limited to anterior one-quarter of body margin; tentacular eyes inside tentacles; few cerebral eyes; color creamy with minute tan speck . . . .
. . . . *Stylochus franciscanus*
— Large, oval to 50 mm long, 25 mm wide; conspicuous, conical nuchal tentacles well back from anterior margin; marginal eyes extending back from anterior end one-quarter to one half

of body length; tentacles with eyes; cerebral eyes numerous in two elongated clusters between tentacles; color buff, with short brown streaks (plate 79A) . . . . *Stylochus tripartitus*
8. With prominent nuchal tentacles 9
— Nuchal tentacles wanting in adult, present in young; adult very large, to 60–100 mm long, 25 mm wide; thick; elongate oval in shape with somewhat pointed anterior end; marginal eyes numerous forward, thinning posteriorly; color dark brown with closely set dark spots . . . .
. . . . *Stylochus atentaculatus*
9. Very large, 50–150 mm long; oval; no posterior notch . . . . .
. . . . 10
— Small, 7 mm long; oval; distinct posterior median notch; cerebral eyes few, in a line or either side of midline forward of tentacles; color buff to brown . . . . *Stylochus exiguus*
10. Elongate oval in shape, narrowing at ends; large, commonly 50–100 mm or more, relatively thin; color gray with inconspicuous darker spots . . . . *Stylochus californicus*
— Broadly oval in shape with conspicuous contractile nuchal tentacles; thick and tough; very large, to 100 mm long and 70 mm wide; color tan or brown, heavily marked with dark brown dashes or spots . . . . *Kaburakia excelsa*
11. With nuchal tentacles. 12
— Without nuchal tentacles; small bosses or small conical elevations may be present 14
12. Small, up to 12 mm long; cuneate (wedge-shaped); expanded anteriorly, tapering to bluntly pointed posterior; pointed nuchal tentacles; four to six eyes in tentacular clusters in and around tentacles; 10–15 cerebral eyes scattered in two longitudinal tracts; color variable, pale brownish yellow or buff, translucent yellow or golden tan; ventral surface white (plate 79D) . . . . . . *Emprosthopharynx gracilis*
— Larger, oval in shape. 13
13. To 40 mm long, thick and firm; nuchal tentacles nipplelike and contractile; tentacular eyes fill the tentacles and occur diagonally before and behind tentacle bases; cerebral eyes in elongated areas broadening anteriorly; color transparent bluish green or light olive with branches of digestive tract showing as zigzag chocolate lines radiating from the from the central main intestine to the periphery. . . . .
. . . . *Pseudoalloioplana californica*
— Similar to *P. californica* in size and shape, but color light brown, speckled, or reticulated, and digestive branches not obvious; prominent nuchal tentacles; 40–90 tentacular eyes around the tentacle bases; cerebral eyes number 60–75 in two groups extending forward between and in front of the tentacles; body firm texture, oval shape to 40 mm or more in length, 25 mm wide . . . . . . *Discosolenia burchami*
14. With a penis stylet 15
— Without a penis stylet. 18
15. With a vaginal duct that exists via an external pore posterior to the female gonopore; shape elongate with rounded anterior and posterior ends; 26 mm long, 7 mm wide; tentacular eyes in compact groups numbering 50–60 each; cerebral eyes in groups of 90–100 loose elongate groups that spread anteriorly . . . . *Copidoplana tripyla*
— Without a vaginal duct. 16
16. With short conical or cylindrical nuchal tentacles; elongate shape, rounded anteriorly then tapering to a blunt posterior end; 25 mm long, 8 mm wide; color light green; tentacular eyes number nine to 13 and are located on the tentacles; cerebral eyes in two clusters of 12–20 eyes . . . . .
. . . . *Triplana viridis*
— Without nuchal tentacles 17

17. Penis armed with straight, relatively short stylet; body elongate with expanded anterior end tapering to bluntly pointed posterior end; to 30 mm long; tentacular eye clusters of up to 13 large eyes, cerebral clusters elongate, each with about 50 small eyes; color light brown or pale tan, dotted with reddish brown spots . . . .
. . . . *Pleioplana inquieta*

— Penis armed with a long, sharp stylet; broad oval shape; 12 mm long, 9 mm wide; tentacular eye clusters elliptically shaped, cerebral eye clusters long and narrow converging anteriorly with a large eye in front of each cluster . . . . . .
. . . . *Pleioplana californica*

18. With highly developed spermiducal bulbs 19
— Without spermiducal bulbs 20

19. With large common genital pore and large, conical unarmed penis; body shape varies from elongate oval to wedge-shaped; to 40 mm long; color highly variable, may be uniform reddish brown with darker streaks outlining pharynx and copulatory apparatus, or mottled dark brown with darker stripe mid-dorsally, or with dark brown to nearly black blotches on nearly white background . . . . . .
. . . . *Notocomplana litoricola*

— With separate, inconspicuous genital pores; penis small; body elongate oval, rounded anteriorly, tapering posteriorly, thick and firm; to 40 mm long, 6 mm wide; eyes in distinct cerebral and tentacular groups of up to 50 eyes each; color speckled gray or light tan mottled with green and brown . . . . *Stylochoplana chloranota*

20. Lang's vesicle short or reduced 21
— Lang's vesicle moderately long to long 24

21. Duct of Lang's vesicle short; small oval Lang's vesicle; shape cuneate; 5 mm long; lacking nuchal tentacles; tentacular eyes four to five in each cluster, four to five cerebral eyes in elongate clusters . . . . . *Parviplana californica*
— Duct of Lang's vesicle moderate to long 22

22. Lang's vesicle short and plump; body widest anteriorly, tapering to pointed tail; to 60 mm long when extended and crawling; tan or pale gray shading to patchy brown mid-dorsally (plate 79F) . . . . *Notocomplana acticola*
— Lang's vesicle small or much reduced 23

23. Small conical penis papilla; elongate shape, rounded anteriorly and bluntly pointed posterior; 18 mm long; lacking nuchal tentacles; color medium tan; 15–25 tentacular eyes in a tight cluster, cerebral eyes in two linear groups . . . . .
. . . . *Notocomplana sciophila*

— Long slender penis papilla; penis exceptionally long and slender; body elongate, not expanded anteriorly, sides nearly parallel along anterior one-third, then tapering gradually to blunt posterior end; to 14 mm long, 4 mm wide; yellowish gray in color . . . . . *Notocomplana saxicola*

24. Lang's vesicle long; broad penis papilla; body broadly oval with a blunt anterior and a slightly rounding posterior; 23 mm long, 12 mm wide; lacking nuchal tentacles; dorsal color a clear translucent white dotted with dark red spots, which are grouped in the central region to form a saddle-shaped blotch; 12 large tentacular eyes in each cluster, 40 eyes in each cerebral eye cluster . . . . *Notocomplana timida*

— Lang's vesicle long and moderately narrow; conical penis papilla; body expanded anteriorly, tapering posteriorly; to 35 mm long, 15 mm wide; up to 30 eyes in each tentacular and cerebral cluster; color pale with pink or red tinge, brown along mid-dorsal region, color faint at margins . . . .
. . . . *Notocomplana rupicola*

25. Upturned folds of anterior margin form pseudotentacles; pharynx ruffled 26
— Marginal tentacles not obvious folds; pharynx tubular . . . . .
. . . . 29

26. Dorsal surface smooth 27
— Dorsal surface with papillae or tubercles; body broad oval; 12 mm long, 8 mm wide; the sucker is slightly posterior to the midpoint; dorsal surface is white or gray with small pinkish or dark red pigmented spots; marginal eyes in two indefinite clusters between the pseudotentacles; two cerebral clusters of 15 eyes each . . . . *Acanthozoon lepidum*

27. Color basically black and white, sometimes also some red 28
— Color not black and white but pale tan or brown with darker brown flecks; oval or rounded, of firm consistency; to 28 mm long; sucker central in position, very large, oval, and much folded (plate 79C) . . . . *Pseudoceros canadensis*

28. Elongate oval with rounded ends and ruffled margin; to 50 mm long; thin, delicate; pronounced mid-dorsal ridge extending most of body length; pseudotentacles broad, flaplike; cerebral eyes in single cluster like an inverted V; eyes on pseudotentacles, sucker well anterior to middle; white with mid-dorsal black stripe along center of ridge, stripe forks anteriorly between pseudotentacles with branches continuing to anterior margin . . . .
. . . . *Pseudoceros luteus*

— Oval, with gracefully ruffled margin; 40–90 mm long; coloration striking, basically white with darker markings, black stripe with tinges of red on mid-dorsal ridge from just behind tentacles to posterior one-third of body, similar stripe encircles entire margin of animal except anterior edge between the pseudotentacles—this marginal band is just inward of edge, leaving outer margin white; elongate spots scattered over entire dorsal surface, mostly black but with some small wine spots and a few indistinct white spots; pseudotentacles with dark spot between them . . . .
. . . . *Pseudoceros montereyensis*

29. Marginal tentacles small or not apparent 30
— Marginal tentacles large 31

30. Mouth and male gonopore open in common to the exterior; body elliptical, to 9 mm long; about 80 small eyes at the base of each tentacle; ground color of dorsum orange with darker shade along mid-dorsal line from eyes to posterior end of midgut—laterally, this color is lighter and, near the margin, alternates with bright yellow in raylike expansions; minute white specks scattered over dorsum
. . . . *Stylostomum lentum*

— Mouth and male gonopore open separately to the exterior; elliptical shape; 8 mm long, 6 mm wide; sucker posterior to middle of the body; 50 large cerebral eyes in irregular clusters . . . . *Acerotisa langi*

31. Tentacles with eyes 32
— Tentacles without eyes, cerebral eyes in two distinct groups; shape a broad oval; 18 mm long, 13 mm wide . . . .
. . . . *Anciliplana graffi*

32. Dorsal surface grayish or greenish white with thin black or red lines 33
— Dorsal colored otherwise 34

33. Dorsal surface grayish white with a pure white border, white mid-dorsal ridge, crisscrossed with black lines usually terminating in red tips; red tips extend into the white border; a few short, scattered, disconnected red streaks; a dark mark at base of the tentacles, distally red, proximally black, and in between is a small black dash; to 20 mm long (plate 79B) . . . . *Eurylepta californica*

— Dorsal surface greenish white with five well-defined red transverse lines and two irregular red longitudinal stripes; body broadly oval, 31 mm long by 20 mm wide; marginal tentacles large and fleshy; small eyes scattered over the tentacles and on the margin between the tentacles; cerebral eyes larger and in two clusters . . . . *Euryleptodes cavicola*

34. Orange-red, yellowish pink, or salmon with minute white specks and pink streak along mid-dorsal line; another color variation is yellowish white with very yellow margin, orange anterior end and numerous white stripes running longitudinally; body oval, thick; sucker small, rounded, somewhat posterior to middle; to 30 mm long . . . . *Eurylepta aurantiaca*

— Striking coloration of alternating black and white longitudinal stripes of varying width with mid-dorsal orange stripe and marginal orange band; black stripes often show orange spots, white stripes sometimes tinged with red; ventral surface white with orange margin; tentacles black and flaring; to 40 mm long . . . . *Prostheceraeus bellostriatus*

List of Species

Only Polycladida are keyed out.

RHABDOCOELA

Alcha evelinae Marcus, 1949. Seaweed and gravel in tide pools at Pacific Grove, including in the "Great Tide Pool" (Karling and Schockaert 1977). Described from Brazil.

Collastoma pacifica Kozloff, 1953. In gut of sipunculan *Themiste pyroides*; see Kozloff 1953, J. Parasitol. 39: 336–340.

Duplacrorhynchus major Schockaert and Karling, 1970. Sandy mudflat under enteromorphoid *Ulva*, Newport (Yaquina Bay), Oregon.

Duplacrorhynchus minor Schockaert and Karling, 1970. Sandy mudflat under enteromorphoid *Ulva*, Tomales Bay and Elkhorn Slough.

Gyratrix hermaphroditus Ehrenberg, 1831. Tide pools, sand, pebbles, algae at Pacific Grove and Elkhorn Slough, and in sand at Yaquina Head, Oregon (Karling and Schockaert 1977). Said to be cosmopolitan.

Gyratrix proaviformis Karling and Schockaert, 1977. Sand with *Phyllospadix*, Boiler Bay, south of Lincoln City, Oregon.

Itaipusa bispina Karling, 1980. Yaquina Head, Oregon, and Pacific Grove, intertidal stones, sand, and the eelgrass *Zostera*.

Itaipusa curvicirra Karling, 1980. General habitat and locations as *I. bispina*.

Neoutelga inermis Karling, 1980. Monterey Marina in fine sand and mud, 1–4 m.

Paraustrorhynchus pacificus Karling and Schockaert, 1977. In the "Great Tide Pool" at Point Pinos, Pacific Grove, among seaweed and pebbles, and from 26 m in fine sand with mud off Fishermen's Wharf, also in Monterey Bay.

Promesostoma dipterostylum Karling, 1967. Washed out from seaweed and stones, from the "Great Tide Pool," Pacific Grove.

Rhinolasius dillonicus Karling, 1980. Intertidal sand with enteromorphoid *Ulva*, Nick's Cove, Tomales Bay. This species was named for the former Pacific Marine Station at Dillon Beach.

Scanorhynchus forcipatus Karling, 1955. Fine sand and mud, Monterey Bay, 2–10 m; also in the North Atlantic. Karling and Schockaert (1977) noted that the material from Monterey Bay may not be identical to the European *S. forcipatus*.

Syndesmis dendrastrorum Stunkard and Corliss, 1951. In intestine of sand dollar *Dendraster*.

Syndisyrinx franciscanus Lehman, 1946 (=*Syndesmis franciscanus*) (plate 81A). Common in intestine of sea urchin *Strongylocentrotus*; good class material (Lehman 1946). See Shinn 1981 Hydrobiologia 84: 155–162 (diet: ingests host intestinal tissue and symbiotic ciliates); Shinn 1983 Ophelia 22: 57–79 (life history).

Utelga montereyensis Karling, 1980. Monterey Marina, in fouling on pilings.

Yaquinaia microrhynchus Schockaert and Karling, 1970. Low-water superficial layer of sandy mudflat under enteromorphoid *Ulva*, Newport (Yaquina Bay) and Boiler Bay in coarse intertidal sand, Oregon.

PROSERIATA

Small, long, narrow, 1–12 mm long and 0.2–0.5 mm wide, worms that are interstitial in sand, mud, shell gravel, and in tide pools. The worms are colorless, translucent or with a brownish dorsal parenchyma.

Archimonocelis coronata Karling, 1966. Tomales Bay (Second Sled Road), tide pools.

Archimonocelis semicircularis Karling, 1966. Pacific Grove (Hopkins Marine Station, tide pools in gravel, stones, seaweed) and Tomales Bay (Second Sled Road).

Archotoplana dillonbeachensis Karling, 1964. Abundant in coarse sand at Dillon Beach, Tomales Bay.

Asilomaria ampullata Karling, 1966. Tomales Bay (Second Sled Road), shell gravel.

Coelogynopora brachystyla Karling, 1966. Point Pinos (Pacific Grove), sandy bottom, 0.5 m.

Coelogynopora paracnida Karling, 1966. Tomales Bay (Second Sled Road); shell gravel.

Coelogynopora tenuiformis Karling, 1966. Point Pinos, in shell gravel and stones with *Polychoerus carmelensis*.

Itaspiella bursituba Karling, 1964. 3 m on sand, Tomales Bay.

Itaspiella bodegae Karling, 1964. Coarse sand, Dillon Beach.

Minona cornupenis Karling, 1966. Hopkins Marine Station, Pacific Grove, sandy beach in subsoil water.

Minona obscura Karling, 1966. Elkhorn Slough (Monterey Bay), in enteromorphoid *Ulva* and mud in slightly brackish water; abundant.

Monocelis cincta Karling, 1966. Point Pinos, abundant in tide pools; Dr. Karling states that *Monocelis cincta* (plate 81B) is the luminous *Monocelis* mentioned in Rickets and Calvin (fourth ed., p. 73). See Martin 1978a, b, c (ciliary gliding).

Monocelis hopkinsi Karling, 1966. In tide pools on the beach close to the Hopkins Marine Station, Pacific Grove.

Monocelis tenella Karling, 1966. Tomales Bay (Lawson's Landing), from sand in a deep tide pool on the mudflat.

Promonotus orthocirrus Karling, 1966. Tomales Bay (Second Sled Road), tide pool.

Serpentiplana doughertyi Karling, 1964. Tomales Bay (Second Sled Road), in shell gravel in tide pool. Dr. Karling indicated a number of his species were from "Bodega Bay," in the larger sense of the outer bight of water between Tomales Bay and Bodega Harbor. Named for Dr. Ellsworth Dougherty.

TRICLADIDA

Nexilis epichitonius Holleman and Hand, 1962. Commensal in mantle cavity of the chiton *Mopalia hindsii*; also with other invertebrates (plate 81C); see Holleman 1972.

Oregoniplana opisthopora Holmquist and Karling, 1972. In sandy substrate of *Zostera* meadows and on rocks in surf zone at low-tide level, Yaquina Head, Oregon.

Pacificides psammophilus Holmquist and Karling, 1972. Interstitial on moderately exposed sandy beaches; near Elkhorn Slough, Monterey Bay (plate 81D).

Procerodes pacifica Hyman, 1954. Described from stranded *Macrocystis* near San Diego collected by the late well-known carcinologist Thomas E. Bowman; reported from Friday Harbor (Holleman 1972), and thus probably occurs in our area.

PROLECITHOPHORA

Small elongate, narrow, 1–2 mm long to 0.2–0.8 mm wide worms that are opaque, yellowish brown as spots or a triangle between the eyes or as transverse or longitudinal stripes or bands. Four eyes or absent. Found in tide pools with seaweed, interstitially in sand, and washed from stones and gravel.

Allostoma amoenum Karling, 1962a. Elkhorn Slough, in wet sand under enteromorphoid *Ulva* on the mudflat, abundant.

Cylindrostoma monotrochum (Graff, 1882). Tomales Bay (Karling, 1962b) and the Atlantic Ocean.

Cylindrostoma triangulum Karling, 1962b. Monterey Bay, intertidal and dredged.

Multipeniata californica Karling and Jondelius, 1995. Hopkins Marine Station, Pacific Grove; gravel under algae.

Plagiostomum abbotti Karling, 1962a. "Great Tide Pool" and Hopkins Marine Station, Pacific Grove, seaweed, stones, gravel. Named for Donald P. Abbott.

Plagiostomum hartmani Karling, 1962a. Monterey Bay, off Hopkins Marine Station in fine sand and detritus, 26 m. Dr. Karling apparently meant to spell the name *hartmanae* (after the polychaetologist Olga Hartman), as he inked in the latter spelling in a reprint that he provided to Ralph Smith.

Plagiostomum hedgpethi Karling, 1962a. Tomales Bay (Second Sled Road, tide pools; in a plankton net with the kelp *Macrocystis*); also at Hopkins Marine Station, Pacific Grove (tide pools with algae and in sand under "wrackbeds"). Named for Joel W. Hedgpeth.

Plagiostomum hymani Karling, 1962a. "Great Tide Pool," Pacific Grove, from seaweed. Evidently Dr. Karling meant to spell this name *hymanae*, after Libbie Hyman, as in a reprint to Ralph Smith, Dr. Karling inked out the *i* and wrote in *ae*.

Plagiostomum langi Karling, 1962a. Tomales Bay (Second Sled Road), tide pools. Named for the Swedish zoologist Karl Lang.

Pseudostomum californicum Karling, 1962b. "Great Tide Pool" in Pacific Grove among stones and gravel with *Polychoerus carmelensis*, among a "mosslike covering" of *Ulva*; Tomales bay in mud-bottom with eelgrass *Zostera* and in tide pools; common.

POLYCLADIDA

ACOTYLEA

Copidoplana tripyla Hyman, 1953. Rare, intertidally in Monterey Bay.

Discosolenia burchami (Heath and McGregor, 1912) (=*Pseudostylochus burchami*). Mainly subtidal but may be found in low intertidal pools.

Emprosthopharynx gracilis Heath and McGregor, 1912 (=*Stylochoplana gracilis*). On blades and stipes of *Macrocystis*, occasionally on pilings.

Hoploplana californica Hyman, 1953. Common on encrusting bryozoans, especially *Celleporaria brunnea*, on pilings.

Kaburakia excelsa Bock, 1925 Under rocks, in holes of rock burrowing bivalves, on boat bottoms, and on pilings around mussels.

Notocomplana acticola (Boone, 1929) (=*Notoplana acticola*). Very common under rocks from high to low tide zone. See Koopowitz et al. 1976 Biol. Bull. 150: 411–425 (behavior).

Notocomplana litoricola (Heath and McGregor, 1912) (=*Freemania litoricola*). Beneath stones below midtide level. See Phillips and Chiarappa 1980 J. Exp. Mar. Biol. Ecol. 47: 179–189 (predation on limpets).

Notocomplana rupicola (Heath and McGregor, 1912) (=*Notoplana rupicola*). Under rocks, low intertidal.

Notocomplana saxicola (Heath and McGregor, 1912) (=*Notoplana saxicola*). On masses of algae in tide pools and on algal holdfasts.

Notocomplana sciophila (Boone, 1929) (=*Notoplana sciophila*). Rare, intertidally in Monterey Bay.

Notocomplana timida (Heath and McGregor, 1912) (=*Notoplana timida*). Rare, rocky intertidal in Monterey Bay.

Parviplana californica Hyman, 1953 (Woodworth, 1894). Rocky intertidal among sponges and bryozoans.

Plehnia caeca Hyman, 1953. Subtidal on muddy substrate Monterey Bay.

Plehnia caeca var. *oculifera* Hyman, 1953. Subtidal on muddy substrate Monterey Bay.

Pleioplana californica (Plehn, 1897) (=*Stylochoplana plehni* [Bock 1913]). Subtidal in Monterey Bay.

Pleioplana inquieta (Heath and McGregor, 1912) (=*Notoplana inquieta*). Found under stones at low water; also associated with *Macrocystis*.

Pseudoalloioplana californica (Heath and McGregor, 1912) (=*Alloioplana californica*). Abundant under boulders on gravel and in crevices at mean tide and above. See Martin 1978b (ciliary gliding).

Stylochoplana chloranota (Boone, 1929) (=*Leptoplana chloranota*). Under stones in low intertidal.

Stylochus atentaculatus Hyman, 1953. In rocky crevices and under stones.

Stylochus californicus Hyman, 1953. Occurs in association with bivalve mollusks, on which it feeds.

Stylochus exiguus Hyman, 1953. In burrows of mudflat dwellers.

Stylochus franciscanus Hyman, 1953. Abundant in fouling community of floating piers and also very low intertidal and subtidally in San Francisco Bay.

Stylochus tripartitus Hyman, 1953. Fouling community of floating piers in San Francisco Bay and on kelp stipes and holdfasts.

Triplana viridis Hyman, 1953 (Freeman, 1933) (=*Phylloplana viridis*). On *Zostera* in sheltered bays and shallow subtidal water; Humboldt Bay to Puget Sound.

COTYLEA

Acanthozoon lepidum (Heath and McGregor, 1912). Under stones and in rock crevices on shore Monterey Bay.

Acerotisa langi Heath and McGregor, 1912. Uncommon, intertidal Monterey Bay.

Anciliplana graffi Heath and McGregor, 1912. Known from Monterey Bay.

Eurylepta aurantiaca Heath and McGregor, 1912. Under stones and crawling about on bottom of tide pools; sluggish, clings tenaciously to rocks.

Eurylepta californica Hyman, 1959. First collected at Cambria by the late David Montgomery, among coralline algae on rocky shores. The reproductive anatomy requires study.

Eurylepta sp. Morris, Abbott and Haderlie, 1980. Dorsal color light yellow with numerous concentric white lines and darker yellow margin. Known from surfaces of seaweed; low intertidal Pacific Grove, Monterey Bay.

Euryleptodes cavicola Heath and McGregor, 1912. Low intertidal loose rocky shore or among holdfasts of seaweeds. The color patterns *of E. californica* and *E. cavicola* are very similar; to distinguish from *E. californica*, the reproductive anatomy needs to be compared.

Euryleptodes pannulus Heath and McGregor, 1912. Broadly oval; tentacles rudimentary or absent; color unknown;. Monterey Bay.

Euryleptodes phyllus Heath and McGregor, 1912. Body elliptical; tentacles present; color unknown; Monterey Bay.

Prostheceraeus bellostriatus Hyman, 1953. On wharf pilings.

Pseudoceros canadensis Hyman, 1953. In semiprotected rocky areas or jetties at low-tide level; Tomales Point and north.

Pseudoceros luteus (Plehn, 1897). In low tide pools.

Pseudoceros montereyensis Hyman, 1953. Under rocks in mid-intertidal, reported only from Pacific Grove.

Stylostomum lentum Heath and McGregor, 1912. Among rocks, low intertidal.

References

Bock, S. 1913. Studien uber Polycladen. Zool. Bidrag. Uppsala, 2: 29–344.

Bock, S. 1925. Papers from Dr. Mortensen's Pacific Expedition 1914–1918. XXVII. Planarians. Pt. IV New stylochids. Vidensk. Medd. Dansk Naturh. Foren. 79: 97–184.

Boone, E. S. 1929. Five new polyclads from the California coast. Ann. Mag. Nat. Hist. (10) 3: 33–46.

Cannon, L. R. G. 1986. Turbellaria of the world. A guide to families & genera. Brisbane: Queensland Museum. 136 pp.

Ching, H. L. 1978. Redescription of a marine flatworm *Pseudoceros canadensis* Hyman, 1953 (Polycladida: Cotylea). Can. J. Zool. 56: 1372–1376.

Faubel, A. 1983. The Polycladida, Turbellaria proposal and establishment of a new system. Part I. The Acotylea. Mitt. Hamb. Zool. Mus. Inst. 80: 17–122.

Faubel, A. 1984. The Polycladida, Turbellaria proposal and establishment of a new system. Part II. The Cotylea. Mitt. Hamb. Zool. Mus. Inst. 81: 189–259.

Freeman, D. 1933. The polyclads of the San Juan region of Puget Sound. Trans. Amer. Micro. Soc. 52: 107–148.

Heath, H. and E. A. McGregor. 1912. New polyclads from Monterey Bay, California. Proc. Acad. Nat. Sci. Philadelphia 64: 455–488.

Holleman, J .J. 1972. Marine turbellarians of the Pacific Coast. I. Proc. Biol. Soc. Washington 85: 405–412.

Holleman, J. J. and C. Hand. 1962. A new species, genus and family of marine flatworms (Turbellaria, Tricladida, Maricola) commensal with mollusks. Veliger 5: 20–22.

Holmquist, C. and T. G. Karling. 1972. Two new species of interstitial marine triclads from the North American Pacific coast, with comments on evolutionary trends and systematics in Tricladida (Turbellaria). Zool. Scripta 1: 175–184.

Hooge M. D. and S. Tyler. 2006. Concordance of molecular and morphological data: The example of the Acoela. Integrative and Comparative Biology 46: 118–124.

Hyman, L. H. 1951. The Invertebrates: Vol. II. Platyhelminthes and Rhynchocoela; the acoelomate Bilateria. New York: McGraw-Hill, 572 pp.

Hyman, L. H. 1953. The polyclad flatworms of the Pacific coast of North America. Bull. Amer. Mus. Nat. Hist. 100: 265–392

Hyman, L. H. 1954. A new marine triclad from the coast of California. Amer. Mus. Novit. 1679, 5 pp.

Hyman, L. H. 1955. The polyclad flatworms of the Pacific Coast of North America: additions and corrections. Amer. Mus. Novitates 1704, 11 pp.

Hyman, L. H. 1959. Some Turbellaria from the coast of California. Amer. Mus. Novit. 1943, 17 pp.

Karling, T. G. 1962a. Marine Turbellaria from the Pacific Coast of North America. I. Plagiostomidae. Arkiv för Zool. 15: 113–141.

Karling, T. G. 1962b. II. Pseudostomidae and Cylindrostomidae. Arkiv för Zool. 15: 181–209.

Karling, T. G. 1964. III. Otoplanidae. Arkiv för Zool. 16: 527–541.

Karling, T. G. 1966. IV. Coelogynoporidae and Monocelididae. Arkiv för Zool. 18: 439–528.

Karling, T. G. 1967. On the genus *Promesostoma* (Turbellaria), with descriptions of four new species from Scandinavia and California. Sarsia 29: 257–268.

Karling, T. G. 1980. Revision of Koinocystididae (Turbellaria). Zoologica Scripta 9: 241–269.

Karling, T. G. and U. Jondelius. 1995. An East-Pacific species of *Multipeniata* Nasonov and three Antarctic *Plagiostomum* species (Platyhelminthes, Prolecithophora). Microfauna Marina 10: 147–158.

Karling T. G. and E. R. Schockaert 1977. Anatomy and systematics of some Polycystididae (Turbellaria, Kalyptorhynchia) from the Pacific and S. Atlantic. Zool. Scripta. 6: 5–19.

Kozloff, E. N. 1953. *Collastoma pacifica* sp. nov., a rhabdocoel turbellarian from the gut of *Dendrostoma pyroides* Chamberlin. J. Parasit. 39: 336–340.

Kozloff, E. N. 1965. New species of acoel turbellarians from the Pacific Coast. Biol. Bull. 129: 151–166.

Lehman, H. E. 1946. A histological study of *Syndisyrinx franciscanus*, gen. et sp. nov., and endoparasitic rhabdocoel of the sea urchin, *Strongylocentrotus franciscanus*. Biol. Bull. 91: 295–311.

MacGinitie, G. E. and N. MacGinitie. 1968 Natural history of marine animals. 2nd ed. New York: McGraw-Hill, 523 pp.

Martin, G. G. 1978a. The duo-gland adhesive system of the archiannelids *Protodrilus* and *Saccocirrus* and the turbellarian *Monocelis*. Zoomorphologie 91: 63–75.

Martin, G. G. 1978b. A new function of rhabdites: mucus production for ciliary gliding. Zoomorphologie 91: 235–248.

Martin, G. G. 1978c. Ciliary gliding in lower invertebrates. Zoomorphologie 91: 249–261.

Morris, R. H., D. P. Abbot, and E. C. Haderlie. 1980. Intertidal invertebrates of California. Stanford Univ. Press, 690 pp.

Plehn, M. 1897. Drei neue Polycladen. Jenaische. Zeitschr. Naturwiss. 31: 90–99.

Prudhoe, S. 1985. A monograph on Polyclad Turbellaria. British Museum (Natural History). Oxford Univ. Press, 259 pp.

Schockaert, E. R. and T. G. Karling. 1970. Three new anatomically remarkable Turbellaria Eukalyptorhynchia from the North American Pacific coast. Arkiv för Zool. 23: 237–253.

Seifarth, W. 2005. Marine flatworms of the world. http://www.rzuser.uni-heidelberg.de/~bu6/.

Sluys, R. 1989. A monograph of the Marine Triclads. Rotterdam: A. A. Balkema. 463 pp.

Stunkard, H. W. and J. O.Corliss. 1951. New species of *Syndesmis* and a revision of the family Umagillidae Wahl, 1910 (Turbellaria: Rhabdocoela). Biol. Bull. 101: 319–334.

Tyler S., S. Schilling, M. Hooge, and L. F. Bush. 2005. Turbellarian taxonomic database. Version 1.4 http://devbio.umesci.maine.edu/styler/turbellaria/.

* = Not in key.

Acoela

MATTHEW D. HOOGE

(Plate 82)

The Acoela is a group of small, acoelomate, hermaphroditic marine worms. Some acoels live in the plankton or on algae and other substrates in intertidal pools, but most species live in the water-filled spaces between grains of sand. Acoels are often nondescript, and distinguishing one species from another requires the preparation of histological sections, or at least the use of a compound microscope. One prominent feature of the acoels is the statocyst, a spherical georeceptor at the anterior end of the

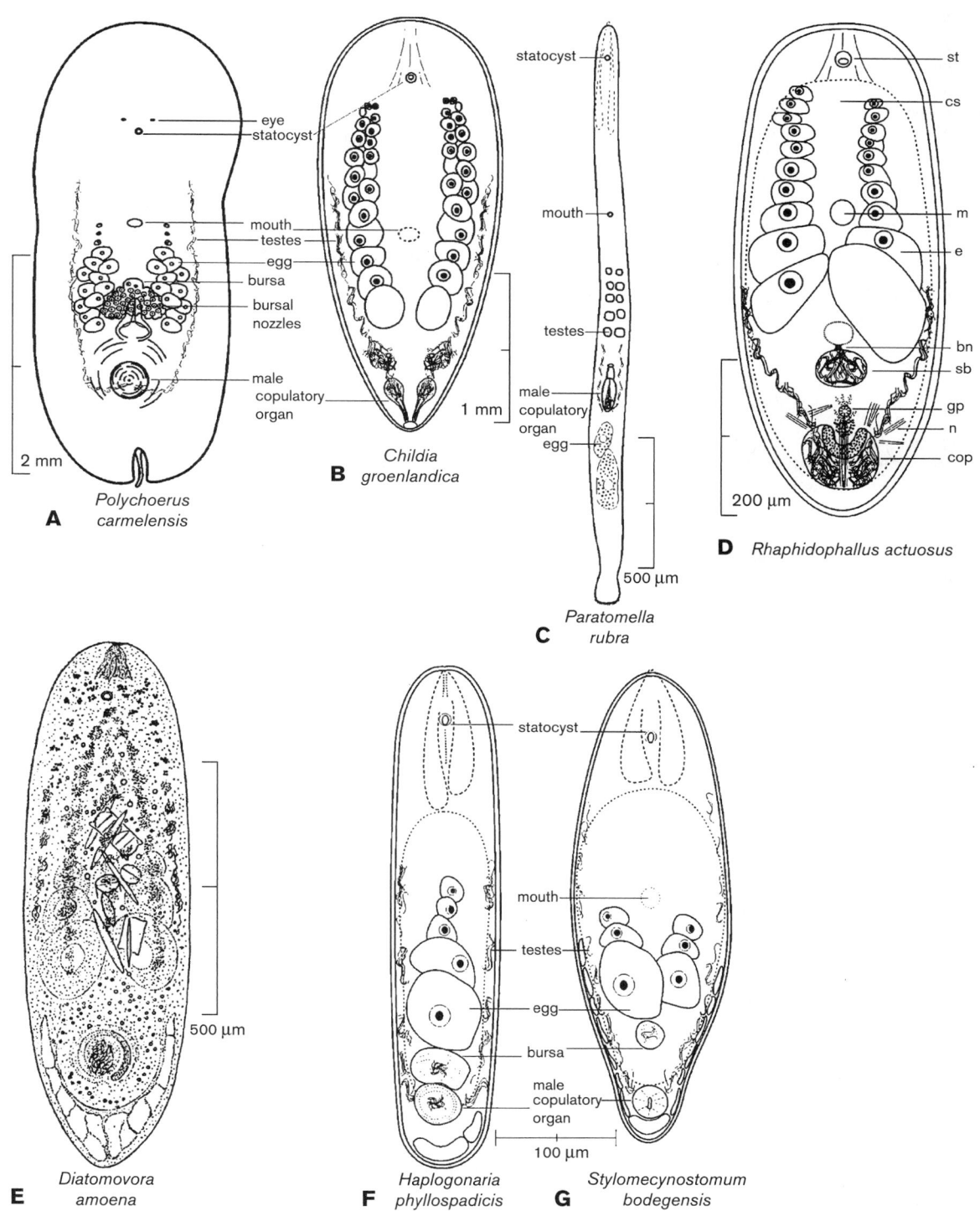

A **Polychoerus carmelensis**

- eye
- statocyst
- mouth
- testes
- egg
- bursa
- bursal nozzles
- male copulatory organ

2 mm

B **Childia groenlandica**

1 mm

C **Paratomella rubra**

- statocyst
- mouth
- testes
- male copulatory organ
- egg

500 μm

D **Rhaphidophallus actuosus**

- st
- cs
- m
- e
- bn
- sb
- gp
- n
- cop

200 μm

E **Diatomovora amoena**

500 μm

F **Haplogonaria phyllospadicis**

- statocyst
- mouth
- testes
- egg
- bursa
- male copulatory organ

G **Stylomecynostomum bodegensis**

100 μm

PLATE 82 Flatworms, acoels (A, after Costello and Costello 1938b; B, after Graff 1911; C, after Crezée 1978, and Rieger and Ott 1971; D, from Hooge and Tyler 2003; E, from Kozloff 1965; F and G, from Hooge and Tyler 2003).

body. Also recognizable are parts of their reproductive systems, including eggs, sperm-filled vesicles, and the posteriorly positioned male copulatory organ. Acoels lack an epithelially lined gut cavity (as found in other flatworms) and instead have syncytial digestive tissue that encloses ingested material in vacuoles.

Clean, well-sorted sand, which is a common sediment type in the highly exposed intertidal zone of the California and Oregon coast, is most often devoid of acoels because most species prefer silty sediment, and this may be a contributing factor to the paucity of known acoel species from this region. Of the more than 330 known species of acoels, 13 species have been positively identified from the Pacific coast of North America (Kozloff 1965, 2000; Hooge and Tyler 2003); six species have been found in California, and one has been identified in Oregon.

List of Species

Childia groenlandica (Levinsen, 1879). Plate 82B. Cosmopolitan species with paired male copulatory organs containing sclerotized penis needles. Found subtidally in San Francisco Bay; see Hyman 1959.

Diatomovora amoena Kozloff, 1965. Plate 82E. From intertidal muddy sand in South Slough, Charleston, Oregon.

Polychoerus carmelensis Costello and Costello, 1938. Plate 82A. Large ovoid-shaped worm from high tide pools and on enteromorphoid *Ulva,* on Monterey Peninsula; abundant in the past but increasingly rare over the past decade (J. Pearse pers. comm.). See Costello and Costello 1938 Ann. Mag. Nat. Hist. (11) 1: 148–155 (description); 1938 Bio. Bull. 75: 85–98 (reproduction); 1939, Bio. Bull. 76: 80–89 (egg laying); Armitage 1961, Pac. Sci. 15: 203–210 (biology); Martin 1978b, c (ciliary gliding).

Rhaphidophallus actuosus Kozloff, 1965. Plate 82D. Cylindrically shaped; penis composed of sclerotized needles; additional needles surround male copulatory organ. From surface mud in Bodega Harbor; see Hooge and Tyler 2003.

Haplogonaria phyllospadicis Hooge and Tyler, 2003. Plate 82F. In sediment underlying the surfgrass *Phyllospadix*, Bodega Bay.

Stylomecynostomum bodegensis Hooge and Tyler, 2003. Plate 82G. Fine-grained sand at the low intertidal level in Campbell Cove, Bodega Harbor.

Paratomella rubra Rieger and Ott, 1971. Plate 82C. Slender worm with bright red coloration. Asexual forms can be found in paratomizing chains of two. Collected from fine-grained sand at the low intertidal level at Dillon Beach. See Crezée 1978, Cah. Biol. Mar. 19: 1–9; Hooge and Tyler 2003.

References

Graff L. v. 1911. Acoela, Rhabdocoela und Alloeocoela des Ostens der Vereinigten Staaten von Amerika. Z. Wiss. Zool. 99: 321–428.

Hooge M. D. 1999. The abundance and horizontal distribution of meiofauna on a northern California Beach. Pacific Science 53: 305–315.

Hooge, M. D. and Tyler, S. 2003. Two new acoels (Acoela, Platyhelminthes) from the central coast of California. Zootaxa 131: 1–14.

Hyman, L. H. 1959. Some Turbellaria from the coast of California. Amer. Mus. Novit. no. 1943: 17 pp.

Kozloff, E. N. 2000. A new genus and five new species of acoel flatworms from the Pacific coast of North America, and resolution of some systematic problems in the families Convolutidae and Otocelididae. Cahiers de Biologie Marine 41: 281–293.

Rieger, R. M. and J. A. Ott. 1971. Gezeitenbedingte Wanderungen von Turbelarien und Nematoden eines Nordadriatischen Sandstrandes. Vie Milieu, Suppl. 22: 425–447.

Fecampiida

ARMAND M. KURIS

(Plate 83)

The Fecampiida are a small but very highly evolved group of parasitic castrators and parasitoids of marine, mostly crustacean, hosts. Although sometimes still considered as enigmatic Turbellaria (Cannon 2005), phylogenetic work (Rohde et al. 1994) recognizes them as a distinctive high-level taxon perhaps unrelated to the other parasitic Platyhelminthes (grouped as the monophyletic Neodermata). The internal anatomy is quite unlike other flatworms. Fecampiids lack a gut, their reproductive systems include paired posterior gonads opening to a posterior genital pore with diffuse vitellaria, and they have a large glandular nidamental organ that secretes the egg cocoon.

Two genera are potentially available in the intertidal and shallow subtidal zones of the Pacific coast of North America. *Kronborgia* are parasites of malacostracan crustaceans including amphipods, isopods, and caridean shrimps. These large worms develop in the haemocoel of their hosts (plate 83A), and then castrate and kill the hosts upon emergence. Sexes are separated, and the mature free-living female worm secretes a cocoon for her eggs; she dies after oviposition. Infective larvae penetrate the cuticle of their host (Christensen and Kanneworff 1965). On the Pacific coast, *K. pugettensis* Shinn and Christensen, 1985, infects the hippolytid shrimp *Heptacarpus kincaidi*. It is bright pink and, when approaching maturity, fills most of the haemocoel of the host. Its cocoons are attached in a tight coil to epibenthic organisms, such as hydroids. An undescribed orange species parasitizes mesopelagic *Gnathophausia* mysids in the vicinity of Isla Guadelupe, Baja California.

Fecampia erythrocephala Girard, 1886, is a parasitoid of juvenile green crabs, *Carcinus maenas* (plate 83B) and in protected outer coast habitats can be a major mortality source for crabs under 12 mm carapace width (Kuris et al. 2002). It also may infect other crustaceans. Adults are rose-colored with bright red heads. These worms are hermaphroditic and, upon emergence, secrete a retort-shaped cocoon on the undersides of rocks (plate 83C). Adult worms fill the cocoons with eggs and then die. Because these worms parasitize very small hosts and small hosts are rarely examined for parasites, these sort of worms are easily overlooked even where common. It has not yet been detected in *Carcinus* on the Pacific coast.

An unidentified fecampid has been reported from the haemocoel of the nudibranch *Aeolidia papillosa.*

References

Cannon, L. 2005. "Turbellaria" (turbellarians). In Marine parasitology, K. Rohde, ed. Collingwood Australia: CSIRO, pp. 47–55.

Christensen, A. M. and B. Kanneworff. 1965. Life history and biology of *Kronborgia amphipodicola* Christensen and Kanneworff (Turbellaria, Neorhabdocoela). Ophelia 2: 237–251.

Kuris, A. M., M. E. Torchin, and K. D. Lafferty. 2002. *Fecampia erythrocephala* rediscovered: prevalence and distribution of a parasitoid of the European shore crab, *Carcinus maenas.* Journal of the Marine Biological Association of the United Kingdom. 82: 955–960.

Rohde, K., K. Luton, P. R. Baverstock, and A. M. Johnson. 1994. The phylogenetic relationship of *Kronborgia* (Platyhelminthes Fecampiida) based on comparison of 18s ribosomal DNA sequences. International Journal for Parasitology 24: 657–669.

Shinn, G. L. and A. M. Christensen. 1985 *Kronborgia pugettensis* sp. nov. (Neorhabdocoela: Fecampiidae), an endoparasitic turbellarian infesting the shrimp *Heptacarpus kincaidi* (Rathbun), with notes on its life-history. Parasitology 91: 431–447.

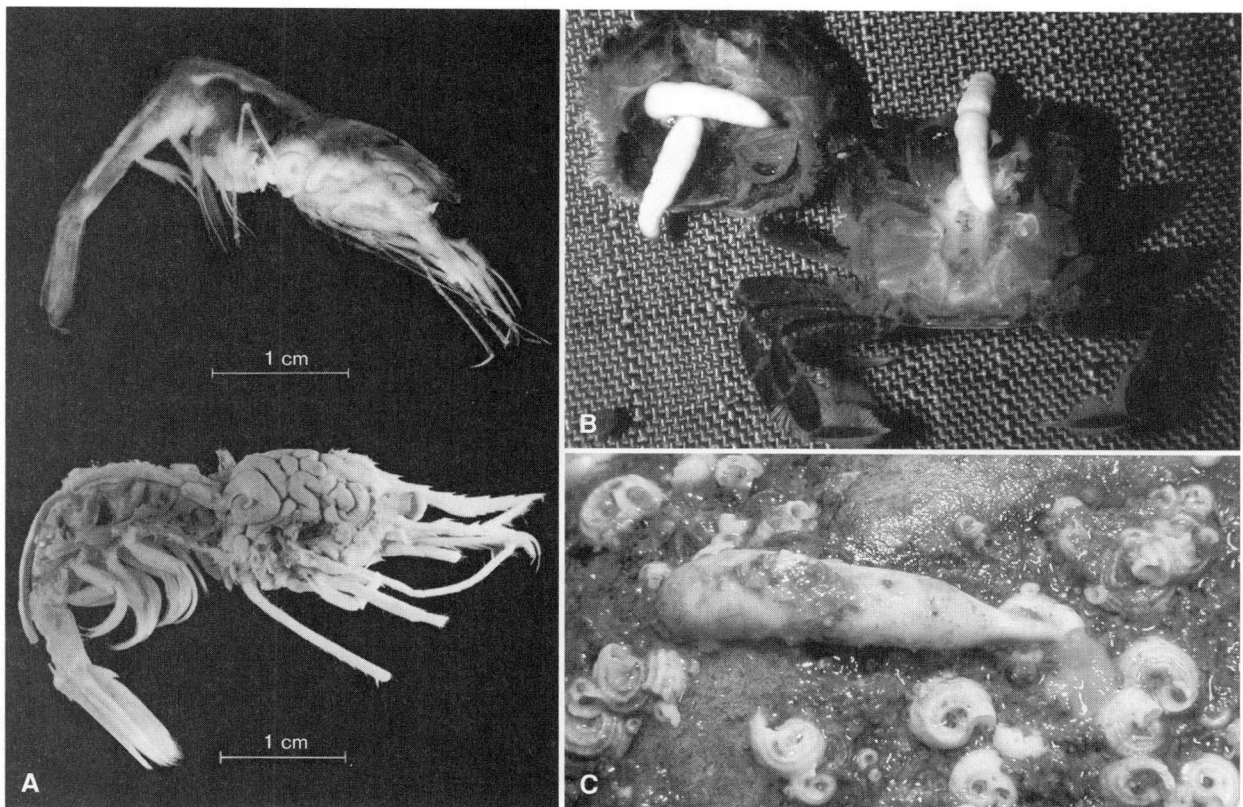

PLATE 83 A, Fecampiida, *Kronborgia caridicola* Kanneworff and Christensen, 1966, in a shrimp; B, *Fecampia erythrocephala* from a juvenile green crab, *Carcinus maenas*, on the Isle of Man; C, egg cocoon of *Fecampia erythrocephala* on the underside of a rock on the Isle of Man (A, from Kanneworff and Christensen 1996, Ophelia 3: 65–80, with permission; B and C, photographs by Kevin Lafferty).

Trematoda

ARMAND M. KURIS

(Plate 84)

The trematodes or flukes include two subclasses, the Aspidobothridea and the Digenea.

Aspidobothridea

Aspidobothrideans are parasites of mollusks, fishes, and turtles. Some have one host (a mollusk) while others have two-host life cycles. Marine species are found in the bile ducts of swell sharks (*Cephaloscyllium ventriosum*) in California and often in the chimaeroid (*Hydrolagus colliei*) in Washington. These are easily recognized, being several centimeters in length and having a long ventral sucker that looks like tractor tread. They have little effect on their hosts. None have been recovered from marine invertebrates on the West Coast of North America. A review of marine forms is available in Rohde (2005).

Digenea

Digenean flukes are abundant and widespread parasites, often causing morbidity and mortality in vertebrate hosts, including many species of interest for fisheries, wildlife conservation, and animal husbandry. Several digenean species parasitize humans and the blood flukes are estimated to infect 200 million people, contributing to perhaps 250,000 fatalities per year.

Adult flukes may be found in a very wide variety of organs in vertebrate hosts, including the gut, lungs, liver, musculature, swim bladders, etc. Adult flukes have two circular suckers, one around the mouth and the other on the ventral surface.

A generalized, typical life cycle is shown in plate 84A. There are many variations on this theme. Digenean life cycles are complex, and most involve three hosts. Eggs pass from the vertebrate final host. The first intermediate host is infected when a miracidium larva penetrates the epidermis of a mollusk, usually a snail, sometimes a bivalve. In this first intermediate host, the trematode asexually multiplies (as sporocysts or rediae, plate 84B) and parasitically castrates (eliminates reproduction of) the molluscan host. Infections usually persist for the rest of the life of the parasitized mollusk, often for several years. Up to 40% of the tissue weight of infected mollusks may be trematode larvae. The impact of larval trematodes on molluscan first intermediate hosts may be substantial with respect to host density, competition, and longevity (Kuris and Lafferty 2005). Ultimately, cercariae leave the host (plate 84C), swim away, and penetrate an appropriate second intermediate host. These vary depending on trematode species.

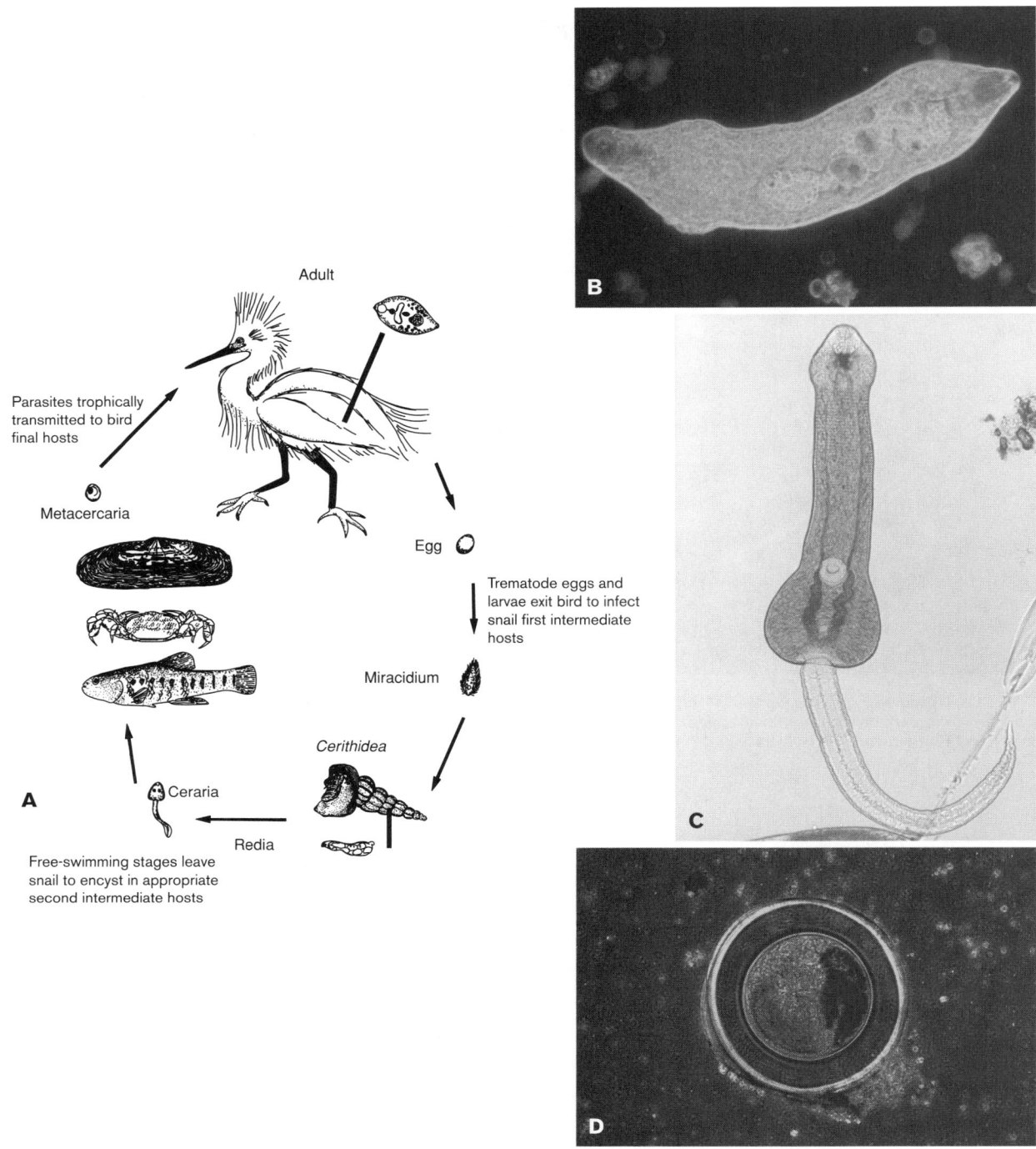

PLATE 84 A, Life cycles of the some of the trematodes using the snail *Cerithidea californica* as a first intermediate host (modified by Ryan Hechinger from K. Lafferty 1997, The ecology of parasites in a salt marsh ecosystem, in N. E. Beckage, ed., Parasites and pathogens: effects on host hormones and behavior, Chapman and Hall); B, Rediae of *Himasthla rhigedana* Dietz 1909 dissected from the snail, *Cerithidea californica;* C, Cercariae of *Himasthla rhigedana* released from the snail *Cerithidea californica;* D, Metacercaria of *Probolocoryphe uca* encysted in a crab (photos for B, C, D by Todd Huspeni).

Common marine second intermediate hosts include crustaceans, polychaetes, mollusks and fishes (Ching 1991). Here, the cercariae encyst as metacercariae (plate 84D) and await transmission to their final hosts via trophic transmission. Common species along the shore are often rich hunting grounds for trematode metacercariae; hosts range from beach hoppers (*Maritrema* sp. in the amphipod *Megalorchestia corniculata* [Ching 1974, sampled at Santa Barbara]) to tide pool cottid fishes (Hall and Pratt 1969, sampled in Oregon) and in perhaps all estuarine fishes in southern and central California. Some carefully studied species behaviorally modify their hosts to increase their likelihood of falling prey to a proper final host. *Euhaplorchis californiensis* Martin 1950 metacercariae increase susceptibility of the California killifish, *Fundulus parvipinnis*, to bird predators by a factor of 10–30 times (Lafferty and Morris 1996). This may be a general phenomenon. Along the California coast, final hosts are often shore birds or teleost fishes.

Some species of snails are much more important as hosts than are others. These include littorine snails and the California horn snail (*Cerithidea californica*), which serves as host for at least 18 species (Martin 1972). Through their parasitic castration effects (Kuris 1990, Lafferty 1993) and their widespread host behavior modification, digenean trematodes may play an important role in intertidal and estuarine ecosystems.

References

Ching, H. L. 1974. Two new species of *Maritrema* (Trematoda: Microphallidae) from the Pacific coast of North America. Can. J. Zool. 52: 865–869.

Ching, H. L. 1991. Lists of larval worms from marine invertebrates of the Pacific Coast of North America. J. Helminthol. Soc Wash. 58: 57–68.

Kuris, A. K. 1990. Guild structure of larval trematodes in molluscan hosts: prevalence, dominance and significance of competition, pp. 69–100. In Parasite communities: patterns and processes. G. W. Esch, A. O. Bush, and J. M. Aho, eds. Chapman and Hall.

Kuris, A. M. and K. D. Lafferty. 2005. Population and community ecology of larval trematodes in molluscan first intermediate hosts. In *Marine Parasitology*. K. Rohde, ed. Australia: CSIRO Publishing.

Lafferty, K. D. 1993. Effects of parasitic castration on growth, reproduction and population dynamics of the marine snail *Cerithidea californica*. Mar. Ecol. Prog. Ser. 96: 229–237.

Lafferty, K. D. and A. K. Morris. 1996. Altered behavior of parasitized killifish increases susceptibility to bird final hosts. Ecology 77: 1390–1397.

Martin, W. E. 1972. An annotated key to the cercariae that develop in the snail *Cerithidea californica*. Bull. S. Calif. Acad. Sci. 71: 39–43.

Rohde, K. 2005. Aspidobothrea. In Marine parasitology. K. Rohde, ed. Australia: CSIRO Publishing.

Monogenea

ARMAND M. KURIS

All Monogenea are parasitic, mostly of fishes. As ectoparasites, they are usually site-specific with various species being located on the skin, fins, gills, mouth, nasal capsules, and cloacas of their hosts (Rohde 1977). Site specificity may be so extreme that some species only occur on a portion of a single gill arch. Among the marine forms, a few species have been recovered from cephalopods, and one, *Udonella caligorum* Johnston, 1835, is commonly found as a hyperparasite on caligoid copepods. All species have a prominent posterior adhesive disk, variously armed with elaborate hooks and suckers. They have a blind bifurcated gut and feed on host tissues. All are hermaphroditic. Long classified with the Trematoda, they are now considered to be more closely related to the Cestoda (Boeger and Kritsky 2001). Hayward (2005) and Whittington (2005) provide an introduction to marine monogenean parasites.

Most monogeneans are typical parasites (macroparasites). Their impact on hosts is intensity-dependent: the more parasites acquired the greater the pathology (Kuris and Lafferty 2000, Lafferty and Kuris 2002). Eggs are released from the uterus, hatch in the water, and a ciliated oncomiracidium larva is the host-location stage. The Gyrodactylidae are a speciose group that have evolved parthenogenesis and viviparity, giving birth to a series of living young. This permits a rapid build-up of large populations on their hosts in an intensity-independent manner. In essence, they are unique helminth pathogens (Kuris and Lafferty 2000). Transfer of adult gyrodactylids by contact is their mode of transmission. Infestations can be fatal and, as for most pathogens, the infestation is ultimately controlled by the host's immune defenses.

Along the coast of California and Oregon, Monogenea are readily recovered from most species of teleosts or elasmobranchs. Most species of Monogenea are highly host-specific (Rohde 1977). Other than gyrodactylids, most only cause significant damage to their hosts at high intensities. Captive fishes in aquaria or fish farms can experience severe, sometimes fatal consequences because otherwise benign forms reach epidemic proportions on enclosed high-density host populations.

Love and Moser (1983) provide a checklist of parasites of California fishes. Readily collected species include *Microcotyle sebastis* (Goto 1894) from the gills of various rockfishes (*Sebastes* spp.) and several species of *Gyrodactylus* from the arrow goby, *Clevelandia ios,* and the long-jawed mudsucker, *Gillichthys mirabilis.*

References

Boeger, W. A. and D. C. Kritsky. 2001. Phylogenetic relationships of the Monogenoidea. In Interrelationships of the Platyhelminthes. D. T. J. Littlewood and R. A. Bray, eds. London: Taylor and Francis, pp. 92–102.

Crane, J. W. 1972. Systematics and new species of marine Monogenea from California. Wasmann Journal of Biology 30: 109–166.

Hayward, C. 2005. Monogenea Polyopisthocotylea (ectoparasitic flukes). In *Marine Parasitology*. K. Rohde, ed. Collingwood, Australia: CSIRO, pp. 55–63.

Kuris, A. M. and K. D. Lafferty. 2000. Parasite-host modeling meets reality: adaptive peaks and their ecological attributes. In Evolutionary biology of host-parasite relationships: theory meets reality. R. Poulin, S. Morand and A. Skorping, ed. Elsevier Science, pp. 9–26.

Lafferty, K. D. and A. M. Kuris. 2002. Trophic strategies, animal diversity and body size. Trends Ecol. Evol. 17: 507–513.

Love, M. S. and M. Moser. 1983. A checklist of parasites of California, Oregon and Washington marine and estuarine fishes. NOAA Tech. Rept. NMFS SSRF-777, 576 pp.

Rohde, K. 1977. Habitat partitioning in Monogenea of marine fishes. Zeit. Parasitenk. 53: 171–182.

Whittington, I. D. 2005. Monogenea Monopisthocotylea (ectoparasitic flukes). In Marine parasitology. K. Rohde, ed. Collingwood, Australia: CSIRO, pp. 63–72.

Cestoda

ARMAND M. KURIS

(Plate 85)

The cestodes, or tapeworms (subclass Eucestoda), are all parasites of the small intestine (or spiral valve of vertebrates). Most

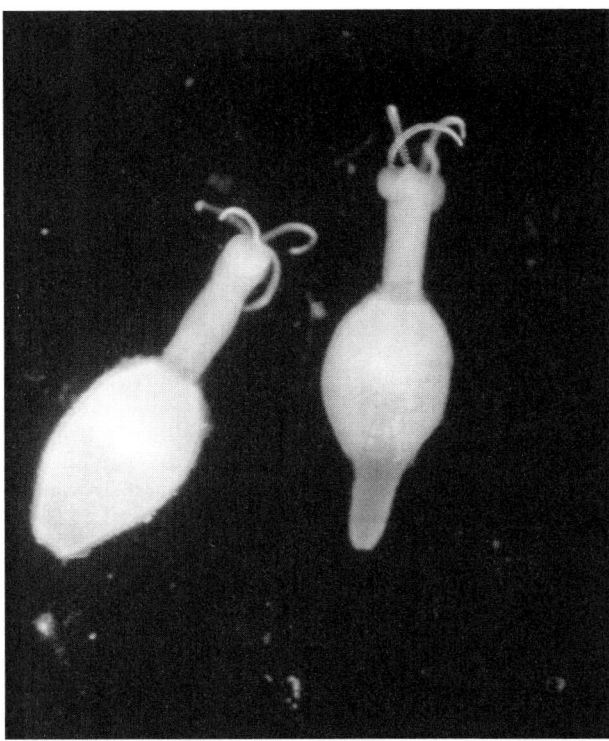

PLATE 85 Excysted larval trypanorhynch tapeworm from the crab *Carcinus maenas* (photograph by Mark Torchin).

tapeworm orders are marine, and most of these involve an elasmobranch as the final host. Life cycles are complex and usually involve three hosts.

For several marine orders of tapeworms, only a few life cycles have been worked out. Thus the following life cycle is generalized and somewhat hypothetical. Eggs pass with the feces of the host and are ingested by a small crustacean, usually a copepod. Depending on the cestode species, this is then typically transmitted to a second intermediate host where the larva migrates to an appropriate tissue, grows, develops an invaginated scolex and encysts (plate 85). The life cycle is completed when the infected second intermediate host is ingested by a vertebrate final host. As with other trophically transmitted parasites, behavior modification of infected hosts has been reported with resultant parasite-increased trophic transmission to the predator final host.

Only a few larval tapeworms have been reported from marine invertebrates in California. For some of these in beach hoppers, the final host is a shore bird (Ching 1991). Others, recovered from crabs such as *Hemigrapsus oregonensis* and from ghost shrimps (*Neotrypaea californiensis*), use elasmobranches as the final hosts, as do the tapeworm larvae found in clams (*Tresus nuttalli, Macoma nasuta, Leukoma staminea* [MacGinitie and MacGinitie 1968, Warner and Katkansky 1969]) and the echiuran, *Urechis caupo* (personal observation).

In contrast to trematodes, there has been very little study of larval marine tapeworms on the Pacific coast.

References

Ching, H. L. 1991. Lists of larval worms from marine invertebrates of the Pacific coast of North America. J. Helminthol. Soc. Wash. 58: 57–68.

MacGinitie, G. E. and N. MacGinitie. 1968. Natural history of marine animals. 2nd ed. New York: McGraw-Hill.

Warner R. W. and S. C. Katkansky. 1969. Infestation of the clam *Protothaca staminea* by two species of tetraphyllidean cestodes (*Echeneibothrium* spp.). J. Invert. Pathol. 13: 129–133.

Nemertea

(Plates 86–90)

PAMELA ROE, JON L. NORENBURG, AND SVETLANA MASLAKOVA

Nemerteans, or ribbon worms (previously known as Rhynchocoela and Nemertini) are soft-bodied, unsegmented worms. About 1,150 nemertean species are recognized (Gibson 1995); 119 benthic forms are reported from U.S. Pacific waters (Crandall and Norenburg 2001) and approximately 50 species are known from intertidal areas of central and northern California and Oregon.

Most nemerteans are free-living benthic marine worms. About 100 species are members of deep-sea pelagic communities, a few species live in damp conditions on land (e.g., *Geonemertes*), a few inhabit fresh water (e.g., *Prostoma*), several live in the mantle cavity of clams (e.g., *Malacobdella*), and a variety of decapod crustacean species host species of the genus *Carcinonemertes*. The primary diagnostic features of nemerteans are the proboscis, usually a long, eversible tubular structure, and the fluid-filled rhynchocoel chamber within which the proboscis resides (dorsal to the digestive tract). Development is via a juvenile-like pelagic (planuliform) larva or benthic encapsulated development, typical of palaeo- and hoplonemerteans, or via a planktonic pilidium larva, typical of heteronemerteans (Norenburg and Stricker 2001).

The traditional higher classification of nemerteans is in flux. Cladistic and molecular studies are likely to significantly modify the classification. The proboscis is unarmed in members of the class Anopla, comprising the orders Palaeonemertea and Heteronemertea, whereas in the order Hoplonemertea (class Enopla), the proboscis is three-chambered and armed with one or many stylets (plate 90A). However, there is strong support for considering Heteronemertea and Hoplonemertea to be sister-groupings (Thollesson and Norenburg 2003). The relationships within Palaeonemertea and to the previous two groupings are still unclear. Therefore, we keep the name Palaeonemertea here and abandon the names Anopla and Enopla. The traditional order Bdellonemertea has been dropped, as its only genus, *Malacobdella,* clearly belongs within the Hoplonemertea Monostilifera (Thollesson and Norenburg 2003).

The proboscis typically is used in food capture; in a few species, it is also used for locomotion, including escape responses (particularly in the terrestrial and some of the supralittoral species). Most nemerteans are predators on polychaetes, various crustaceans or insects, or clams; some include opportunistic necrophagy and detritivory. The proboscis secretes sticky and, in many species, toxic substances that serve to trap, hold, and paralyze the prey. Upon contact with a prey organism, the proboscis is rapidly everted and coiled around the prey, which is swallowed whole; often the mouth is capable of distending several times more than the diameter of the worm itself. Alternatively, the prey is stabbed by the stylet, making a wound through which toxins are delivered; the prey is swallowed whole or digestive enzymes are secreted into the stylet wound and the prey is consumed in suctorial fashion.

Identification of most nemertean species is difficult and time-consuming, usually requiring study of internal anatomy by means of light microscopy on serial sections. Specimens must be thoroughly relaxed (narcotized) prior to fixation. Nemertean species differ widely in their response to various relaxing agents. We have found the most success with magnesium chloride, $MgCl_2$ made isotonic to seawater (approximately 7.5% weight to volume in drinking water). Either the specimen is dropped directly into a 1:1 mixture of 7.5% $MgCl_2$ and seawater, or the 7.5% $MgCl_2$ is added slowly to seawater surrounding the specimen until a 1:1 mixture is achieved. If that fails, we often get good results by using either a 0.15–0.2% emulsion of propylene phenoxetol in seawater (or add 2% emulsion drop-wise to seawater surrounding specimen) or tricaine methanesulfonate (MS-222) crystals added gradually to seawater surrounding the specimen. Relaxed specimens are fixed in buffered 10% formalin (or 10% formalin-seawater) for 24 hours, followed by Bouin or Bouin-Hollande for about 24 hours. After fixation, they are transferred incrementally to 70% ethyl alcohol over a 24-hour period, followed by several changes of 70% ethyl alcohol over a period of days or weeks. The stylet armature of hoplonemerteans contains calcium and will be destroyed or deformed if the specimen is left in fixative too long.

Field identification is reliable for those species having unique color patterns or other unique external features. Some hoplonemertean species can be identified by characteristics of the proboscis armature, such as basis shape, sculpturing of the stylet, or number of accessory stylet sacs. These characteristics are best observed from living specimens. If the animal is sufficiently small and transparent, the proboscis armature can be viewed by compressing the whole animal between a microscope slide and coverslip. For large or opaque hoplonemerteans, it is best to remove the proboscis prior to fixing the specimen, as follows: once the specimen is completely relaxed, cut the animal transversely about two head-lengths from the snout. Pull the proboscis from the cut end of the posterior body part, at least past the bulbous middle chamber, and cut the proboscis from the animal (the proboscis is attached posteriorly in the rhynchocoel by a highly extensible retractor muscle). The proboscis can then be placed in seawater on a microscope slide, with the middle chamber compressed under a coverslip. After observation, the proboscis should be preserved along with the specimen from which it came.

Monographs by Coe (1901, 1904, 1905, 1940) are the most comprehensive taxonomic references for benthic nemerteans of the Pacific coast. Coe (1943), covering U.S. Atlantic species, is useful for West Coast species also having Atlantic distribution. Gibson (1995) is a nomenclatural review of nemertean genera and species. A web site for nemertean research, http://nemertes.si.edu, has available a comprehensive bibliography, searchable by taxon and author, a nomenclator with valid binomens and synonyms, a list of contact information for specialists, digital identification keys, and a comprehensive checklist of U.S. Pacific coast nemerteans (Crandall and Norenburg 2001).

Acknowledgments

We thank Eugene C. Haderlie for information in previous editions of this Manual, thereby making this update less arduous. This material is based upon work supported by the National Science Foundation under Grant No. DEB 9712463.

Glossary

ACCESSORY STYLET SACS or **POUCHES** sacs lateral to the central stylet(s) and basis in the proboscis bulb region of armed proboscis (hoplonemerteans) containing accessory stylets. Plate 90A.

BASIS the structure to which the central stylet(s) in the proboscis of hoplonemerteans is (are) attached. Plate 90A.

CAUDAL CIRRUS small, taillike appendage at the posterior end of some heteronemertean species (plate 87A, 87C, 87E); may be lost during collecting.

CEPHALIC FURROWS pair of lateral, shallow ciliated grooves anterior to or near front of cerebral ganglia and posterior ciliated grooves running obliquely or transversely across the dorsal and ventral surfaces of the head, meeting mid-dorsally and usually mid-ventrally, typically over or in back of the cerebral ganglia. Plates 89A, 89E, 89J, 90H, 90R.

CEPHALIC LOBE front end of body that appears to be the head, but usually the brain is behind this. Plate 89E, 90R, 90AA.

CEPHALIC SLITS deep longitudinal slits or furrows along the anterior-lateral margin in most heteronemerteans, except Baseodiscidae (plate 88E, 88H).

CEREBRAL ORGAN, CEREBRAL SENSORY ORGAN a pair of ciliated canals or pits and a surrounding mass of nervous and glandular tissue, closely associated with the brain; fused to the posterior of the ganglia in Heteronemertea; connected by a nerve in Hoplonemertea, usually in front of, but may be in back of or next to, cerebral ganglia; organs not present in some species. Externally, cerebral organs usually open to the outside at the posterior of the cephalic slits of Heteronemertea, as lateral or sub-lateral pores in the anterior dorso-ventral furrows of Hoplonemertea, or as simple pores on the lateral body surface of Palaeonemertea. When mentioned in the key, look for tiny, circular, lateral holes near the brain.

LATERAL SENSE ORGAN a pair of lateral pits found in a few Palaeonemertea species. Each appears as a small circle well behind the brain on the lateral margin of the body (spot in fifth white band.) Plate 86D.

PROBOSCIS ARMATURE the basis, central stylet(s), and accessory stylets and sacs found at the anterior portion of the bulb region of the proboscis in hoplonemerteans. Plate 90A.

PROBOSCIS CHAMBERS in hoplonemerteans, the proboscis is divided into a long anterior chamber with tall epithelium usually forming distinct papillae, a muscular central chamber or bulb region, and a posterior chamber, often shorter than the anterior chamber and lined with secretory epithelium (plate 90A). A duct passing from the posterior chamber through the bulb region to the base of the central stylet allows posterior chamber secretions (presumably toxins) to bathe the central stylet, thus creating a "poisoned dagger" for use in immobilizing prey.

PROBOSCIS PORE, RHYNCHOSTOME opening through which the proboscis everts, usually at the anterior tip of the body (anteriorly located openings in plate 86B, 86C).

RHYNCHOCOEL the fluid-filled chamber in which the proboscis lies.

RHYNCHODAEUM apical or subapical invagination of the body wall through which the proboscis is everted; also serving as the mouth in Monostilifera species.

STATOCYST structure found in *Ototyphlonemertes*: consists of a cell-lined vesicle (transparent sphere) embedded in the dorsal surface of each ventral ganglion of the brain, just behind the dorsal ganglion. The vesicle contains a statolith consisting either of two to eight easily counted spherical granules (oligogranular, plate 90GG), or it is a spherical aggregation of usually 12 or more such granules (polygranular, plate 90FF); easily seen with low-power microscopy of live worm slightly flattened under coverslip.

STYLET a single tiny, needlelike (monostiliferans) or multiple tacklike (polystiliferans) barbs anchored to the basis in the proboscis of hoplonemerteans (plate 90A, many figures, plates 89 and 90).

Classification

PALAEONEMERTEA Nemerteans with unarmed proboscis and proboscis not differentiated into regions. Mouth below or behind the brain. Typically slender, very soft-bodied and extensile, and not very flattened. Most (all species here) lack ocelli (often present in larvae) and cephalic furrows or slits. If a specimen does not fit any of the more easily differentiated orders, it is probably in this order. Mouth may or may not be conspicuous (plate 86B, 86C); far behind brain in *Procephalothrix* (plate 86G).

HETERONEMERTEA Nemerteans with unarmed proboscis not differentiated into regions. Mouth below or behind the brain and usually conspicuous (plate 88D). Head commonly with cephalic slits (plates 88E, 88H), except for *Baseodiscus* and *Zygeupolia* (among species here). A caudal cirrus (plate 87A, 87C, 87E), if present, is a useful identifying characteristic for *Cerebratulus*, *Micrura*, and *Zygeupolia*.

HOPLONEMERTEA Nemerteans with a more or less three-chambered proboscis, usually armed with stylet(s) (plate 90A). Mouth in front of the brain and, in most hoplonemerteans, combined with the rhynchodaeum, leaving only one visible opening at the anterior end of the body (all species here), called the proboscis pore (often inconspicuous, making the animal appear to have no openings on the head). Head usually with pair of anterior, dorso-ventral cephalic furrows (sites of cerebral sensory organ pores); commonly with posterior, annular cephalic furrow, V-shaped dorsally and ventrally (plates 89A, 89E, 89J, 90H, 90R). One suborder, Monostilifera, is characterized as having a single

PLATE 86 A, *Tubulanus cingulatus*; B, *Tubulanus,* head, ventral view with anteriorly located proboscis pore and posteriorly located mouth; C, *Carinoma mutabilis*, head, ventral view, with anteriorly located proboscis pore and posteriorly located mouth; D, *Tubulanus sexlineatus,* lateral sensory organ appears as a dark spot in the fifth transverse white band; E, *Carinoma mutabilis*; F, *Tubulanus capistratus*; G, *Procephalothrix spiralis*; H, *Tubulanus polymorphus* (A, after Coe 1904; B–E, originals; F and H, after Coe 1901; G, after Coe 1940).

central stylet. However, the genus *Malacobdella*, consisting of commensals in the mantle cavity of clams, is highly modified and lacks armature (Thollesson and Norenburg 2003). Malacobdellans have a posterior sucker and the general appearance of a leech (plate 90B). The suborder Polystilifera, with a padlike basis studded with many minute stylets (mouth and rhynchostome externally separate), is not known from intertidal habitats of California or Oregon.

Key to Nemertea

1. Mouth distinctly ventral, under or behind brain 2
— Mouth and proboscis share pore near tip of snout 32
2. Caudal cirrus present, but may be lost 3
— Caudal cirrus not present or unknown 11

3. Cephalic lobe with longitudinal lateral slits (usually conspicuous) 4
— Without longitudinal cephalic slits, pointed head, no ocelli, intestinal region flattened; body whitish to pale rosy color, translucent in intestinal region; to 8 cm long (plate 87C) . . . . *Zygeupolia rubens*
4. Cephalic lobe wide, often flaring wider than body posteriorly (lanceolate), flattening toward snout; snout relatively pointed; cephalic slits deep; usually no ocelli (inconspicuous, inside cephalic slits, if present); anterior body region cylindrical, strongly muscular (plate 88D) 5
— Cephalic lobe narrow, not flared (not lanceolate) 9
5. Intestinal region moderately to highly flattened (ribbonlike), adapted for swimming, not strongly contractile, usually with thin lateral margins (an ambiguous character); mouth usually large, may be elongate. . . . . . . *Cerebratulus* 6

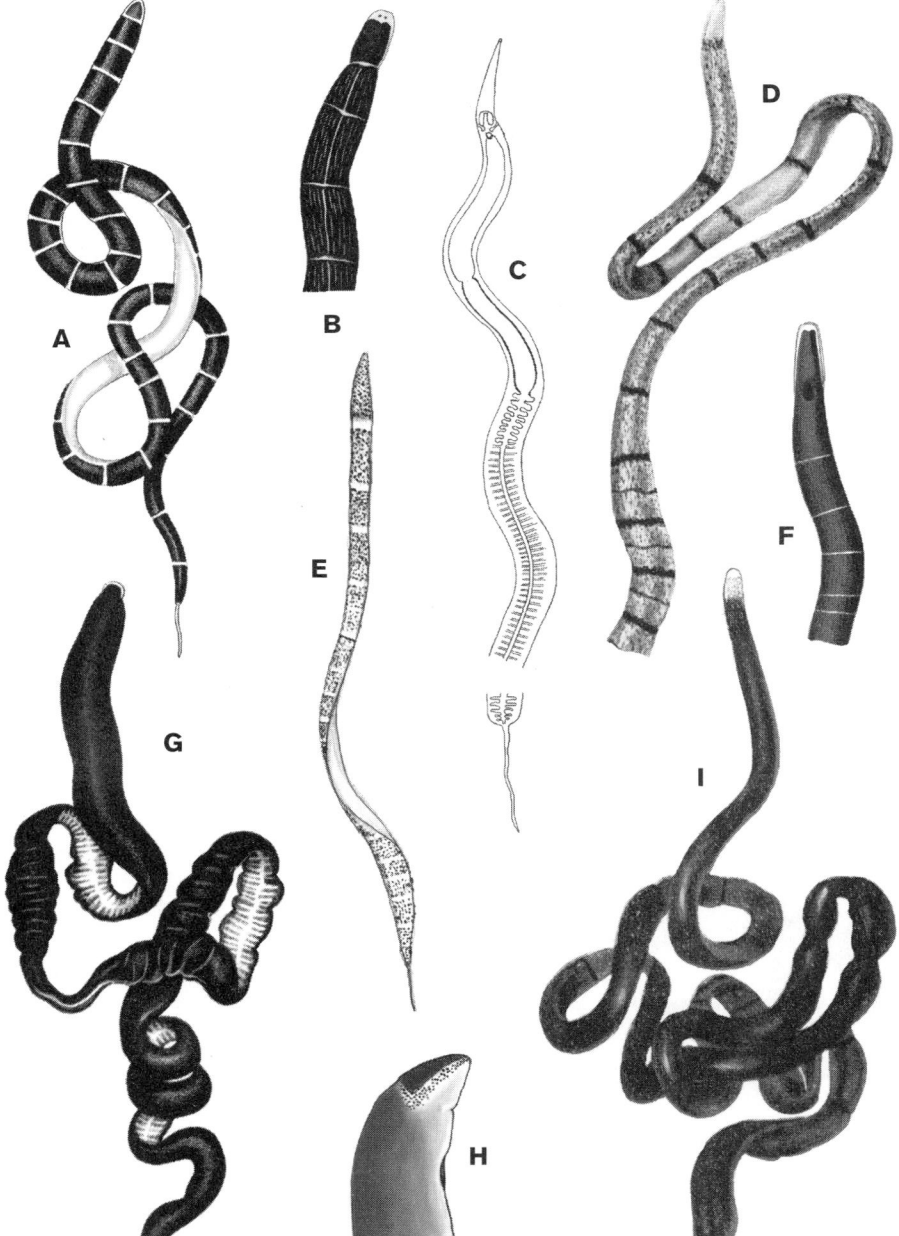

PLATE 87 A, *Micrura verrilli*; B, *Lineus pictifrons*; C, *Zygeupolia rubens*; D, *Euborlasia nigrocincta* var. 1; E, *Micrura coei*; F, *Micrura wilsoni*; G, *Baseodiscus punnetti*; H, *Baseodiscus punnetti*, head, lateral view; I, *Euborlasia nigrocincta* var. 2 (A, after Coe 1901; B, F, after Coe 1905; C, after Coe 1943; D–E, I, after Coe 1940, G–H, after Coe 1904).

— Intestinal region not highly flattened, without thin lateral margins; color ranging from translucent whitish to rosy, with brain showing through as reddish, and intestinal diverticula as salmon to brownish; without white bands or circles; cephalic lobe can be distinctly wider than body, with a pointed snout (cerebratulidlike); intestinal region flattened, but narrow; very long caudal cirrus; no ocelli; to 60 cm long (plate 88E, 88F) . . . . *Micrura alaskensis*

6. Snout with discrete region of white; body with different color. 7

— Cephalic lobe without discrete white coloration. 8

7. Tip of snout with short but conspicuous white region; body blood-red dorsum and ventrum; to 2 m long (plate 88A, 88B). . . . . *Cerebratulus montgomeryi*

— Similar to plate 88A, 88B *Cerebratulus montgomeryi*, but cephalic lobe white for slightly less than length of cephalic

slits; body dorsally and ventrally brown, brownish purple, or blackish; to 30 cm or more . . . .
. . . . *Cerebratulus albifrons*

8. Body slaty brown to grayish or olive, paler ventrally, often with paler to whitish lateral margins; relatively sturdy; very deep cephalic slits; to 1 m long (plate 88C, 88D) . . . .
. . . . *Cerebratulus marginatus*

— Similar to plate 88C, 88D *Cerebratulus marginatus*, but body yellowish to pale rosy salmon, cephalic lobe paler; reddish brain and lateral nerve cords obvious; short caudal cirrus; very fragile; to 15 cm long . . . .
. . . . *Cerebratulus californiensis*

9. Conspicuous narrow white bands, or rings, with entire margins, encircling body at intervals. 10

— Without discrete white bands or circles (transverse "bands," if present, do not have discrete margins); body

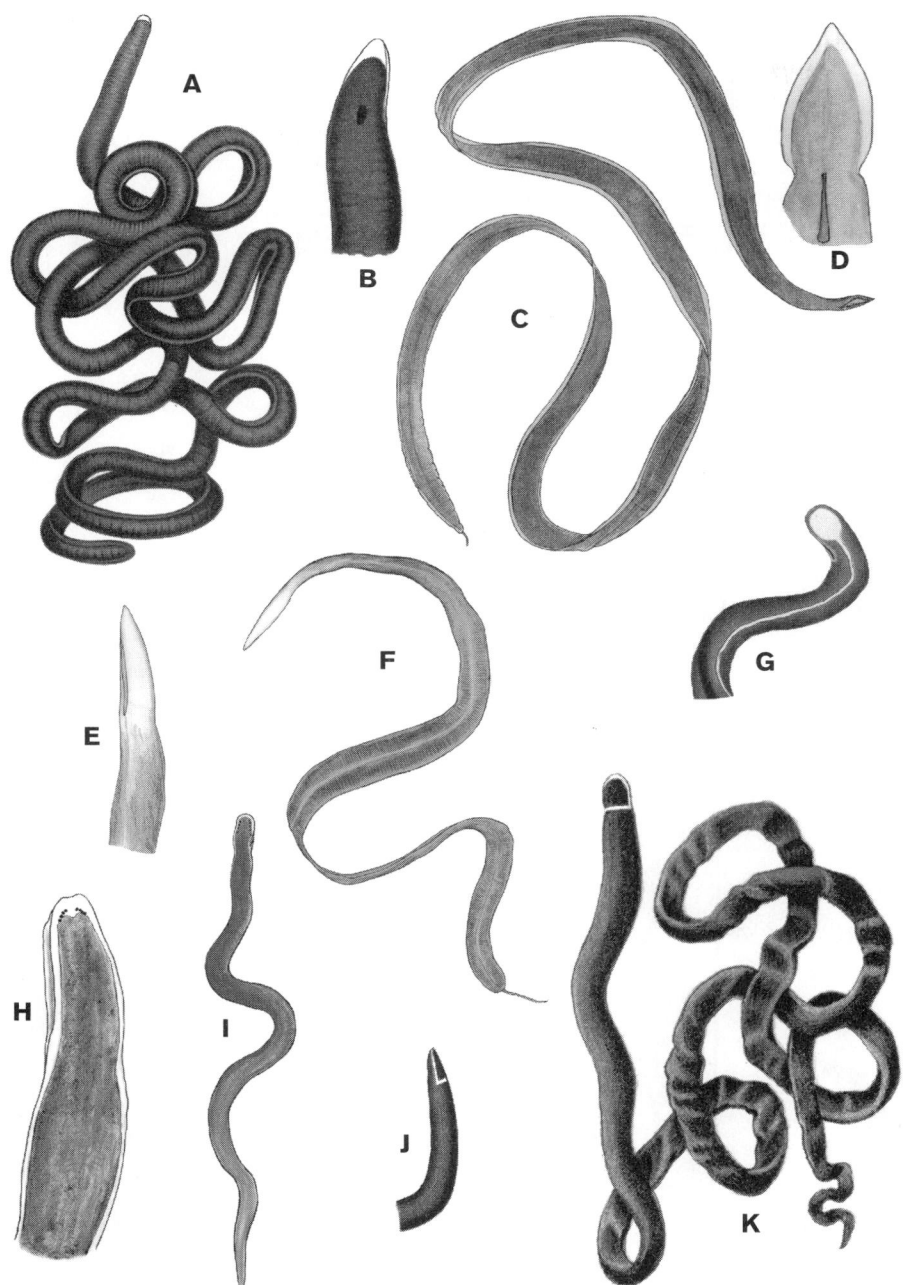

with pale yellowish ground color, dorsally with dense covering of variously sized dark brown to blackish pigment spots, sometimes coalescing into irregular longitudinal stripes that may be separated by transverse unpigmented regions; body cylindrical with strongly flattened ventrum; ocelli present, row of 10–18 inconspicuous ocelli along dorsal and ventral margin of each cephalic furrow; build and live in mucoid tubes; up to 3 cm long (plate 87E) . . . . *Micrura coei*

— Not as above; see *Micrura alaskensis* . . . . back to 5

10. Dorsal half of cephalic lobe clear bright orange, lateral and ventral surfaces white, dorsum with very broad, dark purple medial band transected at intervals by a fine white line, giving the appearance of a series of sharply demarcated rectangles running the length of the body; to 50 cm long (plate 87A) . . . . *Micrura verrilli*

— White margin at tip and along cephalic slits; remainder of body with range of brownish hue, paler ventrally; to 15 cm long (plate 87F) . . . . *Micrura wilsoni*

11. Cephalic lobe with longitudinal lateral slits (e.g., plate 88E, 88H) 12

— Without longitudinal cephalic slits 22

12. Body conspicuously divided into cylindrical foregut region and flattened intestinal region . . . . back to 5

— Body not conspicuously divided into cylindrical foregut and flattened intestinal regions 13

13. Body with discrete pigment patterning 14

— Body without discrete pigment patterning 20

14. Body with one or more obvious white or distinctly pigmented transverse bands or circles with discrete margins 15

— Without transverse bands having discrete margins 18

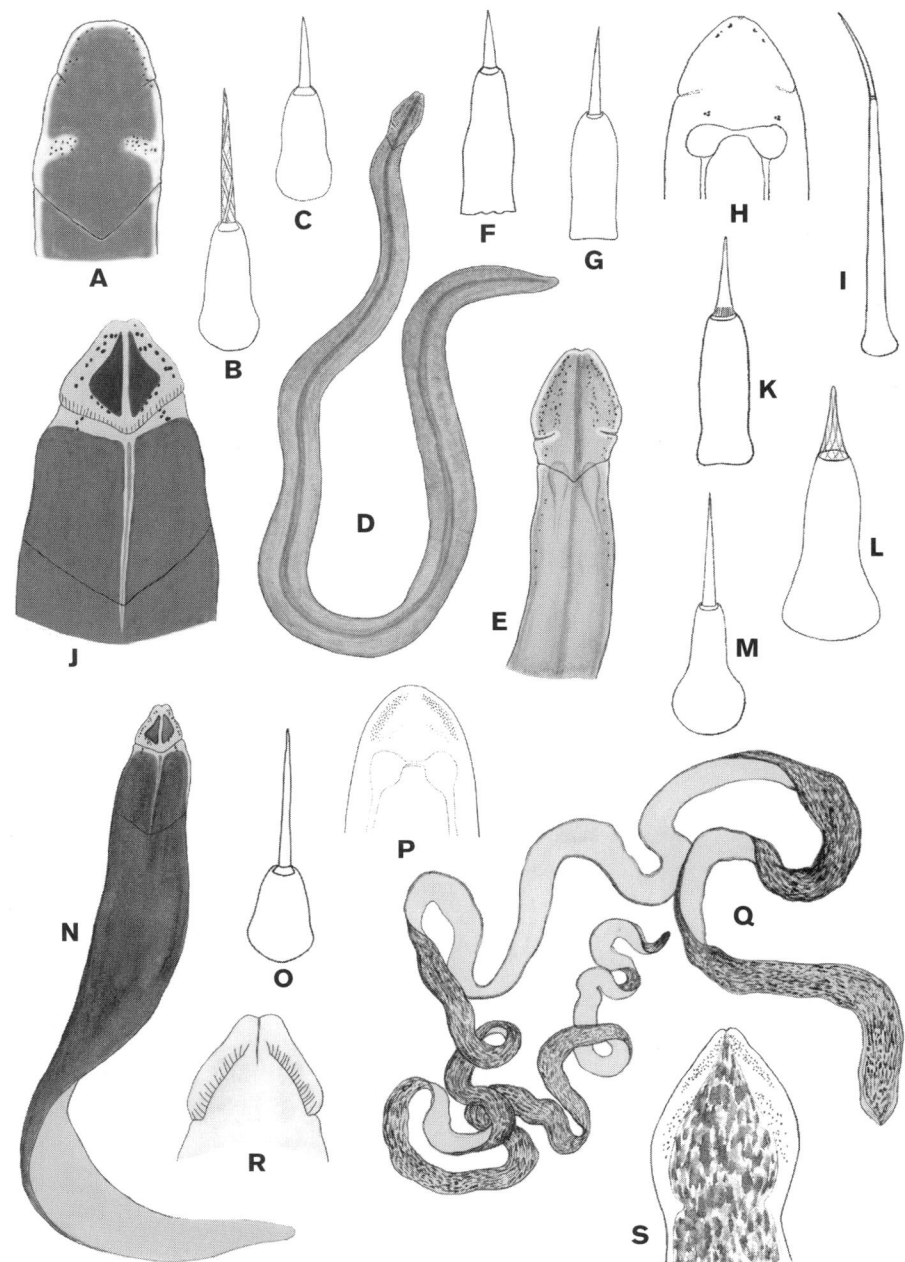

PLATE 89 A, *Paranemertes peregrina*, head, dorsal view; B, *Paranemertes peregrina*, spiral stylet; C, *Paranemertes peregrina*, smooth stylet; D, *Zygonemertes virescens*; E, *Zygonemertes virescens*, anterior end, dorsal view; F, *Zygonemertes virescens*, central stylet; G, *Zygonemertes albida*, central stylet; H, *Paranemertes californica*, head, dorsal view; I, *Emplectonema gracile*, central stylet; J, *Amphiporus bimaculatus*, head, dorsal view; K, *Paranemertes californica*, central stylet; L, *Emplectonema buergeri*, spiral stylet (with six accessory stylet sacs, from Puget Sound); M, *Emplectonema buergeri*, smooth stylet (with two accessory stylet sacs, from Alaska); N, *Amphiporus bimaculatus*; O, *Amphiporus bimaculatus*, central stylet; P, *Emplectonema buergeri*, head, dorsal view, Alaska; Q, *Emplectonema buergeri*, Puget Sound; R, *Amphiporus bimaculatus*, head, ventral view; S, *Emplectonema buergeri*, head, dorsal view (Puget Sound) (A–F, J, L, N–O, Q–S, originals; G, I, P, after Coe 1901; H, K, after Coe 1940; M, after Coe 1905).

15. Transverse bands or rings at intervals along body..... 16
— Single dorsal, transverse, white band connecting posterior ends of white cephalic slits; tip of snout white; remainder of body dark reddish to purplish brown; paler or reddish ventrum; no ocelli; to 40 cm long (plate 88J, 88K)......
.... *Lineus torquatus*
16. Transverse bands or rings paler than ground color 17
— Transverse bands darker than ground color; purplish brown to blackish rings; one variety, with purplish brown dorsum and ventrum with white head speckled with fine red, orange or brown dots; other variety with paler, rosy color with elongated reddish or purplish dots; to 70 cm long, 12 mm wide (plate 87D, 87I) ....
.... *Euborlasia nigrocincta*
17. Whitish transverse bands or rings (see also key item 10; plate 87A) .... *Micrura verrilli*

Or plate 87F......................... *Micrura wilsoni*
— Yellow transverse bands, seven to 15 very fine longitudinal lines of yellow on dorsal surface; body soft and flabby, with longitudinal creases; no ocelli (but may have two surface pigment spots near cephalic tip); to 15 cm long (plate 87B)................................ *Lineus pictifrons*
18. With discrete white or pale yellow dorsomedial stripe; body dark brown; to 20 cm long (plate 88G) ....
.... *Lineus bilineatus*
— With other form of patterning.................... 19
19. Dorsal surface with dense covering of brownish to blackish, variously sized pigment spots, which may coalesce to form irregular longitudinal lines organized into elongate blocks by transverse bands of ground color (plate 87E) .... *Micrura coei*
— Front end of cephalic lobe white, ventral and dorsal,

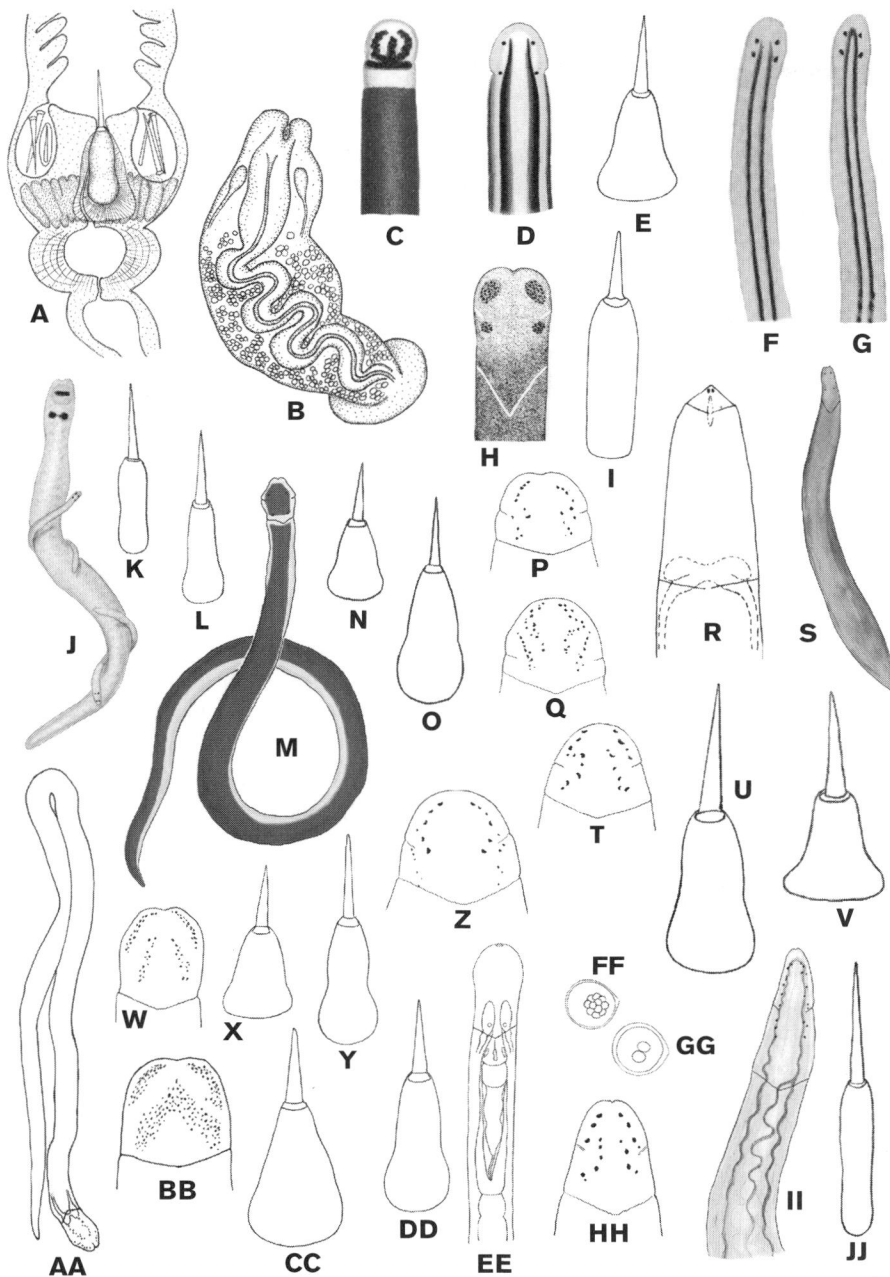

PLATE 90 A, diagram of hoplone-mertean proboscis showing central stylet on basis and accessory stylets and sacs in middle chamber, plus posterior end of anterior chamber and anterior end of posterior chamber; B, *Malacobdella* sp.; C, *Tetrastemma signifer*, head, dorsal view; D, *Tetrastemma quadrilineatum*, head, dorsal view; E, *Tetrastemma quadrilineatum*, central stylet; F, *Nemertopsis gracilis*; G, *Nemertopsis gracilis* var. *bullocki*; H, *Pantinonemertes californiensis*, head, dorsal view; I, *Pantinonemertes californiensis*, central stylet; J, *Tetrastemma albidum*, brooding female with young; K, L, *Tetrastemma albidum*, central stylets; M, *Tetrastemma nigrifrons*; N, O, *Tetrastemma nigrifrons*, central stylets; P, Q, *Amphiporus flavescens*, eye arrangements; R, *Poseidonemertes collaris*, head, dorsal view; S, *Oerstedia dorsalis*; T, *Amphiporus flavescens*, eye arrangement; U, V, *Amphiporus flavescens*, central stylets; W, *Amphiporus imparispinosus*, eye arrangement; X, Y, *Amphiporus imparispinosus*, central stylets; Z, *Amphiporus imparispinosus*, eye arrangement; AA, *Amphiporus imparispinosus*; BB, *Amphiporus formidabilis*, eye arrangement; CC, DD, *Amphiporus formidabilis*, central stylets; EE, *Ototyphlonemertes americana*; FF, *Ototyphlonemertes americana*, polygranular statocyst; GG, *Ototyphlonemertes* sp. oligogranular statocyst; HH, *Amphiporus imparispinosus*, eye arrangement; II, *Amphiporus cruentatus*, head, dorsal view; JJ, *Amphiporus cruentatus*, central stylet. (A–B, M–O, S, W–Y, AA–DD, II, originals; C–E, K–L, P–Q, T–V, Z, HH, JJ, after Coe 1905; F–G, J, after Coe 1940; H–I, after Gibson, Moore, and Crandall, 1982, J. Zool. Lond. 196: 463–474; R, after Roe and Wickham 1984, Proc. Biol. Soc. Wash. 97: 60–70).

sharply delineated from pink to rosy body; four to eight ocelli near tip of snout; to 15 mm long (plate 88H) . . . . *Lineus rubescens*

20. Single row of easily observed ocelli present along each cephalic slit 21
— Without ocelli (see key item 5; plate 88E, 88F) . . . . *Micrura alaskensis*
21. Body contracts by shortening and thickening; cephalic lobe coloration reddish to reddish brown, remainder of body reddish to very dark brown; 1–10 cm (plate 88I) . . . . *Lineus ruber*
— Similar to *Lineus ruber* (plate 88I), but body contracts by coiling when sufficiently irritated; body olive green, brown to reddish, with hair-thin circles of lighter color giving animals a segmented appearance; two to eight ocelli; to 15 cm long . . . . *Ramphogordius sanguineus*

22. Mouth far back of brain, body mostly cylindrical 23
— Mouth under or just behind brain 24
23. Body yellowish to yellowish green with rosy head; weak tendency to contract into corkscrew shape (most reliable external means to distinguish individuals smaller than 15 cm from *P. spiralis*); to 1 m long . . . . *Procephalothrix major*
— Body and head mostly whitish; strong tendency to contract into corkscrew shape; to 15 cm long (plate 86G) . . . . *Procephalothrix spiralis*
24. Ocelli absent 25
— Ocelli present, numerous (40–60 per side); snout with dark brown or blackish dorsal pigment shield with white anterior and lateral margins, white ventral surface; body paler brownish to red dorsum, grading to paler grayish ventrum; short, shallow, oblique (dorso-ventral) cephalic furrows to

sides of brain region mark openings to cerebral organs; to 60 cm long, 10 mm wide (plate 87G, 87H) . . . .
. . . . *Baseodiscus punnetti*
25. Cephalic lobe usually broadly rounded, often exceeding width of body, pointed only when burrowing 26
— Cephalic lobe narrower than body, relatively sharply pointed; intestinal region flattened; mouth close behind brain; caudal cirrus present; to 8 cm long (plate 87C) . . . . *Zygeupolia rubens*
26. Lateral sense organs (see plate 86D) present, usually at least twice as far from snout as brain 27
— Lateral sense organs lacking; no cerebral sensory organs; whitish anteriorly to translucent with brownish intestine posteriorly; intestinal region flattened; to 20 cm long (plate 86C, 86E) . . . . *Carinoma mutabilis*
27. Cerebral sensory organs (therefore, external pores) present; worms often form parchmentlike tubes 28
— Cerebral sensory organs (and external pores) absent; cephalic lobe and body white, more translucent in intestinal region; to 10 cm long (see "Annotated Species List") . . . . *Carinomella lactea*
28. Body of more or less homogenous color, no distinct patterning 29
— Body brown, with white, distinct longitudinal and transverse linear markings; numerous transverse rings 30
29. Body white to translucent; to 25 mm long (see "Annotated Species List") . . . . *Tubulanus pellucidus*
— Body deep orange to reddish orange; to 3 m long (plate 86H) . . . . *Tubulanus polymorphus*
30. With four to six longitudinal lines 31
— With three longitudinal lines, one dorsomedial and one each just below lateral margin of body; to 1 m long (plate 86F) . . . . *Tubulanus capistratus*
31. With four longitudinal lines, one each along lateral margin of body, remaining two dividing dorsum into three equal parts; to 15 cm long (plate 86A) . . . .
. . . . *Tubulanus cingulatus*
— With five or six longitudinal lines, one is median dorsal, pair narrowly spaced lines along each lateral margin of body; the sixth, if present, marks ventral midline; to 1.5 m long (plate 86D) . . . . *Tubulanus sexlineatus*
32. Commensal with crustaceans or clams; no cerebral sense organs 33
— Not commensal with crustaceans or clams 37
33. At base of legs or among appendages of abdomen of crabs; very short proboscis, without accessory stylet sacs; basis about 2.5–3 times length of central stylet; two ocelli 34
— In mantle cavity of clams; posterior sucker, proboscis without stylet, no ocelli (plate 90B) . . . . *Malacobdella* 35
34. On *Pugettia producta, Hemigrapsus oregonensis, H. nudus, Pachygrapsus crassipes, Cancer* spp. except *C. magister*; possibly other brachyuran crabs (see "Annotated Species List" if on nonbrachyuran host); to 6 mm long . . . .
. . . . *Carcinonemertes epialti*
— On *Cancer magister*; to 6 mm long . . . .
. . . . *Carcinonemertes errans*
35. In *Macoma secta* or *M. nasuta*; 11–12 gut undulations in mature specimen; to 20 mm long . . . .
. . . . *Malacobdella macomae*
— Other bivalve host. 36
36. In *Siliqua patula*; 15–16 gut undulations in mature specimen, to 42 mm long . . . . *Malacobdella siliquae*
— In other bivalves (e.g., *Tresus nuttallii, Tresus capax*; see "Annotated Species List"). . . . . *Malacobdella grossa*?

37. Statocysts absent 39
— With a pair of cerebral statocysts, no eyes (except larva), stylet smooth or fluted; in moderately coarse, well-sorted sand . . . . *Ototyphlonemertes* 38
38. Each statolith comprising about 16 granules, stylet with helical fluting, no cerebral sensory organs, proboscis barely reaching past cerebral ganglia; to 10 mm long (plate 90EE, 90FF). . . . . *Ototyphlonemertes americana*
— Each statolith comprising eight granules or fewer, stylet smooth (plate 90GG). . . . . *Ototyphlonemertes* sp.
39. More than two ocelli 40
— With one pair ocelli, conspicuous near tip of snout; snout often retracted into cephalic lobe to level of anterior cephalic grooves, thereby forming a collar at that point (plate 90R). . . . . *Poseidonemertes collaris*
40. With four distinct ocelli, two anterior and two posterior 41
— More than four ocelli 47
41. With distinct color patterning 42
— Without distinct color patterning 45
42. With longitudinal brown stripes 43
— Without longitudinal brown stripes. 44
43. With one pair median stripes; body ground color whitish to pale brown; filiform; ocelli large; up to 15 cm long (var. *bullocki* has a small dark rectangle, a little longer than wide, that unites the anterior ends of the longitudinal stripes); (see plate 90F, 90G for both varieties). . . . . *Nemertopsis gracilis*
— With four stripes, two medial and one along each margin; body ground color whitish; moderately flattened; up to 12 mm long (plate 90D, 90E) . . . .
. . . . *Tetrastemma quadrilineatum*
44. Head white with sharply demarcated dorsal wreath of brown pigment; body reddish brown; red blood corpuscles; up to 25 mm (plate 90C) . . . . *Tetrastemma signifer*
— Head white with variously shaped dorsal shield of deep brown pigment; head with pair of clearly incised furrows on each side of cephalic lobe; body coloration variable, ranging from purplish to brown, sometimes with continuous or broken white dorsomedial line, or reddish with brown flecks or pale brownish dorsomedial band; paler ventrum, often with white longitudinal median band; red blood corpuscles; to 70 mm long (plate 90M–90O) . . . . .
. . . . *Tetrastemma nigrifrons*
45. Body not white 46
— Body and cephalic lobe white to very pale salmon; body somewhat flattened; brain yellowish; ocelli with diffuse corona of pigment; cerebral organs large, reaching next to brain; rhynchocoel to posterior third of body; stylet basis very slender; oviparous or viviparous; to 15 mm long (plate 90J–90L) . . . . *Tetrastemma albidum*
46. Color varies from uniform yellowish to pale salmon to brown; often with darker dorsal mottling, which may include aggregations into transverse rings; ocelli may be difficult to see; body cylindrical; to 15 mm long (plate 90S) . . . . *Oerstedia dorsalis*
— Color predominately greenish or yellowish-green, although immature specimens may be grayish or whitish; body relatively flattened (rather than cylindrical as in (plate 90S) *Oerstedia dorsalis*; small ocelli, to 35 mm long . . . . *Tetrastemma candidum*
47. Ocelli not extending behind cerebral ganglia, but may be dorsal to or above them 49
— Ocelli extend behind cerebral ganglia, along foregut and lateral nerve cords; stylet basis approximately one-and-a-half

to two times length of stylet, with sharply truncated or concave posterior margin; numerous sickle-shaped bodies in epidermis visible with microscopic examination of compressed individuals 48

48. Body pale green to dark bluish green; back end of stylet basis of mature individuals with lobed or serrated appearance; to 20 mm long (plate 89D, 89E, 89F) . . . .
. . . . Zygonemertes virescens
— Body whitish; back end of stylet basis not lobed; to 25 mm long (plate 89G) . . . . Zygonemertes albida

49. Stylets with helical fluting or striation at base 50
— Stylets smooth (but see Emplectonema buergeri) 52

50. Stylets with helical fluting 51
— Stylets with striation at base, basis about twice length of stylet; two, four, or six pouches of accessory stylets, rhynchocoel fluid is red; body rosy to pinkish anteriorly, more gray or pale salmon posteriorly; intestine often green; four clusters of two to three small ocelli each, one pair flanking proboscis opening and one cluster in front of each brain lobe; up to 45 cm long (plate 89H, 89K) . . . .
. . . . Paranemertes californica

51. Brown or purplish brown dorsum, paler ventrum; with unpigmented mask on dorsolateral surface of cephalic lobe over each of the two posterior groups of ocelli; each group with five to 12 ocelli; two pouches of accessory stylets; up to 25 cm long (plate 89A, 89B) . . . .
. . . . Paranemertes peregrina
— Body color fleshy cream to orange-tan throughout; red brain visible; 18–35 ocelli, loosely arranged in four somewhat linear groups along margin of cephalic lobe and extending inward in front of brain; to 11 cm long; central stylet like that of Paranemertes peregrina (plate 89B) . . . . .
. . . . Paranemertes sanjuanensis

52. Rhynchocoel less than half of body length 53
— Rhynchocoel more than half of body length 54

53. Dorsal surface green with whitish, yellowish, or very pale green ventrum; stylets long and slender, and conspicuously curved; basis slender, two to three times length of stylet; up to 50 cm long (plate 89I) . . . .
. . . . Emplectonema gracile
— Dorsal surface brown, often with reddish brown or purple mottling and streaking, with pale yellowish to beige ventrum; width 2-3 times dorso-ventral thickness; stylet short, back end of basis strongly swollen; up to 1 m long (plate 89M, 89P; spiral stylet and 6 accessory sacs in Puget Sound, plate 89L, 89Q, 89S) . . . . Emplectonema buergeri

54. Blood vessels not red. 55
— Blood vessels conspicuous, with red corpuscles; five to 10 ocelli in single row on each side of cephalic lobe; slender stylet basis; rhynchocoel to posterior end of body; up to 25 mm long (plate 90II, 90JJ) . . . . . Amphiporus cruentatus

55. Body whitish, may be tinged with pink. 56
— Body not white or pinkish 57

56. Proboscis with two or three pouches of accessory stylets; brain pale pink to brownish; four groups of ocelli with up to 15 ocelli each, depending on size of individual; up to 50 mm long (plate 90W–90AA, 90HH) . . . .
. . . . Amphiporus imparispinosus
— Proboscis with 6-12 pouches of accessory stylets; brain rosy to red; 4 groups of ocelli with up to 50 ocelli each, depending on size of individual; up to 30 cm (plate 90BB–90DD)
. . . . Amphiporus formidabilis

57. Without pair of distinct, dark pigment patches in front of brain on dorsal surface of cephalic lobe. 58

— Head whitish with two large angular or oval patches of very dark brownish pigment; dorsum of body red, reddish brown, or brownish orange; ventrum rosy, orange, or beige; four clusters with six to 15 relatively large ocelli each and two posterior clusters with 2–5 ocelli each immediately behind the anterior cephalic grooves; stylet about twice length of basis; two or four pouches of accessory stylets; body with flattened dorsum and ventrum, widest at foregut region; swims actively if irritated; up to 15 cm long (plate 89J, 89N, 89O, 89R) . . . .
. . . . Amphiporus bimaculatus

58. Dorsum yellow, but may be whitish; cerebral ganglia yellowish; up to 20 mm long (plate 90P, 90Q, 90T–90V)
. . . . Amphiporus flavescens
— Dorsum dark grayish-green to bluish-green, darker in dorsal midline, shading to lighter greenish-tan laterally; head and ventrum light cream to tan; anterior two groups of ocelli larger and closer to surface than posterior two groups; eye number increases with size, anterior with about five to 100 and posterior with about three to 50 per group; length 20–450 mm (plate 90H, 90I) . . . .
. . . . "Pantinonemertes" californiensis

List of Species

Only West Coast geographic distributions are listed unless broader distribution indicates potential taxonomic problems. Most information is from Coe 1901, 1904, 1905, 1940; Gibson 1995; plus McDermott and Roe 1985, Amer. Zool. 25: 113–125 (feeding) unless otherwise indicated. Identity of some species that were originally described from elsewhere (commonly Europe) are speculative, as no specific comparative work has been done. Our confidence in such identifications varies with the degree to which the known morphological attributes (e.g., color, pattern, armature structure) are unique or diagnostic among all known nemerteans. The type locality is given when the general distribution is reported to extend beyond the Pacific coast of North America. Detailed type information for many of Coe's species can be found in Hochberg and Luniansky (1998, Hydrobiol. 365: 291–300, taxonomy). Some Washington state distribution information is based on Stricker (1996). Distributions for California with Crandall as citation are personal communications in 1999 by F. B. Crandall (National Museum of Natural History). Species noted by an asterisk are not recorded in published literature as occurring or are of uncommon occurrence intertidally along the central or northern California or Oregon coasts. They are listed here for completeness but are not included in the key.

Palaeonemertea

Carinomella lactea Coe, 1905. Central to southern California (type region); also Atlantic Florida. In intertidal sand. Superficially similar to Carinoma mutabilis and Tubulanus pellucidus, but with lateral sense organs, without cerebral sensory organs. For figures, see Coe (1905, figs. 46, 47, 50, 51).

Carinoma mutabilis Griffin, 1898. British Columbia (type region) to Mexico. In intertidal and subtidal sediments. Superficially similar to Carinomella lactea and Tubulanus pellucidus, but without either lateral or cerebral sensory organs.

Procephalothrix major (Coe, 1930) (=Cephalothrix major). Central California to Mexico. In sediments under rocks fully exposed to surf. Large individuals reach a meter in length, width only 2–5 mm. Similar to *Procephalothrix spiralis* (plate 86G), but body usually twisted or knotted, not spirally coiled. *P. major* and *P. spiralis* are morphologically distinguished internally by differences in their nephridia (Coe 1930, Biol. Bull. 58: 203–216). For figure, see Coe (1940, fig. 29).

Procephalothrix spiralis (Coe, 1930) (=*Cephalothrix spiralis*). Alaska to southern California; also northwest Atlantic coast (type region). In mud under intertidal and subtidal rocks. Recognized by spiral coiling of body when contracted.

Tubulanus capistratus (Coe, 1901) (=*Carinella capistrata*). Alaska (type region) and Puget Sound to Monterey, California. Intertidal; in delicate parchment tubes among algae and other surface cover or under rocks.

Tubulanus cingulatus (Coe, 1904) (=*Carinella cingulata*). Bolinas (Crandall) to San Diego. Intertidal and subtidal; on soft sediments.

Tubulanus pellucidus (Coe, 1895) (=*Carinella pellucida*). San Francisco Bay (Crandall) to San Diego; also Atlantic coast (type region). Intertidal; in delicate parchment tubes under rocks or among algae and other surface cover. Superficially similar to *Carinomella lactea* and *Carinoma mutabilis* (plate 86E) but typically much smaller; has both lateral and cerebral sensory organs. For figures, see Coe (1895, Trans. Conn. Acad. Arts Sci. 9: 515–522).

Tubulanus polymorphus Renier, 1804 (=*Carinella polymorpha, C. rubra, C. speciosa*). Aleutian Islands to Monterey, California; also northern Europe, Mediterranean (type region). Intertidal and subtidal; under rocks or among mussels and other surface cover in or adjacent to muddy substrata. Easily recognized by its uniform red-orange color and large size, to 3 m. Body soft and distensible; Roe once held a specimen about 2 m above the substratum, and part of its body still formed a pile on the ground.

Tubulanus sexlineatus (Griffin, 1898) (=*Carinella dinema, C. sexlineata*). Alaska to southern California. Intertidal and subtidal; in tubes among algae and other surface cover on rocks and pilings. Can reach 1.5 m in length.

Heteronemertea

Baseodiscus punnetti (Coe, 1904) (=*Taeniosoma punnetti*). Monterey Bay, California to Mexico. Intertidal and subtidal; among algae and other surface cover on rocks.

Cerebratulus. Cerebratulids are typically large and not strongly contractile in length, often relatively flattened with thin lateral edges throughout the intestinal region; burrowing forms tend to fragment more readily than most other nemerteans; often swim when kept in container of clean seawater. Species without distinctive external markings are difficult to distinguish from each other.

Cerebratulus albifrons Coe, 1901. Alaska to San Diego. Intertidal and subtidal; in mud or under rocks; subtidal off central and southern California. Length 30 cm or more. Similar to *Cerebratulus montgomeryi* (plate 88A, 88B), but cephalic lobe white for greater distance posteriorly. For figures, see Coe (1901, plate 4, figs. 3, 4).

Cerebratulus californiensis Coe, 1905. Washington to Mexico. Intertidal and subtidal burrower in soft sediments. Relatively small, about 10-15 cm length, readily fragments, no distinctive markings. Similar to *Cerebratulus marginatus* (plate 88C, 88D), but with different colors. For figures, see Coe (1940, plate 24, figs. 7–10). We have found a form resembling *C. californiensis* in Elkhorn Slough, but with inconspicuous ocelli in the cephalic slits.

Cerebratulus herculeus Coe, 1901 (=*Cerebratulus latus*). Bering Sea, Alaska (type region), Puget Sound, and southward to southern California (subtidal). Intertidal and subtidal burrower in mud. Enormous in size, often more than 2 m long and 25 mm wide, but capable of graceful swimming and active burrowing. Dark brown to reddish brown, duller ventrally. For figure, see Coe (1904, plate 1, fig. 5).

Cerebratulus longiceps Coe, 1901. Alaska; dredged at Tomales Bay by Corrêa (1964). Lowest intertidal to subtidal. Length to 30 cm, width to 6 mm. Dorsum dark brownish-black or purplish, paler ventrally; much paler on tip of head and borders of cephalic slits; body and head much flattened, head long and pointed anteriorly. For figures, see Coe (1904, plate 5, figs. 4–7).

Cerebratulus marginatus Renier, 1804. Alaska to San Diego; also Japan, western North Atlantic south to Cape Cod, Arctic coasts, Europe south to Madeira, and Mediterranean Sea (type region). Intertidal and subtidal burrower in mud. Length to 1 m.

Cerebratulus montgomeryi Coe, 1901. Alaska to Monterey Bay. Intertidal and subtidal; under stones and in mud.

Cerebratulus occidentalis Coe, 1901. Alaska, British Columbia, and Puget Sound; also San Francisco Bay (Corrêa 1964). Lower intertidal mudflats to predominantly subtidal. Reaches 15-30-plus cm. Dorsum chestnut brown to reddish brown anteriorly, light chocolate brown in intestinal region; brownish flesh color to light chocolate ventrally, with ventromedial ochre stripe; flattened posteriorly. For figure, see Coe (1904, plate 6, fig. 3).

Euborlasia nigrocincta Coe, 1940. California to Chile. Intertidal and subtidal under rocks and in hard clay in both exposed areas and protected bays. Length to 70 cm. Two intergrading color varieties, with many deviations, including inconspicuousness of dark transverse bands, especially in younger individuals.

Lineus bilineatus (Renier, 1804) (=*Cerebratulus bilineatus, Lineus albolineatus, L. bilineata, L. bilineatsu*). Alaska to San Diego; also coasts of Europe, Mediterranean (type region), Black Sea, Madeira, South Africa. Intertidal and subtidal; under and among algae and other surface cover on rocks, often in kelp holdfasts. Feeds on polychaetes.

Lineus flavescens Coe, 1904. Southern California to Gulf of California, with a disjunct northern record in Bodega Harbor (Crandall). Intertidal and subtidal on pilings and rocks, among algae and other surface cover. Ranging from 8–120 mm in length. Pale yellow to deep ochre, sometimes with shadings of orange or green; slender; head narrower than body and often rather pointed and emarginate anteriorly; several small irregular ocelli on each side of tip of head. For figures, see Coe (1904, plate 17, figs. 3, 4).

Lineus pictifrons Coe, 1904. Washington to Mexico. Under and among algae and other surface cover on rocks and piers exposed to surf, in kelp holdfasts, and in bay and harbor mud.

Lineus ruber (Müller, 1774) (see Gibson, 1995 for full synonymy). Alaska to Monterey (but see note below); also Circumpolar, Siberia, North Atlantic coasts of Europe (type region), Greenland to southern New England, Mediterranean and Black Seas, Madeira to South Africa. Mid-intertidal, under rocks on rocky shores. Deposits mucoid egg cocoon; larvae engage in sibling cannibalism; tens of crawl-away juveniles emerge from cocoon (Riser 1974, in Giese and Pearse, Reproduction of Marine Invertebrates, Vol. 1, Academic Press, pp. 359–389; reproduction). Feeds on polychaetes, oligochaetes, crustaceans, littorines, dissolved organics. Junior synonym and accepted as color variety of *Lineus viridis* for many years; reports could refer to either (e.g., Corrêa 1964). Neither senior author has seen *Lineus ruber*

---

* = Not in key.

or *Lineus viridis* (both well known to Norenburg) during many years of collecting along coasts of central California, Puget Sound, and Alaska. Both species have high likelihood of appearing as invasive species (J. T. Carlton reports possible specimens of *L. ruber* in the highly invaded brackish lagoon Lake Merritt, Oakland, in San Francisco Bay). Likely to be confused with reddish variety of *Ramphogordius sanguineus* (=*Lineus vegetus*). Former contracts linearly, latter contracts into spiral form.

Lineus rubescens Coe, 1904. San Francisco Bay (Crandall) to San Diego. Intertidal and subtidal; under or among algae and other surface cover, on rocks and pilings, or under rocks.

Lineus torquatus Coe, 1901. Alaska to San Francisco Bay; also Japan, Kurile Islands, Russia. Intertidal; under rocks and in mud.

*Lineus viridis (Müller, 1774) (see Gibson, 1995, for complete synonymy). Included here only for comparative purposes. Not recorded in Coe 1940 as occurring anywhere on the West Coast. Coe 1901-1905 listed a West Coast occurrence at Annette Island, Alaska, also East Coast of North America, most European coasts, Mediterranean. *Lineus ruber* was considered a junior synonym and color variety of *L. viridis* for many years (with *L. viridis* being more green). See comments above for *L. ruber*. Likely to be confused with greenish variety of *Ramphogordius sanguineus* (=*Lineus vegetus*). Former contracts linearly, latter contracts into spiral form. Also resembles *L. ruber,* but coloration is olive green to dark greenish brown; head may have reddish hue, especially around cerebral ganglia. May occur in mid- to low-intertidal muddy habitats, under rocks or among mussels and other surface cover. Deposits eggs in mucoid cocoon from which approximately 100 or more small crawl-away juveniles emerge.

Micrura alaskensis Coe, 1901 (=*Micrura griffini*). Alaska to Mexico (type region); also Japan. Intertidal; in crevices, under rocks in soft sediments, and in sandy tidal flats.

Micrura coei (Coe, 1905) (=*Micrura pardalis sensu* Coe, 1905, renamed by Gibson, 1995). Monterey Bay to Mexico. Intertidal; in tide pools exposed to surf, on piers, on and under rocks, in mucous tubes.

*Micrura olivaris Coe, 1905. Bodega Head (Crandall) and Monterey Bay. Uncommon, in rock crevices in low intertidal. Pale olive-brown, buff, or grayish ochre, deeper olive in intestinal region, paler median dorsal stripe in esophageal region; six to 12 small black ocelli in irregular row on each side of head. No figure known.

Micrura verrilli Coe, 1901 (=*Lineus striatus*). Alaska to Monterey Bay. Intertidal; among roots of eelgrass, in kelp holdfasts, on rocks and in tide pools, and under rocks in sandy mud.

Micrura wilsoni (Coe, 1904) (=*Lineus wilsoni*). California to Mexico. Intertidal and subtidal; under rocks, in kelp holdfasts, on sandy mud, on piers.

Ramphogordius sanguineus (Rathke, 1799) (=*Lineus vegetus, Myoisophagos sanguineus*). California to Washington State; also cosmopolitan, especially in northern hemisphere (type region, northeast Atlantic coast). Intertidal, especially above mid-intertidal, to high marsh tide pools. Among algae and other surface cover, on rocks exposed to surf, also under rocks on sand and mud; common. Includes several previously described North Atlantic species (Riser 1994, Proc. Biol. Soc. Wash. 107: 548–556; Riser 1998, Hydrobiol. 365: 149–156, taxonomy, morphology). Frequently occurs in clusters, and color resembles either *Lineus viridis* or *Lineus ruber* (plate 88I), but becomes much more slender when fully stretched and coils spirally rather than contracting linearly when disturbed. Generally reproduces by fragmentation (fissipary). Feeds on dead shrimp, minced clams, etc., in laboratory conditions and also on polychaetes, oligochaetes. For figure, see Coe (1940, fig. 13, as *Lineus vegetus*).

Zygeupolia rubens (Coe, 1895) (=*Valencinia rubens, Zygeupolia littoralis*). Monterey Bay to Mexico; also Atlantic coast (type region). Intertidal; soft sediments of bays and harbors.

Hoplonemertea

*Amphiporus angulatus (Müller, 1774). Alaska south to Point Conception, rare south of Puget Sound; also circumpolar, including Greenland (type region), western North Atlantic coast to Cape Cod and western North Pacific coast to Japan. In sandy and cobble situations, especially beneath stones; intertidal and subtidal. Dark or reddish brown or purplish dorsally, with paler margins and a conspicuous, whitish, angular patch on each side of head, continuous with color of ventral body surface. Body large, up to 20 cm, stout; when disturbed, can become greatly thickened anteriorly (Coe 1905). Gibson and Crandall (1989, Zool. Scripta 18: 453–470) declared *Amphiporus angulatus sensu* Müller (1774) as *nomen dubium*, while accepting *A. angulatus sensu* Coe (1901) from Alaska. On the basis of external appearance the latter is indistinguishable from *A. angulatus sensu* Verrill (1892) from New England and eastern Baffin Island (Norenburg pers. obs.), therefore probably also from Greenland, the type locale. For figures, see Coe (1901, fig. 10; plate 6, fig. 4; plate 7, fig. 2; plate 11, fig. 2; plate 13, fig. 3).

Amphiporus bimaculatus Coe, 1901 (=*Nipponnemertes bimaculatus, Collarenemertes bimaculatus*). Alaska to Mexico (type region). Intertidal; on and in rock crevices and pilings, among algae and other surface cover, also under rocks. Subtidal on kelp holdfasts. Specimens from the Sea of Japan called *A. bimaculatus* by Coe are not this species (Crandall pers. comm.). Not *Amphiporus* (Gibson and Crandall 1989, Zool. Scr. 18: 453–470). Often swims when placed in container of clean seawater.

Amphiporus cruentatus Verrill, 1879 (=*Amphiporus leptacanthus* Coe, 1995). Washington to San Diego. Intertidal and subtidal; among algae and other surface cover on rocks and pilings, common in kelp holdfasts.

Amphiporus flavescens Coe, 1905. Monterey Bay to Mexico. Intertidal and subtidal; under rocks and among algae and other surface cover on rocks and pilings. Quite variable in color intensity, number and arrangement of ocelli, and stylet armature.

Amphiporus formidabilis Griffin, 1898 (=*Amphiporus exilis*). Aleutian Islands, Alaska to Puget Sound (type region), south to Monterey Bay; also Japan (Corrêa 1964). Intertidal; among algae and other surface cover on rocks and pilings, under and in rock crevices, often exposed to surf. Often exposed on foggy days. Feeds on amphipods, isopods, and chelipeds of *Petrolisthes* sp. Difficult to distinguish externally from *A. imparispinosus.*

Amphiporus imparispinosus Griffin, 1898. (=*Amphiporus similis, A. leuciodus*). Alaska to Puget Sound (type region), south to Mexico; also Siberia. Upper intertidal to subtidal; under rocks and among algae/coralline algae on rocks and pilings, at exposed and quiet-water sites; common but patchy. Feeds on amphipods. Compare to *A formidabilis*, above. *A. similis* was designated a separate species by Coe (1905) based on specimens having two versus three accessory stylet sacs; designated as *A. imparispinosus* var. *similis* in Coe (1940).

*"Amphiporus" rubellus Coe, 1905. Southern California (type region), may occur farther north. Intertidal to (more commonly) subtidal; among surface cover on wharf pilings and low intertidal rocks. Not an *Amphiporus* (Gibson and Crandall 1989, Zool. Scr. 18: 453–470, taxonomy). Salmon yellowish to pale

* = Not in key.

orange or pale red dorsally, paler and usually grayish ventrally; large pigment cup ocelli in two groups as an irregular anterior row of six to 12 and a posterior cluster of six to eight on each side. For figures, see Coe (1905, plate 1, figs. 11, 12).

Amphiporus similis Coe, 1905. Central California (type region). On pilings in Monterey, among coralline algae at Pacific Grove. Coe (1940, 1943) and others have synonymized this form with *Amphiporus imparispinosus* Griffin, 1898 (Gibson and Crandall 1989, Zool. Scripta 18: 453–470, taxonomy) retained it as a separate *species inquirenda*. Externally, *A. similis* is reported as having two pairs of cephalic grooves, whereas *A. imparispinosus* has a single pair (Crandall). For figures, see Coe (1905, plate 16, figs. 93, 94; plate 22, figs. 152, 153).

Carcinonemertes epialti Coe, 1902. Pacific coast North America. Intertidal and subtidal, at bases of legs and abdomen of several crab species, including *Pugettia producta, Hemigrapsus oregonensis, H. nudus, Pachygrapsus crassipes, Cancer* spp. except *C. magister;* possibly other brachyuran crabs. Common, but patchy by host and region. Feeds on developing crab embryos (Roe 1984, Biol. Bull.167: 426–436; feeding, reproduction). May be a species complex (Wickham and Kuris 1988, Hydrobiologia 156: 23–30; taxonomy). Small size and great transparency make juveniles useful for classroom study. Spiny lobster, *Panulirus interruptus,* harbors *C. wickhami* in southern California (Shields and Kuris 1989, Fishery Bull. U.S. 88: 279–287; taxonomy, morphology).

Carcinonemertes errans Wickham, 1978. Washington (type region), also Alaska to Monterey Bay. Intertidal and subtidal, at bases of legs and abdomen of host, *Cancer magister;* common but patchy. Feeds on developing crab embryos. Migrates from old skeleton to new at time of host molting (Wickham et al. 1984, Biol. Bull. 167: 331–338; biology).

Emplectonema buergeri Coe, 1901 (=*Emplectonema violaceum* Griffin, 1898; not *E. violaceum* Bürger, 1896). Alaska (type region) to Monterey Bay. Intertidal and subtidal; under rocks, among mussels and other surface cover on rocks. Length to 1 m, width of 5 mm. Puget Sound forms with spiral-grooved stylets (Maslakova, Norenburg, pers. obs.), otherwise resemble description of Alaska forms with smooth stylets (Coe 1901, 1905); morphology of California forms (Coe 1940) not known. If differences verified, the forms are likely different species.

Emplectonema gracile (Johnston, 1837) (=*Nemertes gracilis*, plus large list of synonyms for forms not from the western U.S. coast). Alaska to Mexico, also northern Europe (type region) to Mediterranean. Intertidal and subtidal; among mussels and other surface cover on rocks and pilings, both exposed to surf and in muddy situations where barnacles occur; common but patchy; several individuals often together in tangled masses. Feeds on acorn barnacles. Proboscis basis much elongated and stylets curved.

Malacobdella grossa (Müller, 1776) (see Gibson 1995 for full synonymy). Puget Sound to California, also widespread along North Atlantic eastern (type region) and western coasts, and western Pacific coasts. Prior to Kozloff (1991, Can. J. Zool. 69: 1612–1618; taxonomy), all U.S. Pacific coast occurrences of *Malacobdella*, except for *M. minuta*, were designated *M. grossa.* Those from three hosts have been reassigned to new species (see below). The identity of *M. grossa* reported from *Tresus nuttallii* and *T. capax*, from Humboldt Bay north (Haderlie in Morris, Abbott and Haderlie, eds. 1980, *Intertidal Invertebrates of California*, pp. 84–90; general biology) needs to be investigated. Malacobdellids are plankton feeders.

Malacobdella macomae Kozloff, 1991. Oregon and central California. Commensal in clams *Macoma secta* and *M. nasuta.* For figure, see Kozloff (1991, Can. J. Zool. 69: 1612–1618, figs. 10–11).

Malacobdella minuta Coe, 1945. Single specimen found in *Yoldia cooperi* off Point Loma, California. Distribution of host suggests possible occurrence north as far as San Francisco. Distinguished from other species by small size at maturity (5–8 mm), intestine with seven to eight undulations, gonads in single row.

Malacobdella siliquae Kozloff, 1991. Washington and Oregon. Occurs in the razor clam *Siliqua patula.* For figure, see Kozloff (1991, Can. J. Zool. 69: 1612–1618, figs. 1–2).

Nemertopsis gracilis Coe, 1904 (=*Nemertes gracilis* var. *bullocki*). Washington to Mexico. Intertidal and subtidal; among algae and other surface cover on rocks and pilings exposed to surf, also in rocky cracks and under rocks in muddy situations, usually in low intertidal, although Roe found specimens associated with the high-intertidal "*Pantinonemertes" californiensis* at Elkhorn Slough; common. Feeds on the clam *Lasaea adansoni.* Original description for *N. gracilis* (plate 90F) differs from *N. gracilis* var. *bullocki* (plate 90G) only in color pattern. Specimens in China externally identical to those in California, except for minor color differences, designated a new species, *N. bullocki* (Sun 1998, Chin. J. Oceanol. Limnol. 16: 271–279, morphology, taxonomy).

Oerstedia dorsalis (Abildgaard, 1806) (=*Tetrastemma dorsalis, T. dorsale, Oerstedia dorsale*). Washington to Mexico, also both sides of Atlantic from northern Europe (type region) to Madeira and Nova Scotia to south of New England. Low intertidal and subtidal; among algae and other surface cover on rocks and pilings. This species originally described from Europe, where its epidermal coloring and pigmentation can range from a uniform brown to patterned speckling (Envall and Sundberg 1993, J. Zool, Lond. 230: 293–318). Similar to *Tetrastemma candidum,* but more cylindrical in cross-section, and head more bullet-shaped.

Ototyphlonemertes species are very likely to be encountered in the interstitial space of relatively well-sorted, intertidal or subtidal coarse sediments or shell-hash. Specimens can be collected by any of the "wash and decant" techniques (Norenburg 1988, in Higgins and Thiel, *Introduction to the Study of Meiofauna*, Smiths. Inst. Press, pp 287–292; biology); littoral forms often can be persuaded to migrate out of the sediment by simulating low tide—by placing sediment at upper end of a slanted white tray and then dripping seawater over the sediment (Corrêa 1949, Commun. zool. Mus. Hist. nat. Montev. 3: 1–9). Multiple allopatric and sympatric putative species have been found in regions with appropriate habitats that have been well sampled (Norenburg 1988, Hydrobiol. 156: 87–92; taxonomy). The Pacific coast has not been well sampled; known species are *Ototyphlonemertes spiralis* Coe, 1940, intertidal from coarse sand near San Diego; *O. americana* Gerner, 1969 (Helgoländer wiss. Meeresunters. 19: 68–110; taxonomy) from San Juan Island, Puget Sound. Forms that key out to either of the known species have a reasonable probability of not being those species (see Envall and Norenburg 2001, Hydrobiol. 456: 145–163; taxonomy). Two species, one with a smooth stylet and one with a spirally fluted stylet, occur at Half Moon Bay and at San Gregorio Beach (Norenburg pers. obs.); only the second appears to be a described species (see below).

Ototyphlonemertes americana Gerner, 1969. Puget Sound. Several individuals resembling *O. americana* were found mid- to low-tide range at Half Moon Bay and San Gregorio Beach just south of San Francisco (new record).

"*Pantinonemertes" californiensis* Gibson, Moore, and Crandall, 1982. Washington to Huntington Beach, southern California (Crandall). High intertidal and supralittoral; under logs and stones

* = Not in key.

that have been in place a long time, often covered with *Zostera* wrack. Length to 450 mm (Gibson, Moore, and Crandall 1982, J. Zool. London 196: 463–474; distribution, size, taxonomy, morphology). Not *Pantinonemertes* (Norenburg, Crandall pers. obs.; S. Maslakova, 2005, Ph.D. dissertation, George Washington University, Washington, D.C.). Feeds primarily on amphipod *Traskorchestia traskiana*. Most spawn in June–July; spawning can be induced by submerging ripe individuals in seawater for a few minutes, or by exposing submerged individuals to drying conditions (Roe 1993, Hydrobiologia 266: 29–44; feeding, reproduction).

Paranemertes californica Coe, 1904. Monterey Bay to Mexico. Intertidal; burrower in soft sediments in bays and harbors.

Paranemertes peregrina Coe, 1901. Alaska (type region) to Mexico. Intertidal; under rocks and among algae and other surface cover on rocks on wave-impacted shores. Also buried in soft sediments, or crawling on soft sediments when sediments are exposed at low tide; common. Feeds on polychaetes, especially nereids, at low tide (Roe 1976, Biol. Bull. 150: 80–106; feeding, ecology). Coe (1905) attributed two morphotypes to this species. One from California north to Juneau, Alaska, has stylets with spiral fluting, resulting in a braided appearance (plate 89B); with two accessory stylet sacs; dorsal color is purplish brown, tending more toward brown than purple. The second morphotype, described by Coe (1901) from Alaska, has stylets without spiral fluting (plate 89C), often has four accessory stylet sacs, and the body color is more purple than brown.

Paranemertes sanjuanensis Stricker, 1982. False Bay, San Juan Island. Intertidal; crawls on soft sediments at low tide. Beige colored. Stylets with helical fluting, giving stylets a braided appearance similar to that of *P. peregrina* (plate 89B). Roe has seen a few specimens at Bodega Harbor tentatively identified as *P. sanjuanensis* based on similar appearance and spirally fluted stylets. For figures, see Stricker (1982, Zool. Scripta 11: 107–115; taxonomy, morphology).

Poseidonemertes collaris Roe and Wickham, 1984. Bodega Harbor. Intertidal; rapid burrowers in surface sediments, especially sandy muds with moderate amounts of green algae present. When contracting, does not coil or twist, often retracts head to level of anterior grooves, forming a collar. (Roe and Wickham 1984, Proc. Biol. Soc. Wash. 97: 60–70; taxonomy, morphology).

Tetrastemma albidum Coe, 1905 (=*Prosorhochmus albidus*). Monterey Bay to Mexico. Intertidal; among algae and other surface cover on rocks and pilings exposed to surf. Not a *Prosorhochmus* (Norenburg, Maslakova pers. obs.). May be livebearing (Coe 1940).

Tetrastemma candidum (Müller, 1774) (=*Fasciola candida*). Alaska to Mexico, also northeast Atlantic coasts (type region) to Mediterranean, South Africa, and Labrador to south of New England. Intertidal and subtidal; among algae and other surface cover on rocks and piers, and buried in soft sediments. Nondescript; may be a species complex. Compare to *Oerstedia dorsalis*. Will feed on *Artemia* nauplii in laboratory.

Tetrastemma nigrifrons Coe, 1904. Central California (type region), Washington to Costa Rica; also Sea of Japan. Intertidal and subtidal; among algae and other surface cover on rocks and piers; common. Highly variable in both pigmentation and color pattern, four common patterns in geographic range of this manual.

Tetrastemma quadrilineatum Coe, 1904. Monterey Bay to Mexico. Intertidal; among algae and other surface cover on rocks and pilings, also in tide pools.

Tetrastemma reticulatum Coe, 1904. Common at San Diego and San Pedro; may occur north to Monterey Bay (Crandall). Specimens small, only 8–15 mm long and very slender. White with rectangular and longitudinal chocolate brown to light or

reddish brown blocks covering much of dorsum. On pilings and among rocks in lowest intertidal. For figures, see Coe (1904, plate 14, figs. 7, 8; plate 20, figs. 7–9; 1905, plate 2, figs. 16, 17).

Tetrastemma signifer Coe, 1904. Monterey Bay to San Diego. Intertidal and subtidal; among algae and other surface cover on rocks and pilings, and in subtidal kelp holdfasts.

Zygonemertes albida Coe, 1901. British Columbia to Puget Sound, also Monterey Bay (new record). Intertidal and subtidal; among algae and other surface cover on rocks and pilings. Resembles pale, juvenile individuals of *Z. virescens* (see below). Positive identification is only possible with a sexually mature specimen. Sexually mature *Z. albida* remain whitish. Basis without posterior notching.

Zygonemertes thalassina Coe, 1901. Alaska (type region), south to Point Reyes (Crandall). Among algae and surface cover on rocks and among shell hash, low-water and deeper. Uniform pale green, the color of pea soup; basis concave or serrated posteriorly, about three times as long as wide, stubby central stylet, four to five accessory stylets per pouch. For figures, see Coe (1901, fig. 5, plate 2, fig. 5; plate 7, fig. 1; plate 13, fig. 2).

Zygonemertes virescens (Verrill, 1879) (=*Amphiporus virescens*). British Columbia to Mexico; also Atlantic (type region) and Gulf Coasts. Intertidal and subtidal; under rocks, among algae and other surface cover on rocks and pilings, and on soft sediments at low tide. Variable in color (juveniles whitish to pale yellow or green, adults pale to dark hues of green), in number of ocelli (increasing with body size to 100 or more) and in relative sizes of basis and central stylet (basis usually with posterior notching or ruffling, length approximately four times diameter and one and a half to two times length of stylet; two to three accessory stylets per pouch). See above for comparison with *Z. albida*.

References

Coe, W. R. 1901. Papers from the Harriman Alaska Expedition, 20 The nemerteans. Proc. Wash. Acad. Sciences 3: 1–110, pl 1–13.

Coe, W. R. 1904. The nemerteans. Harriman Alaska Expedition 11: 1–220, pl 1–22.

Coe, W. R. 1905. Nemerteans of the west and northwest coasts of America. Bull. Museum Comparative Zool, Harvard 47: 1–319, pl 1–25.

Coe, W. R. 1940. Revision of the nemertean fauna of the Pacific coasts of North, Central and northern South America. Allan Hancock Pacific Expeditions 2: 247–323.

Coe, W. R. 1943. Biology of the nemerteans of the Atlantic Coast of North America. Connecticut Acad. Arts and Sciences 35: 129–328.

Corrêa, D. D. 1964. Nemerteans from California and Oregon. Proc. Cal. Acad. Sci. 31: 515–558.

Crandall, F. B. and J. L. Norenburg. 2001. Checklist of the Nemertean Fauna of the United States. 2nd ed. http://nemertes.si.edu/PDFs/epub2918.pdf. pp. i–ii, 1–36.

Gibson, R. 1995. Nemertean genera and species of the world: an annotated checklist of original names and description citations, synonyms, current taxonomic status, habitats and recorded zoogeographic distribution. J. Nat. Hist. 29: 271–562.

McDermott, J. J. 2001. Status of the Nemertea as prey in marine ecosystems. Hydrobiologia 456: 7–20.

Norenburg, J. L. and S. A. Stricker. 2001. Phylum Nemertea. pp. 163–177 in C. M. Young, M. Rice, and M. A. Sewell, Atlas of marine invertebrate larvae, Academic Press, 656 pp.

Stricker, S. A. 1996. Phylum Nemertea. pp. 94–101 in E. N. Kozloff, Marine invertebrates of the Pacific Northwest, Seattle: Univ. Washington Press, 539 pp.

Thollesson, M. and J. L. Norenburg. 2003. Ribbon worm relationships—A phylogeny of the phylum Nemertea. Proc. Roy. Soc. Lond., B 270: 407–415.

* = Not in key.

Nematoda

(Plates 91–110)

W. DUANE HOPE

Nematodes are distributed universally in marine and estuarine sediments, as well as on epibiotic substrates. Within the past few decades, benthic ecologists have begun to document the relative abundance and diversity of nematodes and their adaptations to physical and chemical factors (e.g., salinity, redox potential, pore-water content, sediment granulometry and organic content, temperature, and various pollutants); assess their feeding habits and food sources; and measure their respiratory rates, biomass, reproductive potential, and so forth. It has been shown that nematodes almost always have the highest population density of any benthic metazoan and have the greatest number of species/unit area, often by an order of magnitude. Their densities range from as much as 0.1 million/m² on rocky shores to 10 million–23 million/m² in muddy estuaries (Platt and Warwick 1980).

Whereas between 4,000 and 5,000 species of marine nematodes have been named, estimates of the total number of marine nematode species range as high as 100,000,000 species (Pearce 1995). This is an estimate of free-living marine nematodes only. Many additional species parasitize various invertebrates (Hope 1977; Hope and Tchesunov 1999; Petter 1980, 1981a and b, 1982a and b, 1983a and b, 1987; Petter and Gourbault 1985; Tchesunov 1988a and b, 1995; Tchesunov and Hope 1997; Tchesunov and Spiridinov 1985, 1993) and vertebrates (Moser, Love, and Sakanari 1983; Dailey 2001).

For the Pacific coast of North America, from Point Barrow, Alaska, to the Mexican border, 243 species have been reported—of which 166 (68%) were reported as new. By contrast, for the entire Atlantic coast of North America, including the coasts of Canada and Gulf of Mexico, 671 species have been reported from near- and off-shore habitats—of which 406 (61%) were species new to science. The difference is attributable not to the fact that the diversity of marine nematodes is less on the Pacific coast but to the greater amount of research that has been done on the Atlantic coast nematode fauna. Of the 243 species known from the Pacific coast, 47 are known from the intertidal zone between the Washington/Oregon border and Point Conception.

Most publications concerned with Pacific coast marine nematodes are faunal or concerned at least in part with taxonomy (Allen and Noffsinger 1978; Allgén 1947; Cobb 1920; Chitwood 1960a and b; Hope 1967a and b; Hope and Yorkoff 1985; Jensen 1989; Jones 1964; Murphy 1962, 1963a, b, and c, 1964a, b, and c, and 1965a, b, and c; Murphy and Jensen 1961; Nelson et al. 1971, 1972; Sharma et al. 1979; Sharma and Vincx 1982; Steiner and Albin 1933; Timm 1970).

Particularly noteworthy is the work of Wieser (1959a) on the marine nematodes of the Puget Sound, Washington area, in which 114 species are described. Of those 76 (67%) were new. Many of these species will probably be discovered on the coasts of Oregon and California. Unfortunately, most of the voucher and type specimens for that investigation no longer exist. Another publication in which a large number of local species has been described is that by Allgén (1947), in which 81 species were reported from the coastal areas of La Jolla and San Diego. Of these 35 (43%) were regarded as new. However, most of Allgén's descriptions of new species and redescriptions of known species are too inadequate to be useful, and his identifications are questionable.

The remaining West Coast publications on marine nematodes are concerned with morphology (Burr and Burr 1975; Burr and Webster 1971; Chitwood and Chitwood 1974; Hope 1969, 1974, and 1982; Hope and Gardiner 1982; Maggenti 1964; Siddiqui and Viglierchio 1970a and b, 1971, and 1977; Wright and Hope 1968; Wright, et al., 1973) and ecology (Chitwood and Murphy 1964; Sharma and Webster 1983; Trotter and Webster 1983; Wieser 1959b; Wing and Clendenning 1971).

Methods of Collection

Most marine nematodes live in the top several centimeters of sediment, so any device that removes at least the surface sediment may be used to collect them. A coring device 2.5 cm in diameter is adequate for quantitative studies. Whether sampling for qualitative or quantitative investigations, it is necessary to take replicate samples because of the patchy distribution of marine nematodes. Sampling methods for meiofauna in general are covered in detail elsewhere (Higgins and Thiel 1988).

A common procedure for extracting nematodes from sediment is to place the sample in a container of seawater with a volume of at least 50 times that of the sediment (plate 91A). The sample should be stirred sufficiently to put the entire

PLATE 91 A, The sediment sample is placed in a container and B, completely suspended by stirring in a volume of sea water at least 50 times greater than that of the sediment. C, After heavier particles settle for about 30 seconds, D, the supernatant is poured onto a sieve with openings not greater than 50 μm. Tilt the screen while pouring so the nematodes and residual detritus are restricted to a small area of the screen. E, The screen is inverted, and the extracted sample is rinsed with seawater into a beaker or other suitable container. The extracted sample is then poured into a counting dish where specimens may be examined and sorted from residues of sediment with the aid of a dissecting microscope. An Irwin loop, small cactus thorn, eyelash, or other small probe is used to sort or transfer specimens to a fixative. Transfer should be done quickly to prevent desiccation (original by Molly Kelly Ryan).

sample in suspension (plate 91B), and after allowing the suspension to stand for 20–30 seconds (plate 91C), the supernatant is poured through a sieve with pores about 50 μm in diameter (plate 91D). The sieve is then rinsed from the back with seawater into a suitable container (plate 91E), be it a vial in which the nematodes will be fixed or a counting dish for live observation with a dissecting microscope. It is desirable to use filtered seawater to avoid contamination of the sample from laboratory seawater systems. This method works espe-cially well when the specific gravity of the particles of sediment is greater than that of the nematodes. However, nematodes have adhesive caudal glands by which they may attach themselves to particles of sediment that may carry the nematodes out of suspension prior to sieving. Therefore, it may be desirable to extract the same sample two or three times to maximize recovery. Subjecting the nematodes to osmotic stress by diluting the seawater suspension with freshwater also minimizes this problem. However, if this procedure is followed, it

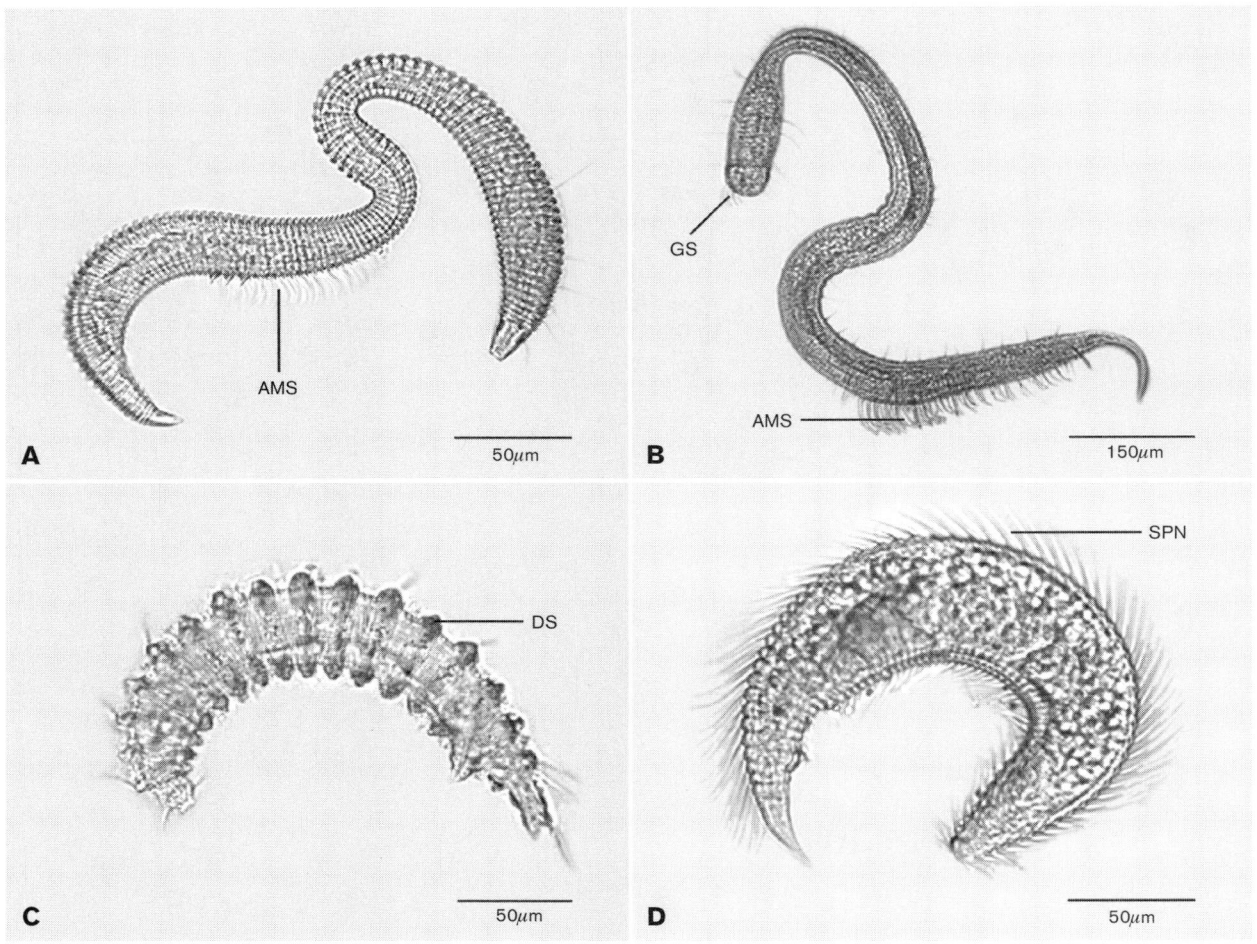

PLATE 92 A, Male of *Epsilonema* sp.; B, male of *Draconema* sp.; C, Male of *Desmoscolex* sp.; D, male of *Greeffiella* sp.

is important to flush the sieve with seawater as quickly as possible so that the nematodes may recover. Even if the nematodes become active once returned to seawater, this procedure may compromise the quality of the specimens for taxonomic study.

There are other techniques for extracting nematodes from sediment, some of which work better when extracting nematodes from sediment containing large amounts of silt or organic detritus. These include the Uhlig's seawater ice method (Uhlig, et al. 1973) or centrifuging samples suspended in silica sols (De Jonge and Bouwman 1977). Sediment trapped within the interstices of the thallus of the seaweeds *Egregia* sp. and *Laminaria* sp. often harbors large nematodes of the family Leptosomatidae. Such nematodes may be collected with fine forceps as the holdfast is broken apart.

Unless the processed sample is very clean sand, it is likely that the extraction will contain considerable detritus from which the nematodes must be sorted by hand with the aid of a dissecting microscope. After stirring the extracted sample, an aliquot is poured or pipetted into a counting dish, and the nematodes are transferred from there to a vial containing a suitable fixative. The transfer should be done quickly to prevent the nematodes from becoming desiccated. Irwin loops (a thin wire folded in half and twisted upon itself leaving a small loop at the end) are satisfactory for this purpose. It will be easier to recognize nematodes in residues of detritus if the sample has been stained with Rose Bengal. However, unstained specimens are preferable for taxonomic purposes. Most marine nematodes are transparent and, for taxonomic purposes, may be studied in detail with a compound microscope without cutting sections or enhancing contrast with stains.

Preparation for Light Microscopy

Taxonomic and/or phylogenetic investigations of marine nematodes require preparations for high-resolution light microscopy. Given that nematodes are transparent, it is possible to study their internal and external morphology from whole mount preparations. The fixative most often used for such preparations is about 7.5% formalin in distilled water for at least 24 hours. Riemann (1988) recommends the use of a mixture of 10% formalin and 0.5% propionic acid in distilled water for the enhanced preservation of nuclei. Although formalin remains the preferred fixative for marine nematodes, it and others have been compared by Timm and Hackney (1969) and Maggenti and Viglierchio (1965).

Anhydrous glycerin is the mounting medium used in preparing permanent mounts. Nematodes are processed into anhydrous glycerin by first transferring them from the fixative into a 2% solution of glycerin in distilled water for a brief period to

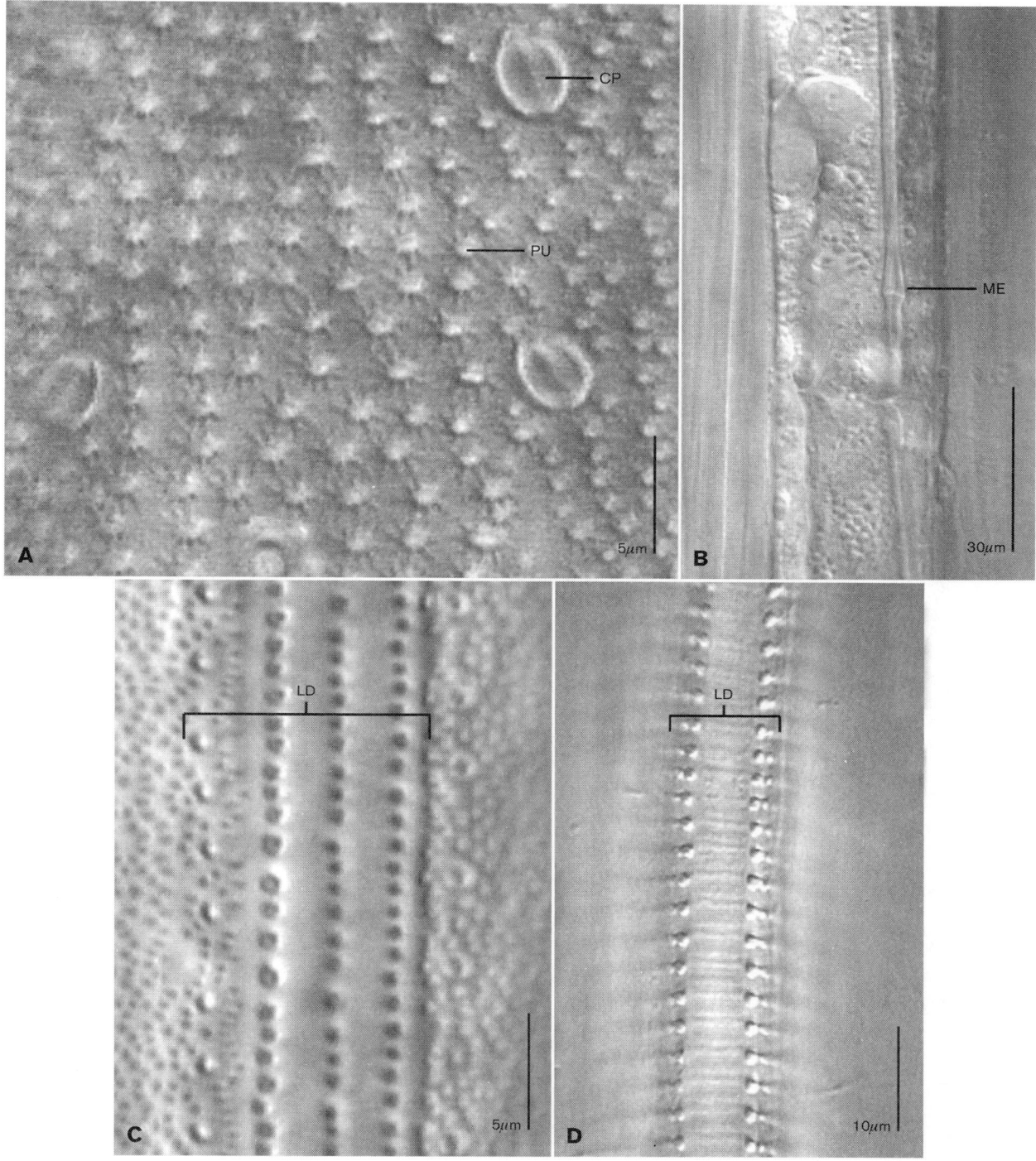

PLATE 93 A, Cuticular punctations with spars and cuticular pores with diagonal aperture in lateral field of *Longicyatholaimus* sp.; B, Metaneme in lateral chord of *Cylicolaimus* sp.; C, lateral differentiation in cuticle of *Mesonchium* sp.; D, lateral differentiation in cuticle of *Spilophorella* sp.

rinse away the fixative, then transferring them to a small container, such as a Bureau of Plant Industry Dish, filled with about 1 ml of the 2% glycerin solution. The dish is partially covered with a coverslip so that evaporation of water from the solution is extended over a period of one week or longer. The specimens will collapse if evaporation is too rapid. After the water has evaporated, the dish and contents are placed in a desiccator for at least one week. The glycerin to be used as the mounting medium should be kept in a desiccator, as well. Specimens may deteriorate if mounted in glycerin containing traces of water.

A small drop of anhydrous glycerin is placed on a slide to which are added the specimen(s) and at least four coverslip supports placed at positions equivalent to about two, four, eight, and 10 on the face of a clock approximately midway to the edge of where the circular coverslip will lie. The volume of

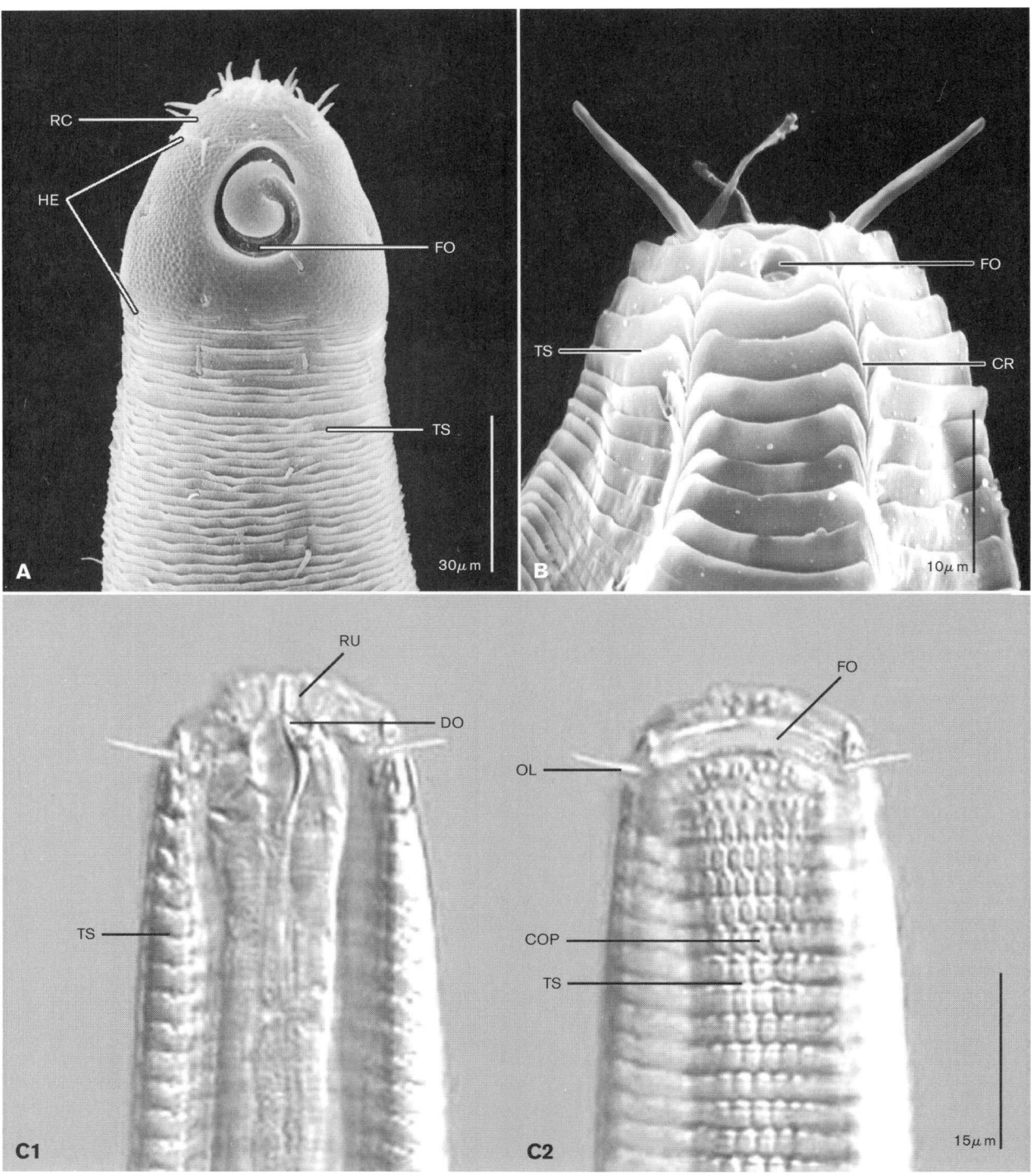

PLATE 94 A, SEM of *Desmodora pilosa* Ditlevsen, 1926, head region in lateral view; B, SEM of *Monoposthia* sp. head region in lateral view; C1, head region of *Euchromadora* sp. in optical section; C2, cuticle of head region and transversely oval fovea of *Euchromadora* sp.

glycerin should be small so that after the coverslip has settled on the supports the glycerin flows to near, but short of, the edge of the coverslip. If the glycerin flows to the edge or beyond, it will not be possible to seal the coverslip later. Specimens will become distorted or destroyed without adequate supports, which is to be avoided not only for obvious reasons, because body measurements that may be necessary later will be unreliable when making identifications. Glass beads, man-

ufactured for use in reflective paint or rods cut from spun glass, are often used as supports. The supports should equal the maximum diameter of the nematode to be mounted. If supports are too thick then specimens may drift to the edge of the mount, especially if the slide happens to be stored in a vertical position. In addition, if the mount is too thick, it may be impossible to focus on specimens with lenses having a small working distance.

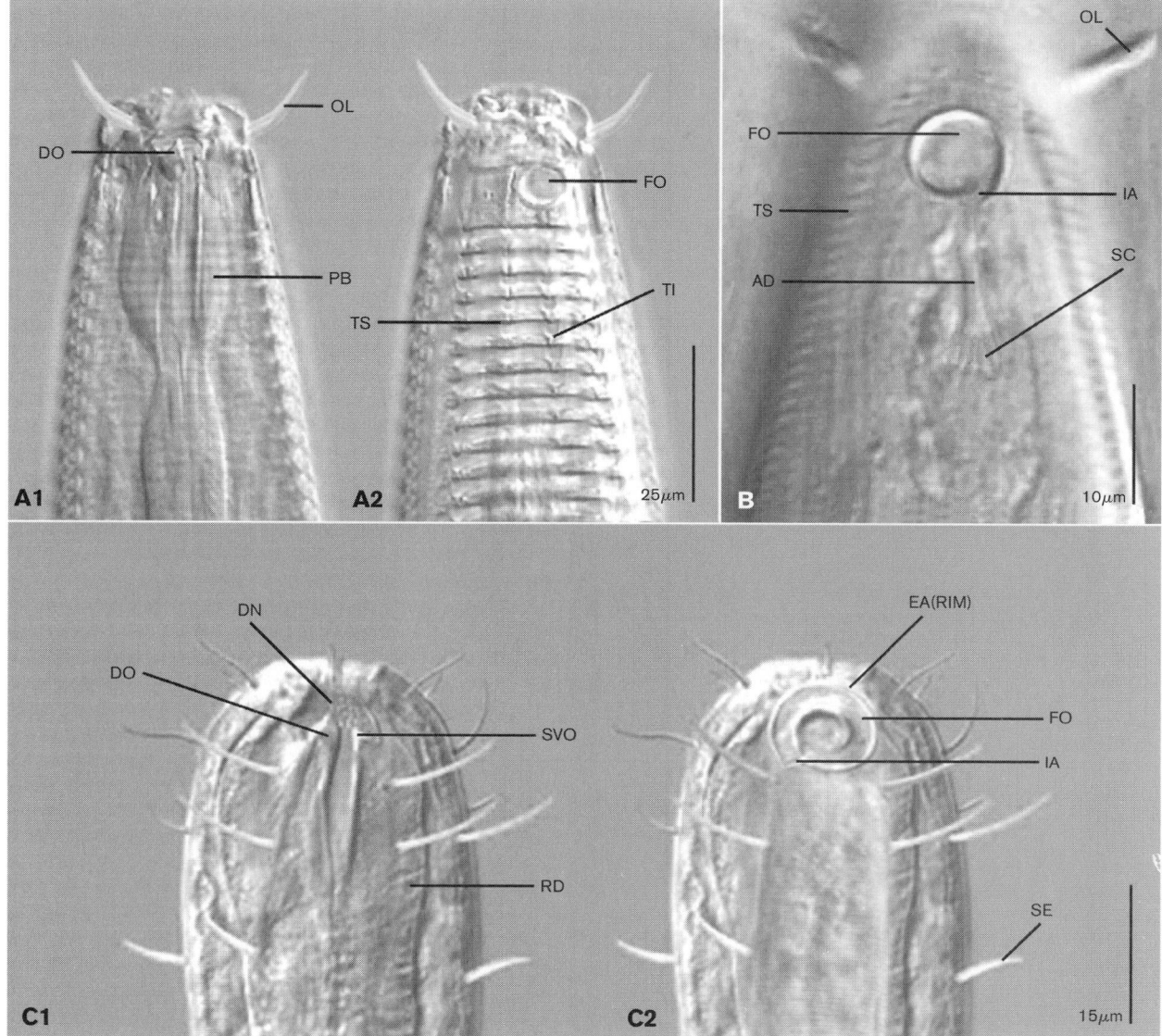

PLATE 95 A1, Head and anterior esophagus of *Nudora* sp. in optical longitudinal section; A2, head and anterior cervical region of *Nudora* sp.; B, amphid of *Plectolaimus* sp. in optical, longitudinal section; C1, head and anterior esophagus of *Sigmophoranema* sp. in optical longitudinal section; C2, head and anterior cervical region of *Sigmophoranema* sp.

The coverslip should be lowered gently onto the drop of glycerin to prevent air from being trapped in the glycerin. Once the coverslip has settled on its supports, the slide must be sealed. Glyceel is the best product for this purpose, and although it is not commercially available at present, a method by which it may be prepared has been published by Bates (1997). Fingernail polish or paraffin may also be used as a sealer.

Although mounts may be prepared using conventional glass slides, it is preferred that nematodes be mounted between two coverslips that in turn are fitted on 1 × 3 inch plastic frames commercially available from Kanto Rika Co. Ltd. The advantage to this system is that a specimen may be examined from both sides. In cases where a publication results from the identification and/or description of specimens, it is imperative that the specimens be deposited in a museum, such as the National Museum of Natural History, or other institutions, where they may be available for future verification, regardless of whether the collections were made as part of faunal, ecological or taxonomic investigations.

Morphology

Nematodes are typically spindle-shaped and vary in length from as much as 4 cm to as little as 0.5 mm and may be extremely thin or robust. That part of the body in which the esophagus (also termed pharynx) occurs is the cervical region, and that from the anal (female; plate 108A; AA) or cloacal (male; plates 97C, 98; CA) aperture posteriorly is the tail. The value "a" in taxonomic descriptions is the body length divided by the body width. Likewise, "b" and "c" are the length of the

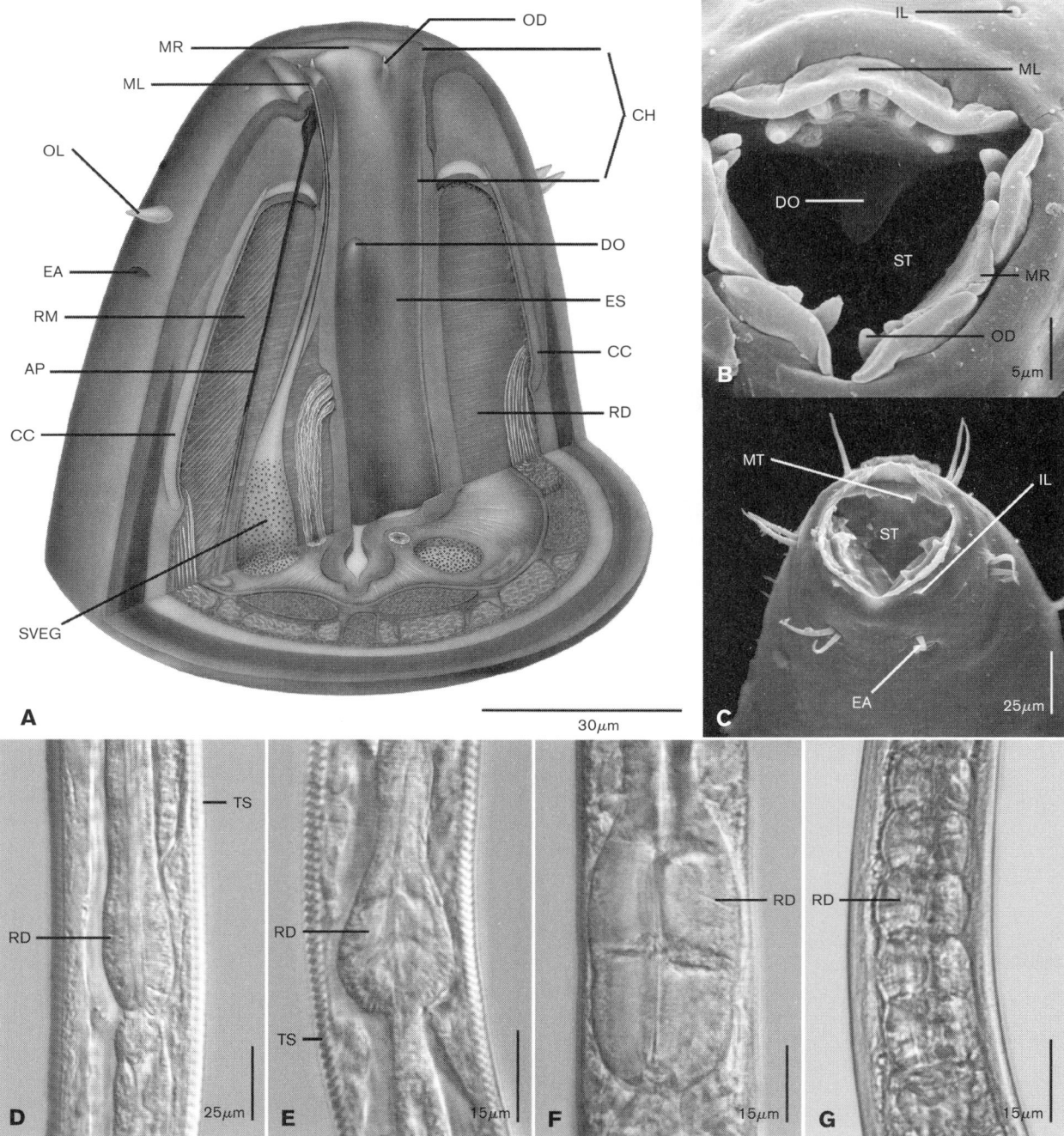

PLATE 96 A, Head region of *Deontostoma californicum* Steiner and Albin, 1933, in cut-away view (from Hope 1982); B, SEM of oral surface of *Deontostoma demani* (Mawson 1956) Hope 1967; C, SEM of oral surface of *Enoplus* sp.; D, posterior end of the esophagus of *Leptolaimus* sp.; E, posterior end of the esophagus of *Paradesmodora* sp.; F, posterior end of the esophagus of *Bolbolaimus* sp.; G, posterior end of the esophagus of *Polygastrophora* sp. (A–G in optical longitudinal section).

body divided by the length of the esophagus and length of tail respectively, and V is the distance from the head to the vulva expressed as a percent of the total body length, including the tail.

Because all marine nematodes are transparent to one degree or another, both external and internal features of their morphology are employed in taxonomic descriptions and assessments of phylogeny. Most investigations of their morphology are based on whole-mount preparations for light microscopy, but studies with both scanning and transmission electron microscopy have greatly advanced our understanding of their structure. Literature concerned with the morphology of nematodes is extensive and cannot be covered in detail here. The following description of morphology is intended to provide only a basic knowledge of structure and terminology for taxonomic purposes. Those interested in a detailed treatment of morphology are referred to Bird and Bird (1991), Chitwood and Chitwood (1974), Maggenti (1981), and Malakhov (1994).

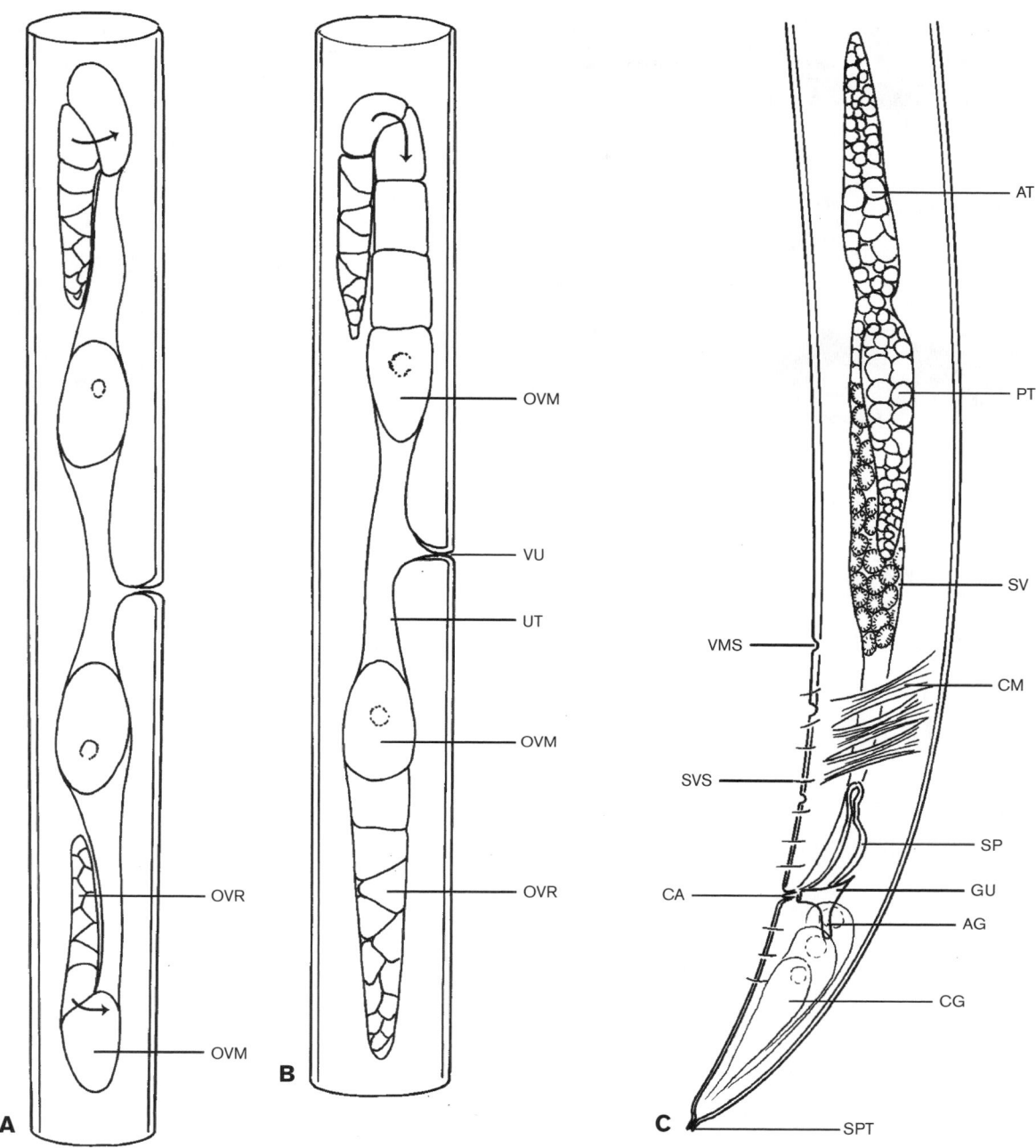

PLATE 97 A, Reproductive system of didelphic female with antidromously reflexed gonoducts; B, reproductive system of didelphic female with anterior (upper) gonoduct homodromously reflexed and posterior gonoduct outstretched; C, male reproductive system with two (diorchic) opposed testes.

Cuticle

The body is covered by a transparent, flexible cuticle, which with the exception of at least a few sensory receptors that all nematodes possess may be devoid of any internal or surface features. Conversely, it may bear such external features as transverse (plates 94A–94C2, 95A2–95B, 96D–96E, 104C–104D, 104G, 105B, 105I; TS) or longitudinal (plate 107E; LS) striae, or intracuticular punctations (plates 93A, 93C, 93D, 104F, 104I, 108D; PU), pores (plates 93A, 105B–105E, 105G; CP),

complex patterns resembling a basket weave (plates 94C2, 104G; COP), or combinations of such features. A complex pattern is homogeneous if it is the same throughout the body length and heterogeneous if the pattern varies from one area of the body to another. These patterns may be interrupted or modified along the lateral region of the body, in which case the lateral field is differentiated. Such modifications may include wider spacing or increased size of the punctations (plates 93C, 105E; LD), arrangement of punctations into longitudinal files (plate 93C; LD), or ladderlike cuticular markings (plate 93D; LD).

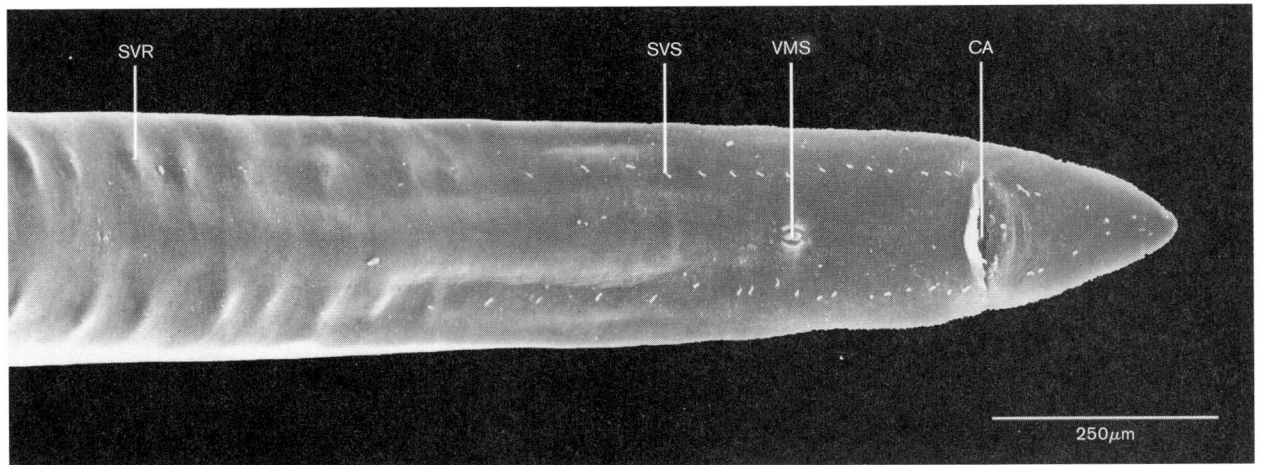

PLATE 98 SEM of the ventral surface of a male of *Deontostoma californicum*.

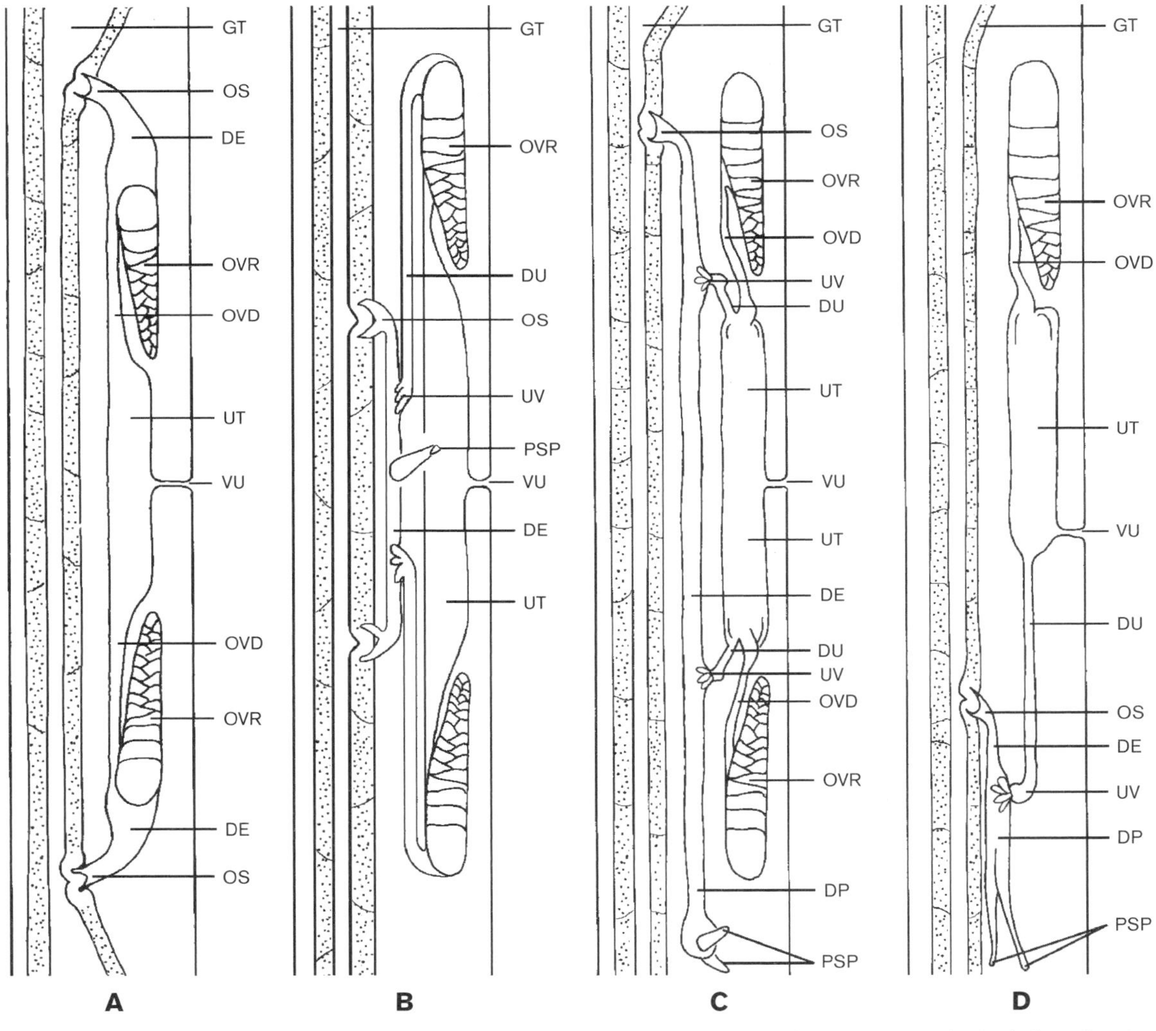

A   B   C   D

PLATE 99 A, Demanian system in a didelphic female with normal insemination by way of the vulva (*Viscosia*; Oncholaimellinae); B, demanian system in a didelphic female with the copulation pore in the vicinity of the vulva (*Meyersia*; Adoncholaiminae); C, complex demanian system in a didelphic female with copulation pores in the posterior body region (*Adoncholaimus*; Adoncholaiminae); D, complex demanian system in monodelphic females with copulation pores in the posterior body region (*Metoncholaimus*, *Oncholaimus*, and *Oncholaimium*; Oncholaiminae) (after Hope 1974).

In cases where striae are deep, the intervals between them are termed annules. In some cases, such as in many members of Desmodoridae and Monoposthiidae, an annule may have an anterior or posterior margin partially overlapping an adjacent annule (plates 94A–94C1, 106F, 107A, 110F; OA), or the annules may be covered by desmens, which are secretions with particulate inclusions and are unique to members of Desmoscolecina (plates 92C, 107B, 107C, 107G–108A; DS). Other members of this suborder have scalelike adornments (plate 107F; SLA) or numerous spines (plates 92D, 107D; SPN). The caudal endring in members of Desmoscolecidae may have a pair of lateral, small circular spots termed phasmata (plates 107H, 108A; PH), the function of which is unknown.

Annules may also have anteriorly or posteriorly directed tines (plates 95A2, 107A; TI) or crests (plates 94B, 110F; CR), each of which may or may not be in register with those of the adjacent annule so as to collectively form longitudinal ridges. When ridges are present, there are usually four or more such ridges with an equal number in both dorsal and ventral sectors of the body. Alae are winglike extensions of the cuticle that, if present, occur along the lateral field on each side of the body. Bursae (plate 100E; CB) are similar to alae but limited to the sublateral or subventral region of the male cloaca where they may aid in copulation and may bear sensory papillae.

The head region of certain members of Enoplida bear a specialized internal layer of cuticle, the cephalic capsule (plates 96A, 100A, 100B, 100D, 100G–I, 101C–F, 101I–102C; CC). The presence of a cephalic capsule, shape of its posterior margin, and presence or absence of locules (plate 100A, 101F, 102A, 102B; LO), and/or fenestrations (plate 102B; FE) are important characters. The cephalic capsule among members of the genera Pseudocella, Thoracostoma, and Triceratonema (Leptosomatidae) has a thickened, internal, ventral ridge termed a tropis (plates 101I–102B; TR). The cephalic capsule is not to be confused with a cephalic helmet, which occurs among members of Desmodoridae (plates 94A, 105I, 106D–106F; HE) and Ceramonematidae (plate 110E–110F; HE). The cuticle of the helmet is often thickened without an internal specialization but devoid of prominent transverse striae that occur throughout the remainder of the body.

Digestive System

The digestive system consists of a stoma (buccal capsule), esophagus (pharynx), midgut, rectum, and anal/cloacal apertures. The buccal capsule may be closed at rest and not differ noticeably from the walls of the esophagus (plates 101C, 102D, 104E, 106D–F, 107B–108A, 109I, 110A, 110E), or it may be fixed open and goblet-shaped (plates 100C, 100E, 100F, 102C, 102F–104A, 104C), or with innumerable other variations in size and shape. The buccal orifice in at least some members of Leptosomatidae is surrounded by dorsal and subventral microlabia (ML) and mandibular ridges (MR), which may or may not be provided with odontia (OD) (plates 96A, 96B). In such cases the buccal orifice is dilated by retrodilator muscles (RM) inserted upon each of the three apodemes (AP) that are in turn connected to a corresponding microlabium (plate 96A). Nematodes with a more spacious stoma often possess thin, foliate lips that may arch over and close the buccal orifice (plates 103H, 103I, 104C, 109B; FOL).

The cheilostome is the anterior part of the buccal capsule, whose depth is the same as the thickness of the cuticle covering the anterior surface of the head (plate 96A; CH), and the remaining posterior part, which may or may not be surrounded by esophageal tissue, is the esophagostome (plate 96A; ES). The cheilostome of all members of the suborder Chromadorina have four, anteriorly directed folds or ribs, termed rugae (plate 94C1, 104F–104I, 105C, 105D; RU), on each of the three buccal walls. The buccal capsule may be unarmed or armed with teeth or mandibles. Teeth situated in the cheilostome are termed odontia (plates 96A, 96B, 108B, 108D; OD), whereas those in the esophagostome are designated onchia (plates 94C1, 95A1, 96A, 96B, 100D, 101E, 102F–102I, 103B–103D, 103I, 104F, 104H–104I, 105C, 105I, 106B, 106G–106I, 107A; ON and DO), although either may be referred to as teeth. Small rasplike teeth, or denticles (plates 102C, 103D–103G, 105D, 106B, 107A, 109F; DN), may be present. Opposable mandibles bearing teeth may also be present, especially among certain members of the order Enoplida (plates 100A, 100C, 100D; MN) and Selachinematidae (plate 105G; MT). Esophageal tissue surrounding the esophagostome may be enlarged to form a pharyngeal bulb (plates 101A, 104H, 105D, 105F, 106B, 106I; PB). Especially in the case of a prominent dorsal tooth, the tooth may be bear a cuticular apodeme to which obliquely or longitudinally oriented muscles of the esophagus are attached (plate 104H, 104I; AP). A stylet (STY) exists in the buccal apparatus of members of Thoracostomopsinae (Thoracostomopsidae: Enoplida) (plate 100B) and Siphonolaimidae (Monhysterida) (plate 109G).

The overall shape of the esophagus is often cylindrical or uniformly widened toward its posterior end (claviform) (plate 96D), or more abruptly enlarged to form a single pyriform to round (plate 96E) or oblong (plate 96F) bulb, or there may be a series of posterior bulbs (plate 96G). The posterior end of the esophagus is connected to the midgut by way of an esophago-intestinal valve (cardium), which may be flat or conical and project into the lumen of the esophagus, or columnar and situated between the posterior end of the esophagus and anterior end of the gut.

Glands

There are several types of cells in marine nematodes that either do, or are assumed to have, a glandular function. Cervical glands (ventral gland; renette) occur widely among marine nematode taxa and is believed to be secretory-excretory. They usually consist of a single cell, but there may be two or more, all of which lie either along the length of the esophagus, usually on the ventral side, or may lie adjacent to the anterior region of the intestine (plates 101A, 102G–103B, 103D, 104F, 104H, 106G, 108B, 108F, 108I, 109I; CRD). The external pore of the cervical gland opens ventrally, usually somewhere along the length of the esophagus (plates 100G, 102G–103B, 104F, 104H, 106H, 108B, 108F, 108I, 109E, 109I; CRP), but may open on the lip region (plate 101A; CRP).

Cells presumed to have a glandular function and/or glandular-sensory function occur in the hypodermal chord and open to the exterior by way of small surface pores or pores through setae and papillae (plate 106D; HG). Some members of Chromadoroidea have in the hypodermal chord glandlike cells, the neck of which passes through an intracuticular ring and opens to the exterior through a slit in a narrow ridge that traverses a shallow depression on the surface of the cuticle (plate 93A; CP) (Wright and Hope 1968). These ring pores (pore complex) are of taxonomic importance, but a thorough investigation of their occurrence among taxa and an assessment of phylogenetic significance is needed.

Glands are believed to be associated with elongate, subventral ambulatory setae (AMS) found exclusively among members of Draconematidae (plates 92B, 106E). Draconematids also have a cluster of glandular setae on the dorsal surface of the anterior end of the head (plates 92B, 106E; GS).

Three caudal glands occur in most marine nematodes, although only two occur in members of Rhabdodemaniidae and in a few species of Monhysteridae (Riemann 1970) and Xyalidae (Lorenzen 1978). It is of taxonomic importance to know in some cases whether the caudal glands (CG) are restricted to the caudal region (plates 103H–103I, 104F, 105C, 106A, 106F, 108I, 110D) or extend anteriorly beyond the level of the rectum (plates 100E, 100G, 101F, 102A), and whether they open to the exterior by way of a common duct at the tip of the tail or through three separate subterminal pores.

The esophagus possesses dorsal and subventral gland cells, the products of which are believed to have a digestive function. Members of Enoplida have either a single dorsal gland cell and one or two gland cells in each subventral sector, whereas most all other marine nematodes have, in addition to a single dorsal gland cell, only one cell in each subventral sector. The aperture by which the dorsal gland opens into the lumen is situated in the anterior region of the esophagus, the wall of the stoma, or through a dorsal tooth in the stoma. The position of the apertures of the subventral glands may also be in the wall of the esophagus, through the stoma wall or through teeth, or on the lip region, as in members of Leptosomatidae (plate 96A; SVEG).

Glands associated with amphids probably have a chemosensory function and are discussed in conjunction with the nervous system.

Nervous System

The various sensory receptors of the peripheral system are of taxonomic interest, especially the sensilla, amphids, ocelli, and metanemes. Sensilla (SE; CRS) are projections of the cuticle having tactile, chemosensory and/or glandular functions. Where present they may be either setiform—that is, relatively long and filamentous (plates 95C2, 101B, 104E, 106F, 108B–108C, 108F, 108I, among others)—or papilliform—that is, relatively short and stout (plates 102D, 108I, 109G, 110A). The number and distribution of sensilla vary considerably among taxa, but all nematodes appear to have had, at least in the primary condition, a circle of two subdorsal, two lateral, and two subventral inner labial sensilla (IL) surrounding the buccal orifice. They may be setiform (plates 100B–100D, 103H, 105D) but are usually papilliform (plates 96B, 96C, 100A, 100E, 100H, 108E, among others) and often not detectable in whole-mount preparations.

Peripheral to it is a second circle of two subdorsal, two lateral, and two subventral outer labial sensilla (plates 94C, 95A–95B, 96A, 100E, 100I, 103D, 103H, 104D, 105B–105D, 108E; OL), and posterior to it a third circle of two subdorsal and two subventral paralabial (PL) sensilla (plates 100B, 100D, 100E, 100I, 103G, 103H, 104A, 108E, among others). Often, the third circle is referred to as cephalic sensilla, but unfortunately that expression is also used to refer collectively to all three circles of sensilla. To avoid this duplicit use of the term, the outer circle of four "cephalic sensilla" are referred to in this work as paralabial sensilla, and cephalic sensilla refers collectively to inner, outer, and paralabial sensilla (plate 100B; CS). When all three circles are separate, it is indicated as such in taxonomic descriptions

as an arrangement of 6 + 6 + 4, and when the outer labial and paralabial are in the same circle as an arrangement of 6 + 10. The outer labial and paralabial sensilla may be either papilliform or setiform.

Additional sensilla, whether papilliform or setiform, may be distributed throughout the body length, limited to cervical and caudal regions, or associated with the vulva of females and cloaca of males where they doubtless mediate copulation.

Amphids (AM), believed to be chemoreceptor organs, are situated laterally, one on each side of the head or anterior neck region. Each amphid consists of a *fusus* (plate 95B; SC), a *canalis* (plate 95B; AD), a *porus canalis* (plate 95B; IA), *fovea* (plate 95B; FO), and, where the *fovea* is partially closed by an external layer of cuticle, the fovea opens to the exterior by way of an external *apertura* (plates 96A, 100C, 100H, among others; EA) (Riemann 1972). In instances where the *fovea* is not partially covered by cuticle, the external *apertura* is nonexistent or may be regarded as the same as the outer rim of the *fovea* (plates 95C2, 108B–108I, 109A–109G, 109I, 110A, 110B, 110D–110F; FO).

The *fusus* is situated within the hypodermis beneath the cuticle. Sensory cilia are rooted in the *fusus* (plate 95B; SC) and may extend by way of the *canalis*, through the *porus canalis* (IA), into the *fovea* (plate 104C; FO). The sensory cilia are embedded in the *corpus gelatum*, a secretion that extends from the *fusus*, through the *canalis*, into the *fovea*, and sometimes beyond the *fovea* as a hyaline cylinder.

The *foveae* are the most important feature of amphids for purposes of systematic research. Structurally, *foveae* are of four types: pocketlike, spiral, circular, and blisterlike.

With few exceptions, members of Enoplida have pocket-shaped amphids. These amphids consist of an intracuticlar cavity with the internal aperture at the posterior end of the *fovea* and an external aperture, usually near the anterior end of the fovea. Both apertures are on the amphid's central axis; thus, the dorsal and ventral halves of each amphid are essentially symmetrical (plates 100C, 100H, 101A, 101C, 101E, 102A, 102C, 102D, 102F, among others). The *fovea* of Halalaiminae is a longitudinal groove on the surface of the cuticle (plate 102E) and is probably a variant of the pocket-shaped amphid. Also, the *fovea* of Rhabdodemaniidae is probably of this type, but is reduced to a simple pore that normally is too small to be observed with light microscopy. In addition, it has an extremely long and sinuous *fusus*.

The *fovea* of spiral amphids is a groove on the external surface of the cuticle. They are usually spiraled ventrally, the predominant type in Chromadorida, Desmodorida, Axonolaimidae, Comesomatidae and Diplopeltidae (Araeolaimida), and Ceramonematidae (Plectida). In the case of these amphids, the internal aperture is on the dorsal side of what would be the amphid's median axis. From there, the *fovea* spirals ventrad as a slight arch (plate 106B), a transverse groove (plate 94C2), an inverted U- or shepherd's crook–shape (plates 106E, 110F), a spiral of one turn (plates 94A, 108F), or of several turns (plate 105A–105H, among others). Less frequently, the *fovea* may be dorsally spiraled, as in *Belbolla, Ditlevsenella, Eurystomina, Megeurystomina,* and *Pareurystomina* (Enchelidiidae) (plate 103D–103G), and rarely in Trefusiidae and Tripyloididae. In these cases, the internal aperture is shifted ventrad and the spiral proceeds from there in a dorsal direction. Whereas the internal aperture is typically at the outside end of the spiral, it is reversed in *Tarvaia* with the internal aperture at the inner end of the spiral. In all spiral amphids, the external aperture is more or less congruent with the rim of the *fovea*.

Among members of Monhysterida, the *fovea* is a circular concavity on the surface of the cuticle (plates 108H–110I). The internal aperture is situated away from the median axis of the *fovea,* so the dorsal and ventral halves of the amphid are not symmetrical. For this reason it is thought that circular amphids may be considered a variant of the spiral amphid. As in the case of spiral amphids, the external aperture is congruent with the rim of the *fovea.*

The last type of amphid is blisterlike and is restricted to the order Desmoscolecida (plate 107B, 107C, 107F–107I). The *fovea* is a cavity in the cuticle and may be round, oval, or triangular in outline. The internal aperture is on the median axis of the amphid, so the dorsal and ventral halves are basically symmetrical. Unlike other amphids, there is no known external aperture. Rather the *fovea* appears to be completely covered by a thin, external layer of cuticle.

Metanemes (plate 93B; ME) are serially arranged filamentous structures in the hypodermal chords of all members of Enoplida as well as those of Tobrilina and Tripylina (Triplonchida). Each metaneme has a spindle-shaped swelling. The spindle is the synaptic junction of a monociliated proprioceptor, which is the anterior half of the metaneme and the dendrite of a neuron in the posterior portion (Hope and Gardiner 1982). Those metanemes that parallel either the dorsal or ventral margin of the hypodermal chord are termed orthometanemes. Those that cross the chord from its dorsal to its ventral margins, or vice versa, but remain within the dorsal and ventral boundaries of the chord are termed loxometanemes type I, and those that cross the chord and extend beyond its margins are termed loxometanemes type II (Lorenzen 1978, 1981). The presence of metanemes is regarded as being apomorphic for Enoplida, and their position and orientation relative to the hypodermal chord are believed to be of importance in the phylogenetic assessment of taxa within Enoplida according to the classification by Lorenzen (1981).

Structures thought to be photoreceptors typically occur on the lateral surface of the anterior region of the esophagus among diverse taxa of marine nematodes. In some cases, only a diffuse area of pigmentation occurs (plates 100A, 101I; OP), whereas in others the pigment is incorporated in a light-shielding cup that partially envelops an innervated lamellar rhabdomere (plates 100G, 101C, 102A, 102B, 108F; OC) on each lateral surface of the anterior esophagus. Conversely, the rhabdomere and pigmented light shield are completely separate from the esophagus in *Araeolaimus elegans* de Man 1888 (Croll et al. 1975) and in *Diplolaimella* sp. (Van de Velde and Coomans 1988). In *Quadricoma cobbi* (Steiner 1916) Filipjev 1922, a pigment spot, presumed to be a photoreceptor, is situated opposite the anterior midgut (Decraemer 1978), and in *Acanthonchus rostratus* Wieser 1959, each ocellus is situated in a tubelike structure dorsolateral to the anterior esophagus (Murphy 1963a).

Reproductive System

Most marine nematodes are dioecious, fertilization is internal, and development is direct, with four juvenile stages preceding the sexually mature adult stage. Males have one (monorchic) or two (diorchic) testes (plate 97C), a seminal vesicle (SV), and, especially in Enoplida, there may be an ejaculatory duct ensheathed in muscle. The male system opens into the rectum, hence the latter is a cloaca. The number of testes, whether in tandem or opposed (plate 97C) when paired, whether outstretched or distally curved back upon itself (reflexed), and on

which side of the gut each is situated are all of taxonomic significance.

A pair of cuticularized spicula (SP), invested with both protractor and retractor muscles, aid in the transfer of sperm. They vary in size and shape from being relatively short and scimitar-shaped (plates 97C, 100A, 101B, 101F–101H, 104A, 106C, among others) to long and needlelike (aciculate) (plates 100E, 100G, 100H, 103B, 103F, 107H, 108C, 109G) with many intermediate lengths and forms. The cuticularized gubernaculum, when present, is in its simplest form a plate situated on the posterior side of the distal ends of the spicula (plates 100F, 100I, 101C; GU). This plate, which is the corpus of the gubernaculum, may bear a dorsally or caudally directed apophysis (plates 106H, 108B, 108D, 108E, 108H, 109A, 109D–109F, 109H, 109I, 110C, 110D, among others; AG) to which are attached protractor muscles that extend from the anterior ends of the spicula to the apophysis and from the latter to the ventral surface of the tail. A series of dorsoventrally oriented copulatory muscles on each side of the posterior body region enable the male to coil around the female during copulation (plates 97C, 100E, 100G, 101D, 101F, 101I–102B, 105C, 110D; CM).

The ventral body surface anterior to the cloaca of males often bears one or a longitudinally arranged series of two or more ventromedian supplements (VMS) believed to be sensory receptors that mediate copulation. These structures may be papilliform (plates 98, 101F, 103F, 103G, 103I, 104H, 105E, 105G, 106I, 109C), setiform minute pores (plate 101I, 101J), cuticular tubules (plates 100A–100D, 100G, 100I, 105C, 110C, 110D), or cuplike (cyathiform) depressions, the latter with (plates 102A, 102B, 103D, 103E, 105A, 105D) or without (plates 97C, 104F) anterior and posterior flanges, to mention but a few. Subventral supplements (SVS) may also occur (plates 98, 100A, 100B, 100I, 101D, 101E, 101F, 101I, 101J, 102C, 103C, 103G, 105C, 108I). They are almost always setiform or papilliform but never tuboid. In some Leptosomatidae they may be situated on mammiform elevations (plates 98, 101D, 101F, 101J, 102B; SVR).

As in the case of males, females may have one (monodelphic) or two (didelphic) gonads (plate 97A, B), but if among marine nematodes there are two, they are always opposed—never in tandem. If they are paired, it is of taxonomic significance as to which side of the gut each is situated. As in the males, the gonoduct may be outstretched (lower end of plate 97B) or reflexed (upper end of plate 97B), but in females two forms of reflexed gonoduct exist. In the case of the homodromous gonad, the ovary (OVR) is simply folded back on itself, very much as in the gonad of males. When an ooctye reaches the bend of the gonoduct, it rounds the bend with its leading edge always remaining in the lead—that is it continues to move in the same direction relative to the gonoduct (arrow in upper end of plate 97B). In antidromous gonads, the end of each ovary that is farthest from the vulva ends blindly, and the entrance to the oviduct lies somewhere along the ovary's midlength (plate 97A). In this case, the oocyte that is next to enter the oviduct must reverse its direction and slide past other less mature oocytes into the oviduct. What had been the trailing edge of the developing oocyte while in the ovary becomes the leading edge as it enters the oviduct (arrows in plate 97A).

The vulva (VU) is always situated on the ventral body surface and usually near mid-body length. In the monodelphic condition, the vulva may be displaced significantly toward the anterior if the gonoduct and ovary are directed toward the posterior (opisthodelphic), or if the gonoduct is directed anteriorly (prodelphic) the vulva may be shifted posteriorly. In Laura-tonematidae, the vulva is either merged with the rectum or situated near it.

Some females in the family Oncholaimidae, particularly those in the subfamilies Oncholaimellinae, Adoncholaiminae, and some Oncholaiminae, possess a demanian system (plate 99). In its most complex condition, it occurs where females are inseminated through puncture wounds (plate 99B–99D; PSP) in the body wall made by the spicula of males (traumatic insemination) rather than by way of the vulva. The punctures usually occur in the posterior or caudal regions, although in some species they may occur in the vicinity of the vulva (plate 99B). A secretion plug and sperm are deposited into the female's body cavity by way of the resulting wound. This process results in the formation of an interstitial channel that eventually reaches and merges with what had been the blind posterior end of the *ductus principalis* (plate 99C, 99D; DP). Upon entering the lumen of the *ductus principalis,* sperm proceed to the *uvette* (plate 99B–99D; UV), where they accumulate and may then pass one at a time through the *ductus uterinus* (plate 99B–99D; DU) into the uterus (UT), where fertilization of ova occurs.

In some didelphic females (plate 99C), there are two *uvettes,* one passing to each of the two branches of the reproductive system. In this case sperm may proceed past the posterior *uvette* through the *ductus entericus* (DE) and enter the anterior branch of the uterus by way of the anterior *uvette* and *ductus uterinus.* Coomans et al. (1988) obtained evidence that excess sperm, be they from the initial or subsequent inseminations, may pass beyond the anterior *uvette* and enter the gut by way of the *osmosium* (OS), where they are presumably digested.

A demanian system also occurs in cases where insemination of females is normal, that is by way of the vulva, such as in *Viscosia* (plate 99A). In this case, there is a single *ductus entericus* extending from each uterus to an *osmosium* at the gut wall, and the demanian system may serve as both a seminal receptacle and a means by which excess sperm are shunted into the digestive system.

Classification

Nematoda are often regarded as a separate phylum (Chitwood 1951; Maggenti 1982), but recent trends have been to regard the Nematoda and Nematomorpha as either subphyla (Coomans 1981) or classes (Andrássy 1976) of the phylum Nemathelminthes. Malakhov (1994), however, presents evidence that the Nematoda and the Gastrotricha are closely allied and gives each the rank of class within Nemathelminthes.

For marine nematodes, almost all of which are in the class Adenophorea, there is disagreement as well concerning postulates of phylogenetic relationships and a classification that best accommodates even the most recent morphological information (Adamson 1987; Andrássy 1976; Maggenti 1982; Inglis 1983, and Lorenzen 1994). Unlike previous classifications that relied heavily on unique combinations of plesiomorphic characters, Lorenzen (1981) based his classification, where possible, on apomorphic characters. Apomorphic characters were not found in all cases, and many families that could not be assigned elsewhere were placed in the paraphyletic suborder Leptolaimina.

The system that is followed for the most part in this section is that of DeLey and Blaxter (2002), who treat nematodes as a separate phylum and have used molecular and morphological data to derive their classification. The author has, however,

deviated from this classification, especially with regard to Enoplina and Ironina. Leptosomatidae and Oxystominidae are assigned to the suborder Ironina in the classifications of Lorenzen (1981) and DeLey and Blaxter (2002). However, in all members of Ironina, except Leptosomatidae and Oxystominidae, buccal teeth are formed in special cells at the anterior end of the esophagus, from whence they become attached to jaws capable of being protruded radially. For this reason, Leptosomatidae and Oxystominidae are reassigned to the paraphyletic suborder Enoplina.

Apomorphic characters are lacking for many taxa. Hence, in following this and any other classification of marine nematodes, elements of the key are often composed of unique combinations of plesiomorphic characters.

Technical terms used in the key are defined in the preceding section on morphology. For most families and/or subfamilies included in this key, there is at least one representative illustration. The illustrations themselves may serve as a general guide for identifications. The pictorial keys of British marine nematodes (Platt and Warwick 1983, 1988; Warwick et al. 1998) may be useful as a guide to families or genera, but the species occurring on our coast are most likely to be different from those in Britain. Tarjan's (1980) dichotomous key to genera may also be helpful, as would Wieser (1959) in identifying some species on the Oregon and northern California coasts. Abebe et al. (2006) also contain keys to marine taxa; their keys to Chromadoridae and Desmodoridae may be helpful supplements to the keys here.

The key to subfamilies that follows is based upon a review of the literature and use of the collections in National Museum of Natural History at the Smithsonian Institution. The computer program DELTA (Dallwitz, Paine, and Zurcher 1993) was used in developing the key.

Acknowledgments

The author is deeply indebted to Rosanne Johnson for translating Russian articles and for critical reviews of the manuscript, to Eunice Pancoast for help with bibliographic assistance and reviews of the manuscript, to Molly Ryan for preparing the illustrations and plates, and to Abbie Yorkoff for helping research literature, reviewing the manuscript, and providing technical assistance with the collections. I also wish to thank Wilfrida Decraemer, D. G. Murphy, Ella Mae Noffsinger, and Reverend R. W. Timm for permission to use their published illustrations.

Key to Subfamilies of Commonly Occurring Marine Nematodes

Keys to the subfamilies of Chromadoridae and Desmodoridae are not included.

1. (*Note multiple choices*) Cuticle usually smooth as observed with light microscopy (obviously striated in Trefusiida and *Oncholaimoides*); metanemes usually present (absent in Trefusiida); cephalic sensilla usually 6 + 10 (6 + 6 + 4 in some Anoplostomatinae, Ironidae and most Trefusiida); fovea of amphid intracuticular and pocketlike; internal aperture situated on median, longitudinal axis of amphid (nonspiral); (dorsally spiraled in some Enchelidiidae; ventrally spiral in most Tripyloididae and some Trefusiidae; obscure in some Enoplinae, Phanodermatidae, Rhabdodemaniinae, and

Pandolaiminae); external aperture of pocketlike amphid usually transverse, slitlike to oval, but may be circular (Oxystomininae) or longitudinally slitlike (Halalaiminae); stoma with or without armament, but never with rugae; esophagus usually cylindrical or claviform, rarely multibulbar; females usually didelphic, amphidelphic, and gonoducts reflexed (monodelphic in some Trefusiida); ejaculatory duct usually muscular (Enoplea) 2

— Cuticle usually with transverse striations, intracuticular punctations, or complex patterns presumably derived from punctations; cephalic sensilla 6 + 6 + 4 with paralabial sensilla usually longer than outer labial sensilla, or 6 + 10 with paralabial sensilla usually shorter than outer labial sensilla; cephalic and somatic sensilla never on peduncles; amphidial fovea spiral or derived from spiral (i.e., looped) transverse slit with internal aperture at dorsal end of slit, circular or transversely oval with internal aperture situated at dorsolateral or ventrolateral side of fovea; stoma usually armed with dorsal and subventral onchia (absent in Selachinematidae, Stilbonematinae, many Epsilonematidae, and some Cyatholaimidae); stoma usually with 12 rugae (apparently absent in all Epsilonematidae and in some Desmodoridae, Draconematidae, Microlaimidae, and Aponchiidae); esophagus usually with posterior bulb; caudal glands not projected anterior to rectum; gonoducts usually reflexed (outstretched in Aponchiidae and Microlaimidae); males diorchic (one anterior testis in Chromadoridae and Desmodoroidea, among others); ejaculatory duct nonmuscular (Chromadorida; Desmodorida) 29

— Cuticle usually striated, with or without prominent setae on peduncles, spines, scales, wartlike protuberances or desmens; if spines present and adherent sediment particles absent, then spines long and in transverse rows; if cuticle with transverse and longitudinal striations, then pseudocoelomocytes often present and esophagus then with muscular anterior part, thin midsection, and glandular posterior region usually with dorsal glands overlapping anterior end of gut; inner and outer labial cephalic sensilla obscure, paralabials setiform, or if outer labials apparent, then often situated on peduncles; amphids vesiculate or circular; stoma very small, unarmed; rugae absent; caudal glands not projected anterior to rectum; female reproductive system didelphic, amphidelphic, and gonoducts outstretched or reflexed; male reproductive system monorchic or diorchic (Desmoscolecida) 43

— Cuticle with or without transverse and longitudinal striations and with or without punctations; cephalic sensilla 6 + 6 + 4 or 6 + 10; amphidial fovea concave and circular, troughlike and looped or spiral; stoma armed or unarmed, but never with rugae; esophagus seldom with posterior bulb; caudal glands not projected anterior to rectum; females monodelphic or didelphic with gonoducts always outstretched; males monorchic or diorchic; ventromedian supplements never tubiform; ejaculatory duct nonmuscular (Monhysterida; Araeolaimida) 46

— Cuticle usually with transverse striations, always without punctations; cephalic sensilla 6 + 6 + 4, or if 6 + 10 then helmet present; fovea spiral, inverted U, or circular; rugae absent; buccal capsule usually unarmed; if buccal capsule armed with onchiostyle, then amphids anterior to paralabial sensilla; if buccal capsule armed with odontia or onchia, then males with tuboid supplements; esophagus with or without mid- and posterior bulbs; caudal glands

not projected anterior to rectum; gonoducts almost always reflexed (outstretched in *Pterygonema* and *Manunema*); females usually didelphic (monodelphic in *Alaimella, Listia,* and *Procamacolaimus*); males monorchic or diorchic; ejaculatory duct nonmuscular; ventromedian supplements often tubiform (Plectida) 58

2. (*Note three choices*) Cuticle smooth (except in Halalaiminae and *Oncholaimoides*); metanemes present; cephalic sensilla usually 6 + 10; outer labial sensilla usually not jointed (jointed in Tripyloidina); cephalic sensilla situated anterior to amphid; amphid nonspiral (i.e., pocketlike or elongated grove with internal aperture always at posterior margin of fovea and situated on lateral line), or unispiral in Tripyloidina; fusus never elongated and sinuous; three caudal glands present, or caudal glands absent (Enoplida) 3

— Cuticle smooth; metanemes present; cephalic sensilla 6 + 6 + 4 or 6 + 10; outer labial sensilla never jointed; amphid poroid and usually not visible with light microscopy; amphidial fusus elongated and sinuous, often difficult to observe; tail with two caudal glands confined to caudal region (Triplonchida) (*Note:* Rhabdodemaniidae and Pandolaimidae are the only truly marine taxa currently assigned to Triplonchida, and only species of those families have porelike amphids 25

— Cuticle finely striated; metanemes absent; cephalic sensilla 6 + 6 + 4 with outer labial sensilla jointed, paralabial sensilla usually posterior to amphid, amphids nonspiral or infrequently spiral, and male and female gonads usually paired (*Rhabdocoma* monodelphic, opisthodelphic). Or cuticle distinctly striated, cephalic sensilla 6 + 10, outer labial sensilla not jointed, paralabial sensilla anterior to amphid; amphidial fovea pocketlike, external aperture slitlike to circular and internal aperture on lateral line; males monorchic and females monodelphic; three caudal glands confined to tail (Trefusiida). 26

3. (*Note multiple choices*) Cephalic sensilla usually 6 + 10, except 6 + 6 + 4 in Oxystominidae; amphid usually pocketlike; cephalic capsule usually present and esophageal tissue usually attached to post-labial cuticle of head; stoma shape variable, but if dolioform then cephalic sensilla more than one-half head diameter in length; stoma unarmed, armed with odontia, onchia, or mandibles, but mandibles never radially everted; dorsal esophageal gland orifice posterior to stoma; caudal glands with at least one caudal gland extended precaudally, rarely confined to tail (as in juveniles and females Enoplidae, Anticomidae, and Halalaiminae); males usually diorchic; females usually didelphic (Enoplina) 4

— Cephalic sensilla 6 + 10; outer labial and paralabial sensilla never jointed; amphids pocket-shaped; cephalic capsule absent; stoma open, spacious, with walls usually more or less parallel (dolioform), but never tubular or with elongate stylet or mandibles (sexually dimorphic males of *Symplocostoma, Calyptronema,* and *Polygastrophora* have reduced stoma with multibulbar esophagus and/or ocelli); stoma usually with onchia; if unarmed then buccal capsule never with longitudinal suture and outer labial sensilla less than one-half head diameter in length; dorsal esophageal gland orifice at level of stoma, usually at distal end of onchium; esophagostome at most only partly enclosed by esophageal musculature (Oncholaimina; Oncholaimoidea) 21

— Cephalic sensilla usually 6 + 10, except 6 + 6 + 4 in *Ironella*; amphids pocket-shaped; cheilostome with radially eversible mandibles; esophagostome narrow, tubular, and

enclosed by longitudinal protractor muscles; males without tubiform ventromedian supplements, except present in *Ironella* (Ironina; Ironoidea; Ironidae) (plate 101A). . . . . . .
. . . . Thalassironinae
— Cephalic sensilla 6 + 10; outer labial sensilla may be jointed; amphids usually dorsally or ventrally unispiral; cephalic capsule absent and esophageal tissue not attached to post-labial, peripheral cuticle of head; onchia often present in posterior stoma; stoma aperture with thin, foliate lips; onchia often present in crescent-shaped chamber in posterior region of stoma; three caudal glands confined to caudal region; females didelphic; males always monorchic (Tripyloidina; Tripyloididae) (plate 103H). . . . . Tripyloidinae
4. Cephalic sensilla 6 + 10 or 6 + 6 + 4; buccal capsule unarmed or armed with odontia and/or onchia, never with mandibles; position of gonads relative to gut variable; ventromedian supplements present but never tubiform (Leptosomatoidea) 5
— Cephalic sensilla 6 + 10; stoma unarmed or armed with mandibles or buccal stylet; anterior and posterior gonads always on left of gut; males usually with single tubiform, ventromedian supplement (Enoploidea) 13
5. Cephalic sensilla 6 + 10; neck region not strongly tapered; amphids usually situated on head region near outer circle of cephalic sensilla; amphidial fovea pocketlike with transverse, slitlike external aperture; cephalic capsule usually present; buccal capsule variable in size and shape and often armed with odontia and/or onchia (Leptosomatidae) 8
— Cephalic sensilla 6 + 6 + 4; neck region usually strongly tapered; amphid usually situated in neck region (near head in *Thalassoalaimus*); amphid either slender, longitudinally elongate groove on surface of cuticle, or pocketlike with oval external aperture; cephalic capsule absent; buccal capsule reduced in size and unarmed; males with or without ventromedian supplements (Oxystominidae) 6
6. External aperture of fovea wide and circular to oval; females didelphic or monodelphic; caudal glands extend anterior to rectum. 7
— External aperture of fovea narrow, longitudinally elongated; females didelphic; caudal glands confined to caudal region (plate 102E) . . . . Halalaiminae
7. Females monodelphic; males without rows of subventral supplements (plate 102D) . . . . Oxystomininae
— Females didelphic; males with two rows of setiform subventral supplements . . . . Paroxystominae
8. Cephalic capsule usually not well-developed and never with deep narrow incisions between lobes or with tropises; tail bluntly conical to conical with cylindrical or flagellate terminus 9
— Cephalic capsule well-developed, usually with one or more tropises; if tropises absent then lens-bearing ocellus and/or odontia present; tail always bluntly conical (plates 101F–102B). . . . . Thoracostomatinae
9. Paralabial sensilla <2 head diameters in length 10
— Paralabial sensilla >2 head diameters in length . . . .
. . . . Barbonematinae
10. Tail conical, almost always with cylindrical or flagellate terminus; stoma unarmed or with odontia and/or onchia or with comblike row of denticles on at least subventral walls of cheilostome 11
— Tail bluntly conical to rounded; stoma unarmed, or at most armed with small dorsal onchium (plate 101C) . . . .
. . . . Leptosomatinae

11. Stoma funnel-shaped, never with anterior comblike row of denticles in cheilostome. 12
— Stoma dolioform with comblike row of anteriorly directed denticles on at least subventral walls in cheilostome (plate 102C). . . . . Cylicolaiminae
12. External aperture of amphid in males never covered by cuticular flap; each subventral wall of esophagostome with one or two onchia (plate 101E). . . . . Synonchinae
— External aperture of amphid in males with anteriorly directed flap with or without bifurcations (flap absent in females); buccal capsule unarmed (plate 101B) . . . .
. . . . Platycominae
13. Stoma unarmed; cephalic capsule with or without tropises
. . . . 17
— Stoma armed with mandibles or stylet; cephalic capsule never with tropises 14
14. Mandibles present; mandibles posteriorly rounded, never branched; mandibular onchia absent; anterior surface of mandibular claws never bearing round to oval cuticular bodies; inner labial sensilla papilliform (Enoplidae) (plates 96C, 100A) . . . . Enoplinae
— Mandibles usually with onchia or with buccal stylet; if mandibles are present, then posteriorly branched and mandibular onchia often present; anterior surface of mandibular claws often with spherical to oval cuticular bodies; inner labial sensilla setiform and robust (Thoracostomopsidae) 15
15. Mandibles present with or without onchia; buccal stylet always absent. 16
— Buccal stylet present (plate 100B). . . . .
. . . . Thoracostomopsinae
16. Distal end of mandible with lateral claws; distal end of mandible without central rounded or toothlike projection (plate 100D). . . . . Enoplolaiminae
— Distal end of mandible without pair of lateral claws; anterior end of mandible with central rounded or toothlike projection (plate 100C) . . . . Trileptiinae
17. Outer labial sensilla less than one head diameter in length; amphids situated near outer ring of cephalic sensilla; buccal capsule not spacious and dolioform; buccal capsule enveloped by esophagus; cervical gland usually present; caudal glands confined to caudal region 19
— Outer labial sensilla one head diameter in length or more; amphids situated well posterior to cephalic sensilla; buccal capsule spacious, dolioform and always unarmed; esophagus enveloping only posterior end of buccal capsule; cervical gland not known (pore may be present); caudal glands extend into precaudal region (Anoplostomatidae)
. . . . 18
18. Amphids not sexually dimorphic; buccal capsule with longitudinal seams; males with copulatory bursa; tubular ventromedian supplement never present (plate 100E) . . . . . .
. . . . Anoplostomatinae
— Amphids sexually dimorphic; buccal capsule without longitudinal seams; males without copulatory bursa; tubular ventromedian supplement present or absent (plate 100F)
. . . . Chaetonematinae
19. Cephalic capsule poorly developed, often indistinct; tropises absent; neck region with distinct grouping of several cervical sensilla, usually arranged in longitudinal series; posterior esophagus never crenate in profile (Anticomidae) (plate 100I). . . . . Anticominae
— Cephalic capsule present; tropises present or absent; neck region without distinct grouping of cervical sensilla; posterior

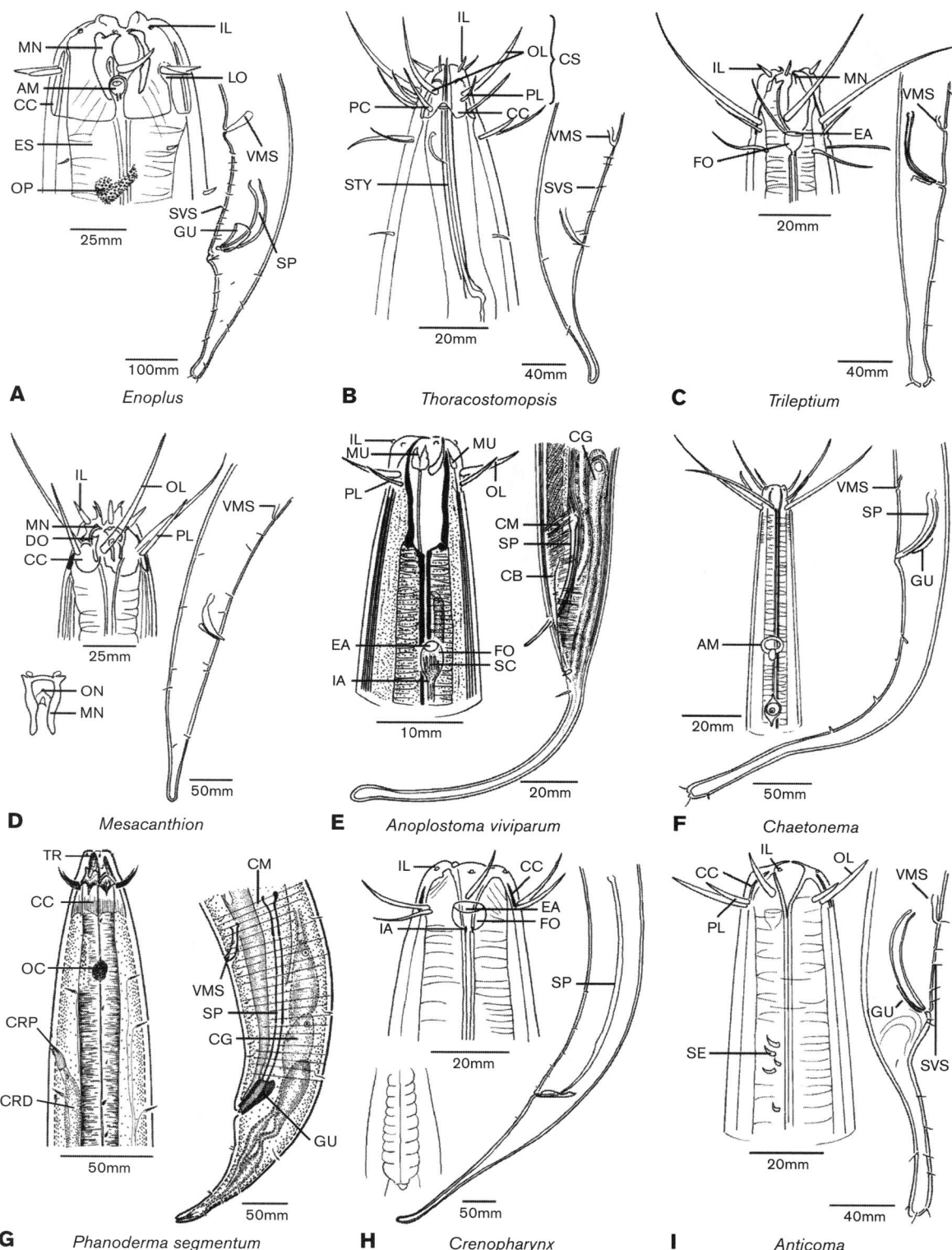

A *Enoplus*

B *Thoracostomopsis*

C *Trileptium*

D *Mesacanthion*

E *Anoplostoma viviparum*

F *Chaetonema*

G *Phanoderma segmentum*

H *Crenopharynx*

I *Anticoma*

PLATE 100 G, From Murphy 1963c with permission from the author and from the Helminthological Society of Washington (HSW).

esophagus crenate in profile (Phanodermatidae) 20

20. Cephalic capsule well-developed and with dorsal and subventral tropises (plate 100G) . . . . Phanodermatinae

— Cephalic capsule narrow, ringlike, and without tropises (plate 100H) . . . . Crenopharynginae

21. Esophagostome not divided by transverse suture or rows

of denticles; buccal capsule always dolioform, usually armed with one dorsal and two (rarely one) subventral onchia, with one subventral onchium larger than the other (onchia reduced or absent in Pelagonematinae and Krampiinae); esophagus always cylindrical to claviform; gubernaculum without apophysis (Oncholaimidae) 22

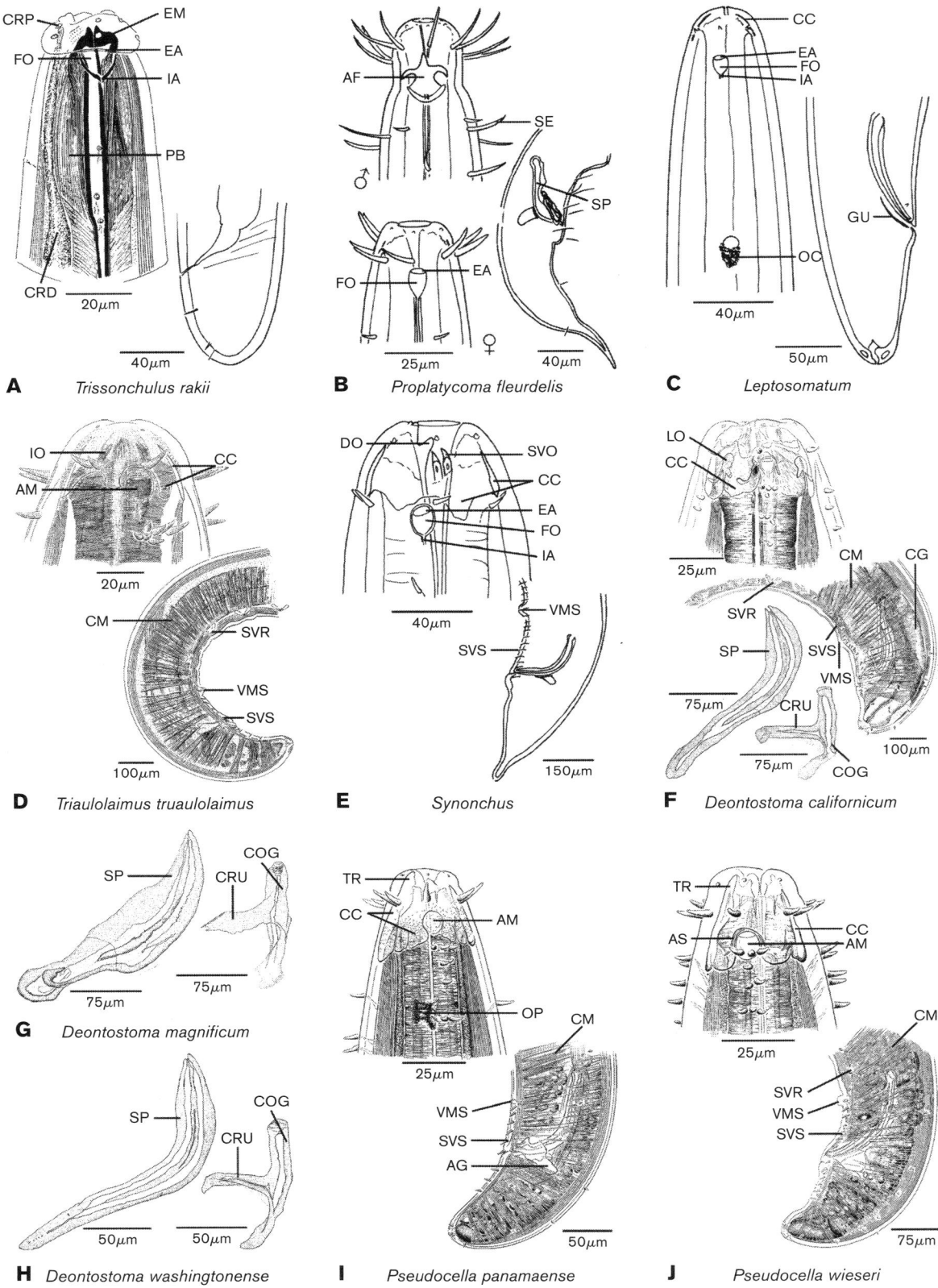

A   *Trissonchulus rakii*

B   *Proplatycoma fleurdelis*

C   *Leptosomatum*

D   *Triaulolaimus truaulolaimus*

E   *Synonchus*

F   *Deontostoma californicum*

G   *Deontostoma magnificum*

H   *Deontostoma washingtonense*

I   *Pseudocella panamaense*

J   *Pseudocella wieseri*

PLATE 101 B, After Hope 1988; D, from Hope 1967a, with permission from the HSW; F–J, from Hope 1967b, with permission from Transactions of the American Microscopical Society (TAMS).

— Esophagostome usually divided by transverse suture or rows of denticles into two or three chambers and usually armed with single, well-developed subventral onchium, or if two or three onchia, then esophagostome always with rows of denticles and gubernaculum with apophysis; if buccal capsule closed and tubular, then ocellus present; esophagus with or without terminal bulbs; ventromedian supplements usually present (Enchelidiidae) (plate 103D–103G) . . . . Enchelidiinae

22. (*Note three choices*) Stoma with all onchia reduced in size or absent; females monodelphic or didelphic; demanian system absent 24
— Stoma with at least one well-developed onchium, but never more than three onchia; females monodelphic or didelphic; demanian system present or absent 23
— Stoma with numerous subventral onchia . . . .
. . . . Octonchinae

23. (*Note three choices*) Stoma with left subventral onchium larger than remaining onchia; females monodelphic; demanian system present (*Oncholaimium, Oncholaimus, Wiesoncholaimus*) or absent (*Prooncholaimus; Pseudoncholaimus*) (plates 102G–103B) . . . . Oncholaiminae
— Stoma with subventral onchia equal in size and both larger than dorsal onchium; females didelphic; demanian system always absent (plate 103C) . . . . Pontonematinae
— Subventral onchia of equal size or right subventral onchium larger; females didelphic; demanian system complex . . . . Adoncholaiminae
— Stoma with right subventral onchium larger than remaining onchia, if others exist; females didelphic (*Oncholaimelloides* monodelphic); demanian system simple (*Oncholaimellus* and *Viscosia*) or absent (*Oncholaimoides*) (plate 102F) . . . . Oncholaimellinae

24. Female reproductive system didelphic . . . .
. . . . Pelagonematinae
— Female reproductive monodelphic . . . . Krampiinae

25. Buccal capsule dolioform and unarmed; tail conical with short cylindrical terminus (Pandolaimidae) (plate 104A)
. . . . Pandolaiminae
— Buccal capsule broadly conical and armed with three pairs of anterior odontia in cheilostome, one onchium on dorsal and on each subventral wall of esophagostome and sometimes additional small denticles at base of esophagostome; tail broad, elongate with bluntly rounded terminus (Rhabdodemaniidae) (plate 103I) . . . . Rhabdodemaniinae

26. Cuticle striated, with or without longitudinal ridges; cephalic sensilla 6 + 6 + 4; outer labial sensilla jointed; amphids pocketlike, rarely dorsally or ventrally spiraled; males monorchic or diorchic; females monodelphic or didelphic; vulva never opens near anal vent or merged with rectum to form cloaca 27
— Cuticle distinctly striated, but never with longitudinal ridges; cephalic sensilla 6 + 10; outer labial sensilla not jointed; amphids pocketlike; females monodelphic, prodelphic; vulva situated near anal vent, or opens into cloaca; males monorchic (Lauratonematidae) (plate 104D) . . . . .
. . . . Lauratonematinae

27. Cuticle without longitudinal ridges; strongly tapered head cone absent; buccal cavity conical or barrel-shaped; gonads usually paired in both sexes (monodelphic and opisthodelphic in *Rhabdocoma*); males often with cervical, ventromedian supplements (Trefusiidae) 28
— Cuticle with thin longitudinal ridges; head cone strongly tapered anteriorly and without cuticular striations; buccal

cavity obscure; males with one anterior testis and females with single posterior ovary (*Porocoma* didelphic); males without cervical, ventromedian supplements (Xennellidae) (plate 104E) . . . . Xennelliinae

28. Buccal cavity small and conical (plate 104B) . . . .
. . . . Trefusiinae
— Buccal cavity barrel-shaped, usually with oval seams in wall of buccal capsule (seams absent in *Africanema*) (plate 104C) . . . . Halanonchinae

29. Cuticle usually striated, always without punctations; helmet present or absent; stoma never with struts (Desmodoroidea; Microlaimoidea) 36
— Cuticle striated, almost always with punctations; if punctations absent then stoma with struts or cuticle with longitudinal files of spines and stoma without armament; helmet always absent (Chromadoroidea) 30

30. Buccal capsule with dorsal tooth and with or without subventral teeth, but never with struts or mandibles 32
— Dorsal tooth always absent; buccal capsule usually with broad anterior and narrower posterior chambers each with longitudinal struts or with elongate mandibles; if struts and mandibles absent then cuticle with longitudinal files of spines (Selachinematidae) 31

31. Buccal capsule with anterior and posterior chambers, each chamber bearing struts or posterior chamber with opposable mandibles; cuticle almost always with transverse rows of punctations (plate 105E–105G) . . . . Selachinematinae
— Buccal capsule without struts or mandibles; cuticle with longitudinal files of spines (plate 105H) . . . .
. . . . Richtersiinae

32. Stout dorsolateral subcephalic seta present near paralabial sensilla; ventromedian supplements cyathiform with prominent, articulated anterior and posterior flanges; spicula with characteristic bend near distal third (Neotonchidae) (plate 105A) . . . . Neotonchinae
— Dorsolateral subcephalic sensilla absent; if ventromedian supplement cyathiform, then without articulated flanges; spicula without bend at distal third 33

33. Amphid commonly oval and transverse, but may be spiral; cephalic sensilla usually 6 + 6 + 4 and six outer labial shorter than four paralabial sensilla; if outer labial and paralabial in single circle, then six outer labial sensilla longer than four paralabials; females didelphic with anterior gonad on right and posterior always on left of gut; males always monorchic (plate 104F–104I) . . . .
. . . . Chromadoridae
— Amphid always multispiral; six outerlabial always longer than four paralabial sensilla; males usually diorchic and females didelphic; anterior gonads may be on right or left, but posterior gonad always on opposite side of anterior gonad (Cyatholaimidae) 34

34. Males with ventromedian supplements 35
— Males without ventromedian supplements (plate 105B)
. . . . Cyatholaiminae

35. Males with tubiform ventromedian supplements (plate 105C) . . . . Paracanthonchinae
— Males with cyathiform ventromedian supplements (plate 105D) . . . . Pomponematinae

36. Body spindle-shaped or epsilon- or S-shaped; if spindle-shaped then cephalic helmet often present; males always monorchic; females didelphic and ovaries antidromous (Desmodoroidea) 37
— Body spindle-shaped; males diorchic or monorchic; helmet absent; females didelphic or monodelphic with outstretched

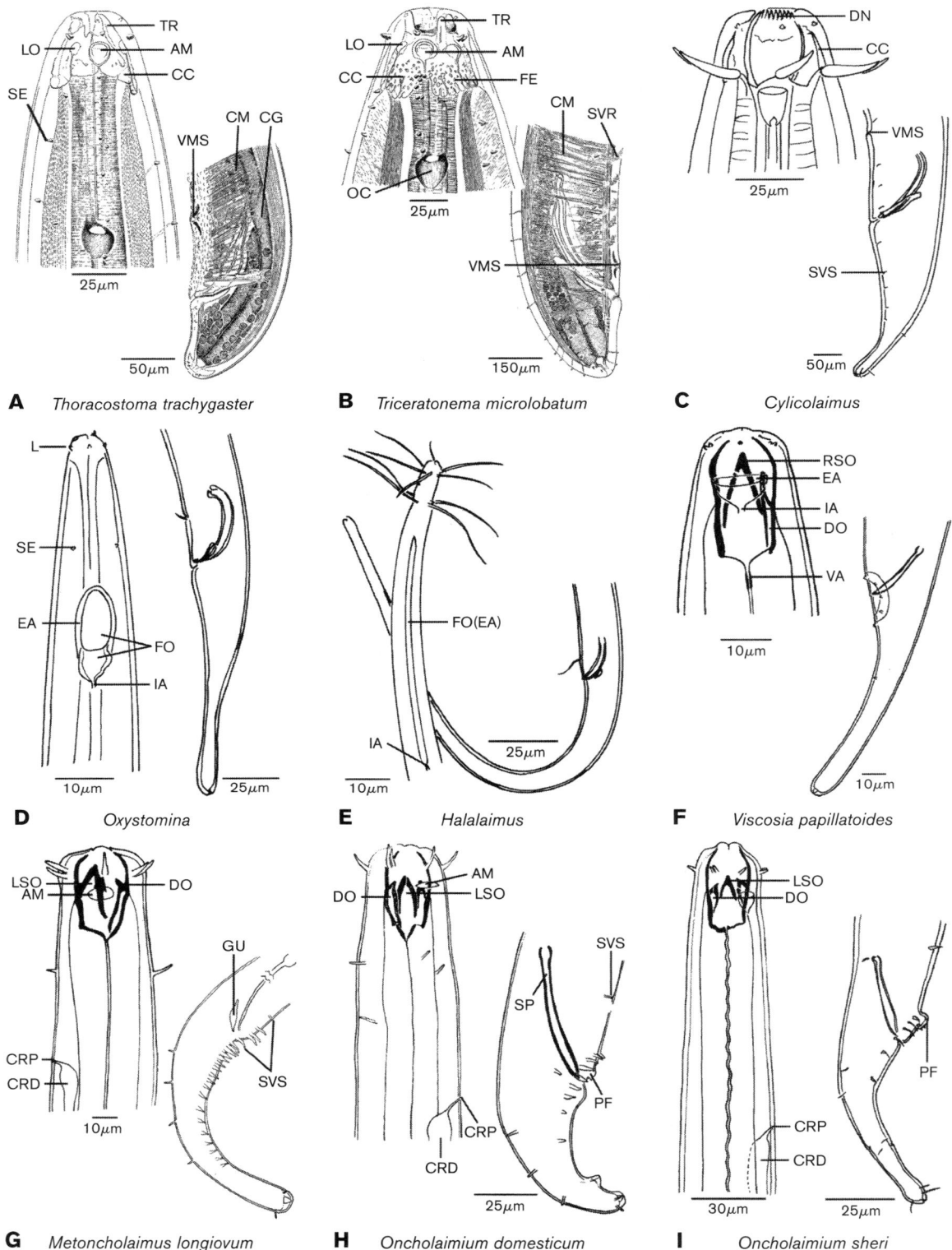

A *Thoracostoma trachygaster*

B *Triceratonema microlobatum*

C *Cylicolaimus*

D *Oxystomina*

E *Halalaimus*

F *Viscosia papillatoides*

G *Metoncholaimus longiovum*

H *Oncholaimium domesticum*

I *Oncholaimium sheri*

PLATE 102 A and B, from Hope 1967b, with permission from TAMS; F–I, from Chitwood 1960b, with permission from TAMS.

ovaries (Microlaimoidea). 41

37. Body usually spindle-shaped, but may have greater width in anterior or posterior body regions than in mid-region, but body not epsilon- or sigmoid-shaped; stilt setae absent (Desmodorinae, Spiriniinae, Pseudonchinae, Stilbonematinae, Molgolaiminae) (plates 105I–106D) . . . .

. . . . Desmodoridae

— Body epsilon- or S-shaped; stilt or adhesive ambulatory setae present on subventral surface of body 38

38. Ovaries situated posterior to dorsal bend of body; stilt setae situated anterior to or in same body region as ovaries (i.e., in posterior region of body); adhesive setae not present

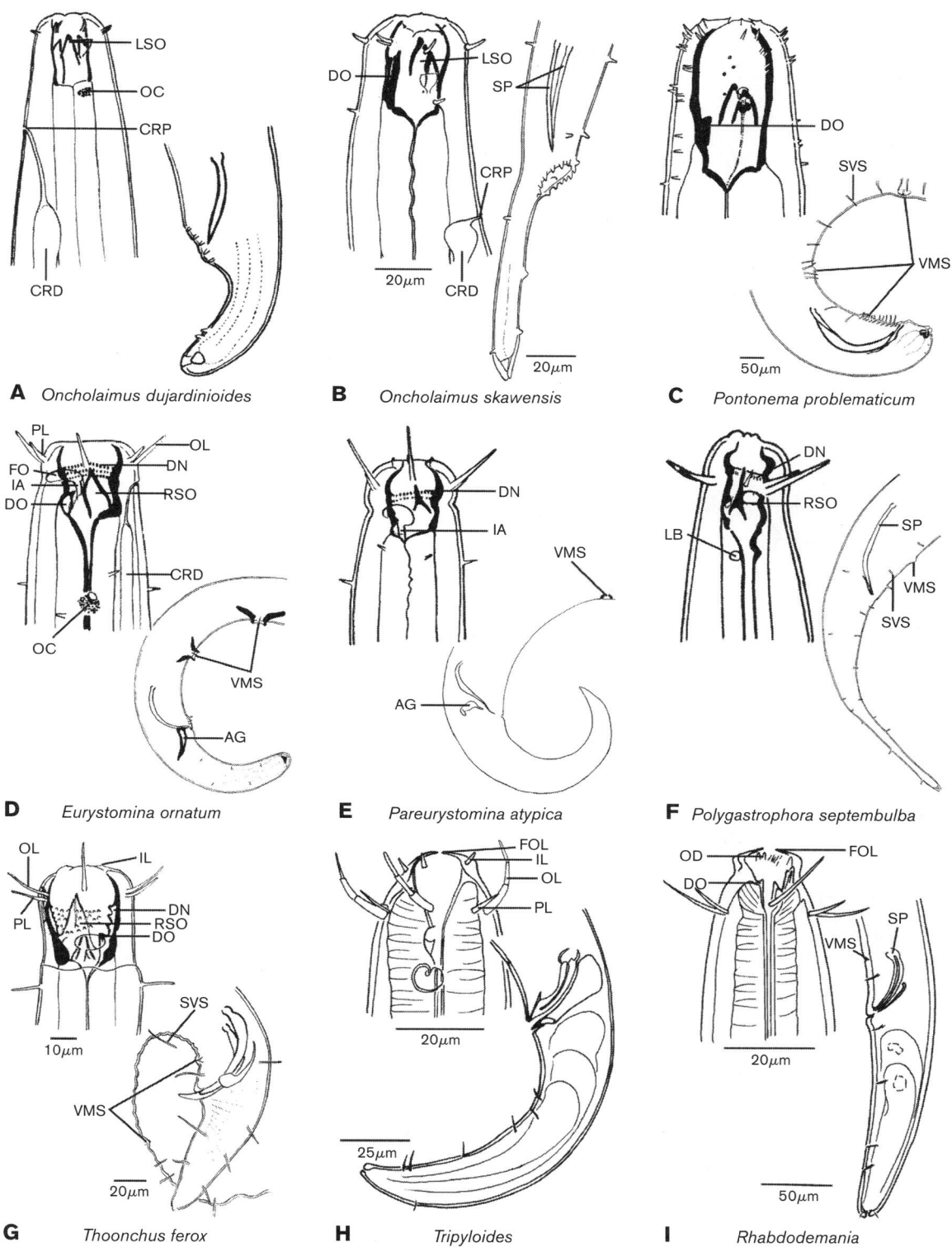

A *Oncholaimus dujardinioides*

B *Oncholaimus skawensis*

C *Pontonema problematicum*

D *Eurystomina ornatum*

E *Pareurystomina atypica*

F *Polygastrophora septembulba*

G *Thoonchus ferox*

H *Tripyloides*

I *Rhabdodemania*

PLATE 103 A–G, from Chitwood 1960b, with permission from TAMS.

on anterior dorsal surface of head (Epsilonematidae) 39

— Ovaries situated anterior to dorsal bend of body (i.e., in mid-region of body; adhesive ambulatory setae situated anterior to region of body bearing ovaries; adhesive setae situated on anterior dorsal surface of head (Draconematidae) 40

39. Thornlike spines not present on dorsal cervical region (plates 92A, 106F). . . . . Epsilonematinae

— Thornlike spines present on anterior dorsal cervical region . . . . Glochinematinae

40. Buccal capsule without teeth; pharynx with mid- and posterior swellings (plates 92B, 106E) . . . . Draconematinae

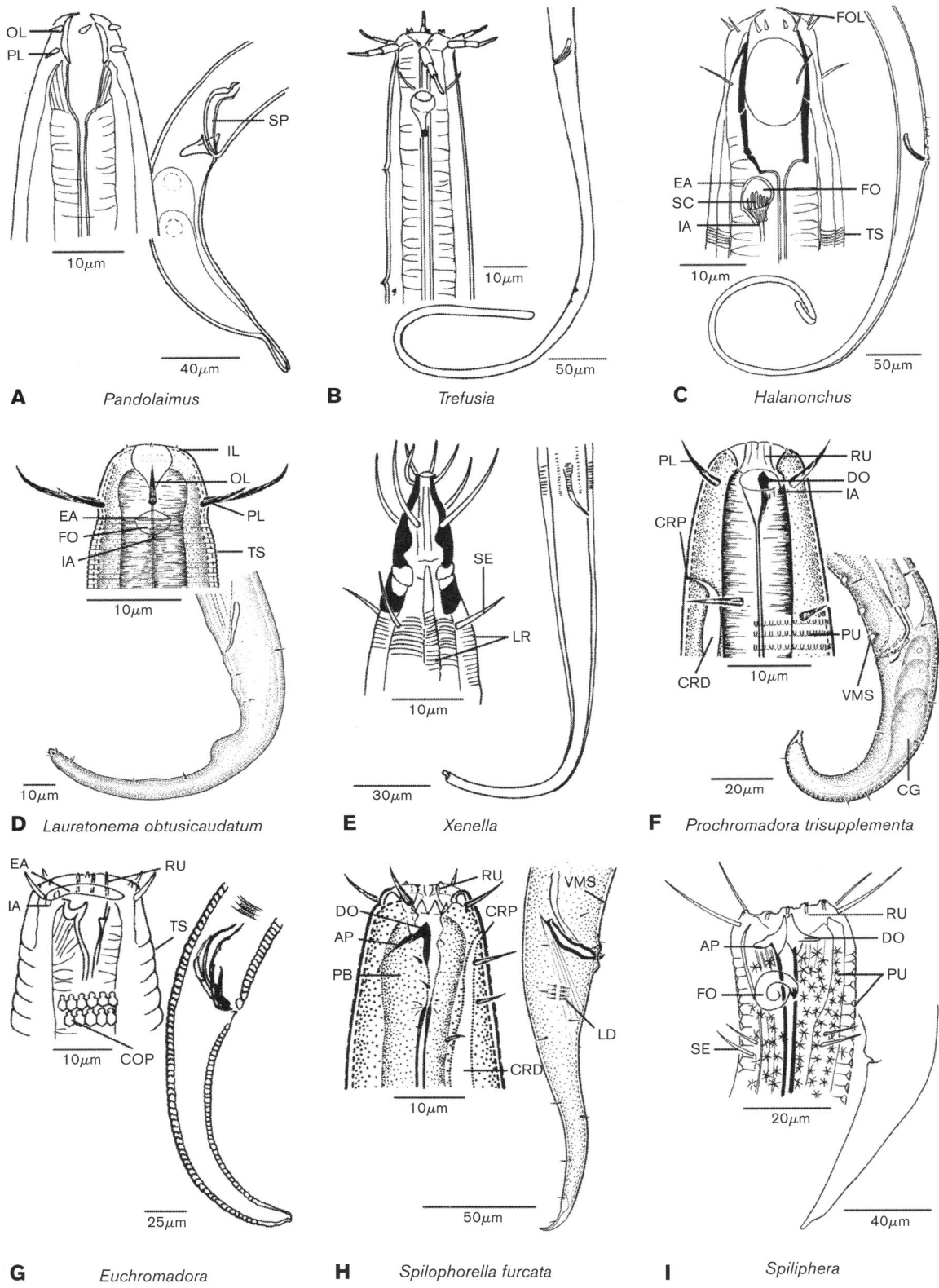

A *Pandolaimus*

B *Trefusia*

C *Halanonchus*

D *Lauratonema obtusicaudatum*

E *Xenella*

F *Prochromadora trisupplementa*

G *Euchromadora*

H *Spilophorella furcata*

I *Spiliphera*

PLATE 104 D, from Murphy and Jensen 1961, with permission from Murphy and HSW; F, from Murphy 1963b and H, from Murphy, 1963c, both with permission from the author and HSW.

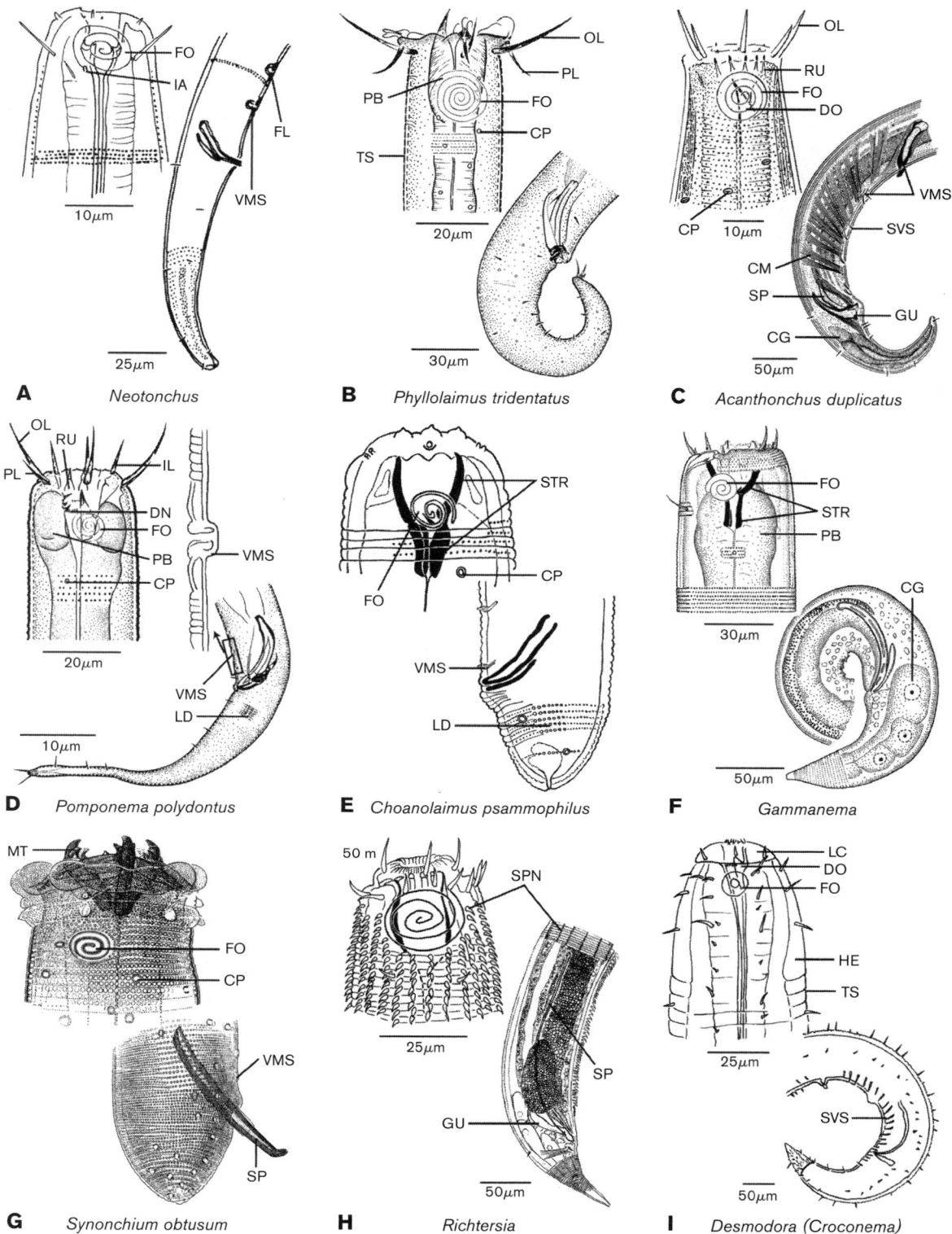

A   *Neotonchus*

B   *Phyllolaimus tridentatus*

C   *Acanthonchus duplicatus*

D   *Pomponema polydontus*

E   *Choanolaimus psammophilus*

F   *Gammanema*

G   *Synonchium obtusum*

H   *Richtersia*

I   *Desmodora (Croconema)*

PLATE 105 B and D, from Murphy 1963b, with permission from the author and HSW; E, from Chitwood 1960a, with permission from Brill Academic Publishers; F, from Murphy 1964b, with permission from the author and HSW; G, from Cobb 1920 (Ventromedian supplement in D, original).

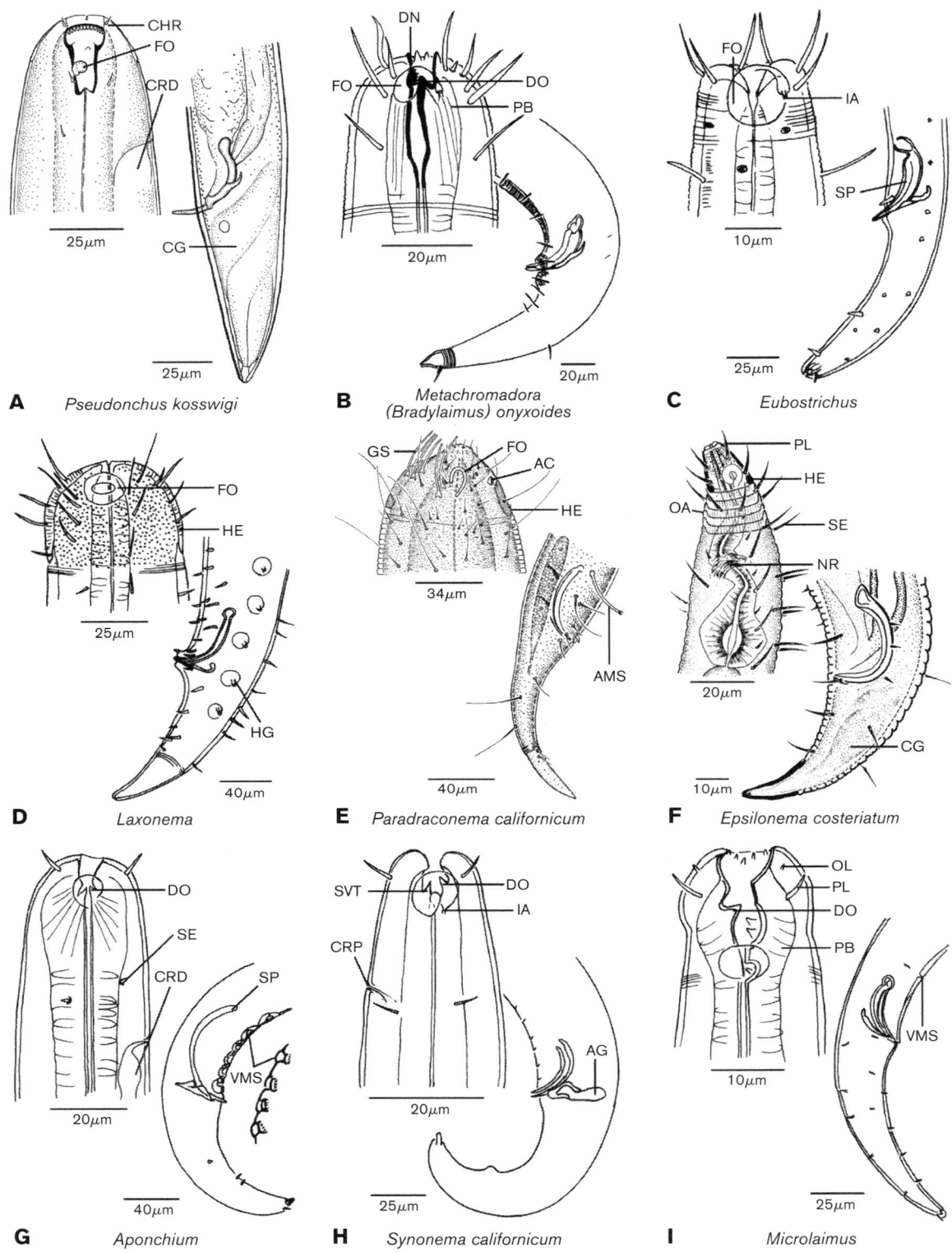

A *Pseudonchus kosswigi*

B *Metachromadora (Bradylaimus) onyxoides*

C *Eubostrichus*

D *Laxonema*

E *Paradraconema californicum*

F *Epsilonema costeriatum*

G *Aponchium*

H *Synonema californicum*

I *Microlaimus*

PLATE 106 A, from Murphy 1964c, with permission from the author and Mitteilungen Hamburgischen Zoologischen Museum und Institut; E, from Allen and Noffsinger 1978, with permission from Noffsinger; F, from Murphy 1963c, with permission from the author and HSW; H, after Jensen 1989.

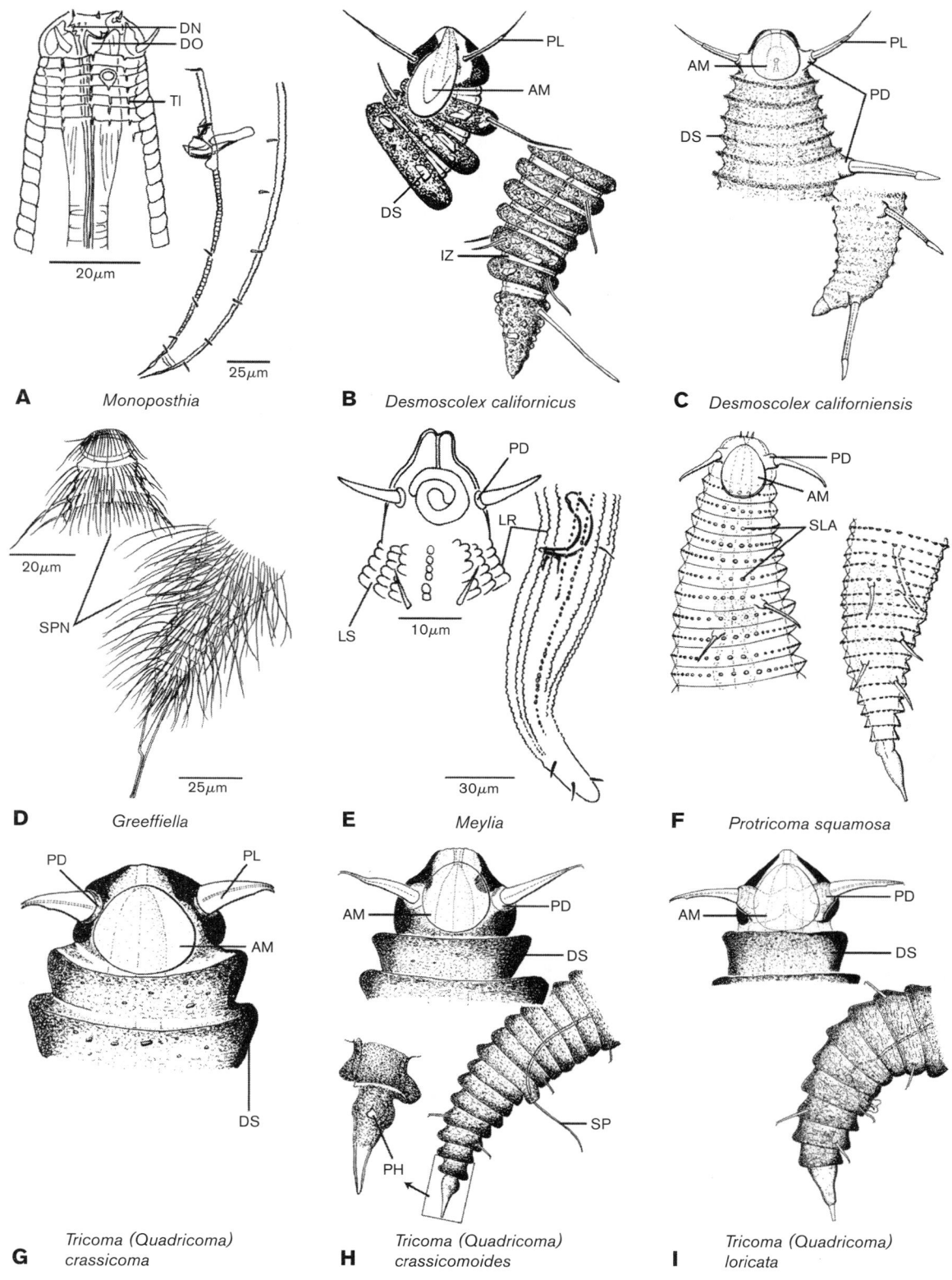

A *Monoposthia*

B *Desmoscolex californicus*

C *Desmoscolex californiensis*

D *Greeffiella*

E *Meylia*

F *Protricoma squamosa*

G *Tricoma (Quadricoma) crassicoma*

H *Tricoma (Quadricoma) crassicomoides*

I *Tricoma (Quadricoma) loricata*

PLATE 107 B, C, F–I, from Timm 1970, with permission from the author; E, after Gerlach 1956.

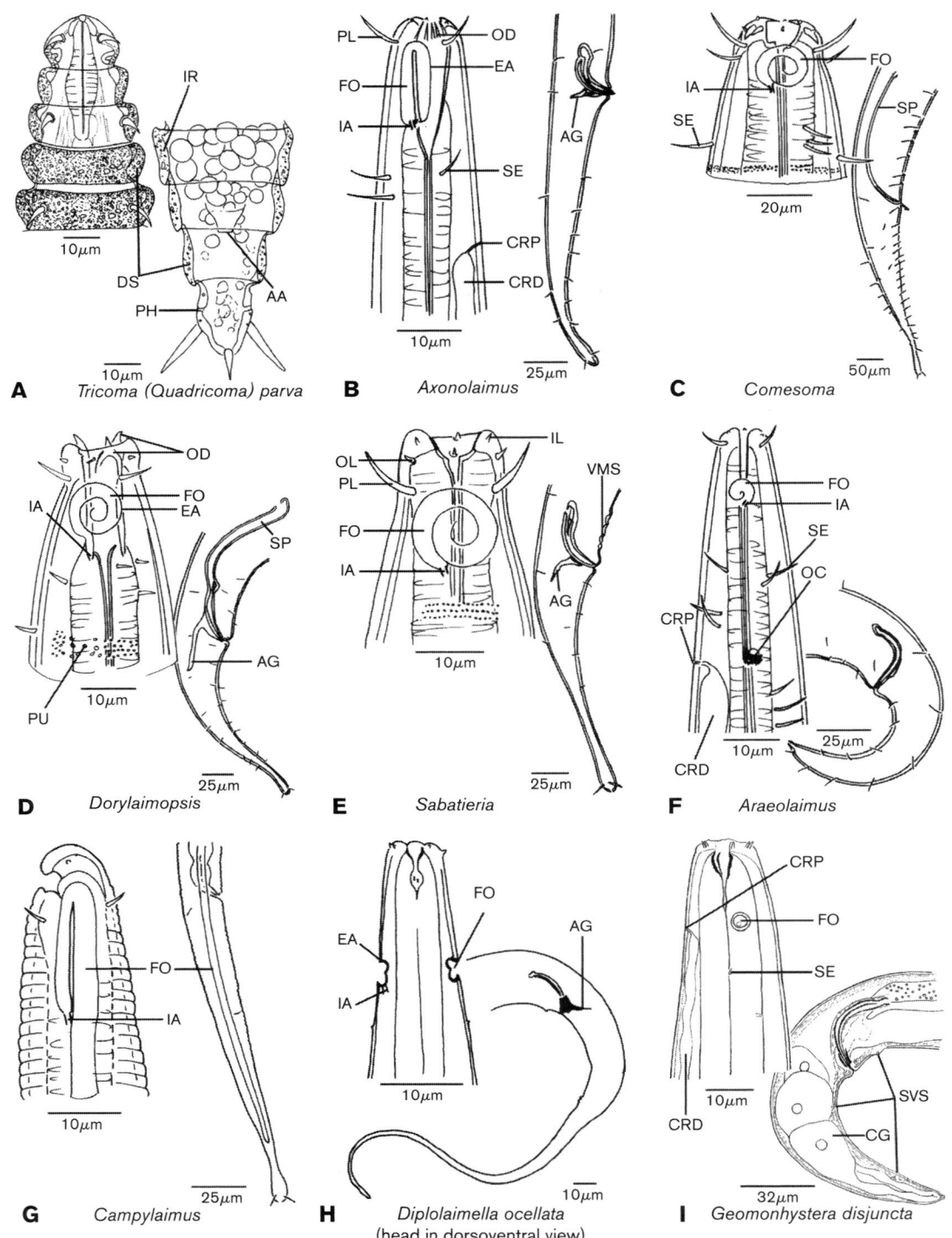

A   *Tricoma (Quadricoma) parva*

B   *Axonolaimus*

C   *Comesoma*

D   *Dorylaimopsis*

E   *Sabatieria*

F   *Araeolaimus*

G   *Campylaimus*

H   *Diplolaimella ocellata*
(head in dorsoventral view)

I   *Geomonhystera disjuncta*

PLATE 108 H and I, from Chitwood and Murphy 1964, with permission from Murphy and TAMS.

— Buccal capsule usually with dorsal tooth; pharynx with posterior swelling, rarely with mid-pharyngeal swelling . . . . Prochaetosomatinae

41. Amphids unispiral; cuticle without distinct overlapping annules and longitudinal ridges; ovaries outstretched; males diorchic or monorchic 42

— Amphid circular; cuticle with distinct overlapping annules and longitudinal files of tines; ovaries antidromous; males diorchic, testes opposed (Monoposthiidae) (plate 107A)

. . . . Monoposthiinae

42. Females didelphic; males diorchic, testes opposed (Microlaimidae) (plate 106I) . . . . Microlaiminae

— Females monodelphic; males monorchic (Aponchiidae) (plate 106F–106H) . . . . Aponchiinae

43. Cuticle striated, ornamented with prominent setae, numerous spines in transverse rows, scales, wartlike protuberances, or desmens; subdorsal and subventral somatic sensilla paired (one on right and left sides in each pair) and

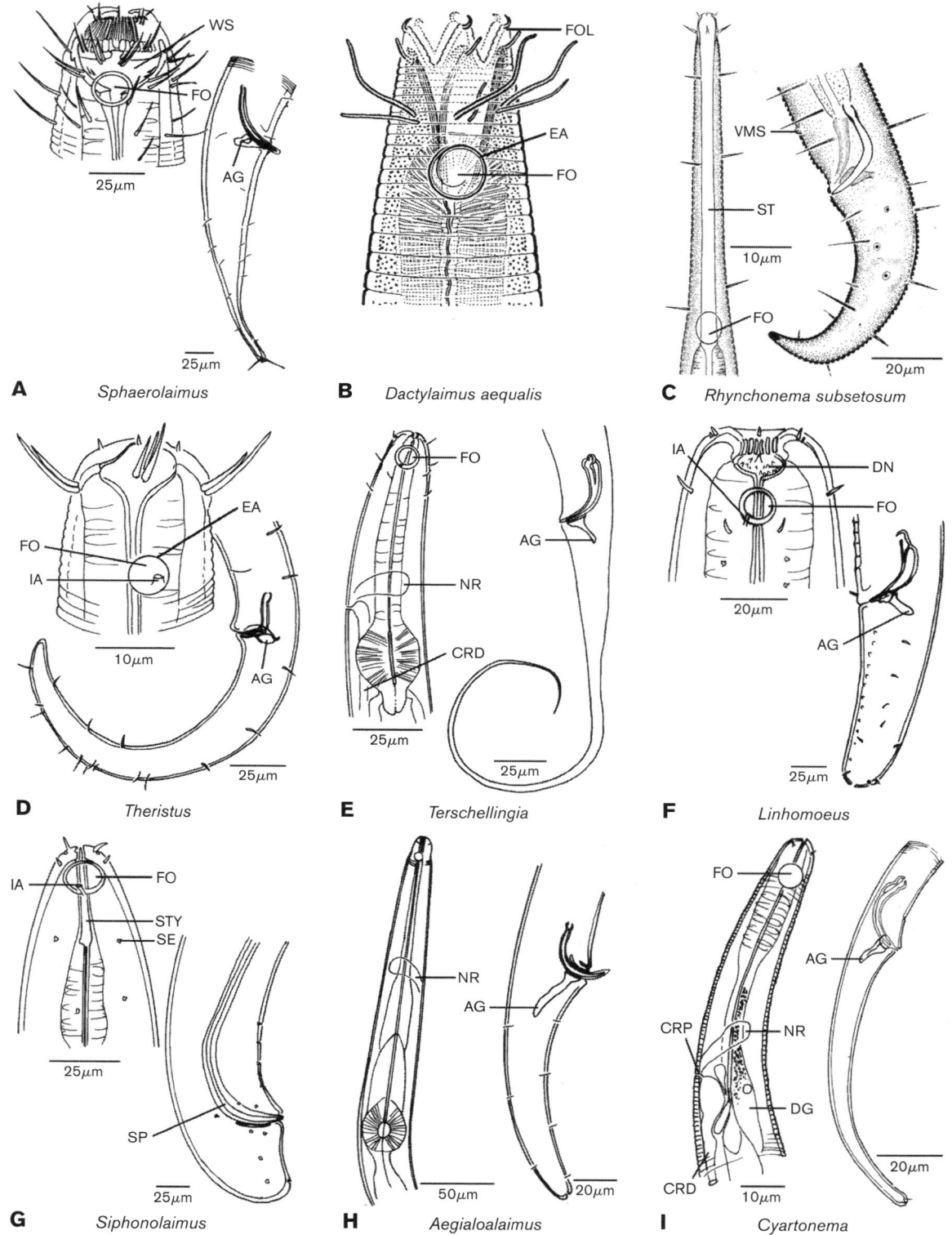

A *Sphaerolaimus*

B *Dactylaimus aequalis*

C *Rhynchonema subsetosum*

D *Theristus*

E *Terschellingia*

F *Linhomoeus*

G *Siphonolaimus*

H *Aegialoalaimus*

I *Cyartonema*

PLATE 109 B, from Cobb 1920; C, from Murphy 1964a, with permission from the author and HSW.

distributed in identifiable patterns (see below); amphids vesiculate, sometimes lobelike, and situated between or posterior to paralabial sensilla; corpus gelatum nonspiral; pseudocoelomocytes absent; male reproductive system monorchic or diorchic; female reproductive system didelphic, amphidelphic, and outstretched (Desmoscolecidae)

....45

— Cuticle without ornamentation; transverse or transverse and longitudinal striae (cobbled; plate 107E) present with or without spines in longitudinal files; amphids circular with indistinct or spiral corpus gelatum; cuticle rarely with thin layer of adherent detritus; subdorsal and subventral

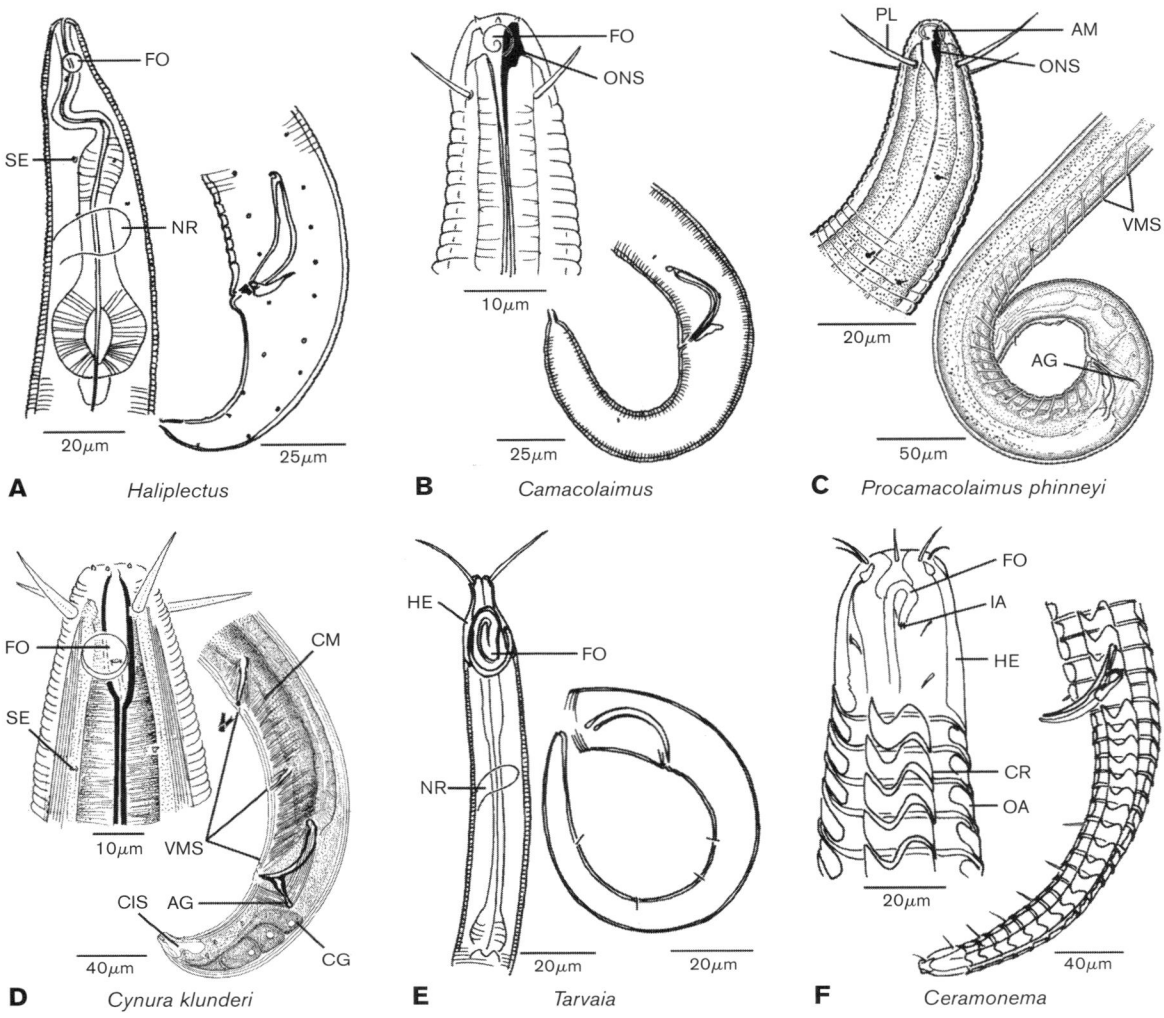

A *Haliplectus*

B *Camacolaimus*

C *Procamacolaimus phinneyi*

D *Cynura klunderi*

E *Tarvaia*

F *Ceramonema*

PLATE 110 C, from Murphy 1964b, with permission from the author and HSW.

somatic sensilla sparse and not paired or distributed in identifiable patterns; esophagus with muscular, cylindrical anterior part, thin mid-section and glandular terminus, often with dorsal esophageal gland overlapping anterior end of gut; pseudocoelomocytes usually present; male reproductive system diorchic; female reproductive system didelphic, amphidelphic, and gonoducts reflexed 44

44. Cervical gland absent; cuticle with transverse striae, with or without longitudinal ridges, and transverse rows or longitudinal files of short spines; lateral field sometimes without striation; cuticle sometimes with thin layer of adherent particles (Meyliidae) (plate 107E) . . . . Meyliinae

— Cervical gland present; cuticle with transverse striations only and without additional ornamentation (Cyartonematidae) (plate 109I) . . . . Cyartonematinae

45. (*Note three choices*) Adults usually with desmens on individual body rings or on specialized papillae; somatic sensilla arranged in desmoscolecid pattern (i.e., subdorsal somatic sensilla more abundant than subventral sensilla), with subdorsals usually on desmens one, three, five, seven, nine, 11, 13, 16, and 17 (the terminal desmen), and subventrally on desmens two, four, six, eight, 10, 12, 14, and 15; first subventral pair situated more laterally than remaining subventral sensilla; last subdorsal pair (ninth) situated on end ring and last subventral pair (eight) on or near anus/cloaca bearing ring; phasmata and pigment spots usually present; males monorchic (plates 92C, 107B–107C) . . . . Desmoscolecinae

— Adults without deposits of secretions and foreign particles on rings, but with numerous spines, papillae, or fine fringes in transverse rows; somatic sensilla, where present, paired and arranged as in Desmoscolecinae (i.e., with more subdorsal than subventral sensilla), and end ring always with pair of subdorsal sensilla; males monorchic (plates 92D, 107D) . . . . Greeffiellinae

— Adults with or without deposits of secretions and foreign particles on rings; somatic sensilla arranged in tricomid pattern (i.e., subdorsal less abundant than subventral sensilla); posterior-most sensilla always subventral, with sensilla never present on terminal desmen; males diorchic (plates 107F–108A) . . . . Tricominae

46. Amphidial fovea circular and concave; gonoducts always outstretched; ventromedian supplements never tubiform (Monhysterida) 52

TABLE 1
Abbreviations Used in Nematoda Plates 91–110

AA: anal aperture
AC: acanthiform seta
AD: amphidial duct
AF: amphidial flap
AG: apophysis of gubernaculum
AM: amphid
AMS: ambulatory setae
AP: apodeme
AS: arched suture
AT: anterior testis
CA: cloacal aperture
CB: copulatory bursa
CC: cephalic capsule
CG: caudal gland
CH: cheilostome
CHR: cheilostome ridge with denticles
CIS: caudal intracuticular shield
CM: copulatory muscle
COG: corpus of gubernaculum
COP: complex punctation
CP: cuticular pore
CR: crest
CRD: cervical gland duct
CRP: cervical gland pore
CRU: crus
CS: cephalic sensilla
CVS: cervical ventromedian supplement
DE: ductus entericus
DG: dorsal esophageal gland
DN: denticles
DO: dorsal onchium
DP: ductus principalis
DS: desmens
DU: ductus uterinus
EA: external aperture of amphid
EM: eversible mandible
ES: esophagostome
FE: fenestrations of cephalic capsule
FL: flanges of ventromedian supplement
FO: fovea of amphid

FOL: foliate lips
GU: gubernaculum
GS: glandular setae
GT: gut
HE: helmet
HG: hypodermal gland
IA: internal aperture of amphid
IL: inner labial sensilla
IO: intraradial odontium
IR: inversion ring where direction of
  annular overlap reverses direction
IZ: interzone
LC: labial cuff
LD: lateral differentiation
LB: lens-like body
LO: locule
LR: longitudinal ridge
LS: longitudinal striae
LSO: left subventral onchium
ME: metaneme
ML: microlabium
MN: mandible
MR: mandibular ridge
MT: mandibular teeth
MU: muniment
NR: nerve ring
OA: overlapping annules
OC: ocellus
OD: odontium
OL: outer labial sensilla
ON: onchium
ONS: onchiostyle
OP: ocellar pigment
OS: osmosium
OVD: oviduct
OVM: ovum
OVR: ovary
PB: pharyngeal bulb
PC: post-cephalic sensilla
PD: peduncle

PF: precloacal flap
PH: phasmata
PL: paralabial sensilla
PSP: point of sperm penetration in
  traumatic insemination
PT: posterior testis
PU: punctation
RC: retractable cuff
RD: radial dilator muscles of the
  esophagus
RM: retrodilator muscle
RSO: right subventral onchium
RU: rugae
SC: sensory cilia of amphid
SE: sensilla (setae)
SLA: scale-like adornments
SP: spiculum
SPN: spines
SPT: spinneret (caudal gland pore)
ST: stoma
STY: stylet
STR: strut
SV: seminal vesicle
SVEG: subventral esophageal gland
SVO: subventral onchium
SVR: subventral supplement on raised
  cuticle
SVS: subventral supplement
SVT: subventral teeth
TI: tine
TR: tropis
TS: transverse striae
UT: uterus
UV: uvette
VA: valve-like apparatus
VMS: ventromedian supplement
VU: vulva
WS: wall of stoma

— Amphidial fovea circular and concave or troughlike and spiral or looped; if fovea circular, then gonoduct antidromous; ventromedian supplements papilliform, setiform, tubiform or absent (Araeolaimida) 47
47. Amphidial fovea looped, or if spiral, then with not more than one and one-quarter turns; cuticle usually without punctations 50
— Fovea spiral with more than two turns; cuticle with punctations (Comesomatidae) 48
48. Spicula not much more than one cloacal-body-diameter long, or if longer, then stoma enveloped with longitudinal protractor muscles and armed with eversible odontia; gubernaculum usually with apophyses 49
— Spicula elongate, more than two cloacal-body diameters; eversible odontia never present; gubernaculum usually without apophyses (plate 108C) . . . . Comesomatinae
49. Stoma elongate and cylindrical, usually with odontia (plate 108D) . . . . Dorylaimopsinae
— Stoma small and cyathiform or indistinct; stoma usually

without odontia or onchia (plate 108E) . . . . . Sabatieriinae
50. Amphidial fovea uni- or multispiral; esophagostome small and cyathiform, or indistinct from esophagus; stoma unarmed 51
— Amphidial fovea looped like inverted U; esophagostome spacious, elongate, and infundibular or cylindrical, often with odontia (Axonolaimidae) (plate 108B) . . . .
. . . . Axonolaiminae
51. Outer labial sensilla papilliform; buccal capsule small and cyathiform or not differentiated from lumen of esophagus; oral aperture sometimes shifted dorsad (Diplopeltidae) (plate 108F–108G) . . . . Diplopeltinae
— Outer labial sensilla setiform; buccal capsule not differentiated from lumen of esophagus; stoma never displaced dorsad (Coninckiidae) . . . . Coninckiinae
52. Cuticle smooth or striated; cephalic sensilla almost always 6 + 10; inner labial sensilla papiliform; fovea of amphids concave and circular; oral rim of cheilostome with thin, foliate flap; buccal cavity funnel-shaped; esophagus cylindrical,

never with posterior bulb; cardium conical or flat and enclosed by anterior end of gut, but never columnar and situated external to gut lumen; females monodelphic, prodelphic (Monhysterina)..........................56
— Cuticle smooth or striated; cephalic sensilla 6 + 6 + 4 or 6 + 10; outer labial longer or shorter than paralabial sensilla; amphid thick-rimmed; cheilostome often thickened and padlike, or lips not evident and buccal cavity with stylet; medial rim of cheilostome never thin and foliate; cardium often columnar, nonmuscular and external to gut lumen; females didelphic (Linhomoeina)...........53
53. Buccal capsule small to large and well-developed, but never degenerate or tuboid and armed with protrusile stylet (Linhomoeidae).......................................54
— Buccal capsule degenerate or tuboid with protrusile stylet at anterior end of esophagus (Siphonolaimidae) (plate 109G)
....Siphonolaiminae
54. Cephalic sensilla 6 + 6 + 4 with four to eight subcephalic sensilla......................................55
— Cephalic sensilla 6 + 6 + 4 without subcephalic sensilla (plate 109E).......................Desmolaiminae
55. Esophagostome variable in shape, but not deep and dolioform (plate 109F)....................Linhomoeinae
— Esophagostome longer than wide and dolioform.......
...............................Eleutherolaiminae
56. (*Note three choices*) Cuticle usually smooth; outer labial about as long as paralabial sensilla; subcephalic sensilla absent; buccal capsule variable in shape, but never dolioform; esophagostome enveloped by esophageal tissue; cervical gland usually present; caudal gland open through common terminal pore; males monorchic; testis always on right side of gut; females monodelphic; ovary always on right of gut; tail without terminal, setiform sensilla (Monhysteridae) (plate 108H–108I).........Monhysterinae
— Cuticle distinctly striated; cephalic sensilla 6 + 10; outer labial sensilla same length as or longer than paralabial sensilla; subcephalic sensilla seldom present; rim of cheilostome foliate; buccal capsule conical; buccal armament usually absent; cervical gland usually absent; caudal glands open through separate, subterminal pores, or not visible; anterior ovary on left of gut; males diorchic or monorchic; anterior testis always on left and posterior, if present, on right of gut (Xyalidae) (plate 109B–109D)
....Xyalinae
— Cuticle striated or smooth; cephalic sensilla 6 + 10; outer labial sensilla always shorter than paralabial sensilla; two lateral and four sublateral groups of subcephalic sensilla present; rim of cheilostome foliate; esophagostome only partially enveloped by esophageal tissue; buccal capsule usually dolioform; cervical gland always present; tail conical and terminally cylindrical; tail with two or three terminal sensilla; caudal glands with separate subterminal pores; males usually diorchic; anterior testis on right of gut or left of gut; posterior testis always on opposite side of anterior (Sphaerolaimidae)..........................57
57. Wall of esophagostome not divided into six plates (plate 109A)...........................Sphaerolaiminae
— Wall of esophagostome divided into plates............
...............................Parasphaerolaiminae
58. Helmet absent; amphid reduced spiral..............61
— Helmet present; amphid inverted U-shaped or reversed spiral (i.e., spiral with internal aperture at center or inner end of spiral).......................................59
59. Cuticle pattern tilelike and amphid inverted U; or cuticle

striated and amphid reversed spiral (Ceramonematoidea)
....60
— Cuticle striated; amphid inverted U-shaped (Diplopeltoididae)...........................Diplopeltoidinae
60. Amphid reversed spiral; cuticle striated; head strongly tapered anteriorly (Tarvaiidae) (plate 110E).......Tarvaiinae
— Amphid U-shaped; cuticle tile-like; head not strongly narrowed anteriorly (Ceramonematidae) (plate 110F)......
...............................Ceramonematinae
61. Cephalic sensilla 6 + 6 + 4, paralabial sensilla setiform; esophagus shape variable, but if with thin anterior region enveloping esophagostome, then never with well-developed posterior bulb; ventromedian supplements almost always tubiform (Leptolaimoidea)..................62
— Cephalic sensilla 6 + 6 + 4, paralabial sensilla papilliform; esophagus anteriorly thin, enveloping elongate esophagostome, then cylindrical with well-developed posterior bulb; posterior bulb with thickened cuticular lining; ventromedian supplements papilliform (Haliplectoidea; Haliplectidae) (plate 110A).....................Haliplectinae
62. Cheilostome without longitudinal ribs..............63
— Cheilostome with six longitudinal ribs (Rhadinematidae)
...............................Rhadinematinae
63. Esophagus with or without muscular posterior bulb, but if present, then lumen of esophagus not extending from oral aperture to bulb as thin, cuticular tube.............64
— Lumen of esophagus extending from oral aperture as thin, cuticular tube, apparently without surrounding tissue, to very muscular posterior bulb (Aegiaolaimidae) (plate 109H)
...............................Aegiaolaiminae
64. Anterior neck region never strongly bent ventrad; amphid spiral, looped like inverted U or circular; if amphids circular, then males with tubiform ventromedian supplements; somatic sensilla never on peduncles; esophagus never with large, pear-shaped posterior bulb; females almost always didelphic, males usually diorchic (Leptolaimidae)......65
— Anterior neck region strongly bent ventrad; amphids circular; somatic sensilla on peduncles; esophagus with large pear-shaped terminal bulb; females monodelphic; males monorchic; ventromedian, precloacal supplements absent (Peresianidae).....................Manunematinae
65. Amphid situated posterior to paralabial sensilla......66
— Amphid situated at level of or anterior to paralabial sensilla (plate 110B–110C)..............Camacolaiminae
66. Esophagostome cylindrical, elongate and spacious; cheilostome with odontia................Anonchinae
— Esophagostome variable, but if cylindrical, elongate, and spacious cheilostome never with odontia (plate 110D)
...............................Leptolaiminae

List of Species

Most of the nematode species listed below are from either the holdfasts of intertidal kelps, such as *Laminaria* and *Egregia*, or from intertidal sediments. Numerous other marine habitats are inhabited by nematodes representing many other species either undescribed or previously unrecognized from this region. Places to explore include mud flats, high-energy sandy beaches, many other rocky intertidal microhabitats (e.g., among sea anemones, in mussel beds, or under other species of algae such the high intertidal mosslike clumps of *Cladophora*, in high intertidal tide pools), fouling communities on pilings and marina floats, marsh channels, and so forth. In estuaries and harbors, particular attention should be made to the possibility of encountering species introduced with oysters, in ship fouling, in ballast water, or by other mechanisms.

Anoplostomatidae

Anoplostoma viviparum (Bastian, 1865) Bütschli, 1874 (=*Symplocostoma viviparum* Bastian, 1865). Aquarium with brackish intertidal sediment from San Francisco Bay near Pinole (Chitwood 1960b: 358). Plate 100E.

Phanodermatidae

Phanoderma segmentum Murphy, 1963 (*lapsus P. segmenta* and *P. segments*, Murphy, 1963c: 249). Intertidal rock scrapings near Depot Bay, north of Devil's Punch Bowl, Oregon (Murphy 1963c: 249). Plate 100G.

Ironidae

Trissonchulus raskii (Chitwood, 1960) Inglis, 1961 (=*Dolicholaimus raskii* Chitwood, 1960). Moss Landing, Monterey Bay (Chitwood 1960b: 352). Plate 101A.

Leptosomatidae

Deontostoma californicum Steiner and Albin, 1933. Known from Pacific Grove (Steiner and Albin 1933: 25); Point Pinos, Pacific Grove, and Shell Beach, La Jolla (Hope 1967b: 323); sediment from holdfasts of *Egregia* in shallow subtidal off Santa Barbara (Hope and Gardiner 1982: 1); sediment in holdfasts of *Laminaria digitata* and *Egregia* spp. on intertidal rocks at Dillon Beach (Croll and Maggenti 1968: 108; Siddiqui and Viglierchio 1971: 264; Hope 1967b: 323; Maggenti 1964: 159; Viglierchio and Siddiqui 1974: 338; Wright et al. 1973: 30). Plate 101F.

Deontostoma magnificum (Timm, 1951) Platonova, 1962: 203 (=*Thoracostoma magnificum* Timm, 1951; *Thoracostoma pacificum* Murphy, 1965a). Sediment from high-tide mussel beds at Haystack Rock, Cannon Beach, Oregon (Murphy 1965b: 106). Plate 101G.

Deontostoma washingtonense (Murphy, 1965) Hope, 1967: 327 (=*Thoracostoma washingtonensis* Murphy, 1965b: 211). Sediment from holdfasts of *Egregia* sp. at Tongue Point; intertidal rocks at Dillon Beach and Point Pinos, Pacific Grove (Hope 1967b: 327). Plate 101H.

Pseudocella panamaense (Allgén, 1947) Wieser, 1953 (=*Thoracostoma panamaense* Allgén, 1947). Sediment in holdfasts of *Egregia* sp.; intertidal sediment at Bolinas Bay (Hope 1967b: 314). Plate 101I.

Pseudocella (Pseudocella) wieseri Hope, 1967 (=*Thoracostoma panamaense* Wieser, 1953 [nec Allgén, 1947]). Sand from holdfasts of *Egregia* on intertidal rocks Dillon Beach and intertidal sediment collected in Bolinas Bay (Hope 1967b: 309). Plate 101J.

Thoracostoma trachygaster Hope, 1967. Holdfasts of *Egregia* at Dillon Beach and Point Pinos, Pacific Grove (Hope 1967b: 317). Plate 102A.

Triceratonema microlobatum (Allgén, 1947) Platonova, 1976 (=*Thoracostoma microlobatum* Allgén, 1947). From holdfasts of *Egregia* on intertidal rocks at Point Pinos, Pacific Grove (Hope 1967b: 321). Plate 102B.

Triaulolaimus triaulolaimus (Hope, 1967) Platonova, 1979 (=*Pseudocella (Corythostoma) triaulolaimus* Hope, 1967). Sediment held by holdfasts of *Egregia menziesii* on intertidal rocks at Dillon Beach (Hope 1967a: 6). Plate 101D.

Oncholaimidae

Viscosia papillatoides Chitwood, 1960. Pinole, San Francisco Bay (Chitwood 1960b: 365). Plate 102F.

Metoncholaimus longiovum Chitwood, 1960. Mud below low-tide mark, Tomales Bay (Chitwood 1960b: 368). Plate 102G.

Oncholaimium domesticum Chitwood and Chitwood, 1938 (=*Oncholaimium oxyuris* Ditlevsen, 1941 *v. domesticus* Chitwood and Chitwood, 1938) Rachor, 1969: 93. Aquarium with sediment from Pinole, San Francisco Bay. Plate 102H.

Oncholaimium sheri Chitwood, 1960. San Francisco Bay (Chitwood 1960b: 364). Plate 102I.

Oncholaimus dujardinioides Chitwood, 1960. Algae on intertidal rocks, Dillon Beach (Chitwood 1960b: 359). Plate 103A.

Oncholaimus skawensis Ditlevsen, 1921. Holdfasts of *Egregia laevigata*, Dillon Beach (Chitwood 1960b: 359). Plate 103B.

Pontonema problematicum Chitwood, 1960. Holdfasts of *Egregia laevigata* near low tide, Dillon Beach (Chitwood 1960b: 370; Maggenti 1964: 159). Plate 103C.

Enchelidiidae

Eurystomina ornatum (Eberth, 1863) Marion, 1870 (=*Enoplus ornatus* Eberth, 1865). Sediment behind rocks at Dillon Beach (Chitwood 1960b: 377). Plate 103D.

Pareurystomina atypica Chitwood, 1960. Sediment in holdfasts of *Egregia laevigata*, Dillon Beach (Chitwood 1960b: 378). Plate 103E.

Polygastrophora septembulba Gerlach, 1954. Aquarium with brackish, intertidal sediment from San Francisco Bay near Pinole (Chitwood 1960b: 374). Plate 103F.

Thoonchus ferox Cobb, 1920. Intertidal sand and tide pools behind rocks at Dillon Beach (Chitwood 1960b: 371). Plate 103G.

Lauratonematidae

Lauratonema obtusicaudatum Murphy and Jensen, 1961. Subtidal sediment from South Slough, Coos Bay. Plate 104D. Murphy and Jensen, 1961: 167. Plate 104D.

Chromadoridae

Prochromadora trisupplementa Murphy, 1963. From epiphytic filamentous algae on stipe of *Egregia menziesii* at Yaquina Head, Newport (Murphy 1963b: 75). Plate 104F.

Spilophorella furcata Murphy, 1963 (*lapsus S. furecata* Murphy, 1963c: 253). From intertidal rock scrapings at Devil's Punch Bowl, Oregon (Murphy 1963c: 249). Plate 104H.

Cyatholaimidae

Phyllolaimus tridentatus Murphy, 1963. Intertidal sand exposed to moderate wave action at Newport, Oregon (Murphy 1963b: 75). Plate 105B.

Acanthonchus duplicatus Wieser, 1959. Sediment associated with holdfasts of *Egregia* at Dillon Beach (Wright and Hope 1968: 1005). Plate 105C.

Pomponema polydontus Murphy, 1963 (*lapsus P. polydonta* Murphy, 1963). Sand from intertidal beach exposed to moderate wave action at Newport, Oregon (Murphy 1963b: 73). Plate 105D.

Selachinematidae

Choanolaimus psammophilus de Man, 1880. About roots of beach spermatophytes, Moss Landing, Monterey Bay (Chitwood 1960a: 57). Plate 105E.

Synonchium obtusum Cobb, 1920. About roots of spermatophytes and in sand of inner bay, Moss Landing, Monterey Bay (Chitwood 1960a: 58). Plate 105G.

Desmodoridae

Pseudonchus kosswigi Murphy, 1964. Beach adjacent to Yaquina Lighthouse, Newport, Oregon (Murphy 1964c: 114). Plate 106A.

Draconematidae

Paradraconema californicum Allen and Noffsinger, 1978. Holdfast of *Laminaria digitata*, Dillon Beach (Allen and Noffsinger 1978: 50). Plate 106E.

Epsilonematidae

Epsilonema costeriatum (Murphy, 1963) Lorenzen, 1973; (=*Notochaetosoma costeriatum* Murphy, 1963) (*lapsus N. costeriata* Murphy, 1963: 250). Holdfast of alga *Costeria costata* at Marine Gardens north of Devil's Punch Bowl, Oregon (Murphy 1963c: 250). Plate 106F.

Aponchiidae

Synonema californicum Jensen, 1989. From holdfasts on rocky bottom in Jenner, near mouth of Russian River (Jensen 1989: 13). Plate 106H.

Desmoscolecidae

Desmoscolex californicus Timm, 1970. Mud at depth of 0.6 m at Miller Park, Tomales Bay (Timm 1970: 19). Plate 107B.

Desmoscolex californiensis Gerlach and Riemann, 1973 (=*Eudesmoscolex californicus* Timm, 1970; *D. californicus* (Timm, 1970) Lorenzen, 1971 is a secondary homonym of *D. californicus* Timm, 1970). *Species inquirenda op* Lorenzen, 1971: 344. Holdfasts of *Laminaria* Dillon Beach (Timm 1970: 33). Plate 107C.

Protricoma squamosa Timm, 1970. Holdfasts, rocks, and mud at Jenner near mouth of Russian River; rock scrapings at Yaquina Head, Oregon (Timm 1970: 36). Plate 107F.

Tricoma (Quadricoma) crassicoma Steiner, 1916. Jenner, Pacific Grove, and Dillon Beach (Timm 1970: 42). Plate 107G.

Tricoma (Quadricoma) crassicomoides Timm, 1970. *Egregia* holdfasts on rocks near Point Pinos, Pacific Grove (Timm 1970: 43). Plate 107H.

Tricoma (Quadricoma) loricata Filipjev, 1922. *Egregia* holdfasts on rocks near Point Pinos, Pacific Grove (Timm 1970: 46). Plate 107I.

Tricoma (Quadricoma) parva Timm, 1970. *Egregia* holdfasts, rocks, and mud, Jenner (Timm 1970: 48). Plate 108A.

Monhysteridae

Diplolaimella (Diplolaimita) ocellata (Bütschli, 1874) Gerlach, 1957 (=*Monhystera ocellata* Bütschli, 1874; *Diplolaimella (Diplolaimita) schneideri* Timm, 1952). Intertidal mud at Pinole, San Francisco Bay and in similar habitat near a freshwater inlet of Tomales Bay (Chitwood and Murphy 1964: 315). Plate 108H.

Geomonhystera disjuncta (Bastian, 1865) Jacobs, 1987 (=*Monhystera disjuncta* Bastian, 1865). Decaying algae at Dillon Beach (Chitwood and Murphy 1964: 319). Plate 108I.

Xyalidae

Dactylaimus aequalis Cobb, 1920: 250. Mud from San Francisco Bay. Plate 109B (juvenile only).

Rhynchonema subsetosum Murphy, 1964 (*lapsus R. subsetosa* Murphy, 1964: 26). Intertidal sand at Governor Patterson Memorial State Park near Waldport, Oregon (Murphy 1964a: 26). Plate 109C.

Leptolaimidae

Procamacolaimus phinneyi (Murphy, 1964) Hope and Tchesunov, 1999 (=*Dagda phinneyi* Murphy, 1964: 192). Intertidal sand on beach at Umpqua Light House State Park, Oregon. Plate 110C.

Cynura klunderi Murphy, 1965. Intertidal sand in small cove subjected to mild wave action at Moss Landing, Monterey Bay (Murphy 1965c: 216). Plate 110D.

References

Abebe, E., W. Traunspurger, and I. Andrássy. 2006. Freshwater nematodes: ecology and taxonomy. CABI Publishing, Cambridge, MA. i–xx and 1–752.

Adamson, M. L. 1987. Phylogenetic analysis of the higher classification of the Nematoda. Can. J. Zool. 65: 1478–1482.

Allen, M. W. and E. M. Noffsinger. 1978. A revision of the marine nematodes of the superfamily Draconematoidea Filipjev 1918 (Nematoda: Draconematina). Univ. Calif. Publs. Zool. 109: 1–133.

Allgén, C. 1947. Papers from Dr. Th. Mortensen's Pacific Expedition 1914–1916. LXXV. West American marine nematodes. Vidensk. Medd Dan. naturhist. Froen. 110: 65–219.

Andrássy, I. 1976. Evolution as a basis for the systematization of nematodes. London: Pitman Publ., 288 pp.

Bates, J. W. 1997. The slide-sealing compound "Glyceel". J. Nema. 29: 565–566.

Bird, A. F. and J. Bird. 1991. The structure of nematodes. 2nd ed. San Diego: Acad. P., 316 pp.

Burr, A. H. and C. Burr 1975. The amphid of the nematode *Oncholaimus vesicarius*: ultrastructural evidence for a dual function as chemoreceptor. J. Ultrastruct. Res. 51: 1–15.

Burr, A. H. and J. M. Webster. 1971. Morphology of the Eyespot and Description of Two Pigment Granules in the Esophageal Muscle of a Marine Nematode, *Oncholaimus vesicarius*. J. Ultrastructure Res. 36: 621–632.

Chitwood, B. G. 1951. North American marine nematodes. Tex. J. Sci. 3: 617–672.

Chitwood, B. G. 1960a. *Choanolaimus psammophilus* J. G. de Man, 1880, rediscovered, and *Synonchium obtusum* N. A. Cobb, 1920, as a natural enemy of plant pathogens. Nematol. (Suppl. II): 55–60.

Chitwood, B. G. 1960b. A preliminary contribution on the marine nemas (Adenophorea) of northern California. Trans. Am. Microsc. Soc. 79: 347–384.

Chitwood, B. G. and M. G. Chitwood. 1938. Zoology–Notes on the "culture" of aquatic nematodes. J. Wash. Acad. Sci. 28: 455–460.

Chitwood, B. G. and M. B. Chitwood. 1974. Introduction to nematology. Baltimore: Univ. Park Press, 334 pp.

Chitwood, B. G. and D. G. Murphy 1964. Observations on two marine monhysterids—their classification, cultivation, and behavior. Trans. Am. Microsc. Soc. 83: 311–329.

Cobb, N. A. 1920. One hundred new nemas (type species of 100 new genera). Contributions to a Science of Nematology 9: 217–343.

Coomans, A. 1981. Aspects of the phylogeny of nematodes. Accad. Naz. Lincei. 49: 161–174.

Coomans, A., D. Verschuren, and R. Vanderhaeghen. 1988. The Demanian System, Traumatic Insemination and Reproductive Strategy in *Oncholaimus oxyuris* Ditlevsen (Nematoda: Oncholaimidae). Zoologica Scripta 17: 15–23.

Croll, N. A., A. A. F. Evans, and J. M. Smith. 1975. Comparative nematode photoreceptors. Comp. Biochem. and Physiol. 51A: 139–143.

Croll, N. A. and A. R. Maggenti. 1968. A peripheral nervous system in Nematoda with a discussion of its functional and phylogenetic significance. Proc. Helminthol. Soc. Wash. 35: 108–115.

Dailey, M. D. 2001. Parasitic Diseases. In: *CRC* Handbook of marine mammal medicine. Dierauf, L. A. and M. D. Gulland. 2nd ed. Washington, DC: CRC Press. 357–379 pp.

Dallwitz, M. J., T. A. Paine, and E. J. Zurcher. 1993. DELTA user's guide. A general system for processing taxonomic descriptions. 4th ed. Canberra, Australia: CSIRO Division of Entomology. 136 pp.

Decraemer, W. 1978. The genus *Quadricoma* Filipjev, 1922 with a redescription of *Q. cobbi* (Steiner, 1916), *Q. crassicomoides* Timm, 1970 and *Q. loricata* Filipjev, 1922 (Nematoda—Desmoscolecida). Contribution No. X on nematodes from the Great Barrier Reef, collected during the Belgian expedition in 1967. Cah. Biol. Mar. 19: 63–89.

De Jonge, V. N. and L. A. Bouwman. 1977. A simple density separation technique for quantitative isolation of meiobenthos using the colloidal silica Ludox-TM. Mar. Biol. 42: 143–148.

DeLey, P. and M. Blaxter. 2002. Systematic Position and Phylogeny. pp 1–30. In The biology of nematodes. D. L. Lee, ed. New York and London: Taylor and Francis. 635 pp.

Gerlach, S. A. 1956. Diagnosen neuer Nematoden aus der Kieler Bucht. Kieler Meeresforschungen, 12 (1): 85–109.

Higgins, R. P. and H. Thiel, eds. 1988. Introduction to the study of meiofauna. Washington, D.C.: Smithson. Inst. P., 488 pp.

Hope, W. D. 1967a. A review of the genus *Pseudocella* Filipjev, 1927 (Nematoda: Leptosomatidae) with a description of *Pseudocella triaulolaimus* n. sp. Proc. Helminth. Soc. Wash. 34: 6–12.

Hope, W. D. 1967b. Free-living marine nematodes of the genera *Pseudocella* Filipjev, 1927, *Thoracostoma* Marion, 1870, and *Deontostoma* Filipjev, 1916 (Nematoda: Leptosomatidae) from the west coast of North America. Trans. Am. Microsc. Soc. 86: 307–334.

Hope, W. D. 1969. Fine structure of the somatic muscles of the free-living marine nematode *Deontostoma californicum* Steiner and Albin, 1933 (Leptosomatidae). Proc. Helminth. Soc. Wash. 36: 10–29.

Hope, W. D. 1974. Nematoda. In Giese, A. C. and J. S. Pearse, eds. Reproduction of marine invertebrates. New York and London: Acad. P., 391–468 pp.

Hope, W. D. 1977. Gutless nematodes of the deep-sea. Mikrofauna Meeresboden 61:307–308.

Hope, W. D. 1982. Structure of head and stoma in the marine nematode genus *Deontostoma* (Enoplida: Leptosomatidae). Smithson. Contrib. Zool. 353: 1–22.

Hope, W. D. 1988. A review of the marine nematode genera *Platycoma* and *Proplatycoma* with a description of *Proplatycoma fleurdelis* (Enoplida: Leptosomatidae). Proceedings of the Biological Society of Washington (3): 693–706.

Hope, W. D. and S. L. Gardiner. 1982. Fine structure of a proprioceptor in the body wall of the marine nematode *Deontostoma californicum* Steiner and Albin, 1933 (Enoplida: Leptosomatidae). Cell and Tissue Res. 225: 1–10.

Hope, W. D. and A. V. Tchesunov. 1999. *Smithsoninema inaequale* n. g. and n. sp. (Nematoda, Leptolaimidae) inhabiting the test of a foraminiferan. Invert. Biol. 118(2): 95–108.

Hope, W. D. and A. Yorkoff. 1985. Nematoda. In W. C. Austin, ed., An annotated checklist of marine invertebrates in the cold temperate northeast Pacific, Vol. 1. Cowichan Bay, BC, Canada: Khoyatan Mar. Lab. 129–152 pp.

Inglis, W. G. 1983. An outline classification of the phylum Nematoda. Aust. J. Zool. 31: 243–255.

Jensen, P. 1989. Revision of Aponchiidae Gerlach, 1963 (Nematoda: Monhysterida), epibiotic nematodes on shells of intertidal epibenthic invertebrates, with description of new species. Phuket Mar. Biol. Cent. Res. Bull. No. 50: 1–24.

Jones, G. F. 1964. Redescription of *Bolbella californica* Allgén, 1951 (Enchelidiidae: Nematoda), with notes on its ecology off southern California. Pac. Sci. 18: 160–165.

Lorenzen, S. 1978. Discovery of stretch receptor organs in nematodes—structure, arrangement and functional analysis. Zool. Scr. 7: 175–178.

Lorenzen, S. 1981. Bau, Anordnung und postembryonale Entwicklung von Metanemen bei Nematoden der Ordnung Enoplida. Veröff. Inst. Meeresforsch. Bremerhaven 19: 89–114.

Lorenzen, S. 1994. The phylogenetic systematics of freeliving nematodes. Vol. 162. Translation by J. Greenwood, H. M. Platt, ed. London: The Ray Soc., 383 pp.

Maggenti, A. R. 1964. Morphology of somatic setae: *Thoracostoma californicum* (Nemata: Enoplidae). Proc. Helminthol. Soc. Wash. 31: 159–166.

Maggenti, A. R. 1981. General nematology. New York: Springer-Verlag, 372 pp.

Maggenti, A. R. 1982. Nemata. In Synopsis and classification of living organisms. S. P. Parker, ed., New York: McGraw Hill, 879–929 pp.

Maggenti, A. R. and D. R. Viglierchio. 1965. Preparation of nematodes for microscopic study–perfusion by vapor phase in killing and fixing. J. Agric. Sci. 36: 435–463.

Malakhov, V. V. 1994. Nematodes. structure, development, classification, and phylogeny. Translation by G. Bentz, W. D. Hope, ed., Washington, DC: Smith. Inst. P., 286 pp.

Moser, M., M. S. Love, and J. Sakanari. 1983. Common parasites of California marine fishes. Cal. Dept. Fish Game. 17 pp.

Murphy, D. G. 1962. Three undescribed nematodes from the coast of Oregon. Limnol. Oceanogr. 7: 386–389.

Murphy, D. G. 1963a. A note on the structure of nematode ocelli. Proc. Helminthol. Soc. Wash. 30: 25–26.

Murphy, D. G. 1963b. A new genus and two new species of nematodes from Newport, Oregon. Proc. Helminthol. Soc. Wash. 30: 73–78.

Murphy, D. G. 1963c. Three new species of marine nematodes from the Pacific near Depot Bay, Oregon. Proc. Helminthol. Soc. Wash. 30: 249–256.

Murphy, D. G. 1964a. *Rhynchonema subsetosa*, a new species of marine nematode, with a note on the genus *Phyllolaimus* Murphy, 1963. Proc. Helminthol. Soc. Wash. 31: 26–28.

Murphy, D. G. 1964b. Free-living marine nematodes, I. *Southerniella youngi*, *Dagda phinneyi*, and *Gammanema smithi*, new species. Proc. Helminthol. Soc. Wash. 31: 190–198.

Murphy, D. G. 1964c. The marine nematode genus *Pseudonchus* Cobb, 1920, with descriptions of *Cheilopseudonchus*, n. g. and *Pseudonchus kosswigi*, n. sp. Mitt. Hamburg. Zool. Mus. Inst. 1964: 113–118.

Murphy, D. G. 1965a. Free-living marine nematodes. II. *Thoracostoma pacifica* n. sp. from the coast of Oregon. Proc. Helminthol. Soc. Wash. 32: 106–109.

Murphy, D. G. 1965b. *Thoracostoma washingtonensis*, n. sp., eine Meeresnematode aus dem pazifischen Küstenbereich vor Washington. Abhandlungen und Verhandlungen des Naturwissenschaftlichen Vereins in Hamburg 9: 211–216.

Murphy, D. G. 1965c. *Cynura klunderi* (Leptolaimidae), a new species of marine nematode. Zool. Anz. 175: 216–222.

Murphy, D. G. and H. J. Jensen. 1961. *Lauratonema obtusicaudatum*, n. sp. (Nemata: Enoploidea), a marine nematode from the coast of Oregon. Proc. Helminthol. Soc. Wash. 28: 167–169.

Nelson, H., J. M. Webster, and A. H. Burr. 1971. A redescription of the nematode *Oncholaimus vesicarius* (Wieser, 1959) and observations on the pigment spots of this species and of *Oncholaimus skawensis* Ditlevsen, 1921. Can. J. Zool. 49: 1193–1197.

Nelson, H., B. Hopper, and J. M. Webster. 1972. *Enoplus anisospiculus*, a new species of marine nematode from the Canadian Pacific coast. Can. J. Zool. 50: 1681–1684.

Pearce, F. 1995. Rockall mud richer than rainforest. New Sci. Sept. 16: 8.

Petter, A. J. 1980. Une nouvelle famille de nématodes parasites d'invertebrés marins, les Benthimermithidae. Ann. Parasit. 55: 209–224.

Petter, A. J. 1981a. Description des mâles d'une nouvelle espèce de nématode marin de la familie des Benthimermithidae. Ann. Parasit., 56: 285–295.

Petter, A. J. 1981b. Description des mâles de trois nouvelles espèces de nématodes de la famille de Benthimermithidae. Bull. Mus. Histoire Nat., Paris, 4e series 3: 455–465.

Petter, A. J. 1982a. *Benthimermis gracilis* n. sp., nouveau mâle de la famille des Benthimermithidae (Nematoda). Bull. Mus. Hist. Nat, Paris, 4e series 4: 71–74.

Petter, A. J. 1982b. Description de deux nouveaux mâles de la famille des Benthimermithidae (Nematoda) de l'Atlantique sud-oriental. Bull. Mus. Histoire Nat., Paris, 4e ser. 4: 397–403.

Petter, A. J. 1983a. Quelques nouvelles espèces du genre *Benthimermis* Petter, 1980 (Benthimermithidae: Nematoda) du Sud de l'Océan Indien. Syst. Parasitol. 5: 2–15.

Petter, A. J. 1983b. Description d'un nouveau genre de Benthimermithidae (Nematoda) présentant des utérus munis de glandes annexes. Ann. Parasitol. Hum. Comp., 58: 177–184.

Petter, A. J. 1987. Quelques nouvelles espèces de femelles du genre *Benthimermis* Petter, 1980 (Benthimermithidae: Nematoda) des grands fonds de la mer de Norvége. Bull. Mus. Hist. Nat., Paris, ser. 9, section A, no. 3: 565–578.

Petter, A. J. and N. Gourbault. 1985. Nématodes abyssaux (campagne Walda du N/O "Jean Charcot"). IV. Des nématodes parasites de nématodes. Bull. Mus. Hist. Nat., Paris, 4e ser. 7: 125–130.

Platonova, T. A. 1962. New species of nematodes of the genus *Pseudocella* Filipjev from the Kurile Islands and Southern Sakhalin. Studies of the Far Eastern Seas of the USSR 8: 200–218. {In Russian}.

Platt, H. M. and R. M. Warwick. 1980. The significance of free-living nematodes to the littoral ecosystem. In The shore environment, Vol. 2. Ecosystems. Price, J. H., D. E. G. Irvine, and W. F. Farnham, eds. London and New York: Acad. P., 729–759 pp.

Platt, H. M. and R. M. Warwick. 1983. Freeliving marine nematodes Part I British Enoplids. Pictorial key to world genera and notes for the identification of British species. Published for the Linnean Society of London and The Estuarine and Brackish-water Sciences Association. Synopses of the British Fauna. Synopses of the British (New Series). Cambr. Univ. P., No 28: i–vii, 210 pp.

Platt, H. M. and R. M. Warwick. 1988. Freeliving marine nematodes Part II British Chromadorids. Pictorial key to world genera and notes for the identification of British species. Published for the Linnean Society of London and The Estuarine and Brackish-water Sciences Association. Synopses of the British Fauna (New Series). Camb. Univ. P., No 38: i–vii, 502 pp.

Rachor, E. 1969. Das de Mansche Organ der Oncholaimidae, eine genitointestinale Verbindung bei Nematoden. Zeitschrift für Morphologie und Ökologie der Tiere 66: 87–166.

Riemann, F. 1970. Das Kiemenlückensystem von Krebsen als Lebensraum der Meiofauna, mit Beschreibung freilebender Nematoden aus Karibischen amphibisch lebenden Decapoden. Veröff. Inst. Meeresforch. Bremerhaven 12: 413–428.

Riemann, F. 1972. Corpus gelatum und ciliäe Strukturen als lichtmikroskopisch sichtbare Bauekenebte des Seitenorgans freilebender Nematoden. Zeitschrift für Morphologie der Tiere, 72: 46–76.

Riemann, F. 1988. Nematoda. In Introduction to the study of meiofauna. Higgins, R. P. and H. Thiel, eds. Washington, D.C.: Smithson. Inst. P., 293–301 pp.

Sharma, J., B. E. Hopper, and J. M. Webster. 1979. Benthic nematodes from the Pacific coast with special reference to the Cyatholaimids. Ann. Soc. R. Zool. Belg. 108(1978): 47–56.

Sharma, J. and M. Vincx. 1982. Cyatholaimidae (Nematoda) from the Canadian Pacific coast. Can. J. Zool. 60: 271–280.

Sharma, J. and J. M. Webster. 1983. The abundance and distribution of free-living nematodes from two Canadian Pacific beaches. Estuar. Coast. Shelf Sci. 16: 217–227.

Siddiqui, I. A. and D. R. Viglierchio. 1970a. Fine structure of photoreceptors in *Deontostoma californicum*. J. Nematology 2: 274–276.

Siddiqui, I. A. and D. R. Viglierchio. 1970b. Ultrastructure of photoreceptors in the marine nematode *Deontostoma californicum*. J. Ultrastruct. Res. 32: 558–571.

Siddiqui, I. A. and D. R. Viglierchio. 1971. Ultrastructure of the lamellated cytoplasmic inclusions in Schwann cells of the marine nematode, *Deontostoma californicum*. J. Nem. 3: 264–275.

Siddiqui, I. A. and D. R. Viglierchio. 1977. Ultrastructure of the anterior body region of marine nematode *Deontostoma californicum*. J. Nem 9: 56–82.

Steiner, G. and F. M. Albin. 1933. On the morphology of *Deontostoma californicum* n. sp. (Leptosomatinae, Nematodes). J. Wash. Acad. Sci. 23: 25–30.

Tarjan, A. C. 1980. An illustrated guide to the marine nematodes. Inst. Food Agric. Sci. Univ. of Florida, 135 pp.

Tchesunov, A. V. 1988a. A case of nematode parasitism in nematodes. A new find and redescription of a rare species *Benthimermis australis* Petter, 1983 (Nematoda: Marimermithida: Benthimermididae) in South Atlantic. Helminthol. 25: 115–128.

Tchesunov, A. V. 1988b. New finds of the Deep-Sea Family Benthimermithidae in South Atlantic with Description of a New Species. Vestn. Zool. 6: 12–22.

Tchesunov, A. V. 1995. New Data on the Anatomy of a Marimermithid Nematode. Nematologica 41: 347.

Tchesunov, A. V. and W. D. Hope. 1997. *Thalassomermis megamphis* n. gen., n. sp. (Mermithidae: Nemata) from the bathyal South Atlantic Ocean. J. Nem. 29: 451–464.

Tchesunov, A. V. and S. E. Spiridinov. 1985. *Australonema euglagiscae* gen. et sp. n. (Nematoda, Marimermithida)—a parasite of a polychaete from Antarctica. Vestn. Zool. 2: 16–21.

Tchesunov, A. V. and S. E. Spiridinov. 1993. *Nematomermis enoplivora* gen. n., sp. n. (Nematoda: Mermithoidea) from marine free-living nematodes, *Enoplus* spp. Russ. J. Nem., 1: 7–16.

Timm, R. W. 1970. A revision of the nematode order Desmoscolecida Filipjev, 1929. Univ. Calif. Publ. Zool. 93: 1–115.

Timm, R. W. and T. Hackney. 1969. Effects of fixation and dehydration procedures on marine nematodes. J. Nem. 1: 146–149.

Trotter, D. and J. M. Webster. 1983. Distribution and abundance of marine nematodes on the kelp *Macrocystis integrifolia*. Mar. Biol. 78: 39–43.

Uhlig, G., H. Thiel, and J. S. Gray. 1973. The quantitative separation of meiofauna. A comparison of methods. Helgol. wiss. Meeresunters. 25: 173–195.

Van de Velde, M. C. and A. Coomans. 1988. Ultrastructure of the photoreceptor of *Diplolaimella* sp. (Nematoda). Tissue Cell 20: 421–429.

Viglierchio, D. R. and I. A. Siddiqui. 1974. Pigments of the ocelli of Antarctic and Pacific marine nematodes. Trans. Am. Microsc. Soc. 93: 338–343.

Warwick, R. M., H. M. Platt, and P. J. Somerfield. 1998. Free-living marine nematodes part III Monhysterids. Synop. British Fauna (New Series) No 53: i–vii, 296 pp.

Wieser, W. 1959a. Free-living nematodes and other small invertebrates of Puget Sound beaches. Univ. Wash. Publ. Biol. 19: 1–179.

Wieser, W. 1959b. The effect of grain size on the distribution of small invertebrates inhabiting the beaches of Puget Sound. Limnol. Oceanogr. 4: 181–194.

Wing, B. L. and K. A. Clendenning. 1971. Kelp surfaces and associated invertebrates. *Macrocystis* surfaces throughout the water column. In The biology of giant kelp beds Macrocystis in California. Wheeler, J. N., ed. Lehre: J. Cramer, 319–339 pp.

Wright, K. A. and W. D. Hope. 1968. Elaborations of the cuticle of *Acanthonchus duplicatus* Wieser, 1959 (Nematoda: Cyatholaimidae) as revealed by light and electron microscopy. Can. J. Zool. 46: 1005–1011.

Wright, K. A., W. D. Hope, and N. O. Jones. 1973. The ultrastructure of the sperm of *Deontostoma californicum*, a free-living marine nematode. Proc. Helminthol. Soc. Wash. 40: 30–36.

Gastrotricha

(Plate 111)

WILLIAM D. HUMMON

Gastrotrichs are an acoelous phylum of about 720 (Hummon 2007) described species that occur in all sorts of aquatic conditions, with about half of these species inhabiting marine and estuarine habitats; they are nearly all free-living ciliary gliders, although a few may occur as epizootics. They are microscopic in size, having a Poisson size distribution that ranges from 70 μm to 4 mm, with a mode of perhaps 200 μm. The marine-estuarine species are mostly interstitial, living amid relatively clean, fine to coarse sands, although some are tolerant of high organic, sulfide or pollution loads, and a few even occur in mud and oozes.

There are two orders of Gastrotricha: the mostly box- to strap-shaped Macrodasyida, with about 265 species in seven families, and the mostly tenpin-shaped Chaetonotida, with 130 species in two suborders and three families. Gastrotricha are D-shaped in cross-section and bear cilia or cirri on their ventral surfaces; many have a naked cuticle, but others have cuticular armature consisting of scales, spines, or warts of various sorts in species-specific patterns. They have a complete digestive system, with a triradiate pharynx and a one-cell-thick gut that acts both as stomach and intestine. Many have cyrtocytic protonephridia. All marine forms have duo-gland adhesive tubes, with two or more on the caudum, from two to many forward beneath the mouth (in macrodasyids and a few chaetonotids), and often several to many arranged along the lateral, dorsal or ventral surfaces. The vast majority are hermaphroditic, although many may reproduce by parthenogenesis. They are easily separated from turbellarians and gnathostomulids by the presence of a terminal mouth and the lack of static organs and lateral/dorsal locomotor cilia.

To begin a study of gastrotrichs, collect moist to wet sand from the lower half of a beach at low tide. Place a spoonful of sand in a small jar, add enough 6–7% $MgCl_2$ to cover the sand, swirl, leave for 10 min, swirl, decant into a small petri dish, add an equal amount of seawater and observe under a dissecting microscope at 40–50× magnification. To study living specimens, transfer them by micropipette, use modeling clay posts beneath the corners of 18 mm square coverslips, and observe under a compound microscope using differential interference contrast optics. Reviews can be found in d'Hondt (1971), Hummon (1982, 2007), Ruppert (1988, 1991), and in various invertebrate texts.

The west coast of North America is vastly understudied, with only eight species (see plate 111) having been published from the intertidal and shallow coastal waters of California (two macrodasyids: Todaro 1995; Hochberg 1998, 1999) and Washington (five macrodasyids and two chaetonotids: Wieser 1957, 1959; Hummon 1966, 1969, 1972).

No gastrotrichs were reported from Oregon until I visited the Oregon Institute of Marine Biology, Coos Bay, during October and November 2001. I surveyed the gastrotrichs of the intertidal Oregon coast. Thirty species in 12 genera were recorded on high-resolution videotape from 25 locations from Ft. Stevens State Park in the north to Brookings in the south. In January 2002, I sampled three locations near Huntington Beach California, where I recorded 10 species in five genera, all of which were included among those from Oregon.

References

d'Hondt, J. L. 1971. Gastrotricha. Ocean. Mar. Biol. Ann. Rev. 9: 141–191.

Hochberg, R. 1998. Postembryonic growth and morphological variability in *Turbanella mustela* (Gastrotricha, Macrodasyida). J. Morphol. 237: 117–126.

Hochberg, R. 1999. Spatiotemporal size-class distribution of *Turbanella mustela* (Gastrotricha: Macrodasyida) on a northern California beach and its effect on tidal suspension. Pacific Sci. 53: 50–60.

Hummon, W. D. 1966. Morphology, life history and significance of the marine gastrotrich, *Chaetonotus testiculophorus* n. sp. Trans. Amer. Microsc. Soc. 85: 450–457.

Hummon, W. D. 1969. *Musellifer sublitoralis*, a new genus and species of Gastrotricha from the San Juan Archipelago, Washington. Trans. Amer. Microsc. Soc. 88: 282–286.

Hummon, W. D. 1972. Distribution of Gastrotricha in a marine Beach of the San Juan Archipelago, Washington. Mar. Biol. 16: 349–355.

Hummon, W. D. 1982. *Gastrotricha*, pp. 857–863 in S. P. Parker, ed., Synopsis and classification of living organisms. McGraw-Hill, New York, Vol. 1, pp. 857–863.

Hummon, W. D., ed. 2007. Global Database for Marine Gastrotricha and Synopsis of Described Species of Gastrotricha. Server at http://132.235.243.28 or www.hummon-nas.biosci.ohiou.edu.

Ruppert, E. E. 1988. *Gastrotricha*, pp. 302–311, in R. P. Higgins and H. Thiel, eds., Introduction to the Study of Meiofauna. Smithsonian Institution Press, Washington, D.C.

Ruppert, E. E. 1991. *Gastrotricha*, pp. 41–109, in F. W. Harrison and E. E. Ruppert, eds., Microscopic anatomy of invertebrates, Wiley-Liss, New York, Vol. 4 Aschelminthes.

PLATE 111 A, *Macrodasys cunctatus* Wieser,
1957: 1—habitus, a composite of dorsal and
ventral views, 2—ventral view of the fore end,
3—internal anatomy of the rear end, as seen
from above; B, *Turbanella cornuta* Remane, 1926:
1—habitus, a dorsal view, 2—ventral view of the
fore end, 3—ventral view of the rear end, 4—juvenile, probably damaged in the mid-pharynx;
C, *Turbanella mustela* Wieser, 1957: habitus, a
composite of dorsal and ventral views; D,
Paraturbanella intermedia Wieser, 1957: 1—habitus, a dorsal view, 2—ventral view of the fore
end, 3—ventral view of the rear end; E,
Tetranchyroderma pugetensis Wieser, 1957: habitus, a ventral view, along with two tetrancres; F,
Paraturbanella solitaria Todaro, 1995: 1—habitus,
a composite of dorsal and ventral views, 2—
ventral view of the fore end, 3—ventral view of
the rear end, 4—internal anatomy of the
foregut region, as seen from below; G,
Halichaetonotus testiculophorus (Hummon, 1966)
[=*Chaetonotus testiculophorus* Hummon, 1966]
new combination: 1—habitus, a dorsal view
(left) and internal anatomy in optical view
(right), 2—ventral ciliary pattern, 3—anterior
portion of tri-radiate pharynx, 4—ventrolateral
spines of the head (above) and midbody (below), which are actually hydrofoil scales that require shifting the species from the genus
Chaetonotus to *Halichaetonotus*, 5—pregnant individual, 6—egg attached to a sand grain, 7—juvenile (key: Ph—pharynx, In—intestine,
Sp—sperm, X—X-organ, Fb—furcal base, Ft—
furcal tip); H, *Musellifer sublitoralis* Hummon,
1969: 1—internal anatomy in optical view, 2—
pincerlike structures that may project from the
mouth, 3, 4, and 5—dorsal scales of the head,
trunk and proximal furcal branch, 6—lateral
view of the way trunk scales imbricate (key: Ph
and In as in plate 111G7, Mz—ciliated muzzle,
Ts—testis, Ov—ovum).

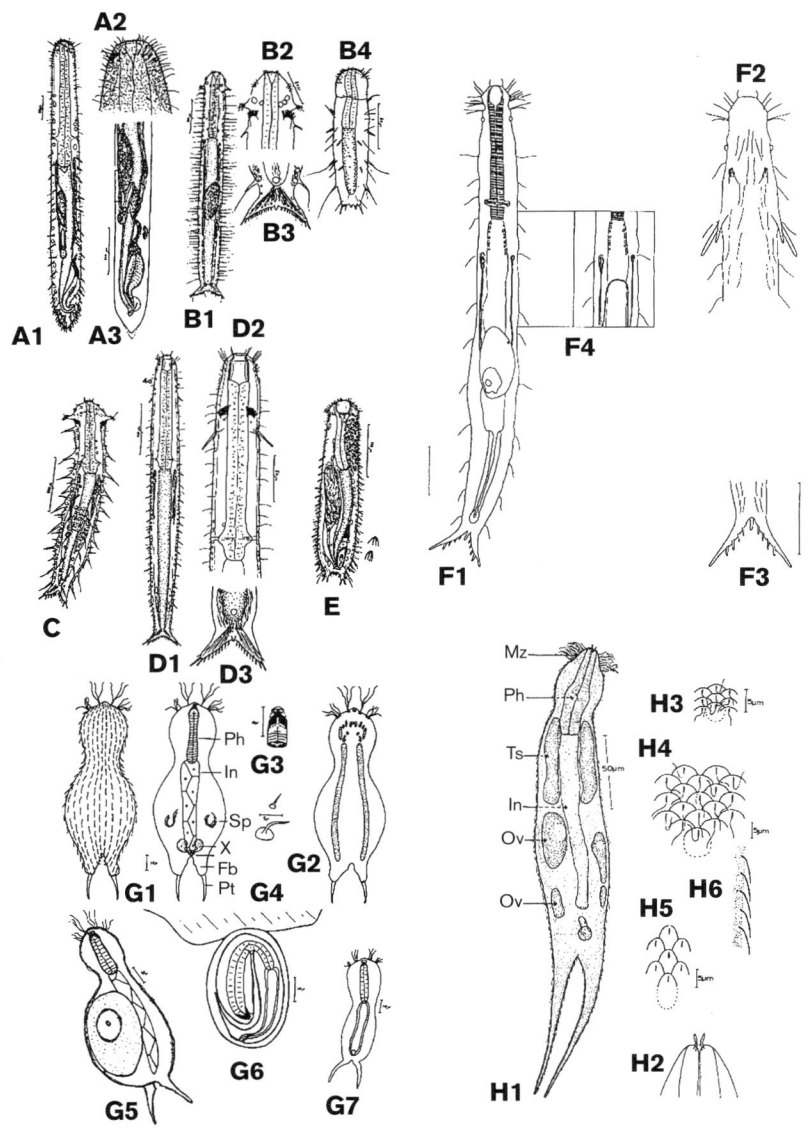

Todaro, M. A. 1995. *Paraturbanella solitaria*, a new psammic species (Gastrotricha: Macrodasyida: Turbanellidae) from the coast of California.
Proc. Biol. Soc. Wash. 108: 553–559.
Todaro, M. A., J. M. Bernhard, and W. D. Hummon. 2000. A new species
of *Urodasys* (Gastrotricha, Macrodasyida) from dysoxic sediments of
the Santa Barbara Basin (California, U.S.A.). Bull. Mar. Sci. 66: 467–476.

Wieser, W. 1957. Gastrotricha Macrodasyiodea from the intertidal of
Puget Sound. Trans. Amer. Microsc. Soc. 76: 372–381.
Wieser, W. 1959. The effect of grain size on the distribution of small in-
vertebrates inhabiting the beaches of Puget Sound. Limnol.
Oceanogr. 4: 181–194.

Kinorhyncha, Loricifera, and Priapulida

(Plate 112)

ROBERT P. HIGGINS

The Kinorhyncha, Loricifera, and Priapulida are three related phyla that are more uncommonly found in the literature than in their habitats. In recent years, they have been included in the Scalidophora (Lemburg 1995). The Kinorhyncha and Priapulida, especially, were once considered part of the "aschelminth" complex, once considered "pseudocoelomates," sometimes included in the Nemathelminthes (especially by European biologists), and, along with the Nematomorpha, considered as constituting the phylum Cephalorhyncha by Malakhov (1980). The majority opinion is that these animals should remain as separate phyla until a more convincing argument can be supported.

All three of these phyla are characterized by a body plan that begins with a spherical eversible head bearing rings of recurved spines or scalids. Centered within this head is a mouth cone which connects to a large muscular pharynx lined with cuticle. In kinorhynchs the mouth cone is protrusible with nine oral styles around the mouth opening, in loriciferans it is telescopic and may have oral styles in some immature stages, and in priapulids it is eversible and covered with numerous rings of teeth.

A neck region is present in all three phyla and often functions as a closing device as well as sensory area. The trunk or abdominal region is covered with a distinctive cuticle and diagnostic cuticular structures relevant to each phylum.

Kinorhyncha

(Plate 112A)

Kinorhynchs or "mud dragons" are marine meiofaunal invertebrates, all less than 1 mm long and easily recognized by their distinctive segmentation. An eversible spherical head, the first of 13 segments, has a terminal opening through which the mouth cone protrudes. The anterior margin of the mouth cone has a series of nine oral styles that may be articulated and are used in feeding. The outer surface of the everted head is covered with a series of recurved scalids. The first ring consists of 10 spinoscalids that, in the process of eversion and withdrawal, function both as sensory and locomotory appendages. A total

of seven rings of scalids are present in the adult, and most of these play some sensory and locomotory role; the last ring of scalids, the trichoscalids, are sensory only. As few as four rings may be found in the first of six juvenile stages. Growth is by molting until the adult stage is reached. All but the trichoscalid ring consist of scalids alternating in multiples of five.

The head may be inverted (withdrawn) into the more heavily cuticularized trunk region and closed off by a series of smaller plates of the neck, segment two. The 11 trunk segments may consist of a complete ring or may be divided into separate plates that vary in their arrangement and are diagnostic at higher taxonomic levels. In cross-section, the trunk segments may be oval, circular or triangular. The trunk segments may have various arrangements of cuticular spines or less prominent setae or secretory tubules.

The cuticle may exhibit a distinctive series of sensory spots (i.e., modified flosculae, a sensory structure found also in Loricifera, Priapulida, and at least some Rotifera and Nematomorpha), secretory pores, muscle scars (i.e., internal muscle attachment areas visible externally), cuticular hairs—with or without perforation sites—and a pectinate fringe of the cuticle at the posterior border of a given segment. All such cuticular structures are of taxonomic importance.

Most kinorhynchs are subtidal, found as deep as 5,000 m. Although found in nearly all kinds of sediments from silt to coarse sand, siliceous to carbonate, and occasionally found in association with other animals such as sponges, bivalves, and bryozoan colonies, most are found in sandy mud. Kinorhynchs have also been found in other parts of the world in the mantle cavities (i.e., gill chambers) of intertidal bivalves, on intertidal mudflats (including the exposed mud of tidal brackish creeks), and in the exposed holdfasts (often with accumulations of sediment) or frond surfaces of macroalgae, suggesting the range of interesting habitats in which the student may search for these animals. With exceptions of some coarse-sand subtidal habitats and the medium- to coarse-sand intertidal beach habitats created by high-energy wave action, most kinorhynchs are found in the upper few centimeters of finer sediments, both subtidal and intertidal.

Kinorhynchs move through sediment by a repetitive everting and inverting of the spherical head, which allows the scalids to gain purchase on the substrate particles and propel

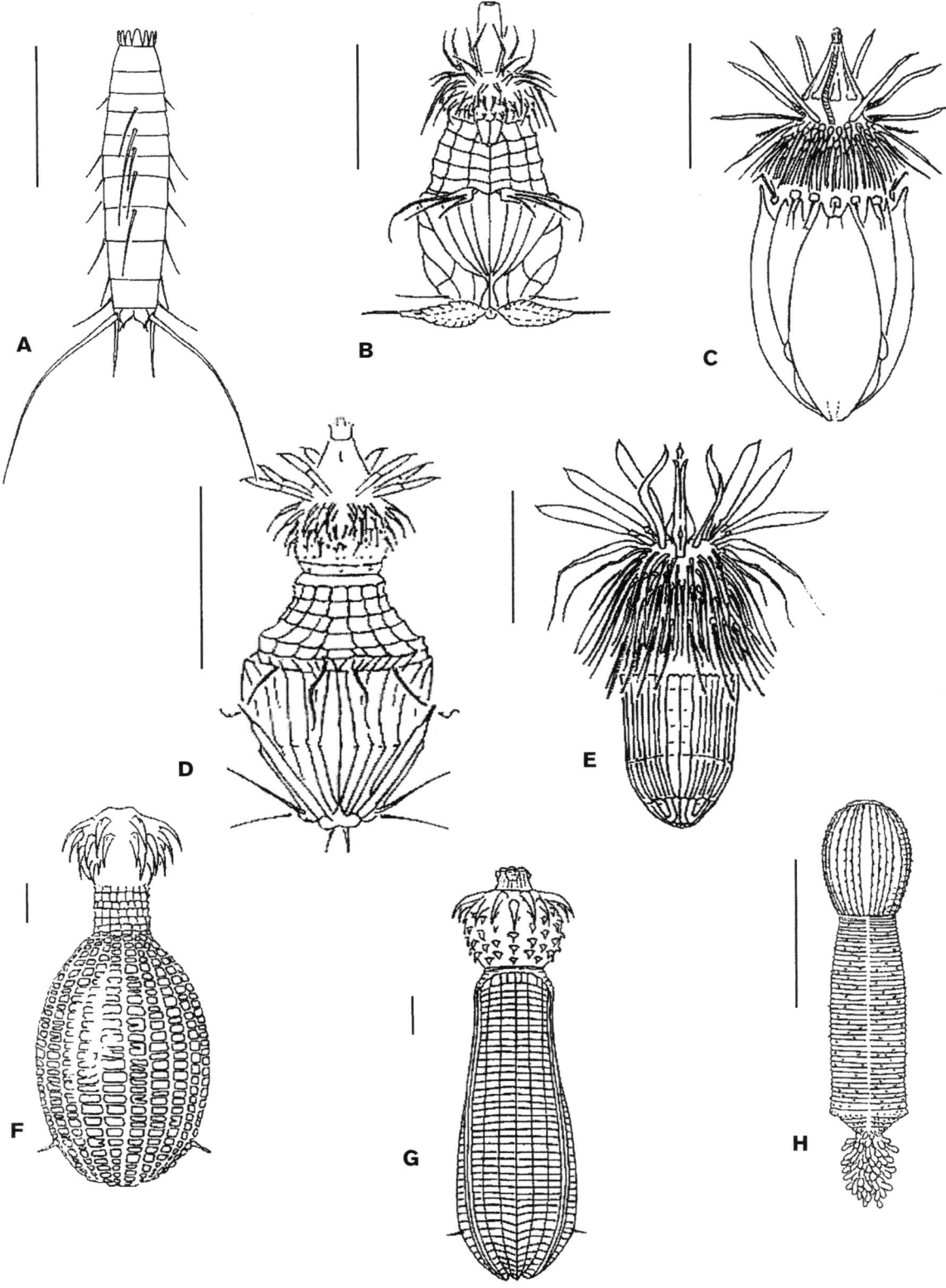

PLATE 112 A, Kinorhyncha, *Cephalorhyncha nybakkeni*, from intertidal coarse sand, dorsal view (after Higgins 1986), vertical scale bar equals 100 μm; B, C, Loricifera, *Nanaloricus mysticus*, Higgins-larva and adult, ventral view (after Kristensen 1983), vertical scale bars equals 100 μm; D, E, Loricifera, *Pliciloricus gracilis*, Higgins-larva and adult, ventral view (after Higgins and Kristensen 1986), vertical scale bars equals 100 μm; F, G, Priapulida, *Priapulus caudatus*, first larval stage and later larval stage, ventral view (after Higgins et al. 1993), scale bar equals 100 μm; H, Priapulida, *Priapulus caudatus*, adult, ventral view (after Storch et al. 1994), scale bar equals 1 mm.

the animal forward through the interstices formed by medium to coarse sand (mesobenthic mode), or through the displacement of finer particulate material (endobenthic mode). When the head is extended, they are able to feed on particulate matter, small algal cells including diatoms, and bacteria by protruding the mouth cone and manipulating the oral styles so as to ingest food.

Only two species of kinorhynch have been reported from the intertidal of the Pacific coast: *Echinoderes kozloffi* Higgins, 1977, from diatom-covered intertidal rock pebble substrate on San Juan Island, Washington, and *Cephalorhyncha nybakkeni* (=*Echinoderes nybakkeni*) (plate 112A) from a high-energy beach at Carmel. A few subtidal species may be encountered intertidally if accidentally transported out of their normal deeper water habitats. These include *Kinorhynchus ilyocryptus* (Higgins, 1961), which is especially abundant below 50 m depth in Monterey Bay (unpublished record based on collections by Nybakken and Higgins in 1985) and which was also found subtidally in Tomales Bay (Higgins 1986). Other possibilities of subtidal species being transported into intertidal habitats include *Pycnophyes sanjuanensis* Higgins, 1961; *P. parasanjuanensis* Adrianov and Higgins, 1996; *Echinoderes pennaki* Higgins, 1960; and *Kinorhynchus cataphractus* (Higgins, 1991), all from Puget Sound, and *Semnoderes pacificus* Higgins, 1967, from subtidal coastal sediment off Redondo Beach, California.

Collecting kinorhynchs intertidally depends on the habitat selected for examination. Looking for kinorhynchs in the holdfasts of macroalgae requires that the holdfasts and associated sediment be carefully isolated and placed in a plastic container containing filtered seawater (filtered to prevent undue contamination from the water column). Fronds of macroalgae or any pieces of macroalgae encountered may be treated similarly. In such instances, the container with the filtered seawater and sample is shaken vigorously to loosen any associated kinorhynchs or other invertebrates, poured through a fine (62 μm) mesh net, and the residue examined under a stereomicroscope with a minimum of 25× magnification.

The same basic technique is also applicable to samples of intertidal bryozoan colonies, sponges, bivalves, and other invertebrates that might harbor kinorhynchs or other meiofaunal organisms. Caution must be taken in drawing conclusions as to whether this is a definitive habitat for whatever is found. For example, bivalves may bring surface sediment and associated meiofauna into the gill chambers, sponges may also take in meiofauna when inhaling water with suspended material originating from the surface of nearby sediment, disturbance within any aquatic sediment-water interface may bring momentarily suspended meiofaunal organisms into contact with bryozoan colonies and these suspended organisms may possibly travel great distances. It has been the author's experience that macroalgae, rolling along subtidal sediment, may also accumulate and variously retain numerous meiofaunal hitchhikers.

Intertidal coarse sediment is best processed by placing samples in a bucket of filtered seawater, stir well, allow a moment for some settling of sediment and immediately decant through a fine mesh sieve and examine its contents. If obtaining live material is not the objective, the use of the freshwater-shock method works very well for kinorhynchs and results in the head being fully extended. Fine material, such as the upper 1–2 cm of intertidal mud, can be collected qualitatively by merely scraping off this layer and processing it. Quantitative methods require the use of a coring device. Usually, a 50 mL plastic syringe with the end cut off to form a small corer used

in sufficient replicate modes will provide adequate material. The processing of fine sediment is a laborious task. There are no satisfactory short-cuts to a careful search, particle by particle.

Samples may be preserved in 10% formalin with Rose Bengal stain added to make the organisms more prominent in the sorting process. Preserved kinorhynchs, especially if immersed for a short period of time in freshwater before preservation, make excellent specimens for study purposes because the osmotic pressure often everts the head region. Sorting live material, although much more interesting, is not necessary unless obtaining living material is the goal. Using the blotting technique described in the meiofauna section is not practical for intertidal mud samples because the kinorhynchs from such habitats are modified so as to not react hydrophobically as in subtidal kinorhynchs. If they were not so adapted, incoming tidal water would lift them from the mud and isolate them in the tenacious surface film, thereby effectively removing them from their habitat.

Loricifera

(Plate 112B–112E)

Loriciferans, like the kinorhynchs, are exclusively marine and meiofaunal, less than 485 μm in length, and poorly known. The first documented loriciferan, an adult, was seen in 1974 (Kristensen 1983) by the current author. Although I recognized that it might be a priapulid larva, I thought it more likely represented an undescribed phylum, but I also recognized the need to have more than one specimen. In the following year Kristensen found a second specimen, a larval stage, but he thought it might be a rotifer or perhaps a larval kinorhynch. After several years of searching Kristensen collected additional specimens, both adults and larvae, off Roscoff, France, and this led us to conclude that, indeed, a new phylum, Loricifera, had to be erected. Kristensen (1983) described the first species, *Nanaloricus mysticus*.

Additional subtidal species have been described or are known from the Atlantic and Gulf Coasts, and one deep-sea species constitutes the only described loriciferan from the Pacific Ocean. Nevertheless, several undescribed species are known from the both the North and South Pacific. Despite the fact that no records of this phylum yet exist within the region covered by this manual, we include Loricifera in order to make the student aware that such animals exist and are likely to be found along the Pacific coast.

Adult loriciferans are 108–485 μm in length (Kristensen 2002). As in the Kinorhyncha and Priapulida, they are bilaterally symmetrical and have a body plan consisting of a spherical eversible head, a neck, and an abdomen (trunk). Centered within the head is a protrusible mouth cone that connects to a large muscular pharynx lined with cuticle, but the mouth cone of loriciferans may have three telescopic sections; a series of six, eight, or 16 oral ridges; and eight cone retractors. The outer surface of the head bears seven to nine rings of sensory/locomotory spines or scalids. In the Loricifera, the anterior ring of scalids consists of eight anteriorly directed clavoscalids, both in adult and larval stages (plate 112B–112E); the remaining rings, some of which may be asymmetrically positioned, are much more complex and consist of >200 specialized appendages called spinoscalids, all directed posteriorly. In nanaloricids, the six dorsal clavoscalids of the male are divided into three units thereby resulting in a total of 20 anteriorly directed appendages. This is the most significant sexually dimorphic character for loriciferans. In addition, in some loriciferans, the pharyngeal tube within the mouth cone has a spiral band of thickened cuticle reminiscent of the Tardigrada. There is a large, bulbous pharynx, sometimes with as many as five rows of three pairs of placoids developed at the distal ends of the cuticle-lined triangular lumen, also reminiscent of the Tardigrada. In Pliciloricidae, the pharynx bulb is located in the mouth cone, but in the Nanaloricidae it is in the neck region. The remaining digestive tract is simply tubular and consists of esophageal, stomach, and rectal areas; the latter, like the pharynx, also is lined with cuticle.

The neck region bears 15–22 specialized appendages called trichoscalids. The neck may be withdrawn partially into the lorica along with the head and, in doing so, closes off the anterior opening of the lorica. The abdomen of the adult of *Nanaloricus mysticus* (plate 112C) has a series of six heavy cuticularized plates that constitute the lorica. In pliciloricids (plate 112E), the abdomen is surrounded by a thinner lorical cuticle with well-defined folds (plicae) as in all loriciferan larvae.

The larvae of loriciferans are called Higgins-larvae. They are 80–385 μm long and have the same body regions as the adults. The mouth cone may be unarmed, as in the nanaloricids (plate 112B), or with six to 12 oral stylets in the pliciloricids (plate 112D). The head has eight anteriorly directed clavoscalids and up to seven rows of specialized spinoscalids whose numbers vary from 68 to 80. The neck of the Higgins-larvae, a distinctive collarlike region in the pliciloricids and a series of distinctive plates in the nanaloricids, is much better defined than in the adults and is more obviously constructed as a closing-apparatus for the withdrawn head. The posterior region of the neck, called the thorax by Kristensen (1991) has three to six rows of plates. The lorica of both nanaloricids and pliciloricids consists of a series of 20–22 longitudinal folds/plates.

Several sensory appendages are present at the anterior limits of the lorica, and two toes with adhesive glands are located more caudally. Details of both external and internal anatomy are provided by Kristensen (1983, 1991) and Higgins and Kristensen (1986).

Loriciferans occur mostly in medium to coarse subtidal sand or sandy mud. So little is known about the Loricifera that one cannot exclude the possibility of finding specimens in intertidal sediment habitats along the Pacific coast either by virtue of it being a natural habitat, or one into which specimens have been transferred by currents and wave action. If seen alive, there should be little doubt about recognizing the animal as a loriciferan; when preserved, specimens, especially Higgins-larvae, may look similar to preserved bdelloid rotifers, not a group well-known to marine biologists. In any event, they are very small and at least 50× magnification must be used to see them in a sample. Any mechanism used to collect meiofauna, subtidally or intertidally, will collect loriciferans if they are present (see meiofauna section). Methods of processing samples and preparing specimens for study may be found in Higgins and Thiel (1988).

Priapulida

(Plate 112F–112H)

The Priapulida consists of 19 living species of wormlike marine invertebrates. Priapulids were the dominant inhabitants of soft sediments preserved in the early and middle Cambrian. Until the first meiofaunal representatives were described from tropical waters (Land 1968), priapulids were known from eight living species, all macroscopic, ranging from 1–20 cm in length, and found only in cold-water habitats throughout the world (a very large priapulid, *Halicryptus higginsi*, up to a half meter in length, was later described from Alaska by Shirley and Storch 1999).

Priapulids, like the kinorhynchs and loriciferans, have a similar body plan insofar that it consists of a spherical evertable head, which contains an evertable mouth cone, a neck region, and an elongate, annulated abdomen or trunk region, sometimes with a caudal appendage. The head (also called an introvert or a proboscis) is covered with rings of scalids, often organized in longitudinal rows in the family Priapulidae. A series of larval stages are found in nearly all known living species of priapulids. Larval stages are all very small, 0.1–1.0 mm in the case of *Priapulus caudatus*. The youngest stage larva of the latter species is ovoid to spherical (plate 112F); subsequent stages of *Priapulus* larvae are dorsoventrally flattened (plate 112G).

Priapulus caudatus Lamarck, 1816 (plate 112H), is the only priapulid reported from the Pacific coast, where it ranges from the Arctic south to at least central California (Shapeero 1962). In our region, it is found in "sticky mud" (as described in the second edition of this manual) in shallow sublittoral waters of Tomales Bay. Both larvae and adults of *P. caudatus* are occasionally found in intertidal mud flats. They tend to be reclusive and often have a patchy distribution. They are commonly found under rocks. In general, priapulids inhabit the upper 20 cm layer of mud.

Most priapulids are found by laborious searching through or by sieving mud samples. Whereas 1 mm mesh sieves are suitable for finding the adults, larvae will pass through; therefore, one must use a finer mesh net or use the bubble-and-blot method of extraction (Higgins 1964). This method involves placing several liters of mud sample in a bucket and adding seawater until one can mix the contents into a souplike slurry. If this slurry is poured into a clean bucket from a height of about a meter, it will have the same effect as introducing a stream of bubbles into the mixture; the bubbles will bring a selection of meiofaunal organisms to the surface where they will be caught in the surface tension. Allow the material to settle for about one minute in order for the sediment to settle away from the surface, thereby leaving a "float" of microorganisms that have been trapped by the surface tension.

Next gently place the lower surface of a piece of copy paper on the surface film and quickly remove it. The floating specimens will be transferred from the surface tension to the paper. Then, with a stream of seawater from a wash-bottle, wash the material from the paper into a fine mesh sieve, directly into a dish for stereomicroscopic examination, or into a collecting bottle for preservation. Repeating the entire process several times may be necessary to accumulate sufficient material. Priapulid larvae are among several meiofaunal taxa that respond to this collecting method. If they are present in the sampled mud, this method will extract them with great efficiency. If one wants to extrude the head from the lorica, immersion for a few minutes in freshwater may help.

References

Adrianov, A. V. and R. P. Higgins. 1996. *Pycnophyes parasanjuanensis*, a new kinorhynch (Kinorhyncha: Homalorhagida: Pycnophyidae) from San Juan Island, Washington, U.S.A. Proc. Biol. Soc. Washington. 109: 236–247.

Higgins, R. P. 1960. A new species of *Echinoderes* (Kinorhyncha) from Puget Sound. Trans. Amer. Microsc. Soc. 79: 85–91.

Higgins, R. P. 1961. Three new homalorhage kinorhynchs from the San Juan Archipelago, Washington. J. Elisha Mitchell Sci. Soc. 77: 81–88.

Higgins, R. P. 1964. A method for meiobenthic invertebrate collection. Amer. Zool. 4: 291.

Higgins, R. P. 1977. Redescription of *Echinoderes dujardini* (Kinorhyncha) with descriptions of closely related species. Smithsonian Contr. Zool. 248: 1–26.

Higgins, R. P. 1986. A new species of *Echinoderes* (Kinorhyncha: Cyclorhagida) from a coarse-sand California beach. Trans. Amer. Microsc. Soc. 105: 266–273.

Higgins, R. P. 1991. *Pycnophyes chukchiensis*, a new homalorhagid kinorhynch from the Arctic Sea. Proc. Biol. Soc. Washington 104: 184–188.

Higgins, R. P. and R. M. Kristensen. 1986. New Loricifera from southeastern United States coastal waters. Smithsonian Contr. Zool. 438: 1–70.

Higgins, R. P. and H. Thiel, eds. 1988. Introduction to the study of meiofauna. Washington, Smithsonian Institution Press, 487 pp.

Kristensen, R. M. 1983. Loricifera, a new phylum with Aschelminthes characters from meiobenthos. Z. Zool. Syst. Evolut.-forsch. 21: 163–180.

Kristensen, R. M. 1991. Loricifera. In M*icroscopic anatomy of invertebrates*. Vol. 4. F. W. Harrison and E. E. Ruppert, eds. Aschelminthes, pp. 351–375. New York: Wiley-Liss.

Kristensen, R. M. 2002. An introduction to Loricifera, Cycliophora, and Micrognathozoa. Integ. Comp. Biol. 42: 641–651.

Land, J. van der. 1968. A new aschelminth, probably related to the Priapulida. *Zool. Meded.* 42: 237–250.

Lemburg, C. 1995. Ultrastructure of the sense organs and receptor cells of the neck and lorica of *Halicryptus spinulosus* larva (Priapulida). Microfauna Marina 10: 7–30

Malakhov, V. V. 1980. Cephalorhyncha, a new type of animal phylum of animal kingdom uniting Priapulida, Kinorhyncha, Gordiacea, and a system of Aschelminthes worms. Zool. Zhurnal 59: 485–499 (in Russian with English summary).

Shapeero, W. L. 1962. The distribution of *Priapulus caudatus* Lam. on the Pacific coast of North America. Amer. Midl. Nat. 68: 237–241.

Shirley, T. C. and V. Storch. 1999. *Halicryptus higginsi* n. sp. (Priapulida)— a giant new species from Barrow, Alaska. Invertebrate Biol. 118: 401–413.

Nematomorpha

(Plate 113)

ARMAND M. KURIS

The Nematomorpha, or horsehair worms, are all protelean parasitoids of arthropods. Most species infect insects as juveniles (Poinar 1991). One genus, *Nectonema,* is marine and attacks decapod crustaceans. All kill their host upon emergence as adults.

Morphologically, nematomorphs are very similar to nematodes. However, they lack the lateral excretory cords, and their gut is greatly reduced and nonfunctional. The parasitic juveniles absorb nutrients through their body surface. Adults do not feed. Their pseudocoel is largely filled with gonads. *Nectonema* adults are distinctive, having rows of lateral cilia that aid in swimming. Sexes are separate, and the free-living adults mate and lay their eggs in the water. Multiple host life cycles and the use of paratenic hosts for some freshwater species have been documented (Hanelt and Janovy 2003). Nematomorphs are famous for inducing a powerful effect on their host's behavior, causing the dying insect to seek water so the adult nematomorph can emerge in its appropriate environment (Thomas et al. 2002).

The Nematomorpha is one of the few phyla never reported from marine environments along the west coast of North America. On the New England coast, palaemonid shrimps are often infected with *Nectonema* (Born 1967). The Atlantic horsehair worm *Nectonema agile* Verrill, 1879, in the body cavity of the common east coast shrimp *Palaemonetes vulgaris* is shown in plate 113. Because very few Pacific coast hermit crabs and caridean shrimps have been even superficially examined for parasites, it is possible that a species of *Nectonema* may yet be recovered from this coast.

References

Born, J. W. 1967. *Palaemonetes vulgaris* (Crustacea Decapoda) as host for the juvenile stage of *Nectonema agile* (Nematomorpha). J. Parasit. 53: 793–794.

Hanelt, B. and J. Janovy. 2003. Spanning the gap: identification of natural paratenic hosts of horsehair worms (Nematomorpha: Gordioidea) by experimental determination of paratenic host specificity. Invert. Biol. 122: 12–18.

Poinar, G. O. 1991. Nematoda and Nematomorpha, in: J. H. Thorpe and A. P. Covich, eds., Ecology and classification of North American freshwater invertebrates. New York: Academic Press.

Thomas, F., A Schmidt-Rhaesa, G. Martin, C. Manu, P. Durand and F. Renaud. 2002. Do hairworms (Nematomorpha) manipulate the water-seeking behavior of their terrestrial hosts? J. Evol. Biol. 15: 356–361.

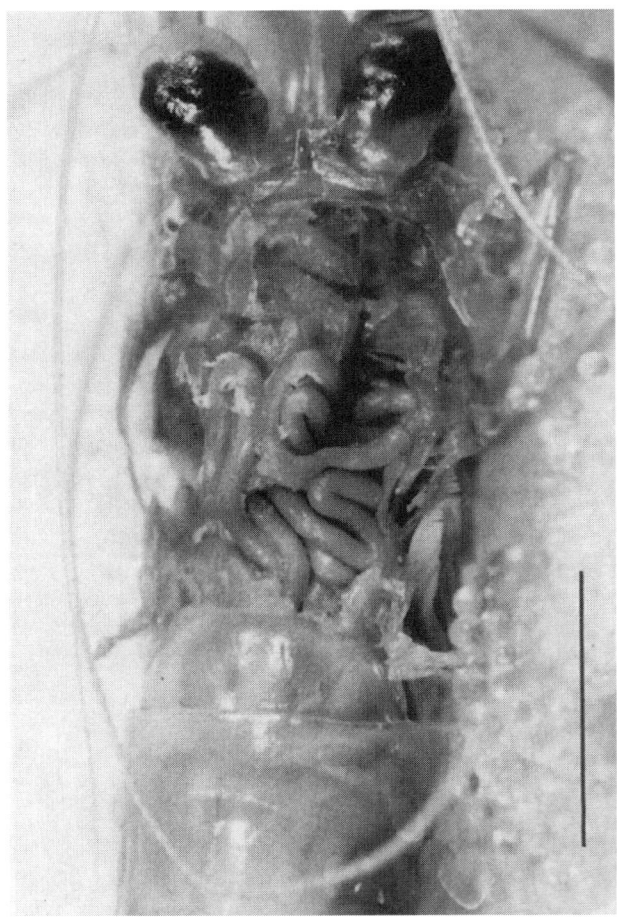

PLATE 113 *Nectonema agile* in the body cavity of the shrimp *Palaemonetes vulgaris* (from J. Bresciani 1991. Nematomorpha, Chapter 5. In Microscopic Anatomy of Invertebrates. Wiley-Liss, Inc., with permission). Scale bar = 0.5 cm.

Acanthocephala

(Plate 114)

ARMAND M. KURIS

The Acanthocephala, or spiny-headed worms, are all parasites. Adults live in the small intestine of teleost fishes and tetrapods. In the gut of their vertebrate hosts, they absorb nutrients through their body surface. They lack a gut and attach to the intestine of the host using an eversible, spiny proboscis, which they can retract into their spacious pseudocoel. Sexes are separate, and eggs are passed with the feces of their host.

Acanthocephalans have a two-host life cycle; using insects or crustaceans as intermediate hosts. In their arthropod intermediate host, the acanthocephalan egg hatches as an acanthor larva. This uses distinctive hooks to penetrate the gut and enter the hemocoel. In the hemocoel, the acanthor loses its hooks and metamorphoses to an acanthella. The acanthella differentiates a proboscis with the adult hooks and gradually transforms to the infective cystacanth larva by invagination of the developing proboscis. In their decapod crustacean intermediate hosts, the cystacanths are often in the posterior portion of the hemocoel near the thoraco-abdominal junction. They are opaque white (sometimes yellow or orange), about 1 mm long, and shaped like a rugby ball. Acanthellae are often present but are more difficult to notice, being translucent, more elongate, flattened, and usually smaller than the cystacanths. To best observe the cystacanths and identify them to species, the cystacanths should be placed in fresh water, which forces them to evert their proboscides (plate 114). Trophic transmission occurs when the infected arthropod is eaten by the vertebrate final host.

Freshwater and terrestrial acanthocephalans are well known to behaviorally modify their intermediate hosts (Moore 1984, 2002). This increases the likelihood that an appropriate vertebrate predator will eat the infected intermediate host. There has been little investigation of marine acanthocephalans for evidence of parasite-increased trophic transmission.

In California, larval acanthocephalans (family Polymorphidae) are commonly found in the hemocoels of the decapods *Emerita analoga, Blepharipoda occidentalis,* and *Hemigrapsus oregonensis.* Adults of these acanthocephalans are in gulls and other shore birds. Lafferty and Gerber (2002) report that *Profilicollis altmani* (Perry 1942) (=*Polymorphus kenti* Van Cleave, 1947) contributed up to 10% of the mortality of California sea otters, presumably a consequence of otters ingesting infected mole crabs. Other species of acanthocephalans parasitize

PLATE 114 Acanthocephala, Cystacanth of *Profilicollis altmani* from the hemocoel of the crab *Hemigrapsus oregonensis*, proboscis everted in fresh water (photograph by Mark Torchin).

marine amphipods and isopods, but these have not been investigated in California. These may serve as hosts for the several species of acanthocephalans reported from teleost fishes along the Pacific coast of North America.

References

Lafferty, K. D. and L. Gerber. 2002. Good medicine for conservation biology: the intersection of epidemiology and conservation theory. Cons. Biol. 16: 593–604.

Moore, J. 1984. Altered behavioral responses in intermediate hosts: an acanthocephalan parasite strategy. Amer. Natur. 123: 572–577.

Moore, J. 2002. Parasites and the behavior of animals. New York: Oxford University Press.

Gnathostomulida

(Plates 115–116)

RICHARD FARRIS

Gnathostomulids (plates 115 and 116) are vermiform organisms with an adult body length between 500 and 1500 μm. The name, meaning "small jaws associated with the mouth," comes from a pair of cuticularized jaws with teeth situated posterior to the mouth opening and within a distinct anterior pharynx. An unpaired basal plate is located just in front of and below the jaws. The pharynx is highly muscular, and in live preparation under a compound microscope, the jaws can frequently be seen snapping open and shut. Three to four pairs of compound sensory bristles, often observed flickering back and forth, are located at the anterior end. The epithelium is monociliated, parenchyma is poorly developed, and a body cavity is lacking. The digestive tract is simple and incomplete.

All adult organisms are hermaphroditic. The male apparatus may include a penis stylet composed of cuticular rods and a muscular penis. Female organs usually consist of a large ovary dorsal to the gut and a bursa complex for sperm storage. Several large eggs may be seen anterior to the bursa. Sperm enter either by epidermal impregnation or, when present, through a vagina. Fertilization is internal and development is direct.

The Gnathostomulida are interstitial organisms living between sand grains in marine environments that are reducing but not totally anoxic. Mean grain size for the substrate is between 150 and 250 μm (Farris 1975). They are thus more likely to be found in sandy beach estuaries rather than mud flats where the silt-clay fraction is much higher. Along the California, Oregon, and Washington coasts, the best place to collect gnathostomulids is among the sediments trapped within the root systems of the surfgrass *Phyllospadix scouleri* on rocky, high wave-energy beaches.

Sediment may be collected by simply pulling the grasses off the rocks and transporting them back to the laboratory. There the sand can be scraped off into a flask and extracted using the magnesium chloride anesthetization-decantation method (Pfannkuche and Thiel 1988). In this procedure, isotonic magnesium chloride is added to the flask containing the sediment. After rotating the mixture several times to ensure thorough mixing, let it stand for eight minutes before agitating one more time, then passing the supernatant through a 62 μm nylon sieve. The sieve is next rinsed with clean seawater and placed in a petri dish with a small amount of seawater for two hours, giving the organisms time to crawl through into the dish. The sieve is then removed, and specimens can be examined in the petri dish using a dissecting microscope.

Often gnathostomulids can be distinguished from other interstitial phyla through a common behavioral characteristic of swimming backward as well as forward. Among interstitial organisms, only the ciliates and a few turbellarians exhibit a similar behavior. For detailed analysis, a wet-mount slide of the organism must be examined using a compound microscope (for more information on extraction and preparation techniques see Higgins and Thiel 1988).

Gnathostomulida were recognized as a phylum in 1969 (Riedl 1969). Its systematic position as either an acoelomate or pseudocoelomate is still debated (Sterrer et al. 1985). More than 100 species have been described in two orders and 18 genera. Only one species has been described from the West Coast (Riedl 1971): *Gnathostomula karlingi* Riedl 1971, was described from low intertidal sediments from Yaquina Head, Newport, Oregon.

Keying to families, genera, and species requires detailed analysis of the jaws and teeth. Identification of the most commonly found suborders can be done using the above described characteristics and extraction technique.

Most Commonly Observed Suborders

Order Bursovaginoidea

Body short (500–900 μm), head well-developed, with paired compound sensory bristles, bursa present.

SUBORDER SCLEROPERALIA

With a cuticular bursa, stylets associated with the penis, sperm small and teardrop-shaped. *Gnathostomula* sp. (plate 115).

SUBORDER CONOPHORALIA

Lacking a cuticular bursa and penis stylets, sperm larger and cone-shaped. *Austrognathia* sp. (plate 116).

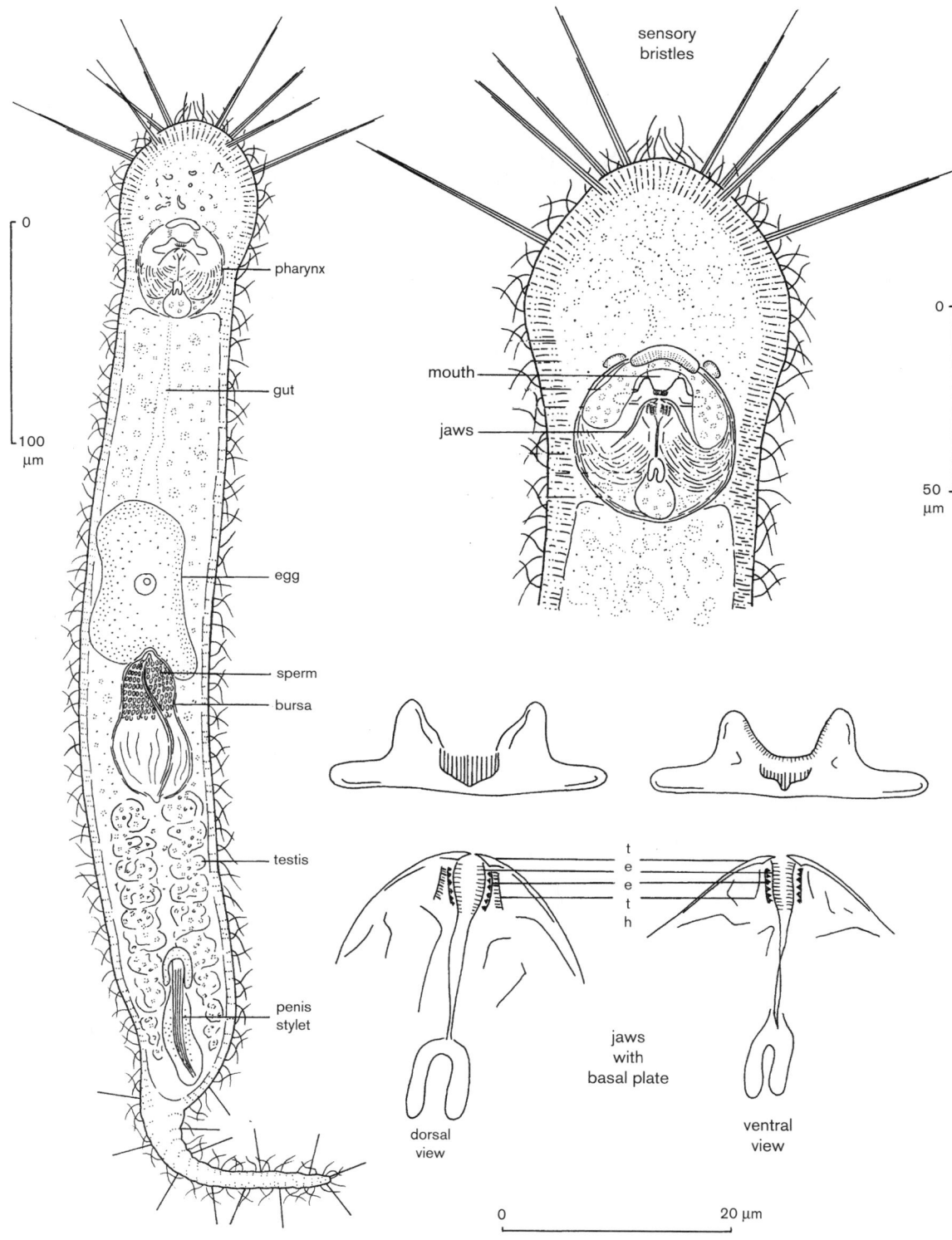

sensory bristles

pharynx

gut

egg

sperm

bursa

testis

penis stylet

mouth

jaws

t
e
e
t
h

jaws
with
basal plate

dorsal
view

ventral
view

0

100
μm

0

50
μm

0   20 μm

PLATE 115 Gnathostomulida, suborder Scleroperalia, *Gnathostomula* sp.

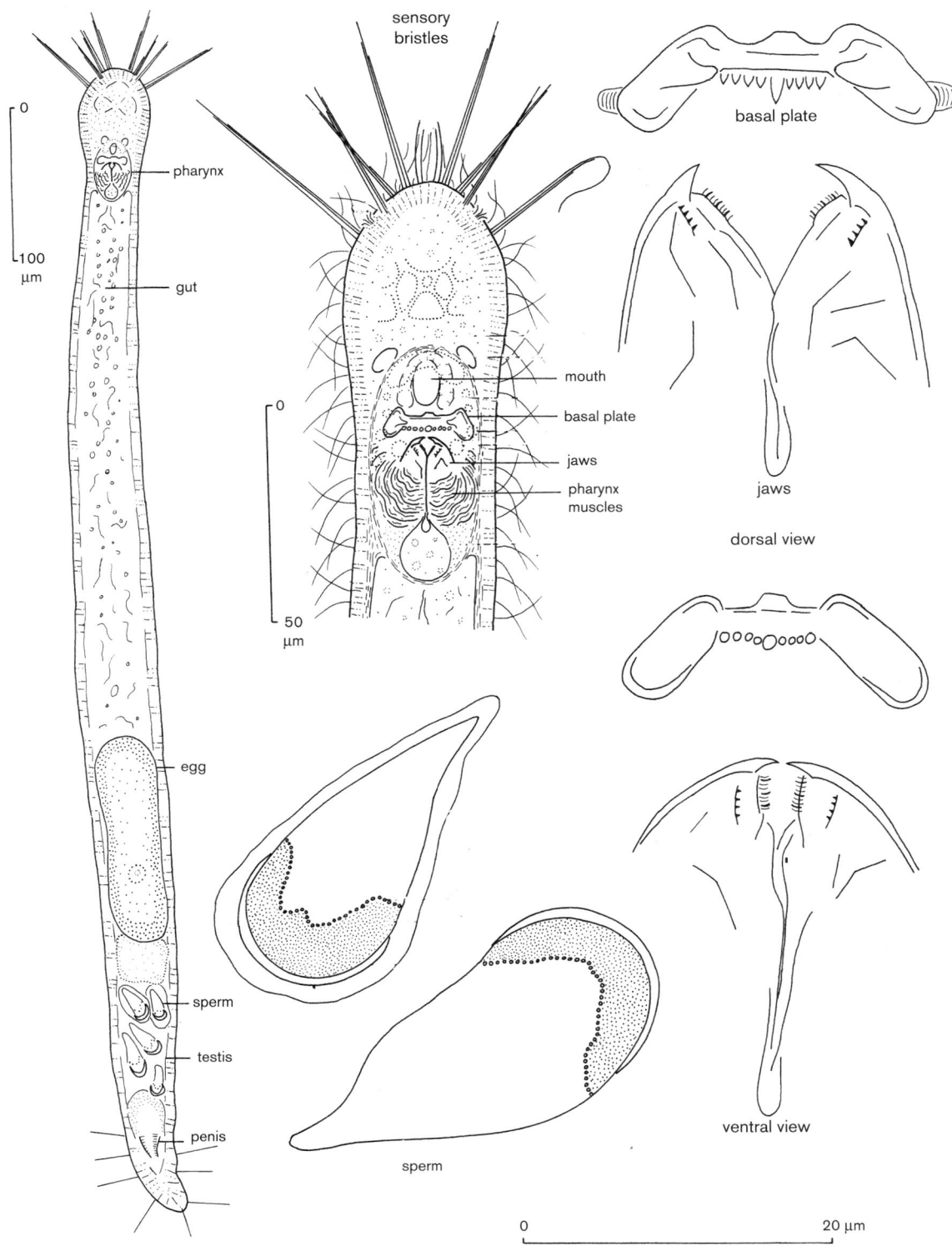

sensory
bristles

0

100
μm

pharynx

gut

0

50
μm

mouth

basal plate

jaws

pharynx
muscles

egg

sperm

testis

penis

sperm

basal plate

jaws

dorsal view

ventral view

0   20 μm

PLATE 116 Gnathostomulida, suborder Conophoralia, *Austrognathia* sp.

References

Bres, M. and R. Higgins. 1992. "Cryptic fauna of marine sand" (video). Produced by American Society of Zoologists. Washington, D.C.: The National Museum of Natural History, Smithsonian Institution.

Farris, R. 1975. On the ecology and systematics of Gnathostomulida from North Carolina and Bermuda. Ph.D. dissertation. Chapel Hill, NC: University of North Carolina.

Farris, R. and D. O'Leary. 1985. Application of videomicroscopy to the study of interstitial fauna. Internationale Revue fur Gesamten der Hydrobiologie 70: 891–895.

Higgins, R. and H. Thiel. 1988. Introduction to the study of meiofauna. Washington, D.C.: Smithsonian Institution Press.

Pfannkuche, O. and H. Thiel. 1988. Sample processing. pp. 134–145, In Introduction to the study of meiofauna. R. Higgins and H. Thiel, eds. Washington, D.C.: Smithsonian Institution Press.

Riedl, R. 1969. Gnathostomulida from America. Science 163:445–452.

Riedl, R. 1971. On the genus Gnathostomula (Gnathostomulida). Internationale Revue fur Gesamten der Hydrobiologie 56: 385–496.

Sterrer, W. 1972. Systematics and Evolution within the Gnathostomulida. Systematic Zoology 21: 151–173.

Sterrer, W., M. Mainitz and R. M. Rieger. 1985. Gnathostomulida: enigmatic as ever. pp. 181–199. In The origins and relationships of lower invertebrates. S. Conway Morris, ed. Oxford: Clarendon Press.

Rotifera

(Plate 117)

STEVEN C. FRADKIN

Rotifers are small, wormlike pseudocoelomate metazoans found worldwide in marine, freshwater, and semiterrestrial habitats. Ranging in size from 100 μm to 2 mm, most rotifers are free-living herbivores, bacteriovores, and predators. Rotifera is a morphologically diverse phylum that exhibits complex life-histories, including a range of reproductive strategies from obligate asexuality to obligate sexuality. Diapause, the ability to arrest development and metabolic activity, is a common life-history strategy employed to survive harsh conditions. Rotifers occupy many intertidal and nearshore marine habitats, including meiobenthic, benthic, and planktonic habitats. Coastal areas under estuarine influence appear to be important areas for many rotifer taxa (Fradkin 2001), which can become very abundant (>1,000 individuals l⁻¹). Only the open ocean is devoid of rotifers, probably due to resource limitation.

The phylum consists of 120 genera, with only a subset known from marine waters. In the Oregonian faunal province, 10 genera from three orders are known, with an unknown number of species. Almost all marine rotifer taxonomy is based on European and North American Atlantic descriptions, with few, if any, descriptions for Pacific species. The present key identifies Californian and Oregonian marine rotifers to the generic level.

Known as "wheeled animalcules" for their prominent anterior "wheel organ" or corona, rotifers have a basic body plan including a head, trunk, and foot. There is substantial deviation from this bauplan throughout the phylum. The head contains the mouth and sense organs (tactile and light-sensitive), and is crowned by a ciliated corona that produces currents used for locomotion and feeding. The trunk cavity is filled with digestive and reproductive organs, in addition to a brain that services organs and muscles via nerves. The digestive tract includes a mastax, esophagus, stomach, and an intestine with flames cells leading to an anus. The mastax is a muscular pharynx containing a complex hard-jaw structure known as the trophi that is composed of mucopolysaccharide.

The structure of the trophi varies considerably between taxa and is widely used as a diagnostic feature in rotifer classification, especially to the species level. Determination of trophi structure requires dissection and mounting prior to examination. See Stemberger (1979) for preparation techniques. In females, the reproductive organ is a germovitellarium composed of the germarium (ovary) and vitellarium that produces yolk. In some species, the trunk and head is encased in a hardened lorica. The lorica is a rigid body wall composed of scleroprotein. The syncytial epidermal layer has no cell boundaries but contains multiple nuclei. The transition from trunk to foot is marked by the anus. The foot, with its associated toes and pedal glands, is used for creeping and attachment to surfaces.

All rotifers have eutylic development, where the body has a constant number of cells, approximately 1,000. After development, growth occurs through cellular expansion rather than cell division. Wallace and Snell (2001) provide a thorough account of general rotifer biology, while Thane-Fenchel (1968) provides a key to Atlantic and Baltic marine rotifer genera.

In coastal California and Oregon, three major taxonomic groups are represented: the digonont orders Seisonidea and Bdelloidea (both in the class Digononta) and the order Ploima (in the class Monogononta).

Seisonidea

This is a monogenic order with paired reproductive organs represented by a single undescribed species, *Seison* sp., on the Pacific coast (Kozloff 1996). *Seison* is 800–1,500 μm and has a reduced corona that is not used in food gathering or locomotion. *Seison* is epizoic on the leptostracan crustacean *Nebalia*, which is found amongst algal wrack on protected muddy substrates. Attachment typically occurs on the thoracic legs and inner carapace surface of *Nebalia*. Up to eight rotifers may be found on an individual *Nebalia*; however, densities are normally between one and three (Ricci et al. 1993). Seisonids are exclusively marine and are thought to be the ancestral rotifer lineage. Seisonids are obligately sexual, with populations composed of both males and females that show no sexual dimorphism. Female reproductive organs consist of only a germarium without a vitellarium. Diapause does not occur in order Seisonidea.

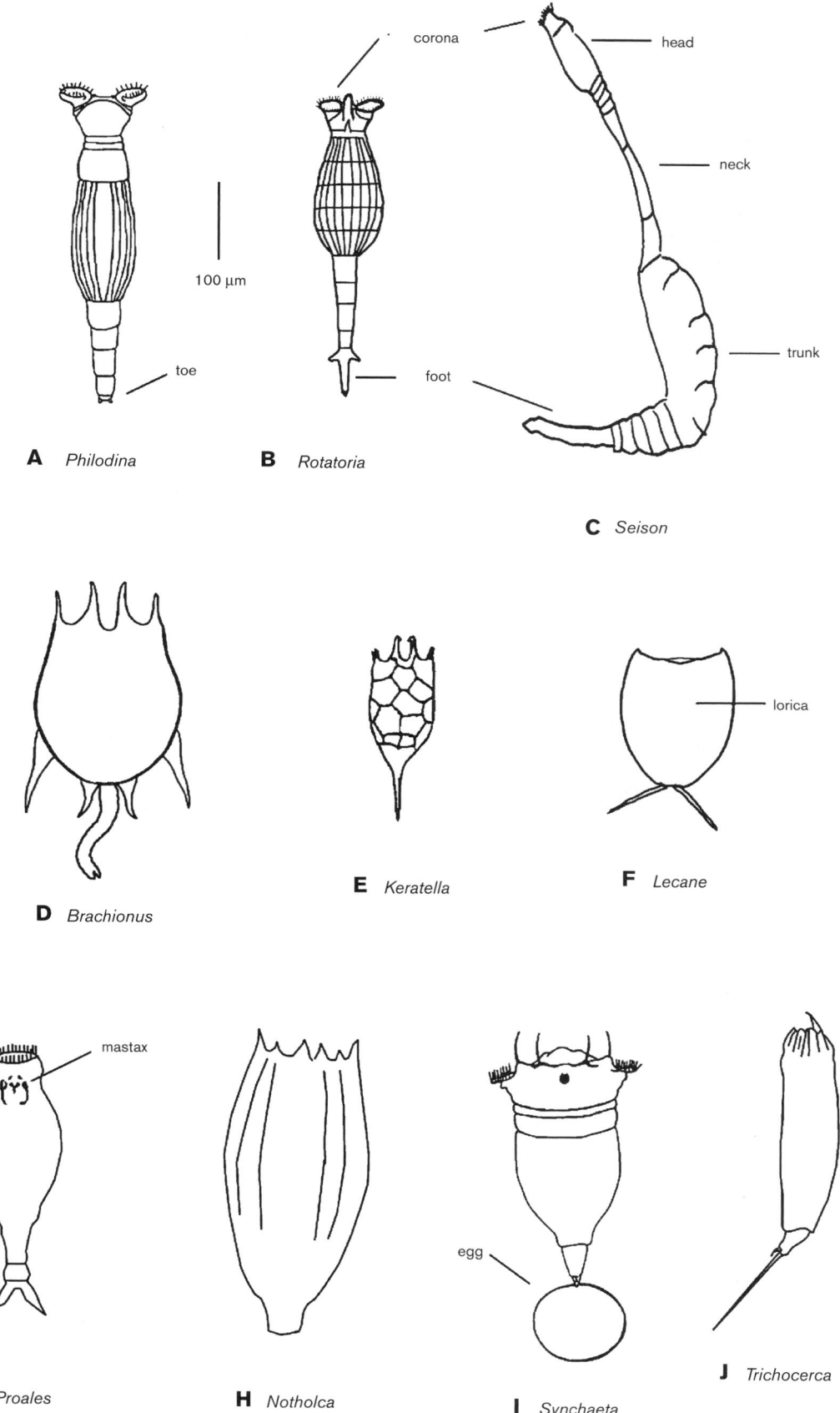

corona

head

neck

trunk

100 μm

toe

foot

A *Philodina*

B *Rotatoria*

C *Seison*

D *Brachionus*

E *Keratella*

F *Lecane*

lorica

mastax

egg

G *Proales*

H *Notholca*

I *Synchaeta*

J *Trichocerca*

PLATE 117 A–J, Examples of rotifer genera of the Oregonian Faunal Province (drawings by S. Fradkin).

Bdelloidea

Bdelloidea is represented on the Pacific coast by a two genera, *Philodinia* and *Rotatoria*. These rotifers have a wormlike vermiform shape and pseudo-segmentation that allows telescoping of the 250 μm body. Both genera are meiobenthic, living in interstitial sandy beaches. Like all bdelloids, populations are exclusively female and reproduction is obligately asexual. Males have never been observed, and genetic evidence suggests a complete lack of sexual reproduction in the order. Females have paired germovitellaria. Diapause in bdelloids occurs through adult anhydrobiosis, where bdelloids are capable of surviving long periods of complete desiccation. Ricci and Melone (2000) provide a comprehensive key of bdelloids.

Ploima

Ploima contains >1,275 species. Unlike class Digononta, class Monogononta (the most diverse group of rotifers, with >100 genera) evolved in fresh water with subsequent reinvasion of marine systems by genera only in the order Ploima. Marine monogonont representatives are considerably less diverse relative to their freshwater counterparts, yet monogononts are still more diverse in marine waters than digononts. On the Pacific coast, Ploima is represented by seven genera that exhibit marked morphological variation. All monogonont rotifers are dioecious, with sexes containing a single gonad. Monogononts reproduce by cyclical parthenogenesis where multiple generations of asexual reproduction are punctuated by periods of sexual reproduction. Populations usually consist of diploid females reproducing asexually. In response to environmental cues of a degrading environment, females produce daughters that in turn produce haploid eggs. If unfertilized, haploid eggs develop into haploid males that have incomplete digestive tracts and contain a single testis that dominates the body cavity. If fertilized, the resultant diploid sexual egg becomes a diapausing "resting" egg, capable of long-term diapause (>20 yrs.). When reproducing asexually, marine monogonont rotifers are capable of high reproductive rates (e.g., 1.4 d^{-1} at 20° C and 0.11 d^{-1} at 14° C, Arndt et al. 1985, Henroth 1983). Ploima genera include *Brachionus*, *Keratella*, *Notholca*, *Lecane*, *Proales*, *Trichocerca*, and *Synchaeta*.

Key to Genera

1. Rotifers with paired ovaries: DIGONONTA 2
— Rotifers with a single ovary: MONOGONONTA 4
2. Robust corona, free swimming, slender wormlike body: Bdelloidea 3
— Reduced corona, sessile epibiont on *Nebalia*: Seisonidea (plate 117C) . . . . *Seison*
3. Rostrum present, viviparous (plate 117B) . . . . *Rotatoria*
— Rostrum absent, oviparous (plate 117A) . . . . *Philodina*
4. Distinct lorica 5
— Without distinct lorica, spindle-shaped body 9
5. Movable posterior foot/toes present. 6
— No posterior foot, posterior spine if present is part of lorica. 8
6. Rigid posterior foot/toes 7
— Single flexible posterior foot; wormlike, retractable foot. dorsoventrally flattened body (plate 117D) . . . . . *Brachionus*
7. Long spinelike foot, cylindrical body (plate 117J) . . . .
. . . . *Trichocerca*
— Two spinelike toes, disk-like body (plate 117F) . . . .
. . . . *Lecane*
8. Dorsal plate with polygonal facet pattern (plate 117E)
. . . . *Keratella*
— Dorsal plate without facets (plate 117H) . . . . *Notholca*
9. Anterolateral ciliated auricles, foot not clearly segment (plate 117I) . . . . *Synchaeta*
— No anterolateral ciliated auricles, foot clearly segmented (plate 117G) . . . . *Proales*

References

Arndt, H, H. J. Kramer, R. Heerkloss, and C. Schröder. 1985. Zwei Methoden zur Bestimmung populationsdynamischer Parameter von Zooplanktern unter laborbedingungen mit ersten Ergebnissen an *Eurytemora affinis* (Copepoda; Calanoida) und *Synchaeta cecilia* (Rotatoria, Monogononta). Wiss. Z. Univ. Rostock, Meeresbiol. Beitr. 34: 17–21.

Fradkin, S. C. 2001. Rotifer distribution in the coastal waters of the northeast Pacific Ocean. Hydrobiologia 446/447: 173–177.

Henroth, L. 1983. Marine pelagic rotifers and tintinnids—important trophic links in the spring plankton community of the Gullmar Fjord, Sweden. J. Plankton Res. 5: 835–846.

Kozloff, E. N. 1996. Marine invertebrates of the Pacific Northwest. University of Washington Press, 539 pp.

Ricci, C. and G. Melone. 2000. Key to the identification of the genera of bdelloid rotifers. Hydrobiologia 418: 73–80.

Ricci, C., G. Melone, and C. Sotgia. 1993. Old and new data on Seisonidea (Rotifera). Hydrobiologia 255/256: 495–511.

Stemberger, R. S. 1979. A guide to the rotifers of the Laurentian Great Lakes. Cincinnati, Ohio: U.S. Environmental Protection Agency, USA, PB80–101280.

Thane-Fenchel, A. 1968. A simple key to the genera of marine and brackish-water rotifers. Ophelia 5: 299–311.

Wallace, R. L. and T. W. Snell. 2001. Rotifera. In Ecology and classification of North American freshwater invertebrates. 2nd ed. pp. 195–254. J. H. Thorp and A. P. Covich, eds. Academic Press.

Kamptozoa (Entoprocta)

(Plates 118–119)

KERSTIN WASSON AND RICHARD N. MARISCAL

Kamptozoans are tiny suspension feeders. Because of their small size and relative obscurity, they are frequently mistaken for stalked protozoans, hydroids, or bryozoans (ectoprocts). Living kamptozoans, however, can be readily distinguished from other groups by the vigorous bending movements of the stalk, reflected in both their scientific name (Greek: Kamptozoa = bending animals) and their common name ("nodding heads"). Further, kamptozoan tentacles are rolled inward when the animal is disturbed; in contrast, bryozoans and hydroids pull their tentacles straight down into a protective housing.

The kamptozoan body plan consists of a basal attachment structure (adhesive disc or stolon), a cylindrical stalk, and a calyx perched atop it (plate 118A–118D). The calyx contains a U-shaped digestive tract, a pair of gonads, protonephridia, and a ganglion. The rim of the cuplike calyx carries a circlet of ciliated tentacles that generate the feeding current and capture particles. All kamptozoans reproduce asexually by budding, forming either clonal aggregations of separate bodies or colonies of interconnected zooids. General reviews of kamptozoan biology can be found in Hyman (1951), Brien (1959), Mariscal (1975), Emschermann (1982), Nielsen and Jespersen (1997), and Wasson (2002).

The phylogenetic position of kamptozoans remains enigmatic. In the past, they have been allied with bryozoans, to which they bear a superficial resemblance. But bryozoans and kamptozoans differ fundamentally in both their adult and larval body plans. Bryozoans are coelomates, while kamptozoans lack a mesodermally lined body cavity. Another conspicuous difference is the position of the anus: it lies within the circlet of tentacles in kamptozoans (hence "entoprocts"), while it lies outside the tentacles in bryozoans (thus "ectoprocts"). Embryology and molecular sequence data place kamptozoans with spiralians, perhaps affiliated with annelids.

Worldwide, about 150 species of kamptozoans have been described; all but one are marine. The kamptozoan fauna of most regions is poorly known, so the true diversity of the group is certainly much greater. There are two major clades of kamptozoans: the Solitaria (loxosomatids) and the Coloniales (barentsiids and pedicellinids).

Loxosomatids are characterized by calycal budding (plate 118A–118B); these buds detach and settle nearby on the substrate, forming clonal aggregations. A muscular suction disc (*Loxosoma*, plate 118A) or a glandular foot or disc (*Loxosomella* and *Loxomespilon*, plate 118B–118C) serves for adhesion. These kamptozoans generally live commensally on other invertebrates that create ciliary feeding or respiratory currents, such as sponges, bryozoans, polychaetes, or sipunculans, and are often host-specific. Most are quite small, unable to generate strong ciliary feeding currents, and seem to benefit from the enhanced circulation of water containing food particles resulting from the ciliary action or movements of their hosts. Although only two loxosomatids have been described from this coast (Soule and Soule 1965), various undescribed species have been found incidentally in California, and a focused search would surely reveal many more.

In colonial kamptozoans, buds form at the base of stalks, rather than on the calyx. These budded zooids remain connected to each other by stolons, forming colonies. Colonial kamptozoans tend to be larger than solitary forms. They also have longer ciliated tentacles and do not rely on the ciliary currents of other invertebrates for food, and they occur on a variety of living and nonliving substrates and are not host-specific. Internally, they have a circulatory organ that spans the calyx-stalk junction and facilitates the transport of materials between these major body regions. Pedicellinids have uniformly muscular stalks (plate 119A–119B); in barentsiids, the stalks are built of wide muscular nodes and narrow rigid rods (plates 118D, 119C–119H). Eight species of colonial kamptozoans are known from the West Coast and have been reviewed by Wasson (1997).

Collection, Examination, and Preservation

Solitary kamptozoans are mostly too small to be sought in the field with the naked eye. Instead, likely host species should be collected and examined in the laboratory with a dissecting microscope and bright lighting so the lit loxosomatids are visible against the darker background. Colonial kamptozoans can be found by laboratory scrutiny of substrates, especially coralline algae and hydroids collected at low tide, or of suspended panels (e.g., deployed in marinas) that have accumulated a fouling community. In the field, some species hang in visible bunches in low intertidal crannies, while others form delicate,

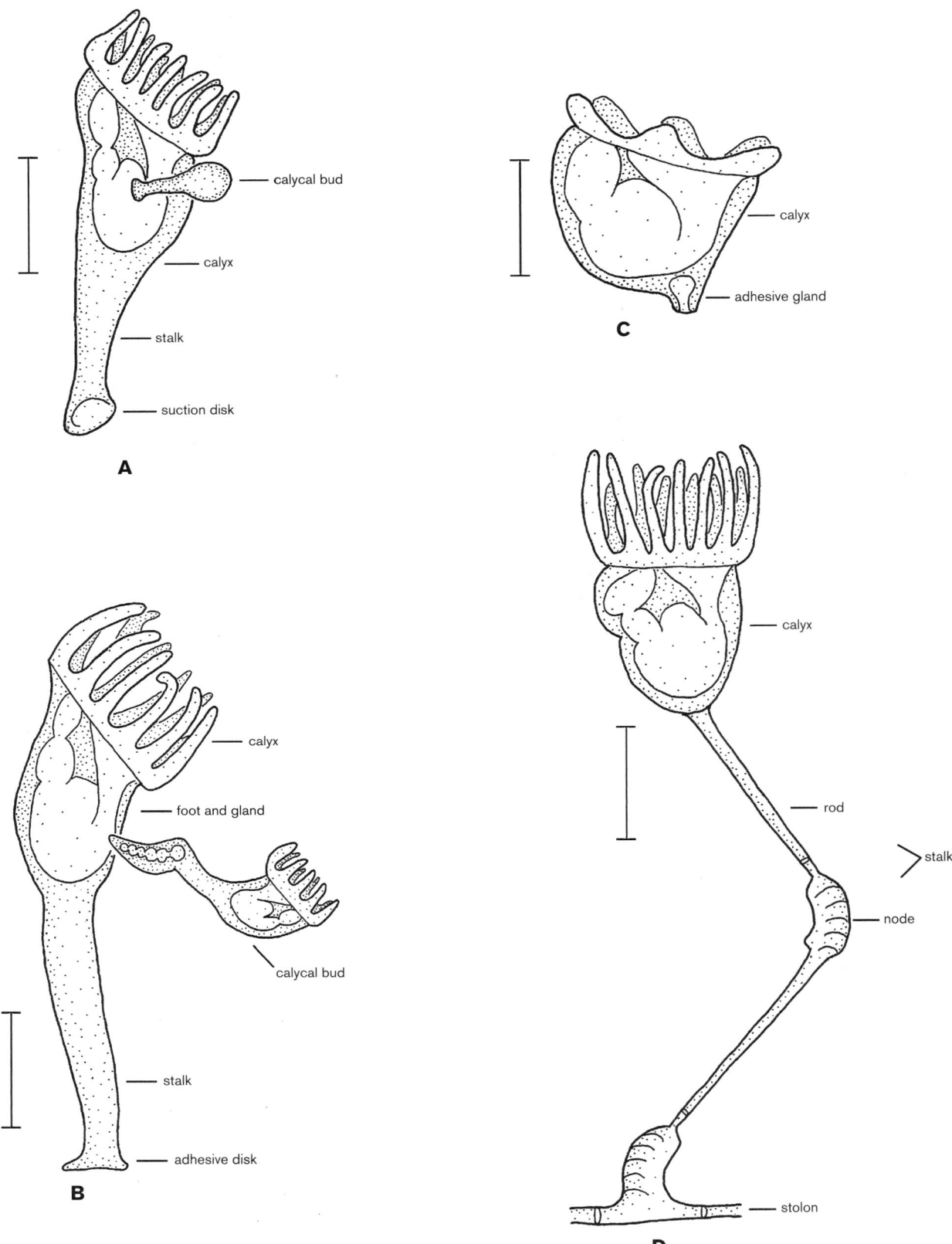

PLATE 118 Generalized members of solitary and colonial genera, shown in left-side view; scale bar = 200 μm in each figure: A, *Loxosoma*; B, *Loxosomella*; C, *Loxomespilon*; D, *Barentsia*.

PLATE 119 Colonial kamptozoans of the Pacific, small portions of colonies drawn to the same scale; scale bar = 2 mm; all calyces are shown in right-side view; stippling in the rods represents rod pores: A, *Myosoma spinosa*; B, *Pedicellina newberryi*; C, *Barentsia benedeni*; D, *Barentsia conferta*; E, *Barentsia discreta*; F, *Barentsia hildegardae*; G, *Barentsia parva*; H, *Barentsia ramosa* (reprinted from the Zoological Journal of the Linnean Society 121:9 1997, by permission of Academic Press).

furry, and dense carpets on floating docks, often intermixed with dense colonies of the bryozoan *Bowerbankia*. To ascertain that colonies are indeed kamptozoans, place them in a little water and stroke them gently; the way the zooids then bend to and fro is unmistakable. Zooids of some species respond to prodding by lying flat against the substrate. Look for potential predators, such as tiny turbellarian flatworms (Canning and Carlton 2000) while watching live *Barentsia* or other colonial kamptozoans under the microscope. In the laboratory, zooids are best examined as whole mounts. Pluck a living zooid from its substrate and place it on a microscope slide in a drop of seawater. Cover it with a coverslip with corners raised on bits of modeling clay so the zooid is not too flattened. Such a live whole mount reveals all the features needed for identification and provides a clear view of the tentacular ciliation, digestive tract, gonads, and larvae (if present) through the thin, transparent body wall. Kamptozoans can be maintained in the laboratory in running or well-oxygenated seawater and will eat cultured phytoplankton; however, only species from bays and harbors will grow luxuriantly.

Kamptozoans can be anaesthetized by the slow addition of 7.5% magnesium chloride solution to seawater and should be fixed in 3–5% formalin; they may be transferred to ethanol later.

Key to Kamptozoa

1. Buds form on the calyx, then detach, resulting in clonal aggregations of separate individuals; musculature continuous between calyx and stalk (plate 118A–118C) . . . . Loxosomatidae 2
— Buds form at the base of stalks, remaining permanently connected by basal stolons, forming colonies; musculature not continuous between calyx and stalk (plates 118D, 119A–119H) 6
2. Stalk absent; attachment to substrate by base of calyx; tentacles greatly reduced to very short, triangular flaps (plate 118C) . . . . *Loxomespilon* sp.
— Stalk present; attachment to substrate by base of stalk 3
3. Basal attachment a muscular suction disc, allowing rapid detachment and reattachment; no permanent attachment to the substrate (plate 118A) . . . . *Loxosoma* spp.
— Basal attachment by secretions from glandular foot in young buds, which adults may retain for slow mobility or lose as they glue themselves permanently to substrate (plate 118B) . . . . *Loxosomella* 4
4. On stomatopod crustaceans 5
— On other substrates (esp. sipunculans, bryozoans, and aphroditid, nephtyid, onuphid, and polynoid polychaetes) . . . . undescribed *Loxosomella* spp.
5. Total height of stalk and calyx 1.0–1.6 mm, stalk about as long as calyx; with 22–34 long tentacles . . . . *Loxosomella macginitieorum*
— Total height 0.8–1.1 mm; stalk shorter than calyx; with 12 short tentacles . . . . *Loxosomella prenanti*
6. Stalk wide, muscular and roughly cylindrical for entire length; stalk not differentiated into nodes and rods (plate 119A, 119B) . . . . Pedicellinidae 7
— Stalk differentiated into short, wide, muscular nodes and long, narrow, rigid, mostly nonmuscular rods (plate 119C–119H) . . . . Barentsiidae 8

7. Stalk and calyx spiny; stalk with diagonal and longitudinal musculature; calyx oriented obliquely on stalk, tilted toward oral side (plate 119A) . . . . *Myosoma spinosa*
— Stalk and calyx without spines; stalk without diagonal musculature; calyx oriented vertically on stalk (plate 119B) . . . . *Pedicellina newberryi*
8. Zooids very small and delicate (total height of stalk and calyx <2 mm); without conspicuous pores on stalk rods 9
— Zooids large and robust (total height >2 mm); with conspicuous pores on stalk rods 10
9. Stalk with many nodes interspersed between short rods; calyx oriented somewhat obliquely on stalk, tilted toward oral side; in bays and harbors (plate 119C) . . . . *Barentsia benedeni*
— Stalk with only one or two nodes interspersed between long rods; calyx oriented vertically on stalk; on open coast (plate 119G) . . . . *Barentsia parva*
10. Stalk with only one node 11
— Stalk with multiple nodes 12
11. Rods short and wide (often >50% width of node), flexible, and pale; pores sparse and only at the base of rods (plate 119D) . . . . *Barentsia conferta*
— Rods long and narrow (often <25% width of node), rigid, and dark brown; many pores along entire length of rod (plate 119E) . . . . *Barentsia discreta*
12. Stalks long (often >10 mm), typically with four to eight nodes; frequent budding at upper stalk nodes resulting in dense, erectly branched, bushy colonies; (plate 119H) . . . . *Barentsia ramosa*
— Stalks shorter (always <10 mm), typically with two to four nodes; budding at upper stalk nodes rare; colonies forming loose, prostrate mats, not dense bushes (plate 119F) . . . . *Barentsia hildegardae*

List of Species

Solitaria

LOXOSOMATIDAE

Loxosoma spp. None have been found in this region so far, but they are likely to occur here on similar hosts as elsewhere in world (e.g., capitellid, maldanid, pectinariid, scalibregmid, and terebellid polychaetes). See Nielsen 1996, Zool. Scripta 25: 61–75 for comprehensive review of genus.

Loxosomella macginitieorum and *L. prenanti* Soule and Soule, 1965. Southern California (Pt. Mugu); subtidal; on the gills of the stomatopod crustacean *Hemisquilla californiensis*; nothing known beyond very brief original description.

Loxosomella spp. Described elsewhere in the world from numerous polychaete families (e.g., aphroditids, chaetopterids, eunicids, maldanids, nephtyids, pectinariids, polynoids, polyodontids), as well as from a variety of other invertebrate hosts (e.g., ascidians, bivalves, bryozoans, crinoids, crustaceans, gorgonians, hydroids, ophiuroids, pterobranchs, scleractinians, sipunculans, sponges). We have found *Loxosomella* on the hosts listed below. Each host probably bears a different species, all probably undescribed. See Nielsen 1989, Entoprocta, Synopses of the British Fauna, E. J. Brill, Leiden for description of European species from similar hosts.

HOSTS OF UNDESCRIBED *LOXOSOMELLA* SPECIES FROM THE PACIFIC COAST

Location of *Loxosomella* on host listed for each.

POLYCHAETES

- under the furry dorsal covering of unidentified aphroditids (dredged from San Juan Archipelago, Washington).
- on the parapodia and under the elytra of various polynoids, including *Eunoe oerstedi* and *Gattyana cirrosa* (dredged from San Juan Archipelago).
- on the branchiae of the onuphids *Diopatra ornata* (from Bodega Bay) and *Onuphis eremita* (dredged from Monterey Bay).
- on the branchiae of unidentified nephtyids (dredged subtidally and collected intertidally in mudflats, San Juan Archipelago).

SIPUNCULANS

- on the bodies of *Golfingia pugettensis* and *Themiste pyroides* (dredged from San Juan Archipelago).

BRYOZOANS

- on colonies of *Dendrobeania lichenoides*, *Hippodiplosia insculpta*, *Schizoporella japonica*, *Tegella aquilirostris*, and other bryozoans (low intertidal, San Juan Island).

Loxomespilon sp. On the branchiae and parapodia of the onuphid polychaete *Diopatra ornata* (dredged from San Juan Archipelago). This genus has only one described species, *L. perezi* Bobin and Prenant, 1953, reported from the Atlantic coast of Europe where it was found associated with aphroditid and sigalionid polychaetes; see Emschermann 1971, Mar. Biol. 9: 51–62. Further investigation is required to determine whether the North American species is *L. perezi* or undescribed.

Coloniales

PEDICELLINIDAE

Myosoma spinosa Robertson, 1900. Southern California (La Jolla) to British Columbia (Queen Charlotte Sound); low intertidal to 50 m; protected and exposed coasts; usually on living substrates (especially red algae, thecate hydroids, chaetopterid polychaete tubes, barentsiid kamptozoans); detailed description, particularly of musculature, in Robertson 1900, Proc. Calif. Acad. Sci. 2: 323–348.

Pedicellina newberryi Wasson, 1997. Southern California (La Jolla) to Alaska (Yakutat); low intertidal to 120 m; protected and exposed coasts; usually on thecate hydroids, bryozoans, barentsiid kamptozoans, or worm tubes.

BARENTSIIDAE

The sexual modes of these six species are described by Wasson, 1997, Biol. Bull. 193: 163–170.

Barentsia benedeni (Foettinger, 1887). Cosmopolitan invasive species, generally limited to bays and harbors; on our coast has been found in Salton Sea, San Francisco Bay, and Coos Bay; forms yellow resting buds (hibernacula) able to withstand extremes of brackish water habitat; morphology investigated by Mariscal 1965, J. Morph. 116: 311–338 (as *B. gracilis*).

Barentsia conferta Wasson, 1997. Central California (Carmel Pt.) to British Columbia (Port Renfrew); low intertidal; exposed coasts; typically on red algae, especially erect coralline algae, but also on hydroids and other substrates.

Barentsia discreta (Busk, 1886). Cosmopolitan; on our coast from Chile (Cape Horn) to southern California (Hueneme); low intertidal to 500 m; protected and exposed coasts; common on serpulid tubes; also on various other substrates; for description of sex determination in Californian colonies see Emschermann 1985, pp. 101–108 in Nielsen & Larwood (eds.), Bryozoa: Ordovician to Recent, Olsen & Olsen, Denmark.

Barentsia hildegardae Wasson, 1997. Oregon (Coos Bay) to British Columbia (Banks Island); low intertidal zone to 80 m; protected and exposed coasts; often on solitary ascidians and chaetopterid polychaete tubes, but also on various other, usually living substrates; preyed upon by nudibranchs *Ancula pacifica* and *Cuthona* sp.; sexual biology described by Wasson 1998, Invert. Biol. 117: 123–128.

Barentsia parva (O'Donoghue and O'Donoghue, 1923). Southern California (Corona del Mar) to British Columbia (Virago Sound); low intertidal to 120 m; protected and exposed coasts; typically on other, larger colonial kamptozoans; also on thecate hydroids, bryozoans, and chaetopterid polychaete tubes.

Barentsia ramosa (Robertson, 1900). Southern California (Santa Rosa Island) to British Columbia (Strait of Georgia); low intertidal to 70 m; exposed coasts; on bare rock in deep crannies or on bryozoans, hydroids, and other animals; preyed upon by nudibranchs *Ancula pacifica* and *Cuthona* sp.; detailed description in Robertson 1900, Proc. Calif. Acad. Sci. 2: 323–348.

References

Brien, P. 1959. Classe des Endoproctes ou Kamptozoaires. In Traité de Zoologie, P-P. Grassé, ed., Vol. 5, pp. 927–1007. Masson et Cie, Paris.

Canning, M. H. and J. T. Carlton. 2000. Predation on kamptozoans (Entoprocta). Invertebrate Biology 119: 386–387.

Emschermann, P. 1972. *Loxokalypus socialis* gen. et sp. nov. (Kamptozoa, Loxokalypodidae fam. nov.), ein neuer Kamptozoentyp aus dem nördlichen Pazifischen Ozean. Ein Vorschlag zur Neufassung der Kamptozoensystematik. Marine Biology 12: 237–254.

Emschermann, P. 1982. Les kamptozoaires. État actuel de nos connaissances sur leur anatomie, leur développement, leur biologie et leur position phylogénétique. Bulletin de la Société Zoologique de France 107: 317–344.

Hyman, L. H. 1951. The pseudocoelomate bilateria—Phylum Entoprocta. In The invertebrates, Vol. 3, L. H. Hyman, pp. 521–554. New York: McGraw-Hill.

Mariscal, R. N. 1975. Entoprocta. In Reproduction of marine invertebrates, Vol. 2, A. C. Giese and J. S. Pearse, eds., pp. 1–41. San Francisco: Academic Press.

Nielsen, C. and A. Jespersen. 1997. Entoprocta. In Microscopic anatomy of invertebrates, Vol. 13, F. W. Harrison, ed., pp. 13–43. New York: Wiley-Liss, Inc.

Soule, D. F. and J. D. Soule. 1965. Two new species of *Loxosomella*, Entoprocta, epizoic on Crustacea. Allan Hancock Foundation Publications, Occasional Paper 29: 1–19.

Wasson, K. 1997. Systematic revision of colonial kamptozoans (entoprocts) of the northeastern Pacific. Zoological Journal of the Linnean Society 121: 1–63.

Wasson, K. 2002. A review of the invertebrate phylum Kamptozoa (Entoprocta) and synopsis of kamptozoan diversity in Australia and New Zealand. Transactions of the Royal Society of South Australia 126: 1–20.

Sipuncula and Echiura

(Plates 120–121)

MARY E. RICE

This section treats two types of marine worms bearing some superficial resemblance in having an apparently unsegmented, stout, cylindrical body and a spacious body cavity or coelom. They were for a century or more (until about the 1940s) called the class Gephyrea of the Annelida, which often also included priapulans. In view of several distinct morphological characters (see below), this concept was discarded (Hyman 1959), and the two groups considered here are commonly placed in the separate phyla Sipuncula and Echiura (Stephen and Edmonds 1972).

Although molecular and developmental studies have suggested close affinities of both groups with the Annelida (Staton 2003; McHugh 1997; Hessling 2003) and, in the case of Echiura, questioned an independent phyletic status, this chapter will retain the more commonly accepted concept of separate phyla for both the sipunculans and echiurans.

Sipuncula

The commonest local sipunculans are known as "peanut worms" because of their appearance when contracted, although the term would not be descriptive for many of the species. The body is usually cylindrical and elongate when relaxed, with a slender anterior part, the introvert (plate 120A1). Upon contraction, this anterior introvert is pulled back into itself, and the worm becomes more ovoid, with the body very firm and turgid. The muscular body wall encases a spacious coelom and a long, coiled intestine. The anus opens dorsally at the base of the introvert (in contrast to its posterior position in the echiurans).

Sipunculans differ from annelids in the lack of segmentation and setae, and from echiurans in the antero-dorsal position of the anus and in other characteristics. The introverted anterior body, which can be extended for feeding, has tentacles that usually surround the terminal mouth. The introvert, often ornamented by spines or hooks, is quite unlike the proboscides of either annelids or echiurans.

Sipunculans are marine or estuarine and commonly, in our area, inhabit rock crevices, empty shells, algal holdfasts, or other protected situations along the open coast. Others occur in firm sand or mud or among eelgrass or surfgrass roots, although they are absent from the shifting sands of exposed beaches. The rocky intertidal forms apparently collect detritus with their tentacles, whereas burrowing forms may directly swallow large quantities of mud and sand.

Identification of sipunculans is facilitated if specimens are well relaxed with tentacles extended prior to fixation. It is best to let specimens relax naturally until tentacles are exposed and, if possible, to anesthetize them in an extended position. A good relaxant for many species is 10% ethanol in seawater. During anesthetization, if the anterior end remains retracted, a slight pressure on the body wall will often cause extension of the introvert. Other relaxants found to be successful for some species include menthol crystals, $MgCl_2$, or refrigeration in cold seawater or freezing. A fixative commonly used is 5% formalin in seawater, followed in one to five days by a transfer to 70% ethyl alcohol for storage. However, the choice of fixative is dependent on intended use of the specimen.

Fisher's monograph (1952) covers local sipunculans, although some of his names have been altered by subsequent workers. Other more comprehensive treatments are those of Stephen and Edmonds (1972) and Cutler (1994).

Key to Sipuncula

1. Tentacles conspicuous when extended, branching (plate 120B) 2
— Tentacles inconspicuous, fingerlike 4
2. Introvert armed with small, black or brown spines; tentacles arising on four stems (plate 120B) . . . .
. . . . *Themiste pyroides*
— Introvert devoid of spines; tentacles arising on six stems
. . . . 3
3. Body fusiform (spindle-shaped) to pyriform (pear-shaped); resembling *T. pyroides*; collar (at base of tentacles) reddish purple (plate 120C) . . . . *Themiste dyscrita*
— Body cylindrical; collar not obviously reddish or purplish (plate 120D) . . . . *Themiste hennahi*
4. Muscles of body wall divided into separate longitudinal bands (as best seen in dissected specimens). 5

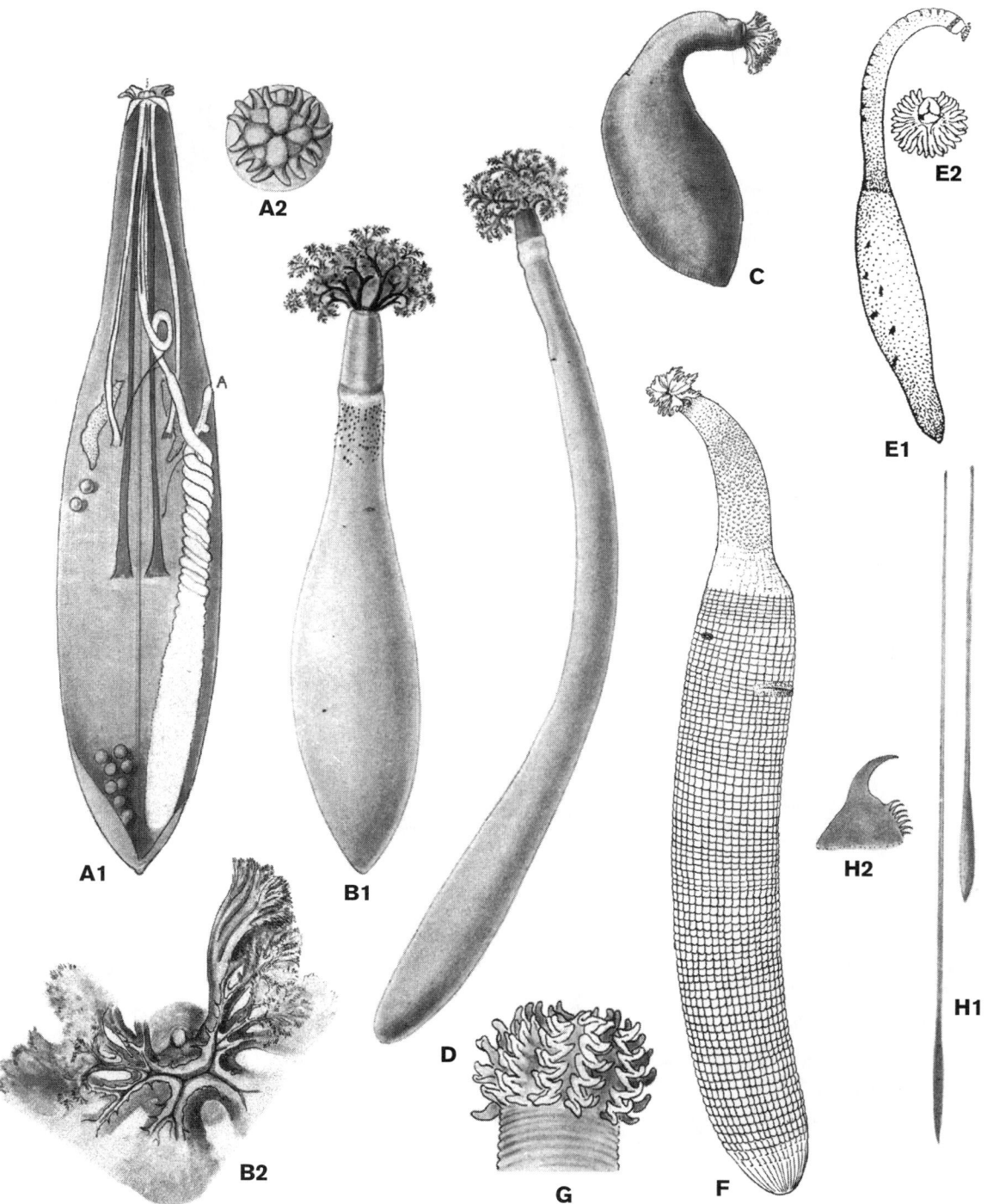

PLATE 120 Unsegmented coelomate worms (not to scale); Sipuncula. A1, *Golfingia margaritacea margaritacea*, dissected specimen; A2, frontal view of tentacles; B1, *Themiste* pyroides; B2, view of mouth, four tentacle stems, and grooves leading to mouth; C, *Themiste dyscrita*, contracted specimen; D, *Themiste hennahi*, formerly *T. zostericola* and *T. perimeces*; E1, *Phascolosoma agassizii*; E2, oral view of extended tentacles; F, *Sipunculus nudus*; G, head and tentacles of *Siphonosoma ingens*; H1, *Apionsoma misakianum*, formerly *Golfingia hespera*, extended and contracted specimens; H2, hook from introvert (A–D, G, H from Fisher 1952; E from *Light's Manual*, 3rd ed., drawn by Carolyn B. Gast; F from Stephen and Edmonds 1972).

— Muscles of body wall without trace of bands 7
5. Adults medium-size (commonly 5–7 cm or up to 12 cm fully extended); skin thick, rough, with prominent papillae largest in anal region and posterior extremity; introvert with anterior rows of small hooks and transverse, brownish bands; body often spotted with black, brown, or pur- ple (plate 120E) . . . . *Phascolosoma agassizii*
— Adults large (12–50 cm extended); skin smooth without prominent papillae on trunk, but may be grooved with longitudinal and circular furrows; no hooks on introvert 6
6. Introvert short, with scalelike papillae and sharply marked

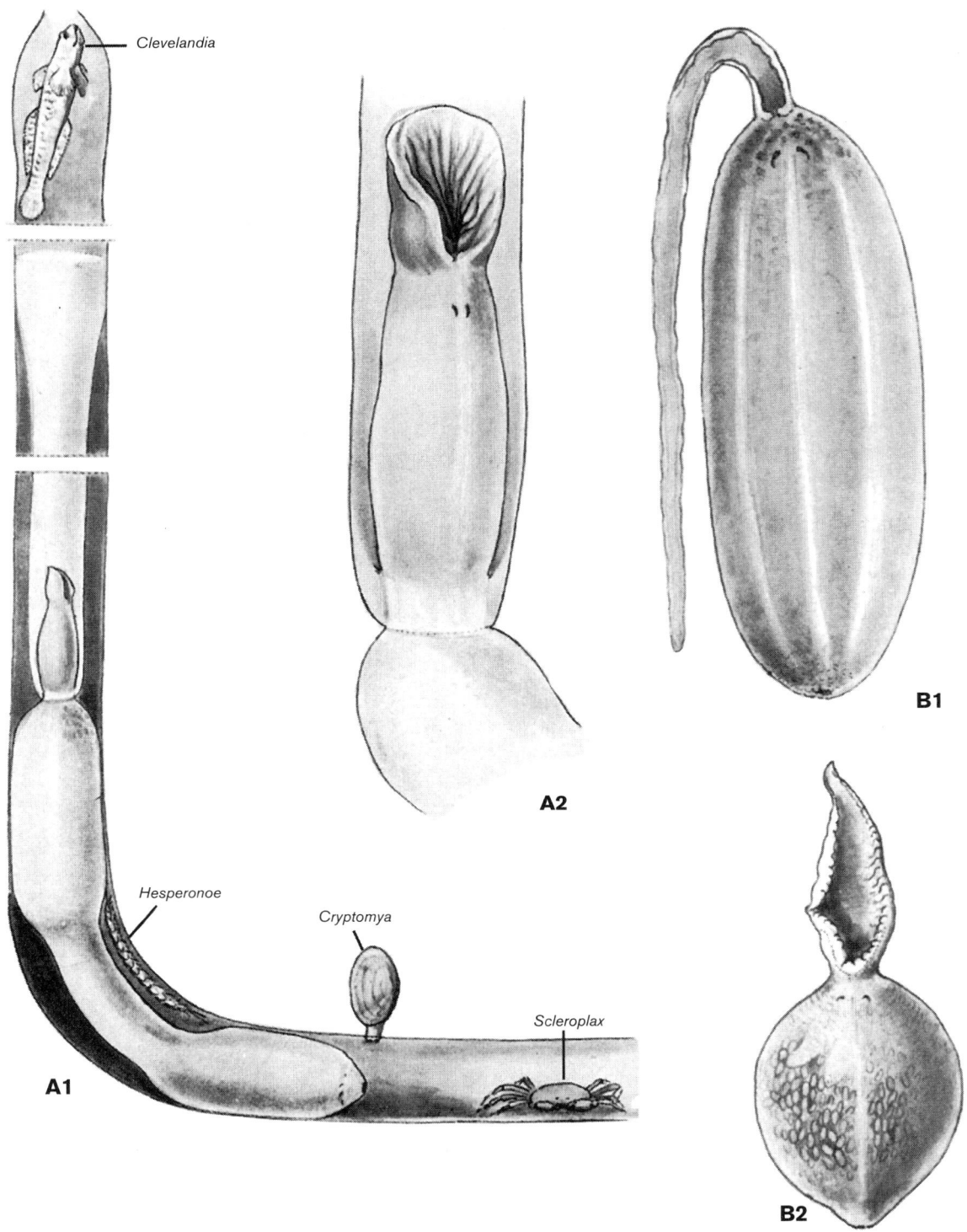

PLATE 121 Unsegmented coelomate worms: Echiura, A1, *Urechis caupo* with mucous feeding net, in burrow with commensals; A2, anterior body, slime tube in place, characteristic pumping posture; B1, *Listriolobus pelodes*, adult; B2, young specimen (all figures after Fisher 1946).

off from body; tentacular fold surrounding mouth with lobulate tentacles (plate 120F) . . . . *Sipunculus nudus*
— Introvert without scalelike papillae, not sharply marked off from body; tentacles arranged in numerous longitudinal series, forming a sort of head (plate 120G) . . . .
. . . . *Siphonosoma ingens*

7. Body small, slender, threadlike; introvert six to eight times length of trunk with rows of small hooks anteriorly (plate 120H) . . . . *Apionsoma misakianum*
— Body small, cylindrical; introvert short, length usually less than trunk, no hooks (plate 120A) . . . .
. . . . *Golfingia margaritacea margaritacea*

List of Species

GOLFINGIIDAE

Golfingia margaritacea margaritacea (Sars, 1851). The curious generic name was coined by E. Ray Lankester to commemorate a holiday spent golfing at St. Andrews in 1885. Many subspecies have been designated for this species, including *G. margaritacea californiensis* Fisher, 1952. In a review of the genus, Cutler and Cutler (1987) describe the variability of characters in the species and conclude that the subspecific designation proposed by Fisher 1952 is not justified. In rock crevices.

THEMISTIDAE

Themiste dyscrita (Fisher, 1952). In sand among rocks and in pholad burrows. This and following species of *Themiste* were formerly *Dendrostomum*.

Themiste hennahi Gray, 1928. Two species of *Themiste* reported in the previous edition, *T. perimeces* (Fisher, 1928) and *T. zostericola* (Chamberlain, 1919), were synonymized with *T. hennahi* by Cutler and Cutler (1988). In mudflats, among *Zostera*, in sand and mud under boulders, low intertidal. See Peebles and Fox 1933 Bull. Scripps Inst. Ocean. 3: 201–224 (structure, function, physiology, behavior).

Themiste pyroides (Chamberlain, 1919). Generally in rock crevices.

PHASCOLOSOMATIDAE

Phascolosoma agassizii Keferstein, 1867. The commonest sipunculan of our intertidal; in mud, rock crevices, shells, holdfasts, and *Mytilus* beds. See Towle and Giese 1967; Rice 1973.

Apionsoma misakianum (Ikeda, 1904). Formerly known as *Golfingia hespera* Fisher, 1952. For a synonymy of *G. misakiana* and *G. hespera* see Cutler 1979; for an explanation of generic change see Gibbs and Cutler 1987.

SIPUNCULIDAE

Siphonosoma ingens (Fisher, 1947). In muddy sand and among *Zostera* roots.

Sipunculus nudus Linnaeus, 1766. Subtidal; occasionally washed ashore.

References

Cutler, E. B. 1979. A reconsideration of the *Golgingia* subgenera *Fisherana* Stephen, *Mitosiphon* Fisher, and *Apionsoma* Sluiter (Sipuncula). Zoological Journal of the Linnean Society 65: 367–384.

Cutler, E. B. 1994. The Sipuncula: their systematics, biology and evolution. Cornell University Press, 453 pp.

Cutler, E. B. and N. J. Cutler. 1987. A revision of the genus *Golfingia* (Sipuncula: Golfingiidae). Proc. Biol. Soc. Wash. 100: 735–761.

Cutler, E. B. and N. J. Cutler. 1988. A revision of the genus *Themiste* (Sipuncula). Proc. Biol. Soc. Wash. 101: 741–766.

Fisher, W. K. 1952. The sipunculid worms of California and Baja California. Proc. U.S. Nat. Mus. 102: 371–450.

Gibbs, P. E. and E. B. Cutler. 1987. A classification of the phylum Sipuncula. Bull. Brit. Mus. Nat. Hist., Zool. 52: 43–58.

Hyman, L. H. 1959. The invertebrates: smaller coelomate groups, Vol. V. McGraw-Hill (Sipunculida, pp. 610–696).

Peebles, F. and D. L. Fox. 1933. The structure, functions, and general reactions of the marine sipunculid worm *Dendrostoma zostericola*. Bull. Scripps Inst. Oceanog. Tech. Ser. 3: 201–224.

Rice, M. E. 1967. A comparative study of the development of *Phascolosoma agassizii*, *Golfingia pugettensis*, and *Themiste pyroides* with a discussion of developmental patterns in the Sipuncula. Ophelia 4: 143–171.

Rice, M. E. 1973. Morphology, behavior, and histogenesis of the pelagosphera larva of *Phascolosoma agassizii* (Sipuncula). Smithson. Contrib. Zool. No.132, 51 pp.

Rice, M. E. and M. Todorovic, editors. 1975. Proc. Internat. Symposium on the Biology of the Sipuncula and Echiura, Vols 1 and 2. Naucno Delo, Belgrade.

Staton, J. L. 2003. Phylogenetic analysis of the mitochondrial cytochrome c oxidase subunit 1 gene from 13 sipunculan genera: intra- and interphylum relationships. Invertebrate Biology 122: 252–264.

Stephen, A. C. 1964. A revision of the classification of the phylum Sipuncula. Ann. Mag. Nat. Hist. Ser. 13, 7: 457–462.

Stephen, A. C. and S. J. Edmonds. 1972. The phyla Sipuncula and Echiura. London, British Museum (Natural History), 528 pp.

Towle, A. and A. C. Giese. 1967. The annual reproductive cycle of the sipunculid *Phascolosoma agassizii*. Physiol. Zool. 40: 229–237.

Echiura

Echiurans resemble annelids in a general way, but are unsegmented as adults. In recent microanatomical studies, findings of repetitive units in developmental stages of the nervous system of the echiuran *Bonellia viridis* have been interpreted as further evidence of a close phylogenetic relationship with the annelids (Hessling 2003). The body of an echiuran is usually sausage-shaped, consisting of a muscular wall surrounding a spacious coelom, and contains a very long, looped intestine, opening by a posterior anus. A characteristic solid, extensible proboscis just anterior to the mouth has given the group the name of "spoon worms," from its shape when contracted. There are usually one or more pairs of setae placed ventrally behind the mouth and one or two posterior rings of setae.

Echiurans are entirely marine. Most burrow in sand or mud, or inhabit rock crevices, empty shells, sand-dollar tests, and pholad burrows, and swallow large quantities of bottom material or lighter detritus gathered by the long proboscis. However, the most common echiuran of this area, *Urechis caupo*, (plate 121A), has the specialized habit of collecting fine-particulate material, including bacteria, in a net of mucus through which it pumps a flow of water, periodically consuming both net and collected food. The natural history of the "inn-keeper" *Urechis*, so called for the various commensals sharing its burrow, is interestingly described by Fisher and MacGinitie (1928). Arp et al. (1992, 1995), Menon and Arp (1994), and Julian et al. (2001) consider *Urechis'* relationship to its burrow environment in general and often sulfide-rich conditions in particular. Suer (1984) reported on growth and spawning of *Urechis* in Bodega Harbor.

Fisher's monograph (1946) has a good general account of the group.

Key to Echiura

— Size large; a ring of conspicuous setae at posterior end of flesh-colored body; preoral proboscis very short; forms permanent burrows in muddy sand; (plate 121A) . . . . *Urechis caupo* Fisher and MacGinitie, 1928

— Size small to medium; body globose, green to gray violet, no posterior setae; proboscis yellow, elongate, soft, easily lost; rare, in mud among *Zostera* and in sandy mud, to off-shore depths (plate 121B) . . . . *Listriolobus pelodes* Fisher, 1946

References

Arp, A. J., J. G. Menon, and D. Julian. 1995. Multiple mechanisms provide tolerance to environmental sulfide in *Urechis caupo*. American Zoologist 35: 132–144.

Arp, A.J., B.M. Hansen, and D. Julian. 1992. The burrow environment and coelomic fluid characteristics of the echiuran worm *Urechis caupo* from three northern California population sites. Marine Biology 113: 613–623.

Fisher, W. K. 1946. Echiuroid worms of the North Pacific Ocean. Proc. U.S. Nat. Mus. 96: 215–292.

Fisher, W. K. and G. E. MacGinitie. 1928a. A new echiuroid worm from California. Ann. Mag. Nat. Hist. (10) 1: 199–204.

Fisher, W. K. and G. E. MacGinitie. 1928b. The natural history of an echiuroid worm (*Urechis*). Ann. Mag. Nat. Hist. (10) 1: 204–213.

Hessling, R. 2003. Novel aspects of the nervous system of *Bonellia viridis* (Echiura) revealed by the combination of immunohistochemistry, confocal laser-scanning microscopy and three-dimensional reconstruction. Hydrobiologia 496: 225–239.

Julian, D., M. L. Chang, J. R. Judd, and A. J. Arp. 2001. Influence of environmental factors on burrow irrigation and oxygen consumption in the mudflat invertebrate *Urechis caupo*. Marine Biology 139: 163–173.

Menon, J. G. and A. J. Arp. 1993. The integument of the marine echiuran worm *Urechis caupo*. Biological Bulletin 185: 440–454.

McHugh, D. 1997. Molecular evidence that echiurans and pogonophorans are derived annelids. Proc. Natl. Acad. Sci. 94: 8006–8009.

Newby, W. W. 1940. The embryology of the echiuroid worm *Urechis caupo*. Mem. Amer. Phil. Soc. 16: 1–213.

Stephen, A. C. and S. J. Edmonds. 1972. The Phyla Sipuncula and Echiura. London: British Museum (Natural History), 528 pp.

Suer, A. L. 1984. Growth and spawning of *Urechis caupo* (Echiura) in Bodega Harbor, California. Marine Biology 78: 275–284.

Tardigrada

(Plates 122–123)

LELAND W. POLLOCK AND ALBERT CARRANZA

In both body form and walking gait, microscopic members of the Tardigrada suggest miniature bears. Indeed, they have been known as "water bears" since their discovery in the eighteenth century. Their size range, typically from 100–500 μm in body length, places them among the meiofaunal community. They inhabit semiaquatic habitats, such as water films surrounding terrestrial plants, as well as fully aquatic settings, including the surfaces of aquatic plants and the network of interstitial spaces between grains of sediment.

Their unique body construction generates discussion regarding their phylogenetic placement. Their microscopic size, large blood-filled pseudocoelom, bulbous muscular pharynx, and bandlike musculature suggest affinity to several aschelminth phyla. However, a metameric body plan, including four pairs of legs terminating in complex claws, a segmented annelid-arthropod style nervous system, and a multilayered cuticle requiring periodic molting for growth, more convincingly supports arthropod affinities.

Largely on the basis of head and foot appendages, the phylum can be divided into three orders: Heterotardigrada (including most marine tardigrades), Mesotardigrada, and Eutardigrada. Heterotardigrades, and especially the primitive, all-marine suborder Arthrotardigrada, are characterized by multiple pairs of cephalic sensory structures and feet ending in distinct toes, claws, or both toes and claws. These features, along with lateral and caudal cuticular projections, are especially useful in distinguishing species. Leg pairs are numbered I–IV, anterior to posterior. In most, cephalic appendages include a single median cirrus, and paired internal, external and lateral cirri. A pair of blunt-tipped clavae are located just medial to the lateral cirri. A pair of posterio-lateral somatic cirri is present in most species, as are either cirri or papillae attached to the shank of each leg pair.

Marine tardigrades of the California and Oregon coasts have received little attention. Two species, one associated with algae (*Bathyechiniscus tetronyx* [now *Stryaconyx sargassi*]) and the other with barnacles (*Echiniscoides sigismundi*) were reported by Mathews (1938) and Schuster and Grigarick (1965), respectively. Pollock's (1989) brief survey of selected beaches added three interstitial species to the two originally identified in California by McGinty (1969). Carranza's (1996) in-depth study of interstitial tardigrades at northern California's Coleman Beach (Sonoma County) added five more species to the list. One-third of the marine species located at the handful of sites explored within the state so far were new to science. This fact should stimulate greater attention to this poorly known but fascinating segment of California and Oregon marine fauna.

Sand-dwelling forms prefer medium to coarse sand-grain size and occupy intertidal beaches from the quarter to high tidal levels, extending from the sand surface to the depth of the low-tide water table. Sand samples should be swirled in a bucket of fresh water to osmotically "shock" tardigrades into releasing their grip on sand grains. Allow the sand to settle, and then decant the water through a piece of plankton netting. Use a spray bottle of seawater to back-rinse trapped specimens from the plankton netting into a small dish of seawater. Examine the specimens using the high magnification of a dissecting microscope. Transfer to a compound microscope will be necessary for accurate identifications.

Key to Marine Tardigrada

1. Fourth pair of legs with toes that bear small terminal claws
   .... Halechiniscidae
2. Fourth pair of legs with toes that lack terminal claws ......
   .... Batillipedidae
3. Fourth pair of legs lack toes but with claws directly attached to foot .... claw-bearing marine Tardigrada

Key to Halechiniscidae

1. Number of primary points associated with caudal appendage:
   A. Caudal appendage with 4-pointed process........4p
   B. Caudal appendage ends in a single point.........1p
   C. Caudal appendage absent x
2. Shape of basal appendage associated with the femur of legs IV:
   A. Legs IV bear triangular basal spikes ....ts

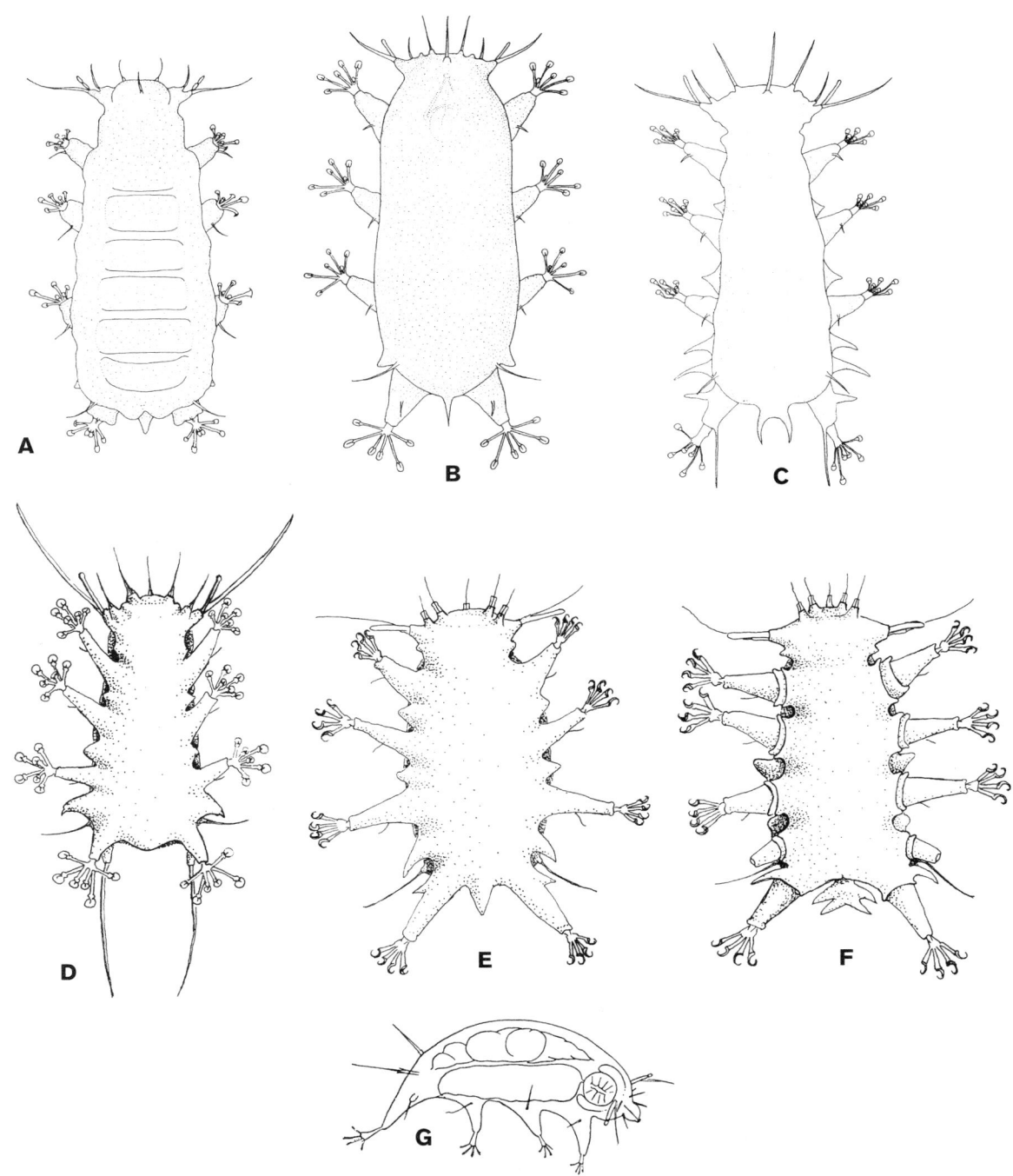

PLATE 122 A, *Batillipes gilmartini*, adult, dorsal view; B, *Batillipes mirus*, adult, dorsal view; C, *Batillipes tridentatus*, adult, dorsal view; D, *Batillipes orientalis*, adult, ventral view; E, *Halechiniscus remanei remanei*, adult, ventral view; F, *Halechiniscus* sp., adult, ventral view; G, *Styraconyx sargassi*, adult, lateral view (A, used with permission from McGinty 1969; B, used with permission from McGinty and Higgins, 1968; C, used with permission from Pollock 1989; D–F, original figures, Carranza; G, from Mathews 1938, as *Bathyechiniscus tetronyx*).

B. Legs IV bear long, branched spines . . . . lb
C. Legs IV bear long, smooth spines . . . . ls
D. Legs IV bear flexible papilla, with or without terminal spine . . . . pa
3. Lateral features of cuticle between legs III and legs IV:
   A. Sharp lateral spikes present between legs III and legs IV
   . . . . ssp

B. Flat-tipped, blunt lateral spikes present between legs III and legs IV . . . . bsp
C. Lateral cuticular features absent between legs III and legs IV. x
4. Number of exposed points on medial claws on legs IV:
   A. One or two exposed points 1
   B. Three exposed points 3

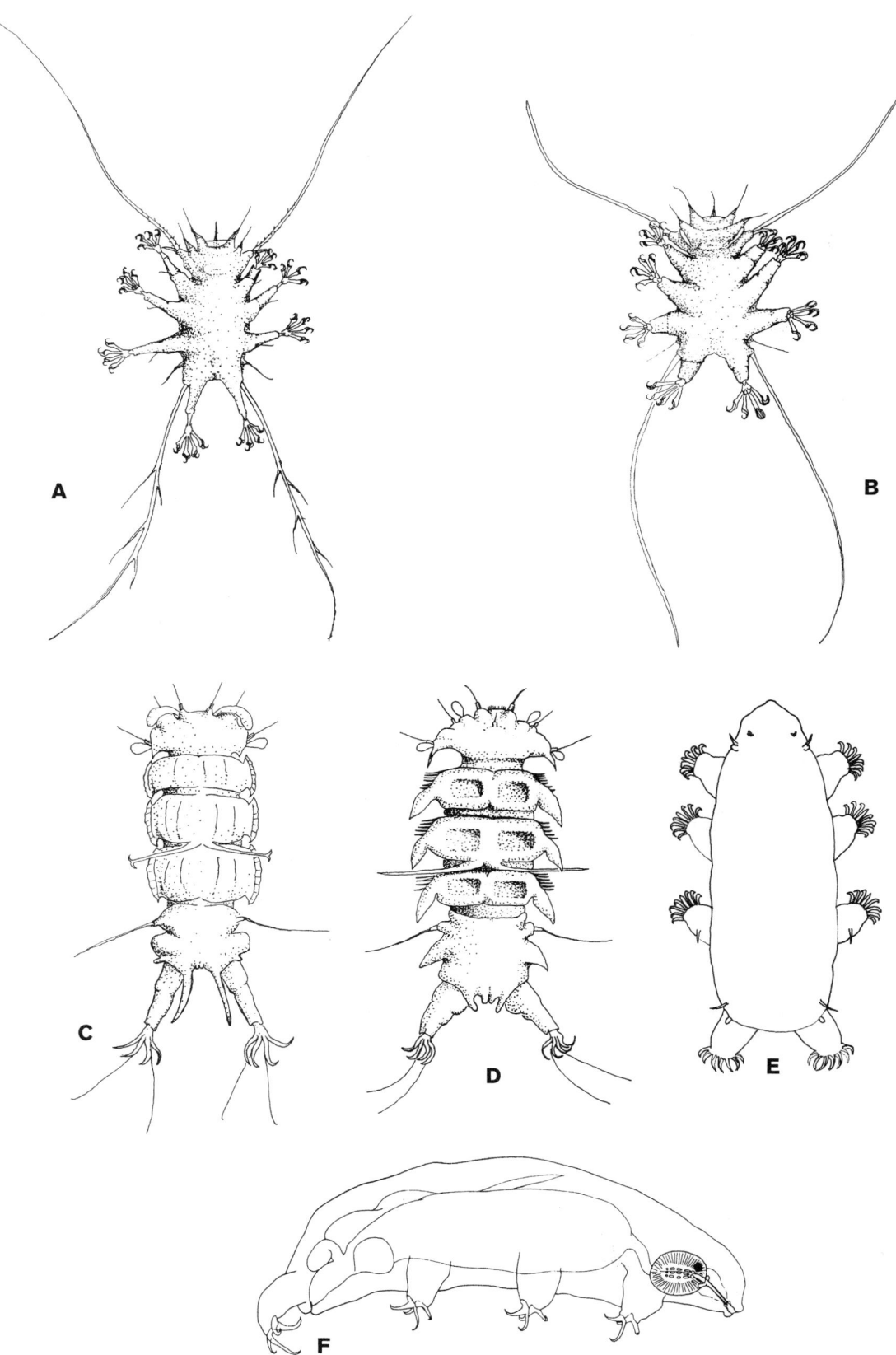

PLATE 123 A, *Tanarctus arborspinosus*, adult, ventral view; B, *Tanarctus ramazzotti*, adult, ventral view; C, *Stygarctus bradypus*, adult, dorsal view; D, *Stygarctus spinifer*, adult, dorsal view; E, *Echiniscoides sigismundi*, adult, dorsal view; F, *Hypsibius geddesi*, adult, lateral view (example of genus; used with permission from Hallas 1971 Steenstrupia 1: 201–206, fig. 4; A–E, original figures, Carranza).

CAUDAL	LEG IV SPINE	LEG III–IV FEATURES	EXPOSED POINTS	
1p	ts	ssp	1	*Halechiniscus remanei remanei* (plate 122E)
4p	ts	bsp	1	*Halechiniscus* sp. (plate 122F)
x	lb	x	1	*Tanarctus arborspinosus* (plate 123A)
x	ls	x	1	*Tanarctus ramazzotti* (plate 123B)
x	pa	x	3	*Styraconyx sargassi* (plate 122G)

Key to Batillipedidae

1. Shape of caudal appendage:
   A. 2 pointed appendage 2p
   B. Single pointed appendage 1p
   C. Caudal appendage absent x
2. Shape of primary clava:
   A. Rod-shaped, uniform width . . . . un
   B. Rod-shaped, inflated tip . . . . in
   C. Short with medial constriction . . . . me
3. Shape of lateral projection(s) between legs III and IV:
   A. Single prominent point 1p
   B. Two prominent points 2p
   C. Single-lobe rounded projection 1r

CAUDAL	CLAVA	PROJECTION	
1p	me	1r	*Batillipes gilmartini* (plate 122A)
1p	un	1p	*Batillipes mirus* (plate 122B)
2p	un	2p	*Batillipes tridentatus* (plate 122C)
X	in	1p	*Batillipes orientalis* (plate 122D)

Key to Claw-bearing Marine Tardigrada

1. Presence of cephalic appendages (spines, clavae):
   A. Cephalic appendages present . . . . ca
   B. Cephalic appendages absent x
2. Claws associated with legs IV:
   A. Four simple claws per foot; middle two claws with projecting hairs 4h
   B. Four branched claws per foot; hairs absent 4b
   C. More than four claws per foot . . . . >4
3. Presence of a pair of caudal spikes:
   A. Pair of caudal spikes present . . . . cs
   B. Caudal spikes absent x

CEPHALIC APPENDAGES	LEG IV CLAWS	CAUDAL SPIKES	
ca	4h	cs	*Stygarctus bradypus* (plate 123C)
ca	4h	x	*Stygarctus spinifer* (plate 123D)
ca	>4	x	*Echiniscoides sigismundi* (plate 123E)
x	4b	x	*Hypsibius* sp. (plate 123F)

List of Species

Heterotardigrada

BATILLIPEDIDAE

Juveniles possess four rather than six toes per foot and may have shorter appendages and modified cuticular features.

Batillipes gilmartini McGinty, 1969. From medium-grain sand, mid- to high-tide levels, at 0–20 cm depth in the beach southwest of the Hopkins Marine Station at Pacific Grove and from Pt. La Jolla.

Batillipes mirus Richters, 1909. Despite cosmopolitan distribution, only juvenile specimens have been reported from superficial fine sand at Cape Meares, Oregon.

Batillipes orientalis Chang and Rho, 1997. Found at mid-tidal level, 30–40 cm in depth, coarse sands of Coleman Beach, Sonoma County (A. Carranza). Californian specimens differ from Korean animals by possessing fourth leg spines half the length of the body.

Batillipes tridentatus Pollock, 1989. Puget Sound to Big Sur. In some beaches, located from 15–45 cm depth, central half of littoral zone; in others, 0–10 cm depth, upper half of littoral zone. Four-toed juveniles with modest caudal papilla and one lateral projection between legs II and IV. They also lack a pointed projection on the femur of leg pair IV in adults.

HALECHINISCIDAE

Juveniles possess two claws per foot and may differ from adults in appendage proportions and the condition of cuticular elaborations.

Halechiniscus remanei Schulz, 1955. From the Mediterranean and the Atlantic and Pacific Oceans. Most abundant between one-quarter and three-quarter tidal heights, from 14–45 cm sand depth.

Halechiniscus sp. 40–50 cm depths, coarse sand; Coleman Beach (A. Carranza).

Styraconyx sargassi Thulin, 1942. Originally identified as *Bathyechiniscus tetronyx* by Mathews (1938); from the marine alga *Dictyota*.

Tanarctus arborspinosus Lindgren, 1971. California specimens at 30–50 cm depth, coarse sand at Coleman Beach (A. Carranza). Long, branched appendages of fourth pair of legs easily broken or lost.

Tanarctus ramazzotti Renaud-Mornant, 1975. Specimens found at all beach heights, usually at depths >20 cm (A. Carranza).

ECHINISCOIDIDAE

Echiniscoides sigismundi Schultze, 1865. Cosmopolitan distribution. Upper intertidal, associated with algae (e.g., enteromorphoid *Ulva* and *Endocladia muricata*) and barnacles. Kristensen and Hallas (1980) list six subspecies, including the notation that "specimens of *E. sigismundi* from the California coast (Schuster et al. 1975) . . . might also be [subspecies] *groenlandicus*, but [were] not sufficiently preserved to allow safe identification." See Crisp and Hobart 1954, Ann. Mag. Nat. Hist. (12) 7: 554–560 (ecological notes).

STYGARCTIDAE

Juveniles possess only the median pair of hair-bearing claws on each foot.

Stygarctus bradypus Schulz, 1951. Cosmopolitan distribution. Spotty distribution all along the Pacific coast. Typically in the upper half of littoral zone, from 25–45 cm sand depth. Caudal spikes and projecting spines on dorsal plates missing in early juvenile stages.

Stygarctus spinifer Hiruta, 1985. Known from Japan, Washington state, and several sites in northern California. Most abundant at three-quarters tidal level, 15–45 cm deep.

Eutardigrada

HYPSIBIIDAE

Hypsibius sp. A single specimen of this eutardigrade genus reported from Coleman Beach (A. Carranza).

References

Carranza, A. 1996. Psammolittoral marine Tardigrada from Coleman Beach, California. Master's of Science Thesis, Department of Biology, Sonoma State University, Rohnert Park, California, 73 pp.

Chang, Y. C. and H. S. Rho. 1997. Two new marine tardigrades of genus *Batillipes* (Heterotardigrada, Batillipedidae) from Korea. Korean Journal of Systematic Zoology 13: 93–102.

Kristensen, R. M. and T. E. Hallas. 1980. The tidal genus *Echiniscoides* and its variability, with erection of Echiniscoididae fam.n. (Tardigrada). Zoologica Scripta 9: 113–127.

Kristensen, R. M. and R. P. Higgins. 1984. Revision of *Styraconyx* (Tardigrada: Halechiniscidae), with descriptions of two new species from Disko Bay, West Greenland. Smithsonian Contributions to Zoology 391: 1–40.

Mathews, G. B. 1938. Tardigrada from North America. American Midland Naturalist 19: 619–627.

McGinty, M. M. 1969. *Batillipes gilmartini*, a new marine tardigrade from a California beach. Pacific Science 23: 394–396.

Pollock, L. W. 1989. Marine interstitial Heterotardigrada from the Pacific coast of the United States, including a description of *Batillipes tridentatus* n. sp. Transactions of the American Microscopical Society 108: 169–189.

Schuster, R. O. and A. A. Grigarick. 1965. Tardigrades from western North America, with emphasis on the fauna of California. Univ. Calif. Publ. Zool. 76: 1–67.

Schuster, R. O., A. A. Grigarick, and E. C. Toftner 1975. Ultrastructure of tardigrade cuticle. Memorie dell'Istituto Italiano di Idrobiologia 32 Supplement: 337–375.

Annelida

(Plates 124–181)

The annelids, or segmented worms, are characterized by an elongated body divided into segments and formed on the plan of a tubular jacket of muscle surrounding a fluid-filled coelom. Although they lack a rigid internal skeleton, annelids can use the hydrostatic pressure of coelomic fluid, acted upon by the muscular body wall, as a "fluid skeleton," aiding in extension of the body and in burrowing. Locomotion in polychaetes is also aided by numerous setae (or chaetae) that project from the sides of the body. The annelid body plan has proved to be extremely plastic and adaptable in an evolutionary sense, so we find a great diversity of form, habitat, and mode of life within the phylum, including deep-sea gutless worms formerly placed in a separate phylum (the Pogonophora).

Older groupings (Polychaeta, Oligochaeta, Hirudinea, and "archiannelids") are now treated in two classes, the Polychaeta (incorporating the archiannelids) and the Clitellata (including the oligochaetes and hirudineans, or leeches). For convenience of identification, we continue to treat the polychaete worms, the marine oligochaetes, and the marine leeches in three separate sections.

Oligochaeta

DAVID G. COOK, RALPH O. BRINKHURST, AND CHRISTER ERSÉUS

(Plate 124)

The Clitellata contains leeches (Hirudinea), the familiar earthworms, and about 15 families of generally small to minute, often obscure, predominantly aquatic worms. The nonleech groups were traditionally referred to as the class Oligochaeta, but DNA-based studies have showed that leeches are derived oligochaetes, which makes Oligochaeta a synonym of Clitellata. In its original sense, Oligochaeta is thus not a natural group, but the collective term "oligochaetes" may still be used for clitellates that are not leeches.

The best-known representatives of the aquatic oligochaetes (also known as "microdriles") are the tubificids. These red, threadlike worms, often sold as petfish food, may occur in great abundance in organically polluted habitats. On the Pacific coast of North America, representatives of three families occur in salt or brackish water: the Tubificidae (now including the former family Naididae), the Enchytraeidae, and the Randiellidae.

Oligochaetes are hermaphroditic, lack parapodia, palps, tentacles, and jaws, and possess comparatively few chaetae (the term "setae" is often used in oligochaete literature and still used in most polychaete literature). Reproduction is confined to a few anterior segments where, at sexual maturity, a glandular clitellum is formed that secretes the cocoon within which one or a few eggs develop directly into juveniles resembling small adults. A microdrile oligochaete consists of a long, thin, segmented cylindrical body with a prostomium, an anterior ventral mouth, and a terminal anus (plate 124A). The gut, ventral nerve cord, and the main dorsal and ventral blood vessels (except in Enchytraeidae) extend the length of the body, but all are modified to some degree in anterior segments. The taxonomically significant alimentary canal begins at the mouth as a narrow, thin-walled esophagus. In the region of the second segment, the esophagus opens into the pharynx, a widened tube with a conspicuous dorsal pharyngeal pad consisting of gland cells; extensions of this pad are the pharyngeal glands (in enchytraeids often called septal glands), which are more or less discrete masses of glandular cells whose long, secretory processes penetrate the pharyngeal wall. In some Enchytraeidae, a pair of tubular peptonephridia (plate 124B) arise from the esophageal wall immediately behind the pharynx; these and the pharyngeal glands probably both serve as a combination of digestive, lubricative, and food-collecting functions.

Other structures are segmentally arranged and include the chaetae, which usually occur in two dorsolateral and two ventrolateral bundles on all segments except the peristomium (segment I) and a pair of excretory organs, the nephridia, which are located ventrolaterally in most postclitellar segments. The reproductive organs in virtually all marine microdriles consist of one pair of testes associated with paired male genitalia; one pair of ovaries associated with a pair of small, ventrolateral, female funnels that conduct eggs to the exterior; and a pair of ectodermal spermathecae that receive and store sperm after copulation. In tubificids and enchytraeids, ovaries and female funnels occur in the segment immediately posterior to the testes segment; that is, they are located in the atrial or male genital segment (plate 124K). For the rare family Randiellidae, with a single species (*Randiella litoralis*) on the Pacific coast, the number and position of gonads are unclear, but *R. litoralis* has

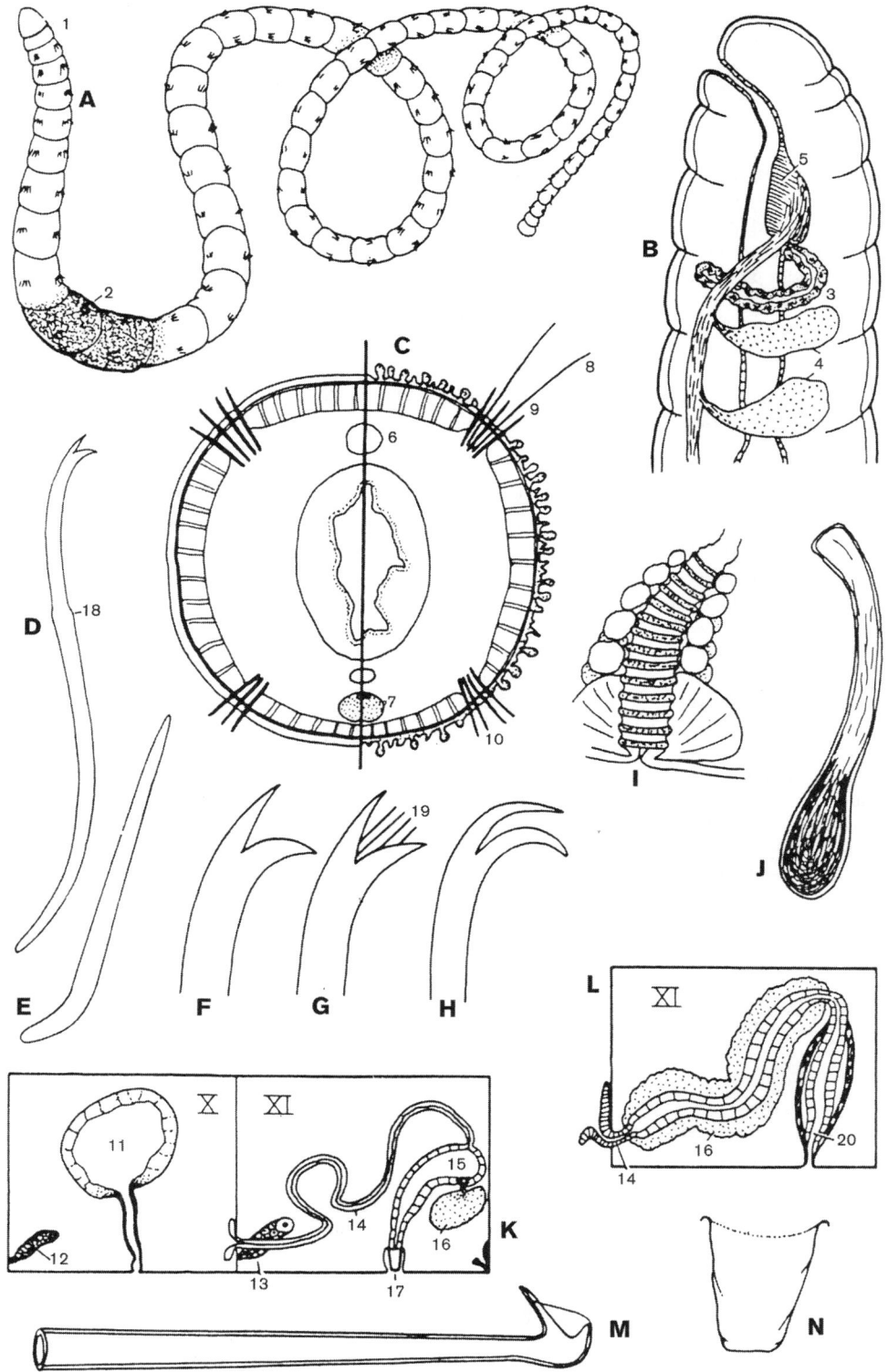

PLATE 124 A, Tubificid oligochaete (from life); B, diagrammatic optical section of anterior end of *Enchytraeus* (Enchytraeidae); C, diagrammatic transverse section through midsection of generalized microdrile (left side) and a tubificid, *Tubificoides* (right side), latter showing papillate body wall; D, sigmoid crotchet chaeta, with nodulus; E, proximally curved (enchytraeid) chaeta, without nodulus; F, distal end of bifid crotchet; G, distal end of pectinate chaeta; H, distal end of chaeta of *Chaetogaster limnaei* (Naidinae, Tubificidae); I, spermathecal duct of a Pacific Northwest *Lumbricillus* (Enchytraeidae; after Tynen 1969); J, sperm bundle of another Pacific Northwest *Lumbricillus* (after Tynen, 1969); K, diagrammatic longitudinal section through genitalia of *Tubificoides* spp. (Tubificidae); L, diagrammatic longitudinal section through male genitalia of the Atlantic *Monopylephorus irroratus* (Tubificidae); M, penis sheath of *Limnodrilus hoffmeisteri* (Tubificidae); N, penis sheath of *Tubificoides* spp. (generalized diagrammatic); 1, prostomium; 2, clitellum; 3, peptonephridium; 4, pharyngeal glands; 5, pharynx; 6, dorsal blood vessel; 7, ventral nerve cord; 8, hair chaeta; 9, dorsal crotchets; 10, ventral chaetae (crotchets); 11, spermatheca; 12, testis; 13, ovary; 14, vas deferens; 15, atrium; 16, prostate gland; 17, penis; 18, nodulus; 19, intermediate teeth; 20, eversible pseudopenis (all by David Cook).

two pairs of male funnels suggesting at least two pairs of testes. Moreover, randiellids have multiple spermathecae.

Taxonomic Criteria

In the microdrile families under consideration, the most important taxonomic characters are the morphology and arrangement of the male genitalia and spermathecae; the number, form, and detailed structure of the chaetae; and, in the Enchytraeidae, the anatomy of the anterior digestive system.

GENITALIA

In the Tubificidae, the male genitalia consist of a pair of funnels, which are on the anterior face of septum X/XI or (in the subfamily Naidinae) in a more anterior position, that collect sperm and conduct them into a pair of narrow, tubular vasa deferentia, which open into the storage and copulatory apparati; the latter consist typically of a pair of muscular atria that each open to the exterior on segment XI (or more anteriorly, in Naidinae) either as a simple pore in the body wall, or, more usually, as some form of modified penis or pseudopenis; true penes often possess sheaths of thickened cuticle around them. The atrium usually bears a stalked, glandular organ, the prostate gland, which probably provides nutrients and lubrication for the sperm prior to copulation. In some tubificid genera, the prostate glands are diffuse masses of cells more or less covering the atria. After copulation, sperm from the sexual partner are stored in spermathecae, which, in tubificids, are located in the segment immediately anterior to that containing the atria. Spermathecae are simple invaginations of the body wall, each consisting of a narrow tubular duct and a voluminous storage ampulla (plate 124K).

The genital system of the Naidinae is basically the same as in the other tubificids, but the various elements are located in segments IV–V to VII–VIII, and if present at all, the prostate glands are diffuse cell masses. The comparative rarity of naidines in a fully sexual condition is explained by the high incidence of asexual reproduction by simple fission in this subfamily.

Enchytraeidae differ considerably from other microdrile families in many respects, including their genital systems; spermathecae, whose ducts often bear gland cells and whose ampullae are often in open communication with the gut, are located in segment V, and the male genitalia are situated in segments XI and XII. Large, more or less tubular male funnels, lying freely in the body cavity of segment XI, connect with the very narrow, often tortuously coiled vasa deferentia, which lie in segment XII. The latter open to the exterior simply on penial papillae or in association with glandular, and often muscular, penial bulbs. Atria, in the tubificid sense, are lacking in this family.

As already noted, the genital system is poorly known for the minute species comprising the Randiellidae, but paired male funnels appear to be present in the posterior ends of both segments X and XI in *R. litoralis*. Sexually mature randiellids typically have three or more pairs of small globular spermathecae, located in segments VII and/or VIII.

In some Tubificidae and all Randiellidae the ventral chaetae of the atrial or spermathecal segments, or both, become modified at sexual maturity as genital (spermathecal and penial) chaetae.

CHAETAE

With a few exceptions, microdriles possess four bundles of chitinous chaetae implanted in the body wall of all segments except the peristomium (segment I). Two basic types occur in the families considered here, namely crotchets and hair chaetae. In the Tubificidae, crotchets are usually slender, sigmoid structures (plate 124D) possessing a more or less median thickening (the node or nodulus); the distal ends may be single-pointed or, more commonly, bifid (forked), and the latter are sometimes further modified by one or a series of fine intermediate teeth between the outer major teeth (pectinate chaetae) (plate 124G). In randiellids, all somatic (nongenital) chaetae are small, single-pointed and barely sigmoid. Enchytraeid crotchets are single-pointed or rounded (with one rare exception) and range in form from slender, sigmoid chaetae with a nodulus to the more usual condition of stout, straight chaetae without nodal thickenings (plate 124E).

Hair chaetae occur only in the dorsal bundles of some tubificids, and consist of an elongate, slender, single-pointed shaft without a nodulus.

OTHER CHARACTERS

The anterior digestive system in Enchytraeidae, especially the arrangement of the pharyngeal glands ("septal glands") and the presence or absence of peptonephridia, are major diagnostic characters and have been covered above. For purposes of species keys, the only other significant character in Tubificidae concerns the presence or absence of large, free cells within the body cavity, known as coelomocytes; these occur in the subfamilies Rhyacodrilinae and Naidinae, often in large numbers, and may then completely fill the coelom of some segments. Coelomocytes may also be abundant in species of Enchytraeidae.

The Naidinae are characterized by anterior segments that are more modified than in other microdrile families. Dorsal chaetae are often absent from the first four or five segments, and sensory structures may be present. The prostomium may bear an elongated tactile extension known as a proboscis, and many naidines possess eyes on the peristomium, which are visible as paired, crescent-shaped masses of opaque pigment granules located at the bases of the photosensitive cells.

CHARACTERS DIFFERENTIATING OLIGOCHAETES FROM OTHER ANNELIDS

Because oligochaetes are often obscure and sometimes unexpected components of marine communities, and because some polychaetes bear a striking superficial resemblance to them, it is useful to note some characters that differentiate oligochaetes from polychaetes. Mature oligochaetes are less difficult than immature ones because their elaborate genitalia and glandular clitellum are unique, albeit that the microdrile clitellum is only one cell thick and less conspicuous than that of earthworms. Immature worms can best be distinguished from those polychaetes with reduced parapodia and small numbers of chaetae, such as some capitellids, by detailed examination of the chaetae: in these polychaetes, the bifid crotchets usually possess a thin, membranelike hood around their distal ends. Unfortunately, the hood is sometimes reduced to a small, membranous keel connecting the lower tooth to the chaetal shaft, and a similar ornamentation is seen also in some tropical marine oligochaetes. With some practice, however, capitellid polychaetes can still be discriminated from tubificids in more subtle ways. First, if red in color, they are more deeply red than any of the latter. Second, a capitellid tends to perform rather slow, peristaltic movements, which in its contracted phase emphasizes the external segmentation (the rings) of the worm, whereas a tubificid (as well as an enchytraeid) normally wriggles vividly and retains a smoother surface when contracted. Finally, capitellids tend to produce more mucus (making them stickier) than marine oligochaetes.

Ecology

Marine and estuarine oligochaetes are intertidal or subtidal benthic animals that live within and feed upon the bottom deposits. The larger species of tubificids and enchytraeids burrow freely in the substrate and appear to ingest sediment rather indiscriminately. The very small species in these families (down to about 0.1 mm in diameter), as well as the randiellids, are meiobenthic (interstitial) worms that inhabit the interstices between substrate particles and feed on fine, organic debris, or sometimes on live microorganisms, such as diatoms. For most species, however, and regardless of worm size, the main target food is probably bacteria. One naidine tubificid, *Chaetogaster limnaei,* is an obligate commensal in the mantle cavity of some bivalves and pulmonate gastropods. Similar to other species of *Chaetogaster,* it is a predator feeding on protozoans, rotifers, and microcrustaceans.

Free burrowing species tend to live in silts and poorly sorted, fine sands. Meiobenthic worms are more restricted to coarser sands, and they are probably more oxygen-demanding than the burrowers. In general, oligochaetes are particularly abundant in areas of organic enrichment, often to the exclusion of other groups because they are able to withstand the low oxygen tensions usually associated with this condition. In the intertidal zone, for example, many enchytraeids are found in or beneath masses of decaying seaweed or under damp wood debris, and tubificids often occur in large numbers under stones that are partly buried in the sediment. Subtidally, in areas of organic pollution, very large populations of a few species of tubificids may occur; for example, in San Francisco Bay, Brinkhurst and Simmons (1968) found that three species of *Tubificoides* formed up to 97.8% of the total bottom fauna at some stations.

Collection, Preservation, and Examination Procedures

Oligochaetes may be extracted from sediments in which they live by any combination of sieving, flotation, or manual-sorting techniques. Intertidal sediments may be collected by digging or scooping material into containers, and any coring device, grab, dredge, or fine-mesh hand net is suitable for obtaining subtidal samples.

Erséus (1994) provides detailed methods for preservation, general observation, and the staining and mounting of specimens. Worms to be preserved, either in the sediment or after extraction, are best fixed in 10% formaldehyde solution for 48 hours, and stored in 70–80% ethyl alcohol. Worms intended for genetic analysis should not be passed through formaldehyde. Fixation in Bouin's solution (a standard mixture of formaldehyde, picric acid, and acetic acid) is also excellent for sections, dissections, and whole mounts, but here too the specimens must be transferred into ethyl alcohol within a maximum of three to four days—otherwise the specimens will become hard and brittle (Erséus 1994).

To identify microdriles, it is necessary to make temporary or permanent microscopical preparations of whole or dissected animals. Staining whole specimens is recommended for critical work. Staining can be done with borax carmine or haematoxylin, but preferred is an alcoholic paracarmine solution (see Erséus 1994 for the staining procedure). Important characters, such as chaetae and the cuticular penis sheaths, can be observed by mounting worms temporarily in glycerol. Glycerol is good for handling large, routine collections and will accept nuclear-stained animals, but it will eventually destroy soft tissues. Therefore, for more critical work, it is recommended that worms should be dehydrated in absolute ethyl alcohol, cleared in xylene or toluene, and permanently mounted in Canada Balsam (or other synthetic mounting media). Erséus (1994) also provides details on the procedure for making such permanent mounts.

Enchytraeidae can be more readily identified from living material; worms should be mounted in a small volume of seawater on a microscope slide and immobilized by gentle pressure of a coverslip. Large, more vigorous individuals can be stopped by mounting them in a drop of gelatin solution that is just ready to set.

Key to Families of Oligochaeta

1. Some or all crotchets bifid, sigmoid; hair chaetae present or absent; modified genital chaetae present or absent; male genitalia (pores) situated in segment V, VI, VII or XI . . . . . .
    . . . . Tubificidae
—   All crotchets conspicuous (rather stout), single-pointed or rounded, straight or curved proximally, or sigmoid; hair and genital chaetae always absent; male genitalia (pores) situated in segment XII . . . . Enchytraeidae
—   All crotchets inconspicuous (slender, maximally 1.5 μm thick), sharply single-pointed, barely sigmoid; somatic hair chaetae absent, but long, distally ornamented genital chaetae present ventrally in segment X; male genitalia (pores) hardly visible, but possibly present in segments XI and XII . . . . Randiellidae

List of Species of Intertidal and Shallow Sublittoral Marine and Estuarine Oligochaeta from Point Conception to the Columbia River

See Baker and Coates (1996) for a key to many of the species noted below and to additional Pacific Northwest species. The primary literature (see "References") should always be consulted when identifying oligochaetes from California and Oregon, especially because a great many species of intertidal and shallow-water oligochaetes (especially intertidal enchytraeids) remain undescribed from the Pacific coast and because unrecorded introduced species from Asia, Australasia, South America, and elsewhere should be expected.

ENCHYTRAEIDAE

Grania incerta Coates and Erséus, 1980. Known from Santa Barbara and British Columbia, so presumably within our range; shallow sublittoral, in sand. See Erséus (1994).

Grania paucispina (Eisen, 1904). Recorded from shallow water in Tomales Bay; see Coates and Erséus (1980).

Lumbricillus santaeclarae Eisen, 1904. Estuarine, under debris and decaying driftwood along the shore; since Eisen's original description, it has only been recorded from estuarine Lake Merritt, Oakland, in San Francisco Bay (J. T. Carlton collector, identified by M. Tynen).

Marionina southerni (Cernosvitov, 1937). Noted from "California" by Coates (1980, p. 1316).

Marionina subterranea (Knöllner, 1935). A very small worm found intertidally in sand near high-tide level, Half Moon Bay, central California (R. E. Mesick), Coos Bay, Oregon (Strehlow, 1982a); probably widely distributed.

Marionina spp. A number of other *Marionina* species occur along our coast. For example, Strehlow (1982a) recorded *M. sjaelandica* Nielsen & Christensen, 1961, *M. vancouverensis* Coates, 1980, and *M. achaeta* Lasserre, 1966, from Coos Bay, Oregon, as well as three undescribed species from Coos Bay (Strehlow's thesis contains full descriptions of these species, but these were never published).

TUBIFICIDAE

Includes the formerly separate family Naididae, now subfamily Naidinae.

Aktedrilus locyi Erséus, 1980. An interstitial species known from California (on an exposed high energy beach in Monterey Bay) and from Oregon (on a protected sandy beach in lower Coos Bay).

Aktedrilus oregonensis Strehlow, 1982. An interstitial species described from Coos Bay, Oregon.

Bathydrilus litoreus Baker, 1983. Described from intertidal habitats in British Columbia and Washington (Baker 1983), but also from subtidal sites off southern California (Erséus 1991); can thus be expected from the littoral along Oregon as well as California.

Chaetogaster diaphanus (Gruithuisen, 1828). Sometimes in brackish water; predatory on smaller animals.

Chaetogaster limnaei von Baer, 1827. A "cosmopolitan" species found in fresh and brackish water; commensal in pulmonate gastropods and freshwater mussels and sphaeriid clams, usually in the mantle cavity.

Ilyodrilus frantzi Brinkhurst, 1965. This taxon is variable, containing worms with and others without hair chaetae; the former being found in estuaries and early on attributed to a separate subspecies, *I. frantzi capillatus* Brinkhurst and Cook, 1966. Kathman and Brinkhurst 1998 (p. 164) suggest that the chaetal variation in *I. frantzi* may be due to changes in salinity. The form without hairs was originally described from Lake Tahoe (Brinkhurst 1965).

Limnodrilus hoffmeisteri Claparede, 1862. A "cosmopolitan" and abundant tubificid, predominantly in fresh water but often in brackish habitats; subtidal.

Limnodriloides barnardi Cook, 1974. To be looked for in subtidal waters (extending into the shallow subtidal) of central California and Oregon; see Erséus (1994).

Limnodriloides monothecus Cook, 1974. This widespread Atlantic species is known from scattered estuaries on the Pacific coast, including San Francisco Bay (Erséus 1982), leading Cohen and Carlton (1995) to suggest that it was not native to the eastern Pacific but was introduced by oystering or shipping activity. Additional plates in Erséus (1990, 1994).

Limnodriloides victoriensis Brinkhurst and Baker, 1979. Known south to Coos Bay, Oregon (Martin Posey collector, identified by R. Brinkhurst).

Nais communis Piguet, 1906. Sometimes in brackish water.

Paranais frici Hrabe, 1941. In brackish water, widely distributed in the world. See Brinkhurst and Coates (1985) for taxonomic treatment.

Paranais litoralis (Müller, 1784). In brackish to fully marine habitats, intertidal and subtidal, widely distributed in the world. See Brinkhurst and Coates 1985.

Peloscolex apectinatus. See *Tubificoides parapectinatus,* below.

Tectidrilus diversus Erséus, 1982. Within our region known subtidally from San Francisco Bay, and in deeper offshore waters along the outer coast. Additional plates in Erséus (1994).

Teneridrilus calvus Erséus and Brinkhurst, 1990 (in Erséus et al. 1990). In the Sacramento-San Joaquin Delta; probably primarily a freshwater species.

Teneridrilus mastix (Brinkhurst, 1978). Freshwater and low salinity regions of estuaries. See Erséus et al. (1990).

Thalassodrilides gurwitschi (Hrabe, 1971). Widely reported in the Mediterranean, Atlantic, and Pacific from coastal muds and locally recorded from Elkhorn Slough in Monterey Bay (Kimberly Heiman, personal communication 2006; identification by Michael Milligan), as well as from the Salton Sea in inland southern California.

Tubificoides bakeri Brinkhurst, 1985. Noted here as possibly within our range in shallow sublittoral, having been recorded from British Columbia in 10 m, but in deeper water off Los Angeles; see Brinkhurst (1985) and Erséus (1994).

Tubificoides brownae Brinkhurst and Baker, 1979 (=*Peloscolex gabriellae* of previous edition and of earlier San Francisco Bay records). Possibly a North Atlantic oligochaete, with records from Coos Bay (Oregon) and San Francisco Bay; treated by Cohen and Carlton (1995) as introduced with oysters or shipping. Reported from Saudi Arabia (Erséus 1985). See also Brinkhurst (1986).

Tubificoides diazi Brinkhurst and Baker, 1979. A possibly Atlantic species introduced to the Pacific Rim; known from Boundary Bay and Coos Bay on the Pacific coast; see Brinkhurst (1986).

Tubificoides fraseri Brinkhurst, 1986. British Columbia to southern California (see Brinkhurst 1986) and also known from Australia and New Zealand.

Tubificoides nerthoides (Brinkhurst, 1965) (=*Peloscolex nerthoides*). Originally reported from Tomales Bay and San Francisco Bay (Point Richmond); subtidal, brackish, and marine; tolerant of organic pollution.

Tubificoides parapectinatus Brinkhurst, 1985. San Francisco Bay. This is the species earlier reported as a part of "*Peloscolex apectinatus*" by Brinkhurst and Simmons (1968) and in the previous edition of this manual.

Tubificoides wasselli Brinkhurst and Baker, 1979. Known on the Pacific coast from Victoria, British Columbia, and San Francisco Bay, and apparently introduced from the Atlantic Ocean; see Brinkhurst (1986).

Varichaetadrilus angustipenis (Brinkhurst and Cook, 1966) (=*Limnodrilus angustipenis*). Noted here because of its abundance in the lower Sacramento-San Joaquin Delta; primarily freshwater; see Erséus et al. 1990.

RANDIELLIDAE

Randiella litoralis Erséus and Strehlow, 1986. Interstitial and intertidal; Oregon. See Erséus and Strehlow (1986).

References

Baker, H. R. 1983. New species of *Bathydrilus* Cook (Oligochaeta, Tubificidae) from British Columbia. Can. J. Zool. 61: 2162–2167.

Baker, H. R. and K. A. Coates. 1996. Class Oligochaeta. pp. 170–178. In: Marine invertebrates of the Pacific Northwest. E. N. Kozloff, ed. University of Washington Press.

Brinkhurst, R. O. 1964. Studies on the North American aquatic Oligochaeta. I. Naididae and Opistocystidae. Proc. Acad. Nat. Sci. Philadelphia 116: 195–230.

Brinkhurst, R. O. 1965. Studies on the North American aquatic Oligochaeta. II. Tubificidae. Proc. Acad. Nat. Sci. Philadelphia 117: 117–172.

Brinkhurst, R. O. 1985. A further contribution to the taxonomy of the genus *Tubificoides* Lastockin (Oligochaeta: Tubificidae). Can. J. Zool. 63: 400–410.

Brinkhurst, R. O. 1986. Taxonomy of the genus *Tubificoides* Lastockin (Oligochaeta, Tubificidae): species with bifid setae. Can. J. Zool. 64: 1270–1279.

Brinkhurst, R. O. and K. A. Coates. 1985. The genus *Paranais* (Oligochaeta: Naididae) in North America. Proc. Biol. Soc. Washington 98: 303–313.

Brinkhurst, R. O. and D. G. Cook. 1966. Studies on the North American aquatic Oligochaeta. III. Lumbriculidae and additional notes and records of other families. Proc. Acad. Nat. Sci. Philadelphia 118: 1–33.

Brinkhurst, R. O. and B. G. M. Jamieson. 1971. Aquatic Oligochaeta of the world. Edinburgh: Oliver & Boyd, 860 pp.

Brinkhurst, R. O. and M. L. Simmons. 1968. The aquatic Oligochaeta of the San Francisco Bay system. Calif. Fish Game 54: 180–194.

Coates, K. 1980. New marine species of *Marionina* and *Enchytraeus* (Oligochaeta, Enchytraeidae) from British Columbia. Can. J. Zool. 58: 1306–1317.

Coates, K. A., and Erséus, C. 1980. Two new species of *Grania* (Oligochaeta, Enchytraeidae) from the Pacific coast of North America. Can. J. Zool. 58: 1037–1041.

Cohen, A. N. and J. T. Carlton. 1995. Biological Study. Nonindigenous Aquatic Species in a United States Estuary: A Case Study of the Biological Invasions of the San Francisco Bay and Delta. A Report for the United States Fish and Wildlife Service, Washington, D.C., and The National Sea Grant College Program, Connecticut Sea Grant, NTIS Report Number PB96-166525, 246 pp. + Appendices.

Cook, D. G. 1974. The systematics and distribution of marine Tubificidae (Annelida, Oligochaeta) in the Bahia de San Quintin, Baja California, with descriptions of five new species. Bull. So. Calif. Acad. Sci. 73: 126–140.

Erséus, C. 1980. Taxonomic studies on the marine genera *Aktedrilus* Knöllner and *Bacescuella* Hrabe (Oligochaeta, Tubificidae), with descriptions of seven new species. Zool. Scr. 9: 97–111.

Erséus, C. 1982. Taxonomic revision of the marine genus *Limnodriloides* (Oligochaeta, Tubificidae). Verh. naturwiss. Ver. Hamburg (New Series) 25: 207–277.

Erséus, C. 1985. Annelida of Saudi Arabia. Marine Tubificidae (Oligochaeta) of the Arabian Gulf coast of Saudi Arabia. Fauna Saudi Arabia 6: 130–154.

Erséus, C. 1990. The marine Tubificidae (Oligochaeta) of the barrier reef ecosystems at Carrie Bow Cay, Belize, and other parts of the Caribbean Sea, with descriptions of twenty-seven new species and revision of *Heterodrilus*, *Thalassodrilides* and *Smithsonidrilus*. Zool. Scr. 19: 243–303.

Erséus, C. 1991. Records of the marine genus *Bathydrilus* (Oligochaeta: Tubificidae) from California, with descriptions of two new species. Proc. Biol. Soc. Washington 104: 622–626.

Erséus, C. 1994. The Oligochaeta. pp. 5–38. In: Taxonomic atlas of the benthic fauna of the Santa Maria Basin and western Santa Barbara Channel. Volume 4. The Annelida Part 1. Oligochaeta and Polychaeta: Phyllodocida (Phyllodocidae to Paralacydoniidae). Santa Barbara, California: Santa Barbara Museum of Natural History.

Erséus, C., J. K. Hiltunen, R. O. Brinkhurst, and D. W. Schloesser. 1990. Redefinition of *Teneridrilus* Holmquist (Oligochaeta, Tubificidae) with description of two new species from North America. Proc. Biol. Soc. Washington 103: 839–846.

Erséus, C. and D. R. Strehlow. 1986. Four new interstitial species of marine Oligochaeta representing a new family. Zool. Scr. 15: 53–60.

Kathman, R. D. and R. O. Brinkhurst. 1998. Guide to the freshwater oligochaetes of North America. College Grove, Tennessee: Aquatic Resources Center, iv + 264 pp.

Strehlow, D. R. 1982a. The relation of tidal height and sediment type to the intertidal distribution of marine oligochaetes in Coos Bay, Oregon. Master of Science thesis, Department of Biology, Oregon State University, Corvallis, Oregon, 82 pp.

Strehlow, D. R. 1982b. *Aktedrilus locyi* Erséus, 1980 and *Aktedrilus oregonensis* n. sp. (Oligochaeta, Tubificidae) from Coos Bay, Oregon, with notes on distribution with tidal height and sediment type. Can. J. Zool. 60: 593–596.

Tynen, M. J. 1969. New Enchytraeidae from the east coast of Vancouver Island. Can. J. Zool. 47: 387–393.

Hirudinida

EUGENE M. BURRESON

(Plates 125–127)

The leeches are highly specialized clitellate annelids. Molecular phylogenetic analyses suggest that leeches are an order of the Clitellata, rather than a separate class. All Hirudinida possess the general characteristics of clitellates in that they are hermaphroditic and have a cocoon-secreting clitellum. They are further characterized by possession of oral and caudal suckers used both for locomotion and feeding, and a coelomic system reduced to a network of sinuses and tubular canals.

There are two recognized suborders of leeches: Arhynchobdellida and Rhynchobdellida. The Arhynchobdellida includes the predaceous Erpobdelliformes and the sanguivorous ("blood-eating") Hirudiniformes, which are freshwater or terrestrial leeches that lack a proboscis and feed with a pharynx that may or may not have jaws. The Rhynchobdellida, which are all aquatic, possess a protrusible proboscis used in feeding on body fluids of their prey.

There are three families of rhynchobdellids: the freshwater Glossiphoniidae that feed on a range of invertebrate and vertebrate taxa; the freshwater and marine Ozobranchidae, parasites on the blood of turtles; and the freshwater and marine Piscicolidae, parasites on the blood of fishes.

The family Piscicolidae is composed of three subfamilies:

- Pontobdellinae are characterized by having two pairs of pulsatile vesicles per urosome segment, but these vesicles are not typically visible externally. The large, tubercled shark leeches are members of the Pontobdellinae.

- Piscicolinae possess one pair of pulsatile vesicles per urosome segment; they are usually, but not always, obvious externally.

- Platybdellinae are characterized by the absence of pulsatile vesicles.

No key to subfamilies is presented here because the presence or absence of pulsatile vesicles cannot be determined reliably without histological sections.

All leeches in the nearshore waters along the northern California and Oregon coasts belong in the family Piscicolidae; thus, all are external parasites of marine or estuarine fishes. Marine leeches are most abundant in polar and temperate seas, although they are not commonly encountered. Worldwide there are approximately 60 genera and well over 100 species of marine piscicolid leeches.

The rocky coastline and estuaries of northern California, Oregon, and Washington have a comparatively rich marine leech fauna on subtidal fishes. The leech fauna of intertidal fishes is still relatively poorly studied, but there is one known species in northern California. Clearly, much work remains to adequately characterize the leech fauna of intertidal fishes.

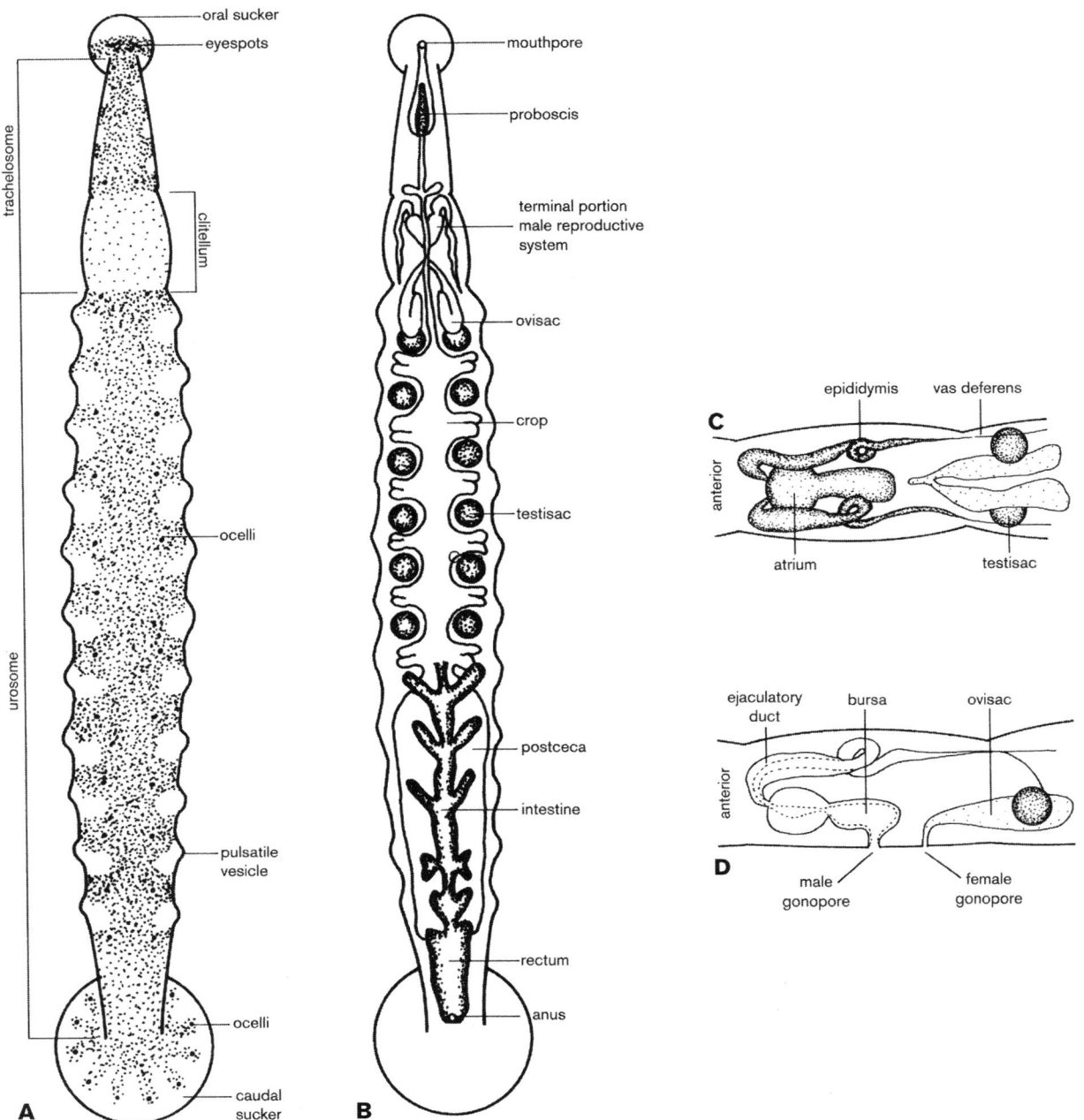

PLATE 125 Hirudinida, Piscicolidae—external and internal morphological terminology of a generalized piscicolid leech: A, dorsal view of entire animal illustrating some external features; B, dorsal view of entire animal illustrating some internal features of the digestive and reproductive systems; C, dorsal view of clitellar region illustrating details of the terminal portion of the male reproductive system and the female reproductive system; D, lateral view of clitellar region illustrating gonopores and the same structures as shown in C.

General Morphology

Mature piscicolid leeches range in size from about 5 mm in total length to well over 15 cm long (the large shark leeches), but most range between 10 mm and 30 mm. Leeches' bodies are divided into two regions: the trachelosome, which is the anterior portion of the body ending at the posterior end of the clitellum, and the urosome, which is the posterior portion of the body between the clitellum and the caudal sucker (plate 125A).

There is an obvious oral and caudal sucker, although the size of both suckers varies considerably among genera. The oral sucker is usually much smaller than the caudal sucker, and it may possess one or two pairs of eyespots or none at all. The caudal sucker is usually wider than the maximum body width and eccentrically attached to the urosome. However, in some species, especially those that tend to attach to crustaceans, the caudal sucker is usually small and terminal. The caudal sucker may possess a ring of punctiform ocelli around the margin (plate 125A).

All leech bodies consist of 34 segments. Externally, each segment is subdivided into superficial rings called annuli. The number of annuli per urosome segment used to be an important taxonomic character, but their use has fallen out of favor with recognition that the number per segment can appear to vary considerably depending on fixation and fullness of the gut.

Pigmentation is highly variable in leeches. Some leeches have little or no pigmentation and appear creamy white. In others, the body may be solidly pigmented, usually as tones of green, yellow, brown, or black, or there may be a pattern of segmental transverse bands or longitudinal stripes. Pigmentation may also occur on both suckers.

There are a variety of external morphological characters that may be present on marine leeches, including leaflike lateral branchiae, papillae, large tubercles, lateral pulsatile vesicles and punctifom ocelli. Pulsatile vesicles are actually part of the internal coelomic system, but they are often visible externally as lateral blisterlike structures. These external features usually repeat segmentally on the urosome and occasionally on the trachelosome. The pattern varies depending on the genus from features on each segment (usually on the middle annulus) to features on each annulus.

The reproductive systems of leeches can be important for identification, but serial sections are usually required for adequate determination. There are usually either five or six pairs of large testisacs in the urosome (plate 125B). If there are only five pairs, it is always the first anterior pair that is lost. The testisacs are connected to paired dorsal vasa deferentia that expand in the clitellar region and enter highly convoluted epididymides. The vasa deferentia continue anteriorly, bend ventrally, and enter the muscular and highly glandular common atrium where the spermatophore is formed (plate 125C, 125D). The atrium is connected to an evertible bursa, which opens through the ventral male gonopore in the clitellum (plate 125D). The female reproductive system consists of paired ovisacs that fuse anteriorly into a common oviduct and open through the ventral female gonopore, which is located directly posterior to the male gonopore (plate 125D). Many variations on this general scheme have evolved to store and transport sperm to the eggs, and these can be important taxonomically (Sawyer 1986). The urosome of leeches is filled with clitellar gland cells that supply nutrients and cocoon material to the external surface of the clitellum during cocoon deposition.

The coelomic system is reduced in leeches and consists of a complex arrangement of segmental tubular canals and sinuses that can only be determined by serial histological sections. The coelomic system is important taxonomically (Sawyer 1986) but will not be discussed more fully here because it will not be used in the keys.

The digestive system of piscicolid leeches consists of a feeding proboscis that is protrusible through the mouthpore, usually centrally located in the oral sucker. The proboscis opens into the esophagus in the trachelosome that merges with the expanded crop in the urosome. There is a dorsal intestine in the posterior portion of the urosome leading to a dorsal anus. Ventrally, beneath the intestine, are large, paired blood-storage sacs termed postceca (plate 125B).

Ecology

Marine leeches may be very host specific, occurring on only a single species of fish, or they may feed on a wide variety of fish hosts. In addition, they may be either temporary or semi-permanent parasites of fishes. Temporary parasites usually leave the host soon after a blood meal, although it is not uncommon for them to remain on the host for a few days. After leaving the fish host, they usually seek a sheltered location among vegetation or under stones or shells to digest the blood. However, some estuarine and marine leeches that inhabit soft substrate areas utilize crustaceans for attachment after leaving the fish host.

Semi-permanent parasites remain on the fish and take successive blood meals, leaving the host only to deposit cocoons. Leeches are hermaphroditic, and, depending on the species, mating may involve copulation through the gonopores or hypodermic implantation of spermatophores anywhere on the body. Mating may occur on or off the fish host, but cocoons are never deposited on the fish. The cocoons of piscicolid leeches, which are usually adhered in clusters to substrate such as vegetation, rocks, shells or even the carapace of live crustaceans, typically contain a single egg. However, some species of *Heptacyclus* may deposit up to five eggs in each cocoon. Cocoons are left unattended, and newly hatched leeches must find a fish host on their own. Young leeches can usually survive for a week or more before their first blood meal. Hatching time is temperature dependent, but some species may aestivate in the cocoon until appropriate environmental conditions reoccur. Leeches usually die after cocoon deposition, but some species may produce successive broods.

Leeches have a strong shadow response that involves extension and active waving to and fro. Many species are capable of prolonged swimming, and when a shadow passes over them, they will detach from the substrate and swim up into the water column. This behavior facilitates finding a new host.

Collection and Preservation

Leeches can be difficult to collect and to identify. They often leave the host after feeding and therefore may go undetected even when abundance is high. They are usually sufficiently large to be detected by the naked eye and occur on the body surface and fins or in the gill cavity or mouth of fishes. Leeches should be collected by gently dislodging the caudal sucker with forceps and placing them in a dish containing water of the temperature and salinity as at the collection site.

Leeches are usually hardy and may live for weeks or longer off the host in a beaker of water. For proper identification, it is important to observe leeches alive and to note as many external characters as possible. It may be necessary to relax leeches in weak alcohol or other narcotizing agents prior to examination. Careful observation should be made of pigmentation color and pattern; number and arrangement of eyes on the oral sucker; ocelli on the body and caudal sucker; number of lateral pulsatile vesicles, if present; and arrangement of papillae, tubercles, or other obvious external characters.

Leeches that have been fixed without prior relaxation are often impossible to identify because they usually contract strongly or curl into a tight ball, making observation of important characters difficult. Leeches relaxed in weak alcohol prior to fixation and preserved in 95% ethanol will usually retain their pigmentation and eyespots for long periods; however, pigmentation usually fades rapidly after fixation in formalin-based fixatives, especially after transfer to 70% ethanol. Thus, leeches that have been fixed in formalin and then stored in alcohol are also often difficult to identify, especially to species, because of the loss of all pigmentation including eyespots. If leeches are to be sectioned, Bouin's fixative is best because it preserves the coelomic system much better than formalin. However, this is usually only possible if a large number of leeches of the same species are available.

Generic determination of leeches that lack obvious external characters may depend on internal anatomy. Unfortunately, most marine leeches are too small to dissect, and stained whole mounts, useful for studies of trematode parasites, do not work well for leeches because the body is full of clitellar gland cells that obscure important structures. Serial sections are usually the only way to determine internal anatomy. Serial transverse sections of the region of the clitellum and the first few segments of the urosome are most useful to document the anatomy of the reproductive systems and the coelomic system. Transverse sections of small leeches involve very small block faces and histology is tedious. All these difficulties combine to make identification of leeches problematic, even for experts.

Key to Marine and Estuarine Hirudinida

The marine leeches of the Pacific coast are relatively well studied except for species that may be collected in the intertidal zone. It is likely that many intertidal leeches remain to be discovered.

The following key includes species that are likely to be found intertidally or near shore in Washington, Oregon, and northern California; it does not include species known only from deep water or species known only from southern California or Canada and Alaska. The key is based on external characters only and will work best if specimens are alive or freshly preserved with pigmentation intact. Leeches that have faded after storage in alcohol are difficult to identify unless they have obvious surface features. Fish host information can also be helpful for species that are host specific.

1. Body with 31 pairs of conspicuous, phylliform lateral projections (branchiae) (plate 126A) . . . . . *Branchellion lobata*
— Body lacking lateral branchiae 2
2. Body with large wartlike tubercles; typically large, more than 30 mm total length, caudal sucker no wider than maximum body width; on elasmobranchs (plate 126B) . . . . *Stibarobdella* sp.
— Body without large wartlike tubercles 3
3. Body with 10 pairs of obvious lateral pulsatile vesicles 4
— Body lacking obvious pulsatile vesicles 5
4. Each annulus of urosome with six large, multispined papillae dorsally and eight or more small papillae ventrally; body very muscular, narrow and cylindrical (plate 126C) . . . . *Orientobdella confluens*
— Body smooth, lacking papillae; body flaccid, tapering at each end; pigmentation uniformly very dark purple, appearing black, except posterior portion of urosome and caudal sucker, which are unpigmented (plate 126D) . . . . *Trachelobdella oregonensis*
5. Oral sucker with one or two pairs of eyespots 6
— Oral sucker lacking eyespots 11
6. Oral sucker with one pair of eyespots; oral sucker larger than caudal sucker and wider than maximum body width; body with segmental small papillae especially obvious on ventral posterior urosome; terminal caudal sucker with 10–12 ocelli; pigmentation uniformly light brown with segmental darker transverse bands on posterior urosome (plate 126E) . . . . *Ostreobdella papillata*
– Oral sucker with two pairs of eyespots 7

7. Caudal sucker lacking ocelli 8
— Caudal sucker with ocelli 9
8. Body lightly pigmented, appearing creamy white with faint reddish-brown transverse bands; caudal sucker unpigmented; eyespots on oral sucker punctiform; one pair punctiform eyespots on second annulus of trachelosome (plate 126F) . . . . *Oceanobdella pallida*
— Body heavily pigmented, uniformly brown but interrupted laterally to form segmental unpigmented areas; oral sucker with two pigment bands, posterior band with two areas of heavy pigmentation obscuring eyespots (plate 127A) . . . . *Pterobdella abditovesiculata*
9. Oral sucker with two pairs of crescent-shaped eyespots and one pair of punctiform eyespots on second annulus of trachelosome 10
— Oral sucker with two pairs of punctiform eyespots and one pair of punctiform eyespots on second annulus of trachelosome; 12–14 ocelli on caudal sucker; pigmentation varies from unpigmented to uniformly reddish- or purplish-brown with segmental clusters of lateral small yellow pigment granules; body up to 15 mm total length (plate 127B) . . . . *Heptacyclus diminutus*
10. Green longitudinal pigment bands interrupted segmentally to form unpigmented areas; 12–14 ocelli on caudal sucker; body up to 30 mm total length (plate 127C) . . . . *Heptacyclus viridus*
— Uniformly reddish-brown pigmentation interrupted laterally by segmental unpigmented areas; oral sucker with reddish-brown pigmentation in the form of a cross with eyespots at the junctures of the pigmentation; 14 ocelli on caudal sucker; small, body only up to 8 mm total length (plate 127D) . . . . *Heptacyclus buthi*
11. Body very small, not over 10 mm total length; lacking eyespots and ocelli; pigmentation consists of faint segmental reddish-brown transverse bands (plate 127E) . . . . *Calliobdella knightjonesi*
— Body up to 30 mm total length; lacking eyespots and ocelli; pigmentation uniformly faint reddish-brown or with transverse bands, with two faint reddish-brown bands on oral sucker and faint reddish-brown bands radiating from end of urosome to edge of sucker; on elasmobranchs (plate 127F) . . . . *Marsipobdella sacculata*

List of Species

PISCICOLIDAE

PONTOBDELLINAE

Stibarobdella sp. On *Myliobatis californicus* in Elkhorn Slough (MacGinitie and MacGinitie 1968, p. 219). Not *S. macrothela*, a parasite of tropical sharks, or *S. loricata* (see Moore 1952), a leech known only from very deep water in the southern hemisphere.

PISCICOLINAE

Branchellion lobata Moore, 1952. Common on rays, especially in Elkhorn Slough (see MacGinitie and MacGinitie 1968, p. 219; Moore 1952); known from *Mustelus henlei*, *Triakis semifasciata*, *Galeorhinus zyopterus*, *Mustelus californicus*, and *Rhinobatos productus* in Monterey Bay; not reported north of Tomales Bay.

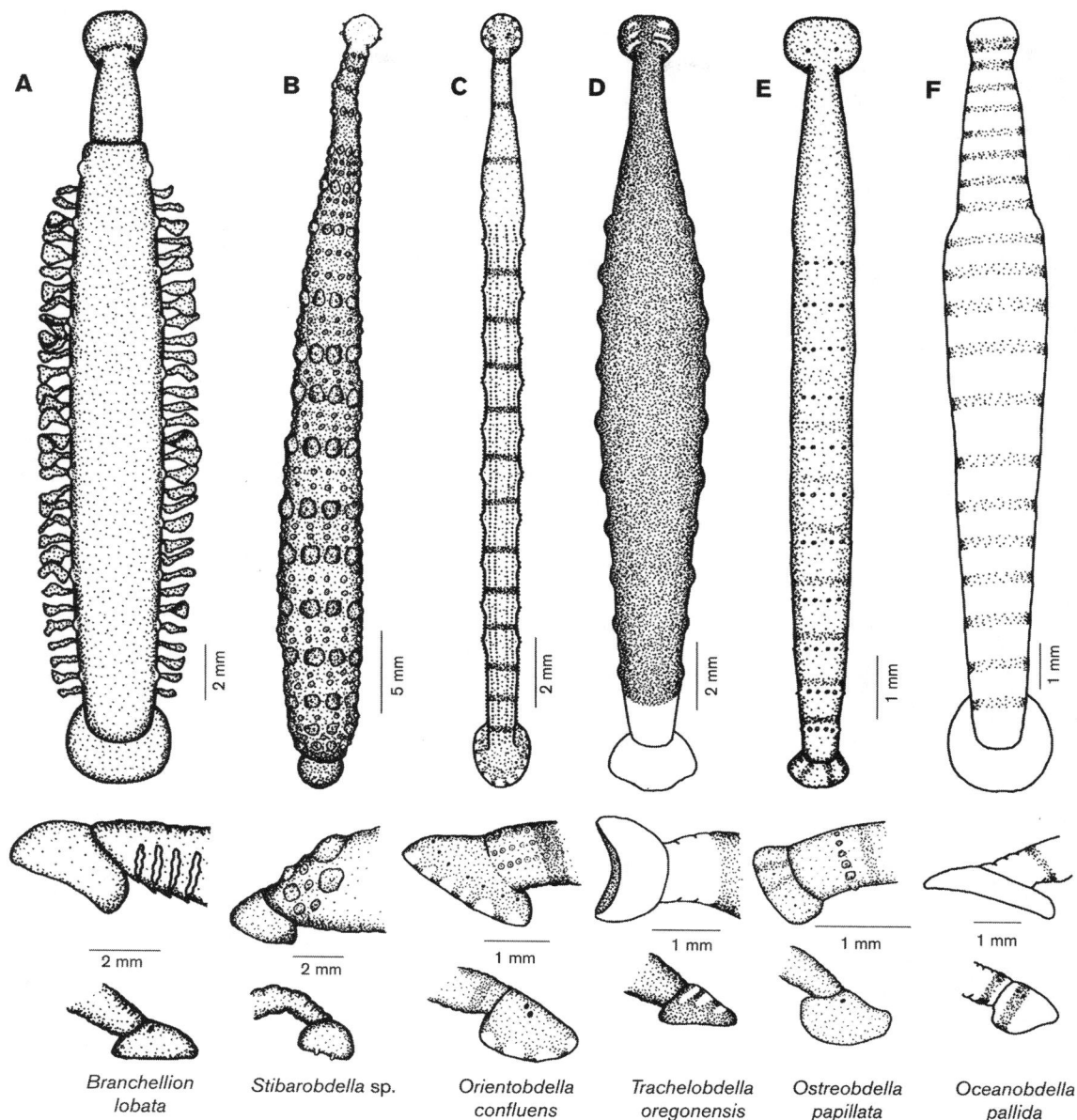

A   B   C   D   E   F

2 mm   5 mm   2 mm   2 mm   1 mm   1 mm

2 mm   2 mm   1 mm   1 mm   1 mm   1 mm

Branchellion lobata   *Stibarobdella* sp.   *Orientobdella confluens*   *Trachelobdella oregonensis*   *Ostreobdella papillata*   *Oceanobdella pallida*

PLATE 126 Hirudinida, Piscicolidae—dorsal view of adult entire animal illustrating body and sucker shape, pigmentation pattern, eyespots and ocelli; for each species: lateral view of caudal sucker (top) and oral sucker (bottom) illustrating pigmentation pattern, eyespots, and ocelli (C, E, from Burreson 1977b, courtesy of Universidad Nacional Autónoma De México; D, from Burreson 1976b, courtesy of Journal of Parasitology; F, from Burreson 1997c, courtesy of Transactions of the American Microscopical Society).

Calliobdella knightjonesi Burreson, 1984. Appears to be a specific parasite on the fins of English sole, *Parophrys vetulus*; known only from the central Oregon coast; rare (see Burreson 1984).

Marsipobdella sacculata Moore, 1952. Reported only from *Raja* spp., including *Raja rhina* and *R. inornata*. San Francisco to San Diego.

Orientobdella confluens Burreson, 1977. Never collected on a fish host; feeds on English sole, *Parophrys vetulus*, and big skate, *Raja binoculata*, in the laboratory (see Burreson 1977b; note that the plates were reversed by the publisher); known from Bodega Bay on the asteroid *Dermasterias imbricata*.

Trachelobdella oregonensis Burreson, 1976. Appears to be a specific parasite of the cabezon, *Scorpaenichthys marmoratus*;

known only from the central Oregon coast (see Burreson 1976b).

PLATYBDELLINAE

Heptacyclus (=*Malmiana*) *viridus* (Burreson, 1977). Appears to be a specific parasite on the pectoral fins of the buffalo sculpin, *Enophrys bison*. Known only from Oregon estuaries, especially Yaquina Bay (see Burreson 1977a).

Heptacyclus diminutus (Burreson, 1977). A common parasite of a number of marine fishes, especially rockfishes of the genus *Sebastes*, the cabezon *Scorpaenichthys marmoratus*, and ling cod *Ophiodon elongatus*; observed in public aquaria at Oregon State

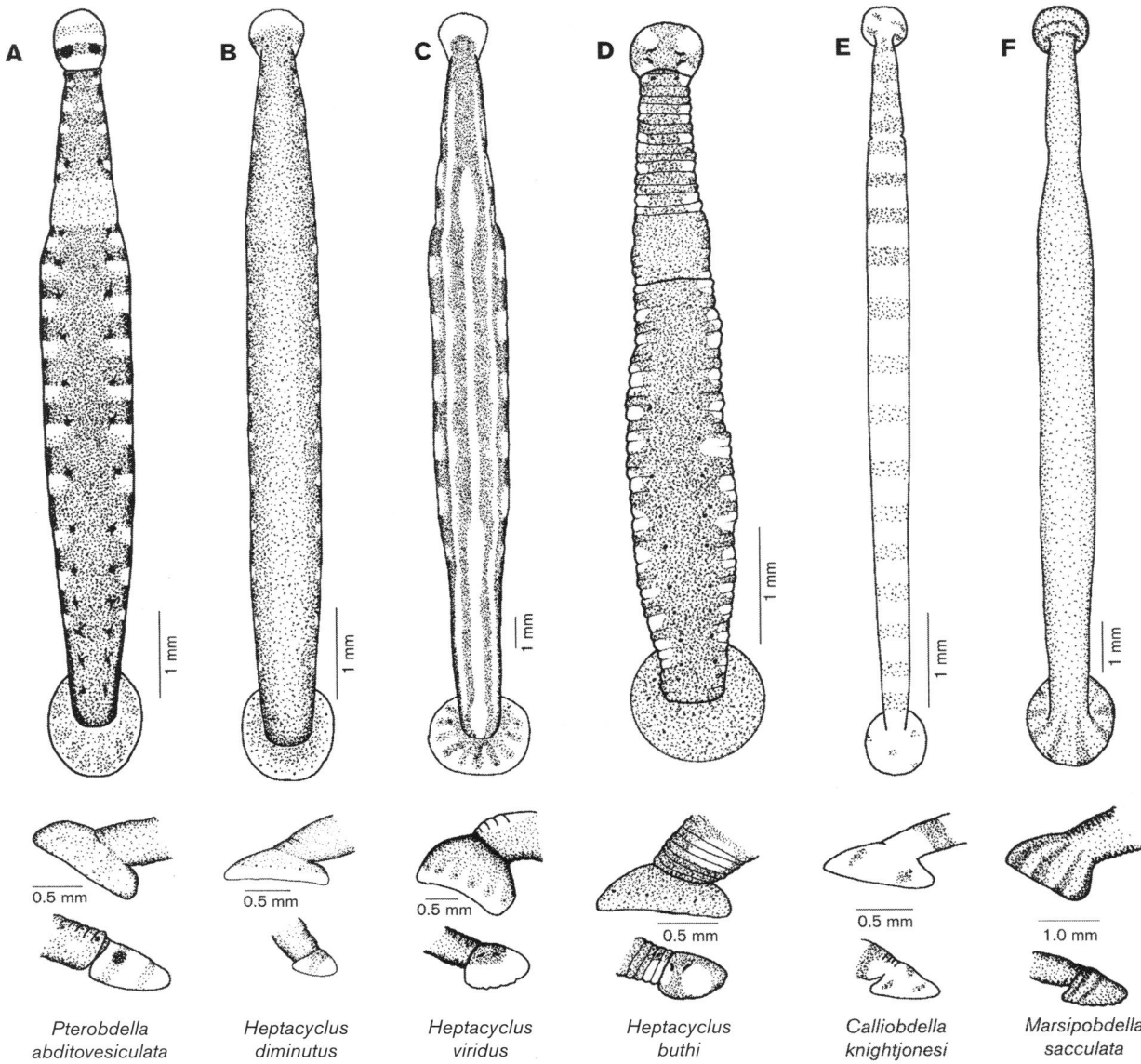

A	B	C	D	E	F
Pterobdella abditovesiculata	Heptacyclus diminutus	Heptacyclus viridus	Heptacyclus buthi	Calliobdella knightjonesi	Marsipobdella sacculata

PLATE 127 Hirudinida, Piscicolidae—dorsal view of adult entire animal illustrating body and sucker shape, pigmentation pattern, eyespots and ocelli; for each species, lateral view of caudal sucker (top) and oral sucker (bottom), illustrating pigmentation pattern, eyespots and ocelli (A, from Burreson 1976a, courtesy of the Journal of Parasitology; B, C, from Burreson 1977a, courtesy of the Journal of Parasitology; D, from Burreson and Kelman 2006, courtesy of the Journal of Parasitology; E, from Burreson 1984, courtesy of the Zoological Journal of the Linnean Society).

University Hatfield Marine Science Center and Vancouver Public Aquarium (see Burreson 1977a).

Heptacyclus buthi (Burreson and Kalman, 2006). The only known intertidal leech from the Pacific coast. Known only from the Bodega Marine Reserve on the tide pool fish *Oligocottus snyderi*, *Oligocottus rubellio*, and *Clinocottus analis* (see Burreson and Kalman 2006).

Oceanobdella pallida Burreson, 1977. Appears to be a specific parasite in the mouth of English sole, *Parophrys vetulus*, along the entire Oregon coast (see Burreson 1977c); also on same host in British Columbia (see Madill 1988).

Ostreobdella papillata Burreson, 1977. Only known hosts are black rockfish, *Sebastes melanops*, in Oregon (see Burreson 1977b; note that the plates were reversed by the publisher) and tiger rockfish, *Sebastes nigrocincta*, in British Columbia (see Madill 1988); one specimen was recovered from a Pacific octopus, *Enteroctopus dofleini*, in Oregon, but it is likely a case of host transfer during feeding by the octopus.

Pterobdella (=*Aestabdella*) *abditovesiculata* (Moore, 1952). Appears to be primarily a parasite of the staghorn sculpin *Leptocottus armatus* in Yaquina Bay, Oregon, and Tomales Bay but also parasitizes other demersal fishes (see Burreson 1976a).

References

Burreson, E. M. 1976a. *Aestabdella* gen. n. (Hirudinea: Piscicolidae) for *Johnassonia abditovesiculata* Moore 1952 and *Ichthyobdella platycephali* Ingram 1957. J. Parasitol. 62: 789–792.

Burreson, E. M. 1976b. *Trachelobdella oregonensis* sp. n. (Hirudinea: Piscicolidae), parasitic on the cabezon, *Scorpaenichthys marmoratus* (Ayres), in Oregon. J. Parasitol. 62: 793–798.

Burreson, E. M. 1977a. Two new species of *Malmiana* (Hirudinea: Piscicolidae) from Oregon coastal waters. J. Parasitol. 63: 130–136.

Burreson, E. M. 1977b. Two new marine leeches (Hirudinea: Piscicolidae) from the west coast of the United States. In Exerta Parasitológica en Memoria Del Doctor Eduardo Caballero Y Caballero, Universidad Nacional Autónoma De México, Inst. Biol. Publ. Espec. 4: 503–512.

Burreson, E. M. 1977c. *Oceanobdella pallida* n. sp. (Hirudinea: Piscicolidae) from the English sole, *Parophrys vetulus*, in Oregon. Trans. Amer. Microsc. Soc. 96: 527–530.

Burreson, E. M. 1984. A new species of marine leech (Hirudinea: Piscicolidae) from the north-eastern Pacific Ocean, parasitic on the English sole, *Parophrys vetulus* Girard. Zool. J. Linn. Soc. 80: 297–301.

Burreson, E. M. 1995. Phylum Annelida: Hirudinea as vectors and disease agents. In P. T. K. Woo (Ed.), *Fish Diseases and Disorders.* Vol. 1. *Protozoan and Metazoan Infections,* Chapter 14. CAB International, pp. 599–629.

Burreson, E. M. and J. E. Kalman. 2006. A new species of *Malmiana* (Oligochaeta: Hirudinida: Piscicolidae) from tidepool fishes in northern California. J. Parasitol. 92: 89–92.

Madill, J. 1988. New Canadian records of leeches (Annelida: Hirudinea) parasitic on fish. Can. Field-Naturalist 102: 685–688.

MacGinitie, G. E. and N. MacGinitie. 1968. Natural history of marine animals. 2nd ed. New York: McGraw-Hill, 523 pp.

Moore, J. P. 1952. New Piscicolidae (leeches) from the Pacific and their anatomy. Occas. Pap. Bernice P. Bishop Mus. 21: 17–44.

Sawyer, R. T. 1986. Leech biology and behaviour. Three Vols. Oxford: Oxford University Press, 1065 pp.

Siddall, M. E., K. Apakupakul, E. M. Burreson, K. A. Coates, C. Erséus, S. R. Gelder, M. Källersjö, and H. Trapido-Rosenthal. 2001. Validating Livanow: Molecular data agree that leeches, branchiobdellidans and *Acanthobdella peledina* form a monophyletic group of oligochaetes. Molec. Phylog. Evol. 21: 346–351.

Williams, J. I. and E. M. Burreson. 2006. Phylogeny of the fish leeches (Oligochaeta, Hirudinida, Piscicolidae) based on nuclear and mitochondrial genes and morphology. Zool. Scripta 35: 627–639.

Polychaeta

JAMES A. BLAKE AND R. EUGENE RUFF

(Plates 128–181)

The dominant class of annelids in the marine environment is the Polychaeta ("many setae"). Polychaetes are numerous, diverse, almost entirely marine, and often constitute a major component of benthic communities. Several attempts to partition the polychaete families into orders were summarized by Blake (1994a), who adopted a classification scheme and arrangement of the families similar to one originally proposed by Pettibone (1982). Subsequently, a comprehensive cladistic analysis of the annelids and in particular polychaetes was published. The monophyly exhibited in this phylogenetic approach was translated into a system of clades within which the families were classified (Rouse and Fauchald 1997).

To simplify an approach for students attempting to use the keys, the arrangement of the families more or less follows that of the previous edition of this manual. However, related families are grouped together, and although the clades proposed by Rouse and Fauchald (1997) are not stated in the classification, they can readily be reconstructed. For students, the family level is the most important category required to understand the different and widely varying morphology of polychaetes.

Thus, it is crucial that users of these keys learn to recognize the families and to work within them. It is for this reason that we have attempted to group families together that are similar in external appearance and habitat.

Several books provide detailed accounts of the wide variety of polychaete morphology. These include Fauchald (1977), Fauchald and Rouse (1997), Glasby et al. (2000), and Rouse and Pleijel (2001). In addition, a series of monographs on California polychaetes by Blake et al. (1994, 1995, 1996, and 2000) provide detailed accounts of most of the families reported in this section, as well as others that may be found in deeper water. All of these works are heavily illustrated and documented, providing students with valuable resources with which to continue their study of polychaetes.

General Morphology

To make clear the basic structure of polychaetes, the morphology of free-living forms is treated first. The structure of free-living polychaetes is perhaps best understood by taking a typical example, describing its structure, and pointing out how other families deviate from this plan.

The body is generally elongated with numerous segments and consists of a **PROSTOMIUM** (anterior cephalic lobe), a **METASTOMIUM** (the following body segments) and a **PYGIDIUM** (the last segment).

The heads of polychaetes are exceedingly diverse. For an example of the head of an "errant" or free-living polychaete, see *Nereis* (plate 128A, 128B). The head, in this genus, consists of a preoral prostomium, provided at its anterior margin with a pair of small **ANTENNAE** and at its sides with paired, fleshy, biarticulated **PALPS.** The prostomium of *Nereis* bears two pairs of eyes; other polychaetes may have one pair or none; in some, numerous eyespots are scattered on the **PERISTOMIUM** or even on the tentacles or sides of the body.

The segment just behind the prostomium is the **PERISTOMIUM**—in *Nereis*, a fusion of two segments. It bears, in *Nereis*, four pairs of **TENTACULAR CIRRI** on short stalks at its anterior margin. Usage of the terms palp, antenna, tentacle, and cirrus varies greatly amongst different families and taxonomists. **ANTENNAE,** unless otherwise specified, are usually dorsal or marginal on the prostomium. **PALPS** are usually associated with the mouth, tend to be lateral or ventral to the prostomium, and border the anterior margin of the mouth. However, certain dorsolateral structures, especially if these are large, elongated, grooved, or prehensile as in the spionids, are frequently called palps. The term **CIRRUS** is usually applied to structures arising dorsally or ventrally on the parapodia, whereas comparable structures on the anterior part of the body, if elongated, may be designated tentacular cirri (or peristomial tentacles). **TENTACLE** is a very general term used to signify any of a variety of elongated sensory or feeding structures, usually on the head.

In most free-living polychaetes the pharyngeal region may be everted to form a **PROBOSCIS,** which in *Nereis* bears stout **JAWS** and small, horny teeth (paragnaths). The proboscis of *Nereis* is divisible into two external regions (plate 128A, 128B): (1) an **ORAL RING** external to the mouth (internal when proboscis is retracted) and divisible into six areas, numbered V–VIII; and (2) a second ring, the **MAXILLARY RING** with stout, horny jaws at its outer end and divisible into six areas, numbered I–IV. Even-numbered areas are paired; odd-numbered

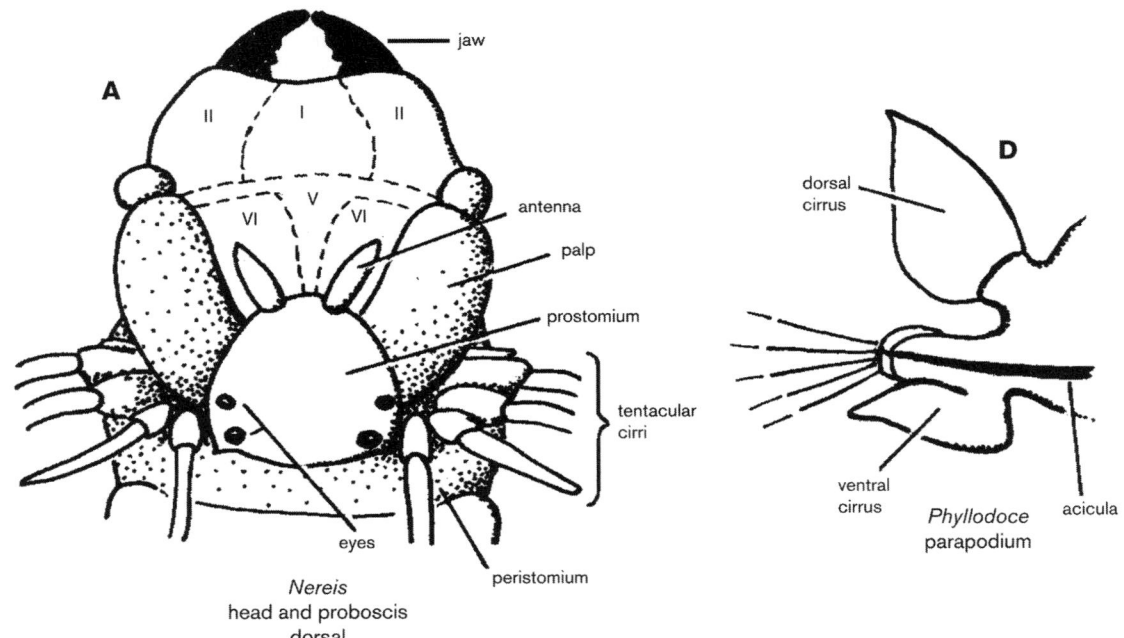

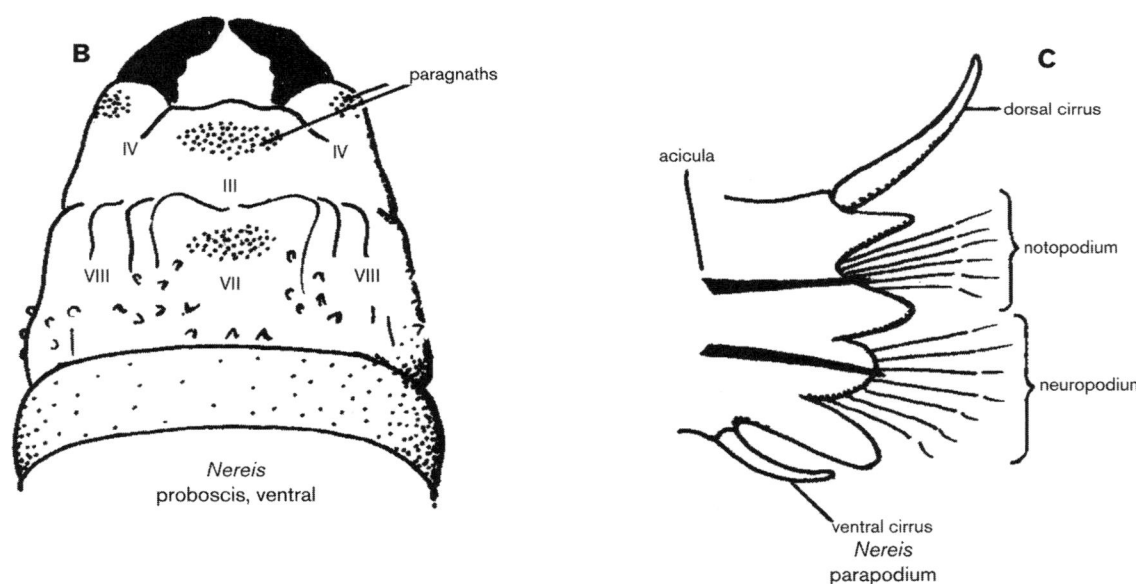

PLATE 128 Polychaeta. Morphology: A, head of *Nereis* with everted proboscis, dorsal; B, same, ventral; C, parapodium, *Nereis*; D, parapodium, *Phyllodoce* (by Blake).

areas lie in the dorsal and ventral midlines. A study of the arrangement of paragnaths is necessary in taxonomic work on nereidids; in other families, the proboscis may be smooth or covered with soft papillae.

The jaw pieces of some polychaetes are complex (e.g., Eunicidae, Oenonidae, Lumbrineridae, Onuphidae, and Dorvillei-

dae) and composed of several parts, each of which may have numerous small teeth (plate 129A, 129B). These are important characteristics in taxonomic work.

The body segments bear conspicuous **PARAPODIA**. The parapodia of *Nereis* are paired locomotor appendages, each of which is composed of an upper lobe, the **NOTOPODIUM**, and a

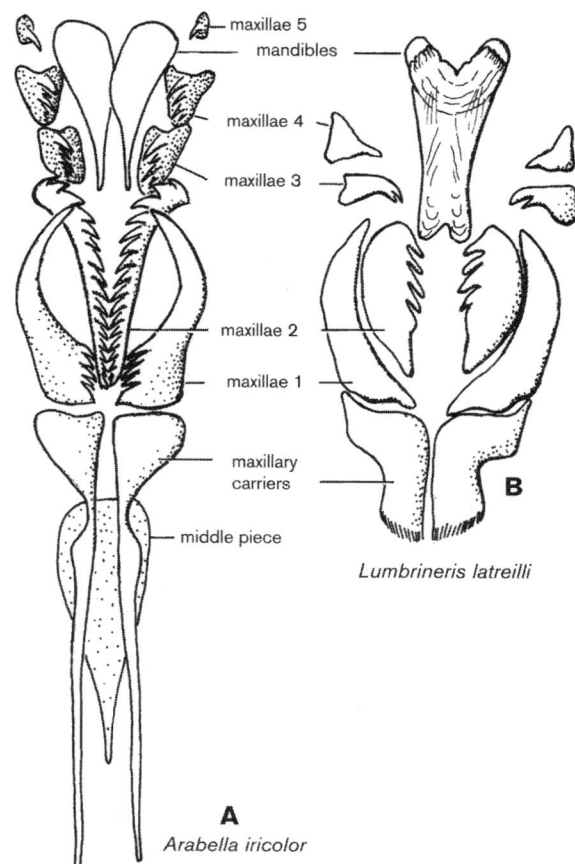

maxillae 5
mandibles
maxillae 4
maxillae 3
maxillae 2
maxillae 1
maxillary carriers
middle piece

B

Lumbrineris latreilli

A

Arabella iricolor

PLATE 129 Polychaeta. Jaw apparatus of some euniciform polychaetes: A, *Arabella iricolor*; B, *Lumbrineris latreilli* (by Blake).

lower **NEUROPODIUM** (plate 128C). Each lobe typically contains a bundle of slender, projecting, chitinous **SETAE**[1] and a larger, dark, internal spine or **ACICULA** (plates 128C, 128D, 130A). The shape, size, number, and position of setae are important in classification. Arising from the bases of the notopodium above and the neuropodium below are often slender, flexible outgrowths—the **DORSAL CIRRUS** (pl., cirri) and **VENTRAL CIRRUS,** respectively. The notopodium consists of a dorsal and middle lobe, between which are an acicula and projecting setae; the neuropodium consists of a neuroacicular lobe, provided with setae and an acicula, and a ventral lobe. In many polychaetes, dorsal gills or **BRANCHIAE,** made conspicuous by red or green blood within, arise from the bases of the parapodia in certain parts of the body.

The first parapodia of *Nereis* are borne on the segment behind the peristomium. The first two pairs of parapodia are **UNI-RAMOUS,** that is, each contains a single acicula and **SETAL FASCICLE** (bundle); all succeeding parapodia are **BIRAMOUS.** In other polychaetes, the parapodia vary: some are only uniramous (notopodium lacking) (plate 128D), others only biramous, some with both types, and some with parapodia greatly reduced or absent (apodous).

1. The Latin *seta,* for "bristle" or "hair," refers to the characteristic feature of polychaetes; the Greek equivalent is *chaeta,* which translates to "long hair" or "mane." *Seta* is used in this section, although *chaeta* is widely used in current polychaete literature.

Setae vary widely in form and furnish precise characteristics for determination of species. Because microscopic examination is necessary, the use of setal characteristics has been avoided where possible in the keys, but it is necessary in some cases. Common setal types and their names should be recognized. Plate 130A–130PP give some idea of the diversity of setal form. We may distinguish **SIMPLE SETAE** (plate 130B–130T, 130CC–130PP) from **COMPOUND** or **JOINTED SETAE** (plate 130U–130BB). Long, slender, simple setae are called **CAPILLARY SETAE.** The tips of setae, whether simple or compound, are either **ENTIRE, BIFID** (plate 130T), or **TRIFID.** If bent like a sickle, they are **FALCATE**; if flattened like an oar blade, they are **LIMBATE** (plate 130I) or **BILIMBATE** (plate 130J). Some simple setae have stubby, bent, usually bifid ends called **HOOKS** or **CROTCHETS** (plate 130CC–130FF). These are usually relatively stout and grade into short, broadened types known as **UNCINI** (plate 130GG–130PP), usually set in close rows, which are especially characteristic of tube-dwelling polychaetes.

Compound setae may be multiarticulate, as in the long bristles of some flabelligerids (plate 130U), but are characteristically two-jointed, composed of a shaft and a blade (plate 130V–130BB). This blade in turn may have various shapes and may itself be a hook (plate 130BB). The blade rests in a notch in the end of the shaft. If the two sides of the notch are equal, it is **HOMOGOMPH** (plate 130AA); if unequal, **HETEROGOMPH** (plate 130Z). Finally setae, either simple or compound, may be embedded at their tips in a clear matrix and are spoken of as hooded. Thus, we may have **SIMPLE HOODED HOOKS** (plate 130T, 130CC) or **COMPOUND HOODED HOOKS** (plate 130BB).

Tube-dwelling or "sedentary" polychaetes depart widely from the body form of predatory or "errant" types. The prostomium and eyes are often reduced, proboscis and jaws usually absent, and the anterior end, especially in families with species dwelling in fixed tubes, greatly elaborated for feeding and respiration. In sabellids and serpulids, the peristomial tentacles form a branchial crown of featherlike "gills," which serves both for feeding and respiration. Cilia pass water between branches of the plumes and transport food, entangled in mucus, down to the mouth. In other forms such as terebellids, the peristomial tentacles are long, filamentous, and prehensile, serving to carry food by ciliary action in a groove running along each filament. Just behind the head, tufted, blood-filled branchiae may be present. Parapodia in tubicolous polychaetes tend to be small and are provided with rows of uncini for gripping the sides of the tube. Special glands may secrete tube-forming material. In serpulids, which form a rigid calcareous tube, one or more peristomial cirri may be modified to a pluglike operculum that can block the tube entrance. The body in tube-dwelling forms is often divisible into a more anterior and specialized **THORAX** and a less specialized posterior **ABDOMEN.** The thorax may bear a **COLLAR** anteriorly, which may be extended back to the posterior end of the thorax as a pair of folds, the **THORACIC MEMBRANES.** The abdomen may be followed by a caudal region. The preceding accounts are of selected free-living and tubicolous types and can convey but a poor idea of the actual diversity of pattern of the numerous annelid families.

Collection and Preservation

Polychaetes are delicate and fragile animals, and special care should be taken when handling them, especially when washing sediment samples through screens. If handled too roughly, the bodies will fragment, or critical parts will be lost, making

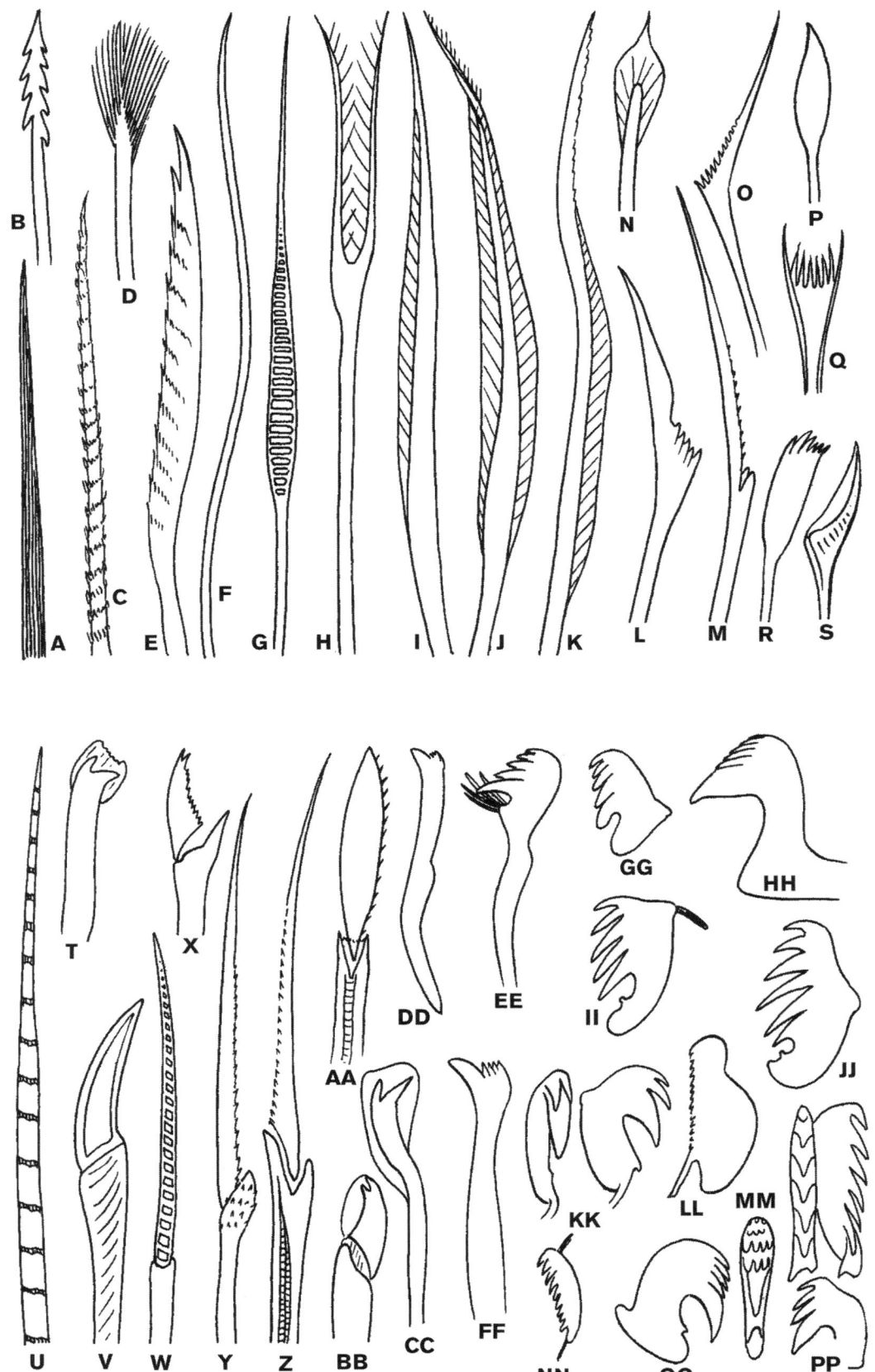

PLATE 130 Polychaeta. Types of setae: A, acicula; B–T, simple setae: B, barbed; C, spinous capillary; D, brushlike; E, serrated; F, simple capillary; G, camerated capillary; H, furcate or forked; I, limbate or winged; J, bilimbate; K, limbate, Serpulidae; L, M, collar setae, Serpulidae; N, spatulate (palea), Sabellidae; O, geniculate or bent, Serpulidae; P, lanceolate or styliform, Chaetopteridae; Q, pectinate, Eunicidae; R, S, paleae, Sabellariidae; T, bifid acicular seta; U, multiarticulate, Flabelligeridae; V–BB, compound setae: V, Flabelligeridae; W, Sigalionidae; X, Syllidae; Y, Phyllodocidae; Z, heterogomph spiniger, Nereididae; AA, homogomph falciger, Nereididae; BB, Eunicidae; CC–PP, simple hooks and uncini: CC, *Polydora*; DD, *Arenicola*; EE, Maldanidae; FF, *Trichobranchus*; GG–PP, uncini: GG, *Serpula*; HH, avicular uncinus of Sabellidae; II–MM, Serpulidae; NN, *Chaetopterus*; OO, *Amphitrite*; PP, Sabellidae (after Fauvel 1923).

identification difficult if not impossible. It is highly recommended that samples be sieved as soon as possible after collection and before the samples are placed in fixative. By doing so, the animals are removed early from the sediments, which, particularly in the case of coarse sands, could damage the specimens by abrasion or crushing if the sediment collected is heavy. There are two types of "sieving." (1) *Elutriation* involves placing the sediment into a container such as a bucket and then washing with a steady flow of water so the lighter sediment and animals are floated off onto the sieve. This ensures that fragile animals are not subject to direct flow of water. (2) *Direct sieving* involves actually sieving sediment onto a screen; however, when doing so, a gentle flow of water should be directed to the underside of the screen, rather than from above. This breaks the force of the water and helps prevent damaging specimens. Large samples should be handled in small portions to avoid having too much material on the screen at one time. Contents of the screen should be gently washed into a jar and kept separate from the sediment residue, which can be resieved at a later time to ensure removal of all specimens. Although these procedures may be more time-consuming than the usual method of dumping samples into formalin at the time of collection and sieving dead material, they have been found to be extremely valuable in ensuring that the specimens are preserved in good condition. The extra time spent in carefully washing sediment samples is minimal compared to the extra time that may be necessary to identify poor material. Damaged specimens may be misidentified or completely lost to counts of total individuals recorded from quantitative samples.

When sorting polychaetes from sediments, samples should be washed through sieves having mesh sizes of at least 0.3–1.0 mm to retain very small species. The 0.3 mm or 300 μm sieve is now considered the lower limit for benthic macrofauna. Meiofauna are considered to be those organisms which pass through a 300 μm sieve and are retained on a mesh of 63 μm.

Specimens may be relaxed prior to preservation, using a solution of 7.5% magnesium chloride or menthol. Fixation of free-living polychaetes is best in formalin of 5–10%, depending on the use of the material; preservation of quantitative benthic samples is usually done in 10% buffered formalin. Tubicolous worms generally have a softer cuticle than errant forms and may alternatively be fixed in Bouin's, Kahle's, or a potassium dichromate fixative, all of which harden the cuticle, making study easier. After a suitable time in the fixative (normally no longer than 24 hours), the worms may be transferred to 70 or 80% ethyl alcohol (ETOH). Specimens intended for molecular analysis require special handling. The simplest type of preservation for molecular analysis is with cold 95–100% ETOH. If −80°C freezers are available, these should be used. For larger specimens, parapodia can be removed and preserved separately for molecular work, leaving the majority of the worm for normal preservation and species confirmation.

During fixation, care should be taken to ensure a straight specimen. Coiled worms are difficult to study and break easily when manipulated. Straightening may be achieved by placing the living worm on a piece of paper towel or filter paper and adding the fixative with a pipette. The same results can be achieved by placing the specimen in a plastic petri dish, adding the fixative a drop at a time, and manipulating the specimen with forceps. If the worm everts its proboscis, this should be held with forceps to prevent inversion until the worm is fixed. After a suitable time in the fixative, worms are preserved in 70–80% ETOH.

Collection and processing of larval and adult polychaetes, such as the sexual stages of syllids taken from the plankton, is best done while the samples are fresh. Samples should be refrigerated as soon as possible after collection and then sorted from fingerbowls with pipettes using a stereomicroscope equipped with fiber optics to avoid heating the water in the dishes. Polychaete larvae are very active and are best removed with pipettes to smaller culture dishes where they can be refrigerated until they can be studied in detail. The study of living polychaete larvae is a rewarding experience. Polychaete larvae often have elegant pigment and ciliary patterns; cilia beat continuously and provide the observer with details of ciliary pattern and distribution. Observation of such larvae is best done using a research quality microscope and a hanging-drop technique in which a coverslip containing a drop of seawater with a confined larva is suspended into in the well of a depression slide. The coverslip is held in place by adding a little seawater to the edges. Heat-absorbing filters or a chilling stage can be used to control temperature on the stage of the microscope to allow sufficient time to make morphological observations or perhaps to take photographs. These and other methods of obtaining and culturing polychaete larvae can be found in Blake (1969, 1975a).

Dissection of Jaws, Mounting of Parapodia, Staining

For taxonomic purposes it is often necessary to dissect out the proboscis to examine jaws or other structures. Parapodia must always be examined under the microscope. The proboscis of a large worm may be dissected out with a pair of fine scissors; smaller worms require a pair of iris scissors, and minute specimens require the use of forceps and a microscalpel. The specimen is held with forceps while the razor-sharp scalpel is used to cut out the proboscis. The jaws of eunicidlike polychaetes may be removed along with the entire pharyngeal apparatus. A medial incision is made dorsally from about setigers 4–8, or wherever the jaws can be seen through the cuticle. The pharynx is then removed with forceps. Musculature is removed or trimmed under a dissecting microscope.

Additional techniques include clearing specimens with Amman's lactophenol as a temporary mounting medium. The mixture consists of 100 g phenol, 100 ml lactic acid, 200 ml glycerin, and 100 ml water. The specimen is mounted on a slide in this mixture and covered with a coverslip. The slide is heated carefully to avoid bubbling. This procedure clears the tissues, making internal sclerotized structures such as jaws and aciculae more visible. Another technique for preparing specimens for observation of small jaw pieces is to clear them in potassium hydroxide (KOH). A 10% solution of KOH is prepared in which the worms are placed for two to three hours. The worms should be checked periodically until they appear to be sufficiently clear for observation of the jaws. These specimens can be transferred back to alcohol for study, usually from a wet mount. Permanent mounts can be made using Euparal® or Permount®, which also clear the specimens. The normal way to prepare such mounts is to dehydrate the specimens in 100% ETOH, followed by a brief immersion in a clearing agent such as toluene or xylene prior to applying the mounting media. Specimens may be mounted directly in Euparal® from 100% ETOH, but each slide must be allowed to dry on a warming tray before storage. Permanent slides of parapodia, jaw pieces, setae, or whole specimens become valuable research tools and important voucher materials for a taxonomist.

The use of methyl green stain has become popular for elucidating differences between species. Staining patterns appear to be species-specific and can provide a quick recognition feature when sorting large amounts of material. Methyl green staining is particularly useful with sphaerodorids, capitellids, cirratulids, maldanids, and sabellids. A saturated solution of the stain is prepared using 70% ETOH. The specimen(s) is then immersed in a few drops of this solution for a few minutes and then placed in clean alcohol to differentiate the pattern. An additional technique includes using Shiralastain, a brownish dye that enhances surface structures such as papillae. Specimens are immersed in the stain and then differentiated in clean alcohol (Petersen 2000).

The Fauna and the Keys

A total of 414 species of polychaetes in 57 families, including nine meiofaunal families, are treated in this section. This represents a 33.5% increase over the number of species (275) that were included in the previous edition of this book (Blake 1975b). This increased number of taxa is due in part to extending the geographic coverage from central and northern California to Oregon, but is mainly due to an increase in our knowledge of the polychaete fauna of the eastern north Pacific. The Spionidae is the largest family, with 50 species. Other large families and approximate numbers of species are: Syllidae, (37); Sabellidae, (23), Phyllodocidae, (21), Cirratulidae (21), Serpulidae (20), Terebellidae (18), Nereididae (17), Capitellidae (15), Polynoidae (14), Ampharetidae (13), and Opheliidae (10).

Although the polychaete fauna of central and northern California and Oregon appears to be large, the northern sector of California and much of Oregon are not as well known as comparable areas in the southern part of California. Microhabitats of the rocky intertidal zones have only recently been explored, bays and estuaries of the central and northern coasts are poorly known, and the interstitial fauna of sand and mud has been essentially ignored. Numerous additional species will be found with increased study.

The keys to Polychaeta begin with a key to families. Because the ability to recognize families is of great practical value to the zoologist, the student should learn early on the common family types. Each family is then considered separately. The families are typically arranged with predatory or crawling forms first and concluding with the tube-dwelling or burrowing forms. This arrangement is considered more desirable than an alphabetical sequence because it places related families together and makes comparisons of plates easier. Families may be located by referring to the page number following the family name in the "Family Key." Under each family are listed the species, and in most cases there are keys to genera and species. For more complete accounts of California polychaetes, the student is referred to Hartman (1968, 1969) and Blake et al. (1994, 1995, 1996, 2000). Additional important general references include Banse and Hobson (1974), Berkeley and Berkeley (1948, 1952), Day (1967), Fauvel (1923, 1927), Hartman (1959, 1965), and Hobson and Banse (1981). For pelagic polychaetes not treated here, see Dales (1957) and Dales and Peter (1972).

Glossary

This glossary includes terms not defined in the text. It is expanded at the request of users of the previous edition and positive response to a detailed glossary published in Blake (1994a). There are so many terms and specialized terminology

associated with individual polychaete families that is it not possible to cover all of these in the brief morphological sections of this section.

ACICULA (pl., aciculae) a stout internal collagenous rod that support each branch of a parapodium; one or several may be present.

ACICULAR SETA a very stout projecting seta homologous with other setae but similar in thickness and shape to an internal acicula.

AILERON accessory jaw plate in the Glyceridae, an imbedded winglike structure.

ANAL CIRRUS (pl., anal cirri) elongated projection(s) from the pygidium.

APICAL TOOTH (TEETH) the smaller denticles or teeth above the main fang (e.g., the hooks of lumbrinerids, spionids, and capitellids).

APINNATE lacking pinnules, smooth (e.g., the smooth branchiae of spionids).

ARISTATE SETA a stout seta with smooth shaft and terminal tuft of one to several hairs.

ARTICULATE jointed.

ASETIGEROUS lacking setae; achaetous.

AURICLE paired appendages of antennal base (family Sigalionidae); also referred to as antennal ctenidia.

AURICULAR ear-shaped.

BIDENTATE with two teeth (setae and jaws).

BIFURCATE with two prongs; bifid.

BIPINNATE a structure formed like a feather with a main stem and two rows of side branches.

BLADE distal portion of a compound seta. (Compare with "shaft.")

BRANCHIAL CROWN a circle of filaments or radioles used for filter feeding and/or respiration; found in sabellids and serpulids.

BUCCAL of or pertaining to the mouth.

CAPILLARY (pl., capillaries) long, slender, tapering, hairlike setae; may be limbate or without sheath.

CAUDAL referring to the tail or posterior region.

CARUNCLE a sensory lobe that is a posterior projection of the prostomium.

CEPHALIC CAGE long, forwardly directed setae that enclose and protect the head.

CEPHALIC RIM a flange encircling the head in maldanids.

CEPHALIC KEEL a median ridge on the prostomium or head in maldanids.

CERATOPHORE the basal joint of an antenna.

CHEVRON V-shaped chitinized jaw piece at the base of the eversible pharynx in some goniadids.

CIRRIFORM shaped like a cirrus; slender, cylindrical, tapering.

CIRROPHORE a basal projection on which a cirrus is mounted.

CIRRUS (pl., cirri) a sensory projection, usually slender and cylindrical; refers to structures on various parts of the body that may or may not be homologous with each other (e.g., the dorsal cirrus is derived from the superior part of the notopodium, the ventral cirrus is derived from the inferior part of the neuropodium; the occipital or nuchal cirrus is found on the posterior part of the prostomium in spionids).

CIRROSTYLE distal joint of cirrus as in some syllids.

CLAVATE club-shaped, with a slender base and inflated tip.

COLLAR an anterior, encircling fold or flap; a rim of tissue (e.g., across the dorsum of setiger 2 in the spionid *Streblospio*, or the tissue encircling the first setiger and covering the base of the branchial crown of sabellids and serpulids).

COLLAR SETAE modified notosetae found in the collar of sabellids and serpulids; taxonomically important.

COMPANION SETAE setae that alternate with modified setae in a fascicle, usually simple limbate capillary setae (e.g., fifth setiger of *Polydora*).

CORDATE, CORDIFORM heart-shaped.

CROOK SETA, CROOKLIKE SETA a stout, curved seta found in the neuropodia of setiger 1 in the spionid *Spiophanes*.

CROTCHET a long-shafted, hooked seta found in many families; usually with two or more teeth, sometimes hooded.

DENTATE with teeth.

DENTICLE a small tooth.

DENTICULATE with small teeth.

DIGITIFORM finger-shaped.

DORSAL TUBERCLE in scale worms, a dorsal swelling on segments with a dorsal cirrus; in same position as elytrophore on segments with elytra.

ELYTRON (pl., elytra) a dorsal, scalelike structure; in scale worms.

ELYTROPHORE a cirrophore bearing an elytron in scale worms.

EPITOKE swimming sexual stage of a polychaete; usually with highly modified body structures.

ESOPHAGEAL CAECA lateral diverticula of the esophagus as in arenicolids.

EVERSIBLE proboscis being extending with the inner part folding outward like a glove; or being everted.

FACIAL TUBERCLE a projecting ridge or lobe on the upper lip below the prostomium; in scale worms.

FALCIGER blades of a compound seta with a stout, curved end. (Compare with "spiniger.")

FASCICLE a setal bundle; a group of similar or differing setae projecting from the tissue as a unit or group.

FELT matted hairs or fine setae, produced by notopodia in Aphroditidae; usually with entangled detritus.

FILIFORM threadlike, slender.

FIMBRIATED with brushlike border.

FOLIACEOUS leaflike.

FURCATE SETA short, fork-, or lyre-shaped setae; in dorvilleids, scalibregmatids, orbiniids, nephtyids, and paraonids.

FUSIFORM spindle-shaped or cigar-shaped.

GENICULATE bent like a knee.

GENITAL HOOK OR SETA modified setae, found in capitellids; thought to be involved in copulation.

GILLS common term for branchiae.

GIZZARD a grinding organ in anterior digestive system in some spionids (*Dipolydora*) and sabellariids.

HARPOON SETA a stout, pointed seta with recurved barbs near the tip; *Laetmonice* (Aphroditidae).

HOOD hyaline envelope or cowl entirely or partially covering the distal end of setae in many families (e.g., spionids, capitellids).

HOOK general term use to refer to a stout-shafted, blunt, often distally curved and dentate seta; smaller hooks arranged in single or double rows are often called uncini.

INFERIOR ventral-most, lower-most.

INTERPARAPODIAL located between or connecting successive parapodia (e.g., interparapodial genital pouches of some spionids [*Laonice*]).

INTERRAMAL located between the dorsal and ventral branches of a single parapodium.

INVOLUTE curved inward, as in the interramal cirri of the nephtyids genus *Aglaophamus*.

JAWS a general term for a set of opposable, chitinized structures found in some polychaete families.

LAMELLA (pl., lamellae) a flattened, sheet- or platelike fleshy structure; a flattened lobe (as pre- and postsetal lamellae).

LANCEOLATE pointed, shaped like a lance.

LAPPET a small, tongue-shaped flap or fleshy process; as in highly reduced pygidial lobes of polydorids, ventral parts of sabellid collars, and lateral extensions of some anterior segments in terebellids.

LIGULE a compressed conical lobe of a parapodium (in nereidids).

LIMBATE SETA a simple seta covered with a transparent hyaline sheath that is visible on one ("unilimbate") or both ("bilimbate") sides of the shaft.

LOBE a parapodial process (e.g., presetal and postsetal lobes).

LONG-HANDLED uncini with a long basal rod or manubrium (terebellids, maldanids, sabellids).

LYRATE furcate; shaped like a lyre; refers to certain types of setae. (See also "furcate.")

MACROGNATH a large paired, black jaw piece on the proboscis of goniadids. (Compare with "micrognath.")

MACROTUBERCLE large, chitinized projection on the elytra of some polynoids. (Compare with "microtubercle.")

MANDIBLE ventral paired, flattened jaw piece found in euniciform families, usually fused along the median line. (See also "maxilla.")

MANUBRIUM a handlelike process or part; refers to the swelling and waistlike constriction seen in the neuropodial hooks of spionids or in the notopodial setae in the abdomen of sabellids.

MAXILLA dorsally attached pharyngeal jaw pieces of euniciform families.

MAXILLARY CARRIER a paired jaw piece supporting the maxillae in euniciform families, with or without a median unpaired supporting piece.

MEDIAL near or toward the midline of the body.

MEDIAN in the midline.

MEMBRANOUS thin, flattened, sheetlike.

MICROGNATH small, black jaw pieces typically arranged around the opening of the proboscis in an arc above and below the macrognaths in the goniadids.

MICROTUBERCLE small chitinized projections on the elytra of some polynoids. (Compare with "macrotubercle.")

MONILIFORM beaded; as in dorsal cirri of some syllids or sequential body segments of some cirratulids.

MULTIARTICULATE with many joints.

NATATORY SETAE special setae that develop during sexual maturity; sometimes broader than normal setae and used for swimming; long and threadlike in some sexually mature cirratulids.

NUCHAL EPAULETTE a raised and elongated sensory organ projecting posterolateral to the prostomium.

NUCHAL ORGAN a sensory organ on the prostomium or extending back from it, usually in the form of a groove or ciliated ridge.

OCCIPITAL pertaining to the posterior part of the prostomium.

OCULAR referring to the eye.

OPERCULUM (pl., opercula) a hard structure used as a stopper or plug in a tube opening; a modified branchia in spirorbins.

OPERCULAR PALEA a setalike structure formed in the thoracic segments of sabellariids, which migrates anteriorly and forms the operculum.

PALEA (pl., paleae) a broad, flattened type of seta.

PALMATE with several digits diverging from a common base, multidigitate; resembling the fronds of a palm (e.g., branchiae in some amphinomids, scalibregmatids).

PECTINATE with a series of projections arranged like the teeth of a comb; refers to setae and branchiae.

PENICILLATE brushlike.

PENNONED tear drop–shaped; refers to the shape of the tip of certain setae (e.g., thoracic notosetae in certain sabellids and some of the modified setae in the fifth setiger in some spionids such as *Pseudopolydora paucibranchiata*).

PHARYNX anterior part of the digestive tract, modified for feeding; sometimes eversible, sometimes also modified for burrowing.

PILOSE covered with very short hairs; referring to some setae.

PINNATE featherlike, with main stem and lateral side branches; side branches may have either digitiform or flattened and platelike pinnules; refers to branchiae (e.g., in spionids and sabellids).

POLYBOSTRICHUS the male sexual stolon of certain species of syllids in which the males and females are dimorphic. (Compare with "sacconereis.")

POSTSETAL posterior to the setae; refers to parapodial lobes or ligules.

PRESETAL anterior to the setae; refers to parapodial lobes or ligules.

PROBOSCIS anterior-most part of the pharynx; epithelial and eversible, simple to branched, or muscularized.

PROBOSCIDEAL ORGANS minute structures, sometimes called papillae, that cover the surface of the proboscis of glycerids; variously ornamented; taxonomically important.

PROVENTRICLE muscularized anterior region of the digestive tract in syllids; posterior to the pharynx.

RADIOLE one of the main radii or tentacles of the branchial crown of a sabellid or serpulid; each radiole normally bears two rows of side branches or pinnules.

RAMOSE branched.

RENIFORM kidney-shaped.

RINGENT SETAE furcate or forked setae with annular serrations on both prongs; in euphrosinids.

RUGOSE rough or lumpy.

SACCONEREIS the female sexual stolon stage of certain syllid species in which the male and female forms are dimorphic. (Compare with "polybostrichus.")

SESQUIRAMOUS a parapodium in which the notopodium is reduced to the base of the dorsal cirrus and acicula. (Compare with "subbiramous.")

SETIGER a seta-bearing segment of the body; also called chaetiger.

SHAFT the proximal portion of a compound seta. (Compare with "blade.")

SIMPLE SETA unjointed seta.

SPATULATE flattened, bladelike seta with blunt end; may bear a terminal mucron (point) in some species.

SPINE stout modified seta found in posterior notopodia of many spionids and orbiniids and also in the fifth setiger of some spionids.

SPINIGER a compound seta whose blade tapers to a fine tip. (Compare with "falciger.")

SUBACICULAR HOOK a simple hooded hook in a position ventral to the acicula in onuphids and eunicids.

SUBBIRAMOUS a parapodium that is neither completely uniramous nor biramous, with the neuropodium well developed and the notopodium reduced in size and bearing very few setae.

SUBTERMINAL almost at the end.

SUBEQUAL approximately equal.

SUPERIOR the dorsalmost of two or more structures.

TENTACLE a slender outgrowth of sensory function emerging from the head (usually called antenna), peristomial segment (also called tentacular cirrus), and anterior body segments (e.g., in cirratulids and cossurids).

TENTACULAR CIRRUS a sensory projection arising either from the peristomium or from associated asetigerous segments, the latter usually considered homologous with dorsal and ventral cirri of normal post-peristomial segments.

TENTACULAR FORMULA a series of letters and numbers used to indicate the arrangement of tentacular cirri and setular cirrus in phyllodocids.

TENTACULOPHORE the basal projection on which a tentacular cirrrus is mounted.

TREPAN the chitinized, anteriorly toothed part of the eversible pharynx of some syllids.

TRIDENTATE with three teeth.

TRUNCATE with the end blunt, not tapering.

TUBERCLE an elongated papilla present in some scale worms.

UNCINUS (pl., uncini) sharply dentate, clawlike setae, often with a square or oval platelike base and several curved teeth, or S-shaped with a broad base and single tooth.

UNCINIGER a segment bearing uncini.

UNIDENTATE distally entire.

ABBREVIATIONS USED ON PLATES

ant,	antenna (or antennae)
br,	branchia (or branchiae)
dCirrus,	dorsal cirrus (or cirri)
dTent,	dorsal tentacle
dTentCirrus,	dorsal tentacular cirrus (or cirri)
fAnt,	frontal antenna
latAnt,	lateral antenna
mAnt,	medial antenna
nuOrg,	nuchal organ
neL,	neuropodial lamella
neS,	neuroseta (or neurosetae)
noL,	notopodial lamella
noS,	notoseta (or notosetae)
notoP,	notopodium
occAnt,	occipital antenna
per,	peristomium
pro,	prostomium
prob,	proboscis
tentCirrus(i),	tentacular cirrus (or cirri)
vCirrus,	ventral cirrus (or cirri)
vTentCirrus,	ventral tentacular cirrus (or cirri)

Key to Families of Polychaeta

1. External segmentation and parapodia not apparent; setae absent. 2
— External segmentation and/or setae present 4
2. Paired antennae on prostomium; body long, slender. 3
— Paired antennae absent; tactile hairs along body; body small, short; interstitial forms (plate 135A, 135B) . . . .
. . . . Dinophilidae and Diurodrilidae
3. Without ventral ciliation; two short, stiff tentacles present, lacking ampullary apparatus at base; cuticle thick, smooth

and iridescent; body turgid, superficially resembling nematode (plate 135H) . . . . Polygordiidae

— With ventral ciliation; two long flexible tentacles present usually with ampullae at base; cuticle thinner, internal segmentation sometimes apparent (plate 135I); body not resembling nematode . . . . Protodrilidae

4. Dorsal surface more or less covered with overlapping elytra (scales) (plate 131A), paleae or felt (plate 131B) 5

— Dorsal surface not covered with elytra, paleae, or felt . . . . .
. . . . 9

5. Dorsal surface more or less concealed by felt (plates 136A, 136B); notosetae may be harpoon-shaped (plate 136C) or held erect over dorsum . . . . Aphroditidae

— Dorsal surface more or less concealed by elytra or paleae; harpoon-shaped setae absent 6

6. Dorsal surface more or less concealed by paleae (plate 141A) . . . . Chrysopetalidae

— Dorsal surface more or less concealed by elytra 7

7. Setae all simple; elytra and dorsal cirri alternate regularly from setiger 4 to about 23; thereafter, every second elytra followed by a dorsal cirrus (plate 137A) . . . . . . Polynoidae

— Compound neurosetae present 8

8. Compound setae with long, slender, articulate blade (plate 130W); 1–3 antennae present (plate 140F) . . . .
. . . . Sigalionidae

— Compound neurosetae falcigerous, with short, unidentate blade (plate 140C); a single medial antenna present (plate 140A) . . . . Pholoidae

9. Parapodia well developed, usually bearing compound setae; setal lobes supported by internal acicula; prostomium usually with sensory appendages (except in Lumbrineridae and Oenonidae); pharynx well developed, muscular, often armed with jaws or teeth 10

— Parapodia reduced, simple setae predominant, aciculae absent; prostomium usually without sensory appendages, often fused with peristomium, which may bear grooved palps, buccal cirri, or a branchial crown; pharynx without jaws or teeth, usually saclike, not muscular. 27

10. Prostomium completely retracted between first parapodia with three pairs of tentacular cirri partially supported by acicula (plate 142A) . . . . Pisionidae

— Prostomium not completely retracted between first parapodia 11

11. Notosetae arranged in transverse rows across dorsum (plate 141F); heavy furcate ringent notosetae present . . . .
. . . . Euphrosinidae

— Notosetae arising from defined fascicles or absent; furcate setae if present; thin, delicate, not heavy 12

12. Dorsal cirri digitiform or filiform, inserted posterior to spreading fascicles of notosetae and tufts of numerous branchiae (plate 141C); prostomium continuing posteriorly as conspicuous caruncle (except genus *Hipponoa*) . . .
. . . . Amphinomidae

— Dorsal cirri absent, reduced to inconspicuous lobe, or enlarged as broad, leaflike or lamellate structure; branchiae, if present, either single, paired, simply branched, or spiraled; never as numerous tufts; caruncle absent 13

13. Dorsal and ventral cirri flattened, leaflike, paddlelike, or globular; prostomium with four frontal antennae (plate 132A) and sometimes a medial one, as well; tentacular cirri 2–4 pairs; parapodia uniramous (parapodia subbiramous in *Notophyllum*); setae compound . . . . Phyllodocidae

— Dorsal and ventral cirri, if present, not leaflike or

globular 14

14. Prostomium conical, annulated, terminating distally in four minute antennae (plate 154A); peristomium fused with prostomium, without tentacular cirri; proboscis large, powerful. 15

— Prostomium otherwise; tentacular cirri present or absent 16

15. Body with parapodia similar throughout, either all uniramous or all biramous; dorsal cirri small, globular; ventral cirri larger, conical; proboscis with four subequal jaws or macrognaths; chevrons never present . . . . Glyceridae

— Body with 2–3 regions: anterior uniramous region; transitional region where notopodia gradually develop (this region may be lacking); and posterior biramous region with noto- and neuropodia well separated; dorsal and ventral cirri conical to fingerlike; proboscis with a pair of dentate macrognaths and few to many micrognaths; chevrons often present (plate 154B) . . . . Goniadidae

16. Body with two or more rows of large spherical capsules or tubercles arranged segmentally; prostomium and tentacular segment indistinct; segments indistinct, except for parapodia. . . . . Sphaerodoridae

— Body without spherical capsules or tubercles; prostomium distinct; segmentation distinct. 17

17. Prostomium with four small frontal antennae (plate 132B); body subrectangular in cross section; biramous parapodia with rami well separated and with long cilia along interramal border; most species with interramal cirrus or branchia; notosetae and neurosetae all simple, arranged in fan-shaped fascicles, with more or less developed presetal and postsetal lamellae; muscular proboscis with a pair of internal jaws. . . . . Nephtyidae

— Prostomium with 0–3, 5, or 7 antennae, but never four 18

18. With an elaborate jaw apparatus consisting of a pair of ventral mandibles and dorsal maxillae consisting of few to numerous paired pieces (plate 129A); with 1–2 asetigerous and apodous tentacular or buccal segments, without tentacular cirri, or with only a single short, dorsolateral pair 19

— Jaws absent or otherwise; with 0–8 pairs of tentacular cirri 23

19. Prostomium simple, conical, or suboval, without antennae or distinct palps (plate 131C); parapodia without dorsal or ventral cirri; first two segments asetigerous and apodous, without tentacular cirri; body smooth, elongate, cylindrical, resembling an earthworm. 20

— Prostomium suboval, with 1–7 antennae, two palps; parapodia with dorsal and ventral cirri; body otherwise. 21

20. Neurosetae consisting of limbate setae with fine tips and hooded hooks (plate 160E); jaw apparatus with two short, broad maxillary carriers, no median piece (plate 129B); eyes absent. . . . . Lumbrineridae

— Neurosetae consisting of limbate setae (plate 161I) with or without projecting acicular setae; without hooks or crotchets; jaw apparatus with two long, slender maxillary carriers plus a median piece (plate 129A); eyes present (plate 131C) or absent . . . . Oenonidae

21. First segment apodous and asetigerous; seven prostomial antennae (five long occipital, two short frontal); paired palps short, globular (plate 157B); tube dwelling . . . .
. . . . Onuphidae

— First two segments apodous and asetigerous. 22

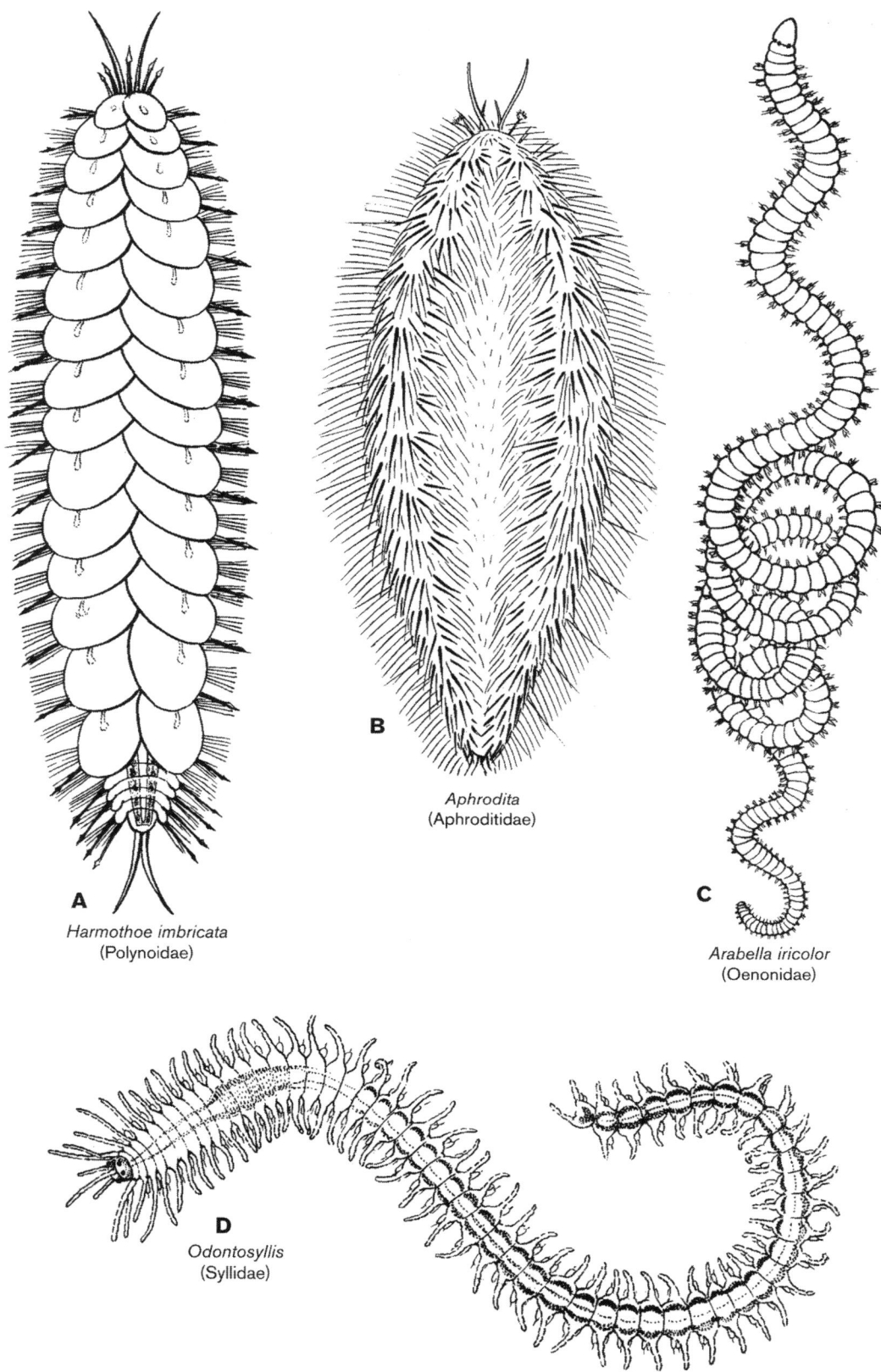

A

Harmothoe imbricata
(Polynoidae)

B

Aphrodita
(Aphroditidae)

C

Arabella iricolor
(Oenonidae)

D

Odontosyllis
(Syllidae)

PLATE 131 Polychaeta. Representative polychaetes: A, *Harmothoe imbricata* (Polynoidae); B, *Aphrodita* (Aphroditidae); C, *Arabella iricolor* (Oenonidae); D, *Odontosyllis* (Syllidae) (after McIntosh).

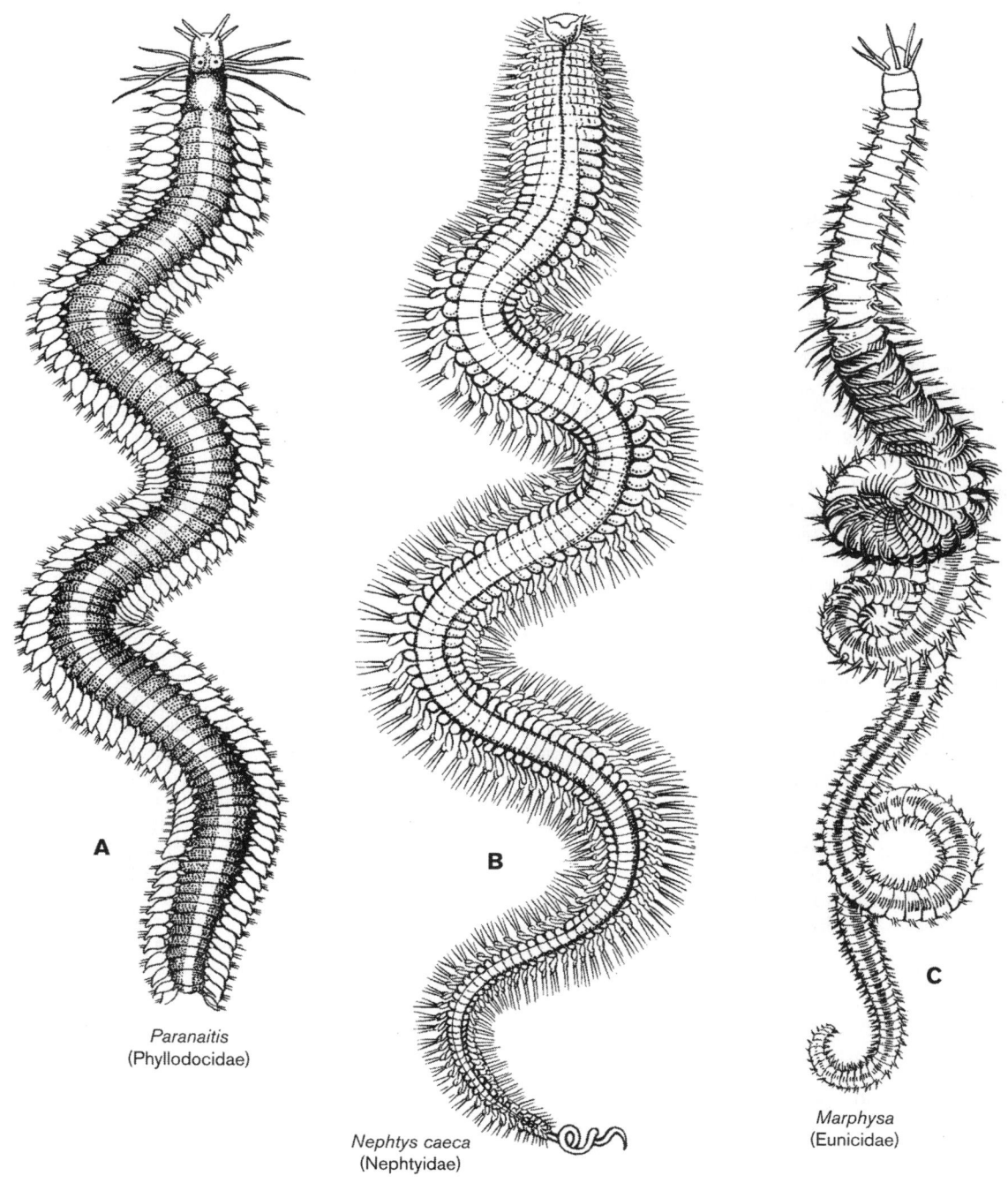

A
Paranaitis
(Phyllodocidae)

B
Nephtys caeca
(Nephtyidae)

C
Marphysa
(Eunicidae)

PLATE 132 Polychaeta. Representative polychaetes: A, *Paranaitis* (Phyllodocidae); B, *Nephtys caeca* (Nephtyidae); C, *Marphysa* (Eunicidae) (after McIntosh).

22. Prostomium with a pair of antennae and a pair of palps (plate 159A); crawlers and burrowers . . . . Dorvilleidae
— Prostomium with 1–5 occipital antennae (plate 132C) and a pair of short, globular ventral palps more or less fused to prostomium; tube dwelling . . . . Eunicidae
23. Body minute, with nine or fewer segments; with zero, two, or three antennae; two palps; without tentacular segment and tentacular cirri; parapodia uniramous with capillary or compound seta in simple fascicle; minute interstitial forms (plate 135C–135E) . . . . Nerillidae
— Body larger, with 15 or more segments; with 1–3 antennae, two palps; with tentacular segment and 1–8 pairs tentacular cirri; parapodia uniramous or biramous; large, free-living forms 24

24. Neurosetae compound (some may have blades secondarily fused to shaft) 25
— Neurosetae and notosetae simple, not compound (notosetae may be stout or hooked); tentacular segment apodous and asetigerous, more or less fused with prostomium, usually with two pairs of tentacular cirri (plate 147E–147F) .. .. Pilargidae
25. Parapodia biramous or subbiramous; notopodia represented at least by internal acicula.................... 26
— Parapodia uniramous (may be biramous in sexual epitokes); tentacular segment apodous and asetigerous, with one to two pairs of tentacular cirri; prostomium suboval with three antennae, two palps (plates 131D, 149A) .... Syllidae
26. Parapodia with varying degrees of development of extra lobes or ligules (plate 128C); prostomium suboval to subpyriform, with two frontal antennae and two biarticulate palps (plate 128A–128B); proboscis with a pair of distal, dentate, hooked jaws; with single apparent tentacular segment bearing three to four pairs of cirri; notosetae compound .... Nereididae
— Parapodia without ligules; prostomium suboval to subquadrangular, with 2–3 antennae (plate 146I); two palps (may be biarticulate); proboscis without jaws or with 2–4 simple teeth; with 1–4 asetigerous tentacular segments and 2–8 pairs of tentacular cirri; notosetae simple or absent .... Hesionidae
27. Body short and stout; posterior end covered ventrally by a chitinized shield; anus surrounded by filamentous gills (plate 134D) .... Sternaspidae
— Body elongate; posterior end not covered by shield; without anal gills 28
28. Anterior end modified by development of frilly membranes, buccal tentacles, or a branchial crown of feathery tentacles around mouth; prostomium often reduced and indistinguishable from buccal segments 49
— Anterior end not greatly modified; prostomium usually well developed and obvious; buccal segment sometimes with parapodia and may bear a pair of palps or a few grooved tentacles 29
29. Buccal segment with tentacles retractile into mouth; tentacles either grooved or papillose (plate 133F) .... Ampharetidae
— Buccal segment otherwise....................... 30
30. Buccal segment with a pair of palps, or several grooved tentacles located on anterior setigers................. 31
— Buccal segment without palps; anterior setigers without grooved tentacles 39
31. Anterior end with a pair of papillose adhesive palps (plate 167C); head flattened and spadelike (plate 167B); branchiae absent; abdominal parapodia with well-developed parapodial lamellae; hooded hooks present .... Magelonidae
— Palps, if present, not papillose or adhesive; head not flattened; branchiae often present..................... 32
32. Body divided into 2–3 distinct regions 33
— Body not divided into distinct regions 35
33. Body divided into three distinct regions; prostomium reduced (plate 167J); peristomium with large lip; setiger 4 with large modified setae; inhabiting distinct thickened tubes .... Chaetopteridae
— Body divided into two distinct regions; prostomium not reduced; peristomium with small lip; setigers 2–4 with stout acicular neurosetae, or acicular Neurosetae absent; tubes simple, not thickened........................... 34

34. With one or two anterior parapodia directed forward, bearing long setae and appearing to form a cage (plate 167H); neuropodia of setigers 2–4 bearing stout acicular setae; following setigers with capillaries; far posterior notopodia with heavy spines, sometimes arranged as rosettes; branchiae absent; hooded hooks absent .... Trochochaetidae
— Anterior parapodia not directed forward; appearing uniramous, but actually subbiramous with notopodia reduced to cirriform process penetrated by acicula; some anterior neuropodia with serrated postsetal lobes; interramal cirrus present on segments 1–7 .... Apistobranchidae
35. Body segments poorly developed; parapodia uniramous if present; setae sometimes entirely lacking; two adhesive pygidial lobes present; two hollow, nongrooved palps present................................... 36
— Body segments well-developed; parapodia biramous; setae always present; pygidial lobes, if present, nonadhesive 37
36. Parapodia lacking; setae usually lacking (plate 135I) .... Protodrilidae
— Parapodia stubby, retractile, containing a single bundle of setae (plate 135L) .... Saccocirridae
37. Neuropodia with compound falcigers; body frequently covered with minute papillae; prostomium with two small palps................................... Acrocirridae
— Setae all simple, never compound 38
38. Anterior end with pair of dorsolateral grooved palps, often long and coiling (plate 133B); anterior margin of prostomium rounded, incised, or with horns; neuropodia and/or notopodia of posterior setigers bear hooded hooks (plate 130CC); none, some, or many segments with paired branchiae................................ Spionidae
— Prostomium usually lacking appendages; anterior margin usually conical, narrowly rounded, first setigerous segment often bearing a pair of large grooved palps or numerous grooved tentacular filaments; numerous long, filamentous branchiae present on several body setigers .... Cirratulidae
39. Multidentate hooks present at least in posterior setigers.. 40
— Multidentate hooks absent; unique serrated setae may be present (i.e., Ctenodrilidae) 42
40. Multidentate hooks with hoods; body resembling an earthworm; body divided into distinct thoracic and abdominal regions, with thoracic setigers having limbate capillary setae and abdominal setigers having long-handled, multidentate hooded hooks; some genera with some thoracic or abdominal segments with both capillaries and hooks..... .... Capitellidae
— Multidentate hooks without hoods; body not resembling an earthworm 41
41. Body segments elongated, with body appearing jointed, but never annulated (plate 134B); branchiae rare....... .... Maldanidae
— Body segments not elongated, always annulated (plate 134A); branchiae present .... Arenicolidae
42. Body minute, usually with 9–12 segments; setae include unique serrated or furcate setae 43
— Body larger, with at least 15 setigers when mature..... 44
43. Unique serrated setae present (plate 171A); reproduces asexually by fragmenting; commonly found in marine aquaria and seawater tables of marine labs; found in fine muds of fouling communities .... Ctenodrilidae

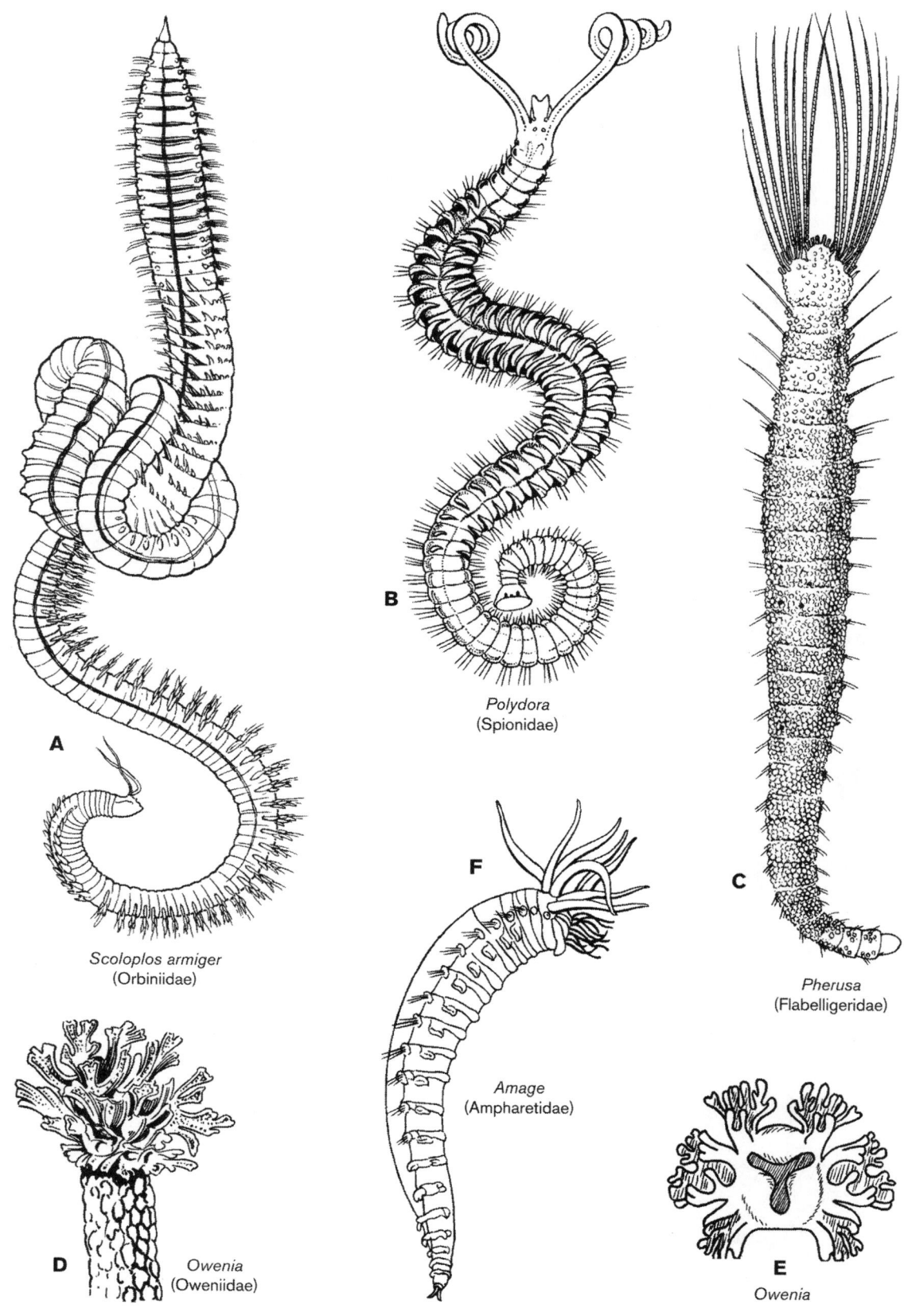

A

Scoloplos armiger
(Orbiniidae)

B

Polydora
(Spionidae)

C

Pherusa
(Flabelligeridae)

D

Owenia
(Oweniidae)

E

Owenia

F

Amage
(Ampharetidae)

PLATE 133 Polychaeta. Representative polychaetes: A, *Scoloplos armiger* (Orbiniidae); B, *Polydora* (Spionidae); C, *Pherusa* (Flabelligeridae); D, E, *Owenia* (Oweniidae); F, *Amage* (Ampharetidae) (after McIntosh).

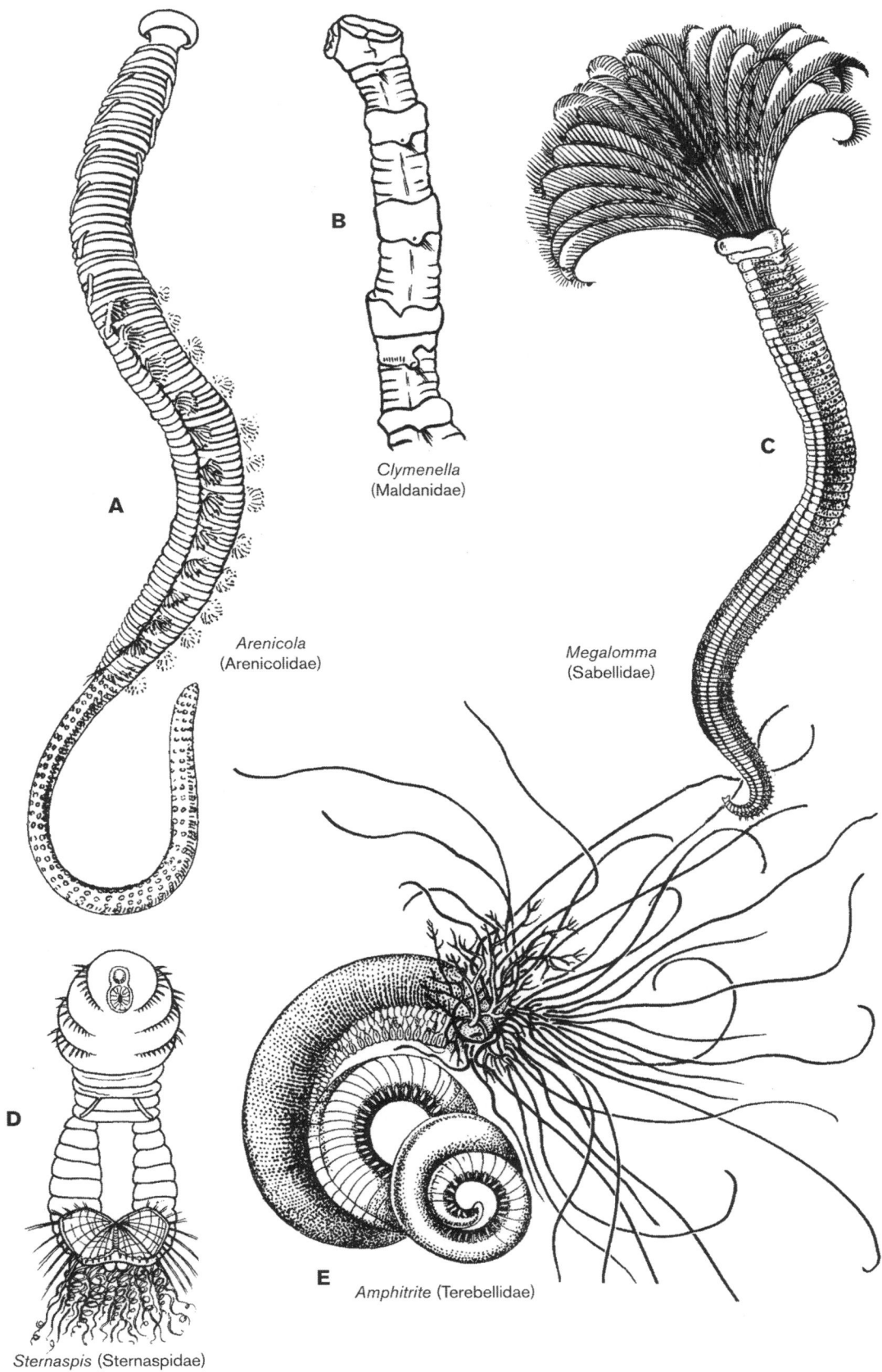

B

Clymenella
(Maldanidae)

A

Arenicola
(Arenicolidae)

C

Megalomma
(Sabellidae)

D

E *Amphitrite* (Terebellidae)

Sternaspis (Sternaspidae)

PLATE 134 Polychaeta. Representative polychaetes: A, *Arenicola* (Arenicolidae); B, *Clymenella* (Maldanidae); C, *Megalomma* (Sabellidae); D, *Sternaspis* (Sternaspidae); E, *Amphitrite* (Terebellidae) (after McIntosh).

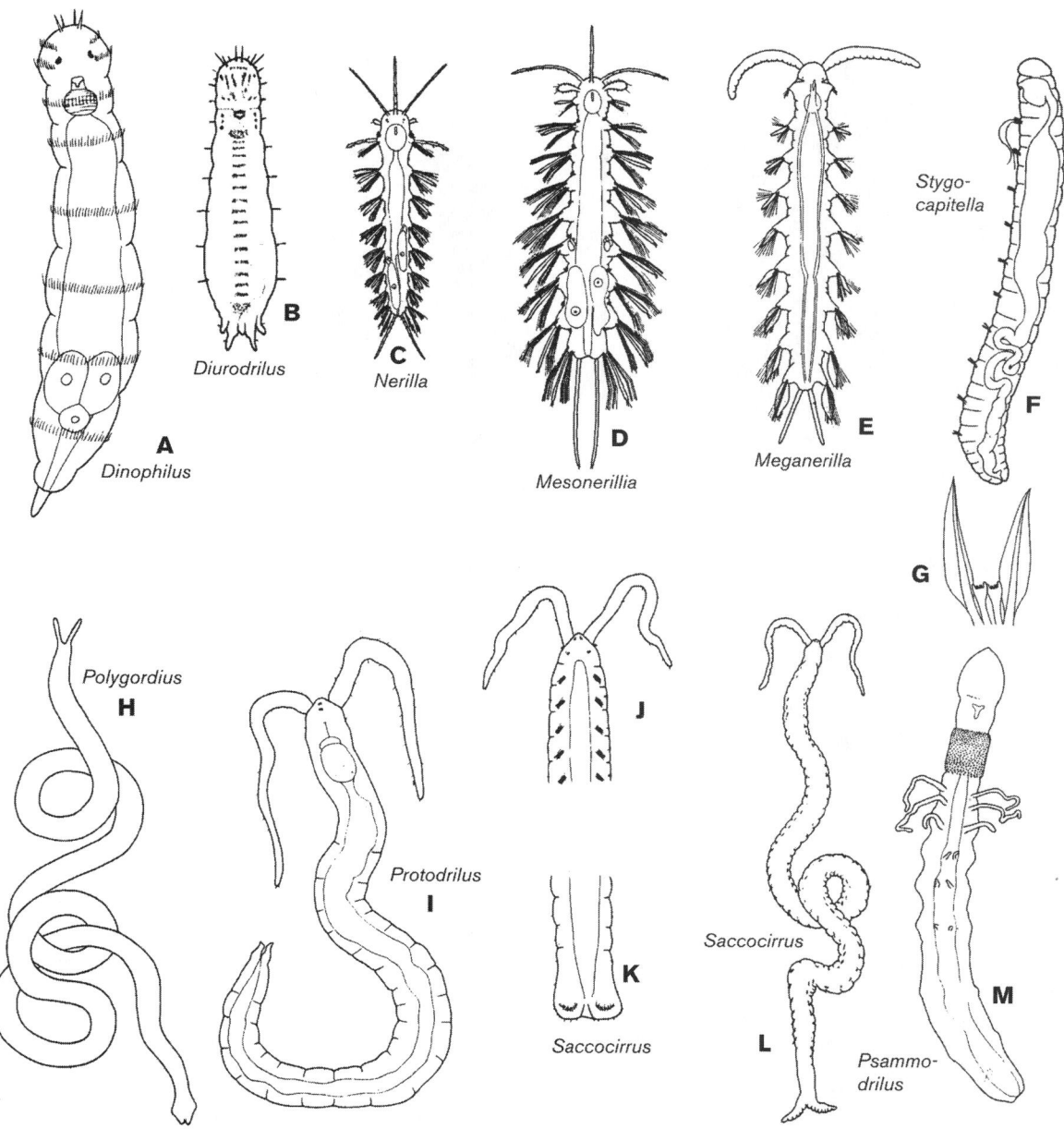

PLATE 135 Polychaeta. Representative meiofaunal polychaetes: A, *Dinophilus* ("Dinophilidae"); B, *Diurodrilus* (Diurodrilidae); C, *Nerilla* (Nerillidae); D, *Mesonerilla* (Nerillidae); E, *Meganerilla* (Nerillidae); F, *Stygocapitella* (Parergodrilidae); G, setae of *Stygocapitella*; H, *Polygordius* (Polygordiidae); I, *Protodrilus* (Protodrilidae); J, K, *Saccocirrus*, anterior and posterior ends (Saccocirridae); L, *Saccocirrus*, whole animal; M, *Psammodrilus* (Psammodrilidae) (after Westheide 1988).

— Unique furcate setae with two central teeth present (plate 135G); found in clean sands, interstitial . . . . Parergodrilidae
44. With a single filiform tentacle or branchia arising from dorsum of an anterior setiger . . . . Cossuridae
— Branchiae if present, in pairs along body segments 45
45. Body sleek, ventral groove often present; segmental eyes sometimes present; prostomium a sharply tapering cone sometimes with terminal palpode; pair of retractile nuchal organs present . . . . Opheliidae

— Body otherwise; segmental eyes absent; prostomium conical, blunt, or lobed, not sharply tapering; nuchal organs not retractile 46
46. Prostomium reduced to simple lobe or retractable, providing simple unadorned appearance to anterior end; body elongate with long anterior segments and short posterior ones; notosetae serrated capillaries; neurosetae very small bidentate hooks in dense fields . . . . Oweniidae (genera *Myriochele* and *Galathowenia*)
— Prostomium distinct, well-developed, extending anteriorly as notched, pointed, or rounded lobe 47

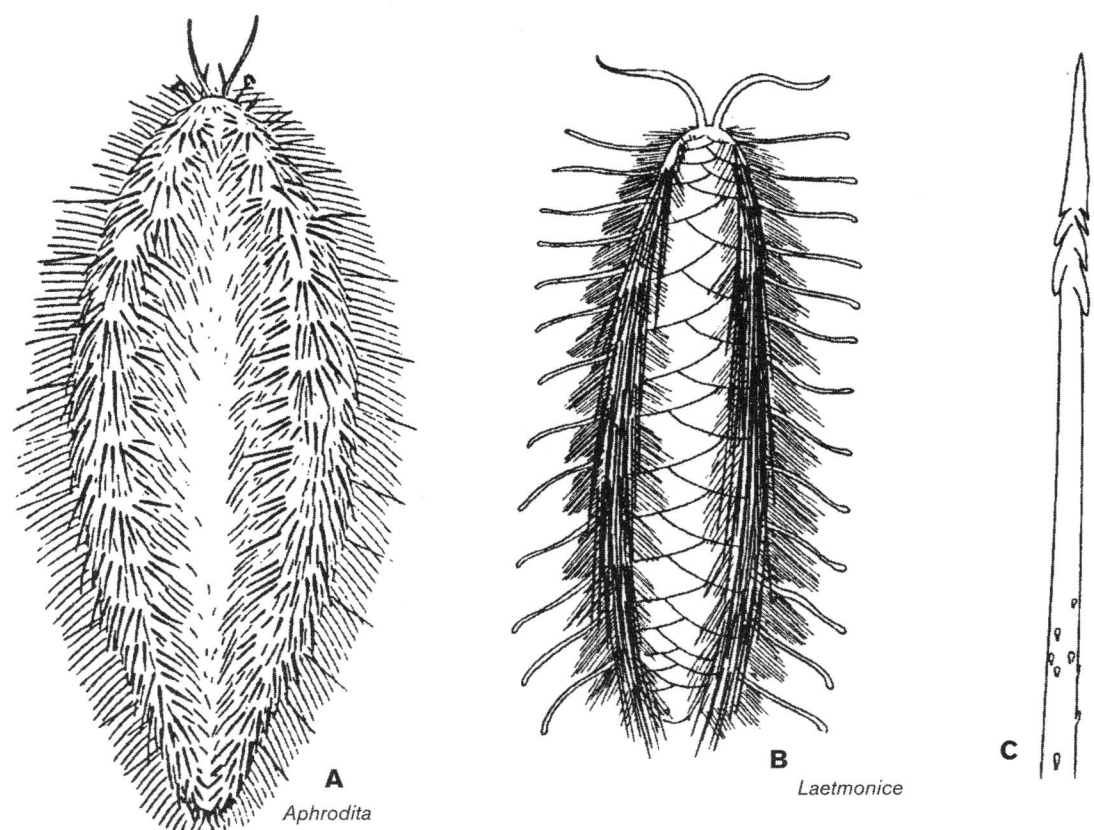

PLATE 136 Polychaeta. Aphroditidae (after McIntosh): A, *Aphrodita*; B; *Laetmonice*; C, harpoon-shaped notoseta of *Laetmonice*.

47. Prostomium T-shaped, notched, or lobed on anterior margin; body swollen anteriorly, often with a rough, areolated appearance (plate 172A); branchiae, if present, branched and restricted to anterior segments . . . . Scalibregmatidae
— Prostomium conical or rounded; body not swollen anteriorly; not rough in appearance; branchiae dorsally directed and distributed over a long body region 48
48. Parapodia with internal acicula, lobes well developed; often elongate; branchiae continuing to posterior end of body; setae including crenulated or camerated capillaries (plate 130G), with or without lyrate setae; flail setae or modified neuropodial uncini; capillaries often arranged in palisades; prostomium without medial antenna (plate 133A) . . . . Orbiniidae
— Parapodia without internal acicula; lobes reduced; branchiae absent from posterior end of body; setae all smooth or faintly striated, not crenulated, lyrate setae present and various forms of modified spines present or absent; prostomium with medial antenna present in some genera (plate 163G); small, threadlike worms . . . . Paraonidae
49. Anterior end terminating in frilled membrane (plate 133D, 133E); body enclosed in tube of closely fitting sand grains; setae including serrated capillaries, and bidentate hooks . . . . Oweniidae (*Owenia*)
— Anterior end with tentacles, palps, or branchial crown of feathery tentacles around mouth 50

50. Multiarticulate compound setae present (plate 130V); capillary setae and uncini, when present appearing annulated; cephalic cage present (plate 133C) or absent; body regions indistinct . . . . Flabelligeridae
— Multiarticulate compound setae absent; body regions well defined. 51
51. Anterior end with paleae present in one or three very distinct rows 52
— Anterior end without setae or paleae, or palaea present as spreading fascicle of heavy spines 53
52. Paleae in a single row (plate 176A); caudal region short and flattened; tube free, conical, formed of close fitting sand grains . . . . Pectinariidae
— Paleae form two to three rows (plate 175A); caudal region long, cylindrical; rigid sand tubes attached to rocks or shells, often in dense colonies . . . . Sabellariidae
53. Anterior end with soft tentacles for deposit feeding; branchiae often present on anterior segments; setal types not inverted in abdominal region 54
— Anterior end with tentacular crown of feathery bipinnate radioles; calcareous operculum present or absent 56
54. Tentacles either grooved or papillose, retractile into mouth (plate 133F). . . . . Ampharetidae
— Tentacles not retractile into mouth, grooved but never papillose (plate 134E) 55
55. Thoracic uncini long-handled (plate 130FF), abdominal

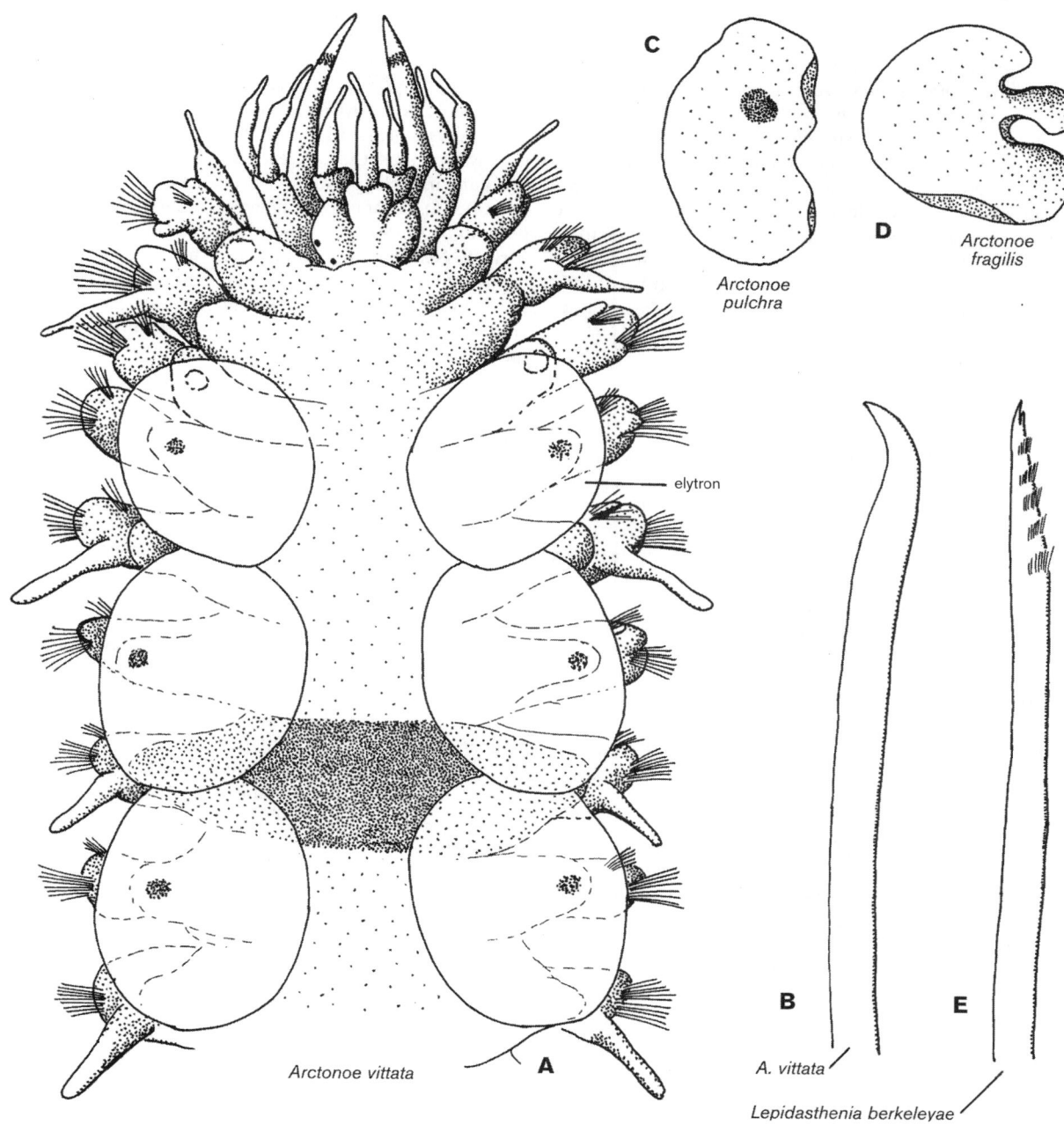

elytron

Arctonoe pulchra

Arctonoe fragilis

Arctonoe vittata A

B
A. vittata

E
Lepidasthenia berkeleyae

PLATE 137 Polychaeta. Polynoidae: A, *Arctonoe vittata*; B, neuroseta; C, *A. pulchra* elytron; D, *A. fragilis*, elytron; E, *Lepidasthenia berkeleyae*, neuroseta. (C, after Johnson).

ones short-handled; some genera with a single branchial trunk . . . . Trichobranchidae
— Both thoracic and abdominal uncini short-handled; sometimes with a posterior prolongation on thoracic uncini (plate 13OOO); branchiae absent or elaborated into numerous forms (e.g., plate 134E) . . . . Terebellidae
56. Tubes leathery or of mucus, sand, or mud; radioles of tentacular crown not modified into operculum (plate 134C) . . . . Sabellidae
— Tubes calcareous; one radiole usually modified into operculum 57
57. Tube irregularly twisted or straight, sometimes coiled near

base; body symmetrical; with more than four thoracic setigers . . . . Serpulidae *sensu stricto*
— Tube coiled into spiral; body asymmetrical; with four or fewer thoracic setigers (plate 181H) . . . .
. . . . Serpulidae, Subfamily Spirorbinae

The Meiofaunal Families

Several families of polychaetes are so small in body size that they are rarely encountered with standard macrofaunal collecting methods. For the most part, these polychaetes are part of the meiofauna, or those organisms that live in sediments but are undersampled even when a 300 μm sieve is used. Most of these polychaetes were formerly referred to a category called the Archiannelida. However, this classification has been abandoned, and it is now recognized that these annelids are true polychaetes, despite several having lost features common to other polychaetes. Meiofaunal polychaetes are poorly known from California but are quite common and diverse (Blake, personal observations). To provide some information to readers that will enable tentative identification and further study, a brief summary of each of the meiofaunal families is presented along with some illustrations of common genera. General accounts of meiofaunal polychaetes may be found in Westheide (1988, 1990), Fauchald and Rouse (1997), and Glasby et al. (2000).

DINOPHILIDAE

Dinophilids include very small, short polychaetes with few segments and a length up to no more than 2 mm. Parapodia, setae, and appendages are absent, except for a short unpaired anal cirrus that may be apparent in species of *Dinophilus* (plate 135A). The prostomium is characterized by stiff sensory cilia and distinct ciliary bands, resembling those found on the larvae of other polychaetes, that are obvious on the prostomium and segments. Some species of *Dinophilus* have dwarf males. The phylogenetic relationships of dinophilids have been investigated by Eibye-Jacobsen and Kristensen (1994), who demonstrated that dinophilids formed a single clade within the Dorvilleidae. Nevertheless, "dinophilids" are distinctive organisms, lacking the complex maxillary apparatus of dorvilleids and, when encountered, will elicit considerable interest. Species of *Dinophilus* often turn up in marine aquaria. Specimens were found on seawater tables at the first author's former laboratory at Dillon Beach. Dinophilids turned up in laboratory cultures in Long Beach (D. J. Reish, personal communication). Important references include Westheide 1967, Helgo. wiss. Meeres.16: 207–215; 1988; Donworth 1986, Zool. Anz. 6: 1–19.

DIURODRILIDAE

Diurodrilids are very small—<0.5 mm long—meiofaunal polychaetes that have a simple prostomium that lacks appendages, a peristomial segment, five indistinct body segments, and a pygidium (plate 135B). A muscular pharynx is present, but jaws are absent. Parapodia and setae are absent. The only appendages are two pygidial cirri, or "toes." The body is enclosed in a cuticle that in some species forms plates in distinct, species-specific patterns. Sensory cilia are present along the body and on the prostomium. These cilia may be arranged in distinct, species-specific patterns. To date, *Diurodrilus* is the only known genus, with six species described. Kristensen and Niilonen 1982, Zool. Scr. 11: 1–12, described the morphology of *Diurodrilus* and established family status for this taxon. No species have been reported from California.

NERILLIDAE

Nerillids are minute polychaetes that live in sandy sediments where they feed on diatoms and bacteria on sand grains. They are about 0.3–2.0 mm long and have no more than seven to nine segments. The prostomium is fused to the peristomium and bears three antennae that may be articulated (plate 135C, 135D) and a pair of short, clavate, ventrolateral palps; or antennae and/or palps may be entirely absent (plate 135E). A stomodaeum has an eversible ventral buccal organ that bears some skeletal elements called "buccal pieces." Parapodia are usually biramous. Cirri are present or absent; setae are simple or compound. The pygidium usually bears two anal cirri (plate 135C–135E). A midventral ciliated groove is present and appears to be the source of the gliding locomotion observed in nerillids. Nerillids have not been reported form from California but should be present. Approximately 15 genera have been defined, including *Nerilla*, *Nerillidium*, *Meosnerilla*, *Meganerilla*, and *Troglochaetus*. See Purschke 1985, Microfauna Mar. 2: 23–60 (pharyngeal morphology); Westheide 1988, 1990.

PARERGODRILIDAE

Parergodrilids are small interstitial worms found in sand beaches. There are two genera: *Stygocapitella*, which is marine, and *Parergodrilus*, which occurs in freshwater. The body of *Stygocapitella* is short and thick; prostomial, parapodial, and pygidial appendages are lacking (plate 135F). Branchiae are absent. The prostomium is bluntly rounded; the peristomium is a complete ring. Retractable nuchal organs are present in *Stygocapitella* and absent in *Parergodrilus*. The stomodaeum has an eversible ventral buccal organ. Aciculae are absent; setae include capillaries and furcate setae of a distinctive form where the two "tynes" are short and blunt (plate 135G) rather than long and whiplike, as in some polychaetes. *Stygocapitella subterranea* Knöller, 1934, has been observed in clean sand beaches in northern California (Blake, unpublished). See Karling 1958, Ark. f. Zool. 11: 307–342 (morphology); Riser 1984, Tane, 30: 339–250 (reproduction, development).

POLYGORDIIDAE

The polygordiids are slender, elongate worms that, although usually considered meiofaunal, often range up to 100 mm long, 1 mm wide, and with more than 180 indistinct segments and thus are part of the macrofauna. The prostomium and peristomium are fused; there are two anterior palps (tentacles); eyes may be present or absent. Paired nuchal organs are present. Parapodia and setae are absent (plate 135H). Because they lack typical polychaete morphology, it is difficult if not impossible to distinguish one species of *Polygordius* from another without resorting to thin sectioning. With histological investigation, details of the stomodaeum, reproductive system, and circulatory system serve to separate species. Specimens of *Polygordius* are widespread in benthic assemblages around North America, where they prefer coarse sediments, but no species have been named in California and Oregon. See Westheide 1990 (general account); Ramey et al. 2006, JMBA 86: 1025–1034 (morphology, n.sp. NE U.S.).

PROTODRILIDAE

Protodrilids are minute interstitial polychaetes that have a pair of flexible palps, distinct from the prostomium. Their bodies are slender, elongate, and range from 2–15 mm long and have

21–77 segments. The palps have internal canals that insert posterior to the brain and two nuchal organs (statocysts), and eyes may be present or absent (plate 135I). The ventral pharynx has a ventral tonguelike extension. The body is segmented, but parapodia and setae are absent. Two pygidial lobes are present, which serve to attach the worms to sand grains. A ventral band of cilia extends along the entire body, and rings or patches of cilia may also be present on the prostomium and other parts of the body. Protodrilids are found in clean sand in intertidal or shallow subtidal areas. It is believed that protodrilids feed on microflora on the surface of sand grains. No species have been reported from California or Oregon; *Protodrilus* has been reported widely elsewhere. See Nordheim 1989, Zool. Scr. 18: 245–268 (systematics, reproductive biology, development); Westheide, 1990 (general account).

*PROTODRILOIDIDAE

Protodriloidids are minute, interstitial worms with slender, flattened bodies that measure up to 13 mm long and have up to 50 segments. They were formerly included in the Protodrilidae but differ in having a pair of palps that are actually extensions of the prostomium, instead of being distinct. Two nuchal organs are present; eyes are absent. The ventral pharynx lacks the ventral tonguelike extension found in Protodrilidae. Parapodia are absent, but setae are present as bidentate hooks. Two species, *Protodrioides symbioticus* (Giard, 1904) and *P. chaetifer* (Remane, 1926) are known but are not reported from California. See Purschke and Jouin 1988, J. Zool. 215: 405–432 (anatomy, ultrastructure of the foregut, phylogeny); Westheide 1990 (general account); Hartmann-Schröder 1996 (descriptions of species).

*PSAMMODRILIDAE

The psammodrilids, represented by a single genus, *Psammodrilus*, include small, interstitial grublike polychaetes that live in coarse intertidal and subtidal sandy sediments. Superficially, they resemble juvenile enteropneusts in that they have an elongated prostomium that is rounded anteriorly and distinctly separated from a peristomium that is formed by two rings, the first of which appears like an enteropneust collar. Antennae and palps are absent; nuchal organs are present or absent. The thoracic segments have slender, elongate notopodial lobes that superficially resemble gills (plate 135M); aciculae are present. The abdominal segments bear multidentate uncini. Pygidial cirri are absent. Sensory cilia are found all over the body and are especially long on the pygidium. No psammodrilids have been reported from California and Oregon. Hobson (1971, Proc. Biol. Soc. Wash. 84: 245–252) recorded *Psammodrilus balanoglossoides* Swedmark, 1953, the type species, from shallow sandy sediments in Cape Cod Bay. Additional specimens were recently collected from Massachusetts Bay in about 30 m. The position of *Psammodrilus* among other polychaetes is not understood. See Swedmark 1955, Arch. Zool. Exp. Gen. 92: 141–219 (morphology and development).

SACCOCIRRIDAE

The saccocirrids are slender, elongate worms with a pair of anterior palps, segments with reduced parapodia, and setae

(plate 135J, 135L). They live in coarse intertidal and subtidal sandy sediments. Individual specimens range up to about 20 mm long and have upwards of 200 segments. The prostomial palps (tentacles) have internal canals that connect posterior to the brain; two nuchal organs and a pair of eyes are present. Parapodia are uniramous, retractile, and bear several different types of furcate simple setae. The pygidium is bilobed and bears adhesive papillae (plate 135K–135L). Saccocirrids are reported to be sexually dimorphic and to have complex reproductive organs. One species, *Saccocirrus sonomacus* Martin, 1977, has been reported from Shell Beach, Sonoma County (see Martin 1977, Trans. Amer. Micros. Soc. 96: 97–103 [morphology]).

The Macrofaunal and Epifaunal Families

APHRODITIDAE

The aphroditids are slow-moving scale worms commonly called sea mice because of a feltlike dorsal surface that is formed by fine, silky setae and entrapped particles. Aphroditids, because of their size, do not occur in dense populations. They are adapted to burrowing in mud or creeping over the sea bottom, where they feed on detritus and sessile or slow-moving animals such as other polychaetes (Day 1967; Fauchald and Jumars 1979). Several species of *Aphrodita* occur in shallow subtidal areas off the California coast; they may be dredged and are occasionally washed ashore. There are two common genera: *Aphrodita* and *Laetmonice* (see Hartman 1968 for key to species; Blake 1995a).

KEY TO APHRODITIDAE

1. Dorsal felt well-developed, completely covering elytra with thick mat; harpoon-shaped notosetae lacking; neurosetae pointed or slightly curved, with or without lateral fringe of soft bristles or basal spur (plate 136A) . . . . . . *Aphrodita*
— Dorsal felt poorly developed, usually not covering elytra (plate 136B); harpoon-shaped notosetae present (plate 136C); neurosetae with lateral fringe of long, stiff bristles and basal spur . . . . *Laetmonice*

LIST OF SPECIES

Aphrodita armifera Moore, 1910. Dredged on silty or rocky bottoms.

Aphrodita castanea Moore, 1910. Dredged on shallow bottoms.

Aphrodita parva Moore, 1905. Dredged on shallow bottoms. The species may be a juvenile of *A. japonica* Marenzeller, 1879. See Blake (1995a).

Aphrodita refulgida Moore, 1910. Dredged on shallow bottoms. Narchi 1969, Veliger 12: 43–52, collected this species from off Tomales Bay and studied a bivalve, *Neaeromya rugifera*, that is attached to the ventral surface of the worm by its byssal threads.

POLYNOIDAE

The family Polynoidae is the largest and most commonly encountered family of scale worms. The approximately 700

* = Not in key.

species in this family are characterized by dorsoventrally flattened bodies, simple setae in both notopodial and neuropodial fascicles, and scales alternating with the dorsal cirri down much of the length of the body. Most species are carnivorous or omnivorous, feeding on small invertebrates, plant fragments, and detritus. Polynoids are slow moving and normally creep along the bottom, hiding in crevices, under rocks, and in algal holdfasts. The dorsum and the elytra are often pigmented with a variety of patterns and colors matching the general background. In addition, the elytral surface is sometimes covered with detritus and epiphytes, making the specimens difficult to detect. A number of polynoids are commensal with other organisms, predominantly echinoderms, mollusks, echiurans, or other polychaetes. In many of the commensal species, the elytra and notopodia are reduced in size with the notosetae fewer in number or absent altogether. Many of these commensal species are pigmented similar to the host organism. Although a few species become quite large (up to 25 cm), the majority of polynoids are only a few centimeters in length. Abbot and Reish (1980) provide detailed summaries of the biology and ecology of species of *Arctonoe*, *Halosydna*, *Harmothoe*, *Hesperonoe*, and *Lepidasthenia*. The most recent review of California Polynoidae is by Ruff (1995).

KEY TO POLYNOIDAE

1. With 12 pairs of elytra; lateral antennae inserted terminally on anterior prolongations of the prostomium. . . . .
. . . . *Lepidonotus squamatus*
— With 15 or more pairs of elytra; lateral antennae inserted ventral to the medial antenna (plates 137A, 138A) 2
2. (*Note three choices*) With 18 pairs of elytra . . . .
. . . . *Halosydna brevisetosa*
— With 15 pairs of elytra 6
— With 20 or more pairs of elytra 3
3. Elytra cover at least two-thirds of the dorsum on anterior segments (plate 137A), those on posterior segments noticeably smaller than those on anterior segments; dorsal cirri and antennae club- or pear-shaped; neuropodial setae falcate, sides smooth (plate 137B) . . . . *Arctonoe* 5
— Elytra reduced in size throughout, those in posterior segments not noticeably smaller than those in anterior segments; dorsal cirri and antennae cirriform (plate 139A, 139C); neuropodial setae serrated (plate 137E) . . . .
. . . . *Lepidasthenia* 4
4. Notopodia short, elytra not covering dorsal midline, ventral neuropodial margin smooth (plate 139B) . . . .
. . . . *Lepidasthenia berkeleyae*
— Notopodia elongate, elytra extending across dorsum; ventral neuropodial margin with row of papillae (plate 139D) . . . . *Lepidasthenia longicirrata*
5. (*Note three choices*) Elytra strongly frilled or folded at external margin (plate 137D); in ambulacral grooves of starfish. . . . . *Arctonoe fragilis*
— Elytral margin smooth; broad, transverse, dark band across anterior segments (plate 137A); in branchial grooves of mollusks, with other invertebrates or free-living . . . .
. . . . *Arctonoe vittata*
— Elytral margin slightly undulate (plate 137C); commensal with holothurians . . . . *Arctonoe pulchra*
6. Notopodial and neuropodial setae each of one kind . . . .
. . . . 8
— Notopodial and neuropodial setae each of two kinds, one stout and the other slender . . . . *Hesperonoe* 7

7. Elytral surface smooth or with very few minute, low papillae (plate 138I); commensal with *Urechis*. . . . .
. . . . *Hesperonoe adventor*
— Elytra with numerous scattered, low papillae (plate 138J); commensal with *Neotrypaea* . . . . . . *Hesperonoe complanata*
8. Some neuropodial setae with bifid tips (plate 138C). . . . .
. . . . 9
— Neuropodial setae entire at the distal end (plate 138H); elytra conspicuously covered with furcated spines . . . . . .
. . . . *Eunoe senta*
9. Anterior eyes far forward, under the prostomial peaks (plate 138F); elytral margin lightly fringed (plate 138G) . . . .
. . . . *Harmothoe imbricata*
— Anterior eyes near the middle of the prostomium, clearly visible dorsally (plate 138A); elytra smooth or strongly fringed 10
10. Elytra strongly fringed along outer margins and with large, hexagonal areas adorned with characteristic spines (plate 138E). . . . . *Harmothoe hirsuta*
— Elytra smooth along outer margins, fringe absent (plate 138B, 138D) . . . . *Malmgreniella* 11
11. Elytra with characteristic dark patch along inner border (plate 138B) . . . . *Malmgreniella macginitiei*
— Elytra with reticulate surface (plate 138D). . . . .
. . . . *Malmgreniella nigralba*

LIST OF SPECIES

See Hartman 1939, 1968; Pettibone 1953; Ruff 1995.

Arctonoe fragilis (Baird, 1863). Commensal in ambulacral grooves of asteroids. This species is characterized by having folded posterior margins on the elytra and vestigial ventral cirri. See Davenport and Hickock 1951, Biol. Bull. 100: 71–83 (physiology of commensalism); Hickock and Davenport 1957, Biol. Bull. 113: 397–406 (behavior and commensalism); Abbott and Reish 1980 (review of ecology, commensalism).

Arctonoe pulchra (Johnson, 1897). A common species commensal with asteroids, holothurians, mollusks, and terebellid polychaetes. The elytra and body of *A. pulchra* are colorful and usually match the coloration found on the host. A dark pigment spot occurs on each elytron, but this is variable. See Davenport and Hickock 1951, Biol. Bull. 100: 71–83 (physiology of commensalism); Dimock and Davenport 1971, Biol. Bull. 141: 472–484 (behavior and host recognition); Abbott and Reish 1980 (review of ecology, commensalism).

Arctonoe vittata (Grube, 1855). Commensal with asteroids, mollusks, or free-living. Common hosts in California include *Cryptochiton stelleri*, *Dermasterias imbricata*, and *Diodora aspera*. The worms often match their hosts' coloration. A dark pigment band typically occurs across the dorsum of segment 8 but is sometimes absent after preservation. See Davenport and Hickock 1951, Biol. Bull. 100: 71–83 (physiology of commensalism); Hickock and Davenport 1957, Biol. Bull. 113: 397–406 (behavior and commensalism); Gerber and Stout 1968, Physiol. Zool. 41: 169–179 (commensalism); Webster 1968, Veliger 11: 121–125 (commensal with *Cryptochiton stelleri*); Wagner, Phillips, Standing, and Hand 1979, J. Exp. Mar. Biol. Ecol. 39: 205–210 (seastar attracted to polychaete); Abbott and Reish 1980 (review of ecology, commensalism).

Eunoe senta (Moore, 1902). A widespread but generally rare species occurring around the North Pacific rim to southern

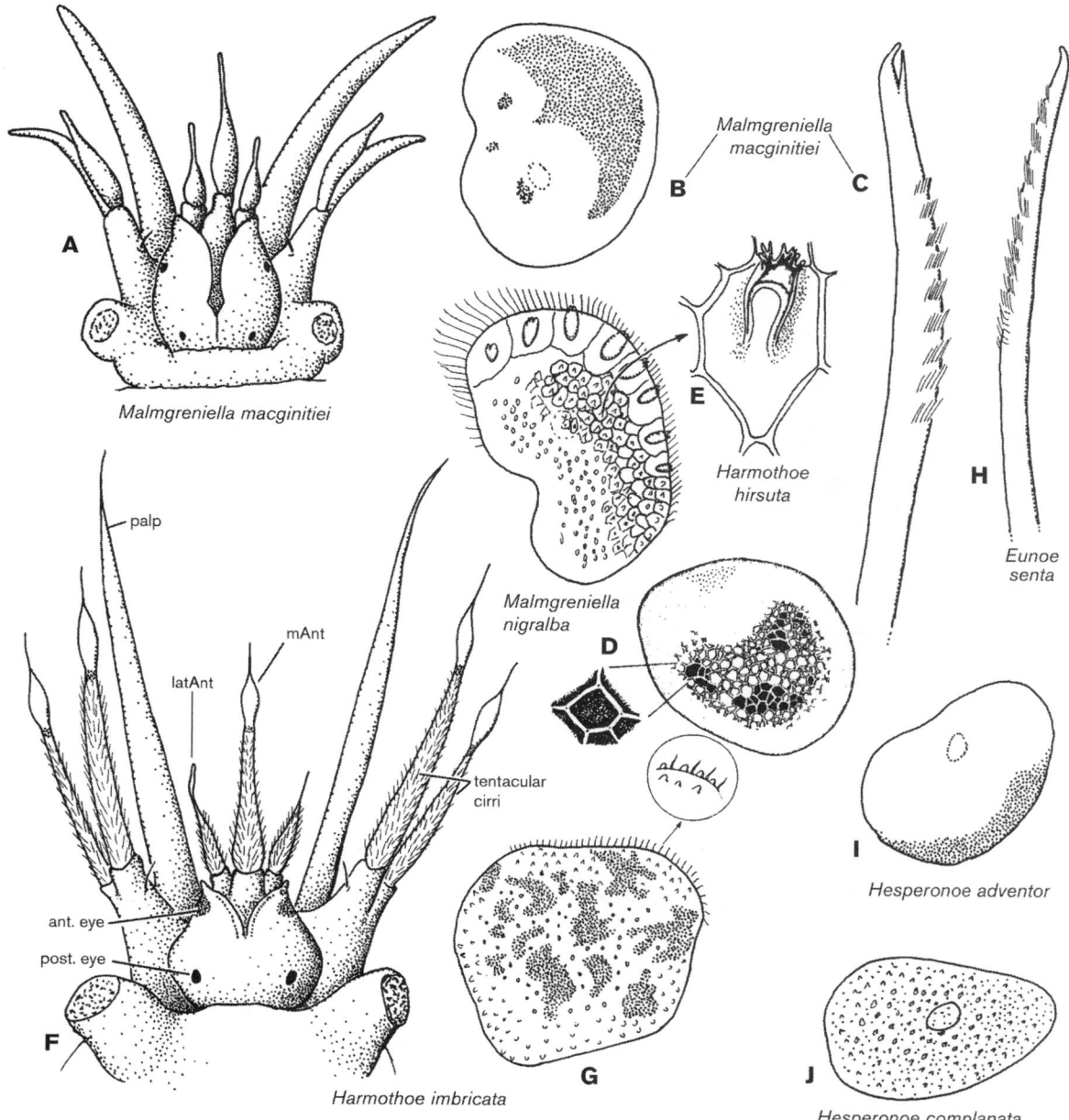

PLATE 138 Polychaeta. Polynoidae: A, *Malmgreniella macginitiei*, elytra omitted; B, elytron; C, neuroseta; D, *Malmgreniella nigralba*, elytron, inset shows detail of surface; E, *Harmothoe hirsuta* elytron, inset shows detail of tubercle; F, *Harmothoe imbricata*; G, elytron, inset shows detail of margin; H, *Eunoe senta*, neuroseta; I, *Hesperonoe adventor*, elytron; J, *H. complanata*, elytron. (D, after Ruff; E, after Johnson; H, after Hartman; I, after Skogsberg).

California. The species occurs on mixed sand and gravel bottoms from 5 m to shelf depths. The body has a shaggy appearance due to the spiny elytra, papillose cirri, and dense fascicles of notosetae. See Ruff 1995.

Halosydna brevisetosa Kinberg, 1855 (=*H. johnsoni* [Darboux, 1899]). The most common scale worm in central and northern California; in rocky habitats; commensal with terebellids or mollusks as well as free-living. See Gaffney 1973, Syst. Zool. 22: 171–175 (intraspefic variation in setae); Blake 1975a (larvae); Ajeska and Nybakken 1976, Veliger 19: 19–26 (commensal with *Milibe*); Abbott and Morris 1980 (review of ecology, reproduction, commensalism).

Harmothoe hirsuta Johnson, 1897. Probably limited to southern California; intertidal to 98 m. The setae, cirri, and elytra of *H. hirsuta* are usually coated with sediment and debris, giving the organism a very shaggy appearance. See Ruff 1995.

Harmothoe imbricata (Linnaeus, 1767). One of the most common scale worms in California. The species utilizes a wide variety of habitats. It is found in the intertidal under rocks, in eelgrass beds, subtidally on rocky, muddy, or sandy substrates, in kelp holdfasts, mussel beds, and old *Sabellaria* reefs. It is found both free-living and as a commensal with echinoderms and other polychaetes. Blake (1975a) recorded egg diameters

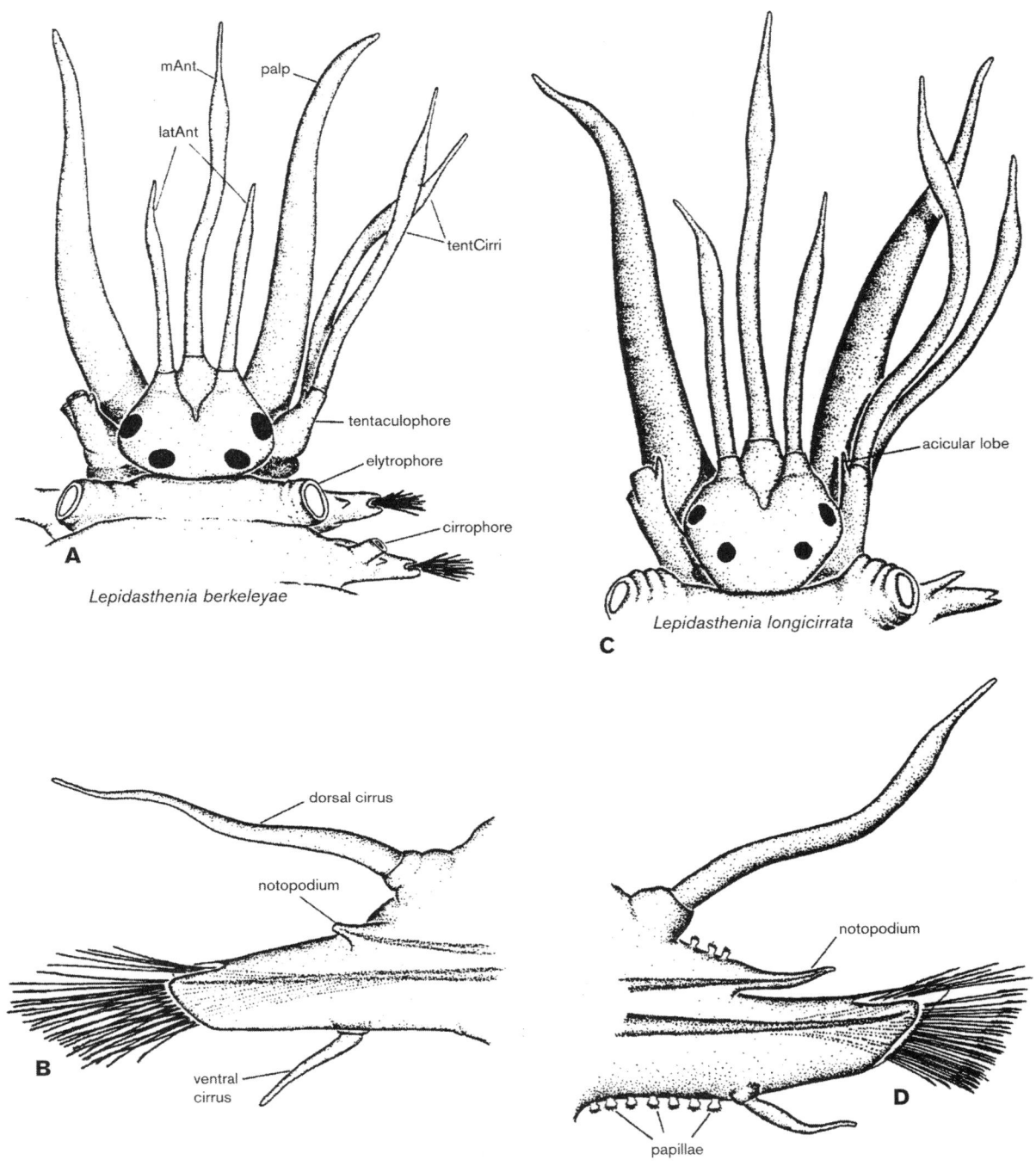

PLATE 139 Polychaeta. Polynoidae: A, *Lepidasthenia berkeleyae*; B, median cirrigerous parapodium, anterior view; C, *Lepidasthenia longicirrata*; D, anterior cirrigerous parapodium, posterior view (all after Ruff).

of 120–123 μm in specimens from Tomales Bay. The eggs are brooded under the elytra throughout much of the year, and after hatching the larvae have a prolonged pelagic life before settling and metamorphosis. In other oceans, however, *H. imbricata* has been found to exhibit a full range of developmental types from direct development under the elytra to free-spawning into seawater (Blake 1975a). This plasticity in mode of reproduction coupled with the diverse habitats in which the species is able to live contribute toward making *H. imbricata*

one of the most widely distributed species of polynoids. See also Daly 1972, Mar. Biol. 12:53–56 (maturation and breeding biology in England); Cazaux 1968, Arch. Zool. Exper. Gen. 109: 477–543 (larval development); Abbott and Reish 1980 (review of ecology, reproduction).

Hesperonoe adventor (Skogsberg, 1928). A gray-green species found in the burrows of the echiuran *Urechis caupo* and *Upogebia pugettensis*. Abbott and Reish 1980 (review of ecology, commensalism).

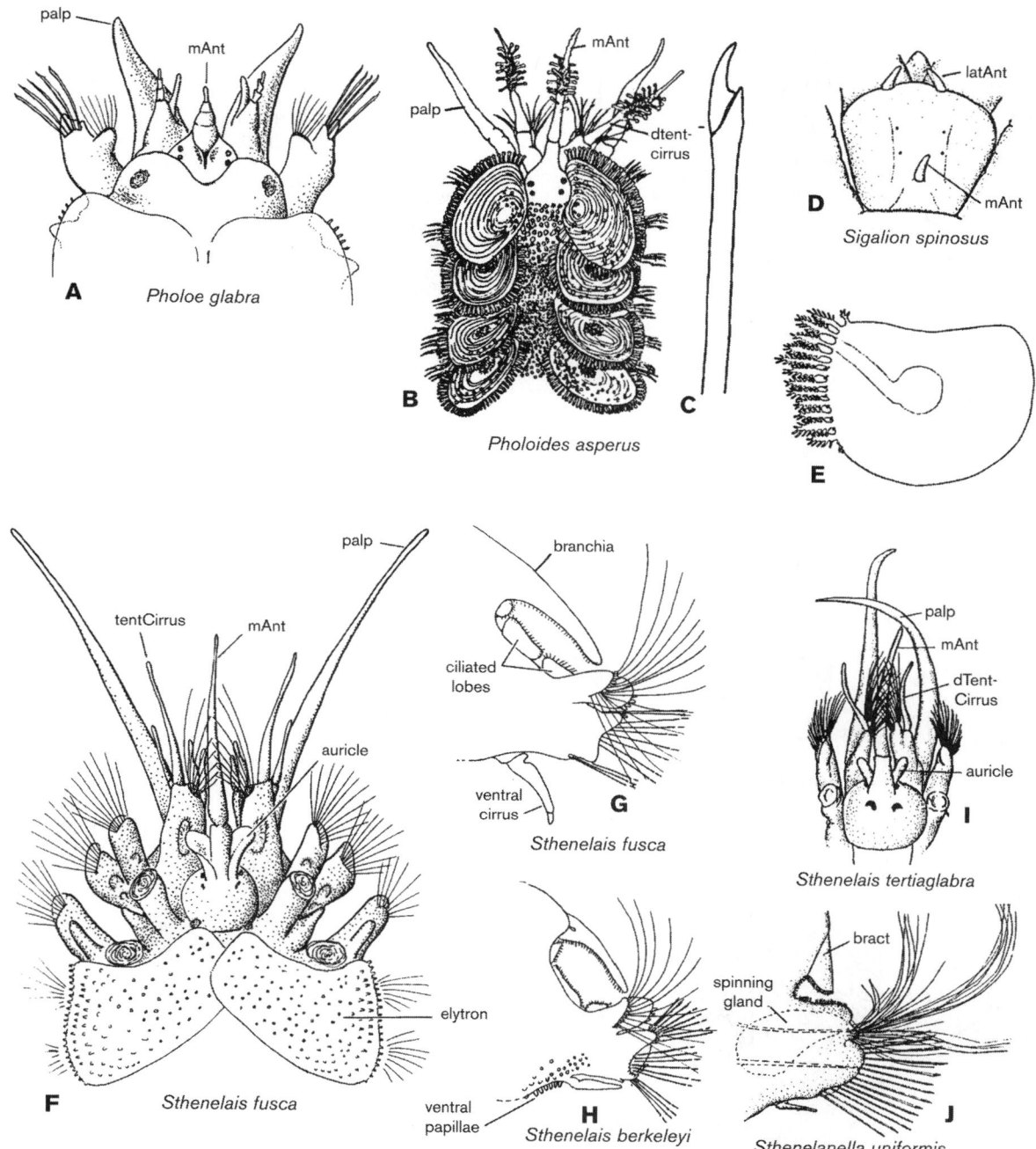

PLATE 140 Polychaeta. Pholoidae: A, *Pholoe glabra*; B, *Pholoides asperus*; C, neuropodial compound falciger; Sigalionidae: D, *Sigalion spinosus*, prostomium, dorsal view; E, elytron; F, *Sthenelais fusca*; G, anterior parapodium, posterior view; H, *Sthenelais berkeleyi*, anterior parapodium, posterior view; I, *Sthenelais tertiaglabra*, anterior end, dorsal view; J, *Sthenelanella uniformis*, anterior parapodium (A, D–E, after Hartman; B, after Johnson; C, G–H, after Pettibone; I, after Moore).

Hesperonoe complanata (Johnson, 1901). A bright yellow-orange species commensal with the ghost shrimp *Neotrypaea californiensis* and with *Urechis caupo*.

Lepidasthenia berkeleyae Pettibone, 1948 (=*L. interrupta* of Blake, 1975b [previous edition], not Marenzeller, 1902). A commensal in the thick mud tubes of the maldanids *Praxillella pacifica* and *Maldanella robusta*; in shallow depths from <50 m to the upper slope. See Ruff 1995.

Lepidasthenia longicirrata Berkeley, 1923. Vancouver Island to southern California. *L. longicirrata* is normally free-living in muds, among rocks, or in shelly substrata; often associated with parchmentlike, sand and shell-covered tubes that the worm apparently constructs; reported as a commensal with maldanid polychaetes; intertidal to 330 m. See Abbott and Reish 1980 (distribution); Ruff 1995.

Lepidonotus squamatus (Linnaeus, 1767). Widely distributed in subarctic and boreal waters from the intertidal to lower slope depths from around the north Pacific rim to northern Mexico; the Atlantic from the east coast of the United States, through Greenland and Iceland, down to the Mediterranean; a free-living

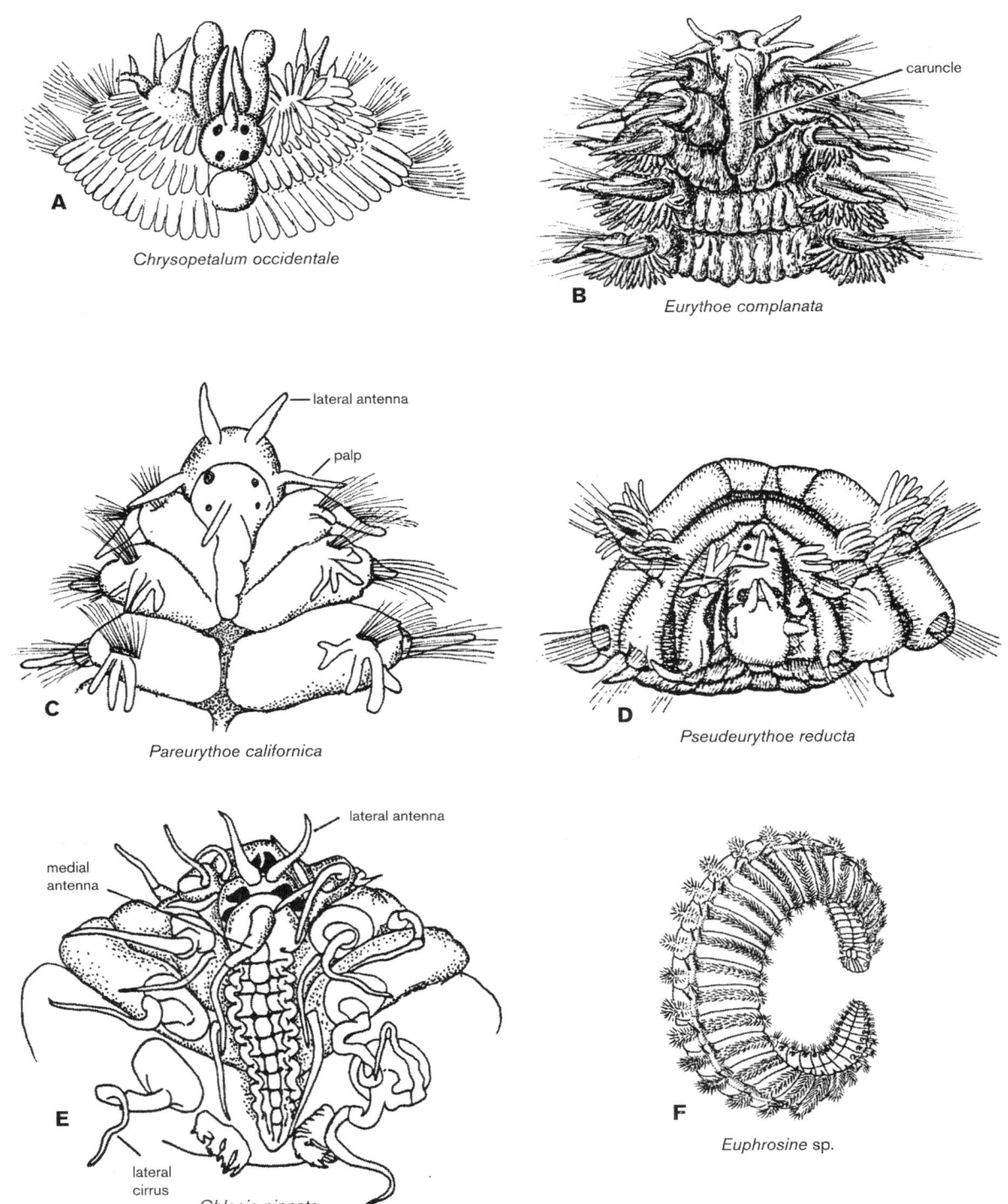

A
Chrysopetalum occidentale

B
caruncle
Eurythoe complanata

C
lateral antenna
palp
Pareurythoe californica

D
Pseudeurythoe reducta

E
medial antenna
lateral antenna
lateral cirrus
Chloeia pinnata

F
Euphrosine sp.

PLATE 141 Polychaeta. Chrysopetalidae: A, *Chrysopetalum occidentale*; Amphinomidae: B, *Eurythoe complanata*; C, *Pareurythoe californica*; D, *Pseudeurythoe reducta*; E, *Chloeia pinnata*; Euphrosinidae: F, *Euphrosine* sp. (A, after Imajima and Hartman; B, after Hartman; C, after Johnson; D, after Kudenov and Blake; E, after Kudenov; F, after McIntosh).

polynoid found on a variety of habitats including kelp holdfasts, under rocks, and among mussel shells or barnacles. The dorsal surface is often covered with marine growths and debris, imparting a scruffy appearance to the worm. When disturbed, it characteristically curls into a tight ball for protection. See Ruff 1995.

Malmgreniella macginitiei Pettibone, 1993 (misidentified as *Harmothoe lunulata* [delle Chiaje, 1841] in previous edition). Free-living or commensal in the burrows of the ghost shrimp *Neotrypaea californiensis*; also in the tubes of the maldanid *Axiothella rubrocincta* and on the arms of the ophiuroid *Amphiodia urtica*. See Ruff 1995.

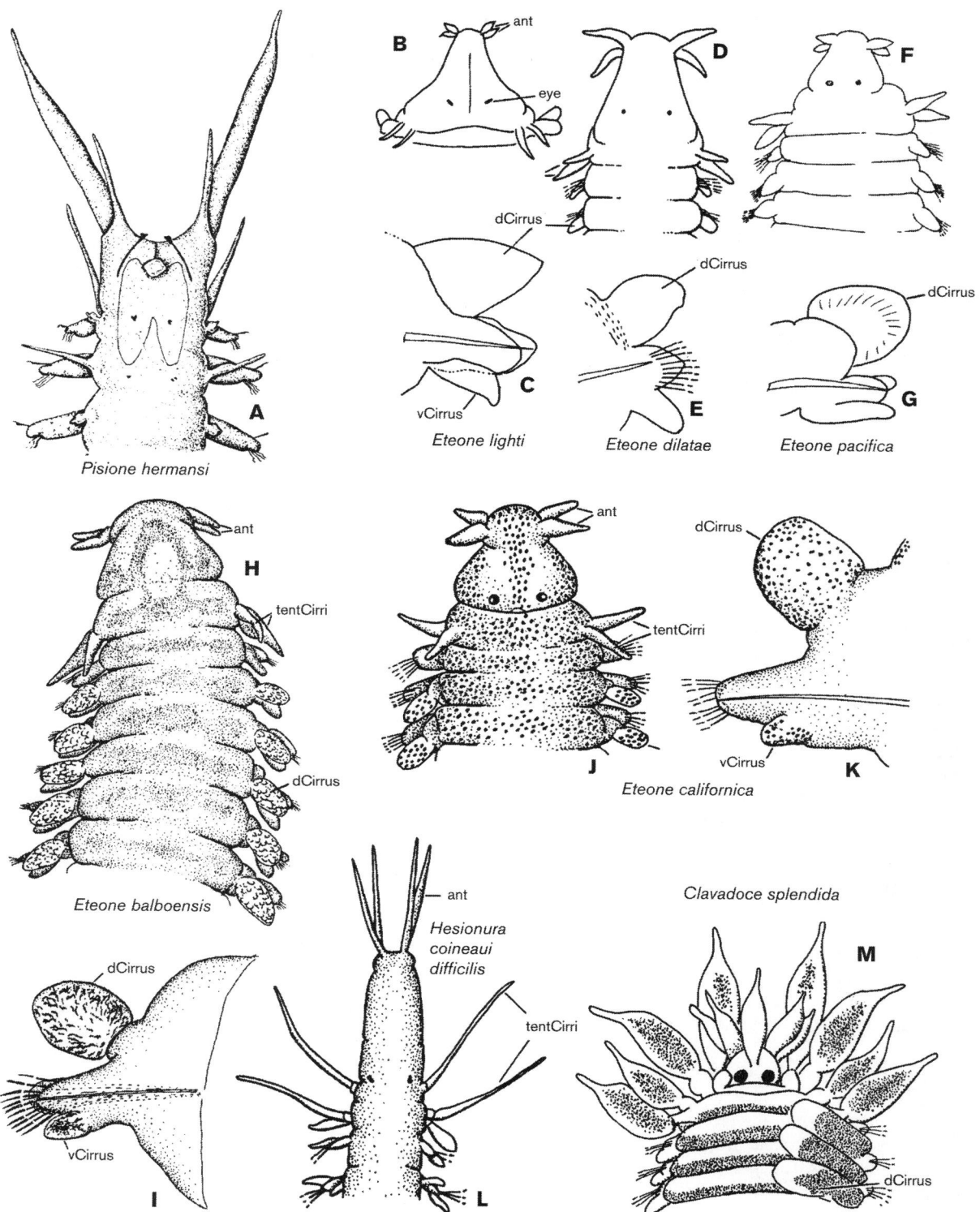

PLATE 142 Polychaeta. Pisionidae: A, *Pisione hermansi*, anterior end, dorsal view; Phyllodocidae: B, *Eteone lighti*; C, parapodium; D, *Eteone dilatae*; E, parapodium; F, *Eteone pacifica*, G, parapodium; H, *Eteone balboensis*; I, parapodium; J, *Eteone californica*; K, parapodium; L, *Hesionura coineaui difficilis*; M, *Clavadoce splendida* (A after Gradek 1991; B–E, G, after Hartman; rest after Blake various).

Malmgreniella nigralba (Berkeley, 1923). In the vertical burrows of the holothurian *Leptosynapta clarki*; prefers coarse gravelly sands from the low intertidal to approximately 100 m. The elytra of this species are distinctly reticulated, although this feature is not always obvious if the black pigmentation has faded. See Ruff 1995.

PHOLOIDAE

The Pholoidae are small, crawling, carnivorous scale worms that are found in algal holdfasts, under rocks, and in crevices among oysters and mussels, and creeping over the surface of muddy sediments littered with shell hash. Some species are interstitial. They are widely distributed and occur from the intertidal zone to great depths. Pholoids have typically been included in the Sigalionidae. Both families include scale worms with compound setae. However, the bodies of pholoids are short, broad, and with relatively few segments; sigalionids have bodies that are long, slender and have numerous segments. Further, pholoids lack branchiae, have the tentaculophores of segment I located medial to the palps, and the blades of the compound setae are short and falcate; in contrast, sigalionids have branchiae, the tentaculophores are located dorsal to the palps, and the blades of the compound setae are typically long and often multiarticulated. Species of *Pholoe* were included in the Sigalionidae in the previous edition.

KEY TO PHOLOIDAE

1. Prostomium oval, bilobed, with medial antenna on ceratophore in anterior notch (plate 140A); with or without lateral antennae; tentaculophores astigerous, with dorsal and ventral tentacular cirri; elytra on segments 2, 4, 5, 7, and continuing on alternate segments to segment 23, then on every segment to end of body; elytra delicate, with short border and surface papillae . . . . *Pholoe*
— Prostomium subrectangular, with medial antenna on ceratophore on anterior border; without lateral antennae (plate 140B); tentaculophores with notosetae and single tentacular cirrus; elytra on segments 2, 4, 5, 7, and continuing on alternate segments to end of body; elytra thick, with concentric rings and numerous long border papillae (plate 140B). . . . . *Pholoides*

LIST OF SPECIES

See Blake 1995b.

Pholoe glabra Hartman, 1961. Misidentified in part as *P. tuberculata* Southern, 1914 in the previous edition. Central California to Mexico on intertidal beaches, but more common in offshore greenish muds. The species is characterized by having a short, triangular facial tubercle, short papillae on the elytra that have several folds or articles, pseudoarticulated joints on the tips of the medial antenna and the dorsal and ventral tentacular cirri, and very few papillae on the surface of the body. Further, the serrations on the blades of the compound setae are very distinct; in related species, these are either absent or very difficult to discern.

Pholoides asperus (Johnson, 1897) (as *Peisidice aspera* in Peisidicidae in the previous edition). Best known from Monterey, where it is found at the low-tide mark crawling over stones or hiding in crevices. The body is short (to 12 mm); prostomium

* = Not in key.

lacks lateral antennae; neuropodia have composite setae; and the elytra have a heavily fimbriated margin. See Johnson 1897, Proc. Calif. Acad. Sci. 1: 153–198; Blake 1975a (larvae, as *Pholoe minuta*), 1995b.

SIGALIONIDAE

The Sigalionidae includes highly active scale worms that have long, narrow bodies with numerous segments and pairs of elytra. Most species are large, carnivorous polychaetes that are also able to burrow rapidly into sediments. These scale worms are widely distributed, with many species being relatively common and familiar. Sigalionids occur from the intertidal to the deep sea. Species of *Pholoe*, included in the Sigalionidae in the previous edition, are now referred to the Pholoidae (see above).

KEY TO SIGALIONIDAE

1. Lateral antennae located on prostomium; antennae small, without ceratophores (plate 140D); elytra with bipinnately branched lateral filaments (plate 140E) . . . . *Sigalion spinosus*
— Lateral antennae, when present, fused to tentacular segment; medial antenna with large ceratophore (plate 140F, 140I) 2
2. Middle body segments with notopodial spinning glands producing long, threadlike setae extending far beyond parapodia (plate 140J) . . . . *Sthenelanella uniformis*
— Notopodial spinning glands absent 3
3. Middle group of compound neurosetae include stout, short-bladed, bidentate falcigers; lateral antennae present or absent 4
— Middle group of compound neurosetae without short-bladed falcigers; lateral antennae absent . . . . *Sthenelais verruculosa*
4. Ventral surface of body and ventral sides of parapodia densely papillated (plate 140H) . . . . . . *Sthenelais berkeleyi*
— Ventral surface smooth or with very few, sparse papillae (plate 140G). 5
5. Stout compound neurosetae of middle segments numbering five or more; first pair of elytra not pigmented, subsequent ones pigmented in various patterns . . . . *Sthenelais fusca*
— Stout compound neurosetae of middle segments numbering only one to two; all elytra semitranslucent, lacking pigmentation, but may be encrusted with rusty colored sediment particles . . . . *Sthenelais tertiaglabra*

LIST OF SPECIES

See Blake 1995c.

Sthenelais fusca Johnson, 1897. An intertidal to shallow-water species, common under rocks, among algal holdfasts, and among rhizomes of *Phyllospadix*. The species is large, up to 130 mm long, 5–7 mm wide, and with about 200 segments. *S. fusca* is distinguished from other eastern Pacific congeners in having many heavy, nonarticulated bifid falcigers in the neuropodia, while at the same time having a smooth, nonpapillated ventral surface. Specimens from Bodega Harbor exhibit a rather elegant body coloration in which the unpigmented first pair of elytra contrasts with the following elytra that are heavily colored with mottled brown patterns. See Pettibone 1971, J. Fish. Res. Bd. Can. 28: 1393–1401.

Sthenelais berkeleyi Pettibone, 1971. Intertidal to shelf depths in sediments with gravel or with shells mixed with mud. A large species, up to 400 mm long, 12 mm wide, and about 270 segments (Pettibone 1971). The details that distinguish *Sthenelais berkeleyi* from the closely related *S. fusca* were defined by Pettibone (1971). The two species are similar in having many heavy, uniarticulated, and bifid falcigerous neurosetae. *S. berkeleyi*, however, has a thickly papillated ventral surface instead of one that is smooth. Further, the bilobed posterior bracts of the neuropodia in *S. berkeleyi* are provided with numerous stylodes on both the upper and lower parts, whereas in *S. fusca*, there are a few stylodes in the upper part and none on the lower part. See Pettibone 1971, J. Fish. Res. Bd. Can. 28: 1393–1401; Pernet 2000, Invert. Biol. 119: 147–151 (setae of the first segment are arranged in the shape of an anterior-dorsal directed tube and thus act as a respiratory snorkel when the worm is buried).

Sthenelais verruculosa Johnson, 1897. In shallow subtidal habitats out to about 100 m. A moderate to large species, at least up to 75 mm long; dull yellow to light tan in color with the elytra semitransparent, with the microtubercles of posterior segments sometimes appearing as orange spots on elytra. *S. verruculosa* differs from *S. fusca, S. berkeleyi,* and *S. tertiaglabra* in completely lacking the stout, short-bladed, and bidentate falcigers in middle and posterior neuropodia. *S. verruculosa* is most readily confused with *S. tertiaglabra,* the latter typically having a rust-colored flocculent material surrounding the microtubercles on the elytra. This substance is lacking in *S. verruculosa.*

Sthenelais tertiaglabra Moore, 1910. A deeper water, offshore species.

Sthenelanella uniformis Moore, 1910. Intertidal to 113 m. *S. uniformis* is readily identified by the mottled red spots on the anterior pairs of elytra and by the long, silky, threadlike notosetae emerging from the notopodial spinning glands. *S. uniformis* inhabits long tubes with thickened walls that are covered with mud or sand and reinforced by the long felt or setal threads produced from spinning glands.

Sigalion spinosus (Hartman, 1939). Shallow subtidal to 119 m in sediments of sand and silt. A large species, up to 100 mm long, 5 mm wide, with about 115 segments. The body is relatively colorless; the elytra are translucent. The bipinnately branched filaments on the lateral margin of the elytra are distinctive.

CHRYSOPETALIDAE

The Chrysopetalidae are fragile, flattened polychaetes that have notosetae modified into large, flattened paleae (plate 141A), providing a golden or coppery appearance to the dorsum. Chryspetalids are active scavengers and carnivores. Most species are small and live in crevices in rock, coral, shell fragments, algal holdfasts, and other cryptic habitats along the shore. Approximately 11 genera and about 45 species are known. Chrysopetalids range from intertidal habitats to the deep sea. Two species, *Chrysopetalum occidentale* and *Paleanotus bellis,* occur commonly along the California coasts.

KEY TO CHRYSOPETALIDAE

1. Paleae of one kind narrow, not longitudinally serrated (plate 141A); color in life pale rust to yellowish . . . . *Chrysopetalum occidentale*

— Paleae of two kinds: dorsal broad ones, longitudinally serrated, and laterally directed narrower ones, not serrated; color in life glistening white or greenish . . . . *Paleanotus bellis*

LIST OF SPECIES

See Hartman 1940.

Chrysopetalum occidentale Johnson, 1897. Rocky habitats; more southern in distribution.

Paleanotus bellis (Johnson, 1897). A small species (to about 3 mm), common among barnacles, bryozoans, and sponges on pilings and elsewhere. See Banse and Hobson 1968; Blake 1975a (larval development).

PISIONIDAE

The pisionids are a rare group of polychaetes occurring in soft to loose substrata. Species of the genus *Pisione* are the most familiar representatives of the family and are noteworthy for their reduced prostomium, lack of antennae, and the possession of a pair of acicula that project forward at an oblique angle in front of the mouth (plate 142A). The proboscis is eversible with marginal papillae and four chitinous jaws. The peristomium is formed of three segments and bears a pair of long, anteriorly directed palps with basal sheaths and two pairs of smaller, biarticulate cirri (plate 142A). Parapodia are uniramous, bearing a superior simple seta and several inferior composite falcigers.

Pisione hermansi Gradek, 1991. Sonoma County, coarse intertidal sands. See Gradek 1991, Trans. Am. Microsc. Soc. 110: 212–225.

AMPHINOMIDAE

Amphinomids are often brilliantly pigmented, sometimes with characteristic color patterns; this is particularly true for *Chloeia* and *Notopygos* species. The bodies of some amphinomids, such as *Chloeia* and *Notopygos,* tend to be grublike and oval in cross-section for a limited number of segments, whereas *Amphinome, Eurythoe, Hermodice, Pareurythoe,* and similar forms are elongate and subrectangular in cross-section and have many more segments (Kudenov 1995a). The head is often hidden and usually bears a caruncle that takes on characteristic shapes. The parapodia bear tufts of stout setae and branched branchiae. In some species, the harpoonlike notosetae may break when handled, can lodge under the skin, and may release a neurotoxin and thus cause local irritation and infection. Such amphinomids are locally called "fire worms." According to Kudenov (1995a), 19 genera and approximately 130 species are known.

KEY TO AMPHINOMIDAE

1. Caruncle completely absent; neurosetae simple hooks. . . . . *Hipponoa gaudichaudi*

— Caruncle well-developed, sometimes with crests and folds (plate 141B–141E) 2

2. Body short, ovoid; first branchiae not larger than following ones; caruncle with numerous folds (plate 141E) . . . . *Chloeia pinnata*

— Body elongated, usually tapering at anterior and posterior ends 3

3. Caruncle small, inconspicuous, stretching maximally through three segments (plate 141C, 141D) 4

— Caruncle large, conspicuous, concealing much of the prostomium, with thick medial ridge, stretching through at least three segments (plate 141B) . . . .
. . . . *Eurythoe complanata*

4. Branchiae present on all segments from second or third; caruncle narrow, elongated, extending posteriorly to middle of second setigerous segment (plate 141C); up to 50 mm long . . . . *Pareurythoe californica*

— Branchiae limited to anterior part of body (plate 141D) . . . . *Pseudeurythoe reducta*

LIST OF SPECIES

See Hartman 1940; Kudenov 1995a.

Chloeia pinnata Moore, 1911. One of the most abundant polychaetes off the coasts of central and southern California, occurring in various sediment types (Kudenov 1995a).

Eurythoe complanata (Pallas, 1766). Under rocks; more common in southern California.

Hipponoa gaudichaudi Audouin and Milne Edwards, 1833. Commensal with floating colonies of the barnacle *Lepas*. See Kudenov 1977, Bull. So. Calif. Acad. Sci. 76: 85–90 (brooding behavior).

Pareurythoe californica (Johnson, 1897). California, intertidal under rocks. See Hartman 1968; Abbott and Reish 1980 (habitats and distribution).

Pseudeurythoe reducta Kudenov and Blake, 1985. Known from a single specimen collected in Elkhorn Slough in the high intertidal zone. The specimen occurred in fine sand where it apparently formed deep burrows into anoxic layers. See Kudenov and Blake 1985, Bull. So. Calif. Acad. Sci. 84: 38–40.

EUPHROSINIDAE

The Euphrosinidae have short, compact bodies, with parapodia forming transverse ridges bearing numerous rows of setae across much of the dorsum (plate 141F). Numerous branched branchiae occur on the dorsum and lateral sides of the body, often hidden by the numerous rows of setae. The prostomium is reduced to a narrow ridge that bears two short lateral antennae and a medial unpaired antenna that is located more posteriorly. The caruncle is well developed and formed of three lobes with paired lateral ciliated ridges. The setae are furcate and sometimes called ringent setae. In California, several species, such as *Euphrosine aurantiaca* Johnson, 1897, *E. arctia* Johnson, 1897, *E. bicirrata* Moore, 1905, and *E. hortensis* Moore, 1905, have been recorded (Hartman 1968; Abbott and Reish 1980; Kudenov 1995b). In life, they are bright red to orange and may occur in algal holdfasts or crevices of rocks. Typically subtidal, they are occasionally washed ashore. See Hartman (1940, 1968) and Kudenov (1995b) for keys to species.

PHYLLODOCIDAE

The phyllodocids are predatory and among the most active and common polychaetes found along the shore. They are most common and conspicuous in shallow-water habitats, especially associated with hard substrata, although some genera are typical of the soft sediments of mud flats and some are found in deep water. Phyllodocids are frequently brightly colored and may have diagnostic pigment patterns. Unfortunately, the pigments fade rapidly in preservatives. One of the most noticeable features of these worms is their production of copious amounts of mucus when disturbed. Their bodies are typically long and slender. The prostomium bears four frontal antennae and sometimes a fifth medial antenna or papilla; eyes may be present or absent; a pair of nuchal organs are present. The proboscis is eversible and usually covered with soft papillae; a few species have hard denticles on the proboscis, but jaws are lacking. Two, three, or four pairs of tentacular cirri may be present on the first one to three segments. The arrangement of these tentacular cirri, the fusion of tentaculate segments, and the presence or absence of setae on these same segments is of major importance at the generic level. A "tentacular formula" has been developed to summarize these characters in which 1 represents a tentacular cirrus; N represents a normal dorsal (D) or ventral (V) lamellar cirrus; and S or 0 indicate the presence or absence of setae. For example, the following formula is typical of the genus *Phyllodoce*: $(0\frac{1}{0} + 0\frac{1}{1}) + S\frac{1}{N}$. This formula describes a worm having three tentacular segments, the first of which has only one elongate tentacular cirrus; the second has no setae but has elongate dorsal and ventral tentacular cirri; and the third segment has setae, a dorsal tentacular cirrus, and a normal ventral cirrus. Parentheses around the formula for the first two segments indicate that those segments are fused. Parapodia are normally uniramous, although genera with subbiramous parapodia are known. Notopodia are represented by a short stalk to which a prominent foliose dorsal cirrus is attached; in genera with subbiramous parapodia, the notopodia have an internal acicula and sometimes a few emergent simple setae. Ventral cirri are smaller and less conspicuous. The shape and size of the dorsal cirrus is an important species-level character. Neurosetae are always compound (plate 143E), with the sculpturing of the shaft sometimes being diagnostic (Eibye-Jacobsen 1991). However, very high magnification (1000×) is required to observe these details. The classification of phyllodocids has changed since the previous edition of this book. Readers are referred to Pleijel (1991) for a review of phylogeny and classification of phyllodocids.

KEY TO PHYLLODOCIDAE

1. With one pair of tentacular cirri on first segment (plates 142L, 143A); total number of tentacular cirri 2–4 pairs . . .
. . . . 6

— With two pairs of tentacular cirri on first segment (plate 142B, 142D, 142F, 142H, 142J) . . . . *Eteone* 2

2. Prostomium wider than long 3

— Prostomium longer than wide, or about as wide as long . .
. . . . 4

3. First segment dorsally reduced (plate 142B); dorsal cirri pointed, nearly triangular in shape (plate 142C); ventral cirri distinctly pointed, directed ventrally . . . .
. . . . *Eteone lighti*

— First segment dorsally entire, well developed (plate 142H); dorsal cirri rounded, apically oval in shape (plate 142I); ventral cirri rounded, not directed ventrally; eyes absent; body and cirri lightly pigmented; ventral cirri small, with narrow basal attachment (plate 142I) . . . .
. . . . *Eteone balboensis*

4. Prostomium semicircular, as long as wide (plate 142J); dorsal cirri as wide as long, (plate 142K); body and cirri darkly pigmented . . . . *Eteone californica*

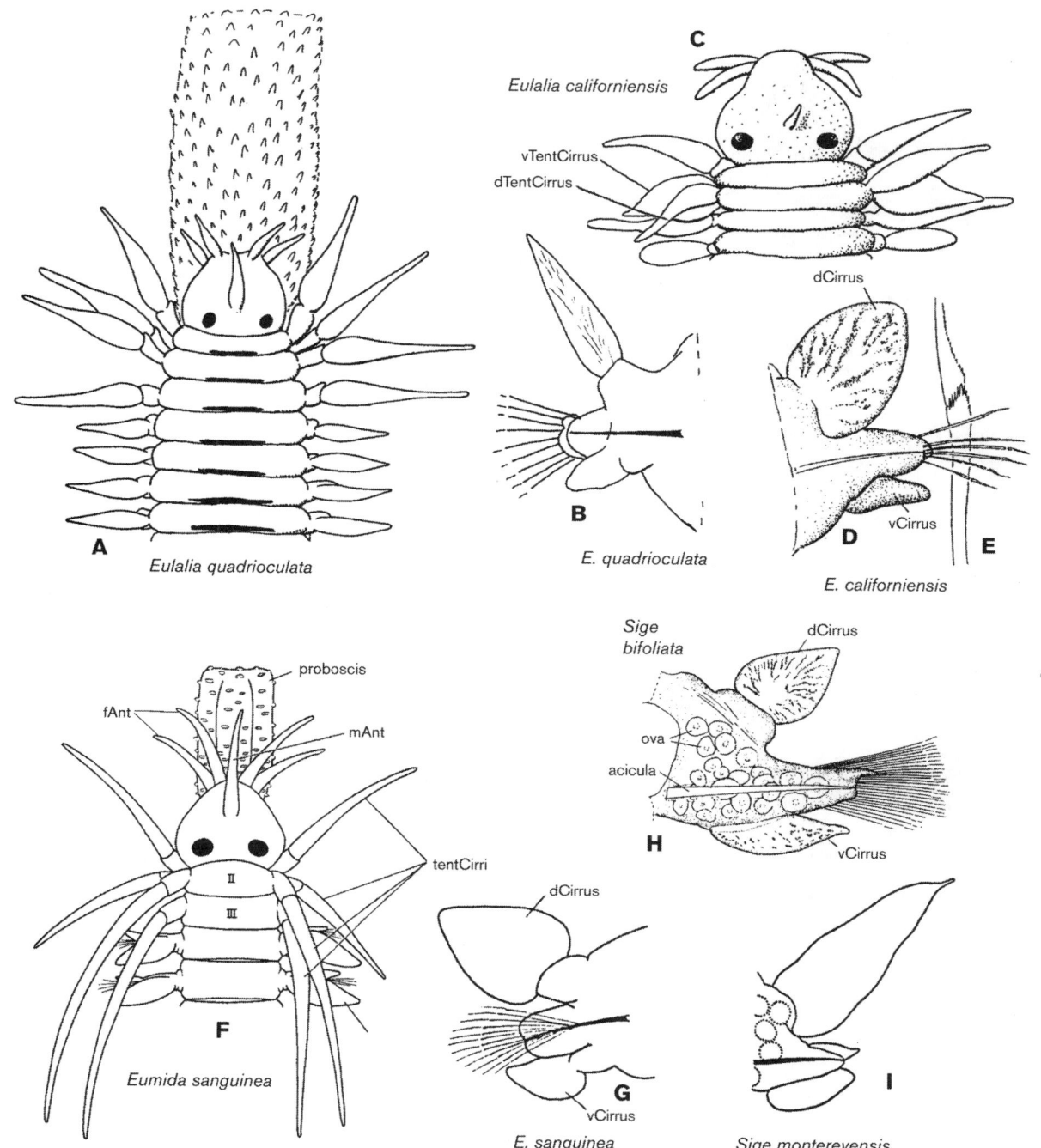

PLATE 143 Polychaeta. Phyllodocidae: A, *Eulalia quadrioculata*; B, parapodium; C, *Eulalia californiensis*; D, parapodium; E, seta; F, *Eumida sanguinea*; G, *E. sanguinea* parapodium; H, *Sige bifoliata*, parapodium; I, *Sige montereyenesis*, parapodium (F–G after Pettibone; rest after Blake various).

— Prostomium longer than wide (plate 142D, 142F); dorsal cirri longer than wide or wider than long; body and cirri either pale, lightly pigmented, or pigment in distinct patterns 5
5. Dorsal cirri subrectangular, slightly longer than wide (plate 142E); body pale green in life with numerous small brown spots; a smaller species, not exceeding 50 mm in length . . . . *Eteone dilatae*
— Dorsal cirri broadly rounded (plate 142G); body bright yellow in life with irregularly spaced black spots; a large

species, to more than 100 mm . . . . *Eteone pacifica*
6. With three pairs of tentacular cirri on first two segments; setae from segment 3 (plate 142L); tentacular formula: 0⅙ + 0⅓ + S⅚; prostomium longer than wide . . . .
. . . . *Hesionura coineaui difficilis*
— With three or four pairs of tentacular cirri on first three segments 7
7. Four frontal and one medial antenna present (plates 142M, 143A) 8
— Four frontal antennae present (plate 144A, 144F); nuchal

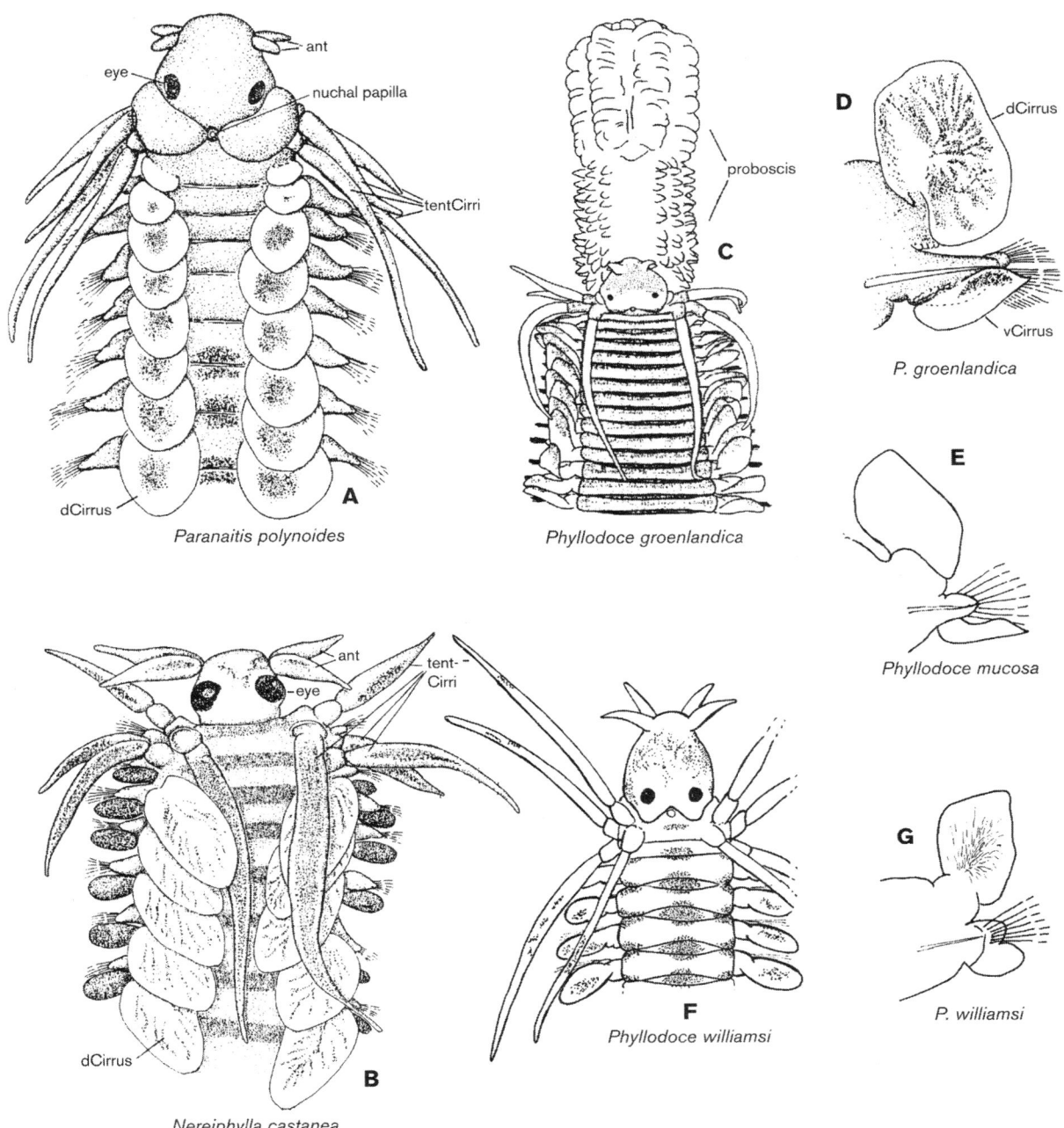

PLATE 144 Polychaeta. Phyllodocidae: A, *Paranaitis polynoides*; B, *Nereiphylla castanea*; C, *Phyllodoce groenlandica*; D, parapodium; E, *Phyllodoce mucosa*, parapodium; F, *Phyllodoce williamsi*; G, parapodium (C, after Pleijel; rest after Blake various).

papilla present or absent 13

8. Antenna and tentacular cirri all large, flattened, club-shaped (plate 142M); ventral cirrus large, reniform, oriented at right angles to acicula . . . . *Clavadoce splendida*

— Antennae and tentacular cirri cirriform, not flattened (plate 143A, 143F); ventral cirrus small, oval to elliptical, oriented in same direction as acicula 9

9. All three tentacular segments distinct dorsally (plate 143A); proboscis thickly papillated . . . . *Eulalia* 10

— First tentacular segment reduced, with first pair tentacular cirri lateral to prostomium (plate 143F); proboscis smooth or only sparsely papillated 11

10. (*Note three choices*) Dorsal cirri of middle parapodia elongate, lanceolate (plate 143B); body greenish, with black transverse stripes in intersegmental grooves (plate 143A) . . . . *Eulalia quadrioculata*

— Dorsal cirri of middle parapodia oval or cordate (plate 143D); antennae all short, none more than one-quarter the length of prostomium (plate 143C); all tentacular cirri short, none extending more than three segments in length; ventral cirri elongate with narrow base (plate 143D); body with two longitudinal rows of pigment derived from lateral segmental pigment . . . . *Eulalia californiensis*

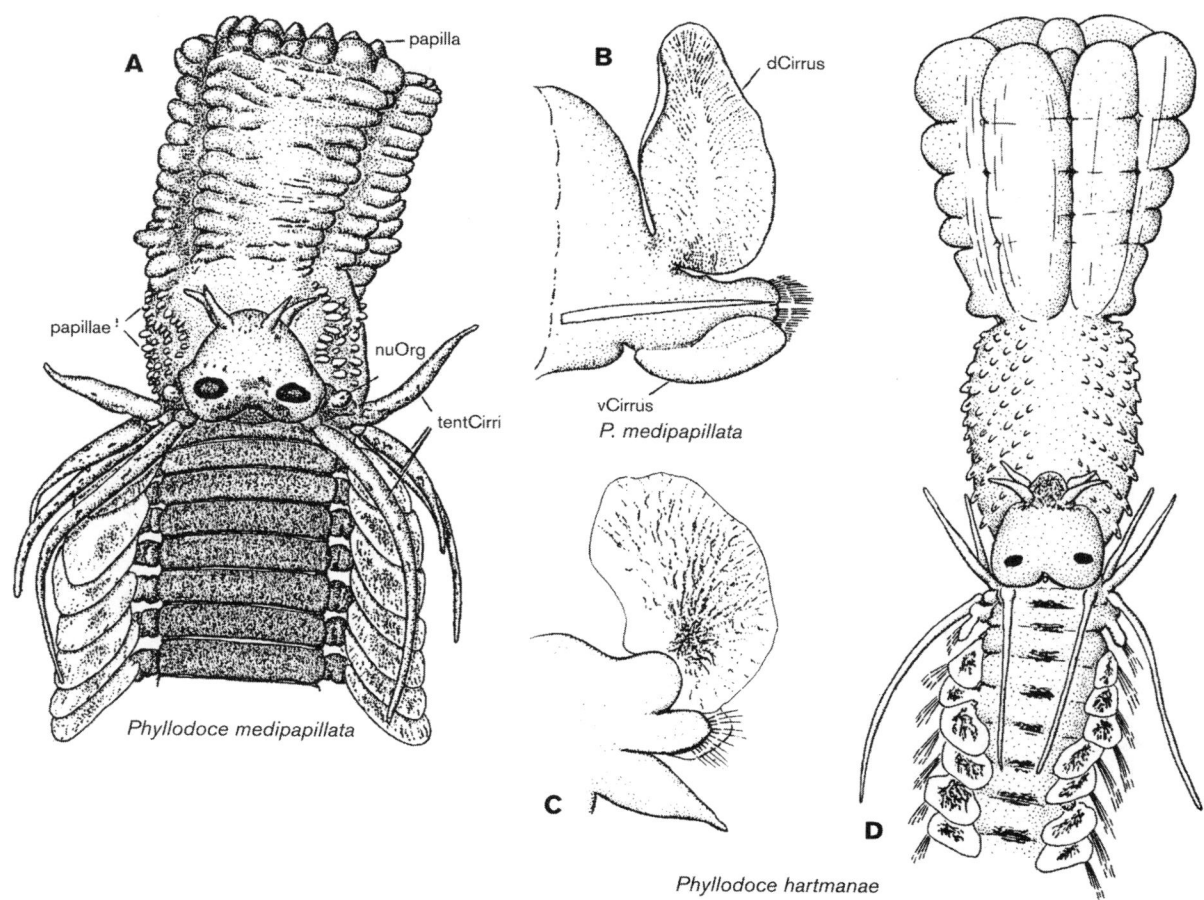

PLATE 145 Polychaeta. Phyllodocidae: A, *Phyllodoce medipapillata,* proboscis everted; B, parapodium; C, *Phyllodoce hartmanae,* parapodium; D, anterior end, dorsal view with proboscis everted (after Blake).

— Dorsal cirri of middle parapodia thickened, elongate, with broad basal attachment; antennae short, all tentacular cirri long, extending more than three segments in length; ventral cirri with broad basal attachment, tapering to narrow tip; body light to dark green including dorsal cirri, brown spots present or absent on dorsum, intersegmental stripes absent . . . . *Eulalia viridis*

11. Neuropodia without superior dorsal lobe (plate 143G). . . . .
. . . . *Eumida sanguinea*

— Neuropodia with superior dorsal lobe (plate 143H, 143I)
. . . . *Sige* 12

12. Dorsal cirri broad, cordate (plate 143H) . . . . . . *Sige bifoliata*

— Dorsal cirri elongate, foliaceous (plate 143I) . . . .
. . . . *Sige montereyensis*

13. Segments 1 to 2 well developed, forming distinct collar around tapering posterior half of prostomium (plate 144A); segments 1 to 2 bearing three pairs tentacular cirri; segment III distinct, with fourth pair tentacular cirri; tentacular formula: $(0\frac{1}{6} + 0\frac{1}{4}) + S\frac{1}{4}$ . . . . *Paranaitis polynoides*

— Segment 1 reduced, not visible dorsally, bearing first pair tentacular cirri lateral to prostomium (plate 144B, 144F); segments 2–3 distinct, bearing three pairs tentacular cirri . . . .
. . . . 14

14. Peristomium oblong or rounded, without nuchal papilla (plate 144B); tentacular formula: $(0\frac{1}{6} + S\frac{1}{1}) + S\frac{1}{4}$ . . . .

. . . . *Nereiphylla castanea*

— Peristomium heart-shaped, with or without nuchal papilla (plate 144F) . . . . *Phyllodoce* 15

15. Neuropodium with elongate superior lobe or protuberance; distal half of proboscis with large, elongate papillae scattered over surface. . . . . *Phyllodoce longipes*

— Neuropodium with smoothly rounded tip, without superior lobe (plate 145B); distal half of proboscis with low rounded or rugose lobes (plates 144C, 145A), conical papillae limited to oral opening 16

16. Setae first present from segment 3 17

— Setae first present from segment 4 19

17. Ventral cirri elongate, narrow, pointed, at least twice as long as setigerous lobe (plate 144E); middorsum with intersegmental brown pigment, sometimes forming longitudinal black band along body . . . . *Phyllodoce mucosa*

— Ventral cirri blunt to pointed; broad, not narrow or elongate, extending only slightly beyond setigerous lobe . . . . .
. . . . 18

18. All anterior segments heavily pigmented (plate 144C); proximal papillae of proboscis all of one type, conical, arranged in six rows of 13–16 papillae on each side; ventral cirri pointed apically, at least half again as long as neuropodial lobe (plate 144D) . . . . *Phyllodoce groenlandica*

— Each body segment with three pigmented intersegmental

areas across dorsum, forming three longitudinal lines along body (plate 144F); proximal papillae of proboscis all of one type, arranged in six rows of nine papillae on a side; ventral cirri thick, fleshy, blunt-tipped, slightly longer than neuropodial lobe (plate 144G) . . . . . . *Phyllodoce williamsi*

19. Body large, robust, with diffuse dark brown pigment dorsally on segments; prostomium wider than long, with lateral nuchal organs present when proboscis everted (plate 145A); proboscis with proximal papillae arranged in six lateral rows; ventral cirri rounded apically (plate 145B) . . . . *Phyllodoce medipapillata*

— Body slender, delicate, with distinct middorsal interseg- mental dark chromatophore (plate 145D); prostomium as wide as long, or slightly longer than wide, nuchal organs never present; ventral cirri apically pointed; proboscis with proximal papillae arranged in numerous spiral rows (plate 145D); ventral cirri elongated, pointed (plate 145C) . . . . *Phyllodoce hartmanae*

LIST OF SPECIES

See Hartman 1936, Univ. Calif. Publ. Zool. 41: 117–132; Plei- jel 1991; Blake 1994b.

Clavadoce splendida Hartman, 1936. In sand flats with *Zostera* and *Ulva* (Hartman, 1936); in algal holdfast, Doran Beach Jetty, Bodega Harbor; known only from Bodega Bay region (Blake 1994b). The species has a mottled coloration that blends with small filamentous algae rendering an effective camouflage.

Eteone californica Hartman, 1936. Widely distributed along the west coast of North America from the intertidal zone to shallow shelf depths, common in bays and estuaries in sandy sediments. A close relative of the Atlantic species, *E. longa,* dif- fering in details of the prostomial shape. Most specimens are covered with numerous small pigment granules.

Eteone balboensis Hartman, 1936. Rare, known only from central and southern California and reported from intertidal and shelf depths (Blake 1994b).

Eteone dilatae Hartman, 1936. Intertidal, more or less re- stricted to central and southern California in clean sand beaches, pale, green in life and characterized by having a long, trapezoidal-shaped prostomium and subrectangular dorsal cirri. See Blake 1975a (larvae).

Eteone lighti Hartman, 1936. Intertidal in estuarine beaches of central and southern California. The species prefers muddy sands and is recognized by having a broad, trapezoidal pros- tomium and triangular-shaped dorsal cirri that are about as wide as long.

Eteone pacifica Hartman, 1936. Intertidal, from British Co- lumbia to central California in silty mud, bright yellow in life, with round black spots laterally and square black spots dorsally. The dorsal cirri are wider than long and distally rounded. See Banse 1972a.

Eulalia quadrioculata Moore, 1906 (=*Eulalia aviculiseta* Hart- man, 1936). One of the most common polychaetes of north- ern California, among mussels and barnacles, in algal holdfasts, and under rocks in debris. The color is a distinctive olive-green on the dorsum with black intersegmental grooves. See Banse 1972a (morphology and synonymy).

Eulalia californiensis (Hartman, 1936) (=*Steggoa californien- sis*). A rare species, apparently limited to central California and ranging from the intertidal to shelf depths on a variety of bot- tom types, including sand, shelly gravel, among crevices in

rocks, and in algal holdfasts. This species has been mistakenly identified as *E. bilineata*.

Eulalia viridis (Linnaeus, 1767). A widely distributed species, probably introduced with oyster culture or ballast water from the western North Atlantic; rare locally; under rocks and in al- gal holdfasts, most similar to *E. quadrioculata,* but readily dis- tinguished by color, pigmentation, and details of the dorsal and ventral cirri. See Pettibone 1963; Hartmann-Schröder 1996.

Eumida sanguinea (Oersted, 1843). A widely distributed species introduced into embayments such as San Francisco Bay, proba- bly from the North Atlantic; may occur among algae and bry- ozoans on rocks.

Hesionura coineaui difficilis (Banse, 1963). Sandy sediments. See Banse 1963, Proc. Biol. Soc. Wash. 76: 197–208; Hartmann- Schröder 1963, Zool. Anz. 171: 204–243.

Nereiphylla castanea (Marenzeller, 1879) (=*Genetyllis cas- tanea*). A widely distributed species, possibly composed of sib- lings. The species is common in algal holdfasts and other microhabitats in rocky areas. Body elegantly pigmented red to orange with dark orange stripes across the dorsum.

Paranaitis polynoides (Moore, 1909). Intertidal to shallow shelf depths in mixed sand and silt. The species is elegantly colored in life with a white body bearing red pigment on the dorsal midline, dorsal cirri, and tentacular cirri; this fades in preser- vative, but the dorsum and parapodia retain dark pigment.

Phyllodoce groenlandica Oersted, 1843 (=*Anaitides groen- landica*). A widespread Arctic-boreal species from the intertidal to shelf and slope depths, distinguished by the broad, pointed ventral cirri.

Phyllodoce hartmanae Blake and Walton, 1977. Shallow sub- tidal sands; characterized by having the proximal papillae on the proboscis in a spiral or oblique pattern. The dorsal surface bears intersegmental black pigment in a row along the dorsum; a similar pigment spot is located on the dorsal cirrus.

Phyllodoce longipes Kinberg, 1866. A widely distributed species: Gulf of Mexico, southeastern United States, Chile, South Africa, northern Europe; Central California, recently found in Tomales Bay and off the airport runway in San Francisco Bay. Low water to shelf depths. See Pleijel 1990: Zool. J. Linn. Soc. 98: 161–184 (systematics); Blake 1994b (morphology).

Phyllodoce medipapillata Moore, 1909 (=*Anaitides medipapil- lata*). A large species found from the intertidal to shelf depths in rocky habitats. This species has setae first present from seg- ment 4; a conspicuous nuchal organ is present just anterior to the first tentacular cirri, body pigment diffuse dark brown. See Abbott and Reish 1980 (distribution, habitat).

Phyllodoce mucosa Oersted, 1843 (=*Anaitides mucosa*). Sand and silt, not common, likely introduced from the western North Atlantic. See Cazaux 1969, Arch. Zool. Exp. Gén. 110: 14–202 (larval development); Pettibone 1963 (morphology).

Phyllodoce williamsi Hartman, 1936 (=*Anaitides williamsi*). Common in mixed sand and mud sediments; lays eggs on blades of *Zostera* on tidal flats. The body is distinctly pigmented with three longitudinal lines on the dorsum and ventrum, bro- ken at the segmental furrows. See Blake 1975a (larvae).

Sige montereyensis Hartman, 1936. Known only as a single specimen from Monterey; in shallow water. The dorsal cirri of this species are long, pointed, and foliaceous.

Sige bifoliata (Moore, 1909) (=*Eumida bifoliata*). Sometimes mistaken for *Eulalia quadrioculata,* this species is common among algal holdfasts and debris in rocky habitats of the Bodega Bay region; also known from sandy-silt sediments and shell bottoms. The species was redescribed by Blake (1994b), who found the medial antenna longer than the frontal antennae.

The broad dorsal cirri readily separate the species from *S. montereyensis*. Intertidal to shallow shelf depths (Blake 1994b).

HESIONIDAE

The hesionids are small- to moderate-size worms with a fragile, dorsoventrally flattened body with the prostomial and parapodial appendages being deciduous. Because of their tendency to fragment, hesionids are often difficult to identify. The prostomium is usually wider than long, oval to quadrangular or pentagonal, and may be incised posteriorly. It typically bears 1–2 pairs of eyes, 2–3 antennae, and two palps. A large, more or less muscular proboscis is present, with the margin often equipped with cilia or papillae. The tentacular segments are fused to the peristomium, ventrally surrounding the mouth, and with each other dorsally. There are 2–8 pairs of tentacular cirri that may be distinctly articulated, wrinkled, or smooth. The ventral tentacular cirri are generally shorter than the corresponding dorsal ones. Parapodia are either biramous or reduced, but never uniramous. The notosetae are simple and include capillaries, heavy spines, furcate setae, and pectinate setae. The neurosetae are compound falcigers or spinigers, sometimes with a few simple neurosetae present. One to four aciculae are present in each parapodial ramus, including the bases of the tentacular cirri. The pygidium is usually small and bears a terminal anal pore and a pair of anal cirri; it may also bear a flat disk or plate (genus *Microphthalmus*).

KEY TO HESIONIDAE

1. Parapodia sesquiramous, with notopodia inconspicuous, reduced to aciculae and bases of dorsal cirri, notosetae entirely absent (plate 146D) 2
— Parapodia biramous or subbiramous, with notopodia bearing distinct setal fascicles 3
2. With eight pairs of tentacular cirri on three visible segments; with three antennae, medial one attached frontally (plate 146C); proboscis with ten papillae; neurosetae modified on several anterior setigers; golden, stout, with distally knobbed shafts and short falcigerous blades . . . . . .
. . . . *Heteropodarke heteromorpha*
— With six pairs of tentacular cirri on three visible segments; with two frontal antennae and two biarticulate palps; proboscis with about 25 papillae; anterior neurosetae not modified; with extra large lobes at bases of parapodia . . . .
. . . . *Micropodarke dubia*
3. Notosetae of one kind, distally serrated, blunt-tipped; dorsal and ventral cirri articulated (plate 146F); distinct brown segmental pigment bands (plate 146E) . . . .
. . . . *Gyptis brunnea*
— Notosetae of two or more kinds 4
4. Palps simple; pygidium disk-shaped (plate 146B); tentacular and parapodial cirri smooth, not inserted on cirrophore (plate 146A); body pale with brown marks along sides of segments and paired spots on the posterior margins of each segment; small interstitial species on sandy beaches . . . . .
. . . . *Microphthalmus sczelkowii*
— Palps biarticulate; pygidium not disk-shaped; tentacular and parapodial cirri smooth to articulated, inserted on cirrophore 5
5. Six pairs tentacular cirri on one visible segment (plate 146I); notosetae first present on setigers 3 or 4; proboscis

with numerous terminal papillae; notopodia reduced almost to a cirrus with inconspicuous setae; body reddish-brown to purple; free-living on muddy bottoms or commensal in ambulacral grooves of starfishes . . . .
. . . . *Ophiodromus pugettensis*
— Eight pairs tentacular cirri on one visible segment (plate 146G); notosetae first present on setigers 4–6; proboscis with few terminal papillae; furcate setae with serrations below shorter tine (plate 146H, visible at 400×); notosetae including smooth spines and furcate setae; longest blades of neurosetae slightly more than twice as long as shortest ones; anterior and posterior eyes subequal . . . .
. . . . *Podarkeopsis glabrus*

LIST OF SPECIES

See Hilbig 1994a; Pleijel 1998.

Gyptis brunnea Hartman, 1961. In shelf depths; larval stages have been taken from plankton in Tomales Bay and reared in the laboratory (Blake 1975a); juveniles grow well in a sand substrate; adults have not been found in the intertidal zone.

Heteropodarke heteromorpha Hartmann-Schröder, 1962. Low water to about 100 m; in sandy sediments, numerous specimens from Bodega Harbor.

Microphthalmus sczelkowii Mecznikow, 1865. Clean sand beaches; Dillon Beach (see Pettibone 1963).

Microphthalmus and *Hesionides* species are small and poorly known from California, and it is likely that several species are present. Both of these genera were shown by Pleijel and Dahlgren 1998, Cladistics, 14: 129–150, to not belong to the Hesionidae. However, because no other family has been suggested for these polychaetes, they are here retained in the Hesionidae; see also Pleijel 1998.

Micropodarke dubia (Hessle, 1925) (=*Micropodarke amemiyai* Okuda, 1938). Low water to 40 m in fine sands; southeastern Alaska to California, identified from Bodega Bay and Morro Bay, California and off Dash Point, Puget Sound; also Yellow Sea and Japan. This species is distinguished by the large lobes that originate near the bases of the parapodia; these reach two-thirds the length of anterior parapodia. See Imajima and Hartman 1964; Banse and Hobson 1968, 1974; Armstrong, Strude, Thom, and Chew 1976, Syesis 9: 277–290 (ecology, Puget Sound, 2–100 individuals m⁻²).

Ophiodromus pugettensis (Johnson, 1901). This species is sometimes referred to the genus *Podarke* and is the most common intertidal hesionid of California and Oregon; on silty bottoms and in ambulacral grooves of asteroids, especially *Patiria miniata*. See Hickock and Davenport 1957, Biol. Bull. 113: 397–406 (behavior and commensalism); Lande and Reish 1968, Bull. So. Calif. Acad. Sci. 67: 104–111; Blake 1975a (larval development); Abbott and Reish 1980; Shaffer 1979, Bio. Bull. 136: 343–355 (feeding); Abbott and Reish 1980 (review).

Podarkeopsis glabrus (Hartman, 1961) (=*Oxydromus arenicolus glabrus* Hartman, 1961). In shallow waters in mixed sand and silt sediments, in Bodega Harbor, and was misidentified as *Gyptis brevipalpa* (=*Podarkeopsis*) in the previous edition (Blake 1975b). See Hilbig 1994a; Blake 1975a (pelagic larvae as *G. brevipalpa*).

PILARGIDAE

The pilargids are a small group of carnivorous or omnivorous polychaetes only rarely encountered in casual collections. Most

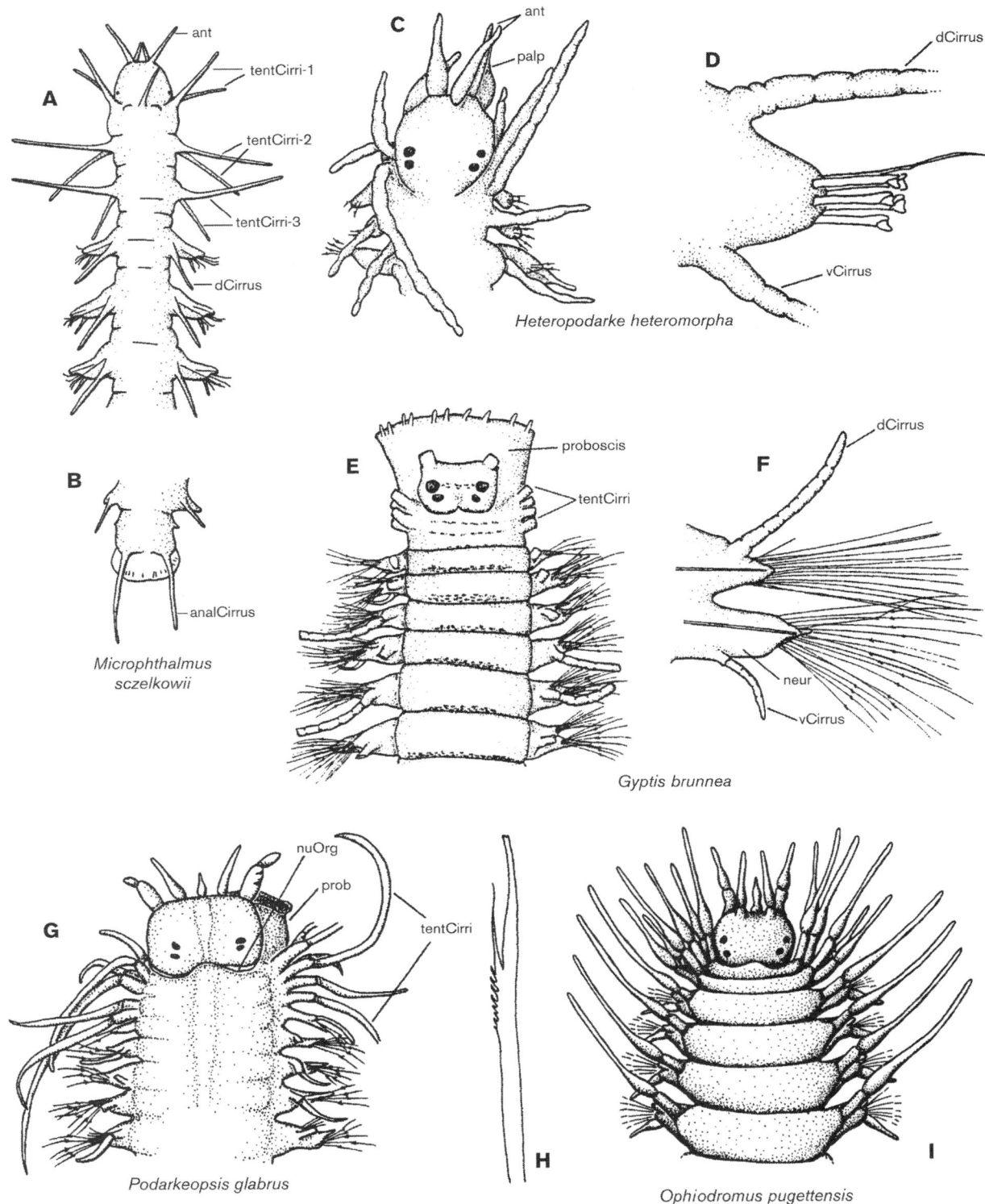

PLATE 146 Polychaeta. Hesionidae: A, *Microphthalmus sczelkowii*; B, posterior end, dorsal view; C, *Heteropodarke heteromorpha*; D, parapodium; E, *Gyptis brunnea*; F, parapodium; G, *Podarkeopsis glabrus*; H, lyrate seta; I, *Ophiodromus pugettensis* (A, B, after Westheide; C, D, after Hartmann-Schröder; E–H, after Hilbig; rest after Blake).

species are subtidal and require careful sieving to remove them from the muds and muddy sands that they inhabit. Pilargids are related to the hesionids and syllids but are easily differentiated from them by the absence of aciculae in tentacular segments, the absence of compound setae, and the unarmed

proboscis. Several of the genera have characteristic recurved notopodial spines, which are shaped like boat hooks.

Pilargids encountered in central California have elongate, somewhat flattened, ribbonlike bodies (e.g., *Sigambra* and *Pilargis*); other genera may be cylindrical. The body surface may

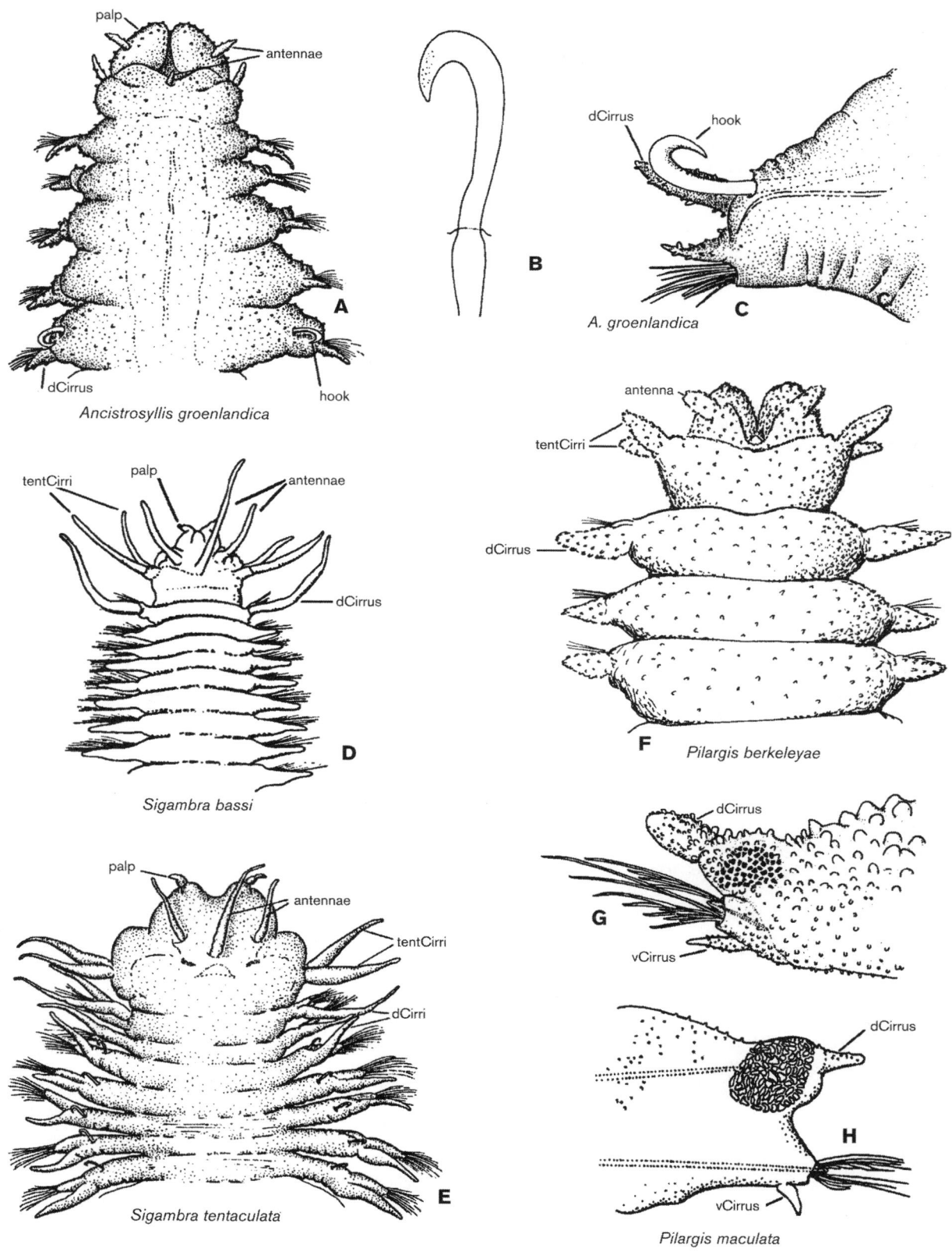

PLATE 147 Polychaeta. Pilargidae: A, *Ancistrosyllis groenlandica*; B, hooked notoseta; C, parapodium showing hooked seta; D, *Sigambra bassi*; E, *Sigambra tentaculata*; F, *Pilargis berkeleyae*; G, parapodium; H, *Pilargis maculata*, parapodium. (D, F–H, after Hartman; rest after Blake).

be smooth or distinctly papillated. The prostomium, tentacular segments, and the first setiger are more or less fused. The prostomium is typically small, inconspicuous, and has no or up to three antennae; eyes are normally lacking. The medial antenna, when present, is on the posterior part of the prostomium. Antennae may be short (e.g., *Pilargis*) or quite long (e.g., *Sigambra*). Palps are biarticulate and usually set off from the prostomium by large palpophores bearing small (or no) palpostyles. The proboscis is unarmed, bulbous, and usually bears a circlet of papillae around the distal opening. The anterior end of the animal often appears enlarged due to the contained proboscis. The tentacular segment is asetigerous and usually bears two pairs of tentacular cirri (rarely absent).

Parapodia are subbiramous, with the notopodia reduced to embedded notoaciculae in the cirrophores of the dorsal cirri. Additional capillary setae, large curved hooks, or straight spines may be present or absent. Neuropodia are somewhat conical with embedded neuroaciculae and simple neurosetae; composite setae are always lacking. Dorsal and ventral cirri are present, with the ventral smaller than the dorsal or both cirri subequal. The pygidium normally bears a terminal anus and a pair of anal cirri.

The most important taxonomic characters include the presence and structure of the notopodial spines, degree of flattening of the body, presence/absence of the antennae and tentacular cirri, and the form of these same structures if present.

KEY TO PILARGIDAE

1. Prostomial antennae and peristomial cirri short (plate 147F); setae short, inconspicuous . . . . *Pilargis* 4
— Prostomial antennae and peristomial cirri long (plate 147A, 147D); setae conspicuous; stout, recurved notosetae present (plate 147B) 2
2. Prostomium reduced, inconspicuous; antennae shorter than palps (plate 147A); tentacular cirri short; dorsal cirri short (plate 147C), with those of setiger 1 similar to or slightly longer than subsequent ones . . . .
. . . . *Ancistrosyllis groenlandica*
— Prostomium larger; antennae longer than palps; tentacular cirri long; dorsal cirri long and slender with those of setiger 1 greatly elongate . . . . *Sigambra* 3
3. Hooked notosetae setae from setigers 11–15; medial antenna twice as long as lateral antennae (plate 147D) . . . . .
. . . . *Sigambra bassi*
— Hooked setae from setigers 3–4; medial antenna subequal to lateral antennae (plate 147E) . . . . . *Sigambra tentaculata*
4. Dorsum heavily papillated; notopodia with subglobular base and terminal clavate process (plate 147F–147G) . . . .
. . . . *Pilargis berkeleyae*
— Dorsum lightly papillated; notopodia with a broad, quadrate base and tapering short cirri (plate 147H) . . . . . .
. . . . *Pilargis maculata*

LIST OF SPECIES

See Hartman 1947b, 1968; Blake 1994c.

Ancistrosyllis groenlandica McIntosh, 1879. Low water to slope depths.

Pilargis berkeleyae Monro, 1933. Intertidal to shelf depths in sandy sediments, this species has a heavily papillated surface.

Pilargis maculata Hartman, 1947. Central California, intertidal to shallow subtidal in cryptic habitats in rocks or other shoreline substrates. Pettibone 1966, Proc. U.S. Natl. Mus. 118: 155–208, regarded *P. berkeleyae* and *P. maculata* as synonyms.

Sigambra bassi (Hartman, 1945). Central and southern California, intertidal to 113 m. *S. bassi* is distinguished from congeners by having the notopodial hooks first present from setigers 11–15 instead of 3–4.

Sigambra tentaculata (Treadwell, 1941). A widely distributed shelf species, sometimes reported from the intertidal; hooked setae from setigers 3–4.

SYLLIDAE

The Syllidae represent one of the most diverse and systematically challenging families of polychaetes. They tend to be small, usually <10 mm long and 1 mm wide, and are especially well represented as epibiota on hard surfaces or on metazoans in intertidal, subtidal, and shelf habitats. Other members of this family are also abundant in soft sediments. Historically, the syllids have been among the most poorly known polychaete families along the California and Oregon coasts. However, Kudenov and Harris (1995) have revised many species and newly described others.

The body of syllids is usually small, short and slender. The integument is generally smooth, but papillae of various shapes and distributions are often present. The prostomium bears 2–3 pairs of eyes. Three antennae are usually present, including an unpaired medial antenna and paired lateral antennae; these may be articulated, wrinkled, or smooth. Paired palps project anteroventrally and are roughly triangular in shape and may be completely fused to one another or free. The palps of *Autolytus* differ in being greatly reduced, not visible dorsally, and projecting ventrally.

The pharynx is an eversible tube with a cuticular lining that is usually armed with a dorsomedial tooth that is more or less subdistal when the pharynx is everted; a distal ring of marginal pharyngeal teeth may accompany the dorsomedial tooth, sometimes collectively forming a crown or "trepan" (as in *Trypanosyllis*). Sometimes the pharynx is described as being unarmed (i.e., pharyngeal teeth are absent), as in *Syllides*. The proventriculus is conspicuous through the body wall, with transversely striated muscles arranged in rows, and connects the anterior pharynx to the posterior intestine.

Two pairs of tentacular cirri are present; *Exogone* and *Sphaerosyllis* have only one pair. Parapodia are uniramous in nonreproductive individuals; reproductive individuals may develop biramous parapodia. Setae include compound falcigers (as in *Typosyllis*), compound spinigers (as in some species of *Exogone* and *Pionosyllis*), or all simple setae (as in *Haplosyllis*). The articulation between the blade and the shaft may be fused. A dorsal and ventral simple seta may also be present among the compound setae. Aciculae are present in all parapodia.

Dorsal cirri are variable and may alternate between long and short and may be articulated or moniliform (as in *Syllis*, *Typosyllis*, or *Trypanosyllis*), smooth (as in *Exogone*, *Odontosyllis*, or *Pionosyllis*), or irregularly wrinkled (as in *Eusyllis* or *Streptosyllis*). Ventral cirri are mostly digitiform and usually do not extend beyond the parapodial lobes; they are entirely absent from the subfamily Autolytinae. The pygidium bears two anal cirri; sometimes with an additional ventromedial cirrus. These may be articulated to smooth.

DEFINITIONS OF SUBFAMILIES

The four subfamilies are characterized according to Kudenov and Harris 1995. See also San Martín 2003.

SUBFAMILY AUTOLYTINAE Body threadlike, 4–20 mm long, 0.5–1 mm wide. Palps partially to completely fused to one another, directed ventrally. Nuchal organs present either as conspicuous epaulettes or lobes. Tentacular cirri numbering two pairs. Antennae, tentacular cirri, and dorsal cirri smooth. Ventral cirri absent. Pharynx long, coiled or sinuous, with a distal circlet of teeth (=trepan).

SUBFAMILY EXOGONINAE Body rather compact, less than 8 mm long, 0.8 mm wide. Palps fused dorsally for at least one-half to all their length. Nuchal organs usually inconspicuous. Tentacular cirri usually numbering 1–2 pairs; may be absent. Antennae, tentacular cirri, and dorsal cirri smooth. Ventral cirri present. Pharynx usually straight, with a middorsal tooth or unarmed.

SUBFAMILY EUSYLLINAE Body usually at least 10 mm long. Palps only fused basally, free from one another distally. Nuchal organs large and conspicuous. Tentacular cirri numbering 1–2 pairs. Dorsal cirri smooth to wrinkled. Ventral cirri present. Pharynx usually straight, with a middorsal tooth with or without a marginal series of teeth, or unarmed.

SUBFAMILY SYLLINAE Body usually at least 10 mm long, sometimes more than 50 mm. Palps usually free basally. Nuchal organs usually small and inconspicuous. Tentacular cirri numbering two pairs. Dorsal cirri strongly articulated or moniliform. Ventral cirri present. Pharynx straight, with a middorsal tooth, with or without distal circlet of teeth (=trepan).

The key to genera and species is modified from Kudenov and Harris (1995).

KEY TO SPECIES AND GENERA OF SYLLIDAE

1. Dorsum of benthic atokous stages with two nuchal epaulettes extending from hind margin of prostomium through first few segments (plate 148G); ventral cirri absent (plate 148C); dorsal cirri smooth (plate 148G); pelagic sexual stages highly modified with distinct body regions and long natatory setae (plate 148A–148B) . . . . Autolytinae 29
— Dorsum without nuchal epaulettes; ventral cirri present; dorsal cirri smooth to articulated; sexual stages, if present, not as highly modified; natatory setae present or absent 2
2. Dorsal cirri smooth; body small, usually <8 mm long; palps fused for at least half their length (plate 149D) . . . . Exogoninae 4
— Dorsal cirri smooth to articulated; body larger, usually >10 mm long; palps not fused or fused only basally 3
3. Palps fused basally (plate 150B); dorsal cirri smooth to wrinkled (plate 150A, 150B) . . . . Eusyllinae 12
— Palps not fused basally (plate 150H); dorsal cirri articulated (plate 150L) . . . . Syllinae 16
4. Two pairs of tentacular cirri (plate 149A); dorsal cirri long and filiform . . . . *Brania brevipharyngea*
— One pair of tentacular cirri (plate 149K, 149L); dorsal cirri papillalike or ovoid 5
5. Dorsal cirri papillalike or ovoid . . . . *Exogone* 6
— Dorsal cirri pyriform or flask-shaped . . . . . *Sphaerosyllis* 10
6. Compound setae with highly modified blades, shafts, and heterogomph joints (plate 149C); dorsal simple setae similar in all segments, or differing between anterior and posterior segments 7
— Compound setae with unmodified blades, shafts, and heterogomph joints (plate 149G, 149H); dorsal simple setae similar in all segments 8
7. Dorsal cirri absent from setiger 2 (plate 149B); blades of compound setae of setigers 1–3 short, deeply bifid (plate 149C); medial antenna only half again as long as lateral antennae; proventriculus from setigers 3–5, with 14–16 rows of muscle cells . . . . *Exogone dwisula*
— Dorsal cirri present on setiger 2 (plate 149D); blades of compound setae of setigers 1–3 long, not bifid, medial antenna 2–3 times longer than lateral antennae; proventriculus extending through four to five segments, with 18–24 rows of muscle cells . . . . *Exogone lourei*
8. Transition between long and short blades of compound falcigers abrubt in anterior setigers, with superior blades 3–4 times times longer than inferior blades within fascicles (plate 149G–149I); medial antenna long, lateral antennae very short (plate 149F) . . . . *Exogone molesta*
— Transition between long and short blades of compound falcigers gradual in all setigers, with superior blades one and a half to three times longer than inferior blades within fascicles; medial antenna short to long, lateral antennae short 9
9. Medial antenna long, extending to setiger 3 (plate 149E); superior blades of compound falcigers three times longer than inferior blades; pharynx in 6–8 segments; all dorsal simple setae distally unidentate . . . . *Exogone acutipalpa*
— Medial antenna shorter, extending to setiger 1 (plate 149J); superior blades of compound falcigers one and a half times longer than inferior blades; pharynx in four segments; all dorsal simple setae distally bidentate . . . . *Exogone breviseta*
10. Dorsal cirri similar to one another; dorsum with dense fields of filiform and elliptical papillae (plate 149K), often encrusted with fine sediments; ventrum lacking papillae; laterally with two pairs of conspicuous papillae per segment, each pair associated with basal anterior and posterior facies of parapodial lobes . . . . *Sphaerosyllis californiensis*
— Dorsal cirri of two forms, not similar to one another 11
11. Dorsum with two conspicuous dorsomedial longitudinal rows of distally knobbed macropapillae, two pairs per segment from setigers 4 or 5, alternating between a large anterior and smaller posterior pair per segment (plate 149L); ventrum with 12 digitiform papillae in four longitudinal rows of three papillae and two additional longitudinal rows of four small, round papillae per segment, altogether forming six longitudinal rows; dorsal cirri flask-shaped anteriorly, cirriform posteriorly . . . . *Sphaerosyllis bilineata*
— Dorsum without obvious dorsomedial longitudinal rows of papillae (plate 149M); ventrum with two pairs of papillae per segment, all arranged in two longitudinal rows; dorsal cirri mammiform in setigers 1–7 or 9 and far posterior segments, digitiform in medial segments from setigers 8–10, or all digitiform from setigers 8–10 . . . . *Sphaerosyllis ranunculus*
12. Eversible pharynx unarmed. . . . . *Syllides*
— Eversible pharynx armed 13

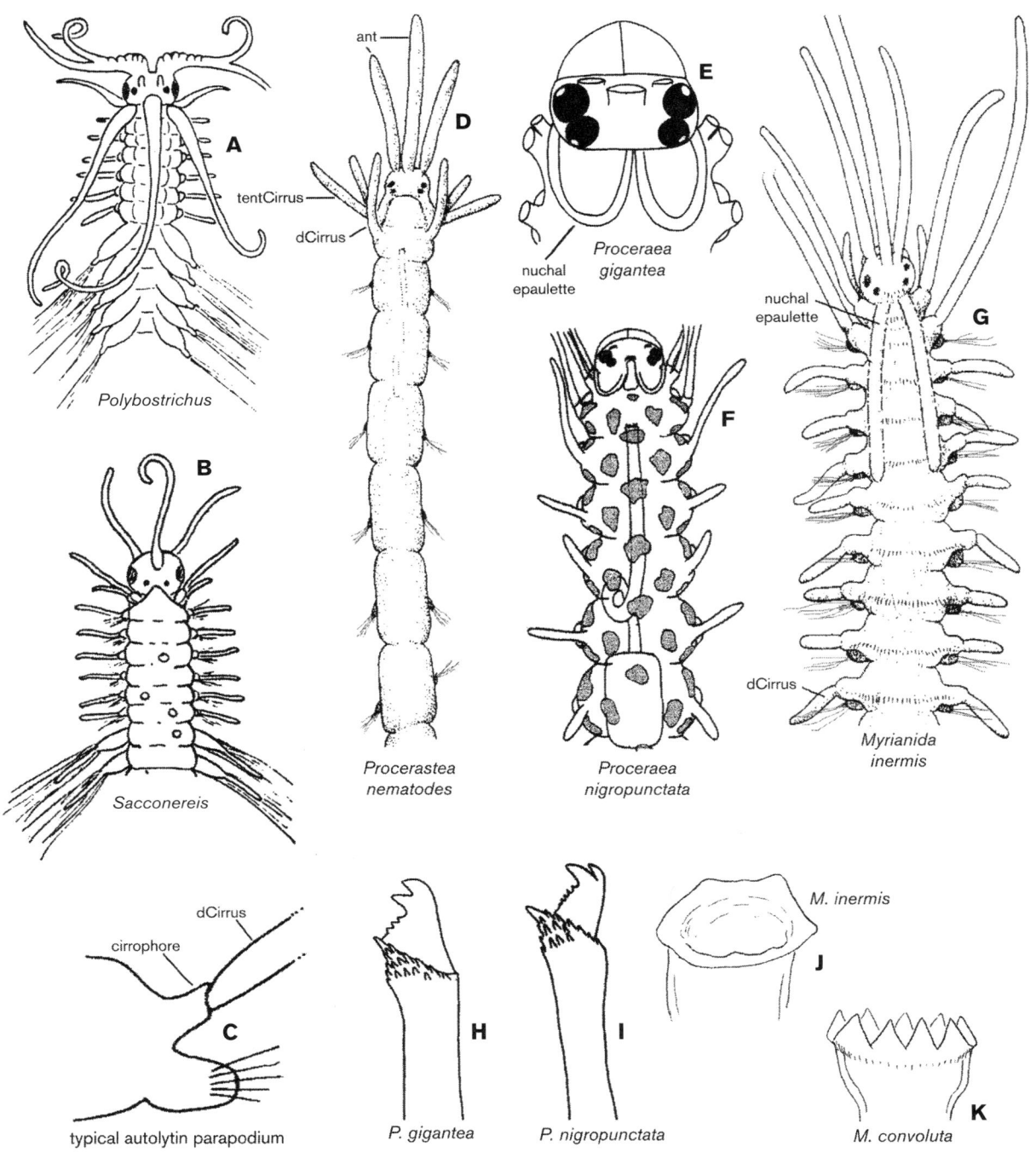

PLATE 148 Polychaeta. Syllidae, subfamily Autolytinae: A, autolytin sexual stage of a male (polybostrichus); B, autolytin sexual stage of a female (sacconereis); C, typical autolytin parapodium; D, *Procerastea nematodes*, dorsal view; E, *Proceraea gigantea*, diagram of anterior end, dorsal view with antennae and cirri cut at bases; F, *Proceraea nigropunctata*, dorsal view, anterior end showing pigment spots; G, *Myrianida inermis*, anterior end, dorsal view; H, *P. gigantea*, compound falciger; I, *P. nigropunctata*, compound falciger; J, *M. inermis*, trepan; K, *M. convoluta*, trepan (D, G, after San Martín; E, F, H, I, after Nygren and Gidholm; J, K, after Nygren; rest after Blake).

13. Prostomium covered posteriorly by nuchal hood (plate 150A); pharynx armed with a distal circlet of smaller curved teeth . . . . *Odontosyllis* 14
— Prostomium not covered posteriorly by nuchal hood (plate 150B); pharynx armed with a single middorsal tooth . . . . *Pionosyllis* 15
14. Dorsal cirri in middle region short, thick, no longer than half the width of the body; dorsum relatively unpigmented . . . . *Odontosyllis parva*
— Dorsal cirri in middle region long, slender and tapering to fine tip; dorsum dark with pale transverse bands (plate 150A) . . . . *Odontosyllis phosphorea*
15. Blades of compound setae at least three times longer than basal width; medial antenna arising from center of

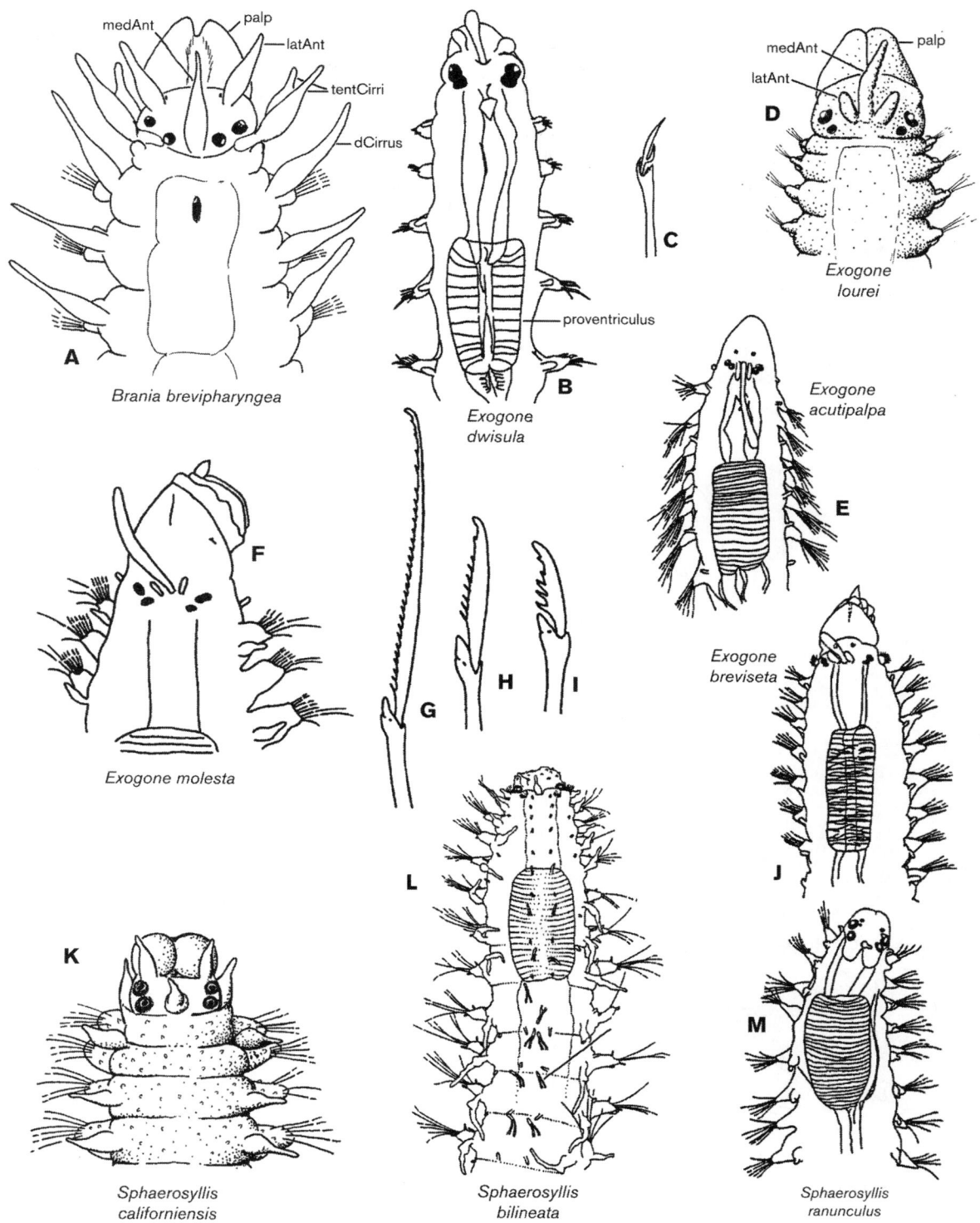

PLATE 149 Polychaeta. Syllidae: A, *Brania brevipharyngea*; B, *Exogone dwisula*; C, compound seta; D, *Exogone lourei*; E, *Exogone acutipalpa*; F, *Exogone molesta*; G–I, compound setae; J, *Exogone breviseta*; K, *Sphaerosyllis californiensis*; L, *Sphaerosyllis bilineata*; M, *Sphaerosyllis ranunculus* (B, C, F–J, L, M, after Kudenov and Harris; rest after Blake).

prostomium . . . . *Pionosyllis gigantea*
— Blades of compound setae short, less than three times longer than wide; medial antenna arising from anterior prostomial margin (plate 150B) . . . .
. . . . *Pionosyllis magnifica*

16. Pharynx with a middorsal tooth 17
— Pharynx with a middorsal tooth and trepan 27
17. Setae include compound falcigers and thick dorsal and ventral simple setae, or pseudocompound setae. . . . . . *Syllis* 18
— Setae include compound falcigers, sometimes with extremely

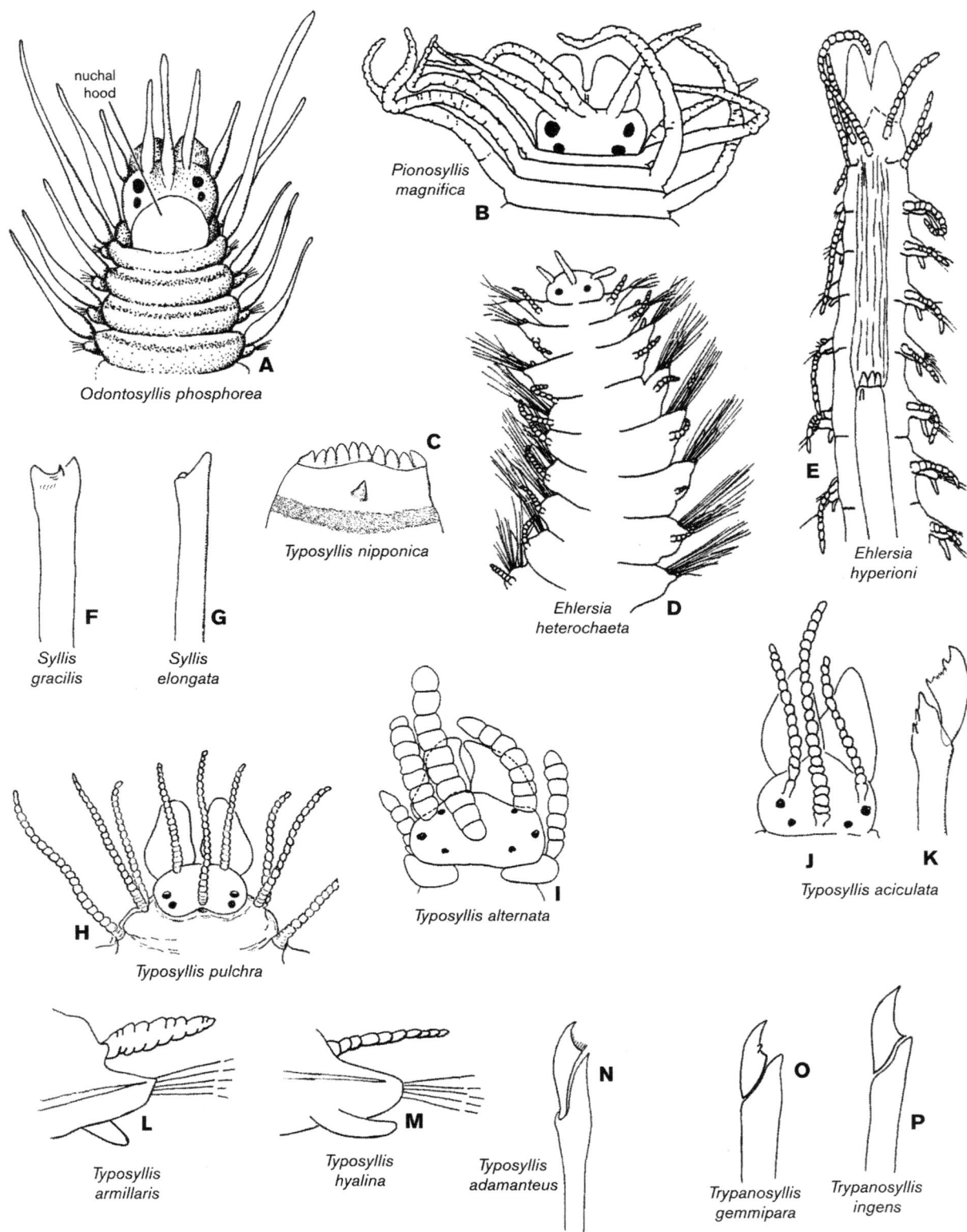

PLATE 150 Polychaeta. Syllidae: A, *Odontosyllis phosphorea*; B, *Pionosyllis magnifica*; C, *Typosyllis nipponica*, pharynx; D, *Ehlersia heterochaeta*; E, *Ehlersia hyperioni*; F, *Syllis gracilis*, Y-shaped seta from median parapodium; G, *Syllis elongata*, acicular seta; H, *Typosyllis pulchra*; I, *Typosyllis alternata*; J, *Typosyllis aciculata*; K, compound falciger; L, *Typosyllis armillaris*, parapodium; M, *Typosyllis hyalina*, parapodium; N, *Typosyllis adamanteus*, compound falciger; O, *Trypanosyllis gemmipara*, compound falciger; P, *Trypanosyllis ingens*, compound falciger (B, after Moore; C, after Imajima; D, E, after Kudenov and Harris; H, after Berkeley and Berkeley; J, K, after Reish; L, M, after Fauvel; rest after Blake).

long spinigerlike blades, and thin dorsal and ventral simple setae 19

18. (*Note three choices*) Middle parapodia with thick, Y-shaped seta (plate 150F) in addition to compound setae . . . .
. . . . *Syllis gracilis*
— Middle parapodia with 2–3 projecting acicular setae (plate 150G) in addition to composite setae . . . . . . *Syllis elongata*
— Middle parapodia with only pseudocompound setae . . . . .
. . . . *Syllis spongiphila*

19. Only compound falcigers present (plate 150N). . . . .
. . . . *Typosyllis* 21
— Compound falcigers and spinigerlike falcigers with extremely long bidentate blades at least four times longer than inferior-most blades within fascicles of midbody segments . . . . *Ehlersia* 20

20. Prostomium with eyes (plate 150D); blades of superior compound falcigers from midbody segments four times longer than inferior blades within fascicles; proventriculus in 6–9 segments with 30–36 rows of muscle cells . . . .
. . . . *Ehlersia heterochaeta*
— Prostomium without eyes (plate 150E); blades of superior compound falcigers from midbody segments 10 times longer than inferior blades within fascicles; proventriculus in 4–7 segments with 40–43 rows of muscle cells . . . . . .
. . . . *Ehlersia hyperioni*

21. Composite falcigers distally entire (plate 150N); paired antennae inserted ventrally . . . . *Typosyllis adamanteus*
— Some or all composite falcigers distally bifid (plate 150K) 22

22. Pharynx with band of black pigment just behind middorsal tooth (plate 150C) . . . . *Typosyllis nipponica*
— Pharynx unpigmented 23

23. Medial antenna inserted at or near the middle of prostomium (plate 150I) 24
— Medial antenna inserted at frontal margin of prostomium 26

24. Paired antennae inserted at frontal margin of prostomium (plate 150H). 25
— Paired antennae inserted in front of anterior eyes (plate 150J). . . . . *Typosyllis aciculata*

25. Antennae and dorsal cirri with numerous articles (plate 150H); dorsal cirri of middle segments with 50–70 articles; prostomium with small dorsal flap . . . . . . *Typosyllis pulchra*
— Antennae and dorsal cirri with relatively few articles (plate 150I); dorsal cirri of middle segments with no more than 25 articles . . . . *Typosyllis alternata*

26. Dorsal cirri thick (plate 150L) . . . . *Typosyllis armillaris*
— Dorsal cirri slender (plate 150M) . . . . . . *Typosyllis hyalina*

27. Palps same length as prostomium; body cylindrical; setae simple. . . . . *Geminosyllis*
— Palps shorter than prostomium; body flattened; setae compound. . . . . *Trypanosyllis* 28

28. Color of dorsum pale, crossed with dark transverse lines; composite falcigers with subdistal tooth (plate 150O) . . . . .
. . . . *Trypanosyllis gemmipara*
— Color ivory to tawny; cirri purplish brown; composite falcigers with smooth cutting edge (plate 150P) . . . .
. . . . *Trypanosyllis ingens*

29. Dorsal cirri present on setiger 1 only (plate 148D) . . . . . .
. . . . *Procerastea nematodes*
— Dorsal cirri present on all setigers (as in plate 148F–148G) 30

30. Cirrophores absent after setiger 1. 31
— Cirrophores present on all setigers. 34

31. Specimens small (≤2.6 mm); blades of compound setae without serrations . . . . *Proceraea penetrans*
— Specimens larger (10–50 mm); blades of compound setae with serrations beneath bidentate tip 32

32. Adult specimens to 50 mm; nuchal epaulettes extending onto setiger 1 (plate 148E); blades of compound setae with large superior tooth, nearly as large as main tooth in middle setigers (plate 148H) . . . . *Proceraea gigantea*
— Adult specimens 10–12 mm; nuchal epaulettes restricted to tentacular segment; blades of compound setae with small superior tooth (plate 148I) 33

33. Body distinctly pigmented, with dorsum bearing seven dark spots arranged in two transverse rows across each segment, two weaker spots present in parapodial lobes, and with a midventral longitudinal brown band; anterior and posterior eyes nearly confluent (plate 148F) . . . .
. . . . *Proceraea nigropunctata*
— Body with two indistinct dark longitudinal bands along sides or without pigment; anterior and posterior eyes separated . . . . *Proceraea cornuta*

34. Dorsal cirri alternating in length; blades of compound setae with large superior tooth (as in plate 148H); living specimens orange with bright yellow middorsal line in life . . . .
. . . . *Epigamia noroi*
— Dorsal cirri uniform in length; superior tooth on compound bidentate setae much smaller than secondary tooth (as in plate 148I); living specimens not brightly colored
. . . . 35

35. Body large (≤30 mm); dorsal cirri approaching body width in length; eyes brown; anterior eyespots not illustrated or described. . . . . *Autolytus varius*
— Body small (≤6.9 mm); dorsal cirri two-thirds of body width or less; eyes red; anterior eyespots present 36

36. Trepan without distinct teeth (plate 148J); cirrophores swollen, inflated, with cirrostyles attached subterminally (plate 148G); nuchal epaulettes extending through setigers 3–5 (plate 148G) . . . . *Myrianida inermis*
— Trepan with 15–16 well-developed teeth (plate 148K); cirrophores small; cirrostyles with terminal attachment; nuchal epaulettes extending maximally through setiger 2. . . . .
. . . . *Myrianida convoluta*.

LIST OF SPECIES

AUTOLYTINAE

The species of *Autolytus* and related genera are not well documented from California. The sexual stages of at least four species were collected from Tomales Bay plankton (Blake, unpublished). Many of the described pelagic sexual forms, including ones observed locally, have not been assigned to named asexual forms. Likewise, some species are only known from the sexual stages. See Gidholm 1965, Zool. Bidrag f. Uppsala 37: 1–44 (reproduction); 1966, Ark. Zool. l9: 157–213 (systematics) of the subfamily Autolytinae. A key to asexual forms of *Autolytus* species for British Columbia and Washington is provided by Banse and Hobson (1974). Nygren (2004) has provided a revision of the Autolytinae. From this paper, we have extracted the following data and developed the key to asexual phases of species reported from California. Readers are referred to Nygren (2004) for details. See also Nygren and Gidholm 2001; San Martín 2003.

Autolytus varius Treadwell, 1914. Intertidal in rocky habitats from San Francisco Bay to British Columbia. Considered as *incertae sedis* by Nygren (2004) since the original description is insufficient and the types are lost.

Epigamia noroi (Imajima and Hartman, 1964). Intertidal and shelf depths, among hydroids, bryozoans and algae. Found off La Jolla in *Phyllospadix* and on a floating dock on Santa Catalina Island; Japan and California.

Myrianida convoluta (Cognetti, 1953). Intertidal and subtidal amongst hydroids, bryozoans and tunicates. Found off Santa Catalina Island on *Laminaria* with hydroids; north Pacific, north Atlantic, and Mediterranean.

Myrianida inermis (Saint-Joseph, 1887). Sublittoral among hydroids, bryozoans, and tunicates. Collected from the Scripps pier and reported from the northeast Pacific from California to Washington.

Proceraea cornuta (Agassiz, 1862). Intertidal to shallow subtidal (~20 m) among algae with bryozoans or hydroids.

Proceraea gigantea Nygren and Gidholm, 2001. Shallow water on stones and in eelgrass, and among hydroids, sponges, and bryozoans. Santa Catalina Island. See Nygren and Gidholm 2001.

Proceraea nigropunctata Nygren and Gidholm, 2001. Shallow water to about 18 m, on stones and shells among algae, sponges, and hydroids; southern California to Washington. See Nygren and Gidholm 2001.

Proceraea penetrans (Wright and Woodwick, 1977). Subtidal, 6–18 m, within blisters on the hydrocoral *Stylaster californicus* (=*Allopora*). Off Santa Catalina Island and Gull Island off Santa Cruz. See Wright and Woodwick 1977, Bull. So. Calif. Acad. Sci. 76: 42–48.

Procerastea nematodes Langerhans, 1884. Found on a floating dock amongst algae and hydroids at Santa Catalina Island; otherwise known to 50 m in the northeast Atlantic on stones and shells.

EXOGONINAE

Brania brevipharyngea Banse, 1972. Intertidal to shelf depths.

Exogone dwisula Kudenov and Harris, 1995. This is probably the *Exogone gemmifera* of earlier editions, falcigers of the first 2–3 setigers have blades that resemble can-openers; intertidal to shelf depths; Canada to southern California. See Kudenov and Harris 1995.

Exogone lourei Berkeley and Berkeley, 1938. The most common species of *Exogone* in central California embayments; the species bears long compound spinigers on setiger 2 that have greatly enlarged shafts; prefers sandy mud; algal holdfasts; see Banse and Hobson 1968; Banse 1972a; Woodin 1974, Ecol. Monogr. 44: 171–187 (life history and ecology).

Exogone molesta (Banse, 1972). A distinctive species having a long medial antenna, long pointed palps, and superior falcigers with long blades; easily confused, however, with *E. acutipalpa*. Shallow shelf and upper slope depths.

Exogone breviseta Kudenov and Harris, 1995. Similar to *E. molesta* and *E. acutipalpa* in having the long medial antenna and palps, but lacking long-bladed superior falcigers; shallow subtidal and shelf depths.

Exogone acutipalpa Kudenov and Harris, 1995. Closely related to *E. molesta*, differing in having an extra pair of eyespots anterior to the antennae, shallow subtidal to shelf depths.

Sphaerosyllis californiensis Hartman, 1966. Records of *S. hystrix* and *S. pirifera* in earlier editions are probably this species;

S. californiensis differs from other local species by having the dorsal cirri all similar to one another instead of being of two types, intertidal to shelf depths; silt and mud sediments; also rocky habitats.

Sphaerosyllis bilineata Kudenov and Harris, 1995. This species was found among the type materials of *S. californiensis* by Kudenov and Harris (1995) and may be widespread in California; readily recognized by having two longitudinal rows of two types of macropapillae along the dorsum of the body, in silty sediments and rocky habitats; intertidal to shelf depths.

Sphaerosyllis ranunculus Kudenov and Harris, 1995. With two types of dorsal cirri, but unlike *S. bilineata*, lacks longitudinal rows of dorsal papillae, low water to shelf depths in mixed sediments; Canada to southern California.

EUSYLLINAE

Odontosyllis parva Berkeley, 1923. Among algae, mussels, bryozoans, and sponges. See Banse 1972a.

Odontosyllis phosphorea Moore, 1909. Common among algal holdfasts and *Zostera*; commonly collected at night with lights; intertidal to shelf depths. See Banse 1972a; Kudenov and Harris 1995.

Pionosyllis gigantea Moore, 1908. Intertidal to shallow subtidal; in mixed and muddy sediments.

Pionosyllis magnifica Moore, 1906. Intertidal to shallow subtidal; in rocky and mixed sediments.

SYLLINAE

*Amblyosyllis sp. Rocky habitat, open coast, associated with siliceous sponges; not confirmed with recent collections.

Ehlersia heterochaeta (Moore, 1909). Widespread in the eastern Pacific, Canada to Mexico, low water to shelf depths in mixed sediments. See Kudenov and Harris 1995.

Ehlersia hyperioni (Dorsey and Phillips, 1987). Similar to *E. heterochaeta* with which it may occur; but differs in lacking eyes and in details of blade lengths of compound falcigers in middle body segments, low water to about 150 m in mixed to fine sediments.

**Haplosyllis spongicola* (Grube, 1855). Among sponges and bryozoans in shallow rocky habitats. All setae simple, terminating in a large fang at right angles to the shaft and one or two small superior teeth. Widely reported throughout the world's oceans.

Syllis elongata (Johnson, 1901). Common among algal holdfasts on rocky shores. See Banse 1972a.

Syllis gracilis Grube, 1840. With algal holdfasts and in crevices in rocky habitats.

Syllis spongiphila Verrill, 1885. Associated with various habitats including sponges, mussel beds, shell surfaces, rocks, and mixed sediments; intertidal to shelf depths. See Kudenov and Harris 1995.

Trypanosyllis gemmipara Johnson, 1901. Among bryozoans, sponges, and tunicates in exposed rocky habitats.

Trypanosyllis ingens Johnson, 1902. Among rocks and algal holdfasts. See Berkeley and Berkeley 1952, J. Fish. Res. Bd. Canada 8: 488–496.

Typosyllis aciculata Treadwell, 1945. Rocky habitats. See Reish 1950, Amer. Mus. Nov. no. 1466: 1–5. (redescription).

Typosyllis adamanteus (Treadwell, 1914) (=*Syllis spenceri* Berkeley and Berkeley, 1938). Among algae and barnacle clumps. See Banse 1972a.

* = Not in key.

Typosyllis alternata (Moore, 1908). In mixed sediments, gravel, and rocks; low water to slope depths.

Typosyllis armillaris (Muller, 1771). A widely reported species.

Typosyllis hyalina (Grube, 1863). In cryptic habitats associated with algae, mussels, and sponges on hard surfaces; intertidal to shallow shelf depths.

Typosyllis nipponica Imajima, 1966. Apparently introduced from Japan; now reported from San Francisco Bay to Los Angeles Harbor. Color green or brown, assuming the color of algae being ingested (L. Harris, personal communication).

Typosyllis pulchra (Berkeley and Berkeley, 1938). Among mussel and algal holdfasts. See Banse 1972a; Heacox, 1980, Pac. Sci. 34: 245–259 (reproduction and larval development).

NEREIDIDAE

Nereidids (known as sand or pile worms) are among the most familiar polychaetes found along the shore. There are numerous species in all habitats from the open coast to estuaries in soft sediments and cryptic habitats. Nereidids are often used in classrooms as examples of "typical" polychaetes and are commonly used as experimental animals. Although the family contains a great number of genera and species with quite variable morphologic features, nereidids are easily recognized by the presence of two palps on the prostomium and four pairs of tentacular cirri arising from the anterolateral corners of the peristomium.

The prostomium bears two antennae (rarely one), two biarticulate palps, and usually two pairs of eyes (plate 128A). The pharynx is an eversible proboscis that bears two fang-shaped, often serrated terminal jaws (plates 128A, 128B). The everted proboscis has a characteristic proximal or oral ring and a distal or maxillary ring; these rings may be equipped with groups of papillae or hardened paragnaths of various sizes, shapes, and distributional patterns that are of great taxonomic importance (plates 128A, 128B). It is therefore often necessary to dissect the proboscis to identify species if it is not everted. The incision should be made ventrally, slightly off to one side to keep median structures intact, from the mouth down to about setigers 6 or 7. Care should be taken not to break the jaws in attempting to spread the often very muscular and rigid proboscideal wall. The paragnaths of the maxillary ring are used for seizing prey; whereas those on the oral ring are mainly used for browsing and burrowing. The peristomium bears four pairs of smooth or moniliform tentacular cirri on the anterior margin. *Cheilonereis* has a greatly expanded peristomium that forms a collar around the prostomium. Parapodia are uniramous in the first two setigers and distinctly biramous in all subsequent setigers. Both parapodial rami possess an often minute acicular lobe and one to three conspicuous lobes above and below the acicula that are called ligules. The proportions of the parapodia, especially the ligules, differ greatly among species and are very reliable taxonomic characters. Because the parapodial shape often changes from anterior to posterior on one animal, an anterior, middle, and posterior parapodium should be examined for accuracy. Setae are generally compound, with the articulation between shaft and blade either symmetrical (homogomph) or asymmetrical (heterogomph; slightly asymmetrical articulations are also called hemigomph). The blades may be distally blunt (falcigerous) or drawn out to a very fine, hairlike tip (spinigerous). The distribution of setal types is more impor-

tant for species identifications than the setal morphology, and is best recorded when parapodial lobes are examined. The pygidium bears two ventrolaterally inserted anal cirri.

KEY TO NEREIDIDAE

1. Distinct asetigerous segment between peristomium and first setiger; paragnaths present or absent 2
— Without asetigerous segment between peristomium and first setiger (plate 151A); paragnaths absent; males with highly modified notosetae (plate 151C). . . . . *Micronereis nanaimoensis*
2. Peristomium enlarged, forming a collar around and under prostomium (plate 152A); commensal with hermit crabs . . . . *Cheilonereis cyclurus*
— Peristomium not so enlarged. 3
3. Posterior notopodia with simple, dark brown, hooked setae (plate 152B); peristomial cirri greatly elongated . . . . *Platynereis bicanaliculata*
— Posterior notopodia without such setae. 4
4. Paragnaths absent from either the maxillary or oral ring 5
— Paragnaths present on both rings 6
5. Paragnaths limited to oral ring (areas VII and VIII) as six brown cones set in transverse row, two cones to each area; neuropodial lobes of posterior setigers large, saclike (plate 152C) . . . . *Eunereis longipes*
— Paragnaths present only on the maxillary ring (area I with none, II with three minute cones, III with one minute cone, IV with three small cones in transverse row); anterior parapodia with large dorsal and ventral lobes (plate 152D) . . . . *Ceratonereis tunicatae*
6. Posterior notopodia with some homogomph falcigers in addition to spinigers (plate 153E) 7
— Posterior notopodia without homogomph falcigers. 11
7. Posterior notopodial lobes greatly elongate, straplike (plate 152E) . . . . *Nereis vexillosa*
— Posterior notopodial lobes otherwise 8
8. Proboscis with many tiny paragnaths over both oral and maxillary rings (plate 152F, 152G) 9
— Proboscis with paragnaths larger and restricted to certain areas 10
9. Jaws with 8–9 teeth (plate 152H); eyes small; inhabits sediments . . . . *Nereis procera*
— Jaws with 3–5 teeth (plate 152I); eyes large; rocky intertidal . . . . *Nereis eakini*
10. (*Note three choices*) Parapodial lobes typically dark (plate 153A); no difference in size of middle and posterior notopodial lobes; proboscis lacks paragnaths in area V . . . . *Nereis pelagica neonigripes*
— Parapodial lobes not dark; notopodial lobe in anterior region (plate 153B) increasing in size to subrectangular lobe about twice as long as broad (plate 153C); body bright green to brown; proboscis lacks paragnaths in area V. . . . . *Nereis grubei*
— Parapodial lobes not dark, posterior notopodial lobes not larger (plate 153D), body pale, dorsum with bars of brown or rust-colored pigment; proboscis with a single paragnath in area V . . . . *Nereis latescens*
11. Area VI of proboscis with transverse paragnaths, body pale with dark, quadrate bars . . . . *Perinereis monterea*
— Area VI of proboscis with conical paragnaths; pigment bars absent 12

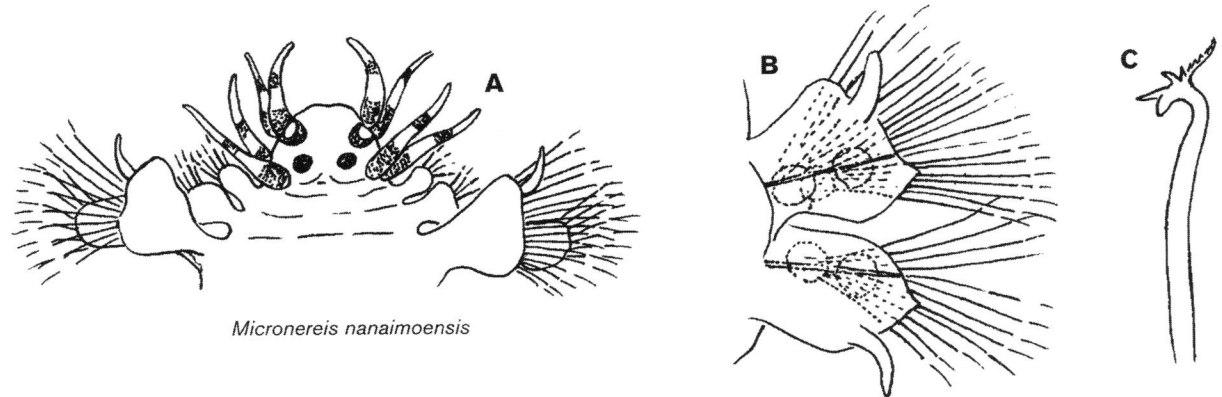

PLATE 151 Polychaeta. Nereididae: A, *Micronereis nanaimoensis*; B, parapodium; C, notoseta of male (all after Berkeley and Berkeley).

12. (*Note three choices*) Posterior notopodial lobe becoming broadly oval, foliose, with minute dorsal cirrus (plate 153F); marine; see note in species list . . . . *Neanthes brandti* and *Neanthes virens*
— Posterior parapodial lobes very elongate (plate 153G); estuarine . . . . *Neanthes succinea*
— Posterior parapodial lobes short (plate 153H); in coastal estuaries, lagoons, and lakes; see note in species list . . . . *Neanthes limnicola*

LIST OF SPECIES

See Hartman 1936, Proc. U.S. Nat. Mus. 83: 467–480; 1940; Hilbig 1994b.

Ceratonereis tunicatae Hartman, 1936. Color in life distinctive, with broad purple bands on dorsum; known only from the intertidal zone of northern California in ascidians.

Cheilonereis cyclurus (Harrington, 1897). Ventral part of the peristomium enlarged, forming a collarlike structure. Commensal with hermit crabs; low intertidal to shallow shelf depths.

Eunereis longipes Hartman, 1936. Prostomium with four unusually large eyes. Reported from northern California, intertidal in rock crevices.

Namanereis pontica (Bobretsky, 1872) (=*Lycastopsis pontica neapolitana* La Greca, 1950). From peat-clay banks in Tomales Bay. See Hartman 1959, Bull. Mar. Sci. 9: 163. This species is a cosmopolitan representative of a subfamily (Namanereinae) of very small worms with reduced nereidid features, inhabiting fresh or brackish water at high intertidal or spray zone levels, and usually overlooked. See Glasby 1999, Rec. Aust. Mus. Suppl. 25: 1–144 (taxonomy, phylogeny, biogeography).

Micronereis nanaimoensis Berkeley and Berkeley, 1953. (=*Phyllodocella bodegae* Fauchald and Belman, 1972). Egg cocoons found attached to eelgrass near the entrance to Tomales Bay and to *Obelia* sp. Bodega Harbor; pelagic larvae released (Blake, unpublished). See Berkeley and Berkeley 1953, J. Fish. Res. Bd. Can. 10: 85–95 (life history).

Neanthes brandti (Malmgren, 1866) and *Neanthes virens* (Sars, 1835). These species are separated with difficulty. The former has many paragnaths on both rings of the proboscis while the second has few. Because of the overlap in geographic range and great similarities in morphology, especially of the parapodia, there is a possibility that these species may not be genetically isolated. Both are known from intertidal areas, including sand and rock. See Abbott and Reish 1980; Fong 1991, J. Exp. Mar. Biol. Ecol. 149: 177–190 (environmental effects on reproduction).

Neanthes limnicola (Johnson, 1901) (=*N. lighti* Harman, 1938). This species is sometimes referred to as *N. diversicolor* (Müller 1776), to which it is closely related. The two species, however, have greatly different reproductive patterns and need not be confused. *Neanthes limnicola* is viviparous and hermaphroditic, whereas *N. diversicolor* of Europe and eastern North America is dioecious and has free-living demersal larvae. See Smith 1950, J. Morph. 87: 417–465 (embryology); Pettibone 1963; Fong and Pearse 1992a, Mar. Biol. 112: 81–89; 1992b, Biol. Bull. 182: 289–297 (photoperiodicity in life history); Fong and Garthwaite 1994, Mar. Biol. 118: 463–470 (genetics and life history). Inhabits brackish or even freshwater of estuarine streams, coastal lagoons, and Lake Merced (San Francisco).

Neanthes succinea (Frey and Leuckart, 1847). One of the most common nereidids in North America. The species is easily recognized by the large posterior notopodial lobes that bear a small distal cirrus; sandy mud to muddy sediments of bays.

Nereis eakini Hartman, 1936. This species is characterized by having numerous paragnaths on the oral ring; known only from central California in intertidal rocky habitats.

Nereis grubei (Kinberg, 1866) (=*N. mediator* Chamberlin, 1918). Among algae on rocky shores; intertidal to low water. See Reish 1954, Occ. Pap. no. 14, Allan Hancock Found. (life history, ecology); Schroeder 1967, Biol. Bull. 133: 426–437 (metamorphosis and eiptikous setae); 1968, Pac. Sci. 22: 476–481 (life history); Abbott and Reish 1980 (review).

Nereis latescens Chamberlin, 1919. Intertidal to shallow depths, common in algal holdfasts and among rocks and debris.

Nereis natans Hartman, 1936. Perhaps a sexual stage of another central California species in this list.

Nereis pelagica Linneaus, 1758 (includes *N. pelagica neonigripes* Hartman, 1936). A widely distributed species from intertidal to upper slope depths in rocky habitats.

Nereis procera Ehlers, 1868. Intertidal to upper slope depths in silt, sand, and on rocks.

Nereis vexillosa Grube, 1851. This species is easily recognized by the green color and the greatly elongate, narrow notopodial lobes bearing a terminal cirrus found in posterior setigers. An eastern Pacific species, intertidal on the open coast with

* = Not in key.

PLATE 152 Polychaeta. Nereididae: A, *Cheilonereis cyclurus*; B, *Platynereis bicanaliculata*, posterior notopodial hook; C, *Eunereis longipes*, posterior parapodium; D, *Ceratonereis tunicatae*, anterior parapodium; E, *Nereis vexillosa*, posterior parapodium; F, *Nereis procera*, ventral view with proboscis everted; G, dorsal view, with proboscis everted; H; jaw showing number of teeth; I, *Nereis eakini*, jaw.

mussels and barnacles, also on pilings. See Johnson 1943, Biol. Bull. 84: 106–114 (life history); Woodin 1977, Mar. Biol. 44: 39–42 (algal gardening behavior); Wilson 1980, J. Exp. Mar. Biol. Ecol. 46: 73–80 (interaction between terebellid and nereidid).

Perinereis monterea (Chamberlin, 1918). Rocky habitats; rare.

Platynereis bicanaliculata (Baird, 1863). Abundant in protected rocky habitats among algal holdfasts; builds tubes and aggregates; with *Phyllospadix*. See Woodin 1974, Ecol. Monogr.

44: 171–187 (life history and ecology); Woodin 1977, Mar. Biol. 44: 39–42 (algal gardening behavior); Blake 1975a (larval development); Abbott and Reish 1980.

SPHAERODORIDAE

The Sphaerodoridae is a relatively small family of deposit-feeding polychaetes that typically occur in offshore shelf and slope

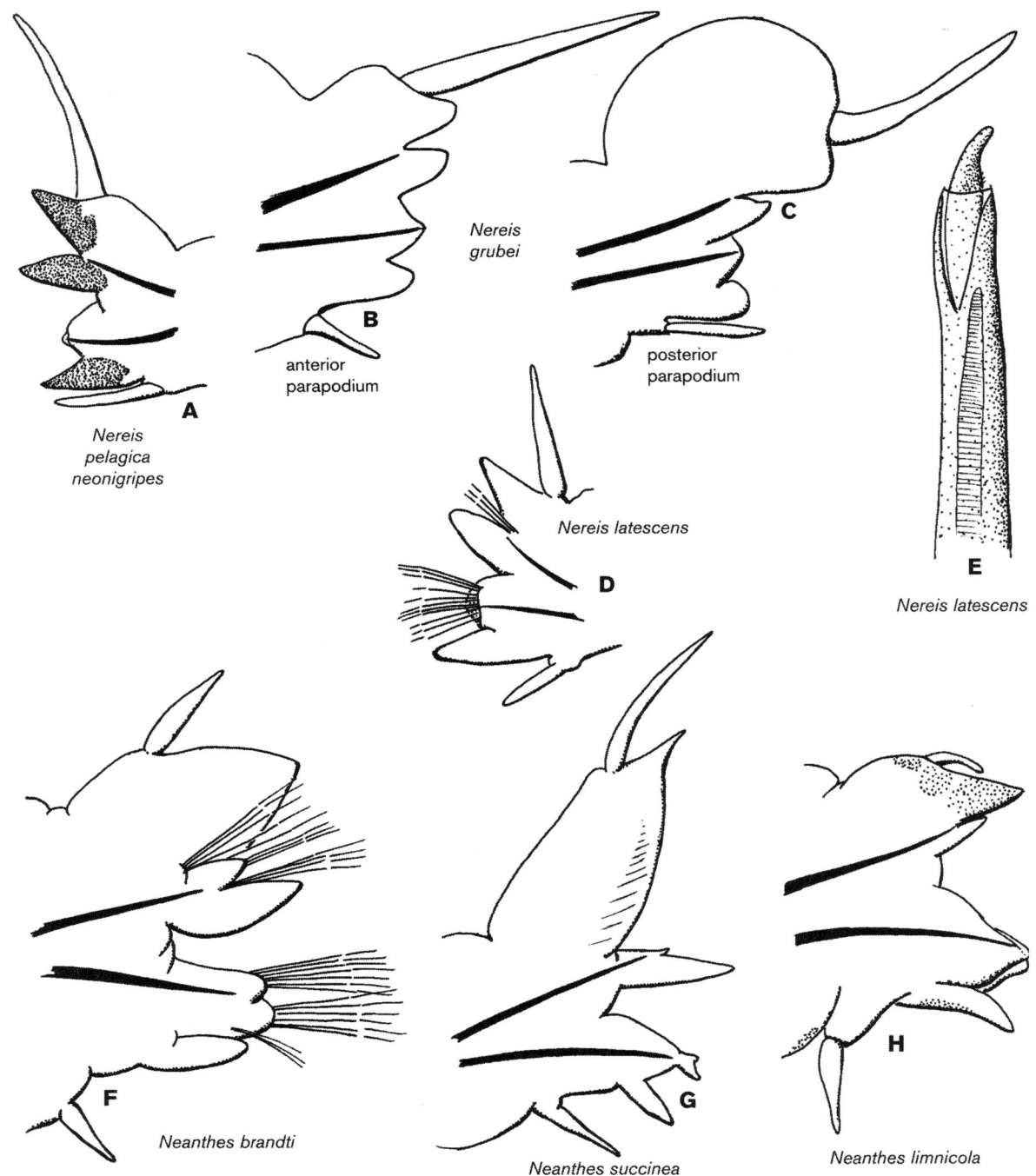

PLATE 153 Polychaeta. Nereididae: A, *Nereis pelagica neonigripes*, posterior parapodium; B, *Nereis grubei*, anterior parapodium; C, posterior parapodium; D, *Nereis latescens*, posterior parapodium; E, homogomph falciger; F, *Nereis brandti*, posterior parapodium; G, *Neanthes succinea*, posterior parapodium; H, *Neanthes limnicola*, posterior parapodium (all after Blake).

depths. They are commonly called "bubble worms" because they have two or more rows of inflated capsules on the dorsum. In addition, the body is typically ornamented with additional epidermal tubercles or papillae on the dorsal and ventral surfaces. The bodies of sphaerodorids are either short and grublike or long and slender. The prostomium bears two or three pairs of lateral antennae and a single medial antenna. There is one pair of peristomial cirri. Dorsal cirri are absent; ventral cirri are present. Setae are either simple or compound; simple curved hooks may be present in anterior segments. See Fauchald 1974, J. Nat. Hist. 8: 257–289 (revision of the family); Kudenov 1994 (California species).

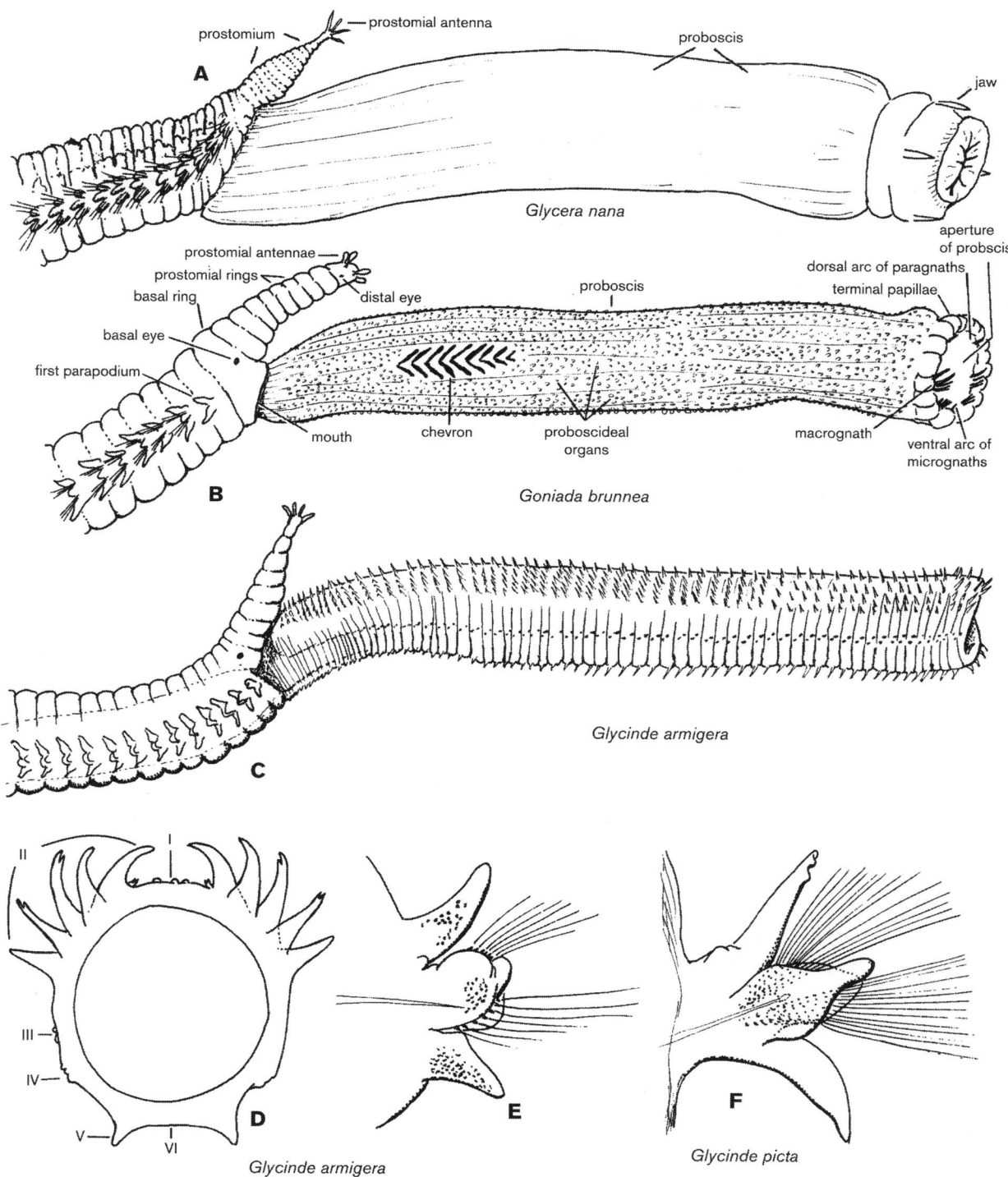

PLATE 154 Polychaeta. Glyceridae and Goniadidae: A, *Glycera nana*, proboscis everted; B, *Goniada brunnea*, proboscis everted; C, *Glycinde armigera*, proboscis everted; D, cross-section through middle of everted proboscis showing distribution of proboscideal armature; E, anterior parapodium; F, *Glycinde picta*, anterior parapodium (B–D, F, after Hartman; rest after Blake).

GONIADIDAE

The goniadids are easily recognizable by the elongate, conical, annulated prostomium and the slender, long body divided into two to three distinct regions: 1) an anterior region with uniramous parapodia and 2) a wider posterior region carrying bira-

mous parapodia. In contrast, the closely related glycerids do not exhibit distinct body regions. To more carefully distinguish goniadids from glycerids, an examination of the parapodia and proboscis is often required. The proboscis of goniadids is distinctive; very long, cylindrical, and equipped with a variety of terminal jaws (macro- and micrognaths) arranged in a more or less complete circlet and covered with proboscideal organs of very different and sometimes striking appearances. The proboscideal armature may also include chevrons, which are dark, V-shaped jaw pieces arranged in two lateral rows close to the base of the proboscis. Unlike glycerids, goniadids rarely evert their proboscis when dying in the fixative; proboscideal organs and macro- and micrognaths are therefore only visible after dissection. The chevrons can sometimes be seen through the body wall, unless the specimen is too large or heavily pigmented. The proboscis of goniadids can be extremely long, so a longitudinal cut to open the pharynx should reach back about 50 segments to reveal the terminal jaws. For the observation of the proboscideal organs, it may be necessary to examine an entire cross-section of the proboscis because shapes may vary greatly from dorsal to ventral.

KEY TO GONIADIDAE

1. Proboscis with soft papillae over the surface and with a set of chevrons on each side at base (plate 154B) . . . . *Goniada brunnea*
— Proboscis with numerous hard, yellow, chitinized spines (plate 154C); chevrons absent . . . . *Glycinde* 2
2. Dorsal cirrus incised near the tip (plate 154F); proboscideal organs of area V with unique duck foot–shaped papillae; ventral micrognaths present . . . . *Glycinde picta*
— Dorsal cirrus entire (plate 154E); proboscideal organs of area V conspicuous, conical papillae resembling those of area II (plate 154D) . . . . *Glycinde armigera*

LIST OF SPECIES

See Hartman 1940, 1950; Hilbig 1994c; Böggemann 2005 (revision of family).

Glycinde armigera Moore, 1911. Intertidal to upper slope depths in sandy mud sediments. See Blake 1975a (larvae).

Glycinde picta Berkeley, 1927 (=*G. polygnatha* Hartman, 1950). Intertidal to shallow shelf depths; in sandy mud. See Blake 1975a (juvenile morphology); Böggemann 2005 (revision of family).

Goniada brunnea Treadwell, 1906. Intertidal to outer shelf depths; in sandy mud and mixed sediments; low intertidal to deep water.

GLYCERIDAE

The glycerids, commonly called bloodworms, can reach considerable size (up to 800 mm long). They are easily recognized by their pointed, annulated prostomium and numerous and very crowded segments. To separate the glycerids from the closely related goniadids, the parapodia and proboscis must be examined (see below), although goniadids are typically more cylindrical and slender than glycerids and have bodies that are divided into 2–3 distinct regions instead of having the body segments consistent from the anterior to posterior end. The prostomium of glycerids is conical, elongate, more or less distinctly annulated, and usually tapers to a very fine tip. The terminal ring bears two pairs of small, typically biarticulate antennae and sometimes a pair of minute eyes. The basal ring is fused with the peristomium and is often distinctly wider and longer than the other rings. A pair of more conspicuous eyes may be present, and the posterior margin bears a pair of

nuchal organs. The very long and muscular proboscis is densely covered with small, transparent papillae called proboscideal organs that occur in various shapes ranging from spherical to cylindrical and slender. These proboscideal organs also bear a variety of ridges whose form and number are highly diagnostic but unfortunately are best seen with the scanning electron microscope. When the proboscis is fully everted, four dark, chitinous terminal jaws are visible, each consisting of a hook-shaped fang and a rodlike or V-shaped support, the aileron. The branchiae are important for species identification and, when present, are protrusions of the parapodia. They may be simple and digitiform or blister- or saclike in appearance, or they may be branched. Some species have retractile branchiae that may only be discernible, when retracted, as an area of somewhat loose, very thin epidermis on the parapodial wall.

KEY TO GLYCERIDAE

1. Parapodia uniramous; setae all compound 7
— Parapodia biramous; notosetae simple, neurosetae compound 2
2. Parapodia with one postsetal lobe; branchiae absent 3
— Parapodia with two postsetal lobes; branchiae present 4
3. Parapodia with two presetal lobes throughout; proboscideal organs without ridges (plate 155A) . . . . *Glycera nana*
— Parapodia with one presetal lobe in posterior setigers; proboscideal organs with transverse ridges (plate 155B) . . . . *Glycera tenuis*
4. Parapodia with blisterlike branchiae on posterodorsal sides (plate 155C) . . . . *Glycera robusta*
— Parapodia with long fingerlike branchiae 5
5. Branchiae retractile, emerging from posterodorsal side of the parapodia, forming branched lobes (plate 155D) . . . . *Glycera americana*
— Branchiae not retractile, not branched 6
6. Parapodia with a single branchia located above setal lobe (plate 155E) . . . . *Glycera macrobranchia*
— Parapodia with two fingerlike branchiae, one located above and other below the setal lobe (plate 155F) . . . . *Glycera dibranchiata*
7. Parapodial lobes short, wider than long; color in life light green (plate 155G) . . . . *Hemipodia californiensis*
— Parapodial lobes longer than wide; color in life dull or bright red (plate 155H) . . . . *Hemipodia simplex*

LIST OF SPECIES

The family was revised by Böggemann (2002), resulting in synonymies that create numerous widely distributed species. These synonymies contrast with the results of O'Connor (1986), who suggested that there were many sibling species as yet undescribed. Some of Böggemann's synonymies are incorporated, others are not. The names used in the previous edition are indicated as synonyms where relevant. Readers should expect this issue to be revisited in the future. A thoughtful review of Böggemann (2002) together with comments on California glycerids is presented in *SCAMIT Newletter*, 2002, Vol. 21(7): 1–6 + Annotated Table. See Hartman 1940, 1950; O'Connor 1986; Hilbig 1994d; Böggemann 2002.

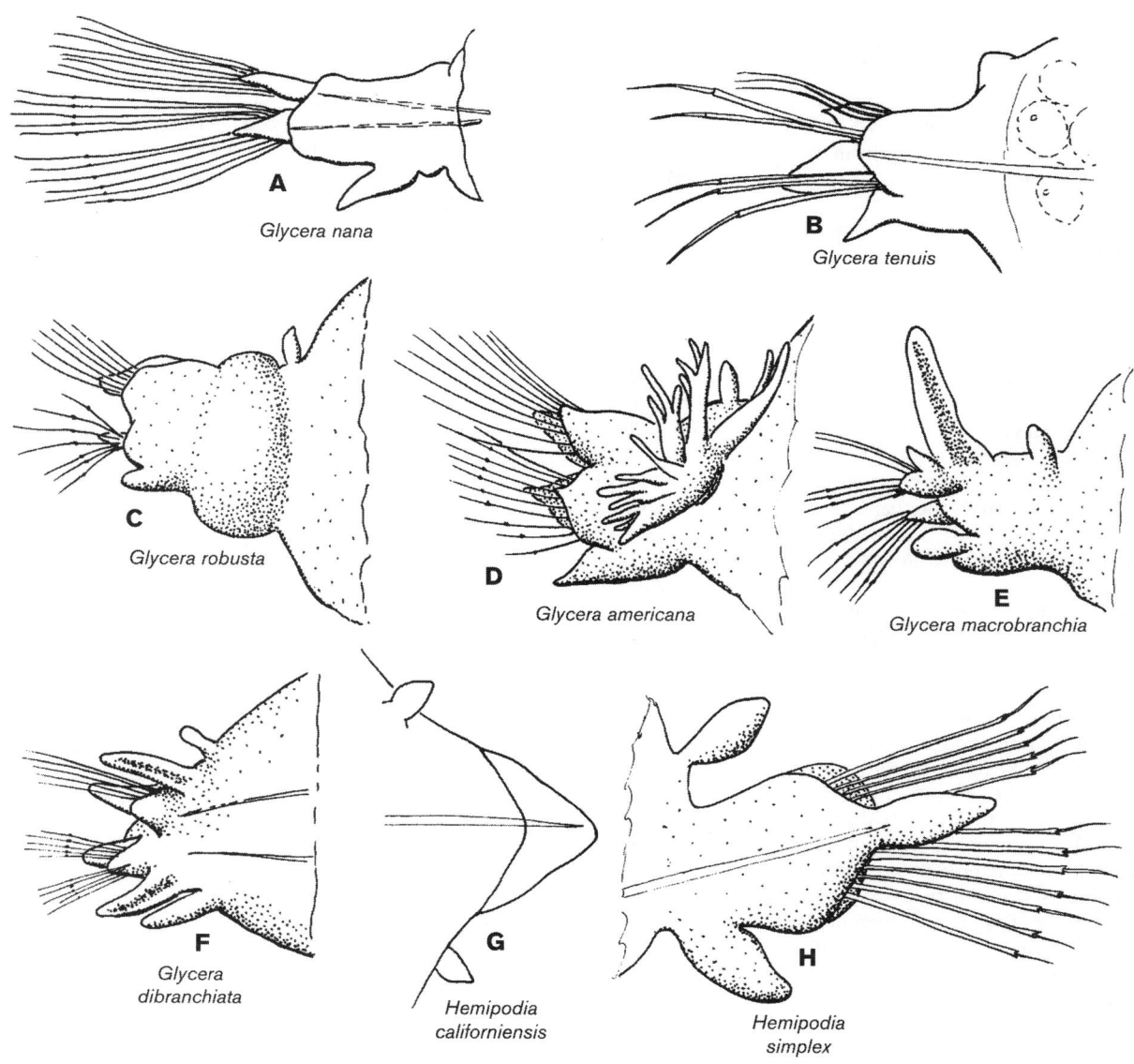

PLATE 155 Polychaeta. Glycerid parapodia: A, *Glycera nana*; B, *Glycera tenuis*; C, *Glycera robusta*; D, *Glycera americana*; E, *Glycera macrobranchia*; F, *Glycera dibranchiata*; G, *Hemipodia californiensis*; H, *Hemipodia simplex* (A, B, G after Hartman; rest after Blake).

Glycera americana Leidy, 1855. Intertidal to 120 m in sandy mud sediments. See Abbott and Reish 1980.

Glycera dibranchiata Ehlers, 1868. In sandy mud to mud; intertidal to shelf depths. A well-known bait-worm that is harvested commercially in Maine and eastern Canada, packed in *Fucus* and shipped via air freight to all coasts of the United States. See Klawe and Dickie 1957, Bull. Fish. Res. Bd. Can. 115: 1–37 (biology, and use as a baitworm).

Glycera macrobranchia Moore, 1911 (=*G. convoluta* Keferstein, 1862 in previous edition). Sandy sediments; intertidal to midshelf depths.

Glycera nana Johnson, 1901 (=*G. capitata* of previous edition). Hilbig (1994d) referred local records of *G. capitata* Oersted, 1843, to *G. nana* Johnson, thus supporting the suggestion of O'Connor (1986) that *G. capitata* included a complex of closely related species. However, Böggemann (2002) defined *G. capitata* more broadly and includes *G. nana* in the synonymy. This synonymy is not supported here. A comparison of *G. cap-*

itata specimens from the northeastern United States and elsewhere with shallow and deep-water California material clearly indicates that there are at least two different North American species. *G. nana* occurs in mixed sand and muddy sediments; intertidal to slope depths.

Glycera robusta Ehlers, 1868. Intertidal to outer-shelf depths in sand and cobble sediments.

Glycera tenuis Hartman, 1944. Intertidal to shelf depths in clean sands. Böggemann (2002) referred this species to the cosmopolitan *Glycera oxycephala* Ehlers, 1887. However, *G. tenuis* is here retained as a distinct species because it has a single presetal lobe and 13–16 ridges on the proboscideal organs, whereas *G. oxcephala* has two presetal lobes and 9–10 ridges on the proboscideal organs. Böggemann (2002) defines *G. oxycephala* with 5–20 ridges on the proboscideal organs but apparently does not account for the differences in numbers of presetal lobes. See Hartman 1944b; Blake 1975a (larvae); Böggemann 2002 (taxonomy, morphology).

Hemipodia californiensis (Hartman, 1938). Muddy sediments of estuaries. According to Böggemann (2002), the well-known generic name *Hemipodus* Quatrefages, 1866, is superseded by *Hemipodia* Kinberg, 1865.

Hemipodia simplex (Grube, 1857) (=*G. borealis* Johnson, 1901). Intertidal to shallow-shelf depths in sands to sandy muds. See Blake 1975a (larvae); Böggemann 2002 (taxonomy, morphology).

NEPHTYIDAE

Nephtyids are among the more conspicuous and familiar polychaetes found in marine sediment. They are long, strongly muscular, with a cylindrical anterior region that contains the proboscis and a tapering middle and posterior region that is roughly rectangular in cross-section. Although the smallest species are <10 mm long when reaching sexual maturity, other species can grow to considerable size, reaching 20 cm in length and 1 cm in width. The most characteristic feature of nephtyids is the shape of the parapodia. The widely separated rami and the usually dense, fan-shaped setal fascicles pointing diagonally outward from the "corners" of each segment are characteristic for all nephtyids. Nephtyids preserve well and are easy to identify because taxonomically important characters remain intact; only the posterior end or the anal cirrus will occasionally break off during fixation. Prostomial shape is defined very early during postlarval development, so it can be used as a reliable taxonomic character in the identification of all stages of development. Other reliable taxonomic characters include the numbers and arrangement of papillae on the anterior of the everted pharynx, and especially the form of the interramal cirrus. This structure is sometimes called a branchia and is inserted just underneath the dorsal cirrus; it is present on almost all parapodia, starting from setigers 3, 4, or between 5 and 10 and lacking on the last 10–20 setigers. Generally, interramal cirri are slender and digitiform, but they may be basally inflated or foliaceous; they are either short and straight, or long and curved outward (recurved) or inward (involute).

KEY TO NEPHTYIDAE

1. Interramal cirri curved inward, with free end directed toward parapodial wall (involute) . . . . *Aglaophamus*
— Interramal cirri curved outward, with free end directed away from parapodium (recurved) (plate 156B) . . . .
. . . . *Nephtys* 2
2. Interramal cirri from setiger 3; prostomium with spread-eagle pigment pattern (plate 156C); first few segments pale dorsally; inhabits clean, sandy beaches . . . .
. . . . *Nephtys californiensis*
— Interramal cirri from setigers 4–6 3
3. Interramal cirri begin on setiger 4; prostomium and first few segments with characteristic dark pigment pattern on dorsal side (plate 156A); inhabits muddy sand flats . . . . .
. . . . *Nephtys caecoides*
— Interramal cirri begin on setigers 5–6 4
4. Posterior prostomial antennae bifurcate (plate 156D); interramal cirri begin on setiger 5; subdermal eyes present at level of setiger 3 (plate 156E); small species . . . .
. . . . *Nephtys cornuta*
— Posterior prostomial antennae not bifurcate; interramal cirri begin on setigers 5–6; subdermal eyes not present in adults; large species . . . . *Nephtys caeca*

LIST OF SPECIES

See Hartman 1938, Proc. U.S. Nat. Mus. 85: 143–158; Hartman 1940, 1950; Hilbig 1994e; Lovell 1997.

Aglaophamus spp. Several species of this genus occur in shelf and slope sediments. See Hilbig 1994e.

Nephtys caeca (Fabricius, 1780). Sandy sediments; rare, possibly introduced from the eastern United States.

Nephtys caecoides Hartman, 1938. Specimens of this species have distinctive but variable pigmentation on the prostomium and some anterior setigers; in sandy mud sediments of bays, lagoons, and the open shelf; intertidal to shelf depths. See Clark and Haderlie 1962, J. Anim. Ecol. 31: 339–357.

Nephtys californiensis Hartman, 1938. Similar to *N. caecoides*, but differing in having a more limited anterior dorsal pigment pattern that includes a sort of V-shaped pigment structure across the prostomium, sometimes with a central reddish spot. Intertidal in clean sandy beaches; very common, see Clark and Haderlie, above; also reported from shelf depths (Hilbig 1994e). See Abbott and Reish 1980 (review of biology).

Nephtys cornuta Berkeley and Berkeley, 1945. The synonymy of this species was established by Hilbig (1994e). The bifid ventral or posterior antennae, although variable in appearance, are highly distinctive. This species includes the subspecies *N. cornuta franciscana* Clark and Jones, 1955, from San Francisco Bay and *Aglaophamus neotenus* Noyes, 1980, the latter originally described from New England. The species is small, rarely exceeding 15 mm and usually much smaller. It retains juvenile features including the presence of subdermal eyes at about the level of setiger 3; these are often seen in postlarval forms of related species. Intertidal to outer shelf depths; in coastal embayments, the species occurs in muddy sediments where dense populations sometimes develop.

Nephtys ferruginea Hartman, 1940. Similar to *N. californiensis*, but occurring in subtidal depths (16–450 m) in mixed sediments and differing chiefly in details of the dorsal pigment pattern, numbers of rows of subdistal papillae on the proboscis, and by having (instead of lacking) an unpaired middorsal papilla on the proboscis. See Lovell 1997.

Nephtys parva Clark and Jones, 1955. A small species originally described from muddy sediments of San Francisco Bay and included in the previous edition (Blake 1975b) is not a valid taxon. However, there is a difference of opinion as to what it actually represents. *N. parva* was referred to *N. caecoides* by Hilbig (1994e) following examination of a paratype from the NMNH collections. Subsequently, Lovell (1997) referred the species to *N. cornuta* Berkeley and Berkeley following examination of the holotype. Given that Clark and Jones also described a subspecies of *N. cornuta*, *N. cornuta franciscana*, in the same paper with *N. parva*, it seems inconceivable that a careful morphologist such as the late Dr. Meredith Jones would have mistaken two so very different species, as is evident in the published descriptions of *N. cornuta franciscana* and *N. parva*. It is more likely that the archived types were miscurated because the published description of *N. parva* is more like *N. caecoides* than *N. cornuta*. See Clark and Jones 1955, J. Wash. Acad. Sci. 45: 143–146.

ONUPHIDAE

Onuphids are midsize to large worms with dorsoventrally flattened bodies. These worms have a distinctive prostomium

* = Not in key.

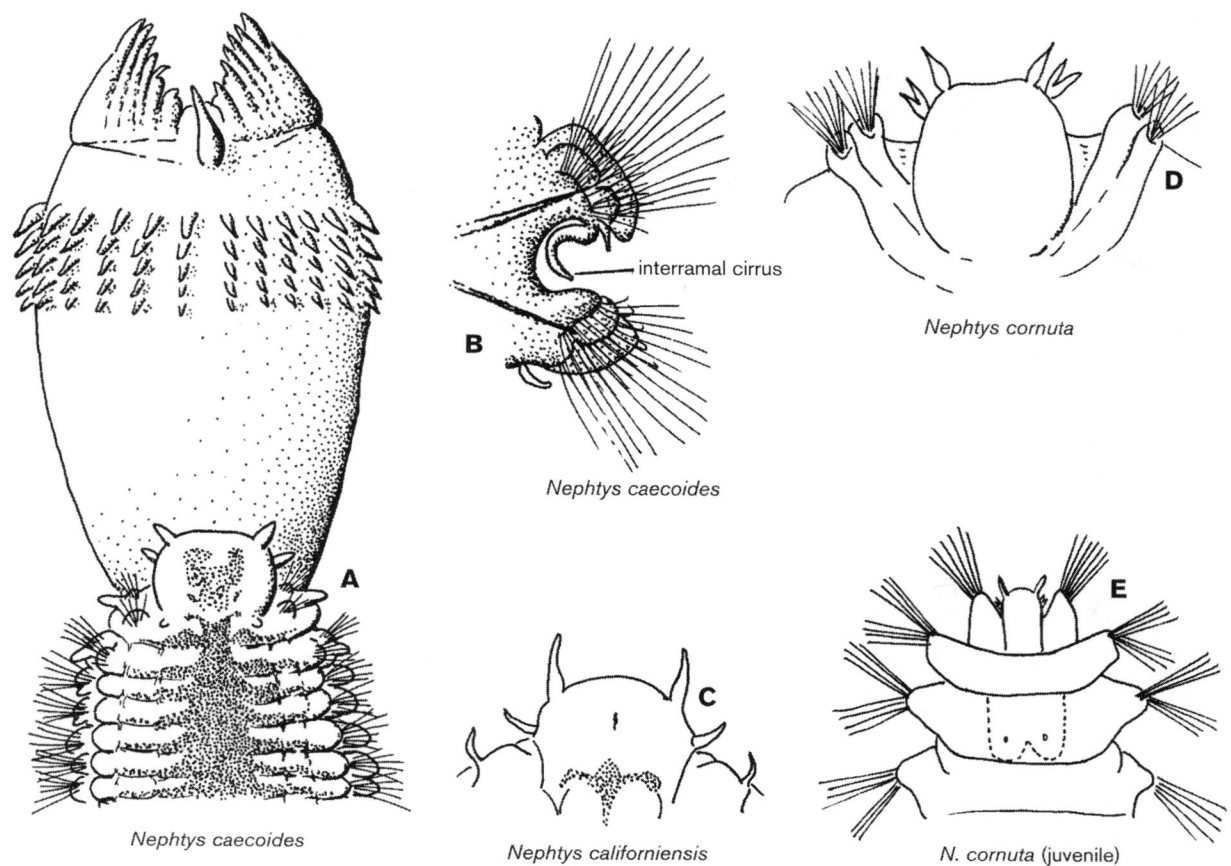

interramal cirrus

B

Nephtys caecoides

Nephtys cornuta

D

Nephtys caecoides

Nephtys californiensis

C

N. cornuta (juvenile)

E

PLATE 156 Polychaeta. Nephtyidae: A, *Nephtys caecoides*, dorsal view with proboscis everted; B, 30th parapodium, anterior view; C, *Nephtys californiensis*; D, *Nephtys cornuta*, dorsal view of adult; E, same, juvenile (C, after Hartman; D, after Berkeley; E, after Clark and Jones; rest after Blake).

bearing five occipital antennae, two frontal palps, and two labial palps. This arrangement of antennae and palps is consistent in all onuphids, in contrast to the morphologically similar eunicids where the number of antennae varies from 1–5. The first few parapodia are often specialized and, in combination with the head appendages, make the Onuphidae a distinctive and spectacular family of polychaetes. Onuphids occur in sandy sediments, low intertidal to subtidal.

KEY TO ONUPHIDAE

1. Peristomium without tentacular cirri . . . .
   . . . . *Hyalinoecia juvenalis*
— Peristomium with tentacular cirri (plate 157B) 2
2. Branchial filaments spiraled on the main stalk (plate 157A)
   . . . . *Diopatra* 3
— Branchial filaments simple, cirriform (plate 157D, 157G) 4
3. Pectinate setae with numerous small, short teeth . . . .
   . . . . *Diopatra ornata*
— Pectinate setae with 7–8 large, long teeth . . . .
   . . . . *Diopatra splendidissima*
4. Branchiae first present on setiger 1 . . . . *Onuphis* 5
— Branchiae first present from posterior setigers . . . .
   . . . . *Mooreonuphis stigmatis*
5. Branchiae on setigers 10–80 thick (plate 157D); subacicular hooks usually begin on setiger 9 (plate 157C); pseudo-

compound hooks usually tridentate, but third tooth may be very small, usually on setigers 1–4 (plate 157E, 157F) . . . . *Onuphis elegans*
— All branchiae slender (plate 157G); subacicular hooks usually begin on setigers 11–14; pseudocompound hooks tridentate . . . . *Onuphis iridescens*

LIST OF SPECIES

See Hartman 1944a; Hobson 1971; Fauchald 1968, 1982; Hilbig 1995a.

Diopatra ornata Moore, 1911. Central California to Canada in low water; forming erect, parchmentlike tubes to which are attached pieces of shell and algae. See Hartman 1944a; 1968; Fauchald 1968; Abbott and Reish 1980.

Diopatra splendidissima Kinberg, 1865. Southern California to Ecuador; intertidal to 30 m in protected embayments forming erect, parchmentlike tubes to which are attached small pieces of shell and sticks, not as ragged in appearance as tube of *D. ornata*. See Hartman 1944a, 1968; Fauchald 1968; Abbott and Reish 1980.

Hyalinoecia juvenilis Moore, 1911. These onuphids are commonly called "soda straw" worms because of their transparent cylindrical tubes; offshore in shelf sediments, 10 m to the shelf break.

Mooreonuphis stigmatis (Treadwell, 1922) (=*Onuphis stigmatis*; =*Nothria stigmatis* in previous edition). Intertidal to about

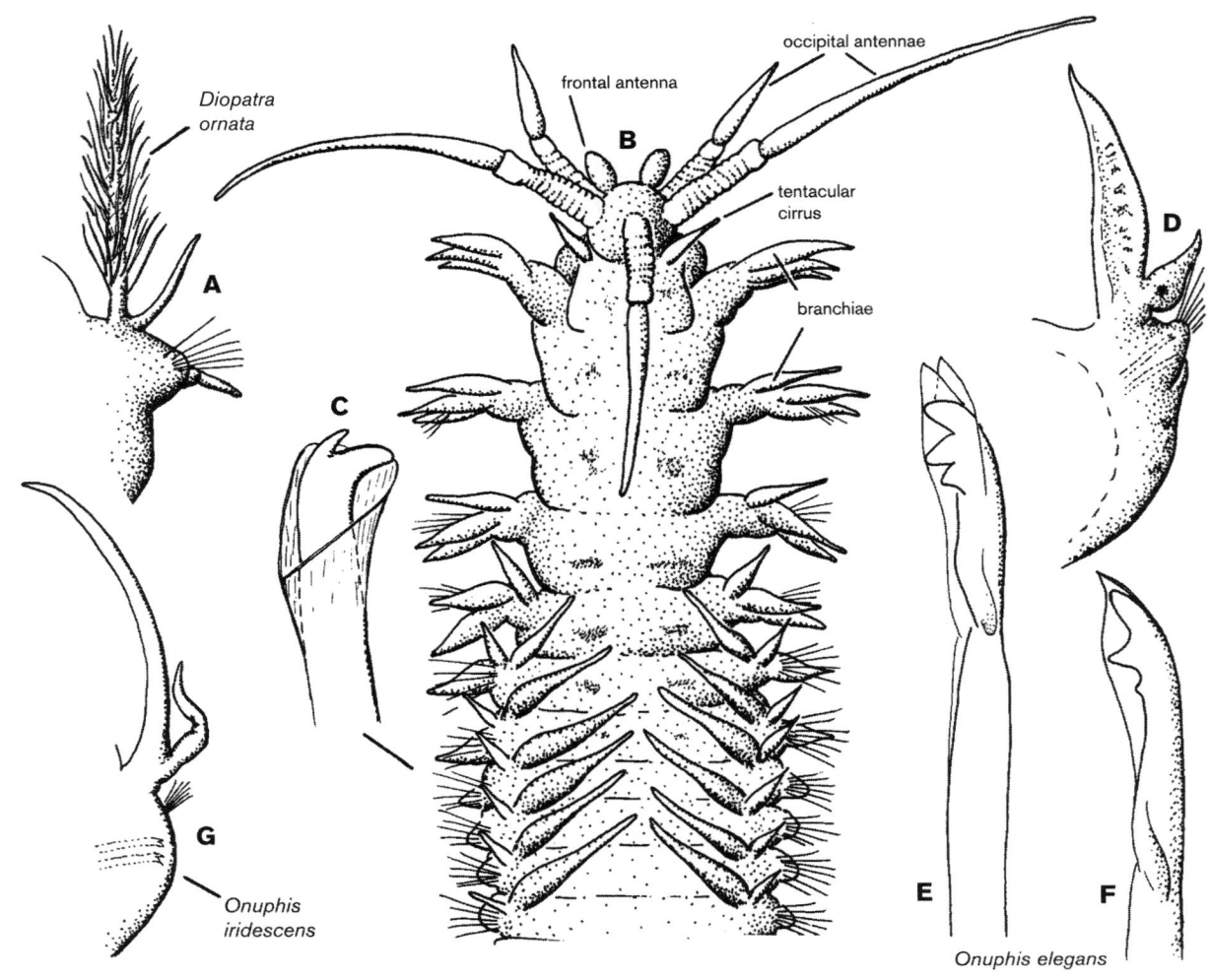

PLATE 157 Polychaeta. Onuphidae: A, *Diopatra ornata*, anterior parapodium; B, *Onuphis elegans*; C, subacicular hook from setiger 60; D, 16th parapodium; E, tridentate pseudocompound hook from setiger 2; F, tridentate pseudocompound hook from setiger 2 with small third tooth; G, *Onuphis iridescens*, 25th parapodium (G, after Hobson; rest after Blake).

40 m in sandy sediments; not forming permanent tubes. Hartman 1944a, 1968; Fauchald 1982 (taxonomy).

Onuphis elegans (Johnson, 1901) (=*Nothria elegans*). A common species in California embayments; prefers sandy sediments. See Blake 1975a (larval development, results suggest that post larval and juvenile setae and setal patterns are different than in adults). British Columbia to southern California, intertidal to 23 m.

Onuphis iridescens (Johnson, 1901) (=*Nothria iridescens*). Similar in morphology to *O. elegans,* but preferring mixed sand and silt sediments; intertidal to slope depths.

*Onuphis intermediates (*sensu* Hobson, 1971). Specimens intermediate between *O. elegans* and *O. iridescens* were reported by Hobson (1971). Hilbig (1995a) supported Hobson's suggestion that intermediates were in the process of speciation and may be isolated by biological characters. Central California, intertidal to about 75 m.

*Paradiopatra parva (Moore, 1911). An offshore species, in silty sediments, 10–300 m. Distinguished by the heads of the pseudocompound hooks, which are drawn out to long acute points. See Hilbig 1995a.

* = Not in key.

EUNICIDAE

The eunicids are among the more spectacular polychaetes because of their considerable size and vivid coloration. They are not only one of the largest polychaete families in terms of number of species, but also one of the few with a fossil record that reaches back to the Cretaceous (fossils are also known from most other euniciform families, although more scattered than for the eunicids). The largest known polychaete, reported by Fauvel (1923) as being about 3 m long, is a species of *Eunice.* The only polychaetes eaten regularly by people are swarming species of the genus *Palola* caught in regions of the South Pacific. The family is important ecologically because they break down coral rock. Eunicids are most closely related to the Onuphidae but can be distinguished from the latter by the presence of five antennae (but sometimes only one or three) arising from very short, cylindrical ceratophores; in contrast, onuphids always have seven prostomial appendages, including five antennae that arise from long, usually annulated ceratophores, and two frontal appendages referred to as frontal antennae or frontal palps. Eunicids are not quite as iridescent as onuphids but are often more colorful and typically

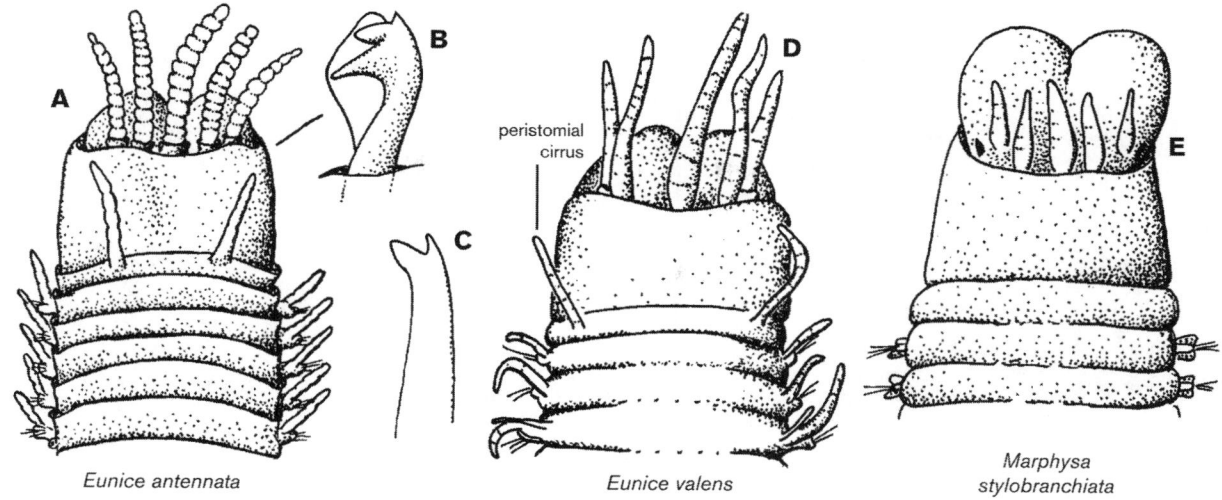

PLATE 158 Polychaeta. Eunicidae: A, *Eunice antennata*; B, subacicular hook; C, *Eunice valens*, subacicular hook; D, *Eunice valens*, anterior end; E, *Marphysa stylobranchiata* (C, D, after Fauchald; rest after Blake).

have larger, more delicate gills. In addition, eunicids build fragile, parchmentlike tubes, if any, whereas many onuphids are known for their hard, quill-like or very tough fibrous tubes.

KEY TO EUNICIDAE

1. With three prostomial antennae . . . . *Lysidice*
— With five prostomial antennae (plate 158A) 2
2. Peristomial cirri present (plate 158D) 4
— Peristomial cirri absent (plate 158E) . . . . *Marphysa* 3
3. Branchiae with 4–7 palmate filaments; compound setae with long pointed blades . . . .
. . . . *Marphysa sanguinea*
— Branchiae simple throughout; compound setae falcate with bidentate tips . . . . *Marphysa stylobranchiata*
4. Subacicular hooks (plate 158B, 158C) and pectinate setae present . . . . *Eunice* 5
— Subacicular hooks and pectinate setae absent . . . . . *Palola*
5. Prostomial antennae moniliform (plate 158A); branchiae from setigers 5–6 . . . . *Eunice antennata*
— Prostomial antennae with faint articulations (plate 158D), not moniliform; branchiae from setiger 3 . . . .
. . . . *Eunice valens*

LIST OF SPECIES

See Fauchald 1970, 1992; Hartman 1944a; Orensanz 1990; Hilbig 1995b.

Eunice antennata (Savigny in Lamarck, 1818). A rocky intertidal and shallow subtidal; rare. A widely reported species, Fauchald (1992) published a redescription based on material from the Gulf of Suez, near the type locality. Our local specimens appear to be the same, but detailed comparison of the maxillae is required to confirm the identity.

Eunice valens (Chamberlin, 1919). Intertidal to low water. See Fauchald 1969, Smiths. Contr. Zool. 6: 10–12, plate 5; 1992;

Åkesson 1967, Acta Zool. 48: 141–192 (embryology, as *E. kobiensis*).

Marphysa sanguinea (Montague, 1815). Intertidal to shallow subtidal; shaley rocks and hard-packed clays; circumtropical and subtropical; southern California and Mexico, recently introduced to San Francisco Bay. See Hartman 1944a; Cohen and Carlton 1995.

Marphysa stylobranchiata Moore, 1909. Rocky habitats, in crevices; intertidal to the continental slope. A specimen identified as *Lysidice* sp. in the previous edition was probably a juvenile of this species.

*Palola paloloides (Moore, 1909). Known from shallow rocky habitats in southern California.

DORVILLEIDAE

The dorvilleids are mostly small, only a few millimeters long; most species have been discovered and described in the last quarter of the twentieth century when finer mesh screens were used. Some species of *Ophryotrocha* have been investigated extensively because of their role as pollution indicators, but they are poorly known in California. Quantitative sampling of the deep-sea floor has resulted in a wealth of new genera and species of unexpected morphological diversity. Among the Eunicida, the dorvilleids are considered phylogenetically the oldest group because they are the only extant family with a ctenognath jaw apparatus (i.e., the maxillae consist of several rows of small denticles rather than 4–6 pairs of larger pieces [Hilbig 1995c]).

KEY TO DORVILLEIDAE

1. Parapodia subbiramous, with elongate dorsal cirri (=notopodia) and enclosed notoacicula (plate 159C) 3
— Parapodia uniramous, without elongate dorsal cirri (=notopodia) and notoacicula (plate 159F) 2

* = Not in key.

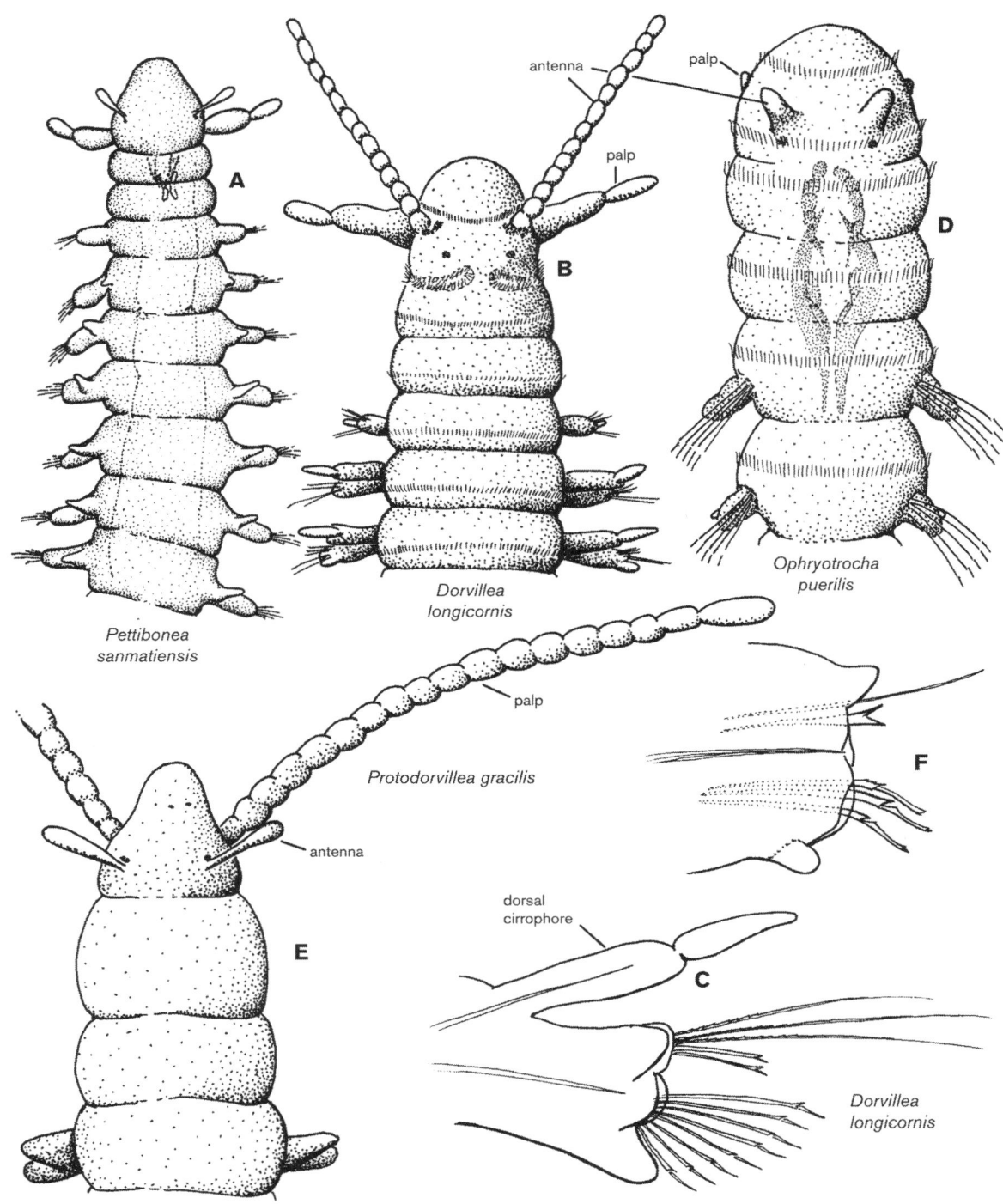

PLATE 159 Polychaeta. Dorvilleidae: A, *Pettiboneia sanmatiensis*; B, *Dorvillea longicornis*; C, parapodium; D, *Ophryotrocha puerilis*; E, *Protodorvillea gracilis*; F, parapodium (F, after Hartman; rest after Blake).

2. Animal minute (2–8 mm); prostomium with minute, inconspicuous antennae (plate 159D) . . . . . *Ophryotrocha puerilis*
— Animals larger; prostomium with short, smooth antennae, palps long, moniliform (plate 159E) . . . .
. . . . *Protodorvillea gracilis*
3. Maxillae with 7–8 rows of denticles; antennae short, smooth; palps biarticulate (plate 159A); furcate setae present . . . . *Pettiboneia* 4

— Maxillae with four rows of denticles; antennae articulated to moniliform; palps biarticulate to moniliform furcate setae present or absent 5
4. Notopodia from setiger 2; two asetigerous peristomial rings visible dorsally (plate 159A) . . . .
. . . . *Pettiboneia sanmatiensis*
— Notopodia from setiger 3; two asetigerous peristomial rings fused dorsally . . . . *Pettiboneia pugettensis*

5. Furcate setae if present, with short tynes; maxillae with maxillary carriers and both superior and inferior basal plates . . . . *Dorvillea* 6
— Furcate setae if present, with long, slender tynes; maxillae without inferior basal plates . . . . *Parougia caeca*
6. Prostomium with a prominent nuchal papilla; palps longer than antennae; in life with alternating bands of white and pink on dorsum . . . . *Dorvillea moniloceras*
— Prostomium without nuchal papilla; antennae longer than palps (plate 159B); without alternating bands of color on dorsum . . . . *Dorvillea longicornis*

LIST OF SPECIES

See Hartman 1944a; Fauchald, 1970; Jumars 1974, Zool. J., Linn. Soc. 54: 101–135; Hilbig and Blake 1991, Zool. Scr. 20: 147–183; Hilbig 1995c (California offshore species).

Dorvillea moniloceras (Moore, 1909). An elegantly colored species with alternating transverse bands of pink and white on the dorsum. Among worms and tunicates on floats and pilings in coastal embayments; rocky intertidal to low water. See Abbott and Reish 1980 (distribution, feeding behavior).

Dorvillea longicornis (delle Chiaje, 1828) (=*D. articulata* of Hartman's Atlas, 1968, p. 817; =*D. rudolphi* in 3rd ed. of *Light's Manual*). In harbors, in mud on floats and pilings, tolerant of semipolluted waters. See Richards 1967, Mar. Biol. 1: 124–133 (reproduction and development).

Ophryotrocha puerilis Claparéde and Mecznikow, 1869. In mud and detrital masses in harbors; a contaminant in aquaria; cosmopolitan, principally in southern California.

Ophryotrocha spp. Most species of *Ophryotrocha* are meiofaunal; numerous species have been reported from elsewhere in North America, but the genus is poorly known in California.

Parougia caeca (Webster and Benedict, 1884). An East Coast species, possibly introduced to California. Reported from southern California (SCAMIT, 2001).

Pettiboneia pugettensis (Armstrong and Jumars, 1978). Originally described from Puget Sound, later reported from British Columbia; intertidal to 11–12 m; specimens have been identified from the east side of San Francisco Bay in shallow water. See Blake and Hilbig 1990, Bull. So. Calif. Acad. Sci. 89: 109–114.

Pettiboneia sanmatiensis Orensanz, 1973. Originally described from Uruguay, this species has been found in intertidal sand flats in Tomales Bay, California. The species will likely be found more widely once fine mesh sieves are used to separate organisms from sediment. See Blake 1979, Bull. So. Cal. Acad. Sci. 78: 136–140.

Protodorvillea gracilis (Hartman, 1938). Sand or muddy sand sediments; intertidal and shelf depths. See Hobson 1971.

LUMBRINERIDAE

Lumbrinerids are long, cylindrical, burrowing worms that normally lack palps, antennae, and eyes. Despite the relatively featureless morphology, a large number of species have been described. For example, the genus *Lumbrineris sensu lato* contains about 200 species. Lumbrinerids can be distinguished from the similar appearing Oenonidae (=Arabellidae) by the shape of the prostomium. The prostomium is always rounded in side view in lumbrinerids and dorsoventrally flattened in oenonids. Dissection of the pharynx will

*= Not in key.

also reveal that the two families have very different jaw morphology.

The morphological homogeneity of the lumbrinerids has made attempts to classify them at the generic level difficult. Most of the commonly known lumbrinerid species were described in the nineteenth century, and although the majority of these species are still valid, the genera have been revised several times. The most current emphasis in lumbrinerid systematics has been a shift to the less readily accessible jaw or maxillary characters. The generic revisions by Orensanz 1990, Antarctic Res. Ser. 52: 1–183, and Frame 1992, Proc. Biol. Soc. Wash. 105: 185–218, are the most recent. Other important works are by Hartman (1944a), Fauchald (1970), and Hilbig (1995d).

KEY TO LUMBRINERIDAE

1. Compound hooded hooks present in some anterior setigers (plate 160H) 5
— Hooded hooks all simple (plate 160E) 2
2. Hooded hooks present from setiger 1 3
— Hooded hooks first present from setiger 7; posterior postsetal lobes elongated and erect (plate 160C) . . . . *Scoletoma erecta*
3. Posterior postsetal lobes elongated 4
— Posterior postsetal lobes not elongated (plate 160B); hooded hooks in middle setigers with numerous small teeth; each segment encircled with broad, brown pigmented band (plate 160A) . . . . *Scoletoma zonata*
4. Both pre- and postsetal lobes elongated in posterior setigers (plate 160F) . . . . *Eranno lagunae*
— Presetal lobes not prolonged in any setiger, postsetal lobes thick, elongated in posterior setigers only (plate 106D) . . . . *Scoletoma luti*
5. Aciculae black (plate 160L) 6
— Aciculae yellow or amber 7
6. Posterior pre- and postsetal lobes prolonged (plate 160J); prostomium narrow, conical, tapering to pointed tip (plate 160I) . . . . *Lumbrineris californiensis*
— Posterior postsetal lobes not prolonged (plate 160L); prostomium tapering to bluntly rounded anterior tip (plate 160K) . . . . *Lumbrineris japonica*
7. Posterior postsetal lobes elongated 8
— Posterior postsetal lobes not elongated (plate 160G); maxillae III with two teeth, maxillae IV with one tooth . . . . *Lumbrineris latreilli*
8. Prostomium conical, tapering to narrow anterior tip; both pre- and postsetal lobes prolonged in posterior setigers; maxillae III and IV each with a single tooth . . . . *Lumbrineris cruzensis*
— Prostomium as wide as long, globular, broadly rounded on anterior margin (plate 160M); presetal lobes not prolonged in any setiger; maxillae III with three teeth; maxillae IV with two teeth (plate 160N) . . . . *Lumbrineris inflata*

LIST OF SPECIES

See Hartman 1944a; Fauchald 1970; Hilbig 1995d.

Eranno lagunae (Fauchald, 1970). Shallow subtidal to shelf depths. See Hilbig 1995d.

Lumbrineris californiensis Hartman, 1944. In sandy and muddy sediments, low water to slope depths.

Lumbrineris cruzensis Hartman, 1944. Low water to shelf depths, in mixed sediments.

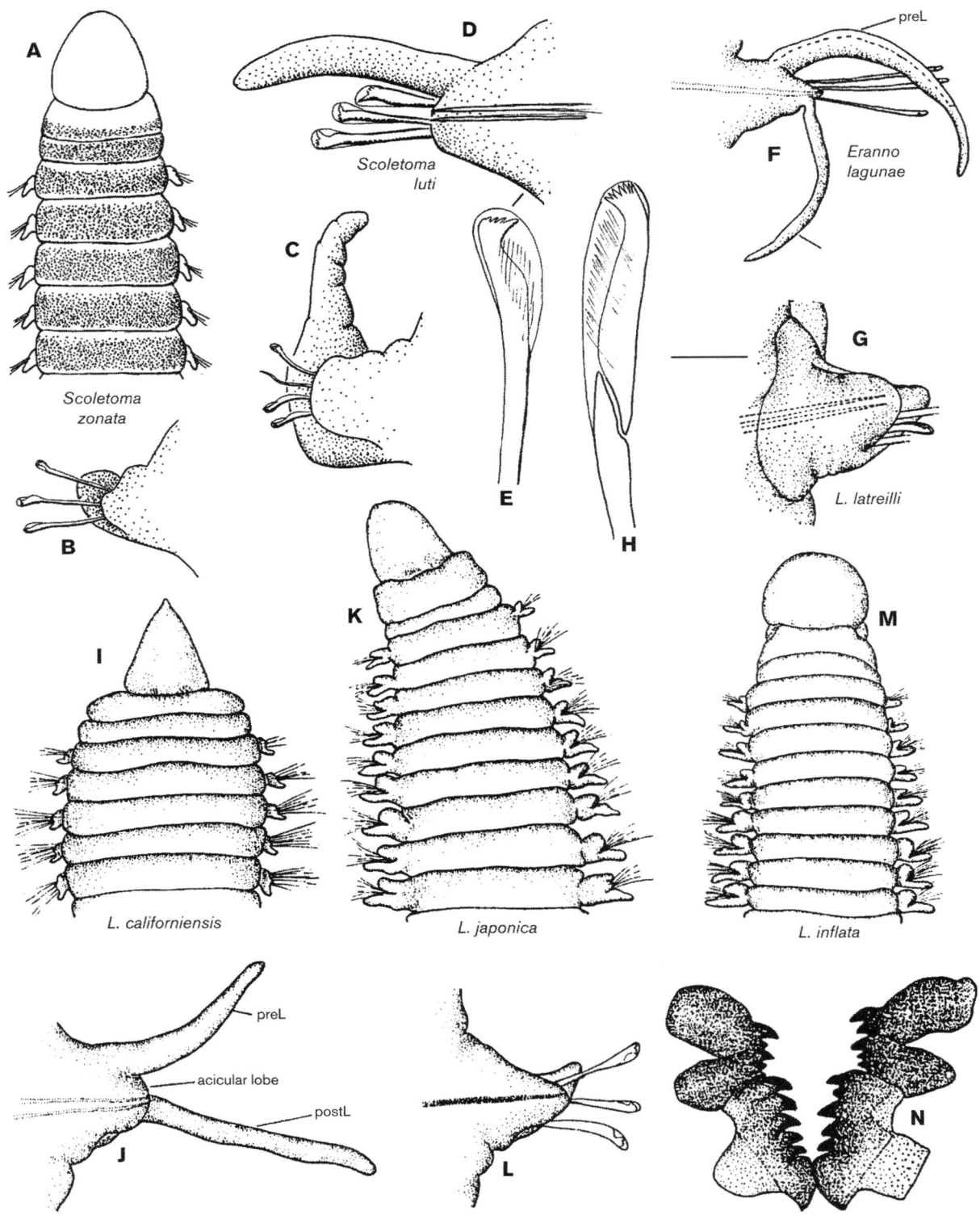

PLATE 160 Polychaeta. Lumbrineridae: A, *Scoletoma zonata*; B, posterior parapodium; C, *Scoletoma erecta*, posterior parapodium; D, *Scoletoma luti*, posterior parapodium; E, simple hooded hook; F, *Eranno lagunae*, posterior parapodium; G, *Lumbrineris latreilli*, posterior parapodium; H, compound hooded hook; I, *L. californiensis*; J, posterior parapodium; K, *L. japonica*; L, posterior parapodium; M, *L. inflata*; N, maxillary apparatus (I–L, after Hilbig; rest after Blake).

Scoletoma erecta (Moore, 1904). Intertidal to low water in sandy mud sediments.

Lumbrineris inflata Moore, 1911. Intertidal to shelf depths in coarse sediments. See Woodin 1974, Ecol. Monogr. 44: 171–187 (life history and ecology).

Lumbrineris japonica (Marenzeller, 1879). Intertidal to about 75 m, in mixed sediments.

Lumbrineris latreilli Audouin and Milne Edwards, 1834. A widely distributed, cosmopolitan species, intertidal to slope depths in mud, sand, eelgrass beds, under rocks. See Hartman 1944a (observations on reproduction); Pettibone 1963 (development summary).

Scoletoma zonata (Johnson, 1901) (=*Lumbrineris zonata*). The most common intertidal lumbrinerid in northern California and Oregon; common in sand flats of Tomales Bay and Bodega Harbor; sand or mixed sand-mud sediments.

Scoletoma luti (Berkeley and Berkeley, 1945) (=*Lumbrineris tetraura* of previous edition). Southeast Alaska to central California, Chile; intertidal to shelf depths in muddy sediments.

OENONIDAE (=ARABELLIDAE)

The oenonids superficially resemble lumbrinerids, but differ in several characters. The most obvious difference is the complete absence of hooded hooks; limbate capillaries are the only setae present. In many species, these limbate setae are strongly geniculate and serrated or covered with surficial spines. Many oenonids have a characteristically thin and wiry body; the cuticle is very rigid and often strikingly iridescent. The prostomium of oenonids is generally dorsoventrally depressed, whereas the typical lumbrinerid prostomium is conical or even globular. The most important difference between lumbrinerids and oenonids, from an evolutionary and systematic point of view, is the arrangement of the jaw apparatus (see Hilbig 1995e). Oenonids are distributed worldwide, but are never abundant; some species are parasitic, spending at least part of their life histories in the bodies of other polychaetes, bivalves, and echiurans. The following key is modified from Hilbig (1995e).

KEY TO OENONIDAE

1. Setae include limbate capillaries and thick, projecting acicular spines in middle and posterior parapodia (plate 161E) . . . . *Drilonereis* 2
— All setae limbate; thick acicular spines absent (plate 161I, 161K) . . . . *Arabella* 4
2. Acicular spines present from setiger 1; anterior parapodia minute, barely visible dorsally, with short pre- and postsetal lobes; postsetal lobes (plate 161A, 161B) gradually turning digitiform in middle setigers, presetal lobes turning digitiform in posterior setigers and becoming subequal to postsetal lobes, parapodia thus distinctly bilabiate in posterior third of body; MI with 1–5 basal teeth, MII with 3–5 teeth, MIII and MIV with one tooth and sometimes additional minute teeth . . . . *Drilonereis longa*
— Acicular spines first present in setiger 10 or later; anterior parapodia readily visible dorsally (plate 161C); parapodia not distinctly bilabiate in posterior setigers; postsetal lobes papillalike or conical, as long as parapodial stem at most 3
3. Mandibles large (plate 161G); parapodia smaller in anterior setigers than in posterior (plate 161D), but proportions

of parapodial lobes not changing; acicular spines first present in setigers 10–13; MI with 3–4 basal teeth (large specimens up to seven), MII 6–9, MIII 1–5, MIV 1–2, MV one (occasionally absent) (plate 161F) . . . . . *Drilonereis falcata*
— Mandibles absent; parapodia conspicuous throughout (plate 161C), with digitiform postsetal lobes in posterior setigers; acicular spines first present in setigers 25–30, projecting about as far as postsetal lobes; MI basally smooth . . . . *Drilonereis nuda*
4. Postsetal lobes in posterior setigers digitiform, very long, directed dorsally (plate 161K); upper supra-acicular and subacicular setae smooth, lower supra-acicular setae with serrated wings; eyespots absent or obscure in adult; color in life pale gray or with three dorsal rows of dark bluish spots and with an iridescent epithelium . . . . *Arabella semimaculata*
— Postsetal lobes of posterior setigers prominent (plate 161J), but not directed dorsally; surficial structures of setae different 5
5. All setae with serrated wings (plate 161I); MI with unidentate tips on both sides, MIII and MIV with about 4–5 conical teeth, alternating large and small; prostomium usually with four eyespots in a transverse row (plate 161H); color in life dark red to green and highly iridescent . . . . *Arabella iricolor*
— Some setae with smooth wings; left MI with bidentate tip 6
6. Subacicular setae smooth in middle and posterior setigers, supra-acicular setae serrated; MIII and MIV with about 4–5 conical teeth, alternating large and small (plate 161M); prostomium with four eyes in transverse row along posterior margin (plate 161L); color in alcohol pale to light brown, with broad band of reddish pigment. . . . . *Arabella protomutans*
— All setae smooth in anterior setigers, serrated in middle and posterior setigers; MIII and MIV with about ten needlelike, equal teeth, giving comblike shape to maxillae (plate 161N); eyes absent; color in alcohol copper to purplish brown, highly iridescent . . . . *Arabella pectinata*

LIST OF SPECIES

See Hartman 1944a; Fauchald 1970; Hilbig 1995e.

Arabella iricolor (Montagu, 1804). A widely distributed, cosmopolitan species. According to Hilbig (1995e), *A. iricolor,* while one of the most familiar species of *Arabella* in California, has likely been confused with *A. pectinata* and *A. protomutans.* All three species are similar superficially, but *A. iricolor* is the only one to have serrated pectinate setae. Common in rocky habitats, crevices, oyster and mussel beds; with algae and debris; also in mixed sediments where it burrows deeply; intertidal to 90 m. See Abbott and Reish 1980 (habitats, distribution).

Arabella pectinata Fauchald, 1970. Hilbig (1995e) provided the first records of this species from California based on specimens from Tomales Bay. *A. pectinata* has most likely been confused with *A. iricolor,* but differs in that the setae are all smooth instead of serrated; the third and fourth maxillae have about ten needlelike teeth of equal size instead of five conical teeth of alternating size. Central California to western Mexico; intertidal to shallow subtidal in mixed sediments.

Arabella protomutans Orensanz, 1990. Hilbig (1995e) provided the first record of this species from western North America based on specimens from rocky habitats at Tomales Point, Dillon Beach, and Avila. *A. protomutans* is closely related and

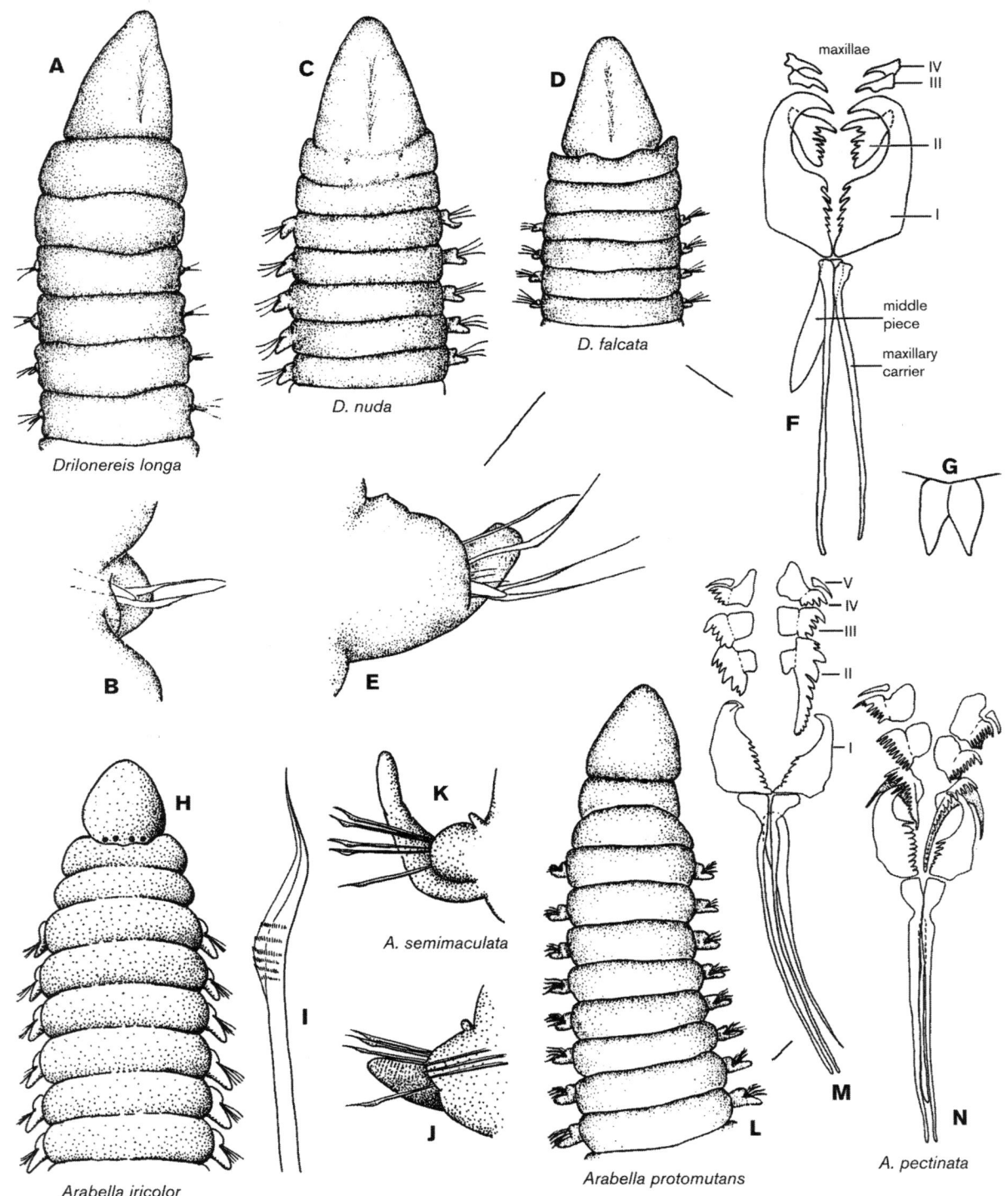

PLATE 161 Polychaeta. Oenonidae: A, *Drilonereis longa*, B, parapodium; C, *Drilonereis nuda*; D, *Drilonereis falcata*, E, parapodium; F, maxillary apparatus; G, mandibles; H, *Arabella iricolor*; I, simple seta; J, posterior parapodium; K, *A. semimaculata*, posterior parapodium; L, *A. protomutans*; M, maxillary apparatus; N, *A. pectinata*, maxillary apparatus (A–E, L–N, after Hilbig; rest after Blake).

intermediate to both *A. iricolor* and *A. pectinata* in the nature of the maxillary apparatus, with MI resembling that of *A. pectinata* and the irregular dentition of MIII and MIV being similar to *A. iricolor*. The species is most readily recognized by the smooth subacicular setae in posterior parapodia; if only anterior fragments are available, the pharynx should be dissected to examine the first and third maxillae. Central California, Patagonia; in rocks and mixed sediments, intertidal to shallow subtidal.

Arabella semimaculata (Moore, 1911). Easily distinguished by the long, dorsally directed postsetal lobes in posterior setigers; in rocks and mixed sediments; intertidal to 80 m.

Drilonereis falcata Moore, 1911. A widespread, free-living species; central California; in mixed sediments; intertidal to 350 m.

Drilonereis longa Webster, 1879. Intertidal to slope depths.

Drilonereis nuda Moore, 1909. Rocky bottoms with kelp, in tide pools; intertidal to about 100 m.

ORBINIIDAE

The orbiniids are burrowing deposit feeders that do not make permanent tubes. They are common in tidal flats and sometimes attain dense populations in sediments that are of mixed sands and muds. Their bodies are distinctive, with a well-developed and firm anterior region and a fragile and ragged-appearing posterior abdominal region bearing long, dorsally erect branchiae and parapodial lobes. Orbiniids are readily recognized by their body form and characteristic camerated and crenulated setae.

KEY TO ORBINIIDAE

1. Anterior margin of prostomium broadly rounded (plate 162B) . . . . *Naineris dendritica*
— Anterior margin of prostomium pointed (plate 162A) 2
2. Thoracic region with ventral papillae (plate 162A) . . . . *Orbinia johnsoni*
— Thoracic region without ventral papillae 3
3. Thoracic neuropodia with slender pointed setae only (plate 162E) . . . . *Leitoscoloplos pugettensis*
— Thoracic neuropodia with blunt spines (plate 162D) and pointed setae 4
4. Neuropodia of posterior thoracic setigers with 2 postsetal lobes and an additional subpodial lobe (plate 162F) . . . . *Scoloplos armiger*
— Posterior thoracic neuropodia with a single postsetal lobe; subpodial lobe absent (plate 162C) . . . . . *Scoloplos acmeceps*

LIST OF SPECIES

See Hartman 1957; Pettibone 1957; Blake 1996a.

Leitoscoloplos pugettensis (Pettibone, 1957) (=*Haploscoloplos elongatus* of previous edition). The most common orbiniid in bays and estuaries in California and Oregon, also in subtidal sediments in offshore areas; inhabits sands and muds; intertidal to 220 m. Reproduction and larval development were described by Blake (1980). The species deposits cocoons in the sediment, with larvae emerging for a short, lecithotrophic phase in the plankton. See Parkinson 1978, Pac. Sci. 32: 149–155 (feeding, burrowing, ecology).

Naineris dendritica (Kinberg, 1867). The most common orbiniid in rocky intertidal areas; found in algal holdfasts, bryozoan masses, and debris under and between rocks; intertidal to shallow subtidal. Specimens reported as *Protoaricia* sp. in the previous edition are juveniles of *N. dendritica* (Blake, unpublished).

Orbinia johnsoni (Moore, 1909). Intertidal to shallow subtidal in sandy sediments of protected beaches.

Scoloplos acmeceps Chamberlin, 1919. In mud, algal holdfasts, roots of *Zostera*; intertidal to shallow subtidal depths. The lecithotrophic larval development was described by Blake (1980).

Scoloplos armiger (Müller, 1776). A cosmopolitan species in mixed sediment types; rare in northern California; largely subtidal.

PARAONIDAE

The paraonids are small, elongate burrowing worms that rarely exceed 20 mm in length. They have bodies that are somewhat expanded and flattened anteriorly, narrowing to numerous posterior segments. The bodies of some species are tightly spiraled after preservation, resembling a corkscrew, suggesting that they maintain this position in the sediments. Paraonids have no palps and are considered to be subsurface deposit feeders; they have only a simple, soft proboscis. The number of taxa has increased in recent years through the use of finer mesh screens and the increased exploration of continental shelf and slope depths, where they are most abundant. The intertidal and shallow water fauna of California is poorly known. A few common species known to occur in shallow-water embayments in California are cited here. Other intertidal species are to be expected with careful inspection of marine sediments. Readers are referred to Blake (1996b) for a comprehensive review of California paraonids.

KEY TO SELECTED PARAONID GENERA

1. Prostomium with medial antenna (plate 163A, 163G) 2
— Prostomium without medial antenna (plate 163D, 163E) 3
2. Posterior neuropodia with modified setae (plate 163H, 163I) . . . . *Aricidea*
— Posterior notopodia with modified setae (plate 163B, 163C) . . . . *Cirrophorus*
3. Posterior neuropodia with modified setae (plate 163F) . . . . *Levinsenia*
— Neuropodia without modified setae . . . . *Paraonella*

LIST OF SPECIES

See Hartman 1944c, 1957; Hobson 1972, Proc. Biol. Soc. Wash. 85: 549–556; Strelzov 1973/1979; Blake 1996b.

Aricidea catherinae Laubier, 1967. A widespread species and one of the most familiar paraonids in North America. The species is recognized by the shape of the medial antenna that becomes thickest in the middle, then tapers to a fine tip (plate 163G); most common in shallow embayments in sandy mud, but ranges to shelf depths. *Aricidea sueica* Eliason, 1920, a European species listed in the previous edition, was misidentified and may have been *A. catherinae*.

Aricidea lopezi Berkeley and Berkeley, 1956. *A. lopezi* is superficially similar to *A. catherinae* in the shape of the antenna, but this is longer, extending back to setigers 3–6 instead of 1–3 (plate 163J). Other differences include the dorsal cirri on setigers 1 and 2 that are cirriform instead of short and stubby, and the modified spines have an arista arising from the convex side of the curved tip instead of from the tip (plate 163I). Shallow subtidal sediments to shelf depths.

Aricidea ramosa Annenkova, 1934. Recognized by the branched medial antenna; common in fine sands and coarse silt sediments; low water to slope depths. There has been discussion among local taxonomists regarding differences in specimens having two to three branches on the antenna and specimens having many branches. In a sample from slope depths, one of us (Blake) found specimens with two lateral

* = Not in key.

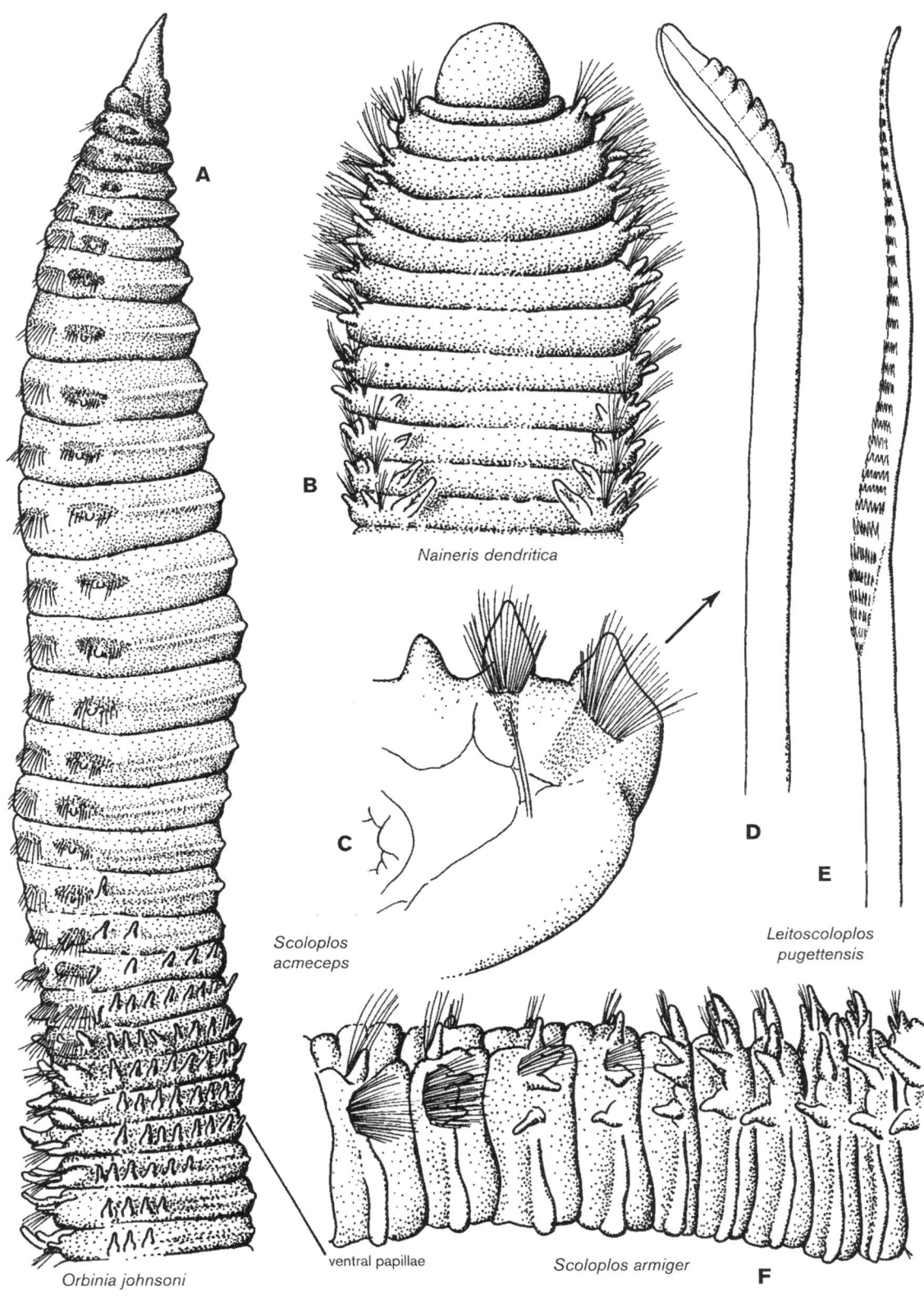

Naineris dendritica

Scoloplos acmeceps

Leitoscoloplos pugettensis

Orbinia johnsoni

ventral papillae   *Scoloplos armiger*

PLATE 162 Polychaeta. Orbiniidae: A, *Orbinia johnsoni*; B, *Naineris dendritica*; C, *Scoloplos acmeceps*, 18th parapodium; D, thoracic neuroseta; E, *Leitoscoloplos pugettensis*, thoracic neuroseta; F, *Scoloplos armiger*, lateral view of transitional region, setigers 16–24 (A, C–F, after Hartman; B, after Blake).

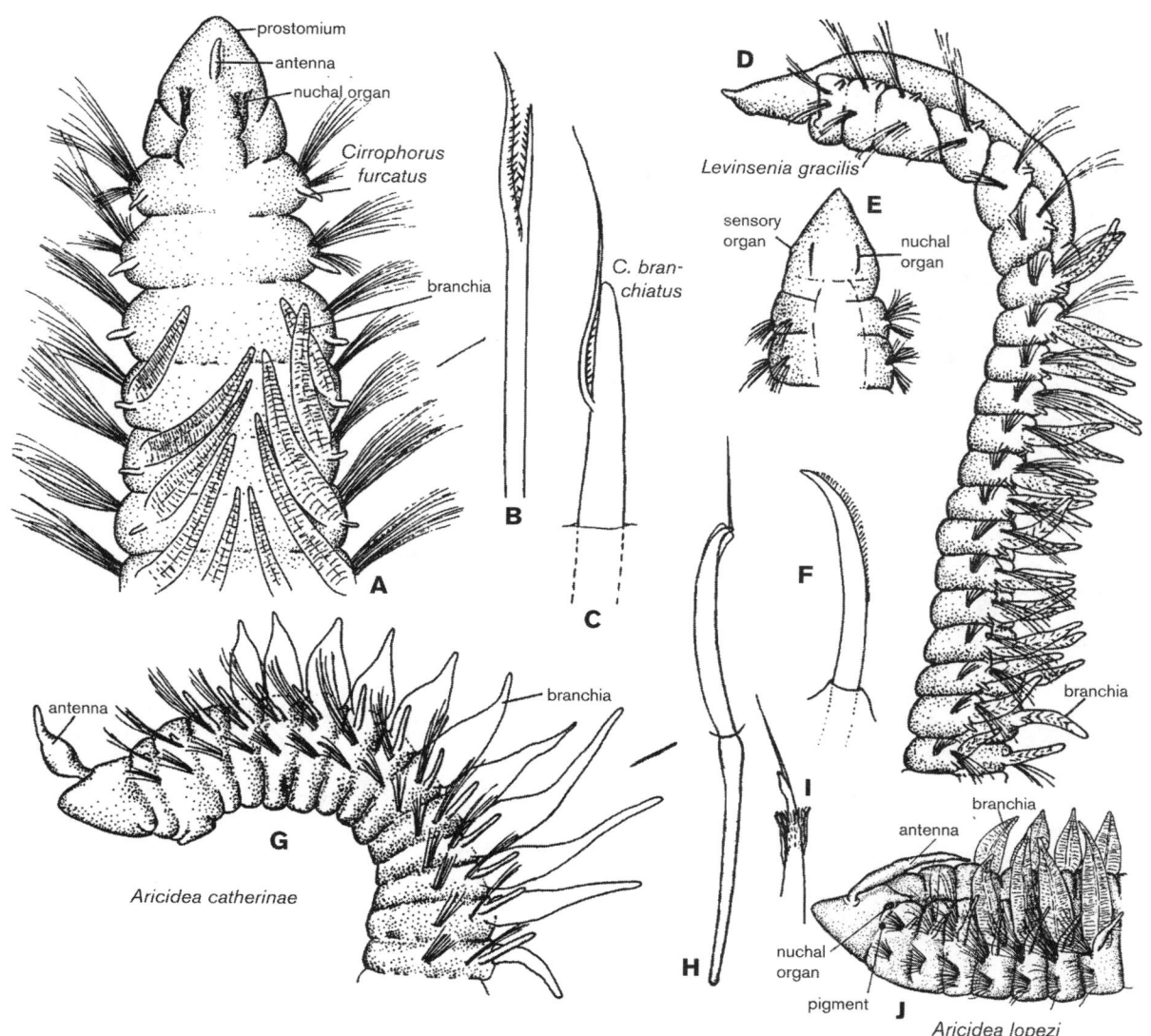

PLATE 163 Polychaeta. Paraonidae: A, *Cirrophorus furcatus*; B, furcate notoseta; C, *Cirrophorus branchiatus*, furcate notoseta; D, *Levinsenia gracilis*, anterior end, lateral view; E, anterior end, dorsal view; F, posterior neuroseta; G, *Aricidea catherinae*, anterior end, lateral, view; H, posterior neuroseta; I, *Aricidea lopezi*, posterior neuroseta; J, anterior end, lateral view (B, after Hartman; C, after Strelzov; G–H, after Laubier; rest after Blake).

branches and other specimens with six lateral branches in the same samples; all other characters among these specimens were appropriate for this species. However, specimens from shallow and deep water need to be carefully compared to ensure that a single species is involved. See Strelzov 1973/1979 (taxonomy); Blake 1996b.

Cirrophorus branchiatus Ehlers, 1908. Readily recognized by the form of its modified notopodial spines, where a long, thin lateral filament arises on a large blunt-tipped spine (plate 163C); in fine sediments; low water to slope depths.

Cirrophorus furcatus (Hartman, 1957). Readily recognized by its rusty brown pigmentation and the lyrate setae (plate 163B) that occur in the notopodia. California, low water to slope depths.

Levinsenia gracilis (Tauber, 1879) (=*Paraonis gracilis*). A small, threadlike species (plate 163D), only collected when fine mesh sieves are used to separate the fauna from sediments. Branchiae begin on setiger 7 (plate 163D). The species is one of the most widespread of the Paraonidae, occurring in silt and mud; shallow subtidal to slope depths. In California, *L. gracilis* may be confused with *L. oculata* (Hartman, 1957).

Paraonella platybranchia (Hartman, 1961). Intertidal in sand beaches (collected at Dillon Beach, 1975); shallow subtidal; Columbia River to southern California.

SPIONIDAE

The Spionidae is among the largest and most common polychaete families in benthic communities. Spionids occur in a wide variety of habitats from the intertidal to the deep sea. They are readily recognized by their anterior ends, which carry a pair of long prehensile palps. Spionids sometimes form dense assemblages with individuals extending their palps from burrows or tubes to filter particles from the water; in other

* = Not in key.

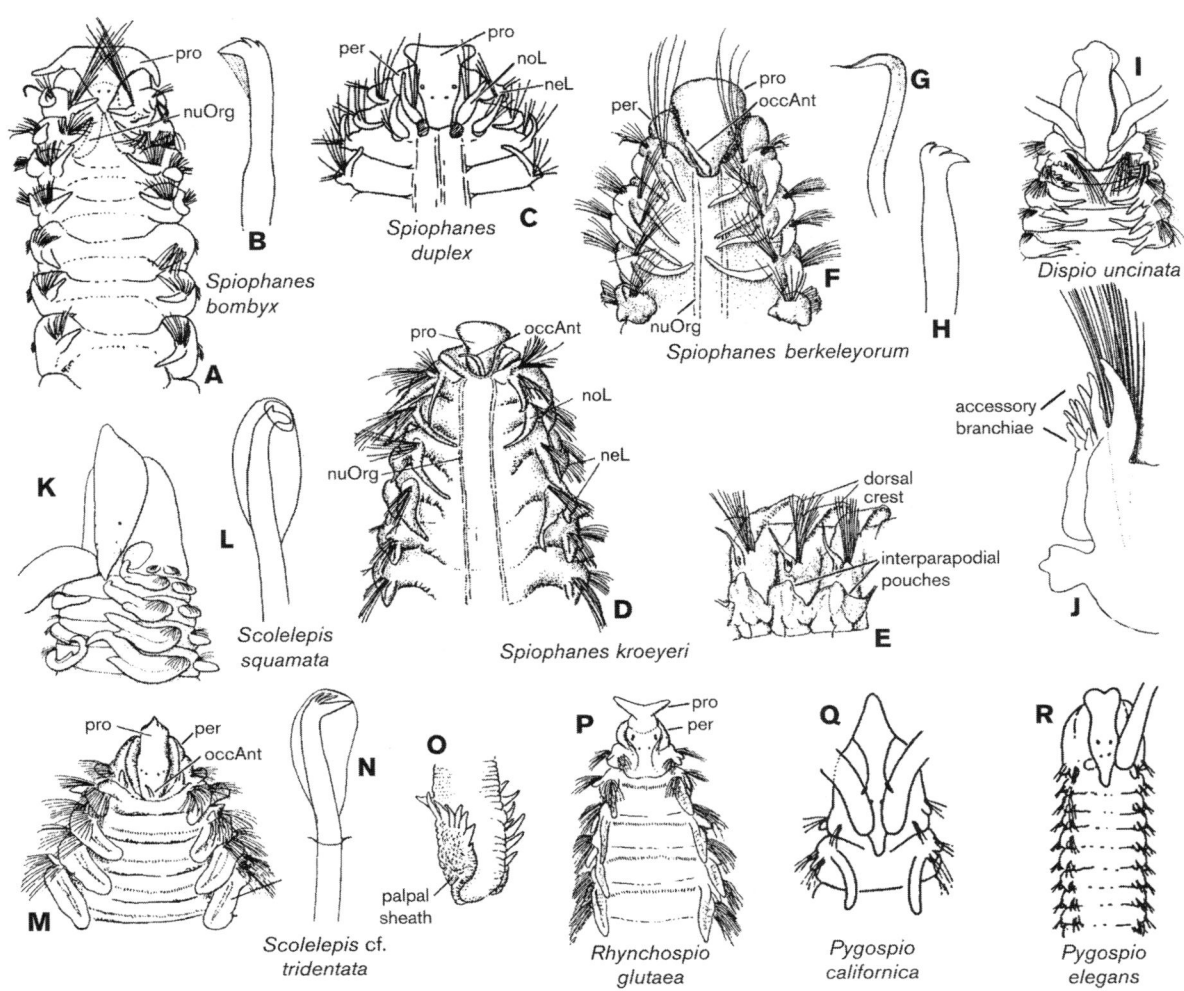

PLATE 164 Polychaeta. Spionidae: A, *Spiophanes bombyx*; B, neuropodial hook; C, *Spiophanes duplex*; D, *Spiophanes kroeyeri*; E, lateral view of setigers 27–29; F, *Spiophanes berkeleyorum*; G, crooklike neuroseta from setiger 1; H, neuropodial hook; I, *Dispio uncinata*; J, parapodium showing accessory branchiae; K, *Scolelepis squamata*; L, neuropodial hooded hook; M, *Scolelepis* cf. *tridentata*; N, neuropodial hooded hook; O, base of palp showing palpal sheath; P, *Rhynchospio glutaea*; Q, *Pygospio californica*; R, *Pygospio elegans* (A, B, D, E, G, H, N–P, after Imajima; C, I, J, after Hartman; K, after Hartmann-Schröder; rest after Blake).

situations, the worms are surface deposit feeders and use the palps to sweep the sediment surface. Some spionids, such as *Polydora*, bore into calcareous substrates and are sometimes considered pests by the shellfish industry. A few species are opportunistic, occupying environments that are disturbed or organically enriched. Such species have life history patterns that allow them to populate available areas rapidly. In California waters, there are at least 19 genera and approximately 90 known spionid species. A comprehensive review of the biology and systematics of the California spionids has been published by Blake (1996c).

KEY TO SPIONIDAE

1. Branchiae absent (plate 164A, 164C); setiger 1 with 1–2 large curved neuropodial hooks (plate 164G) in addition to normal capillaries . . . . *Spiophanes* 2
— Branchiae present; setiger 1 without large neuropodial hooks (plate 164M) 5
2. Prostomium triangular, with long frontal horns; paired nuchal organs extending to setigers 3–4, segmental organs not present (plate 164A); neuropodial hooks with hood below main fang (plate 164B), with 1–2 apical teeth; occipital antenna absent; typically shallow water, inhabiting sand . . . . *Spiophanes* cf. *bombyx*
— Prostomium T-shaped, bell-shaped, or subtriangular (plate 164C, 164F), without long frontal horns; neuropodial hooks without hood (plate 164H), with 3–4 apical teeth; occipital antenna present or absent 3
3. Occipital antenna absent; prostomium T-shaped; nuchal organs absent (plate 164C) . . . . *Spiophanes duplex*
— Occipital antenna present; prostomium bell-shaped to subtriangular; nuchal organs paired, ciliated grooves extending posteriorly over numerous segments (plate 164D, 164F) 4
4. Interparapodial pouches and ciliated dorsal ridges present on anterior and middle body segments (plate 164E) . . . . *Spiophanes* cf. *kroeyeri*
— Interparapodial pouches and ciliated dorsal ridges absent . . . . *Spiophanes berkeleyorum*

5. Setiger 5 modified, with specialized setae 21
— Setiger 5 not modified 6
6. Prostomium pointed (plate 164I, 164K, 164M), sometimes appearing conical with rounded apex in contracted specimens 7
— Prostomium broadly rounded or incised, or with lateral or frontal horns on anterior margin 10
7. Branchiae from setiger 1; notosetae of setiger 1 long, thin (plate 164I); accessory branchiae present (plate 164J) . . . . .
. . . . Dispio uncinata
— Branchiae from setiger 2 (plate 164M); notosetae of setiger 1 present or absent, if present, not long, thin; accessory branchiae absent . . . . Scolelepis 8
8. Branchiae and dorsal notopodial lamellae basally fused (plate 164K); hooded hooks with 2–3 apical teeth; hooded hooks with more or less straight shaft bearing short, falcate main fang surmounted by 0–2 small apical teeth (plate 164L); notosetae present on setiger 1; without palpal sheath . . . . Scolelepis (Scolelepis) squamata
— Branchiae and dorsal notopodial lamellae fused for most of length, free at tips (plate 164M); hooded hooks with 0–2 apical teeth; hooded hooks with curved shaft bearing long main fang oriented at 90–120° angle to shaft, bearing 2–6, sharply pointed apical teeth (plate 164N); notosetae present or absent on setiger 1; palps with membranous basal sheath (plate 164O) 9
9. Notosetae present on setiger 1; neuropodial hooded hooks from setigers 15–16 . . . .
. . . . Scolelepis (Parascolelepis) cf. tridentata
— Notosetae absent on setiger 1; neuropodial hooded hooks from setigers 13–27 . . . . . . Scolelepis (Parascolelepis) texana
10. Prostomium with laterally directed horns (plate 164P) . . . . .
. . . . Rhynchospio glutaea
— Prostomium truncate, rounded, or incised on anterior margin, without lateral or frontal horns 11
11. Branchiae limited to middle and posterior setigers except for single pair on setiger 2 in sexually mature males (plate 164Q) . . . . Pygospio 12
— Branchiae from setigers 1 or 2, continuing posteriorly for variable number of setigers 13
12. Prostomium bilobed on anterior margin; branchiae from setigers 11–13; neuropodial hooded hooks from setigers 8–9 (plate 164R) . . . . Pygospio elegans
— Prostomium conical on anterior margin (plate 164Q); branchiae from setiger 19; neuropodial hooks from setiger 23 . . . . Pygospio californica
13. Branchiae concentrated in anterior setigers, 1–22 pairs (plate 165B, 165C), absent posteriorly 14
— Branchiae present over most of body length 18
14. Branchiae present from setiger 1, no more than three-pairs 15
— Branchiae present from setiger 2, with four or more pairs 16
15. Branchiae, one pair; setiger 2 with raised dorsal collar (plate 165A) . . . . Streblospio benedicti
— Branchiae, three pairs, each with flattened bifoliate platelike pinnules; with transverse ridge or membrane between branchial bases of setiger 1 (plate 165B) . . . .
. . . . Paraprionospio pinnata
16. Branchiae entirely apinnate; pinnate branchiae entirely absent (plate 165C) . . . . Prionospio lighti
— Branchiae include both apinnate and pinnate types in different combinations 17

17. Branchiae with first three pairs apinnate and fourth pair with platelike pinnules (plate 165D, 165E) . . . .
. . . . Apoprionospio pygmaea
— Branchiae with first and fourth pairs pinnate and second and third pairs apinnate (plate 165F) . . . .
. . . . Prionospio steenstrupi
18. Branchiae from setiger 1 19
— Branchiae from setiger 2 20
19. Prostomium narrow, blunt to weakly incised on anterior margin (plate 165J); hooded hooks in neuropodia only; branchiae basally free from notopodial lamellae; pygidium with no more than four anal cirri . . . . Spio
— Prostomium bell-shaped, broadly rounded on anterior margin (plate 165I); hooded hooks present in both neuro- and notopodia; branchiae basally fused to notopodial lamellae; pygidium with numerous cirri . . . .
. . . . Marenzelleria viridis
20. Prostomium broad, bluntly rounded, or flattened on anterior margin (plate 165G); branchiae free from dorsal lamellae; interparapodial lateral pouches present between some parapodia (plate 165H) . . . . Laonice cirrata
— Prostomium narrow, entire or incised on anterior margin; branchiae basally or entirely fused to dorsal lamellae; interparapodial lateral pouches absent . . . . Microspio
21. Branchiae from setiger 2 22
— Branchiae from setigers 6–12 28
22. Modified spines of setiger 5 of two types, one with expanded end bearing cusps or bristles (plates 166C, 166D, 166G), second simple, falcate (plate 166B) . . . .
. . . . Boccardia 23
— Modified spines of setiger 5 of one type, simple, falcate, with bilimbate companion setae . . . . Boccardiella 26
23. Major spines of setiger 5: (1) falcate (plate 166B) and (2) tridentate (plate 166C); branchiae anterior to setiger 5 small, inconspicuous (plate 166A) . . . . . Boccardia tricuspa
— Major spines of setiger 5: (1) falcate and (2) bristle-topped (plate 166D, 166G); branchiae anterior to setiger 5 larger, conspicuous 24
24. Notosetae absent on setiger 1, acicular setae present on posterior notopodia (plate 166E); bristle-topped spines of setiger 5 with small accessory tooth (plate 166D) . . . .
. . . . Boccardia berkeleyorum
— Notosetae present on setiger 1; acicular setae absent in posterior notopodia; bristle-topped spines of setiger 5 lack an accessory tooth 25
25. Notosetae of setiger 1 long, directed forward in fan-shaped fascicle (plate 166F) . . . . Boccardia columbiana
— Notosetae of setiger 1 short . . . . Boccardia proboscidea
26. Posterior notopodia with large, recurved boat hook–shaped spines (plate 166H); pygidium with paired cirri or lappets
. . . . 27
— Posterior notopodia with capillaries, spines absent; pygidium saucerlike . . . . Boccardiella truncata
27. Pygidium with two broad ventral lappets, each with a short process (plate 166I); bores into calcareous substrates, sometimes found in mud . . . . Boccardiella hamata
— Pygidium with two long anal cirri (plate 166J); inhabits soft sediments in areas of very low salinity, sometimes nearly freshwater . . . . Boccardiella ligerica
28. Setiger 5 slightly to moderately modified, usually with well-developed parapodia; two types of major spines on setiger 5: (1) simple, acicular, or falcate and (2) pennoned, with both types usually arranged in U- or J-shaped double

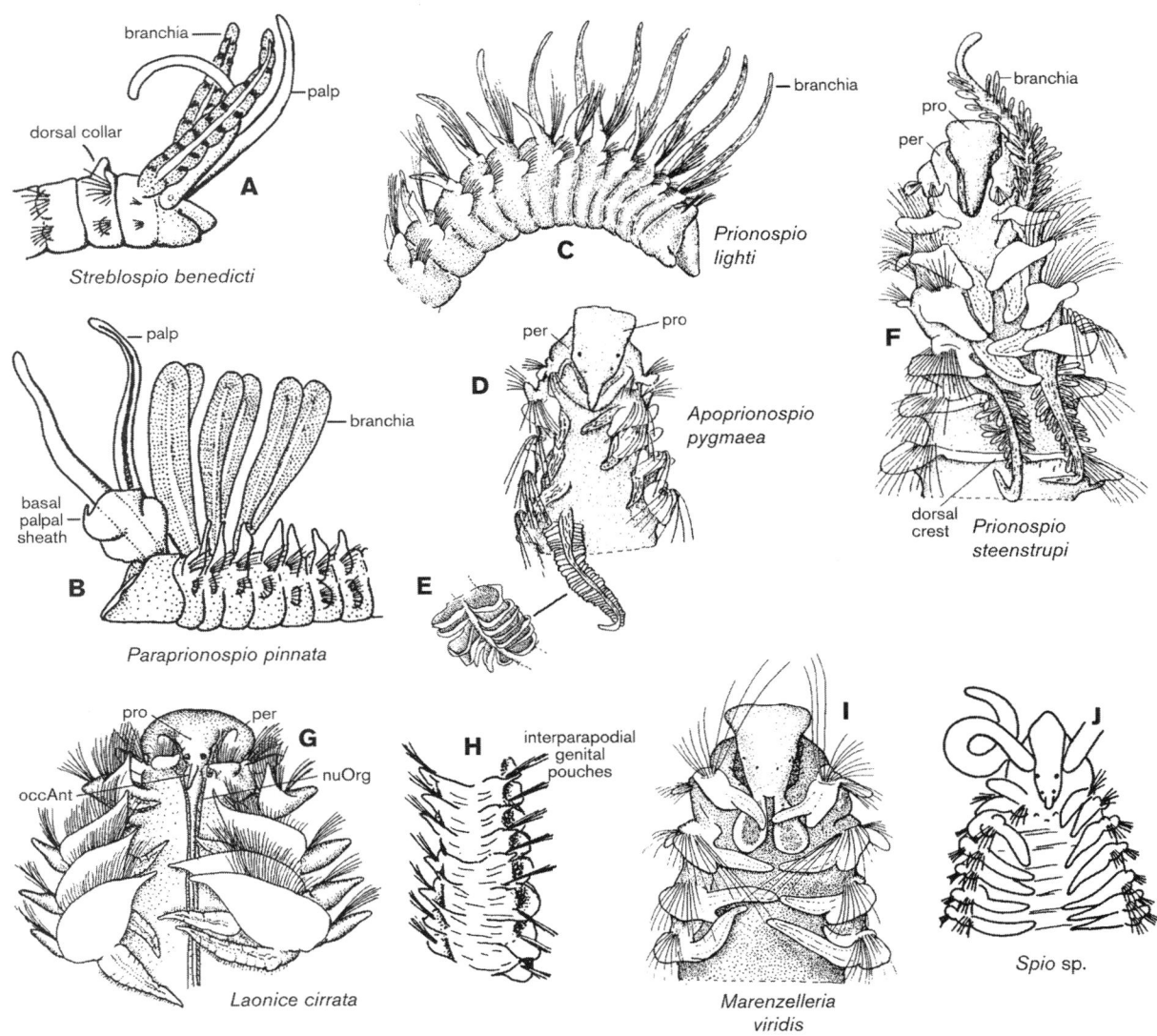

PLATE 165 Polychaeta. Spionidae: A, *Streblospio benedicti*; B, *Paraprionospio pinnata*; C, *Prionospio lighti*; D, *Apoprionospio pygmaea*; E, detail of branchia showing platelike structures; F, *Prionospio steenstrupi*; G, *Laonice cirrata*; H, dorsolateral view of some middle body segments showing interparapodial genital pouches; I, *Marenzelleria viridis*; J, *Spio* sp. (A, B, J, after Hartman; D–F, I, after Maciolek; G, after Berkeley and Berkeley; rest after Blake).

row; hooded hooks with secondary tooth closely applied to main fang . . . . *Pseudopolydora* 29

— Setiger 5 greatly modified; major spines of one or two types in curved row, not in U- or J-shaped rows; hooded hooks with prominent angle between teeth 30

29. Prostomium rounded on anterior margin (plate 166K); pygidium saucerlike, with smooth margin (plate 166L); small (about 4–6 mm long); in life, palps with reflective yellow pigment spots . . . . *Pseudopolydora paucibranchiata*

— Prostomium bifid on anterior margin (plate 166M); pygidium cup-shaped with two dorsal processes (plate 166N); larger (up to 28 mm); exhibits intersegmental black pigment on anterior segments . . . . *Pseudopolydora kempi*

30. Modified spines of setiger 5 of one type, variously shaped, with or without companion setae 31

— Modified spines of setiger 5 of two types, (1) with expanded end, (2) falcate; one or both types usually tips covered or cloaked with bristles . . . . *Carazziella calafia*

31. Hooded hooks with constriction and manubrium on shaft, and with main fang at a more or less right angle to shaft and wide angle with apical tooth; notosetae absent on setiger 1; anterior part of digestive tract never interrupted by gizzardlike structure (plate 166P) . . . . *Polydora* 32

— Hooded hooks with smooth, curved shafts with main fang directed apically, forming wide angle with shaft and reduced angle with apical tooth (plate 166EE); notosetae present on setiger 1; anterior part of digestive tract sometimes interrupted by enlarged, thickened gizzardlike structure . . . . *Dipolydora* 41

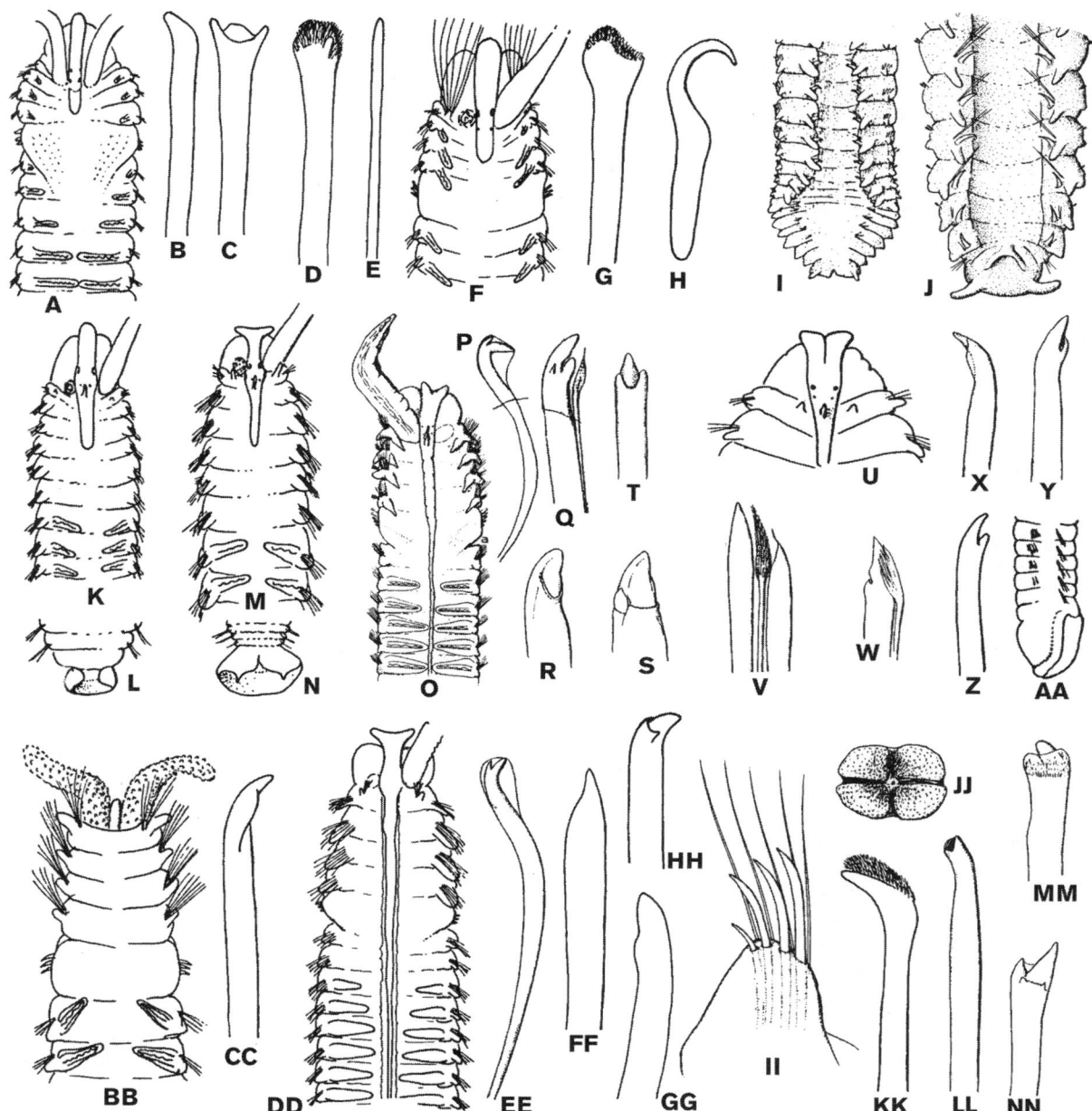

PLATE 166 Polychaeta. Spionidae (*Polydora*-complex): A, *Boccardia tricuspa*; B–C, major spines from setiger 5; D, *Boccardia berkeleyorum*, major spine from setiger 5; E, acicular spine from posterior notopodium; F, *Boccardia columbiana*; G, major spine from setiger 5; H, *Boccardiella hamata*, posterior notopodial hook; I, posterior end, dorsal view; J, *Boccardiella ligerica*, posterior end, dorsal view; K, *Pseudopolydora paucibranchiata*; L, posterior end; M, *Pseudopolydora kempi*; N, posterior end; O, *Polydora bioccipitalis*; P, hooded hook; Q, major spine from setiger 5; R, *Polydora alloporis*, major spine from setiger 5; S, *Polydora narica*; T, *Polydora spongicola*; U, *Polydora nuchalis*; V, major spines and companion seta from setiger 5; W, *Polydora cornuta*, major spine and companion seta from setiger 5; X–Y, *Polydora websteri*, major spines from setiger 5; Z, *Dipolydora pygidialis*, major spine from setiger 5; AA, posterior end; BB, *Polydora commensalis*; CC, major spine from setiger 5; DD, *Dipolydora elegantissima*; EE, hooded hook; FF, major spine from setiger 5; GG, *Dipolydora socialis*, major spine from setiger 5; HH, *Dipolydora bidentata*, major spine from setiger 5; II, *Dipolydora bifurcata*, posterior setiger showing notopodial spines; JJ, pygidium; KK, *Dipolydora brachycephala*, major spine from setiger 5; LL, *Dipolydora quadrilobata*, major spine from setiger 5; MM–NN, *Dipolydora armata*, major spines from setiger 5 (A–C, F–G, U, after Woodwick; rest after Blake).

— Major spines of setiger 5 either simple, without accessory structures (plate 166V), or with subterminal tooth (plate 166W) or thin flange sometimes oriented obliquely between curved tip of spine and main shaft (plate 166X) 36

34. Major spines of setiger 5 with distinct subdistal concavity surrounded by shelf or collar, without lateral teeth or flanges (plate 166R); commensal with alloporan corals . . . . *Polydora alloporis*

— Major spines of setiger 5 without distinct subdistal concavity, except for space formed behind projecting transverse collar or shelf; distinct lateral teeth or collar may appear lobed in lateral view; commensal with sponges or free living 35

35. Prostomium weakly incised along anterior margin, sometimes turned ventrally, appearing entire from dorsum, with narrow caruncle extending to end of setiger 2; major spines with large, prominent subterminal collar, appearing as two large teeth in lateral view (plate 166T); commensal with sponges . . . . *Polydora spongicola*

— Prostomium blunt along anterior margin, with large swollen caruncle extending posteriorly to setiger 4; major spines of setiger 5 with subterminal shelf and distinct lateral tooth (plate 166S); known only from soft sediments associated with ampharetids. . . . . *Polydora narica*

36. Prostomium with occipital antenna (plate 166U) 37
— Prostomium without occipital antenna 38

37. Major spines of setiger 5 simple, without accessory tooth; companion setae of setiger 5 with expanded, plumose tips (plate 166V); anterior pair of eyes spaced more widely than posterior pair . . . . *Polydora nuchalis*

— Major spines of setiger 5 falcate, with distinct subterminal tooth, sometimes with additional narrow keel between tooth and tip of spine; companion setae of setiger 5 thin, flattened, feathery, closely adhering to major spines (plate 166W); anterior and posterior pairs of eyes spaced equidistant from one another, forming a square . . . . *Polydora cornuta*

38. Prostomium incised along anterior margin 39
— Prostomium entire along anterior margin 40

39. Modified spines of setiger 5 with obliquely curved flange extending between straight shaft and curved tip (plate 166X–166Y); pygidium disklike, anus just below narrow dorsal notch; bores in mollusk shells . . . . . . *Polydora websteri*

— Modified spines of setiger 5 with small, triangular accessory tooth; pygidium disklike, with shallow, broad dorsal notch, and anus more toward center of disk; forms dense populations on mud flats or on bottom of boats and piers . . . . *Polydora limicola*

40. Major spines of setiger 5 with large accessory tooth (plate 166Z); caruncle narrow, extending to end of setiger 2 pygidium scoop-shaped (plate 166AA); palps without pigment bands; bores in mollusk shells, with hermit crabs, and with encrusting bryozoans and coralline algae . . . . *Polydora pygidialis*

— Major spines of setiger 5 with large accessory flange, similar to that of *P. websteri*, but larger; caruncle extending to middle of setiger 3 or to setiger 4; pygidium disklike, with dorsal notch and anus in center of disk; palps with distinct black pigmented bands; borer in bivalve shells and gastropods occupied by hermit crabs. . . . . . . *Polydora brevipalpa*

41. Hooded hooks from setiger 7 43
— Hooded hooks from setigers 10–17 42

42. Prostomium recessed into setiger 1, rounded to weakly incised anteriorly, caruncle absent (plate 166BB); palps very short; modified spines of setiger 5 with long, lateral sheath (plate 166CC); pygidium with 4–14 small, narrow lobes; commensal with hermit crabs . . . . *Dipolydora commensalis*

— Prostomium bifurcate along anterior margin, caruncle extending well beyond setiger 5 (plate 166DD); palps very long; modified spines of setiger 5 with long lateral sheath (plate 166FF); pygidium with four nearly equal lobes; bores into bivalve shells and gastropod shells occupied by hermit crabs . . . . *Dipolydora elegantissima*

43. Major spines of setiger 5 simple, falcate, with subterminal boss, accessory teeth, flanges and collars absent; notopodial postsetal lamellae of setiger 1 short, conical to triangular (plate 166GG); posterior notopodia with a few short and long capillaries . . . . *Dipolydora socialis*

— Major spines of setiger 5 with accessory teeth, flanges, collars, often cloaked with bristles 44

44. Major spines of setiger 5 with collar on convex side of falcate tip (plate 166HH); posterior notopodia with packets of small glistening needles; hooded hooks bidentate in anterior and middle setigers, unidentate in posterior setigers; branchiae from setiger 8 . . . . *Dipolydora bidentata*

— Major spines of setiger 5 with various accessory teeth, flanges, or collars not restricted to convex side of spine. 45

45. Modified spines of setiger 5 with large accessory tooth on concave side and small spur on convex side, tip not covered with bristles; posterior notopodial spines present or absent, if present not arranged in rosette 46

— Modified spines of setiger 5 always covered with fine bristles, with or without accessory teeth, teeth if present, sometimes joined by collar or cowling (plate 166KK–166MM); posterior notopodial spines present, forming projecting rosette 47

46. Posterior notopodia with 3–4 recurved spines (plate 166II); pygidium with four nearly equal lobes (plate 166JJ); branchiae from setiger 8 . . . . *Dipolydora bifurcata*

— Posterior notopodia with long capillaries, spines absent; pygidium disklike, with dorsal gap; branchiae from setiger 9 . . . . *Dipolydora giardi*

47. Major spines of setiger 5 with large, flattened, beaklike curved end bearing large and distinctive crest of bristles (plate 166KK); companion setae reduced to thin, simple capillaries; prostomium distinctly notched along anterior margin . . . . *Dipolydora brachycephala*

— Major spines of setiger 5 with two teeth, some bristles between teeth or partially covering spine 48

48. Major spines of setiger 5 bidentate, with two large but unequal teeth, each connected by lateral hood or cowling on one side and thin shelf on other side; cowling covered with fine bristles (plate 166MM); eyes absent; pygidium cuffshaped, with broad dorsal notch (plate 166NN); bores into calcareous substrates . . . . *Dipolydora armata*

— Major spines of setiger 5 bidentate, with two nearly equal teeth between which arise a tuft of fine bristles (plate 166LL); four eyes present, arranged in transverse row; pygidium with four equal lobes; builds tubes in soft sediments . . . . *Dipolydora quadrilobata*

LIST OF SPECIES

See Hartman 1936, 1941a, 1969; Blake 1969, 1971; 1996c; Blake and Evans 1973; Blake and Woodwick 1971, 1972; Light 1977, 1978.

Apoprionospio pygmaea (Hartman, 1961). Intertidal to shallow subtidal depths in silty mud; also known from Gulf of Mexico and the eastern United States. See Foster 1969, Proc. Biol. Soc. Wash. 82: 381–400; 1971, Stud. Fauna Curaçao Carib. Isl. 36: 1–183; Light 1978; Hannon 1981, Limnol. and Oceanogr., 26: 159–171 (larval settlement, Montery Bay); 1985, Zool J. Linn. Soc., 84: 325–383; Blake 1996c.

Boccardia berkeleyorum Blake and Woodwick, 1971. Central California, intertidal, bores in coralline algae, hermit crab shells, and the jingle shell *Pododesmus*. See Blake and Woodwick 1971.

Boccardia columbiana Berkeley, 1927. Bores in mollusk shells and coralline algae. See Woodwick 1963a (morphology and ecology); Blake and Arnofsky 1999 (larval development); Blake 2006 (larval development).

Boccardia proboscidea Hartman, 1940. High intertidal rock pools, in crevices; sandy mudflats; occasionally in debris in rocky intertidal; introduced into Australia. See Woodwick 1963a (morphology and ecology); Woodwick 1977, Hartman Memorial Volume, Allan Hancock Found., pp. 347–371 (larval development); Blake and Kudenov 1981, Mar. Ecol. Prog. Ser. 6: 175–182 (larval biology); Gibson 1997, Invert. Biol. 116: 213–226 (larval biology); Gibson and Gibson 2004, Evolution 58: 2704–2717 (larval development as a model for evolution of poecilogony in spionids).

Boccardia tricuspa (Hartman, 1939). Bores in mollusk shells and coralline algae. See Woodwick 1963b (morphology and ecology); Blake 2006 (pelagic larvae).

Boccardiella hamata (Webster, 1879) (=*Boccardia uncata* Berkeley, 1927). Intertidal as a borer into hermit crab shells and bivalves; may also build tubes in mud flats. See Blake 1966, Bull. So. Calif. Acad. Sci. 65: 176–184 (morphology and ecology); Dean and Blake 1966, Biol. Bull. 130: 316–330 (life history and larval development).

Boccardiella ligerica (Ferronière, 1898). A widely distributed species, opportunistic in sediments with waters of very low salinity; redescribed by Blake and Woodwick 1971, later reported from San Francisco Bay by Light (1978) and in freshwater streams in southern California by Kudenov 1983, Bull. So. Calif. Acad. Sci. 82: 144–146; Cohen and Carlton 1995.

Boccardiella truncata Hartman, 1936. Known only from Moss Beach, San Mateo County in northern California, intertidal in sandstone reefs. See Hartman 1936, 1968.

Carazziella calafia Blake, 1979. Soft sediments; intertidal to about 90 m. See Blake 1979, Proc. Biol. Soc. Wash. 92: 466–481 (morphology); 1996c (morphology, ecology); Blake and Arnofsky 1999 (pelagic larvae); Blake 2006 (pelagic larvae).

Dipolydora armata (Langerhans, 1880). A widespread, warm-water species associated with coralline algal crusts, living coral, and coral rock; rare in temperate latitudes; intertidal to shallow subtidal; southern California. See Blake 1996c; Radashevsky, and Nogueira 2003, J. Mar. Biol. Ass. U. K. 83: 375–384 (reproduction, larvae, asexual reproduction).

Dipolydora bidentata (Zachs, 1933). (=*Polydora convexa* Blake and Woodwick, 1972). California, western Pacific. Intertidal to shallow subtidal, bores into hermit crab shells and shells of living bivalves and gastropods. See Blake and Woodwick 1972; Radashevsky 1993 (morphology, synonyms, habitats, distribution).

Dipolydora bifurcata Blake, 1981. A borer in coralline algae; known only from northern California in outer coast tide pools. See Blake 1981, Bull. So. Calif. Acad. Sci. 80: 32–35.

Dipolydora brachycephala (Hartman, 1936) (=*Polydora brachycephala*). Intertidal to shallow shelf depths in soft sediments. This species is very similar to *D. caulleryi* (Mesnil, 1897) but differs in morphology of the larvae. See Blake 2006 (larval development).

Dipolydora commensalis (Andrews, 1891) (=*Polydora commensalis*). A well-known commensal with hermit crabs. Intertidal on both coasts of North America. See Hatfield 1965, Biol. Bull. 128: 356–368 (larval development); Blake 1969 (larvae); Radashevsky 1993 (morphology, synonyms, habitats, distribution); Williams and McDermott 1997, Invert. Biol. 116: 237–247 (feeding biology).

Dipolydora elegantissima (Blake and Woodwick, 1972) (=*Polydora elegantissima*). Bores in hermit crab and Pismo clam (*Tivela*) shells. See Blake and Woodwick 1972.

Dipolydora giardi (Mesnil, 1896) (=*Polydora giardi*). This small species is widely distributed around North America, Europe, Australia, and New Zealand. Intertidal to about 200 m and bores in calcareous substrata. In California, the species bores into coralline algae and mollusk shells, especially those of the abalone *Haliotis rufescens*. See Day and Blake 1979, Biol. Bull. 156: 20–30 (larval development); Radashevsky and Petersen 2005, Zootaxa 1086: 25–36 (morphology and distribution).

Dipolydora quadrilobata (Jacobi, 1883) (=*Polydora quadrilobata*). Widely distributed in the northern hemisphere, intertidal to shallow subtidal in soft sediments; probably introduced into San Francisco Bay. See Light 1977; Blake 1969 (larval development); Radashevsky 1993 (morphology, synonyms, habitats, distribution).

Dipolydora socialis (Schmarda, 1861) (=*Polydora socialis*). Widely distributed in North America, intertidal to shallow subtidal in fine sands to coarse silts; also algal holdfasts, occasionally boring into mollusk shells as a nestler in mud pockets. See Blake 1969 (larvae); 1971, 1996c (morphology).

Dispio uncinata Hartman, 1951. A widespread species reported locally from Tomales Bay to southern California, intertidal to shallow subtidal in mixed sediments; reported from San Francisco Bay by Light (1977). See Blake and Arnofsky 1999 (larval development); Blake 2006 (larval development).

Laonice cirrata (Sars, 1851) (=*Laonice foliata* Moore, 1923 has been determined to be a synonym of *L. cirrata* Blake, unpublished). A widely distributed cosmopolitan species; low water to slope depths in soft sediments.

Marenzelleria viridis (Verrill, 1873) (=*Scolecolepides viridis*). This species is native to the northeastern United States, but has been introduced into northern Europe and most recently San Francisco Bay (Cohen and Carlton 1995). The species is capable of building dense populations in soft sediments from the intertidal to shallow subtidal. The adults and larvae are tolerant of very low salinities. See George 1966, Biol. Bull. 130: 76–93 (reproduction and larval development); Maciolek 1984, Proc. 1st Int. Polychaete Conf., Sydney, Linn. Soc. NSW, pp. 48–62 (systematics, distribution); Cohen and Carlton 1995 (introduced into San Francisco Bay); Blake and Arnofsky 1999 (review of reproduction).

Microspio pigmentata (Reish, 1959). Central and southern California to Mexico, shallow subtidal to outer shelf depths. Characterized by having a distinctive pigment spot in the center of the prostomium, spinous posterior notosetae, and multidentate hooded hooks. See Maciolek 1990, J. Nat. Hist. 24: 1109–1141 (revision of the genera *Spio* and *Microspio*); Blake 1996c (keys and descriptions of California species).

Paraprionospio pinnata (Ehlers, 1901) (=*Prionospio pinnata*). Widely distributed in North and South America in silty mud, low water to shelf depths. See Dauer 1985, Mar. Biol. 85:

* = Not in key.

143–151 (feeding morphology and behavior); Blake and Arnofsky 1999 (pelagic larvae).

Polydora alloporis Light, 1970. Subtidal, boring in *Stylaster californicus* (=*Allopora*) and creating distinctive parallel holes; also intertidal in *Stylantheca porphyra*. See Light 1970, Proc. Calif. Acad. Sci. 37: 459–471.

Polydora bioccipitalis Blake and Woodwick, 1972. Intertidal and shallow subtidal in hermit crab shells; southern California, Chile. See Blake and Woodwick 1972; Blake 1996c.

Polydora brevipalpa Zachs, 1933 (=*Polydora variegata* Imajima and Sato, 1984 from Japan). Intertidal to shallow subtidal in various calcareous habitats including bivalves and gastropod shells with or without hermit crabs; similar to *P. websteri*, but noted to be distinct in surveys of intertidal cryptic habitats in California (Blake and Woodwick, unpublished). See Radashevsky 1993 (morphology, synonyms, habitats, distribution).

Polydora cornuta Bosc, 1802 (=*Polydora ligni* Webster, 1879). The type-species of *Polydora;* widely distributed in North America, common on mud flats of bays and estuaries; possibly introduced multiple times. An extensive literature exists on the biology and ecology of this species. See Blake 1969 (larval development); Rice and Simon 1980, Ophelia 19: 79–115 (population variability); Blake 1971; Blake and Maciolek 1987, Bull. Biol. Soc. Wash. 7: 11–15 (neotype designation); Cohen and Carlton 1995 (introduced into San Francisco Bay); Radashevsky 2005, Zootaxa 1064: 1–24 (adult and larval morphology, global distribution).

Polydora limicola Annenkova, 1934. Forms large aggregations on rocks, wharves, and ship bottoms; possibly introduced into California. See Blake 1996c for review.

Polydora narica Light, 1969. Subtidal off Monterey, California; associated with an unidentified ampharetid.

Polydora nuchalis Woodwick, 1953. Intertidal in mudflats of estuaries and bays. See Woodwick 1953, J. Wash. Acad. Sci. 43: 381–383 and 1960, Pac. Sci. 14: 122–128 (larval development).

Polydora pygidialis Blake and Woodwick, 1972. Intertidal as a borer in mollusk shells and ectoprocts. See Blake and Woodwick 1972; Blake 1996c.

Polydora spongicola Berkeley and Berkeley, 1950. Commensal with sponges. See Woodwick 1963b (morphology, ecology); Radashevsky 1993 (morphology, synonyms, habitats, distribution).

Polydora websteri Hartman, 1943. Bores in mollusk shells and other calcareous materials. See Blake 1969 (larval development); 1971 (morphology and ecology).

Prionospio dubia Day, 1961. A widespread but not overly common species; shallow subtidal to shelf and slope depths in soft sediments. See Blake 1996c for review.

Prionospio lighti Maciolek, 1985 (=previous eastern Pacific records of *P. cirrifera*). Intertidal to shelf depths. See Maciolek 1985, Zool. J. Linn. Soc. 84: 325–383 (morphology and distribution); Blake 1996c.

Prionospio steenstrupi Malmgren, 1867. A widespread northern hemispheric species, sometimes forming dense populations in shallow subtidal sediments. See Maciolek 1985, Zool. J. Linn. Soc. 84: 325–383 (morphology and distribution).

Pseudopolydora kempi (Southern, 1921). Intertidal in sandy mud sediments of bays and estuaries. It has generally assumed that *P. kempi* has been introduced from the western Pacific (Cohen and Carlton 1995). However, studies of the reproduction and larval development from India (the type locality) by Srikrishnadhas and Ramamoorthi 1977, Proc. Symp. Warm Water Zooplankton, Spec. Pub. UNESCO/NIO: 617–677; the Sea of Japan by Radashevsky 1985, Mar. Biol. Vladivostok No.

2: 39–46; and central California by Blake and Woodwick 1975, Biol. Bull. 149: 109–127 suggest otherwise. California populations exhibit an extreme form of adelphophagia where highly modified eggs serve as nurse cells for a few developing larvae that are brooded in capsules until released to the plankton at about the 15-setiger stage. A short pelagic period precedes settlement and metamorphosis. In contrast, developing larvae from India and the Sea of Japan lack nurse cells and are released as planktotrophic larvae at the three-setiger stage where they spend a prolonged period in the plankton. There are differences in larval pigmentation patterns among the three populations, as well, suggesting that they might not all represent the same species.

Pseudopolydora paucibranchiata (Okuda, 1937). Puget Sound to southern California. Intertidal on mud flats of bays and estuaries. In Tomales Bay, *P. paucibranchiata* may be the dominant spionid polychaete on many sand flats, preferring finer sediments than *P. kempi*. Like *P. kempi*, *P. paucibranchiata* has been regarded as an introduction from the western Pacific (Cohen and Carlton 1995). The reproduction and larval development have been studied by Blake and Woodwick 1975, Biol. Bull. 149: 109–127, from central California and by Radashevsky 1983, Mar. Biol. Vladivostok No. 2: 38–46. In both reports, the planktotrophic larvae were released at the three-setiger stage. Although similar, there are differences in larval pigmentation and ciliary patterns between larvae reported in these two studies. Clearly, further investigation of the systematics, biology, and ecology of both *P. kempi* and *P. paucibranchiata* are required relative to whether they are introduced. See Radashevsky 1993 (morphology, synonyms, ecology, and distribution).

Pygospio californica Hartman, 1936. Known only from central California in high intertidal sand flats. See Blake 2006 (larval development).

Pygospio elegans Claparède, 1863. Locally found with the sabellid *Chone minuta* in sand and tube aggregations in rocky intertidal areas; sand flats, high intertidal. See Blake 2006 (review of reproduction, larval development, and asexual reproduction).

Rhynchospio glutaea (Ehlers, 1897) (=*R. arenincola* Hartman, 1936). A widespread cosmopolitan species, intertidal to shallow shelf depths in sand and mud sediments. See Banse 1964, Proc. Biol. Soc. Wash. 76: 203–204; Blake 1996c.

Scolelepis (*Scolelepis*) *squamata* (Müller, 1806) (=*Nerinides acuta* [Treadwell, 1914]). On protected beaches of clean sand sediments. See Hartman 1941a (morphology, reproduction).

Scolelepis (*Parascolelepis*) *texana* Foster, 1971. First reported from central California by Blake (1996c), the species is also reported from the Gulf of Mexico and the eastern United States; mixed sediments.

Scolelepis (*Parascolelepis*) cf. *tridentata* Southern, 1914 (=*Nerinides tridentata*). Specimens near this species have been taken from intertidal sandy muds in Tomales Bay and Bodega Bay. This species may be the same as *S. yamaguchii* Imajima. See Blake and Arnofsky 1999 (larval development); Blake 2006 (larval development).

Spio filicornis (Müller, 1776). Species of the genus *Spio* are relatively uncommon in the eastern Pacific (Blake 1996c). *S. filicornis* has been reported from the east coast of Vancouver Island by Berkeley and Berkeley 1952, but is not so well known in California and Oregon. Additional keys to eastern Pacific species are provided by Hobson and Banse 1981 (British Columbia and Washington) and Blake 1996c (California). See

* = Not in key.

Maciolek 1990, J. Nat. Hist. 24: 1109–1141 (revision of the genera *Spio* and *Microspio*).

Spiophanes berkeleyorum Pettibone, 1962. Low intertidal to shelf depths in mixed sand and mud sediments. See Blake 1996c.

Spiophanes cf. *bombyx* (Claparède, 1870). A reportedly widespread species, but further study will likely reveal local speciation; intertidal to shelf depths in sandy sediments. See Blake 2006 (pelagic larvae, reports differences in larval pigment patterns among widespread populations).

Spiophanes duplex (Chamberlin, 1919) (=*S. missionensis* Hartman, 1941). Sandy mud sediments, intertidal to shelf depths. See Blake 2006 (larval development).

Spiophanes kroeyeri Grube, 1860 (=*S. fimbriata* Moore, 1923). Low water to shelf and slope depths in mixed sand and mud. *S. kroeyeri* is one of the dominant polychaetes in slope depths off California. See Blake 1996c.

Streblospio benedicti Webster, 1879. One of the most common and familiar polychaetes in intertidal mud flats of estuaries and tributaries. Considered introduced by Cohen and Carlton (1995). See Levin 1984, Biol. Bull. 166: 494–508 (larval development from three coasts of North America); Blake and Arnofsky 1999 (review of development); Blake 2006 (review of development).

MAGELONIDAE

The Magelonidae are represented by a single genus, *Magelona,* with more than 50 described species. They are characterized as having slender, threadlike bodies with a distinctive, flattened, and sometimes spatulate head region that is modified to assist burrowing through sandy sediments. The body of magelonids is slender and divided into two distinct regions, including a thoracic or anterior region consisting of prostomium, peristomium, and nine setigers and an abdominal region with a variable number of segments. The pharynx is eversible and a pair of long, prehensile, papillate palps arise from the prostomium-peristomium that are used to collect detrital particles.

The principal diagnostic characters important in differentiating species of *Magelona* include (1) the presence or absence of frontal horns on the prostomium, (2) relative dimensions of the prostomium, (3) presence or absence of dorsal median lobes on thoracic notopodia, (4) presence and location of lateral pouches between abdominal segments, (5) modifications to thoracic setae of setiger 9, (6) morphology of the abdominal hooded hooks, (7) presence of abdominal medial lamellae, and (8) the presence/absence and form of interlamellae on abdominal parapodia.

KEY TO MAGELONIDAE

1. Prostomium with frontal horns (plate 167A, 167B) 2
— Prostomium rounded anteriorly, without frontal horns (plate 167C, 167F) 3
2. Hooded hooks of one type; without modified setae in setiger 9 . . . . *Magelona berkeleyi*
— Hooded hooks of two types, including normal long-shafted ones and a single short one located at base of lateral lamellae; with modified setae in setiger 9 . . . .
. . . . *Magelona hartmanae*
3. Bidentate hooded hooks present . . . .
. . . . *Magelona californica*
— Tridentate hooded hooks present 4
4. Lateral pouches present between successive parapodia of abdominal segments (plate 167C); modified setae of setiger

9 of two types, one limbate (plate 167E), the other mucronate (plate 167D); hooded hooks of one type . . . .
. . . . *Magelona sacculata*
— Lateral pouches absent; modified setae of setiger 9 limbate (plate 167G); hooded hooks of two types, including normal long-shafted ones and a single short one located at base of lateral lamellae. . . . . *Magelona pitelkai*

LIST OF SPECIES

See Hartman 1944b; Blake, 1996d.

Magelona berkeleyi Jones, 1971. Shallow subtidal to outer shelf depths. See Jones 1971, J. Fish. Res. Bd. Can. 28: 1445–1454 (morphology).

Magelona californica Hartman, 1944. Southern California embayments in the intertidal zone. See Hartman 1961, 1969.

Magelona hartmanae Jones, 1978. Subtidal.

Magelona pitelkai Hartman, 1944. The most common magelonid in central and northern California; clean sand to sandy mud. See Jones 1978. Proc. Bio. Soc. Wash. 91: 336–363 (redescription, distribution); Blake 2006 (pelagic larvae).

Magelona sacculata Hartman, 1961. Fine sands; rare. See Blake 2006 (pelagic larvae).

TROCHOCHAETIDAE (=DISOMIDAE)

Trochochaeta (=*Disoma*) *franciscanum* (Hartman, 1947) (plate 167H) is known from shallow mud in San Francisco Bay and other embayments up and down the West Coast; larvae have been taken in plankton from Tomales Bay. See Hartman 1947, J. Wash. Acad. Sci. 37: 160–169; Banse and Hobson 1968; Blake and Arnofsky 1999 (larval development); Blake 2006 (larval development). Pettibone (1976) synonymized *T. franciscanum* with the well-known Atlantic species, *T. multisetosa* (Oersted 1844); however, Blake 2006 reports morphological differences in pelagic larvae of California specimens and those described for *T. multisetosa.*

CHAETOPTERIDAE

The chaetopterids are highly modified tubicolous polychaetes that are adapted to pump water through their tubes as part of suspension feeding activities. Their bodies are elongate and fragile, composed of numerous segments, and divided into three distinct regions. Sizes range from 7 mm to nearly 300 mm in length. Some chaetopterids are widely distributed on sandy and muddy bottoms, occurring from the intertidal to the deep sea. They may form solitary tubes or sometimes colonies where the tubes compose dense mats; at least one species has branched tubes. Some species are well known as experimental animals in developmental and behavioral research. Recent studies have demonstrated that a sibling species complex is included in taxa formerly referred to *Spiochaetopterus costarum.*

KEY TO CHAETOPTERIDAE

1. Anterior end with a pair of palps and a pair of tentacular cirri arising from the first setigerous segment (plate 167I); middle region with 4–12 segments; prostomium with eyes; white ventral glandular shield from setigers 6–12; tube

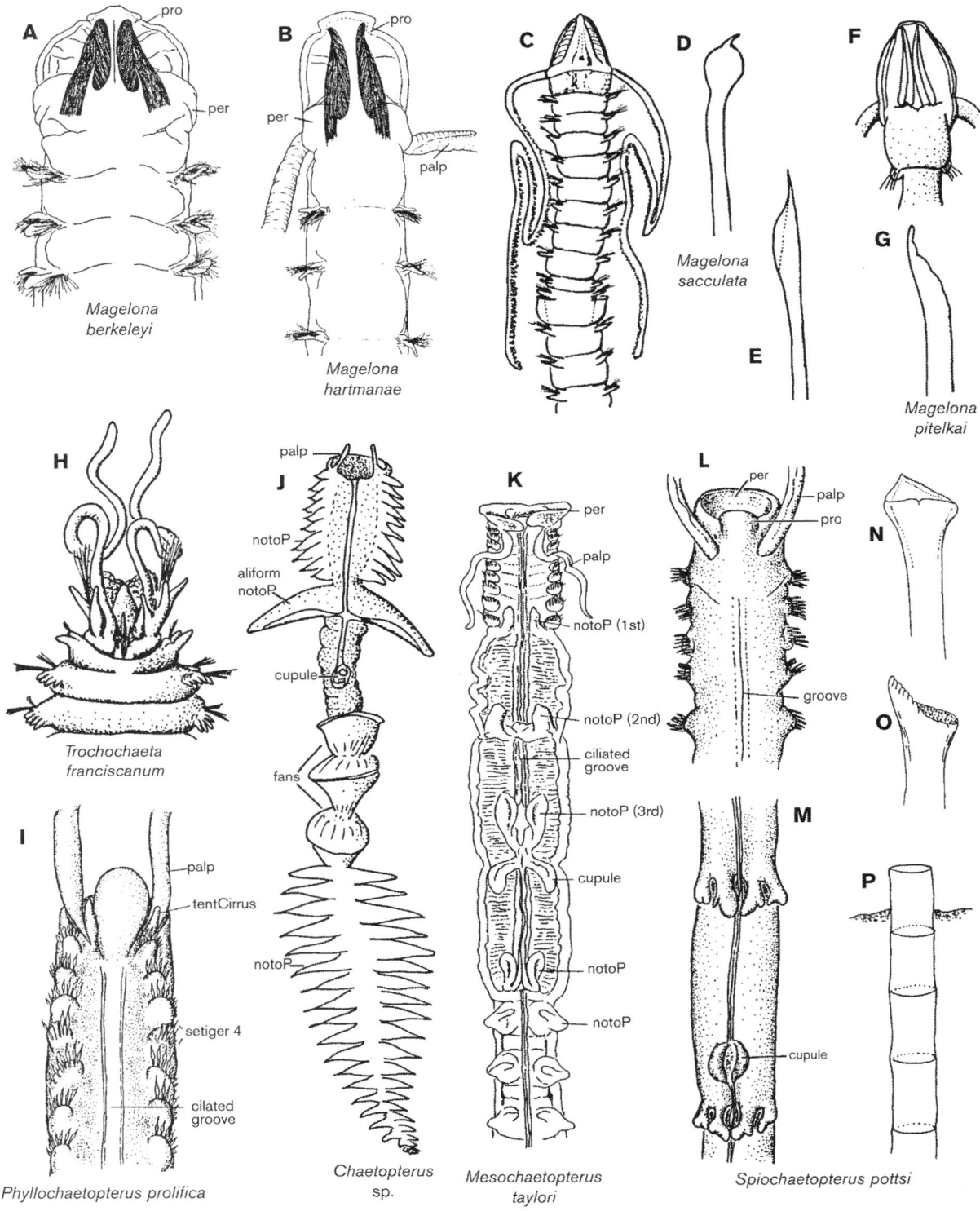

PLATE 167 Polychaeta. Magelonidae, Trochochaetidae, and Chaetopteridae: A, *Magelona berkeleyi*; B, *Magelona hartmanae*; C, *Magelona sacculata*, D, mucronate seta from setiger 9; E, limbate seta from setiger 9; F, *Magelona pitelkai*; G, limbate seta from setiger 9; H, *Trochochaeta franciscanum*; I, *Phyllochaetopterus prolifica*; J, *Chaetopterus* sp.; K, *Mesochaetopterus taylori*; L, *Spiochaetopterus pottsi*; M, middle region of body; N–O, spines from setiger 4; P, tube. (A–B, after Jones; C–H, after Hartman; J, after MacGinitie; K, after Potts; rest after Blake).

usually branched; asexual buds and regenerates commonly present . . . . *Phyllochaetopterus prolifica*
— Anterior end with a pair of palps; tentacular cirri absent 2
2. Palps long, prehensile, longer than anterior region (plate 167K, 167L); notopodia of middle region with 1–3 lobes, never fused across dorsum 3
— Palps shorter than anterior region (plate 167J); middle region with five segments, with first bearing winglike notopodia, remainder formed into paddles; peristomium a broad collar; tube large, U-shaped, sometimes covered with sand grains . . . . *Chaetopterus* spp.
3. Middle region with unilobed, conical, fleshy or fingerlike notopodia (plate 167K); setiger 4 with several stout modified setae; tube usually straight or J-shaped, covered with sand or shell . . . . *Mesochaetopterus taylori*
— Middle region with bi- or trilobed notopodia (plate 167M); setiger 4 usually with only a single stout modified seta (plate 167N, 167O); tube distinctly annulated, semitransparent, straight or twisted (plate 167P) . . . .
. . . . *Spiochaetopterus pottsi*

LIST OF SPECIES

See Barnes 1965, Biol. Bull. 129: 217–233 (feeding and tube-building); Blake 1996e (review).

Chaetopterus spp. (=*Chaetopterus variopedatus* of previous edition). Two types of species are present. One forms U-shaped tubes in sand flats of bays and estuaries, and the other is a smaller species that attaches to wharf pilings. Taxonomic studies required to define these species have not been performed. The mucus-bag feeding mechanism was described by MacGinitie 1939, Biol. Bull. 77: 115–118. See Abbott and Reish 1980.

Mesochaetopterus taylori Potts, 1914. Intertidal in sand flats, usually associated with *Zostera*, reported from Dillon Beach and Humboldt Bay; juveniles were reported from the Santa Maria Basin by Blake 1996e. See Abbott and Reish 1980; Blake 2006 (pelagic larvae).

Phyllochaetopterus prolifica Potts, 1914. Western Canada to southern California. Intertidal to about 100 m; common; rocky habitats. See Abbott and Reish 1980.

Spiochaetopterus pottsi (E. Berkeley, 1927) (=*Spiochaetopterus costarum* in the previous edition and in Blake 1996e; =*Telepsavus costarum* in Hartman 1969). British Columbia to southern California. Low water to shelf depths; in fine sands. See Gitay 1969, Sarsia 37: 9–20 (taxonomy); Blake 1996e (review); Bhaud 2003, Hydrobiologia 496: 279–287 (larvae and adults of *Spiochaetopterus*); Blake 2006 (pelagic larvae).

APISTOBRANCHIDAE

The apistobranchids are elongate, fragile, deposit-feeding spioniform polychaetes that occur in shallow shelf depths to the lower slope. The anterior end has a pair of grooved spionidlike palps that are attached to a "head" that consists of the fused prostomium and peristomium. The parapodia are subbiramous with the notopodia reduced to clavate erect ciliated structures that superficially resemble branchiae. These notopodia are supported by internal aciculae, a feature unusual in sedentary tubicolous polychaetes. Some anterior postsetal lamellae are fimbriated and interramal cirri are present that provide a superficial resemblance to orbiniids. Setae are simple. The taxonomic history is very confused. According to Blake (1996f), there may be up to five valid species. The common eastern Pacific species is *Apistobranchus ornatus* Hartman, 1965, which has been collected in shallow subtidal habitats and slope depths. See Blake 1996f (review of morphology, taxonomic history, and California records).

CIRRATULIDAE

Cirratulids are deposit-feeding polychaetes that burrow or crawl through the substratum, although at least one genus, *Dodecaceria*, is able to bore into calcareous structures. In organically enriched sediment in estuaries, cirratulids often reach high population densities. They typically bury their bodies just below the surface with their long branchiae and tentacles visible at the surface. Cirratulids have a frilly appearance because they usually bear numerous pairs of branchiae along the length of their bodies. Important recognition characters include the shape of the prostomium, placement of the tentacles on the peristomium or anterior setigers, enlargement or swelling in some thoracic setigers, presence or absence of moniliform abdominal segments, form of the posterior portion of the body, presence or absence of spines or hooked setae, and the patterns imparted by methyl green stain.

KEY TO CIRRATULIDAE

1. Anterior end with a single pair of long, grooved dorsal tentacles or tentacular filaments 2
— Anterior end with two groups of numerous tentacular filaments (plate 169A, 169D) 13
2. Setae all smooth or denticulate capillaries; modified acicular spines or hooks absent 3
— Setae include smooth capillaries and modified setae such as acicular spines, hooks, or knoblike setae 5
3. Capillary setae smooth, sometimes with fibrils splayed or spread out along edge . . . . *Aphelochaeta* 4
— Capillaries include both smooth setae and ones with broad, basally flattened blades having fine denticles along one edge . . . . *Monticellina*
4. Expanded anterior region with segments not crowded; methyl green producing prominent ventral transverse bands in thoracic region extending up lateral sides to anterior edge of notopodia (plate 168B); middle body segments strongly moniliform; prostomium as wide as long; body slender, fragile . . . . *Aphelochaeta monilaris*
— Body elongate, thin throughout (plate 168C, 168D), not expanded anteriorly, without moniliform segments; prostomium at least twice as long as wide (plate 168C) . . . . .
. . . . *Aphelochaeta elongata*
5. Segmental branchiae limited to few anterior setigers (plate 168A); modified setae usually spatulate, with distal excavation, spoon-shaped; constructs calcareous tube masses; color in life dark green to black; in alcohol usually black; sometimes regenerating anterior or posterior sections of body . . . . *Dodecaceria fewkesi*
— Segmental branchiae numerous, extending over anterior and middle parts of body (plate 168E, 168J), sometimes to posterior end; modified setae never with excavate tip; body color sometimes dark, but usually light tan or brown in alcohol; regenerates rare 6
6. (*Note three choices*) Modified setae distally entire acicular spines (plate 168I, 168L), rarely with one to two additional

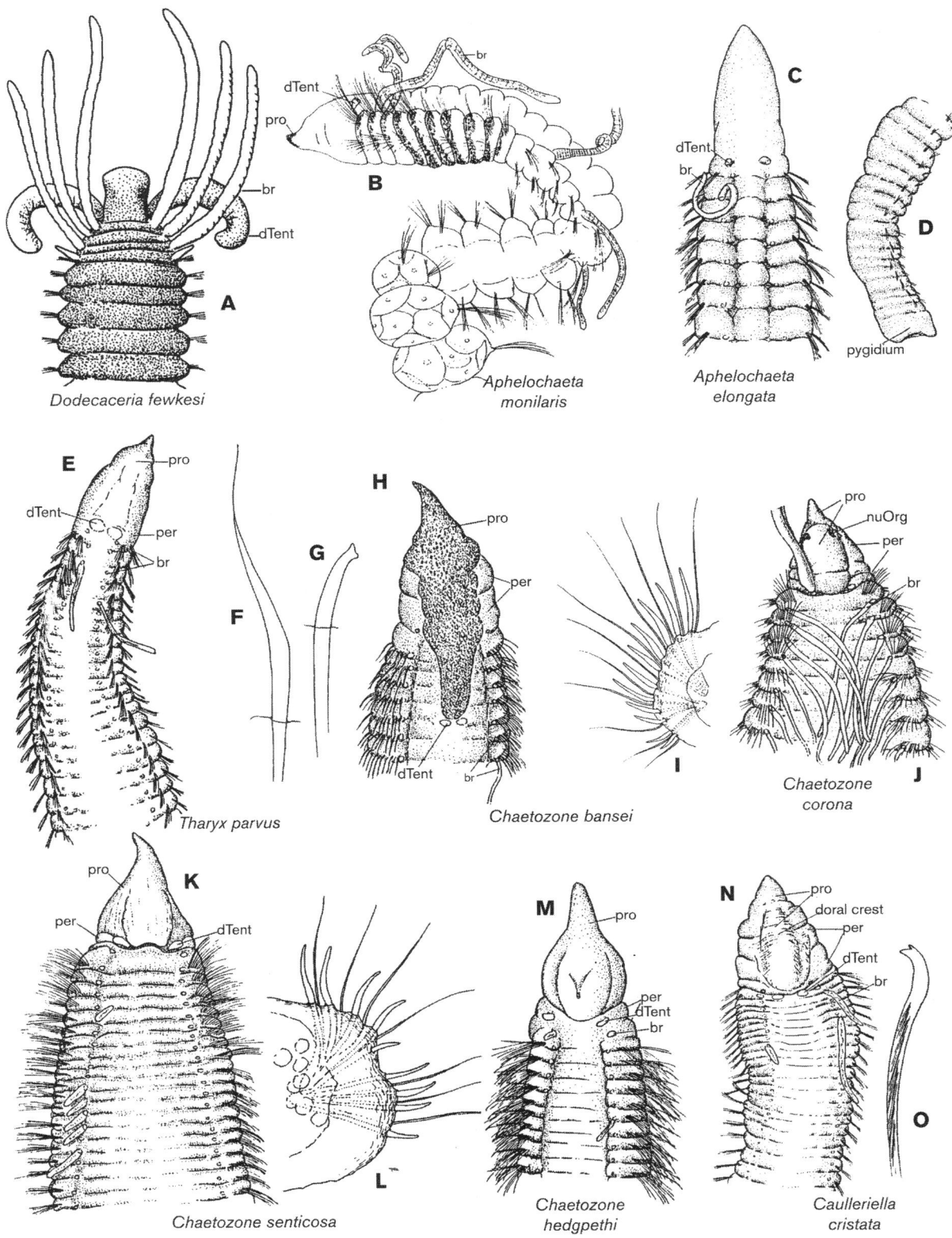

PLATE 168 Polychaeta. Cirratulidae: A, *Dodecaceria fewkesi*; B, *Aphelochaeta monilaris*; C, *Aphelochaeta elongata*; D, posterior end; E, *Tharyx parvus*; F, capillary seta; G, hooked spine from posterior setiger; H, *Chaetozone bansei*; I, posterior setiger showing cincture of spines and capillary setae; J, *Chaetozone corona*; K, *Chaetozone senticosa*; L, posterior setiger showing posterior spines and capillaries; M, *Chaetozone hedgpethi*; N, *Caulleriella cristata*; O, hooked seta (rest after Blake).

bidentate spines; spines of posterior setigers usually arranged as cinctures, providing bristly armature . . . .
. . . . *Chaetozone* 7
— Modified setae bidentate crotchetlike hooks (plate 168O), not arranged as cinctures in posterior segments . . . .
. . . . *Caulleriella* 10
— Modified setae with irregular knoblike or subbidentate tips (plate 168G), usually geniculate and grading into similarly shaped capillaries (plate 168F), never arranged in cinctures
. . . . *Tharyx parvus*
7. Paired dorsal tentacles first present from peristomium or from setiger 1 8
— Paired dorsal tentacles first present from setiger 4–7 (plate 168H) (may occur more anteriorly in small specimens); pretentacular region glandular, staining intensely with methyl green . . . . *Chaetozone bansei*
8. Neuropodial acicular spines first present from setiger 1; spines of posterior segments forming partial cinctures with 10–12 spines on a side; peristomium with dorsal medial crest continuing posteriorly over dorsum of setiger 1; prostomium with paired, pigmented nuchal organs (plate 168J) . . . . *Chaetozone corona*
— Neuropodial acicular spines first present from middle body segments (setigers 65–100+). 9
9. Peristomium with one short, asetigerous annulation (plate 168K); prostomium triangular; first pair of branchiae arising on anterior half of setiger 1; partial cinctures of posterior end with 9–10 spines (plate 168L); prostomium and peristomium not staining with methyl green . . . .
. . . . *Chaetozone senticosa*
— Peristomium with two large annulations (plate 168M); prostomium pear-shaped; first pair of branchiae arising from second peristomial annulation; methyl green staining forming light stripes around some anterior setigers . . . .
. . . . *Chaetozone hedgpethi*
10. Bidentate neuropodial hooks from setiger 1 11
— Bidentate neuropodial hooks from setigers 20–30 or later 12
11. Bidentate hooks without accompanying capillaries; peristomium with three asetigerous annulations and overlying dorsal crest (plate 168N); dorsal tentacles arising from posterior margin of setiger 1; body staining weakly yet uniformly with methyl green, no pattern evident . . . .
. . . . *Caulleriella cristata*
— Bidentate hooks with accompanying capillaries; peristomium with two asetigerous annulations, without dorsal crest; dorsal tentacles arising from anterior margin of setiger 1; staining pattern weak, but darker in intersegmental grooves . . . . *Caulleriella pacifica*
12. Prostomium triangular, not elongate, eyes absent; bidentate hooks from setigers 17–30, accompanied by capillaries only in anterior setigers . . . . *Caulleriella hamata*
— Prostomium elongate, narrow, pointed apically, with large conspicuous eyes; bidentate hooks from setigers 80–86, accompanied by capillaries throughout . . . .
. . . . *Caulleriella apicula*
13. Setae all capillaries . . . . *Protocirrineris socialis*
— Setae include capillaries and acicular spines (plate 169C)
. . . . 14
14. Branchiae first present from same setiger as tentacular filaments; eyes present (plate 169B) . . . . *Cirratulus* 15
— Branchiae arise from segments anterior to tentacular filaments; eyes absent 16

15. Prostomium broadly rounded anteriorly (plate 169B); with 5–6 red eyes in two transverse rows; may reproduce asexually . . . . *Cirratulus dillonensis*
— Prostomium wedge-shaped or triangular, tapering to narrow, rounded anterior tip; with 2–3 black eyes in two transverse rows . . . . *Cirratulus spectabilis*
16. Tentacular filaments arise from 2–5 anterior segments (plate 169G); branchiae shift from just dorsal to notopodium in anterior setigers toward middorsum of body in middle setigers (plate 169H) . . . . *Timarete perbranchiata*
— Branchial filaments arise from 1–3 anterior segments (plate 169D); branchiae remaining just dorsal to notopodium, not shifting to middorsal location . . . . *Cirriformia* 17
17. Neuroacicular spines from setigers 40–45; notoacicular spines from setigers 70–80; acicular spines dark brown to brassy in color; tentacular filaments arising from setiger 5 in adults (plate 169D); methyl green producing distinctive bands encircling most body segments and with distinctive patterns on prostomium and peristomium; inhabits rocky intertidal habitats . . . . *Cirriformia spirabrancha*
— Neuroacicular spines from setiger 80 or more posteriorly; notoacicular spines from about setiger 100 (plate 169F); acicular spines light brown to yellow in color; tentacular filaments arising from setiger 6 in adults (plate 169E); methyl green producing dark pattern on dorsum anterior to tentacular cirri and weak dorsal bands on some anterior setigers; inhabits intertidal mud flats . . . .
. . . . *Cirriformia moorei*

LIST OF SPECIES

Cirratulids are among the most abundant polychaetes in infaunal benthic communities, yet they are poorly known. Blake (1996g) described 47 species from eastern Pacific localities, yet many of these are very limited in range and habitat. Although the seemingly restricted distributions are in part due to limited collecting, it does appear that most cirratulids have restricted habitats and sediment requirements. Additional species of *Aphelochaeta*, *Chaetozone*[†] and *Cirratulus* are to be expected in the study area. Species included here are those most likely to be encountered in intertidal and nearshore collecting. Readers are referred to Blake (1996g) for more extensive keys and descriptions.

Aphelochaeta monilaris (Hartman, 1960). This species is highly distinctive with its expanded anterior and posterior ends and moniliform middle-body segments. The moniliform nature of the segments is best developed in sexually mature specimens; atokous specimens are less distinctive. In shallow subtidal embayments to slope depths. See Blake 1996g.

Aphelochaeta elongata Blake, 1996. This species differs from other eastern Pacific congeners in having a long, narrow head region and narrow, nonexpanded posterior end. See Blake 1996g.

*Aphelochaeta spp. Additional species of *Aphelochaeta* occur in our area. In Blake's (1996g) review of eastern Pacific cirratulids, *A. multifilis* (Moore, 1909) (=*Tharyx mulitfilis* of previous edition)

---

* = Not in key.

[†]*Chaetozone setosa* Malmgren, 1867, the type species, does not occur in California. This species is restricted to the Arctic. The records in the previous edition of this book and Hartman (1969) were misidentifications of various species. Eastern Pacific records included in Blake 1996g refer to as yet undescribed species.

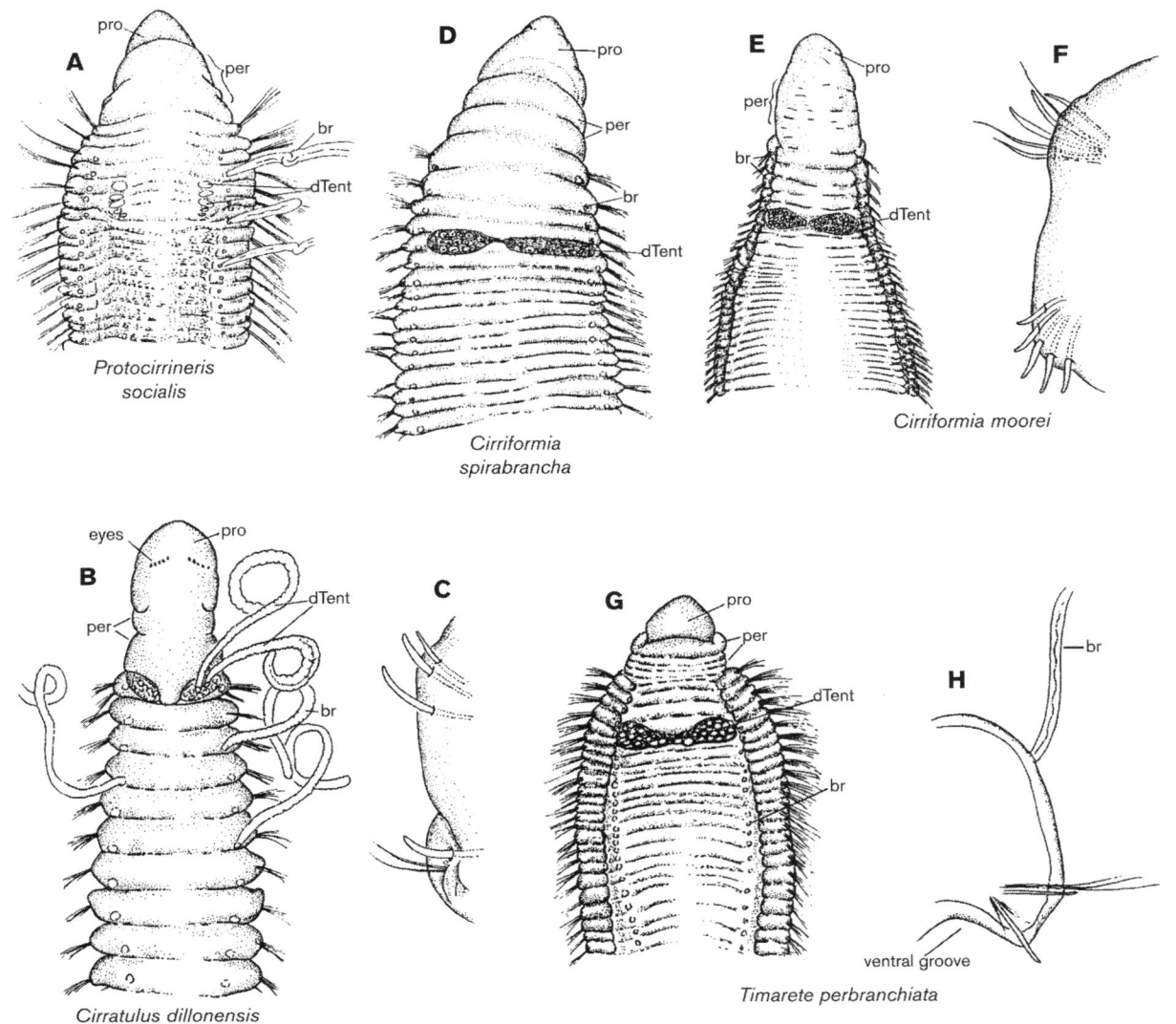

PLATE 169 Polychaeta. Cirratulidae: A, *Protocirrineris socialis*; B, *Cirratulus dillonensis*; C, posterior parapodium showing distribution of spines in noto- and neuropodia; D, *Cirriformia spirabrancha*; E, *Cirriformia moorei*; F, middle body parapodium showing noto- and neurosetae; G, *Timarete perbranchiata*; H, posterior parapodium (after Blake).

was restricted to southern California. Earlier records of this species in central and northern California may be *A. glandaria* Blake, 1996, *A. williamsae* Blake, 1996, or other species as yet undescribed. See Blake 1996g.

Caulleriella apicula Blake, 1996. Originally described from shallow water in southern California, one of us (Ruff) has recently identified a specimen from Santa Cruz, Monterey County.

Caulleriella cristata Blake, 1996. Neuropodial hooks from setiger 1. A rocky intertidal species occurring on the outer coast in algal holdfasts and under coralline algae. See Blake 1996g.

Caulleriella hamata (Hartman, 1948). Earlier California records of this species were referred to other species by Blake (1996g). An intertidal to shallow-water species known from Alaska to Washington; California records are not confirmed. See Blake 1996g.

Caulleriella pacifica Berkeley, 1929. Puget Sound to Monterey Bay; intertidal to shallow subtidal. See Blake 1996g (morphology, systematics, distribution).

Chaetozone bansei Blake, 1996. An unusual species in which the paired dorsal tentacles are shifted posteriorly to the end

of a dorsal ridge over setigers 4–7; neuropodial spines first present from setiger 28–30. The species has been collected from shallow-water shifting sands off the Golden Gate. See Blake 1996g.

Chaetozone corona Berkeley and Berkeley, 1941. This species has paired pigmented nuchal organs (previously thought to be eyes) and neuropodial acicular spines first present from setiger 1; probably limited to southern California embayments. See Blake 1996g.

Chaetozone hedgpethi Blake, 1996. Acicular spines first present from middle body segments; prostomium pear-shaped; peristomium with two short asetigerous segments. Intertidal to shallow subtidal; Tomales Bay and other embayments in fine sand. See Blake 1996g.

Chaetozone senticosa Blake, 1996. Acicular spines first present in middle body segments; prostomium and peristomium fused, forming triangular "head." Bodega Harbor in shallow water sediments, where it is one of the dominant infaunal species. See Blake 1996g.

Cirriformia moorei Blake, 1996. *Cirriformia moorei* has been misidentified as *C. spirabrancha* for many years, but it is one of the most common polychaetes found in the low intertidal zone of mudflats of estuaries and embayments in California (Blake 1996g). The species is often associated with beds of the eelgrass, *Zostera marina*. The worms live in sulfide-rich muds and respire through branchiae that they project into clear, overlying water. The species feeds at or just below the sediment/water interface (Ronan 1978, Paleobiology 3: 389–403). The red branchiae are clearly visible on the surface of the sediment at the bottoms of tidal pools. When populations of the species are dense, the numerous branchiae look like thousands of red worms on the surface (Blake 1996g). An excellent account of the feeding and burrowing mechanisms of *C. grandis*, a closely related East Coast species is that of Shull and Yasuda 2001, J. Mar. Res. 59: 453–473. Central and southern California in estuaries, intertidal to about 25 m. See Blake 1975, Trans. Amer. Microsc. Soc. 94: 179–188 (larval development as *C. spirabrancha*); Abbott and Reish 1980 (biology as *C. spirabrancha*).

Cirriformia spirabrancha (Moore, 1904). Misidentified in most publications as the common low intertidal mudflat species occurring in northern and central California. This form has been determined to be a separate species, *C. moorei* (see above). *C. spirabrancha* resides in crevices under rocks in tide pools in the rocky intertidal on semiexposed shores. See Blake 1996g.

Cirratulus dillonensis Blake, 1996. Known only from Dillon Beach, in cryptic habitats in the rocky intertidal. The species is characterized by having a narrow, elongate prostomium; two peristomial annulations where the second is twice as long as the first; and neuroacicular spines from setigers 7–8. Methyl green stain imparts an elegant pattern of transverse bands all along the body of this species that may be diagnostic. A central band of numerous stained speckles completely encircles each segment while the intersegmental areas are unstained; each tentacular filament and branchiae are covered with numerous small stained speckles. The prostomium and peristomium are also covered with small speckles; the dorsum of the peristomium does not stain. Blake (1996g) reported evidence of asexual reproduction.

Cirratulus spectabilis (Kinberg, 1866). Intertidal in rocks; a cryptic species with few records; one specimen taken from rocky tide pools in Cayucos suggests it may be more common than records indicate. See Blake 1996g.

Cirratulus spp. Blake (1996g) reviewed *Cirratulus* records and the confusing taxonomic history of the genus from the eastern Pacific. He produced a key and descriptions for five species and reorganized many of the records based on an examination of type specimens. Apart from *C. dillonensis* and a single specimen of *C. spectabilis* at Cayucos, none of the other species are confirmed from the study area of this section, although it is likely that one or more will be found with further collection and study. Previous records of *C. cirratus* in the eastern Pacific refer to other species. See Blake 1996g.

Dodecaceria fewkesi Berkeley and Berkeley, 1954. Common in rocky habitats, forming calcareous masses; the more cosmopolitan *D. concharum* Oersted has been reported in our fauna, but despite examination of hundreds of specimens of *Dodecaceria* from central California, none have been seen that can be referred to *D. concharum*. See Berkeley and Berkeley 1954. J. Fish. Res. Bd. Can. 11: 326–334 (life history); Abbott and Reish 1980.

Monticellina spp. No species of *Monticellina* are known from intertidal or shallow waters in our area of study; however, Blake (1996g) reported six species from the eastern Pacific, largely from offshore sediments.

Protocirrineris socialis Blake, 1996. Rocky intertidal habitats, perhaps subtidal. See Blake 1996g.

Timarete perbranchiata (Chamberlin, 1918). Under intertidal rocks covered with algae. The species is characterized by having the black neuropodial acicular spines from setigers 17–20, notopodial spines from setigers 31–33, the tentacular filaments arising in two dense groups separated by a distinct dorsal gap, and the branchiae number 3–5 per side on the posttentacular segments. Yoshiyama and Darling 1982, Env. Biol. Fish. 7: 39–45, reported that the tentacles of this species (as *Cirriformia luxuriosa*) were distasteful to fish in the rocky intertidal of central California. See Abbott and Reish 1980 (biology and ecology as *Cirriformia luxuriosa*); Blake 1996g (morphology, distribution).

Timarete luxuriosa (Moore, 1904). (=*Cirriformia luxuriosa* in the previous edition). After an examination of the holotype, Blake (1996g) referred the species to *Timarete* and restricted the distribution to southern California. Northern California records of this species are *T. perbranchiata*.

Tharyx parvus E. Berkeley, 1929. Species of *Tharyx* have knob-tipped spines in posterior parapodia. *T. parvus* has only a few such spines in posterior setigers. British Columbia; California, San Francisco Bay; shallow water. See Blake 1996g.

COSSURIDAE

The cossurids are small, burrowing worms that occur in mixed sandy and muddy sediments from shallow waters to the deep sea. They resemble cirratulids but are easily recognized by the single middorsal branchial filament that arises from one of the anterior segments. The prostomium is conical, without appendages or eyes, with a more or less distinct transverse prostomial furrow. The peristomium consists of a single apodous and asetigerous segment. The body is divided into a thorax and abdomen. The abdominal segments are usually cylindrical and often beaded. A single dorsomedial filamentous gill arises from an anterior segment (setigers 2–5). Aciculae are absent and parapodial lobes are poorly developed. Setae are usually all capillaries with bent and abruptly tapering to acute tips, sometimes appearing hirsute, limbate, or serrated. The pygidium bears three anal cirri and sometimes numerous shorter papillae. See Hilbig (1996) for a detailed account of California Cossuridae.

KEY TO COSSURIDAE

1. Dorsal branchial filament arising from segment 2 (plate 170A); pygidium with up to 20 digitiform processes and three anal cirri (plate 170B); methyl green stain solid through setiger 2, remainder of thorax with dorso- and ventrolateral circlets or irregular groups of single cells (plate 170D) . . . . *Cossura pygodactylata*
— Dorsal branchial filament arising from anterior or posterior margin of segment 3; digitiform pygidial processes absent (low papillae may be present). 2
2. Thorax consisting of 19–21 segments; methyl green staining pattern including large dorso- and ventrolateral patches (plate 170C), middorsum and midventrum not staining except for last few segments; in shelf and slope depths. . . . . *Cossura rostrata*
— Thorax with >20 segments. 3

* = Not in key.

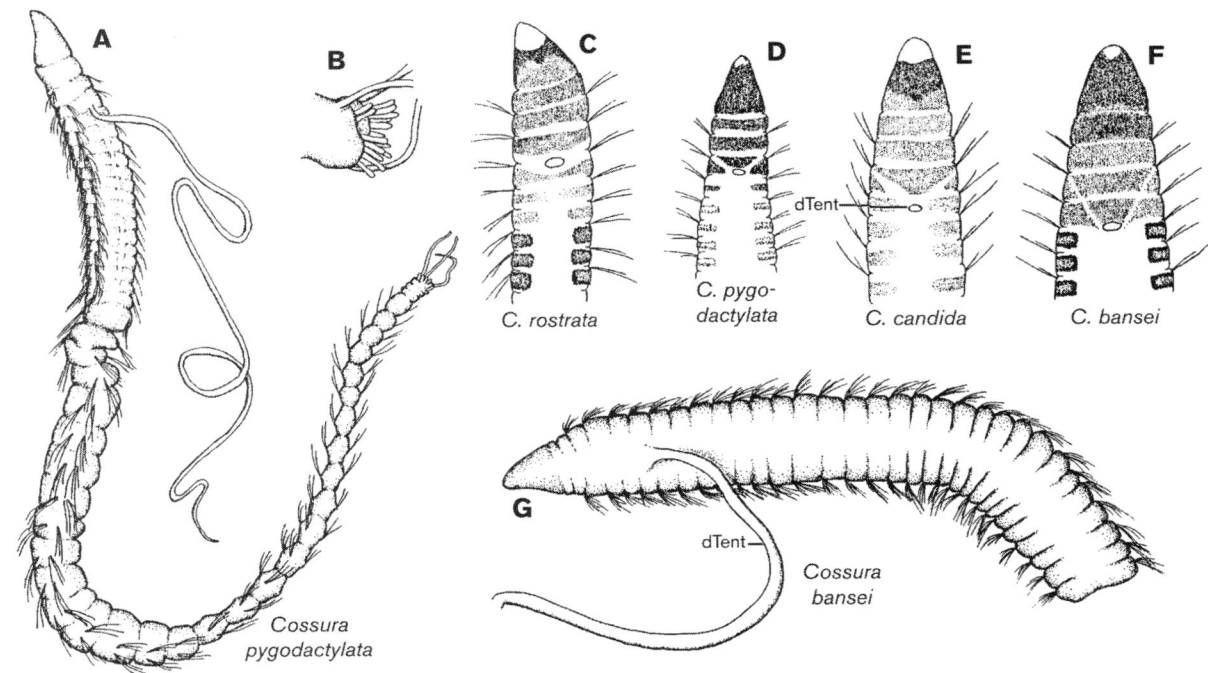

PLATE 170 Polychaeta. Cossuridae: A, *Cossura pygodactylata,* entire animal; B, posterior end, showing pygidial cirri and anal papillae; C, *C. rostrata,* showing methyl green staining pattern; D, *C. pygodactylata,* showing methyl green staining pattern; E, *C. candida,* showing methyl green staining pattern; F, *C. bansei,* showing methyl green staining pattern; G, *Cossura bansei* dorsal view of anterior end (all after Hilbig).

3. Large species, width to nearly 1 mm; thorax with 24–35 segments, with deep middorsal groove in postbranchial segments; methyl green staining solid through segment 2, reduced to lateral patches between setal fascicles in postbranchial thorax (plate 170E); posterior abdomen with three rows of segmental ventral patches . . . . *Cossura candida*

— Small species, width to 0.5 mm; thorax with 20–31 segments; abdomen with narrow, flattened segments similar to thorax; methyl green staining pattern including large dorso- and ventrolateral patches on thorax and small ones on abdomen; no lateral stain between setal fascicles; insertion of branchial filament very far back on setiger 3 (plate 170F, 170G) . . . . *Cossura bansei*

LIST OF SPECIES

See Hilbig 1996.

Cossura bansei Hilbig, 1996. The species occurs in mixed sand and silt, 18–160 m.

Cossura candida Hartman, 1955. In mud and mixed sand and mud sediments, 11–2,400 m.

Cossura pygodactylata Jones, 1956. Originally described from San Francisco Bay, off Point Richmond; now widely reported from Washington to central California; Japan; Cape Hatteras to South Carolina; western France, Bay of Biscay; in sand and silt sediments from the lowest intertidal zones to slope depths. See Bachelet and Laubier 1994, Mem. Mus. Nat. d'hist. Nat. 162: 355–369 (postlarval development).

Cossura rostrata Fauchald, 1972. A common species in northern California embayments, with densities up to 2,227 m^{-2} (Hilbig 1996; Blake, unpublished data); 6–3,348 m.

CTENODRILIDAE

Ctenodrilus serratus (Schmidt, 1857) (plate 171B) is cosmopolitan in distribution. It is inconspicuous, and occurs in the intertidal in clumps of debris; also commonly found in seawater systems of marine laboratories. The species is small (2.5–6.0 mm in length); purplish with a speckled surface (plate 171B); serrated setae are characteristic (plate 171A); rapid asexual reproduction is prevalent. See Hartman 1944c.

FLABELLIGERIDAE

The flabelligerids are highly modified, sedentary, surface deposit feeders that are commonly called "cage worms" because of the characteristic long, anteriorly directed setae found in many genera. The body is elongate, thickened, and usually somewhat fusiform in shape. The cuticle is often ornamented with numerous short to long mucus-secreting epithelial papillae. Silt and sand particles often adhere to the mucoid secretions, providing some species with a covering of particles through which setae and papillae protrude. Other species secrete a thick, transparent mucous sheath. The prostomium and peristomium retract into the first few setigers, sometimes making it difficult to study critical cephalic structures. Two long, grooved prostomial palps are present and used in feeding. The peristomium bears eight or more branchiae arising from a membrane or hood. Parapodia are reduced. Notosetae are capillaries, with those of setigers 1–6 often elongated and directed anteriorly forming the cephalic cage. Neurosetae include distinct cross-barred capillaries, similar to notosetae, simple or bidentate, compound or pseudocompound hooks, or various combinations of these

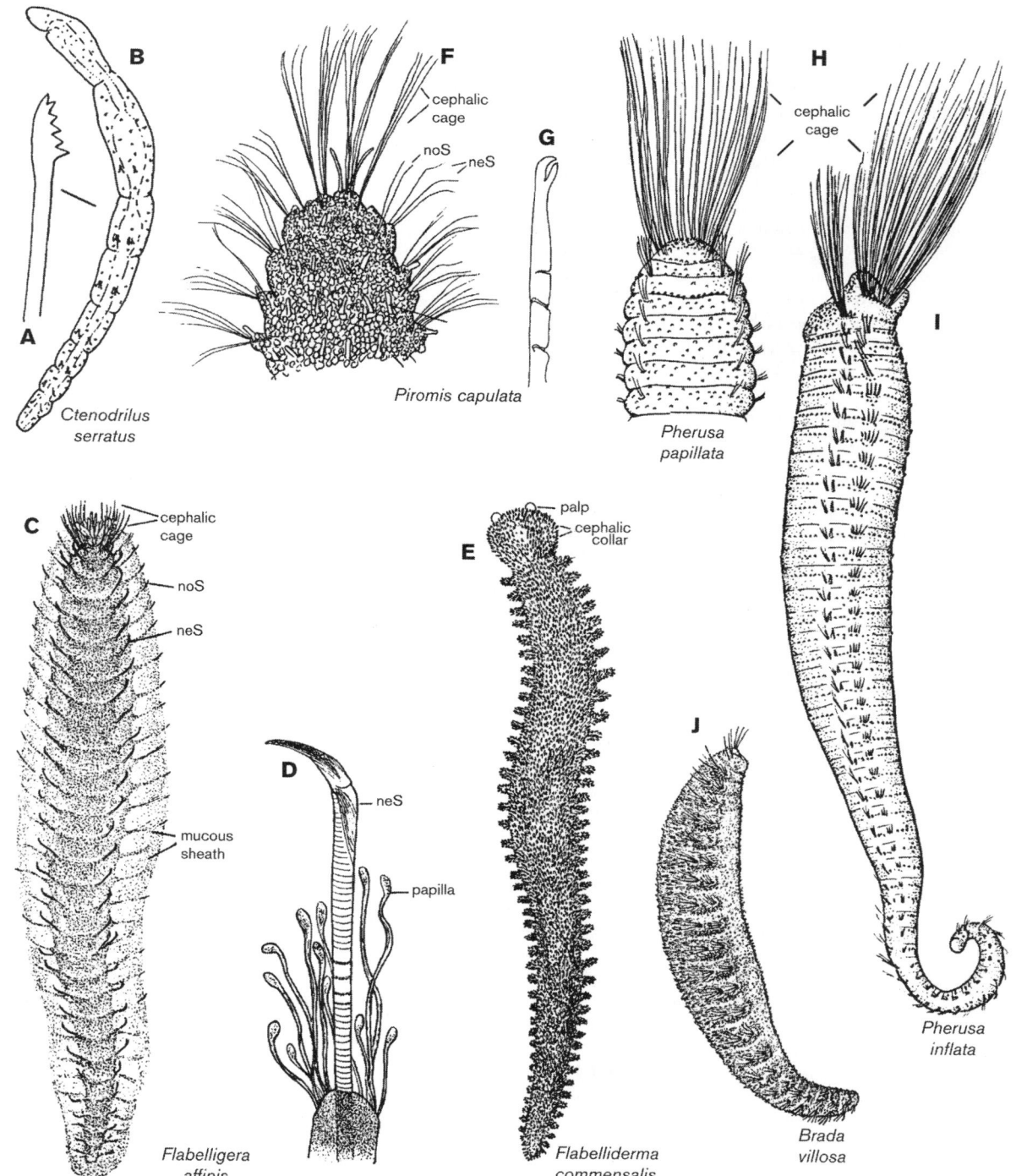

PLATE 171 Polychaeta. Ctenodrilidae and Flabelligeridae: A, *Ctenodrilus serratus*, seta; B, whole animal, lateral view; C, *Flabelligera affinis*; D, neuroseta and papillae; E, *Flabelliderma commensalis*; F, *Piromis capulata*; G, neuroseta; H, *Pherusa papillata*; I, *Pherusa inflata*; J, *Brada villosa* (A–B, E, I, after Hartman; C–D, J, after Støp-Bowitz; rest after Blake).

setae. If at all possible, flabelligerids should be examined alive for an unobstructed view of prostomial and peristomial features.

KEY TO FLABELLIGERIDAE

1. Body covered with a transparent, mucilaginous sheath (plate 171C) through which papillae and setae protrude (plate 171D) . . . . *Flabelligera affinis*

— Body covered with papillae, sand grains, or silt, not a transparent sheath. 2

2. Neurosetae from segments posterior to cephalic cage include bidentate falcigers (plate 171G); dorsum with tips of numerous papillae emerging between sand grains on each segment (plate 171F); branchial filaments thick, limited to four pairs . . . . *Piromis capulata*

— Neurosetae from segments posterior to cephalic cage all unidentate 3
3. Cephalic cage absent or poorly developed, with a few long noto- and neurosetae present on setigers 1–2; postcephalic neurosetae long, pointed, cross-barred; body surface covered with elongate papillae, usually with sand grains at base (plate 171J) . . . . *Brada villosa*
— Cephalic cage well developed; postcephalic neurosetae short, curved falcigers 4
4. Cephalic cage formed of numerous long, anteriorly directed setae from first 1–4 segments; body usually with some adhering sand grains . . . . *Pherusa* 5
— Cephalic cage covered with dorsal collar or hood, with setae best seen ventrally; dorsum densely covered with papillae, usually lacking adhering sand grains (plate 171E); commensal with sea urchins . . . .
. . . . *Flabelliderma commensalis*
5. Papillae mostly limited to anterior margins of segments (plate 171I); dorsum of anterior end oblique, densely covered with sand grains. . . . . *Pherusa inflata*
— Papillae mostly concentrated on dorsal surface of some anterior setigers, not limited to anterior margins (plate 171H); dorsum of anterior end not oblique; papillae sometimes lightly covered with silt or sand . . . . *Pherusa papillata*

LIST OF SPECIES

Hartman 1961, 1969; Blake 2000a.

Brada villosa (Rathke, 1843). Widespread in the northern hemisphere, intertidal to slope depths. There has been considerable discussion regarding synonyms of this species. Pettibone 1954, Proc. U.S. Nat. Mus. 103: 203–256, and Blake 2000a reviewed the literature and morphology attributed to *B. villosa* and other reported species of *Brada*. There is an unusual pattern of body surface abrasion in *B. villosa* that affects the number and appearance of papillae on the surface of the body. As the worms grow and age, they accumulate sand particles that form abrasive encrustations that erode the papillae so that eventually some specimens have a body surface that is rugose and nodular instead of papillate. This process needs to be further investigated. It is possible that proposed North American synonyms of *B. villosa* might actually represent valid species.

Flabelliderma commensalis (Moore, 1909). Rocky intertidal; free-living or commensal on the surface of the purple sea urchin *Strongylocentrotus purpuratus*; not known north of Monterey Bay. See Spies 1973, J. Morph. 139: 465–490 (circulatory system); 1977, Hartman Memorial Volume, Allan Hancock Found., pp. 323–345 (larval development); Light 1978, Proc. Biol. Soc. Wash. 91: 681–690 (taxonomy).

Flabelligera affinis Sars, 1829 (=*Flabelligera infundibularis* Johnson, 1901). Widespread in the northern hemisphere, Alaska to southern California. Low intertidal to slope depths in muddy bottoms; mainly subtidal.

Piromis capulata (Moore, 1909). Intertidal to slope depths; in coarse sediments; worms with sand grains adhering tightly to their body. See Abbott and Reish 1980.

Pherusa inflata (Treadwell, 1914). Intertidal rocky habitats, associated with sponges and shells. See Hartman 1959, Pac. Sci. 6: 71–74.

Pherusa papillata (Johnson, 1901). Intertidal to shelf depths in mixed sediments of sand, rock, and silt. See Abbott and Reish 1980.

ACROCIRRIDAE

The Acrocirridae include a small group of polychaetes believed to be deposit feeders. There are three known genera that have elongate slender bodies with numerous segments. The prostomium is reduced and bears a pair of palps on the frontal margin; eyes are present or absent. The pharynx is unarmed, and a small buccal sac is only partially eversible. Branchiae are present or absent, when present there are 1–6 pairs. The body is often covered with numerous papillae. Setae include long, hollow-looking neuropodial compound falcigers. Acrocirrids were removed from the Cirratulidae by Banse 1969, J. Fish. Res. Bd. Can. 26: 2595–2620. Acrocirrids occur from the intertidal to the deep sea but are not commonly encountered, and, as a result, little is known about their biology. There are four genera: *Acrocirrus*, *Macrochaeta*, *Chauvinelia*, and *Flabelligella*. An additional genus, *Flabelligena* does not appear to be necessary and needs to be reviewed. The two former genera are known from intertidal and shelf depths; others are only known from slope and abyssal depths. *Acrocirrus heterochaetus* has been reported from shallow shelf depths in California by Blake et al. 2000.

SCALIBREGMATIDAE

Scalibregmatids are burrowing infaunal deposit feeding worms that are widely distributed but not commonly collected. They range from the intertidal to the deep sea, with most species occurring deeper than 1,000 m. Their bodies have a rugged appearance because the cuticle is usually areolated with up to six annulae per segment. Their bodies are elongate and usually inflated anteriorly or sometimes throughout, giving a maggot shape. They typically have a prostomium that is broad on the anterior margin and often bears frontal horns (plate 172A). Scalibregmatids have relatively simple, biramous parapodia that bear simple setae, some of which are lyrate; others are spines that are sometimes large and conspicuous. Representatives of this family are characteristic of subtidal muddy sediments and are rarely encountered in the intertidal. An exception in California is *Hyboscolex pacificus* (Moore, 1909) (=*Oncoscolex pacificus*), which may occasionally be taken in the low rocky intertidal. *Scalibregma californicum* Blake, 2000, is the most common scalibregmatid in shelf and slope depths. California species of Scalibregmatidae were reviewed by Blake (2000b).

OPHELIIDAE

Opheliids are burrowing, infaunal, deposit-feeding worms that are among the better known and familiar polychaete families in near-coastal marine habitats. Opheliid species are typically elongate, slender worms with pointed prostomia. Their bodies are cylindrical and often very sleek looking; however, species of *Travisia* are thick, bulky, and grublike. Opheliids are well adapted for rapid burrowing and have a variety of sensory adaptations including lateral segmental eyespots, sensory pits, or papillae. Parapodia are biramous with small, inconspicuous rami provided with simple capillary setae. Dorsal cirri are absent; ventral cirri may be present. Branchiae are absent or present; if present, they are sometimes branched. The pygidium is often elongate and tubular with species-specific arrangements of papillae and cirri. Blake (2000c) reviewed the Opheliidae from California.

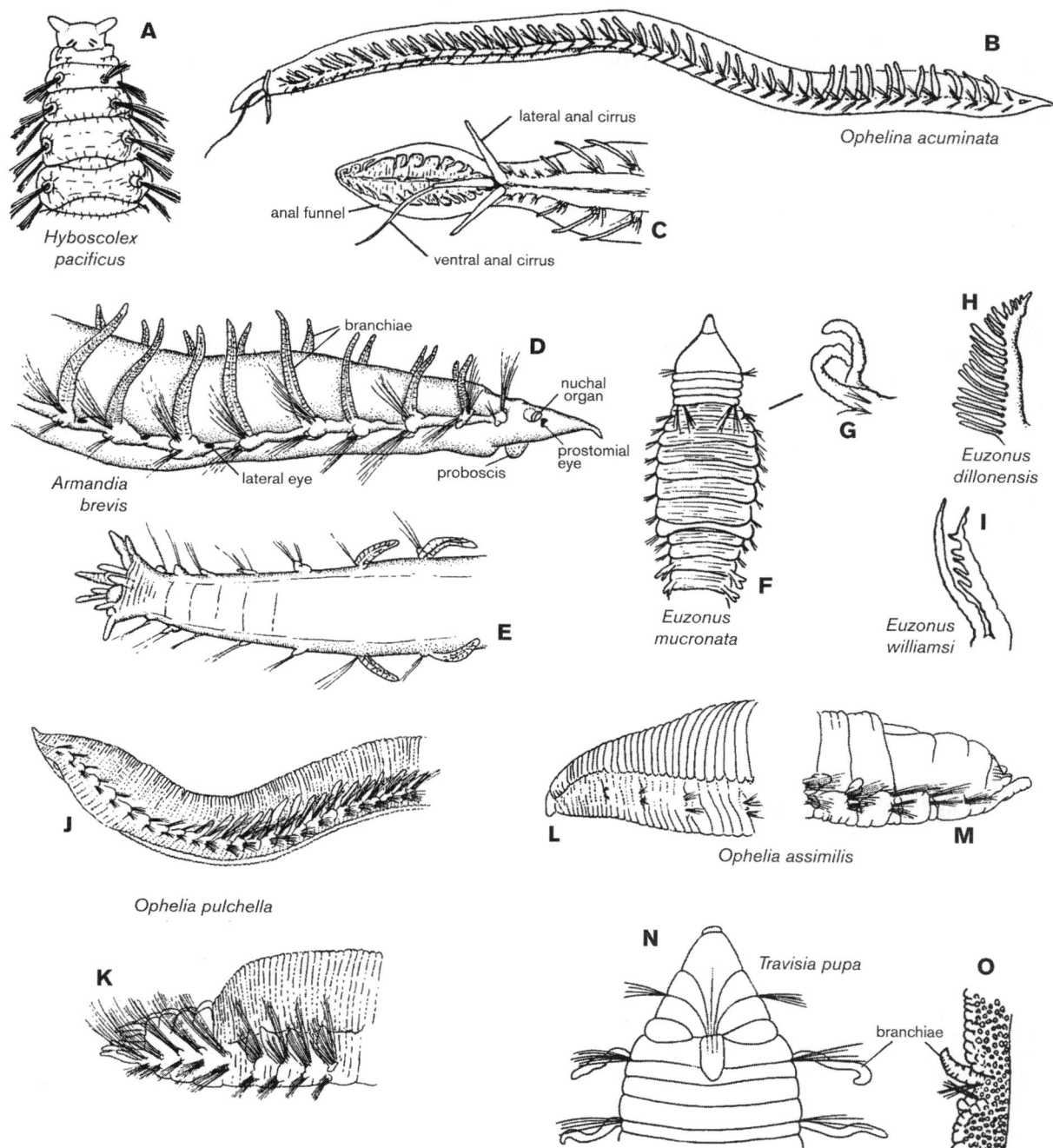

PLATE 172 Polychaeta. Scalibregmatidae and Opheliidae: A, *Hyboscolex pacificus*; B, *Ophelina acuminata*; C, posterior end; D, *Armandia brevis*; E, posterior end; F, *Euzonus mucronata*; G, branchiae; H, *Euzonus dillonensis*, branchia; I, *Euzonus williamsi*, branchiae; J, *Ophelia pulchella*; K, posterior end; L, *Ophelia assimilis*; M, posterior end; N, *Travisia pupa*; O, parapodium showing branchiae and setae (A, after Imajima; B–C, after Støp-Bowitz; F, after Treadwell; G–I, L–N, after Hartman; O, after Hobson and Banse; rest after Blake).

KEY TO OPHELIIDAE

1. Body elongate, generally streamlined, with distinct ventral groove extending along body (plate 172B) 2
— Body large, thick, grublike, without ventral groove (plate 172N) 9
2. Ventral groove extending throughout length of body (plate 172B) 3
— Ventral groove absent in anterior setigers, limited to posterior region. 5

3. Lateral parapodial eyespots present (plate 172D) 4
— Lateral parapodial eyespots absent; branchiae present; anal funnel an expanded scoop, opening ventrally, with a single midventral cirrus and two shorter lateral cirri (plate 172C) . . . . *Ophelina acuminata*
4. Branchiae present (plate 172D); anal funnel with long, unpaired cirrus and eight shorter papillae (plate 172E). . . . . *Armandia brevis*
— Branchiae absent; anal funnel with only small papillae . . . . *Polyophthalmus pictus*

5. Body with three distinct regions; anterior two setigers set off from thorax by constricted ring (plate 172F) 6
—   Body with two regions, expanded anteriorly, but without constricted ring 8
6. Branchiae with a single branch, each with long lateral pinnules (plate 172H) . . . . *Euzonus dillonensis*
—   Branchiae with 2–3 branches, with or without lateral pinnules 7
7. Branchiae with two branches, without lateral pinnules (plate 172G) . . . . *Euzonus mucronata*
—   Branchiae with 2–3 branches, with lateral pinnules (plate 172I) . . . . *Euzonus williamsi*
8. Body with nine anterior, abranchiate setigers (plate 172J); body with 38 setigers; with five posterior branchial setigers (plate 172K) . . . . *Ophelia pulchella*
—   Body with 10 anterior, abranchiate setigers (plate 172L); body with 33 setigers; with three posterior branchial setigers (plate 172M) . . . . *Ophelia assimilis*
9. Posterior parapodia with enlarged, pointed, triangular parapodial lobes; vesicles or beads on body surface uniform in size; body with 46 setigers . . . . *Travisia gigas*
—   Posterior parapodia without enlarged parapodial lobes (plate 172O); vesicles or beads on body surface not uniform, those of posterior side of annuli larger, wartlike in appearance; with 25–27 setigers . . . . *Travisia pupa*

LIST OF SPECIES

Armandia brevis (Moore, 1906) (=*A. bioculata* Hartman, 1938). Intertidal to shallow shelf depths in sandy mud and silt. See Woodin 1974, Ecol. Monogr. 44: 171–187 (life history and ecology); Hermans 1978 (larval development and metamorphosis); Hannan 1981, Limnol. Oceanogr. 26: 159–171 (larval settlement patterns).

Euzonus dillonensis (Hartman, 1938). Dillon Beach, California in clean sand beaches.

Euzonus mucronata (Treadwell, 1914) (=*Thoracophelia mucronata*). In clean sand beaches. See Dales 1952, Biol. Bull. 102: 232–252 (ecology, larval development); McConnaughey and Fox 1949, Univ. Calif. Publ. Zool. 47: 319–340.

Euzonus williamsi (Hartman, 1938). Dillon Beach, California in clean sand beaches.

Ophelia assimilis Tebble, 1953. In sand beaches.

Ophelia pulchella Tebble, 1953. In sandy mud sediments; shallow subtidal. See Tebble 1953, Ann. Mag. Nat. Hist. (12) 6: 361–368 (review of *Ophelia*).

Ophelina acuminata Oersted, 1843 (=*Ammotrypane aulogaster* Rathke, 1843). Muddy sediments; low intertidal to shelf and slope depths; widespread in the northern hemisphere.

Polyophthalmus pictus (Dujardin, 1839). Intertidal in rocky habitats associated with algae.

Travisia gigas Hartman, 1938. In sandy mudflats.

Travisia pupa Moore, 1906. Shelf depths.

STERNASPIDAE

Sternaspis fossor Stimpson, 1854, is a small, swollen worm with an ovoid shape (plate 134D). It has very few segments and burrows head first into fine, silty sediments. Hard anal plates cover the entrance to the burrow allowing the terminal branchial filaments to extend into the water. It is usually sub-

tidal and may be taken with shallow grabs. *S. fossor* is widely distributed around North America from 10–400 m. A recent review of California collections by Petersen (2000) suggests that local specimens might be an undescribed species.

CAPITELLIDAE

Capitellids are burrowing, earthwormlike polychaetes that have a short, conical or pointed prostomium and an elongated body divided into two distinct regions. There are no prostomial or peristomial appendages, and branchiae are rare. The anterior or thoracic region typically bears capillary setae, whereas the posterior or abdominal region bears hooded hooks. The numbers of segments in these regions, the detail and initiation of the hooks, and presence or absence of other structures, such as branchiae, nuchal organs, or genital hooks, have served to establish a highly complex generic hierarchy. Capitellids are among the most commonly encountered polychaetes in marine soft sediments. They are often the dominant components of benthic infaunal communities especially in sediments that are organically enriched. Because of their accessibility and importance in sedimentary environments, capitellids have been the subject of numerous ecological studies. Blake (2000d) reviewed the systematics and biology of California Capitellidae.

KEY TO CAPITELLIDAE

1. Thorax with eight setigers; capillary setae entirely absent, setae hooded hooks; first two setigers 3–4 times as long as setigers 3–8; transition from thorax to abdomen abrupt . . . . *Amastigos acutus*
—   Thorax with nine or more setigers; capillary setae present 2
2. Thorax with nine setigers, genital spines present in setigers 8–9 (plate 173A, 173B) . . . . *Capitella* species complex
—   Thorax with 10 or more setigers, genital spines absent 3
3. Thorax with 10 setigers; setigers 1–4 with capillaries, 5–10 with hooks . . . . *Mediomastus* 4
—   Thorax with 11 or more setigers. 6
4. Prostomium acutely pointed (plate 173C); some thoracic capillaries limbate (plate 173D), others short, paddlelike with smooth tips (plate 173E) . . . . *Mediomastus acutus*
—   Prostomium small, conical (plate 173F); thoracic capillaries limbate, with distinct wings, not spatulate 2
5. Posterior abdominal notopodial capillaries present (plate 173G); thoracic hooks long, with straight shaft, no constriction; abdominal hooks shorter, with distinct shoulder and constriction on shaft; body rarely exceeding 14 mm in length, usually smaller . . . . *Mediomastus ambiseta*
—   Posterior abdominal notopodial capillaries absent; thoracic and abdominal capillaries more or less similar, both with long, straight shaft and weakly developed shoulder and constriction; body larger, up to 35 mm long. . . . . *Mediomastus californiensis*
6. Thorax with 11 setigers. 7
—   Thorax with 12 or more setigers. 12
7. Thoracic setae all capillaries . . . . *Notomastus* 8
—   Thoracic setae include capillaries in setigers 1–5, and with hooks in 6–11 (plate 173P) . . . . *Heteromastus* 11

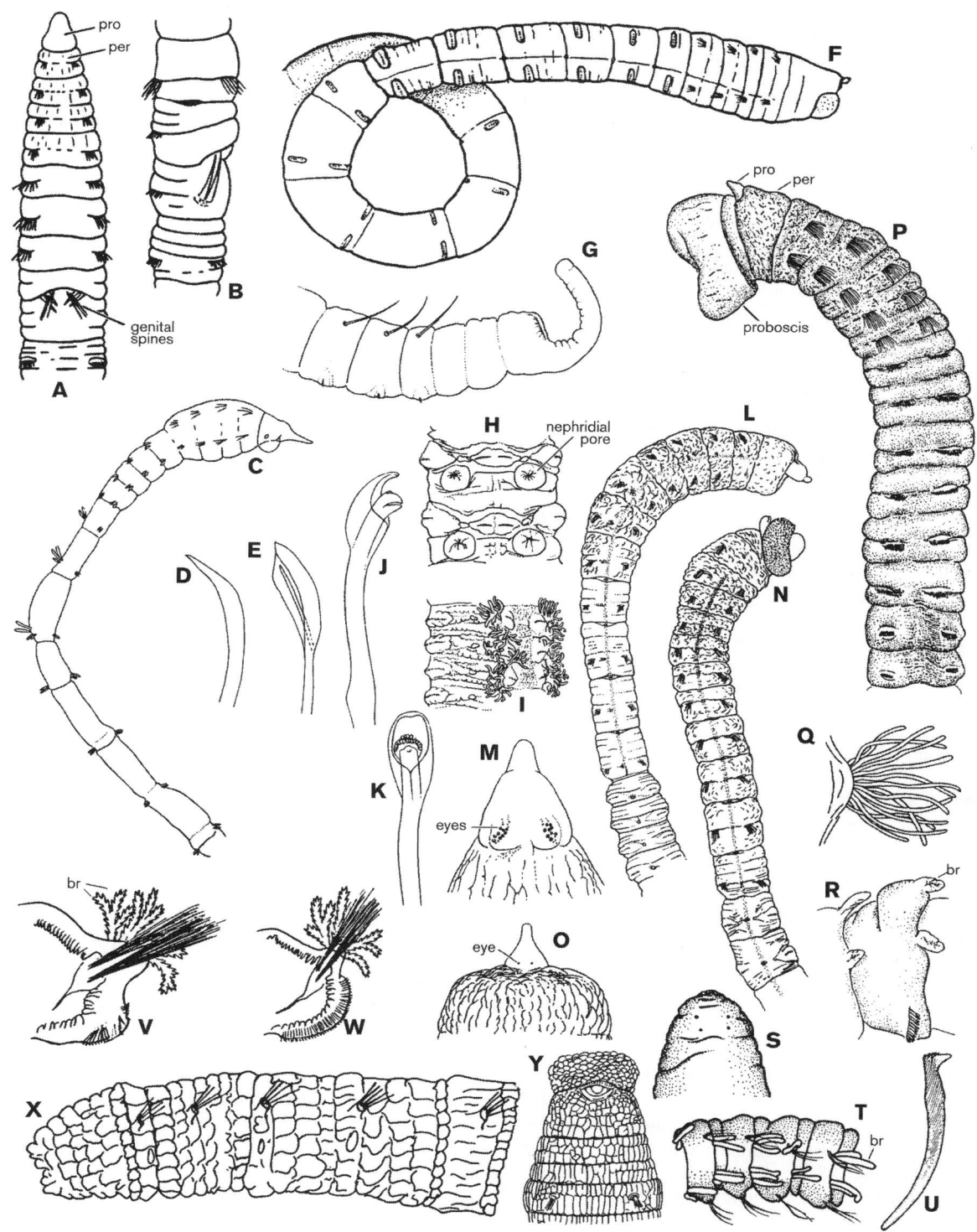

PLATE 173 Polychaeta. Capitellidae and Arenicolidae: A, *Capitella capitata*; B, transitional region showing genital spines; C, *Mediomastus acutus*; D, limbate thoracic capillary seta; E, modified thoracic notoseta; F, *Mediomastus californiensis*; G, *Mediomastus ambiseta*, posterior end; H, *Notomastus magnus*, anterior abdominal segments showing nephridial openings; I, far posterior segments showing branchiae; J–K, abdominal hooded hooks; L, *Notomastus tenuis*; M, prostomium and anterior part of peristomium, showing eyes; N, *Notomastus hemipodus*; O, prostomium and peristomium, dorsal view; P, *Heteromastus filobranchus*; Q, detail of branchiae from posterior setiger; R, *Heteromastus filiformis*, far posterior setiger showing branchiae; S, *Branchiomaldane simplex*, anterior end, dorsal view; T, posterior setigers showing branchiae; U, hooked seta or uncinus; V, *Abarenicola*, branchial segment showing ventral extent of the neurosetae; W, *Arenicola*, branchial segment showing ventral extent of the neurosetae; X, *Abarenicola pacifica*; Y, *Arenicola cristata* (A–B, R, after Berkeley and Berkeley; H–O, after Hartman; C–D, after Warren; S–U, after Fournier and Barrie; Y, after Ashworth; rest after Blake).

8. Nephridial pores large, conspicuous middorsally on abdominal segments (plate 173H); absent in thoracic segments; hood of abdominal hooks greatly enlarged, voluminous (plate 173J, 173K) 9
— Nephridia absent in abdominal segments, present or absent in thoracic segments; hood of abdominal hooks inflated, but not voluminous 10
9. Branchiae multiple, branched, in tufts (plate 173I) . . . . . .
. . . . *Notomastus magnus*
— Branchiae single, long, filamentous . . . .
. . . . *Notomastus latericeus*
10. Anterior thoracic segments gradually decrease in width from anterior to posterior, not changing abruptly in size (plate 173L); prostomium with paired groups of numerous eyespots (plate 173M); methyl green staining body more or less uniformly, without striking patterns . . . .
. . . . *Notomastus tenuis*
— First 3–5 thoracic setigers expanded, followed by narrower segments (plate 173N); prostomium with a single pair of minute eyespots (plate 173O); methyl green producing distinct patterns especially in abdominal segments where pair of longitudinal midventral bands extend to posterior end
. . . . *Notomastus hemipodus*
11. Branchiae with multiple filaments, from about setigers 30–50 (plate 173Q); body robust . . . .
. . . . *Heteromastus filobranchus*
— Branchiae small, single, not with multiple filaments (plate 173R), from setiger 100 or more posteriorly; body slender, fragile . . . . *Heteromastus filiformis*
12. Thorax with 12 setigers, all with capillaries; abdominal notopodial hooks of posterior setigers change to acicular spines; pygidium forms anal funnel with two ventral lobes; branchiae absent . . . . *Scyphoproctus oculatus*
— Thorax with 15 or more setigers 13
13. Thorax with 15 setigers 14
— Thorax with 17–18 setigers, all with capillaries . . . .
. . . . *Anotomastus gordiodes*
14. Thoracic segments relatively smooth; branchiae with 2–3 lobes . . . . *Dasybranchus glabrus*
— Thoracic segments with deep lines; branchiae large, branched . . . . *Dasybranchus lumbricoides*

LIST OF SPECIES

See Hartman 1947a, 1969; Blake 2000d.

Amastigos acutus Piltz, 1977. Intertidal sandy beaches, southern California.

Anotomastus gordiodes (Moore, 1909). The large number of thoracic setigers readily differentiates this species from other eastern Pacific capitellids.

Capitella species complex. Several morphologically very similar species are common in estuarine mud flats of the eastern Pacific; systematic studies necessary to distinguish these species from the type species *C. capitata* (Fabricius, 1780) from Greenland have not been carried out. See Grassle and Grassle 1977, Science 192: 567–569 (sibling species in *Capitella*); Grassle, Gelfman, and Mills 1987, Bull. Biol. Soc. Wash. 7: 77–88 (karyotypes of *Capitella* sibling species); Eckelbarger and Grassle 1987, Mar. Biol. 95: 415–429 (reproductive and larval morphology of *Capitella* sibling species); Blake 2000d (review of systematics and biology); Abbott and Reish 1980 (review).

Dasybranchus glabrus Moore, 1909. Intertidal to shallow subtidal in fine sands.

Dasybranchus lumbricoides Grube, 1878. A widespread species; intertidal in mud.

Heteromastus filiformis (Claparède, 1864). Cosmopolitan in fine silty sediments; common. See Fredette, T. J. 1982, Proc. Biol. Soc. Wash. 95: 194–197 (developmental morphology); Shaffer, P. L. 1983, Neth. J. Sea Res. 17: 106–125 (population ecology).

Heteromastus filobranchus Berkeley and Berkeley, 1932. Shallow subtidal to shelf depths.

Mediomastus acutus Hartman, 1969. Shallow subtidal in fine silty sands.

Mediomastus ambiseta (Hartman, 1947). Intertidal to shelf depths.

Mediomastus californiensis Hartman, 1944. Intertidal to shelf depths.

Notomastus hemipodus Hartman, 1945. Recorded from all three coasts of North America, intertidal to shelf depths in fine sands and silts.

Notomastus latericeus Sars, 1851. This species has been reported from worldwide localities and identified from monitoring programs in southern California (SCAMIT 2001). Descriptions in Hartmann-Schröder (1996) from northern Europe and Ewing (1984) from the Gulf of Mexico.

Notomastus magnus Hartman, 1947. Intertidal to shelf depths in sandy muds.

Notomastus tenuis Moore, 1909. Intertidal to shelf depths in sandy muds.

Scyphoproctus oculatus Reish, 1959. Southern California in shallow, nearshore sediments.

ARENICOLIDAE

Arenicolids ("lug" worms) are large, burrowing worms that form U-shaped tubes in mudflats of estuaries. They are particle feeders and obtain these by pumping water through their burrows. Typically, the external evidence of an arenicolid burrow is a funnellike depression on the surface, near which is a fecal-castings mound that denotes the other end. The body of arenicolids is cylindrical and separated into two or three distinct regions: a prebranchial region with about six setigers, a branchial region, and a caudal region that is often asetigerous. Dissection is necessary to identify *Abarenicola*.

KEY TO ARENICOLIDAE

1. Body slender; branchiae first present from posterior setigers (setiger 18 or later); branchiae simple with 1–3 in a tuft (plate 173T); prostomium bluntly rounded on anterior margin, with four small eyespots (plate 173S); uncini with long handles (plate 173U) . . . . . . *Branchiomaldane simplex*
— Body thick, robust (plate 173X–173Y); branchiae first present from setiger 7; branchiae fine filaments in thick tufts or clusters; prostomium with 2–3 lobes 2
2. Neuropodia of branchial segments long, extending ventrally to near the midline (plate 173W); with a single pair of esophageal caeca . . . . *Arenicola* 3
— Neuropodia of branchial segments short throughout and widely separated ventrally (plate 173V); with more than one pair of esophageal caeca . . . . *Abarenicola* 4
3. Branchiae, 11 pairs; color in life, light yellow with a purplish cast . . . . *Arenicola brasiliensis*
— Branchiae, 16–18 pairs; color in life rich dark green anteriorly and brown posteriorly . . . . *Arenicola cristata*

4. Nephridial pores present setigers 5–9, each uncovered; esophageal caeca include one large anterior pair and 3–6 smaller pairs . . . . *Abarenicola pacifica*
— Nephridial pores present setigers 5–9, each covered with a hood; esophageal caeca include one large anterior pair and seven to nine smaller pairs . . . .
. . . . *Abarenicola claparedii oceanica*

LIST OF SPECIES

See Wells 1962, Proc. Zool. Soc. London 138: 331–353.

Abarenicola claparedii oceanica Healy and Wells, 1959 (=*A. vagabunda oceanica*). Sand flats; Humboldt Bay and north. See Wells 1963, Syst. Assoc. Publ. 5: 79–98.

Abarenicola pacifica Healy and Wells, 1959. Sand flats; Humboldt Bay and north. See Healy and Wells 1959, Proc. Zool. Soc. Lond. 133: 315–335; Hobson 1976, Biol. Bull. 133: 343–354 (feeding biology); Swainbanks 1981, J. Sed. Pet. 51: 1137–1145 (sediment reworking).

Arenicola brasiliensis Nonato, 1958. Reported locally from San Francisco Bay; intertidal in sand. See Abbott and Reish 1980 (ecology, reproduction).

Arenicola cristata Stimpson, 1856. Humboldt Bay to southern California, intertidal in sand flats.

Branchiomaldane simplex Berkeley and Berkeley, 1932 (=*B. vincentii* of the previous edition). British Columbia to southern California. Intertidal to shallow subtidal, in fine sand, rocky beaches, recorded from among rhizomes of *Phyllospadix*. See Fournier and Barrie 1987, Bull. Biol. Soc. Wash. 7: 97–107.

OWENIIDAE

Oweniids are slender, fragile polychaetes that inhabit closely adhering tubes formed of a mucous matrix with overlapping sand grains or shell fragments. They are surface deposit feeders and occur from the intertidal to the deep sea. They sometimes form dense tube mats. The body is elongate, rigid, and more or less divided into anterior and posterior regions; it is truncated anteriorly, tapered posteriorly. The prostomium and peristomium are fused to form a truncate head. The mouth is terminal and surrounded by peristomial lips, which are variously developed. One genus bears terminal palplike extensions. Parapodia are reduced. Setae include serrated capillaries in notopodia and dense fields of small, long-shafted hooks in neuropodia. Oweniids are distinguished from one another on the basis of head structures, number of thoracic setigers, shape and orientation of the neuropodial hooks, and shape and number of pygidial lobes. Most species occur in offshore sediments; a few, mostly of the genus *Owenia*, occur in sandy sediments, sometimes in intertidal beaches. The following key is intended to identify two common species of *Owenia* and three of the genera that occur offshore. Readers are referred to Blake (2000e) for keys and descriptions of California Oweniidae.

KEY TO OWENIIDAE

1. Head bearing tentacular crown of small, branched gill-like structures . . . . *Owenia* 2
— Head without tentacular crown or with pair of long palplike structures 3

2. Tentacular crown distinctly pigmented (plate 174A); peristomial collar pigmented; prostomial eyes absent; posterior end with dorsal groove . . . . *Owenia collaris*
— Tentacular crown and peristomial collar not pigmented (plate 174B); prostomial eyes present; posterior end without groove . . . . *Owenia johnsoni*

3. Head with pair of long, anteriorly directed palplike structures; uncini with main fang surmounted by small, closely adhering tooth . . . . *Myriowenia californiensis*
— Head without appendages; uncini with two large teeth, sometimes in tandem 4

4. Mouth terminal, continued ventrally as elongate oral slit surrounded by thin, membranous lips . . . . . *Galathowenia*
— Mouth terminal or shifted ventrally, without elongate oral slit and overlapping oral lips . . . . *Myriochele*

LIST OF SPECIES

See Blake 2000e.

Galathowenia spp. Blake (2000e) reports on three offshore California species that are largely restricted to fine-grained sediments.

Myriochele spp. Blake (2000e) reports on three offshore California species. Of these, *M. striolata* Blake may be expected in nearshore sediments.

Myriowenia californiensis Hartman, 1960. This unusual oweniid occurs offshore in mixed sediments.

Owenia collaris Hartman, 1955 (formerly referred to *O. fusiformis* delle Chiaje, 1841). Common in shallow, sandy sediments. See Fager 1964, Science 143: 356–359; Eckman et al. 1981, J. Mar. Res. 39: 361–374 (ecology).

Owenia johnsoni Blake, 2000. Central and northern California in shallow, sandy sediments.

MALDANIDAE

Maldanids, or bamboo worms, are elongate, often very large infaunal deposit feeders. The body is cylindrical, with a relatively few number of segments that are each greatly elongated. The prostomium is small and poorly defined as such, but with the larger peristomium forms a distinctive looking head region that often appears to be superficially truncated. The peristomium sometimes bears a dorsal crest and may be surrounded by a cephalic plate with a raised rim or ridge. The mouth is ventral. A simple, papillated proboscis is eversible. Parapodia are reduced but bear notopodial capillaries and neuropodial rostrate hooks; neurosetae of setigers 1–3 may be modified to form spines. Up to 10 posterior preanal segments may be asetigerous. The pygidium may be conical, truncate, funnel-shaped, or petaloid, and the anus may be dorsal, terminal, or recessed into the funnel. Most maldanids are head-down deposit feeders, forming deep feeding voids and then defecating onto mounds on the surface.

KEY TO MALDANIDAE

1. Pygidium a slanting plate without anal cirri (plate 174J), with a dorsal anal opening; neurosetae absent on setiger 1 5
— Pygidium a terminal plate encircled by anal cirri (plate 174C, 174H); neurosetae present on setiger 1 2

2. Neurosetae all rostrate hooks (plate 174D); body green with red bands on segments 4–8 (plate 174C) . . . .
. . . . *Axiothella rubrocincta*

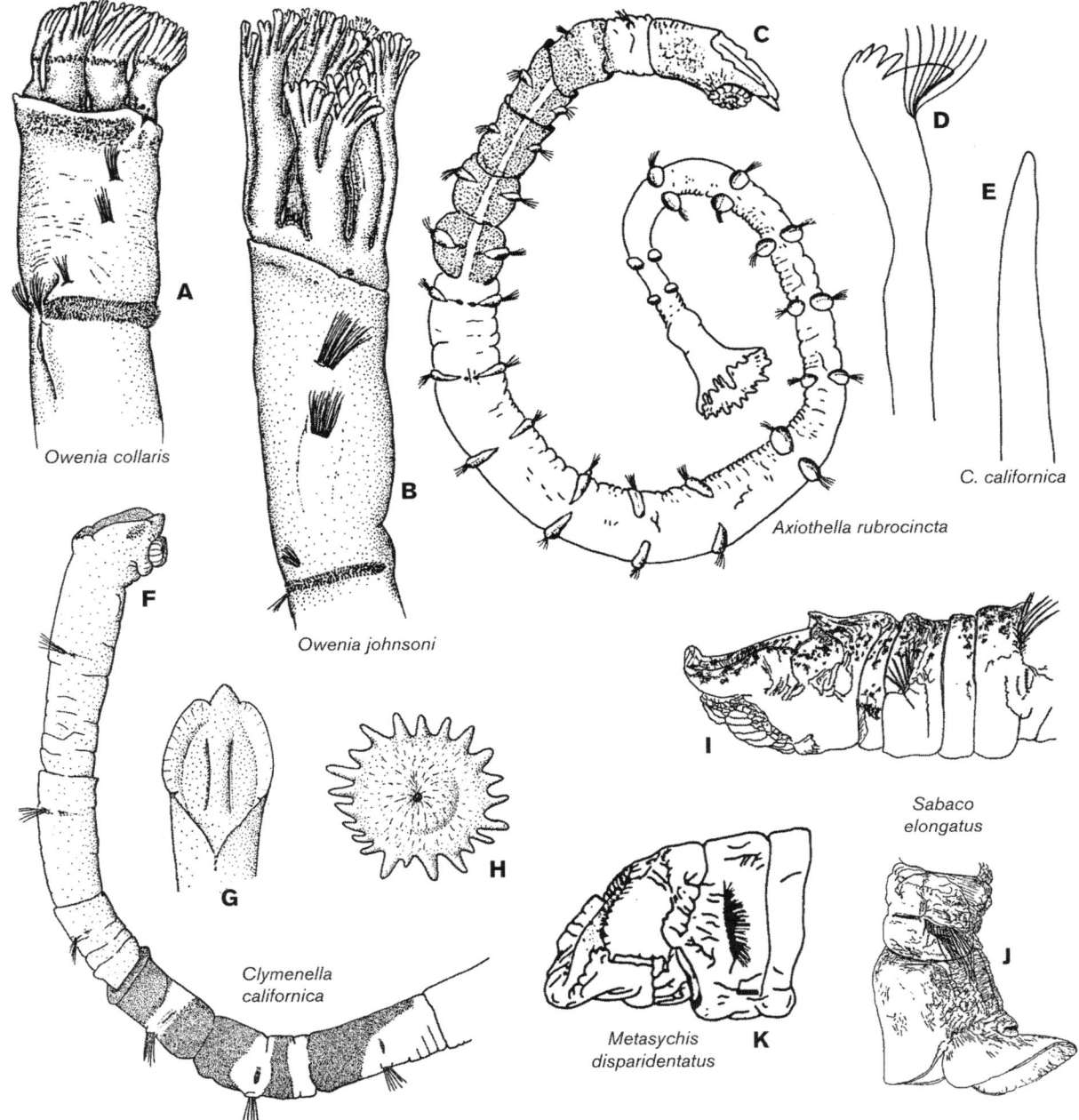

PLATE 174 Polychaeta. Oweniidae and Maldanidae: A, *Owenia collaris*; B, *Owenia johnsoni*; C, *Axiothella rubrocincta*; D, neuropodial rostrate hook; E, *Clymenella californica*, neuropodial spine from anterior setiger; F, anterior end, lateral view; G, prostomium and peristomium, dorsal view; H, pygidium; I, *Sabaco elongatus*; J, last setiger and pygidium; K, *Metasychis disparidentatus* (C, after Spies; I–K, after Light; rest after Blake).

— Neurosetae of first 3–4 setigers thick acicular spines (plate 174E), with those of subsequent setigers rostrate hooks. 3

3. Collar present on anterior margin of setiger 4 (plate 174F) . . . . *Clymenella* 4
— Segments without collars . . . . *Praxillella pacifica*

4. Head with well-developed, raised margin (plate 174F, 174G); 22–27 setigers; mudflats . . . . *Clymenella californica*
— Head without raised margin; 21 setigers; shaley rocks . . . . *Clymenella complanata*

5. Collar on setiger 1 entire, encircling body (plate 174I); lateral cephalic lobes with low, entire rim; nuchal organs very small, crescentic, and isolated from cephalic rim; asetigerous preanal segments absent (plate 174J) . . . . *Sabaco elongatus*
— Collar on setiger 1 ventrally produced, not encircling body (plate 174K); lateral cephalic rim low, crenulated, sculptured; nuchal organs J-shaped, connected to cephalic rim; with one asetigerous preanal segment . . . . *Metasychis disparidentatus*

See Hartman 1969; Imajima and Shiraki 1982.

Axiothella rubrocincta (Johnson, 1901). Fine sand sediments; very common in bays and estuaries. See Woodin 1974, Ecol. Monogr. 44: 171–187 (life history and ecology); Kudenov 1977, Zool. J. Linn. Soc. 60: 95–109 (functional morphology of feeding); Kudenov 1978: J. Exp. Mar. Biol. Ecol. 31: 209–221 (feeding ecology); Kudenov 1982, Mar. Biol. 70: 181–186 (sediment reworking); Spies 1969, MS thesis, University of the Pacific, Stockton (functional morphology); Weinberg 1979, Mar. Eco. Prog. Ser. 1: 301–314 (spionid distribution in *Axiothella* patches).

Clymenella complanata Hartman, 1969. In crevices of shale rocks in San Mateo County.

Clymenella californica Blake and Kudenov, 1974. Tomales Bay in muds at extreme low intertidal. See Blake and Kudenov 1974, Bull. So. Calif. Acad. Sci. 73: 48–50; Kudenov 1977, Zool. J. Linn. Soc. 60: 95–109 (functional morphology of feeding).

*Clymenella torquata (Leidy, 1855). A well-known North Atlantic species was reported from British Columbia by Banse 1981, Can. J. Fish. Aquat. Sci. 38: 633–637; see also Hobson and Banse 1981. Banse suggests that this species may have been introduced via oyster culture and that it might be more widely distributed along the eastern Pacific. The species is similar to *Axiothella rubrocincta* in having only rostrate uncini in neuropodia, thus differing from the two California species that have spines in some anterior neuropodia.

Metasychis disparidentatus (Moore, 1904). This species has been found in Drakes Bay and Monterey Bay in relatively shallow shelf depths. It has been reported from the intertidal in British Columbia, but not California. See Imajima and Shiraki 1982.

Praxillella pacifica Berkeley, 1929 (=*P. affiinis pacifica*). Mud of harbors and bays. See Banse and Hobson 1968; Imajima and Shiraki 1982; Kudenov 1977, Zool. J. Linn. Soc. 60: 95–109 (functional morphology of feeding).

Sabaco elongatus (Verrill, 1873) (=*Asychis amphiglypta* and *A. elongata* of authors). San Francisco Bay, introduced from the Atlantic. See Light 1974, Proc. Biol. Soc. Wash. 87: 175–184; Light 1991, Ophelia Suppl. 5: 133–146 (revision of subfamily Maldaninae); Cohen and Carlton 1995.

SABELLARIIDAE

Sabellariids, although relatively small, often form dense colonies and establish large reefs along the shore or on subtidal rocks. These reefs grow so large that they have been known to modify shipping channels along some coastlines. Sabellariids are permanently encased in a sandy tube attached either singly to a rock or as part of the large reeflike colonies. The bodies of sabellariids are divided into four regions: (1) an anterior end formed into an opercular stalk consisting of fused anterior larval segments and the thorax; (2) a parathoracic region consisting of three to four segments, (3) an abdominal region, and (4) a tubular asetigerous far posterior region. The opercular stalk bears a crown that consists of circles of paleae projecting anteriorly. The degree of fusion of

* = Not in key.

the operculum and the kinds and arrangements of the paleae are of major taxonomic importance in sabellariids. The most important work on sabellariids is by the late David Kirtley (1994). This monograph is global in its coverage and provides keys, descriptions, and illustrations to all known species. Kirtley recognized two subfamilies and 12 genera among the sabellariids. However, this classification was not fully justified in the original publication and to date has not been subjected to a rigorous cladistic analysis. For this section, the genus *Sabellaria* is defined and used in a broad sense, and thus there are no changes at the generic level from the previous edition. The following key is intended to identify local species and, as such, does not follow the hierarchical arrangement of Kirtley (1994).

KEY TO SABELLARIIDAE

1. Opercular paleae form a black cone (plate 175E); outer opercular paleae with a distal plume (plate 175D); colonies often form massive reefs......*Phragmatopoma californica*
— Operculum formed by diverging yellow spines (plate 175A) 2
2. Opercular paleae form two visible rows (plate 175I); outer opercular paleae pinnate (plate 175J).................
....*Idanthyrsus saxicavus*
— Opercular paleae form three visible rows; outer opercular paleae are broad plates with some type of terminal spine or arista ....*Sabellaria* 3
3. Outer opercular paleae terminate in a flat plate, with a distal spinose arista (plate 175B–175C); opercular stalk with black speckles (plate 175A).......*Sabellaria cementarium*
— Outer opercular paleae and stalk otherwise...........4
4. (*Note three choices*) Outer opercular paleae with longest spine marginally serrated (plate 175F).................
....*Sabellaria spinulosa*
— Outer opercular paleae with longest spine a smooth spike (plate 175G).......................*Sabellaria nanella*
— Outer opercular paleae with longest spine a plumed arista (plate 175H) ....*Sabellaria gracilis*

LIST OF SPECIES

See Hartman 1944c; Kirtley 1994.

Idanthyrsus saxicavus (Baird, 1863). (=*Idanthyrsus ornamentatus* Chamberlin, 1919). Rocky habitats, not forming large colonies; low intertidal to about 150 m.

Phragmatopoma californica (Fewkes, 1889). Constructs large colonies of tubes on rocks; may form massive reefs; intertidal to 230 m. See Dales 1952, Quart. J. Micr. Sci. 93: 435–452 (ecology); Roy 1974, Bull. So. Calif. Acad. Sci. 73: 117–125 (tube dwelling behavior); Woodin 1977, Mar. Biol. 44: 39–42 (algal gardening behavior); Eckelbarger 1977, Bull. Mar. Sci. 27: 241–255 (larval development); Jensen and Morse 1984, J. Exp. Mar. Biol. Ecol. 83: 107–126 (larval recruitment); Pawlik et al. 1991, Science 251: 421–424 (settlement of larvae); Abbott and Reish 1980 (review).

Sabellaria cementarium Moore, 1906 (=*Neosabellaria cementarium* sensu Kirtley, 1994). Rocky habitats; tubes of sand and gravel, may be solitary or colonial, often forms reefs; common. See Smith and Chia 1985a, Can. J. Zool. 63: 1037–1049 (larval development and metamorphosis); 1985b, Can. J. Zool. 63: 2852–2866 (metamorphosis); Abbott and Reish 1980 (review).

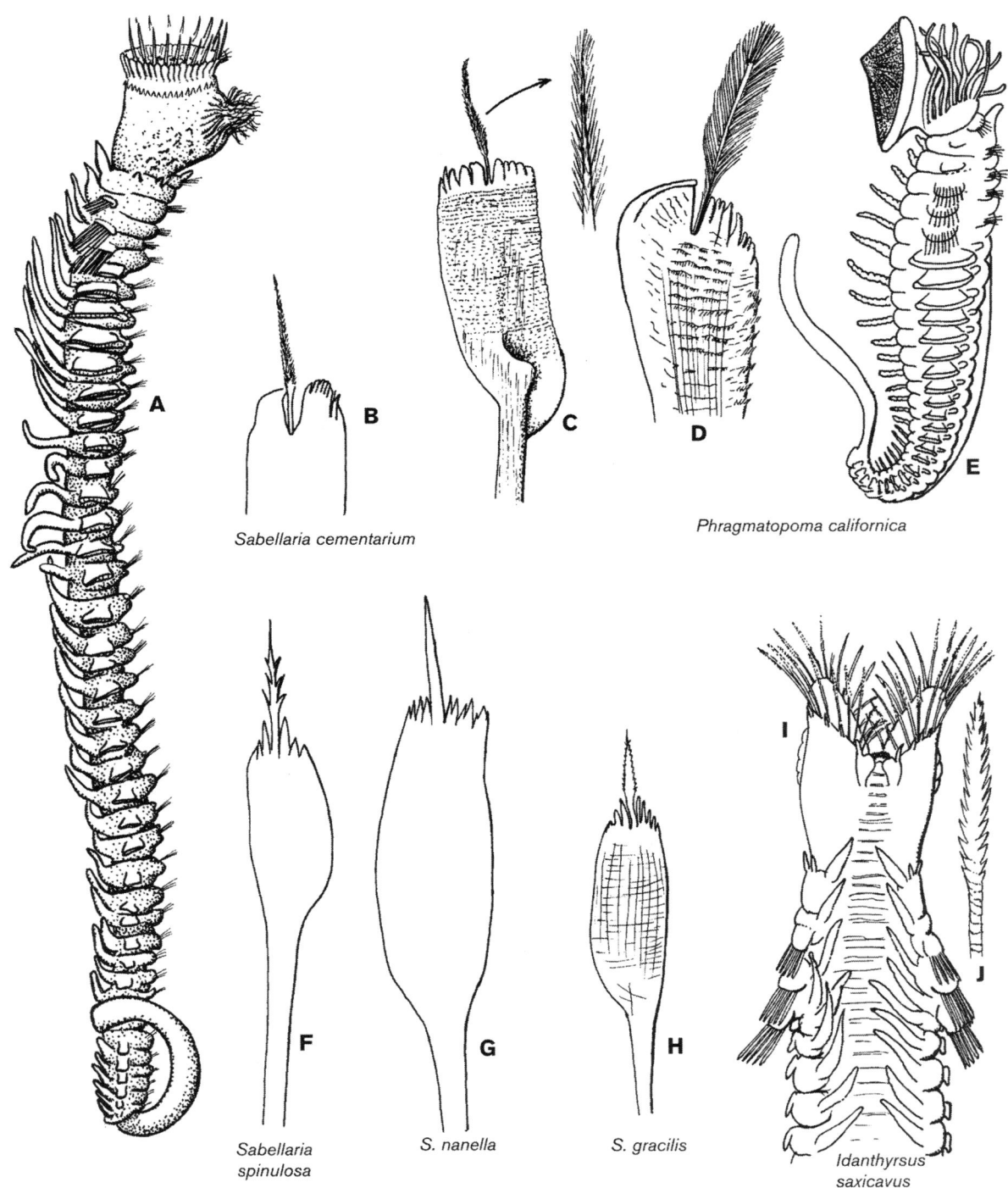

PLATE 175 Polychaeta. Sabellariidae: A, *Sabellaria cementarium*; B–C, outer opercular paleae; D, *Phragmatopoma californica*, outer opercular palea; E, whole animal, lateral view; F, *Sabellaria spinulosa*, outer opercular palea; G, *S. nanella*, outer opercular palea; H, *S. gracilis*, outer opercular palea; I, *Idanthyrsus saxicavus*; J, outer opercular palea (A, C, after Okuda; D, G, H, after Hartman; rest after Blake).

Sabellaria gracilis Hartman, 1944. Rocky habitats; tubes made of fine sand, colonies smaller, more delicate, seldom conspicuous; common; intertidal to about 45 m.

Sabellaria nanella Chamberlin, 1919. San Francisco; rare, probably introduced from South America.

Sabellaria spinulosa Leuckart, 1849. Collected by Olga Hartman in the 1930s in San Francisco Bay but not known since;

probably introduced as a fouling organism; a common European species.

PECTINARIIDAE

Pectinariids, the "ice-cream cone" worms, are permanently encased in a tapered, sandy tube that is open at both ends. The

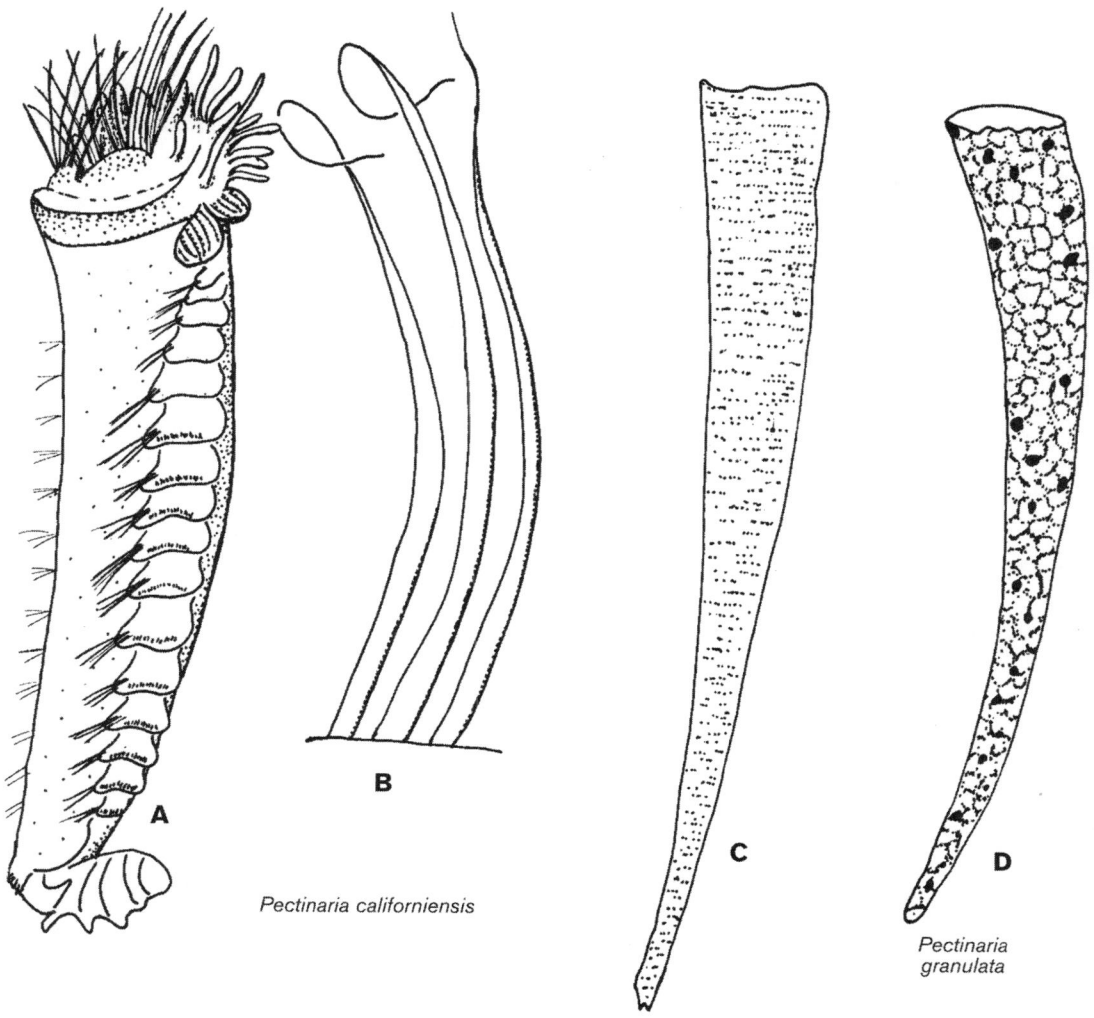

PLATE 176 Polychaeta. Pectinariidae: A, *Pectinaria californiensis*; B, cephalic spines; C, tube; D, *Pectinaria granulata*, tube (after Blake).

worms lie buried in the sediment and feed head down, forming a void into which the head of the worm projects. The tube with the worm is oriented such that only the narrow posterior tip projects above the sediment. Through this opening at the surface, a respiratory current carries water down the tube, over the posterior end of the worm and along the body to the gills. The head of the worm is provided with a thickened operculum armed with a row of flattened setae or paleae. The body is divided into two regions: (1) an anterior region with biramous parapodia and (2) a posterior region with well-developed neuropodia and notopodia either reduced or absent altogether. The tube structure is often delicate and diagnostic in shape and composition. Depending on which species, the tube is formed of sand grains, sponge spicules, foraminiferan shells, or even shell fragments. Readers are referred to Hartman (1941b), which is still the most important paper on eastern Pacific pectinariids.

KEY TO PECTINARIIDAE

1. Cephalic spines long, tapering to fine tips (plate 176A, 176B); tubes straight, formed of fine, reddish brown sand grains (plate 176C) . . . . *Pectinaria californiensis*

— Cephalic spines short and blunt; tubes curved and formed of coarse, black and white sand grains (plate 176D) . . . . *Pectinaria granulata*

LIST OF SPECIES

See Hartman 1941b.
   Pectinaria californiensis Hartman, 1941. Shelf depths; in sand. See Abbott and Reish 1980 (biology).
   Pectinaria granulata (Linnaeus, 1767) (=*Cistenides brevicoma* [Johnson, 1901]). Shallow subtidal in gravel and coarse sand.

AMPHARETIDAE

Ampharetids are small- to medium-size polychaetes that build tubes in soft sediment. They are somewhat difficult to identify because almost all diagnostic characters are soft-body features; with a few exceptions, the only hard structures, the setae, do not differ much even among genera. The body of ampharetids is divided into two regions: (1) an anterior region with biramous parapodia, and (2) a posterior region with well-developed neuropodia, with notopodia either reduced or absent. The most distinctive features of ampharetids are associated with the anterior end. The prostomium ranges from simple to more complex with lateral folds and glandular ridges. Two to four pairs of branchiae may be smooth or pinnate. The first notopodia are sometimes enlarged, directed anteriorly, and bear specialized, golden-colored paleae. Capillary setae are only present in the thorax; neuropodial uncini occur in both the thorax and abdomen. The uncini are small, platelike structures that occur in rows and have one to three rows of teeth above a small rostral point or basal prow. The Ampharetidae superficially resemble the Terebellidae, but can be distinguished by having branchiae that are simple rather than palmate or dendritically branched, branchiae that arise in a straight row across the dorsum of a single segment rather than in a segmental fashion, and the tentacles originating from the mouth or buccal cavity may be retracted rather than emerging freely from the ventral side of the prostomium.

KEY TO AMPHARETIDAE

1. Branchiae of two kinds, one pinnate, the other smooth (plate 177A); thorax with 15 setigers . . . .
. . . . *Schistocomus hiltoni*
— Branchiae all smooth (plate 177C) 2
2. Three pairs of branchiae present (check for scars if missing) 3
— Four pairs of branchiae present 4
3. Thorax with nine uncinigers; last thoracic notopodia elevated (plate 177B); paleae present; anterior part of body expanded . . . . *Mugga wahrbergi*
— Thorax with 11 uncinigers; last thoracic notopodia not elevated; paleae absent; body elongate, slender; prostomium scoop-shaped (plate 177C) . . . .
. . . . *Glyphanostomum pallescens*
4. Segment 3 with curved dorsal hooks (plate 177D), segment 6 with postbranchial membrane across dorsum; segments 3–6 fused, forming lateral collar bearing three fascicles of needlelike neurosetae; abdomen with 40–50 segments . . . .
. . . . *Melinna oculata*
— Postbranchial membrane and nuchal hooks absent; segments 3–6 free or only partially fused, not forming collar; neurosetae absent from first six segments; abdomen usually with 10–15 segments 5
5. Thorax with 12 uncinigers 6
— Thorax with 13 or more uncinigers 10
6. Paleae absent; abdomen with 14 segments; abdominal dorsal cirri much longer than neuropodial lobes . . . .
. . . . *Asabellides lineata*
— Paleae present (plate 177E) 7
7. Buccal (oral) tentacles smooth; with glandular ridge across dorsum of unciniger 8 (fifth to last thoracic setiger) (plate 177F) . . . . *Anobothrus gracilis*
— Buccal (oral) tentacles pinnate; dorsal glandular ridge absent 8
8. Ventral side of upper lip with band of eyespots (plate 177G); paleae short, brassy, indistinctly limbate; segments 1 and 2 separate . . . .*Ampharete labrops*
— Ventral side of upper lip without eyespots; paleae otherwise; segments 1 and 2 fused ventrally to form lower lip 9
9. Paleae slender, tapering gradually (plate 177I); branchiae separated by distinct median gap (plate 177H); no methyl

green stain posterior to branchiae; pygidium with a circlet of about 15 equally long anal cirri . . . .
. . . . *Ampharete acutifrons*
— Paleae wide, tapering abruptly (plate 177K); branchiae not separated by median gap (plate 177J); methyl green stains a few segments posterior to branchiae; pygidium with two lateral anal cirri and about 6–8 low, rounded lobes (not cirri) . . . . *Ampharete finmarchica*
10. Thorax with 13 uncinigers; lower lip large, scoop-shaped, crenulated (plate 177L); branchiae all similar, cylindrical
. . . . *Lysippe labiata*
— Thorax with 14 uncinigers; lower lip not enlarged or crenulated 11
11. All branchiae similar, cylindrical; segment 3 with thin, long setae . . . . *Hobsonia florida*
— Inner branchiae enlarged, foliaceous, with scalloped anterior margin, others cylindrical (plate 177M); segment 3 with broad palea with mucronate tips . . . .
. . . . *Amphicteis scaphobranchiata*

LIST OF SPECIES

See Hilbig 2000a.

Ampharete acutifrons (Grube, 1860). A widely distributed species; low water to slope depths.

Ampharete finmarchica (Sars, 1864). Widely distributed in the northern hemisphere; intertidal to slope depths.

Ampharete labrops Hartman, 1961. Widespread in California, fine sands. Lecithotrophic larvae with a short pelagic life have been obtained from fertilizations of sperm and eggs taken from males and females collected in Bodega Bay (Blake, unpublished).

Amphicteis scaphobranchiata Moore, 1906. Low water to slope depths.

Anobothrus gracilis (Malmgren, 1866). Low water to slope depths.

Asabellides lineata (Berkeley and Berkeley, 1943). Low water to shelf depths.

Glyphanostomum pallescens (Théel, 1879). Low water to slope depths.

Hobsonia florida (Hartman, 1951) (=*Amphicteis gunneri floridus* Hartman, 1951; *Hypaniola grayi* Pettibone, 1953). Intertidal in estuarine mud flats and marshes; western Canada to Oregon; also New England, Gulf of Mexico. See Zottoli 1974, Trans. Am. Microsc. Soc. 93: 78–89 (reproduction and development); Fauchald and Jumars 1979 (feeding); Banse 1979, Can. J. Zool. 57: 1543–1552 (morphology, systematics, distribution).

Lysippe labiata Malmgren, 1866. Widespread in the northern hemisphere; Alaska to California; intertidal to upper slope depths.

Melinna oculata Hartman, 1969. Low water to shelf depths.

Mugga wahrbergi Eliason, 1955. Low water to slope depths.

Schistocomus hiltoni Chamberlin, 1919. In sand- and debris-covered tubes under rocks and in crevices on open coast.

TEREBELLIDAE

The terebellids are tube-dwelling or burrowing polychaetes that have very soft, fragile bodies. They are readily recognized in the field by their long, prehensile tentacles that

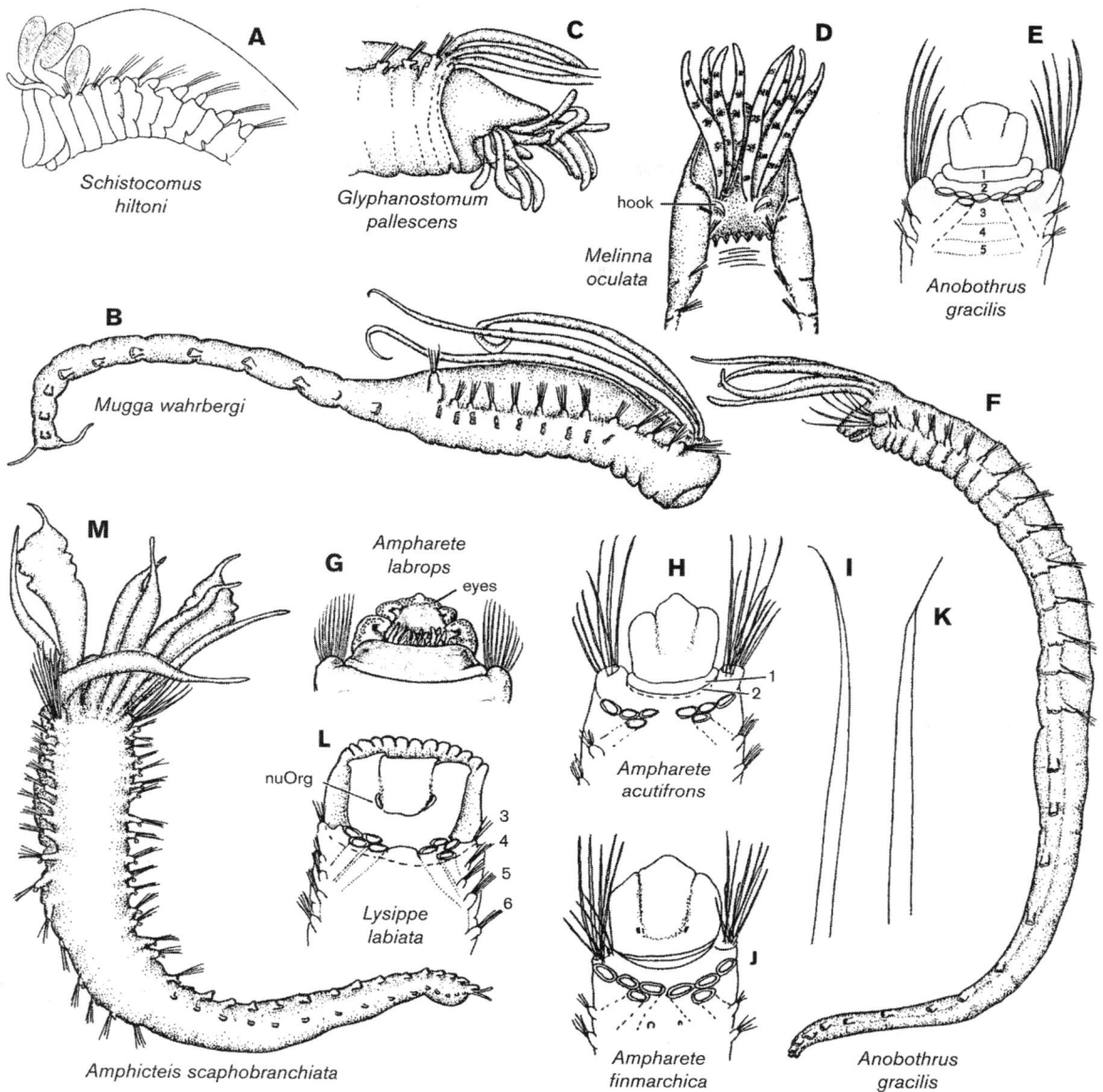

PLATE 177 Polychaeta. Ampharetidae: A, *Schistocomus hiltoni*, anterior end, left lateral view; B, *Mugga wahrbergi*, entire animal, right lateral view; C, *Glyphanostomum pallescens*, anterior end, right lateral view; D, *Melinna oculata*, anterior end, dorsal view; E, *Anobothrus gracilis*, anterior end, dorsal view; F, whole animal, left lateral view; G, *Amphrete labrops*, anterior end, ventral view; H, *Ampharete acutifrons*; anterior end, dorsal view; I, palea; J, *Ampharete finmarchica*, anterior end, dorsal view; K, palea; L, *Lysippe labiata*, anterior end, dorsal view; M, *Amphicteis scaphobranchiata*, entire animal, dorsal view (A, G, after Hartman; rest after Hilbig).

tend to spread out over the sediment while they feed. The tentacles contract into a tangled mass and often break off following preservation. The largest species, such as *Neoamphitrite robusta,* occur in intertidal or shallow-water habitats, whereas smaller species are more typical in deep water. Terebellids are often difficult to identify because their soft bodies do not preserve well. In contrast to ampharetids, terebellids never have paleae, their feeding tentacles originate from a tentacular membrane and cannot be retracted, and their branchiae are often branched or tufted and arranged segmentally.

The prostomium is reduced, fused to the peristomium, and bears a tentacular membrane with tentacles. The tentacular membrane is variable in shape and is sometimes expanded into a foliaceous lobe; eyespots may be present or absent. The first two to three segments are asetigerous, but may bear branchiae and sometimes lateral lappets. Parapodia are simple. Notosetae are usually limbate capillaries, but other unique types are also known. Neurosetae are typically avicular uncini. The shape and arrangement of the uncini is diagnostic at the subfamily level and important in species identification. The most important review of California terebellids is by Hilbig (2000b).

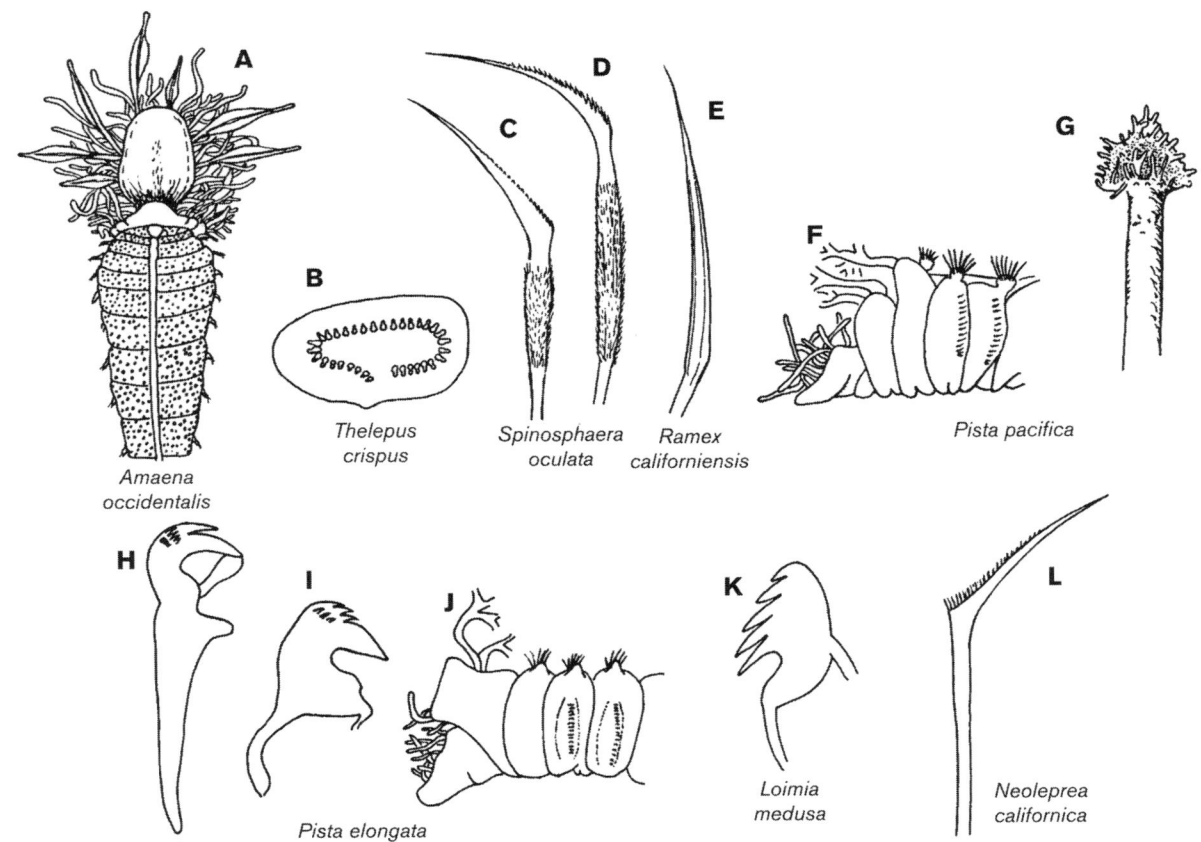

PLATE 178 Polychaeta. Terebellidae: A, *Amaeana occidentalis*; B, *Thelepus crispus*, arrangement of thoracic uncini; C–D, *Spinosphaera oculata*, thoracic notosetae; E, *Ramex californiensis*, thoracic notoseta; F, *Pista pacifica*; G, tube; H, *Pista elongata*, long-shafted thoracic uncinus; I, short uncinus; J, anterior end, lateral view; K, *Loimia medusa*, thoracic uncinus; L, *Neoleprea californica*, thoracic notoseta (all after Hartman).

KEY TO TEREBELLIDAE

1. Thoracic neuropodia and uncini absent; prostomial lobe greatly prolonged as a long flap with three lobes (plate 178A) . . . . *Amaeana occidentalis*
— Thoracic uncini present 2
2. (*Note three choices*) Thoracic uncini arranged in single rows 3
— Thoracic uncini arranged in double or alternating rows from setigers 7–11 4
— Thoracic uncini arranged in single rows on setigers 2–5, then in a depressed ring on succeeding thoracic setigers (plate 178B); notosetae begin on second branchial segment; branchiae are tufts of unbranched coiling filaments . . . . *Thelepus crispus*
3. With three pairs of branchiae; notosetae begin on first branchial segment . . . . *Streblosoma crassibranchia*
— Branchiae absent; notosetae begin on second segment . . . . *Polycirrus* spp.
4. Branchiae absent; peristomium with transverse rows of minute eyespots; thoracic notosetae include long and short curved spines with flaring denticulate tips and thickened shafts bearing numerous spinelets (plate 178C, 178D); with more than 30 thoracic setigers . . . . . . *Spinosphaera oculata*
— Branchiae present 5

5. With one pair of slightly branched branchiae; peristomium with many small eyespots behind tentacular bases; notosetae limbate, without denticles; with 13 thoracic setigers; color in life light red (plate 178E) . . . . *Ramex californiensis*
— With two or three pairs of branchiae 6
6. With two pairs of branchiae 7
— With three pairs of branchiae 8
7. Anterior branchiae larger; ventral scutes (scalelike structures) prominent, continuous through 16 setigers, thereafter reduced; thoracic notosetae with smooth shafts and denticulate tips (plate 178L); with 23–28 or more thoracic setigers . . . . *Neoleprea californica*
— Posterior branchiae larger; lateral lappets on segments 1–3 and smaller auricular lobes on segments 4–6; thoracic notosetae limbate, smooth; with 17 thoracic setigers . . . . *Pista agassizi*
8. Anterior end with conspicuous lateral lappets (plate 178F) 9
— Anterior end without lateral lappets 11
9. Uncini of first few setigers avicular with long shafts (plate 178H), remaining uncini with short shafts (plate 178I) 10
— Uncini of first few setigers with short base and pectinate with 5–6 teeth in a row (plate 178K) . . . . . . *Loimia medusa*

10. Large lappets on second branchial segment (plate 178J); tube with reticulated top; under rocks . . . . . . *Pista elongata*
— Large lappets on second and third branchial segments (plate 178F); tube with large, hoodlike, overlapping membrane (plate 178G); sandy mud flats . . . . *Pista pacifica*
11. Color in life green with light tan tentacles and deep reddish brown branchiae; notosetae with smooth tips; numerous small dark eyespots behind tentacular bases; tubes sandy . . . . *Eupolymnia heterobranchia*
— Color in life brown with numerous white tentacles and red branchiae; notosetae with denticulate tips; eyes absent; tubes muddy . . . . *Neoamphitrite robusta*

LIST OF SPECIES

See Hilbig 2000b.

Amaeana occidentalis (Hartman, 1944). Mud flats of bays and estuaries; intertidal to shallow shelf depths. See Hartman 1944b.

Eupolymnia heterobranchia (Johnson, 1901) (=*Eupolymnia crescentis* Chamberlin, 1919). Sandy mud sediments of bays and estuaries; in eelgrass beds; intertidal to low water. See Banse 1980, Can. J. Fish. Aquat. Sci. 37: 20–40; Wilson 1980: J. Exp. Mar. Biol. Ecol. 46: 73–80 (interaction between terebellid and *Nereis vexillosa*).

Lanice conchilega (Pallas, 1766). Low intertidal to 200 m.

Loimia medusa (Savigny, 1818). Intertidal to shallow shelf depths in sandy mud sediments. See Hilbig 2000b.

Nicolea zostericola (Oersted, 1844). Collected from the Bay Farm Borrow Area in San Francisco Bay; possibly introduced from the east coast. See Hartmann-Schröder 1996.

Neoamphitrite robusta (Johnson, 1901). In mud tubes under rocks. See Brown and Ellis 1971, J. Fish, Res. Bd. Can. 28: 1433–1435 (tube-building, feeding).

Neoleprea californica Moore, 1904 (=*Terebella californica*). Rocky habitats; algal holdfasts; Oregon to southern California. See Banse 1980, Can. J. Fish. Aquat. Sci. 37: 20–40.

Pista agassizi Hilbig, 2000. This species was named to replace *P. brevibranchia* Chamberlin, 1919, which is a homonym of *P. brevibranchia* Caullery, 1915. The original specimen that was collected by A. Agassizi from Mendocino in northern California formed the basis for the redescription by Hilbig (2000b). Hilbig indicated that the species was collected in southern California, but Mendicino, the type locality, is in northern California. The habitat is unknown.

Pista elongata Moore, 1909. In rocky habitats, forms muddy tubes in crevices and under rocks; often has commensal scale worms and pea crabs in the burrows. See Abbott and Reish 1980.

Pista pacifica Berkeley and Berkeley, 1942. Sandy mud sediments; forms characteristic tubes that extend deep into the substratum; the scale worm *Halosydna brevisetosa* may be a commensal in the tubes. See Hartman 1944b; Abbott and Reish 1980.

Polycirrus californicus Moore, 1909. Intertidal to shelf depths in sandy and rocky areas; has a distinctive elongated postsetal lobe. See Hilbig 2000b.

Polycirrus spp. Common in rocky habitats and also in mud in tires and other habitats on floats and docks. Several species may be present.

Proclea graffi (Langerhans, 1884). Intertidal to shelf depths.

Ramex californiensis Hartman, 1944. Rocky habitats on exposed outer coast. See Hartman 1944b; Blake, J. A., 1991, Bull. Mar. Sci. 448(460 (larval development).

Spinosphaera oculata Hartman, 1944. Known only from northern California; rocky habitats on exposed outer coast; intertidal. See Hartman 1944b.

Spinosphaera pacifica Hessle, 1917. Japan and Central California, low water to shelf depths.

Streblosoma crassibranchia Treadwell, 1914. Intertidal to basin depths in sandy mud; creates a long, sand-encrusted tube that is closely coiled through much of its length; central and southern California.

Thelepus crispus Johnson, 1901. Tubes attached to undersides of rocks in exposed rocky habitats; one of the most common intertidal terebellids; scale worms sometimes occur as commensals in the burrows. See Abbott and Reish 1980.

Thelepus setosus (Quatrefages, 1865). Tubes sandy, distinctive, with the upper end terminating in a triangular overlapping hood bearing many thin threads of cemented particles; the scale worm *Halosydna brevisetosa* may occur as a commensal in the tubes. See Abbott and Reish 1980 (biology).

TRICHOBRANCHIDAE

The trichobranchids are terebelliform polychaetes, sometimes confused with species of Terebellidae. As with all terebelliforms, there are two body regions: (1) a thorax with biramous parapodia, and (2) an abdominal region with uniramous parapodia. Trichobranchids are distinguished from terebellids by having long-handled thoracic hooks instead of avicular uncini. In addition, the anterior thorax of some species has lateral lappets that continue across the ventrum that results in a telescoped appearance to the anterior part of the body. Three genera are reported from California by Hilbig (2000c): *Artacamella*, *Octobranchus*, and *Terebellides*. For the most part, trichobranchids are subtidal in shelf and slope depths and are not commonly encountered. At least one species, *Artacamella hancocki* Hartman, 1955, is known from relatively shallow depths off central and southern California; no local species are known from the intertidal. Hilbig (2000c) provides an important summary of trichobranchids from California.

SABELLIDAE

The sabellids, or feather-duster worms, are characterized by having the prostomium modified into an often colorful radiolar crown and a peristomium that is usually modified into an anterior collar. Species range in size from a millimeter up to half a meter long. The body is divided into a relatively short thoracic region (typically eight setigers) and an abdominal region that can range from a few to over a hundred setigers. The first thoracic setiger is usually uniramous, while the following thoracic segments have limbate or bilimbate capillary setae in the notopodia and uncini in the neuropodia. These setal types are inverted at the junction of the thoracic and abdomen. A ciliated fecal groove runs from the ventral anus along the ventral abdomen, swinging around the body at the point of setal inversion and continuing along the dorsal thorax to the peristomial collar.

* = Not in key.

Sabellids are found at all depths throughout the world's oceans and are common in the intertidal. Nearly all species build tubes, using the radioles to combine particles with mucus to construct temporary burrows in muddy silts, or building permanent tubes in rocky crevices and against wharves, pilings, and other hard substrates. A few genera (e.g., *Manayunkia*) have successfully moved into brackish and fresh water. Other species live in close association with sponges and colonial ascidians, or burrow into coral reef environments. There are nearly 500 valid species in the family, and a number of species are encountered throughout the central California region yet to be described.

KEY TO SABELLIDAE

1. Abdominal uncini in tori nearly encircling body 2
— Abdominal uncini in short tori 3
2. Thorax with eight setigers (plate 179R); abdominal uncini with several apical teeth . . . . *Myxicola infundibulum*
— Thorax with 1–4 setigers; abdominal uncini with a single apical tooth . . . . *Myxicola aesthetica*
3. Thoracic uncini acicular, with gently curved handles beneath toothed heads (plate 179M) 4
— Thoracic uncini avicular (Z-shaped) beneath toothed heads (plate 179N), with or without handles 13
4. Specimens small (<10 mm); abdomen with only three setigers 5
— Specimens small or large; abdomen with seven or more setigers 9
5. Two pairs of radioles; pygidial eyes absent (plate 179A) . . . . *Manayunkia speciosa*
— Three pairs of radioles; pygidial eyes present. 6
6. Length to about 8 mm (including radioles); collar not apparent; peristomium more than three times longer than first setiger . . . . *Fabricinuda limnicola*
— Length to about 3 mm; collar present; peristomium not exceeding twice first setiger length 7
7. Level, membranous collar encircling base of radioles; ventral filamentous appendages present; inferior pseudospatulate notosetae absent . . . . *Fabriciola* cf. *berkeleyi*
— Collar thick, only obvious on ventral side; ventral filamentous appendage absent; inferior pseudospatulate notosetae present (plate 179C) 8
8. Ventral collar as a broad lobe (plate 179B); inferior thoracic pseudospatulate notosetae present in setigers 3–7; dorsal lips well-developed, triangular . . . . *Fabricia stellaris*
— Ventral collar as a tonguelike lobe; inferior thoracic pseudospatulate notosetae restricted to setigers 3–5; dorsal lips low, rounded . . . . *Novafabricia brunnea*
9. Ventrum strongly flattened through posteriormost 10 segments (plate 179Q) . . . . *Euchone limnicola*
— Posterior setigers circular in cross-section, not ventrally flattened. 10
10. Small specimens (<5 mm); inferior paleate thoracic notosetae absent; thoracic uncini with unequal teeth above main fang (similar to plate 179M); abdominal uncini raspshaped, with poorly developed breast (plate 179I) . . . . *Amphicorina gracilis*
— Small to large specimens (5–60+ mm); inferior paleate thoracic notosetae present; thoracic uncini with crown of subequal teeth; abdominal uncini with breast well-developed (plate 179J) 11

11. Paleate notosetae distally rounded (plate 179K); large specimens with up to 15 radiolar pairs . . . . *Chone mollis*
— Paleate notosetae distally pointed (plate 179L) 12
12. Specimens small (10 mm); up to seven radiolar pairs. . . . . *Chone minuta*
— Specimens larger (30 mm); up to 10 radiolar pairs . . . . *Chone gracilis*
13. Thoracic uncini with handles (plate 179N) 14
— Thoracic uncini without handles (as in plate 179O) . . . . *Laonome* sp.
14. Companion seta with toothed head and thin distal mucro (plate 179D); radiolar eyes absent 15
— Companion seta with membranous distal structure (plate 179E); compound radiolar eyes present 16
15. Collar extending above short fused base of branchial crown (plate 179P) . . . . *Demonax medius*
— Fused base of branchial crown elongated and not entire overlapped by collar. . . . . *Demonax rugosus*
16. Eyes restricted to distal tips of dorsalmost radioles (plate 179G). . . . . *Megalomma splendida*
— Eyes aligned along proximal axis of radioles 17
17. Eyes paired along radioles and masked by transverse, deep maroon pigment bands (plate 179H). . . . . *Bispira* sp.
— Eyes usually conspicuous, in single series along radioles. 18
18. Radioles dichotomously branched (plate 179F). . . . . *Schizobranchia insignis*
— Radioles usually not divided (occasionally with single bifurcation in large *Eudistylia*) 19
19. Radioles arranged in crescents; outer margins with sharpedged ridges and appearing quadrangular; eyes absent on dorsalmost pair 20
— Radioles arranged in spirals (large specimens only); outer margins rounded; eyes present on dorsalmost pair 22
20. Dorsal margins of branchial base cleft; radioles with 7–12 eyespots in each row (plate 179S) . . . . *Pseudopotamilla occelata*
— Dorsal margins of branchial base entire (similar to plate 179T); radioles with 0–5 eyespots in each row 21
21. Specimens to 25 mm; uncini in last two thoracic tori reduced in number . . . . *Pseudopotamilla socialis*
— Specimens to 58 mm; uncinal counts in last two thoracic tori not reduced . . . . *Pseudopotamilla* sp.
22. Dorsal margin of branchial base with deep cleft (similar to plate 179S). . . . . *Eudistylia polymorpha*
— Dorsal margin of branchial base without cleft (plate 179T) . . . . *Eudistylia vancouveri*

LIST OF SPECIES

See Banse 1972b, 1979; Fitzhugh 1989, 1990a–b, 1992, 1993; Hartman 1944b, 1951b, 1969; Knight-Jones and Perkins 1998; Perkins 1984.

Amphicorina gracilis (Hartman, 1969) (=*Oriopsis gracilis*). Rocky intertidal, forming tube clusters with sponges and tunicates in sandy tide pools. Body with red-brown longitudinal stripes ventrolaterally.

Bispira sp. Rocky and mixed bottoms; tube membranous covered with muddy sand and debris. Includes *Sabella crassicornis* from the previous edition and may include several closely related species in San Francisco Bay. See Knight-Jones and Perkins 1998; Abbott and Reish 1980 (biology as *Sabella crassicornis*).

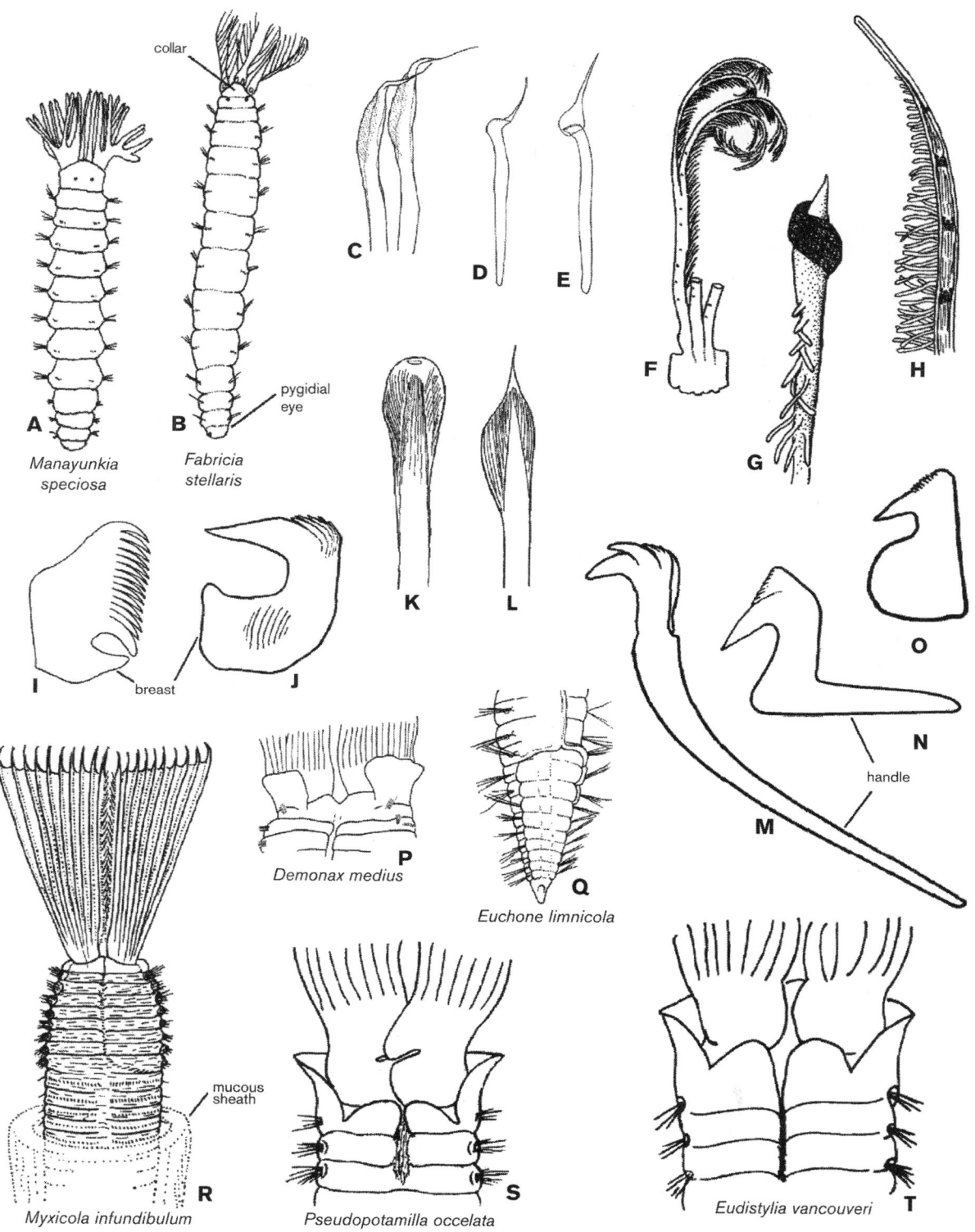

PLATE 179 Polychaeta. Sabellidae: A, *Manayunkia speciosa*, ventral view; B, *Fabricia stellaris*, ventral view; C, pseudospatulate setae, setiger 4; D, *Demonax rugosus*, companion seta; E, *Pseudopotamilla intermedia*, companion seta; F, *Schizobranchia insignis*, radiole; G, *Megalomma splendida*, tip of radiole with eye; H, *Bispira* sp., distal radiole; I, *Amphicorina gracilis*, abdominal uncinus; J, *Chone minuta*, abdominal uncinus; K, *C. mollis*, thoracic paleate seta; L, *C. minuta*, thoracic paleate seta; M, general sabellid acicular thoracic neuropodial uncinus; N, general sabellid avicular thoracic neuropodial uncinus with handle; O, same, without handle; P, *Demonax medius*, collar region, dorsal view; Q, *Euchone limnicola*, posterior, ventral view; R, *Myxicola infundibulum*, anterior end surrounded by mucus sheath, dorsal view; S, *Pseudopotamilla occelata*, dorsal view; T, *Eudistylia vancouveri*, dorsal view (B–C, O, after Fitzhugh; D, after Perkins; E, after Moore; F, H, J–L, N, Q, after Hartman; I, P, after Banse; R, after Day; S after Berkeley and Berkeley).

Chone gracilis Moore, 1906. Sandy mud sediments with rocks and broken shells; tube nearly transparent; normally subtidal.

Chone minuta Hartman, 1944. Rocky intertidal and shallow subtidal, in algal holdfasts and among compound ascidians; builds close-fitting tubes covered with edge-mounted shell fragments that accumulate sand and form mounds in the intertidal. Listed as *C. ecaudata* in the previous edition.

Chone mollis (Bush, 1905). Common in sand-mud sediments; mucoid tube covered with fine sand. See Bonar 1972, J. Exp. Mar. Biol. Ecol. 9: 1–18 (tube building, feeding).

Demonax medius (Bush, 1905) (=*Sabella media*). Rocky habitats, in crevices in flexible, leathery, sand-coated tubes. Branchial crown red-brown mottled with white.

Demonax rugosus (Moore, 1904). Associated with pilings and hard substrata; in thick, translucent, sand-coated tubes similar to those of *D. medius*.

Euchone limnicola Reish, 1960. Estuarine, often in great abundance in sandy muds; constructs straight, upright tubes of fine-grained silts.

Eudistylia polymorpha (Johnson, 1901). Rocky habitats, also in harbors on floats and wharves. Branchial crown yellowish-tan to orange; tubes thick, parchmentlike, basally attached to hard substratum. May be confused with *E. vancouveri* (see below). See Abbott and Reish 1980 (biology).

Eudistylia vancouveri (Kinberg, 1867). Common on floats and wharves, also in sandy mudflats. Branchial crown with transverse bands of dark green and maroon; tubes constructed of thick solidified mucus coated with sand grains, often forming masses and intermixed with *Schizobranchia insignis*. Specimens exhibiting characteristics of both *E. polymorpha* and *E. vancouveri* have been noted, suggesting possibility of hybridization. Found with the symbiotic copepod *Gastrodelphys dalesi* at Tomales Point; see Dudley 1964, Amer. Mus. Nov. no. 2194.

Fabricinuda limnicola (Hartman, 1951). In estuarine mudflats; closely adhering mucoid tube coated with silt particles.

Fabriciola cf. *berkeleyi* (Banse, 1956). In estuarine mudflats and among sponges and ascidians; tubes composed of fine detritus, free in the mud but attached along their length to the sponges or tunicates.

Fabricia stellaris (Müller, 1774) (=*F. sabella* [Ehrenberg, 1836]; see Fitzhugh 1990). In estuarine muds; often among algae and barnacles on pilings. See Lewis 1968, J. Linn. Soc. London 47: 515–526 (ecology).

Laonome sp. Fitzhugh. An undescribed species found in great abundance on some floating docks in San Francisco Bay, normally in areas with lowered salinity (K. F. Fitzhugh, pers. comm.). Previously recognized as *Potamilla* sp. by Cohen and Carlton 1995 (introduced species in San Francisco Bay).

Manayunkia speciosa Leidy, 1858. In freshwater streams, canals, and lakes in silty sand. Body olive green with brown crown. Mucoid mud tubes attached to fixed objects at one end, sometimes branched. See Pettibone 1953, Biol. Bull. 105: 149–153 (morphology); Cohen and Carlton 1995 (introduced species in San Francisco Bay).

Megalomma splendida (Moore, 1905). Rocky bottoms, rare intertidal, proximal radioles with several transverse, red bars.

Myxicola aesthetica (Claparède, 1870). Tube thick, mucilaginous; worms often found in masses attached to ropes or other sunken objects; <50 mm long.

Myxicola infundibulum (Renier, 1804). Tube thick, mucilaginous, positioned vertically in soft sediments; often abundant across large areas; cosmopolitan; body yellow-orange with violet-brown radiolar tips; up to 200 mm long. See Abbott and Reish 1980 (biology).

Novafabricia brunnea (Hartman, 1969) (=*Fabricia brunnea*). Rocky intertidal, in sand bound by holdfasts of red and green algae and in interstices of *Phragmatopoma* tubes; branchial crown and first few segments often dark brown, becoming cream-colored in posterior region.

Pseudopotamilla sp. Rocky or mixed bottoms; rare. Body orange to red, with dark wine-colored radioles. Tube horny, transparent or covered with fine sand grains. Listed in previous edition as *Pseudopotamilla intermedia* Moore, 1905, but differs in details from type specimen that was dredged at over 1600 m depth off southeast Alaska and that appears to be regenerating anteriorly (L. Harris, pers. comm.).

Pseudopotamilla occelata Moore, 1905. Rocky habitats, open coast; common; branchial crown with red-brown transverse bands; horny, translucent tubes often in masses, occasionally with small, commensal, two-tentacled hydroid *Proboscidactyla* on rims. See Abbott and Reish 1980 (biology as *Potamilla occelata*).

Pseudopotamilla socialis Hartman, 1944. Rocky habitats, open coast; associated with sponges and colonial tunicates, body pale or tinged with black; slender opaque tubes covered with sand and shell fragments.

Schizobranchia insignis Bush, 1905. Rocky habitats; in harbors on floats and wharves; common; radioles variable in color between specimens, ranging from orange and tan through dark pink to maroon, and occasionally green-tan; body fawn-colored; often found in clumps of thick, sand-coated, solidified mucus tubes intermixed with *Eudistylia vancouveri*.

SERPULIDAE

The serpulids are a distinctive group of sessile polychaetes that construct permanent calcareous tubes. The worms have a feathery crown of radioles, one of which is usually modified into an operculum that acts as a plug when the individual withdraws into its tube. The body has a distinct thorax with a thoracic membrane continuous with the peristomial collar and a long abdominal region. Many authors use the subfamilies Serpulinae, Filograninae, and Spirorbinae (ten Hove 1984; Rouse and Pleijel 2001). We will use a more conservative classification of two subfamilies, Serpulinae with tubes elongate and the Spirorbinae with tubes coiled either in a dextral or sinistral direction.

SERPULINAE

In the Serpulinae the thorax is symmetrical and most often has seven setigers. The first setiger is usually uniramous with notopodial "collar setae," which may be ornamented with spines or serrations. Subsequent thoracic setigers have limbate or geniculate capillary notosetae and neuropodial rasp-shaped uncini. Setal inversion occurs at the end of the thorax, and the abdomen is equipped with neuropodial capillary setae and notopodial uncini similar to those in the thorax.

Secretions from glands behind the ventral collar enable serpulids to precipitate calcium carbonate from seawater (Hedley 1956, Quart. J. Microsc. Sci. 97: 421–427). This material is mixed with a mucopolysaccharide to construct the tubes, which are most often attached to hard substrates in an irregular pattern, or intertwined in aggregations. Features of taxonomic importance include the shape and ornamentation of the operculum and peduncle, details of the collar setae, and aspects of the calcareous tube.

KEY TO SERPULINAE

1. Three pairs of radioles; radiolar tips swollen but not distinctly modified as opercula; thorax separated from abdomen by long asetigerous section; slender tubes often massed (plate 180I) . . . . *Salmacina tribranchiata*
— 10 or more pairs of radioles; operculum large and on bare peduncle; thorax and abdomen not greatly separated 2
2. Operculum without terminal spines 3
— Operculum with distal processes 4
3. Operculum with about 26 ribs; three large knobs at base of funnel; peduncle offset (plate 180E) . . . .
. . . . *Crucigera zygophora*
— Operculum with about 100 ribs; peduncle directly beneath funnel, with slight distal constriction (plate 180F) . . . . . .
. . . . *Serpula columbiana*
4. Opercular peduncle distally produced into pair of membranous lateral wings (plate 180K). . . . .
. . . . *Spirobranchus spinosus*
— Peduncular wings absent 5
5. Operculum egg-shaped, prolonged distally into single long, curved, black spine (plate 180L) . . . .
. . . . *Vermiliopsis multiannulata*
— Operculum funnel-shaped with numerous distal spines
. . . . 6
6. Operculum an elongate funnel edged with rows of dark, chitinous spines (plate 180G, 180H); collar setae simple serrated blades; tubes with peristomes (plate 180J) . . . . . .
. . . . *Ficopomatus enigmaticus*
— Operculum distally expanded, marginally crenulated, with crown of chitinized hooks; collar setae with knobs or spines at base of blade; tubes without peristomes. 7
7. Distal opercular spines smooth, directed inward (plate 180A); collar setae with smooth blade and pair of basal knobs (plate 180C) . . . . *Hydroides gracilis*
— Distal opercular spines directed outward (plate 180B), each with lateral projections; collar setae with small basal spines separated from finely serrated blade by conspicuous gap (plate 180D) . . . . *Hydroides elegans*

LIST OF SPECIES

See Bailey-Brock and Hartman 1987; Hartman 1969; Johnson 1901; Moore 1923; Hobson and Banse 1981.

Crucigera zygophora (Johnson, 1901). Rocky intertidal habitats. Crown red at base and broadly barred distally; thick tube usually fully attached to hard surface.

Ficopomatus enigmaticus (Fauvel, 1923) (=*Mercierella enigmatica*). Abundant in San Francisco Bay and Elkhorn Slough in quiet, brackish areas where it sometimes forms massive aggregations of white, ringed tubes; an introduced cosmopolitan fouling species; crown green and brown. See ten Hove and Weerdenburg 1978, Biol. Bull. 154: 96–120; Abbott and Reish 1980 (review of biology).

*Hydroides diramphus Mörch, 1863 (=*Hydroides lunifer* Claparède, 1870). A southern and Baja California (and cosmopolitan subtropical) fouling species. See Bastida-Zavala and ten Hove

* = Not in key.

2003, Beaufortia 53: 67–110 (systematics of Eastern Pacific *Hydroides*).

Hydroides elegans (Haswell, 1883) (=*Hydroides pacificus* Hartman, 1969; *H. norvegicus* of authors). An introduced fouling species in harbors and marinas of southern California; white, erect tubes, slightly angular and crossed by obscure growth lines; sometimes massed. See Bastida-Zavala and ten Hove 2003 (above).

Hydroides gracilis (Bush, 1905) (=*Eupomatus gracilis*). On rocks and gastropod shells; Monterey Bay and south; crown white to grayish yellow, banded with black; solitary tubes irregularly coiled and longitudinally ridged. See Bastida-Zavala and ten Hove 2003 (above).

Salmacina tribranchiata (Moore, 1923). Sheltered rocky habitats. Crown red-orange; slender tubes white, cylindrical, often closely massed. See Abbott and Reish 1980 (review of biology).

Serpula columbiana Johnson, 1901 (=*Serpula vermicularis* of previous edition). In harbors on floats and wharves and attached to hard surfaces along exposed shores; crown white with several crimson or violet bands; irregularly coiled white tubes mostly attached to substrate, often with raised distal ends. See Kupriyanova 1999, Ophelia, 50: 21–34 (taxonomy); Abbott and Reish 1980 (review of biology as *S. vermicularis*).

Spirobranchus spinosus Moore, 1923. Usually found subtidally on hard substrates in massed tubes, but occasionally as solitary individuals in the intertidal; crown with purple-black and white bands; tubes white with a high median keel, often pitted and overgrown.

Vermiliopsis multiannulata (Moore, 1923). Rocky habitats; tube white, fully attached to substratum, with a serrated median longitudinal keel.

SPIRORBINAE

The spirorbins are fanworms that construct small, usually tightly coiled calcareous tubes. They are characterized by having shortened, asymmetrical thoracic regions that are normally only three or four segments in length. The first setiger is uniramous with limbate or modified "collar setae" in the notopodia. In addition, the notosetae of the third setiger are sometimes modified as "sickle setae" with serrated distal tips.

Spirorbins are commonly found attached to rocks, shells, pilings, and kelp fronds where they use their radiolar crowns for respiration and feeding. The animals are only a few millimeters in length, however, and identification usually requires careful observation with the compound microscope. Features of taxonomic importance include the shape and ornamentation of the tube, the positioning of and structures associated with the embryos, the shape and ornamentation of the opercula, and details of the setae and uncini. The direction of tube coiling is important to note, although some species may coil in either direction.

KEY TO SPIRORBINAE

1. Tube coiling dextral (counterclockwise). 2
— Tube coiling sinistral (clockwise) 7
2. Tube glassy, transparent, with three heavy longitudinal ridges (plate 181I); red pigment of living worm visible through tube walls . . . . *Paradexiospira vitrea* (in part)

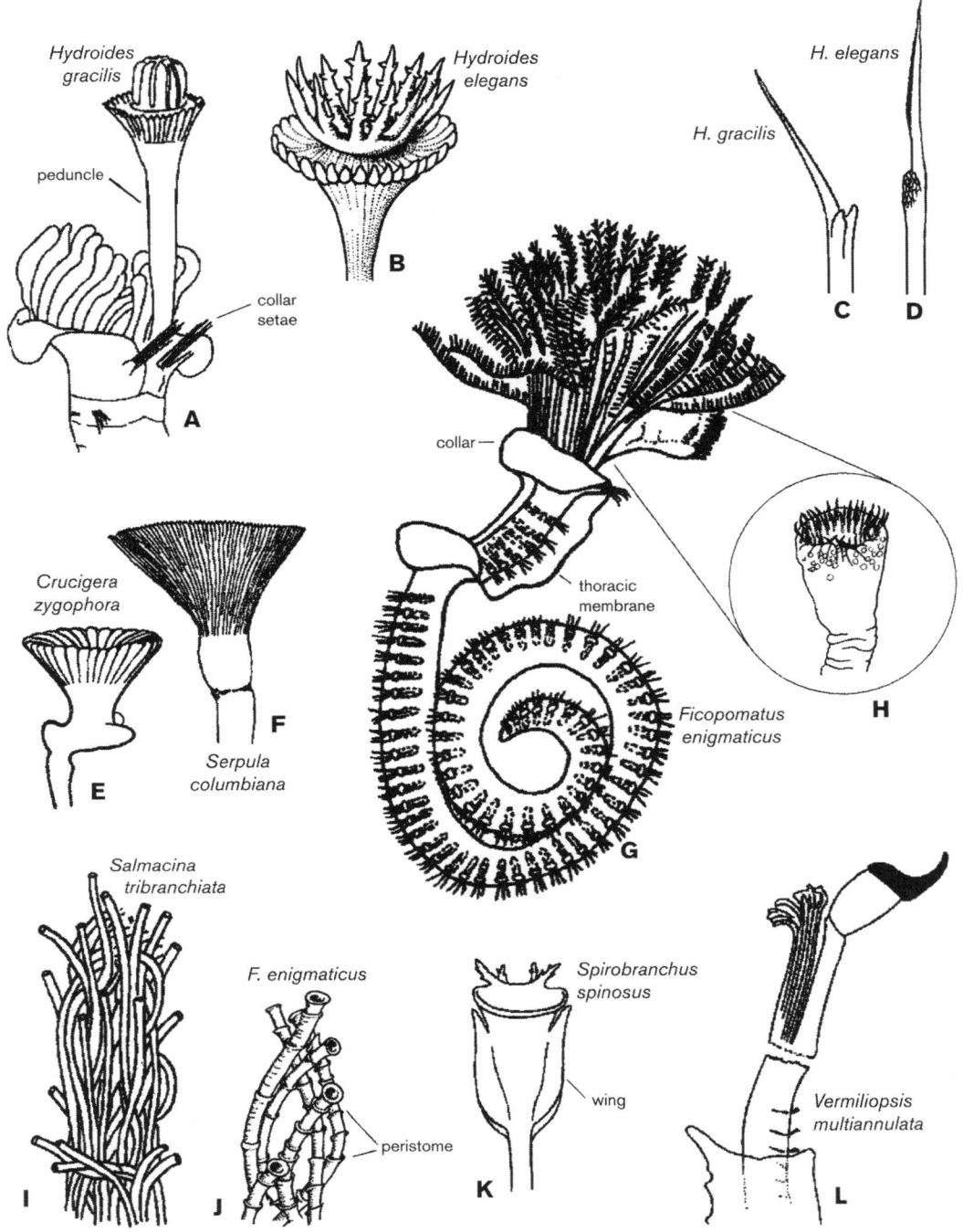

PLATE 180 Polychaeta. Serpulidae, Subfamily Serpulinae: A, *Hydroides gracilis*, anterior end; B, *H. elegans*, operculum; C, *H. gracilis*, collar seta; D, *H. elegans*, collar seta; E, *Crucigera zygophora*, operculum; F, *Serpula columbiana*, operculum; G, *Ficopomatus enigmaticus*, whole animal; H, detail of operculum; I, *Salmacina tribranchiata*, tube aggregation; J, *F. enigmaticus*, tubes; K, *Spirobranchus spinosus*, operculum and peduncle; L, *Vermiliopsis multiannulata*, crown. (A, C, I, K, and L, after Hartman; B, after Bailey-Brock and Hartman; D, after Bianchi; E–F, after Johnson; H, after Hartmann-Schröder; J, after Day; rest after Blake).

— Tube translucent or opaque, with or without ridges; worm not visible 3
3. Tube smooth, thin-walled and round in cross section (as in plate 181J); embryos stuck directly on inside of tube wall 4
— Tube thicker, opaque, usually with longitudinal ridging and sloping sides (as in plate 181K); if in tube, embryos not

attached directly to walls 5
4. Collar setae sharply geniculate, without cross striations; tube porcellaneous, pressed entirely against substratum (plate 181B) . . . . *Circeis armoricana* (in part)
— Collar setae obliquely geniculate, with cross striations (plate 181C); distal portion of translucent tube usually coiling upward (plate 181J) . . . . *Circeis spirillum*

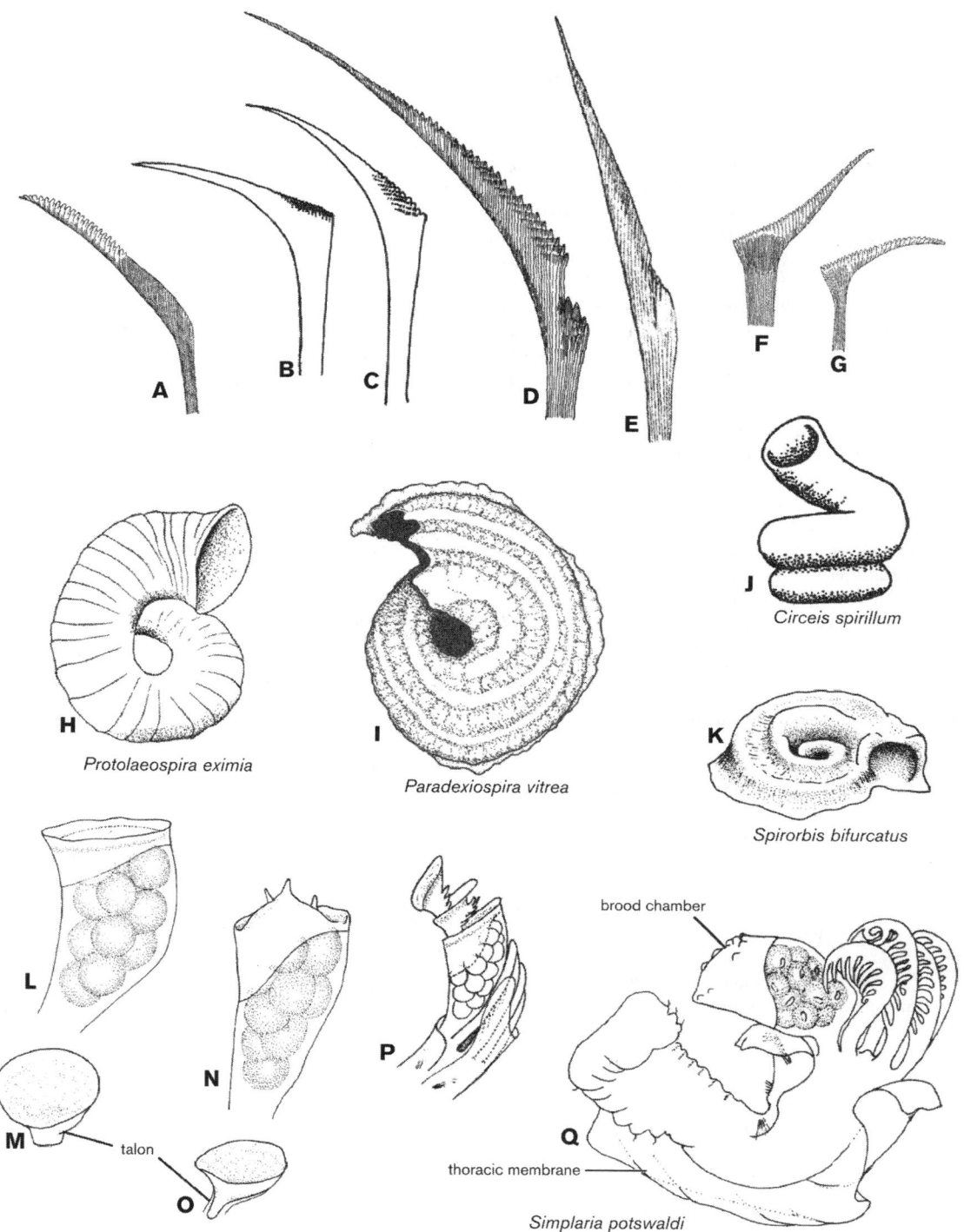

PLATE 181 Polychaeta. Serpulidae, Subfamily Spirorbinae: A, *Pileolaria marginata*, sickle seta; B, *Circeis armoricana*, collar seta; C, *C. spirillum*, collar seta; D, *Simplaria potswaldi*, collar seta; E, *Bushiella abnormis*, collar seta; F, *Spirorbis bifurcatus*, abdominal neuroseta; G, *S. marioni*, abdominal neuroseta; H, *Protolaeospira eximia*, tube; I, *Paradexiospira vitrea*, dextral tube; J, *Circeis spirillum*, tube; K, *Spirorbis bifurcatus*, sinistral tube; L, *Pileolaria marginata*, brood chamber; M, operculum; N, *Pileolaria lateralis*, brood chamber; O, operculum; P, *Bushiella abnormis*, opercula; Q, *Simplaria potswaldi*, whole animal dorsal view (A–G, J–Q, after Knight-Jones; H, after Vine; I, after Crisp).

5. Embryos incubated in inverted cuticular cup (similar to plate 181L); collar setae simple, without fin and blade construction (similar to plate 181C); sickle setae absent. . . . .
. . . . *Neodexiospira pseudocorrugata*
—   Embryos in string attached by single thread to inside of tube wall; collar setae with distinct gap between blade and fin (plate 181D); sickle setae present (plate 181A). 6

6. Subquadrangular tube to 2 mm (plate 181K); abdominal setae with very wide distal shafts (plate 181F) . . . .
. . . . *Spirorbis bifurcatus* (in part)

— Subquadrangular tube usually not exceeding 1 mm; abdominal setae with narrow distal shafts (plate 181G) . . . . *Spirorbis marioni*

7. Collar setae simple or with indistinct gap between blade and basal fin (plate 181E); cross striations absent 8
— Collar setae with cross striations and coarsely toothed fin separated from blade by distinct gap (plate 181D) 9

8. Tube usually with longitudinal ridges; sickle setae present (as in plate 181A); collar setae with indistinct gap between blade and basal fin (plate 181E); stacked opercula often present above brood chamber (plate 181P) . . . . *Bushiella abnormis*
— Tube smooth and porcellaneous; sickle setae absent; collar setae geniculate and without basal fin (plate 181B); single operculum present (similar to plate 181M); embryos brooded in tube . . . . *Circeis armoricana* (in part)

9. Embryos not associated with operculum 10
— Embryos brooded in opercular chamber 12

10. Tube transparent, with three heavy longitudinal ridges (plate 181I); embryos attached directly to tube wall . . . . *Paradexiospira vitrea* (in part)
— Opaque tube masking inhabitant; embryos not directly attached to inside of tube walls 11

11. Tube with transverse growth rings (plate 181H); embryos in sac arising near base of opercular stalk; thorax with three tori . . . . *Protolaeospira eximia*
— Tube with longitudinal ridges; embryos on thread attached to tube wall; thorax with two tori . . . . *Spirorbis bifurcatus* (in part)

12. Sickle setae (plate 181A) present in third thoracic setiger 13
— Sickle setae absent 14

13. Tube usually with distinct transverse growth rings; rim of brood chamber symmetrical, without spines (plate 181L); operculum with short, blunt, eccentric talon (plate 181M) . . . . *Pileolaria marginata*
— Tube usually smooth with indistinct growth rings; rim of brood chamber asymmetric, often with spines (plate 181N); operculum with elongated, bluntly pointed talon (plate 181O) . . . . *Pileolaria lateralis*

14. Tube coil up to 3 mm in diameter, with irregular transverse growth rings; thoracic membrane extending posteriorly and forming large "cloak" (plate 181Q); brood chamber relatively smooth distally . . . . *Simplaria potswaldi*
— Tube coil up to 2 mm, without ridges; greatly enlarged thoracic membrane absent; brood chamber often with distal spines (similar to plate 181N) . . . . . *Simplaria pseudomilitaris*

LIST OF SPECIES

See Bush 1905; E. Knight-Jones et al. 1995; P. Knight-Jones 1978, 1984; P. Knight-Jones et al. 1979; Vine 1977.

Bushiella abnormis (Bush, 1905). On shaded areas of intertidal rocks and stones; coiling always sinistral; often found associated with *Paradexiospira vitrea*.

Circeis armoricana Saint-Joseph, 1894. Attaches to shells and serpulid tubes, and particularly to algae; widespread in arctic and boreal waters; coiling usually dextral, occasionally sinistral; often associated with *Pileolaria marginata*.

Circeis spirillum (Linnaeus, 1758) (=*Spirorbis spirillum*). Sublittoral on bryozoans and hydroids; coiling always dextral.

Neodexiospira pseudocorrugata (Bush, 1905). On intertidal rocks, stones, and shells; coiling always dextral; found sparsely intermixed with *Spirorbis bifurcatus* or *Simplaria potswaldi*.

Paradexiospira vitrea (Fabricius, 1780). On shaded areas of intertidal stones, rocks and shells; coiling usually dextral but occasionally sinistral; sometimes found associated with *Bushiella abnormis*, *Protolaeospira eximia*, *Simplaria potswaldi*, or *Spirorbis bifurcatus*. See Crisp et al. 1967, J. Mar. Biol. Assoc. U.K. 47: 511–521 (ecology).

Pileolaria lateralis Knight-Jones, 1978. Found mostly on small algae, including *Crytopleura ruprechtiana*, *Rhodymenia pacifica* and *Chondracanthus corymbiferus* in the Monterey area; coiling always sinistral.

Pileolaria marginata Knight-Jones, 1978. On algae and kelp, including *Macrocystis pyrifera*; coiling always sinistral; often abundant and found associated with *Circeis armoricana*.

Protolaeospira eximia (Bush, 1905). Attaches to hard surfaces including wood, shells, and stones. Coiling always sinistral; found intermixed with *Simplaria potswaldi* or *Circeis armoricana*.

Simplaria potswaldi Knight-Jones, 1978. On shaded areas of wood and stones; coiling always sinistral; often associated with *Paradexiospira vitrea*, *Spirorbis bifurcatus*, *Bushiella abnormis*, and occasionally with *Neodexiospira pseudocorrugata*.

Simplaria pseudomilitaris (Thiriot-Quièvreux, 1965). On shaded areas of intertidal stones and hard substrata; coiling always sinistral; often associated with *Spirorbis bifurcatus* and *S. marioni*.

Spirorbis bifurcatus Knight-Jones, 1978. On shaded areas of intertidal rocks and stones; coiling both dextral and sinistral; often found with *Paradexiospira vitrea*, *Simplaria potswaldi*, or *S. pseudomilitaris*.

Spirorbis marioni (Caullery & Mesnil, 1897). Normally occurring in more tropical waters, but occasionally found under rocks in and around ship harbors; coiling always dextral; noted at Bodega Head intermixed with *Protolaeospira eximia*, *Simplaria pseudomilitaris*, and the sinistrally coiled *Paradexiospira vitrea* (Knight-Jones, unpublished data).

ACKNOWLEDGMENTS

In compiling this section, we have drawn extensively from published literature—especially the many monographs included in the four volumes on polychaetes published as part of the *Taxonomic Atlas of the Santa Maria Basin and Western Santa Barbara Channel* (Blake et al. 1994, 1995, 1996, 2000). To that end, we are grateful for the contributions of Jerry Kudenov, Leslie Harris, Mary Petersen, and Brigitte Hilbig, whose excellent keys, well-written descriptions, illustrations, and attention to detail have made it possible to orient ourselves in the compilation of this very complex fauna. We also thank Paul Valentich Scott of the Santa Barbara Museum for permission to cite this work extensively. The following individuals are thanked for providing specimens and contributing information: Kirk Fitzhugh, Leslie Harris, Phyllis K. Knight-Jones, E. Wyn Knight-Jones, and Thomas H. Perkins. Nancy J. Maciolek and Isabelle P. Williams read and commented on drafts of this section. Preparation of this section is based upon work supported by the National Science Foundation under Grant No. DEB-0118693 (PEET) to James A. Blake, University of Massachusetts, Boston.

References

Abbott, D. P. and D. J. Reish. 1980. Polychaeta: The marine annelid worms, pp. 448–489. In Intertidal invertebrates of California. Morris, R. H., D. P. Abbott, and E.C. Haderlie, eds. Stanford, CA: Stanford University Press, 690 pp.

Bailey-Brock, J. H. and O. Hartman. 1987. Class Polychaeta. In Reef and shore fauna of Hawaii, Section 2: Platyhelminthes through Phoronida, and Section 3: Sipuncula through Annelida. D. M. Devaney and L. G. Eldredge, eds. Bishop Museum Special Publication 64(2&3): 1–461.

Banse. 1972a. On some species of Phyllodocidae, Syllidae, Nephtyidae, Goniadidae, Apistobranchidae, and Spionidae (Polychaeta) from the northeast Pacific Ocean. Pac. Sci. 26: 191–222.

Banse, K. 1972b Redescription of some species of Chone Kroeyer (Sabellidae, Polychaeta). Fishery Bull. 70: 459–495.

Banse, K. 1979. Sabellidae (Polychaeta) principally from the northeast Pacific Ocean. J. Fish. Res. Bd. Can. 36(8): 869–882.

Banse, K. and K. D. Hobson 1968. Benthic polychaetes from Puget Sound, Washington, with remarks on four other species. Proc. U.S. Nat. Mus. 125: 1–53.

Banse, K. and K. D. Hobson. 1974. Benthic errantiate polychaetes of British Columbia and Washington. Bull. Fish Res. Board Canada 185: 1–111.

Berkeley, E. and C. Berkeley. 1948. Canadian Pacific fauna 9. Annelida 9b(1). Polychaeta Errantia. Fish. Res. Bd. Canada, pp. 1–100.

Berkeley, E. and C. Berkeley. 1952. Canadian Pacific fauna 9. Annelida 9b(2). Polychaeta Sedentaria. Fish. Res. Bd. of Canada, pp. 1–139. (See Pettibone 1967, Proc. U.S. Natl. Mus. 119: 1–23, for review of the Berkeley type material.)

Blake, J. A. 1969. Reproduction and larval development of Polydora from northern New England (Polychaeta: Spionidae). Ophelia 7: 1–63.

Blake, J. A. 1971. Revision of the genus Polydora from the east coast of North America (Polychaeta: Spionidae). Smithson. Contrib. Zool. 75: 1–32.

Blake, J. A. 1975a. The larval development of Polychaeta from the northern California coast. III. Eighteen species of Errantia. Ophelia 14: 23–84.

Blake, J. A. 1975b. Phylum Annelida: Class Polychaeta. In Light's manual, intertidal invertebrates of the central California Coast. R. I. Smith and J. A. Carlton, eds. University of California Press, pp. 151–243.

Blake, J. A. 1980. The larval development of Polychaeta from the Northern California coast. IV. Leitoscoloplos pugettensis and Scoloplos acmeceps (Family Orbiniidae). Ophelia 19: 1–18.

Blake, J. A. 1994a. Chapter 3. Introduction to the Polychaeta. In Taxonomic atlas of the benthic fauna of the Santa Maria Basin and western Santa Barbara Channel. Vol. 4. J. A. Blake and B. Hilbig, eds. Santa Barbara, CA: Santa Barbara Museum of Natural History, pp. 39–114.

Blake, J. A. 1994b. Chapter 4. Family Phyllodocidae Savigny, 1818. In Taxonomic atlas of the benthic fauna of the Santa Maria Basin and western Santa Barbara Channel. Vol. 4. J. A. Blake and B. Hilbig, eds. Santa Barbara, CA: Santa Barbara Museum of Natural History, pp. 115–187.

Blake, J. A. 1994c. Chapter 10. Family Pilargidae Saint Joseph, 1899. In Taxonomic atlas of the benthic fauna of the Santa Maria Basin and western Santa Barbara Channel. Vol. 4. J. A. Blake and B. Hilbig, eds. Santa Barbara, CA: Santa Barbara Museum of Natural History, pp. 275–299.

Blake, J. A. 1995a. Chapter 2. Family Aphroditidae Malmgren, 1867. In Taxonomic atlas of the Santa Maria Basin and western Santa Barbara Channel. Vol. 5. Annelida Part 2. Polychaeta: Phyllodocida (Syllidae and Scale Bearing Families), Amphinomida, and Eunicida. J. A. Blake, B. Hilbig, P. H. Scott, eds. Santa Barbara, CA: Santa Barbara Museum of Natural History, pp. 99–104.

Blake, J. A. 1995b. Chapter 5. Family Pholoidae Kinberg, 1858. In Taxonomic atlas of the Santa Maria Basin and western Santa Barbara Channel. Vol. 5. Annelida Part 2. Polychaeta: Phyllodocida (Syllidae and Scale Bearing Families), Amphinomida, and Eunicida. J. A. Blake, B. Hilbig, P. H. Scott, eds. Santa Barbara, CA: Santa Barbara Museum of Natural History, pp. 175–188.

Blake, J. A. 1995c. Chapter 6. Family Sigalionidae. In Taxonomic atlas of the Santa Maria Basin and western Santa Barbara Channel. Vol. 5. Annelida Part 2. Polychaeta: Phyllodocida (Syllidae and Scale Bearing Families), Amphinomida, and Eunicida. J. A. Blake, B. Hilbig, P. H. Scott, eds. Santa Barbara, CA: Santa Barbara Museum of Natural History, pp. 189–206.

Blake, J. A. 1996a. Chapter 1. Family Orbiniidae. In Taxonomic atlas of the Santa Maria Basin and western Santa Barbara Channel. Vol. 6. Annelida Part 3. Polychaeta: Orbiniidae to Cossuridae. J. A. Blake, B. Hilbig, P. H. Scott, eds. Santa Barbara, CA: Santa Barbara Museum of Natural History, pp. 1–26.

Blake, J. A. 1996b. Chapter 2. Family Paraonidae. In Taxonomic atlas of the Santa Maria Basin and western Santa Barbara Channel. Vol. 6. Annelida Part 3. Polychaeta: Orbiniidae to Cossuridae. J. A. Blake, B.

Hilbig, P. H. Scott, eds. Santa Barbara, CA: Santa Barbara Museum of Natural History, pp. 27–70.

Blake, J.A. 1996c. Chapter 4. Family Spionidae Grube, 1850, including a review of the genera and species from California and a revision of the genus Polydora Bosc, 1802. In Taxonomic atlas of the Santa Maria Basin and western Santa Barbara Channel. Vol. 6. Annelida Part 3. Polychaeta: Orbiniidae to Cossuridae. J. A. Blake, B. Hilbig, P. H. Scott, eds. Santa Barbara, CA: Santa Barbara Museum of Natural History, pp. 81–223.

Blake, J. A. 1996d. Chapter 7. Family Magelonidae. In Taxonomic atlas of the Santa Maria Basin and western Santa Barbara Channel. Vol. 6. Annelida Part 3. Polychaeta: Orbiniidae to Cossuridae. J. A. Blake, B. Hilbig, P. H. Scott, eds. Santa Barbara, CA: Santa Barbara Museum of Natural History, pp. 253–262.

Blake, J. A. 1996e. Chapter 6. Family Chaetopteridae. In Taxonomic atlas of the Santa Maria Basin and western Santa Barbara Channel. Vol. 6. Annelida Part 3. Polychaeta: Orbiniidae to Cossuridae. J. A. Blake, B. Hilbig, P. H. Scott, eds. Santa Barbara, CA: Santa Barbara Museum of Natural History, pp. 233–251.

Blake, J. A. 1996f. Chapter 3. Family Apistobranchidae. In Taxonomic atlas of the Santa Maria Basin and western Santa Barbara Channel. Vol. 6. Annelida Part 3. Polychaeta: Orbiniidae to Cossuridae. J. A. Blake, B. Hilbig, P. H. Scott, eds. Santa Barbara, CA: Santa Barbara Museum of Natural History, pp. 71–79.

Blake, J. A. 1996g. Chapter 8. Family Cirratulidae. In Taxonomic atlas of the Santa Maria Basin and western Santa Barbara Channel. Vol. 6. Annelida Part 3. Polychaeta: Orbiniidae to Cossuridae. J. A. Blake, B. Hilbig, P. H. Scott, eds. Santa Barbara, CA: Santa Barbara Museum of Natural History, pp. 263–384.

Blake, J. A. 2000a. Chapter 1. Family Flabelligeridae Saint Joseph, 1894. In Taxonomic atlas of the Santa Maria Basin and western Santa Barbara Channel. Vol. 7. Annelida Part 4. Polychaeta: Flabelligeridae to Sternaspidae. J. A. Blake, B. Hilbig, P. V. Scott, eds. Santa Barbara, CA: Santa Barbara Museum of Natural History, pp. 1–23.

Blake, J. A. 2000b Chapter 6. Family Scalibregmatidae Malmgren, 1867. In Taxonomic atlas of the Santa Maria Basin and western Santa Barbara Channel. Vol. 7. Annelida Part 4. Polychaeta: Flabelligeridae to Sternaspidae. J. A. Blake, B. Hilbig, P. V. Scott, eds. Santa Barbara, CA: Santa Barbara Museum of Natural History, pp. 129–144.

Blake, J. A. 2000c. Chapter 7. Family Opheliidae Malmgren, 1867. In Taxonomic atlas of the Santa Maria Basin and western Santa Barbara Channel. Vol. 7. Annelida Part 4. Polychaeta: Flabelligeridae to Sternaspidae. J. A. Blake, B. Hilbig, P. V. Scott, eds. Santa Barbara, CA: Santa Barbara Museum of Natural History, pp. 145–168.

Blake, J. A. 2000d. Chapter 4. Family Capitellidae Grube, 1862. In Taxonomic atlas of the Santa Maria Basin and western Santa Barbara Channel. Vol. 7. Annelida Part 4. Polychaeta: Flabelligeridae to Sternaspidae. J. A. Blake, B. Hilbig, P. V. Scott, eds. Santa Barbara, CA: Santa Barbara Museum of Natural History, pp. 47–96.

Blake, J. A. 2000e. Chapter 5. Family Oweniidae Rioja, 1917. In Taxonomic atlas of the Santa Maria Basin and western Santa Barbara Channel. Vol. 7. Annelida Part 4. Polychaeta: Flabelligeridae to Sternaspidae. J. A. Blake, B. Hilbig, P. V. Scott, eds. Santa Barbara, CA: Santa Barbara Museum of Natural History, pp. 97–127.

Blake, J. A. 2006. Spionida. Chapter 13. In Reproductive Biology and Phylogeny of Annelida. Vol. 4. G. W. Rouse and F. Pleijel, eds. G. M. Jamieson, series ed. Enfield, NH, USA; Plymouth, U.K.: Science Publishers, Inc., pp. 565–638.

Blake, J. A. and P. A. Arnofsky. 1999. Reproduction and larval development of the spioniform Polychaeta with application to systematics and phylogeny. In Reproductive strategies and developmental patterns in annelids. A. W. C. Dorresteijn and W. Westheide, eds. Hydrobiologia 402: 57–106.

Blake, J. A. and J. W. Evans 1973. Polydora and related genera as borers in mollusk shells and other calcareous substrates. Veliger 15: 23–249.

Blake, J.A. and B. Hilbig (eds.). 1994. Taxonomic atlas of the Santa Maria Basin and western Santa Barbara Channel. Vol. 4. Annelida Part 1. Oligochaeta and Polychaeta: Phyllodocida (Phyllodocidae to Paralacydoniidae). Santa Barbara, CA: Santa Barbara Museum of Natural History, pp. 1–379.

Blake, J. A., B. Hilbig, and P. H. Scott (eds.). 1995. Taxonomic atlas of the Santa Maria Basin and western Santa Barbara Channel. Vol. 5. Annelida Part 2. Polychaeta: Phyllodocida (Syllidae and Scale Bearing Families). Santa Barbara, CA: Santa Barbara Museum of Natural History, pp. 1–378.

Blake, J. A., B. Hilbig, and P. H. Scott (eds.). 1996. Taxonomic atlas of the Santa Maria Basin and western Santa Barbara Channel. Vol. 6. Annelida Part 3. Polychaeta: Orbiniidae to Cossuridae. Santa Barbara, CA: Santa Barbara Museum of Natural History, pp. 1–418.

Blake, J. A., B. Hilbig, and P. V. Scott (eds.). 2000. Taxonomic atlas of the Santa Maria Basin and western Santa Barbara Channel. Vol. 7. Annelida Part 4. Polychaeta: Flabelligeridae to Sternaspidae. Santa Barbara, CA: Santa Barbara Museum of Natural History, pp. 1–348.

Blake, J. A. and K. H. Woodwick. 1971. A review of the genus *Boccardia* Carazzi (Polychaeta: Spionidae) with descriptions of two new species. Bull. So. Calif. Acad. Sci. 70: 31–42.

Blake, J. A. and K. H. Woodwick. 1972. New species of *Polydora* from the coast of California (Polychaeta: Spionidae). Bull. So.. Calif. Acad. Sci. 70: 72–79.

Böggemann, M. 2002. Revision of the Glyceridae Grube 1850 (Annelida: Polychaeta). Abh. Senckenberg. Naturforsch. Ges. 555: 1–249.

Böggemann, M. 2005. Revision of the Goniadidae (Annelida, Polychaeta). Abhandlungen des Naturwissenschaftlichen Vereins in Hamburg 39: 1–354.

Bush, K. J. 1905. Tubicolous annelids of the tribes Sabellides and Serpulides from the Pacific Ocean. Harriman Alaska Exped., N.Y. 12: 169–355.

Cohen, A. N. and J. T. Carlton, 1995. Nonindigenous aquatic species in a United States Estuary: A case study of the biological invasions of the San Francisco Bay and Delta. Report prepared for the U.S. Fish and Wildlife Service, Washington, D.C. and The National Sea Grant Program, Connecticut Sea Grant (NOAA Grant No. NA36RG0467). 241 pp.

Dales, R. P. 1957. Pelagic polychaetes of the Pacific Ocean. Bull. Scripps Inst. Oceanogr. 7: 99–168.

Dales, R. P. and G. Peter, 1972. A synopsis of the pelagic Polychaeta. J. Nat. Hist. 6: 55–92.

Day, J. H. 1967. A monograph on the Polychaeta of southern Africa Brit. Mus. Nat. Hist. London Publ. no. 656: 1–878, 2 volumes.

Eibye-Jacobsen, D. 1991. Observations on setal morphology in Phyllodocidae (Polychaeta: Annelida), with some taxonomic considerations. Bull. Mar. Sci. 48: 530–543.

Eibye-Jacobsen, D. and R. M. Kristensen. 1994. A new genus and species of Dorvilleidae (Annelida, Polychaeta) from Bermuda, with a phylogenetic analysis of Dorvilleidae, Iphitimidae, and Dinophilidae. Zool. Scr. 23: 107–131.

Ewing, R. M. 1984. Chapter 14. Capitellidae. In Taxonomic guide to the polychaetes of the northern Gulf of Mexico. Vol. 2. J. M. Uebelacker and P. G. Johnson, eds. Barry A. Vittor & Associates, Inc. Mobil, 47 pp.

Fauchald, K. 1968. Onuphidae (Polychaeta) from western Mexico. Allan Hancock Monographs in Marine Biology 3: 1–82.

Fauchald, K. 1970. Polychaetous annelids of the families Eunicidae, Lumbrineridae, Iphitimidae, Arabellidae, Lysaretidae, and Dorvilleidae from western Mexico. Allan Hancock Monogr. Mar. Biol. 5: 1–335.

Fauchald, K. 1977. The polychaete worms, definitions and keys to the orders, families, and genera. Natural History Museum of Los Angeles County, Science Series 28: 1–188.

Fauchald, K. 1982. Revision of *Onuphis, Nothria,* and *Paradiopatra* (Polychaeta: Onuphidae) based upon type material. Smithsonian Contributions to Zoology 356: 1–109.

Fauchald, K. 1992. A review of the genus *Eunice* (Polychaeta Eunicidae) based upon type material. Smithsonian Contributions to Zoology 523: 1–422.

Fauchald, K. and P. A. Jumars. 1979. The diet of worms: a study of polychaete feeding guilds. Oceanography and Marine Biology Annual Review 17: 193–284.

Fauchald, K. G. Rouse. 1997. Polychaete systematics: Past and present. Zool. Scr. 26: 71–138.

Fauvel, P. 1923. Polychètes errantes. Faune de France 5: 1–488.

Fauvel, P. 1927. Polychètes sédentaires. Addenda aux Errantes, Archiannélides, Myzostomaires. Faune de France 16: 1–494.

Fitzhugh, K. 1989. A systematic revision of the Sabellidae-Caobangiidae-Sabellongidae complex (Annelida: Polychaeta). Bull. Am. Mus. Nat. Hist. 192: 1–104.

Fitzhugh, K. 1990a. A revision of the genus *Fabricia* Blainville, 1828 (Polychaeta: Sabellidae). Sarsia 75: 1–16.

Fitzhugh, K. 1990b. *Fabricinuda,* a new genus of Fabriciinae (Polychaeta: Sabellidae). Proc. Biol. Soc. Wash. 103(1): 161–178.

Fitzhugh, K. 1992. Species of *Fabriciola* Friedrich, 1939 (Polychaeta: Sabellidae: Fabriciinae), from the California coast. Pacific Science 46(1): 68–76.

Fitzhugh, K. 1993. *Novafabricia brunnea* (Hartman, 1969), new combination, with an update on relationships among Fabriciinae taxa (Polychaeta: Sabellidae). Contr. Sci. 438: 1–12.

Glasby, C. J., P. A. Hutchings, K. Fauchald, H. Paxton, G. W. Rouse, C. Watson Russell, and R. S. Wilson. 2000. 1. Class Polychaeta. In Polychaetes and allies: The Southern Synthesis. Fauna of Australia. Vol. 4. P. L. Beasley, G. J. B. Ross, and C. J. Glasby, eds. Polychaeta, Myzostomida, Pogonophora, Echiura, Sipuncula. CSIRO Publishing. Melbourne. xii + 465 pp.

Hartman, O. 1936. New species of Spionidae (Annelida: Polychaeta) from the coast of California. Univ. Calif. Pub. Zool. 41: 45–52.

Hartman, O. 1939. Polychaetous annelids. Part I. Aphroditidae to Pisionidae. Allan Hancock Pac. Exped. 71–156.

Hartman, O. 1940. Polychaetous annelids. Part II. Chrysopetalidae to Goniadidae. Allan Hancock Pac. Exped. 7: 173–287.

Hartman, O. 1941a. Polychaetous annelids. Part III. Spionidae. Some contributions to the biology and life history of Spionidae from California. Allan Hancock Pac. Exped. 7: 289–323.

Hartman, O. 1941b. Polychaetous annelids. Part IV. Pectinariidae. Allan Hancock Pac Exped. 7: 325–345.

Hartman, O. 1944a. Polychaetous annelids. Part V. Eunicea. Allan Hancock Pac. Exped. 10: 1–238.

Hartman, O. 1944b. Polychaetous annelids from California including the descriptions of two new genera and nine new species. Allan Hancock Pac. Exped. 10: 239–310.

Hartman, O. 1944c. Polychaetous annelids. Part VI. Paraonidae, Magelonidae, Longosomidae, Ctenodrilidae, and Sabellariidae. Allan Hancock Pac. Exped. 10: 311–390.

Hartman, O. 1947a. Polychaetous annelids. Part VII. Capitellidae. Allan Hancock Pac Exped. 10: 391–482.

Hartman, O. 1947b. Polychaetous annelids. Part VIII. Pilargidae. Allan Hancock Pac. Exped. 10: 483–524.

Hartman, O. 1950. Goniadidae, Glyceridae, and Nephtyidae. Allan Hancock Pac. Exped. 15: 1–182.

Hartman, O. 1951a. Literature of the Polychaetous annelids. Los Angeles: privately published, 290 pp.

Hartman, O. 1951b. Fabricinae (Feather-duster Polychaetous annelids) in the Pacific. Pac. Sci. 5: 379–391.

Hartman, O. 1957. Orbiniidae, Apistobranchidae, Paraonidae, and Longosomidae. Allan Hancock Pac. Exped. 15: 211–394.

Hartman, O. 1959. Catalogue of the Polychaetous annelids of the world. Parts I and II. Allan Hancock Found. Pub. Occ. Pap. 23: 1–628.

Hartman, O. 1961. Polychaetous annelids from California. Allan Hancock Pac. Exped. 25: 1–226.

Hartman, O. 1965. Catalogue of the Polychaetous annelids of the world. Supplement 1960–1965 and Index, Allan Hancock Found. Publ., Occ. Pap. 23: 197 pp.

Hartman, O. 1968. Atlas of the Errantiate Polychaetous annelids from California. Los Angeles: Allan Hancock Found. Univ. South. Calif., 828 pp.

Hartman, O. 1969. Atlas of the Sedentariate Polychaetous annelids of California. Los Angeles: Allan Hancock Found. Univ. South. Calif., 812 pp.

Hartmann-Schröder, G. 1996. Annelida, Borstenwürmer, Polychaeta. Die Tierwelt Deutschlands 58: 1–648.

Hermans, C. O. 1978. Metamorphosis in the opheliid polychaete *Armandia brevis.* In Settlement and metamorphosis of marine invertebrate larvae. F. Chia and M. E. Rice, eds. pp. 113–126. New York: Elsevier.

Hilbig, B. 1994a. Chapter 9. Family Hesionidae Sars, 1862. In Taxonomic atlas of the benthic fauna of the Santa Maria Basin and western Santa Barbara Channel. Vol. 4. J. A. Blake and B. Hilbig, eds. Santa Barbara, CA: Santa Barbara Museum of Natural History, pp. 243–269.

Hilbig B. 1994b. Chapter 12. Family Nereididae Johnson, 1845. In Blake, J.A. and B. Hilbig (eds.), Taxonomic atlas of the benthic fauna of the Santa Maria Basin and western Santa Barbara Channel. Vol. 4. Santa Barbara Museum of Natural History, pp. 301–327.

Hilbig, B. 1994c. Chapter 7. Family Goniadidae Kinberg, 1866. In Taxonomic atlas of the benthic fauna of the Santa Maria Basin and western Santa Barbara Channel. Vol. 4. J. A. Blake and B. Hilbig, eds. Santa Barbara, CA: Santa Barbara Museum of Natural History, pp. 215–230.

Hilbig, B. 1994d. Chapter 6. Family Glyceridae Grube, 1850. In Blake, J.A. and B. Hilbig (eds.), Taxonomic atlas of the benthic fauna of the Santa Maria Basin and western Santa Barbara Channel. Vol. 4. Santa Barbara Museum of Natural History, pp. 197–214.

Hilbig, B. 1994e. Chapter 13. Family Nephtyidae Grube, 1850. In Taxonomic atlas of the benthic fauna of the Santa Maria Basin and western Santa Barbara Channel. Vol. 4. J. A. Blake and B. Hilbig, eds. Santa Barbara, CA: Santa Barbara Museum of Natural History, pp. 329–362.

Hilbig, B. 1995a. Family Onuphidae Kinberg, 1865. In Taxonomic atlas of the Santa Maria Basin and western Santa Barbara Channel. Vol. 5. Annelida Part 2. Polychaeta: Phyllodocida (Syllidae and Scale Bearing Families), Amphinomida, and Eunicida. J. A. Blake, B. Hilbig, and P. H. Scott, eds. Santa Barbara, CA: Santa Barbara Museum of Natural History, pp. 229–262.

Hilbig, B. 1995b. Family Eunicidae Savigny, 1818. In Taxonomic atlas of the Santa Maria Basin and western Santa Barbara Channel. Vol. 5. Annelida Part 2. Polychaeta: Phyllodocida (Syllidae and Scale Bearing Families), Amphinomida, and Eunicida. J. A. Blake, B. Hilbig, and P. H. Scott, eds. Santa Barbara, CA: Santa Barbara Museum of Natural History, pp. 263–278.

Hilbig, B. 1995c. Family Dorvilleidae Chamberlin, 1919. In Taxonomic atlas of the Santa Maria Basin and western Santa Barbara Channel. Vol. 5. Annelida Part 2. Polychaeta: Phyllodocida (Syllidae and Scale Bearing Families), Amphinomida, and Eunicida. J. A. Blake, B. Hilbig, and P. H. Scott, eds. Santa Barbara, CA: Santa Barbara Museum of Natural History, pp. 341–364.

Hilbig, B. 1995d. Family Lumbrineridae Malmgren, 1867, emended Orensanz 1990. In Taxonomic atlas of the Santa Maria Basin and western Santa Barbara Channel. Vol. 5. Annelida Part 2. Polychaeta: Phyllodocida (Syllidae and Scale Bearing Families), Amphinomida, and Eunicida. J. A. Blake, B. Hilbig, and P. H. Scott, eds. Santa Barbara, CA: Santa Barbara Museum of Natural History, pp. 279–313.

Hilbig, B. 1995e. Family Oenonidae Kinberg, 1865, emended Orensanz, 1990. In Taxonomic atlas of the Santa Maria Basin and western Santa Barbara Channel. Vol. 5. Annelida Part 2. Polychaeta: Phyllodocida (Syllidae and Scale Bearing Families), Amphinomida, and Eunicida. J. A. Blake, B. Hilbig, and P. H. Scott, eds. Santa Barbara, CA: Santa Barbara Museum of Natural History, pp. 315–339.

Hilbig, B. 1996. Family Cossuridae Day, 1963. In Taxonomic atlas of the Santa Maria Basin and western Santa Barbara Channel. Vol. 6. Annelida Part 3. Polychaeta: Orbiniidae to Cossuridae. J. A. Blake, B. Hilbig, and P. H. Scott, eds. Santa Barbara, CA: Santa Barbara Museum of Natural History, pp. 385–404.

Hilbig, B. 2000a. Family Ampharetidae Malmgren, 1867. In Taxonomic atlas of the Santa Maria Basin and western Santa Barbara Channel. Vol. 7. Annelida Part 4. Polychaeta: Flabelligeridae to Sternaspidae. J. A. Blake, B. Hilbig, and P. V. Scott, eds. Santa Barbara, CA: Santa Barbara Museum of Natural History, pp. 169–230.

Hilbig, B. 2000b. Family Terebellidae Grube, 1851. In Taxonomic atlas of the Santa Maria Basin and western Santa Barbara Channel. Vol. 7. Annelida Part 4. Polychaeta: Flabelligeridae to Sternaspidae. J. A. Blake, B. Hilbig, and P. V. Scott, eds. Santa Barbara, CA: Santa Barbara Museum of Natural History, pp. 231–294.

Hilbig, B. 2000c. Family Trichobranchidae Malmgren, 1866. In Taxonomic atlas of the Santa Maria Basin and western Santa Barbara Channel. Vol. 7. Annelida Part 4. Polychaeta: Flabelligeridae to Sternaspidae. J. A. Blake, B. Hilbig, and P. V. Scott, eds. Santa Barbara, CA: Santa Barbara Museum of Natural History, pp. 295–309.

Hobson, K. D. 1971. Some polychaetes of the superfamily Eunicea from the North Pacific and North Atlantic oceans. Proc. Biol. Soc. Wash. 83: 527–544.

Hobson, K. D. and K. Banse. 1981. Sedentariate and archiannelid polychaetes of British Columbia and Washington. Bull. Fish Res. Board Canada 209: 1–144.

Imajima, M. and O. Hartman. 1964. Polychaetous annelids of Japan. Allan Hancock Found., Pub. Occ. Pap. 26: 1–462 (2 vols.).

Imajima, M. and Y. Shiraki. 1982. Maldanidae (Annelida: Polychaeta) from Japan. Bull. Nat. Sci. Mus. 8(2): 7–88.

Johnson, H. P. 1901. The Polychaeta of the Puget Sound region. Proc. Boston Soc. Nat. Hist. 29: 381–437.

Kirtley, D. W. 1994. A review and taxonomic revision of the family Sabellariidae Johnston, 1865 (Annelida: Polychaeta). Science Series No. 1, Sabecon Press. Vero Beach, Florida, 223 pp.

Knight-Jones, E. W., P. Knight-Jones, and A. Nelson-Smith. 1995. Annelids (Phylum Annelida). In Handbook of the marine fauna of north-West Europe, P. J. Hayward and J. S. Ryland, eds. Oxford, England: Oxford University Press, 800 pp.

Knight-Jones, P. 1978. New Spirorbidae (Polychaeta: Sedentaria) from the East Pacific, Atlantic, Indian, and Southern Oceans. Zool. J. Linn. Soc. 64: 201–240.

Knight-Jones, P. 1984. A new species of Protoleodora (Spirorbidae: Polychaeta) from Eastern U.S.S.R., with a brief revision of related genera. Zool. J. Linn. Soc. 80: 109–120.

Knight-Jones, P., E. W. Knight-Jones, and R. P. Dales. 1979. Spirorbidae (Polychaeta: Sedentaria) from Alaska to Panama. J. Zool. Soc. London 189: 419–458.

Knight-Jones, P. and T. H. Perkins. 1998. A revision of Sabella, Bispira, and Stylomma (Polychaeta: Sabellidae). Zool. J. Linn. Soc. 123: 385–467.

Kudenov, J. D. 1994. Chapter 8. Family Sphaerodoridae Malmgren, 1867. In Taxonomic atlas of the benthic fauna of the Santa Maria Basin and western Santa Barbara Channel. Vol. 4. J. A. Blake and B. Hilbig, eds. Santa Barbara, CA: Santa Barbara Museum of Natural History, pp. 223–234.

Kudenov, J. D. 1995a. Family Amphinomidae Lamarck, 1818. In Taxonomic atlas of the Santa Maria Basin and western Santa Barbara Channel. Vol. 5. Annelida Part 2. Polychaeta: Phyllodocida (Syllidae and Scale Bearing Families), Amphinomida, and Eunicida. J. A. Blake, B. Hilbig, and P. H. Scott, eds. Santa Barbara, CA: Santa Barbara Museum of Natural History, pp. 207–215.

Kudenov, J. D. 1995b. Family Euphrosinidae Williams, 1851. In Taxonomic atlas of the Santa Maria Basin and western Santa Barbara Channel. Vol. 5. Annelida Part 2. Polychaeta: Phyllodocida (Syllidae and Scale Bearing Families), Amphinomida, and Eunicida. J. A. Blake, B. Hilbig, and P. H. Scott, eds. Santa Barbara, CA: Santa Barbara Museum of Natural History, pp. 217–228.

Kudenov, J. D. and L. H. Harris. 1995. Family Syllidae Grube, 1850. In Taxonomic atlas of the Santa Maria Basin and western Santa Barbara Channel. Vol. 5. Annelida Part 2. Polychaeta: Phyllodocida (Syllidae and Scale Bearing Families), Amphinomida, and Eunicida. J. A. Blake, B. Hilbig, and P. H. Scott, eds. Santa Barbara, CA: Santa Barbara Museum of Natural History, pp. 1–97.

Light, W. J. 1977. Spionidae (Annelida: Polychaeta) from San Francisco Bay, California: A revised list with nomenclatural changes, new records, and comments on related species from the northeastern Pacific Ocean. Proc. Biol. Soc. Wash. 90: 66–88.

Light, W. J. 1978. Spionidae Polychaeta Annelida. In Invertebrates of the San Francisco Bay estuary system. W. L. Lee, ed. California Academy of Sciences. Pacific Grove, CA: Boxwood Press, 211 pp.

Lovell, L. L. 1997. A review of six species of Nephtys (Cuvier, 1817) (Nephtyidae: Polychaeta) described from the eastern Pacific. Bull. Mar. Sci. 60: 350–363.

Moore, J. P. 1923. The polychaetous annelids dredged by the U.S.S. Albatross off the coast of southern California in 1904. Spionidae to Sabellariidae. Proc. Acad. Nat. Sci. Philadelphia 75: 179–259.

Nygren, A. 2004. Revision of Autolytinae (Syllidae: Polychaeta). Zootaxa 680: 1–314.

Nygren, A. and L. Gidholm. 2001. Three new species of Proceraea (Polychaeta: Syllidae: Autolytinae) from Brazil and the United States, with a synopsis of all Proceraealike taxa. Ophelia 54: 177–191.

O'Connor, B. D. S. 1986. The Glyceridae (Polychaeta) of the North Atlantic and Mediterranean, with descriptions of two new species. Jour. Nat. Hist. 21: 167–189.

Orensanz, J. M. 1990. The eunicimorph polychaete annelids from Antarctic and Subantarctic seas. With an addenda to the Eunicemorpha of Argentina, Chile, New Zealand, Australia, and southern Indian Ocean. Biology of Antarctic Seas XXI. Antarctic Res. Ser. 52: 1–183.

Perkins, T. H. 1984. Revision of Demonax Kinberg, Hypsicomus Grube, and Notaulax Tauber, with a review of Megalomma Johansson from Florida (Polychaeta: Sabellidae). Proc. Biol. Soc. Wash. 97(2): 285–368.

Petersen, M. H. 2000. Family Sternaspidae Carus, 1863, including a review of described species and comments on some points of confusion. In Taxonomic atlas of the Santa Maria Basin and western Santa Barbara Channel. Vol. 7. Annelida Part 4. Polychaeta: Flabelligeridae to Sternaspidae. J. A. Blake, B. Hilbig, and P. V. Scott, eds. Santa Barbara, CA: Santa Barbara Museum of Natural History, pp. 311–336.

Pettibone, M. H. 1953. Some scale-bearing polychaetes of Puget Sound and adjacent waters. University of Washington Press, 89 pp.

Pettibone, M. H. 1957. North American genera of the family Orbiniidae (Annelida Polychaeta), with descriptions of new species. J. Wash. Acad. Sci. 47: 159–167.

Pettibone, M. H. 1963. Marine polychaete worms of the New England region. l. Aphroditidae through Trochochaetidae. Bull. U.S. Nat. Mus. 227: 1–356.

Pettibone, M. H. 1976. Contribution to the polychaete family Trochochaetidae Pettibone. Smiths. Contr.. Zool. 230: 1–21.

Pettibone, M. H. 1982. Classification of Polychaeta. In Parker, S.P. (ed.), Synopsis of living organisms. McGraw-Hill 2: 3–43.

Pleijel, F. 1991. Phylogeny and classification of the Phyllodocidae. Zool. Scr. 20: 225–261.

Pleijel, F. 1998. Phylogeny and classification of Hesionidae. Zool. Scr. 27: 89–163.

Radashevsky, V. I. 1993. Revision of the genus Polydora and related genera from the north west Pacific (Polychaeta: Spionidae). Pub. Seto Mar. Biol. Lab. 36: 1–60.

Rouse, G. W. and K. Fauchald. 1997. Cladistics and polychaetes. Zool. Scr. 24: 269–301.

Rouse, G. W. and F. Pleijel. 2001. Polychaetes. Oxford University Press, Inc., New York. 354 pp.

Ruff, R. E. 1995. Family Polynoidae Malmgren, 1867. In Taxonomic atlas of the Santa Maria Basin and western Santa Barbara Channel. Vol. 5. Annelida Part 2. Polychaeta: Phyllodocida (Syllidae and Scale Bearing Families), Amphinomida, and Eunicida. J. A. Blake, B. Hilbig, and P. H. Scott, eds. Santa Barbara, CA: Santa Barbara Museum of Natural History, pp. 105–166.

San Martín, G. 2003. Annelida Polycaheta II. Syllidae. In *Fauna Iberica*. Vol. 21. M. A. Ramos et al., eds. Madrid: Museo Nacional de Ciencia Naturales, CSIC, 554 pp.

SCAMIT, 2001. A taxonomic listing of soft bottom macro- and mega-invertebrates from infaunal and epifaunal monitoring programs in the Southern California Bight. Edition 4. San Pedro, CA: The Southern California Association of Marine Invertebrate Taxonomists, 196 pp.

Strelzov, V. E. 1973. Polychaete worms of the family Paraonidae Cerruti, 1909—Polychaeta Sedentaria. Akad. Leningrad: Nauk S.S.S.R., 170 pp. [translated 1979]

ten Hove, H. A. 1984. Toward a phylogeny in serpulids (Annelida; Polychaeta). In P. A. Hutchings (ed.), Proceedings of the First International Polychaete Conference, Sydney, Australia, July 1983. The Linnaean Society of New South Wales, pp. 181–196.

Vine, P. J. 1977. The marine fauna of New Zealand: Spirorbinae (Polychaeta: Serpulidae). Mem. N. Z. Oceanogr. Inst. 68: 1–68.

Westheide, W. 1988. Polychaeta. In Introduction to the study of meiofauna. R. P. Higgins and H. Thiel, eds. Washington, D.C.: Smithsonian Institution Press, pp. 332–344.

Westheide, W. 1990. Polychaetes: Interstitial Families. Keys and notes for the identification of the species. Universal Book Services, Oegstgeest, 152 pp.

Woodwick, K. H. 1963a. Comparison of *Boccardia columbiana* Berkeley and *Boccardia proboscidea* Hartman (Annelida, Polychaeta). Bull. So. Calif. Acad. Sci. 62: 132–139.

Woodwick, K. H. 1963b: Taxonomic revision of two polydorid species (Annelida, Polychaeta, Spionidae). Proc. Biol. Soc. Wash. 76: 209–216.

Arthropoda

(Plates 182–348)

Introduction

JOEL W. MARTIN

Arthropods are the dominant group of metazoans on earth, and this has been the case for hundreds of millions of years. This dominance is evident in terms of the number of described species, global biomass, ecological import, and just about any other category one can imagine. Indeed, even if our comparison extended to all forms of life instead of being restricted to metazoans, the only exception would be for the poorly understood groups of "microbes" and especially for phage (viruses that attack prokaryotes), whose incredible diversity is just beginning to be appreciated (Rohwer and Edwards 2002; Rohwer 2003).

For metazoans, in terms of the numbers of described species, the insects rule. Of the estimated 1.75 million described species of plants and animals on Earth (Gleich et al. 2002; Brusca and Brusca 2003; Chapman 2005), more than 1.33 million are metazoans; more than half of the metazoans are insects, with the vast majority of those being beetles. Arthropods dominate not only in species but also in numbers; who could begin to estimate the number of ants? (But others have made the same claim for the nauplius larval stage of crustaceans; see Martin and Davis 2001: 2, citing Fryer 1987).

Arthropods also include other extremely significant groups. Foremost among these in terms of species numbers are the incredibly diverse chelicerate groups (including not only spiders, but also ticks, mites, scorpions, xiphosurans, opilionids, and many other lesser-known groups), the crustaceans (often described as the marine equivalent of the insects in terms of ecological and numerical dominance), and the myriapods (millipedes and centipedes). Apart from insects, chelicerates, crustaceans, and myriapods, a host of other chitin-enclosed creatures exist that may, or may not, be related (see Brusca and Brusca 2003). In terms of marine metazoans, mention also must be made of the extinct trilobites, yet another arthropod group that included what were, at one time, major players in the world oceans.

With so much of the planet's biodiversity contained within the Arthropoda, one might think that the group is well studied, with major relationships well resolved. However, our current understanding of arthropod phylogeny and the relationships among (and within) the many groups of arthropods is far from it (see Table 1). Are arthropods part of a larger clade that includes all other "molting" organisms, the "Ecdysozoa" hypothesis (Aguinaldo et al. 1997; Mallatt et al. 2004)? Or is the older "Articulata" hypothesis (uniting annelids and arthropods) more accurate (Scholtz 2002)? Do the various arthropod groups comprise a monophyletic group (the currently favored hypothesis), or has the combination of a bilateral segmented body, pronounced tagmosis, well-developed exoskeleton (and related modifications), and other features (Brusca and Brusca 2003: 475) arisen more than once such that there are several "arthropodous" phyla rather than a single phylum? Did crustaceans give rise to insects, such that the world's most numerous creatures are actually just an outgrowth of one branch of the crustaceans (Brusca 2000; Regier et al. 2005)? These are just some of the seemingly persistent questions about arthropods and their relationships, many of which are now being addressed with molecular methodology (Giribet et al. 2001; Regier and Schultz 1997, 1998a, b, 2001; Regier et al. 2005, and papers cited therein) and with studies of evolutionary development (Averof and Akam 1993, 1995a, b). Far from being a well-studied, well-resolved group, the arthropods continue to present questions and problems that at times seem nowhere close to being resolved.

Arthropods share several key features, most obvious of which perhaps are the well-developed exoskeleton (necessitating growth by ecdysis and all that goes with that process), obvious segmentation, high degree of tagmatization (specialization of body regions), ventral nerve cord, more or less open circulatory system, and (usually) compound paired eyes. They have traditionally been grouped or allied with other "cuticle-bearing" groups, such as tardigrades (Garey et al. 1996) and onychophorans (and indeed some workers treat these three groups together as the "Panarthropoda"), and most workers have assumed a relationship to, if not a descent from, the annelids. A traditional approach (Brusca and Brusca 2003) would recognize five major arthropodan subphyla: Trilobitomorpha, Crustacea, Hexapoda (the insects and their kin), Myriapoda, and Cheliceriformes. But although the monophyly of the arthropods as a whole is today widely accepted, the exact nature of the relationships among these five groups remains unclear. Some compelling molecular evidence indicates that hexapods and crustaceans are more closely related to each other than to other arthropod subphyla (Friedrich and Tautz 1995; Boore et al. 1995, 1998; Giribet et al. 2001, 2005; Regier and Schultz 2001;

TABLE 1

Classification of the Phylum Arthropoda Used in this Manual

Subphylum Crustacea

  Class Branchiopoda

    Subclass Sarsostraca

      Order Anostraca (fairy shrimp)

    Subclass Phyllopoda

      Order Notostraca (tadpole shrimp)

      Order Diplostraca (water fleas and clam shrimp)

  Class Cephalocarida (cephalocarids)

  Class Thecostraca

    Infraclass Cirripedia (barnacles)

  Class Branchiura (fish lice)

  Class Mystacocarida (mystacocarids)

  Class Copepoda (copepods)

  Class Ostracoda (ostracodes)

  Class Malacostraca (crabs, krill, pill bugs, shrimp, and relatives)

    Subclass Eumalacostraca

      Superorder Eucarida

        Order Decapoda (crabs, shrimp, and relatives)

        Order Euphausiacea (krill)

      Superorder Peracarida

        Order Mysidacea (opossum shrimp, mysids)

        Order Cumacea (cumaceans)

        Order Isopoda (pillbugs and sowbugs)

        Order Tanaidacea (tanaidaceans)

        Order Amphipoda (amphipods, scuds, beach hoppers)

    Subclass Hoplocarida

      Order Stomatopoda (mantis shrimp)

    Subclass Phyllocarida

      Order Leptostraca (leptostracans)

Subphylum Chelicerata

  Class Pycnogonida (sea spiders)

  Class Arachnida

    Order Acari (mites)

    Order Pseudoscorpiones (pseudoscorpions)

Subphylum Hexapoda

  Class Insecta

    Order Archaeognatha (jumping bristletails)

    Order Collembola (springtails)

    Order Hemiptera (water boatmen)

    Order Diptera (flies and midges)

    Order Dermaptera (earwigs)

    Order Coleoptera (beetles)

Subphylum Myriapoda

  Class Chilopoda (centipedes)

Regier et al. 2005), which results in the introduction of terms such as "Pancrustacea," "Tetraconata," or "Cormogonida" to accommodate some of the newly recognized clades (reviewed by Giribet et al. 2005), although this has not been rigorously tested using morphological data and modern phylogenetic methods.

Ironically, although the name "Mandibulata" (encompassing insects, crustaceans, and myriapods) that appeared in the last edition of *Light's Manual* is today almost never used, we have to some extent come full circle in that there does appear to be a close relationship between crustaceans and insects. The former grouping Atelocerata, uniting insects and myriapods, is now—like Mandibulata and Uniramia—only of historical interest. Placement of the myriapods with either the "Pancrustacea" or with chelicerates (sometimes called the "Paradoxopoda hypothesis"; Mallatt et al. 2004) remains one of the most vexing questions.

In the following sections, crustaceans, as the most important group of marine arthropods, are treated in some detail. Remaining arthropod chapters cover the chelicerates and selected hexapods and myriapods. Although chelicerates and insects are unquestionably dominant in terrestrial ecosystems, they take a back seat to the crustaceans in the marine world. Yet even here, the presence of terrestrial arthropods (in addition to the dominant crustaceans) is not only felt but integral to the balanced marine shallow water ecosystems of California and Oregon.

Crustacea

Although insects clearly are the most numerically dominant group of animals on earth, a different accolade must go to the crustaceans. As noted by Martin and Davis (2001) in their introduction to an updated classification of the Crustacea (a publication on which I rely heavily in this brief overview), "no group of plants or animals on the planet exhibits the range of morphological diversity seen among the extant Crustacea. This morphological diversity . . . is what makes the study of crustaceans so exciting. Yet it is also what makes deciphering the phylogeny of the group and ordering them into some sort of coherent classification so difficult." Crustaceans exhibit so many different shapes, sizes, and appendages; are found in so many habitats; and occupy so many ecological roles that their phylogeny and classification have proved elusive for more than 200 years.

The number of described extant Crustacea species is approximately 68,000 (Brusca and Brusca 2003); the actual number of species is, of course, far more (see Martin and Davis 2006). Among arthropod groups, this places crustaceans third in number of described species, following insects and chelicerates. But in terms of morphological diversity, they are unrivaled.

Relative to crustacean morphological diversity, Martin and Davis (2001, and references therein) state that "the known size of crabs now ranges from a maximum leg span of approximately 4 m in the giant Japanese spider crab *Macrocheira kaempferi,* and a maximum carapace width of 46 cm in the giant Tasmanian crab *Pseudocarcinus gigas,* to a minimum of 1.5 mm across the carapace for a mature ovigerous female pinnotherid, *Nannotheres moorei,* the smallest known species of crab. An ovigerous hermit crab (probably genus *Pygmaeopagurus*) with a shield length of only 0.76 mm taken from dredge samples in the Seychelles might hold the record for decapods, and of course much smaller crustaceans exist. Tantulocarids, recently discovered parasites found on other deep-sea crustaceans, are so small that they are sometimes found attached to the aesthetascs of the antennule of copepods; the total body length of *Stygotantulus stocki* is only 94 μm from tip of rostrum to end of caudal rami. In terms of biomass, that of the Antarctic krill *Euphausia superba* has been estimated at 500 million tons at any given time, probably surpassing the biomass of any other group of metazoans. In terms of sheer numbers, the crustacean nauplius has been called 'the most abundant type of multicellular animal on earth' (Fryer 1987). Crustaceans have been found in virtually every imaginable habitat, have been mistaken for molluscs, worms, and other distantly related animals, and continue to defy our attempts to force them into convenient taxonomic groupings."

Martin and Davis (2001) go on to say, "the inclusion of pentastomids among the Crustacea takes the known morphological diversity and lifestyle extremes of the Crustacea—already far greater than for any other taxon on earth—to new heights. How many other predominantly marine invertebrate taxa can claim to have representatives living in the respiratory passages of crocodilians, reindeer, and lions?"

It will come as no surprise that in a group this diverse, there is no universally agreed-upon phylogenetic scheme or classification. The most recent attempt to organize all extant crustacean families by Martin and Davis (2001), although more of a traditional classification than a true phylogenetic analysis based on new data, has enjoyed fairly wide use and acceptance (Brusca and Brusca 2003; McLaughlin et al. 2005) and serves as a starting point for organizing our thoughts and discussions about the group. In that classification, the Crustacea—treated collectively as a subphylum of the arthropods—is partitioned into six extant classes.

Branchiopods, the majority of which are inhabitants of freshwater ephemeral systems, are posited as the sister group to all other extant crustaceans (a point not without controversy). Branchiopods include the Anostraca (fairy shrimp), Notostraca (tadpole shrimp), "cladocerans" (a possibly unnatural assemblage containing a diverse group sometimes referred to as "water fleas"), and the older name "Conchostraca" (an artificial grouping that contained, at one time, all of the "clam shrimp" families).

On the other main branch of the Crustacea, according to the Martin and Davis classification, would be the Cephalocarida, Remipedia (small, vermiform crustaceans known primarily from tropical and subtropical anchialine cave systems), Maxillopoda, Ostracoda (considered part of the Maxillopoda by some workers), and Malacostraca (containing not only the decapods, but also the diverse amphipods and isopods, making the mala-

MYSTACOCARIDA

Mystacocarids are tiny interstitial (meiobenthic) crustaceans usually less than a millimeter long. The elongate, cylindrical body is divided into an anterior cephalon, which bears the antennules, antennae, and three pairs of mouthparts, and a posterior trunk composed of 10 segments, the first of which is usually reduced and bears an appendage often termed the maxilliped. Following the tenth trunk segment is a telson that bears furcae (sometimes called uropods). They were first discovered in intertidal sandy beaches of southern New England in 1939. James Nybakken reports mystacocarids from a sandy beach at Moss Landing, on the shores of Monterey Bay, in the meiofauna section of this manual. No mystacocarids have been identified to species on the Pacific coast, and their distribution and biology in our region remain unknown.—*JTC and JWM*

costracans by far the largest crustacean group in terms of number of species, size of species, and ecological importance).

Of the above groups, I regret most our maintaining the Maxillopoda as a natural monophyletic clade in the Martin and Davis compendium, as nearly all lines of evidence indicate—and nearly all major workers on the group now concede—that it is not (Regier et al. 2005; Giribet et al. 2005; see also Martin and Davis 2001, for a discussion of maxillopod unity). Thus, a more realistic breakdown of the extant Crustacea, abolishing the Maxillopoda and elevating the constituent taxa, would recognize as valid (monophyletic) groups the Branchiopoda, Remipedia, Cephalocarida, Thecostraca, Tantulocarida, Branchiura, Pentastomida, Mystacocarida, Copepoda, Ostracoda, and Malacostraca (the latter containing the Phyllocarida [Leptostraca], Hoplocarida [Stomatopoda], and Eumalacostraca). As noted above, the Hexapoda (insects) also may be nested within Crustacea, although we do not take that step here.

Crustaceans stand apart from other arthropods in their possession of a head that, at least primitively, seems to have been composed of five fused body somites and is followed by a postcephalic segmented trunk (often divisible into a thorax and an abdomen), two pairs of preoral appendages (antennules and antennae), limbs that are or were biramous (though this condition often has been lost, and flattened, phyllopodous limbs also are widespread), a distinctive nauplius larval stage (bypassed in many extant groups), and gills or other cuticular structures for aquatic gas exchange (Brusca and Brusca 2003).

In the following sections, we have attempted to summarize what is known and, to some degree, indicate what is still unknown about the crustaceans of the central California to Oregon coast. Tremendous progress has been made in our understanding of this fauna in the 30 years following the publication of the last edition of *Light's Manual* (1975). But this progress has been uneven, and the observant reader will note that although some sections are greatly expanded and others are completely new (meaning these groups were not deemed sufficiently important or sufficiently known for inclusion in the previous edition), other groups remain more or less unchanged. What follows are treatments of the cephalocarids, mystacocarids, branchiopods, ostracodes, copepods, branchiurans, cirripedes, leptostracans, mysidaceans, cumaceans, isopods, tanaidaceans, amphipods (now expanded into four sections by eight different authors), hoplocarids, euphausiaceans, and decapods (now including contributions from three authors).

References

Aguinaldo, A. M., J. M. Turbeville, L. S. Linford, M. C. Rivera, J. R. Garey, R. A. Raffe, and J. A. Lake. 1997. Evidence for a clade of nematodes, arthropods, and other moulting animals. Nature 387: 489–493.

Averof, M., and M. Akam. 1993. Hom/Hox genes of *Artemia*: implications for the origin of insect and crustacean body plans. Current Biology 3: 73–78.

Averof, M., and M. Akam. 1995a. Hox genes and the diversification of insect and crustacean body plans. Nature 376: 420–423.

Averof, M., and M. Akam. 1995b. Insect-crustacean relationships: insights from comparative developmental and molecular studies. Philosophical Transactions of the Royal Society of London B 347: 293–303.

Boore, J. L., T. M. Collins, D. Stanton, L. L. Daehler, and W. M. Brown. 1995. Deducing arthropod phylogeny from mitochondrial DNA rearrangements. Nature 376: 163–165.

Boore, J. L., D. Lavrov, and W. M. Brown. 1998. Gene translocation links insects and crustaceans. Nature 392: 667–668.

Brusca, R. C. 2000. Unraveling the history of arthropod biodiversification. In: Our Unknown Planet: Recent Discoveries and the Future. Proceedings of the 45th Annual Systematics Symposium of the Missouri Botanical Garden. Annals of the Missouri Botanical Garden 87: 13–25.

Brusca, R. C., and G. J. Brusca. 2003. Invertebrates. Second Edition. Sunderland, MA: Sinauer Associates, Inc., 936 pp.

Chapman, A. D. 2005. Numbers of living species in Australia and the world. Australian Government, Department of the Environment and Heritage, Australian Biological Resources Study. 60 pp.

Friedrich, M., and D. Tautz. 1995. Ribosomal DNA phylogeny of the major extant arthropod classes and the evolution of myriapods. Nature 376: 165–167.

Fryer, G. 1987. Quantitative and qualitative: numbers and reality in the study of living organisms. Freshwater Biology 17: 177–189.

Garey, J. R., M. Krotec, D. R. Nelson, and J. Brooks. 1996. Molecular analysis supports a tardigrade-arthropod association. Invertebrate Biology 115: 79–88.

Giribet, G., G. D. Edgecomb, and W. C. Wheeler. 2001. Arthropod phylogeny based on eight molecular loci and morphology. Nature 413: 157–161.

Giribet, G., S. Richter, G. D. Edgecombe and W. C. Wheeler. 2005. The position of crustaceans within the Arthropoda—evidence from nine molecular loci and morphology. In Crustacea and Arthropod Relationships. S. Koenemann and R. Jenner, eds. Crustacean Issues 16: 307–352.

Gleich, M., D. Maxeiner, M. Miersch, and F. Nicolay (English edition translated by S. Rendall). 2002. Life Counts: Cataloging Life On Earth. Grove Press, Atlantic Monthly Press. 288 pp.

Mallatt, J. M., J. R. Garey, and J. W. Shultz. 2004. Ecdysozoan phylogeny and Bayesian inference: first use of nearly complete 28S and 18S rRNA gene sequences to classify the arthropods and their kin. Molecular Phylogenetics and Evolution 31: 178–191.

Martin, J. W., and G. E. Davis. 2001. An Updated Classification of the Recent Crustacea. Natural History Museum of Los Angeles County, Science Series No. 39: 1–124.

Martin, J. W., and G. E. Davis. 2006. Historical trends in crustacean systematics. Crustaceana 79: 1347–1368.

McLaughlin, P. A., D. K. Camp, M. V. Angel, et al. 2005. Common and Scientific Names of Aquatic Invertebrates from the United States and Canada: Crustaceans. American Fisheries Society Special Publication 31: 1–545.

Regier, J. C., and J. W. Shultz. 1997. Molecular phylogeny of the major arthropod groups indicates polyphyly of crustaceans and a new hypothesis for the origin of hexapods. Molecular Biology and Evolution 14: 902–913.

Regier, J. C., and J. W. Shultz. 1998a. Resolving arthropod phylogeny using multiple nuclear genes. American Zoologist 37: 102A.

Regier, J. C., and J. W. Shultz. 1998b. Molecular phylogeny of arthropods and the significance of the Cambrian "explosion" for molecular systematics. American Zoologist 38: 918–928.

Regier, J. C., and J. W. Shultz. 2001. Elongation factor-2: a useful gene for arthropod phylogenetics. Molecular Phylogenetics and Evolution 20: 136–148.

Regier, J. C., J. W. Shultz, and R. E. Kambic. 2005. Pancrustacean phylogeny: hexapods are terrestrial crustaceans and maxillopods are not monophyletic. Proceedings of the Royal Society B, 272: 395–401.

Rohwer, F. 2003. Global phage diversity. Cell 113: 141.

Rohwer, F., and R. Edwards. 2002. The phage proteomic tree: a genome-based taxonomy for phage. Journal of Bacteriology 184: 4529–4535.

Scholtz, G. 2002. The Articulata hypothesis—or what is a segment? Organisms Diversity and Evolution 2: 197–215.

Cephalocarida

ROBERT HESSLER

Cephalocarids, the most primitive living crustaceans, are tiny (about 3 mm long) subtidal crustaceans with a horseshoe-shaped head, a thorax of nine segments, and an abdomen of 11 segments (including the telson); only the thorax bears appendages, which are triramous (rather than uni- or biramous as in most other crustaceans). The telson bears a pair of large processes forming a caudal fork (furca). They are benthic, with direct, gradual development from brooded eggs. No one knows their precise habitat with respect to the bottom.

Lightiella serendipita Jones, 1961, has been found in shallow water on muddy sand in San Francisco Bay. One specimen of an unidentified cephalocarid has been collected in Anaheim Bay in southern California (Reish et al., 1975). They are surely more widely distributed.

References

Carpucino, M., et al. 2006. A new species of the genus *Lightiella*: the first record of Cephalocarida (Crustacea) in Europe. Zoological Journal of the Linnean Society 148: 209–220.

Jones, M. L. 1961. *Lightiella serendipita* gen. nov., sp. nov., a cephalocarid from San Francisco Bay, California. Crustaceana 3: 31–46.

Reish, D. J., T. J. Kauwling, and T. C. Schreiber. 1975. Annotated checklist of the marine invertebrates of Anaheim Bay. In, E. D. Lane and C. W. Hill, eds., The marine resources of Anaheim Bay. California Department of Fish and Game, Fish Bulletin 165: 41–55.

Sanders, H. L. 1955. The Cephalocarida, a new subclass of Crustacea from Long Island Sound. Proceedings of the National Academy of Sciences 41: 61–66.

Sanders, H. L. 1957. The Cephalocarida and crustacean phylogeny. Syst. Zool. 6: 112–129.

Sanders, H. L. 1963. Cephalocarida. Functional morphology, larval development, comparative external anatomy. Memoirs of the Connecticut Academy of Arts and Sciences 15: 1–80.

Branchiopoda

DENTON BELK

(Plates 182–183)

Branchiopods are a diverse group of crustaceans unambiguously united by possession of a complex post-mandibular, filter-feeding apparatus with a sternal food groove (Walossek 1993). At the level of common names, branchiopods were classically divided on the basis of size into the *large* branchiopods and *small* branchiopods. The large branchiopods were grouped into three orders: Anostraca (fairy shrimp), Conchostraca (clam shrimp), and Notostraca (tadpole shrimp). The famous Norwegian biologist G. O. Sars (1867) once united these three groups in a suborder he named the Phyllopoda, a name that was long abandoned in formal taxonomy; to this day, the large branchiopods are often called phyllopods. The small branchiopods belong to the Cladocera (water fleas). Martin and Davis (2001) resurrected the name Phyllopoda as a subclass to

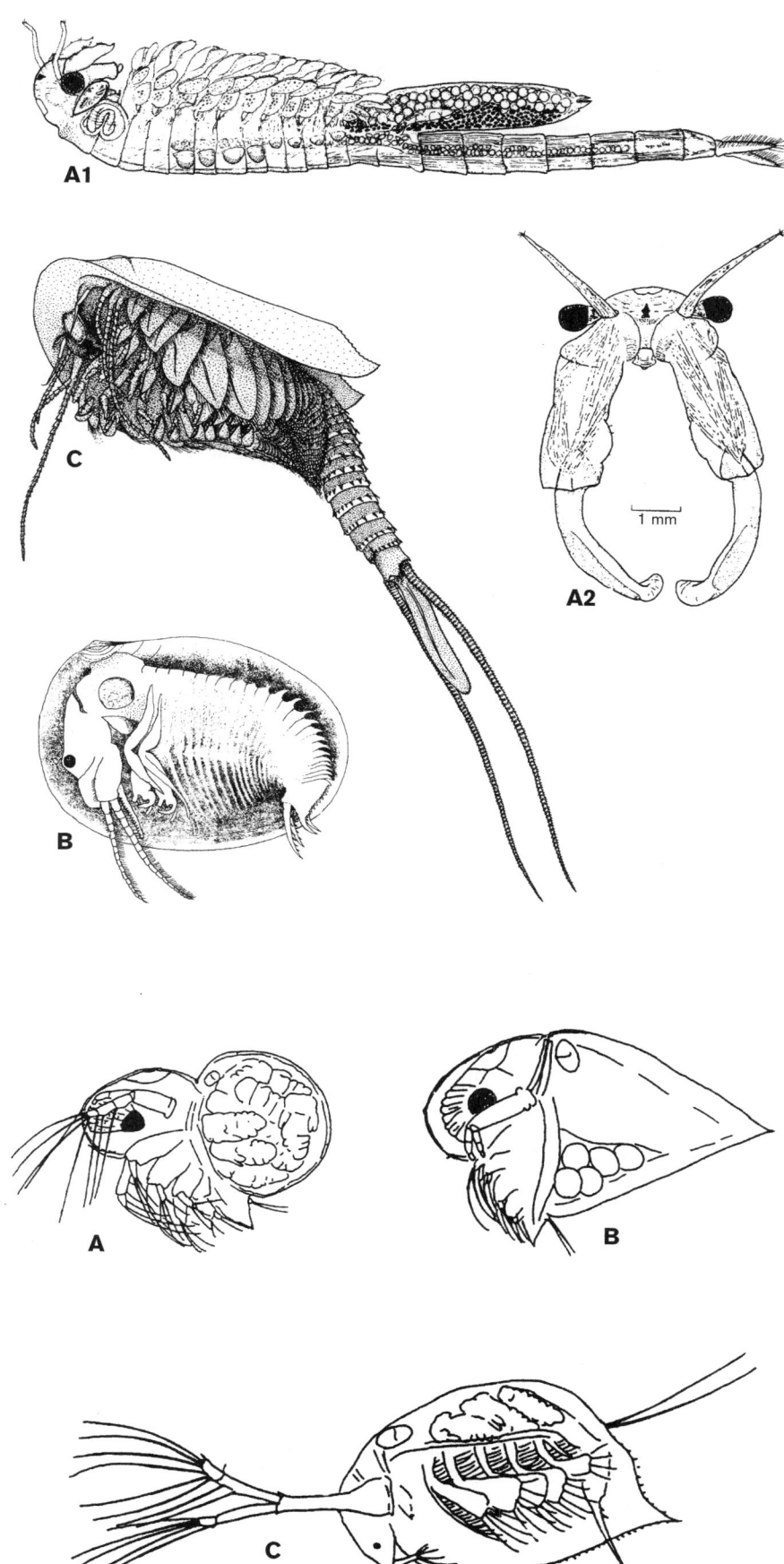

PLATE 182 A, Anostraca—A1, lateral view of a female *Branchinecta lindahli;* A2, anterior view of the head of a male *B. lindahli;* B, Diplostraca—left lateral view of a male *Cyzicus californicus* with the left half of the carapace removed to expose the body of the clam shrimp; C, Notostraca—ventrolateral view of *Lepidurus packardi* (A, from Lynch 1964; reprinted courtesy of American Midland Naturalist; B, redrawn from Packard 1883; C, courtesy of Joel W. Martin).

1 mm

PLATE 183 A, Diplostraca—principal morphotype of *Podon* and *Pleopis;* B, Diplostraca—principal morphotype of *Evadne* and *Pseudevadne;* C, Diplostraca—*Penilia avirostris* (modified from Egloff et al. 1997, courtesy of Academic Press).

encompass the notostracans, conchostracans, and cladocerans. In turn, the cladocerans and former "conchostracans" are grouped together in the order Diplostraca; which group, however, may be polyphyletic (Martin and Davis 2001).

The Branchiopoda is an ancient group that most likely originated in marine waters. The oldest known branchiopod is an anostracan from Upper Cambrian marine deposits in Sweden (Walossek 1993). Kerfoot and Lynch (1987) suggest that the evolution of bony fishes forced large branchiopods from the sea and freshwater inland lakes into temporary pools and hypersaline lakes. They also argue that the evolving dominance of fish predation in fresh waters shaped the morphology, adaptive biology, and evolution of the Cladocera.

For a detailed treatment of branchiopod anatomy, consult Martin (1992).

Anostraca

Fairy shrimps (plate 182A) have an elongate and segmented body with their head bearing stalked compound eyes distinctly set off from a thorax consisting of 11 leg-bearing segments and two fused genital segments. The seven-segmented abdomen has flukelike caudal rami called cercopods extending from the last segment. Adults range mostly from 6 mm–25 mm, but there is one species, the raptorial predator *Branchinecta gigas* Lynch, 1937, that does not reach sexual maturity until it is around 50 mm. I observed one female that measured over 17 cm. Anostracans have a thin, flexible exoskeleton and do not have a carapace. Fairy shrimps swim ventral side up except when they feed by scraping surfaces to collect algae and other organic materials. They also feed by nonselectively filtering particles from the water as they swim. Males use their large second antennae to grasp females during mating.

Anostracans inhabit inland waters ranging from hypersaline to those almost devoid of dissolved substances. The two species most likely encountered near the coastline are *Artemia franciscana* Kellogg, 1906 (=*Artemia salina* of earlier West Coast literature) and *Branchinecta lindahli* Packard, 1883. *Artemia franciscana* lives in seawater evaporation pools used for commercial salt production around San Francisco Bay and other bays. *Branchinecta lindahli* are common in rain-filled ephemeral pools just inland from San Francisco Bay. They occupy a number of these habitats in the San Francisco National Wildlife Refuge. A wealth of information about anostracans and a key to the species you are likely to encounter near the coastline of California and Oregon may be found in Eriksen and Belk (1999).

Diplostraca

Adult "conchostracans" (plate 182B) range in size from about 2 mm–20 mm. The whole body of clam shrimps is enclosed in a bivalved carapace to which it attaches by transverse adductor muscles and other tissues arising from the second maxillary somite. Their compound eyes are internalized in the head and situated close to one another or fused into one functional unit in some. The trunk contains 10–32 segments, each bearing a pair of legs. In males, the first one or two pairs of legs are modified as claspers for grasping the female carapace during mating. Modes of reproduction include obligate sexual, mixed sexual and female self-fertilization, and unisexual (Sassaman 1995). Clam shrimps extend their biramous second antennae outside the carapace and use them for swimming, burrowing,

adhering to surfaces, and as aids in seizing females. To identify conchostracans, consult Martin and Belk (1988) and Belk (1989) along with the key by Mattox (1959). When making identifications, Sassaman's comment (1995) that "classification of clam shrimps is controversial at all levels ranging from the diagnosis of species to that of orders" should be kept in mind.

Mature water fleas (plate 183), or cladocerans, mostly range in size from 0.2 mm–3 mm, but females of *Leptodora kindtii,* an almost transparent cladoceran that feeds on plankton in Holarctic lakes, grow to 18 mm. The carapace encloses only the thorax and abdomen. There are four to six pairs of legs on the thorax. Except for two freshwater genera, all species have a single black compound eye located medially within their heads. Most cladocerans use their biramous antennae to swim. Others use their antennae, legs, and abdomen singly or in various combinations for crawling, scrambling, levering, and burrowing. Reproduction generally alternates between parthenogenetic and sexual. As a result, populations are composed mostly of females. Males have their first pair of thoracic legs modified to include copulatory hooks. In most females, a brood chamber occupies the space between the dorsal surface of their body and the inner surface of the carapace. In some, the carapace is reduced to a brood chamber on the animal's back.

Only eight of the roughly 400 species of Cladocera are truly neritic or oceanic species. The others are found in inland waters ranging from fresh to saline. One of the marine species is *Penilia avirostris,* a member of the family Sididae (plate 183C). *Penilia* is readily distinguished from the other seven marine species by having six pairs of phyllopodial legs enclosed by a carapace. The others, all members of the family Podonidae, have their carapace reduced to a dorsal brood chamber and have four pairs of exposed, elongate, and prehensile legs useful in raptorial feeding (plate 183A–183B). Numerous inland water cladocerans regularly wash into estuaries.

To get started with an identification of these displaced species, consult Dodson and Frey (2001). They present a very useful discussion of cladoceran biology and a key to the North American genera. For an excellent introduction to the marine cladocerans and their reproductive biology, see Egloff, Fofonoff, and Onbé (1997). They also present in their Table 3 the number of setae on the exopods of the thoracic legs of the seven marine species of the family Podonidae. This information allows definitive identification of these species and is included in the following section.

List of Cladoceran Species

Evadne nordmanni (Lovén, 1836). Number of exopodite setae on thoracopods I–2, II–2, III–1, IV–1; Monterey Bay and North American West Coast (Baker, 1938, Proc. Cal. Acad. Sci. 23: 311–365). See Dye and Onbé 1993, Bull. Plankton Soc. Japan 40: 67–69 (feeding).

Evadne spinifera (Müller, 1868). Number of exopodite setae on thoracopods I–2, II–2, III–2, IV–1; Monterey Bay and southern California coast (Baker 1938, Proc. Cal. Acad. Sci. 23: 311–365).

Penilia avirostris Dana, 1849. Off La Jolla (personal communication M. M. Mullin to Takashi Onbé); a warm–water species whose most northern record is the Bering Strait (Onbé et al. 1996, Proc. NIPR Sym. Polar Biol. 9: 141–152). See Uye and Onbé 1993, Bull. Plankton Soc. Japan 40: 67–69 (feeding).

Pleopis polyphemoides (Leuckart, 1859) (=*Podon polyphemoides*). Number of exopodite setae on thoracopods I–3, II–3, III–3, IV–2; Monterey Bay and North American West Coast (Baker 1938, Proc. Cal. Acad. Sci. 23: 311–365).

Pleopis schmackeri (Poppe, 1889). Number of exopodite setae on thoracopods I–4, II–4, III–4, IV–2; not yet known from the North American West Coast (T. Onbé, personal communication).

Podon intermedius Liljeborg, 1853. Number of exopodite setae on thoracopods I–2, II–1, III–1, IV–2; also not yet known from the North American West Coast (T. Onbé, personal communication).

Podon leuckarti Sars, 1862. Number of exopodite setae on thoracopods I–1, II–1, III–1, IV–2; Northern Bering Sea, Chukchi Sea, Bering Strait (Onbé et al. 1996, Proc. NIPR Symp. Polar Biol. 9: 141–152).

Pseudevadne tergestina Claus, 1877. Number of exopodite setae on thoracopods I–2, II–3, III–3, VI–1; Monterey Bay and North American West Coast (Baker 1938, Proc. Cal. Acad. Sci. 23: 311–365, as *Evadne tergestina*). See Uye and Onbé 1993, Bull. Plankton Soc. Japan 40: 67–69 (feeding).

Notostraca

Notostracans (plate 182C) range in size from about 2 cm–10 cm. They have an elongate body covered dorsally at the anterior end of the animal by a broad shieldlike carapace giving them a tadpolelike appearance. The head is incorporated into the anterior portion of the carapace. Their compound eyes are internalized and located on top of the head. Such positioning goes along with the generally bottom–dwelling nature of these crustaceans. The exoskeleton of the thorax and abdomen appears segmented; however, these "segments" are best referred to as "body-rings" because they do not always accurately reflect underlying segmentation (Linder 1952). The first 11 leg-bearing body rings make up the thorax, with the genital openings on the eleventh. The abdomen includes both an anterior leg-bearing section and a posterior legless section. There is variation within each species in the number of body rings in the abdomen and in the numbers of body rings having and not having legs. Also, the number of legs per body ring in the abdomen is variable. The abdomen ends in a telson from which extends a pair of long, thin, cylindrical, and multiarticulate caudal rami. This telson identifies membership in the genus *Triops*, one of the two genera in the order Notostraca. An elongate platelike process extends from the dorsal edge of the telson between the caudal rami in members of the other genus, *Lepidurus*.

Notostracans are omnivores that readily capture and eat fairy shrimps, small fishes, and other small animals. Notostracan reproduction includes obligate sexual, mixed sexual and female self-fertilization, and unisexual modes. Both North American species of *Triops*, *T. longicaudatus* Leconte, 1846, and *T. newberryi* (Packard, 1871), represent complexes of bisexual and unisexual populations (Sassaman et al. 1997).

Tadpole shrimps live in ephemeral inland waters ranging from fresh to alkaline or even brackish. Temporary rain-pools at the San Francisco National Wildlife Refuge and surrounding areas on San Francisco Bay are home to *Lepidurus packardi* Simon, 1886. *Lepidurus packardi* is protected under the U.S. Endangered Species Act and cannot be collected without a federal permit. It is the only notostracan so protected.

Identification of notostracans to species is often a difficult process that ends in an uncertain conclusion. Use the key in Linder (1959) and then consult Lynch (1966), Lynch (1972),

Saunders (1980), and King and Hanner (1998) if you are working with a species of *Lepidurus,* or Sassaman et al. (1997) if you have a species of *Triops*.

References

Belk, D. 1989. Identification of species in the conchostracan genus *Eulimnadia* by egg shell morphology. Journal of Crustacean Biology 9: 115–125.

Dodson, S. I., and D. G. Frey. 2001. Cladocera and other Branchiopoda, pp. 849–913. In: Ecology and classification of North American freshwater invertebrates. 2nd ed. J. H. Thorp and A. P. Covich, eds. San Diego: Academic Press.

Egloff, D. A., P. W. Fofonoff, and T. Onbé. 1997. Reproductive biology of marine cladocerans. Advances in Marine Biology 31: 79–167.

Eriksen, C. H., and D. Belk. 1999. Fairy shrimps of California's puddles, pools, and playas. Mad River Press, 196 pp.

Kerfoot, W. C., and M. Lynch. 1987. Branchiopoda communities: associations with planktivorous fish in space and time, pp. 367–378. In: Predation. Direct and indirect impacts on aquatic communities. W. C. Kerfoot and A. Sih, eds. Hanover, NH: University Press of New England.

King, J. L., and R. Hanner. 1998. Cryptic species in a "living fossil" lineage: taxonomic and phylogenetic relationships within the genus *Lepidurus* (Crustacea: Notostraca) in North America. Molecular Phylogenetics and Evolution 10: 23–36.

Linder, F. 1952. Contributions to the morphology and taxonomy of the Branchiopoda Notostraca, with special reference to the North American species. Proceedings of the United States National Museum 102: 1–69.

Linder, F. 1959. Notostraca, pp. 572–576. In: Fresh-water biology. 2nd ed. W. T. Edmondson, ed. New York: John Wiley and Sons.

Lynch, J. E. 1966. *Lepidurus lemmoni* Holmes: a redescription with notes on variation and distribution. Transactions of the American Microscopical Society 85: 181–192.

Lynch, J. E. 1972. *Lepidurus couesii* Packard (Notostraca) redescribed with a discussion of specific characters in the genus. Crustaceana 23: 43–49.

Martin, J. W. 1992. Branchiopoda, pp. 25–224. In: Microscopic anatomy of invertebrates. Volume 9, Crustacea. F. W. Harrison and A. G. Humes, eds. New York: Wiley-Liss.

Martin, J. W., and D. Belk. 1988. Review of the clam shrimp family Lynceidae Stebbing, 1902 (Branchiopoda: Conchostraca), in the Americas. Journal of Crustacean Biology 8: 451–482.

Martin, J. W., and G. E. Davis. 2001. An Updated Classification of the Recent Crustacea. Natural History Museum of Los Angeles County, Science Series No. 39: 1–124.

Mattox, N. T. 1959. Conchostraca, pp. 577–586. In: Fresh-water biology. 2nd ed. W. T. Edmondson, ed. New York: John Wiley and Sons.

Sars, G. O. 1867. Histoire naturelle des Crustacés d'eau douce Norvége. C. Johnson, 145 pp.

Sassaman, C. 1995. Sex determination and evolution of unisexuality in the Conchostraca. Hydrobiologia 212: 169–179.

Sassaman, C., M. A. Simovich, and M. Fugate. 1997. Reproductive isolation and genetic differentiation in North American species of *Triops* (Crustacea: Branchiopoda: Notostraca). Hydrobiologia 359: 125–147.

Saunders, J. F. 1980. A redescription of *Lepidurus bilobatus* Packard (Crustacea: Notostraca). Transactions of the American Microscopical Society 99: 179–186.

Walossek, D. 1993. The Upper Cambrian *Rehbachiella* and the phylogeny of Branchiopoda and Crustacea. Fossils and Strata 32: 1–202.

Ostracoda

ANNE C. COHEN, DAWN E. PETERSON, AND ROSALIE F. MADDOCKS

(Plates 184–196)

The Ostracoda are a large and important class of small bivalved crustaceans. Because of their microscopic size (0.20 mm–2 mm, rarely to 32 mm), they are not conspicuous animals, but they are abundant and diverse in most aquatic ecosystems if sampled correctly. There are probably at least 25,000 extant species, of which roughly 12,000 have been described (3,000 freshwater,

9,500 marine). Ostracoda have the best fossil record of any arthropod group: the carapace is usually well impregnated with calcite (calcium carbonate), so dead and molted carapaces accumulate abundantly as fossils in both modern sediments and ancient sedimentary rocks.

Ostracodes are found worldwide in the ocean from intertidal to abyssal and hadal depths (7,000 m), in coastal estuaries and marshes, in most fresh waters, and in a few terrestrial habitats. In tropical coral reef ecosystems and in deep-sea faunas, the careful collector may expect to find several hundred species, whereas high-latitude and stressed environments (e.g., coastal marshes, salt lakes, and ephemeral ponds) may yield only a few species. Diversity in the littoral zone varies with the habitat but may be surprisingly high; 58 species of interstitial and meiofaunal ostracodes were collected from intertidal sandy beaches of Anglesey, North Wales; 40 podocopid species from intertidal habitats of the Kurile Islands; and 20 species in a single sample of intertidal mussels near Seto, Japan (see "References").

In seasonally freshwater rock pools in the high intertidal to supratidal zone and in the freshwater lens above the halocline of anchialine (cavern and fracture) systems, certain freshwater (some salt-tolerant) species of Cyprididae may thrive; such populations are usually parthenogenetic and are probably dispersed by wind or birds. On foggy and rainy shores, the collector can inspect water-holding plants, such as mosses and bromeliads, and leaf litter for rare semiterrestrial ostracodes.

Although ostracodes are sometimes abundant, they are poorly known from the California and Oregon intertidal, which hosts more species than are reported here. The faunas of coastal estuaries, deltas, and lagoons of the accretionary coasts of North America (especially of the southern Atlantic and Gulf of Mexico) have been more intensively studied than those of erosional coasts, such as the rocky shores of central and northern California and Oregon. Users of this manual who collect from algae, mussel, and oyster beds; rocky tide pools; pocket beaches; seagrass; silty sand; and other appropriate habitats should expect to discover many additional unreported, and even some undescribed, species and more than a few taxonomic tangles.

This section lists the ostracode species known from the littoral and immediate sublittoral (to a depth of a few meters) within the geographic range of this handbook (from Pt. Conception through central and northern California and Oregon). The list of 32 Podocopa and 10 Myodocopa includes those species reported in previous literature that are considered valid, as well as new records for this region (specimens identified by the authors). Generic assignments have been updated for many species, and some species are left in open nomenclature because they are new or not yet firmly identified.

Of the 10 myodocopid ostracodes known from our region, three are also found both north and south of our region, three south only, one is a probable European introduction, and three have not yet been reported beyond our boundaries. Some of the seven additional myodocopids that have been reported from other localities on the West Coast may yet be found within our region, particularly two reported both north and south of our region (see the "List of Species").

Because of the cold California Current, the intertidal podocopid ostracodes are closely related to those of British Columbia, Alaska, the Aleutian Islands, the Kurile Islands, and northern Japan. Alaskan faunas are poorly known. Probably most of the 32 species described from Vancouver Island range south into California. The intertidal faunas of rocky shores at these latitudes are dominated by phytal dwellers,

especially species of Cytheridae, Hemicytheridae, Leptocytheridae, Loxoconchidae, Microcytheridae, Paradoxostomatidae and Xestoleberididae. Because of their small size and fragility and difficulties of identification, several of these families are greatly under-represented in the California literature, although doubtless they are abundant on California shores (see "References").

By contrast, the littoral and sublittoral faunas of southern California (south of Pt. Conception) include many podocopids in common with coastal Mexico and Central America. A greater research effort off southern California has yielded a more diverse ostracode fauna (littoral and sublittoral) with many podocopid species in common with coastal Mexico and Central America. Many of these warmer water species are also well represented in late Cenozoic fossil assemblages of California.

Most ostracodes are benthic, crawling over or burrowing beneath the sediment surface, through the interstices of shelly sands and gravels, over rocks and plants, or through microalgae. Some are demersal plankton, swimming for short distances usually just above the sea floor to feed or mate (often at night). Fewer are planktonic in the open ocean. Ostracodes include detritivores, scavengers, herbivores, suspension-feeders, predators, commensals (of crustaceans, polychaetes, sponges, and echinoderms) and a single fish parasite. In turn, some are parasitized by even more minute copepods, isopods, nematodes, and ciliates.

Ostracodes are important food for fish and benthic invertebrates. Their carapace remains are consistently included in inventories of gut contents of fish, although not necessarily as dominant or selected components of the diet. The biotic interactions of most ostracodes with other invertebrates and fish are not well documented, and collectors are urged to watch for evidence of such relationships.

Some myodocopid and unrelated halocyprid ostracodes produce bioluminescence to deter predators. The luciferin and luciferase are exuded from glands in the upper lip (some Cypridinidae) or from epidermal glands around the margin of the carapace (some planktonic Halocyprididae). *Porichthys notatus*, the California midshipman fish, has been shown to be bioluminescent only in the southern part of its range where it can feed on the bioluminescent California myodocopid *Vargula tsujii*. Otherwise nonbioluminescent midshipman fish from Puget Sound were able to bioluminesce after being fed bioluminescent ostracodes. Additionally, a speciose Caribbean group of myodocopids has males capable of performing spectacular species-specific bioluminescent mating displays somewhat similar to those of fireflies.

General Morphology

Ostracodes can withdraw all limbs inside the enveloping bivalved **CARAPACE**, the margins of which may lock to form a tight seal. (There are reports of ostracode survival after passage through the guts of fish. A few may survive hours or even days of emersion, which has facilitated geographic dispersal to isolated coastal lagoons and terrestrial waters by hitchhiking on migratory waterfowl.) Ostracode valves are closed by **CENTRAL ADDUCTOR MUSCLES** connecting the inner dermis of each valve through the enclosed limb-bearing part of the body. The valves are either nudged apart by appendages or opened hydraulically, as there is no complementary set of opening muscles, nor is there any functional equivalent to the bivalve mollusk ligament. The ostracode carapace varies from smooth to highly ornamented but never has the series of fine concentric "growth lines" (indicating previous molts) present on the bivalved carapaces of Conchostraca. It is perforated by numerous, minute NORMAL PORE CANALS with chemosensory and tactile sensilla, which communicate with nerve cells in the underlying epidermis. Ecophenotypic variability of carapace structure is common in podocopid species populations of the coastal zone, especially in certain families of ornamented Cytheroidea, in response to fluctuations of salinity, substrate, organic carbon, and other poorly understood environmental parameters.

The exoskeleton of the soft body consists of a somewhat rigid head capsule (with labrum [upper lip] and MOUTH) and a flexible cuticular integument over the posterior region. The endoskeleton is a loose array of internal chitinous SCLERITES. The demarcation between the cephalon and thorax is indistinct, and there is argument about whether it has segmental or functional significance. The abdomen is either missing or not differentiated from the thorax; the posterior body (TRUNK) of most ostracodes appears under light microscopy to be unsegmented. In a few taxa, external transverse folds fringed with spines or small setae (most visible with scanning electron micrography [SEM]), and sometimes an internal chitinous framework, probably indicate vestigial trunk segmentation. Vestiges of as many as 10 or perhaps 11 segments (plus a telson) occur in the Platycopida (and fewer in the podocopid *Saipanetta*). In the Myodocopa, obvious dorsal trunk segmentation is found in a few Cladocopina (up to four segments) and a few Thaumatocypridoidea (probably more).

There are five to seven paired limbs (six to eight if copulatory limbs included), all more or less different in structure and function, and their homologies with other Crustacea are difficult to trace. Ostracoda never have six or more serially repeated pairs of trunk limbs, as in the superficially similar barnacle "cypris" larvae or in Conchostraca. The male and female external genitalia are thought, at least in some cases (particularly Myodocopa), to be a modified last (usually eighth) pair of limbs. Certain perhaps sensory structures have also been interpreted as the remnants of limbs. The body terminates in a pair of caudal FURCAE, which in Myodocopa is posterior to the anus but in Podocopa is anterior to the anus.

Ostracodes have no larval stage and no metamorphosis but hatch from the egg as a juvenile with at least three functional limbs and the enveloping bivalved carapace (less calcified than adult). Juveniles resemble reduced adults. During a fixed series (usually five to eight) of molts, they grow larger and acquire additional limbs; limbs also become more complex, more setose, and, in some cases, change function. Carapace details and primary and secondary sexual characteristics may begin to take shape in late instars but are not fully developed until the adult (terminal) instar. Because molting requires abandonment of the calcified part of the carapace cuticle, the hinge structures and marginal details are fully generated only during the final molt of Podocopa. Fossil assemblages in modern and ancient sediments include numerous molted valves and carapaces in addition to adult specimens. Juveniles are difficult to identify, so collectors should strive to recover large populations with complete growth series and adults of both sexes.

Sexes are separate, and fertilization is internal. Parthenogenesis is common in some freshwater lineages but has not been conclusively verified in marine dwellers. Many ingenious adaptations have evolved to enable copulation in spite of the enveloping carapace. Sexual dimorphism occurs in the internal soft anatomy (testes, ovaries, and associated receptacles, reservoirs and pumps), external genitalia (especially complex in males), appendages (with sensory, clasping and egg-cleaning modifications), DOMICILIUM (in brooders), and external size, shape and ornament of the carapace (sometimes very dimorphic). The details of sexual dimorphism are essential for identification at all taxonomic levels. One sex is sometimes easier to identify than the other.

Identification of Ostracoda has been difficult for nonspecialists. Dissection and examination must be performed under a light microscope. Few keys exist, and none should be relied upon as substitutes for consulting the primary taxonomic literature of many countries and languages. Secondary faunal compilations exist for a few other regions, but there are none for marine ostracodes of North America (except papers on Myodocopida of Western North Atlantic).

Considerable information about ostracode specialists and their research can be found on the Web site of the International Research Group on Ostracoda (www.uh.edu/~rmaddock/IRGO/cypris.html; http://userpage.fu-ber/in.de/~palaeont/irgo/ostracoda.html), with links to other sites, such as Cypris, the International Ostracoda Newsletter and Cohen's tabular key (illustrated). A Web search on "Ostracoda" will turn up numerous other sites, many beautifully illustrated. Southern California fauna are listed at www.scamit.org/taxonomic_tools.htm.

How to Collect and Prepare Ostracodes for Identification

Look for ostracodes in most intertidal habitats—even freshwater pools in the highest zone. Use the same collecting equipment and methods that you would use for other tiny crustaceans. Collect samples of algae, pitted rocks, oysters, or mussels into buckets or stout plastic bags and then extract the hidden tiny animals. Examine these samples directly under a dissecting microscope and remove the ostracodes with fine forceps (the flexible ones are good) or an eyedropper (for swimming ones).

For stubbornly hidden ostracodes, a wash may help. Try adding some fresh water or a little alcohol to the seawater and shaking or soaking the algae (etc.) until more animals emerge. You can wash and strain the sample water through a fine-meshed net or screen (holes not more than 0.5 mm for myodocopids, as small as 0.02 mm for the tiniest interstitial podocopids) and flush the inverted contents with a squirt bottle into a small dish for observation. Live ones often play possum at first but are fun to watch when they open their valves and move around after a few minutes. Picking and observing lively ostracodes from a sorting dish may be easier if you slow them down by adding a little anesthetic such as $MgCl_2$ (50% 0.36M) for a few minutes; revive them by returning them to plain seawater.

To hunt for ostracodes in bottom sediment and seagrass, drag a fine-meshed net (or homemade hand dredge) along the surface of the bottom, not digging in more than just a few centimeters. Cohen's favorite dredge is one that Todd Oakley made by attaching pantyhose to a frame. Flush seawater through the net (small running seawater hoses work) to remove fine sediment before flushing net contents into a sorting dish and picking ostracodes from the coarser sediment (stirring the sediment and quickly decanting mainly liquid may produce many but not all of the ostracodes). Some myodocopids are attracted to traps baited with a little fish or a light stick and anchored overnight. Some male myodocopids are attracted to a bright light suspended from a dark dock, and they then can be scooped up with nets. Preserve and store your labeled ostracode samples in 70% EtOH until you are ready to identify them. Take field notes.

Start the identification process by examining a sample under a dissecting microscope and sorting the ostracodes into groups of similar ones. Gently transfer an ostracode into a little glycerol in a labeled depression slide (use alcohol instead if you are preparing for SEM or molecular studies) and examine it under a microscope. Fine flexible forceps, an eyedropper (don't use much liquid), mouth pipette, brush, needles, or tiny wire loop (using a fine probe in your other hand as a pusher) are useful. Examine the ostracode under a compound microscope and perhaps sketch it (camera lucida helpful). Use a coverslip for higher powers. With the slide under the dissecting microscope, remove one or both carapace valves. Hold the ostracode in place with one tool (try the tiny wire loop) while you slide a sharp tool (fine dissecting needle or single edged razor blade) between the valve and the body to sever the central adductor muscle. Pry the valve off the body and cut the weak dorsal body connection to the valves. Examine the specimen again under the compound microscope. You may be able to identify the subclass or order of ostracode at this point.

If you are unfamiliar with ostracodes, sketch the limb positions and shapes before removing limbs (with fine needles) under the dissecting microscope. Cohen uses insect minuten pins glued inside the tip of narrow wooden sticks or fine pipette tips. Retain the valves and body in the glycerin slide, but with a needle move each limb to a permanent slide; cover both slides with a coverslip. Cohen makes permanent slides with polyvinyl lactophenol (this is very poisonous, so use care). When the limb slide is fully dry, the limbs can be examined and drawn under high power with the compound microscope.

The valves of myodocopans will weaken only somewhat over time in glycerin. Both valves and dissected limbs can also be stored in a vial of alcohol stoppered with cotton and placed in a larger vial that also contains a permanently inked label. Cotton stoppered vials should be stored in a larger and well-sealed jar of alcohol, perhaps with labeled vials of similar ostracodes. However, the valves of podocopans will decalcify in alcohol and glycerin. For storage, they should be dried on paper micropaleontological slides. SEM reveals many important minute characters (particularly of the carapace and copulatory limbs) not otherwise visible (see "References"). Additional voucher specimens (preserved in alcohol) of species used for SEM, molecular and ecological studies should always be retained and deposited in a museum collection as verification of your identifications.

Classification of Living Ostracoda

Early classifications of living Ostracoda were based mostly on soft anatomy, whereas classifications of fossils were based on carapace features. Older classifications are obsolete, and none of the newer syntheses is universally regarded as authoritative. Since the mid-twentieth century, in an effort to reconcile the discrepancies, some paleontologists began to study living faunas, and the advent of SEM revolutionized the description of carapace and appendage structure. Myodocopa and Podocopa differ noticeably from each other (see tabular key below), and

some molecular analyses suggest that they may actually not form a monophyletic clade. The classification of Podocopa remains more uneven than that of Myodocopa.

Glossary

This list contains abbreviations used in the plates and gives illustrative examples from among the plates.

Terms used herein (and in many other ostracode publications) are defined (some with figure references).

1ST ANTENNA, ANTENNULA, A1 1st (cephalic) pair of limbs. Plates 184, 190A, 191D, 191E.

2ND ANTENNA, ANTENNA, A2 2nd (cephalic) pair of limbs. Plates 184, 191B, 191D, 191E.

3RD (CEPHALIC) LIMB see "mandible (md)." Plate 184.

4TH LIMB (1ST MAXILLA, MAXILLULA, 4) 4th (cephalic) limb, labeled "maxilla" in many papers; use care in reading literature descriptions. Plates 184, 192D, 193F.

5TH LIMB (2ND MAXILLA, 5) 5th limb, labeled "maxilla" in many papers; controversially considered to be and labeled "1st trunk leg," "1st thoracic limb," or "maxilliped" by some podocopid specialists; use care in reading literature descriptions. Plates 184, 191D, 191E, 192F.

6TH LIMB (6) 6th limb (and 1st trunk/thoracic limb); controversially considered to be and labeled "2nd trunk leg" by some podocopid specialists; use care in reading descriptions. Plates 184, 191D, 191E, 192G.

7TH LIMB (7) 7th limb (and 2nd trunk/thoracic limb); controversially considered to be and labeled "3rd trunk limb" by some podocopid specialists; use care in reading descriptions. Plates 184, 191D, 191E, 192H.

8TH LIMB (8) see "copulatory limb (CP)."

ADDUCTOR MUSCLE SCAR PATTERN (AMS) rather central pattern of scars (raised or depressed) indicating where central adductor muscles attach to valves; not always clearly visible. Plates 189A–189G, 189I, 189J, 190K, 190L.

ARTICLE see "podomere."

BASIS distal part of divided protopod (separated from coxa by suture). Plates 185C, 188A–188D.

BELLONCI ORGAN (BO) organ projecting from forehead or naupliar (medial) eye of most Myodocopa. Plates 184B, 185B.

BRANCHIAL PLATE (VIBRATORY PLATE, BP) flat setose epipod, (at least in Myodocopa) used to circulate water through the domicilium. Plates 184C, 184D, 192D.

BROOD POUCH an expanded region of the female domicilium, usually posterior, for protecting developing eggs and instars.

BRUSH-SHAPED ORGANS small pair of lobes bearing numerous fine setae; found between the fifth limbs in males of most Cytherocopina and some Cypridocopina; might represent vestigial appendages. Plates 190F, 194D.

CALCIFIED INNER LAMELLA (CIL) calcified part of the inner (medial) lamella of valve margin. Plate 189F.

CARAPACE lateral outfolds of dorsal epithelium and cuticle, bivalved and enveloping the entire body. Plates 184D, 187, 195, 196.

CAUDAL PROCESS (KEEL, K) posterior extension of carapace. Plates 184A, 186G, 186H.

COPULATORY LIMBS (CP) male limbs modified for copulation, at least in most Myodocopa, usually the 8th limb pair (see also "hemipenes"). Plate 184A.

COXA proximal part of divided protopod (separated from "basis" by suture; see "protopod"). Plates 185C, 188A–188D.

DOMICILIUM the volume between the two valves of the carapace and occupied by the soft body and limbs.

ENDITE medial extension (tooth, lobe, masticatory process) of protopod, generally used in feeding (see also "masticatory process"). Plate 188A.

ENDOPOD/ENDOPODITE (EN) in distally biramous crustacean limb, the medial ramus. Plates 184C, 184D, 185C, 190B.

EPIPOD (EP) extension (usually setose flat lobe) of lateral protopod; in ostracodes (at least in Myodocopa) it often forms a branchial plate used to circulate water. Plate 184A, 184B.

EXOPOD/EXOPODITE (EX) in distally biramous crustacean limb, the lateral ramus. Plates 184, 185C, 190B.

FREE MARGIN the perimeter of the valves (exclusive of the hinge region) where valves are in contact.

FURCA(E) (F) terminal (may be ventral in position) body extensions. Plates 184, 185D, 185H, 186F, 188E, 191E.

FUSED ZONE see "zone of concrescence."

HEMIPENES (HP) podocopan term for paired male copulatory appendages. Plates 184C, 191D, 193H.

HINGE (H) complementary interlocking articulatory structures on inner dorsal margin of valves. Basic types mentioned in key:

   ADONT single element hinge, bar on one valve fits into groove of other valve

   MERODONT three-element hinge, with anterior and posterior teeth (or sockets) separated by median bar (or groove)

   LOPHODONT a merodont hinge in which all elements are smooth

   PENTODONT a five element hinge, in which the anterior and posterior region of the median element are differentiated, either crenulate or smooth (family Pectocytheridae only)

   HOLOMERODONT a merodont hinge in which all elements are crenulate and negative on one valve, positive on the other

   ANTIMERODONT a merodont hinge in which the median element is reversed (negative or positive) from the terminal elements

   ENTOMODONT a four-element hinge, with the anterior region of the median bar or groove differentiated, usually with all elements crenulate

   GONGYLODONT a crenulate hinge with a systematic increase in size of teeth and decrease in size of sockets from anterior to posterior in the right valve, left valve complementary (family Loxoconchidae only)

   AMPHIDONT a modified entomodont hinge in which the four elements are well differentiated (families Hemicytheridae, Trachyleberididae)

   HOLAMPHIDONT an amphidont hinge in which all elements are smooth

   HEMIAMPHIDONT an amphidont hinge in which the posterior element is crenulate and the others are smooth

   SCHIZODONT entomodont hinge in which the anterior and anteromedian elements are bifid, and the posterior element is bifid or crenulate (Family Schizocytheridae only). Plate 191F.

INCISUR (INCISURE, ROSTRAL INCISUR, I) an anterior indentation (ranging from slight to deep and even slitlike) of the valves of the carapace; it acts like an oar-lock for the rowing antennae of many swimming Myodocopa. Plates 184A, 186G.

INNER LAMELLA inner layer of epidermis and cuticle of the dorsal body fold forming the valve, entirely uncalcified or calcified only around the free margin.

INNER MARGIN (IM) the inner or proximal edge of the calcified inner carapace lamella, forming an abrupt line or shelf in dead or fossil valves of Podocopida. Plates 189A, 189H, 190L.

KEEL (K) posterior extension of carapace (also called "caudal process"). Plates 184A, 184B, 185A, 186G, 186H.

LATERAL EYE (LE) paired compound eyes of most Myodocopida. Plates 184A, 184B, 185A, 185F.

LINE OF CONCRESCENCE (LC) the proximal edge of the zone of concrescence. Plates 189I, 190L.

MANDIBLE third limb (third head limb). Plates 184A–184D, 192E, 193G.

MAXILLULA, "MAXILLA" see "4th and 5th limbs."

MASTICATORY PROCESS see "endite."

NAUPLIAR EYE (MEDIAL EYE, NE) single medial anterior eye, normally composed of three eye cups, has reflecting tapetal layer. Plates 184, 185B.

NORMAL PORE CANALS (NPC) numerous perforations through valve with chemosensory and tactile sensilla, which communicate with nerve cells in the underlying epidermis; may be simple (open) or covered with sieve plates. Plate 189I.

OUTER LAMELLA the outer layer of epidermis and cuticle of the dorsal body fold forming the valve; may be calcified or uncalcified.

PALP term used for a podocopan armlike jointed portion (endopod or exopod plus basis) of limb, either directed forward and used for food manipulation, or directed backward and used for clasping.

PODOMERE (ARTICLE) segment of jointed crustacean limb separated from other segments by articulated joint or inflexible suture.

PROTOPOD (PROTOPODITE, PRO) basal podomere(s) of crustacean limb (plate 184B), often divided into two (sometimes three) segments: the coxa attaching the limb to the body and the more distal basis. Plate 188A, 188B.

RADIAL PORE CANALS (RPC) modified normal pores located at free margins of valves, housing nerve filaments that lead from tactile sensilla to nerve cells in the underlying epidermis. Plate 189H.

ROSTRUM anterior projection of valves of many myodocopans, overhangs incisur. Plates 184A, 184B, 186E, 186G.

SCLERITE internal chitinous struts supporting body and limbs. Plate 188C.

SEM scanning electron micrography. Plates 195, 196.

SPINNERET GLAND (SG) gland (connected to 2nd antenna) secreting adhesive thread. Plates 184D, 190B, 191D.

VALVE right or left half of bivalved ostracode carapace (sometimes erroneously referred to as "shell").

VESTIBULE (VESTIBULUM, V) in Podocopa, a marginal cavity between the outer valve lamella and calcified part of the inner valve lamella, extending from the line of concrescence to the inner margin, housing epidermal, glandular, and sometimes reproductive tissues. Plates 189I, 190L.

VIBRATORY PLATE (BRANCHIAL PLATE, BP) flat plate (with marginal setae) located proximally and laterally on limb (an epipod in Myodocopa, but podocopan homologies uncertain); used to circulate water. Plates 184C, 184D, 192D.

"WALKING LEGS" posterior limbs that are long, slender, jointed, directed ventrally, and used for locomotion. Plates 184C, 184D, 191D, 191E.

ZENKER'S ORGAN (Z) muscular sperm pump (ejaculatory pump) located dorsally within male body (Cypridocopina) or interlamellar cavity (Sigilliocopina).

ZONE OF CONCRESCENCE (FUSED ZONE, ZC) a clear marginal band extending from the free margin to the line of concrescence, formed by a flangelike outgrowth of the outer lamella and traversed by radial pore canals.

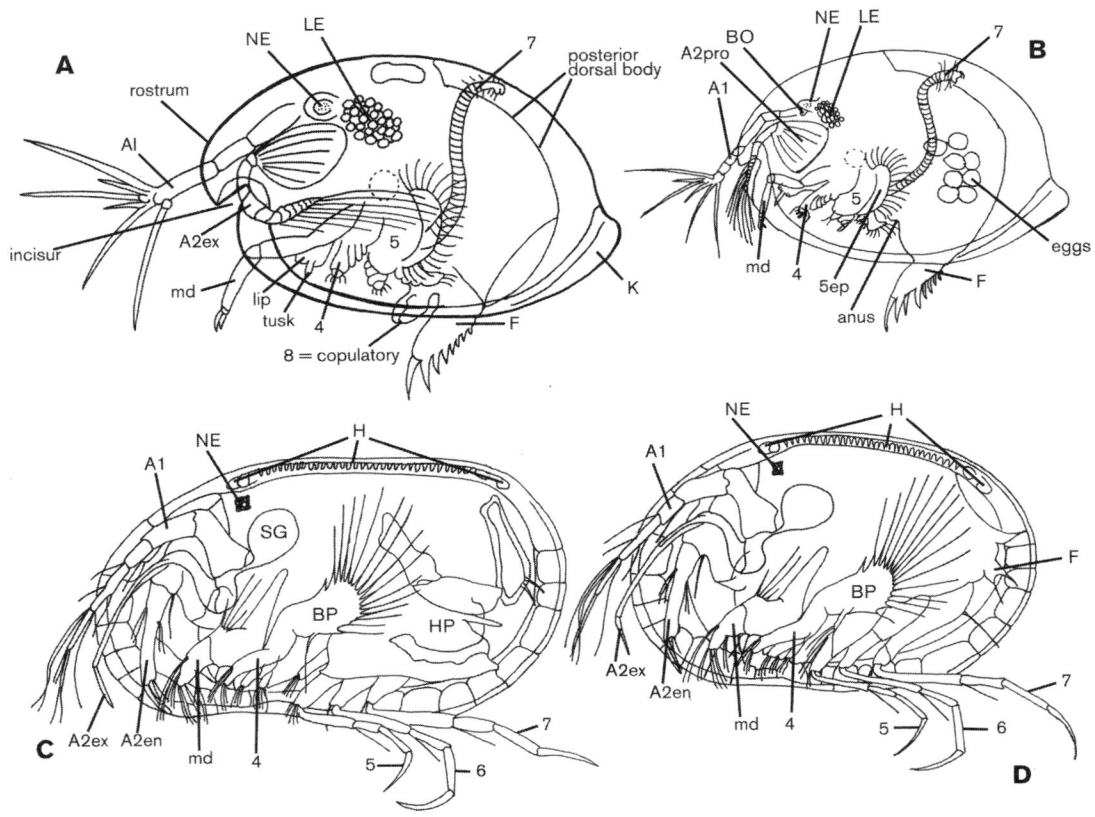

PLATE 184 Ostracoda, diagrammatic comparison of Myodocopa and Podocopa. Myodocopida (Cypridinidae) after Morin 1986: A, male; B, female; Podocopida (Cytheroidea), after Athersuch, et al. 1989: C, male; D, female (redrawn by Ginny Allen).

Key to the Ostracoda

TABULAR KEYS TO MYODOCOPA

ANNE C. COHEN

The three tabular keys on the following pages cover all Myodocopa from subclass to the family level (along with just the subclass Podocopa). A dichotomous key by Rosalie Maddocks further identifies Podocopa from orders to families (along with just the subclass Myodocopa). Because of the chance of encountering unreported species, taxonomic literature should be consulted if identification is attempted beyond the family level.

Key to Podocopa (and Subclass only of Myodocopa)

ROSALIE F. MADDOCKS

1. Carapace elongate oblong to almost circular; with or without rostrum and incisur; valves usually not overlapping; 2nd antenna with large muscular protopod; long exopod with nine podomeres (rarely less) with long setae (often used for swimming); smaller (often much smaller) endopod with one to three podomeres, usually conspicuously dimorphic in adults; large lamellar furcae posterior to anus; with or without lateral eyes and Bellonci organ (plates 184A, 184B, 185–188) . . . . Subclass Myodocopa

— Carapace ovoid, inflated-subtriangular, oblong elongate, or compressed; no rostrum or incisur; valves overlap around free margin; 2nd antenna geniculate, pediform; with very small exopod with no more than two podomeres; much larger propulsive endopod with up to four podomeres; variable furca anterior to anus; no lateral eyes or Bellonci organ (plates 184C, 184D, 189–196) . . . . . . Subclass Podocopa 2

2. Carapace size, shape diverse; left (sometimes right) valve usually overlaps right valve. Adductor muscle scar pattern varied (but includes rows) (plates 189A, 190K, 191D); seven pairs of limbs (including at least one walking leg) (plates 184C, 184D, 191D, 191E) plus large male hemipenes (plates 184C, 191D, 192I, 193H, 194H); 2nd antennal exopod greatly reduced (plates 184C, 184D, 190B); 4th limb (and usually mandible) without lateral setal comb (plates 184C, 184D, 192D, 192E), furcae mostly reduced (plates 184C, 184D, 191E), posterior body only rarely transversely ridged (plates 184C, 184D, 189–196) . . . .
. . . . Order Podocopida 3

— Carapace oblong, laterally compressed, right valve overlapping left all around; adductor muscle scar pattern biserial; six pairs of limbs (no walking legs) plus male large hemipenes; 2nd antenna with large setose exopod (two podomeres); 4th limb (and mandible) with setal combs; furcae large, lamellar, with stout claws; posterior body transversely ridged; marine, not intertidal . . . .
. . . . Order Platycopida

Key to the Suborders of Podocopida

3. Carapace ovoid to elongate, usually smooth (or punctate); adductor muscle scars an aggregate or rosette of numerous scars; calcified inner lamella narrow 4

— Carapace form and ornament diverse; adductor muscle scars varied, but distinctive arrangements of a smaller number of discrete scars (see descriptions further in key); calcified inner lamella relatively broad (plate 189F) 5

4. Carapace elongate-ovoid, symmetrical; right valve overlaps left; anterior end narrower, posterior inflated as brood chamber; adductor muscle scar pattern of six to 12 in compact rosette; hinge adont; mandibular palp with lateral setal comb; posterior body smooth; furcae generally absent; males rare; freshwater to slightly oligohaline; global . . . .
. . . . Suborder Darwinulocopina

— Carapace inflated-ovoid, asymmetrical; left valve strongly overlapping right; adductor muscle scar pattern an aggregate of 20–35 scars; hinge merodont; mandible without setal comb; posterior body transversely ridged; large lamellar furca with fairly stout claws; males with spirally muscularized Zenker's organs; marine, always rare, circumtropical in reef habitats, probably cryptic and interstitial . . . . Suborder Sigilliocopina

5. Carapace diverse, usually smooth (or slightly punctate) (plate 195A, 195B); hinge usually adont (otherwise macrocyprid described in couplet 7a); adductor muscle scar pattern of five to nine discrete scars in a distinctive arrangement (plate 189); 5th limb endopod a small walking leg, or reduced in females and modified as a recurved clasper in males; 7th limb recurved over posterior body, modified distally for cleaning; some males with brush-shaped organ; testes and ovaries sometimes extend into interlamellar cavity of valves; Zenker's organs in dorsal region of male body; many marine and most freshwater ostracodes, some parthenogenetic (plates 189, 195A, 195B) . . . . Suborder Cypridocopina 7

— Carapace shape varied, often ornate, with complex hinges and marginal details; adductor muscle scar pattern of four to nine discrete scars in distinctive arrangement (see next couplets); 5th, 6th, and 7th limbs are walking legs (plates 184C, 184D, 191D, 191E); paired brush-shaped organ in males (plate 190F); testes and ovaries not extending into interlamellar cavity of valves; no Zenker's organs (plates 184C, 184D, 190–194, 195C–195J, 196). 6

6. Carapace with characteristic bairdioid shape (plate 195H), smooth or punctate; left valve overlaps right; adductor muscle scar pattern of three pairs of scars in anterior vertical column plus posterior pair of scars; 2nd antennal exopod a minute scale with up to three small setae; furcae small, rodlike with small setae; one to two balls of sediment usually visible in dorsal region of midgut; marine, global, especially diverse in reef habitats but represented in most subtidal faunas from all depths and sediment types . . . . Suborder Bairdiocopina

— Carapace extremely varied; left valve usually slightly overlaps right; adductor muscle scar pattern typically a vertical row of four (rarely, five) small scars (rarely some of the scars divided) (plates 190K, 190L, 191A, 191D–191F, 192J, 193J, 193K, 194I–194K); hinge adont, merodont, amphidont, schizodont (plate 191F), or other specialized condition; sexual dimorphism may be conspicuous; 2nd antennal exopod modified as a hollow spinning seta connected to spinneret gland (plates 184D, 190B, 191D); furcae reduced to two tiny setae (plate 191E); gut contents not usually

identifiable; marine and nonmarine; global, extremely abundant and diverse in all aquatic habitats, accounting for more than 90% of all living Ostracoda, especially dominant in brackish water, intertidal and shallow sublittoral faunas (plates 184C, 184D, 190–194, 195C–195G, 195I, 195J, 196) . . . . Suborder Cytherocopina 9

Key to the Superfamilies of Cypridocopina

7. Carapace elongate; right valve overlaps left; hinge sinuous, consisting of five elements, terminal elements crenulated; adductor muscle scar pattern a compact circular spot with three discrete upper scars and a lower group of indistinct scars; female 5th limb jointed; 6th limb with one long and two shorter terminal claws; testes and ovaries do not enter interlamellar valve cavity; Zenker's organ with numerous chitinous spikes not arranged in rosettes. One to two balls of sediment visible in hindgut. Global, marine benthos, represented in all sublittoral and deeper habitats but usually uncommon . . . . Superfamily Macrocypridoidea

— Carapace shape and overlap variable; hinge adont; adductor muscle scar pattern of five to six discrete scars (anterior column of three scars and slightly lower posterior column of two to three scars); female 5th limb unjointed, 6th limb with one terminal claw; ovaries and testes or vas deferens partly housed within interlamellar valve cavity; Zenker's organ without chitinous spikes or with spikes arranged in rosettes; gut contents unrecognizable 8

8. Carapace shape and overlap varied; usually smooth; adductor muscle scar pattern of five scars in two columns; mandible with weak simple teeth; 7th limb with three short terminal setae, none recurved, no pincer; Zenker's organ without chitinous spikes; global, marine; swimming, crawling and burrowing; represented in all sublittoral and deeper habitats. . . . . Superfamily Pontocypridoidea

— Carapace varied, usually smooth (but not always; e.g., plate 195A, 195B); left valve usually slightly overlaps right; adductor muscle scar pattern (plate 189) of one cap (dorsal) scar plus about five scars in two columns; mandible with strong trifid teeth; 7th limb with long recurved pectinate seta and one to two short or long terminal setae or modified as pincer; Zenker's organ with chitinous spikes usually grouped in rosettes; global, highly diverse; marine, brackish and freshwater, found in all aquatic habitats, dominant in terrestrial waters; also well represented in littoral and shallow-sublittoral habitats of tropical regions, coastal lagoons, and estuaries, and in anchialine cave faunas (plates 189, 195A, 195B) . . . .
. . . . Superfamily Cypridoidea, Family Cyprididae 9

Key to the Families of Cytheroidea Known to Be Present in the California and Oregon Littoral

This key is valid only for the species included in the species list of this section.

9. Carapace smooth or very faintly pitted (plates 190K, 190L, 192J, 193J, 193K, 194J, 194K) 10

— Carapace conspicuously (except *Acuminocythere*, plate 196C) pitted, ridged, or otherwise ornamented (plates 195C–195G, 195I, 195J, 196) 14

TABLE 2

Tabular Key to Myodocopa (and Subclass only of Podocopa)

Character	Subclass Podocopa	Subclass Myodocopa	Order Myodocopida	Order Halocyprida
Carapace (valve)				
Adult length (approximate)	0.1 mm–8 mm	0.1 mm–32 mm	Most 1 mm–3 (but up to 32) mm	0.1 mm–3 mm
Shape (lateral outline) (plates 184–187, 189–196)	Ovoid, inflated-subtriangular, oblong or elongate, **not circular**	Elongate oblong to almost circular	Elongate-ovoid or ovoid, few rather circular	Varied, elongate sub-quadrate to almost circular
Anterior rostrum (plates 184–187)	**Without** (projection in some Cypridoidea)	**With or without**	Most	Many, but not most
Anterior incisur (notch) (plates 184, 190–194)	**Without**	**With or without**	Most with (usually not above midheight), but not obvious on more circular valves	With (often above mid-height) or without (more circular valves)
Ventral margin	Straight, sinuous (partly concave) or slightly convex	Usually convex, some straight, none concave		
Posterior keel (project-ing caudal process) (plates 184–186)	n.g.	Yes or no	Present in many	None, except some with posterodorsal corner-shaped process or spine
Anterior inner valve surface	n.g.	Setose or not	**With at least 1 seta**	Without setae
Body (plate 184)				
Medial eye	With or without	With or without	Present in most	Absent
lateral eye	**Absent**	**Present or absent**	**Present in most**	**Absent**
Bellonci organ	Absent	Usually present	Usually present	Usually present
Bifurcate?		Bifurcate or not	Not bifurcate	Many bifurcate
Posterodorsal body	Few with some segmentation	Few with some segmentation	Body smooth; some with gills or pouches	A few with some segmentation; no gills or pouches
Furca (plates 184, 185D, 185H, 186F, 188E, 191E)				
Furcal position	**Anterior to anus**	**Posterior to anus**		
Furcal shape	Varied, most not plate-like with claws, (no strongly sclerotized plates)	**Always strongly sclerotized flat plate with claws**		
No. of Limbs (including male copulatory limb)	7–8 (plus rudimentary brush organ in many)	6–8	8	6–8
2nd Antenna				
Best developed ramus	**Endopod (plates 191D, 191E, 192B**	**Exopod (plate 184)**		
Number of endopod articles	3–4	1–3	1–3	2–3
Male endopod	No clasper, but some dimorphic	Most with clasper	Many with clasper (usually reflexed)	Many with hooklike clasper
Number of exopod articles	1–2 (or seta only)	**9 (rarely less)**	9 (rarely less)	9

TABLE 2 (continued)

Character	Subclass Podocopa	Subclass Myodocopa	Order Myodocopida	Order Halocyprida

Mandible (plates 185C, 188A–188D, 190G, 192E, 193G, 194B)

Character	Subclass Podocopa	Subclass Myodocopa	Order Myodocopida	Order Halocyprida
Sexual dimorphism	n.g.	Some	Dimorphic in Sarsielloidea	Apparently none
Proximal setose branchial plate	Many Podocopida	None		
Lateral setose comb	Platycopida only	None		
Proximal endite (on coxa)	n.g.	Usually present, varied	**Usually present and lobelike**	**Present as a cusped tooth**
Basis endite	n.g.	Present in some, varied	1 family with lobe, none with large tooth	**Many with cusped tooth**
Endopod: 1st article setal position	n.g.	V and/or D setae	V setae only	D (and sometimes V) setae
Exopod	n.g.	Present in some = lobe distally inserted on basis with endopod	Present in most (1 article) or absent	Present (1 article) or absent

4th Limb (plates 184, 190D, 190E, 192D, 193E)

Character	Subclass Podocopa	Subclass Myodocopa	Order Myodocopida	Order Halocyprida
Sexual dimorphism	n.g.	Present in some	Reduced only in Sarsielloidea males	Apparently none
Proximal setose branchial plate	Present in Podocopida, absent in Platycopida	Absent (?except in Cylindroleberididae)	Possibly = bare proximal lobe in Cylindroleberididae	Absent
Lateral setose comb	Platycopida only	Present in some	Cylindroleberididae only	Probably absent
Exopod	n.g.	Many	Usually present = 1 article	Present (1–2 articles) or not
Endopod	Podocopida jointed	Jointed	2 articles	2–3 articles

5th Limb (plates 184, 190H, 191D, 191E, 192F, 193C, 194E)

Character	Subclass Podocopa	Subclass Myodocopa	Order Myodocopida	Order Halocyprida
Sexual dimorphism	Often, some males with hooklike clasper	Some	Limb reduced in male Sarsielloidea; Cyclasteropinae males with process	Rare
Proximal setose branchial plate	Well-developed to absent	Present = epipod		
Limb shape	Leglike or reduced	Varied	**Compact feeding limb**	Leglike
With setose comb	No	Some	Cylindroleberididae only	No
Teeth or short claws	n.g.	Present or absent	Many with strong non-terminal teeth or claws	Some with terminal claw (only)
Endopod	Present	Present	Rami unclear	Present and jointed
Exopod	Absent	present in few	Rami unclear	Polycopina only

6th Limb (plates 184, 190I, 191D, 191E, 193D, 194F)

Character	Subclass Podocopa	Subclass Myodocopa	Order Myodocopida	Order Halocyprida
Limb present	Present	Present or absent	Present	Present or absent
Sexual dimorphism	Platycopida	Some	None or slight	Some (*Conchoecia*)
Proximal setose branchial plate	Well-developed to absent	Present in some = epipod	**Absent**	**Present in some = epipod**
Limb shape	Leglike or reduced	Varied	**Short, flat, poorly jointed**	Leglike or absent

(continued)

TABLE 2 (continued)

Character	Subclass Podocopa	Subclass Myodocopa	Order Myodocopida	Order Halocyprida
Endopod	Present	Present	Rami unclear	Present
Exopod	Absent	Present in a few	Rami unclear	Present in some (1 article)

7th Limb (plates 190J, 191D, 191E, 192H, 193E, 194G)

Limb present	Present or absent	Present or absent	Usually present	Present or absent
Sexual dimorphism	n.g.	Some	Limb reduced or absent in some males	None
Proximal setose branchial plate	Reduced if present	Absent		
Limb size	n.g.	Varied	**Usually long**	**Reduced or absent**
Limb shape	Leglike, absent, or a jointed (but not annu-lated) cleaning limb	Not leglike	**Long wormlike annulated cleaning limb**	**Lobe with few setae or limb absent**

Copulatory (8th) Limb or Hemipenis (plates 184, 191D, 192I, 193H, 194H)

Paired or single	Paired	Paired or single	Paired	**Single**
Branches	n.g.	Branched in many	3 lobes	Single or 2 branches

NOTE: Bold = major character difference; D = dorsal; V = ventral; n.g. = character not given for Podocopa.

10. Adductor muscle scar pattern of five scars in vertical row (plate 190K, 190L); 1st antenna (plate 190A) with eight podomeres; basal podomere of 5th limb (plate 190H) has small branchial plate with setae (plate 190) . . . . Bythocytheridae

— Adductor muscle scar of four scars in vertical row (plates 191A, 191D, 191E, 192J, 193J, 193K, 194I–194K); 1st antenna with no more than six podomeres (plate 192A); basal podomere of 5th limb without branchial plate (plate 193C) 11

11. In lateral view, anterior end of carapace more narrowly rounded than posterior end (plates 191A, 191D, 191E, 193J, 193K, 194I, 194J) 12

— In lateral view, anterior end of carapace usually more broadly rounded than posterior end (plates 192J, 194K) 13

12. Carapace fragile, elongate, laterally compressed (plates 191A, 191D, 191E, 193J, 193K); inconspicuous sexual dimorphism; hinge lophodont; rather oval frontal scar (plate 191A); normal pore canals inconspicuous, funnel-type; no eye scar; mouth region (plate 193I) extended as cone with sucking disk; mandibles and 4th limbs greatly reduced and transformed to hypodermic-like apparatus (plates 191A, 191D, 191E, 193) . . . . Paradoxostomatidae

— Carapace strongly calcified, inflated, egg-shaped and ventrally flattened; conspicuous sexual dimorphism; hinge merodont; frontal scar V-, J-, or W-shaped; normal pore canals conspicuous, sieve-type; eye scar present behind eye spot; mouth region and mouth parts normal (plate 194I, 194J) . . . . Xestoleberididae

13. In lateral view, anterior carapace margin broadly rounded, posterior margin slightly more narrowly rounded, not cau-date; ventral margin sinuate; exterior smooth; deep vestibules; hinge lophodont (plate 194K) . . . . Cytheromatidae

— In lateral view, anterior carapace margin obliquely rounded, posterior margin angulate to caudate; ventral margin nearly straight; exterior ribbed and pitted, with subtle ventrolateral alar thickenings, vestibules shallow or absent; hinge merodont (plate 192J) . . . . Cytheruridae

14. Carapace subrhomboidal to subhexagonal in lateral outline; with lightly pitted ornament (plates 191B, 191C, 196B) 15

— Carapace subovate or subquadrate in lateral outline; with lightly to heavily pitted to reticulate ornament (plates 191F, 195C–195G, 195I, 195J, 196A, 196C–196J) 16

15. Carapace subrhomboidal in lateral outline, with caudal process located above midheight; hinge gongylodont; normal pore canals simple (plate 196B). . . . . Loxoconchidae

— Carapace subhexagonal in lateral outline, with more or less distinct posterior angle at about midheight; lightly pitted ornament with faint radial ridges; hinge antimerodont; sieve-type and simple normal pore canals (plate 191B, 191C) . . . . Cytheridae

16. Hinge amphidont 17

— Hinge merodont or pentodont. 18

17. Carapace elongate-oblong in lateral view, with concave posterodorsal element and slightly angulate posteroventral region; hinge hemiamphidont to holamphidont; adductor muscle scar pattern with upper one to two scars divided, plus two to three frontal scars; anterior radial pore canals very numerous; internal chitinous supports (sclerites) present in knee joints of walking legs (plates 195C–195G, 196E–196J) . . . . Hemicytheridae

TABLE 3

Tabular Key to Suborders and Superfamiliies of Order Halocyprida

Character	Suborder Polycopina Family Polycopidae	Suborder Halocypridina	Family Thaumatocyprididae	Family Halocyprididae
Carapace (valve)				
Adult size (length)	0.1 mm–1.1 mm	0.4 mm–3.2 mm	0.4 mm–2.5 mm	0.75 mm–3.3 mm
Shape (outline)	Ovoid or almost circular	Elongate-ovate to ovate, few rather circular	Almost circular	Usually somewhat sub-quadrate to oblong, less often rather ovate (*Deeveya*)
Rostrum	Rare (only *Pontopolycope*)	Present in many	Present in none	Usually conspicuous
Incisur	**Present in none** (except overhang in *Pontopolycope*)	**Present in many**	**Absent, but anteroventral margin straight or slightly concave and delimited by anterior and anteroventral processes**	**Usually conspicuous (often above midheight)** (slight in *Deeveya*)
Dorsal margin	Convex	Convex or straight	Convex or straight	Usually rather straight, sometimes convex
Posterodorsal margin	Without keel	Without keel	With or without long spinelike process	Often with spine or corner-shaped glandular process
Body				
Bellonci organ	Usually bifurcate and Formed as two setae	Bifurcate or not, not formed as setae	Not bifurcate; sometimes absent	Bifurcate or not, often sexually dimorphic; sometimes absent
Posterodorsal body	Few with some segmentation	Few with some segmentation	Some with traces of segmentation	Smooth
Furca	**With short triangular process between at least longest claws**	**No triangular processes** (few with a minute blunt peg) **between claws**	**Anterior and ventral edges forming right angle, claws on anterior margin longer and articulated at base, claws on ventral margin unarticulated**	**All claws articulated at base** (Deeveyinae with minute glandular peg between two claws)
No. of Limbs (including male copulatory limb)	6	8	8	8
1st Antenna				
No. of articles	3–5 (inferred 8)	3–8	7–8	3–8
Bend in limb	None	Vanes	Between first and second article	Distal (*Deeveya* also bent between first and second)
Distal filament pad	Absent	Present in some	Absent	Euconchoecinae only

(continued)

TABLE 3 (continued)

Character	Suborder Polycopina Family Polycopidae	Suborder Halocypridina 	Family Thaumatocyprididae	Family Halocyprididae
Mandible				
Coxal endite	Cusped tooth	Cusped tooth		
Basis endite	**No tooth**	**Cusped tooth**		
Exopod	Present (one article)	Absent		
Endopod—number of articles	Two articles	Jointed (number of articles unclear)		
Fourth Limb				
Exopod	Present (one to two articles)	Absent		
Endopod	Two to three articles	Two articles		
Fifth Limb				
Endopod—number of articles	One article	Jointed (number of articles unclear)		
Exopod	One article	None		
Sixth Limb: presence				
	Absent	**Present**		
Endopod	NA	Jointed		
Exopod	NA	Present in few (one article), usually absent	Present (one article—except *Danielopolina*)	Present only in Deeveyinae (one article)
Seventh Limb: presence				
	Absent	Reduced (or absent), with one to three setae	Unjointed	One to two articles
Male Copulatory Limb				
Branched or not	Two parts	One to two parts	Two parts	One to two parts

NOTE: Key differences are indicated in boldface.

— Carapace (plates 191F, 196C, 196D) elongate-ovate to subtriangular in lateral view, without concave or angulate elements; hinge schizodont (plate 191F); adductor muscle scar pattern with no divided scars (plate 191F); few anterior radial pore canals; knee joints of walking legs without internal supports . . . . Schizocytheridae

18. Carapace elongate-subquadrate in lateral view; coarsely pitted to reticulate; simple frontal scar; few radial pore canals; deep vestibules; hinge pentodont; egg care not known; male 6th and 7th limbs symmetrical (plate 196A) . . . . . .
. . . . Pectocytheridae

— Carapace elongate-oval in lateral view; surface smooth to weakly pitted; frontal scar V- to J-shaped; numerous radial pore canals; vestibules shallow or absent; females brood young in domicilium; male right and left 6th and 7th limbs asymmetrical (plate 195I, 195J) . . . . Cytherideidae

List of Species

MYODOCOPA

ANNE C. COHEN

Myodocopa are mostly subtidal and few have yet been collected intertidally within the geographic range of this book; most listed below have been reported slightly subtidal near shore. AC = Anne C. Cohen; RFM = Rosalie F. Maddocks; DP = Dawn Peterson; identified specimens of many of the following species are in the California Academy of Sciences (San Francisco).

MYODOCOPIDA

To identify juvenile instars see Kornicker and Harrison-Nelson 1999, Smithsonian Cont. Zool. 602: 1–55 (see pp. 32–36) and

TABLE 4

Tabular Key to Families of the Order Myodocopida

Character	Subfamily Cypridinidoidea Cypridinidae	Subfamily Cylindroleberoidea Cylindroleberididae	Subfamily Sarsielloidea Philomedidae	Rutidermatidae	Sarsiellidae
Carapace (valves]					
Outline round (except for keel)	None (plates 184A, 184B, 185A)	None (plates 185E–G; 187A)	None (plate 186A, 186B, 186D, 186E)	None (plate 187B, 187C)	**Female and juvenile Sarsielli-nae round or round except for keel (plate 187G, 187H)**
Incisur (anterior slot or indentation)	Deep (1 exception) (plate 184A, 184B)	Shallow (plate 187A) to deep (plate 185E) or **slitlike (Cylin droleberidinae) (plate 185F, 185G)**	Minute to deep (plate 186A, 186B, 186D, 186E) (*Pleoschisma* = none)	Shallow (plate 187B, 187C)	Minute (plate 186H) to shallow (plate 186G)
Smooth at low magnification?	Usually	Cylindroleberidinae, Cyclasteropinae	Some	None completely smooth	None completely smooth
Ornamented with nodes, ridges, etc.	Few	Asteropteroninae only	Many	Most	All
Posterior vertical row of external hairs	None except male *Codonocera*	All males (except Asteropteroninae)	None	None	None
Male more elongate?	Some	Some	Yes	Yes	Yes
Body					
Bellonci organ	Short or long	Usually long	Long	Long	Long
Upper lip with valvular nozzles?	**Present in all (many nozzles) (plate 185B)**	Absent	**Present (few nozzles) or absent**	Absent	Absent
Posterodorsal body	Most smooth (plate 184A, 184B); some males with row of large lobes	**Usually 7–8 (some fewer) dorsal over-lapping flat leaf-like gills**	Smooth	Smooth	Smooth
Furca					
1st claw with basal suture	Yes (plates 184A, 184B, 185D)	Yes (plate 185H)	Yes (plate 186F)	Yes	**No (plate 188E)**
Other claws without basal suture	Sometimes	No	No	No	Sometimes
1st Antenna					
3rd and 4th articles	Separated by suture	Usually separated by complete suture	Separated by suture	Sometimes fused	Fused
Male 5th article	Not reduced	Usually little reduction	Wedged ventrally between 4th and 6th	Wedged ventrally between 4th and 6th	Wedged ventrally between 4th and 6th
Male seta of 5th article with bush of long filaments?	No or slight (some with bushes on setae of terminal article)	Cyclasteropinae, most Cylindrole-berididae and Asteropteroninae	Yes	Yes	Yes

(continued)

TABLE 4 (continued)

Character	Subfamily Cypridinidoidea Cypridinidae	Subfamily Cylindroleberoidea Cylindroleberididae	Subfamily Sarsielloidea Philomedidae	Rutidermatidae	Sarsiellidae
Males with terminal filaments with suckers?	Yes	No	No	No	No
7th article with clawlike a-seta	None	All Cylindroleberidinae, most Cyclasteropinae, some Asteropteroninae	None	None	None
2nd Antenna					
Protopod with seta?	With minute distal seta	Most with minute distal seta	No seta	No seta	No seta
Male endopod a clasper?	Not in most, but in some	Always	Always	Always	In most
Mandible					
Coxal endite shape	Spiny lobe, rarely absent	**Long, scythe-shaped branched, spiny**	Spiny lobe, some long, reduced in male	Spiny lobe (plate 188A), reduced in male (plate 188B)	Absent or small spiny lobe, reduced in male
Basis endite	Absent	**Lobe with setae**	Absent	Absent	Absent
Exopod	Short, 2 setae	Short, 2 setae	Short, 2 setae	Absent or reduced, 0–?2 setae	Absent or reduced, 0–1 setae
Endopodial claws form	**Terminal group (plate 185C)**	**Terminal group**	**Terminal group (plate 186C)**	**Pincer in females and juveniles (except Metaschismatinae)**	**3-pronged rake in female and juvenile Sarsiellinae (plate 188C)**
Endopodial claws—number and position	**All terminal: (2-) 3 relatively short & all on 3rd article (plate 185C)**	**All terminal: (1-) 3 relatively long & all on 3rd article**	**All terminal:** females & juveniles with (2-) 3 relatively long claws (plate 186C); males with 1–3 claws and all on 3rd article	**Not all terminal in females & juveniles: claws form pincers (plate 188A) (except *Metaschisma*), with 1 (-2) claw on 2nd article and 1 claw on 3rd; males with 1 terminal claw (plate 188B)**	**Not all terminal in females & juveniles: usually with 1 (Dantyinae 0–3) claw on each of 3 endopodial articles (plate 188C); males with 1–3 terminal claws only (or plus weaker proximal claws) (plate 188D)**
4th Limb (plate 184A, 184B)					
Reduced and weak in males?	No	No	**Yes**	**Yes**	**Yes**
With setal comb	No	**Yes, on protopod**	No	No	No
With basal epipod?	No	Bare lobe ?= epipod	No	No	No
Endopod—claws on 2nd article (terminal)	At least 2 claws, several setae, many pectinate	**Reduced; no claws**	Females and juveniles with 2 claws and various setae	Females and juveniles with at least 2 claws, several/few setae; *Metaschisma* with 3	**Females and juveniles with terminal row of 5 stout claws (3–5 triangular) and 1 subterminal long thin seta on each side of claw row**

(continued)

TABLE 4 (continued)

Character	Subfamily Cypridinidoidea Cypridinidae	Subfamily Cylindroleberoidea Cylindroleberididae	Subfamily Sarsielloidea Philomedidae	Rutidermatidae	Sarsiellidae
5th Limb (plate 184A, 184B)					
Reduced, weak in males?	No	No	Yes	Yes	Yes
With setal comb	No	**Yes**	No	No	No
With teeth or claws	**Row of claws (usually 6) + short peg on 1st exopodial article**	No teeth or claws	**1 big squarish or elongate tooth on 2nd article, smaller teeth on 1st article**	**1 big squarish tooth on 2nd article, smaller teeth on 1st article**	**Sarsiellinae none; Dantyinae with big squarish tooth on 2nd article, small tooth on 1st article**
6th Limb (plate 184A, 184B)					
With well developed endites?	4	All reduced	4	4	1–4
7th Limb (plate 184A, 184B)					
Reduced in males?	No	No	In a few	No	Usually
Terminus	Peg or tooth opposite comb	Opposing combs	Peg(s) opposite comb	Peg(s) or comb opposite comb	Small opposing combs
Male 8th Limb (plate 184A)					
Copulatory limb	Big, complex, 3 lobes (1 jointed) with internal sclerites	Small in most males, 3 lobes, unjointed, no sclerites	Short to elongate, 3 lobes, some/all with sclerites	Small elongate with 3 lobes	Small elongate with 3 lobes, with distal sclerotized hook

NOTE: Key differences are indicated in boldface.

Kornicker 1992, Smithsoman Cont. Zool. 531: 1–243. For information on seven Northwestern American coastal species beyond range of this book, see Juday 1907, Univ. Calif. Publ. Zool. 3: 135–156; Lie, 1968, Fiskedirektoratets Skrifter, serie HavUnders¢kelser, Bergen 14: 229–556 (identifications unverified, specimens not extant); Lie and Evans 1973, Marine Biol. 21: 122–126 (identifications unverified, specimens not extant); Poulsen 1965, Dana Report 65: 1–484; Kornicker and Myers 1981, Smithsonian Cont. Zool. 334: 1–35; Chess and Hobson 1997, NOAA Tech. Memorandum NMFS-SWFSC-243, U.S. Dept. Com.; and website of SCAMIT (South. Calif. Assoc. Mar. Invert. Taxonomists): http://www.scamit.org/taxonomic_tools. htm or http://www. scamit.org/index.htm) (plates 184A, 184B, 185–188).

CYPRIDINIDAE

For summary and phylogenetic analysis see Cohen, A. C. and J. Morin 2003. Sexual morphology, reproduction and bioluminescence in Ostracoda. In Bridging the Gap: Trends in the Ostracode Biological and Geological Sciences. L. E. Park and A. J. Smith, eds. The Paleontological Society Papers 9. New Haven: Yale University Press.

Vargula tsujii Kornicker and Baker 1977, Proc. Biol. Soc. Washington 90: 218–231 (= *Vargula americana* [Müller, 1890] of Hobson and Chess 1976, Fisheries Bull. 74: 567–598). Benthic in sand, planktonic, 3 m–931 m, Baja California to south of Monterey Bay, not reported further north. Planktonic occurrence probably only as nocturnal demersal plankton: see Hammer 1981, Mar. Biol. 62: 275–280; Hobson and Chess 1976 (also fish predation); Chess and Hobson 1997, NOAA Tech. Memorandum NMFSSWFSC-243, U.S. Dept. Com. (also fish predation); Stepien and Brusca 1985, Mar. Ecol. Prog. Ser. 25: 91–105 (also fish predation). Bioluminescence in midshipman fish is related to predation on *V. tsujii*: see Thompson, Nafpaktitis, and Tsuji 1988, Mar. Biol. 98: 7–13; 1988, Comp. Biochem. Physiol. A, 89: 203–210; Thompson and Tsuji 1989, Mar. Biol. 102: 161–165; Thompson et al. 1998, J. Exp. Biol. 137: 39–52. For more on bioluminescence see Huvard 1990, Acta Zoologica 71: 217–223; 1993, Comp. Biochem. Physiol. A, 104: 333–338, J. Morph. 218: 181–193 (plate 185A–185D).

CYLINDROLEBERIDIDAE

For keys see Kornicker 1981, Smithsonian Cont. Zool. 319: 1–548 (subfamilies of Cylindroleberididae, genera of Cylasteropinae,

Asteropteroninae; also general morphology of Myodocopida); Kornicker 1986, Smithsonian Cont. Zool. 425: 1–139 (some genera of Cylindroleberidinae). To identify juvenile instars see Kornicker 1992, Smithsonian Cont. Zool. 531: 1–243.

ASTEROPTERONINAE

Asteropella slatteryi Kornicker, 1981. Benthic in sand mixed with silt and clay in harbor (often in kelp bed) at Half Moon Bay; Moss Landing, Monterey Bay and Pilar Point Harbor, Half Moon Bay; 1.8 m–37 m. See Kornicker and Harrison-Nelson 1997, Smithsonian Cont. Zool. 593: 1–53 (description, distribution, ecology); Chess and Hobson 1997, NOAA Tech. Memorandum NMFS-SWFSC-243 (Catalina Is., benthic in sediment by day, nocturnal plankton, fish predation) (plate 187A).

CYCLASTEROPINAE

Leuroleberis sharpei Kornicker, 1981 (=*Cylindroleberis lobianci* Müller, 1894 of Sharpe 1908, Proc. U.S. Nat. Mus. 35: 399-430 [not =*C. lobianci* Müller]; =*Cycloleberis lobiancoi* of Hobson and Chess, 1976). Depth 2 m–146 m, planktonic (probably only as demersal plankton at night) and benthic from Gulf of California to Monterey Bay, possibly to Alaska; in Monterey Bay off Sunset Beach, Kaiser Trestle, Watsonville outfall, and Pajaro River, and also off South Jetty and northwest of harbor entrance at Moss Landing. For fish predation, benthic and planktonic habitats at Catalina Id., see Hobson and Chess 1976, Fish. Bull. 74: 567–598; Chess and Hobson 1997, NOAA Tech. Memorandum NMFS-SWFSC-243 (plate 185E).

CYLINDROLEBERIDINAE

Postasterope barnesi (Baker, 1978) (=*Parasterope barnesi* Baker 1978, Crustaceana 35: 139–141; =*Parasterope* sp. of Tuel et al. 1976, Biol. Survey Pillar Point Harbor, El Granada, Calif., Rep. Mar. Ecol. Inst., Redwood City, CA). Benthic in sand, 1.8 m–210 m; San Diego to Point Conception; Pillar Point Harbor, Half Moon Bay; also males collected at night in subtidal benthic light trap on silty sea grass near Spud Pt. Marina Bodega Harbor, CAS, AC. See Kornicker and Harrison-Nelson, 1997, Smithsonian Cont. Zool. 593: 1–53 (description, distribution, ecology) (plate 185F–185H)

PHILOMEDIDAE

For keys, see Kornicker 1981, Smithsonian Cont. Zool. 332: 1–16 (most genera of Philomedinae); Kornicker 1989, Smithsonian Cont. Zool. 467: 1–134 (genera of Pseudophilomedinae).
Euphilomedes carcharodonta (Smith 1952) (=*Philomedes carcharodonta* Smith 1952, J. Fish. Res. Board Canada 9: 16–41; ="myodocopid" in Tomales Bay, of Kornicker 1977, in Aspects Ecol. Zoogeog. Recent Fossil Ostracoda: 159–173 [pers. comm. Kornicker]). Benthic in 5 m–180 m from La Jolla to Ganges Harbor, British Columbia, including Monterey Bay, Half Moon Bay, Tomales Bay, and Bodega Bay. See Kornicker and Harrison-Nelson 1997, Smithsonian Cont. Zool. 593: 1–53 (description, distribution, ecology); Poulsen 1962, Dana Report 57 (description); Baker 1974, Texas J, Science 25: 131–132 (abundant

on S. Calif. shelf); Oliver et al. 1980, Fish. Bull. 78: 437–454 (physical ecology, Monterey Bay) (plate 186A–186C).
Euphilomedes longiseta (Juday, 1907) (=*Philomedes longiseta* Juday 1907, Univ. Calif. Publ. Zool. 3: 135–156; =*P. longiseta*, of Lucas, 1952, Contr. Can. Biol. Fish., Stud. Biol. Stat. Canada, n.s. 6: 398–416). Surface plankton tows off San Diego (Juday, 1907); benthic sand, 6 m –9 m off Monterey Bay (Oliver et al. 1980, Fish. Bull. 78: 437–454); 20 m–90 m, near Vancouver Island (Lucas, 1952).
Euphilomedes morini Kornicker and Harrison-Nelson, 1997 (=*Philomedes longiseta* of Tuel et al. 1976, Biol. Survey Pillar Point Harbor). Benthic in 1.8 m on sand, Pillar Point Harbor, Half Moon Bay. See Kornicker and Harrison-Nelson 1997, Smithsonian Cont. Zool. 593: 1–53 (description, distribution, ecology) (plate 186D–186F).
Zeugophilomedes oblonga (Juday, 1907) (=*Euphilomedes oblonga*, of Oliver et al. 1980, Fishery Bull. 78: 437–454). Surface plankton off San Diego Bay and San Pedro (Juday 1907, Univ. Calif. Publ. Zool. 3: 135–156); benthic, 9 m–14 m, subtidal high energy beach, Monterey Bay (Oliver et al. 1980); see Kornicker and Harrison-Nelson 1997, Smithsonian Cont. Zool. 593: 1–53.

RUTIDERMATIDAE

For bibliography giving synonymy, distribution, biology for all Rutidermatidae as of 1986, see Cohen and Kornicker 1986, Smithsonian Cont. Zool. 449: 1–11. For keys see Kornicker and Myers 1981, Smithsonian Cont. Zool. 334: 1–35 (S. Calif.); Kornicker 1983, Smithsonian Cont. Zool. 371: 1–86 (part of Rutidermatidae); Kornicker 1992, Smithsonian Cont. Zool. 531: 1–243 (juvenile instars of *Rutiderma*). For subfamily Metaschismatinae see Kornicker 1994, Smithsonian Cont. Zool. 553: 1–200.
Rutiderma apex Kornicker and Nelson, 1997 (=*Rutiderma* sp. of Tuel et al. 1976, Biol. Survey Pillar Point Harbor, El Granada CA, Rep. Marine Ecol. Inst.). Benthic on sand, sandy silt, and silty clay, 1.8 m–11 m, Pillar Point Harbor, Half Moon Bay, Tomales Bay, and Dark Gulch, Mendocino County; also benthic on silty sea grass near Spud Pt. Marina, Bodega Harbor (males scooped from bottom, depth <1 m, and at night in subtidal benthic light trap) CAS, AC. See Kornicker and Harrison-Nelson 1997, Smithsonian Cont. Zool. 593: 1–53 (description, distribution, ecology) (plates 187B, 187C, 188A, 188B).

SARSIELLIDAE

For keys to subfamilies of Sarsiellidae, genera of Sarsiellinae, see Kornicker 1986, Smithsonian Cont. Zool. 415: 1–217; Kornicker 1991, Smithsonian Cont. Zool. 505: 1–140. Kornicker, L. S. and F. E. Caraion. For key to genera of subfamily Dantyinae see Kornicker and Caraion 1980, Smithsonian Cont. Zool. 309: 1–27.

SARSIELLINAE

Eusarsiella zostericola (Cushman, 1906) (=*Sarsiella zostericola* Cushman, 1906; =*Sarsiella tricostata* Jones, 1958). Benthic at <1 m–3 m, off Pt. Richmond, San Francisco Bay; shallow shores of Atlantic bays from Maine to Chesapeake Bay; also coastal lagoons of Texas, and near Essex, England. The species was introduced to San Francisco Bay, the Gulf of Mexico, and England with oysters (*Crassostrea virginica*) transplanted from the Atlantic coast. See Kornicker and Wise 1962, Crustaceana

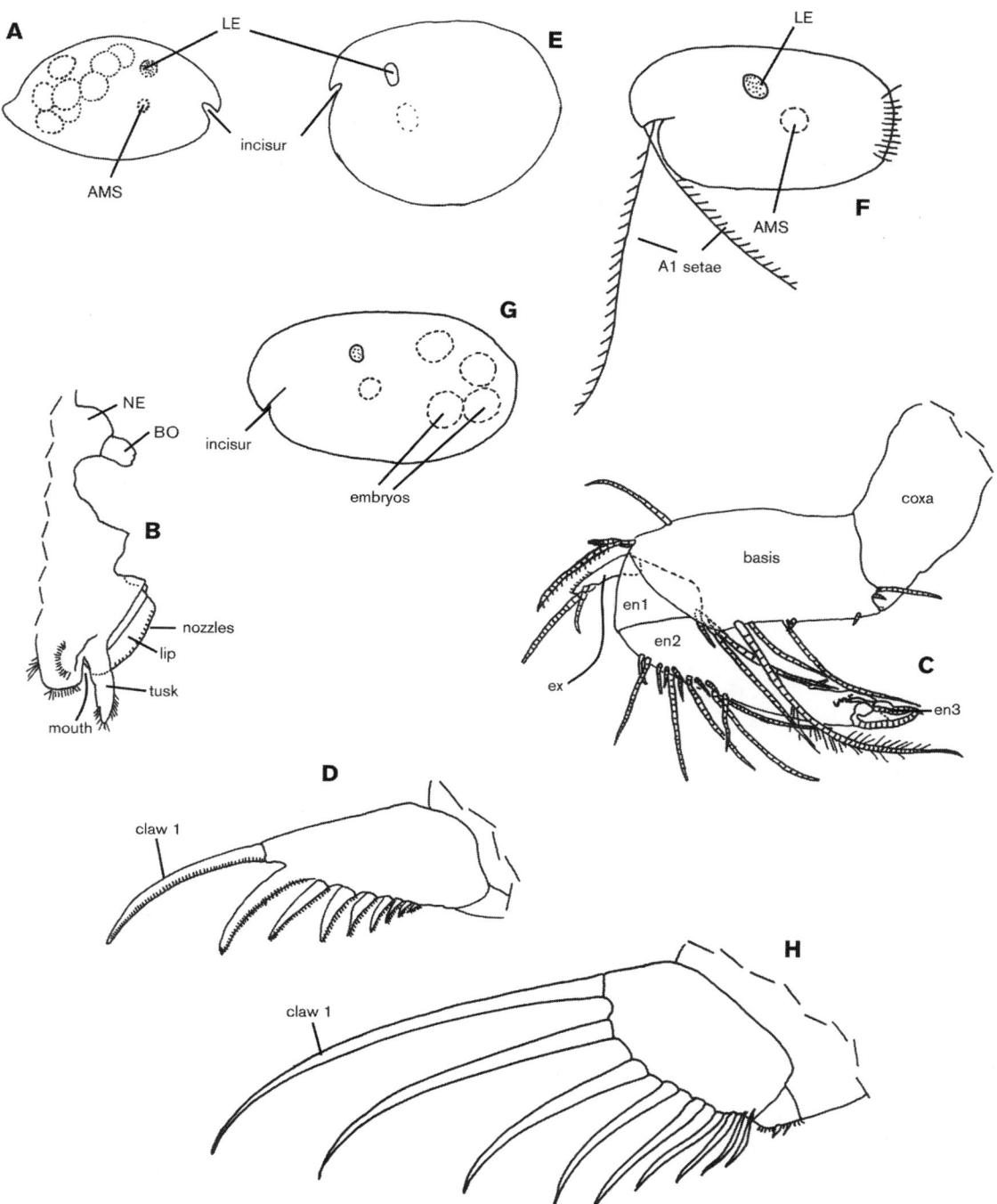

PLATE 185 Myodocopida, Cypridinidae and Cylindroleberididae; Cypridinidae: *Vargula tsujii*, after Kornicker and Baker, 1977: A, female showing deep anterior incisur, interior brooded embryos; B, anterior of body showing Bellonci organ attached to naupliar eye, upper lip (with nozzles and tusk); C, male mandible (medial view; coxal endite not shown, but showing relative shortness of terminal claws); D, left male furca showing 2nd claw without basal suture, other claws with sutures; Cylindroleberididae: E. *Leuroleberis sharpei* after Kornicker, 1981: female, showing circular shape and deep anterior incisur; *Postasterope barnesi* after Kornicker and Harrison-Nelson 1997: F, male, showing deep slitlike anterior incisur, long male setae of 1st antennae, male posterior hair row on valve; G, female, showing slitlike incisur, interior embryos; H, left furca showing all claws with basal suture (all but E, redrawn by Ginny Allen; E by A. Cohen).

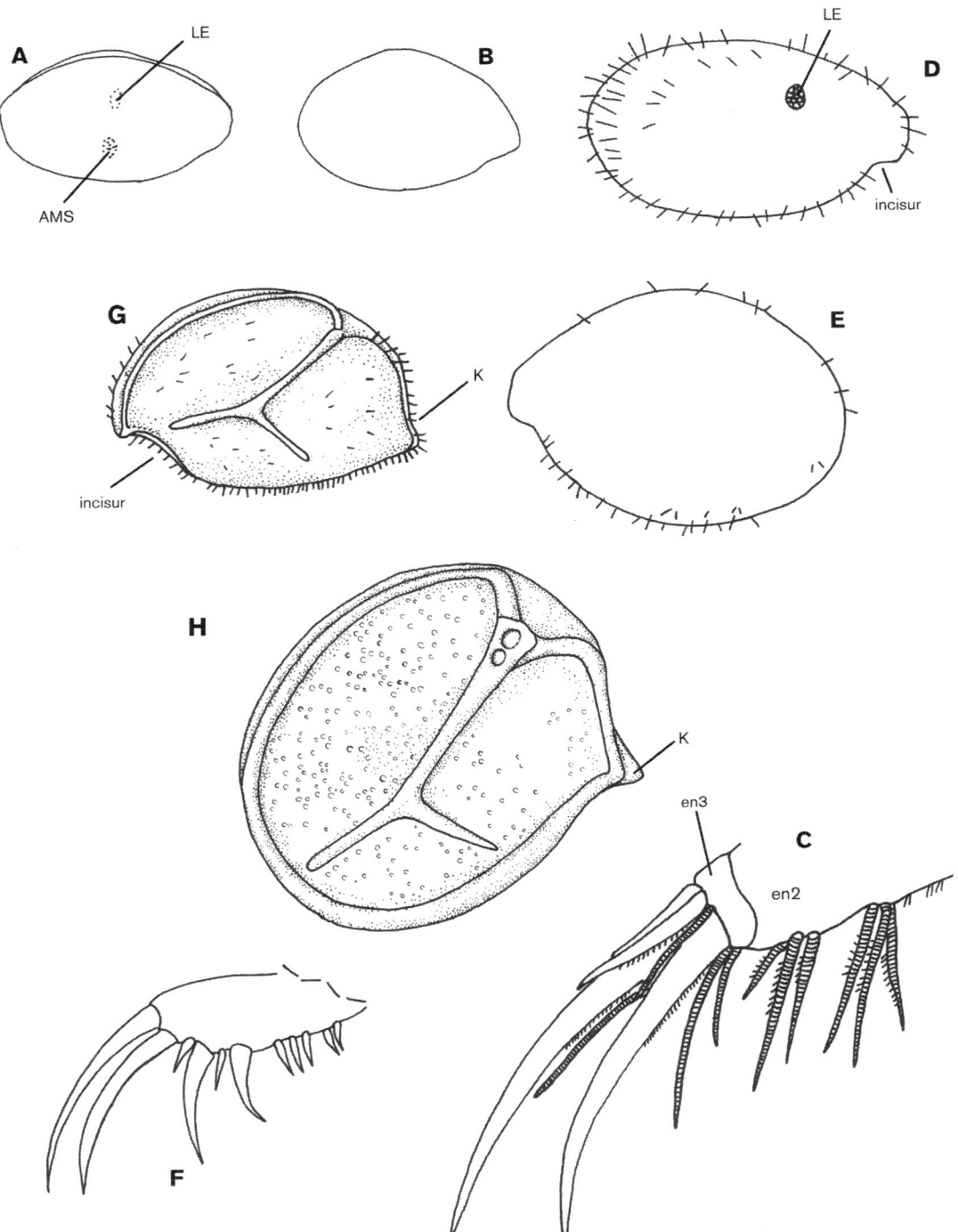

PLATE 186 Myodocopida, Philomedidae, Sarsiellidae; Philomedidae: *Euphilomedes carcharodonta*, after Poulsen 1962: A, male (more elongate valve, larger more visible lateral eye than female); B, female; C, female mandible tip, showing long terminal claws; *Euphilomedes morini,* after Kornicker and Harrison-Nelson, 1997: D, male; E, female; F, male left furca; Sarsiellidae: *Eusarsiella zostericola*, after Kornicker 1967, anterior of valves to left: G, male (with overhanging anterior rostrum); H, female (with mostly circular shape, no rostrum) (all redrawn by Ginny Allen).

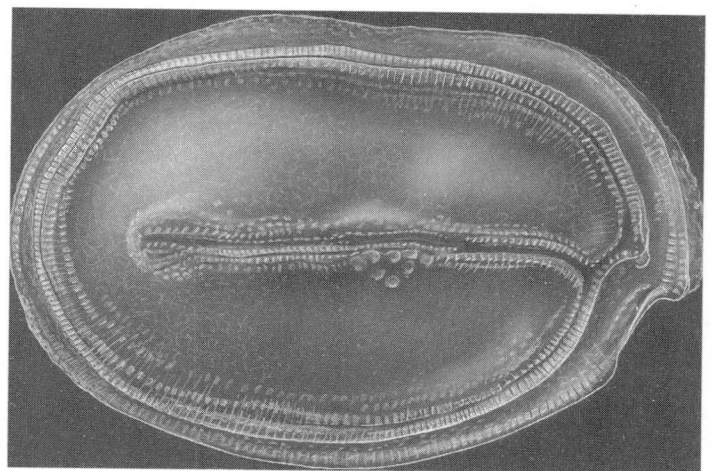

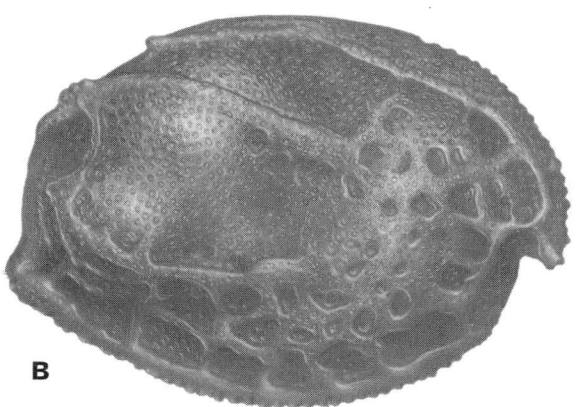

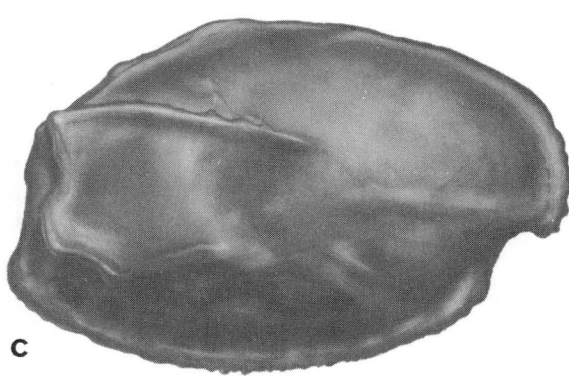

PLATE 187 Myodocopida, Cylindroleberididae and Rutidermatidae, showing ornate valves. Cylindroleberididae: A, *Asteropella slatteryi* female, from Kornicker, 1981, showing shallow incisur; Rutidermatidae: *Rutiderma apex,* from Kornicker and Harrison-Nelson, 1997: B, female; C, male (more elongate and less ornate than female) (drawn by Carolyn Gast, loaned by Smithsonian Institution).

4: 59–74 (description); Kornicker 1975, Bull. Amer. Paleont. 65: 129–139 (spread with oysters); Kornicker and Harrison-Nelson 1997, Smithsonian Cont. Zool. 593: 1–53 (distribution, and references therein); Bamber 1987, J. Micropaleo. 6: 57–62 (life history in Great Britain) (plates 186G, 186H, 188C–188E).

PODOCOPA

ROSALIE F. MADDOCKS AND DAWN PETERSON

PODOCOPIDA

BAIRDIOCOPINA

BAIRDIIDAE

Neonesidea sp. In sand dredged from shore, Whaler's Cove, Point Lobos (plate 195H).

CYPRIDOCOPINA

CYPRIDIDAE

Cypridopsis vidua (O. F. Müller, 1786). Females and juveniles in high intertidal/supratidal tide pools with freshwater influx, Bodega Bay, RFM. This parthenogenetic species is abundant worldwide in permanent and temporary fresh waters. See Sars 1925, An Account Crustacea Norway (Ostracoda) 9: 1–277; Tressler, 1959, in Fresh-water Biology, Ward, Whipple and Edmonson, eds.: 657–734); Delorme 1970, Can. J. Zool. 48: 253–266) (plate 189C–189E).

Herpetocypris reptans (Baird, 1835). Females in high intertidal/supratidal tide pools with freshwater influx, Bodega Bay, RFM. Parthenogenetic, common in fresh waters of Europe and North America. See above references (plate 189F–189H).

Heterocypris salina (Brady, 1868) (=*Cyprinotus salinus*). Empty valves in roots of eelgrass in brackish water inlet, depth <10 cm, Richardson Bay, Marin County, CA, DP. Females and juveniles in high intertidal/supratidal pools with freshwater influx, Bodega Bay, RFM. This parthenogenetic, halophile, somewhat variable species is very widely reported in brackish waters of Europe, Russia, the Caspian Sea and North America. See above references and Delorme 1970, Can. J. Zool. 48: 153–168; Swain (1999, Fossil Nonmarine Ostracoda U.S.). See also Ganning 1967, Helgoländer wiss. Meeresunters. 15: 27–40 (ecology of rockpool populations of Scandinavia) (plate 189A, 189B, 189I–189L).

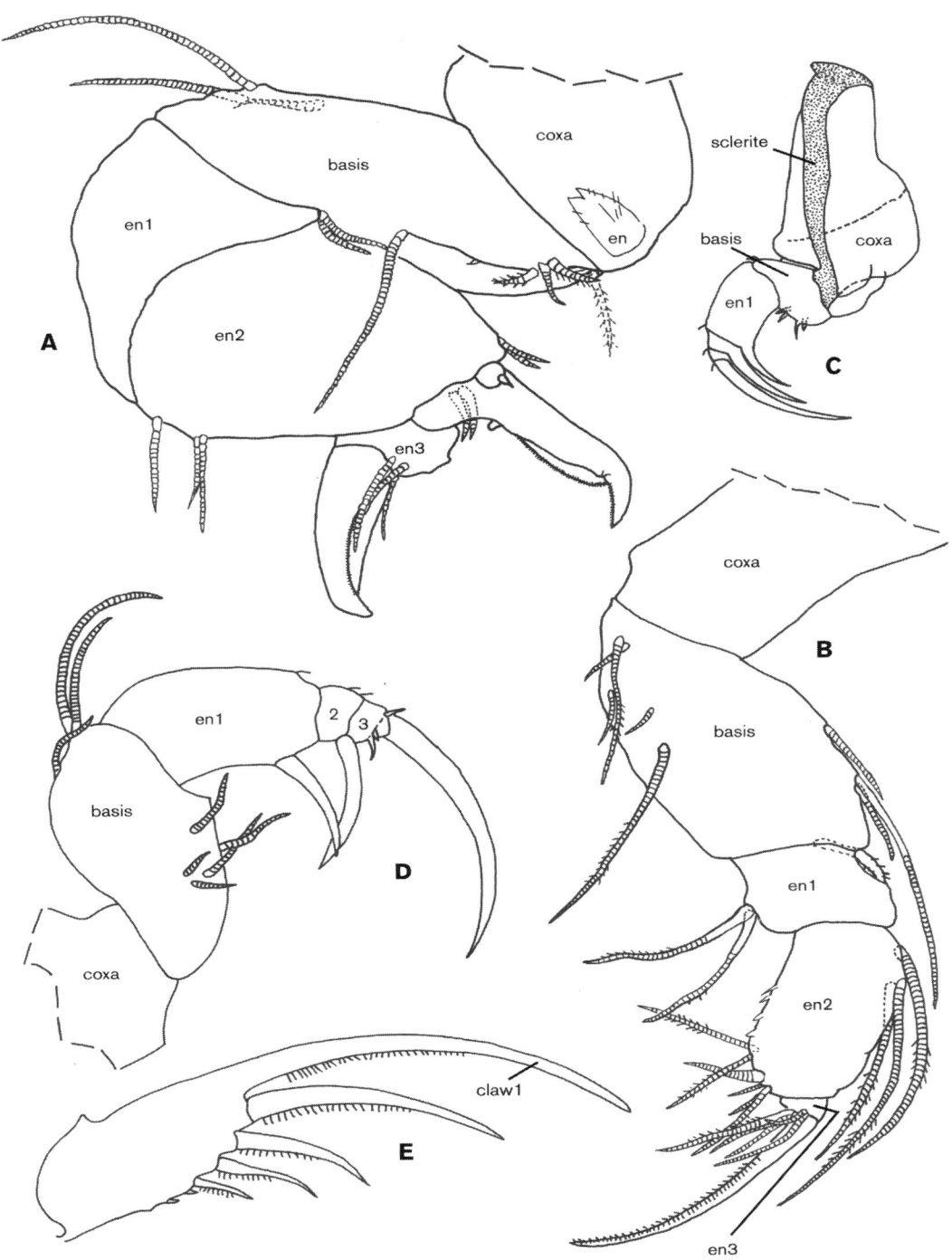

PLATE 188 Myodocopida, Rutidermatidae and Sarsiellidae; Rutidermatidae: *Rutiderma apex,* after Kornicker and Harrison-Nelson, 1997: A, female mandible (showing terminal pincer-claws), medial view; B, male mandible (no pincer), medial view; Sarsiellidae: *Eusarsiella zostericola*: C, female mandible (showing one long ventral claw on each endopodial podomere forming rake), medial view; D, male mandible (long claw only on terminal podomere), medial view; E, male right furca showing 1st (longest) claw without basal suture, other claws with sutures (C, D, after Kornicker 1967; E, after Kornicker and Wise 1962; redrawn by Ginny Allen).

Sarscypridopsis aculeata (Costa, 1847) (=*Cypridopsis aculeata*). Empty valves in roots of eelgrass in brackish water inlet, depth <10 cm, Richardson Bay, Marin County, DP. It has been widely reported from Europe, Iceland, the Americas, central Asia and Africa. See above references and Delorme 1970, Can. J. Zool. 48: 253–266; Forester and Brouwers 1985, J. Paleont. 59: 344–369 (plate 195A, 195B).

CYTHEROCOPINA

BYTHOCYTHERIDAE

Sclerochilus sp. On intertidal mud and fine sand flat near Sandpiper Restaurant, northeast Bodega Harbor, RFM (plate 190A–190L).

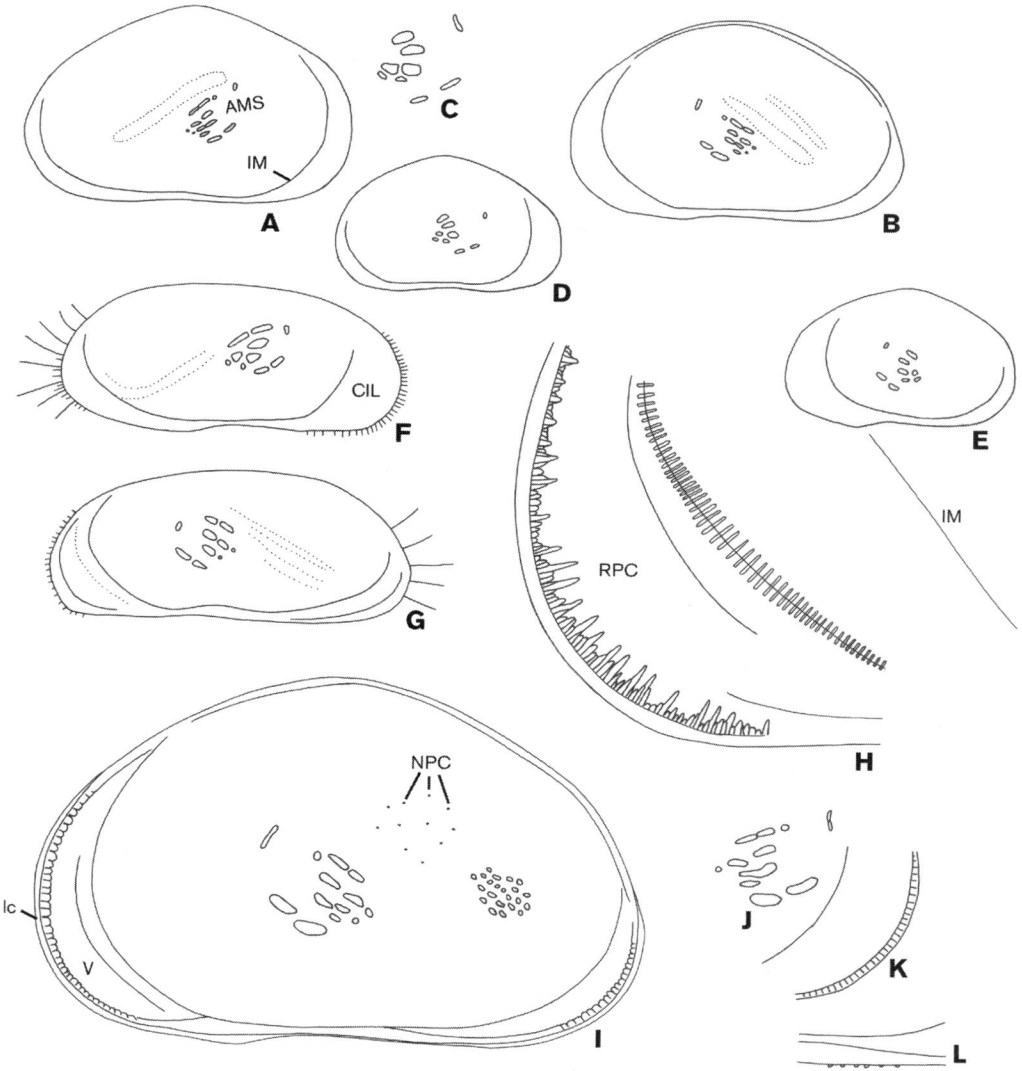

PLATE 189 Podocopida: Freshwater Cyprididae: *Heterocypris salina* (Brady, 1868), female specimen RFM 3413F (A, B, I–L): A, B, exteriors of right and left valves with traces of ovaries, X60; I, exterior of left valve, normal pore canals and granular micro-ornament indicated as dots and ovals, X105; J, muscle scar pattern of right valve, X105; K, anterior margin of right valve with radial pore canals, X105; L, anteroventral margin of right valve with minute tuberculae, X175; *Cypridopsis vidua* (O. F. Müller, 1786), female specimen RFM 3411F (C–E): C, muscle scar pattern of right valve, X105; D, E, exteriors of right and left valve, X60; *Herpetocypris reptans* (Baird, 1835), female specimen RFM 3412F (F–H): F, G, exteriors of right and left valves, X35; H, anterior margin of left valve with flange, radial pore canals, outer list, striate inner list, and inner margin, X175 (drawn by R. Maddocks).

CYTHERIDAE

Cythere alveolivalva Smith, 1952 (=*Cythere uncifalcata* Smith, 1952; *Cythere* sp. A of Valentine, 1976). This species is common in coastal regions of the northern Kurile Islands, Kamchatka Peninsula, and Alaska to California; also fossil (Pleistocene and Holocene) in Oregon and California (Tsukagoshi and Ikeya 1987) (plate 191B, 191C).

Cythere valentinei Tsukagoshi and Ikeya, 1987 (=*Cythere* sp. B of Valentine, 1976; *Cythere* cf. *C. lutea* and *Cythere alveolivalva* (part) of Swain 1969, in Neale (ed.), Taxon. Morph. Ecol. Rec. Ostracoda; *Cythere lutea* of Swain and Gilby 1974, Micropaleo.

20: 257–352). On intertidal mud and fine sand flat in Bodega Harbor, RFM. Valentine (1976) reported range in offshore waters as Point Reyes to southern Vancouver Island; also fossil (Pleistocene of Cape Blanco, Oregon).

Cythere sp. of Tsukagoshi and Ikeya, 1987 (=*Cythere alveolivalva* (part) of Swain 1969, in Neale (ed.), Taxon., Morph. Ecol. Rec. Ostracoda, and Swain and Gilby 1974, Micropaleo. 20: 257–352; *Cythere* sp. aff. *C. lutea* of Swain and Gilby 1974; *Cythere maia* of Valentine, 1976). Tide pool at Point Piedras Blancas, CA. Valentine (1976) reported range in offshore waters as Baja California to Point Piedras Blancas, CA; also fossil (Pliocene-Pleistocene).

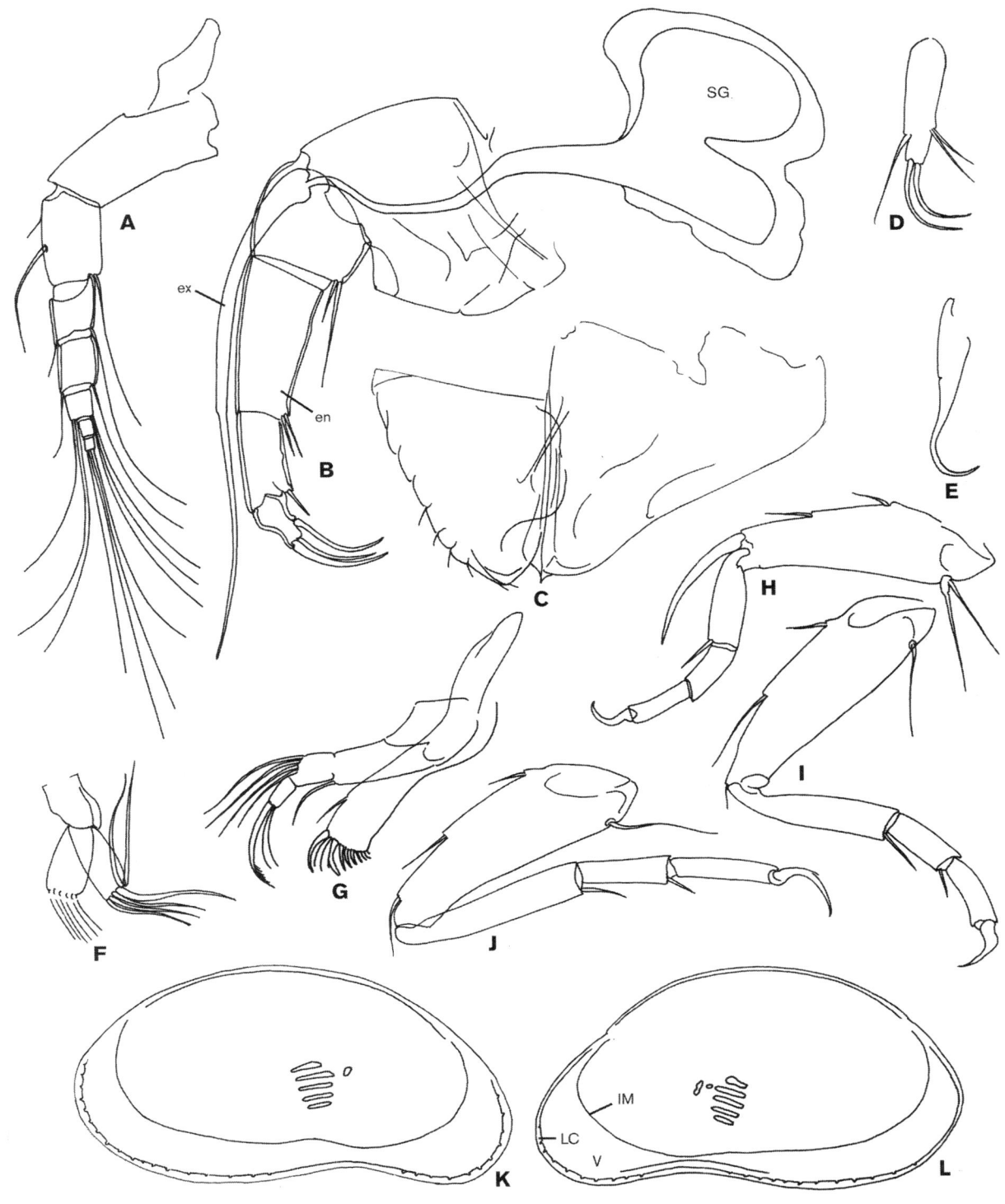

PLATE 190 Podocopida: Bythocytheridae: *Sclerochilus* sp.: female specimen RFM 3408F (A–E, G–L): A, first antenna; B, second antenna with spinnaret gland; C, mouth region with forehead, upper and lower lip; D, E, palp and masticatory process of maxillula (fourth limb); G, mandible; H–J, fifth, sixth and seventh limbs; all X430; K, L, exteriors of right and left valves, X120; F, male specimen RFM 3403M, brush-shaped organ, X430 (drawn by R. Maddocks).

CYTHERIDEIDAE

Cyprideis beaconensis (LeRoy, 1943). Shallow water among algae bloom, salt marsh, Skaggs Island Naval Station, Sonoma Co., CA, DP; Lake Merritt, Oakland; widely distributed in estuaries and marshes along the Pacific coast from southern Chile to British Columbia, also Midway and Oahu Islands in the North Pacific; probably dispersed on migratory waterfowl (Sandberg 1964, Stockholm Contribs. Geology 7: 1–178; Sandberg and Plusquellec 1974, Geoscience and Man 6: 1–26) (plate 195I, 195J).

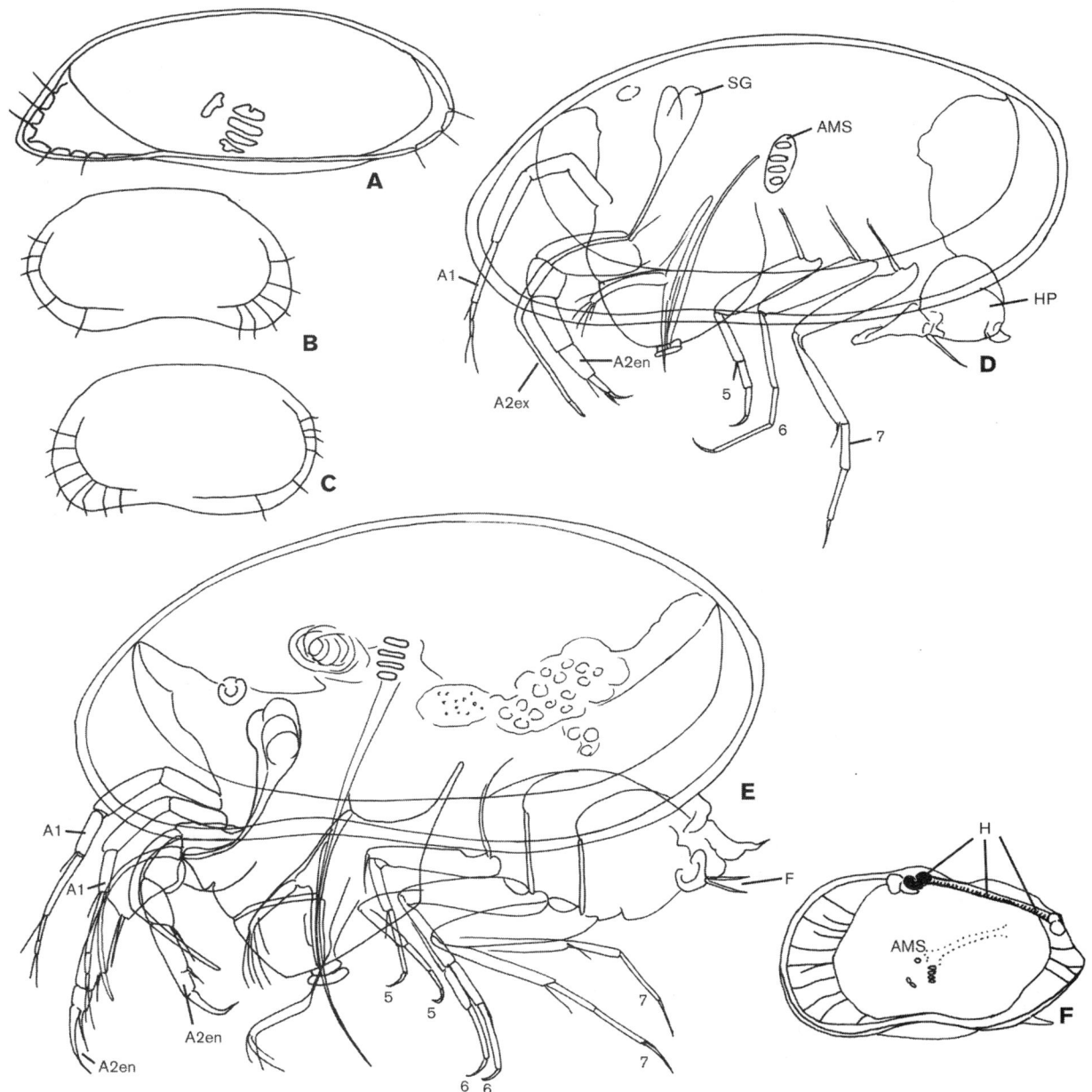

PLATE 191 Podocopida: Cytheridae, Paradoxostomatidae, Schizocytheridae. Paradoxostomatidae: A, *Acetabulostoma californica,* exterior of male left valve, X80, after Watling, 1972. Cytheridae: B, C,, *Cythere alveolivalva* Smith, 1952, exteriors of male right and left valves, X35, after Smith, 1952; Paradoxostomatidae: *Paradoxostoma striungulum* Smith, 1952 (D, E): D, entire male specimen RFM 3400M, as seen from left side in transmitted light; E, entire female specimen RFM 3401F, as seen from left side in transmitted light; with postmortem extrusion of body from carapace; all X175; Schizocytheridae: F, *Spinileberis hyalina,* interior of right valve, X65, redrawn from Watling 1970 (drawn by R. Maddocks).

CYTHEROMATIDAE

Paracytheroma similis Skogsberg, 1950 (Calif. Acad. Sci., Proc. 26: 483–505). On sand and rocks, 15 m in Monterey Bay, near Pacific Grove (plate 194K).

CYTHERURIDAE

Howeina sp. aff. *H. camptocytheroidea* Hanai, 1957. On intertidal mud and fine sand flat in Bodega Harbor, RFM (plate 192A–192J).

HEMICYTHERIDAE

Ambostracon glaucum (Skogsberg, 1928) (=*Cythereis glaucum* Skogsberg 1928, Calif. Acad. Sci., Occ. Pap. 15: 1–143). On holdfasts of algae in Carmel Bay; also on calcareous algae in tide pool just outside Hopkins Marine Station, Pacific Grove; Swain and Gilby (1974) reported it from 12 m in Bodega Bay; from 23 m in Morro Bay, and from 11 m–26 m in Coos Bay (Micropaleo. 20: 257–352). Valentine (1976) gave the range in offshore waters as Baja California to Pt. Piedras Blancas (Micropaleo. 20: 257–352) (plate 195C, 195D).

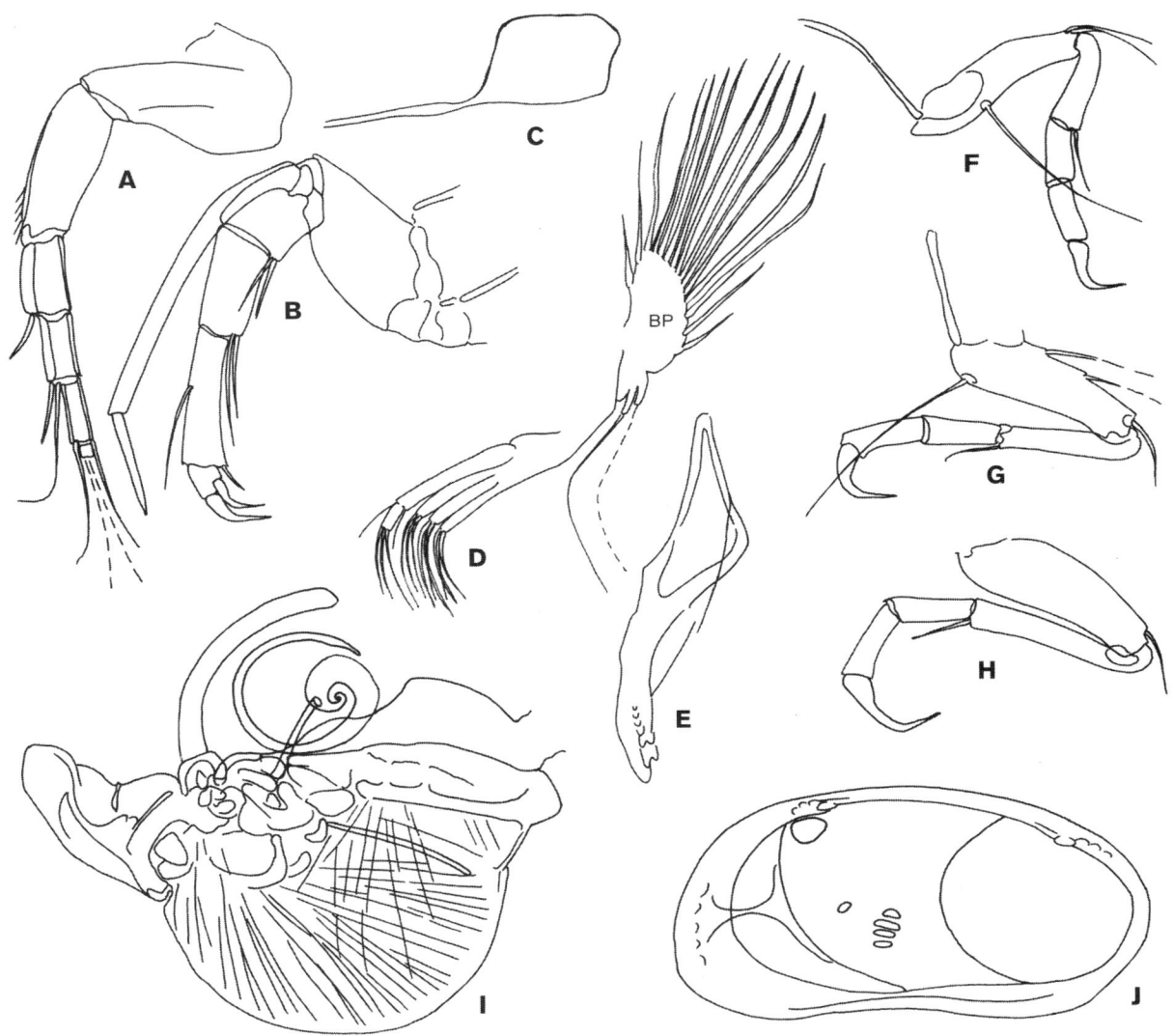

PLATE 192 Podocopida: Cytheruridae: *Howeina* sp. aff. *H. camptocytheroidea* Hanai, 1957, male specimen RFM 3403M: A, first antenna; B, second antenna; C, spinnaret gland; D, maxillula (fourth limb); E, mandible; F–H, fifth, sixth, and seventh limbs; I, hemipenis, all X430; J, exterior of left valve, X175 (drawn by R. Maddocks).

Aurila laeviculoidea Swain and Gilby, 1974. In sand in 10 m off Crescent City.

Aurila lincolnensis (LeRoy, 1943). Among shore rocks at Rockaway Beach, San Mateo Co., DP; also on intertidal mud and fine sand flats in Bodega Harbor; Swain and Gilby (1974) reported it from sand in 12 m in Bodega Bay, RFM, and in 10 m off Crescent City (Micropaleo. 20: 257–352). The species is widely reported off southern California and Mexico. Valentine (1976) gave the range in offshore waters as Baja California to Cape Flattery (plate 196E, 196F).

Aurila montereyensis (Skogsberg, 1928) (=*Cythereis montereyensis*, Skogsberg, 1928). On calcareous algae, among roots of eelgrass, and on holdfasts of *Macrocystis* in 2 m, in tide pools at Carmel Bay; Bodega Harbor, in fairly dense seagrass, silt with fine sand, RFM. Valentine (1976) gave the latitudinal range in offshore waters as north of Pt. Conception to near Cape Alava, Washington (plate 195G).

Aurila sp. aff. *A. corniculata* (Okubo, 1980). McNears Beach, San Francisco Bay, *A. corniculata* is reported from coastal waters of Japan, Korea, and the South Kurile Islands (Schornikov and Tsareva 1995, Mitt. Hamburg. Zoolog. Mus. Inst. 92: 237–253) (plate 196I, 196J).

Aurila sp. of Swain and Gilby, 1974 (Micropaleo. 20: 257–352). In sand in 10 m in Bodega Bay. It is likely that additional species of *Aurila* and related genera occur in the intertidal zone within the range of this manual (see Valentine 1976).

Radimella aurita (Skogsberg, 1928) (=*Cythereis aurita* Skogsberg 1928, Calif. Acad. Sci., Occ. Pap. 15: 1–143). On calcareous algae in tide pool and on holdfasts of *Macrocystis*, outside Hopkins Marine Station on Monterey Bay, near Pacific Grove; also on kelp near shore and on eelgrass in Carmel Bay. Valentine (1976) reported the range in offshore waters as Baja California to north of Santa Cruz); also fossil (plate 195E, 195F).

Radimella? pacifica (Skogsberg, 1928) (=*Cythereis pacifica* Skogsberg 1928, Calif. Acad. Sci., Occ. Pap. 15: 1–143). On calcareous algae in tide pool and on holdfasts of *Macrocystis* in 2

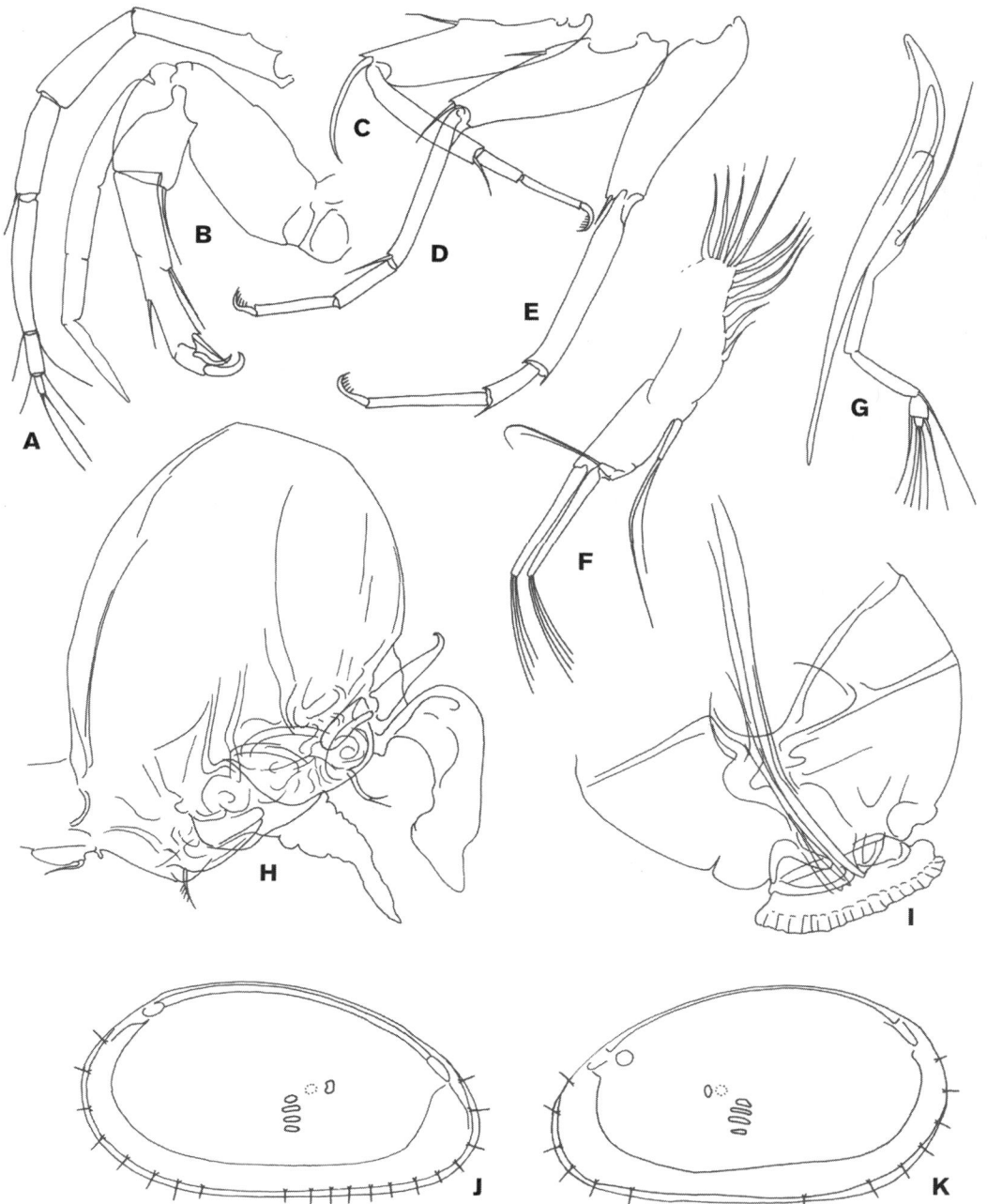

PLATE 193 Podocopida: Paradoxostomatidae: *Paradoxostoma* sp., male specimen RFM 3407M: A, first antenna; B, second antenna; C–E, fifth, sixth, and seventh limbs; F, maxillula (fourth limb); G, mandible; H, hemipenis; I, mouth region with forehead, upper lip, oral sucking disk, and mandible within esophagus; all X430; J, K, exteriors of right and left valves, X175 (drawn by R. Maddocks).

m, outside Hopkins Marine Station near Pacific Grove; also on kelp near shore and on eelgrass in Carmel Bay. Valentine (1976) reported the range in offshore waters as Baja California to south of Cape Vizcaino; also fossil.

Robustaurila jollaensis (LeRoy, 1943) (=*Aurila jollaensis*, of Swain and Gilby 1974, Micropaleo. 20: 257–352). Bodega Bay in 12 m and off Crescent City in 10 m. Valentine (1976) reported the range in offshore waters as Baja California to Cape Flattery; also fossil (plate 196G, 196H).

LOXOCONCHIDAE

Cytheromorpha sp. Empty valves in brackish water on medium-grained sand and rocks, in less than 1 m, McNears Beach, Marin Co.; in brackish water on muddy sand, 10 cm, Richardson Bay, Marin Co.; in brackish water on mudflat among floating algae, in 5 cm, Bolinas Bay, Bodega Bay, DP.

Loxoconcha lenticulata LeRoy, 1943. Empty valves in brackish water on medium-grained sand and rocks, <1 m, McNears

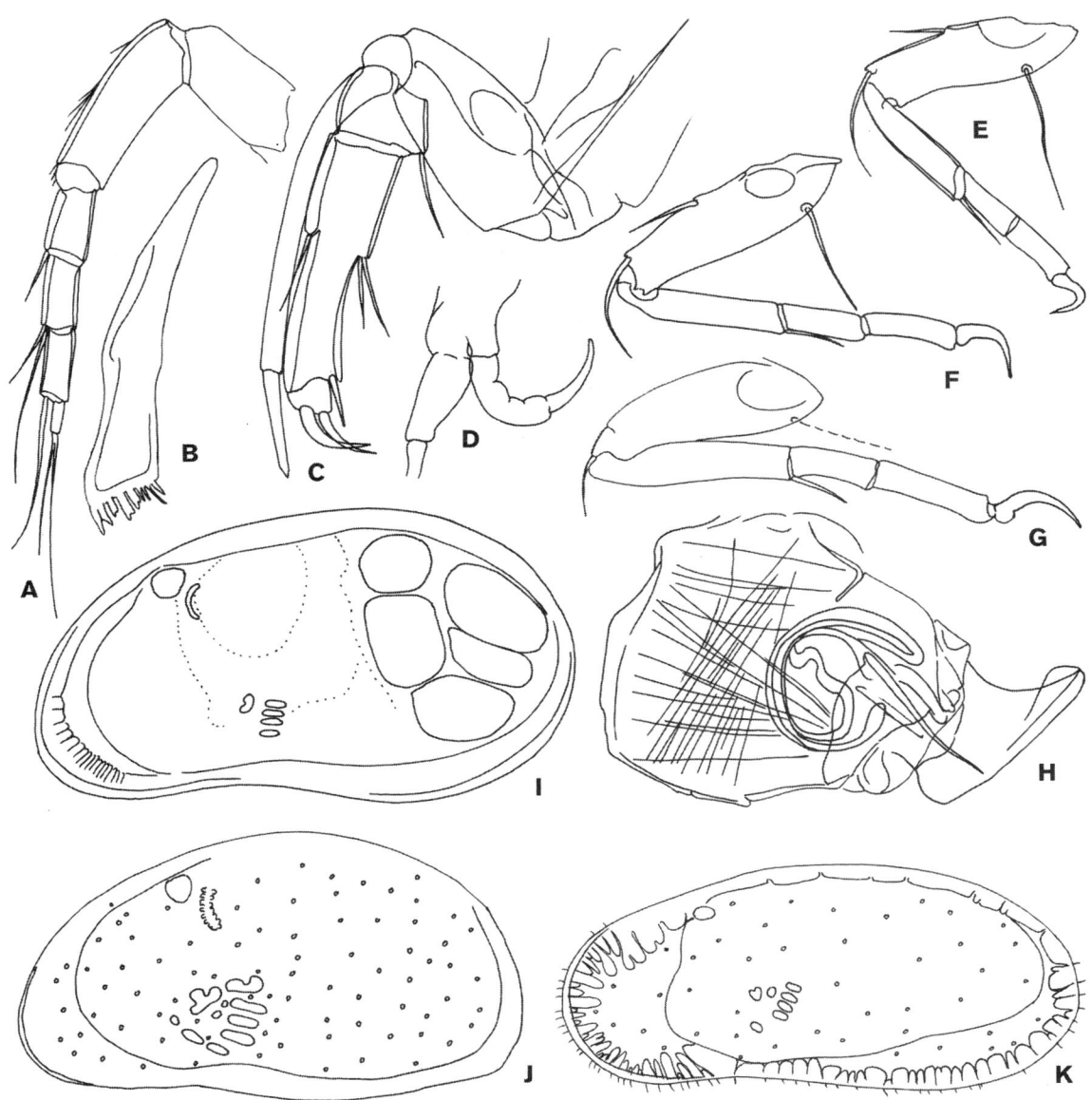

PLATE 194 Podocopida: Cytheromatidae and Xestoleberididae. Xestoleberididae: A–I, *Xestoleberis* sp., male specimen RFM 3402M (A–H); A, first antenna; B, mandibular base; C, antenna; D, brush-shaped appendage; E–G, fifth, sixth, and seventh limbs; H, hemipenis; all X430; I, left exterior of female specimen RFM 3410F, with developing eggs in domiciliary brood pouch, pigmentation and patch pattern indicated by dots, X175; J, *Xestoleberis hopkinsi* male left valve exterior, X120; Cytheromatidae: K, *Paracytheroma similis* male left valve exterior, X100 (J, redrawn from Skogsberg, 1950, pl. 29, fig. 1; K, redrawn from Skogsberg, 1950; drawn by R. Maddocks).

Beach, Marin Co., DP. Valentine (1976) reported the range in offshore waters as northernmost Baja California to Pt. Conception; also fossil (plate 196B).

PARADOXOSTOMATIDAE

Acetabulostoma californica Watling 1972 (Proc. Biol. Soc. Washington 85: 481–488). Low intertidal on exposed side of Tomales Point, Marin Co., in the zone of the red alga *Corallina gracilis*. Most species of *Acetabulostoma* are parasitic on gammarid amphipods, but the host (if any) of *A. californica* is unknown (plate 191A).

Paradoxostoma striungulum Smith, 1952. Empty valves in brackish water on medium-grained sand and rocks, less than 1 m, McNears Beach, Marin Co., DP; living on intertidal fouled eelgrass (*Zostera marina*) blades in the South Slough National Estuarine Research Reserve, Coos Bay (J. T. Carlton, collector), CAS, RFM. Described from Departure Bay, British Colombia, on *Obelia* near water surface (Smith 1952, J. Fish. Res. Board Canada 9: 16–41) (plate 191D, 191E).

Paradoxostoma sp. One male living on mud and fine sand flat exposed at low tide, Bodega Harbor, RFM. It is likely that additional species of *Paradoxostoma* and related genera occur within the range of this book (plate 193A–193K).

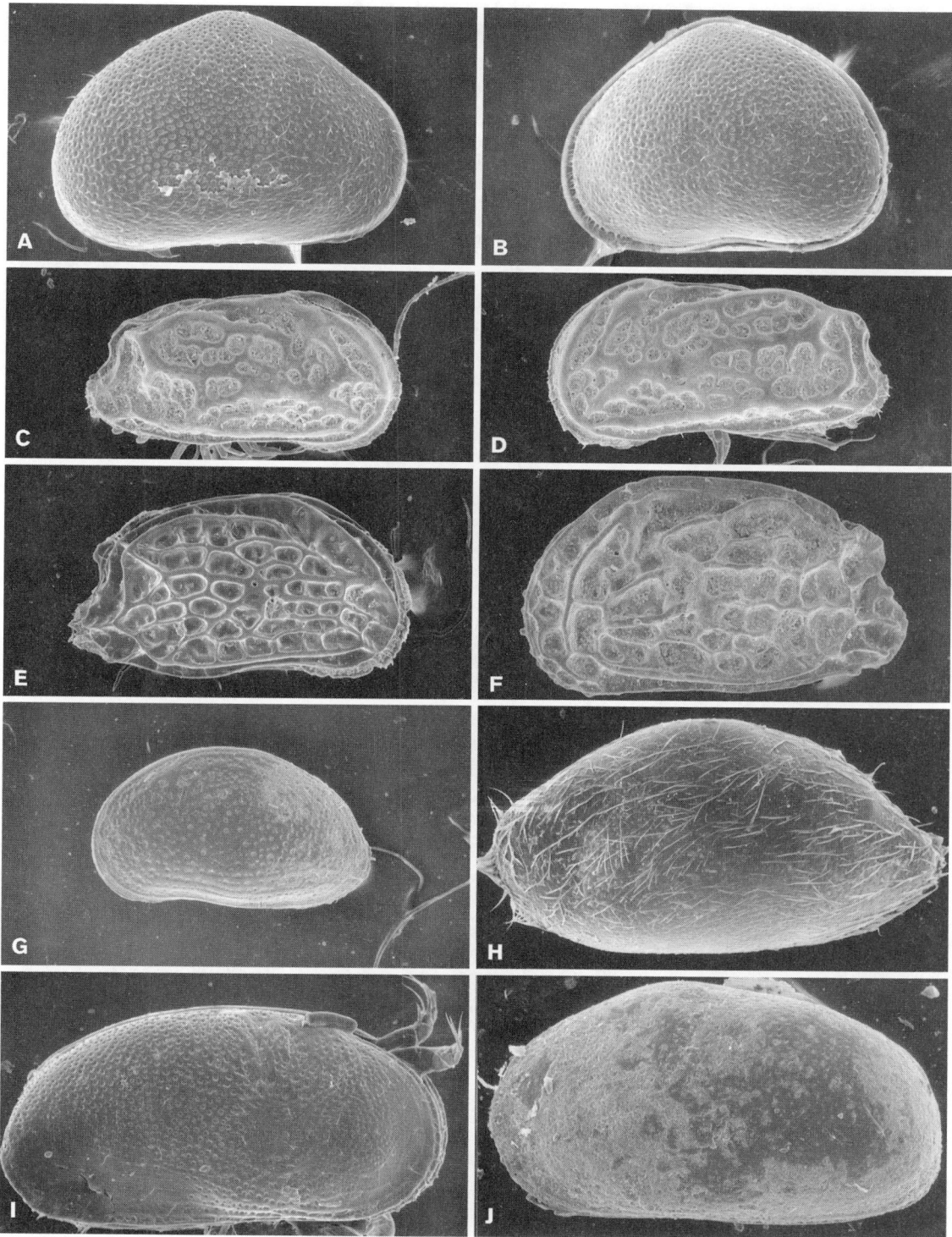

PLATE 195 Podocopida: Bairdiidae, Cyprididae, Cytherideidae, Hemicytheridae. A, B, Cyprididae: *Sarscypridopsis aculeata* (Costa, 1847), CAS 121631, right and left exteriors of female carapaces, X67; Hemicytheridae: *Ambostracon glaucum* (Skogsberg, 1928), CAS 121629 (C, D): C, right exterior of female; D, left exterior of male, X67; E, F, *Radimella aurita* (Skogsberg, 1928), CAS 121627: E, exterior of male (?) right valve; F, exterior of female (?) left valve; both X67; G, *Aurila montereyensis* (Skogsberg, 1928), CAS 121622, exterior of female (?) left valve, X67; Bairdiidae: H, *Neonesidea* sp., CAS 120522, left exterior of carapace, X67; Cytherideidae: I, J: *Cyprideis beaconensis* (LeRoy, 1943), CAS 121619: I, right exterior of male carapace; J, left exterior of female carapace with heavy coating of microbial slime; both X67 (SEMs by D. Peterson).

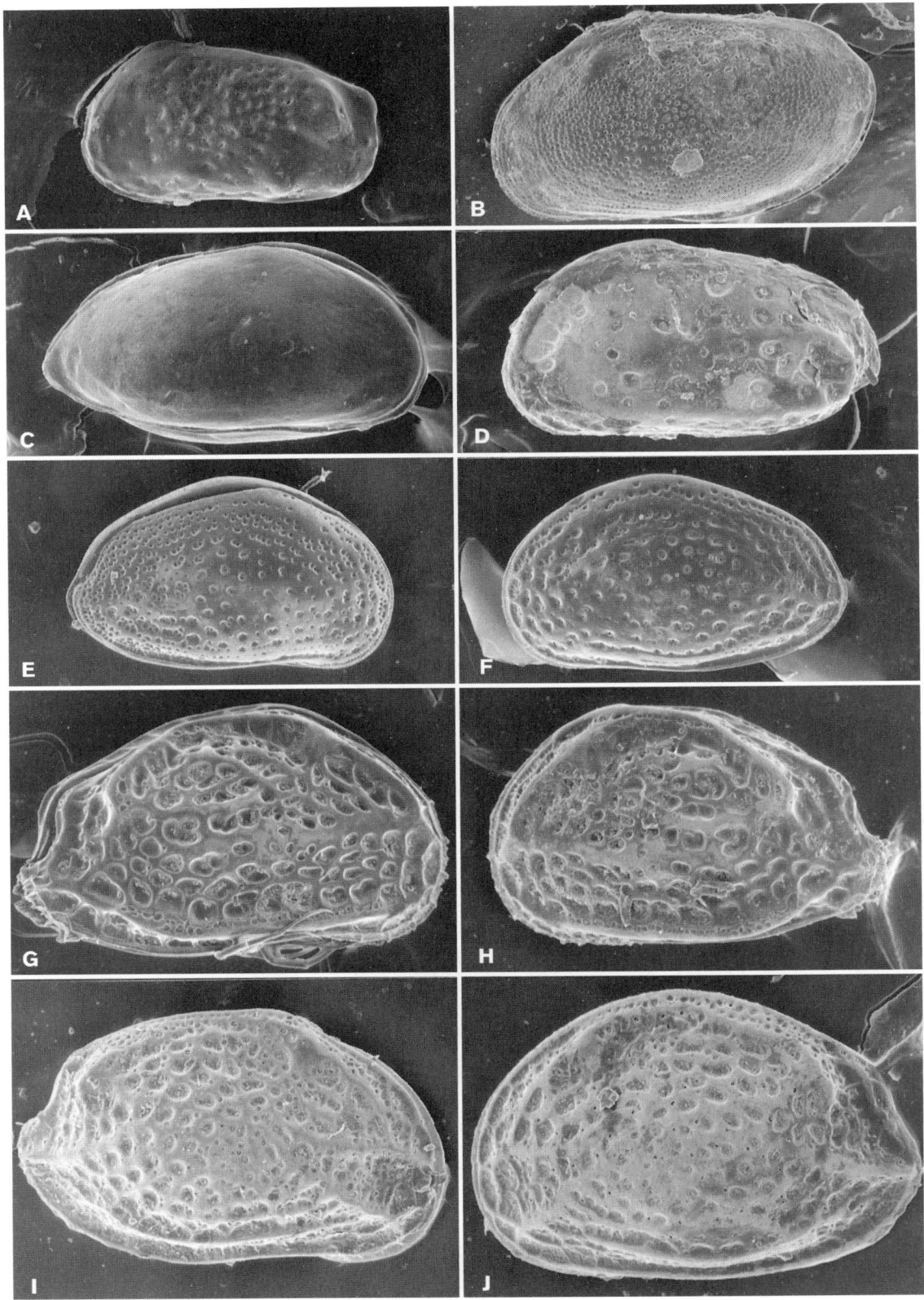

PLATE 196 Podocopida: Hemicytheridae, Loxoconchidae, Pectocytheridae, Schizocytheridae. Pectocytheridae: A, *Pectocythere parkerae* Swain and Gilby, 1974, CAS 121626, exterior of left valve, X100; Loxoconchidae: B, *Loxoconcha lenticulata* (LeRoy, 1943), UMPC 12220, exterior of male left valve, X100; Schizocytheridae: C, *Acuminocythere crescentensis* Swain and Gilby, 1974, UMPC 12096, right exterior of female carapace, X100; D, *Acuminocythere* sp. of Swain and Gilby, 1974, UMPC 12080, exterior of left valve, X100; Hemicytheridae: E, F, *Aurila lincolnensis* (LeRoy, 1943), CAS 121620, right and left exteriors of carapaces, X100; G, H, *Robustaurila jollaensis* (LeRoy, 1943), CAS 1212630, right and left exteriors of carapaces, X100; I, J, *Aurila* aff. *A. corniculata* (Okubo, 1980), CAS 121628, exteriors of right and left valves, X100 (SEMs by D. Peterson).

PECTOCYTHERIDAE

Pectocythere parkerae Swain and Gilby, 1974 (=*Munseyella* sp. B of Valentine, 1976). On sand in 11 m–36 m in Bodega Bay, off Crescent City, and in Coos Bay. Valentine (1976) reported the range in offshore waters as San Diego to Cape Lookout, OR (plate 196A).

Pectocythere tomalensis Watling, 1970 (Crustaceana 19: 251–263). In 6 m on sandy bottom, White Gulch, Tomales Bay; also on sand in 12 m in Bodega Bay (Swain and Gilby 1974, Micropaleo. 20: 257–352); Valentine (1976) reported the range in offshore waters as Monterey Bay to Cape Flattery.

SCHIZOCYTHERIDAE

Acuminocythere crescentensis Swain and Gilby, 1974 (=*"Paijenborchella"* sp. A of Valentine, 1976). Morro Bay and off Crescent City and Coos Bay in 10 m–26 m. Valentine (1976) reported the range in offshore waters as north of Point Conception to Cape Flattery (plate 196C).

Acuminocythere sp. of Swain and Gilby, 1974 (=*"Paijenborchella"* sp. B of Valentine, 1976). Empty carapace in mud and fine sand, flat exposed at low tide, in Bodega Harbor, RFM. Valentine (1976) reported the range in offshore waters as Pt. Buchon to Little River, CA (plate 196D).

Spinileberis hyalina Watling, 1970. On bottom silt and clay, 2 m, in Tomales Bay. See Watling 1970, Crustaceana 19: 251–263 (plate 191F).

XESTOLEBERIDIDAE

Xestoleberis hopkinsi Skogsberg, 1950. On holdfasts of algae in rocky tide pool full of brown algae, just outside the Hopkins Marine Station, Monterey Bay (plate 194J).

Xestoleberis sp. On mud and fine sand flat exposed at low tide, Bodega Harbor, RFM. It is likely that additional species of *Xestoleberis* occur within our range (plate 194A–194I).

ACKNOWLEDGMENTS

We thank Todd Oakley, Jeff Spees, and James T. Carlton for assistance and for providing new collections of Ostracoda. We also thank Frederick M. Swain for the loan of specimens described by Swain and Gilby (1974) from the University of Minnesota Paleontological Collections; Louis Kornicker and Elizabeth Nelson for literature and information regarding myodocopans and for the loan of the half-tone figures (drawn by Carolyn Gast) in plate 187; Ginny Allen for redrawing many myodocopid figures; and the Bodega Marine Laboratory for providing lab space for Cohen and Oakley during collections and for other assistance.

References

GENERAL REFERENCES ON OSTRACODA

Athersuch, J., D. J. Horne, and J. E. Whittaker. 1989. Marine and Brackish Water Ostracods, Synopses of the British Fauna (New Series) (D. M. Kermack and R. S. K. Barnes, eds.) 43: 1–343. New York: E. J. Brill.

Cohen, A. C., J. W. Martin, and L. S. Kornicker. 1998. Homology of Holocene ostracode biramous appendages with those of other crustaceans: the protopod, epipod, exopod and endopod. Lethaia 31: 251–265.

Cohen, A. C., and J. G. Morin. 1990. Patterns of reproduction in ostracodes; a review. Journal of Crustacean Biology 10: 84–211.

Ellis and Messina Catalogue of Ostracoda. Micropaleontology Press, American Museum of Natural History. For online access go to: http://www.micropress.org.

Hartmann, G. 1966, 1967, 1968, 1975, 1989. Ostracoda. In Dr. H. G. Bronns Klassen und Ordnungen des Tierreichs, Fünfter Band: Arthropoda, I. Abteilung, 2. Buch, IV. Teil, Lieferungen 1–5: 1–1067.

Hartmann, G., and H. S. Puri 1974. Summary of neontological and paleontological classification of Ostracoda. Mitteilungen aus dem Hamburgischen Zoologischen Museum und Institut 70: 7–73.

Horne, D., A. C. Cohen, and K. Martens. 2002. Taxonomy, morphology and biology of Quarternary and living Ostracoda. In The Ostracoda: Applications in Quaternary Research, J. Holmes and A. Chivas, eds. AGU Geophysical Monograph Series 131: 5–36.

Kaesler, R. L. 1987. Superclass Crustacea. In Fossil invertebrates: 241-258. R. S. Boardman, A. H. Cheetham, and A. J. Rowell, eds. London: Blackwell Scientific Publs.

Kempf, E. K. 1980. Index and Bibliography of Nonmarine Ostracoda 1, Index A. Geologisches Institut der Universitaet zu Koeln, Sonderveroeffentlichungen, no. 35 (and later volumes in this series).

Kempf, E. K. 1986. Index and Bibliography of Marine Ostracoda, Vol. 1 Index A. Geologisches Institut der Universitaet zu Koeln, Sonderveroeffentlichungen, no. 50 (and later volumes in this series).

Maddocks, R. F. 1982. Ostracoda. In The biology of crustacea, vol. I: systematics, the fossil record and biogeography: 221–239, L. G. Abele, ed. New York: Academic Press.

Maddocks, R. F. 1992. Ostracoda. In Microscopic anatomy of invertebrates 9: crustacea. F. W. Harrison and A. G. Humes, eds. 415–442. New York: Wiley-Liss, Inc.

Martens, K., ed. 1998. Sex and Parthenogenesis. Evolutionary ecology of reproductive modes in non-marine ostracods. Leiden: Backhuys Publ., 336 pp.

Moore, R. C., ed. 1961. Treatise on Invertebrate Paleontology, Part Q Arthropoda 3 Crustacea Ostracoda. Lawrence: Geological Society of America and University of Kansas Press, 442 pp.

Morin, J. G., and A. C. Cohen. 1991. Bioluminescent displays, courtship, and reproduction in ostracodes. In Crustacean Sexual Biology, R. Bauer and J. Martin eds: 1–16. New York: Columbia University Press.

Morkhoven, F. P. C. M. 1962, 1963. Post-Palaeozoic Ostracoda, Their Morphology, Taxonomy and Economic Use 1, 2. Amsterdam: Elsevier, 204 and 478 pp.

Oakley, T., and C. Cunningham. 2002. Molecular phylogenetic evidence for the independent evolutionary origin of an arthropod compound eye. Proceedings of the National Academy of Science 99: 1426–1430.

Smith, R. J., T. Kamiya, and D. J. Horne. 2006. Living males of the "ancient asexual" Darwinulidae (Ostracoda: Crustacea). Proceedings of the Royal Society B: 10 pp.

Spears, T., and Abele, L. G. 1998. Crustacean phylogeny inferred from 18S rDNA. In Arthropod relationships, systematics association special volume series 55, Fortey, R. A. and R. H. Thomas, eds. 169–187. London: Chapman & Hall.

Wingstrand, K. G. 1988. Comparative spermatology of the Crustacea Entomostraca. 2. Subclass Ostracoda. Biologiske Skrifter 32: 1–149.

Some Additional References (but not cited in Faunal List):

OSTRACODA OF NORTHWEST AMERICA

Benson, R. H. 1959. Ecology of recent ostracodes of the Todos Santos Bay region, Baja California, Mexico. University of Kansas Paleontological Contributions, Arthropoda, Article 1: 1–80.

Benson, R. H., and R. L. Kaesler. 1963. Recent marine and lagoonal ostracodes from the Estero de Tastiota region, Sonora, Mexico (northeastern Gulf of California). University of Kansas Paleontological Contributions, Arthropoda, Article 3: 1–34.

Brouwers, E. M. 1983. Occurrence and distribution chart of ostracodes from the northeastern Gulf of Alaska. U.S. Geological Survey, Miscellaneous Field Studies Map MF-1518, Pamphlet, pp. 1–13.

Brouwers, E. M. 1988. Paleobathymetry on the continental shelf based on examples using ostracods from the Gulf of Alaska. In Ostracoda in the earth sciences. P. De Deckker, J.-P. Colin and J.-P. Peypouquet, eds.: pp. 55–76. Amsterdam: Elsevier.

Brouwers, E. M. 1990. Systematic paleontology of Quaternary ostracode assemblages from the Gulf of Alaska, Part 1. Families Cytherellidae, Bairdiidae, Cytheridae, Leptocytheridae, Limnocytheridae, Eucytheridae, Krithidae, Cushmanideidae. U.S. Geological Survey Professional Paper 1510: 1–43.

Brouwers, E. M. 1993. Systematic paleontology of Quaternary ostracode assemblages from the Gulf of Alaska, part 2. Families Trachyleberididae, Hemicytheridae, Loxoconchidae, Paracytherideidae. U.S. Geological Survey Professional Paper 1531: 1–47.

Ishizaki, K., and F. J. Gunther. 1974. Ostracoda of the Family Cytheruridae from the Gulf of Panama. Science Reports of the Tohoku University, Sendai, Second Series (Geology) 45: 1–50.

Ishizaki, K., and F. J. Gunther. 1976. Ostracoda of the Family Loxoconchidae from the Gulf of Panama. Science Reports of the Tohoku University, Sendai, Second Series (Geology) 46: 11–26.

McKenzie, K. G. 1965. Myodocopid Ostracoda (Cypridinacea) from Scammon Lagoon, Baja California. Crustaceana 9: 57–70.

McKenzie, K. G., and F. M. Swain. 1967. Recent Ostracoda from Scammon Lagoon, Baja California. Journal of Paleontology 41: 281–305.

Swain, F. M., and J. M. Gilby. 1967. Recent Ostracoda from Corinto Bay, western Nicaragua, and their relationship to some other assemblages of the Pacific Coast. Journal of Paleontology 41: 306–334.

Valentine, P. C. 1976. Zoogeography of Holocene Ostracoda off western North America and paleoclimatic implications. United States Geological Survey Professional Paper 916, 47 pp.

EXAMPLES OF DIVERSITY IN LITTORAL ZONES

Schornikov, E. I. 1974. Kizucheniuo ostrakod (Crustacea, Ostracoda) litorali Kuril'ckix ostrovov: 137–214. Rastitel'n'ii i Zhivotn'ii Mir Litorali Kuril'skix Ostrovov (in Russian).

Schornikov, E. I. 1975. Ostracod fauna of the intertidal zone in the vicinity of the Seto Marine Biological Laboratory. Publications Seto Marine Biological Laboratory 22: 1–30.

Williams, R. 1969. Ecology of the Ostracoda from selected marine intertidal localities on the coast of Anglesey. In The taxonomy, morphology and ecology of recent ostracoda, J. W. Neale, ed.: pp. 299–329. Edinburgh: Oliver and Boyd.

PREPARATION OF SPECIMENS

Cohen, A. C., and J. G. Morin. 1986. Three new luminescent ostracodes of the genus Vargula from the San Blas region of Panama. Contributions in Science, Natural History Museum of Los Angeles County, 373: 1–23 (anesthetics, etc.).

Cohen, A. C., and J. G. Morin. 1997. External Anatomy of the Female Genital (Eighth) Limbs and the Setose Openings in Myodocopid Ostracodes (Cypridinidae). Acta Zoologica 78: 85–96 (SEM, and references therein).

Copepoda

JEFFERY R. CORDELL

Free-Living Copepoda (Orders Calanoida, Cyclopoida, and Harpacticoida)

(Plates 197–206)

Copepods have been compared to insects, because, like their terrestrial counterparts, they have successfully occupied an astounding diversity of habitats and modes of life (see Huys and Boxshall 1991 for a summary of copepod habitats). In terms of their importance in marine food webs, the role of copepods cannot be overstated. By way of their conversion of detritus and phytoplankton into animal biomass, copepods often form the first link between primary and secondary consumers. Many commercially important fish, including herring, anchovies, and rockfish feed on planktonic copepods during some or all of their life history stages. Benthic and epibenthic harpacticoid copepods often dominate the diets of flatfish and several species of Pacific salmon during their early life histories.

Given their importance in nearshore and estuarine food webs, one would expect to find a number of ecological and taxonomic studies of copepods of coastal California and Oregon. However, this is not the case. Even though coastal and estuarine ecosystems have undergone and continue to undergo rapid changes and increasing stress due to land use, urbanization, and introduced species, there have been no long-term studies of plankton or meiobenthic dynamics in this region. Likewise, taxonomic compendia of copepods for the Pacific coast are few. Esterly (1905, 1906, 1911, 1924) reported on marine plankton of San Francisco Bay and the San Diego region, Dawson and Knatz (1980) published a list and keys of the planktonic copepods of San Pedro Bay, and Gardner and Szabo (1982) give keys and an annotated bibliography for pelagic marine copepods of British Columbia. Lang (1965) produced what remains one of the finest treatments of a local fauna for harpacticoid copepods from interstitial waters and tide pools near the Hopkins Marine Station in Pacific Grove and the now-gone Pacific Marine Station (formerly located in Dillon Beach). He described 98 species of harpacticoids, of which 81 were new.

Thus the taxonomy of the copepod faunas of many California nearshore habitats remains little-studied, including those from salt marshes, estuaries, seagrass beds, sandy beaches, mudflats, rocky shores, fouling communities, and the shallow subtidal. Many species remain undescribed, and an unknown number of introduced taxa remain undetected. Great care must be exercised in attempting identification of copepods from these habitats. Unfortunately, the primary taxonomic literature usually must be used to identify the most common intertidal copepods, the Harpacticoida, to genus or species levels. However, there are several publications that have keys to higher taxonomic levels. Huys et al. (1996) provide a key to world harpacticoid families, and Boxshall and Halsey's An Introduction to Copepod Diversity (2004) contains keys to families of marine planktonic, marine benthic, and fish parasite copepods, and to genera of many copepod families. The latter book is also an excellent starting point for anyone wanting to study copepod classification.

EXTERNAL STRUCTURE OF THREE ORDERS OF FREE-LIVING COPEPODA

Of the nine copepod orders currently recognized, three—Calanoida, Cyclopoida, and Harpacticoida—contain most of the free-living individuals likely to be found in nearshore habitats. A fourth order, the Poecilostomatoida, has been placed into the Cyclopoida (Boxshall and Halsey 2004).

Copepods have developed two basic body plans, gymnoplean and podoplean, which are defined by the position of the major body articulation between prosome and urosome (plate 197). In the gymnoplean plan (Order Calanoida), this articulation is behind the fifth pedigerous somite (plate 197A), and in the podoplean plan (Orders Cyclopoida and Harpacticoida), it is between the fourth and fifth pedigerous segments (plate 197B). The prosome consists of the cephalosome that bears the first six pairs of appendages together with three or four free prosomites, which are sometimes referred to as the metasome and bear four or five pairs of swimming legs (plate 197A, 197B). In most harpacticoids, the somite bearing the first pair of legs is also fused to the cephalosome and together they form the cephalothorax. The urosome is composed of a segment bearing reduced fifth legs (except in calanoids) followed by remaining body segments (plate 197A, 197B). In most free-living copepods, the first two urosome segments are fused in the female to form a genital double somite, resulting in fewer urosome segments in females as compared with males.

The cephalosome bears six pairs of appendages (seven if a cephalothorax) (plate 197C): first antennae (A1), second antennae (A2), mandibles (Md), maxillules (Mxl), maxillae (Mx), and maxillipeds (Mpd).

The first antennae are usually short in harpacticoids and a few cyclopoids (e.g., *Corycaeus*) (nine or fewer segments), intermediate length in free-living cyclopoids (eight to 26 segments and usually shorter than the prosome), and long in calanoids (18–27 segments, usually exceeding the prosome). Except in some cyclopoids, one or both first antennae in the male are modified for grasping the female during mate guarding and copulation. Such geniculation consists of one or several segments being slightly to greatly swollen, and the first antenna being articulated such that it forms a grasping organ (plates 197C, 198A).

The second antennae are diagnostic in separating the free-living suborders of Copepoda. In cyclopoids, the exopod of the second antenna is either absent or reduced such that the appendage appears uniramous (plate 198C). In harpacticoids, the antennary exopod is usually small, consisting of one to four segments (plate 198D), but may rarely have up to eight segments or be absent. In calanoids, the exopod is large and multisegmented (plate 198E).

· Copepod swimming legs (pereiopods, P1–P4) in their basic form consist of a biramous limb with an exopod and endopod (plate 198F). Each limb is united into a pair by an intercoxal sclerite. In free-living calanoids, cyclopoids, and a few harpacticoid families, the P1 has retained the basic unmodified form and is similar to the rest of the swimming legs. However, in the vast majority of the harpacticoids, the endopod and/or exopod of P1 is extensively modified, often into a prehensile limb (plate 198F). For this reason the P1 is of great taxonomic importance in the suborder Harpacticoida.

The fifth pair of legs (P5) are also helpful for separating the free-living copepod suborders. In calanoids, the fifth legs are sexually dimorphic, often complexly modified in the male; in the female the fifth legs are often reduced or absent (plate 198G). In free-living cyclopoids, the endopod of the fifth leg is absent, and the exopod is usually reduced to a single segment (plate 198H). In all but a few harpacticoids, the fifth legs are biramous and leaflike, with articles of the inner ramus fused into a single baseoendopod, and a single-segmented exopod (plate 198I).

The terminal segment of the copepod body is the anal somite, which bears the anus either terminally or dorsally. The caudal rami, which are attached to the anal somite, are unsegmented lobes that are armed with setae of various sizes and lengths (plate 197A, 197B).

DEVELOPMENT

After hatching from the egg, copepods undergo six naupliar stages, followed by six copepodite stages (plate 199). Dahms (1990) extensively describes harpacticoid nauplii and summarizes the pertinent literature. Nauplii have three pairs of developed appendages: the antennae, antennules, and mandible. In the naupliar stage, these appendages are used for both feeding and locomotion. Most harpacticoid nauplii are ovate and flattened and associated with the bottom. Nauplii of cyclopoids and calanoids are usually not flat and are free swimming (plate 199B). The sixth nauplius molts into the first copepodite stage, which is similar in shape to the adult (plate 199C).

The first copepodite stage (CI) usually has five somites and two pairs of pereiopods. With each successive molt, a somite is added and further addition and development of the pereiopods takes place such that by CV there are nine somites and all pairs of legs are present. The molt to the adult (CVI) adds one more somite and the pereiopods and genital segment become fully developed. It is important to distinguish between CV and adult when identifying copepods to avoid making erroneous species determinations based on characters that are not fully formed.

COMMON COPEPOD HABITATS

PLANKTON

Planktonic copepods can be generally divided into three assemblages that may be encountered near shore. These assemblages are not fixed in time and space, and some components of each may mix with the others.

First, a number of pelagic taxa typical of nearshore marine waters can often be collected along the outer coast or near the entrances to bays and estuaries, and some of them have reproducing populations in embayments (plate 200A–200F). Common taxa include the calanoids *Calanus pacificus, Pseudocalanus* spp., *Clausocalanus* spp., *Ctenocalanus vanus, Epilabidocera longipedata, Metridia pacifica, Acartia tonsa, Tortanus discaudatus, Centropages abdominalis*, the cyclopoid *Corycaeus anglicus*, and several species of the cyclopoid genus *Oithona*. Most of the taxa encountered from nearshore waters can be identified using the keys of Dawson and Knatz (1980) and Gardner and Szabo (1982).

Second, the plankton of bays and marine-influenced parts of estuaries are dominated by another group of copepods (plate 201A–201F). Usually, the dominant copepods in these conditions are calanoids of the genus *Acartia*, subgenus *Acartiura*. Several species have been recorded in California bays, including *A. hudsonica* and *A. omorii*, but there are also undescribed species present (Bradford 1976). The calanoids *Acartia (Acanthacartia) californiensis, A. (A.) tonsa*, and *Paracalanus* sp. are usually abundant. Another calanoid, *Eurytemora americana*, can be very abundant in the water column near eelgrass beds and in estuaries. Two species of cosmopolitan planktonic harpacticoids, *Euterpina acutifrons* and *Microsetella rosea*, are also abundant in marine embayments. The cyclopoid *Oithona davisae* and the calanoids *Tortanus dextrilobatus* and *Pseudodiaptomus marinus* have been introduced from Asia and successfully established themselves in San Francisco Bay; they are abundant there. The latter species also occurs in other California bays. A native species of *Pseudodiaptomus, P. euryhalinus*, is often abundant in salt marsh ponds in San Francisco Bay and southern California.

The third assemblage of planktonic copepods to be found in nearshore waters occurs in oligohaline and tidal fresh reaches of estuaries (plate 202A–202F). Until the 1970s and 1980s the only numerous copepods in these habitats were the calanoid *Eurytemora affinis* and the harpacticoids *Coullana canadensis* and *Pseudobradya* sp. However, this assemblage has changed

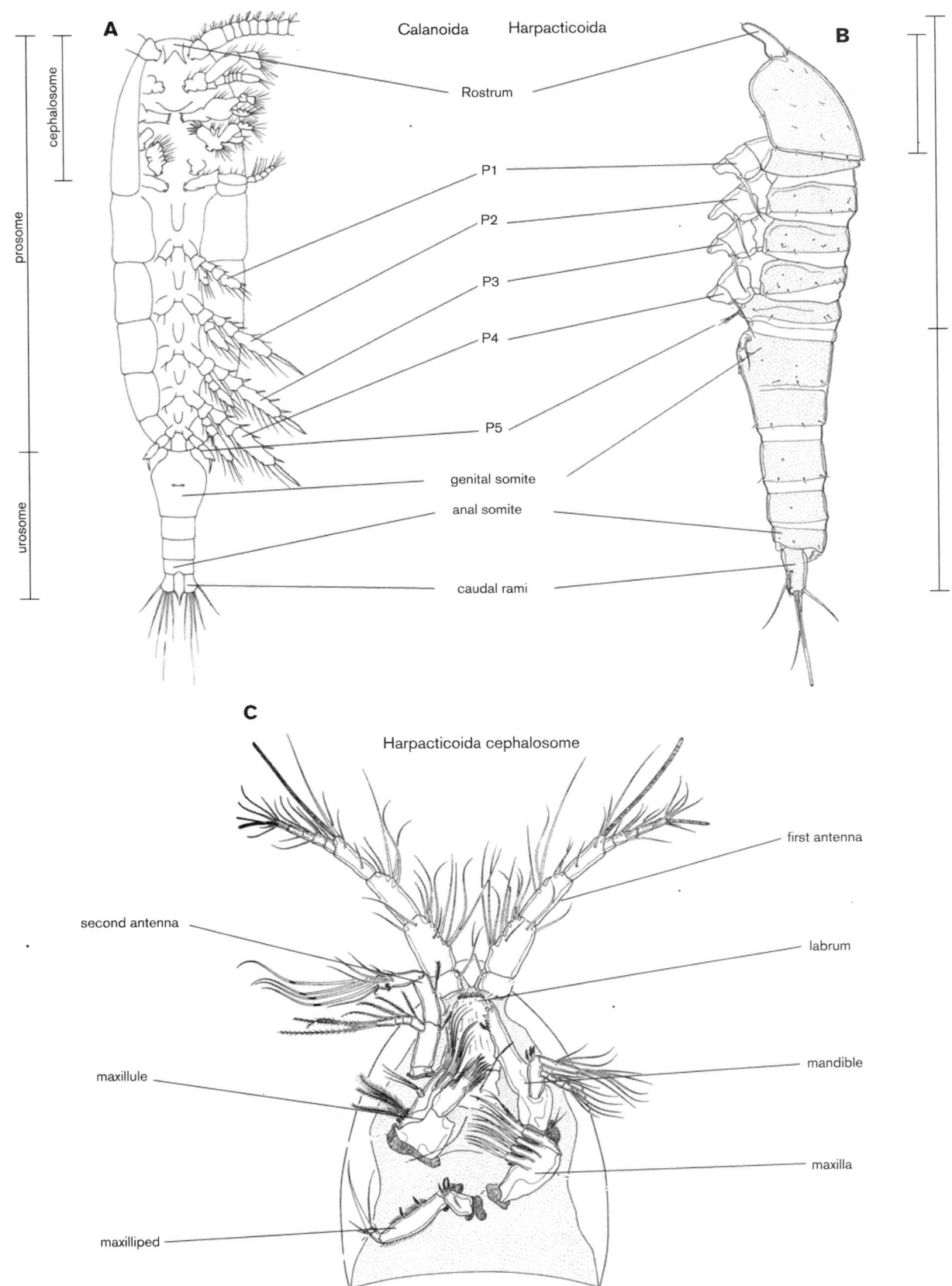

PLATE 197 Copepoda: A, body plan of Calanoida and B, Harpacticoida, abbreviations: P1–4, first through fourth swimming legs; P5, fifth leg; C, cephalosome of Harpacticoida, appendages shown on only one side (A. from Gardner and Szabo 1982; B and C, from Huys et al. 1996).

FIRST ANTENNA

Calanoida

Harpacticoida

A

B

♂

♀

SECOND ANTENNA

Cyclopoida

Harpacticoida

Calanoida

C

D

E

FIFTH LEGS (♂ TOP ♀ BOTTOM)

Cyclopoida

Calanoida

Harpacticoida

HARPACTICOIDA FIRST AID AND SWIMMING LEGS

sclerite

endopod

exopod

G

H

I

F

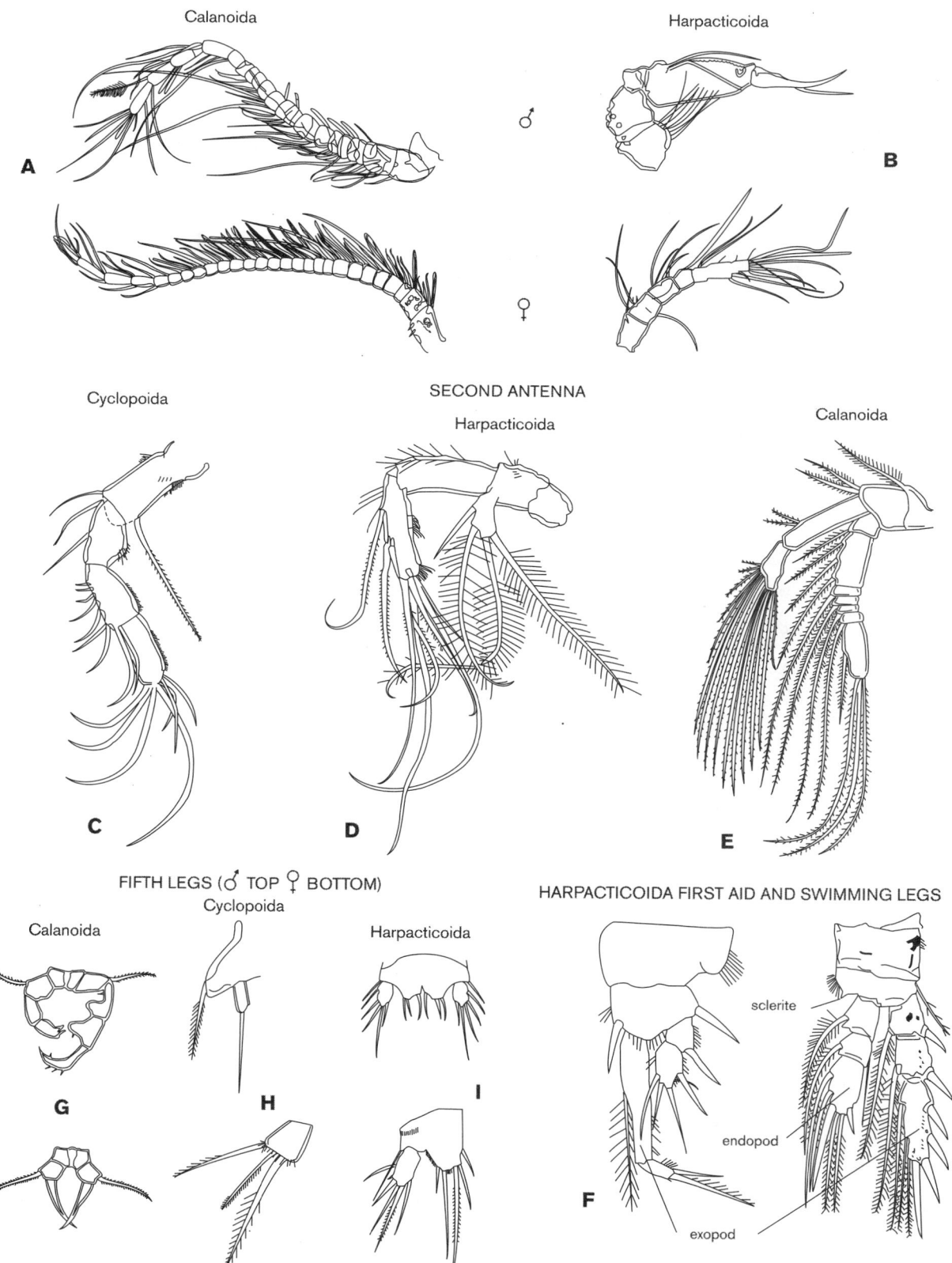

PLATE 198 Copepoda: examples of first antennae of male and female of A, Calanoida and B, Harpacticoida; examples of second antennae of C, Cyclopoida, D, Harpacticoida, E, and Calanoida; example of swimming and first leg of F, Harpacticoida; G, examples of male and female fifth legs of G, Calanoida, H, Cyclopoida, I, and Harpacticoida (A, from Huys and Boxshall 1991; B, from Huys et al. 1996; C–E, G, H, from Boxshall and Halsey 2004; F, from Lang 1965 (left) and Huys et al. 1996 (right); H, I, modified from Lang 1965).

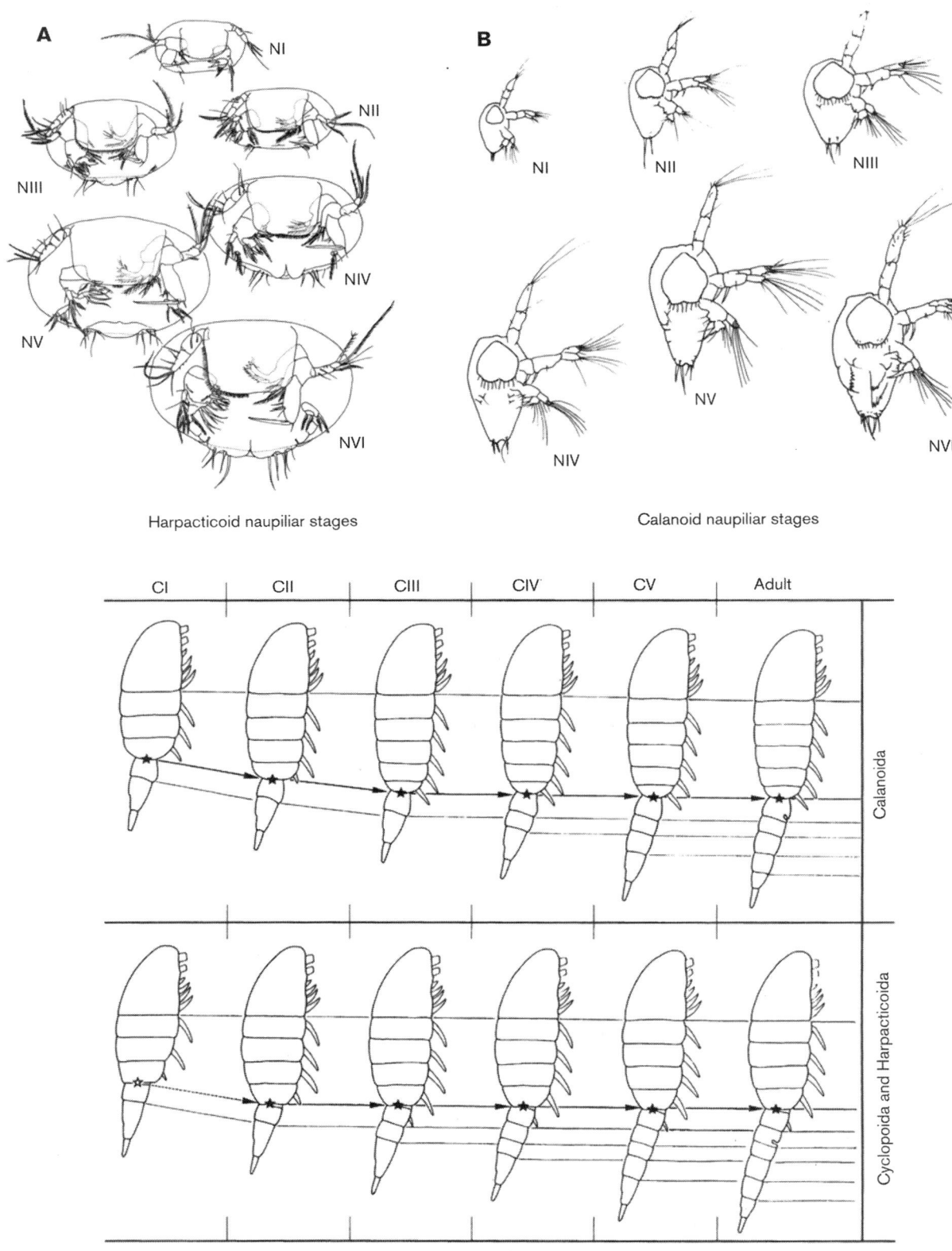

A

NI

NII

NIII

NIV

NV

NVI

Harpacticoid naupiliar stages

B

NI

NII

NIII

NIV

NV

NVI

Calanoid naupiliar stages

CI   CII   CIII   CIV   CV   Adult

Calanoida

Cyclopoida and Harpacticoida

C   Copepoda copepodite stages

PLATE 199 Copepoda: developmental stages of copepods: A, naupliar stages of Harpacticoida and B, Calanoida; C, copepodite stages of Calanoida, Cyclopoida, and Harpacticoida (A, from Dahms 1990; B, from Gardner and Szabo 1982; C, from Boltkovskoy 1999).

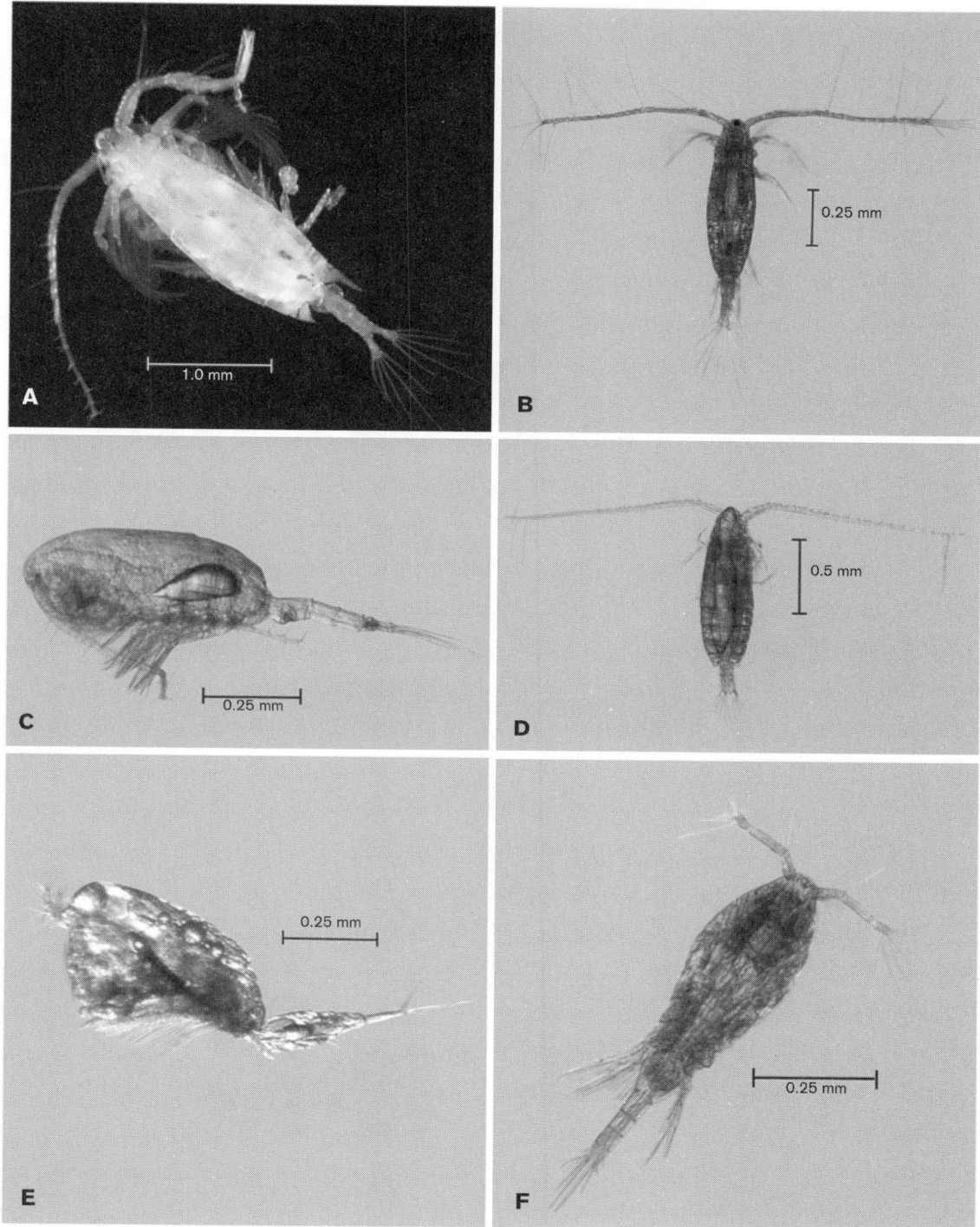

PLATE 200 Copepoda: examples of pelagic copepods characteristic of nearshore marine plankton; Calanoida: A, *Epilabidocera longipedata*, male; B, *Acartia longiremis*; C, *Pseudocalanus* sp.; D, *Calanus* sp.; Cyclopoida: E, *Corycaeus anglicus*; F, *Oncaea* sp.

drastically due to the introduction and establishment of ballast water-introduced Asian species. In the Sacramento-San Joaquin estuary, these include the calanoids *Pseudodiaptomus forbesi*, *Sinocalanus doerrii*, and *Acartiella sinensis,* and the cyclopoids *Limnoithona sinensis*, and *L. tetraspina*. The establishment of these nonnative copepods in the Sacramento-San Joaquin estuary has been accompanied by a decline and shift in seasonality in the previously dominant *E. affinis*. The Columbia River and numerous smaller coastal estuaries in Oregon and Washington have also been invaded by *S. doerrii*, *P. forbesi*, and another species of *Pseudodiaptomus*, *P. inopinus*; these species can dominate the mesozooplankton in brackish and tidal-fresh parts of invaded estuaries.

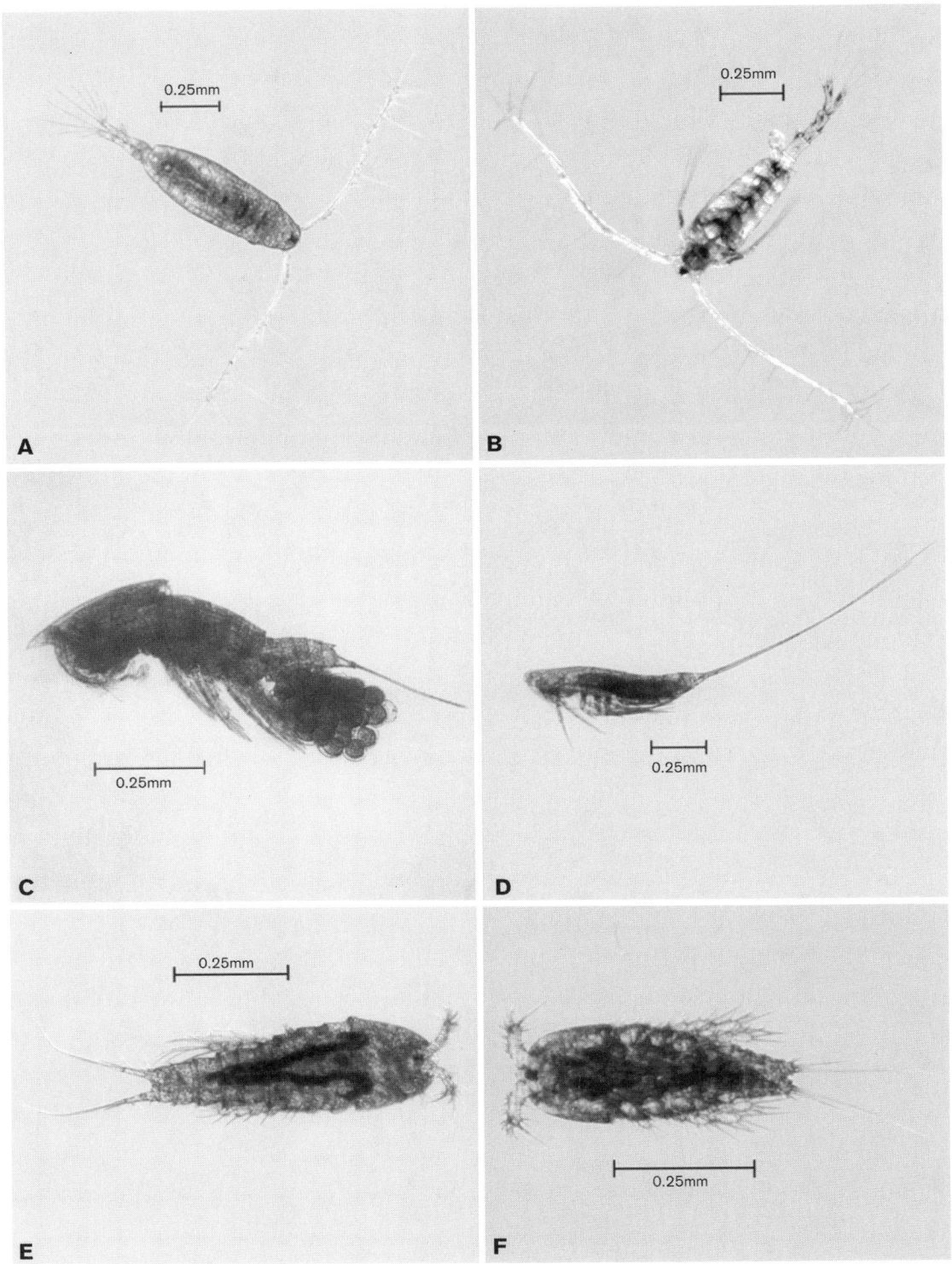

PLATE 201 Copepoda: examples of copepods common in plankton of bays and estuaries; Calanoida: A, *Acartia (Acartiura)* sp.; B, *Tortanus discaudatus*, male; Harpacticoida: C, *Euterpina acutifrons*; D, *Microsetella rosea*; E, *Leimia vaga*; F, *Microarthridion littorale*.

EPIBENTHOS AND EPIPHYTON

Copepod assemblages that are closely associated with dock epifauna, eelgrass, algae, and microphyton and detritus at the sediment-water interface are dominated by harpacticoids, although one family of cyclopoids, the Cyclopinidae, can also be very abundant. They are easily collected by sweeping with a 253 µm or finer mesh hand net. The number of species likely to be encountered is large (see systematic list), and the likelihood of a large number of undescribed species or introduced species makes presenting a species key unrealistic at this time. However, many taxa can be identified

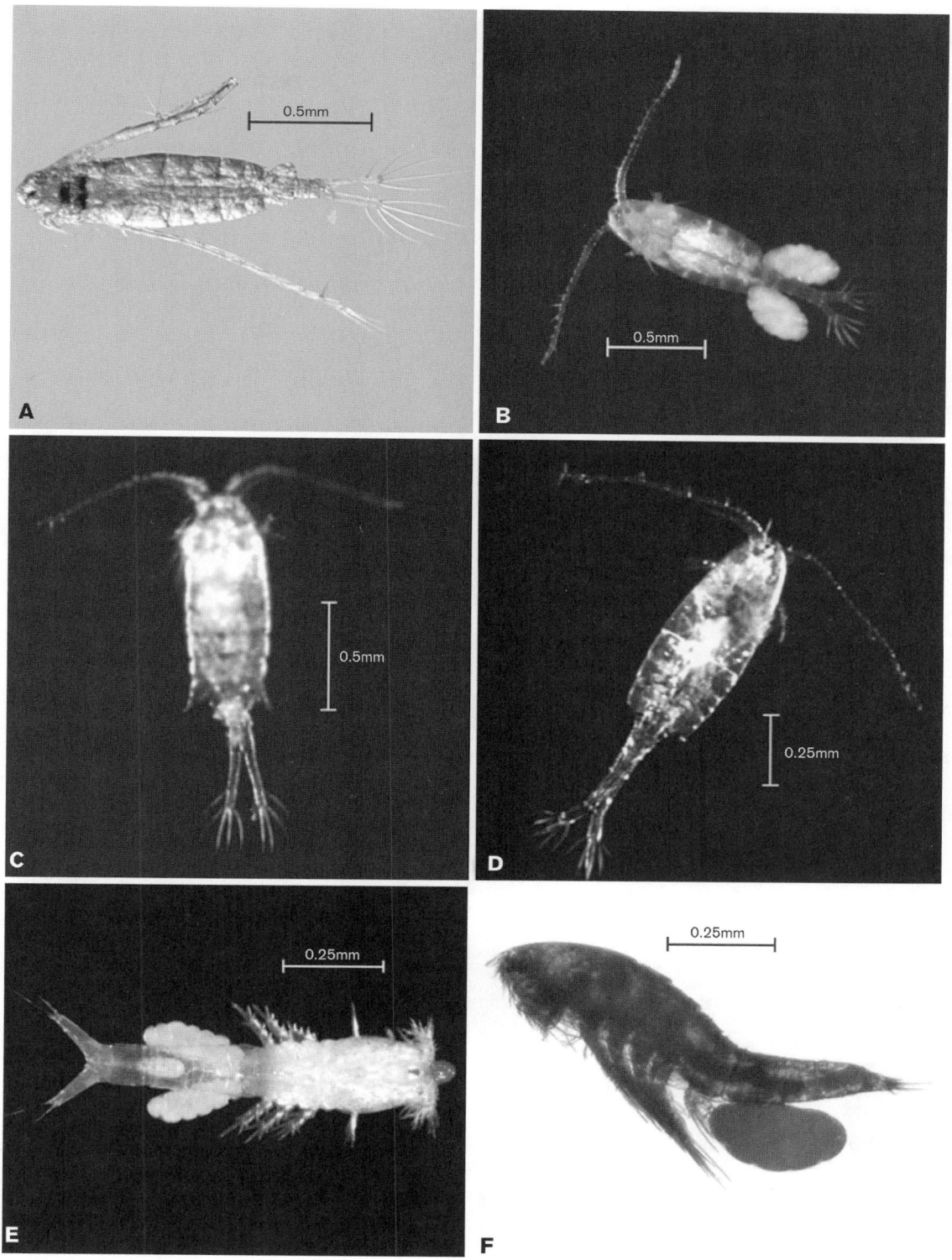

PLATE 202 Copepoda: examples of copepods common in oligohaline and tidal-freshwater parts of estuaries; Calanoida: A, *Sinocalanus doerrii*, male; B, *Pseudodiaptomus forbesi*; C, *Eurytemora affinis*; D, *Pseudodiaptomus inopinus*; Harpacticoida: E, *Coullana canadensis*; F, *Pseudobradya* sp.

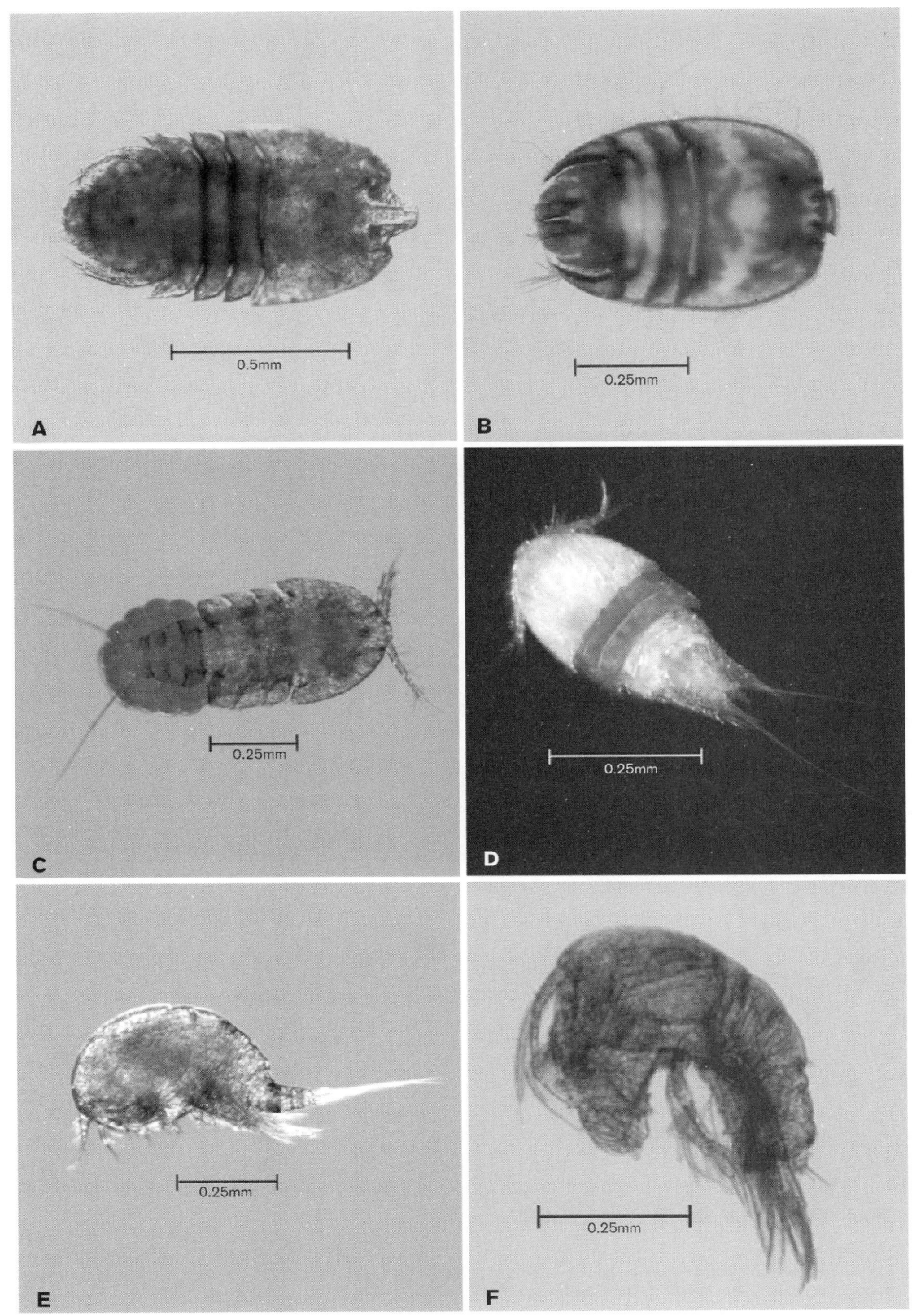

PLATE 203 Copepoda: examples of copepods associated with epibenthic/epiphytic habitats; Harpacticoida: A, family Peltidiidae; B, *Porcellidium* sp.; C, *Zaus caeruleus*; D, *Scutellidium* sp. (male); E, *Diarthrodes* sp.; F, family Tegastidae.

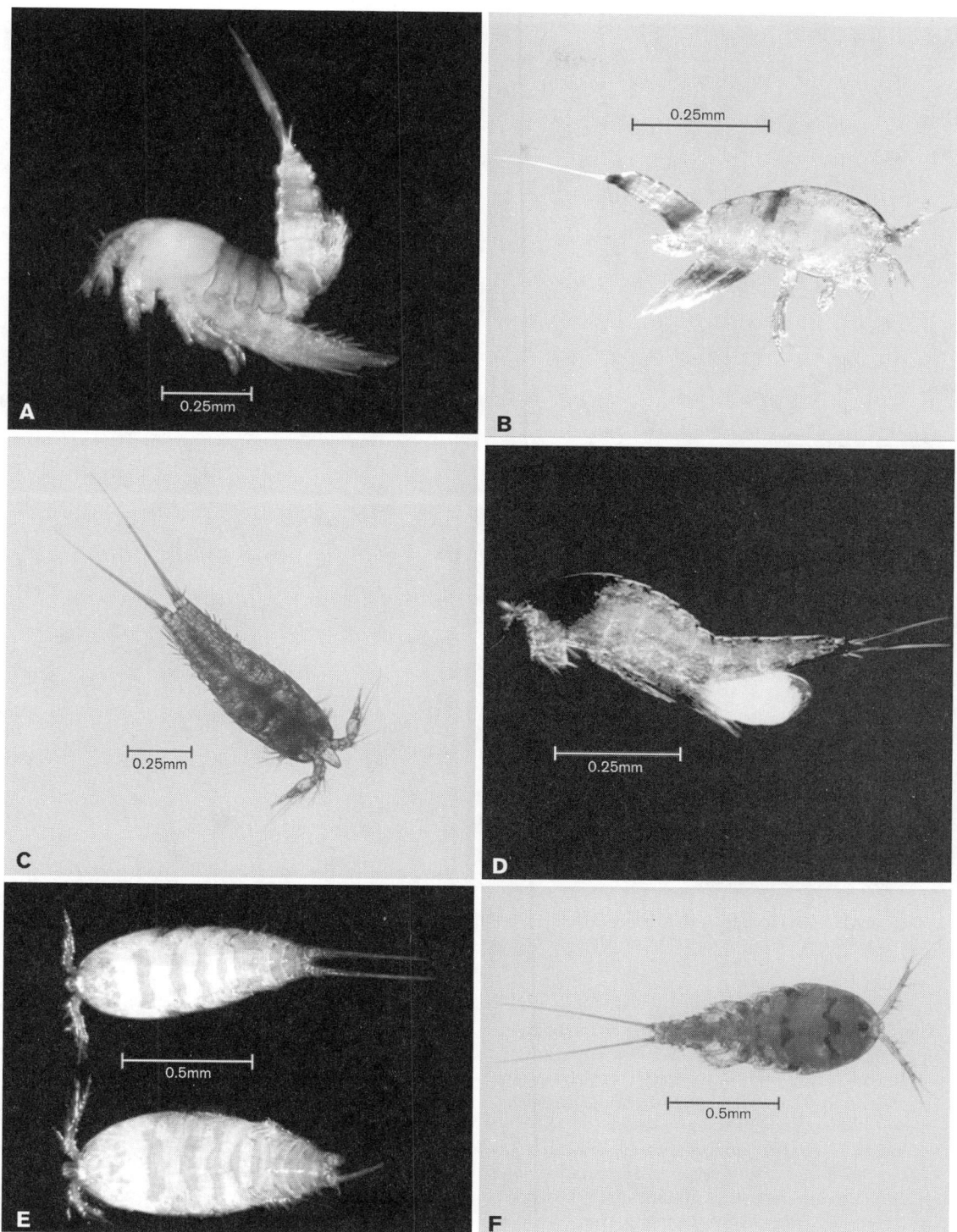

PLATE 204 Copepoda: examples of copepods associated with epibenthic/epiphytic habitats, continued; Harpacticoida: A, *Amphias-copsis cinctus*; B, *Diosaccus spinatus*; C, *Amonardia perturbata* (male); D, *Ectinosoma melaniceps*; E, *Harpacticus uniremis* (male, top; female, bottom); F, *Tisbe* sp.

to genus and species using Lang (1948, 1965) and the other taxonomic resources listed at the end of this section.

Several families of harpacticoids are either exclusively affili-ated with epibenthic and epiphytic habitats or have large num-bers of species associated with these habitats (plates 204A–205D). These include Diosaccidae, Harpacticidae, Laophonti-dae, Parastenheliidae, Peltidiidae, Porcellidiidae, Tegastidae, Thalestridae, and Tisbidae. These harpacticoids all have devel-oped morphological characters in adaptation to life on or near the substratum. The most common of these adaptations is the development of modified first legs into prehensile appendages that allow the animals to grasp and move about heterogeneous

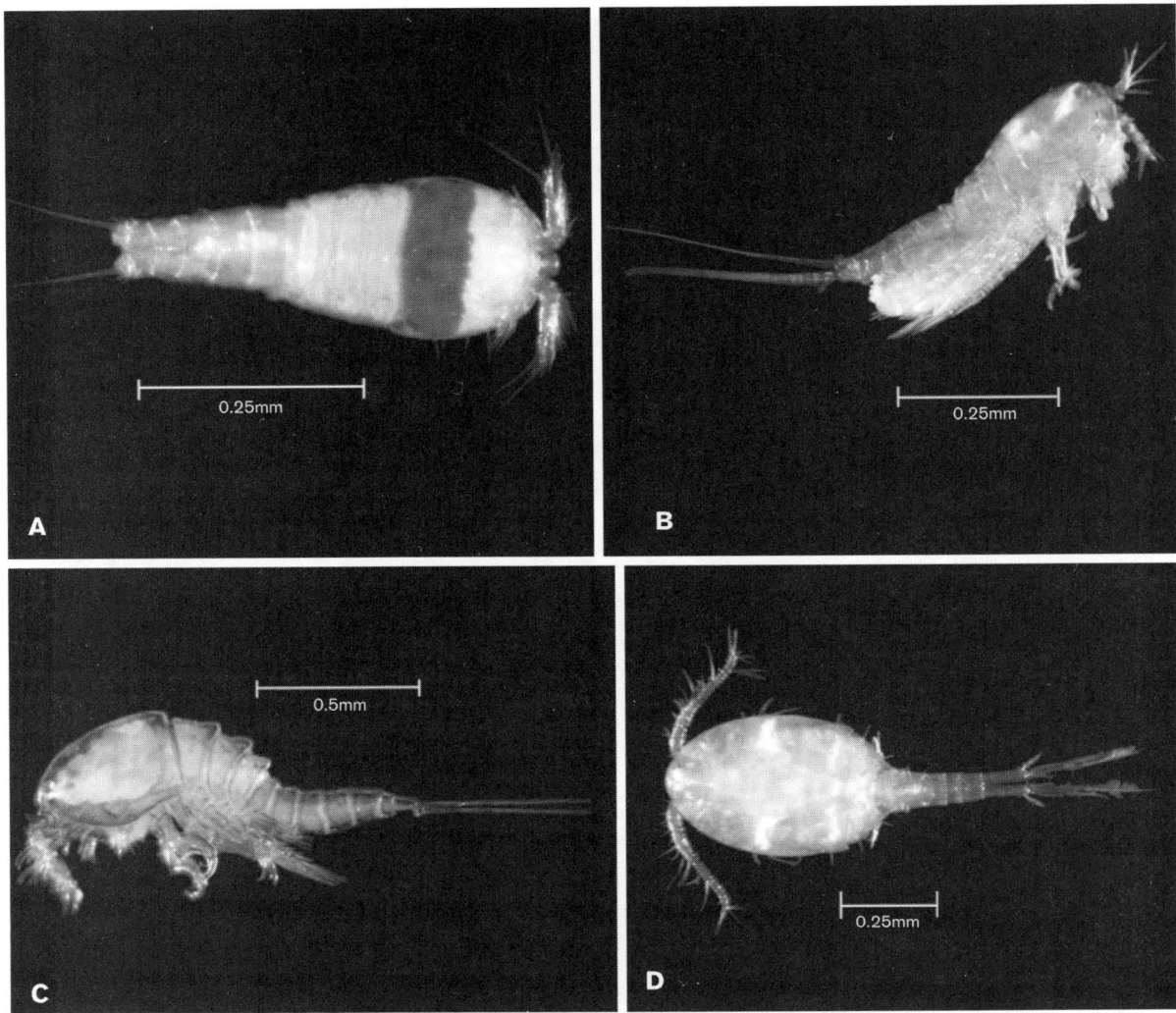

PLATE 205 Copepoda: examples of copepods associated with epibenthic/epiphytic habitats, continued; Harpacticoida: A, *Paradactylopodia* sp. (male); B, *Parathalestris californica*; C, *Thalestris longimana* (male); Cyclopoida: D, family Cyclopinidae.

surfaces. Another such adaptation is flattening of the body, either dorsoventrally (e.g., Porcellidiidae, Peltidiidae) or antero-laterally (Tegastidae) so the animals can attach themselves to the leaves of seagrasses and algae. One thalestrid genus, *Diarthrodes*, is a "leaf miner," boring tunnels through blades of algae.

Several harpacticoids in this group warrant special mention because of their prominence or ecological importance. One of the most abundant inhabitants of supralittoral splash pools is *Tigriopus californicus*, a member of the family Harpacticidae. It can be observed as numerous small red specks in the water column and on the bottom. On many algae and eelgrass blades, the bright purple or maroon genus *Porcellidium* can be seen without magnification crawling on the plants, often near leaf junctions. In this genus, the cephalic appendages are modified to form an effective suction device, and *Porcellidium* occurs even in areas with significant wave energy. Net sweepings of algae will yield several large and colorful harpacticoids. The diosaccid *Amphiascopsis cinctus* has both red and blue color variants and is prominent in eelgrass beds and algae on docks. Another large brightly colored diosaccid is *Diosaccus spinatus*, which occurs in the late summer among

eelgrass epiphytes and is lemon yellow with several prominent red stripes.

The largest epibenthic/epiphytic harpacticoid likely to be encountered in California is the thalestrid *Eudactylopus latipes typica*, which is common in eelgrass beds of San Francisco Bay. Two other genera, *Harpacticus* (family Harpacticidae) and *Tisbe* (family Tisbidae) are epibenthic but not always strictly associated with plants or epiphytes. These two taxa together can become very abundant in the early spring, reaching combined densities of over 1 million per square meter. From Puget Sound north these two taxa often comprise the principal food source for juvenile pink and chum salmon.

INFAUNA

After nematodes, harpacticoid copepods are the most numerous benthic metazoans in terms of biomass. In intertidal habitats, benthic harpacticoids either live in the interstices between sand grains or burrow through the sediment. In sandy sediments, vermiform body types are very common

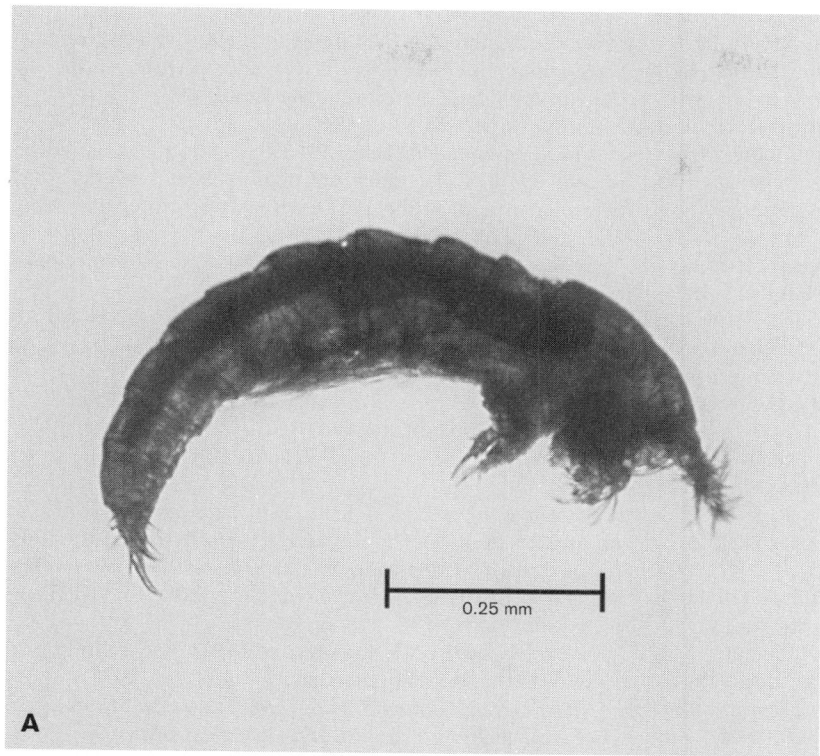

0.25 mm

A

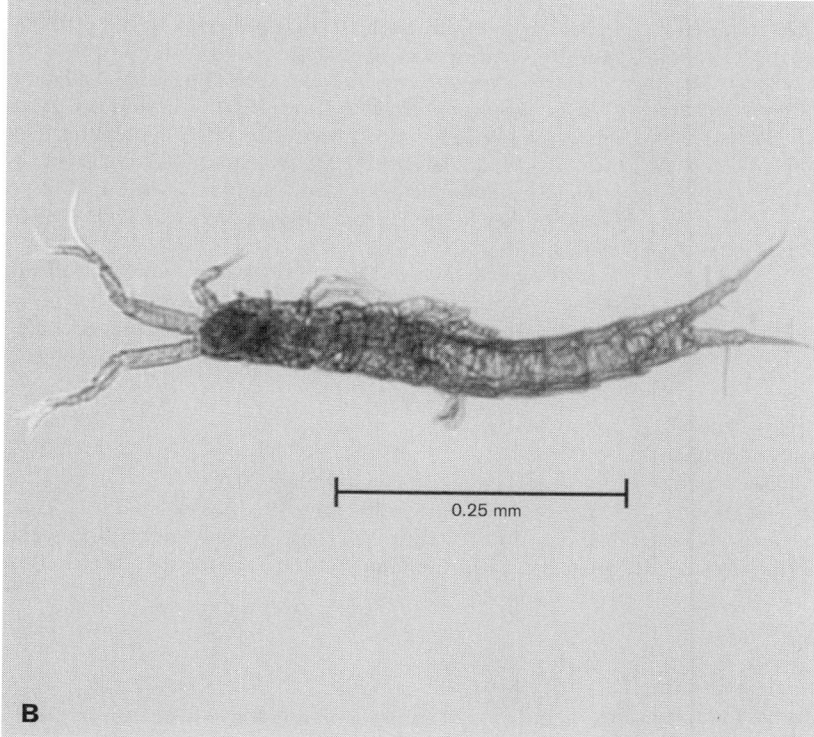

0.25 mm

B

PLATE 206 Copepoda: examples of infaunal benthic copepods; Harpacticoida: A, *Huntemannia jadensis*; B, family Leptastacidae (male).

(e.g., Leptastacidae, Cyllindropsyllidae), but more robust forms that burrow through or crawl over the sand surface also occur (e.g., *Huntemannia*, Tachidiidae) (plate 206A, 206B). These types can be collected by digging in the sand and filtering the water that collects in the resulting hole. In finer mud substrates, burrowing and sediment surface types predominate (e.g., Cletodidae).

LIST OF FREE-LIVING HARPACTICOIDA

For the most part, harpacticoids listed below were taken from the taxa found by Lang (1965), Kask et al. (1982), Watkins (1983), Webb (1983), and from collections made by the author in Puget Sound and in coastal marine and estuarine habitats of Washington, Oregon, and California. Information for the annotations also came from these sources, from Lang (1948), and from the author's observations. This list should be regarded as partial, because many habitats remain unsampled. Also for this reason, any species may occur in more habitats/regions than listed. In many cases, the only record for a species is from Lang's (1965) original collections, mainly from near Hopkins Marine Station, Pacific Grove (abbreviated as HMS in the species list), and Pacific Marine Station, Dillon Beach, California (DB).

Chappuis (1958) recorded 76 (mainly European) species of harpacticoids from Puget Sound, Washington, but Lang (1965) refuted most of these, and they are not included in the list. Similarly, Crandell (1966) listed 57 mainly northern European species in Yaquina Bay, Oregon, that were identified just prior to Lang's (1965) regional monograph and are also not listed unless also recorded elsewhere within the region.

The symbol + indicates that taxon has not been recorded from California or Oregon, but occurs in Washington and/or southern British Columbia. Also listed are several species of algae-dwelling harpacticoids that were found near La Jolla in southern California (Gunnill 1982).

The symbol † indicates that additional undescribed or unidentified species have been observed by the author, for that genus, within the region covered. This designation includes species in the genera *Tegastes, Dactylopusia, Psyllocamptus, Mesochra, and Itunella* that were described in a PhD thesis (Watkins 1983) but have not been published. Lang (1948) summarized harpacticoid taxonomic literature and gave keys to families, genera, and species. Bodin (1997) gives a complete listing of all additional descriptions, new species, revisions, and taxonomic publications that occurred after Lang's monograph. Huys et al. (1996) provide a key to the marine and brackish water harpacticoid families. Where keys published subsequent to Lang's 1948 monograph exist, they are cited for the families and genera listed.

HARPACTICOIDA

POLYARTHRA

CANUELLIDAE

Key to genera: Huys et al. 1996, Boxshall and Halsey 2004.

Coullana canadensis (Willey, 1923). Usually dominant epibenthic and planktonic copepod in oligohaline to tidal fresh reaches of large rivers; also common in smaller estuaries. Description of male and redescription of female (as *Scottolana canadensis*): Coull 1972, Crustaceana 22: 209–214.

LONGIPEDIIDAE

Longipedia americana Wells, 1980. Lower intertidal into the subtidal, in medium sand and/or mixtures of sand and mud. Abundant in sediment around eelgrass, *Zostera marina*, and often occurs in near-bottom plankton net samples. Other species in this genus may be present. Key to species: Wells 1980, Zool. J. Linn. Soc. 70: 103–189.

OLIGOARTHRA

AMEIRIDAE

Key to genera: Boxshall and Halsey 2004.

Ameira longipes Boeck, 1864.[†] Geographically widespread and common in high intertidal pools to shallow subtidal eelgrass beds where microalgal epiphytes and detritus occur. Redescription: Lang (1965) from near HMS.

Ameira parvuloides Lang, 1965.

Ameira parasimulans Lang, 1965. This and previous species uncommon in collections from tidal pools and eelgrass; may be affiliated with sediment. Described from interstitial shell and sand at DB.

Ameira minuta Boeck, 1865. Sediments and leaves in eelgrass and subtidal mud and gravel.

Ameiropsis longicornis Sars, 1907.[+] Occurs in the North Atlantic and Mediterranean and found in shallow subtidal sand in southern B.C. (Kask et al. 1982).

Interleptomesochra reducta Lang, 1965. Interstitial fine shell sand. Key to species: Lang 1965.

Nitokra typica Boeck, 1865.[†] Eelgrass and sandy mud in estuaries.

Nitokra spinipes Boeck, 1865. Estuarine marshes and tide pools with freshwater input. Lang (1965) described the California variant of this widespread species as *N. s. armata*, but Wells and Rao [1987, Mem. Zool. Surv. India 16: 1–385] rejected subspecies of *N. spinipes*.

Nitokra lacustris (Schmankevitch, 1875). Low-salinity and fresh coastal and inland marshes.

Nitokra affinis californica Lang, 1965. Described from a tide pool with shell sand, stones, and algae near HMS.

Praeleptomesochra similis Lang, 1965. Tidal pools, shell sand, stones, algae. Key to species: Lang 1965, Pesce 1981, Bull. Zool. Mus. Univ. Amsterdam 8: 69–72.

Pseudoleptomesochra typica Lang, 1965. Interstitial shell sand.

Psyllocamptus triarticulatus Lang, 1965.[†] This and the previous species were described from tidal pools containing shell sand, stones, and algae off HMS. Watkins (1983) collected another unnamed species in Mendocino and San Luis Obispo Counties. Key to species: Lang 1965, Ceccherelli 1988, Vie Milieu 38: 155–171.

Psyllocamptus minutus Sars, 1911.[+] Shallow subtidal sand and cobble.

Sarsameira sp.[+] Intertidal mud in estuaries. Key to species: Reidenauer and Thistle 1983, Trans. Am. Micros. Soc. 102: 105–112.

ANCORABOLIDAE[†]

Key to genera: Boxshall and Halsey 2004.

Laophontodes hedgpethi Lang, 1965. Associated with *Laminaria* and other algal holdfasts. In New England, this and other genera in the family also associated with colonial invertebrates.

CANTHOCAMPTIDAE

See Bodin (1997) for notes on revisions of this family. Key to genera of fresh and brackish water: Boxshall and Halsey 2004.

Itunella muelleri (Gagern, 1922).[†] Described from Europe, this species was reported from the Santa Rosa creek estuary by Watkins (1983), who also reported an undescribed species from supralittoral pools.

Leimia vaga Willey, 1923.[+] Benthic and epibenthic; often abundant in estuarine mudflats and tidal channels. Described from New Brunswick.

Mesochra pygmaea (Claus, 1863).[†] Usually affiliated with vegetation; one of the most abundant and widespread harpacticoids in eelgrass along the Pacific coast; also recorded from brackish water. Key to species: Hamond 1971, Austral. J. Zool. Suppl. 7: 1–32, Fiers and Rutledge 1990, Bull. Inst. R. Sci. Nat. Belg. Biol. 60: 105–125.

Mesochra rapiens (Schmeil, 1894).[+] Mud and sand in oligohaline and tidal freshwater reaches of estuaries.

Mesochra alaskana M.S. Wilson, 1958.[+] Same as previous species, in more saline regions.

Nannomesochra arupinensis (Brian, 1925). Widely distributed in warm marine and brackish waters; in California, found in *Enteromorpha* in brackish pools (Watkins 1983).

Cletocamptus deborahdexterae Gomez et al., 2004. Salton Sea. The genus has also been reported (as *C. deitersi* Schmankewitsch, 1875) from coastal brackish and inland saline environments, including desert springs in California. However, the taxon *"C. deitersi"* may comprise a number of morphologically similar but genetically distinct species (Gomez et al. 2004, J. Nat. Hist. 38: 2669–2732. Revision of the genus: Fleeger 1980, Trans. Am. Micros. Soc. 99: 25–31).

CLETODIDAE[†]

Key to the genera: Boxshall and Halsey 2004.

Acrenydrosoma karlingi Lang, 1965. Tidal pools with coarse sand. Unidentified species which may be *A. macalli* Schizas and Shirley 1994 [Crustaceana 67: 331–340] occurs in lower intertidal/shallow subtidal mud and fine sand with detritus.

Cletodes hartmannae Lang, 1965. Described from tidal pools with shell sand, stones, and algae near HMS. Key to species: Lang 1965, Hamond 1973, Mem. Queensland Mus. 16: 471–483, Fiers 1991, Beaufortia 42: 13–47.

Enhydrosoma hopkinsi Lang, 1965. Described from tidal pools with shell sand, stones, and algae near HMS. Key to species: Thistle 1980, Trans. Am. Micro. Soc. 99: 384–397, Bell and Kern 1983, Bull. Mar. Sci. 33: 899–904.

Kollerua breviarticulatum Shen and Tai, 1964.[+] Intertidal and shallow subtidal eelgrass, mud, and sand. Described from freshwater in south China and may be an introduced species. New generic combination by Gee 1994 Sarsia 79: 83–107, for *Enhydrosoma breviarticulatum*.

Limnocletodes behningi Borutzky, 1926.[+] Common in estuaries near river mouths, and abundant in intertidal mud near vegetation such as *Spartina* spp. Revision of the genus: Wells 1971, J. Nat. Hist. 5: 507–520.

Strongylacron buchholzi (Boeck, 1872).[+] Mud and sandy mud in estuaries. New generic combination by Gee and Huys 1996, Sarsia 81: 161–191 for *Enhydrosoma buchholzi*.

Stylicletodes verisimilis Lang, 1965. Described from Monterey Bay in fine sand and detritus at about 26 m depth. Key to species: Bodin 1968, Mém. Mus. Natn. Hist. Nat. 55: 1–107.

CLYTEMNESTRIDAE

Key to the genera and species: Boxshall and Halsey 2004.

Clytemnestra rostrata (Brady 1883).

Clytemnestra scutellata Dana 1848. This and the previous species are planktonic and widespread throughout tropical and temperate oceanic waters down to 60 m but are sometimes collected near shore.

DACTYLOPUSIIDAE

Elevated from subfamily of the Thalestridae to family rank by Willen (2000).

Dactylopusia crassipes Lang, 1965.[†] Very abundant in estuaries where salinity is slightly lowered and also in a variety of coastal vegetated and unvegetated habitats. Important prey for juvenile salmon.

Dactylopusia glacialis (Sars, 1909). Common but usually not abundant in microalgae and other vegetation; also occurs in the North Atlantic and Arctic.

Dactylopusia paratisboides Lang, 1965. Microalgae and epiphytes on eelgrass and other substrates.

Dactylopusia tisboides (Claus, 1863).[+] A widely geographically distributed species (see Bodin 1964, Rec. Trav. Sta. Mar. Endoume 51: 107–183) that has been found in southern British Columbia and Puget Sound.

Dactylopusia vulgaris inornata Lang, 1965. Abundant in epiphytes on eelgrass and macroalgae. Important in diets of juvenile salmon and flatfish.

Diarthrodes cystoecus Fahrenbach, 1954.[†] See Fahrenbach 1954, J. Wash. Acad. Sci. 44: 326–329; 1962, La Cellule 62: 301–376 for a redescription and complete details of the biology of this algal leaf-mining species. It is common inside the intertidal red alga *Halosaccion*.

Diarthrodes dissimilis Lang, 1965.

Diarthrodes unisetosus Lang, 1965. This and the previous species were described from algae at DB and were also found by Watkins (1983) in intertidal vegetated habitats.

DANIELSSENIIDAE

Key to genera: Boxshall and Halsey 2004.

Danielssenia typica Boeck, 1872. Intertidal and shallow subtidal sand and sandy mud; sediment in eelgrass. Some previous records are probably *D. reducta*. See Gee 1988, Zool. Scr. 17: 39–53 and Huys and Gee 1993, Bull. Nat. Hist. Mus. 59: 45–81 for revisions and keys to species.

D. reducta Gee, 1988.[+] Sediments around eelgrass.

DARCYTHOMPSONIIDAE

Key to genera: Huys et al. 1996, Boxshall and Halsey 2004.

Leptocaris armatus Lang, 1965. Associated with algae and/or algal detritus. Key to species: Lang 1965, Fleeger and Clark 1980, Northeast Gulf Sci. 3: 53–59.

Leptocaris doughertyi Lang, 1965.

Leptocaris pori Lang, 1965. This and the previous species were described from interstitial fine shell sand in Monterey Bay.

Leptocaris sp.[+] Kask et al. 1982 found an unidentified *Leptocaris* species in intertidal sandy mud of several British Columbia estuaries. Another, possibly the same species is associated with woody and other plant debris in Puget Sound and coastal estuaries.

ECTINOSOMATIDAE

Key to genera: Huys et al. 1996, Boxshall and Halsey 2004.

Arenosetella kaiseri Lang, 1965. Described from interstitial shell sand off HMS. Key to species: Lang 1965, McLachlan and Moore 1978, Ann. S. Afr. Mus. 76: 191–211, Bodin 1979, Vie Milieu 27: 311–357.

Bradya cladiofera Lang, 1965. Described from off HMS in fine sand with detritus, at about 26 m depth.

Bradyellopsis foliatus Watkins, 1987. Collected infrequently by Watkins (1983) by plankton net from lower intertidal/shallow subtidal sites dominated by macroalgae. Key to species: Watkins 1987, J. Crust. Biol. 7: 380–395.

Ectinosoma melaniceps Boeck, 1865.[†] Cosmopolitan species; very common in lower intertidal marine habitats rich in detritus and in eelgrass; less common in brackish water. Redescription: Lang 1965, Pallares 1970, Physis 30: 255–282. Key to species: Lang 1965.

Ectinosoma breviarticulatum Lang, 1965.

Ectinosoma paranormani Lang, 1965. This and the previous species were described from off HMS in a tidal pool with sand and algae.

Ectinosoma californicum Lang, 1965. Described from rinsing of algae collected off HMS.

Halectinosoma kunzi Lang, 1965.[†] Described from off HMS in a tidal pool with shell and gravel.

Halectinosoma longisetosum Lang, 1965. Described from off HMS in a tidal pool with sand and gravel.

Halectinosoma ornatum Lang, 1965. Described from rinsings of algae from tide pool at Point Pinos, Monterey Bay.

Halectinosoma similidistinctum Lang, 1965. Described from off HMS in fine sand and detritus at about 26 m depth. Key to species: Lang 1965, Clément and Moore 1995, Zool. J. Linn. Soc. 114: 247–306.

Halectinosoma unicum Lang, 1965. Described from off HMS in fine sand with some detritus at about 7 m depth.

Hastigerella abbotti Lang, 1965. Described from intertidal interstitial shell sand off HMS. Key to species: Lang 1965, McLachlan and Moore, 1978, Ann. S. Afr. Mus. 76: 191–211.

Pseudobradya cornuta Lang, 1965.[†] Key to species: Lang 1965.

Pseudobradya crassipes Lang, 1965.

Pseudobradya pectinifera Lang, 1965. This and the previous two species were described from off HMS in fine sand and detritus at 26 m depth.

Pseudobradya pulchera Lang, 1965. Described from rinsings of algae collected at DB.

Pseudobradya lanceta Coull, 1986.[+] Sediments around eelgrass beds in Washington and southern British Columbia.

EUTERPINIDAE

Euterpina acutifrons (Dana, 1848). Planktonic, cosmopolitan between 66°N and 40°S. See Boxshall 1979, Bull. Brit. Mus. (Nat. Hist.) Zool. 35: 201–264 for redescription and distributional notes.

HAMONDIIDAE

Key to the genera and species: Boxshall and Halsey 2004.

Ambunguipes rufocincta (Brady, 1880).[+] Uncommon on eelgrass and macroalgae in Puget Sound and Hood Canal, WA. Comb. nov. for *Rhynchothalestris rufocincta* (Brady, 1880). See Huys 1990, Zool. J. Linn. Soc. 99: 51–115.

HARPACTICIDAE

Key to genera: Huys et al. (1996), Boxshall and Halsey 2004.

Harpacticella paradoxa (Brehm, 1924). *Harpacticella* is a brackish- and freshwater genus, previously known only from Asia. *H. paradoxa*, which was described from China and also occurs in Japan, has been found in the Klamath (CA) and Samish (WA) river estuaries and in the estuary and reservoirs of the Columbia river. It was probably introduced from Asia to the NE Pacific.

Harpacticus compressus Frost, 1967.[+†] Lower intertidal/shallow subtidal mud and sand with microalgae, epiphytes on eelgrass and macroalgae, epiphyton on docks. Described from Kodiak Island.

Harpacticus compsonyx Monard, 1926.[+] A widely distributed warm-water species that was found by Kask (1982) on subtidal eelgrass in southern British Columbia.

Harpacticus pacificus Lang, 1965. Common in epiphytes on macroalgae, eelgrass, and rocks. Part of a species group containing the European species *H. obscurus*, *H. littoralis*, and *H. giesbrechti*, which are difficult to separate. Given presence of other unidentified Pacific coast species of this group, care should be taken in assigning names.

Harpacticus septentrionalis Klie, 1939. Lower intertidal/shallow subtidal algae and eelgrass. Also in northern Europe and Japan.

Harpacticus spinulosus Lang, 1965. Described from rinsings of algae collected at DB. In Puget Sound and southern British Columbia, very abundant in sandy sediments around eelgrass and in unvegetated sand, where it is eaten by juvenile flatfish.

Harpacticus uniremis Krøyer, 1842. Mid-intertidal to 20 m on a variety of habitats. Circumboreal distribution, occurring as far south as La Jolla, where it occurred on subtidal algae (Gunnill 1982). From Alaska through Washington, this species is abundant and forms a principal prey resource for outmigrating juvenile pink and chum salmon.

Perissocope biarticulatus Watkins, 1987. Described from CA in sandy sediments near macroalgae. Key to species: Watkins 1987, J. Crust. Biol. 7: 380–393.

Tigriopus californicus (Baker, 1912). Easily collected and recognized harpacticoid common all along the coast in supralittoral pools. Key to species: Bradford 1967, Trans. Roy. Soc. N.Z. Zool. 10: 51–59, Itô 1969, J. Fac. Sci. Hokkaido Univ. 17: 58–77. See Powlik 1999, J. Mar. Biol. Assoc. U.K. 79: 85–92 (ecology, and references therein); Burton 1997 Evolution 51: 993–998 and Edmands 2001 Mol. Ecol. 10: 1743–1750 (genetics, and references therein).

Zaus aurelii Poppe, 1884.[†] Very common on eelgrass and macroalgae. Itô 1980, Publ. Seto Mar. Biol. Lab. 25: 51–77 redescribed this species from Kodiak Island, Alaska. Lang (1948) and Itô (1980) considered this and *Z. caeruleus* Campbell, 1929 described from British Columbia, to be synonymous.

Zaus robustus Itô, 1974.[+] Described from Japan and reported by Kask et al. (1982) from dock epiphyton at the Pacific Biological Station, Nanaimo.

Zaus spinatus hopkinsi Lang, 1965. Eelgrass, algae, and rock habitats from upper intertidal to shallow subtidal.

Zausodes septimus Lang 1965. Key to species: Lang 1965.

Zausodes sextus Lang 1965. This and the previous species were described from off HMS at 7 m.

HUNTEMANNIIDAE

Key to the genera: Boxshall and Halsey 2004.

Huntemannia jadensis Poppe, 1884. Common all along the coast in medium to coarse sand, from marine to tidal-fresh waters.

Nannopus palustris Brady, 1880. Estuaries, in brackish and freshwater mud and sand.

LAOPHONTIDAE

Lang (1965) was unable to construct a key to the genera, and addition of numerous genera since 1965 will require that the family be revised before a reliable generic key can be created.

Arenolaophonte stygia Lang, 1965. Described from interstitial intertidal coarse sand near HMS. Watkins (1983) reported another apparently undescribed species collected from sediment trapped in *Phyllospadix* turf.

Echinolaophonte horrida (Norman, 1876). One specimen was found by Kask et al. (1982) on subtidal sand in southern British Columbia. Key to species: Lang 1965.

Echinolaophonte armiger briani Lang, 1965. A cosmopolitan species; holdfasts and epiphytes of algae and eelgrass.

Esola sp. A single specimen was collected near Little River, Mendocino County by Watkins (1983).

Heterolaophonte discophora (Willey, 1929).[†] Macro- and microalgal and detritus-rich habitats. Boreal on both coasts of the U.S.; in the Pacific, occurs from AK to Monterey Bay and in Japan. Redescription: Lang 1965, Itô 1974, J. Fac. Sci. Hokkaido Univ. 19: 546–640. Key to species: Lang 1965. Species of the 'quinquespinosa' group were removed to new genus *Quinquelaophonte* by Wells, et al. 1982, N.Z. J. Zool. 9: 151–184.

Heterolaophonte hamondi Hicks, 1975.[+] Common in intertidal eelgrass and sandy mud with detritus in Puget Sound; also southern B.C. (Kask et al. 1982).

Heterolaophonte longisetigera (Klie, 1950).[+] Common in vegetated and detritus-rich habitats in Puget Sound and southern British Columbia.

Heterolaophonte mendax (Klie, 1939).[+] One specimen found by Kask et al. (1982) in intertidal sandy mud at Nanaimo.

Heterolaophonte variabilis Lang, 1965. Lower intertidal algae; very common on eelgrass epiphytes all along the coast.

Laophonte acutirostris Lang, 1965. Described from a tide pool containing coarse shell sand, near HMS.

Laophonte cornuta Phillipi, 1840.[†] Cosmopolitan; associated with algae and common all along the coast. Redescription: Itô 1968, J. Fac. Sci. Hokkaido Univ. 16: 369–381.

Laophonte elongata Boeck, 1872.[+] Specimens common in Puget Sound very similar to *L. elongata triarticulata* Monard, 1928. Watkins (1983) also found a single specimen of a species resembling but distinct from *L. elongata* in northern CA.

Laophonte sp. (*inopinata* group). Specimens similar to but distinct from *L. inopinata* T. Scott, 1892 occur in tide pools near macroalgae (Watkins 1983) and also in epiphytes and microalgae in Puget Sound.

Laophonte inornata A. Scott, 1902.[+] In southern British Columbia and Washington in intertidal and shallow subtidal sand in sheltered embayments.

Onychocamptus mohammed (Blanchard and Richard, 1891). Occurs over a wide geographical range and a variety of habitats. Intertidal and shallow subtidal in oligohaline and tidal-fresh regions of the Sacramento/San Joaquin Delta and other estuaries along the Pacific coast and in reservoirs of the Columbia River.

Paralaophonte asellopsiformis Lang, 1965. Described from a tide pool containing shell sand, near HMS. Lang (1965) remarked that this species was very similar to the mostly sand-dwelling genus *Asellopsis*, and described it as a "true sand-dweller." Key to species: Lang 1965; Wells et al. 1982, N.Z. J. Zool. 9: 151–184.

Paralaophonte congenera (Sars, 1908).[+] Intertidal and shallow subtidal mud and sandy mud bays in Washington and southern British Columbia.

Paralaophonte pacifica Lang, 1965. Common from southern British Columbia to southern California in epiphytes and microalgae. A subspecies, *P. pacifica galapagoensis* Mielke 1981, Mikrofauna Meeresbodens 52: 1–134 was described from the Galapagos Islands.

Paralaophonte perplexa (T. Scott, 1898).[+] Commonly co-occurs with *P. pacifica* in southern British Columbia and Puget Sound; also occurs in North Atlantic and Arctic.

Paralaophonte subterranea Lang, 1965. Described from interstices from sand near DB; also collected by Lang 1965 near HMS.

Paronychocamptus proprius Lang, 1965. Described from a tide pool with shell sand, stones, and algae near HMS.

Paronychocamptus sp. This species of *Paronychocamptus*, similar to *P. huntsmani* (Willey, 1923) occurs near the mouth of the Columbia River.

Pseudonychocamptus paraproximus Lang 1965. Described from a Point Pinos, Monterey Bay tide pool, in fine sand. Common in Puget Sound shallow subtidal. Considered synonymous with *P. proximus* (Sars, 1908) by Mielke 1975, Mikrofauna Meeresbodens 52: 1–134. Key to species: Lang 1965, Ceccherelli 1988, Vie Milieu 38: 155–171.

Pseudonychocamptus spinifer Lang, 1965. Described from a tide pool with shell sand, stones, and algae near HMS. Common in Puget Sound around eelgrass and in lower intertidal and shallow subtidal sandy mud.

Quinquelaophonte capillata (C. B. Wilson, 1932).[+] Mainly Atlantic species found by Kask et al. (1982) in intertidal sandy mud and shallow subtidal sand in southern British Columbia; associated with eelgrass in Padilla Bay, northern Washington. Redescription: Coull 1976, Trans. Am. Micros. Soc. 95: 35–45.

Quinquelaophonte longifurcata (Lang, 1965). Described from a tide pool with shell sand near HMS.

LEPTASTACIDAE[†]

Key to genera: Huys 1992, Boxshall and Halsey 2004.

Arenopontia dillonbeachia Lang, 1965. Described from fine sand at about 3 m depth at DB; also found in Monterey Bay and in Japan by Itô 1969, J. Fac. Sci. Hokkaido Univ. 17: 58–77 who described the male. Key to species: Lang 1965; Bodiou and Colomines 1986, Crustaceana 57: 288–294.

Cerconeotes constrictus (Lang, 1965). Described as a species of *Leptastacus* from fine sand in a DB tide pool; also found year around in intertidal sandy mud in Nanaimo estuary, British Columbia (Kask et al. 1982). Key to species: Huys 1992, Med. Kon. Acad. Wetensch., Lett. Sch. Kunst. Belg., 54: 21–196.

Paraleptastacus spinicauda (T. and A. Scott, 1895). Reported from sand around eelgrass near the mouth of the Fraser River, British Columbia by Webb 1983.

Sextonis incurvatus (Lang, 1965). Described as a species of *Leptastacus* from fine sand in Tomales Bay. Key to species: Huys 1992, Med. Kon. Acad. Wetensch., Lett. Sch. Kunst. Belg. 54: 21–196.

LOURINIIDAE

Lourinia armatus (Claus, 1866). Geographically widespread in warm waters; found on algae near La Jolla by Gunnill (1982). Watkins (1983) reported collecting a single specimen of another, undescribed species near Piedras Blancas Point, CA.

MIRACIIDAE

The four genera previously included in the Miraciidae have been included together with the genera formerly placed in the Diosaccidae, in the family Miraciidae, which has priority (summarized by Boxshall and Halsey 2004). Key to genera (of Diosaccidae): Lang 1965.

Amonardia normani (Brady, 1872).[+] Eelgrass and macro- and microalgae. Key to species (males only): Lang 1965.

Amonardia perturbata Lang, 1965. Often very abundant on eelgrass leaves and macroalgae.

Amphiascoides dimorphus Lang, 1965.[†] Described from intertidal shell sand at DB. Key to species: Lang 1965.

Amphiascoides lancisetiger Lang, 1965. Described from tidal pool with shell sand, stones, and algae near HMS.

Amphiascoides petkovskii Lang, 1965. Intertidal and shallow subtidal sandy mud, including around eelgrass.

Amphiascoides subdebilis (Willey, 1935).[+] In sediments around intertidal and subtidal eelgrass and unvegetated sand and mud flats. Nearly cosmopolitan distribution.

Amphiascopsis cinctus (Claus, 1866). Intertidal sand and mud with microalgae, intertidal algae, epiphyton on docks. Cosmopolitan.

Amphiascus minutus (Claus, 1863).[†] Intertidal eelgrass and sandy mud, and a variety of mid- to low rocky intertidal habitats.

Amphiascus parvus Sars, 1906. European species recorded from at least one site in California (Watkins 1983) and abundant in eelgrass beds of coastal embayments in Oregon and Washington.

Amphiascus undosus Lang, 1965. Lower intertidal habitats. In Oregon and Washington, an abundant species on eelgrass in estuaries.

Bulbamphiascus imus (Brady, 1872).[†] Uncommon in low intertidal/shallow subtidal sand, except in areas of high organic loading or pollution, where it can dominate. Key to species: Dinet 1971, Tethys 2: 747–762.

Diosaccus spinatus Campbell, 1929.[†] A prominent bright yellow species on macroalgae and eelgrass. In Washington and British Columbia, this species is rare in the spring but abundant from late summer into the winter. Key to species: Lang 1965.

Paramphiascella xiphophora Lang, 1965. Described from off HMS in 27 m of water on a bottom of fine sand, detritus, and algae. Key to species: Marcotte 1974, Zool. J. Linn. Soc. 55: 65–82.

Paramphiascopsis ekmani Lang, 1965. Tide pools in mid- to lower intertidal. Key to species: Lang 1965, Hicks 1986, New Zealand. J. Nat. Hist. 20: 389–397.

Robertgurneya diversa Lang, 1965.[†]

Robertgurneya hopkinsi Lang, 1965. This and the previous species were described from off HMS at 26 m depth on a bottom of fine sand, detritus, and algae; both also occur intertidally. Species similar to *R. diversa* and *R. hopkinsi*, as well as several unidentified species have been found in intertidal sand in Puget Sound. Key to species: Lang 1965, Hamond 1973, J. Nat. Hist. 7: 65–76.

Robertsonia knoxi (Thompson and A. Scott, 1903). Sheltered sand beaches and lower intertidal/shallow subtidal sand and sandy mud with detritus or eelgrass. Key to species: Hamond 1973, Rec. Aust. Mus. Sydney 28: 421–435, Fiers 1996, Bull. Mar. Sci. 58: 117–130.

Robertsonia propinqua (T. Scott, 1893).[+] Abundant in the autumn and winter in subtidal sand and sediments in eelgrass.

Schizopera knabeni Lang, 1965.[†] Sediments around estuarine eelgrass; extends upstream into oligohaline tidal reaches of estuaries. Co-occurs with one or more unidentified species in Puget Sound. Key to species: Lang 1965, Apostolov 1973, Acta Mus. Maced. Sc. Nat., Skopje 13: 81–107.

Schizopera sp. Watkins (1983) found an unidentified species in a salt marsh at Chorro Creek, San Luis Obispo County.

Stenhelia (Stenhelia) asetosa Thistle and Coull, 1979.[†] Key to species of the subgenus: Thistle and Coull 1979, Zool. J. Linn. Soc. 66: 63–72.

Stenhelia (Stenhelia) peniculata Lang, 1965. Described from off HMS, at 26 m depth on a bottom of fine sand and detritus but is also abundant intertidally. Unlike most of its congeners, this species occurs in algae- and seagrass-rich habitats.

Stenhelia (Delavalia) latipes Lang, 1965. Described from same habitat as previous species. Key to species of the subgenus: Coull 1976, Zool. J. Linn. Soc. 59: 353–364.

Stenhelia (Delavalia) latioperculata Itô, 1981.[+] Described from sandy bottom at 26 m off Hokkaido, Japan. Abundant in intertidal/shallow subtidal sediments near eelgrass and *Spartina* in Washington and British Columbia.

Stenhelia (Delavalia) longicaudata longicaudata Boeck, 1872.[+] Described from the North Atlantic, this species was found by Kask et al. (1982) in subtidal mud in Departure Bay, British Columbia.

Stenhelia (Delavalia) longipilosa Lang, 1965. Described from same habitat as *S. (S.) peniculata*.

Stenhelia (Delavalia) oblonga Lang, 1965. Intertidal mud flats and salt marshes.

Typhlamphiascus pectinifer Lang, 1965.[†] Described from Monterey Bay tidal pools containing shell sand, stones, and algae. Occurs in a variety of lower intertidal/shallow subtidal habitats in Puget Sound.

Typhlamphiascus unisetosus Lang, 1965. Described from same habitat as the previous species.

NORMANELLIDAE

Normanella bolini Lang, 1965.[†] Key to species: Lang 1965; review of the genus, Lee and Huys 1999, Cah. Biol. Mar. 40: 203–262.

Normanella confluens Lang, 1965. This and the previous species were described from a tide pool with shell sand, stones, and algae near HMS.

ORTHOPSYLLIDAE

Key to genera: Boxshall and Halsey 2004.

Orthopsyllus linearis illgi (Chappuis, 1958). Described from Puget Sound, where it is abundant in intertidal and shallow subtidal sand beaches; redescribed by Lang (1965) from tide pools near HMS.

PARAMESOCHRIDAE

Key to genera: Boxshall and Halsey 2004.

Apodopsyllus vermiculiformis Lang, 1965.[†] Described from fine sand in a tide pool at Point Pinos, Monterey Bay. Key to species: Lang 1965, Coull and Hogue 1978, Trans. Amer. Micros. Soc. 97: 149–159.

Scottopsyllus (Scottopsyllus) pararobertsoni Lang, 1965. Described from a tide pool with shell sand, stones, and algae near HMS. Key to species: Lang 1965.

PARASTENHELIIDAE

Parastenhelia hornelli Thompson and A. Scott, 1903.[+] Common in microalgae and epiphytes in a variety of intertidal and shallow subtidal habitats in southern B.C. and Puget Sound. Key to species: Wells et al. 1982, N.Z. J. Zool., 9: 151–184.

Parastenhelia spinosa (Fischer, 1860). A cosmopolitan species recorded in vegetated habitats from southern British Columbia to La Jolla, CA.

PELTIDIIDAE

Key to genera: Huys et al. 1996, Boxshall and Halsey 2004. All members of this family are strongly dorsoventrally compressed and occur mainly on macroalgae and eelgrass blades.

Alteutha langi Monk, 1941.[†] Described from La Jolla. Watkins (1983) found two other unidentified species from the California intertidal.

Alteuthella sp. Watkins (1983) found an unidentified species near Piedras Blancas point.

Eupelte setacauda Monk, 1941.[†] Key to species: Hicks 1982, South Africa. Zool. J. Linn. Soc. 75: 49–90.

Eupelte simile (Monk, 1941). This and the previous species were described from La Jolla. Watkins (1983) reported two (one as *Paralteutha*) apparently undescribed species from Mendocino County intertidal.

Peltidium sp.[†] Unidentified specimens have been found in coastal bays of Washington and reported from Mendocino County intertidal by Watkins (1983).

PORCELLIDIIDAE

This family is also dorsoventrally flattened. Key to genera: Boxshall and Halsey 2004.

Porcellidium fimbriatum Claus, 1863.[†] Found by Gunnill (1982) on algae near La Jolla. Kask et al. (1982) reported *P.* cf. *fimbriatum* from subtidal eelgrass and gravel in southern British Columbia.

P. viride (Phillipi, 1840). Puget Sound to La Jolla on eelgrass and macroalgae.

PSEUDOTACHIDIIDAE

Elevated from subfamily of the Thalestridae to family rank by Willen (2000). Partial key to genera: Huys et al. 1996.

Dactylopodella sp.[†] Isolated specimens found in California (Watkins 1983) and Puget Sound. Key to species: Hicks 1989, Nat. Mus. N.Z. Rec. 3: 101–117.

Idomene purpurocincta Norman and T. Scott, 1905.[†] On algae and eelgrass. Supplemental description of specimens from California: Lang 1965.

RHIZOTRICHIDAE

Rhizothrix sp.[+] Common in lower intertidal/shallow subtidal sandy habitats in Puget Sound. Key to species: Bodin 1979, Vie Milieu 27: 311–357.

TACHIDIIDAE

Key to genera and species: Boxshall and Halsey 2004.

Geeopsis incisipes (Klie, 1913).[+] Common in European salt marshes, and also found in emergent marshes in Puget Sound estuaries.

Microarthridion littorale (Poppe, 1881).[†] Euryhaline; mud and muddy sand in estuaries and salt marshes; also occurs in epibenthic plankton. Probably a complex of several species. Key to species in Bodin, P. 1970, Tethys 2: 385–486.

Tachidius discipes Giesbrecht, 1881. Common in oligohaline and tidal freshwater portions of estuaries in sand and mud and in epibenthic plankton.

Neotachidius triangularis Shen and Tai, 1963. One of the most abundant harpacticoids in marine-influenced tidal channels of coastal estuaries; also in eelgrass beds. Described from the Pearl River Delta, South China, and may be introduced. Two new species have been described from Korea, and it is possible that records of *N. triangularis* in the Northeast Pacific may be one of these species or an as yet undescribed species. Table distinguishing species in Huys et al. 2005, Zool. J. Linn. Soc. 143: 133–159.

TEGASTIDAE

All members of this family are strongly anterolaterally compressed and have extremely hard integument. They are often associated with epibionts and colonial invertebrates Key to genera: Huys et al. 1996, Boxshall and Halsey 2004.

Tegastes perforatus Lang, 1965.[†] This and following species described from males only from 7 m depth off HMS; Watkins 1983 figured female of this and females and males of three undescribed species in intertidal pools and recorded a fourth undescribed species.

Syngastes serratus Lang, 1965.

TETRAGONICIPITIDAE

Key to genera: Huys 1995, Hydrobiologia 308: 23–28, Boxshall and Halsey 2004.

Phyllopodopsyllus parabradyi Lang, 1965. Described from near HML in a tide pool with shell sand, stones, and algae. Key to species: Lang 1965, Coull 1973, Proc. Biol. Soc. Wash. 86: 9–24.

Phyllopodopsyllus borutzkyi Lang, 1965. Described from tide pool algae, DB.

Paraschizopera trifida (Yeatman, 1980).[+] Comb. nov. Huys, 1995, Hydrobiologia 308: 23–28, for *Diagoniceps trifidus* Yeatman, 1980 described from Puget Sound. Huys also gives a key to species.

THALESTRIDAE

Key to genera of each subfamily: Boxshall and Halsey 2004.

Eudactylopus latipes typica Sewell, 1940. Eelgrass and macroalgae. Redescription: Lang 1965.

Paradactylopodia serrata Lang, 1965.[†] Mid- to lower intertidal on sand and muddy sand.

Parathalestris bulbiseta Lang, 1965.[†] Southern British Columbia to La Jolla. Key to species: Lang 1965.

Parathalestris californica Lang, 1965. This and previous species are usually associated with macroalgae but can also occur on unvegetated sand.

Parathalestris jacksoni (T. Scott, 1898).[+] Occurs in the North Atlantic and Arctic and also in Alaska and southern British Columbia.

Parathalestris verrucosa Itô, 1970.[+] Described from Japan, this species was also reported from epiphyton in southern British Columbia by Kask et al. (1982).

Phyllothalestris mysis (Claus, 1863).[†] Algae rich lower intertidal habitats.

Rhynchothalestris helgolandica (Claus, 1863). Lower intertidal near algae; also occurs in Puget Sound. Redescription: Huys 1990, Zool. J. Linn. Soc. 99: 51–115.

Thalestris longimana Claus, 1863. Described from Europe; abundant on algae in Puget Sound.

TISBIDAE

Key to genera: Boxshall and Halsey 2004.

Scutellidium arthuri Poppe, 1884. On eelgrass and algae blades. Recorded from southern British Columbia to southern California, and in Japan. Redescription: Lang 1965. Key to species: Branch 1975, Ann. S. Afr. Mus. 66: 221–232.

S. hippolytes (Kroyer, 1863). Widely distributed on Atlantic coast in same habitats as above species. Recorded from southern British Columbia to La Jolla.

Tisbe cf. *furcata* (Baird, 1837).[†] Eelgrass, detrital, and microalgal habitats. Due to many closely related "sibling" species, identifications and distributions are difficult to establish for this genus. There are up to six unidentified/undescribed species in the northeastern pacific (Kask et al. 1982). *T.* cf. *furcata* appears to be cosmopolitan. It is an important prey item for juvenile salmon and other small fish.

Tachidiella parva Lang, 1965. Described from off HMS in fine sand and detritus at about 26 m depth. This is a *species incerta* because there is a contradiction between the Lang's key and his diagnosis of the species: the key indicates five setae on the exopodite of the fifth leg, while the description and drawing mention only four. Review of genus, Lee and Huys 1999, Zoosystema 21: 419–444.

REFERENCES FOR FREE-LIVING COPEPODS

Bodin, P. 1997. Catalogue of the new marine harpacticoid copepods. Documents de Travail de l'Institut Royal des Sciences Naturelles de Belgique No. 89, 304 pp.

Boltovskoy, D. ed. 1999. South Atlantic Zooplankton. Backhuys Publishers, Leiden, 1706 pp.

Boxshall, G. A., and S. H. Halsey. 2004. An Introduction to Copepod Diversity. The Ray Society, London. No. 166, 966 pp.

Bradford, J. 1976. Partial revision of the *Acartia* subgenus *Acartiura* (Copepoda: Calanoida: Acartiidae). N.Z. Journal of Marine and Freshwater Research 10: 159–202.

Chappuis P. A. 1958. Harpacticoïdes psammiques marins des environs de Seattle (Washington, U.S.A.). Vie Milieu 8: 409–422.

Crandell, G. F. 1966. Seasonal and spatial distribution of harpacticoid copepods in relation to salinity and temperature in Yaquina Bay, Oregon. PhD Thesis, Oregon State University, Corvallis, Oregon, 137 pp.

Dahms, H.-U. 1990. Naupliar development of Harpacticoida (Crustacea, Copepoda) and its significance for phylogenetic systematics. Microfauna Marina 6: 169–272.

Dawson, J. K., and G. Knatz. 1980. Illustrated key to the planktonic copepods of San Pedro Bay. Technical Reports of the Allan Hancock Foundation No. 2. The Allan Hancock Foundation and the Institute for Marine and Coastal Studies, University of Southern California, Los Angeles. 1–106.

Esterly, C. O. 1905. The pelagic Copepoda of the San Diego Region. University of California Publications in Zoology 2: 113–233.

Esterly, C. O. 1906. Additions to the copepod fauna of the San Diego Region. University of California Publications in Zoology 3(5): 53–92.

Esterly, C. O. 1911. Third report on the Copepoda of the San Diego Region. University of California Publications in Zoology 6: 313–352.

Esterly, C. O. 1924. The free-swimming Copepoda of San Francisco Bay. University of California Publications in Zoology 26: 81–129.

Gardner, G. A., and I. Szabo. 1982. British Columbia Pelagic marine Copepoda: an identification manual and annotated bibliography. Can. Spec. Publ. Fish. Aquat. Sci. 62: 536 pp.

Gunnill, F. C. 1982. Effects of plant size and distribution on the numbers of invertebrate species and individuals inhabiting the brown alga *Pelvetia fastigiata*. Marine Biology 69: 263–280.

Huys, R., J. M. Gee, C. G. Moore, and R. Hamond. 1996. Marine and brackish water harpacticoid copepods. Part One. Synopses of the British Fauna (New Series). The Linnaean Society of London and the Estuarine and Coastal Sciences Association, 352 pp.

Huys, R., and Boxshall, G. A. 1991. Copepod Evolution. The Ray Society, London. No. 159, 468 pp.

Kask, B. A., J. R. Sibert, and B. Windecker. 1982. A check list of marine and brackish water harpacticoid copepods from the Nanaimo estuary, southwestern British Columbia. Syesis, 15: 25–38.

Lang, K. 1948. Monographie der Harpacticiden 2 vols. Lund, Håkan Ohlsson's Bøktryckeri. Stockholm, Nordiska Bøkhandeln, 1–1682.

Lang, K. 1965. Copepoda Harpacticoidea from the Californian Pacific coast. Kungliga Svenska Vetenskapsakademiens Handlingar (4) 10: 1–560.

Owre, H. B., and M. Foyo. 1967. Copepods of the Florida Current, with illustrated keys to genera and species. Fauna Caribaea 1. Crustacea, Part 1, Copepoda, 137 pp.

Watkins, R. L. 1983. Harpacticoida (Crustacea: Copepoda) from the California Coast. PhD Thesis, Arizona State University, 312 pp.

Webb, D. G. 1983. Predation by juvenile salmonids on harpacticoid copepods in a shallow subtidal seagrass bed: effects on copepod community structure and dynamics. PhD Dissertation, University of British Columbia, 246 pp.

Willen, E. 2000. Phylogeny of the Thalestridimorpha Lang, 1944 (Crustacea, Copepoda). Cuvier Verlag, Göttingen, 233 pp.

Commensal and Parasitic Copepoda (Orders Cyclopoida, Siphonostomatoida, and Monstrilloida)

(Plates 207–211)

Commensal and parasitic copepods exhibit enormous morphological diversity. They range in form from those that resemble free-living copepods to those that are transformed to such a degree that it is hard to recognize them as crustaceans. Two orders—Cyclopoida and Siphonostomatoida—contain the majority of copepod species associated with marine invertebrates and fishes. Many of these species were previously included in the Poecilostomatoida until this order was placed in the Cyclopoida by Boxshall and Halsey (2004). These authors present a useful list of animal phyla and/or classes along with the types of symbiotic copepods associated with each group and an extensive bibliography. They also give keys to families of copepods parasitic on marine fishes and to genus and/or species within each copepod family. For each family they also detail host species or groups (if known) for individual copepod genera and/or species.

Symbiotic siphonostomatoids can be easily distinguished from cyclopoids in most cases by the presence of an oral cone, formed by the labrum and labium, enclosing the styletlike mandible (plate 207A).

COPEPODS ASSOCIATED WITH NEARSHORE FISHES

Ectoparasites most often encountered on marine fishes are sea lice of the families Caligidae and Trebiidae. Most of the caligid species are in the genera *Caligus* and *Lepeophtheirus*. Two species, *C. clemensi* and *L. salmonis* (plate 207B) cause serious disease in farmed salmon. Trebiids are similar in appearance to caligids and occur on several nearshore sharks and rays. Three families, the Bomolochidae (plate 207C), Chondracanthidae (plate 207D), and Lernaeopodidae (plate 208A) are associated mainly with the nasal cavities and gills of a wide variety of nearshore fishes.

Copepods in the family Penellidae (plate 208B) have a life cycle involving two different hosts. On the first host, which is a fish or a pelagic gastropod, development to the sexually mature adult takes place on the gills. After mating, the fertilized female leaves the first host and finds the second host (a fish or marine mammal), where it embeds itself and metamorphoses into a much larger saclike form. Often the most visible part of penellids are the long paired egg sacs that stream out from the host's integument. Similarly, in the family Sphyriidae (plate 208C) adult females are highly modified, with the cephalothorax embedded in the host and the rest of the body outside the host.

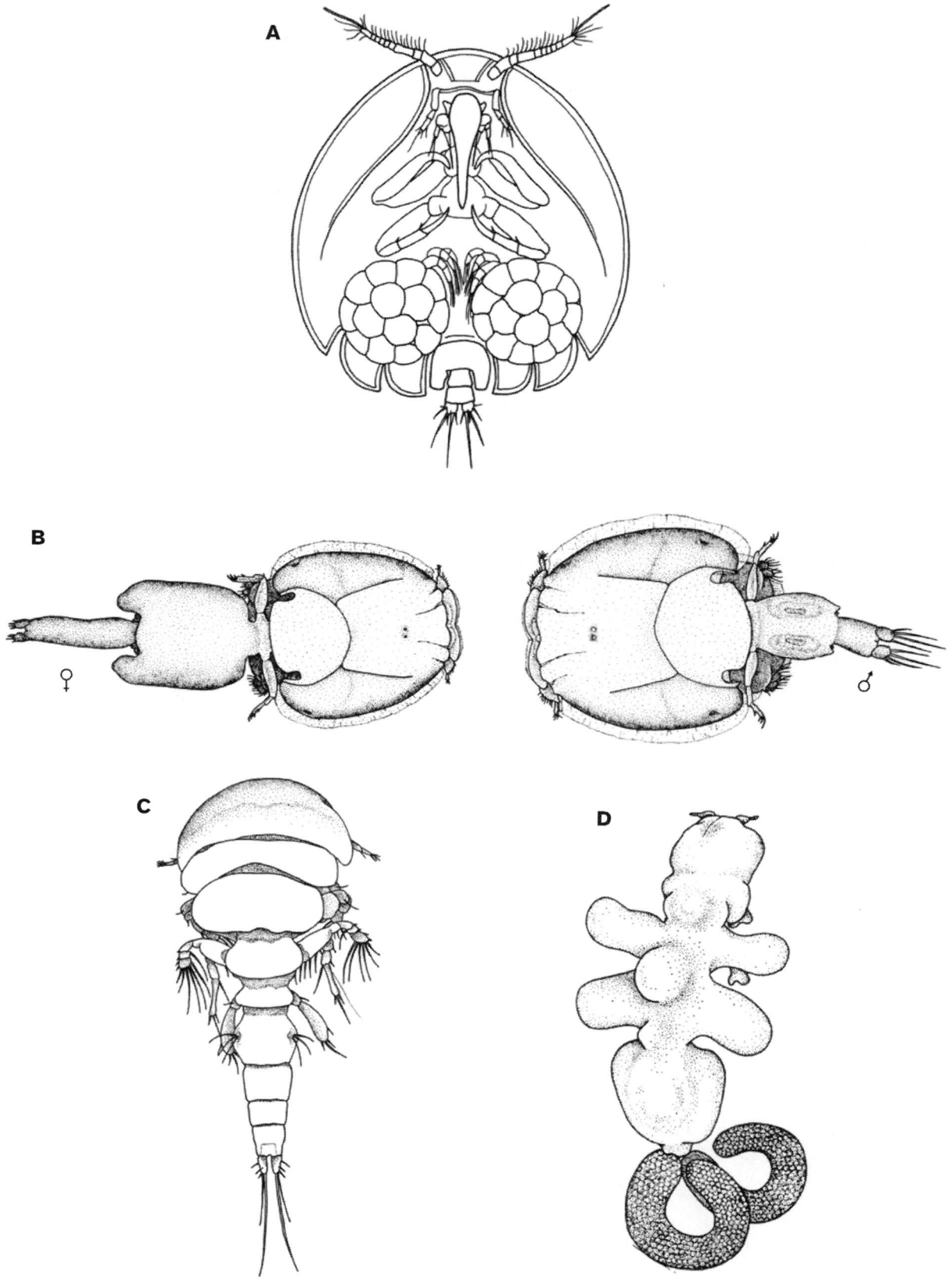

PLATE 207 Symbiotic Copepoda: A, example of the morphology of a siphonostomatoid copepod, *Artotrogus orbicularus* Boeck, 1859 (Artotrogidae); B, *Lepeophtheirus salmonis* (Caligidae); C, *Holobomolochus venustus* (Bomolochidae); D, *Chondracanthus narium* (Chondracanthidae) (A, from Boxshall and Halsey 2004; B, from Kabata 1979; C, from Kabata 1971; D, from Kabata 1969).

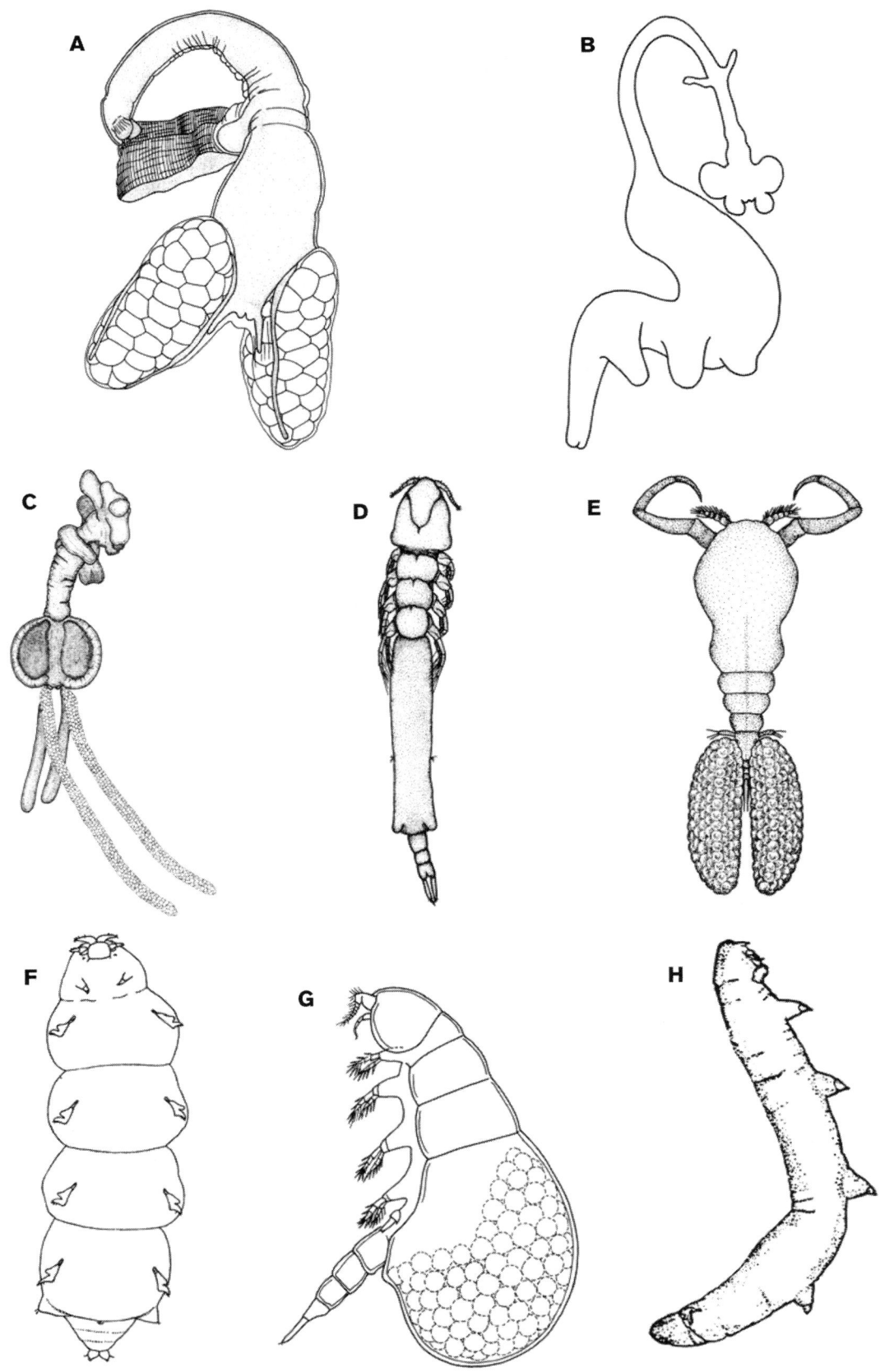

PLATE 208 Symbiotic Copepoda: A, *Parabrachiella microsoma* (Lernaeopodidae); B, *Haemobaphes* sp. (Penellidae); C, *Norkus cladocephalus* (Sphyriidae); D, *Kroeyerina deborahae* (Kroyeriidae); E, *Ergasilus nanus* (Ergasilidae); F, *Haplostoma albicatum* (Botryllophilidae); G, *Doropygus psyllus* (Notodelphyidae); H, *Enteropsis abbotti* (Enteropsidae) (A, from Dojiri 1981; B and G, from Boxshall and Halsey 2004; C, from Dojiri and Deets 1988; E, from Kabata 1979; F, from Ooishi and Illg 1977; H, from Illg and Dudley 1980.

In California and Oregon, the family Kroyeriidae is represented by the genus *Kroeyerina* (plate 208D), which occurs in the gill and nasal lamellae of elasmobranchs. In fresh waters, copepod fish parasites are dominated by the family Ergasilidae, which occur on the gills, nostrils, fins, and body surfaces of their hosts. Ergasilids are uncommon in marine fishes, but several species of *Ergasilus* occur on estuarine fishes (plate 208E).

COPEPODS ASSOCIATED WITH INVERTEBRATES

UROCHORDATA

Solitary and colonial tunicates are often inhabited by one or more copepod species, and they can often be seen through the transparent body wall inside their hosts. Copepod symbionts of tunicates are relatively well described for the northeastern Pacific. The families most often encountered in ascidians include Botryllophilidae (plate 208F), Notodelphyidae (plate 208G), Enteropsidae (plate 208H), and Ascidicolidae (plate 209A).

ECHINODERMATA

In other regions, echinoderms are hosts to a large diversity and at times very high densities of symbiotic copepods from 13 families. Few species have been described from the Pacific coast, and additional species and records are to be expected. *Caribeopsyllus amphiodiae* (Thaumatopsyllidae) (plate 209B) and several species of *Cancerilla* (Cancerillidae) (plate 209C) occur in the stomach and as external parasites, respectively, of some California brittle stars. Based on studies from other regions, it is probable that additional copepod associates of echinoderms from the families Asterocheridae, Pseudanthessiidae, Nanaspididae, Synaptiphilidae, and Lichomolgidae will be found in the northeastern Pacific.

MOLLUSCA

Copepod families most often found symbiotic in bivalve mollusks in the northeastern Pacific include the Anthessiidae, Myicolidae, Mytilicolidae, and Lichomolgidae. *Anthessius fitchi* (Anthessiidae) (plate 209D) occurs in the boring clams *Chaceia ovoidea* and *Zirfaea pilsbryi*. *Pseudomyicola spinosus* (Myicolidae) (plate 209E) is extremely widespread geographically and locally infests the mantle cavity of *Mytilus* spp. Likewise, *Mytilicola orientalis* (Mytilicolidae) (plate 209F) is a widespread intestinal parasite that occurs in the Pacific oyster *Crassostrea gigas* and the bivalve *Saxidomus giganteus*. The most speciose family of bivalve copepod symbionts in this region is the Lichomolgidae, containing at least six species of *Herrmannella* (plate 210A) that are found in a variety of clams and scallops.

Compared with bivalves, gastropods have relatively few copepod symbionts. Several species of *Anthessius* occur on limpets, cephalaspideans, and anaspideans. Splanchnotrophidae are highly transformed and live inside nudibranchs and sacoglossans. They are usually noticed only because paired egg sacs protrude through the body wall of the host. In addition to a number of undescribed species in this family, several species of *Ismaila* (plate 210B) have been described from California nudibranchs.

CRUSTACEA

The Nicothoidae is a widespread and speciose family that parasitizes peracarids and other crustaceans, often as egg mimics. *Hansenulus trebax* occurs in the brood chambers of *Neomysis mercedis* and several other mysids, where it consumes the host's eggs (plate 210C). Another copepod family often found symbiotic with crustaceans is the Clausidiidae, which has two genera (*Hemicyclops* and *Clausidium*—plate 210D, 210E) commonly found on the surface and in the burrows of callianassid shrimp.

ANNELIDA

Polychaetes appear to be the main hosts for copepods of the order Monstrilloida, which has a single family, the Monstrillidae (plate 211A). Adult monstrilloids are planktonic and do not feed. After hatching, nauplii locate a host, burrow into it, and transform into a saclike body. The last copepodid stage leaves the host and molts into the adult. Most of the copepod polychaete symbionts described from the northeastern Pacific are from the cyclopoid family Gastrodelphyidae (plate 211B) and are external associates of sabellids. *Spiophanicola spinulosus* (Spiophanicolidae) (plate 211C) is an external parasite of the polychaete genus *Spiophanes*. In the Atlantic, the scale worm *Harmothoe imbricata* is parasitized by at least five species in the family Herpyllobiidae (plate 211D), and this family may also occur on polynoids in the Pacific.

BRYOZOA

In other regions, species from two closely related siphonostomatoid families, Artotrogidae (plate 207A) and Asterocheridae (plate 211E) are commonly associated with bryozoan colonies. Although few if any species from these families have been described from the northeast Pacific, several possibly undescribed species in these families occur with bryozoans in the Puget Sound region.

PLATYHELMINTHES

Only two copepods are known as associates of flatworms, and one, *Pseudanthessius latus* (Pseudanthessiidae) (plate 211F), was described from the turbellarian *Kaburakia excelsa*.

CNIDARIA

The Rhynchomolgidae is the largest family of copepods associated with cnidarians. On the Pacific coast, five species of *Doridicola* (Rhynchomolgidae) (plate 211G) have been described, and occur on a variety of anemones and the sea pen *Ptilosarcus gurneyi*. As with bryozoans, many species of Asterocheridae and Artotrogidae have been found associated with cnidarian hosts elsewhere, and more may be expected in the northeastern Pacific.

LIST OF SOME COMMENSAL AND PARASITIC COPEPODA

The following list includes symbiotic copepods associated with invertebrates and fishes that occur in intertidal and shallow subtidal marine and estuarine habitats. Species associated only with deeper subtidal or pelagic hosts (e.g., most rockfishes, scombrids) are not included.

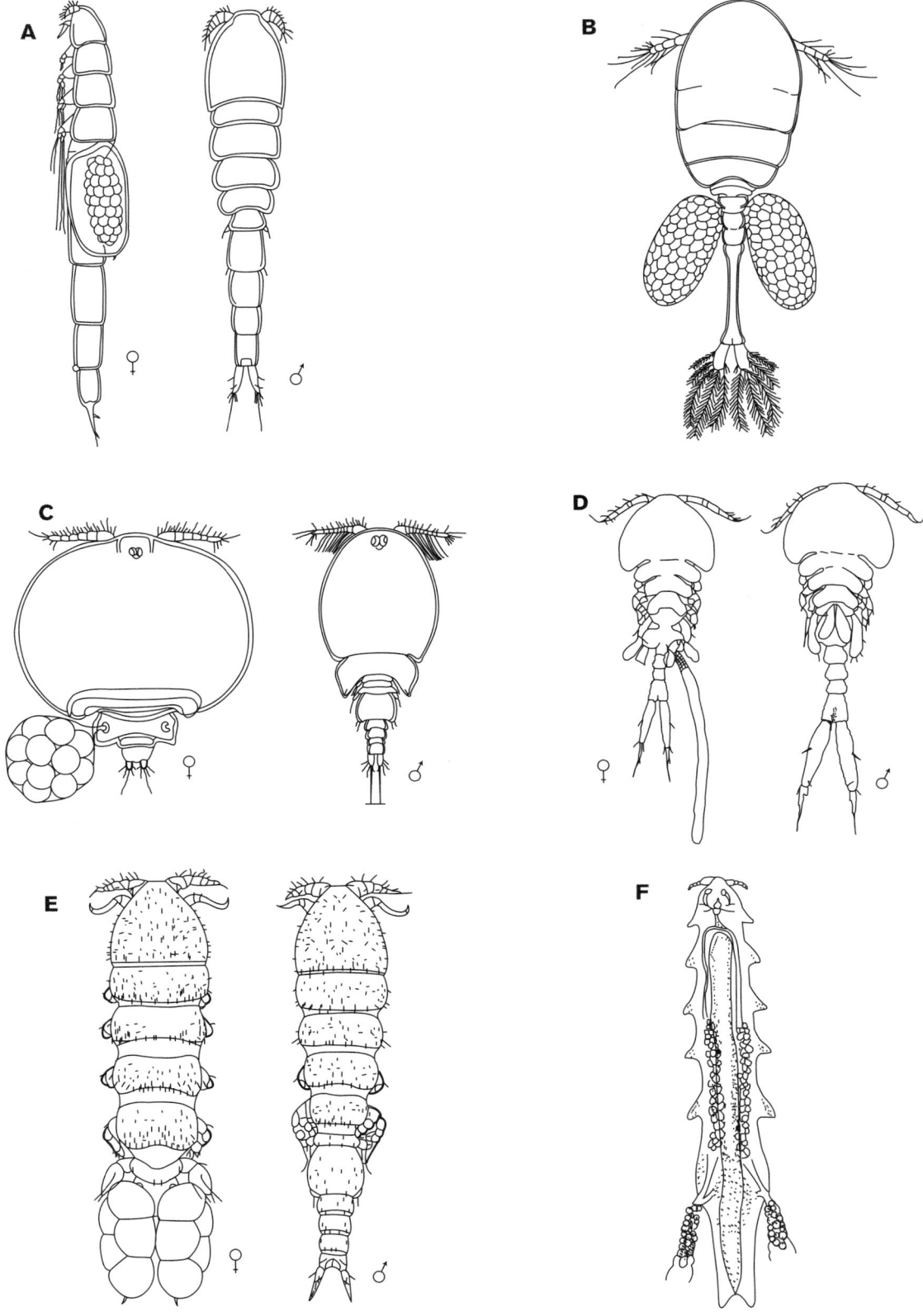

PLATE 209 Symbiotic Copepoda: A, *Ascidocola rosea* (Ascidocolidae); B, *Caribeopsyllus amphiodiae* (Thaumatopsyllidae); C, *Cancerilla tubulata* (Cancerillidae); D, *Anthessius fitchi* (Anthessiidae); E, *Pseudomyicola spinosus* (Myicolidae); F, *Mytilicola orientalis* (Mytilicolidae) (A and C, from Boxshall and Halsey 2004; B, from Ho et al. 2003; D, from Illg 1960; E, from Humes 1968; F, from Grizel 1985).

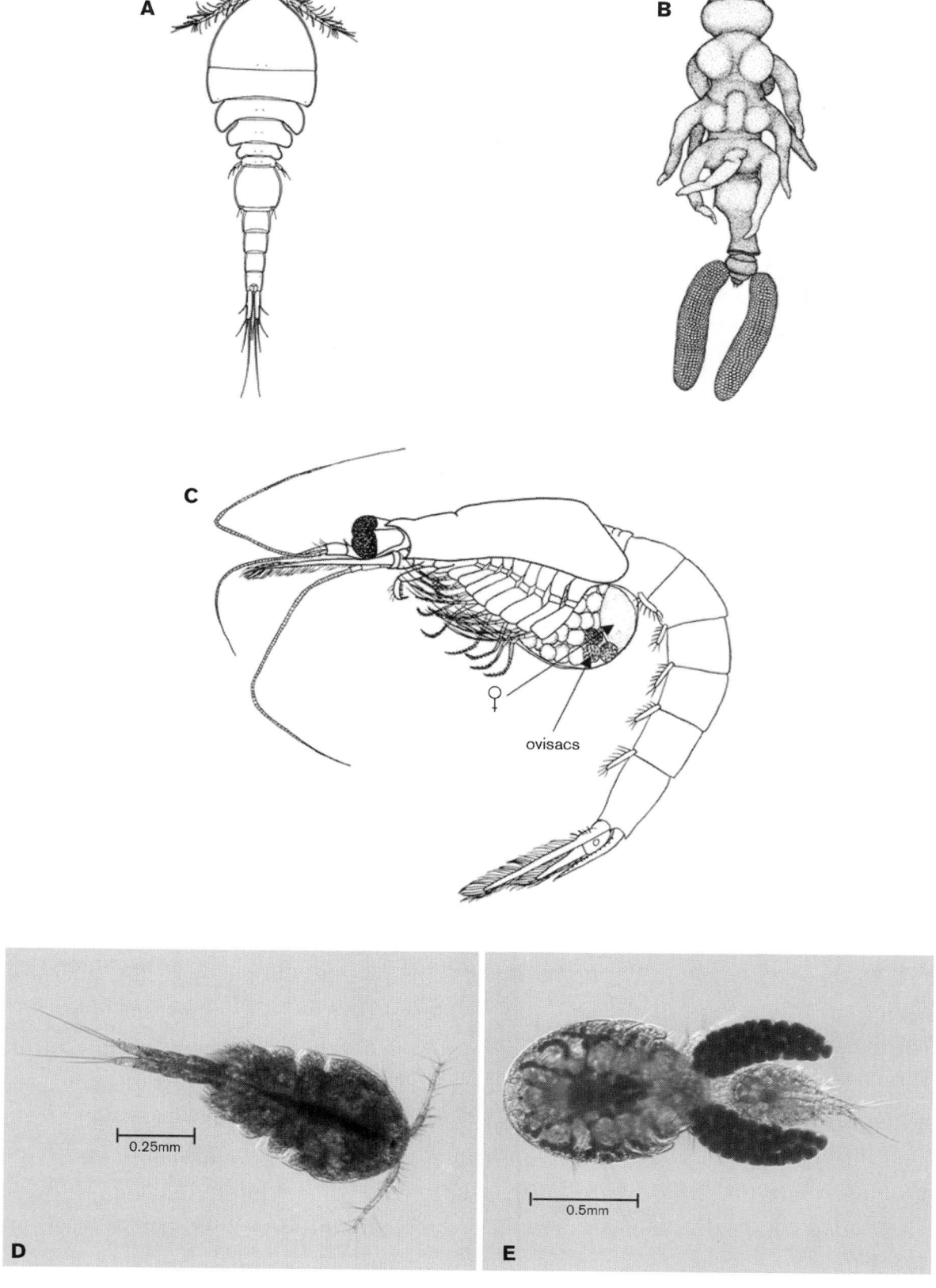

PLATE 210 Symbiotic Copepoda: A, *Herrmannella saxidomi* (Lichomolgidae); B, *Ismaila occulta* (Splanchnotrophidae); C, *Hansenulus trebax* (Nicothoidae) female and ovisacs in the marsupium of *Neomysis mercedis*; D, *Hemicyclops subadhaerens* (Clausidiidae); E, Female *Clausidium vancouverense* (Clausidiidae) with attached male (A, from Humes and Stock 1973; B, from Ho 1981; C, from Daly and Damkaer 1986).

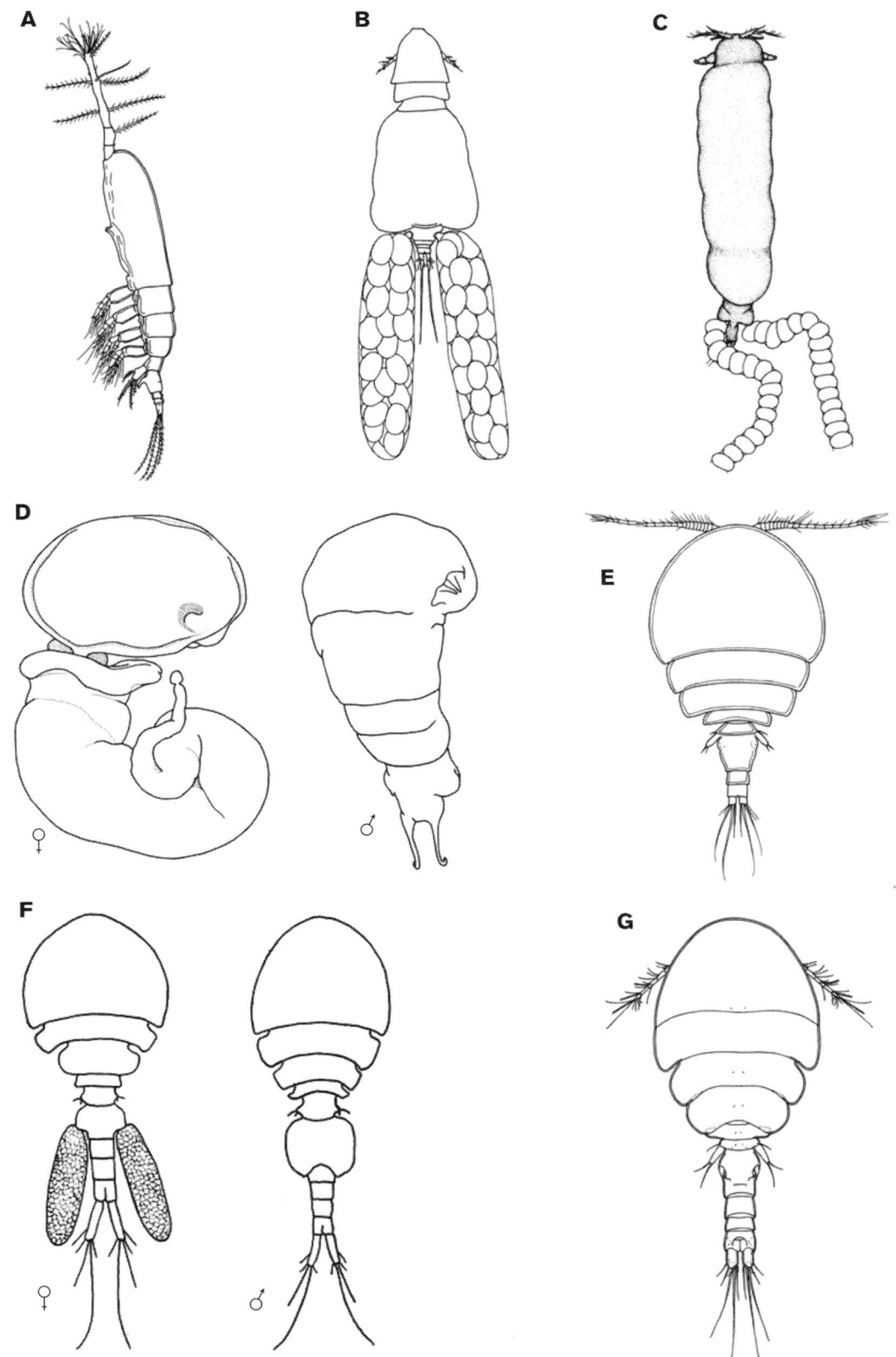

PLATE 211 Symbiotic Copepoda: A, *Monstrilla longicornis* (Monstrillidae); B, *Gastrodelphys fernaldi* (Gastrodelphyidae); C, *Spiophanicola spinulosus* (Spiophanicolidae); D, *Eurysilenium oblongum* (Herpyllobiidae); E, *Asterocheres boecki* (Asterocheridae); F, *Pseudanthessius latus* (Pseudanthessiidae), antennules not shown; G, *Doridicola ptilosarci* (Rynchomolgidae) (A, [modified], B, C, and D, from Boxshall and Halsey 2004; B, from Ho 1984; E, from Illg 1950; F, from Humes and Stock 1973).

HARPACTICOIDA

LAOPHONTIDAE

Namakosiramia californiensis Ho & Perkins, 1977. On the sea cucumber *Parastichopus parvimensis*.

THALESTRIDAE

Donsiella limnoriae Stephensen, 1939. On the ventral side of wood-boring isopods, genus *Limnoria*.

CYCLOPOIDA

ANTHESSIIDAE

Anthessius fitchi Illg, 1960. In the bivalves *Chaceia ovoidea* and *Zirfaea pilsbryi*.

Anthessius lighti Illg, 1960. On the anaspidean mollusk *Aplysia californica*. Named for S. F. Light, the author of the first edition of this manual.

Anthessius navanacis Illg, 1960. On the cephalaspidean mollusk *Navanax inermis*.

Anthessius nortoni Illg, 1960. On the keyhole limpet, *Diodora aspera*.

ASCIDOCOLIDAE

Ascidocola rosea Thorell, 1859. Geographically widely distributed, in ascidians.

BOMOLOCHIDAE

Bomolochus constrictus (Cressey & Collette, 1970). Described from the Pacific topsmelt *Atherinops affinis*.

Bomolochus cuneatus Fraser, 1920. On the gills of Embiotocidae (surf perches), the threespine stickleback *Gasterosteus aculeatus*, the California grunion, *Leuresthes tenuis*, and the Pacific herring *Clupea harengus pallasi*.

Bomolochus longicaudus Cressey, 1969. In the gill cavity of the barred sand bass *Paralabrax nebulifer*.

Bomolochus paucus Cressey and Dojiri, 1984. On the white croaker *Genyonemus lineatus*.

Bomolochus pectinatus Stock, 1955. Parasitic on the California grunion.

Holobomolochus embiotocae Hanan, 1976. In the nasal cavity of Embiotocidae.

Holobomolochus prolixus (Cressey, 1969). In the gill cavity of the staghorn sculpin *Leptocottus armatus*.

Holobomolochus spinulus (Cressey, 1969). In the gill cavity of Cottidae (sculpins) and the California scorpionfish *Scorpaena guttata*.

Holobomolochus wilsoni Cressey & Cressey, 1985. Described from the gills of the garibaldi *Hypsypops rubicundus*.

Holobomolochus venustus Kabata, 1971. In the nasal cavity of the cabezon *Scorpaenichthys marmorata*.

BOTRYLLOPHILIDAE (IN COMPOUND ASCIDIANS)

Botryllophilus abbotti Ooishi & Illg, 1989. In *Eudistoma* sp.

Haplosaccus elongatus Ooishi & Illg, 1977. In *Aplidium arenatum*.

Haplostoma albicatum Ooishi & Illg, 1977. In *Distaplia occidentalis*.

Haplostoma minutum Ooishi & Illg, 1977. In *Aplidium arenatum*.

Haplostoma dentatum Ooishi & Illg, 1977. In *Cystodytes lobatus*.

Haplostoma setiferum Ooishi & Illg, 1977. In *Eudistoma ritteri*.

Haplostoma elegans Ooishi & Illg, 1977. In *Aplidium propinquum*.

Haplostomella dubia Ooishi & Illg, 1977. In *Aplidium arenatum*.

Haplostomella distincta Ooishi & Illg, 1977. In *Aplidium arenatum*.

Haplostomella oceanica Ooishi & Illg, 1977. In *Eudistoma ritteri*.

Haplostomella reducta Ooishi & Illg, 1977. In *Distaplia occidentalis*.

Haplostomides bellus Ooishi & Illg, 1977. In *Ritterella aequalisiphonis*.

CHONDRACANTHIDAE

Acanthochondria deltoidea (Fraser, 1920). In the gill cavity of Hexagrammidae (greenlings).

Acanthochondria. dojirii Kabata, 1984. In the gill cavity of the English sole *Parophrys vetulus*.

Acanthochondria fraseri Ho, 1972. On the c-o sole *Pleuronichthys coenosus*.

Acanthochondria hoi Kalman, 2003. In the gill cavity of the California halibut, *Paralichthys californicus*.

Acanthochondria holocephalarum Kabata, 1968. On the claspers of male ratfish, *Hydrolagus colliei*.

Acanthochondria rectangularis (Fraser, 1920). On the English sole.

Acanthochondria vancouverensis Kabata, 1984. In the nasal capsules of the cabezon.

Auchenochondria lobosa Dojiri & Perkins, 1979. In the oral cavity of the California halibut.

Chondracanthus gracilis Fraser, 1920. In the gill cavity of Hexagrammidae and the cabezon.

Chondracanthus heterostichi Ho, 1972. On the giant kelpfish *Heterostichus rostratus*.

Chondracanthus irregularis Fraser, 1920. On the gills and in the gill cavity of the buffalo sculpin *Enophrys bison* and great sculpin *Myoxocephalus polycanthocephalus*.

Chondracanthus narium Kabata, 1969. In the nasal cavity of the lingcod *Ophiodon elongatus*.

Chondracanthus pinguis Wilson, 1912. In the gill cavity of Hexagrammidae, cabezon, Scorpaenidae (rockfishes), and Stichaeidae (pricklebacks).

Chondracanthus pusillus Kabata, 1968. On the gills and in the gill cavity of Stichaeidae and the penpoint gunnel *Apodichthys flavidus*.

Heterochondria atypica Ho, 1972. On the gills of the lingcod and the senorita *Oxyjulis californica*.

Pseudodiocus scorpaenus Ho, 1972. On the California scorpionfish.

CLAUSIDIIDAE

Clausidium vancouverense (Haddon, 1912). In the branchial chamber and on the exoskeleton of the ghost shrimps *Neotrypaea californiensis* and *N. gigas* and the mud shrimp *Upogebia pugettensis*.

Hemicylops subadhaerens Gooding, 1960. In burrows of *Neotrypaea californiensis*.

Hemicylops thysanotus Wilson, 1935. Occurs with *Clausidium vancouverense* and also on the nudibranch *Hermissenda crassicornis*.

ENTEROPSIDAE

Enterocola laticeps Illg & Dudley, 1980. In the simple ascidian *Styela gibbsii*.

Enteropsis abbotti Illg & Dudley, 1980. In the simple ascidian *Styela* sp.

Enteropsis capitulatus Illg & Dudley, 1980. In the simple ascidian *Boltenia villosa*

Enteropsis minor Illg & Dudley, 1980. In the colonial ascidian *Metandrocarpa taylori*.

Enteropsis superbus Illg & Dudley, 1980. In the simple ascidian *Pyura haustor*.

ERGASILIDAE

Ergasilus lizae Krøyer, 1863. On the shiner perch *Cymatogaster aggregata*.

Ergasilus turgidus Fraser, 1920. On the roughback sculpin *Chitonotus pugetensis*.

GASTRODELPHYIDAE

Gastrodelphys dalesi (Green, 1961). External associate of the sabellid polychaete *Eudistylia polymorpha*.

Gastrodelphys fernaldi Dudley, 1964. External associate of the sabellid polychaete *Bispira* sp.

Sabellacheres gracilis G. O. Sars, 1862. External associate of the sabellid polychaete *Myxicola infundibulum*.

Sabellacheres illgi Dudley, 1964. External associate of the sabellid polychaetes *Branchiomma burrardum, Distylidia rugosa, and Pseudopotamilla occelata*.

GONOPHYSEMA-GROUP

This group comprises four highly transformed genera that have not been formally placed in any family.

Arthurhumesia canadiensis Bresciani & Lopez-Gonzalez, 2001. Inside zooids of the compound ascidian *Aplidium solidum*.

HERPYLLOBIIDAE

Eurysilenium oblongum Hansen, 1887. This and the following species are parasites of the scale worm *Harmothoe imbricata*.

Eurysilenium truncatum M. Sars, 1870.

Herpyllobius arcticus Steenstrup & Lützen, 1861.

Herpyllobius haddoni Lützen, 1964.

Herpyllobius polynoes (Krøyer, 1863).

LICHOMOLOGIDAE

Herrmannella bullata Humes & Stock, 1973. Associated with the scallops *Chlamys hastata hericia* and *C. rubida*.

Herrmannella columbiae (Thompson, 1897). In the bivalves *Tresus nuttallii* and *Callithaca tenerrima*.

Herrmannella kodiakensis Humes, 1995. In the bivalves *Saxidomus giganteus* and *Leukoma staminea*.

Herrmannella panopeae (Illg, 1949). In the bivalve *Panopea abrupta*.

Herrmannella saxidomi (Illg, 1949). In the bivalve *Saxidomus nuttalli*.

Herrmannella tivelae (Illg, 1949). In the Pismo clam *Tivela stultorum*.

Modiolicola gracilis C. B. Wilson, 1935. On the gill filaments of the mussel *Mytilus californianus*.

MYICOLIDAE

Myicola ostreae Hoshina & Sugiara, 1953. Attached to the gills of the Pacific oyster *Crassostrea gigas*.

Pseudomyicola spinosus (Raffaele & Monticelli, 1885). Parasitic in the mantle cavities of more than 50 species of bivalves including *Mytilus* spp.

MYTILICOLIDAE

Mytilicola orientalis Mori, 1935. Intestinal parasite of the Pacific oyster *Crassostrea gigas* and the bivalve *Saxidomus giganteus*.

NOTODELPHYIDAE

Doropygus bayeri Illg, 1958. In the simple ascidian *Pyura haustor*.

Doropygus fernaldi Illg, 1958. In the simple ascidian *Boltenia villosa*.

Doropygus mohri Illg, 1958. In the simple ascidian *Styela gibbsii*.

Doropygus seclusis Illg, 1958. In the simple ascidian *Chelyosoma productum*.

Doropygopsis longicauda (C. W. S. Aurivillius, 1882). In the simple Ascidian *Ascidia paratropa*.

Notodelphys affinis Illg, 1958. In the simple ascidians *Ascidia paratropa* and *Corella willmeriana*.

Pholeterides furtiva Illg, 1958. In the compound ascidian *Aplidium californicum*.

Pygodelphys aquilonaris Illg, 1958. In at least eight species of simple ascidians.

Pythodelphys acruris Dudley & Solomon, 1966. In the compound ascidian *Eudistoma psammion*.

Scolecodes huntsmani (Henderson, 1931). In the simple ascidians *Boltenia villosa, Pyura haustor*, and *Styela gibbsii*.

PHILICHTHYIDAE

Colobomatus embiotocae Hanan, 1976. In the nasal cavity of the fish family Embiotocidae.

PHILOBLENNIDAE

Philoblenna bupulda Ho & Kim, 1992. Parasitic in the gastropod *Fusitriton oregonensis* (in Korea).

PSEUDANTHESSIIDAE

Pseudanthessius latus Illg, 1950. On various turbellarians.

RHYNCHOMOLGIDAE

Doridicola confinis (Humes, 1982). This and the following three species are found on the column and tentacles of the sea anemones *Anthopleura, Epiactis*, and *Urticina*.

Doridicola pertinax (Humes, 1982).

Doridicola sunnivae (Humes, 1982).

Doridicola turmalis (Humes, 1982).

Doridicola ptilosarci Humes & Stock, 1973. On the surface and in folds of the sea pen *Ptilosarcus gurneyi*.

SPIOPHANICOLIDAE

Spiophanicola spinulosus Ho, 1984. External parasite of the polychaetes *Spiophanes berkeleyorum, S. duplex*, and *S. kroeyeri*.

SPLANCHNOTROPHIDAE

Unidentified species in this family have been found in the nudibranch genera *Dendronotus, Dirona, Eubranchus,* and *Triopha* in the northeastern Pacific region.

Ismaila occulta Ho, 1981. Parasitic in the cerata of the nudibranch *Dendronotus iris.*

Ismaila besicki Ho, 1989. In the nudibranch *Flabellina trophina.*

TAENIACANTHIDAE

Taeniacanthodes haakeri Ho, 1972. On the body surface of the shiner perch.

Taeniastrotos californiensis Cressey, 1969. On the barred sand bass.

TEREDICOLA-GROUP

Boxshall and Halsey (2004) were unable to place the genera in this group in to a family.

Teredicola typica C. B. Wilson, 1942. In the mantle cavity of the shipworm *Lyrodus pedicellatus.*

THAUMATOPSYLLIDAE

Caribeopsyllus amphiodiae Ho, Dojiri, Hendler, & Deets, 2003. In the stomach of the brittle star *Amphiodia urtica.*

SIPHONOSTOMATOIDA

ASTEROCHERIDAE

Asterocheres lilljeborgi Boeck, 1859. Associated with the asteroid *Henricia leviuscula.*

Asterocheres rubrum (Campbell, 1939). Described from plankton; host unknown.

ARTOTROGIDAE

Unidentified species from this family have been observed associated with bryozoans.

CALIGIDAE

Found on the body surfaces of fish.

Caligus bonito Wilson, 1905. On the northern anchovy *Engraulis mordax.*

Caligus clemensi Parker & Margolis, 1964. On the Pacific herring, threespine stickleback, Hexagrammidae, Pacific salmon, and the ratfish.

Caligus elongatus Nordmann, 1832. On Pacific herring.

Caligus hobsoni Cressey, 1969. On Embiotocidae and cabezon.

Caligus klawei Shiino, 1959. On the northern anchovy.

Caligus serratus Shiino, 1965. On the topsmelt and the jack smelt *Athineropsis californiensis.*

Lepeophtheirus bifidus Fraser, 1920. On rock sole *Lepidopsetta bilineata* and the buffalo sculpin.

Lepeophtheirus bifurcatus Wilson, 1905. On the sharpnose surf perch *Phanerodon atripes.*

Lepeophtheirus breviventris Fraser, 1920. In the mouth of the lingcod.

Lepeophtheirus hospitalis Fraser, 1920. On Pleuronectidae (flatfishes) and the buffalo sculpin.

Lepeophtheirus kareii Yamaguti, 1936. On flatfishes.

Lepeophtheirus oblitus Kabata, 1973. On the body and in the gill cavity of Hexagrammidae and Scorpaenidae.

Lepeophtheirus parvicruris Fraser, 1920. On the starry flounder *Platichthys stellatus.*

Lepeophtheirus parviventris Wilson, 1905. On the body surface and fins of numerous Cottidae (sculpins) and the big skate *Raja binoculata.*

Lepeophtheirus parvus Wilson, 1908. On the pile perch *Rhacochilus vacca.*

Lepeophtheirus pravipes Wilson, 1912. On the big skate and lingcod.

Lepeophtheirus salmonis (Krøyer, 1837). On Pacific salmon, the white sturgeon *Acipenser transmontanus,* the surf smelt *Ammodytes hexapterus,* and Hexagrammidae.

Lepeophtheirus spatha Dojiri and Brantley, 1991. On the California halibut.

CANCERILLIDAE

Cancerilla durbanensis Stephenson, 1933. This and the following species are external parasites of the brittle star *Amphipholis squamata.*

Cancerilla neozelandica Stephensen, 1933.

Cancerilla tubulata, Dalyell, 1851.

EUDACTYLINIDAE

Bariaka pamelae (Laubier, Maillard, & Oliver, 1966). Described from San Francisco Bay, on the gill bar of the sevengill shark *Notorhynchus maculosus.*

Eudactylina acanthii Scott, 1901. On the gills of the spiny dogfish *Squalus acanthias.*

Eudactylinodes keratophagus Deets & Benz, 1986. Described from the California horn shark *Heterodontus francisci.*

KROYERIIDAE

Kroeyerina deborahae Deets, 1987. Described from the California shovelnose guitarfish *Rhinobatos productus.*

LERNAEOPODIDAE

Clavella embiotocae Dojiri, 1981. Attached to the gill filaments of Embiotocidae.

C. parva Wilson, 1912. On the fins of Cottidae, Embiotocidae, Hexagrammidae, and Scorpaenidae.

Lernaeopodina pacifica Kabata & Gussev, 1966. On the gills and in the buccal cavity of Rajidae (skates).

Dendrapta cameroni longiklavata Kabata & Gussev, 1966. On the gills and in the buccal cavity of Rajidae.

Naobranchia occidentalis Wilson, 1915. On the gills of rock sole, Pacific sand dab, and various Cottidae and Scorpaenidae.

Nectobranchia indivisa Fraser, 1920. On the gills of Pleuronectidae (flatfish).

Parabrachiella nitida (Wilson, 1915). In the buccal cavity of the Pacific sanddab *Citharychthys sordidus.*

Parabrachiella microsoma (Dojiri, 1981). On the gill filaments of the white croaker and Pacific sanddab.

Parabrachiella paralichthyos (Piasecki, 1993). In the buccal cavity of the California halibut.

Schistobranchia tertia Kabata, 1970. On the gills and in the buccal cavity of Rajidae.

NICOTHOIDAE

Hansenulus trebax Heron & Damkaer, 1986. Brood pouch parasite in the mysids *Neomysis mercedis, Alienacanthomysis macropsis, Proneomysis wailesi,* and *Xenacanthomysis pseudomacropsis.*

PANDARIDAE

Perissopus oblongus (Wilson, 1908). On the body surface of the bat ray *Myliobatis californica.*

PENELLIDAE

Haemobaphes diceraus Wilson, 1917. On the gills of the shiner perch and the eulachon *Thaleichthys pacificus.*

Haemobaphes disphaerocephalus Grabda, 1976. In the gill cavity of the eulachon and on farmed Atlantic salmon in net pens in the Pacific Northwest.

Haemobaphes intermedius Kabata, 1967. On the gills of Cottidae.

Peniculus fissipes Wilson, 1917. Embedded in the fins of the topsmelt and Embiotocidae.

Phrixocephalus cincinnatus Wilson, 1908. Eye parasite of the Pacific sanddab.

SPHYRIIDAE

Norkus cladocephalus Dojiri & Deets, 1988. Described from the hemibranchs of the shovelnose guitarfish.

TREBIIDAE

Trebius heterodonti Deets & Dojiri, 1989. External parasite on the California horn shark.

Trebius caudatus Krøyer, 1838. External parasite on the spiny dogfish.

Trebius latifurcatus Wilson, 1921. External parasite on the rays *Gymnura marmorata, Urolophus halleri, Platyrhinoides triseriata, Raja inornata,* and *Torpedo californica,* the California bat ray, the guitarfish, and the angel shark *Squatina californica.*

MONSTRILLOIDA

The adult and first nauplius occur in the plankton; other juvenile stages are highly transformed endoparasites of invertebrates.

Monstrilla canadensis McMurrich, 1917.
Monstrilla helgolandica Claus, 1863.
Monstrilla longiremis Giesbrecht, 1892.
Monstrilla spinosa Park, 1967.
Monstrilla wandelii Sehensen, 1913.

REFERENCES (COMMENSAL AND PARASITIC COPEPODS)

Boxshall, G. A. 1988. A review of the copepod endoparasites of brittle stars (Ophiuroida). Bull. Br. Mus. Nat. Hist (Zool) 54: 261–270.

Cressey, R. F. 1969. Five new parasitic copepods from California inshore fish. Proc. Biol. Soc. Wash. 82: 409–428.

Cressey, R. F. and H. B. Cressey. 1985. *Holobomolochus* (Copepoda: Bomolochidae) redefined, with descriptions of three new species from the eastern Pacific. J. Crust. Biol. 5: 717–727.

Deets, G. B. 1987. Phylogenetic analysis and revision of Kroeyerina Wilson, 1932 (Siphonostomatoida: Kroyeriidae), copepods parasitic on chondrichthyans, with descriptions of four new species and the erection of a new genus, *Prokroyeria.* Can. J. Zool. 65: 2121–2148.

Dojiri, M. 1981. Copepods of the families Lernaeopodidae and Naobranchiidae parasitic on fishes from Southern California in shore waters. J. Crust. Biol. 1: 251–264.

Dojiri, M., and R. F. Cressey. 1987. Revision of the Taeniacanthidae (Copepoda: Poecilostomatoida) parasitic on fishes and sea urchins. Smithson. Contr. Zool. No. 447: i–iv, 1–250.

Dojiri, M., and G. B. Deets. 1988. *Norkus cladocephalus,* new genus, new species (Siphonostomatoida: Sphyriidae), a copepod parasitic on an elasmobranch from southern California waters, with a phylogenetic analysis of the Sphyriidae. J. Crust. Biol. 8: 679–687.

Dudley, P. L. 1964. Some gastrodelphyid copepods from the Pacific coast of North America. Amer. Mus. Novit., No. 2194. 51 pp.

Dudley, P. L. 1966. Development and systematics of some Pacific marine symbiotic copepods. A study of the Notodelphyidae, associates of ascidians. Univ. Wash. Publ. Biol., 21, 282 pp.

Fraser, C. M. 1920. Copepods parasitic on fish from the Vancouver Island region. Proc. Trans. R. Soc. Can. (3) 13 (Section 5): 45–67.

Gooding, R. V. 1960. North and South American copepods of the genus *Hemicyclops* (Cyclopoida: Clausidiidae). Proc. U.S. Nat. Mus. 112: 159–195.

Gotto, V. 2004. Commensal and Parasitic Copepods Associated with Marine Invertebrates. Synopses of the British Fauna (New Series), J. H. Crothers and P. J. Hayward eds., No. 46 (second edition), 1–350.

Grizel, H. 1985. *Mytilicola orientalis* Mori, parasitism. Parasitose à *Mytilicola orientalis* Mori. Fiches D'Identification des Maladies et Parasies des Poissons, Crustacés et Mollusques 20: 1–4.

Grygier, M. J. 1995. Annotated chronological bibliography of Monstrilloida (Crustacea: Copepoda). Galaxea 12: 1–82.

Heron, G. A. 1986. A new nicothoid copepod parasitic on mysids from northwestern North America. J. Crust. Biol. 6: 652–665.

Ho, J.-S. 1970. Revision of the genera of the Chondracanthidae, a copepod family parasitic on marine fishes. Beaufortia 17: 105–218.

Ho, J.-S. 1972. Four new parasitic copepods of the family Chondracanthidae from California inshore. Proc. Biol. Soc. Wash. 85: 523–540.

Ho, J.-S. 1981. *Ismaila occulta,* A new species of poecilostomatoid copepod parasitic in a dendronotid nudibranch from California. J. Crust. Biol. 1: 130–136.

Ho, J.-S. 1984. New family of poecilostomatoid copepods (Spiophanicolidae) parasitic on polychaetes from southern California, with a phylogenetic analysis of nereicoliform families. J. Crust. Biol. 4: 134–146.

Ho, J.-S. 1992. Phylogenetic analysis of the Myicolidae, a family of poecilostome copepods chiefly parasitic in marine bivalve mollusks. Acta Zool. Taiwan 3: 67–77.

Ho, J. S., M. Dojiri, G. Hendler, and G. B. Deets. 2003. A New Species of Copepoda (Thaumatopsyllidae) symbiotic with a brittle star from California, U.S.A., and designation of a new order Thaumatopsylloida. J. Crust. Biol. 23: 582–594.

Humes, A. G. 1968. The cyclopoid copepod *Pseudomyicola spinosus* (Raffaele & Monticelli) from marine pelecypods, chiefly in Bermuda and the West Indies. Beaufortia 178: 203–226.

Humes, A. G. 1980. A review of the copepods associated with holothurians including new species from the Indo-Pacific. Beaufortia 30: 31–123.

Humes, A. G. 1986. *Myicola metisiensis* (Copepoda: Poecilostomatoida), a parasite of the bivalve *Mya arenaria* in eastern Canada, redefinition of the Myicolidae, and diagnosis of the Anthessiidae n. fam. Can. J. Zool. 64: 1021–1033.

Humes, A. G. 1986. Synopsis of copepods associated with asteroid echinoderms, including new species from the Moluccas. J. Nat. Hist. 20: 981–1020.

Humes, A. G. 1982. A review of Copepoda associated with sea anemones and anemonelike forms (Cnidaria, Anthozoa). Trans. Amer. Phil. Soc. 72: 1–120.

Humes, A. G., and G. A. Boxshall. 1996. A revision of the lichomolgoid complex (Copepoda: Poecilostomatoida), with the recognition of six new families. J. Nat. Hist. 30: 175–227.

Humes, A. G., and J. H. Stock. 1973. A revision of the family Lichomolgidae Kossman, 1877, cyclopoid copepods mainly associated with marine invertebrates. Smithson. Contr. Zool., No. 127, 368 pp.

Huys, R. 1988. On the identity of the Namakosiramiidae Ho & Perkins 1977 (Crustacea, Copepoda), including a review of harpacticoid associates of Echinodermata. J. Nat. Hist. 22: 1517–1532.

Huys, R. 2001. Splanchnotrophid systematics: a case of polyphyly and taxonomic myopia. J. Crust. Biol. 21: 106–156.

Huys, R., and G. A. Boxshall. 1991. Copepod Evolution. The Ray Society, London, 468 pp.

Illg, P. L. 1949. A review of the copepod genus *Paranthessius* Claus. Proc. U.S. Nat. Mus. 99: 391–428.

Illg, P. 1950. A new copepod, *Pseudanthessius latus* (Cyclopoida: Lichomolgidae) commensal with a marine flatworm. J. Wash. Acad. Sci. 40: 129–133.

Illg, P. L. 1958. North American copepods of the family Notodelphyidae. Proc. U.S. Nat. Mus. 107: 463–649.

Illg, P. L. 1960. Marine copepods of the genus *Anthessius* from the northeastern Pacific Ocean. Pac. Sci. 14: 337–372.

Illg, P. L., and P. L. Dudley. 1980. The family Ascidicolidae and its subfamilies (Copepoda, Cyclopoida), with descriptions of new species. Mem. Mus. Nat. Hist. Nat. Paris, Zool.117: 1–192.

Kabata, Z. 1971. Four Bomolochidae (Copepoda) from fishes of British Columbia. J. Fish. Res. Bd. Canada 28: 1563–1572.

Kabata, Z. 1969. *Chondracanthus narium* sp. n. (Copepoda: Chondracanthidae), a parasite of the nasal cavities of *Ophiodon elongatus* (Pisces: Teleostei) in British Columbia. J. Fish. Res. Bd. Can. 26: 3043–3047.

Kabata, Z. 1967. The genus *Haemobaphes* (Copepoda: Lernaeoceridae) in the waters of British Columbia. Can. J. Zool. 45: 853–875.

Kabata, Z. 1970. Some Lernaeopodidae (Copepoda) from fishes of British Columbia J. Fish. Res. Bd. Canada 27: 865–885.

Kabata, Z. 1973. The species of *Lepeophtheirus* (Copepoda: Caligidae) from fishes of British Columbia. J. Fish. Res. Bd. Canada 30: 729–759.

Kabata, Z. 1979. Parasitic Copepoda of British fishes. The Ray Society, London, Monograph 152, 468 pp.

Kabata, Z. 1984. A contribution to the knowledge of Chondracanthidae (Copepoda: Poecilostomatoida) parasitic on fishes of British Columbia. Can. J. Zool. 62: 1703–1713.

Love, M.S., and M. Moser. 1983. A checklist of parasites of California, Oregon, and Washington marine and estuarine fishes. NOAA Tech. Rep. NMFS SSRF-777, 576 pp.

Margolis, l., and J. R. Arthur. 1979. Synopsis of the parasites of fishes of Canada. Bull. Fish. Res. Bd. Canada, no. 199, 269 pp.

Ooishi, S. 1991. North Pacific copepods (Cyclopoida: Ascidicolidae) associated mostly with compound ascidians. In: Proceedings of the Fourth International Conference on Copepoda. Bull. Plankton Soc. Japan (Special Volume): 49–68.

Ooishi, S., and P. L. Illg. 1977. Haplostominae (Copepoda, Cyclopoida) associated with compound ascidians from the San Juan Archipelago and vicinity. Spec. Publ. Seto Mar. Biol. Lab. ser. V. 154 pp.

Wilson, C. B. 1905. North American parasitic copepods belonging to the family Caligidae. Part 1. Caliginae. Proc. U.S. Nat. Mus. 28: 479–672.

Wilson, C. B. 1907. North American parasitic copepods belonging to the family Caligidae. Part 2. Trebinae and Euryphorinae. Proc. U.S. Nat. Mus. 31: 669–720.

Wilson, C. B. 1907. North American parasitic copepods belonging to the family Caligidae. Parts 3 and 4. A revision of the Pandaridae and the Cecropinae. Proc. U.S. Nat. Mus. 31: 669–720.

Wilson, C. B. 1908. North American parasitic copepods. A list of those found upon the fishes of the Pacific coast, with descriptions of new genera and species. Proc. U.S. Nat. Mus. 33: 323–490.

Wilson, C. B. 1915. North American parasitic copepods belonging to the Lernaeopodidae, with a revision of the entire family. Proc. U.S Nat. Mus. 47: 565–579.

Wilson, C. B. 1935. Parasitic Copepoda from the Pacific coast. Amer. Midl. Nat.16: 776–797.

Yamaguti, S. 1963. Parasitic Copepoda and Branchiura of fishes. Interscience Publishers, New York and London, 1104 pp.

Branchiura

ARMAND M. KURIS

(Plate 212)

The Branchiura, or fish lice, are superficially similar to copepods. However, they have compound eyes, their antennae are reduced or absent, the maxillules are usually modified suckers,

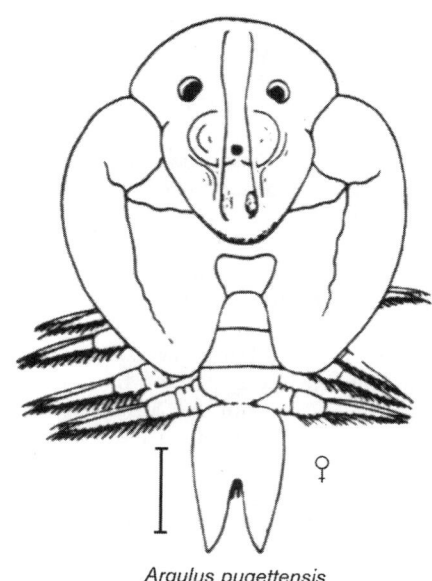

Argulus pugettensis

PLATE 212 Branchiura; *Argulus pugettensis,* scale bar = 1 mm (after Wilson 1912).

and the mandibles are highly modified for piercing and sucking (plate 212). Also, in contrast to copepods, Branchiura usually have direct development, and their eggs are attached to the substrate. All are micropredators of teleosts or amphibians, feeding on blood or tissue fluids, moving from host to host.

Three species have been reported from the Pacific coast of North America: *Argulus borealis* Wilson, 1912, from rockfish and flatfish, *A. pugettensis* Dana, 1852, from surf perch and rockfish and *A. melanostictus* Wilson, 1935, from the grunion *Leuresthes tenuis.*

There have been few investigations of fish lice from the California and Oregon coasts. Love and Moser (1983) provide host records. Perhaps the best known is the grunion parasite (Olson 1972). Elsewhere, as fish farms have expanded, fish lice have become very serious infectious pests. Their abundance builds up on the high-density host populations that are sustained in these penned facilities. There are numerous studies of their population ecology and their control under these artificial circumstances (see for example Hakalahti and Valtonen 2003).

References

Hakalahti, T., and E. T. Valtonen. 2003. Population structure and recruitment of the ectoparasite *Argulus coregoni* Thorell (Crustacea: Branchiura) on a fish farm. Parasitology 127: 79–85.

Love, M. S., and M. Moser. 1983. A checklist of parasites of California, Oregon and Washington marine and estuarine fishes. NOAA Tech Rep NMFS SSRF-777: 1–576.

Olson, A. C. 1972. *Argulus melanostictus* and other parasitic crustaceans on the California grunion, *Leuresthes tenuis* (Osteichthyes Atherinidae). J. Parasit. 58: 1201–1204.

Cirripedia

WILLIAM A. NEWMAN

(Plates 213–216)

Barnacles are some of the most conspicuous and well-known organisms of the seashore. Cirripeds are specialized, attached crustaceans that first appear in the early Paleozoic, and those called barnacles generally use feathery thoracic limbs (**CIRRI**) in feeding. They usually hatch from brooded eggs as setose-feeding **NAUPLIUS** larvae (plate 214D), the last of six stages of which metamorphoses into a nonfeeing **CYPRID** larva (plate 214E). Nauplii and cyprids from the field can be identified to species genetically, if not always by morphological characters (Newman and Ross 2001). The cyprid finds a place to settle before metamorphosing into a juvenile, and the resulting adult remains attached for life. Most adults are planktotrophic setose feeders, but one superorder, the Rhizocephala, is wholly parasitic and so highly modified that members are only recognizable as crustaceans by their larval stages. More often than not, such parasites represent ancient clades whose ancestors within the Crustacea as well as within the Cirripedia are revealed by molecular genetics. The Cirripedia includes three superorders and representatives of each occur in central California (Newman 1979, Newman and Abbott 1980).

The superorder **RHIZOCEPHALA**, as mentioned above, encompasses the most specialized parasites among the crustaceans. Many species start life as a typical, albeit nonfeeding nauplius, which, after four stages, metamorphoses into a typical cyprid larva. The female cyprid searches out a host, usually a recently molted brachyuran or anomuran decapod, and attaches by one or both of its antennae to the surface of a gill or at the base of a seta. It then undergoes a typical metamorphosis in which the legs and compound eyes are cast. But the remaining tissue is extruded through an antenna or the mouth field, sometimes by way of a kentrogon, into the host. There it becomes the **INTERNA**, a body that migrates around the digestive tract and sends out nutritive rootlets that pervade the host's body. Eventually a globular reproductive body, the **EXTERNA**, erupts, usually beneath or on the abdomen of the host. It contains the ovaries and develops a cavity in which to brood the eggs that in turn has a cavity or cavities to house males. The brood cavity has an aperture to the outside where the male cyprids settle and metamorphose into wormlike males that migrate into one or two special pockets inside to await laying of the eggs. The solitary, globose externa of *Heterosaccus californicus* is fairly common protruding from beneath the abdomen of older kelp crabs, *Pugettia producta,* and a small percentage of hermit crabs bear the gregarious, fingerlike externae of *Peltogasterella gracilis* on their abdomen (see Reinhard 1944, Høeg 1992).

The second superorder, the **ACROTHORACICA**, contains relatively small, largely shell-less forms found burrowing in calcareous substrates, such as mollusk shells, coral, and limestone. The superorder reaches its greatest diversity in coral seas. Locally, a minute apygophoran, *Trypetesa lateralis,* makes its burrows in the shells of *Chlorostoma* and occasionally other gastropods occupied by female hermit crabs. The female cyprid settles inside such occupied shells and excavates the small burrow with a slit-like opening through which it can feed and the male cyprid can enter. The male cyprid locates and enters a burrow, where it attaches close to or on the female before metamorphosing into a gutless sac ("dwarf male") containing little more than a testis. The female lays the eggs in her mantle cavity, where a dwarf male fertilizes them. Naupliar stages are generally passed though in the eggs which hatch as cyprid larvae destine to become males or females. By and large, the cyprid crawls rather than swims in search of an established female if destined to be a male, or an appropriate place in the gastropod shell inhabited by a hermit crab, if female. Female **APYGOPHORA** have a blind gut and reduced cirri and mouthparts and feed largely on the eggs of the host hermit crab (Tomlinson 1969, Williams and Boyko 2006).

The third or principal superorder, the **THORACICA**, encompasses the stalked and sessile barnacles, and representatives are common in marine and estuarine waters along the California coast. Their shell remains are so numerous in some situations that Charles Darwin suggested that the present geological period might be known in the fossil record as the "Age of Barnacles." The orders and family-group taxa to which these forms belong are taken up below.

In a primitive stalked barnacle, such as the goose barnacle *Lepas* (plate 213A1), the body is divided into two regions: the capitulum and peduncle, or stalk. The former is armored with five calcareous plates: a pair of large **SCUTA** (sing., **SCUTUM**), a pair of **TERGA** (sing., **TERGUM**), and a narrow median **CARINA** along the dorsal margin. Together, the terga and scuta close the aperture through which the cirri are extended during feeding. In the stalked barnacle *Pollicipes* (plate 213B), a sixth plate, the **ROSTRUM**, and numerous accessory platelets have been added around the basal margin of the capitulum. These plates, and some plates of the accessory whorls, are homologous with the **OPERCULUM** and **PARIETAL** or wall plates of sessile barnacles. It is the terga and scuta of a pedunculate ancestor that have became separated from the wall and specialized to form the operculum, and this development allowed the wall to evolve into a very rigid structure (Darwin 1854).

The Balanomorpha, the sessile or acorn barnacles, are largely divided between the relatively primitive superfamily, Chthamaloidea, and the Balanoidea, and the nomenclature of the plates was largely established by Darwin (1854). Species of balanoid genera such as *Balanus* and *Amphibalanus* (plate 213C, 213E), have three calcareous regions: (1) a generally flat calcareous basis, attached to the substratum; (2) a wall formed of immovable articulated plates (parietes); and (3) a movable portion, the opercular valves, closing the aperture and usually consisting of the terga (singular, tergum) and scuta (singular, scutum) (plate 214A, 214B). The wall of these genera consists of six parietal plates, each known as a compartment (plate 213E). The unpaired plates are the carina (plural, carinae), furthest from the head (posterodorsal), and the rostrum (close to anterior), a compound plate in these genera as the paired rostrolateral plates are fused to it. The first carinolaterals are the pair overlapped by the compound rostrum, and the second carinolaterals, which appear by replication of the first during ontogeny, are overlapped by the carina (Ross and Newman 1996).

There are also six plates in chthamaloids, such as *Chthamalus* (plate 213D), but the rostrum is free, and by this and some other characters they are more generalized than the balanoids. In addition to this, the most obvious difference is that the second carinolaterals have failed to develop in *Chthamalus*. Thus, *Chthamalus* is also six-plated, but the rostrum is overlapped rather than underlapped by the adjacent plates (plate 213D–213E), and that can be a good way to tell most chthamaloids from balanoids, especially when relatively young. There are only four plates in *Tetraclita* because not only do the carinolaterals fail to replicate during ontogeny, but the rostrum is compound. Although the maximum number of wall plates in the local sessile barnacles is six, it is still eight in such relatively primitive intertidal forms as *Catomerus* found in southeast Australia and Tasmania. The primitive condition for the basis was membranous rather than calcareous, as it is in *Tetraclita, Chthamalus,* and in some balanoid species, such as *Semibalanus,* but its being membranous also can be secondary.

The exposed median triangular portion of each compartment is the **PARIES** (plural, **PARIETES**). The edge of a compartment that is overlapped by an adjacent compartment is an **ALA**

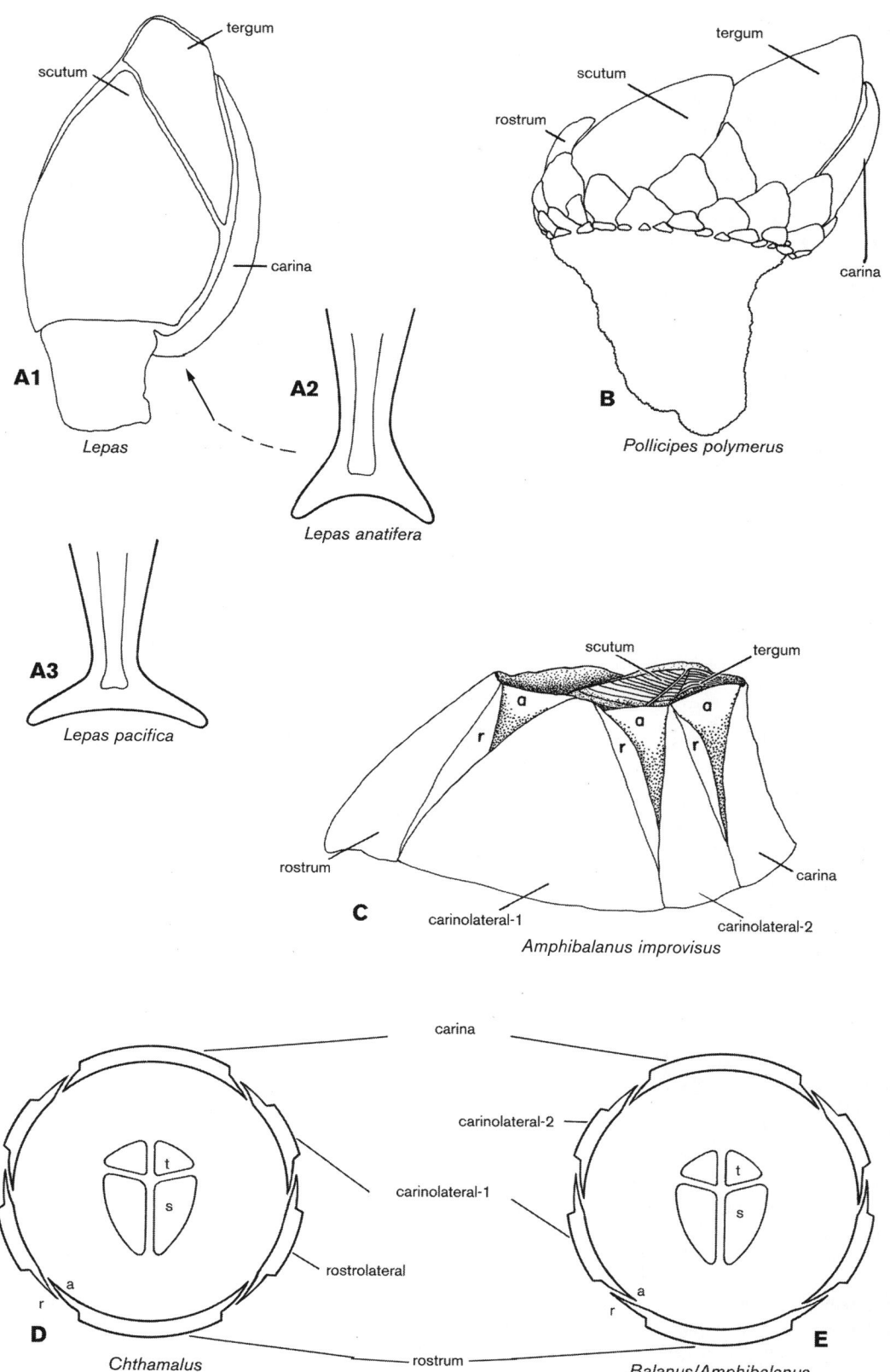

PLATE 213 Types of barnacles and terminology of plates: A1, goose barnacle (*Lepas*) with five plates, from right; A2, base of carina of *L. anatifera*; A3, of *L. pacifica*; B, leaf barnacle (*Pollicipes polymerus*) with additional plates; C, sessile barnacle (*Amphibalanus improvisus*), showing how radii overlap alae; D, E, schematic cross-sections to show arrangement of plates of *Chthamalus* and *Balanus* (r = radius, a = ala, t = tergum, s = scutum) (Newman, original).

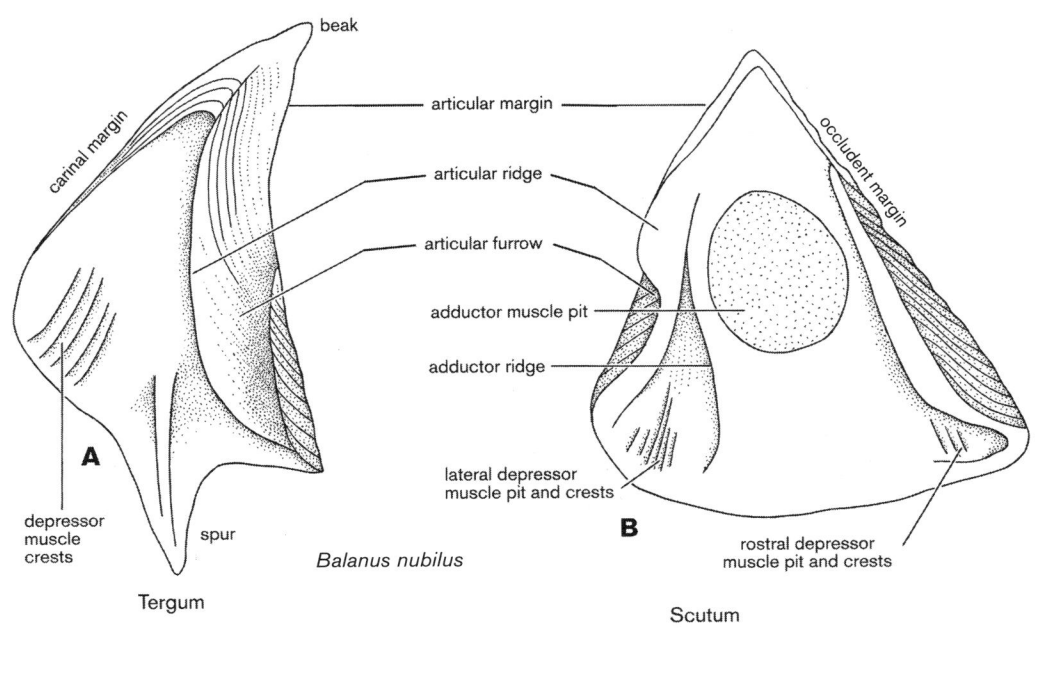

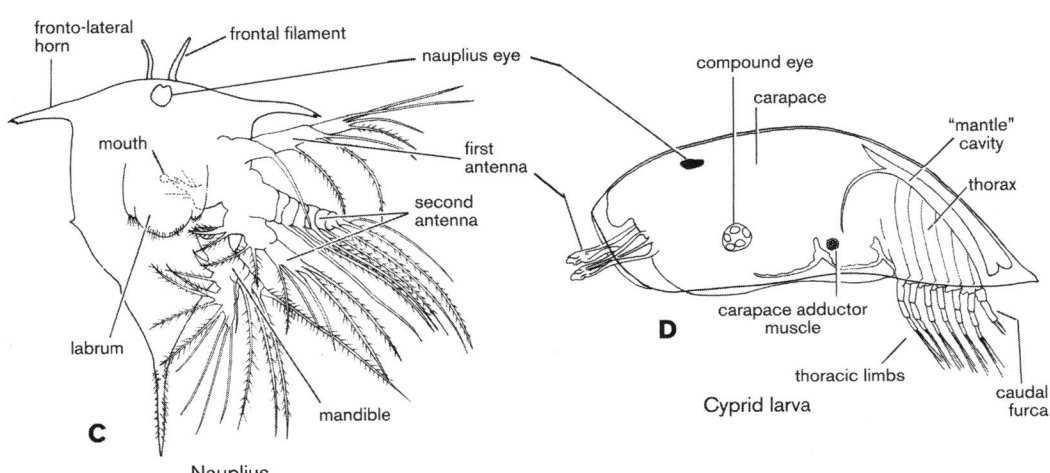

PLATE 214 Terminology of interior of opercular valves of *Balanus nubilus*: A, right tergum; B, right scutum; the beak is composed of hardened shell exposed by erosion, and the terga of *Semibalanus cariosus, B. nubilus, Menesiniella aquila,* and *Tetraclita rubescens* are "beaked"; Larval forms: C, second stage nauplius of *Amphibalanus amphitrite* seen from below, right appendages deleted; D, cyprid of *Trypetesa* seen from left, modified after Kuhnert (1934) (all but D, Newman, original).

(plural, **ALAE**). If the overlapping edge is marked by a distinct change in the direction of growth lines, it is known as a **RADIUS** (plural, **RADII**). Radii can form on the overlapping portion of the compartments in all but the most primitive sessile barnacles, and although a few highly advanced forms, such as *Tetraclita,* have lost them, their configuration can be useful in identifying species. While primitively solid, as in *Chthamalus* and lower balanoids like *Hesperibalanus,* the parietes become permeated by **LONGITUDINAL CANALS** or tubes in higher forms, and in addition the radii of *Megabalanus* are permeated by **TRANSVERSE CANALS**. However such specializations can be lost; for example, while juveniles of *Balanus glandula* have poorly developed longitudinal canals, the wall becomes solid as individuals grow.

The operculum of generalized acorn barnacles guards the orifice of the wall. It is formed by the paired terga and scuta (plates 213C, 214A–214B) of each side articulated together, and the slit between them guards the aperture through which the cirri are extended when feeding. Muscles controlling the operculum include the (1) adductor running between the scuta of each pair that closes the aperture, and (2) three pairs of depressor muscles that control the position of the operculum in the orifice of the shell. The interior of the valves are marked by distinctive ridges, furrows, and pits, and most of these are related to the origins and insertions of these muscles and/or the evolution of the tergal spur. Similarly, the exteriors of the valves in some species are ornamented in various distinctive ways, and these characteristics are also useful in identification.

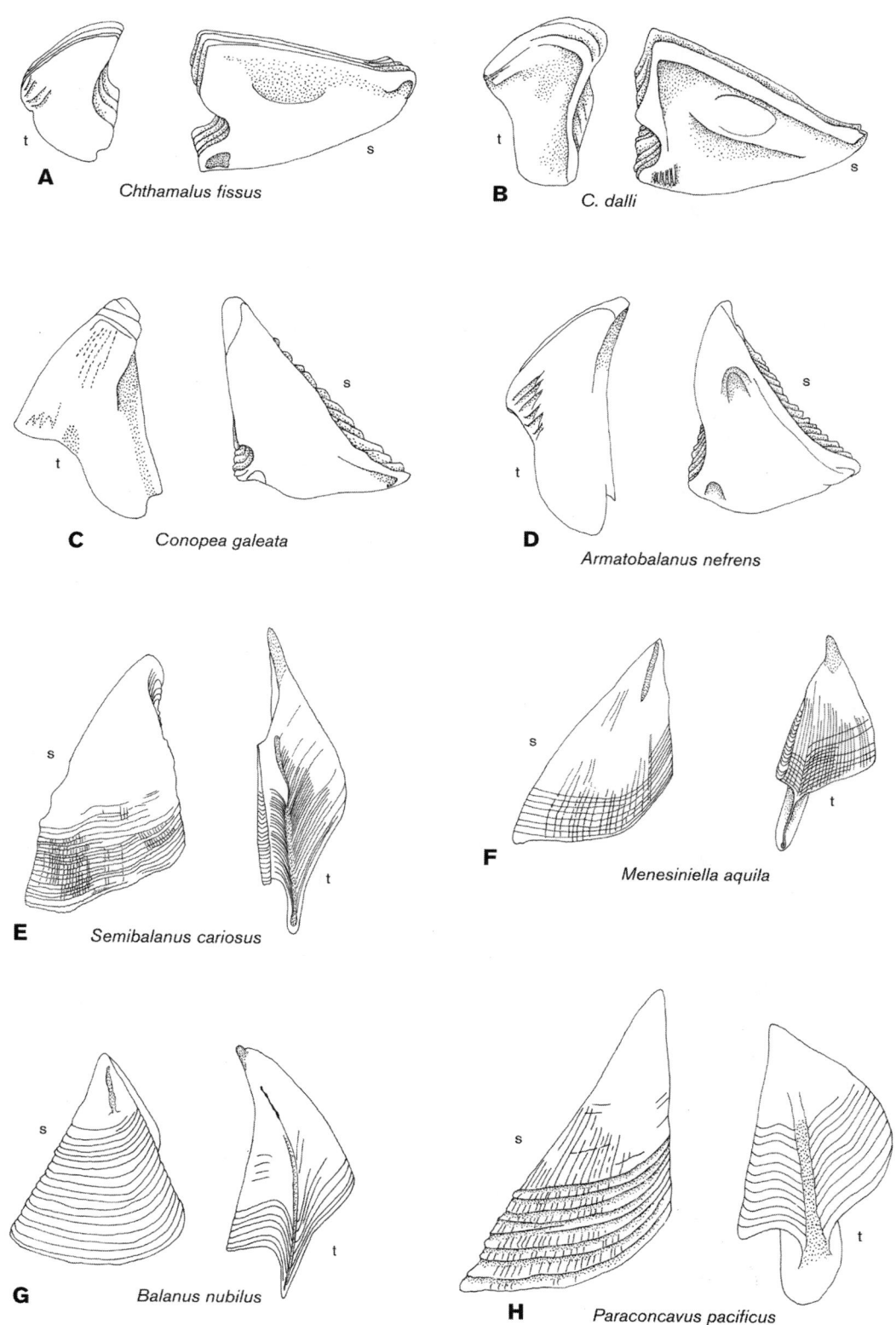

A   *Chthamalus fissus*

B   *C. dalli*

C   *Conopea galeata*

D   *Armatobalanus nefrens*

E   *Semibalanus cariosus*

F   *Menesiniella aquila*

G   *Balanus nubilus*

H   *Paraconcavus pacificus*

PLATE 215 A–D, right-hand terga and scuta of sessile barnacles, interior; E–H, exterior (Newman, original).

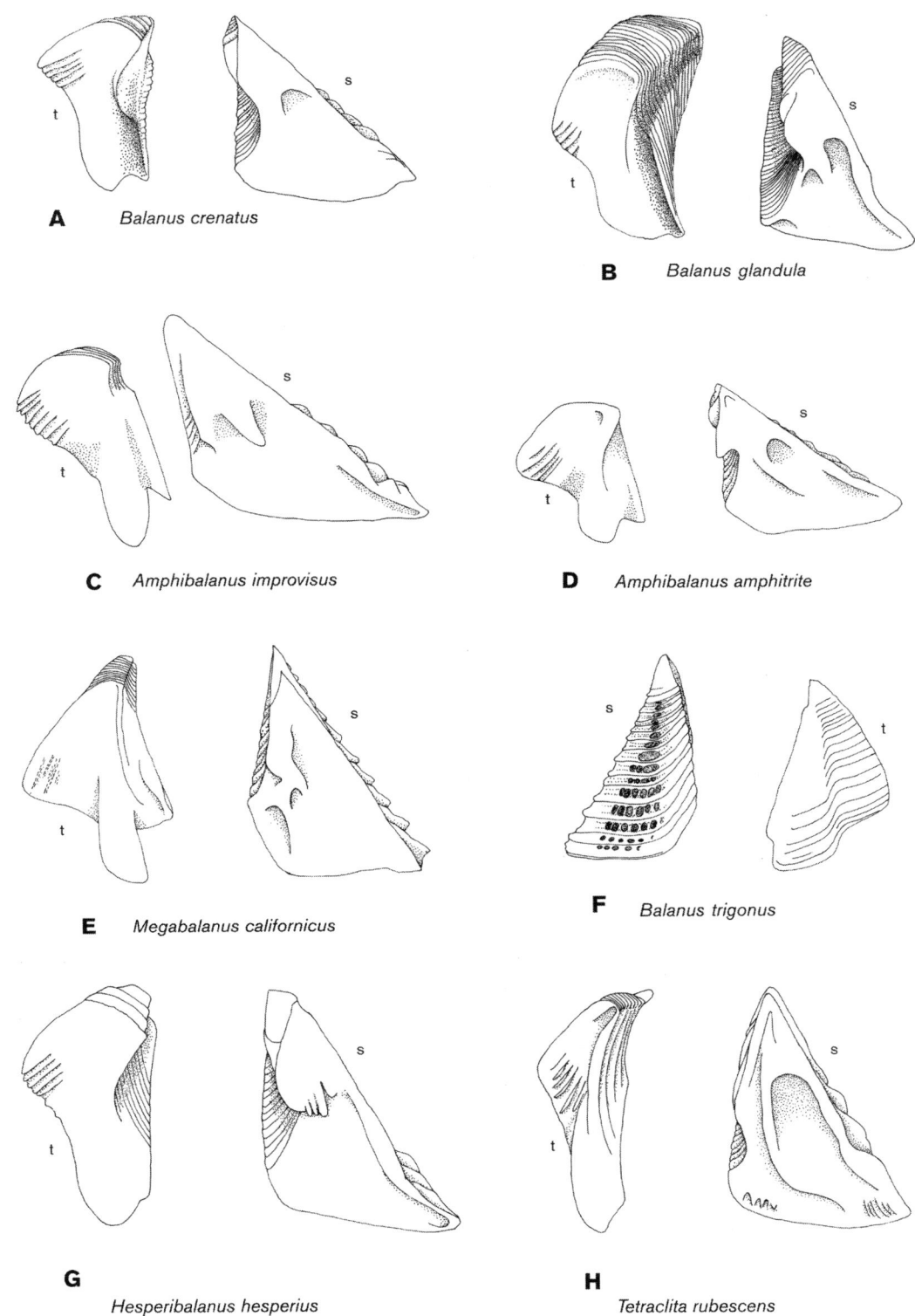

A *Balanus crenatus*

B *Balanus glandula*

C *Amphibalanus improvisus*

D *Amphibalanus amphitrite*

E *Megabalanus californicus*

F *Balanus trigonus*

G *Hesperibalanus hesperius*

H *Tetraclita rubescens*

PLATE 216 Right-hand terga and scuta of sessile barnacles: F, exterior; A–E, G, H, interior (Newman, original).

As noted above, sexes in the Rhizocephala and Acrothoracica are separate. Thoracicans, on the other hand, are primarily cross-fertilized hermaphrodites; eggs are laid in the mantle cavity of the adult, where they are cross-fertilized and develop into larvae. Although a few species develop directly into and release cyprid larvae, most produce nauplii, which are released into the plankton where they develop into cyprid larvae.

For the identification of barnacles found on whales that may be washed ashore, see Newman and Abbott 1980.

Key to Thoracic Cirripedia

1. Stalked forms . . . . Lepadomorpha 2
— Sessile forms . . . . Balanomorpha 7
2. Capitular plates more than five in number, surrounded basally by whorl of imbricating plates; exposed intertidal on rocks and artificial structures (plate 213B) . . . .
. . . . Pollicipes polymerus
— Capitular plates five or less in number, without basal whorl of imbricate plates 3
3. Capitular plates five in number 4
— Capitular plates two in number, reduced to a pair of Y-shaped scuta, on scyphomedusae . . . . Alepas pacifica
4. Capitular plates completely covering capitulum (plate 213A1) . . . . Lepas 5
— Capitular plates reduced to narrow slips; occurring in gill chambers of some crabs and the California spiny lobster
. . . . Octolasmis californiana
5. Capitulum laterally compressed; individuals always attached to floating objects, not forming floats of their own 6
— Capitulum more or less globular; plates thin, papery; individuals forming floats of their own, initial attachment to feathers, Velella, or floats of adult . . . . . Dosima fascicularis
6. Plates thin but not papery, dark pigment of hypodermis showing through scuta; base of carina laterally expanded like whale flukes (plate 213A3) . . . . Lepas pacifica
— Plates thick, pigment of hypodermic not showing through scuta except in very small specimens; base of carina forming a blunt fork (plate 213A1, 213A2) . . . . Lepas anatifera
7. Wall composed of four plates, permeated by numerous longitudinal tubes that filled with calcareous reddish material exposed externally by erosion; interior of opercular valves heavily sculptured (plate 216H) . . . . . . Tetraclita rubescens
— Wall composed of six plates 8
8. Rostrum overlapped by adjacent plates (rostrolaterals) (plate 213D); small, brown, or gray-green 9
— Rostrum overlapping adjacent plates (carinolaterals-1) (plate 213E) 10
9. Scutum with long, strong, adductor ridge, with lateral depressor muscle crests (plate 215B); specialized setae at ends of second cirri finely bipectinate . . . . Chthamalus dalli
— Scutum with short, strong, adductor ridge, without lateral depressor muscle crests (plate 215A); specialized setae at ends of second cirri coarsely bipectinate . . . . . Chthamalus fissus
10. Wall solid, not permeated by longitudinal tubes (except for very young B. glandula) 11
— Wall permeated by longitudinal tubes 14
11. Free-living, on wide variety of substrates 12
— Living buried in gorgonians and hydrocorals 13
12. Wall white, ribbed (immature individuals may have small, irregular unfilled tubes in the wall); scutum with pit on either of adductor ridge (plate 216B) . . . .
. . . . Balanus glandula

— Wall white, smooth or weakly ribbed; scutum with callus forming several ridges extending from the articular ridge to the adductor ridge (plate 216G) . . . .
. . . . Hesperibalanus hesperius
13. In gorgonians; basis boat-shaped, tergum epically truncated (plate 215C) . . . . Conopea galeata
— In hydrocorals; basis more or less flat, tergum apically pointed (plate 215D) . . . . Armatobalanus nefrens
14. Wall permeated by a single row of more or less uniformly spaced tubes 15
— Wall permeated by many irregularly spaced tubes not in a single row, outer surface white, irregularly ribbed, having thatched appearance; tergum with beak (plate 215E) . . . .
. . . . Semibalanus cariosus
15. Exterior of wall white, without colored markings 18
— Exterior of wall ornamented by conspicuous reddish or maroon, more or less longitudinal markings 16
16. Radii solid, not permeated by transverse tubes 17
— Radii permeated by transverse tubes; exterior of scutum without longitudinal striations; tergum with faint depressor muscle crests (plate 216E) . . . . Megabalanus californicus
17. (Note three choices) Exterior of scutum marked by one or more longitudinal rows of pits; tergal spur nearly half width of basal margin (plate 216F) . . . . . . Balanus trigonus
— Exterior of scutum with numerous longitudinal striations; tergal spur about one-third width of basal margin (plate 215H) . . . . Paraconcavus pacificus
— Exterior of scutum marked by simple transverse growth lines, without pits or longitudinal striations; tergal spur about one-third width of basal margin; in bays and harbors (plate 216D) . . . . Amphibalanus amphitrite
18. Tergum beaked (plate 215F–215G) 19
— Tergum not beaked 20
19. Aperture of shell relatively small, not flaring; exterior of scutum and tergum marked by longitudinal striations (plate 215F) . . . . Menesiniella aquila
— Aperture of shell relatively large, flaring; exterior of scutum and tergum without longitudinal striations (plates 214A, 214B, 215G) . . . . Balanus nubilus
20. Scutum without adductor ridge; tergal spur wider than long, occupying half of basal margin (plate 216A); anterior surface of intermediate articles of cirrus clothed with fine spinules . . . . Balanus crenatus
— Scutum with strong adductor ridge; tergal spur longer than wide, occupying less than half of basal margin; only in brackish water (plates 213C, 216C) . . . .
. . . . Amphibalanus improvisus

List of Species

ACROTHORACICA

APYGOPHORA

*Trypetesa lateralis Tomlinson, 1953. Burrowing inside gastropod shells inhabited by female hermit crabs, an egg predator and a short-range endemic on the Pacific coast of North America. See Tomlinson 1953, J. Wash. Acad. Sci. 43: 373–381; 1955, J. Morph. 96: 97–114 (morphology); Newman and Abbott 1980; Williams and Boyko 2006.

* = Not in key.

RHIZOCEPHALA

KENTROGONIDA

Heterosaccus californicus Boschma, 1933. Occurs on spider crabs; solitary, castrates host, host males are feminized, externa mimics egg mass of host and is cared for accordingly. See Newman and Abbott 1980; Høeg 1992.

Peltogasterella gracilis (Boschma, 1927). Externa occur gregariously on hermit crabs. See Yanagimachi 1961, Crustaceana 2: 183–186 (life cycle); Newman and Abbott 1980.

THORACICA

LEPADOMORPHA

LEPADIDAE

Pelagic goose barnacles; see Newman and Abbott 1980.

Alepas pacifica Pilsbry, 1907. Attached to bell of scyphomedusae, usually *Phacellophora.*

Dosima fascicularis (Ellis and Solander, 1786) (=*Lepas fascicularis*). Like other goose barnacles, but capable of forming gas-filled float of it own. Often blue (see Fox and Crozier 1967 Experientia 23: 12–14). Sometimes found attached to the by-the-wind-sailor *Velella* (Knudsen 1963, Bull. So. Calif. Acad. Sci. 62: 130–131).

Lepas anatifera Linnaeus, 1758. On floating timber, tar balls, and other objects. Look for the small neustonic crab *Planes* among *Lepas* populations washed ashore.

Lepas pacifica Henry, 1940. As *L. anatifera* but apparently an Eastern Pacific endemic. Common on driftwood, milk cartons, tar balls, kelp stipes, and other substrates washed ashore.

POECILASMATIDAE

Octolasmis californiana Newman, 1960. A southern species, ranging north to Monterey Bay; inhabiting gill chambers of crabs and spiny lobsters. See Newman 1960 Veliger 3: 9–11; Newman and Abbott 1980; Jeffries and Voris 1996, Raffles Bull. Zool. 44: 575–592 (bibliography); Alvarez et al. 2003 J. Crustacean Biol. 23: 758–764 (cypris larva); Blomsterberg et al. 2004, J. Morph. 260: 154–164 (growth and molting).

SCALPELLOMORPHA

Pollicipes polymerus Sowerby, 1833 (=*Mitella polymerus*). Leaf barnacle, have also been called gooseneck barnacles. Frequently forming dense stands with the mussel *Mytilus californianus* on mid-tidal rocks exposed to heavy wave action. Terga and scuta occasionally exfoliated. Look for camouflaged specimens of the limpet *Lottia digitalis* on the plates of this barnacle. Abbott (1987) presents sketches of the external and internal anatomy and nauplius larva of *Pollicipes* from Monterey Bay. See Barnes and Reese 1959, Proc. Zool. Soc. Lond. 132: 569–585 (feeding) and 1960, J. Anim. Ecol. 29: 169–185 (behavior, ecology); Howard and Scott 1959, Science 129: 716–718 (predaceous feeding); Hilgard 1960, Biol. Bull. 119: 169–188 (reproduction); Burnett 1972 J. Morph. 136: 79–107 (circulatory system); Fyhn et al. 1972, J. Exp. Biol. 57: 83–102 (physiology, including desiccation tolerance); Petersen et al.

1974 J. Exp. Biol. 61: 309–320 (respiration); Newman and Abbott 1980; Simberg 1981, Biol. Bull. 160: 31–42 (reproduction); Lewis 1975, Mar. Biol. 32: 141–153 (reproduction); Lewis 1981, Crustaceana 41: 14–20 (feeding; in high energy systems *Pollicipes* extend their cirri in a fixed position to feed, but juvenile *Pollicipes* in calm water in the laboratory undertake cirral beating); Lewis and Chia 1981, Can. J. Zool. 59: 893–901 (growth, fecundity, reproductive biology in Puget Sound); Page 1983, J. Exp. Mar. Biol. Ecol. 69: 189–202 (energetics); Chaffee and Lewis 1988, J. Exp. Mar. Biol. Ecol. 124: 145–162 (stalk growth); Hoffman 1988, Pac. Sci. 42: 154–159 and 1989, J. Exp. Mar. Biol. Ecol. 125: 83–98 (settlement, growth, recruitment); van Syoc 1995, pp. 197–312, in: Schram and Høeg, New Frontiers in Barnacle Evolution (A. A. Balkema, Rotterdam) (genetics and global history of genus); Miner 2002, Invert. Biol. 121: 158–162 (genetics along California coast).

BALANOMORPHA

CHTHAMALOIDEA

Chthamalus dalli Pilsbry, 1916. High-intertidal northern species, generally with and above *Balanus glandula.* See Blower and Roughgarden 1988, Oecologia 75: 512–515 (parasitic castration by isopod *Hemioniscus*); Wares and Castaneda 2005, J. Mar. Biol. Assoc. U.K. 85: 327–331 (genetics, distribution).

Chthamalus fissus Darwin, 1854 (=*C. microtretus* Cornwall, 1937. based on specimens with a narrow, slitlike orifice). Southern species ranging north to below San Francisco; occurring with *C. dalli* in central California from which it can be distinguished by specialized spines on its cirri. See Hines (1978, 1979, reproduction); Newman and Abbott 1980; Shanks 1986, Oecologia 69: 420–428 (ecology); Blower and Roughgarden 1988, Oecologia 75: 512–515 (parasitic castration by isopod *Hemioniscus*); Wares and Castaneda 2005, J. Mar. Biol. Assoc. U.K. 85: 327–331 (genetics, distribution).

TETRACLITOIDEA

Tetraclita rubescens Darwin, 1854 (=*Tetraclita squamosa rubescens;* =*Tetraclita rubescens elegans*). Commonly intertidal, singly, or in clusters; peltate white "variety" *elegans* on subtidal crabs and mollusk shells; prior to the 1970s reported as rare north of San Francisco, but now occurring at least north to Cape Mendocino, California (Connolly and Roughgarden 1998, Cal. Fish Game 84: 182–183). See also Hines (1978, 1979, reproduction); Newman and Abbott 1980; Miller and Roughgarden 1994 (larvae). Abbott (1987) presents sketches of the internal anatomy of *Tetraclita* from Monterey Bay.

BALANOIDEA

ARCHAEOBALANIDAE

Armatobalanus nefrens (Zullo, 1963) (=*Balanus nefrens*). Subtidal southern species ranging north to Monterey Bay, embedded in the hydrocorals *Stylaster californicus* and *Errinopora pourtalesii*; associated with but not restricted to coral from the low intertidal to 80 m.

Conopea galeata (Linnaeus, 1771) (=*Balanus galeatus*). Southern form extending as far north as Monterey Bay, occurring on gorgonians; only shore barnacle in this region with complemental males. See Newman and Abbott 1980.

Hesperibalanus hesperius (Pilsbry, 1916) (=*Balanus hesperius laevidomus*). Subtidal northern species ranging south to San Francisco Bay, frequently on crabs and mollusk shells; low intertidal in laminarian holdfasts, etc. See Newman and Abbott 1980.

Semibalanus cariosus (Pallas, 1788) (=*Balanus cariosus*). Common intertidal northern species ranging to somewhat south of San Francisco Bay. See Gwilliam 1976, Biol. Bull. 151: 141–160 (shadow reflex, and earlier papers cited therein); Spight 1981 (settlement); Strathmann et al. 1981 (settlement).

BALANIDAE

Balanus crenatus Bruguière, 1789. Low intertidal and subtidal, northern species found in protected situations, can be distinguished from worn or otherwise similar-looking *Balanus glandula* by presence of fine spinules on anterior faces of intermediate articles of cirrus IV. On a wide variety of substrates; also in fouling communities. See Newman and Abbott 1980.

Balanus glandula Darwin, 1854. Common; intertidal, outer coast, and in bays; forms dense stands with and below *Chthamalus*. Abbott (1987) presents sketches of the external and internal anatomy of *B. glandula*, as well as sketches of action of muscles in opercular closure. This is the first barnacle species to show ecophenotypic plasticity in its cirri relative to high versus low flow regimes (Marchinko 2003, Evolution 57: 1281–1290). This is the only native California barnacle that has been introduced elsewhere in the world: it is abundant along the shores of Argentina and has been introduced to Japan as well. See Barnes and Barnes 1956, Pac. Sci. 10: 415–422 (biology); Newman 1967 (physiology, behavior); Connell 1970, Ecol. Monogr. 40: 49–78 (ecology); Hines (1978, 1979, reproduction); Wu and Levings 1979, Mar. Biol. 54: 83–89 (energy flow, population dynamics); Wu 1980, Can. J. Zool. 58: 559–566, and 1981, Can. J. Zool. 59: 890–892 (breeding and aggregation); Strathmann et al. 1981 (settlement, dispersal); Spight 1981 (settlement); Palmer et al. Mar. Biol. 67: 51–55 (predator evasion behavior); Gaines et al. 1985, Oecologia 67: 267–171 (larval concentration, settlement); Gaines and Roughgarden 1985, Proc. Natl. Acad. Sci. 82: 3707–3711 (settlement and community ecology); Blower and Roughgarden 1988 (parasitic castration by isopod *Hemioniscus*); Judge et al., 1988, J. Exp. Mar. Biol. Ecol. 119: 235–251 (recruitment); Dill and Gillett 1991, J. Exp. Mar. Biol. Ecol. 153: 115–127 ("economic logic" of hiding behavior, induced by shadows); Lohse 1993 (substrate ecology); Sanford and Menge 2001, Mar. Ecol. Prog. Ser. 209: 143–157 (growth); Sotka et al. 2004, Mol. Ecol. 13: 2143–2156, and Wares and Cunningham 2005, Biol. Bull. 208: 60–68 (genetics). There are numerous additional papers on the community ecology of this species.

Balanus nubilus Darwin, 1854 (=*B. flos* Pilsbry, 1907). Often misspelled *nubilis*, following an error by Pilsbry (1916); Darwin's original spelling was *nubilus*. Very large; low intertidal on pilings and rocks, as well on marina floats. See Newman and Abbott 1980.

Balanus trigonus Darwin, 1854. Low intertidal and subtidal southern species extending north to Monterey Bay. See Newman and Abbott 1980.

Amphibalanus amphitrite (Darwin, 1854) (=*Balanus amphitrite*). The genus *Amphibalanus* was established by Pitombo (2004). A southern hemisphere species introduced to the Pacific coast; generally restricted to warmer portions of bays and harbors from San Francisco Bay and south; low intertidal and subtidal. A common fouling organism. Abundant in the Salton Sea, where some beaches are primarily composed of the shells

of this barnacle; see Raimondi 1992, Biol. Bull. 182: 210–220 (adult plasticity and rapid larval evolution in the Salton Sea); Simpson and Hurlbert 1998, Hydrobiologia 381: 179–190 (salinity effects on growth, mortality, and shell strength in the Salton Sea); see also Newman 1967 (physiology, behavior, San Francisco Bay); Zullo et al. 1972 (introduction to California); Henry and McLaughlin 1975; Newman and Abbott 1980.

*Amphibalanus eburneus (Gould, 1841) (=*Balanus eburneus*). Introduced to southern California (Colorado Lagoon, Long Beach, J. T. Carlton, 2000, pers. comm.). An estuarine barnacle (and an often abundant fouling organism) from the East Coast now widely distributed around the world largely in tropical and subtropical waters, long known from Mazatlán on the west coast of Mexico and Guaymas in the Gulf as well as the Hawaiian Islands. To be watched for in bays and harbors north of Point Conception. Similar in appearance to *A. improvisus*, with which it may co-occur and from which it can be distinguished by a scutum with longitudinal striae and a tergum with a deeply concave rather than relatively straight margin between the spur and the depressor muscle crests. See Henry and McLaughlin (1975) and Zullo (1979) for illustrations.

Amphibalanus improvisus (Darwin, 1854) (=*Balanus improvisus*). Introduced from the Atlantic coast of North America and first found in the Port of San Francisco in the 1850s soon after the arrival of the first Gold Rush ships from New England (Carlton and Zullo 1969). Low intertidal and subtidal, an often very abundant fouling organism, in brackish water of bays on shells, pilings, etc. On occasion, it has settled in the entirely freshwater Delta Mendota Canal south of Tracy, in the midst of the farmlands of central California. See Newman 1967 (physiology, behavior, San Francisco Bay); Zullo et al. 1972 (occurrence in fresh water in California); Henry and McLaughlin 1975; Newman and Abbott 1980. There is an extensive Atlantic literature on this species.

Menesiniella aquila (Pilsbry, 1907) (=*Balanus aquila*). Large, common in low intertidal and subtidal on pilings and rocks in bays; rare north of San Francisco. See Newman and Abbott 1980.

Paraconcavus pacificus (Pilsbry, 1916) (=*Balanus pacificus*). Subtidal southern form extending north to Monterey Bay; usually on other organisms or manmade structures. See Giltay 1934, Bull. Mus. Roy. Hist. Natur. Belgique 10: 1–7 (on sand dollar *Dendraster*); Newman and Abbott 1980.

Megabalanus californicus (Pilsbry, 1916) (=*Balanus tintinnabulum californicus*). Low intertidal and subtidal species, often found in large clusters on buoys and pilings; scarce north of Monterey Bay, but occasionally occurring north of San Francisco, such as along the Sonoma County coast and buoys north of Humboldt Bay. See Zullo 1968, Occ. Pap. Calif. Acad. Sci. 70, 3 pp. (records north of San Francisco); Henry and McLaughlin 1986; Miller and Roughgarden 1994 (larvae); Newman and McConnaughey 1987, Pac. Sci. 41: 31–36 (El Niño waifs).

References

Abbott, D. P. 1987. Observing marine invertebrates. Drawings from the laboratory. Edited by G. H. Hilgard. Stanford University Press, Stanford, 380 pp.

Anderson, D. T. 1994. Barnacles. Structure, function, development, and evolution. London: Chapman and Hall, 376 pp.

Carlton, J. T., and V. A. Zullo. 1969. Early records of the barnacle *Balanus improvisus* Darwin from the Pacific coast of North America. Occ. Pap. Calif. Acad. Sci. 75: 1–6.

Connolly, S. R., and J. Roughgarden. 1999. Increased recruitment of northeast Pacific barnacles during the 1997 El Niño. Limnol. Oceanogr. 44: 466–469.

* = Not in key.

Cornwall, I. E. 1952. The barnacles of California (Cirripedia). Wasmann J. Biol. 9: 311–346 (dated fall 1951, but published 1952).

Cornwall, I. E. 1955. Canadian Pacific Fauna. 10. Arthropoda. 10e. Cirripedia. Fisheries Research Board of Canada, 49 pp.

Crisp, D. J., and P. S. Meadows. 1962. The chemical basis of gregariousness in cirripedes. Proc. Roy. Soc. Lond. B, 156: 500–520.

Darwin, C. 1852. A Monograph of the subclass Cirripedia. Part I, Lepadidae; 1854, Part II, Balanidae. Ray Society, London.

Gruvel, A. 1905. Monographie des Cirrhipedes. Paris: Masson, 472 pp. (reprinted, A. Asher, 1965).

Henry, D. P. 1942. Studies on the sessile Cirripedia of the Pacific coast of North America. Univ. Wash. Publ. Oceanogr. 4: 95–134.

Henry, D. P. 1965. Unique occurrence of complemental males in a sessile barnacle. Nature 207: 1107–1108.

Henry, D. P., and P. A. McLaughlin. 1975. The barnacles of the *Balanus amphitrite* complex (Cirripedia, Thoracica). Zoologische Verhandelingen 141, 254 pp.

Henry, D. P., and P. A. McLaughlin. 1986. The Recent species of *Megabalanus* (Cirripedia: Balanomorpha) with special emphasis on *Balanus tintinnabulum* (Linnaeus) sensu lato. Zoologische Verhandelingen 235, 69 pp.

Hines, A. H. 1978. Reproduction in three species of intertidal barnacles from central California. Biol. Bull. 154: 262–281.

Hines, A. H. 1979. The comparative reproductive ecology of three species of intertidal barnacles, pp. 213–234 in: S. E. Stancyk, ed., Reproductive ecology of marine invertebrates. Columbia, SC: University of South Carolina Press.

Høeg, J. T. 1992. Rhizocephala, pp. 313–345. In *Microscopical Anatomy of Invertebrates*, Volume 9. F. W. Harrison and A. G. Humes, eds. New York: Wiley-Liss, New York.

Høeg, J. T. 1995. The biology and life cycle of the Rhizocephala (Cirripedia). J. Mar. Biol. Assoc. U.K. 75: 517–550.

Miller, K. M., and J. Roughgarden. 1994. Descriptions of the larvae of *Tetraclita rubescens* and *Megabalanus californicus* with a comparison of the common barnacle larvae of the central California coast. Journal of Crustacean Biology 14: 579–600.

Newman, W. A. 1967. On physiology and behavior of estuarine barnacles. Proc. Symposium on Crustacea, Part III, pp. 1038-1066, Mar. Biol. Soc. India.

Newman, W. A. 1979. California transition zone: significance of short-range endemics, pp. 399–416. In, Historical biogeography, plate tectonics, and the changing environment. J. Gray and A. J. Boucot, eds. Corvallis, OR: Oregon State University Press.

Newman, W. A., and D. P. Abbott. 1980. Cirripedia, pp. 504–535. In Intertidal invertebrates of California. R. H. Morris, D. P. Abbott, and E. C. Haderlie, eds. Stanford University Press.

Newman, W. A., and A. Ross. 1976. Revision of the balanomorph barnacles, including a catalog of the species. San Diego Soc. Nat. Hist. Memoir 9, 108 pp.

Newman, W. A., and A. Ross. 2001. Prospectus on larval cirriped setation formulae, revisited. Journal of Crustacean Biology 21: 56–77.

Newman, W. A., V. Zullo, and T. H. Withers. 1969. Cirripedia. In Treatise on Invertebrate Paleontology, R. C. Moore, ed., Part R. Arthropoda. 4 (1): R206-295. Geol. Soc. Amer. and Univ. Kansas Press.

Perez-Losada, M. J., J. T. Høeg, and K. A. Crandall. 2004. Unraveling the evolutionary radiation of the thoracican barnacles using molecular and morphological evidence: A comparison of several divergence time estimation approaches. Systematic Biology 53: 244–264.

Pilsbry, H. A. 1916. The sessile barnacles (Cirripedia) contained in the collections of the U.S. National Museum; including a monograph of the American species. Bull. U.S. Nat. Mus. 93, 366 pp.

Pitombo, F. B. 2004. Phylogenetic analysis of the Balanidae (Cirripedia, Balanomorpha). Zoologica Scripta 33: 261–276.

Reinhard, E. C. 1944. Rhizocephalan parasites of hermit crabs from the Northwest Pacific. Jour. Wash. Acad. Sci. 34: 49–58.

Reischman, P. G. 1959. Rhizocephala of the genus *Peltogasterella* from the coast of the State of Washington and the Bering Sea. Koninkl. Nederl. Akad. Wetensch. Proc. (C) 62: 409–435.

Ross, A., and W. A. Newman. 1996. A unique experiment in four-platedness by a Miocene barnacle (Cirripedia: Balanidae) that Darwin considered improbable. Journal of Crustacean Biology 16: 663–668.

Spight, T. M. 1981. Settlement patterns of two rocky shore barnacles. Ecosynthesis 1: 93–120.

Standing, J. D. 1980. Common inshore barnacle cyprids of the Oregonian faunal province (Crustacea: Cirripedia). Proc. Biol. Soc. Wash. 93: 1184–1203.

Strathmann, R. R., E. S. Branscomb, and K. Vedder. 1981. Fatal errors in set as a cost of dispersal and the influence of intertidal flora on set of barnacles. Oecologia 48: 13–18.

Tomlinson, J. T. 1969. The burrowing barnacles (Cirripedia: order Acrothoracica) Bull. U.S. Nat. Mus. 296, 162 pp.

Williams J. D. and Boyko C. B. 2006. A new species of the burrowing barnacle genus *Tomlinsonia* Turquier, 1985 (Crustacea, Cirripedia, Acrothoracica, Apygophora, Trypetesidae) in hermit crab shells from the Philippines with the description of a new parasite species of *Hemioniscus* Buchholz, 1866 (Crustacea, Isopoda, Cryptoniscoidea, Hemioniscidae). Zoosystema 28: 285–305.

Zullo, V. A. 1963. A review of the subgenus *Armatobalanus* Hoek (Cirripedia: Thoracica), with the description of a new species from the California coast. Ann. Mag. Nat. Hist. (13) 6: 587–594.

Zullo, V. A. 1979. Marine Flora and Fauna of the Northeastern United States. Arthropoda: Cirripedia. NOAA Technical Report NMFS Circular 425, 31 pp.

Zullo, V. A. 1992. Revision of the balanid barnacle genus *Concavus* Newman, 1982, with the description of a new subfamily, two new genera and eight new species. Journal of Paleontology 27: 1–46.

Zullo, V. A., D. B. Beach, and J.T. Carlton. 1972. New barnacle records (Cirripedia, Thoracica). Proc. Calif. Acad. Sci. (4) 39: 65–74.

Leptostraca

TODD A. HANEY, JOEL W. MARTIN, AND ERIC W. VETTER

(Plates 217–219)

Leptostracans ("thin-shelled shrimp") are relatively small, broadly distributed, marine crustaceans. Leptostraca are known from a wide variety of habitats from estuaries to the deep sea, including pelagic waters, tropical anchialine caves, coral reefs, seagrass beds, and mudflats (Haney and Martin 2000). Many species remain undescribed around the world, and "cosmopolitan" species are not likely, given that all species brood their eggs and that eclosion is postlarval.

Leptostraca have a hinged rostrum, folded carapace, phyllopodous thoracic appendages, and eight abdominal segments (plate 217). Most are 5 mm–15 mm in length, with females often larger than males. A thin carapace covers all eight thoracic segments, as well as the first three or more abdominal segments. Superficially the carapace appears to be bivalved, but it is merely folded and possesses no hinge. In this regard, the carapace is more similar to that of some clam shrimp (Spinicaudata) than to the hinged valves of ostracods.

One of the most notable features of leptostracans is the hinged rostrum. In all other malacostracan crustaceans (except stomatopods), the rostrum, if present, is fixed. The rostrum is a thin flap of cuticle that extends from the anterior carapace margin. It is most broad proximally, where its base is defined by its articulation with the carapace, and it tapers to a rounded distal end. The rostrum covers the base of the eyestalks, and, in most species, its ventral surface bears a rostral keel that is positioned between the eyestalks.

Eye morphology varies considerably, so much so that eyestalk morphology is a key feature for generic identification (see Walker-Smith and Poore 2001: fig. 1). The eyestalks of *Nebalia* are typically shorter than the rostrum, and most species of the genus have eyes with a substantial number of ommatidia and photopigments that cause the eyes to appear dark to bright red. There is some variation of eye morphology within this large genus as well; for instance, small protuberances or invaginations of the eyestalk have been used as taxonomic characters (Dahl 1990; Vetter 1996c; Haney et al. 2001).

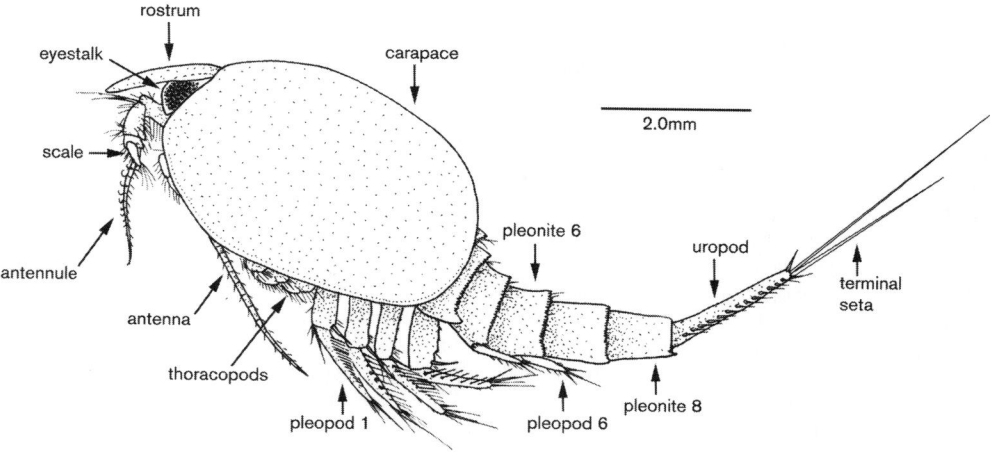

PLATE 217 Leptostracan morphology.

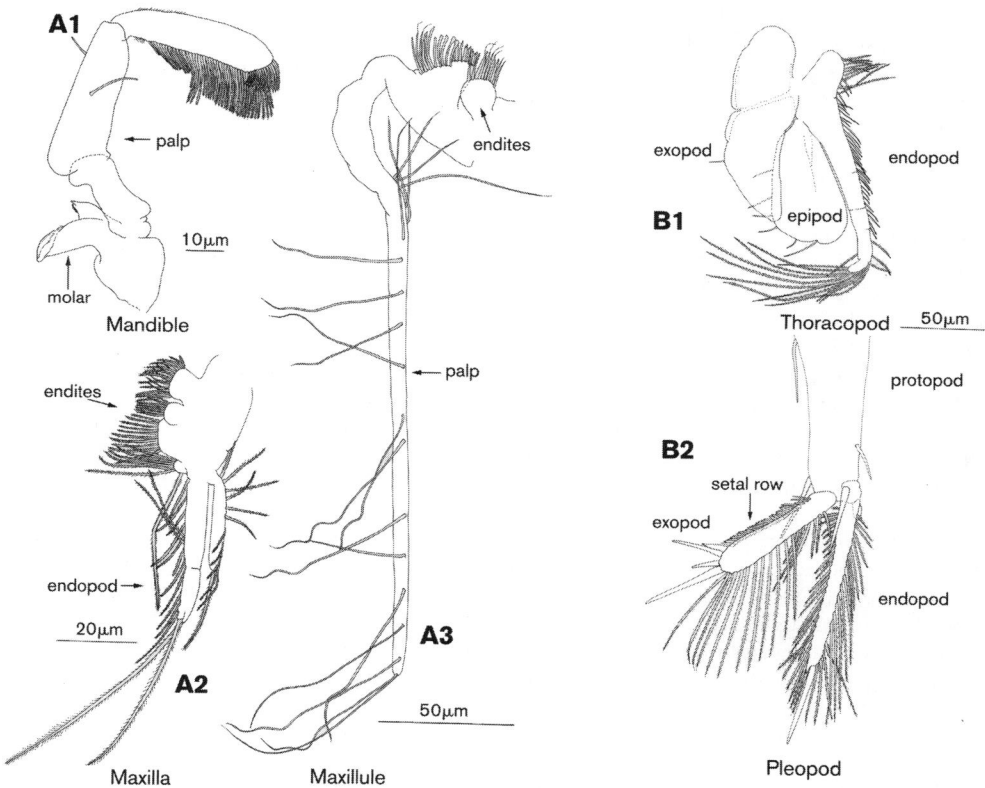

PLATE 218 A, Leptostracan mouthparts: A1, mandible; A2, maxilla; A3, maxillule; B, limbs: B1, thoracopod; B2, pleopod.

The first and second pairs of antennae (antennules and antennae, respectively) vary considerably among genera, but their morphology is fairly well conserved within each genus. The antennular peduncle is composed of the four basalmost segments, with the fourth article giving rise to an ovate article called the antennular scale and a multisegmented flagellum. The most striking variation in the morphology of the antennule is sexual dimorphism: in females, the antennular flagellum is composed of multiple articles of similar size that are generally distinct and arranged linearly; in males, the flagellum is highly curved.

The antenna possesses a peduncle of three articles, the first two of which each bear a distal spine in the Nebaliidae. Different patterns of spination and setal arrangements on the third article of this appendage have been used as diagnostic features. Sexual dimorphism of the antennal flagellum is not as pronounced, although the antennal flagellum of male leptostracans is often much longer than that of the females in, for example, *Nebalia*.

Leptostracans possess the three pairs of mouthparts shared by all Crustacea (plate 218A); these are the mandibles and the first and second pairs of maxillae (maxillules and maxillae, respectively). The mandible bears a molar process and an anteriorly directed palp composed of three segments. The terminal article of the palp bears rows of plumose setae along its distal and inferior margin, most likely used to assist in the movement of food particles toward the mouth. The maxillule has two endites and an elongate maxillulary palp, which extends back nearly the length of the carapace. Given its position between the carapace and the thoracic limbs and the serrate setae that it bears, the maxillulary palp probably serves to prevent the area beneath the carapace from becoming fouled by debris. The maxilla bears an endopod and exopod (except in *Nebaliopsis*) and three endites. As with those of the maxillule, each maxillary endite bears one or more dense rows of setae.

The thoracic limbs (thoracopods) are biramous and often described as foliaceous, although these limbs are not as markedly flattened among the members of the Paranebaliidae as they are in other nebaliaceans. In all Leptostraca, the thoracic limb is composed of an endopod, which gives rise to both an exopod and an epipod (plate 218B). The length and ornamentation of these appendages varies both among taxa and developmental stages. For example, the distal articles of the endopods bear long, plumose setae in ovigerous females, but such setae are shed in post-ovigerous stages (Dahl 1985). The thoracopods differ depending on the segment on which they are born; most notably, the first and eighth thoracopods are typically more slender than thoracopods two through seven.

In addition to eight thoracopods, leptostracans possess six abdominal appendages (pleopods) (plate 218B). Because the thoracopods are mostly or entirely concealed by the carapace, the most clearly visible limbs are the first (anteriormost) four pairs of pleopods; they are robust, biramous limbs and do not share the foliaceous morphology of the thoracopods. The last two (fifth and sixth) pairs of pleopods are short and uniramous. The fifth pleopod consists of two segments. The sixth pleopod is most often single-segmented and the smallest of the abdominal appendages; it is two-segmented in the Nebaliopsididae and the genus *Nebaliella*.

The abdomen consists of seven abdominal somites, or pleonites. In most species the carapace conceals the first (anteriormost) three pleonites. The fourth pleonite is slightly expanded along its ventro-lateral margin, producing an epimeron unique to this pleonite (see Haney and Martin 2000: fig. 8a). The entire posterior margin of pleonites one through six bears dentition that ranges from blunt to highly acute teeth (see Martin et al. 1996: figs. 3f, 14c). Although the shape of such teeth varies slightly with respect to position on the individual, this character exhibits little intraspecific variation and, hence, appears to be of taxonomic value.

The terminal abdominal somite most commonly has been referred to as the telson; however, both Sharov (1966) and Bowman (1971) argued against this, claiming the terminal segment to represent a true abdominal somite rather than a telson. We agree with this interpretation; the leptostracan "telson" is simply an eighth pleonite. Bowman (1971) referred to this terminal body segment as an anal somite, as it bears a terminal anus. The postero-ventral surface of the anal somite is characterized by two broad, subtriangular processes; these processes are called anal plates, as it is the deep invagination between them that marks the site of the anus. The anal somite differs conspicuously from the first six pleonites in two other regards: it lacks posterior-margin dentition, and it bears a single pair of large appendages. The appendages have been referred to as caudal furcae but are better described as uropods given the arguments of Sharov (1966) and Bowman (1971). The uropod bears a row of simple setae along its lateral margin, a row of plumose setae medially, and one or more elongate setae terminally.

Life History and Ecology

Leptostracans are sexually dimorphic, perhaps the most conspicuous difference between the two sexes being antennule morphology. In *Nebalia*, the antennule of females is nearly straight and posteriorly directed; the male antennule is curved forward. Additionally, in many species, including *N. daytoni* Vetter (plate 219B), the antennal flagellum of the male is significantly longer than that of the female, often trailing beyond the posterior-most end of the animal. In *N. daytoni*, the eyestalk of the male is proportionally larger than that of the female, but significant dimorphism in this feature is uncommon. As with most malacostracans, the male gonopore is on the eighth thoracic somite, whereas that of the female is on the sixth somite.

The size and shape of the carapace is also sexually dimorphic; males tend to be more laterally compressed and be more shallow-bodied along their dorso-ventral axis. This difference is in part associated with the role of females in brooding. Leptostracans exhibit direct development and possess no free-swimming larval stage. Unlike the marsupium of isopods and amphipods, which is formed by two or more overlapping brood plates (oostegites), female leptostracans possess a brooding "basket." The developing ova are held beneath the thoracic segments, between the valves of the carapace, and they are bound ventrally by many elongate setae that arise from the enlarged distal articles of the thoracopods and intertwine to form the basket. Such setae are not present in the pre- and post-ovigerous female (see Dahl 1985, figs. 6–10). Populations of leptostracans tend to be dominated by females. In collections of *Nebalia gerkenae* from Elkhorn Slough (Monterey County), for instance, Gerken (1995) found that adult males generally represented <15% of the individuals.

Although leptostracans have been collected from oligotrophic environments (e.g., Vetter, 1996a–c), they are most often associated with areas rich with organic material. *Nebalia* sometimes occurs in prodigious numbers. *Nebalia hessleri* occurs in densities ranging from 700,000 to 2 million animals per square meter at the head of the Scripps Submarine Canyon, densities that exceed any previously reported for a macroscopic marine invertebrate (Vetter 1994). Similar reports suggest that dense aggregations of leptostracans might be common. Conlan and Ellis (1979) estimated densities of *Nebalia* to reach 2,462 individuals per square meter at log-handling sites in British Columbia, and Gerken (1995) estimated densities of as many as 1,856 individuals per liter for *Nebalia gerkenae* (as *Nebalia pugettensis*) in Elkhorn Slough. The significance of such dense aggregations is not yet well understood, but some studies provide estimates of the significant contribution of leptostracan populations to secondary productivity (Rainer and Unsworth 1991; Vetter 1995, 1996b, 1998) suggesting that these animals warrant further study of their role in the benthos.

The studies of Gerken (1995) and Vetter (1996a) are among a handful on leptostracan population dynamics and reproduction; both reported clear seasonal cycles in *Nebalia*. They also cultured *Nebalia* and reported locomotory, reproductive, and feeding behaviors. Of note is that *Nebalia* appears to be able to survive in fetid, hypoxic environments where many Crustacea that otherwise might compete with *Nebalia* do not occur (Okey 2003).

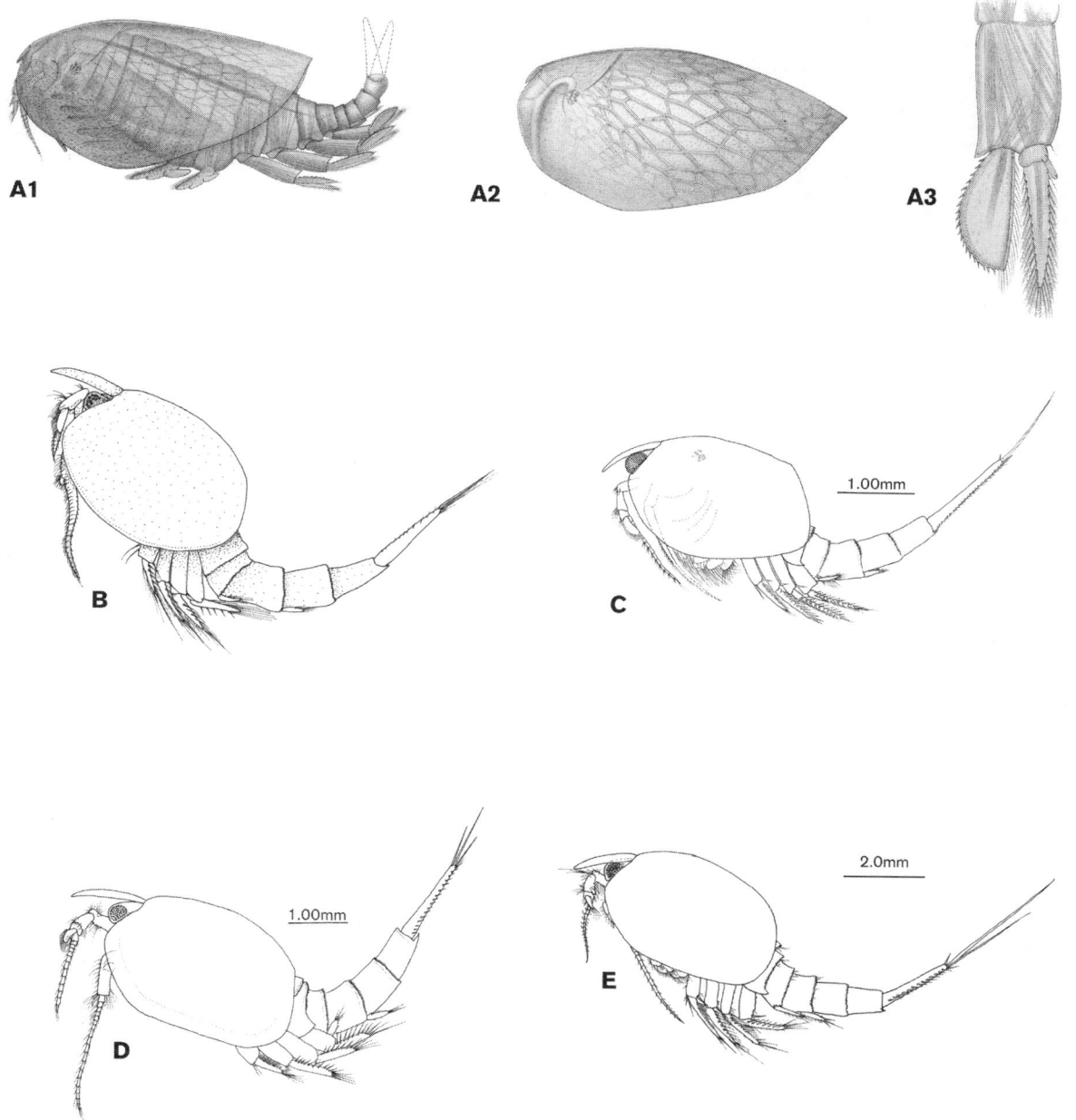

PLATE 219 A, *Nebaliopsis* (modified from Sars, 1887): A1, whole animal (female), left side; A2, carapace; A3, pleopod 3; B, *Nebalia daytoni* (male) Vetter, 1996, from La Jolla; C, *Nebalia hessleri* (female) Martin et al., 1996, from La Jolla; D, *Nebalia gerkenae* (female) Haney and Martin, 2000, from Monterey Bay, with permission of the Biological Society of Washington; E, *Nebalia kensleyi* (female) Haney and Martin, 2005, from Tomales Bay.

Cannon (1927) provided a detailed description of the filter-feeding mechanism of leptostracans. Predation has also been documented (Martin et al. 1996). Nishimura and Hamabe (1964) noted that "in certain localities at least nebaliaceans are distributed so densely as to swarm enormously to and devour such large-sized dead animals as sharks till these are finally disintegrated in a rather short time." Some species are able to switch between feeding modes.

This potential for tremendous density, their obvious ecological importance, and the out-of-date and poorly known taxonomy makes the group surely one of the most obvious of the many "gaps" in our knowledge of West Coast marine invertebrates.

The Fauna of California and Oregon

For crustaceans that often occur in great numbers within the intertidal zone, the Leptostraca of the Pacific coast are not well known. The first published records of *Nebalia* from the West Coast were those of Packard (1883) in Puget Sound and LaFollette (1914) for several specimens collected from a kelp holdfast at Laguna Beach. For many years, it was assumed by most authors that all leptostracans along the West Coast were *Nebalia pugettensis* (Clark, 1932). This species is unrecognizable, and the name has been abandoned (Martin et al. 1996). New species of *Nebalia* have been described, and more remain to be named to accommodate what was previously referred to as *N. pugettensis*. A nebaliid reported as nonnative in San Francisco Bay (Ruiz et al. 2000) remains unidentified.

Because of the variability of leptostracans, and the differences between the sexes, the use of scanning electron microscopy has been of great value in the identification and description of new species. The few species described to date from the West Coast are recognized by readily observable morphological characters (e.g., the eyes in *N. daytoni*). The keys permit the identification of these taxa with a light microscope and without the need for dissection.

Collection and Preservation

Leptostracans are commonly encountered in areas of fine-grained sediments that are rich with organic matter. In the intertidal zone, individual leptostracans can be collected with a small hand net or spoon. They are most easily noticed at the waterline during low tide. On mudflats or other areas of relatively fine-grained sediments, if the collector disturbs small pools of water found in depressions or permits water to rush beneath rocks as they are overturned, often several leptostracans become trapped at the surface. Indeed, they briefly "flail" at the surface, and the regular flexion of their abdomen is a behavior that permits them to be quickly distinguished from similarly sized and faster-swimming crustaceans, such as amphipods, with which they are often found.

When leptostracans are abundant, a simple bulk collection of any algal matter with which they are associated often yields many specimens. Algae can be washed vigorously in a bucket and the wash run through a 250 μm to 500 μm sieve for collection of the animals. From subtidal environments, leptostracans often are retrieved from benthic grab samples.

Light and baited traps also work well. An inexpensive light or baited trap can be constructed in any number of ways and should work well as long as consideration is given to the size of the entrance(s) to the trap and to the period of exposure. A diving flashlight or chemical light stick can be placed in a small, weighted bucket and covered with a lid in which narrow (<1cm) slits have been cut to permit animals to enter. Such a trap generally yields good results if exposed for a minimum of one hour. Traps baited with meat or other attractants require much longer exposures and can often be left overnight; however, the return can diminish with time if the animals are able to escape the trap after the bait is consumed.

Specimens should be fixed in 75%–95% ethanol, which will also permit genetic analysis. Preservation in solutions lower than ~50% ethanol permits degradation of tissues; very high concentrations of these solutions will tend to make tissues more hardened and brittle. For studies of ultrastructure, exposing the animals to low-concentration alcohol solutions or other relaxants (e.g., magnesium chloride) prior to fixation helps to avoid wrinkling or other damage to soft tissues.

ACKNOWLEDGMENTS

We acknowledge grants from the National Science Foundation's PEET initiative and the Biotic Surveys and Inventories program. For their attention to leptostracans in the field, we thank Don Cadien, Lisa Haney, Leslie Harris, Regina Wetzer, David Jacobs, Amy Poopatanapong, Chris Winchell, Tony Phillip, Craig Staude, Mary Wicksten, and Todd Zimmerman.

Key to Species

Note that there are undescribed species along the coast.

1. Carapace large, concealing thoracic segments and at least first six pleonites; carapace with fine dentition along postero-dorsal margin; uropods broad, paddlelike; holopelagic (plate 219A) . . . . *Nebaliopsis*
— Carapace not extending beyond pleonite 5; carapace postero-dorsal margin even, lacking dentition and strongly emarginate (indented); uropods tapering evenly from base to tip; benthic 2
2. Eyestalk distally truncate, with slight dorsal and ventral protrusion of corneal surface; eyestalk and supraorbital spine subequal in length; article 2 of mandibular palp with long, plumose subterminal seta (plate 219B) . . . .
. . . . *Nebalia daytoni*
— Eyes not distally truncate, without dorsal and ventral corneal protrusions; eyestalk much longer than supraorbital spine; article 2 of mandibular palp lacking elongate, plumose subterminal seta 3
3. Body length typically >6.0 mm, reaching 15.0 mm; peduncular article 4 of antennule with four or five robust apical setae; posterior margins of pleonites 5–7 with highly acute teeth (plate 219C) . . . . *Nebalia hessleri*
— Body length typically <6.0 mm; peduncular article 4 of antennule with two or less robust apical setae; posterior margins of pleonites 5–7 with subacute or blunt-ended teeth
. . . . 4
4. Eyes with pigment on distal third of eyestalk; posterior margins of pleonites 5–7 with subacute teeth; terminal seta shorter than uropod (plate 219D) . . . . . . *Nebalia gerkenae*
— Eyes with pigment on distal two-thirds of eyestalk; posterior margins of pleonites 5–7 with distally rounded teeth; terminal seta 1.7× length of uropod (plate 219E) . . . . . .
. . . . *Nebalia kensleyi*

List of Species

NEBALIIDAE

Nebalia daytoni Vetter, 1996. Oligotrophic sands, 8 mm–33 mm, off La Jolla; occurring farther north. See Vetter 1996, Mar. Ecol. Prog. Ser. 137: 83–93 (enrichment and population cycles).

Nebalia gerkenae Haney and Martin, 2000. Surface mats of *Gracilaria* and *Ulva*, as well as beneath small rocks partly embedded in mud; high intertidal; Bennett Slough, an arm of Elkhorn Slough, Monterey Bay.

Nebalia kensleyi Haney and Martin, 2005. Intertidal mats of *Ulva* in Tomales Bay.

*Nebalia sp. A widespread species in the Pacific Northwest; the figure here is of a specimen close to the original collecting site of *N. pugettensis*; some diagnostic figures of the species were noted by Martin et al. 1996.

Nebalia hessleri Martin, Vetter, and Cash-Clark, 1996. Deeper water off the southern California coast; noted here on the possibility it may be discovered in shallower water.

*Nebalia "*pugettensis* Clark, 1932" (=*Epinebalia pugettensis*). Martin et al. (1996) declared this species a *nomen nudum*, until it can be determined that there is only one species in the immediate vicinity of Clark's original collecting site at Friday Harbor.

* = Not in key.

NEBALIOPSIDIDAE

Nebaliopsis sp. Deeper water off the California coast.

References

Bowman, T. E. 1971. The case of the nonubiquitous telson and the fraudulent furca. Crustaceana 21: 165–175.

Cannon, H. G. 1927. On the feeding mechanism of *Nebalia bipes*. Transactions of the Royal Society of Edinburgh 55: 355–369.

Clark, A. E. 1932. *Nebalia Caboti*, n. sp., with observations on other Nebaliacea. Transactions of the Royal Society of Canada 26: 217–235.

Conlan, K. E., and D. V. Ellis. 1979. Effects of wood waste on sand-bed benthos. Marine Pollution Bulletin 10: 262–267.

Dahl, E. 1985. Crustacea Leptostraca, principles of taxonomy and a revision of European shelf species. Sarsia 70: 135–165.

Dahl, E. 1987. Malacostraca maltreated—the case of the Phyllocarida. Journal of Crustacean Biology 7: 721–726.

Dahl, E. 1990. Records of *Nebalia* (Crustacea Leptostraca) from the Southern Hemisphere—a critical review. Bulletin of the British Museum of Natural History (Zoology) 56: 73–91.

Gerken, S. 1995. The population ecology of the leptostracan crustacean, *Nebalia pugettensis* (Clark, 1932), at Elkhorn Slough, California. M.S. thesis, University of California, Santa Cruz, 53 pp.

Haney, T. A., R. R. Hessler, and J. W. Martin. 2001. *Nebalia schizophthalma*, a new species of leptostracan (Crustacea: Malacostraca) from deep waters off the eastern United States. Journal of Crustacean Biology 21: 192–201.

Haney, T. A., and J. W. Martin. 2000. *Nebalia gerkenae*, a new species of leptostracan (Crustacea, Phyllocarida) from the Bennett Slough region of Monterey Bay, California. Proceedings of the Biological Society of Washington 113: 996–1014.

Haney, T. A., and J. W. Martin. 2005. *Nebalia kensleyi*, a new species of leptostracan (Crustacea: Phyllocarida) from Tomales Bay, California. Proceedings of the Biological Society of Washington 118: 3–20.

LaFollette, R. 1914. A *Nebalia* from Laguna Beach. Journal of Entomology and Zoology 6: 204–206.

Martin, J. W., E. W. Vetter, and C. E. Cash-Clark. 1996. Description, external morphology, and natural history observations of *Nebalia hessleri*, new species (Phyllocarida: Leptostraca), from southern California, with a key to the extant families and genera of the Leptostraca. Journal of Crustacean Biology 16: 347–372.

Nishimura, S., and M. Hamabe. 1964. A case of economical damage done by *Nebalia*. Publications of the Seto Marine Biological Laboratory 12: 173–175.

Okey, T. A. 2003. Macrobenthic colonist guilds and renegades in Monterey Canyon (U.S.A.) drift algae: partitioning multidimensions. Ecological Monographs 73: 415–440.

Packard, A. S. 1883. A monograph of the phyllopod Crustacea of North America, with remarks on the order Phyllocarida. F. V. Hayden ed., in Twelfth Annual Report of the U.S. Geological and Geographical Survey of the Territories of Wyoming and Idaho: a report of the progress of the exploration of Wyoming and Idaho for the year 1878, Part I. (Washington, D.C.: Government Printing Office), pp. 295–497.

Rainer, S. F., and P. Unsworth. 1991. Ecology and production of *Nebalia* sp. (Crustacea: Leptostraca) in a shallow-water seagrass community. Australian Journal of Marine and Freshwater Research 42: 53–68.

Ruiz, G. M., P. W. Fofonoff, J. T. Carlton, M. J. Wonham, and A. H. Hines. 2000. Invasion of coastal marine communities in North America: apparent patterns, processes, and biases. Annual Review of Ecology and Systematics 31: 481–531.

Sharov, A. G. 1966. Basic Arthropodan Stock with Special Reference to Insects. (Oxford: Pergamon Press), 271 pp.

Vetter, E. W. 1994. Hotspots of benthic production. Nature 372: 47.

Vetter, E. W. 1995. Detritus-based patches of high secondary production in the nearshore benthos. Marine Ecology Progress Series 120: 251–262.

Vetter, E. W. 1996a. Life-history patterns of two Southern California *Nebalia* species (Crustacea, Leptostraca)—the failure of form to predict function. Marine Biology 127: 131–141.

Vetter, E. W. 1996b. Secondary production of a Southern California *Nebalia* (Crustacea, Leptostraca). Marine Ecology Progress Series 137: 95–101.

Vetter, E. W. 1996c. *Nebalia daytoni*, n. sp., a leptostracan from southern California (Phyllocarida). Crustaceana 69: 379–386.

Vetter, E. W. 1998. Population dynamics of a dense assemblage of marine detritivores. Journal of Experimental Marine Biology and Ecology 226: 131–161.

Walker-Smith, G. K., and G. C. B. Poore. 2001. A phylogeny of the Leptostraca (Crustacea) with keys to the families and genera. Memoirs of Museum Victoria 58: 383–410.

Mysidacea

RICHARD F. MODLIN

(Plates 220–223)

Mysids are inconspicuous, but abundant and diverse, shrimp-like peracaridan crustaceans. They are rarely collected if conventional sampling techniques are utilized because of the variety of microhabitats they occupy. Generally, many coastal species inhabit the sediment surface or shoal within the water column contiguous with, or a few meters above, the bottom. Other species are truly planktonic and exist in the water column for their entire life cycle. Others opt for more cryptic habitats such as the interstices of sponges, corals, discarded gastropod shells, and holdfasts of macroalgae. Some even establish symbiotic relationship with sponges, anemones and scyphozoans. Consequently, sampling mysid populations usually requires a variety of specialized collecting techniques.

Except for *Mysis relicta*, which is strictly freshwater and may be endemic, or introduced into, some deep-water and glacial lakes of northern California and Oregon, mysids are marine crustaceans. However, because of its adaptations to estuarine habitats, *Neomysis mercedis* occurs in some freshwater tributaries to the Sacramento-San Joaquin Estuary, as well as those of other estuarine systems along the northwestern coast of North America.

Taxonomically, the Mysidacea are divided into two suborders: the Lophogastrida and Mysida. Lophogastrids are large, pelagic, deep-ocean mysids very rarely collected in coastal environments, while species of the Mysida dominate most coastal habitats from intertidal to pelagic environs and occupy both benthic and planktonic habitats. The two suborders are easily distinguished. Specimens of Lophogastrida have branchial gills extending from the base of some thoracic legs, well-developed, unmodified biramous pleopods in both sexes, and lack a statocyst on the endopods of the uropod (plate 220A, 220B). Statocysts on the endopods of the uropod are obvious in the Mysida, branchia are absent, and pleopods of females are rudimentary (uniramous and platelike) or absent in the Mysida (plate 220C, 220D). Lophogastrida are not treated here; Kathman et al. (1986) provide taxonomic information on Lophogastrida reported along the northwestern coast of the United States.

Juvenile and mature females are easily separated from males by the presence of an obvious brood pouch (marsupium) (plate 220C). The brood pouch is composed of a number of large paired, interlocking plates (oostegites) ballooning posteriorly from ventral surface of the thorax. The number of oostegites can be important in the identification of some mysid species.

Males, juvenile and mature, are easily identified by a pair of penes located at the junction of the thorax and abdomen on the ventral surface. Also, depending on species, male third, fourth, and/or fifth pleopods are sexually dimorphic to enable the possible transfer of spermatophores. Modified males pleopods are an important character for taxonomic diagnosis. The magnitude of male pleopod dimorphism in some mysid taxa varies between juveniles and mature individuals of the

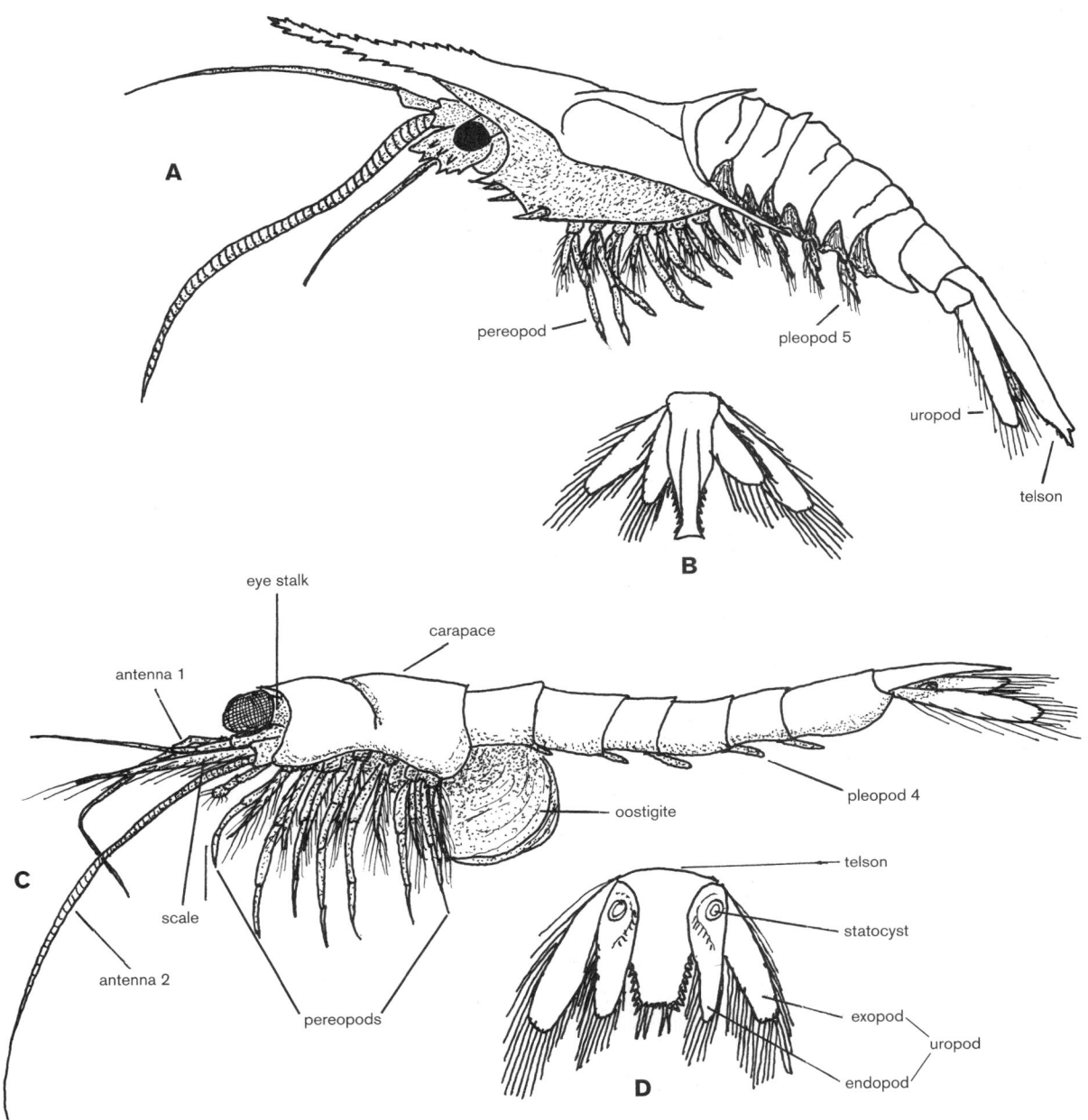

PLATE 220 A, order Mysidacea, suborder Lophogasterida, *Gnathophausia* sp.; B, telson-uropod complex, dorsal view, of *Gnathophausia* sp.; C, order Mysidacea, suborder Mysida, generalize mysid species; D, telson-uropod complex, ventral view, of species in suborder Mysida (figures are not to scale).

same species (plate 221H, 221I). This variation can sometimes lead to misidentification.

Taxonomy of the Mysida, except in some limited cases, is well established. Their identification is based on a variety of obvious, and for the most part invariable, morphological characteristics. Consequently, with the use of a hand lens, mysids collected in the field can usually be identified to genus. Microscopic examination and limited dissection enable easy identification of a mysid to species.

Identification of the species considered in this section is based primarily on variations in the morphology of telson, uropods, male pleopods, and antennal scales. Except for the male pleopods, the design of the other key characteristics does not vary between the sexes. For information on Mysidacea

south, of our region the reader should examine the taxonomic atlas edited by Blake and Scott (1997), and for coastal areas to the north, Daly and Holmquist (1986) and Kathman et al. (1986). The only comprehensive global treatment of mysid biology and taxonomy is the older summary of Mauchline (1980).

Key to Mysidacea

1. Eye stalk length normal (plate 220A, 220C) 2
— Eye stalk length elongated (plate 221A) . . . .
. . . . *Alienacanthomysis macropsis*
2. Lateral margins of telson completely or partially armed with spines, but if partial, a group of spines also located

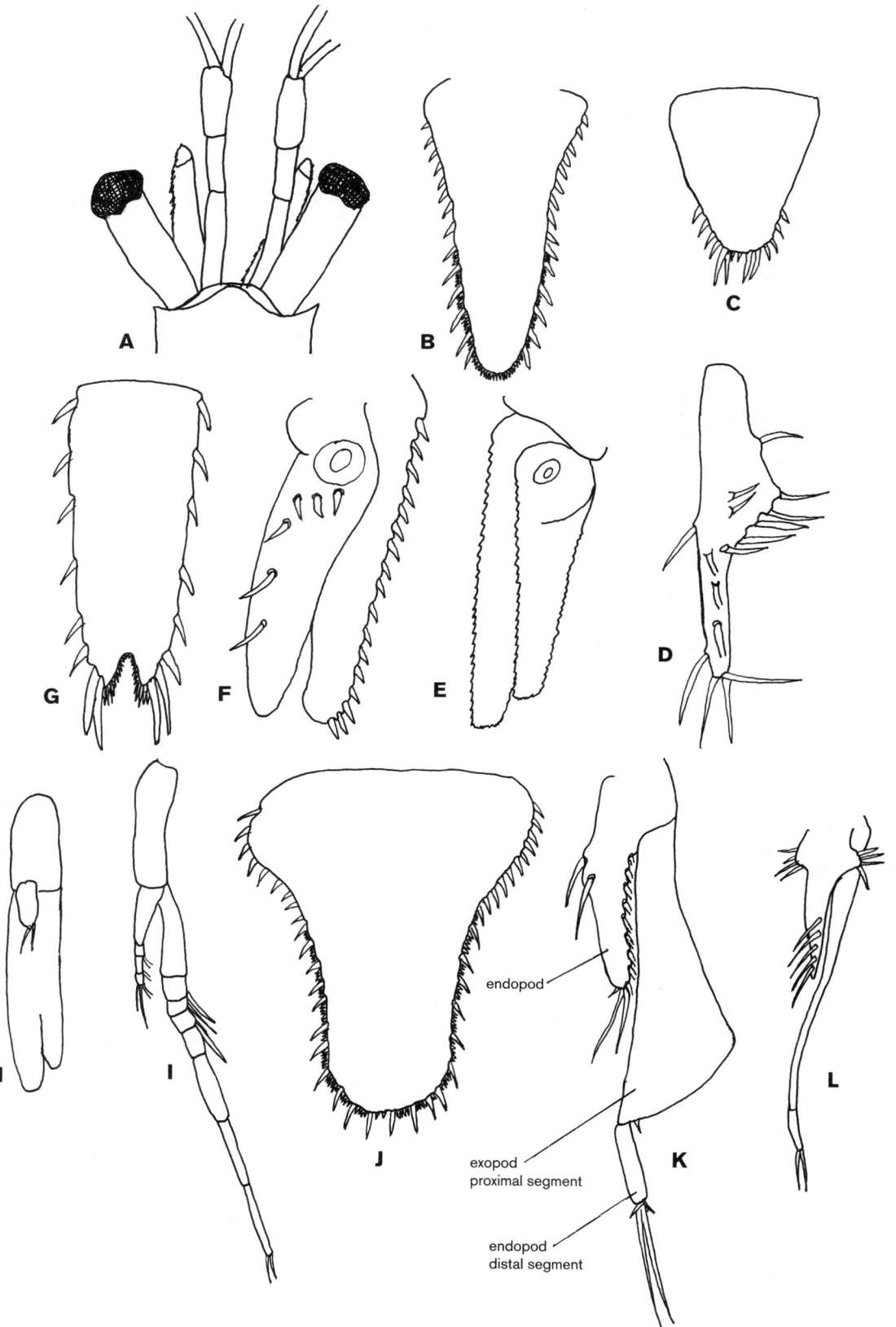

endopod

exopod
proximal segment

endopod
distal segment

PLATE 221 A, *Alienacanthomysis macropsis*, anterio-dorsal view; B, *Alienacanthomysis macropsis*, telson; C, *Deltamysis holmquistae*, telson; D, *Deltamysis holmquistae*, male pleopod 5; E, *Deltamysis holmquistae*, uropodal exopod and endopod; F, *Archeomysis grebnitzkii*, uropodal exopod and endopod; G, *Archeomysis grebnitzkii*, telson; H, *Archeomysis grebnitzkii*, pleopod 3 of immature male; I, *Archeomysis grebnitzkii*, pleopod 3 of mature male; J, *Hippacanthomysis platypoda*, telson; K, *Hippacanthomysis platypoda*, male pleopod 4; L, generalized pleopod 4 of *Neomysis* spp. and *Acanthomyis* spp. (figures are not to scale).

proximally (plates 221B, 223A) 3
— Lateral margins of telson armed with spines only in distal
half (plate 221C). . . . . *Deltamysis holmquistae*
3. Exopod of uropods without spines along lateral margin
(plate 221E) 4
— Exopod of uropods with spines along lateral margin (plate
221F) . . . . *Archaeomysis grebnitzkii*
4. Exopod of male fourth pleopod two segmented, cylindrical in cross section (plate 221L) 5
— Exopod of male fourth pleopod with proximal segment flattened, bladelike, distal segment cylindrical (plate 221K)
. . . . *Hippacanthomysis platypoda*
5. Abdominal segments smooth, without furrows or folds
. . . . 7
— Abdominal segments with furrows or folds, some may be
faint and/or disconnected (plate 222A–222D). 6
6. Telson sharply triangular with longest marginal spines in
distal half, apex with one pair of minute spines and one
pair of long spines (plate 222F) . . . .
. . . . *Exacanthomysis davisi*
— Telson broadly triangular with longest lateral spines along
entire margin, apex with one pair of minute spines and
two pair of long spines (plate 222E) . . . .
. . . . *Holmesimysis costata*
7. Distal tip of antennal scale sharply pointed (plate 222G)
. . . . 8
— Distal tip of antennal scale rounded (plate 222H) . . . . . .
. . . . 10
8. Length of telson greater than two times width measured
proximally at broadest interval 9
— Length of telson two times or less than width measured
proximally at broadest interval (plate 222I) . . . .
. . . . *Neomysis mercedis*
9. Lateral spines on margin of telson 25 or less (plate 222J)
. . . . *Neomysis rayii*
— Lateral spines on margin of telson >25 (plate 222K) . . . .
. . . . *Neomysis kadiakensis*
10. Lateral margins of telson completely armed with spines
(plate 223C, 223F, 223I) 12
— Spination along lateral margins of telson interrupted proximally (plate 223A, 223B) 11
11. Endopod of uropod with four to five spines in vicinity of
statocyst (plate 223D). . . . . *Acanthomysis californica*
— Endopod of uropod with two spines in vicinity of statocyst
(plate 223E) . . . . *Hyperacanthomysis longirostris*
12. Margin of telson armed with long spines interspersed with
minute spines (plate 223F, 223I) 13
— Margin of telson armed with 35–40 subequal (of about the
same length) spines. . . . . *Columbiaemysis ignota*
13. Endopod of uropod with single spine in vicinity of statocyst (plate 223H). . . . . *Acanthomysis aspera*
— Endopod of uropod with 4 spines in vicinity of statocyst
. . . . *Acanthomysis hwanhaiensis*

List of Species

Acanthomysis californica Murano and Chess, 1987. A probable mid-water species collected offshore of the Big Sur region
at 119 m (at a site 143 m deep) (Murano and Chess 1987); specimens occasionally collected off shallow exposed beaches.

Hyperacanthomysis longirostris Ii, 1936 (=*Acanthomysis bowmani* Modlin and Orsi, 1997; see Fukuoka and Murano 2000,

Plankton Biol. Ecol. 47: 122–128). An exotic Asian species
collected in Suisun Bay of the Sacramento–San Joaquin
Estuarine system. In low-salinity waters.

Acanthomysis aspera Ii, 1964. An exotic brackish-water Japanese species established in the delta region of the Sacramento–
San Joaquin Estuary (Modlin and Orsi 1997).

Acanthomysis hwanhaiensis Ii, 1964. This exotic Korean
species has been collected in the Sacramento–San Joaquin Estuary delta (Modlin and Orsi 2000).

Alienacanthomysis macropsis (Tattersall, 1932) (=*Neomysis
macropsis* Tattersall, 1932a, *Acanthomysis macropsis*). Uncommon; San Francisco Bay to Alaska. Closely related to *Acanthomysis pseudomacropsis* (Tattersall), which inhabits eastern
Pacific coastal waters from Alaska to Japan (Banner 1948, Tattersall 1951, Ii 1964). Some mature females of *A. macropsis* from
the Columbia River Estuary harbor the ectoparasitic copepod
Hansenulus trebax Heron and Damkaer, 1986, in their marsupium (Daly and Damkaer 1986).

Archaeomysis grebnitzkii Czerniavsky 1882 (=*Archaeomysis
maculata* Holmes, 1894; *Callomysis maculata*; *Bowmaniella banneri* Bacescu, 1968). Common off sandy and gravelly beaches
from central California northward.

Columbiaemysis ignota Holmquist, 1982 (=*Acanthomysis
brunnea* Murano and Chess, 1987; see Fukuoka and Murano,
2001). Freshly caught specimens are rich brown in color,
closely resembling their habitat of the brown macroalgae *Laminaria* and *Nereocystis* (Murano and Chess 1987). A coastal
species reported from British Columbia and from Albion Cove,
Mendocino County.

Deltamysis holmquistae Bowman and Orsi, 1992. Collected in
the Sacramento–San Joaquin Estuary; tolerant of low salinity
waters, 1.1‰–2.2‰ (Bowman and Orsi 1992).

Exacanthomysis davisi (Banner, 1948) (=*Acanthomysis davisi*).
From deeper coastal bays and inlets from northern California
to British Columbia.

Hippacanthomysis platypoda Murano and Chess, 1987. Males
of fresh specimens are dark brown, while females tend to be
green; common in the vicinity of coarse sandy bottoms, moving in a manner resembling the swimming behavior of sea
horses. Known from Albion and Mendocino Coves, California.

Holmesimysis costata (Holmes, 1900) (=*Neomysis sculpta* Tattersall, 1933; =*Acanthomysis costata*). Littoral species common
under *Macrocystis* fronds and other kelp species forming
canopies on or near water surface (Turpen et al. 1994); bays and
inlets from central California north.

Neomysis japonica Nakazawa, 1910. This Japanese species was
first collected in San Francisco Bay in 2004 (Petaluma River Boat
Basin, and probably more widely distributed; John Chapman,
personal communication); it appears to occupy a similar habitat as the native *N. mercedis*. Easily separated from *N. mercedis*
by the following characteristics: antennal scale 10 times as long
as broad, distal tip articulated; telson broadly triangular, length
2.5 times width measured proximally at broadest expanse, lateral margins armed with 40 or more uniformly small, regularly
spaced spines. See Tattersall 1951 (illustrations); Nakazawa
1910, Annotationes Zoologicae Japonenses 7: 247–261.

Neomysis kadiakensis Ortmann, 1908. Common in deep-water bays and inlets from San Francisco Bay north to British Columbia.

Neomysis mercedis Holmes, 1897 (=*Neomysis awatschensis* of
Banner; *N. intermedia* of Simmons et al. 1974, Calif. Fish Game
60: 23–25 and 60: 211–212, and Simmons and Knight 1975,

* = Not in key.

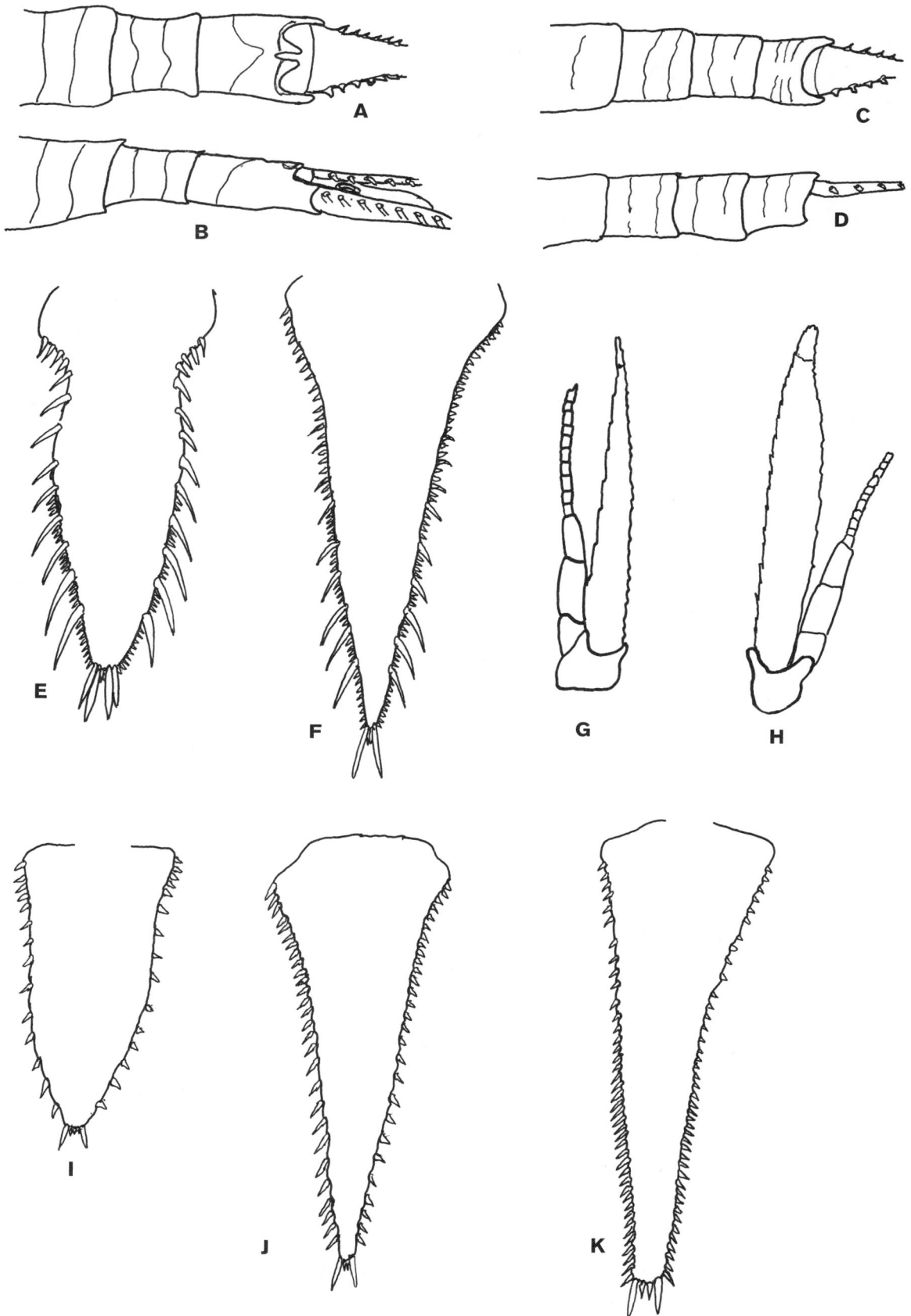

PLATE 222 A, *Holmesimysis costata*, abdominal segments 4–6, dorsal view; B, *Holmesimysis costata*, abdominal segments 4–6, lateral view; C, *Exacanthomysis davisi*, abdominal segments 3–6, dorsal view; D, *Exacanthomysis davisi*, abdominal segments 3–6, lateral view; E, *Holmesimysis costata*, telson; F, *Exacanthomysis davisi*, telson; G, *Neomysis* spp., antenna 2 and scale; H, *Acanthomysis* spp., antenna 2 and scale; I, *Neomysis mercedis*, telson; J, *Neomysis rayii*, telson; K, *Neomysis kadiakensis*, telson (figures are not to scale).

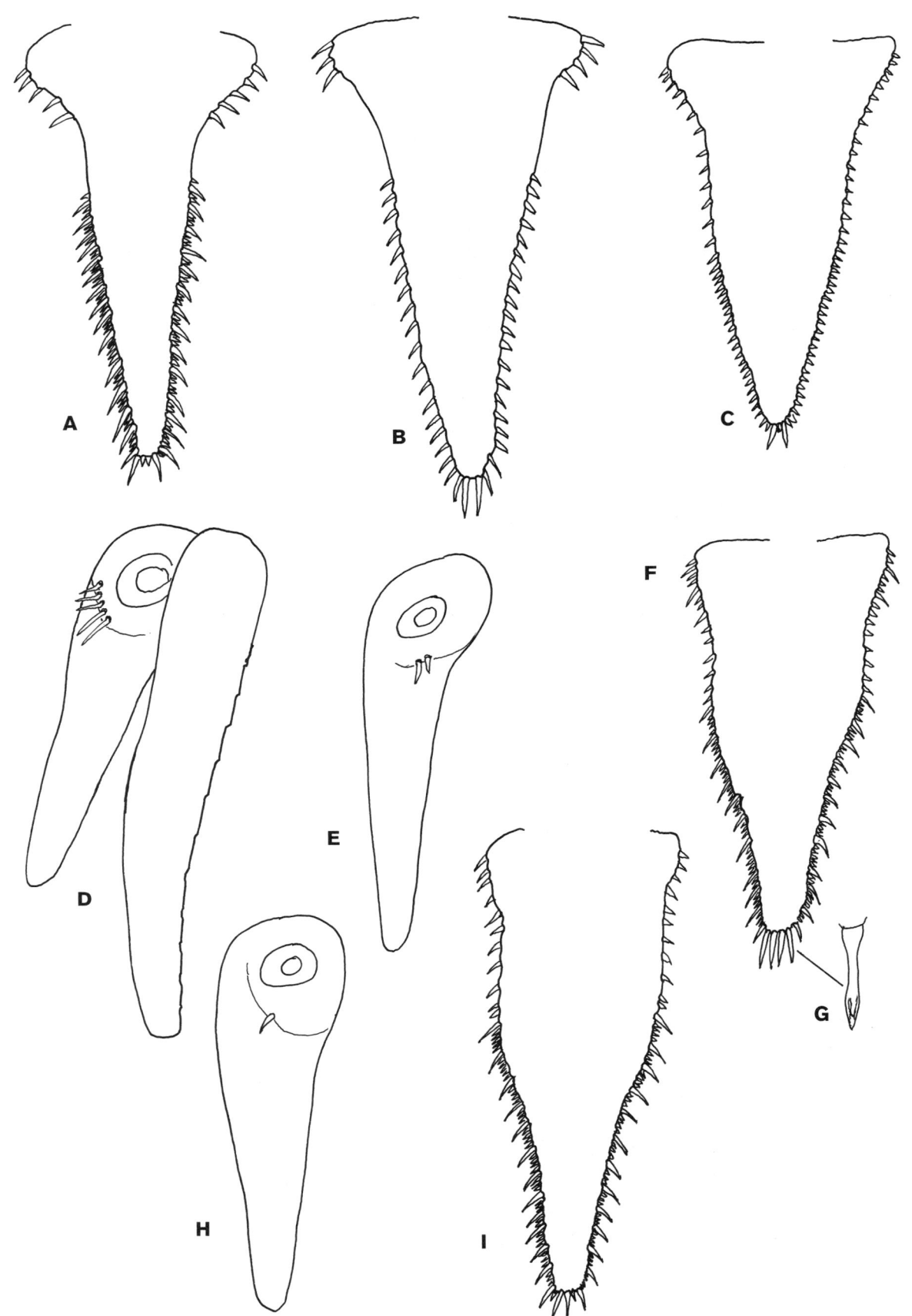

PLATE 223 A, *Acanthomysis californica*, telson; B, *Hyperacanthomysis longirostris*, telson; C, *Columbiaemysis ignota*, telson; D, *Acanthomysis californica*, uropodal exopod and endopod; E, *Hyperacanthomysis longirostris*, uropodal endopod; F, *Acanthomysis aspera*, telson; G, *Acanthomysis aspera*, design of terminal spines on apex of telson; H, *Acanthomysis aspera*, uropodal endopod; I, *Acanthomysis hwanhaiensis*, telson (figures are not to scale).

Comp. Biochem. Physiol. (A) 50: 181–193, the latter on respiration as related to salinity and temperature). Euryhaline; one of the most abundant species inhabiting shallow coastal estuarine bays and inlets, from brackish to fresh water, throughout our entire area. In some of its endemic habitats probably being replaced by exotic mysid species (Bowman and Orsi 1992; Modlin and Orsi 1997). *Neomysis mercedis* is omnivorous, feeding both on phytoplankton and zooplankton, with its diet selective and varying with life stage, season, and prey availability (Siegfried and Kopache 1980, Biol. Bull. 159: 193–205; Murtaugh 1981 Ecology 62: 894–900, 1983, Can. J. Fish. Aquat. Sci. 40: 1968–1974), and on meiobenthic harpacticoid copepods (Johnston and Lasenby 1982, Can. J. Zool. 60: 813–824). In some locations, mature females may harbor the ectoparasitic copepod *Hansenulus trebax* Heron and Damkaer, 1986, in their marsupium (Daly and Damkaer 1986).

Neomysis rayii (Murdoch, 1885) (=*Neomysis franciscorum* Holmes, 1900; *Neomysis toion* Derzhavin, 1913). A more offshore species, also collected in deeper bays and inlets, from San Francisco Bay north across the Pacific rim to northeast Asia.

References

Banner, A. H. 1948. A taxonomic study of the Mysidacea and Euphausiacea (Crustacea) of the northeastern Pacific, Part II. Mysidacea, from tribe Mysini through subfamily Mysidellinae. Transactions of the Royal Canadian Institute 27: 65–124.

Blake, J. A., and P. H. Scott (eds.). 1997. Taxonomic atlas of the benthic fauna of the Santa Maria Basin and western Santa Barbara Channel, Volume 10, The Arthropoda, The Crustacea Part 1. Santa Barbara: Santa Barbara Museum of Natural History, 151 pp.

Bowman, T. E., and J. J. Orsi. 1992. *Deltamysis holmquistae*, a new genus and species of Mysidacea from the Sacramento-San Joaquin Estuary of California (Mysidae: Mysinae: Heteromysini). Proceeding of the Biological Society of Washington 105: 733–742.

Daly, K. L., and D. M. Damkaer. 1986. Population dynamics and distribution of *Neomysis mercedis* and *Alienacanthomysis macropsis* (Crustacea: Mysidacea) in relation to the parasitic copepod *Hansenulus trebax* in the Columbia River Estuary. Journal of Crustacean Biology 6: 840–857.

Daly, K. L., and C. Holmquist. 1986. A key to the Mysidacea of the Pacific Northwest. Can. J. Zool. 64: 1201–1210.

Fukuoka, K., and M. Murano. 2001. *Telacanthomysis*, a new genus, for *Acanthomysis columbiae*, and redescription of *Columbiaemysis ignota* (Crustacea: Mysidacea: Mysidae). Proceedings of the Biological Society of Washington 114: 197–206.

Heron, G. A., and D. M. Damkaer. 1986. A new nicthoid copepod parasitic on mysids from northwestern North America. Journal of Crustacean Biology 6: 652–665.

Holmes, S. J. 1900. California Stalk-eyed Crustacea. Occasional Papers of the California Academy of Sciences VII: 1–262 + 4 plates.

Holmquist, C. 1973. Taxonomy, distribution and ecology of the three species *Neomysis intermedia* (Czerniavsky), *N. awatschensis* (Brandt) and *N. mercedis* Holmes (Crustacea, Mysidacea). Zoologische Jahrbucher, Abteilung fur Systematik, Okologie und Geographie der Tiere 100: 197–222.

Holmquist, C. 1975. A revision of the species *Archaeomysis grebnitzkii* Czerniavsky and *A. maculata* (Holmes) (Crustacea, Mysidacea). Zoologische Jahrbucher, Abteilung fur Systematik, Okologie und Geographie der Tiere 102: 51–71.

Holmquist, C. 1979. *Mysis costata* Holmes, 1900, and its relations (Crustacea, Mysidacea). Zoologische Jahrbucher, Abteilung fur Systematik, Okologie und Geographie der Tiere 106: 471–499.

Holmquist, C. 1980. *Xenacanthomysis*—a new genus for the species known as *Acanthomysis pseudomacropsis* (W. M. Tattersall, 1933) (Crustacea, Mysidacea). Zoologische Jahrbucher, Abteilung fur Systematik, Okologie und Geographie der Tiere 107: 501–510.

Holmquist, C. 1981a. The Genus *Acanthomysis* Czerniavsky, 1882 (Crustacea, Mysidacea). Zoologische Jahrbucher, Abteilung fur Systematik, Okologie und Geographie der Tiere 108: 386–415.

Holmquist, C. 1981b. *Exacanthomysis* gen. nov., another detachment from the genus *Acanthomysis* Czerniavsky (Crustacea, Mysidacea). Zoologische Jahrbucher, Abteilung fur Systematik, Okologie und Geographie der Tiere 108: 247–263.

Ii, N. 1964. Fauna Japonica Mysidacea. Biogeographical Society of Japan, Tokyo, 610 pp.

Kathman, R. D., W. C. Austin, J. C. Saltman, and J. D. Fulton. 1986. Identification manual to the Mysidacea and Euphausiacea of the Northeast Pacific. Canadian Special Publication of Fisheries and Aquatic Sciences 93, Department of Fisheries and Game, Ottawa, Canada, 401 pp.

Mauchline, J. 1980. The Biology of Mysids, Part I. Yonge (eds.), Advances in marine biology, Vol. 18. J. H. S. Blaxter, F. S. Russell, and M. Yonge, eds. New York: Academic Press, pp. 1–369.

Modlin, R. F., and J. J. Orsi. 1997. *Acanthomysis bowmani*, a new species, and *A. aspera* Ii, Mysidacea newly reported from the Sacramento–San Joaquin Estuary, California (Crustacea: Mysidae). Proceeding of the Biological Society of Washington 110: 439–446.

Modlin, R. F., and J. J. Orsi. 2000. Range extension of *Acanthomysis hwanhaiensis* Ii, 1964, to the San Francisco estuary, California, and notes on its description (Crustacea: Mysidacea). Proceedings of the Biological Society of Washington 113: 690–695.

Murano, M., and J. R. Chess. 1987. Four new mysids from California coastal waters. Journal of Crustacean Biology 7: 182–197.

Siegfried, C.A., and M. E. Kopache. 1980. Feeding of *Neomysis mercedis* (Holmes). Biological Bulletin 159: 193–205.

Tattersall, W. M. 1932a. Contributions to a knowledge of the Mysidacea of California, I. On a collection of Mysidae from La Jolla, California. University of California Publications in Zoology 37: 301–314.

Tattersall, W. M. 1932b. Contributions to a knowledge of the Mysidacea of California, II. The Mysidacea collected during the survey of San Francisco Bay by the *U.S.S. Albatross* in 1914. University of California Publications in Zoology 37: 315–347.

Tattersall, W. M. 1951. A review of the Mysidacea of the United States National Museum. United States National Museum Bulletin 201, 292 pp. (and supplement by A. H. Banner, 1954, Proc. U.S. Natl. Mus. 103: 575–583).

Turpen, S., J. W. Hunt, B. S. Anderson, and J. S. Pearse. 1994. Population structure, growth, and fecundity of the kelp forest mysid *Holmesimysis costata* in Monterey Bay, California. Journal of Crustacean Biology 14: 657–664.

Cumacea

LES WATLING

(Plates 224–230)

Cumaceans are small crustaceans, generally ranging in size from 1 mm to 1 cm; however, a few species, such as the Arctic *Diastylis goodsiri*, may be 3 cm or more in length. The Cumacea currently contains more than 1,200 species worldwide. Of these, 49 species are known from the Pacific coast of the United States. Many of the habitats where cumaceans are likely to be found, such as estuaries, shallow embayments, beaches, tidal flats, and the inner continental shelf, have not yet had their cumacean fauna documented. In contrast, the far less diverse cumacean fauna of the northeastern United States is completely known (Watling 1979).

Cumaceans are malacostracans, distinguished by the following combination of features: the carapace covers the first three or four, or rarely six, thoracic somites; the anterior margin of the carapace is extended in front of the head as pseudorostral lobes; the telson may be present, reduced, or incorporated into the last abdominal somite (pleonite); the eyes are united dorsally in all but a very few genera; the second antennae lack an exopod; and pleopods are absent in females (with the exception of one deep-sea species) and often reduced in number or absent in males (plate 224).

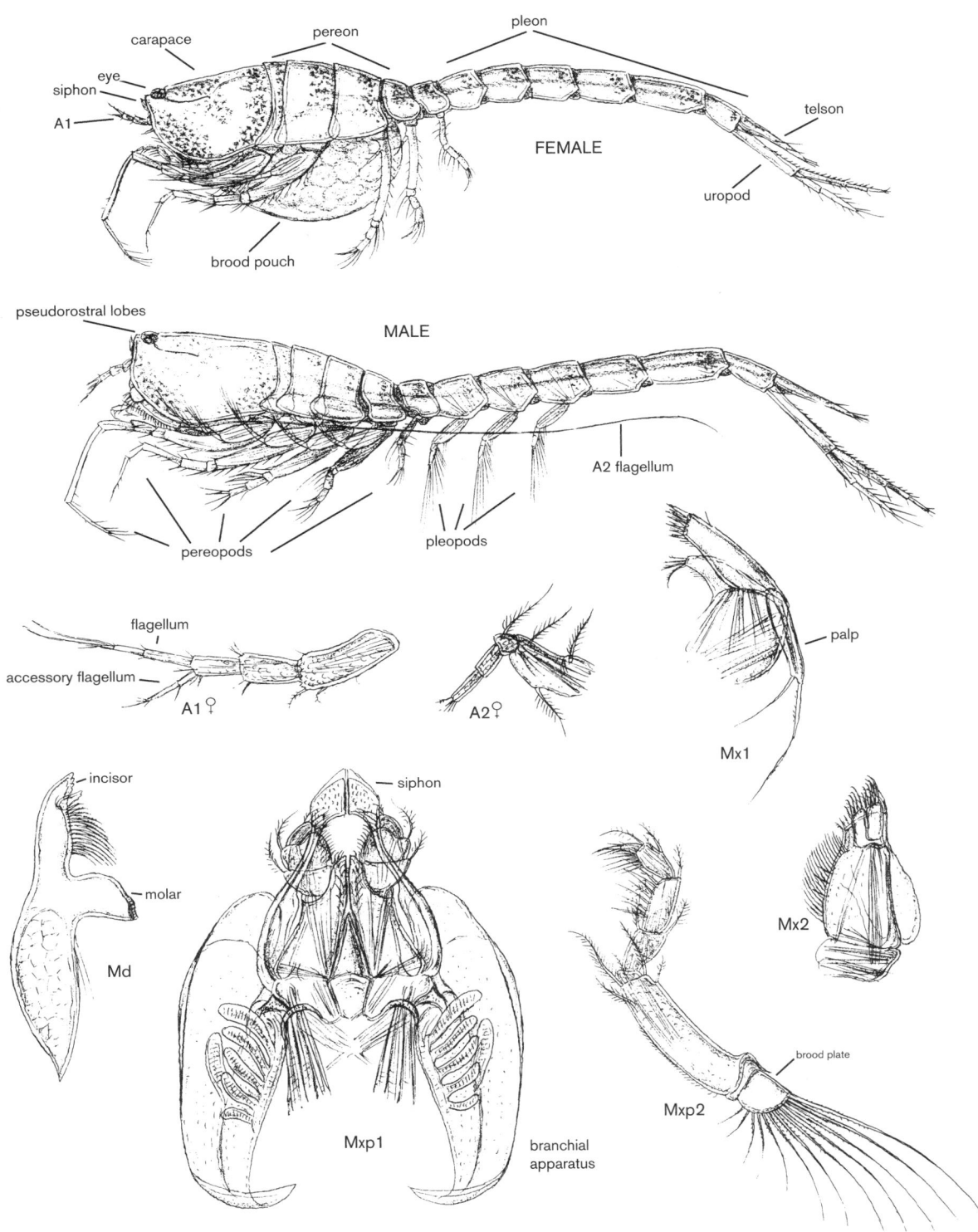

PLATE 224 Basic cumacean morphology: A1, antenna 1; A2, antenna 2; Md, mandible; Mx1, first maxilla; Mx2, second maxilla; Mxp1, first maxilliped (from Sars, 1899–1900).

The cumacean body is externally divided into carapace, thorax, and abdominal regions (plate 224). Externally the body can be divided into three regions, the carapace (which covers the head and first three thoracic somites), the pereon (consisting of the remaining thoracic somites), and the abdomen or pleon. The pereon usually consists of five somites, but fewer may be visible depending on the extent of the carapace. The abdomen always contains six somites, of which pleonite 5 is usually the longest. An articulated telson may or may not be present terminally.

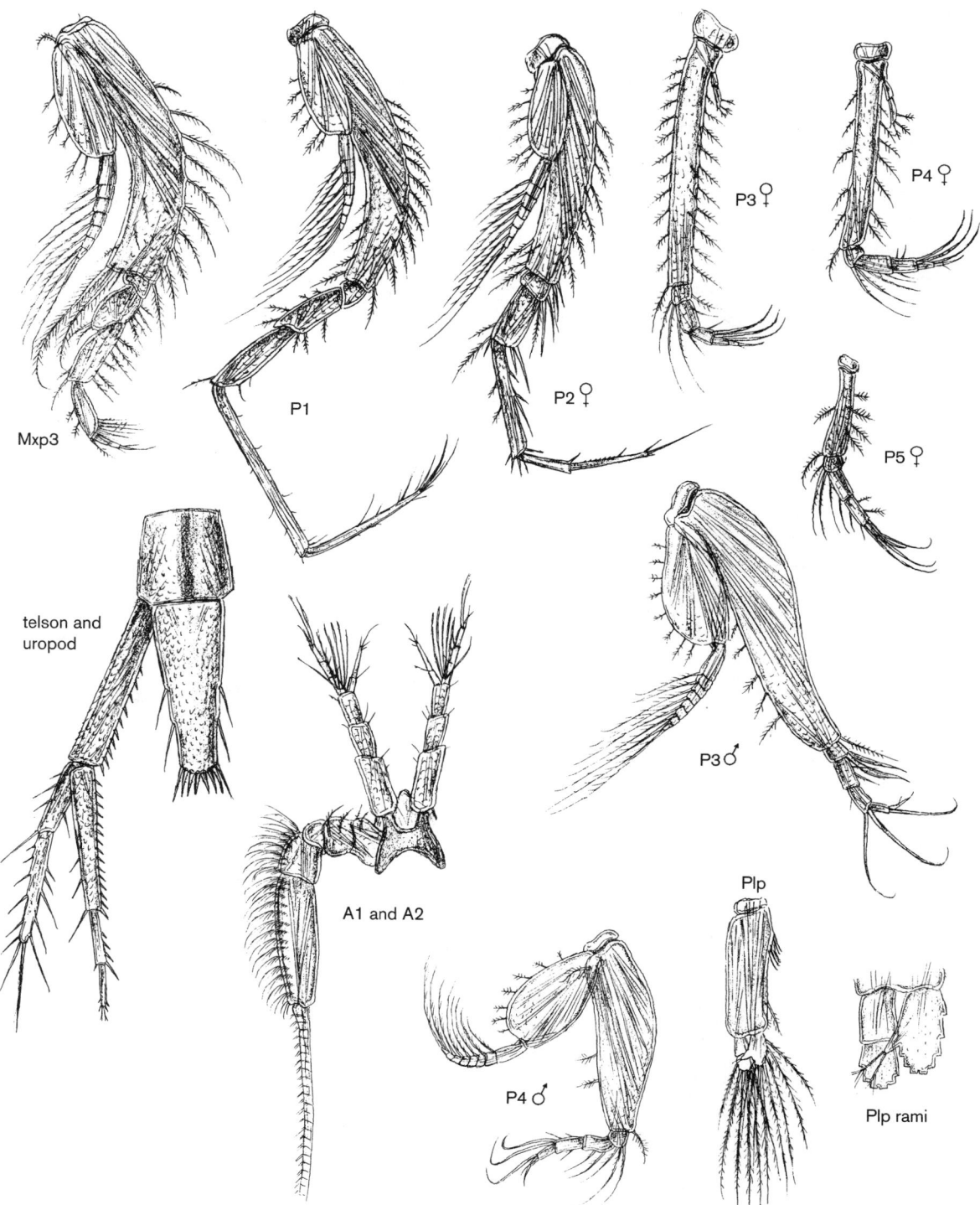

Mxp3

P1

P2 ♀

P3 ♀

P4 ♀

P5 ♀

P3 ♂

telson and uropod

A1 and A2

P4 ♂

Plp

Plp rami

PLATE 225 Basic cumacean morphology (continued): A1, antenna 1; A2, antenna 2; Mxp3, third maxilliped; P1–P5, pereopods 1 through 5; Plp, pleopod (from Sars, 1899–1900).

The carapace is expanded ventrally and laterally to form a branchial chamber. Each side of the carapace is produced anteriorly in the form of pseudorostral lobes that meet, but are not fused, in front of the head, forming a pseudorostrum (plate 224). Reaching to the end of, or projecting beyond, the pseudorostrum are the tips of the branchial epipods of the first maxilliped, which together form the branchial siphon, or exhalant canal, for the respiratory current. The

pseudorostrum may be directed anteriorly at various angles, or be completely reflexed such that the branchial opening is dorsal.

As in all crustaceans the head bears five pairs of appendages, viz., the first and second pairs of antennae, mandibles, and first and second pairs of maxillae.

The first antenna (plate 224, A1) consists of a three-articulate peduncle, the distal-most of which bears two rami, a main

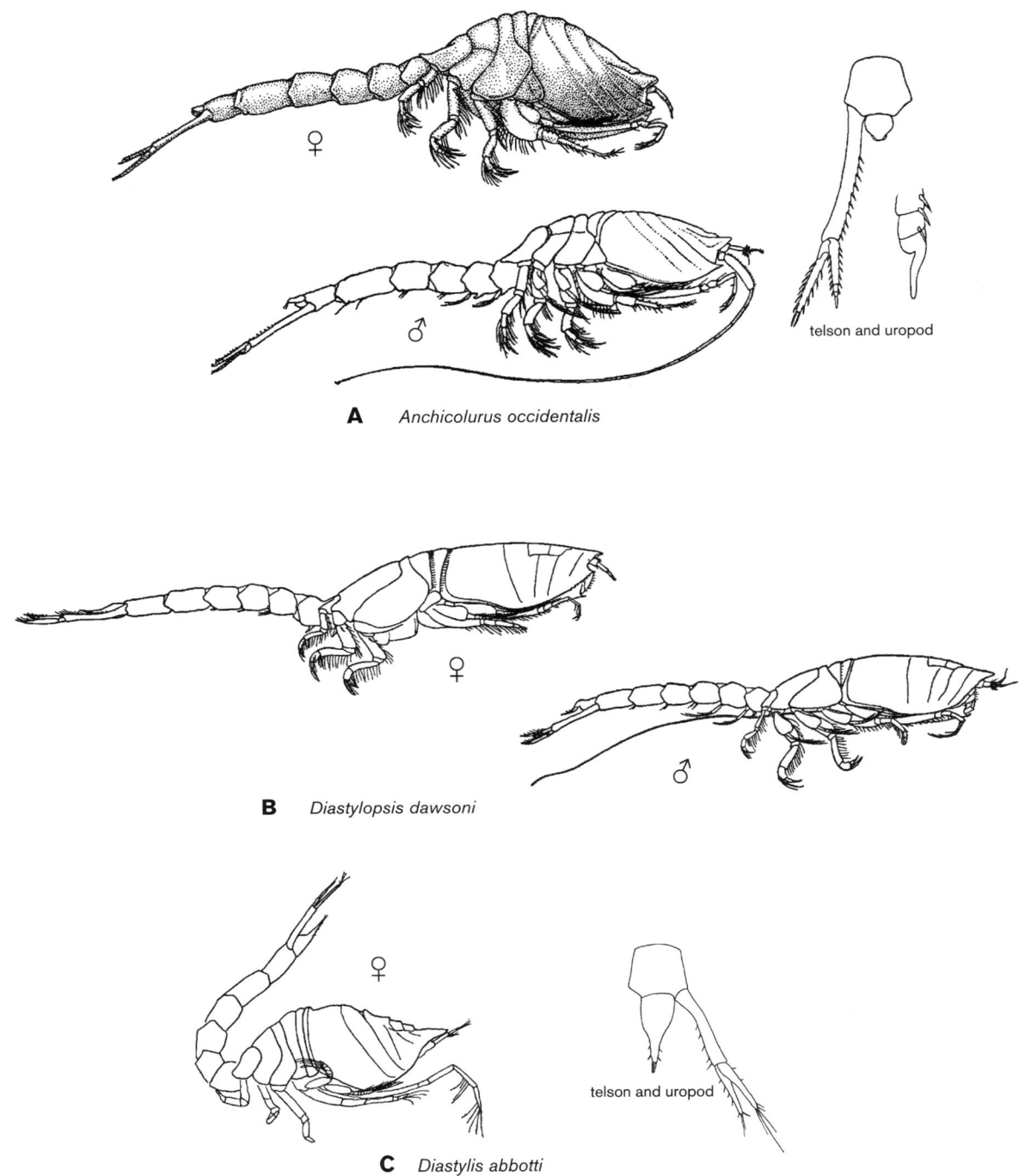

A *Anchicolurus occidentalis*

telson and uropod

B *Diastylopsis dawsoni*

C *Diastylis abbotti*

telson and uropod

PLATE 226 A, *Anchicolurus occidentalis*; B, *Diastylopsis dawsoni*; C, *Diastylis abbotti* (drawings not to scale).

flagellum of two to six articles, and an accessory flagellum (when present) of one to four articles. In several families, the accessory flagellum consists of a single article ranging in size from minute to nearly as long as the main flagellum. The main flagellum is often festooned with long sensory setae, especially in males of the genus *Leptostylis*.

The second antenna (plate 224, A2) is generally rudimentary in the female, whereas in the male it typically consists of a five-articulate peduncle bearing a very long multiarticulate flagellum. Peduncle articles four and five bear a strong brush of sensory setae, and peduncle article five is much longer than ar-

ticle four (plate 225). In the immature male, the setal brush and elongate flagellum are not present; instead the second antenna is shaped like an elongate club.

In cumaceans the mouth appendages (mandible, first and second maxillae, first and second maxillipeds) are well enclosed by the carapace fold. The morphology of these appendages does not need to be known for routine identifications, so they will not be dealt with here. The outermost of the mouth appendages is the third maxilliped. It is also the third thoracic appendage.

The third maxilliped is the most leglike of the three pairs of maxillipeds (plate 225, Mxp3). It consists of an elongate

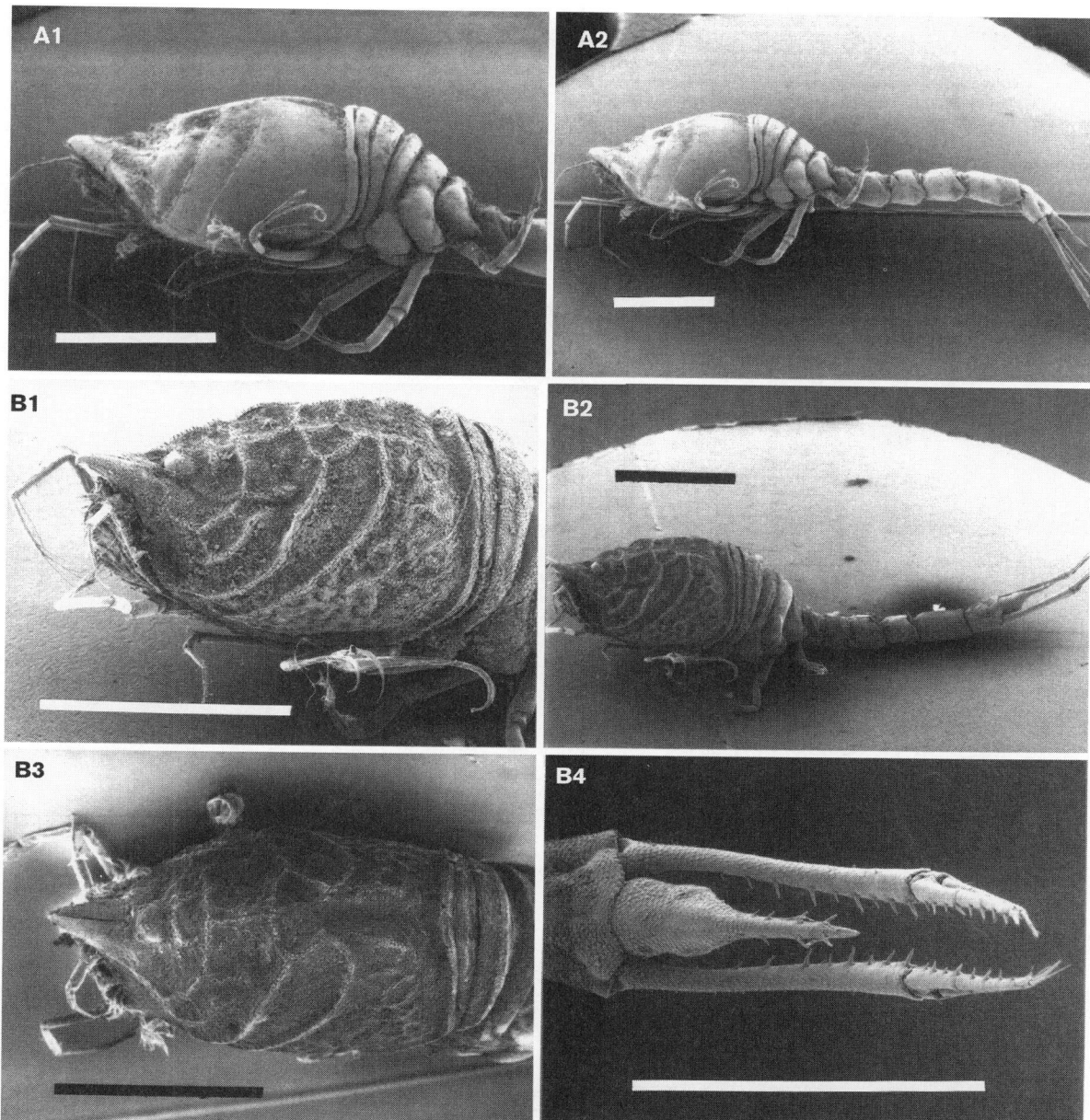

PLATE 227 A, *Diastylis pellucida*; B, *Diastylis santamariensis*; scale bars = 1 mm (photographs from Watling and McCann 1997).

basis and five-articulate endopod and usually possesses an exopod. The shape of this appendage ranges from very elongate to broad and operculate. Whether truly opercular or not, the third maxilliped usually covers the ventral aspect of the mouth field.

Pereopod 1 is the first true ambulatory appendage, but is, in fact the fourth thoracic appendage (plate 225, P1). Its structure is much the same as for the third maxilliped, but the endopod is generally more elongate. An exopod is usually present. The remaining pereopods decrease in length and robustness posteriorly (plate 225, P3, P4, P5); exopods may or may not be present, and if present, may be very small, on pereopods 2–4. An exopod is never present on pereopod 5.

The appendages of abdominal somites 1–5 are known as pleopods and are present (with a single exception in a deep-sea species) only in males (plate 225, Plp). Depending on the fam-ily, there may be one to five pairs of pleopods, or they may be absent altogether.

The last pair of abdominal appendages are the uropods. They consist of a uniarticulate peduncle bearing two rami, the endopod, and exopod (plate 225). The exopod is always two-articulate, but the endopod may consist of one to three articles.

The cumacean body terminates with a telson, on the ventral side of which is located the anus and anal valves. In the Bodotriidae, Leuconidae, and Nannastacidae, the telson is very short and is fused to the sixth pleonite; in these families the telson is said to be absent. Freely articulated telsons of varying length can be found in the other families (plate 225).

Cumaceans can be found in all sedimentary habitats, from mud to sand, but they are usually more diverse in sandy areas. They also occur in a range of salinities and are found at all ocean depths. Along our coast, the best areas for obtaining cumaceans

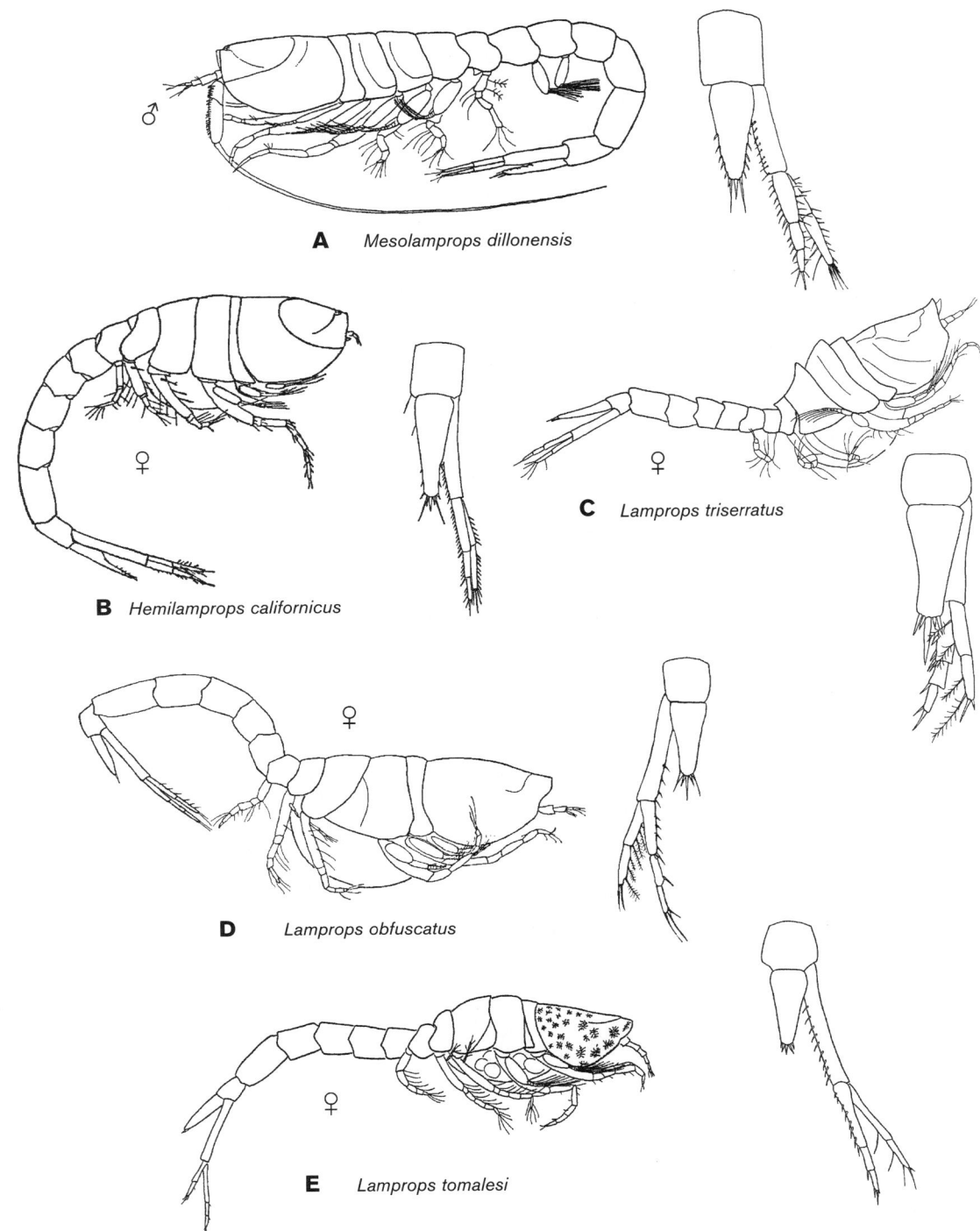

A *Mesolamprops dillonensis*

B *Hemilamprops californicus*

C *Lamprops triserratus*

D *Lamprops obfuscatus*

E *Lamprops tomalesi*

PLATE 228 A, *Mesolamprops dillonensis*; B, *Hemilamprops californicus*; C, *Lamprops triserratus*; D, *Lamprops obfuscatus*; E, *Lamprops tomalesi* (drawings not to scale).

are the shallow offshore sand bars and most sandy to muddy-sand subtidal areas where currents are not too strong. Cumaceans can be collected by many devices, including box cores, grabs, and dredges lined with a fine mesh. On sandy or grassy bottoms, an effective method for obtaining large numbers of cumaceans is to drag a plankton net with a weight attached about a meter in front of the net. As the weight drags along the bottom cumaceans are stirred out of the sediment and caught by the net.

Key to the Families of Cumacea

1. With freely articulated telson 2
— Without freely articulated telson 3
2. Telson with zero or two terminal setae . . . . . . Diastylidae
— Telson with three or more terminal setae . . . .
. . . . Lampropidae
3. Uropod endopod uniarticulate . . . . Nannastacidae
— Uropod endopod two-articulate 4

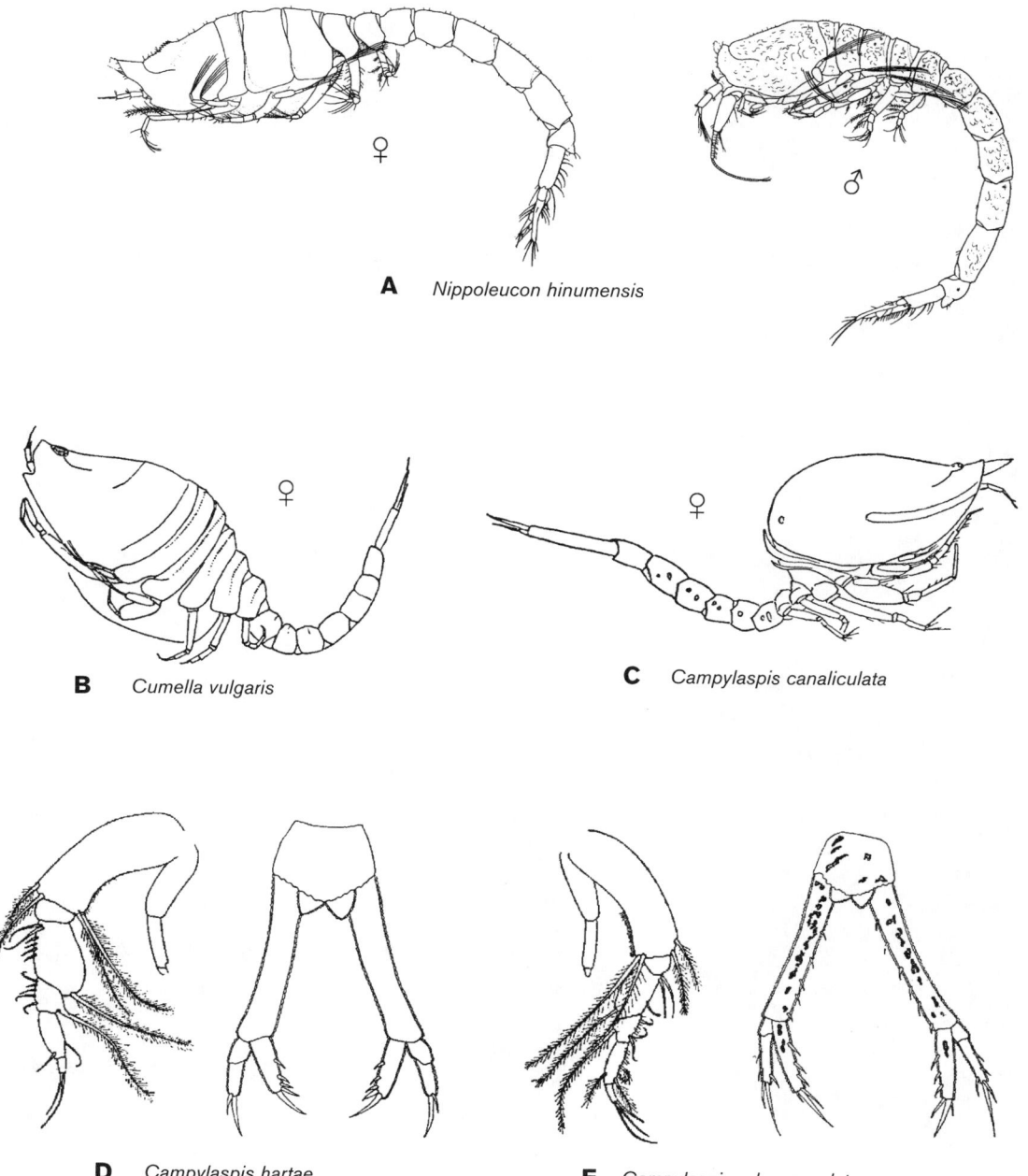

A *Nippoleucon hinumensis*

♀

♂

B *Cumella vulgaris*

♀

C *Campylaspis canaliculata*

♀

D *Campylaspis hartae*

E *Campylaspis rubromaculata*

PLATE 229 A, *Nippoleucon hinumensis*; B, *Cumella vulgaris*; C, *Campylaspis canaliculata*; D, pereopod 2 and uropods of *Campylaspis hartae*; E, *Campylaspis rubromaculata* (drawings not to scale).

4. Male with zero or two pairs of pleopods, females with exopods on pereopods 1–3 . . . . Leuconidae
— Male with five pairs of pleopods; females with exopods on pereopod 1 only . . . . . Bodotriidae, Subfamily Bodotriinae

Key to Species with Free Telson

1. Telson very short, less than one-third the length of uropod peduncles (plate 226A) . . . . *Anchicolurus occidentalis*
— Telson more than half the length of uropod peduncle, or longer than peduncle 2
2. Telson bearing two terminal setae 3

— Telson bearing more than two terminal setae 6
3. Body with pereonite 4 greatly enlarged, especially as seen in dorsal view (plate 226B) . . . . *Diastylopsis dawsoni*
— Body with pereonite 4 no more than twice length of pereonite 3 as seen in dorsal view. 4
4. Uropod exopod slightly longer than peduncle; telson slightly shorter than uropod peduncle (plate 226C) . . . . .
. . . . *Diastylis abbotti*
— Uropod exopod about half the length of peduncle; telson half to three-quarters the length of uropod peduncle 5
5. Carapace with two oblique lines; telson about half the length of uropod peduncle (plate 227A) . . . .
. . . . *Diastylis pellucida*

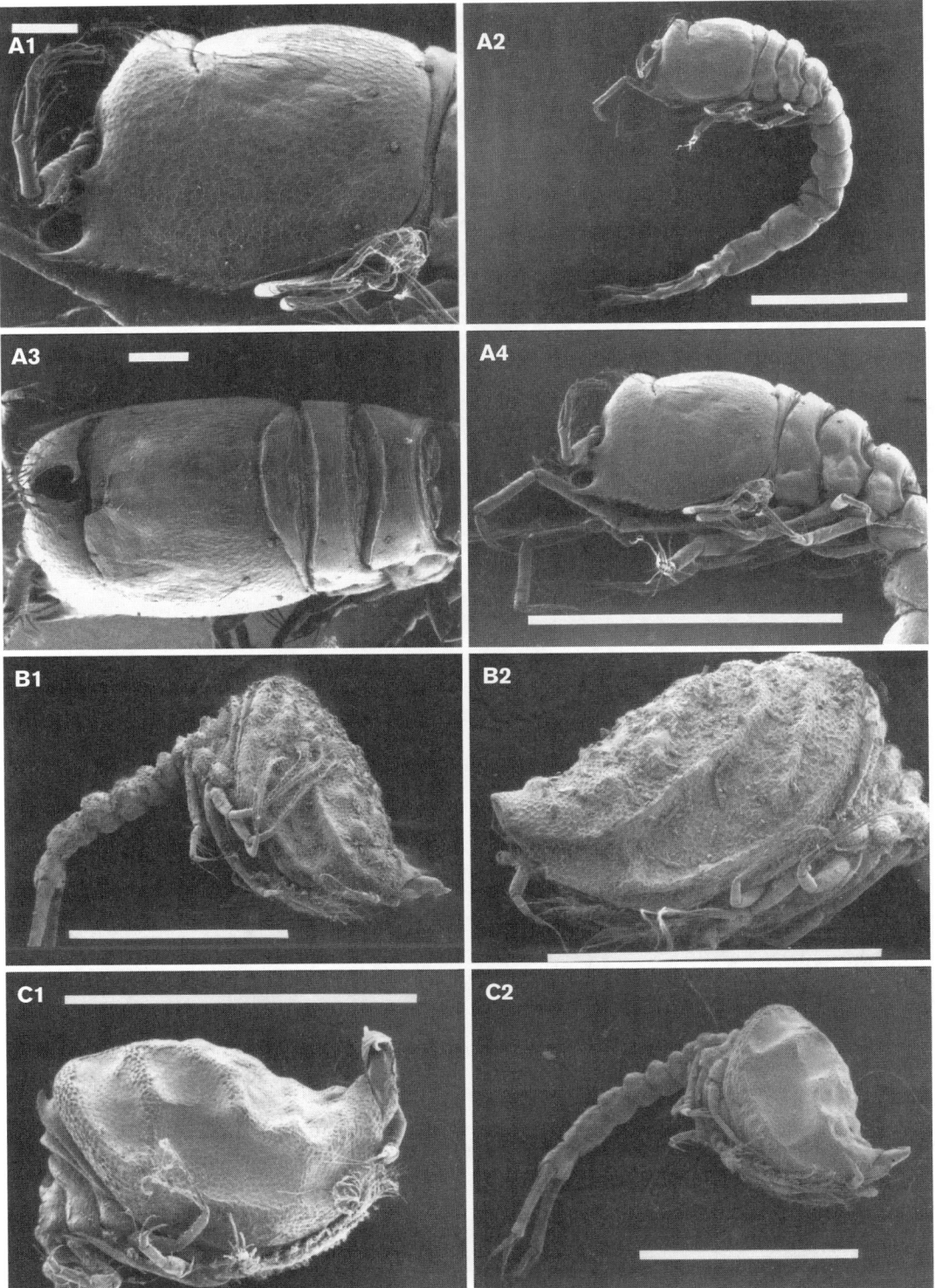

PLATE 230 A, *Eudorella pacifica*; B, *Campylaspis rubromaculata*; C, *Campylaspis hartae*; scale bars = 1 mm (photographs from Watling and McCann 1997).

— Carapace surface very rough and with several oblique lines; telson about three-quarters the length uropod peduncle (plates 227B) . . . . *Diastylis santamariensis*
6. Uropod exopod equal to or longer than endopod (plate 228A). . . . . *Mesolamprops dillonensis*
— Uropod exopod clearly shorter than endopod. 7

7. Carapace with oblique lateral ridges or falcate lateral ridge appearing as faint lines 8
— Carapace without any ridges 9
8. Carapace with falcate lateral ridge (plate 228B). . . . . *Hemilamprops californicus*
— Carapace with oblique lateral ridges (plate 228C) . . . . . .

.... *Lamprops triserratus*

9. Uropod exopod equal in length to proximal two articles of endopod (plate 228D) .... *Lamprops obfuscatus*
— Uropod exopod extends beyond end of second article of endopod (plate 228E) .... *Lamprops tomalesi*

Key to Species with No Free Telson

1. Uropod endopod composed of two distinct articles 2
— Uropod endopod uniarticulate 3
2. First antenna with conspicuous "elbow"; carapace with large tooth at anteroventral corner; first pereonite very narrow (plate 230A) .... *Eudorella pacifica*
— First antenna without elbow; carapace anteroventral corner rounded; first pereonite wide enough to be easily seen in lateral view (plate 229A) ....
.... *Nippoleucon hinumensis*
3. Carapace extended posteriorly, overhanging first few pereonites so that pereonites 1 and 2 are much narrower than pereonites 3–5 4
— Carapace not overhanging pereon, pereonites 1 and 2 same length as pereonites 3–5 (plate 229B) ..... *Cumella vulgaris*
4. Carapace smooth, female with a small groove extending posteriorly from anterior margin (plate 229C) ....
.... *Campylaspis canaliculata*
— Carapace with bumps or large ridges 5
5. Carapace with series of large ridges, no bumps or tubercles (plates 229D, 230C) .... *Campylaspis hartae*
— Carapace with series of tubercles, some organized into shallow ridges; carapace, legs and uropods with many pigment spots (plates 229E, 230B) ..... *Campylaspis rubromaculata*

List of Species

LAMPROPIDAE

Hemilamprops californicus Zimmer, 1936.
Lamprops tomalesi Gladfelter, 1975.
Lamprops obfuscatus (Gladfelter, 1975) (=*Diastylis obfuscata*). This species and *L. triserratus* were assigned by Gladfelter to the genus *Diastylis*, which belongs to a different family.
Lamprops quadriplicata Smith, 1879.
Lamprops triserratus (Gladfelter, 1975) (=*Diastylis triserrata*).
Mesolamprops dillonensis Gladfelter, 1975. This species may be the same as *Hemilamprops californicus*.

DIASTYLIDAE

Anchicolurus occidentalis (Calman, 1912).
Diastylopsis dawsoni Smith, 1880. One of the most common shallow-water species.
Diastylis abbotti Gladfelter, 1975.
Diastylis santamariensis Watling and McCann, 1997.
Diastylis pellucida Hart, 1930.

LEUCONIDAE

Eudorella pacifica Hart, 1930.
Nippoleucon hinumensis (Gamô, 1967) (=*Hemileucon hinumensis*). This species was introduced from Japan in ballast water and

occurs along much of the coast in estuaries and bays; it is particularly common, for example, in San Francisco Bay and Coos Bay.

NANNASTACIDAE

Campylaspis canaliculata Zimmer, 1936.
Campylaspis rubromaculata Lie, 1971 (=*C. nodulosa* Lie, 1969).
Campylaspis hartae Lie, 1969.
Cumella vulgaris Hart, 1930.

References

Calman, W. T. 1912. The Crustacea of the Order Cumacea in the collection of the United States National Museum. Proceedings of the U.S. National Museum 41: 603–676.

Gladfelter, W. B. 1975. Quantitative distribution of shallow-water cumaceans from the vicinity of Dillon Beach, California, with descriptions of five new species. Crustaceana 29: 241–251.

Hart, J. F. L. 1930. Some Cumacea of the Vancouver Island region. Contributions to Canadian Biology and Fisheries 6: 1–8.

Lie, U. 1969. Cumacea from Puget Sound and off the Northwestern coast of Washington, with descriptions of two new species. Crustaceana 17: 19–30.

Lie, U. 1971. Additional Cumacea from Washington, U.S.A., with description of a new species. Crustaceana 21: 33–36.

Sars, G. O. 1899–1900. Cumacea. An Account of the Crustacea of Norway. Volume 3. Christiania, Bergen, Norway.

Watling, L. 1979. Marine fauna and flora of the Northeastern United States: Cumacea. National Marine Fisheries Service, Circular 423, 22 pp.

Watling, L., and L. D. McCann. 1997. Cumacea, pp. 121–180. In Taxonomic atlas of the benthic fauna of the Santa Maria Basin and western Santa Barbara Channel. Volume 11: The Crustacea Part 2—The Isopoda, Cumacea and Tanaidacea. J. A. Blake and P. H. Scott, eds. Santa Barbara: Santa Barbara Museum of Natural History, Santa Barbara, California, 278 pp.

Isopoda

RICHARD C. BRUSCA, VÂNIA R. COELHO, AND STEFANO TAITI

(Plates 231–252)

Isopods are often common and important members of many marine habitats. They can be distinguished from other peracarids, and other crustaceans in general, by the following combination of characteristics:

1. Body usually flattened (except in Anthuridea and Phreatoicidea).
2. Head (cephalon) compact, with unstalked compound eyes, two pairs of antennae (first pair minute in Oniscidea), and mouthparts comprising a pair of mandibles, two pairs of maxillae (maxillules and maxillae), and one pair of maxillipeds.
3. A long thorax of eight thoracomeres, the first (and also the second in Gnathiidea) fused with the head and bearing the maxillipeds, the remaining seven (called pereonites) being free and collectively comprising a body division called the pereon.
4. Seven pairs of uniramous legs (pereopods), all more or less alike (hence, "iso-pod"), except Gnathiidea, which have only five pairs of walking legs.
5. Appendages never chelate (i.e., the subterminal article, or propodus, is not modified into "hand" that works with the terminal article, or dactyl, as a true claw).

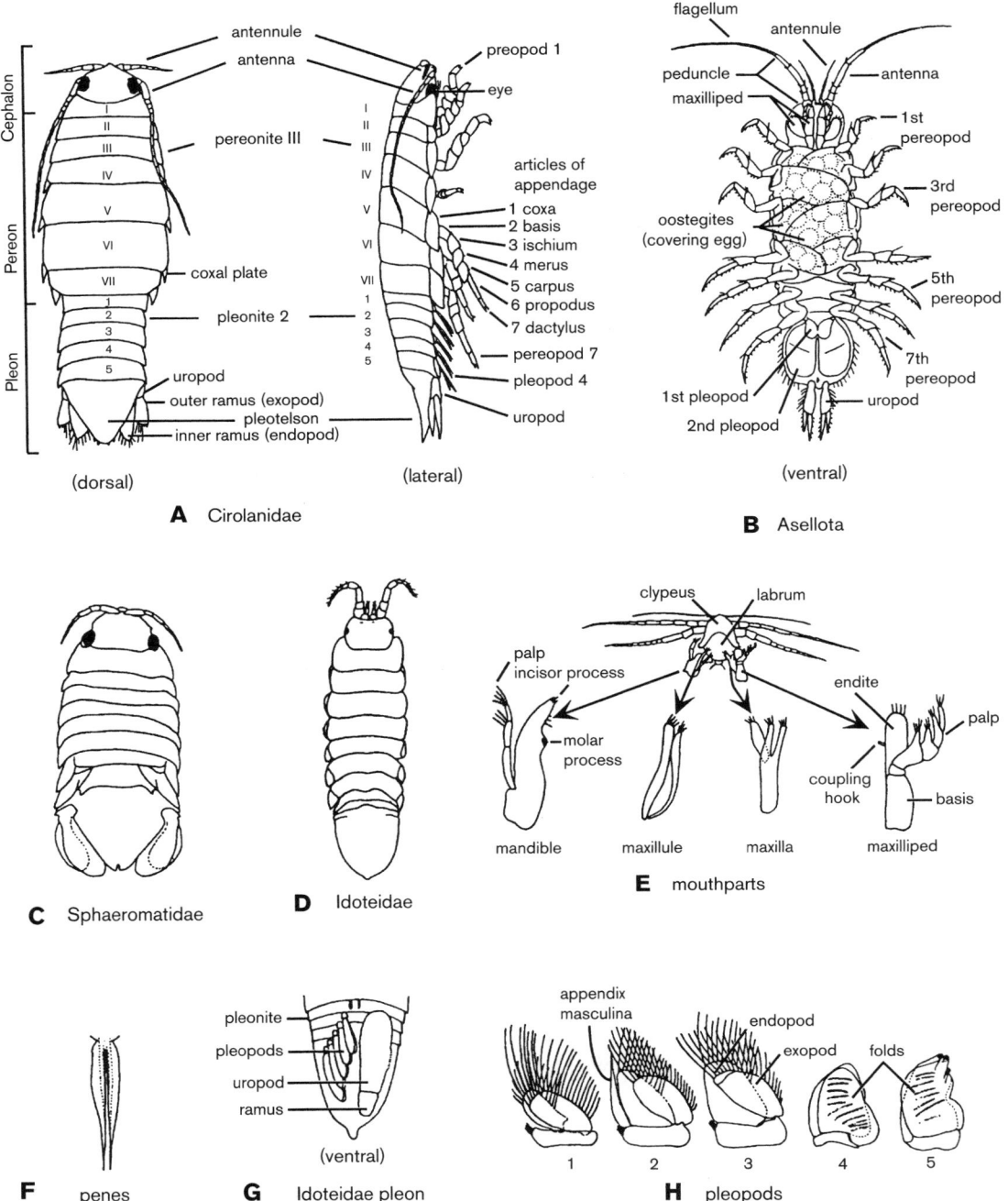

PLATE 231 Isopoda. Isopod anatomy in representative groups: A, Cirolanidae; B, Asellota; C, Sphaeromatidae; D, Idoteidae; E, generalized mouthparts; F, penes; G, pleon of Valvifera (ventral view); H, Generalized pleopods (after Van Name 1936; Menzies and Frankenberg 1966; Menzies and Glynn 1968).

6. A relatively short abdomen (pleon) composed of six somites (pleonites), at least one of which is always fused to the terminal anal plate (telson) to form a pleotelson.

7. Six pairs of biramous pleonal appendages, including five pairs of platelike respiratory/natatory pleopods and a single pair of fanlike or sticklike, uniarticulate (unjointed) uropods.

8. Heart located primarily in the pleon.

9. Biphasic molting (i.e., posterior half of body molts before anterior half).

For general isopod morphology, see plate 231.

All isopods possess one of two fundamental morphologies, being "short-tailed" or "long-tailed" (Brusca and Wilson 1991). In the more primitive, short-tailed isopods, the telsonic region is very small, positioning the anus and uropods terminally or subterminally on the pleotelson (Phreatoicidea, Asellota, Microcerberidea, Oniscidea, Calabozoidea). The more highly derived long-tailed isopods have the telsonic region greatly elongated, thus shifting the anus and uropods to a subterminal position on the pleotelson (Flabellifera, Anthuridea, Gnathiidea, Epicaridea, Valvifera).

Isopods can be sexed in several ways. If oöstegites, or a marsupium, are present, one is obviously examining a female. The openings of the oviducts in females (near the base of the legs on the fifth pereonite) are difficult to observe. If oöstegites are absent, males can be distinguished by the presence of paired penes on the sternum of pereonite 7 (or pleonite 1) or appendices masculinae (sing. appendix masculina) on the endopods of the second pleopods. Absence of penes, appendices masculinae, and oöstegites indicates the specimen is either a nongravid female or a juvenile that has not yet developed secondary sexual features.

Isopods are a large, diverse order with 10 named suborders, all but two (Phreatoicidea and Calabozoidea) of which occur in California and Oregon. They are found in all seas and at all depths, in fresh and brackish waters and on land (the Oniscidea). The approximately 10,000 species are more or less equally split between marine and terrestrial/freshwater environments.

Several general guides to marine isopods of Pacific North and Middle America have been published. These include: Richardson (1905) (still a valuable reference, although obviously out of date), Schultz (1969), Brusca (1980) and Brusca et al. (2004, keys to common Gulf of California species), Brusca and Iverson (1985, the only summary treatment available for the tropical eastern Pacific region), and Wetzer et al. (1997). Kensley and Schotte (1989) is also a useful reference, especially for keys to higher taxa. Key citations to the original literature are provided in this section, and the history of Pacific isopodology (with a complete bibliography) can be found in Wetzer et al. (1997).

The California marine isopod fauna (native and introduced) numbers approximately 200 named species in eight suborders. The keys treat species occurring primarily in the intertidal and supralittoral zones for all of the California and Oregon coasts, plus the commonly encountered fish parasites of the family Cymothoidae.

In the sea, isopods compare in ecological importance to the related Amphipoda and Tanaidacea, notably as intermediate links in food chains. They typically predominate (numerically), along with tanaids, bivalves, and polychaetes, in soft bottom sediment samples from continental shelves. On some tropical coasts, isopods may constitute the majority of prey items consumed by rocky-shore fishes. In the Arctic region, they are one of the primary food items of gray whales. Intertidal isopods are predominantly benthic and cryptic, living under rocks, in crevices, empty shells and worm tubes, and among sessile and sedentary organisms, such as algae, sponges, hydroids, ectoprocts, mussels, urchins, barnacles, and ascidians.

Some isopods burrow in natural substrates including mud, sand, soft rocks, and driftwood, and some burrowers, such as *Limnoria* (the gribbles) and *Sphaeroma,* do extensive damage to pilings and wooden boats. In the tropics, some species of *Sphaeroma* burrow into mangroves, weakening the prop roots and causing them to break more easily, which typically stimulates the growth of multiple new rootlets, leading to the classic stairstep structure of red mangrove prop roots (Perry and Brusca 1989). Several species are important scavengers on shore wrack or dead animals (e.g., *Ligia, Tylos*). Cirolanids, corallanids, and tridentellids are voracious carnivores, functioning both as predators and scavengers. Epicarideans are all parasites on other crustaceans, cymothoids are all parasites on fishes, and aegids are "temporary parasites" (or "micropredators") on fishes. Some invertebrate parasites, notably acanthocephalans, use isopods as intermediate hosts.

Identification of isopods often requires dissection and microscopic examination of appendages and other structures us-ing fine-pointed "jewelers" forceps under a binocular dissecting microscope. Dissected parts may be mounted on microscope slides in glycerin or a more permanent medium for observation under a compound microscope.

Key to the Suborders of Isopoda

1. With five pairs of pereopods (thoracomere 2 entirely fused to cephalon, with its appendages modified as pylopods and functioning as a second pair of maxillipeds; thoracomere 8 reduced, without legs); adult males with mandibles grossly enlarged, forcepslike, projecting in front of head; adult females without mandibles . . . . Gnathiidea
— With seven pairs of pereopods (thoracomere 2 not fused with cephalon, with one pair of maxillipeds and seven pairs of pereopods); males without projecting, forcepslike mandibles; females with mandibles 2
2. Adults obligate parasites on other crustaceans; bilateral symmetry reduced or lost in females; male a small bilaterally symmetrical symbiont living on the body of the female; antennae (antennae 2) vestigial; antennules (antennae 1) reduced to three or fewer articles; without maxillules (maxillae 1) . . . . Epicaridea
— Not obligate parasites on other crustaceans; bilateral symmetry retained in both sexes; male not as above; antennae never vestigial; antennules variable; usually with maxillules 3
3. Body cylindrical or tubular in cross-section, but often appearing laterally compressed (amphipodlike) due to ventrally elongated abdominal pleura; with distinct row of filter setae along medial margin of maxilla (maxilla 2); penes located on coxae of male pereopod 7; apex of pleotelson curves dorsally; pleonite 5 elongate, markedly longer than any other pleonites (known only from the southern hemisphere and India) . . . . Phreatoicidea
— Body variable, but not appearing laterally compressed as above; without row of filter setae along medial margin of maxilla; penes on sternum of male pereonite 7 (or on sternum of pleonite 1); apex of pleotelson does not curve dorsally; pleonite 5 rarely elongate (markedly longer than other pleonites only in Limnoriidae) 4
4. Terrestrial; antennules vestigial, minute; pleon always of five free pleonites, plus the pleotelson . . . . Oniscidea
— Aquatic; antennules normal, or if reduced not minute; pleon variable, with or without fused pleonites 5
5. Anus and articulating base of uropods positioned terminally (or subterminally) on pleotelson; uropods styliform 6
— Anus and articulating base of uropods positioned at base of pleotelson; uropods flattened 8
6. With lateral coxal plates; antenna peduncle 5-articulate; maxillipeds without coupling setae; penes of male arise from articulation between pereonite 7 and pleonite 1; mandible without palp; pleopodal exopods broad and opercular to the thick tumescent endopods; female pleopod 1 present . . . . Calabozoidea
— Without lateral coxal plates (pereopodal coxae small); antenna peduncle 6-articulate; maxillipeds with or without coupling setae; penes of male arise on sternum of pereonite 7; mandible with palp; pleopods not as above; female pleopod 1 absent 7
7. Minute, usually <3 mm long; long and slender, length about six times width; antenna peduncle without a scale;

antennule reduced, peduncle indistinguishable from flagellum; maxilliped without coupling setae on endite; female pleopod 2 biramous; male pleopod 2 endopod not geniculate; interstitial . . . . Microcerberidea
— Rarely minute, usually >4 mm long; body not elongate (length less than six times width); antenna peduncle usually with a scale; antennule rarely reduced, peduncle and flagellum distinct; maxilliped almost always with coupling setae on endite; female pleopod 2 uniramous; male pleopod 2 endopod large and geniculate; rarely interstitial . . . . . .
. . . . Asellota
8. Body elongate, length usually more than six times width; uropodal exopod curving dorsally over pleotelson; coxae of maxillipeds fused to head (i.e., not freely articulating); mandible with lamina dentata in lieu of spine row and lacinia mobilis (lamina dentata, spine row and lacinia mobilis lacking in Paranthuridae); maxillule an elongate stylet with apical hooks or serrate margin; maxilla vestigial and fused with paragnath (or absent) . . . . Anthuridea
— Body not markedly elongate, length usually less than four times width; uropodal exopod not curving over pleotelson; coxae of maxillipeds not fused to head; mandible without lamina dentata; maxillule variable; maxilla well developed, never fused with paragnath 9
9. Uropods modified as a pair of ventral opercula covering the entire pleopodal chamber; males with penes arising on sternum of pleonite 1, or on articulation between pereonite 7 and pleonite 1; mandibular molar process a stout, flattened grinding structure . . . . Valvifera
— Uropods not modified as ventral opercula covering pleopods, but positioned laterally; males with penes arising on sternum of pereonite 7; mandibular molar process usually a thin, bladelike, cutting structure, or absent (flattened only in Sphaeromatidae). . . . . Flabellifera

ANTHURIDEA

Key general references: Menzies 1951; Menzies and Barnard 1959; Negoescu and Wägele 1984; Poore 1984; Kensley and Schotte 1989; Cadien and Brusca 1993; Wetzer and Brusca 1997.

Anthurideans are long, slender, subcylindrical isopods, with a length usually six to 15 times the width. The pereonites are mostly longer than wide (in contrast to most isopods, in which the reverse is true), and the dorsum often bears distinctive ridges, grooves, or chromatophore patterns. Distinct coxal plates are rarely evident. The pleonites are often fused in various combinations, and pleonite 6 usually has its line of fusion with the telson demarcated by a deep dorsal groove. The first antennae are short (except in males of some species), as are the second antennae. The mandibles lack a distinct lacinia mobilis or spine row, instead usually having a dentate lobe or plate (the "lamina dentata"). The outer ramus of the maxillule is a slender stylet with terminal spines; the maxillae are rudimentary. The maxillipeds are more or less fused to the head and lack coupling setae on the endites.

Anthurideans are thought to be primarily carnivores, feeding on small invertebrates. Most inhabit littoral or shallow shelf environments, although some deep benthic (and some freshwater) species are also known. Many are known to be protogynous sequential hermaphrodites, and males have not yet been reported for several species. Fewer than 600 species of anthurideans have been named, but many remain undescribed.

Four families of Anthuridea are currently recognized, distinguished primarily by characters of the mouthparts and pleon: Hyssuridae, Antheluridae, Anthuridae, Paranthuridae—the latter two occur in California waters.

1. Mouthparts styletlike, adapted for piercing and sucking, forming a conelike structure; mandible usually with smooth incisor, no molar process or lamina dentate; pleonites 1–6 usually with distinct sutures . . . .
. . . . Paranthuridae 2
— Mouthparts adapted for cutting and chewing; mandible usually with molar process, lamina dentate and toothed incisor; all or most pleonites usually fused . . . . Anthuridae 4
2. Seven pairs of pereopods; pereonite seven not minute (plate 232D) . . . . Paranthura elegans

Note: Paranthura japonica (plate 252E) is a recently introduced species found in fouling communities in bays and estuaries; see species list.

— Six pairs of pereopods; pereonite seven minute, <20% length of pereonite 6 3
3. Pleon slightly longer than pereonites 6+7 (plate 232B) . . . . Colanthura bruscai
— Pleon slightly shorter than pereonites 6+7 . . . .
. . . . Califanthura squamosissima
4. Maxilliped of four articles (at least three free); no pigmentation pattern on pereonites (plate 232A) . . . .
. . . . Cyathura munda
— Maxilliped of five articles; pereonites 1–6 each with a rectangular outline of pigment, characteristically discontinuous on each segment, and segment 7 with posterior transverse pigmentation (plate 232C) . . . .
. . . . Mesanthura occidentalis

ASELLOTA

Key general references: Richardson 1905; Menzies 1951, 1952; Menzies and Barnard 1959; Kussakin 1988; George and Strömberg 1968; Wilson 1994, 1997; Wilson and Wägele 1994.

Asellotans are easily recognized by the following combination of features: uropods terminal and styliform; pleonites 4–5, and often pleonite 3, fused to pleotelson, creating an enlarged terminal piece; pleonite 1, 2, or 3 forming an operculum over the more posterior pleopods; male pleopods 2 with specialized copulatory apparatus consisting of an enlarged protopod, a geniculate (kneelike) endopod, and typically a well-muscled exopod; pereonites without coxal plates.

The Asellota are one of the most diverse groups of isopods, comprising about 25% of all marine species. They are most successful and diverse in the deep sea. Thirty-eight species of Asellota, in nine families, are known from California waters; 18 (in four families) occur in California's intertidal region.

1. Eyes on lateral, peduncle-like projections; terminal article (dactylus) of pereopods 2–7 with two claws 2
— Eyes (if present) dorsolateral on head, not pedunculate; dactylus of pereopods 2–7 with two or three claws 3
2. Pleotelson somewhat pear-shaped; uropods greatly reduced, barely visible dorsally . . . . Munnidae 4
— Pleotelson broad, shieldlike; uropods short but clearly visible in dorsal view . . . . Santiidae

Note: Only one species of this family, Santia hirsuta (plate 235A) is known from California.

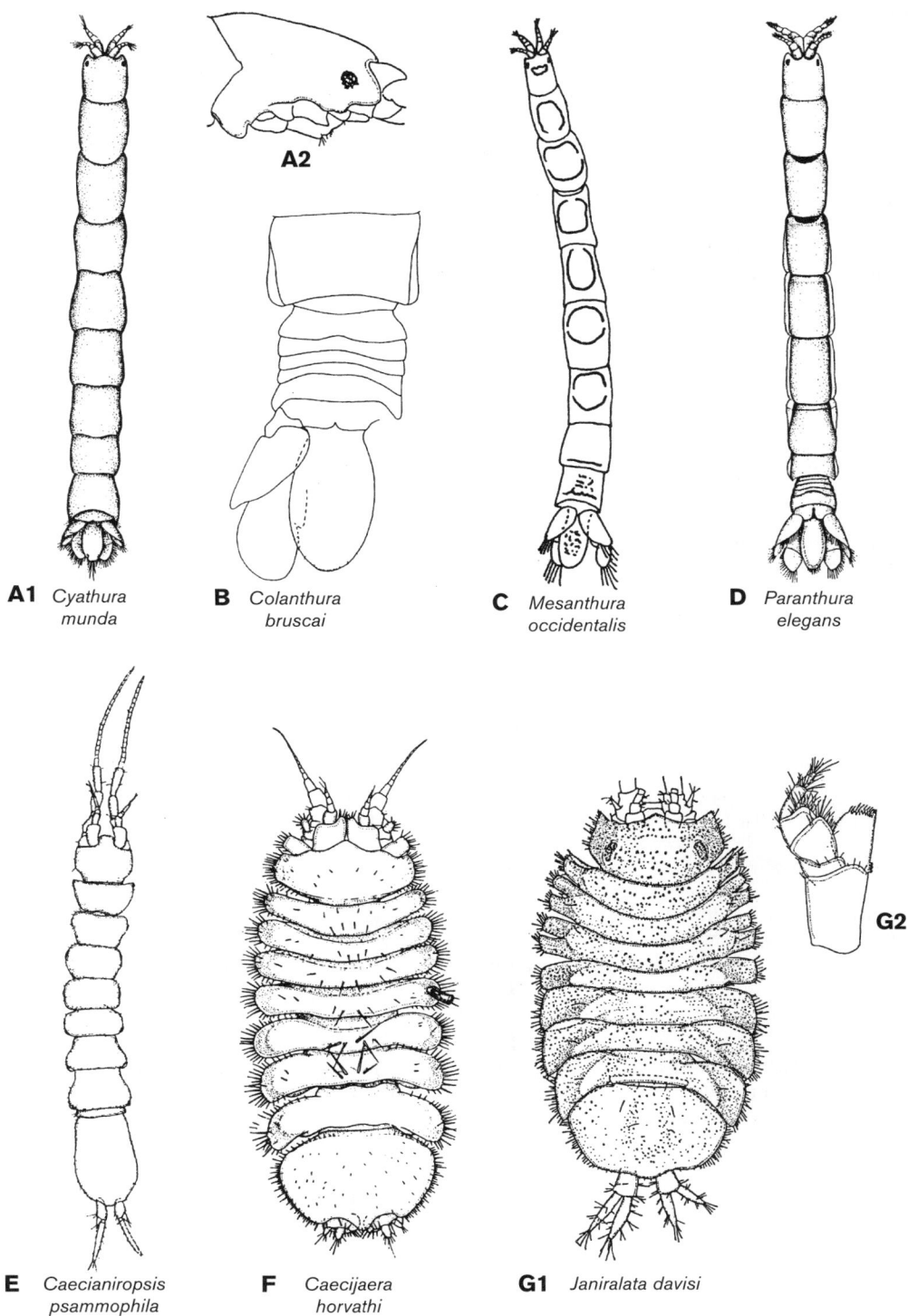

A1 *Cyathura munda*

A2

B *Colanthura bruscai*

C *Mesanthura occidentalis*

D *Paranthura elegans*

E *Caecianiropsis psammophila*

F *Caecijaera horvathi*

G1 *Janiralata davisi*

G2

PLATE 232 Isopoda. Anthuridea: A, *Cyathura munda*, A1, whole animal, A2, detail of the head in lateral view; B, *Colanthura bruscai*, pleon; C, *Mesanthura occidentalis*; D, *Paranthura elegans*; E–G, Asellota: E, *Caecianiropsis psammophila*; F, *Caecijaera horvathi*; G, *Janiralata davisi*, G1, whole animal, G2, maxilliped (after Menzies 1951A; Menzies and Petit 1956; Menzies and Barnard 1959; Poore 1984; Wetzer and Brusca 1997).

3. Both pairs of antennae small, flagella lacking or rudimentary; antenna articles of peduncle dilated; uropods short, inserted in subterminal excavations of pleotelson, not extending much beyond its posterior margin, if at all . . . . Joeropsididae 7
— Antennae long with multiarticulate flagella (caution: often broken off); antenna articles of peduncle not dilated; uropods well developed. . . . . Janiridae 8

Note: *Iais californica* (plate 252A), a tiny commensal species that lives on the underside of its host isopod *Sphaeroma quoianum,* is often common and may become disassociated from its host in samples.

4. Uropods minute, without serrate distal margin; male first pleopods with apices tapering to tip (plate 235E) . . . . *Uromunna ubiquita*
— Uropods not minute, with serrate distal margin; male first pleopods with apices laterally expanded 5
5. Uropods without large acute spinelike protuberances on distal margin; dentate suburopodal shelf visible in dorsal view (plate 235C). . . . . *Munna halei*
— Uropods with large acute spinelike protuberances on distal margin; no dentate suburopodal shelf visible in dorsal view. 6
6. Pleotelson broad (length about 0.8 times width); body stout (length about 1.7 times width) (plate 235D) . . . . *Munna stephenseni*
— Pleotelson narrow (length about 1.6 times width); body relatively elongate (length about 2.6 times width) (plate 235B) . . . . *Munna chromatocephala*
7. Pleotelson with five to seven spines on each lateral border (plate 234D) . . . . *Joeropsis dubia dubia*
— Pleotelson with three spines on each lateral border (plate 234E). *Joeropsis dubia paucispinis*
8. Eyes lacking. 9
— Eyes present. 10
9. Body not elongate, length less than three times width; not a minute interstitial species (plate 232F) . . . . *Caecijaera horvathi*
— Body elongate, length about six times width; minute (<2 mm long) interstitial species (plate 232E). . . . . *Caecianiropsis psammophila*
10. Propodus (next to last article) of first pereopod with conspicuous serrated margin on proximal third of ventral margin; basal three articles of maxillipedal palp as wide as endite. . . . . *Janiralata* 11
— Propodus of first pereopod with proximal third of inferior border smooth; maxillipedal palp with second and third articles much wider than endite . . . . *Ianiropsis* 12
11. Pleotelson with distinct, medially curved, spinelike posterolateral angles (plate 233A) . . . . . *Janiralata occidentalis*
— Pleotelson with posterolateral angles evenly curved, lacking distinct angles or spinelike processes (plate 232G) . . . . *Janiralata davisi*
12. Lateral borders of pleotelson with spinelike serrations. 13
— Lateral borders of pleotelson spineless (fine setae may be present) 15
13. Pleotelson with four to seven spinelike serrations on each side; lateral apices of first male pleopod not directed abruptly posteriorly (plate 233B) . . . . . . *Ianiropsis analoga*
— Pleotelson with two to three spinelike serrations on each side; lateral apices of first male pleopod directed abruptly posteriorly 14

14. Pleotelson with two spinelike serrations on each side (plate 233D). . . . . *Ianiropsis epilittoralis*
— Pleotelson with three spinelike serrations on each side (plate 234C). . . . . *Ianiropsis tridens*
15. Uropods half or less length of pleotelson. 16
— Uropods considerably exceeding half pleotelson length 17
16. Pleotelson with distinct posterolateral angles lateral to uropod insertions (plate 233C) . . . . *Ianiropsis derjugini*
— Pleotelson lacking posterolateral angles lateral to uropod insertions (plate 234A). . . . . *Ianiropsis minuta*
17. Uropods exceeding length of pleotelson; lateral apices of first male pleopod bifurcate (plate 234B). . . . . *Ianiropsis montereyensis*
— Uropods not exceeding pleotelson length; lateral apices of first male pleopod not bifurcate (plate 233E) . . . . *Ianiropsis kincaidi*

EPICARIDEA

Key general references: Richardson 1905; Shiino 1964; Markham 1974, 1977.

Epicarideans are ectoparasites of other crustaceans (malacostracans, ostracodes, copepods, and cirripeds). Females are usually greatly distorted, being little more than an egg sac in some species. Males are symmetrical but minute and live on the body of the female. Eyes are usually present in males, but reduced or absent in females. The antennules (first antennae) are very reduced, usually of only two or three articles; a 3-articulate peduncle is generally apparent only in juvenile stages. The antennae (second antennae) are vestigial in adults. The mouthparts are reduced, forming a suctorial cone with a pair of piercing stylets formed from the mandibles; a mandibular palp is absent. The maxillules and maxillae are reduced or absent.

There are no good references for the Epicaridea as a whole, although Strömberg (1971) reviews the embryology (including that of several California species), and Jay (1989) cites several other papers containing general information. The California fauna is poorly known, both taxonomically and biologically. About 700 species of epicarideans have been described worldwide in 11 families. Three of these families are represented in California waters by 16 species, six of which occur in the intertidal region and are included in the key.

Species in the family Bopyridae retain complete, or nearly complete, body segmentation, and usually have six or seven pereopods on one side but far fewer on the other. The sides of the pleonites are often produced as large lateral plates (epimeres) that resemble pleopods. Adult bopyrids are parasites either on the abdomen or in the branchial chamber of decapod crustaceans. In branchial parasites, the female attaches ventrally to the host's branchiostegite, inducing a bulge in the host's carapace. Males are much smaller and usually found on the ventral side of the pleon of the female isopod. Females brood many small eggs in an oöstegial brood pouch that hatch as a free-swimming epicaridium stage. The epicaridium attaches to an intermediate host, a calanoid copepod. Once on the copepod, the isopod molts into a microniscus stage and then into the cryptoniscus stage. The cryptoniscus detaches from the copepod, is free-swimming, and eventually attaches to the definitive host. All species are probably sequential hermaphrodites. About 500 species have been described worldwide.

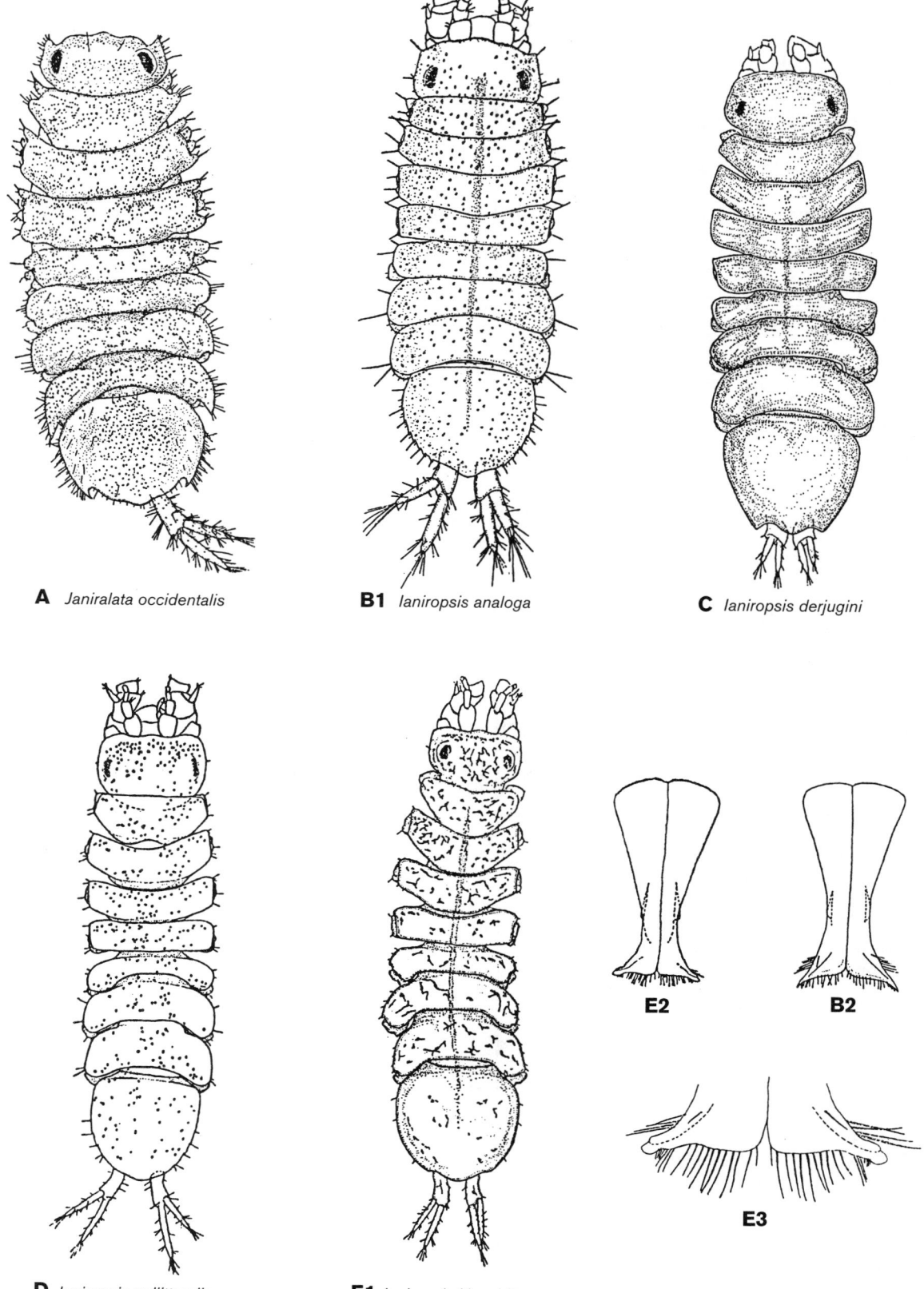

A *Janiralata occidentalis*

B1 *Ianiropsis analoga*

C *Ianiropsis derjugini*

D *Ianiropsis epilittoralis*

E1 *Ianiropsis kincaidi*

E2

B2

E3

PLATE 233 Isopoda. Asellota: A, *Janiralata occidentalis*; B, *Ianiropsis analoga*, B1, whole animal, B2, male first pleopods; C, *Ianiropsis derjugini*; D, *Ianiropsis epilittoralis*; E, *Ianiropsis kincaidi*, E1, whole animal, E2, male first pleopods, E3, detail of the distal part of the male first pleopods (after Menzies 1951A, 1952).

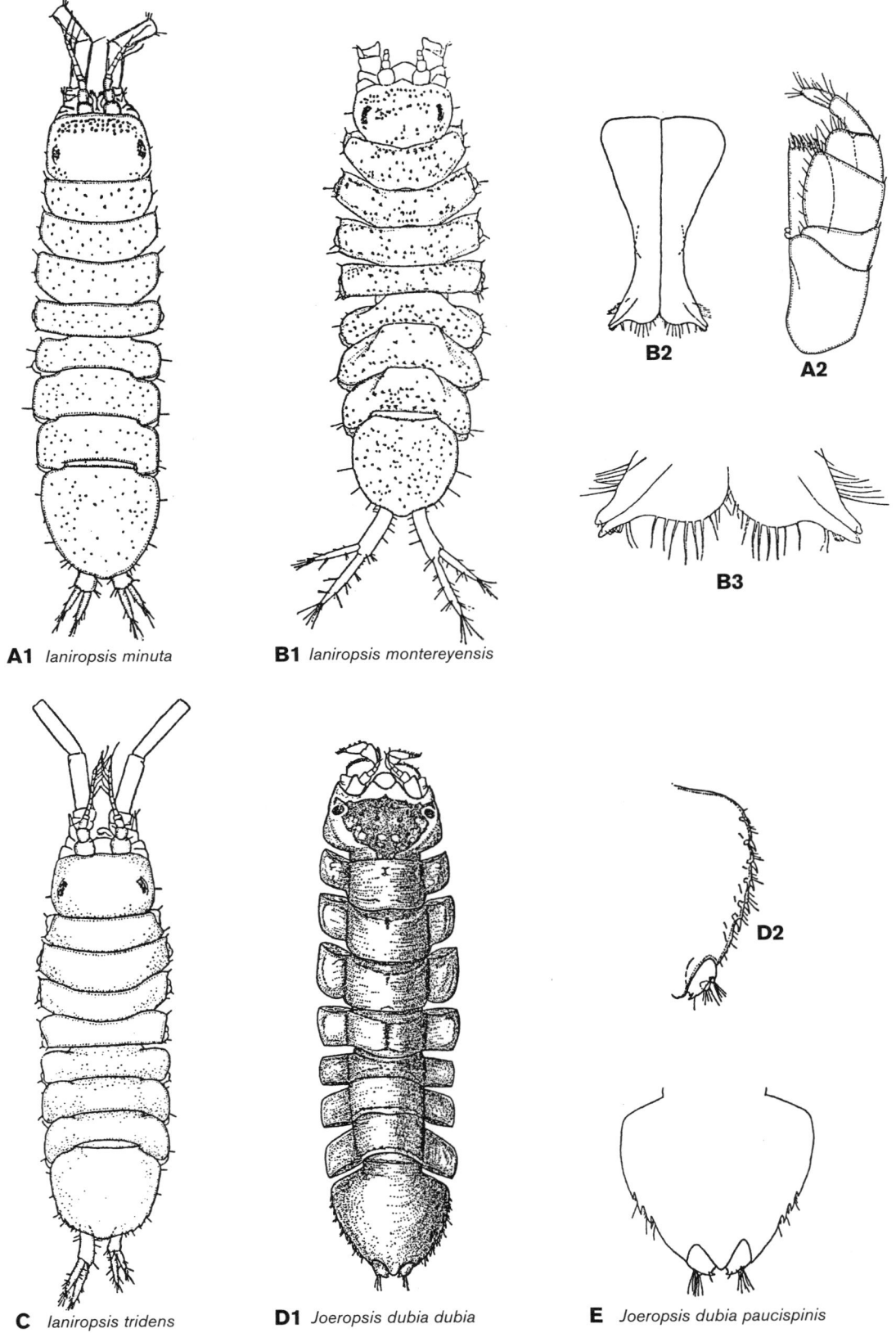

A1 *Ianiropsis minuta*

B1 *Ianiropsis montereyensis*

B2

A2

B3

C *Ianiropsis tridens*

D1 *Joeropsis dubia dubia*

D2

E *Joeropsis dubia paucispinis*

PLATE 234 Isopoda. Asellota: A, *Ianiropsis minuta*, A1, whole animal, A2, maxilliped; B, *Ianiropsis montereyensis*, B1, whole animal, B2, male first pleopods, B3, detail of the distal part of the male first pleopods; C, *Ianiropsis tridens*; D, *Joeropsis dubia dubia*, D1, whole animal, D2, detail of the lateral margin of the pleotelson; E, *Joeropsis dubia paucispinis*, detail of pleotelson (after Menzies 1951A, 1952).

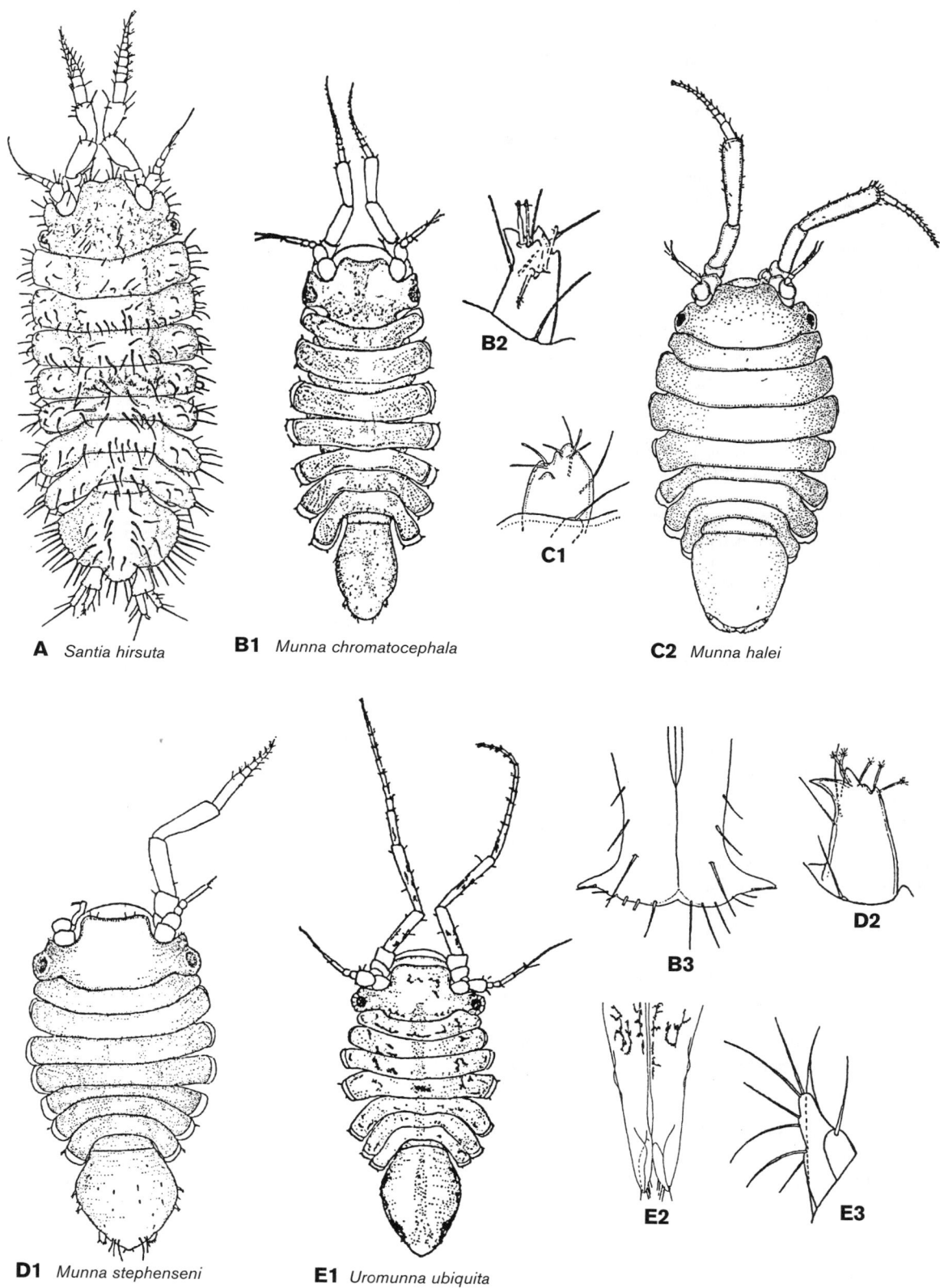

A *Santia hirsuta*

B1 *Munna chromatocephala*

C2 *Munna halei*

D1 *Munna stephenseni*

E1 *Uromunna ubiquita*

PLATE 235 Isopoda. Asellota: A, *Santia hirsuta*; B, *Munna chromatocephala*, B1, whole animal, B2, uropod, B3, male first pleopods; C, *Munna halei*, C1, uropod, C2, whole animal; D, *Munna stephenseni*, D1, whole animal, D2, uropod; E, *Uromunna ubiquita*, E1, whole animal, E2, male first pleopods, E3, uropod (after Menzies 1951A, 1952).

Species in the family Entoniscidae are internal parasites of crabs and shrimps. Females are usually modified beyond recognition, with the marsupium grossly inflated and in some cases extending dorsally over the head. Males and mancas, however, are less distorted with a flattened body, complete segmentation, and pereopods. Mature females are surrounded by a host response sheath, with an external communication to the environment via a small hole or furrow in the carapace of the hosts. Most are parasitic castrators, and in some cases entoniscids can feminize male hosts. Good references on the biology of this family include: Giard (1887), Giard and Bonnier (1887), Veillet (1945), and Reinhard (1956). The following key is based mainly on adult females.

1. Female without segmentation, simply an egg sac; antennae and mouthparts absent. . . . . Hemioniscidae

   Note: One California species, *Hemioniscus balani*, parasitic in barnacles of the genera *Balanus* and *Chthamalus*.

— Female with distinct or weak segmentation; not simply an egg sac; antennae and mouthparts present, although may be greatly reduced 2

2. Body of female without indication of rigid exoskeleton, seemingly undifferentiated, but body divisions and segmentation present; pereonites expanded laterally into thin plates; maxillipeds are the only recognizable mouthparts; pereopods stubby or absent; endoparasites in body cavity of decapod crustaceans . . . . Entoniscidae

   Note: One California species, *Portunion conformis*, plate 237A, in body cavity of the crab *Hemigrapsus* spp.

— Female distinctly segmented; pereonites not expanded laterally into thin plates; mouthparts rudimentary; pereopods prehensile, seven present on one side, but all except first may be absent on the other side; parasites of branchial cavity or on pleopods of decapod crustaceans . . . .
. . . . Bopyridae 3

3. Pleon with lateral plates (epimeres or pleural lamellae) elongate, those of female fringed with long, branched processes, those of male without such digitations; in branchial cavity of ghost shrimps of the genus *Neotrypaea* (plate 236B) . . . . *Ione cornuta*

— Pleon in both sexes with pleural lamellae rudimentary or absent (caution: do not confuse lateral biramous pleopods with pleural lamellae) 4

4. Female pleopods not prominent, relatively short, not noticeable in dorsal view; in branchial chamber of the snapping shrimp *Synalpheus lockingtoni* and *Alpheopsis equidactylus* (plate 236A) . . . . *Bopyriscus calmani*

— Female pleopods prominent, long, visible in dorsal view
. . . . 5

5. Pleopods biramous, with narrow branches arising from a peduncle or stem, extending laterally from narrow pleon; among pleopods of the mud shrimp *Upogebia pugettensis* (plate 236D). . . . . *Phyllodurus abdominalis*

   Note: Compare to the recently recognized *Orthione griffensis* (plate 252B), now abundant along the coast in the mud shrimp *Upogebia* (see species list).

— Pleopods biramous, lanceolate, not arising from a peduncle, extending posteriorly from pleon; in branchial chamber of the pelagic galatheid crab *Pleuroncodes planipes* (plate 236C). . . . . *Munidion pleuroncodis*

FLABELLIFERA

Key general references: Stimpson 1857; Richardson 1899, 1905, 1909; Holmes and Gay 1909; Hatch 1947; Menzies 1962; Menzies and Barnard 1959; Schultz 1969; Brusca 1981, 1989; Bruce et al. 1982; Bruce 1986, 1990, 1993; Harrison and Ellis 1991; Brusca and Wilson 1991; Brusca et al. 1995; Wetzer and Brusca 1997.

Flabellifera comprise a large paraphyletic assemblage of families defined more by the absence of certain features than by any unique attributes. The eyes are usually large and well-developed but are reduced or absent in cave and deep-sea species. The mouthparts are usually robust, adapted for cutting and grinding, or occasionally for piercing. Both the maxillules and maxillae are biramous. The pereopods are usually subsimilar, but in Serolidae, and some Cirolanidae and Sphaeromatidae, the anterior pairs may be subchelate/prehensile. The pleon comprises one to five free segments, plus the pleotelson. The uropods arise laterally, usually forming a distinct tailfan with the pleotelson.

With more than 3,000 described species, the Flabellifera is the second largest isopod suborder, represented in California by seven families, three of which (Anuropidae, Excorallanidae, Serolidae) have not been reported north of Point Conception. Because of the great diversity of this suborder, it is more convenient to key the families first, and then the species in each family.

Key to Families

1. Uropods greatly reduced, with very small, often clawlike exopod; body less than 4 mm long; burrowing in wood or algal holdfasts . . . . Limnoriidae
— Uropods not greatly reduced; body rarely <3 mm long; rarely burrowing in wood or algae (a few species of Sphaeromatidae burrow into coastal wood structures, but they are large animals) 2

2. Pleon composed of three or fewer dorsally visible free pleonites, plus the pleotelson. 3
— Pleon composed of four or five dorsally visible free pleonites, plus the pleotelson. 4

3. Pleon composed of three dorsally visible free (complete) pleonites, plus pleotelson; cephalon fused medially with first pereonite; body strongly depressed and expanded laterally; pereonite 7 tergite incomplete or absent; antennae set very close together; frontal lamina reduced to a small triangular plate visible only by pushing aside antennal bases; pleopods 1–3 small and natatory, basis elongated; pleopods 4–5 large, broadly ovate, suboperculiform . . . . . . Serolidae
— Pleon composed of one or two dorsally visible free (complete) pleonites plus pleotelson; cephalon not fused with first pereonite (except in *Ancinus* and *Bathycopea*); body convex dorsally, not strongly depressed; pereonite 7 tergite complete; antennae not set close together; frontal lamina large and distinct; pleopods subequal, of modest size, basis not elongated; pleopods 4–5 ovate but not operculiform . . . . Sphaeromatidae

4. All pereopods prehensile (dactyli longer than propodi); antennae reduced, without clear distinction between peduncle and flagellum; maxillipedal palp two-articulate. . . . . . .
. . . . Cymothoidae
— At least pereopods 4–7 ambulatory (dactyli not longer than propodi); antennae not as above, with clear distinction between peduncle and flagellum; maxillipedal palp of two to five articles. 5

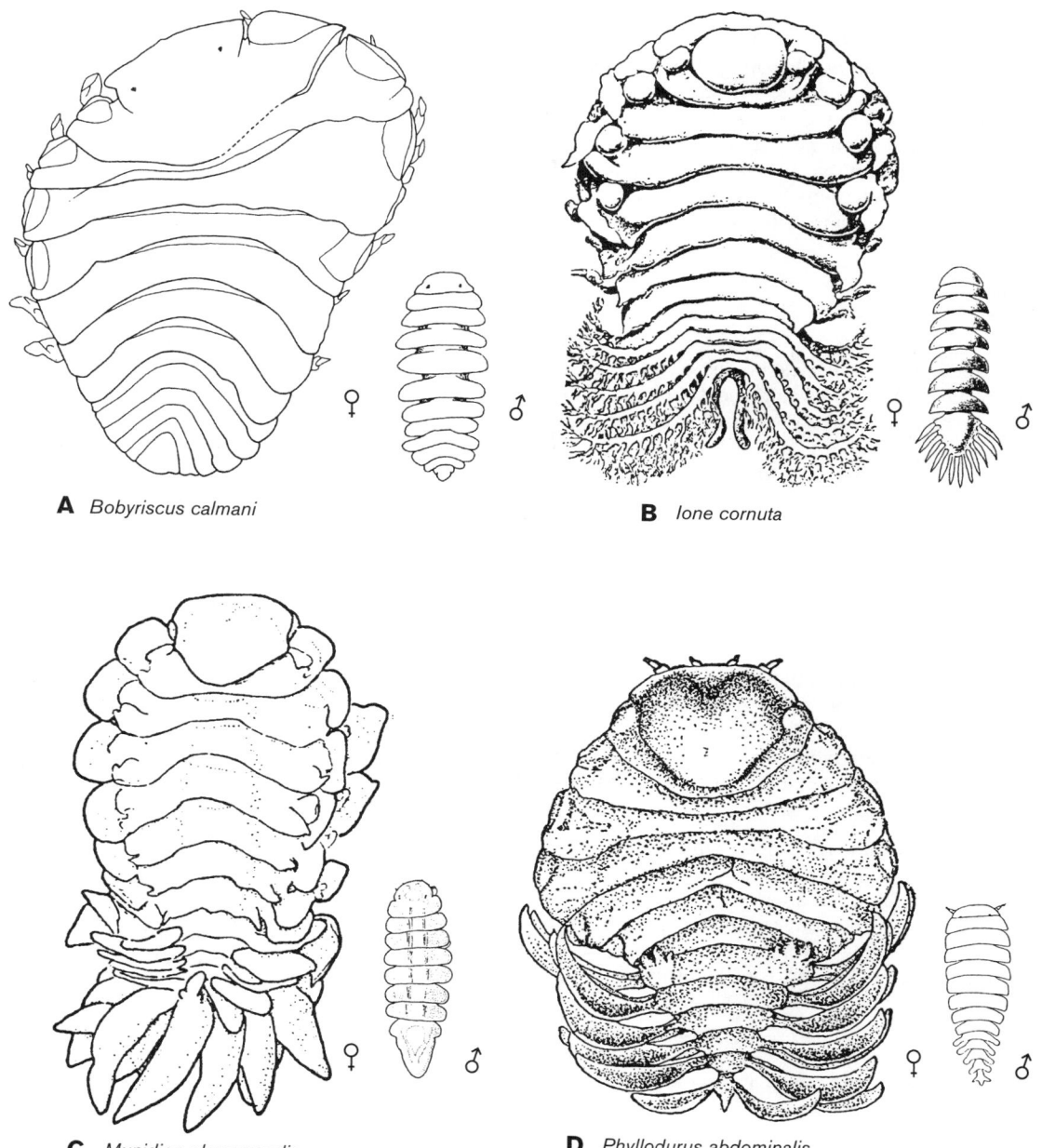

A *Bobyriscus calmani*

B *Ione cornuta*

C *Munidion pleuroncodis*

D *Phyllodurus abdominalis*

PLATE 236 Isopoda. Epicaridea: A, *Bobyriscus calmani*; B, *Ione cornuta*; C, *Munidion pleuroncodis*; D, *Phyllodurus abdominalis* (after Richardson 1905; Markham 1975; Sassaman et al. 1984).

5. Pereopods 1–3 strongly prehensile (dactyli longer than propodi); maxillipeds and maxillules and maxillae with stout, curved, apical setae; lacinia and molar process of mandible reduced or absent; maxilla reduced to a single slender stylet . . . . Aegidae
— Pereopods 1–3 weakly prehensile at best; maxillipeds without stout, curved setae; mandible with or without lacinia and molar process; maxilla not a slender stylet. 6
6. Mandible with distinct lacinia and large bladelike molar process; mandibular incisor generally broad, three-dentate; maxillule lateral (outer) lobe often with several (10–14) stout spines, never styletlike or falcate; maxilla well-developed; pereopods 1–3 not prehensile (dactyli not longer than propodi) . . . . Cirolanidae

— Mandible with lacinia and molar process greatly reduced, vestigial, or absent; mandibular incisor narrow; maxillule lateral (outer) lobe simple and falcate; maxilla reduced; pereopods 1–3 weakly prehensile or ambulatory. . . . .
. . . . Corallanidae

AEGIDAE

Aegids are cirolanidlike, with the smooth dorsal surface either vaulted or flattened. The maxillipedal palp is of two, three, or five articles, the terminal ones with stout acute setae ("spines"). The mandible is elongate, with a narrow incisor and reduced or vestigial molar process. Coxal plates of pereonites 2–6 are large and distinct. Pereopods 1–3 are prehensile (i.e., the dactyli are as long or longer than the propodi and strongly curved); pereopods 4–7 are ambulatory. The family Aegidae comprises six genera. All are temporary parasites on marine fishes. Adults engorge themselves with food (presumably blood) from their hosts, then dislodge and sit on the bottom to digest their meal. Nine species, in two genera, have been reported from Pacific North America, six of which inhabit California waters. However, only a single species occurs in the intertidal zone, *Rocinela signata* (plate 237B).

CIROLANIDAE

Cirolanids have sleek symmetrical bodies, two to 6.5 times longer than wide, with well-developed coxal plates on pereonites 2–7. The mandible has a broad tridentate incisor and a spinose bladelike molar process. The maxillipedal palp typically is five-articulate, and the articles never have hooked or curved setae or spines. All pereopods are ambulatory, although legs 1–3 tend towards a grasping form, with well-developed dactyli. The uropods form a tail fan with the pleotelson.

Cirolanids are all carnivores, either predatory or scavenging. A number of species are known to attack sick or weakened fish, or fish trapped in fishing nets, and some are capable of stripping a fish to the bones in a matter of hours. Stepien and Brusca (1985, cited in the species list at *Cirolana diminuta*) review this phenomenon and describe the behavior from Catalina Island. This large family includes 55 genera. Eight species (in six genera) are known from California waters, six of which occur intertidally.

1. Antennule peduncle article 1 longer than articles 2 or 3; antennule article 2 arising at right angle to article 1; maxilliped endite barely reaching (or extending barely beyond) first palp article; maxilliped endite without coupling setae; antennae long, extending beyond pereonite 7; lateral margins of pleonite 5 not encompassed by pleonite 4 . . . . *Eurydice*

   Note: One species in California: E. caudata, plate 238B.

— Antennule peduncle article 2 or 3 longest; antennule article 2 not arising at right angle to article 1; maxilliped endite extending well beyond first palp article, usually to distal margin of second palp article; maxilliped endite with coupling setae; antennae length variable; lateral margins of pleonite 5 variable 2
2. Antennule peduncle article 2 or 3 longest; clypeus projecting ventrally 3
— Antennule peduncle article 3 always longer than 1 or 2; clypeus short, broad, flat, and sessile, not projecting ventrally . . . . *Cirolana* 4
3. Prominent rostral process, apically spatulate, separating antennules . . . . *Excirolana* 5
— Without prominent rostral process . . . . *Eurylana*

   Note: One species, E. arcuata, plate 238A.

4. Uropodal rami without apical notch; rostrum meets but does not overlap frontal lamina; antennule peduncle articles 1 and 2 not fused (plate 237D) . . . . *Cirolana harfordi*
— Both uropodal rami with apical notch; rostrum overlaps frontal lamina; antennule peduncle articles 1 and 2 fused (plate 237C) . . . . *Cirolana diminuta*

5. Pleotelson broadly rounded and crenulate posteriorly; antennule peduncle with articles 2 and 3 subequal in length (plate 238C) . . . . *Excirolana linguifrons*
— Pleotelson obtusely rounded and acuminate posteriorly; antennule peduncle with article 3 longer than article 2 (plate 238D) . . . . *Excirolana chiltoni*

CORALLANIDAE

Corallanids resemble cirolanids but are even more highly modified as predators. Characteristic features of the family include very large eyes, absence of an endite on the maxilliped, large falcate apical setae on the lateral lobes of the maxillules (often tended by subapical accessory setae), vestigial uniramous maxillae, and frequently a heavily ornamented dorsum beset with setae, spines, tubercles or carinae (especially in males). There are always five free pleonites. The first three pairs of pereopods are often grasping (dactylus as long or longer than the propodus).

Corallanidae is a small group, with six genera and about 70 species. The family is largely confined to tropical and subtropical shallow-water marine habitats, although some brackish and freshwater species are known. Many species are common on coral reefs (hence the name). Because they are often found attached to large prey, such as fishes, rays, turtles, or shrimps, they are sometimes called parasites, but they are actually predators. Two species in the large New World genus *Excorallana* occur in our intertidal region. Both can be collected using night lights over rocky bottoms. *E. tricornis occidentalis*, at least in Costa Rican waters, has nocturnal mass-migrations into the water column, perhaps preying on other microcrustaceans (Guzman et al. 1988, Bull. Mar. Sci. 43: 77–87). Two species belonging to the closely related family Tridentellidae are easily mistaken for corallanids; *Tridentella glutacantha* and *T. quinicornis* both occur in shallow subtidal rocky regions of California's offshore islands (see Delaney and Brusca 1985, J. Crust. Biol. 5: 728–742; Wetzer and Brusca 1997).

1. Head of male ornamented with three tubercles; pereonites 4–7 with row of tubercles on posterior margin; pleotelson not densely covered by setae (plate 238E) . . . . *Excorallana tricornis occidentalis*
— Head of male not ornamented with tubercles; pereonites 4–7 without row of tubercles on posterior margin; pleotelson densely covered by setae (plate 238F) . . . . *Excorallana truncata*

CYMOTHOIDAE

Cymothoids resemble cirolanids and corallanids but are modified for a parasitic lifestyle—all are fish parasites. The definitive features of the family are that all seven pairs of pereopods are prehensile (with long, strongly recurved dactyli as long as or longer than the propodi) and the maxillipedal endite lacks coupling setae. Overall, the mouth appendages are highly modified for the parasitic lifestyle. The maxillipeds are reduced to small palps of two or three articles, the maxillules are modified as slender uniarticulate stylets lying adjacent to one another to facilitate transfer of the host's blood to the mouth, and the maxillae are reduced to small bilobed appendages. All of these mouth appendages bear stout, curved, terminal, or subterminal spinelike setae that serve to hold the buccal region strongly affixed to the flesh of the host fish. All cymothoid species are probably

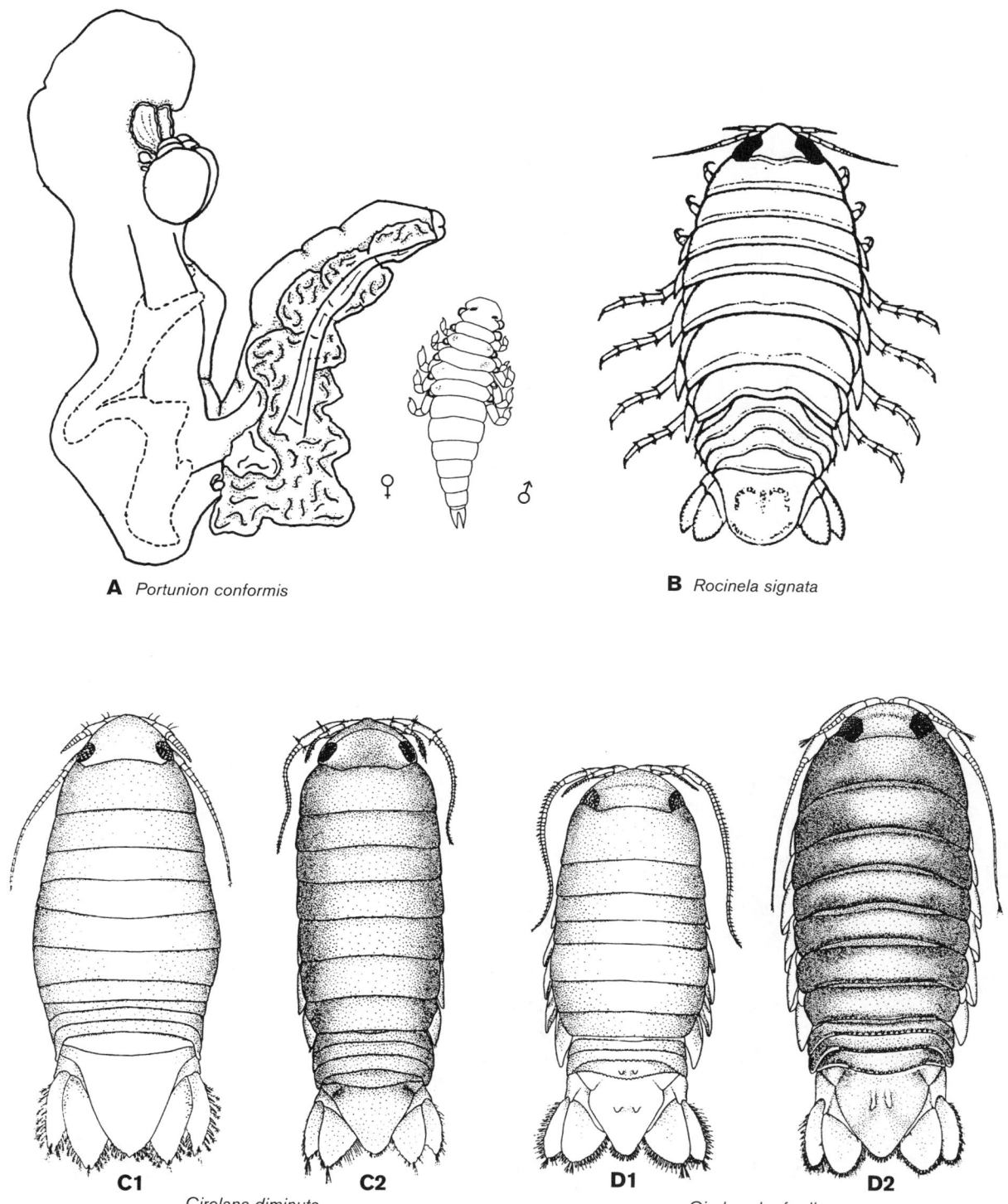

A **Portunion conformis**

B **Rocinela signata**

C1 C2
Cirolana diminuta

D1 D2
Cirolana harfordi

PLATE 237 Isopoda. Epicaridea: A, *Portunion conformis*; B-D, Flabellifera: B, *Rocinela signata*; C, *Cirolana diminuta*, C1, postmanca, C2, adult male; D, *Cirolana harfordi*, D1, male, D2, a different male morphotype (after Richardson 1905; after Schiödte and Meinert; Muscatine 1956; Brusca et al. 1995).

protandric hermaphrodites, first maturing into males and later transforming into females (unless retained in the male stage by the presence of a female already in place on the host fish).

Cymothoids are parasites on marine or freshwater fishes, and they are commonly found on sport and commercial fishes, such as mullet, jacks, groupers, flounder, perch, anchovies, and many others. Although they are not intertidal species, they are often seen by sport fishers and researchers, Most species attach either epidermally, in the gill chamber, or in the buccal region. However, species in some genera actually burrow beneath the skin where they live in a pocket or capsule formed within the musculature of the host (e.g., *Artystone, Riggia, Ichthyoxenus,*

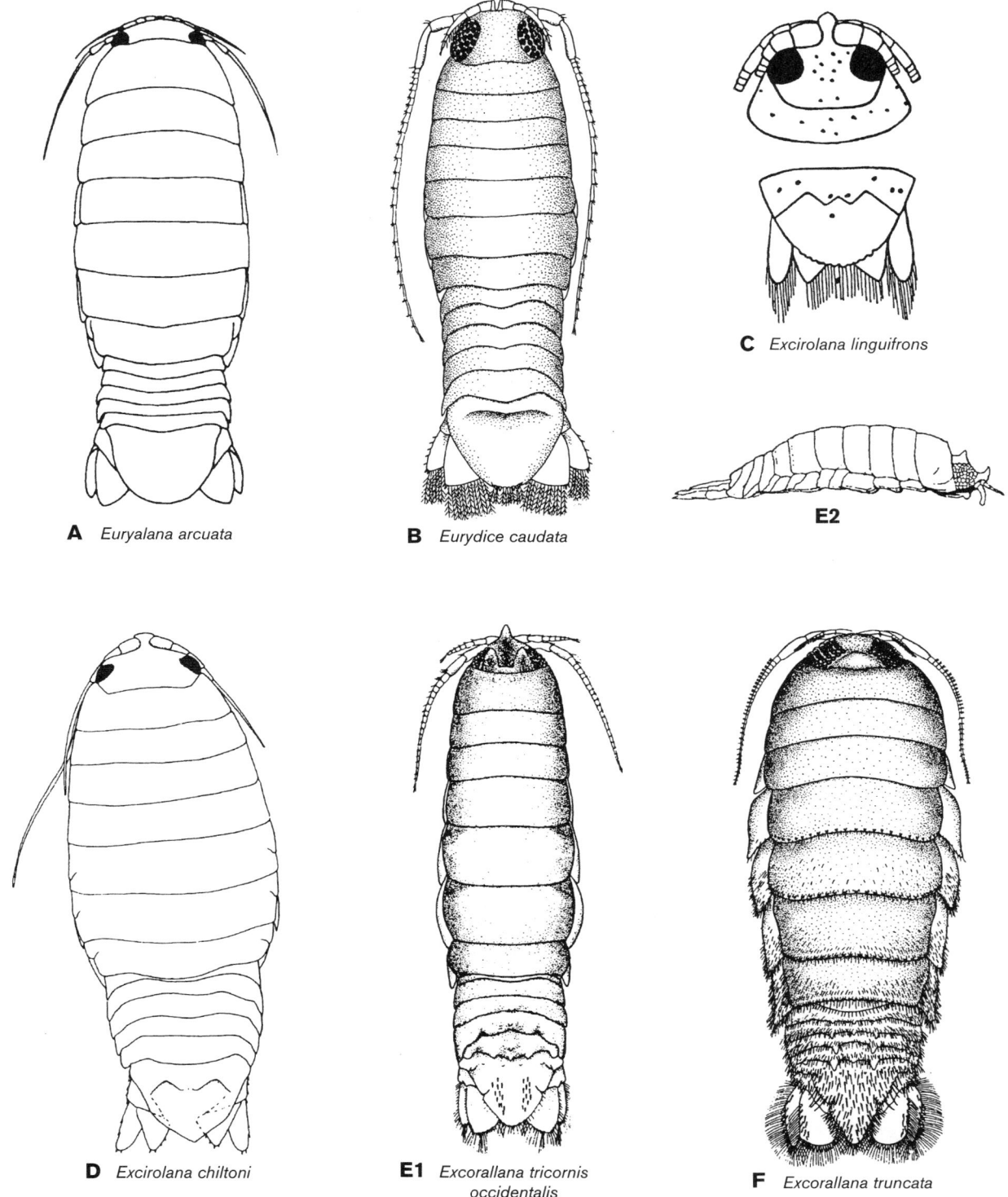

A *Euryalana arcuata*

B *Eurydice caudata*

C *Excirolana linguifrons*

E2

D *Excirolana chiltoni*

E1 *Excorallana tricornis occidentalis*

F *Excorallana truncata*

PLATE 238 Isopoda. Flabellifera: A, *Eurylana arcuata*; B, *Eurydice caudata*; C, *Excirolana linguifrons*, cephalon, pleotelson; D, *Excirolana chiltoni*; E, *Excorallana tricornis occidentalis*, E1, whole animal in dorsal view, E2, whole animal in lateral view; F, *Excorallana truncata* (after Richardson 1899, 1905, after Hansen; Bruce and Jones 1981; Jansen 1981; Delaney 1982, 1984; Brusca et al. 1995).

Ourozeuktes). Aside from some localized damage, in most cases cymothoids do not appear to create a great hardship for their hosts. Host-parasite specificity varies between genera, being high in some (e.g., *Cymothoa, Idusa, Mothocya*) and low in others (e.g., *Anilocra, Nerocila, Livoneca, Elthusa*). The only known case of a parasite functionally replacing a host organ occurs in *Cymothoa exigua,* a species that sucks so much blood from its host fish's tongue that the tongue atrophies and is destroyed, but the isopod remains attached to the remaining tongue stub where the host uses it as a replacement tongue for food manipulation (Brusca and Gilligan 1983). An extensive radiation of cymothoid genera and species has taken place in the freshwater rivers of the Amazon Basin, and to a lesser extent central Africa and southeast Asia. Forty-three nominate genera and more than 400 species of cymothoids exist, but the taxonomy of this family is very poorly understood. Seven species, in five genera, are known from California waters.

1. Posterior margin of cephalon trisinuate; pleon not immersed in pereon 2
— Posterior margin of cephalon not trisinuate; pleon partially immersed in pereon 3
2. Cephalon not immersed in pereonite 1; uropods generally extend beyond posterior border of pleotelson, and clearly visible in dorsal view . . . . Nerocila

Note: One species in California, N. acuminata, plate 240C.

— Cephalon somewhat immersed in pereonite 1; uropods barely or not extending beyond posterior border of pleotelson and typically held concealed under the pleotelson (not visible in dorsal view) . . . . Enispa

Note: One species in California, E. convexa, plate 239C.

3. Basal articles of antennules not expanded and touching 4
— Basal articles of antennules expanded and touching or nearly touching . . . . Ceratothoa 5
4. Antennule longer than antenna . . . . Mothocya

Note: One species in California, M. rosea (plate 239D).

— Antennule shorter than antenna . . . . Elthusa 6
5. Pereopods 4–7 not carinate; posterior margin of pleonite 5 smooth, not trisinuate; labrum with free margin wavy, with wide medial notch (plate 239B) . . . . Ceratothoa gilberti
— Pereopods 4–7 carinate; posterior margin of pleonite 5 trisinuate (except in occasional males); labrum with free margin broadly excavate, without medial notch (plate 239A) . . . . Ceratothoa gaudichaudii
6. Pleotelson in adult female nearly twice as broad as long; eyes medium-size and widely separated; anterior border of head broadly rounded or truncate; antenna of 10–11 articles; juveniles with diffuse dark pigmentation on uropodal exopod and anterolateral areas of pleotelson (plate 240B) . . . . Elthusa vulgaris
— Pleotelson in adult female about as broad as long; eyes large, close-set medially; anterior border of head strongly produced, apically blunt; antenna of eight to nine articles; juveniles with pigment granules concentrated in melanophores, lacking distinct color pattern (plate 240A) . . . . Elthusa californica

LIMNORIIDAE

Limnoriids are a cosmopolitan family of wood and algae-boring isopods (the marine gribbles), distinguished by their minute size (4 mm or less in length), wood/algae boring habits, and several unique anatomical features: the head is set off from the pereon and freely rotates, the mandible incisor process lacks teeth and instead forms a projecting rasp-and-file device used to work wood, the mandibular molar process is absent, the basis of the maxillipeds is elongated and waisted, and the uropods are greatly reduced, with a minute often clawlike exopod.

More than 70 species, in three genera (Limnoria, Lynseia, Paralimnoria), have been described. Four species are known from California waters, one of which is an algal borer (L. algarum) and can be most easily found in the holdfasts of large brown algae such as Macrocystis, Egregia, Laminaria, Postelsia, and Nereocystis. The others infest marine woods, such as pier pilings, docks, boats, driftwood, etc.

1. Incisor process of mandibles simple, lacking rasp or file; algal holdfast borers (plate 240D) . . . . Limnoria algarum
— Incisor of right mandible with filelike ridges, that of left with rasplike sclerotized plates; wood borers 2
2. Dorsal surface of pleotelson with a median Y-shaped keel at base; lateral and posterior borders of pleotelson smooth (plate 241B) . . . . Limnoria lignorum
— Dorsal surface of pleotelson with symmetrically arranged tubercles anteriorly; lateral and posterior borders of pleotelson smooth or tuberculate 3
3. Four anterior tubercles on pleotelson; posterior and lateral margins of pleotelson not tuberculate (plate 241A) . . . . Limnoria quadripunctata
— Three anterior tubercles on pleotelson; posterior and lateral borders of pleotelson tuberculate (plate 241C) . . . . Limnoria tripunctata

SEROLIDAE

Serolids are quickly recognized by their broadly ovate, very thin, flattened bodies with broadly expanded coxal plates. The head is deeply immersed in the pereon. Some species are quite large (to 80 mm). The mandible lacks a molar process, and the maxilliped lacks coupling setae on the endite. Pereonite 1 is fused dorsally with the cephalon and encompasses it laterally. Pereopod 1 of both sexes and pereopod 2 of most adult males are subchelate, with the dactylus folding back upon an inflated propodus.

Serolidae is a cold-water family, primarily distributed in the southern hemisphere. Deep-sea species often have reduced eyes or are blind. They are carnivores, scavengers, or omnivores. Heteroserolis carinata, which ranges form southern California to the Gulf of California, is the only California species (plate 241D). It burrows just under the sediment surface, from the low intertidal zone to about 100 m depth.

SPHAEROMATIDAE

Sphaeromatid isopods can be recognized by their compact, convex bodies, usually capable of rolling into a ball (conglobation); by their pleon which is consolidated into two or three divisions; and by their lateral uropods in which the endopod is rigidly fused to the basal article and the exopod (if present) is movable. In their ability to conglobate, sphaeromatids resemble certain terrestrial isopods called "pillbugs"—a striking example of parallel evolution. Identification of genera and species is often difficult because of marked sexual dimorphism. Hence it is advisable, when making determinations, to have a representative sample including adults of both sexes. Twenty-five species of sphaeromatids, in 10 genera, have been described from California waters, 12 of which occur intertidally and are included in the following key. Some workers place Ancinus, Bathycopea, and Tecticeps in separate families, while others recognize various subfamilies. However, the relationships of the sphaeromatid genera have yet to be analyzed phylogenetically and such taxonomic opinions are based largely on intuition.

Sphaeromatids are primitive flabelliferans with herbivorous habits. The molar process of the mandible is a broad, ovate

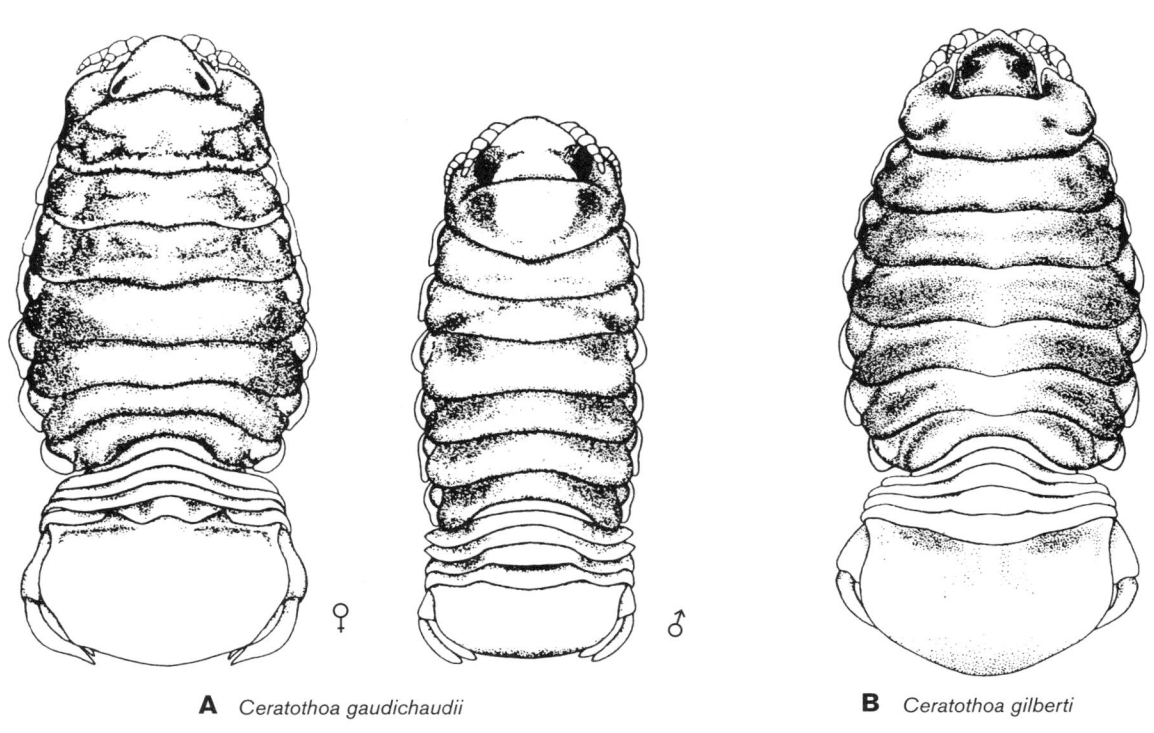

A *Ceratothoa gaudichaudii*

B *Ceratothoa gilberti*

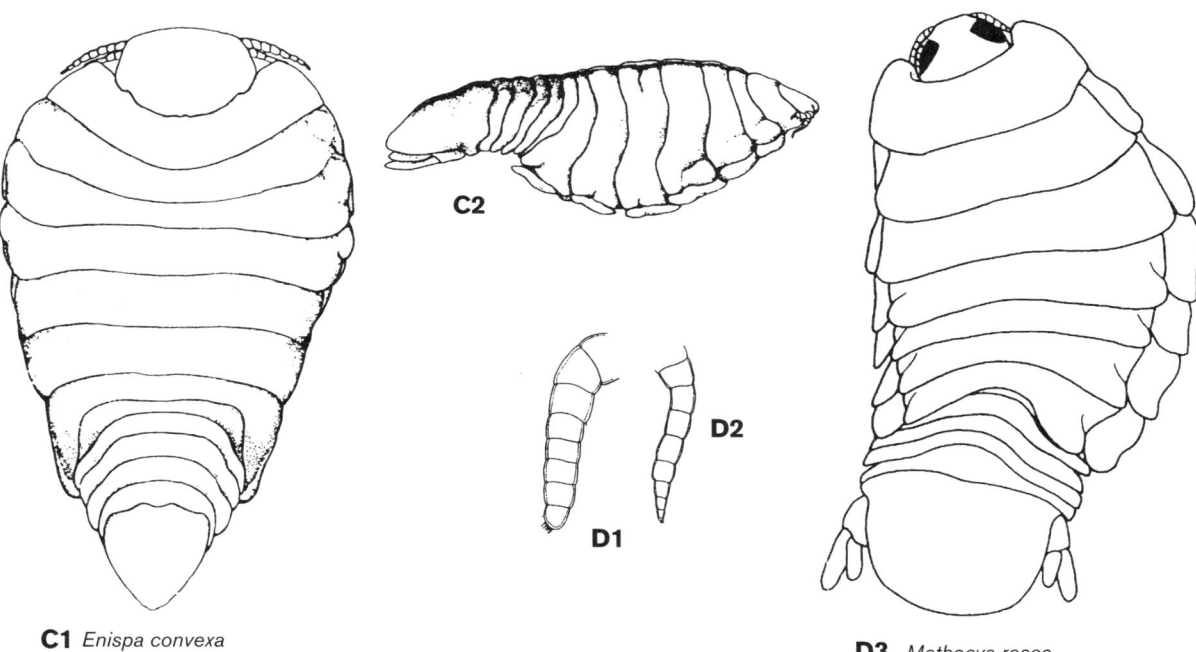

C1 *Enispa convexa*

D3 *Mothocya rosea*

PLATE 239 Isopoda. Flabellifera: A, *Ceratothoa gaudichaudii*; B, *Ceratothoa gilberti*; C, *Enispa convexa*, C1, whole animal in dorsal view, C2, whole animal in lateral view; D, *Mothocya rosea*, D1, antenna 1, D2, antenna 2, D3, whole animal (after Brusca 1981; Bruce 1986).

grinding structure used to chew algae or other plant material. Smaller species probably feed by scraping diatoms and detritus off sand grains. *Paracerceis sculpta*, a subtropical species that finds its way north to southern California, is unique in that it is possesses three distinct male morphs (designated alpha, beta, and gamma males). Alpha males are large, with a distinct mor-

phology typical of other members of the genus; beta males mimic females; gamma males mimic juveniles. The advantage of the beta and gamma males is thought to be in allowing them to sneak into the harem, protected by a single alpha male, to inseminate females (see Shuster et al. citations under *Paracerceis* in species list). In the Sea of Cortez, harems most

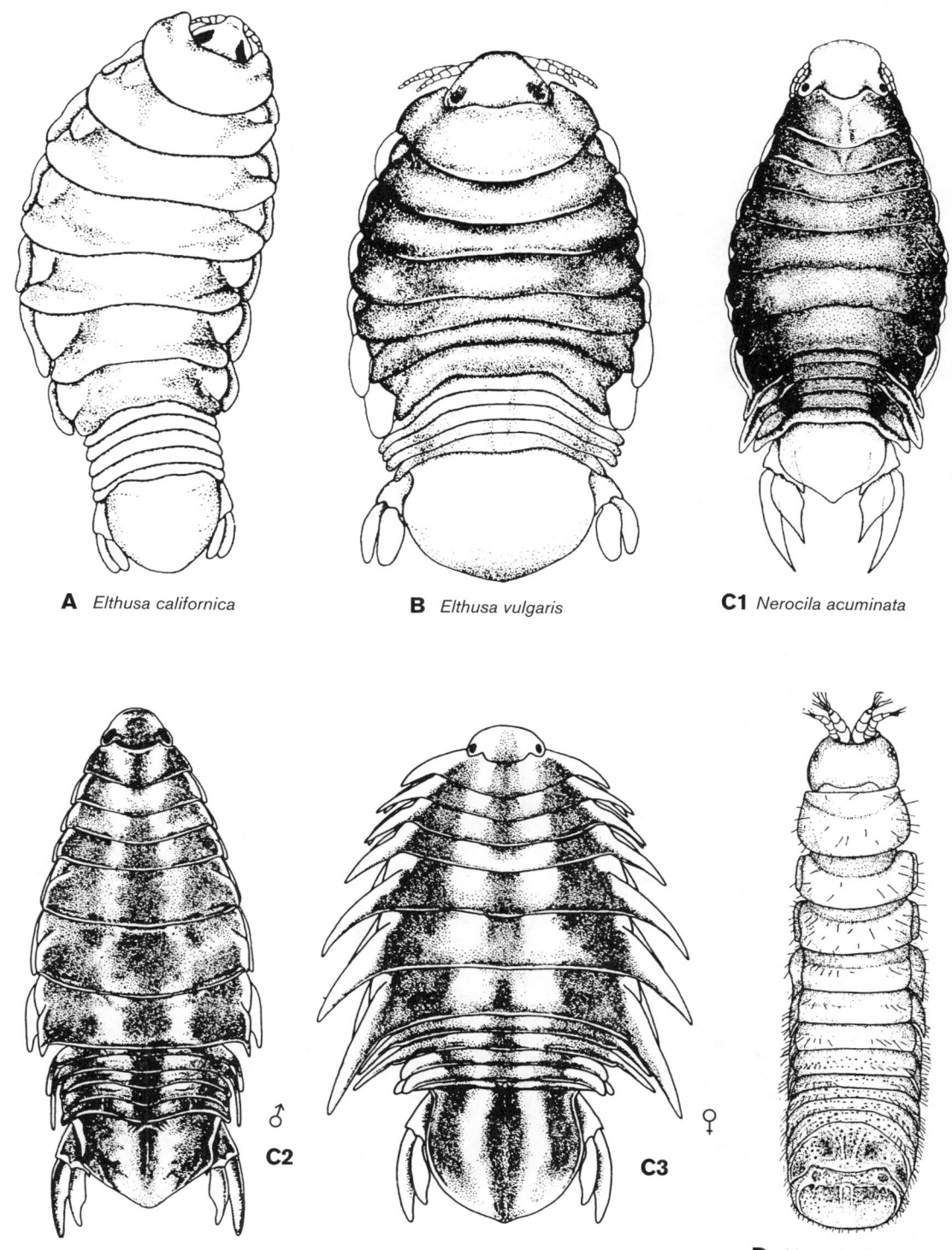

A *Elthusa californica*   **B** *Elthusa vulgaris*   **C1** *Nerocila acuminata* ♀

C2 ♂   **C3** ♀

D *Limnoria algarum*

PLATE 240 Isopoda. Flabellifera: A, *Elthusa californica*; B, *Elthusa vulgaris*; C, *Nerocila acuminata*, C1, acuminata form, female, C2, acuminata form, male, C3, aster form, female; D, *Limnoria algarum* (after Menzies 1957; Brusca 1981).

commonly form in calcareous sponges; the natural history of California populations of *P. sculpta* has not been studied.

Note: A sometimes common sandy beach and surf zone isopod is the distinctive *Tecticeps convexus*, with a broad oval, flattened body, about 15 mm long. Morris, Abbott, and Haderlie (1980, *Intertidal Invertebrates of California*) present a color photo. See further notes in species list.

1. Pereopod 1 prehensile; uropod lacking exopod (plate 241E) . . . . *Ancinus*

Note: One species in California, *A. granulatus*.

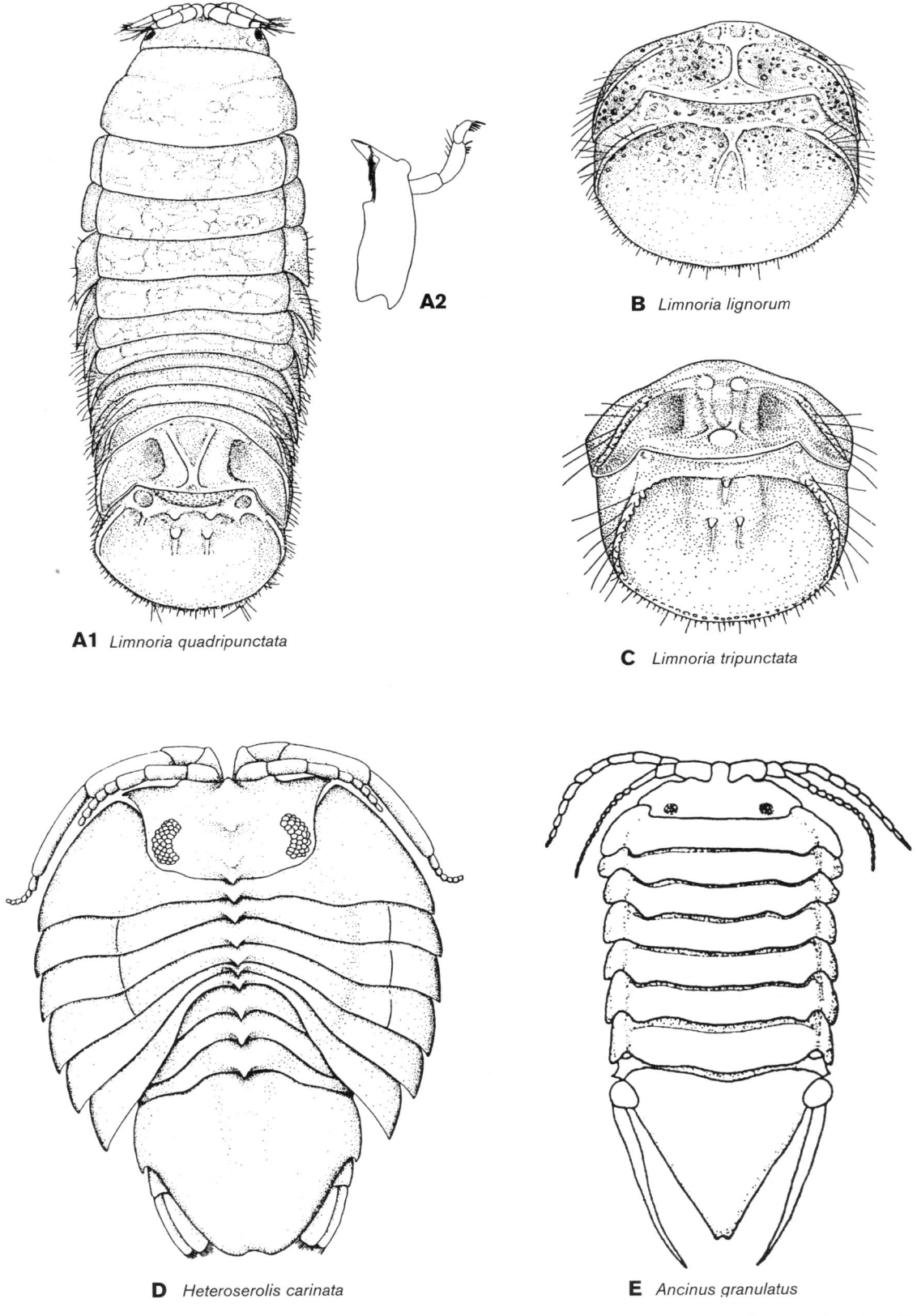

A1 *Limnoria quadripunctata*

B *Limnoria lignorum*

C *Limnoria tripunctata*

D *Heteroserolis carinata*

E *Ancinus granulatus*

PLATE 241 Isopoda. Flabellifera: A, *Limnoria quadripunctata*, A1, whole animal, A2, mandible; B, *Limnoria lignorum*, pleotelson; C, *Limnoria tripunctata*, pleotelson; D, *Heteroserolis carinata*; E, *Ancinus granulatus* (after Menzies 1957; Trask 1970; Wetzer and Brusca 1997).

— Pereopod 1 ambulatory; uropod with exopod 2
2. Pleopods 4 and 5 lacking pleats . . . . . Gnorimosphaeroma 3
— Pleopods 4 and 5 with pleats on endopods 4
3. First article of peduncles of right and left antennae touching each other (plate 243B) . . . .
. . . . Gnorimosphaeroma noblei
— First article of peduncles of right and left antennae not touching each other (plate 243C) . . . .
. . . . Gnorimosphaeroma oregonense

Note: Two additional species of Gnorimosphaeroma occur in our region: G. insulare, common intertidally in some areas, such as estuarine reaches of San Francisco Bay, and G. rayi, intertidal in Tomales Bay. In G. insulare (plate 252C1), only the first two pleonites of second (first visible) pleonal division form its lateral margin; in G. oregonense (plate 243C), all three pleonites comprising the second (first visible) pleonal division reach and form its lateral margin. In G. rayi (plate 252C2), the basis (first free joint) of pereopod 1 has a tuft of seven to nine setae, and the sternal crest of the ischium (second free joint) has two to three setae; in G. oregonense (plate 252C3) the basis of pereopod 1 has only 1 seta, and the ischium has rows of long setae.

4. Pleopod 4 and 5 with branchial pleats on both rami 5
— Pleopod 4 and 5 with branchial pleats on endopods only
. . . . 6
5. Uropods lamellar in females, endopod reduced and exopod elongate-cylindrical in males; ovigerous females with four pairs of oöstegites . . . . Paracerceis 7
— Uropods lamellar in both sexes; ovigerous females lacking oöstegites 8
6. Uropodal exopod with serrate outer margin . . . .
. . . . Sphaeroma 12
— Uropodal exopod with smooth or lightly crenulate outer margin . . . . Exosphaeroma 13
7. Male uropods with spines; female pleotelson stout, with four tubercles (plate 242F) . . . . Paracerceis cordata
— Male uropods without spines; female pleotelson elongate, with three tubercles (plate 243A) . . . . . . Paracerceis sculpta
8. Frontal margin of head produced as a quadrangular process; first two articles of antennules dilated (plate 242E) . . . . . .
. . . . Dynamenella dilatata
— Frontal margin of head not produced; articles of antennules not dilated 9
9. Uropod rami with crenulate margin (at least in males) (plate 242A) . . . . Paradella dianae
— Uropod rami without crenulate margin 10
10. Pleotelson with many tubercles (plate 242B) . . . .
. . . . Dynamenella sheareri
— Pleotelson without tubercles 11
11. Pleotelson with many ridges; uropod rami of similar length (plate 242C) . . . . Dynamenella benedicti
— Pleotelson smooth; uropod with exopod (outer ramus) longer than endopod (inner ramus) (plate 242D) . . . .
. . . . Dynamenella glabra
12. Pleotelson with many rows of tubercles, posterior extremity without prominent transverse elevation (plate 244B) . . . . Sphaeroma walkeri
— Pleotelson with two rows of tubercules, posterior extremity with prominent transverse elevation (plate 244C) . . . . . .
. . . . Sphaeroma quoianum

Note: Pseudosphaeroma campbellenis is a recently introduced, small New Zealand sphaeromatid that may be confused with S. walkeri and S. quoianum. Often light green in color, P. campbellensis

has a narrow, upturned granulated pleotelson; there are two tubercles on pleonite 5, and four tubercles on the anterior portion of the pleotelson; the uropods are smooth and rounded (A. Cohen, J. T. Carlton, and J. Chapman, personal communication).

13. Pleotelson and uropods relatively small; posterior margin of pleotelson rounded (plate 244A) . . . .
. . . . Exosphaeroma inornata
— Pleotelson and uropods very large; posterior margin of pleotelson acuminate (plate 243D) . . . .
. . . . Exosphaeroma amplicauda

GNATHIIDEA

Key general references: Monod 1926; Menzies and Barnard 1959; Menzies 1962; Schultz 1966; Brusca 1989; Cohen and Poore 1994; Wetzer and Brusca 1997.

Gnathiids are quickly recognized by the presence of only six free pereonites and five pairs of pereopods, the first pereonite being fused to the cephalon (with its appendages functioning as a second pair of maxillipeds, or pylopods) and the seventh pereonite being greatly reduced and without legs. The pleon is abruptly narrower than the pereon, always with five free pleonites (plus the pleotelson). Adult males have broad flattened heads with grossly enlarged mandibles that project in the front. Females have small narrow heads and no mandibles at all. In both sexes the eyes are well developed and frequently on short processes (ocular lobes). The embryos are incubated internally, distending the entire body cavity and displacing the internal organs.

Gnathiids occur from the littoral zone to the deep sea, and they are often numerous in shallow soft-bottom benthic samples. Adults probably do not feed and are often found in association with sponges. Adults are benthic, but the juvenile stage, called "praniza," is a temporary parasite on marine fishes. Praniza are good swimmers, whereas adults have only limited swimming capabilities. Females and juveniles cannot be identified, and the taxonomy of this suborder is based entirely on males. About 10 genera and 125 species, in a single family (Gnathiidae), have been described worldwide. Eight species have been found in California waters, all but G. steveni (plate 244D) being subtidal. Only two species have been reported from north of Point Conception, Gnathia tridens and Caecognathia crenulatifrons, the latter in subtidal waters in Monterey Bay. For a key to all known California species see Wetzer and Brusca (1997).

MICROCERBERIDEA

Key general reference: Wägele et al. 1995.

Being tiny (<2 mm in length) and cryptic, members of this suborder are overlooked by most collectors. Microcerberids resemble anthurid isopods in having an elongate body and subchelate first pereopods. However, they are most closely related to the Asellota, with which they share the terminal styliform uropods and many other features. An asellote species, Caecianiropsis psammophila, also lives interstitially in intertidal sands of central California and shows the same adaptations to this habitat as microcerberids—i.e., elongation, small size, and loss of eyes and pigmentation. Only one species of microcerberid has been reported from California waters, Coxicerberus abbotti, known from the interstitial environment in the Monterey Bay area (plate 244E).

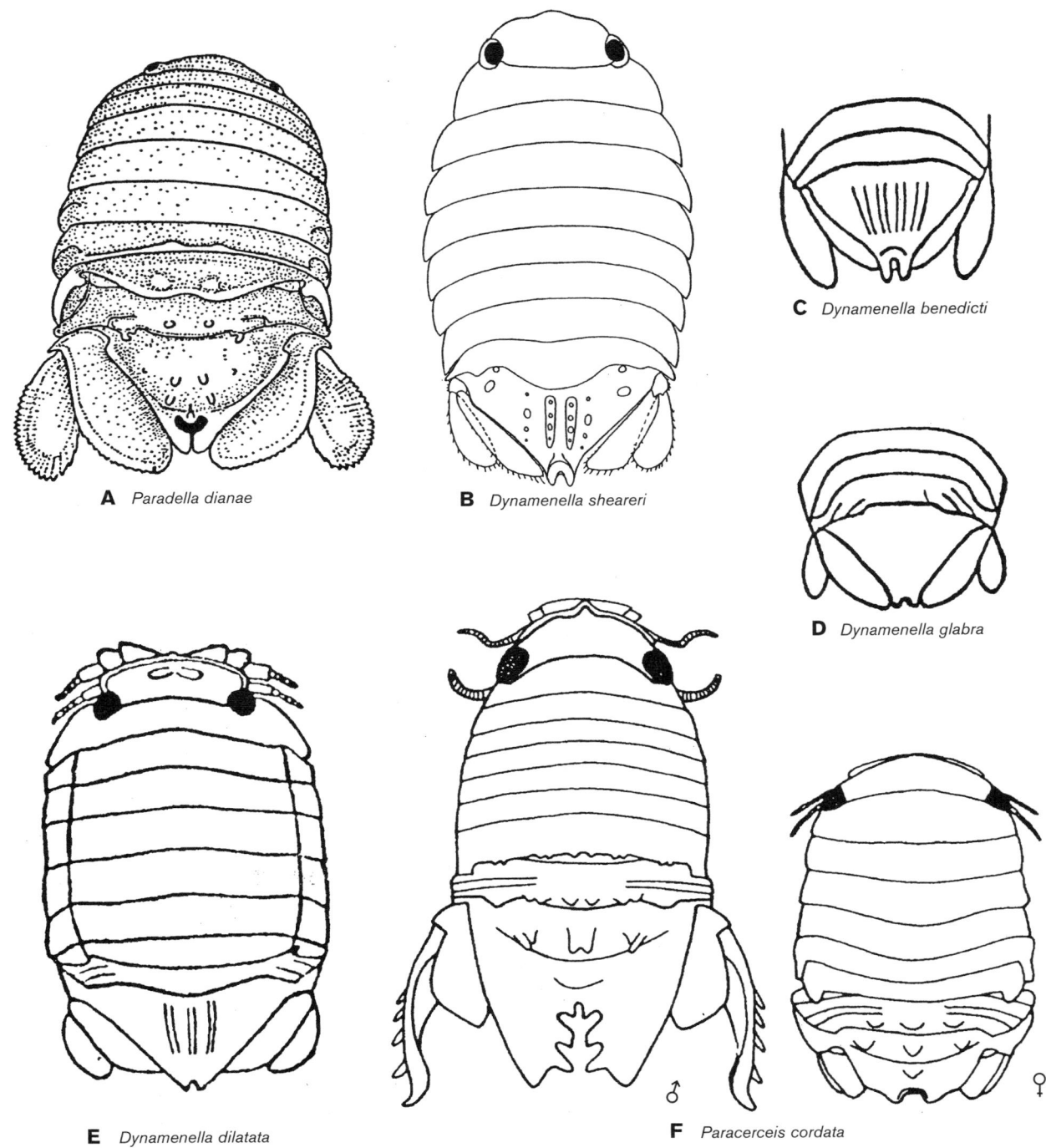

A *Paradella dianae*

B *Dynamenella sheareri*

C *Dynamenella benedicti*

D *Dynamenella glabra*

E *Dynamenella dilatata*

F *Paracerceis cordata*

PLATE 242 Isopoda. Flabellifera: A, *Paradella dianae*; B, *Dynamenella sheareri*; C, *Dynamenella benedicti*, pleotelson; D, *Dynamenella glabra*, pleotelson; E, *Dynamenella dilatata*; F, *Paracerceis cordata* (after Richardson 1899; Menzies 1962; George and Strömberg 1968; Schultz 1969).

VALVIFERA

Key general references: Stimpson 1857; Richardson 1905.

Valviferans are distinguished by the unique opercular uropods that form hinged doors ("valves") covering the pleopods. Additional features that aid in recognition are the well-developed coxal plates, often partly fused pleonites, absence of mandibular palps (except in the southern hemisphere family Holognathidae), and the penes of males arising from pleonite 1, or on the articulation of pleonite 1 and pereonite 7 (rather than on the thorax, as in all other marine isopods).

Three families and 34 species are represented in our waters. *Mesidotea entomon*, an offshore circum-Arctic species, is reported to occur as far south as Pacific Grove and is the only representative of the Chaetiliidae in California. Twenty-one species in the families Arcturidae and Idoteidae occur in our intertidal region.

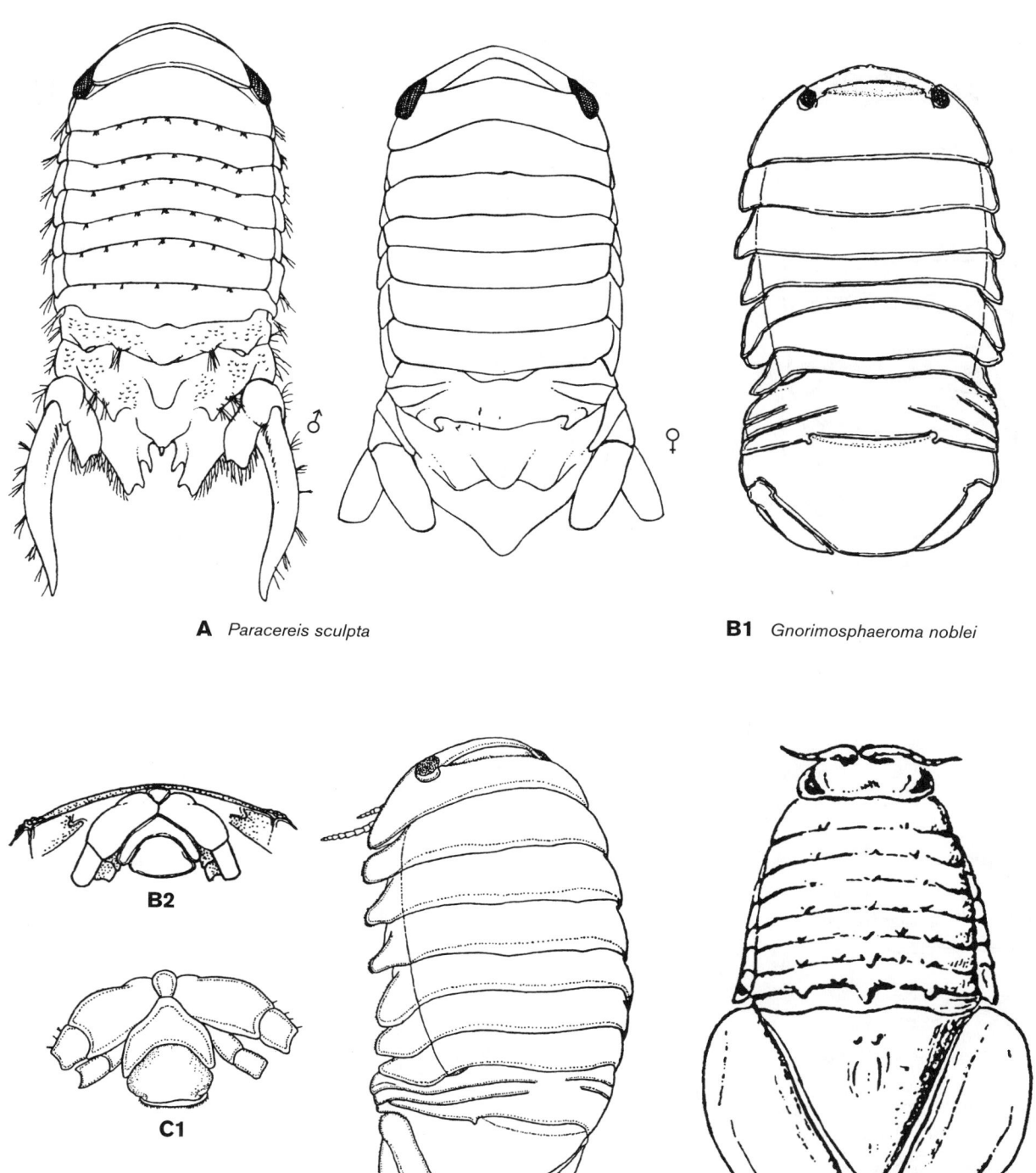

A *Paracereis sculpta*

B1 *Gnorimosphaeroma noblei*

B2

C1

C2 *Gnorimosphaeroma oregonense*

D *Exosphaeroma amplicauda*

PLATE 243 Isopoda. Flabellifera: A, *Paracerceis sculpta*; B, *Gnorimosphaeroma noblei*, B1, whole animal, B2, frontal view of cephalon; C, *Gnorimosphaeroma oregonense*, C1, frontal view of cephalon, C2, whole animal; D, *Exosphaeroma amplicauda* (after Stimpson 1857; Menzies 1954A; Brusca 1980).

1. Body narrow, subcylindrical; anterior four pereopods unlike posterior three, being smaller, setose, and nonambulatory; head fused with first pereonite, leaving six free pereonites . . . . Arcturidae: *Idarcturus*

   Note: One intertidal species in California and Oregon, *I. hedgpethi*, plate 244F.

— Body dorsoventrally depressed; pereopods subsimilar and ambulatory; seven free pereonites . . . . Idoteidae 2

2. Pleon composed of three complete pleonites and one incomplete pleonite (represented by a pair of lateral suture lines), plus pleotelson. . . . . *Cleantioides*

   Note: One species in California and Oregon, *C. occidentalis,* plate 244G.

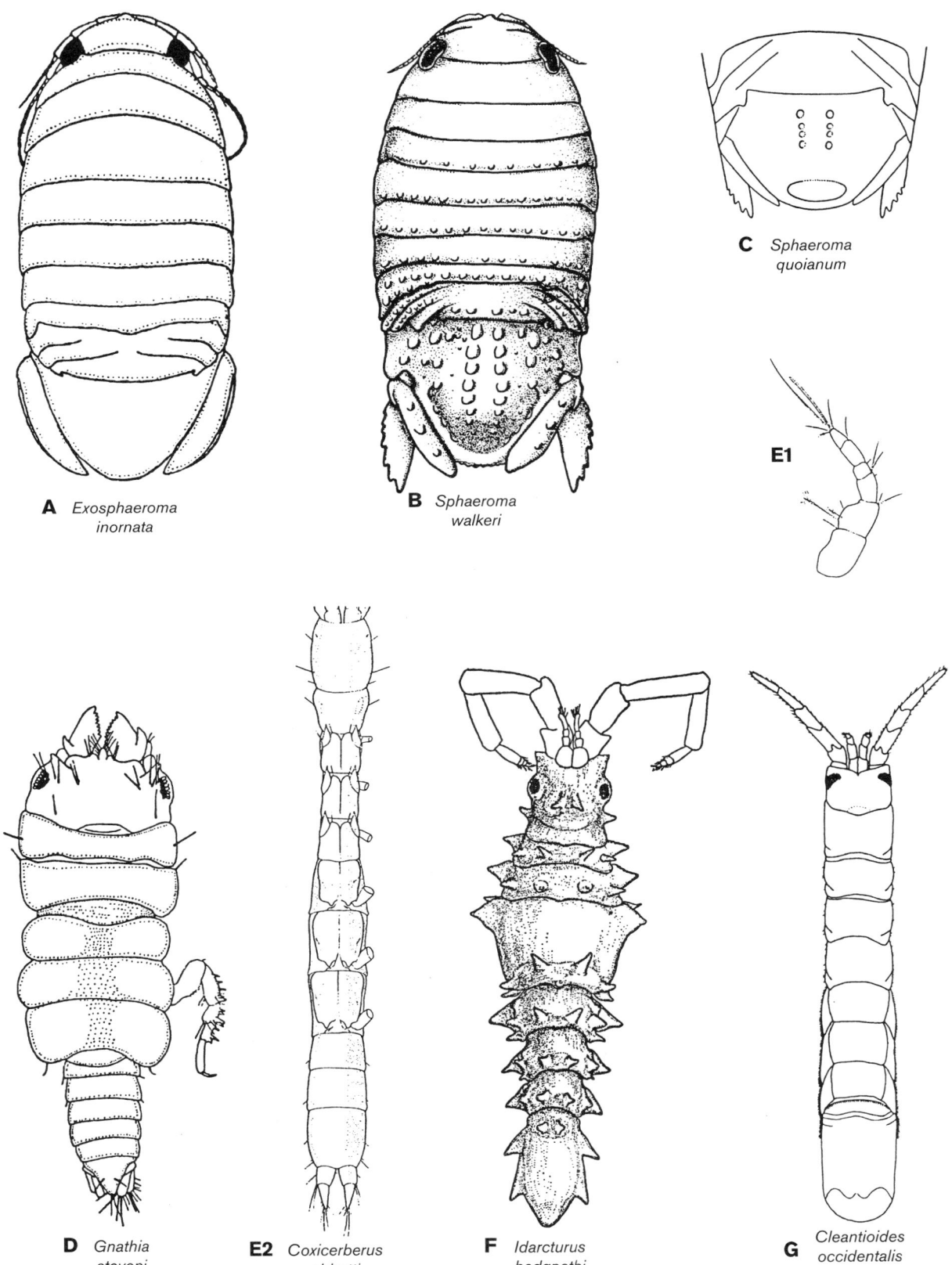

A *Exosphaeroma inornata*

B *Sphaeroma walkeri*

C *Sphaeroma quoianum*

E1

D *Gnathia steveni*

E2 *Coxicerberus abbotti*

F *Idarcturus hedgpethi*

G *Cleantioides occidentalis*

PLATE 244 Isopoda. Flabellifera: A, *Exosphaeroma inornata*; B, *Sphaeroma walkeri*; C, *Sphaeroma quoianum*, pleotelson; D, Gnathiidea: D, *Gnathia steveni*; E, Microcerberidea: E, *Coxicerberus abbotti*, E1, antenna 1, E2, whole animal; F–G, Valvifera: F, *Idarcturus hedgpethi*; G, *Cleantioides occidentalis* (after Richardson 1905; Menzies 1951A, 1962; Dow 1958; Lang 1960; Brusca and Wallerstein 1979; Kensley and Schotte 1989).

— Pleon with less than three complete pleonites 3
3. Pleon composed of a single segment, with or without incomplete suture lines 4
— Pleon composed of two complete pleonites and one incomplete pleonite . . . . *Idotea* 6
4. Antenna with multiarticulate flagellum; pleon with one pair of incomplete suture lines 5
— Antenna with single clavate flagellar article; pleon usually without suture lines . . . . *Erichsonella* 15
5. Maxillipedal palp of four articles . . . . *Colidotea* 16
— Maxillipedal palp of three articles . . . . *Synidotea* 17
6. Maxillipedal palp of four articles 7
— Maxillipedal palp of five articles 9
7. Pleotelson posterior margin concave (plate 246D) . . . . . .
. . . . *Idotea rufescens*
— Pleotelson posterior margin not concave 8
8. Pleotelson posterior margin with strong median process, triangular in shape and with rounded apex (plate 246B)
. . . . *Idotea ochotensis*
— Pleotelson posterior margin without strong median process (plate 247B) . . . . *Idotea urotoma*
9. Eyes transversely (dorsoventrally) elongate; maxilliped with one, two or three coupling setae (plate 246F) . . . . .
. . . . *Idotea stenops*
— Eyes not transversely elongate; maxilliped with one coupling seta 10
10. Posterior border of pleotelson strongly concave (plate 246C) . . . . *Idotea resecata*
— Posterior border of pleotelson not concave 11
11. Pleonite 1 with acute lateral borders 12
— Pleonite 1 without acute lateral borders 13
12. Eyes reniform; anterior margin of pereonite 1 encompassing cephalon (plate 247A) . . . . *Idotea wosnesenskii*
— Eyes rectangular; anterior margin of pereonite 1 not encompassing cephalon (plate 246E) . . . . *Idotea schmitti*
13. Pleotelson with median posterior projection 14
— Pleotelson without median posterior projection (plate 245F) . . . . *Idotea kirchanskii*
14. Eyes circular; pleotelson median posterior projection long (plate 245E) . . . . *Idotea aculeata*
— Eyes with straight anterior and convex posterior border; pleotelson median posterior projection short (plate 246A) . . . . *I. montereyensis*
15. Body not elongated (length about 3 times width) (plate 245D) . . . . *Erichsonella pseudoculata*
— Body elongate (length about 7.4 times width) (plate 245C)
. . . . *Erichsonella crenulata*
16. Posterior margin of pleotelson rounded; body relatively stout (length about 2.6 times width); antenna not, or barely, reaching pereonite 2; body dark purple or dark red (fading to bluish-gray in alcohol); commensal on sea urchins (plate 245A) (*Strongylocentrotus*) . . . . *Colidotea rostrata*
— Posterior margin of pleotelson triangular-shaped; body elongate (length about 5.5 times width); antenna reaching pereonite 3 or 4; body brown to brownish-green; not commensal on sea urchins (usually in brown algae) (plate 245B)
. . . . *Colidotea findleyi*
17. Body smooth; head without preocular horns or other projections (plate 247E) . . . . *Synidotea harfordi*

Note: *Synidotea laticauda* (plate 252D) is a very abundant isopod in fouling communities of San Francisco Bay, living on hydroids on floats, buoys, and pilings. The frontal margin of the head is transverse or slightly concave with a slight median excavation; in *S. harfordi*,

the frontal margin of the head is transverse or slightly convex, with no median excavation. In *S. laticauda*, the pleon is less than one-third longer (in midline) than the greatest width; in *S. harfordi*, the pleon is at least one-third longer than broad.

— Body with tuberculations, carinae or bumps; head with preocular horns or other processes 18
18. Pereon lacking tubercles (plate 247D) . . . .
. . . . *Synidotea consolidata*
— Pereon with tubercles 19
19. Preocular horns project forward (plate 247G) . . . .
. . . . *Synidotea ritteri*
— Preocular horns project laterally 20
20. Lateral borders of first four pereonites acute; each pereonite with a transverse row of three pointed tubercles (plate 247F) . . . . *Synidotea pettiboneae*
— Lateral borders of second, third and fourth pereonites blunt; pereonites with many small tubercles (plate 247C)
. . . . *Synidotea berolzheimeri*

ONISCIDEA (TERRESTRIAL, MARITIME ISOPODS)

Key general references: Van Name 1936 and 1940, 1942 supplements; Mulaik and Mulaik 1942; Garthwaite et al. 1985, 1992; Garthwaite 1992; Leistikow and Wägele 1999.

The Oniscidea (formerly "Oniscoidea") are the only group of crustaceans fully adapted to live on land. They are distinguished by: extreme reduction (to one to three articles) of the antennules; endopods of male pleopod 1 and/or 2 elongate, styliform, specialized as a copulatory apparatus; and, presence of a complex water-conducting system (Hoese 1981, 1982 a, b). In species best adapted to terrestrial life (e.g., Porcellionidae, Armadillidiidae, Armadillidae) the exopods of pleopods 1–2 or 1–5 bear respiratory structures, called pseudotracheae or "lungs." Terrestrial isopods possess general body morphologies correlated to their ecological strategies and behaviour, and can be grouped in three main categories (Schmalfuss 1984): the runners, with an elongate, slightly convex body and long pereopods; the clingers, with a flat broad body and short strong pereopods; and the rollers, with a highly convex body able to roll up into a ball (pillbugs).

With almost 4,000 described species, Oniscidea is the largest isopod suborder. They occur in any kind of terrestrial habitat, from littoral to high mountains, from forests to deserts. In our region, 22 species, in 10 families, occur in littoral biotopes, but only species of *Ligia*, *Tylos*, *Littorophiloscia*, the Detonidae, and the Alloniscidae are typical inhabitants of the eulittoral zone.

The key and species list include all the strictly littoral oniscid species, some of which have wide distributions or have been introduced to North America, and some of which occur on both coasts.

1. Uropods ventral, hidden by pleotelson and not visible in dorsal view of the animal (plate 248A) . . . .
. . . . Tylidae *Tylos punctatus*
— Uropods terminal, clearly visible in dorsal view 2
2. Flagellum of antenna with more than 10 articles; eye with more than 50 ommatidia . . . . Ligiidae 3
— Flagellum of antenna with two to seven articles; eye with <30 ommatidia, or eyes absent 6
3. Pleotelson with posterolateral projections; uropod with insertion of exopod and endopod at the same level . . . .
. . . . *Ligia* 4

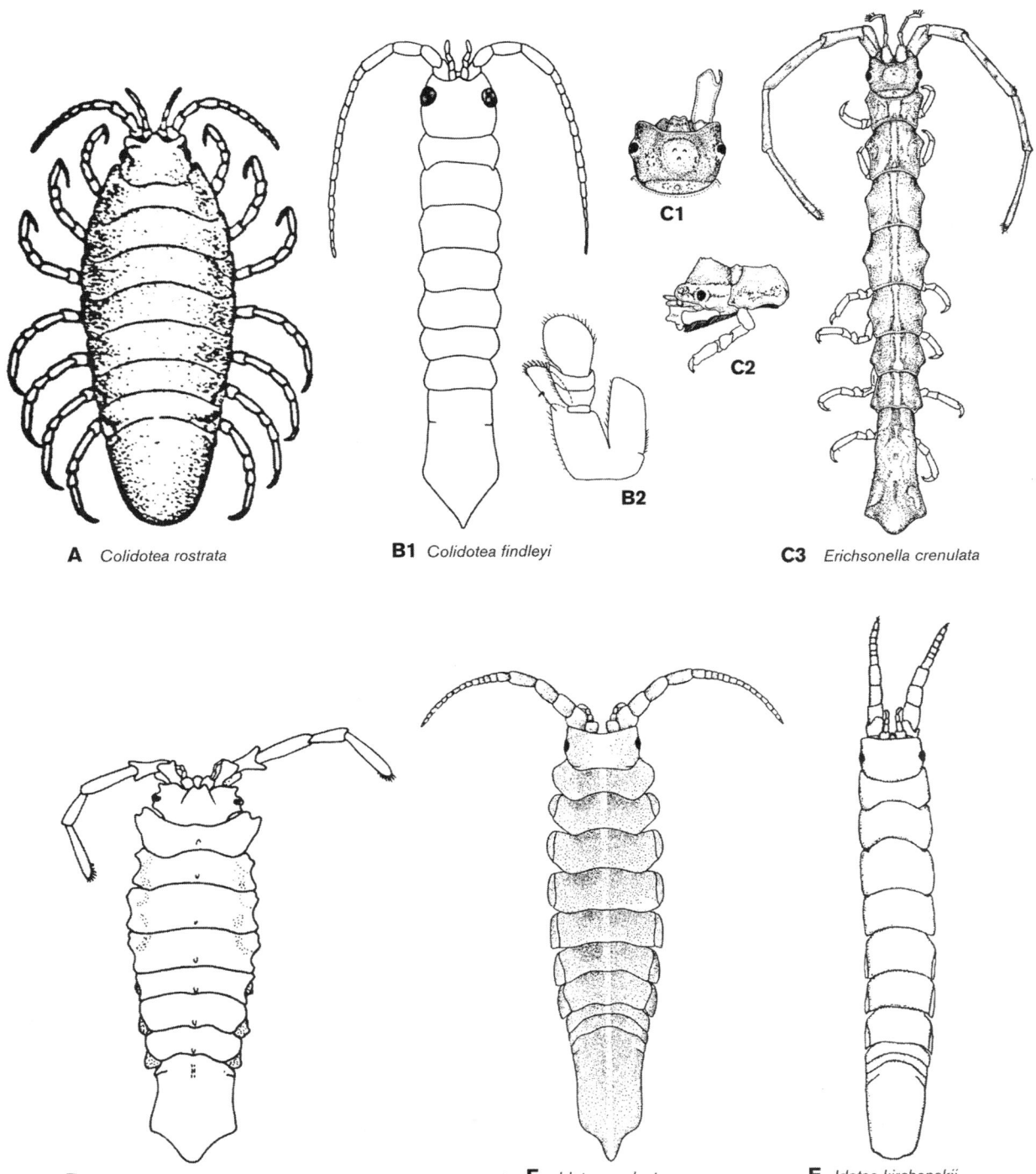

PLATE 245 Isopoda. Valvifera: A, *Colidotea rostrata*; B, *Colidotea findleyi*, B1, whole animal, B2, maxilliped; C, *Erichsonella crenulata*, C1, dorsal view of cephalon, C2, lateral view of cephalon, C3, whole animal; D, *Erichsonella pseudoculata*; E, *Idotea aculeata*; F, *Idotea kirchanskii* (after Benedict 1898; Stafford 1913; Menzies 1950a; Schultz 1969; Miller and Lee 1970; Brusca and Wallerstein 1977).

— Pleotelson without posterolateral projections; uropod with insertion of exopod distinctly proximal to that of endopod . . . . *Ligidium* 5

4. Distance between eyes equal to length of one eye; peduncle of uropod several times longer than broad (plate 248B) . . . . *Ligia occidentalis*

— Distance between eyes equal to twice length of one eye; peduncle of uropod about as broad as long (plate 248C) . . . . *Ligia pallasii*

5. Surface of body smooth and shiny; eye ovoid, far from posterior margin of cephalon; endopod of second male pleopod with rounded apex (plate 248D). . . . . *Ligidium gracile*

— Surface of body rough with sparse scales; eye subtriangular, almost reaching posterior margin of cephalon; endo-

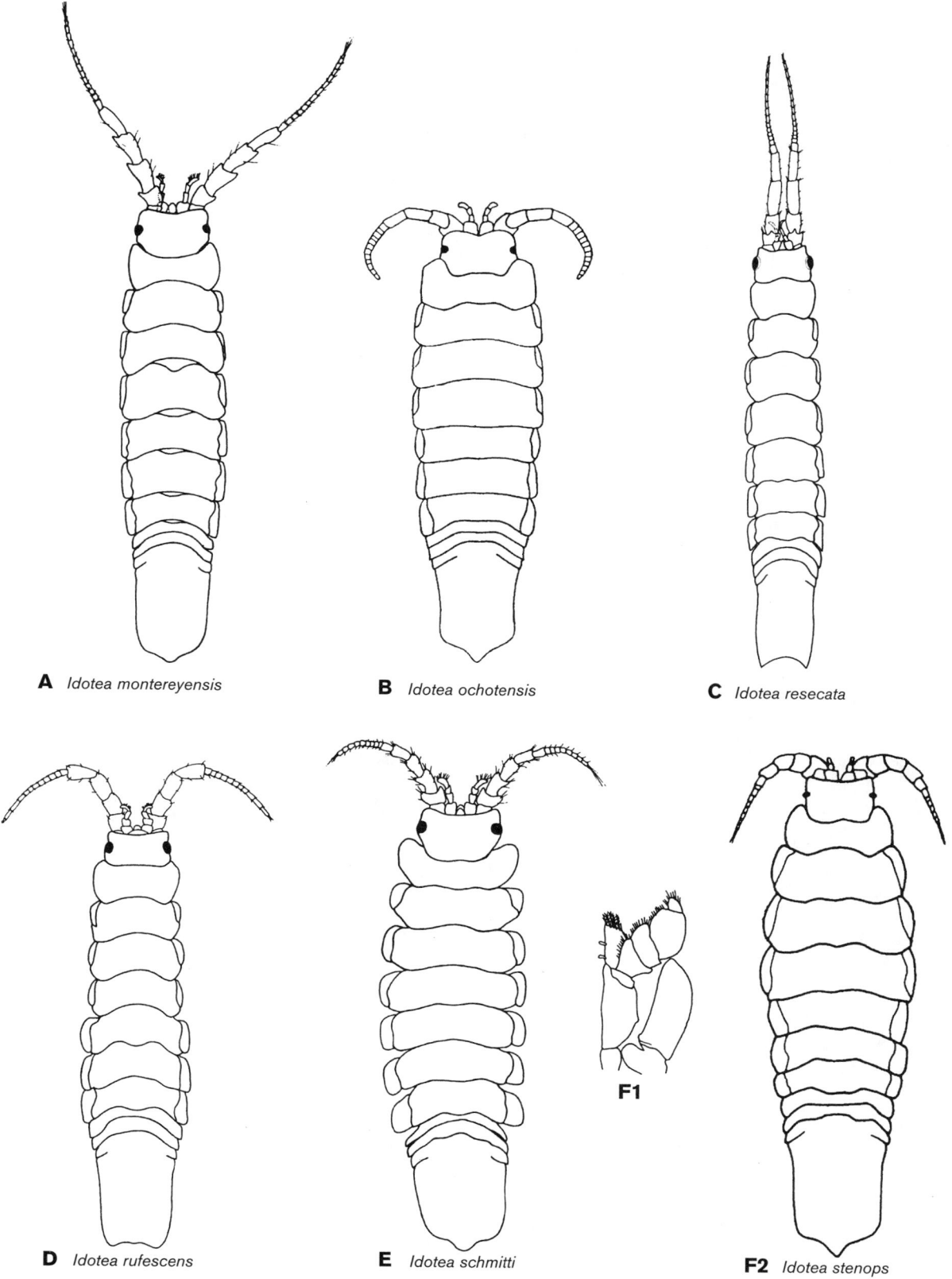

A *Idotea montereyensis*

B *Idotea ochotensis*

C *Idotea resecata*

D *Idotea rufescens*

E *Idotea schmitti*

F1

F2 *Idotea stenops*

PLATE 246 Isopoda. Valvifera: A, *Idotea montereyensis*; B, *Idotea ochotensis*; C, *Idotea resecata*; D, *Idotea rufescens*; E, *Idotea schmitti*; F, *Idotea stenops*, F1, maxilliped, F2, whole animal (after Richardson 1905; Schultz 1969; Rafi and Laubitz 1990).

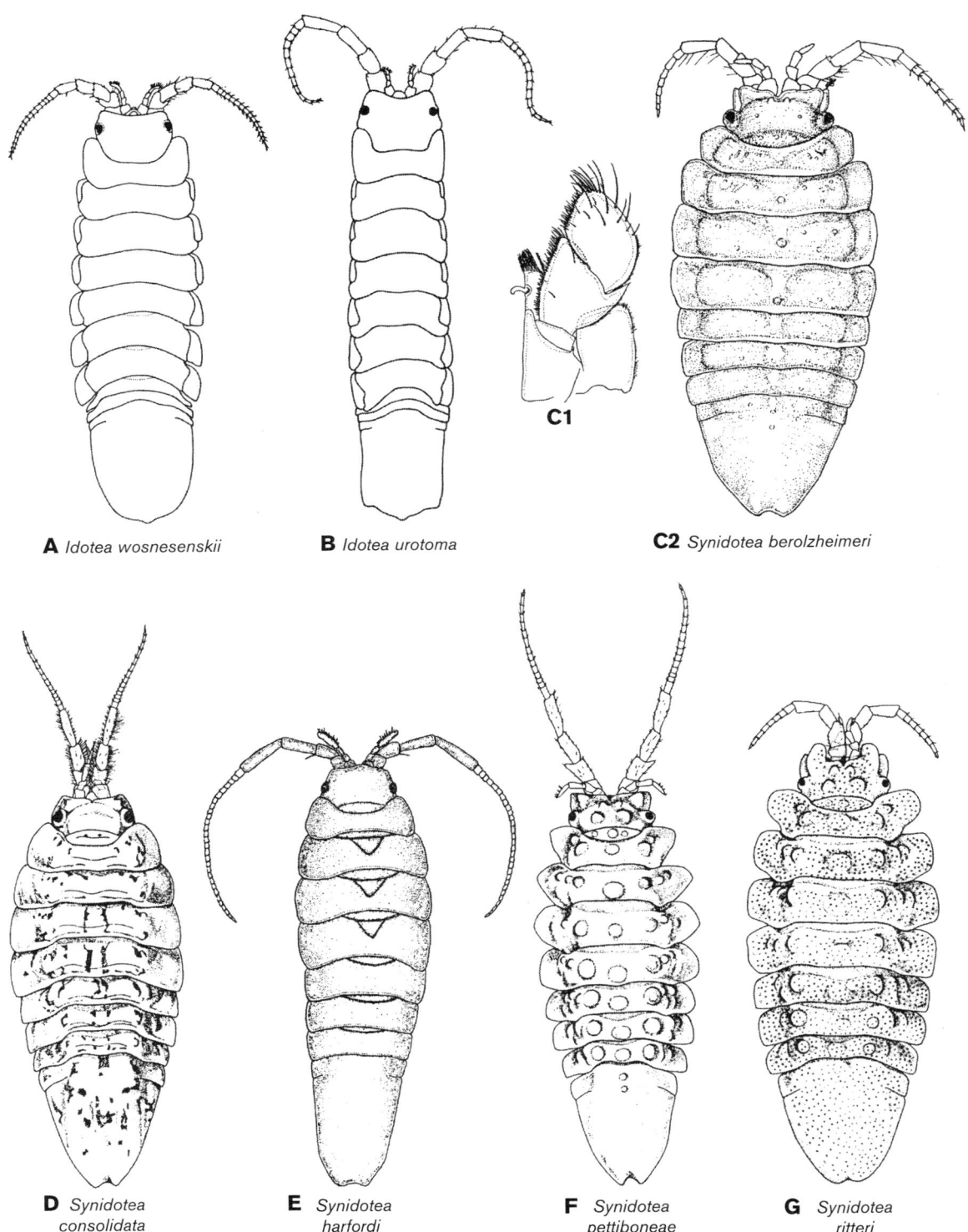

A *Idotea wosnesenskii*

B *Idotea urotoma*

C1

C2 *Synidotea berolzheimeri*

D *Synidotea consolidata*

E *Synidotea harfordi*

F *Synidotea pettiboneae*

G *Synidotea ritteri*

PLATE 247 Isopoda. Valvifera: A, *Idotea wosnesenskii*; B, *Idotea urotoma*; C, *Synidotea berolzheimeri*, C1, maxilliped, C2, whole animal; D, *Synidotea consolidata*; E, *Synidotea harfordi*; F, *Synidotea pettiboneae*; G, *Synidotea ritteri* (after Menzies and Miller 1972; Rafi and Laubitz 1990).

pod of second male pleopod with pointed apex (plate 248E) . . . . *Ligidium latum*
6. Flagellum of antenna tapering to a point, with articles distinguishable only in micropreparations. . . . .
. . . . Trichoniscidae 7

— Flagellum of antenna with two to four clearly distinct articles. 8
7. Flagellum of antenna of three minute articles; eye consisting of a single black ommatidium (plate 249A). . . . .
. . . . *Haplophthalmus danicus*

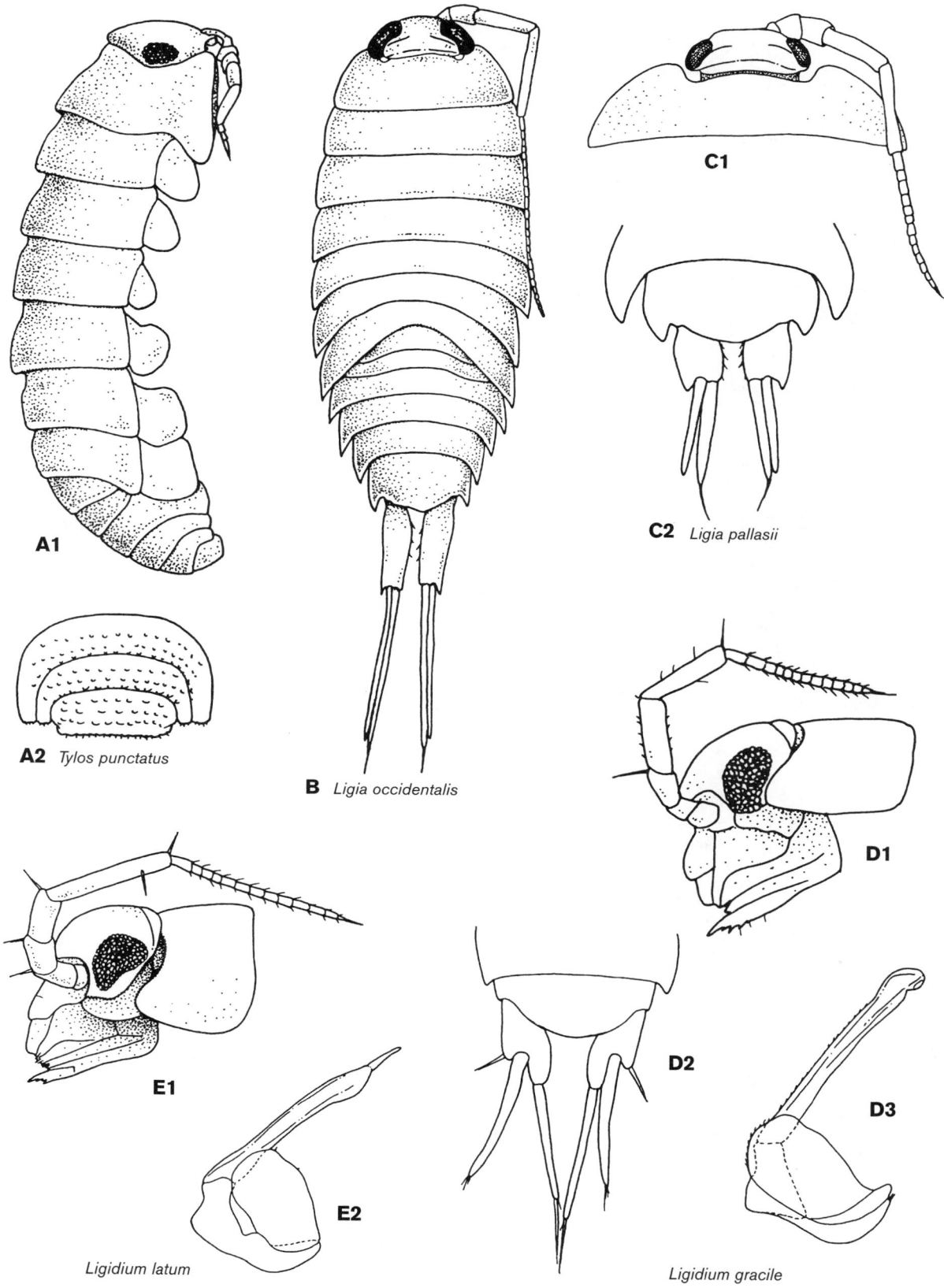

A1

A2 *Tylos punctatus*

B *Ligia occidentalis*

C1

C2 *Ligia pallasii*

D1

D2

D3

E1

E2

Ligidium latum

Ligidium gracile

PLATE 248 Isopoda. Oniscidea: A, *Tylos punctatus*, A1, lateral view of whole animal, A2, fourth and fifth pleonite and pleotelson; B, *Ligia occidentalis*; C, *Ligia pallasii*, C1, cephalon and first pereonite, C2, fifth pleonite, pleotelson, and uropods; D, *Ligidium gracile*, D1, lateral view of cephalon and first pereonite, D2, fifth pleonite, pleotelson, and uropods, D3, second male pleopod; E, *Ligidium latum*, E1, lateral view of cephalon and first pereonite, E2, second male pleopod.

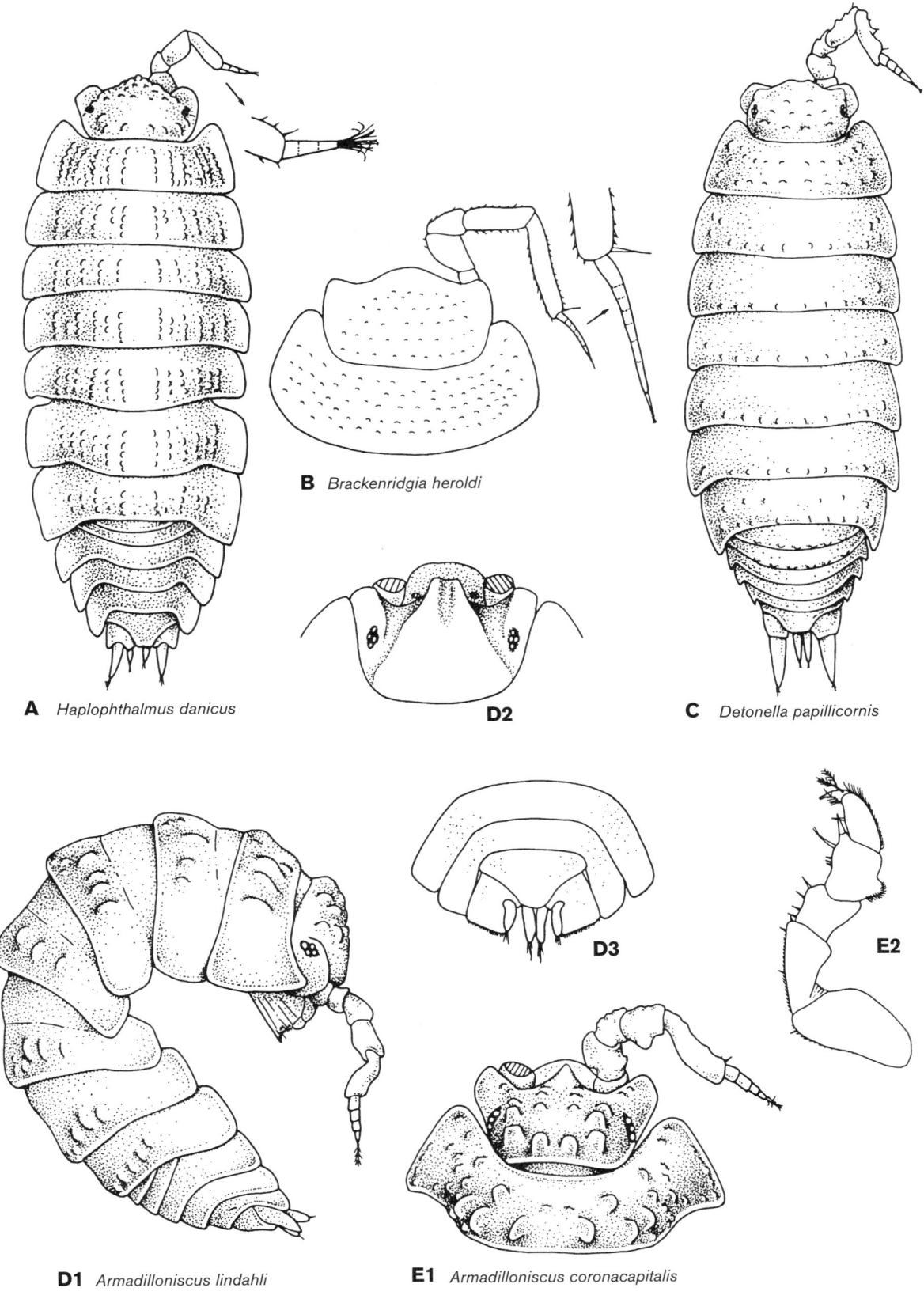

A *Haplophthalmus danicus*

B *Brackenridgia heroldi*

C *Detonella papillicornis*

D2

D3

E2

D1 *Armadilloniscus lindahli*

E1 *Armadilloniscus coronacapitalis*

PLATE 249 Isopoda. Oniscidea: A, *Haplophthalmus danicus*; B, *Brackenridgia heroldi*, cephalon and first pereonite; C, *Detonella papillicornis*; D, *Armadilloniscus lindahli*, D1, lateral view of whole animal, D2, dorsal view of cephalon, D3, fourth and fifth pleonite, pleotelson, and uropods; E, *Armadilloniscus coronacapitalis*, E1, female cephalon and first pereonite, E2, male seventh pereopod.

— Flagellum of antenna of six or seven minute articles; eyes lacking (plate 249B) . . . . *Brackenridgia heroldi*

8. Flagellum of antenna with four articles . . . .
. . . . Detonidae 9

— Flagellum of antenna with two or three articles 12

9. Uropods with peduncle subcylindrical, exopod inserted terminally and distinctly protruding from body outline (plate 249C) . . . . *Detonella papillicornis*

— Uropods with peduncle lamellar, exopod inserted on medial margin and not protruding from body outline 10

10. Body markedly convex and capable of rolling into a ball; cephalon with median lobe truncate (plate 249D) . . . . . .
. . . . *Armadilloniscus lindahli*

— Body not markedly convex and incapable of rolling into a ball; cephalon with median lobe pointed 11

11. Penultimate article of peduncle of antenna with spurlike process on lateral margin; dorsal body surface of adult female covered with conspicuous tubercles; seventh male pereopod with a strong spine caudally directed and a rounded lobe on carpus (plate 249E) . . . .
. . . . *Armadilloniscus coronacapitalis*

— Penultimate article of peduncle of antenna without spurlike process on lateral margin; dorsal body surface rough with low, rounded tubercles; seventh male pereopod without spine and lobe on carpus (plate 250A) . . . .
. . . . *Armadilloniscus holmesi*

12. Flagellum of antenna with three articles 13

— Flagellum of antenna with two articles 15

13. Cephalon with cone-shaped lateral lobes protruding frontwards; pleon not abruptly narrower than pereon . . . .
. . . . Alloniscidae *Alloniscus* 14

— Cephalon without cone-shaped lateral lobes; pleon abruptly narrower than pereon (plate 250D) . . . .
. . . . Halophilosciidae *Littorophiloscia richardsonae*

14. Peduncle of uropod with posterolateral margin produced, rounded (plate 250C) . . . . *Alloniscus mirabilis*

— Peduncle of uropod with posterolateral margin not produced, oblique (plate 250B) . . . . *Alloniscus perconvexus*

15. Body moderately convex, unable to roll into a ball; uropod subcylindrical, distictly protruding backwards compared with pleotelson tip 16

— Body very convex, able to roll into a ball; uropod flattened, reaching pleotelson tip 21

16. Dorsal surface of body covered with fine but distinct scales; first article of flagellum of antenna distinctly shorter than second . . . . Platyarthridae 17

— Dorsal surface of body with no distinctly visible scales; first article of flagellum of antenna as long or longer than second . . . . Porcellionidae 18

17. Eyes with about 10 ommatidia; pleotelson tip reaching distal margin of uropodal peduncle (plate 250E) . . . .
. . . . *Niambia capensis*

— Eyes lacking; pleotelson much shorter than uropodal peduncle (plate 250F) . . . . *Platyarthrus aiasensis*

18. Cephalon with a V-shaped suprantennal line; pereonite 1 with regularly convex posterior margin (plate 251A) . . . . .
. . . . *Porcellionides floria*

— Cephalon with no suprantennal line; pereonite 1 with posterior margin concave at sides . . . . *Porcellio* 19

19. Pleotelson with a rounded apex (plate 251B) . . . .
. . . . *Porcellio dilatatus*

— Pleotelson with an acute apex 20

20. Dorsal surface of body granulated; posterior margin of first pereonite distinctly concave at sides (plate 251C) . . . . . .

. . . . *Porcellio scaber*

— Dorsal surface of body smooth; posterior margin of first pereonite slightly concave at sides (plate 251D) . . . .
. . . . *Porcellio laevis*

21. Cephalon with a triangular frontal scutellum; eyes with 20–25; posterolateral corner of first pereonite entire; uropod with large flattened exopod filling gap between pleotelson and fifth pleonite (plate 251E) . . . .
. . . . Armadillidiidae *Armadillidium vulgare*

— Cephalon with no triangular frontal scutellum; eyes with four to eight ommatidia; posterolateral corner of first pereonite cleft; uropod with large flattened peduncle filling gap between pleotelson and fifth pleonite, exopod minute inserted dorsally (plate 251F) . . . .
. . . . Armadillidae *Venezillo microphthalmus*

List of Species

ANTHURIDEA

ANTHURIDAE

*Other anthurids may be present in our region; Ernest Iverson noted (1974) a small (2mm) bright orange anthurid in empty spirorbid tubes, with its telson and uropods modified to form an operculum to close the tube, in the Bodega Bay region.

Cyathura munda Menzies, 1951. Marin County and south; low intertidal to 132 m; common in kelp holdfasts (e.g., *Egregia* and *Laminaria*) and on surfgrass (*Phyllospadix*). See Wetzer and Brusca 1997.

Mesanthura occidentalis Menzies and Barnard, 1959. Point Conception and south; intertidal to 20 m on kelp and rocks.

PARANTHURIDAE

Califanthura squamosissima (Menzies, 1951). Dillon Beach and south; shallow subtidal to 142 m, muddy or sandy sediments and kelp beds; *Macrocystis* holdfasts.

Colanthura bruscai Poore, 1984. San Clemente and south; intertidal to 27 m.

*Paranthura algicola Nunomura, 1978. A questionable species, not distinguishable by its description; possibly *P. elegans*. Reported by Nunomura (1978) from California, but no specific locality provided.

Paranthura elegans Menzies, 1951. Marin County and south; intertidal to 55 m on algal mats, mud bottoms, pier pilings, rocky low intertidal in holdfasts of *Laminaria* and *Macrocystis*, among coralline algae, and other habitats. See Wetzer and Brusca 1997.

*Paranthura japonica Richardson, 1909. An introduced Asian species found in fouling communities in San Francisco Bay and in marinas in southern California (John Chapman, personal communication).

*Paranthura linearis Boone, 1923. A *nomen nudum*. Reported by Boone from Laguna Beach.

ASELLOTA

ASELLIDAE

Asellus tomalensis Harford, 1877. Central California and north; shallow subtidal in fresh and brackish water.

* = Not in key.

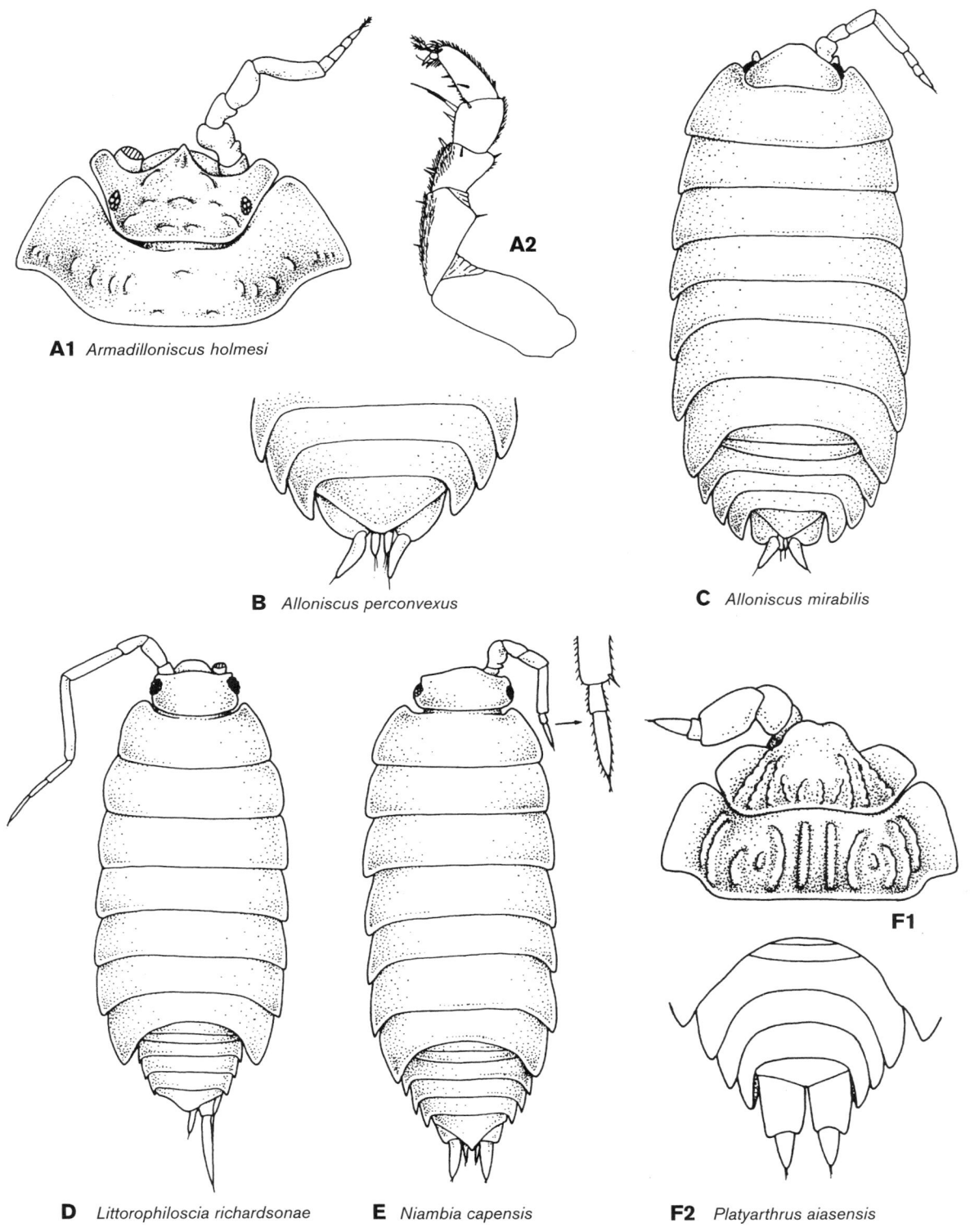

A1 *Armadilloniscus holmesi*

A2

B *Alloniscus perconvexus*

C *Alloniscus mirabilis*

D *Littorophiloscia richardsonae*

E *Niambia capensis*

F1

F2 *Platyarthrus aiasensis*

PLATE 250 Isopoda. Oniscidea: A, *Armadilloniscus holmesi*, A1, cephalon and first pereonite, A2, male seventh pereopod; B, *Alloniscus percon-vexus*, third to fifth pleonite, pleotelson and uropods; C, *Alloniscus mirabilis*; D, *Littorophiloscia richardsonae*; E, *Niambia capensis*; F, *Platyarthrus aiasensis*, F1, cephalon and first pereonite, F2, pleon, pleotelson and uropods.

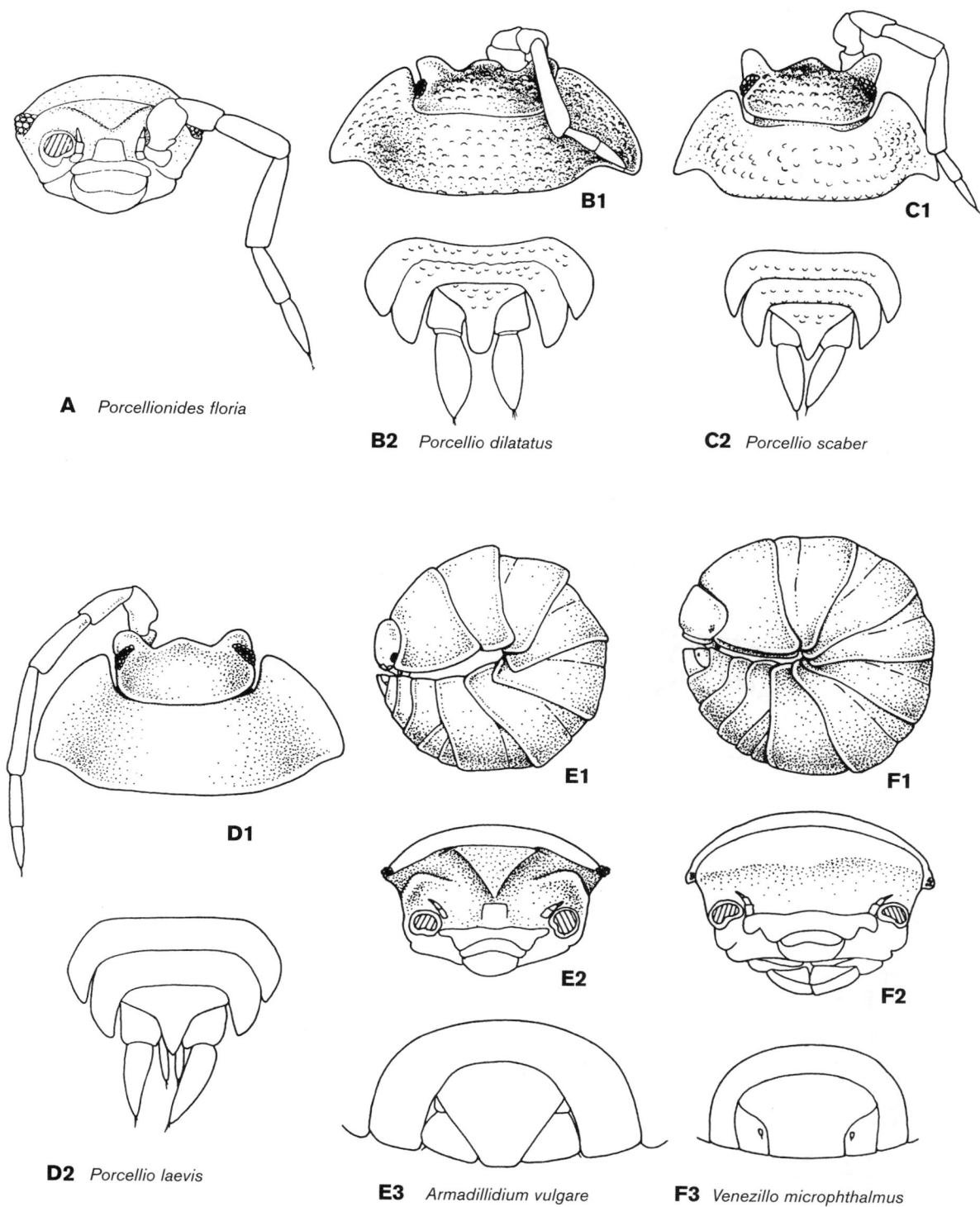

A *Porcellionides floria*

B1

B2 *Porcellio dilatatus*

C1

C2 *Porcellio scaber*

D1

D2 *Porcellio laevis*

E1

E2

E3 *Armadillidium vulgare*

F1

F2

F3 *Venezillo microphthalmus*

PLATE 251 Isopoda. Oniscidea: A, *Porcellionides floria*, frontal view of cephalon; B, *Porcellio dilatatus*, B1, cephalon and first pereonite, B2, fourth and fifth pleonite, pleotelson, and uropods; C, *Porcellio scaber*, C1, cephalon and first pereonite, C2, fourth and fifth pleonite, pleotelson, and uropods; D, *Porcellio laevis*, D1, cephalon and first pereonite, D2, fourth and fifth pleonite, pleotelson, and uropods; E, *Armadillidium vulgare*, E1, lateral view of whole animal, E2, frontal view of cephalon, E3, fifth pleonite, pleotelson, and uropods; F, *Venezillo microphthalmus*, F1, lateral view of whole animal, F2, frontal view of cephalon, F3, fifth pleonite, pleotelson, and uropods (F3 after Arcangeli 1932).

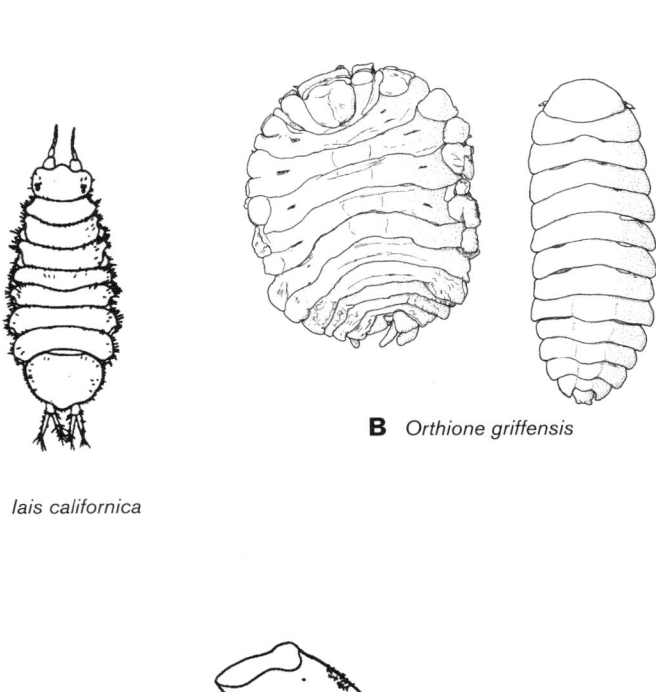

B *Orthione griffensis*

pleonite 3

C1 *Gnorimosphaeroma insulare*

A *Iais californica*

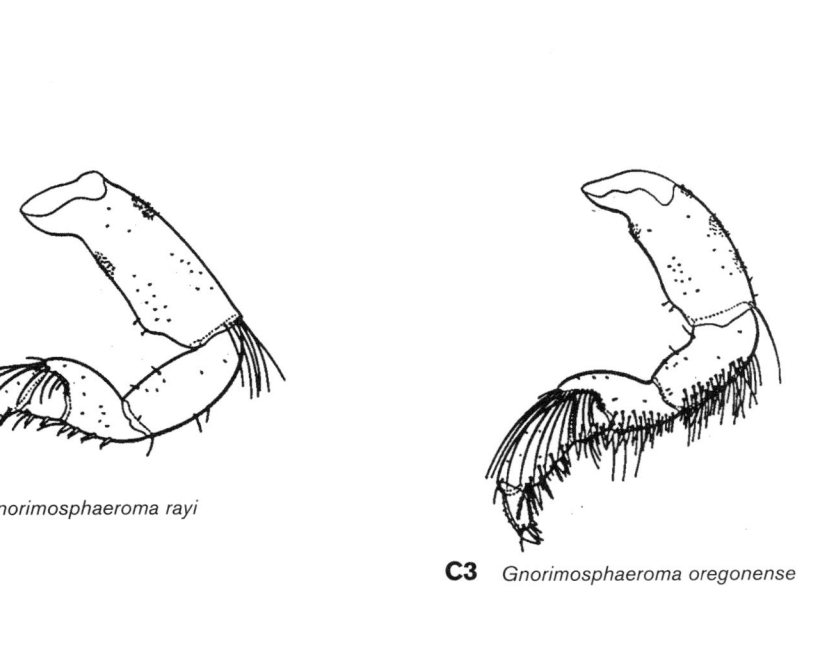

C2 *Gnorimosphaeroma rayi*

C3 *Gnorimosphaeroma oregonense*

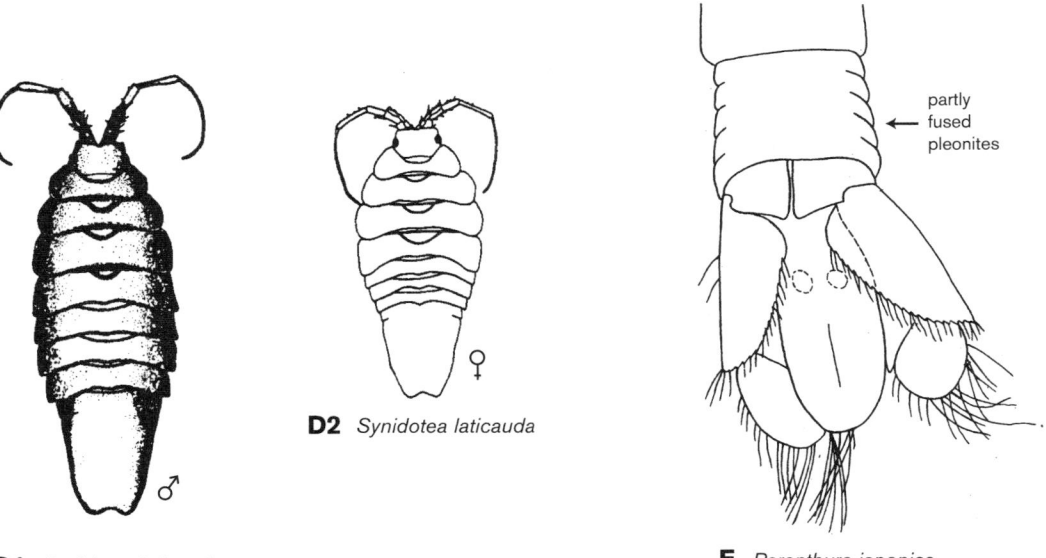

partly
fused
pleonites

D2 *Synidotea laticauda*

D1 *Synidotea laticauda*

E *Paranthura japonica*

PLATE 252 Isopoda. A, *Iais californica*; B, *Orthione griffensis*; C1, *Gnorimosphaeroma insulare*, C2, *Gnorimosphaeroma rayi*, pereopod 1 (from Japan); C3, *Gnorimosphaeroma oregonense*, pereopod 1 (from American Pacific coast); D1, D2, *Synidotea laticauda*, male and female; E, *Paranthura japonica* (A, after Menzies and Barnard 1951; B, from Markham 2004; C1, after Menzies 1954; C2, C3, from Hoestlandt 1975; D, from Richardson 1909.

JANIRIDAE

See Wilson and Wägele 1994, Invert. Taxon. 8: 683–747 (review of family); Kussakin 1962, Trudy Zool. Inst. Akad. Nauk SSSR 30: 17–65 (in Russian; janirids of the seas of U.S.S.R.).

Caecianiropsis psammophila Menzies and Pettit, 1956. Tomales Bluff at Tomales Point (Marin County) and Asilomar (Monterey County); interstitial, buried in sand. See Menzies and Pettit 1956, Proc. U.S. Nat. Mus. 106 (3376): 441–446 (description).

Caecijaera horvathi Menzies, 1951. Hawaii and southern California; intertidal, living in burrows excavated in wood by the isopod *Limnoria*.

*Iais californica (Richardson, 1904). Introduced from Australia or New Zealand with its host isopod *Sphaeroma quoianum*; in bays and estuaries. See Menzies and Barnard 1951, Bull. So. Calif. Acad. Sci. 50: 136–151; Rotramel, 1972.

Ianiropsis analoga Menzies, 1952. Marin County and north; intertidal under rocks or in *Laminaria* holdfasts. Hatch (1947) misidentified this species in Washington as the European *J. maculosa*; Carvacho's (1981) distribution for *Janira maculosa* (Washington State) is based on Hatch, and therefore incorrect.

Ianiropsis derjugini (Gurjanova, 1933) (=*Ianiropsis kincaidi derjugini*). Monterey County and north; intertidal under rocks covered by algae. See Miller 1968.

Ianiropsis epilittoralis Menzies, 1952. Marin County to San Luis Obispo County; on green filamentous algae in high intertidal (Iverson 1974).

Ianiropsis kincaidi (Richardson, 1904) (=*Ianiropsis pugettensis* Hatch, 1947). Monterey County and north; intertidal.

Ianiropsis minuta Menzies, 1952. Marin County; intertidal under rocks or sand.

Ianiropsis montereyensis Menzies, 1952. Marin to Monterey Counties; intertidal to shallow subtidal, under rocks or in *Macrocystis* holdfasts.

Ianiropsis tridens Menzies, 1952. San Juan Island to Monterey County; northern Chile; intertidal, on algae and occasionally found in sponges.

Janiralata davisi Menzies, 1951. Carmel Cove, Monterey County, low intertidal under rocks.

Janiralata occidentalis (Walker, 1898). Washington to Orange County; intertidal under rocks.

*Janiralata triangulata (Richardson, 1899). Monterey Bay; shallow water.

JOEROPSIDIDAE

Joeropsis dubia dubia (Menzies, 1951) formerly *Jaeropsis*. Dillon Beach, Marin County and south; low intertidal to 100 m; on algal holdfasts, bryozoans, tunicates, hydroids, barnacles and under rocks. See Miller 1968.

Joeropsis dubia paucispinis (Menzies, 1951). Marin County; intertidal to 116 m. See Miller 1968.

*Joeropsis lobata (Richardson, 1899). Monterey Bay; shallow water.

MUNNIDAE

See Kussakin, 1962, Trudy Zool. Inst. Akad. Nauk S.S.S.R. 30: 66–109 (in Russian; munnids of the seas of U.S.S.R.).

Munna chromatocephala Menzies, 1952. Central California and north; intertidal on red algae and among incrusting organisms on rocks.

Munna halei Menzies, 1952. Cape Arago (Oregon) to San Luis Obispo; intertidal under rocks, in *Macrocystis* holdfast, and among spines of the purple sea urchin *Strongylocentrotus purpuratus*. See Harty 1979, Bull. So. Calif. Acad. Sci., 78: 196–199 (occurrence and behavior on urchins at Cape Arago: when trapped under urchin's spines, the isopod remains still until the spines become erect and the isopod can crawl away; when held by the urchin's pedicellaria, the isopod is eventually freed and moves away apparently unharmed).

Munna stephenseni Gurjanova, 1933. Central California and north; intertidal to 18 m.

Uromunna ubiquita (Menzies, 1952) (=*Munna minuta* Hansen in Hatch, 1947). Intertidal to shallow subtidal; reported among colonies of the tube-building worm *Owenia* at La Jolla by Fager (1964, Science 143: 356–359).

PARAMUNNIDAE

*Munnogonium tillerae (Menzies and Barnard, 1959) (=*Munnogonium waldronensis* George and Strömberg, 1968; =*Munnogonium erratum* [Schultz, 1964]). Central to southern California; 5 m–150 m; see Bowman and Schultz 1974, Proc. Biol. Soc. Wash. 87: 265–272 (redescription); Wilson 1997.

SANTIIDAE

Santia hirsuta (Menzies, 1951) (=*Antias hirsutus*). Tomales Bluff at Tomales Point, Marin County; intertidal in rock and sand between coralline and laminarian algal zones.

EPICARIDEA

BOPYRIDAE

*Aporobopyrus muguensis Shiino, 1964. Bodega Bay (Milton Miller) and south; 10 m–12 m; in branchial chamber of porcelain crab *Pachycheles rudis*.

*Aporobopyrus oviformis Shiino, 1934. Seto, Japan and Mugu Pier at Point Mugu; 10 m–12 m; in branchial chamber of porcelain crab *Pachycheles pubescens* in California.

*Argeia pugettensis Dana, 1853 (=*Argeia pauperata* Stimpson, 1857; =*Argeia calmani* Bonnier, 1900; =*Argeia pingi* Yu, 1935). Branchial parasites on crangonid shrimps, 32 m–188 m. See Jay 1989, Amer. Midl. Nat., 121: 68–77 (parasitism on *Crangon franciscorum*).

*Asymmetrione ambodistorta Markham, 1985. Southern California, 3 m infesting the hermit crab *Isocheles pilosus*. See Markham 1985, Bull. So. Calif. Acad. Sci. 84: 104–108 (description).

Bopyriscus calmani (Richardson, 1905) (=*Bopyrella macginitiei* Shiino, 1964). Southern and central California, intertidal to 9 m on branchial chamber of the snapping shrimp *Synalpheus lockingtoni* and *Alpheopsis equidactylus*. See Sassaman et al. 1984, Proc. Biol. Soc. Wash. 97: 645–654. (biology, taxonomy).

Ione cornuta Bate, 1864 (=*Ione brevicauda* Bonier, 1900). San Francisco and north; intertidal to shallow water in branchial chamber of ghost shrimps (on *C. longimana* in the eastern Pacific and *N. japonica* in the western Pacific).

*Munidion pleuroncodis Markham, 1975. Central California and south; known to infest only the pelagic red galatheid *Pleuroncodes planipes,* which occurs in California only during warm

* = Not in key.

years when the host moves north from the tropical eastern Pacific. Offshore storms occasionally move *P. planipes* ashore where they are beached. See Markham 1975, Bull. Mar. Sci. 25: 422–441 (systematics); Wetzer and Brusca 1997.

Orthione griffensis Markham, 2004. Abundant on the mud shrimp *Upogebia pugettensis* in Oregon. See Markham 2004, Proc. Biol. Soc. Wash. 117: 186–198 (description).

Phyllodurus abdominalis Stimpson, 1857. Intertidal among pleopods of mud shrimp *Upogebia pugettensis* (female is posterior to first pair of large pleopods, small male roves; A. Kuris, observations). See Markham 1977, Proc. Biol. Soc. Wash. 90: 813–818 (systematics).

Schizobopyrina striata (Nierstrasz and Brender à Brandis, 1929). Shallow water on shrimps *Hippolyte californiensis* (in San Diego Bay) and on *Thor algicola* (in Gulf of California).

ENTONISCIDAE

Portunion conformis Muscatine, 1956. San Francisco to Marin County; intertidal. An endoparasitic castrator in the crabs *Hemigrapsus oregonensis* and *H. nudus*. See Muscatine 1956, J. Wash. Acad. Sci. 46: 122–126; Piltz, 1969, Bull. So. Calif. Acad. Sci. 68: 257–259; Kuris et al. 1980, Parasitology 80: 211–232 (host defensive mechanisms sometimes kill female *Portunion*); Shields and Kuris 1985, J. Invert. Path. 45: 122–124 (ectopic infections of host).

CABIROPIDAE (FORMERLY AS CABIROPSIDAE)

Cabirops montereyensis Sassaman, 1985. Monterey Bay. Shallow water on marsupium of the isopod *Aporobopyrus muguensis* (which in turn lives in the branchial cavity of the porcelain crab *Pachycheles*). See Sassaman 1988, Proc. Biol. Soc. Wash. 98: 778–789; 1992, Proc. Biol. Soc. Wash. 105: 575–584 (description of mature female and epicaridium larva).

*Undescribed cabiropid. An undescribed species occurs in the isopod *Tecticeps convexus* at Horseshoe Cove at Bodega Head (A. Kuris, unpublished observations).

HEMIONISCIDAE

Hemioniscus balani Buchholz, 1866. European species apparently introduced throughout the world; in eastern Pacific from Alaska (Coyle and Mueller 1981, Sarsia 66: 7–18) to Baja California. Parasitic in intertidal barnacles (see Blower and Roughgarden 1988, Oecologia 75: 512–515). This species has also been assigned to *Cryptothir*, *Cryptothiria*, and *Cryptoniscus*.

FLABELLIFERA

AEGIDAE

Aega (Aega) lecontii (Dana, 1854). Central and southern California. Offshore; taken from fish or from soft bottoms. See Brusca 1983, Allan Hancock Fdn. Monogr. Mar. Biol. no. 12, 39 pp. (systematics).

Rocinela signata Schiödte and Meinert, 1879 (=*Rocinela aries* Schiödte and Meinert, 1879).

Los Angeles to Ecuador; also in tropical western Atlantic; intertidal to 68 m; common, taken from fish or from soft bot-

toms. See Brusca and France, 1982, Zool. J. Linn. Soc. 106: 231–275 (systematics).

CIROLANIDAE

See Brusca and Ninos 1978, Proc. Biol. Soc. Wash. 91: 379–385; key to California species; Bruce and Jones 1981; Brusca et al. 1995.

Cirolana diminuta Menzies, 1962. Point Conception and south; intertidal to 50 m; easily confused with the tropical *C. parva. C. harfordi* var. *spongicola* Stafford, 1912, is probably this species. *C. diminuta* attack nearshore fishes in southern California, perhaps attacking fish initially injured by carnivorous ostracodes (Stepien and Brusca 1985, Mar. Ecol. Prog. Ser. 25: 91–105.

Cirolana harfordi (Lockington, 1877). Abundant in mussel beds on rocky shores, where they may occur in densities of thousands per square meter; intertidal to shallow subtidal. See Brusca 1966 (salinity and humidity tolerance); Johnson 1976, Mar. Biol. 36: 343–350 (biology, population dynamics); 1976, Mar. Biol. 36: 351–357 (population energetics). Abbott (1987) presents extensive sketches of external and internal anatomy based upon material from Pacific Grove (Monterey Bay).

Eurydice caudata Richardson, 1899 (=*E. branchuropus* Menzies and Barnard, 1959). San Diego and south; intertidal to 160 m.

Eurylana arcuata (Hale, 1925) (=*Cirolana arcuata*). Introduced to San Francisco Bay; occurs in New Zealand, Australia, and west coast of South America; intertidal to shallow subtidal. See Bowman et al. 1981, J. Crustacean Biol. 1: 545–557 (introduction).

Excirolana chiltoni (Richardson, 1905) (=*E. kincaidi* [Hatch, 1947]; =*E. vancouverensis* [Fee, 1926]; =*E. japonica* Richardson, 1912). Intertidal on sandy beaches. See Enright 1965, Science 147: 864–867; 1971, J. Comp. Physiol. 75: 332–346; 1972, J. Comp. Physiol. 77: 141–162; 1976, J. Comp. Physiol. 107: 13–37 (all, tidal rhythms); Klapow 1972, Biol. Bull. 143: 568–591 (molting and reproductive cycles); Iverson 1974.

Excirolana linguifrons (Richardson, 1899). Monterey Bay to southern California; intertidal on sandy beaches. See Connors et al. 1981, Auk 98: 49–64 (preyed upon by sanderlings, Bodega Bay area).

CORALLANIDAE

See Bruce et al. 1982.

Excorallana tricornis occidentalis Richardson, 1905. Southern California to Panama; intertidal to 138 m on rocks, sandy beaches, and in mangrove habitats. See Delaney 1993, Bull. So. Calif. Acad. Sci. 92: 64–69 (cuticle).

Excorallana truncata (Richardson, 1899) (=*E. kathyae* Menzies, 1962). Point Conception and south; intertidal to 183 m. See Delaney 1982, J. Crust. Biol. 2: 273–280; 1984, Bull. Mar. Sci. 34: 1–20 (systematics).

CYMOTHOIDAE

See Brusca 1981; Brusca and Gilligan 1983; Bruce 1986, 1990.

Ceratothoa gaudichaudii (H. Milne Edwards, 1840). Southern California (rare) to Cape Horn and around to southern Patagonia. Found on many species of pelagic fishes.

Elthusa californica (Schioedte and Meinert, 1884) (=*Livoneca californica*; misspelled as *Lironeca*). On dwarf surfperch (*Micrometrus minimus*), shiner surfperch (*Cymatogaster aggregata*), surf smelt (*Hypomesus preitiosus*), topsmelt (*Atherinops affinis*), arrow goby (*Clevelandia ios*), and California killifish (*Fundulus parvipinnis*). See Waugh et al. 1989, Bull. So. Calif. Acad. Sci. 88: 33–39 (incidence of infestation on fish in Bodega Harbor).

Elthusa vulgaris (Stimpson, 1857) (=*Livoneca vulgaris*). In gill chambers of a wide variety of fishes. See Brusca, 1978, Occ. Paps. Allan Hancock Fdn. n. ser. 2: 1–19 (biology and systematics).

Enispa convexa (Richardson, 1905) (=*Livoneca convexa*). San Diego and south, but rare in California; a tropical species. Found in gill chambers of Pacific bumper (*Chloroscombrus orqueta*), pompanos (*Trachinotus rhodopus* or *T. paitensis*), and *Serranus* sp.

Mothocya rosea Bruce, 1986. San Diego and south; found in *Hyporhampus rosea* and *H. snyderi*.

Nerocila acuminata Schioedte and Meinert, 1881 (=*Nerocila californica* Schioedte and Meinert, 1881). Southern California and south; parasite of many fish species. See Brusca 1978, Crustaceana 34: 141–154 (biology).

LIMNORIIDAE

See Cookson 1991, Mem. Mus. Victoria 52: 137–262 (systematics, including treatments of *L. lignorum*, *L. tripunctata*, and *L. quadripunctata*); Menzies 1954, Bull. Mus. Comp. Zool. Harvard 112: 364–388 (reproduction); Menzies 1957, Bull. Mar. Sci. Gulf and Caribbean 7: 101–200 (systematics).

Limnoria algarum Menzies, 1957. Oregon to southern California, intertidal to 15 m. In holdfasts of *Macrocystis*, *Egregia*, *Laminaria*, *Postelsia*, *Nereocystis*, *Sargassum* and *Pelagophycus*.

Limnoria lignorum (Rathke, 1799). Temperate and boreal northern hemisphere distribution; south to Point Arena on the Pacific coast; intertidal to 20 m.; wood borer. See Cookson 1991, above.

Limnoria quadripunctata Holthuis, 1949. Widespread cool temperate distribution; central to southern California; intertidal to 30 m; wood borer. See Cookson 1991, above.

Limnoria tripunctata Menzies, 1951. Temperate and tropical locations around the world; on our coast from at least Oregon south; intertidal to 7 m; wood borer. See Menzies 1951, Bull. So. Calif. Acad. Sci. 50: 86–88; Cookson 1991 (above); Johnson and Menzies 1956, Biol. Bull. 110: 54–68 (migratory behavior); Beckman and Menzies 1960 Bio. Bull. 118: 9–16 (reproductive temperature and geographic range).

SEROLIDAE

Heteroserolis carinata (Lockington, 1877) (=*Serolis carinata*). Southern California and south; intertidal to 98 m. on soft bottoms. See Wetzer and Brusca 1997.

SPHAEROMATIDAE

See Harrison and Ellis 1991; Bruce 1993.

Ancinus granulatus Holmes and Gay, 1909 (=*A. seticomvus* Trask, 1970). Santa Barbara and south; intertidal to 10 m. *Ancinus* and *Bathycopea* are placed in the family Ancinidae by Bruce (1993) and some other workers. See Trask 1970, Bull. So. Calif. Acad. Sci. 69: 145–149.

Clianella elegans Boone, 1923. *Nomen dubium*. La Jolla and San Pedro.

Dynamene tuberculosa Richardson, 1899. Shallow water.

Dynamenella benedicti (Richardson, 1899). Monterey Bay; intertidal.

Dynamenella conica Boone, 1923. *Species inquirenda*. San Francisco to Monterey Bay; intertidal.

Dynamenella dilatata (Richardson, 1899). Monterey Bay; intertidal.

Dynamenella glabra (Richardson, 1899). Oregon to San Diego; intertidal.

Dynamenella sheareri (Hatch, 1947). Intertidal to shallow subtidal.

Exosphaeroma amplicauda (Stimpson, 1857). Intertidal under rocks and stones; see Rees 1975, Mar. Biol. 30: 21–25 (habitat; competition with *Gnorimosphaeroma oregonense*).

Exosphaeroma aphrodita Boone, 1923. *Nomen dubium*. La Jolla.

Exosphaeroma inornata Dow, 1958 (=*E. media* George and Strömberg, 1968). Northern California to Los Angeles; intertidal and shallow subtidal in holdfasts of kelp *Macrocystis*. See Dow 1958, Bull. So. Calif. Acad. Sci. 57: 93–97; Iverson 1974; Iverson 1978, J. Fish. Res. Bd. Can. 35: 1381–1384.

Exosphaeroma octoncum (Richardson, 1897). Monterey to Marin County; shallow water. See Iverson 1974.

Exosphaeroma rhomburum (Richardson, 1899). Monterey Bay; shallow water.

Gnorimosphaeroma insulare (Van Name, 1940) (=*G. oregonensis lutea* Menzies, 1954; =*G. lutea* Menzies, 1954). Fresh and brackish water, in shallow estuaries and lagoons, including Lake Merced on the San Francisco Peninsula. See Menzies 1954 (below); Eriksen 1968 Crustaceana 14: 1–12 (ecology); Riegel 1959, Biol. Bull. 116: 272–284 (osmoregulation) and 1959, Biol. Bull. 117: 154–162 (physiology, ecology, taxonomy); Hoestlandt 1973, Arch. Zool. Exper. Gen. 114: 349–395, and 1977, Crustaceana 32: 35–54 (taxonomy); Iverson 1974.

Gnorimosphaeroma noblei Menzies 1954. Central California; high intertidal under rocks. See Menzies 1954, Amer. Mus. Novitates 1683, 24 pp. (review of *Gnorimosphaeroma* species); Iverson 1974.

Gnorimosphaeroma oregonense (Dana, 1853). Formerly spelled *G. oregonensis*. San Francisco Bay and north; intertidal to 24 m; brackish to salt water. See Menzies 1954 (above); Riegel 1959 Biol. Bull. 116: 272–284 (osmoregulation), and 1959, Biol. Bull. 117: 154–162 (physiology, ecology, taxonomy); Eriksen 1968, Crustaceana 14: 1–12 (ecology); Hoestlandt 1970, C.R. Acad. Sci. Hebd. Seances Acad. Sci. (D) 270: 2124–2125 (polychromatism); Rees 1975, Mar. Biol. 30: 21–25 (habitat; competition with *Exosphaeroma amplicauda*); Standing and Beatty 1978, Can. J. Zool. 56: 2004–2014 (humidity behavior and reception); Brook et al. 1994, Biol. Bull. 187: 99–111 (protogynous sex change); Zimmer et al. 2002 Mar. Biol. 140: 1207–1213 (cellulose digestion and phenol oxidation).

Gnorimosphaeroma rayi Hoestlandt, 1969. Japan, eastern Siberia, Hawaii, and Tomales Bay; shallow water; introduced with Japanese oysters planted in Tomales Bay. See Hoestlandt 1969, C.R. Acad. Sci. Paris 268: 325–327; Hoestlandt 1973 (cited above); Hoestlandt 1975, Publ. Seto Mar. Biol. Lab. 22: 31–46 (occurrence on Pacific coast).

Paracerceis cordata (Richardson, 1899). Intertidal to shallow subtidal, on pink coralline algae and kelp holdfasts (Lee and Miller 1980).

Paracerceis sculpta (Holmes, 1904). Southern California and south. Widely introduced around the world by shipping. Intertidal to shallow subtidal. Males with harems occurring in calcareous sponges. See Miller 1968; Shuster and Wade 1991, Nature 350: 606–610; Shuster 1989, Evolution 43: 1683–1698; Shuster 1992, Behavior 121: 231–258; Shuster and Sassaman

---

* = Not in key.

1997, Nature 388: 373–377 (all reproduction, genetics, in forms of this species); Shuster 1987, J. Crust. Biol. 7: 318–327 (three discrete male morphs); Shuster 1992, J. Exp. Mar. Biol. Ecol. 165: 75–89 (use of artificial sponges as breeding habit).

Paradella dianae (Menzies, 1962). Southern California to Bahía de San Quintín; intertidal to shallow subtidal. See Iverson 1974.

*Pseudosphaeroma campbellenis Chilton, 1909. An introduced New Zealand species common in fouling communities in brackish water of Coos Bay (Oregon), San Francisco Bay, and other estuaries.

Sphaeroma quoianum H. Milne Edwards, 1840 (commonly spelled as *S. quoyanum,* an unnecessary correction of the original spelling; =*S. pentodon* Richardson, 1904). Intertidal to shallow subtidal in wood, mud and soft rock borer. Introduced to western North America in the late 1800s on ships from Australia (see Rotramel 1972; Carlton 1979; Carlton and Iverson 1981). See also Talley et al. 2001, Mar. Biol. 138: 561–573 (habitat utilization and alteration in California salt marshes).

Sphaeroma walkeri Stebbing, 1905. A western Pacific and Indian Ocean species introduced to southern California. See Carlton and Iverson 1981, J. Nat. Hist. 15: 31–48 (introduction to California).

*Tecticeps convexus Richardson, 1899. Oregon to Point Conception. Intertidal to 9 m; common at times on the sandy beaches in the intertidal surf zone, as in Sonoma County, where they match in color the sediment of the beach they are on. *T. convexus* has an additional broad range of defensive mechanisms, including the ability to fold in half while protruding its sharp uropods, and, when disturbed, to emit a cucumberlike smell, all suggestive of predation pressure (J. T. Carlton). Placed in the family Tecticipididae by Iverson (1982), Bruce (1993), and other workers. See Iverson 1974.

GNATHIIDEA

GNATHIIDAE

See Cohen and Poore 1994.

Gnathia steveni Menzies, 1962. Redondo Beach to northwestern Baja California; intertidal.

MICROCERBERIDEA

MICROCERBERIDAE

Coxicerberus abbotti (Lang, 1960) (=*Microcerberus abbotti*), central California. Interstitial; intertidal. See Lang 1960, Arkiv for Zool. 13: 493–510 (description). Abbott (1987) presents a sketch of a specimen from the sandy beach in front of the Agassiz Laboratory at the Hopkins Marine Station in Pacific Grove.

VALVIFERA

ARCTURIDAE

Idarcturus hedgpethi Menzies, 1951. Tomales Bluff at Tomales Point, Marin County; collected by Joel Hedgpeth in low intertidal on hydroids.

* = Not in key.

CHAETILIIDAE

Mesidotea entomon (Linnaeus, 1767) (=*Saduria entomon*). Circumpolar, on our coast south to Pacific Grove. Intertidal in the northern part of its range, to 30 m in the south.

HOLOGNATHIDAE

See Poore 1990.

Cleantioides occidentalis (Richardson, 1899). Southern California and south; intertidal to 50 m. See Kensley and Kaufman 1978, Proc. Biol. Soc. Wash. 91: 658–665 (genus description); Brusca and Wallerstein 1979.

IDOTEIDAE

See Menzies 1950, Wasmann J. Biol. 8: 155–195 (*Idotea* of northern California); Brusca and Wallerstein 1977; Brusca and Wallerstein 1979, Proc. Biol. Soc. Wash. 92: 253–271 (both, idoteids of the Gulf of California); Brusca and Wallerstein 1979, Bull. Biol. Soc. Wash. 3: 67–105 (idoteid zoogeography); Wallerstein and Brusca 1982, J. Biogeogr. 9: 135–190 (fish predation and role in zoogeography and evolution); Brusca 1984, Trans. San Diego Soc. Nat. Hist. 20: 99–134 (phylogeny, evolution, biogeography of idoteids); Rafi and Laubitz 1970 Can. J. Zool. 68: 2649–2687 (idoteids of northeast Pacific); Poore and Lew Ton 1993, Invert. Taxon. 7: 197–278 (idoteids of Australia and New Zealand)).

Colidotea findleyi Brusca and Wallerstein, 1977. San Diego and south; intertidal to at least 1 m; common on the brown algae *Sargassum*. See Brusca and Wallerstein 1977; Brusca 1983, Trans. San Diego Soc. Nat. Hist. 20: 69–79 (evolution).

Colidotea rostrata (Benedict, 1898). Northern California (rare) and south; commensal of sea urchin *Strongylocentrotus*. See Brusca 1983 (above); Stebbins 1988 J. Crust. Biol. 8: 539–547 (natural history, behavior); 1988, J. Exp. Mar. Biol. Ecol. 124: 97–113 (urchins as refuge from fish predation); 1989, Mar. Biol. 101: 329–337 (population dynamics and reproductive biology in southern California); Delaney 1993, Bull. So. Calif. Acad. Sci. 92: 64–69 (cuticle).

Erichsonella crenulata Menzies, 1950. Southern California (Newport Bay); intertidal to shallow subtidal; on eelgrass *Zostera*.

Erichsonella pseudoculata Boone, 1923 (=*Ronalea pseudoculata*). Point Conception to the Mexican border. Intertidal to 18 m. See Menzies and Bowman 1956, Proc. U.S. Natl. Mus. 106: 339–343 (redescription).

Idotea aculeata (Stafford, 1913) (=*Pentidotea aculeata*). Intertidal on various habitats, including pink-colored individuals matching *Melobesia* encrusting on the surfgrass *Phyllospadix* (D. Carlton, Horseshoe Cove, Bodega Head).

Idotea fewkesi Richardson, 1905. Shallow water.

Idotea kirchanskii Miller and Lee, 1970. Oregon and south; bright green on the green surfgrass *Phyllospadix* and like *I. aculeata* also occasionally matching the pink epiphytic alga *Melobesia*. See Miller and Lee 1970, Proc. Biol. Soc. Wash. 82: 789–798 (description).

Idotea metallica Bosc, 1802. A rare tropical species occasionally occurring in southern California and Gulf of California during warm years; pelagic, attached to floating seaweed. Cosmotropical.

Idotea montereyensis (Maloney, 1933) (=*Pentidotea montereyensis*; =*Idotea gracillima* (Dana) of Richardson, 1905, and Schultz, 1969). Common on surfgrass *Phyllospadix*. See Lee 1966, Comp.

Biochem. Physiol. 18: 17–36; 1966, Ecology 47: 930–941; 1972, J. Exp. Mar. Biol. Ecol. 8: 201–215 (all, pigmentation, color change, ecology); Iverson 1974; Lee and Miller 1980.

Idotea ochotensis Brandt, 1851. Northern California and north; intertidal to 36 m.

Idotea resecata Stimpson, 1857. Intertidal; frequently found living in kelp (e.g., *Macrocystis*, *Egregia*) and eelgrass (*Zostera*). Consumes seeds of the surfgrass *Phyllospadix* (Holbrook et al. 2000, Mar. Biol. 136: 739–747); see also Menzies and Waidzunas 1948, Bio. Bull. 95: 107–113 (postembryonic growth); Miller 1968; Lee and Gilchrist 1972, J. Exp. Mar. Biol. Ecol. 10: 1–27 (coloration and ecology); Iverson 1974; Brusca and Wallerstein 1977 (description, range); Lee and Miller 1980; Alexander 1988, J. Exp. Biol. 138: 37–49; and Alexander and Chen 1990, J. Crustacean Biol. 10: 406–412 (both, swimming behavior); preyed upon in southern California kelp beds by the labrid fish *Oxyjulis californica*; when released from regulation by this fish, *I. resecata* "multiplies rapidly and destroys the kelp canopy" (Bernstein and Jung 1979, Eco. Mono. 49: 335–355). Abbott (1987) presents sketches of internal and external anatomy based upon material from *Macrocystis* kelp beds in Monterey Bay.

Idotea rufescens Fee, 1926. Intertidal to 82 m, on algae. Possibly a synonym of *I. resecata*. See Iverson 1974; Wetzer and Brusca 1997.

Idotea schmitti (Menzies, 1950). (=*Pentidotea schmitti*; =*Pentidotea whitei* Stimpson of Richardson, 1905). Intertidal to shallow subtidal. See Iverson 1974.

Idotea stenops Benedict, 1898. Intertidal to shallow subtidal. See Miller 1968; Iverson 1974; see Brusca and Wallerstein 1977 (description, range). Abbott (1987) presents sketches of internal and external anatomy based upon material from Point Pinos (Monterey Bay).

Idotea urotoma Stimpson, 1864 (=*Cleantis heathii* Richardson, 1899; =*Idotea rectilinea* Lockington, 1877). Intertidal to shallow subtidal; see Brusca and Wallerstein 1977 (description, range).

Idotea wosnesenskii Brandt, 1851 (=*Idotea hirtipes* Dana, 1853; =*Idotea oregonensis* Dana, 1853). San Francisco and north; one anomalous record from La Paz (Baja California). Intertidal to shallow subtidal. Named for the famous Russian naturalist Ilya G. Voznesenskii. See Brusca 1966 (salinity and humidity tolerance); Miller 1968; Brusca and Wallerstein 1977 (description, range); Alexander 1988, J. Exp. Biol. 138: 37–49; Alexander and Chen 1990, J. Crustacean Biol. 10: 406–412 (both, swimming behavior); Zimmer et al. 2002, Mar. Biol. 140: 1207–1213 (cellulose digestion; cannot oxidize dietary phenolics, despite feeding on seaweeds rich in phenols).

Synidotea berolzheimeri Menzies and Miller, 1972. Central California (San Luis Obispo to Sonoma Counties); intertidal, on hydroid *Aglaophenia*. See Menzies and Miller 1972, Smithsonian Contr. Zool. 102, 33 pp. for review of the genus *Synidotea*.

Synidotea consolidata (Stimpson, 1856) (=*Synidotea macginitiei* Maloney, 1933). Central California and north; intertidal to 20 m. This species has been confused in the literature with the very similar circumarctic *Synidotea bicuspida* (Owen 1839).

Synidotea harfordi Benedict, 1897. Oregon and south; introduced to Japan. Intertidal to shallow subtidal. See Brusca and Wallerstein (1979).

*Synidotea laticauda Benedict, 1897. Abundant in fouling communities on floats and buoys in San Francisco Bay and also in Willapa Bay, Washington. Poore (1996, J. Crust. Biol. 16:

384–394) retained the name *S. laticauda*, while Chapman and Carlton (1991, J. Crust. Biol. 11: 386–400; 1994, J. Crust. Biol. 14: 700–714) indicate that this species is introduced and a synonym of the Japanese *Synidotea laevidorsalis* (Miers, 1881). See also Miller 1968.

Synidotea pettiboneae Hatch, 1947. Central California and north; intertidal on hydroids and bryozoans.

Synidotea ritteri Richardson, 1904. Alaska (Coyle and Mueller 1981, Sarsia 66: 7–18) to north of San Francisco; intertidal.

ONISCIDEA

See Miller (1938) and Brusca (1966) for aspects of biology and ecology of maritime isopods of the San Francisco Bay and Dillon Beach areas, respectively.

ARMADILLIDAE

Venezillo microphthalmus (Arcangeli, 1932). Southern and central California.

ARMADILLIDIIDAE

Armadillidium vulgare (Latreille, 1804). Cosmopolitan species of Mediterranean origin.

LIGIIDAE

Ligia occidentalis Dana, 1853. Oregon and south on rocky shores. See Armitage 1960, Crustaceana 1: 193–207 (chromatophores); Wilson 1970, Bio. Bull. 138: 96–108 (osmoregulation); Lee and Miller 1980.

Ligia pallasii Brandt, 1833. Santa Cruz and north; rocky shores on open coast. Principal food is encrusting diatoms, insect larvae, algae, and "occasional members of the same species" (Carefoot 1973, Mar. Biol. 18: 228–236). See also Wilson 1970, Bio. Bull. 138: 96–108 (osmoregulation); Carefoot 1973, Mar. Biol. 18: 302–311 (growth, reproduction, life cycle), and 1979, Crustaceana 36: 209–214 (habitat of young); Lee and Miller 1980; Zimmer et al. 2001, Mar. Biol. 138: 955–963 (possesses high numbers of microbial symbionts in hepatopancreatic caeca, which contribute to digestive processes); Zimmer et al. 2002, Mar. Biol. 140: 1207–1213 (cellulose digestion and phenol oxidation).

Ligidium gracile (Dana, 1856). Riparian.

Ligidium latum Jackson, 1923. San Francisco Bay area to Santa Barbara County; riparian.

PHILOSCIIDAE

Littorophiloscia richardsonae (Holmes and Gay, 1909). Littoral species common in marshes, along bays and estuaries. See Taiti and Ferrara 1986, J. Nat. Hist. 20: 1347–1380 (systematics).

PLATYARTHRIDAE

Niambia capensis (Dollfus, 1895) (=*Porcellio littorina* Miller, 1936). Introduced from southern Africa; supralittoral and riparian. See Miller 1936. Univ. Calif. Publ. Zool. 41: 165–172 (descriptions).

* = Not in key.

Platyarthrus aiasensis Legrand, 1953. Introduced; western Mediterranean/Atlantic; known in the United States from southern California and Texas. A myrmecophile (sharing the nests of ants).

PORCELLIONIDAE

Porcellio dilatatus Brandt, 1833 (=*Porcellio spinicornis occidentalis* Miller, 1936). Introduced from Europe. See Miller 1936, Univ. Calif. Publ. Zool. 41: 165–172 (descriptions).

Porcellio laevis Latreille, 1804. A cosmopolitan introduced species of Mediterranean origin. Synanthropic. See Miller 1936, Univ. Calif. Publ. Zool. 41: 165–172 (description).

Porcellio scaber Latreille, 1804 (=*Porcellio scaber americanus* Arcangeli, 1932). A cosmopolitan species of European origin. See Miller 1936, Univ. Calif. Publ. Zool. 41: 165–172 (descriptions).

Porcellionides floria Garthwaite and Sassaman, 1985. Southern and western United States and Baja California; very similar to the cosmopolitan synanthropic *Porcellionides pruinosus* (Brandt, 1833), which is present in the United States but does not seem to occur on the Pacific coast (Garthwaite and Sassaman 1985, J. Crust. Biol. 5: 539–555).

ALLONISCIDAE

See Menzies 1950, Proc. Calif. Acad. Sci. (4), 26: 467–481, on California *Armadilloniscus*; Schultz 1972, Proc. Biol. Soc. Wash. 84: 477–488 (systematics); Garthwaite et al. 1992.

Alloniscus mirabilis (Stuxberg, 1875) (=*Alloniscus cornutus* Budde-Lund, 1885). San Mateo County to Magdalena Bay; littoral halophilic species common on sandy beaches above high-tide line, where it borrows in sand under driftwood. See Schultz 1984, Crustaceana 47; 149–167 (systematics).

Alloniscus perconvexus Dana, 1856. A littoral halophilic species common on sandy beaches above high-tide line, where it borrows in sand under driftwood. See Lee and Miller 1980; Schultz 1984 (above).

DETONIDAE

Armadilloniscus coronacapitalis Menzies, 1950. Marin County to San Miguel and Anacapa Islands. A littoral halophilic species.

Armadilloniscus holmesi Arcangeli, 1933 (=*Actoniscus tuberculatus* Holmes and Gay, 1909, a preoccupied name). A littoral halophilic species found in marshes, bays, and estuaries under rocks and driftwood.

Armadilloniscus lindahli (Richardson, 1905). Marin County (Tomales Bay) and south; a littoral halophilic species. Schultz (1972, Proc. Biol. Soc. Wash. 84: 477–488) notes that this species is unique among West Coast *Armadilloniscus* in being capable of rolling into a ball like a pillbug.

Detonella papillicornis (Richardson, 1904). San Francisco Bay and north; a littoral halophilic species common under rocks above high tide line. See Garthwaite 1988, Bull. So. Calif. Acad. Sci. 87: 46–47 (occurrence in Bolinas Lagoon, California).

TRICHONISCIDAE

Brackenridgia heroldi (Arcangeli, 1932). Central and southern California.

Haplophthalmus danicus Budde-Lund, 1885. Cosmopolitan.

TYLIDAE

Tylos punctatus Holmes and Gay, 1909. Southern California and south; a littoral halophilic species restricted to sandy beaches where it burrows above the most recent high-tide line during the day and is active on surface at night (Hays 1977, Pac. Sci. 31: 165–186). See also Hamner et al. 1968, Anim. Behav. 16: 405–409 (orientation), and 1969, Ecology 50: 442–453 (behavior, life history); Schultz 1970, Crustaceana 19: 297–305 (systematics); Hayes 1974, Ecology 55: 838–847, and 1977, Pac. Sci. 31: 165–186 (ecology); Holanov and Hendrickson 1980, J. Exp. Mar. Biol. Ecol. 46: 81–88 (burrowing).

References

Abbott, D. P. 1987. Observing marine invertebrates. Drawings from the laboratory. G. H. Hilgard, ed. Stanford, CA: Stanford University Press, 380 pp.

Arcangeli, A. 1932. Isopodi terrestri raccolti dal Prof. Silvestri nel Nord-America. Boll. Lab. Zool. gen. agr. Portici 26: 121–141.

Bruce, N. L. 1986. Revision of the isopod crustacean genus *Mothocya* Costa, in Hope, 1851 (Cymothoidae: Flabellifera), parasitic on marine fishes. J. Nat. Hist. 20: 1089–1192.

Bruce, N. L. 1990. The genera *Catoessa, Elthusa, Enispa, Ichthyoxenus, Idusa, Livoneca* and *Norileca* n. gen. (Isopoda, Cymothoidae), crustacean parasites of marine fishes, with descriptions of eastern Australian species. Rec. Australian Mus. 42: 247–300.

Bruce, N. L. 1993. Two new genera of marine isopod crustaceans (Flabellifera: Sphaeromatidae) from southern Australia, with a reappraisal of the Sphaeromatidae. Invertebrate Taxon. 7: 151–171.

Bruce, N. L., and D. A. Jones. 1981. The systematics and ecology of some cirolanid isopods from southern Japan. J. Nat. Hist. 15: 67–85.

Bruce, N. L., R. C. Brusca, and P. M. Delaney. 1982. The status of the isopod families Corallanidae Hansen, 1890 and Excorallanidae Stebbing, 1904 (Flabellifera). J. Crustacean Biol. 2: 464–468.

Brusca, G. J. 1966. Studies on the salinity and humidity tolerance of five species of isopods in a transition from marine to terrestrial life. Bull. So. Calif. Acad. Sci. 65: 146–154.

Brusca, R. C. 1980. Common intertidal invertebrates of the Gulf of California. 2nd ed. Tucson, AZ: Univ. Arizona Press.

Brusca, R. C. 1981. A monograph on the Isopoda Cymothoidae (Crustacea) of the Eastern Pacific. Zool. J. Linn. Soc. 73: 117–199.

Brusca, R. C. 1989. Provisional keys to the genera *Cirolana, Gnathia,* and *Limnoria* known from California waters. SCAMIT [Southern California Association of Marine Invertebrate Taxonomists] Newsletter 8: 17–21.

Brusca, R. C., and M. R. Gilligan. 1983. Tongue replacement in a marine fish (*Lutjanus guttatus*) by a parasitic isopod (Crustacea: Isopoda). Copeia 3: 813–816.

Brusca, R. C., and E. W. Iverson. 1985. A guide to the marine isopod Crustacea of Pacific Costa Rica. Rev. Biol. Trop. 33 (Suppl. 1): 1–77.

Brusca, R. C., E. Kimrey, and W. Moore. 2004. Invertebrates [of the Northern Gulf of California]. Pp. 35–107. In Seashore guide to the northern Gulf of California. R. C. Brusca, E. Kimrey, and W. Moore, eds. Tuscon, AZ: Arizona-Sonora Desert Museum.

Brusca, R. C., and M. Ninos. 1978. The status of *Cirolana californiensis* Schultz and *Cirolana deminuta* Menzies and George, with a key to the California species of *Cirolana* (Isopoda: Cirolanidae). Proc. Biol. Soc. Wash. 91: 379–385.

Brusca, R. C., and B. R. Wallerstein. 1977. The marine isopod Crustacea of the Gulf of California. I. Family Idoteidae. Amer. Mus. Novitates 2634: 1–17.

Brusca, R. C., and B. R. Wallerstein. 1979. The marine isopod crustaceans of the Gulf of California. II. Idoteidae. New genus, new species, new records, and comments on the morphology, taxonomy and evolution within the family. Proc. Biol. Soc. Wash. 92: 253–271.

Brusca, R. C., and G. D. F. Wilson. 1991. A phylogenetic analysis of the Isopoda with some classificatory recommendations. Mem. Queensland Mus. 31: 143–204.

Brusca, R. C., R. Wetzer, and S. France. 1995. Cirolanidae (Crustacea; Isopoda; Flabellifera) of the tropical eastern Pacific. Proc. San Diego Nat. Hist. Soc., No. 30, 96 pp.

Cadien, D., and R. C. Brusca. 1993. Anthuridean isopods (Crustacea) of California and the temperate northeast Pacific. SCAMIT [Southern

California Association of Marine Invertebrate Taxonomists] Newsletter 12: 1–26.

Carlton, J. T. 1979. Introduced invertebrates of San Francisco Bay. Pp. 427–444. In San Francisco Bay: the urbanized estuary. T. J. Conomos, ed. San Francisco: California Academy of Sciences.

Carvacho, A. 1981. Le genre Janira Leach, avec description d'une nouvelle espèce (Isopoda, Asellota). Crustaceana 41: 131–142.

Cohen, B. J., and G. C. B. Poore. 1994. Phylogeny and biogeography of the Gnathiidae (Crustacea: Isopoda) with description of new genera and species, most from southeastern Australia. Mem. Mus. Victoria 54: 271–397.

Garthwaite, R. 1992. Oniscidea (Isopoda) of the San Francisco Bay Area. Proc. Calif. Acad. Sci. 47: 303–328.

Garthwaite, R., F. G. Hochberg, and C. Sassaman. 1985. The occurrence and distribution of terrestrial isopods (Oniscoidea) on Santa Cruz Island with preliminary data for the other California islands. Bull. So. Calif. Acad. Sci. 84: 23–37.

Garthwaite, R., R. Lawson, and S. Taiti. 1992. Morphological and genetic relationships among four species of Armadilloniscus Uljanin, 1875 (Isopoda: Oniscidea: Scyphacidae). J. Nat. Hist. 26: 327–338.

Giard, A. 1887. Fragments biologiques. VII. Sur les Danalia, genre de Cryptonisciens parasites des Sacculines. Bull. Biol. Fr. Belgique (2) 18: 47–53.

Giard, A., and J. Bonnier. 1887. Contributions a l'etude des bopyriens. Trav. Inst. Zool. Lille Lab. Mar. Wimereux 5: 1–272.

George, R. Y., and J. O. Strömberg. 1968. Some new species and new records of marine isopods from San Juan Archipelago, Washington, U.S.A. Crustaceana 14: 225–254.

Harrison, K., and J. P. Ellis. 1991. The genera of the Sphaeromatidae (Crustacea: Isopoda). A key and distributional list. Invert. Taxon. 5: 915–952.

Hatch, M. H. 1947. The Chelifera and Isopoda of Washington and adjacent regions. Univ. Wash. Publ. Biol. 10: 155–274.

Hoese, B. 1981. Morphologie und Funktion des Wasserleitungssystems der terrestrischen Isopoden (Crustacea, Isopoda, Oniscoidea). Zoomorphology 98: 135–167.

Hoese, B. 1982a. Der Ligia-Typ des Wasserleitungssystems bei den terrestrischen Isopoden und seine Entwicklung in der Familie Ligiidae (Crustacea, Isopoda, Oniscoidea). Zool. Jb. (Anat.) 108: 225–261.

Hoese, B. 1982b. Morphologie und Evolution der Lungen bei den terrestrischen Isopoden (Crustacea, Isopoda, Oniscoidea). Zool. Jb. (Anat.) 197: 396–422.

Holmes, S., and M. E. Gay. 1909. Four new species of isopods from the coast of California. Proc. U. S. Nat. Mus. 36: 375–379.

Iverson, E. W. 1974. Range extensions for some California marine isopod crustaceans. Bull. Soc. Calif. Acad. Sci. 73: 164–169.

Iverson, E. 1982. Revision of the isopod family Sphaeromatidae (Crustaccea: Isopoda: Flabellifera) I. Subfamily names with diagnoses and key. Journal of Crustacean Biology 2: 248–254.

Jay C. V. 1989. Prevalence, size and fecundity of the parasitic isopod Argeia pugettensis on its host shrimp Crangon francisorum. American Midland Naturalist 121: 68–77.

Kensley, B., and R. C. Brusca (eds). 2001. Isopod systematics and evolution. Crustacean Issues 13, A. A. Balkema, Rotterdam, 365 pp.

Kensley, B., and M. Schotte. 1989. Guide to marine isopod crustaceans of the Caribbean. Washington, D.C.: Smithsonian Institution Press, 308 pp.

Kussakin, O. G. 1979. Marine and brackish isopods (Isopoda) of cold and temperate waters of the northern hemisphere. Volume 1. Suborder Flabellifera. (In Russian.) Opred. Faune S.S.S.R. Akad. Nauk 122: 1–470.

Kussakin, O. G. 1982. Marine and brackish isopods (Isopoda) of cold and temperate waters of the northern hemisphere. Volume 2. Suborder Anthuridea, Microcerberidea, Valvifera, Tyloidea. (In Russian.) Opred. Faune S.S.S.R. Akad. Nauk 131, 461 pp.

Kussakin, O. G. 1988. Marine and brackish isopods (Isopoda) of cold and temperate waters of the northern hemisphere. Volume 3. Suborder Asellota. Part 1. Families Janiridae, Santiidae, Dendrotionidae, Munnidae, Paramunnidae, Haplomunnidae, Mesosignidae, Haploniscidae, Mictosomatidae, Ischnomesidae. (In Russian.) Opred. Faune S.S.S.R. 152, 500 pp.

Lee, W. L., and M. A. Miller. 1980. Isopoda and Tanaidacea: the isopods and allies, pp. 536–558. In Intertidal invertebrates of California. R. H. Morris, D. P. Abbott, and E. C. Haderlie, eds. Stanford, CA: Stanford University Press, 690 pp.

Leistikow, A., and J. W. Wägele. 1999. Checklist of the terrestrial isopods of the new world (Crustacea, Isopoda, Oniscidea). Revta Bras. Zool. 16: 1–72.

Markham, J. C. 1974. Parasitic bopyrid isopods of the amphi-American genus Stegophryxus Thompson with the description of a new species from California. Bull. So. Calif. Acad. Sci. 73: 33–41.

Markham, J. C. 1977. Description of a new western Atlantic species of Argeia Dana with a proposed new subfamily for this and related genera (Crustacea: Isopoda: Bopyridae). Zoologische Mededelingen 52: 107–123.

Menzies, R. J. 1951. New marine isopods, chiefly from Northern California, with notes on related forms. Proc. U.S. Nat. Mus. 101: 105–156.

Menzies, R. J. 1952. Some marine asellote isopods from Northern California, with descriptions of nine new species. Proc. U.S. Nat. Mus. 102: 117–159.

Menzies, R. J. 1962. The marine isopod fauna of Bahía de San Quintin, Baja California, Mexico. Pac. Nat. 3: 337–348.

Menzies, R. J., and J. L. Barnard. 1959. Marine Isopoda on coastal shelf bottoms of Southern California: systematics and ecology. Pac. Nat. 1: 3–35.

Menzies, R. J., and D. Frankenberg. 1966. Handbook on the common marine isopod Crustacea of Georgia. Univ. Georgia Press, Athens, 93 pp.

Menzies, R. J., and P. W. Glynn. 1968. The common marine isopod Crustacea of Puerto Rico: a handbook for marine biologists. Stud. Fauna Curacao Other Caribb. Is. 27 (104): 1–133.

Miller, M. A. 1938. Comparative ecological studies of the terrestrial isopod Crustacea of the San Francisco Bay region. Univ. Calif. Publ. Zool. 43: 113–142.

Miller, M. A. 1968. Isopoda and Tanaidacea from buoys in coastal waters of the continental United States, Hawaii, and the Bahamas (Crustacea). Proc. U.S. Nat. Mus. 125: 1–53.

Monod, T. 1926. Les Gnathiidae. Essai monographique (morphologie, biologie, systematique). Mem. Soc. Sci. Nat. Maroc. 12: 1–667.

Mulaik, S., and D. Mulaik. 1942. New species and records of American terrestrial isopods. Bull. Univ. Utah 32: 1–23.

Negoescu, I., and J. W. Wägele. 1984. World list of the anthuridean isopods (Crustacea, Isopoda, Anthuridea). Trav. Mus. Hist. Nat. "GR Antipa" XXV: 99–146.

Perry, D. M., and R. C. Brusca. 1989. Effects of the root-boring isopod Sphaeroma peruvianum on red mangrove forests. Mar. Ecol. Prog. Ser. 57: 287–292.

Poore, G. C. B. 1984. Colanthura, Califanthura, Cruranthura and Cruregens, related genera of the Paranthuridae (Crustacea: Isopoda). J. Nat. Hist. 18: 697–715.

Poore, G. C. B. 1990. The Holognathidae (Crustacea: Isopoda: Valvifera) expanded and redefined on the basis of body-plan. Invert. Taxon. 4: 55–80.

Reinhard, E. G. 1956. Parasitological reviews. Parasitic castration of Crustacea. Exp. Parasit. 5: 79–107.

Richardson, H. R. 1899. Key to the isopods of the Pacific coast of North America, with descriptions of twenty-two new species. Proc. U.S. Nat. Mus. 21: 815–869.

Richardson, H. 1905. A monograph on the isopods of North America. Bull. U.S. Nat. Mus. 54: 727 pp.

Richardson, H. 1909. Isopods collected in the northwest Pacific by the U.S. Bureau of Fisheries Steamer "Albatross" in 1906. Proc. U.S. Nat. Mus. 37: 75–129.

Rotramel, G. 1972. Iais californica and Sphaeroma quoyanum, two symbiotic isopods introduced to California (Isopoda, Janiridae and Sphaeromatidae). Crustaceana, Suppl. 3: 193–197.

Schmalfuss, H. 1984. Eco-morphological strategies in terrestrial isopods. Symp. zool. Soc. Lond. 53: 49–63.

Schultz, G. A. 1966. Marine isopods of the submarine canyons of the southern California shelf. Allan Hancock Pac. Exp. 27: 1–56.

Schultz, G. A. 1969. How to know the marine isopod crustaceans. Dubuque: W. C. Brown, 359 pp.

Shiino, S. M. 1964. On three bopyrid isopods from California. Rept. Fac. Fish. Pref. Univ. Mie 5: 19–25.

Stafford, B. E. 1913. Studies in Laguna Beach Isopoda, IIB. J. Ent. Zool. 5: 182–188.

Stimpson, W. 1857. The Crustacea and Echinodermata of the Pacific shores of North America. Boston J. Nat. Hist. 6: 503–513.

Strömberg, J. O. 1971. Contribution to the embryology of bopyrid isopods with special reference to Bopyroides, Hemiarthrus, and Pseudione (Isopoda, Epicaridea). Sarsia 47: 1–46.

Van Name, W. G. 1936. The American land and fresh-water isopod Crustacea. Bull. Amer. Mus. Nat. Hist. 71: 1–535 (and Supplements, 1940, 77: 109–142 and 1942, 80: 299–329).

Veillet, A. 1945. Recherches sur le parasitisme des crabes et des galathees par les Rhizocephales et les Epicarides. Ann. Inst. Oceanogr. Monaco, 22: 193–341.

Wägele, J. W., N. J. Voelz and J. Vaun McArthur. 1995. Older than the Atlantic Ocean: discovery of a fresh-water *Microcerberus* (Isopoda) in North America and erection of *Coxicerberus*, new genus. J. Crustacean Biol. 15: 733–745.

Wetzer, R., H. G. Kuck, P. Baez, R. C. Brusca, and L. M. Jurkevics. 1991. Catalog of the Isopod Crustacea type collection of the Natural History Museum of Los Angeles County. Nat. Hist. Mus. Los Angeles Co., Tech. Rpt. No. 3: 1–59.

Wetzer, R., and R. C. Brusca. 1997. The Order Isopoda. Descriptions of the Species of the Suborders Anthuridea, Epicaridea, Flabellifera, Gnathiidea and Valvifera, pp. 9–58. In: Taxonomic atlas of the benthic fauna of the Santa Maria Basin and Western Santa Barbara Channel. Vol. 11. The Crustacea, Part 2: *Isopoda, Cumacea* and *Tanaidacea*. J.A. Blake and P. H. Scott, eds. Santa Barbara: Santa Barbara Museum Natural History.

Wetzer, R., R. C. Brusca, and G. D. F. Wilson. 1997. The Order Isopoda. Introduction to the Marine Isopoda, pp. 1–8. In: Taxonomic atlas of the benthic fauna of the Santa Maria Basin and Western Santa Barbara Channel. Vol. 11. The Crustacea, Part 2: *Isopoda, Cumacea* and *Tanaidacea*. J. A. Blake and P. H. Scott, eds. Santa Barbara: Santa Barbara Museum of Natural History.

Wilson, G. D. F. 1994. A phylogenetic analysis of the isopod family Janiridae (Crustacea). Invert. Taxon. 8: 749–766.

Wilson, G. D. F. 1997. The suborder Asellota, pp. 59–120. In: Taxonomic atlas of the benthic fauna of the Santa Maria Basin and Western santa barbara channel. Vol. 11. The Crustacea, Part 2: *Isopoda, Cumacea* and *Tanaidacea*. J. A. Blake and P. H. Scott, eds. Santa Barbara: Santa Barbara Museum Natural History.

Wilson, G. D. F. and J. =W Wägele. 1994. A systematic review of the family Janiridae (Isopoda, Asellota). Invertebrate Toxonomy 8. 683–747.

Tanaidacea

ANDREW N. COHEN

(Plate 253)

Tanaids are small, mostly marine creatures that look like tiny lobsters with elongate bodies a few millimeters in length and conspicuous claws that they hold in front of their heads. Some species are typically found on hydroids, bryozoans, coralline algae, barnacles, or other epibenthic organisms, and sometimes in fouling communities on floats or pilings, while other species occur on mud. Most live either in tunnels or in mucous tubes cemented together from particles of detritus, where they often appear with their head and claws poking out. Some members of the family *Pagurapseudidae* live coiled inside tiny snail shells with their claws protruding, like minute hermit crabs.

The tanaid body is subcylindrical or flattened dorsoventrally and is divided into three sections (plate 253A): a **CEPHALOTHO-RAX** (a small carapace consisting of the cephalon fused with the first two thoracic segments), which typically bears a pair of compound eyes, two pairs of antennae, mouth parts (including paired mandibles, first and second maxillae, and maxillipeds), and a pair of clawed appendages (**CHELIPEDS**); a **PEREON** consisting of six segments or **PEREONITES** (thoracic segments 3–8), each of which bears a pair of legs (**PEREOPODS**); and a short abdomen or **PLEON**, with two to three free segments (**PLEONITES**), plus a terminal **PLEOTELSON** (the telson fused with the 6th pleonite). The pleon usually bears a series of up to five pairs of flattened, two-branched **PLEOPODS** and a pair of caudal appendages called **UROPODS**. Tanaids differ from isopods in having six rather than seven pereonites, at least one jointed uropod branch, and (with few exceptions) a pair of pincers or true chelae on the chelipeds, where these are simple or subchelate in isopods.

The young are brooded in the female's brood pouch (**MAR-SUPIUM**), and emerge as epibenthic juveniles called **MANCAS**. The marsupium is formed on the underside of the pereon from thin plates (**OÖSTEGITES**) that project from the basal segments of one or more pairs of legs. The sexes are often dissimilar, and in some species different types of males may develop either from mancas or secondarily from females. Males of highly dimorphic species can generally be distinguished from females by their more strongly developed chelipeds, often bearing large and sometimes grotesque chelae; longer first antennae with more flagellar segments, which bear sensory setae (**AES-THETASCS**); larger eyes; and in some genera, fused or vestigial mouth parts.

Lang (1956) divided the tanaids into two suborders, the Monokonophora with a single, small genital cone (the penial process at the end of the sperm duct) between the last pair of legs, and the Dikonophora with two. Sieg (1980) proposed an arrangement with three suborders, the Apseudomorpha (corresponding to the Monokonophora), the Tanaidomorpha, and the Neotanaidomorpha (together corresponding to the Dikonophora), which is followed here. Only the Apseudomorpha and Tanaidomorpha are represented by species in this key.

ACKNOWLEDGMENTS

My thanks to Richard Heard, Don Cadien, and Jim Carlton for their very helpful comments on this section.

Key to Tanaidacea

1. First antenna with two-branched flagellum (plate 253B2); pleopods sometimes lacking; mandible with three-articled palp (plate 253H1); marsupium in females formed by four pairs of oostegites; not tube dwellers . . . . Apseudomorpha 2

— First antenna with unbranched flagellum; mandible without palp (plate 253H2); pleopods always present; marsupium in females formed by one or four pairs of oostegite; tube dwellers . . . . Tanaidomorpha 3

2. Five pleonites plus pleotelson; pleon coiled or asymmetrical; pereopods cylindrical; first pereopod more than twice the length of pereopods 2–5; lives in tiny snail shells . . . . *Pagurotanais* sp.

— Two free pleonites (pleonites 3–5 fused with pleotelson) plus a sharply triangular pleotelson with three dorsal swellings, each bearing a few spines; pleon straight and symmetrical; pereopods somewhat flattened and stout; first pereopod little longer than pereopods 2–5 (plate 253B) . . . . *Synapseudes intumescens*

3. Five pleonites plus pleotelson; five pairs of pleopods; four pairs of oostegites; uropods two-branched, though the outer branch may be inconspicuous; usually found on mud (plate 253A, 253G) . . . . *Leptochelia* sp.

— Three to five pleonites plus pleotelson; three pairs of pleopods (one may be rudimentary); one pair of oostegites modified into ovisacs on the fifth pair of pereopods; uropods unbranched; usually found on hard substrates . . . . Tanaidae 4

4. Three pleonites plus pleotelson; two functional pairs and one rudimentary pair of pleopods (plate 253C) . . . . *Pancolus californiensis*

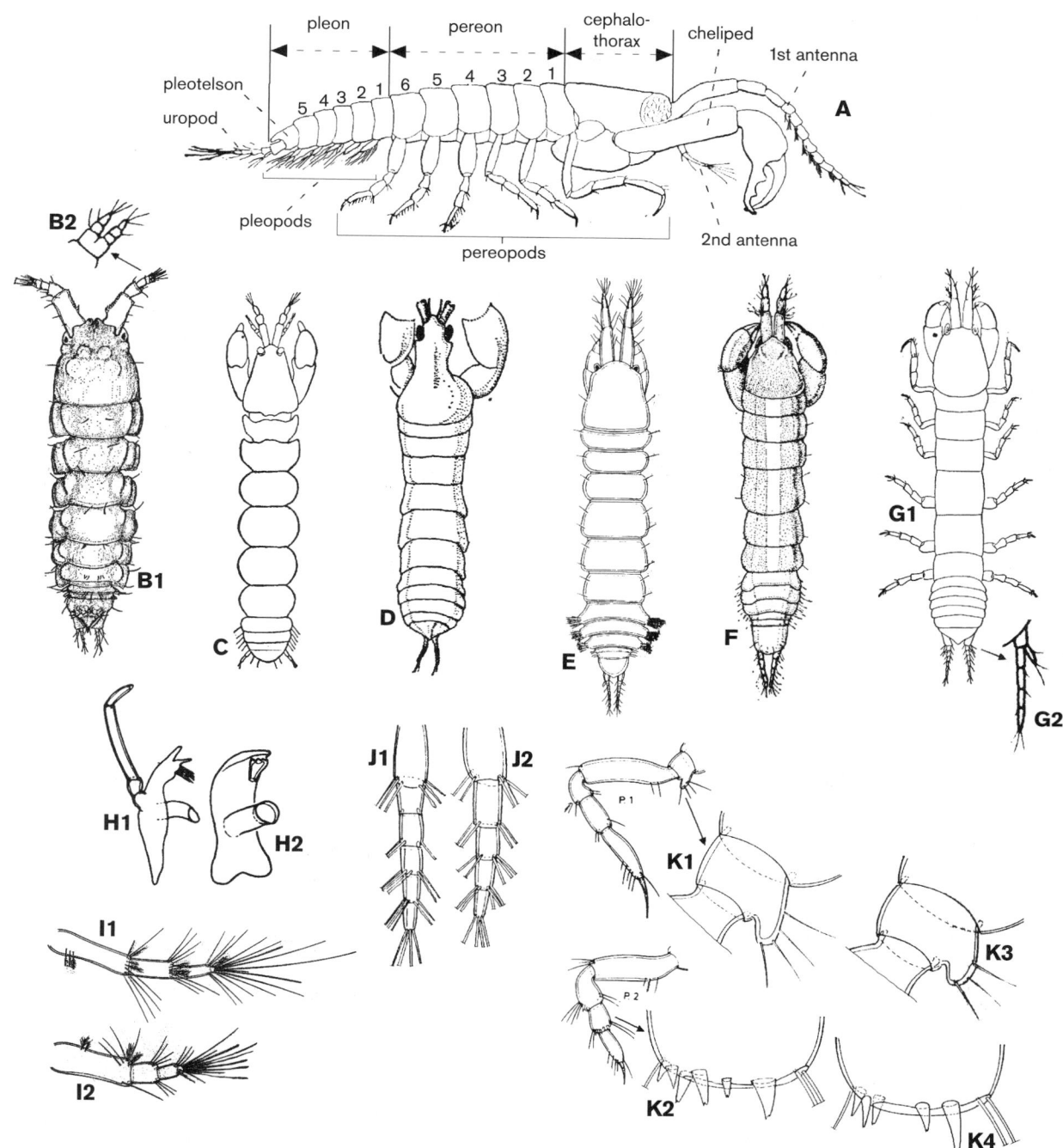

PLATE 253 A, *Leptochelia* sp., male, specimen from New England, from Richardson, 1905b (from Harger), as *L. savignyi*; B1, *Synapseudes intumescens*, from Menzies, 1949; B2, flagellum of 1st antenna, from Miller, 1968; C, *Pancolus californiensis*, modified after Richardson, 1905a, to show three pleonites; D, *Sinelobus* sp., damaged male, from Miller, 1968 (as *Tanais* sp.); E, *Anatanais pseudonormani*, from Sieg and Winn, 1981; F, *Zeuxo normani*, from Miller, 1968; G, *Leptochelia* sp., G1, female, specimen from New England, from Richardson, 1905b (from Harger), as *L. savignyi*; G2, uropod, from Holdich and Jones, 1983; H1, Apseudomorpha mandible, with 3-articled palp; H2, Tanaidomorpha mandible, without palp, from Sieg and Winn, 1979; I, 1st antenna, I1, of *Anatanais pseudonormani*, from Sieg and Winn, 1981; I2, of *Zeuxo normani*, from Sieg, 1980; J, uropod, J1, of *Zeuxo normani*; J2, of *Zeuxo paranormani*, from Sieg and Winn, 1981; K, *Zeuxo normani*, K1, coxa of 1st pereopod; K2, distal end of carpus of 2nd pereopod, lateral view; *Zeuxo paranormani*, K3, coxa of 1st pereopod; K4, distal end of carpus of 2nd pereopod, lateral view, from Sieg and Winn, 1981 (figures from Sieg and Winn [1979, 1981] used with permission of the Biological Society of Washington; figure from Holdich and Jones [1983] used with permission of Cambridge University Press).

— Four pleonites plus pleotelson; three pairs of functional pleopods on pleonites 1–3; complete transverse dorsal rows of setae on first two to three pleonites; male cephalon strongly narrowed toward anterior (plate 253D) . . . .
. . . . *Sinelobus* sp.
— Five pleonites plus pleotelson; three pairs of functional pleopods on pleonites 1–3 5
5. First article of the first antenna twice the length of the second article (plate 253E, 253I1) . . . .
. . . . *Anatanais pseudonormani*
— First article of the first antenna two and a half to three times the length of the second article (plate 253I2). 6
6. Adult uropods with six articles (plate 253J1); coxa of first pereopod with longer protuberance (plate 253K1); distal end of carpus of second pereopod with four lateral and two medial spines (plate 253K2, 253E) . . . . . . *Zeuxo normani*
— Adult uropods with five articles (plate 253J2); coxa of first pereopod with shorter protuberance (plate 253K3); distal end of carpus of second pereopod with three lateral and two medial spines (plate 253K4) . . . . . *Zeuxo paranormani*

List of Species

APSEUDOMORPHA

PAGURAPSEUDIDAE

Pagurotanais sp. (=*Pagurapseudes* of previous west coast literature). *Pagurotanais* species, which are adapted for occupying snail shells in the manner of hermit crabs, have rarely been reported in the northeastern Pacific. Menzies (1953) described *P. laevis* from kelp holdfasts in 4–6 m off Santa Catalina Island, and from 91–93 m off Guadalupe Island, Mexico; Howard (1952) noted a species (identified as *Pagurapseudes* sp. by Menzies) living in a caecid snail shell in Concepcion Bay, Baja, California. Lee and Miller (1980) reported females of an undescribed species, similar to but not *P. laevis,* at Pacific Grove in Monterey Bay "occupying small snail shells among holdfasts of red algae, low intertidal zone on rocky shores protected from strong surf"; apparently the same species was collected from red algae in the lower intertidal at Hopkins Marine Station in 1967 "in shells of *Barleeia* and other gastropods" and illustrated in Abbott (1987).

METAPSEUDIDAE

Synapseudes intumescens Menzies, 1949. Fairly common on exposed rocky shores in the low intertidal from Sonoma County to Guadalupe Island, Mexico, occasionally down to 66 m. Reported on a variety of substrates including the holdfasts and lower blades of brown and red algae, arborescent bryozoans, tunicates, the dorsal surface of the seastar *Patiria*, abalone shells, in *Mytilus* beds, and on and under rocks (Menzies 1949, 1953; Lee and Miller 1980; Abbott 1987).

TANAIDOMORPHA

TANAIDAE

Anatanais pseudonormani Sieg, 1980. See Sieg and Winn 1981. A sublittoral southern California species (as shallow as

13 m in Scorpio Harbor on Santa Cruz Island), but included here so it may be watched for over a broader area.

Pancolus californiensis Richardson, 1905. Habitats include the sand held underneath cushionlike clumps of *Cladophora* in the high intertidal and the holdfasts of sea palms (*Postelsia*) in the low intertidal, from central to southern California; records in the Columbia River estuary and Puget Sound require confirmation. Richardson (1905a, b) described and illustrated this species as having two pleonites, but Lang (1950, 1961) re-examined the type material and provided photographs that clearly show three pleonites.

Sinelobus sp. (=*Tanais* sp. of previous editions). Introduced species in fouling in bays and estuaries. Although Sieg (1980) assigned this species to *S. stanfordi* (Richardson, 1901) (=*Tanais stanfordi*) based on an illustration of a damaged specimen, it is apparently not that species. *S. stanfordi* has two dorsal, separated, curved rows of setae on each of the first two pleonites, rather than continuous transverse rows.

Zeuxo normani (Richardson, 1905) (=*Anatanais normani*, =*Tanais normani*). On bryozoans, hydroids, and red (especially coralline) algae, from British Columbia to southern California, and Japan (Hatch 1947; Miller 1968; Sieg 1980; Sieg and Winn 1981).

Zeuxo paranormani Sieg, 1980. Sieg (1980) determined that part of Richardson's *Zeuxo normani* type material consisted of this very similar species. Reported in Humboldt Bay, Monterey Bay, and southern California.

LEPTOCHELIIDAE

Leptochelia spp. Usually white, sometimes greenish or tinged with orange, abundant on mudflats and among algae in pools. *Leptochelia* species, all under the name *Leptochelia dubia* (Krøyer, 1842), which was first described from Brazil, have been reported from sandy intertidal flats in Puget Sound (where they prey on sand dollar larvae; Highsmith 1983, Ecology 63: 329–337; see also Highsmith 1983, Ecology 64: 719–726, sex reversal and fighting behavior in Puget Sound), from deep water (to nearly 600 m) off southern California (Dojiri and Sieg 1997), and from bay mud and occasionally in fouling communities elsewhere. These doubtless represent more than one species, possibly including both native and introduced taxa. Globally, the name *L. dubia* has been variously restricted to tropical-subtropical populations around the world (as reviewed by Miller 1968) or used for all temperate to tropical populations of this species group (Sieg 1986). The name *Leptochelia savignyi* (Krøyer, 1842), has been used as both a junior and a senior synonym of *L. dubia*; if the two are the same, *L. savignyi* is the "older" name, with page priority.

LEPTOGNATHIIDAE

There are a few unconfirmed reports in shallow water in central California of the usually deepwater genus *Leptognathia*. However, the Leptognathiidae has been revised (Larsen and Wilson 2002), and it is unclear to which genus or family these records should now be referred.

References

Abbott, D. P. 1987. Observing marine invertebrates. Drawings from the laboratory. Edited by Galen Howard Hilgard. Stanford University Press, 380 pp.

Dojiri, M., and J. Sieg, 1997. The Tanaidacea, pp. 181–278. In: J. A. Blake and P. H. Scott, Taxonomic atlas of the benthic fauna of the Santa Maria Basin and western Santa Barbara Channel. 11. The Crustacea. Part 2 The *Isopoda, Cumacea* and *Tanaidacea*. Santa Barbara Museum of Natural History, Santa Barbara, California.

Hatch, M. H. 1947. The Chelifera and Isopoda of Washington and adjacent regions. Univ. Wash. Publ. Biol. 10: 155–274.

Holdich, D. M., and J. A. Jones. 1983. Tanaids: keys and notes for the identification of the species. New York: Cambridge University Press.

Howard, A. D. 1952. Molluscan shells occupied by tanaids. Nautilus 65: 74–75.

Lang, K. 1950. The genus *Pancolus* Richardson and some remarks on *Paratanais euelpis* Barnard (Tanaidacea). Arkiv. for Zool. 1: 357–360.

Lang, K. 1956. Neotanaidae nov. fam., with some remarks on the phylogeny of the Tanaidacea. Arkiv. for Zool. 9: 469–475.

Lang, K. 1961. Further notes on *Pancolus californiensis* Richardson. Arkiv. for Zool. 13: 573–577.

Larsen, K. and G. D. F. Wilson. 2002. Tanaidacean phylogeny, the first step: the superfamily Paratanaidoidea. J. Zool. Syst. Evol. Res. 40: 205–222.

Lee, W. L., and M. A. Miller. 1980. Isopoda and Tanaidacea: the isopods and allies. In Intertidal invertebrates of California. pp. 536–558. R. H. Morris, D. P. Abbott, and E. C. Haderlie, eds. pp. 536-558. Stanford, CA: Stanford University Press, 690 pp.

Menzies, R. J. 1949. A new species of Apseudid crustacean of the genus *Synapseudes* from northern California (Tanaidacea). Proc. U.S. Natl. Mus. 99: 509–515.

Menzies, R. J. 1953. The Apseudid Chelifera of the eastern tropical and north temperate Pacific Ocean. Bull. Mus. Comp. Zool. 107: 443–496.

Miller, M. A. 1940. The isopod Crustacea of the Hawaiian Islands (Chelifera and Valvifera). Occ. Pap. Bernice P. Bishop Mus. 15, no. 26, pp. 299–321.

Miller, M. A. 1968. Isopoda and Tanaidacea from buoys in coastal waters of the continental United States, Hawaii, and the Bahamas (Crustacea). Proc. U.S. Natl. Mus. 125: 1–53.

Richardson, H. 1905a. Descriptions of a new genus of Isopoda belonging to the family Tanaidae and of a new species of *Tanais*, both from Monterey Bay, California. Proc. U.S. Natl. Mus. 28: 367–370.

Richardson, H. 1905b. A monograph on the isopods of North America. Washington, D.C. Smithsonian Institution, 727 pp.

Sieg, J. 1980. Taxonomische Monographie der Tanaidae Dana, 1849 (Crustacea: Tanaidacea). Abhandlungen Senckenbergische Naturforschende Gesellschaft 537: 1–267.

Sieg, J. 1986. Distribution of the Tanaidacea: Synopsis of the known data and suggestions on possible distribution patterns, pp. 165–193. In: Crustacean Issues, vol. 4, Crustacean Biogeography, F. R. Schram, ed., Balkema, Rotterdam, The Netherlands.

Sieg, J., and R. N. Winn. 1979. Keys to suborders and families of Tanaidacea (Crustacea). Proc. Biol. Soc. Wash. 91: 840–846.

Sieg, J., and R. N. Winn. 1981. The Tanaidae (Crustacea; Tanaidacea) of California, with a key to the world genera. Proc. Biol. Soc. Wash. 94: 315–343.

Amphipoda

(Plate 254)

The Amphipoda have been divided into the suborders Gammaridea, Caprellidea, Cyamidea, Hyperiidea and Ingolfiellidea (Schram 1986, Crustacea. Oxford University Press, New York). However, Myers and Lowry (2003) regard the caprellids, or skeleton shrimps, and the cyamids, or whale lice, as families Caprellidae and Cyamidae. These distinctive groups are covered in separate sections in this manual, for ease of recognition and identification.

The Caprellidae (plate 254A) occur on solid surfaces and are strictly marine or estuarine. The Cyamidae are ectoparasites of cetaceans and are occasionally found on beached whales and dolphins (plate 254B). The Hyperiidea (plate 254C) are parasites and commensals of marine macrozooplankton and are exclusively pelagic. Hyperiids are occasionally discovered free swimming intertidally or in shallow-water plankton tows, or are found attached beneath or embedded in the bells of stranded medusae or salps. The Gammaridea (scuds, landhoppers, and beachhoppers) (plate 254E) are the most abundant and familiar amphipods. They occur in pelagic and benthic habitats of fresh, brackish, and marine waters, the supralittoral fringe of the seashore, and in a few damp terrestrial habitats and are difficult to overlook. The wormlike, 2-mm-long interstitial Ingofiellidea (plate 254D) has not been reported from the eastern Pacific, but they may slip through standard sieves and their interstitial habitats are poorly sampled.

Key to Amphipoda

1. Gills not exceeding three pairs, female oöstegites not exceeding two pairs; pleon and urosome (abdomen) vestigial and pereonite 1 fused to head 2
— Gills and oöstegites exceeding three pairs, abdomen and abdominal appendages well developed; head and pereonite 1 separate 3
2. Body segments tubular, legs with moderate hooks, free living (plate 254A) . . . . Caprellidae
— Body segments loosely separated, legs powerful with sharp hooks, parasites of cetaceans (plate 254B) . . . . . Cyamidae
3. Urosome with only two segments; palps of maxillipeds absent; eyes usually cover most of head but can be tiny; entirely pelagic (plate 254C) . . . . Hyperiidea
— Urosome with three segments; palps of maxillipeds present 4
4. Pleopods leaflike, vestigial, or absent; movable compound claw of gnathopods formed of articles 6 and 7 together; body vermiform; without coxal and epimeral plates; entirely interstitial (unreported from the northeast Pacific) (plate 254D) . . . . Ingolfiellidea
— Pleopods well developed, with few exceptions; dactyls of gnathopods formed by article 7 alone (plate 254E) . . . . Gammaridea

Gammaridea

JOHN W. CHAPMAN

(Plates 255–304)

The ubiquitous and abundant gammaridean amphipods are critically important in marine and estuarine shallow-water ecosystems of the northeast Pacific and warrant reliable, workable guides to the species. The numerical abundances and species and life-history diversities of the Gammaridea exceed all other eucaridan or peracaridan orders. Gammaridean amphipods are one of the most common aquatic taxa. The taxonomy and systematics of marine eastern Pacific species have greatly advanced since 1975, but many undescribed species occur in the region and little more than the names of most described species are known. The lack of research is disproportionate to these species' importance in ecosystems that are of great interest to humans.

Gammaridean amphipods are critical food sources of whales, fish, and birds, (Moore et al. 2003, McCurdy et al. 2005, Schneider and Harrington 1981) and are highly sensitive to environmental alterations (Conlan 1994, Zajac et al. 2003). All amphipods care for their offspring for extended periods (Jones 1971, Shillaker and Moore 1987). Some change sex (Lowry and Stoddart 1986); others attract, hold, and defend mates (Borowsky1983, 1984, 1985; Conlan 1989, 1995a) and

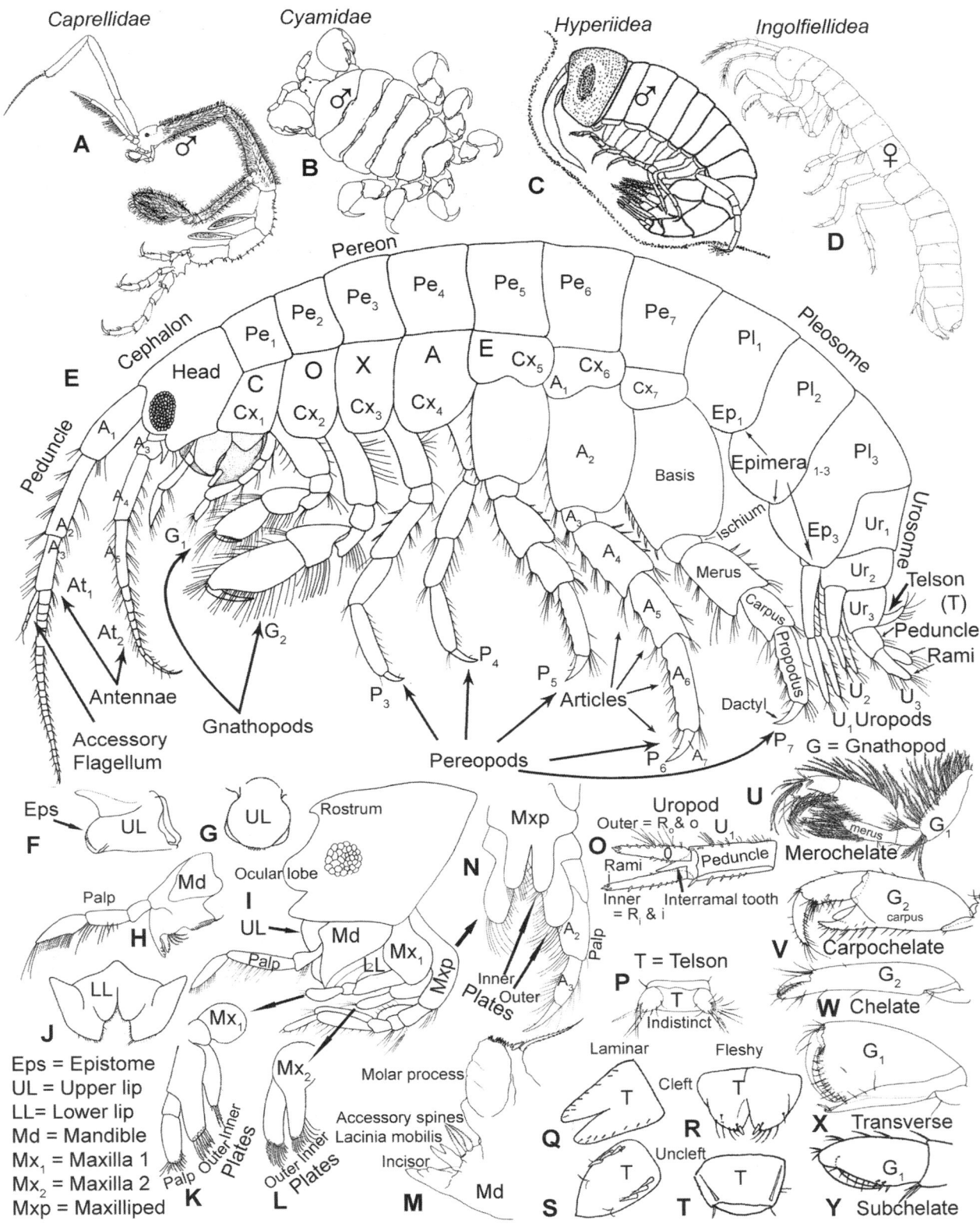

PLATE 254 Amphipoda. A, Caprellidae—*Caprella mutica*; B, Cyamidae—*Cyamus scammoni*; C, Hyperiidea—*Hyperoche medusarum* (Müller, 1776) in situ; D, Ingolfiellidea—*Ingolfiella fuscina* Dojiri and Seig, 1987; E, Gammaridea—generalized body; F, G, generalized upper lip; H, generalized mandible; I, generalized head; J, generalized lower lip; K, generalized maxilla 1; L, generalized maxilla 2; M, *Polycheria* mandible; N, generalized maxilliped; O, uropod 1, *Paragrubia uncinata*; P, telson, *Eohaustorius*; Q, telson, *Batea lobata*; R, telson, *Parallorchestes leblondi*; S, telson, *Stenothoe estacola*; T, telson, *Paracorophium* sp.; U, gnathopod 1, *Aoroides secundus*; V, gnathopod 2, *Ericthonius brasiliensis*; W, gnathopod 1, *Americhelidium shoemakeri*; X, gnathopod 2, *Americhelidium rectipalmum*; Y, gnathopod 1, *Stenothoe valida* (figures modified from: Barnard 1953,1962c, 1965, 1975; Barnard and Karaman 1991a, 1991b; Bousfield 1973; Bousfield and Chevrier 1996; Bousfield and Hendrycks 2002; Bousfield and Kendall 1994; Doiji and Sieg 1987, Flores and Brusca 1975; Gurjanova 1938; Margolis et al. 2000; Todd Miller, personal communication; and Platovoet et al. 1995).

territories (Connell 1963). Some use chemicals for defense (Hay et al. 1987, Hay et al. 1990) or are repelled by defensive chemicals (Hay et al. 1988, 1990; Cronin and Hay 1996a, 1996b). Some species undergo risky long-distance migrations (Chess 1979, Mills 1967, Watkin 1941), and others exploit, imitate, parasitize, eat (Crane 1969, Cartwright and Behrens 1980, Goddard, Skogsberg, and Vansell 1928), displace, or attack other invertebrates and fish (Bousfield 1987, Wilhelm and Schindler 1999); burrow in wood (Barnard 1955c) or macroalgae (Conlan and Chess 1992); or alter sediment dynamics in estuaries (Olafsson and Persson 1986). Lysianassid amphipods are adapted for engorgement and are among the most important carrion feeders in the sea (Dahl 1979, Thurston 1990). Some amphipods die of unknown diseases (Pelletier and Chapman 1996), which they may introduce with them by humans to new areas (Slothouber Galbreath et al. 2004).

Genus-level variation within most families occurs in geographical patterns that correspond with the Cretaceous continental divisions. However, these evolutionary patterns have been and continue to be obscured by human introductions of shallow-water amphipods among continents. One in 10 of the eastern Pacific shallow-water species treated herein are likely introductions.

Morphologies for stridulation (to make harsh sounds) occur among Isaeidae, Melitidae, and perhaps Phoxocephalidae and are likely adaptations for attracting mates. Their sounds have not been recorded and their adaptive values have not been resolved. Most amphipod species are beautiful (flamboyant) in color and form, but color pictures of live amphipods are seldom published and only a few are posted on the internet.

All coastal marine and estuarine gammaridean amphipod families, genera, and species from the Columbia River to Point Conception reported from <10 m depths are included in the keys or listed. A few species known only from depths >10 m or slightly north or south of the region are listed and indicated by asterisks. These latter species are recognized from limited taxonomical or geographical information and thus, although not clearly within the geographical region, cannot be reliably discounted. Few of these latter species are included in the keys.

The first edition of *Light's Manual* (Light 1941) covered 47 gammaridean species; the second edition Barnard (1954d) covered approximately 61, and the third edition (Barnard 1975) included 141 species. This section includes 351 species. The exponential increase in species (including nearly half of the species discovered or resolved since the 1970s) is due largely to the contributions of Bousfield, Conlan, Hendrycks, and coworkers at the Canadian Museum of Nature. The many species and additional character variations and types of taxonomic characters that Bousfield and his colleagues discovered reveal the paramount importance of distinguishing interspecific and intraspecific variation. The taxonomic outlines for this section rely extensively on their contributions. General treatments of gammaridean amphipod taxonomy and systematics also include Barnard and Barnard (1983a, b), Barnard and Karaman (1991a, b), Bellan-Santini (1999), Bousfield (2001), Myers and Lowry (2003), Serejo (2004) and Staude (1997).

Additional guides to amphipod taxonomic literature of the eastern Pacific include bibliographies (SCAMIT 2001) and Internet postings, such as the "Amphipod Newsletter."

Although intended to be comprehensive within the region, these keys are not reliable outside of their specified geographic,

bathymetric, or ecological boundaries. Barnard's (1975) emphasis on durable and external morphology is followed with "natural" dichotomies sacrificed when artificial distinctions are more apparent, where family, genus, or species relationships remain poorly resolved, where difficult dissections or magnifications of greater than 40x can be avoided, or where characters are fragile or difficult to observe or to define. Occasional notes in the species lists are to assist with identifications, indicate pitfalls, or provoke interest.

The Gammaridea (plate 254E, center) (legged order) have a clearly defined **CEPHALON (HEAD)**, a thorax or **PEREON** of seven freely articulated segments, (**PEREONITES = PE**$_{1-7}$), a six-segmented **ABDOMEN (PLEON)** of three (**PLEONITES = PL**$_{1-3}$) (**PLEOSOME**) and three (**UROSOMITES = UR**$_{1-3}$) (**UROSOME**) and a **TELSON (=T)**. The **TELSON** is a flap over the anus attached to pleonite 6 (urosomite 3). Most gammarideans are laterally flattened. Each pleonite has a pair of **PLEOPODS** (swimmerets). The pleopods are complex and seldom illustrated. The lower lateral edges of the pleonites that extend below the body are **EPIMERA (=EP**$_{1-3}$). The urosomites of the **UROSOME** each bear rigid, lateral, posterior projecting **UROPODS (=U**$_{1-3}$). The uropods usually consist of a basal **PEDUNCLE** and one or two distal **RAMI**.

The appendages of the head (plate 254E, 254I center left) are, in order from anterior to posterior: **ANTENNA 1, ANTENNA 2 (=AT**$_{1-2}$) (plate 254E, center left), **UPPER LIP (UL)** (plate 254F, 254G, bottom left), **MANDIBLE (MD)** (plate 254H, 254M), **LOWER LIP (LL)** (plate 254J), **MAXILLA 1 (MX**$_1$) (plate 254K), **MAXILLA 2 (MX**$_2$) (plate 254L), and the **MAXILLIPED (MXP)** (plate 254N).

ANTENNA 1 (plate 254E, center left) is composed of a three-article peduncle and a **FLAGELLUM** of variable article numbers. An **ACCESSORY FLAGELLUM** extends from the distal medial surface of the third peduncle article and is prominent on many species and families but is also minute (requiring high magnification to observe) or absent in other species and families.

ANTENNA 2 (plate 254E, center left) consists of five peduncular articles and a flagellum of variable article numbers. The morphologies of the peduncle articles and the flagellum and the distributions and morphologies of their spines and sensory organs (calceoli) are important taxonomic characters (Steele and Steele 1993, Bousfield 2001). Count peduncle articles from the fifth backward because articles 1–3 are usually more difficult to distinguish than the difference between peduncle and flagellum.

The **EPISTOME (EPS)** (plate 254F, bottom left) is a cephalic sclerite attached to of the anterior upper lip and usually fused to the upper lip when it is large. The epistome is taxonomically important mainly in Phoxocephalidae and Lysianassidae. However, the epistome varies greatly in other families and perhaps the most spectacular is the noselike extension of *Proboscinotus loquax* (plate 256B) for which this species is named.

The **UPPER LIP** (plate 254F, 254G, bottom left) is variable among species and families but seldom used for taxonomic purposes unless it has a well-developed epistome. Najnidae and Pleustidae have asymmetrical upper lips.

The **MANDIBLE** (plate 254H, 254M) has a **MOLAR** (plate 254M), which can be large and triturative, reduced, or vestigial and without a grinding surface. The molar often bears a large pinnate seta (illustrated in plate 254M). The apex of the mandible is the **INCISOR** (plate 254M), which usually projects as a series of teeth. Above the incisor is the **LACINIA MOBILIS** (plate 254M), a spinelike movable appendage of the medial mandibular edge that consists of variable numbers of individual or fused articulated spines (see E. Dahl and R. T. Hessler, 1982, Zool. J. Linnean Soc. 74: 133–146 on origin, function and phylogeny). The lacinia mobilis (plate 254M) varies laterally in Phoxocephalidae and Pleustidae and is used for species distinctions among pleustids (Bousfield and Hendrycks 1994a, 1994b; Hendrycks and Bousfield 2004). A row of accessory spines usually occur above the lacinia mobilis (plate 254M). The mandible also usually bears a triarticulate **PALP** (plate 254H, 254I). The mandibular palp is sometimes only one or two articles and is lacking in Dogielinotidae, Eophliantidae, Hyalidae, Hyalellidae, Najnidae, Phliantidae and Talitridae and variably present in Dexaminidae and Synopiidae.

The **LOWER LIP** (plate 254J) lies behind the mandibles and in front of the first pair of maxillae. The lower lip can be difficult to remove without damage. (see instructions for dissection below) but is a particularly important character for distinguishing ampithoid and pleustid species and genera and for distinguishing eusirids from pleustids.

The **MAXILLA 1** (plate 254K) have an inner and an outer plate and a palp of two or more articles. The inner plate is not closely contiguous with the outer plate and can be overlooked or lost during dissection when it remains partially attached to the lower lip. The outer plate bears heavy distal spines. The palp is occasionally reduced to one article or is absent.

The **MAXILLA 2** (plate 254L) are two simple, setose plates and lack a palp.

The **MAXILLIPED** (plate 254N) usually covers the other mouth parts from below and is therefore usually removed first in mouth part dissections. The maxillipeds are fused at the base and appear as a single branched appendage. Each branch has an inner and an outer plate and a palp that is usually composed of four articles. The palp is occasionally reduced to three articles (absent in hyperiids), and the plates are often severely reduced in size. The distal palp article is usually pointed and referred to as a dactyl.

The amphipod **ROSTRUM** (plate 254I) is variable and can extend over or between the first antennae, or it can be greatly reduced or absent. Some amphipods are blind. Most have lateral eyes of one to hundreds of ommatidia. The ocular lobe (plate 254I) of many Gammaroidea, including Hadzioidea, Eusiroidea, and Corophioidea, extends over the second antenna and usually forms a ventral antennal sinus notch. Oedicerotidae and Synopiidae ommatidia merge dorsally into a single eye. Argissidae eyes each consist of four lenticular facets and Ampeliscidae eyes consist of two corneal lenses.

Most uropods (=U_{1-3}) (plate 254E, right; 254O) are composed of a **PEDUNCLE** and one or two **RAMI** (plate 254O). The inner and outer rami are labelled in illustrations as R_i and R_o respectively, or simply "$_i$" and "$_o$." The rami vary from bare or spinose stubs, to short ornamented appendages with fine teeth and denticles, to tubular, to long and spinose, lanceolate (spear-shaped), and to foliose (leaf-shaped). Many species have a large peduncular tooth extending between the rami (plate 254O) of the first or second uropods or other large spines that are important taxonomic characters. Most rami are of a single article, but on uropod 3, outer rami of two articles are characteristic of Lysianassidae, Liljeborgiidae, Gammaridae, Hadzioidea, Melitidae, and Stenothoidae.

General **TELSON** (=T) morphologies (plate 254P–254T, bottom center) are, respectively: **INDISTINCT** (plate 254P), **LAMINAR CLEFT** (plate 254Q), **FLESHY CLEFT** (plate 254R), **LAMINAR UNCLEFT** (plate 254S), and **FLESHY UNCLEFT** (plate 254T). Fleshy

and laminar telsons range greatly in form between **CLEFT** (split into two lobes) and **UNCLEFT** (fused into a single piece). **LAMINAR** telsons are dorsoventrally thin (flattened). Some laminar telsons (Pleustidae) have a ventral keel that creates a thickened appearance from a lateral view even though the edges and apex are laminar. Fleshy telsons are at least one-third as thick as they are wide or long. The tonguelike telsons of some Stenothoidae (plate 254S) seem to fall in between, but they are mostly longer and wider than thick. Haustoriidae telsons (plate 254P) are referred to as indistinct because the widely separated lobes are not clearly fleshy or laminar.

"Amphipod" means "double feet." The amphipod thorax (**PEREON**) (plate 254E, center) bears seven pairs of walking legs (**PEREOPODS** = P_{1-7}). Naming systems for amphipod legs (pereopods) vary. The first two pereopods are adapted primarily for feeding, defense, cleaning, and reproductive activities. Evolution of gnathopod morphology for feeding, mating, and defense is unlikely to have occurred in response to the same selection processes as pereopods 3–7, which are used for attachment, mobility, and nest or tube construction. The reference to pereopods 1 and 2 as **GNATHOPODS** (=G_{1-2}) and continuing the sequence with **PEREOPODS** 3–7 (=P_{3-7}) is used here (plate 254E, center; plate 254U–254Y). However, reference to all walking legs as pereopods 1–7 is also correct. Beware of amphipod descriptions previous to the 1980s that commonly refer to gnathopods 1 and 2 and then to pereopods 2–7 as pereopods 1, 2, 3, 4, and 5.

Each pereopod has seven **ARTICLES** (=A_{1-7}) (plate 254E, center) ("joint" is an inappropriate term). These seven articles are, respectively, the **COXA**, **BASIS**, **ISCHIUM**, **MERUS**, **CARPUS**, **PROPODUS**, and **DACTYL** (plate 254E, center). The **COXAE** (=CX_{1-7}) (plate 254E, center) is pereopod article 1 and often expands from its attachment point on the body downward to cover remaining parts of the pereopod. The coxae are normally the most conspicuous articles of the pereopods. Reference to the coxae and dactyls and then to article numbers 2–6, rather than **MERUS–PROPODUS**, is common usage, but exceptions are numerous in this key and elsewhere when descriptions require fine details or distinctions.

Details of the prehensile gnathopod morphology are critical in amphipod taxonomy. **MEROCHELATE** (plate 254U, lower right) refers to the condition in which articles 5 (carpus), 6 (propodus), and 7 (dactyl), respectively, fold around the merus (article 4) to become prehensile. **CARPOCHELATE** (plate 254V) refers to the condition in which articles 6 (propodus) and 7 (dactyl) fold around the carpus (article 5) to become prehensile. The posterior edge of gnathopod article 6 that is overlapped by the dactyl is commonly referred to as a **HAND** or **PALM**. **CHELATE** (plate 254W) refers to the dactyl (article 7) closing onto an extended finger of the palm at >90°. **TRANSVERSE** (plate 254X) applies to pereopods and gnathopods on which the dactyl closes against a palm at a 90° angle. **SUBCHELATE** (plate 254Y) is the condition where the dactyl closes against the palm at less than a 90° angle. **SIMPLE** refers to the condition in which the dactyl does not fold onto article 6 and is thus not prehensile. Thus, most pereopods (P_{3-7}) (plate 254E, center) are simple.

Sexual differences are distinct in many species, with males bearing enlarged, heavily prehensile gnathopods; extra large, long, and densely setose antennae; large and densely setose uropod 3; powerful pleopods; and, occasionally, large eyes. However, many species have few external sexual differences. Mature males have a minute pair of **PENIAL PROCESSES** (penes) that hang ventrally from the pereonite 7 between the coxae. Penes often bear tiny spines that are difficult to see and, after the seventh pereopods are removed, can be confused with broken ends of tendons. The penes can also be confused with gills (plate 255MM), which are attached to the coxae and vary greatly in shape and size among amphipod families. Among appendages of the pereopods, if they break off easily, they are probably gills.

Breeding females bear up to six laminate brood plates (OÖSTEGITES) (plate 255MM, 255NN) in the space between the bases of coxae 2 and 5. The oöstegites form a pericardium (marsupium or brood chamber) that encases the eggs and for which peracaridans are named. The oöstegites are attached to the coxae medially and can be confused with the gills but are clearly distinguished by their long, interleaving, pinnate setae on mature females.

COLLECTION AND DISSECTION

The great diversity and superficially similar morphologies of gammaridean amphipods could make their taxonomy appear difficult, but they don't. Sharp forceps, sometimes probes tipped with insect pins or fine sewing needles, a stereomicroscope with magnifications ranging between 6x and 40x, patience, and interest to learn the simple anatomy, follow instructions, and learn from mistakes are all that is needed to identify amphipods.

Dissections of mouth parts under high magnification can appear daunting, but preparation is more important than skill. The top half of 50-mm petri dishes are good containers for dissections. Glass is preferable. Detach the base of transmitting stereoscopes that do not have hand rests or otherwise provide a platform level with the specimens that will allow palms to rest on a surface and stabilize both hands for the fine manipulations that are needed. Replace or cover clear glass microscope stages with black or dark blue covers to prevent transmitted light and maximize reflected light. Adjust the light source to maximize unobstructed illumination of the specimen without reflection from overlying liquid surface.

A compound microscope with 100x–1,000X magnifications (and transmitted light) is necessary to observe tiny appendages or fine anatomical characters. Prepare mounts of these characters under the stereoscope using the equipment above. Tiny parts are easy to lose. A slide placed off center on the petri dish with a centered drop of glycerin will allow continuous observation of the transfer of dissected parts directly from the dissecting dish into the glycerin drop.

Begin with numerous large, mature, unbroken specimens of a single species bearing both pairs of antennae, all three pairs of uropods, a telson, both pairs of gnathopods, and at least the first three articles of all other legs. Previously identified species, if available, limit the need for guessing at the conditions of difficult-to-find or -observe characters. Readily identifiable gammaridean amphipods that are easy to find include beach hoppers (Talitridae) of open coastal beaches; the large green Ampithoidae of docks and floats in bays and estuaries, and Corophiidae and Aoridae of estuary mudflats and fouling communities are usually large enough for observations of basic morphology and anatomy under low magnification.

Amphipods occur nearly everywhere that permanent water occurs, and they are easy to find by washing aquatic sediments or plants on a 0.5 mm or 1.0 mm mesh sieve or in a section of plastic window screening or by sweeping with a dip net. Spread the washed material in a shallow pan and sort out the animals using light forceps or plastic eyedropper (pipette).

Freezing kills painlessly and does not ruin amphipods for later use if they are preserved in 70% alcohol within a few hours. Suitable preservation for morphological analyses includes fixing the animals in 10% formalin for a few days before permanent storage in 70% alcohol. Preserve the specimens directly in alcohol if they are to be used for molecular genetics analyses.

For dissection, immerse a specimen in 70% alcohol that is sufficiently deep for manipulations using forceps without distorting the liquid surface directly over the specimen. The best light is usually by reflection from the sides of the specimen rather than transmitted from beneath. Two lamps from different angles are better than one. Cool fiber optics lamps are better than direct, hot light from tungsten lamps.

Examine coxa 1 to determine whether it is significantly smaller than or hidden by coxa 2, or nearly as large as coxa 2 and freely visible. Tilt and rotate the amphipod and adjust the light(s) to provide maximum lateral illumination and contrast of plate and segment edges for these observations. Count the coxae to ensure that all seven are being observed.

Manipulate the telson to determine its fleshy or laminar condition. A laminar telson is freely articulate at its base. Remove the urosome and mount it dorsal side up on a depression slide filled with glycerin overlain with a coverslip. Note whether the urosome consists of three separate segments or has one or two fused segments. The rami of uropod 3 are often lost during preservation. Check the mounted urosome and count the rami of the uropods (usually three pairs). Damaged uropods are especially common among Iphimediidae, Megaluropidae, Oedicerotidae, Pleustidae, Eusiroidea, and a few genera of the Podoceridae. Some Gammaridae and Hyalidae have extremely short or inconspicuous inner rami. Remove uropod 3 if necessary and mount it on slide for observation under 100x magnification or more. A sclerotic socket usually remains to mark the presence of a ramus that has been lost.

SPINES and SETAE are the ends of a range of homologous structures. Setae are highly flexible and can be bent in the middle without breaking. Spines are thickened setae that are less flexible and can break when bent. Subtle differences in the placement and arrangement of spines and setae are becoming critical characteristics for distinguishing species in many families. Whether the distal spines of uropod 1 are apicomedial (between the rami) or apicolateral (lateral to the rami), for example, are critical taxonomic characters for distinguishing genera of Phoxocephalidae.

Examine antenna 1 for the presence of an accessory flagellum on the distal medial corner of peduncle article 3 and its condition, if it is present. If an accessory flagellum is not obvious, tiny accessory flagella are readily observed by mounting antenna 1 in glycerin on a slide with a bit of clay or a few grains of sand under a thin coverslip. The sand or clay on the slide allow movement of the coverslip, which can be used to roll appendages to suitable angles for observation. The glycerin slide mounts also work for other small appendages requiring high magnification observations.

Make a slide for each appendage. Hold the body with a dissecting pin or forceps and remove pereopod 5, including the coxa, by grasping deeply into the basal musculature with forceps. Place the pereopod on a drop of glycerin with the outside up. Add the coverslip. Repeat this process for pereopods 4 and 3 and for gnathopods 2 and 1 in order. Remove and mount the urosome. Label the slides as they are produced using a grease pencil or tape labels.

Study the slide of the urosome (telson and urosomites 1–3) and note whether the urosomites are fused, the proportionate lengths of the urosomites, the numbers of uropods, the number of rami on the uropods, the relative lengths of the uropods, and the lengths of the rami relative to the peduncles of the uropods.

Examine the general condition of gnathopods 1 and 2. Proceed in the key as far as is possible with these observations. Make additional slides as needed, or fully dissect the amphipod into its component parts. (About 20 slides are required to mount each major character of a specimen.) Observations on the head and **EPIMERA** (the ventral, lateral, posterior sides of the pleonites) can be difficult using a stereomicroscope. Remove the pleopods for a clearer view of the epimera. Closely observe the head, noting the general outline, the shape of the ocular lobe and any anterior or ventral incisions before removing the antennae or any mouth parts.

Test the amphipod for shrinkage in glycerin before mounting parts in critical dissections. Use a slow-drip method for an hour to replace the alcohol preservative with glycerin if the amphipod develops significant "frost," or air bubbles. Hyalidae are especially sensitive to glycerin.

Spear the head with a needle or clutch it dorsally with forceps in a dish of alcohol to remove mouth parts. Right-handers should hold the left side down with the left forceps or needle and point the mouth parts toward 12 o'clock. Mouth parts are easy to dissect from back to front. Grab the maxilliped across its base with the right forceps to pull it off. Mount it in a drop of glycerin with the curved posterior side upward and without separating the lobes or palps. Follow this procedure to remove and mount maxilla 1 and 2.

The mandibles are heavily sclerotized, solidly attached, or somewhat twisted, can be difficult to remove. If possible, note the presence or absence of the mandibular palps before dissecting the mandibles because they are easily lost during dissection. Use extreme care not to grab the mandible near the molar, incisor, or palp because these characters are readily shattered or broken away. Rotate the mandible outward with slight pressure of the forceps to identify the medial molar before grasping heavily. Remove the mandible by grabbing deeply into and tearing out the fleshy and flexible tissue immediately *behind* it and place it in a puddle of glycerin on a slide. This mass will often include the right and left maxilla 1 and lower lip attached together. Tease away the lower lip, leaving the inner plates of the maxillae attached to their outer plates. Separate and mount the maxillae 1 in glycerin under a coverslip. Mount each mandible in glycerin with sand grains under a coverslip. The sand will allow the mandibles to be rolled over and properly oriented for observation. Label the left and right mandibles.

A clear view of the epistome is possible from the lateral front of the head. Pull the first and second antennae forward and up. The epistome reaches forward beyond the mouth part bundle and can be extended into a significant tooth, spike, or cusp. Care is needed not to confuse the **EPISTOMAL SPIKE** with the **MANDIBULAR PALPS**; the latter are flexible and setose whereas the epistomal spike is solid, smooth, and fixed. An epistomal spike may also be confused with the lateral **EXCRETORY SPOUTS** or **ENSIFORM PROCESSES** projecting from the ventral side of the second antennae peduncle article 2 of phoxocephalids.

AIDS TO IDENTIFICATION

Characters referred to in the keys are usually illustrated in the introduction, family key, or keys to species. Families are clustered in the illustrated keys into similar groups where whole body pictures of each genus are attempted. "Flipping" can be a good way to quickly search among taxa, and the plates are ordered, in part, for this purpose. Use keys forward and backward from known species to test or "verify" identifications (and the key). Read the ecology, natural history, and identification notes in the species lists for more hints on their identities. Most anatomical characters referred to in the keys are clustered in the illustrations of plate 254 to allow quick access to explanations of morphologies referred to in the keys. Mark or copy plate 254 for continuous reference to anatomical notes.

First identifications should begin with large mature specimens in good condition, free of debris or damaged appendages. Return to a mature specimen of the opposite sex or the sex appropriate for the particular key to check critical characters. Test identifications further by reference to any additional relevant literature. Keys interpret nature from incomplete knowledge. Even the best keys can be wrong, incomplete, or unclear. Many northeast Pacific amphipod species remain to be identified or fully described. New species continue to be introduced. Identifications of species using a single key can provide only a first level of confidence.

Specific identifications are increasingly reliable as they are based on increasing numbers of characters and include more biogeographic, ecological, and natural history information. Comparisons with original taxonomic descriptions or with type specimens provide increasingly confident identifications. Character distinctions should also account for variations due to size, reproductive development, and age.

Specific differences are nearly all based on adult morphologies and often on only one sex. The opposite sex often is too poorly described for specific distinctions. Groups emphasizing males usually provide a lower proportion of specimens suitable for identifications. Species are keyed out twice in some cases when sexual characters are critical in the taxonomy and where sexual dimorphism is known.

TABULAR KEY TO FAMILIES

In addition to flaws in the keys, damaged specimens (from gut contents in particular) often lack critical anatomical features, preventing progress directly through dichotomous keys. The gammaridean families and suborders are therefore distinguished additionally in a tabular key and followed by notes on distinctive external and readily observed internal characters. The auxiliary information on the shapes of heads, gnathopods, telsons, and third uropods (and quick-to-observe internal characters) are useful for identifying families "at a glance," or to test questionable endpoints. Gammaridean families and suborders are arranged in Table 5 by telson shape and condition, then by the number and shapes of the rami of uropod 3, and then by external similarity. Use the tabular key and notes to check the dichotomous keys (and vice versa) and broaden searches for family placements of specimens missing critical morphological characters. The following notes, in order of family or suborder, include salient characters that would not readily fit in the tabular key.

TAXONOMIC NOTES BY FAMILY AND SUPERFAMILY

AMPELISCIDAE tiny eyes when present, two separated dorsal frontal lenses when present, massive head, pleated gills, build pocket-shaped silt tubes.

TABLE 5

Tabular Key for Gammaridea

Family/Suborder	Plates	Telson Shape	Telson Cleft	U3 No.	Rami Shape	Urosomites	Oöst. setae	Gnathopods Type G1	Gnathopods Type G2	Gnathopods Largest size	Pereopods P6 & P7 Smlrty	Pereopods P6 & P7 Largest	Acc. flag.	Mandible Molar	Mandible Palp
Talitridae	303–304	Flesh	UnCl	1	Stubby	CmprUr2&3	Curl	Smpl–Tnsv	SbC	G2	S	P7	0	+	−
Phliantidae	258	Flesh	UnCl	0	—	CmprUr2&3	Curl	Smpl	Smpl	=	D	P7	0	−	−
Eophliantidae	258	Fused	—	0	—	FsdUr2&3	Curl?	Smpl	Smpl	=	S	=	0	−	−
Hyalidae	273	Flesh	Cleft	1	Stubby	CmprUr2&3	Curl	Tnsv–Chlt	SbC	G2	~S	P7	0	−	−
Hyalellidae	272	Flesh	UnCl	1–2	Stubby	CmprUr2&3	Curl	Tnsv–Chlt	SbC	G2	~S	P7	0	−	−
Najnidae	271	Flesh	UnCl	1	Stubby	CmprUr2&3	Curl	SbC–Tnsv	SbC–Tnsv	=	S	P7	0	−	−
Dogielinotidae	271	Flesh	UnCl	1	Stubby	CmprUr2&3	Curl	SbC	SbC	G2	D	P7	0	−	−
Podoceridae	259	Flesh	UnCl	0	—	LongUr1	Stra	Smpl–SbC	SbC	G2	S	=,P6	0–~1	+	+
Cheluridae	260	Flesh	UnCl	2	Leaf	Sep	Stra	Tnsv	Tnsv–Chlt	G2	S	~P7	0	+	+
Corophiidae	269–270	Flesh	UnCl	1–2	Stubby	Sep, Fsd	Stra	SbC	Smpl	G2	D	P7	0	+	+
Aoridae	262	Flesh	UnCl	1–2	Stubby	Sep	Stra	SbC–Mero	Mero	G1	S	P7	1	+	+
Isaeidae	263–264	Flesh	UnCl	1–2	Stubby	Sep, FUr2&3	Stra	Smpl–SbC	SbC–Tnsv	G2	=–D	P7	1–3+	+	+
Ischyroceridae	267–268	Flesh	UnCl	1–2	ShMd	Sep	Stra	SbC	SbC–Carp	G2	~S	P7	0	+	+
Ampithoidae	265–266	Flesh	UnCl	2	ShMd	Sep	Stra	Tnsv–Chlt	SbC	G2	S	P7	0–3+	+	+
Lysianassoidea	286–287	Lmnr	UnCl–Cleft	0–2	NrSpn–Lnc	Sep	Stra	Smpl–Chlt	Chlt	G1	=–D	=–P6	0–3+	−–+	+
Stenothoidae	275	Lmnr	UnCl	1	Tblr	Sep	Stra	Smpl–SbC	SbC	G2	=–D	P6	0	−–+	+
Leucothoidae	282	Lmnr	UnCl	2	NrSpn	Sep	Stra	Carp	SbC	G2	S	P7	0	−–+	−–+
Amphilochidae	274	Lmnr	UnCl	2	NrSpn	Sep	Stra	Smpl–Tnsv	Smpl–Tnsv	G2	S	P7	0	−.5	+
Stegocephaloidea	283	Lmnr	UnCl	2	NrSpn	Sep	Stra	Chlt	Smpl, Tnsv	G2	S	=	0	−–+	+
Dexaminidae	279	Lmnr	Cleft	2	NrSpn	FsdUr2&3	Stra	SbC	Tnsv	=	=–D	=	0	−–+	−,+
Ampeliscidae	276–278	Lmnr	Cleft	2	Lnc	FsdUr2&3	Stra	Smpl	Smpl	=	D	P6	0	+	+
Argissidae	284	Lmnr	Cleft	2	Lnc	Sep	Stra	Smpl	Smpl	G2	~S	P6	0	+	+
Megaluropidae	285	Lmnr	Cleft	2	Leaf	Sep	Stra	Smpl	Smpl	G2	D	P7	2~2	−+	+
Oedicerotidae	280	Lmnr	UnCl	2	NrSpn	Sep	Stra	SbC–Tnsv	SbC–Chlt	G1	D	P7	2~2	+	+
Synopiidae	281	Lmnr	Cleft	2	Lnc	Sep	Stra	Smpl	Smpl	=	~S	P7	0	+	+
Phoxocephalidae	289–292	Lmnr	Cleft	2	Lnc	Sep	Stra	SbC–~Chlt	SbC–~Chlt	G2	D	~P6	3+	+	−–+
Urothoidae	288	Lmnr	Cleft	2	Lnc	Sep	Stra	Smpl	SbC	=	S	P6	3+	+	+
Haustoriidae	261	Sep	Cleft	2	Lnc	Sep	Stra	Smpl	Chlt	G2	D	P6	3+	+	+
Pontoporeiidae	293	Lmnr	Cleft	2	Lnc	Sep	Stra	SbC	Tnsv	G1	D	P6	3+	+	+
Liljeborgiidae	298	Lmnr	Cleft	2	Lnc	Sep	Stra	SbC	SbC	G2	S	P6	3+	+	+
Pleustidae	294–295	Lmnr	UnCl	2	NrSpn	Sep	Stra	SbC	SbC	G2	S	P7	2–3+	−.5	+
Eusiriodea	296–297	Lmnr	UnCl–Cleft	2	NrSpn	Sep	Stra	Red, SbC	SbC	=–G2	S	=	0–1	−–+	+
Hadzioidea	300–301	Lmnr	Cleft	2	Lnc	Sep	Stra	SbC	SbC–Chlt	G2	~S	~P7	0–1	+	+
Gammaroidea	299	Lmnr	Cleft	2	Lnc	Sep	Stra	SbC	SbC	~=	S	=	3+	+	+
Crangonyctidae	302	Lmnr	Cleft	2	Lnc	Sep	Stra	SbC	SbC	G2	S	P6	3+	+	+

NOTE: Family-Superfamily Key including; Plate number(s); Telson "Shape" (Lmnr, Flesh, Fused and Sep respectively: laminar, fleshy, fused to urosome, and separate) and "Cleft" (Cleft and Uncl respectively: cleft and uncleft); Uropod 3 "(U3)" rami number (0, 1, and 2) and "Rami Shape" (Stubby, ShMd, NrSpn, Lnc and Leaf, respectively: stubby, short and modified [bearing denticles, teeth or hooks], narrow with dorsal spine rows, lance shaped with lateral spines, and leaf shaped); Urosomite (CmprUr2&3, LongU1, FsdUr2&3, Sep, Fsd indicate respectively: compressed or shortened urosomites 2 and 3, long urosomite 1, fused urosomites 2 and 3, separate urosomites, and all urosomites fused); "oöstegite setae" (Curled and Stra respectively: curled and straight); "Gnathopods" "G1" and "G2", "Type" (Smpl, SbC, Tnsv, Chlt, Red), G1 and G2 are respectively: simple, subchelate, transverse, chelate, reduced) and "Largest" size (G1 or G2 and =); "pereopods" 6 and 7 "P6 & P7" similarity "Smlrty" (S, D, –S respectively: closely similar, different, nearly similar) and "Largest" pereopod 6 or 7 (respectively: P6, P7, =); "Acc. flag." accessory flagellum article numbers (0, 1, 2 and, 3+ are respectively: absent, a tiny difficult to observe article, one or two prominent articles, and multiple prominent articles) and; "Mandible" "molar" and "palp" (+, – and .5 indicate respectively: present, absent and greatly reduced). Character qualifications are: dash "–" indicates a range of conditions between two states, comma "," indicates exclusive condition states, tilde "~" indicates poor resolution of the condition and an empty underline "__" indicates irrelevance of the character.

AMPHILOCHIDAE coxa 1 obscured, gnathopods 1 and 2 and article 5 extending along article 6, telson distally acute, uropod 2 not extending as far as uropods 1 and 3.

AMPITHOIDAE uropod 3 outer ramus with large hooks, lower lip with distinctive prominent inner lobes.

AORIDAE gnathopod 1 basket-shaped and merochelate or carpochelate and larger than gnathopod 2, uropod 3 biramous except for uniramous *Grandidierella*.

ARGISSIDAE coxa 1–4 ventrally rounded, coxa 3 shorter than coxa 2 and 4, eyes consist of 4 elements or are absent, male urosomite 2 bearing a dorsal tooth.

CHELURIDAE urosomite 1 bearing a giant dorsal tooth, uropods 1–3 rami grossly different, lives in wood chambers.

COROPHIIDAE basket-shaped gnathopod 2, suspension feeders, sediment tube builders.

CRANGONYCTIDAE lacks ventral antenna sinus or notch, uropod 3 inner ramus short, freshwater or low salinity estuary.

DEXAMINIDAE pereopods 3–7 dactyls short, pleated gills.

DOGIELINOTIDAE fossorial, uropods 1 and 2 peduncles lined dorsally with long spines, pereopods 5 and 6 lined with straight stout digging spines, pereopod 6 and 7 dactyls straight.

EOPHLIANTIDAE head blunt, body antlike, mature specimens unknown.

EUSIROIDEA rostrum pointed or minute, rostrum inserted between antenna 1 peduncle first articles, ventral antenna sinus notched (except *Batea*), dorsal urosome spineless.

GAMMAROIDEA urosomites 1–3 dorsally spinose, gnathopods 1 and 2 palm lined with thick short spines, uropod outer ramus minutely biarticulate, ventral antenna sinus notch lacking.

HADZIOIDEA male gnathopod 2 larger than gnathopod 1, ventral antenna sinus notched (except *Maera*, *Netamelita*, and *Quadrimaera*), urosome dorsally spineless (although teeth or setae may occur), uropod 3 rami short or long, uniarticulate or biarticulate.

HAUSTORIIDAE fossorial, blind, rostrum lacking, antenna 2 peduncle article 4 laterally expanded, gnathopod and pereopod dactyls minute or absent.

HYALELLIDAE maxilla 1 palp lacking or consisting of a single inconspicuous article.

HYALIDAE maxilla 1 palp of 2 articles, male gnathopod 1 dactyl modified for clasping female.

ISAEIDAE gnathopod 2 larger than 1, uropod 3 usually with 2 short rami and accessory flagellum usually present.

ISCHYROCERIDAE uropod 3 outer ramus with denticles or teeth, telsons diverse.

LEUCOTHOIDAE gnathopod 2 article 5 extending along article 6, telson distally acute. Liljeborgiidae: mandibular molar reduced, telson lobes distally notched, commensal with polychaetes and echiuroids.

LILJEBORGIIDAE mandibular molar reduced, telson lobes distally notched; commensal with polychaetes and echiurans.

LYSIANASSOIDEA antenna 1 article 1 swollen, gnathopod 2 hand mitten-shaped and article 3 longer than wide.

MEGALUROPIDAE coxae 1–4 ventrally rounded, coxa 3 shorter than coxa 2 and 4, eyes large.

NAJNIDAE down-turned antennae, rounded forehead, kelp burrower.

OEDICEROTIDAE eyes dorsal, rostrum helmet-shaped, gnathopods 1 and 2 article 5 variously extended along article 6, pereopod 7 is 50% longer than pereopod 6 and with long straight dactyl.

PHLIANTIDAE rostrum spatulate, body isopodlike and dorsoventrally flattened.

PHOXOCEPHALIDAE fossorial, rostrum shieldlike, antenna 2 spinose, antenna 2 article 4 narrow or expanded.

PLEUSTIDAE rostrum massive to minute, lacking a ventral antenna sinus, telson notch ventrally keeled.

PODOCERIDAE telson spinose, pleopods powerful, eyes bulge laterally.

PONTOPOREIIDAE fossorial, rostrum lacking, head lacking ventral antenna sinus notch, sternal gills, fresh to brackish waters.

STEGOCEPHALOIDEA epimeron 3 ornate, coxa 1 ventrally acute, rostrum large decurved (*Iphimediidae* gnathopod 2 with long lysianassoid article 3).

STENOTHOIDAE coxa 3 and 4 massive, coxa 1 small, obscured, distal telson bluntly acute, tongue-like.

SYNOPIIDAE rostrum helmet-shaped, eyes fused dorsally, tiny accessory lateral eyes in species of the region, telson large, urosomites 1 and 2 with dorsal tooth.

TALITRIDAE terrestrial, antenna 1 tiny.

UROTHOIDAE fossorial, head with extended ventral anterior edge, rostrum lacking, antenna 2 peduncle article 4 narrow.

KEY TO FAMILIES AND SUPERFAMILIES

Families of the region occurring only in deep water, offshore, or only to the north or south of the region, and thus not included, are Lafystiidae, Melphidippidae, Parampithoidae, Pardaliscidae, and Stilipedidae.

1. Telson fleshy, thick, short, or minute not readily articulated at base, uncleft or cleft (plate 255A–255F); telson lobes broadly separate (plate 255G) or telson indistinct (plate 255H); rami of uropod 3 (if present) shorter than peduncle (with numerous exceptions) 2

— Telson flat, laminar, and moveable, uncleft or deeply cleft, always distinct, never both uncleft and fleshy (plate 255I–255N); rami of uropod 3 always present and usually longer than the peduncle 18

2. Antenna 1 much shorter than antenna 2, and no longer than the head (plate 255O); telson with 10 or more irregularly distributed stout spines (plate 255A) and pereopods particularly heavy, terrestrial or semiterrestrial . . . . Talitridae (plates 303–304)

— Antenna 1 of similar size or larger than antenna 2 or significantly longer than the head (plate 255P); telson with six or less irregularly spaced stout spines (not counting long spines or setae) entirely aquatic or intertidal 3

3. Uropod 3 indistinct or absent 6 (plate 255F, 255H, 255Q) 4

— Uropod 3 large and readily visible (plate 255R–255V) . . 6

4. Telson fused to urosome and urosomites 2 and 3 fused (plate 255H), body tubular, ant- or tanaidaceanlike, burrows into kelp . . . . Eophliantidae (plate 258)

— Telson separate from urosome, not ant- or tanaidaceanlike (plate 255F, 255Q–255V), body laterally or dorsoventrally flattened. 5

5. Urosome less than twice as long as deep (plate 255Q); rostrum spatulate (plate 255W). . . . . Phliantidae (plate 258)

— Urosome more than twice as long as deep (plate 255F2); body laterally compressed; rostrum an evenly rounded bulge, small or absent. . . . . Podoceridae (plate 259)

6. Pleonite 3 with immense posteriorly projecting dorsal tooth, uropod 2 peduncle greatly expanded, uropods 2 and 3 enormous and urosomites 1–3 fused (plate 255P) . . . . Cheluridae (plate 260)

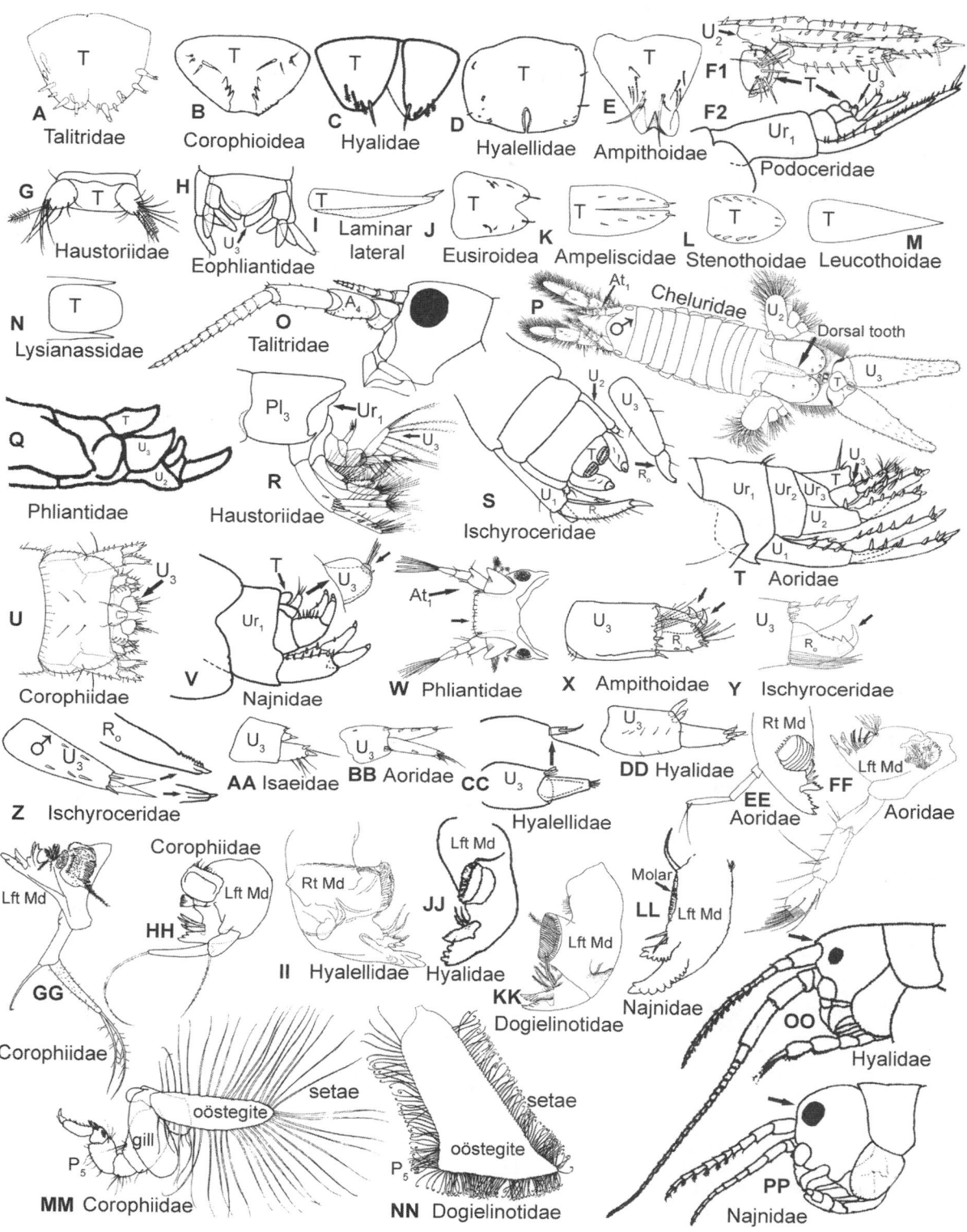

PLATE 255 Family Key. A, Telson, *Orchestia gammarellus*; B, telson, *Americorophium brevis*; C, telson, *Apohyale anceps*; D, telson, *Allorchestes bellabella*; E, telson, *Ampithoe aptos*; F, urosome, *Podocerus cristatus*; G, telson, *Eohaustorius*; H, urosome, *Lignophliantis pyrifera*; I, telson, generalized; J, telson, *Oligochinus lighti*; K, telson, *Ampelisca*; L, telson, stenothoid; M, telson, *Leucothoe*; N, telson, *Lysianassa*; O, head, *Orchestia georgiana*; P, body, *Chelura terebrans*; Q, urosome with uropod 1 removed, *Pariphinotus seclusus*; R, urosome, *Eohaustorius washingtonianus*; S, urosome, *Cerapus tubularis*; T, urosome, *Bemlos concavus*; U, urosome, *Laticorophium baconi*; V, urosome, *Carinonajna kitamati*; W, head and rostrum dorsal, *Pariphinotus escabrosus*; X, uropod 3, *Ampithoe plumulosa*; Y, uropod 3, *Jassa falcata*; Z, uropod 3, *Ischyrocerus pelagops*; AA, uropod 3, *Cheiriphotis megacheles*; BB, uropod 3, *Columbaora cyclocoxa*; CC, uropod 3, *Parallorchestes bellabella*; DD, uropod 3, *Prohyale frequens*; EE, mandible, *Aoroides exilis*; FF, mandible, *Grandidierella japonica*; GG, mandible, *Corophium alienense*; HH, mandible, *Americorophium spinicorne*; II, mandible, *Allorchestes angusta*; JJ, mandible, *Parallorchestes americana*; KK, mandible, *Proboscinotus loquax*; LL, mandible, *Carinonajna bicarinata*; MM, oöstegite, pereopod 5, *Americorophium salmonis*; NN, oöstegite, pereopod 5, *Proboscinotus loquax*; OO, head, *Ptilohyale longipalpa*; PP, head, *Carinonajna bicarinata*. (figures modified from Barnard 1950, 1954a, 1962a, 1962c, 1965, 1967a, 1969a, 1972c, 1975, 1979a; Bousfield 1958a, 1961a, 1973; Bousfield and Conlan 1982; Bousfield and Hendrycks 2002; Bousfield and Hoover 1995, 1997; Bousfield and Marcoux 2004; Chapman 1988; Chapman and Dorman 1975 ; Conlan 1990; Conlan and Bousfield 1982a, 1982b; and Shoemaker 1933, 1934).

— Pleonite 3 without posteriorly projecting dorsal tooth; uropod 2 without greatly expanded peduncle (plate 255F2, 255Q–255V). 7

7. Uropod 3 with 2 prominent short or long rami that may not be equal (plate 255X–255BB). 8
— Uropod 3 with 1 ramus only or with inner ramus minute, scalelike or otherwise indistinct and difficult to observe (plate 255S, 255U–255V, 255CC, 255DD) 12

8. Urosome decurved, nearly ventral to pleonite 3 (plate 255R); telson lobes widely separated (plate 255G) . . . . Haustoriidae (plate 261)
— Urosome normally aligned with (not ventral to) pleonite 3 (plate 255S–255V); telson lobes adjacent or fused (plate 255B–255E) 9

9. Uropod 3 rami structurally similar and the outer ramus with setae or with short, straight spines but not hooks or denticles (plate 255AA, 255BB). 10
— Uropod 3 rami different in structure and the outer ramus bearing conspicuous hooks or denticles (plate 255X–255Z) 11

10. Male gnathopod 1 larger than gnathopod 2 . . . . Aoridae (plate 262)
— Male gnathopod 1 smaller than gnathopod 2 . . . . Isaeidae (plates 263–264)

11. Outer ramus of uropod 3 stout, with two heavy, hooked spines and inner ramus flat and apically setose (plate 255X) . . . . Ampithoidae (plates 265–266)
— Outer ramus of uropod 3 apically stout, bearing a single large hook (plate 255Y) or relatively slender, outer ramus either denticulate (plate 255Z) or unornamented (an exception is *Cerapus* [plate 255S], which lacks an inner ramus) . . . . Ischyroceridae (plates 267–268)

12. Combined lengths of urosomites 2 and 3 greater than one-half of urosomite 1 (plate 255T) or urosomites 1–3 fused (plate 255U); mandibular palp present (plate 255EE–255HH); oöstegites lined with evenly curved or straight setae (plate 255MM). 13
— Urosomites 2 and 3 combined lengths less than one half of urosomite 1 (plate 255V); mandibular palp absent (plate 255II–255LL); oöstegites lined with distally curled setae (plate 255NN) 15

13. Male gnathopod 1 or gnathopod 2 carpochelate (plate 254V). 14
— Male and female gnathopod 2 basket-shaped, merochelate, or simple, ventrally lined with long pinnate setae (plate 254U) and larger than gnathopod 1 . . . . Corophiidae (plates 269–270)

14. Gnathopod 1 carpochelate (*Grandidierella japonica*) . . . . Aoridae (with Corophiidae) (plate 262)
— Gnathopod 2 carpochelate (*Cerapus*) . . . . Ischyroceridae (plates 267–268)

15. Head anteriorly square and antenna 1 insertion dorsal to the eye (plate 255OO); molar prominent (plate 255JJ–255KK); uropod 3 ramus short but readily apparent (plate 255CC, 255DD). 16
— Head anteriorly decurved, insertions of antenna 1 ventral to the eye (plate 255PP); mandibular molar an indistinct flat plate (plate 255LL); uropod 3 ramus tiny (plate 255V) . . . . Najnidae (plate 271)

16. Pereopods 2–7 fossorial, pereopod 6 articles 4–6 densely lined with long, straight setae and with straight dactyl (plate 256A); epistome of upper lip proboscoid (noselike) (plate 256B); uropod 1 peduncle lined with long spines and rami without spines (plate 256C) . . . . Dogielinotidae (plate 271)

— Pereopods 2–7, articles 4–7 with sparse, short setae and with curved dactyls (plate 256D); uropod 1 rami and lateral peduncle with short stout spines (plate 256E); epistome reduced (plate 256F) 17

17. Telson uncleft or cleft less than one-third of total length (plate 255D); maxilla 1 palp extremely reduced or absent, not extending to distal plate end (plate 256G) . . . . Hyalellidae (plate 272)
— Telson cleft more than one half of the entire length into subtriangular lobes (plate 255C); maxilla 1 palp extending to distal end of outer plate (plate 256H, 256I) . . . . Hyalidae (plate 273)

18. Coxa 1 tiny and obscured by coxa 2 and coxa 2–4 enlarged or immense (arrows, plate 256J, 256K) 19
— Coxa 1 at least half as large as coxa 2 and coxa 2–4 not greatly enlarged or immense (plate 256L, 256M) 22

19. Rostrum inserted between first antennae (plate 256N); telson laminar and deeply cleft (plate 254Q); gnathopod 1 vestigial . . . . Eusiroidea (Bateidae) (plate 296)
— Rostrum vestigial (plate 256K) or extended over first antenna (plate 256M) but not inserted between the antennae; telson uncleft (plate 256L–256M). 20

20. Gnathopod 1 carpochelate (plate 256O, 256P) . . . . Leucothoidae (plate 282)
— Gnathopod 1 simple, transverse or subchelate but not carpochelate 21

21. Uropod 3 biramous, rami of a single article, uropod 2 not reaching distal end of uropod 3 and telson acute (plate 256Q); gnathopods 1 and 2 article 5 extending to the posterior palm edge of article 6 (plates 254X, 256R) . . . . Amphilochidae (plate 274)
— Uropod 3 with a single biarticulate ramus (plate 256T); gnathopod 2 article 5 not extending along posterior edge of the palm of article 6 (plate 256S); telson evenly rounded or bluntly acute posteriorly (plate 255L) and uropod 2 extending as far as distal uropods 1 and 4 . . . . Stenothoidae (plate 275)

22. Urosomites 2 and 3 fused together (plate 256L, 256U). 23
— Urosomites separate (plate 256J, 256K, 256M) 24

23. Head as long as combined lengths of pereonites 1–3, pereopod 3 and 4 dactyls longer than combined articles 5 and 6; pereopods 6 and 7 dissimilar, eyes tiny, consisting of one dorsal lateral and one anteroventral cuticular lens (plate 256L) . . . . Ampeliscidae (plates 276–278)
— Head shorter than the combined lengths of pereonites 1 and 3; pereopods 2 and 3 dactyls shorter than combined articles 5 and 6; pereopods 6 and 7 similar; eyes with numerous ommatidia (plate 256V). . . . . Dexaminidae (plate 279)

24. Eyes coalesced into a single dorsal anterior mass on a strongly decurved, usually helmet shaped, rostrum (plate 256W, 256X) 25
— Eyes lateral, rostrum present or absent (plate 256M, 256N, 256Y–256AA). 26

25. Telson emarginated (plate 256BB) or evenly rounded; urosome dorsally unarmed and telson not extending beyond peduncle of uropod 3 (plate 256CC); gnathopod 1 article 6 stout (plate 254X); accessory eyes lacking (plate 256W) . . . . Oedicerotidae (plate 280)
— Telson deeply cleft (plate 256DD); telson extending beyond peduncle of uropod 3 and urosomites 1 and 2 dorsally toothed (plate 256EE); gnathopod 1 article 6 weak (plate 256FF); accessory flagellum prominent and multiarticulated (plate 256GG) . . . . Synopiidae (plate 281)

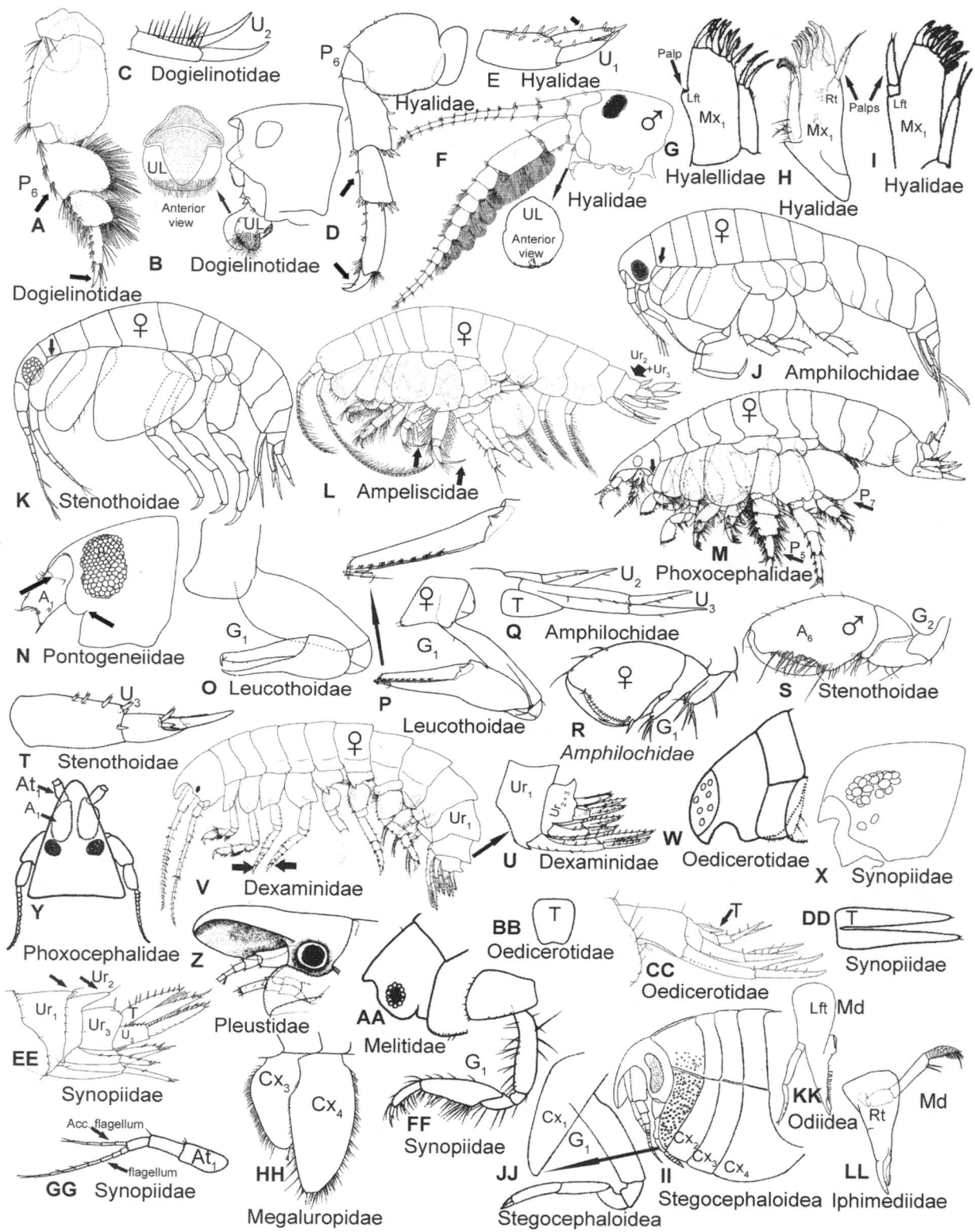

PLATE 256 Family Key. A, pereopod 6, B, head and upper lip, C, uropod 1, *Proboscinotus loquax*; D, pereopod 6, *Apohyale anceps*; E, uropod 1, *Apohyale anceps*; F, head and upper lip (UL = *Apohyale pugettensis*), *Ptilohyale littoralis*; G, maxilla 1, *Allorchestes rickeri*; H, maxilla 1, *Apohyale anceps*; I, maxilla 1, *Parallorchestes cowani*; J, body, *Gitana calitemplado*; K, body, *Stenula modosa*; L, body, *Ampelisca milleri*; M, body, *Eobrolgus chumashi*; N, head, *Pontogeneia rostrata*; O, gnathopod 1, *Leucothoe alata*; P, gnathopod 1, *Leucothoides pacifica*; Q, telson and uropods 1 and 2, *Apolochus littoralis*; R, gnathopod 1, *Apolochus barnardi*; S, gnathopod 2, *Stenothoe estacola*; T, uropod 3, *Stenula incola*; U, urosome, V, body, *Atylus levidensus*; W, head, *Americhelidium shoemakeri*; X, head, *Metatiron tropakis*; Y, head, *Foxiphalus obtusidens*; Z, rostrum, *Thorlaksonius grandirostris*; AA, head, *Elasmopus antennatus*; BB, telson, *Monoculodes emarginatus*; CC, urosome, *Americhelidium micropleon*; DD, telson, *Tiron biocellata*; EE, urosome, *Metatiron tropakis*; FF, gnathopod 1, *Tiron biocellata*; GG, antenna 1, *Tiron biocellata*; HH, coxae 3 and 4, *Gibberosus myersi*; II, head and 3 pereonites, *Cryptodius kelleri*; JJ, gnathopod 1, *Cryptodius kelleri*; KK, left mandible, *Cryptodius kelleri*; LL, right mandible, *Coboldus hedgpethi* (figures modified from: Barnard 1960a, 1962a, 1962b, 1962c, 1962e, 1967a, 1969a, 1972b, 1972c, 1977, 1979a; Bousfield and Chevrier 1996; Bousfield and Kendall 1994; Bousfield and Hendrycks 1994a, 2002; Dickinson 1982; Gurjanova 1951; Hendrycks and Bousfield 2001; Jarrett and Bousfield 1994b; Moore 1992; and Nagata 1965a).

26. Gnathopod 1 carpochelate (plate 256O, 256P) . . . .
. . . . Leucothoidae (plate 282)
— Gnathopod 1 not carpochelate. 27
27. Coxa 4 is 50% longer than coxa 3 or more and ventrally
rounded (plate 256HH) 29
— Coxae 3 and 4 within 30% of the same length
and ventrally square or acute, not rounded (plate
256II) 28
28. Coxa 1 ventrally acute (plate 256JJ); mandibles needlelike
with molars weak or lacking (plate 256KK, 256LL) . . . . . .
. . . . Stegocephaloidea (plate 283)
— Coxa 1 ventrally square or rounded (plate 257A); mandible
not as above 30
29. Eye round and of four distinctive ommatidia (plate 257B);
uropod 3 rami posteriorly acute (plate 257C); mandibular
palp slight with article 3 longest (plate 257D). . . . .
. . . . Argissidae (plate 284)
— Eye variously shaped, multifaceted (plate 257E); uropod 3
rami foliate (plate 257F) (this appendage is often lost);
mandible palp stout and article 2 is longest (plate 257G);
mandible with triturative molar (plate 257H) . . . .
. . . . Megaluropidae (plate 285)
30. Antenna 1 article 1 depth usually half or more of the
length (plate 257I, 257J); gnathopod 2 article 3 at least
1.5 times longer than wide and article 6 mitten-shaped
with dactyl minutely transverse (plate 257K); body usually
white, compact, shiny and densely calcified . . . .
. . . . Lysianassoidea (plates 286–287)
— Antenna 1 article 1 longer than deep (plates 256M, 256N,
257N, 257Q); gnathopod 2 article 3 less than 1.2 times
longer than wide, article 6 not mitten–shaped and dactyl
prominent (plate 257L, 257M) 31
31. Fossorial—dense long stout lateral spines lining antennae
2 (plates 256M 257N, 257O) and lining pereopod 5 articles
4–6 (plates 256M, 257N, 257P). 33
— Nonfossorial—antennae 2 and pereopod 5 articles 4–6 not
lined with long dense stout lateral spines (plate 257Q, 257R)
. . . . 32
32. Accessory flagellum two articles or more and apparent at
magnifications of 40x or less (plate 257Q, 257R); all telsons
cleft and with prominent distal setae or spines (plate 257S,
257T) 37
— Accessory flagellum absent or a tiny article apparent only
at magnifications >40x (plate 257U–257W); telsons deeply
cleft or evenly rounded and with few or no prominent distal setae or spines (plate 257X, 257Y) 35
33. Pereopods 6 and 7 similar in length and form, ventral
cephalic margin extended (see arrow; plate 257N) entirely
marine . . . . Urothoidae (plate 288)
— Pereopod 7 different in form and at least 40% shorter than
pereopod 6 (plate 256M); ventral cephalic margin reduced
(see arrows; plates 256M, 257Z) 34
34. Rostrum extended and visorlike (plate 256M, 256Y) entirely marine or high-salinity estuary. . . . .
. . . . Phoxocephalidae (plates 289–292)
— Rostrum minute; entirely freshwater or low-salinity estuary (plate 257Z) . . . . Pontoporeiidae (plate 293)
35. Telson evenly rounded (plate 257X) or emarginate (plate
255J). 36
— Telson deeply cleft (plate 257Y) . . . .
. . . . Eusiroidea (Pontogeneiidae) (plates 296–297)
36. Ventral antennal sinus without a notch (plate 257W); upper lip ventrally bilobed (plate 257AA, 257BB); lower lip
with inwardly tilting pillow shaped inner and outer lobes

(except for *Anomalosymtes coxalis*) (plate 257CC) and with
short mandibular extensions (arrows, plate 257DD, 257EE)
. . . . Pleustidae (plates 294–295)
— Ventral antennal sinus with a notch (plate 256N);
lower lip ventrally convex (plate 257FF); inner and
outer lobes of lower lip not pillow shaped and bearing
large extensions of the outer lobes (plate 257GG,
257HH). . . . . . Eusiroidea (Calliopidae) (plates 296–297)
37. Pereopods 5–7 dactyls small and straight (plate 257Q);
uropod 3 rami sharply pointed distally, nearly equal in
length and lined with single thick spines (plate 257II,
257JJ); molar reduced (plate 257KK). . . . .
. . . . Liljeborgiidae (plate 298)
— Pereopods 5–7 dactyls stout and slightly curved (plate
257R); uropod 3 rami with thick spines in clusters or inner
ramus greatly reduced (plate 257LL–257NN); molar
prominent (plate 257OO, 257PP). 38
38. Urosome with dorsal clusters of large stout spines or setae
(plate 257QQ) . . . . Gammaroidea (plate 299)
— Urosome dorsum bare (plate 257P) or variously toothed
(plate 257RR) but without clusters of spines 39
39. Head with an inferior antennal sinus (plate 257R, 257SS,
257TT); accessory flagellum of three or more segments
(plate 257R) . . . . Hadzioidea (plates 300–301)
— Head lacking inferior antennal sinus and accessory flagellum of two segments (plate 257UU). . . . .
. . . . Crangonyctidae (plate 302)

LISTS OF GAMMARIDEA SPECIES BY FAMILY

Species lists include author, notes, species lengths and depth
ranges. Species lengths are a crude index of size based on the
distance from the distal end of the head to the posterior edge
of the telson and are usually of the largest specimens reported.
Species preceded by an asterisk are not in the key or are out
of the range of the region.

EOPHILANTIDAE

Eophliantidae are kelp burrowers of the eastern Pacific and
southern hemisphere. This rare, antlike species is the only
member of the family with a fused telson. Urosomites 2 and 3
are also fused. Reproductive individuals are unknown. The
name *Lignophliantis* indicates an eophliantid with lignin in its
gut (Barnard 1969a: 104).

KEY TO EOPHILIANTIDAE

1. Tiny tubular body, lacking accessory flagellum, with sparse
body setae and spines and pereonites lacking a ventral flange
(plate 258A); mandible lacking palp or molar (plate 258B);
uropod 3 consisting of a peduncle only (plate 255H) . . . . .
. . . . *Lignophliantis pyrifera*

LIST OF SPECIES

Lignophliantis pyrifera Barnard, 1969a. Bores into haptera of
the kelp *Macrocystis pyrifera*. A lack of records is likely due to
low probability of retention on standard 0.5 mm mesh collecting sieves normally used and the difficulty of recognizing
such a small, unusual amphipod in nearshore algae samples;
1.4 mm; intertidal—3 m.

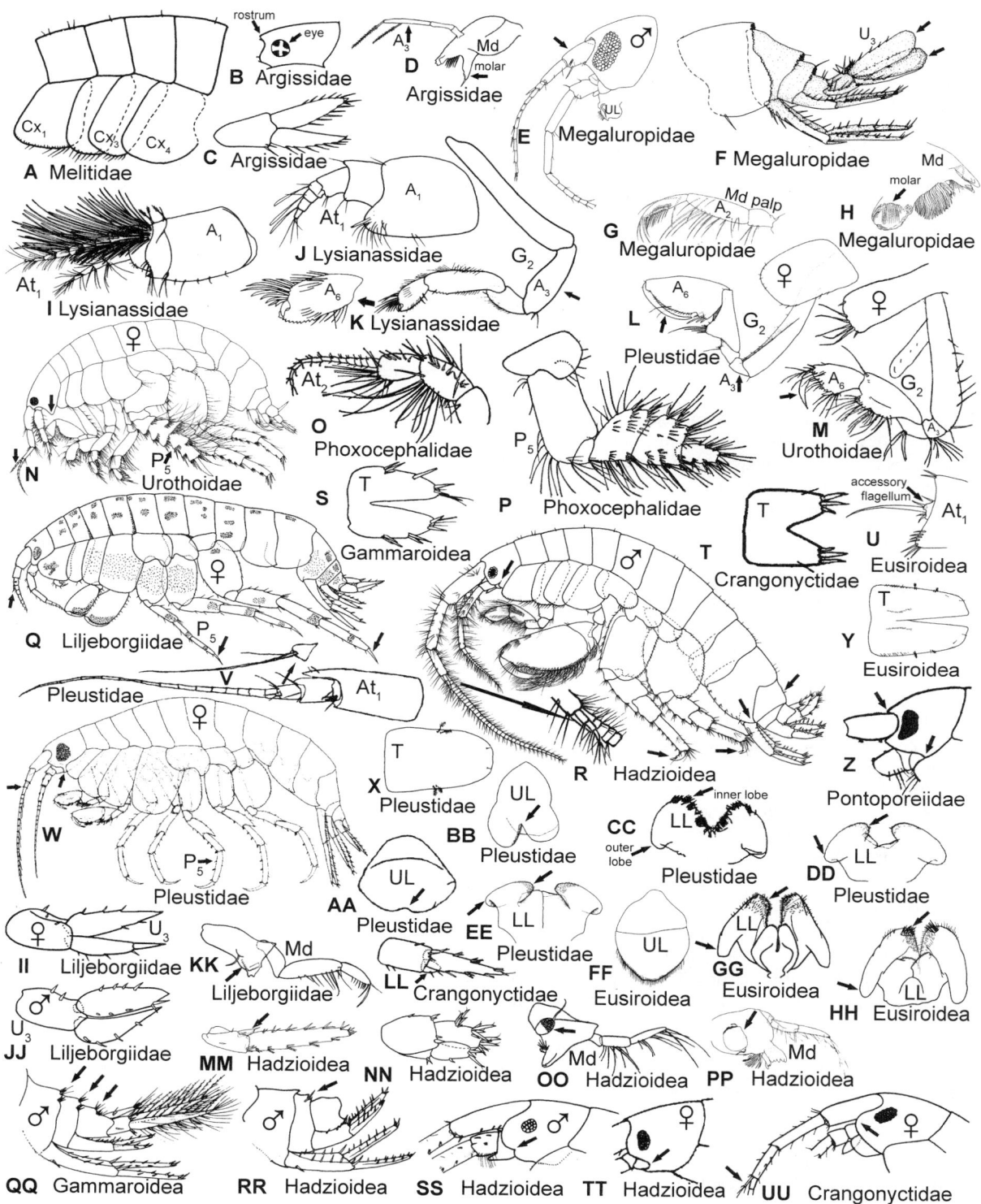

PLATE 257 Family Key. A, coxae 1-4, *Elasmopus antennatus*; B, head, C, telson, D, mandible, *Argissa hamatipes*; E, head, F, urosome, G, mandibular palp, H, molar, *Gibberosus myersi*; I, antenna 1, *Macronassa pariter*; J, antenna 1, *Ocosingo borlus*; K, gnathopod 2, *Macronassa macromer*; L, gnathopod 2, *Pleusirus secorrus*; M, gnathopod 2, *Urothoe varvarini*; N, body, *Urothoe marina*; O, antenna 2, *Grandifoxus grandis*; P, pereopod 5, *Grandifoxus grandis*; Q, body, *Listriella diffusa*; R, body, *Elasmopus antennatus*; S, telson, *Anisogammarus pugettensis*; T, telson, *Crangonyx pseudogracilis*; U, accessory flagellum, *Oligochinus lighti*; V, antenna 1 and accessory flagellum, *Anomalosymtes coxalis*; W, body, *Kamptopleustes coquillus*; X, telson, *Chromopleustes lineatus*; Y, telson, *Pontogeneia rostrata*; Z, head, *Monoporeia affinis*; AA, upper lip, *Anomalosymtes coxalis*; BB, upper lip, *Chromopleustes lineatus*; CC, lower lip, *Anomalosymtes coxalis*; DD, upper lip, *Holopleustes aequipes*; EE, lower lip, *Chromopleustes lineatus*; FF, upper lip, *Oligochinus lighti*; GG, lower lip, *Accedomoera vagor*; HH, lower lip, *Paracalliopiella pratti*; II, female uropod 3, *Listriella diffusa*; JJ, male uropod 3, *Listriella diffusa*; KK, mandible, *Listriella melanica*; LL, uropod 3, *Crangonyx pseudogracilis*; MM, uropod 3, *Melita nitida*; NN, uropod 3, *Elasmopus antennatus*; OO, mandible, *Maera similis*; PP, mandible, *Megamoera dentata*; QQ, urosome, *Gammarus daiberi*; RR, urosome, *Desdimelita microdentata*; SS, head, *Melita nitida*; TT, head, *Maera jerrica*; UU, head, *Crangonyx pseudogracilis* (figures modified from: Barnard 1954a, 1959a, 1959b,1960a, 1962b, 1969a, 1969b, 1979a; Bousfield 1958b, 1973; Bousfield and Hendrycks 1995b; Gurjanova 1953; Hendrycks and Bousfield 2004; Jarrett and Bousfield 1996; Krapp-Schickel and Jarrett 2000; Lincoln 1979; McKinney 1980; Segerstråle 1937; and Thomas and Barnard 1986).

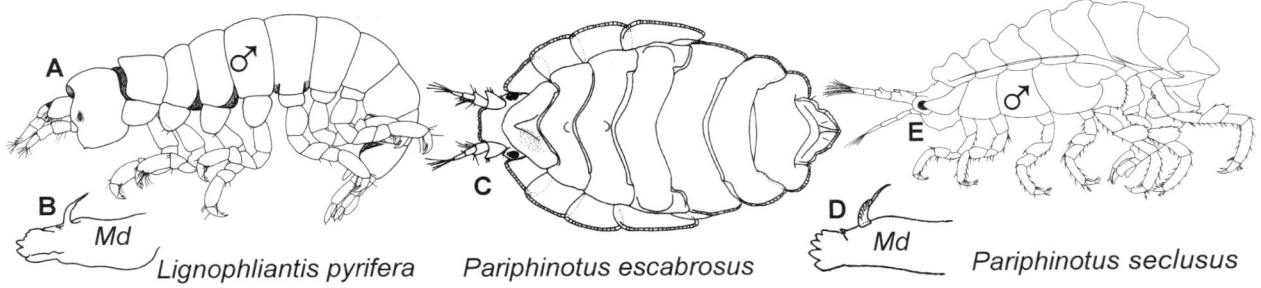

A
♂
B
Md
Lignophliantis pyrifera

C

D
Md
Pariphinotus escabrosus

E
♂
Pariphinotus seclusus

PLATE 258 Eophliantidae and Phliantidae. A, B, *Lignophliantis pyrifera*; C, E, *Pariphinotus escabrosus*; D, *Pariphinotus seclusus* (figures modified from: *Pariphinotus escabrosus, Pariphinotus seclusus* of: Barnard 1969a, 1979a, and Shoemaker 1933).

PHLIANTIDAE

Phliantidae look more like isopods than amphipods, with their simple or barely subchelate gnathopods, square rostrum, dorsoventrally flattened calcified body, short antennae, splayed coxae, and the lack of a third uropod.

KEY TO PHLIANTIDAE

1. Body broad, dorsoventrally flattened (plate 258C, sex not known); mandibule lacking palp and molar (plate 258D); coxae splayed (plate 258C, 258E); rostrum square and distally annulated (plate 255W) . . . . . *Pariphinotus escabrosus*

LIST OF SPECIES

Pariphinotus escabrosus (Barnard, 1962b) (=*Heterophlias*). Moderately abundant under rock substrata, in kelp *Macrocystis* holdfasts, rare in surfgrass *Phyllospadix*. *P. escabrosus* was initially misidentified as *Pariphinotus seclusus* Shoemaker, 1933, from the Dry Tortugas, Florida, which does not occur in the Pacific; 3.8 mm; intertidal—16 m.

PODOCERIDAE

Podoceridae have an extended urosomite 1, minute or absent uropod 3, and fleshy, entire telsons. The delicate antennae, pereopods, and pleopods of preserved specimens are usually missing. Some species are brilliantly pigmented and occur in highly visible locations (Goddard 1984). An unidentified *Podocerus* of Oregon (probably in the *P.* "*cristatus*" group) appears to be a Batesian mimic of *Flabellina trilineata* (Goddard 1984, Shells and Sea Life 16: 220–222). The particularly long urosomite 1 of males may be an adaptation for their powerful pleopods, which are used for pelagic swimming in search of mates (Conlan 1991, Hydrobiologia 223: 255–282). The broad geographic ranges of many podocerids are likely due to human introductions or to poorly resolved species definitions.

KEY TO PODOCERIDAE

1. Urosomites 1–3 separate and antenna 1 shorter than antenna 2 (plate 259A, 259B); uropod 3 minute (plate 259B) 2
— Urosomites 2 and 3 fused (plate 259C); antenna 1 as long or longer than antenna 2, uropod 3 absent (plate 259D) 4

2. Pleonites with raised carina (plate 259E) . . . . *Podocerus* "*cristatus*"
— Pleonites without raised carina (plate 259B) 3
3. Male gnathopod 2 article 4 extended forward (plate 259F) . . . . *Podocerus spongicolus*
— Male gnathopod 2 article 4 not greatly extended forward (plate 259G) . . . . *Podocerus brasiliensis*
4. Pereopods 3 and 4 article 2 expanded and pereopods 5–7 lengths <1.5 times length of pereopods 3–4 (plate 259H) 6
— Pereopods 3 and 4 article 2 narrow and pereopods 5–7 greater than two times lengths of pereopods 3–4 (plate 259D) 5
5. Pereopod 7 article 5 shorter than article 6 (plate 259D); gnathopod 2 dactyls denticulate, uropods 1 and 2 lateral edges with spines (plate 259B) . . . . *Dulichia rhabdoplastis*
— Pereopod 7 article 5 longer than article 6, gnathopod 2 palm without teeth, uropods 1 and two lateral edges without spines (not illustrated) . . . . *Dulichia* sp.
6. Male and female coxa 1 with anteriorly directed spine (plate 259H, 259I), eyes small and within lateral cephalic bulge, accessory flagellum of one article . . . . *Dyopedos arcticus*
— Female coxa 1 with out anteriorly directed spine (plate 259J), eyes large, usually with light outer ring, accessory flagellum with three articles . . . . . . *Dyopedos monacanthus*

LIST OF SPECIES

Dulichia rhabdoplastis McClosky, 1970. Commensal on sea urchin *Strongylocentrotus franciscanus*; 25 mm; intertidal—25 m.
Dulichia sp. Soft benthos in San Francisco Bay (Presidio Yacht Club, Sausalito, collected 2003). Possibly an introduced undescribed species, not commensal on echinoderms; 6 mm; shallow subtidal—3 m.
Dyopedos arcticus (Murdoch 1885). Pan Arctic; in Pacific from Pt. Barrow to southern California; to 20 mm; 3 m–410 m.
Dyopedos monacanthus (Metzger, 1875). Pan Arctic; in Pacific in Northern California, often clinging to algae, hydroids, and bryozoans, on sand-gravel, to silt clay; 8 mm; 12 m–217 m.
Podocerus brasiliensis (Dana, 1853). Cosmopolitan in tropical and warm temperate seas and likely an introduction in California harbors; 8 mm; intertidal—12 m.
Podocerus "*cristatus*" (Thompson, 1879). A likely species complex reported widely from warm temperate waters; on our coast from southern California to Magdalena Bay among Sertularidae, *Boltenia*, and seaweeds, on mud and gravel, on corals and

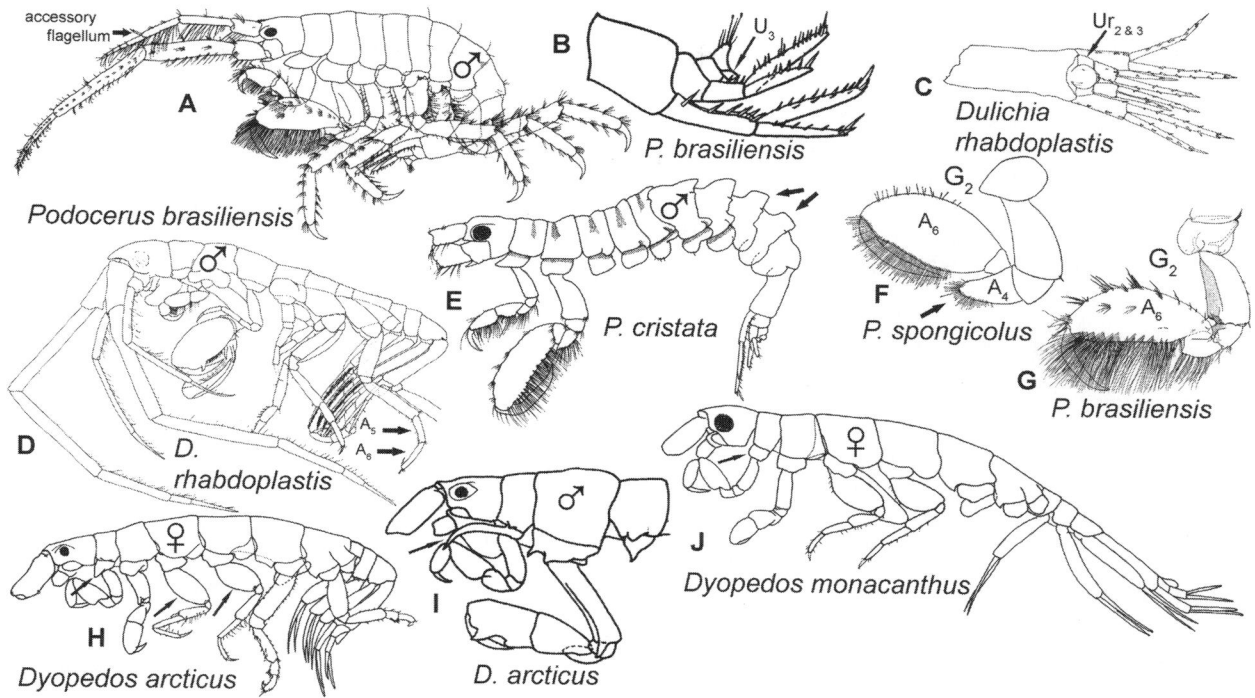

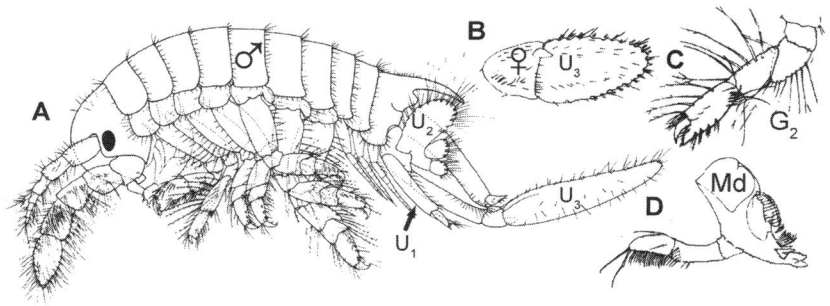

PLATE 259 Podoceridae. C, D, *Dulichia rhabdoplastis*; H, I, *Dyopedos arcticus*; J, *Dyopedos monacanthus*; A, B, G, *Podocerus brasiliensis*; E, *Podocerus cristata*; F, *Podocerus spongicolus* (figures modified from Alderman 1936; Barnard 1962a, 1970; Laubitz 1977; and McClosky 1970).

PLATE 260 Cheluridae. A-D, *Chelura terebrans* (figures modified from Barnard 1950, and Bousfield 1973).

among anemones; some species or populations are possible Batesian mimics; 6 mm; 7 m–100 m.

Podocerus spongicolus Alderman, 1936. In sponges; poorly known; 6 mm; intertidal—4m.

CHELURIDAE

Cheluridae invade holes that the isopod *Limnoria* make in wood; the chelurids then enlarge the holes into galleries by scraping furrows in the soft grains (Barnard 1955, Essays in the Natural Sciences in Honor of Captain Allan Hancock, pp. 87–98, Los Angeles: Univ. So. Calif.). *Chelura terebrans* is the only chelurid in this region and is distinguished from all other species by its completely fused urosome, enormous uropods, and dorsally spiked pleonite 3.

KEY TO CHELURIDAE

1. Body dorsally depressed and cylindrical (plates 255P, 260A); sexual dimorphism in the third uropod (plate 260A,

260B); gnathopods weak and chelate (plate 260C); mandible with large palp and molar (plate 260D) . . . . . .
. . . . *Chelura terebrans*

LIST OF SPECIES

Chelura terebrans (Philippi, 1839). An introduced cosmopolitan mid-latitude wood-boring species that was not reported from the eastern Pacific until the 1950s (Barnard, 1950, 1952). Warm, high salinity protected areas of bays and estuaries; 6 mm; shallow subtidal and intertidal.

HAUSTORIIDAE

Haustoriidae are blind, unpigmented, and fossorial and are most abundant in clean, fine marine, or estuarine sands where they swim and burrow upside down. Haustoriid burrows result in distinctive, punctate indents and shiny marks on wet sand surfaces of swash zones and high intertidal sand pools.

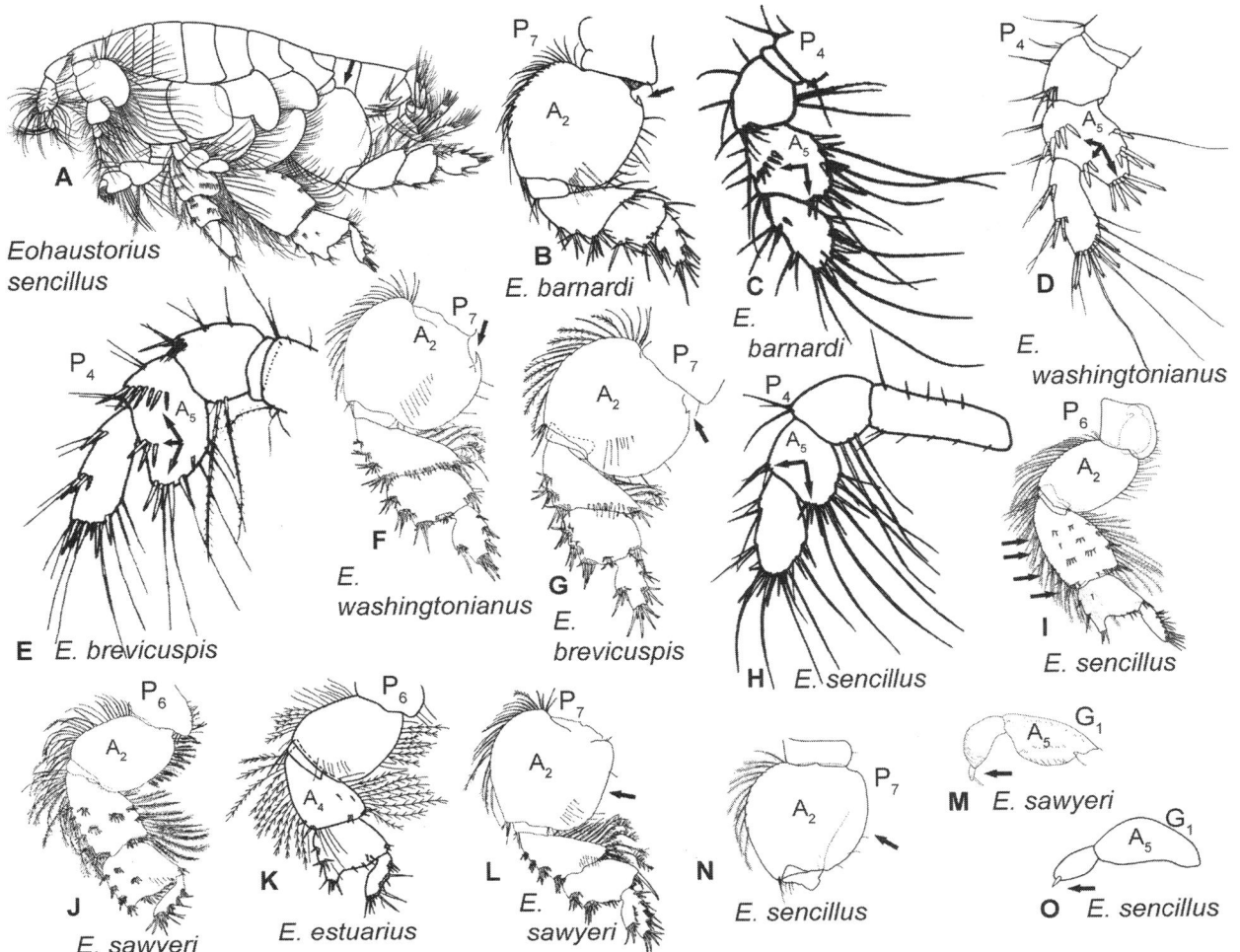

PLATE 261 Haustoriidae. B, C, *Eohaustorius barnardi*; E, G, *Eohaustorius brevicuspis*; K, *Eohaustorius estuarius*; A, H, I, N, O, *Eohaustorius sencillus*; J, L, M, *Eohaustorius sawyeri*; D, F, *Eohaustorius washingtonianus* (figures modified from Barnard 1957a, 1962e; Bousfield and Hoover 1995; Bosworth 1973; and Thorsteinson 1941).

See Jones 1977, Wasmann J. Biol. 21: 114–149 (seasonal occurrence).

KEY TO HAUSTORIIDAE

1. Pereopod 7 article 2 lacking posterior dorsal cusp (plate 261A) 4
— Pereopod 7 article 2 with sharp posterior dorsal cusp (plate 261B) 2
2. Pereopod 4 article 5 width and length nearly equal, bearing only two ventral spine fascicles and extending only slightly posteriorly (plate 261C) . . . . *Eohaustorius barnardi*
— Pereopod 4 article 5 width nearly twice the length, bearing three ventral spine fascicles, and greatly extending posteriorly (plate 261D) 3
3. Pereopod 7 posterior dorsal cusp height twice its base (plate 261F) . . . . *Eohaustorius washingtonianus*
— Pereopod 7 posterior dorsal cusp short, height equal to its base (plate 261G) . . . . *Eohaustorius brevicuspis*
4. Pereopod 4 article 5 width equal to length (not extending posteriorly) and with two ventral spine fascicles (plate 261H) 5

— Pereopod 4 article 5 width nearly twice length, with three ventral spine fascicles and extending posteriorly (plate 261E) . . . . *Eohaustorius brevicuspis*
5. Pereopod 6 article 4 long, nearly as wide at mid-length as at distal end and with three or more sets of spine fascicle rows (plate 261I, 261J) 6
— Pereopod 6 article 4 narrower at mid-length and with less than three sets of spine fascicle rows (plate 261K) . . . . *Eohaustorius estuarius*
6. Pereopod 7 article 2 posterior edge nearly straight (plate 261L), dactyl of gnathopod 1 small (plate 261M) . . . . *Eohaustorius sawyeri*
— Pereopod 7 article 2 posterior edge evenly rounded (plate 261N), dactyl of gnathopod 1 missing or fused with article 6 (plate 261O) . . . . *Eohaustorius sencillus*

LIST OF SPECIES

Eohaustorius barnardi (Barnard 1957). Fine sand marine beaches; 5 mm; 5 m–20 m.

Eohaustorius brevicuspis Bosworth, 1973. *E. brevicuspis* and *E. washingtonianus* are distinguished only by the variable posterior

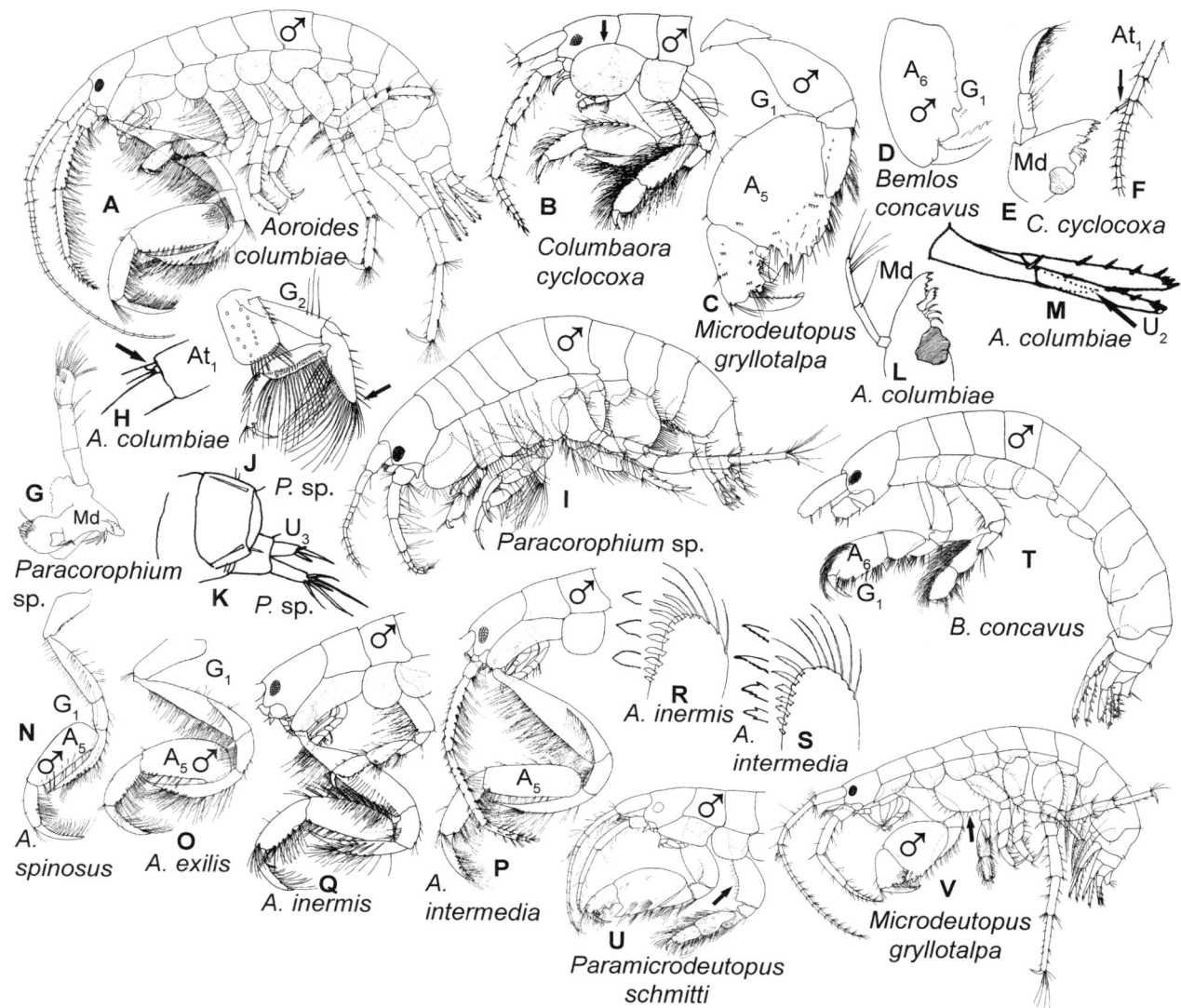

PLATE 262 Aoridae. A, H, L, M, *Aoroides columbiae*; O, *Aoroides exilis*; P, R, *Aoroides inermis*; Q, S, *Aoroides intermedia*; N, *Aoroides spinosus*; D, T, *Bemlos concavus*; B, E, F, *Columbaora cyclocoxa*; C, V, *Microdeutopus gryllotalpa*; U, *Paramicrodeutopus schmitti*; G, I, J, K, *Paracorophium* sp. (figures modified from Bousfield 1973; Chapman and Dorman 1975; Chapman, personal observation; Conlan and Bousfield 1982b; Miller, personal observation; and Shoemaker 1942).

cusp of pereopod 7 and therefore are not clearly distinct species; 5 mm; intertidal—1 m.

Eohaustorius estuarius Bosworth, 1973. Estuarine sands; an important species for toxicity tests (Hecht and Boese, 2002, Envir. Toxicol. and Chem. 21: 816–819); 1 mm–5 mm; intertidal—7 m.

Eohaustorius sawyeri Bosworth, 1973. Fine sands; not clearly distinct from *E. sencillus*; 4 mm; intertidal—1 m.

Eohaustorius sencillus Barnard, 1962. Fine sand beaches and sandy mud benthos; 4 mm; intertidal—30 m.

Eohaustorius washingtonianus (Thorsteinson 1941). Sandy and muddy sand sediments; 6 mm; intertidal—18 m.

AORIDAE

Aoridae, including the corophiid *Paracorophium* keyed here, are tube-building suspension feeders and occur over a wide range of depths in marine and estuarine rocky, fouling, and soft ben-

thic communities. Urosome articles are separate and uropod 3 is usually biramous. Rostrum short or absent, eyes small or large, ocular lobe large and rounded. Antenna 1 usually long but often lost in preservation. Male gnathopod 1 larger than gnathopod 2 in most genera. Telson entire and fleshy. Pereopod 7 long, extending further than pereopod 6. Aorid taxonomy is not reliable for females. At least four nonnative aorids occur within this region. More arrivals and new discoveries are expected in busy international ports and oystering bays.

KEY TO AORIDAE

1. Male gnathopod 1 merochelate (plate 262A, 262B) 2
— Male gnathopod 1 carpochelate (plate 262C) or subchelate (plate 262D) 9
2. Male coxa 1 greatly inflated (plate 262B); mandibular palp article 3 falcate and densely setose (plate 262E); antenna 1 accessory flagellum conspicuous but only as long as first

article of flagellum (plate 262F) ....
.... *Columbaora cyclocoxa*
— Male coxa 1 not inflated (plate 262A); mandibular palp article 3 not falcate or densely setose (plate 262G, 262L); antenna 1 accessory flagellum minute (plate 262H) or absent (view under 100×) 3
3. Coxae 2–4 deeper than wide (plate 262I); article 4 of gnathopod 2 distally blunt (plate 262J); uropod 3 short (plate 262K); mandibular palp stout (plate 262G)........
.... *Paracorophium* sp.
— Coxae 2–4 not deeper than wide (plate 262A); distal end of gnathopod 1 article 4 sharply pointed (plate 262A, 262B); uropod 3 long (plate 262A); mandibular palp slender, article 3 longer than article 2 (plate 262L) 4
4. Uropod 2, peduncle lacking ventral interramal spine.....
.... *Aoroides secundus*
— Uropod 2, peduncle with prominent ventral interramal spine (plate 262M) 5
5. Male gnathopod 1, article 5 without anterior setae bundles, and width >1.4 times width of article 2 (plate 262A, 262N, 262O).. 6
— Male gnathopod 1 article 5 with anterior setae bundles and article 5 and article 2 widths nearly equal (plate 262P, 262Q).. 8
6. Gnathopod 1, anterior and lateral edges of article 2 densely setose and hind margin of article 2 bare (plate 262A) .....
.... *Aoroides columbiae*
— Gnathopod 1, anterior and lateral edges of article 2 sparsely setose and posterior edge of article 2 with setae (plate 262N, 262O) 7
7. Gnathopod 1, anterior edge of article 3 with sparse setae (plate 262N) .... *Aoroides spinosus*
— Gnathopod 1, anterior edge of article 3 with dense setae (plate 262O)....................... *Aoroides exilis*
8. Gnathopod 1, article 5 anterior margin densely setose (plate 262Q), thick spines of inner edge of inner plate of maxilliped nearly smooth (plate 262R) .... *Aoroides inermis*
— Gnathopod 1, article 5 anterior margin sparsely setose (plate 262P), thick spines of inner edge of inner plate of maxilliped serrate (plate 262S) .... *Aoroides intermedia*
9. Male gnathopod 1 subchelate (plate 262D, 262T).......
.... *Bemlos concavus*
— Male gnathopod 1 carpochelate (plate 262C) 10
10. Article 2 of male gnathopods 1 and 2 not expanded (plate 262U).................... *Paramicrodeutopus schmitti*
— Article 2 of male gnathopods 1 and 2 both expanded (plate 262C, 262V) .... *Microdeutopus gryllotalpa*

LIST OF SPECIES

Aoroides columbiae Walker, 1898. Abundant in subtidal fouling communities of rocks, pilings and floats; 6 mm; intertidal—>100 m.

Aoroides exilis Conlan and Bousfield, 1982b. Among algae and sponges under stones and on sand and gravel beaches of open coasts and protected waters; 6 mm; intertidal—50 m.

Aoroides inermis Conlan and Bousfield, 1982b. High-salinity sand and rock surfaces of exposed and protected waters; 6.5 mm; intertidal—148 m.

Aoroides intermedia Conlan and Bousfield, 1982b. 6 mm; intertidal—63 m.

Aoroides secundus Gurjanova, 1938. An Asian species introduced probably by ships to the Pacific coast where it occurs on floats and docks of central San Francisco Bay and southern California harbors; 3.5 mm; intertidal—2 m.

Aoroides spinosus Conlan and Bousfield, 1982b. Low intertidal and subtidal; on various substrata, but especially with algae and among debris; not known south of Coos Bay; 7 mm; intertidal—45 m.

Bemlos concavus (Stout, 1913). Stony bottoms, surf exposed bedrock, *Phyllospadix*, kelp, *Corallina*; 6 mm; intertidal—3 m.

Columbaora cyclocoxa Conlan and Bousfield, 1982b. Under boulders and among *Laminaria* on exposed algal-covered rocky beaches; 7 mm; intertidal—10 m.

Microdeutopus gryllotalpa Costa, 1853. Introduced, a well-known western Atlantic and Mediterranean species of shallow estuaries found on the intertidal mud flats of Humboldt Bay since the 1980s (Boyd et al. 2002); 10 mm; to 150 m in Atlantic.

Paramicrodeutopus schmitti (Shoemaker, 1942). Rocky surf-washed beaches among *Phyllospadix* and red algae; 5 mm; intertidal—43 m.

LIST OF SPECIES

Paracorophium sp. An introduced intertidal mudflat species of northern Humboldt Bay, possibly from South America, included here because of its biramous uropod 3, collected and illustrated by Todd Miller; 4 mm; intertidal—2 m.

ISAEIDAE

Isaeidae are entirely marine suspension feeders that build tubes or occupy empty shells and occur at a wide depth range. Male gnathopod 1 is smaller than gnathopod 2. *Photis* males bear conspicuous stridulation ridges on the lateral face of gnathopod article 2 and medial ventral edge of coxa 2. Rostrum short or absent, eyes small or large, ocular lobe prominent and pointed. Pereopod 7 longer than pereopod 6. The common loss of pereopods and antennae in preservation can greatly complicate identifications. Urosome articles are separate except for *Chevalia*. Uropod 3 is biramous and the telson is entire. The taxonomy is reliable for males only.

KEY TO ISAEIDAE

1. Uropod 3 inner ramus less than half as long as outer ramus and scale- or platelike (plate 263A, 263B) 2
— Uropod 3 inner ramus more than half as long as outer ramus (plate 263C)................................. 5
2. Gnathopod 2 article 5 of males less than one-third as large as article 6 (plate 263D) 3
— Gnathopod 2 article 5 of males more than half as large as article 6 (plate 263E) 4
3. Antenna 1, accessory flagellum a tiny nub (plate 263F) (view at 100×); coxa 3 deeper than pereonite 3 (plate 263G)....... 8
— Antenna 1, accessory flagellum multiarticulated, and coxa 3 shallower than pereonite 3 (plate 263D); large teeth on palm of male gnathopod 2 vary from three to five (adults lose inner ramus of uropod 3). ..... *Cheiriphotis megacheles*
4. Gnathopod 2 article 5 broader than article 6 (plate 263E); gnathopods 1 and 2 palms transverse and greatly overlapped by dactyls (plate 263H, 263I) ....
.... *Cheirimedeia zotea*
— Gnathopod 2 article 5 and 6 approximately equal in width (plate 263J); gnathopods 1 and 2 palms oblique and not

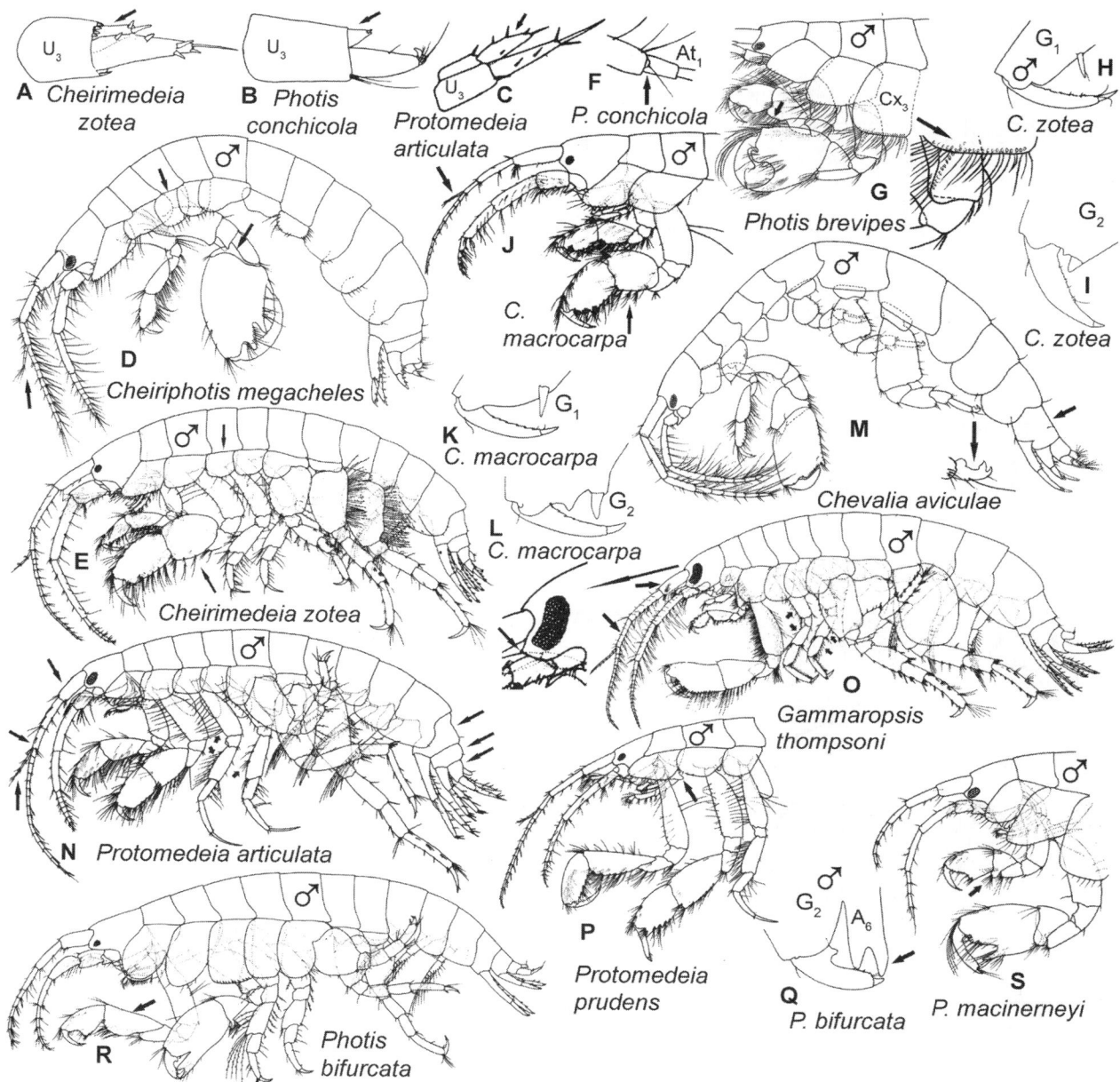

PLATE 263 Isaeidae. J, K, L, *Cheirimedeia macrocarpa*; A, E, H, I, *Cheirimedeia zotea*; D, *Cheiriphotis megacheles*; M, *Chevalia aviculae*; O, *Gammaropsis thompsoni*; Q, R, *Photis bifurcata*; G, *Photis brevipes*; B, F, *Photis conchicola*; S, *Photis macinerneyi*; C, N, *Protomedeia articulata*; P, *Protomedeia prudens* (figures modified from Barnard 1962a; and Conlan 1983).

greatly overlapped by dactyls (plate 263K, 263L) . . . .
. . . . *Cheirimedeia macrocarpa*
5. Urosomites 1 and 2 coalesced and pereopod 5–7 with heavy gripping dactyl (plate 263M) . . . . *Chevalia aviculae*
— Urosomites 1 and 2 separate, accessory flagellum of two or more articles, usually conspicuous (plate 263N) 6
6. Antenna 1 article 3 shorter than article 1, pereopods 3 and 4 anterior margins of articles 2 and 4 strongly setose, male ocular lobes distally rounded (plate 263N) 7
— Antenna 1 article 3 as long as article 1 or longer, pereopods 3 and 4 anterior margins of articles 2 and 4 weakly setose, ocular lobes distally pointed (plate 263O) 13
7. Gnathopods 1 and 2 palms more than half as long as dactyls and coxa 1 without a posterior tooth (plate

263N) . . . . *Protomedeia articulata*
— Gnathopods 1 and 2 palms less than half as long as dactyls and coxa 1 with a posterior tooth (plate 263P) . . . .
. . . . *Protomedeia prudens*
8. Gnathopod 2 with two teeth defining the palm process of article 6 (plate 263Q); gnathopod 1 article 5 nearly three times as long as wide (plate 263R) . . . .
. . . . *Photis bifurcata*
— Gnathopod 2 with a single tooth defining the palmer process of article 6 and gnathopod 1 article 5 less than twice as long as wide (plate 263S) 9
9. Gnathopod 1 article 5 posterior margin short, less than one-third the length of the anterior margin (plate 263S) . . . . .
. . . . *Photis macinerneyi*

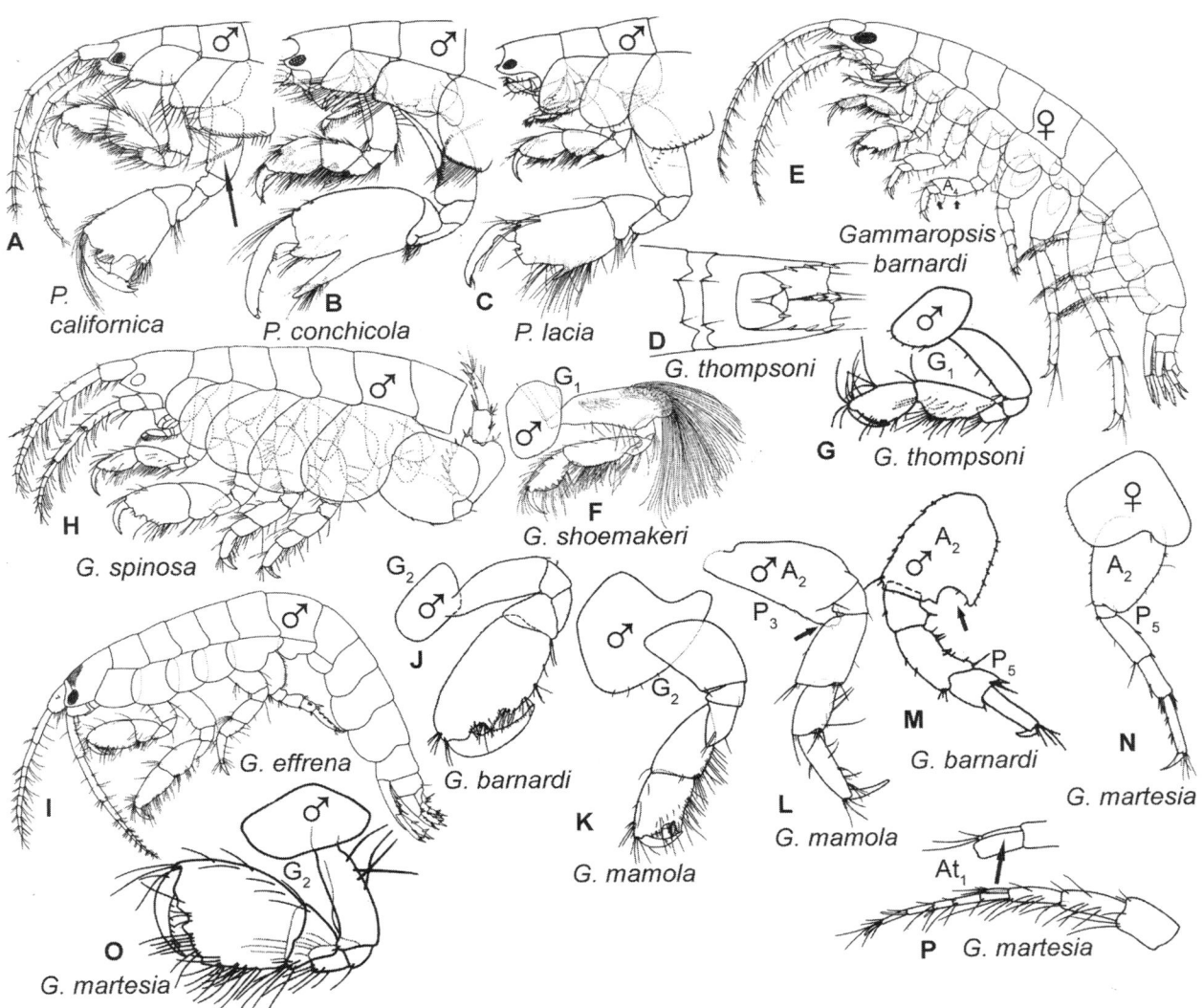

PLATE 264 Isaeidae. E, J, M, *Gammaropsis barnardi*; I, *Gammaropsis effrena*; K, L, *Gammaropsis mamola*; N–P, *Gammaropsis martesia*; F, *Gammaropsis shoemakeri*; H, *Gammaropsis spinosa*; D, G, *Gammaropsis thompsoni*; A, *Photis californica*; B, *Photis conchicola*; C, *Photis lacia* (figures modified from Barnard 1959b, 1962a, 1969a; Conlan 1983; Kudrajaskov and Tzvetkova 1975; and Shoemaker 1942).

— Gnathopod 1 article 5 posterior margin extended, more than one-third the length of the anterior margin (plate 263G) 10

10. Palmar excavation deeply rounded (plates 263G, 264A) 11

— Palmar excavation sharply incised (plate 264B) 12

11. Inner margin of gnathopod 2 dactyl without a large protrusion (plate 264A) . . . . *Photis californica*

— Inner margin of gnathopod 2 dactyl with a large protrusion (plate 263G) . . . . *Photis brevipes*

12. Dactyl of gnathopod 2 extending past the defining palmar tooth of article 6 (plate 264B); lives in small snail shells attached by mucus to algae on rocky coasts . . . . *Photis conchicola*

— Dactyl of gnathopod 2 not extending past the defining palmar tooth of article 6 (plate 264C) . . . . *Photis lacia*

13. Urosome of males dorsally cusped (plate 264D); coxa 7 greatly expanded posteriorly, pereopods 3 and 4 articles 4 and 5 subequal (plate 263O) 14

— Urosome of both sexes dorsally smooth, pereopod 7 coxa short, pereopods 3 and 4 articles 5 half to three-quarters the length of article 4 (plate 264E) 15

14. Male gnathopod 1 posterior distal corner of article 2 expanded and densely covered with setae (plate 264F) . . . . *Gammaropsis shoemakeri*

— Gnathopod 1 posterior distal corner of article 2 unexpanded and without dense cover of setae (plate 264G) . . . . *Gammaropsis thompsoni*

15. Gnathopod 2 (both sexes) posterior margin of gnathopod 2 article 5 more than one-third the length of article 6 (plate 264H, 264I) 16

— Gnathopod 2 (both sexes) posterior margin of gnathopod 2 article 5 less than one-fifth the length of article 6 (plate 264E, 264J) 18

16. Male coxa 2 posteriorly straight or only slightly concave (plate 264H, 264I); pereopod 3 anteriodistal article 2 not expanded 17

— Male coxa 2 posteriorly lobed (plate 264K); pereopod 3 anteriodistal article 2 expanded (plate 264L) . . . . *Gammaropsis mamola*

17. Antenna 1 article 1 twice as thick as article 2, accessory flagellum tiny and of two articles, articles 2 and 4 of pereopod 5 normal, head pigmented, ocular lobes rounded (plate 264I) . . . . *Gammaropsis effrena*
— Antenna 1 article 1 only slightly thicker than article 2, accessory flagellum prominent and of three articles, pereopod 5 articles 2 and 4 of thick, article 4 posterior lined with spines, head unpigmented, ocular lobes pointed (plate 264H) . . . . *Gammaropsis spinosa*
18. Male pereopod 5 article 2 posterior ventral edge deeply notched (plate 264M); gnathopod 2 article 6 half as wide as long (plate 264J); accessory flagellum a microscopic button (not shown) (plate 264E) . . . . . *Gammaropsis barnardi*
— Male pereopod 5 article 2 posterior ventral edge evenly rounded (plate 264N); gnathopod 2 article 6 two-thirds as wide as long (plate 264O); accessory flagellum as long as the first article of the flagellum (plate 264P) . . . .
. . . . *Gammaropsis martesia*

LIST OF SPECIES

Cheirimedeia macrocarpa Bulytscheva, 1952. In brackish to full marine waters on semiprotected sand flats; possibly introduced; 5 mm; intertidal.

Cheirimedeia zotea (Barnard, 1962) (=*Protomedeia zotea*). In mixed mud and sand sediments; 5 mm; intertidal—113 m.

Cheiriphotis megacheles (Giles, 1885). Abundant among *Phyllospadix* and *Silvetia* and under rocks in California; also reported widely from the warmer Pacific and Indian Oceans. Cryptogenic, possible species complex; 3 mm; intertidal—16 m.

Chevalia aviculae (Walker, 1898). Reported also in the Indian Ocean, South Africa, and the Caribbean Sea; cryptogenic; soft benthos; 4 mm; intertidal—35 m.

Gammaropsis barnardi (Kudriaschov and Tzvetkova, 1975) (=*Podoceropsis barnardi*). In mixed rock sediments and sand; 5 mm; intertidal—17 m.

Gammaropsis effrena (Barnard, 1964). Among *Phyllospadix*, algae, and polychaete tubes in rocky areas; 3.7 mm; intertidal.

Gammaropsis mamola (Barnard, 1962). Among algae holdfasts and on hard surfaces including submerged logs. 4 mm; 3 m–25 m.

Gammaropsis martesia (Barnard, 1964a). Among *Phyllospadix*, tunicates, and sponges; 3 mm; intertidal—84 m.

Gammaropsis shoemakeri Conlan, 1983. Among kelp and hydroids; 5.5 mm; intertidal—27 m.

Gammaropsis spinosa (Shoemaker, 1942). Among algae, sponges, and polychaete tubes; 3.5 mm; intertidal—27 m.

Gammaropsis thompsoni (Walker, 1898). Among encrusting animals and in algal holdfasts; 11.5 mm; intertidal—27 m.

Photis bifurcata Barnard, 1962. Usually on soft sediments; 4 mm; low water—109 m.

Photis brevipes Shoemaker, 1942. In various sediments but especially sand; 7 mm; low water—289 m.

Photis californica Stout, 1913. Among *Phyllospadix* and on open coast rocky shores; 6 mm; low intertidal—147 m.

Photis conchicola Alderman, 1936. On rocky beaches with algae and surfgrass, often paguridlike, living in empty gastropod shell; 5.5 mm; intertidal—42 m.

Photis lacia Barnard, 1962a. In sandy sediments of exposed coasts; 3.3 mm; low intertidal—40 m.

Photis macinerneyi Conlan, 1983. Sandy substrates of exposed and protected marine coasts; 4.3 mm; low intertidal—40 m.

Protomedeia articulata Barnard, 1962. In soft sediments; 8 mm; low intertidal to deep subtidal

Protomedeia prudens Barnard, 1966. In soft sediments; 7.5 mm; intertidal—400 m.

AMPITHOIDAE

Ampithoidae are herbivores that build nests of algae or burrow into kelp stipes and commonly attain the same color as the algae they inhabit. The third uropods and rami are short, with two (occasionally one) distinctive stout hook spines on the outer ramus. Taxonomy emphasizes males.

KEY TO AMPITHOIDAE

1. Pereopods 3 and 4 article 2 strongly inflated, width more than three-fourths of the width of the coxa (plate 265A); gnathopod 1 palm transverse (plate 265B) 9
— Pereopods 3 and 4 article 2 width less than one-half of the width of the coxa (plate 265C); gnathopod 1 palm subchelate (plate 265D) 2
2. Antenna 1 accessory flagellum multiarticulated (plate 265C); uropods 1 and 2 with distal ventral spinose process projecting below the rami (plate 265E) . . . .
. . . . *Paragrubia uncinata*
— Antenna 1 accessory flagellum vestigial or absent (plate 265F); uropods 1 and 2 with distal ventral spinose process small or absent (plate 265G). 3
3. Gnathopod 1 posterior lobe of article 5 long, more than 40% of the length of the entire article (plate 265F, 265H). . . . . .
. . . . 4
— Gnathopod 1 posterior lobe of article 5 short, <40% of the length of the entire article (plate 265I) 6
4. Antenna 2 peduncle 5 and flagellum with dense plumose setae (plate 265H); male gnathopod 1 article 5 shorter than article 6 (plate 265H); male gnathopod 2 palm slightly oblique (plate 265J); epimeron 3 hind margin evenly rounded (plate 265K). . . . . *Ampithoe plumulosa*
— Antenna 2 lacking dense plumose setae, gnathopod 1 article 5 as long or longer than article 6, male gnathopod 2 palm transverse or produced forward (plate 265F); epimeron 3 posterior ventral corner with intersecting ridge and angular or slightly notched (plate 265F) 5
5. Male gnathopod 2 palm produced forward (plate 265F); epimeron 3 posterior ventral corner with small notch at the end of the intersecting ridge (plate 265F); lower lip lobes widely separated (plate 265L). . . . . *Ampithoe lacertosa*
— Male gnathopod 2 palm transverse and bearing square tooth (plate 265M); epimeron 3 posterior ventral corner without a notch at the end of the intersecting ridge (plate 265N); lower lip lobes separated by narrow gap (plate 265O) . . . . *Ampithoe valida*
6. Apex of telson with two enlarged, lobed "rabbit ear" folds (plate 265P); pereopod 5 article 5 less than half as long as article 6 (plate 265S) . . . . *Ampithoe aptos*
— Apex of telson with two minute lateral knobs (plate 265R); pereopod 5 article 5 more than half as long as article 6 (plate 265Q). 7
7. Male gnathopod 2 palm sharply incised to form a large pointed tooth (plate 266A); antenna 2 slightly shorter than antenna 1 (plate 266A); antenna 2 setose and with flagellum shorter than combined articles 4 and 5 (plate 266A)
. . . . *Ampithoe sectimanus*
— Male gnathopod 2 palm roundly incised to form short, blunt tooth (plate 266B); antenna 2 longer than antenna 1 (plate 266B), antenna 2 weakly setose and with flagellum

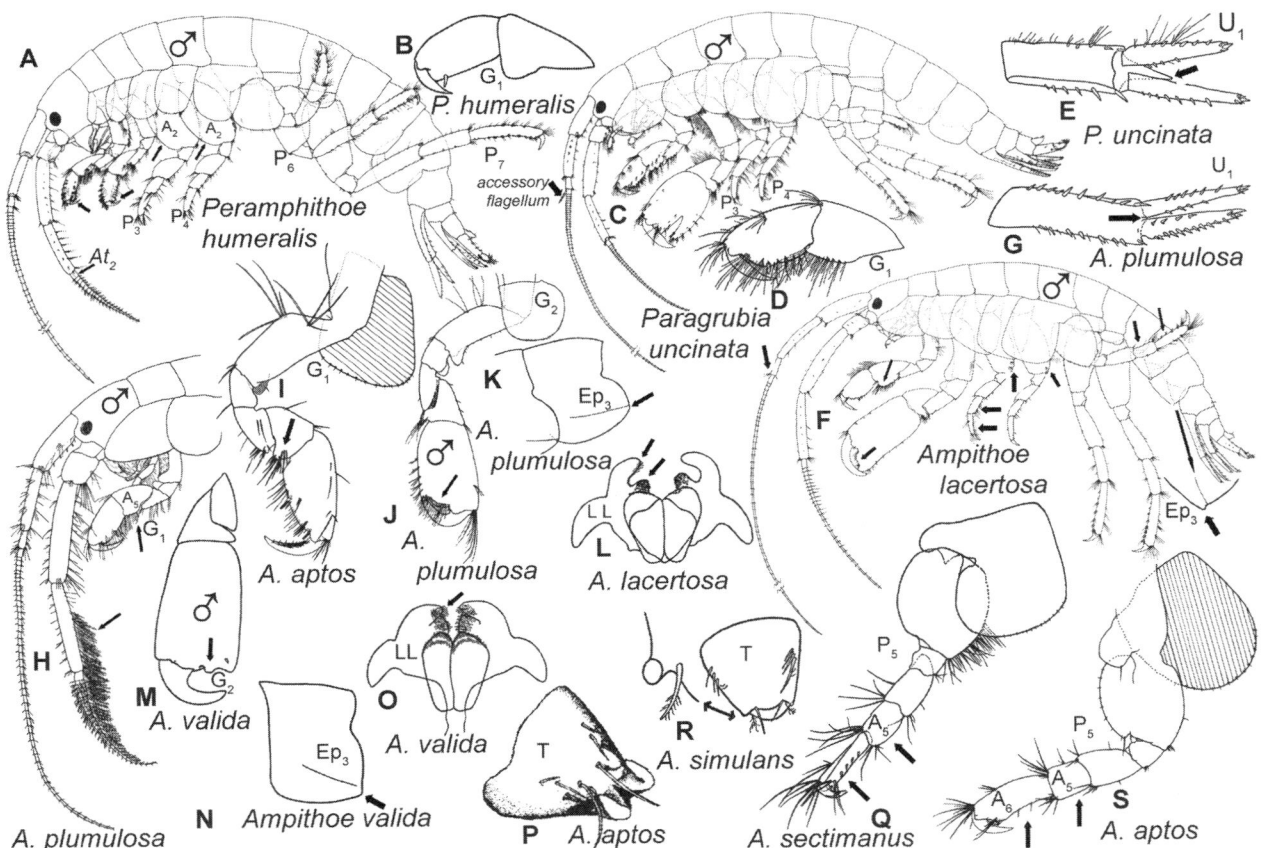

PLATE 265 Ampithoidae. I, P, S, *Ampithoe aptos*; F, L, *Ampithoe lacertosa*; G, H, J, K, *Ampithoe plumulosa*; Q, *Ampithoe sectimanus*; R, *Ampithoe simulans*; M–O, *Ampithoe valida*; C–E, *Paragrubia uncinata*; A, B, *Peramphithoe humeralis* (figures modified from Barnard 1952b, 1965, 1969a; Conlan and Bousfield 1982a; and Shoemaker 1938a).

as long as peduncular articles 4 and 5 (plate 266B). 8

8. Male gnathopod 1 articles 2 anterior edge lined with plumose setae (plate 266B); mandibular palp article 3 distal seta row marked by angle at inner proximal margin (plate 266C); epimeron 3 posterior ventral corner evenly rounded (plate 266D) . . . . *Ampithoe dalli*

— Male gnathopod 1 article 2 anterior edge bare (plate 266E); mandibular palp article 3 distal seta row rounding evenly into inner proximal margin (plate 266F); epimeron 3 posterior ventral corner notched (plate 266G) . . . .
. . . . *Ampithoe simulans*

9. Male gnathopod 2 article 6 less than twice as thick as gnathopod 1 article 6 (plates 265A, 266H, 266I). 10

— Male gnathopod 2 article 6 more than twice as thick as gnathopod 1 article 6 (plate 266J–266L) 12

10. Antenna 2 flagellum proximal articles fused into one article longer than wide (plate 266I) . . . .
. . . . *Peramphithoe stypotrupetes*

— Antenna 2 flagellum proximal articles separate and not longer than wide (plate 265A) 11

11. Pereopod 7 more than 1.5 times as long as pereopod 6 (plate 265A); gnathopod 2 (both sexes) palm transverse; and article 5 equal to or longer than article 6 (plate 266M) . . . .
. . . . *Peramphithoe humeralis*

— Pereopod 7 less than 1.2 times length of pereopod 6; and gnathopod 2 (both sexes) palm oblique with article 5 length less than article 6 (plate 266H) . . . .
. . . . *Peramphithoe mea*

12. Male gnathopod 2 palm well defined, extending about half the length of posterior edge of article 6 (plate 266K); antenna 2 article 5 shorter than article 4 (plate 266N) . . . . .
. . . . *Peramphithoe lindbergi*

— Male gnathopod 2 palm poorly defined and extending more than half length of article 6; antenna 2 article 4 length approximately equal to article 5 (plate 266L) 13

13. Lower lip lateral and medial lobes projecting equally (plate 266O); antenna 2 flagellum article 1 nearly 2 times longer than more distal articles (plate 266L) . . . .
. . . . *Peramphithoe plea*

— Lower lip lateral lobes projecting further than medial lobes (plate 266P); antenna 2 flagellum article 1 less than two times wider than more distal articles (plate 266Q) . . . . . .
. . . . *Peramphithoe tea*

LIST OF SPECIES

Ampithoe aptos (Barnard, 1969) (=*Pleonexes aptos*). Algal covered bottoms where it is scarce; 7 mm; intertidal.

*Ampithoe corallina Stout, 1913. Southern California; possible *nomen nudum*.

Ampithoe dalli Shoemaker, 1938. Boreal, south to Cape Arago on exposed and protected beaches, in tide pools, under rocks and log fouling organisms, in 10–34‰ salinity. Females ovigerous March to August; 20 mm; intertidal—10 m.

* = Not in key.

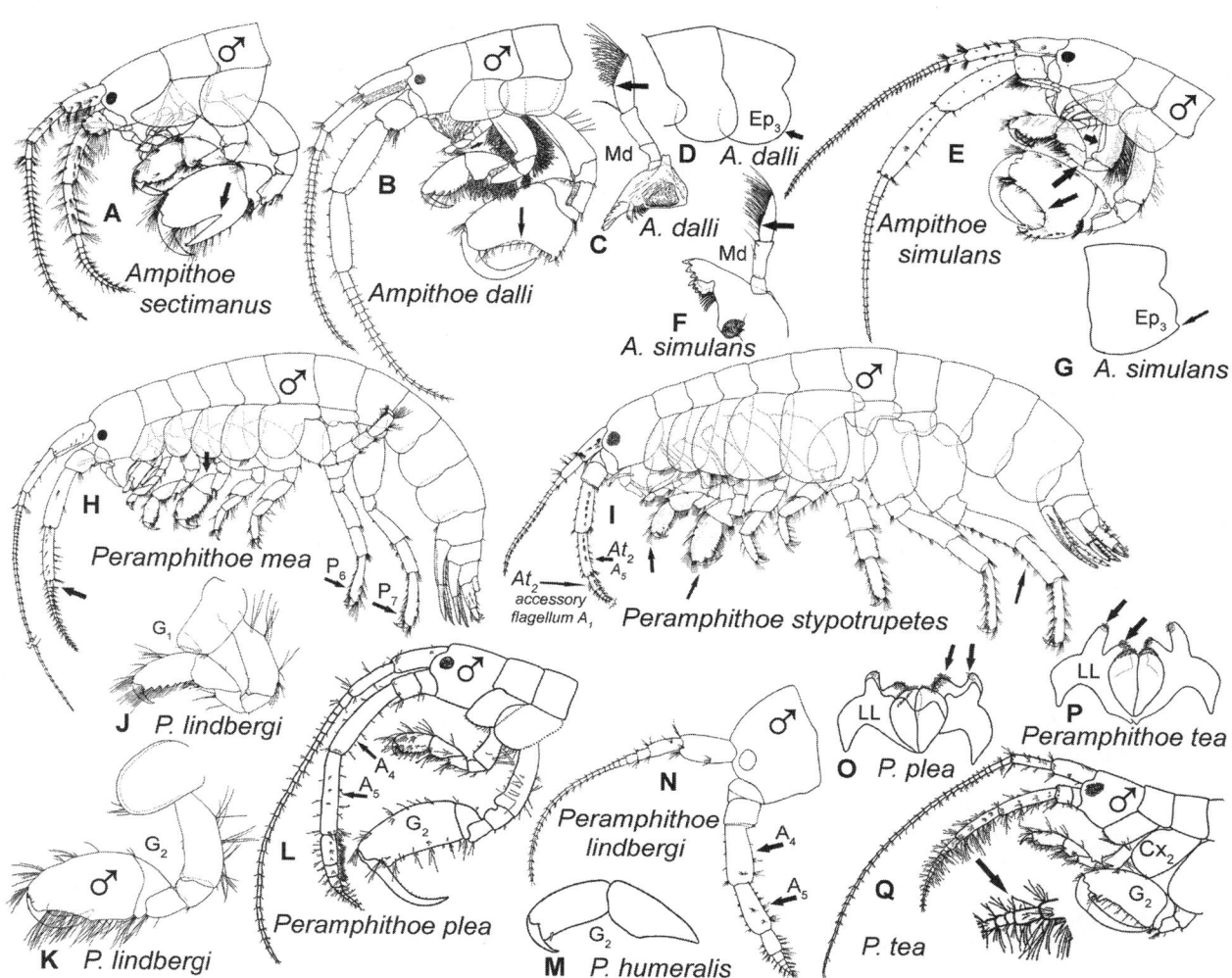

PLATE 266 Ampithoidae. B, C, D, *Ampithoe dalli*; A, *Ampithoe sectimanus*; E–G, *Ampithoe simulans*; M, *Peramphithoe humeralis*; J, K, N, *Peramphithoe lindbergi*; H, *Peramphithoe mea*; L, O, *Peramphithoe plea*; I, *Peramphithoe stypotrupetes*; P, Q, *Peramphithoe tea* (figures modified from Barnard 1952b, 1965; Conlan and Bousfield 1982a; Conlan and Chess 1992; and Shoemaker 1938a).

Ampithoe lacertosa (Bate, 1858). Among algae, gravel, or woody debris and on pilings and floats of estuaries; also protected open coasts; heavily speckled with diffuse spots. See Heller 1968, MSc thesis, Univ. Washington 132 pp. (biology and development); 24 mm; intertidal—11 m.

*Ampithoe longimana (Smith, 1873). North Atlantic, introduced to southern California, may receive protection from predators by accumulating toxins from algae it ingests (Hay et al. 1990, Ecology 71: 733–743); 10 mm; intertidal—10 m.

Ampithoe plumulosa Shoemaker, 1938. Eastern Pacific and western Atlantic; common on algae and *Mytilus* beds; origins unclear, a likely introduction or misidentified elsewhere in the world; 16 mm; intertidal—15 m.

*Ampithoe pollex Kunkel, 1910. Northeast Pacific records unclear due to poor description of type populations; possibly introduced to southern California; 5.5 mm; intertidal.

*Ampithoe ramondi Audoin, 1828. Cosmopolitan at latitudes <45°; not reported north of Point Conception but may appear to the north with climate warming; in diverse algae; 12 mm; intertidal—32 m.

Ampithoe sectimanus Conlan and Bousfield, 1982. High salinity exposed rocky coasts among algae, females ovigerous May to August; 12.5 mm; intertidal.

Ampithoe simulans (Alderman, 1936). Among algae and *Phyllospadix* of open and semiprotected coasts, occasionally in brackish water; 30 mm; intertidal—4 m.

Ampithoe valida Smith, 1873. Abundant among green algae and in fouling communities of pilings floats, docks, and on mudflats of estuaries in Europe, eastern and western United States, Japan, Argentina; a likely Atlantic species introduced to the Pacific coast. See Alonso et al. 1995, Oebalia 21: 77–91 (seasonal population changes); Pardali et al. 2000, Mar. Ecol. Prog. Ser. 196: 207–219 (biology, ecology in Portugal); Borowsky 1983, Mar. Biol. 77:257–263 (tube building and reproductive ecology); 12.5 mm; intertidal—30 m.

Paragrubia uncinata (Stout, 1912) (=*Cymadusa uncinata*). Rolls blades of kelp *Macrocystis pyrifera* into cigar-shaped tubes, occurs also among *Phyllospadix*; 35 mm; 4 m–27 m.

*Peramphithoe eoa (Bruggen, 1907). Sea of Japan, northeast Pacific records of this species and its distinction from *P. mea* are unclear; 10 mm; intertidal—90 m.

Peramphithoe humeralis (Stimpson, 1864). This very large amphipod (like *Paragrubia*) makes nests in *Alaria* or *Macrocystis* by curling the fronds into a tube in which the young

* = Not in key.

may remain in a colony for several instars after emerging from the female oötangium. The upper walls of the tube are consumed by adults and their juveniles. Reproduction June to August. See Jones 1971, pp. 343–367, in W. North, ed., The biology of giant kelp beds (*Macrocystis*) in California. Nova Hedwigia 32 (general biology). Conlan and Bousfield (1982a) consider the South African *Peramphithoe humeralis* (see Griffiths 1979) to be a different species, although it may live in *Macrocystis* there also; up to 53 mm; low intertidal—18 m.

Peramphithoe lindbergi (Gurjanova, 1938). Boreal south to Corona del Mar, among eelgrass and algal holdfasts, ovigerous June to September; 12.5 mm; intertidal—18 m.

Peramphithoe mea (Gurjanova, 1938). Boreal, south possibly to Coos Bay, Oregon, or southern California; southern populations of eastern Pacific *P. mea*, *P. plea*, and *P. tea* are not clearly distinguished; among eelgrass; 22 mm; rarely intertidal—60 m.

Peramphithoe plea (Barnard, 1965). Among kelp holdfasts on exposed coasts; 12.5 mm; intertidal—17 m.

Peramphithoe stypotrupetes Conlan and Chess, 1992. Burrows into *Eisenia* and *Laminaria* stipes, cohorts remain and graze on the stipe's interior; 21 mm; shallow subtidal—10 m.

Peramphithoe tea (Barnard, 1965). Among algae of exposed and semiprotected high salinity areas, distinction from *P. plea* unclear, ovigerous May to August; 12 mm; intertidal—67 m.

ISCHYROCERIDAE

Ischyroceridae construct tubes on hard surfaces in areas of high water velocity and include many of the most common amphipods of fouling communities. *Jassa* males and probably males of all other genera use gnathopod 2 for mate guarding, combat, and display, while "sneaker" males obtain mates as paedomorphs (Kurdzie and Knowles 2002, Roy. Soc. 269: 1749–1754). Male gnathopod 2 larger than gnathopod 1. Intraspecific variation in male secondary sex characters among mating systems (Conlan 1989, 1991, 1995a, 1995b) and the adaptive variations in mating systems with environmental conditions complicate the taxonomy of Ischyroceridae based on male secondary sex characters. Uropod 3 bearing short rami, the outer ramus bearing single large hook spines in *Jassa* and *Microjassa* and comblike fused spines among *Ischyrocerus*. Uropod 3 of *Ericthonius* and *Cerapus* bearing a single ramus.

KEY TO ISCHROCERIDAE

1. Male gnathopod 2 carpochelate, coxa 4 longer than deep (plate 267A, 267B); telson extremely short and covered with dorsally directed spines (plate 267C, 267D); uropod 3 uniramous (plate 267C, 267E) 2
— Male gnathopod 2 subchelate, coxa 4 as deep or deeper than long (plate 267F); telson without dense dorsally directed spines (plate 267G–267I); uropod 3 biramous (plate 267J). 4
2. Rostrum acute, antenna 1 article 1 swollen (plate 267A); uropod 2 reduced, bearing single vestigial ramus (plate 267C) . . . . *Cerapus* spp.
— Rostrum absent, antenna 1 article 1 only slightly thicker than article 2 and uropod 2 with two normal rami (plate 267B) 3
3. Male gnathopod 1 article 2 with dorsal posterior protrusion (plate 267K); male gnathopod 2 article 5 apically bifid and

coxa 2 bearing stridulating ridges (plate 267L) . . . .
. . . . *Ericthonius brasiliensis*
— Male gnathopod 2 article 5 with a simple apical tooth; male gnathopod 1 article 2 without dorsal posterior protrusion (plate 267B) and coxa 2 without stridulating ridges . . . . *Ericthonius rubricornis*
4. Peduncle of uropod 1 with lateral row of plumose setae (plate 267F, 267M) . . . . *Ruffojassa angularis*
— Peduncle of uropod 1 bearing only short, stout spines, without lateral plumose setae (plate 267N) 5
5. Coxa 5 anterior and posterior lobes approximately equal (plate 267N). 6
— Coxa 5 anterior at least three times larger than the posterior lobe (plate 267O) 7
6. Male gnathopod 2 article 6 with swollen dorsoanterior margin (plate 267N) . . . . *Microjassa litotes*
— Male gnathopod 2 article 6 with evenly rounded dorsoanterior margin (plate 267P) . . . . *Microjassa barnardi*
7. Uropod 3 rami blunt, outer ramus bearing irregular teeth proximal to a single heavy distal hooked spine rami (plate 267Q). 9
— Uropod 3 rami pointed, outer ramus bearing small straight distal spine or no spine, usually lined with microscopic denticles (plate 267R) (confirm at 100–400x) 8
8. Eye diameter less than one-fifth of the depth of head (plate 267S); male gnathopod 2 palm concave (plate 267T). . . . .
. . . . *Ischyrocerus anguipes*
— Eye diameter more than one-fourth of head, male gnathopod 2 palm straight (plate 267O) . . . .
. . . . *Ischyrocerus pelagops*
9. Gnathopod 2 anterolateral margin of article 2 with a row of setae (plates 267U, 268A) 11
— Gnathopod 2 article 2 without anteriolateral setae (plate 268B) 10
10. Uropod 1, posterioventral interramal spine less than one-eighth length of outer ramus (plate 268C). . . . . *Jassa shawi*
— Uropod 1, posterioventral interramal spine more than one-third length of outer ramus (plate 268D) (*J. shawi* and *J. falcata* are poorly distinguished species) . . . . . *Jassa falcata*
11. Male gnathopod 1 article 5 without anteriodistal seta at the junction of article 6 (plate 268E) . . . . *Jassa staudei*
— Male gnathopod 1 article 5 with one or more anteriodistal seta at the junction of article 6 (plate 268F–268H) 12
12. Anteriodistal seta of gnathopod 1 tiny and slightly lateral (plate 268F) . . . . *Jassa marmorata*
— Anterior distal seta of gnathopod 1 article 5 long, slightly medial or dorsal (plate 268G, 268H) 13
13. Uropod 1 ventral distal peduncle spine more than one-fourth of the length of the shortest ramus (plate 268D). 14
— Uropod 1 ventral distal peduncle spine less than one-eighth of the length of the shortest ramus (plate 268C); antenna 2 anterior article 5 distal half densely setose, gnathopod 2 article 2 anterior setae dense and article 6 thumb small (plate 268I); female gnathopod 2 palm concave (plate 268J) . . . . *Jassa borowskyae*
14. Tip of telson bearing apical setae as well as lateral setae (plate 268K) and gnathopod 2 of large-thumbed male with defining spines on a ledge (plate 268L) . . . . . *Jassa morinoi*
— Tip of telson without apical setae (plate 268M); gnathopod 2 of large-thumbed male with defining spines absent (plate 268A) or not in a deep ledge (plate 268N). 15
15. Male gnathopod 2 thumb tip angled and bearing defining spines (plate 268N); female gnathopod 1 palm (not illustrated) and male evenly convex (plate 268G) . . . .

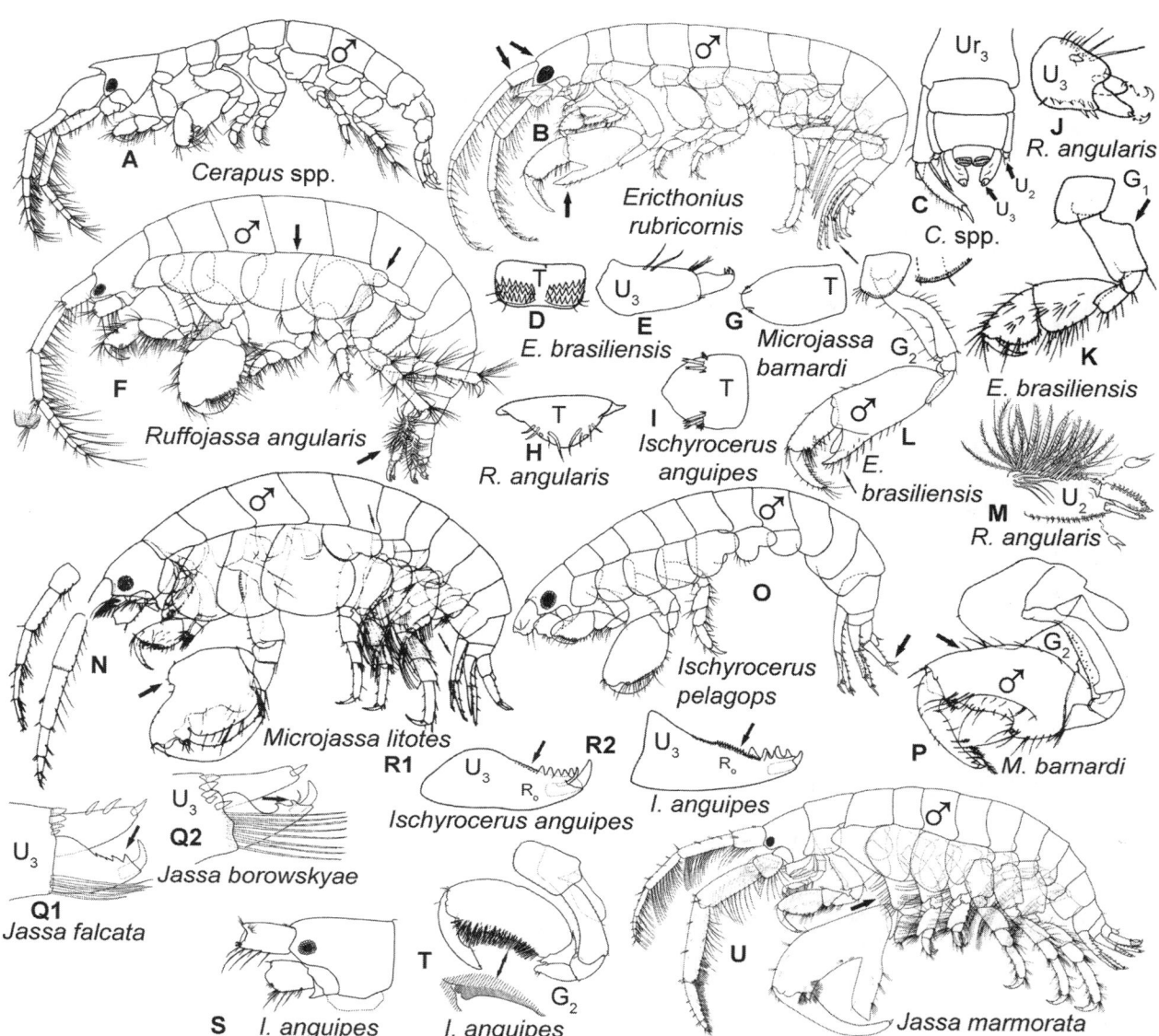

PLATE 267 Ischyroceridae. A, C, *Cerapus* spp.; B, *Ericthonius rubricornis*; D, E, K, L, *Ericthonius brasiliensis*; I, R–T, *Ischyrocerus anguipes*; O, *Ischyrocerus pelagops*; Q2, *Jassa borowskyae*; Q1, *Jassa falcata*; U, *Jassa marmorata*; G, P, *Microjassa barnardi*; N, *Microjassa litotes*; F, H, J, M, *Ruffojassa angularis* (figures modified from Bousfield 1973; Barnard 1962a, 1969a ; and Conlan 1990, 1995b).

. . . . *Jassa carltoni*
— Gnathopod 2 thumb tip straight, defining spines absent (plate 268A) and gnathopod 1 palm female (not illustrated) and male straight or concave (plate 268O) . . . .
. . . . *Jassa slatteryi*

LIST OF SPECIES

Cerapus spp. Referred to previously in the eastern Pacific as the Atlantic species *Cerapus tubularis* Say, 1817, but Pacific taxa likely represent one or more undescribed native species; build thick, pliable, striped, cylindrical tubes open at both ends; 3.2 mm; intertidal and shallow subtidal.

Ericthonius brasiliensis (Dana, 1853). Taxonomy poorly resolved: open coast populations (in habitats such as *Phyllospadix*) and harbor populations (likely introduced) probably represent different species; exhibits territorial behavior (Con-

nell 1964, Res. Pop. Ecol. 87: 87–101); 6.5 mm; intertidal—300 m.

Ericthonius rubricornis (Stimpson, 1853) (=*Ericthonius hunteri*). Amphiboreal, forming mats of muddy tubes on diverse substrata; shallow water populations may be introduced; possibly a hypermale of *E. brasiliensis* (Myers and McGrath 1984, J. Mar. Biol. Assoc. UK 64: 379–400) or part of a species complex; 9 mm; intertidal—235 m.

Ischyrocerus anguipes (Kroyer, 1838). Boreal-temperate North Atlantic and eastern Pacific, tube-building on various substrata, a likely species complex with origin of shallow water harbor species uncertain; 12 mm; intertidal—326 m.

Ischyrocerus parvus Stout, 1913. Possibly *I. anguipes*; 3 mm; rocky intertidal.

Ischyrocerus pelagops Barnard, 1962. Fine gray sands; 4.5 mm; intertidal—24 m.

* = Not in key.

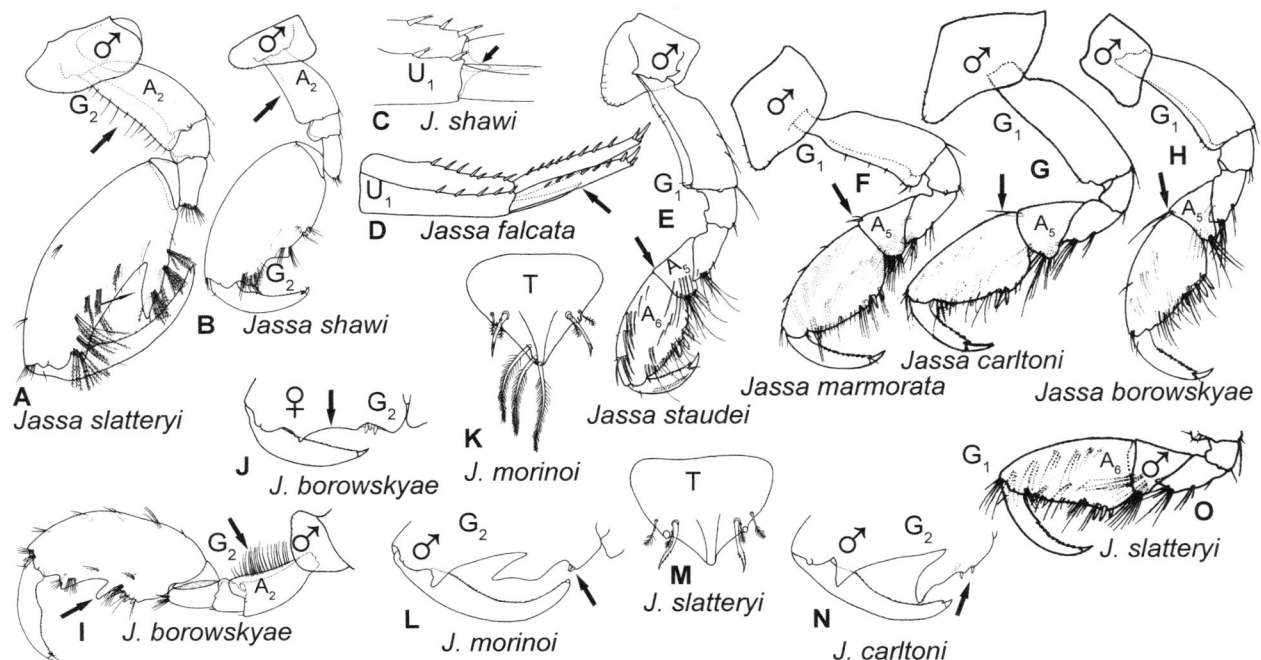

PLATE 268 Ischyroceridae. H, I, J, *Jassa borowskyae*; G, N, *Jassa carltoni*; D, *Jassa falcata*; F, *Jassa marmorata*; K, L, *Jassa morinoi*; B, C, *Jassa shawi*; A, M, O, *Jassa slatteryi*; E, *Jassa staudei* (figures modified from Barnard 1962a, 1969a; and Conlan 1990).

Ischyrocerus sp. A Barnard, 1969. Possibly *I. pelagops*; 3.8 mm; rocky intertidal.

Ischyrocerus sp. B Barnard, 1969. Possibly *I. anguipes*; 3.4 mm; intertidal.

Jassa borowskyae Conlan, 1990. California, Siberia, Sea of Japan, exposed rocky shores on algae and surfgrass; 7.7 mm; low intertidal—20 m.

Jassa carltoni Conlan, 1990. Southern California in *Phyllospadix*, named in honor of James T. Carlton (of Light and Smith's Manual); difficult to distinguish from *J. morinoi* or *J. slatteryi*; 3.5 mm; intertidal.

Jassa falcata (Montagu, 1808). Most shallow water mid-latitude marine *Jassa* of the world were referred to as *J. falcata* prior to the work of Conlan (1990); presently recognized only in European harbors, but not clearly absent elsewhere; 7 mm; low intertidal—40 m.

Jassa marmorata Holmes, 1903. Introduced cosmopolitan marine and estuarine species, found in fouling communities on floats and pilings in harbors of Asia, Europe, New England, and the northeast Pacific; 7 mm; low intertidal—30 m.

Jassa morinoi Conlan, 1990. North Pacific, Atlantic, and Mediterranean; a likely introduced species, on rocks and algae; 6 mm; low intertidal—7 m.

Jassa shawi Conlan, 1990. On hard substrata and sponges; 7 mm; low intertidal.

Jassa slatteryi Conlan, 1990. On algae and hydroids; Ecology (Jeong et al. 2007, J. Crust. Biol. 27[1]:65–70); 5.5 mm; low intertidal.

Jassa staudei Conlan, 1990. On rocks and algae; 11.4 mm; low intertidal—82 m.

Microjassa barnardi Conlan, 1995b. On algal holdfasts and rocks; 2.5 mm; intertidal—52 m.

Microjassa litotes Barnard, 1954. On algal holdfasts; 3.5 mm; intertidal—157 m.

* = Not in key.

Ruffojassa angularis Shoemaker, 1942b. A southern species and a likely introduction that occurs as far north as Carmel; also reported from Madagascar, Hawaiian Islands; 3.5 mm; shallow subtidal—30 m.

COROPHIIDAE

Corophiidae build U-shaped tubes in soft sediments or on hard surfaces. Morphological variations in the male rostrums and massive peduncle of the second antennae of most species allow field identifications. Most females can be reliably identified to species. Telson fleshy and entire, outer lobes of lower lip entire, article 5 of pereopods 3–6 short and reniform, urosomites 1–3 fused or separate and similar in length, uropod 3 with one ramus which can bear multiple articulate setae or spines, gnathopod 2 article 5 of most corophiids is fused over a broad suture to article 4 and lined posteriorly with long, pinnate setae that form a basket used for suspension feeding. Also keyed here is the aorid *Grandidierella japonica* because of its similarity to the corophiids.

KEY TO COROPHIIDAE

1. Male gnathopod 1 massive and carpochelate (plate 269A); male and female uropod 3 ramus spinose, more than three times longer than wide and round in cross-section (plate 269B); pereopod 7 only slightly longer than pereopod 6 (plate 269C) (Aoridae) . . . . *Grandidierella japonica*
—   Male gnathopod 1 relatively small; pereopod 6 half as long as pereopod 7 (plate 269D); uropod 3 ramus oval and dorsoventrally flattened (plate 269E–269G) (Corophiidae) 2

2. Urosomites separate (plate 269D–269G) 3
—   Urosomites fused (plate 269F–269G) (gently clean dorsal urosome with fine needle or brush if unclear). 13

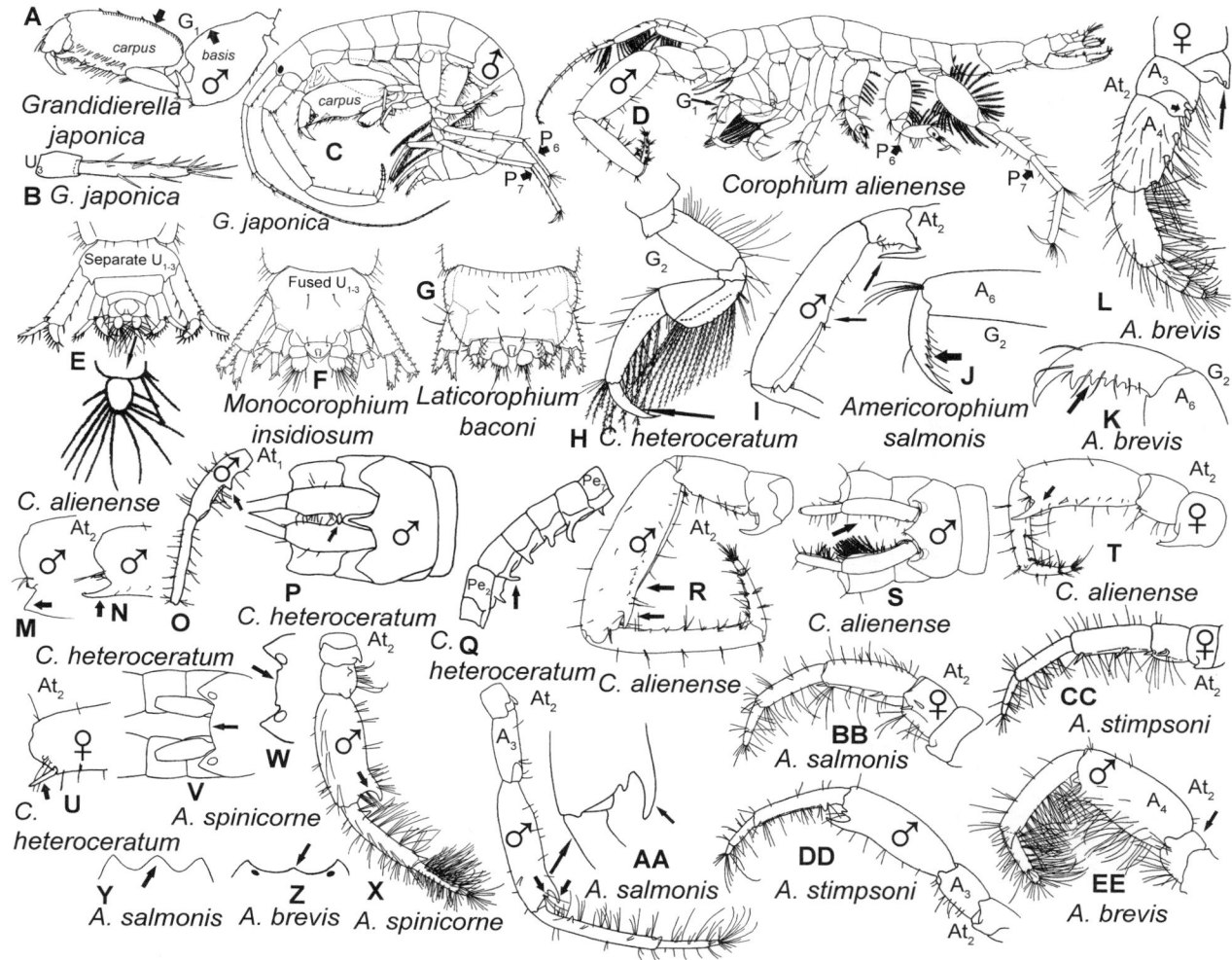

PLATE 269 Corophiidae. K, L, Z, EE, *Americorophium brevis*; J, Y, AA, BB, *Americorophium salmonis*; V–X, *Americorophium spinicorne*; CC, DD, *Americorophium stimpsoni*; D, E, R–T, *Corophium alienense*; H, I, M–Q, U, *Corophium heteroceratum*; A–C, *Grandidierella japonica*; G, *Laticorophium baconi*; F, *Monocorophium insidiosum* (figures modified from Barnard 1954a; Faith Cole, personal communication; Chapman and Dorman 1975; Hiryama 1984; Nagata 1965b; Chapman 1988; and Shoemaker 1934, 1947, 1949).

3. Gnathopod 2 dactyl posterior edge smooth (plate 269H) (apparent at 40×–50×); antenna 2 article 2 excretory spout more than half as long as article 1 (plate 269I) 4
— Gnathopod 2 dactyl posterior edge toothed (plate 269J, 269K); antenna 2 article 2 excretory spout less than half as long as article 1 (plate 269L) 7
4. Male (lacking brood lamellae) 5
— Female (bearing brood lamellae) 6
5. Antenna 2 article 4 with a single denticle on medial edge (plate 269I) and variably pointed or truncated distal tooth (plate 269M, 269N) ; antenna 1 article 1 inner edge with 1 (sometimes 2), medial tooth (plate 269O, 269P); pereonites 2–7 with ventral projections (plate 269Q) . . . .
. . . . *Corophium heteroceratum*
— Antenna 2 with multiple denticles lining ventral medial edges of articles 4 and 5 and bearing a pointed distal tooth on article 4 (plate 269R); antenna 1 inner edge of article 1 without tooth (plate 269S); pereonites 2–7 without ventral projections. . . . . *Corophium alienense*
6. Antenna 2 article 4 with stout distal medial tooth (plate 269T) . . . . *Corophium alienense*
— Antenna 2 article 4 with a stout distal medial spine and no

tooth (plate 269U) . . . . *Corophium heteroceratum*
7. Rostrum broadly rounded (plate 269V) or flat (plate 269W); antenna 2 with a single prominent distal tooth on article 4 (plate 269X) . . . . *Corophium spinicorne*
— Rostrum narrowly rounded (plate 269Y) or pointed (plate 269Z); antenna 2 article 4 of males with one prominent and one accessory distal ventral tooth (plate 269AA) and female antenna 2 article 4 with single distal spine and without distal teeth (plate 269BB, 269CC) 8
8. Male (lacking brood lamellae) 9
— Female (bearing brood lamellae) 11
9. Antenna 2 with few setae, article 3 longer than wide (plate 269AA, 269DD) 10
— Antenna 2 setose, article 3 half as long as wide (plate 269EE). . . . . *Americorophium brevis*
10. Antenna 1 article 1 dorsally more than twice as long as wide (plate 270A) and with a ventral tooth (plate 270B) . . . . . .
. . . . *Americorophium stimpsoni*
— Antenna 1 article 1 broadly expanded laterally (plate 270C) and without a ventral tooth (plate 270D) . . . .
. . . . *Americorophium salmonis*

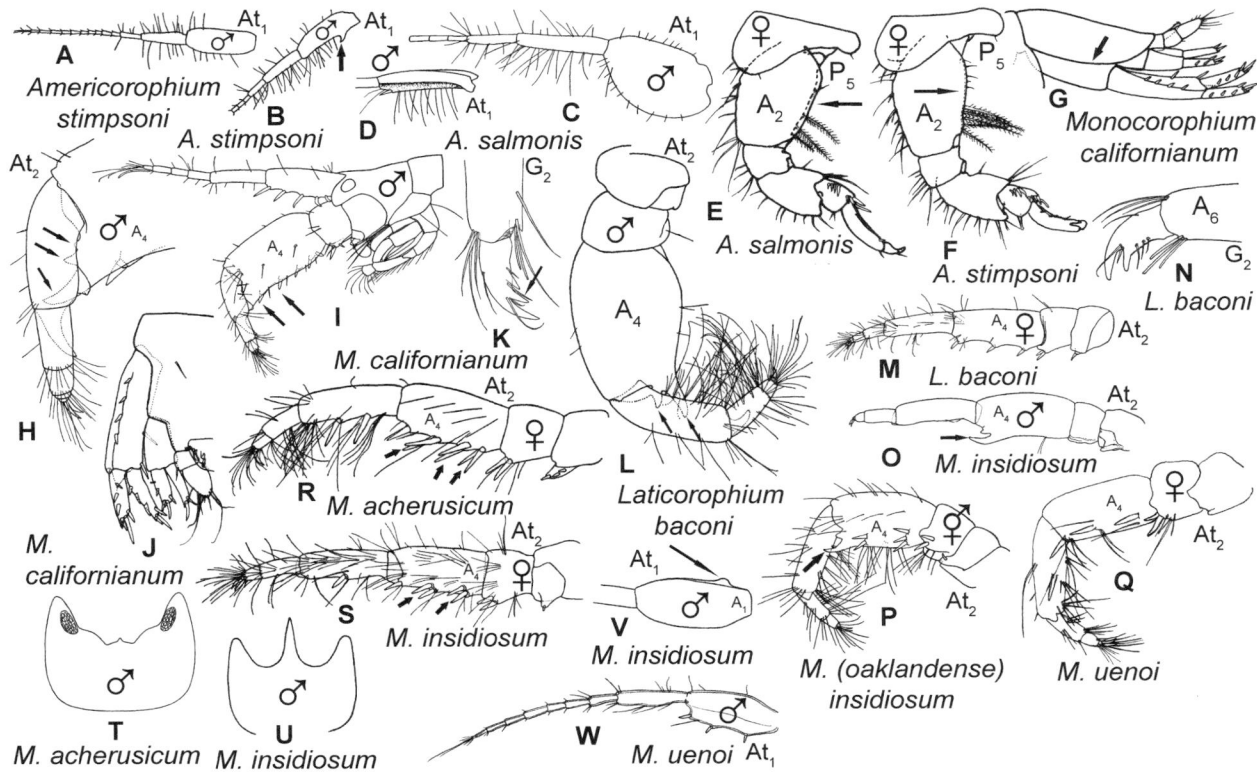

PLATE 270 Corophiidae. C–E, *Americorophium salmonis*; A, B, F, *Americorophium stimpsoni*; L, M, *Laticorophium baconi*; R, T, *Monocorophium acherusicum*; G–K, *Monocorophium californianum*; O, S–U, V, *Monocorophium insidiosum*; P, *Monocorophium (oaklandense) insidiosum*; Q, W, *Monocorophium uenoi* (figures modified from Bousfield and Hoover 1997; Shoemaker 1934, 1947, 1949; and Stephensen 1932).

11. Antenna 2, setose, peduncle article 3 with three ventral spines and article 4 with two pairs of ventral spines (plate 269L) . . . . *Americorophium brevis*
— Antenna 2 not setose, articles 3 and 4 each with two ventral spines (plate 269BB, 269CC) (female *A. salmonis* and *A. stimpsoni* may be indistinguishable) 12
12. Female pereopod 5 article 2 posterior edge faintly concave and with sharp lateral edge (plate 270E) . . . . *Americorophium salmonis*
— Female pereopod 5 article 2 straight or slightly convex posteriorly and with rounded posterolateral edge (plate 270F) . . . . *Americorophium stimpsoni*
13. Uropod 1 inserted ventral to dorsolateral ridge of urosome (plates 269G, 270G) 14
— Uropod 1 inserted laterally and urosome without dorsolateral ridge (plate 269F) 15
14. Antenna 2 article 4 with one large and two small distal medial teeth (plate 270H) and lined on ventral medial edge with four to five stout spines (plate 270I); uropod 2 length 1.5 times uropod 3 (plate 270J); gnathopod 2 dactyl with three teeth (plate 270K). . . . . *Monocorophium californianum*
— Antenna 2 article 4 with two large distal medial teeth and two or less ventromedial spines (plate 270L); female antenna 2 article 5 without spines on medial ventral edge or a distal medial tooth on article 4 (plate 270M); gnathopod 2 dactyl with two teeth (plate 270N); uropods 2 and 3 lengths equal (plate 269G) . . . . *Laticorophium baconi*
15. Antenna 2 article 4 with large distal medial tooth (plate 270O, 270P). 18
— Antenna 2 article 4 without a distal medial tooth (plate 270Q–270S) 16

16. Antenna 2 article 4 lined ventrally with single stout spines in tandem and article 5 with single ventral spine (plate 270Q). . . . . *Monocorophium uenoi*
— Antenna 2 article 4 lined ventrally with pairs of stout spines (plate 270R, 270S) 17
17. Antenna 2 article 4 with three ventral pairs of stout spines and article 5 with two single spines (plate 270R) . . . . *Monocorophium acherusicum*
— Antenna 2 article 4 with two ventral pairs of stout spines and article 5 with a single spine (plate 270S) . . . . *Monocorophium insidiosum*
18. Antenna 2 article 4 lined with ventral triads or pairs of spines and with a distal medial tooth (plate 270P) . . . . *Monocorophium (oaklandense) insidiosum*
— Antenna 2 article 4 without ventral spines (plate 270O) 19
19. Rostrum short, not extending past ocular lobes (plate 270T). *Monocorophium acherusicum*
— Rostrum long, extending past ocular lobes (plate 270U) 20
20. Antenna 1 article 1 with medial protrusion (plate 270V) . . . . *Monocorophium insidiosum*
— Antenna 1 article 1 without medial protrusion (plate 270W) . . . . *Monocorophium uenoi*

LIST OF SPECIES

Americorophium brevis (Shoemaker, 1949) (=*Corophium brevis*). Previously ranging from Prince William Sound to San Francisco Bay (extinct in San Francisco Bay, its type locality; next nearest population is Humboldt Bay); predominantly in marine fouling communities, but also soft benthos of estuaries; 6 mm; intertidal—35 m.

Americorophium salmonis (Stimpson, 1857) (=*Corophium salmonis*). Southern Alaska to Humboldt Bay, high salinity estuary to freshwater on muddy bottoms; probably introduced far up Columbia River; critical prey of juvenile salmon (Bottom and Jones 1990, Prog. Oceanogr. 25: 243–270); 7 mm; intertidal—10 m.

Americorophium spinicorne (Stimpson, 1857) (=*Corophium spinicorne*). Vancouver Island to San Louis Obispo, estuarine and freshwater. Introduced to upper Putah Creek, California, and upper Columbia River (Lester and Clark 2002, West. N. American Nat. 62: 230–233). Status and taxonomy of southern populations unclear. Tubes almost exclusively attached to hard surfaces. The long article 3 of male *A. salmonis* and *A. stimpsoni* antenna 2 allow distinctions of these species from *A. spinicorne* in the field, which has a nearly square article 3; 7 mm; intertidal—20 m. See Aldrich 1961, Proc. Acad. Natl. Sci. Phil. 113: 21–28 (ecology); Eriksen 1968, Crustaceana 14: 1–12 (ecology).

Americorophium stimpsoni (Shoemaker, 1941) (=*Corophium stimpsoni*). Historically from Mendocino County south to Santa Cruz Island, estuarine and freshwater, exclusively in soft benthos. A potentially threatened species not found in recent decades outside of the San Francisco Bay Delta east of Carquinez Strait; 6 mm; intertidal—10 m.

Corophium alienense Chapman 1988. San Francisco Bay, Tomales Bay, Los Angeles Harbor, introduced from Asia during the Vietnam War, also in China (Ren, 1995, Studia Marina Sinica 10: 267–271, as *Corophium dentalium*), occasionally in high pools reaching temperatures of 30°C; 6.5 mm; intertidal—3 m.

Corophium heteroceratum Yu, 1938. San Francisco Bay and Los Angeles Harbor, morphologically plastic, introduced, probably from the Yellow Sea, estuarine and marine; 9 mm; shallow subtidal—10 m.

*Crassicorophium bonellii (Milne Edwards 1830) (=*Corophium bonellii*). A "bipolar" (Bousfield 1973) cold water marine parthenogenic morphotype transferred around the world by humans; Arctic, North Atlantic, Falkland Islands, Chile. Not formally reported from the northeast Pacific, but the proposed differences between *C. bonellii*, *M. acherusicum*, and *M. insidiosum* are gnathopod 2 dactyl teeth numbers and antenna spine patterns that are too variable for species distinctions. *Crassicorophium bonellii* is possibly a parthenogenic form of one or both species, but see Myers et al. 1989, J. Mar. Biol. Assoc. U.K., 69: 319–321 (a presumed male); 6 mm; intertidal—18 m.

Laticorophium baconi (Shoemaker, 1934) (=*Corophium baconi*). On benthos off coastal shelf in California and among marine float fouling communities; 4 mm; intertidal—55 m.

Monocorophium acherusicum (Costa 1857) (=*Corophium acherusicum*). Cosmopolitan marine, introduced from North Atlantic by shipping and other means to all protected marine coasts between 50° north and 50° south latitude; abundant in float fouling communities and estuary soft benthos. *Crassicorophium bonellii* is indistinguishable from *M. acherusicum*; 4.5 mm; intertidal—10 m.

Monocorophium californianum (Shoemaker, 1934) (=*Corophium californianum*). Marine rocky and sandy bottoms; 3.5 mm; intertidal—100 m. Extremely rare.

*Monocorophium carlottensis Bousfield and Hoover, 1997. Marine fouling communities, northern species (Prince William Sound to Puget Sound); 4.2 mm; low intertidal—10 m.

Monocorophium insidiosum (Crawford, 1937) (=*Corophium insidiosum*). Cosmopolitan marine and estuarine, introduced

from North Atlantic; high frequencies of an undescribed nicothoid copepod egg predator occur among summer Puget Sound populations; 4.5 mm; intertidal—10 m.

Monocorophium oaklandense (Shoemaker, 1949) (=*Corophium oaklandense*). The occasional appearance of this morphotype in pure lab cultures of *M. insidiosum* suggests that *M. oaklandense* is a triploid intersex and thus a synonym of *M. insidiosum*; 5 mm; intertidal—2 m.

Monocorophium uenoi (Stephensen, 1932) (=*Corophium uenoi*). Sea of Japan, South China Sea, introduced to California; 5 mm; intertidal to 24 m.

AORIDAE

Grandidierella japonica Stephensen 1938. Keyed here with corophiids due to the uniramous uropod 3. The distinctive gnathopod, green eggs, and black head permit recognition of females and wandering males in the field. Preserved specimens are readily confused with *Microdeutopus gryllotalpa*, which has a biramous uropod 3. The mature male gnathopod 1 (plate 269A) basis is expanded forward and bears onto anterior ridges of the carpus (see arrows in figure) in an apparent adaptation for stridulation. This Japanese species ranges from the Fraser River to Bahia de San Quintin and also occurs in Hawaii, England, and Australia in fine muds of estuarine flats. *G. japonica* feeds on epiphytes, suspended partiles, and detritus and is a facultative cannibal and amphipod predator. See Bay et al. 1989, Environ. Toxicol. Chem. 8: 1191–1200 (toxicology); Greenstein and Tiefenthaler 1997, Bull. So. Calif. Acad. Sci. 96: 34–41 (reproduction and population dynamics in Newport Bay); 13 mm; high intertidal—10 m.

NAJNIDAE

Najnidae are algivores that burrow into and form galls in the stipes and holdfasts of intertidal and shallow subtidal macrophytes including *Alaria*, *Egregia*, *Macrocystis*, and *Lessoniopsis*. The najnid molar is a uniquely thickened surface on the mandible and the palp is reduced or absent (plate 255LL), and the sharply produced posterior coxa 4 (plate 271A) is characteristic of the family. Sexual dimorphism is weak. All *Carinonajna* were previously recognized as *Najna conciliorum* Derzhaven 1937, occurring on both Asian and western North American coasts (Barnard 1962c). Barnard (1979a) distinguished the North American populations (as *Najna kitamati*) from the Asian *N. conciliorum* by their rounded rather than square third epimeron, their longer maxilliped palp dactyl, and their minute ramus of uropod 3. Bousfield (1981) and Bousfield and Marcoux (2004) erected the North American *Carinonajna* based on the above distinctions and the occurrence of a lateral carina on the urosome and pleonite 1 of the North American forms. Bousfield and Marcoux (2004) added 10 species to *Carinonajna*. However, the morphological variations proposed to distinguish these *Carinonajna* species (eye size, lacinia mobilis tooth numbers, gnathopod palm spine numbers and sizes, the spine numbers on uropod 1 and 2 rami, and dorsal urosomal carination) are unclear in illustrations and the descriptions and do not fully reveal species differences.

KEY TO NAJNIDAE

1. Gnathopods 1 and 2 article 6 posteriodistal corner with one large and one tiny medial spine (plate 271B, 271C); epimeron

---

* = Not in key.

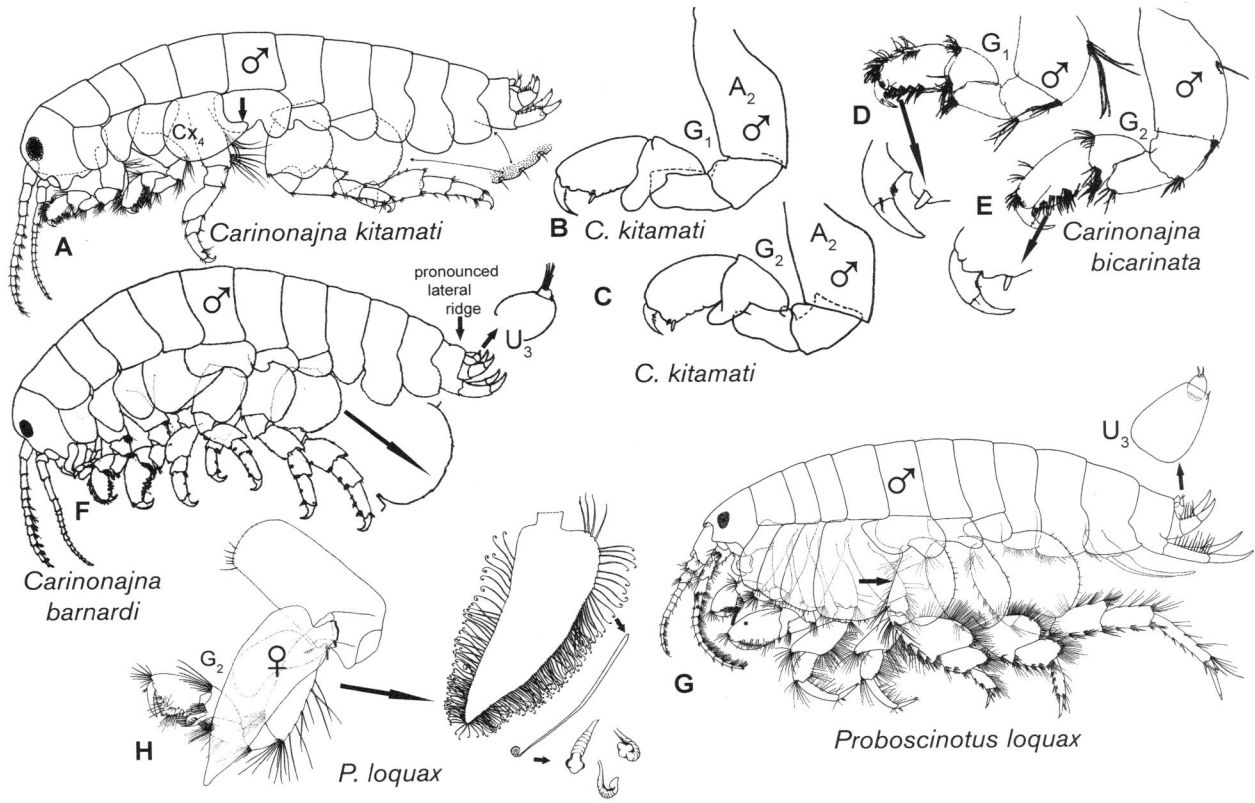

PLATE 271 Najnidae and Dogielinotidae. A–C, *Carinonajna kitamati*; D, E, *Carinonajna bicarinata*; F, *Carinonajna barnardi*; G, H, *Proboscinotus loquax* (figures modified from Barnard 1962c, 1967a; and Bousfield and Marcoux 2004).

3, pereopod 7 posterior pereopod 7 basis and epimeron 3 strongly crenulate (plate 271A) (*C. kitamati* subgroup). . . . . *Carinonajna kitamati, C. lessoniophila, C. bispinosus*
— Posteriodistal corner of gnathopod 2 article 6 bearing a single stout spine (plate 271D, 271E); posterior edge of epimeron 3 and pereopod 7 basis weakly crenulate (plate 271F) (*C. bicarinata* subgroup) . . . . *Carinonajna barnardi, C. longimana, C. carli,* and *C. bicarinata*

LIST OF SPECIES

Carinonajna barnardi (Bousfield 1981). 9.5 mm; intertidal—10 m.

Carinonajna bicarinata (Bousfield, 1981). In *Phyllospadix* and *Laminaria* holdfasts; 8.5 mm; intertidal—10 m.

Carinonajna bispinosa Bousfield and Marcoux, 2004. 7.5 mm; intertidal—10 m.

Carinonajna carli Bousfield and Marcoux, 2004. *Phyllospadix,* boulders and gravel, *Hedophyllum*; 8.2 mm; intertidal—10 m.

Carinonajna kitamati (Barnard, 1979) (=*Najna ?consiliorum*). Among *Egregia* and rarely *Postelsia* and *Macrocystis*; 8 mm; intertidal—17 m.

Carinonajna lessoniophila (Bousfield, 1981). From galls in stipes of *Lessoniopsis littoralis*; 9.2 mm; intertidal—10 m.

Carinonajna longimana (Bousfield, 1981). On *Hedophyllum, Laminaria,* and in *Phyllospadix* root mass communities on semi-protected beaches; 5.5 mm; intertidal—1 m.

DOGIELINOTIDAE

Dogielinotidae superficially resemble other fossorial families of the region (Phoxocephalidae, Urothoidae, and Pontoporeiidae); however, the reduced urosomites 2 and 3, reduced mandibular palp (plate 255KK), single, reduced ramus of uropod 3 (plate 271G), lack of an accessory flagellum and remarkable distally curled setae of the oöstegites (plate 271H) indicate their talitrid origins along with the Najnidae, Hyalellidae, and Hyalidae.

KEY TO DOGIELINOTIDAE

1. Surf-zone sand burrowing (fossorial) lacking accessory flagellum (plate 271G); posterior coxa 4 not produced, uropods 1 and 2 rami bare and uropod 3 with minute ramus (plate 271G); oöstegites lined with distally curled setae (plates 255NN, 271H); epistome proboscoid shaped (plate 256B) . . . . *Proboscinotus loquax*

LIST OF SPECIES

Proboscinotus loquax (Barnard, 1967d). The talking nose amphipod—from the root meaning of *proboscis* "nose" (due to the noselike epistome) (plate 256B) and the root meaning of *loquax* "talk." Open coast fine and coarse sand beaches. The restricted range, Washington coast of the Juan de Fuca Straits to Clam Beach, Eureka, in northern California of this abundant, distinctive species is unusual among native species. Its

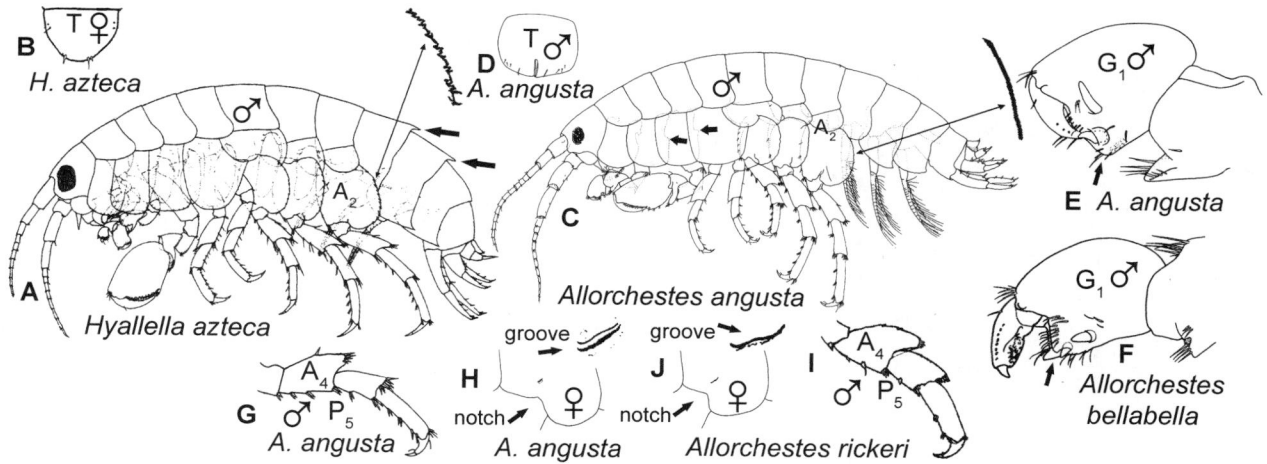

PLATE 272 Hyalellidae. A, B, *Hyalella azteca*; C–E, G, H, *Allorchestes angusta*; F, *Allorchestes bellabella*; I, J, *Allorchestes rickeri* (figures modified from Barnard 1979a and Hendrycks and Bousfield 2001).

endemic status in North America should be examined. An important prey of shorebirds in the region. See Hughes 1982, Mar. Biol. 71: 167–175 for population biology. Open sandy beaches; 8 mm; intertidal.

HYALELLIDAE

Hyalellidae are herbivorous talitroideans closely related to Hyalidae and Najnidae that live in coarse sand and rock-cobble areas and among aquatic plants. The hyalellids are relatively helpless out of water, while the allorchestids hop and otherwise move quickly when exposed. The first male gnathopods are modified for clasping to the highly modified female ventral pereonite 2 and dorsal coxa 2 (Hendrycks and Bousfield 2001) (plate 272H, 272J).

KEY TO HYALELLIDAE

1. Primarily freshwater, pereopod 7 article 2 posterior edge serrate (plate 272A); telson uncleft (plate 272B) . . . .
. . . . *Hyalella azteca*
— Marine and estuarine, pereopod 7 article 2 posterior edge smooth or slightly crenulate but not serrate (plate 272C), telson slightly cleft (plate 272D) 2
2. Male gnathopod 1 dactyl not inflated, 5 times longer than wide, palm nearly straight, and broadly square posteriorly (plate 272E) 3
— Male gnathopod 1 dactyl inflated, half as wide as long, palm deeply incised and sharply angular posteriorly (plate 272F) . . . . *Allorchestes bellabella*
3. Pereopod 5, article 4 width one-half of the length (plate 272G); female coxa 2 anteriodistal preamplexing notch broadly obtuse (plate 272H) . . . . *Allorchestes angusta*
— Pereopod 5, article 4 width two-thirds of the length (plate 272I); preamplexing notch nearly at right angle (plate 272J) . . . . *Allorchestes rickeri*

LIST OF SPECIES

Allorchestes angusta Dana, 1856. Japanese records refer to *A. malleola* (Stebbing 1899); 10 mm; intertidal—1 m.

Allorchestes bellabella Barnard, 1974. Marine to estuarine, sometimes planktonic; 13 mm; intertidal—7 m.

Allorchestes rickeri Hendrycks and Bousfield, 2001. Open coast and semiprotected sand and rock beaches; 6 mm; intertidal.

Hyalella azteca (Saussure, 1858). A mostly freshwater species with low-salinity populations in upper estuaries, coastal lakes, rivers, and barrier beach lagoons to the tree line; likely species complex (Hogg et al. 1998), but also with many likely introduced populations. The illustration of *H. azteca* from San Francisco Bay in Toft et al. (2002) is of *Hyalella montezuma* Cole and Watkins, 1977, from Montezuma Well, Arizona, and not of San Francisco Bay material; 5 mm; intertidal—20 m.

HYALIDAE

Hyalidae are intertidal marine and estuarine herbivores with entirely cleft fleshy telsons and greatly reduced urosomite 2. Hyalids hop and otherwise move quickly when exposed. The first male gnathopods are modified for clasping to the highly modified female ventral pereonite 2 and dorsal coxa 2 (Bousfield and Hendrycks 2002) (plate 273T).

KEY TO HYALIDAE

1. Uropod 3 with scalelike inner ramus (plate 273A); maxilla 1 palp consisting of two articles (plate 273B) 2
— Uropod 3 with single ramus (plate 273C); maxilla 1 palp consisting of a single article (plate 273D) 4
2. Antenna 2 peduncle length less than length from anterior head to posterior pereonite 1 and faint dorsal carination on pereonites 1–6 (plate 273E) . . . .
. . . . *Parallorchestes americana*
— Antenna 2 peduncle length greater than length from anterior head to posterior pereonite 1 and faint dorsal carination on pereonites 1–5 only or absent entirely (plate 273H) 3
3. Peduncle of antenna 2 length equal to distance from anterior of head to pereonite 2 and carinations on pereonites absent (plate 273G) . . . . *Parallorchestes cowani*
— Peduncle of antenna 2 length greater than distance from anterior head to pereonite 2, faint carination on pereonites

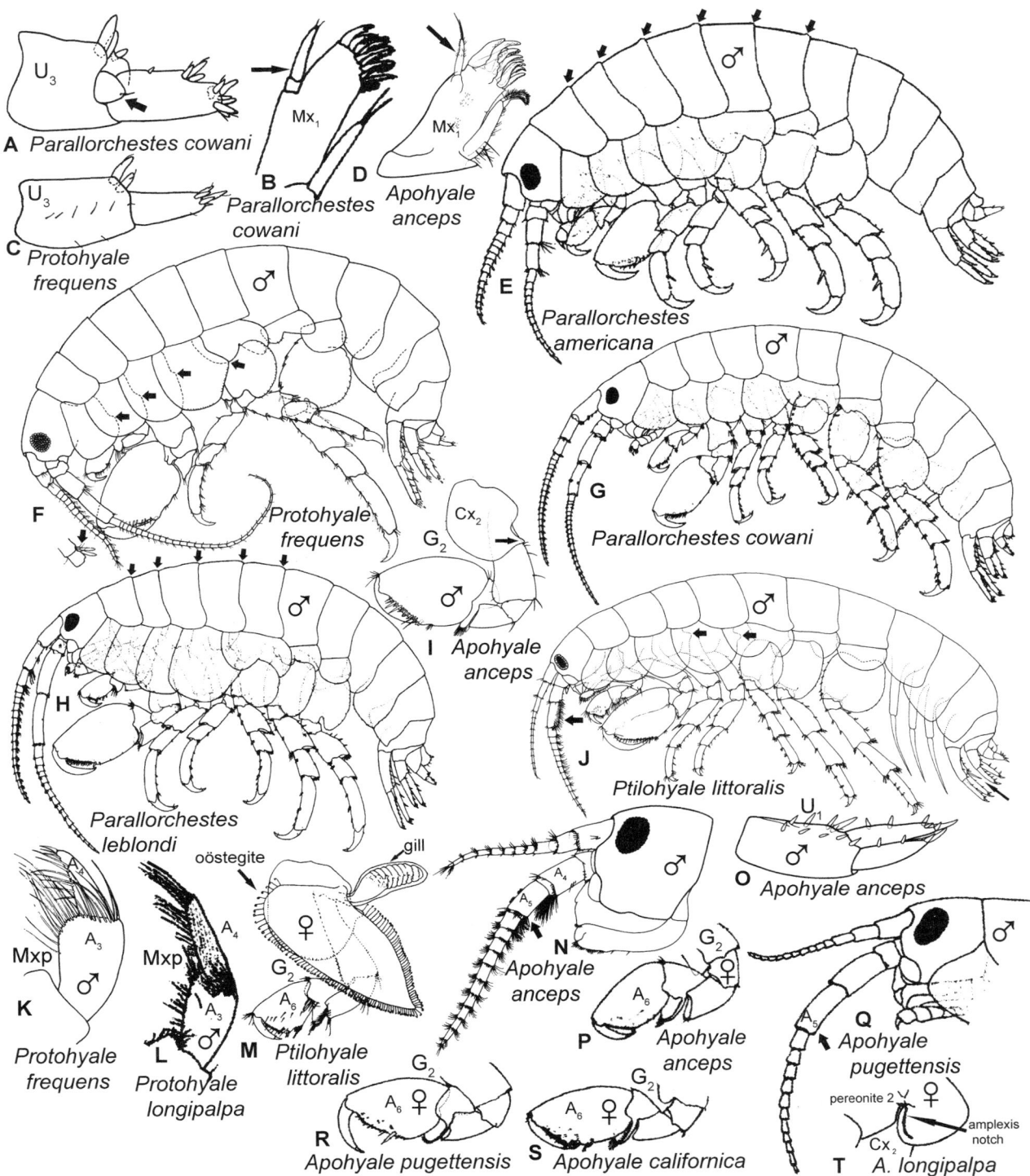

PLATE 273 Hyalidae. D, I, N–P, *Apohyale anceps*; S, *Apohyale californica*; Q, R, *Apohyale pugettensis*; E, *Parallorchestes americana*; A, B, G, *Parallorchestes cowani*; H, *Parallorchestes leblondi*; C, F, K, *Protohyale frequens*; L, *Protohyale longipalpa* (T, amplexis notch); J, M, *Ptilohyale littoralis* (figures modified from Barnard 1952b, 1962c, 1969a; Bousfield 1973; and Bousfield and Hendrycks 2002).

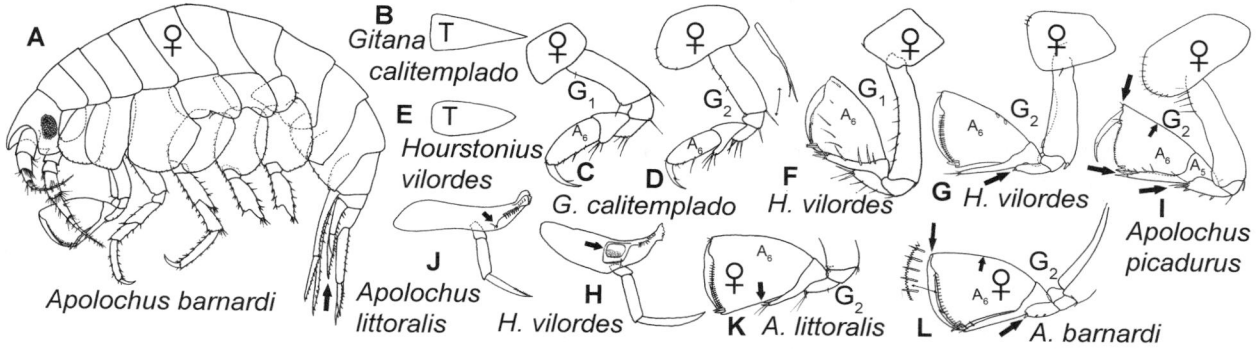

PLATE 274 Amphilochidae. A, L, *Apolochus barnardi*; J, K, *Aplolochus litoralis*; I, *Apolochus picadurus*; B–D, *Gitana calitemplado*; E–H, *Hourstonius vilordes* (figures modified from Barnard 1962c and Hoover and Bousfield 2001).

— Antenna 2, peduncle article 5 lacking distal setae (plate 273N); uropod 1 lacking prominent distal spine (plate 273O); female gnathopod 2 article 6 with parallel anterior and posterior edges (plate 273P). 7
7. Antenna 2 peduncle article 5 with proximal ventral setae (plate 273N) . . . . *Apohyale anceps*
— Antenna 2 peduncle article 5 without ventral setae (plate 273Q) 8
8. Female gnathopod 2 palm longer than posterior edge (plate 273R) . . . . *Apohyale pugettensis*
— Female gnathopod 2 palm shorter than posterior edge (plate 273S). . . . . *Apohyale californica*

LIST OF SPECIES

Apohyale anceps (Barnard, 1969a) (=*Hyale anceps*). Abundant on wave-dashed turf platforms and under cobbles and in *Silvetia* of the rocky open coast; 12–18 mm; intertidal.

Apohyale californica (Barnard, 1969a) (=*Hyale grandicornis californica*). Abundant on wave-dashed turf platforms and under cobbles and in *Silvetia* of the rocky open coast. Whether *A. californica* and *A. pugettensis* are different species or size-related morphologies of a single species is unclear; 6 mm–12 mm; intertidal.

Apohyale pugettensis (Dana, 1853) (=*Hyale pugettensis*). Frequent in nearly freshwater open coast spray pools and above high-water level along bedrock shores; 18 mm; intertidal.

*Hyale seminuda (Stimpson 1856). Stimpson noted that this species occurred "on seaweed and among barnacles on piles, stones, etc. at half tide in San Francisco Harbor." The identity of *H. seminuda* is unclear; no additional records have appeared since the original report; 13 mm; intertidal.

Parallorchestes americana Bousfield and Hendrycks, 2002. Specific distinctions in carination among *Parallorchestes* of the northeast Pacific are unclear and awaiting analyses of sexual, allometric, and meristic variation. Commonly free swimming in intertidal areas of surf-exposed costs; 7.5 mm; intertidal.

Parallorchestes cowani Bousfield and Hendrycks, 2002 (=*Allorchestes ochotensis* in part). Free swimming or associated with brown algae and *Phyllospadix* at low water on exposed and semiprotected rocky coasts; 13 mm; intertidal.

Parallorchestes leblondi Bousfield and Hendrycks, 2002. Exposed sandy and rocky beaches at low water level; 11 mm; intertidal.

Prohyale frequens (Stout, 1913) (=*Hyale frequens*). Characters distinguishing all other *Prohyale* of the region (*P. jarrettae, P. oclairi,*

* = Not in key.

and *P. setucornis*) from *P. frequens* (Bousfield and Hendrycks 2002) are indistinct when adjusted for size. One of the most abundant intertidal amphipods and particularly abundant among *Phyllospadix* roots and coralline algae of open and semiprotected coasts; 8 mm; intertidal—6 m.

Protohyale longipalpa Bousfield and Hendrycks 2002. Among algae in semiprotected areas; 8.5 mm; intertidal—1 m.

Ptilohyale littoralis (Stimpson, 1853) (=*Hyale plumulosa, Ptilohyale plumulosa, Hyale crassicorne, Ptilohyale litoralis*). Male head shown in plate 256F. A probable solid ballast introduction between the northwest Atlantic and northwest Pacific and Australia. Protected shores in salt marshes among *Spartina* and fucoids, stones, or high-tide, low-salinity pools; 8 mm; intertidal.

AMPHILOCHIDAE

Seldom observed alive, Amphilochidae are small colorful leucothoideanlike amphipods commensal with sea fans, hydroids, and other sessile marine invertebrates. They are distinguished by prominent, decurved rostrums, projecting article 5 of gnathopod 2 along the posterior edge of article 6, round or oval eyes with darkly pigmented centers surrounded by pale ommatidia, laminate uncleft acute telsons and second uropods that do not extend as far as uropods 1 and 3 (plates 256J, 274A).

KEY TO AMPHILOCHIDAE

1. Telson 2.3 times as long as wide, lateral edges straight and distally acute (plate 274B); gnathopods 1 and 2 weak, palm of article 6 indistinct (plate 274C, 274D). . . . .
. . . . *Gitana calitemplado*
— Telson only twice as long as wide, distally blunt and lateral edges convex (plate 274E); gnathopods 1 and 2 article 6 with distinct palm separated from the posterior edge by a corner (plate 274F, 274G). 2
2. Gnathopod 2 article 4 with single large spine at apex (plate 274G); mandibular molar triturative (plate 274H) . . . . . .
. . . . *Hourstonius vilordes*
— Gnathopod 2 article 4 with multiple spines at apex and posterior edge (plate 274I); mandibular molar vestigial (plate 274J, arrow). 3
3. Gnathopod 2 article 5 not projecting half way along posterior edge of article 6 (plate 274K) . . . .
. . . . *Apolochus littoralis*
— Gnathopod 2 article 5 projecting more than halfway along posterior edge of article 6 (plate 274I) 4

4. Gnathopod 2 anterior margin nearly straight and project-ing over the dactyl hinge; distal margin of article 4 bear-ing one spine and one seta (plate 274I).............. .... *Apolochus picadurus*
— Gnathopod 2 anterior margin curved outward and not pro-jecting over dactyl hinge; distal margin of article 4 bearing one spine only (plate 274L) .... *Apolochus barnardi*

LIST OF SPECIES

Apolochus barnardi Hoover and Bousfield, 2001 (=*Amphilochus neapolitanus*). In *Phyllospadix* and *Egregia* root masses and among coralline algae; 2.5 mm; intertidal—6 m.

Apolochus littoralis (Stout, 1912) (=*Amphilochus littoralis*). Low intertidal rocks and shell and among coralline algae; 2.3 mm; intertidal—2 m.

Apolochus picadurus (Barnard, 1962c) (=*Amphilochus picadurus*). Mud and rock bottoms; 2.7 mm; 2 m–6 m.

Gitana calitemplado Barnard, 1962c. A rare shallow water species of bays and protected coasts. Whole body illustration plate 256J; 2.0 mm; 9 m–27 m.

Hourstonius vilordes (Barnard, 1962c) (=*Gitanopsis vilordes*). From rocks and *Egregia*; 3.0 mm; intertidal—4 m.

STENOTHOIDAE

Stenothoidae are commensals and probable parasites or micro-predators on hydroids. Some species, including *Stenothoe valida*, are beautifully pigmented. "Steno" and "tho" mean narrow and quick, but stenothoids are fat and are not remarkably quick. Their massive coxae 2–4 cover all appendages (plate 275A), al-lowing rapid identification of the family. Undescribed species may occur in this region but are obscured by the poor taxo-nomic resolution of existing species. Concepts of stenothoid genera are based on the degree of fusion and reduction of mouth parts, which are delicate and easily broken or lost in dissections. Fusion or separation of articles can be difficult to determine (Barnard 1962c) and intraspecific variation in mouth part mor-phology is unknown. External morphology is emphasized here, but mouth part morphology may be more reliable.

KEY TO STENOTHOIDAE

1. Article 2 of pereopod 6 linear, thin (plate 275A, 275B)..... 2
— Article 2 of pereopod 6 expanded (plate 275C, 275D) 4
2. Mandibular palp 1 articulate or absent (no published il-lustrations)..................... *Stenothoides burbanki*
— Mandibular palp 2–3 articulate (plate 275E, 275F) 3
3. Article 5 of gnathopod 1 twice as long as article 6 (plate 275G); mandibular palp large and 3 articulate (plate 275E) .... *Mesometopa esmarki*
— Articles 5 and 6 of gnathopod 1 equal in length (plate 275H); mandibular palp minute and 2 articulate (plate 275F)........................... *Mesometopa sinuata*
4. Gnathopod palm shallowly concave, with distal notch and large tooth, densely setose and lacking a proximal defin-ing tooth (plate 275I); maxilla 1 palp of two articles (plate 275J); mandible lacking palp (plate 275K)............. .... *Stenothoe valida*
— Gnathopod 2 palm with relatively few setae, a proximal defining tooth and a small distal hinge tooth (plate 275L–275N); mandible with palp (plate 275O, 275P); max-

illa 1 palp of one article (plate 275Q, 275R) (difficult, re-quiring dissection and mounting on a slide for 100x mag-nification observation) 5
5. Telson with four stout spines (plate 275S); mandibular palp of two articles (plate 275O) .... *Metopa cistella*
— Telson lacking stout spines (plate 275T, 275U); mandibu-lar palp of one article (plate 275P)...................6
6. Gnathopod 1 article 5 longer than article 6 (plate 275V); pereopod 7 article 4 extending less than half of the length of article 5 (plate 275W)............... *Stenula modosa*
— Gnathopod 1 article 5 length equal to article 6 (plate 275Y); pereopod 7 article 4 extending the length of article 5 (plate 275X) .... *Stenula incola*

LIST OF SPECIES

Mesometopa esmarki (Boeck, 1872). Boeck's description and the only record of this species are based upon a specimen from central California, perhaps from San Francisco Bay. Only the incomplete illustrations reproduced herein were published. The long article 5 of gnathopod 1 (plate 275G) was reported to be of a male but is characteristic of females among stenothoids; 5 mm; intertidal.

Mesometopa sinuata Shoemaker, 1964. Coos Bay to Monterey Bay (holotype collected from a boat bottom in Monterey Bay); the description is based on a male. Whether a female could be distinguished from *M. esmarki* is unclear; 4 mm; intertidal.

Metopa cistella Barnard, 1969. Commensal with anemones, hydroids and sea pens; 2.3 mm; low intertidal to deep subtidal.

Stenothoe estacola Barnard, 1962c. Pt. Conception and south, associated with the worm *Phragmatopoma*. Expanded posterior basis of pereopod 6, six stout spines on dorsal telson, but lack-ing extended setose palm of *S. valida*; 3.0 mm; intertidal.

Stenothoe valida (Dana, 1852). Cosmopolitan in marine bays and harbors of temperate latitudes; transported by human activity. Hydroid predator or commensal; 5 mm; shallow subtidal—10 m.

Stenothoides burbanki Barnard, 1969a. Among tunicates and sponges, algal turf, *Egregia* and *Laminaria* holdfasts; scarce. Ex-cept for lacking a mandibular palp, not distinguished from *Me-sometopa sinuata*; 3 mm; intertidal—3 m.

Stenula incola Barnard, 1969a. Not clearly distinguished from *S. modosa* morphologically. Sex-based variation in *Metopa cis-tella* gnathopod morphology (Barnard 1969) closely matches the differences between *S. incola*, described entirely from a male specimen, and *S. modosa*, described from a female speci-men. Occurring in algal turf; 3 mm; intertidal.

Stenula modosa Barnard, 1962c. Mud bottoms. Body shown in plate 256K. Distinguished from *S. incola* primarily by eco-logical differences; 2 mm; subtidal—92 m.

AMPELISCIDAE

Ampeliscidae build pocket-shaped tubes with a single opening in fine sand or mud bottoms and feed by sweeping in suspended particulates using their antennae. Urosomites 2–3 are fused; the head is longer than deep and lacks a rostrum. The eyes, when present, consist of dorsal frontal lenses with anterior pairs of om-matidia. *Byblis* and *Haploops* are predominantly deep-sea species. Pelagic phase males have long antennae, larger pleosomes, broad setose uropod 3 rami, and larger dorsal carina on uro-somites 1. The taxonomy is based on female morphology.

* = Not in key.

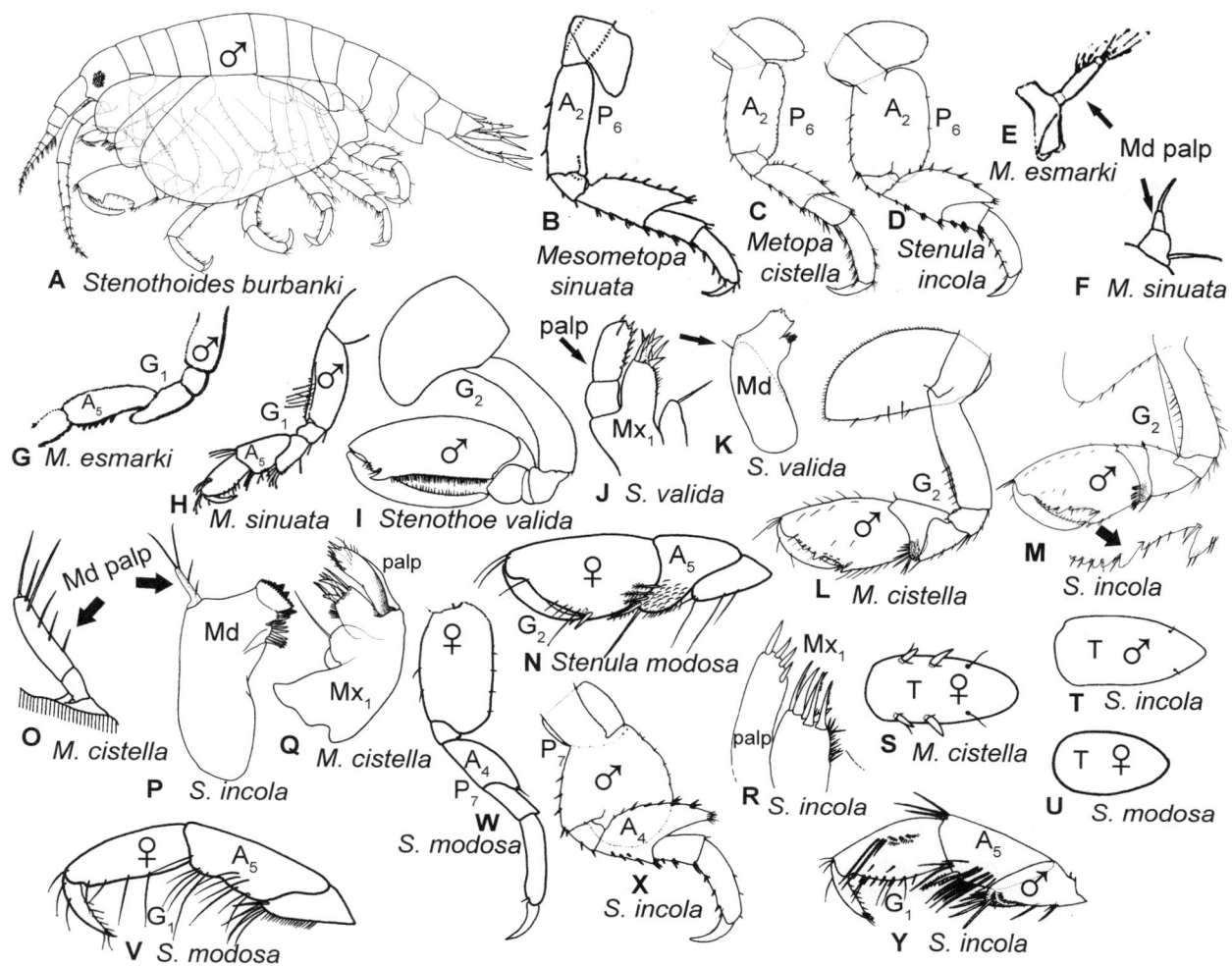

PLATE 275 Stenothoidae. E, G, *Mesometopa esmarki*; B, F, H, *Mesometopa sinuata*; C, L, O, Q, S, *Metopa cistella*; I–K, *Stenothoe valida*; A, *Stenothoides burbanki*; D, M, P, R, T, X, Y, *Stenula incola*; N, U–W, *Stenula modosa* (figures modified from Barnard 1953, 1962c, 1969a; Boeck 1872; and Shoemaker 1964).

KEY TO AMPELISCIDAE

1. Pereopod 7 article 2 with roundly expanded posterior edge and more than twice as wide as article 3 (plate 276A) 2
— Pereopod 7 article 2 with anterior and posterior edges nearly parallel and less than twice as wide as article 3 (plate 276B) . . . . *Haploops*
2. Pereopod 7 article 2 anterior edge of posterioventral lobe without setae at junction with article 3 and lobe extending ventrally (plate 276C); uropod 3 rami facing edges smooth or evenly serrate (plate 276D) 3
— Pereopod 7 article 2 anterior ventral edge bearing setae at junction with article 3 and lobe extending obliquely (plate 276E1); uropod 3 facing edges of rami unevenly serrate (plate 276E2) 20
3. Pleonite 3 posteriodistal corner produced into large or small acute tooth (plate 276A) 4
— Pleonite 3 posteriodistal corner square or rounded but not produced (plate 276F) 12
4. Pereopod 7 article 5 anterior margin notched and article 4 posterior lobe broad, extending more than two-thirds the length of segment 5 (plate 276G); uropod 1 not reaching beyond midpoint of uropod 2 ramus (plate 276A); telson dorsal surface with long spines (plate 276H) 5
— Pereopod 7 article 5 without anterior notch and article 4 posterior lobe acute, extending less than two-thirds the length of article 5 (plate 276I); uropod 1 reaching beyond midpoint of uropod 2 ramus (plate 276J); telson dorsal surface with short spines (plate 276K) 6
5. Epimeron 3 posterior ventral tooth minute (plate 276L); head ventral edge slightly concave (plate 276M), uropod 1 rami extending to middle of uropod 2 rami (plate 276L). . . . . *Ampelisca indentata*
— Epimeron 3 posterior ventral tooth distinct (plate 276A); head ventral edge straight or convex (plate 276N), uropod 1 rami not extending to the middle of uropod 2 rami (plate 276A) . . . . *Ampelisca pugetica*
6. Uropod 2 outer ramus lacking subapical spine and pleonite 3 posterior margin evenly concave (plate 276J) . . . . *Ampelisca hancocki*
— Uropod 2 outer ramus with long subapical spine and pleonite 3 posterior margin sinuate (plate 276O) 7
7. Head lower front margin deeply concave and parallel with upper margin (plate 276P) 8
— Head lower front margin convex or only slightly concave

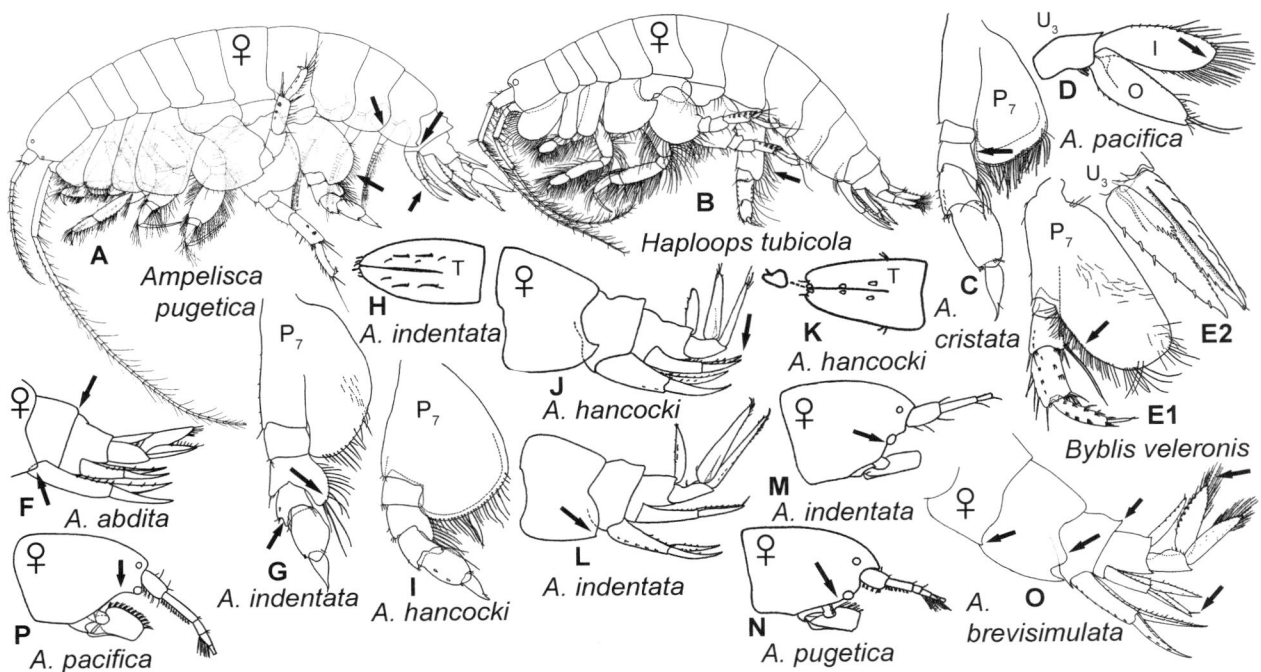

PLATE 276 Ampeliscidae. F, *Ampelisca abdita*; O, *Ampelisca brevisimulata*; C, *Ampelisca cristata*; I–K, *Ampelisca hancocki*; G, H, L, M, *Ampelisca indentata*; D, P, *Ampelisca pacifica*; A, N, *Ampelisca pugetica*; E1, E2, *Byblis veleronis*; B, *Haploops tubicola* (figures modified from Barnard 1954b, 1966a; Dickinson 1982; and Mills 1964).

but never parallel with upper margin (plate 277Q).... 10
8. Uropod 3 inner ramus distally rounded (plate 276D); pleonite 2 posterioventral corner without a tooth .... *Ampelisca pacifica*
— Uropod 3 inner ramus distally pointed and pleonite 2 posterioventral corner with a tooth (plate 276O) 9
9. Pleonite 3 posterior produced into a large rounded process extending posteriorly to or beyond the ventral tooth; uropod 1 peduncle dorsolateral edge evenly curved without expansion; urosomal carina massive and not laminar (plate 276O); telson apex lobes narrow and lacking a notch (plate 277B) .... *Ampelisca brevisimulata*
— Pleonite 3 posterior edge weakly convex, not extending posteriorly beyond the ventral tooth; uropod 1 peduncle proximolateral edge expanded dorsally and urosomal carina laminar (plate 277C); telson apex lobes with a notch (plate 277D)..................... *Ampelisca cristata*
10. Telson dorsal surface with short, blunt spines aligned in a median row (plate 277E) (this species may not occur in this region: see *A. careyi* and *A. unsocalae*) .... *Ampelisca macrocephala*
— Telson dorsal surface with long slender spines irregularly positioned over the lobes (plate 277F)............. 11
11. Antenna 1 insertion below dorsal head margin, forming a forehead, head lower anterior slightly concave, antenna 1 reaching end of peduncle of antenna 2 (plate 277A); telson lobes with apical notch laterally facing (plate 277F) (distinctions between *A. caryei* and *A. unsocalae* questioned by Watling 1997: 140)............... *Ampelisca careyi*
— Antenna 1 insertion at dorsal head margin precluding a forehead, head lower anterior straight, without a concavity, antenna 1 not reaching end of peduncle of antenna 2 (plate 277G) telson lobes with apical notch facing slightly medial (plate 277H)............. *Ampelisca unsocalae*

12. Pereopod 7 article 3 equal to or longer than article 4 and article 2 basal lobe not extending past article 3 (plate 277I); mandibular palp article 2 with inflated edges curved (plate 277J)... 13
— Pereopod 7 article 3 shorter than article 4 and article 2 basal lobe extending past article 3 (plate 276C, 276I); mandibular palp article 2 straight, parallel sided (plate 277K) 14
13. Pereopod 5 article 2 posterioventral lobe enlarged (plate 277L); pereopod 7 article 2 posterior edge expanded (plate 277M) .... *Ampelisca abdita*
— Pereopod 5 article 2 posterioventral lobe straight and indistinct (plate 277N); pereopod 7 article 2 posterior edge almost evenly rounded, not expanded (plates 256L, 277I) .... *Ampelisca milleri*
14. Uropod 2 outer ramus bearing subapical spine (plate 277O); telson lobes middle dorsal surface bearing one or two diagonal rows of five to nine setae (plate 277P); head ventral margin concave (plate 277Q)................. .... *Ampelisca venetiensis*
— Uropod 2 outer ramus not bearing subapical spine (plate 277R); telson lobes middle dorsal surface bearing various arrangements of single setae (plate 277S); head ventral margin faintly incised or convex (plate 277T)........ 15
15. Urosomite 1 produced dorsally into a prominent carina or bump (plate 277C, 277R); female uropod 3 inner ramus inner edge smooth or only faintly serrate (plate 277U)..... 16
— Urosomite 1 evenly rounded not produced into prominent carina or bump (plate 277V); female uropod 3 inner ramus inner edge serrate (plate 277W).................... 18
16. Uropod 1 rami similar in length and longer or equal in length to the peduncle (plate 277R, 277X), telson lobes without a medial projection (plate 277Y) 17

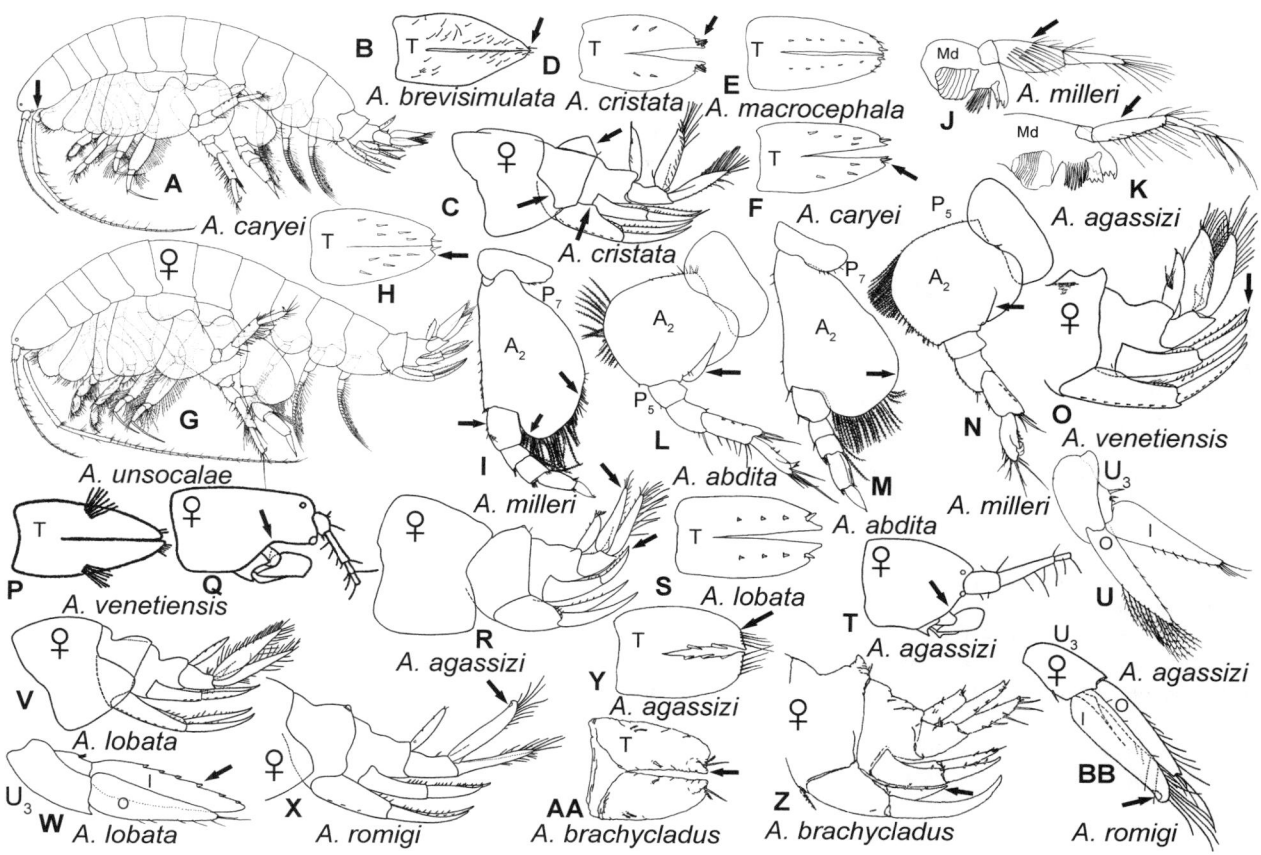

PLATE 277 Ampeliscidae. L, M, *Ampelisca abdita*; K, R, T, U, Y, *Ampelisca agassizi*; B, *Ampelisca brevisimulata*; Z, AA, *Ampelisca brachycladus*; A, F, *Ampelisca careyi*; C, D, *Ampelisca cristata*; S, V, W, *Ampelisca lobata*; E, *Ampelisca macrocephala*; I, J, N, *Ampelisca milleri*; X, BB, *Ampelisca romigi*; G, H, *Ampelisca unsocalae*; O–Q, *Ampelisca venetiensis* (figures modified from: Barnard 1954b, 1960b; Chapman 1988; Dickinson 1982; and Roney 1990).

— Uropod 1 inner ramus less than one-half as long as the outer ramus and shorter than the peduncle (plate 277Z); telson lobes with a medial projection (plate 277AA). . . . . *Ampelisca brachycladus*

17. Uropod 1 rami and peduncle equal in length (plate 277X); telson lobes bare on medial edge; antenna 1 reaching distal end of antenna 2 peduncle, uropod 3 inner ramus uncinate (plate 277BB). . . . . *Ampelisca romigi*

— Uropod 1 rami longer than peduncle (plate 277R) and uropod 3 inner ramus sharply lanceolate (plate 277U); telson lobes with four to five medial spines (plate 277Y); antenna 1 not reaching distal end of antenna 2 peduncle . . . . *Ampelisca agassizi*

18. Pereopod 7 article 5 lacking spine bearing notch on anterior margin (plate 278A); uropod 1 inner margin of outer ramus with spines (plate 278B); uropod 3 inner edge of inner ramus lined with evenly spaced serrations bearing inserted spines (plate 277W) . . . . *Ampelisca lobata*

— Pereopod 7 article 5 with spine bearing notch on anterior margin (plate 278C); uropod 1 inner margin outer ramus without spines (plate 278D); uropod 3 inner edge of inner ramus lined with smaller, unevenly spaced serrations that are without inserted spines (plate 278E, 278F) 19

19. Uropod 3 inner and outer rami lengths equal and with tiny serrations on inner edge of inner ramus (plate 278E, note arrow); coxa 1 expanding only slightly distally (plate 278G) . . . . *Ampelisca fageri*

— Female uropod 3 inner ramus longer than outer ramus and inner edge of inner ramus with medium-size serrations (plate 278F); coxa 1 expanding distally (plate 278H) (Pacific records are probably *A. fageri* variants). . . . . *Ampelisca schellenbergi*

20. Distal peduncle of uropod 1 reaching distal end of uropod 2 peduncle, coxae 2–3 posterodistal corner evenly truncated with posterior edge as long as anterior edge and coxa 1 with straight anterior edge and about equal in length to coxa 2 (plate 278I). . . . . *Byblis millsi*

— Distal peduncle of uropod 1 reaching less than two-thirds of the length of uropod 2 peduncle (not shown); coxae 2–3 posterodistal corners obliquely truncated with anterior edges longer than posterior edges and coxa 1 with concave anterior edge and longer than coxa 2 (plate 278J) . . . . *Byblis veleronis*

LIST OF SPECIES

Ampelisca abdita Mills, 1967 (=*A. milleri* of earlier San Francisco Bay literature, not of Barnard, 1954b). An estuarine species native to and characteristic of the North American Atlantic coast, and introduced to central California (San Francisco and Tomales Bays); see Mills 1967, J. Fish. Res. Bd. Can. 24: 305–355 (biology, ecology); Chapman 1988, J. Crust. Biol. 8: 364–382 (introduced status). Whole body illustration shown in plate 256L; 8 mm; intertidal to 15 m.

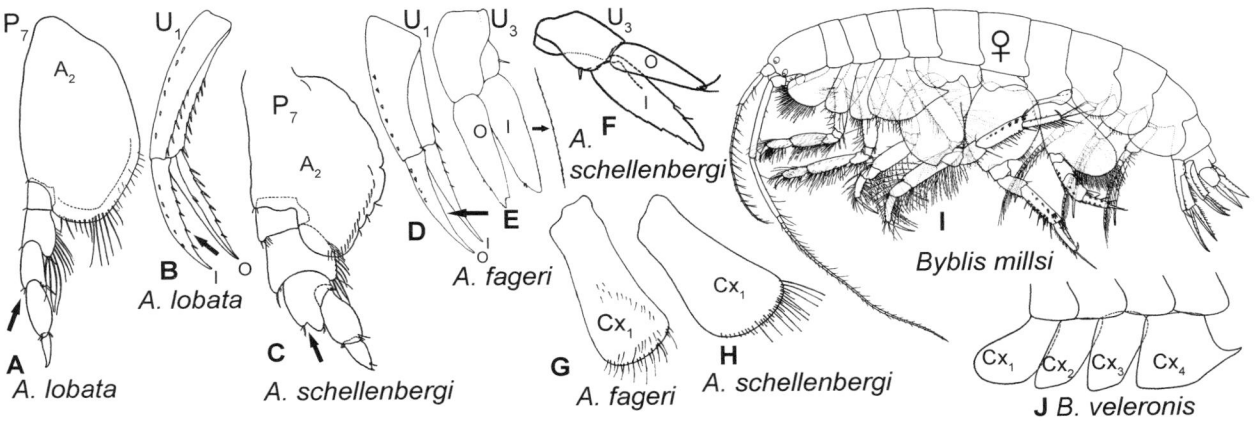

PLATE 278 Ampeliscidae. D, E, G, *Ampelisca fageri*; A, B, *Ampelisca lobata*; C, F, H, *Ampelisca schellenbergi*; I, *Byblis millsi*; J, *Byblis veleronis* (figures modified from Barnard 1954b, 1967b; and Dickinson 1982, 1983).

Ampelisca agassizi (Judd, 1896). Western Atlantic and Eastern Pacific, cold temperate to tropical, and probably more than one species; 7.5 mm; 5 m–450 m.

*Ampelisca brachycladus Roney, 1990. Southern, not reported in the region of this book, but likely to be confused with *A. agassizi*; 10 m–50 m.

Ampelisca brevisimulata Barnard, 1954b. 9 mm; 4 m–456 m.

Ampelisca careyi Dickinson 1982. A possible variant of *A. unsocalae*; 12 mm; intertidal—200 m.

Ampelisca cristata Holmes, 1908. In coarse sand; 14 mm; intertidal—152 m.

Ampelisca fageri Dickinson, 1982. Mixed bottom areas of sand and boulders; 8 mm; intertidal—40 m.

Ampelisca hancocki Barnard, 1954b. Fine sand and silt; 6.5 mm; 9 m–200 m.

*Ampelisca indentata Barnard, 1954b. Point Conception and south; not clearly within the region but could be confused with small *A. pugetica*; 5 mm; 33 m–98 m.

Ampelisca lobata Holmes, 1908, 7 mm, shallow subtidal to 591 m.

*Ampelisca macrocephala (Liljeborg 1852). Boreal; 9 mm; 10 m–280 m.

Ampelisca milleri Barnard 1954b. Central California (Gulf of the Farallones) to Ecuador, and the Galapagos Islands, a native marine species earlier confused with the estuarine *A. abdita* in San Francisco Bay; 6 mm; intertidal—187 m.

Ampelisca pacifica Holmes, 1908. Monterey Bay and south; 12 mm; 5 m–1,821 m.

Ampelisca pugetica Stimpson, 1864. 8.5 mm; intertidal—255 m.

Ampelisca romigi Barnard, 1954b. Monterey Bay and south; 10 mm; 3 m–508 m in coarse sand and gravel.

*Ampelisca schellenbergi Shoemaker, 1933. A species recorded by this name from tropical and boreal seas. Records in our region are probably *A. fageri* variants; 7.6 mm; intertidal—46 m.

Ampelisca unsocalae Barnard, 1960b. 9 mm; 50 m–1,700 m.

*Ampelisca venetiensis, Shoemaker, 1916. Venice, California and south, but distribution poorly resolved; 18 mm; intertidal—84 m.

Byblis millsi Dickinson, 1983. 10 mm; intertidal—100 m.

Byblis veleronis Barnard, 1954b. 14 mm; 5 m–422 m.

Haploops spp. Key to species in Dickinson 1983. A deep-water genus with two confirmed species in the region (*H. lodo* and *H. tubicola*, 18 mm).

* = Not in key.

DEXAMINIDAE

Dexaminidae have fused urosomites 2 and 3 biramous uropod 3 and variable length rostrums. The inferior antennal sinus is small or lacking, the gnathopods weak or simple, the telson is laminar and deeply cleft, and coxae 1–4 are deep or shallow and, in common with Ampeliscidae, have deeply pleated gills. *Guernea*, *Paradexamine*, and *Polycheria* lack mandibular palps.

KEY TO DEXAMINIDAE

1. Pereopods 3–7 fully prehensile, pereopods 5–7 article 2 narrow, coxae 1–4 longer than deep and rostrum indistinct (plate 279A); feeds upside down from inside tunicate *Amaroucium* colonies (plate 279B) . . . . . *Polycheria osborni*
— Pereopods 3–7 simple (not prehensile), pereopods 5–7 article 2 expanded, coxae 1–4 deeper than long and rostrum distinct (plate 279C) 2
2. Pleonites 1–3 bearing lateral carina in addition to dorsal medial carina (plate 279C) . . . .*Paradexamine* sp.
— Pleonites 1–3 lacking lateral carina, dorsal carina present or absent (plate 279D). 3
3. Dorsal anteriorly directed tooth of urosomite 1 and deeply beveled, (possibly adapted for interlocking urosomites 2 and 3 with urosomite 1) and coxa 5 larger than coxa 4 (plate 279D) . . . .*Guernea reduncans*
— Dorsal tooth of urosomite 1 absent or posteriorly directed, pereopod 5 coxa smaller than coxa 4 (plate 279E, 279F) 4
4. Coxa 4 posteriorly expanded, urosomite 1 and fused urosomites 2 and 3 with a sharp mid dorsal tooth preceded by a deep notch (plate 279E) . . . .*Atylus tridens*
— Coxa 4 lacking posterior expansion and urosomites with rounded mid dorsal teeth that lack an anterior notch (plate 279F) 5
5. Telson wider than long (plate 279G); eye longer than the rostrum (plate 279F) . . . .*Atylus georgianus*
— Telson longer than wide (plate 279H); eye shorter than the rostrum (plate 279I) . . . .*Atylus levidensus*

LIST OF SPECIES

Atylus species were previously in Atylidae.

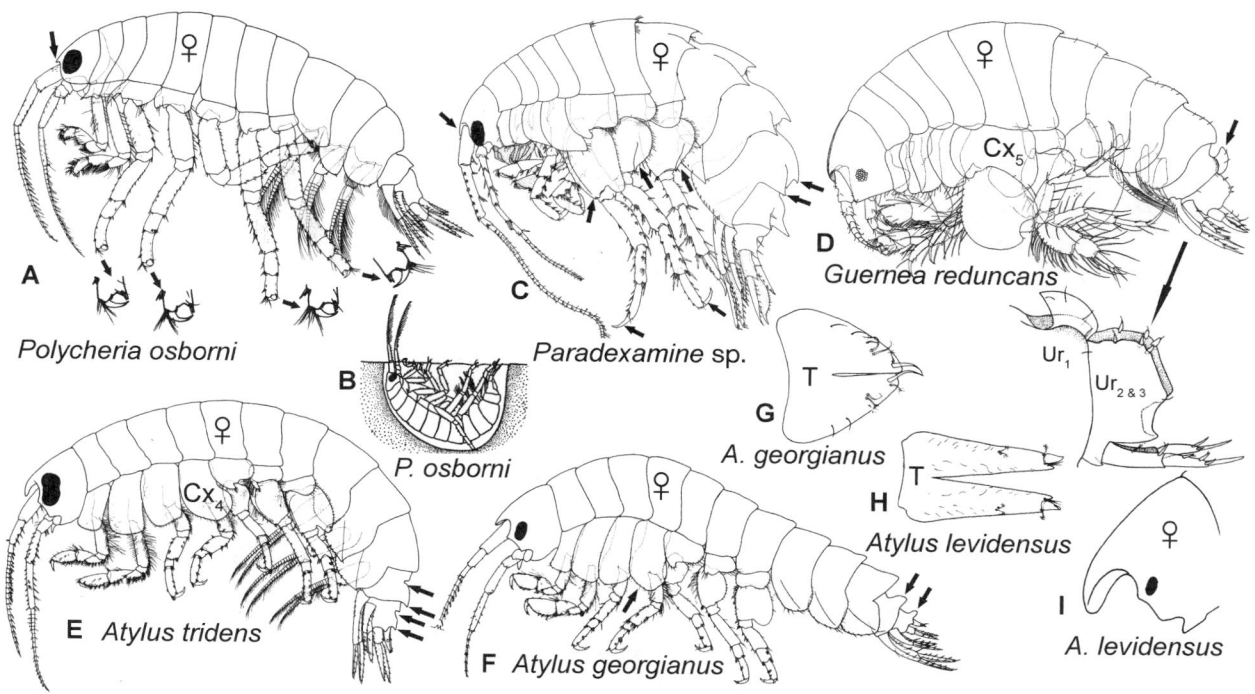

PLATE 279 Dexaminidae. F, G, *Atylus georgianus*; H, I, *Atylus levidensus*; E, *Atylus tridens*; D, *Guernea reduncans*; C, *Paradexamine* sp.; A, B, *Polycheria osborni* (figures modified from Barnard 1972a; Barnard and Karaman 1991a; Bousfield and Kendall 1994; and Skogsberg and Vansell 1928).

Atylus georgianus Bousfield and Kendall, 1994. Subtidal sand and eelgrass; 8 mm; intertidal—3 m.

Atylus levidensus Barnard, 1956. Various sediments but especially in sand; 12 mm; intertidal—3 m.

Atylus tridens (Alderman, 1936). Sand, eelgrass, and rocky bottoms, occasionally pelagic; 10 mm; intertidal—6 m.

Guernea reduncans (Barnard, 1957b) (=*Dexamonica reduncans*). In fine sand and green mud; 2.5 mm; subtidal—180 m.

Paradexamine sp. Introduced, occurs in high-salinity fouling communities of Los Angeles, Long Beach Harbor and San Francisco Bay; 5 mm, intertidal—3 m. The illustration of the Australian *P. frinsdorfi* Sheard, 1938 is provided as an example of the genus.

Polycheria osborni (Calman, 1898). *Polycheria* makes pits in compound tunicate tests (especially *Aplidium californicum*); they live upside down in the pits (which they can open and close), with their legs extended and pleopods propelling water forward along the body, while feeding on suspended particles. Broods up to 80 young in early summer. Intertidal rocky shores; 5.8 mm.

OEDICEROTIDAE

Oedicerotidae have weakly fossorial pereopods and burrow into fine sand or mud where most species are probably predators on meiofauna. The eyes are dorsally coalesced into a single mass on the decurved rostrum. The telson is laminar and entire, and pereopod 7 is more than half again as long as pereopod 6. Oedicerotidae lack an accessory flagellum. Mandibular palps are present and molars are prominent or reduced. The taxonomy is based on females.

KEY TO OEDICEROTIDAE

1. Gnathopod 2 chelate (plate 280A); mandibular molar reduced (plate 280B) 2

— Gnathopod 2 subchelate (plate 280C); mandibular molar prominent (plate 280D) 6

2. Pereopod 5, article 5 approximately half as long as article 6 (plate 280E) .... *Eochelidium* cf. *miraculum*

— Pereopod 5 article 5 equal to or longer than article 6 (plate 280F) 3

3. Uropod 3 half as long as uropod 1 (plate 280F) .... *Americhelidium micropleon*

— Uropod 3 more than two-thirds as long as uropod 1 (plate 280G) 4

4. Gnathopod 1 palm of article 6 transverse (plate 280H), gnathopod 2 dactyl nearly one-third of the length of article 6 (plate 280I) .... *Americhelidium rectipalmum*

— Gnathopod 1 palm of article 6 is slightly oblique, verging toward subchelate (plate 280J); gnathopod 2 dactyl less than one-fourth of the length of article 6 (plate 280K) 5

5. Epimeron 2 posterior ventral corner produced (plate 280L) .... *Americhelidium shoemakeri*

— Epimeron 2 posterior ventral corner square (plate 280M) .... *Americhelidium pectinatum*

6. Gnathopod 2 article 5 not extending over the posterior edge of article 6 (plate 280C); head as long as first four pereonites combined (plate 280N) .... *Westwoodilla tone*

— Gnathopod 2 article 5 extending over more than half of the posterior edge of article 6 (plate 280O); head shorter than first four pereonites combined (plate 280P) 7

7. Pereopod 4 article 4 anteriorly expanded (plate 280Q) pereopod 7 article 2 posterior ventral corner expanded (plate 280R); pleonite 2 posterior ventral corner strongly produced and rostrum anterior curved to less than 60° angle of the head dorsum (plate 280R) .... *Pacifoculodes spinipes*

— Pereopod 4 article 4 parallel sided (plate 280P, 280S); pleonite 2 posterior ventral corner square or rounded (plate

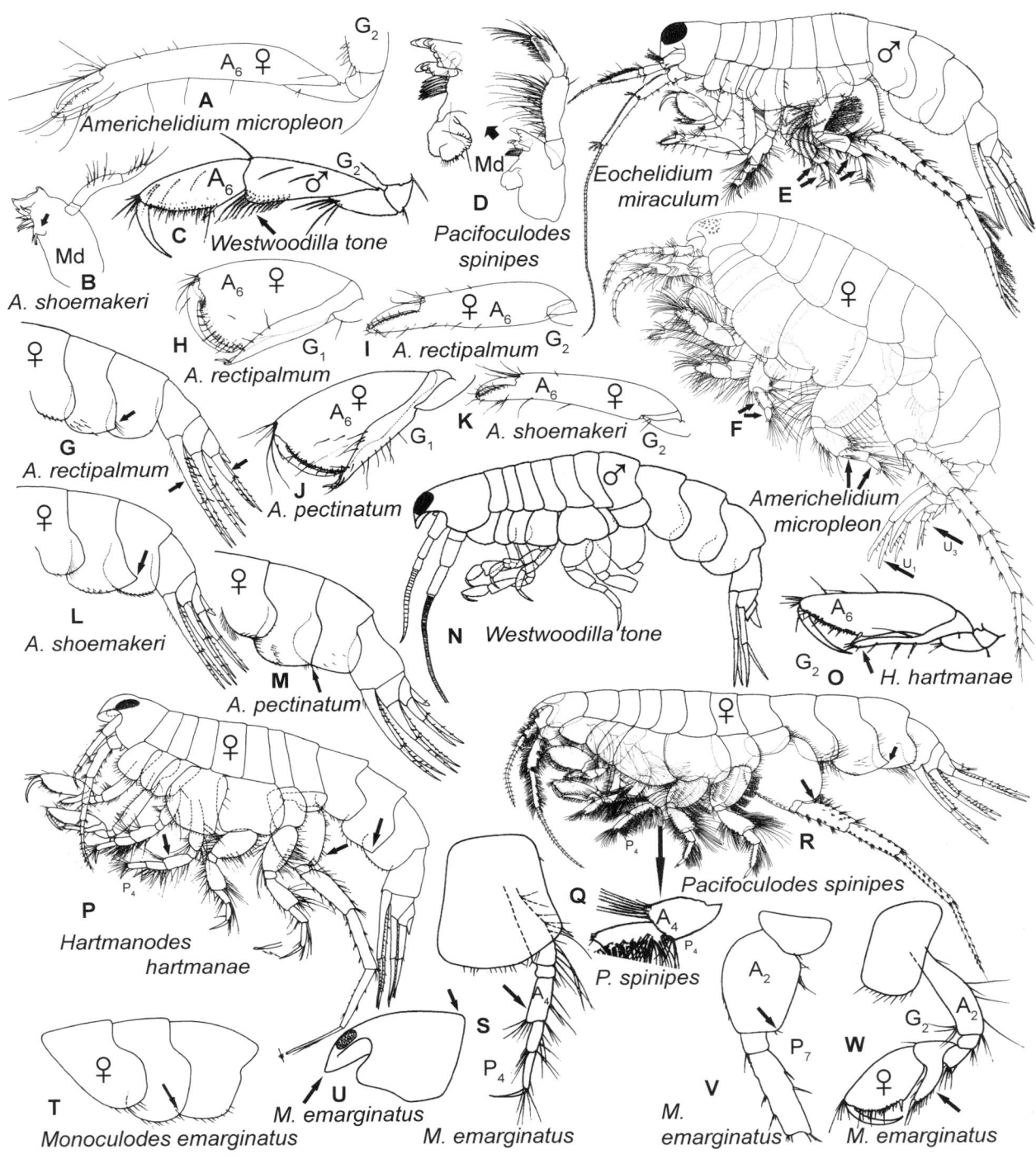

PLATE 280 Oedicerotidae. A, F, *Americhelidium micropleon*; J, M, *Americhelidium pectinatum*; G–I, *Americhelidium rectipalmum*; B, K, L, *Americhelidium shoemakeri*; E, *Eochelidium* sp. cf. *E. miraculum*; O, P, *Hartmanodes hartmanae*; S–W, *Monoculodes emarginatus*; D, Q, R, *Pacifoculodes spinipes*; C, N, *Westwoodilla tone* (figures modified from Barnard 1962d, 1977; Bousfield and Chevrier 1996; Imbach 1967; Jansen 2002; and Mills 1962).

280P, 280T); rostrum anterior curved to 80° or greater angle with the head dorsum (plate 280U); pereopod 7, posterior ventral corner of article 2 not expanded ventrally (plate 280P, 280V) 8

8. Gnathopod 2 extension of article 5 less than 60% total length of article 6 (plate 280W); pleonite 2 posterior ventral corner rounded (plate 280T) . . . . *Monoculodes emarginatus*

— Gnathopod 2 extension of article 5 more than 60% length

of article 6 (plate 280O); pleonite 2 posterior ventral corner square (plate 280P) . . . . *Hartmanodes hartmanae*

LIST OF SPECIES

Americhelidium micropleon (Barnard, 1977) (=*Synchelidium micropleon*). On open fine sand beaches, not reported north of Dillon Beach; 3.5 mm; intertidal.

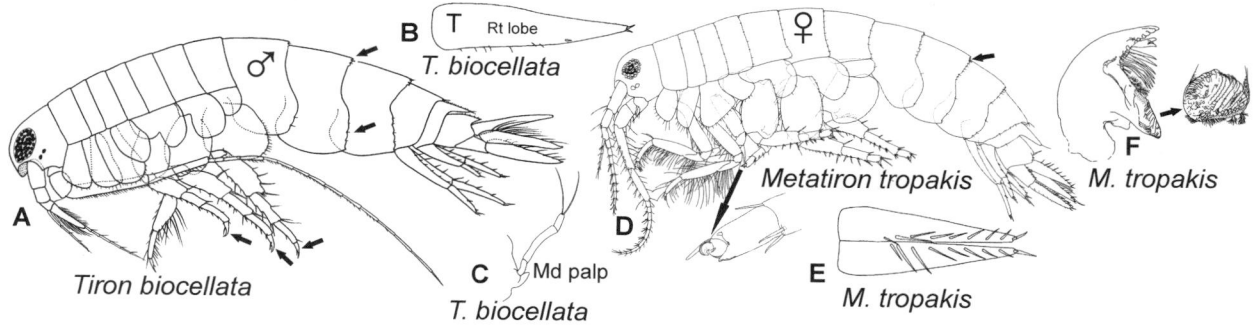

PLATE 281 Synopiidae. D–F, *Metatiron tropakis*; A–C, *Tiron biocellata* (figures modified from Barnard 1962b, 1972b).

Americhelidium pectinatum Bousfield and Chevrier, 1996. Shallow sandy sediments; 4 mm; intertidal—50 m.

Americhelidium rectipalmum (Mills, 1962) (=*Synchelidium rectipalmum*). Clean sand bottoms; 6 mm; intertidal—90 m.

Americhelidium shoemakeri (Mills, 1962) (=*Synchelidium shoemakeri*). Clean sand bottoms; 4.5 mm; intertidal—180 m.

Eochelidium sp. cf. *E. miraculum* (Imbach, 1967). The identity of this species is unclear. In fine silty sediments of warm, high-salinity harbor areas (Puget Sound, San Francisco Bay, and Los Angeles Harbor); a likely ballast water introduction from Asia; 5 mm; intertidal—10 m.

Hartmanodes hartmanae (Barnard, 1962d) (=*Monoculodes hartmanae*). Common in rocky intertidal and shallow water communities and scarce below 37 m; 5 mm; intertidal—146 m.

Monoculodes emarginatus Barnard, 1962d. Fine mud and sand, may not be a shallow water species, difficult to distinguish from *M. perditus* Barnard 1966; 4.5 mm; 10 m–200 m.

Pacifoculodes spinipes Mills, 1962 (=*Monoculodes spinipes*). Surf zone of open sandy beaches; 11 mm; intertidal—98 m.

Westwoodilla tone Jansen, 2002. Fine mud and sand, usually offshore; broad-bodied species; previously confused with the North Atlantic *W. caecula* (Bate, 1857) (see: Jansen 2002, Steenstrupia 27(1): 83–136); 8 mm; shallow subtidal—500 m.

SYNOPIIDAE

Synopiidae (Tironidae) have helmet or plough-shaped heads, prominent and multiarticulate accessory flagella, similar length pereopods 6 and 7 (relative to Oedicerotidae), and strong sexual dimorphism. Collected in both benthic grabs and vertical plankton hauls, and some species having peculiar grasping dactyls (see *Metatiron*). Synopiidae may be commensals of benthic or epibenthic invertebrates. The family remains poorly studied in the region. Gnathopods of *Tiron* and *Metatiron* are weakly subchelate or simple, and they bear two accessory eyes near the base of antenna 2. *Garosyrrhoe* and *Syrrhoe*, occurring near our region, have transverse gnathopods and lack accessory eyes.

KEY TO SYNOPIIDAE

1. Pleonites 1–3 crenulate dorsally and on medioposterior edges and dactyls normal and strong (plate 281A); telson lacking dorsal spines (plates 256DD, 281B); mandibular palp present (plate 281C) (condition of the molar has not been reported) . . . . *Tiron biocellata*

— Pleonites crenulate only on dorsal third, dactyls stubby and twisted (plate 281D); telson lobes lined with large medial spines (plate 281E); mandibular palp absent and molar large (plate 281F) . . . . *Metatiron tropakis*

LIST OF SPECIES

Garosyrrhoe bigarra (Barnard 1962b). Southern California; 4.5 mm; sublittoral.

Metatiron tropakis (Barnard, 1972b). In sand, both coasts of North America and the Caribbean; a species complex; 6 mm; 3 m–357 m.

Syrrhoe crenulata Göes, 1866. 10 mm; off of Oregon in deeper water (40 m–200 m).

Syrrhoe longifrons Shoemaker, 1964. Vancouver Island; 10 mm, "shallow waters."

Tiron biocellata Barnard 1962b. Rock bottoms associated with the worms *Diopatra* and *Nothria*. Caribbean records are probably of another species; 4.6 mm; shallow subtidal—180 m.

LEUCOTHOIDAE

Leucothoidae are commensals of tunicates identifiable to family by their carpochelate gnathopods. Previously recognized Anamixidae appear to be nonfeeding super-male leucothoid morphs referred to as "anamorph" stages with mandibles and maxillae replaced by a ventral keel (Thomas and Barnard 1983) and "anamixid" females were previously recognized as *Leucothoides*. The Leucothoidae are thus likely to be sequential hermaphrodites. Distinctions among families, genera, species, and morphs (and the biogeography) in this group are difficult to resolve without greater knowledge of their life histories and development patterns. See Thomas 1997 Rec. Australian Mus. 49: 35–98.

KEY TO LEUCOTHOIDAE

1. Gnathopod 2 dactyl reaching distal process of article 5 (plate 282A) 2

— Gnathopod 2 dactyl not reaching extended process of article 5 (plate 282D, 282E); mandible (plate 282B) and maxilliped (plate 282C) fully formed . . . . *Leucothoides pacifica*

2. Mandibles and maxilliped replaced by a ventral keel (males only) (plate 282F) . . . .

* = Not in key.

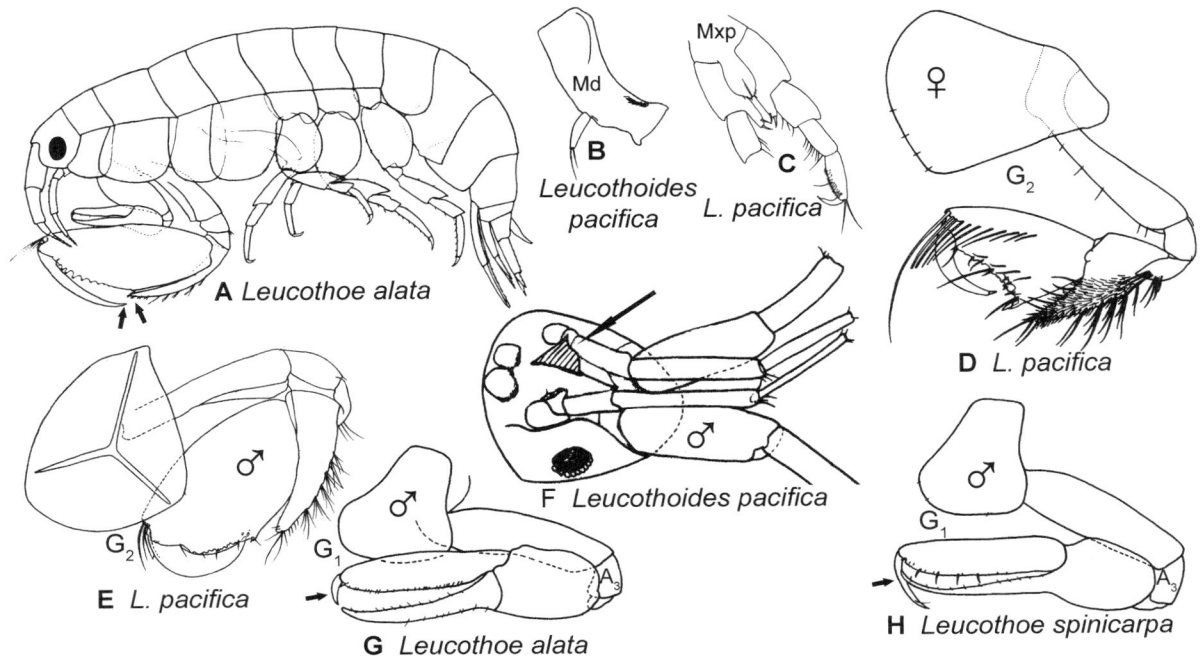

PLATE 282 Leucothoidae. A, G, *Leucothoe alata* (male?); H, *Leucothoe spinicarpa*; B–F, *Leucothoides pacifica* (figures modified from Barnard 1955b, 1962c, 1969a and Nagata 1965a).

. . . . . . *Leucothoides pacifica* (previously *Anamixis lindsleyi*)
— Mandible and maxilliped fully formed (plate 282B, 282C)
. . . . 3
3. Male gnathopod 1 dactyl shorter than article 3 (plate
282G) . . . . *Leucothoe alata*
— Male gnathopod 1 dactyl 1.5 times longer than article 3
(plate 282H) . . . . *Leucothoe spinicarpa*

LIST OF SPECIES

Leucothoides pacifica (Barnard 1955b) (=*A. lindsleyi, Anamixis pacifica*). In sponges and tunicates, especially on pilings. *Leucothoides* could be a morph of *Leucothoe*; 2.8 mm; intertidal—8 m.

Leucothoe alata Barnard, 1959b (=*Anamixis pacifica*). Introduced parasite and commensal of tunicates in San Francisco Bay, and recorded in "open-sea" algal bottoms (Barnard 1962C); Japan; sex (plate 282A) not given; to 12.3 mm; intertidal—18 m.

Leucothoe spinicarpa (Abildgaard, 1789). Introduced cosmopolitan marine parasite and commensal of tunicates "widely distributed from subarctic waters to south temperate regions; perhaps universally distributed" (Barnard 1962C) and thus likely to consist of multiple species; 3 mm; intertidal—10 m.

STEGOCEPHALOIDEA

Stegocephaloidea (Iphimediidae and Odiidae) are distinguished primarily by mouth part morphology because the body sculpturing and ornamentations are often similar among genera. The cone-shaped mouth bundle, needle-shaped mandibles, and weak molars (plate 256KK–256LL) are probable adaptations for predation on coelenterates, sponges, and bryozoans (Coleman and Barnard 1991, Moore and Rainbow 1992). Most species appear to be protandrous hermaphrodites or parthenogenic (Moore 1992).

KEY TO STEGOCEPHALOIDEA

1. Pleonites 1–3 dorsal and lateral teeth reduced except for a single rounded dorsal tooth on pleonite 3 (plate 283A); telson with convergent lateral edges and bluntly acute (plate 283B); gnathopod 2 article 6 expanding distally (plate 283C). . . . . *Cryptodius kelleri*
— Pleonites 1–3 with pairs of teeth at the posterior margins and dorsally (plate 283D, 283E); gnathopod 2 article 6 distally narrow (plate 283F); telson with nearly parallel lateral edges and broad posterior excavation (plate 283G, 283H) 2
2. Telson posterior excavation deep (plate 283G); maxilla 1 palp of two articles (plate 283I) . . . . *Iphimedia rickettsi*
— Telson posterior excavation shallow (plate 283H); maxilla 1 palp of one article (plate 283J) . . . . . *Coboldus hedgpethi*

LIST OF SPECIES

IPHIMEDIIDAE

Coboldus hedgpethi (Barnard, 1969a) (=*Panoploea* in Iphimediidae, previously Acanthonotozomatidae). In mixed sediments (especially cobbles) among algae and on harbor pilings; 4.5 mm; 1 m–82 m.

Iphimedia rickettsi (Shoemaker, 1931) (=*Panoploea rickettsi*). Rocky substrata, especially in holdfasts of kelps and coralline algae, possibly commensal; 8 mm; low intertidal—60 m.

LIST OF SPECIES

Cryptodius kelleri (Bruggen, 1907) (=*Odius kelleri*). Boreal northern Pacific, rocky substrata, especially among algae; 5 mm; intertidal—90 m.

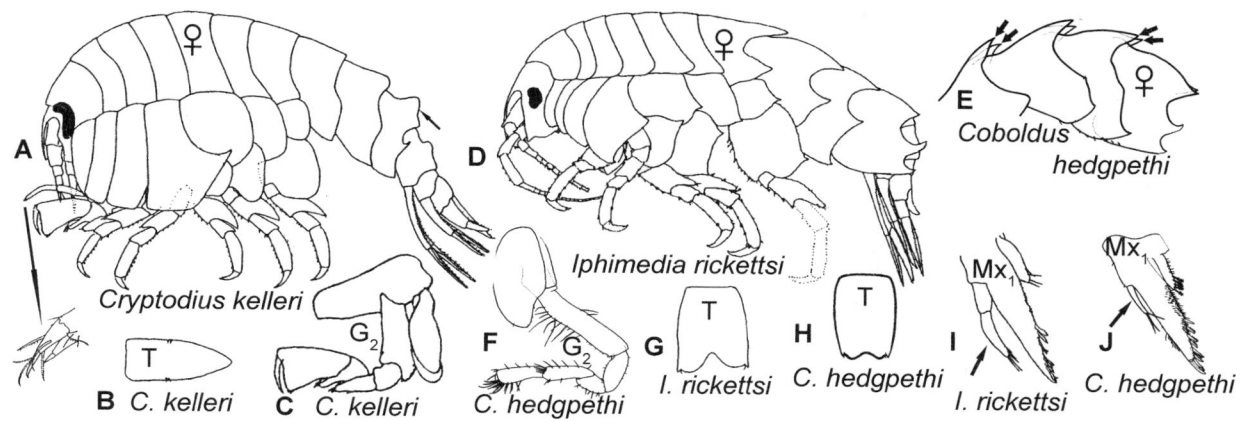

PLATE 283 Stegocephaloidea. E, F, H, J, *Coboldus hedgpethi*; A–C, *Cryptodius kelleri*; D, G, I, *Iphimedia rickettsi* (figures modified from: Barnard 1969a and Moore 1992).

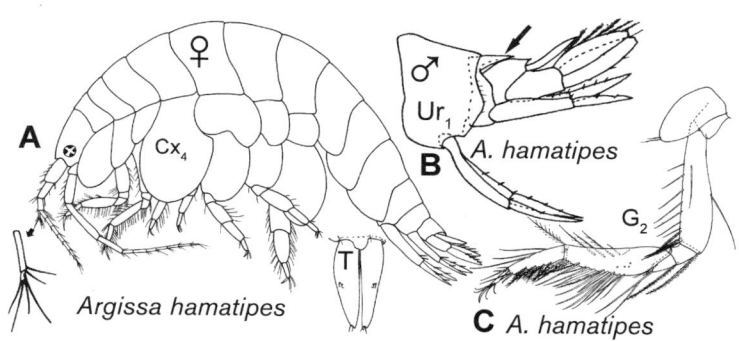

PLATE 284 Argissidae. A–C, *Argissa hamatipes* (figures modified from Barnard 1969b and Hirayama 1983).

ARGISSIDAE

Argissidae of this region include a single cosmopolitan species distinguished by the combination of long coxae 1 and 4, feeble gnathopods, telson deeply cleft and posteriorly expanded pereopod 7 article 2. Eyes (when present) consist of four visual elements. The Pacific species is unlikely to be the same as *A. hamatipes* of the North Atlantic. The unusual coxa of Argissidae may allow upside-down burrowing and feeding similar to Megaluropidae, which Argissidae resemble.

KEY TO ARGISSIDAE

1. Eye of four visual elements, coxa 4 longer than coxa 3, female urosomite 1 evenly rounded (plate 284A); male urosomite 1 with a large tooth (plate 284B); gnathopod 2 simple (plate 284C) . . . . *Argissa hamatipes*

LIST OF SPECIES

Argissa hamatipes (Norman, 1869). Cosmopolitan in mud, sand and rock benthos. A likely complex of species of which there are shallow and deep-water members. Males of eastern Pacific populations have carina on dorsal urosomites 1–3 (no illustrations published); 4.5 mm; 4 m–1,096 m.

MEGALUROPIDAE

Megaluropidae feed upside down at the sand episurface. Their unusual coxae allow dorsal extension of pereopods 3–4 from the upside down position; the long, flexible pereopods 5–7 can quickly dig into well-sorted sediments leaving only a hole at the sand surface that is maintained by the legs and leaf like uropods (Barnard et al., 1988, Crustaceana Suppl. 13: 234–244). Terminal pelagic males have large eyes and antenna 2 with long flagellum and setal tufts on the peduncle, a "distinctive" gnathopod 2 and large pleon.

KEY TO MEGALUROPIDAE

1. Gnathopod 2 article 4 distally produced (plate 285A, 285B); male rostrum short and ocular lobe of head bearing a sharp angle (plate 285C); accessory flagellum of two articles (plate 285D) 2
— Gnathopod 2 article 4 not distally produced (plate 285E); male rostrum long, ocular lobe lacking sharp angle, accessory flagellum of one article (plate 285F) . . . . . . *Resupinus*
2. Epimeron 3 posteriorly serrate (plate 285A, 285G) . . . . . .
. . . . *Gibberosus myersi*
— Epimeron 3 posteriorly smooth (plate 285H) . . . .
. . . . *Gibberosus devaneyi*

LIST OF SPECIES

Resupinus sp. Shallow-water tropical genus (Thomas and Barnard 1986) included for reference due to complex geography and taxonomy of *Gibberosus*.
Gibberosus myersi (McKinney, 1980) (=*Megaluropus longimerus* Barnard 1962b). Cryptogenic: in Atlantic from Caribbean to

* = Not in key.

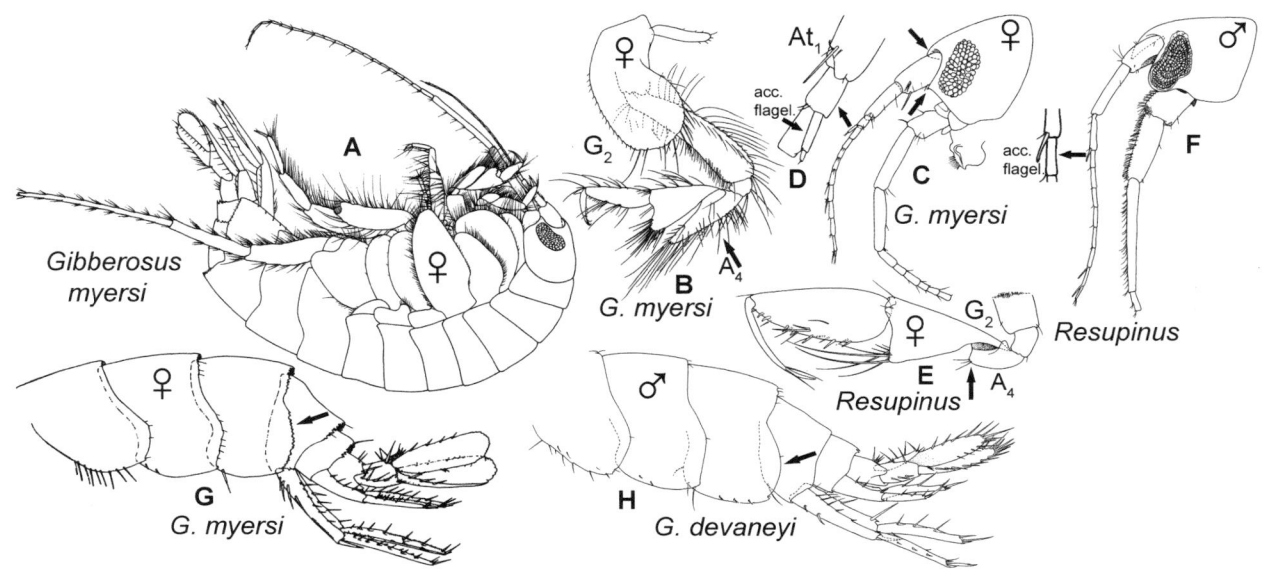

PLATE 285 Megaluropidae. H, *Gibberosus devaneyi*; A–D, G, *Gibberosus myersi*; E, F, *Resupinus* (figures modified from Barnard 1962b; McKinney 1980; Thomas and Barnard 1986).

North Carolina and in the eastern Pacific from Peru to British Columbia; not likely to be a single species; eastern Pacific forms on sand bottoms, among *Phyllospadix* and *Silvetia,* and occasionally among *Anthopleura elegantissima*; 4 mm; intertidal—27 m.

Gibberosus devaneyi Thomas and Barnard 1986. La Jolla and south but may occur in the southern end of our region; 2–3 mm; intertidal sand beaches.

LYSIANASSOIDEA

Lysianassoidea (Aristidae, Lysianassidae, Opisidae, Uristidae) are predators, scavengers, commensals, and parasites (Dahl, 1979; Conlan 1994). The mitten-shaped article 6 and long article 3 of gnathopod 2 and stubby article 1 of antenna 2 are distinctive characteristics of the infraorder. Telsons of lysianassids range from flat and deeply cleft to entire and stubby.

KEY TO LYSIANASSOIDEA

1. Uropod 3 consisting of peduncle only (plate 286A); pereopod 7 article 2 greatly expanded (plate 286B); body may be covered with scales and fuzz; one to three pleonites plus the first urosomite forming erect peaks (plate 286B); or body smooth, with only the third pleonite forming peaks (plate 286C) (secondary phase male) . . . . . *Ocosingo borlus*
— Uropod 3 with two rami (plate 286D); body not as above
. . . . 2
2. Gnathopod 1 chelate (plate 286E) 3
— Gnathopod 1 simple (plate 286F), or subchelate, or parachelate (plate 286G), maximum extension of thumb not beyond dactyl hinge 4
3. Gnathopod 1 inner margins of dactyl and palm nearly parallel (plate 286E, 286H); telson entire (plate 286I); eye usually present . . . . *Prachynella lodo*
— Dactyl and palm of gnathopod 1 outlining a circular gap

with the dactyl and thumb touching only at the tips (plate 286J); telson deeply cleft (plate 286K), eyes always present
. . . . *Opisa tridentata*
4. Telson entire (plate 286L) 5
— Telson cleft (plate 286M) 10
5. Telson distally concave (plate 286L) 6
— Telson distally convex (plate 286N) 8
6. Female antenna 1 article 3 half as long as wide and epistome not extending past upper lip (plate 286O); outer ramus of uropod 3 of two prominent articles (plate 286P) . . . .
. . . . *Dissiminassa dissimilis*
— Female antenna 1 article 3 less than one-third as long as wide (plate 286Q); and epistome extending past upper lip (plate 286R); outer ramus of uropod 3 with a minute distal article (plate 286S) 7
7. Female third pleonal epimeron a quadrate plate (plate 286T) . . . . *Aruga holmesi*
— Female third pleonal epimeron posteriorly concave (plate 286U) . . . . *Aruga oculata*
8. Posterior edge of telson with two stout spines (plate 286N); outer ramus of uropod 3 of two articles (plate 286V); gnathopod 1 palm transverse 7 (plate 286G) . . . .
. . . . *Orchomenella recondita*
— Posterior edge of telson without stout spines (plate 286W, 286X); uropod 3 outer ramus of a single article (plate 286Y); gnathopod 1 simple (plate 286Z) 9
9. Uropod 3 inner ramus over half as large as outer ramus (plate 286Y); anterior cephalic lobe and posterior margins of pereopod 7 crenulate (plate 286AA); mandibular molar large (plate 286BB) . . . . *Macronassa macromera*
— Uropod 3 inner ramus <30% as large as outer ramus (plate 286CC); anterior cephalic lobe and posterior margins of pereopod 7 smooth (plate 286DD, 286EE); mandibular molar small and fuzzy (plate 286FF) . . . .
. . . . *Macronassa pariter*
10. Ventral posterior of pleonal epimeron 3 faintly or strongly hooked (plate 287A, 287B); mandibular palp even with molar (plate 287C) 14

*= Not in key.

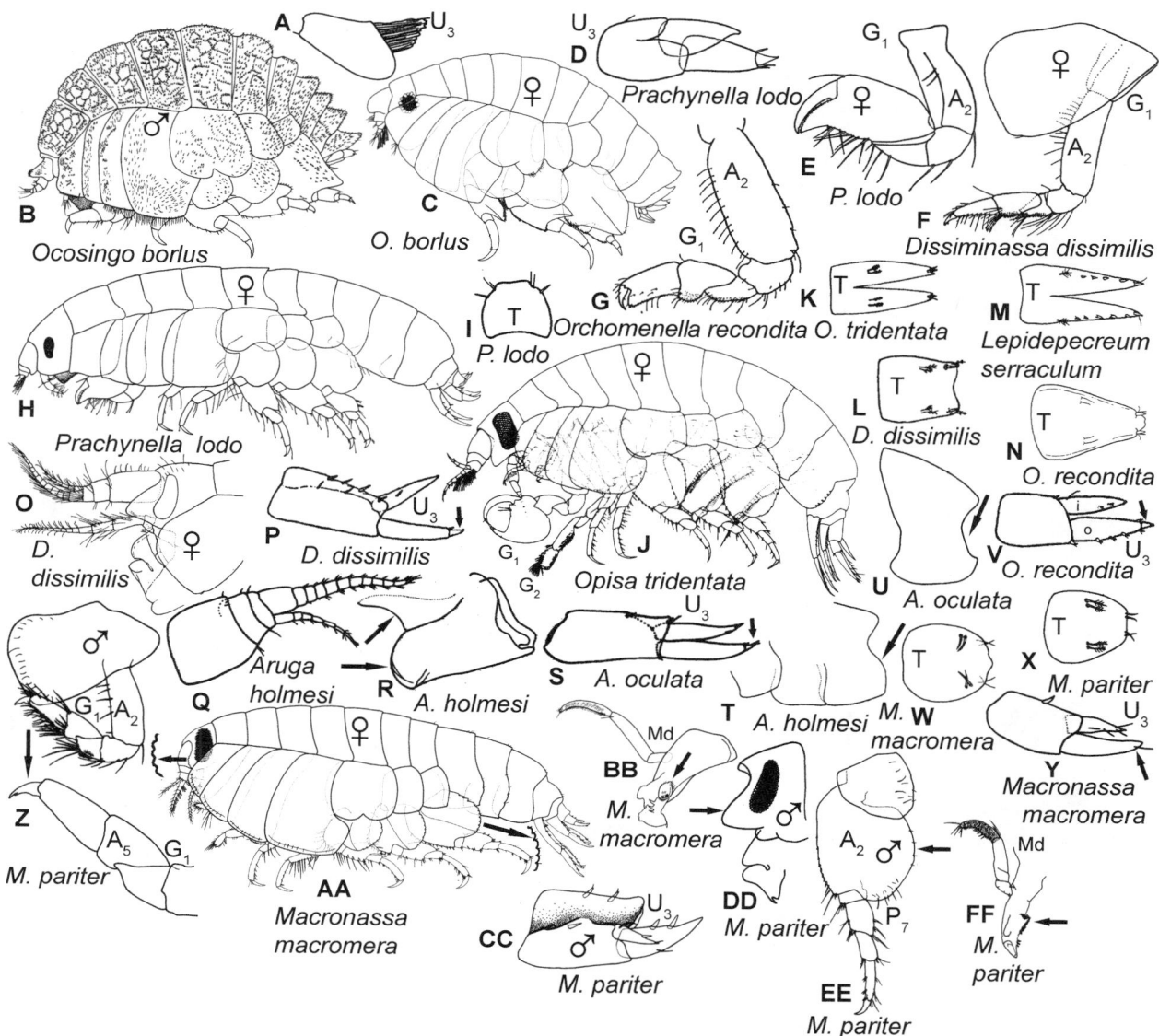

PLATE 286 Lysianassoidea. Q, R, T, *Aruga holmesi*; S, U, *Aruga oculata*; F, L, O, P, *Dissiminassa dissimilis*; M, *Lepidepecreum serraculum*; W, Y, AA, BB, *Macronassa macromera*; X, Z, CC–FF, *Macronassa pariter*; G, N, V, *Orchomenella recondita*; A–C, *Ocosingo borlus*; J, K, *Opisa tridentata*; D, E, H, I, *Prachynella lodo* (figures modified from Barnard 1955a, 1964b, 1967c, 1969a; Bousfield 1987; Dalkey 1998; Shoemaker 1942; and Stasek 1958).

— Ventral posterior of third pleonal epimeron square, rounded, not hooked (plate 287D); mandibular palp proximal to molar (plates 286BB, 286FF, 287E) 11
11. Urosomite 1 dorsally carinate, overlapping urosomite 2 (plate 287D). 12
— Urosomite 1 rounded, not overlapping urosomite 2 (plate 287F) . . . . *Orchomene minutus*
12. Dorsal pereonites and pleonites rounded (plate 287D); anterior extension of antenna 1 article 1 over article 2 slight (plate 287G). 13
— Dorsal pereonites and pleonites carinate or sharply extending posteriorly (plate 287H); large projection on dorsal antenna 1, article 1 (plate 287I) . . . . *Lepidepecreum gurjanovae*
13. Third pleonal epimeron extension acuminate (plate 287J) . . . . *Orchomene pacifica*
— Third pleonal epimeron square (plate 287D). . . . . *Lepidepecreum serraculum*
14. Coxa 1 tapering distally, smaller and partially hidden by coxa 2 (plate 287K); uropod 3 outer ramus article 2 narrower

than article 1 (plate 287L). . . . . *Aristias veleronis*
— Coxa 1 not tapering distally, of similar size and not obscured by coxa 2 and uropod 3 outer ramus article 2 expanding evenly to width of 1 (plate 287A) 15
15. Coxa 1 anterior concave (plate 287A); and upper lip greatly extending beyond the epistome (not shown) 16
— Coxa 1 anterior convex (plate 287M); upper lip not extending to or extending only slightly beyond the epistome (plate 287N). 17
16. Epimeron 2 ventral posterior corner square (plate 287O); uropod 2 weakly constricted (not shown), outer ramus with enlarged spines (plate 287P); possibly a new species . . . . *Anonyx* cf. *lilljeborgi*
— Epimeron 2 ventral posterior corner produced into a sharp tooth (plate 287Q); uropod 2 distal rami with small spines (plate 287R) . . . . *Anonyx* cf. *nugax*
17. Pereopod 7 longer than pereopod 6 (plate 287M) . . . . *Psammonyx longimerus*
— Pereopod 7 shorter than pereopod 6 (plate 287S) 18

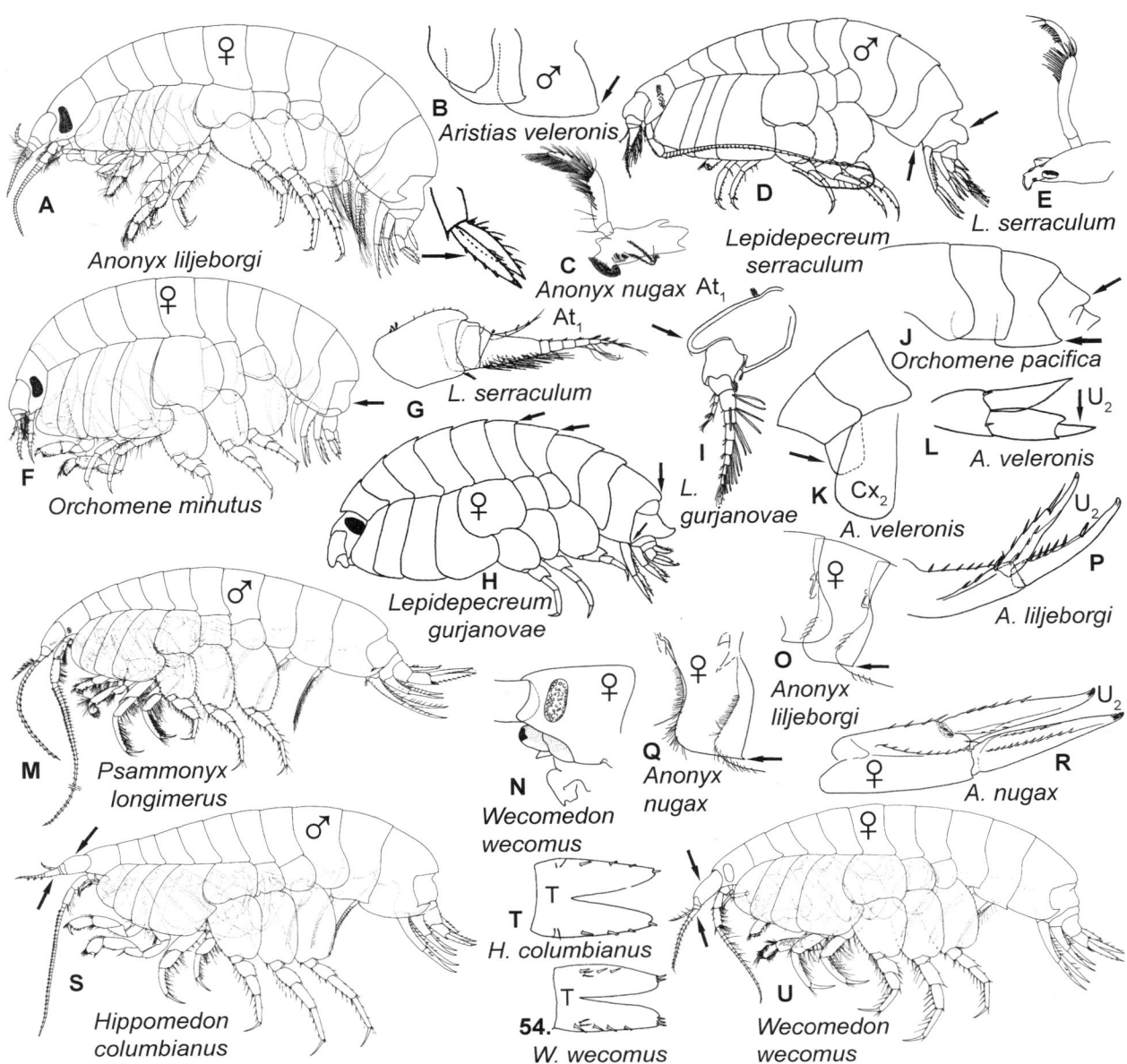

PLATE 287 Lysianassoidea. A, O, P, *Anonyx lilljeborgi*; C, Q, R, *Anonyx nugax*; B, K, L, *Aristias veleronis*; S, T, *Hippomedon columbianus*; H, I, *Lepidepecreum gurjanovae*; D, E, G, *Lepidepecreum serraculum*; F, *Orchomene minutus*; J, *Orchomene pacifica*; M, *Psammonyx longimerus*; N, U, V, *Wecomedon wecomus* (figures modified from Barnard 1964c, 1971; Bousfield 1973; Dalkey 1998; Hurley 1963; Jarrett and Bousfield 1982; Lincoln 1979; and Steel and Brunel 1968).

18. Antenna 1 flagellum article 1 more than half as long as peduncle article 1 (plate 287S); tips of telson pointed and bearing one to two spines (plate 287T) . . . .
. . . . *Hippomedon columbianus*
— Antenna 1 flagellum article 1 less than half as long as article 1 (plate 287U); tips of telson blunt and bearing two or more spines (plate 287V) . . . . *Wecomedon wecomus*

LIST OF SPECIES

ARISTIDAE

Aristias veleronis Hurley, 1963. A likely commensal with brachiopods, sponges, and ascidians. Possible synonym of *A. pacificus* Schellenberg, 1936; *Aristias* sp. A (1985, SCAMIT

Newsletter 3[10]) in the sponge *Staurocalyptus* may also be this species; 6 mm; intertidal—18 m.

LYSIANASSIDAE

Aruga holmesi Barnard, 1955a. Soft sediment; Washington, California, Gulf of Mexico, Western Florida; perhaps two species, one in each ocean; 11.5 mm; intertidal—183 m.

Aruga oculata Holmes, 1908. Most common in shallow soft benthos of California; 15 mm; 1 m–457 m.

Dissiminassa dissimilis (Stout, 1913) (=*Lysianassa dissimilis*). Tomales Bay and south among *Macrocystis* holdfasts, *Aplidium* sp., loose rocks, *Phyllospadix* and coralline algae; 6 mm; intertidal—73 m.

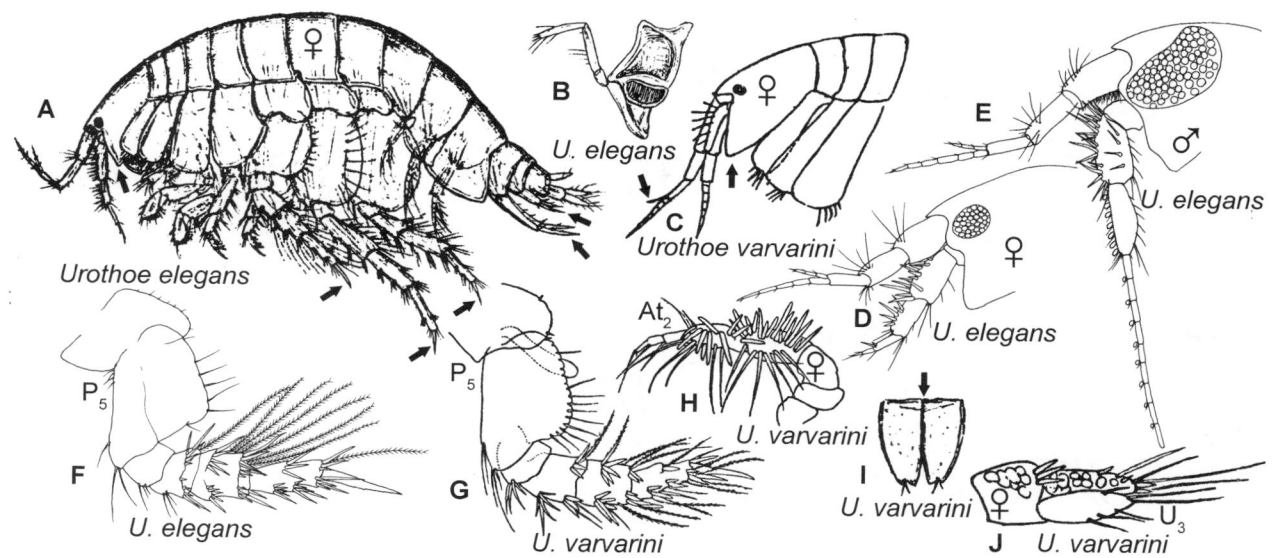

PLATE 288 Urothoidae. A, B, D–F, *Urothoe elegans*; C, G–J, *Urothoe varvarini* (figures modified from Gurjanova 1953; Lincoln 1979; and Sars 1895).

Hippomedon columbianus Jarrett and Bousfield, 1982. Soft benthos, epimeral notch not apparent in specimens <3 mm; 4.8 mm; 4 m–320 m.

Lepidepecreum gurjanovae Hurley, 1963. Sex of illlustrated specimen not given. Three forms occur (1) in Carmel and Goleta, 0 m–3 m, (2) in southern California shelf, 15 m–135 m, and (3) from southern California to British Columbia, being the typical 3-mm form described by Hurley, 1963 (see Barnard 1969a: 175); intertidal–1,720 m; tiny (to 3 mm).

Lepidepecreum serraculum Dalkey, 1998. A shallow-water species of the *L. gurjanovae*—complex, fine sandy silt to coarse red sand off open coasts; also in harbors; 3 mm at Mexico-United States border ranging to 6 mm in Canada; intertidal—150 m.

Macronassa macromera (Shoemaker, 1916) (=*Lysianassa macromera*). Abundant in high-energy intertidal environments among *Egregia* holdfasts and *Anthopleura elegantissima*; 5 mm.

Macronassa pariter (Barnard, 1969a) (=*Lysianassa pariter*). Among sponges and tunicates, Cayucos and south; 5.7 mm; intertidal.

Ocosingo borlus Barnard, 1964c (=*Fresnillo fimbriatus* [Barnard, 1969]). A sequential hermaphrodite; the secondary phase male was renamed *F. fimbriatus* (see Lowry and Stoddart 1983, 1986); 2 mm; intertidal—180 m.

Orchomene minutus (Kroyer, 1846). A boreal species found south to Oregon and south to the Gulf of St. Lawrence in the Atlantic; 11 mm; intertidal—547 m.

Orchomene pacifica (Gurjanova, 1938). Japan Sea, coastal shelf of southern California; 5 mm; 3 m–421 m.

Orchomenella recondita (Stasek, 1958) (=*Allogaussia recondita*; =*Orchomene recondita*). Commensal in the gut of the sea anemone *Anthopleura elegantissima,* Oregon to Santa Cruz Island; intertidal; with global warming, should be watched for north of Oregon; 4 mm. See De Broyer and Vader 1990; Beaufortia 41: 31–38 (biology).

Prachynella lodo Barnard, 1964b. A southern species found as far north as Monterey Bay; also reported from Sea of Japan; 5.8 mm; 10 m–439 m.

Psammonyx longimerus Jarrett and Bousfield, 1982. Sandy sediments; 14 mm; intertidal—200 m.

Wecomedon wecomus (Barnard, 1971). Soft sandy sediments; 13 mm; intertidal—100 m.

OPISIDAE

Opisa tridentata Hurley, 1963. Fish gill parasite; 8 mm; 17 m–183 m.

URISTIDAE

Anonyx cf. *lilljeborgi* Boeck, 1871b. Soft sediments; another boreal species assumed to occur from the Gulf of Alaska to Mexico on our coast (southern populations should be re-examined), and from Nova Scotia to Delaware in the Atlantic, but perhaps representing a species complex; 11 mm; intertidal—1,015 m.

Anonyx cf. *nugax* (Phipps, 1774). Panboreal, south to California, perhaps representing a species complex but also perhaps misreported; up to 42 mm; 4 m–1,184 m.

UROTHOIDAE

Urothoidae of our region are probably undescribed but have been variously assigned to *Urothoe varvarini* and *U. elegans*. Urothoids live in fine subtidal sediments over a large depth range and are likely meiofaunal predators. Although reported mostly from deep water, this obscure, low-density group could be overlooked in shallow marine benthic habitats. The description and illustrations here are composites from previous reports in which specimens from the region were compared with Sars' (1895) and Lincoln's (1979) illustrations of *U. elegans* and with Gurjanova's (1953) illustrations of *U. varvarini*.

KEY TO UROTHOIDAE

1. Large dactyls on pereopods 5–7; pereopod 7 article 2 oval (plate 288A); prominent mandibular palp and molar (plate 288B); weak rostrum, broad ventral extensions of the head and prominent accessory flagellum (plate 288A, 288C–D); small subchelate gnathopods (plate 257L); spinose fossorial pereopods 5 (plate 288F, 288G) and

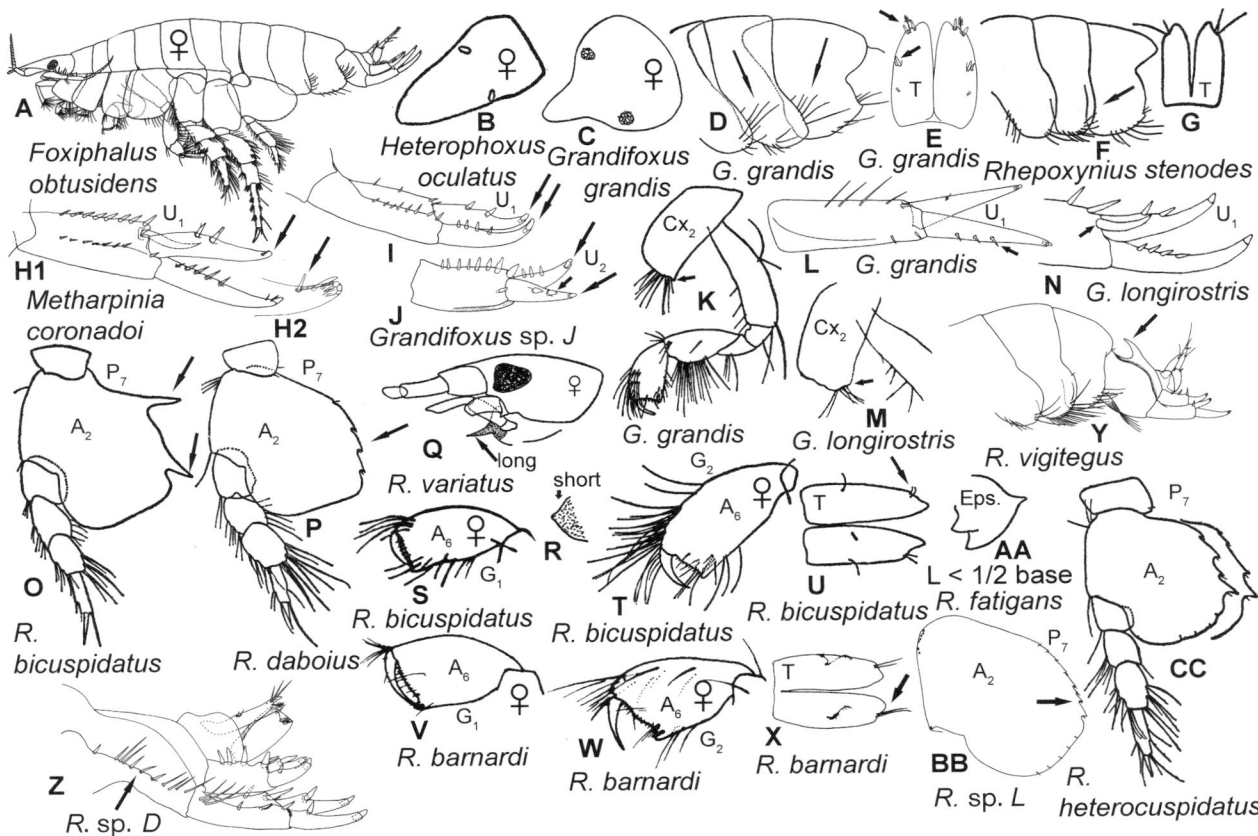

PLATE 289 Phoxocephalidae. A, *Foxiphalus obtusidens*; C–E, K, L, *Grandifoxus grandis*; M, N, *Grandifoxus longirostris*; I, J, *Grandifoxus* sp.; B, *Heterophoxus oculatus*; H, *Metharpinia coronodoi* (expanded tip of outer ramus); AA, *Rhepoxynius fatigans*; V–X, *Rhepoxynius barnardi*; O, R–U, *Rhepoxynius bicuspidatus*; P, *Rhepoxynius daboius*; CC, *Rhepoxynius heterocuspidatus*; Z, *Rhepoxynius* sp. D; BB, *Rhepoxynius* sp. L; F, G, *Rhepoxynius stenodes*; Q, *Rhepoxynius variatus*; Y, *Rhepoxynius vigitegus* (figures modified from Barnard 1960a, 1971, 1980a, 1980b; Barnard and Barnard 1982a;, Coyle 1982; Gurjanova 1938; and Jarrett and Bousfield 1994a).

antenna 2 (plate 288H); straight rami of uropods 1 and 2 (plate 288A); laminar telson cleft to the base (plate 288I), and biramous uropod 3 (plate 288J) . . . . *Urothoe* sp.

LIST OF SPECIES

Urothoe elegans Bate 1857. Cited in the eastern Pacific but not clearly present; mud benthos; 6 mm; shallow subtidal—shelf depths.

Urothoe varvarini Gurjanova 1953. Rare in mud benthic samples; 5 mm; 5 m–1,292 m.

PHOXOCEPHALIDAE

Phoxocephalidae, "spiny heads," are the most diverse and abundant sand- and mud-burrowing marine crustaceans of the 1 mm–10 mm range in coastal soft bottoms after ostracodes (Barnard 1960a). Phoxocephalidae variously resemble Dogielonotidae, Haustoriidae, Urothoidae, Gammaridae, and Pontoporeiidae, but are distinguished readily from these taxa by their shieldlike pointed rostrums. Sexual dimorphism occurs in eye development, uropod 3, gnathopods, and antennae. Phoxocephalid taxonomy is reliable only for females and rests on

* = Not in key.

untested assumptions of the invariance of characters. Phoxocephalidae are predators of meiofauna and invertebrate larvae (Oliver et al. 1982, Mar. Ecol. Prog. Ser. 7: 179–184; Oakden 1984, J. Crust. Biol. 4: 233–247). They live a year or more (Kemp et al., 1985, J. Crust. Biol. 5: 449–464) and are used extensively in aquatic toxicology due to their great sensitivity to pollutants (Robinson et al. 1988, Environ. Tox. Chem. 7: 953–959).

KEY TO PHOXOCEPALIDAE

1. Rostrum unconstricted, lateral edges straight or slightly convex (plate 289A, 289B) 22
   Rostrum constricted, lateral edges concave (plate 289C) 2
2. Epimera 1 and 2 posterior edges with numerous long setae (plate 289D); telson with distal and medial dorsal spines (plate 289E) 3
— Epimera 1 and 2 without posterior setae (plate 289F); telson with only distal spines (plate 289G) 6
3. Uropods 1 and 2 with tiny, subapical supernumerary spines on one or more rami and with subapical spines poorly developed (plate 289H); examine this character under a minimum of 50x magnification . . . . *Metharpinia* spp.
— Uropods 1 and 2 without tiny, subapical supernumerary spines on any rami and subapical ramal spines well developed (plate 289I, 289J) 4

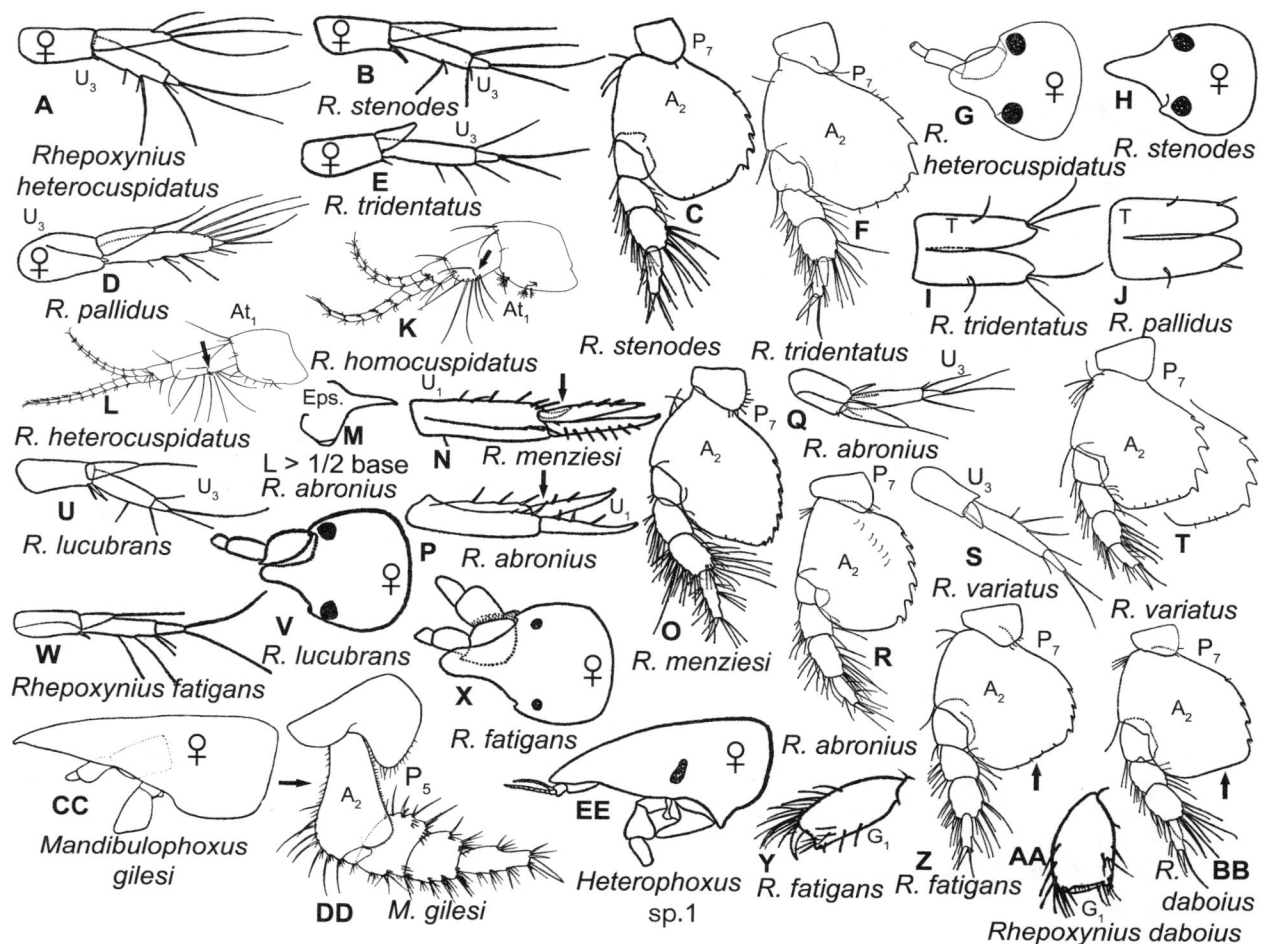

PLATE 290 Phoxocephalidae. EE, *Heterophoxus* sp. 1; CC, DD, *Mandibulophoxus gilesi*; M, P, R, *Rhepoxynius abronius*; AA, BB, *Rhepoxynius daboius*; W–Z, *Rhepoxynius fatigans*; A, G, L, *Rhepoxynius heterocuspidatus*; K, *Rhepoxynius homocuspidatus*; U, V, *Rhepoxynius lucubrans*; N, M, *Rhepoxynius menziesi*; D, J, *Rhepoxynius pallidus*; B, C, H, *Rhepoxynius stenodes*; E, F, I, *Rhepoxynius tridentatus*; S, T, *Rhepoxynius variatus* (figures modified from Barnard 1954a, 1957c, 1960a; and Barnard and Barnard 1982a).

4. Coxae 1–3 with distinct posterioventral tooth (plate 289K); uropod 1 inner distal peduncle lacking large displaced spine (plate 289L) 5

— Coxae 1–3 posteriorly rounded (plate 289M); uropod 3 with large distal medial spine (plate 289N) ....
.... *Grandifoxus longirostris*

5. Uropods 1 and 2 outer ramus spines small (plate 289L) ....
.... *Grandifoxus grandis*

— Uropods 1 and 2 (U_2 not illustrated) outer ramus spines thick, rhomboid (plate 289J); coxae 1 setae narrowly spread (not illustrated)................. *Grandifoxus* sp. J

6. Pereopod 7 article 2 posterior edge with two prominent spurs (plate 289O)............................... 7

— Pereopod 7 article 2 posterior edge with three or more large spurs (plate 289P) 9

7. Epistome prominent and pointed (plate 289Q) (*R.* sp. A not illustrated but similar to *R. variatus*).................
.... *Rhepoxynius* sp. A

— Epistome blunt, inconspicuous (plate 289R) 8

8. Female gnathopods 1 and 2 article 6 anterior and posterior edges parallel at distal end (plate 289S, 289T); telson with short distal setae (plate 289U) ....
.... *Rhepoxynius bicuspidatus*

— Female gnathopods 1 and 2 article 6 expanding at distal ends (plate 289V, 289W); telson with long distal setae (plate 289X) .... *Rhepoxynius barnardi*

9. Urosome dorsal surface smooth, without a tooth (plate 289A)... 10

— Urosome dorsal surface bearing conspicuous anteriorly reverting tooth (plate 289Y)......... *Rhepoxynius vigitegus*

10. Urosome with lateral spine row (plate 289Z)...........
.... *Rhepoxynius* sp. D

— Urosome bare, without lateral spine row (plate 289A).....
.... 11

11. Epistome rounded, lacking anterior cusp............. 12

— Epistome anterior cusp pointed (long or short) (plate 289AA).. 17

12. Pereopod 7, article 2 posterior edge with six or more small teeth (plate 289BB) 16

— Pereopod 7, article 2 posterior with five or less prominent teeth (plate 289CC)................................ 13

13. Female uropod 3 inner ramus more than two-thirds length of outer ramus (plate 290A, 290B); pereopod 7 article 2 posterior edge with more than three teeth (plates 289CC, 290C).. 14

— Female uropod 3 inner ramus length one-half or less of outer ramus (plate 290D, 290E); pereopod 7 article 2 with three large posterior teeth (plate 290F) 15

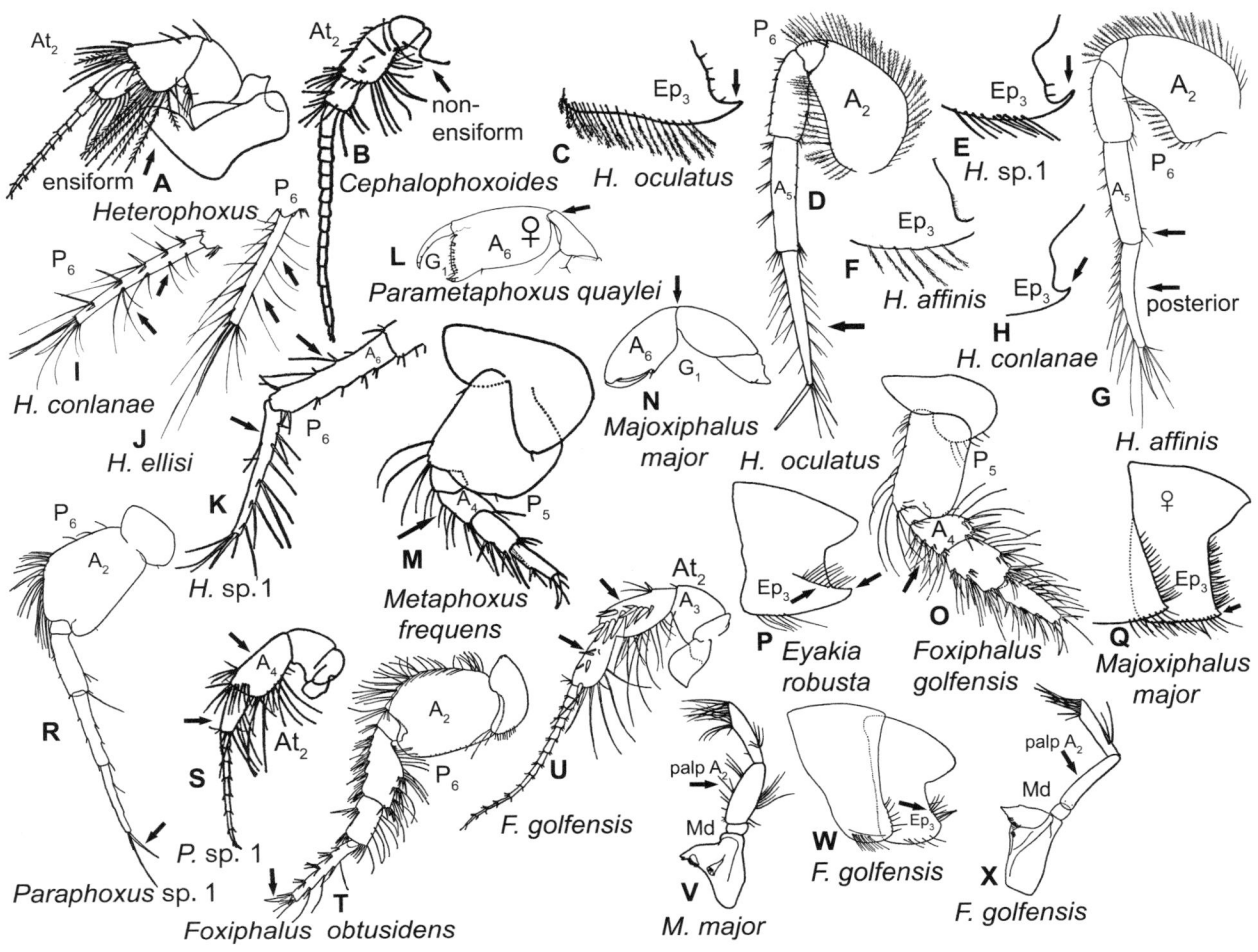

PLATE 291 Phoxocephalidae. B, *Cephalophoxoides homilis*; P, *Eyakia robusta*; O, U, W, X, *Foxiphalus golfensis*; T, *Foxiphalus obtusidens*; A, *Heterophoxus*; F, G, *Heterophoxus affinis*; H, I, *Heterophoxus conlanae*; J, *Heterophoxus ellisi*; C, D, *Heterophoxus oculatus* (original drawing incomplete); E, K, *Heterophoxus* sp. 1; N, Q, V, *Majoxiphalus major*; M, *Metaphoxus frequens*; L, *Parametaphoxus quaylei*; R, S, *Paraphoxus* sp. 1 (figures modified from Barnard 1960a; Holmes 1908; and Jarrett and Bousfield 1994b).

14. Pereopod 7 article 2 with four to five asymmetrical posterior serrations on article 2 (plate 289CC); female rostrum broad, with base greater than one-half of the head width (plate 290G) . . . . *Rhepoxynius heterocuspidatus*
— Pereopod 7 article 2 with four symmetrical posterior teeth (plate 290C); rostrum narrow, with base less than one-third of the head width (plate 290H) . . . . . *Rhepoxynius stenodes*

15. Female uropod 3 inner ramus length one-third of outer ramus (plate 290E); male telson medial setae long (plate 290I) . . . . *Rhepoxynius tridentatus*
— Female uropod 3 inner ramus length one-half of outer ramus (plate 290D); male telson medial setae short (plate 290J) . . . . *Rhepoxynius pallidus*

16. Antenna 1, article 2 lateral marginal setae extending to apex (plate 290K) . . . . *Rhepoxynius homocuspidatus*
— Antenna 1, article 2 lateral marginal setae extending less than one-third of the distance to the apex (plate 290L) (illustrations of R. sp. L not published; R. heterocuspidatus is a similar example) . . . . *Rhepoxynius* sp. L

17. Epistome cusp long, extending >1.5 times the width of its base (plate 290M) 18
— Epistome cusp short, extending approximately equal to width of its base (plate 289AA) 20

18. Female uropod 1 peduncle with large displaced dorsomedial spine (plate 290N); pereopod 7 article 2 with eight small posterior teeth (plate 290O) . . . . *Rhepoxynius menziesi*
— Female uropod 1 peduncle without large displaced dorsomedial spine (plate 290P); pereopod 7 article 2 with less than eight posterior teeth 19

19. Female uropod 3 inner ramus length greater than one-half of the outer ramus (plate 290Q); pereopod 7 article 2 with five to six posterior teeth (plate 290R) . . . .
. . . . *Rhepoxynius abronius*
— Female uropod 3 inner ramus length less than one-half of the outer ramus (plate 290S); pereopod 7 article 2 with three to five (usually four) variably sized teeth lining posterior edge (plate 290T) . . . . *Rhepoxynius variatus*

20. Female uropod 3 inner ramus length about one-half of the outer ramus length (plate 290U); eyes large (plate 290V) . . . . *Rhepoxynius lucubrans*
— Female uropod 3 inner ramus length less than one-third of the outer ramus (plate 290W); eyes small (plate 290X) 21

21. Gnathopod 1 article 6 narrow (plate 290Y); pereopod 7 article 2 ventral edge slightly beveled, not straight (plate 290Z) . . . . *Rhepoxynius fatigans*

— Gnathopod 1 article 6 broad (plate 290AA); pereopod 7 article 2 ventral edge straight (plate 290BB) . . . . *Rhepoxynius daboius*

22. Pigmented eyes absent (plate 290CC) (check also *Heterophoxus oculatus* if antenna 2 ensiform process is longer than wide); pereopod 5 article 2 slightly concave and expanding distally (plate 290DD) . . . . *Mandibulophoxus gilesi*

— Pigmented eyes present (plate 290EE); pereopod 5 article 2 convex and parallel-sided or slightly narrowing distally 23

23. Antenna 2 with ensiform process longer than wide (plate 291A) 24

— Antenna 2 ensiform process absent or wider than long (plate 291B) 28

24. Epimeron 3 tooth weakly upturned or straight (plate 291C); pereopod 6, article 6 with posterior setae (plate 291D) 25

— Epimeron 3 tooth strongly upturned (plate 291E, 291F); pereopod 6, article 6 without posterior setae (plate 291G) 27

25. Pleonite 3 with more than 14 ventral, plumose setae and posterioventral tooth as thick at the base as length (plate 291C) . . . . *Heterophoxus oculatus*

— Pleonite 3 with less than 14 ventral, plumose setae and with posterioventral tooth longer than thick (plate 291H) 26

26. Pereopod 6 article 6 posterior edge lined with doubly or triply inserted setae (plate 291I); female pereopod 7 with small coxal gill . . . . *Heterophoxus conlanae*

— Pereopod 6 article 6 posterior edge lined with singly inserted setae (plate 291J); female pereopod 7 lacking coxal gill . . . . *Heterophoxus ellisi*

27. Pereopod 6 article 5 posterior margin with three setae pairs plus a single seta and article 6 with one mid-posterior seta (plate 291K) . . . . *Heterophoxus* sp. 1

— Pereopod 6 articles 5 and 6 with posterior setae only on extreme distal ends (plate 291G) . . . . . . *Heterophoxus affinis*

28. Female gnathopod 1 article 5 attachment to article 6 constricted (plate 291L); pereopod 5, article 4 deeper than wide (plate 291M) 40

— Female gnathopod 1 article 5 attachment to article 6 normal, unconstricted (plate 291N); pereopod 5, article 4 wider than deep (plate 291O) 29

29. Epimeron 3 with large posterioventral tooth and an oblique row of facial setae (plate 291P) . . . . *Eyakia robusta*

— Epimeron 3 posterioventrally square or rounded and without an oblique row of facial setae (plate 291Q) 30

30. Pereopod 6 dactyl long and thin (plate 291R); antenna 2, articles 4 and 5 without facial spine clusters (plate 291S) . . . . *Paraphoxus* sp. 1

— Pereopod 6 dactyl shorter and relatively stout (plate 291T); antenna 2, articles 4 and 5 with facial spine clusters (plate 291U) 31

31. Epimeron 3 posterior edge lined with 20 or more setae (plate 291Q); mandibular palp second article swollen (plate 291V) . . . . *Majoxiphalus major*

— Epimeron 3 posterior edge lined with 15 or less setae (plate 291W); mandibular palp article 2 linear with parallel sides (plate 291X) 32

32. Uropod 1 peduncle with large displaced lateral or medial distal spine (plate 292A) (variable in *Eobrolgus*, plate 292L) 34

— Uropod 1 peduncle without large displaced lateral or medial distal spine (plate 292B) (variable in *Eobrolgus*, plate 292L) 33

33. Pereopod 7 article 2 ventral edge slightly crenulated and lined with long setae (plate 292C) . . . . *Foxiphalus golfensis*

— Pereopod 7 article 2 ventral edge smooth and without long setae (plate 292D) . . . . *Foxiphalus falciformis*

34. Epistome produced (plate 292E) 35

— Epistome unproduced (plate 292F, 292G) 36

35. Epistome cusp longer than width of base (plate 292H, 292I) . . . . *Foxiphalus similis*

— Epistome length no greater than width of base (plate 292E) . . . . *Foxiphalus cognatus*

36. Telson lobes with dorsolateral spines (plate 292J) 37

— Telson lobes without dorsolateral spines (plate 292K) 38

37. Uropod 1 inner ramus apical nail flexible, articulate (flex the inner distal nail with a fine needle to make this observation) (plate 292A, right arrow) . . . . *Foxiphalus obtusidens*

— Uropod 1 inner ramus apical nail immersed, rigid (plate 292M) . . . . *Foxiphalus xiximeus*

38. Female rostrum lateral edges straight or slightly concave (plate 292N); two stout distal spines on the inner maxilliped palp (plate 292O) . . . . *Foxiphalus aleuti*

— Female rostrum lateral edges slightly convex (plate 292P); one stout distal spine on the inner maxilliped palp (plate 292Q) 39

39. Epimeron 3 bearing one to two ventral setae (plate 292R); outer plate of maxilla 1 with 11 spines (incompletely illustrated) (plate 292S) . . . . *Eobrolgus chumashi*

— Epimeron 3 bearing a single ventral seta (plate 292T); outer plate of maxilla 1 with nine spines (not illustrated) . . . . *Eobrolgus spinosus*

40. Gnathopods 1 and 2 sixth articles similar in shape and length (plate 292U, 292V); mandibular molar triturative (plate 292W) . . . . *Cephalophoxoides homilis*

— Gnathopod 1 article 6 longer than gnathopod 2 article 6 (plate 292X, 292Y); mandibular molar weak, nontriturative (plate 292Z) 41

41. Gnathopod 1 palm not extending beyond anterior distal corner and posterior article 5 of gnathopods 1 and 2 overlapped by articles 6 and 4 (plate 292X, 292Y) . . . . *Metaphoxus frequens*

— Gnathopod 2 weakly chelate, palm extending beyond anterior distal corner, and posterior lobe of article 5 of gnathopod 1 free of articles 6 and 4 (plate 292AA, 292BB) 42

42. Pereopod 5 coxa extending less than 40% of article 2 length (plate 292CC); pereopod 6 article 2 ventral posterior corner extended and rounded (plate 292CC) . . . . *Parametaphoxus quaylei*

— Pereopod 5 coxa extending more than 50% of article 2 length (plate 292DD); pereopod 6 article 2 posterior ventral corner square (plate 292EE) . . . . *Parametaphoxus* sp.

LIST OF SPECIES

Cephalophoxoides homilis (Barnard 1960a) (=*Phoxocephalus homilis*). Intertidal eelgrass beds (Dean and Jewitt 2001, Ecol. Appl. 11: 1456–1471) and soft benthos; 4.3 mm; intertidal—250 m.

Eobrolgus chumashi Barnard and Barnard, 1981. Marine, estuary, muddy sands; body (plate 256M). *Eobrolgus* are not

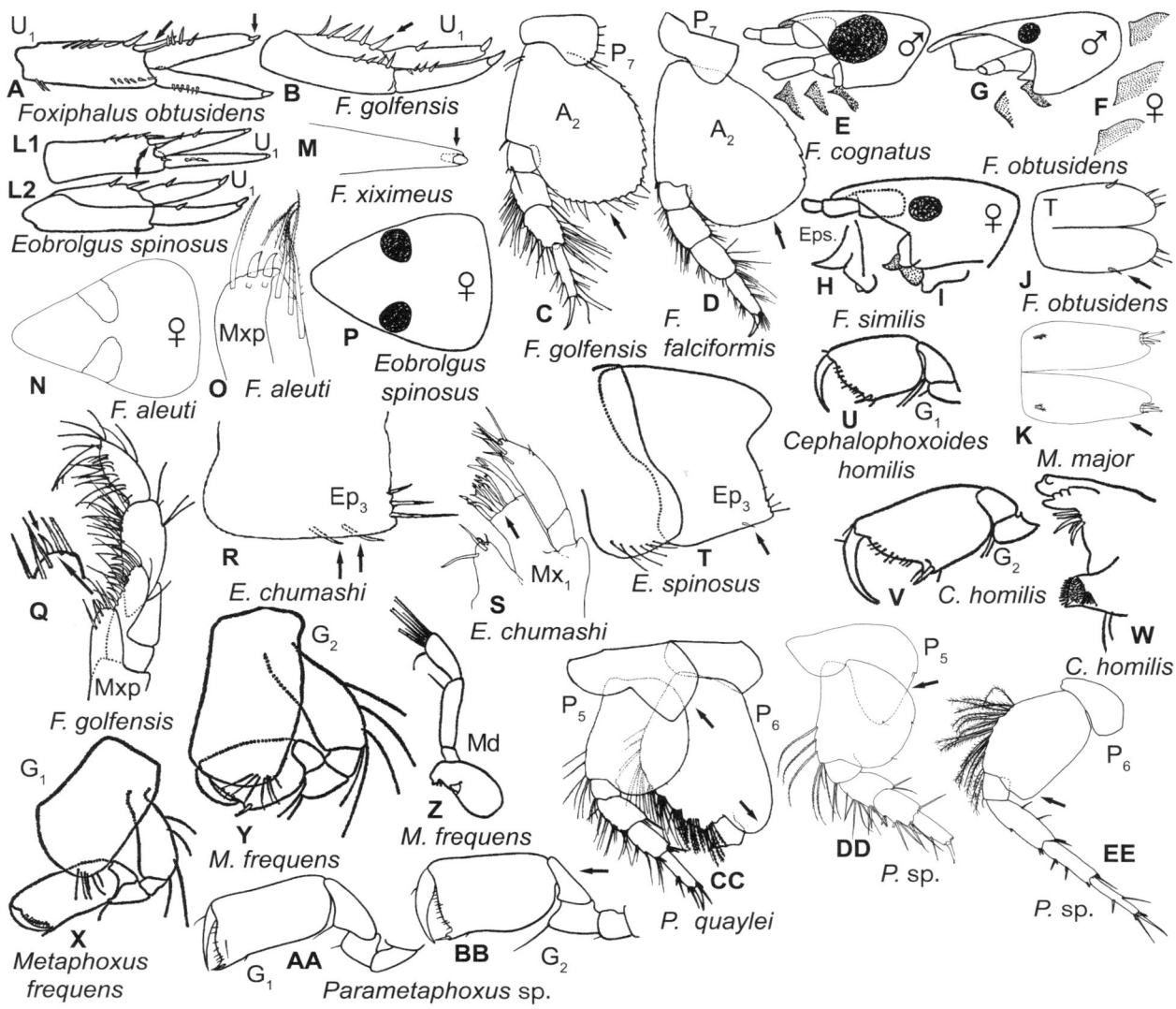

PLATE 292 Phoxocephalidae. U–W, *Cephalophoxoides homilis*; R, S, *Eobrolgus chumashi*; L, P, T, *Eobrolgus spinosus*; N, O, *Foxiphalus aleuti*; E, *Foxiphalus cognatus* (epistomes); D, *Foxiphalus falciformis*; B, C, Q, *Foxiphalus golfensis* (distal inner plate); E, F, G (epistomes); A, F, G, J, *Foxiphalus obtusidens*; H, I, *Foxiphalus similis*; M, *Foxiphalus xiximeus*; K, *Majoxiphalus major*; X–Z, *Metaphoxus frequens*; CC, *Parametaphoxus quaylei*; AA, BB, DD, EE, *Parametaphoxus* sp. (figures modified from Alderman 1936; Barnard 1960a, 1964a; Barnard and Barnard 1982a, 1982b; Chapman, personal communication; and Jarrett and Bousfield 1994a, 1994b).

distinguished morphologically from *Foxiphalus falciformis*; 4.5 mm; intertidal—11 m.

Eobrolgus spinosus (Holmes, 1903). A possible introduction from the Northwest Atlantic; "study on hybridization [with *E. chumashi*] is warranted" (Barnard and Barnard, 1981). Displaced uropod 1 spine on uropod 1 variably present. Estuarine, muddy sand; 4 mm; intertidal.

Eyakia robusta. (Homes, 1908). Associated with brittle stars, occasional surface swimmer of neritic zone; Alaska population is possibly a different species (Jarrett and Bousfield 1994a Amphipacifica 1: 89); 6.5 mm–15 mm; intertidal—320 m.

Foxiphalus aleuti Barnard and Barnard, 1982b. In sand; 9 mm; subtidal—110 m.

Foxiphalus cognatus (Barnard, 1960). In coarse shell and sand; 5 mm; intertidal—324 m.

Foxiphalus falciformis Jarrett and Bousfield, 1994a. Fine marine sands; doubtfully distinguished from *Eobrolgus* and *F. golfensis* by pereopod 7 and minute differences in mandibles; 8 mm; intertidal.

Foxiphalus golfensis Barnard and Barnard, 1982b. Oregon and south; 9.1 mm; intertidal—91 m.

Foxiphalus obtusidens (Alderman, 1936). Common in sand tide pools; 5.5 mm–15 mm; intertidal—210 m.

Foxiphalus similis (Barnard, 1960a). Surf-protected fine sands; 5 mm; sublittoral—324 m.

Foxiphalus xiximeus Barnard and Barnard, 1982. May not be distinct from *F. obtusidens*; medium surf-exposed beaches in sand; 8 mm; low intertidal—20 m.

Grandifoxus grandis (Stimpson, 1856) (=*Paraphoxus milleri*). Often in reduced salinities; 9.5 mm–14 mm; intertidal.

Grandifoxus longirostris (Gurjanova, 1938). In sand, largely subtidal; 8 mm; 10 m–90 m.

Grandifoxus sp. Barnard 1980a. Pacific Grove, from a "senile" incompletely described 14.6 mm male, may not be distinct from *G. grandis*; intertidal sands.

Heterophoxus affinis (Holmes, 1908). In fine sand to mud, the deep-water populations may include *Heterophoxus* sp. 1

of Jarrett and Bousfield 1994b; 9 mm; shallow subtidal to 600+ m.

Heterophoxus conlanae Jarrett and Bousfield, 1994b. Not clearly distinguished from *H. oculatus*; 8 mm; intertidal—40 m.

Heterophoxus ellisi Jarrett and Bousfield 1994b. Fine sands and mud; 7 mm; intertidal—155 m.

Heterophoxus oculatus (Holmes, 1908). In fine sands, eye loss occurs in deeper populations and is not accompanied by other character differences (Cadien 2002, SCAMIT 21[2]: 7); 9 mm; 10 m–120 m.

Heterophoxus sp. 1 Jarrett and Bousfield 1994b. Southern California, not clearly distinguished from *H. affinis*; fine sediments; 7 mm; 90 m–360 m.

Majoxiphalus major (Barnard, 1960a). *Majoxiphalus* is poorly distinguished from *Foxiphalus*; differences may be size- or age-related; 6.5 mm–17.5 mm; intertidal—91 m.

Mandibulophoxus gilesi Barnard, 1957c. Fine sands; 6 mm; shallow subtidal—14 m.

Metaphoxus frequens Barnard, 1960a. Fine sands and muddy sand; 3.5 mm; intertidal—496 m.

Metharpinia coronadoi Barnard, 1980a. Southern California, posterioventral corner of pleonite 3 produced into a large hook, muddy sand; 7 mm; 18 m–43 m.

Metharpinia jonesi (Barnard, 1963). Southern, pleonite 3 without posterioventral hook; 3.8 mm; intertidal—18 m.

Parametaphoxus sp. Chapman (undescribed). Southern California and south; 3.5 mm; intertidal—170 m.

Parametaphoxus quaylei Jarrett and Bousfield 1994b. Fine sand and mud, Washington and north; 2.8 mm; 25 m–100 m. *Parametaphoxus* sp. Chapman and *P. quaylei* may occur in our region.

Paraphoxus spp. Barnard 1960a. Shallow water *Paraphoxus* reported north and south of the region in mixed sediments and mud but not confirmed in the region (Barnard 1979b); 3 mm–5 mm; shallow subtidal to 2,800 m.

Rhepoxynius abronius (Barnard, 1960a). Abundant inshore and subtidally at the high salinity mouths of estuaries, mostly in surf-protected localities, in sand to below 50 m. An important species for toxicity bioassays (Ambrose 1984, J. Exp. Mar. Biol. Ecol. 80: 67–75 (behavior); DeWitt et al. 1988, Mar. Envir. Res. 25: 99–124 (sediment features, toxicity); Swartz 1986, Mar. Envir. Res. 18: 133–153 (toxicity); 5.5 mm; shallow subtidal—90 m.

Rhepoxynius barnardi Jarrett and Bousfield 1994a. Sand habitats, a possible synonym of *R. bicuspidatus*; 4 mm; intertidal—59 m.

Rhepoxynius bicuspidatus (Barnard, 1960a). Fine sand and sandy mud, a low proportion of specimens of this species has three spurs on article 2 of one or both seventh pereopods; 4.5 mm; 8 m–475 m.

Rhepoxynius boreovariatus Jarrett and Bousfield 1994a. Northern; 4.5 mm; intertidal—40 m.

Rhepoxynius daboius (Barnard, 1960a). Sandy mud, a probable synonym of *R. fatigans*; 4 mm; intertidal—813 m.

Rhepoxynius fatigans (Barnard, 1960a). Sandy mud; 4 mm; intertidal—330 m.

Rhepoxynius heterocuspidatus (Barnard, 1960a). 4.8 mm, intertidal—146 m. This species, *Rhrepoxynius* sp. C, *R. stenodes* and the following three species occur south of this region.

Rhepoxynius homocuspidatus (Barnard and Barnard, 1982a). 3.5 mm; intertidal—64 m.

Rhepoxynius lucubrans (Barnard, 1960a). 5.3 mm; intertidal—91 m.

Rhepoxynius menziesi (Barnard and Barnard, 1982a). 7 mm; intertidal—22 m.

Rhepoxynius pallidus (Barnard, 1960). British Columbia and Washington; possible synonym of *R. tridentatus* ; 6 mm; intertidal—40 m.

Rhepoxynius sp. A SCAMIT, 1987. Sand benthos; length not known; <20 m.

Rhepoxynius sp. C Barnard and Barnard 1982a. Sand; 4.3 mm; intertidal—15 m.

Rhepoxynius sp. D Barnard and Barnard 1982a. Southern California, a possible morph of *R. menziesi*; 8 mm; intertidal—27 m.

Rhepoxynius sp. L Barnard and Barnard 1982a. Dillon Beach, fine sand; epistome is assumed to be rounded since it combines "characters of both *R. heterocuspidatus* and *R. homocuspidatus*" (Barnard and Barnard 1982a); a lack of illustrations of the epistome and antenna 1 leave the placement of this species uncertain; 5.7 mm; intertidal—2 m.

Rhepoxynius stenodes (Barnard, 1960a). Muddy sand; 3.5 mm; 2 m–374 m.

Rhepoxynius tridentatus (Barnard, 1954a). Mud and sand; 5 mm, intertidal—89 m.

Rhepoxynius variatus (Barnard, 1960a). Muddy sands; number and relative sizes of teeth on posterior pereopod 7 are variable; 5 mm; intertidal—89 m.

Rhepoxynius vigitegus (Barnard,1971). Sandy mud; 4.5 mm; shallow subtidal to 30 m.

PONTOPOREIIDAE

Pontoporeiidae are represented in the eastern Pacific by *Diporeia erythrophthalma* and *Monoporeia* sp. (see species list below). The figures are of *Monoporeia affinis* for identifying the genus. American pontoporeiids were long assumed to be "glacial marine relicts" dispersed over North America and Eurasia during the Pleistocene deglaciation by marine inundations of coastal regions that trapped brackish water species in freshening ponds and then forced them inland (e.g., Segerstråle 1976, Dadswell 1974). However molecular (Väinölä and Varvio 1989) and morphological data (Bousfield 1989) indicate that speciation among these "relicts" has occurred, a pattern expected from long isolation among distant populations rather than recent arrivals. Pontoporeiidae mate pelagically (Bousfield 1989), and male antennae can be twice as long as their bodies. Pontoporeiidae differ from Gammaridae by lacking pereopod 7 coxal gills, from Phoxocephalidae by lacking a rostrum, and from Urothoidae by dissimilar pereopods 6 and 7.

KEY TO PONTOPOREIIDAE

1. Urosome evenly rounded, lateral head lobe slightly acute, eyes dark, prominent accessory flagellum (plate 293A); gnathopod 1 article 5 posterior edge length greater than half the anterior article length (plate 293B); gnathopod 2 article 6 narrowest distally and with pinnate distal setae (plate 293C); coxa 5 lobes ventral projection equal (plate 293D); sternal gills on pereonites 2–5 (arrows) (plate 293A, 293E); telson as wide as long cleft two-thirds of its length (plate 293F) and; uropod 3 outer ramus with a tiny distal nail (plate 293G) . . . . *Monoporeia* sp.

* = Not in key.

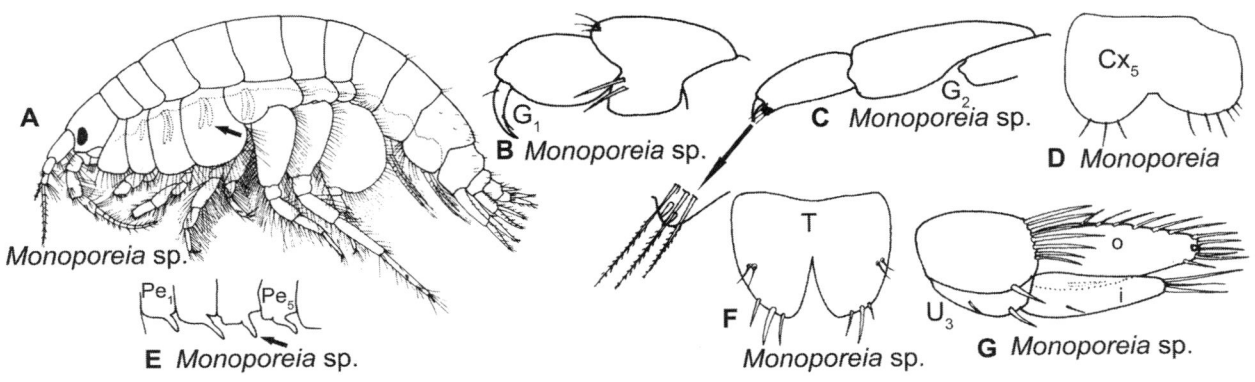

PLATE 293 Pontoporeiidae. A–G, *Monoporeia* sp. (figures modified from Bousfield 1989).

LIST OF SPECIES

*Diporeia erythrophthalma (Waldron 1953). Named for its red eyes, known only from freshwater Lake Washington and the only other pontoporeiid of the eastern Pacific; 5 mm; 0 m–50 m.

Monoporeia sp. Restricted to the low-salinity benthos of the lower Columbia River where it is common, and from a single male collected in August 2004 from low-salinity benthos of Yaquina Bay, Oregon. Cryptogenic (historical occurrence in the region unclear, but unknown elsewhere); 8 mm; intertidal—20 m.

PLEUSTIDAE

Pleustidae are commensals, egg predators, and microparasites of other invertebrates and are common in fouling communities. The left mandible morphology is used for taxonomy because the right lacinia mobilis is greatly reduced or missing in many species. *Thorlaksonius* may be Batesian mimics of snails, while the bright colors of *Chromopleustes* may be for warning or Mullerian mimicry. Males have relatively larger gnathopods and smaller, narrower bodies, but sexual dimorphism is weak. The distinctive and beautiful pigmentation of pleustids is lost in preservation. The loss of pigment and the emphasis placed on the left mandible morphology to define species increases the difficulty of distinguishing pleustids of the region. Genera and species erected without complete notes on the presence or absence of the mandibular molar are doubtful; in particular, all characters proposed to distinguish *Gnathopleustes*, *Incisocalliope*, and *Trachypleustes* vary uniformly among the taxa or with size and thus do not yet reveal significant differences.

KEY TO PLEUSTIDAE

1. Gnathopods 1 and 2 article 5 distally truncate and broadly attached to article 6 (plate 294A, 294B); pereopods 3 and 4 dactyls more than one-third of the length of article 6 and simple (plate 294A) or short and notched (plate 294C) . . . 2
— Gnathopods 1 and 2 article 5 distally produced and narrowly attached to article 6, eusiridlike and pereopod 3 and 4 dactyls less than one-third of the length of article 6 and simple (plate 294D) . . . . *Pleusirus secorrus*
2. Mandibular molar reduced, nontriturative (plate 294E), antenna 1 article 1 without anterior projections (plate 294A) 3
— Mandibular molar fully developed, triturative (plate 294F);

and antenna 1 article 1 with anterior ventral or dorsal projections (except for *Heteropleustes setosus*) (plate 294G). 18
3. Rostrum massive, extending beyond antenna 1 peduncle article 1, dorsal pleonites1–3 weakly or strongly carinate (plate 294A) 4
— Rostrum moderate or indistinct, extending less than length of antenna 1 peduncle article 1 and dorsal pereonites 1–4 smooth, not carinate or ridged (plate 294D) 8
4. Rostrum apex strongly deflexed (80°–90°) with nearly straight lower margin (plate 294A). 5
— Rostrum apex at <80° angle to dorsum, lower margin convex (plate 294H) 7
5. Coxa 7 not laterally ridged or sharply pointed posteriorly (plate 294A) . . . . *Thorlaksonius brevirostris*
— Coxa 7 laterally ridged or sharply pointed posteriorly (plate 294I). 6
6. Coxa 7 posteriorly pointed but not laterally ridged (plate 294I) . . . . *Thorlaksonius depressus*
— Coxa 7 posteriorly pointed bluntly and laterally ridged (plate 294J) . . . . *Thorlaksonius subcarinatus*
7. Coxa 5 with large lateral ridge and pointed behind (plate 294H). . . . . *Thorlaksonius borealis*
— Coxa 5 with reduced lateral ridge and obtuse behind (plate 294K) . . . . *Thorlaksonius grandirostris*
8. Antenna 1 and 2 length less than one-third of the total body length; gnathopods 1 and 2 article 6 palms shorter than posterior margin (plate 294L) 9
— Antennae 1 and 2 and more than one-third of the total body length (not shown); gnathopods 1 and 2 palms equal to or longer than article 6 posterior margin (plate 294M) 11
9. Pereopods 3–7 dactyls less than one-third the length of article 6 (plate 294N) and distally notched (plate 294C), coxa 1 smaller than coxa 2 (plate 294N) . . . . *Dactylopleustes echinoides*
— Pereopods 3–7 dactyls more than one-third of the length of article 6 and unnotched; coxa 1 approximately equal to coxa 2 (plate 294L) 10
10. Gnathopods 1 and 2 article 6 width about equal to of posterior margin length (plate 294L). . . . . *Micropleustes nautilus*
— Gnathopods 1 and 2 article 6 width about 60% of posterior margin length (plate 294O) . . . . . . *Micropleustes nautiloides*
11. Pereopod 4 article 6 swollen and lined posteriorly with large stout spines (plate 294P) . . . . *Commensipleustes commensalis*

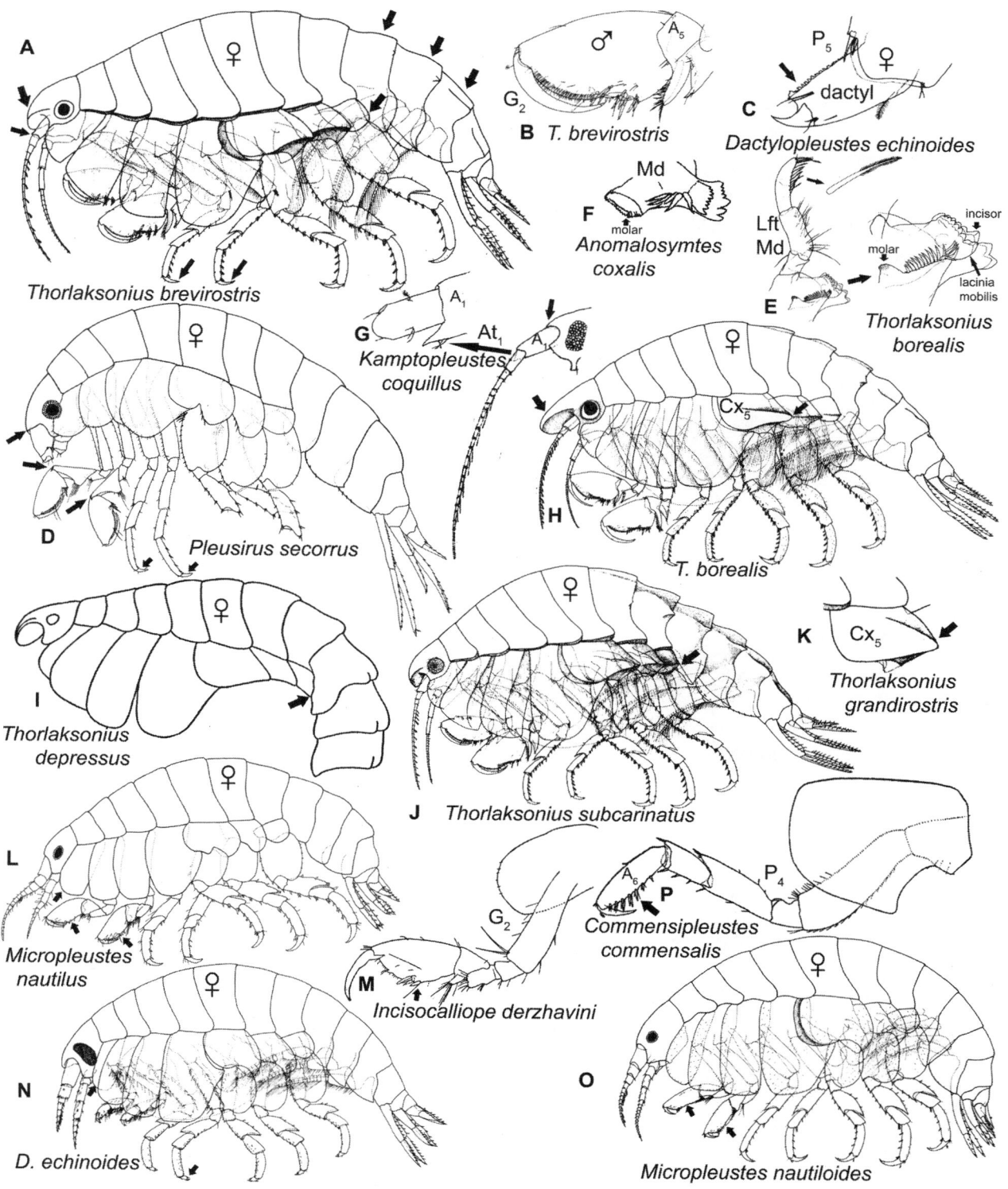

PLATE 294 Pleustidae. F, *Anomalosymtes coxalis*; P, *Commensipleustes commensalis*; C, N, *Dactylopleustes echinoides*; M, *Incisocalliope derzhavini*; G, *Kamptopleustes coquillus*; O, *Micropleustes nautiloides*; L, *Micropleustes nautilus*; D, *Pleusirus secorrus*; E, H, *Thorlaksonius borealis*; A, B, *Thorlaksonius brevirostris*; I, *Thorlaksonius depressus*; K, *Thorlaksonius grandirostris*; J, *Thorlaksonius subcarinatus* (figures modified from Alderman 1936; Barnard 1969a; Bousfield and Hendrycks 1994b, 1995b; Hendrycks and Bousfield 2004; and Shoemaker 1952).

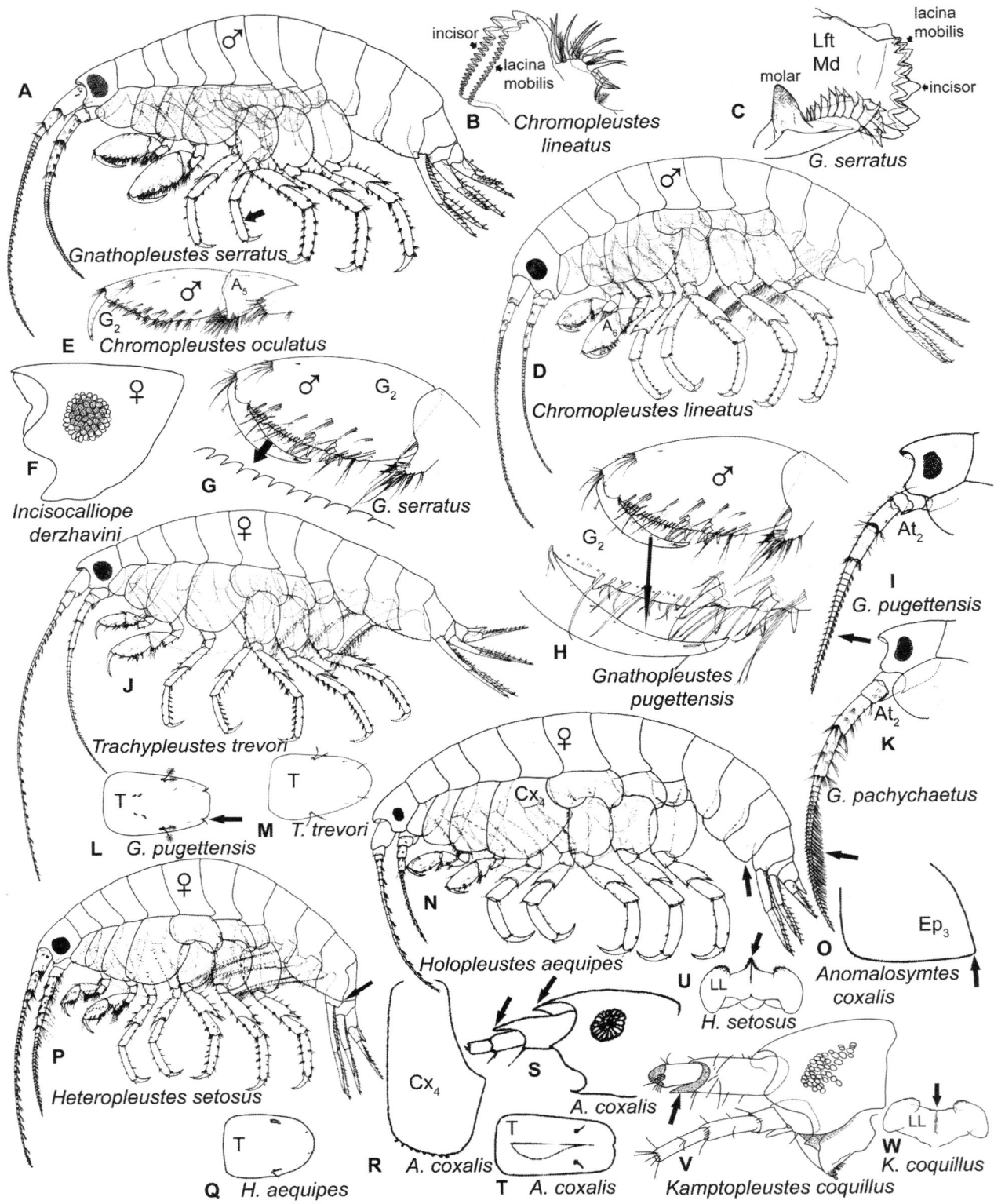

PLATE 295 Pleustidae. O, R–T, *Anomalosymtes coxalis*; B, D, *Chromopleustes lineatus*; E, *Chromopleustes oculatus*; K, *Gnathopleustes pachychaetus*; H, I, L, *Gnathopleustes pugettensis*; A, C, G, *Gnathopleustes serratus*; P, U, *Heteropleustes setosus*; N, Q, *Holopleustes aequipes*; F, *Incisocalliope derzhavini*; V, W, *Kamptopleustes coquillus*; J, M, *Trachypleustes trevori* (figures modified from Barnard 1971; Chapman 1988; Bousfield and Hendrycks 1995b; and Hendrycks and Bousfield 2004).

— Pereopod 4 article 6 of uniform width and lined posteriorly with long setae or short spines but not stout large stout spines (plate 295A)........................12

12. Lacinia mobilis of left mandible with 17–50 teeth (plate 295B), live specimens brilliantly pigmented 13
— Lacinia mobilis of left mandible with 10 or less teeth (plate 295C), live pigmentation unknown...............14

13. Male gnathopod 2 article 5 less than half the length of article 6 (plate 295D)............ *Chromopleustes lineatus*
— Male and female gnathopod 2 article 5 more than half as long as article 6 (plate 295E)..........*Chromopleustes oculatus*

14. Anterior edge of eye convex (oval), protected bays and estuaries (plate 295F) ....*Incisocalliope derzhavini*
— Anterior edge of eye concave or flat (bean-shaped), marine (plate 295A)...................................15

15. Gnathopod dactyls serrate (plate 295G).............
...........................*Gnathopleustes serratus*
— Gnathopod dactyls smooth (plate 295H)...........16

16. Antenna 2 flagellum sparsely setose (plate 295I, 295J).....
....17
— Antenna 2 flagellum densely setose (plate 295K) ....
............................*Gnathopleustes pachychaetus*

17. Telson posterior edge truncate and with lateral setae set into tiny notches (plate 295L) ....
........................*Gnathopleustes pugettensis*
— Telson posterior evenly rounded and without distal notches (plate 295M) ....*Trachypleustes trevori*

18. Pleon epimera 3 posterior ventral corner blunt or rounded, not produced (plate 295N, 295O) 19
— Epimera 3 posterior ventral corner sharply acute (plate 295P)...20

19. Coxa 4 nearly as wide as long, ocular lobe rounded or obtuse, and anterior of antenna 1 article 1 blunt or bluntly produced (plate 295N); telson distally rounded (plate 295Q) ....*Holopleustes aequipes*
— Coxa 4 nearly twice as deep as long (plate 295R), antenna 1 peduncle dorsal anterior article 1 and ocular lobe both sharply produced (plate 295S); telson distally blunt or notched (plate 295T) (note ventral keel) ....
.......................*Anomalosymtes coxalis*

20. Antenna 1 article 1 bearing an acute dorsal projection and ocular lobe evenly rounded (plate 295P); lower lip with acute medial process (plate 295U) ....*Heteropleustes setosus*
— Antenna 1 article 1 bearing an acute ventral projection (plate 295V); ocular lobes acute (plate 257W, 295V); medial junction of lower lip flat (plate 295W)...........
.......................*Kamptopleustes coquillus*

LIST OF SPECIES

Anomalosymtes coxalis Hendrycks and Bousfield 2004. Natural history and ecology unknown. Lack of a ventral antenna 2 sinus in common with Pleustidae, but the lower lip lacks pillow-shaped inner lobes and resembles Eusiroidea. Mandibular palp present and molar triturative; 3 mm; shallow subtidal—25 m.

Chromopleustes (=*Parapleustes*) *lineatus* (Bousfield, 1985). A commensal and possible egg predator of echinoderms in rocky habitats. Four to five bright yellow and brown longitudinal body stripes (Bousfield, 1985, Rotunda 18: 30–36); 9 mm; shallow subtidal—17 m.

Chromopleustes (=*Parapleustes*) *oculatus* (Holmes, 1908) (=*Parapleustes oculatus*). Predator of the sea cucumber *Cucumaria miniata* (Chen and Norton, 2005, Abstracts, Estuarine

Research Federation Annual Meeting, Norfolk, VA), and also associated with the brittle star *Amphiodia urtica* (Barnard and Given, 1960, Pac. Nat. 1: 46). Not clearly distinguished from *Heteropleustes setosus* or *Pleusymptes pacifica*; 11 mm; intertidal—2 m or more.

Commensipleustes (=*Parapleustes*) *commensalis* (Shoemaker, 1952). Commensal and possible lobster egg predator. Bousfield and Hendrycks (1995a) give northern records on sponges, indicating plasticity in the species or taxonomic complications; 5.5 mm; intertidal—50 m.

Dactylopleustes echinoides Bousfield and Hendrycks 1995b. Commensal or egg predator of sea urchins; 3.3 mm; intertidal—2 m.

Gnathopleustes pachychaetus Bousfield and Hendrycks, 1995b. Rocky intertidal (to 2 m) among algae; 7 mm.

Gnathopleustes pugettensis (Dana, 1853) (=*Parapleustes pugettensis*). Rocky and soft benthos. See also *Trachypleustes trevori*; 6 mm; intertidal—140 m.

Gnathopleustes serratus Bousfield and Hendrycks 1995b. Under intertidal boulders, associated with sessile invertebrates; 10 mm.

Heteropleustes setosus Hendrycks and Bousfield 2004. Associated with sponges; 6.7 mm; intertidal.

Holopleustes aequipes Hendrycks and Bousfield 2004. Open-coast sand and algae; 3.3 mm; intertidal—2 m.

Incisocalliope bairdi Hendrycks and Bousfield 2004 (=*Parapleustes bairdi* of Barnard, 1956). Soft benthos, probably associated with hydroids or bryozoans; could be misidentified *Gnathopleustes*; 5.5 mm; intertidal—140 m.

Incisocalliope derzhavini (Gurjanova, 1938) (=*Parapleustes derzhavini*). Introduced Asian species in protected bays, harbors and estuaries; may include *I. newportensis*; 4 mm; shallow subtidal—2 m.

Incisocalliope newportensis Barnard 1959c (=*Parapleustes newportensis*). Bays and estuaries among fouling organisms of floats and pilings and of doubtful distinction from *I. derzhavini*; 5 mm; intertidal—2 m.

Kamptopleustes coquillus (Barnard, 1971) (=*Pleusymptes coquillus*). Whole body illustration plate 257W; on mud and sandy mud; 3.8 mm; 3 m–200 m.

Micropleustes nautiloides Bousfield and Hendrycks, 1995b (=*Parapleustes nautiloides*). In coralline algae and *Phyllospadix* mats; 2.9 mm; intertidal.

Micropleustes nautilus (Barnard, 1969a) (=*Parapleustes nautilus*). In exposed rocky shore algal mats, sponges and among *Phyllospadix*; 3.4 mm; intertidal—3 m.

Pleusirus secorrus Barnard, 1969a. Low intertidal and subtidal cobbles; 3.8 mm; intertidal—25 m.

Pleusymptes subglaber Barnard and Given 1960 (=*Sympleustes subglaber*). Recorded from San Clemente sublittoral, but the unknown condition of the *P. subglaber* mandibular molar and distal ventral extension of antenna 1 article prevent confidence in its distinction from species of either *Chromopleustes* or other *Pleusymptes*; 4 mm; 9 m or less to 110 m.

Thorlaksonius borealis Bousfield and Hendrycks, 1994b. Occurring in offshore fouling communities of hard substrate; 11 mm; intertidal—10 m.

Thorlaksonius brevirostris Bousfield and Hendrycks, 1994b. Among algae on rocks; 7 mm; intertidal—35 m.

Thorlaksonius depressus (Alderman, 1936) (=*Pleustes depressus*). Among algae on rocks and among *Phyllospadix*. Mimics snails (Carter and Behrens, 1980, Veliger 22: 376–377); 8.5 mm; intertidal—4 m.

* = Not in key.

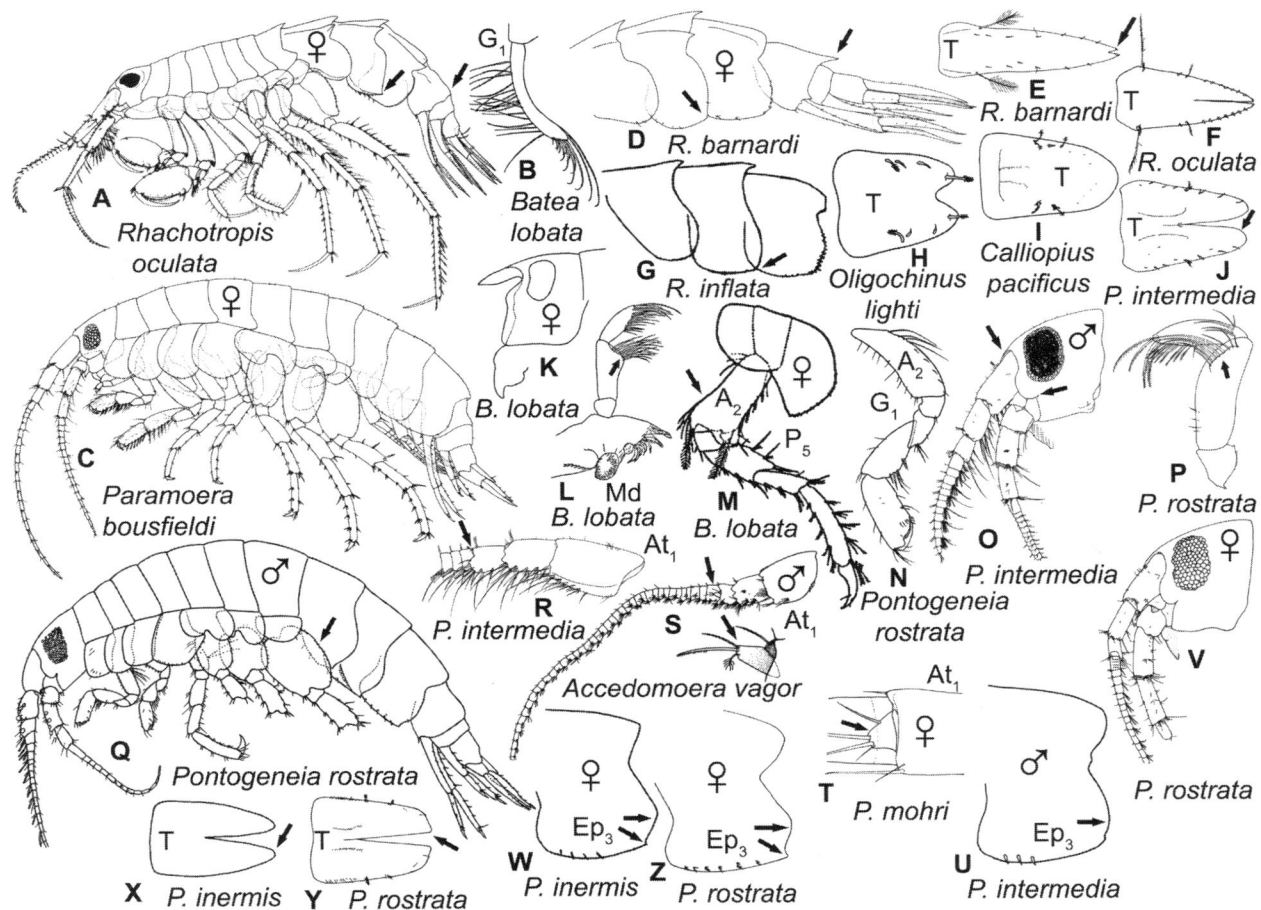

PLATE 296 Eusiroidea. S, *Accedomoera vagor*; B, K, L, M, *Batea lobata*; I, *Calliopius pacificus*; H, *Oligochinus lighti*; C, *Paramoera bousfieldi*; W, X, *Pontogeneia inermis*; T, *Paramoera mohri*; J, O, R, *Pontogeneia intermedia*; N, P, Q, V, Y, Z, *Pontogeneia rostrata*; D, E, *Rhachotropis barnardi*; G, *Rhachotropis inflata*; A, F, *Rhachotropis oculata* (figures modified from Barnard 1952b, 1964c, 1969a, 1971, 1979; Bousfield 1973; Bousfield and Hendrycks 1995a, 1997; Sars 1895; Shoemaker 1926; and Staude 1995).

Thorlaksonius grandirostris Bousfield and Hendrycks, 1994b. On rocks with seagrass, probably mimics a snail; 6 mm; intertidal—2 m.

*Thorlaksonius platypus (Barnard and Given, 1960) (=*Pleustes platypa*). Pt. Conception and south; but not clearly distinct from the more northern *T. grandirostris*. On various macrophytes. Imitates a snail (Crane 1969, Veliger 12: 200; Field 1974. Pacific Science 28: 439–447); 8.5 mm; 3 m–100 m.

Thorlaksonius subcarinatus Bousfield and Hendrycks, 1994b. On rocks and algae; 9.5 mm; intertidal—25 m.

Trachypleustes trevori Bousfield and Hendrycks, 1995b. Associated with sponges and tunicates under rocks of exposed coasts. Distinction from *Gnathopleustes pugettensis* is mainly the smaller gnathopods which are described largely from males; 5 mm; intertidal.

EUSIROIDEA

Eusiroidea (Bateidae, Calliopiidae, Eusiridae, Pontogeneiidae) include a broad range of morphological diversity (which is particularly apparent in the shapes of telsons—laminar cleft or entire—and in the absolute and relative sizes of first and second gnathopods), which makes this group difficult to distinguish

* = Not in key.

from other families. The Bateidae, with vestigial gnathopod 1 and coxa 1 obscured by coxa 2, are among the most extreme of gnathopod morphotypes. Accessory flagellum either a tiny button or absent. Rostrum variable, urosomites separate, third uropod biramous. Eusiroidea are free living, with well-developed molars and mandibular palps. *Pontogeneia* and *Paramoera* occasionally swarm in intertidal pools and eusirids are extremely abundant in hard bottom nearshore marine communities.

KEY TO EUSIROIDEA

1. Gnathopods powerful and subchelate with dactyls reaching more than two-thirds of the length of article 6, dorsal urosomites 1 and 2 and posterior ventral epimeron 3 toothed, dactyls of pereopods 10–20 times as long as broad (plate 296A) 2
— Gnathopod 1 vestigial, (plate 296B), or normally subchelate (plate 296C); if dactyl present, not reaching one-half of the length of article 6 (plate 296C) 4
2. Urosomite 1 dorsally toothed, epimeron 2 nearly square (plate 296D); distal notch of telson small (plate 296E). . . . .
. . . . *Rhachotropis barnardi*
— Urosomite 1 not dorsally toothed (plate 296A); epimeron 2 produced or rounded; telson cleft more than one-third of its length (plate 296F). 3

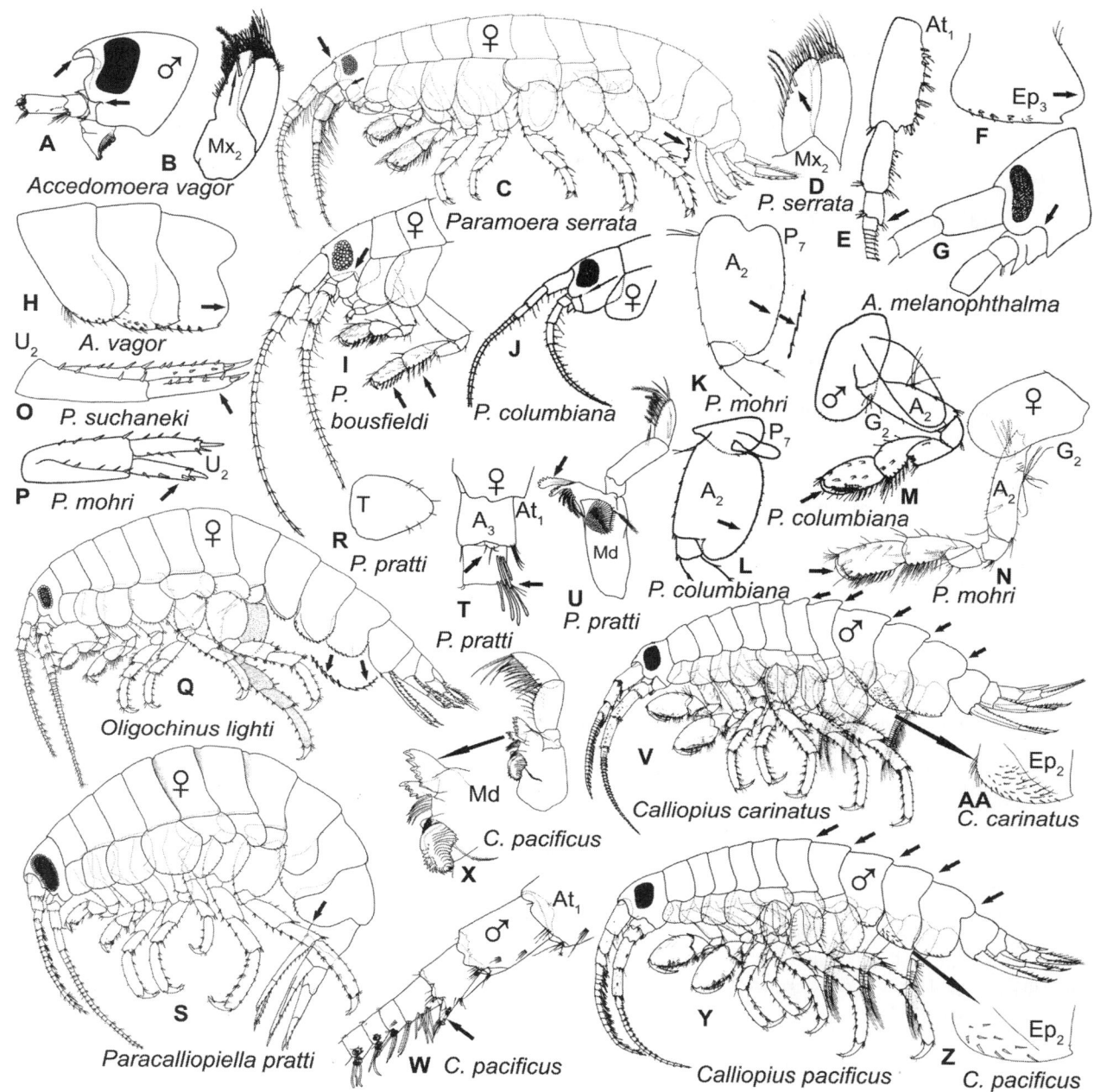

PLATE 297 Eusiroidea. E–G, *Accedomoera melanophthalma*; A, B, H, *Accedomoera vagor*; V, AA, *Calliopius carinatus*; W–Z, *Calliopius pacificus*; Q, *Oligochinus lighti*; R–U, *Paracalliopiella pratti*; I, *Paramoera bousfieldi*; J, L, M, *Paramoera columbiana*; K, N, P, *Paramoera mohri*; C, D, *Paramoera serrata*; O, *Paramoera suchaneki* (figures modified from Barnard 1952b, 1969a; Bousfield 1958a; Bousfield and Hendrycks 1997; Gurjanova 1938; and Staude 1995).

epimeron 3 slight (plate 296U). . . . . *Pontogeneia intermedia*

— Rostrum reaching more than one-third of the length of antenna 1 article 1 (plate 296V); posterior extension of epimeron 3 prominent (plate 296W). 8

8. Apices of telson rounded (plate 296X); posterior margin of epimeron 3 strongly convex and with obtuse ventral corner (plate 296W). . . . . *Pontogeneia inermis*

— Apices of telson angled (plate 296Y); posterior margins of epimeron 3 with moderate convex posterior edge and weak ventral tooth (plate 296Z). . . . . *Pontogeneia rostrata*

9. Rostrum prominent (plate 297A); inner plate of maxilla 2 with one medial spine (plate 297B). 10

— Rostrum indistinct or absent (plate 297C); inner plate of maxilla 2 with multiple medial spines (plate 297D). 11

10. Antenna 1, peduncular article 3 with ventral distal tooth (plate 297E); epimeron 3 posterior edge greatly expanded beyond ventral corner (plate 297F); ventral antennal sinus evenly rounded (plate 297G). . . . .
. . . . *Accedomoera melanophthalma*

— Antenna 1, peduncular article 3 without a ventral distal tooth (plate 296S); epimeron 3 posterior edge only slightly expanded beyond ventral corner (plate 297H); ventral antennal lobe notched (plate 297A). . . . . *Accedomoera vagor*

11. Deep cleft separating ocular lobe and second antenna sinus and female gnathopod 2 article 5 shorter than or equal to article 6 (plate 297C). 12

— Shallow cleft separating ocular lobe and antenna 2 sinus and female gnathopod 2 article 5 as long as article 6 (plate 297I). . . . . *Paramoera bousfieldi*

12. Anterioventral region of head below the antennal notch extending nearly even with the ocular lobe and posterior edges of pereopod 7 article 2 strongly serrate (plate 297C)
. . . . *Paramoera serrata*

— Anterioventral region of head posterior to ocular lobe (plate 297J); posterior edges of pereopod 7 article 2 weakly serrate (plate 297K) or smooth (plate 297L). 13

13. Male gnathopod 2 palm one-half of the length of article 6 and subchelate (plate 297M). . . . . . *Paramoera columbiana*

— Male gnathopod 2 palm less than one-third of the length of article 6 and oblique (plate 297N). 14

14. Female uropod 2 outer ramus equal to or longer than inner ramus (plate 297O). . . . . *Paramoera suchaneki*

— Female uropod 2 outer ramus shorter than inner ramus (plate 297P). . . . . *Paramoera mohri*

15. Telson, posteriorly concave (plate 296H); pleonal epimeron 3, posteriorly serrate and with ventral spines (plate 297Q)
. . . . *Oligochinus lighti*

— Telson, posteriorly convex and evenly rounded (plates 296I, 297R); epimeron 3, without serrations (plate 297S)
. . . . 16

16. Pleonite 3 posteriorly rounded (plate 297S); antenna 1 article 3 without ventromedial extension (plate 297T); lacinia mobilis of mandible greatly extended (plate 297U). . . . . .
. . . . *Paracalliopiella pratti*

— Pleonite 3 posterior ventrally square and with a minute tooth (plate 297V); antenna 1 article 3 ventromedially extended (plate 297W); lacinia mobilis of mandible normal (plate 297X). 17

17. Dorsal pereonites 5–7 and pleonites 1 and 2 not carinate (plate 297Y); pleon plate 2 with facial setae in two to three submarginal rows (plate 297Z). . . . . *Calliopius pacificus*

— Dorsal pereonites 5–7 and pleonites 1 and 2 carinate (plate 297V); pleon plate 2 with facial setae in five to seven submarginal rows (plate 297AA). . . . . *Calliopius carinatus*

LIST OF SPECIES

BATEIDAE

Batea lobata Shoemaker, 1926. Inshore sand and mud bottoms and pier pilings; 6 mm; intertidal—8 m.

CALLIOPIIDAE

Calliopius carinatus Bousfield and Hendrycks, 1997. Common in surf-swash zone, mainly along rocky shores, marine to mesohaline inshore waters; 9 mm; intertidal.

Calliopius pacificus Bousfield and Hendrycks, 1997. Dominant in inshore waters of bays and estuaries, apparently moderately euryhaline, among submerged plants and algae and on floats; 15 mm; intertidal to shallow depths.

Oligochinus lighti Barnard, 1969a. In the cobble–*Silvetia*–*Phyllospadix* zone; among the most abundant amphipods in *Mastocarpus papillatus* and *Endocladia muricata* of high and middle intertidal where they feed on epiphytic algae; see Johnson 1975, pp. 559–587 in Gates and Schmerl, eds., Perspectives of biophysical ecology, Springer Verlag (ecology); named in honor of the founder of this book, Sol Felty Light, 1886–1947; 11.5 mm.

Paracalliopiella pratti (Barnard 1954a). Low intertidal and subtidal on algae, mixed sediment, and seagrass. Known only from Alaska, Puget Sound, and Fossil Point in Coos Bay, Oregon, the latter collected from the introduced Japanese seaweed *Sargassum muticum*; 5 mm; intertidal—2 m.

EUSIRIDAE

Rhachotropis barnardi Bousfield and Hendrycks 1995a. Deep subtidal on fine sediment and probably also pelagic; abundance in shallow waters unclear; 4 mm; 17 m–350 m.

Rhachotropis inflata (Sars, G. O., 1883). On fine sediments and pelagic, circum-Arctic; occurrence in shallow waters possible; 8 mm; 10 m–154 m.

Rhachotropis oculata (Hansen,1888). Pan-arctic, swimming, planktivorous; south to southern California; occurrence in shallow waters unclear; 10 mm; 18 m–274 m.

PONTOGENEIIDAE

Accedomoera melanophthalma (Gurjanova, 1938). On mixed algae and sediments of boreal western Pacific to California, but Eastern Pacific occurrences poorly documented; 8 mm; intertidal—80 m.

Accedomoera vagor Barnard, 1969a. On algae in exposed rocky areas; 7.5 mm, intertidal—2 m.

Paramoera bousfieldi Staude, 1986. Gravel of brackish, stream mouths or intertidal freshwater seeps; 4.5 mm.

Paramoera columbiana Bousfield, 1958. Estuary, in gravel and other mixed sediments; 11 mm; intertidal.

Paramoera mohri Barnard, 1952b. Marine rocky open coasts; 6.5 mm, intertidal—10 m.

Paramoera serrata Staude, 1995. Marine, sand, and mixed sediments; 6 mm; low intertidal.

Paramoera suchaneki Staude, 1986. Marine, gravel, cobble and mussel beds; 13 mm; intertidal.

Pontogeneia inermis (Kroyer, 1838). Pan boreal in northern hemisphere, eastern Pacific identification uncertain; mixed sediments, possible echinoderm and coelenterate commensal

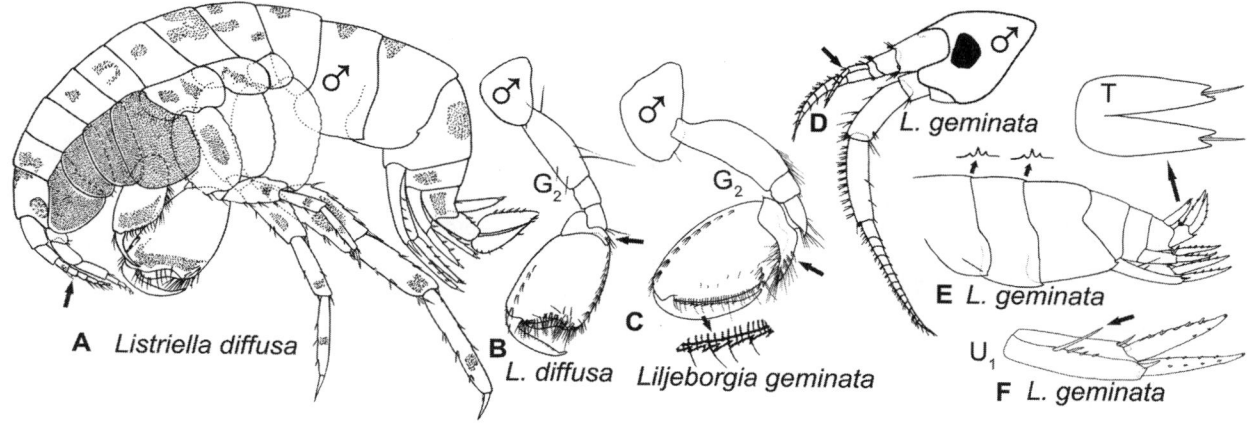

PLATE 298 Liljeborgiidae. C–F, *Liljeborgia geminata*; A–B, *Listriella diffusa* (figures modified from Barnard 1959a, 1969a).

and also common in nocturnal plankton samples; 4.5 mm; intertidal—220 m.

Pontogeneia intermedia Gurjanova, 1938. Intertidal and shallow subtidal on algae and various rocky sediments occurring also Japan and eastern Russia; 7.5 mm; intertidal.

Pontogeneia rostrata Gurjanova, 1938. On algae and mixed sediments; 6.5 mm; shallow subtidal and low intertidal.

Pontogeneia sp. A shallow subtidal undescribed purple species that occurs among *Strongylocentrotus purpuratus* spines (Harty 1979, p. 198, concerning observations at Cape Arago, Oregon, in: Bull. So. Calif. Acad. Sci. 78: 196–199); 7 mm; shallow subtidal.

LILJEBORGIIDAE

Liljeborgiidae have tiny rostrums, short antenna 2 relative to antenna 1, prominent accessory flagellum, poorly developed mandibular molars (plate 257KK), large gnathopods, and distally notched, laminar telsons that are cleft to the base. Pigmentation of many *Listriella* remains partially intact in preservation. Liljeborgiidae are likely commensals in tubes and burrows of large subtidal invertebrates, including polychaetes and echiuroids.

KEY TO LILJEBORGIIDAE

1. Gnathopods and often body pigmented urosome smooth and accessory flagellum of two articles (plates 257Q, 298A); gnathopod 2 article 5 posterior lobe not extending behind article 6 and dactyl posterior edges minutely serrate or smooth (plate 298B) . . . . *Listriella diffusa*
— Gnathopods and body not pigmented, gnathopod 2 article 5 posterior lobe extending behind article 6 and dactyls deeply serrate (plate 298C); accessory flagellum multiarticulate (plate 298D); pleonites 1 and 2 minutely toothed dorsally (plate 298E); uropod 1 peduncle with long lateral spines (plate 298F) . . . . *Liljeborgia ? geminata*

LIST OF SPECIES

Liljeborgia geminata Barnard 1969a. Of a poorly distinguished species complex (Barnard 1969a) occurring in the Atlantic and Pacific Oceans; on floats and pilings of southern California harbors and shallow coastal waters in rhizomes of *Macrocystis pyrifera*. Multiple long spines on peduncles of uropods 1 and 2 (plate 298F) are characteristic of this species; 8.7 mm; 1 m–70 m.

Listriella albina Barnard 1959a. Oregon to Baja California, a shallow warm-water species in its southern range and found at great depths in its northern range (Barnard 1971); 7.5 mm; 16 m–721 m.

Listriella diffusa Barnard 1959a. Shallow subtidal to 23 m in sandy sediments, possibly a commensal with large tube building polychaetes. Additional species of *Listriella* reported from southern California (Barnard 1959a) are expected in our region. Whole female body illustration plate 257Q. The tooth-like structures on the inner lobes of the lower lip (functions unknown) occur also in Melitidae; 3.5 mm; 3 m–172 m.

Listriella goleta Barnard 1959a. Oregon to Baja California, a shallow warm-water species in its southern range found at great depths in its northern range; 3.5 mm, 16 m–721 m.

GAMMAROIDEA

Gammaroidea (Anisogammaridae and Gammaridae) are free-living, benthic, and epibenthic omnivores and zooplankton predators that range widely in shallow marine shores, estuaries, tidal creeks, and low-elevation rivers and lakes. Sexual dimorphism is weak. Whether native Gammaridae occur in the region is unclear.

KEY TO GAMMAROIDEA

1. Male gnathopod palms nearly transverse, lined with thick peg spines and gnathopod dactyls thick (plate 299A–299C); gills with accessory lobes (plate 299D) 2
— Gnathopod palms oblique, lined with simple spines and dactyls slender (plate 299E); gills normal, without accessory lobes (plate 299F) 6
2. Urosomite 2 with prominent median tooth (plate 299G); uropod 3 inner ramus >60% length of outer ramus (plate 299H) . . . . *Anisogammarus pugettensis*
— Urosomite 2 without prominent median tooth (plate 299A, 299I); uropod 3 inner ramus less than 30% length of outer ramus (plate 299J) 3
3. Pleon segments dorsum bare (plate 299A) or with a few tiny setae (plate 299I) . . . . *Eogammarus confervicolus*

* = Not in key.

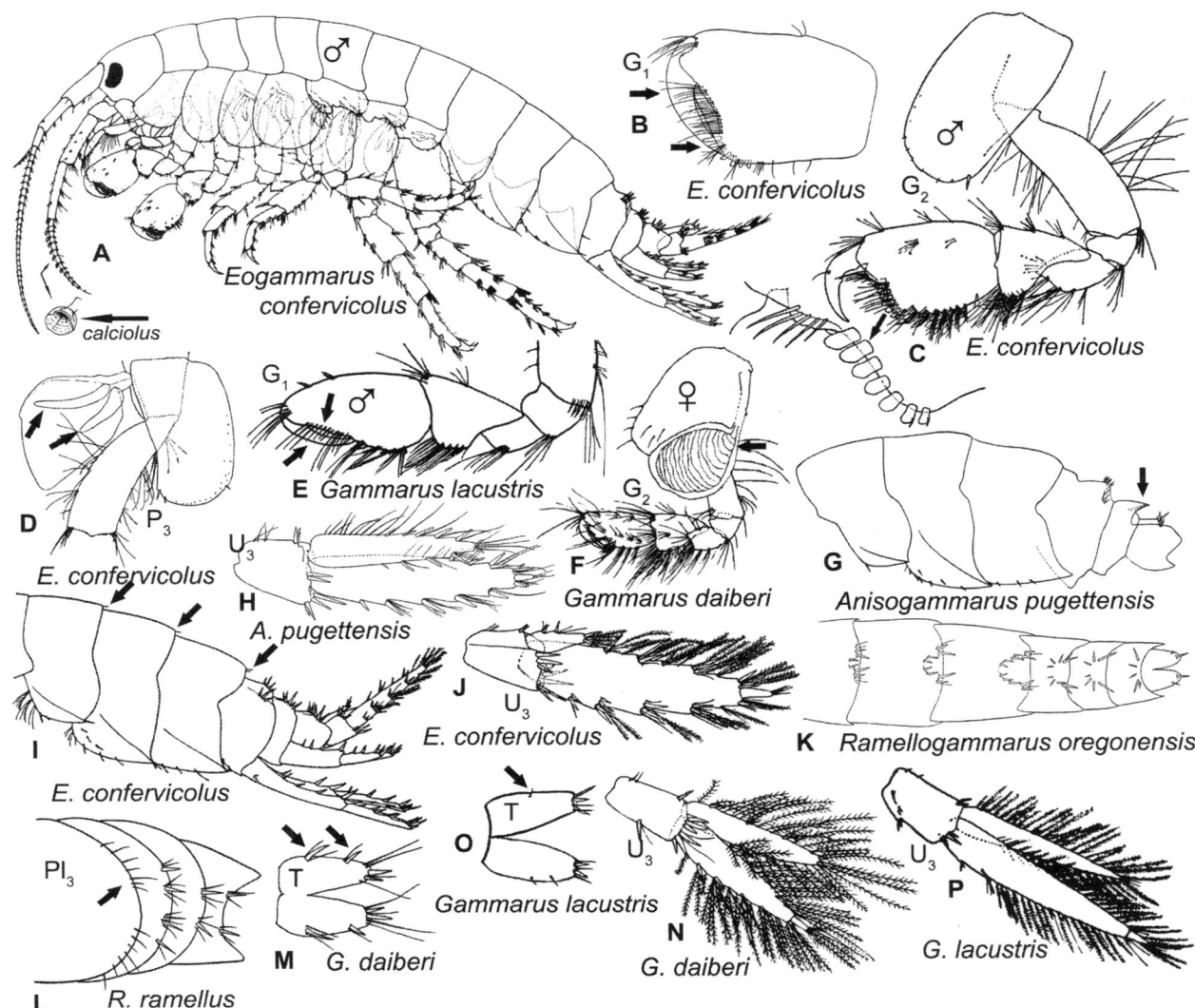

PLATE 299 Gammaroidea. G, H, *Anisogammarus pugettensis*; A–D, I, J, *Eogammarus confervicolus*; F, M, N, *Gammarus daiberi*; E, O, P, *Gammarus lacustris*; K, *Ramellogammarus oregonensis*; L, *Ramellogammarus ramellus* (figures modified from Barnard 1954a; Bousfield 1958b, 1973, 1979; Shoemaker 1944, 1964; and Weckel 1907).

— Pleon segments (one or more) with numerous, conspicuous, dorsal spines or setae (plate 299K, 299L) 4

4. Pleonite 3 with few or no setae and numerous stout spines (plate 299K) 5

— Pleonite 3 with numerous long setae and few stout spines (plate 299L) . . . . *Ramellogammarus ramellus*

5. Pleonites 1–3 with few setae and stout spines anterior to posterior pleonite margins (plate 299K) . . . . *Ramellogammarus oregonensis*

— Pleonites 1–3 with few setae and stout spines only on posterior edges . . . . *Ramellogammarus* spp.

6. Telson with two lateral bundles of prominent spines and seta (plate 299M); inner ramus extending to distal end of the first article of the outer ramus article 1 (plate 299N) . . . . *Gammarus daiberi*

— Telson with one or no lateral spine bundles and few seta (plate 299O); inner ramus not reaching distal end of the first article of the outer ramus article 1 (plate 299P) . . . . *Gammarus lacustris*

LIST OF SPECIES

ANISOGAMMARIDAE

Anisogammarus pugettensis (Dana, 1853). Marshes and low-salinity estuaries; high tolerance of low oxygen (Waldichuck and Bousfield 1962, J. Fish. Res. Bd. Canada 19: 1163–1165); 17 mm; subtidal to intertidal.

Eogammarus confervicolus (Stimpson, 1856) (=*E. oclairi* Bousfield, 1979). Estuarine, intertidal, and subtidal; various substrata but especially associated with sedges, eelgrass, algae, and wood chips; calceoli (plate 299A) are chemosensory organs of ecological and taxonomic interest (Stanhope et al. 1992, J. Chem. Ecol.18: 1871–1887); 19 mm; subtidal to intertidal. The major character separating *E. oclairi* and *E. confervicolus* (two distal telson lobe spines instead of one) is size dependent: *E. oclairi* are large (19 mm) *E. confervicolus*, and are thus synonyms.

*Eogammarus oclairi Bousfield 1979. See *E. confervicolus*.

* = Not in key.

Ramellogammarus columbianus Bousfield and Morino, 1992. Freshwater, occurring in pebble and stone bottoms in moss or woody detritus often at the mouths of medium-size streams flowing into protected bays; 13 mm; intertidal.

Ramellogammarus littoralis Bousfield and Morino, 1992. Freshwater, occurring in pebble and stone bottoms in moss or woody detritus often at the mouths of medium-size streams flowing into protected bays; 9.5 mm; intertidal.

Ramellogammarus oregonensis (Shoemaker 1944) (=*Anisogammarus oregonensis, Eogammarus oregonensis*). Known only from extreme low salinities and freshwater of Big Creek and mouth of D River, Lincoln County, Oregon, and Siltcoos River, Lane County, Oregon; 10 mm; subtidal to intertidal.

Ramellogammarus ramellus (Weckel, 1907) (=*Gammarus ramellus, Anisogammarus ramellus* and *Eogammarus ramellus*). Low-salinity and freshwater marshes and stream mouths among coarse sand, stones, and wood debris, a morphologically variable, poorly described species or species complex; 13 mm; subtidal to intertidal.

Ramellogammarus spp. Several freshwater species from aquatic plants, coarse gravel and benthos of the lower Columbia River and coastal river mouths in up to 2% salinities (Bousfield and Morino 1992, Cont. Nat. Sci. 17: 1–22)

GAMMARIDAE

Gammarus daiberi Bousfield 1969. Ballast water introduction from eastern North America to 0–15% salinity areas of San Francisco Bay and Delta, benthic and semipelagic; 12.5 mm; subtidal to intertidal.

Gammarus tigrinus Sexton 1939. A benthic and pelagic species, introduced to the North Sea from eastern North America with ballast water, a likely invader of the intertidal Pacific coast and estuaries; referred to in Europe as a "scourge" due to its likely replacement of native gammaroid species (see Dielman and Pinkster 1977, Bull. Zool. Mus. Univ. Amsterdam 6: 21–29; Pinkster et al. 1977, Crustaceana Suppl. 4: 91–105); morphology and ecology are similar to *G. daiberi*, 1–25% salinity; 14 mm; low intertidal to shallow subtidal.

Gammarus lacustris (Sars, 1863). Filamentous algae in weed and rush margins of hard-water lakes and ponds of American Pacific coastal alpine, rare in tidal waters of rivers, West Coast distribution and taxonomy unclear and may be present along the Pacific coast south of 45° N (Barnard and Barnard 1983: 81); important zooplankton predator in lakes (Wilhelm and Schindler 1999, Can. J. Fish. Aquat. Sci. 56: 1401–1408); 15 mm; intertidal river mouths, low elevation lakes and streams.

HADZIOIDEA

Hadzioidea (Hadziidae and Melitidae) occur in marine and estuarine benthic fouling communities. The Hadzioidea have large accessory flagella, short antenna 2 relative to antenna 1, waxy cuticles and greater lateral body compression than most Gammaridea. The only Hadziidae of the region is marine. Estuarine Melitidae may overlap with Crangonyctidae in low-salinity environments. The diversity of secondary sex characters in Melitidae, ranging from the enormous male gnathopods of *Dulichia*, probably adapted for competition for females, to the stridulating anatomy in *Melita* perhaps to attract males, indicate broad variation in mating behaviors in the family.

* = Not in key.

KEY TO HADZIOIDEA

1. Uropod 3 inner ramus less than a fifth as long as the outer ramus (plate 300A) 2
— Rami of uropod 3 similar in length (plate 300B) 12
2. Male and female gnathopod 2 article 6 equal to or smaller than article 5 and coxa 2–3 longer than deep (plate 300C) . . . . *Netamelita cortada*
— Male and female gnathopod 2 article 6 larger than article 5 and coxa 2–3 deeper than long (plate 300D) 3
3. Pleosome segments 1–3 with a central dorsal tooth plus accessory lateral teeth (plate 300E) 4
— Pleosome segments 1–3 without dorsal teeth (plate 300D) or with only dorsal lateral teeth (plate 300F) 6
4. Male gnathopod 2 article 6 immense and chelate (plate 300G) . . . . *Dulichiella spinosa*
— Male gnathopod 2 large but subchelate and not immense (plate 300H) 5
5. Gnathopod 2 dactyl distally blunt and without dense anterior setae (plate 300H); dorsal pleonite 1 with multiple lateral teeth (plate 300I) . . . . *Megamoera subtener*
— Gnathopod 2 dactyl distally pointed and covered anteriorly with setae (plate 300J); pleonite 1 with single lateral teeth (plate 300E) . . . . *Megamoera dentata*
6. Male gnathopod 1 article 6 and dactyl highly modified (plate 300K); distinct from simple female dactyl (plate 300L) 7
— Male gnathopod 1 article 6 and dactyl normally subchelate, not modified or distinct from simple dactyl of female (plate 300M) 10
7. Urosomite 1 posterior edge with a distinct dorsal medial tooth (plate 300D); condition of female coxa 5 not reported . . . . *Melita sulca*
— Urosomite 1 posterior edge without a medial tooth (plate 300F, 300N); female coxa 5 ventrally extended (plate 300O) 8
8. Urosomite 3 dorsally bare and pleonal epimeron 3 of both sexes weakly toothed or square (plate 300N) . . . . *Melita nitida*
— Urosomite 2 with dorsal lateral teeth (plate 300F, 300P); posterior edge of pleonal epimeron 3 of both sexes sharply toothed (plate 300Q, 300R) 9
9. Urosomite 2 with widely separate pairs of dorsal lateral teeth (plate 300P); male gnathopod 1 dactyl overlapped by article 6 less than half its length (plate 300S) . . . . *Melita oregonensis*
— Urosomite 2 with only two closely spaced dorsolateral teeth (plate 300F); male gnathopod 1 dactyl overlapped more than half by article 6 (plate 300K) . . . . *Melita rylovae*
10. Urosomite 1 with 3–5 dorsal lateral teeth (plate 301A) . . . . *Desdimelita californica*
— Urosomite 1 with a single dorsal tooth (plate 301C) 11
11. Pereopods 5–7 dactyl lengths greater than 4 times width (plate 301C) . . . . *Desdimelita desdichada*
— Pereopods 5–7 dactyls lengths <3 times width (plate 301D) . . . . *Desdimelita microdentata*
12. Pleon epimeron 3 with multiple irregular posterior teeth (plate 300T); mandibular palp article 3 less than one third of the length of article 2 (plate 300V) . . . . *Ceradocus spinicauda*
— Pleon epimeron 3 rounded (plate 300U) or with one ventral posterior tooth (plate 301O); mandibular palp article 3 greater than two thirds of the length of article 2 (plate 301F) 13

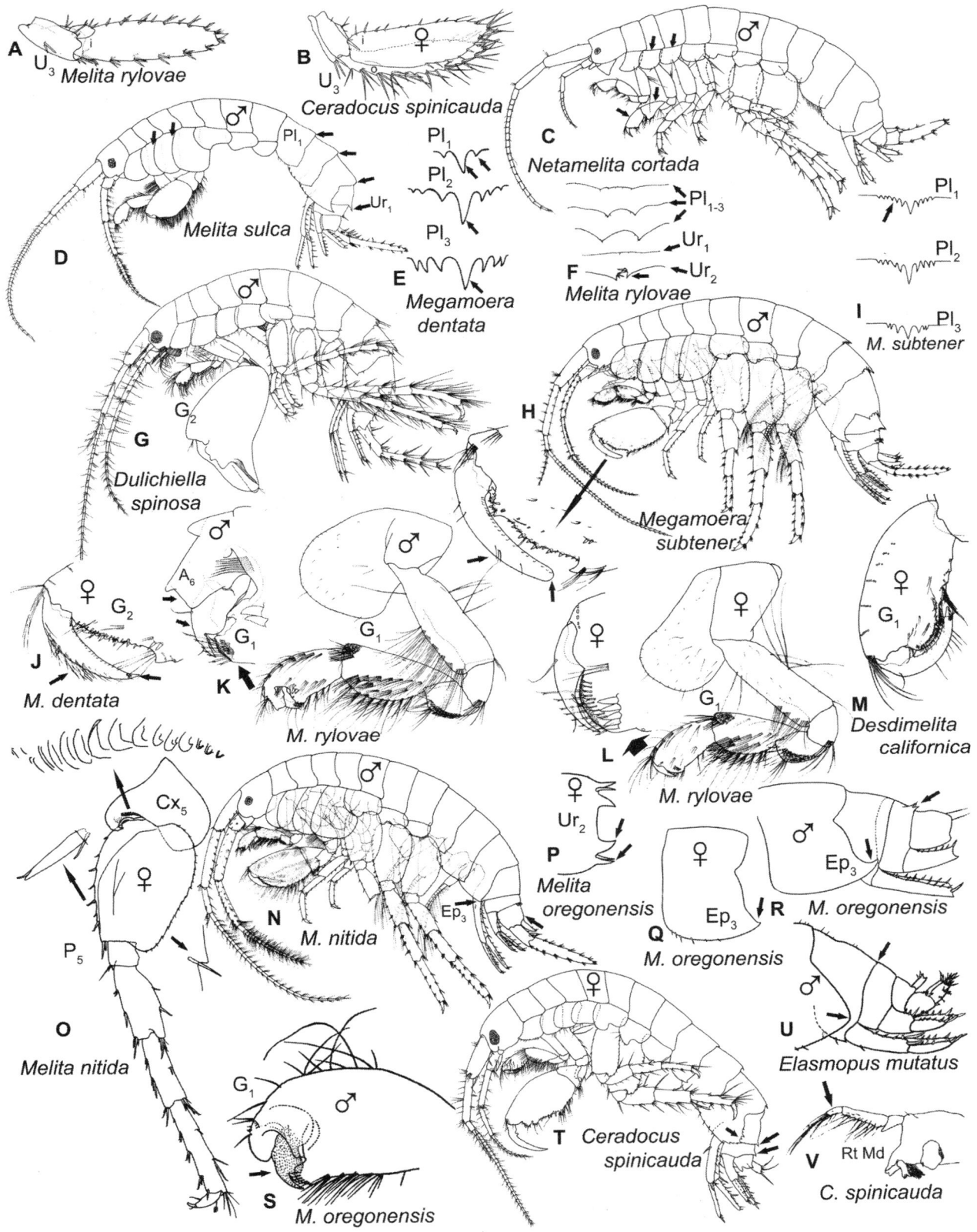

PLATE 300 Hadzioidea. B, T, V, *Ceradocus spinicauda*; M, *Desdimelita californica*; G, *Dulichiella spinosa*; U, *Elasmopus mutatus*; E, J, *Megamoera dentata*; H, I, *Megamoera subtener*; N, O, *Melita nitida*; P–S, *Melita oregonensis*; A, F, K, L, *Melita rylovae*; D, *Melita sulca*; C, *Netamelita cortada* (figures modified from Barnard 1954a, 1962b, 1969a, 1970; Chapman 1988; Jarrett and Bousfield 1996; Krapp-Schickel and Jarrett 2000; and Yamato 1987, 1988).

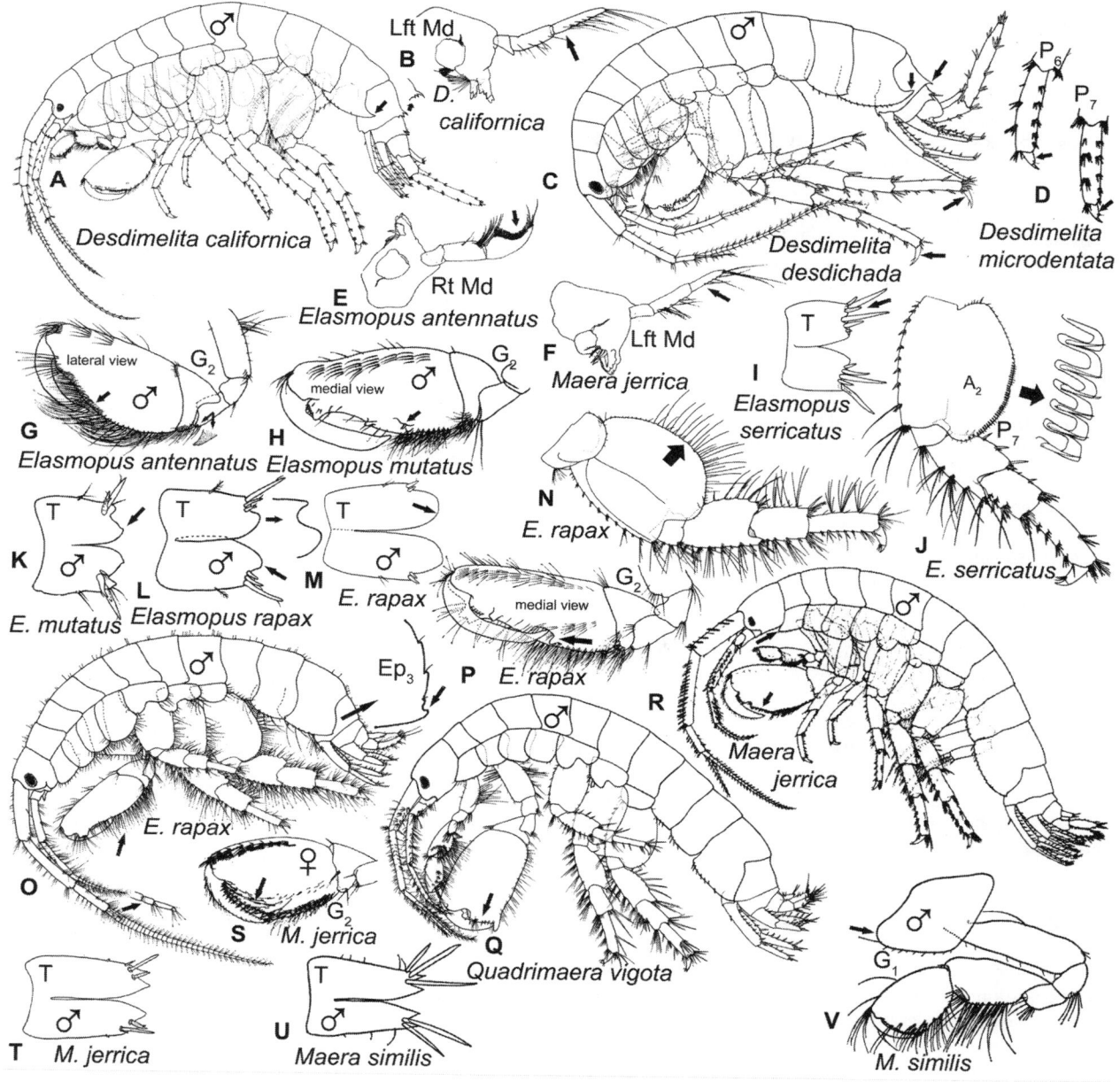

PLATE 301 Hadzioidea. A, B, *Desdimelita californica*; C, *Desdimelita desdichada*; D, *Desdimelita microdentata*; E, G, *Elasmopus antennatus*; H, K, *Elasmopus mutatus*; L–P, *Elasmopus rapax*; I, J, *Elasmopus serricatus*; F, R–T, *Maera jerrica*; U, V, *Maera similis*; Q, *Quadrimaera vigota* (figures modified from Barnard 1954a, 1959b, 1962b, 1969b, 1979a and Krapp-Schickel and Jarrett 2000).

13. Mandibular palp article 3 falcate and with dense comblike setae (plate 301E) 14
— Mandibular palp ordinary and sparsely setose (plate 301F) 17
14. Male gnathopod palm without defining proximal processes (plate 301G) (examine from lateral and medial. These processes are nearly transparent and can be obscured by dense setae) . . . . *Elasmopus antennatus*
— Male gnathopod palm with defining proximal processes (plate 301H) 15
15. Telson distally truncate and spinose (plate 301I); posterior pereopod 5 deeply serrated (plate 301J) . . . . *Elasmopus serricatus*

— Telson less spinose, distally incised and with medial extensions of the lobes (plate 301K–301M); posterior pereopod 5 weakly serrate (plate 301N) 16
16. Third pleonal epimera posterior ventrally rounded (plate 300U); male gnathopod 2 article 6 proximal palmar tooth reduced and palm bearing few setae (plate 301H); medial lobe of telson bluntly acute (plate 301K) . . . . *Elasmopus mutatus*
— Third pleonal epimera posterior ventrally square or with a small tooth (plate 301O); male gnathopod 2 with a defining hinge process at the proximal medial corner of article 6; palm (plate 301P) with dense setae (plate 301O, 301P) . . . . *Elasmopus rapax*

17. Male gnathopod 2 nearly transverse, 80–90° angle of article 6 distal posterior corner (plate 301Q). . . . .
. . . . *Quadrimaera vigota*
— Male and female gnathopod 2 subchelate, with palm angle >100° (plate 301R, 301S) 18
18. Telson distal spines less than one-third of the telson length (plate 301T); coxa 1 anterior acute (plate 301R) . . . .
. . . . *Maera jerrica*
— Telson distal spines greater than one-half of the telson length (plate 301U); coxa 1 anterior rounded (plate 301V) . . . . . .
. . . . *Maera similis*

LIST OF SPECIES

HADZIIDAE

Netamelita cortada Barnard, 1962b. Tunicate colonies at base of *Phyllospadix* beds; 3.5 mm; intertidal—20 m.

MELITIDAE

Ceradocus spinicauda (Holmes, 1908). Intertidal algae among cobbles; 12 mm; 3 m–218 m.

Desdimelita californica (Alderman, 1936). Among cobbles to fine sediments; 10 mm; intertidal—10 m.

Desdimelita desdichada (Barnard, 1962b) (=*Melita desdichada*). Soft sediments; 9 mm; 10 m–108 m.

Desdimelita microdentata Jarrett and Bousfield, 1996. 11 mm; 1 m–35 m.

Dulichiella spinosa Stout, 1912 (=*Melita appendiculata*). A rocky intertidal semitropical species reported north of Pt. Conception only once (Bousfield and Jarrett 1996, Amphipacifica 2[2]: 13); not clearly distinct from tropicopolitan *Dulichiella appendiculata* (Say, 1818); 4.5 mm; intertidal—3 m.

Elasmopus antennatus (Stout, 1913). Distinguished from *E. mutatus* by its acute rather than round posterior epimeron 3 (plate 257R); among *Phyllospadix* and algae bottoms; 10.5 mm; intertidal—18 m.

Elasmopus mutatus Barnard, 1962b. Open rocky coast among algae turf. Allometric distinctions between *E. mutatus* and the larger *E. rapax* are unclear; 7.5 mm; intertidal.

Elasmopus rapax Costa, 1853. Cosmopolitan in latitudes below 45° and restricted to enclosed bays. Introduced in California. Variation in telson morphology with size is apparent in male telsons of 7.5 mm *E. mutatus* (plate 301K) and 8 mm and 11 mm *E. rapax* (plate 301L, 301M); to 11 mm; intertidal—100 m.

Elasmopus serricatus Barnard, 1969b. Among *Egregia*, *Phyllospadix* and coralline algae. Poorly distinguished from southern Californian *Elasmopus holgurus* Barnard, 1962b. 8 mm; intertidal.

*Maera grossimana (Montagu, 1808). The northeast Pacific record of this North Atlantic species (Bousfield 2001) is uncertain; 10 mm.

Maera jerrica Krapp-Schickel and Jarrett, 2000 (=*Maera inaequipes* Barnard, 1954a). Among intertidal algae and in soft offshore sediments; 14 mm; intertidal—135 m.

Maera similis Stout, 1913. Soft benthos of estuaries and coastal waters. 9 mm, intertidal—221 m.

Megamoera dentata (Kroyer, 1842). Cosmopolitan in Arctic to cold temperate northern hemisphere oceans, on rocky and sedimentary bottoms; to 28 mm; intertidal—672 m.

Megamoera subtener (Stimpson, 1864) In coarse gravel and shell, under stones and kelp; 12 mm; intertidal—10 m.

Melita nitida Smith, 1874. Estuarine, abundant among algae and hydroids. Introduced probably from the northwest Atlantic, but also indistinguishable from the Asian *Melita setiflagellata* Yamato 1987 and therefore may be introduced to or from Asia. See Borowsky et al. 1997, J. Exp. Mar. Biol. Ecol. 214: 85–95 (reproductive morphology and physiology in polluted estuarine sediments); 12 mm; intertidal—10 m.

Melita oregonensis Barnard, 1954a. Rocky shores; 12 mm, intertidal.

Melita sulca (Stout, 1913). Condition of female coxa 5 not described. Harbors and among cobbles and algae holdfasts of open coasts; 12 mm; intertidal—101 m.

Melita rylovae Bulycheva, 1955. Introduced to San Francisco Bay from Asia and also found in ballast water samples collected in Australia where it is also introduced (Williams et al. 1996, Est. Coastal Shelf Sci. 26: 409–420); in fouling communities of docks and floats; 7.5 mm; 1 m–10 m.

Quadrimaera vigota (Barnard, 1969a) (=*Maera vigota*). Abundant under cobbles, on sponges and tunicates and rarely on algal holdfasts; dark pink; 8.5 mm; intertidal.

CRANGONYCTIDAE

Crangonyctidae have two segment accessory flagella (plate 257UU) with small terminal articles, dorsally smooth urosomes and lack a ventral antenna sinus. They are distinguished from the Hadzioidea also by having pereopod 6 longer than pereopod 7. Sexual dimorphism is reduced. As crangonyctids, *Crangonyx pseudogracilis* and *C. floridanus* share a biramous uropod 3 with reduced inner ramus (plate 302A) singly inserted spine rows on lateral article 6 of gnathopods 1 and 2 (plate 302B, 302C), eyes, and above ground occurrence; but morphological distinctions between the two species are not clear, with pleon tooth development and ventral comb setae of the male uropod 2 outer ramus being variable and of uncertain significance. The low-salinity occurrences of *Crangonyx pseudogracilis* and *C. floridanus* are unique among the almost exclusively freshwater Crangonyctidae.

KEY TO CRANGONYCTIDAE

1. Pleon epimera teeth reduced (plate 302D); male uropod 2 outer ramus slightly decurved and lined ventrally with tiny comb spines (plate 302E) . . . . *Crangonyx pseudogracilis*
— Pleon epimera teeth large (plate 302F); male uropod 2 outer ramus straight and lined ventrally with large comb spines (plate 302G). . . . . *Crangonyx floridanus*

LIST OF SPECIES

Crangonyx floridanus Bousfield, 1963. Endemic to sloughs, swamps, caves, and ponds along the U.S. Gulf Coast and possibly introduced to San Francisco Bay. The specific identity of *C. floridanus* in the San Francisco Bay Delta (Toft et al. 2002) is unclear since the associated illustration in the report is of a previously published figure of *Crangonyx forbesi* (Hubricht and Mackin 1940); 6 mm; intertidal—10 m.

Crangonyx pseudogracilis Bousfield, 1958. Occurring in aquatic vegetation in still or slow flowing, organically polluted, low salinity waters. Introduced to western North America and Japan (Zhang 1997) and Europe, where it spread secondarily from Great Britain to Ireland possibly in aquarium

* = Not in key.

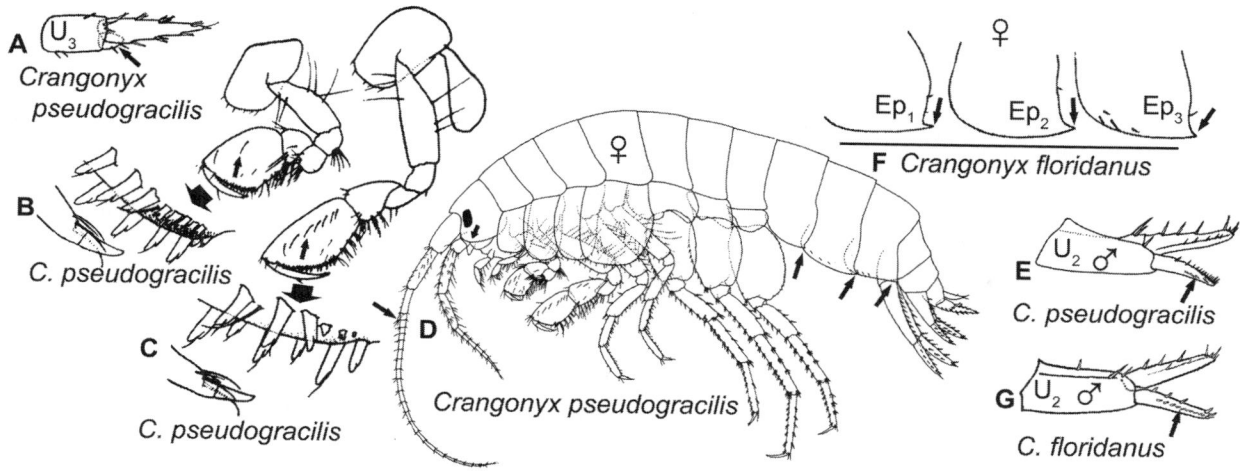

PLATE 302 Crangonyctidae. F, G, *Crangonyx floridanus*; A–E, *Crangonyx pseudogracilis* (figures modified from Bousfield 1958b, 1963, 1973).

plants (Costello 1993, Crustaceana 65: 287–299). Whether *C. floridanus* or *C. pseudogracilis* occur in San Francisco Bay should be more than an academic concern. *Crangonyx pseudogracilis* is declining in some areas of its native central and eastern North American range in the presence of invading introduced species (Beckett et al. 1997). In contrast, native European amphipod populations decline in the presence of the invading *C. pseudogracilis* which are largely unaffected by native parasites (MacNeil et al. 2003). However, the microsporidian parasite *Fibrillanosema crangonycis* of *C. pseudogracilis* appears to be transmitted with the host in invasion events and then vertically transmitted to native European amphipod hosts (Slothouber Galbeath et al. 2004). Vertical transmission combined with host sex ratio distortion may enhance host invasion success through increased rates of population growth and declines of potential native competitors (MacNeil et al. 2003). *C. pseudogracilis* also occurs in Oregon (Bousfield 1961b) and southern California (Bottoroff et al. 2003); 5 mm; intertidal—10 m.

TALITRIDAE

EDWARD L. BOUSFIELD

Talitrids comprise mainly beach hoppers, common at night on damp sand beaches, where they feed upon seaweeds cast up by the tide. Fresh beach wrack may contain purely aquatic amphipods, but their death is rapid in the air, whereas beach hoppers survive well out of water. Because the patterns and colors by which they may be identified in life are lost in preservatives, a morphological key to the Talitridae precedes a key to *Megalorchestia* based largely on color (Bowers 1963).

Although entirely terrestrial talitrids (land hoppers) are not native to our area, the student and professional zoologist will encounter southern hemisphere species introduced in urban and agriculture environments in California. Abundant, for example, under *Eucalyptus* and other leaf litter in Golden Gate Park (and other parks) in the City of San Francisco is the introduced Australian leafhopper *Arcitalitrus sylvaticus* (Haswell, 1880).

KEY TO TALITRIDAE GENERA

From Bousfield 1982.

1. Male gnathopod 1 simple, article 6 more than twice as long as wide (plate 303A1, 303C3, 303F1, 303H1); pere-

opods and uropods stout, with large spines; pleopod peduncles laterally spinose (plate 303A2); burrowers . . . . . .
. . . . *Megalorchestia*
— Male gnathopod 1 transverse, article 6 less than twice as long as wide (plate 303B1, 303E1); appendages slender, with small spines; pleopod peduncles with few or no lateral spines (plate 303B2); under debris or rocks 2
2. Male antenna 2 thick (plate 303E6) and sexual dimorphism is strong; male gnathopod 2 dactyl sinuate (plate 303E2)
. . . . *Transorchestia*
— Male antenna 2 slender (plate 303B1, 303G1) and sexual dimorphism is weak; male gnathopod 2 dactyl evenly curved (plate 303B1) 3
3. Uropod 1 with distolateral spine (plate 303G2); telson longer than broad, with single dorsolateral spines (plate 303G3); female gnathopod 1 articles 5 and 6 posteriorly swollen, dactyl not exceeding palm (303G4); brood plate setae simple (i.e., plate 255MM) . . . . *Paciforchestia*
— Uropod 1 lacking distolateral spine (plate 303B1, 303E4); telson broader than long, with groups of dorsal and marginal spines (plate 303B3); female gnathopod 1 segments 5 and 6 not swollen posteriorly and dactyl slightly exceeding palm (plate 303B4); brood plate setae hook-tipped (i.e., plate 255NN) . . . . *Traskorchestia*

KEY TO TALITRIDAE

1. Male gnathopod 1 transverse, dactyl not or barely overlapping palm (plate 303B1, 303E1, 303G4); pereopod 7 longer than 6 (plate 303B1); uropod 3, ramus narrowing distally and shorter than peduncle (plate 303E3) 2
— Male and female gnathopod 1 simple (both sexes), dactyl strong, heavy (fossorial) (plate 303A1, 303C3, 303D1, 303F1, 303H1); pereopod 6 longer than 7 (plate 303A1); uropod 3 ramus distally broad and as long as peduncle (plate 303C1) 4
2. Pereopods 3 and 4 slender, article 5 about equal to article 6 length and width (plate 303E5); male gnathopod 2 palm slightly concave and dactyl sinuate (plate 303E2); male antenna 2 peduncle thick (plate 303E6) . . . .
. . . . *Transorchestia enigmatica*
— Pereopods 3 and 4 article 5 shorter and thicker than segment 6 (303B1); male gnathopod 2 palm evenly convex,

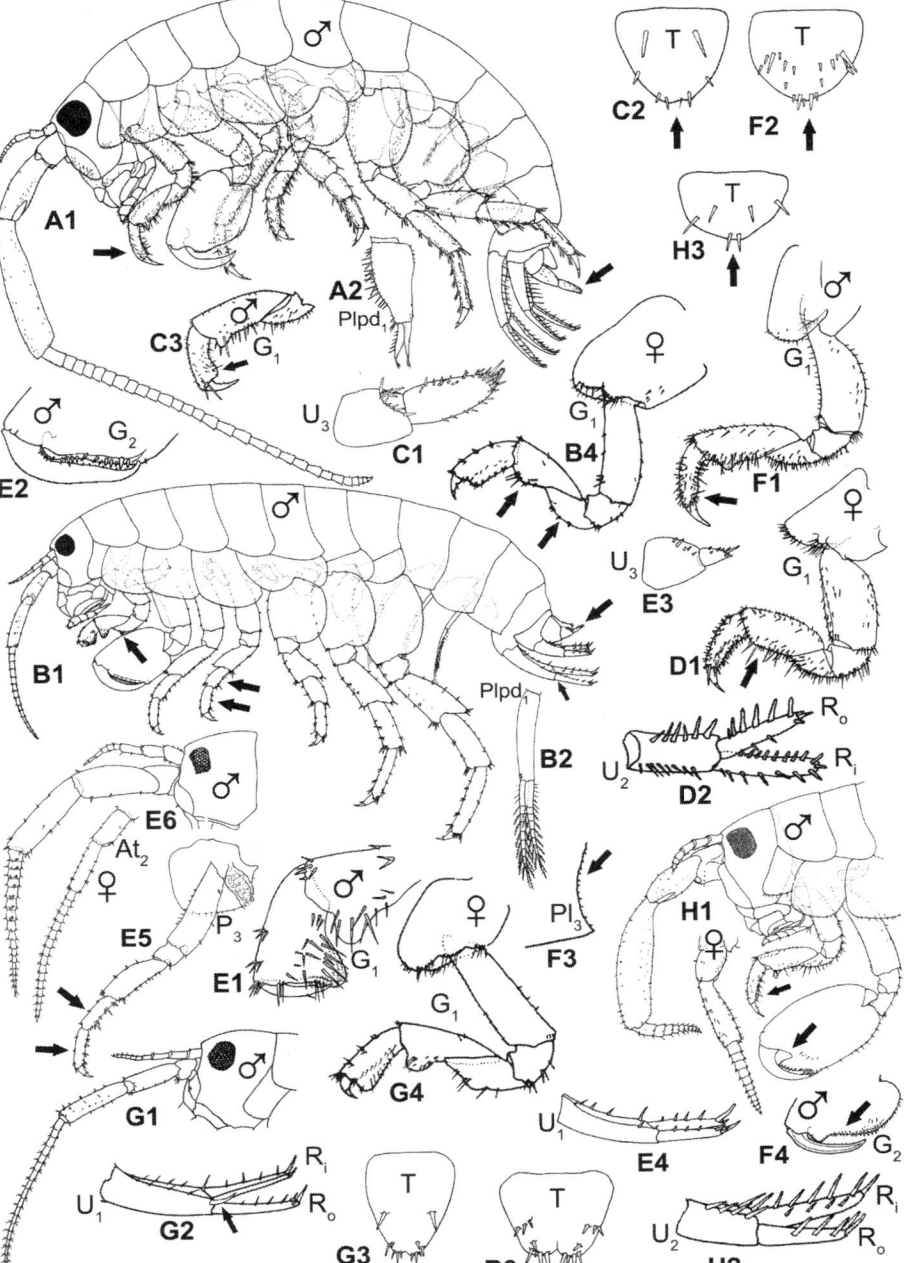

PLATE 303 Talitridae. H, *Megalorchestia benedicti*; A, *Megalorchestia californiana*; D, *Megalorchestia pugettensis*; F, *Megalorchestia corniculata*; C, *Megalorchestia californiana*; G, *Paciforchestia klawei*; E, *Transorchestia enigmatica*; B, *Traskorchestia traskiana* (figures modified from: Bousfield 1961a; Bousfield 1982, Nat. Mus. Canada, Publ. Biol. Oceangr. 11, 73 p.; Bousfield and Carlton 1967. Bull. Sth. Calif. Acad. Sci. 66: 277–283).

dactyl evenly curved and male antenna 2 peduncle relatively short and thin (plate 303B1) 3

3. Pleopods weak, rami 4–6 segmented; male gnathopod 1 article 4 lacking posterior translucent process ("blister") . . .
. . . . *Traskorchestia georgiana*

— Pleopods strong, rami 7–10 segmented (plate 303B2); male gnathopod 1 article 4 with small, posterior, translucent process (plate 303B1, arrow) . . . .
. . . . *Traskorchestia traskiana*

4. Uropod 2, inner and outer margins of outer ramus bearing spines (plate 303D2); flagellum of male antenna 2 as long or longer than peduncle (plate 303A1) 5

— Uropod 2, only outer margin of outer ramus bearing spines (plate 303H2); flagellum of antenna 2 shorter than peduncle (plate 303H1). 6

5. Posterior margins of pleonites with numerous small spines (plate 303F3); female gnathopod 1 article 5 with posterior translucent process ("blister"); pleopod rami less than one half of the peduncle length . . . .
. . . . *Megalorchestia californiana*

— Posterior margins of pleonites without spines; female gnathopod 1 without translucent process on article 5 (plate 303D1, arrow); pleopod rami half to three fourths as long as peduncles. . . . . *Megalorchestia columbiana*

6. Telson with shallow distal notch (plate 303C2); anteroventral margin of pleonite 1 with 1–7 spines; male gnathopod 1 article 6 with posterior distal expansion (plate 303C3, arrow) . . . . *Megalorchestia pugettensis*

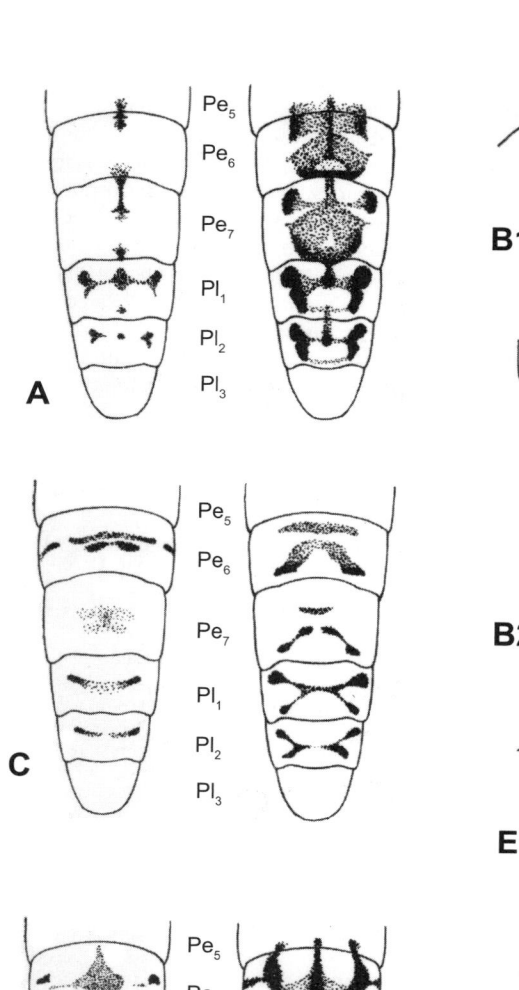

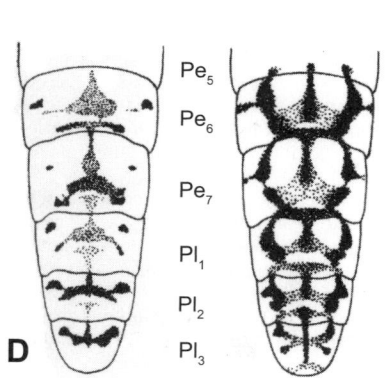

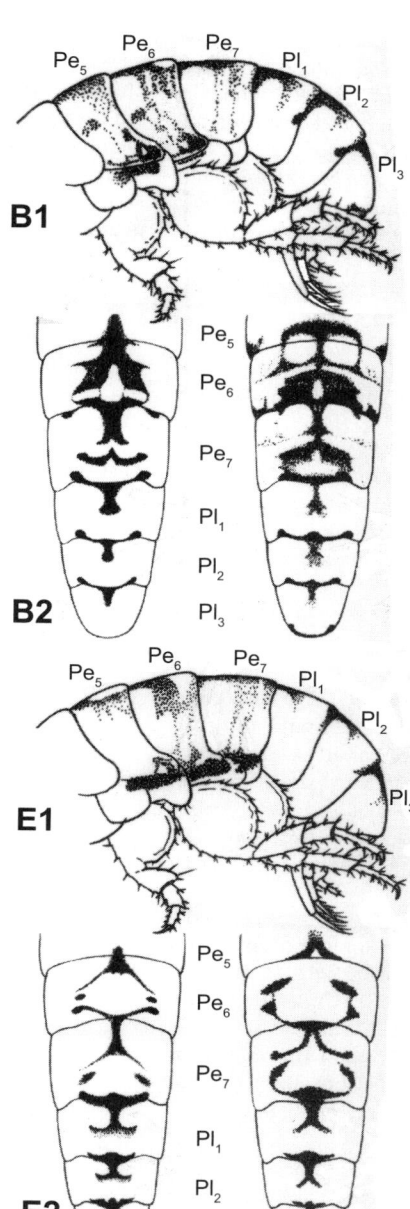

PLATE 304 Talitridae. Color patterns of *Megalorchestia*: dorsal and lateral views; paired figures show extent of pattern variation. A, *M. californiana*; B, *M. corniculata*; C, *M. columbiana*; D, *M. benedicti*; E, *M. pugettensis* (figures after Bowers 1963, Pac. Sci. 17:315–320).

— Telson without distal notch (entire) (plate 303F2, 303H3); anteroventral margin of pleonite 1 without spines; male gnathopod 1 article 6 without posterior distal expansion (plate 303F1, 303H1, arrows) 7

7. Posterior pleonite edges with 10 or more small spines (plate 303F2); male gnathopod 2 shallowly concave (plate 303F4) . . . . *Megalorchestia corniculata*

— Posterior pleonite edges with five or less spines; male gnathopod 2 deeply incised (plate 303H1, arrow) . . . . *Megalorchestia benedicti*

A FIELD (COLOR-PATTERN) KEY TO *MEGALORCHESTIA*

DARL E. BOWERS

(Plate 304)

1. Mature *Megalorchestia* 2

— *Megalorchestia* immature or not distinguished in couplets 2a through 4a below 5

2. Antenna 2 when folded reaching back to or past middle of body; flagellum longer than peduncle 3

— Antenna 2 when folded not reaching middle of body; flagellum shorter than peduncle 4

3. Color of antennae 2 rosy red . . . . *Megalorchestia californiana*

— Color of antennae 2 bluish white . . . . *Megalorchestia columbiana*

4. Color of antennae 2 usually salmon pink . . . . *Megalorchestia corniculata*

— Color of antennae 2 otherwise 5

5. Dorsal pigment pattern containing "butterfly" design 6

— Dorsal pigment pattern containing T-shaped figures; the lower limb of the T may be faint or missing 8

6. Mid-dorsal line absent; "butterfly" spots are flattened . . . . *Megalorchestia columbiana*

— Mid-dorsal line present 7

7. No markings on third pleonite; sides of body relatively free of pigment marks . . . . *Megalorchestia californiana*
— Markings on third pleonite; sides of body blotched in checkerboard pattern . . . . *Megalorchestia benedicti*
8. Two diffuse spots on lateral pereonites 5–7 . . . . *Megalorchestia corniculata*
— Three discrete spots on lateral pereonites 5–7 . . . . *Megalorchestia pugettensis*

LIST OF SPECIES

EDWARD L. BOUSFIELD

Species of *Megalorchestia* were formerly *Orchestoidea*, and species of *Paciforchestia*, *Transorchestia*, and *Traskorchestia* were formerly in *Orchestia*.

Megalorchestia benedicti (Shoemaker, 1930). Common on fine-sand beaches with *M. californiana*; 8 mm.

Megalorchestia californiana (Brand, 1851). Large and common, high up on wide, exposed beaches of fine sand; digs burrows of elliptical cross-section. May have parasitic mites (see note under *M. corniculata*); 23 mm.

Megalorchestia columbiana (Bousfield, 1958). On coarse-sand beaches with little seaweed. See Bowers 1964, Ecology 45: 677–696 (ecology); 22 mm.

Megalorchestia corniculata (Stout, 1913). Large and common, on steep, protected beaches with coarse sand and considerable seaweed; burrow nearly circular in cross-section. May be infested on their ventral surface with parasitic mites. See Craig 1971, Anim. Behav. 19: 368–374 and 1973, Anim. Behav. 21: 699–706 (lunar orientation); Craig 1973, Mar. Biol. 23: 101–109 (ecology). See Bowers 1964, Ecology 45: 677–696 (ecology); 21 mm.

*Megalorchestia minor (Bousfield, 1957). A southern species occurring north to San Simeon, just north of Point Conception, on surf-exposed flat sand beaches; see Bousfield, 1982; 15 mm.

Megalorchestia pugettensis (Dana, 1853). Under debris on coarse-sand beaches with little seaweed; 17 mm.

*Paciforchestia klawei (Bousfield, 1961). Known from British Columbia and from southern and Baja California; to be expected within our range. Under debris on protected coarse-sand and pebble beaches; 14.5 mm.

Transorchestia enigmatica Bousfield and Carlton, 1967. Described as a new species from the estuarine Lake Merritt, in Oakland, in San Francisco Bay, this amphipod is a member of the *T. chiliensis* species group, known from Chile and New Zealand (although the *enigmatica* clade remains unknown from either region). It was introduced in solid ballast from the southern hemisphere, perhaps by sailing ships carrying lumber from California to Valparaiso or Iquique, and returning in ballast, which was known to then be dumped into the Oakland Estuary near Lake Merritt. Under debris on sandy beaches; 15 mm.

Traskorchestia georgiana (Bousfield, 1958). In the drift line of protected stony and pebbly beaches, on sand with windrows of *Zostera* and *Sargassum* and usually co-occurring with *T. traskiana*. Possibly sexually dimorphic uropod 1; 13.5 mm.

Traskorchestia traskiana (Stimpson, 1857). On rocky beaches, occasionally on sandy beaches with algae; under debris and boards in salt marshes. See Page, 1979, Crustaceana 37: 247–252 (antennal growth); Busath, 1980, pp. 395–401 in Power ed, The California Islands Santa Barbara Mus. Natl. Hist. (genetics); Koch 1980, Crustaceana 57: 295–303 (behavior); Koch 1990, Crustaceana 59: 35–52 (population biology); 17 mm.

* = Not in key.

ACKNOWLEDGMENTS

Many improvements were provided by SCAMIT (Southern California Association of Marine Invertebrate Taxonomists) members, E. L. Bousfield, K. E. Conlan and E. A. Hendrycks (Canadian Museum of Nature), D. Cadien, S. McCormick, D. Pasko, and W. C. Fields. Assistance was provided by Amy Chapman and Carol Cole. Nicole Rudel assembled many plates. L. Weber and G. Boehlert allowed space at the OSU, HMSC. J. Webster, J. Mullens and S. Gilmont, Guin library, located difficult references. V. Gertseva translated Russian texts. The Gammaridea section is dedicated to J. Laurens Barnard, author of the two previous versions (Barnard 1954d. 1975), a patient, generous mentor and an inspiration to every amphipod taxonomist (Bousfield and Staude 1994, Thomas 1992, Rothman 1993).

REFERENCES

Alderman, A. L. 1936. Some new and little known amphipods of California. University of California Publications in Zoology 41: 53–74.

Barnard, J. L. 1950. The occurrence of *Chelura terebrans* Philippi in Los Angeles and San Francisco Harbors. Bulletin of the Southern California Academy of Science. 49: 90–97.

Barnard, J. L. 1952. Some Amphipoda from central California. Wasmann Journal of Biology 10: 9–36.

Barnard, J. L. 1953. On two new amphipod records from Los Angeles Harbor. Bulletin of the Southern California Academy of Science 52: 83–87.

Barnard, J. L. 1954a. Marine Amphipoda of Oregon. Oregon State Monographs 8: 1–103.

Barnard, J. L. 1954b. Amphipoda of the family Ampeliscidae collected in the eastern Pacific Ocean by the Velero III and Velero IV, Allan Hancock Pacific Expeditions 18: 1–137.

Barnard, J. L. 1954c. A new species of *Microjassa* (Amphipoda) from Los Angeles Harbor. Bulletin of the Southern California Academy of Science 53: 127–130.

Barnard, J. L. 1954d. Amphipoda, pp. 155–167 in R. I. Smith et al., (eds.) Intertidal Invertebrates of the Central California Coast. University of California Press, Berkeley, 446 pp.

Barnard, J. L. 1955a. Notes on the amphipod genus *Aruga* with the description of a new species. Bulletin of the Southern California Academy of Science 54: 97–103.

Barnard, J. L. 1955b. Two new spongicolous amphipods (Crustacea) from California. Pacific Science 9: 26–30.

Barnard, J. L. 1956. Two rare amphipods from California with notes on the genus *Atylus*. Bulletin of the Southern California Academy of Science 55: 35–43.

Barnard, J. L. 1957a. A new genus of haustoriid amphipod from the northeastern Pacific Ocean and the southern distribution of *Urothoe varvarini* Gurjanova. Bulletin of the Southern California Academy of Sciences 56: 81–84.

Barnard, J. L. 1957b. A new genus of dexaminid amphipod (marine Crustacea) from California. Bulletin of the Southern California Academy of Sciences 56: 130–132.

Barnard, J. L. 1957c. A new genus of phoxocephalid Amphipoda (Crustacea) from Africa, India, and California. Annals and the Magazine of Natural History 10: 432–438.

Barnard, J. L. 1958. Revisionary notes on the Phoxocephalidae (Amphipoda), with a key to the genera. Pacific Science 12: 146–151.

Barnard, J. L. 1959a. Liljeborgiid amphipods of southern California coastal bottoms, with a revision of the family. Pacific Naturalist 1: 12–28.

Barnard, J. L. 1959b. Part II. Estuarine Amphipoda, pp. 13–69, In Ecology of Amphipoda and Polychaeta of Newport Bay, California. J. L. Barnard and D. J. Reish, eds. Allan Hancock Foundation Publications Occasional Papers 21, 106 pp.

Barnard, J. L. 1960a. The amphipod family Phoxocephalidae in the eastern Pacific Ocean, with analyses of other species and notes for a revision of the family. Allan Hancock Pacific Expeditions 18: 175–375.

Barnard, J. L. 1960b. New bathyal and sublittoral ampeliscid amphipods from California, with an illustrated key to *Ampelisca*. Pacific Naturalist 1: 1–36.

Barnard, J. L. 1962a. Benthic marine Amphipoda of southern California: Families Aoridae, Photidae, Ischyroceridae, Corophiidae, Podoceridae. Pacific Naturalist 3: 1–72.

Barnard, J. L. 1962b. Benthic marine Amphipoda of southern California: Families Tironidae to Gammaridae. Pacific Naturalist 3: 73–115.

Barnard, J. L. 1962c. Benthic marine Amphipoda of southern California: Families Amphilochidae, Leucothoidae, Stenothoidae, Argissidae, Hyalidae. Pacific Naturalist 3: 116–163.

Barnard, J. L. 1962d. Benthic marine Amphipoda of southern California: Family Oedicerotidae. Pacific Naturalist 3: 351–371.

Barnard, J. L. 1962e. A new species of sand-burrowing marine Amphipoda from California. Bulletin of the Southern California Academy of Science 61: 249–252.

Barnard, J. L. 1964a. Deep-sea Amphipoda (Crustacea) collected by the R/V Vema in the eastern Pacific Ocean and the Caribbean and Mediterranean Seas. Bulletin of the American Museum of Natural History 127: 3–46.

Barnard, J. L. 1964b. Los anfípodos bentónicos marinos de la costa occidental de Baja California. Revista de la Sociedad Mexicana Historia Natural 24: 205–274.

Barnard, J. L. 1964c. Marine Amphipoda of Bahia de San Quintin, Baja California. Pacific Naturalist 4: 55–139.

Barnard, J. L. 1965. Marine Amphipoda of the family Ampithoidae from southern California. Proceedings of the U.S. National Museum 118: 1–46.

Barnard, J. L. 1966. Benthic Amphipoda of Monterey Bay, California. Proceedings of the U.S. National Museum 119: 1–41.

Barnard, J. L. 1967a. New and old dogielinotid marine Amphipoda. Crustaceana 13: 281–291.

Barnard, J. L. 1967b. New species and records of Pacific Ampeliscidae (Crustacea: Amphipoda). Proceedings of the U.S. National Museum 121: 1–20.

Barnard, J. L. 1967c. Bathyal and abyssal gammaridean Amphipoda of Cedros Trench, Baja California. United States National Museum Bulletin 260, 204 pp.

Barnard, J. L. 1969a. Gammaridean Amphipoda of the rocky intertidal of California: Monterey Bay to La Jolla. Bulletin of the U.S. National Museum 258: 1–230.

Barnard, J. L. 1969b. The families and genera of marine gammaridean Amphipoda. Bulletin of the U.S. National Museum 271: 1–535.

Barnard, J. L. 1970. Sublittoral Gammaridea (Amphipoda) of the Hawaiian Islands. Smithsonian Contributions to Zoology 34: 1–286.

Barnard, J. L. 1971. Gammaridean Amphipoda from a deep-sea transect off Oregon. Smithsonian Contributions to Zoology 61: 1–86.

Barnard, J. L. 1972a. Gammaridean Amphipoda of Australia, Part I. Smithsonian Contributions to Zoology 103: 1–333.

Barnard, J. L. 1972b. A review of the family Synipiidae (=Tironidae), mainly distributed in the deep sea (Crustacea: Amphipoda). Smithsonian Contributions to Zoology 124: 1–194.

Barnard, J. L. 1972c. The marine fauna of New Zealand: Algae-living littoral Gammaridea (Crustacea, Amphipoda). New Zealand Oceanographic Institute Memoirs 62: 1–216.

Barnard, J. L. 1975. Crustacea, Amphipoda: Gammaridea, pp. 313–366. In Light's manual: intertidal invertebrates of the Central California Coast. R. I. Smith and J. T. Carlton, eds. University of California Press, Berkeley.

Barnard, J. L. 1977. A new species of Synchelidium (Crustacea, Amphipoda) from sand beaches in California. Proceedings of the Biological Society of Washington 90: 877–883.

Barnard, J. L. 1979a. Littoral gammaridean Amphipoda from the Gulf of California and the Galapagos Islands. Smithsonian Contributions to Zoology 271: 1–149.

Barnard, J. L. 1979b. Revision of American species of the marine amphipod genus Paraphoxus (Gammaridea: Phoxocephalidae). Proceedings of the Biological Society of Washington 92: 368–379.

Barnard, J. L. 1980a. The genus Grandifoxus (Crustacea: Amphipoda: Phoxocephalidae) from the northeastern Pacific Ocean. Proceedings of the Biological Society of Washington 93: 490–514.

Barnard, J. L. 1980b. Revision of Metharpina and Microphoxus (marine phoxocephalid Amphipoda of the Americas). Proceedings of the Biological Society of Washington 93: 104–135.

Barnard, J. L., and C. M. Barnard. 1981. The amphipod genera, Eobrolgus and Eyakia (Crustacea: Phoxocephalidae) in the Pacific Ocean. Proceedings of the Biological Society of Washington 94: 295–313.

Barnard, J. L., and C. M. Barnard. 1982a. The genus Rhepoxynius (Crustacea: Amphipoda: Phoxocephalidae) in American Seas. Smithsonian Contributions to Zoology 357, 149 pp.

Barnard, J. L., and C. M. Barnard. 1982b. Revision of Foxiphalus and Eobrolgus (Crustacea: Amphipoda: Phoxocephalidae) from American oceans. Smithsonian Contributions to Zoology 372: 1–35.

Barnard, J. L., and C. M. Barnard. 1983a. Freshwater Amphipoda of the World I. Evolutionary Patterns 1. Hayfield Associates, Mt. Vernon, VA, pp. 1–359.

Barnard, J. L., and C. M. Barnard. 1983b. Freshwater Amphipoda of the World II. Handbook and Bibliography, 2, Hayfield Associates, Mt. Vernon, VA, pp. 359–830.

Barnard, J. L., and R. R. Given. 1960. Common pleustid amphipods of southern California, with a projected revision of the family. Pacific Naturalist 1: 37–48.

Barnard, J. L., and G. S. Karaman. 1990a. The families and genera of marine gammaridean Amphipoda (except marine gammaroids). Part 1, Records of the Australian Museum 13: 1–417.

Barnard, J. L., and G. S. Karaman. 1990b. The families and genera of marine gammaridean Amphipoda (except marine gammaroids). Part 2. Records of the Australian Museum 13: 419–866.

Beckett, D. C., P. A. Lewis, and J. H. Green. 1997. Where have all the Crangonyx gone? The disappearance of the amphipod Crangonyx pseudogracilis, and subsequent appearance of Gammarus nr. fasciatus, in the Ohio River. American Midland Naturalist 139: 201–209.

Bellan-Santini, D. 1999. Ordre des Amphipodes (Amphipoda Latreille, 1816). In: Bacescu, M. et al. Treatise on zoology: anatomy, systematics, biology: 7. Crustaceans: 3A. Peracarida. Mémoires de l'Institut océanographique, Monaco 19: 93–176.

Boeck, A. 1872. Bidrag til Californiens Amphipodefauna. Forhandlinger i Videnskskabs -Selskabet i Christiana, 1871, 22 pp.

Borowsky, B. 1983. Reproductive behavior of three tube-building peracarid crustaceans: the amphipods Jassa falcata and Ampithoe valida and the tanaid Tanais cavolinii. Marine Biology 77: 257–263.

Borowsky, B. 1984. The use of males' gnathopods during precopulation in some gammaridean Amphipoda. Crustaceana 47: 245–250.

Borowsky, B. 1985. Differences in reproductive behavior between two male morphs of the amphipod crustacean Jassa falcata Montagu. Physiological Zoology 58: 497–502.

Bosworth, W. S. 1973. Three new species of Eohaustorius (Amphipoda, Haustoriidae) from the Oregon coast. Crustaceana 25: 253–260.

Bottoroff, R. L., B. A. Hamill, and W. I. Hamill. 2003. Records of the exotic freshwater amphipod, Crangonyx pseudogracilis, in San Luis Obispo County, California. California Fish and Game 89: 197–200.

Bousfield, E. L. 1957. Notes on the amphipod genus Orchestoidea on the Pacific Coast of North America. Bulletin of the Southern California Academy of Science 56: 119–129.

Bousfield, E. L. 1958a. Distributional ecology of the terrestrial Talitridae (Crustacea: Amphipoda) of Canada. Proceedings of the 10th International Congress of Entomology 1: 883–898.

Bousfield, E. L. 1958b. Fresh-water amphipod crustaceans of glaciated North America. Canadian Field Naturalist 72: 55–113.

Bousfield, E. L. 1961a. New records of beach hoppers (Crustacea: Amphipoda) from the coast of California. National Museum of Canada Bulletin 172, Contributions to Zoology, 1959: 1–12.

Bousfield, E. L. 1961b. New records of fresh-water amphipod crustaceans from Oregon. National Museum of Canada, Natural History Papers, 12: 1–7.

Bousfield, E. L. 1963. New freshwater amphipod crustaceans from Florida. National Museum of Canada, Natural History Papers 18: 1–9.

Bousfield, E. L. 1969. New records of Gammarus (Crustacea: Amphipoda) from the middle Atlantic region. Chesapeake Science 10: 1–17.

Bousfield, E. L. 1973. Shallow-water gammaridean Amphipoda of New England. Cornell University Press, Ithaca, NY, 312 pp.

Bousfield, E. L. 1979. The amphipod superfamily Gammaroidea in the northeastern Pacific region: Systematics and distributional ecology. Bulletin of the Biological Society of Washington 3: 297–357.

Bousfield, E. L. 1987. Amphipod parasites of fishes of Canada. Canadian Bulletin of Fisheries and Aquatic Sciences 217: 1–37.

Bousfield, E. L. 1989. Revised morphological relationships within the amphipod genera Pontoporeia and Gammaracanthus and the "glacial relict" significance of their postglacial distributions. Canadian Journal of Fisheries and Aquatic Sciences 46: 1714–1725.

Bousfield, E. L. 2001. An updated commentary on phyletic classification of the amphipod Crustacea and its application to the North American fauna. Amphipacifica 3(1): 49–119.

Bousfield, E. L., and A. Chevrier. 1996. The amphipod family Oedicerotidae on the Pacific coast of North America. Part 1. The Monoculodes and Synchelidium generic complexes: Systematics and distributional ecology. Amphipacifica 2(2): 75–148.

Bousfield, E. L., and E. A. Hendrycks. 1994a. A revision of family Pleustidae (Crustacea: Amphipoda: Leucothoidea) Part I. Systematics and biography of component subfamilies. Amphipacifica, 1(1): 17–57.

Bousfield, E. L., and E. A. Hendrycks. 1994b. The amphipod superfamily Leucothoidea on the Pacific coast of North America. Family Pleustidae: Subfamily Pleustinae. Systematics and biogeography. Amphipacifica 1(2): 3–69.

Bousfield, E. L., and E. A. Hendrycks. 1995a. The amphipod superfamily Eusiroidea in the North American Pacific region. I. Family Eusiridae: systematics and distributional ecology. Amphipacifica 1(4): 3–59.

Bousfield, E. L., and E. A. Hendrycks. 1995b. The amphipod family Pleustidae on the Pacific coast of North America. Part III. Subfamilies Parapleustinae, Dactylopleustinae, and Pleusirinae: Systematics and distributional ecology. Amphipacifica 2(1): 65–133.

Bousfield, E. L., and E. A. Hendrycks. 1997. The amphipod superfamily Eusiroidea in the North American Pacific region. II. Calliopiidae. Systematics and distributional ecology. Amphipacifica 2(3): 3–66.

Bousfield, E. L., and E. A. Hendrycks. 2002. The talitroidean amphipod family Hyalidae revised, with emphasis on the North Pacific fauna: systematics and distributional ecology. Amphipacifica 3(3): 17–134.

Bousfield, E. L., and P. M. Hoover. 1995. The amphipod superfamily Pontoporeioidea on the Pacific Coast of North America. I. Family Haustoriidae. Genus *Eohaustorius* J. L. Barnard: Systematics and distributional ecology. Amphipacifica 2(1): 35–64.

Bousfield, E. L., and P. M. Hoover. 1997. The amphipod superfamily Corophioidea on the Pacific Coast of North America. Part V. Family Corophiinae, new subfamily. Systematics and distributional ecology. Amphipacifica 2(3): 67–139.

Bousfield, E. L., and J. A. Kendall. 1994. The amphipod superfamily Dexaminoidea on the North American Pacific coast; Families Atylidae and Dexaminidae: Systematics and distributional ecology. Amphipacifica 1(3): 3–66.

Bousfield, E. L., and P. Marcoux. 2004. Talitroidean amphipod family Najnidae in the North Pacific region: systematics and distributional ecology. Amphipacifica 3(4): 3–44.

Bousfield, E. L., and H. Morino. 1992. The amphipod genus *Ramellogammarus* (Amphipoda: Anisogammaridae) on the Pacific coast of North America. Contributions of the Royal British Columbia Museum 17: 1–22.

Bousfield, E. L., and C. P. Staude. 1994. The impact of J. L. Barnard on North American Pacific amphipod research: A tribute. Amphipacifica 1(1): 3–16.

Boyd, M. J., T. J. Mulligan, and F. J. Shaughnessy. 2002. Non-indigenous Marine Species of Humboldt Bay, California. Report to the California Department of Fish and Game. Humboldt State University, 118 pp.

Chapman, J. W. 1988. Invasions of the northeast Pacific by Asian and Atlantic gammaridean amphipods crustaceans, including a new species of *Corophium*. Journal of Crustacean Biology 8: 364–382.

Chapman, J. W., and J. A. Dorman. 1975. Diagnosis, systematics and notes on *Grandidierella japonica* (Amphipoda: Gammaridea) and its introduction to the Pacific coast of the United States. Bulletin of the Southern California Academy of Science 74: 104–108.

Chess, J. R. 1979. High densities of benthic amphipods related to upwelling on the northern California coast. Coastal Oceanography and Climatology News 1: 31.

Cole, G. A., and R. L. Watkins. 1977. *Hyalella montezuma*, a new species (Crustacea: Amphipoda) from Montezuma Well, Arizona. Hydrobiologia 52: 175–184.

Coleman, C. O., and J. L. Barnard. 1991. Revision of Iphimediidae and similar families (Amphipoda, Gammaridea). Proceedings of the Biological Society of Washington 104: 253–268.

Conlan, K. E. 1982. Revision of the gammaridean amphipod family Ampithoidae using numerical analytical methods. Canadian Journal of Zoology 60: 2015–2027.

Conlan, K. E. 1983. The amphipod superfamily Corophioidea in the northeastern Pacific region. 3. Family Isaeidae: Systematics and distributional ecology. Publications in Natural Sciences, National Museums of Canada 4: 1–75.

Conlan, K. 1989. Delayed reproduction and adult dimorphism in males of the amphipod genus *Jassa* (Corophioidea: Ischyroceridae): an explanation for systematic confusion. Journal of Crustacean Biology 9: 601–625.

Conlan, K. E. 1990. Revision of the crustacean amphipod genus *Jassa* Leach (Corophioidea: Ischyroceridae). Canadian Journal of Zoology 68: 2031–2075.

Conlan, K. E. 1991. Precopulatory mating behaviour and sexual dimorphism in the amphipod Crustacea. Hydrobiologia 223: 255–282.

Conlan, K. E. 1994. Amphipod crustaceans and environmental disturbance: a review. Journal of Natural History 28: 519–554.

Conlan, K. E. 1995a. Thumb evolution in the amphipod genus *Microjassa* Stebbing (Corophiidea: Ischyroceridae). Journal of Crustacean Biology 15: 693–702.

Conlan, K. E. 1995b. Thumbing doesn't always make the genus. Revision of *Microjassa* Stebbing (Crustacea : Amphipoda: Ischyroceridae). Bulletin of Marine Science 57: 333–377.

Conlan, K. E., and E. L. Bousfield. 1982a. The amphipod superfamily Corophioidea in the northeastern Pacific region. Family Ampithoidae: systematics and distributional ecology. National Museum of Natural Sciences (Ottawa). Publications in Biological Oceanography 10: 41–75.

Conlan, K. E., and E. L. Bousfield. 1982b. The amphipod superfamily Corophioidea in the northeastern Pacific Region. Family Aoridae: Systematics and distributional ecology. National Museum of Natural Sciences (Ottawa), Publications in Biological Oceanography 10: 77–101.

Conlan, K. E., and J. R. Chess. 1992. Phylogeny and ecology of a kelp-boring amphipod, *Parampithoe stypotrupetes*, a new species (Corophioidea: Amphithoidae). Journal of Crustacean Biology 12: 410–422.

Connell, J. H. 1963. Territorial behavior and dispersion in some marine invertebrates. Research in Population Ecology 5: 87–101.

Costello, M. J. 1993. Biogeography of alien amphipods occurring in Ireland, and interactions with native species. Crustaceana 65: 287–299.

Coyle, K. O. 1982. The amphipod genus *Grandifoxus* in Alaska. Journal of Crustacean Biology 2: 430–450.

Cronin, G., and M. E. Hay. 1996a. Susceptibility to herbivores depends on recent history of both the plant and animal. Ecology 77: 1531–1543.

Cronin, G., and M. E. Hay. 1996b. Induction of seaweed chemical defenses by amphipod grazing. Ecology 77: 2287–2301.

Dadswell, M. J. 1974. Distribution, ecology, and postglacial dispersal of certain crustaceans and fishes in eastern North America. National Museum of Natural Sciences Publications in Zoology 11: 110 pp.

Dahl, E. 1979. Deep-sea carrion feeding amphipods: Evolutionary patterns in niche adaptation. Oikos 33: 167–175.

Dalkey, A. 1998. A new species of amphipod (Crustacea: Amphipoda: Lysianassoidea) from the Pacific Coast of North America. Proceedings of the Biological Society of Washington 111: 621–626.

Dickinson, J. J. 1982. The systematics and distributional ecology of the family Ampeliscidae (Amphipoda: Gammaridea) in the northeastern Pacific Region I. The genus *Ampelisca*. National Museum of Natural Sciences (Ottawa). Publications in Biological Oceanography 10: 1–39.

Dickinson, J. J. 1983. The systematics and distributional ecology of the superfamily Ampeliscoidea (Amphipoda: Gammaridea) in the northeastern Pacific region. II. The genera *Byblis* and *Haploops*. Publications in Natural Sciences, National Museum of Natural Sciences, Canada 1: 1–38.

Dojiri, M., and J. Sieg. 1987. *Ingolfiella fuscina*, new species (Crustacea: Amphipoda) from the Gulf of Mexico and the Atlantic coast of North America, and partial redescription of *I. atlantisi* Mills, 1967. Proceedings of the Biological Society of Washington 100: 494–505.

Flores, M., and G. J. Brusca. 1975. Observations on two species of hyperiid amphipods associated with the ctenophore *Pleurobrachia bachei*. Bulletin of the Southern California Academy of Sciences 74: 10–15.

Griffiths, C. L. 1979. A redescription of the kelp curler *Ampithoe humeralis* (Crustacea, Amphipoda) from South Africa and its relationship to *Macropisthopous*. Annals of the South African Museum 79: 131–138.

Gurjanova, E. F. 1938. Amphipoda. Gammaroidea of Siaukhu Bay and Sudzuhke Bay (Japan Sea). Reports of the Japan Sea Hydrobiological Expedition of the Zoological Institute of the Academy of Sciences USSR in 1934, 1: 241–404.

Gurjanova, E. F. 1951. Bokoplavi moreii SSSR i sopredelnikh vod (Amphipoda: Gammaridea). Akad. Nauk SSSR, Moskow 41: 1–1029.

Gurjanova, E. F. 1953. Novye dopolneija k dal'nevostochnoi fauna morskik bokoplavov. Akademiia Nauk SSSR. Trudy Zoologicheskogo Institute 13: 216–241.

Gurjanova, E. 1962. Bokoplavy severnoi chasti Tixogo Okeana (Amphipoda-Gammaridea) chast' 1, Akad. Nauk SSSR, 74: 1–440.

Haertel, L., and C. Osterberg. 1967. Ecology of zooplankton benthos and fishes in the Columbia River Estuary. Ecology, 48: 459–472.

Hay, M. E., J. E. Duffy, C. A. Pfister, and W. Fenical. 1987. Chemical defense against different marine herbivores: are amphipods insect equivalents? Ecology 68: 1567–1580.

Hay, M. E., J. E. Duffy, and W. Fenical. 1990. Host-plant specialization decreases predation on a marine amphipod: an herbivore in plant's clothing. Ecology 71: 733–743.

Hay, M. E., J. E. Duffy, W. Fenical, and K. Gustafson. 1988. Chemical defense in the seaweed *Dictyopteris delicatula*: differential effects against reef fishes and amphipods. Marine Ecology Progress Series 48: 185–192.

Hendrycks, E. A., and E. L. Bousfield. 2001. The amphipod genus *Allorchestes* in the North Pacific region: systematics and distributional ecology. Amphipacifica 3(2): 3–38.

Hendrycks, E. A., and E. L. Bousfield. 2004. The amphipod family Pleustidae (mainly subfamilies Mesopleustinae, Neopleustinae, Pleusymtinae, and Stenopleustinae) from the Pacific coast of North America: systematics and distributional ecology. Amphipacifica 3(4): 45–113.

Hirayama, A. 1983. Taxonomic studies on the shallow-water gammaridean Amphipoda of West Kyushu, Japan. I. Acanthonotozomatidae, Ampeliscidae, Amphithoidae, Amphilochidae, Argissidae, Atylidae, and Colomastigidae. Publications of the Seto Marine Biological Laboratory 28: 75–150.

Hirayama, A. 1984. Taxonomic studies on the shallow water gammaridean Amphipoda of West Kyushu, Japan. II. Corophiidae. Publications of the Seto Marine Biological Laboratory 29: 1–92.

Hogg, I. D., C. Larose, Y. Delafontaine, and K. G. Doe. 1998. Genetic evidence for a *Hyalella* species complex within the Great Lakes St Lawrence River drainage basin: implications for ecotoxicology and conservation biology. Canadian Journal of Zoology 76: 1134–1140.

Holmes, S. J. 1908. The Amphipoda collected by the U.S. Bureau of Fisheries steamer "*Albatross*" off the west coast of North America, in 1903 and 1904, with descriptions of a new family and several new genera and species. Proceedings of the U.S. National Museum 35: 489–543.

Hoover, P. M., and E. L. Bousfield. 2001. The amphipod superfamily Leucothoidea on the Pacific coast of North America: Family Amphilochidae: systematics and distributional ecology. Amphipacifica 3(1): 3–28.

Hubricht, L., and J. G. Mackin. 1940. Descriptions of nine new species of fresh-water amphipod crustaceans with notes and new localities for other species. American Midland Naturalist 23: 187–218.

Hurley, D. E. 1963. Amphipods of the family Lysianassidae from the west coast of North and Central America. Allan Hancock Foundation Publications Occasional Paper 25: 1–160.

Imbach, M. C. 1967. Gammaridean Amphipoda from the South China Sea. Naga Report 4: 39–167.

Jarrett, N. E., and E. L. Bousfield. 1982. Studies on the amphipod family Lysianassidae in the Northeastern Pacific region. *Hippomedon*: and related genera. National Museum of Natural Sciences (Ottawa). Publications in Biological Oceanography 10: 103–128.

Jarrett, N. E., and E. L. Bousfield. 1994a. The amphipod superfamily Phoxocephaloidea on the Pacific coast of North America. Family Phoxocephalidae. Part 1. Metharpiniinae, new subfamily. Amphipacifica 1: 58–140.

Jarrett, N. E., and E. L. Bousfield. 1994b. The amphipod superfamily Phoxocephaloidea on the Pacific Coast of North America. Family Phoxocephalidae. Part II. Subfamilies Pontharpiniinae, Brolginae, Phoxocephalinae, and Harpiniinae. Systematics and distributional ecology. Amphipacifica 1(2): 71–150.

Jarrett, N. E., and E. L. Bousfield. 1996. The amphipod superfamily Hadzioidea on the Pacific Coast of North America: Family Melitidae. Part I. The *Melita* group: systematics and distribution ecology. Amphipacifica 2(2): 3–74.

Krapp-Schickel, T., and N. Jarrett. 2000. The amphipod family Melitidae on the Pacific coast of North America. Part II. The *Maera-Ceradocus* complex. Amphipacifica 2(4): 23–61.

Kudrjaschov, V. A., and N. L. Tzvetkova. 1975. New and rare species off Amphipoda (Gammaridea) from the coastal waters of the South Sakhalin. Zoologicheskii Zhurnal 54: 1306–1315 (in Russian).

Laubitz, D. R. 1977. A revision of the genera *Dulichia* Kroyer and *Paradulichia* Boeck (Amphipoda, Podoceridae). Canadian Journal of Zoology 55: 942–982.

Light, S. F. 1941. Amphipoda, pp. 93–104, in S. F. Light, Laboratory and Field Text in Invertebrate Zoology, Associated Students Store, University of California, Berkeley, 232 pp.

Lincoln, R. J. 1979. British marine Amphipoda: Gammaridea. British Museum (Natural History), London, 658 pp.

Lowry, J. K., and H. E. Stoddart. 1983. The shallow-water gammaridean Amphipoda of the subantarctic islands of New Zealand and Australia: Lysianassoid. Journal of the Royal Society of New Zealand 13: 279–394.

MacNeil C., J. T. Dick, M. J. Hatcher, R. S. Terry, J. E. Smith, and A. M. Dunn. 2003. Parasite-mediated predation between native and invasive amphipods. Proceedings in Biological Science 270: 1309–1314.

McCarthy, J. E. 1973. The distribution, substrate selection and sediment displacement of *Corophium salmonis* (Stimpson) and *Corophium spinicorne* (Stimpson) on the coast of Oregon, PhD, Oregon State University, 61 pp.

McCloskey, L. R. 1970. A new species of *Dulichia* (Amphipoda, Podoceridae) commensal with a sea urchin. Pacific Science 24: 90–98.

McCurdy, D. G., M. R. Forbes, S. P. Logan, D. Lancaster, and S. I. Mautner. 2005. Foraging and impacts by benthic fish on the intertidal amphipod *Corophium volutator*. Journal of Crustacean Biology 25: 558–564.

McKinney, L. D. 1980. Four new and unusual amphipods from the Gulf of Mexico and Caribbean Sea. Proceedings of the Biological Society of Washington 93: 83–103.

Mills, E. L. 1961. Amphipod crustaceans of the Pacific coast of Canada, I. Family Atylidae. Bulletin of the National Museum of Canada 172: 13–33.

Mills, E. L. 1962. Amphipod crustaceans of the Pacific coast of Canada. II. Family Oedicerotidae. National Museum of Canada Natural History Papers 15: 1–21.

Mills, E. L. 1964. *Ampelisca abdita*, a new amphipod crustacean from eastern North America. Canadian Journal of Zoology 42: 559–575.

Mills, E. L. 1967. A reexamination of some species of *Ampelisca* (Crustacea: Amphipoda) from the east coast of North America. Canadian Journal of Zoology 45: 635–652.

Moore, P. G. 1992. A study of the amphipods from the superfamily Stegocephaloidea Dana, 1852 from the northeastern Pacific region: systematics and distributional ecology. Journal of Natural History 26: 905–936.

Moore, P. G., and P. S. Rainbow. 1992. Aspects of the biology of iron, copper and other metals in relation to feeding in *Andaneixis abyssi*, with notes on *Andaniopsis nordlandica* and *Stegocephalus inflatus* (Amphipoda: Stegocephalidae), from Norwegian waters. Sarsia 76: 215–225.

Moore, S. E., J. M. Grebmeier, and J. R. Davies. 2003. Gray whale distribution relative to forage habitat in the northern Bering Sea: current conditions and retrospective summary. Canadian Journal of Zoology 81: 734–742.

Mouritsen, K. N., and K. T. Jensen. 1997. Parasite transmission between soft-bottom invertebrates: temperature mediated infection rates and mortality in *Corophium volutator*. Marine Ecology Progress Series 151: 123–134.

Myers, A. A., and J. K. Lowry. 2003. A phylogeny and a new classification of the Corophiidea Leach 1814 (Amphipoda). Journal of Crustacean Biology 23: 443–485.

Nagata, K. 1965a. Studies of marine gammaridean Amphipoda of the Seto Inland Sea, I. Publications of the Seto Marine Biology Laboratory 13: 131–170.

Nagata, K. 1965b. Studies of marine gammaridean Amphipoda of the Seto Inland Sea, III. Publications of the Seto Marine Biology Laboratory 13: 191–326.

Olafsson, E. B., and L.-E. Persson. 1986. The interaction between *Nereis diversicolor* O. F. Müller and *Corophium volutator* Pallas as a structuring force in a shallow brackish sediment. Journal of Experimental Marine Biology and Ecology 103: 103–117.

Pelletier, J., and J. W. Chapman. 1996. Application of antibiotics to cultures of the gammaridean amphipod *Corophium spinicorne* Stimpson, 1857. Journal of Crustacean Biology 16: 291–294.

Roney, J. D. 1990. A new species of marine amphipod (Gammaridea: Ampeliscidae) from the sublittoral of southern California. Bulletin of the Southern California Academy of Sciences 89: 124–129.

Rothman, P. L. 1993. New families, genera and species of amphipod crustaceans described by J. L. Barnard (1928–1991). Journal of Natural History 27: 743–780.

Sars, G. O. 1895. Amphipoda, An account of the Crustacea of Norway with short descriptions and figures of all the species, 711 pp.

Schneider, D. C., and B. A. Harrington. 1981. Timing of shorebird migration in relation to prey depletion. Auk 98: 801–811.

Segerstråle, S. G. 1937. Susien uber die Bodentierwelt in sudfinnlandischen kustengewassern III. Zur moorphologie und biologie des ampipoden *Pontoporeia affinis*, nebst einer revision der *Pontoporeia*-systematic. Societas Scientiarum Fennica. Commentationes Biologicae 7: 1–183.

Segerstråle, S. G. 1976. Postglacial lakes and the dispersal of glacial relicts. Commentationes Biologicae 83: 1–15.

Serejo, C. S. 2004. Cladistic revision of talitroidean amphipods (Crustacea, Gammaridea), with a proposal of a new classification. Zoologica Scripta 33: 551–586.

Shillaker, R. O., and P. G. Moore. 1987. The biology of brooding in the amphipods *Lembos websteri* Bate and *Corophium bonelli* Milne Edwards. Journal of Experimental Marine Biology and Ecology 110: 113–132.

Shoemaker, C. R. 1926. Amphipods of the family Bateidae in the collection of the United States Museum. Proceedings of the United States National Museum 68: 1–26.

Shoemaker, C. R. 1931. A new species of amphipod crustacean (Acanthonotozomatidae) from California and notes on *Euystheus tenuicornis*. Proceedings of the United States National Museum 78: 1–8.

Shoemaker, C. R. 1933. Two new genera and six new species of Amphipoda from Tortugas. Carnegie Institution of Washington Publication 435: 245–256.

Shoemaker, C. R. 1934a. Two new species of *Corophium* from the west coast of America. Journal of the Washington Academy of Sciences 24: 356–360.

Shoemaker, C. R. 1938a. Three new species of the amphipod genus *Ampithoe* from the west coast of America. Journal of the Washington Academy of Sciences 28: 15–25.

Shoemaker, C. R. 1938b. Two new species of amphipod crustaceans from the east coast of the United States. Journal of the Washington Academy of Sciences 28: 326–332.

Shoemaker, C. R. 1941. A new genus and a new species of Amphipoda from the Pacific coast of North America. Proceedings of the Biological Society of Washington 54: 183–186.

Shoemaker, C. L. 1942. Amphipod crustaceans collected on the Presidential Cruise of 1938. Smithsonian Miscellaneous Collections 101: 1–52.

Shoemaker, C. R. 1944. Description of a new species of Amphipoda of the genus *Anisogammarus* from Oregon. Journal of the Washington Academy of Sciences 34: 89–93.

Shoemaker, C. R. 1947. Further notes on the amphipod genus *Corophium* from the east coast of America. Journal of the Washington Academy of Sciences 37: 47–63.

Shoemaker, C. L. 1949. The amphipod genus *Corophium* on the west coast of America. Journal of the Washington Academy of Science 39: 66–82.

Shoemaker, C. R. 1952. A new species of commensal amphipod from a spiny lobster. Proceedings of the United States National Museum 102: 231–233.

Shoemaker, C. L. 1955a. Amphipoda collected at the Arctic Laboratory, Office of Naval Research, Point Barrow, Alaska, by G. E. MacGinitie. Smithsonian Miscellaneous Collections 128: 1–78.

Shoemaker, C. L. 1955b. Notes on the amphipod crustacean *Maeroides thompsoni* Walker. Journal of the Washington Academy of Science 45: 59.

Shoemaker, C. R. 1964. Seven new amphipods from the west coast of North America with notes on some unusual species. Proceedings of the United States National Museum 115: 391.

Skogsberg, T., and G. H. Vansell. 1928. Structure and behavior of the amphipod, *Polycheria osborni*. Proceedings of the California Academy of Sciences 17: 267–295.

Slothouber Galbreath J. G., J. E. Smith, T. S. Terry, J. J. Becnel, and A. M. Dunn. 2004. Invasion success of *Fibrillanosema crangonycis*, n. sp., n. g., a novel vertically transmitted microsporidian parasite from the invasive amphipod host *Crangonyx pseudogracilis*. International Journal for Parasitology 34: 235–244.

Stasek, C. R. 1958. A new species of *Allogaussia* (Amphipoda, Lysianassidae) found living within the gastrovascular cavity of the sea-anemone *Anthopleura elegantissima*. Journal of the Washington Academy of Science 48: 119–126.

Staude, C. P. 1995. The amphipod genus *Paramoera* Meirs (Gammaridea: Eusiroidea: Pontogeneiidae) in the eastern North Pacific. Amphipacifica 1(4): 61–102.

Staude, C. P. 1997. Phylum Arthropoda: Subphylum Crustacea: Class Malacostraca: Order Amphipoda, pp. 346–391. In *Marine Invertebrates of the Pacific Northwest*. E. N. Kozloff and L. H. Price, eds. Seattle: University of Washington Press.

Stebbing, T. R. R. 1906. Amphipoda I. Gammaridea. Das Tierreich 21, 806 pp.

Steele, D. H. and P. Brunel. 1968. Amphipoda of the Atlantic and Arctic coasts of North America: *Anonyx* (Lysianassidae). Journal of the Fisheries Research Board of Canada 25: 943–1060.

Steele, V. J., and D. H. Steele. 1993. Presence of two types of calceoli on *Gammarellus angulosus* (Amphipoda: Gammaridea). Journal of Crustacean Biology 13: 538–543.

Stephensen, K. 1932. Some new amphipods from Japan. Annotationes Zoologicae Japonenses 13: 487–501.

Stimpson, W. 1857. On the Crustacea and Echinodermata of the Pacific shores of North America. Boston Journal of Natural History 6: 444–532.

Thomas, J. D. 1992. J. Laurens Barnard (1928–1991), Journal of Crustacean Biology 12: 324–326.

Thomas, J. D., and J. L. Barnard. 1983. Transformation of the *Leucothoides* morph to the *Anamixis* morph (Amphipoda). Journal of Crustacean Biology 3: 154–157.

Thomas, J. D., and J. L. Barnard. 1986. New genera and species of the *Megaluropus* group (Amphipoda, Megaluropidae) from American seas. Bulletin of Marine Science 38: 442–476.

Thorsteinson, E. D. 1941. New or noteworthy amphipods from the North Pacific coast. University of Washington Publications in Oceanography 4: 50–96.

Thurston, M. H. 1990. Abyssal necrophagous amphipods (Crustacea: Amphipoda) in the northeast and tropical Atlantic Ocean. Progress in Oceanography 24: 257–274.

Toft, J. D., J. R. Cordell, and W. C. Fields. 2002. New records off crustaceans (Amphipoda, Isopoda) in the Sacramento/San Joaquin Delta, California, and application of criteria for introduced species. Journal of Crustacean Biology 22: 190–200.

Väinölä, R., and S. Varvio. 1989. Molecular divergence and evolutionary relationships in *Pontoporeia* (Crustacea: Amphipoda). Canadian Journal of Fisheries and Aquatic Science 46: 1705–1713.

Waldron, K. D. 1953. A new subspecies of *Pontoporeia affinis* in Lake Washington with a description of its morphology and life cycle. Masters thesis, University of Washington, Seattle, 123 pp.

Watkin, E. E. 1941. Observations on the night tidal migrant Crustacea of Kames Bay. Journal of the Marine Biological Association of the United Kingdom 25: 81–96.

Watling, L. 1997. 3. The families Ampeliscidae, Amphilochidae, Liljeborgiidae, and Pleustidae, pp. 137–175. In Taxonomic atlas of the benthic fauna of the Santa Maria basin and the western Santa Barbara channel, Volume 12, The Crustacea Part 3. J. A. Blake, L. Watling, and P. V. Scott, eds. The Amphipoda, Santa Barbara Museum of Natural History, 251 pp.

Weckel, A. L. 1907. The freshwater Amphipoda of North America. Proceedings of the United States National Museum 32: 25–58.

Yamato, S. 1987. Four intertidal species of the genus *Melita* (Crustacea: Amphipoda) from Japanese waters, including descriptions of two new species. Publications of the Seto Marine Biology Laboratory 32: 275–302.

Yamato, S. 1988. Two new species of the genus *Melita* (Crustacea: Amphipoda) from brackish waters in Japan. Publications of the Seto Marine Biology Laboratory 33: 79–95.

Zajac, R. N., R. S. Lewis, L. J. Poppe, D. C. Twichell, J. V. Millstone, and M. L. DiGiacomo-Cohen. 2003. Responses of infaunal populations to benthoscape structure and the potential importance of transition zones. Limnology and Oceanography 48: 829–842.

Zhang, J. 1997. Systematics of the freshwater amphipod genus Crangonyx (Crangonyctidae) in North America. PhD thesis, Old Dominion University, 361 pp.

CAPRELLIDAE

LES WATLING AND JAMES T. CARLTON

(Plates 305–311)

Caprellids, or "skeleton shrimp," are remarkable crustaceans in which one can easily invest many hours of profitable observation, watching their feeding behavior, inter- or intraspecific interactions, and sheer gymnastics, as they cling and climb, often perfectly camouflaged, on hydroids, bryozoans, or other substrates. Much remains to be learned about their biology, ecology, and distribution along the Pacific coast, and some relatively common intertidal species remain known from hardly more than their original descriptions. Especially overlooked—or mistaken for juveniles—are those species that are only a few millimeters in length as adults. The patient student working with living substrates, a comfortable chair, and a good microscope will be richly rewarded.

Caprellids, often looking somewhat skeletonlike because of their elongated segments and thus sticklike nature, differ from other amphipods in their overall body shape and the strong reduction of abdominal somites (plate 305). In many species, legs 3 and 4 may also be reduced to a minute article or two, such that only coxal gills are present. The body is usually slender and cylindrical, and the head is generally completely or partly fused with the first pereonite. As a result, the head appears to possess an additional pair of appendages behind the maxillipeds, but these are actually the first pair of gnathopods. The mouth appendages are typical for amphipods in general, but the mandible may be modified through reduction of the palp. All the legs of caprellids, unless reduced, are modified for clinging or grasping. There are generally two pairs of coxal gills, on pereopods 3 and 4, but occasionally a gill is present on pereopod 2 or absent from pereopod 4.

Öostegites, or brood plates, are present in mature females on pereopods 3 and 4 only. When present the öostegites may obscure the presence of minute legs or gills, so those specimens need to be examined carefully. The abdomen is always reduced and generally consists of one or two somites bearing very reduced appendages or simple lobes. In a few cases the details of these lobes can be important in determining the identity of a specimen.

Until the revision of caprellid families by Laubitz (1993), the taxonomy of caprellids had remained virtually unchanged since the early seminal work of Mayer (1882, 1890, 1903). McCain (1970) and Vassilenko (1974) proposed slightly different familial arrangements for the caprellids. However, the characters used were basically those established by Mayer along with the addition of some mandibular features. Laubitz (1993) reduced in significance a few of Mayer's characters, incorporated McCain's mandible features, and added characters from maxilla 1 and the lower lip. While Laubitz (1993) states that her familial divisions are "both preliminary and speculative," they augment those established by Vassilenko (1974). Laubitz (1993) and Takeuchi (1999), however, suggested that caprellids, as presently constituted, are polyphyletic, perhaps originating twice from different gammarid ancestors. Myers and Lowry (2003) conducted a detailed phylogenetic analysis of all corophiidean amphipods, with caprellids included. They showed that caprellids were, as has been suspected for a long time, merely highly modified members of the suborder Corophiidea and did not constitute a separate suborder of amphipods. We have chosen to adopt the Myers and Lowry (2003) arrangement of families and subfamilies that resulted from their phylogenetic analysis in preference to the higher level schemes of Vassilenko, Laubitz, or Takeuchi.

Caprellids are clingers, and so can be found on almost any substratum that is erect above the bottom, including parts of other invertebrates. For the most part, caprellids use their mandibles to scrape microalgae from the surface of the substratum to which they cling. In other cases, they may feed on particles suspended in the water or, rarely, are predaceous (e.g., Caine 1980).

Collecting of caprellids is made somewhat easy by their tendency to aggregate on certain substrates. Rarely does a hydroid or bryozoan colony produce just one specimen. In Japanese and eastern American waters, caprellids have been observed to engage in extended parental care in which the young are either carried on the female's appendages or always in close proximity to the female (Aoki and Kikuchi 1991; Thiel 1997). Similar examples may be found in littoral California and Oregon waters with careful collecting and observing.

The following key is for species found in littoral to shallow waters only and works best for males (which are often much more elongate than females), although an attempt has been made to keep the key as gender-neutral as possible. When keying out caprellids, care must be taken to not let your attention be drawn to certain obvious features, such as body spines and tubercles. Unless explicitly noted, body spination or tuberculation can be a quite variable character and may not be very useful for species discrimination. To facilitate identification with a dissection microscope, other very useful characters, such as those of the mouth appendages or the abdomen, which require higher magnification, have not been used in this key, but should be consulted in the primary literature.

Guides to the Californian and Oregonian caprellid fauna include Laubitz (1970), Martin (1977), and Marelli (1981). Jessen (1969) is an excellent source of habitat and general ecological information on many Pacific coast species. Given the number of introduced caprellids now present in the bays and estuaries of the American Pacific coast, including *Caprella drepanochir, C. equilibra, C. mutica, C. penantis,* and *C. scaura,* further species, especially harbor-dwelling Asian taxa, should be expected. The monographs of Ishitaro Arimoto (1976) and Stella Vassilenko (1974) should be consulted in this regard, as well as more recent summaries and papers, such as those by Takeuchi (1999) and Guerra-Garcia and Takeuchi (2003).

Deeper-water caprellids can be identified using Laubitz (1970) and Watling (1995).

ACKNOWLEDGMENTS

We are very grateful to Marvin Peterson Jessen (formerly with Grand View College, Des Moines, Iowa) for providing us with and allowing us to use the original artwork from his unpublished PhD dissertation. We thank Suzanna Stoike for bringing Jessen's dissertation to our attention.

KEY TO CAPRELLIDS

In part from Martin 1977.

1. Pereonites 3 and 4 with minute, rudimentary pereopods; mandible with palp............................2
— Pereonites 3 and 4 bearing only gills, and öostegites in females; mandible without palp....................7
2. Pereonite 5 with rudimentary pereopods (plate 306B).....
..................................*Mayerella banksia*
— Pereonite 5 leg similar to legs on pereonites 6 and 7.....3
3. Gills present on pereonites 3 and 4.................4
— Gills present on pereonite 2 as well as on pereonites 3 and 4...6
4. Antenna 2 without swimming setae; head with spine or knob (plate 306A)................*Deutella californica*
— Antenna 2 with swimming setae; head smooth 5
5. Pereonites 3 and 4 with median lateral projections over the gills; antenna 2 flagellum slender with long swimming setae; pereopods 5–7 articles slender (plate 306D)........
..................................*Tritella pilimana*
— Pereonites 3 and 4 without median lateral projection over the gills; antenna 2 flagellum stout with short setae; pereopods 5–7 articles stout (plate 306C)......*Tritella laevis*
6. Abdomen five-segmented and with uropods (does not look like typical caprellid); pereonites 5 and 6 short and stout (plate 310A).....................*Cercops compactus*

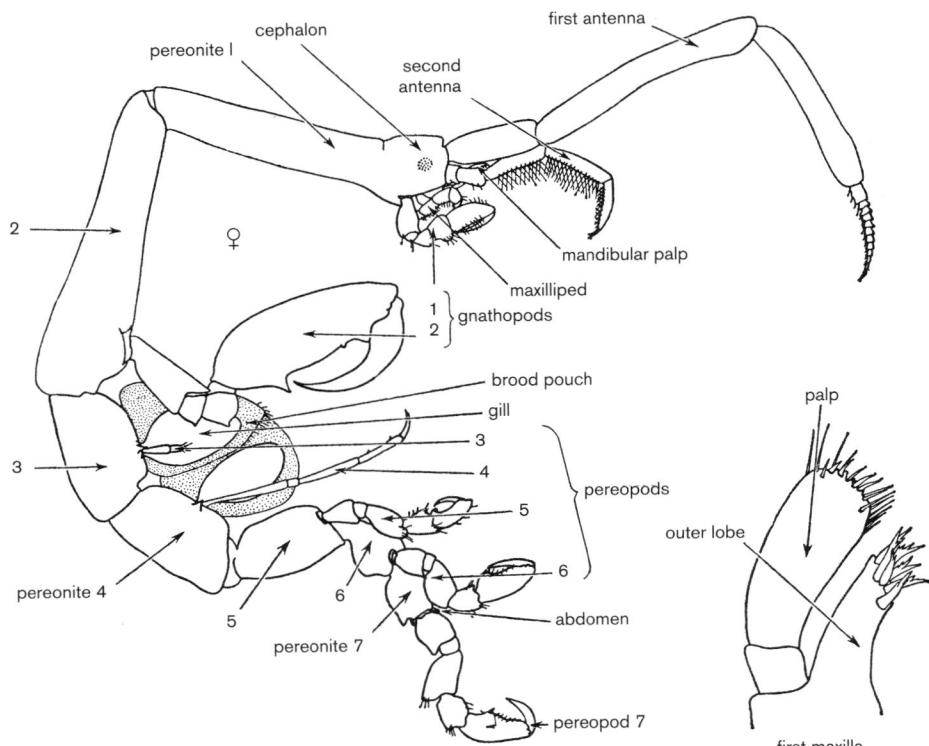

PLATE 305 Basic morphology of a caprellid amphipod (from McCain 1968).

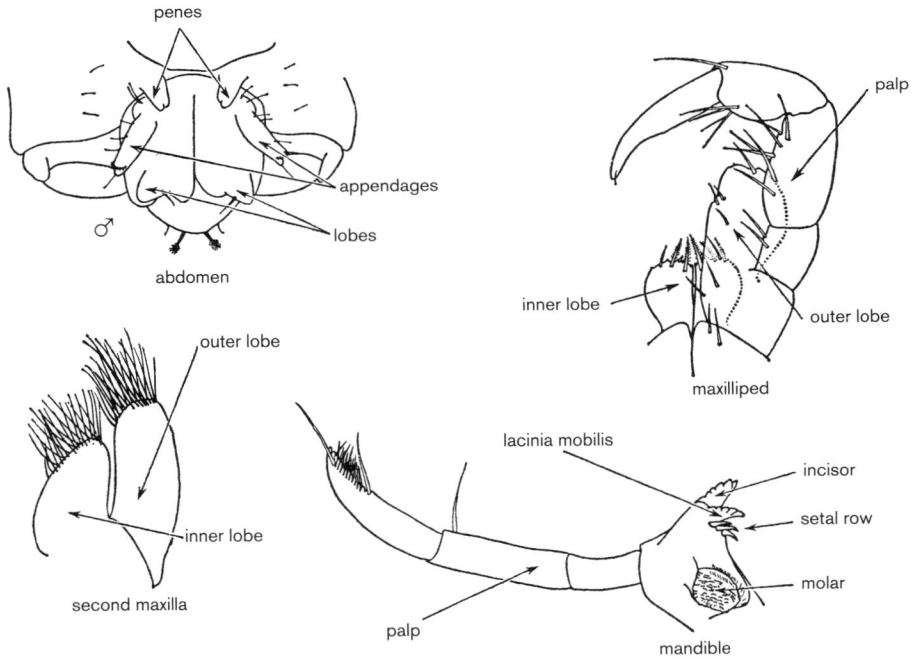

— Abdomen minute; pereonites 5 and 6 long and slender (plate 310B) . . . . *Perotripus brevis*

7. Body short, compact, stout; antenna 1 peduncle articles not much longer than wide (plate 307A). . . . . *Caprella greenleyi*

— Body may be shortened, but not excessively stout; antenna 1 peduncle articles much longer than wide 8

8. Head with paired dorsal spines or tubercles. 9

— Head with 1 anteriorly directed or dorsal spine or without spine 11

9. Head spines appear more as small tubercles; gnathopod 2 merus anteroventral corner smoothly rounded (plate 307B). . . . . *Caprella ferrea*

— Head spines sharp, directed anteriorly; gnathopod 2 merus

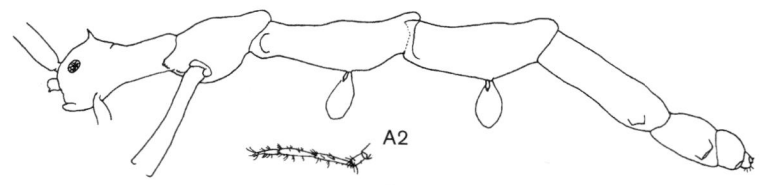

A *Deutella californica*

A2

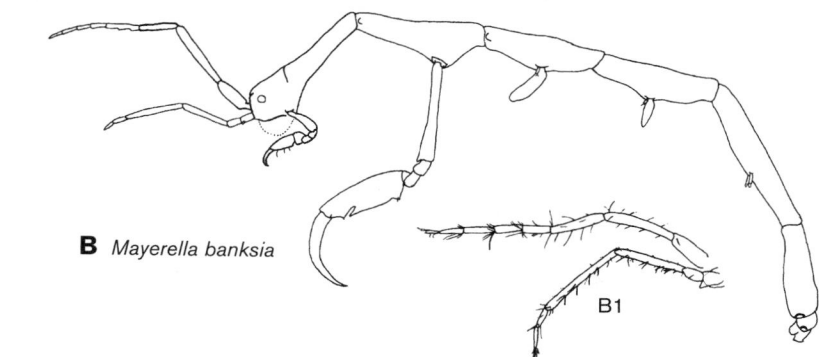

B *Mayerella banksia*

B1

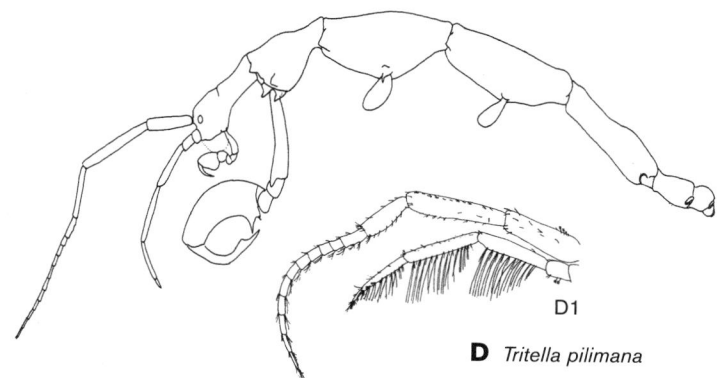

C *Tritella laevis*

C1

PLATE 306 A, *Deutella californica*,
A2, antenna 2; B, *Mayerella banksia*,
B1, antenna 1-2; C, *Tritella laevis*, C1,
antenna 1-2; D, *Tritella pilimana*,
D1, antenna 1-2; drawings not to
scale; (from Laubitz 1970).

D *Tritella pilimana*

D1

anteroventral corner with stout spinelike seta or produced into spinous projection 10

10. Male antenna 1 peduncle articles very setose and stout, flagellum with fewer than 20 articles; pereonites 3 and 4 in both sexes with several pairs of spines (plate 307C) . . . . *Caprella kennerlyi*

— Male antenna 1 peduncle elongate and not heavily setose, flagellum with at least 21 articles; pereonites 3 and 4 in male smooth, in female with few pairs of spines (plate 307D) . . . . *Caprella anomala*

11. Head smooth, without anteriorly directed spine, or with minute dorsal spine. 12

— Head with one large, anteriorly-directed spine 17

12. Pereonites 3, 4, and 5 in males with several to many sharp spines (plate 310C1); note that females may have spines on anterior pereonites, including on cephalon (plate 310C2) . . . . *Caprella mutica*

— Pereonites 3, 4, and 5 in males with no spines, or few blunt spines. 13

13. Pereonites 5–7 in males with one or two small blunt spines (plate 310D); spines may be extend forward onto cephalon in some males; females may be much spinier, with spines

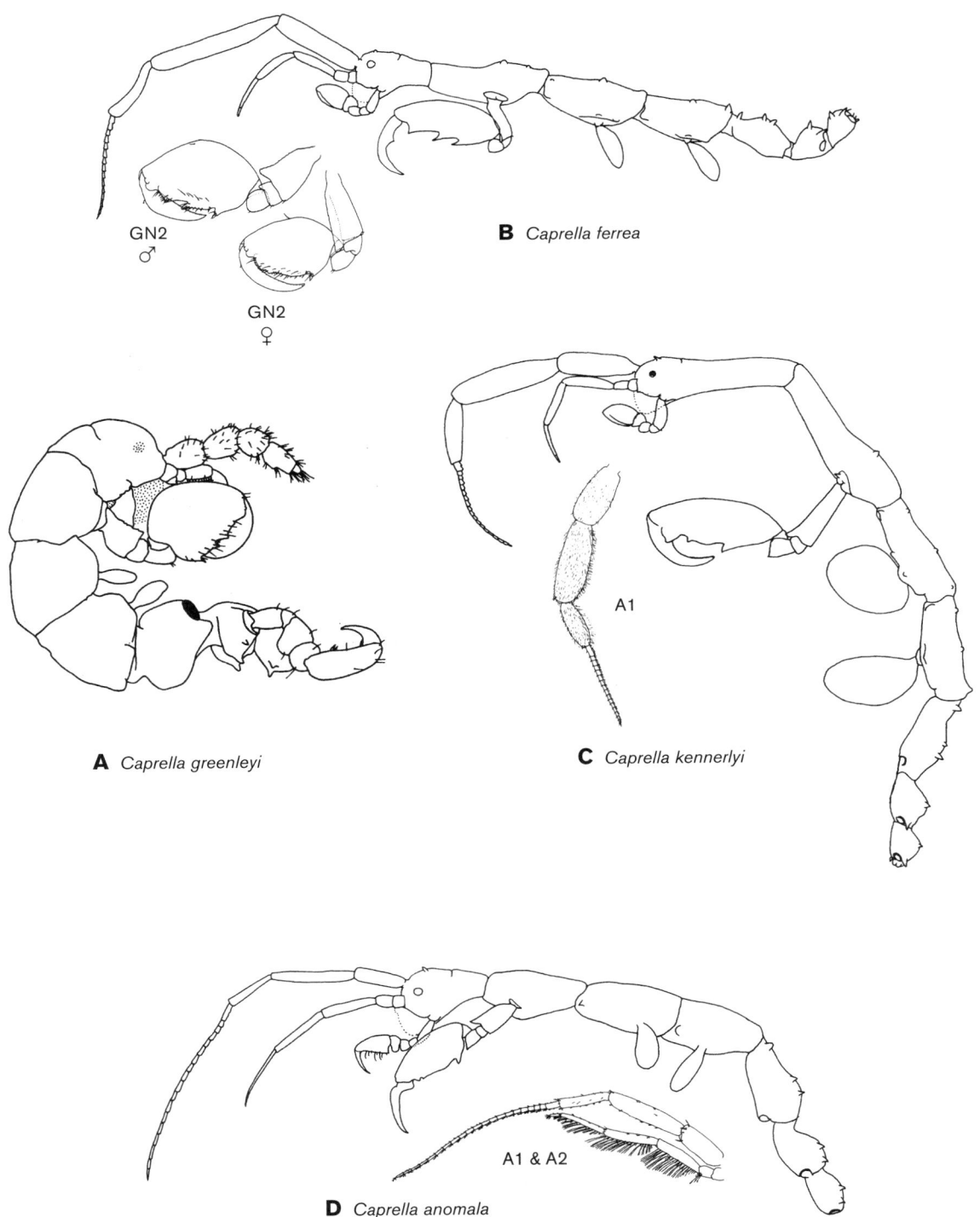

GN2
♂

GN2
♀

B *Caprella ferrea*

A *Caprella greenleyi*

A1

C *Caprella kennerlyi*

A1 & A2

D *Caprella anomala*

PLATE 307 A, *Caprella greenleyi*; B, *Caprella ferrea*; C, *Caprella kennerlyi*; D, *Caprella anomala*; drawings not to scale; A1 = antenna 1, A2 = antenna 2, GN1 = gnathopod 1, GN2 = gnathopod 2 (A, from McCain 1969; rest, Laubitz 1970).

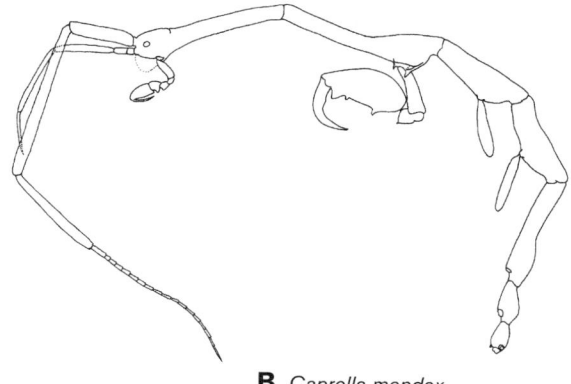

B *Caprella mendax*

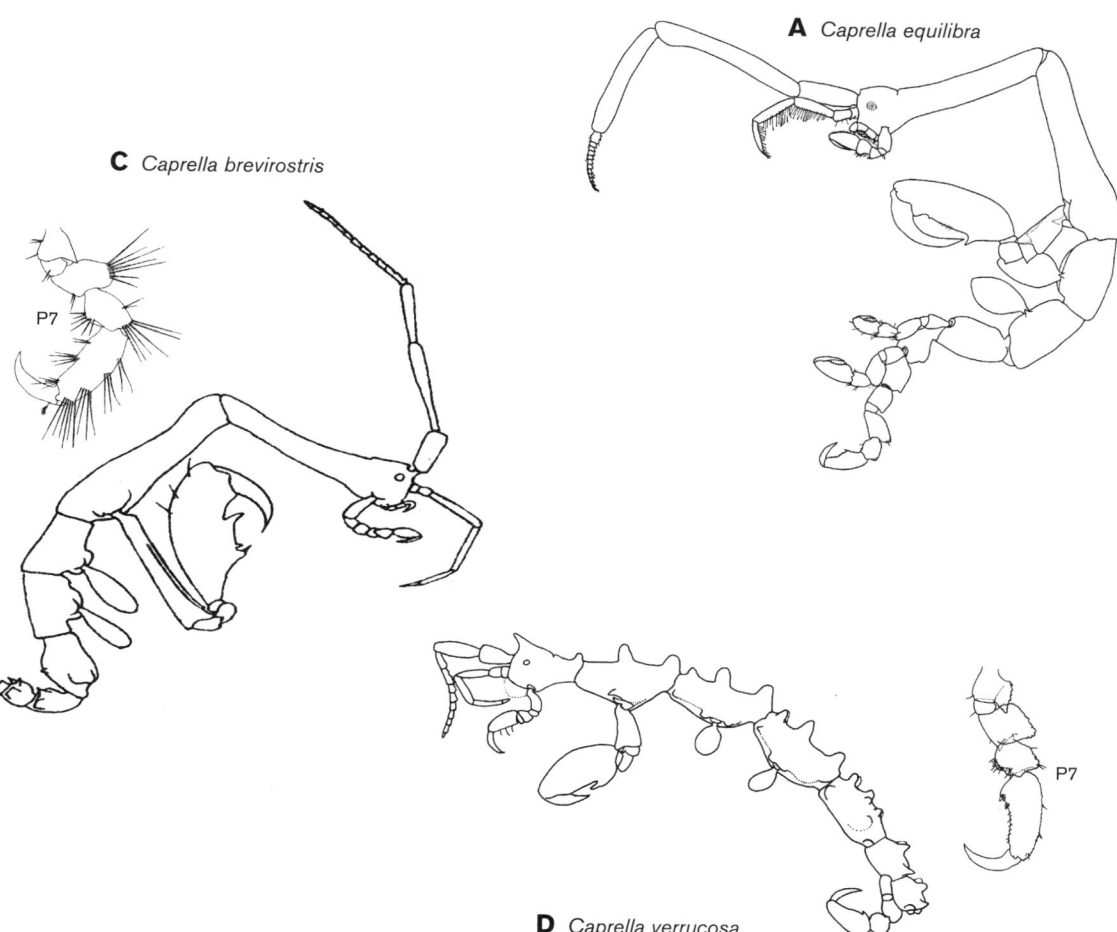

A *Caprella equilibra*

C *Caprella brevirostris*

P7

D *Caprella verrucosa*

P7

PLATE 308 A, *Caprella equilibra*; B, *Caprella mendax*; C, *Caprella brevirostris*; D, *Caprella verrucosa*; drawings not to scale; P7 = pereopod 7 (A, McCain 1968; B, D, from Laubitz 1970; C, from Mayer 1903).

17. Propodus of pereopods lacking heavy, "grasping" setae (plate 308C) . . . . *Caprella brevirostris*
— Propodus of pereopods with grasping setae 18
18. Body tubercles large, wide at base and at tips (plate 308D) . . . . *Caprella verrucosa*
— Body tubercles may be numerous but are low and narrow sharply toward tips or body tubercles absent 19
19. Head spine originating from dorsal margin of head. 20

— Head spine originating from anterior margin of head 24
20. Head spine long and narrow, anteriorly curving, head otherwise smooth 21
— Head with small tubercles and dorsally directed spine with wide base 23
21. Pereonite 5 with one dorsal spine in subadult and adult male; antenna 1 with 18–20 flagellar articles; adult male gnathopod

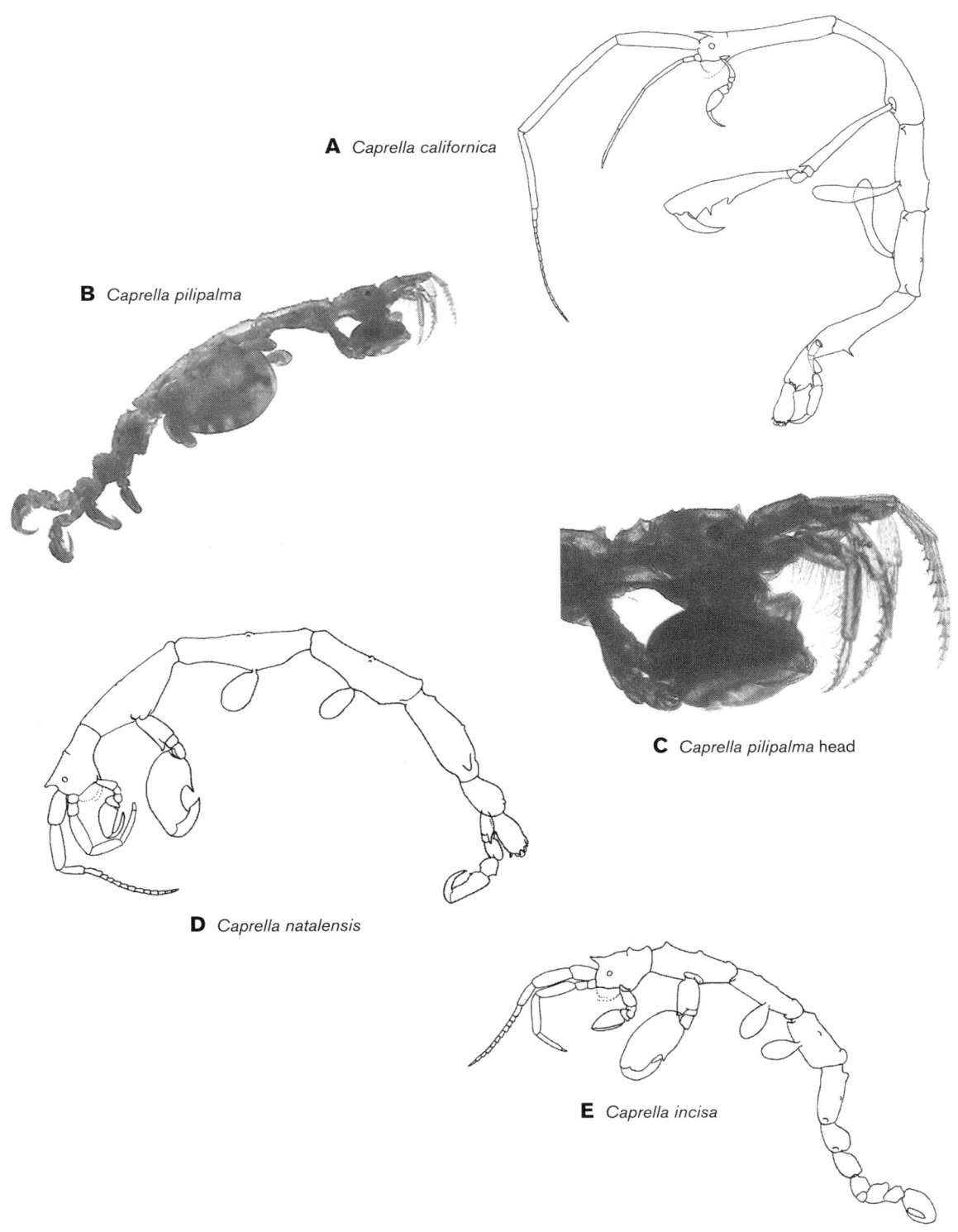

A *Caprella californica*

B *Caprella pilipalma*

C *Caprella pilipalma* head

D *Caprella natalensis*

E *Caprella incisa*

PLATE 309 A, *Caprella californica*; B, *Caprella pilipalma*; C, *Caprella pilipalma* head; D, *Caprella natalensis*; E, *Caprella incisa*; drawings and photos not to scale (A, D, E, from Laubitz 1970; B, photograph by L. Watling).

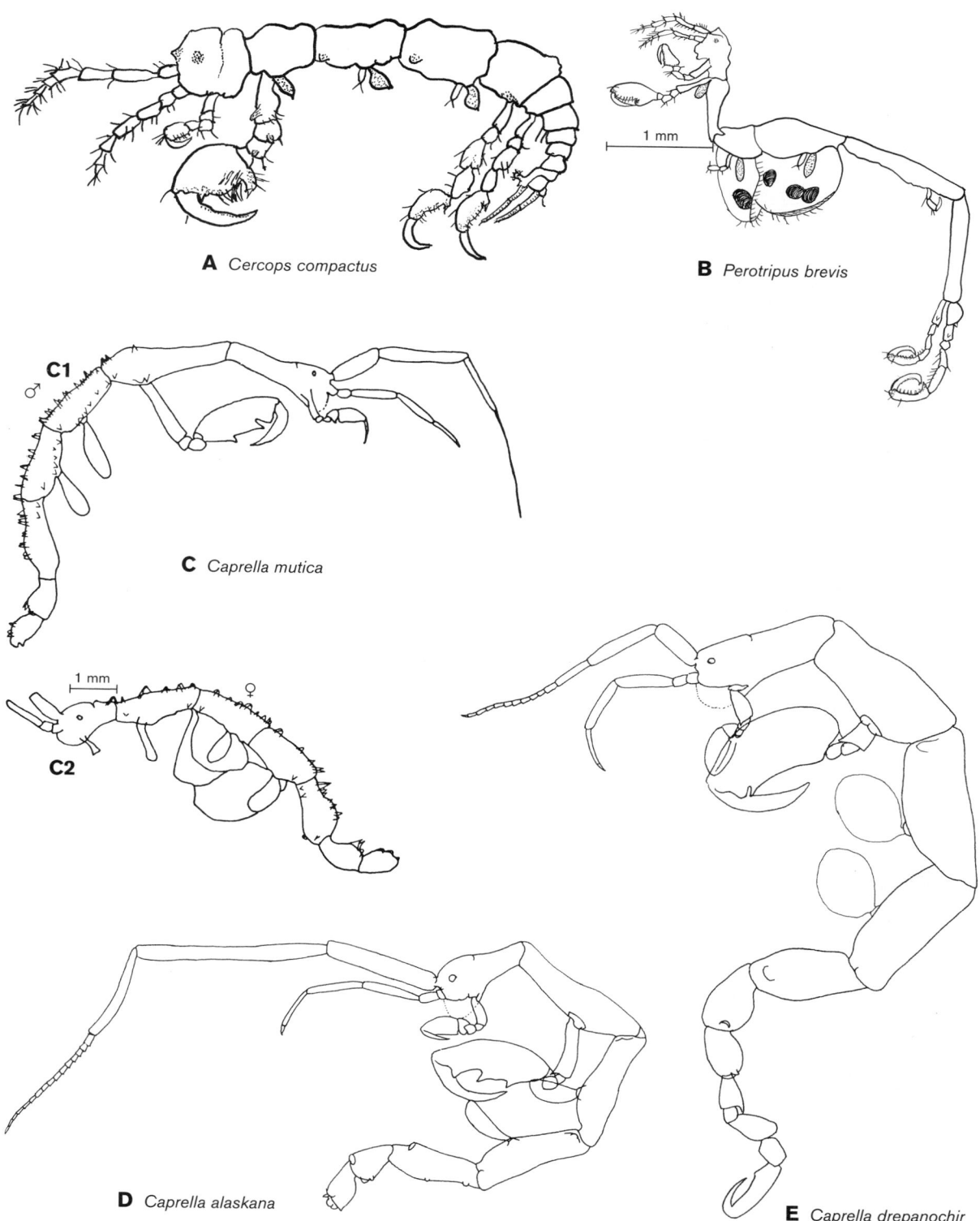

A *Cercops compactus*

B *Perotripus brevis*

1 mm

C1

C *Caprella mutica*

1 mm

C2

D *Caprella alaskana*

E *Caprella drepanochir*

PLATE 310 A, *Cercops compactus*, male; B, *Perotripus brevis*, female; C, *Caprella mutica*, C1, male (California); C2, female (California); D, *Caprella alaskana*, subadult male; E, *Caprella drepanochir*, male; scale bars = 1 mm (A, B, from Jessen 1969; C1, from Marelli 1981; C2, from Martin 1977; D, E, Laubitz 1970).

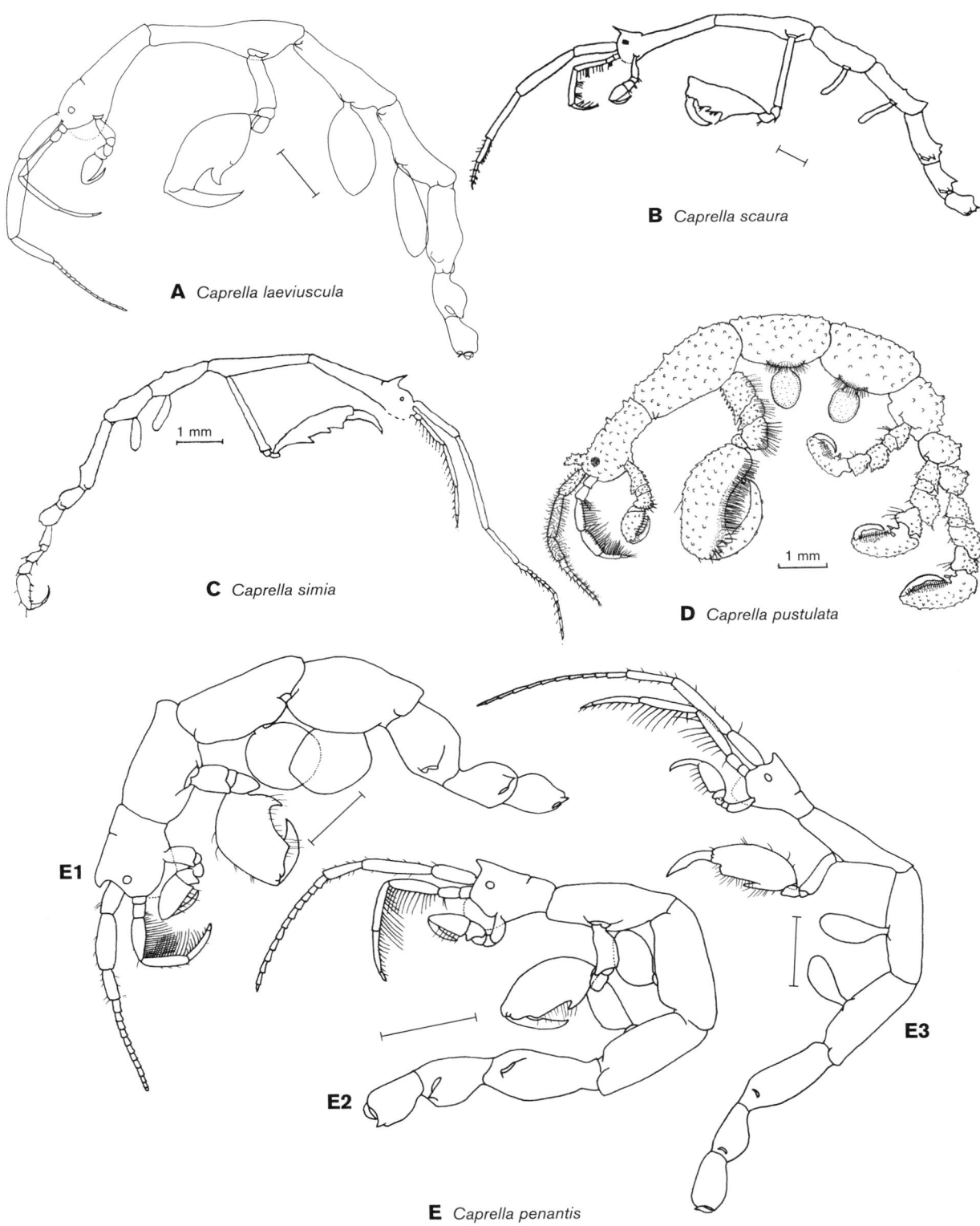

A *Caprella laeviuscula*

B *Caprella scaura*

C *Caprella simia*

1 mm

D *Caprella pustulata*

1 mm

E1

E2

E3

E *Caprella penantis*

PLATE 311 A, *Caprella laeviuscula*, male; B, *Caprella scaura*, male (from California); C, *Caprella simia*, adult male (from Japan); D, *Caprella pustulata*, male; E, *Caprella penantis*, males: E1, from Florida, E2, from Cape Cod, Massachusetts, E3, from Prince Edward Island, Canada, to show variation in robustness; scale bars = 1 mm (A, from Laubitz 1970; B, from Marelli 1981; C, from Arimoto 1976; D, Jessen 1969; E, from Laubitz 1972).

24. Body with tuberculations on all pereonites; boundary between head and pereonite 1 marked with groove and ridge on anterior margin of pereonite 1 (plate 309E) . . . . . *Caprella incisa*
— Body with tuberculations on posterior pereonites; boundary between head and pereonite 1 marked with simple or faint groove 25
25. Pereonite 5 usually shorter than pereonite 6 plus 7 (plate 311E; see also discussion in species list) . . . .
. . . . *Caprella penantis*
— Pereonite 5 usually longer than pereonite 6 plus 7 (plate 309D; see also discussion in species list) . . . .
. . . . *Caprella natalensis*

LIST OF SPECIES

CAPRELLIDAE

CAPRELLINAE

Caprella alaskana Mayer, 1903. San Francisco Bay (Marelli, 1981) and north, intertidal on bryozoans and sabellid worm tubes.

Caprella andreae Mayer, 1890. A high-seas pelagic species found on drifting objects; recorded in the North Pacific and should be looked for on oceanic debris washed ashore in the Pacific Northwest. Illustrated in McCain (1968).

Caprella angusta. See *Caprella natalensis.*

Caprella anomala (Mayer, 1903) (=*Metacaprella anomala*). Intertidal on hydroids; sublittoral to 100 m. Mori (1999) argued for the submergence of the genus *Metacaprella* into *Caprella*; it is retained, however, as a full genus by Myers and Lowry (2003).

Caprella brevirostris Mayer, 1903. On hydroids, algae, and abalone shells; see Martin (1977) for redescription and additional figures.

Caprella californica Stimpson, 1856. In bays (on algae, eelgrass, and hydroids) and on outer coast on coralline algae; sublittoral in kelp beds. See Keith 1969, Crustaceana 16: 119–124 (omnivorous diet); Keith 1971, Pac. Sci. 25: 387–394 (substrate preference: prefers bryozoan *Bugula neritina* over algae *Ulva* and *Polysiphonia*).

Caprella carina Mayer, 1903. Jessen (1969) reported this boreal species washed ashore on a sandy beach near Coos Bay. Illustrated in Laubitz (1972).

Caprella drepanochir Mayer, 1890. Largely known from Asian and subarctic Alaskan waters, and also occurring throughout Coos Bay, Oregon, in fouling communities (Rudy and Rudy 1985; S. Stoike, field collections, 2005), to where it was probably introduced in ship fouling from Japan; to be looked for in other bays and harbors along the coast.

Caprella equilibra Say, 1818. Originally described from South Carolina, and said to occur in many habitats world-over. Often common along the coast in intertidal and shallow waters on hydroids, bryozoans, and ascidians, on wharf pilings and marina floats. Part of its distribution is doubtless due to transport in ship fouling: Mayer (1903) records a specimen from Hong Kong collected "off ships bottoms," and several specimens from ships and buoys in the Atlantic. *C. equilibra* is also reported from 100 m–145 m off southern California (Watling 1995). This species may consist of both a harbor- and bay-dwelling taxon that has been distributed globally in ship-fouling, as well as cryptic species (that may be distinguishable only genetically) that occur on the open rocky shore, or in the deep sea. Martin (1977) reports that intersex individuals occur in San Francisco Bay. See Keith 1969, Crustaceana 16: 119–124 (omnivorous diet); Keith 1971, Pac. Sci. 25: 387–394 (substrate preference).

Caprella ferrea Mayer, 1903 (=*Metacaprella ferrea*). Intertidal on algae, hydroids, bryozoans, sabellid tubes, and other substrates. Martin (1977) noted that specimens from northern California "differed from Laubitz's (1970) description in that peduncular segments 2 and 3 of antenna 1 were heavily setose."

Caprella gracilior Mayer, 1903. Keyed out in the previous edition of this manual, but a generally sublittoral and deeper-water species (in Monterey Bay, for example, in depths of more than 1,700 m). McCain and Steinberg (1970) record it from Tomales Bay, but this may have a transcription error from Dougherty and Steinberg (1953), where no such record is noted. See Caine 1978, Biol. Bull. 155: 288–296 (observations on the seastar *Luidia foliata*, Puget Sound)

Caprella greenleyi McCain, 1969. Originally taken on the seastar *Henricia* in Boiler Bay, Oregon, but a rare habitat (McCain 1969; Martin 1977); on coralline algae and bryozoans. See Martin (1977) for supplemental description and plates; a small species, ranging as adults from 1.5 m–3.6 mm in length.

Caprella incisa Mayer, 1903. On hydroids (such as *Aglaophenia* and campanulariids), bryozoans, coralline algae, and kelp holdfasts.

Caprella kennerlyi (Stimpson, 1864) (=*Metacaprella kennerlyi*). Common all along coast on open shores, intertidal and subtidal, on many substrates, and to be expected in marine fouling communities, having been first described from the bottom of a revenue cutter at Port Townsend, Washington. Jensen (1969) noted that it was "particularly unnoticeable on *Plumularia lagenifera*, where the caprellid body resembles closely a hydrocladium branching from the main axis of the hydroid." Ricketts et al. (1985) describe this species as "prettily pink-banded." See Martin 1977 (habitat diversity); Caine 1978, Biol. Bull. 155: 288–296 (observations on behavior, Puget Sound).

Caprella laeviuscula Mayer, 1903. On many substrates (hydroids, bryozoans, compound ascidians, algae, eelgrass); Martin (1977) notes the occasional presence on the animals of hydroids and the epiphytic diatom *Isthmia*. See Caine 1980, Mar. Biol. 56: 327–335 (ecology, Puget Sound); 1991, J. Crust. Biol. 11: 56–63 (reproductive behavior and sexual dimorphism, Puget Sound).

Caprella mendax Mayer, 1903. Predominately sublittoral throughout range but may extend into shallower water; close to *C. equilibra*, and still listed as a synonym of this species by some workers (Stoddart and Lowry 2003).

Caprella mutica Schurin, 1935 (=*Caprella acanthogaster humboldtiensis* Martin, 1977, described from Humboldt Bay; =*Caprella macho* Platvoet, de Bruyne, and Meyling, 1995, from the Netherlands, in both cases based upon nonnative populations). A distinctive Asian species introduced to the Pacific coast with shipping or oysters, and often very abundant on hydroids in fouling communities on floats and pilings. See Marelli (1981) for synonymy and detailed description.

Caprella natalensis Mayer, 1903 (=*Caprella angusta* in Dougherty and Steinberg, 1953, and in Laubitz, 1970, not of Mayer, 1903; =*Caprella uniforma* LaFollette, 1915). A probable global species complex, this morphospecies is recorded from many habitats worldwide. On California and Oregon shores, intertidal on hydroids, algae, the surfgrass *Phyllospadix*, bryozoans, and so forth, as well as in deeper southern California waters (Watling 1995).

Caprella penantis Leach, 1814. In her Atlantic monograph, Laubitz (1972) reported the "true *C. penantis*" from Monterey Bay (but without specific location, habitat, or date of collection). John McCain, in the previous edition of this manual (1975), speculated that Monterey Bay "may be the northern limit of *C. penantis* and the southern of *C. natalensis*" (but the

* = Not in key.

latter was subsequently reported from deeper water off southern California by Watling [1995]). *C. penantis* is widely known throughout the Atlantic basin, with scattered records through the Pacific Ocean, and may represent an introduction to the California coast. *C. penantis* and *C. natalensis* "are very similar in general appearance" (Laubitz 1972); in addition to differences noted in the key between the species in the ratio of the length of pereonite segments, Laubitz illustrates differences in the propodus palm of gnathopod 2 and in the abdomen of males and females of both species. She also notes that *C. penantis* "tends to be stouter than *C. natalensis* and, particularly in mature specimens, to have very obvious pleural development not present in *C. natalensis*"; the Monterey Bay material reported by Laubitz showed "the typical stout body and strong pleural development . . . and [were] obviously more setose than *C. natalensis*, particularly the adult males." *C. penantis* is well-known to vary considerably in overall morphology and robustness; Laubitz (1972) illustrated a latitudinal gradient in stoutness along the Atlantic coast of North America (plate 311E, herein). See Bynum (1980, Est. Coastal Mar. Sci. 10: 225–237) and Caine, 1989, Crustaceana 9: 425–431) relative to the relationship of robustness to degree of wave exposure.

*Caprella pilidigitata Laubitz, 1970. Intertidal and subtidal in British Columbia and occurring in deeper waters in southern California (Watling 1995). To be looked for in intertidal and shallow waters along the intervening coast. Laubitz (1970) noted that *C. pilidigitata* is similar to *C. mendax* and *C. equilibra*; it is distinguished from *C. mendax* by a setose gnathopod 2 dactylus (it is not setose in *mendax*) and by the absence of a lateral spine at the base of gnathopod 2 (there is a small lateral spine in this position in *mendax*); it is distinguished from *C. equilibra* by the absence of spines on the sides of pereonites 3, 4, and 5 (present in *C. equilibra*); in addition, pereonite 4 is longer than pereonite 5 in *C. pilidigitata*, but equal to or less than the length of pereonite 5 in *C. equilibra*. Illustrated in Laubitz (1970).

Caprella pilipalma Dougherty and Steinberg, 1953. Monterey Peninsula; to be looked for elsewhere. This species has never been illustrated and needs a thorough redescription. Dougherty and Steinberg (1953) noted that *C. natalensis*, *C. incisa*, *C. verrucosa*, and *C. pilipalma* may all occur together on the same hydroids.

Caprella pustulata Laubitz, 1970. A northern species occurring at least as far south as the Coos Bay, Oregon, region, and first noted there by Jessen (1969). See also Martin (1977). On hydroids, bryozoans, and sabellid worm tubes.

Caprella scaura Templeton, 1836. A possible species complex reminiscent of *C. equilibra* and *C. natalensis*; first described from the Indian Ocean, and since reported widely from the Atlantic and Pacific (McCain 1968; Foster et al. 2004). May owe part of its global distribution to shipping. Introduced to the California coast; known from San Francisco Bay and Elkhorn Slough (Marelli 1981). In fouling communities (see Ren and Zhang 1996).

Caprella simia Mayer, 1903. A Japanese species introduced to southern California harbors (Cohen et al. 2005; identification by John Chapman). To be watched for in central California and perhaps further north as well.

Caprella uniforma. Keyed out in the previous edition; see *C. natalensis*.

Caprella verrucosa Boeck, 1871. On open coast and in bays, on hydroids, bryozoans, coralline algae. See Marelli (1981) for morphological variations.

Deutella californica Mayer, 1890. Open rocky shore and on pilings of marine wharfs such as at Monterey, on many substrates; also deeper water off southern California (Watling 1995). See Martin 1977 (habitat diversity); Caine, 1980, Mar. Biol. 56: 327–335 (ecology, Puget Sound).

Mayerella banksia Laubitz, 1970. Primarily sublittoral; see Laubitz 1970 and Watling 1995.

Tritella laevis Mayer, 1903. Intertidal on many substrates (hydroids, algae, bryozoans, coralline algae [e.g., *Odonthalia*], sponges); reported at 88 m in Monterey Bay.

Tritella pilimana Mayer, 1890. Widely occurring on open intertidal coast on many substrates, in bays on eelgrass with hydroids, subtidally on crabpots from 9 m and in deeper water to 145 m off southern California. Martin (1977) reports that intersex specimens exist.

*Tritella tenuissima Dougherty and Steinberg, 1953. Keyed out in previous edition, but a species of deeper offshore waters of central and southern California. McCain (1968) noted that this species "should probably be transferred to *Triliropus*" based on lack of swimming setae on antenna 2 and because pereopod 5 is inserted near the midlength on pereonite 5.

PARACERCOPINAE

Cercops compactus Laubitz, 1970. Open rocky shore, on alga *Plocamium*, coralline algae, bryozoans, hydroids. A small species, reaching 3.8 mm in length.

PHTISICINAE

Perotripus brevis (LaFollette, 1915). Intertidal on hydroids and other substrates on open coast; also known from the sublittoral in southern California (Watling 1995). A small species, under 5 mm (Martin's [1977] largest male was 2.7 mm). See Caine, 1978, Biol. Bull. 155: 288–296 (on tubes of *Phyllochaetopterus prolifica* in very low intertidal, Puget Sound; harpacticoid copepods are major prey).

REFERENCES

Aoki, M., and T. Kikuchi. 1991. Two types of maternal care for juveniles observed in *Caprella monoceros* Mayer, 1890 and *Caprella decipiens* Mayer, 1890 (Amphipoda: Caprellidae). Hydrobiologia 223: 229–237.

Arimoto, I. 1976. Taxonomic studies of caprellids (Crustacea, Amphipoda, Caprellidae) found in the Japanese and adjacent waters. Seto Marine Biological Laboratory, Special Publications III, 229 pp.

Caine, E.A. 1980. Ecology of two littoral species of caprellid amphipods (Crustacea) from Washington, USA. Marine Biology 56: 327–335.

Cohen, A. N., L. H. Harris, B. L. Bingham, J. T. Carlton, J. W. Chapman, C. C. Lambert, G. Lambert, J. C. Ljubenkov, S. N. Murray, L. C. Rao, K. Reardon, and E. Schwindt. 2005. Rapid Assessment Survey for exotic organisms in southern California bays and harbors, and abundance in port and non-port areas. Biological Invasions 7: 995–1002.

Dougherty, E. C., and J. E. Steinberg. 1953. Notes on the skeleton shrimps (Crustacea: Caprellidae) of California. Proceedings of the Biological Society of Washington 66: 39–50.

Foster, J. M., R. W. Heard, and D. M. Knott. 2004. Northern range extensions for *Caprella scaura* Templeton, 1836 (Crustacea: Amphipoda: Caprellidae) on the Florida Gulf coast and in South Carolina. Gulf and Caribbean Research 16: 65–69.

Guerra-Garcia, J. M., and I. Takeuchi. 2003. The Caprellidea (Malacostraca: Amphipoda) from Mirs Bay, Hong Kong, with the description of a new genus and two new species. Journal of Crustacean Biology 23: 154–168. (*Caprella hirayamai*, new species, similar to *Caprella californica* and *C. scaura*).

Jensen, M. P. 1969. The ecology and taxonomy of the Caprellidae (Order: Amphipoda; Suborder: Caprellidea) of the Coos Bay, Oregon, area. PhD dissertation, Department of Entomology, University of Minnesota, 248 pp.

* = Not in key.

LaFollette, R. 1915. Caprellidae from Laguna Beach. Part II. Journal of Entomology and Zoology, Pomona College, California 7: 55–63.

Laubitz, D. R. 1970. Studies on the Caprellidae (Crustacea, Amphipoda) of the American North Pacific. National Museum of Natural Sciences, Canada, Publications in Biological Oceanography, No. 1, pp. 1–89.

Laubitz, D. R. 1972. The Caprellidae (Crustacea, Amphipoda) of Atlantic and Arctic Canada. Natl. Mus. Can. Publs. Biol. Ocean. No. 4, 82 pp.

Laubitz, D.R. 1993. Caprellidea (Crustacea: Amphipoda): towards a new synthesis. Journal of Natural History 27: 965–976.

Marelli, D. C. 1981. New records for Caprellidae in California. Proc. Biol. Soc. Wash. 94: 654–662.

Martin, D. M. 1977. A survey of the family Caprellidae (Crustacea, Amphipoda) from selected sites along the northern California coast. Bull. So. Calif. Acad. Sci. 76: 146–167.

Mayer, P. 1882. Die Caprelliden des Golfes von Neapel und der angrenzenden Meeres-Abschnitte. Ein Monographie. Fauna und Flora des Golfes von Neapel 6: 1–201.

Mayer, P. 1890. Die Caprelliden des Golfes von Neapel. Nachtrag zur Monographie derselben. Fauna und Flora des Golfes von Neapel 17: 1–157.

Mayer, P. 1903. Die Caprellidae der Siboga-Expedition. Siboga Expedition 34: 1–160.

McCain, J. C. 1968. The Caprellidae (Crustacea: Amphipoda) of the western North Atlantic. Bulletin of the United States National Museum 278: 1–147.

McCain, J. C. 1969. A new species of caprellid (Crustacea: Amphipoda) from Oregon. Proceedings of the Biological Society of Washington 82: 507–510.

McCain, J. C. 1970. Familial taxa within the Caprellidae (Crustacea: Amphipoda). Proceedings of the Biological Society of Washington 82: 837–842.

McCain, J. C., and J. E. Steinberg. 1970. Crustaceorum Catalogus. Part 2. Amphipoda I, Caprellidea I, Fam. Caprellidae. Dr. W. Junk N. V., Den Haag, Netherlands, 78 pp.

Mori, A. 1999. *Caprella kuroshio*, a new species (Crustacea: Amphipoda: Caprellidae), with a redescription of *Caprella cicur* Mayer, 1903, and an evaluation of the genus *Metacaprella* Mayer, 1903. Proc. Biol. Soc. Wash. 112: 722–738.

Myers, A. A., and J. K. Lowry. 2003. A phylogeny and new classification of the Corophiidea Leach, 1814 (Amphipoda). Journal of Crustacean Biology 23: 443–485.

Ren, X., and Ch. Zhang. 1996. (Fouling Amphipoda (Crustacea) from Dayawan, Guangdong province, China (South China Sea). Institute of Oceanology (China Academy of Sciences) 1: 58–78 (in Chinese).

Ricketts, E. F., J. Calvin, J. W. Hedgpeth, and D.W. Phillips. 1985. Between Pacific Tides. 5th ed. Stanford, CA: Stanford University Press.

Rudy, P., and L. Rudy. 1985. Oregon estuarine invertebrates. Supplement. On deposit in the Library, University of Oregon Institute of Marine Biology, Charleston, Oregon.

Stoddart, H. E., and J. K. Lowry. 2003. Zoological catalogue of Australia. Crustacea: Malacostraca: Peracarida: Amphipoda, Cumacea, Mysidacea. Volume 19.2B. Victoria, Australia: CSIRO Publishing.

Takeuchi, I. 1999. Checklist and bibliography of the Caprellidae (Crustacea, Amphipoda) from Japanese waters. Otsuchi Marine Science 24, 5–17.

Thiel, M. 1997. Another caprellid amphipod with extended parental care: *Aeginina longicornis*. Journal of Crustacean Biology 17: 275–278.

Vassilenko, S. V. 1974. Caprellids (skeleton shrimps) of the seas of the USSR and adjacent waters. Opredeleliteli po Faune SSSR 107: 1–287. [In Russian]

Watling, L. 1995. The Suborder Caprellidea. Taxonomic Atlas of the Benthic Fauna of the Santa Maria Basin and Western Santa Barbara Channel. Volume 12, Part 3, pp. 223–240. Santa Barbara, CA: Special Publications of the Santa Barbara Museum of Natural History.

CYAMIDAE

JOEL W. MARTIN AND TODD A. HANEY

(Plate 312)

The family Cyamidae is a relatively species-poor group of crustaceans, all of which live in obligate symbiotic association with marine cetaceans (Laubitz 1982; Martin and Heyning 1999; Haney 1999). There are seven genera and 28 species (Martin and Heyning 1999; Margolis et al. 2000). Of those species of

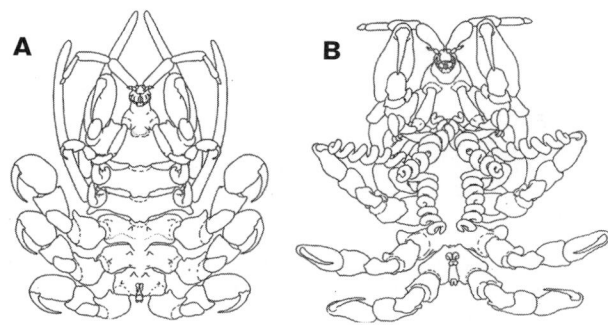

PLATE 312 Ventral aspects of two male cyamids from the gray whale: A, *Cyamus ceti*; B, *Cyamus scammoni* (Haney, original).

cetaceans that most often harbor cyamids, the gray whale (*Eschrichtius robustus*) is by far the most likely to strand on California beaches, with the humpback whale (*Megaptera novaeangliae*) a distant second (J. Heyning, personal communication).

Three species of cyamids are known from gray whales: *Cyamus scammoni* Dall, 1872; *Cyamus kessleri* Brandt, 1872; and *Cyamus ceti* (Linnaeus, 1758). *Cyamus scammoni* is the most commonly encountered species and is immediately distinguished from other cyamids by the branched and tightly coiled gills (plate 312). *Cyamus scammoni* is also the largest of all cyamids, with males and females reaching body lengths of 27 mm and 16 mm, respectively (Leung 1976). *Cyamus scammoni* and *C. ceti* usually attach to the head region of their gray whale host, where as *C. kessleri* typically is found near the anus and genital valves.

The second species, *Cyamus kessleri,* was redescribed by Hurley and Mohr (1957). This species can also be relatively large (up to 21 mm). It is most easily recognized by its narrow body and elongate, uniramous gills. *Cyamus ceti,* which is also known to be associated with the bowhead whale (*Balaena mysticetus*), is smaller and apparently less abundant than the other two species in collections from the gray whale. From humpback whales, Martin and Heyning (1999) list only two cyamid species: *Cyamus boopis* Lütken, 1870, and *Cyamus erraticus* Roussel de Vauzème, 1834. It is possible that the two published records of *Cyamus erraticus* from the humpback whale can be referred to *Cyamus boopis*, given the similar appearances of individuals of these two species. On both gray and humpback whales, respectively, cyamids are commonly found associated with the parasitic barnacles *Cryptolepas rachianecti* and *Conchoderma auritum*. They are otherwise typically restricted to sites on the host animal such as folds of skin, scars, and unhealed wounds, presumably where they are protected from the water currents.

In addition to gill morphology and location, cuticular processes (spines) and morphology of the mouthparts are important taxonomic characters in this family. The above species can be reliably identified without dissection; Leung's (1967) key is particularly helpful. Few other cyamids have been reported from the coast of California and Oregon. Leung (1965) reported an unidentified species of *Cyamus* from Baird's beaked whale (*Berardius bairdii*) from San Francisco Bay and the Farallon Islands. Only two other species have been recorded from California: *Isocyamus kogiae* Sedlak-Weinstein, 1992, from the pygmy sperm whale (*Kogia breviceps*) (Martin and Heyning 1999), and *Syncyamus aequus* Lincoln and Hurley, 1981, from the striped dolphin (*Stenella coeruleoalba*) (Haney et al. 2004).

See Martin and Heyning (1999) for a list of known cyamids and their cetacean hosts.

This section was partially supported by NSF PEET grant DEB 9978193 to J. Martin and D. Jacobs.

REFERENCES

Haney, T. A. 1999. A phylogenetic analysis of the whale-lice (Amphipoda: Cyamidae). University of Charleston, South Carolina, MS thesis, 362 pp.

Haney, T. A., O. De Almeida, and M. S. S. Reis. 2004. A new species of cyamid (Crustacea: Amphipoda) from a stranded cetacean in southern Bahia, Brazil. Bulletin of Marine Science 75: 409–421.

Hurley, D. E., and J. L. Mohr. 1957. On whale-lice (Amphipoda: Cyamidae) from the California gray whale, Eschrichtius glaucas. Journal of Parasitology 43: 352–357.

Laubitz, D. R. 1982. Caprellidea. pp. 292–293. In Synopsis and classification of living organisms. S. P. Parker, ed. New York: McGraw-Hill, Inc.

Leung, Y.-M. 1965. A collection of whale-lice (Cyamidae, Amphipoda). Bulletin of the Southern California Academy of Sciences 64: 132–143.

Leung, Y.-M. 1967. An illustrated key to the species of whale-lice (Amphipoda, Cyamidae), ectoparasites of Cetacea, with a guide to the literature. Crustaceana 12: 279–291.

Leung, Y. -M. 1976. Life cycle of Cyamus scammoni (Amphipoda, Cyamidae), ectoparasite of gray whale, with a remark on the associated species. Scientific Reports of the Whales Research Institute 28: 153–160.

Margolis, L., T. E. McDonald, and E. L. Bousfield. 2000. The whale-lice (Amphipoda: Cyamidae) of the northeastern Pacific region. Amphipacifica 2: 63–117.

Martin, J. W., and J. E. Heyning. 1999. First record of Isocyamus kogiae Sedlak-Weinstein, 1992 (Crustacea, Amphipoda, Cyamidae) from the eastern Pacific, with comments on morphological characters, a key to the genera of the Cyamidae, and a checklist of the cyamids and their hosts. Bulletin of the Southern California Academy of Sciences 98: 26–38.

Hyperiidea

BERTHA E. LAVANIEGOS

Hyperiids are amphipods adapted to pelagic life. Their abundance tends to be low relative to other planktonic crustaceans such as copepods or euphausiids. Hyperiids can be distinguished by the morphology of mouth parts and limbs. Some hyperiids have large compound eyes or inflated bodies; others are elongate or have long antennae. Most hyperiid amphipods live as parasitoids in association with gelatinous zooplankton hosts. Cutting-edge chelae are characteristic of hyperiid species associated with gelatinous organisms such as salps, hydromedusae, siphonophores, and ctenophores. They use these chelae to penetrate the gelatinous tissue of the host. Phronima produce barrellike houses from the bodies of salps, or from large nectophores of siphonophores such as Rosacea cymbiformis.

In some regions of the California Current, aggregations of hyperiid amphipods can play an important role in food webs. For example, Vibilia australis has been found as a significant prey item in stomachs of the chinook (Oncorhynchus tshawytscha) and coho salmon (O. kisutch) (Schabetsberger et al. 2003).

Hyperiids reproduce sexually, but the proportion of females is higher than males. Males of many rare species remain unknown. Despite low abundances, the group is diverse. Up to 80 species have been recorded off southern California (Lavaniegos and Ohman 1999) and about 60 species off central California (Lavaniegos and Ohman, unpublished). The species most frequently found in central California are Primno brevidens, Themisto pacifica (=Parathemisto pacifica), Vibilia armata, Paraphronima gracilis, Tryphana malmi, Phronimopsis spinifera, and Phronima sedentaria. Abundant summer species off the Oregon coast include Themisto pacifica, Paraphronima gracilis, Streetsia challengeri, Tryphana malmi, Hyperia medusarum, Primno macropa, and Hyperoche medusarum (Lorz and Pearcy 1975). A figure of the later species is included in the introductory plate to amphipods (plate 254C).

The most complete identification key of hyperiid amphipods is provided by Vinogradov et al. (1996). Brusca (1981) provides a key to hyperiids along our coast.

REFERENCES

Brusca, G. J. 1981. Annotated keys to the Hyperiidea (Crustacea: Amphipoda) of North American coastal waters. Allan Hancock Foundation, Technical Report 5, 1–76.

Lavaniegos B. E., and M. D. Ohman. 1999. Hyperiid amphipods as indicators of climate change in the California Current, pp. 489–509. In Crustaceans and the biodiversity crisis. Vol. I. F. R. Schram, J. C. von Vaupel Klein, eds.. Brill, Leiden.

Lorz H. V. and Pearcy W. G. 1975. Distribution of hyperiid amphipods off the Oregon coast. J. Fish. Res. Board Can. 32: 1448–1447.

Schabetsberger R., C. A. Morgan, R. D. Brodeur, C. L. Potts, W. T. Peterson, and R. L. Emmett, 2003. Prey selectivity and diel feeding chronology of juvenile chinook (Oncorhynchus tshawytscha) and coho (O. kisutch) salmon in the Columbia River plume. Fisheries Oceanography 12: 523–540.

Vinogradov, M. E., A. F. Volkov, and T. N. Semenova. 1996. Hyperiid amphipods of the world oceans. Science Publishers Inc., New Delhi, 632 pp.

Stomatopoda

ROY L. CALDWELL

(Plate 313)

The only living representatives of the Hoplocarida are the stomatopods, an order of marine predators characterized by a pair of greatly enlarged second maxillipeds, or raptorial appendages, used to capture and process prey. These appendages are analogous to the predatory front legs of the praying mantis. This gives the group their common name of "mantis shrimp."

Functionally, stomatopods can be divided into two groups—spearers and smashers—depending on the type of raptorial appendage they possess (Caldwell and Dingle 1976). Spearers have a thin dactylus armed with two or more spines that are used to impale relatively unarmored prey such as shrimp and fish. In smashers, the heel of the dactylus is enlarged and heavily calcified. This weapon is used as a powerful hammer to smash the shells of armored prey such as snails and crabs. There are more than 450 species of stomatopods, the great majority of which occur in tropical and subtropical shallow coastal seas. However, four species of stomatopod have been recorded from Californian waters including two, Hemisquilla californiensis and Pseudosquillopsis marmorata, that may occasionally be found in our range.

List of Species

Hemisquilla californiensis Stephenson, 1967 (=H. ensigera californiensis, elevated to specific rank by Ahyong, 2001; =H. stylifera of the older California literature). The largest of all smashers, this colorful mantis shrimp can reach 30 cm in length. Adults exhibit an overall yellow-brown body color; telson and distal segments of raptorial appendages greenish yellow to bright yellow; distal segments of antennules, maxillipeds, walking legs and pleopods blue; distal segment of uropods deep blue with red setae fringe, antennal scales bluish at base brownish yellow distally with pink setae; adult males have red polarized patches on their carapace. Postlarvae recruit from the plankton at under 35 mm and burrow in soft

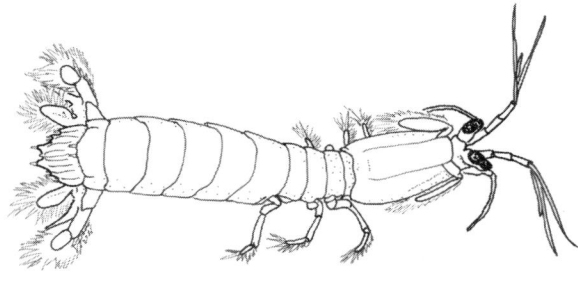

PLATE 313 Stomatopoda. Juvenile of *Pseudosquilliopsis marmorata*, Monterey, California (drawn from life by Christine Huffard).

sediments. *Hemisquilla californiensis* has been recorded at depths from 4 m to more than 100 m, but it typically is found at 7 m–20 m (Hendrickx and Salgado-Barragan 1989). Burrows are simple blind tunnels that extend into the substrate at an angle. A large adult will have a burrow nearly 2 m long and 8 cm–10 cm in diameter. Males may decorate the entrance to their burrow with shells and stones; the burrow entrances of females are unadorned. Adults may forage away from their burrow during the day and can travel 50 m or more, occasionally moving into the low rocky intertidal to collect prey (Basch and Engle 1989). When disturbed, individuals may emit a low frequency hum produced by vibrating the carapace. Breeding occurs in the spring when large numbers of males may be captured in trawls. MacGinitie and MacGinitie (1968) report a single haul of more than 200 males taken at 20 m off Ventura County. Occasionally such catches are sold commercially and appear in local seafood markets. The strike of this species can inflict serious injury and care should be taken when handling adults. *Hemisquilla californiensis* can be recognized by the subglobular cornea of the eye set obliquely on the stalk. The midband of the eyes is made up of six rows of ommatidia. The raptorial claw dactylus is unarmed. *Hemisquilla californiensis* is found from Point Conception south to Golfo de Chirique, Panama. During El Nino years, larvae have been reported as far north as Monterey Bay.

Pseudosquillopsis marmorata Lockington, 1877 (referred to as *Pseudosquilla lessonii* in Schmitt 1940, and Ricketts and Calvin 1968). This is the one spearing stomatopod likely to be encountered in the low intertidal. Postlarvae recruit at 25 mm–29 mm (Manning 1969), and adults may reach 145 mm. The body is a mottled burnt orange to brown and in adults, the distal segments of the uropods are deep pink with purple setae. The telson also has a purple caste and the setae of the antennal scales and pleopods rose purple. Little is known of the biology of this species, but adults construct burrows, often under the edges of rocks. Postlarvae and juveniles may occupy cavities in rubble. *Pseudosquillopsis marmorata* can be recognized by the cornea of the eye being bilobed with the midband consisting of three rows of ommatidia. The raptorial claw dactylus is armed with two teeth. *Pseudosquillopsis marmorata* is normally found from south of Point Conception down the Mexican coast and in the Galapagos Islands. Individuals may occasionally recruit further north. A juvenile has been collected from rubble at 6 m off Monastery Beach in Monterey Bay (plate 313), and a 124 mm adult female was taken from a commercial oyster bed in Tomales Bay.

Nannosquilla anomala Manning, 1967. This is a small spearer that has been reported from San Clemente Island living in vertical burrows in sandy bottoms at a depth of 5 m–23 m. Maximum reported size is 41 mm. It can be recognized by its small eyes with the subglobular cornea set obliquely on the stalk. The midband consists of six rows of ommatidia. The raptorial claw dactylus is armed with 10–14 teeth.

Schmittius politus Manning, 1972. A small spearer found from Monterey Bay to Punta Abreojos, Mexico, at depths of 12 m–185 m, adults reach a maximum size is 60 mm. It can be recognized by its bilobed eye with the midband consisting of two rows of ommatidia. The raptorial claw dactylus is armed with four teeth.

References

Ahyong, S. T. 2001. Revision of the Australian Stomatopod Crustacea. Records of the Australian Museum, Supplement 26; 1–326.

Basch, L. V., and J. M. Engle. 1989. Aspects of the ecology and behavior of the stomatopod *Hemisquilla ensigera californiensis* (Gonodactyloidea: Hemisquillidae). Biology of Stomatopods. E. A. Ferrero (ed.) Selected Symposia and Monographs U.A. I. 3, Mucchi, Modena, pp. 199–212.

Caldwell, R. L., and H. Dingle. 1976. Stomatopods. Sci. Amer. 234: 80–89.

Hendrickx, M. E., and J. Salgado-Barragan. 1989. Ecology and fishery of stomatopods in the Gulf of California, Mexico.). Biology of Stomatopods. E. A. Ferrero, ed. Selected Symposia and Monographs U.A. I. 3, Mucchi, Modena, pp. 241–249.

MacGinitie, G. E. and N. MacGinitie. 1968. Natural history of marine animals. Second Edition. McGraw-Hill Book Co., New York, 523 pp.

Manning, R. 1967. *Nannosquilla anomala*, a new stomatopod crustacean from California. Proc. Biol. Soc. Wash. 80: 147–150.

Manning, R. 1969. The postlarvae and juvenile stages of two species of *Pseudosquillopsis* (Crustacea, Stomatopoda) from the Eastern Pacific region. Proc. Biol. Soc. Wash. 82: 525–37.

Manning, R. 1972. Notes on some stomatopod crustaceans from Peru. Proc. Biol. Soc. Wash. 85: 297–308.

Ricketts, E. F., and J. Calvin. 1968. Between Pacific tides. 4th ed. Revised by J. W. Hedgpeth. Stanford, CA: Stanford University Press, 614 pp.

Schmitt, W. 1940. The stomatopods of the west coast of America based on collections made by the Allan Hancock Expeditions, 1933–1938. Allan Hancock Pacific Exped. 5: 129–225.

Stephenson, W. 1967. A comparison of Australian and American specimens of *Hemisquilla ensigera* (Owen, 1832) (Crustacea: Stomatopoda). Proc. U.S. Nat. Mus. 120: 1–18.

Eucarida

Euphausiacea

LANGDON QUETIN AND ROBIN ROSS

Euphausiids are shrimplike but have biramous thoracic and abdominal appendages, and all thoracic appendages are of a similar form. Decapods, in contrast, have the first three thoracic appendages greatly reduced in size and modified as maxillipeds. In euphausiids, the carapace does not cover the branched gills whereas in decapods the gills are enclosed in branchial chambers. All euphausiids are marine and planktonic, generally less than 35 mm in total length, and have swimming abilities that allow control of both horizontal and vertical location under some conditions. The word *euphausia* is derived from the Greek words *eu-* for good and true and *–phausia* for shining or light-emitting, because bioluminescence is common to this order. In many parts of the world's oceans, euphausiids are an important food for baleen whales, fish, seals, and seabirds. The term "krill" has become synonymous with euphausiids. Norwegian whalers first used krill to describe the swarming "little fish" (euphausiids) in whale feeding grounds (Brinton et al. 1999).

Of the 32 species whose distributions include the northern Californian and Oregonian coast, three are abundant enough in the nearshore California Current System (CCS) to merit mention; all three serve as an important food source for fish, birds, and whales (Brinton et al. 1999). In the CCS, the frequently co-occurring *Euphausia pacifica* Hansen, 1911, and *Nyctiphanes simplex* Hansen, 1911, have large round eyes and lack a rostrum. Although *E. pacifica* can be larger (adults are 11 mm–25 mm vs. 8 mm–16 mm for *N. simplex*), the presence of a leaflet on the first segment of the first antenna of *N. simplex* is a distinguishing characteristic. *E. pacifica* is the primary cool-water species in the CCS, whereas *N. simplex* is a subtropical coastal species that can dominate the communities in southern parts of the CCS.

The third species, *Thysanöessa spinifera* Holmes, 1900, a neritic species that is most abundant at nearshore stations, also has a large, essentially round eye, but the rostrum is long, very acute and reaches past the eye. The adults are 16 mm–25 mm in the CCS. *T. spinifera* ranges from the southeastern Bering Sea to as far as mid-Baja California during particularly cool springtimes (Brinton et al. 1999). This euphausiid is often infected by a dajid isopod parasitic castrator.

Regional distribution and abundance of several of these euphausiids appear to be tied to the ENSO cycle. For example, *N. simplex* has been reported as far north as 46°N (Brodeur 1986) during an extreme warming ENSO episode. The low abundance of *T. spinifera* off Oregon during El Niño may be due to a reversal of the shelf current, moving the adults off shore (Feinberg and Peterson 2003).

Daytime surface swarms have been observed in all three species. Swarms of *T. spinifera* have been observed from San Diego north to Oregon, including Monterey, San Francisco, and Tomales Bays (Brinton et al. 1999), and swarms of *E. pacifica* have been observed off Monterey. Although records of surface swarms of *N. simplex* are primarily from the southern parts of its range in the Gulf of California (Gendron 1992), such swarms may occur throughout its range, which extends into the Monterey/San Francisco/Tomales Bay region during warming episodes.

Mauchline and Fisher (1969) provide an older, but still useful, summary of euphausiid biology and ecology.

REFERENCES

Brinton, E., M. D. Ohman, A. W. Townsend, M. D. Knight, and A. L. Bridgeman. 1999. Euphausiids of the World Ocean. World Biodiversity Database. CD-ROM Series. Springer-Verlag, UNESCO.

Brodeur, R. D. 1986. Northward displacement of the euphausiid *Nyctiphanes simplex* Hansen to Oregon and Washington waters following the El Niño event of 1982–1983. Journal of Crustacean Biology 6: 686–692.

Feinberg, L. R., and W. T. Peterson. 2003. Variability in duration and intensity of euphausiid spawning off central Oregon, 1996–2001. Progress in Oceanography 57: 363–379.

Gendron, D. 1992. Population structure of daytime surface swarms of *Nyctiphanes simplex* (Crustacea: Euphausiacea) in the Gulf of California, Mexico. Marine Ecology Progress Series 87: 1–6.

Mauchline, J., and L. R. Fisher. 1969. The biology of euphausiids. Advances in Marine Biology 7: 1–454.

Decapoda

(Plates 314–326)

Members of this large and diverse order have five pairs of thoracic limbs, developed for walking, grasping, or swimming, and three pairs of maxillipeds. Decapods include the familiar shrimps, prawns, lobsters, crayfishes, and crabs, as well as others less familiar; the group is diverse and presently divided into a number of suborders and infraorders:

Suborder Dendrobranchiata: primitive prawns; no local intertidal species
Suborder Pleocyemata
   Infraorder Caridea: shrimps and prawns
   Infraorder Palinura: spiny lobsters and slipper lobsters, no local intertidal species
   Infraorder Astacidea: lobsters and crayfish, locally represented only by freshwater species
   Infraorder Brachyura: true crabs
   Infraorder Thalassinidea: mud and ghost shrimps
   Infraorder Anomura: diverse forms, including:
      Superfamily Paguroidea: hermit crabs and stone crabs
      Superfamily Galatheoidea: porcelain crabs and pelagic galatheids ("red crabs")
      Superfamily Hippoidea: sand crabs

Of these, the Caridea, Thalassinidea, Anomura, and Brachyura are treated in detail below.

BIOLOGY

ARMAND M. KURIS AND PATRICIA S. SADEGHIAN

Decapod crustaceans are relatively large and sturdy, well suited for field observations and study of living or freshly killed specimens. Recognition of life-history features will provide many clues for physiological, ecological, and behavioral discoveries. This section provides a protocol for efficient observation of crustacean field biology. Details of life history have not been recorded for most of our local species. Thus the primary intention of this protocol is not to describe well-known features, but to direct attention to topics where interesting discoveries may be made.

SEXING

All local species of decapods attach their eggs to the pleopods of the female for embryonic development. These **OVIGEROUS FEMALES** provide the quickest means for sexual determination, but not all females breed at the same time or in all seasons.

The Caridea are only weakly sexually dimorphic; one must look for the presence (males) or absence (females) of an **APPENDIX MASCULINA** on the second pair of pleopods (plate 315C1, 315C2).

Brachyura are markedly sexually dimorphic. Males have relatively large chelae and narrow, often triangular abdomens (plate 315A1); first and second pleopods are specialized for copulation (plate 315E1, 315E2); third and fourth pleopods are absent. Females have wide, flaplike abdomens (plate 315A3–315A5) and four pairs of pleopods with long setae for egg attachment.

Anomura have a variety of sexually dimorphic features. In general, males have fewer pairs of relatively small, slender, or inflexible pleopods. Local female hermit crabs (*Pagurus*) have four pairs of pleopods, males have three (plate 315F1, 315F2). Stone crabs (*Hapalogaster*, *Oedignathus*) are quickly sexed, as only females show signs of segmentation on the left side of

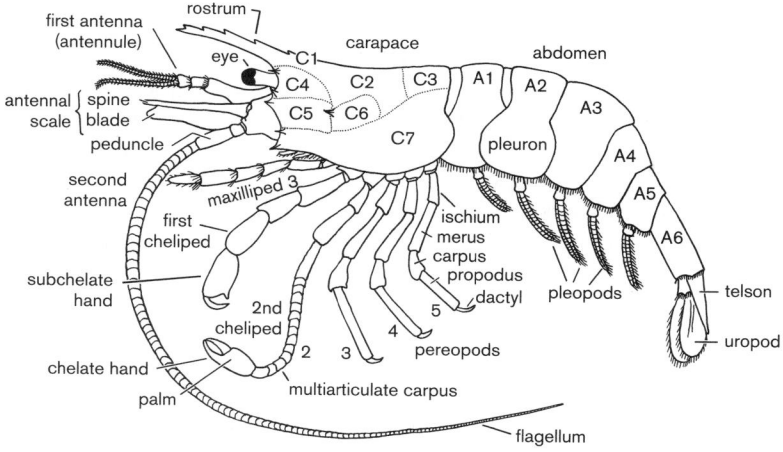

A GENERALIZED CARIDEAN

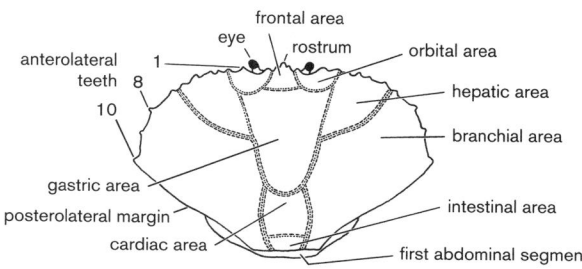

B DORSAL VIEW OF BRACHYURAN

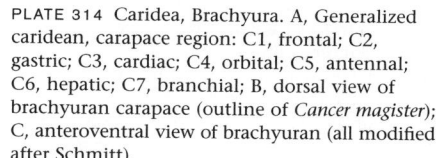

PLATE 314 Caridea, Brachyura. A, Generalized caridean, carapace region: C1, frontal; C2, gastric; C3, cardiac; C4, orbital; C5, antennal; C6, hepatic; C7, branchial; B, dorsal view of brachyuran carapace (outline of *Cancer magister*); C, anteroventral view of brachyuran (all modified after Schmitt).

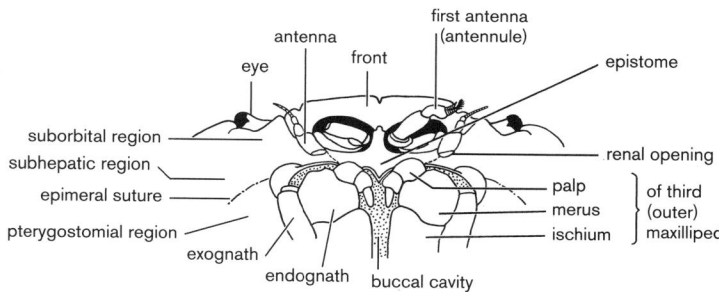

C ANTEROVENTRAL VIEW OF BRACHYURAN

their soft, asymmetrical abdomens. Female porcelain crabs (*Petrolisthes*) have two pairs of pleopods; males have one (plate 315G1, 315G2). Sand crab (*Emerita*) females have three pairs of pleopods; males have none. Ghost and mud shrimp (*Neotrypaea* and *Upogebia*) males also have fewer and/or smaller and more slender pleopods than do females (plate 315H, 315I). The function of these appendages remains to be studied. However, *Neotrypaea* is most easily sexed by noting the presence of the large major chela of the males, often twice as long as the minor chela; in females, the major chela exceeds the minor by <50% in length.

Size is another sexually dimorphic feature. Adult female sand crabs (*Emerita*) and most pea crabs (*Pinnixa*, *Fabia*) are much larger than males. Most female shrimp are also larger, but the size ranges of the sexes often overlap. Sizes are similar for both sexes of hermit crabs. In most of the remaining Brachyura, Thalassinidea, and Anomura, males are larger than

females, again with considerable overlap. These patterns probably reflect differences in mating systems.

MATURATION

Upon reaching sexual maturity, most if not all decapods undergo a **MOLT OF PUBERTY** to attain their adult morphology. This is most easily seen in female brachyurans, in which the width of the abdomen relative to other parts of the crab greatly increases at the molt of puberty (plate 315A4, 315A5). In all decapods, more subtle changes in the relative size of the chelae, abdominal width, presence of setae on pleopods and the abdominal margin, and shape of the genital opening can usually be detected in at least one sex. Relative growth techniques (Teissier 1960, Hartnoll 1985, Vogel 1988, Clayton 1990) are very useful in maturation studies. The relative size of

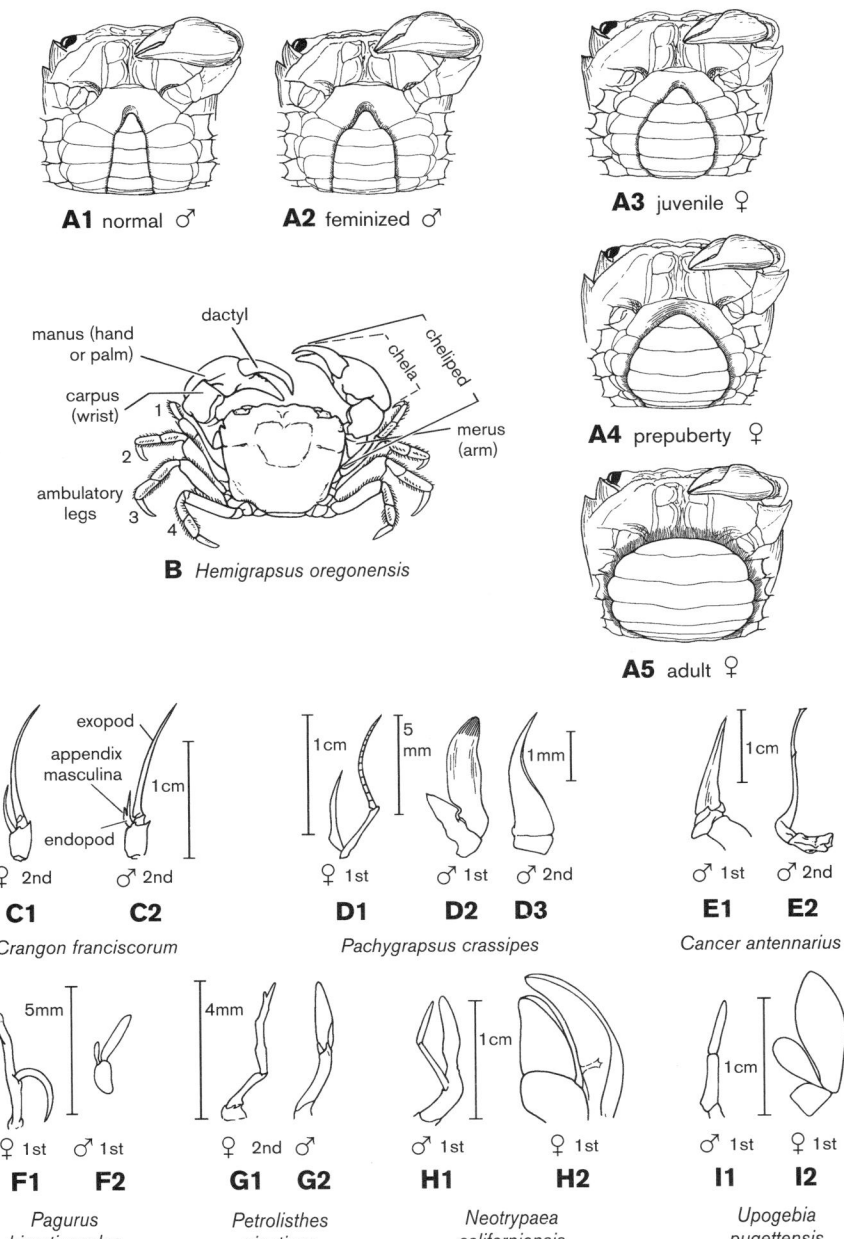

PLATE 315 Caridea, Brachyura, Anomura. A, Abdominal shape and chela length of *Hemigrapsus oregonensis* in various stages of maturation and parasitism by the entoniscid isopod parasitic castrator *Portunion conformis*; all crabs 10.3 mm–10.8 mm carapace width: A1, normal male, A2, male feminized by *Portunion*, A3, juvenile female, A4, prepuberty female, A5, adult female; B, *H. oregonensis*, as labeled (redrawn from Joel Hedgpeth by E. Reid); C–I, sexual dimorphism of decapod pleopods, sex and number as labeled: C, *Crangon*: C1, left, posterior face, C2, 50 mm total length; D, *Pachygrapsus*: D1, 20 mm carapace width, left, anterior face, D2, D3, left, posterior face; E, *Cancer*: 150 mm carapace width, left, anterior face; F, *Pagurus*: left, F1, 8 mm carapace length; G, *Petrolisthes*: left, anterior face, G1, 15 mm carapace length; H, *Neotrypaea*: H1, posterior face, H2, 100 mm total length, anterior face; I, *Upogebia*: posterior face, I2, 100 mm total length (A, by Emily Reid, except A4, drawn by Dottie McLaren; rest from A. Kuris, drawn by E. Reid).

claws is also an aid to recognition of social systems. Males with particularly large claws are generally dominant, and males of different sizes and claw development may adopt very different behavioral reproductive strategies (Kuris et al. 1987).

A related phenomenon is the acquisition of **BREEDING DRESS** prior to oviposition in many shrimp and possibly other decapods. Female shrimp acquire longer setae on their pleopods at the preceding molt. Shorter setae appear at molts preceding nonovigerous instars.

REPRODUCTION

Ovaries of decapod Crustacea lie dorsal to the other organs in the carapace and anterior part of the abdomen (entirely abdominal for hermit crabs and ghost and mud shrimps). They can read-

ily be seen through the transparent cuticle of most shrimps. Corresponding observations can also be made on typically opaque crabs by looking through the flexible **ARTHRODIAL MEMBRANES** connecting the ventral surfaces of thorax and abdomen. Ovarian color varies as does egg color, being red (*Pugettia*), purple (*Pachygrapsus*), brown (*Fabia*), green (*Crangon*), orange (*Emerita*), or magenta (*Petrolisthes*). As the ovary matures and grows, pigmentation becomes more intense. Pigments in the yolk are complexed with vitellogenin proteins.

Relative size of the ripening ovary may be judged by its relation to certain morphological markers, such as sutures separating abdominal segments. Recently laid eggs lie in a gelatinous mass on the pleopods. Within a day, an outer membrane forms around each egg and a thin strand attaches it to pleopodal setae. Young eggs are evenly pigmented. As the yolk is gradually displaced by the growing embryo, a transparent area appears in the egg; in advanced embryos, eyespots and larval chromatophores may also be recognized. Broods about to hatch often have a grayish cast since all the brightly colored yolk has been absorbed. Information about the reproductive state of male decapods is relatively difficult to obtain, dissection and microscopic examination often being required. Most crustaceans oviposit a single brood in an instar. Some species regularly oviposit more than once in an instar, sometimes stripping old egg shells from the egg-bearing setae leaving the setae a glistening gold color.

MOLTING

Discontinuity of growth is one of the most singular aspects of the lives of arthropods. **ECDYSIS**—the actual shedding of the old skin, the **EXUVIA**—and the accompanying increase in size take only a few minutes. However, the entire interval between molts represents a dynamic, cyclical process. Following ecdysis, the animal is soft. In this **POSTMOLT** period, the cuticle gradually hardens as different areas of the exoskeleton calcify sequentially. A series of animals of a given species must be compared to detect this hardening sequence. Postmolt ends with the deposition of a thin, membranous layer. This may be detected by carefully cracking the carapace. If a shiny membrane holds the cracked pieces together, the animal has passed into the "intermolt" period.

The **PREMOLT** period, or preparation for the coming molt, is signaled by the separation of the epidermis from the old cuticle. A mitotic burst, expanding the number of epidermal cells, precedes the deposition of new cuticular structures. First new setae are organized using the previous cuticle as a template. After the new setae are organized, pre-exuvial layers are secreted over the general body surface. These early premolt stages may be recognized in intact carideans and in ghost and mud shrimps by observing setal formation along the margins of the uropods, telson, or antennal scales. Dissection of the transparent mouth parts is usually necessary to see this condition in more heavily calcified decapods. In late premolt, decalcification of the old exoskeleton may be detected if gentle pressure along the epimeral suture (plate 314C) causes it to crack. Molting is imminent if the epimeral suture of a crab, or the dorsal thoracic-abdominal suture of a shrimp, is visibly split. Passano (1960), Skinner (1985), and Chang et al. (1993) provide a good summary of the molt cycle.

Exuviae may be distinguished from the empty remains of a dead animal by the absence of pigment from the corneas of the eyestalks of an exuvia. Discovery of a fragile intact exuvia suggests that a very soft, recently molted animal may be hiding nearby. The previous owner of the exuvia can be certified by matching details of the pigment patterns of the exuvia and the soft animal. Comparison of the soft animal and its exuvia will demonstrate the growth increment per molt for animals of that size.

REGENERATION

Decapods are able to cast off (**AUTOTOMIZE**) their limbs under duress and then regenerate the appendages at subsequent molts. Autotomy is readily demonstrated by squeezing basal segments of an appendage. A specific muscle is stimulated that slices through a cuticular apodeme, severing the limb. This is not a haphazard process. The autotomized limb is always severed at a preformed breakage plane. Upon autotomy, a flap of skin closes over the severed limb base so that scarcely a drop of blood is lost. The regenerating limb forms in a bud that protrudes from the stump of the autotomized limb. Limb buds are transparent in early regenerative stages. Only when the animal passes into premolt does the bud become pigmented (the new cuticle is being deposited on the regenerating appendage). Limb buds take various shapes depending on the species and the limb lost. They are generally compact, conserving space. This interesting developmental variation has not been much studied. Recently regenerated appendages are smaller than normal limbs, but this size discrepancy is no longer apparent after a second or third molt. In some species the regenerated appendage always has a distinctive appearance. Since spider crabs (e.g., *Pugettia*) and purse crabs (e.g., *Randallia*) cease molting after the molt of puberty, a calcified cap is secreted over stumps of limbs lost in their terminal instar. Presence of such a cap verifies that the animal is an adult. Frequency of missing appendages is high among porcelain crabs, true crabs, and *Betaeus* shrimps; it is low for hermit crabs and rare for most local carideans.

BEHAVIOR

All copulating pairs merit careful study, noting molt stages, reproductive states, and relative sizes. Some species appear to form long-term pair bonds; *Alpheus*, *Betaeus*, and *Pachycheles* are examples (see MacGinitie 1937). Several species of majid crabs form seasonal pods in deeper water with males gathering around mounds of many females. *Loxorhynchus grandis* is the best-studied example (Culver and Kuris 2001). Crabs migrate long distances to join these pods. Examining mating behavior in other majids would be interesting.

Hermit crabs spend much time and energy procuring and retaining their snail shell resource. Competition for shells is associated with complex behavioral signaling. Dominance among species and by large crabs over smaller individuals has often been noted. Some species of hermit crabs have strong preferences for certain species of snail shells, while others may be more concerned with size of the shell resource. Further study of the kinds of snail shells that hermit crabs occupy, how they fit, and the relative sizes of the hermits and their shells will increase our understanding of the use of their limited shell resource.

In local decapods, burrowing ranges from the brief escape behavior of some caridean shrimps to the construction of deep, complex, permanent burrows by ghost and mud shrimps. *Crangon* often settles into shallow depressions leaving only its eyes and antennae exposed. Sand crabs burrow backward as waves recede on sandy beaches. Their long, setose, second antennae are then extended to feed in the moving sand. Ghost shrimps (*Neotrypaea*) build poorly defined burrows in muddy sand; mud shrimps (*Upogebia*) construct permanent U-shaped burrows with strong walls cemented by mucous secretions. *Upogebia* burrows may have enlarged sections, side chambers and two or three openings.

Some decapods display strong tidal rhythms. *Emerita* moves up and down the beach with the incoming and outgoing tides. Diel rhythms are less obvious, but at least *Pachygrapsus* is distinctly nocturnal, foraging at night and tightly wedged into crevices by day. Many rhythmic behavior patterns remain to be observed and described. Many species of majid crabs are decorators, attaching fragments of algae, hydroids, bryozoans, and other encrusting organisms to specialized hooked setae on their carapaces (Wicksten 1980, 1978).

KEYS TO DECAPOD CRUSTACEA

ARMAND M. KURIS, PATRICIA S. SADEGHIAN, AND JAMES T. CARLTON

Schmitt's classic monograph, *Marine Decapod Crustacea of California* (1921), and Jensen's (1995) photographic guide are the most useful references for local decapods. Revisions to major groups (families) are given in the list of decapod species below. Terms used in the following keys are illustrated in plates 314A–C and 315B. The key to *Betaeus* is adapted from the work of Josephine Hart (1964, Proc. U.S. Natl. Mus. 115: 431–466), and the key to *Petrolisthes* from the work of Janet Haig (1960, Allan Hancock Pac. Exped. 24, 440 pp.).

KEY TO MAJOR DECAPOD GROUPS

1. Abdomen shrimplike, with well-developed tail fan. 2
— Abdomen small and folded under carapace or soft, reduced, usually asymmetrical. 4
2. Body generally laterally compressed, shrimplike in form; side plates (pleura) of second abdominal segment overlap those of first; abdomen usually with a sharp bend, third pair of legs chelate. Caridea (Key A)
— Third pair of legs not chelate, abdomen dorsoventrally flattened, lateral margins of second abdominal segment not overlapping first segment. 3
3. Chelipeds large and strong usually symmetrical, carapace cylindrical; freshwater. . . . . Astacura

  See Riegel, 1959 (Calif. Fish Game 45: 29–50). Along the coast, two introduced crayfish, *Pacifastacus leniusculus* (Dana, 1852) and *Procambarus clarkii* (Girard, 1852), occur in shallow muddy sloughs, irrigation ditches, lakes, streams, and at the heads of estuaries. An endemic species, *Pacifastacus nigrescens* (Stimpson, 1859), was present in the San Francisco Bay Area in the 19th century, but may have been replaced by *P. leniusculus*. Epizoic branchiobdellid worms may be found on the body and gills.

— Chelipeds variable usually asymmetrical, sometimes subchelate. Carapace flattened, burrowing shrimp. . . . .
. . . . Thalassinidea (Key C)
4. Abdomen small, folded under thorax, symmetrical; uropods absent; last pair of legs not markedly reduced; antennae between eyes. . . . . Brachyura (Key B)
— Abdomen usually asymmetrical and/or reduced; uropods present or absent; last (fourth) pair of legs almost always reduced, folded up behind bases of preceding pair; posterior sternite of thorax not fused to others; antennae external to eyes. . . . . Anomura (Key C)

KEY A: CARIDEA

1. Rostrum absent or very short, without dorsal teeth. 2
— Rostrum present, distinct, usually well developed and spinose. 18
2. Rostrum very short, dorsally flattened; eyes free (not covered by carapace); hands subchelate. 3
— Rostrum absent or very small and spinelike; eyes free or covered by carapace; hands chelate. 9
3. Carapace with two median gastric spines. . . . .
. . . . *Mesocrangon munitella*
— Carapace with none or one median gastric spines. 4
4. No gastric spine; rostrum narrow, tip pointed (plate 319G) curving strongly downward; telson shorter than uropods;

first (antepenultimate) article of third maxilliped (see plate 316A) broadly expanded. . . . . *Lissocrangon stylirostris*
— One gastric spine; rostrum relatively broad, tip round, straight; telson equal to or longer than uropods; first article of third maxilliped narrow, not dilated. . . . . *Crangon* 5
5. Finger of hand (dactyl of chela) turned down almost parallel (180°) to hand (plate 319A); an acute spine on posterodorsal corner of fifth abdominal segment; inner flagellum of first antenna more than two times as long as outer flagellum (plate 316F). . . . . *Crangon franciscorum*
— Finger of hand at a 45° angle, or less, to hand; no spine on posterodorsal corner of fifth abdominal segment; inner flagellum on first antenna distinctly less than two times as long as outer flagellum. 6
6. Flagella of first antenna equal in length; length of antennal scale about equal to or less than two times width; spine of antennal scale not exceeding blade (plate 319E); anterodistal corner of first pleopod without spine, color mottled, including orange and magenta blotches. . . . . . .
. . . . *Crangon handi*
— Inner flagellum of first antenna distinctly longer than outer flagellum; antennal scale length always greater than two times width; spine of antennal scale almost always distinctly exceeding blade (common exception is *nigricauda*); anterodistal corner of first pleopod with a spine color speckled. 7
7. Tip of telson without three small spines, flanking each side (but with a single small spine on each side slightly proximal to tip); dorsum of sixth abdominal segment smooth; without a distinct row of small setae; antennae as long or longer than body; finger of hand at about a 30° angle to hand (plate 319F) (in living specimens, always distinguished by one prominent circular spot [blue center with concentric black and yellow rings] on side of sixth abdominal segment; this often fading in preservative). . . . . *Crangon nigromaculata*
— Tip of telson with three spines on each side; dorsum of sixth abdominal segment slightly grooved, with a distinct row of central setae (may be worn) (living specimens never with colored spot on sixth abdominal segment). 8
8. Antennal scale blade tip narrow, spine long, much exceeding blade (plate 319B); scale greater than two-thirds length of carapace; finger of hand at about 45° angle to hand; antennae about two-thirds body length. . . . .
. . . . *Crangon alaskensis*
— Antennal scale blade tip broad, spine generally short, hardly exceeding blade (plate 319C); scale about two-thirds length of carapace; finger of hand tending toward transverse, at about 30° angle to hand (plate 319D); antennae from two-thirds body length to as long as body (a variable species). . . . . *Crangon nigricauda*
9. Eyes free, chelae not powerfully developed; rostrum very small, reduced to a small spine on frontal margin; three teeth on carapace behind rostral spine, the median the largest; one very prominent supraorbital spine extending beyond anterior margin of carapace; antennal scale broad, subrectangular. . . . . *Lebbeus lagunae*
— Eyes covered by carapace; one or both chelae powerfully developed. 10
10. Rostrum present; one chela greatly enlarged and complex, with dactyl above. 11
— Rostrum absent, one chela usually and about equal, inverted so that dactyls are below (plate 317I–317M); see note in species list. . . . . *Betaeus* 13
11. Pteryostomian spine at anterolateral margin of the carapace,

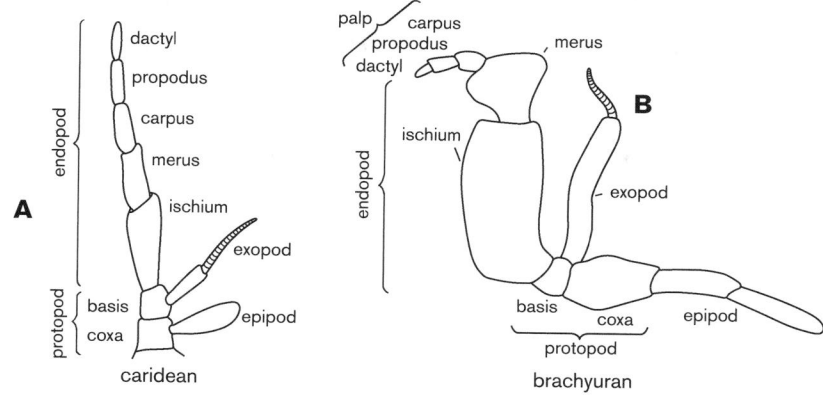

TYPICAL THIRD MAXILLIPEDS

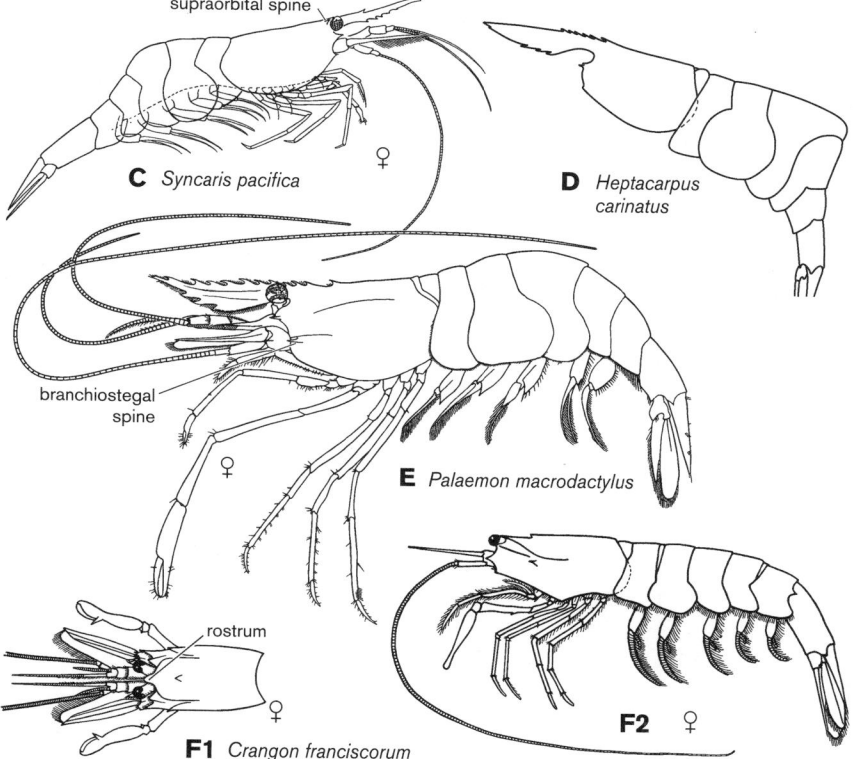

PLATE 316 Caridea and Brachyura. Typical third maxillipeds of caridean, A and brachyuran, B (C, after Hedgpeth 1968; D, after Holmes 1900; E, after Newman 1963; F, after Schmitt 1921).

dactyls of legs bifid yellowish green with red spots . . . . . .
. . . . *Synalpheus lockingtoni*
— Anterolateral margin of the carapace smooth dactyls of legs simple . . . . *Alpheus* 12
12. Tips of walking legs 3–5 simple (plate 317A2), claws mottled orange, yellow, white, body red. . . . . . . *Alpheus bellimanus*
— Tip of walking legs 3–5 with two spines (plate 317A1), claws with black blotches and spots . . . . *Alpheus clamator*
13. Dactyls of ambulatory legs slender and simple 14
— Dactyls of legs stout and bifid 15
14. Chelae with fingers longer than palm (plate 317F, 317K)
. . . . *Betaeus longidactylus*
— Chelae with fingers not longer than palm (plate 317B, 317E, 317L, 317M) . . . . *Betaeus harrimani*
15. Front of carapace rounded, not emarginated (notched) (plate 317G). *Betaeus macginitieae*
— Front emarginated (plate 317C, 317D, 317H) 16

16. Emargination shallow (plate 317D); telson with posterolateral spines small or absent. . . . . *Betaeus harfordi*
— Emargination deep (plate 317C, 317H); telson with well-developed posterolateral spines 17
17. Peduncle of antennule less than one-half carapace length; lower inner ridge of merus of cheliped with long bristles, upper ridge ending in sharp tooth, chelae three times as long as wide, with fingers subequal to palm length (plate 317C, 317J). . . . . *Betaeus gracilis*
— Antennular peduncle approximately equal to carapace length; merus of cheliped with lower inner ridge usually tuberculate, upper ridge with tuft of hairs, not ending in sharp tooth; chela twice as long as wide, with fingers longer than palm (plate 317H, 317I) . . . . *Betaeus setosus*
18. Both legs of first pair simple; second pair of legs very unequal, both with multiarticulate carpus; medium- to large-size, to about 13 cm or more . . . . *Pandalus danae*

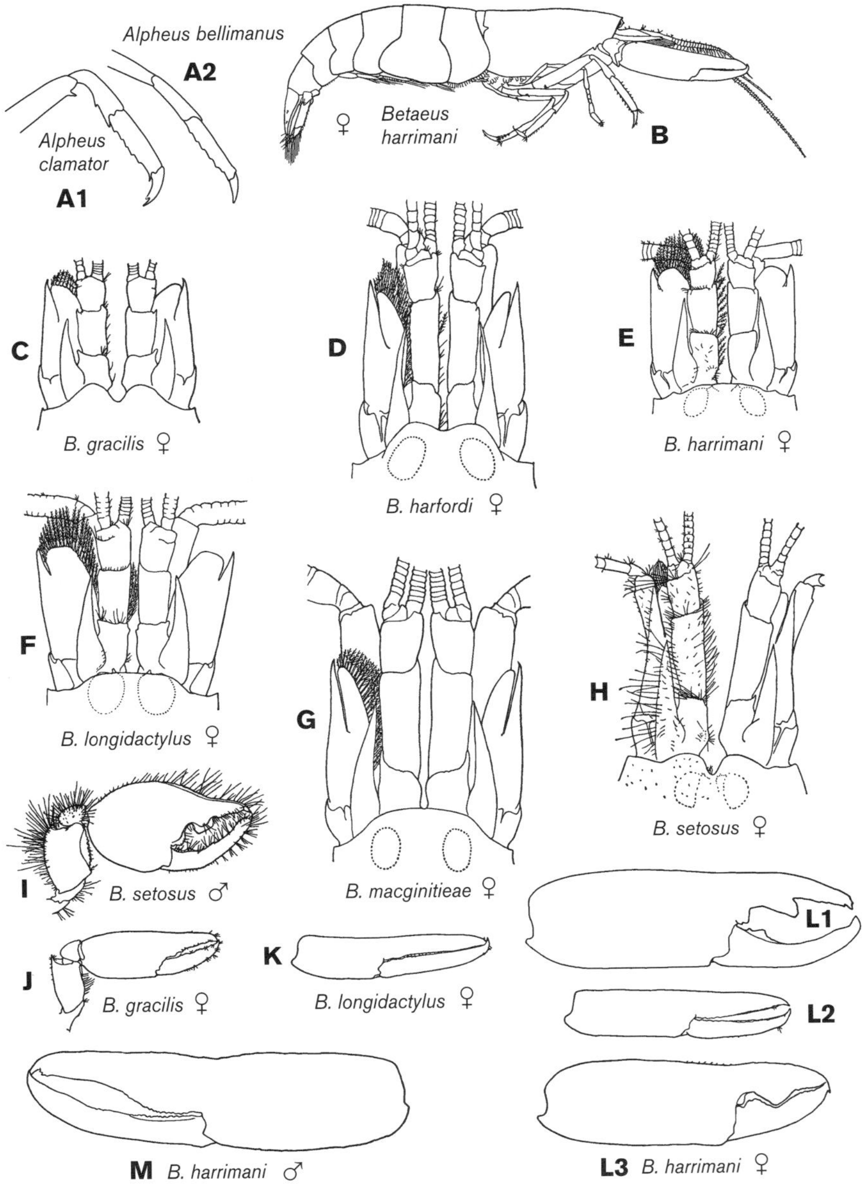

PLATE 317 Caridea. A, walking legs of *Alpheus*; *Betaeus*: B, adult female; C–H, females, frontal region, dorsal; I, male right cheliped; J, female right cheliped; K, female right chela; L, variations in female right chela; M, male left chela (A, after Word and Charwat 1976 [SCCWRP Natantia], redrawn by Dottie McLaren; rest after Hart 1964).

— Both legs of first pair chelate (chelae small); second pair equal or nearly so 19
19. Carpus of second legs not annulated 20
— Carpus of second legs annulated 23
20. Rostrum with one to two dorsal teeth; supraorbital spine present (plates 316C, 318I); in freshwater streams . . . . . .
. . . . *Syncaris pacifica*
— Rostrum with four or more dorsal teeth, supraorbital spine absent, in brackish or marine waters 21
21. Abdominal tergites smooth, lacking ridges, rostrum with at least eight dorsal teeth, at least three teeth behind the orbit, length of dactyl of second walking leg short, less than half the length of the propodus (plate 316E) . . . . . .
. . . . *Palaemon macrodactylus*
— Abdominal tergites with dorsal ridges, no orbital teeth behind the orbit, length of dactyl of second walking leg long, more than half the length of the propodus. . . . .
. . . . *Exopalaemon* 22
22. Dactyl of claw long, more than half the length of the

propodus . . . . *Exopalaemon carinicauda*
— Dactyl of claw short, less than half the length of the propodus . . . . *Exopalaemon modestus*
23. Carpus of second legs with three articles (plate 318H2); colors usually bright green, sometimes brown and red . . . . .
. . . . *Hippolyte* 24
— Carpus of second legs with seven articles; color green, red brown, mottled, or various. 25
24. Rostrum ending in two points (plate 318H1) . . . .
. . . . *Hippolyte californiensis*
— Rostrum ending in three points (plate 318G) . . . .
. . . . *Hippolyte clarki*
25. Carapace with more than 20 segments . . . .
. . . . *Lysmata californica*
— Carapace with seven segments. 26
26. With two to three small, supraorbital spines in a longitudinal series; rostrum high, leaflike, with three dorsal teeth bearing serrate margins, third maxilliped with small exopod; body opaque in life . . . . *Spirontocaris prionota*

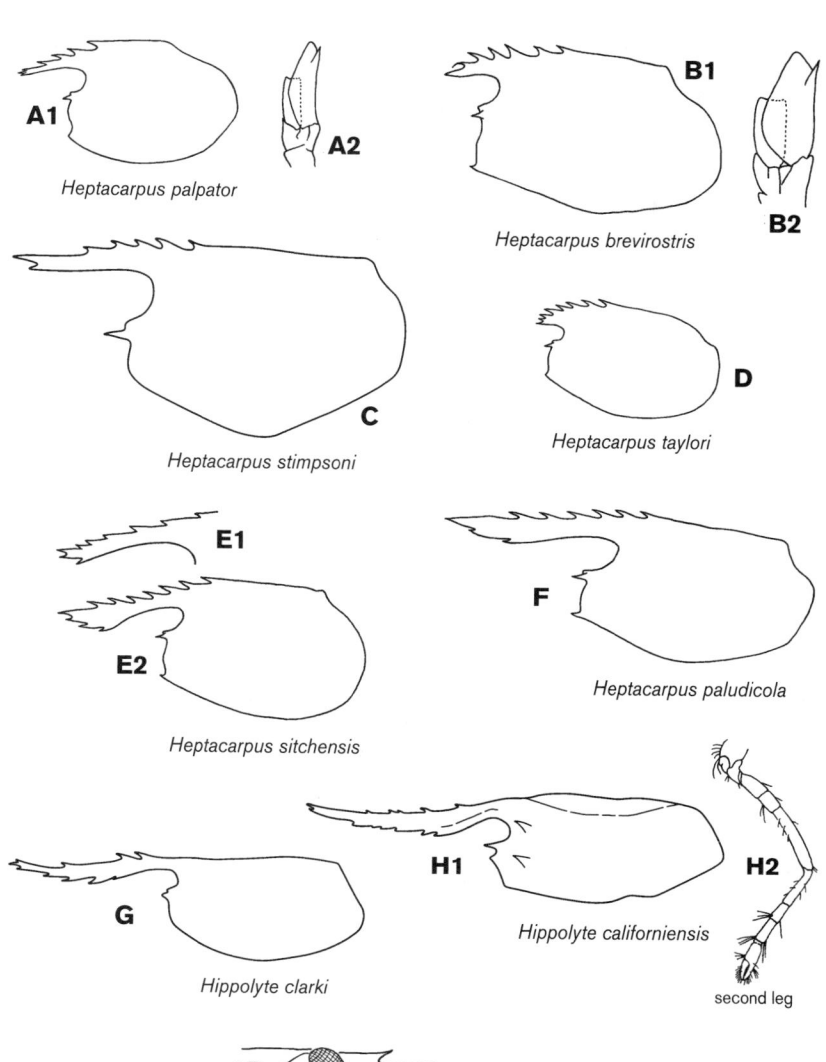

Heptacarpus palpator

Heptacarpus brevirostris

Heptacarpus stimpsoni

Heptacarpus taylori

Heptacarpus sitchensis

Heptacarpus paludicola

Hippolyte clarki

Hippolyte californiensis

second leg

Syncaris pacifica

PLATE 318 Caridea. A–I, outline views of carapaces and antennal scales, as labeled; E, rostral variations; H2, female second right leg (A, B, D–F, after Holmes 1900; C, G, after Word and Charwat 1976 [SCCWRP Natantia], redrawn by Dottie McLaren; E1, after Schmitt; H2, after Chace 1951; I, after Hedgpeth 1968).

— No supraorbital spines; rostral teeth various, not as above; third maxilliped without exopod . . . . *Heptacarpus* 27
27. Rostrum (length measured from posterior margin of orbit to tip) generally as long as or longer than rest of carapace 28
— Rostrum generally shorter than rest of carapace 30
28. Anterior half of rostrum with some dorsal teeth 29
— Anterior half of rostrum lacking dorsal teeth 31
29. Rostral teeth in mature specimens 4–8 dorsally and 1–5 below 4–8/1–5 (plate 318F), subadults with fewer teeth, spine at the anterolateral margin of the carapace (pterygostomian spine) present; uniform green with broken, red brown stripes on carapace (do not confuse with *H. sitchensis*, the rostrum of which does not reach the end of the antennal scale). . . . . *Heptacarpus paludicola*
— Rostral teeth 5 dorsally and 6–7 below 5/6–7, spine at anterolateral margin of carapace (pterygostomian spine) absent . . . . *Heptacarpus franciscanus*
30. Sixth abdominal segment less than two times as long as wide; rostrum deep, one-fourth as deep as long; rostral teeth 4–6/4–6 (plate 316D); epipod present on third max-

illiped color highly variable, often matches algal substrate . . . . *Heptacarpus carinatus*
— Sixth abdominal segment elongate, more than two times as long as wide; rostrum very narrow; rostral teeth 4/4–5; epipod absent on third maxilliped; transparent with a long red line . . . . *Heptacarpus tenuissimus*
31. Rostrum elongate, generally reaching beyond middle of antennal scale, but not to end; rostral teeth in mature specimens 6–7 dorsally and 2–4 below (6–7/2–7) (plates 318E), subadults with fewer teeth, greenish, translucent, with oblique red bands on carapace and crimson bars on legs (do not confuse with *H. paludicola*, the rostrum of which is larger and reaches to or beyond end of antennal scale) . . . . *Heptacarpus sitchensis*
— Rostrum short, generally no reaching middle of antennal scale 32
32. Rostrum not reaching as far as cornea of eye; rostral teeth 5–7/0 (plate 318D); anteriormost rostral teeth often above and slightly behind, rather than well behind, tip; color highly variable, including red-brown, greenish with white carapace, or mottled colors . . . . *Heptacarpus taylori*

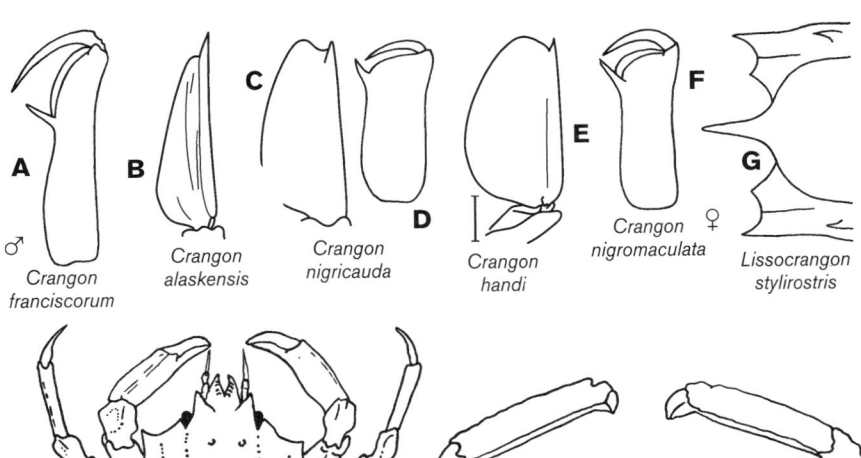

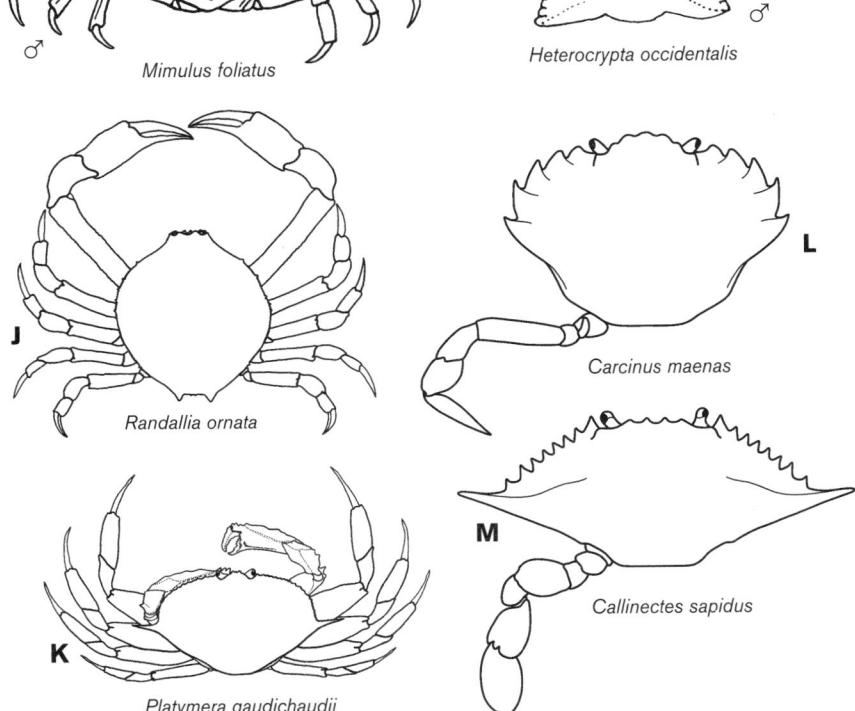

PLATE 319 Caridea and Brachyura. A–C, E, antennal scales; D, F, chela; G, dorsal anterior view of carapace (A, B, D, F, after Rathbun 1904, redrawn by Emily Reid; E, after Kuris and Carlton, 1977, redrawn by Dottie McLaren; C, G, after Holmes 1900, redrawn by Reid; H–K, after Schmitt; L, M, A. Kuris, drawn by Dottie McLaren).

— Rostrum reaching as far as or farther than cornea 33

33. Rostrum reaching beyond first segment of peduncle of antennule; in male may only overlap second antennular segment; rostral teeth 5–8/1–3 (plate 318C); dactyls of ambulatory legs long and slender, about one-third to one-half length of propodus . . . . *Heptacarpus stimpsoni*

— Rostrum not reaching beyond first segment of antennular peduncle; dactyls of legs short and stout, not long and slender 34

34. Epipods on first two pereiopods . . . .
. . . . *Heptacarpus pugettensis*

— Epipods on first three pereiopods 35

35. Antennal scale equal to or shorter than telson; rostral teeth 5–6/0 (plate 318B). . . . . *Heptacarpus brevirostris*

— Antennal scale distinctly longer than telson; rostral teeth 5–6/0–1 (plate 318A). . . . . *Heptacarpus palpator*

KEY B: BRACHYURA

1. Carapace round, with two prominent posterior spines, or ovate with large, straight, lateral spines and more than 12 teeth on anterolateral margin ; mouth field triangular, narrow in front 2

— Carapace nearly square, triangular, ovate, or round; if round, without spines on posterior margin and, if bearing long lateral spines, then with not more than 10 anterolateral teeth; mouth field square. 3

2. Carapace round with two short, prominent spines posteriorly (plate 319J); color white, often with purple patches . . . . *Randallia ornata*

— Carapace ovate with pronounced lateral spines and about 15 small, anterolateral teeth (plate 319K); reddish . . . . . .
. . . . *Platymera gaudichaudii*

3. Carapace nearly square; sides approximately parallel; anterior edge nearly transverse; eyes at anterolateral corners 4
— Carapace triangular, oval, or nearly round, not square; sides not parallel 8
4. Carapace margin smooth, without teeth . . . . Planes cyaneus
— Carapace margin with teeth 5
5. Carapace about as long as wide, frontal margin toothed, outer margin of claws hairy, claws white-tipped . . . . Eriocheir sinensis
— Carapace considerably broader than long, frontal margin without teeth, outer margin of claws not hairy, claws brown-tipped. 6
6. Carapace with transverse flat ridges; strongest laterally; two teeth on anterolateral margin; surface blackish green with numerous red or purple transverse lines . . . . Pachygrapsus crassipes
— Carapace smooth; three teeth on anterolateral margin; without transverse lines . . . . Hemigrapsus 7
7. Color red, purple, or whitish; no hair on legs; chelipeds red-spotted (plate 320A) a green morph lacking spots is fairly common in the northern part of the range . . . . Hemigrapsus nudus
— Color dull brownish green; legs hairy; chelipeds without red spots (plate 315A, 315B) (see note in species list to distinguish young Hemigrapsus). . . . . Hemigrapsus oregonensis
8. Body narrow anteriorly; rostrum single or bifid 9
— Body broad anteriorly; rostrum usually reduced or absent 21
9. Rostrum single. 10
— Rostrum bifid. 12
10. Chelipeds short and stout; carapace broadly pyriform (pear-shaped), with tubercles and fine hairs; short, prominent, spinelike tubercle on first abdominal segment. 11
— Chelipeds much longer and heavier than ambulatory legs; carapace broadly triangular (plate 319I). . . . . Heterocrypta occidentalis
11. Carapace pear-shaped with large tubercles, curved spine behind eye . . . . Pyromaia tuberculata
— Carapace triangular, no spine near eye . . . . Podochela hemphilli
12. Carapace about as broad as long with lateral margins markedly flattened and produced, leaflike; surface smooth, usually encrusted with sponges, bryozoans, etc. (plate 319H). . . . . Mimulus foliatus
— Carapace longer than broad; lateral margins not flattened and produced; surface smooth or rough, sometimes encrusted, obscuring carapace 13
13. Posterolateral margin of carapace without spines 14
— Prominent posterolateral projections. 18
14. Small crabs <5 cm wide, rostrum straight. 15
— Large crab to 25 cm wide, rostrum curved down . . . . Loxorhynchus 17
15. Rostrum two flat plates, claws usually with orange markings, often encrusted with sponges . . . . . . Scyra acutifrons
— Rostrum pointed 16
16. Rostrum short, carapace almost circular, tan to orange, often encrusted with sponges, legs slender . . . . Herbstia parvifrons
— Rostrum long, carapace pear-shaped, very small (to 15 mm), sponges mostly on stocky walking legs . . . .

. . . . Pelia tumida
17. Very large crabs to 25 cm, rostral spines curve strongly down, carapace inflated, covered with small spines and tubercles in juveniles, worn smooth in older terminal adults . . . . Loxorhynchus grandis
— Large crabs to 12 cm, rostral spines slightly decurved, carapace triangular with few large blunt tubercles (plate 320E) . . . . Loxorhynchus crispatus
18. Rostrum consisting of two long, very slender spines; preorbital spine (internal to eye) absent; postorbital spine (external to eye) prominent, slender and acute, far from eye . . . . Oregonia gracilis
— Rostrum otherwise, preorbital spine present or absent. . . . . Pugettia 19
19. Surface of carapace smooth; distance between eyes less than one-third width of carapace in adult specimens (plate 320C) . . . . Pugettia producta
— Carapace tuberculate or spiny; distance between eyes about half the greatest width of carapace. 20
20. Carapace distinctly broader posteriorly; anterolateral teeth narrow, laterally directed; legs moderately long and slender; merus of cheliped with a few tubercles dorsally, not carinate, fingers of claw white. . . . . Pugettia richii
— Carapace not expanded posteriorly; anterolateral teeth broad, anteriorly directed; legs relatively short; merus of cheliped with irregularly dentate keel dorsally, fingers of claw with orange tips . . . . Pugettia gracilis
21. Front (area between eyes) either five-toothed or divided by median notch; carapace hard, anterolateral margin toothed; free living 22
— Front area entire (with the exception of some species of Pinnixa, see below); carapace often membranous, frequently rounded or may be much wider than long; carapace margin not toothed but may have anterolateral acute tubercles or conical spines; commensals in polychaete tubes, molluscan mantle cavities, sea cucumber cloacas, echiuran, and polychaete, bivalve and ghost- and mud-shrimp burrows. . . . . Pinnotheridae (see key, below)
22. Front five-toothed; carapace broadly oval; antennules fold back longitudinally . . . . Cancer 23
— Front area with four or fewer teeth or divided by median notch; antennules fold back longitudinally, transversely, or obliquely 30
23. Carapace widest at seventh or eighth tooth, 12–13 teeth, small, to 55 mm . . . . Cancer oregonensis
— Carapace widest at ninth or tenth tooth, nine to 11 teeth 24
24. Dactyl of cheliped black and spiny . . . . . . Cancer branneri
— Dactyl of cheliped smooth 25
25. Front area markedly produced beyond outer orbital angles forming five nearly equal teeth; fingers of chelipeds black tipped (plate 320D); adults uniformly brick red above, young often brightly and variably colored with spots or stripes . . . . Cancer productus
— Front not markedly produced, with five unequal teeth, color of young not variable, usually similar to adults. 26
26. Carapace widest at tenth anterolateral tooth (first anterolateral tooth is external to eye)(plate 314B); fingers of chelipeds not black-tipped. . . . . Cancer magister
— Carapace widest at eighth or ninth tooth 27
27. Carapace widest at eighth tooth; tenth, and eleventh teeth distinct; teeth with entire edges, curving forward;

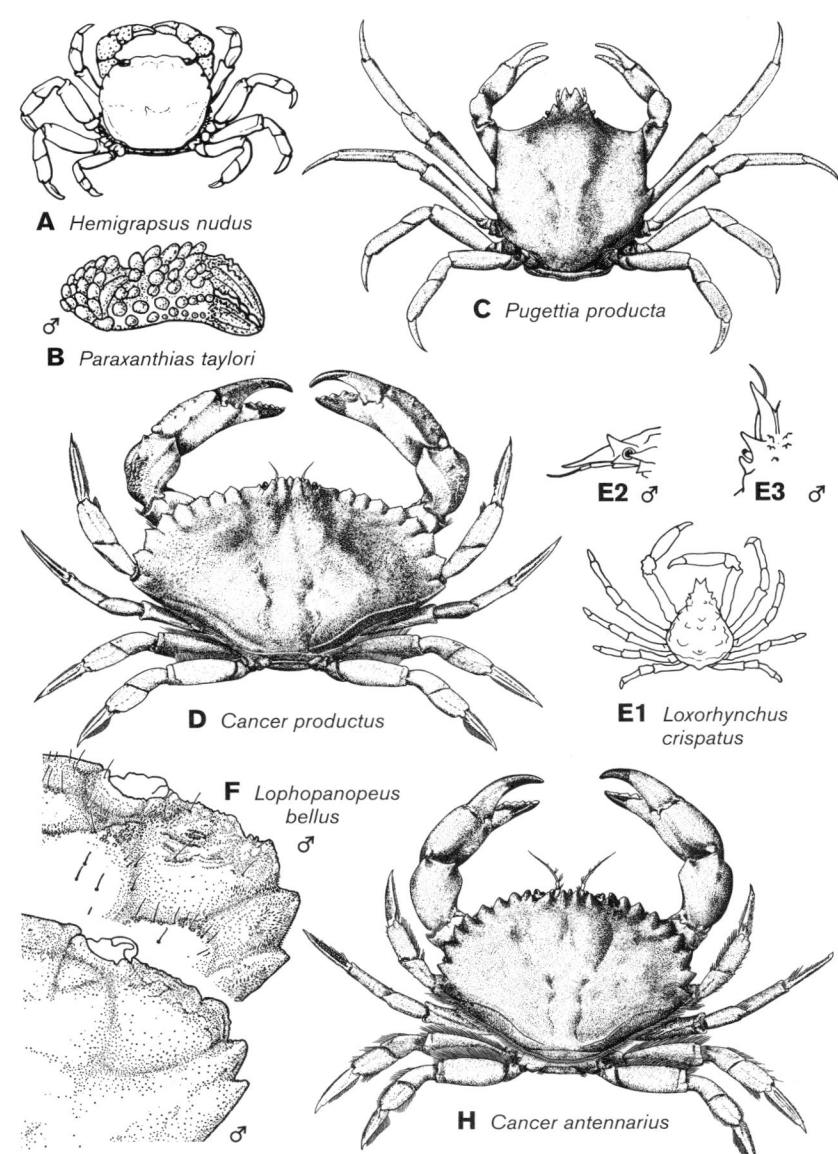

A *Hemigrapsus nudus*

♂

B *Paraxanthias taylori*

C *Pugettia producta*

PLATE 320 Brachyura. B, male chela; F–G, male, dorsal right side of carapace (A, after Hedgpeth 1962; E1, after Rathbun 1925; F, G, after Menzies 1954; rest after Schmitt).

E2 ♂   **E3** ♂

E1 *Loxorhynchus crispatus*

D *Cancer productus*

F *Lophopanopeus bellus* ♂

♂

G *Lophopanopeus leucomanus*

H *Cancer antennarius*

red-spotted beneath; black on fingers of chelipeds (plate 320H) . . . . *Cancer antennarius*
— Carapace generally widest at 9th tooth; not red-spotted beneath 28
28. Upper surface of carapace hairy (pubescent); teeth sharp, curving, with entire edges . . . . *Cancer jordani*
— Carapace smooth, not hairy (glabrous); teeth blunt, with serrate posterior edges 29
29. Fingers of chelipeds white-tipped; merus of third (outer) maxilliped (plate 316B) rounded anteriorly. . . . . . *Cancer gracilis*
— Fingers black-tipped; merus of outer maxilliped truncate anteriorly . . . . *Cancer anthonyi*
30. Carapace subcircular with 6 large lateral spines, hairy, attenules fold back longitudinally, antennal flagellae long and hairy . . . . *Telmessus cheiragonus*
— Carapace shape and spines otherwise, antennules fold back transversely or obliquely, antennal flagellae shorter not hairy 31

31. Front with teeth, carapace with a prominent lateral spine, legs sometimes flattened for swimming 32
— Front divided by a median notch 33
32. Fourth pereiopod a flattened paddle for swimming, carapace with a long sharp lateral spine (plate 319M), claws blue . . . . *Callinectes sapidus*
— Fourth pereiopod somewhat flattened, carapace with five large teeth pointing forward (plate 319L), body and claws mottled green . . . . *Carcinus maenas*
33. Chelipeds with numerous, prominent, rounded tubercles (plate 320B); legs hairy . . . . *Paraxanthias taylori*
— Chelipeds otherwise 34
34. Fingers whitish, in brackish water . . . .
. . . . *Rhithropanopeus harrisii*
— Fingers black; not in brackish water 35
35. Anterolateral margin with eight to 10 small, subequal, acute teeth; carapace broadly oval . . . .
. . . . *Cycloxanthops novemdentatus*

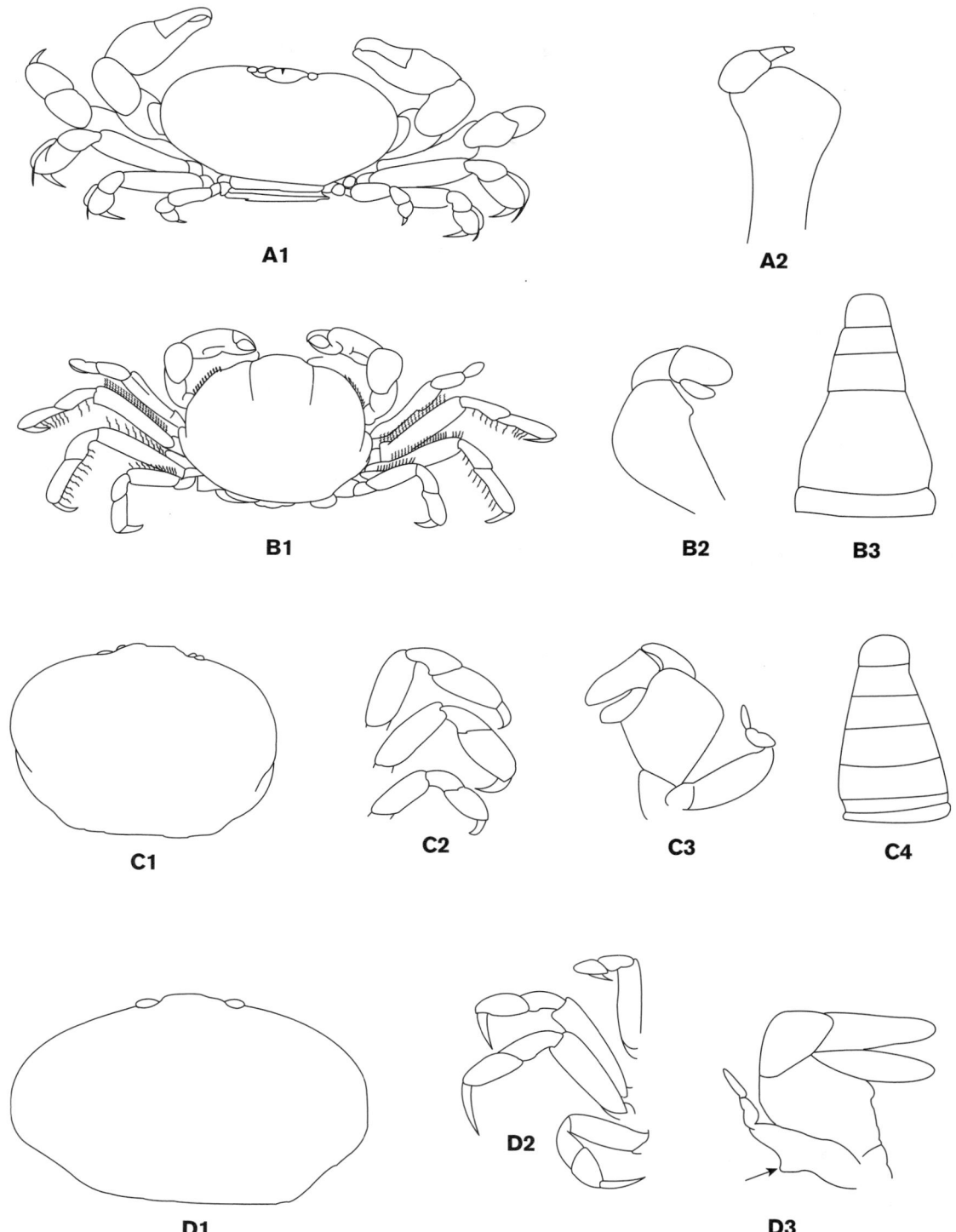

PLATE 321 Brachyura. Pinnotheridae. MXP3 = third maxilliped; WL = walking legs; arrow indicates the external lobe of the exopod of MXP3. *Parapinnixa affinis*: A1, dorsal view; A2, MXP3; *Fabia subquadrata*: B1, dorsal view; B2, MXP3; B3, male abdomen; *Opisthopus transversus*: C1, dorsal view; C2, WL2–WL4; C3, MXP3; C4, male abdomen; *Scleroplax granulata*: D1, carapace dorsal view; D2, WL1–WL4; D3, MXP3 (B1, modified from Bonfil et al. 1992, Ciencias Marinas 18: 37–56). Not to scale.

— Anterolateral margin with 3 prominent, subequal teeth (plate 320F, 320G). . . . . *Lophopanopeus* 36
36. Carapace with distal segments of ambulatory legs hairy (plate 320F) . . . . *Lophopanopeus bellus*
— Carapace and ambulatory legs smooth, not pubescent (plate 320G). *Lophopanopeus leucomanus*

KEY TO PINNOTHERIDAE

ERNESTO CAMPOS

1. First pair of walking legs (WL) stouter and longer than the others (plate 321A1); dactylus of the third maxilliped (MXP3) inserted distally on the propodus (plate 321A2) . . . .

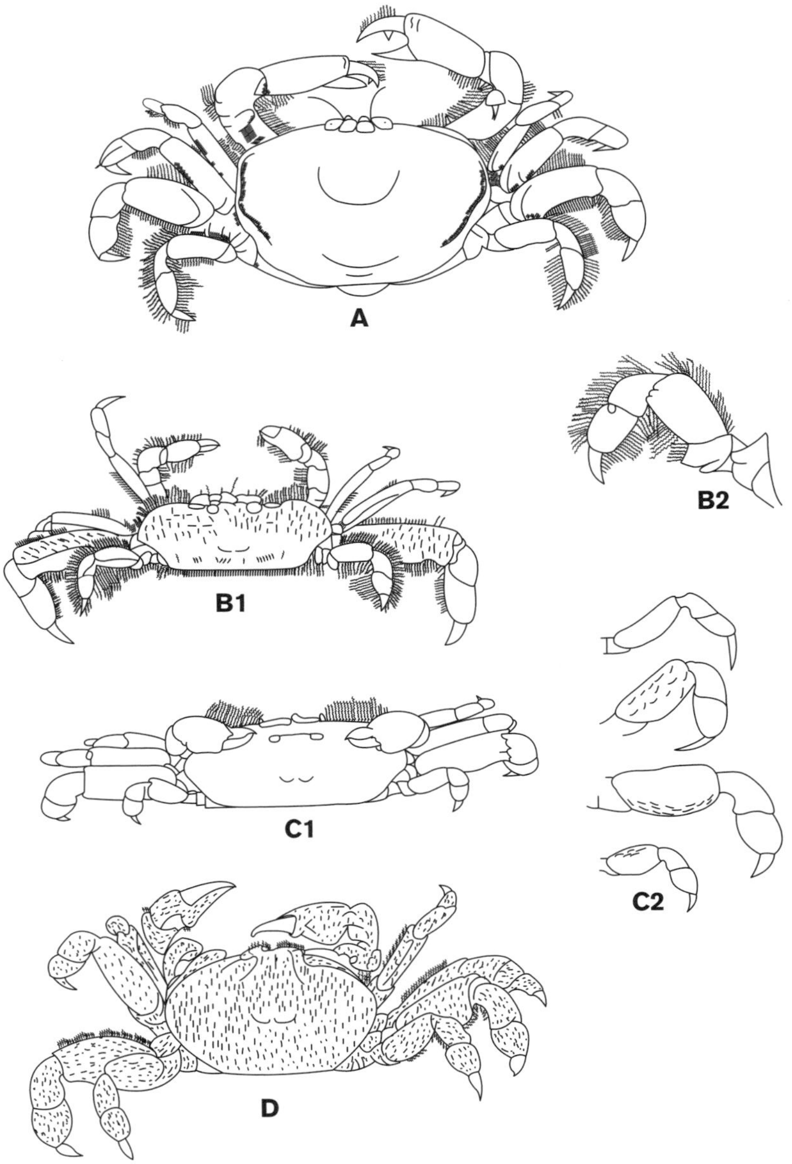

PLATE 322 Brachyura. Pinnotheridae. WL = walking legs. *Pinnixa barnharti*: A, dorsal view; *P. longipes*: B1, dorsal view; B2, WL5, arrows indicate tubercles of the ischium; *P. tubicola*: C1, dorsal view; C2, WL1–WL4; *P. tomentosa*: D, dorsal view. (all after Zmarzly 1992, J. Crust. Biol. 12: 677–713). Scale (mm): A = 2; B1 = 2.5; B2 = 1; C1, D = 5; C2 = 2.

. . . . *Parapinnixa affinis*
— Second or third pair of WL longer than the others (plates 321B1, 322A); dactylus of MXP3 inserted on the ventral margin of the propodus (plate 321B2, 321C3, 321D3) 2
2. Carapace transverse, wider than long (plates 321D1, 322A); third pair of WL longest than the others (plates 321D2, 322C1); exopod of the MXP3 with a lobe on the external margin (plate 321D3, arrow) 4
— Carapace suborbicular or subquadrate (plate 321B1, 321C1); second pair of WL longest than the others (plate 321B1, 321C2); external margin of the exopod without a lobe (plate 321C3). 3
3. Female: carapace subquadrate, whitish, with two longitudinal sulci arising from the upper margin of orbit and extending as far as gastric region (plate 321B1); male: carapace porcelainlike; anterolateral carapace margin with a fringe or hairlike setae; abdominal somites 2–4 fused (plate 321B3). . . . . *Fabia subquadrata*

— Female: carapace suborbicular, red-spotted to green, without longitudinal sulci (plate 321C1); male: carapace red-spotted to green, anterolateral carapace margin without a fringe of hairlike setae; abdominal somites and telson well separated (plate 321C4) . . . . *Opisthopus transversus*
4. Third pair of WL slightly longer than the others (plate 321D1); WL slender and somewhat rounded; carapace hard, dorsally convex and often granulated (plate 321D2) . . . . *Scleroplax granulata*
— Third pair of WL distinctly longer and larger than the others (plate 322B1, 322C2); WL flattened; carapace variable, but if dorsally hard and convex it is smooth (plate 322A) 5
5. Carapace strongly convex and calcified, 1.5 times wider than long (plate 322A). . . . . *Pinnixa barnharti*
— Carapace flat or slightly convex, not strongly calcified, >1.5 times wider than long 6
6. Dactylus shorter than propodus on WL3 (plate 322C2) 7

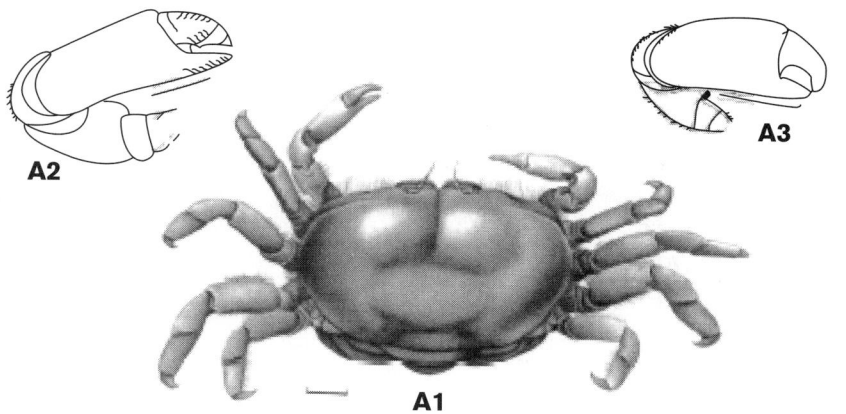

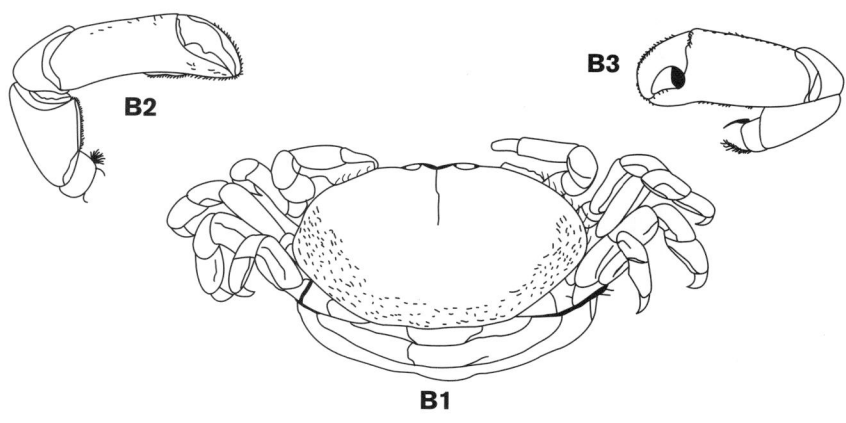

PLATE 323 Brachyura. Pinnotheridae. WL = walking legs. *Pinnixa faba*: A1, dorsal view; A2, female cheliped; A3, male cheliped; *P. littoralis*: B1, dorsal view; B2, female cheliped; B3, male cheliped; *P. weymouthi*: C1, carapace, dorsal view; C2, WL1–WL4 (C, after Zmarzly 1992). Scale (mm): A1 = 4; A2, B3 and C1 = 2; A3, B1 = 5; B2 = 3; C2 = 1.

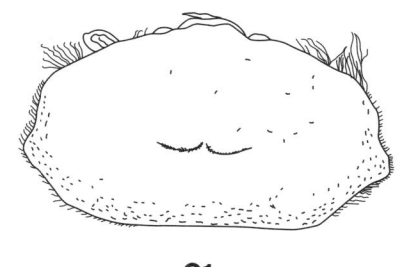

— Dactylus subequal to or exceeding length of propodus on WL3 (plates 323C2, 324A1) 11
7. Distal tip of dactylus of WL4 falling short of or just reaching to distal end of merus of WL3 when both legs extended (plate 322B1, 322C1) 8
— Distal tip of dactylus of WL4 reaching beyond distal end of merus of WL3 when both legs extended (plate 322D) 9
8. Posteroventral margin of ischium of WL4 with two or three large tubercles (plate 322B2 arrow); margins of WL4 with long setal fringe (plate 322B1) . . . . *Pinnixa longipes*
— Posteroventral margin of ischium of WL4 without tubercles; WL4 without long setal fringe (plate 322C1) . . . . *Pinnixa tubicola*
9. Carapace and legs covered with short coarse setae (plate 322D); ventral margin of propodus of WL3 bicarinate, the

carinae granulate or serrate; dactylus of WL3 spinous and slightly curved . . . . *Pinnixa tomentosa*
— Carapace and legs without short coarse setae; ventral margin of propodus of WL3 without carinae; dactylus of WL3 without spines and strongly curved 10
10. Female: carapace oblong (plate 323A1), fingers of cheliped not gaping when tightly closed (plate 323A2); male: fixed finger of chela straight relative to line defined by ventral margin of palm (plate 323A3); inner margin of dactylus of chela with single blunt triangular tooth (plate 323A3) . . . . *Pinnixa faba*
— Female: carapace pointed at sides (plate 323B1); fingers of cheliped gaping when tightly closed (plate 323B2); male: fixed finger of chela slightly deflexed relative to line defined by ventral margin of palm (plate 323B3); inner margin of dactylus of chela toothless (plate 323B3) . . . . *Pinnixa littoralis*

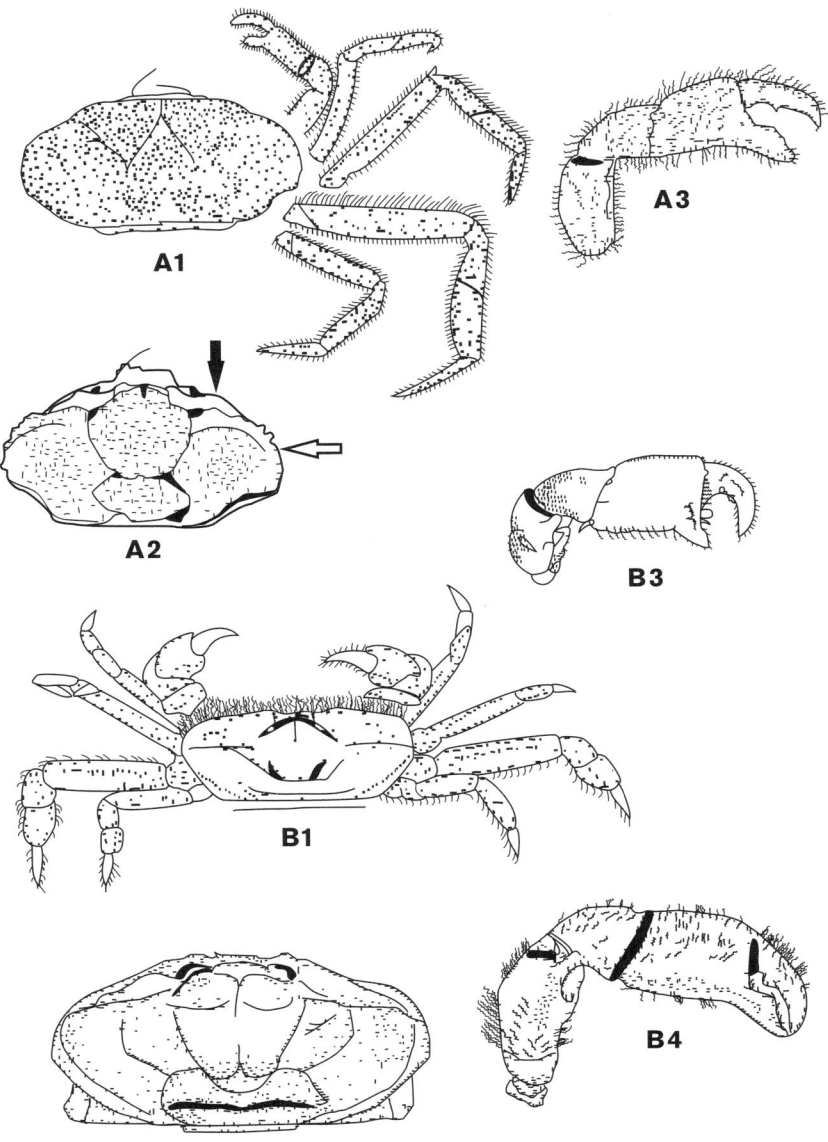

PLATE 324 Brachyura. Pinnotheridae. Closed arrows indicate the subhepatic tooth; open arrows indicate the granulate or serrate anterolateral ridge. *Pinnixa scamit*: A1, A2, female and male, dorsal view; A3, female cheliped; *P. occidentalis*: B1, B2, male and juvenile male, dorsal view; B3, B4, male and female cheliped (A1, A3, after Martin and Zmarzly, 1994; A2, after Campos et al, 1998; B1, B2, B3, B4, after Zmarzly 1992). Scale (mm): A1, A2 = 2; A3 = 1; B1 = 4; B2, B3, B4 = 1.

11. Anterolateral area of carapace smooth and round (plate 323C1) . . . . *Pinnixa weymouthi*

— Anterolateral area of carapace with granulate or serrate ridge (plates 324A2, open arrow, 325A1, 325A2, 325B1, 325B2) 12

12. Fixed finger of chela deflexed (plate 324A3, 324B3, 324B4) 13

— Fixed finger of chela straight or curving upward (plate 325A3, 325B3, 325B4) 14

13. Carapace (plate 324A1, 324A2) with a well-developed granular cardiac ridge; larger, acute, slightly curved teeth along the anterolateral margin of carapace; a well-developed subhepatic tooth (plate 324A1, 324A2, closed arrows); length of propodus of WL3 at least 2.5 times width . . . . *Pinnixa scamit*

— Carapace (plate 324B1, 324B2) with an acute sometimes bilobate cardiac ridge; granulated ridge along the anterolateral margin of carapace; no traces of subhepatic tooth; length of propodus of WL3 1.5–2 times width . . . . *Pinnixa occidentalis*

14. Anterior face of chela of male and female with prominent line of densely packed granules forming ridge just above ventral margin, running most of length of propodus (plate 325A3, 325A4); females often with a second row of granules medially on anterior face (plate 325A3) . . . . *Pinnixa franciscana*

— Anterior face of chela of mature male entirely smooth, without granules (plate 325B4); female and immature males with line of coarse granules just above ventral margin of propodus and scattered granules over rest of propodus, without a second row of granules medially on anterior face (plate 325B3) . . . . *Pinnixa schmitti*

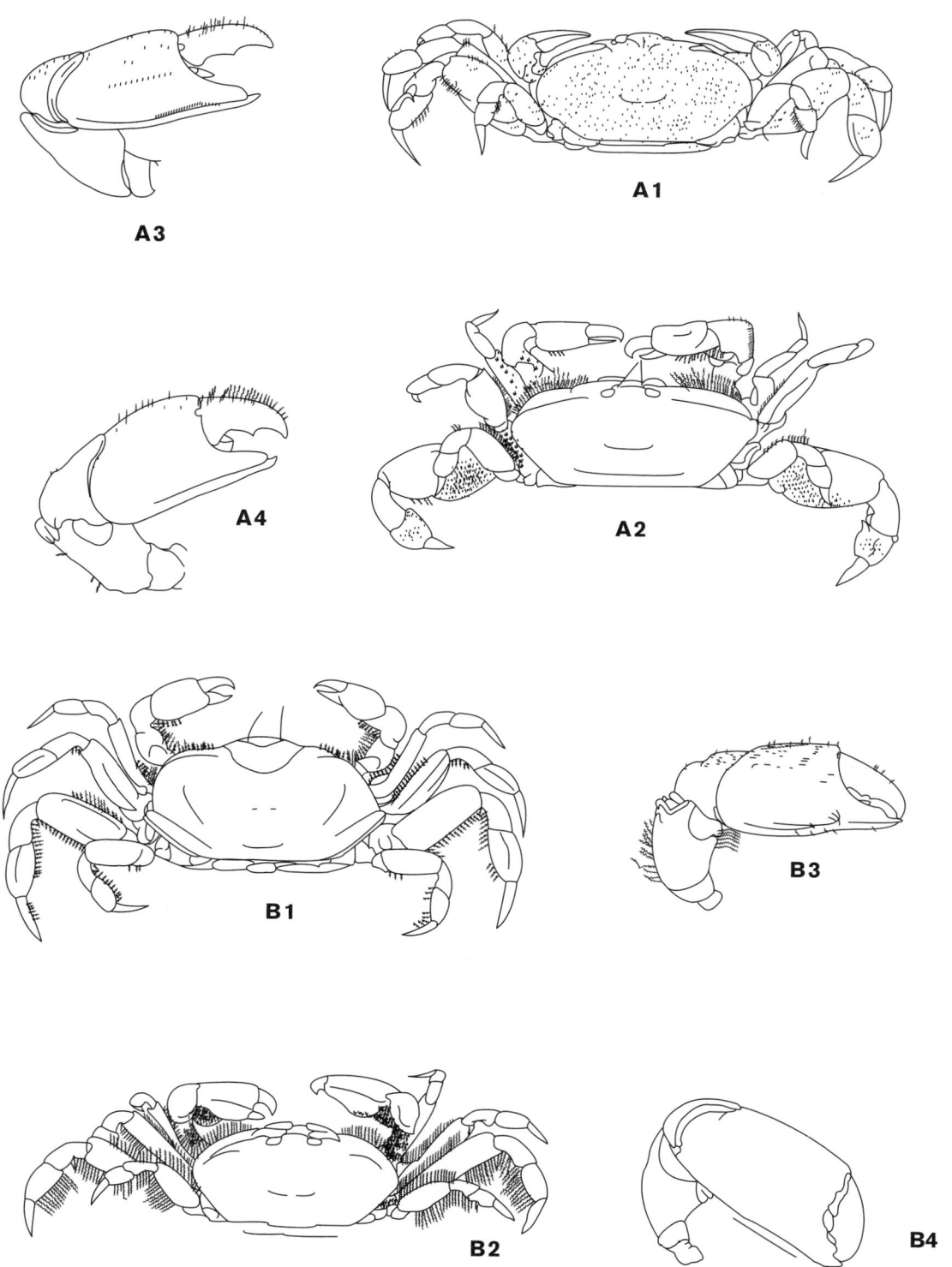

PLATE 325 Brachyura. Pinnotheridae. *Pinnixa franciscana*: A1, A2, female and male dorsal view; A3, A4, female and male cheliped; *P. schmitti*: B1, B2, female and male dorsal view; B3, B4, female and male cheliped (all after Zmarzly 1992). Scale (mm): A1, A2 = 5; A3, A4, B1, B3, B4 = 1; B2 = 3.

KEY C: ANOMURA AND THALASSINIDEA

1. Abdomen reflexed 2
— Abdomen not reflexed, may be twisted 18
2. Abdomen not folded against thorax, body shrimplike, legs greatly elongate, slender . . . . *Pleuroncodes planipes*

— Abdomen folded against thorax, body crablike 3
3. Body egg-shaped; second to fourth legs with last joint curved and flattened; on sandy beaches 4
— Body not egg-shaped; 2nd to 4th legs with last joint ending in sharp, pointed dactyl; in rocky areas 6

4. First pair of legs without claws; carapace without sharp spines along anterolateral margin (plate 80); color gray. . . . .
. . . . *Emerita analoga*
— First pair of legs chelate or subchelate 5
5. Carapace and chelipeds with several sharp spines, claws chelate; large to 60 mm, white, not iridescent . . . .
. . . . *Blepharipoda occidentalis*
— Chelipeds smooth, carapace with small spine, claws subchelate, small to 20 mm, color iridescent white . . . .
. . . . *Lepidopa californica*
6. Uropods absent; carapace as long as, or longer than, broad; abdomen thick and fleshy 7
— Uropods absent; carapace wider than long; abdomen small and flattened. 10
— Uropods present; carapace nearly round in outline, abdomen folded against body 11
7. Carapace completely covering claws, legs, and body. . . . .
. . . . *Cryptolithodes sitchensis*
— Claws and legs visible dorsally 8
8. Claws, legs, and carapace with large spines, carapace triangular, adults brown, juveniles variable in color. . . . . . .
. . . . *Phyllolithodes papillosus*
— Claws, legs, and carapace smooth 9
9. A large smooth foramen formed by the lateral margin of the carpus of the claw against the first walking leg, reddish brown to tan. . . . . *Lopholithodes foraminatus*
— Claws without smooth concavity on carpus, brightly colored, red, orange, yellow . . . . *Lopholithodes mandtii*
10. Legs and carapace hairy, flattened . . . .
. . . . *Hapalogaster cavicauda*
— Legs and carapace roughly tuberculate, not hairy, legs nearly cylindrical. . . . . *Oedignathus inermis*
11. Body and chelae thick; chelae unequal and tuberculate or granular; carpus of chelipeds as long as broad . . . .
. . . . *Pachycheles* 12
— Body and chelae flattened; chelae equal or nearly so, smooth; carpus of chelipeds longer than broad . . . .
. . . . *Petrolisthes* 14
12. Telson with five plates (plate 326B2), lacking small plate at anterior margin of each lateral plate 13
— Telson with 7 plates, small plate at anterior margin of each lateral plate (plate 326C) (sometimes missing in females)
. . . . *Pachycheles pubescens*
13. Chelipeds with long, scattered hairs, carpus of cheliped with a broad triangular lobe on inner margin. . . . .
. . . . *Pachycheles rudis*
— Chelipeds with dense coat of short soft hairs, carpus of cheliped with a toothed lobe on inner margin, occurs in sponges. . . . . *Pachycheles holosericus*
14. Carpus of chelipeds more than twice as long as wide. . . . .
. . . . 15
— Carpus twice as long as wide or less. 16
15. Carapace covered with short, transverse, hairy striations and large, flattened tubercles; carpus about 2.5 times as long as wide; distal portion of maxillipeds bright orange red . . . . *Petrolisthes rathbunae*
— Carapace nearly smooth posteriorly, often granular anteriorly, never with hairy striations; carpus a little over twice to nearly three times as long as wide; outer edge of palp of maxilliped blue (see note in species list) . . . .
. . . . *Petrolisthes manimaculis*
16. Carpus without lobe on anterior margin, margins subparallel; outer edge of palp of maxilliped arthrodial membrane of cheliped bright blue (see note in species list) . . . .
. . . . *Petrolisthes eriomerus*
— Anterior margin of carpus with a distinct proximal lobe, margins of carpus converging distally from their highest points; palp of maxilliped and arthrodial membrane of dactyl of cheliped orange red (plate 326A) 17
17. Carpus of chelipeds with an anterior lobe about one-quarter the length of the carpus, excluding lobe carpus margins almost parallel, carpus with dense covering of hair, body color usually light brown . . . . *Petrolisthes cabrilloi*
— Carpus of chelipeds with long anterior lobe extending more than .25 the length of the carpus, carpus margins excluding lobe converge distally, carpus smooth without hairs . . . . *Petrolisthes cinctipes*
18. Burrowing in mud or sand; abdomen symmetrical, extended, externally segmented; ghost and mud shrimps. 19
— Living in snail shells or worm tubes, abdomen soft, hermit crabs. 22
19. First pair of legs approximately equal and subchelate, other legs simple; eyestalks cylindrical, corneas terminal; four pairs of fanlike pleopods; body hairy, often bluish. . . . . . .
. . . . *Upogebia pugettensis*
— First pair of walking legs very unequal and chelate, second pair chelate; eyestalks flattened, corneas dorsal; three pairs of fanlike pleopods; body smooth, whitish to reddish. . . . .
. . . . *Neotrypaea* 20
20. Median rostral tooth sharply pointed . . . .
. . . . *Neotrypaea gigas*
— Rostrum blunt. 21
21. Tip of eyestalk pointed, color orange to pinkish. . . . .
. . . . *Neotrypaea californiensis*
— Tip of eyestalk blunt, rounded, color white. . . . .
. . . . *Neotrypaea biffari*
22. Abdomen straight, living in worm tubes; right claw slightly larger than left, body color light, claws tipped with orange, legs with orange-brown bands . . . .
. . . . *Discorsopagurus schmitti*
— Abdomen twisted, asymmetrical, usually in snail shells, color various 23
23. Left cheliped equal to or larger than right; outer maxillipeds approximated at base. 24
— Right cheliped larger than left, outer maxillipeds widely separated at base . . . . *Pagurus* 26

Note: The key for *Pagurus* species is primarily based on color; for preserved specimens also consult McLaughlin and Fisher (1974) and Kozloff (1987).

24. No appendages on anterior abdominal segments, fourth legs subchelate, body whitish with median brown stripe, claws and legs with blue blotches . . . . *Isocheles pilosus*
— Paired pleopods present on first abdominal segment, fourth legs not chelate . . . . *Paguristes* 25
25. Eyestalks stout, less than three-quarters the width of hard portion of carapace, propodus of claw wide, only one-fifth longer than wide, outer margin strongly convex, antennae with sparse hairs, reddish, legs with blue spots. . . . .
. . . . *Paguristes bakeri*
— Eyestalks slender, at least as long as hard portion of carapace is wide, propodus of claw narrow, at least one-third larger than wide, outer margin almost straight, antennae with a row of long setae, orange to brown, obscured by golden setae . . . . *Paguristes ulreyi*
26. Minor (left) cheliped with flat dorsal surface, dactyl of walking legs 2 and 3 twisted, with two brown stripes 27

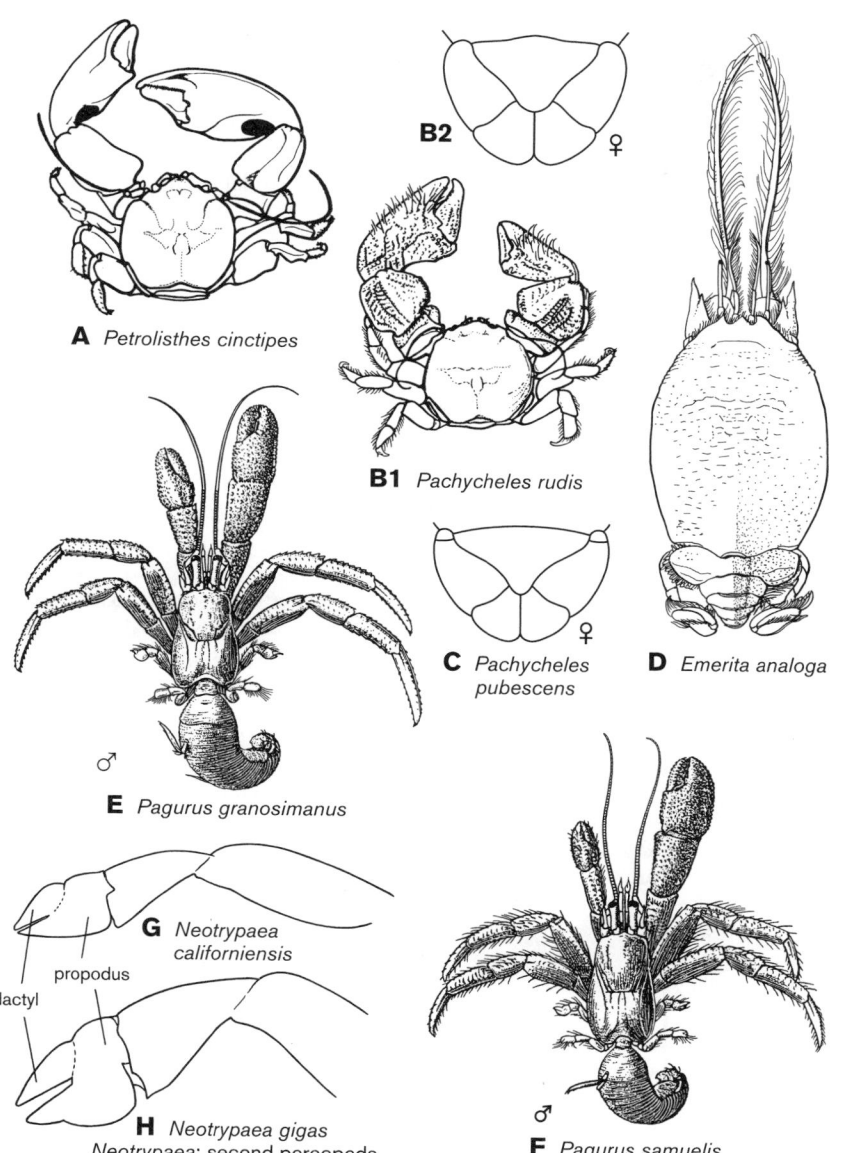

B2 ♀

A *Petrolisthes cinctipes*

B1 *Pachycheles rudis*

C *Pachycheles pubescens* ♀

D *Emerita analoga*

PLATE 326 Anomura. B2, C, female telsons; E, young male; G, H, second right pereopods (A, B1, D, after Hedgpeth; G, H, Rogene Thompson, redrawn by Emily Reid; rest after Schmitt 1921).

E *Pagurus granosimanus* ♂

G *Neotrypaea californiensis*

dactyl   propodus

H *Neotrypaea gigas*
Neotrypaea: second pereopods

F *Pagurus samuelis* ♂

— Minor (left) cheliped with rounded dorsal surface, dactyl of walking legs straight 28

27. Dactyl and propodus with more spines in dorsal than on ventral surface, cornea of eye yellowish green, claws with red stripe . . . . *Pagurus ochotensis*

— Spines distributed rather equally on dorsal and ventral surfaces of claw, cornea black, claws and legs with orange bands . . . . *Pagurus armatus*

28. Merus of major (right) cheliped without prominent tubercules, antennae irregularly banded, chelipeds brown with gray tubercules . . . . *Pagurus quaylei*

— Merus of major (right) cheliped with one to two prominent tubercules. 29

29. Legs not banded, antennae orange, body color uniform dark green stippled with blue or white granulations, major chelae evenly granulated, rostrum rounded (plate 326E). . . . .
. . . . *Pagurus granosimanus*

— Legs banded, color various, major chelipeds not evenly granulated 30

30. Eyes with yellow circles, antennae red, overall body color deep red, dactyl of walking legs with white spot at tip, rostrum triangular, acute; carpus of major cheliped flat perpendicular face. . . . . *Pagurus hemphilli*

— Color not deep red, carpus otherwise 31

31. Body color pale, chelipeds red and spiny, walking legs light blue with red spots and bands, antennae light translucent orange, rostrum rounded . . . . *Pagurus beringanus*

— Body color dark, major cheliped with tubercules, granules or spines, rostrum rounded or acute 32

32. Antennae red, walking legs with bright blue dactyls and bands on distal portion of propodus (white bands and dactyls on small crabs), body and claw color green, rostrum triangular, acute, chelae with tubercles, hard carapace longer than wide (plate 326F) . . . . *Pagurus samuelis*

— Antennae banded or orange, body color brownish, hard carapace about as long as wide. 33

33. Antennae banded green and white, propodus with white distal band and blunt tips, small crabs with white bands

on chelipeds and other walking legs, rostrum triangular, acute, chelipeds with tubercles and granules. . . . .
. . . . *Pagurus hirsutiusculus*
— Antennae pale orange, no blue color on walking legs, chelipeds with orange tips and rows of spines, rostrum rounded, small crabs, carapace <10 mm . . . .
. . . . *Pagurus caurinus*

LIST OF SPECIES FOR DECAPODA

Jensen (1995) provides good photographs and considerable information on habitat, geographic range, color and behavior of many of these species. McLaughlin et al. 2005 provide a comprehensive species list.

PLEOCYEMATA

CARIDEA

CRANGONIDAE

See Zarenkov (1965, Zoologicheskii Zhurnal 44: 1761–1775, in Russian) for an older review of group in general and aspects of evolution and biology; Kuris and Carlton (1977, Biol. Bull. 153: 540–559) for relative growth analyses and taxonomy of west coast species. These shrimp are parasitized by bopyrid isopods, such as *Argeia pugettensis*, which induce a large, asymmetrical swelling of the gill chamber.

Crangon alaskensis Lockington, 1877. Common, in shallow water of bays on soft bottoms; should not be confused with *Crangon nigricauda*, a larger species.

Crangon franciscorum Stimpson, 1859. In bays on mud bottoms, common to abundant; also offshore in deeper waters. See Israel 1936, Calif. Fish Game, Bull. 46 (life history, biology, fishery).

Crangon handi Kuris & Carlton, 1977. Coarse sand of coves and surge channels, sometimes among surfgrass along rocky outer coast, matching substrate in color. Most *Crangon* spp. have salt-and-pepper pattern for crypsis on mud or sand. *C. handi* is cryptic over gravelly substrates, and its beautiful large splotches of color include orange and magenta. See Kuris and Carlton 1977, Biol. Bull. 153: 540–559.

Crangon nigricauda Stimpson, 1856. Common to abundant in bays and offshore in deeper waters; among eelgrass, rocks, and on sand bottoms. See Israel 1936, above.

Crangon nigromaculata Lockington, 1877. On mud and sand bottoms.

Lissocrangon stylirostris (Holmes, 1900). Common in surf zone of semiprotected sandy beaches; sublittoral on sandy-rocky bottoms.

Mesocrangon munitella Walker, 1898. Recorded by Schmitt from shallow rocky bottoms in San Francisco Bay.

ALPHEIDAE

See Ache and Case, 1969, Physiol. Zool. 42: 361–371 (*Betaeus* spp., antennular chemoreception).

Alpheus spp. "Pistol shrimps" injure prey by percussion, producing a high-speed jet of water by clicking dactyl of chela against palm; the snapping sound is caused by the implosion of a cavitation bubble in the jet (Versulis et al. 2000, Science 289: 2114–2117). Very low rocky intertidal, in sponges, kelp holdfasts,

old pholad bore holes. See Kim and Abele 1988, Smith. Contrib. Zool. 454, 119 pp. (taxonomy of Eastern Pacific species).

Alpheus bellimanus Lockington, 1877. Very low intertidal, under rocks, and in holdfasts.

Alpheus clamator Lockington, 1877. A southern species, common in low tide pools, burrows under rocks, paired.

Betaeus gracilis Hart, 1964. Intertidal on rocky shores; kelp holdfasts. See Hart (1964, Proc. U.S. Natl. Mus. 115: 431–466) for a detailed treatment of this genus. Size, shape, and dentition of chelae vary with age, sex, and extent of regeneration and are not reliable systematic characters.

Betaeus harfordi (Kingsley, 1878). Commensal in mantle cavities of abalones *(Haliotis* spp.); leaves host readily and often missed when abalones are collected.

Betaeus harrimani Rathbun, 1904. In burrows of *Upogebia* and *Neotrypaea* on mudflats.

Betaeus longidactylus Lockington, 1877. Intertidal on rocky shores; kelp, kelp holdfasts, and in eelgrass; also recorded in southern California from *Urechis* and *Upogebia* burrows.

Betaeus macginitieae Hart, 1964. Occur in pairs under sea urchins *(Strongylocentrotus* spp.).

Betaeus setosus Hart, 1964. Under rocks and in algae on semiprotected rocky coasts; in kelp holdfasts (especially *Laminaria*) and surfgrass *(Phyllospadix)* roots; on pilings. Small, symbiotic with pairs of *Pachycheles rudis*, underneath crabs, translucent, easily missed.

Synalpheus lockingtoni (Coutière, 1909). Rare in central and northern California; more common to the south. Records include collections on wharf piles at Santa Cruz, and in Elkhorn Slough.

PANDALIDAE

Pandalus danae Stimpson, 1857. Dock shrimp; sublittoral, but occasional in shallow water near harbor channels and over eelgrass *(Zostera)* beds.

ATYIDAE

Syncaris pacifica (Holmes, 1895). Formally designated an endangered species in California and restricted to freshwater streams of Marin, Sonoma, and Napa counties; see Hedgpeth 1968, Intern. Revue Ges. Hydrobiol. 53: 511–524; Martin and Wicksten 2004, J. Crust. Biol. 24: 447–462. Its southern counterpart, *Syncaris pasadenae* (Kingsley, 1896), is extinct; its type locality is said to be underneath the site of the Rose Bowl in Pasadena.

PALAEMONIDAE

See Wicksten 1989, Bull. So. Calif. Acad. Sci. 88: 11–20 (key to species of Eastern Pacific).

Exopalaemon carinicauda (Holthuis, 1950). A Korean and Chinese species first collected in San Francisco Bay in 1993. Introduced either in ballast water or as live bait. See Wicksten 1997, Calif. Fish Game 83: 43–44.

Exopalaemon modestus (Heller, 1862). An Asian species introduced to the Columbia River mouth in ballast water in the 1990s. See Emmett et al. 2002, Biol. Invas. 4: 447–450 (introduction to West Coast).

Palaemon macrodactylus Rathbun, 1902. An Asian species introduced to San Francisco Bay in ballast water in the early 1950s and now found from southern California to Oregon, especially in brackish water, where it may be abundant along floats, wharf pilings, and in algae; see Newman 1963, Crustaceana 5: 199–132; Born, 1968, Bio. Bull. 134: 235–241 (osmoregulation); Little 1969, Crustaceana 17: 69–87 (larval development).

HIPPOLYTIDAE

See Wicksten 1990, Fishery Bulletin 88: 587–598 (key to Eastern Pacific species).

Heptacarpus brevirostris (Dana, 1852). In low-intertidal pools in algae and under rocks; on floats, pilings; sublittoral on algae and on rocky bottoms.

Heptacarpus carinatus Holmes, 1900. Low intertidal; in algae and surfgrass, often matching color of algae.

Heptacarpus franciscanus (Schmitt, 1921). Recorded from shallow water of San Francisco Bay over sandy and rocky bottoms.

Heptacarpus palpator (Owen, 1839). Low intertidal to sublittoral; in tidepools and under rocks to the sublittoral; also on wharf piles. See Wicksten 1986, Bull. So. Calif. Acad. Sci. 85: 46–55 (redescription).

Heptacarpus paludicola Holmes, 1900. In *Zostera* beds and on algae, such as *Ulva*, in shallow pools on mudflats; on wharf pilings, floats; also in mid-intertidal pools of rocky coast; common to abundant.

Heptacarpus pugettensis Jensen, 1983. Very low intertidal; aggregate under boulders.

Heptacarpus sitchensis (Brandt, 1851) (=*H. pictus* (Stimpson, 1871). Common to abundant in middle and lower tide pools of rocky coasts, to the sublittoral; also in *Zostera* beds, and on floats. See Wicksten et al. 1996, Crustaceana 69: 71–75 (taxonomy).

Heptacarpus stimpsoni Holthius, 1947. Low intertidal to sublittoral on soft bottoms.

Heptacarpus taylori (Stimpson, 1857). Mid- to low intertidal under rocks and clumps of bryozoans, sponges; in algae; wharf pilings.

Heptacarpus tenuissimus Holmes, 1900. Low intertidal; soft bottoms.

Hippolyte californiensis Holmes, 1895. California green shrimp; locally common in bays on *Zostera*, matching its color and oriented longitudinally along blade; see Chace 1951, J. Wash. Acad. Sci. 41: 35–39 (taxonomy); Barry 1974, Mar. Biol. 26: 261–270 (habitat selection).

Hippolyte clarki Chace, 1951 Similar to *H. californiensis*, low intertidal in eelgrass beds.

Lebbeus lagunae (Schmitt, 1921). Rare, in rocky pools; a southern species.

Lysmata californica (Stimpson, 1866). Low intertidal; rocky shores, may behave as a cleaner, feeding on fish parasites. A southern species, more common in central California and Oregon after El Niños.

Spirontocaris prionota (Stimpson, 1864). Uncommon, underneath rocks in tide pools.

BRACHYURA

Some of the traditional families of true crabs have been split into groups (superfamilies) of related families. These include the Majidae, now superfamily Majoidea (Epialtidae, Inachidae, Inachoididae, Oregoniidae, Pisidae); the Xanthidae, now superfamily Xanthoidea (Xanthidae, Panopeidae); and the Grapsidae, now superfamily Grapsoidea (Grapsidae, and Varunidae).

LEUCOSIIDAE

Randallia ornata (Randall, 1840). Subtidal; rarely intertidal, sandy substrates; occasionally washed inshore. Frequently infested with corkscrew nemertean *Carcinonemertes*. See Sadeghian and Kuris 2001, Hydrobiologia 456: 59–63 (nemertean egg predator).

CALAPPIDAE

Platymera gaudichaudii (Milne-Edwards, 1837). Sublittoral; occasional specimens in harbors released from fishing boats.

GRAPSIDAE

Pachygrapsus crassipes Randall, 1840. Lined shore crab; ubiquitous in upper intertidal of rocky areas of bays as well as rocky outer coast, abundant; in burrows in *Salicornia* marshes; active, aggressive, nocturnal; see Hiatt 1948, Pac. Sci. 2: 135–213 (biology); Bovbjerg 1960, Ecology 41: 668–672 (behavioral ecology), 1960 Pac. Sci. 14: 421–422 (courtship behavior); Willason 1981, Mar. Biol. 64: 125–133 (salt march ecology).

Planes cyaneus Dana, 1852. Flotsam crab; pelagic; occasionally washed ashore on drift logs with goose barnacle *Lepas*, and also associated with sea turtles; see Chace 1951, Proc. U.S. Nat. Mus. 101: 65–103 (taxonomy); Wicksten and Behrens 2000, SCAMIT Newsletter 19(5): 7 (California record).

VARUNIDAE

Eriocheir sinensis Edwards, 1853. Chinese mitten crab; introduced from Asia in the 1990s, this large catadromous species is abundant in the rivers and sloughs draining into the San Francisco Bay. In the brackish water of the Petaluma River, small juveniles are common among the tubes of the tubeworm *Ficopomatus enigmaticus*. It is much larger than native grapsids and varunids. *E. sinensis* is amphibious; during migration, it can be an abundant pest in low-lying areas, ditches, and canals, clogging pipes, undermining banks and levees, and entering houses. This edible species is highly valued in some Asian cuisines. In Asia, it is an intermediate host for the human lung fluke, an important pathogen. See Cohen and Carlton 1997, Pac. Sci. 51: 1–11, and Rudnick et al. 2005, Bio. Invasions 7: 333–350 (introduction to California).

Hemigrapsus nudus (Dana, 1851). Purple shore crab; mid-intertidal of semiprotected and protected rocky coasts and bays, locally abundant; prefers coarse sand to gravel substrates overlain with large rock cover; sluggish. Sometimes infected with the parasitic castrator entoniscid isopod, *Portunion conformis*. Uncommon south of Morro Bay. See Morris et al. 1996, Physiol. Zool., 69: 864–886 (air breathing); McGaw 2003, Bio. Bull. 204: 38–49 (behavioral thermoregulation); Keppel and Scrosati 2004, Animal Behav. 67: 915–920 (prey avoidance).

Hemigrapsus oregonensis (Dana, 1851). Green shore crab; mid and low intertidal of bays under rock cover overlying mud or muddy sand, abundant; sometimes exposed and active over large areas of mudflats; sublittoral populations in shallow water of bays with profuse *Ulva* cover; small populations along protected outer coast under rocks over mud; also in burrows in *Salicornia* marshes; moderately active. The entoniscid isopod, *Portunion conformis*, a parasitic castrator, often attains infection rates above 40%. In turn, *P. conformis* suffers frequent mortality from a picornavirus (Kuris et al. 1979, 1980). Both species of *Hemigrapsus* and also *Pachygrapsus crassipes* can be infested with the egg predator *Carcinonemertes epialti*. At times, brood mortality is substantial (Shields and Kuris 1988). These shore crabs are also intermediate hosts for a variety of trematode metacercariae, larval trypanorhynch tapeworms, larval *Polymorphus* acanthocephalans, and larval *Ascarophis* nematodes. Very small *Hemigrapsus nudus* may be distinguished from very small *Hemigrapsus oregonensis* by a combination of the following characters: in *H. oregonensis* there is a marked frontal notch, in *H. nudus* a shallow depression; in *H. oregonensis* the lateral spines are sharp and clearly set out, in *H. nudus* they are not sharp, nor as clearly separated from the side; the dactyls of ambulatory legs 1–3 are long in *H. oregonensis*, shorter in *H. nudus;* the dactyl of leg 4 is quite flat in *H. nudus*, rounded in *H. oregonensis*. See Lindberg 1980, Crustaceana 39: 263–281 (behavior); Willason 1981, Mar. Biol. 64: 125–133 (salt marsh ecology).

PARTHENOPIDAE

Heterocrypta occidentalis (Dana, 1854). Sublittoral; sandy bottoms; rare in intertidal.

INACHOIDIDAE

Podochela hemphilli (Lockington, 1877). Low intertidal and subtidal; wharf pilings. Decorators, particularly on first walking legs.
Pyromaia tuberculata (Lockington, 1877). Sublittoral on wharf pilings; often encrusted with sponges and algae; common in shallow dredge hauls in San Francisco Bay. Introduced in recent years to Asia, New Zealand, and Australia.

PISIDAE

Herbstia parvifrons (Randall, 1840) Rare; sponge-encrusted under stones; retreats into crevices, low intertidal. Monterey Bay and north.
Loxorhynchus crispatus Stimpson, 1857. Moss crab; sublittoral; low intertidal on semiprotected rocky coasts in crevices; often heavily decorated with hydroids, sponges and algae. See Wicksten 1977, Calif. Fish Game 63: 122–124 (feeding), 1978, Trans Amer. Micr. Soc. 97: 217–220, 1979, Crustaceana 5 Suppl: 37–46, and 1980, Sci. Amer. 242: 146–154 (decorating).
Loxorhynchus grandis Stimpson, 1857. Sheep crab; a southern species, subtidal, occasionally lower intertidal, very large, males reach 24 cm in length, 11 kg in weight; females are smaller. Subtidal breeding pods in the spring include hundreds of females with many males gathered at periphery. See Culver and Kuris (2001) for information on biology and mating. Hairy outer surfaces wear away in adult terminal molt phase males, revealing blue-green tubercules. Egg masses with the nemertean egg predator, *Carcinonemertes*; supports a fishery in southern California.
Pelia tumida (Lockington, 1877). A southern species; under stones, low-intertidal zone on rocky shores.
Scyra acutifrons Dana, 1851. Uncommon in low intertidal of semiprotected rocky coasts, often encrusted; rare; south of Monterey.

OREGONIIDAE

Oregonia gracilis Dana, 1851. Occasional on wharf pilings and in *Zostera*; usually sublittoral and generally northern; boreal,

also occurring in Japan. Very effective decorator; sexual dimorphism pronounced.

EPIALTIDAE

Mimulus foliatus Stimpson, 1860. Low intertidal of rocky coast; among algae, under rocks, often encrusted with sponges or bryozoans.
Pugettia gracilis Dana, 1851. In low, rocky intertidal; in bays among *Zostera*.
Pugettia producta (Randall, 1840). Northern kelp crab; low intertidal and sublittoral of protected and semiprotected rocky coasts, in kelp beds and other macro-algae and on jetties, wharf pilings; pods of aggregated females sometimes reported subtidally; adults often encrusted with barnacles, bryozoans, and sponges; a lively and aggressive spider crab with a strong pinch. Occasionally parasitized by rhizocephalan barnacles, egg masses with commensal turbellarians and rarely the egg predator nemertean *Carcinonemertes epialti*. The similar *Taliepus nuttallii* occurs south of Point Conception.
Pugettia richii Dana, 1851. Low intertidal among corallines and other algae; often encrusted with hydroids and coralline algae.

PINNOTHERIDAE

The pea crabs, symbionts with annelids, mollusks, sea cucumbers, and in crustacean and echiuran burrows; see Schmitt et al. 1973, Crustaceorum Catalogus. Dr. W. Junk B. V., The Hague, The Netherlands, pp. 1–160; Rathbun 1918, U.S. Nat. Mus. 97: 1–461; Schmitt 1921.
Fabia subquadrata Dana 1851. Taxonomy and distribution see Campos, 1996, J. Nat. Hist. 34: 1157–1178. In bivalve mollusks, especially *Mytilus californianus*. The females undergo several postplanktonic stages (prehard, hard to posthard IV) before they become a large, soft-shelled, ovigerous female (=posthard V). The small hard-shelled (hard stage) males may move between hosts and like female in hard stage are able to swim; see Pearce 1966, Pac. Sci. 20: 3–35.
Opisthopus transversus Rathbun, 1893. This species appears to be non-host specific; symbiont in the mantle cavity of mollusks and the cloaca of sea cucumbers; see Campos et al. 1992, Proc. Biol. Soc. Wash. 105: 753–759; Campos et al. 1998, Proc. Biol. Soc. Wash. 111: 372–381 (taxonomy and distribution); Hopkins and Scanland 1964, So. Calif. Acad. Sci. Bull 63: 85–88 (hosts).
Parapinnixa affinis Holmes. 1940. In tubes of polychaetes *Terebella californica* and *Loimia;* see Glassell 1933, Trans. San Diego Soc. Nat. Hist. 7: 319–330; Berkeley and Berkeley 1941, So. Calif. Acad. Sci. Bull. 40: 16–60; Campos et al. 1992, Proc. Biol. Soc. Wash. 105: 753–759.
Pinnixa barnharti Rathbun, 1918. A southern species; an obligated symbiont of the sea cucumber *Caudina arenicola*. See Campos et al. 1998, Proc. Biol. Soc. Wash. 111: 372–381 (taxonomy and distribution).
Pinnixa faba (Dana, 1851). Predominantly a symbiont of clams; it prefers the gaper clams *Tresus capax* and *T. nuttallii;* see Pearce 1966, pp. 565–589, in H. Barnes ed. Some contemporary studies in marine science. Allen & Unwin, London.
Pinnixa franciscana Rathbun, 1918. Adults recorded from the burrows of *Urechis caupo*, and the ghost shrimps *Neotrypaea californiensis, N. gigas* and *Upogebia pugettensis*; juveniles inhabit the tubes of polychaetes; see Garth and Abbott (1980).

Pinnixa littoralis Holmes, 1894. In the mantle cavity of clams; prefers the gaper clam *Tresus capax*; see Pearce 1965, Veliger 7: 166–170; Campos-González 1986, Veliger 29: 238–239.

Pinnixa longipes (Lockington, 1877). Common in sandy sediments in tubes of the polychaete worms *Axiothella rubrocincta*, *Pectinaria californiensis*, and *Pista elongata*, and occasionally in burrows of *Urechis caupo*; see Garth and Abbott (1980).

Pinnixa occidentalis Rathbun, 1893. A northern species, in burrows of the echiuran *Echiurus* sp., and free-living; may represent a species complex; see Zmarzly 1992, J. Crust. Biol. 12: 677–713; Martin and Zmarzly 1994, Proc. Biol. Soc. Wash. 107: 354–359; Campos et al. 1998, Proc. Biol. Soc. Wash. 111: 372–381.

Pinnixa scamit Martin and Zmarzly, 1994. See Martin and Zmarzly 1994, Proc. Biol. Soc. Wash. 107: 354–359 (taxonomy and distribution); Campos et al. 1998 Proc. Biol. Soc. Wash. 111: 372–381.

Pinnixa schmitti Rathbun, 1918. Adults in the burrows of *Urechis caupo* and the ghost shrimps *Neotrypaea californiensis, N. gigas* and *Upogebia* spp.; see Zmarzly 1992, J. Crust. Biol. 12: 677–713 (taxonomy and distribution); Garth and Abbott (1980).

Pinnixa tomentosa Lockington, 1876. In tubes of chaetopterid, onuphid and terebellid polychaete worms; see Scanland and Hopkins 1978, Proc. Biol. Soc. Wash. 91: 636–641.

Pinnixa tubicola Holmes, 1894. Heterosexual pairs occur in tubes of large polychaete worms, particularly terebellids and chaetopterids; see Zmarzly 1992, J. Crust. Biol. 12: 677–713.

Pinnixa weymouthi Rathbun, 1918. Nothing is known of the biology or symbiotic relationships of this species.

Scleroplax granulata Rathbun, 1893. Common in burrows of the echiuran *Urechis caupo* and the ghost shrimps *Neotrypaea californiensis, N. gigas, Upogebia pugettensis*, and *U. macginitieorum*; see Garth and Abbott 1980; Campos 2006, Zootaxa, 1344: 33–41 (systematics and distribution).

CANCRIDAE

Juvenile *Cancer* species under 20 mm are not readily distinguished using the key. See Schmitt (1921) for a key to the small specimens. See Nations 1975 Los Angeles Co. Mus. Natur. Hist. Sci. Bull. 23: 1–104, Schweitzer and Feldmann (2000) (systematics, biogeography, fossil record).

Cancer antennarius Stimpson, 1856. Pacific rock crab; lower intertidal, common in subtidal; partially imbedded in sand among rocks; protected and semiprotected coast, as well as in bays; often encrusted; common. Iphitimid polychaetes recorded from branchial cavity in southern California (Pilger 1972, Bull. So. Calif. Acad. Sci 70: 84–87). With *Cancer anthonyi* and *Cancer productus* it supports a rock crab fishery in southern California.

Cancer anthonyi Rathbun, 1897. Yellow rock crab; low intertidal; under rocks, common in subtidal; in bays.

Cancer branneri (Rathbun, 1926) (=*Cancer gibbosulus* [De Haan, 1835]). A small species (reaching about 35 mm in width) that may be mistaken for young *Cancer antennarius* (the granules on the carapace in *branneri* are in scattered groups; in *antennarius* crowded); rare in intertidal; in bays on shelly gravel; subtidal.

Cancer gracilis Dana, 1852. Graceful crab; intertidal to sublittoral on sandy shores; megalops and post-larval instars phoretic on scyphozoan medusae.

Cancer jordani Rathbun, 1900. Hairy rock crab; low intertidal and subtidal in bays; uncommon, under rocks and in holdfasts.

Cancer magister Dana, 1852. Dungeness crab; generally offshore on sandy bottoms; occasionally inshore, juveniles in bay and estuary nurseries; support an important fishery see Armstrong et al. 1995, Fish. Bull. 93: 456–470; Hobbs et al. 1992, Can. J. Fish. Aq. Sci. 49: 1379–1388; Paul et al. 2002, Univ. Alaska Sea Grant Coll. Rpt.; Wild et al. 1983, Calif. Dept. Fish and Game Fish Bull.; suffers substantial brood mortality from the symbiotic nemertean egg predator, *Carcinonemertes errans* (see Wickham 1978, Proc. Biol. Soc. Wash. 91:197–202).

Cancer oregonensis (Dana, 1852). Lower intertidal; semiprotected rocky coast, under well-embedded rocks; rare south of Oregon.

Cancer productus Randall, 1840. Red rock crab; under rocks of semiprotected outer coast; also in bays, under rocks or partly buried in sand and mud; active nocturnally; common. Juveniles highly variable in color and pattern (from white to red and brown with spots and stripes or vermiculations; all gradually grow towards a uniform brick red color through successive molts. See Boulding and LaBarbera 1986, Biol. Bull. 171: 538–547 (repeated claw pressure (loading) at the same location on shells facilitates predation on the clam *Leukoma staminea*); Robles et al. 1989, J. Nat. Hist. 23: 1041–1049 (diel variation in intertidal foraging).

CHEIRAGONIDAE

Telmessus cheiragonus (Tilesius, 1815). Helmet crab; northern, subtidal, rarely low intertidal.

PORTUNIDAE

Callinectes sapidus Rathbun, 1896. The Atlantic blue crab; low intertidal to shallow subtidal; will swim in the water column; very aggressive. Occasional specimens are found in San Francisco Bay. May be confused with *Callinectes bellicosus, C. arcuatus,* or *Portunus xantusii*, southern species which may reach central California when there is a strong El Niño.

Carcinus maenas (Linnaeus, 1758). European green or shore crab; low intertidal and shallow subtidal; introduced to San Francisco Bay in early 1990s, population exploded and geographic range extended north to British Columbia in just 10 years; a voracious predator, it has caused substantial declines in the abundance of native crabs and clams in Bodega Harbor; see Grosholz and Ruiz 1995; Grosholz 2005, Proc. Natl. Acad. Sci. 102: 1088–1091 (fisheries and aquaculture). The native symbiotic nemertean egg predator, *Carcinonemertes epialti*, now also infests green crabs (Torchin et al. 1996, J. Parasitol. 83: 449–453), threatens fisheries and aquaculture. Young crabs are variable in color and gradually grow toward the greenish adult color through successive molts. Rapidly growing crabs are yellow or green on the underside, while slow growing crabs are orange to red. See Cohen et al. 1995, Mar. Biol. 122: 225–237 (introduction to California); Lafferty and Kuris 1996, Ecol. 77: 1889–2000 (potential for biological control); Jensen et al. 2002, Mar. Ecol. Prog. Ser. 225: 251–262 (competition with *Hemigrapsus*); Behrens, Yamada, and Hunt 2000, Dreissena 11: 1–7 (introduction to Pacific Northwest); Carlton and Cohen 2003, J. Biogeog. 30: 1809–1820 (global distribution).

XANTHIDAE

See Knudsen 1957, Bull. So. Calif. Acad. Sci. 56: 133–142 (molting); 1959, Wasmann J. Biol. 17: 9–104 (autotomy and regeneration); 1959, Ecology 40: 113–115 (shell formation and growth); 1960, Ecol. Monogr. 30: 16–185 (ecology).

Cycloxanthops novemdentatus (Lockington, 1877). Low intertidal under rocks in gravel and shell substrate; usually rare north of Point Conception, locally common south of Monterey. Active and aggressive for a xanthid. See Knudsen 1960, Bull. So. Calif. Acad. Sci. 59: 1–8 (life cycle).

Paraxanthias taylori (Stimpson, 1860). Lower intertidal; protected outer coast, under well-impacted rocks; rare north of Point Conception. See Knudsen 1959, Bull. So. Calif. Acad. Sci. 58: 138–145 (life history).

PANOPEIDAE

Like most xanthids, panopeids are slow-moving, inactive crabs that "play dead" when handled. For taxonomy of *Lophopanopeus* species, see Menzies 1948.

Lophopanopeus bellus (Stimpson, 1860). Intertidal under rocks; stones of protected and unprotected coast; see Menzies 1948, Allan Hancock Found. Publs. Occ. Pap. 4, 45 pp. (taxonomy). See Knudsen 1959, Bull So. Calif. Acad. Sci. 58: 57–64 (life cycle).

Lophopanopeus leucomanus Lockington, 1876. Intertidal in coarse sand under rocks and in surfgrass roots; see Menzies 1948, above. See Knudsen 1958, Bull So. Calif. Acad. Sci. 57: 51–59 (life cycle).

Rhithropanopeus harrisii (Gould, 1841). Mud crab; introduced from Atlantic coast; common to abundant in sloughs, estuarine habitats with mud banks in San Francisco Bay, as well as Coos Bay and other estuaries in Oregon.

ANOMURA, GALATHEOIDEA

GALATHEIDAE

Pleuroncodes planipes Stimpson, 1860. Pelagic red crab; sometimes beached in vast swarms from Monterey south, more common in El Niño years; for occurrence, biology, and fisheries, see Kato 1974, Mar. Fish. Rev. 36: 1–9; Gomez, G. J. and Sanchez 1997, Bull. Marine Sci. 61: 305–326.

ANOMURA, HIPPOIDEA (MOLE AND SAND CRABS)

HIPPIDAE

Emerita analoga (Stimpson, 1857). Pacific sand crab; intertidal of exposed sandy beaches; abundant but distribution patchy; moves up and down beach with tidal cycle, burrowing to depth of several centimeters when tide is out; regularly found south of Oregon; larvae long-lived in some years they settling in great numbers on outer beaches north to Vancouver Island and sometimes even Alaska. Commonly serving as an intermediate host for bird acanthocephalans, (*Polymorphus* spp.). Important food source for shorebirds. See Dugan et al. 2000 (see above); Jaramillo et al. 2000 Mar Ecol-PSZNI 21: 113–127 (abundance, population structure, burrowing rate); Barron et al. 1999, Bull. Environ. Contam. Toxicol. 62: 469–475 (sensitivity to weathered oil).

ALBUNEIDAE

Lepidopa californica Efford, 1971. A southern California mole crab; low intertidal and subtidal of sandy beaches; filter feeders. See Dugan et al. 2000 (above) and Boyko 2002 (above).

BLEPHARIPODIDAE

Blepharipoda occidentalis Randall, 1840. Spiny mole crab; low intertidal; more common sublittorally; exposed sandy beaches; filter feeders. Important sea-otter food. The small clam *Mysella pedroana* is commonly attached in the gill chamber (Carpenter 2005, Nautilus 119: 105–108). See Knight 1968, Proc. Calif. Acad. Sci. (4) 35: 337–370 (larval development, distribution, ecology.); Dugan et al. 2000, J. Exp. Mar. Biol. Ecol. 255: 229–245 (burrowing abilities and swash behavior); Kreuder et al. 2003, J. Widl. Dis. 39: 495–509 (parasites); Boyko 2002, Bull. Amer. Mus. Natl. Hist. 272 (systematics, literature).

LITHODIDAE

Cryptolithodes sitchensis Brandt, 1853. Umbrella crab; low intertidal to subtidal; in crevices, on sponges; algae; color widely variable.

Hapalogaster cavicauda Stimpson, 1859. Low intertidal of protected rocky coast; under rocks and in deep crevices; uncommon.

Oedignathus inermis (Stimpson, 1860). Exposed and semiprotected rocky coasts; deep in old pholad bore holes, sea urchin holes, or on rock crevices; uncommon, infrequently seen because habitat is inaccessible.

Phyllolithodes papillosus Brandt, 1849. Juveniles rare under low intertidal rocks, adults subtidal.

Lopholithodes foraminatus (Stimpson, 1859). Box crab; very low intertidal; juveniles rare under rocks.

Lopholithodes mandtii Brandt, 1845. Very low intertidal; juveniles under rocks.

PORCELLANIDAE

Porcelain crabs are filter feeders; they readily autotomize their claws and legs and are positively thigmotactic; see Haig (1960, Allan Hancock Pac. Exped. 24, 440 pp.) for detailed descriptions and systematics; Stillman and Somero 2000, Physiol. Biochem. Zool. 73: 200–208 (physiology); Stillman and Somero 1996, J. Exp. Biol. 199: 1845–1855 (morphology); Stillman and Reeb 2001, Mol. Phylo. Ecol. 19: 236–245 (molecular phylogenetics).

Pachycheles holosericus Schmitt, 1921. A southern species; low intertidal, embedded in sponges.

Pachycheles pubescens Holmes, 1900. Low intertidal; rocky areas.

Pachycheles rudis Stimpson, 1859. Low intertidal; semiprotected rocky coast; adults live in permanent pairs, often trapped in old pholad bore holes, and in concavities of *Laminaria* and *Egregia* holdfasts, also under rocks, on wharf pilings; a bopyrid isopod (*Aporobopyrus muguensis*) may occur in branchial cavity reducing fecundity by about 50% (Van Wyk 1982, Parasitol. 85: 459–473).

Petrolisthes cabrilloi Glassell, 1945. Southern species; under rocks and cobble habitats and in mussel beds. Frequently infected with the rhizocephalan barnacle parasitic castrator, *Lernaeodiscus porcellanae* (Høeg and Lutzen 1995, Ocean. Mar. Biol. Ann. Rev. 33: 427–485). See Kropp 1981, Crustaceana 40: 307–310 (deposit feeding).

Petrolisthes cinctipes (Randall, 1840). Mid- and upper intertidal of exposed protected; semiprotected rocky coast, under rocks in mussel beds; often abundant, replaced to the south by *P. cabrilloi*, see Wicksten 1973, Bull. So. Calif. Acad. Sci. 72:

161–163 (feeding); Donahue 2004, Mar. Ecol. Prog. Ser. 2004, Mar. Ecol. Prog. Ser. 267: 196–207 (competition).

Petrolisthes eriomerus Stimpson, 1871. Mid-intertidal of protected rocky coasts; bays, under rocks over gravel substrates; also in eelgrass and kelp holdfast can occur in sandier habitats than *P. cinctipes*; the rhizocephalan *Lernaeodiscus porcellanae* has been reported from specimens from southern California.

Petrolisthes manimaculis Classell, 1945. Low intertidal under rocks; females and juveniles often closely resemble *P. eriomerus*.

Petrolisthes rathbunae Schmitt, 1921. A southern species, under stones, rarely subtidal, low intertidal under rocks and in crevices.

ANOMURA, PAGUROIDEA (HERMIT CRABS)

See McLaughlin and Fisher (1974) for systematics.

DIOGENIDAE (LEFT-HANDED HERMIT CRABS)

Isocheles pilosus (Holmes, 1900). Moon snail hermit; found in low intertidal in sand on semiprotected beaches; often in moon snail shells; more common in subtidal.

Paguristes ulreyi Schmitt, 1921. Furry hermit; low intertidal to subtidal; orange to brown covered with golden hairs.

Paguristes bakeri (Holmes, 1900). In quiet waters over sand or mud; subtidal, rarely intertidal. Dark reddish brown, often in moon snail shells.

PAGURIDAE (RIGHT-HANDED HERMIT CRABS)

See Elwood and Stewart 1985, Anim. Behav. 33: 620–627 (behavior of European hermit crab *Pagurus bernhardus*); Hazlett 1981, Ann. Rev. Ecol. Syst. 12: 1–22 (behavioral ecology), Rittschof 1980, J. Chem. Ecol. 6: 103–118 (chemical attraction to simulated gastropod predation sites), Osorno et al. 1998, J. Exp. Mar. Biol. Ecol. 222: 163–173 (shell selection). Shells inhabited by hermit crabs usually become encrusted with a crab-associated biota (discussed in sections on intertidal parasites and commensals). See Williams and McDermott (2004) J. Exp. Mar. Biol. Ecol. 305: 1–128 for a review of these associations.

Discorsopagurus schmitti (Stevens, 1925) Low intertidal and subtidal; in broken or attached worm tubes.

Pagurus armatus (Dana, 1851). Low intertidal to subtidal; often in moonsnail shells encrusted with hydroids.

Pagurus beringanus (Benedict, 1892). Low intertidal on rock jetties; sublittoral.

Pagurus caurinus Hart, 1971. Rare; a northern species found in protected waters.

Pagurus granosimanus (Stimpson, 1859). Exposed and semiprotected outer coast; lower intertidal pools; common.

Pagurus hemphilli (Benedict, 1892). Mid to low intertidal.

Pagurus hirsutiusculus (Dana, 1851). Mid intertidal of rocky coast; common; in bays, under rock cover; tide pools, coarse sand to gravel substrates; uses a variety of shells, frequently unable to fully withdraw into its shell.

Pagurus ochotensis Brandt, 1851. Low intertidal to subtidal over sandy or softer bottoms; often in moon snail shells.

Pagurus quaylei Hart, 1971. In gravelly areas; shallow water; see Hart 1971, J. Fish. Res. Bd. Canada 28: 1527–1544.

Pagurus samuelis (Stimpson, 1857). Rocky coasts; mid to lower tidepools, abundant, usually in turban shells, behaviorally dominant in shell exchanges to *P. hirsutiusculus*; occasionally in coarse substrates in bays.

THALASSINIDEA

UPOGEBIIDAE (MUD SHRIMP)

Upogebia pugettensis (Dana, 1852). Blue mud shrimp; mid- to lower intertidal of bays; and occasionally on outer coast in protected areas, such as at Cape Arago, Oregon. Locally common; builds D- or Y-shaped burrows, replaced south of Pt. Conception by *Upogebia macginitieorum* Williams, 1986, a white, less robust species (see Williams 1986, Mem. San Diego Soc. Natl. Hist. 14: 1–60 for taxonomy, morphology); firm-walled burrows in mud or muddy sand; commensals include *Betaeus, Hesperonoe*, the clam *Cryptomya californica*, pinnotherids, copepods, and the phoronid *Phoronis pallida*; the isopod *Phyllodurus abdominalis* and the clam *Neaeromya rugifera* may both occur on the abdomen; the parasitic castrator isopod, *Orthione griffenis* Markham 2004, can attain high prevalence; see MacGinitie 1930, Amer. Midl. Nat. (10) 6: 36–44 (natural history); Powell 1974, Univ. Calif. Publ. Zool. 102: 1–41 (gut morphology of *Upogebia* and *Neotrypaea*); Griffen et al. 2004, Mar. Ecol. Prog. Ser. 269: 223–236 (bioturbation); Santagata 2004, Biol. Bull. 207: 103-115 (behavioral cues).

CALLIANASSIDAE (GHOST SHRIMP)

See Manning and Felder 1991, Proc. Biol. Soc. Wash. 104: 764–792 (taxonomic revision of family, including the new genus *Neotrypaea*). See Sakai 2005, *Callianassoidea of the World* (Decapoda, Thalassinidea). Crustaceana, Monographs Volume 4, 200 pp.

Neotrypaea biffari (Holthuis, 1991). Tidepool ghost shrimp; pools under rubble on outer coast; paired, usually in turn with a pair of blind gobies, *Typhlogobius californiensis*.

Neotrypaea californiensis (Dana, 1854) (=*Callianassa californiensis*) Bay ghost shrimp; burrowing in mud or sand of upper to mid-intertidal in bays; often covering large areas of intertidal flats; locally abundant; burrows with poorly defined walls; commensals include shrimp *Betaeus* spp., polynoid worms, various pinnotherid crabs, copepods (*Hemicyclops* and *Clausidium*), and the goby *Clevelandia ios*; the parasitic castrator bopyrid isopod, *Ione*, may occur in gill chamber. See Hoffman 1981, Pac. Sci. 35: 211–216 (association with *Clevelandia*); Labadie and Palmer 1996, J. Zool. 240: 659–675 (claw dimorphism); Feldman et al. 1997, Mar. Ecol. Prog. Ser: 150: 121–136 (recruitment), Lau et al. 2002, Microbial Ecol. 43: 455–466 (digestive bacteria). Bioturbation by ghost shrimps is a very serious problem for oyster mariculture; sediment suspended by these shrimp alters water quality and can foul oyster gills (Dumbauld et al. 2004, pp. 53–61, and DeWitt et al. 2004, pp. 107–118, both in: Proc. Symp. Ecology Large Bioturbators in Tidal Flats and Shallow Sublittoral Sediments. Nagasaki University).

Neotrypaea gigas (Dana, 1852); Giant ghost shrimp; low to subtidal; rare, burrowing in sand; builds deep burrows. More common south of Pt. Conception. The rostral distinction between *N. gigas* and *N. californiensis* is noted in the keys; a further character that distinguishes these species is the nature of the second pereopod: in *N. californiensis*, the propodus and dactyl of the second pereopod is approximately equal (plate 326G), whereas in *N. gigas*, the propodus of the second pereopod is curved and wider than the dactyl (plate 326H).

REFERENCES

Boyko, C. B. 2002. A worldwide revision of the Recent and fossil sand crabs of the Albuneidae Stimpson and Blepharipodidae, new family (Crustacea: Decapoda: Anomura: Hippoidea). Bulletin of the American Museum of Natural History, number 272, 396 pp.

Chace, F. A. 1951. The grass shrimp of the genus *Hippolyte* from the west coast of North America. J. Wash. Acad. Sci. 41: 35–39.

Chang E. S., M. J. Bruce, and S. L. Tamone 1993. Regulation of crustacean molting: a multi-hormonal system. Am. Zool. 33: 324–329.

Clayton, D. A. 1990. Crustacean allometric growth: A case of caution. Crustaceana 58: 270–290.

Culver, C. S., and A. M. Kuris. 2001. Sheep Crab. pp 115–117. In California living marine resources: a status report. W. S. Leet, C. M. Dewees, R. Klingbiel, and E.J. Larsen, eds. Univ. Calif. ANR Publ. #SG01-11.

DeWitt, T. H., A. F. D'Andrea, C. A. Brown, B. D. Griffen, and P. M. Eldridge. 2004. Impact of burrowing shrimp populations on nitrogen cycling and water quality in western North American temperate estuaries, pp. 107–118. In: Proceedings of the Symposium on Ecology of Large Bioturbators in Tidal Flats and Shallow Sublittoral Sediments—from individual behavior to their role as ecosystem engineers. Nagasaki University.

Dumbauld, B., K. Feldman and D. Armstrong. 2004. A comparison of the ecology and effects of two species of thalassinidean shrimps on oyster aquaculture operations in the eastern North Pacific, pp. 53–61. In: Proceedings of the Symposium on Ecology of Large Bioturbators in Tidal Flats and Shallow Sublittoral Sediments from individual behavior to their role as ecosystem engineers. Nagasaki University.

Elwood R.W., and A. Stewart 1985. The timing of decisions during shell investigation by the hermit crab, *Pagurus bernhardus*. Anim. Behav. 33: 620–627.

Haig, J. 1960. The Porcellanidae (Crustacean, Anomura) of the eastern Pacific. Allan Hancock Pac. Exped. 24, 440 pp.

Hart, J. F. L. 1964. Shrimps of the genus *Betaeus* on the Pacific coast of North America with descriptions of three new species. Proc. U.S. Nat. Mus. 115: 431–466.

Hartnoll, R. G. 1985. Growth, pp. 111–196. In: The Biology of Crustacea v. 2. L.G. Abele (ed.)

Hazlett, B. A. 1981. The behavior ecology of hermit crabs. Ann. Rev. Ecol. Syst. 12: 1–22.

Høeg, J. T., and J. Lutzen. 1995. Life cycle and reproduction of the Cirripedia Rhizocephala. Oceanography and Marine Biology Annual Review 33: 427–85.

Iguchi and Ikeda. 2004. Vertical distribution, population structure and life history of *Thysanoessa longipes* (Crustacea: Euphausiacea) around Yamato Rise, central Japan Sea. J. Plankton Res 26: 1015–1023.

Jensen, G. C. 1995. Pacific Coast Crabs and Shrimps. Sea Challengers, Monterey, 87 pp.

Kozloff, E. 1987. Marine Invertebrates of the Pacific Northwest. University of Washington Press: Seattle and London, 511 pp.

Kuris, A. M., and J. T. Carlton. 1977. Description of a new species, *Crangon handi*, and new genus, *Lissocrangon*, of crangonid shrimps (Crustacea: Caridea) from the California coast, with notes on adaptation in body shape and coloration. Biol. Bull. 153: 540–559.

Kuris A. M., G. O. Poinar, R. Hess, and T. J. Morris. 1979. Virus particles in an internal parasite, *Portunion conformis* (Crustacea: Isopoda: Entoniscidae), and its marine crab host, *Hemigrapsus oregonensis*. J. Invert. Path. 34: 26–31.

Kuris, A. M., G. O. Poinar, and R. Hess. 1980. Mortality of the internal isopodan parasitic castrator, *Portunion conformis* (Epicaridea, Entoniscidae), in the shore crab *Hemigrapsus oregonensis* with a description of the host response. Parasitology 80: 211–232.

Kuris, A. M, Z. Ra'anan, A. Sagi, and D. Cohen. 1987. Morphotypic differentiation of male Malaysian giant prawns, *Macrobrachium rosenbergii*. J. Crust. Biol. 7: 219–237.

MacGinitie, G. E. 1937. Notes on the natural history of several marine Crustacea. Amer. Midl. Nat. 18: 1031–1037.

Martin, J. W., and G. E. Davis. 2001. An updated classification of the recent crustacea. Natural History Museum of Los Angeles County Science Series 39: 1–124.

McLaughlin, J., and L. R. Fisher. 1974. The hermit crabs (Crustacea: Decapoda: Paguridea) of northwestern North America. Zool. Verhandel. No. 130, 396 pp.

McLaughlin, J. et al. 2005. Common and Scientific Names of Aquatic Invertebrates from the United States and Canada: Crustaceans. American Fisheries Society Special Publication 31: 1–325.

Passano, L. M. 1960. Molting and its control. In The physiology of crustacea, T. H. Waterman, ed., Vol. 1, Chap. 15, pp. 473–536. Academic Press.

Rittschof, D. 1980. Chemical attraction of hermit crabs and other attendants to simulated gastropod predation sites. J. Chem. Ecol. 6: 103–118.

Schmitt, W. L. 1921. The Marine Decapod Crustacea of California. Univ. Calif. Publ. Zool. 23: 470 pp.

Schweitzer, C. E., and R. M. Feldman. 2000. Re-evaluation of Cancridae Latreille, 1802 (Decapoda: Brachyura) including three new genera and three new species. Contrib. Zool. 69: 1–36.

Shields, J. D., and A. M. Kuris. 1988. Temporal variation in abundance of the egg predator *Carcinonemertes epialti* (Nemertea) and its effect on egg mortality of its host, the shore crab, *Hemigrapsus oregonensis*. Hydrobiologia 156: 31–38.

Skinner D. M. 1985. Molting and Regeneration In D. E. Bliss and L. H. Mantel (eds.), The Biology of Crustacea, pp. 43–146. Academic Press, New York.

Teissier, G. 1960. Relative growth. In The Physiology of Crustacea, T. H. Waterman, ed., Academic Press. Vol. 1, Chap. 16, pp. 537–560.

Torchin M. E., K. D. Lafferty, and A. M. Kuris. 1996. Infestation of an introduced host, the European green crab, *Carcinus maenas*, by a native symbiotic nemertean egg predator, *Carcinonemertes epialti*. J. Parasitol. 83: 449–453.

Versulis, M., B. Schmitz, A. von der Heydt, and D. Lohse. 2000. How snapping shrimp snap: through cavitating bubbles. Science 289: 2114–2117.

Vogel, S. 1988. Life's Devices: The Physical World of Animal and Plants. Princeton Univ. Press. Princeton, New Jersey.

Wicksten, M. K. 1978. Attachment of decorating materials in *Loxorhynchus crispatus* (Brachyura: Majidae). Trans. Amer. Micr. Soc. 97: 217–220.

Wicksten, M. K. 1980. Decorator crabs. Sci. Amer. 242: 146–154.

Wicksten, M. K., and M.D. Behrens 2000. New record of the pelagic crab *Planes cyaneus* in California (Brachyura: Grapsidae). SCAMIT newsletter 19(5): 7.

Pycnogonida

C. ALLAN CHILD AND JOEL W. HEDGPETH

(Plates 327–332)

The Class Pycnogonida are exclusively marine invertebrates found in all oceans at all depths. Identification of most genera usually relies on the presence or absence of appendages or reduction of their segment numbers. They have a central linear body or trunk that often has visible segmentation and contains the internal circulatory system, the nervous system, and the gut, parts of which extend into each leg. The trunk has four paired lateral processes (rarely five or six pairs in non-California and non-Oregon species) each carrying a leg.

The anterior or cephalic segment of the trunk carries most of the other appendages: a dorsal ocular tubercle (sometimes lacking) with four eyes or blind and a pair of chelifores that consists of one, two, or three segments and are fully chelate to entirely lacking among the wealth of more than 1,200 species. The cephalic segment also has a pair of tactile palps that have from one to nine segments or are absent; the first pair of lateral processes; and an anterior suctorial proboscis that varies greatly in shape and size.

A pair of ventral ovigers, which are unique to pycnogonids, are also born on the cephalic segment. The ovigers consist of 10 segments, less in some genera, and are believed to be modified legs that males use to carry eggs passed by the females. The ovigers sometimes have a distal strigilis (a shepherd's crook) used in some species by both sexes to clean appendages. The strigilis is armed with compound or simple spines and a terminal claw or no claw among the various genera. There are usually eight legs, which carry internal gonads along with gut diverticula. Each leg consists of eight segments: three short proximal coxae, a femur, two tibiae, a shorter tarsus, and a propodus. The propodus has a terminal claw and usually a pair of lateral auxiliary claws. The femur contains a cement gland in males only, exiting

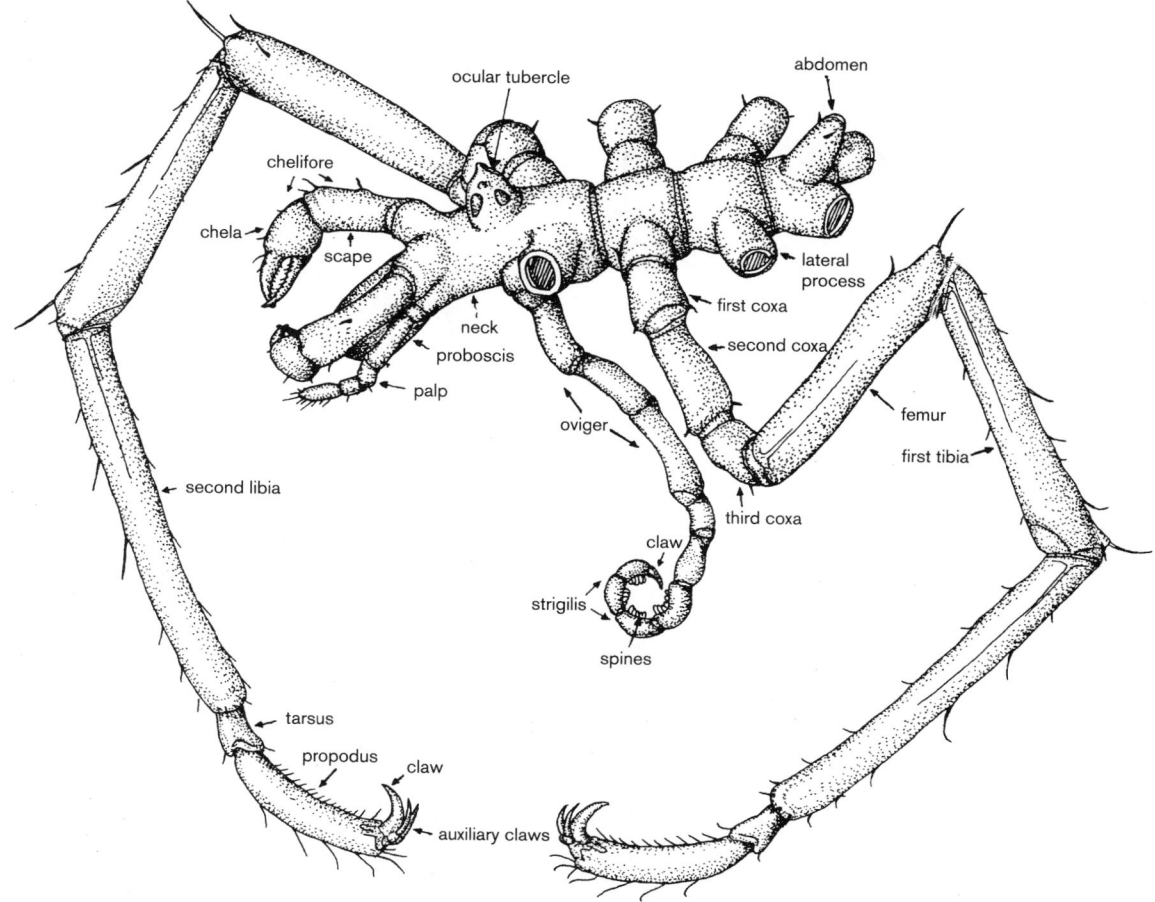

PLATE 327 A diagrammatic pycnogonid (after Child).

through a single or multiple tube, pore, or sieve plate and used to cement the eggs together. The trunk has a small posterior abdomen that contains nothing but the anus.

Identification of most pycnogonids does not require dissection or fancy preparation. A low-power dissecting microscope is usually all that is required to identify almost all California species. The few diagnostic appendage pairs are located on the cephalic segment. Their presence or absence can be determined under a microscope's low power. The number of appendage segments and claw lengths might require slightly higher magnification. Specimens should be stored in the same preservatives as for Crustacea.

Our shallow waters are known to contain five of the nine families, 12 of the 80 genera, and 29 of the more than 1,200 known species of Pycnogonida. Several of the following species are found in the sublittoral as well as littoral habitats. There are many more pycnogonid species and genera found in deeper waters off California and Oregon.

Ammotheidae

This is the most heterogeneous family among the pycnogonids; the ammotheids contain more species and genera than any other family. Chelifores are usually present (see generalized pycnogonid, plate 327) in each species. They can be fully chelate, with chelae atrophied, or with chelae lacking or with-

out chelifores entirely. Their palps are one- to 10-segmented. Ovigers are nine- to 10-segmented, found in both sexes but larger in males, and all known local species lack a functional strigilis. Six genera are known in our shallow waters.

ACHELIA

Genus size small to very small, leg span usually <10 mm; trunk circular or ovoid; lateral processes compact, touching or nearly so, often with distal tubercles and a few setae or spines. Ocular tubercle usually low, eyes present, usually prominent as befits shallow-water species. Abdomen often carried horizontally. Chelifore scapes of one segment; chelae almost always atrophied, without fingers. Palps with seven to nine segments, usually eight, with four or five short distal segments. Ovigers with denticulate or plain spines on pseudostrigilis (merely a curve in the strigilis), without a terminal claw. Leg segments short, propodus with strong heel spines, robust claw and usually long auxiliary claws. One dorsodistal cement gland pore. Seven known species along this coast.

KEY TO ACHELIA SPECIES

1. Chelae atrophied to fingerless bumps in adults. 2
— Adult chelae fully formed (plate 328A). . . . . Achelia chelata

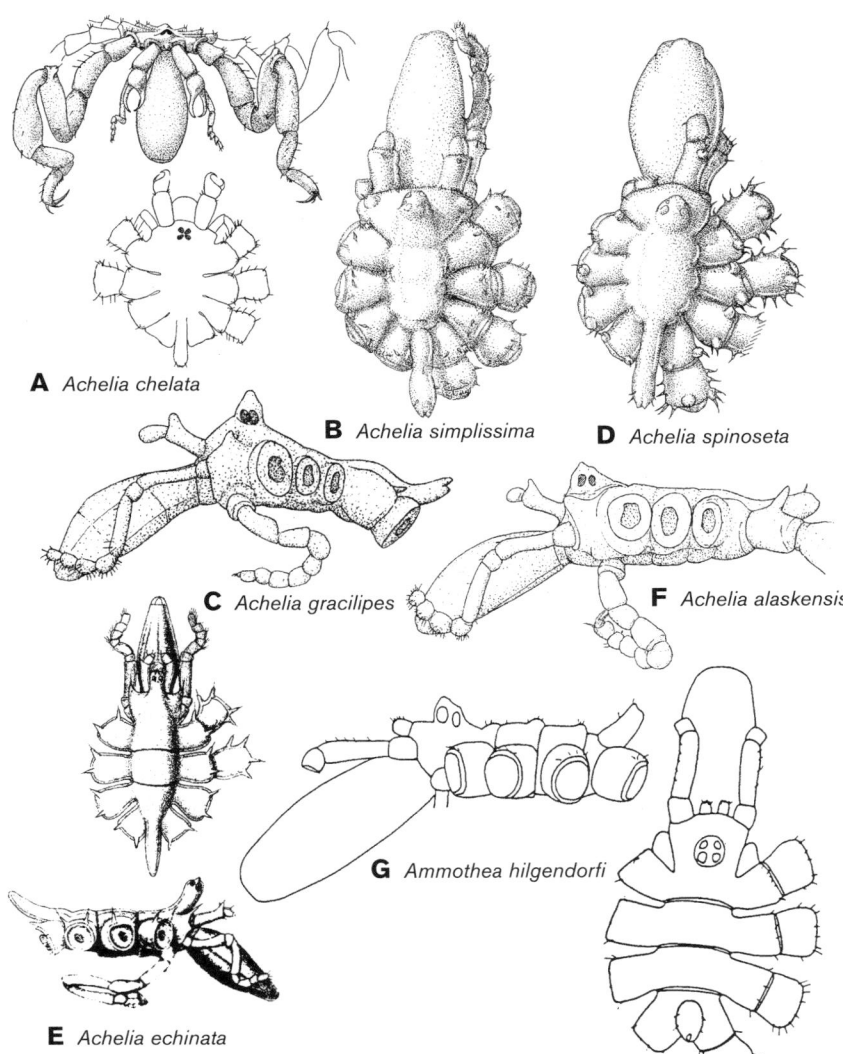

PLATE 328 A, *Achelia chelata*; B, *Achelia simplissima*; C, *Achelia gracilipes*; D, *Achelia spinoseta*; E, *Achelia echinata*; F, *Achelia alaskensis*; G, *Ammothea hilgendorfi* (A, after Hedgpeth; B, D, after Child; C, F, after Cole; E, after Sars; G, after Nakamura).

A *Achelia chelata*

B *Achelia simplissima*

D *Achelia spinoseta*

C *Achelia gracilipes*

F *Achelia alaskensis*

E *Achelia echinata*

G *Ammothea hilgendorfi*

2. With conspicuous dorsal tubercles on distal lateral processes, chelifores, and/or first coxae; palps eight-segmented. 3
— Without conspicuous tubercles on lateral processes, chelifores, coxae; palps seven-segmented (plate 328B) . . . . *Achelia simplissima*
3. First coxae with paired or several tubercles all shorter than segment diameter 4
— First coxae with single conspicuous dorsal tubercle longer than segment diameter and two to three short lateral tubercles (plate 328C) . . . . *Achelia gracilipes*
4. Distal palp segments 5, 6, and 7 ovoid, without projections 5
— Distal palp segments 5, 6, and 7 with ventral projections, eighth segment club-shaped (plate 328D) . . . . *Achelia spinoseta*
5. Lateral processes with two pointed dorsolateral tubercles, first and second coxae with four to five pointed tubercles each; ocular tubercle taller than wide (plate 328E) . . . . *Achelia echinata*
— Lateral processes with tiny tubercles, first coxae with two, second coxae with no dorsal tubercles; ocular tubercle low, shorter than wide (plate 328F) . . . . *Achelia alaskensis*

AMMOTHEA

There is a single known California species, *Ammothea hilgendorfi* (plate 328G). Trunk without dorsomedian tubercles, fully segmented, posterior segment sometimes faint, ocular tubercle a low broad cone with large eyes, proboscis a long oval; lateral processes well separated, glabrous; chelifores one-segmented, short, chelae entirely lacking; palps nine-segmented, distal short segments oval; ovigers 10-segmented, without strigilis, distal three segments carried anaxially on seventh segment, with few tiny spines; legs moderately long, with few short spines, tarsus very short, propodus with heel, four to five large heel spines; claw robust, auxiliary claws well curved, half main claw length. Cement gland a pore on dorsodistal swelling proximal to tip of femora.

AMMOTHELLA

Genus: Trunk with or without dorsomedian tubercles or other adornment; lateral processes often with large spines or dorsodistal tubercles; appendages often with long tubular or pointed spines; chelifore scapes with two segments, chelae atrophied

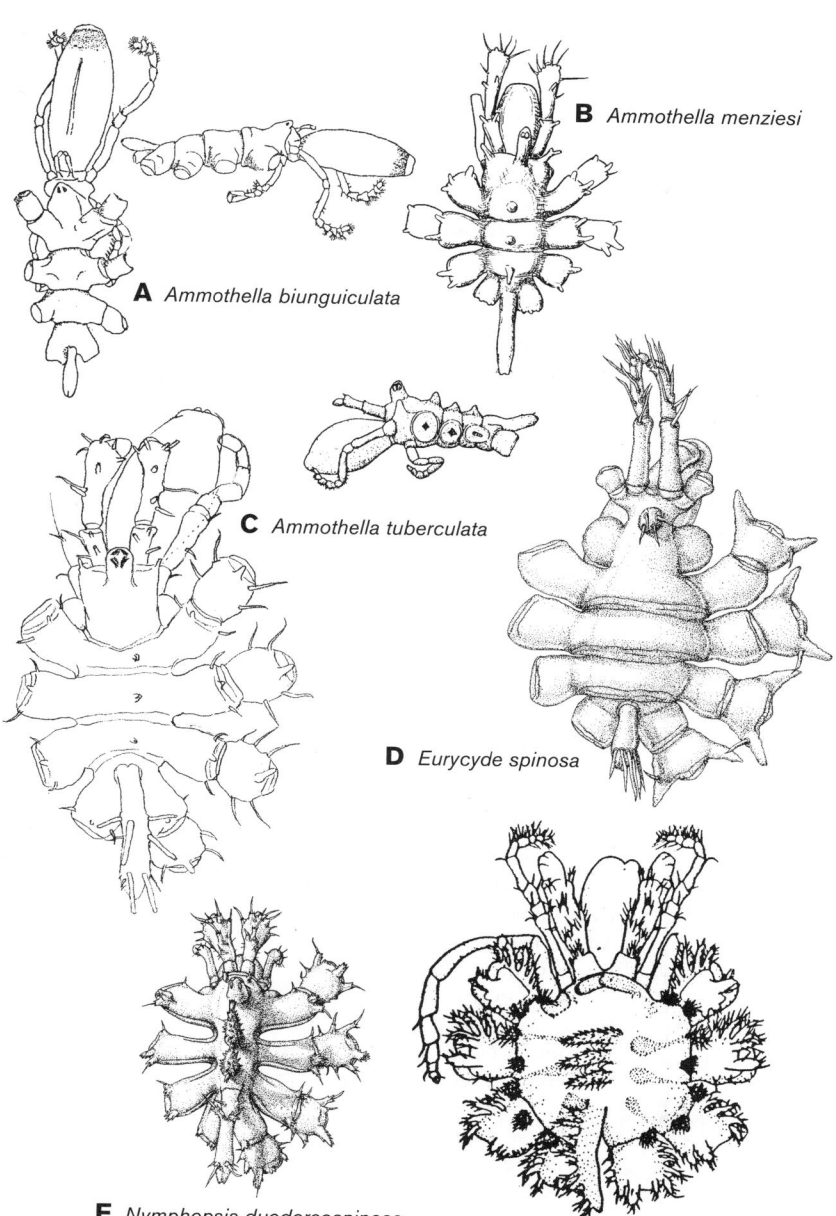

B *Ammothella menziesi*

A *Ammothella biunguiculata*

C *Ammothella tuberculata*

D *Eurycyde spinosa*

PLATE 329 A, *Ammothella biunguiculata*; B, *Ammothella menziesi*; C, *Ammothella tuberculata*; D, *Eurycyde spinosa*; E, *Nymphopsis spinosissima*; F, *Nymphopsis duodorsospinosa* (A, D, E, F, after Child; B, Hedgpeth, original; C, after Cole).

F *Nymphopsis duodorsospinosa*

E *Nymphosis spinosissima*

into bumps; palps nine-segmented; ovigers 10-segmented, without strigilis, distal segments with denticulate spines; legs usually setose, sometimes heavily, tarsus short, propodus usually quite long, with large heel spines; claw robust, auxiliaries long. Cement gland outlet usually a long dorsodistal tube. Four species.

KEY TO *AMMOTHELLA* SPECIES

1. Trunk with spines or glabrous, without dorsomedian tubercles 2
— Trunk with large or small dorsomedian tubercles 3
2. Trunk with one to two dorsomedian spines per segment; trunk and lateral processes lacking tubular spines; ocular tubercle and abdomen long, erect; chelifore scape first segment shorter than second; palp distal three segments slender, twice longer than diameters; oviger usually with denticulate spines on two distal segments only; cement gland opening on a long robust dorsodistal tube . . . .
. . . . *Ammothella spinifera*
— Species much like *Ammothea*. Trunk without adornment, fully segmented, ocular tubercle a low cone, eyes small, abdomen moderately short; chelifores short, tiny; palp fifth to eighth segments with small ventral extensions; ovigers distal four segments with one to three denticulate spines each; legs typical except for propodus, which is straight, lacks larger heel and heel spines, bears robust auxiliary claws but lacks main claw; cement gland pore inconspicuous (plate 329A) . . . . *Ammothella biunguiculata*
3. Trunk, lateral processes with short, rounded dorsal tubercles wider than tall, lateral processes crowded, touching; ocular tubercle little taller than wide; proboscis very wide, bulbous; chelifores short, scapes of equal length (plate 329C) . . . . *Ammothella tuberculata*

— Trunk with three tall, slender tubercles pointed anterior or posterior, lateral processes crowded, narrowly separated, with long or short distal tubercles; ocular tubercle three times taller than wide; proboscis narrow, ovoid; chelifores long, second segment longest (plate 329B) . . . . *Ammothella menziesi*

EURYCYDE

There is a single known California species, *Eurycyde spinosa* (plate 329D). Trunk fully segmented, segment posteriors swollen, lateral processes crowded, almost touching. Ocular tubercle and abdomen very short, with group of spines toward tips of each. Proboscis of two sections, a short basal cylinder and distal slender pyriform process. Chelifore scapes slender, two-segmented; palps 10-segmented; ovigers with weak strigilis having a single row of denticulate spines and short terminal claw. Legs with very long spines each bearing spinules, tarsus short, propodus without major spines, main claw short, robust, auxiliaries lacking. Cement gland a bulge with distal tube proximal on side of femora.

NYMPHOPSIS

Description of the two known California species: tuberculate, all tubercles with lateral and distal spines; trunk with dorsomedian tubercles; ocular tubercle, chelifore scapes, abdomen all tall; chelifore scapes two-segmented, chelae atrophied, carried within trumpet-shaped scape tip; palps nine-segmented; ovigers 10-segmented, without strigilis, with few distal denticulate spines; legs short, with tall tubercles on both tibiae, claw robust, auxiliaries minute.

— Trunk with three tall dorsomedian tubercles; tibiae with fields of crowded dorsal tubercles (plate 329E) . . . . *Nymphopsis spinosissima*
— Trunk with two tall dorsomedian tubercles; legs with few tall tubercles on tibiae (plate 329F) . . . . *Nymphopsis duodorsospinosa*

TANYSTYLUM

Genus: trunk very tiny, unsegmented, lateral processes crowded, touching, forming circular shape, usually with small dorsolateral tubercles; proboscis usually short, tapering distally. Chelifores usually one-segmented stumps, chelae usually lacking. Palps four to seven segmented, short; oviger 10-segmented, with few denticulate or plain spines, without strigilis, most species with distal three oviger segments anaxial to enlarged seventh. Legs short, robust, major segments with dorsal bulges, tarsus short, propodus well curved, with heel spines, main claw robust, auxiliaries present. Cement gland outlet a dorsodistal pore or tiny tube. Three species are represented in the following key.

Two *Tanystylum* species are not keyed below. *T. duospinum* is noted in the species list. *T. nudum* Hilton, 1939, was inadequately described and cannot be identified with certainty. Its type specimen is lost.

KEY TO *TANYSTYLUM* SPECIES

1. Chelifores lacking any form of chelae, scape a one-segmented stump 2

— Chelifores with chelae retained as knobs; short lateral processes with large dorsal tubercle matching three on first coxae; proboscis narrow, pyriform; palp seven-segmented, short; leg segments elongate; cement gland tube on dorsodistal tubercle (plate 330A) . . . . *Tanystylum intermedium*

2. Trunk very compact, without tubercles, abdomen erect, placed toward anterior, not extending as far as first coxa, proboscis short, distally rounded; palps four-segmented; oviger with few plain setae, without anaxial placement of distal segments; legs typical, auxiliary claws half main claw length (plate 330B) . . . . *Tanystylum occidentalis*
— Trunk less compact, with low paired lateral process tubercles, abdomen with basal bulge, carried more horizontally, extending to first coxae, proboscis longer, tapered, styliform; palps six-segmented; oviger distal three segments placed anaxially on seventh; legs more robust, stout, auxiliary claws very short, only 0.2 main claw length (plate 330C) . . . . *Tanystylum californicum*

Rhynchothoracidae

This is a family with a single genus of peculiar species, some of which have repeatedly been found in beach sand, sometimes at considerable depth (1+ m). Their structure is usually adapted for this interstitial form of living in that most species (but not all) lack protruding tubercles and major spines and setae. Perhaps the few species with long tubercles have a different habitat. Several species have a low eye tubercle (some lack both tubercle and eyes) and dorsomedian tubercles that slightly protrude but apparently do not hinder progress among sand grains. All species are among the smallest pycnogonids and rarely have leg spans greater than 4–5 mm.

The proboscis extends out in the same flat plane as the trunk and usually has three anterior lips rather than a circular oral surface. Chelifores are lacking in adults. The short palps have three to five segments and originate on lateral extensions of the cephalic segment, which in the past were often considered an additional segment. The 10-segmented ovigers are very reduced in size, lack a strigilis, and have a peculiar large terminal segment bearing a row of tiny spines opposing a terminal claw that is carried laterally. Leg segments are usually short with the propodus longer than the second tibia. The tarsus is short, propodus well curved, claw robust, and with or without auxiliaries.

RHYNCHOTHORAX

There is one species in our range, *Rhynchothorax philopsammum* (plate 330D) Trunk dorsally compressed, fully segmented, lateral processes shorter than their diameters, with small anterior and posterior tubercles. Cephalic segment with small paired tubercles at anterior and lateral corners; ocular tubercle and eyes lacking; proboscis ovoid, tapering distally; abdomen short, a truncate cone; palps four-segmented with low dorsodistal tubercle on third, fourth upturned. Oviger distal segments with one to two lightly denticulate spines, terminal segment carried anaxially, terminal claw well curved, not as long as segment diameter. Cement gland outlet on legs unknown; in other species, where known, it is a single ventral tube or pore.

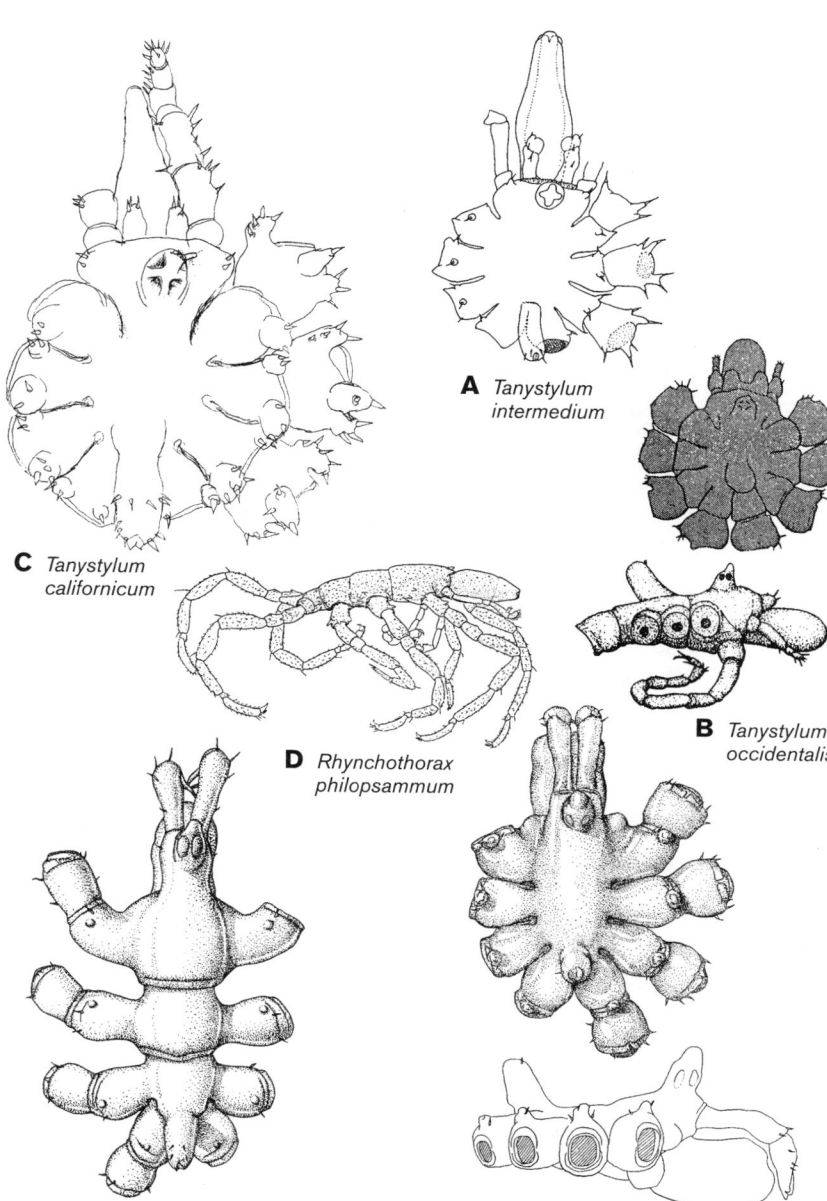

A *Tanystylum intermedium*

C *Tanystylum californicum*

D *Rhynchothorax philopsammum*

B *Tanystylum occidentalis*

F *Anoplodactylus californicus*

E *Anoplodactylus compactus*

PLATE 330 A, *Tanystylum intermedium*; B, *Tanystylum occidentalis*; C, *Tanystylum californicum*; D, *Rhynchothorax philopsammum*; E, *Anoplodactylus compactus*; F, *Anoplodactylus californicus* (A, after Stock; B, after Cole; C, Child, original; D, after Hedgpeth; E, F, after Child).

Phoxichilidiidae

A large family of mostly shallow-water species living for the most part in tropical-temperate habitats with few found in cold or deep waters. The trunks of species in this family have no tubercles or other decoration, are seldom fully segmented, and the lateral processes sometimes have rounded or conical dorsodistal tubercles. This family's additional characters include chelifores with full but small chelae; palps entirely lacking but species sometimes have palp "buds" on the first lateral process anteriors. They have ovigers carried only by males. The ovigers consist of six segments, although a few have only five. The leg has a short tarsus and a propodus with larger heel spines and often a variable length cutting lamina of tiny fused spines on the distal sole.

Species of the genus *Anoplodactylus* have very tiny auxiliary claws on proximal sides of the main claw, while species of the genus *Phoxichilidium* have larger distal auxiliaries that are presumably functional, unlike the tiny lateral form. The cement gland opening is dorsal and prominent, consists of one or more tubes, swollen or flat cups, slits, or pores of different sizes and shapes.

ANOPLODACTYLUS, PHOXICHILIDIUM

Anoplodactylus contains the majority of species in the family and can be separated from the only other California genus, *Phoxichilidium*, by close examination of the propodus and its claws. The auxiliary claws in this genus are tiny, difficult to discern, and sometimes are lacking entirely.

KEY TO *ANOPLODACTYLUS* AND *PHOXICHILIDIUM* SPECIES

1. Trunk with closely crowded lateral processes; neck and proboscis short, robust; palp buds appear as bumps lateral

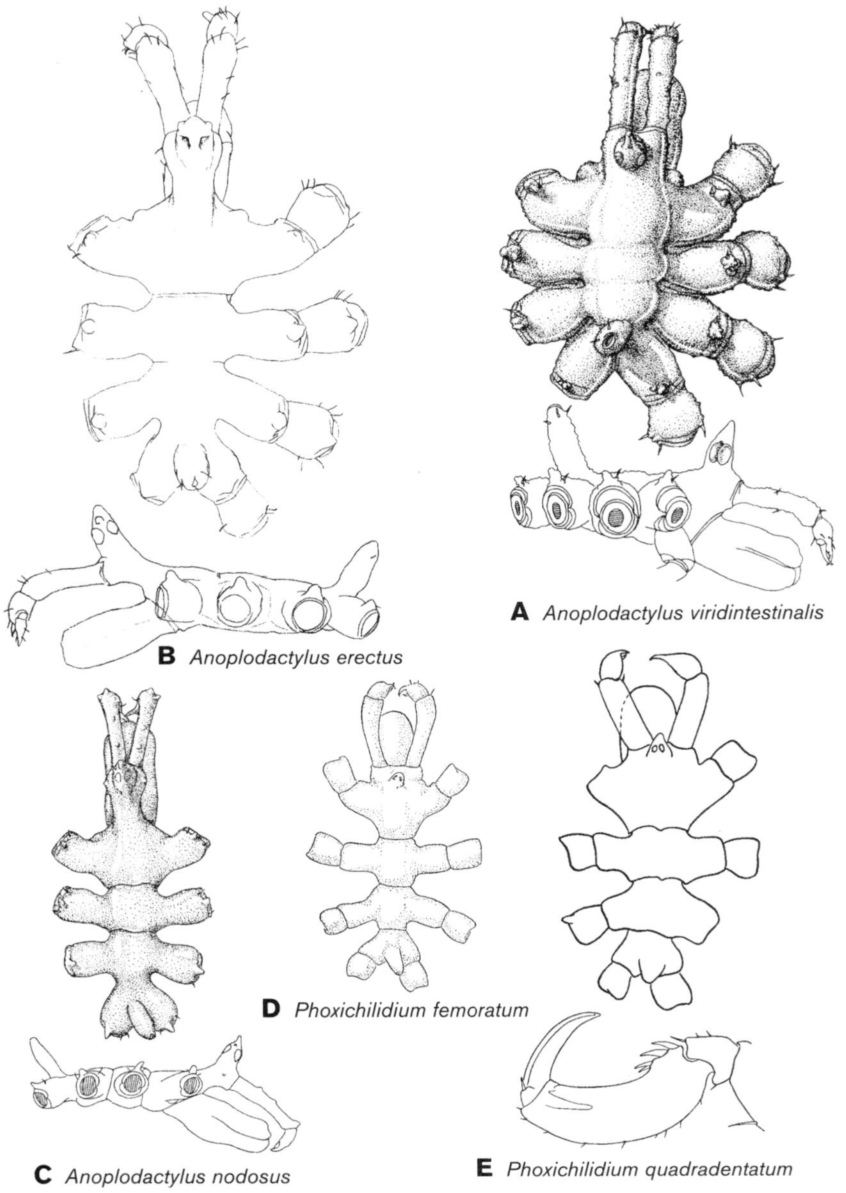

PLATE 331 A, *Anoplodactylus viridintestinalis*; B, *Anoplodactylus erectus*; C, *Anoplodactylus nodosus*; D, *Phoxichilidium femoratum*; E, *Phoxichilidium quadradentatum* (A, C, after Child; B, Child, original; D, E, after Hedgpeth).

A *Anoplodactylus viridintestinalis*

B *Anoplodactylus erectus*

C *Anoplodactylus nodosus*

D *Phoxichilidium femoratum*

E *Phoxichilidium quadradentatum*

to neck; propodal lamina as long or slightly less than entire sole, with large heel 2
— Trunk elongate, lateral processes well separated; neck moderately long, slender; palp buds lacking or not evident; propodal lamina very short, to less than half length of sole, heel small or lacking 3
2. Trunk ovoid in dorsal view, lateral processes less than twice longer than wide, each with broad rounded tubercle, ocular tubercle distally rounded, height equal to that of abdomen; oviger first segment twice wider than distal segments; cement gland a tiny pore on slightly raised bump, auxiliary claws robust (plate 330E). . . . . . . *Anoplodactylus compactus*
— Trunk round in dorsal view, length of lateral processes twice their diameters, each with narrow conical tubercle, ocular tubercle with narrow distal cone, abdomen taller than cone; oviger first segment little wider than distal segments; cement gland a small truncate cone, auxiliary claws usually lacking or minute (plate 331A) . . . .
. . . . *Anoplodactylus viridintestinalis*

3. Lateral process tubercles as tall or taller than wide; female proboscis without ventral adornment; second coxae with ventral tubercle shorter than segment diameter or lacking 4
— Lateral process tubercles low, wider than tall; female proboscis with paired proximoventral alar processes; second coxae ventrodistal tubercle longer than segment diameter; femur with distal tubercle (plate 330F). . . . . . *Anoplodactylus californicus*
4. Lateral processes separated by less than their diameters; ocular tubercle with narrow apical cone; proboscis swollen at midlength; legs with many low bumps or nodes, second coxae lack ventral tubercle (plate 331C) . . . .
. . . . *Anoplodactylus nodosus*
— Lateral processes separated by their diameters or more; ocular tubercle rounded at tip; proboscis cylindrical; legs smooth, second coxae of male with ventral tubercle shorter than coxal diameter (plate 331B) . . . . *Anoplodactylus erectus*
— Four to five major heel spines alternate laterally, auxiliary claws as long or slightly longer than main claw diameter at

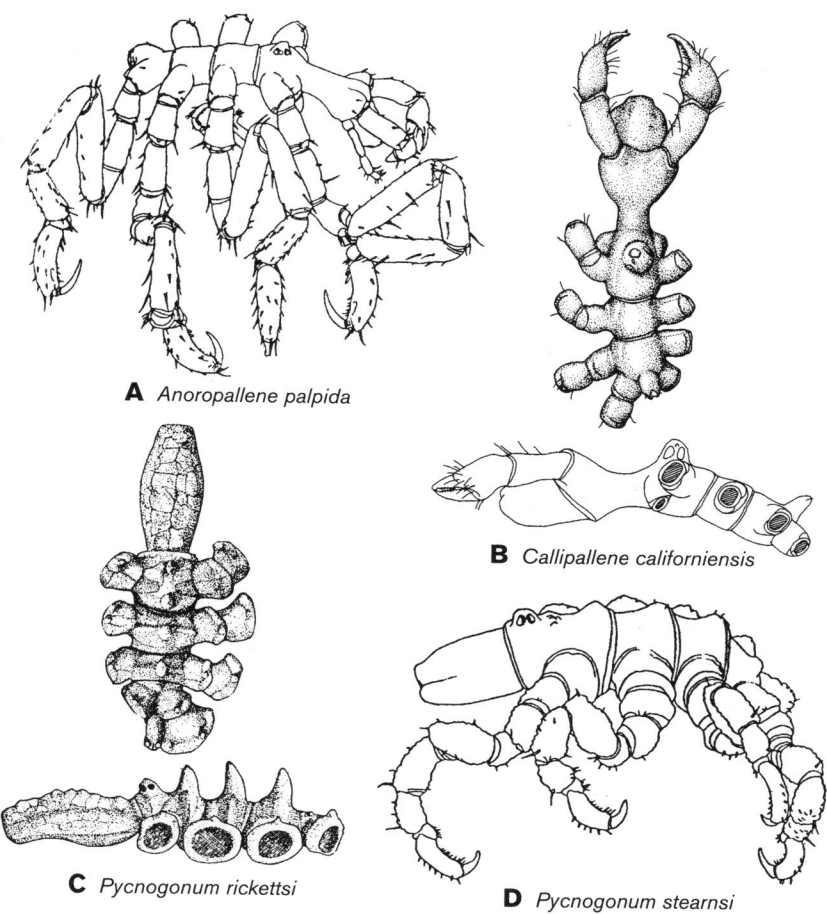

A *Anoropallene palpida*

B *Callipallene californiensis*

C *Pycnogonum rickettsi*

D *Pycnogonum stearnsi*

PLATE 332 A, *Anoropallene palpida*; B, *Callipallene californiensis*; C, *Pycnogonum rickettsi*; D, *Pycnogonum stearnsi* (A, B, D, after Child; C, after Schmitt).

their point of insertion; abdomen at least twice as long as its maximum diameter; lateral processes separated by more than their diameters (plate 331D) . . . . . *Phoxichilidium femoratum*

— Two major heel spines in single row, auxiliary claws very short, about half main claw diameter; abdomen only slightly longer than its diameter; lateral processes separated by less than their diameters (plate 331E) . . . .
. . . . *Phoxichilidium quadradentatum*

Callipallenidae

This family is top heavy with more than 20 genera, only three of which occur in our waters. Family members have a fully segmented trunk, chelifores with large functional chelae, most of which have teeth, and nine- to 10-segmented ovigers in both sexes. The genera of this family often lack palps or have palps of reduced segment numbers only in the males. Sometimes the palp is represented by a bump having a single blunt segment, but not in any recorded California genera. The legs are variously short or long, usually do not have tubercles, and more of the genera lack auxiliary claws than have them. Cement glands are usually difficult to discern but sometimes are found as tiny ventral pores or tubes.

ANOROPALLENE

One local species, *Anoropallene palpida* (plate 332A). Trunk fully segmented, lateral processes well separated, without adorn-

ment; proboscis short, tapering distally to narrow oral surface; chelifore scapes one-segmented, chelae with long fingers and teeth; palps four-segmented, little longer than proboscis, few long distal setae; ovigers 10-segmented, fifth segment longest, with distal lateral apophysis in males, strigilis with short denticulate spines, without claw; legs short, segments robust, tarsus very short, propodus short, with stout heel having two major spines bearing serrate anterior edges, claw short, without auxiliaries. Cement gland with several ventral tubes.

CALLIPALLENE

One species in our range, *Callipallene californiensis* (plate 332B). The genus is similar in habitus to *Anoropallene*, except that *Callipallene* species all lack palps in any form and have auxiliary claws. Ocular tubercle rounded at apex; chelae with seven to 10 well-formed blunt teeth; oviger bases large, round, crowding posterior of short neck; leg segments moderately short, propodus with many short endal and distal setae, main claw less than half as long as propodus.

Pycnogonidae

This is probably the most advanced family of the pycnogonids in terms of reduction of appendage segments or complete loss of the appendage itself. The approximately 60 species of *Pycnogonum* lack chelifores, palps, and have small ovigers with reduced segment numbers in the male only or they lack ovigers entirely. Some species' integument is finely reticulate with the entire animal embraced with many fine lines of darker pigment. Other species have pebbled integument without reticulation. All species have the proboscis and abdomen carried horizontally, have short leg segments, and have short auxiliary claws or lack them entirely. Most cement gland outlets have not been described, but scant evidence places them ventrally on a few species. *Pycnogonum* species are often found at the base of anemones on which they presumably feed.

KEY TO *PYCNOGONUM* SPECIES

— Integument fully reticulate with conspicuous lines; trunk with three conical mediandorsal tubercles taller than low ocular tubercle; proboscis with proximoventral swelling and uneven dorsal surface with bumps; lateral processes, first coxae, femora, and first tibiae with dorsodistal nodes or bumps; oviger nine-segmented, with small terminal claw, few tiny simple spines on distal segments; propodus hardly curved, claw short, without auxiliaries (plate 332C). . . . . . .
. . . . *Pycnogonum rickettsi*
— Integument pebbled, without reticulations; trunk with swellings anterior to segmentation lines, with small dorsal tubercles; proboscis barrel-shaped, without bumps, swellings; lateral process tubercles low, inconspicuous; ovigers nine-segmented, with large terminal claw; legs without tubercles, femur with proximoventral swelling, propodus tapering distally, curved, without auxiliaries (plate 332D). *Pycnogonum stearnsi*

LIST OF SPECIES

AMMOTHEIDAE

Achelia alaskensis (Cole, 1904) (=*Ammothea nudiuscula* Hall, 1913). Described by Cole from Alaska, it occurs as far south as San Francisco; also in Japan, Korea, and Russian far east, mostly in the intertidal; also in bays, tolerating reduced salinities.

Achelia chelata (Hilton, 1939) (=*Ammothea chelata*; *Ammothea euchelata* Hedgpeth, 1940 [redescription and plates]). Distribution very limited with a few intertidal localities confined to central California, including Moss Beach, and on the bryozoan *Bugula* at Pescadero. Also in mussel beds and, in winter, in *Mytilus californianus* (Benson and Chivers 1960 Veliger 3: 16–18).

Achelia echinata (Hodge, 1864). This far-ranging species was first collected in Europe and later in both Atlantic and Pacific shallows. It is extremely variable and has several subspecific names.

Achelia gracilipes (Cole, 1904) (=*Ammothea gracilipes*). A few shallow records from San Francisco to British Columbia.

Achelia simplissima (Hilton, 1939). Most *Achelia* species, including this one, were designated as *Ammothea* until *Achelia* came into general use by the 1940s. This species is rare and only known by two syntypes from the central California coast. Redescribed by Child, 1996, Proc. Bio. Soc.Wash. 190: 679–681.

Achelia spinoseta (Hilton, 1939). Known from only a unique type collected in shallows. Redescribed by Child, 1996, Proc. Biol. Soc. Wash. 109: 681–684.

Ammothea hilgendorfi (Bohm, 1879) (=*Corniger hilgendorfi*; *Lecythorhynchus hilgendorfi*; *L. marginatus* Cole, 1904). This species is common along shores and shallows of the Pacific Rim

from California to Japan, China, and to the Society Islands. Among hydroids and in sheltered crevices; one of the characteristic species of the central California intertidal. See Russell and Hedgpeth, 1990.

Ammothella biunguiculata (Dohrn, 1881). This variable species was given three subspecific names over many years that are no longer valid. It occupies subtidal habitats from California to Hawaii and Australia.

Ammothella menziesi Hedgpeth, 1951. A rare species with only two records north of San Francisco.

Ammothella spinifera Cole, 1904. This rather common species is known from southern California shores to Ecuador, the Caribbean, and Brazil. It is one of a few trans-Panamanian species known.

Ammothella tuberculata Cole, 1904. This is one of only a few *Ammothella* species with dorsal trunk tubercles. Known from British Columbia to southern California shallows in a restricted distribution; the most common pycnogonid of the surfgrass *Phyllospadix* holdfasts.

Eurycyde spinosa Hilton, 1916. It has been known from southern California in most of its records but was lately collected afar in the Galapagos. Redescription: Child, 1992, Smiths. Contrib. Zool. 526: 17.

Nymphopsis spinosissima (Hall, 1912) (=*Ammothella spinosissima*). Known only from the intertidal of the southern California coast, it is easily recognized by its three tall spinose trunk tubercles.

Nymphopsis duodorsospinosa Hilton, 1942. There are only two tall spinose trunk tubercles on this species, and the leg tubercles are clumped. It is known from South Carolina and the Gulf of Mexico to California and to the Galapagos Islands, mostly intertidal.

Tanystylum californicum Hilton, 1939. Infrequently collected; on the hydroid *Aglaophenia*; known only from central and southern California. Its ocular tubercle arises on a mound and it has very tiny chelifore stumps.

*Tanystylum duospinum Hilton, 1939. Similar to *T. californicum*, but smaller and less pigmented; palp 5-segmented; *T. californicum* abdomen about as long as last pair of lateral processes, while *T. duospinum* abdomen is longer than last pair of processes; see Child, 1996; Russell and Hedgpeth, 1990. On hydroids; larvae ectoparasitic on hydroid *Orthopyxis everta*.

Tanystylum intermedium Cole, 1904. Known from Monterey Bay to Chile and the Galapagos in shallow depths, this is the only California species to retain chelifore stumps.

Tanystylum occidentalis (Cole, 1904) (=*Clotenia occidentalis*). This rare species has a clean rounded appearance and its horizontal abdomen originates from a trunk swelling. Found in littoral habitats from Oregon to southern California.

RHYNCHOTHORACIDAE

Rhynchothorax philopsammum Hedgpeth, 1951. This frequently recorded species has a distribution that is almost pantemperate/pantropical. It was described from the intertidal of central California. It is very tiny and has slender legs spanning about 3 mm–4 mm.

PHOXICHILIDIIDAE

Anoplodactylus californicus Hall, 1912 (=*Anoplodactylus portus* Calman, 1927; *A. robustus* Hilton, 1939; *A. carvalhoi* Marcus,

* = Not in key.

1940; *A. projectus* Hilton, 1942). The species has a pantropical/pantemperate range and is one of the few species of the genus with female ventral proboscis outgrowths of unknown use.

Anoplodactylus compactus (Hilton, 1939) (=*Phoxichilidium compactum; Halosoma compactum* Marcus, 1940). This tiny rare form has crowded lateral processes and a short proboscis. It has only been taken in three localities south of San Francisco. Figures and redescription: Child 1975, Proc. Bio. Soc. Wash., 88: 191–193.

Anoplodactylus erectus Cole, 1904. Occurs around the North Pacific Rim and at several Pacific island groups in littoral and shallow localities.

Anoplodactylus nodosus Hilton, 1939. This species has been found only once at Santa Catalina Island but is easily recognized by its many leg outgrowths. Redescription and figures: Child 1975, Proc. Bio. Soc. Wash. 88: 193–196.

Anoplodactylus viridintestinalis (Cole, 1904) (=*Halosoma viridintestinalis*). Common from central California to Panama in shallow depths. It is another species with crowded lateral processes and almost circular trunk dorsally. Common in Tomales Bay, where it may be the most abundant and characteristic sea spider of shallow, sheltered water, conspicuous by virtue of its bright green intestines that branch out to the legs.

Phoxichilidium femoratum (Rathke, 1790) (=*Phoxichilidium tubulariae* Lebour, 1945). This often-taken species is distributed from Europe to Canada and from Los Angeles to Alaska and the Russian far east in littoral depths or mostly deeper.

*Phoxichilidium parvum Hilton, 1939. Santa Cruz and Japan. See Child 1975.

Phoxichilidium quadradentatum Hilton, 1942. Often collected from nearshore buoys in Alaska and northern California: in the second (1954) edition of this manual, one of us noted that "more than 10,000 specimens" of this species were collected on buoys near the Golden Gate Bridge in fouling surveys of the 1940s. It has extremely short auxiliary claws that are sometimes difficult to see.

CALLIPALLENIDAE

Anoropallene palpida (Hilton, 1939) (=*Palene [sic] palpida; Oropallene palpida* Hilton 1942; *O. heterodenta* Hilton, 1942; *Anoropallene crenispina* Stock, 1956). This species is very similar to several *Nymphon* species, except that it has four rather than five palp segments and its abdomen points down at an angle. Shallow water from California to Peru.

Callipallene californiensis (Hall, 1913) (=*Pallene californiensis; Callipallene solicitatus* Child, 1979). Unlike the previous species, this genus has no palps but does have prominent auxiliary claws; from California to Chile in shallow water.

PYCNOGONIDAE

Pycnogonum rickettsi Schmitt, 1934. This species has relatively few records, all from the central California coast, subtidal to intertidal, from wharf pilings, anemones, and hydroids. It can be readily separated from the following species by its fine brown reticulations on a lighter integument and its very large dorsal trunk tubercles.

Pycnogonum stearnsi Ives, 1892. This species boasts many records from the California coast (and Mexico) and around the North Pacific rim to the northern Kurile Islands, all in shallow depths. It has small, low dorsal trunk tubercles and similar low distal lateral process tubercles. It lacks reticulation. Often on *An-*

thopleura, Metridium, and *Aglaophenia.* Both *P. rickettsi* and *P. stearnsi* occur sympatrically at Duxbury Reef on the same species of sea anemones but have not been found on the same individual host.

References

Child, C. A. 1975. The Pycnogonida types of William A. Hilton, I. Phoxichilidiidae. Proceedings of the Biological Society of Washington 88: 189–209.

Child, C. A. 1979. Shallow-Water Pycnogonida of the Isthmus of Panama and the Coasts of Middle America. Smithsonian Contributions to Zoology 293: 1–86.

Child, C. A. 1992a. Shallow-Water Pycnogonida of the Gulf of Mexico. Memoirs of the Hourglass Cruises 9: 1–86.

Child, C. A. 1992b. Pycnogonida of the Southeast Pacific Biological Oceanographic Project (SEPBOP). Smithsonian Contributions to Zoology 526: 1–43.

Child, C. A. 1996. The Pycnogonida types of William A. Hilton, II. The remaining undescribed species. Proceedings of the Biological Society of Washington 109: 677–686.

Cole, L. J. 1904. Pycnogonida of the west coast of North America. Harriman Alaska Expedition 10: 249–298.

Hedgpeth, J. W. 1940. A new pycnogonid from Pescadero, Calif., and distributional notes on other species. Journal of the Washington Academy of Sciences 30: 84–87.

Hedgpeth, J. W. 1951. Pycnogonids from Dillon Beach and vicinity, California, with descriptions of two new species. Wasmann Journal of Biology 9: 105–117.

Hilton, W. A. 1939. A preliminary list of pycnognids [sic] from the shores of California. Journal of Entomology and Zoology of Pomona College 31: 72–74.

Hilton, W. A. 1942. Pycnogonids from the Pacific. Family Phoxichilidiidae Sars, 1891. Journal of Entomology and Zoology of Pomona College 34: 71–74.

Russell, D. J., and J. W. Hedgpeth. 1990. Host utilization during ontogeny by two pycnogonid species (*Tanystylum duospinum*) and *Ammothea hilgendorfi* parasitic on the hydroid *Eucopella everta* (Coelenterata: Campanulariidae). Bijdragen tot de Dierkunde 60: 215–224.

Schmitt, W. L. 1934. Notes on certain pycnogonids including descriptions of two new species of *Pycnogonum.* Journal of the Washington Academy of Sciences 24: 61–70.

Ziegler, A. C. 1960. Annotated list of Pycnogonida collected near Bolinas, California. Veliger 3: 19–22.

Arachnida

Marine and maritime arachnids include representatives of the mites (Acari) and the spiders (Araneae). W. G. Evans (1980, in *Intertidal Invertebrates of California*) notes the presence of the small linphyiid spider *Spirembolus mundus* Chamberlin and Ives, 1933, on the high intertidal shore on rock surfaces on the green algae *Ulva* and under debris on sand (it also occurs in inland situations and along freshwater creeks). A large number of species of spiders are to be expected in supralittoral habitats and other nearshore environments, especially in salt marshes. We treat the marine mites, below.

Acari

IRWIN M. NEWELL AND ILSE BARTSCH

(Plates 333 and 334)

Intertidal mites include representatives of the four major suborders of Acari, but most marine mites are in a single family, the Halacaridae, of the superfamily Halacaroidea in the suborder Prostigmata. This family has been unusually successful in

* = Not in key.

its evolutionary adaptation to numerous marine niches. Halacaridae probably evolved in the sea from several semiaquatic lines and should be regarded as polyphyletic. They have also invaded fresh water. Some species are phytophagous, others are predators, and several have developed truly parasitic habits. Krantz (1973) reported upon predatory halacarid mites in the genera *Agauopsis, Halacarus,* and *Halacarellus,* from intertidal mussel beds in Oregon; Krantz (1976) reported upon arenicolous species, and MacQuitty (1983, 1984) has reported on marine halacaroids from California.

An abundant and easily observed supralittoral and high intertidal mite, reaching 3 mm–4 mm in length, is the bright red "velvet mite," *Neomolgus littoralis* (Linnaeus, 1758) in the family Bdellidae, reported widely in the North Atlantic and North Pacific Oceans. These tiny predators that feed on flies and other prey may be seen actively moving on rock surfaces and on splash zone lichens and are also common under small stones, rocks, and beach wrack. A photograph of *Neomolgus* may be found in Evans (1980), and Abbott (1987) presents a sketch of the external anatomy of *Neomolgus* from Monterey Bay.

The other major suborders with intertidal representatives are the Mesostigmata, Astigmata, and Oribatida (Cryptostigmata). One common species of the Mesostigmata is the eviphidid *Thinoseius orchestoideae* (Hall, 1912) (=*Gammaridacarus brevisternalis* Canaris, 1962), which attaches to the undersides of the amphipod beach hoppers *Traskorchestia* and *Megalorchestia,* and preys upon rhabtidid nematodes that also reside on the amphipods (Canaris 1962; Kitron 1980; Rigby 1996a, b).

Other genera of the Mesostigmata occur free-living; they are mostly predatory. The Astigmata are represented by the Hyadesioidea, with small weakly sclerotized forms. Hyadesiids may locally be very abundant, especially among green algae in tide pools. The Oribatida are dark colored mites, the adults being heavily sclerotized. Representatives of the superfamily Ameronothroidea are often aquatic or semiaquatic; on the seashore, they are mostly herbivorous and may be found in large numbers, generally restricted to the upper intertidal.

Of the marine genera of Halacaridae, at least 14 are known from the North Pacific (Newell 1975), and one, *Thalassacarus,* is known only from this region. Halacaridae occupy a great number of marine habitats, even to depths of more than 5,000 m (Newell 1971). Nevertheless, the ecological distribution of any given species is probably fairly restricted. For example, *Isobactrus* spp. are usually confined to brackish tide pools or estuaries; *Rhombognathus* spp. are rarely encountered subtidally, and never below the euphotic zone; *Scaptognathus, Anomalohalacarus,* and *Actacarus* are interstitial in coarse sand.

Halacaridae range from 0.18 mm–2 mm in body length. They are usually abundant: a liter of coralline algae may contain hundreds of individuals and up to 15 species. Despite their small size they are a conspicuous and omnipresent element, well worth a few hours of the student's time. Ecological studies are of particular importance, since little is known of the actual niches occupied by the various species.

Halacaridae are easily collected by placing algae, gravel, barnacles, mussels, and other substrates in seawater, anesthetizing for about 10 minutes with chloroform, and washing the substrate vigorously with either salt or fresh water. Washings can be preserved in 65% alcohol. If the mites are to be observed alive, the chloroform treatment should be greatly reduced or eliminated, and a vigorous jet of tap water should be used to separate the mites from the substrate. They should then be returned to seawater for further study. Intertidal mites tolerate immersion in fresh water for one to three hours, but longer periods are usually fatal.

Temporary mounts of Halacaridae can be made in Berlese fluid or Hoyer's modification of it. They may also be cleared with lactic acid and transferred to 15% glycerine in water, which is then allowed to evaporate slowly while the mites are being examined. For permanent mounts the mites should be cleared with enzymes and mounted in Hyrax or glycerine, following procedures outlined by Newell (1947).

The interested student should then begin with the classic works of Newell cited below, updated by Krantz (1973, 1976), MacQuitty (1983, 1984), and Bartsch (2004).

The following key to genera is based on adults only. Males are distinguishable by the phorotype (plate 333E), the organ that produces the spermatophore (spermatopositor). Probably all species of Halacaridae utilize spermatophores. There is a six-legged larva in the life cycle, followed by one, two, or three nymphal instars, depending on the genus. Protonymphs have one, deutonymphs two, and tritonymphs (known in *Isobactrus*) have three pairs of provisory genital acetabula, but no genitalia. Protonymphs also have the femur of leg IV undivided.

KEY TO GENERA OF MARINE HALACARIDAE OF THE EASTERN NORTH PACIFIC

1. Insertions of palpi lateral to rostrum; trochanters separated by an interval appreciably greater than their width, so they are largely or fully visible in ventral view; abundant to rare, on various substrates, occasionally interstitial in coarse sand (plate 333A) 2
— Insertion of palpi dorsal to rostrum; trochanters separated by an interval less than their width, so they are largely concealed in ventral view; palpi very long to short; generally rare, often interstitial in coarse sand (plate 333B–333D) 12

2. Middle piece of claw articulating directly with tip of the tarsus (note that oil immersion may be necessary at first to interpret this important character); color in life yellow, brown, or rarely green or green black; predaceous or parasitic (plate 334M) 3
— Middle piece of claw articulating with an intermediate sclerite, the carpite, and this in turn articulates freely with (plate 334K) or is a flexible extension of (plate 334L), the tip of the tarsus (note that some species of *Agauopsis* have a minute, carpitelike structure at the tip of the tarsus, but it appears to be rigid, rodlike extension of the tarsus); color in life green to black; phytophagous; not living under conditions precluding algal growth 11

3. Patella of palp with a seta, variable in position and form (plate 334E–334F); note that in *Copidognathus pseudosetosus* and related species, there is a sharp spine here, but there is no alveolus (socket) and it is not a seta (plate 334G) 5
— Patella of palp lacking a seta (plates 333A, 334G) 4
4. Tarsi of legs bowed; median claw massive, thicker than lateral claws; slow-moving forms, adapted for clinging to hydroids or bryozoans (plate 333F) . . . . *Bradyagaue* Newell, 1971
— Tarsi of legs not bowed, but straight; median claw minute, not as thick as lateral claws (plate 333G) . . . . *Copidognathus* Trouessart, 1888

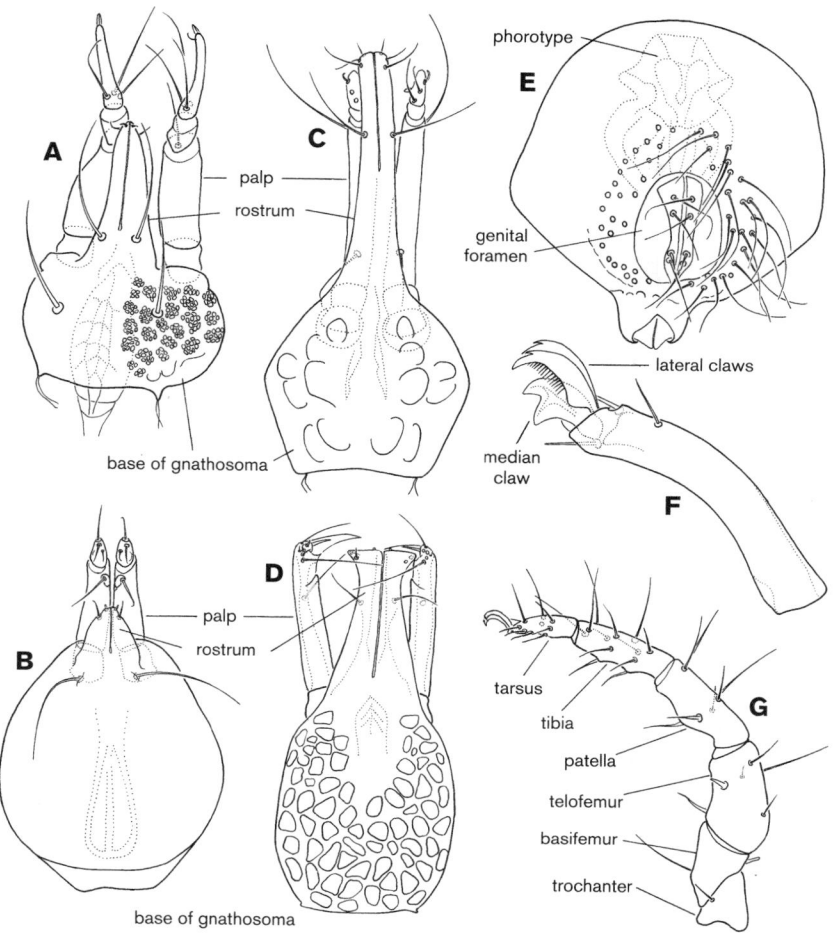

PLATE 333 Halacaridae. A, *Copidognathus curtus* Hall, gnathosoma, ventral; B, *Simognathus* sp., gnathosoma, ventral; C, *Lohmannella falcata* (Hodge), gnathosoma, ventral; D, *Scaptognathus* sp., gnathosoma, ventral; E, *Copidognathus curtus* Hall, genitoanal plate, male, phorotype in dotted line; F, *Bradyagaue bradypus* Newell, tarsus of leg III, showing curvature and massive median claw; G, *Halacarus frontiporus* Newell, leg I, showing segmentation (Newell, original).

5. Patella of legs relatively long, nearly as long as either the telofemur or the tibia (plate 333G); femur of palp with two setae (plate 334E); usually rare 6
— Patella of legs distinctly shorter than telofemur or tibia; femur of palp with only one seta 7
6. Idiosoma (body, exclusive of gnathosoma or "capitulum" slender, very flexible in life, modified for moving freely and quickly through interstices in sand, posterior dorsal plate often divided into right and left halves . . . .
. . . . *Anomalohalacarus* Newell, 1949
— Idiosoma neither slender nor flexible, not modified for interstitial life; posterior dorsal plate either absent, or (if present) not divided into right and left halves . . . .
. . . . *Halacarus* Gosse, 1855
7. Ocular plates large, readily visible on dorsal surface; habitat variable, but not normally interstitial in coarse sand (plate 334A–334C). 8
— Ocular plates very small, at sides of idiosoma (body), often easily overlooked; normally interstitial in coarse sand and under boulders . . . . *Actacarus* Schulz, 1936
8. Leg I rakelike in appearance, with a row of several very heavy peg setae, along anterior ("medial") margin; palpi very short, straight (plate 334D). . . . .
. . . . *Agauopsis* Viets, 1927
— Leg I not rakelike, although some heavy setae may be present ventrally or anteroventrally; palpi longer 9
9. Ocular plates with a thick, taillike extension (cauda), reaching nearly to insertions of legs IV (plate 334C);

cheliceral tarsus with two massive teeth on basal half of dorsal margin, minutely denticulate in distal margin (plate 334H) . . . . *Thalassacarus* Newell, 1949
— Ocular plates without such a cauda, not reaching beyond level of insertions of legs III 10
10. Tarsus of chelicera minutely denticulate throughout (plate 334I) . . . . *Halacarellus* Viets, 1927

Note: (As *Thalassarachna* in previous edition, but there are no Pacific records of this genus)

— Tarsus of chelicera with a few (five to seven) coarse teeth along dorsal margin (plate 334J) . . . .
. . . . *Agaue* Lohmann, 1889
11. Each ocular plate with two setae; with three or more setae on or near lateral margin of body, between insertions of leg II and III; carpite straight, stiff, rodlike (plate 334K); gnathosoma readily visible in dorsal view, projecting anteriorly or anteroventrally; usually abundant (plate 334A) . . . .
. . . . *Rhombognathus* Trouessart, 1888
— Ocular plates without setae, a few setae free in the striated, membranous cuticle (plate 334B); with only one seta on lateral margin of body between insertions of legs II and III; gnathosoma directed ventrally so it is concealed in dorsal view (undistorted specimens) by the overhanging anterior dorsal plate (AD); carpite flexible, curved, monoliform (plate 334L); generally in brackish water . . . .
. . . . *Isobactrus* Newell, 1947

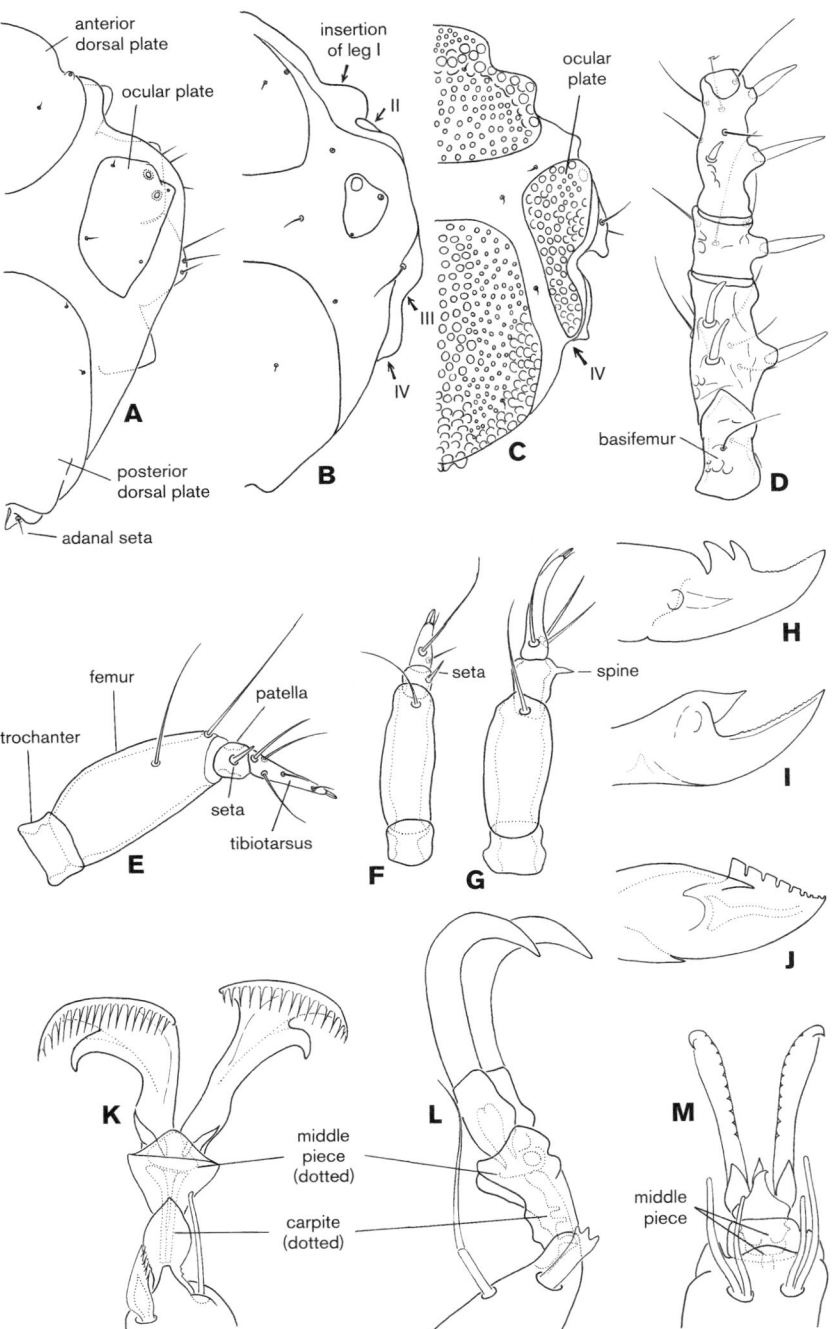

PLATE 334 Halacaridae. A, *Rhombognathus* sp., dorsum, right side; B, *Isobactrus* sp., dorsum, right side; C, *Thalassacarus commatops* Newell, dorsum, right side; D, *Agauopsis productus* Newell, basifemur-tibia I, right side, ventral view; E, *Halacarus frontiporus* Newell, left palp, anterior (="medial") view; F, *Agauopsis* sp., right palp, dorsal view; G, *Copidognathus pseudosetosus* Newell, left palp, dorsal view, showing spine (not a seta) on patella of palp; H, *Thalassacarus commatops* Newell, tarsus of chelicera, side view; I, *Halacarellus capuzinus* (Lohmann), tarsus of chelicera, side view; J, *Agaue longiseta* Newell, tarsus of chelicera, side view; K, *Rhombognathus* sp., ambulacrum and tip of tarsus III, left side, ventral view, showing rodlike carpite; L, *Isobactrus* sp., ambulacrum III, showing moniliform carpite; M, *Copidognathus curtus* Hall, ambulacrum II, ventral view (carpite absent); note that in figures K and L, middle piece and carpite (where present) are shown as dotted outlines and are surrounded by thin, membranous cuticle (Newell, original).

12. Tip of rostrum flared at end; palpi with an exceptionally heavy spiniform seta at tip (plate 333D) . . . .
. . . . *Scaptognathus* Trouessart, 1889
— Tip of rostrum narrowly or bluntly rounded at end, not flared; rostrum long and slender, or short and thick; palpi with only minute setae at tip (plate 333B, 333C) 13
13. Palpi long, slender, extending to or only slightly beyond the end of the long rostrum; rostrum parallel-sided throughout most of length; claws of tarsus I similar in form to those on tarsi II–IV (plate 333C) . . . .
. . . .*Lohmannella* Trouessart, 1901
— Palpi shorter, but extending well beyond the tip of the short, thick, subtriangular rostrum; median claw of tarsus

I grossly enlarged, median and lateral claws markedly different in form from those of tarsi II–IV (plate 333B) . . . . .
. . . .*Simognathus* Trouessart, 1889

REFERENCES

Abbott, D. P. 1987. Observing marine invertebrates. Drawings from the laboratory. Edited by G. H. Hilgard. Stanford, CA: Stanford University Press, 380 pp.

Bartsch, I. 1997. *Thalassarachna* and *Halacarellus* (Halacaridae: Acari): two separate genera. J. Nat. Hist. 31: 1223–1236.

Bartsch, I. 2004. Geographical and ecological distribution of marine halacarid genera and species. Exp. Appl. Acarol. 34: 37–58. (Lists of genera, number of species, and geographical distribution).

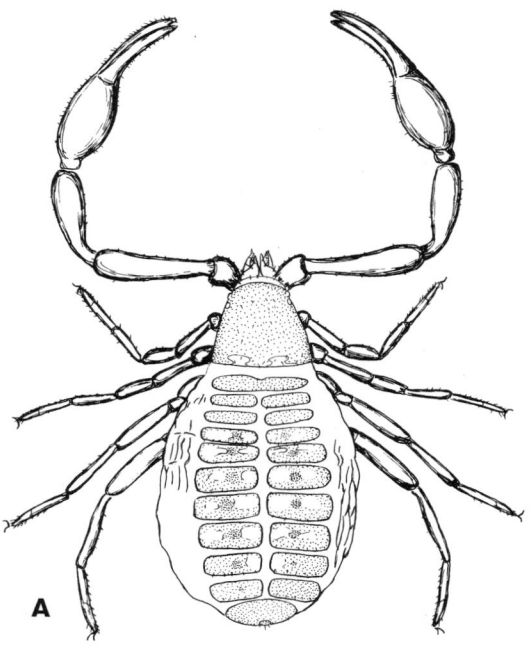

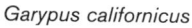
Garypus californicus

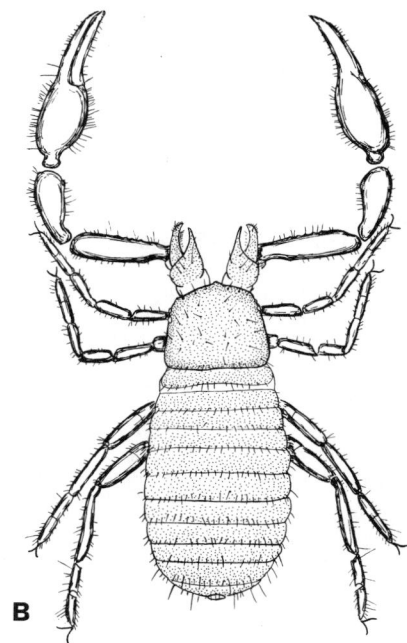
Halobisium occidentale

PLATE 335 Pseudoscorpions. A, *Garypus californicus*; B, *Halobisium occidentale* (both drawn by R. O. Schuster, from 3rd ed.).

Canaris, A. G. 1962. A new genus and species of mite (Laelaptidae) from *Orchestoidea californiana* (Gammaridea). J. Parasitology 48: 467–469.

Evans, G. O. 1963. The systematic position of *Gammaridacarus brevisternalis* Canaris (Acari: Mesostigmata). Ann. Mag. Nat. Hist. (13) 5: 395–399.

Evans, W. G. 1980. Insecta, Chilopoda, and Arachnida: Insects and allies, pp. 641–658. In Intertidal invertebrates of California. R. H. Morris, D. P. Abbott, and E. C. Haderlie, eds. Stanford, CA: Stanford University Press, 666 pp.

Kitron, U. D. 1980. The pattern of infestation of the beach-hopper amphipod *Orchestoidea corniculata*, by a parasitic mite. Parasitology 81: 235–249.

Krantz, G. W. 1973. Four new predatory species of Halacaridae (Acari: Prostigmata) from Oregon with remarks on their distribution in the intertidal mussel habitat (Pelecypoda: Mytilidae). Ann. Ent. Soc. Amer. 66: 979–985.

Krantz, G. W. 1976. Arenicolous Halacaridae from the intertidal zone of Schooner Creek, Oregon (Acari: Prostigmata). Acarologia 18: 241–258.

MacQuitty, M. 1983. Description of a new species of marine mite, *Agauopsis filirostris* (Acari: Halacaroidea) from southern California. Acarologia 24: 59–64.

MacQuitty, M. 1984. The marine Halacaroidea from California. J. Nat. Hist. 18: 527–554.

Newell, I. M. 1947. A systematic and ecological study of the Halacaridae of eastern North America. Bull. Bingham Oceanogr. Coll. 10: 1–232.

Newell, I. M. 1952. Further studies on Alaskan Halacaridae (Acari). Amer. Mus. Novitates 1536: 1–56.

Newell, I. M. 1971. Halacaridae (Acari) collected during Cruise 17 of the R/V Anton Bruun in the southeastern Pacific Ocean. Anton Bruun Report No. 8: 1–58.

Newell, I. M. 1984. Antarctic Halacaroidea. Antarctic Res. Ser. 40: 1–284.

Proches, S., and D. J. Marshall. 2001. Global distribution patterns of non–halacarid marine intertidal mites: implications for their origins in marine habitats. J. Biogeography 28: 47–58.

Pugh, P. J. A., and P. E. King. 1988. Acari of the British supralittoral. J. Nat. Hist. 22: 107–122.

Rigby, M.C. 1996. The epibionts of beach hoppers (Crustacea: Talitridae) of the North American Pacific coast. J. Natl. Hist. 30: 1329–1336.

Rigby, M.C. 1996. Association of a juvenile phoretic uropodid mite with the beach hopper *Traskorchestia traskiana* (Stimpson, 1857) (Crustacea: Talitridae). J. Nat. Hist. 30: 1617–1624.

Pseudoscorpiones

VINCENT F. LEE

(Plate 335)

Pseudoscorpions, or chelonethids, are minute (1 mm–5 mm) animals resembling tiny scorpions, but with a flattened, disc-shaped, or elongated posterior body, relatively enormous pincers, and tailless. They are active hunters, and in woodlands are found in leaf mold and under bark. Two species are often encountered along the shore. *Garypus californicus* Banks, 1909 (plate 335A), occurs under driftwood and beach wrack on sandy beaches above high-tide mark from northern Baja California to northern California. Donald P. Abbott (1987) provides a sketch of a specimen of *G. californicus,* with details of the prosoma, from under rocks on the beach at Hopkins Marine Station. Lighton and Joos (2002a, b) provide one of the few studies on this species.

Halobisium occidentale Beier, 1931 (plate 335B), occurs intertidally in the mud and under logs and rocks in salt marshes (particularly in flats of the pickleweed *Salicornia*) and in crevices of intertidal rocks and under cobblestones on the open coast from central California to Alaska (Schulte 1976).

Initially described from a single female specimen collected in 1927 "in the rubble beneath a log above the high-tide line" at Cannon Beach, Oregon, *Parobisium hesperum* (Chamberlin 1930) is also found under driftwood. The only other record, if correct, appears to be one from an inland locality (Dunsmuir) in California.

Collecting pseudoscorpions in the marine environment is fairly straightforward. Turning over driftwood, kelp, and other debris on sandy beaches can yield the common *G. californicus.* Prying crevices in driftwood is not very productive, but crevices of rocks might yield *H. occidentale,* although they are more common in marshes and estuaries. Here, by digging into the exposed mud at the base of the pickleweed, one can find them in their silken chambers.

For keys to genera, see Chamberlin (1931) and Muchmore (1990).

REFERENCES

Abbott, D. P. 1987. Observing marine invertebrates. Drawings from the laboratory. Edited by Galen Howard Hilgard. Stanford University Press, 380 pp.

Chamberlin, J. C. 1930. A synoptic classification of the false scorpions or chela-spinners, with a report on a cosmopolitan collection of the same. Part II. The Diplosphyronida (Arachnida-Chelonethida). Ann. Mag. Nat. Hist. (10) 5: 1–48.

Chamberlin, J. C. 1931. The arachnid order Chelonethida. Stanford Univ. Pub. Biol. Sci., 7: 1–284.

Lee, V. F. 1979. The maritime pseudoscorpions of Baja California, México. Occas. Pap. Calif. Acad. Sci. 131, 38 pp.

Lighton, J. R. B., and B. Joos. 2002a. Discontinuous gas exchange in a tracheate arthropod, the pseudoscorpion *Garypus californicus*: Occurrence, characteristics and temperature dependence. J. Insect Sci., 2, No. 23, 1–4.

Lighton, J. R. B., and B. Joos. 2002b. Discontinuous gas exchange in the pseudoscorpion *Garypus californicus* is regulated by hypoxia, not hypercapnia. Physiol. Biochem. Zool. 75: 345–349.

Muchmore, W. B. 1990. Pseudoscorpionida, pp. 503–527. In Soil biology guide. D. L. Dindal, ed. New York, Chichester, Brisbane, Toronto, Singapore: John Wiley & Sons.

Schulte, G. 1976. Littoralzonierung von Pseudoskorpionen an der nordamerikanischen Pazifikküste (Arachnida: Pseudoscorpiones: Neobisiidae, Garypidae). Entomol. Germ. 3: 119–124.

Weygoldt, P. 1969. The Biology of Pseudoscorpions. Harvard University Press, 145 pp.

Insecta

Orders of Intertidal Insects

HOWELL V. DALY

Few features of insects are as striking as the difference in the diversity of insects and other terrestrial arthropods among terrestrial, freshwater, and marine faunas. Nearly three-quarters of the earth's animal species are insects, but only about 3% of insect species are aquatic, and a fraction of these are marine or intertidal.

Judging by the tracheate respiratory system and impermeable cuticle, insects evolved as terrestrial arthropods and colonized aquatic habitats only secondarily. Some continued to use oxygen in air, while others developed various devices to obtain oxygen dissolved in water (Eriksen et al. 1996). Five orders (Odonata, Ephemeroptera, Plecoptera, Megaloptera, Trichoptera) and many entire families in other orders are now restricted almost exclusively to fresh water. Many other insects in groups that are normally terrestrial also live in fresh water as herbivores on aquatic plants or as predators or parasites of aquatic organisms.

In the marine environment, a surprising number of insects have been recorded, including parasites of other insects and of marine mammals. Insects as individual organisms are not scarce on the coasts or the surface of the ocean; indeed, some species are exceedingly abundant. Cheng and Frank (1993) list 21 orders from pelagic, coastal, intertidal, mangrove and other tropical/subtropical brackish waters, and salt-marsh and other temperate brackish waters. Almost all species in marine environments belong to families that are found elsewhere in freshwater or terrestrial habitats. Of well over 700 families of insects, only two families are exclusively marine: the rare but worldwide, intertidal coral treader (Hermatobatidae, order Hemiptera; one genus and nine species; Cheng 1977; Foster 1989) and the marine caddis-

flies of New Zealand and Australia (Chathamiidae, order Trichoptera; two genera and four species; Cheng and Frank 1993).

Why is the largest group of animals virtually absent from the largest habitat? Cheng (1976), Norris (1991), Wallace and Anderson (1996), and Usinger (1957), among others, provide discussions about this fascinating question. Regarding the intertidal, Hinton (1977) presented the argument that, on a world basis, the coastline measured in miles is much smaller than the miles of freshwater rivers and streams. Consequently, the intertidal habitat is actually relatively more diverse in species than fresh water.

Insects have succeeded in living under certain physical and chemical aspects of the intertidal environment. Examples are tidal submergence (species in several orders including the widespread collembolan *Anurida maritima* [Hypogastruridae, order Collembola], whose presence on the Pacific coast remains uncertain, the bug *Aepophilus bonairei* (Saldidae, order Hemiptera) in England, and the coral treader mentioned above); life cycle synchronized with tides (the midge *Clunio marinus* [Chironomidae, order Diptera]); wave action (larvae of midges of the widespread genus *Clunio* [Chironomidae, order Diptera] and the barnacle-eating larvae of the Pacific coast fly *Oedoparena glauca* [Dryomyzidae, order Diptera] live here, but species richness of populations of shore flies [Ephydridae, order Diptera] is reduced in areas of violent wave action); and salinity (especially salt-marsh mosquitoes, e.g., *Aedes taeniorhynchus* [Culicidae, order Diptera], shore flies, water boatmen *Trichocorixa* [Corixidae, order Hemiptera], and, in Australia, the intertidal rockpool caddisfly larva of *Philanisus* [Chathamiidae, order Trichoptera]).

Of various reasons why insects have not dominated the intertidal, Hinton (1977) supported the view that the physical violence of intertidal areas tends to exclude insects (see also Steinly 1986). Hynes (1984) deemed this implausible and argued in favor of competitive exclusion. While insects were evolving on land in the Paleozoic Era, they were prevented from colonizing the marine environment by already well-established invertebrates. To this may be added the observation of Vincent Resh (personal communication) that the average body size of benthic invertebrates in the intertidal is larger than the size of the comparable fauna in fresh water, hence the marine benthic fauna is a decided threat to the survival of insects.

Offshore on the surface of the ocean are five species of pelagic water striders, *Halobates* (Gerridae, Hemiptera), that occur far from land in the Atlantic, Pacific, and Indian Oceans. Beneath the surface, larvae of the midge *Chironomus oceanicus* have been dredged from 36 m and another midge, *Pontomyia* sp., from 30 m (Chironomidae, order Diptera).

However, insects are entirely absent in deeper waters of the ocean. Usinger (1957) noted that, except for some chironomids, insects also are rarely successful in colonizing deep freshwater lakes. He proposed that the most limiting factor is probably the inability of insects to respire indefinitely beneath the water's surface. As adults, nearly all insects have access to oxygen in air. Perhaps the energy needed for their reproductive physiology depends on the rich supply of oxygen in the atmosphere. Eriksen et al. (1996) describe several problems involved in obtaining oxygen dissolved in water among which the amount dissolved, even in cold saturated, water is only 0.01% of the amount available in air. Two exceptional freshwater insects complete their lives entirely submerged. The bug *Aphelocheirus* (Aphelocheiridae, order Hemiptera) completes its life history in cold water by the use of a plastron (Hinton 1976). The peculiar stonefly *Capnia lacustria* (Capniidae, order Plecoptera; first reported as *Utacapnia* sp. by Jewett, 1963; Baumann, personal communication) in Lake Tahoe exists indefinitely in deep water. Possibly other stoneflies, such as *Baikaloperla* in Lake Baikal, complete their life cycles in the lacustrian environment (Zapekina-Dulkeit and Zhiltzolva 1973; Baumann, personal communication).

The insects discussed in this manual are limited to species regularly living part or all of their lives in the intertidal zone of the outer coast. Only the water strider, *Halobates sericeus* Eschscholtz (Gerridae, order Hemiptera), inhabits the open ocean near our shores. It occurs in warm, offshore waters 50 miles or more from the coast and as far as 40° north latitude. *Halobates* eggs have been found glued to floating material at sea. Our species is one of the true pelagic forms that are taken nearshore only after severe storms. Zooplankton (e.g., copepods and euphausiids) trapped at the sea surface is their usual food. Cheng (1985) provides a review of the biology of *Halobates*.

The rich insect faunas of the beaches above the tide, coastal dunes, salt marshes, and estuaries are beyond the scope of this treatment. Insects from these areas and inland may occur by accident in the intertidal zone. Honeybees, for example, may fly offshore from inland colonies and are commonly found in the surf. Unless an insect is clearly an adult resident of brackish tidal pools, intertidal rocks, or beaches, readers are advised to identify order and family with the aid of general keys for adult insects (Johnson and Triplehorn 2004; Daly et al. 1998), adult and larval aquatic insects (Merritt and Cummins 1996), or larval insects (Stehr 1987, 1991). Further information on the biologies of marine insects is provided by Benedetti (1973), Cheng (1976), Cheng and Frank (1993), and Evans (1980).

ACKNOWLEDGMENTS

I wish to thank R. Baumann, L. Cheng, K. Christiansen, V. Resh, H. Sturm, and D. S. White for information and comments provided during the preparation of this section.

KEY TO THE ORDERS OF ADULT INTERTIDAL INSECTS

1. Wings absent; insects often jump to escape capture 2
— Wings present, though sometimes concealed by leathery front wings (elytra), or reduced to small, articulated lobes; insects usually fly or run to escape capture 3
2. Hump-backed insects about 1 cm in length; eyes very large and meeting along the midline; long filamentous antennae; three long filaments at the tip of the abdomen (a long median filament and two shorter, lateral cerci); abdomen with more than six segments . . . .
. . . . jumping bristletails, order Archaeognatha
— Small, soft-bodied, insects; body length usually 1 mm–5 mm in length; often with short antennae; abdomen with only six segments; first abdominal segment with short, thick, ventral projection (ventral tube); many species with a distally bifurcate jumping appendage (furcula) on the ventral side of the fourth abdominal segment; furcula normally folded forward (furcula reduced in Anurida maritima and often absent in littoral species) . . . .
. . . . springtails, order Collembola

Note: (the phylogenetic position of Collembola amongst the hexapods is unclear)

3. Front wings overlapping apically when at rest, the basal part divided by converging sutures which form a triangle on the dorsum; brackish pools . . . .
. . . . water boatmen, Corixidae, order Hemiptera
— Front wings not as above: either entirely free and membranous, reduced to vestigial stumps, or covering the dorsum without overlapping 4
4. Front wings free and membranous or reduced to vestigial stumps; hind wings highly modified as halteres . . . .
. . . . flies and midges, order Diptera
— Front wings leathery, completely or sometimes only partly covering dorsum, concealing hind wings when at rest 5
5. Tip of abdomen with strongly sclerotized forcep-shaped cerci . . . . earwigs, order Dermaptera
— Tip of abdomen without forcep-shaped cerci . . . .
. . . . beetles, order Coleoptera

REFERENCES

Benedetti, R. 1973. Notes on the biology of *Neomachilis halophila* on a California sandy beach (Thysanura: Machilidae). Pan-Pacific Entomologist 49: 246–249.

Cheng, L. 1976. Marine insects. Amsterdam: North-Holland Publishing Company, 581 pp.

Cheng, L. 1977. The elusive sea bug *Hermatobates* (Heteroptera). Pan-Pacific Entomologist 53: 87–97.

Cheng, L. 1985. Biology of *Halobates* (Heteroptera: Gerridae). Ann. Rev. Entomol. 30: 111–135.

Cheng, L., and J. H. Frank. 1993. Marine insects and their reproduction. Oceanography and Marine Biology: An Annual Review 31: 479–506.

Daly, H. V., J. T. Doyen, and A. H. Purcell III. 1998. Introduction to insect biology and diversity. 2nd ed. New York: Oxford University Press, 680 pp.

Eriksen, C. H., V. H. Resh, and G. A. Lamberti. 1996. Aquatic insect respiration, pp. 29–40. In An introduction to the aquatic insects of North America, 3rd ed. W. Merritt and K. W. Cummins, eds. Dubuque, IA: Kendall/Hunt Publishing Co.

Evans, W. G. 1980. Insecta, Chilopoda, and Arachnida: insects and allies, pp. 189–200. In Intertidal invertebrates of California. R. H. Morris, D. P. Abbott, and E. C. Haderlie, eds, Stanford, CA: Stanford University Press.

Foster, W. A. 1989. Zonation, behavior and morphology of the intertidal coral-treader *Hermatobates* (Hemiptera: Hermatobatidae) in the south-west Pacific. Zoological Journal of the Linnaean Society 96: 87–105.

Hinton, H. E. 1976. Plastron respiration in bugs and beetles. Journal of Insect Physiology 22: 1529–1550.

Hinton, H. E. 1977. Enabling mechanisms. Proceedings of the XV International Congress of Entomology, Washington, D.C., 1976, pp. 71–83.

Hynes, H. B. N. 1984. The relationships between taxonomy and ecology of aquatic insects, pp. 9–23. In The ecology of aquatic insects. V. H. Resh and D. M. Rosenberg, eds. New York: Praeger.

Jewett, S. G., Jr. 1963. A stonefly aquatic in the adult stage. Science 139: 484–485.

Johnson, N. F., and C. A. Triplehorn. 2004. Borror and DeLong's An introduction to the study of insects. 7th ed. Pacific Grove, CA: Brooks/Cole, 864 pp.

Merritt, R. W., and K. W., Cummins, eds. 1996. An introduction to the aquatic insects of North America. 3rd ed. Dubuque, IA: Kendall/Hunt Publishing Company, 862 pp.

Norris, K. R. 1991. General biology, pp. 68–108. In The insects of Australia, vol. 1, Commonwealth Scientific and Industrial Research Organisation. Ithaca, NY: Cornell University Press.

Stehr, F. W. 1987. Immature insects. Volume 1. Dubuque, IA: Kendall/Hunt Publishing Company, 754 pp.

Stehr, F. W. 1991. Immature insects. Vol. 2. Dubuque, IA: Kendall/Hunt Publishing Company, 975 pp.

Steinly, B. A. 1986. Violent wave action and the exclusion of Ephydridae (Diptera) from marine temperate intertidal and freshwater beach habitats. Proceedings of the Entomological Society of Washington 88: 427–437.

Usinger, R. L. 1957. Marine insects, pp. 1177–1182. In Treatise on marine ecology and paleoecology. Vol. 1. J. W. Hedgpeth, ed. Ecology, Geological Society of America Memoir 67.

Wallace, J. B., and N. H. Anderson. 1996. Habitat, life history, and behavioral adaptations of aquatic insects, pp. 41–73. An introduction

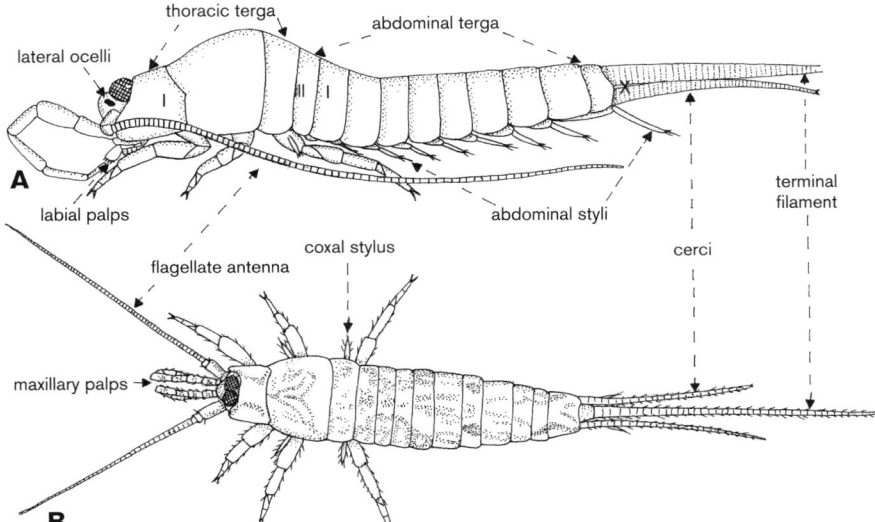

PLATE 336 General structure of Archaeognatha, semidiagrammatic: A, lateral view; B, dorsal view; color pattern of dorsal scales implied; abdominal segment XI is hidden under segment X (Helmut Sturm, original).

to the aquatic insects of North America. 3rd ed. R. W. Merritt and K. W. Cummins, eds. Dubuque, IA: Kendall/Hunt Publishing Co.

Zapekina-Dulkeit, J. I., and L. A. Zhiltzolva. 1973. A new genus of stoneflies (Plecoptera) from Lake Baikal. Entomol. Obozr. 52: 340–346. (English translation in Entomol. Review)

Archaeognatha

HELMUT STURM

(Plate 336)

Archaeognatha, or bristletails, are primitively wingless insects generally <20 mm long as adults (plate 336). All species have a unique jumping mechanism; the tergites, abdominal coxites, and caudal appendages are scaled, and the compound eyes are large and contiguous. Along the western beaches of North America, the halophilous *Neomachilis halophila* Silvestri, 1911 (adults 12–13 mm in length), occurs in the high intertidal and supralittoral zone. Benedetti (1973) working in Pacific Grove on the shores of Monterey Bay, "found it most abundant just on the seaward side of the last terrestrial vegetation in areas covered with rocks too high on the beach to be disturbed by most high tides, and especially among rocks piled in such a manner that a small space occurs between rock and sand." It also occurs a short distance inland, where it may be found on the bark of trees and on stones some 20 m–30 m from the shore, as at Baker Beach in San Francisco (H. Sturm, personal observations).

On the beaches of San Francisco and in southern California, females of *Neomachilis* are generally much more abundant than males; however, Benedetti (1973) reported that males and females were in approximately equal numbers at Pacific Grove. Benedetti found that *Neomachilis* was largely nocturnal, with the greatest activity occurring just before dawn. *Neomachilis* eats unicellular green algae, apparently derived from grazing lichens, as well as yeast and pine pollen. Similarly, Willem (1924) found that the European maritime archaeognathid *Petrobius maritimus* feeds on the unicellular alga *Pleurococcus* growing on gravel and sand grains.

Near or along the beach zone on the Pacific coast are at least three additional bristletails. *Pedetontus californicus* Silvestri, 1911, may co-occur with *Neomachilis* not far from the shore

(the two species co-occur, for example, near Baker Beach). Occurring closer to the high-tide line are *Petridiobius canadensis* Sturm, 2001 (on the Queen Charlotte Islands), and *Petridiobius arcticus* (Paclt, 1970) (on the rocky shores of southern Alaska) (Sturm 2001, Sturm and Bowser, 2004).

Neomachilis, *Pedetontus*, and *Petridiobius* are in the family Machilidae and specifically in the subfamily Petrobiinae, a common feature of which is the absence of scales on the flagellum of the antennae.

REFERENCES

Benedetti, R. 1973. Notes on the biology of *Neomachilis halophila* on a California sandy beach (Thysanura: Machilidae). Pan-Pacific Entomologist 49: 246–249.

Sturm, H. 2001. Possibilities and problems of morphological taxonomy shown by North American representatives of the subgenus *Pedetontus* s. str. and *Petridiobius canadensis* (Archaeognatha, Machilidae, Petrobiinae). Mitteilungen aus dem Museum für Naturkunde in Berlin, Deutsche Entomologische Zeitschrift 48: 3–21.

Sturm, H., and M. Bowser. 2004. Notes on some Archaeognatha (Insecta, Apterygota) from extreme localities and a complementary description of *Petridiobius* (*P.*) *arcticus* (Paclt, 1970). Entomologische Mitteilungen aus dem Zoologischen Museum Hamburg 14: 197–203.

Willem, V. 1924. Observations sur *Machilis maritima*. Bulletin Biologique de la France et de la Belgique 58: 306–320.

Collembola

KENNETH CHRISTIANSEN AND PETER BELLINGER

(Plates 337–343)

There have been many studies of littoral Collembola in Europe, beginning with the work of Laboulbéne and Moniez in the last century and continuing to the present (i.e., Sterzynska and Ehrnsberger 1997), and there are scattered records and descriptions from many parts of the globe. Records from North America are comparatively few, and mainly from the East Coast.

Christiansen and Bellinger (1988) summarized what was known of the fauna north of Panama and recorded nine species from the Pacific coast. Since then, seven more species have been described and nine recorded from the Pacific coasts of Mexico and Nicaragua, but none from farther north. Collembola are

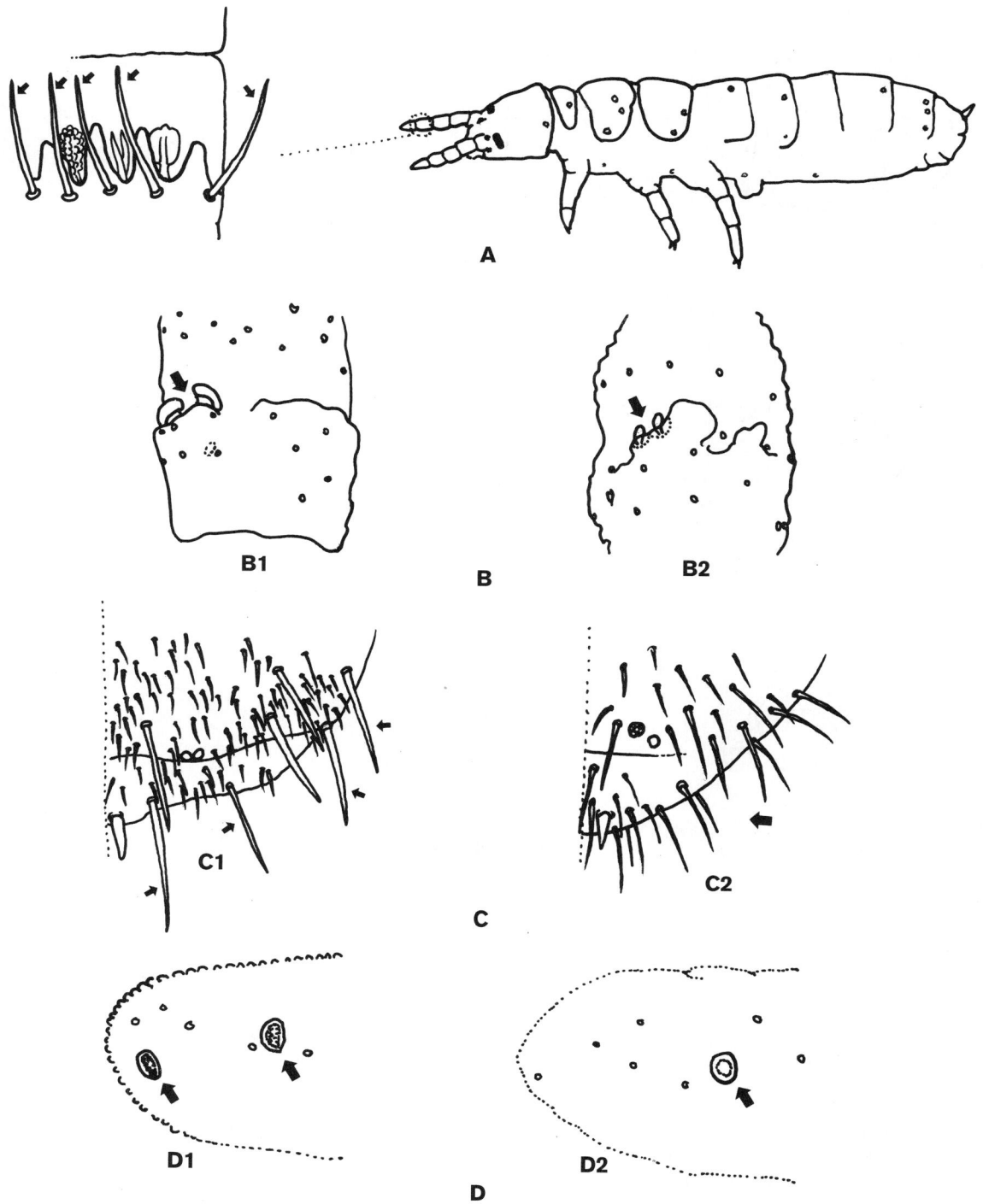

PLATE 337 All figures with setae omitted except where significant in key; small circles represent seta bases: A, semi-diagrammatic typical *Onychiurus* showing variations in structure of sense clubs on third antennal segment sense organ; B, third antennal segment sense organs: B1, *Tullbergia yosii*; B2, *Willemia persimilis*; C, half of posterior abdomen: C1, *Onychiurus dentatus*; C2, *Onychiurus lagunensis*; D, left half of dorsum of first thoracic segment: D1, *Onychiurus hoguei*; D2, *Onychiurus debilis* (B, after Bonet 1945; C, after Palacios-Vargas and Diaz 1996; D, after Palacios-Vargas in litt.).

both small and often cryptic and easily overlooked by collectors in the marine littoral. As a result they are poorly known in North America, and we think it quite likely that some of the species described from Mexican coasts may yet be found in the Oregonian faunal province; accordingly, they are included in the key.

Two habitats in the Pacific marine environment not commonly examined for Collembola may yield many interesting forms. The interstitial sand fauna, which has not been investigated on our Pacific coast for Collembola, includes many species, some highly specialized (see Thibaud and Christian 1998 for a summary of the world fauna). In addition, Collembola have

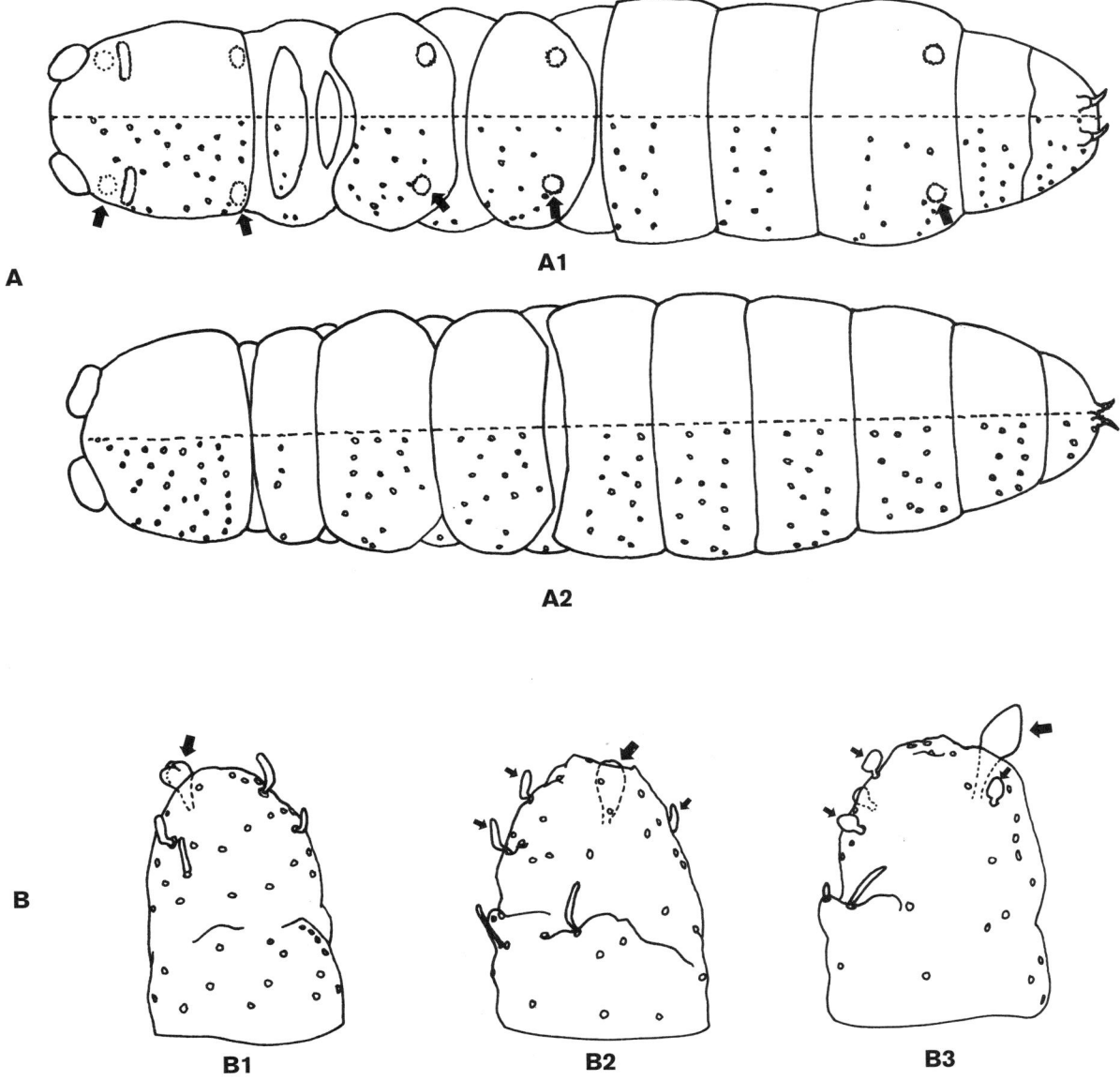

PLATE 338 A, Semi-diagrammatic dorsum of: A1, *Tullbergia yosii* showing pseudocelli, A2, *Willemia bellingeri*; B, apex of antenna showing swollen setae: B1, *Willemia bellingeri*, B2, *W. persimilis*, B3, *W. arenicola* (A, B1, B3, after Palacios-Vargas and Vazquez 1989).

been found in bottom deposits under as much as 20 m of water in the Black Sea (da Gama 1966) and the Mediterranean (Jacquemart and Jacques 1980) and have been collected from deep waters in Pearl Harbor (Bellinger, unpublished). Careful examination of these habitats might add significantly to the local species list.

We omit from this key genera and species recorded only from tropical or subtropical regions (including the Gulf of California), which are very unlikely to be found in the Oregonian coastal province. We include the genus *Anurida* because of old records of *A. maritima* (Gervais) from British Columbia and San Diego; recent descriptions of other littoral species make the identity of these records uncertain. The genera *Isotogastrura* and *Spinactaletes* are widespread and might eventually be found in our area. In addition to typical littoral species, many eurytopic Collembola, especially in the families Hypogastruridae and Isotomidae, have been recorded from wrack and interstitial sand habitats in other parts of the world. Inclusion of these would have made the key impossibly large; specimens not appearing to key out should be checked with the Collembola of North America (Christiansen and Bellinger 1998) or Dindal's *Soil Biology Guide* (1990).

Basic collembolan anatomy and biology are covered in these works and in Hopkin's *Biology of Springtails* (1997).

KEY TO GENERA AND SPECIES OF LITTORAL COLLEMBOLA

Those species and genera marked * are not yet known north of Mexico.

1. Eyes, furcula, and pigment all lacking 2
— Eyes present; pigment and furcula usually present 9
2. Third antennal segment with four or five guard papillae (plate 337A) 3

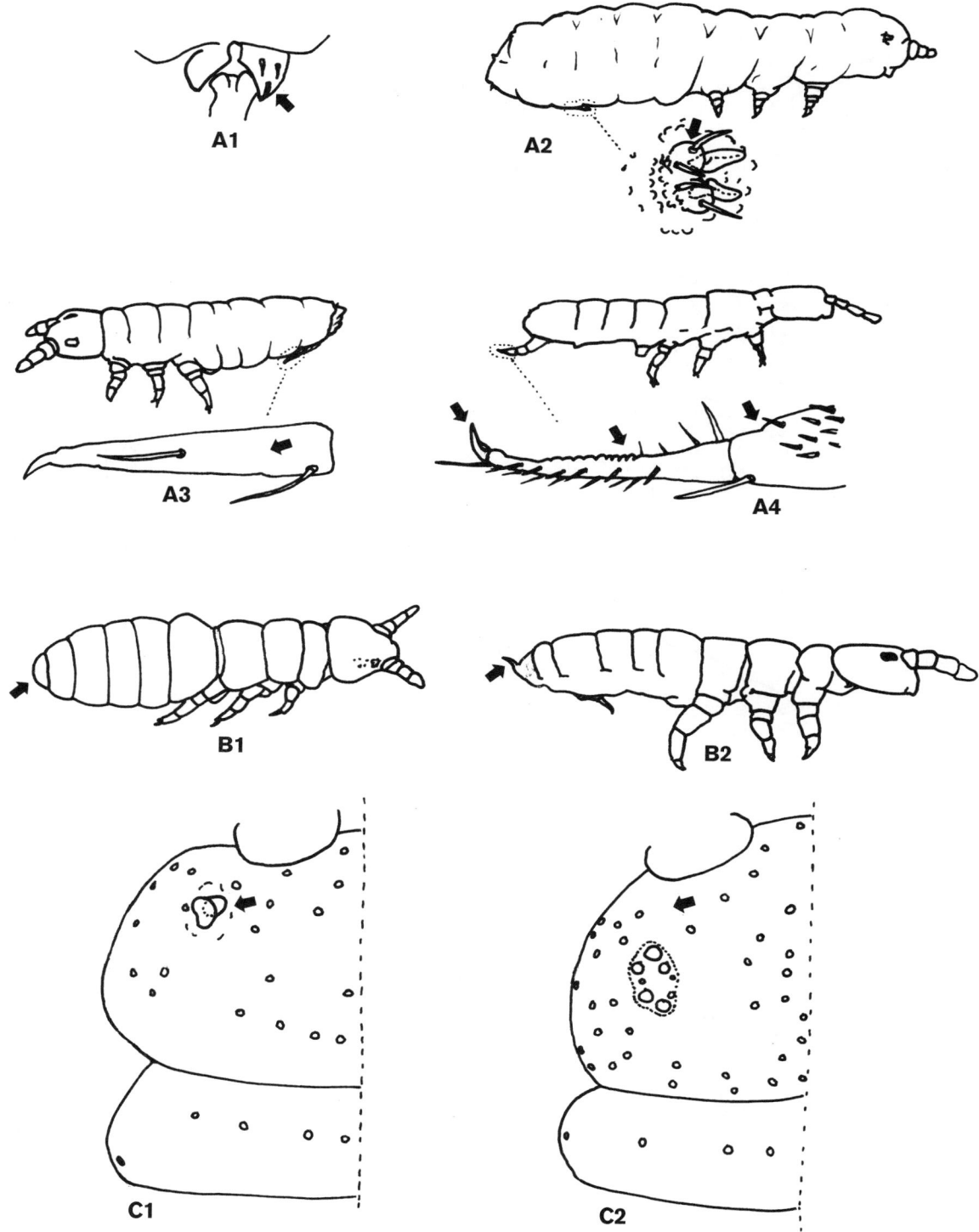

PLATE 339 A, Furcula and habitus: A1, furcula of *Friesea fara*, A2–A4, furcula and habitus; A2, *Xenylla pseudomaritima*; A3, *Pseudostachia xicoana*; A4, *Folsomina onychiurina*; B, Habitus: B1, *Anurida*; B2, *Friesea*; C, left side of head: C1, *Pseudostachia xicoana*; C2, *Paraxenylla lapazana* (A, B, C1, after Palacios-Vargas and Najt 1985; C2, after Palacios-Vargas and Vazquez 1989).

— Third antennal segment without guard papillae (plate 337B) 6
3. Body with stiff spinelike setae among normal setae (plate 337C1; see also plate 337A . . . .
. . . . *Onychiurus (Onychiurus) dentatus*
— Body with only normal setae (plate 337C2) 4

4. First thoracic segment without pseudocelli (see also plate 337A and 337C2). . . . *Onychiurus (Protaphorura) lagunensis**
— First thoracic segment with pseudocelli (plate 337D, see also 337A) 5
5. First thoracic segment with 2 + 2 pseudocelli (plate 337D1) . . . . *Onychiurus (Protaphorura) hoguei**

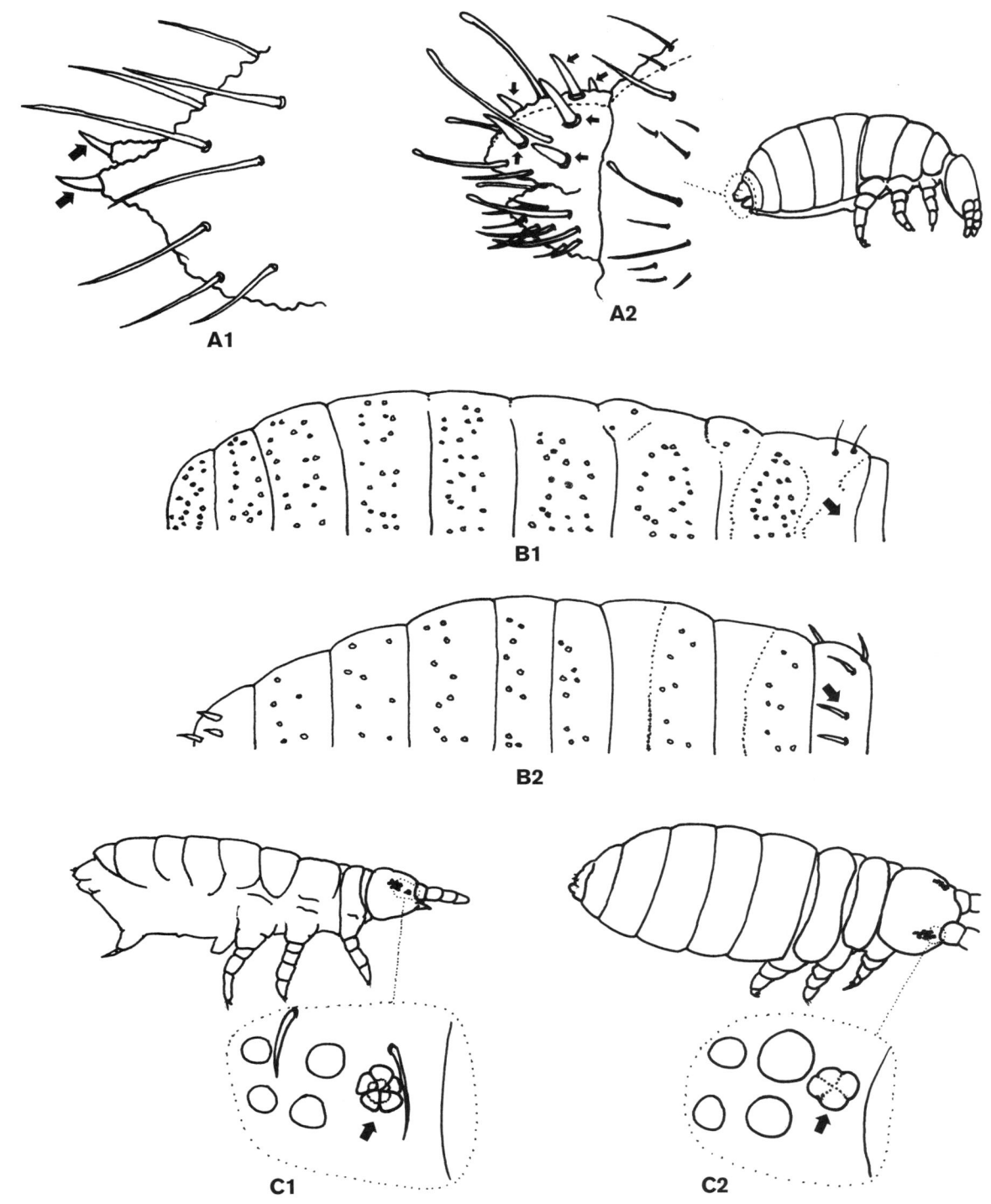

PLATE 340 A, Anal spines: A1, *Xenylla pseudomaritima*; A2, Anal spines and habitus *Friesea fara*; B, semi-diagrammatic dorsum of body, setae shown on only on first thoracic segment: B1, *Isogastrura ahuizotli*; B2, *Oudemansia georgia*; C, habitus and post antennal organ: C1, *Pseudachorutes americanus*; C2, *Hypogastrura (Schoettella) distincta* (B, after Palacios-Vargas and Thibaud 1998; C, after Denis 1931).

— First thoracic segment with 1 + 1 pseudocelli (plate 337D2) . . . . *Onychiurus (Protaphorura) debilis*
6. Pseudocelli present (plate 338A1; see also plate 337B1) . . . . *Tullbergia (Mesaphorura) yosii*
— Pseudocelli absent (plate 338A2) 7
7. Fourth antennal segment apical organ trilobed (plate 338B2). . . . . *Willemia bellingeri**

— Fourth antennal segment apical organ single (plate 338B2, 338B3) 8
8. Fourth antennal sensilla cylindrical (plate 338B2; see also plate 337B2). *Willemia persimilis**
— Fourth antennal sensilla spherical (plate 338B3). . . . . *Willemia arenicola**
9. Mucro reduced, lacking mucro (plate 339A1) or with

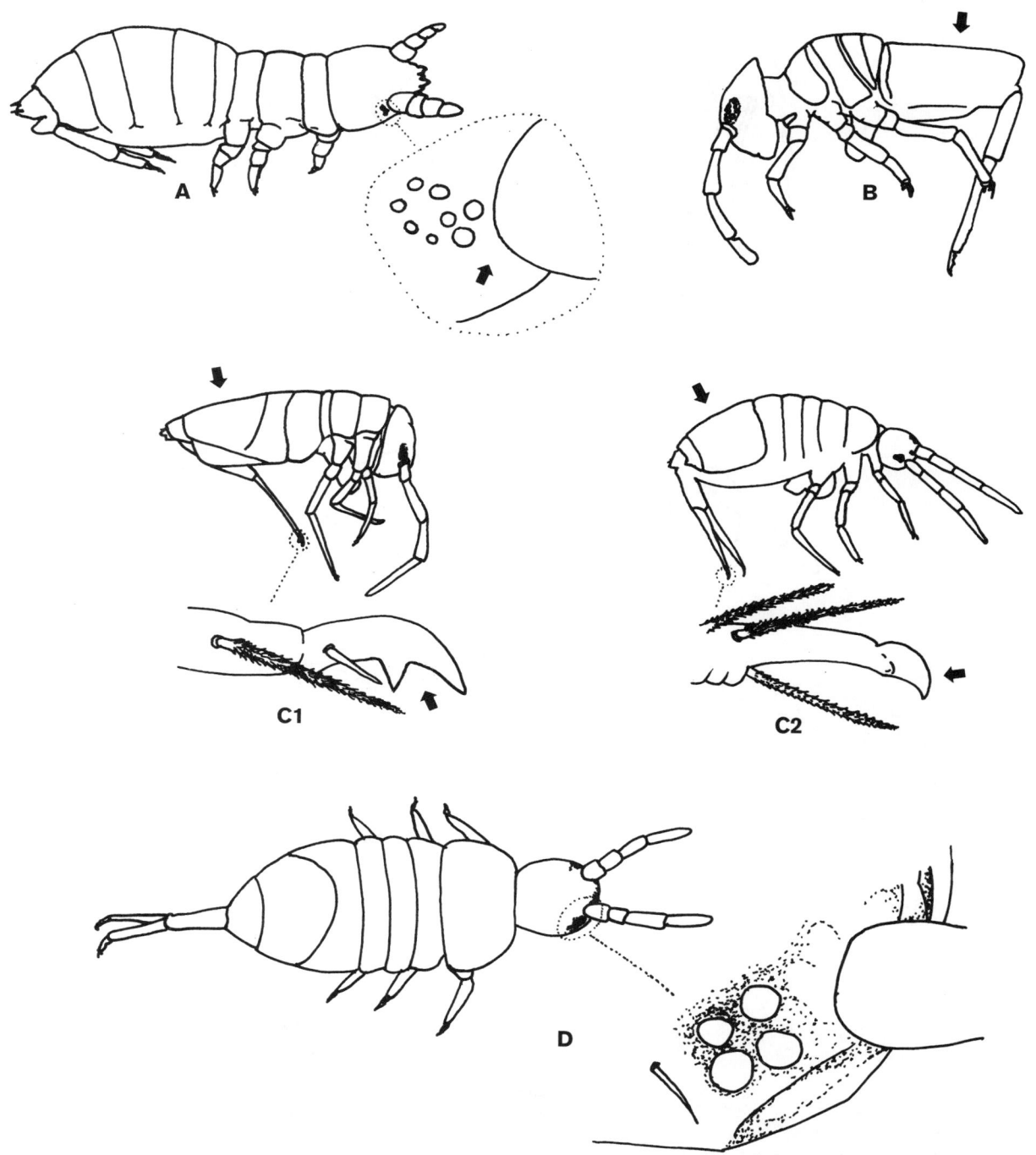

PLATE 341 A, Habitus and detail of eye region, *Oudemansia georgia*; B, Habitus, *Spinactaletes*; C, habitus and mucro: C1, *Entomobrya*; C2, *Seira*; D, habitus and eyes of *Pseudosinella lahainensis* (B, after Soto-Adames 1988).

dens-manubrium fused (plate 339A2, 339A3) or part or all of furcula absent 10
— Mucro, dens, and manubrium all clear (plate 338A4) . . . . .
. . . . 14
10. Anal spines lacking (plate 339B1) 11
— Anal spines present (plate 339B2) 13
11. Furcula absent (plate 339B1) . . . . *Anurida*
— Furcula reduced but present (plate 339A2, 339A3) 12
12. Post antennal organ present (plate 339C1; see also plate 339A3) . . . . *Pseudostachia**

— Post antennal organ absent (plate 339C2) . . . .
. . . . *Paraxenylla lapazana**
13. Anal spines: two (plate 340A1; also plate 339B2) . . . .
. . . . *Xenylla pseudomaritima*
— Anal spines: six (plate 340A2; see also plate 339A1) . . . . .
. . . . *Friesea fara*
14. Prothorax much reduced, not bearing setae (plate 340B1)
. . . . 17
— Prothorax bearing setae (plate 340B2) 15
15. Postantennal organ present (plate 340C) 16

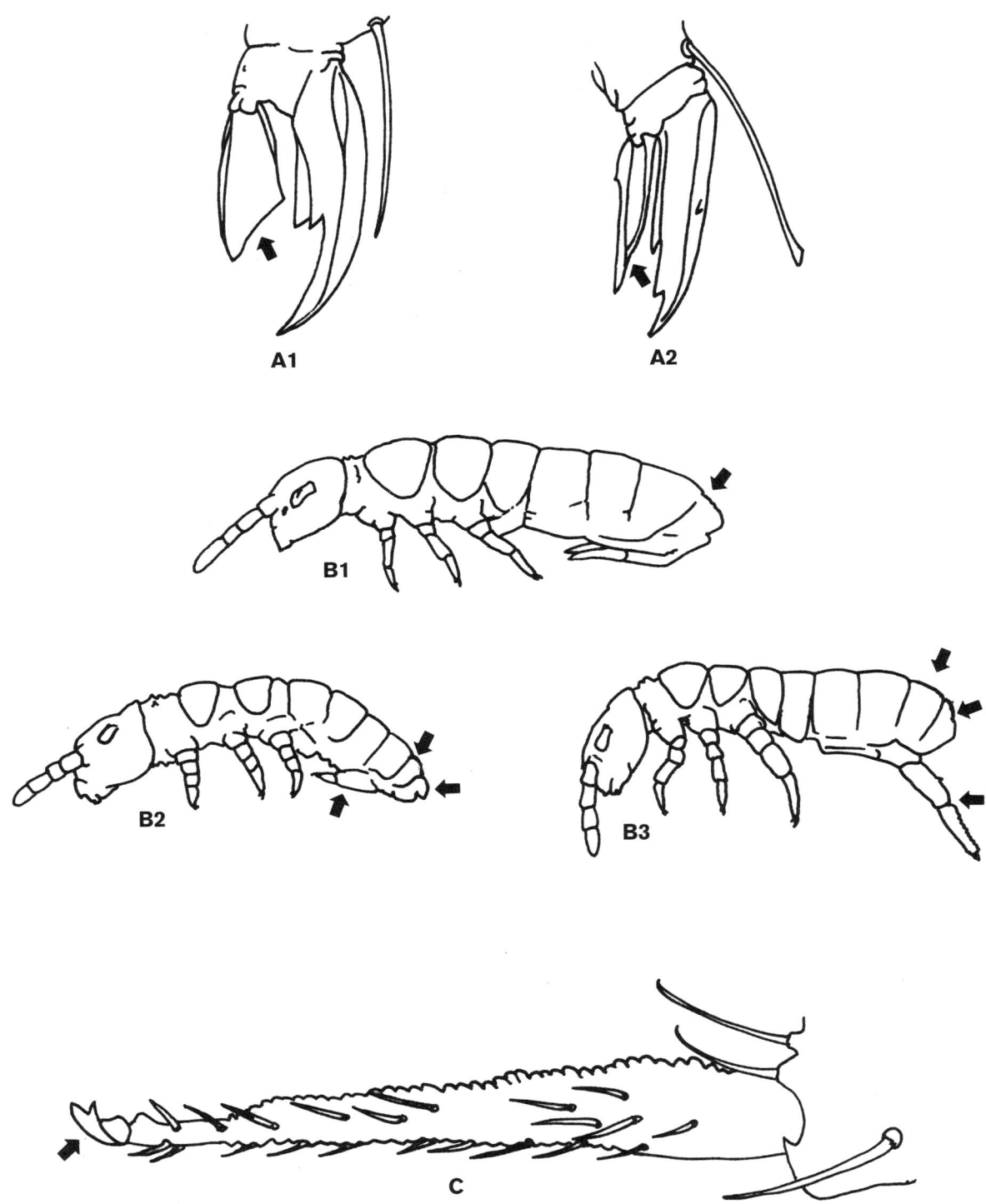

PLATE 342 A, Foot complex: A1, *Entomobrya* (*Mesentotoma*) *laguna*; A2, *Entomobrya* (*Entomobrya*) *arula*; B, habitus: B1, *Cryptopygus thermophilus*; B2, *Isotogastrura* (after Fjellberg 1994); B3, *Archisotoma besselsi*; C, Mucro and dens of *Cryptopygus thermophilus*.

— Postantennal organ lacking (plate 341A; also plate 340B2)
. . . . *Oudemansia georgia*
16. Postantennal organ with five to eight clear lobes (plate 340C1) . . . . *Pseudachorutes americanus*
— Postantennal organ with four unclear lobes (plate 340C2)
. . . . *Hypogastrura* (*Schoettella*) *distincta**

17. With last three abdominal segments fused (plates 339A4, 341B) 18
— With at least abdominal segment 4 separate from other (plate 341A, 341C, 341D) 19
18. Eyeless and unpigmented (plate 339A4) . . . .
. . . . *Folsomina onychiurina*

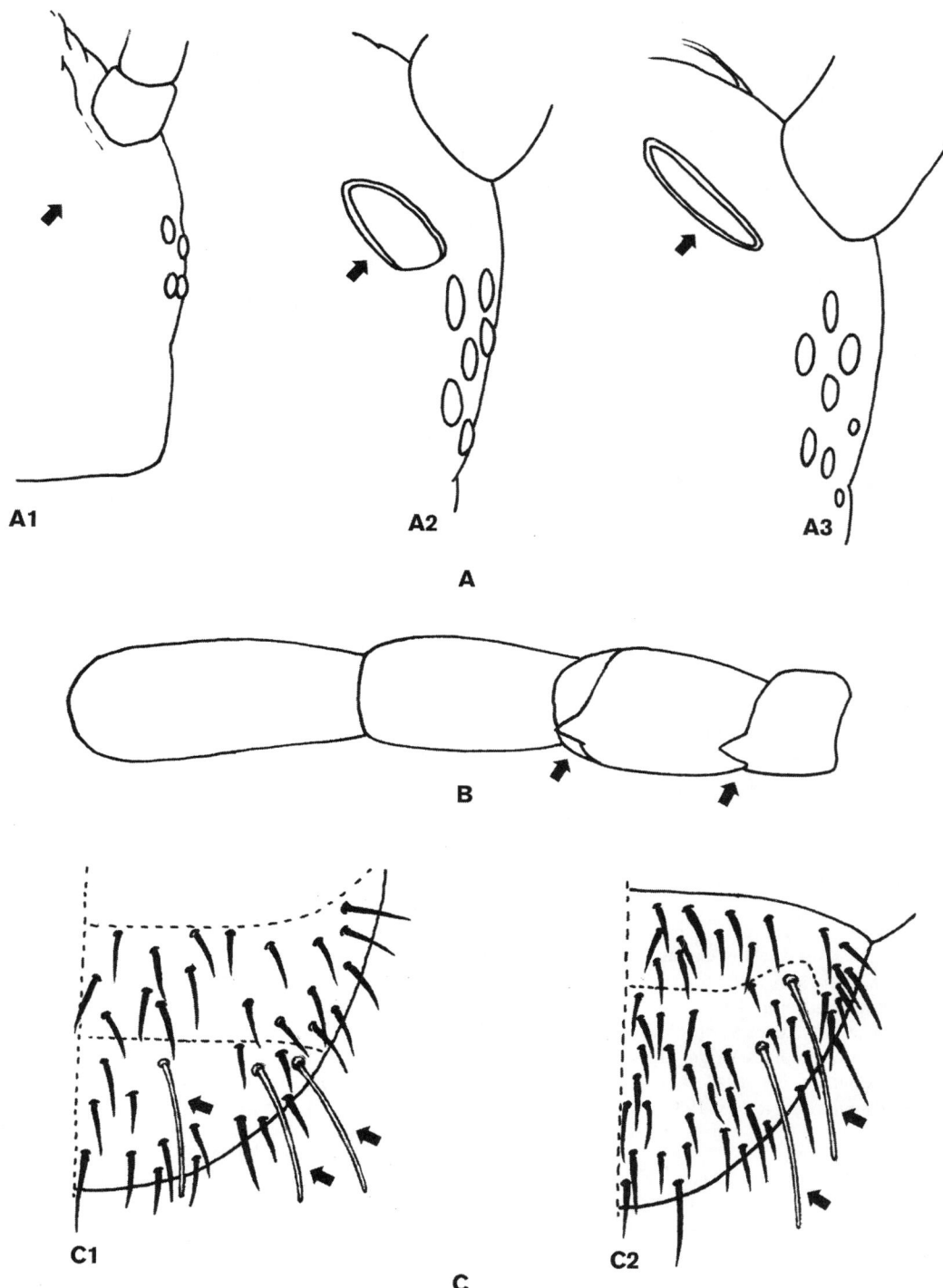

PLATE 343 A, Eyes and P.A.O. region: A1, *Isotogastrura* (after Fjellberg 1995); A2, *Isotoma (Halisotoma) marisca*; A3, *Archisotoma besselsi*; B, antenna of *Isotoma (Halisotoma) marisca*; C, Dorsum of right half of end of abdomen: C1, *Archisotoma besselsi*; C2, *Archisotoma interstitialis*.

— With eyes and pigment (plate 341B)....... *Spinactaletes* *
19. With fourth abdominal segment greater than three times as long as third (plate 341C)..................... 20
— With fourth abdominal segment less than two times as long as third (plate 342B) 23
20. With mucro falcate (plate 341C2)............... *Seira*

— Mucro bidentate (plate 341C1)................... 21
21. Eyes 4 + 4 or fewer (plate 341D), body with scales.....
.... *Pseudosinella*
— Eyes 8 + 8, body without scales................... 22
22. Unguiculus truncate (plate 341A1; also plate 341C1)....
.... *Entomobrya (Mesentotoma) laguna*

— Unguiculus acuminate (plate 342A2; also plate 341C1) . . . . *Entomobrya (Entomobrya) arula*

23. Abdominal segments 5 and 6 fused (plate 342B1), mucro bidentate (plate 342C) . . . . *Cryptopygus*

— Abdominal segments 5 and 6 separate (plate 342B2, 342B3), mucro varied but never bidentate. 24

24. Furcula short (plate 342B2), postantennal organ absent (plate 343A1; also plate 340B1) . . . . *Isotogastrura**

— Furcula long (plate 342B3), postantennal organ present (plate 343A2, 343A3) 25

25. First and second antennal segments with apical projecting angles (plate 343B; also plate 343A2) . . . . *Isotoma (Halisotoma) marisca*

— First and second antennal segments without such 26

26. Sixth abdominal segment with 3 + 3 bothriotricha (plate 343C1, 342B3) . . . . *Archisotoma besselsi*

— Sixth abdominal segment with 2 + 2 bothriotricha (plate 343C2) . . . . *Archisotoma interstitialis*

LIST OF SPECIES

Archisotoma besselsi (Packard, 1877).
Archisotoma interstitialis Delamare, 1954.
Entomobrya (s.s.) arula Christiansen and Bellinger, 1980.
Entomobrya (s.s.) laguna Bacon, 1913.
Folsomina onychiurina Denis, 1931.
Friesea fara Christiansen and Bellinger, 1974.
Isotoma (Halisotoma) marisca Christiansen and Bellinger, 1988.
Mesaphorura yosii Rusek, 1967.
Onychiurus (s.s.) dentatus (Folsom, 1902).
Onychiurus (Protaphorura) debilis (Moniez, 1889).
Onychiurus (Protaphorura) hoguei (Palacios-Vargas and Diaz, 1996).
Onychiurus (Protaphorura) lagunensis (Palacios-Vargas and Diaz, 1996).
Oudemansia georgia Christiansen and Bellinger, 1988.
Paraxenylla lapazana Palacios-Vargas and Vázquez, 1989.
Pseudachorutes americanus Stach, 1949.
Schoettella distincta Denis, 1931.
Willemia arenicola Palacios-Vargas and Vázquez, 1989.
Willemia bellingeri Palacios-Vargas and Vázquez, 1989.
Willemia persimilis Bonet, 1945.
Xenylla pseudomaritima James, 1933.

REFERENCES

Christiansen, K., and P. Bellinger. 1988. Marine littoral Collembola of North and Central America. Bull. Marine Sci. 42: 215–245.

Christiansen, K., and P. Bellinger. 1998. The Collembola of North America north of the Rio Grande. 2nd ed. 1518 pp. Grinnell, Iowa: Grinnell College.

Dindal, D. L., ed. 1990. *Soil Biology Guide.* 1349 pp. New York: John Wiley.

da Gama, M. M. 1966. Notes taxonomiques sur quelques espèces de collemboles. Mem. Estud. Mus. Zool. Univ. Coimbra 295: 1–21.

Hopkin, S. P. 1997. Biology of the Springtails. 330 pp. Oxford University Press.

Jacquemart, S., and J.-M. Jacques. 1980. À propos d'un collembole entomobryen à la fois marin t désertique. Annls Soc. R. Zool. Belg. 109: 9–18.

Sterzynska, M., and R. Ehrnsberger. 1997. Marine algae wrack Collembola of European coasts. Abh. Ber. Naturkundemus. Goerlitz 69: 165–178.

Thibaud, J.-M., and E. Christian. 1998. Biodiversity of interstitial Collembola in sand environments. European J. Soil Biol. 33: 123–127.

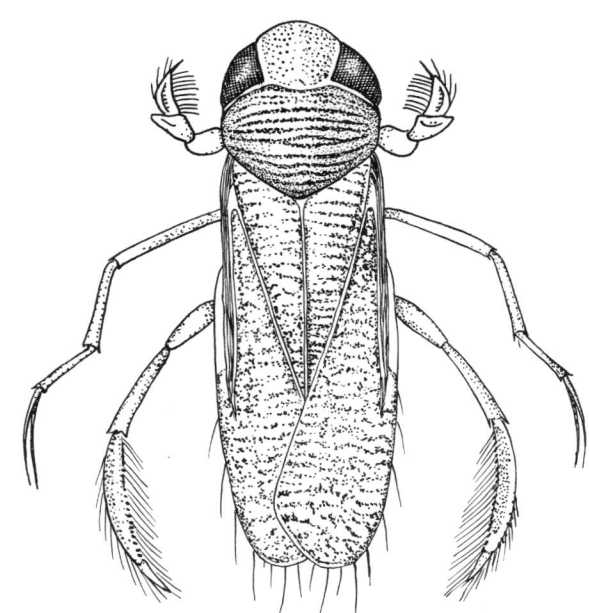

Trichocorixa reticulata

PLATE 344 Hemiptera. *Trichocorixa reticulata* (from Usinger 1963, redrawn by Emily Reid).

Hemiptera

HOWELL V. DALY

(Plate 344)

Water boatmen, *Trichocorixa* (family Corixidae), are characteristic of brackish pools throughout the world and can tolerate salinities considerably above that of the sea (Hutchinson 1931). They are also found with the brine shrimp, *Artemia franciscana,* and brine flies, *Ephydra gracilis,* in the brine pools of southern San Francisco Bay. Our species of *Trichocorixa* in saline waters are distinguished as follows: *T. reticulata* (Guérin-Meneville, 1857) (plate 344) has the width of the interocular space at its narrowest point distinctly exceeding the width of the eye along the hind margin as seen from above; *T. verticalis* (Fieber, 1851) has the interocular space subequal to or less than the width of the hind margin of the eye.

Trichocorixa reticulata occurs in great abundance in tide pools and salt marshes along the entire coast of California (Menke 1979). The coastal populations of *T. verticalis* have been assigned to the subspecies *T. verticalis californica* Sailer, 1948, and may occur with *T. reticulata* in the San Francisco area and Humboldt Bay region. Also along the coast in saline waters are two species of another genus, *Corisella decolor* (Uhler, 1871) and *C. inscripta* (Uhler, 1894), that are distinguished from *Trichocorixa* by having shorter and more numerous hairs on the hemelytra. See Menke (1979) for a key to species.

REFERENCES

Hutchinson, G. E. 1931. On the occurrence of *Trichocorixa* Kirkaldy (Corixidae, Hemiptera-Heteroptera) in salt water and its zoogeographical significance. American Naturalist 65: 573–574.

Menke, A. S. 1979. The semiaquatic and aquatic Hemiptera of California (Heteroptera: Hemiptera). Bulletin of the California Insect Survey 21: 1–166.

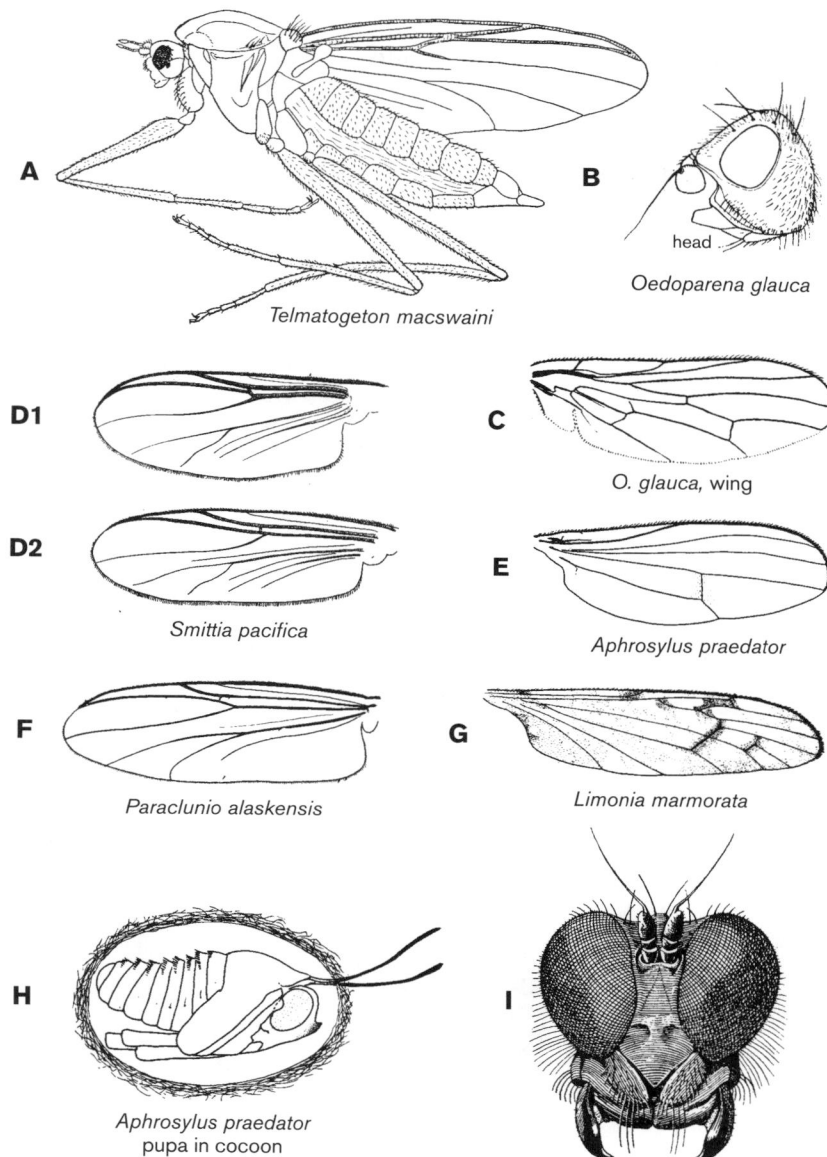

PLATE 345 A, *Telmatogeton macswaini*, adult; B, *Oedoparena glauca*, head; C, *O. glauca*, wing; D, *Smittia pacifica*, wing; D1, male; D2, female; E, *Aphrosylus praedator*, wing; F, *Paraclunio alaskensis*, wing; G, *Limonia marmorata*, wing; H, *A. praedator*, pupa in cocoon; I, *Melanderia mandibulata*, head, note that apparent "mandible" is a lobe of the labellum (A, Wirth 1949; B, C, Curran 1934; D–H, after Saunders 1928; I, Snodgrass 1922).

Diptera

EVERT I. SCHLINGER

(Plates 345–346)

About 20 species of truly marine flies are often abundant elements of the central California and Oregon intertidal zones (many more species, listed at the end of this section, are also found in or near the shore). Some of these marine flies, such as the Chironomidae and Dolichopodidae, are specialized members of taxa that are well adapted to fresh water; others are primarily terrestrial, such as the Limoniidae and Dryomyzidae. The Canacidae appears to have more worldwide saline species than freshwater species. The student is referred to *A Catalog of the Diptera of America North of Mexico* (Stone et al. 1965), *The Flies of Western North America* (Cole and Schlinger 1969), and the *Manual of Nearctic Diptera* (McAlpine et al. 1981, 1989; McAlpine 1987, 1989) for Diptera terminology, history, classification, references, and synonymy of the species and other taxa discussed here.

KEY TO SUBORDERS AND FAMILIES COMMON IN THE COASTAL INTERTIDAL ZONE

1. Antenna of six or more segments; palpus often elongated, four- or five-segmented; calypters (squamae) absent; abdomen often without much hair and rarely with bristles (plates 345A, 345D, 345F, 345G, 346A–346L) . . . .
. . . . Suborder Nematocera 2
— Antenna of three segments, the terminal segment often annulated or bearing a style or arista; palpus short, one- or two-segmented; calypters present or absent; abdomen often with both hair and bristles (plate 345B–345C, 345E, 345H, 345I) . . . . Suborder Brachycera 3
2. Mesonotum with V-shaped suture, starting on each side in front of wing base and pointed in middle part at scutellum; at least nine wing veins reach wing margin; discal cell present (plate 345G) . . . . Limoniidae
— Mesonotum with transverse suture; less than nine wing veins reach wing margins; discal cell absent (plate 345D,

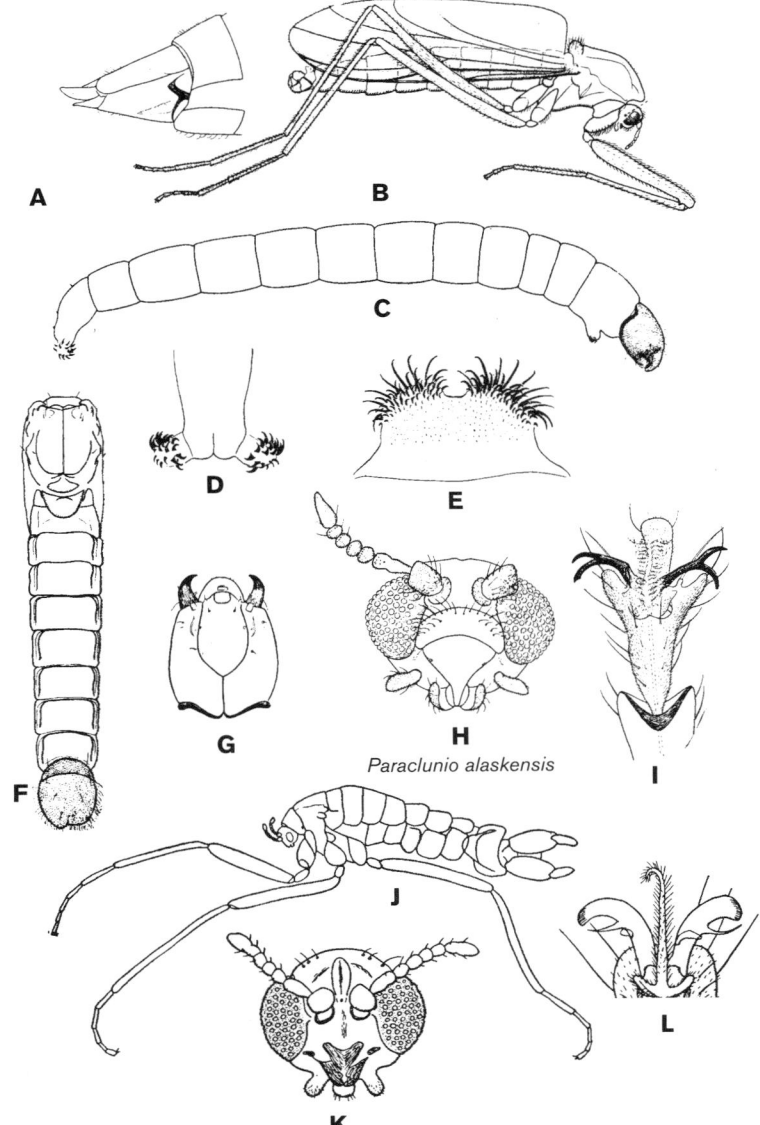

PLATE 346 A–I, *Paraclunio alaskensis*: A, female, tip of abdomen; B, male; C, larva; D, anal pseudopods of larva; E, prothoracic pseudopods of larva; F, pupa; G, larval head, dorsum; H, female, head; I, male, last tarsal segment; J–L, *Tethymyia aptena*: J, adult; K, head; L, male, fore-tarsal claws (A–I, Saunders 1928; J–L, Wirth 1949).

Paraclunio alaskensis

Tethymyia aptena

345F) .... Chironomidae

3. Anal cell much longer than second basal cell and narrowed or closed at wing margins 4

— Anal cell as long as or shorter than second basal cell, or absent; when present it ends or is closed some distance from wing margin 5

4. Head with depressed region at vertex; proboscis pointed, modified for piercing .... Asilidae

— Head without depression at vertex; proboscis not modified for piercing 5

5. Wing with five posterior cells; antennae usually thick ... . Therevidae

— Wings with three or four posterior cells; antennae thin .. ..Bombyliidae*

6. Antennae not enclosed above with a crescent-shaped frontal suture; antennal arista usually terminal. 6

— Antennae enclosed above with a crescent-shaped frontal suture; antennal arista usually dorsal................7

7. Discal cell usually separate from second basal cell; anterior

crossvein located beyond basal fourth of wing ....
.... Empididae*

— Discal cell confluent with second basal cell; anterior crossvein located within basal fifth of wing. ....
.... Dolichopodidae

8. Second antennal segment with longitudinal seam along upper outer edge; lower calypter large 8

— Second antennal segment without such seam; lower calypter undeveloped or vestigial 9

9. Hypopleura with one or more rows of vertical bristles; anal vein usually does not reach margin of wing...........
.... Sarcophagidae*

— Hypopleura without bristles; anal vein reaches wing margin..................................Anthomyiidae*

10. Costa not fractured near end of subcosta or near end of humeral crossvein 10

*Families of flies encountered in the intertidal zone, but are never truly aquatic. Common genera and species of these families are listed at the end of the Diptera section.

— Costa fractured at end of subcosta, or at end of humeral crossvein or both. 12
11. Thorax flattened; last tarsal segment flat and enlarged. . . . .
. . . . Coelopidae*
— Thorax convex; last tarsal segment not flattened 11
12. Mouth rim raised; palpus with apical bristle; antennae close together at their bases (plate 345B, 345C) . . . .
. . . . Dryomyzidae
— Mouth rim not raised; palpus without apical bristle; antennae separated by upper ridge of face. . . . .
. . . . Helcomyzidae*
13. Costa broken only at end of the subcosta 13
— Costa broken at end of subcosta and humeral crossvein
. . . . Ephydridae*
14. Oral vibrissae present at vibrissal angle; preapical tibial bristles present . . . . Heleomyzidae*
— Oral vibrissae absent although one or more pairs of genal bristles may be present; tibiae without preapical bristles
. . . . Canacidae

LIMONIIDAE

A truly marine crane fly, *Limonia (Idioglochina) marmorata* (plate 345G), occurs on the California coast (Alexander 1967). The immature stages were first discovered by Saunders (1928) on Vancouver Island, living in filamentous algae under *Fucus* and among barnacles, small mussels, and other marine animals. Most commonly the larvae are found in the bright green *Ulva*, which cover rocks a foot or two below high-water mark. The larvae are most often observed feeding on the vegetation from the end of their loosely woven silken tubes. The larvae overwinter in these tubes, and adults emerge from May to August. This is the only marine crane fly in our region, and the adults swarm over seaweeds or rest on vertical or overhanging rock sides (Evans 1980).

CHIRONOMIDAE

This family of midge flies appears to be the most adaptive group of dipterans inhabiting the coastal marine area, where no less than eight species representing five genera are found. These species may be wingless and small (1 mm), or fully winged and up to 7 mm. Larvae of most species are known and have been found feeding in algal mats of species of *Ulva* (Saunders 1928 and Wirth 1949). Some species are rather abundant in areas of sewage outfalls and *Paraclunio alaskensis* may be an important species in consuming algae that develop massively in such areas (see Cheesemen and Preissler 1972; Morley and Ring 1972).

KEY TO INTERTIDAL MIDGES

1. Pronotal lobes widely separated; antennae not plumose . . .
. . . . 2
— Pronotum not, or only slightly, notched anteriorly on median line; male antennae usually plumose 6
2. Hind tarsus with second segment not longer than third . . .
. . . . 3

*Families of flies encountered in the intertidal zone, but are never truly aquatic. Common genera and species of these families are listed at the end of the Diptera section.

— Hind tarsus with second segment longer than third 4
3. Wings straplike, reaching to fourth segment of abdomen, halters present . . . . *Eretmoptera browni*
— Wings vestigial, not reaching to abdomen, halters vestigial (plate 346J–346L) . . . . *Tethymyia aptena*
4. Hairs on legs strong and sometimes with scales, front legs of male modified; femora swollen 5
— Hairs of legs weak; legs unmodified (plate 345A) . . . .
. . . . *Telmatogeton macswaini*
5. Tarsal claws slender, those of male deeply cleft; tibiae with rows of strong bristly hairs (plates 345F, 346A–346I) . . . .
. . . . *Paraclunio alaskensis*
— Tarsal claws flattened and broadened, those of male shallowly cleft; tibiae with rows of hairs mostly replaced by scales . . . . *Paraclunio trilobatus*
6. Wings milky white with yellowish basal area; male antenna with eight or 13 flagellar segments 7
— Wings milky white throughout; male antenna with 12 flagellar segments . . . . *Thalassosmittia marina*
7. Maxillary palpus four-segmented; male antenna with eight flagellar segments . . . . *Thalassosmittia clavicornis*
— Maxillary palpus five-segmented; male antenna with 13 flagellar segments (plate 345D) . . . . . . *Thalassosmittia pacifica*

DOLICHOPODIDAE

There are two genera and six species of truly marine Dolichopodidae flies in California and Oregon. Biological information on both is scant, but immature stages and their habits are known for *Paraphrosylus*. Adults of *P. praedator* (wing, plate 345E) are predaceous upon both larvae and adult midges (*Smittia*) on Vancouver Island, according to Saunders (1928). The larvae are likely also predaceous, but their true feeding preferences remain unknown. They inhabit the same mats of filamentous algae as do the midges. The pupae are formed in rounded, elliptical cocoons loosely spread in the algae (plate 345H). *Melanderia mandibulata* and *M. crepuscula* are both found in our area (Arnaud 1958). The adults are remarkable in having the lobes of the labellum developed into "mandiblelike" structures (plate 345I) possibly used in predatory activities.

KEY TO INTERTIDAL DOLICHOPODIDAE

1. Mouthparts appearing "mandiblelike" (plate 345I) 2
— Mouthparts without "mandibles" 3
2. Front femur of male on inner side near base with dense tuft of eight to 10 bristles on a very slight protuberance. . . . .
. . . . *Melanderia mandibulata*
— Front femur of male without tuft of bristles on inner side near base, but with long, black hair . . . . . . *Melanderia crepuscula*
3. Wing clear without distinct black spot over hind crossvein 4
— Wing with distinct black spot over hind crossvein . . . . . .
. . . . *Paraphrosylus direptor*
4. Antennal arista bare 5
— Arista pubescent . . . . *Paraphrosylus grassator*
5. Wing with second, third, and fourth longitudinal veins greatly broadened; fore and middle tibiae with delicate black hairs on lower surface . . . . *Paraphrosylus wirthi*
— Wing without such broadened veins; fore and middle tibiae without such hairs (plate 345E) . . . .
. . . . *Paraphrosylus praedator*

CANACIDAE

Little is known biologically about these small shore flies, but Wirth (1951) indicated that some species breed preferably on tide-covered rocks that are clothed with *Ulva*. Three species are known from this region.

KEY TO INTERTIDAL CANACIDAE

1. Mesofrons with on pair of bristles, just outside the ocellars 2
— Mesofrons with several pairs of long, interfrontal bristles . . . . *Canace aldrichi*
2. Four genal bristles; anterior notopleural bristles present but small . . . . *Canaceoides nudatus*
— Three genal bristles; anterior notopleural bristles absent . . . . *Nocticanace arnaudi*

DRYOMYZIDAE

One species, *Oedoparena glauca* (plate 345B, 345C), of this small family is a predator during its larval stages on the barnacle *Balanus glandula* and species of *Chthamalus*. The adults apparently feed upon diatoms and are never observed away from the *Endocladia-Balanus* association. After mating, males often remain on the back of the female while she lays eggs in and around the mouth of a large barnacle. Large larvae move from barnacle to barnacle and take several months to become adults. Death of the barnacle may be due indirectly to starvation because the larval feeding action seems to restrict the barnacles' cirri from operating properly and thus impairs their feeding. The adult fly is lead-colored with grayish hyaline wings and yellow halters. See Knudsen 1968; Burger et al. 1980.

LIST OF INTERTIDAL DIPTERA INCLUDED IN KEYS

See Cole and Schlinger, 1969.

LIMONIIDAE (CRANE FLIES)

Limonia (Idioglochina) marmorata (Osten-Sacken, 1861) (=*Dicranomyia signipennis* Coquillett, 1905). See Evans 1980 (review and photograph); Robles and Cubit 1981; Robles 1982 (ecology).

CHIRONOMIDAE, ORTHOCLADIINAE (CLUNIONINAE OF AUTHORS) (MIDGES)

METRIOCNEMINI

Eretmoptera browni (Kellogg, 1900).
Metriocnemus yaquina Cranston and Judd, 1987. Splash zone rock pools, Yaquina Head, Oregon.
Tethymyia aptena Wirth, 1949. See Evans 1980 (review and photograph).
Thalassosmittia clavicornis (Saunders, 1928).
Thalassosmittia marina (Saunders, 1928) (=*Saunderia marinus*). See Robles and Cubit 1981; Robles 1982 (ecology).
Thalassosmittia pacifica (Saunders, 1928).

TELMATOGETONINI

Paraclunio alaskensis (Coquillett 1900). See Evans 1980; Robles and Cubit 1981; Robles, 1982 (ecology).

Paraclunio trilobatus Kieffer, 1911. See Evans 1980 (review and photograph); Robles and Cubit 1981; Robles 1982 (ecology).
Telmatogeton macswaini Wirth, 1949.

DOLICHOPODIDAE

Asyndetus spp.
Melanderia crepuscula Arnaud, 1958.
Melanderia mandibulata Aldrich, 1922.
Paraphrosylus direptor Wheeler, 1897.
Paraphrosylus grassator Wheeler, 1897.
Paraphrosylus praedator Wheeler, 1897. See Evans 1980 (review and photograph).
Paraphrosylus wirthi Harmston, 1951.
Thambemyia borealis (Takagi, 1965) (=*Conchopus borealis*). Asian species introduced to San Francisco Bay. See Masunaga et al. 1999, Entomological Science 2: 399–404.

CANACIDAE

Canace aldrichi Cresson, 1936.
Canaceoides nudatus (Cresson, 1926). See Evans 1980 (notes and photograph).
Nocticanace arnaudi Wirth, 1954. See Robles and Cubit 1981; Robles 1982 (ecology).

DRYOMYZIDAE

Oedoparena glauca (Coquillett, 1900). Predatory as larvae upon barnacles (Knudsen 1968; Burger et al. 1980).

LIST AND DIAGNOSES OF REPRESENTATIVE INTERTIDAL DIPTERA NOT IN KEYS

EMPIDIDAE

Small, winged, or brachypterous, predaceous flies found in shoreline sandy areas.
Chersodromia cana Melander, 1945 (*Coloboneura* and *Thinodromia* of authors).
Chersodromia inchoata (Melander, 1906).
Chersodromia insignita Melander, 1945.
Chersodromia magacetes Melander, 1945.
Parathalassius aldrichi Melander, 1906.
Parathalassius melanderi Cole, 1912.

ASILIDAE

Medium to large, predaceous "robber flies," found on sand and occasionally close to ocean water line.
Laphystia actius (Melander, 1923).
Stichopogon coquilletti (Bezzi, 1910).

BOMBYLIIDAE

Small to large flies similar to Therevidae, but antennae generally thinner. Adults feed on nectar; immatures are parasitic on other insects. *Acreophthiria maculipennis* (Cole, 1923). In introduced ice plants just above the shore (N. Evenhuis, personal communication, 2007).

THEREVIDAE

Small to large flies similar to Asilidae, but adults are not predaceous. Larvae are predaceous on several sand-dwelling insects, particularly in dune areas.

Acrosathe pacifica (Cole, 1923).
Chromolepida bella Cole, 1923.
Pherocera spp.
Thereva hirticeps Loew, 1874.
Tabudamima melanophleba (Loew, 1876).

EPHYDRIDAE

Mostly small flies with large "mouths," often called brine flies or shore flies.

No particular species are noted, but taxa likely to be found intertidally close to freshwater areas may be in such genera as *Ephydra, Hostis, Parydra, Scatella, Dimecoenia, Scatophila,* and *Lipochaeta.*

COELOPIDAE

Small, flattened "kelp flies," sometimes common on drying seaweed; their larvae feed on decaying kelp.

Coelopa (Neocoelopa) vanduzeei Cresson, 1914. In lower beach wrack at Pacific Grove (Kompfner 1974). See Evans 1980 (review and photograph).

HELCOMYZIDAE (DRYOMYZIDAE OF AUTHORS)

Large, sea-beach flies that develop in rotting seaweed. Although not yet known from California, they occur from Oregon north and may be found in California.

HELEOMYZIDAE

Small to large flies rarely encountered in the intertidal zone; several species of one sand-colored genus are an exception.

Anorostoma grande Darlington, 1908.
Anorostoma maculatum Darlington, 1908.
Anorostoma wilcoxi Curran, 1933.

ANTHOMYIIDAE

Medium-size flies whose larvae are entomophagous on kelp-fly larvae; see Huckett 1971.

Fucellia antennata Stein, 1910.
Fucellia assimilis Malloch, 1918.
Fucellia costalis Stein, 1910. In higher beach wrack at Pacific Grove (Kompfner 1974).
Fucellia fucorum (Fallen, 1819).
Fucellia pacifica Malloch, 1923.
Fucellia rufitibia Stein, 1910. In middle and upper beach wrack at Pacific Grove (Kompfner 1974).
Fucellia separata Stein, 1910.
Fucellia thinobia (Thomson, 1869).

SARCOPHAGIDAE

Large scavenger flies that may occasionally enter the intertidal zone.

SPHAEROCERIDAE

Thoracochaeta johnsoni (Spuler, 1925) (=*Leptocera johnsoni*). Reported by Kompfner (1974) in all levels of beach wrack at Pacific Grove.

ACKNOWLEDGMENTS

I thank my colleague Neal Evenhuis (Bishop Museum, Hawaii) for reviewing this chapter and for suggesting additional taxa to be treated.

REFERENCES

(Additional overall references may be located in Cole and Schlinger 1969).

Alexander, C. P. 1967. The crane flies of California. Bull. Calif. Ins. Survey, 8: 1–269.

Arnaud, P. H. 1958. A synopsis of the genus Melanderia Aldrich (Diptera: Dolichopodidae). Proc. Ent. Soc. Wash. 60: 179–186.

Axtell, R. C. 1976. Coastal horse flies and deer flies. In Marine insects. L. Cheng, ed. Chapt. 15, pp. 415–436. Amsterdam: North Holland Publishing Co.

Burger, J. F., J. R. Anderson, and M. F. Knudsen. 1980. The habits and life history of *Oedoparena glauca* (Diptera: Dryomyzidae), a predator of barnacles. Proc. Ent. Soc. Wash. 82: 360–377.

Byers, G. W. 1996. Tipulidae. In An introduction to the aquatic insects of North America. 3rd ed. R. W. Merritt and K. W. Cummins eds., Chap. 23. pp. 549–570. Dubuque, IA: Kendall/Hunt Publ.

Cheesemen, D. T., and P. Preissler. 1972. Larval distribution of *Paraclunio alaskensis* at Point Pinos sewage outfall, Monterey County, California. Pan-Pacific Ent. 48: 204–207.

Cheng, L. 1976. Marine insects. Amsterdam: North-Holland Publishing Co. 581 pp.

Coffman, W. P., and L. C. Ferrington, Jr. 1996. Chironomidae. In An introduction to the aquatic insects of North America. R. W. Merritt and K. W. Cummins, eds. Chapt. 26, pp. 635–754 Dubuque, IA: Kendall/Hunt Publ.

Cole, F. R., and E. I. Schlinger. 1969. The flies of Western North America. Berkeley: University of California Press, 693 pp.

Courtney, G. W., R. W. Merritt, H. J. Teskey, B. A. Foote. 1996. Larvae of aquatic Diptera. In An introduction to the aquatic insects of North America. 3rd ed. R. W. Merritt and K. W. Cummins, eds. Chapt. 22, pt. 1, pp. 484–514, Dubuque, IA: Kendall/Hunt Publ.

Dobson, T. 1976. Seaweed flies (Diptera: Coelopidae, etc.), In Marine insects. L. Cheng, ed. Chapt. 16, pp. 447–462. Amsterdam: North Holland Publishing Co.

Evans, W. G. 1980. Insecta, Chilopoda, and Arachnida: insects and allies. In Intertidal invertebrates of California. R. H. Morris, D. P. Abbott, and E. C. Haderlie, eds. Stanford, CA: Stanford Univ. Press.

Foster, W. A., and J. E. Treherne. 1976. Insects of saltmarshes: problems and adaptations. In Marine insects. L. Cheng, ed. Chapt. 2, pp. 5–35. Amsterdam: North Holland Publishing Co.

Hashimoto, H. 1976. Non-biting midges of marine habitats (Diptera: Chironomidae). In Marine insects. L. Cheng, ed. Chapt. 14, pp. 377–412. Amsterdam: North Holland Publishing Co.

Hinton, H. E. 1967. Spiracular gills in the marine fly *Aphrosylus* and their relation to the respiratory horns of other Dolichopodidae. J. Mar. Biol. Assoc. U.K., 47: 485–497.

Hinton, H. E. 1976. Respiratory adaptations of marine insects. In *Marine Insects*. L. Cheng, ed. Chapt. 3, pp. 43–76. Amsterdam: North Holland Publishing Co.

Huckett, H. C. 1971. The Anthomyiidae of California, exclusive of the subfamily Scatophaginae (Diptera). Bull. Calif. Ins. Survey 12: 1–121.

Irwin, M. E., and L. Lyneborg. 1981. The genera of Nearctic Therevidae. Ill. Nat. Hist. Survey Bulletin 32: 194–277.

Knudsen, M. 1968. The biology and life history of *Oedoparena glauca* (Diptera: Dryomyzidae), a predator of barnacles. Master's Thesis, Parasitology, University of California, Berkeley.

Kompfner, H. 1974. Larvae and pupae of some wrack dipterans on a California beach (Diptera: Coelopidae, Anthomyiidae, Sphaeroceridae). Pan-Pac. Entomol. 50: 44–52.

Lonely, J. R. 1976. Biting midges of mangrove swamps and saltmarshes (Diptera: Ceratopogonidae). In Marine insects. L. Cheng, ed. Chapt. 13, pp. 335–365. Amsterdam: North Holland Publishing Co.

McAlpine, J. F. ed. 1987. Manual of Nearctic Diptera: Volume 2. Biosystematics Research Institute Ottawa, Ontario. Research Branch Agriculture Canada. Monograph No. 28. pp. 675–1332.

McAlpine, J. F., et al. 1981. Manual of Nearctic Diptera: Volume 1. Biosystematics Research Institute Ottawa, Ontario. Research Branch Agriculture Canada. Monograph No. 27, 674 pp.

McAlpine, J. F., et. al. 1989. Manual of Nearctic Diptera: Volume 3. Biosystematics Research Institute Ottawa, Ontario. Research Branch Agriculture Canada. Monograph No. 32. pp.1333–1581.

Mathis, W. N. 1993. A revision of the shore-fly genera *Hostis* Cresson and *Paratissa* Coquillett (Diptera: Ephydridae). Proc. Ent. Soc. Wash. 95: 21–47.

Merritt, R. W., and K. W. Cummins, eds. 1984. An introduction to the aquatic insects of North America. 2nd Ed. Dubuque, IA: Kendall/Hunt Publishing Co., 722 pp.

Merritt, R. W., D. W. Webb, and E. I. Schlinger. 1984b Aquatic Diptera, Part Two: Adults of Aquatic Diptera. In An introduction to the aquatic insects of North America. 2nd ed. R. W Merritt and K. W. Cummins, eds. Chapt. 22, pp. 467–490. Dubuque, IA: Kendall/Hunt Publishing Co.

Merritt, R. W., D. W. Webb, and E. I. Schlinger. 1996. Pupae and Adults of Aquatic Diptera. In An introduction to the aquatic insects of North America. 3rd ed. R. W Merritt and K. W. Cummins, eds. Chapt. 22, pt. 2, pp. 515–548. Dubuque, IA: Kendall/Hunt Publishing Co.

Morley, R. L., and R. A. Ring. 1972. The intertidal Chironomidae (Diptera) of British Columbia I. Keys to their life stages; II. Life history and population dynamics. Can. Ent. 104: 1093–1121.

Mundie, J. H. 1957. The ecology of Chironomidae in storage reservoirs. Trans. Ent. Soc. London, 109: 149–232.

Neumann, D. 1976. Adaptations of chironomids to intertidal environments. Ann. Rev. Ent. 21: 387–414.

Nicolson, S. W. 1972. Osmoregulation in larvae of the New Zealand saltwater mosquito, *Opifex fuscus* Hutton (Diptera: Culicidae) J. Entomol. (A) 47: 101–108.

O'Meara, G.F. 1976. Saltmarsh mosquitoes (Diptera: Culicidae). In Marine insects. L. Cheng, ed. Chapt. 12, pp. 303–327. Amsterdam: North Holland Publishing Co.

Ring, R. A. 1989. Intertidal Chironomidae of B.C. Canada. Acta. Biol. Oecol. Hungary 3: 275–288.

Robles, C. 1982. Disturbance and predation in an assemblage of herbivorous Diptera and algae on rocky shores. Oecologia 54: 23–31.

Robles, C. D., and J. Cubit. 1981. Influence of biotic factors in an upper intertidal community: dipteran larvae grazing on algae. Ecology 62: 1536–1547.

Saunders, L. G. 1928. Some marine insects of the Pacific Coast of Canada. Ann. Ent. Soc. Amer. 21: 521–545.

Saunders, L. G. 1930. The early stages of *Geranomyia unicolor* Haliday, a marine tipulid. Entom. Mon. Mag. 66: 185–187.

Schlinger, E. I. 1975. Diptera. In Lights manual: intertidal invertebrates of the Central California Coast. 3rd ed. R. I. Smith and J. T. Carlton eds., pp. 436–446. University of California Press, Berkeley.

Simpson, K. W. 1976. Shore flies and brine flies (Diptera: Ephydridae). In Marine insects. L. Cheng, ed. Chapt. 17, pp. 465–492. Amsterdam: North Holland Publishing Co.

Smith, M. E. 1952. Immature stages of the marine fly *Hypocharassus pruinosus* Wh., with review of the biology of immature Dolichopodidae. Am. Midl. Nat. 48: 421–432.

Stone, A. et al. 1965. A Catalog of the Diptera of America north of Mexico. Washington, D.C.: Agricultural Handbook 276, USDA, 1969 pp.

Usinger, R. L. ed. 1956. Aquatic insects of California. Berkeley: University of California Press, 508 pp.

Wheeler, W. M. 1897. A genus of maritime Dolichopodidae new to America. Proc. Calif. Acad. Sci. 3: 145–152.

Williams, F. X. 1938. Biological Studies in Hawaiian water loving insects. Part III. Diptera or flies. A. Ephydridae and Anthomyiidae. Proc. Hawaii Ent. Soc. 10: 85–119.

Wirth, W. W. 1949. A revision of the Clunionine midges with descriptions of a new genus and four new species (Diptera: Tendipedidae). Univ. Calif. Publ. Entomol. 8: 151–182.

Wirth, W. W. 1951. A revision of the dipterous family Canaceidae. Occ. Pap. B. P. Bishop Mus. 20: 245–275.

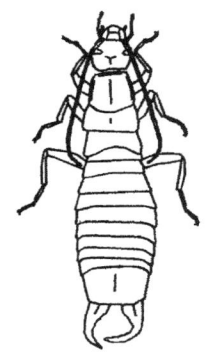

PLATE 347 *Anisolabis maritima* (drawn by H. V. Daly).

Dermaptera

HOWELL V. DALY

(Plate 347)

The maritime or seaside earwig, *Anisolabis maritima* (Bonelli, 1832) (family Carcinophoridae), is found on seashores throughout the world. Along our coast, the species occurs from British Columbia (Guppy 1950) to California (Langston 1974). The first records in the San Francisco Bay area were in 1935 (Langston and Powell 1975). Individuals of this large, shiny black or brown insect (plate 347) are usually found at the high-tide level. During the day, the earwigs hide in crevices, waiting until night to feed on other insects. The pincerlike cerci of the female are nearly straight, while those of the male are strongly bent.

REFERENCES

Guppy, R. 1950. Biology of *Anisolabis maritima* (Géné) the seaside earwig, on Vancouver Island (Dermaptera, Labiidae). Proceedings of the Entomological Society of British Columbia 46:14–18.

Langston, R. L. 1974. The maritime earwig in California (Dermaptera: Carcinophoridae). Pan-Pacific Entomologist 50: 28–34.

Langston, R. L., and J. A. Powell 1975. The earwigs of California (order Dermaptera). Bulletin of the California Insect Survey 20: 1–25.

Coleoptera

DAVID WHITE AND AMANDA NELSON

(Plate 348)

The Order Coleoptera contains more than 350,000 described species worldwide, with perhaps another 1,500 new species being described each year. More than 24,000 species have been described for North America alone. Although beetles are found in just about every terrestrial and freshwater aquatic habitat, relatively few species are associated with intertidal habitats and even fewer enter into fully marine environments (White and Brigham 1996).

The majority of the aquatic beetle families commonly found in freshwater habitats (e.g., Dytiscidae, Elmidae, Dryopidae, Hydrophilidae) have no intertidal representatives or are represented by only a few species (Doyen 1975, 1976). Most commonly found in marine and intertidal habitats are representatives of primarily substrate-dwelling and largely terrestrial families. These are families that also may occupy the littoral zones of lakes and streams (e.g., Carabidae, Staphylinidae). Included here are taxa that spend their life cycles associated with and are restricted to some aspect of the marine environment. Not included are those that periodically visit sandy beaches such as tiger beetles (Cicindelinae), a wide variety of other ground beetles (Carabidae), or those that may occasionally/accidentally enter brackish water (Haliplidae, Dytiscidae, and Gyrinidae).

Worldwide, at least 12 families and more than 90 genera have marine representatives, two-thirds of which are rove beetles (Staphylinidae). The majority of the described species exist along the cold and temperate intertidal zones of western North America and western Europe. Other species have been found along the southern coast of South America and islands of the Indian and Pacific Oceans (Jäch 1998). Species present along the Pacific coast of North America are often widely distributed and may be dispersed by shoreline currents. Undoubtedly, more species will be found along this and other coasts as investigations continue.

Coleoptera are unique among the holometabolus aquatic insects in having the larvae and adults occupying the same habitats and occasionally sharing similar food. Intertidal species often have larvae that mature quickly, with the adults being present the year around. Adults of some marine genera are wingless or flightless, including the carabid *Thalassotrechus*, the salpingid *Aegialites*, and the curculionid *Emphyastes*, while many of the staphylinids are active fliers. Night-active fliers may be attracted to streetlights, porch lights, and building windows, occasionally in large numbers. A seemingly common event for beetles is mass flights or swarms, often during the day and usually for no reason that we can easily determine. Evans (1980) observed such swarms that were flying parallel to the Monterey and Marin County beaches, which lasted an hour or more and contained several species, particularly staphylinids.

Beetles occupy three primary intertidal habitats—rock crevices, open beaches/mud flats, and decaying seaweed wrack—but some may be found under any object that provides shelter. Species living in rock crevices (e.g., *Thalassotrechus* and *Endeodes*) exist throughout the intertidal zone, persisting in trapped air pockets at high tide and crawling about at low tide (Doyen 1976). Species of open beaches and mud flats (e.g., a number of staphylinids) are often very active, moving with the tides. Beetles associated with seaweed wrack are more common at the high tidal zone, feeding on the decaying algae and associated crustaceans or fly eggs and larvae. Intertidal beetles in the families Carabidae, Staphylinidae, Histeridae, and Hydrophilidae are largely predators, while species in the remaining families feed primarily on decaying algae, fungus, and organic matter.

Reviews of the intertidal and marine Coleoptera of the North American Pacific shoreline include Leech and Chandler (1956), Doyen (1975, 1976), Moore and Legner (1976—Staphylinidae), and Evans (1980). Jäch (1998) provides an annotated checklist of riparian and littoral beetles of the world. Community level studies of California beaches have been summarized by Lavoie (1985) and Dugan et al. (2000, 2003). Useful general references on aquatic Coleoptera include Young (1954), Brown (1972), McCafferty (1981) Brigham (1982), White (1982), and White and Brigham (1996). The two-volume set, *American Beetles*, is an indispensable reference to all the coleopteran families and genera and was widely used in revising this section (Arnett and Thomas 2001; Arnett et al. 2002).

KEY TO FAMILIES AND GENERA OF ADULT INTERTIDAL COLEOPTERA

Here we provide a key to the genera of adults likely to be encountered in the intertidal zone. References to species keys for adults and larvae (where known) can be found in Arnett and Thomas (2001) and Arnett et al. (2002). Although it is rather dated, Boving and Craighead (1930) is useful for Coleoptera larvae, particularly to the family level. With minor exceptions, there are no species level keys to larval Coleoptera, and generic level keys are often problematic. Keys to the Staphylinidae genera follow Moore and Legner (1976). Some of the more common intertidal adult staphylinids can be keyed to species using Doyen (1975), but Newton et al. (2001) should be consulted if there are questions.

KEY TO ADULT INTERTIDAL COLEOPTERA

1. Mesothoracic wings absent, antennae with three or fewer segments, tarsi with a single segment (plate 348B2, 348G2) . . . . Larvae (not keyed)
— Mesothoracic wings present, often covering entire abdomen (plate 348G1) or sometimes just the base (plate 348D), antennae with at least four segments, tarsi with at least three segments (plate 348A) . . . . Adults 2
2. Head usually produced anteriorly as a beak-like rostrum . . . . Curculionidae 3
— Head not produced as a rostrum 4
3. Front tibia prolonged beyond the articulation of tarsus into long, flattened paddle; hind tibia markedly expanded at apex, wider than maximum width hind femur . . . . *Emphyastes*
— Front tibia not prolonged beyond the articulation of tarsus into long, flattened paddle; hind tibia not markedly expanded at apex, not wider than maximum width hind femur . . . . *Thalasselephas*
4. Elytra truncate, exposing two or more abdominal terga (plate 348B1, 348C, 348E, 348F) 5
— Elytra entire, usually exposing only one abdominal tergite 7
5. Elytra exposing two abdominal segments, antennae geniculate (strongly bent), head much narrower than pronotum, all tibia usually dilated . . . . Histeridae *Neopachylopus*
— Elytra exposing more than two abdominal terga 6
6. Prothorax and abdomen with yellow or orange protusible vesicles (plate 348D) . . . . Melyridae *Endeodes*
— Prothorax and abdomen without protusible vesicles (plate 348C) . . . . Staphylinidae 18
7. First abdominal sternite completely divided by hind coxal cavities . . . . Carabidae 8
— First visible abdominal segment transversely complete 11
8. Middle coxal cavities not entirely closed by sterna . . . . *Dyschiriodes*
— Middle coxal cavities entirely closed by sterna 9
9. Elytral margin without internal plica toward apex; antennomeres 1–2 with only a ring of long setae toward apex . . . . *Anisodactylus*
— Elytral margin with internal plica toward apex, antennomeres 2–3 with ring of long setae towards apex only 10
10. Head with frontal grooves broad, shallow, not extended to plane of posterior margin of eyes; dorsal surface of hind tarsomeres each with a single pair of setae . . . . . *Thalassotrechus*
— Head with frontal grooves narrow, deep, not extended to plane of posterior margin of eyes; dorsal surface of hind tarsomeres each with numerous setae . . . . *Trechus*
11. Antennae terminating in an abrupt, globular or elongate club (plate 348A) 12

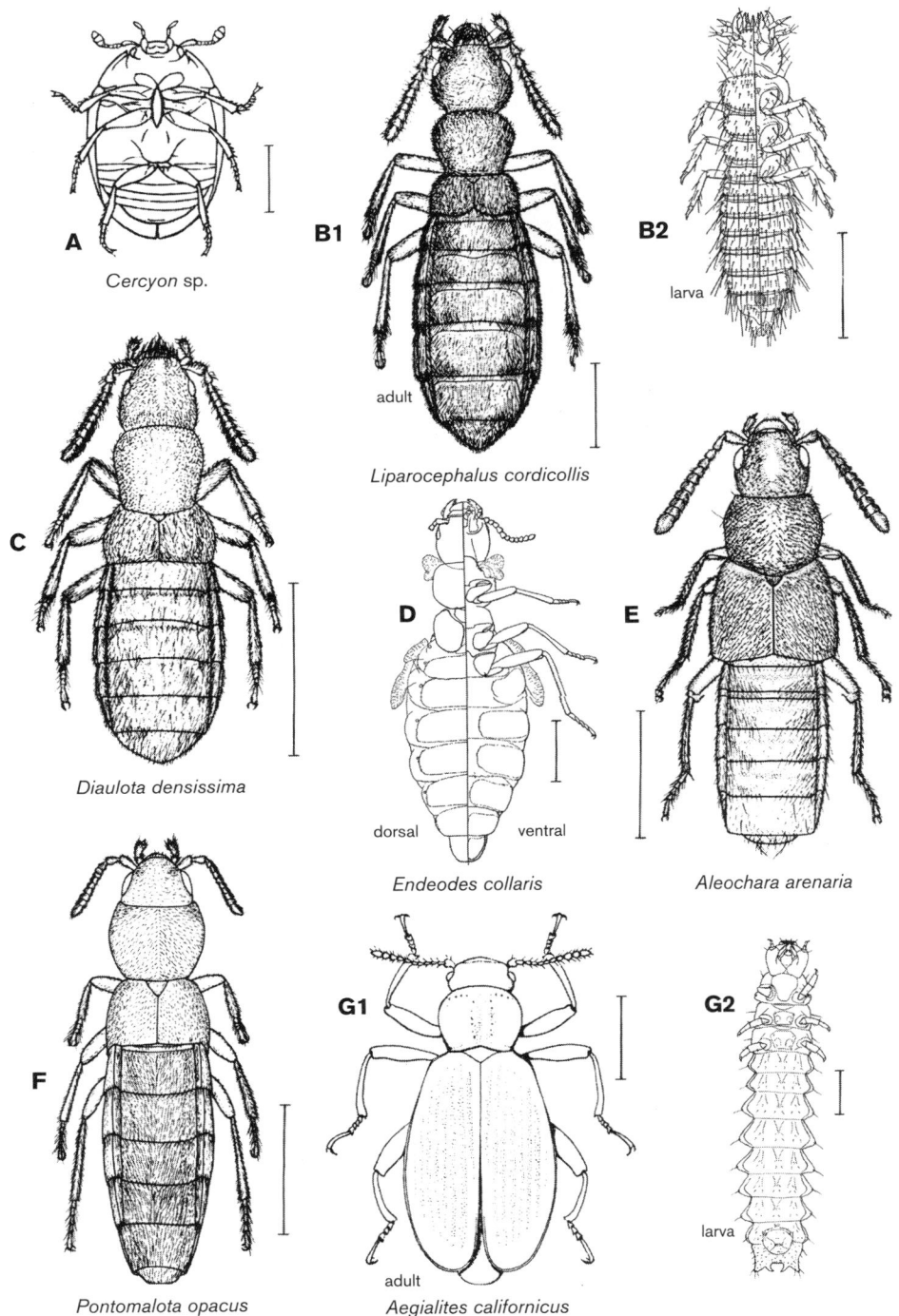

PLATE 348 Coleoptera: Scale bars = 1 mm. A, *Cercyon* sp., ventral; B1, *Liparocephalus cordicollis*, adult; B2, larva, dorsal surface on left, ventral on right; C, *Diaulota densissima*, adult, dorsal; D, *Endeodes collaris*, adult, dorsal surface on left, ventral on right; E, *Aleochara arenaria*, adult, dorsal; F, *Pontomalota opacus*, adult, dorsal; G1, *Aegialites californicus*, adult, dorsal; G2, larva, ventral (A, Mulsant 1844; B1, E, F (latter as *P. luctuosa*) Hatch 1957; B2, Chamberlin and Ferris 1929; C, Hatch 1947; D, Blackwelder 1932; G1, Spilman 1967; G2, Wickham 1904; note figures F and G are Pacific Northwest species).

— Antennae slender or very short and thick with basal segment enlarged 14
12. Antennal club with two segments, prominent frontal ridge . . . . Rhizophagidae *Phyconomus*
— Antennal club with three or more segments 13
13. Abdomen with six to seven visible sternites, antennal club with five segments . . . . Hydraenidae *Neochthebius*
— Abdomen with five visible sternites, antennal club with three segments (plate 348A). . . . . . Hydrophilidae *Cercyon*
14. Tarsal formula 5-5-5 . . . . Limnichidae *Throscinus*
— Tarsal formula 5-5-4 or less. 15
15. Tarsal formula 5-5-4 16

— Tarsal formula <5-5-4 . . . . Ptiliidae *Motschulskium*
16. Hind coxae usually separated by more than coxal width, basal two abdominal segments fused . . . .
. . . . Salpingidae *Aegialites*
— Hind coxae usually separated by much less than coxal width, eyes notched anteriorly . . . .
. . . . Tenebrionidae 17
17. Anterior tibiae flat, spinose, adapted for digging . . . .
. . . . *Phaleria*
— Anterior tibiae slender . . . . *Epantius*
18. Antennae extremely slender and filamentous . . . .
. . . . *Giulianium*
— Antennae not extremely slender and filamentous 19
19. Antennal fossae located on surface of the head between the anterior margins of the eyes 21
— Antennal fossae located at the front or side margins of the head 20
20. Second sternite complete . . . . *Bledius*
— Second sternite absent or rudimentary 22
21. Superior and inferior lateral lines or pronotum widely separated at anterior angles 33
— Superior and inferior lateral lines or pronotum united at anterior angle 34
22. Outer lobe of maxilla entirely corneous, without pubescence on inner side . . . . *Bryothinusa*
— Outer lobe of maxilla partly membranous, with long pubescence on inner side 23
23. Posterior tarsus four-segmented . . . . *Diglotta*
— Posterior tarsus four-segmented 24
24. Middle tarsus four-segmented 26
— Middle tarsus five-segmented 25
25. Anterior tarsus four-segmented 31
— Anterior tarsus five-segmented (plate 348E) . . . . . *Aleochara*
26. Anterior and middle tibiae spinose . . . . *Thinusa*
— Anterior and middle tibiae not spinose 27
27. Tergites not impressed at base (plate 348B1) . . . .
. . . . *Liparocephalus*
— At least three tergites impressed at base 28
28. Fifth tergite impressed at base . . . . *Amblopusa*
— Fifth tergite not impressed at base 29
29. Fourth tergite is impressed at base . . . . *Bryobiota*
— Fourth tergite is not impressed at base (plate 348C) 30
30. Mentum triangular, apex with a very deep and acute V-shaped emargination; abdomen parallel-sided, not distinctly broader at segments V–VI than at base . . . . . *Paramblopusa*
— Mentum trapezoidal, apex with a very broad and deep U-shaped emargination; abdomen oval, distinctly broader at segments V–VI than at base . . . . *Diaulota*
31. Pubescence of disc of pronotum longitudinal (plate 348F)
. . . . *Pontomalata*
— Pubescence of disc of pronotum transverse 32
32. Middle coxal cavities margined by a carina . . . . . *Halobrecta*
— Middle coxal cavities not margined . . . . *Tarphiota*
33. Elytral suture overlapping, yellow and brown variegated
. . . . *Thinopinus*
— Elytral suture not overlapping, dark brown . . . . . *Hadrotes*
34. Superior lateral line of prothorax deflexed in front so that the large lateral setigerous puncture is removed from it by at least three times the width of the puncture . . . .
. . . . *Cafius*
— Superior lateral line of prothorax is not deflexed in front so that the large lateral setigerous puncture is on it or separated from it by at most by the width of the puncture . .
. . . . *Philonthus*

LIST OF SPECIES

The following list contains 12 families, 33 genera, and 56 species. We have been conservative in including only species found along the California to Washington coast and where we could verify the species in more than one literature source. Undoubtedly, we may have missed some synonymies and overlooked potential species, especially in the Staphylinidae. Any decaying organic matter (e.g., beach wrack) can attract a great variety of insects that simply may be incidental to the habitat. This possibly is the case for a number of beetles noted by Lavoie (1985) that we have not included in the key or annotations. We simply list these here: Anabeidae species; Anthicidae (*Amblyderus obesus* Casey); Curculionidae (*Curculio* sp.); Histeridae (*Hypocaccus (Baeckmenniolus) gandens* Leconte, *Hypocaccus bigemminus* LeConte, *Euspilotus scissus* LeConte); Limnebiidae (*Octhebius rectua* LeConte); Mordellidae (*Mordella marginata* Melsheimer); Nitidulidae (*Carpophilus* sp.); Staphylinidae (*Omalium algarum* Casey, *Proteinus* sp.); Tenebrionidae (*Eleodes* sp.).

ADEPHAGA

CARABIDAE (GROUND BEETLES)

This is an extremely large family with more than 40,000 species described worldwide. Given the number of species associated with freshwater littoral zones, marshes, and swamps, it is surprising that more are not adapted to intertidal zones. Many carabids are predators or omnivores feeding on dead and decaying organic matter, which may also be true for the intertidal species.

Thalassotrechus barbarae (Horn, 1892). Rarely seen; in deep rock cracks. Evans (1976, 1977, 1980) has reviewed the behavior, circadian and circatidal rhythms, and variation and distribution. Adults emerge from crevices at dusk and retreat at dawn or during a full moon or when the tide floods. Adults are active all year, and all life-cycle stages can be found year-round.

Thalassotrechus nigripennis Van Dyke, 1918. This is most likely a regional variey of *T. barbarae* (Leech and Chandler 1956).

Dyschiriodes marinus (LeConte, 1852) (=*Dyschirius marinus*, 1852). In seaweed piles on sandy beaches. Evans (1980) notes that this sand-burrower is usually associated with *Bledius monstratus*. Both larvae and adults are predators on insects and crustaceans associated with decaying kelp (Evans 1980).

Trechus ovipennis Motschulsky, 1845. Uncommon, found in crevices and under loose stones. Adults are nocturnal and have predator/scavenger tendencies (Evans 1980).

Anisodactylus californicus (LeConte, 1857) (=*Stenocellus californicus* DeJong). Found in salt marshes.

POLYPHAGA

HISTERIDAE (CLOWN BEETLES)

Of the approximately 3,900 species and 330 genera found worldwide, most are small, black, shiny predators on insect eggs and larvae associated with decaying plant and fecal matter. The genus *Hypocaccus* occurs on sand bars both in marine and river environments worldwide with at least two species along the beaches of California (Lavoie 1985). Only the monotypic genus *Neopachylopus* is commonly found in the intertidal zone.

Neopachylopus sulcifrons (Mannerheim, 1843). Carniverous on insects, associated with decaying matter, feeds on collembolla associated with decaying kelp (Evans 1980).

HYDRAENIDAE (MINUTE MOSS BEETLES)

Hydraenidae is a small family (about 1,200 species world wide) of generally tiny beetles that live in mosses, organic matter, and gravel usually at the margins of freshwater habitats. A single species of hydraenid has been found in the intertidal zone.

Neochthebius vandykei (Knisch, 1924) (=*Ochthebius vandykei*). Common in crevices and the high-intertidal zone on rocky shores. Adults occur in large numbers and are detritus feeders, while larvae are carnivores and may eat other hydraenid larvae (Evans 1980).

HYDROPHILIDAE (WATER SCAVENGER BEETLES)

There are more than 160 genera and 2,800 species worldwide. Freshwater hydrophilids may be very diverse and abundant, particularly in small ponds. The common name "water scavenger beetles" is a bit of a misnomer because most of the larvae are predators, and many are terrestrial, including the subfamily Sphaeridiinae to which the two species *Cercyon* belong.

Cercyon fimbriatum Mannerheim, 1852. High-intertidal zones, on rocky shores. Adults feed on saprophytic fungi and bacteria, while larvae eat soft-skinned, slow-moving animals, such as annelids (Evans 1980).

Cercyon lunigerum Mannerheim, 1853. Has similar habits to *C. fimbriatum*. Evans (1980) notes that adults are frequently observed swarming in large numbers with other beetles, such as *Cafius luteipennis*, of the "decaying kelp community."

STAPHYLINIDAE (ROVE BEETLES)

Rove beetles comprise the second largest family of Coleoptera with more than 46,000 species in more than 3,200 genera. Adults are recognized by their greatly abbreviated elytra and generally soft bodies. Although adults of most species are active fliers, moving away from tides and waves, others are flightless, such as *Thinopinus*. Feeding habits include detritivores, herbivores, and predators. More than 60 genera are known from intertidal zones worldwide, of which at least 17 occur along the Pacific coast. A number of genera are endemic to the Pacific coast from Baja California north to Alaska, e.g., *Bryothinusa, Diaulota, Liparocephalus, Hadrotes, Pontomalota, Tarphiota, Thinopinus,* and *Thinusa*. Species of *Aleochara, Cafius,* and *Hadrotes* feed on the eggs and larvae of flies that are laid in decaying algae and other vegetation. *Bledius* feeds on algae. *Thinopinus* is nocturnal and feeds on small arthropods. *Philonthus* feeds on decaying organic matter. Feeding habits are less well known for *Amblopusa, Bryobiota, Bryothinusa, Diglotta, Diaulota, Liparocephalus, Pontomalota, Tarphiota,* and *Thinusa*, but probably most are predators of fly eggs and small insects and crustaceans.

Aleochara sulcicollis Mannerheim, 1843. In decaying vegetation.

Aleochara arenaria (Casey, 1893) (=*Emplenota arenaria* Maklin 1853). In decaying vegetation.

Amblopusa brevipes Casey, 1893.

Bledius fenyesi Bernhauer and Schubert, 1911.

Bledius monstratus Casey, 1889. Relatively uncommon, in burrows in wrack-line sand. Evans (1980) notes that this beetle is often associated in sand burrows with the carabid *Dyschiriodes marinus*.

Bryobiota bicolor (Casey, 1885). On or under decaying kelp (Evans 1980).

Bryobiota giulianii (Moore, 1978).

Bryothinusa catalinae Casey, 1904. Common on and under rocks and boulders, on sand along intertidal reefs (Moore and Orth 1979).

Cafius canescens (Maklin, 1852). In decaying vegetation.

Cafius decipiens (Horn, 1863). On or under decaying kelp in the high- to middle- intertidal zones on sandy beach (Evans, 1980).

Cafius luteipennis Horn, 1884. Evans (1980) notes that this species is "found under freshly cast-up kelp in company with fly maggots" or as a member of the "decomposing kelp community." *Cafius* adults and larvae prey on fly (*Fucellia*) larvae and pupae, amphipods, and other *Cafius* species (Evans 1980).

Cafius lithocharinus (LeConte, 1863). Common around decaying kelp in the high- to middle- intertidal zones on sandy beach (Evans 1980).

Cafius opacus (LeConte, 1863). Lives among decaying vegetation, especially kelp.

Cafius seminitens Horn, 1884. This voracious predator is common on decomposing kelp in high-intertidal zone on sandy beaches (Evans 1980).

Cafius sulcicollis (LeConte, 1863). Adults and larvae are predators among decaying kelp.

Cafius nauticus Fairmaire, 1849. Found among decomposing kelp.

Diaulota densissima Casey, 1893. In crevices, among algae and barnacles, high and mid intertidal. Glynn (1965) found it abundant in high-intertidal *Endocladia* beds. They also may occur in air pockets below barnacles (Jones 1968, Evans 1980).

Diaulota fulviventris Moore, 1956. Found in protected crevices or among algae and barnacles, high- and middle-intertidal zones on rocky shores (Evans 1980).

Diaulota vandykei (Moore, 1956). Often with *Diaulota densissima* and *Liparocephalus cordicollis*; larvae found under dense growths of the green alga *Ulva* (Jones 1968, Evans 1980).

Diaulota harteri Moore, 1956. Lives among crevices on rocky shores.

Diglotta pacifica Fenyes, 1921.

Diglotta legneri Moore and Orth, 1979.

Giulianium campbelli Moore, 1976. Found under stones and boards below the high-tide mark (Ahn and Ashe 1999).

Giulianium newtoni Ahn and Ashe, 1999. Found below the high-tide mark, under debris (Ahn and Ashe 1999).

Hadrotes crassus (Mannerheim, 1846). In decaying vegetation. Nocturnal predators of crustaceans and insects. Moore (1964) has described the larvae.

Halobrecta algophila (Fenyes, 1909) (=*Atheta algophila* Fenyes, 1920). A marine littoral species (Gusarov 2004).

Liparocephalus brevipennis Mäklin, 1853.

Liparocephalus cordicollis LeConte, 1880. Larvae, pupae, and adults all occur in the mid-intertidal; pupae enclosed in silken cocoons in rock crevices. Evans (1980) notes that at Pacific Grove the larvae are found in the holdfasts of the feather-boat kelp *Egregia*, while the adults may be under the red alga *Mazzaella flaccida* on rock surfaces (see also Jones 1968).

Paramblopusa borealis (Casey, 1906) (=*Amblopusa borealis*).

Philonthus nudus Sharp, 1874.

Pontomalota opacus (LeConte, 1863). High- and middle-intertidal zones on sandy beaches. Adults and larvae are predators on dead crustaceans, such as amphipods (Evans 1980). Synonyms for this species include *Phytosus opaca* LeConte, 1863, *Pontomalota californica* Casey, 1885, *Pontomalota nigriceps* Casey, 1885, *Pontomalota luctuosa* Casey, 1911, and *Pontomalota bakeri* Bernhauer, 1912 (see Ahn and Ashe 1992).

Pontomalota terminalia Ahn and Ashe, 1992.

Tarphiota geniculata (Mäklin, 1852).

Tarphiota pallidipes (Casey, 1893).

Thinopinus pictus LeConte, 1852. A nocturnal species feeding on small arthropods. Burrows in wet sand. A voracious sit-and-wait predator consuming talitrid amphipods, beach isopods, and other prey. Its behavior and ecology have been examined by Craig (1970) and Richards (1982, 1984). Also see Evans (1980).

Thinusa maritima (Casey, 1885).

Thinusa fletcheri Casey, 1906.

LIMNICHIDAE (MINUTE MARSH-LOVING BEETLES)

This small family of about 250 species worldwide is usually associated with vegetation and debris along streams (Brown 1972). The adults of *Throscinus* can be found on intertidal mudflats. The larvae are burrowers in moist soils and probably feed on algae and perhaps decaying organic matter.

Throscinus crotchii LeConte, 1874. Intertidal mudflats (Leech and Chandler 1956).

MELYRIDAE (SOFT-WINGED FLOWER BEETLES)

This large (6,000+ species and 300+ genera) primarily terrestrial family feeds on pollen and other plant materials, but some species are known to feed on decaying organic matter while a few other are predacious. Only one intertidal genus, *Endeodes*, is present along the Pacific coast. Of the eight species in the genus, three are known from Baja California, and the remaining five given here are endemic to the California coast.

Endeodes basalis (LeConte, 1852). Species of *Endeodes* are found in rock crevices and under high intertidal and supratidal debris and driftwood.

Endeodes blaisdelli Moore, 1954. Found under debris on beaches and on rock surfaces (Evans 1980).

Endeodes collaris (LeConte, 1852). Larvae in crevices above high-tide line; adults on rock surfaces at low tide; see Evans (1980).

Endeodes rugiceps Blackwelder, 1932. Found on rock surfaces and under beach debris (Evans 1980).

Endeodes insularis Blackwelder, 1932. Common under beach debris (Evans 1980).

PTILIIDAE (FEATHERWING BEETLES)

Ptillids represent a small family of about 70 genera and 550 species worldwide. Most species are terrestrial and associated with moist organic debris, feeding on fungus. The monotypic genus *Motschulskium* is known only from the West Coast.

Motschulskium sinuatocolle Matthews, 1872. Found on wrack or under decaying kelp, high to middle intertidal zones on sandy beach (Evans 1980).

SALPINGIDAE (NARROW-WAISTED BARK BEETLES)

Salpingidae is a poorly known terrestrial family of about 300 species worldwide. The family common name may not be too appropriate as many are not associated with bark. The genus

Aegialites is present in rock crevices in the intertidal zone and may feed on organic matter and aquatic mites.

Aegialites fuchsii Horn,1892. Found in rock crevices.

Aegialites subopacus (Van Dyke, 1918). Active throughout year in rock crevices of mid-intertidal; larvae and pupae occur during summer months.

Aegialites californicus (Motschulsky, 1845). The species name is based upon a mistake; this beetle occurs from Oregon north to Alaska.

TENEBRIONIDAE (DARKLING BEETLES)

About 19,000 species and 2,000 genera worldwide, darkling beetles are primarily terrestrial feeding on organic matter in a wide variety of habitats. *Phaleria rotundata*, "the kelp beetle," and *Epantius obscurus* are common inhabitants of the intertidal zone.

Phaleria rotundata LeConte, 1851. A common component of the kelp wrack community from Baja California to central California. The larvae and adults feed on decomposing kelp. Under dried kelp by day, high-intertidal zone on sandy beaches (Evans 1980).

Epantius obscurus (LeConte, 1851) (=*Eulabis obscurus*). Under wrack in the high-intertidal zone. These slow-moving scavengers occasionally are found above high-water mark on sandy beaches (Evans 1980).

RHIZOPHAGIDAE (ROOT-EATING BEETLES)

This small primarily terrestrial family (about 220 species worldwide) lives in decaying organic matter, but some species live in ant nests. The primary food resource is fungus. The monotypic *Phyconomus marinus* is known only from California coasts.

Phyconomus marinus LeConte. Under algae and driftwood in the high intertidal.

CURCULIONIDAE (WEEVILS)

With about 60,000 species worldwide and 3,000 species in North America, weevils feed on living and decaying plant material, and several are associated with aquatic plants. Many species have been imported for use in aquatic weed control. Two species, *Emphyastes fucicola* and *Phycocoetes testaceus* are locally abundant along the California to Alaska coast.

Emphyastes fucicola Mannerheim, 1852. Burrowing under decaying vegetation. Adults burrow up to 30 cm in sandy beaches at Pacific Grove (Evans 1980). Larvae eat decaying kelp below the surface of the water.

Thalasselephas testaceus (LeConte, 1876). (=*Phycocoetes testaceus*) Found in the high intertidal zone on sandy beaches and rocky shores. Adults are detritus feeders, are found in groups, and occasionally wash up on shore (Evans 1980).

REFERENCES

Ahn, K. J., and J. S. Ashe. 1992. Revision of the intertidal aleocharine genus *Pontomalota* Casey (Coleoptera: Staphylinidae) with a discussion of its phylogenetic relationships. Entomol. Scand. 23: 347–359

Ahn, K. J., and J. S. Ashe. 1999. Two new species of *Giulianium* Moore from the Pacific Coasts of Alaska and California (Coleoptera: Staphylinidae: Omaliinae). Pan–Pacific Entomologist. 75(3): 159–164.

Arnett, R. H. 1960. The beetles of the United States. Catholic University of America Press, Washington, D.C., 1048 pp.

Arnett, R. H., and M. C. Thomas, eds. 2001. American beetles, Volume 1, Archostemata, Myxophaga, Adephaga, Polyphaga: Stapyliniformia. CRC Press, 443 pp.

Arnett, R. H., M. C. Thomas, P. E. Skelley, and J. H. Frank, eds. 2002. American beetles, Volume 2, Polyphaga: Scarabaeoidea through Curculionoidea. CRC Press, 861 pp.

Boving, A. G., and F. C. Craighead. 1930. An illustrated synopsis of the principal larval forms of the order Coleoptera. J. Ent. Soc. Amer. 11: 1–351.

Brigham, W. U. 1982. Aquatic Coleoptera. In Aquatic insects and oliochaetes of North and South Carolina. A. R. Brigham, W. U. Brigham, and A. Gnilka, eds. Mahomet, Ill.: Midwest Aquatic Enterprises, 837 pp.

Brown, H. P. 1972. Aquatic dryopid beetles (Coleoptera) of the United States. Biota of freshwater ecosystems identification manual no. 6. Wat. Poll Conf. Res. Ser., EPA, Washington, D.C., 82 pp.

Craig, P. C. 1970. The behavior and distribution of the intertidal sand beetle, Thinopinus pictus (Coleoptera: Staphylinidae). Ecology 51: 1012–1017.

Doyen, J. T. 1975. Intertidal insects: Order Coleoptera. In Light's manual of intertidal invertebrates of the Central California Coast. 3rd ed. R. I. Smith and J. T. Carlton, eds. University of California Press, Berkeley, 721 pp.

Doyen, J. T. 1976. Marine beetles (Coleoptera excluding Staphylinidae). In Marine Insects. L. Chang, ed. Amsterdam: North Holland Publ. Co., 581. pp.

Dugan, J. E., D. M. Hubbard, D. L. Martin, J. M. Engle, D. M. Richards, G. E. Davis, K. D. Lafferty, and R. F. Ambrose. 2000. Macrofauna communities of exposed sandy beaches on the southern California mainland and Channel Islands. Fifth California Islands Symposium, OCS Study, MMS 99-0038: 339–346.

Dugan, J. E., D. M. Hubbard, M. D. McCrary, and M. O. Pierson. 2003. The response of macrofauna communities on exposed sandy beaches of southern California. Estuarine Coastal Shelf Sci. 58S: 133–148.

Evans, W. G. 1976. Circadian and circatidal locomotory rhythms in the intertidal beetle Thalassotrechus barbarae (Horn): Carabidae. J. Exper. Mar. Biol. Ecol. 22: 79–90.

Evans, W. G. 1977. Geographic variation, distribution and taxonomic status of the intertidal insect Thalassotrechus barbarae (Horn) (Coleoptera: Carabidae). Quaest. Entomol. 13: 83–90.

Evans, W. G. 1980. Insecta, Chilopoda, and Arachnida: insects and allies. In Intertidal invertebrates of California. R. H. Morris, D. P. Abbott, and E. C. Haderlie, eds. Stanford, CA: Stanford Univ. Press.

Glynn, P. W. 1965. Community composition, structure, and interrelationships in the marine intertidal Endocladia muricata—Balanus glandula association in Monterey Bay, California. Beaufortia 12, 198 pp.

Gusarov, V. I. 2004. A revision of the Nearctic species of the genus Halobrecta Thomson, 1858 (Coleoptera: Staphylinidae: Aleocharinae) with notes on some Palaearctic species of the genus. Zootaxa 746: 1–25.

Jäch, M. A. 1998. Annotated check list of aquatic and riparian/littoral beetle families of the world (Coleoptera). In M. A. Jäch and L. Ji, eds. Water Beetles of China 2: 25–42.

Jones, T. W. 1968. The zonal distribution of three species of Staphylinidae in the rocky intertidal zone of California. Pan-Pac. Entomol. 44: 203–210.

Lavoie, D. 1985. Population dynamics and ecology of beach wrack macroinvertebrates of the central California coast. Bulletin of the Southern California Academy of Sciences 84: 1–22.

Leech, H. B., and H. P. Chandler. 1956. Aquatic Coleoptera. In Aquatic insects of California. R. L. Usinger, ed. University of California Press, Berkeley, 508 pp.

McCafferty, W. P. 1981. Aquatic entomology. Science Books International, Boston, 448 pp.

Moore, I. 1964. The larva of Hadrotes crassus (Mannerheim) (Coleoptera: Staphylinidae). Trans. San Diego. Soc. Natural Hist. 13: 309–312.

Moore, I., and E. F. Legner. 1976. Intertidal rove beetles (Coleoptera: Staphylinidae). In Marine insects. L. Chang, ed. North Holland Publ. Co., Amsterdam. 581, pp.

Newton, A. E., M. K. Thayer, J. S. Ashe, and D. S. Chandler. 2001. Staphylinidae. In American beetles, Volume 1, Archostemata, Myxophaga, Adephaga, Polyphaga: Stapyliniformia. R. H. Arnett and M. C. Thomas, eds. CRC Press, 443 pp.

Richards, L. J. 1982. Prey selection by an intertidal beetle: field test of an optimal diet model. Oecologia 55: 325–332.

Richards, L. J. 1984. Field studies of foraging behaviour of an intertidal beetle. Ecological Entomology 9: 189–194.

White, D. S. 1982. Elmidae. In Aquatic insects and oligochaetes of North and South Carolina. A. R. Brigham, W. U. Brigham, and A. Gnilka, eds. Midwest Aquatic Enterprises, Mahomet, Ill, 837 pp.

White, D. S., and W. U. Brigham. 1996. Aquatic Coleoptera. In An introduction to the aquatic insects of North America. R. W. Merritt and K. W. Cummins, eds. Kendall/Hunt Publ. Co, 862 pp.

Young, F. N. 1954. The water beetles of Florida. University of Florida Press, Gainesville.

Chilopoda

RICHARD HOFFMAN AND JAMES T. CARLTON

A rewarding enterprise awaits the student of maritime centipedes, whose diversity, distribution, biology, and ecology along most shorelines of the world, including the Pacific coast, remain essentially unknown.

Of the four arthropod classes traditionally denoted by the name Myriapoda, only species of the class Chilopoda are known to be facultative or obligate halophiles. Chilopods (the only arthropods in which the first pair of legs is modified into poison fangs) are dispersed among five orders (Scutigeromorpha, Lithobiomorpha, Scolopendromorpha, Craterostigmomorpha, and Geophilomorpha). Of these, only species of the last order are adapted to littoral habitats and are able to endure prolonged submersion in the intertidal zone, although there is every indication that many scolopendromorphs tolerate at least some exposure to seawater.

Geophilomorphs are typically very long and slender animals, usually yellowish in color, the body rather threadlike in appearance, ranging in size from about 5 mm to more than 200 mm, with from 33 to about 170 body segments. None of the known species has any trace of optic organs, and the vast majority of species are edaphobites or corticoles. All species are carnivorous, feeding on any small organisms they can overcome. Structure of the mouth parts is a primary basis for classification at both the generic and family level, and the minute size of many species imposes severe constraints on their identification. Even specialists have described new species in the wrong family owing to faulty observation of these tiny structures.

Most littoral geophilomorphs are members of the family Schendylidae, a primarily pantropical group. On the Pacific coast, Evans (1980) has noted that the presence of Nyctunguis heathii (Chamberlin, 1909) as being "fairly common in some areas in crevices during day or on surfaces of rocks at night during low tide," on high-intertidal rocky shores, noting its occurrence specifically at Monterey. It was first described from below the surface in an Indian shell mound, near Cypress Point, at Monterey (no exact locality was given by Chamberlin, although some shell mounds are on the cliff edges). Evans (1980, plate 189, fig. 28.1) presented a photograph of a 45-mm-long centipede identified as Nyctunguis heathii, collected at Monterey. Pearse et al. (1987, p. 568) also published a photograph of a 50-mm-long marine centipede, identified as Nyctunguis heathii; this is a photograph of a specimen probably collected on the rocky shore in Santa Cruz (John Pearse, personal communication, 2005). The identification of any schendylid should be approached with caution, however, and identify of Chamberlin's species with intertidal centipedes may bear re-examination.

The family Geophilidae is represented on the California shore by two or more species. Lionyx hedgpethi Chamberlin, 1960, was collected in the intertidal of Tomales Bay, and is described as a species of about 25 mm in length. A robust unidentified geophilid, not Lionyx hedgpethi, and about 50 mm in length, occurs along the shore of Bodega Harbor (collected in the summer of 1980 by Ralph I. Smith, under stones at the strandline, south of the yacht club). Other centipedes may occur in our region: Brusca and Brusca (1978) note that in the Humboldt County region of northern California, "We have, on numerous occasions, seen a small red centipede inhabiting the vertical rock faces in the spray zone, originally pointed out . . . by the late Fred Telonicher."

Further south, *Pectiniunguis* is represented in marine habitats in Baja California as well as the coast of South American and the West Indies.

The role of these or other species as intertidal predators remains unknown. *Strigamia maritima* of British shores preys on barnacles and periwinkles, and Evans (1980) speculated that *N. heathii* may prey on the intertidal insect and mite fauna. Ecophysiology of other species has been studied in the European fauna, among which *Strigamia maritima* and *Hydroschendyla submarina* are the best known, and summarized by Lewis (l981), the best source of information on Chilopoda.

References

Brusca, G. J., and R. C. Brusca. 1978. A naturalist's seashore guide. Common marine life of the northern California coast and adjacent shores. Eureka, CA: Mad River Press, 205 pp.

Chamberlin, R. V. 1960. A new marine centipede from the California littoral. Proceedings of the Biological Society of Washington 73: 99–102.

Evans, W. G. 1980. Insecta, Chilopoda, and Arachnida: Insects and Allies, pp. 641–658. In Intertidal invertebrates of California. R. H. Morris, D. P. Abbott, and E. C. Haderlie, eds. Stanford, CA: Stanford University Press, Stanford, California, 690 pp.

Hoffman, R. L. 1980. Chilopoda, pp. 681–688, S. P. Parker (ed.): Synopsis and classification of living organisms, vol. 2, pp. 1–1232. New York: McGraw Hil.

Lewis, J. G. E. 1981. The biology of centipedes. Cambridge University Press, Cambridge, London, New York. 432 pp.

Pearse, V., J. Pearse, M. Buchsbaum, and R. Buchsbaum. 1987. Living invertebrates. Blackwell Scientific Publ., Palo Alto, California, and Boxwood Press, Pacific Grove, California, 848 pp.

Mollusca

(Plates 349–434)

Introduction

DAVID R. LINDBERG

The Mollusca are diverse in body form and size, ranging from giant squids over 20 m in length to adult body sizes of only 500 microns. They are often considered to be the second largest phylum next to Arthropoda, with about 200,000 living species, of which about 75,000 have been named. An additional 35,000 fossils have also been named, making the Mollusca one of the better known invertebrate groups. They exhibit a great range of physiological, behavioral, and ecological adaptations and have an excellent fossil record extending back some 560 million years to the earliest Cambrian. Three major classes—Gastropoda (snails, slugs, limpets), Bivalvia (scallops, clams, oysters, mussels) and Cephalopoda (squid, cuttlefish, octopuses, nautilus)—are recognized, as well as four to five minor living classes—Aplacophora (spicule worms, which are sometimes divided into two separate classes), Polyplacophora (chitons), Scaphopoda (tusk shells), and Monoplacophora (a small group of deep sea limpets).

The majority of mollusks are marine, but large numbers also occupy freshwater and terrestrial habitats. Many nonmarine taxa are in jeopardy as a result of human activities, and there are more recorded extinctions of terrestrial mollusks than of birds and mammals combined. Mollusks are extremely diverse in their food habits, ranging from grazers and browsers on many different biotic substrates to suspension feeders, predators, and parasites. Many are economically important as food, cultural objects, and hosts for human parasites and pests. References to mollusks pervade our legends, our literature, and even our everyday speech (e.g., "snail-mail").

There is no consensus as to the identity of the sister taxon of the Mollusca. What is clear is that mollusks reside among the Lophotrochozoa (mollusks, annelids, brachiopods, bryozoans, and phoronids), but the relationships among these taxa are not clearly delineated. Traditionally, the Annelida have been considered as the sister taxon of the Mollusca by most workers. The mollusks and annelids share several characters, including the trochophore larvae, anteriorly positioned ferrous oxide structures as teeth and jaws, and a cross-configuration of micromeres during early development. Surprisingly, there is little molecular evidence to test the hypothesis of the Annelida as the sister taxa of the Mollusca. However, as a cautionary note,

the inability to clearly identify a sister taxon of the Mollusca may result from a burst of rapid speciation in the Cambrian within the lophotrochozoan ancestor.

The molluscan body plan typically consists of four body components: (1) a head with tentacles and eyes, (2) a ventral muscular foot, (3) a dorsal visceral mass, and (4) the enveloping mantle that secretes the shell. A space between the covering mantle and the side of the foot forms a mantle (or pallial groove). In most mollusks, the groove deepens in the posterior region to form a cavity that contains a pair of gills or ctenidia, as well as openings of the rectum, paired renal organs, and gonads. In the Mollusca, the coelom is represented by the pericardium, kidneys, and gonads. The ancestral mollusks were likely benthic forms living on hard surfaces and feeding by rasping their food from the substrate by means of a unique organ, the radula, which is found in all classes of mollusks except for the Bivalvia. In the Bivalvia, the ctenidia have been highly modified for the filtration of microscopic food material from the water, and they have lost all semblance of a head. Further details are discussed in relation to the various classes and orders.

General References on Mollusca

Beesley, P. L., G. J. B. Ross, and A. Wells, eds. 1998. Mollusca: The Southern Synthesis. Parts A & B. Fauna of Australia, vol. 5. CSIRO Publishing, Melbourne. (Although focused on Australia fauna, this compendium provides excellent introduction to molluscan groups and most taxa to the family level).

Broadhead, T. W. ed. 1985. Mollusks. Notes for a Short Course, University of Tennessee, Department of Geological Sciences, Studies in Geology 13. (Covers both fossil and living mollusks.)

Hyman, L. H. 1967. The Invertebrates: Mollusca I, Vol. VI. McGraw-Hill, 792 pp. (Covers Polyplacophora, Gastropoda, Aplacophora, and Monoplacophora.)

Keen, M. 1971. Sea Shells of Tropical West America. Stanford University Press, 1064 pp. (Valuable on a supraspecific level for the central California coast.)

Keen, M. and E. Coan 1974. Marine Molluscan Genera of Western North America: an illustrated key. 2nd ed. Stanford University Press, 208 pp.

Lydeard C. and D. R. Lindberg, eds. 2003. Molecular Systematics and Phylogeography of Mollusks. Smithsonian Institution Press, Washington, D.C.

McLean, J. H. 1978. Marine Shells of Southern California. Los Angeles County Mus. Nat. Hist. Science Series 24, Revised Ed. (Covers Polyplacophora, Gastropoda, Bivalvia, and Scaphopoda.)

Morton, J. E. 1967. Molluscs. 4th ed. Hutchinson University Library, 244 pp.

Taylor, J. D., ed. 1996. Origin and evolutionary radiation of the Mollusca. Oxford: Oxford University Press.

Wilbur, K. M. and C. M. Yonge, eds. 1964, 1966. Physiology of Mollusca. Vols. I and II. New York: Academic Press.

Wilbur, K. ed. (1983–1988). The Mollusca. Vols. 1–12. New York: Academic Press.

Yonge, C. M. 1960. General characters of Mollusca. In Treatise on invertebrate paleontology. Part I. Mollusca 1. Moore RC ed., pp. I3–I36. Geological Society of America, Inc. and University of Kansas Press, Lawrence.

Aplacophora

AMELIE H. SCHELTEMA

Aplacophorans are primarily mollusks of the continental shelf and deep sea, but they have occasionally been found interstitially at lowest tide levels and even on well-fouled rock jetties where they feed on hydrozoans. When found, they are unmistakable. Both taxa (Neomeniomorpha or Solenogastres, the neomenioids; Chaetodermomorpha or Caudofoveata, the chaetoderms) are vermiform and covered by a cuticle invested with innumerable, shiny, aragonite spines or scales homologous to the scales on the mantle of chitons (hence the name Aculifera for both). Neomenioids creep on a foot, a simple ciliated ridge within a narrow pedal groove that runs as a ventral line from just below the mouth to the posterior. Chaetoderms are burrowers and have no foot; they are distinguished by a cuticular shield around the mouth with which they dig.

Both groups are mainly carnivorous; neomenioids feed on Cnidaria (e.g., hydrozoans), and chaetoderms on Foraminifera or other small organisms. Their molluscan affinities are shown by presence of a radula (in most); a posterior mantle cavity containing, in the chaetoderms, paired ctenidia; and by a ladderlike nervous system much like that in chitons.

Subtidally, aplacophorans are not uncommon off the Pacific coast, with at least 20 species described between 22 m and 1,830 m (see Heath 1911; Scheltema 1998). An organism from intertidal mud-sand in Bodega Harbor (collected by G. Ruiz) was tentatively identified by Ralph Smith as an aplacophoran. Subtidal or low intertidal meiofaunal investigations are likely to turn up neomenioids such as Meiomenia swedmarki Morse, 1979. The only caution in identifying an organism as an aplacophoran is a superficial resemblance to some kinorhynchs with a very spiny cuticle.

References

Heath, H. 1911. The Solenogastres. Reports on the scientific results of the expedition to the tropical Pacific. . . by the "Albatross." Memoirs of the Museum of Comparative Zoology 45: 1–179.

Morse, M. P. 1979. Meiomenia swedmarki gen. et sp. n., a new interstitial solenogaster from Washington, USA. Zoologica Scripta 8: 249–253.

Scheltema, A. H. 1998. Aplacophora. pp. 3–47. In Taxonomic atlas of the benthic fauna of the Santa Maria Basin and the Western Santa Barbara Channel. Vol. 8. P. V. Scott and J. A. Blake, eds. The Mollusca Part I. Santa Barbara, CA: Santa Barbara Museum of Natural History.

Scaphopoda

RONALD L. SHIMEK

(Plate 349)

Scaphopods are infaunal predatory mollusks with a tubular, curved shell that has openings at both ends. They are morphologically well-known (see Shimek and Steiner 1997 for a review). The large aperture is functionally anterior, the concave portion of the shell is dorsal, and the convex portion of the shell is ventral. Scaphopods are bilaterally symmetrical and completely surrounded by the shell, which encloses the mantle cavity. The body is suspended from the functionally dorsal part of the shell by the mantle and surrounded laterally and ventrally by the mantle cavity. Ctenidia are lacking, but the mantle cavity typically bears prominent ciliated respiratory ridges.

The two scaphopod orders, the Dentaliida and the Gadilida, differ significantly in shell shape and soft-part morphology (Palmer 1974, Starobogatov 1974, Steiner 1992a, Shimek and Steiner 1997). Generally, the Dentaliids have a longitudinally ribbed or smooth, "unpolished" shell that is largest at the anterior aperture. Gadilids typically lack sculpture and have a polished shell, with the widest portion of the shell some distance behind the aperture (Shimek 1989).

The foot, which extends from the anterior aperture during burrowing, retracts by bending into the shell in the dentaliids and by introverting within itself in the Gadilids (Shimek and Steiner 1997). While the dentaliid foot bears a pair of lateral lobes slightly proximal to the end of the foot, the Gadilid foot terminates in a fringed disk (Steiner 1992b). Several hundred specialized feeding tentacles, captacula, originate lateral to the proboscis or buccal tube base (Morton 1959, Dinamani 1964, Gainey 1972, Poon 1987, Shimek 1988, Shimek and Steiner 1997).

Most illustrations erroneously depict scaphopods with the apical aperture extending from the sediment. Although this posture occasionally occurs, particularly in spawning animals, most Californian scaphopods spend most of their lives completely buried into sediments. Gadila aberrans and Rhabdus rectius (plate 349) can both be found commonly at least 30 cm under the sediment-water interface (Shimek 1990). The natural history and ecological interactions of some Northeastern Pacific scaphopods are relatively well-known. Some, such as Rhabdus rectius, are generalist predators, although most species are selective predators on specific foraminiferans (Poon 1987, Shimek 1990).

PLATE 349 A, Gadila aberrans; B, Rhabdus rectius (original).

At least 13 scaphopod species may be found in the sediments off the central California coast (Emerson 1978, Baxter 1987, Shimek and Moreno 1996, Shimek 1997, 1998). A number of these are from very deep water and are unlikely to be collected at depths <1,000 m. Most of the rest are potentially collectable in depths of 30 m or less, and some may be found in quite shallow waters (Keen 1971). Some are habitat specialists, but others are true generalists. None, however, are likely to be found intertidally except as beach drift or shells with hermit crabs in them. Scaphopods are preyed upon by fishes and crabs and may be important components of their diets in some localities (Shimek 1990).

Until recently, scaphopod nomenclature was simple: most dentaliids were assigned to *Dentalium,* and most gadilids to *Cadulus* (Emerson 1952, 1962). Morphological, systematic and taxonomic advances have resulted in many nomenclatural changes, which are reflected here (Steiner 1992a, Shimek 1998).

List of Species

A key to most of the central California species is found in Shimek 1998.

DENTILIIDA

DENTALIIDAE

Antalis pretiosum (Sowerby, 1860) (=*Dentalium pretiosum*). In exposed habitats with significant wave action. Collected on the Pacific Northwest coast and traded by Native Americans throughout the Northern Plains. Robust shell; length to 6 cm.

Dentalium neohexogonum Sharp and Pilsbry in Pilsbry and Sharp, 1897.

Dentalium vallicolens Raymond, 1904. Moderate depths, uncommon or absent in shallow water. Faint longitudinal sculpture on the apical half of the shell.

Fissidentalium spp. Three species, all below depths of 300 m.

RHABDIDAE

Rhabdus rectius (Carpenter, 1864) (=*Dentalium rectius*). Found in shallow silty sands; generalist carnivore, also eats sediment and fecal pellets. Shell thin, straight, fragile when dry; length may exceed 10 cm, diameter to about 6 mm.

GADILIDA

GADILIDAE

Gadila perpusillus (Sowerby in Broderip and Sowerby, 1832) (=*Cadulus perpusillus*). Replaces *Gadila aberrans* in southern California. Slender, length to 10 mm; aperture constricted.

Gadila aberrans (Whiteaves, 1887) (=*Cadulus fusiformis*). Found in clean sands from central California north; eats foraminiferans. Slender, length to 10 mm; very slight apertural constriction.

Siphonodentalium quadrifissatum (Pilsbry and Sharp, 1898). Found from central California to central Washington. Apical aperture with four lobes.

Cadulus californicus Pilsbry and Sharp, 1898. Cadulid, to 15 mm long; in silty sands from Southern California to Vancouver Island. Opaque, very highly polished shell.

Cadulus tolmiei Dall, 1897. Cadulid, to 15 mm long; in silt along northwest coast, reported from California; similar to *Cadulus californicus*, but shell thinner, more hyaline, translucent. Eats foraminiferans in the genus *Uvigerina*.

PULSELLIDAE

Pulsellum salishorum Marshall, 1980. Habitat generalist, up to 7 mm long, looks like a small dentaliid, but the foot introverts.

References

Baxter, R. 1987. Mollusks of Alaska. Bayside, CA: Shells and Sea Life Publications, 163 pp.

Dinamani, P. 1964. Feeding in *Dentalium conspicuum*. Proceedings of the Malacological Society of London 36: 1–5.

Emerson, W. K. 1952. Generic and subgeneric names in the molluscan class Scaphopoda. Journal of the Washington Academy of Sciences 42: 296–303.

Emerson, W. K. 1962. A classification of the scaphopod mollusks. Journal of Paleontology 36: 76–80.

Emerson, W. K. 1978. Two new Eastern Pacific species of *Cadulus*, with remarks on the classification of the scaphopod mollusks. Nautilus 92: 117–123.

Gainey, Jr., L. F. 1972. The use of the foot and the captacula in the feeding in *Dentalium*. Veliger 15: 29–34.

Keen, A. M. 1971. Sea shells of tropical West America. Stanford, CA: Stanford University Press, pp. 883–891.

Morton, J. E. 1959. The habits and feeding organs of *Dentalium entalis*. Journal of the Marine Biological Association of the United Kingdom 38: 225–238.

Palmer, C. P. 1974. A supraspecific classification of the Scaphopod Mollusca. Veliger 17: 115–123.

Poon, Perry A. 1987. The diet and feeding behavior of *Cadulus tolmiei* Dall, 1897 (Scaphopoda: Siphonodentalioida). The Nautilus 101: 88–92.

Shimek, R. L. 1988. The functional morphology of scaphopod captacula. The Veliger 30: 213–221.

Shimek, R. L. 1989. Shell morphometrics and systematics: A revision of the slender, shallow-water *Cadulus* of the Northeastern Pacific (Scaphopoda: Gadilida). Veliger 32: 233–246.

Shimek, R. L. 1990. Diet and habitat utilization in a Northeastern Pacific Ocean scaphopod assemblage. American Malacological Bulletin 7: 147–169.

Shimek, R. L. 1997. A new species of Eastern Pacific *Fissidentalium* (Mollusca: Scaphopoda) with a symbiotic sea anemone. Veliger 40: 178–191.

Shimek, R. L. 1998. Scaphopoda. In Taxonomic atlas of the benthic fauna of the Santa Maria Basin and Western Santa Barbara Channel. Volume 9. Mollusca. Blake, J. A., A. L. Lissner, and P. H. Scott, eds. Santa Barbara, CA: Santa Barbara Natural History Museum.

Shimek, R. L. and G. Moreno. 1996. A new species of Eastern Pacific *Fissidentalium* (Mollusca: Scaphopoda). Veliger 39: 71–82.

Shimek, R. L. and G. Steiner. 1997. Scaphopoda. In Mollusca II. Microscopic anatomy of the invertebrates. Volume 6B: 719–781. Harrison F., and A. J. Kohn, eds. New York: Wiley-Liss Inc.

Starobogatov, Ya. I. 1974. Ksenokowhii i ikh znachenive dlya filogenii i sistemy nekotorykh klass mollyuskov (Xenochonchias and their bearing on the phylogeny and systematics of some molluscan classes). Paleontologicheskii Zhurnal 1974(1): 3–18. Translated in: Paleontological Journal (American Geological Institute) 8: 1–13.

Steiner, G. 1992a. Phylogeny and classification of Scaphopoda. Journal of Molluscan Studies 58: 385–400.

Steiner, G. 1992b. The organisation of the pedal musculature and its connection to the dorsoventral musculature in Scaphopoda. Journal of Molluscan Studies 58: 181–197.

Steiner, G., and Kabat, A. R. 2004. Catalogue of species-group names of Recent and fossil Scaphopoda (Mollusca). Zoosystema 26: 549–726.

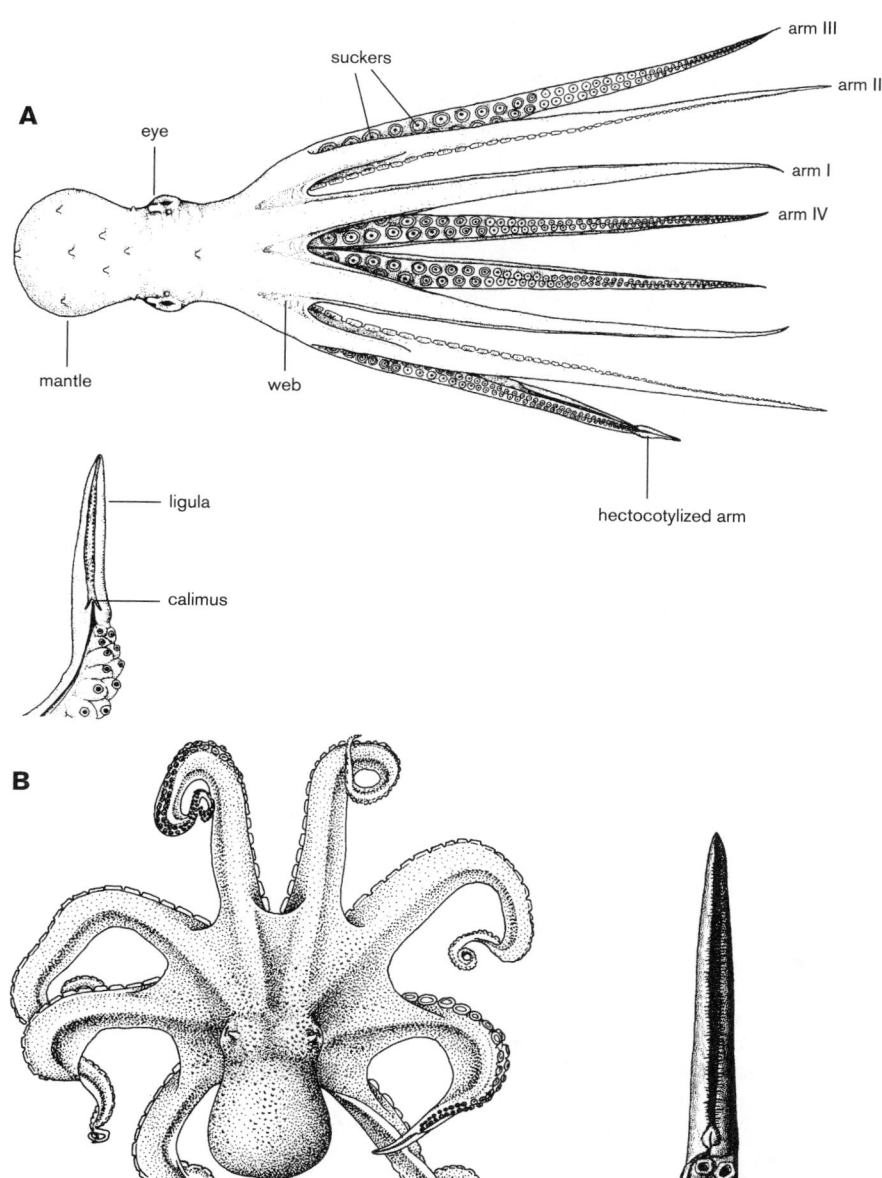

PLATE 350 Octopods with details of hectocotylized arms: A, *"Octopus" rubescens* with labeled anatomy (Atlas); B, *Enteroctopus dofleini* (FAO) (A, from Hochberg (1998); B, from Roper et al. (1984), with permissions).

Cephalopoda

F. G. HOCHBERG AND DANNA JOY SHULMAN

(Plates 350–351)

The octopuses, squids, cuttlefishes, and relatives that make up the class Cephalopoda are unique among mollusks in a number of ways. Their highly developed nervous systems and correspondingly complex behaviors have fascinated observers for centuries. Who cannot be intrigued by stories of octopuses escaping the most cunningly sealed aquaria to steal crabs from neighboring tanks? However, the escape artistry of octopuses is due as much to their lack of a shell as to their cranial capacity.

Although ancestral cephalopods such as the famously fossilized ammonites wore heavy external shells, only a single extant cephalopod genus, *Nautilus,* has retained this character. In the subclass Coleoidea, which includes all other cephalopods, the shell is either reduced to an internal structure (the cuttlefish's cuttlebone, the squid's pen or the octopus' stylets) or lost entirely, as in some octopuses.

Shell loss allowed cephalopods to become swift, agile predators in a variety of marine environments, but it also revealed their soft, tasty mantle to a host of potential predators. A variety of defenses evolved in response to this threat. All coleoids (with the exception of deep-sea species that have secondarily lost their ink sac) can squirt out a cloud of dark black or brown ink to discourage predators. Sometimes the ink seems to resemble the animal itself, possibly serving as a decoy during escape; in other situations, it is thought to confuse, blind, or chemically irritate a predator. But inking is hardly the only cephalopod defense.

Members of the Order Octopoda are mostly nocturnal and fond of hiding in dens, which range from rocky crevices to gastropod shells to kelp holdfasts, and can often be identified by the nearby carapaces of the octopus' prey. The octopuses themselves are difficult to find, for in addition to their reclusive tendencies, they are masters of camouflage, quick to match any background. Their swift color and pattern changes are made possible by a network of pigment sacs (chromatophores), reflective platelets (iridophores), and refractive platelets (leucophores) in the skin, as well as superficial muscles that permit texture control.

Squids, in the Order Teuthida, also possess chromatophores, iridophores, and leucophores, but in lower density than octopuses, and their camouflage abilities pale in comparison. As neritic or oceanic animals, they have less use for these skills; rather, they are built for speed. The squid's elongate mantle sports a pair of fins, which the intertidal and subtidal octopuses lack. Their swimming is powered by jet propulsion, as they draw water into the mantle cavity and expel it through a specialized siphon. Octopuses can also swim by jet propulsion, but they prefer to crawl along the seafloor on their eight arms. These arms are used to grasp their prey, often crustaceans or gastropods. The eight arms of squid are generally shorter, and the squid instead use two long tentacles to catch their meals of fishes and, not infrequently, other squids.

The cannibalistic tendency of squids is well-known to fishermen, who will sometimes hook an animal, only to pull in the line and find it has been half-eaten by its fellows on the way up. This method of line fishing is commonly used to catch *Dosidicus gigas,* the Jumbo or Humboldt squid, which supports a sport fishery in California and artesanal fisheries in Mexico. *Doryteuthis opalescens,* long known as *Loligo opalescens,* the common market squid, although smaller than *Dosidicus* by several feet, is subject to a commercial fishery throughout its range, as is the Giant Pacific Octopus, *Enteroctopus dofleini.*

However, despite the commercial importance of a number of cephalopod species, much of their biology and behavior is still poorly understood. Humboldt squid have been seen stranding in large numbers on the coast of southern California, a phenomenon that baffles scientists and laypeople alike. Whether humans are directly responsible through water pollution, or the squid are suffering from indirect effects of climate change, or the behavior is in some way a natural part of their life cycle remains unknown.

As squid tend to swim in offshore waters, intertidal explorers may find evidence of these cephalopods limited to fish markets and beach strandings. Evidence of cephalopods whose range lies beyond the scope of this manual can also be found in the flotsam at times. A notable example is the shell of the paper nautilus, *Argonauta argo.* The females of this unusual octopod make a small, thin coiled shell for use as an eggcase, as well as a mobile home.

Argonauts, like many cephalopods, exhibit pronounced sexual dimorphism—the males are a tenth the size of the females and make no shells. The egg brooding behavior of female argonauts is an octopod trait; the eggs of squids and sepiolids are not attended by either parent. Hatchlings receive no parental care in most species, as the majority of cephalopods are semelparous and die after spawning or brooding their eggs.

Cephalopods are home to a variety of parasites, including the vermiform phylum Dicyemida, which occurs exclusively in the renal appendages of cuttlefishes, sepiolids, squids, and octopuses. The nature of the relationship between dicyemids and their cephalopod hosts is unknown.

Collection and Preservation

Care must be taken in handling cephalopods, not only for the sake of the animal, whose skin is easily scraped and torn, but also for the sake of the collector, whose skin is in turn vulnerable to the animal's sharp beak. All cephalopods have parrotlike beaks, and most octopuses also have salivary glands that produce venom. None of the local species are known to be life-threatening, but a bite can be extremely painful.

For ease of future identification and studies, cephalopods should be preserved with a minimum of contraction or distortion of body parts. Specimens should be relaxed under anaesthesia, by addition of ethyl alcohol, or, more archaically, magnesium sulfate. The fully relaxed animal can be preserved in 4% formaldehyde, which should be flushed into the mantle chamber with a syringe. Larger animals should have full-strength formalin injected into the viscera.

Key to Orders

1. Ten circumoral appendages (eight arms plus two tentacles); arms with stalked suckers and/or hooks; body rounded to elongate with fins; with an internal shell (gladius or pen) 2
— Eight circumoral appendages; arms with sessile suckers; body ovoid, not elongate . . . .
. . . . Order Octopoda (five species; see list)
2. Eye membrane absent; tentacles retractile into small pockets between arms 3 and 4; mantle rounded, free all around, mantle lengths (ML) <50 mm; fins semicircular, almost as long as mantle, with broad free lobe; arms short, circular in transverse section, lengths subequal; dorsal arms hectocotylized . . . . Order Sepiolioidea

Note: Represented in our fauna by *Rossia pacifica* (plate 351D; see species list).

— Eye covered by transparent membrane; tentacle not retractile into pockets; tentacle clubs narrow and without fixing apparatus; mantle elongate and tapered, attached to body, mantle lengths typically >100 mm; fins triangular, less than half mantle length, attached along entire length; arms long, angular in transverse section, lengths unequal, ventral pair long and broad; left ventral arm of male hectocotylized (plate 351C) . . . .
. . . . Order Teuthoidea (*Doryteuthis opalescens*)

Note: Two additional teuthoids may be encountered, as noted below.

List of Species

SEPIOLIOIDEA

SEPIOLIDAE

Rossia pacifica Berry, 1911. Plate 351D. Body size small, mature females larger than males (ML to 55 mm in females and to 35 mm in males); fins small, round; eggs large (4 mm–5 mm),

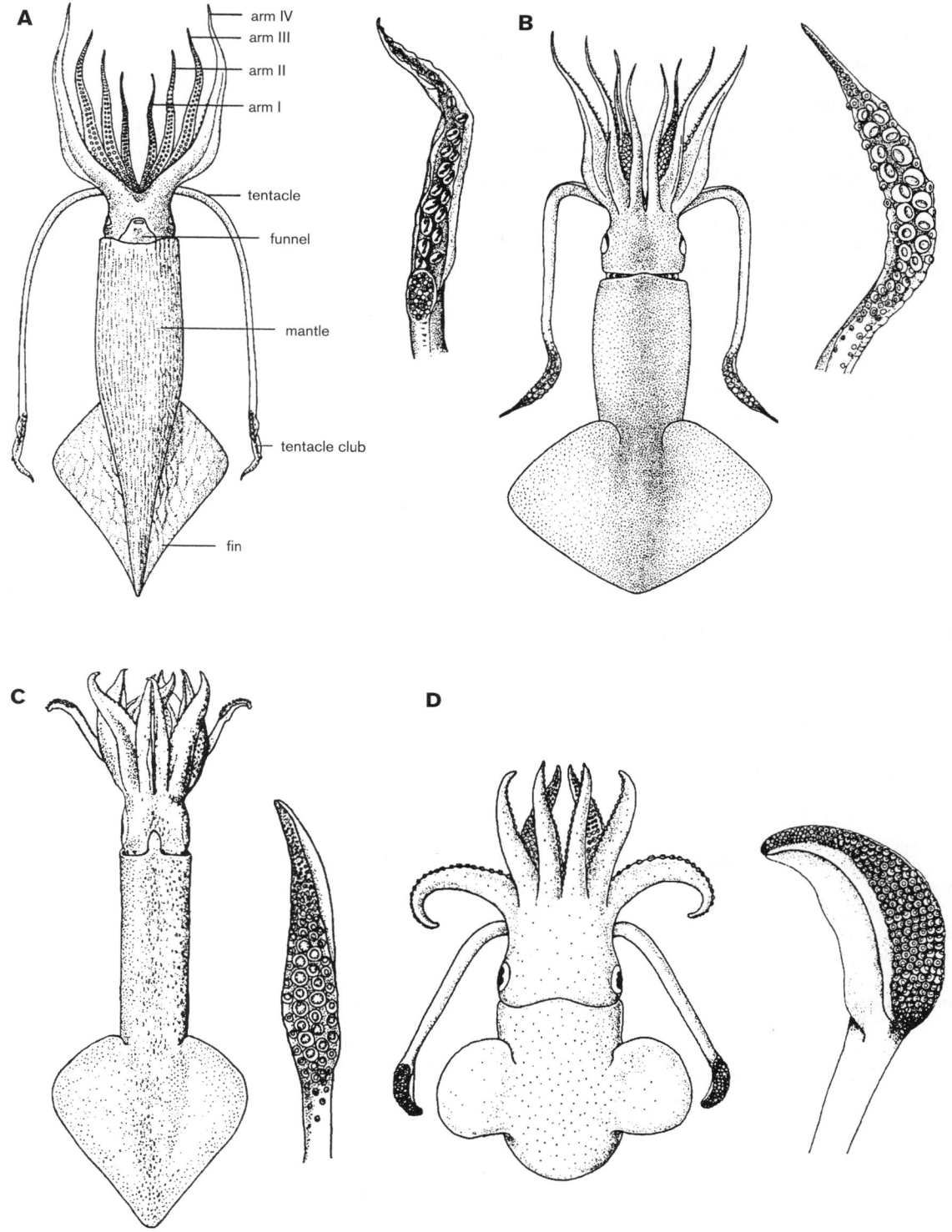

A
- arm IV
- arm III
- arm II
- arm I
- tentacle
- funnel
- mantle
- tentacle club
- fin

B

C

D

PLATE 351 Decapods with details of tentacle clubs: A, *Moroteuthis robustus* with labeled anatomy; B, *Dosidicus gigas*; C, *Doryteuthis opalescens*; D, *Rossia pacifica* (all from Roper et al. 1984).

laid in capsules that are attached in small groups to objects on the bottom. Depth 10 m–250 m; typically neritic in shallow coastal waters, where they live on sand and mud bottoms. Taxonomy of this species in the North Pacific has not been stabilized. See Anderson and Shimek 1993 Veliger 37: 17–19 (egg masses).

TEUTHOIDEA

LOLIGINIDAE

Doryteuthis opalescens (Berry, 1911) [=*Loligo opalescens*; formerly *L. stearnsii* Hemphill, 1892]. Plate 351C. Body medium size, mature males larger than females (ML to 190 mm in males and to 170 mm in females); eggs small (2.3 mm) laid in cylindrical capsules anchored in soft substrates. Depth range not known; common in near shore (neritic waters); seasonally aggregate in shallow water (15 m–35 m) to mate and spawn. See Fields 1965, Calif. Dept. Fish Game Fish Bull. 131: 1–108 (morphology, development, biology); Hixon 1983, pp. 95–114 in Boyle, ed., Cephalopod Life Cycles, Vol. I, Academic Press; Yang et al. 1986, Fish Bull. 84: 771–798 (growth, behavior); Wing and Mercer 1990, Veliger 33: 238–240 (in Alaska); Recksiek and Frey, eds. 1978, Calif. Dept. Fish Game Fish Bull. 169: 1–185 (monograph on biology, oceanography, acoustics); McGowan 1954, Calif. Fish Game 40: 47–54 (spawning).

OMMASTREPHIDAE

Dosidicus gigas (d'Orbigny, 1835) [=*Ommastrephes gigas*; formerly *Ommastrephes giganteus* Gray, 1849; *D. eschrichti* Steenstrup, 1857; *D. steenstrupi* Pfeffer, 1884]. Plate 351B. Body size large (adult ML to 150 cm); surface of mantle smooth; fins large, about one-half length of mantle, sagittate; tentacle clubs with suckers only; eggs small (1.0 mm), spawning habits unknown. Often washed ashore in large numbers from Baja California to Washington. Depth 0 m–1,200 m; oceanic species known to migrate diurnally.

ONYCHOTEUTHIDAE

Moroteuthis robustus (Verrill, 1876) [=*Onykia robusta*; *Ommastrephes robustus*; formerly *M. japonica* (Taki, 1964); *M. pacifica* Okutani, 1983]. Plate 351A. Body size large (adult ML to 230 cm); surface of mantle with numerous fine longitudinal ridges; fins large, occupy more than one-half length of mantle, sagittate; tentacle clubs with 15–18 pairs of hooks in two characteristic rows; eggs small (1.0 mm), spawning habits unknown. Depths of capture range from 100 m–600 m; pelagic in coastal waters; occasionally washed ashore or caught by trawl fishermen. See Pattie 1968, Fish. Res. Papers Wash. Dept. Fisheries 3: 47–50; Green 1989, Calif. Fish Game 75: 241–243 (strandings in British Columbia); van Hyning and Magill 1964, Res. Briefs Fish Comm. Oregon 10: 67–68 (off Oregon); Tsuchiya and Okutani 1991, Bull. Mar. Sci. 49: 137–147 (growth); Hochberg 1974, Tabulata 7: 83–85 (southern California records).

OCTOPODA

OCTOPODIDAE

Enteroctopus dofleini (Wülker, 1910) [=*Octopus dofleini*; *Paroctopus dofleini*; formerly *O. punctatus* Gabb, 1862; *O. hongkongensis* Hoyle, 1885; *O. gilbertianus* Berry, 1912]. Plate 350B. Body size large (adult ML to 350 mm); dark ocelli absent; gills with 12–15 lamellae; enlarged suckers present on all arms of mature males; copulatory organ (ligula) of males very long (16%–18% of arm length); spawned eggs small (6 mm–6.5 mm), laid in festoons; hatchlings planktonic. Depth 0 m–1,500 m; intertidal in northern part of range; inhabit substrates littered with rocks and boulders. See Pickford 1964, Bull. Bingham Ocean. Coll. 19: 1–70; Hartwick 1983, pp. 277–291 in Boyle, ed., Cephalopod Life Cycles, Vol. 1. Academic Press.

"*Octopus*" *rubescens* Berry, 1953. Plate 350A. Body size medium (adult ML to 100 mm); dark ocelli absent; gills with 11–13 lamellae; enlarged suckers present on all but ventral arms in mature males; spawned eggs small (3 mm–4 mm), laid in festoons; hatchlings planktonic. Depth 0 m–300 m; common in the intertidal from northern California to Alaska; live in rocky inshore areas and on sand/mud bottoms offshore.

Octopus bimaculoides Pickford and McConnaughey, 1949. Body size medium (adult ML to 200 mm); two dark ocelli present, each with necklace-like iridescent blue ring; gills with 8–10 lamellae; enlarged suckers on lateral arms of mature males; copulatory organ (ligula) of males very small (2%–4% of arm length); spawned eggs large (10 mm–17 mm), laid in festoons; hatchlings benthonic. A more southern species, occurring from San Simeon south. Depth 0 m–25 m; common in the intertidal; live on mudflats or in protected holes and crevices on rocky substrates. See Forsythe and Hanlon 1988, Malacologia 29: 41–55 (biology).

Octopus bimaculatus Verrill, 1883. Similar in size, general appearance and characteristic features to preceding species but typically larger and with longer arms; two dark ocelli present, each with iridescent blue ring with radiating spokes; spawned eggs small (2 mm–4 mm), laid in festoons; hatchlings planktonic. A southern species, from Point Conception south. Depth 0 m–50 m; common in the low intertidal in Mexico; typically associated with rocky substrates. See Ambrose 1981, Veliger 24: 139–146 (development); 1984, J. Exp. Mar. Biol. Ecol. 77: 29–44 (feeding); 1988, Malacologia 29: 23–29 (population dynamics).

"*Octopus*" *micropyrsus* Berry, 1953. Body size very small (adult ML less than 30 mm); gills with 5–6 lamellae; 1 or 2 enlarged suckers on arms 1–3 of both mature males and females; copulatory organ (ligula) moderately long (7%–15% of arm length); spawned eggs large (10 mm–12 mm), attached singly to substrate in small clusters; hatchlings benthonic. A short range transition endemic that often is common in the intertidal in the northern part of range which extends from Point Conception, California to Pt. Eugenia and the islands off Baja California, Mexico. Depth 0 m–20 m; typically found in kelp holdfasts, piddock holes or in empty gastropod shells. See Haaker 1985, Shells and Sea Life 17: 39–40 (photos).

References

Berry, S. S. 1912. A review of the cephalopods of western North America. Bulletin of the Bureau of Fisheries 30 (1910): 267–336.

Berry, S. S. 1953. Preliminary diagnosis of six west American species of *Octopus*. Leaflets in Malacology 1: 51–58.

Hanlon, R. T., and J. B. Messenger. 1996. Cephalopod Behaviour. Cambridge: Cambridge University Press, 232 pp.

Hochberg, F. G. 1998. Class Cephalopoda, pp. 175–236. In Taxonomic atlas of the benthic fauna of the Santa Maria Basin and the Western Santa Barbara Channel. Vol. 8. The Mollusca Part 1. P. Valentich Scott and J. A. Blake, eds. Santa Barbara, CA: Santa Barbara Museum of Natural History.

Hochberg, F. G. and W. G. Fields. 1980. Cephalopoda: The squids and octopuses, pp. 429–444. In Intertidal invertebrates of California. R. H. Morris, D. P. Abbott, E. C. Haderlie, eds. Stanford, CA: Stanford University Press.

Lang, M. A., and F. G. Hochberg, eds. 1997. Proceedings of the Workshop on The Fishery and Market Potential of Octopus in California. Washington, D.C.: Smithsonian Institution, 192 pp.

Phillips, J. B. 1933a. Description of a giant squid taken at Monterey, with notes on other squid taken off the California coast. California Fish and Game 19: 128–136.

Phillips, J. B. 1933b. Octopi of California. California Fish and Game 20: 20–29.

Phillips, J. B. 1961. Two unusual cephalopods taken near Monterey. California Fish and Game 47: 416–417.

Pickford, G. E., and B. H. McConnaughey. 1949. The *Octopus bimaculatus* problem: a study in sibling species. Bulletin of the Bingham Oceanographic Collection 12: 1–66.

Roper. C. F. E., M. J. Sweeney, C. E. Naun. 1984. FAO species catalogue. Volume 3. Cephalopods of the world. An annotated and illustrated guide to species of interest to fisheries. FAO Fish Synopsis 125(3): 1–277.

Verrill, A. E. 1883. Descriptions of two species of *Octopus* from California. Bulletin of the Museum of Comparative Zoology 11: 117–123.

Winkler, L. R., and L.M. Ashley. 1954. The anatomy of the common octopus of northern Washington. Walla Walla College Publications in Biological Science 10: 1–30.

Polyplacophora

DOUGLAS J. EERNISSE, ROGER N. CLARK, AND ANTHONY DRAEGER

(Plates 352–354)

Chitons are conspicuous in intertidal and shallow subtidal habitats along much of the Pacific coast of North America, where they are often abundant and ecologically important members of the community (Dethier and Duggins 1984; Duggins and Dethier 1985). Indeed, the Pacific coast supports both an unusually high diversity of species and the largest-bodied chiton species in the world.

This diversity was relatively well known when the noted malacologist Allyn G. Smith wrote the chiton key in the previous edition of this manual. Nevertheless, many changes have occurred in our understanding of the diversity of the chiton fauna, nomenclatural advances have been introduced, and we have added more species that, although largely subtidal, find their upper limits in the lower intertidal zone. Most chitons found within 15 m (a depth readily accessible by scuba) may also be expected to occur occasionally in the intertidal. The number of recognized species has also increased due to morphological and molecular studies (see especially the worldwide monograph series by P. Kaas and R. A. Van Belle [1985–1994], and the publications of A. J. Ferreira, R. N. Clark, and D. J. Eernisse).

The following key and species list include some reassignments of genera and the revival of some older nominal species rescued from synonymy. Those that are higher-level changes are based on phylogenetic studies (D. J. Eernisse, unpublished; R. P. Kelly and D. J. Eernisse, unpublished; see also Kelly and Eernisse, 2007; Kelly et al., 2007), which have extended earlier worldwide phylogenetic (Okusu et al. 2003) and morphological analyses (review by Eernisse and Reynolds 1994; see also Buckland-Nicks 1995; Sirenko 1993; 1997; 2006).

Chitons are exclusively marine and relatively conservative in appearance and life styles. All chitons normally have eight shells, or **VALVES**, embedded in a tough but flexible mantle referred to as the **GIRDLE** (plate 352A, 352B). Rare specimens may have six, seven, or nine valves (Roth 1966, Veliger 9: 249–250). Chitons cling to rocks or other hard substrates with their muscular broad foot. Their anterior mouth is separated from the foot, but chitons lack a true head—a condition typical of mollusks except for gastropods and cephalopods. Alongside the foot are paired rows of interlocking **CTENIDIA** (referred to here as **GILLS**). Noting the length and position of each gill row and whether the size of gills decreases toward the posterior anus can aid in identification.

Most chitons (members of order Chitonida), including most in this key, have an **INTERSPACE** between the posterior ends of the left and right gill rows, and each gill row extends at least halfway to the anterior end of the groove alongside the foot. With this arrangement, each gill row functionally divides this pallial groove between the foot and girdle into outer inhalant and inner exhalant spaces because the gills have interlocking cilia, hanging curtainlike from the roof of each pallial groove. Cilia on each gill power water through the row and eject it at surprising velocities past the anus, allowing these chitons to have effective aquatic respiration despite their firm attachment to hard substrates (Yonge 1939).

When chitons are exposed during low tide, when oxygen is more abundant, they have a large surface area of gills with which to respire in air by direct diffusion, provided they are able to keep their gills moist. Chitons of the suborder Acanthochitonina (including Lepidochitonidae and Mopaliidae in this key) have an **ABANAL** gill arrangement in which the largest gill in each gill row is the most posterior. In contrast, members of the suborder Chitonina (Chaetopleuridae and Ischnochitonidae in this key) have an **ADANAL** gill arrangement with the largest gill away from the posterior end of the gill row.

Members of the mostly deep-water order Lepidopleurida (Leptochitonidae in this key, =Lepidopleuridae) are most readily distinguished by their posterior gill arrangement. As in Chitonina, their gill rows are adanal but they do not have an interspace. Instead, the left and right gill rows form a nearly continuous U-shaped arrangement adjacent to the anus. The respiratory mantle cavity, including all the gills, is restricted to the posterior one-third of the animal, resulting in a different and probably primitive functional arrangement with implications not well studied by Yonge (1939) or subsequent authors.

Chitons sense their surroundings with numerous sensory organs distributed on their girdle and across the upper surface of their valves. The presence of these shell organs, called **ESTHETES** (or **AESTHETES**), in the upper layer of valves known as the **TEGMENTUM** is unique to chitons among mollusks. Elsewhere (especially certain genera in tropical seas), esthetes are impressively modified as shell eyes (ocelli) large enough to be visible to the naked eye. Many Pacific coast species have photosensory esthetes, among those used for other sensory functions. Chitons also have many sensory organs among their diverse girdle ornamentation (Leise and Cloney 1982).

It is relatively easy to learn to recognize most chiton genera, whereas distinguishing species within some genera can be quite challenging. The shape of the girdle and the various structures on it provide many of the clues to species identification. Within chitons the girdle shape varies from merely a flexible skirt surrounding the valves, through various degrees of intrusion between the valves, to a covering completely enclosing the valves. The elements on the dorsal girdle surface (bare girdle, granules, scales, spicules, spines, fleshy bristles, or setae) are even more varied and useful for distinguishing species (plate 352C–352H). Closer examination is required to reveal finer diagnostic girdle element features: the organization of scales variously ranges from **IMBRICATING** (i.e., shingled and overlapping) (plate 352E, 352F), to scattered without apparent order, to scales sculptured with microscopic bumps and ridges; the distribution of spines (or setae) varies from scattered to specifically located at the valve sutures, and from individual structures to several structures gathered into tufts, and the setae usually bear further species-specific elaborations of the form of spicules and bristles, which often ornament individual setae. In particular, seta features are the most reliable morphological clues for identifying the 17 species of the most diverse genus in this key: *Mopalia* (plates 353, 354). These setae distinctions have been corroborated with molecular sampling (Kelly et al. 2007; R. P. Kelly and D. J. Eernisse, unpublished).

There are three structures of diagnostic importance on setae. First, setae emerge from a follicle in the girdle as a central supporting shaft, and this can have or lack a dorsal groove.

Second, there can be thinner flexible bristles borne on the shaft and attached either in the groove or in a matrix adhering to the shaft. Third, setae can have or lack rigid, sharp, fracturable mineral spicules, and these spicules can either be located directly on the shaft or be mounted on the end of short to long bristles. If one searches for bristles that are intact, and those found are carefully examined, this will reveal that most species of *Mopalia* have bristles with a spicule at their tip. Likewise, most species lack spicules or bristles on the ventral surfaces of the setae.

Setae are subject to erosion, fouling, and malformation. The setae chosen as models (plates 353, 354) reflect our experience with typical variation in setae due to erosion, and extremely high or low levels of erosion could lead to setae that differ from our key descriptions and drawings. For example, some species have setae with long shafts that we suspect are typically worn clean of bristles and spicules, but exceptionally uneroded setae might have bristles or spicules clear to the tip. Similarly, we have used the proportion of the length of the setal shaft versus the length of valve 5 tegmentum to distinguish some species, but these distinctions might not work well for the occasional animal subject to exceptionally high or low erosion. Such challenges can partly be avoided by examining a selection of setae from different regions of each animal's dorsal girdle surface.

Some environments generate biological and sediment fouling of the setae and valves, which can impede identification. Fouled preserved specimens can be cleaned with needle-pointed forceps and cautious brushing with fine-bristled brushes a few millimeters wide. For field identification of living animals, the jet from a pump-spray bottle filled with seawater aids in dislodging enough material to facilitate identification of familiar species.

The details of the bristles, spicules, and shaft of the setae are minute and are best viewed with a magnification of 50x or higher. With experience, a hand lens will usually suffice to identify species. However, very small *Mopalia* remain challenging: their setae often differ from the adult form.

Chiton valves are typically divided into regions, more pronounced in species with heavier sculpturing patterns, and these partly reflect the radiating or longitudinal rows of esthete sensory organs. Valves are of three types: the anterior or **HEAD** valve, six **INTERMEDIATE** valves, and posterior or **TAIL** valve. The dorsal surface of an intermediate valve can have as many as three distinctive symmetrical regions of sculpturing. The median longitudinal ridge is called the **JUGUM** (or **JUGAL RIDGE**), and the area along the ridge is referred to as the **JUGAL AREA** only if it is set off with distinctive sculpturing. Most chitons with a distinctive jugal area are more southern in California (e.g., *Acanthochitona* spp.), but *Oldroydia percrassa* is a local (albeit rare) exception. The jugum can be sharp-angled in chitons with a high profile or rounded when chitons are flat and broad. The apices of the valve can have or lack a pointed beak. On either side of the jugum is the **CENTRAL** area, extending to paired triangular **LATERAL** areas. The anterior portion of the tail valve has sculpturing similar to the central areas, often with longitudinal riblets or latticelike sculpturing. The posterior part of the tail valve has sculpturing like the lateral areas, often with radiating rows of **RIBS**, finer **RIBLETS**, or discrete nodules. The apex of the tail valve, called the **MUCRO** (or beak), requires special notice. In lateral view, chitons differ in the position of the mucro and in whether the **POST-MUCRONAL SLOPE** (from the mucro to the posterior shell margin) is concave, straight, convex, or even bulging.

The different patterns of pitting, ribbing, nodules, and growth lines alone are seldom sufficient to enable correct identifications. These seemingly fundamental aspects of the skeletal structure can display intraspecific variations in both the number and magnitude of features ornamenting the valves, as well as interspecific similarities in structure. Being aware of some causes for this variability is helpful. This variability results from the nature of the valve's growth, from environmental insults and from genetic variability within a species. Except for the posterior portion of the tail valve, valves grow primarily from their anterior and lateral edges, with the number of sculpturing elements (nodules, ribs, etc.) increasing as the animal grows. Valves are also subject to environmental factors from simple erosion and breakage to damage from encrusting organisms. Even individuals within a species of similar size and apparently pristine sculpture can show enough variability so valve sculpturing is not by itself sufficient for identification.

Although in this key we have largely avoided using characteristics that can only be viewed in disarticulated specimens, important additional characters may often include normally hidden features of the valves. For some species of similar appearance, knowing to search for these normally hidden valve features could be the most efficient route to positive identification. A preserved chiton can be disarticulated by slow heating in a beaker, starting with cold water and a monolayer of KOH pellets. The individual valves can then be carefully separated and rinsed. This will reveal that the valves have an upper exposed, and often colored, layer (the tegmentum) overlaying a thicker, often porous (or solid) intermediate layer, and an inner, porcelain **ARTICULAMENTUM** layer. In some species, the color of the articulamentum varies away from white and can help distinguish between species of similar appearance. In all but some ancient fossil chitons, the articulamentum layer extends anteriorly beneath the preceding valve as paired semi-circular to angular **EAVES** (or **SUTURAL LAMINAE** or **APOPHYSES**). The proportions of the tegmentum versus the eaves, as well as variances in the profile of the anterior and posterior margins of the valves, can also be of taxonomic value. In all but the most phylogenetically basal living chitons (e.g., *Leptochiton* spp.), this layer also extends laterally from the intermediate valves or distally from the terminal valves as **INSERTION PLATES** to anchor the valves firmly in the girdle. These can often be exposed without complete disarticulation by temporarily teasing the girdle tissue away from the valves at their dorsal margin.

Most chitons have **SLITS** in the insertion plates, which correspond to the innervation of the radiating rows of esthetes in the tegmental layer. Some keys to chitons (including the one in the previous edition of this volume) list a **SLIT FORMULA** expressing the number, or range of numbers, of slits observed in the head, each side of an intermediate, and tail valves, respectively. Their omission here reflects our opinion that these are not generally necessary or informative for species-level identifications, besides requiring disarticulation to observe.

Two features that are apparent at first glance are coloration and body proportions. These turn out to be of only modest utility for identifying species. Coloration and pattern can be striking in many of the chiton species. Unfortunately coloration and pattern are also strikingly variable within most species, and it is the exceptional case where color is diagnostic. The body proportions of length to width to height do provide clues, but the ratio of these proportions is not constant between species for some genera. Allometry, or shape change with size,

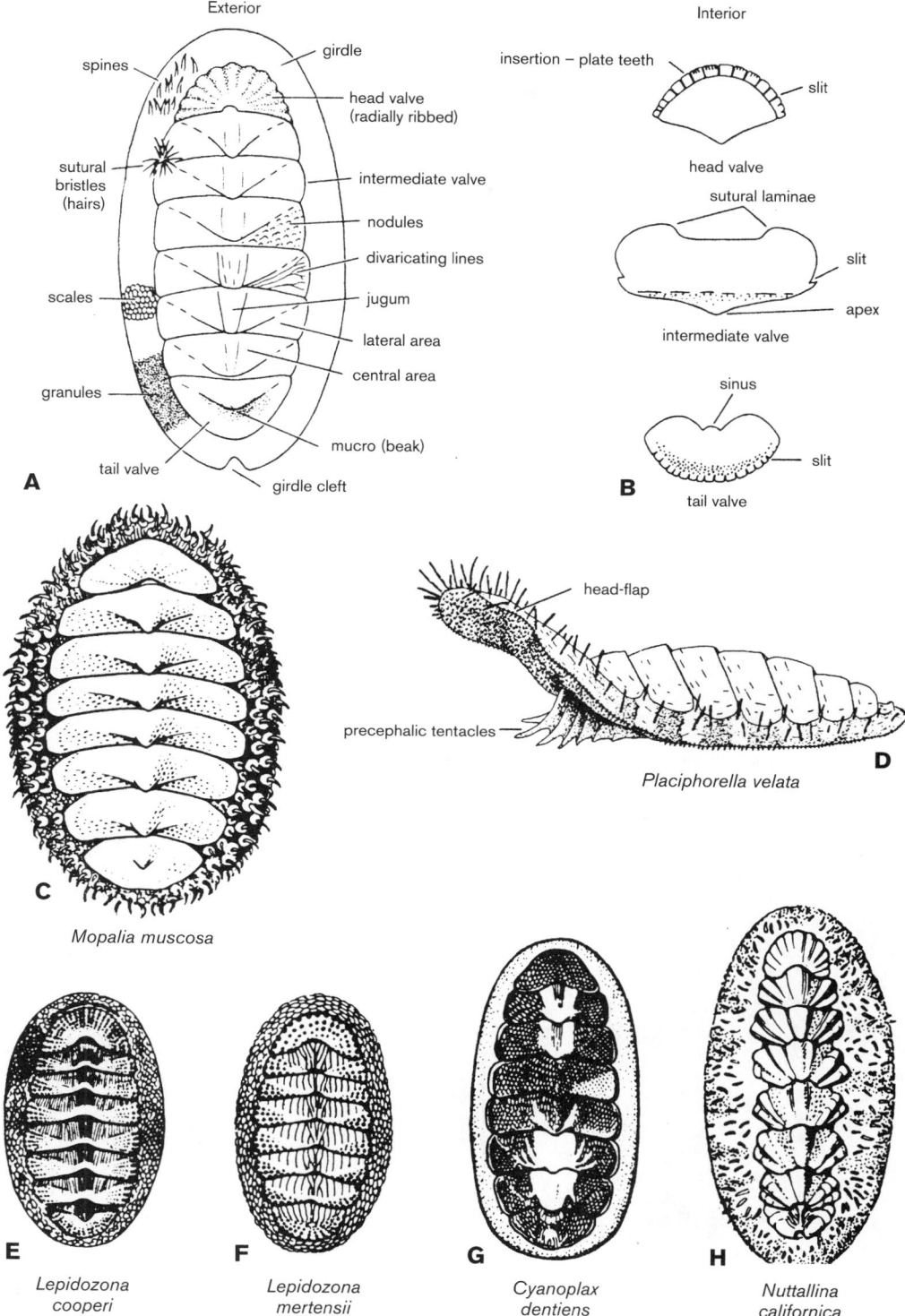

PLATE 352 Chitons: A, diagrammatic chiton showing girdle and shell ornamentation; B, terminology of valves; C, *Mopalia muscosa;* D, *Placiphorella velata;* E–H, other representative chitons (A, B, redrawn by Emily Reid after Yonge (1960); D, McLean, 1962; not to scale).

is another confounding factor as the proportions are often quite different for smaller chitons of any species. As a chiton grows, its cross-sectional profile tends to change from flattened to more peaked. In some species, the change is only a mild increase in proportional height, but in other species the change can be from a flat juvenile cross section to a nearly circular adult cross section. The outline of many species will also change from a rounded oval in juveniles to more elongated in large specimens. Finally, the addition of sculpturing can intensify as a chiton reaches adult size. For example, members of

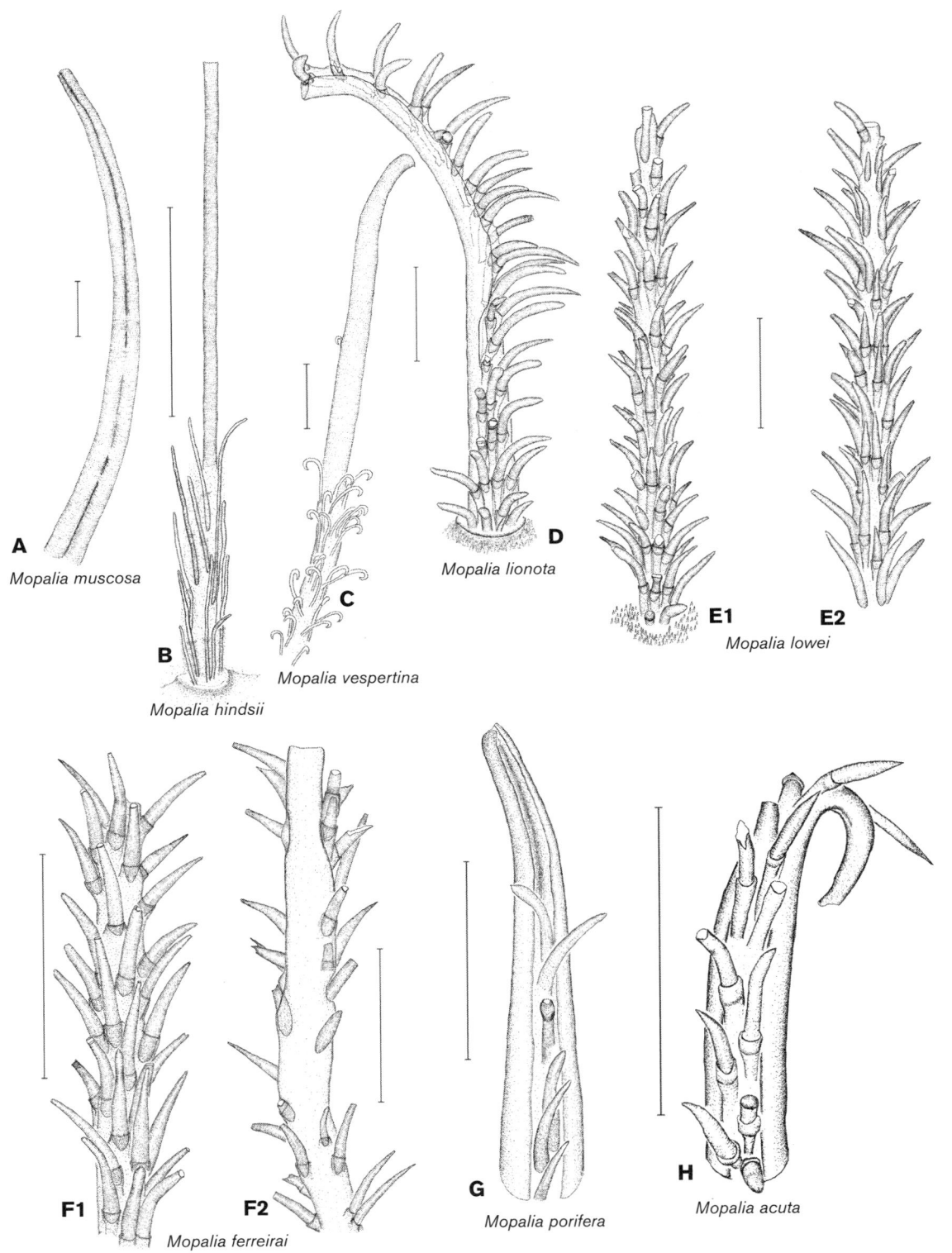

A
Mopalia muscosa

B
Mopalia hindsii

C
Mopalia vespertina

D
Mopalia lionota

E1   E2
Mopalia lowei

F1   F2
Mopalia ferreirai

G
Mopalia porifera

H
Mopalia acuta

PLATE 353 *Mopalia* setae, species as labeled (original artwork by Anthony Draeger).

Callistochiton of similar length can vary dramatically in the prominence of their ribs and bulging tail valves, and juveniles barely exhibit these sculpturing features. For these reasons, we have tried to use characteristics that are evident regardless of the chiton's age. However, this key generally describes adult animals, and juveniles can be challenging to identify.

Chitons feed with a ribbon of teeth, or **RADULA**. Radular properties are relatively conservative within chitons compared to the tremendous variation found in gastropods. Chitons typically have 17 teeth in each row and up to hundreds of rows of teeth. The main (**MAJOR LATERAL** or **SECOND LATERAL**) paired teeth are the primary working teeth and, in chitons, are always

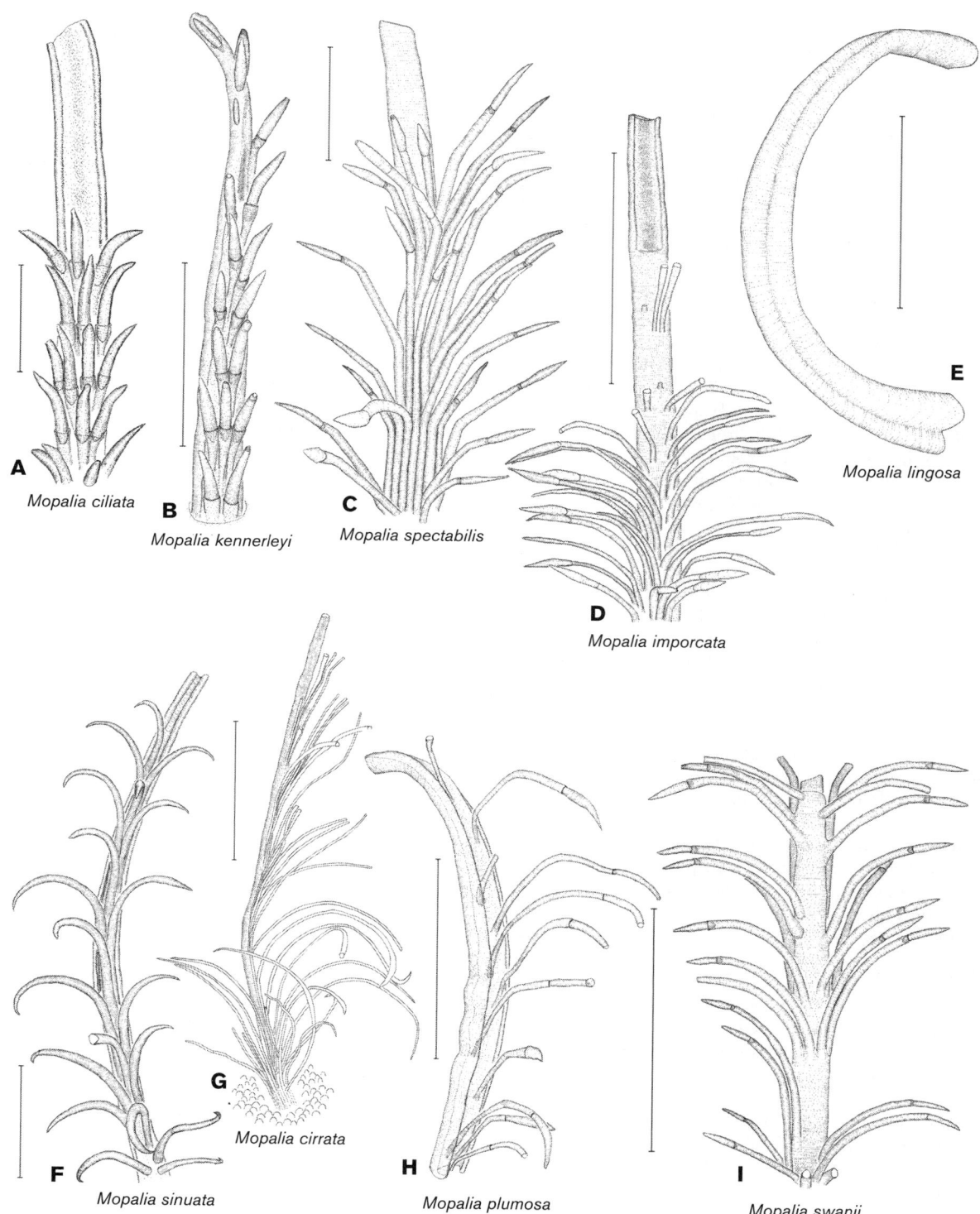

A *Mopalia ciliata*

B *Mopalia kennerleyi*

C *Mopalia spectabilis*

D *Mopalia imporcata*

E *Mopalia lingosa*

F *Mopalia sinuata*

G *Mopalia cirrata*

H *Mopalia plumosa*

I *Mopalia swanii*

PLATE 354 *Mopalia* setae, species as labeled (original artwork by Anthony Draeger).

covered with black shiny magnetite, an iron mineral harder than stainless steel. The number or shape of their cusp(s) varies somewhat with species, partly reflecting variation in their diet.

Like limpets, many chiton species scrape diatoms or other microscopic algae off rock surfaces. Some chitons are more specialized. For example, certain *Tonicella* species feed on en-

crusting coralline algae. Other, especially large, chitons will take bites of fleshy algae. One local shallow-water chiton, *Placiphorella velata* (plate 352D), is an ambush predator, trapping small prey such as amphipods with a rapid lowering of its unusually large anterior girdle, modified as an extended head flap, but its radula is only somewhat shorter than most, without much tooth specialization. Radular characters are not employed in this key because central to northern California and Oregon chitons mostly belong to a few families that show relatively little radular variation, and those not belonging to these families are more easily distinguished by external features such as girdle ornamentation, gill arrangement, or valve sculpturing patterns.

Most chitons can be quite difficult to remove from rocks without injury, especially if they have been alerted to a threat, as could occur from a mere passing shadow, their being touched, or from being uncovered by the turning of a rock. Typically, a chiton can be safely removed from a rock by surprising it with a sudden "dig" with a pointed tool directed into the rock under the foot. Alternatively, especially for chitons found underneath rocks with a smooth undersurface, one can wait for the chiton to start crawling (in either direction) at which point one can more easily slide it laterally and dislodge it. A few species (e.g., *Callistochiton* spp.) will readily detach themselves and drop, curled up, from an overturned rock, often before they are noticed. Once chitons are removed from their attachment, most (but not all) can effectively roll up into a ball, remaining tightly in this position until, conceivably, a wave has rolled them to a new safe position, where they can unroll, extend their foot for a new attachment, and escape.

It is important to flatten specimens before they are placed in preservative. It is much easier to study a flat specimen than one that has curled when put into the preserving fluid. Fortunately, this can be accomplished with a little extra effort and patience. Attempting to force curled chitons flat is futile, but if they are left undisturbed in clean aerated seawater they eventually unroll and attach themselves to the sides of the container, from which they can be swiftly slid, placed on a "chiton stick," and immediately bound into place. The chiton stick may be a tongue depressor or any similar flat strip of wood, and its width should be wider than that of the chiton. Plastic flats can be used in place of wood, but plastic is less porous than wood and thus tends to prevent the penetration of alcohol to the chiton's undersurface. Strips of nylon (i.e., pantyhose) are porous to ethanol and perform well for binding the chiton firmly to the stick, especially if the ends of the stick have been broken to expose a jagged edge that will naturally catch the nylon to hold it in place.

To produce a correctly preserved and flattened specimen, orient each chiton with its long axis parallel to that of the stick and tightly stretch the nylon strips along the longitudinal axis of its body, securing the chiton flat against the stick. With small specimens, the nylon strip can effectively cover several valves; with larger specimens, it may be necessary to use wider strips or several wraps per valve. Once wrapped, the chiton can simply be dropped into ethanol and unwrapped after it is no longer responsive. Methods that call for placing the chiton in hot water to relax the specimen before dehydration should be avoided because they could damage the specimen's DNA, but a magnesium chloride solution that is isotonic with seawater is an effective relaxant.

Alcohol is a generally useful preservative agent for chitons, with other preservatives useful for particular cases, such as initial buffered formaldehyde or (better) cold buffered glutaraldehyde fixation for the study of internal anatomy or gametes. Ethanol is preferable to isopropanol primarily because the former is better for potential DNA extraction, and for this same reason an initial high percentage of ethanol (95%–100%) is generally preferred to more dilute solutions. However, some workers have reported that subsequent trans-

fer and storage in a more dilute concentration (70%–80%) will help avoid the girdle elements becoming brittle. Likewise, although we have found it generally unnecessary or even less desirable for specimens stored in alcohol, others (see Berry, 1961; Burghardt and Burghardt, 1969; Hanselman, 1970) have advocated adding 1%–2% of glycerol to improve the flexibility of the specimen, or even higher amounts if the specimen will be stored dry. Glycerol dry preservation is not trustworthy for molecular study, but specimens for morphological comparisons might be more conveniently handled and housed when dry. There are successful collections produced this way. The recipes and procedures involved with glycerin preservation are more complicated than for ethanol storage, and anyone interested in this preservation mode should first examine several recipes to determine the most appropriate.

Unfortunately, neither method of preservation retains a flawlessly natural appearance. Color is the aspect that most conspicuously suffers in many but not all species from either preservation method, and photographs of the live or freshly fixed specimens are currently the only recourse if accurate records of coloration are important. Lamb and Hanby (2005) provide excellent images of living representatives of most of the species in this key.

Key to the Intertidal and Shallow Water Chitons from Central California to Oregon

1. Valves completely concealed by thick, red-brown, velvety girdle (juveniles <10 mm in length are yellow and have the tiny apices of valves exposed); length often exceeds 20 cm . . . . *Cryptochiton stelleri*
— Valves exposed 2
2. Girdle black, leathery, covering about two-thirds of each valve, exposed areas roughly diamond shaped; length up to 13 cm . . . . *Katharina tunicata*
— Girdle otherwise and not covering most of valves 3
3. Gill rows without an interspace and restricted to posterior one-third of animal; insertion plates absent 4
— Gill rows with an interspace, and extending to the anterior at least halfway along each side of the foot; insertion plates present and visible when valves are disarticulated 6
4. Girdle extends more than halfway into the sutures between valves and has conspicuous spicules; valves thick; jugum raised, anteriorly extended relative to forward edge of a valve's tegmentum; sculpture thickened and indistinct; sculpture on all valves increasingly inflated with increasing body size; color dull brown tinged toward tan or gray; rare in intertidal; length to 3.4 cm . . . . *Oldroydia percrassa*
— Girdle not extending noticeably between the valves, girdle spicules present or absent; valves thin, jugal area not raised or extended toward anterior; sculpture (if present) of radial and longitudinal rows of granules; length <2 cm 5
5. Color of valves brown with darker specks; profile varying from flattened in small specimens to conspicuously arched in large animals; color of foot cream to only slightly pinkish; dorsal girdle surface with numerous scattered spicules much longer than other girdle elements . . . . *Leptochiton nexus*
— Color of valves cream, tan, or orange; valves curved and relatively high in profile; color of foot reddish or "liver colored," dorsal girdle surface without scattered long spicules

6. Tail valve conspicuously longer than head valve 7
— Tail valve smaller or about equal in length to the head valve. 8

7. Foot of living animal cream; valves whitish or tan, streaked or speckled with brown or greenish; girdle minutely granular (like fine sand paper), at 80x girdle scales with 10–14 laterally oriented ridges visible; can exceed 8 cm . . . .
. . . . *Stenoplax heathiana*
— Foot of living animal orange (except in juveniles <1.5 cm); valves purple and reddish tones with brown or white speckles; girdle scales very fine, giving girdle leathery appearance; at 80x, these scales lack visible sculpture; can exceed 8 cm . . . . *Stenoplax fallax*

8. Dorsal girdle covered with closely packed conspicuous scales 9
— Dorsal girdle appears fleshy or gritty, with or without tiny stout corpuscles, more elongate spicules, or setae (hairs or bristles) 16

9. End valves and lateral areas of intermediate valves with very prominent radial ribs (one to two on lateral areas)
. . . . 10
— End valves and lateral areas of intermediate valves with at most riblets or rows of tubercles, pustules or granules (two to many on lateral areas) 12

10. Animal small, not exceeding 8 mm, oval in outline; central areas with even, netlike reticulation; color yellowish or tan (very rare) . . . . *Callistochiton connellyi*
— Animal larger, to 3 cm or more; elongated in outline; central areas with longitudinal ribs crossed by finer cross ribs
. . . . 11

11. Head valve with nine to 10 ribs, lateral areas of intermediate valves with two ribs topped by rounded pustules; tail valve inflated, in adults posterior ribbed area very swollen and "rolled" forward (juveniles are much flatter with much less prominent sculpturing); color tan, brown, or cream . . . . *Callistochiton palmulatus*
— Head valve with seven ribs, lateral areas of intermediate valves with a single heavy rib; tail valve with pointed, posterior mucro (juveniles are much flatter with more central mucro and much less prominent sculpture) . . . .
. . . . *Callistochiton crassicostatus*

12. Central and radial areas of valves granular or with weak riblets. 13
— Central areas prominently ribbed (usually with both longitudinal and lateral cross ribs); radial areas with ribs, or rows of pustules or nodules 14

13. Central areas with numerous fine longitudinal riblets; when animal flat, total length of all valves about 2.5x width of valve 4 tegmentum, body elongate; coloration monochrome olive, slate grey, or (rarely) sky blue; length to 4.5 cm . . . . *Lepidozona regularis*
— Central areas granular, lacking longitudinal riblets; when animals flat, total valve length about 2x the width of valve 4 tegmentum, body broadly oval; color pattern highly variable, colors green, black, blue, white; length to 3 cm . . . .
. . . . *Lepidozona radians*

14. Dorsal girdle scales smooth or with only very faint ribbing; scattered scales conspicuously mammillated; radial areas with rows of pustules or nodules; color variable, often with reddish or purplish tones, but also with uniform or mottled orange, with or without symmetrical anterior and posterior curved markings, or maculated with white, yellow, tan, or brown tones; length to 5 cm. . . . . . . *Lepidozona mertensii*

— Dorsal girdle scales conspicuously ribbed and never mammillated. 15

15. Radial areas with rows of pustules or nodules; color uniform gray-green or occasionally brown; length to about 4 cm . . . . *Lepidozona cooperi*
— Radial areas with two to six ribs or riblets, with considerable variation in strength of the sculpture; color and pattern extremely variable; length to 2.5 cm . . . .
. . . . *Lepidozona scrobiculata*

16. Girdle fleshy, often sandy or leathery in appearance, lacking any noticeable setae or spicules 17
— Girdle bearing spines, spicules, or setae. 27

17. Valves orange or pinkish with alternating colored lines (in life, some lines may be brilliant blue) 18
— Valves not orange or pinkish, and lacking alternating colored lines. 21

18. Head valve with dark maroon or black lines extending the full length of the lined pattern (with similar markings on intermediate valves) 19
— Head valve lacking or with only partial dark maroon or black lines 20

19. Lines more or less longitudinal, with concentric alternating colored lines on head valve forming a distinctive "arrowhead" or "arch" shape, but without zigzags; slope of tail valve usually straight or slightly convex; length to 5 cm . . . .
. . . . *Tonicella lineata*
— Lines with zigzag patterns, with concentric zigzags on head valve; slope of tail valve concave; length to 5 cm. . . . . . .
. . . . *Tonicella lokii*

20. Lines on central areas of intermediate valves expanding to large flamelike markings, concentric lines on head valve similar in color to valve background color and difficult to see; slope of tail valve concave; 30 × 18 μm dorsal girdle scales crowded to touching at 60×; length to 1.5 cm (rare in intertidal) . . . . *Tonicella venusta*
— Lines not expanding into flamelike markings, head valve usually with conspicuous concentric zigzag lines in two colors; slope of tail valve variable; 20 × 10 μm dorsal girdle scales dispersed (not touching) at 60×; length to 2.5 cm
. . . . *Tonicella undocaerulea*

21. Uneroded portions of valves have generally regular granular sculpturing except on their lateral edges, where somewhat larger wartlike pustules are superimposed between the more regular granules, best viewed at 40× when valves are dry 22
— Uneroded portions of valves have regular granular sculpturing extending to the extreme margins of lateral valves, where there are no superimposed wartlike pustules . . . . .
. . . . 23

22. Intermediate valves scarcely if at all beaked (or often eroded where beak would be); color usually olive to dark green, occasionally maroon, often with black sub-apical stripes; gills extending entire length of foot, 28–35 plumes per side; length to 4 cm . . . . *Cyanoplax hartwegii*
— Valves prominently beaked; color green with white, black, and/or blue specks or streaks; found only in sea caves or under *Nuttallina californica*; gills extending about 80% of foot length, about 11–20 plumes per side; length rarely exceeding 1.2 cm. *Cyanoplax caverna*

23. Head valve with prominent radial ribs; valves pale yellowish or tan; found exclusively on the holdfasts of *Macrocystis pyrifera*; length to 2.5 cm (rare) . . . . . . *Cyanoplax lowei*
— Head valve lacking radial ribs, color and habitat not as above 24

24. Slope of tail valve distinctly concave; shell interior white
. . . . 25
— Slope of tail valve more or less straight to somewhat convex; shell interior blue 26
25. Valves typically rusty orange with ultramarine irregular spots, but portions of one or more valves can be solid in color; body outline oval-elongate; length rarely exceeding 1.4 cm . . . . *Cyanoplax berryana*
— Valves variable in color, often tan, black, green, or orange with white or black speckles; body outline oval; length to 1.5 cm . . . . *Cyanoplax keepiana*
26. Valves uniformly dark brown to black or can have a few white streaks, but central areas often eroded to underlying bluish shell layer; mucro nearly terminal, slope convex; girdle uniformly dark brown to black, like the valves, rough appearing because corpuscules vary in length; restricted to midintertidal, length to 1.6 cm . . . . . . *Cyanoplax thomasi*
— Valves extremely variable in color and pattern, often reddish, brown, or greenish tones with blue-green, whitish, or dark specks, spots, or streaks; mucro subcentral to central, slope more or less straight; girdle with light and dark banding or uniformly dark, with corpuscules of similar length; mid- to especially low intertidal; length occasionally >2.0 cm but more normally to 1.6 cm . . . . *Cyanoplax dentiens*
27. Valves with longitudinal and radial rows of well-spaced rounded granules; girdle with scattered fine glassy spicules visible under magnification; color often orange, reddish, green, yellow, or brown; tail valve typically black or dark brown with white central stripe; length to 1.8 cm . . . . . .
. . . . *Chaetopleura gemma*
— Valves and girdle not as above, girdle bearing spines or flexible setae 28
28. Girdle covered densely with stout, sharp spines 29
— Girdle bearing various forms of simple or complex setae
. . . . 31
29. Valves (when not eroded) dark brown or black, sometimes with white subjugal stripes; gills extending nearly 100% of foot length; length to 5 cm . . . . *Nuttallina californica*
— Valves not black, gills extending 80% or less of foot length
. . . . 30
30. Exposed portion of intermediate valves about twice as wide as long; color brown, white, yellow, or pink tones; gills extending about 75%–80% of foot length; length to 3.5 cm
. . . . *Nuttallina fluxa*
— Exposed portion of intermediate valves about as long as wide; color brown or pink-orange tones; gills extending less than 75% of foot length; length to 3.5 cm . . . . *Nuttallina* sp.
31. Girdle greatly expanded anteriorly; setae scaled; valves very wide and much shorter than wide, variously colored with brown, pink, blue, tan, and green; length to 5 cm . . . . . .
. . . . *Placiphorella velata*
— Girdle not expanded anteriorly, setae not scaled; valves not short and wide. 32
32. Valves smooth appearing, microgranular; sculpture, if present, only in the form of longitudinal ribs on central areas that are especially conspicuous when valves are dry; largest setae located chiefly near valve junctures and around posterior girdle, setae can be scarce with girdle nearly bare; girdle never with posterior slit 33
— Valves variously sculptured; head valve often with 10 radial ribs, including two flanking each posterior margin of the valve and the other eight corresponding at the valve margin to slits in the hidden eaves (articulamentum) anchoring the valves in the girdle; setae simple to complex, scattered

or profuse; girdle with or without posterior slit 34
33. Valves microscopically granular; lateral areas not defined; color red, orange, or green speckled with blue, pale green and gray; setae with single row of recurved bristles (often restricted to area around posterior two valves); length usually <2 cm . . . . *Dendrochiton flectens*
— Lateral areas of intermediate valves raised; central areas with longitudinal ribs most conspicuous when valves are dry; setae plumose; coloration highly variable, often speckled with pink, green, gray, orange and/or brown; length usually <1.2 cm. . . . . *Dendrochiton thamnoporus*
34. Setae abundant, coarse, dark brown, simple stiff bristles; valves coarsely sculptured, dark brown, sometimes with gray mottling or white subjugal stripes; animals shorter than 1.5 cm can have spicules in the setal groove; body length to 8 cm (plate 353A) . . . . *Mopalia muscosa*
— Setae not as above 35
35. Setae extremely fine, hairlike 36
— Setae not fine and hairlike 37
36. Central area of valves with fine longitudinal riblets or lirae cut by growth lines; lateral areas granulose; setae profuse, very fine, often bearing short, extremely fine bristles (plate 353B); common in exposed rocky intertidal; length can exceed 6 cm. . . . . *Mopalia hindsii*
— Central areas with netlike reticulation (sometimes obsolete); lateral areas faintly granulose to smooth, bounded by a row of weak, spaced, rounded granules; setae fine, bearing short, strongly recurved bristles; rare and only subtidal in California; length can exceed 6 cm (plate 353C) . . . . .
. . . . *Mopalia vespertina*
37. Setae with white or yellow-tinged spicules, which are located directly on the shaft or are mounted on a short bristle that is usually less than twice the length of the spicule (plates 353D–353H, 354A, 354B) 38
— Setae either with long, soft, and flexible chitinous bristles that are usually tipped with a minute calcareous spicule, or setae recurved and devoid of bristles or spicules (plate 354C–354I) 44
38. Setae as long or longer than valve 5 tegmentum, thick, bearing numerous long, white or yellow-tinged spicules, with or without stalks. 39
— Setae one-half to three-quarters as long as valve 5 tegmentum, most spicules close to the shaft but those in curved portions of the setae can be supported away from shaft on bristles 40
39. Jugal areas smooth (or nearly smooth); setae with two (occasionally three to four) rows of curved, sharply pointed white spicules arising from the upper side (plate 353D); color grayish green, sometimes with black or rarely red or white subjugal markings; length usually <2 cm . . . .
. . . . *Mopalia lionota*
— Jugal areas not smooth; setae bearing sharply pointed white or yellow tinged spicules entirely encircling the shaft; color greenish tones, with dark (rarely colorful) markings; length usually <3 cm (plate 353E) . . . .
. . . . *Mopalia lowei*
40. Setae without dorsal groove, short (to about 2 mm), bearing five irregular rows of spicules, three along the dorsal side, and one on each edge; ventral surface of setae bare; valves variously colored and patterned with rose, lavender, blue, orange, white, and gray (plate 353F). . . . .
. . . . *Mopalia ferreirai*
— Setae with trough-shaped dorsal groove, but trough can be shallow and setae straplike, bearing one to four longitudinal

rows of spicules 41

41. Setae trough shaped, sparsely distributed in two or three alternating rows and especially prominent at valve sutures; setae bear a single (occasionally double) row of slender spicules (plate 353G); valve color green with black subjugal stripes or markings; length to 2.5 cm. . . . .
. . . . *Mopalia porifera*
— Setae not restricted to two or three alternating rows, more numerous; spicules in two to three (or four) rows; valve color various 42

42. Posterior girdle notch absent; unbroken setae tapered, tip width one-half or less of base width; somewhat recurved; seta length generally less than one-half the length of the valve 5 tegmentum; proximal part of setae with two alternating rows of short-stalked, stout spicules, distal portions usually bare; color pattern highly variable, often streaked, speckled or mottled with red, purple, green, pink, gray, and blue; length rarely exceeding 3 cm (plate 353H) . . . .
. . . . *Mopalia acuta*
— Posterior girdle notch present; unbroken setae not tapering strongly 43

43. Setae straplike or a broad shallow trough, bearing three (or four) rows of sharply pointed stalked white spicules along the dorsal side, stalks about one to 1.5 times the length of the spicules; spicules usually restricted to the proximal one-third of shaft; color and pattern highly variable, often green, with red, orange, blue, or white markings; length to 6 cm (plate 354A). *Mopalia ciliata*
— Setae slender, trough shaped or cylindrical, bearing two rows of slender stalked white spicules; on cylindrical setae, stalks and spicules can lie entirely within the setal groove; on curved setae, stalks can diverge from the groove and rise as rows of spicule-tipped bristles, the bristles about two (or three) times the length of the spicules; color and pattern variable, often green tones with reddish, black, yellow, or white markings; length to 6 cm (plate 354B) . . . .
. . . . *Mopalia kennerleyi*

44. Setae with four or more rows of bristles. 45
— Bristles absent or in one to three rows on setae 46

45. Setae with four or more conspicuous rows of coarse bristles, proximal half of bristles clinging to the setae in crowded parallel alignment often obscuring the setal shaft, distal half of bristles angling away from setae at acute angles, tending to form tiers; head valve with radial rows of pustules, lateral area of valves set off by an indistinct row of pustules; valves light green or olive, with red flecks and brilliant blue zigzag lines, sometimes valves are partially or completely suffused with orange (plate 354C) . . . .
. . . . *Mopalia spectabilis*
— Setae with slender, usually recurved, bristles in numerous indistinct rows, setal shaft visible between bristles, bristles angled away from the setal shaft at their attachment point; head valve heavily corded with prominent, usually nodulose, radial ribbing and lateral areas set off by similar prominent ribbing; valve color generally beige/brown but varies from white to pastel orange; length usually <1.5 cm (plate 354D) . . . . *Mopalia imporcata*

46. Setae lacking bristles, recurved, trough shaped (plate 354E; sparse, slender bristles exist on chitons <2.5 cm); valve sculpture smooth except for subdued rhomboid pitting, more pronounced in larger individuals; color pattern streaked or feathered with gray-brown and green, or white and black with occasional burgundy accents; length to 7 cm. . . . . *Mopalia lignosa*

— Setae with one to three rows of conspicuous bristles. 47

47. Bristles on setae robust with base of bristles roughly one-third the width of setae, recurved, tapering to spiculeless point, setae one to two times as long as valve 5 and bearing two to three bristle rows; valves smooth, with reticulation of deep pitting; radial ribs distinct, smooth or faintly beaded, rib of posterior edge of intermediate valves obsolete; color mottled with red, dark brown, white, or blue-green; length to 2 cm (plate 354F). . . . . . . *Mopalia sinuata*
— Bristles on setae slender, minimally tapering. 48

48. Setae with abundant long (often curled) filamentous bristles in a single row within the dorsal groove; setae two to three times as long as valve 5; central areas with strong longitudinal ribbing and weaker cross ribbing, radial ribs prominent, nodulose, posterior edge of valves with longitudinally elongate nodules; color mottled with brown, white, and dark green; length to about 2 cm (plate 354G)
. . . . *Mopalia cirrata*
— Setae generally with two rows of bristles, placed laterally along the setal shaft; radial ribbing and longitudinal ribs of central area reduced or obsolete. 49

49. Jugal sculpture of distinct regular pitting, similar to that of adjacent central area; lateral one-third of anterior edge of valve 5 tegmentum angles away from posterior edge of valve 4 at 10–20 degrees; setae usually profuse, about one-half to one time as long as tegmentum of valve 5, dorsal groove either broad and bearing two irregular alternating rows of long bristles or narrow with bristles in a single line; color gray-white, pale blue, or burgundy, maculated with dark brown or black; length usually <3.5 cm (plate 354H)
. . . . *Mopalia plumosa*
— Central area sculpture of indistinct to distinct regular pitting, diminishing to fine longitudinal ridges at the jugum; lateral anterior one-third of valve 5 tegmentum edge angles away from posterior edge of valve 4 at 30–40 degrees; setae very fine, less than half length of valve 5 tegmentum, often sparsely scattered, bearing two rows of fine, often opposite, laterally diverging, curved bristles; color and pattern variable, with one or more valve portions often colored uniformly or marked with orange, green, yellow, brown, and other colors; length to 6 cm (plate 354I) . . . .
. . . . *Mopalia swanii*

List of Species

Unless otherwise indicated, habitats range from middle and low intertidal into the subtidal at depths of up to about 15 m (about 50 ft). The following list also includes species that are more typical of warm temperate southern California that are not in the key, even though occasional reports exist for central California. With relatively few exceptions, the above key will also serve adequately for the Pacific Northwest, the outer Channel Islands, and the northern Baja coast. The faunas of the Channel Islands and of the cold upwelling region south of Ensenada, near Punta Banda and Punta Santo Tomás, resemble the fauna of central California more than they do southern California, although there is an interesting mix of the two faunal elements.

The following classification to families represents a conservative estimate of phylogenetic affinities based on mitochondrial DNA sequence comparisons (unpublished research by DJE and collaborators). The composition of Lepidochitonidae and Mopaliidae is very robust in these molecular analyses but differs substantially from recent classifications. For simplicity's sake, we have elected to either not use subgenus names or to elevate them to generic status, despite the possibility that their recognition might reflect an accurate phylogenetic pattern of nesting.

The late Donald Abbott (1987) presented sketches of the anatomy and functional morphology of *Nuttallina californica, Mopalia muscosa, Katharina tunicata, Cyanoplax hartwegii,* and *Lepidozona mertensii* from the Monterey Peninsula.

LEPTOCHITONIDAE

Leptochiton rugatus (Pilsbry, 1892). Occasional individuals can be found under rocks well submerged into sand or mud, especially those in shallow warm mid-zone pools near the shore; subtidally especially at 8 m–12 m, or within kelp holdfasts. The red foot and gills are from tissue hemoglobins (Eernisse et al. 1988).

Leptochiton nexus Carpenter, 1864. Rarely collected and primarily subtidal. This species is not rare but has the cryptic habit of living on the side and top surfaces of rocks that are well covered by sand.

Oldroydia percrassa (Dall, 1894) [alternatively placed in separate family, Protochitonidae, by Sirenko (1997), together with an undescribed West Coast *Deshayesiella* sp. from >15 m (R. N. Clark and B. Sirenko, unpublished)]. Rare, under rocks resting on soft substrate but not silt (5 m–10 m) and from granitic ridge under rocks resting on a mixture of course gravel and finer sediment (22 m–24 m).

Hanleyella oldroydi (Dall, 1919). Subtidal, >15 m.

CHAETOPLEURIDAE

Chaetopleura gemma Dall, 1879 (assigned to subgenus *Pallochiton* by Kaas and Van Belle, 1985–1994, volume 3). Common on top and sides of rocks throughout Monterey Peninsula kelp forest down to 10 m.

ISCHNOCHITONIDAE

Stenoplax fallax (Carpenter in Pilsbry, 1892). Primarily subtidal. Along Monterey Peninsula, juveniles <1.5 cm are often under thin layers of sediment on top of rocks, while adults are buried below the sand line along the sides of rocks.

Stenoplax heathiana Berry, 1946 (assigned to subgenus *Stenoradsia* by Kaas and Van Belle, 1985–1994, volume 3). Intertidal down to 7 m under rocks well submerged in sand. Named for Harold Heath who, as a Stanford University professor at Hopkins Marine Station, pioneered the study of California chitons, including an extensive cell lineage study of this species (Heath, 1899). This species is unusual in spawning a sticky egg mass from which crawl-away larvae emerge (Haderlie and Abbott, 1980). Look for the tiny commensal snail *Vitrinella oldroydi* in the mantle cavity. See Andrus and Legard 1975 (habitat); Linsenmeyer 1975, Veliger 18 Supplement: 83–86 (behavior); Putman 1990, Veliger 33: 372–374 (diet).

Stenoplax conspicua (Pilsbry, 1892) (assigned to subgenus *Stenoradsia* by Kaas and Van Belle, 1985–1994, volume 3). Rare north of southern California, where it is common under rocks in a similar habitat to *S. heathiana*. May be preyed upon by octopus, which drill small holes through the plates (Pilson and Taylor 1961, Science 134: 1366–1368). Abbott and Haderlie (1980) note that tiny snails in the genera *Teinostoma* and *Vitrinella* may occur under the girdle.

Lepidozona cooperi (Dall, 1879) (=*Ischnochiton cooperi*). Most common from the low intertidal to 8 m, under rocks and hidden beneath sediment deposits on rocky surfaces. For a review of the genus, see Ferreira (1978).

Lepidozona radians (Pilsbry, 1892) (=*Ischnochiton radians*) Recognized as distinct herein; formerly (e.g., Ferreira, 1978) considered a synonym of the somewhat more northern (Alaska to Washington) and more uniformly tan-colored or reddish *L. interstincta* (Gould, 1852) (=*Ischnochiton interstinctus*). *L. radians* is highly variable in its coloration pattern and is found at shallower depths and in somewhat more exposed habitats, and its range is from southeastern Alaska to northern Baja. Molecular distinctions have also been found (DJE and R. P. Kelly, unpublished). Occasional in the intertidal but most common between 5 m–13 m (ranging deeper) under rocks and hidden beneath sediment deposits on rocky surfaces.

Lepidozona mertensii (Middendorff, 1847). Common in the intertidal to about 8 m, but ranging deeper, on bottom and sides of rocks. See Helfman 1968, Veliger 10: 290–291 (ctenostome bryozoan *Farella elongata* on ventral surface of girdle).

Lepidozona pectinulata (Carpenter *in* Pilsbry, 1893) [=*L. californiensis* (Berry, 1931)]. Rare north of southern California, where it is common under rocks in the low intertidal.

Lepidozona regularis (Carpenter, 1855) (=*Ischnochiton regularis*). Assigned to subgenus *Tripoplax* by Kaas and Van Belle, 1985–1994, volume 4. Relatively rare, sometimes occurring under smooth cobbles in high energy shores.

Lepidozona retiporosa (Carpenter, 1864). Rare in <15 m.

Lepidozona scabricostata (Carpenter, 1864). Rare in <15 m and not likely north of southern California.

Lepidozona scrobiculata (von Middendorff, 1847) [=*Lepidozona sinudentata* (Carpenter in Pilsbry, 1892)]. Most common from 5 m–10 m, under rocks and shells on sand.

Lepidozona willetti (Berry, 1917). Rare in <15 m.

Callistochiton connellyi Willett, 1937. Apparently a rare small species, known to occur in the intertidal. See Ferreira (1979) for a review of the genus.

Callistochiton crassicostatus Pilsbry, 1893. Especially common under rocks in shallow subtidal habitats.

Callistochiton palmulatus Dall, 1879. Especially common under rocks in sandy to silty shallow subtidal habitats; juveniles lack the bulging terminal valves typical of adults.

LEPIDOCHITONIDAE

Cyanoplax berryana (Eernisse, 1986) (=*Lepidochitona berryana*). Especially common in sandy flat shelves on the top and sides of rocks at 0 m–3 m; not known north of San Mateo County.

Cyanoplax caverna (Eernisse, 1986) (=*Lepidochitona caverna*). A small hermaphroditic chiton that normally appears to self-fertilize its brooded embryos (or is parthenogenetic), only locally common and with limited known range between Santa Cruz and San Luis Obispo Counties (Eernisse 1988). Sometimes found nestled in the pallial groove of the larger *Nuttallina californica*, even while brooding, creating Russian doll–like layers of nested chitons (Gomez 1975, Veliger 18 Supplement: 28–29 mistakenly as *C. dentiens*; Eernisse 1986).

Cyanoplax dentiens (Gould, 1846) (=*Lepidochitona dentiens*). Very common species, especially from the low intertidal to about 1 m, on the top and sides of rocky outcrops and boulders between central California and Alaska. Often overlooked because of its small size and cryptically variable color patterns.

* = Not in key.

This species is easy to confuse with other members of the genus (Eernisse 1986; 1988). See Piercy 1987 (habitat, feeding).

Cyanoplax hartwegii (Carpenter, 1855) (=*Lepidochitona hartwegii*). Common under the rockweed *Silvetia compressa* as well as in mid-intertidal tide pools, from Santa Cruz to northern Baja California. See DeBevoise (predation by seastars and crabs); Lyman (behavior); Robb (diet), Andrus and Legard (habitat), McGill (osmotic stress), and Connor (ecology), all in Veliger 18 Supplement, 1975.

Cyanoplax keepiana (Berry, 1948) (=*Lepidochitona keepiana*). Found in warm protected pools under small stones, only rarely observed north of Cayucos.

Cyanoplax lowei (Pilsbry, 1918) (=*Cyanoplax fackenthallae* Berry, 1919). Found exclusively amongst the holdfasts of the giant kelp, *Macrocystis pyrifera*, but rarely collected.

*Cyanoplax cryptica (Kues, 1974). Found exclusively on the southern sea palm kelp, *Eisenia arborea*. Originally proposed as subspecies of *Cyanoplax dentiens*; not known north of Catalina Island.

Cyanoplax thomasi (Pilsbry, 1898) (=*Lepidochitona thomasi*; =*Nuttallina thomasi*). A brooder with separate sexes, only locally abundant in mid-intertidal rocky cracks or under barnacle hummocks; known from the Monterey Peninsula to the southern Big Sur coastline. Closely related to the Pacific Northwest to southeastern Alaska *C. fernaldi* (Eernisse, 1986), which reproductively resembles *C. caverna* in being a selfing (or parthenogenetic) hermaphroditic brooder.

Nuttallina californica (Reeve, 1847). Extremely common mid-intertidal species. Rare north of central California or south of Point Conception, California, but does occur as far south as northern Baja California. See Moore (predation by gulls), Nishi (feeding), Robbins (respiration), Andrus and Legard (habitat), Gomez (association with *Cyanoplax*), Linsenmeyer (behavior), Piper (physiology), Simonsen (osmotic stress), all in Veliger 18 Supplement, 1975.

Nuttallina fluxa (Carpenter, 1864) [=*Nuttallina scabra* (Reeve, 1847), see Piper 1984]. Rare north of southern California, where it is common in mid- to low intertidal habitats, including home depressions when the substrate is sandstone.

Nuttallina sp. of Piper, 1984. More common in southern California, but does occur at central California localities with sandstone shelves, where it forms home depressions in the low intertidal.

MOPALIIDAE

Placiphorella velata Dall, 1879. Can entrap small prey beneath anterior girdle flap (McLean 1962, Proc. Malacol. Soc. London 35: 23–26). Occasionally found in the intertidal but more common at 5 m–10 m on sides and bottoms of rocks.

*Placiphorella mirabilis Clark, 1994. Greater than 15 m. See Clark 1994, Veliger 37: 290–311.

Katharina tunicata (Wood, 1815). Occurs with *Nuttallina californica* in central California; lives among corallines and mussels on exposed rocks from the Big Sur coastline to Alaska. See Giese et al. 1959 (reproduction); Tucker and Giese 1959 (shell repair); Nimitz and Giese 1964, Quart. J. Micr. Sci. 105: 481–495 and Lawrence and Giese 1969, Physiol. Zool. 42: 353–360 (both, chemical changes in reproduction and nutrition); Himmelman 1978, J. Exp. Mar. Biol. Ecol. 31: 27–41 (reproduction); Piercy 1987 (habitat, feeding); Stebbins 1988, Veliger 30: 351–357 (population structure, tenacity); Rostal and Simpson 1988, Veliger 31: 120–126 (salinity); Dethier and Duggins 1984, Amer.

Nat. 124: 205–219 (ecology); Markel and DeWreede 1998, Mar. Ecol. Prog. Ser. 166: 151–161 (impact on kelp *Hedophyllum*).

Tonicella lineata (Wood, 1815). Much rarer than the next species in the central California intertidal but not uncommon at 3 m–8 m and by far the most common intertidal and shallow subtidal member of the genus from northern California to Alaska. Feeds on the upper layer of persistent coralline crustose algae, keeping other organisms from attaching. See Piercy 1987 (habitat, feeding); Clark 1999 (for discussion of literature prior to 1999 and proper species attributions).

Tonicella lokii Clark, 1999. The most common of four lined chiton species in the intertidal of central California; formerly confused with the previous species.

Tonicella undocaerulea Sirenko, 1973. Rare in the intertidal in central California; most common on top and sides of rocks at 12 m–17 m; our species is probably not the same as the one originally described from the northwestern Pacific, based on mitochondrial DNA distinctions (DJE, unpublished; see also Clark 1999).

Tonicella venusta Clark, 1999. Most common on top and sides of rocks at 13 m–18 m.

Cryptochiton stelleri (von Middendorff, 1847). A northern species found south to Monterey; intertidal throughout much of its range, but more commonly subtidal from 3 m–13 m around Monterey Peninsula. Occasionally found south to the Channel Islands, although it has been found in Native American middens from cold upwelling regions of northern Baja California (Emerson, 1956). See Heath 1897, Proc. Acad. Nat. Sci. Phil. 1897: 299–302 (juvenile morphology); Okuda 1947, J. Fac. Sci. Hokkaido Univ. Zool. 9: 267–275 (postlarval development); Tucker and Giese 1959 (shell repair); Tucker and Giese 1962, J. Exp. Zool. 150: 33–43 (reproduction); MacGinitie and MacGinitie 1968, Veliger 11: 59–61 (food, growth, age, external cleaning); Webster 1968, Veliger 11: 121–125 (commensals; Palmer and Frank 1974, Veliger 16: 301–304 (growth); McDermid 1981 Veliger 23: 317–320 (association with epizoic red alga *Pleonosporium*). Talmadge (1975, Veliger 17: 414) reported that the carnivorous snail *Ocinebrina lurida* makes pits on the dorsal surface of *C. stelleri*, rasping down to the flesh under the valves.

Dendrochiton flectens (Carpenter, 1864). [=*Basiliochiton heathii* (Pilsbry, 1898)] Mostly subtidal, 5 m–10 m, common on all sides of rocks, occasionally in low intertidal.

Dendrochiton thamnoporus (Berry, 1911). Common on the Monterey Peninsula on top and sides of rocks from 4 m–15 m; rare in low intertidal.

Mopalia acuta (Carpenter, 1855). Formerly confused with *M. plumosa*, in part (see below). Lowest intertidal to subtidal under rocks and shells on sand and beneath the sand line on larger rocks. Most abundant on Monterey Peninsula at 5 m–13 m.

Mopalia ciliata (Sowerby, 1840). Locally common in low intertidal, under overhangs and in crevices to about 10 m on all sides of rocks; rare north of Monterey Bay. See Fitzgerald 1975, Veliger 18 Supplement: 37–39 (movement, phototactic responses); Piercy 1987 (habitat, feeding).

Mopalia cirrata Berry, 1919. Subtidal in California.

*Mopalia egretta Berry, 1919. Rare in central California; >15 m.

Mopalia ferreirai Clark, 1991. Subtidal in California at 5 m–15 m, on top and sides of rocks.

Mopalia hindsii (Sowerby in Reeve, 1847). Most common in the mid intertidal to 2 m on exposed coasts, often found deep in crevices or on the walls of sea caves. See Giese et al. 1959 (reproduction); Tucker and Giese 1959 (shell repair); Andrus and Legard 1975 (habitat); Himmelman 1980 (reproduction,

* = Not in key.

British Columbia); Piercy 1987 (habitat, feeding); Rostal and Simpson 1988, Veliger 31: 120–126 (salinity).

Mopalia imporcata Carpenter, 1865. Subtidal in California, especially at about 8 m–12 m, or apparently somewhat deeper in canyons, but occurs in the intertidal further north.

Mopalia kennerleyi Carpenter, 1864. Recognized as distinct herein; formerly considered a synonym of *Mopalia ciliata;* rare south of San Francisco Bay. Himmelman 1980 (reproduction, British Columbia, as *M. ciliata*).

Mopalia lignosa (Gould, 1846). Common under rocks in intertidal. Around Monterey Peninsula, populations extend below 10 m in the kelp forests. See Fulton 1975 (diet), Watanabe and Cox (reproduction), Andrus and Legard (habitat), Lebsack (physiology), Linsenmeyer (behavior), all in Veliger 18 Supplement, 1975; Himmelman 1980 (reproduction, British Columbia).

Mopalia lionota Pilsbry, 1918. Of the many species of *Mopalia* with dense setae, this is probably the most heavily ornamented. Most common from the low intertidal to about 3 m, especially in the granite and sand channel habitat in Monterey.

Mopalia lowei Pilsbry, 1918. Subtidal in California, especially from 5 m–10 m on all sides of rocks.

Mopalia muscosa (Gould, 1846). A familiar high- to low-intertidal chiton often covered with algae. Its stiff setae and oval shape distinguish it from the superficially similar but narrower members of *Nuttallina,* which also differ in bearing spines on close inspection. See Fitzgerald (movement, phototactic responses), Smith (behavior), Watanabe and Cox (reproduction), Andrus and Legard (habitat), and Westersund (movement), all in Veliger 18 Supplement, 1975; Monroe and Boolootian 1965, Bull. So. Calif. Acad. Sci. 64: 223–228 (reproduction); Himmelman 1980 (reproduction, British Columbia); Leise 1984, Zoomorphology 104: 337–343 (metamorphosis); Piercy 1987 (habitat, feeding). See also Barnawell (1960), who found that *Mopalia muscosa, M. ciliata,* and *M. hindsii* include bryozoans, hydroids, and barnacles in their diets.

*Mopalia phorminx Berry, 1919. Greater than 15 m.

Mopalia plumosa Carpenter in Pilsbry, 1893. Recognized as distinct herein; formerly considered a synonym of *Mopalia acuta.* Ranges from low intertidal to 7 m in Monterey Bay.

Mopalia porifera Pilsbry, 1893. More common in northern Baja California.

Mopalia sinuata Carpenter, 1864. Subtidal in California, most common on the upper surfaces of rocks, from 8 m downward.

Mopalia spectabilis Cowan and Cowan, 1977. Subtidal in California, under rocks at 7 m–12 m. See Cowan and Cowan 1977, Syesis 10: 45–52.

Mopalia swanii Carpenter, 1864. Rare south of Oregon.

Mopalia vespertina (Gould, 1852) (=*Mopalia laevior* Pilsbry, 1918). Includes *M. hindsii recurvans* Barnawell, 1960. Rare in central California, usually on sides and top of rocks at 3 m–15 m.

References

Abbott, D. P. 1987. Observing marine invertebrates. G. H. Hilgard, ed. Stanford, CA: Stanford University Press, 380 pp.

Andrus, J. K., and W. B. Legard. 1975. Description of the habitats of several intertidal chitons (Mollusca: Polyplacophora) found along the Monterey Peninsula of central California. Veliger 18 (supplement): 3–8.

Barnawell, E. B. 1960. The carnivorous habit among the Polyplacophora. Veliger 2: 85–88.

Berry, S. S. 1917, 1919. Notes on West American chitons—I and II. Proc. Calif. Acad. Sci. (4) 7: 229–248 and 9: 1–36.

Berry, S. S. 1961. Chitons, their collection and preservation, pp. 44–49. In How to collect shells. 2nd ed. American Malacological Union.

Buckland-Nicks, J. 1995. Ultrastructure of sperm and sperm-egg interaction in Aculifera: implications for molluscan phylogeny. Mémoires du Muséum national d'Histoire naturelle 166: 129–153.

Burghardt, G. E., and L. E. Burghardt. 1969. A collector's guide to west coast chitons. San Francisco Aquarium Society, Special Publication 4, 45 pp.

Clark R. N. 1991. A new species of *Mopalia* (Polyplacophora: Mopaliidae) from the northeast Pacific. Veliger 34: 309–313.

Clark R. N. 1999. The *Tonicella lineata* (Wood, 1815) species complex (Polyplacophora: Tonicellidae), with descriptions of two new species. American Malacological Bulletin 15: 33–46.

Clark R. N. 2004. On the identity of von Middendorff's *Chiton sitchensis* and *Chiton scrobiculatus.* Festivus 36: 49–52.

Dethier, M. N., and D. O. Duggins. 1984. An "indirect commensalisms" between marine herbivores and the importance of competitive hierarchies. Am. Nat. 124:205–219.

Duggins, D. O., and M. N. Dethier. 1985. Experimental studies on herbivory and algal competition in a low intertidal habitat. Oecologia 67: 183–191.

Eernisse, D. J. 1986. The genus *Lepidochitona* Gray, 1821 (Mollusca: Polyplacophora) in the northeastern Pacific Ocean (Oregonian and Californian provinces). Zoologische Verhandelingen 228: 3–52.

Eernisse, D. J. 1988. Reproductive patterns in six species of *Lepidochitona* (Mollusca: Polyplacophora) from the Pacific coast of North America. Biological Bulletin 174: 287–302.

Eernisse, D. J. 1998. Class Polyplacophora, pp. 49–73. In Taxonomic atlas of the benthic fauna of the Santa Maria Basin and the Western Santa Barbara Channel. Volume 8. The Mollusca, Part 1: Aplacophora, Polyplacophora, Scaphopoda, Bivalvia and Cephalopoda. P. V. Scott and J. A. Blake, eds. Santa Barbara, CA: Santa Barbara Museum of Natural History.

Eernisse, D. J., and P. D. Reynolds. 1994. Chapter 3. Polyplacophora, pp. 56–110. In Microscopic anatomy of invertebrates, Volume 5, Mollusca 1. New York: Wiley-Liss.

Eernisse, D. J., N. B. Terwilliger, and R. C. Terwilliger. 1988. The red foot of a lepidopleurid chiton: Evidence for tissue hemoglobins. Veliger 30: 244–247.

Emerson, W. K. 1956a. Upwelling and associated marine life along Pacific Baja California, Mexico. Journal of Paleontology 30: 393–397.

Ferreira, A. J. 1978. The genus *Lepidozona* (Mollusca: Polyplacophora) in the temperate eastern Pacific, Baja California to Alaska, with the description of a new species. Veliger 21: 19–44.

Ferreira, A. J. 1979. The genus *Callistochiton* Dall, 1879 (Mollusca: Polyplacophora) in the eastern Pacific, with the description of a new species. Veliger 21: 444–466.

Ferreira, A. J. 1982. The family Lepidochitonidae Iredale, 1914 (Mollusca: Polyplacophora) in the northeastern Pacific. Veliger 25: 93–138.

Giese, A. C., J. S. Tucker, and R. A. Boolootian 1959. Annual reproductive cycles of the chitons *Katharina tunicata* and *Mopalia hindsii.* Biol. Bull. 117: 81–88.

Haderlie, E. C., and D. P. Abbott. 1980. Polyplachophora: the chitons, pp. 412–428. In Intertidal invertebrates of California. Morris, R. H., D. P. Abbott, and E. C. Haderlie, eds. Stanford, CA: Stanford University Press.

Hanselman, G. A. 1970. Preparation of chitons for the collector's cabinet. Of Sea and Shore 1: 17–22.

Heath, H. 1899. The development of *Ischnochiton.* Zool. Jahrb., Abt. Anat. Ontog. Tiere 12: 567–656.

Himmelman, J. H. 1980. Reproductive cycle patterns in the chiton genus *Mopalia* (Polyplacophora). Nautilus 94: 39–49.

Hyman, L. H. 1967. The Invertebrates: Mollusca I, Vol. VI. McGraw-Hill, pp. 70–142.

Kaas, P., and R. A. Van Belle, 1985–1994. Monograph of living chitons. Vols. 1–5. E. J. Brill/Dr W. Backhuys, Leiden.

Kelly, R. P., and D. J. Eernisse. 2007. Southern hospitality: A latitudinal gradient in gene flow in the marine environment. Evolution, 61.

Kelly. R. P., I. N. Sarkar, D. J. Eernisse, and R. Desalle. 2007. DNA barcoding using chitons (genus *Mopalia*). Molecular Ecology Notes 7.

Lamb, A., and B. P. Hanby. 2005. Marine life of the Pacific Northwest: a photographic encyclopedia of invertebrates, seaweeds and selected fishes. Madeira Park, B.C., Canada: Harbour Publishing.

Leise, E. M., and R. A. Cloney. 1982. Chiton integument: ultrastructure of the sensory hairs of *Mopalia muscosa* (Mollusca: Polyplacophora). Cell and Tissue Research 223: 43–59.

Lowenstam, H. A. 1962. Magnetite in denticle capping in Recent chitons (Polyplacophora.) Bull. Geol. Soc. Amer. 73: 435–438.

* = Not in key.

Okusu, A., E. Schwabe, D. J. Eernisse, and G. Giribet. 2003. Towards a phylogeny of chitons (Mollusca: Polyplacophora) based on combined analysis of five molecular loci. Organisms Diversity and Evolution 3: 281–302.

Omelich, P. 1967. The behavioral role and the structure of the aesthetes of chitons. Veliger 10: 77–82.

Piercy, R. D. 1987. Habitat and food preferences in six Eastern Pacific chiton species (Mollusca: Polyplacophora). Veliger 29: 388–393.

Pilsbry, H. A. 1892–1894. Polyplacophora (Chitons). Manual of Conchology 14, 350 pp.; 15, 133 pp.

Piper, S. C. 1984. Biology of the marine intertidal mollusc Nuttallina, with special reference to vertical zonation, taxonomy and biogeography (electrophoresis, growth, movement). Ph.D. Dissertation, University of California, San Diego, 698 pp.

Sirenko, B. I. 1993. Revision of the system of the order Chitonida (Mollusca: Polyplacophora) on the basis of correlation between the type of gills arrangement and the shape of the chorion processes. Ruthenica 3: 93–117.

Sirenko, B. I. 1997. The importance of the development of articulamentum for taxonomy of chitons (Mollusca, Polyplacophora). Ruthenica 7: 1–24.

Sirenko, B. 2006. New outlook on the system of chitons (Mollusca: Polyplacophora). Venus 65: 27–49.

Smith, A. G. 1960. Amphineura. In Treatise on invertebrate paleontology. Part I, Mollusca 1, pp. 41–76. R. C. Moore, ed. Univ. Kansas Press and Geol. Soc. Amer.

Smith, A. G. 1966. The larval development of chitons (Amphineura). Proc. Calif. Acad. Sci. (4) 32: 433–446.

Smith, A.G. 1977. Rectification of West Coast chiton nomenclature (Mollusca: Polyplacophora). Veliger 19: 215–258.

Strathmann, M., and D. J. Eernisse. 1987. Phylum Mollusca, Class Polyplacophora, pp. 205–219 in The Friday Harbor Labs Handbook of Marine Invertebrate Embryology. Seattle: Univ. of Wash. Press.

Thorpe, S. R., Jr. 1962. A preliminary report on spawning and related phenomena in California chitons. Veliger 4: 202–210.

Tomlinson, J., D. Reilly, and R. Ballering. 1980. Magnetic radular teeth and geomagnetic responses in chitons. Veliger 23: 167–170.

Tucker, J. S., and A. C. Giese. 1959. Shell repair in chitons. Biol. Bull. 116: 318–322.

Yonge, C. M. 1939. On the mantle cavity and its contained organs in the Loricata (Placophora). Quart. J. Micro. Sci. 81: 367–390.

Yonge, C. M. 1960. General Characters of Mollusca, pp. 3–36. In Treatise on invertebrate paleontology. Part I. Mollusca 1. R.C. Moore, ed. New York: Geol. Soc. Amer.; Lawrence: University of Kansas Press.

Gastropoda

Shelled Gastropoda

JAMES H. McLEAN

(Plates 355–373)

The gastropods are the largest class of mollusks and exhibit enormous diversity in form and habitat. Limpets, top shells, abalone shells, periwinkles, slipper shells, and whelks are well known to observers of tide pool animals. The beauty of many gastropod shells, especially from tropical regions, has long made them favored objects for collections. Our relatively advanced knowledge of the taxonomy of the gastropods is in large part due to the interest of amateur shell collectors.

This section deals with those gastropods with external shells that occur between Oregon and Point Conception, California, other than the patellogastropod limpets (see separate text by David Lindberg), all species of Littorina (separate text by David Reid) and pelagic gastropods (separate text by Roger Seapy and Carol Lalli). As in the 1975 text, shelled opisthobranchs are included, which are also treated separately by Gosliner and Williams.*

Gastropods possess a muscular foot for creeping or burrowing, a head with sensory tentacles and eyes, and a characteristic rasping radula (absent in some). As in all mollusks, the mantle secretes the shell and provides, in the pallial cavity, a shelter for the gills (CTENIDIA). A hallmark of the gastropods that sets them apart from other mollusks is the phenomenon of TORSION, which occurs early in development. Torsion consists of a 180° counterclockwise rotation of the visceral mass upon the head and foot; the result is that the mantle cavity, ctenidia, and anus, which were originally at the rear, come to lie just above the head. Torsion in its fullest expression characterizes the prosobranch grade (meaning front gills), in which the ctenidia lie anteriorly and the nervous system is twisted into a crude figure 8 (the STREPTONEUROUS condition). Other groups of gastropods have tended to modify the extreme effects of torsion, one change being a straightening out of the nervous system to the EUTHYNEUROUS condition. Euthyneury has been attained in two ways: in opisthobranchs, the body has "unwound" itself in DETORSION; in the pulmonates, the body has retained much of its torsion, but the central nervous system has straightened out by condensation into a ring of ganglia around the esophagus.

Torsion is not the same thing as the coiling of the shell and visceral hump of most gastropods. Coiling serves to strengthen the shell, but it is lost in limpetlike gastropods, land slugs, and nudibranchs, which as adults have reduced or lost the shell and flattened the visceral hump. Coiling is not unique to the gastropods; it is also found in the cephalopod Nautilus and many extinct, shelled cephalopods.

CLASSIFICATION

Higher classification of gastropods has undergone fundamental changes in the 30 years since the 1975 publication of the last edition of Light's Manual. In that work, the prevailing classification was followed in which the divisions for the class Gastropoda were the subclasses Prosobranchia, Opisthobranchia, and Pulmonata; the prosobranchs were further subdivided into the orders Archaeogastropoda, Mesogastropoda, and Neogastropoda. That classification scheme is now considered to have been based on recognition of grades of complexity. The basic hypotheses of gastropod phylogeny have been greatly altered by the application of cladistic methodology and molecular genetics (see Introduction to Mollusca).

The classification system adopted here was introduced by consensus during the 1990s, in papers by Lindberg, Haszprunar, and Ponder and other authors who preceded the publication of the two mollusk volumes for the monumental Fauna of Australia (Beesley et al., 1998). A general phylogeny for Mollusca was presented in simplified form and reiterated by Lindberg, Ponder and Haszprunar (2004) in their section on Mollusca for the Tree of Life volume.

The two major divisions for the class Gastropoda are the subclasses EOGASTROPODA (represented by the living Patellogastropoda) and ORTHOGASTROPODA (containing all other gastropods), now placed within five monophyletic clades. These five groups are the superorders: VETIGASTROPODA, NERITOGASTROPODA, COCCULINIDA, CAENOGASTROPODA, and HETEROBRANCHIA.

The eogastropod superorder Patellogastropoda is treated separately by Lindberg. Two of the five orthogastropod superorders are not represented in the intertidal of Oregon and central California: the Neritogastropoda, which are mostly tropical, and the Cocculinida, which occur offshore in deep water.

*This section is revised from 1975 text by James T. Carlton and Barry Roth.

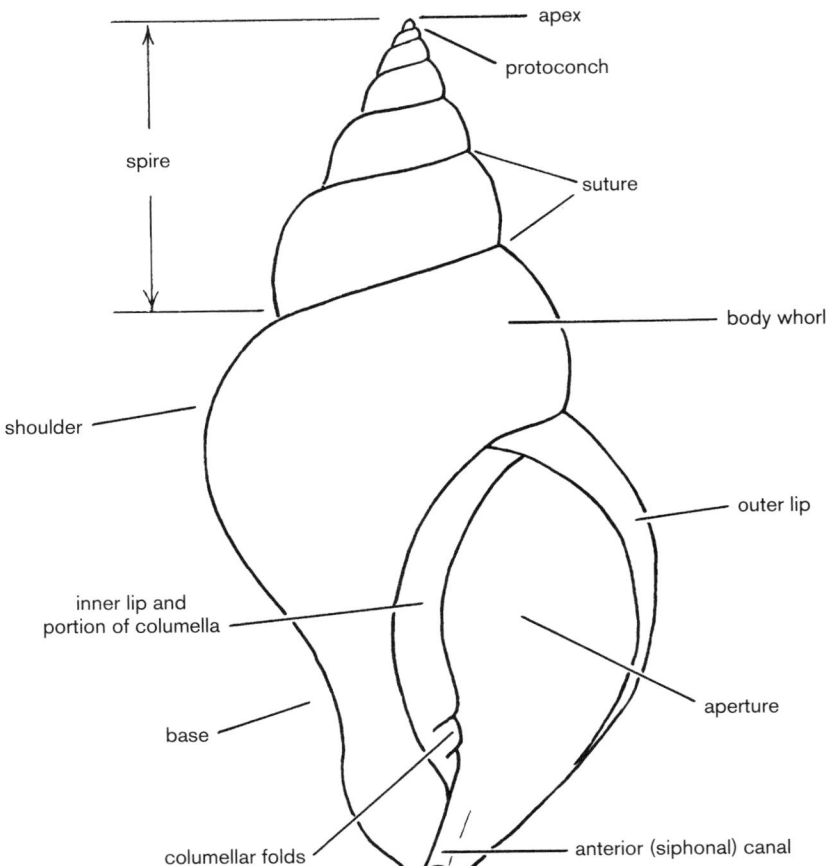

PLATE 355 Generalized gastropod shell.

Classification at the family and superfamily level remains much the same as in the 1975 edition of *Light's Manual* and follows the usage by Ponder and others in the *Fauna of Australia* (1998). That work should be consulted for greater detail at all levels of higher classification to the family and subfamily level. The families and superfamilies not represented in shallow water from central California to Oregon are omitted in the outline of classification that follows.

After the text for this work was completed, a revised classification of gastropods was provided as Part 2 of Bouchet and Rocroi (2005). The nomenclator of Part 1 provides a long-needed standard source for authorship and dates at the family, subfamily, and superfamily level. It was too late to fully adopt the new classification, which introduced a few changes to the content of superfamilies. The content of families has remained essentially unchanged in the revised classification of Bouchet and Rocroi.

VETIGASTROPODA

Shell coiled or limpet-shaped; with nacreous or non-nacreous shells; ctenidia bipectinate (with cilia-bearing filaments on both sides of gill axis); operculum usually present at maturity (except in limpet groups). Those groups with paired ctenidia have the shell with a slit or holes for excurrent flow of water. Groups that have lost the right ctenidium have also lost the slit or holes in the shell. The radula is rhipidoglossate (many teeth in row), used for grazing on algae or diatoms, but some groups are carnivores on sponges and other sessile invertebrates. Sexes are separate, and most are broadcast spawners. The superfamilies

included here were previously considered to comprise the primitive prosobranch order Archaeogastropoda, along with the excluded "docoglossate" limpets, which are now known as the Patellogastropoda. Superfamilies treated here are: Fissurelloidea, Pleurotomarioidea, and Trochoidea.

CAENOGASTROPODA

Shell never nacreous, operculum usually present at maturity (except in those of limpet form), ctenidium single, monopectinate, fused with roof of mantle cavity. Radula with seven teeth per row (taenioglossate condition), simplified to the fanglike ptenoglossate condition in some parasitic members, or reduced to three teeth per row (rachiglossate condition), or lost. There are substantial modifications in shell form correlated with diverse kinds of feeding. Shell form ranges from groups with simple apertures to groups with siphonal canals, including some that are mucus- and filter-feeders (calyptraeids and vermetids), carnivores on sessile invertebrates (velutinoids), shell drillers of bivalves (naticids), sponge feeders (triphoroids), and ectoparasites (epitoniids and eulimids). Most have the males with a cephalic penis, and the females produce egg capsules that allow for direct development or release of veliger larvae. Some are protandrous hermaprodites, changing from males to females (calyptraeids).

This is the largest monophyletic clade of gastropods; this name has long been in use to include all groups that were previously regarded as mesogastropod and neogastropod prosobranchs. Among this large group, only the neogastropods (rachiglossate or toxoglossate radula) are now regarded as a distinct monophyletic clade within the Caenogastropoda. Superfamilies treated here are: Cerithioidea, Littorinoidea, Cingulopsoidea, Rissoidea, Vanikoroidea, Calyptraeoidea, Vermetoidea, Velutinoidea, Naticoidea, Triphoroidea, Epitonioidea, and Eulimoidea.

NEOGASTROPODA

Shell with siphonal canal to shield the incurrent siphon formed by mantle edge; predaceous feeding by means of long, extensible proboscis with mouth and radula at the tip; heightened development of sensory osphradium for detection of living prey; some scavenging of fresh dead prey (Nassariidae). Radula with three teeth per row (rachiglossate condition), or single hollow tooth in Conoidea, modified for injection of venom (toxoglossate condition). Superfamilies treated here are the Muricoidea and Conoidea.

HETEROBRANCHIA

This clade now includes the large orders Opisthobranchia and Pulmonata, as well as some basal members, which previously had been considered to be mesogastropod prosobranchs. Members with external shells have a heterostrophic protoconch, in which the coiling direction of the mature whorls changes abruptly. All are hermaphroditic and engage in reciprocal copulation. There are various kinds of development. The basal groups (here represented by the first three superfamilies) are operculate, by which they differ from opisthobranchs, in which the operculum is lacking. Superfamilies treated here are Rissoelloidea, Omalogyroidea, and Pyramidelloidea.

OPISTHOBRANCHIA

Opisthobranchs are characterized by full detorsion, which places the gills on the right side or rear. They are highly diverse, the best known are shell-less nudibranchs or sea-slugs (treated in a separate section by McDonald). Several clades (suborder Cephalaspidea and Notaspidea) retain a shell in the adult; only these two suborders are keyed in this section (see separate section for further details about anatomy and biology in these two groups).

CEPHALASPIDEA

Shell retained at maturity, body with a broad head-shield. Includes carnivores, herbivores, and detritus feeders. They are included here for comparison of the shells with those of caenogastropods. Superfamilies treated here are Acteonoidea, Philinoidea, Diaphanoidea, Haminoidea, and Bulloidea. Shelled species of cephalaspideans are also here treated by Gosliner and Williams.

NOTASPIDEA

Shell limpet-shaped, body much larger than shell. Superfamily treated here: Umbraculoidea. See also Gosliner and Williams.

PULMONATA

The vascularized lining of the mantle cavity serves as a lung; ctenidia are lost. Most members are terrestrial or freshwater; only a few families are marine or live in brackish water. Marine pulmonates are primarily tropical; there are only a few species in temperate waters.

BASOMMATOPHORA

Eyes at the base of the cephalic tentacles. Herbivores and filter feeders. Superfamilies treated here: Siphonarioidea, Ellobioidea, Trimusculoidea.

TERMINOLOGY

Many of the terms used in the key are illustrated in plate 355, which shows a generalized gastropod shell. The gastropod shell is essentially a spirally coiled tube of increasing diameter; each turn of the spiral is a WHORL. Coiling may take place in a single plane (PLANISPIRAL coiling) or, more frequently, in descending stages encircling a projected COILING AXIS. The spiral trace of juncture with preceding or succeeding whorls is the SUTURE, which may be weakly or strong IMPRESSED. The APEX of the shell is formed by the early growth stage, the PROTOCONCH, together with the early TELEOCONCH whorls, forms the APICAL WHORLS. The apical and subsequent whorls, except the final whorl, form the SPIRE. The FINAL WHORL, also called the BODY WHORL, terminates at the APERTURE, which may be marked by a thickening of the INNER and OUTER LIPS. The base of the shell near the coiling axis forms the COLUMELLA (or PILLAR), which may be marked by COLUMELLAR FOLDS, and the outer lip may bear DENTICLES or elongate LIRATIONS.

Shell profile of coiled gastropods ranges from AURIFORM (broadly inflated), TROCHIFORM (shell diameter greater than shell height), FUSIFORM (with siphonal canal or notch), or long and slender. Trochiform shells may have a rounded aperture (COMPLETE PERISTOME) or an aperture interrupted above the columella (incomplete peristome); there may be a hollow indentation, an UMBILICUS, on the base of the shell that coincides with the coiling axis. The upper part of the aperture above the columella is the PARIETAL area, which may produce a PARIETAL SHIELD that thickens the body whorl to the left of the aperture. The area of greatest breadth in a coiled shell is termed the PERIPHERY of the whorls. The whorl profile may be ROUNDED, ANGULATE, or SHOULDERED.

Most gastropods have a separate, usually uncalcified, chitinous OPERCULUM, which is attached to the upper part of the foot in an active snail, but occludes part or all of the aperture when the animal withdraws into its shell. Opercular shapes include MULTISPIRAL, if the surface is evenly coiled (as in trochids), PAUCISPIRAL if unevenly coiled (littorinids and naticids), or LEAF-SHAPED if the operculum is uncoiled and has its nucleus toward the anterior edge (most neogastropods). Calcified opercula (Turbinidae and Tricoliidae and in some members of Naticidae) are enveloped by the foot and have a pattern unlike that of the inner side.

The surface of gastropod shell may be smooth or variously sculptured. Surface features are described as AXIAL if oriented in the direction of the coiling axis, SPIRAL if following the direction of the growing whorl. Features of axial sculpture are RIBS, NODES, and LAMELLAE (thin, raised, platelike structures); features of spiral sculpture are CORDS, THREADS, or INCISED STRIAE. A netlike sculpture composed of equally strong radial and concentric ribs is termed RETICULATE or CANCELLATE. On limpet shells, sculpture that extends from the apex of the shell outward toward the margin is termed **RADIAL**; sculpture that parallels the margin is termed **CONCENTRIC**.

Some gastropods, notably certain Muricidae, periodically develop prominent axial thickenings that may be ornamented with scales or spines; these thickenings are called **VARICES** (singular **VARIX**) and represent resting periods during which shell length does not increase, instead increasing the axial calcification at the **VARIX**, which may become the final lip in species that have **DETERMINATE** growth. Fine axial **GROWTH LINES** may traverse the whorls; these represent traces of the former position of the aperture. A thin exterior layer of organic material, at times textured or embellished, forms the **PERIOSTRACUM**, which may be removed by wear, exposing the calcareous shell layer beneath. Shells of marine gastropods have at least two layers, a pearly **NACREOUS** layer (if present) is the innermost layer; shell pigments are usually confined to the outermost layer.

The **PROTOCONCH**, at the shell apex (formerly called the nuclear whorls), consists of the embryonic whorls, which directly represents the apical view of the larval shell (if it is retained at later growth stages). If the protoconch has numerous microscopic whorls (**MULTISPIRAL** condition), it represents the entire **VELIGER** shell of the free-swimming larval stage; if it has few whorls and the tip is relatively large and off-center, it is called **PAUCISPIRAL** and is indicative of direct development without a free-swimming larval stage.

The apical whorls of mature specimens may be worn away if the animal lives in exposed habitats, as frequently occurs in the intertidal zone; in such cases, calcareous deposits form plugs from within that seal the apical whorls so that the shell continues to be fully protective. In some species, it is unusual to find a fully mature shell with an intact protoconch, so for such species it is necessary to examine growth stages to observe protoconch morphology. Some shells of living animals may be so heavily encrusted with epizoic bryozoa or algae that the encrustations have to be chipped away to reveal the surface sculpture. Shells of species that live at greater depths are least affected by encrusting organisms.

Direction of coiling is usually to the right (**DEXTRAL**) or rarely to the left (**SINISTRAL**), when viewed with the apex uppermost and the aperture in view. If the aperture lies to the right of the coiling axis, the shell is dextral; if to the left, it is sinistral. With regard to the living animal, the apex of a shell is posterior, its base anterior. The coiling axis, however, does not coincide with the long axis of the extended animal, because the spire projects posteriorly and to the right.

Height or length of a coiled gastropod shell are the same and should be measured along the projected coiling axis. In limpets, length indicates the longest dimension. Width or breadth is measured at right angles to height.

Shells that are cap-shaped have arisen several times, but all are known collectively as limpets. In calyptraeid and hipponicid limpets, the early coiling stage is retained in the adult shell, but in patellogastropod limpets and some fissurellid limpets, the coiled stage may be limited to the protoconch, which is seldom retained at later growth stages. In most limpet shells, the anterior and posterior may be determined from the muscle scar, which is open anteriorly corresponding to the head and opening of the mantle cavity.

COLLECTION AND PRESERVATION

Examination of particular habitats will reveal many species that would otherwise be overlooked. Species living on algae or marine grasses can be collected with a hand net or strainer. Dark overhanging ledges or deep crevices may hide nocturnal, negatively phototactic species. The expanded mantle of *Trivia* and *Lamellaria* match the color and shape of ascidians they prey upon; apertures of hermit crab shells may be examined for *Crepidula* and sponges for triphorids and cerithiopsids; the base and column of sea anemones are often fruitful hunting grounds for epitoniids; echinoderms should be examined for eulimids. Mollusks, sedentary polychaetes, and other invertebrates may be examined for ectoparasitic pyramidellids.

Microgastropods may be collected by vigorous shaking or washing of algae, and the holdfasts or roots of the surfgrass *Phyllospadix* and the eelgrass *Zostera*. Other methods are the shaking of small rocks in plastic bags or buckets and the direct sampling of sand and gravel sediments that accumulate under rocks. If the entire sample is placed in a basin of seawater, living specimens will emerge from the sediments and crawl up the sides of the container. Sand, gravel, or detritus collected in such ways may be screened to separate the size fractions and preserved in alcohol or dried, the microgastropods later to be sorted out under low magnification.

Minimal damage to the habitat should always be the rule in collecting; overturned rocks should be replaced in their original positions, and only small amounts of algae or grasses should be taken from any one area. Large collections of a single species from one area should be avoided.

Snails may be relaxed with the bodies extended, either for dissection or subsequent fixation, in a 7% solution of $MgCl_2 \cdot 6H_2O$ in fresh water, which is isotonic with seawater and may be used to partially or completely replace seawater. Another method is to sprinkle menthol crystals on the surface of a shallow dish of seawater, or by adding one or two drops of propylene phenoxetol in 200 ml–400 ml of seawater. It is usually necessary to crack the shells of living specimens in a vise so the columellar muscle may be separated from the shell for a dissection or to ensure that preservatives can reach the gonads. Bodies may be preserved in 75% ethyl or isopropyl alcohol, or in 95% ethanol for genetic studies; the use of formalin will leach the shell surface and make the tissues unfit for purposes of molecular genetics.

Specimens for collections or retained as vouchers should be preserved both wet and dry. For long-term preservation, microshells are best preserved dry because fluid preservation will eventually damage the shell.

THE FAUNA AND THE KEYS

The relatively low sea surface temperatures in central California and Oregon are indicative of a temperate, cool water faunal province, known to molluscan workers as the Oregonian faunal province. The Californian faunal province prevails in the subtropical conditions to the south of Point Conception, where summer sea surface temperatures are higher. Many northern species extend south to Point Conception, and a number of more southern species do not extend north of Point Conception. Toward the north, the transition to the boreal, cold water conditions is not as sharply defined; many of the species of central California have distributions that extend to Alaska. For entry to the literature on faunal provinces, see Roy et al. (1998).

The intertidal, shelled gastropod fauna of the region is generally well known. The majority of species were described before 1875, with P. P. Carpenter, W. H. Dall, and A. A. Gould naming most of the species treated here. However, some well-known taxa have long masked the presence of cryptic species. Two common species (*Littorina plena* and *Nucella ostrina*) were recently distinguished from *Littorina scutulata* and *Nucella emarginata,* respectively, based on anatomical distinctions that are also reflected in finer details of shell morphology (see species lists for further details). *Nucella analoga* is here distinguished from *N. canaliculata,* although further work is needed to confirm this. More such cases are to be expected.

Except for microgastropods, few new species have been found in our intertidal fauna after 1920. Northern range extensions of southern species during warm water years, new records of microgastropods, and newly introduced species are to be expected.

A surprisingly large number of species have been introduced to the northeastern Pacific, most arriving from the North Atlantic or northwestern Pacific with oyster culture, including *Batillaria attramentaria, Littorina littorea, Crepidula plana, Ocenebrellus inornatus, Urosalpinx cinerea, Ilyanassa obsoleta,* and *Busycotypus canaliculatus. Littorina saxatilis* is a recent arrival introduced in seaweed packed with bait worms from Maine.

The keys to species include most of the intertidal shelled gastropods occurring between the Columbia River and Point Conception. Certain species common in southern California or Washington have not been included, and users of the key are cautioned against attempting to apply it outside its intended range. Worn, eroded, or juvenile specimens with immature lips will not key out or will key out incorrectly. Sizes given in the keys are average sizes.

For most species of shelled gastropods, the intertidal zone represents an upper limit of the available marine habitat in which conditions are not severe for very long and the amount of exposure to air is minimal. For such species, most members of a local population live and reproduce in a much more extensive sublittoral region. There is only a very small group of species for which sublittoral occurrences are not generally known. Of the species treated here, the exclusively intertidal species are: most Patellogastropoda, *Fissurella volcano, Homalopoma baculum, Chlorostoma funebralis, Lirularia succincta, Cerithidea californica, Batillaria attramentaria,* all *Littorina,* *Assiminea californica, Fartulum orcutti,* all *Acanthinucella,* all *Nucella, Lirabuccinum dirum, Ilyanassa obsoleta, Myosotella myosotis,* and *Trimusculus reticulatus.*

There were 106 gastropod species keyed and illustrated by Carlton and Roth (1975), including 17 species of patellogastropods and three species of *Littorina.* Here I key and illustrate 170 species (not counting patellogastropods and *Littorina*), which amounts to the inclusion of about 88 species not previously treated. This increase is due to the inclusion of microgastropods and the addition of some species that are more likely to occur in the shallow sublittoral than in the intertidal zone. Shells of offshore species are frequently inhabited by hermit crabs in shallow water.

The key to the families is an artificial key based on shell characters intended to reach the correct family in the shortest number of steps. Some families have species with such a range of shell morphology that the family for certain genera has had to be keyed more than once. Families and the keys to species within families are arranged in the now current systematic order so that their position in the classification can be kept in mind. Notes on superfamilies are given only if there is more than one family in the superfamily treated here. General information for each family precedes the keys to the species for each family.

CHANGES TO NAMES

An extensive number of changes to names at the generic, specific, and in some cases at the family level, compared to the 1975 edition, are evident here. In many cases, the changes are necessitated by the need to recognize the recent work of specialists. Some of the changes have already been introduced in an account that treated offshore gastropods of southern California (McLean 1996). Some species have until now been known under broadly defined genera, and their more correct assignment to more narrowly defined subgenera have been recognized in current taxonomic works but ignored in faunal guides. In such cases, I make changes that are elevations of long-established subgenera to full genera, recognizing that our species differ in significant ways from the familiar genera, for which the type species may be from a far distant faunal province. Examples of such changes introduced here are the replacement of *Tegula* by *Chlorostoma* and the replacement of *Olivella* by *Callianax.* Another reason to avoid the usage of subgenera is that modern systematists have not wanted to imply relationships not demonstrated by cladistic analysis. All changes and the reasons for making them are noted in the text.

This effort is part of a larger taxonomic manual and full revision of the northeastern Pacific marine gastropods, comparable to the book on the northeastern Pacific bivalves of Coan, Scott, and Bernard (2000). Changes in taxonomy are justified in greater detail in the forthcoming revisions and taxonomic guides (northern and southern) to the shelled gastropods of the northeastern Pacific.

MAJOR SOURCES

For detailed treatments of the morphology, biology, and classification of gastropods at the family and superfamily level, the current standard is provided in the gastropod volume of *Fauna of Australia* (Beesley et al. 1999), with separate contributions by authors in their group of specialization. Excellent accounts and drawings of gastropod anatomy are provided in the revised edition of *British Prosobranch Molluscs* (Fretter and Graham 1994). For shell character terminology see Cox (1960). See the general introduction to Mollusca for other useful works on mollusks.

Marine gastropods from Washington and the Pacific Northwest were treated by Kozloff (1987); those of southern California by McLean (1978); and some deeper-water species of southern California by McLean (1996). D. P. Abbott and Haderlie (1980) provided color illustrations and accounts of biology for some of the common intertidal gastropods of both northern and southern California.

Basic works on taxonomy are Dall's (1921) distributional checklist, Oldroyd's (1927) copies of original descriptions of species listed by Dall, and R. T. Abbott's *American Seashells* (1974); common and scientific names were updated by Turgeon (2nd ed., 1998). Palmer's (1958) treatment of species described by P. P. Carpenter is an extremely useful compilation; Boss, Rosewater, and Ruhoff (1968) and Johnson (1964) have provided listings of the taxa of W. H. Dall and A. A. Gould respectively.

KEY TO FAMILIES

Based on shell characters.
1. Shell of limpet form 2
— Shell spirally coiled or tubular 8

2. Shell with apical foramen or interior septum 3
— Shell not with apical foramen or interior septum 4
3. Shell with apical foramen . . . . Fissuerellidae
— Shell with interior septum . . . . Calyptraeidae
4. Apex anterior, above opening of muscle scar . . . .
. . . . Patellogastropoda
— Apex central or posterior 5
5. Shell colored 6
— Shell white under periostracum 7
6. Periostracum thick, extending beyond shell margin . . . . .
. . . . Tylodinidae
— Periostracum thin, extending only to shell margin . . . . . .
. . . . Siphonariidae
7. Apex posterior . . . . Hipponicidae
— Apex central . . . . Trimusculidae
8. Shell tubular 9
— Shell with regular coiling 10
9. Shell a curved tube . . . . Caecidae
— Shell an irregular tube . . . . Vermetidae
10. Shell with slit-band and single foramen or with row of tubular holes 11
— Shell not with single foramen or row of tubular holes . . . . .
. . . . 12
11. Shell large, inflated, nacreous, with row of tubular holes
. . . . Haliotidae
— Shell minute, white, non-nacreous, slit band and single
hole . . . . Scissurellidae
12. Shell interior nacreous 13
— Shell interior not nacreous 15
13. Operculum multispiral with short growing edge . . . .
. . . . Trochidae
— Operculum with long growing edge, calcareous externally,
or multispiral with calcareous beads 14
14. Operculum multispiral with calcareous beads . . . .
. . . . Liotiidae
— Operculum calcareous on outer surface . . . . . Turbinidae
15. Operculum calcareous . . . . Tricoliidae
— Operculum corneous 16
16. Aperture lacking siphonal notch 17
— Aperture with siphonal canal or siphonal notch 38
17. Aperture round or oval 18
— Aperture elongate 54
18. Protoconch not heterostrophic 19
— Protoconch heterostrophic, immersed or at angle to body
whorls 52
19. Shell low-spired, few whorls 20
— Shell slender and high-spired, numerous whorls 34
20. Shell white 21
— Shell colored 25
21. Height greater than breadth 22
— Breadth greater than height 24
22. Sculpture of axial ribs . . . . Rissoidae (part)
— Sculpture smooth 23
23. Shell transparent, under 1 mm . . . . Rissoellidae
— Shell translucent, over 3 mm in height . . . . . Hydrobiidae
24. Spire moderately high, with strong spiral cords . . . .
. . . . Skeneidae
— Spire low, profile discoidal . . . . Vitrinellidae
25. Shell small to large (over 5 mm) 26
— Shell minute (under 4 mm) 30
26. Final whorl greatly inflated . . . . Velutinidae
— Final whorl not greatly inflated 27
27. Peritreme interrupted, shell thin, purple . . . .

. . . . Janthinidae (treated in separate section)
— Peritreme entire, shell not purple 28
28. Small to medium (height to 15 mm), suture impressed
. . . . 29
— Large (height to 35 mm), suture not impressed . . . .
. . . . Naticidae
29. Shell with strong spiral sculpture . . . .
. . . . Littorinidae (Littorininae)
— Shell lacking strong spiral sculpture . . . .
. . . . Littorinidae (Lacuninae)
30. Inner lip set off from columella by shelf . . . .
. . . . Anabathridae
— Inner lip set off from columella by shelf 31
31. Shell surface smooth 32
— Shell surface sculptured . . . . Rissoidae (part)
32. Shell narrowly umbilicate . . . . Cingulopsidae
— Umbilicus lacking 33
33. Inner lip narrow . . . . Barleeidae
— Inner lip broad, forming callus . . . . Assimineidae
34. Shell white 35
— Shell brown colored 36
35. Aperture circular, surface not glossy, with axial sculpture
. . . . Epitoniidae
— Aperture narrow posteriorly, surface glossy, no axial sculpture
. . . . Eulimidae
36. Small (under 15 mm), outer not flaring . . . . . . Cerithiidae
— Large (over 20 mm), outer lip flaring 37
37. Lacking varices . . . . Batillariidae
— With varices . . . . Potamididae
38. Aperture long and narrow, more than two-thirds length
of shell 39
— Aperture short, less than one-half length of shell 41
39. Lacking denticles on inner and outer lip . . . . Conidae
— With denticles on inner or outer lip 40
40. Lip dentition wrapping across inner and outer lip . . . . . .
. . . . Triviidae
— Lip dentition set back from edge of inner and outer lip
. . . . Cystiscidae
41. Shell small (under 12 mm in height), profile tall and
narrow 42
— Shell moderately large to large (not extremely slender)
. . . . 43
42. Shell dextral . . . . Cerithiopsidae
— Shell sinistral . . . . Triphoridae (also dextral Metaxia)
43. Shell with shiny black periostracum and strong columellar
placations . . . . Mitridae
— Shell not with shiny black periostracum and strong
columellar placations 44
44. Siphonal canal a short notch 45
— Siphonal canal moderately long to very long 48
45. Shell surface smooth, glossy . . . . Olividae
— Shell surface with periostracum 46
46. Base with shallow groove . . . . Nassariidae
— Base lacking shallow groove 47
47. Sculpture not coarsely clathrate . . . . Columbellidae
— Sculpture coarsely clathrate . . . . Turridae (part)
48. Shell sculpture imbricated (scaly) . . . . Muricidae
— Shell sculpture not imbricated 49
49. Shell with posterior anal sinus . . . . Turridae (part)
— Shell lacking posterior anal sinus 50
50. Very large (over 90 mm), spire low, canal long . . . .
. . . . Melongenidae
— Not large (under 60 mm), spire high 51

51. Suture not strongly impressed . . . . Buccinidae
— Suture strongly impressed. . . . . Fasciolariidae
52. Shell discoidal. . . . . Omalogyridae
— Shell not discoidal. 53
53. Umbilicus narrow or lacking . . . . Pyramidellidae
— Umbilicus broad. . . . . Amathinidae
54. Tip of spire recessed 55
— Tip of spire projecting. 57
55. Lip not extending above apex . . . . Diaphanidae
— Lip flaring above apex. 56
56. Shell large (to 50 mm), color pattern mottled. . . . . Bullidae
— Shell relatively small, pale green (under 25 mm) . . . .
. . . . Haminoeidae
57. Aperture length about one-half length of shell . . . .
. . . . Ellobiidae
— Aperture length at least two-thirds length of shell. 58
58. Shell with dark bands and spiral rows of pits . . . .
. . . . Acteonidae
— Shell lacking dark bands and rows of pits . . . .
. . . . Cylichnidae (Actocininae)

KEYS TO SPECIES BY FAMILY

EOGASTROPODA

PATELLOGASTROPODA

See the section on Patellogastropoda by Lindberg.

ORTHOGASTROPODA

VETIGASTROPODA

SUPERFAMILY FISSURELLOIDEA, FAMILY FISSURELLIDAE

Fissurellid limpets, the keyhole limpets, have non-nacreous shells in which the operculum is lacking in the adult. Bipectinate ctenidia are paired and of nearly equal size; there is an anterior notch or hole in the shell, corresponding to the excurrent hole or notch in the mantle. Epipodial tentacles are stubby and in a single row on foot sides. The mantle folds are capable of expansion to envelop the shell, head, and foot; shells of some genera are normally fully enveloped. Left kidney greatly is reduced. The radula is rhipidoglossate, outer lateral tooth greatly enlarged. Many species are carnivorous grazers. Broadcast spawners, lecithotrophic development. See Hickman, 1995, Gastropod volume of Fauna of Australia, p. 669 (general features); See McLean and Geiger 1998, Nat. Hist. Mus. L. A. Co., Cont. Sci., 475: 1–32 (phylogeny).

1. Apical hole large, widely oval; animal much larger than shell, the mantle nearly covering shell 2
— Apical hole small, either circular or elongate, mantle not extending over shell 3
2. Shell small (to 16 mm); margin set off on inner side by a broad, shallow, encircling groove; ends slightly elevated, shell buff color with radiating brown or gray bands; mantle variously colored (red, orange, lemon yellow, gray, brown) (plate 356C) . . . .
. . . . Fissurellidea bimaculata
— Shell very large (to 13 cm); inner margin crenulate, lacking groove; shell buff color, without radiating bands;

mantle black or gray (plate 356D) . . . Megathura crenulata
3. Internal apical callus truncate posteriorly; with concentric sculpture 4
— Callus rounded posteriorly, not sharply truncate; radiating ribs of varying sizes, or faint radial striae; no concentric sculpture; shell pink with red–brown or black rays; foot yellow, mantle red-striped, length to 25 mm (plate 356E). . . . .
. . . . Fissurella volcano
4. Shell outline oval, apical hole round, length to 35 mm (plate 356A) . . . . Diodora aspera
— Shell outline elongate, apical hole oval, length to 20 mm (plate 356B). . . . . Diodora arnoldi

Diodora arnoldi McLean, 1966. Mostly sublittoral, occasionally washed ashore; differs from *D. aspera* in smaller size, nearly parallel sides, and oval rather than round apical hole.

Diodora aspera (Rathke, 1833). Low intertidal zone under rocks, in crevices; diet includes encrusting bryozoans (Gonor, 1968, Veliger 11: 134); commensal polychaete *Arctonoe vittata* often in mantle cavity (Dimock and Dimock 1969, Veliger 12: 65–68). See Margolin 1964, Animal Behav. 12: 187–194 (escape response); Pernet 1997, Veliger 40: 77–83 (development).

Fissurellidea bimaculata (Dall, 1871) (=*Megatebennus bimaculatus*). Low intertidal zone to sublittoral, feeding on compound ascidians. See Ghiselin et al. 1975, Veliger 18: 40–43 (feeding); McLean 1984, Amer. Malac. Bull. 2: 21–34 (taxonomy).

Fissurella volcano Reeve, 1849. Rocky intertidal zone only, on and under coralline–encrusted rocks. See McLean 1984, Cont. Sci., L. A. Co. Mus. Nat. Hist., 354: 1–70 (taxonomy).

*Lucapinella callomarginata (Dall, 1871). A southern species feeding on sponges, Morro Bay south, could move northward. See Miller 1968, Veliger, 11: 130–134 (feeding).

Megathura crenulata (G. B. Sowerby I, 1825). Giant keyhole limpet; low intertidal zone to sublittoral, in rocky areas; pea-crab *Opisthopus transversus* among commensals. See Beninger et. al. 2001, J. Shellfish Res. 20: 301–307 (reproduction).

PLEUROTOMARIODEA

Pleurotomarioideans are vetigastropods with paired ctenidia and coiled shells; the right ctenidium is reduced as a result of space reduction toward the columella of the coiled shell. Shell with slit or series of holes. Two families are treated here: Haliotidae and Scissurellidae.

FAMILY HALIOTIDAE

Shell very large, with nacreous interior layer; apical whorls of such low profile and expansion of final whorl so extensive that shell form is limpetlike; right shell muscle greatly enlarged, producing large clamping foot; mantle cavity shortened and displaced to left side; ctenidia paired, unequal in size, left the largest. Excurrent openings in shell a series of open holes along left side, formed at aperture and sealed at later stages. Epipodium well developed, producing fluted lobes and numerous tentacles. Radula rhipidoglossate. Broadcast spawners, with lecithotrophic veliger stage.

Halitotids, or abalones are large-shelled species used for sport and commercial fisheries to such an extent that there are strict

* = Not in key.

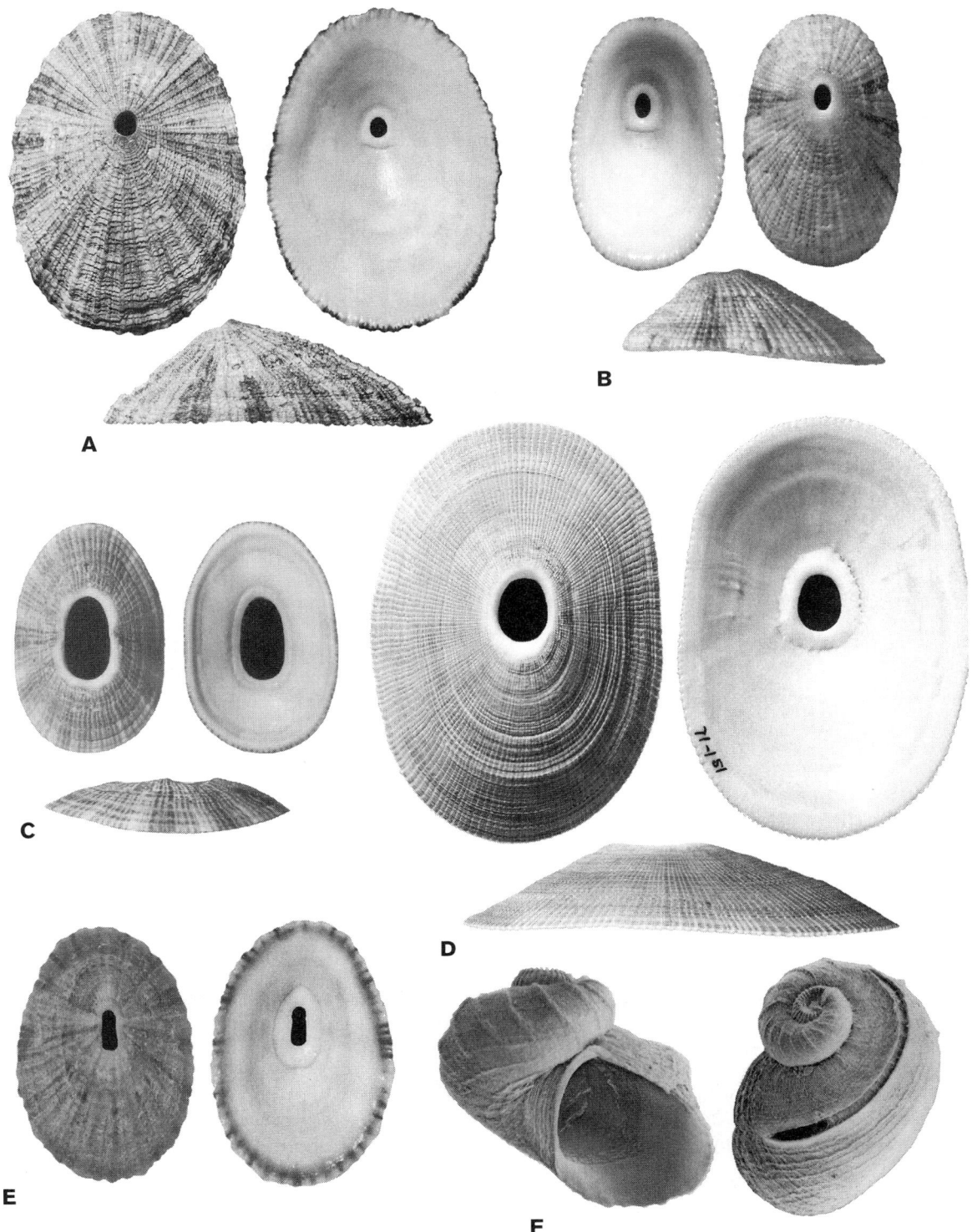

PLATE 356 Fissurellidae and Scissurellidae: A, *Diodora aspera* (three views), length 34.6 mm; B, *Diodora arnoldi* (three views), length 17.7 mm; C, *Fissurellidea bimaculata* (three views), length 19.2 mm; D, *Megathura crenulata* (three views), length 109 mm; E, *Fissurella volcano* (two views), length 20.8 mm; F, *Sinezona rimuloides* (two views), height 0.8 mm.

regulations for size and seasonal limits. There is a large collection of literature on the biology of West Coast species. See Owen, McLean, and Meyer 1971, L.A. Co. Mus., Bull. 9 (hybridization); Geiger 1998, Nautilus, 11: 85–116 (taxonomy of family). For status of aquaculture see McBride 1998, J. Shellfish Res. 17: 593–600.

1. Shell greenish black to dark blue; holes round and flush with surface, which is nearly smooth; muscle scar may be present in older specimens; length to 12 cm (plate 357D) . . . . *Haliotis cracherodii*
— Shell not black or dark blue; holes oval, raised above shell surface 2

PLATE 357 Haliotidae: A, *Haliotis rufescens* (two views), length 218 mm; B, *Haliotis kamtschatkana* (two views), length 57.4 mm; C, *Haliotis walallensis* (two views), length 108 mm; D, *Haliotis cracherodii* (two views), length 114 mm.

2. Sculpture of low, rounded, spiral ridges crossed by closely spaced, raised striations; shallow, indistinct groove along shell margin; shell flat, elongated; brick red with white, blue, and green mottling; no muscle scar; length to 11 cm (plate 357C). . . . . *Haliotis walallensis*
— Sculpture irregular, lumpy, undulating; shell moderately deep, not flat. 3
3. Shallow, broad channel parallel to shell edge; shell thin, mottled green-brown or red-brown with scattered blue and white areas; no muscle scar; length to 9 cm (plate 357B) . . . . *Haliotis kamtschatkana*
— Lacking broad channel; shell dull brick red, may have light color bands (pink, white, green); muscle scar prominent; usually covered with algae, barnacles, and other organisms; length to 22 cm (plate 357A) . . . . *Haliotis rufescens*

Haliotis cracherodii Leach, 1814. Black abalone; mid- to low intertidal; in crevices or under large rocks. See Douros 1987, J. Exp. Mar. Biol. Ecol. 108: 1–14 (stacking behavior); Haaker et al. 1995, J. Shellfish Res. 14: 519–525 (growth); Schiel et al. 2006, J. Exp. Mar. Biol. Ecol. 331: 158–172 (genetics, withering syndrome disease).

Haliotis kamtschatkana Jonas, 1845. Pinto abalone; specimens from south of Marin County may show characters approaching the subspecies *H. k. assimilis* Dall, 1878, found south of Point Conception, which has a higher, more rounded, less corrugated shell. Sublittoral, shells occasionally washed ashore. See Paul et al. 1977, Veliger 19: 303–309 (feeding, growth).

Haliotis rufescens Swainson, 1822. Red abalone; primarily sublittoral, extending to low intertidal in areas of considerable wave action in northern California. See Olsen 1968, Biol. Bull. 134: 139–147 (banding patterns); Haaker et al. 1998, J. Shellfish Res. 17: 747–753 (growth).

Haliotis walallensis Stearns, 1899. Chiefly sublittoral, rarely intertidal. Stohler (1975, Veliger 17: 250, figs. 1–4) provided notes on the righting response.

FAMILY SCISSURELLIDAE

Scissurellids, the minute slit shells, non-nacreous white shells; either with open slit or elongate foramen; animal with paired ctenidia; operculum present. See Geiger 2003, Molluscan Res. 23: 21–83 (phylogeny).

1. Shell minute (0.8 mm height); white, turbinate; three rapidly enlarging whorls; sculpture of axial folds on upper part of whorl, spiral cords on base; with elongate foramen in outer lip near aperture (plate 356F) . . . . *Sinezona rimuloides*

Sinezona rimuloides (Carpenter, 1865). Interstitial in sand and gravel. See McLean 1967, Veliger 9: 404–419 (taxonomy, distribution).

TROCHOIDEA

Trochoideans are vetigastropods in which the left ctenidium remains and the right one is lost; there is no slit or hole in the shell; excurrent flow takes place at the shell edge on the right side. Cephalic tentacles have sensory structures; epipodium well developed with long sensory tentacles; neck lobes assist in moving water through the mantle cavity. Families differ in structure of the operculum and whether it is enveloped by the foot. Shells of most groups are nacreous, but nacre is lost in some groups. The radula is rhipidoglossate and provides distinctions useful in classification. See Hickman and McLean 1990, Nat. Hist. Mus. L.A. Co., Sci. Ser. 35, 1–169 (systematics and subfamilial classification of Turbinidae and Trochidae). For an alternative in high-level classification see Bouchet and Warén (2005, Malacologia 47: 245). Some changes to family level classification are made here. Families treated here are Liotiidae, Turbinidae, Tricoliidae, Trochidae, and Skeneidae.

FAMILY LIOTIIDAE

Genera of the family Liotiidae have rounded apertures and a fine lamellar shell surface. The operculum is multispiral, and its outer surface has fine calcareous beads, formed at the long growing edge. Unlike the Turbinidae, the outer surface of the operculum is not enveloped by the foot. This family is better represented in tropical waters. Prior to Hickman and McLean 1990 (above), this was regarded as a full family; for a time it was regarded as a subfamily of Turbinidae, but it is here returned to family-level status.

1. Shell white, spire low, umbilicus deep, aperture circular, interior nacreous; operculum with fine calcareous beads; sculpture cancellate, of deep square pits formed by strong spiral and axial ribs; small, height to 3 mm (plate 358A) . . . . *Liotia fenestrata*

Liotia fenestrata Carpenter, 1864. In gravel under rocks. The biology of this species has not been studied.

FAMILY TURBINIDAE

The turbinids resemble trochids in having nacreous interiors, low profiles, and broad apertures, differing from trochids in having a solid calcareous operculum enveloped by the foot and thickened on its outer side. Two subfamilies are represented: the Turbininae are large-shelled, with mostly tropical species, and the Colloniinae, with small shells, for which little is known of the biology. See Hickman and McLean, 1990 Nat. Hist. Mus. L.A. County, Sci. ser. 35, 1–169 (systematics and subfamilial classification of Turbinidae and Trochidae).

1. Shell medium to large (25 mm-75 mm in height), conical; brick red; sculpture of diagonal folds and small, rounded nodes; periphery angulate or rounded; base with strong spiral cords (plate 358F) . . . . *Pomaulax gibberosa*
— Shell small, not more than 10 mm in height, with columellar denticles and some spiral sculpture 2
2. Small (to 3.5 mm in height), with coarse spiral and axial ribbing; white with pink dots on spiral cords (plate 358E) . . . . *Homalopoma radiatum*
— Sculpture spiral only 3
3. Small (to 4.5 mm in height), with few prominent spiral cords only; color white, pink, or red (plate 358D) . . . . *Homalopoma paucicostatum*
— Spiral cords numerous. 4
4. Small (to 4.8 mm in height), globose; nearly smooth with faint, incised, spiral grooves; gray to reddish gray or brown (plate 358B). *Homalopoma baculum*
— Small (to 9 mm in height), numerous rounded spiral cords over body whorl and base; often purple or red; juveniles grayish; highly variable in color (plate 358C) . . . . *Homalopoma luridum*

Homalopoma baculum (Carpenter, 1864). Restricted to mid-intertidal zone under rocks. Subfamily Colloniinae.

Homalopoma luridum (Dall, 1885) (=*H. carpenteri* Pilsbry, 1888). Rocky intertidal to sublittoral.

Homalopoma paucicostatum (Dall, 1871). Under rocks; offshore under kelp in gravel and shell bottoms.

Homalopoma radiatum (Dall, 1918) (=*H. fenestratum* Dall, 1919). Found with *H. paucicostatum* from low intertidal to sublittoral, under rocks.

Pomaulax gibberosa (Dillwyn, 1817) (=*Astraea inaequalis* Martyn, 1784). More common at sublittoral depths, occurring in shallower water toward the north. Beach shells common at Pacific Grove. The genus *Pomaulax* Gray, 1850, previously a subgenus of *Astraea* Röding, 1798, is characterized by the nearly smooth outer surface of the operculum. It is here regarded as a full genus in the subfamily Turbininae.

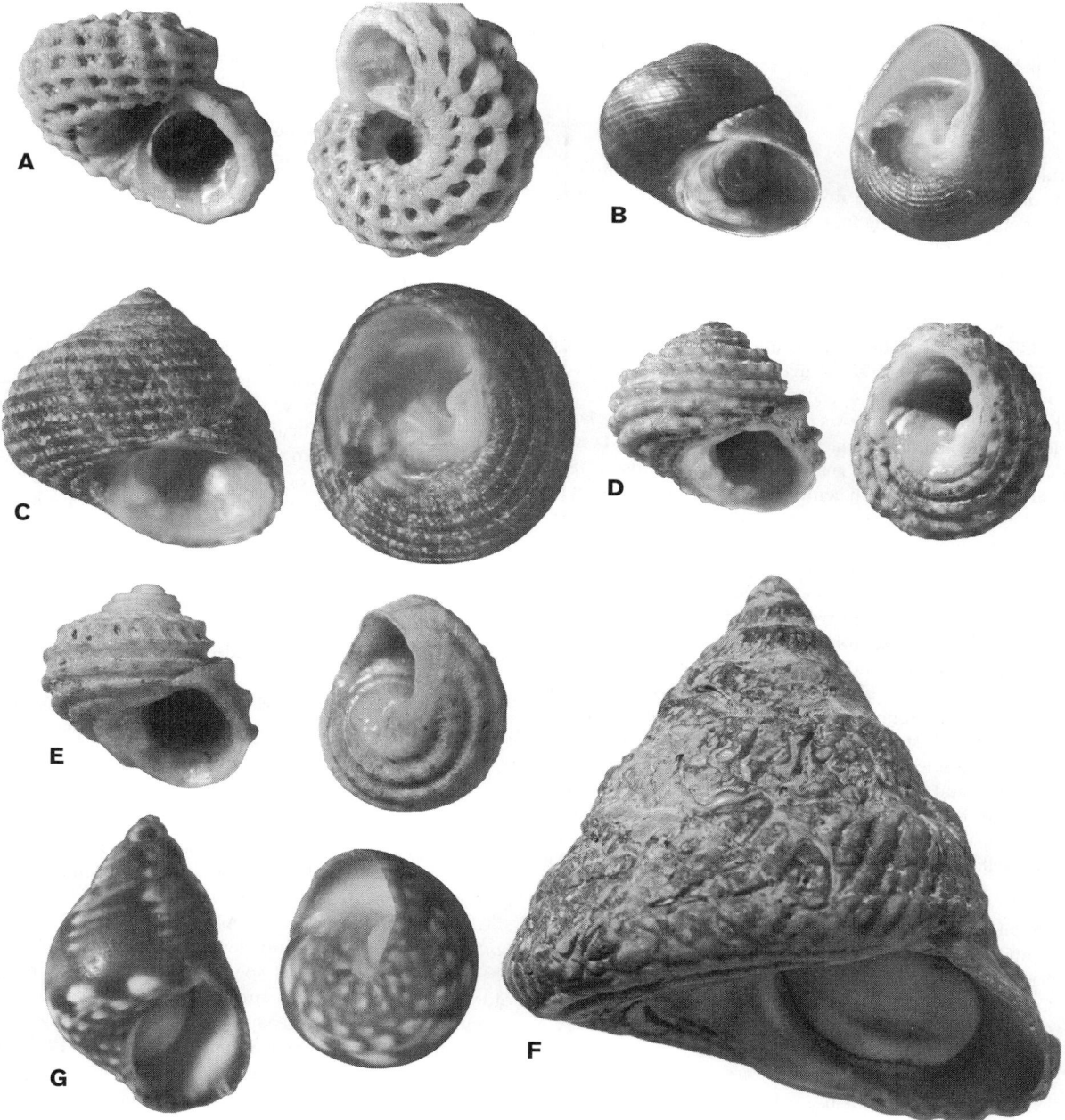

PLATE 358 Liotiidae, Turbinidae (subfamilies Colloniinae and Turbininae) and Tricoliidae: A, *Liotia fenestrata* (two views), height 3.3 mm; B, *Homalopoma baculum* (two views), height 4.8 mm; C, *Homalopoma luridum* (two views), height 8.4 mm; D, *Homalopoma pauci-costatum* (two views), height 4.5 mm; E, *Homalopoma radiatum* (two views), height 3.5 mm; F, *Pomaulax gibberosa*, height 60 mm; G, *Eu-lithidium pulloides* (two views), height 6.0 mm.

FAMILY TRICOLIIDAE

Species in this family are characterized by their lack of interior nacre, small size, rounded whorls of high profile, with mottled color patterns. One genus occurs in the eastern Pacific; it was previously treated as *Tricolia* in the family Phasianellidae, until the group was separated at a higher level from *Phasianella* in the Phasianellidae and treated in the subfamily Tricoliinae. See Hickman and McLean 1990, Nat. Hist. Mus. L.A. Co., Sci. Ser. 35, 1–169 (systematics and subfamilial classification of Turbinidae and Trochidae). The genus *Eulithidium* Pilsbry, 1898, which until then had been regarded as a subgenus of the Indo-

Pacific genus *Tricolia*, was raised by Hickman and McLean (above) to a full genus, based on radular and opercular differences.

1. Small (to 6 mm in height), ovate, of high profile; sculpture lacking, surface smooth; shell mottled red and white, occasionally with brown blotches on periphery; thin, green periostracum; sea-green calcareous operculum with linear grooves perpendicular to outer edge (plate 358G) . . . . *Eulithidium pulloides*

Eulithidium pulloides (Carpenter, 1865) (=*Tricolia pulloides*). In gravel, under rocks, or associated with surfgrass or algae.

Females have substantially larger shells than males. See Mooers, 1981, Veliger 24: 103–108 (feeding, sexual dimorphism and reproduction).

FAMILY TROCHIDAE

Trochidae, the top shells, have nacreous interiors, differing from turbinids in having a non-calcified, multispiral operculum not enveloped by the foot. The operculum further differs from that of turbinids in having a short growing edge. They have elaborate sensory tentacles on the epipodium; most are herbivorous. See Hickman and McLean 1990, Nat. Hist. Mus. L.A. Co., Sci. Ser. 35, 169 pp (external anatomy, radula, systematics and subfamilial classification of Turbinidae and Trochidae); Hickman 1992, Veliger 35: 245–272 (review of reproduction).

The subfamily Teguliinae is represented by relatively large-shelled species previously placed in the genus *Tegula* Lesson, 1835, but here placed in the genus *Chlorostoma* Swainson, 1840 (a raising of subgenus to full genus), necessary because the type species of *Tegula* is the tropical species *elegans* Lesson, 1835, from Panama, which has columellar morphology unlike that of temperate species of *Chlorostoma*. This distinction has long been recognized by Japanese authors. The biology of *Chlorostoma* species (as *Tegula*) has been well studied: see Riedman et al. 1981, Veliger 24: 97–102 (zonation); Hellberg 1998, Evolution 52: 1311–1324 (genetics and speciation). The radula of the tegulines resembles that of turbinids leading to reconsiderations of the affinity of this group (Bouchet and Warén 2005, Malacologia, 47: 245). *Calliostoma* species in the subfamily Calliostomatinae have the external anatomy modified to feed on hydroids (see Hickman and McLean 1990, above). The small-shelled *Lirularia* species have long gill filaments indicative of filter feeding (see Hickman and McLean 1990, above), but species may graze as well; the biology of *Lirularia* species is much in need of study.

1. Umbilicus closed 2
— Umbilicus open 8
2. Base of columella not having strong nodes 3
— Columella with one or two nodes emerging from umbilical region 7
3. Spiral cords beaded 4
— Spiral cords not beaded 5
4. Whorls flat-sided, with beaded spiral cords; shell golden yellow to yellowish brown with bright purple band adjacent to columella; height to 25 mm (plate 359B) . . . .
. . . . *Calliostoma annulatum*
— Whorls shouldered, yellow-brown with irregular mottling; fine cords strongly beaded on early whorls; height to 14 mm (plate 359E) . . . . *Calliostoma supragranosum*
5. Spiral cords fine; yellow-orange with darker markings at base; height to 24 mm (plate 359F) . . . . . . *Calliostoma gloriosum*
— Spiral cords raised, with darker color in interspaces . . . .
. . . . 6
6. Whorls flat-sided, yellowish tan to white or buff, with prominent revolving cords, cords paler in color than interspaces; blue stain next to columella; height to 35 mm (plate 359C) . . . . *Calliostoma canaliculatum*
— Whorls rounded, chocolate brown, with narrow, light tan spiral cords; nacre blue; height to 25 mm (plate 359D)
. . . . *Calliostoma ligatum*
7. Shell purplish black to black; scaly band below suture; mature specimens with two teeth on columella (lower tooth

occasionally worn); height to 35 mm (plate 360A) . . . . . .
. . . . *Chlorostoma funebralis*
— Shell brown or orange brown; no scaly subsutural band; one tooth on columella; height to 30 mm (plate 360B)
. . . . *Chlorostoma brunnea*
8. Shell relatively large, over 9 mm in height 9
— Shell relatively small, under 7 mm in height 11
9. Uniformly red-brown to orange; subconical; height to 10 mm (plate 359A) . . . . *Pupillaria salmonea*
— Shell brown (not reddish); height more than 20 mm . . . . . .
. . . . 10
10. Whorls flat-sided, base angulate; top of inner lip receding into aperture; umbilicus narrow, defined by a strong spiral cord terminating in node; brown; height to 30 mm (plate 360C) . . . . *Chlorostoma montereyi*
— Whorls rounded, base rounded, top of inner lip produced into flange on apertural side of umbilicus; no strong spiral cord defining umbilicus; brown or gray, at times with orange, white, or brown spots on periphery, height to 30 mm (plate 360D) . . . . *Promartynia pulligo*
11. With narrow, raised axial lamellae; height to 5 mm (plate 360F) . . . . *Lirularia parcipicta*
— Lacking narrow, raised axial lamellae 12
12. Base inflated, with a shallow, spiral channel 13
— Base without channel; basal cords strong, few; periphery rounded; height to 7 mm (plate 360E) . . . . . . *Lirularia* sp.
13. Grayish brown, base rounded, spiral cords broad, low; height to 4 mm (plate 360G) . . . . *Lirularia succincta*
— Mottled brown and yellow, base angulate, spiral cords narrow; height to 4.5 mm (plate 360H) . . . . *Lirularia discors*

Chlorostoma brunnea (Philippi, 1848) (=*Tegula brunnea*). Brown turban, occurs lower than *T. funebralis* and on offshore kelp beds near surface. For trochid species on kelp see Lowry et al. 1974, Biol. Bull. 147: 386–396; see Watanabe 1983, J. Exp. Mar. Biol. Ecol. 71: 257–270 (anti-predator defense against *Pisaster* and *Pycnopodia* in *Chlorostoma* spp.).

Chlorostoma funebralis (A. Adams, 1855) (=*Tegula funebralis*). Black turban; midtide levels, avoiding exposed outer-coast habitats; occasional specimens umbilicate. Very tall, older specimens on low-energy flat reefs. See Veliger 6, Suppl. 82 pp. (1964) for papers on ecology, biology; Frank 1975 (and earlier papers), Mar. Biol. 31: 181–192; Paine, 1971, Limnol. Oceanogr. 16: 86–98 (population, energy flow); Moran 1997, Mar. Biol. 128: 107–114 (spawning and larval development).

Chlorostoma montereyi (Kiener, 1850) (=*Tegula montereyi*). Low intertidal zone and on offshore kelp beds.

Promartynia pulligo (Gmelin, 1791). This species is better known as *Tegula pulligo* but is here placed in the genus *Promartynia* Dall, 1909, which previously had been regarded as a subgenus. This genus completely lacks the denticles at the base of the columella that characterize *Chlorostoma* species. Uncommon in low intertidal zone and on offshore kelp beds.

Pupillaria salmonea (Carpenter, 1864). On and under surfaces of rocks, low intertidal zone. *Pupillaria* Dall, 1909, has previously been considered a subgenus of *Margarites* Gray, 1847; it is here raised to generic level, characterized by its strong spiral sculpture and flat, offset base.

*Pupillaria rhodia Dall, 1921. Offshore rocky bottoms (illustrated by Abbott, 1974).

* = Not in key.

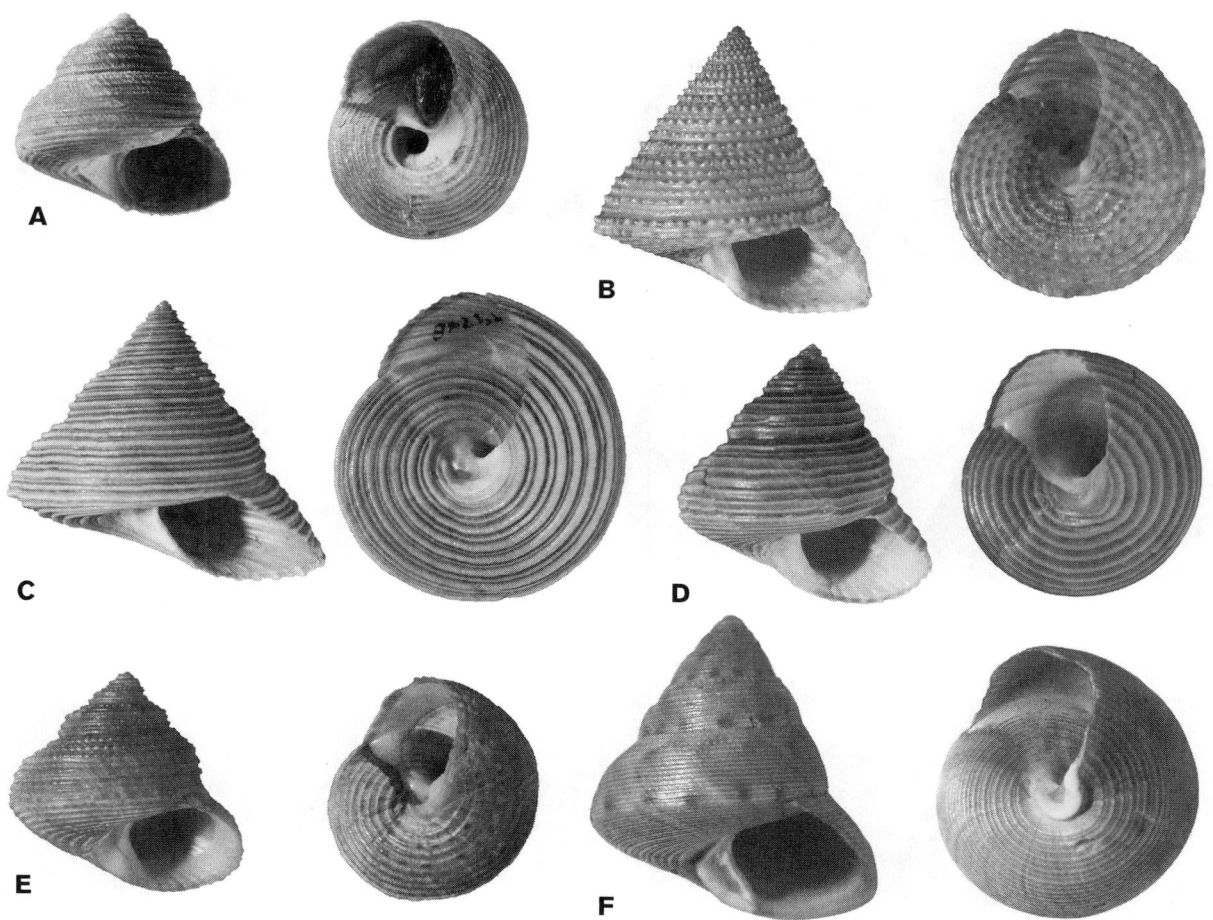

PLATE 359 Trochidae (subfamilies Margaritinae and Calliostomatinae): A, *Pupillaria salmonea* (two views), height 8.9 mm; B, *Calliostoma annulatum* (two views), height 22.8 mm; C, *Calliostoma canaliculatum* (two views), height 34.2 mm; D, *Calliostoma ligatum* (two views), height 22.0 mm; E, *Calliostoma supragranosum* (two views), height 9.0 mm; F, *Calliostoma gloriosum* (two views), height 23.7 mm.

Calliostoma annulatum (Lightfoot, 1786). With *C. canaliculatum* and *C. ligatum* on offshore *Macrocystis* stands and in low rocky intertidal zone. See Perron 1978, Veliger 18: 52–54 (feeding on hydroids).

Calliostoma canaliculatum (Lightfoot, 1786). Largest specimens occur on *Macrocystis*.

Calliostoma ligatum (Gould, 1849) (=*C. costatum* Martyn, 1784). See Holyoak 1988, Veliger 30: 369–371 (spawning and larval development).

Calliostoma supragranosum Carpenter, 1864 (=*C. splendens* Carpenter, 1864). Generally sublittoral, but shells occasionally washed ashore and occupied by hermit crabs.

Calliostoma gloriosum Dall, 1871. Sublittoral rocky bottoms; worn shells on shore.

*Calliostoma tricolor Gabb, 1865. Primarily a southern species, rare in central California. See photo in McLean (1978).

*Halistylus pupoideus (Carpenter, 1864). Small, slender, living on gravel bottoms offshore. See photo in Hickman and McLean 1990, above.

*Norrisia norrisi (Sowerby, 1838). A well-known and distinctive top shell with a bright red animal, occurring south of Monterey Bay, but noted here as a species that may finds its way further north with coastal warming (and a representative of po-

* = Not in key.

tentially many other southern California snails that may do the same). The shell is quite solid, broader than high, with smooth, rounded whorls and a deep umbilicus; it is chestnut brown in color, black near the umbilicus, and the columella is tinged with green. The operculum is spirally tufted. May reach 45 mm in height and 55 mm in diameter. The species is common on kelp and other brown algae.

Lirularia sp. An undescribed species on mudflats, on hard substrates, algae, *Zostera*, particularly in Tomales Bay and Bodega Harbor. In the previous edition this was incorrectly identified as the more northern species *L. funiculata* (Carpenter, 1864).

Lirularia succincta (Carpenter, 1864) (=*Margarites succinctus*). Low intertidal zone; common on gravel, under loose rocks.

Lirularia discors McLean, 1984 (Veliger, 26: 237). Occurs with *L. succincta*, but living in shallow sublittoral where the two occur together.

Lirularia parcipicta (Carpenter, 1864) (=*Margarites parcipictus*). Chiefly sublittoral among rocks, in gravel, among algae.

FAMILY SKENEIDAE

Skeneids have minute, non-nacreous, white shells of low profile, with the umbilicus usually open; the radula is

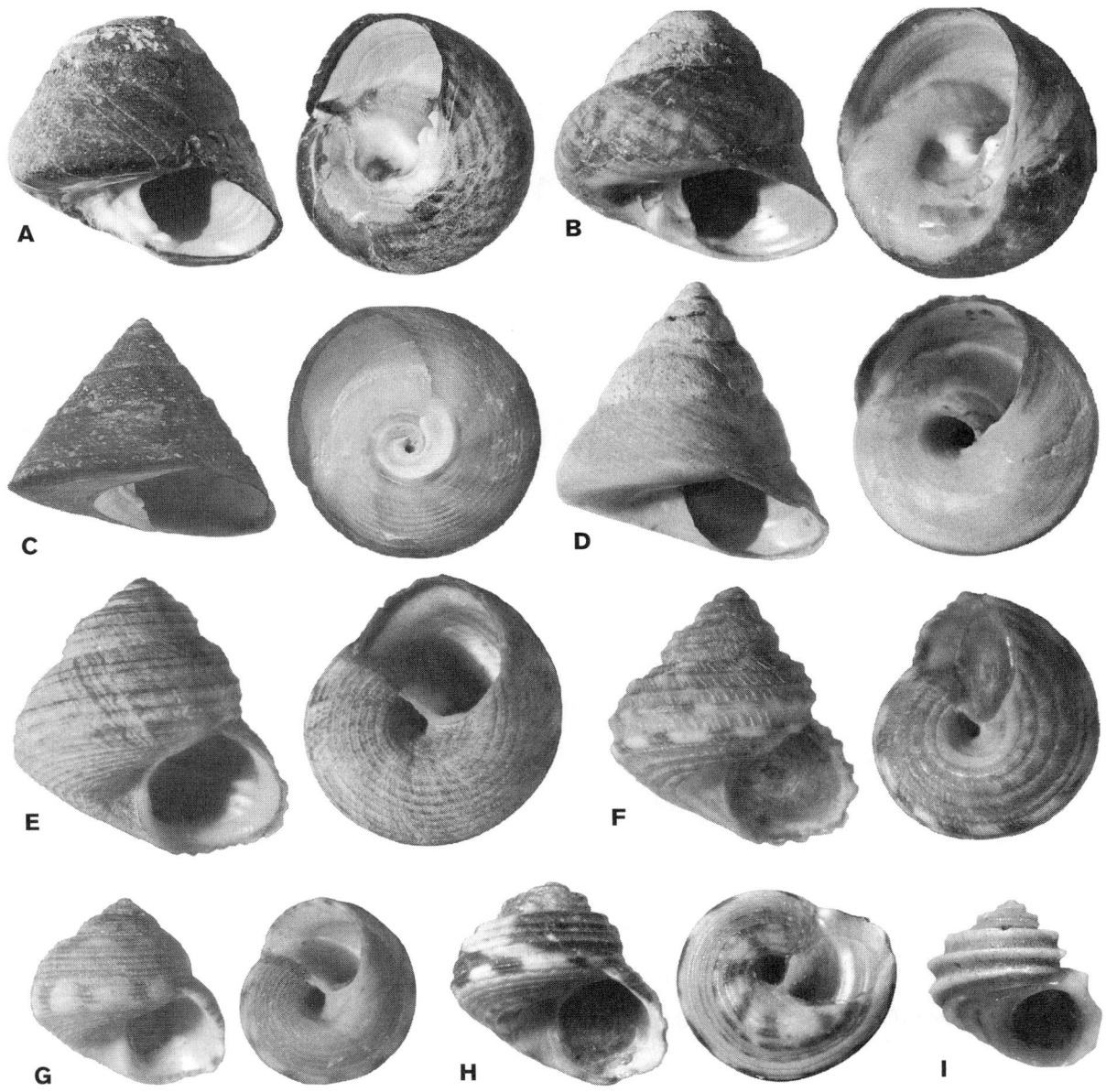

PLATE 360 Trochidae (subfamilies Tegulinae and Lirulariinae) and Skeneidae: A, *Chlorostoma funebralis* (two views), height 24.0 mm; B, *Chlorostoma brunnea* (two views), height 31.5 mm; C, *Chlorostoma montereyi* (two views), height 26 mm; D, *Promartynia pulligo* (two views), height 30.3 mm; E, *Lirularia* sp. (two views), height 6.8 mm; F, *Lirularia parcipicta* (two views), height 5.0 mm; G, *Lirularia succincta* (two views), height 4.0 mm; H, *Lirularia discors* (two views), height 4.3 mm; I, *Parviturbo acuticostatus*, height 2.7 mm.

rhipidoglossate and the operculum multispiral. This family is highly diverse in Australia and New Zealand; see Marshall 1988, J. Nat. Hist. 22: 949–1004 (taxonomy); also Hickman and McLean 1990, Nat. Hist. Mus. L.A. Co., Sci. Ser. 35, 1–169 (systematic position); Hickman, 1998, Gastropod volume, Fauna of Australia, pp. 690–391 (general review).

1. White, interior non-nacreous, profile with moderately high spire, three projecting cords on body whorl and three on base, fine axial lamellae; height 2.5 mm (plate 360I) . . . . *Parviturbo acuticostatus*

Parviturbo acuticostatus (Carpenter, 1864). Intertidal and rocky sublittoral. There is a *Homalopoma* species living off-shore, *H. mimicum* LaFollette 1976, Veliger 19: 68–77, that closely resembles this but has a strong columellar denticle and the thick calcareous operculum of *Homalopoma*.

CAENOGASTROPODA

CERITHIOIDEA

Cerithioideans are mostly slender shells in which the reproductive anatomy of all included families is primitive in having open pallial gonoducts and males that lack a copulatory organ; spermatophores are produced. Detritus feeders or browsers; the radula has seven teeth per row and is taenioglossate. Families differ in shell form, details of anatomy and opercular morphology. Families treated here are Cerithiidae, Batillariidae, and Potamididae. See Houbrick 1988, Mal. Rev., Suppl. 4: 88–128 (phylogeny); Simone 2001, Arquivos Zoologia 36: 147–263 (phylogeny).

FAMILY CERITHIIDAE

(Plate 361)

Cerithiidae have slender shells of numerous whorls; the operculum is multispiral or paucispiral. Large-shelled members of the subfamily Cerithiinae are primarily tropical; northeastern Pacific species are small-shelled members of the subfamily Bittiinae with a short siphonal canal that is not strongly notched. Protoconchs are paucispiral; development is non-planktotrophic. Species were previously assigned to *Bittium* subgenera, a European group. *S. eschrichtii* is placed in the monotypic genus *Stylidium* based on its large size and anatomic distinctions; all other smaller species are retained in *Lirobittium* (Houbrick, 1993, Malacologia 35: 262–314). Little is known of their biology other than notes on *S. eschrichtii* in Strathmann (1987).

1. Shell relatively large, height to 16 mm, no axial sculpture, spiral cords separated by deep grooves, surface with light brown maculations (plate 361A) . . . . . *Stylidium eschrichtii*
—   Shell under 12 mm in length, with axial sculpture in early whorls 2
2. With strongly beaded cancellate sculpture on all whorls 3
—   Spiral sculpture dominant on final whorl 4
3. Sculpture of strong beads and square pits, formed by axial ribs and two strong spiral cords per whorl, base with deep channel; height to 9 mm (plate 361B) . . . . *Lirobittium interfossum*
—   Shell with three strongly beaded spiral cords per whorl and two projecting cords at base; height to 11 mm (plate 361C) . . . . *Lirobittium purpureum*
4. Spiral cords of final whorl broad and projecting, separated by interspaces of about same width as cords (plate 361D) . . . . *Lirobittium latifilosum*
—   Spiral cords of final whorl broad and low, not separated by broad interspaces. 5
5. Broad spiral cords of last whorl with upper edge slightly projecting (plate 361E). . . . . *Lirobittium attenuatum*
—   Broad spiral cords of last whorl slightly inflated and separated by narrow grooves (plate 361F) . . . . *Lirobittium esuriens*

Lirobittium attenuatum (Carpenter, 1864) (=*Bittium attenuatum*). In sand, gravel, under rocks and in surfgrass holdfasts.
Lirobittium esuriens (Carpenter, 1864). Previously regarded as a synonym of *S. eschrichtii*, but is here recognized as a smaller species with prominent axial ribs on the base separated by narrow grooves.
Lirobittium interfossum (Carpenter, 1864) (=*Bittium interfossim*). In gravel, under algae; uncommon intertidally.
Lirobittium latifilosum (Bartsch, 1911) (=*Bittium latifilosum*). In gravel, under algae; uncommon intertidally.
Lirobittium purpureum (Carpenter, 1864) (=*Bittium purpureum*). Algae, surfgrass holdfasts, in sand.
Stylidium eschrichtii (Middendorff, 1849) (=*Bittium eschrichtii*, *B. e. montereyense* Bartsch, 1907). In clean, coarse sand among rocks. Often inhabited by the shell dwelling amphipod *Photis conchicola*.

FAMILY POTAMIDIDAE

Potamidids, like the batillariids, are mud-snail detritivores and also have a multispiral operculum. The family differs from Batillariidae in sperm structure and the radula. *Cerithidea* differs from *Batillaria* in the radula and in having a pallial eye on the mantle edge of the siphon. A single species is now living in California; a southern species *C. fuscata* Gould, 1845, from San Diego Bay is now considered extinct (see Carlton 1993, Amer. Zool. 33: 499–509). See Houbrick 1984, Amer. Malac. Bull. 2:1–20 (systematics); Healy and Wells 1998, pp. 724–727, in Gastropod volume of Fauna of Australia (biology and classification). Populations are often heavily infected with cercarial parasites (see Armitage 2001, S. Calif. Acad. Sci., Bull. 100: 51–58).

1. Shell dark brown; suture moderately impressed; axial ribs low; a few rounded varices on lower whorls; aperture with inflated lip; operculum multispiral; height 25 mm–30 mm (plate 361G) . . . . *Cerithidea californica*

Cerithidea californica (Haldeman, 1840). California horn snail. In bays, estuaries, on mud, in aggregations, under debris. Populations in San Francisco Bay are endangered; they now live in high intertidal marshes, having been displaced by the invasive *Ilyanassa obsoleta* (Race 1981, Veliger 24: 18–27). See papers by Byers cited under *Batillaria attramentaria*, showing that populations of *Cerithidea* have reduced or been replaced by the invasive *Batillaria attramentaria*. See also Bright 1958, Bull. So. Calif. Acad. Sci. 57: 127–139 and *ibid.* 1960, 59: 9–18 (morphology).

FAMILY BATILLARIIDAE

Batillaria species are slender mud-snails, detritivores, feeding by ingesting large quantities of mud. The operculum is multispiral. The family differs from Potamididae most strikingly in its sperm structure (see Healy and Wells 1998, gastropod volume of Fauna of Australia, pp. 720–721). Well represented in the northwestern Pacific by at least four species, a single species has been introduced with Japanese oyster seed and is well established in Elkhorn Slough and Tomales Bay, and other estuaries north to British Columbia.

1. Profile tall, suture not deeply impressed; aperture projecting, anal sinus pinched, anterior canal short; shell graybrown, often with white band below suture; spiral sculpture of strong cords and low axial ribs; axial ribs fading out on lower whorls; operculum multispiral; height 24 mm–28 mm (plate 361H) . . . . *Batillaria attramentaria*

Batillaria attramentaria (G. B. Sowerby I, 1855) (=*B. cumingi* Crosse, 1862; =*B. zonalis* of authors). In dense intertidal aggregations in bays, on soft mud. There is now a large literature on this species. See Whitlach 1974, Veliger 17: 47–55 (ecology); Whitlach and Obrebski 1980, Mar. Biol. 58: 219–225 (feeding); Driscoll 1972, Veliger 14: 375–386 (functional morphology); Yamada and Sankurathri 1977, Veliger 10: 179 (development). This species has gradually displaced the native *Cerithidea californica* in bays in which the two species occur together; see Byers 1999, Biolog. Invasions, 1: 339–352; Byers 2000, J. Exp. Mar. Biol. Ecol. 248: 133–150; Byers 2000, Ecol. 81: 1225–1239; Byers and Goldwasser 2001, Ecology 82: 1330–1343.

PLATE 361. Cerithiidae (subfamily Bittiinae), Potamididae, Batillariidae, and Littorinidae (subfamily Lacuninae). A, *Stylidium eschrichtii*, height 14.2 mm; B, *Lirobittium interfossum*, height 8.4 mm; C, *Lirobittium purpureum*, height 10.2 mm; D, *Lirobittium latifilosum*, height 9.7 mm; E, *Lirobittium attenuatum*, height 8.3 mm; F, *Lirobittium esuriens*, height 9.5 mm; G, *Cerithidea californica*, height 26.4 mm; H, *Batillaria attramentaria* (two specimens), height 24.2 mm, height 27.2 mm; I, *Lacuna porrecta*, height 10 mm; J, *Lacuna marmorata*, height 6.0 mm; K, *Lacuna unifasciata*, height 5.5 mm; L, *Lacuna variegata*, height 6.5 mm.

LITTINOROIDEA

Littorinoids are relatively small-shelled with rounded peristome, lacking anterior or posterior canals or grooves; penis present behind the right cephalic tentacle in males; females have a closed oviduct with complex spiral structure. Some lesser known families not represented in the eastern Pacific are mentioned by Reid 1998, p. 737 (Gastropod vol., Fauna of Australia).

FAMILY LITTORINIDAE, SUBFAMILY LITTORININAE

See the section on *Littorina* by Reid.

FAMILY LITTORINIDAE, SUBFAMILY LACUNINAE

Five genera have been included in this subfamily, but only *Lacuna* is represented in the eastern Pacific. Shells are small, umbilicate, with rounded aperture and narrow to broad columellar shelf. *Lacuna* resembles *Littorina* in shell form but is placed in the subfamily Lacuninae, differing in having a thinner shell, the animal having a pair of small tentacles behind the operculum. Most species occur on algae or sea grasses, on which they browse directly or feed upon epiphytic diatoms. All eastern Pacific species have been assigned to the subgenus *Epheria,* in which the protoconch is multispiral, indicative of a planktotrophic larval stage, in contrast to species of the subgenus *Lacuna,* for which the protoconch is paucispiral, indicative of non-planktotrophic development. See Reid 1989, Phil. Trans. Roy. Soc. London B, 324: 1–110 (morphology, classification, phylogeny). Padilla et al. 1996, J. Molluscan Stud. 62: 275–280, studied radular production in eastern Pacific species. Padilla 1998, Veliger 41: 201–204, showed that the shape of the radular teeth is plastic, changing in individuals transferred between eelgrass and kelp, and that this response is triggered by both diet and environment (Padilla 2001, Evol. Ecol. Res. 3: 15–25). Breeding experiments have confirmed the *L. marmorata* and *L. unifasciata* are distinct species (Langan-Cranford and Pearse 1995, J. Exp. Mar. Biol. 186: 17–31). Population dynamics have been studied by Martel and Chia 1991, Mar. Biol. 110: 237–247, defense against seastars by Fishlyn and Phillips, 1980, Biol. Bull. 158: 34–48, feeding preferences by Van Alstyne, et al. 2001. Mar. Biol. 139: 201–210, and larval behavior by Martel and Diffenbach 1993, Mar. Ecol. Prog. Ser. 99: 215–220.

1. Shell profile relatively slender .2
—   Shell profile relatively broad. 3
2. Suture shallow, with dark brown line at sharply angulate base; columellar shelf narrow; height 5 mm–7 mm (plate 361K) . . . . *Lacuna unifasciata*
—   Suture deep, usually with brown markings of back-directed chevrons; columellar shelf relatively broad; height 8 mm (plate 361L) . . . . *Lacuna variegata*
3. Shell profile depressed, surface smooth with fine spiral striae; light brown, marbled with white, especially at periphery; aperture often with white band showing on interior of aperture; columellar shelf broad; height 6 mm–7 mm (plate 361J) . . . . *Lacuna marmorata*
—   Large, globose; surface wrinkled with fine, wavy, spiral striae; solid brown color, generally lacking a white band in aperture; columellar shelf broad; height to 12 mm (plate 361I) . . . . *Lacuna porrecta*

Lacuna marmorata Dall, 1919. Common; intertidal on rocks, algae, on surfgrass *Phyllospadix.* Eaten by the sea star *Leptasterias,* to which it responds by leaping off the surfgrass blade and lowering itself by a mucous thread into the water below, which it then climbs up again. See Fishlyn and Philips 1980, above.

Lacuna porrecta Carpenter, 1864. On algae and eelgrass.

Lacuna unifasciata Carpenter, 1857. Common, generally Monterey Bay south, in kelp beds, eelgrass, algae. Reported also from Bodega Harbor (D. Padilla, pers. comm. to J. T. Carlton 2003).

Lacuna variegata Carpenter, 1864. On algae, or grasses in bays; more common to north in Puget Sound.

SUPERFAMILY CINGULOPSOIDEA, FAMILY CINGULOPSIDAE

Shells of cingulopsids resemble those of rissooideans, but there are major anatomical differences, particularly in males being aphallate. The operculum has a peg, as in the barleeids. Anatomy was first studied by Fretter and Patil 1958, Proc. Mal. Soc. London, 33: 114–126. See Ponder and Yoo 1980. Rec. Aust. Mus. 33: 1–88 (anatomy, classification); and Ponder and De Keyser 1998: 741, Gastropod volume of Fauna of Australia.

1. Shell brown, whorls rounded, suture deep; umbilicus deep, narrow; height to 1.4 mm (plate 362A) . . . . *Mistostigma* sp.

Mistostigma sp. Rocky intertidal and sublittoral near brown algae. Common in sediment samples but yet to be collected alive.

RISSOIDEA

Rissoideans are minute caenogastropods with round or oval apertures; reproduction in all families involves a cephalic penis. The radula is taenioglossate, as in other basal caenogastropods. Families are defined chiefly on anatomical distinctions and opercular differences. The highly speciose rissooideans occur in marine and fresh water, and there are terrestrial groups. Many families have smooth-surfaced shells, but there are a few groups with strong sculpture. See Ponder and De Keyser 1998, pp. 745–746 (in Gastropod volume of Fauna of Australia). There are many more families than those treated here. Families represented in central California and Oregon are Barleeidae, Anabathridae, Rissoidae, Hydrobiidae, Assimineidae, Caecidae, and Vitrinellidae.

FAMILY BARLEEIDAE

(Plate 362)

Barleeids are rissooideans with smooth, dark brown shells and an opercular peg on the inner side. A single genus *Barleeia* occurs in the northeastern Pacific. They are common in the rocky intertidal and sublittoral, often occurring on algae. Biology of the northeastern Pacific species has not been studied. See Ponder 1983, Rec. Aust. Mus. 35: 231–281 (taxonomy).

1. Shell over 3 mm in length, with sharp basal keel, whorls nearly flat-sided. 2
—   Shell under 3 mm, slight basal angulation; whorls somewhat rounded; color variable, translucent yellow-white to dark brown (plate 362C) . . . . *Barleeia haliotiphila*
2. Large and broad, length 4 mm–5 mm, surface shiny, with color mottling (plate 362B) . . . . *Barleeia acuta*
—   Shell length to 3.3 mm, slender, dark brown (plate 362D) . . . . *Barleeia oldroydi*

Barleeia acuta (Carpenter, 1864) [=*Diala acuta;* =*Barleeia marmorea* (Carpenter, 1864); =*B. dalli* Bartsch, 1920]. Characterized by its relatively large size.

Barleeia haliotiphila Carpenter, 1864 (=*B. oldroydi* Bartsch, 1920). Common among algae, rocks, gravel, sand; also in kelp holdfasts, and reported from high intertidal *Endocladia/Balanus* zone.

Barleeia oldroydi Bartsch, 1920. Resembles *B. acuta* but smaller and more slender, of more common occurrence in rocky sublittoral.

PLATE 362 Cingulopsidae, Barleeidae, Anabathridae, Rissoidae, Hydrobiidae, and Assimineidae: A, *Mistostigma* sp., height 1.4 mm; B, *Barleeia acuta*, height 3.8 mm; C, *Barleeia haliotiphila*, height 2.0 mm; D, *Barleeia oldroydi*, height 3.1 mm; E, *Amphithalamus tenuis*, height 1.0 mm; F, *Alvania compacta*, height 2.1 mm; G, *Alvania almo*, height 1.8 mm; H, *Alvinia aequisculpta*, height 2.8 mm; I, *Alvinia purpurea*, height 2.3 mm; J, *Onoba carpenteri*, height 2.6 mm; K, *Onoba dinora*, height 1.8 mm; L, *Tryonia imitator*, height 3.6 mm; M, *Assiminea californica*, height 2.9 mm.

FAMILY ANABATHRIDAE

Anabathrids are minute rissooideans having an opercular peg, like barleeids, but much smaller and with a thick columellar shelf. Two species occur in southern California, but only one species extends into central California.

See Ponder 1988, Mal. Rev., Suppl. 4: 129–166 (anatomy and classification).

1. Inner lip offset from columella by a shelf; slightly over 1 mm in height (plate 362E) . . . . *Amphithalamus tenuis*

Amphithalamus tenuis Bartsch, 1911. In sand and gravel and on algae.

FAMILY RISSOIDAE

Rissoids are minute rissooideans with a paucispiral operculum that lacks an opercular peg. Most have both axial and spiral sculpture. Feeding on diatoms and microalgal films. This is a very large family with numerous genera and species worldwide. As used here, *Alvinia* differs from *Alvania* in having a protoconch with strong spiral cords. Generic assignments follow Ponder (1985, Rec. Aust. Mus., Suppl. 4, 221 pp.), who reviewed the genera worldwide.

1. Shell tan to brown 2
— Shell white or yellowish white 5
2. Sharply clathrate, with narrow projecting axial and spiral sculpture 3
— Finely clathrate, axial and spiral sculpture not strongly projecting 4
3. Coarsely clathrate with two spiral cords at periphery; protoconch with spiral sculpture; height 2.3 mm (plate 362I) . . . . *Alvinia purpurea*
— Finely clathrate, with four or more spiral cords at periphery; protoconch with fine clathrate sculpture; height 2.6 mm (plate 362J) . . . . *Onoba carpenteri*
4. Suture shallow, with narrow low axial ribs; protoconch smooth; height 2.1 mm (plate 362F) . . . . . *Alvania compacta*
— Suture deeper, no axial sculpture, protoconch smooth; height 1.8 mm (plate 362K) . . . . *Onoba dinora*
5. Shell relatively large, three spiral cords; protoconch with spiral sculpture; height 3.3 mm (plate 362H) . . . . *Alvinia aequisculpta*
— Shell relatively small, three spiral cords; protoconch smooth; height 1.5 mm (plate 362G) . . . . *Alvania almo*

Alvania compacta (Carpenter, 1864); *Alvania almo* Bartsch, 1911; *Alvinia aequisculpta* (Keep, 1887); *Alvinia purpurea* Dall, 1871; *Onoba carpenteri* (Weinkauff, 1885); *Onoba dinora* (Bartsch, 1917). All are microgastropods that can be collected by screening algae and eelgrass and by taking sand and gravel samples for later examination.

FAMILY HYDROBIIDAE

Hydrobiid species usually have smooth sculpture. They are primarily a freshwater group of rissooideans, with only a few species occurring in lagoons and brackish water estuaries. See Kabat and Hershler 1993, Smithsonian Contr. Zool., 547. 1–94 (classification); Hershler and Ponder 1998, Smithsonian Contr. Zool., 600: 1–55 (taxonomy).

1. Shell white, smooth except for fine spiral sculpture, suture deeply impressed; height 4 mm–5 mm (plate 362L) . . . . *Tryonia imitator*

Tryonia imitator (Pilsbry, 1899). Muddy bottoms in shallow bays, now restricted to only a few brackish-water localities in central California, including Elkhorn Slough; a victim of extensive estuarine modification and destruction. See Taylor 1966, Malacologia 4: 53 (taxonomy); Kellogg 1980, Calif. Dept. Fish & Game, Special Pub. 80, 23 pp. (ecology and status).

FAMILY ASSIMINEIDAE

Assimineids are small-shelled rissooideans, characterized the thickened callus on the columella; they live only at the upper limits of the tide in bays and estuaries. The group is speciose in the western Pacific, but there is only one broad-ranging northeastern Pacific species. See Ponder and de Keyzer 1998, pp. 756–758, in Gastropod volume of Fauna of Australia (biology, classification).

1. Inner lip with small, thickened callus; whorls rounded, convex; shell smooth, stoutly conical; glossy brown; height 3 mm (plate 362M) . . . . *Assiminea californica*

Assiminea californica (Tryon, 1865) [=*A. translucens* (Carpenter, 1866)]. Abundant in *Salicornia* marshes on mud, under debris. See Fowler 1980, Veliger, 23: 163–166 (reproduction); Berman and Carlton 1991, J. Exp. Mar. Biol. Ecol. 150: 267–281 (ecology, diet).

FAMILY CAECIDAE

Caecids are rissooideans with minute, tubular shells that are slightly curved and have a plug that seals the posterior end; the operculum is multispiral. Caecids grow by discarding entire earlier growth stages and forming a succession of new plugs. The initial spiral protoconch is seen only in the first grown stage. Living interstitially in sand and gravel. Much is known about the group, but taxonomic work has been of local scope; there are no worldwide revisions. See Bandel 1996, Mitt. Geol.-Paleont. Inst. Univ. Hamburg 79: 53–115 (fossil record); Ponder 1998, pp. 761–773 (Gastropod volume of Fauna of Australia); Absalão and Pizzini 2002, Archiv. Moll. 131: 167–183 (classification).

1. Shell surface with distinct rings and pointed plug 2
— Shell surface nearly smooth, plug rounded to pointed 3
2. With 30–40 closely set rings; length to 3 mm (plate 363A) . . . . *Caecum californicum*
— With 28–24 rings and interspaces between; length 2.5 mm (plate 363B) . . . . *Caecum dalli*
3. Plug rounded, aperture drawn out; length to 2.3 mm (plate 363D) . . . . *Fartulum orcutti*
— Plug angulate to rounded, aperture not drawn out; length to 3.8 mm (plate 363C) . . . . *Fartulum occidentale*

Caecum californicum Dall, 1885. Interstitial in sand, gravel, especially near roots of surfgrass *Phyllospadix*.
Caecum dalli Bartsch, 1920. Lower intertidal; more abundant in sublittoral.
Fartulum occidentale (Bartsch, 1920) (=*Fartulum hemphilli* Bartsch, 1920). Intertidal and sublittoral.
Fartulum orcutti (Dall, 1885). Among debris under stones at high tide zone.
Micranellum crebricinctum (Carpenter, 1864). Large (5 mm–6 mm), with numerous rings. Offshore, sandy bottoms. See photo in McLean (1978).

FAMILY VITRINELLIDAE

Vitrinellids are minute rissooideans with depressed spires and multispiral opercula, best represented in tropical and subtropical

* = Not in key.

PLATE 363 Caecidae, Vitrinellidae, and Hipponicidae: A, *Caecum californicum*, length 3.0 mm; B, *Caecum dalli*, length 2.5 mm; C, *Fartulum occidentale*, length 3.3 mm; D, *Fartulum orcutti*, length 2.2 mm; E, *Vitrinella oldroydi* (three views), diameter 2.7 mm; F, *Antisabia panamensis*, diameter 20 mm (two views); G, *Hipponix tumens* (two views), diameter 13.5 mm.

regions. They differ from skeneids in having a taenioglossate rather than rhipidoglossate radula. Many species have never been collected alive, but at least some species are known to live commensally with other invertebrates. No work has been done on the biology of any eastern Pacific species. See Bieler and Mikkelsen 1988 (Nautilus 102: 1–29) for work on western Atlantic species.

1. Lenticular, minute (to 2.7 mm diameter), white; whorls and aperture rounded; umbilicus broadly open; shell often eroded, with apex missing but replaced by plug from within (plate 363E) . . . . *Vitrinella oldroydi*

Vitrinella oldroydi Bartsch, 1907. Commensal in mantle cavity (foot groove) of the chitons *Stenoplax heathiana* and *S. conspicua*.

VANIKOROIDEA

Two families are included, the limpet family Hipponicidae, and the Vanikoridae, the latter with coiled shells and represented at sublittoral depths in California.

FAMILY HIPPONICIDAE

Hipponicids are sedentary limpets that retain a coiled early stage; they obtain food with their extensible snout. Most species coat their site of attachment with a shelly plate; loose shells show a prominent horseshoe-shaped muscle scar. See Ponder, *in* Gastropod volume of Fauna of Australia, 1998: 770–771 (biology and taxonomy).

1. Sculpture of flat, concentric lamellae bearing fine radial striae under periostracum; apex low, subcentral or near margin; diameter to 25 mm (plate 363F). . . . .
. . . . *Antisabia panamensis*
— Sculpture of strong radial ridges with weaker concentric sculpture; periostracum with fine hairs; apex elevated, overhanging posterior margin; diameter to 15 mm (plate 363G). . . . . *Hipponix tumens*

Antisabia panamensis (C. B. Adams, 1852) [=*Hipponix cranioides* Carpenter, 1864; =*H. antiquatus* of authors, not of Linnaeus, 1767]. A single species of this genus is thought to occur throughout the eastern Pacific, extending from tropical to temperate regions. In colonies on under-surfaces of rocks at low tide and in the sublittoral. Use of *Antisabia* follows Ponder 1998 (above). See Yonge 1953, Proc. Calif. Acad. Sci. Ser. 4, 28: 1–24, and 1960, 31: 111–119 (anatomy, biology, ecology).

Hipponix tumens Carpenter, 1864. Low intertidal, in rock crevices. This is comparable to the Caribbean species *H. subrufus* (Lamarck, 1819).

SUPERFAMILY CALYPTROIDEA, FAMILY
CALYPTRAEIDAE

(Plate 364)

Calyptraeids, the slipper limpets, are sedentary limpets, with coiled early whorls, the mature shell characterized by a shell septum that separates the visceral organs from the broad, posteriorly extended foot. The irregular edge of the shell conforms to the area of attachment. Elongate ctenidial filaments function like bivalves gills, collecting food particles that are shunted to a food groove leading to the mouth. Protandrous hermaphrodites, changing from male to female with size increase; many with differing reproductive modes. Worldwide species of *Crepidula* were reviewed by Hoagland, 1977, Malacologia 16: 352–420. See Collin 2003a, Biol. J. Linn. Soc., 78: 541–593 (phylogeny), and Collin 2003b, Syst. Biol. 52: 618–640 (phylogeny); Collin 2003c, Mar. Ecol. Prog. Ser. 247: 103–122 (global development patterns). There is a very large literature on the biology of *Crepidula* cited by these authors.

1. With paired shell muscles, septum extending anteriorly on both sides 2
— Not with paired muscles, septum not extending anteriorly on both sides. 3
2. Shell interior dark brown; septum extending anteriorly on both sides; apex overhanging shell margin; length to 20 mm (plate 364E) . . . . *Garnotia adunca*
— Shell interior light brown; septum extending anteriorly on both sides, apex close to shell margin; length to 30 mm (plate 364F). . . . . *Garnotia norissiarum*
3. Septum extending forward on left side (ventral view) but not on right side, right side with muscle scar, reddish brown; shell length under 10 mm (plate 364G) . . . .
. . . . *Crepidula convexa*
— Septum notched at left side 4
4. Septum deeply notched at left side and with raised medial fold, shell nearly circular in outline; mottled or radially striped brown and white; length to 20 mm (plate 364H)
. . . . *Crepipatella lingulata*
— Septum with shallow notch at left side, septum sinuous with raised medial fold. 5
5. Apex strongly turned to one side and united with margin of shell; shell with brown blotches or interrupted, wavy,

chestnut-colored markings; length to 35 mm (plate 364A)
. . . . *Crepidula fornicata*
— Apex at shell edge, shell white 6
6. Thick, shaggy, golden brown periostracum; shell planar, relatively thick, often broadly oval; length to 40 mm (plate 364B) . . . . *Crepidula nummaria*
— Thin, shiny brown periostracum, shell form planar . . . . .
. . . . 7
7. Shell relatively thin; shape variable, may occur as foliated, elongate shells in pholad holes or smooth, very thin, concave specimens in hermit-crab shells (plate 364C1) or as low, white shells on undersides of rocks; length to 25 mm (plate 364C2) . . . . *Crepidula perforans*
— Shell thin, flattened, profile broad; living on inner surface of bivalve shells, in bays; length to 30 mm (plate 364D). . . . .
. . . . *Crepidula plana*

Crepidula convexa Say, 1822 (=*C. glauca* Say, 1822). Introduced with Atlantic oysters; in San Francisco Bay, often on shells including hermit crab–occupied *Ilyanassa obsoleta*. See Franz and Hendler 1970, Univ. Conn. Occ. Pap. (Biol. Sci. Ser.) 1: 281–289 (taxonomy); Hendler and Franz 1971, Biol. Bull. 141: 514–526 (reproductive biology and population dynamics).

Crepidula fornicata Linnaeus, 1758. Introduced to Puget Sound with Atlantic oysters; may occasionally occur in central California and Oregon; shape highly variable; occurs in characteristic stacks with male on top.

Crepidula nummaria Gould, 1846 (=*C. nivea* of authors). Also compare *C. plana* and *C. perforans*. Low intertidal zone of outer coast, under rocks, occasionally in abandoned pholad holes.

Crepidula perforans (Valenciennes, 1846). Compare with *C. nummaria* (which possesses a shaggy golden brown periostracum) and *C. plana* (which occurs in San Francisco Bay); in abandoned pholad holes, hermit crab shells, and under rocks along open rocky coast.

Crepidula plana Say, 1822. Introduced with Atlantic oysters; in San Francisco Bay on rocks and often (as concave specimens) in hermit crab shells. *C. perforans* occurs on the open coast. See Collin 2000, Can. J. Zool. 78: 1500–1514 (taxonomy).

Crepipatella lingulata (Gould, 1846). Contrary to Hoagland (1977), this is not the same as *Crepipatella dorsata*, a tropical eastern Pacific species in which the detachment of the septum on the left side is not as deep. On rocks, shells, intertidal to offshore depths. See Collin 2000, Veliger 43: 24–33 (reproduction).

Garnotia adunca (G. B. Sowerby I, 1825). Common on shells of larger snails such as *Chlorostoma* and *Calliostoma*. See Vermeij et al. 1987, Nautilus 101: 69–74 (gastropod hosts); Collin 2000, Veliger 43: 23–33 (reproduction). This is the type species of *Garnotia* Gray, 1867, which is here recognized as a full genus because there are shell muscles on both sides and the septum projects anteriorly on both sides.

Garnotia norissiarum (Williamson, 1905). In southern California occurs chiefly on the teguline trochid *Norrisia norrisi*; here reported on the teguline trochid *Promartynia pulligo* at Carmel. See G. MacGinitie and N. MacGinitie, 1964, Veliger 7: 34 (reproduction); Warner et al. 1996. J. Exp. Mar. Biol. Ecol. 204: 155–167 (social control of sex change); Hobday and Riser 1998 J. Exp. Mar. Biol. Ecol. 225: 139–154 (movement and reproductive potential).

SUPERFAMILY VERMETOIDEA, FAMILY VERMETIDAE

Vermetid shells are attached to a hard substratum, uncoiling with growth; teleoconch whorls irregular, solitary or forming clumps of wormlike tubes. The tubes of vermetid snails are readily

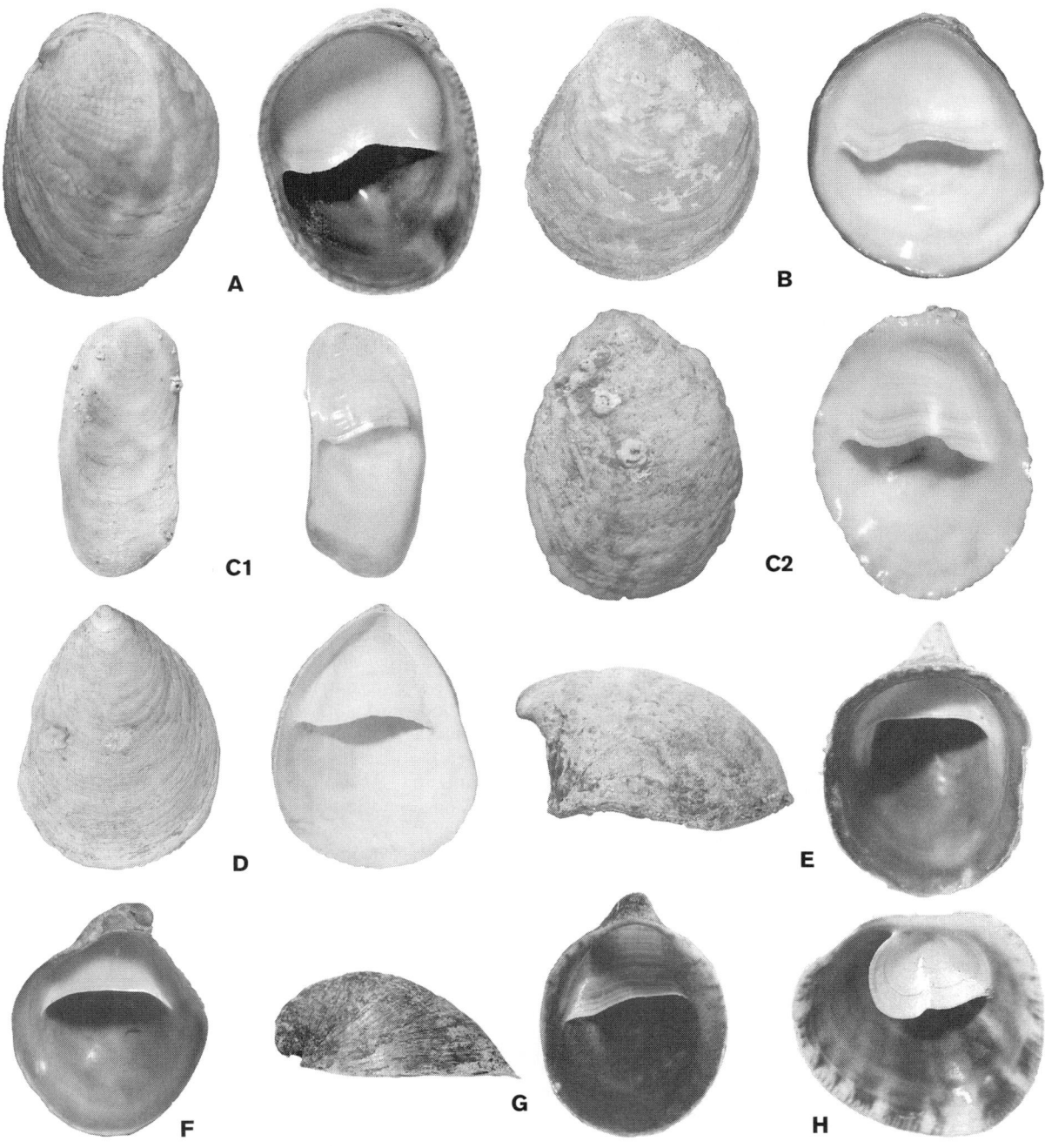

PLATE 364 Calyptraeidae: A, *Crepidula fornicata*, length 33.5 mm (two views); B, *Crepidula nummaria* (two views), length 38.7 mm; C1, *Crepidula perforans* (two views, shell aperture form), length 20 mm; C2, *Crepidula perforans* (two views), length 21.6 mm; D, *Crepidula plana* (two views), length 25.7 mm; E, *Garnotia adunca* (two views), diameter 19 mm; F, *Garnotia norissiarum*, length 30.1 mm; G, *Crepidula convexa* (two views), length 8 mm; H, *Crepipatella lingulata*, diameter 21 mm.

distinguished from the tubes of serpulid worms: vermetids have a three-layered shell, are glossy and white (often tinged with brown) within, and begin with a spirally coiled embryonic shell; tube worms have a two-layered shell, are dull and lusterless within, and begin with a single noncoiled tubular chamber. Some vermetids feed by casting out a mucous net, which traps planktonic particles on which they feed. See Keen 1961, Bull. Brit. Mus. (Nat. Hist.) 7: 181–213 (taxonomy). See Hadfield and Hopper, 1980, Mar. Biol. 57: 315–325 (spermatophores); Hadfield and Iaea 1989, Bull. Mar. Sci. 45: 377–386 (larvae).

1. Solitary; shell coiled, flat, with a strong dorsal ridge and scaly longitudinal ribs; tube diameter 1.5 mm–3 mm; often on abalones, corroding a channel into shell surface (plate 365A) . . . . *Dendropoma lituella*
— Gregarious, generally occurring in clumps or masses (except *Serpulorbis squamigerus* in central California; see below) 2
2. Tube diameter about 2 mm; sculpture of diagonal wrinkles on early portion of tube, projecting tubes smooth; with internal spiral thread on columella in medial whorls;

occasional isolated individuals tightly coiled (plate 365D)
. . . . *Petaloconchus montereyensis*
— Tube diameter much >2 mm 3
3. Lacking operculum; shell relatively large, diameter to about 12 mm; tube scaly to wrinkled, longitudinally ribbed, not embedded in substratum (plate 365C) . . . . .
. . . . *Serpulorbis squamigerus*
— With operculum; tube diameter to 15 mm, similar to above in size and sculpture, but initial whorls embedded in substratum (if on a mollusk shell) and later whorls somewhat corroded where one tube crosses another (plate 365B) . . . .
. . . . *Dendropoma rastrum*

Dendropoma lituella (Mörch, 1861) (=*Spiroglyphus lituellus*). Often found embedded in abalone shells; also on other shells and rocks.

Dendropoma rastrum (Mörch, 1861). May occur in clusters on soft rock or on shells, such as abalones. Has been confused with *Serpulorbis squamigerus* with which it may occur in the same cluster. The sculpture is similar to that of *S. squamigerus*, but *D. rastrum* possesses an operculum and shows slight corrosion where one tube crosses another.

Petaloconchus montereyensis Dall, 1919. Under rocks in low intertidal in areas of heavy but broken wave action; possibly unique among gastropods in periodic production of a new, and molting of the old, operculum. See Hadfield 1970, Veliger 12: 301–309 (anatomy, ecology, biology).

Serpulorbis squamigerus (Carpenter, 1857) (=*Aletes squamigerus*). Twisted masses found south of Point Conception, generally found only as individuals in central California; on rocks, shells, pilings. See Hadfield, above. Larvae settle on bryozoans (see Osman 1987, J. Exp. Mar. Biol. Ecol. 111: 267–284).

VELUNTINOIDEA

Two families comprise the Velutinoidea, the Triviidae and the Velutinidae, which seem not to have shell features in common, although some members of both families have mantle margins that envelop the shell. However, all members are grazing carnivores on ascidians and both groups have the double-walled echinospira larvae; see Wilson 1998, p. 787 (in Gastropod volume, Fauna of Australia). Both families in the superfamily are represented in central California.

FAMILY TRIVIIDAE

Triviids, called the false cowries, have small shells that somewhat resemble cowries in having the outer lip of the mature shell inturned and bearing elongate denticles on both the inner and outer lips. Shells are either strongly ridged (subfamily Triviinae) or smooth (subfamily Eratoinae). Shells lack periostracum and are enveloped by the mantle, which can expand over the entire shell and retract when the animal is disturbed; there is no operculum. No work has been done on Californian species, but they have been studied from other areas of the world. See Gosliner and Liltved 1987, Zool. J. Linn. Soc. 90: 207–254 (general features); Wilson 1998, p. 787 (in Gastropod volume, Fauna of Australia).

1. Aperture slotlike, running full length of shell; shell dark purple brown; about the shape and size of a coffee bean; numerous transverse ridges extending from a dorsal longitudinal furrow to aperture; length to 10 mm (plate 365E)
. . . . *Trivia californiana*

— Shell red to gray dorsally, glossy; inverted pear-shaped
. . . . 2
2. Outer lip with seven to 10 denticles; color purple red dorsally; length to about 16 mm (plate 365G) . . . .
. . . . *Erato vitellina*
— Outer lip with about 12 or more denticles; gray to orange-brown or reddish brown dorsally; length to about 8 mm (plate 365F) . . . . *Erato columbella*

Trivia californiana (Gray, 1827). The coffee-bean shell, more common in subtidal zone, associated with ascidians.

Erato columbella Menke, 1847. Under rocks, low intertidal to offshore, associated with and feeding upon ascidians.

Erato vitellina Hinds, 1844. As above, may be encountered in beach drift. Northern limit is recorded as Bodega Bay, where, along with *Trivia*, it is common in beach drift at Bodega Head.

FAMILY VELUTINIDAE

Velutinidae have broadly inflated shells of low profile. There are two subfamilies: the Velutininae, with strongly developed periostracum and external shells, and the Lamellariinae, with thin periostracum and enveloped shells. There are radular distinctions between the subfamilies; one genus *Marsenina* presents a problem because the shell is like that of Lamellariinae whereas the radula is like that of Velutininae. Members of both subfamilies are found in association with ascidians, on which they feed and deposit their egg masses. The Lamellariinae are often overlooked because of their remarkable similarity in color and texture to the host. See Behrens 1980, Veliger 22: 323–339 (taxonomy of Lamellariinae); Fretter and Graham 1981, J. Moll. Stud., Suppl. 9, 285–262 (biology).

1. Shell with thick brown periostracum 2
— Periostracum very thin, tan 3
2. Periostracum with spiral rows of fine bristles; spire depressed; diameter to 6 mm (plate 365I) . . . . . . *Velutina* sp.
— Periostracum nearly smooth, with irregular axial ridges; spire projecting; diameter to 24 mm (plate 365H) . . . . . .
. . . . *Velutina prolongata*
3. Shell internal, fully enveloped by mantle, translucent white 4
— Mantle with dorsal pore that opens to expose translucent white shell 5
4. Body with flat areas separated by angular ridges, shell surface not maleated; diameter 5 mm (plate 366C) . . . .
. . . . *Marseniopsis sharonae*
— Body bulbous, shell surface malleated (like hammered metal); diameter 17 mm (plate 366D) . . . . *Lamellaria diegoensis*
5. Body surface finely pitted; diameter 6 mm (plate 366B)
. . . . *Marsenina stearnsii*
— Body surface warty; diameter 6 mm (plate 366A) . . . .
. . . . *Marsenina rhombica*

Velutina prolongata Carpenter, 1865. Generally sublittoral; rare in intertidal zone.

Velutina sp. [=*V. velutina* (Müller, 1776), of authors]. Intertidal; may be common under rocks, in crevices. This is much smaller and has less projecting apical whorls than the more northern species *V. velutina*.

Marsenina rhombica (Dall, 1871). The two species of *Marsenina* cannot be told apart on shell characters.

PLATE 365 Vermetidae, Triviidae, and Velutinidae (subfamily Velutininae): A, *Dendropoma lituella*, diameter of individual 16 mm; B, *Dendropoma rastrum*, diameter 64 mm; C, *Serpulorbis squamigerus*, diameter of clump 38 mm; D, *Petaloconchus montereyensis*, diameter of clump 20 mm; E, *Trivia californiana* (two views), length 9.8 mm; F, *Erato columbella*, height 6.2 mm; G, *Erato vitellina*, height 13.7 mm; H, *Velutina prolongata* (two views), diameter 24 mm; I, *Velutina* sp. (two views), diameter 5.7 mm.

Marsenina stearnsii (Dall, 1871) [=*Lamellaria stearnsi*]. See Ghiselin 1964, Veliger 6: 123–124 (on ascidian *Trididemnum opacum*).

Lamellaria diegoensis Dall, 1885. Scarce, on ascidians. See Lambert 1980, Veliger 22: 340–344 (feeding on ascidian *Cystodytes*).

Marseniopsis sharonae (Willett, 1939). Scarce, on ascidians.

SUPERFAMILY NATICOIDEA, FAMILY NATICIDAE

Naticid moon snails have low profiles; shells lack surface sculpture and are fully enveloped by the mantle; the foot has a propodium, which enables the animal to plow through soft bottoms. They prey upon bivalves, which they drill, forming

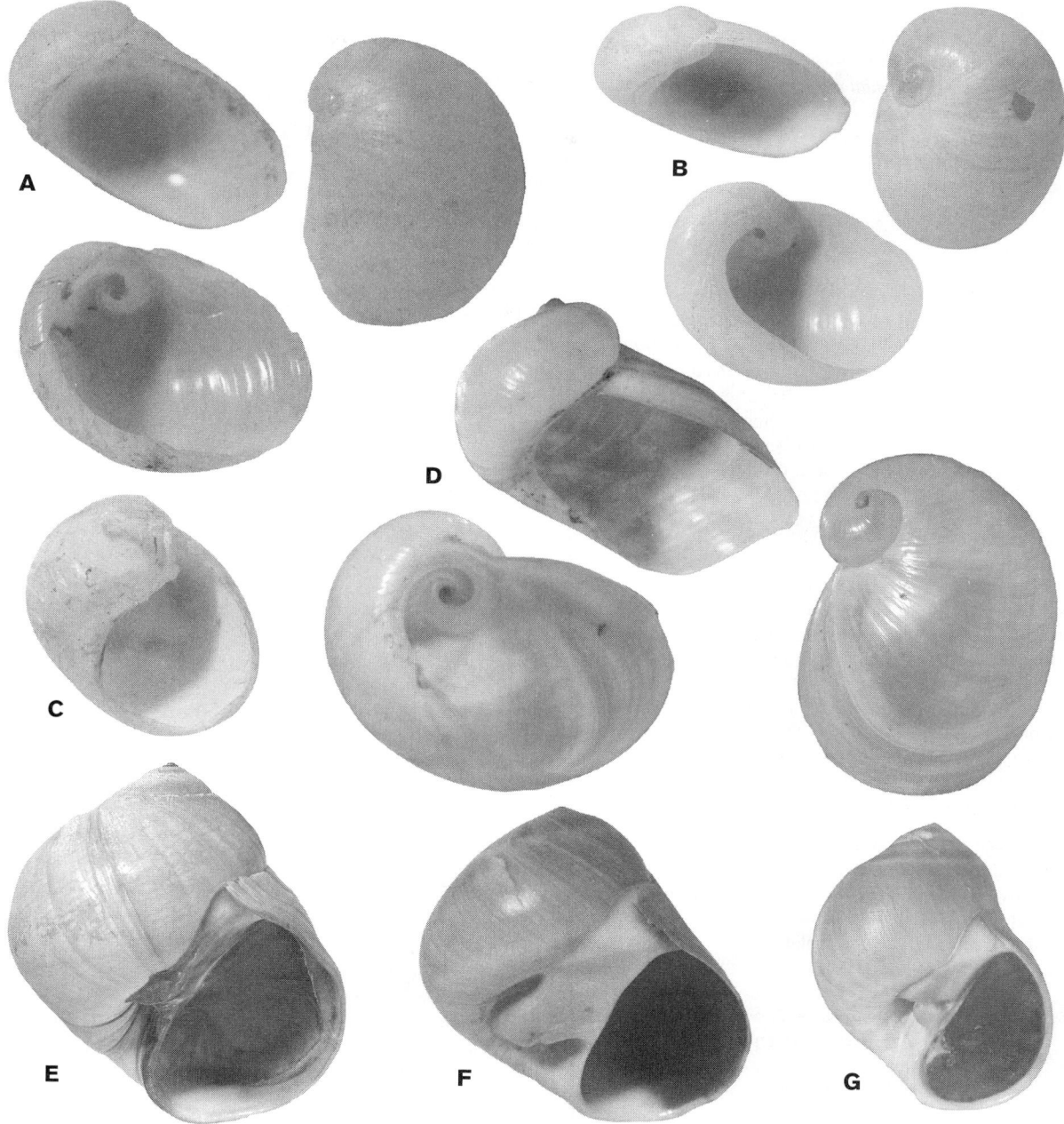

PLATE 366 Velutininae (subfamily Lamellariinae) and Naticidae: A, *Marsenina rhombica* (three views), diameter 11.3 mm; B, *Marsenina stearnsii* (three views), diameter, 6.2 mm; C, *Marseniopsis sharonae*, diameter 5.5 mm; D, *Lamellaria diegoensis* (three views), diameter 17 mm; E, *Euspira lewisii*, height 65 mm; F, *Glossaulax reclusianus*, height 38.8 mm; G, *Glossaulax altus*, height 33.6 mm.

rounded, beveled holes. Drilling involves the accessory boring organ of the proboscis. Radula is taenioglossate, with seven teeth per row. Egg masses form "sand collars" in which the gelatinous capsules are embedded. Former subgenera of the genus *Polinices* are now recognized as genera (for taxonomy see Kabat 1991, Bull. Mus. Comp. Zool., Harvard Univ. 152: 417–499).

1. Large (adult 12.5 cm–15 cm), heavy, globose shell with a large, broadly rounded body whorl; umbilicus deep, upper part covered by a wide columellar callus; shoulder weakly angulate with shallow sulcus below (plate 366E) . . . . *Euspira lewisii*

— Smaller (30 mm–40 mm), umbilicus partially blocked by tongue of callus 2
2. Profile broad, callus tongue white; height 40 mm (plate 366F) . . . . *Glossaulax reclusianus*
— Profile more slender, callus tongue brown, two-lobed; height 40 mm (plate 366G) . . . . *Glossaulax altus*

Euspira lewisii (Gould, 1847) (=*Polinices lewisii*). Largest living moonsnail. On mud and sand flats in bays, lagoons, *Zostera* flats, also soft bottoms offshore. A voracious consumer of infaunal bivalves, including *Macoma*. The older generic name *Polinices* is from Greek, meaning "many victories." See Bernard

1968, Nautilus 82: 1–3 (sexual dimorphism); Reid and Friesen 1980, Veliger 23: 25–34 (digestive system).

Glossaulax altus (Arnold, 1903) (=*Polinices altus*). Separation of this species from *Glossaulax reclusianus* is controversial, but I consider it to be distinct.

Glossaulax reclusianus (Deshayes, 1839) (=*Polinices reclusianus*). More common in southern California, but known from Bodega Harbor since the 1970s.

TRIPHOROIDEA

Triphoroideans have small, slender shells, assigned to two families, the Cerithiopsidae, which are usually dextral and the Triphoridae, which are usually sinistral. Both families are aphallate and feed on sponges. Protoconchs are important for generic-level classification in both families. The radula is usually ptenoglossate, with repeating elements, departing from the taenioglossate condition with seven teeth per row. Both families are represented in central California.

FAMILY CERITHIOPSIDAE

Cerithiopsids are dextral and have a short, siphonal canal open, notched; the final lip is not flared. Sponge feeders, but biology of eastern Pacific species has not been studied, and there has been no recent taxonomic work for the region subsequent to that of Bartsch 1911, Proc. U.S. Nat. Mus. 40: 327–367. Cerithiopsids should not be confused with the cerithiid genera *Lirobittium* and *Stylidium*, shells of which do not have the siphonal notch of cerithiopsids. The cerithiopsid species treated here also differ in having more protoconch whorls than those of the latter two genera.

1. Relatively large (to 10 mm), sculpture of raised spiral cords with minute axial threads in interspaces; brown, flat-sided, relatively large (plate 367A) . . . . *Seila montereyensis*
— Relatively small (to 3 mm); sculpture clathrate, with strongly projecting beads (plate 367B) . . . . *Cerithiopsis berryi*

Seila montereyensis Bartsch, 1907. More common offshore; for taxonomy see DuShane and Draper 1975, Veliger 17: 335–345.

Cerithiopsis berryi Bartsch, 1911. Shells occur in sediment samples, but living specimens should be found on sponge colonies.

FAMILY TRIPHORIDAE

Triphorids are usually sinistral (the subfamily Metaxiinae is an exception) and have a flaring final lip, a pinched posterior sinus (which forms a tube in some genera), and a short anterior canal that is usually sealed to form a tube. The family Triphoridae includes the dextral subfamily Metaxiinae and the sinistral subfamily Triphorinae. Like the Cerithiopsidae, they are sponge feeders. See Marshall 1983, Rec. Aust. Mus., Suppl. 2: 1–119 (taxonomy).

1. Shell sinistral, small (to about 6 mm), slightly convex, with three beaded cords per whorl, the middle cord arising as the shell attains its full size; anterior canal short, closed (plate 367C) . . . . *Triphora pedroana*
— Shell dextral (to about 6 mm), whorls convex, sutures deeply impressed; with weak ribs, four spiral cords per whorl bearing elongate beads; base concave (plate 367D) . . . . *Metaxia convexa*

Metaxia convexa (Carpenter, 1857) (=*M. diadema* Bartsch, 1907). Low tide to sublittoral, gravel, rocks.

Triphora pedroana Bartsch 1907 (=*T. montereyensis* Bartsch, 1907). In sand, gravel, rubble; associated with sponges.

EPITONIOIDEA

Members of the families that comprise the superfamily Epitonioidea are predators or parasites on coelenterates and have a ptenoglossate radula of numerous fanglike teeth that expand over the odontophore. Purple dye is secreted by the hypobranchial gland. Two families are represented in the superfamily Epitonioidea, the Janthinidae, which are treated as part of a separate section on pelagic gastropods, and the Epitoniidae.

FAMILY JANTHINIDAE

See treatment of janthinids in Pelagic Gastropoda by Seapy and Lalli.

FAMILY EPITONIIDAE

Epitoniids, the wentletrap shells, generally have white shells of high profile and impressed suture, usually with many prominent axial ribs or bladelike lamellae; protoconchs are multispiral. The aperture is circular and the operculum paucispiral; purple dye is released by living specimens. They feed on sea anemones by everting the proboscis and biting chunks of the prey; protandrous hermaphrodites, lacking the penis; planktotrophic veligers emerge from egg capsules formed in strings. For biology see Smith 1998, pp. 814–817 (Gastropod volume of Fauna of Australia). See DuShane 1979, Veliger 22: 91–134 (taxonomy of West Coast species).

1. Base not set off by a spiral keel; axial sculpture of thin, sharp lamellae, continuous from whorl to whorl 2
— Base set off by a spiral keel; axial lamellae thick 3
2. Relatively large (height to 35 mm), lacking brown line (plate 367E) . . . . *Epitonium indianorum*
— Relatively small (height to 15 mm) fresh specimens with a characteristic purplish or brown line below suture (plate 367F) . . . . *Epitonium tinctum*
3. Small (height to 10 mm); axial ribs acute; spiral keel strong, projecting; ribs not continuing over onto shell base (plate 367G) . . . . *Opalia montereyensis*
— Large (height to 35 mm); axial ribs broadly rounded; spiral keel low to obscure; about every third rib stronger, continuing over keel onto base of shell (plate 367H) . . . . *Opalia borealis*

Epitonium tinctum (Carpenter, 1864) (=*Nitidiscala tincta*). In sand at base of sea anemones *Anthopleura elegantissima* and *A. xanthogrammica*; feeds at high tide upon tips of anemone tentacles (Hochberg 1971, Ann. Rept. West. Soc. Malacologists 4: 22–23). See also Smith 1977, Veliger 19: 331–340 (chemoreception); Salo 1977, Veliger 20: 168–172 (feeding); Resch and Breyer 1983, Veliger 26: 37–40 (northern and southern populations); Collin 2000, Veliger 43: 302–313 (anatomy and development).

Epitonium indianorum (Carpenter, 1864) (=*Nitisdiscala indanorum*). More common offshore, occasionally at low intertidal zone; hermit-crab shells occur.

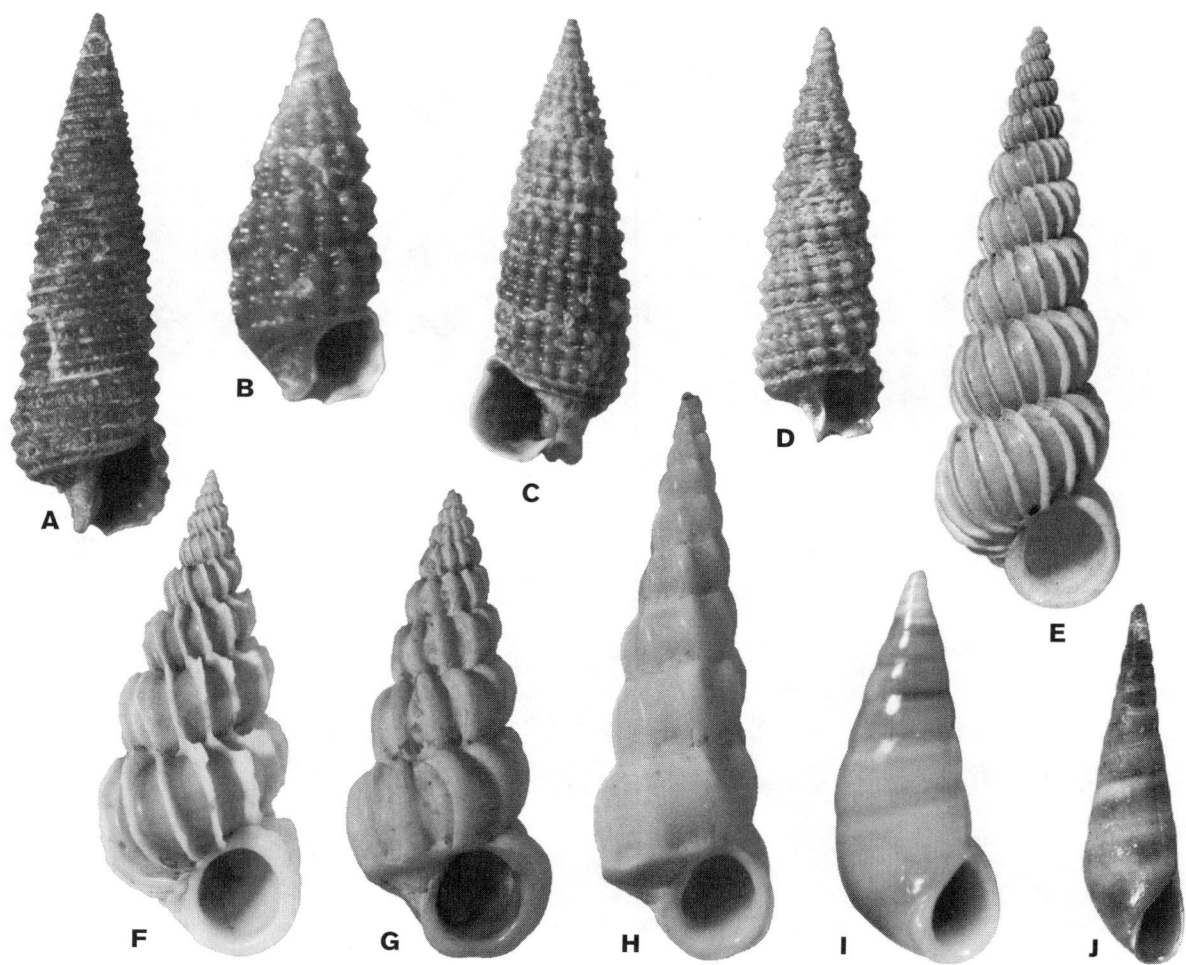

PLATE 367 Cerithiopsidae, Triphoridae, Epitoniidae, and Eulimidae: A, *Seila montereyensis*, height 11.2 mm; B, *Cerithiopsis berryi*, height 2.6 mm; C, *Triphora pedroana*, height 6.1 mm; D, *Metaxia convexa*, height 5.3 mm; E, *Epitonium indianorum*, height 37.8 mm; F, *Epitonium tinctum*, height 15.2 mm; G, *Opalia montereyensis*, height 11.5 mm; H, *Opalia borealis*, height 34.4 mm; I, *Melanella thersites*, height 7.0 mm; J, *Polygireulima rutila*, height 6.8 mm.

*Epitonium hindsii (Carpenter, 1856) (=*Nitidiscala hindsii*; *E. cooperi* Strong, 1930). Very deep suture; on sandy or muddy bottoms offshore. For illustrations see DuShane 1979, Veliger 22, figs. 32–35.

Opalia borealis Keep, 1881 (=*O. chacei* Strong, 1937). In sand at the base of the sea anemone *Anthopleura xanthogrammica*; feeds at high tide upon the anemone by inserting its proboscis directly in the column.

Opalia montereyensis (Dall, 1907). Largely sublittoral; both species also occur occasionally as hermit crab shells intertidally.

SUPERFAMILY EULIMOIDEA, FAMILY EULIMIDAE

Eulimids have small, slender, glossy shells. All are suctorial parasitic on echinoderms; most lack a radula. Species of mobile genera have an operculum, others are permanently attached to hosts and lack opercula. Sexes are separate and males have a cephalic penis. Individuals of mobile species away from hosts are thought to be either resting between feedings or searching for their echinoderm host. See Warén

* = Not in key.

1984, J. Moll. Stud., Suppl. 13, 1–96 (taxonomic revision of family and review of genera worldwide).

1. Shell sturdy, white, profile broad, whorls irregular with moderately deep suture; height to 7 mm (plate 367I) . . . .
    . . . . *Melanella thersites*
—   Shell thin, transparent white, profile slender, whorls even, flat-sided; height to 7 mm (plate 367J) . . . .
    . . . . *Polygireulima rutila*

Melanella thersites (Carpenter, 1864) (=*Balcis thersites*). Occurs on small holothurians at low tide and in the rocky sublittoral, although it is more often found away from its host.

Polygireulima rutila (Carpenter, 1864) (=*Balcis rutila*). Occurs on ventral surfaces of seastars of various species, in the immediate sublittoral zone to moderate depths. The change in the generic name was discussed by McLean (1996: 74).

NEOGASTROPODA
MURICOIDEA

Muricoidean families generally have fusiform shells with elongate apertures and siphonal canals. The radula is rachiglossate,

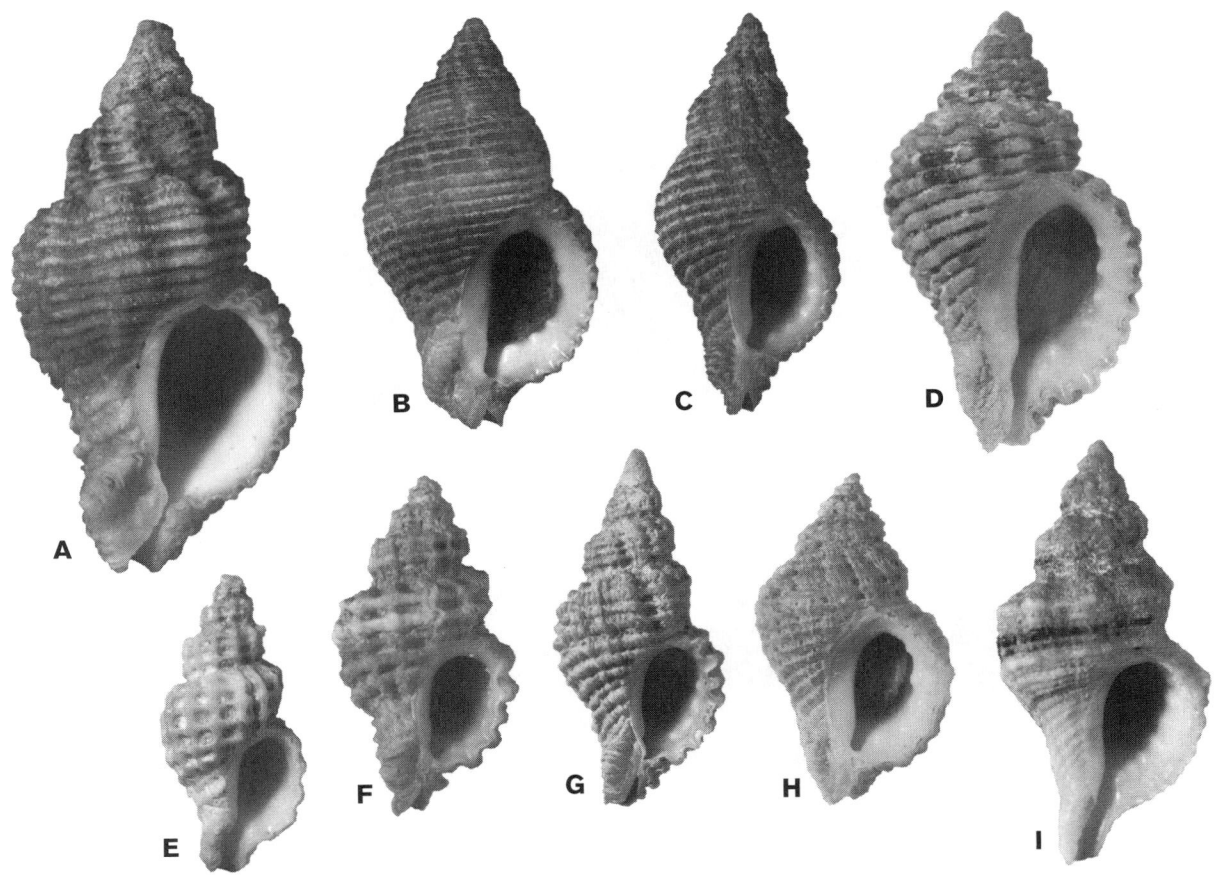

PLATE 368 Muricidae: A, *Ocinebrina aspera*, height 24.0 mm; B, *Ocinebrina lurida*, height 18.5 mm; C, *Ocinebrina munda*, height 17.2 mm; D, *Ocinebrina circumtexta*, height 19.8 mm; E, *Ocinebrina minor*, height 9.7 mm; F, *Ocinebrina atropurpurea*, height 10.9 mm; G, *Ocinebrina interfossa*, height 13.4 mm; H, *Ocinebrina gracillima*, height 12.2 mm; I, *Ocinebrina subangulata*, height 19.0 mm.

with three teeth per row. Most are carnivores, employing an extendable, eversible proboscis to reach food from a distance; digestive system with paired salivary glands, anterior gut with gland of Leiblein, posterior gut with anal gland. See Ponder 1998, pp. 819–820 (Gastropod volume, Fauna of Australia). Only some families are represented in shallow water of central California; those treated here are Muricidae, Columbellidae, Nassariidae, Buccinidae, Fasciolariidae, Melongenidae, Olividae, Cystiscidae, and Mitridae.

FAMILY MURICIDAE

Muricids have elaborate sculpture, both axial and spiral, some forming prominent varices bearing spines or winglike projections. Muricids are carnivorous gastropods, most of which are capable of drilling into the shell of their bivalve or barnacle prey, using both the radula and a special structure called the accessory boring organ. Some (*Acanthinucella, Ceratostoma*) have a labial spine that is used as a wedge to open bivalves or mussels for insertion of proboscis (see Perry 1985, Mar. Biol. 88: 51–58). Egg capsules are attached to rocks. This is a large family that is well represented in tropical waters; it is the most speciose caenogastropod family in shallow waters of central California.

All muricids treated here are members of the subfamily Ocenebrinae, characterized by the fine lamellar sculpture. The genus *Ocinebrina* is well represented but the biology of eastern Pacific species has been little studied; species were formerly placed in *Ocenebra* (see McLean 1996: 80). Species of the genus *Acanthinucella* were formerly placed in *Acanthina* (see Vermeij 1993, Cont. Tert. Quat. Geol. 30: 22). There is a large literature on the genus *Nucella*, species of which are prominent intertidal predators on mussels and barnacles. For evolutionary history of *Nucella*, see Collins et al. 1996, Evol. 50: 2287–2304; and Marko 1998, Evol. 52: 757–774.

1. Siphonal canal usually sealed may be open in some species
    2
—   Siphonal canal open 11
2. Three or more varices per whorl. 3
—   Lacking varices on body whorls but mature lip thickened
    4
3. Three prominent varices per whorl; with projecting tooth on outer lip near base; surface smooth, lustrous; height to 65 mm (plate 369A) . . . . *Ceratostoma foliatum*
—   More than three varices per whorl; shell surface dull, texture chalky; sculpture of alternating large and small spiral cords; height to 40 mm (plate 369B) . . . .
    . . . . *Ocinebrellus inornatus*
4. Spiral sculpture equal in strength to axial sculpture . . . . .
    5
—   Spiral sculpture overriding strong axial ribs 6
5. Profile slender, shell white; height to 10 mm (plate 368E)
    . . . . *Ocinebrina minor*

PLATE 369 Muricidae: A, *Ceratostoma foliatum*, height 62.3 mm; B, *Ocinebrellus inornatus,* height 40.2 mm; C, *Nucella lamellosa*, height 60.6 mm; D, *Urosalpinx cinerea*, height 33.2 mm; E, *Acanthinucella punctulata*, height 26.7 mm; F, *Acanthinucella spirata*, height 27.3 mm; G, *Nucella ostrina*, height 26.7 mm; H, *Nucella emarginata*, height 35.4 mm; I, *Nucella canaliculata*, height 28.4 mm; J, *Nucella analoga analoga*, height 32.7 mm; K, *Nucella analoga compressa*, height 35.1 mm.

— Profile broad, shell brown with yellow rib at periphery; height to 12 mm (plate 368F) . . . .
. . . . *Ocinebrina atropurpurea*

6. Shell large (20 mm–25 mm) 7

— Shell smaller (12 mm–15 mm) 8

7. Color orange, suture deeply impressed; height to 25 mm (plate 368A) . . . . *Ocinebrina aspera*

— Color black, white, or gray, canal usually not sealed; height to 20 mm (plate 368D) . . . . *Ocinebrina circumtexta*

8. Suture weakly impressed, profile broad 9

— Suture strongly impressed, profile slender 10

9. Dark orange, without varix behind outer lip; height to 20 mm (plate 368B) . . . . *Ocinebrina lurida*

— Light orange, small with low varix just behind outer lip; height to 12 mm (plate 368H) . . . . . *Ocinebrina gracillima*

10. Spiral ribs alternately large and small; grayish brown; height to 15 mm, sometimes larger; canal may not be sealed (plate 368G) . . . . *Ocinebrina interfossa*

— Reddish orange; height to 17 mm (plate 368C) . . . .
. . . . *Ocinebrina munda*

11. Outer lip with a projecting tooth near base (immature specimens may lack tooth); color pattern of revolving, interrupted, brown bands. 12
— Outer lip lacking projecting tooth 13

12. Shoulder rounded or weakly angulate; spire low; height to 27 mm (plate 369E) . . . . *Acanthinucella punctulata*
— Prominent keel at shoulder; spire produced; height to 28 mm (plate 369F). . . . . *Acanthinucella spirata*

13. Canal short 14
— Canal moderately long 19

14. Sculpture of closely spaced spiral cords 15
— Sculpture of broad, irregular spiral cords 17

15. Interspaces between cords deeply channeled; height to 30 mm (plate 369I) . . . . *Nucella canaliculata*
— Interspaces between cords with narrow channels 16

16. Spiral cords narrow; height to 35 mm (plate 369J) . . . . . .
. . . . *Nucella analoga analoga*
— Spiral cords broad, interspaces narrow; height to 35 mm (plate 369K) . . . . *Nucella analoga compressa*

17. Shell with irregularly nodulose, often well-separated, spiral cords (some populations with shells nearly smooth); columella excavated; umbilicus closed. 18
— Sculpture various: nearly smooth, or with prominent axial lamellae, or with spiral cords and weaker, irregular, axial swellings; umbilicus small, sometimes closed; height to 60 mm (plate 369C). . . . . *Nucella lamellosa*

18. Spire relatively high, parietal nub of aperture lacking, height to 27 mm; egg capsules vase shaped with long neck (plate 369G) . . . . *Nucella ostrina*
— Spire relatively low, spiral cords with irregular nodes, aperture with parietal nub, height to 35 mm; egg capsules cylindrical, with short neck and flared distally (plate 369H)
. . . . *Nucella emarginata*

19. Axial sculpture strong across shoulder; numerous fine spiral cords, in older specimens most pronounced between axial ribs; canal short, constricted; shell color variable, gray, yellow-brown, aperture purplish; height to 35 mm (plate 369D). . . . . *Urosalpinx cinerea*
— Axial sculpture of seven to nine strong ribs, sharply angled at shoulder of whorl; spiral sculpture reduced to fine threads; whitish, often with brown spiral line at periphery; height to 20 mm (plate 368I). . . . . *Ocenebrina subangulata*

Ceratostoma foliatum (Gmelin, 1791) (=*Purpura foliata*). In low intertidal of semiprotected outer coast and sublittoral. A large species, which may develop prominent varices, especially when living sublittorally. See Kent 1981, Nautilus 95: 38–42 (feeding); Carefoot and Donovan 1995, Biol. Bull. 189: 59–68 and Donovan et al. 1999, J. Exp. Mar. Biol. Ecol. 236: 235–251 (functional significance of shell sculpture).

Ocinebrellus inornatus (Récluz, 1851). [=*Ceratostoma inornatum*; =*Ocenebra japonica* (Dunker, 1869)]. Introduced from Japan with oysters in Tomales Bay, but presence there variable; now common in Puget Sound.

Ocinebrina aspera (Baird, 1863). Here distinguished from *O. lurida* by its larger size and more deeply channeled interspaced between the spiral cords.

Ocinebrina atropurpurea Carpenter, 1865 [=*Ocenebra atropurpurea*; =*O. clathrata* (Dall, 1919)]. Often as hermit-crab shells intertidally.

Ocinebrina circumtexta Stearns, 1871 (=*Ocenebra circumtexta*). Common mid-intertidal species.

Ocinebrina gracillima Stearns, 1871 (=*Ocenebra gracillima*). Under rocks; more common in southern California.

Ocinebrina interfossa Carpenter, 1864 (=*Ocenebra interfossa*). Common under alga-covered rocks.

Ocinebrina lurida (Middendorff, 1848) (=*Ocenebra lurida*). Common on and under rocks. See Palmer, 1988, Veliger 31: 192–302 (biology).

Ocinebrina minor (Dall, 1919). Here removed from synonymy of *O. atropurpurea*.

Ocinebrina munda (Dall, 1892). Here separated from *O. lurida*, differing in its more slender profile. Where sympatric with *O. lurida*, this occurs in sublittoral zone.

Ocinebrina subangulata (Stearns, 1873). This species retains an open siphonal canal and is thereby unlike other members of the genus; its affinity is unsettled.

Ocinebrina sclera (Dall, 1919). A more northern species, much larger than *O. aspera*, and living sublittorally in central California.

Urosalpinx cinerea (Say, 1822). The Atlantic oyster drill, introduced with oysters; may be common among oysters and barnacles in estuaries. See Franz 1971, Biol. Bull. 140: 63–72 (biology); Carriker, 1969, Amer. Zool. 9: 917–933 (drilling).

Acanthinucella punctulata (G. B. Sowerby I, 1825) (=*Acanthina punctulata*). Upper intertidal zone on rocks; moving downward during breeding season. See Sleder 1981, Veliger 24: 172–180 (biology).

Acanthinucella spirata (Blainville, 1832) (=*Acanthina spirata*). Where the unicorn snails *A. punctulata* and *A. spirata* occur together, as at Monterey, *spirata* generally is in the lower, and *punctulata* in the upper intertidal. Feeds on barnacles and mollusks. See Gianniny and Geary 1992, Veliger 35: 195–204 (shell form in genus); Perry 1985, Mar. Biol. 88: 51–58 (function of spine in opening barnacle prey).

Nucella analoga analoga (Forbes, 1852) (previously misidentified as *N. canaliculata*). This name had been considered a synonym of *N. canaliculata* but represents a species that has incised interspaces rather than deeply channeled interspaces. This is common in Oregon and northern California living exposed among mussels on which it feeds. Its range overlaps with that of *N. canaliculata* from Ketchikan, Alaska to Fidalgo Island, Washington. This must have been the species studied by Sanford et al. 2003, Science 300: 1135–1136. See McLean 2006, Festivus 38: 17–20 (systematics).

Nucella analoga compressa (Dall, 1915). This is a southern subspecies of *N. analoga* that occurs sparsely in the exposed mussel zone in Monterey County, south of Pacific Grove. See McLean 2006, Festivus 38: 17–20 (systematics).

Nucella canaliculata (Duclos, 1832). This is a northern species, occurring from Alaska to Puget Sound; it is common in the barnacle zone of waters exposed to tidal currents rather than exposed to strong wave action in the mussel zone.

Nucella emarginata (Deshayes, 1839). Common at upper tide levels. This has more southern distribution, compared to that of *N. ostrina*. The profile is lower than that of *N. ostrina*. See West 1986, Ecology, 67: 798–809 (prey selection); Wayne 1987, Veliger 30: 138–147 (defensive behavior of prey *Mytilus*).

Nucella ostrina (Gould, 1852). Marko 1998, Evolution 52: 757–774, and earlier papers by other authors cited therein, showed that northern populations formerly thought to represent *N. emarginata*, should take the name *N. ostrina*, based on genetic evidence as well as distinctions in the shells. The two species overlap in distribution between Half Moon Bay

* = Not in key.

and Point Conception, California; Marko et al. 2003, Veliger 46: 77–85, noted that in the region of overlap, *N. ostrina* occurs on wave-swept shores and *N. emarginata* is found within embayments such as Monterey Bay, Half Moon Bay, and Morro Bay. This has a higher profile than that of *N. emarginata*.

Nucella lamellosa (Gmelin, 1791). Low tide on rocks, often in protected bays; highly variable sculpture, smoother specimens occurring in more exposed situations, delicate and prominent lamellae developing in protected areas; see Spight 1973, J. Expt. Mar. Ecol. 13: 215–228; 1974, Ecology 55: 712–729.

FAMILY COLUMBELLIDAE

Columbellids are small to minute in size with short siphonal canals, the outer lip may be denticulate and the columella with small plications, or these may be lacking. The radula is rachiglossate. Feeding is varied, some are carnivorous, and some feed on algal films or detritus (see Wilson 1998, pp. 827–829, Gastropod volume, Fauna of Australia). This is a large family with numerous genera and species in tropical regions. There are few papers on the biology of northeastern Pacific species. For taxonomy see Radwin 1977a, Veliger 19: 403–417; 1977b, Veliger 20: 119–133, 1978, Veliger 20: 328–344; Guralnick and de Maintenon 1997, J. Molluscan Stud., 63: 65–77 (radular formation); de Maintenon 1999, Invert. Biol. 118: 258–288 (phylogeny). *Amphissa* and *Alia* have multispiral protoconchs indicative of planktonic development; *Astyris* has a paucispiral protoconch, which is indicative of lecithotrophic development.

1. With incised spiral grooves restricted to base of shell; weak fold on columella separating aperture from short siphonal canal 2
— Spiral sculpture not restricted to base of shell 4
2. Protoconch paucispiral; shell small, slender, with a chevron pattern of thin revolving, brown lines on a yellow-brown background (not an irregular pattern of wavy longitudinal lines), relatively small; height 4.5 mm (plate 370C) . . . . *Astyris aurantiaca*
— Protoconch multispiral; color tan to yellow-brown or variously mottled 3
3. Shoulder of body whorl varying from smooth to strongly keeled (keel usually lighter in color than rest of shell); whorls somewhat inflated; outer lip sinuous; periostracum smooth; color usually yellow-brown to dark brown, at times with white and darker brown mottling; height to 10 mm (plate 370E) . . . . *Alia carinata*
— Slender, whorls nearly flat-sided; periostracum forming thin, projecting, axial blades in living animals; color usually tan, sometimes darker, may show fine white dots; height to 7 mm (plate 370D) . . . . *Alia tuberosa*
4. Shell large (to 18 mm in height), columella with fine plications (plate 370A) . . . . *Amphissa columbiana*
— Shell smaller (to 13 mm in height), variable in color, columella lacking plications (plate 370B) . . . . *Amphissa versicolor*

Amphissa columbiana Dall, 1916. On algae-covered rocks; under rocks in sand and gravel. See Kent 1981, Veliger 23: 275–276 (behavior).

Amphissa versicolor Dall, 1871. Common in rocky intertidal and sublittoral zones. Highly variable in color pattern.

Alia carinata (Hinds, 1844) (=*Mitrella carinata*; =*Mitrella gausapata* of authors). Common on algae and on rocks. See

Carter and Behrens 1980, Veliger 22: 376–377 (mimicry by amphipod); Bergman et al. 1983 Veliger 26: 116–118 (variation and shell repair); Tupen 1999, Veliger 42: 249–259 (variation in shell form and color).

Alia tuberosa (Carpenter, 1864) (=*Mitrella tuberosa*). In sand and gravel at low tide and in beach drift; more common sublittorally.

Astyris aurantiaca (Dall, 1871) (=*Mitrella aurantiaca*). Chiefly sublittoral; uncommon in low intertidal zone among rocks and algae.

FAMILY BUCCINIDAE

Buccinid whelks are medium to large-size, with siphonal canals, and extendable proboscis; they are scavengers or feed on living prey. Numerous genera and species occur in cold northern waters, but there are only two species occurring in shallow water in California.

1. Body whorl with distinct spiral sculpture; low, rounded axial ribs on spire; columella arched, glossy; canal short, twisted; color dull gray or brownish purple; sculpture may be obscured by growths of purple coralline algae (plate 370F) . . . . *Lirabuccinum dirum*

Lirabuccinum dirum (Reeve, 1846) (=*Searlesia dira*). Dire whelk; chiefly intertidal, on coralline-encrusted rocks and among gravel and rocks in crevices; abundant further north, but of spotty occurrence in central California. See Lloyd 1973, Ann. Rept. Western Soc. Malac. 5: 32 (biology, feeding); Vermeij 1991, Veliger 34: 264–271 (taxonomy).

*Kelletia kelletii (Forbes, 1852). Kellet's whelk; this large-shelled species common in southern California has since the 1980s occurred offshore in Monterey Bay; see Herlinger 1981, Veliger, 24: 78 (Monterey record); Rosenthal 1979, Veliger 12: 319–324 (reproduction); Lonhart and Tupen 2001, Bull. S. Calif. Acad. Sci., 100: 238–248 (northern occurrence); Zacherl et al. 2003, J. Biogeog. 30: 913–924 (northern occurrence). See photo in McLean 1978.

FAMILY NASSARIIDAE

The Nassariidae, the mud snails, are scavengers occurring on mud or sand. Like other neogastropods, the radula is rachiglossate. Shells are recognized as nassariids by a prominent furrow at the base of the shell. There are posterior pedal tentacles. Most eastern Pacific species produce egg capsules and have planktotrophic larval stages, as indicated by their multispiral protoconchs. See Demond, 1952, Pac. Sci. 6: 300–317 (review of northeastern Pacific species); Cernohorsky 1984, Bull. Auckland Inst. Mus. 14: 1–359 (taxonomy at subgeneric level); Harasewych 1998, Gastropod volume of Fauna of Australia, pp. 829–831 (general features); Hassl 2000, J. Paleont., 74: 839–852 (phylogeny).

Until now, most New World species had been placed in the genus *Nassarius*; however, that genus is based on a large-shelled Indo-Pacific-type species with a broad columellar shield, representing a group unlike any New World species. Here I raise to full generic level two taxa previously regarded as subgenera: *Caesia* H. and A. Adams, 1853, for large-shelled northeastern Pacific species with broad parietal callus and the lip thickened from the inner side, and the broadly distributed *Hima* Leach, 1852, for slender species with a narrow parietal callus and with the final lip thickened from both sides.

* = Not in key.

PLATE 370 Columbellidae, Buccinidae, Nassariidae, Fasciolariidae, and Melongenidae: A, *Amphissa columbiana*, height 16.7 mm; B, *Amphissa versicolor*, height 11.3 mm; C, *Astyris aurantiaca*, height 4.5 mm; D, *Alia tuberosa*, height 6.5 mm; E, *Alia carinata*, height 9.8 mm; F, *Lirabuccinum dirum*, height 38.6 mm; G, *Ilyanassa obsoleta*, height 30 mm; H, *Caesia fossata*, height 47 mm; I, *Hima cooperi*, height 15.8 mm; J, *Hima mendica*, height 16.5 mm; K, *Harfordia harfordii*, height 55 mm; L, *Harfordia* sp., height 41.6 mm; M, *Aptyxis luteopictus*, height 24.1 mm; N, *Busycotypus canaliculatus*, height 103 mm.

1. With a distinct revolving furrow around base 2
— Lacking a distinct furrow; with columellar fold at base; shell sculpture of revolving, weakly beaded cords crossed by growth lines and oblique axial ribs; apex often eroded; aperture black-glazed; shell dark brown to black, often with adherent detritus and algae; height to 30 mm (plate 370G) . . . . *Ilyanassa obsoleta*
2. Shell relatively large (to 50 mm), broad, with orange callus spreading over parietal area; periphery rounded to carinate; axial sculpture of widely spaced axial ribs, limited to upper part of body whorl (plate 370H) . . . . *Caesia fossata*
— Shell small (to 20 mm), generally slender, without parietal callus; periphery rounded; axial sculpture of rounded ribs 3
3. Axial ribs few, broadly spaced; height to 17 mm (plate 370I) . . . . *Hima cooperi*
— Axial ribs more numerous, closely spaced; height to 18 mm (plate 370J) . . . . *Hima mendica*

Ilyanassa obsoleta (Say, 1822) (=*Nassarius obsoletus*). Introduced from Atlantic coast; an omnivore and deposit feeder, very abundant on San Francisco Bay mudflats. Only American nassariid having a crystalline style. Differs from most other genera because it lacks the deep basal groove, caudal cirri, bifurcated foot, and other features expected in the family. There is an extensive literature on Atlantic coast populations. See Curtis and Hurd 1981, Veliger 24: 91–96 (crystalline style); Scheltema 1964, Ches. Sci. 5: 161–166 (feeding and growth).

Caesia fossatus (Gould, 1850) (=*Nassarius fossatus*). Common on mud and sand in bays, estuaries; see MacGinitie 1931, Ann. Mag. Nat. Hist. (10) 8: 258–261 (egg–laying).

Caesia perpinguis (Hinds, 1844) (=*Nassarius perpinguis*). Sublittoral on sandy bottoms; with a narrow shelf below suture and fine cancellate sculpture with beaded axial ridges. See photo in McLean 1978.

Caesia rhinetes (Berry, 1953) (=*Nassarius rhinetes*). An uncommon, large-shelled species occurring offshore. See photo in McLean 1996.

Hima mendica (Gould, 1849) (=*Nassarius mendicus*). Common in sand, mud, on rocks, of open coast and in bays.

Hima cooperi (Forbes, 1852) (=*Nassarius cooperi*). Differs from *H. mendica* in having fewer axial ribs. Usually regarded as a variant of *H. mendica*, but the consistency of shell form in certain populations leaves the question open, a matter for further investigation.

Hima fratercula (Dunker, 1862) (=*Nassarius fraterculus*). An introduced northwestern Pacific species well established in Puget Sound and southern British Columbia; can be expected to extend its distribution to central California. See photo in Abbott 1974.

FAMILY FASCIOLARIIDAE

Fasciolariids are carnivorous neogastropods with long siphonal canals. The group is mostly tropical; members of subfamily Fusininae occur in California. Aside from radular and anatomical differences, they differ from buccinids in having the animal red-colored. Genera previously regarded as subgenera of *Fusinus* are here recognized as full genera, distinguished from *Fusinus* Rafinesque, 1815, a tropical group for which the type species has a very long canal. Biology of the California species has not been studied.

* = Not in key.

1. Small (height to 26 mm); strong spiral cords continuous over strongly projecting axial ribs; canal short; coloration dark brown, white spiral band at periphery (plate 370M) . . . . *Aptyxis luteopictus*
— Larger, strong spiral cords continuous over strongly projecting axial ribs; canal; shell coloration uniformly brown 2
2. Large, to 55 mm in shell height; profile broad, canal relatively short, not strongly constricted (plate 370K) . . . . *Harfordia harfordii*
— To 45 mm in height; profile slender, canal more constricted (plate 370L) . . . . *Harfordia* sp.

Aptyxis luteopictus (Dall, 1877) (=*Fusinus luteopictus*). Low intertidal to sublittoral, on and under rocks of protected coast; Monterey Bay and south.

Harfordia harfordii (Stearns, 1871) (=*Fusinus harfordii*). Intertidal in partially exposed rocky areas in northern California.

Harfordia sp. Rocky sublittoral; shells occupied by hermit crabs in shallow water.

FAMILY MELONGENIDAE

Melongenids are large-shelled whelks, represented in shallow, mostly tropical waters. The subfamily Busyconinae is well represented along the Atlantic and Gulf of Mexico coasts of the United States. See Hollister 1959, Paleontol. Amer., 4: 59–126 (generic revision); for Atlantic species of Busyconinae see also Abbott (1974: 222–223); Kosyan and Kantor 2004, Ruthenica 14: 9–36 (anatomy and phylogeny of family).

1. No axial ribs; outer lip smooth; shell relatively thin; large, height to 160 mm; suture strongly channeled; shoulder keeled; long, slightly curved canal; with a yellow-brown, feltlike, hairy periostracum, often partly worn off (plate 370N) . . . . *Busycotypus canaliculatus*

Busycotypus canaliculatus (Linnaeus, 1758) (=*Busycon canaliculatum*). Channeled whelk, introduced from Atlantic; sublittorally and in mud at low tide in San Francisco Bay; distinctive strings of large egg capsules are commonly washed ashore. See Stohler 1962, Veliger 4: 211–212 (occurrence); Rohrkasse and Atema 2002, Biol. Bull. 203: 235–236 (chemoreception).

FAMILY OLIVIDAE

The mostly tropical family Olividae is represented in California and more northern regions by the genus *Callianax* H. and A. Adams, 1853, in the subfamily Olivellinae. Until now, *Callianax* had been regarded as a subgenus of the tropical genus *Olivella* Swainson, 1831; here it is regarded as a full genus because it significantly differs in having an operculum, not having multiple columellar folds, and not having columellar callus that extends posteriorly toward the shell apex. All members of the family are carnivores or scavengers of shallow, sandy bottoms. The glossy shell is partially enveloped by the mantle and the propodium is plow-shaped for burrowing. See Olsson 1956, Proc. Acad. Nat. Sci. Phil. 108: 155–225 (taxonomy); Kantor 1991, Ruthenica 1: 17–52 (anatomy and phylogeny).

1. Shell to about 30 mm in length, broad, and robust; variously colored, from almost all white to a black-gray, often violet at base; base offset with a dark line; columellar callus

relatively strong; fold at base of columella often with several incised spiral lines (plate 371A) . . . . *Callianax biplicata*
— Shell smaller (under 20 mm) 2
2. Shell stout and chunky; often with brown, longitudinal, zigzag lines on a brownish buff, gray, or olive-gray background; occasionally with a red-brown spot beside fold at base of columella; height to 14 mm (plate 371B) . . . . *Callianax pycna*
— Shell oblong and slender; may have brown longitudinal lines, color generally gray-brown to tan with faint purplish brown maculations near suture; height to 20 mm (plate 371C) . . . . *Callianax baetica*

Callianax biplicata (G. B. Sowerby I, 1825) (=*Olivella biplicata*). The purple olive; common intertidally, burrowing in clean sand of sloping, protected beaches and offshore of more exposed beaches. See Edwards 1968, Veliger 10: 297–304 (reproduction) and 1969, Veliger 11: 326–333 (predators); Stohler 1969, Veliger 11: 259–267 (growth); Hickman and Lipps 1983, J. Foraminiferal Res., 13: 198–114 (feeding on foraminifera).

Callianax baetica (Carpenter, 1864) (=*Olivella baetica*). More common offshore, but occasionally found in intertidal zone.

Callianax pycna Berry, 1935 (=*Olivella pycna*). Least abundant species, more likely found offshore.

FAMILY CYSTISCIDAE

Cystiscids have minute, glossy white shells; shells are fully enveloped by the brightly colored mantle; found in gravel, among coralline algae and in surfgrass holdfasts in the low intertidal. This group was first separated at the family level from Marginellidae by Coovert and Coovert (1995, Nautilus 109: 43–110). See also Coan and Roth 1966, Veliger 8: 276–299 (taxonomy).

1. Anterior of shell not notched 2
— Anterior of shell with siphonal notch visible in dorsal view; columella with two folds, outer lip dentate; height less than 4 mm (plate 371D) . . . . *Gibberula subtrigona*
2. Spire concealed by extension of aperture; columella with four folds, outer lip finely dentate; height 2 mm–3 mm (plate 371E) . . . . *Granulina margaritula*
— Spire low but not concealed; outer lip smooth within; height to 4 mm–5.5 mm (plate 371F) . . . . . . *Plesiocystiscus jewettii*

Plesiocystiscus jewettii (Carpenter, 1857) (=*Cystiscus jewettii*). The genus was proposed by Coovert and Coovert (see above). In gravel of tide pools.

Gibberula subtrigona (Carpenter, 1864) (=*Granula subtrigona*). In gravel of tide pools.

Granulina margaritula (Carpenter, 1857) [=*Cypraeolina pyriformis* (Carpenter, 1864)]. Common in gravel of tide pools, and on the undersurfaces of medium-size rocks in the low intertidal; when disturbed, extensively protrudes a brightly-colored, mucus-covered mantle (J. T. Carlton, personal observations).

FAMILY MITRIDAE

Mitrids are characterized by a long aperture, short canal and pillar with strong plications. Most species of Mitridae are characteristic of tropical habitats; the Californian species is

an exception for its subtropical to temperate distribution. All members of the family feed on sipunculans. See Cernohorsky 1970, Bull. Auckland Inst. & Mus. 8, 1–190 (taxonomy); Ponder 1972, Malacologia 11: 295–342 (anatomy).

1. Shell large, whorls flat-sided, with shiny black periostracum bearing rows of fine pits; base with strong spiral cords; columella with three strong plications, body color white; height to 50 mm (plate 371G) . . . . *Mitra idae*

Mitra idae Melville, 1893. Rocky sublittoral; occasional as hermit crab shells in intertidal, Bodega Head and south. See Cate 1968, Veliger 10: 247–252 (mating); Chess and Rosenthal 1971, Veliger 14: 172–176 (reproduction); Fukuyama and Nybakken 1983, Veliger 26: 96–100 (feeding); West 1990, Bull. Mar. Sci., 46: 761–779 (feeding).

CONOIDEA

This neogastropod superfamily is characterized by the poison gland and the toxoglossate radular tooth, a single, hollow tooth used singly and injected with venom from the poison gland. Two families are treated here: Turridae and Conidae.

FAMILY TURRIDAE

Turrids are characterized by the posterior anal notch ("turrid notch") at or near the suture. All have a poison gland; the more advanced members have a single, hollow, harpoonlike radular tooth with which venom is delivered to the prey, consisting mostly of polychaete worms. This is a highly diverse family; numerous genera and species occur in deep water. Shallow-water species treated here are in the advanced group with the hollow tooth, with the exception of *Pseudomelatoma* in subfamily Pseudomelatominae, a more primitive group with a rachiglossate radula. Most species treated here are small-shelled and easily overlooked, except for the large-shelled genera *Pseudomelatoma* and *Ophiodermella*. Authors have not been in agreement over classification. See McLean 1971, Veliger 14: 114–130 (classification); Taylor et al. 1993, Bull. Nat. Hist. Mus., London 59: 125–170 (phylogeny); Rosenberg 1998, Amer. Malac. Bull. 14: 219–228 (phylogeny).

1. Shell relatively large for group, over 20 mm in length 2
— Shell relatively small, under 10 mm in length 3
2. With strong light-colored nodes at periphery; rusty brown or yellow-brown to blackish; height to 25 mm (plate 371H) . . . . *Pseudomelatoma torosa*
— Lacking peripheral nodes; with finely cancellate early sculpture; light colored, with darker axial markings; height to 25 mm (plate 371I) . . . . *Ophiodermella inermis*
3. Anal notch deep, at suture, bordered by thickened callus; with about 15 strong axial ridges per whorl, crossed by spiral ribs; height to 8 mm (plate 371L) . . . . *Clathurella canfeldi*
— Anal notch shallow 4
4. Sculpture of strong spiral cords, axial sculpture weak; purple-brown, sometimes mottled with white; height to 7 mm (plate 371K) . . . . *Cymakra gracilior*
— Sculpture coarsely clathrate 5
5. With three spiral cords per whorl, axial sculpture strongly projecting, overridden by narrow spiral cords; height to 9 mm (plate 371N) . . . . *Perimangelia interfossa*

PLATE 371 Olividae (subfamily Olivellinae), Cystiscidae, Mitridae, Turridae, and Conidae: A, *Callianax biplicata*, height 27 mm; B, *Callianax pycna*, height 13.6 mm; C, *Callianax baetica*, height 19 mm; D, *Gibberula subtrigona*, height 3.4 mm; E, *Granulina margaritula*, height 3.0 mm; F, *Plesiocystiscus jewettii*, height 5.3 mm; G, *Mitra idae*, height 37.5 mm; H, *Pseudomelatoma torosa*, height 23.0 mm; I, *Ophiodermella inermis*, height 23.8 mm; J, *Cymakra aspera*, height 6.0 mm; K, *Cymakra gracilior*, height 6.5 mm; L, *Clathurella canfieldi*, height 7.4 mm; M, *Clathromangelia fuscoligata*, height 9.5 mm; N, *Perimangelia interfossa*, height 8.5 mm; O, *Conus californicus*, height 28 mm.

— With two spiral cords per whorl. 6

6. Coarsely cancellate, white with brown spiral banding; height to 8 mm (plate 371M). . . . . *Clathromangelia fuscoligata*
— Brown; axial and spiral ribs equally spaced, strongly beaded at intersections, producing squarish cancellations; height to 5 mm (plate 371J) . . . . *Cymakra aspera*

Pseudomelatoma torosa (Carpenter, 1865). Intertidal rocky areas. See Kantor, 1988, Apex (Société Belge de Malacologie) 3: 1–19 (anatomy). Unlike other species treated here, in its membership in a primitive group not having a toxoglossate radula.
    Perimangelia interfossa (Carpenter, 1864) (=*Clathromangelia interfossa*) (=*Mangelia interlirata* Stearns, 1871). Rocky intertidal and sublittoral. See McLean 1999, Nautilus 114: 101 (generic proposal of *Perimangelia*).
    Clathromangelia fuscoligata (Dall, 1871). Rocky intertidal and sublittoral, scarce.
    Clathurella canfieldi Dall, 1871. In sand among surfgrass roots.
    Cymakra aspera (Carpenter, 1864) (=*Mitromorpha aspera*). Rocky intertidal and sublittoral.
    Cymakra gracilior (Tryon, 1884) (=*Mitromorpha gracilior*). Rocky areas of low intertidal.
    Ophiodermella inermis (Reeve, 1843) [=*Ophiodermella ophioderma* (Dall, 1909)]. Seldom found intertidally, except for hermit crab shells. See Shimek 1983, Malacologia 23: 281–313 (biology).

FAMILY CONIDAE

Although Conidae are speciose in tropical waters, a single species of *Conus* occurs in California. Many genera or subgenera have been proposed but current authors do not agree on how to subdivide the family; some authors consider that there is a single genus in the family. The toxoglossate radula employs a single hollow tooth to inject venom into the prey; some tropical species have become specialized fish or mollusk feeders, but most species prey upon polychaete worms. There is a very large literature on feeding and venom specializations for tropical species. See Kohn 1998: 852–854 (Gastropod Volume, Fauna of Australia).

1. Shell obconic (inversely conical), spire dome-shaped; dull gray, tan, or gray-brown with fine spiral markings of brown under a heavy, dark brown periostracum (removed for illustration to show color pattern); height to 35 mm (plate 371O). . . . . *Conus californicus*

Conus californicus Reeve, 1844. Low intertidal in rock crevices or sand pockets; offshore on sand and rock bottoms; diverse diet of worms, mollusks, crustaceans. See Saunders and Wolfson 1961, Veliger 3: 73–76; Kohn 1966, Ecology 47: 1041–1043 (feeding); Stewart and Gilly 2005, Biol. Bull. 209: 146–153 (feeding on prickleback fishes).

HETEROBRANCHIA

See definition under "Classification" above.

SUPERFAMILY RISSOELLOIDEA, FAMILY RISSOELLIDAE

Rissoellids have minute, transparent shells that lack sculpture, with impressed suture, characteristic radula and operculum. Eastern Pacific species previously placed in this family have been misidentified. See Fretter and Graham 1978, J. Molluscan Stud., Suppl. 6, 153–241; Wise 1998, Nautilus, 111: 13–21 (anatomy and systematics).

1. Whorls rounded, profile higher than broad, suture impressed, with umbilical chink; height about 1 mm (plate 372A) . . . . *Rissoella* sp.

Rissoella sp. From tide pool micromollusk sampling at Carmel, Monterey County.

OMALOGYROIDEA

FAMILY OMALOGYRIDAE

Omalogyridae are very small heterobranch gastropods with a nearly planispiral shell form. See Fretter 1948. J. Mar. Biol. Assoc. U.K. 27: 597–632 (biology of the European *O. atomus*); Bieler and Mikkelson 1998, Nautilus 111: 1–112 (species in Florida).

1. Minute (about 1 mm diameter), planorboid, of about two and a half regularly increasing whorls; spire depressed and base broadly umbilicate; shell smooth, lacking axial sculpture, translucent, with thin, brown periostracum and lighter radial markings (plate 372B). . . . . *Omalogyra* sp.

Omalogyra sp. Minute, in low, rocky intertidal on algae on which egg capsules are deposited.

PYRAMIDELLOIDEA

Pyramidelloidean are ectoparasitic snails that include a range of shell forms, including limpets (in the family Amathinidae). Aperture oval, lacking siphonal canal; operculum present. Protoconch heterostrophic. Eyes are at the bases of broad based cephalic tentacles, the snout on the mentum, or propodium; with acrembolic proboscis. Hermaphroditic. See Ponder and DeKeyzer 1998, p. 865 (Fauna of Australia). Two families are treated: Pyramidellidae and Amathinidae

FAMILY PYRAMIDELLIDAE

Pyramidellids are small to minute ectoparasitic snails that extract body fluids from their molluscan or other invertebrate hosts by means of a long acrembolic proboscis and piercing stylet; the radula is lacking. An operculum is present and there is usually a columellar plication, at least in genera of low profile. The protoconch is heterostrophic—the coiling direction of the adult whorls changes abruptly from that of the early whorls. Species are hermaphroditic; some produces spermatophores for transfer of sperm. Although hosts of few species from the West Coast are known, there is a large literature on pyramidellid biology for other faunal regions. See Fretter and Graham 1949, J. Mar. Biol. Assoc. U.K. 28: 493–532 (anatomy, biology); Wise 1996, Malacologia 37: 443–451 (biology, phylogeny).
    There are two main groups in central California, the variously sculptured, few-whorled Odostominae, which have a columellar plication, and the many-whorled, tall and slender Turbonillinae. Hosts are known for some odostomines, but little is known about the biology of Turbonillinae, which are usually not found associated with hosts. Although many genera were originally proposed as subgenera of *Odostomia* or *Turbonilla*, current authors agree in recognizing a large number of full

PLATE 372 Rissoellidae, Omalogyridae, and Pyramidellidae (subfamilies Odostomiinae and Turbonillinae), and Amathinidae: A, *Rissoella* sp., height 1.0 mm; B, *Omalogyra* sp., diameter 0.6 mm; C, *Ividella navisa*, height 3.3 mm; D, *Boonea lucca*, height 2.9 mm; E, *Boonea oregonensis*, height 3.2 mm; F, *Aartsenia pupiformis*, height 8.4 mm; G, *Brachystomia angularis*, height 5.6 mm; H, *Evalea tenuisculpta*, height 5.4 mm; I, *Odetta fetella*, height 4.3 mm; J, *Turbonilla victoriana*, height 7.0 mm; K, *Bartschella laminata*, height 7.7 mm; L, *Pyrgiscus tenuicula*, height 6.8 mm; M, *Turbonilla cayucosensis*, height 9.0 mm; N, *Iselica ovoidea*, height 7.5 mm.

genera; see Schander, et al. 1999, Boll. Malac. 34: 145–166 (listing of genera). Some authors, starting with Robertson (1978, Biol. Bull. 155: 360–382) have attempted to base generic definitions for certain odostomines on traits relating to spermatophores, but McLean (2002, Amer. Malac. Soc., Charleston, Abstracts), argued that shell characters are best suited for classification in all odostomiine genera. Many species have not been collected alive. Numerous poorly known species occur in the eastern Pacific; species of the region were monographed by Dall and Bartsch (1909, Bull. U.S. Nat. Mus. 68: 1–258), but it is now apparent that our species have been greatly overnamed, which has hampered their study.

1. Shell slender, with many whorls; axial and spiral sculpture various, columellar plication lacking 8
— Shell broadly ovate to conic, with single columellar plication, of few whorls 2
2. Sculpture both axial and spiral 3
— Axial sculpture lacking 5
3. Suture deeply impressed, whorls shouldered; sculpture broadly clathrate; height 3.3 mm (plate 372C) . . . .
. . . . Ividella navisa
— Whorls rounded, axial and spiral sculpture forming beads at intersections 4
4. Profile broad; height 2.9 mm (plate 372D) . . . .
. . . . Boonea lucca
— Profile slender; height 3.2 mm (plate 372E) . . . .
. . . . Boonea oregonensis
5. Whorls evenly rounded, base also rounded 6
— Whorls weakly rounded, base subangulate 7
6. Relatively large (to 8 mm), often shouldered (plate 372F)
. . . . Aartsenia pupiformis
— Smaller (to 5 mm), with fine spiral cords (plate 372H) . . . . .
. . . . Evalea tenuisculpta
7. Spiral sculpture deeply incised; height 4.3 mm (plate 372I)
. . . . Odetta fetella
— Spiral sculpture of faint incisions; height 5.6 mm (plate 372G) . . . . Brachystomia angularis
8. Sculpture axial only 9
— Sculpture both axial and spiral 10
9. Gray, axial ribs extending across base (plate 372J) . . . . . .
. . . . Turbonilla victoriana
— White, axial ribs not extending across base (plate 372M)
. . . . Turbonilla cayucosensis
10. Axial ribs extending whorl to whorl (plate 372K) . . . .
. . . . Bartschella laminata
— Axial ribs strong only on upper part of whorl (plate 372L)
. . . . Pyrgiscus tenuicula

Ividella navisa (Dall & Bartsch, 1907).
Boonea lucca (Dall & Bartsch, 1909).
Boonea oregonensis (Dall & Bartsch, 1907).
Aartsenia pupiformis (Carpenter, 1865) (=O. nota Dall and Bartsch, 1909). In roots of surfgrass Phyllospadix.
Brachystomia angularis (Dall & Bartsch, 1907). Whorls flat-sided, base angulate.
Evalea tenuisculpta (Carpenter, 1864). On red abalone.
*Odetta bisuturalis (Say, 1812) (=Boonea bisuturalis, Menestho bisuturalis). An Atlantic species introduced to San Francisco Bay in the days of commercial oyster culture (and to be looked for elsewhere along the coast). Found associated with Cerithidea californica and Ilyanassa obsoleta, and to be expected with other species.

* = Not in key.

Odetta fetella (Dall & Bartsch, 1909). Occurs on oysters in lagoons.
Turbonilla victoriana Dall & Bartsch 1907. In gravel samples, tide pools.
Bartschella laminata (Carpenter, 1864). A southern species extending to Monterey.
Pyrgiscus tenuicula (Gould, 1853). A southern species extending to Monterey.
Turbonilla cayucosensis (Willett, 1929). In gravel in tide pools.

FAMILY AMATHINIDAE

The family Amathinidae was proposed by Ponder 1987 (Asian Mar. Biol. 4: 1–34) as a pyramidelloidean family that includes the coiled genus Iselica, as well as some limpet-shaped genera, including Cyclothyca, in the tropical eastern Pacific. Members of the family are suctorial parasites on mollusks.

1. Shell broadly ovate, with strong spiral ribs and thin axial lamellae; white with brown periostracum; height to 8 mm (plate 372N) . . . . Iselica ovoidea

Iselica ovoidea (Gould, 1853) [=I. fenestrata (Carpenter, 1864)]. Among Mytilus beds in bays; also associated with other invertebrates; in mud, gravel.

ORDER OPISTHOBRANCHIA, SUBORDER CEPHALASPIDEA

SUPERFAMILY ACTEONOIDEA, FAMILY ACTEONIDAE

Acteonids are usually considered as primitive cephalaspideans opisthobranchs, but were considered by Mikkelsen 1996 (Malacologia 37: 375–442) as transitional between heterobranchs and opisthobranchs.

1. Columella with one fold; shell with incised, pitted, spiral striations, or grooves; white, with two spiral, gray-black bands on body whorl; height 16 mm (plate 373A) . . . . . .
. . . . Rictaxis punctocaelatus

Rictaxis punctocaelatus (Carpenter, 1864) (=Acteon punctocaelatus). Of sporadic occurrence on mud- and sand flats in bays.

SUPERFAMILY PHILINOIDEA, FAMILY CYLICHNIDAE

The family Cylichnidae includes two subfamilies, the Cylichninae, in which there are species of Cylichna occurring offshore, and the Acteocininae, with the genus Acteocina, in which species may occur in shallower water. Feeding on foraminifera. For the systematics of Acteocininae, based on Atlantic species, see Mikkelsen and Mikkelsen 1984, Veliger 27: 164–192.

1. Suture not deeply channeled 2
— Suture deeply channeled 3
2. Shoulder strongly keeled; strong axial striations on upper one-half of whorl; small, to about 6 mm (plate 373B) . . . . .
. . . . Acteocina harpa
— Shoulder subangulate, shell sides nearly parallel, with fine brown periostracum; small, height 4 mm (plate 373C) . . . . .
. . . . Acteocina inculta
3. Shoulder subangulate, spire projecting, columellar fold projecting, sculpture of numerous, spiral striations, reflected by brown periostracum; height to 22 mm (plate 373D) . . . . Acteocina culcitella

— Shoulder rounded, spire recessed, columellar fold weakly developed, with pattern of fine brown lines; height to 14 mm (plate 373E) . . . . *Acteocina cerealis*

Acteocina harpa (Dall, 1871) (=*Retusa harpa*). In sand, gravel, and mud, low intertidal to offshore.

Acteocina inculta (Gould, 1855). Monterey Bay south; common in mud of marsh channels, bays, lagoons.

Acteocina culcitella (Gould, 1853). Sporadically common in bays and lagoons, on sand and mud. See Shonman and Nybakken 1978, Veliger 21: 120–126 (feeding).

Acteocina cerealis (Gould, 1853) [=*A. eximia* (Baird, 1863)]. More common offshore, but reported on intertidal flats in Bodega Harbor, by Gosliner 1979, Nautilus 93: 85–92.

Cylichna attonsa Carpenter, 1864. On soft bottoms offshore. See photo in Palmer (1958).

Cylichna diegensis (Dall, 1919). On soft bottoms offshore.

SUPERFAMILY DIAPHANOIDEA, FAMILY DIAPHANIDAE

The genus *Diaphana* was revised by Schiötte 1999, Steenstrupia, 24: 77–140.

1. Shell small (to about 4 mm in height), umbilicate, thin, translucent, smooth, with weak axial growth lines; three whorls, with globular nucleus (plate 373F) . . . .
. . . . *Diaphana californica*

Diaphana californica Dall, 1919. Uncommon intertidally on algae; sublittorally in sand and kelp holdfasts.

SUPERFAMILY HAMINEOIDEA, FAMILY HAMINOEIDAE

The Haminoeidae (formerly Atyidae) have thin shells with involute spires.

1. Upper portion of body whorl tapered, with a shallow constriction; height 24 mm (plate 373H) . . . .
. . . . *Haminoea virescens*
— Upper portion of body whorl relatively broad 2
2. Body whorl slightly elongate; height 14 mm (plate 373G)
. . . . *Haminoea vesicula*
— Body whorl more shortened; height 14 mm (plate 373I)
. . . . *Haminoea japonica*

Haminoea vesicula (Gould, 1855). Sporadically abundant among algae *Ulva* and *Polysiphonia*, in sloughs, lagoons, bay mudflats.

Haminoea virescens (G. B. Sowerby I, 1833). Occasional in higher tide pools of open-coast, rocky areas.

Haminoea japonica (Pilsbry, 1895) (=*H. callidegenita* Gibson and Chia, 1989). Gibson and Chia's species (Can. J. Zool. 67: 914–922) represented an unrecognized invasion of a Japanese species in the northeastern Pacific, where this species is now broadly established.

SUPERFAMILY BULLOIDEA, FAMILY BULLIDAE

Bullids are relatively large-shelled and are known as bubble shells, a tropical group except for *B. gouldiana*. Feeding on algae. See Willan 1978, J. Malac. Soc Aust. 4: 57–68 (taxonomy).

1. Aperture broadly rounded anteriorly, not flaring; shell pink-gray to brown with cloudy maculations bordered by white on their left edges; height to 50 mm (plate 373J) . . . .
. . . . *Bulla gouldiana*

Bulla gouldiana Pilsbry, 1893. A southern species, occasional in central California; on mudflats in lagoons, bays, estuaries. See Robles 1975, Veliger 17: 278–291 (reproductive system).

NOTASPIDEA

SUPERFAMILY UMBRACULOIDEA, FAMILY TYLODINIDAE

The Tylodinidae (formerly included in the Umbraculidae) are unlike most other notaspideans in having an external shell of limpet form. Feeding is on sponges. See Willan 1987, Amer. Malac. Bull. 5: 215–241 (phylogeny).

1. Shell thin, with central apex, thick brown periostracum extending beyond edge of shell; muscle scar horseshoe-shaped, opening on right side of shell, length to 35 mm (plate 373K) . . . . *Tylodina fungina*

Tylodina fungina Gabb, 1865. Southern (Cayucos, south); found on yellow sponges, which it closely resembles.

ORDER PULMONATA, SUBORDER BASOMMATOPHORA

SUPERFAMILY SIPHONARIOIDEA, FAMILY SIPHONARIIDAE

Siphonariids are pulmonate limpets, mostly occurring in tropical intertidal zones. Only the cosmopolitan genus *Williamia* extends into temperate waters and extends from the intertidal into the sublittoral zone. See Marshall 1981, N. Z. J. Zool. 8: 487–492 (taxonomy of *Williamia* in western Pacific).

1. Shell smooth, thin, waxy, orange or red-brown with translucent, lighter-colored rays and thin periostracum; apex hooked, one-third distance from posterior end; muscle scar opening on right side, length to 10 mm (plate 373L) . . . . *Williamia peltoides*

Williamia peltoides (Carpenter, 1864) [=*W. vernalis* (Dall, 1870)]. In protected low intertidal to sublittoral, under rocks, on coralline algae and on coralline-covered shells of *Chlorostoma*, *Pomaulax*, etc. See Yonge 1960, Proc. Calif. Acad. Sci. (4) 31: 111–119 (biology); McLean 1998, Veliger 41: 243–248 (taxonomy); Collin 2000, Nautilus 114: 117–119 (development).

Williamia subspiralis (Carpenter, 1864). Of higher profile, occurring in rocky sublittoral, Point Sur and south. See McLean 1998, for photo.

SUPERFAMILY ELLOBIOIDEA, FAMILY ELLOBIIDAE

Ellobiidae (formerly Melampidae) are marine pulmonates with sturdy shells, living at high-tide lines in salt marsh habitats. See Martins 1996, Malacologia 37: 163–332 (taxonomy).

1. Three columellar folds (third may be weakly expressed); spire elevated; color variable, brown or brown-purple to yellow; juveniles with small periostracal hairs; height 6 mm (plate 373M) . . . . *Myosotella myosotis*

Myosotella myosotis (Draparnaud, 1801) [=*Ovatella myosotis*; =*Phytia setifer* (Cooper, 1872)]. In *Salicornia* marshes, often very

* = Not in key.

PLATE 373 Acteonidae, Cylichnidae, Diaphanidae, Haminoeidae, Bullidae, Tylodinidae, Siphonariidae, Ellobiidae, and Trimusculidae: A, *Rictaxis punctocaelatus*, height 16 mm; B, *Acteocina harpa*, height 5.5 mm; C, *Acteocina inculta*, height 4.0 mm; D, *Acteocina culcitella*, height 20 mm; E, *Acteocina cerealis*, height 13.4 mm; F, *Diaphana californica*, height 2.8 mm; G, *Haminoea vesicula*, height 23.7 mm; H, *Haminoea virescens*, height 15.3 mm; I, *Haminoea japonica*, height 14.2 mm; J, *Bulla gouldiana*, height 50 mm; K, *Tylodina fungina* (three views), length 32 mm; L, *Williamia peltoides*, length 8.5 mm (two views); M, *Myosotella myosotis*, height 3.7 mm; N, *Trimusculus reticulatus* (two views), length 14 mm.

abundant on mud, under debris, and in crevices of old docks and pilings, at highest levels of spring tides. See Berman and Carlton 1991, J. Exp. Mar. Biol. Ecol. 150: 267–281 (ecology, diet). Introduced from the Atlantic Ocean.

SUPERFAMILY TRIMUSCULOIDEA, FAMILY TRIMUSCULIDAE

The pulmonate limpet family Trimusculidae (formerly Gadinidae) contains a single genus of worldwide occurrence. Hermaphroditic, with the opening of mantle cavity and muscle scar on right side.

1. Shell white, without periostracum, nearly circular; apex central; sculpture reticulate (plate 373N). . . . . *Trimusculus reticulatus*

Trimusculus reticulatus (Sowerby, 1835) (=*Gadinia reticulata*). In groups on roofs of caves and under overhanging ledges in low intertidal; also in abandoned pholad holes. See Yonge, 1958, Proc. Malacol. Soc. London 33: 31–37, and 1960, Proc. Calif. Acad. Sci. 31: 111–119 (biology, ecology); Walsby 1975, Veliger 18: 139–145 (feeding); Haddock 1989, Veliger 32: 403–405 (existence in subtidal air pockets).

ACKNOWLEDGMENTS

This effort has built upon the work of James T. Carlton and Barry Roth, the two previous coauthors for the section on shelled gastropods for the 1975 edition. I have used extensive portions of their text with slight changes and some substantial additions. Here I thank David R. Lindberg for advice with the section on the revised system of classification. I thank Jim Carlton, Lindsey Groves and Gene Coan for critical commentary on the manuscript.

All illustrations of shells are from my original photographs; negatives for each image were scanned and improved in Photoshop by my imaging assistant Michelle Schwengel. Her work on this project as well as the work toward my books in preparation on northeastern Pacific gastropods has been supported in part by the Packard Foundation and in part by a gift from Twila Bratcher.

REFERENCES

Abbott, D. P., and E. C. Haderlie. 1980. Chapter 13. Prosobranchia: Marine snails, pp. 230–307. In Intertidal invertebrates of California. Morris, R. H., D. P. Abbott, and E. C. Haderlie, Stanford University Press.

Abbott, R. T. 1974. American Seashells, 2nd edition. The Marine Mollusca of the Atlantic and Pacific coasts of North America. New York: Van Nostrand Reinhold, 663 pp., 24 pls.

Beesley, P. L., G. J. B. Ross, and A. Wells, eds. 1998. Mollusca: The Southern Synthesis. Fauna of Australia. Vol. 5. CSIRO Publishing: Melbourne. Part A, xvi, 1–653. Part B, viii, 663–1234. [Abbreviated citations in the text cite the author and date, e.g., Ponder, 1998, F of A]

Boss, K. J., J. Rosewater, and F. A. Ruhoff. 1968. The zoological taxa of William Healey Dall. United States National Museum, Bulletin 287: 1–427.

Bouchet, P., and J-P. Rocroi. 2005. Classification and nomenclator of gastropod families. Malacologia, 47(1–2): 1–397. Part 1. Nomenclator of Family-Group Names (Bouchet & Rocroi). Part 2. Working Classification of the Gastropoda (Bouchet, Fryda, Hausdorf, Ponder, Valdés, and Warén).

Carlton, J. T., and B. Roth. 1975. Phylum Mollusca: Shelled gastropods, pp. 467–514 In Light's manual: intertidal invertebrates of the Central California Coast. 3rd ed. R. I. Smith and J. T. Carlton, eds. Berkeley, CA: University of California Press.

Coan, E. V., P. V. Scott, and F. R. Bernard. 2000. Bivalve seashells of western North America; Marine bivalve mollusks from Arctic Alaska to Baja California. Santa Barbara Museum of Natural History, Monographs no. 2, vii + 764 pp.

Cox, L. R. 1960. Gastropoda, General Characteristics of Gastropoda. pp. 84–169. In Treatise on Invertebrate Paleontology, Univ. Kansas Press and Geol. Soc. Amer., Part l, Mollusca 1. R. C. Moore, ed.

Dall, W. H. 1921. Summary of the marine shellbearing mollusks of the northwest coast of America, from San Diego, California, to the Polar Sea, mostly contained in the collection of the United States National Museum, with illustrations of hitherto unfigured species. United States National Museum, Bulletin 112, 217 pp., 22 pls.

Fretter, V., and A. Graham. 1994. British Prosobranch Molluscs. Their Functional Anatomy and Ecology. Revised and Updated Edition. London: Ray Society, 820 pp.

Johnson, R. I. 1964. The Recent Mollusca of Augustus Addison Gould. United States National Museum, Bulletin 239: 1–182.

Kozloff, E. N. 1987. Marine Invertebrates of the Pacific Northwest. University of Washington Press, Seattle, vi + 511 pp.

Lindberg, D.R., W. F. Ponder, and G. Haszprunar. 2004. The Mollusca: Relationships and patterns from their first half-billion years. In Assembling the tree of life. J. Cracraft and M. J. Donoghue, eds. pp. 252–278. New York: Oxford University Press.

McLean, J. H. 1978. Marine shells of southern California, Revised edition. Natural History Museum of Los Angeles County, Science Series, no. 24, 1–104.

McLean, J. H. 1996. The Prosobranchia. pp. i–vii, 1–160. In Taxonomic atlas of the benthic fauna of the Santa Maria Basin and Western Santa Barbara Channel. Volume 9. The Mollusca, Part 2—The Gastropoda. P. H. Scott, J. A. Blake and A. L. Lissner, eds.

Oldroyd, I. S. 1927. The marine shells of the west coast of North America. Stanford University Publications, University Series, Geological Sciences, vol. 2, part I, pp. 1–298, pls. 1–29; part II, pp. 299–604, pls. 30–72; part III, pp. 605–941, pls. 73–l08.

Palmer, K. V. W. 1958. Type specimens of marine Mollusca described by P. P. Carpenter from the west coast (San Diego to British Columbia). Geological Society of America, Memoir 76: 376 pp.

Ponder, W. F., and D. R. Lindberg. 1997. Towards a phylogeny of gastropod molluscs: an analysis using morphological characters. Zoological Journal of the Linnean Society of London 119: 83–265.

Roy, K., D. Jablonski, J. W. Valentine, and G. Rosenberg. 1998. Marine latitudinal diversity gradients: Tests of causal hypotheses. Proceedings of the National Academy of Sciences 95: 3699–3702.

Strathmann, M. F., ed. 1987. Reproduction and development of marine invertebrates of the northern Pacific coast. Seattle: University of Washington Press, xii + 670 pp.

Turgeon, D. D., ed., with contributions by J. F. Quinn, Jr., A. E. Bogan, E. V. Coan, F. G. Hochberg, W. G. Lyons, P. M. Mikkelsen, R. J. Neves, C. F. E. Roper, G. Rosenberg, B. Roth, A. Scheltema, F. G. Thompson, M. Vecchione, and J. D. Williams. 1998. Common and scientific names of aquatic invertebrates from the United States and Canada: Mollusks, 2nd ed. American Fisheries Society Special Publication , 26, ix + 526 pp.

Patellogastropoda

DAVID R. LINDBERG

(Plates 374–377)

The California and Oregon patellogastropod fauna consists of more than 23 species found in diverse niches in intertidal and nearshore habitats. Its potential as a model system for ecological, behavioral, physiological, and evolutionary studies has been exploited numerous times over the past 75 years (see Abbott and Haderlie 1980, Carlton 1981, Ricketts et al. 1985, and references therein). Recent molecular phylogenies (Clabaugh 1997, Simison 2000, Begovic 2004) have produced more robust estimates of relationships than previously available, which will undoubtedly promote a new round of investigations parsed by phylogenetic pattern.

The California fauna is composed of two distinct clades—the Acmaeidae and Lottiidae (Lindberg 1998). The Acmaeidae are mostly deep-water species associated with unique habitats such as waterlogged wood and low oxygen environments. In the northeastern Pacific Ocean. Two species are found in shallower waters: *Acmaea funiculata* and *A. mitra*. The Lottiidae are broadly distributed along the Pacific Rim and is composed of several subclades recognizable by morphology and biogeography, in addition to molecular distinctness. The Lottiidae include all of the remaining central and northern California and Oregon species. In central California, northern and southern groups meet and transition, producing one of the highest diversities of patellogastropods found anywhere in the world. On a single rocky shore in central California, as many as 16 species of acmaeid and lottiid limpets can be found.

Molecular work has revealed that previously used radula, shell structure, and gill characters are often not congruent with each other or with the molecular groupings. However, often maligned shell characters such as shell sculpture and pigmentation patterns are congruent with molecular groupings. Ecophenotypic variation is common in several California lottiid subgroups, and unnamed species remain numerous. These include cryptic species and taxa misidentified as ecophenotypes of other species. However, there are also examples in which long distinguished putative taxa are identical for observed molecular markers. Subgroups are often latitudinally distinct and likely reflect speciation events related to Pleistocene climate events.

This mosaic of niche and geographical sister taxa relationships seen in the molecular phylogenies suggest complex interactions possibly involving several different isolating mechanisms. This explanation for the populating of the California patellogastropod fauna is not new. Avery R. G. Test (1946), who treated the limpets for Professor S. F. Light in the first edition of this book, observed that, "It has become apparent that in the genus *Acmaea* [=*Lottia*], in addition to the process of speciation based on geographical isolation, there has been, perhaps even more frequently, speciation from eurytopic ancestors by the process of ecological isolation and selection." Problems in species identification remain in the Lottiidae, and additional taxa are likely to be discovered as molecular markers for more individuals are obtained and compared.

In the past few decades, the generic names *Collisella*, *Notoacmea*, *Tectura*, *Niveotectura*, *Macclintockia*, and *Discurria* have all been applied to limpets in our fauna. Grant (1937) was the first worker to assign some of the northeastern Pacific "acmaeids" to *Notoacmea*, which she considered as a subgenus of *Acmaea*. Fritchman (1961) adopted Grant's classification and published subgeneric assignments for many of the northeastern Pacific species. Mclean (1966), in a major systematic revision of the northeastern Pacific limpet fauna, also used *Notoacmea* as a subgenus and then later (Mclean 1969) considered *Notoacmea* as a full genus. Lindberg (1981) followed McLean's classification, but later (Lindberg 1986) synonymized *Collisella* with *Lottia* and replaced *Notoacmea* with *Tectura*. With the advent of molecular techniques in the early 1990s, it became apparent that there was substantial convergence in many of the characters that were being used to delineate and delimit many of these generic groupings. Based on this knowledge, I have chosen to adopt a more conservative approach to limpet nomenclature then previously done and use only the genera *Acmaea* and *Lottia* to encompass the species treated here.

The late Harry Fritchman (1961–1962, Veliger, 3–4) studied the reproductive cycles of many of our common local species, and Thomas Wolcott (1973, Bio. Bull. 145: 389–422) produced an important study on the physiological ecology and zonation of central California limpets. Avery R. G. Test produced a classic paper in 1945 on the distributional ecology of our common species (Ecology 26: 395–405). A set of student-produced papers, edited by the late Donald P. Abbott and colleagues, on Monterey Bay limpets from a summer course at Hopkins Marine Station (Veliger 11, Supplement, 112 pp., 1968) stimulated many subsequent biological and ecological studies, many of them published in *The Veliger* as well. Abbott and Haderlie (1980) provided detailed summaries of the biology and ecology of a number of common species.

KEY TO SPECIES

Note: Some species key out more than once.

1. Shell entirely white or white with irregular brown radial markings; lateral teeth of radula approximately equal in size and shape (plate 374C) (shell commonly encrusted with coralline algae) 2
—   Shell colored; lateral teeth of radula unequal in size and shape (plate 375H) (may or may not be encrusted with coralline algae) 4
2. Shell white or white with sparse brown rays and/or small brown apical spot; apex subcentral; aperture oval or compressed laterally; radula with uncini near base of third lateral teeth (plate 374D) . . . . *Lottia triangularis*
—   Shell entirely white, lacks apical spot; apex central; aperture oval; radula without uncini near base of third lateral teeth (cf. plate 375E, 375H) 3
3. With radial sculpture (plate 374A) . . . . *Acmaea funiculata*
—   Without radial sculpture (plate 374B) . . . . . *Acmaea mitra*
4. Sides of shell parallel or nearly so; shell compressed, aperture narrow, oblong. 5
—   Sides of shell not parallel, aperture oval 9
5. Ends of shell curved upward; nearly smooth with obscure, low, radial ribbing at margin; exterior brown, interior bluish, with brown apical stain; on *Laminaria* and *Pterygophora* stipes in low intertidal (plate 374G) . . . . . .
    . . . . *Lottia instabilis*
—   Ends of shell not curved upward 6
6. Shell three to four times longer than wide; yellow to brown; second and third lateral teeth with straight, broad cutting surfaces (plate 375D); occurring on the seagrasses *Zostera* and *Phyllospadix* spp. 7
—   Shell less than three times longer than wide; exterior dark brown, lustrous, smooth, with fine radial sculpture; occurring on *Egregia* 8
7. Shell color light yellow with brown-red chevron markings; sculpture of concentric growth lines; on intertidal eel grass *Zostera marina* (plate 375A) . . . . "*Lottia*" *depicta*
—   Shell color light to dark brown with small, rounded, radial ribs; on surfgrass *Phyllospadix* (plate 375C) . . . .
    . . . . "*Lottia*"*paleacea*
8. Interior of shell rich dark brown; third lateral teeth with prominent lateral extension (plate 375E, 375F). . . . .
    . . . . *Lottia insessa*
—   Interior gray to light brown, usually with brown apical stain; edge of third lateral teeth sigmoidal; externally resembling *L. insessa* (plate 375H, 375J) (see species list). . .
    . . . . *Lottia pelta*
9. Shell pink, mottled with white streaks and white and yellow brown dots; thin, elevated, small (to about 8 mm); smooth, or with fine radial ribs; (plate 374E) . . . .
    . . . . *Lottia rosacea*
—   Color otherwise 10

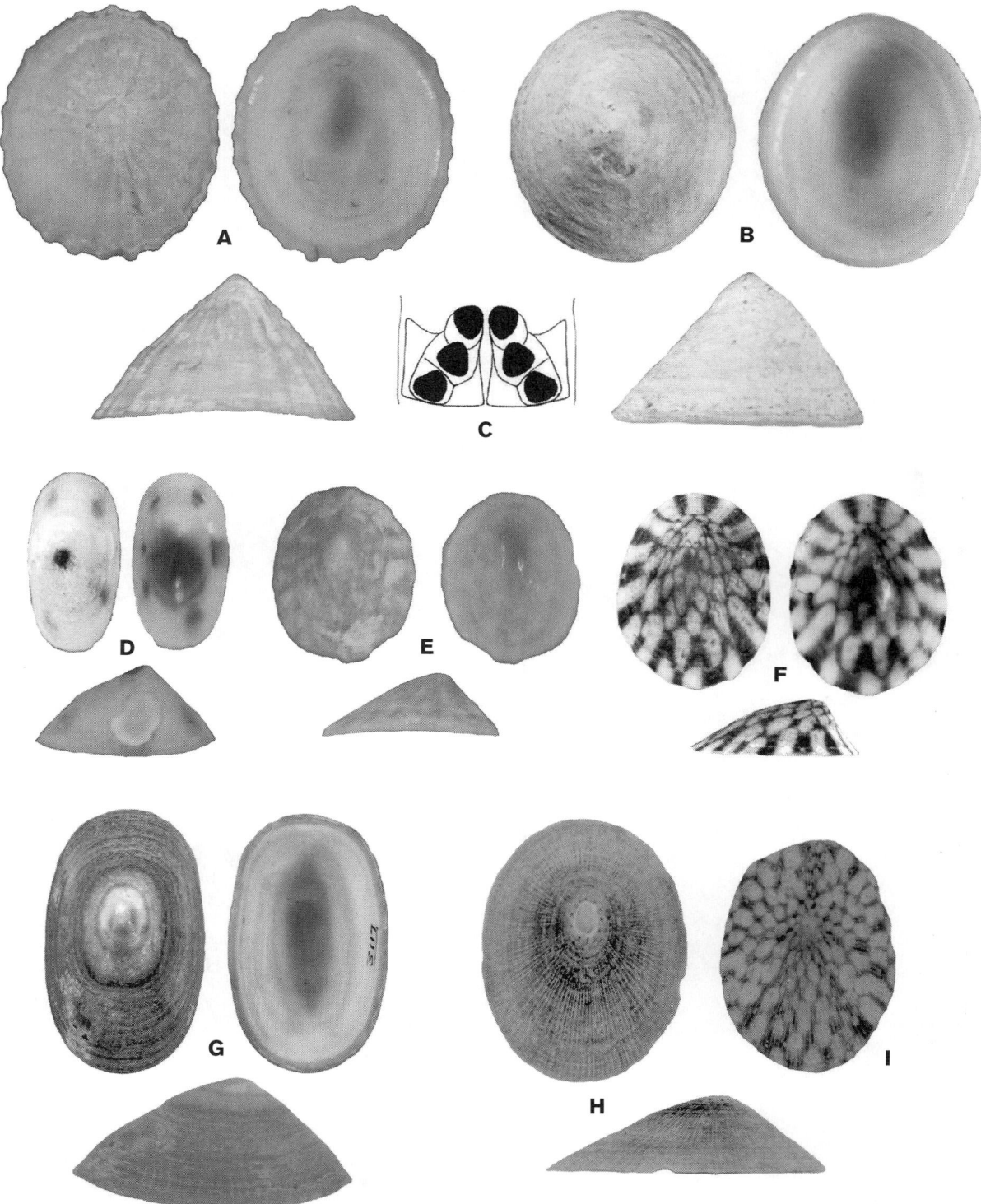

PLATE 374 A, *Acmaea funiculata,* length 20 mm; B, *Acmaea mitra,* length 30 mm; C, radular row of *Acmaea mitra;* D, *Lottia triangularis,* length 5 mm; E, *Lottia rosacea,* length 5 mm; F, *Lottia* sp. from *Haliotis* spp. length 7 mm; G, *Lottia instabilis* (kelp form), length 25 mm; H, *Lottia instabilis* (solid form), length 15 mm; I, *Lottia instabilis* (tessellate form), length 10 mm; sizes are typical of central California specimens.

10. Apex positioned in the anterior quarter of the shell, may overhang the edge of the shell 11
— Apex position subcentral to anterior third of the shell. 13
11. Shell long-oval, low, large (to about 100 mm) and heavy; maculated brown and white; shell often eroded, or small

(<25 mm) blue-black with concentric growth lines; inner margin dark brown, intermediate area black with prominent, owl-shaped muscle scar at center; sides of foot and head black to gray (plate 377A) . . . . *Lottia gigantea*
— Shell oval, moderate profile, medium size (to about 25 mm); shell white to brown with tessellate markings; intermediate

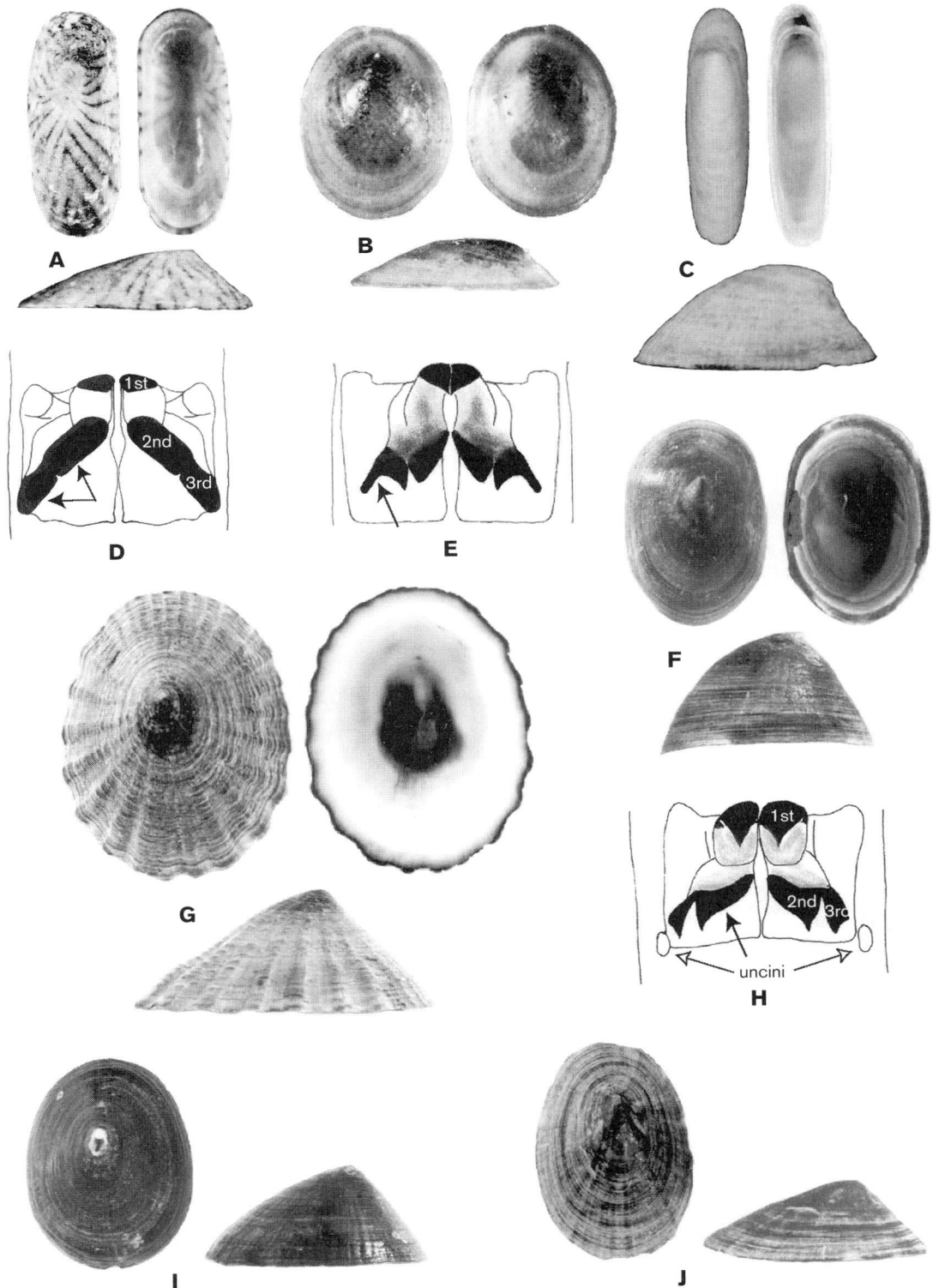

PLATE 375 A, *"Lottia" depicta,* length 8 mm; B, *"Lottia" depicta* (oval form), length 8 mm; C, *"Lottia" paleacea,* length 8 mm; D, radular row of *"Lottia" depicta,* note broad, flat cusps; E, radular row of *Lottia insessa,* note extensions on 3rd lateral teeth; F, *Lottia insessa,* length 12 mm; G, *Lottia pelta* (rock form), length 35 mm; H, radular row of *Lottia pelta,* note uncini and sigmodial shaped 3rd lateral teeth; I, *Lottia pelta* (*Mytilus* form), length 15 mm; J, *Lottia pelta* (kelp form), length 10 mm; sizes are typical of central California specimens.

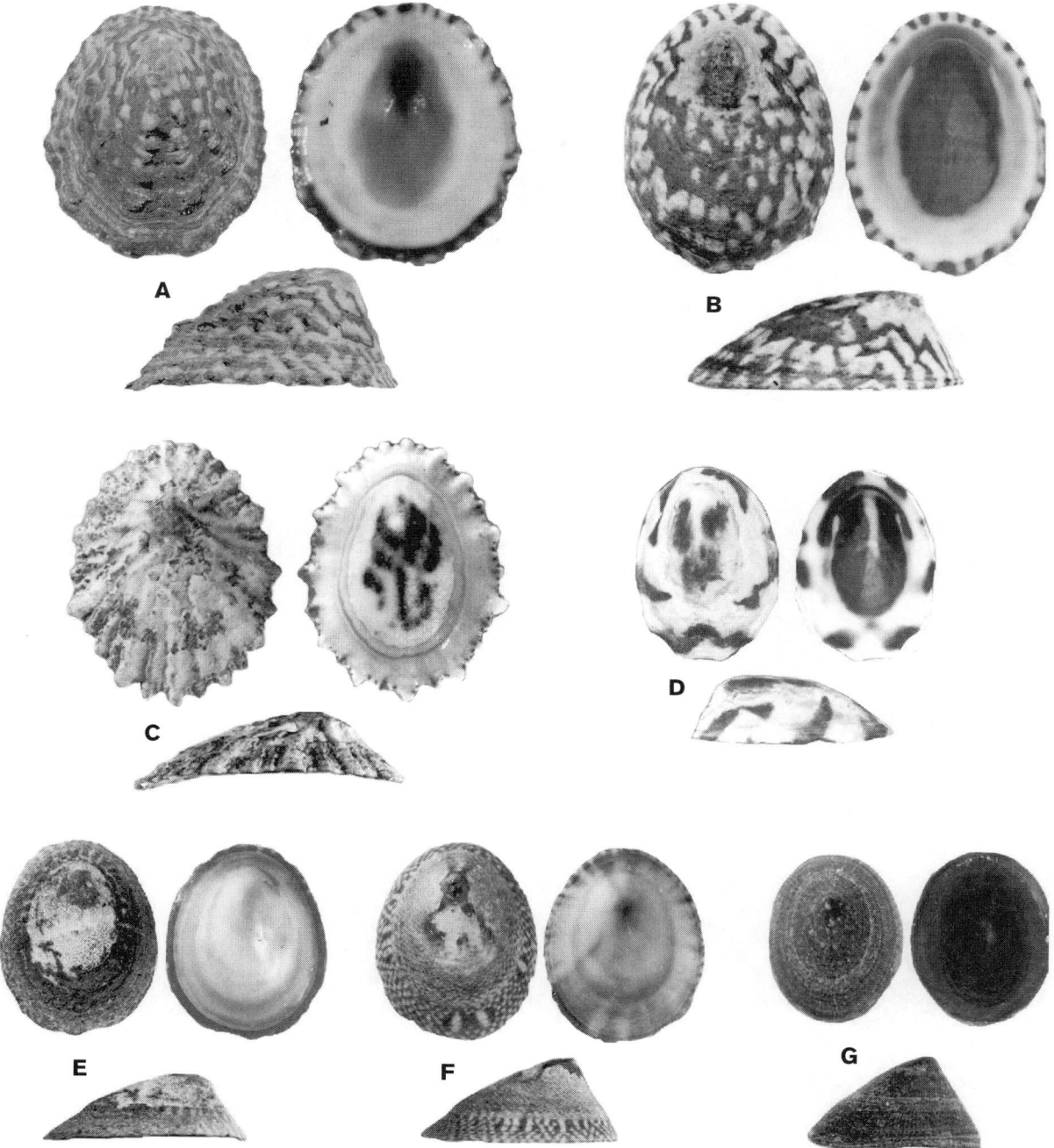

PLATE 376 A, *Lottia digitalis* (rock form), length 20 mm; B, *Lottia austrodigitalis* (rock form), length 20 mm; C, *Lottia scabra*, length 20 mm; D, *Lottia digitalis* (*Pollicipes* form), length 8 mm; E, F, *Lottia paradigitalis*, length 10 mm; G, *Lottia asmi*, length 5 mm; sizes are typical of central California specimens.

area white to blue-white 12
12. Ribs triangular in profile, usually light colored (often with darker-colored spines) and tessellate interspaces between ribs; ribs project strongly in all directions, forming strong scalloped margin; apical region often covered with callus; animal with black spots on head and sides of foot (plate 376C) . . . . *Lottia scabra*
— Ribs rounded in profile, usually not lighter than interspaces; posterior margin sometimes scalloped; anterior slope generally concave, ribs strongest on posterior slope,

may be absent at anterior end; animal lacks dark spots on head and sides of foot (plate 376A) . . . .
. . . . *Lottia digitalis*
— Ribs not pronounced (Monterey Bay southward) (plate 376B) (see species list) . . . . *Lottia austrodigitalis*
13. Exterior shell surface sculpted with coarse radial ribs; sometimes eroded, with ribs visible only at margin of shell . . . . .
. . . . 14
— Exterior shell surface sculpted with concentric growth lines, fine radial ribs or striations 15

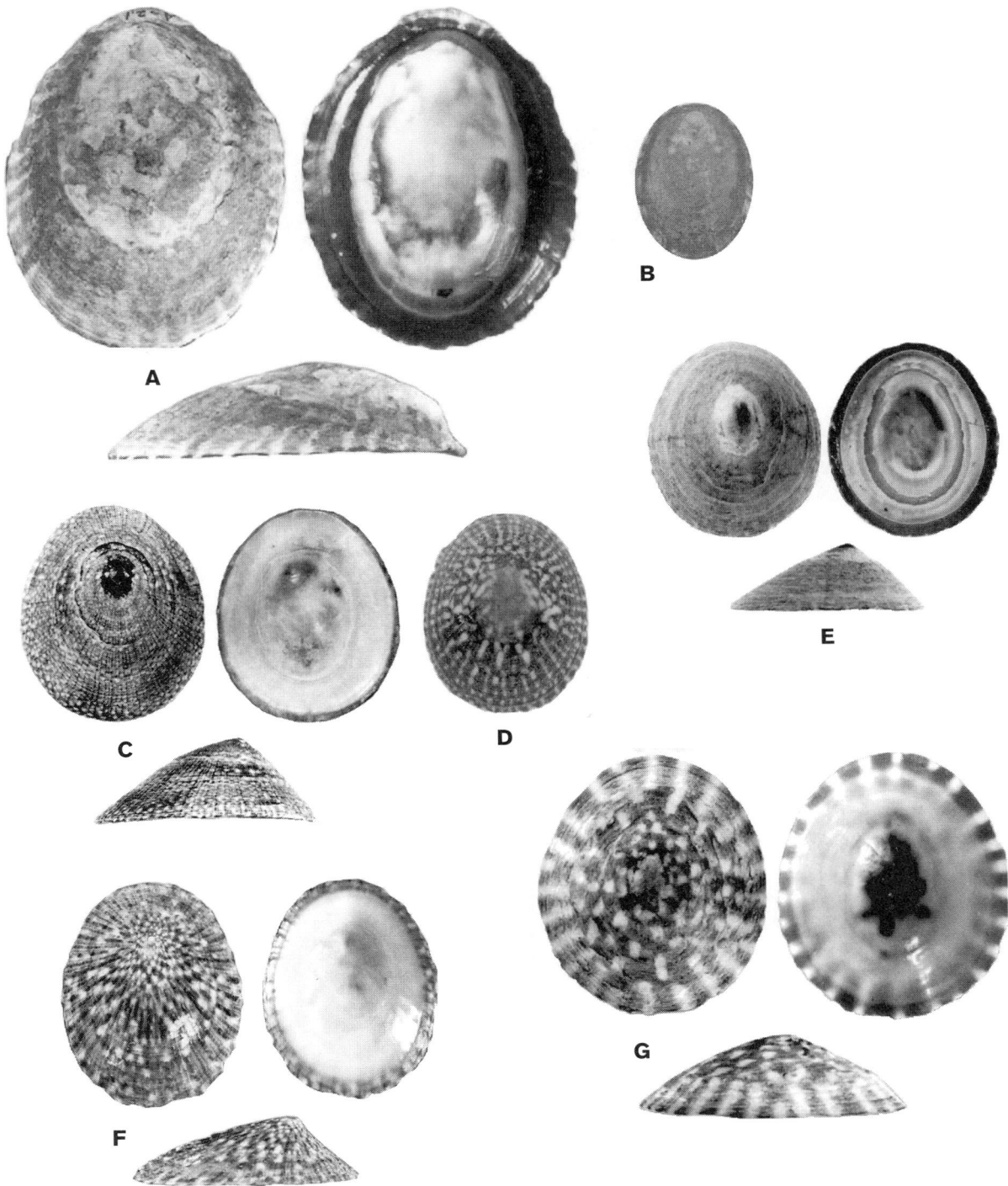

PLATE 377 A, *Lottia gigantea* (rock form), length 60 mm; B, *Lottia gigantea* (*Mytilus* form), length 10 mm; C, *Lottia persona*, length 25 mm; D, *Lottia persona* (northern form), length 25 mm; E, *Lottia fenestrata*, length 20 mm; F, *Lottia limatula*, length 25 mm; G, *Lottia scutum,* length 35 mm; sizes are typical of central California specimens.

14. Rib usually light-colored (often with darker-colored spines) and darker interspaces; ribs projecting strongly in all directions, forming strong scalloped margin; apical region often covered with callus; animal with black spots on head and sides of foot (plate 376C) . . . . *Lottia scabra*
— Ribs broad and generally equally developed on all slopes, may appear knobby; color various, brown, green, or green-black, checkered with white tessellations or peripheral rays and bands of white (plate 375G). . . . . *Lottia pelta*

15. Interior intermediate area dark (dark gray or blue-brown) 16
— Intermediate area blue-white to white. 17

16. Black to dark gray-brown inside and out; small (to about 10 mm), elevated, exterior often eroded, generally with fine radial striae visible at least at margin; generally on turban snail *Chlorostoma funebralis* (but not the only limpet

on this *Chlorostoma*) (plate 376G)...........*Lottia asmi*

— Intermediate area between shell margin and apex suffused with brown; shell conical with round aperture and weak radial sculpture; olive to gray with small, white tessellations, sometimes drawn out around aperture; apex often eroded to brown (plate 377E)...........*Lottia fenestrata*

17. With fine, imbricate (scaly) radial ribbing; margin serrate (with sawlike notching); shell low or elevated; color buff yellow, with fine darker mottlings, or green brown with white tessellations or bands; sides of foot and head black to gray (plate 377F)...................*Lottia limatula*

— Without imbricate radial ribbing..................18

18. Shell profile low, height usually less than one-third of shell length.......................................19

— Shell profile elevated usually greater than one-third of shell length.......................................20

19. Shell thick, large (to about 50 mm); apex subcentral; sculpture coarser, of flat-topped ridges, color pattern of variable spotting, typically arrayed into irregular rays; top of head and mantle tentacles brown (plate 377G).........*Lottia scutum*

— Shell thin, small to medium (to about 30 mm); animal white and not pigmented........................21

20. Shell elongate oval, medium size (to 30 mm); sculpture of straight threadlike radial riblets color variable, either tessellate pattern of oval white spots on brown background (plate 374I) or solid color (yellow white to red-brown, buff) (plate 374H).......................*Lottia instabilis*

— Shell oval (<10 mm); white, yellow or translucent with fine yellow to red-brown markings.................22

21. Markings red-brown in a reticulating network; sculpture of numerous threadlike riblets; on the shells of living *Haliotis* spp. (plate 374F) (see species list)..........*Lottia* sp.

— Shell color light yellow with yellow-brown chevron markings; sculpture of concentric growth lines; on subtidal eelgrass *Zostera marina* (plate 375B)........*"Lottia" depicta*

22. Shell large, to 50 mm; with fine, regular riblets; olivaceous green with scattered white tessellations arranged into two lateral rays, anterior markings translucent; apex eroding to brown (plate 377C, 377D)...............*Lottia persona*

— Shell lacking scattered white tessellations arranged into two lateral rays................................23

23. Shell surface with few fine, raised, radial lines (sometimes obsolete); color variable, typically with fine white spots drawn out as white streaks at margin; apex eroding to white because of lack of brown apical stain (plate 376E, 376F).........................*Lottia paradigitalis*

— Shell surface with fine, irregular, radial ribbing; apex erodes to brown; shell color dark brown to black (plate 375I), with or without scattered white dots and rays; brown apical stain present; in mussel beds and associated with the algae *Pelvetia, Egregia, Postelsia, Laminaria,* and *Alaria* (plate 375J) ....*Lottia pelta*

LIST OF SPECIES

ACMAEIDAE

Acmaea funiculata Carpenter, 1864 (=*Niveotectura funiculata*). Typically found below the thermocline at 30 m–60 m on the central California and Oregon coasts; closely resembles *A. mitra* but has a ribbed shell often covered by bryozoans; shells occur in beach drift, especially in the Monterey Bay area.

Acmaea mitra Rathke, 1833. Dunce cap limpet; low intertidal and subtidal living and feeding on encrusting coralline algae *Lithothamnion* and *Lithophyllum;* the worn bright white shells are common in beach drift, frequently revealing the distinctive burrows of a polydorid worm.

LOTTIIDAE

Lottia gigantea Sowerby, 1834. Owl limpet; the largest of the California limpets (100+ mm). A territorial, protandric species found on exposed open coasts. Smaller specimens, common on *Mytilus californianus,* are more elongate, dark blue-black in color, and shell sculpture consists only of concentric growth lines. Common in the middle and high intertidal zone when human predation is minimal; this limpet is thus readily observed at low tide on the protected Asilomar shores of Monterey Bay. See Stimson 1970, Ecology 51: 113–118 and 1973, Ecology 54: 1020–1030 (territorial behavior and ecology); Lindberg and Wright 1985, Veliger 27: 261–265 (sex change patterns); Shanks and Wright 1986, Oecologia 69: 420–428 (shell damage by wave-borne rocks); Wright and Shanks 1993, J. Exp. Mar. Biol. Ecol. 166: 217–229 (territorial behavior).

Lottia limatula (Carpenter, 1864) (=*Collisella limatula*). File limpet; specimens from southern California often have two wide posterior-lateral rays of tessellations. Those from estuarine habitats (e.g., Tomales Bay) and the high intertidal often have elevated shells, and were formerly considered the subspecies *moerchii* Dall, 1878. The imbrications on the riblets distinguish this species from all other northeastern Pacific lottiids. See Eaton 1968, Veliger 11, Suppl. 5–12 (activity and feeding); Wells 1980, J. Exp. Mar. Biol. Ecol. 48: 151–168 (activity, predation).

Lottia pelta (Rathke, 1833) (=*Collisella pelta*). Shield limpet; this highly variable species is actually a group of related taxa that could be individually recognized at the species level given their molecular distance (Begovic 2003). Similar ecophenotypic forms are present in several of these sibling taxa, further complicating identification. Those living on rock substrates in the mid-intertidal zone tend to be heavily ribbed and brown to brown-green in color. Specimens from algal fronds (*Egregia, Postelsia, Laminaria,* and *Alaria* species) have apertures that fit the host algae and typically lack ribbing (plate 375J) (these smooth and ribbed phenotypes were given various trinomial and quadrinomial designations in the 19th century). Color patterns are similar to the host algae. Specimens from *Mytilus* aggregations also lack ribbing and are blue-black in color. See Craig 1968, Veliger 11 Suppl., 13–19 (activity, feeding); Sorensen and Lindberg 1991, J. Exp. Mar. Biol. Ecol. 154: 123–136 (predation by oystercatchers).

Lottia insessa (Hinds 1842) (=*Discurria insessa*). Low intertidal, usually on the stipe or holdfast of the alga *Egregia menziesii,* on which it feeds, by excavating deep pits. See Black 1976, Ecology 57: 265–277 (grazing); Choat and Black 1979, J. Exp. Mar. Biol. Ecol. 41: 25–50 (life history).

Lottia digitalis (Rathke, 1833) (=*Collisella digitalis*). Finger or ribbed limpet; high intertidal; commonly infected by the marine fungus *Pharcidia balani,* which erodes sculpture and alters shell color and patterns. A *situs* form occurs on the plates of the stalked barnacle *Pollicipes polymerus;* this form is white with black markings that closely resemble the barnacle's plate sutures. At low tide, forming characteristic dense clusters in shaded areas. See Collins 1976, Veliger 19: 199–203 and 1977, Veliger 20: 43–48 (ecology, biology); Lindberg and Pearse 1990,

J. Exp. Mar. Biol. Ecol. 140: 173–185 (experiments on shell color and morphology); Hobday 1995, J. Exp. Mar. Biol. Ecol. 189: 29–45 (ecology of body size variation); Kay 2002, Mar. Biol. 141: 467–477 (recruitment); Kay and Emlet 2002, Invert. Biol. 12: 11–24 (reproduction, larvae).

Lottia austrodigitalis (Murphy, 1978) (=*Collisella austrodigitalis*). Co-occurs with *Lottia digitalis* in central California; shells typically smoother with weaker ribs. Also has *Pollicipes polymerus* form. A high intertidal species first discovered by protein electrophoresis. See Murphy 1976, Bio. Bull. 155: 193–206.

Lottia paradigitalis (Fritchman, 1960) (=*Lottia strigatella* of authors, not of Carpenter, 1864; =*Collisella paradigitalis*). There has been substantial confusion surrounding this species. Light (1941) considered it to be a hybrid between *Lottia pelta* and *L. digitalis*. Fritchman (1960) described this species after a careful study of material between Vancouver Island and Ensenada. Simison and Lindberg (2003, Veliger, 46: 1–19) further documented the distinctness and range of this species using molecular techniques. Common on vertical surfaces and on shells of other mollusks, including turban snails and *Mytilus* spp. High and middle intertidal. See Dixon 1981, Veliger 24: 181–184 (larval settlement).

Lottia asmi (Middendorff, 1847) (=*Collisella asmi*). Specimens living on *Chlorostoma*, its most common habitat, are slightly laterally compressed and have a high profile. Specimens on *Mytilus* are flatter than those on *Chlorostoma* but color is the same. Specimens from rock substrates are also flatter and lack the dark pigmentation of *Chlorostoma* and *Mytilus* specimens; instead, they are gray-brown and may have white tessellations. Low and mid-intertidal zone. See Lindberg and Pearse 1990, J. Exp. Mar. Biol. Ecol. 140: 173–185 (experiments on shell color and morphology); Lindberg 1990, Veliger 33: 375–383 (movement patterns); Kay and Emlet 2002, Invert. Biol. 12: 11–24 (reproduction, larvae).

Lottia triangularis (Carpenter, 1846) (=*Collisella triangularis*). Two forms occur: an oval form associated with encrusting coralline algae, and a compressed form on the stipes of upright corallines. Also sometimes on trochid gastropods in the low intertidal. Low intertidal to subtidal.

Lottia scabra (Gould, 1846) (=*Collisella scabra, Macclintockia scabra*). Rough limpet; specimens larger than 10 mm usually home to a specific site on the substrate; the shell margin conforms to the site or a home depression in form. Commonly occurs on shells of *Lottia gigantea*. Middle intertidal on horizontal surfaces. See Collins 1976, Veliger 19: 199–203 and 1977, Veliger 20: 43–48 (ecology, biology); Wells 1980, J. Exp. Mar. Biol. Ecol. 48: 151–168 (activity, homing); Lindberg and Dwyer 1983, Veliger 25: 229–234 (home depressions); Geller 1991, J. Exp. Mar. Biol. Ecol. 150: 1–17 (grazing ecology); Sanders et al. 1991, Physiol. Zool. 64: 1471–1489 (stress protein responses); Gilman 2007, Veliger 48: 235–242 (shell microstructure).

Lottia scutum (Rathke, 1833) (=*Notoacmea scutum*). Plate limpet; apex position changes with size in this species: small specimens of <10 mm in length have more anterior apices than larger specimens. Middle to low intertidal. Small amphipods of the genus *Hyale* may occur under the shell margin in the mantle cavity. See Phillips 1981, Mar. Biol. 64: 95–103 (life-history).

Lottia instabilis (Gould, 1846) (=*Acmaea ochracea* Dall, 1871; =*Collisella instabilis*). *Lottia instabilis* is a polytypic species with at least five ecological variants or forms. The most common forms are either solid colors (pale yellow to rose red) (plate 374H) or tessellate (brown to gray with symmetrical markings)

(plate 374I). The tessellate form predominates in the southern portion of the range, while solid colors predominate in the northern portion; northern specimens also tend to be larger. Both of these forms are found on bare rock substrates. On trochid gastropods (including those occupied by hermit crabs), this species is tan to red-brown in color and is typically marked with white. The sides of the shell are parallel. This species also commonly occurs on the stipes of the brown algae *Laminaria setchellii* and *Pterygophora californica* (plate 374G). Depending on how early in its ontogeny *Lottia instabilis* resides on these algae determines whether the apex of the shell is tessellate, and specimens that directly settle on these algae typically lack the tessellate apex. See Lindberg 1979, Nautilus 92: 50–56 (variation).

Lottia sp. This species was first considered to be a coralline alga ecophenotype of *Lottia pelta* commonly found on the shells of living *Haliotis rufescens* and *H. kamtschatkana* (Lindberg 1981); however, it is an undescribed species (based on molecular analysis). Range includes central and northern California.

Lottia ochracea (Dall, 1871) (=*Collisella ochracea*). See *Lottia instabilis*.

Lottia persona (Rathke, 1833) (=*Notoacmea persona, Tectura persona*). Concentration of larger white markings into symmetrical lateral rays give this limpet the common name, mask limpet. In specimens from the northern part of the distribution these markings are commonly drawn out into short rays (plate 377D). Holding the shell up to a bright light source reveals the anterior markings to be translucent, which the limpet uses to assess ambient light conditions. A nocturnal species, abundant in the high and middle intertidal zones in caves, cracks, and crevices during the day. See Lindberg et al. 1975, Veliger 17: 383–386 and 1982, Veliger 25: 173–174 (light reception).

Lottia fenestrata (Reeve, 1855) (=*Notoacmea fenestrata*). Middle to low intertidal on bare rocks embedded in sand or mud; less commonly in aggregations or individually on rock benches. See Blanchard and Feder 2000, Veliger 43: 289–301 (distribution, reproduction, growth, in Alaska).

Lottia rosacea (Carpenter, 1846) (=*Acmaea rosacea, Tectura rosacea*). Molecular analyses show that this species is actually a group of related taxa that could be individually recognized at the specific level; however, there are no known shell or radular characters that delineate the different lineages. Almost exclusively subtidal to 30 m; closely associated with coralline algae.

The following two seagrass-associated species are more closely related to lottiid species found in the New World tropics than they are to the above species. This remoteness was first noted using molecular characters by Clabaugh (1997), and their placement relative to tropical taxa corroborated by Simison (2000). Here they are nominally placed in the genus Lottia pending revision of the tropical New World Lottiidae.

"*Lottia*" *paleacea* (Gould, 1853) (=*Notoacmea paleacea, Tectura paleacea*). Surfgrass limpet; occurs and feeds on the surfgrass *Phyllospadix torreyi* and *P. scouleri* in northern California and Oregon. Low intertidal zone. Fishlyn and Phillips (1980, Biol. Bull. 158: 34–48) reported that this limpet incorporates the phenolic compounds of the surfgrass into its shell, a case of "chemical camouflage" that then reduces the ability of the seastar *Leptasterias* to detect it as prey. Has a small notch on the right anterior margin and sometimes on the left side as well (Debby Carlton, personal communication); these notches permit water flow for respiration when the limpet is firmly attached to the seagrass blade.

"Lottia" depicta (Hinds 1842) (=Notoacmea depicta, Tectura depicta). Eelgrass limpet; a southern species occurring on the eelgrass Zostera marina in estuaries and offshore; offshore specimens tend to be more oval reflecting the wider blade width of Zostera in these habitats (plate 375B). Northern range extended by El Niño events to Monterey Bay. See Lindberg 1982 Bull. So. Calif. Acad. Sci. 81: 87–95 (morphometrics, taxonomy); Zimmerman et al. 1996, Oecologia 107: 560–567 (herbivory).

REFERENCES

Abbott, D. P., and E. C. Haderlie. 1980. Chapter 13. Prosobranchia: Marine snails, pp. 230–307. In Intertidal invertebrates of California. Morris, R. H., D. P. Abbott, and E. C. Haderlie, Stanford University Press.

Begovic, E. 2004. Population structuring mechanisms and diversification patterns in the Patellogastropoda of the North Pacific. PhD dissertation, Department of Integrative Biology, University of California at Berkeley, 291 pp.

Carlton, J. T. 1981. Bibliography of Pacific coast Acmaeidae, pp. 106–119. In D. R. Lindberg, Acmaeidae, Gastropoda, MOLLUSCA, Invertebrates of the San Francisco Bay Estuary System. The Boxwood Press, Pacific Grove, California, 122 pp.

Clabaugh, J. P. 1997. Molecular phylogenetics of eastern Pacific limpets (Patellogastropoda: Lottiidae): rapid evolution in the intertidal? MA thesis, Biology Department, University of California, Los Angeles, 52 pp.

Fritchman, H. K., III. 1961. A study of the reproductive cycles in the California Acmaeidae (Gastropoda). Part I. Veliger 3:57–63.

Grant, A. R. 1937. A systematic revision of the genus Acmaea Eschscholtz, including consideration of ecology and speciation. Ph.D. Dissertation, Department of Zoology, University of California at Berkeley, 432 pp.

Lindberg, D. R. 1981. Invertebrates of the San Francisco Bay Estuary System: Mollusca, Family Acmaeidae, The Boxwood Press, Pacific Grove, CA, 120 pp.

Lindberg, D. R. 1986. Name changes in the Acmaeidae. Veliger 29: 142–148.

Lindberg, D. R. 1998. Order Patellogastropoda, pp. 639–652. In P. L. Beesley, G. J. B. Ross, and A. Wells, eds. Mollusca: The Southern Synthesis. Part B. Fauna of Australia. Volume 5. CSIRO Publishing, Melbourne.

McLean, J. H. 1966. West American prosobranch Gastropoda: superfamilies Patellacea, Pleurotomariacea, Fissurellacea. Ph.D. Dissertation, Biology, Stanford University, Stanford, 255 pp.

McLean, J. H. 1969. Marine shells of southern California. Los Angeles County Museum of Natural History, Science Series 24, Zoology 11, 104 pp.

Ricketts, E. F., J. Calvin, and J. W. Hedgpeth, revised by D. W. Phillips. Between Pacific Tides, Fifth Edition. 1985. Stanford University Press, xxvi + 652 pp.

Simison, W. B. 2000. Evolution and phylogeography of New World gastropod faunas. Ph.D. Dissertation, Department of Integrative Biology, University of California at Berkeley. 202 pp.

Test, A. R. G. 1946. Speciation in limpets of the genus Acmaea. Contributions of the Laboratory of Vertebrate Biology, University of Michigan 31: 1–24.

Littorina

DAVID REID

(Plates 378–379)

Although they are among the most common intertidal snails and ecologically important, the identification of Littorina species poses a challenge. A remarkable attribute of the group is the great variability in shell shape, size, and color. Shell variation is often correlated with habitat, so habitat-specific shell forms (ecotypes) can usefully be recognized, although there is intergradation between them. At least in the nonplanktotrophic species (with limited gene flow), these ecotypes are believed to result mainly from genetic adaptation caused by natural selection; for example, crab predators common on sheltered shores select for large, strong shells, whereas on exposed shores the danger of dislodgement by waves selects for thin, small shells with a large aperture accommodating a large foot. There is also some evidence for direct ecophenotypic effects on shell form.

As a result of this variability, and because there are few distinguishing features in the shells, identification requires care. Accurate field identification based only on shell features can be achieved with experience and knowledge of local variation, but identification should be confirmed by examination of the diagnostic characters of the reproductive system (penis and pallial oviduct). This is easily done, for these structures are readily seen on the outside of the animal when the shell is crushed and removed, and internal dissection is not required. Further, the penis can be examined even in living male animals; it is an organ up to one-third of the length of the shell, situated just behind the eye at the base of the right head tentacle, visible when the shell of a crawling animal is gently raised. Special attention should be paid to the proportions of the wrinkled basal region and the smooth distal filament of the penis and to the number of flask-shaped "mammilliform penial glands" that are visible externally as raised papillae along the posterior edge of the penis (plate 379A). Females lack a penis and develop a pale wrinkled skin patch (the ovipositor) in the corresponding position. A complication is that in a few species, notably L. sitkana, the males shed and regrow the penis following the breeding season.

The critical anatomical feature of the females is the pallial oviduct, a glandular structure producing the layers of the egg capsule, which is visible as a complex looping spiral structure (plate 379K), embedded in the mantle wall adjacent to the columellar muscle. The pallial oviduct becomes smaller outside the reproductive season but does not disappear. In the corresponding position, males possess only a simple glandular groove, the prostate gland, with no spiral loops.

For a detailed account of the taxonomy and systematics of all Littorina species, including descriptions and reviews of shell variation, reproductive anatomy, radula, distribution, and habitat, as well as evolutionary history, see the monograph by Reid (1996). The general anatomy of Littorina has been described by Fretter and Graham (1994). There is a large literature on the life history, population genetics, and ecology of littorinids, as reviewed by McQuaid (1996a, b). The proceedings of symposia on littorinid biology have been published in volumes 193 (1990), 309 (1995) and 378 (1998) of the journal Hydrobiologia. Behrens Yamada (1992) has described the comparative ecology of the four most common Pacific Coast Littorina species (L. sitkana, L. keenae, L. scutulata and L. plena), and Strathmann (1987) their spawn and development.

KEY TO SPECIES

1. Length of aperture (measured parallel to coiling axis of shell) less than half total shell length; shell smooth or with fine spiral grooves, but not strong spiral ribs; shell often with marbled, spotted, checkerboard or banded pattern 2
— Length of aperture half or more than half total shell length; shell with strong spiral ribs or fine spiral grooves (sometimes eroded smooth) 4
2. Shell delicate, easily crushed between fingers; length to 12 mm; shell color pale brown, often with darker spiral bands or pale dashes; habitat salt marsh or sheltered, muddy rocks; southern limit Humboldt Bay (plate 378J) . . . .
. . . . Littorina subrotundata (saltmarsh ecotype)

PLATE 378 A, B, *Littorina scutulata*; C, D, *L. plena*; E, *L. keenae*; F, *L. littorea* (from Maine); G, *L. saxatilis* (from San Francisco Bay); H, I, *L. sitkana* (ribbed and smooth forms); J, *L. subrotundata* (saltmarsh ecotype); K, *L. subrotundata* (barnacle ecotype). (B, C, E, F, I–K, from Reid 1996; rest, D. Reid).

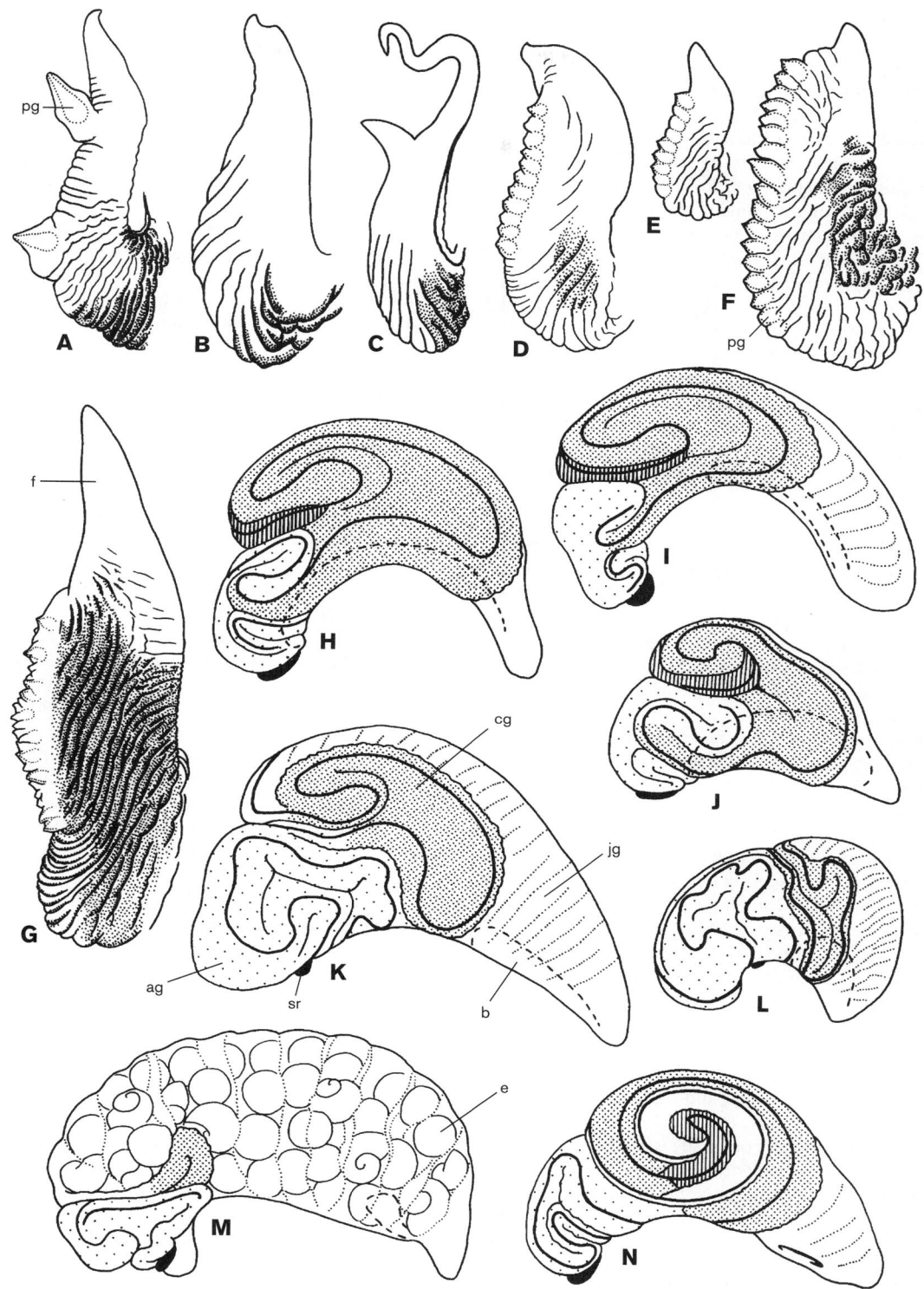

PLATE 379 *Littorina* penes (A–G) and pallial oviducts (H–N): A, N, *L. keenae*; B, H, *L. scutulata*; C, I, *L. plena*; D, M, *L. saxatilis* (from Maine); E, L, *L. subrotundata*; F, K, *L. sitkana*; G, J, *L. littorea* (G, from Portugal; J from Maine); not to scale; abbreviations and shading: ag, albumen gland (sparse stipple); b, copulatory bursa (dashed outline; visible only by dissection); cg, capsule gland (differentiated into opaque pink and translucent reddish portions; dense stipple and vertical hatching respectively); e, embryos in brood pouch; f, penial filament; jg, septate swollen jelly gland; pg, mammilliform penial gland; sr, seminal receptacle (black); shading on penes is black pigmentation (from Reid 1996).

— Shell solid; shell with predominantly marbled, spotted or checkerboard pattern, or black; habitat high intertidal zone on rocky shores 3

3. Shell pattern of fine pale marbling or small spots on brown, olive, or black background (plate 378D), or all black; pale stripe usually present within base of aperture (plate 378C); length to 19 mm; penis with lateral glandular projection halfway along its length, continuing beyond as a long, narrow (sometimes coiled) filament (plate 379C); copulatory bursa extends only halfway beneath coiled glandular portion of pallial oviduct (plate 379I); each cephalic tentacle with black pigmented line along its length, or entirely black . . . . *Littorina plena*

— Shell pattern of coarse white spots or checkerboard pattern (rarely two broad white bands) on brown, olive, or black background (plate 378A, 378B); usually no pale stripe within aperture; length to 18 mm; penis gradually tapering, with small terminal bifurcation (plate 379B); copulatory bursa extends far back beneath coiled glandular portion of pallial oviduct, almost to seminal receptacle (plate 379H); cephalic tentacles with transverse black bands . . . . *Littorina scutulata*

4. Shell smooth, with fine spiral grooves (often eroded); conspicuous white stripe within base of brown aperture; flattened parietal area (smoothly eroded by animal) adjacent to columella (plate 378E); length to 23 mm; shell color black or brown with irregular white flecks, but often eroded; penis with two (rarely one or three) large mammilliform penial glands (plate 379A); multispiral pallial oviduct with enlarged, septate jelly gland (plate 379N) . . . . *Littorina keenae*

— Shell smooth, finely grooved or coarsely ribbed; no basal white stripe within aperture; no eroded parietal area; shell color variable; penis with three to 42 small mammilliform penial glands 5

5. Shell with 20–40 fine spiral grooves (plate 378F); length to 30 mm; color brown or blackish with eight to 25 narrow dark spiral lines; columella white, lined pattern within aperture; penis with tapering terminal filament 30%–50% total length, and 10–42 small mammilliform penial glands on posterior edge of base (plate 379G); pallial oviduct of female has large albumen and capsule glands, and terminal portion (jelly gland) is not enlarged or swollen (plate 379J); introduced to San Francisco Bay (from Atlantic coast) . . . . *Littorina littorea*

— Shell smooth, or with <10 fine spiral grooves, or with coarse spiral ribs; color variable; penis with terminal filament 10%–30% total length, and three to 25 small mammilliform penial glands (plate 379D–379F); terminal portion of pallial oviduct of female is a swollen jelly gland or enlarged brood pouch with embryos 6

6. Terminal portion of pallial oviduct of female is a thinwalled, septate brood pouch, containing numerous shelled embryos 400–650 μm diameter (i.e., ovoviviparous development), albumen and capsule glands small (plate 379M); shell highly variable, but commonly with aperture half of total shell length, nine to 14 spiral ribs (plate 378G), and color often pale brown or cream, frequently with pale flecks; length to 17 mm; introduced population (from Atlantic coast) in San Francisco Bay . . . . . . *Littorina saxatilis*

— Terminal portion of pallial oviduct of female is a thickwalled, septate, jelly gland, producing gelatinous matrix of benthic egg masses (i.e., oviparous development), albumen and capsule glands large (plate 379K, 379L); shell variable, but frequently blackish brown or with spiral color bands 7

7. Shell usually with eight to 11 coarse spiral ribs on last whorl (plate 378H), and then with fine spiral microstriae in spaces between ribs (but sometimes smooth or ribbed only on base, plate 378I); length to 21 mm; shell color usually blackish brown (occasionally white, yellow, orange, or banded); unequivocal separation from *L. subrotundata* based on details of capsule gland of pallial oviduct (plate 379K); habitat sheltered rocky shores; southern limit Cape Arago, Oregon . . . . *Littorina sitkana*

— Shell smooth, or with up to 10 fine spiral grooves, or with nine to 10 coarse spiral ribs on last whorl (plate 378K), but without spiral microstriae in spaces between ribs; length to 7 mm; shell color blackish brown, often with one to eight pale or white spiral bands (occasionally white or orange); unequivocal separation from *L. sitkana* based on details of capsule gland of pallial oviduct (plate 379L); habitat among mussels, barnacles and *Mazzaella* on wave-exposed rocky shores (for saltmarsh form, see key item 2 above); southern limit Cape Arago, Oregon . . . . *Littorina subrotundata* (barnacle ecotype)

LIST OF SPECIES

Littorina keenae Rosewater, 1978. [=*L. planaxis* (Philippi, 1847)]. Abundant on open coasts on bare rock in upper intertidal; Baja California to Coos Bay, Oregon. Diet black lichens, epilithic and endolithic microalgae. Spawn is a free gelatinous egg mass from which pelagic biconvex capsules (to 400 μm diameter) containing single eggs are released; development planktotrophic. Distinguished from *L. scutulata* or *L. plena*, with which it co-occurs, by eroded parietal area beside columella and by lower spire; penis and pallial oviduct are diagnostic. This is the basal species of *Littorina* (Williams, Reid and Littlewood 2003). See Bock and Johnson 1967, Veliger 10: 42–54 (zonation); Behrens Yamada 1977, Mar. Biol. 39: 61–65 (range limit); Schmitt 1979, Mar. Biol. 50: 359–366 (spawn); Behrens Yamada 1992 (ecology, diet); Schmitt 1993, in Hochberg, ed., Third California Islands Symposium, 257–271, Santa Barbara Mus. (demography); Reid 1996 (systematics, review).

Littorina littorea (Linnaeus, 1758). Records from San Francisco Bay are introductions from East Coast. Tide pools on sheltered rocky shores, also sand- and mudflats. Grazes on a wide range of algae, especially small, ephemeral macroalgae. A large species, recognized by smooth, onion-shaped profile with relatively indistinct sutures between whorls, fine spiral ribs and pattern of brown lines. See Carlton 1992, J. Shellfish Res. 11: 489–505 (West Coast records); Reid 1996 (systematics, review).

Littorina plena Gould, 1849. Included with *L. scutulata* from 1864 to 1979. Common throughout intertidal, among algae, mussels, and barnacles, on moderately exposed and sheltered shores, also salt marshes and in estuaries; Baja California to Kodiak Island, Alaska. Diet not known in detail, probably resembles that of *L. scutulata*. Spawn of pelagic biconcave capsules, usually with two rims of equal diameter (to 1,340 μm diameter) containing two to 47 eggs; development planktotrophic. Rediscovered as a species distinct from the closely similar *L. scutulata* (see below) using penis and egg capsule shape (Murray 1979). Only reproductive anatomy is 100% diagnostic for morphological identification: penis with long terminal filament and lateral projection halfway along its length (plate 379C); copulatory bursa of female extends only halfway beneath coiled glandular portion of pallial oviduct (this character requires microdissection; plate 379I); but presence of broad unbroken black stripe along cephalic tentacle (or tentacle entirely black) gives accurate identification in more than 96% of cases. Useful shell characters are presence of pale basal band within aperture, relatively finely spotted shell pattern, and stronger incised lines (at least 80% accuracy). See Murray 1979, Veliger, 21: 469–474 (anatomical identification); Murray 1982, Veliger 24: 233–238 (discriminant analysis of shells); Mastro, Chow, and Hedgecock 1982, Veliger 24: 239–246 (allozyme analysis); Strathmann 1987 (spawn); Chow 1987a, Veliger 29: 359–366 (discriminant analysis of shells); Chow 1987b, J. Exp. Mar. Biol. Ecol. 110: 69–89 (growth, ecology); Chow 1989, J. Exp. Mar. Biol. Ecol. 130: 147–165 (competition, diet); Geller 1991, J. Exp. Mar. Biol. Ecol. 150: 1–17 (ecology); Reid 1996 (systematics, review); Rugh 1997, Veliger 40: 350–357 (shell morphology); Kyle and Boulding 2000, Mar. Biol. 137: 835–845 (molecular genetic population structure); Hamilton and Heithaus 2001, Proc. R. Soc. London Ser. B 268: 2585–2588 (predation, behavior); Hohenlohe and Boulding 2001, J. Shellfish Res. 20: 453–457 (molecular and morphological distinction from *L. scutulata*); Hohenlohe 2002, Invert. Biol. 121: 25–37 (spawn and life history); Hohenlohe 2003a, b, Veliger 46: 162–168, 211–219 (local and geographical distribution).

Littorina scutulata Gould, 1849. Not separated from *L. plena* from 1864 to 1979, so references to *L. scutulata* during this period may have included both species. Common from mid- to high intertidal, among algae and barnacles, on wave-exposed and protected coasts, but absent from the most sheltered habitats, estuaries, and salt marshes; Baja California to Sitka, Alaska. Diet mainly epilithic and endolithic microalgae and diatom films, but also includes lichens, various macroalgae and plant litter. Spawn of pelagic biconvex capsules with two rims of unequal diameter (to 1,000 μm diameter) containing one to 11 eggs; development planktotrophic. For morphological separation from closely similar *L. plena* (see above) only reproductive anatomy is 100% diagnostic: penis gradually tapering, with small terminal bifurcation (plate 379B); copulatory bursa of female extends far back beneath coiled glandular portion of pallial oviduct, almost to seminal receptacle (this character requires microdissection; plate 379H); but presence of transverse black bands on cephalic tentacles gives at least 96% accuracy of identification. Useful shell characters are absence of pale basal band within aperture, relatively coarsely spotted shell pattern, and weaker sculpture (at least 80% accuracy). See Murray 1979, Veliger, 21: 469–474 (anatomical identification); Jensen 1981, Veliger 23: 333–338 (diet); Murray 1982, Veliger, 24: 233–238 (discriminant analysis of shells); Mastro, Chow, and Hedgecock 1982, Veliger 24: 239–246 (allozyme analysis); d'Antonio 1985, J. Exp. Mar. Biol. Ecol. 86: 197–218 (diet); Strathmann 1987 (spawn); Chow 1987a, Veliger, 29: 359–366 (discriminant analysis of shells); Chow 1987b, J. Exp. Mar. Biol. Ecol. 110: 69–89 (growth, ecology); Sacchi and Voltolina 1987, Atti Soc. Ital. Sci. Nat. Mus. Civ. Stor. Nat. Milano 128: 209–234 (ecology); Voltolina and Sacchi 1990, Hydrobiologia 193: 147–154 (diet); Behrens Yamada 1992 (ecology, diet); Behrens Yamada and Boulding 1996, J. Exp. Mar. Biol. Ecol. 204: 59–83 (predation, zonation); Reid 1996 (systematics, review); Rugh 1997, Veliger 40: 350–357 (shell morphology); Boulding and Harper 1998, Hydrobiologia 378: 105–114 (habitat); Boulding, Holst, and Pilon 1999, J. Exp. Mar. Biol. Ecol. 232: 217–239 (predation); Kyle and Boulding 2000, Mar. Biol. 137: 835–845 (molecular genetic population structure); Rochette and Dill 2000, J. Exp. Mar. Biol. Ecol. 253: 165–191 (predation, behavior); Hamilton and Heithaus 2001, Proc. R. Soc. London Ser. B 268: 2585–2588 (predation, behavior); Hohenlohe and Boulding

2001, J. Shellfish Res. 20: 453–457 (molecular and morphological distinction from *L. plena*); Hohenlohe 2002, Invert. Biol. 12: 25–37 (spawn and life history); Hohenlohe 2003a, b, Veliger 46: 162–168, 211–219 (local and geographical distribution).

Littorina saxatilis (Olivi, 1792). [=*L. rudis* (Maton, 1797); =*L. tenebrosa* (Montagu, 1803)]. Introduced from East Coast; several reproducing populations established in San Francisco Bay, in upper eulittoral, among barnacles; in North Atlantic, it occupies a wide habitat range, from littoral fringe to shallow sublittoral, on wave-exposed and sheltered rocky shores, in lagoons and salt marshes. Diet includes a broad range of diatoms, blue-green algae, ephemeral filamentous and foliose macroalgae, and macrophyte litter. Development ovoviviparous; crawling juveniles (to 650 μm diameter) released from brood pouch. West Coast shells could be confused with those of *L. sitkana* and *L. subrotundata*: microstriae present between spiral ribs; spire relatively taller than in most *L. sitkana*; coloration variable but commonly pale brown or cream, with pale flecks (West Coast populations have not been seen with the dark blackish brown, often banded or lined, shells frequent in the other two species, although these color forms do occur in some Atlantic populations). Diagnostic feature is presence of shelled embryos in brood pouch of pallial oviduct (plate 379M; contrast swollen jelly gland in corresponding position in *L. sitkana* and *L. subrotundata*); penes similar in these three species. So far, *L. saxatilis* has not been found within the geographical range of *L. sitkana* and *L. subrotundata* (northern California, Oregon and further north). Potentially, this introduction could pose a threat to these native species if it spreads northwards, because in the Atlantic *L. saxatilis* is extremely abundant in a wide range of intertidal habitats and is a rapid colonist. In the Atlantic, the shells are highly variable (1.2 mm–25.8 mm; smooth or ribbed, globular to tall), with characteristic ecotypes in different habitats (Reid 1996); barnacle and saltmarsh ecotypes are almost identical to corresponding ecotypes of *L. subrotundata*. See Reid 1996 (systematics, review); Carlton and Cohen 1998, Veliger 41: 333–338 (West Coast records).

Littorina sitkana Philippi, 1846. [=*L. kurila* Middendorff, 1848; =*L. atkana* (Dall, 1886)]. Only clearly distinguished from barnacle ecotype of *L. subrotundata* since 1991. Common on protected rocky and boulder shores, mainly in upper intertidal, in crevices and among macroalgae; rarely on wave-exposed shores or in salt marshes; Cape Arago, Oregon, to Alaska and northwest Pacific. Diet includes a broad range of microalgae, macroalgae, epiphytes, black lichens, and plant litter. Spawn a benthic gelatinous mass up to 37 mm long, often in a much larger communal mass, deposited under stones or on seaweed, containing eggs in individual capsules (to 1,000 μm diameter); development nonplanktotrophic with intracapsular metamorphosis. Over the entire geographical range there is considerable variation in shell features, from the extremes of saltmarsh ecotype (smooth, tall spire, thin shell) to wave-exposed ecotype (low spire, large aperture); in Oregon and Washington, most shells are of a moderate ecotype, but sculpture ranges from smooth to strongly ribbed (plate 378H, 378I; smooth shells are produced under conditions of fast growth), and coloration is also variable. Small shells from open coasts and salt-marsh form are hard to distinguish from corresponding ecotypes of *L. subrotundata*, but *L. sitkana* is rare in these habitats. Distinguishing features of the shell are microstriae between spiral ribs (but both absent in smooth shells), and absence of black-lined color pattern that is common in *L. subrotundata*. Compare also with introduced *L. saxatilis* (see above). Diagnostic character is shape of capsule gland of pallial oviduct (plate 379K); penes of these three species are similar. See Behrens 1972, Veliger 15: 129–132 (zonation); Behrens Yamada 1977, Mar. Biol. 39: 61–65 (range limit); McCormack 1982, Oecologia 54: 177–183 (zonation); Voltolina and Sacchi 1990, Hydrobiologia 193: 147–154 (diet); Reid and Golikov 1991, Nautilus 105: 7–15 (identification); Behrens Yamada 1992 (ecology, diet); Boulding and Van Alstyne 1993, J. Exp. Mar. Biol. Ecol. 169: 139–166 (habitat, selection, predation, growth); Boulding, Buckland-Nicks, and Van Alstyne 1993, Veliger 36: 43–68 (morphology, identification, genetics); Behrens Yamada and Boulding 1996, J. Exp. Mar. Biol. Ecol. 204: 59–83 (predation, zonation); Reid 1996 (systematics, review); Behrens Yamada, Navarrete and Needham 1998, J. Exp. Mar. Biol. Ecol. 220: 213–226 (predation, growth, behavior); Boulding, Holst, and Pilon 1999, J. Exp. Mar. Biol. Ecol. 232: 217–239 (predation); Jones and Boulding 1999, J. Exp. Mar. Biol. Ecol. 242: 149–177 (behavior); Kyle and Boulding 2000, Mar. Biol. 137: 835–845 (molecular genetic population structure); Rochette and Dill 2000, J. Exp. Mar. Biol. Ecol. 253: 165–191 (predation, behavior); Boulding, Pakes, and Kamel 2001, J. Shellfish Res. 20: 403–409 (predation); Hamilton and Heithaus 2001, Proc. R. Soc. London Ser. B 268: 2585–2588 (predation, behavior); Rochette, Dunmall, and Dill 2003, J. Sea Res. 49: 119–132 (population biology).

Littorina subrotundata (Carpenter, 1864). [=*Algamorda subrotundata*; =*L. (Algamorda) newcombiana* (Hemphill, 1877); =*L. kurila* of authors, not Middendorff, 1848]. Current concept of this species only established since 1991; previously the name was restricted to the salt-marsh ecotype, whereas shells from rocky shores were included with *L. sitkana.* Widely distributed in northern Pacific, occupying a range of intertidal habitats. At southeastern extreme of its range (northern California, Oregon, Washington) it is restricted to two contrasting habitats: on salt-marsh plants (saltmarsh ecotype) and in mid- to upper intertidal on moderately to strongly wave-exposed rocky shores, where it is found (barnacle ecotype) among barnacles, mussels and beds of red alga *Mazzaella.* It is virtually absent from protected rocky shores, probably as a result of intense predation by crabs on these intermediate habitats; here it is replaced by *L. sitkana,* which, with a thicker and larger shell, is more resistant to crabs. Salt-marsh ecotype from Humboldt Bay northward; barnacle ecotype from Cape Arago, Oregon, northward; both are of restricted distribution, but not (as has been suggested) in danger of extinction. Diet includes diatoms, enteromorphine *Ulva, Mazzaella* tips, and epithelial cells of salt-marsh plants. Spawn a benthic gelatinous mass up to 3 mm long, deposited on salt-marsh plants, hard mud, seaweed or barnacles, containing eggs in individual capsules (to 960 μm diameter); development nonplanktotrophic with intracapsular metamorphosis. Over the entire geographical range there is great variation in shell size, shape, sculpture, and color; shells can be informally classified into moderate, wave-exposed, barnacle and salt-marsh ecotypes (Reid 1996), of which only the last two occur in California and Oregon. Barnacle ecotype (plate 378K) is <7 mm, globular, variable in color (black, white, banded, lined, or orange) and sculpture (ribbed or smooth); shells are difficult to distinguish from small *L. sitkana* (see above), but if ribbed then grooves between ribs lack microstriae. Salt-marsh ecotype (plate 378J) up to 12 mm, delicate, tall-spired, smooth, or with few incised lines, pale brown with dark spiral lines or pale dashes; shells could be confused with *L. saxatilis* (see above). Diagnostic character is shape of capsule gland of pallial oviduct (plate 379L); penes of these three species are similar. Conspecificity of saltmarsh and barnacle ecotypes has been confirmed by molecular genetics. See MacDonald 1969a, Ecol.

Monogr. 39: 33–60 (ecology); MacDonald 1969b, Veliger 11: 399–405 (ecology); Boulding 1990, Hydrobiologia 193: 41–52 (natural selection; as 'Littorina sp.'); Berman and Carlton 1991, J. Exp. Mar. Biol. Ecol. 150: 267–281 (ecology, diet); Reid and Golikov 1991, Nautilus 105: 7–15 (identification); Boulding and Van Alstyne 1993, J. Exp. Mar. Biol. Ecol. 169: 139–166 (habitat, selection, predation, growth; as 'Littorina sp.'); Boulding, Buckland-Nicks, and Van Alstyne, 1993, Veliger 36: 43–68 (morphology, identification, genetics; as 'Littorina sp.' and 'L. kurila'); Boulding and Hay 1993, Evolution 47: 576–592 (ecophenotypic and genetic components of shell shape); Reid 1996 (systematics, review); Boulding and Harper 1998, Hydrobiologia 378: 105–114 (habitat); Kyle and Boulding 1998, Proc. R. Soc. London 265: 303–308 (molecular genetics); Boulding Holst, and Pilon 1999, J. Exp. Mar. Biol. Ecol. 232: 217–239 (predation); Kyle and Boulding 2000, Mar. Biol. 137: 835–845 (molecular genetic population structure); Tie, Boulding, and Naish 2000, Mol. Ecol. 9: 107–118 (microsatellite markers).

REFERENCES

Behrens Yamada, S. 1992. Niche relationships of northeastern Pacific littorines. In J. Grahame, P. J. Mill and D. G. Reid, eds., Proceedings of the Third International Symposium on Littorinid Biology, 281–291. Malacological Society of London.

Fretter, V., and A. Graham. 1994. British Prosobranch Molluscs. Their Functional Anatomy and Ecology. Revised and Updated Edition. London: Ray Society, 820 pp.

McQuaid, C. D. 1996a. Biology of the gastropod family Littorinidae. I. Evolutionary aspects. Oceanogr. Mar. Biol. Ann. Rev. 34: 233–262.

McQuaid, C. D. 1996b. Biology of the gastropod family Littorinidae. II. Role in the ecology of intertidal and shallow marine ecosystems. Oceanogr. Mar. Biol. Ann. Rev. 34: 263–302.

Reid, D. G. 1989. The comparative morphology, phylogeny and evolution of the gastropod family Littorinidae. Phil. Trans. R. Soc. Lond. Ser. B, 324: 1–110.

Reid, D. G. 1996. Systematics and evolution of *Littorina.* London: Ray Society, 463 pp.

Reid, D. G., E. Rumbak, and R. H. Thomas. 1996. DNA, morphology and fossils: phylogeny and evolutionary rates of the gastropod genus *Littorina.* Philosophical Transactions of the Royal Society of London, Series B, 351: 877–895.

Strathmann, M. F., ed. 1987. Reproduction and development of marine invertebrates of the northern Pacific coast. Seattle: University of Washington Press, xii + 670 pp.

Williams, S. T., D. G. Reid, and D. T. J. Littlewood. 2003. A molecular phylogeny of the Littorininae (Gastropoda: Littorinidae): unequal evolutionary rates, morphological parallelism and biogeography of the Southern Ocean. Mol. Phyl. Evol. 28: 60–86.

Pelagic Gastropoda
(Heteropods, Pteropods, and Janthinids)

ROGER R. SEAPY AND CAROL M. LALLI

(Plates 380–385)

Holopelagic gastropod snails live out their life histories in the open ocean, although many species may be present in nearshore waters. They are distributed from tropical to polar latitudes of the world's oceans but are represented by a mere handful (about 150) of the approximately 40,000 described species of marine gastropods. The microscopic species may be collected in plankton tows, and pseudoconchs, shells, and bodies of macroscopic species may occasionally wash ashore. Unfortunately, many of the holopelagic species are not widely known. Their biology and ecology were reviewed by Lalli and Gilmer (1989), and photographs and brief characterizations of a number of West Coast taxa were included in Wrobel and Mills (1998).

Those species of pelagic gastropods normally encountered off the West Coast are associated with the cold, southward-flowing waters of the broad California Current. Occasionally, however, species that commonly occur in the warm waters off Baja California and farther south can be transported northward in a narrow tongue of water adjacent to the coast. This transport can occur by one of two mechanisms. The first of these is the Davidson Countercurrent, which develops regularly during the winter months in response to a weakening of the northerly winds. Southerly winds can then develop to produce a strong, northward flow that potentially can transport planktonic organisms as far north as British Columbia. The second mechanism is associated with periodic El Niño/Southern Oscillation (ENSO) events. At such times warm, southern waters flow northward along the coast and, as with the Davidson Countercurrent, displace California Current waters offshore and carry warm water species northward, far beyond their normal range limits off Baja California. For the groups discussed here, those species associated with warm, southern waters that potentially can be carried north of Point Conception during intrusion events are referred to as "warm-water species" and, with the exception of the janthinids, are not included in the keys. Taxonomic references that can be consulted for their identification are included. Determination of which species to include as warm-water taxa was based on the distributional records of McGowan (1967, 1968) from the southern portion of the California Current and Cummings and Seapy (2003) and Seapy (unpublished) from southern California waters.

A common adaptation in shelled pelagic gastropods is reduction in the size and weight of the calcareous shell. As a result, these shells will usually dissolve if stored in a preservative that has not been buffered and maintained at the pH of seawater. Fixation and preservation of shell-bearing specimens can best be accomplished using 5%–10% formalin solution in seawater, buffered to pH 8.3 using sodium borate (borax). If only the shell remains, the appropriate method of long-term storage is to rinse the shells thoroughly in fresh water, dry them, and place them in acid-free glass tubes or gelatin capsules.

The pelagic snails include representatives from the Caenogastropoda (janthinids and heteropods) and the Heterobranchia (nudibranchs and pteropods). With the exception of *Fiona pinnata*, pelagic nudibranchs (*Phylliroë*, *Cephalopyge*, and *Glaucus*) have not been recorded from waters north of Point Conception to our knowledge. *Fiona pinnata* is widespread in the northern hemisphere and is associated with the pelagic cnidarians *Velella* and *Porpita* and with goose barnacles *Lepas* growing on floating debris (Behrens and Hermosillo, 2005). It is occasionally washed ashore with *Lepas* (e.g., Marcus 1961, MacFarland 1966, Holleman 1972) and is included in the key to Nudibranchs and Sacoglossans. In addition, Wrobel and Mills (1998) stated that *Phylliroë* and *Glaucus* are carried northward only occasionally into southern California waters. Thus, we have not included the pelagic nudibranchs here.

The most commonly encountered planktonic gastropods are the shelled veliger larvae of benthic species, many of which are of the same size as the microscopic shelled heteropods and euthecosomatous pteropods. Veliger larvae of all gastropods are distinguished by having a ciliated, bi- or multi-lobed velum used in swimming and, in planktotrophic forms, food collection. The veliger shells of benthic prosobranchs are frequently ornamented and usually exhibit dextral coiling, while those of benthic opisthobranchs and other benthic heterobranchs are sinistrally coiled and usually unornamented. An illustrated key to gastropod larvae found off Oregon and Washington is provided by Goddard (2001). The shells of atlantid heteropods are dextrally coiled, whereas those of euthecosomes are either sinistrally coiled or uncoiled; the shells of both heteropods and euthecosomes tend to be thin, transparent, and unornamented.

In this section, we have attempted to give brief introductions to the heteropods, pteropods (euthecosomes, pseudothecosomes and gymnosomes), and janthinids; describe some of their adaptations to a pelagic existence; and provide a series of keys and illustrations that will facilitate their identification.

KEY TO HETEROPODS, PTEROPODS, AND JANTHINIDS

1. Pair of well-developed eyes present; foot modified as a single, laterally flattened swimming fin (plates 380A, 381A1, 381B). . . . . Heteropoda
—   Without well-developed eyes; foot not modified as a single, laterally flattened swimming fin 2
2. Foot flat, solelike; violet-colored, dextrally (clockwise) coiled shell present (plate 385C, 385D) . . . . Janthinidae (*Janthina*)
—   Foot modified as paired swimming wings (plates 382A1, 384A1) or as a single, fused and flattened wingplate (plate 383) . . . . (Pteropoda) 3
3. Paired swimming wings and external calcareous shell present. . . . . Euthecosomata
—   Broad, fused wingplate present; shell present or absent . . . . Pseudothecosomata
—   Paired swimming wings present; body streamlined; shell absent . . . . Gymnosomata

HETEROPODA

The name Heteropoda (*hetero-* meaning "other"; *-poda* for "foot") refers to the modification of the foot as a laterally flattened **SWIMMING FIN**. Heteropods swim in an upsidedown orientation, with their swimming fin directed upward (plate 381A1, 381B). They are carnivores that locate their prey visually with a pair of large, image-forming **EYES**, consisting of a distal spherical lens, a middle pigmented region, and a proximal ribbonlike retina. Prey are captured with a protrusible **RADULA** (with elongate and distally hooked teeth), located in the **BUCCAL MASS** at the end of a mobile and extensible **PROBOSCIS**. In the microscopic atlantids, the prey is held with the **FIN SUCKER** and pieces of tissue are torn off and ingested using the radula. In the macroscopic carinariids and pterotracheids, however, the fin sucker has not been observed to manipulate prey; instead, prey are captured and ingested whole. Heteropods feed on a variety of zooplanktonic taxa (reviewed by Lalli and Gilmer 1989). The prey of atlantids includes shelled pteropods, other atlantids, copepods, and chaetognaths, while carinariids feed on salps, doliolids, chaetognaths, siphonophores, and crustaceans (mainly copepods and euphausids). The feeding habits of pterotracheids, however, are poorly known.

Like many other shallow water zooplankters, the shells and bodies (except for the opaque eyes and viscera) of heteropods are transparent. All heteropods are dioecious and show sexual dimorphism (e.g., the penis and penial appendage in males; absence of a swimming fin sucker and presence of cuticular spines anterior to the eyes in female pterotracheids; and, radular differences in some atlantids). Fertilized eggs are laid in egg strings that break free from the female in most species. Development leads to a free-swimming, planktotrophic veliger larva, characterized by four or six long, slender velar lobes. The veligers of each species typically have distinctive larval shells.

The three families of heteropods (Atlantidae, Carinariidae, and Pterotracheidae) illustrate a hypothetical evolutionary sequence involving a change from microscopic to macroscopic size, with a reduction in shell size or loss of the shell. In the microscopic atlantids (average size about 2 mm–4 mm; maximal size about 10 mm), the animal can retract into its **DEXTRALLY** (clockwise) **COILED SHELL** and seal the shell aperture with a thin, chitinous **OPERCULUM** (plate 380A1). The carinariids and pterotracheids are macroscopic, and the viscera and gonads are compacted into a **VISCERAL NUCLEUS** (plate 381A1, 381B), which is closely associated with the heart and gills. Their transparent bodies are elongate in the anterior-posterior axis, cylindrical (in most species), and muscular. In both groups the body is divided into three regions (**PROBOSCIS, TRUNK,** and **TAIL**), and an operculum is lacking. In the carinariids, the shell only covers the visceral nucleus (plate 381A1), and in most species, the shell is laterally compressed and triangular in side view, with a keel along the anterior margin (plate 381A2). Pterotracheids lack a shell, and the visceral mass receives protection by being recessed to varying degrees into the cylindrical body (plate 381B). Also, their bodies are more elongate and streamlined than the carinariids, and they are the fastest swimmers among the heteropods.

Heteropods are usually preserved in a formalin-seawater solution while the animals are still alive. As a result, the bodies are often observed to be fully retracted into the shell in the atlantids and contracted to varied degrees in the carinariids and pterotracheids. For the atlantids, examination of important taxonomic features (e.g., eye and operculum morphologies) can only be accomplished in animals that are extended outside the shell by relaxation prior to preservation or by destroying the shell to expose these structures. Relaxation of heteropods can be accomplished using MS222 or dilute magnesium chloride or magnesium sulfate solution in sea water.

Thirty-four species of heteropods are recognized in the world's oceans (Richter and Seapy 1999). Most of these are tropical to subtropical and cosmopolitan in distribution, although two temperate species in the North Pacific (*Carinaria japonica* and *Atlanta californiensis*) are noteworthy for their restriction to the Transition Zone Faunal Province of the North Pacific (McGowan 1971, Seapy and Richter 1993).

Except for two species of pterotracheids that are mesopelagic and undergo diel vertical migration (Pafort-van Iersel 1983), all other species are epipelagic. Many of the epipelagic species undergo limited diel vertical migrations (Seapy 1990).

KEY TO HETEROPODS

1. Shell present; microscopic or macroscopic size 2
— Shell lacking; body macroscopic, with proboscis, trunk, and tail (lacking in *Firoloida*) regions; swimming fin located about midway between eyes and visceral mass; visceral mass partially to mostly imbedded in gelatinous, cylindrical body (plate 381B) . . . . Pterotracheidae

Note: Pterotracheids have not been recorded north of Point Conception. However, four species (*Firoloida desmaresti, Pterotrachea coronata, P. hippocampus,* and *P. scutata*) were reported from the southern

portion of the California Current by McGowan (1967), and three (*F. desmaresti, P. coronata,* and *P. scutata*) have been collected off southern California (Cummings and Seapy 2003; Seapy, unpublished). The presence in *F. desmaresti* of tentacles anterior to the eyes and a terminal visceral nucleus (and concomitant absence of a tail) distinguish this genus from *Pterotrachea* spp. For identification of the species of *Pterotrachea* see Seapy (1985, Malacologia 26: 125–135).

2. Microscopic size (most species less than about 5 mm; maximum about 10 mm); animal can retract into the shell; shell laterally compressed . . . . Atlantidae

Note: *Atlanta californiensis* (plate 380B). Shell moderately small (maximal diameter 3.5 mm), consisting of up to four and two-thirds whorls; shell spire low conical and surface sculpture lacking; shell surface clear to pigmented, with clear to light purple sutures; keel base orange to red-brown. See Seapy and Richter (1993) for species description and comparison with morphologically similar species. Additional species in the genus *Atlanta* either have been recorded from the southern portion of the California Current or from adjacent waters. These include six species (*A. gaudichaudi, A. inflata, A. lesueuri, A. peroni, A. inclinata,* and *A. turriculata*) from the California Current off Baja California (McGowan (1967). McGowan and Fraundorf (1966, Limnol. Ocean. 11:456–469) also recorded *A. fusca* from waters to the southeast of Cabo San Lucas. Since the mid-1960s, the species composition of the genus *Atlanta* has changed substantially (see Richter and Seapy 1999). Three species described in the early 1970s (*A. echinogyra, A. plana,* and *A. tokiokai*) were recorded from the Gulf of California by Seapy and Skoglund (2001, Festivus 33: 33–44), and they undoubtedly also occur in the southern portion of the California Current. Last, *A. oligogyra* was considered to be synonymous with *A. lesueuri* from the late 1940s to the mid 1970s but was reestablished as a valid species by Richter (1974). Like the above three species, it was recorded from the Gulf of California (Seapy and Skoglund (2001, Festivus 33:33–44) and probably occurs here as well. Except for *A. gaudichaudi,* all of these species were reported from Hawaiian waters by Seapy (1990, Malacologia 32: 107–130), and the interested reader is referred to this paper for their identification.

— Macroscopic size, with proboscis, trunk and tail regions; shell only covers the stalked visceral mass and is either external and cap-shaped (*Carinaria*), external and flattened (*Pterosoma*), or internal and greatly reduced (*Cardiapoda*); swimming fin located beneath or slightly anterior to the visceral mass . . . . Carinariidae

Note: *Carinaria japonica* (plate 381A) Maximal body length 105 mm (from waters off northern Baja California) (Seapy, 1980, Mar. Biol. 60: 137–146). Shell cap-shaped, triangular in lateral profile, and with a well developed keel along the anterior edge. Tail with a prominent dorsal crest.

LIST OF HETEROPODS

PTEROTRACHEOIDA (=HETEROPODA)

Atlantidae

Atlanta californiensis Seapy and Richter, 1993. Color photograph in Wrobel and Mills (1998: fig. 125). Abundant in southern California waters, with seasonal maxima in late summer (Cummings and Seapy 2003). Males exhibit periodic neustonic swarming that hypothetically is associated with mating in this

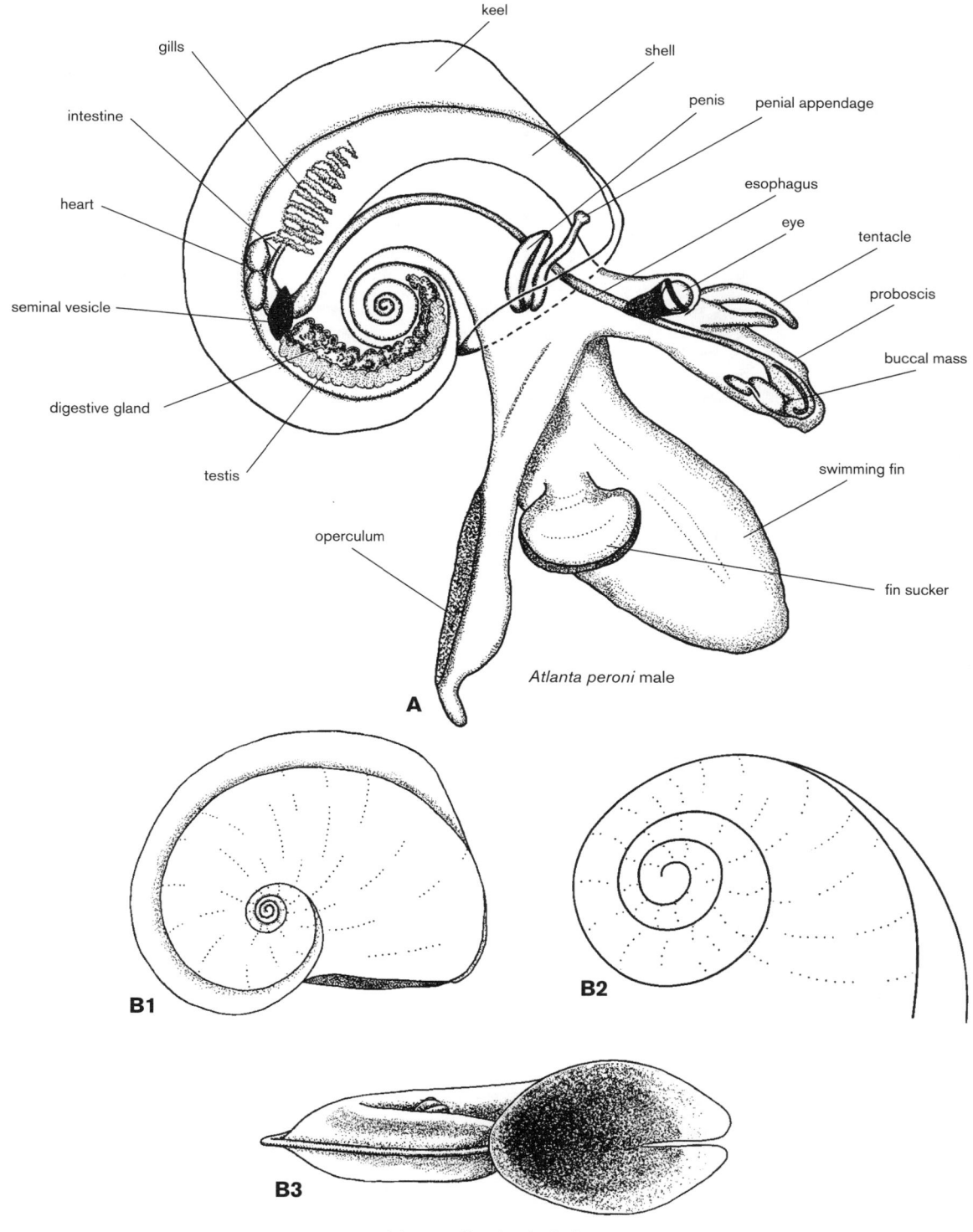

keel

gills

shell

intestine

penis

penial appendage

esophagus

heart

eye

tentacle

seminal vesicle

proboscis

buccal mass

digestive gland

testis

swimming fin

operculum

fin sucker

Atlanta peroni male

A

B1

B2

B3

Atlanta californiensis shell

PLATE 380 A, *Atlanta peroni,* male from right side; B. *Atlanta californiensis,* B1, shell from right, B2, shell spire, B3, shell from aperture (ventral).

and several other species of atlantids (Seapy and Richter 1993). Member of Transition Zone Faunal Province.

Atlanta echinogyra Richter, 1972. Color photograph in Seapy (1990, Malacologia, 32: 107–130, fig. 5E). Warm-water species.

Atlanta fusca Souleyet, 1852. Warm-water species.

Atlanta gaudichaudi Souleyet, 1852. Color photographs by Newman and Flowers (1996: 210) in Fossa and Nilsen, Korallenriff-Aquarium, Vol. 5, Birgit Schmettkamp Verlag, Bornheim. Warm-water species.

Atlanta inclinata Souleyet, 1852. Although reported by McGowan (1967) from CalCOFI samples, an unknown number of

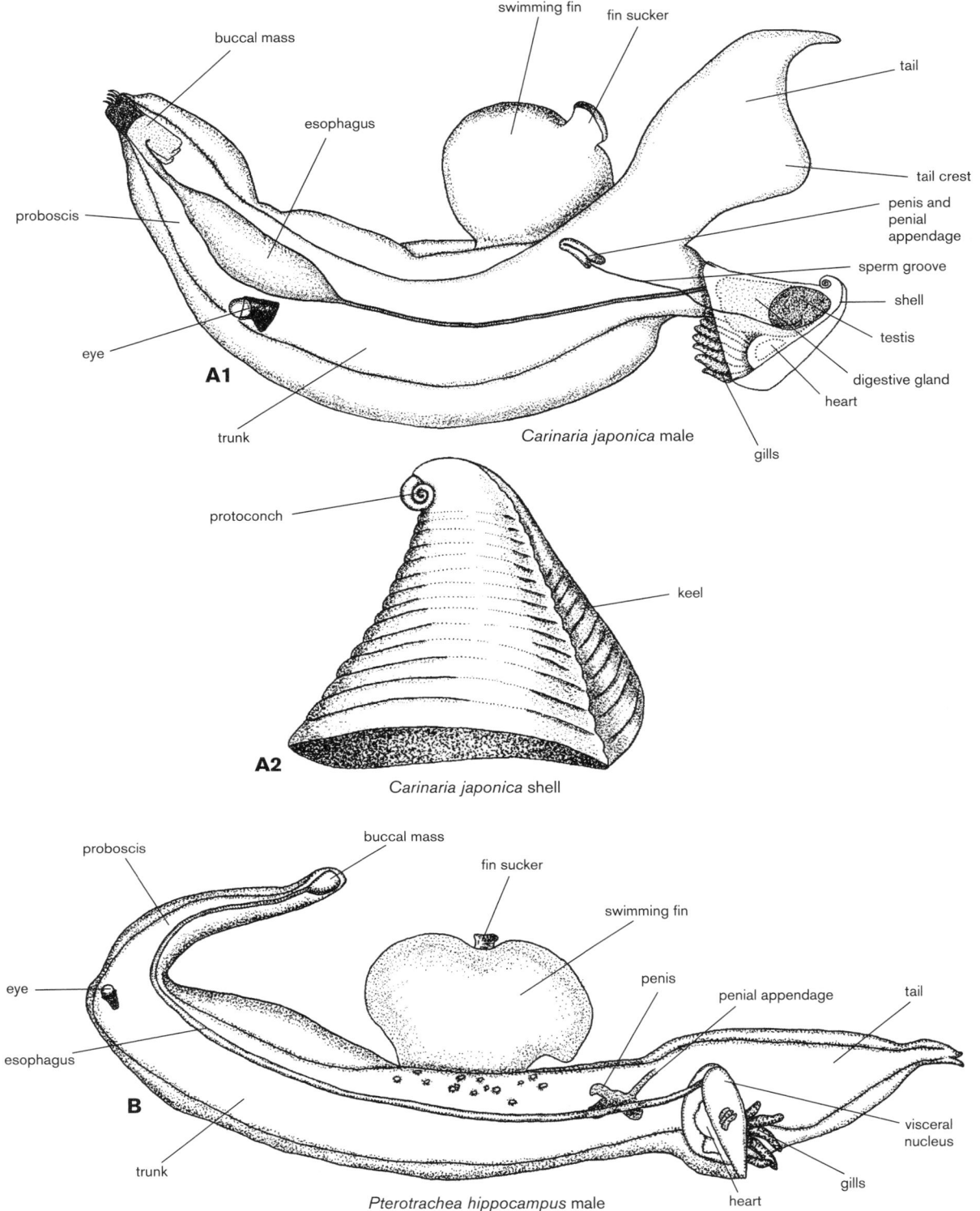

PLATE 381 A, *Carinaria japonica*, A1, male from right side, A2, shell from right side; B, *Pterotrachea hippocampus*, male from right side.

specimens assigned to this species most probably belong instead to *A. tokiokai* (see discussion below). Warm-water species.

Atlanta inflata Souleyet, 1852. Color photograph in Seapy (1990, Malacologia 32: 107–130, fig. 5F). Warm-water species.

Atlanta lesueuri Souleyet, 1852. Color photograph in Seapy (1990, Malacologia 32: 107–130, fig. 5B). Warm-water species.

Atlanta oligogyra Tesch, 1908. This species was not reported from the California Current region by McGowan (1967) undoubtedly because it was not generally recognized as a valid species at the time. In 1949, Tesch (Dana Rept. 34: 1–54) concluded that *A. oligogyra* was a junior synonym of *A. lesueuri*. However, Richter in 1974 ("Meteor" Forschung-Ergebnisse, (D),

17: 55–78) and 1986 (Archiv. Moll. 117: 19–31) cited distinctive morphological differences between *A. lesueuri* and *A. oligogyra* that justified their status as separate species. Since the two species co-occur in moderately high abundances in Hawaiian waters (Seapy, 1990, Malacologia 32: 107–130), it is reasonable to expect that they also would occur together in the southern portion of the California Current. Warm-water species.

Atlanta peroni Lesueur, 1817. Color photograph in Lalli and Gilmer (1989: color fig. 2). Recorded rarely off southern California (Cummings and Seapy 2003). Warm-water species.

Atlanta plana Richter, 1972. Color photograph in Newman and Flowers (1996: 211) in Fossa and Nilsen, Korallenriff-Aquarium, Vol. 5, Birgit Schmettkamp Verlag, Bornheim. Warm-water species.

Atlanta tokiokai van der Spoel and Troost, 1972. Color photograph in Seapy (1990, Malacologia 32: 107–130, fig. 5D). Prior to the evaluation of those species of atlantids with tilted spires by Richter (1990, Archiv. Moll. 119: 259–275), the taxonomic status of this species was uncertain. Richter showed that *A. tokiokai* is a valid species and shares many characteristics with *A. inclinata* Souleyet, 1852. Of the two species, only *A. tokiokai* was identified from Hawaiian waters by Seapy (1990, Malacologia 32: 107–130). In the Gulf of California, four specimens of *A. tokiokai* were identified from among the samples examined, while only one *A. inclinata* was found (Seapy and Skoglund, 2001, Festivus 33: 33–44). Thus, it is highly probable that *A. tokiokai* was present in the CalCOFI samples examined by McGowan (1967). Warm-water species.

Atlanta turriculata d'Orbigny, 1836. Color photograph in Seapy (1990, Malacologia 32: 107–130, fig. 5C). Warm-water species.

Carinariidae

Carinaria japonica Okutani, 1955. Color photograph in Wrobel and Mills (1998: fig. 126). Abundance varies seasonally in the California Current (Dales, 1953, Proc. Zool. Soc. Lond. 122(IV): 1007–1015; McGowan, 1967) and in waters off southern California (Seapy, 1974, Mar. Biol. 24: 243–250). The larva and larval shell were described by Seapy and Thiriot-Quiévreux (1994, Veliger 37: 336–343). Diet and prey preferences were studied from southern California waters by Seapy (1980, Mar. Biol. 60: 137–146). Member of Transition Zone Faunal Province.

Pterosoma planum Lesson, 1827. Color photograph in Lalli and Gilmer (1989: color fig. 4). Warm-water species.

Pterotracheidae

Firoloida desmaresti Lesueur, 1817. Color photograph in Wrobel and Mills (1998: fig. 127). Collected rarely off southern California (Cummings and Seapy 2003). Warm-water species.

Pterotrachea coronata Niebuhr (ms. Forskål, 1775). Color photograph in Wrobel and Mills (1998: fig. 128). Captured occasionally in southern California waters (Cummings and Seapy 2003). Warm-water species.

Pterotrachea hippocampus Philippi, 1836 (=*P. minuta* Bonnevie, 1920). Until recently, both *P. hippocampus* and *P. minuta* were widely recognized as valid species. Both were reported by Dales (1953, Proc. Zool. Soc. Lond. 122(IV): 1007–1015) and McGowan (1967) from the California Current. In a review of the genus *Pterotrachea* from Hawaiian waters (Seapy, 1985, Malacologia 26: 125–135), all specimens that could be tentatively identified as *P. minuta* were shown instead to be young

P. hippocampus. Subsequently, Seapy (2000, J. Moll. Stud. 66: 99–117) substantiated these findings based upon animals from the North Atlantic and concluded that *P. minuta* is not a valid species. Warm-water species.

Pterotrachea scutata Gegenbaur, 1855. Collected rarely off southern California (Seapy, unpublished). Warm-water species.

THECOSOMATA

The Thecosomata constitutes one of two clades of holopelagic gastropods commonly referred to as "pteropods," the other group being the Gymnosomata (see below). The name refers to the fact that they have swimming wings derived from the foot (*ptero-* meaning wings; -*pod* referring to the foot). There are two groups of thecosomatous pteropods: the Euthecosomata and Pseudothecosomata.

Euthecosomes possess a thin, external, calcareous shell that is **SINISTRALLY** (counter-clockwise) **COILED** in the more primitive species (Family Limacinidae; plate 382A, 382B) and **UNCOILED** in the other species (Family Cavoliniidae; plate 382C, 382D). In near-surface waters off California and Oregon, the shell size of euthecosomes likely to be encountered ranges from about 1 mm to 3 mm in diameter in limacinids and up to 20 mm in length in cavoliniids. The morphology of the shell is a distinguishing feature used in identification of euthecosome species. The paired **SWIMMING WINGS** lie dorsally and laterally to the flat **MEDIAN FOOTLOBE** (plate 382A1). Smaller **LATERAL FOOTLOBES** border either side of the mouth. The head is not well defined; there are one or two small **TENTACLES**. In species of *Cavolinia*, large extensions of mantle lining project from the shells of living animals; these **MANTLE APPENDAGES** function in buoyancy regulation and in feeding. *Cavolinia* species also produce a transparent, gelatinous **TEMPORARY PSEUDOCONCH** (or "false shell") that covers the true shell and is jettisoned when the animal is disturbed; thus it is almost never found with animals that have been collected in nets.

Pseudothecosomes differ from euthecosomes in having the swimming wings fused into a single **WINGPLATE** (plate 383A–383C). Most also have a well-developed **PROBOSCIS** formed by elongation of part of the footlobes. Only one genus (*Peraclis*) in this group has an external, sinistrally coiled calcareous shell that typically has a prolonged rostrum and displays various types of sculpture. All other pseudothecosomes lack an external shell, but they may have an internal, slipper-shaped **PSEUDOCONCH** that encases the viscera (plate 383A), although its attachment to the body is tenuous. Empty pseudoconchs and/or detached animals are sometimes found in plankton tows or washed up on beaches.

There are about 50 species of thecosomatous pteropods; the majority are found in tropical and subtropical waters, and most species are oceanic and epipelagic (van der Spoel 1967, 1976; Bé and Gilmer 1977; van der Spoel et al. 1997). Diel vertical migration is common among the group (e.g., Wormuth 1981), and shallow-water plankton tows taken at night tend to capture more individuals and more species. All euthecosomes and pseudothecosomes are ciliary-mucus feeders that use an external mucous web many times the size of their bodies to capture bacteria, phytoplankton, and/or small zooplankton (Gilmer 1972, 1990; Gilmer and Harbison 1986). Thecosomes are protandrous; they mature and function first as males, then as females (for a review of reproduction, see Lalli and Gilmer 1989). Most lay free-floating egg masses from which veliger larvae hatch. Only a few species retain fertilized eggs until hatching. Bandel and Hemleben (1995) provide descriptions of embryonic development for many thecosome species.

KEY TO THECOSOMES

1. Shell present; animal with a pair of swimming wings 2
— Shell or pseudoconch present; animal with a single, flattened wingplate 3
— No shell or pseudoconch; wingplate present and bearing two ciliated tentacles on the postero-lateral edges . . . . Desmopteridae (Pseudothecosomata)

Note: *Desmopterus pacificus* (plate 383C). Body length to 2 mm; diameter of wingplate to 5 mm. Wingplate tentacles short, not extending beyond wingplate border.

Desmopterus papilio. Body length to 2 mm; wingplate span to 6 mm. Wing tentacles long, trailing beyond wingplate border. It is not clear whether these two species of *Desmopterus* are present off California. The description and figure of *D. pacificus* by Essenberg (1919, Univ. Calif. Publ. Zool. 19: 85–88) clearly indicate very short wingplate tentacles, while the photograph in Wrobel and Mills (1998: fig. 136) is unquestionably of *D. papilio* with long wing tentacles. In other respects, the anatomy of the two species is very similar. McGowan reported (1967) and illustrated (1968: plate 20) *D. pacificus* as the only species of *Desmopterus* from the California Current. Also, only *D. pacificus* has been collected from southern California waters (Cummings and Seapy, unpublished).

2. Shell coiled sinistrally, lacking surface sculpture . . . . Limacinidae (Euthecosomata)

Note: *Limacina helicina* (plate 382A). Shell thin and transparent, with a low spire and a conspicuous open umbilicus. Whorls expand regularly. Striations may be present. Shell diameter to 3 mm–5 mm (larger in northern parts of range).

Limacina inflata (plate 382B). Shell with depressed spire and large body whorl. Umbilicus present. In adults, a thickened rib is present along the outer part of the body whorl; this rib projects over the aperture like a beak. Maximal shell diameter <1.5 mm.

Three other species of *Limacina* (*L. bulimoides, L. lesueuri,* and *L. trochiformis*) were reported from the southern portion of the California Current (McGowan 1967, 1968). See Bé and Gilmer (1977), van der Spoel (1967), or van der Spoel et al. (1997) for identification.

— Shell bilaterally symmetrical, ranging in shape from straight and pointed, to pyramidal, to globose, to conical . . . . Cavoliniidae (Euthecosomata)

Note: *Clio pyramidata* (plate 382C). Shell of pyramidal shape, with a sharply pointed posterior tip. Triangular in cross-section, with a flat ventral surface and a strongly arched dorsal surface. Shell length to about 20 mm.

Creseis virgula (plate 382D). Shell conical, straight or curved dorsally, with no ornamentation. Aperture rounded. Shell length <9 mm.

Eleven other cavoliniid species have been reported from the southern portion of the California Current (McGowan 1967, 1968); these are included in the "List of Thecosomes." All are epipelagic, warm-water species. See Bé and Gilmer (1977), van der Spoel (1967), or van der Spoel et al. (1997) for identification.

3. Shell coiled sinistrally; with reticulate or spiral ornamentation; columella prolonged into a rostrum . . . . Peraclididae (Pseudothecosomata)

Note: Three species of *Peraclis* (*P. apicifulva, P. bispinosa,* and *P. reticulata*) have been recorded from California waters (McGowan 1967, 1968). All species are mesopelagic or bathypelagic and are not described here. See van der Spoel (1976) or van der Spoel et al. (1997) for identification.

— Shell absent; gelatinous pseudoconch present . . . . Cymbuliidae (Pseudothecosomata)

Note: *Corolla spectabilis* (plate 383A). Body of animal encased in a rounded, ovate, gelatinous pseudoconch studded with many distinct tubercles. Pseudoconch length to 40 mm. Wingplate broadly rounded, reaching 80 mm in diameter; 12 mucous glands are present on each anterio-lateral margin. Proboscis short and broad.

Gleba cordata (plate 383B). Pseudoconch flattened with few tubercles, length to 45 mm. Wingplate diameter to 60 mm. Five to six large mucous glands present on the lateral margins of the wingplate. Proboscis long and narrow, capable of great extension, and free from the wingplate for most of its length.

LIST OF SPECIES

CAVOLINIOIDEA (=EUTHECOSOMATA)

Limacinidae

Limacina bulimoides (d'Orbigny, 1836). Recorded in low numbers off southern California (Cummings and Seapy 2003). Warm-water species.

Limacina helicina (Phipps, 1774). Color photographs in Lalli and Gilmer (1989: color fig. 6) and Wrobel and Mills (1998: fig. 130). Reproduction and development described by Paranjape (1968, Veliger 10: 322–326). In situ feeding observations reported by Gilmer and Harbison (1986) and Gilmer (1990). The most abundant species of euthecosome off southern California, with maximal numbers in summer and fall (Cummings and Seapy 2003). Common in polar and cold temperate waters of both hemispheres.

Limacina inflata (d'Orbigny, 1836). Reproduction and development described by Lalli and Wells (1973, Bull Mar. Sci. 23: 933–941; 1978, J. Zool. 186: 95–108). The third most abundant species of euthecosome off southern California (after *L. helicina* and *Creseis virgula*), with seasonal maxima in winter and early spring (Cummings and Seapy 2003). Cosmopolitan and abundant in tropical and subtropical waters, but extending to about 45°N.

Limacina lesueuri (d'Orbigny, 1836). Warm-water species.

Limacina trochiformis (d'Orbigny, 1836). Warm-water species.

Cavoliniidae

Cavoliniinae

Cavolinia inflexa (Lesueur, 1813). Color photograph in Gilmer and Harbison (1986: fig. 3a). Low numbers off southern California, with seasonal maximum in spring (Cummings and Seapy 2003). Warm-water species.

Cavolinia tridentata (Niebuhr, 1775). Color photographs in Lalli and Gilmer (1989: color fig. 7) and in Wrobel and Mills (1998: fig. 132). Rare off southern California (Cummings and Seapy 2003). Warm-water species.

Cavolinia uncinata (Rang, 1829). Warm-water species.

Diacria quadridentata (de Blainville, 1821). Warm-water species.

Diacria trispinosa (de Blainville, 1821). Color photograph in Lalli and Gilmer (1989: color fig. 8). Rare off southern California (Cummings and Seapy 2003). Warm-water species.

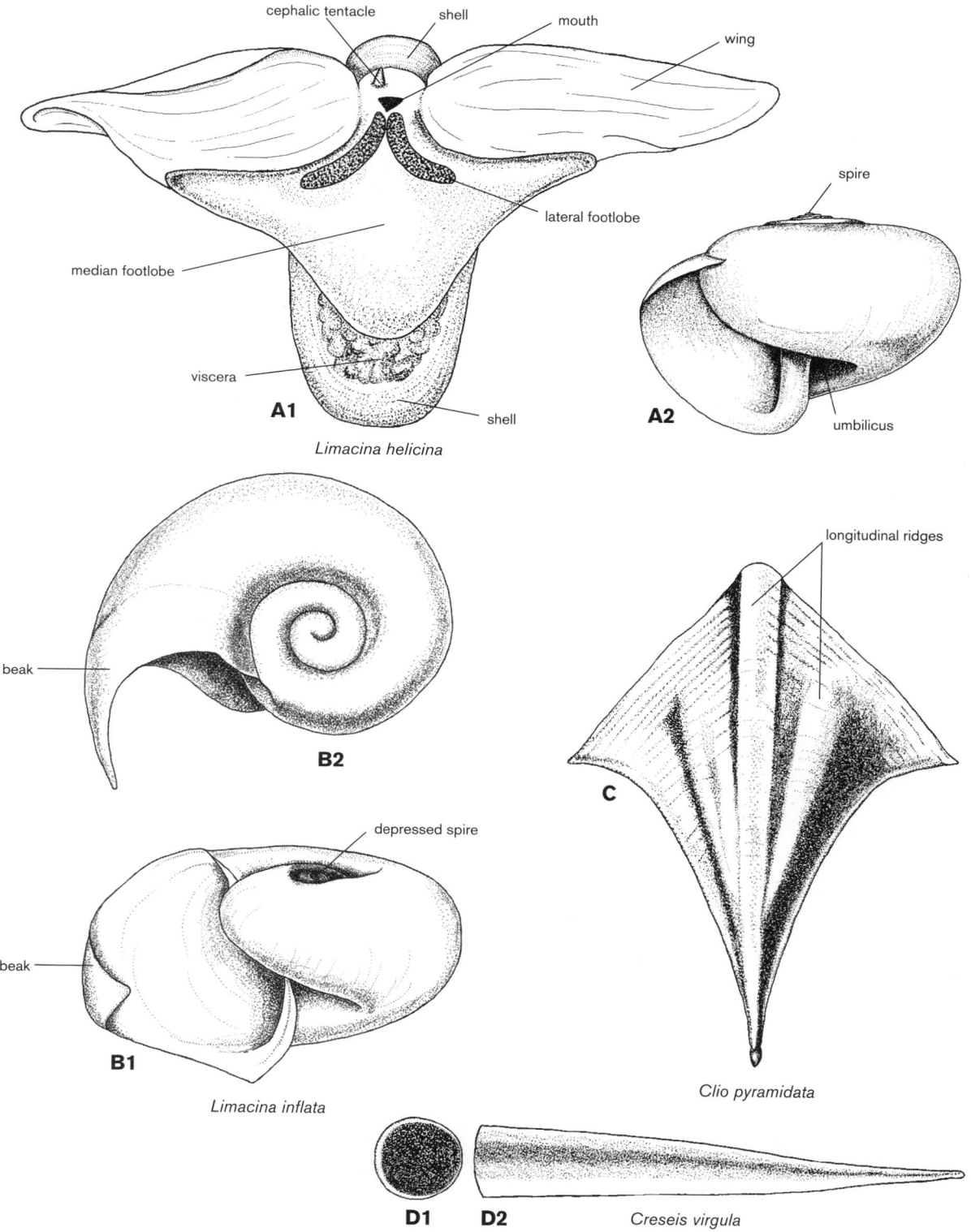

cephalic tentacle
shell
mouth
wing
lateral footlobe
median footlobe
viscera
A1
shell
Limacina helicina

spire
A2
umbilicus

beak
B2

longitudinal ridges
C

depressed spire
beak
B1
Limacina inflata

Clio pyramidata

D1 **D2** *Creseis virgula*

PLATE 382 A, *Limacina helicina*, A1, ventral view of animal, A2, shell from aperture; B, *Limacina inflata*, B1, shell from aperture, B2, shell from left side; C, *Clio pyramidata*, dorsal view of shell; D, *Creseis virgula*, D1, shell from aperture, D2, shell from side.

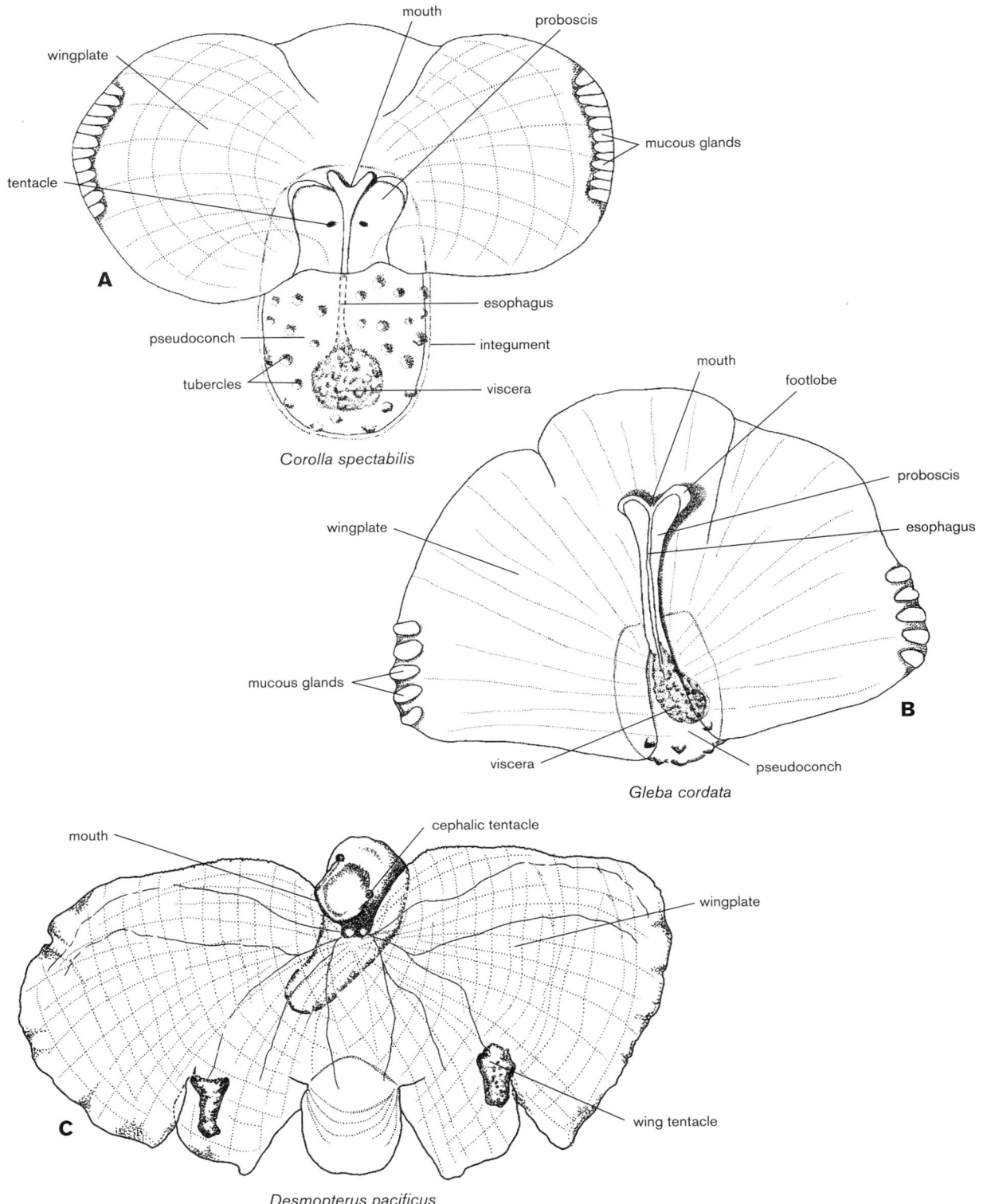

A

Corolla spectabilis

wingplate
mouth
proboscis
mucous glands
tentacle
esophagus
pseudoconch
integument
tubercles
viscera

B

Gleba cordata

mouth
footlobe
proboscis
esophagus
wingplate
mucous glands
viscera
pseudoconch

C

Desmopterus pacificus

mouth
cephalic tentacle
wingplate
wing tentacle

PLATE 383 A, *Corolla spectabilis*, ventral view of animal; B, *Gleba cordata*, ventral view of animal; C, *Desmopterus pacificus*, ventral view of animal.

Clioinae

Clio cuspidata (Bosc, 1802). Uncommon off southern California (Cummings and Seapy 2003). Warm-water species.

Clio pyramidata Linnaeus, 1767. Color photograph in Wrobel and Mills (1998: fig. 131). Feeding with a mucous web described by Gilmer and Harbison (1986). Relatively common in temperate, oceanic waters; migrates vertically between mesopelagic and epipelagic depths. Moderately abundant off southern California throughout the year (Cummings and Seapy 2003). McGowan (1968) reported several varieties present throughout the California Current. Only one variety (*C. pyramidata lanceolata*) was identified by Cummings and Seapy from southern California waters.

Clio recurva (Children, 1823) (=*C. balantium* Pelseneer, 1888). Warm-water species.

Creseis acicula (Rang, 1828). Warm-water species.

Creseis virgula (Rang, 1828). Color photograph in Wrobel and Mills (1998: fig. 133). Feeding observations by Gilmer and Harbison (1986); embryonic development described by Bandel and Hemleben (1995). Abundant in southern California waters, with a seasonal maximum in the summer (Cummings and Seapy 2003). Three varieties in the California Current (McGowan 1968), although only one (*C. virgula virgula*) was identified off southern California by Cummings and Seapy. Cosmopolitan warm-water species.

Hyalocylis striata (Rang, 1828). Warm-water species.

Styliola subula (Quoy and Gaimard, 1827). Warm-water species.

Cuvierininae

Cuvierina columnella (Rang, 1827). Rare off southern California (Cummings and Seapy 2003). Warm-water species.

CYMBULIOIDEA (=PSEUDOTHECOSOMATA)

Peraclididae

Peraclis apicifulva Meisenheimer, 1906. Meso- or bathypelagic species.

Peraclis bispinosa Pelseneer, 1888. Meso- or bathypelagic species. Identification of California species in doubt (van der Spoel, 1976).

Peraclis reticulata (d'Orbigny, 1836). Color photograph in Lalli and Gilmer (1989: color fig. 9). Mesopelagic species.

Cymbuliidae

Corolla spectabilis Dall, 1871. Color photograph in Wrobel and Mills (1998: fig. 134). Feeding on plankton described by Gilmer (1972). Relatively common in the California Current. Reported in the eastern Pacific between 20°N and 45°N. Pseudoconchs occasionally found washed up on beaches.

Gleba cordata Niebuhr, 1776. Color photograph in Wrobel and Mills (1998: fig. 135). Feeding described by Gilmer (1972). A warm-water species (van der Spoel 1976) reported to range northward into central California waters (Wrobel and Mills 1998).

For a comparison of the cymbuliid genera, see Rampal (1996, Beaufortia 46: 1–10).

Desmopteridae

Desmopterus pacificus Essenberg, 1919. Relatively common in the California Current (McGowan 1967, 1968) and off southern California (Cummings and Seapy, unpublished). Biology unknown.

Desmopterus papilio Chun 1889. Distributed in tropical/subtropical waters throughout the Atlantic and Indian Oceans and penetrating into the western Pacific (van der Spoel 1976). First reported in the eastern North Pacific off central California by Wrobel and Mills (1998). Biology unknown.

GYMNOSOMATA

The gymnosomes are fast-swimming carnivores. Like the thecosomes, they have swimming wings derived from the molluscan foot, and thus both groups are referred to as "pteropods." In other respects, the two groups are very different anatomically and biologically, and they are not close phylogenetic relatives.

All adult gymnosomes lack a shell; the name refers to their naked (*gymno-*) body (*-some*). They also lack a mantle and mantle cavity. The body is either streamlined or baglike in shape. In Pacific waters, adult body lengths range from <5 mm to about 30 mm. The **PAIRED SWIMMING WINGS** (plates 384, 385A, 385B) are the most conspicuous distinguishing feature. The body usually consists of a well-defined **HEAD** with two pairs of tentacles; the larger **ANTERIOR TENTACLES** are located at the tip of the head, and the inconspicuous **POSTERIOR TENTACLES** are set in depressions on the dorsal surface. The remnant of the molluscan foot is found ventrally between the wings and usually consists of two **LATERAL FOOTLOBES** and a **MEDIAN, POSTERIOR FOOTLOBE** (plate 384A1, 384B). The integument may be transparent or colored from the presence of chromatophores in some species. **GILLS**, if present, are located laterally in the midbody region and/or at the posterior tip of the body (plate 384B); they range from simple, thin, unpigmented areas to fringed structures.

Gymnosomes are feeding specialists, usually eating only specific thecosome species (see review by Lalli and Gilmer 1989). This specificity has been accompanied by an elaborate development of unique structures in the feeding apparatus, or **BUCCAL ORGANS,** and identification of gymnosome species depends largely on anatomical differences in these structures. Many species have special eversible tentacles used to capture prey; these include **BUCCAL CONES** (=**CEPHALOCONI**) with a papillate surface (plate 384A1, 384A2), or **SUCKER-BEARING ARMS** (=**ACETABULIFEROUS ARMS**) with either stalked or sessile suckers. Some species have a large eversible **PROBOSCIS** used to extract thecosome prey from their shells. Most species have paired **HOOK SACS** (plates 384A1, 385A), which are unique to this group; the sacs are located on either side of the **RADULA** and contain a variable number of chitinous hooks used to extract prey from shells. Some gymnosomes also have a spinous **JAW.**

Gymnosomes are simultaneous hermaphrodites that lay floating egg masses. Veliger larvae with simple uncoiled shells hatch within a few days. They later metamorphose, losing the shell and becoming **POLYTROCHOUS LARVAE** (plate 384A3) characterized by three bands of cilia encircling the head, mid-body, and posterior tip. The final metamorphosis to the adult form occurs slowly, with development of the wings, concomitant loss of ciliary bands, and elongation of the body.

Gymnosomes are rarely found in nearshore plankton tows, and they are very poorly known in the geographic area under consideration. The great majority of specimens have been captured in plankton nets and preserved without prior relaxation in an anesthetic, resulting in highly contracted specimens in which almost all of the distinguishing anatomical features are hidden. These shell-less opisthobranchs are notoriously difficult to identify because of their small size and because many of the unique morphological structures are normally retracted within the head of non-feeding animals or are almost always retracted in preserved specimens. Identification thus requires the use of special relaxation methods with living animals, or dissections of preserved specimens.

When working with living gymnosomes, eversion of buccal structures can sometimes be achieved by the gradual addition of small quantities of any of the following chemicals to a small amount of seawater containing the animals: urethane crystals, menthol crystals, MS222, or dilute magnesium chloride or magnesium sulfate solution. Once an animal has everted its feeding apparatus and no longer responds to gentle probing, it can be preserved in a 5%–10% formalin solution. If working with preserved, highly contracted specimens, it may be preferable to render the animal transparent before attempting dissection. This can be done by transferring the specimen to 70% alcohol and then moving it to 95% and 100% alcohol for several hours in each solution until all water is removed. It can then be cleared by immersion in an oil, such as oil of wintergreen or oil of cloves. This technique will usually reveal the presence and location of suckers, hooks, and radula when the animal is viewed under a dissecting microscope. Very small individuals that have been cleared in oil can also be mounted on a slide, directly in a mounting medium such as Permount or balsam.

KEY TO GYMNOSOMES

The following key has been designed to include genera that have been reported from the northern and central sections of the California Current (McGowan 1968) or from waters off northern California and Oregon. Those few species that have been identified and that may be present in inshore surface waters are described in more detail. New or previously unrecorded gymnosomes will not key out correctly; it is suggested that problematic species be checked against complete descriptions in van der Spoel (1976) and van der Spoel et al. (1997).

1. Buccal cones present (plate 384A, 384A2); no posterior or lateral gills 2
— Sucker-bearing arms present; posterior gill, lateral gill, or both present (plate 384B) 3
— Buccal cones and sucker-bearing arms absent 4
2. Two pairs of buccal cones . . . . Paraclione

Note: Paraclione (=Clionina) longicaudata. Small, slender body pointed posteriorly; length <10 mm. Chromatophores on body and wings. Small shallow hook sacs with <16 hooks per sac.

— Three pairs of buccal cones . . . . Clione

Note: Clione limacina (plate 384A). Slender, streamlined body tapering posteriorly; maximum body length about 30 mm in California Current region. Red pigmentation on buccal cones and posterior body. Viscera not extending to posterior end in adults. Transparent integument; no chromatophores; no gills. Large hook sacs, each containing approximately 30 hooks in adults.

3. Two short lateral and one median sucker-bearing arms (the laterals may be lacking if suckers are attached directly to the buccal cavity); small hook sacs; lateral gill generally present, posterior gill mostly absent . . . .
. . . . Pneumodermopsis
— Two well-developed lateral sucker-bearing arms, but median arm usually lacking or greatly reduced; long, voluminous hook sacs; lateral and posterior gills both present (plate 384B) and usually fringed . . . . Pneumoderma

Note: Pneumoderma atlanticum (Oken, 1815) pacificum (Dall, 1871) (plate 384B). Elongate body, with clearly defined head; body length to about 25 mm. Integument contains chromatophores that pro-

duce a purplish-brown coloration in living animals. Lateral arms with about 50 small suckers each. Large footlobes; median lobe long and pointed. Small lateral gill; posterior gill has four radiating crests with simple fringes.

4. Long retractile proboscis present; head small relative to body (plate 385B) 5
— Proboscis absent; large head occupying at least half of the total body length . . . . Thliptodon

Note: Thliptodon sp. (plate 385A). Body ovate to cylindrical; body length to ca. 25 mm. The expanded head contains a pair of gullet bladders (unique to this genus) and large hook sacs. Wings long, narrow at base and wide distally. Transparent integument; no gills. No buccal cones or sucker-bearing arms. Footlobes small.

5. Extremely long proboscis, extending twice as long as the body when evaginated; no lateral gill . . . . Cliopsis

Note: Cliopsis krohni (plate 385B). Body rounded, not streamlined; length to 40 mm. Head small relative to body; small anterior tentacles. Hexagonal posterior gill in adults only; no lateral gill. Lateral footlobes in the shape of a horseshoe; median footlobe greatly reduced or absent. Shallow hook sacs and radula present.

— Lateral gill present, encircling all or most of the mid-body . . . . Pruvotella

LIST OF SPECIES

CLIONOIDEA

CLIONIDAE

Clioninae
Clione limacina (Phipps, 1774). Color photographs in Lalli and Gilmer (1989: color fig. 12), Gilmer and Lalli (1990, Amer. Malacol. Bull. 8: 67–75; fig. 1), and Wrobel and Mills (1998: fig. 140). The most abundant species of gymnosome in the northern hemisphere, widely distributed throughout Arctic, Subarctic, and northern temperate waters of the Pacific and Atlantic oceans. Preys exclusively on the thecosomes Limacina helicina (in the subarctic and cold temperate North Pacific) and Limacina retroversa (in the temperate North Atlantic) (Lalli 1970, J. Exp. Mar. Biol. Ecol. 4: 101–118; Conover and Lalli 1972, J. Exp. Mar. Biol. Ecol. 9: 279–302). Reproduction and development reviewed in Lalli and Gilmer (1989).
Paraclione (=Clionina) longicaudata (Souleyet, 1852). Uncommon in area; extends northward into the Gulf of Alaska (van der Spoel 1987, Biol. Oceanogr. 5: 29–42). Biology unknown.

Thliptodontinae
Thliptodon sp. The color photograph in Wrobel and Mills (1998: fig. 141) of Thliptodon sp. is of an unidentified species, described as rare and poorly known; recorded from both deep and surface waters off California. In the eastern Pacific, T. diaphanus was reported by van der Spoel (1987, Biol. Ocean. 5: 29–42) to be abundant in the Gulf of Alaska.

Cliopsidae
Cliopsis krohni Troschel, 1854. Color photograph in Wrobel and Mills (1998: fig. 139). This species feeds on the pseudothecosome Corolla. It is widely distributed between 50°N and 50°S in all oceans.
Pruvotella sp. Uncommon in area.

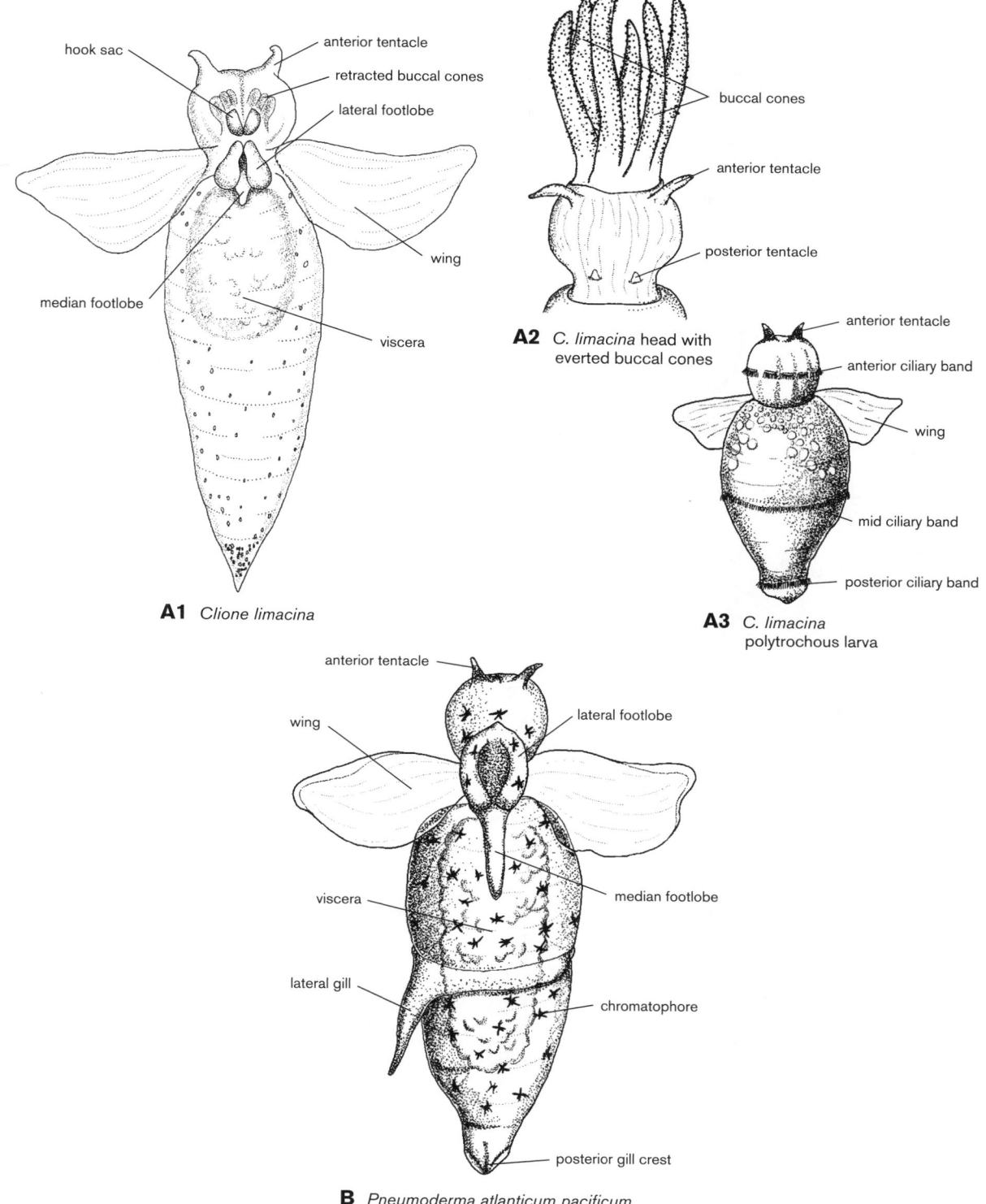

A1 *Clione limacina*

hook sac

anterior tentacle

retracted buccal cones

lateral footlobe

median footlobe

wing

viscera

buccal cones

anterior tentacle

posterior tentacle

A2 *C. limacina* head with everted buccal cones

anterior tentacle

anterior ciliary band

wing

mid ciliary band

posterior ciliary band

A3 *C. limacina* polytrochous larva

anterior tentacle

wing

lateral footlobe

viscera

median footlobe

lateral gill

chromatophore

posterior gill crest

B *Pneumoderma atlanticum pacificum*

PLATE 384 A, *Clione limacina*, A1, ventral view of animal, A2, view of head with buccal cones everted, A3, polytrochous larva with three ciliary bands; B, *Pneumoderma atlanticum pacificum*, ventral view of animal.

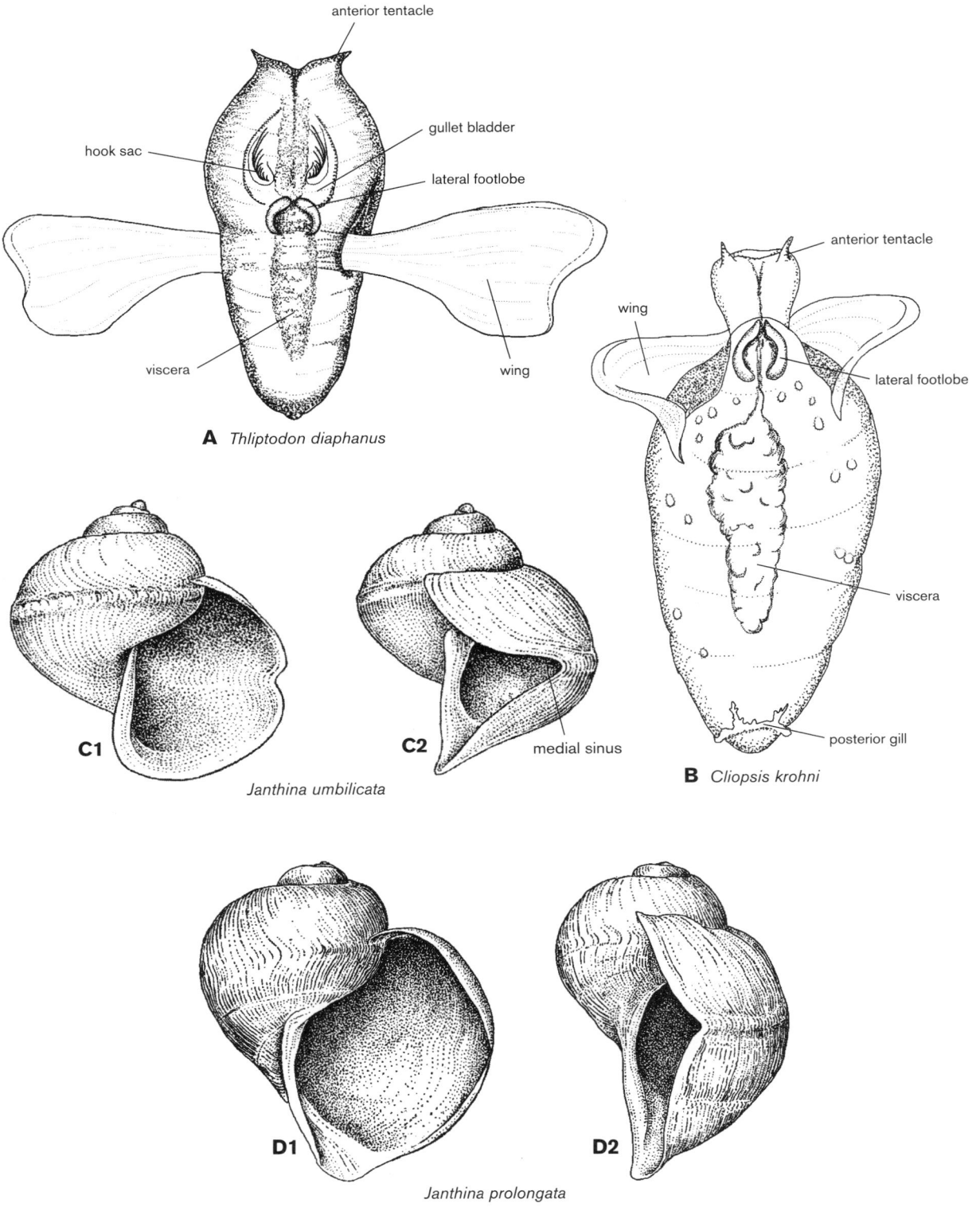

anterior tentacle

gullet bladder

hook sac

lateral footlobe

viscera

wing

A *Thliptodon diaphanus*

anterior tentacle

wing

lateral footlobe

viscera

posterior gill

B *Cliopsis krohni*

C1

C2

medial sinus

Janthina umbilicata

D1

D2

Janthina prolongata

PLATE 385 A, *Thliptodon diaphanus*; B, *Cliopsis krohni*, ventral view of animal; C, *Janthina umbilicata*: side view of shell oriented (C1), with the aperture outward and (C2), with the apertural lip outward; D, *Janthina prolongata*: side view of shell oriented (D1) with the aperture outward and (D2) with the apertural lip outward.

Pneumodermatidae

Pneumoderma atlanticum (Oken, 1815) *pacificum* (Dall, 1871). Color photograph in Wrobel and Mills (1998: fig. 137). Recorded off the west coast between 35°N and 45°N. Biology unknown.

Pneumodermopsis sp. Eastern Pacific representatives of this genus have not been identified to species. The color photograph in Wrobel and Mills (1998: fig. 138) is of an Atlantic species that is not known to occur off California.

JANTHINIDAE

Among the pelagic snails, members of the ptenoglossate family Janthinidae are least modified for an open ocean existence. They have a **DEXTRALLY COILED SHELL** and a flat, **SOLELIKE FOOT** used to produce a **BUBBLE RAFT** that floats on the ocean surface. The animals are normally oriented in an upside-down position with the foot attached to the underside of the float. To build or enlarge the float, the foot is extended above the water surface and a bubble of air is trapped and encased in mucus (secreted by the foot). The mucus dries and each air bubble is added to the float, which attains a size of 13 cm long by 2 cm wide in *Janthina prolongata* (Laursen, 1953). Janthinids are not able to swim; if they become detached from their float or the prey upon which they are feeding, they will sink. They can be carried great distances by surface currents and are occasionally blown onshore where they may be collected by beachcombers.

Despite the retention of a basic ptenoglossate morphology, the janthinids exhibit a number of morphological and reproductive adaptations to a pleustonic life style (reviewed in Lalli and Gilmer 1989). The wall thickness of the shell is much reduced, resulting in a very light shell. The radula is modified by loss of the central tooth and elongation of the lateral teeth, which are hook-shaped terminally to facilitate feeding on the tissues of their pelagic cnidarian prey, e.g., *Velella*, *Porpita*, and *Physalia*. Eyes, statocysts and an operculum are lacking in the adults.

The body and shell in *Janthina* are violet to blue in color, which approximates that of the surface waters and presumably serves as camouflage against potential predators both from above and below the ocean surface. Last, the reproductive biology of the janthinids shows substantial modifications from their ptenoglossate relatives. They have been characterized variously as hermaphroditic (with individuals containing both male and female systems) and protandrous (with small individuals being males and large ones females). Both males and hermaphrodites lack a copulatory organ. Instead, they have two different types of sperm; small eupyrene ones (that fertilize eggs) and large, oligopyrene ones (that function to transport the eupyrene sperm to the female gonopore).

Only one species (*J. janthina*) releases free-swimming veliger larvae directly into the sea. The remaining species produce egg capsules, which they attach to the underside of the bubble raft and from which veliger larvae are hatched. The veligers are planktotrophic and possess eyespots, statoliths, and an operculum; however, these structures are lost at metamorphosis.

The family Janthinidae consists of two genera, *Janthina* and *Recluzia*. The former genus includes five species (Laursen 1953). A total of 11 species belonging to the genus *Recluzia* were reported in the review by van der Spoel et al. (1997), eight of which were considered dubious, leaving perhaps three valid species. In addition to the uncertainty of the species composition of *Recluzia*, very little is known about the biology of these animals (discussed in Lalli and Gilmer 1989). The janthinids found most commonly in West Coast waters (Seapy, unpub-lished) are *Janthina umbilicata* and *J. prolongata* (the latter species is a distant second in abundance to the former one). The majority of records for these two species are from southern California waters, although the ranges of both extend well northward into Oregon and Washington waters. Two other species, *J. exigua* and *J. janthina*, have been collected only in limited numbers from southern California (Seapy, unpublished) and are not included here.

The interested reader is referred to the monograph on *Janthina* by Laursen (1953); it includes a review of their biology and resolves the previous taxonomic chaos in the genus, reducing the number of valid species from 60 to five. The biology and ecology of the group are discussed by Lalli and Gilmer (1989).

KEY TO *JANTHINA*

1. Shell small (maximal height = 9 mm); apertural lip with a deep, V-shaped medial sinus; surface of shell finely striated, with striae on each whorl curving inward strongly and converging at the midline of each whorl (plate 385C) . . . . *Janthina umbilicata*
— Shell large (maximal height = 39 mm); apertural lip with a very shallow medial sinus; surface of shell irregularly striated, with striae on each whorl curving inward weakly at the midline of the whorl (plate 385D). . . . . *Janthina prolongata*

LIST OF SPECIES

EPITONIOIDEA

Janthinidae

Janthina prolongata Blainville, 1822. Color photographs in Lalli and Gilmer (1989: color fig. 1), Nybakken and Bertness (2004, Marine biology: an ecological approach, sixth ed. San Francisco: Benjamin Cummings Pearson: fig. 2.20) Wrobel and Mills (1998: fig. 124). *Velella* and *Porpita* prey detected at distances up to 10 cm away (Bieri 1966, Publ. Seto Mar. Biol. Lab. 14: 161–170). Warm-water species.

Janthina umbilicata d'Orbigny, 1840. Egg capsules and lateral radular teeth figured in Laursen (1953). Warm-water species.

REFERENCES

Bandel, K. and C. Hemleben. 1995. Observations on the ontogeny of thecosomatous pteropods (holoplanktic Gastropoda) in the southern Red Sea and from Bermuda. Mar. Biol. 124: 225–243.

Bé, A. W. H., and R. W. Gilmer. 1977. A zoogeographic and taxonomic review of euthecosomatous Pteropoda, pp. 733-808. In: A. T. S. Ramsay (ed.), Oceanic micropaleontology, Vol. I, London: Academic Press.

Behrens, D. W., and A. Hermosillo. 2005. Eastern Pacific nudibranchs. Sea Challengers: Monterey, California. 137 pp.

Cummings, F. A., and R. R. Seapy. 2003. Seasonal abundances of euthecosomatous pteropods and heteropods from waters overlying San Pedro Basin, California. Veliger 46: 305–313.

Gilmer, R. W. 1972. Free-floating mucus webs: a novel feeding adaptation for the open ocean. Science 176: 1239–1240.

Gilmer, R. W. 1990. In situ observations of feeding behavior of thecosome pteropod molluscs. Amer. Malacol. Bull. 8: 53–59.

Gilmer, R. W., and G. R. Harbison. 1986. Morphology and field behavior of pteropod molluscs: feeding methods in the families Cavoliniidae, Limaciniae and Peraclididae (Gastropoda: Thecosomata). Mar. Biol. 91: 47–57.

Goddard, J. H. R. 2001. Mollusca: Gastropoda, pp. 86–128. In: A. L. Shanks (ed.), An Identification guide to the larval marine invertebrates of the

Pacific Northwest. Oregon State University Press: Corvallis, Oregon, 314 pp.

Holleman, J. J. 1972. Observations on growth, feeding, reproduction and development in the opisthobranch, *Fiona pinnata* (Eschscholtz). Veliger 15: 142–146.

Lalli, C. M., and R. W. Gilmer. 1989. Pelagic snails: the biology of holoplanktonic gastropod mollusks. Stanford, California: Stanford University Press, 259 pp.

Laursen, D. 1953. The genus *Ianthina*: a monograph. Dana Rept. 38: 1–40.

MacFarland, F. M. 1966. Studies of opisthobranchiate mollusks of the Pacific Coast of North America. Memoirs of the California Academy of Sciences, Vol. VI. San Francisco, California: California Academy of Sciences, 546 pp.

Marcus, E. 1961. Opisthobranch mollusks from California. Veliger 3 (Suppl.): 1–85.

McGowan, J. A. 1967. Distributional atlas of pelagic molluscs in the California Current region. Calif. Coop. Ocean. Fish. Invest. Atlas 6, 218 pp.

McGowan, J. A. 1968. The Thecosomata and Gymnosomata of California. Veliger 3 (Suppl.): 103–129.

McGowan, J. A. 1971. Oceanographic biogeography of the Pacific, pp. 3–74. In The micropaleontology of oceans. B. M. Funnell and W. R. Riedel, eds. Cambridge: University Press.

Pafort-van Iersel, T. 1983. Distribution and variation of Carinariidae and Pterotracheidae (Heteropoda, Gastropoda) of the Amsterdam Mid North Atlantic Plankton Expedition 1980. Beaufortia 33: 73–96.

Richter, G. 1974. Die Heteropoden der "Meteor" -Expedition in den Indischen Ozean, 1964/65. "Meteor" Forschungsergebnisse (D) 17: 55–78

Richter, G., and R. R. Seapy. 1999. Heteropoda, pp. 621–647. In Zooplankton of the South Atlantic Ocean. D. Boltovskoy, ed. Leiden: Backhuys.

Seapy, R. R. 1990. Patterns of vertical distribution in epipelagic heteropod molluscs off Hawaii. Mar. Ecol. Prog. Ser. 60: 235–246.

Seapy, R. R., and G. Richter. 1993. *Atlanta californiensis*, a new species of atlantid heteropod (Mollusca: Gastropoda) from the California Current. Veliger 36: 389–398.

Spoel, S. van der. 1967. Euthecosomata, a Group with Remarkable Developmental Stages (Gastropoda, Pteropoda). Gorinchem: J. Noordiujn, 375 pp.

Spoel, S. van der. 1976. Pseudothecosomata, Gymnosomata and Heteropoda (Gastropoda). Utrecht: Bohn, Scheltema, and Holkema, 484 pp.

Spoel, S. van der, L. Newman, and K. W. Estep. 1997. Pelagic molluscs of the world. World Biodiversity Database, CD-ROM Series. Expert Center for Taxonomic Identification (ETI), Amsterdam and Paris: UNESCO.

Wormuth, J. H. 1981. Vertical distributions and diel migrations of Euthecosomata in the northwest Sargasso Sea. Deep-Sea Res. 28: 1493–1515.

Wrobel, D., and C. Mills. 1998. (Revised Printing, 2003). Pacific Coast pelagic invertebrates. A guide to the common gelatinous animals. Monterey: Sea Challengers and Monterey Bay Aquarium, 112 pp.

Opisthobranchia Clades and Onchidiacea

GARY C. WILLIAMS

(Plate 386)

Since the mid-1970s, a plethora of interest by various authors has resulted in substantially revised views concerning our concepts of gastropod (including opisthobranch) phylogeny (Haszprunar 1988, Bieler 1992, Mikkelsen 1996, Valdés and Camacho-García 2004). This has resulted both from detailed comparisons of comparative anatomy and morphology and from phylogenetic methodology (cladistics) and molecular systematics. Under these revised concepts, the traditional taxon Cephalaspidea is considered paraphyletic: it does not represent a natural grouping, but rather an assemblage that does not contain all the descendents of a common ancestor. As a consequence, three families (Acteonidae, Bullidae, and Ringiculidae) have been

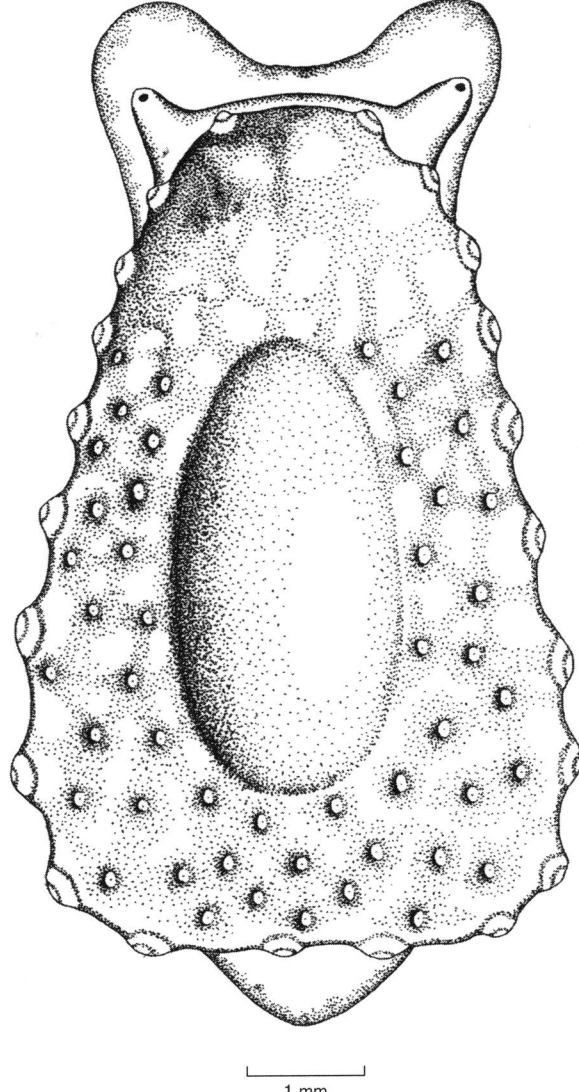

1 mm

PLATE 386 *Onchidella borealis* (Drawing by Gary C. Williams).

removed from the traditional grouping, but disagreement still exists concerning the evolutionary relationships and nomenclature of these and other taxa. Despite these revisions and the present state of controversy, the name "Cephalaspidea" is retained here as one of convenience due to superficial similarities in morphology and life habits and to reflect some degree of historical consistency.

The following sections provide keys to Cephalaspidea and Acteonacea, Anaspidea, Notaspidea, and Nudibranchia and Sacoglossa. Also provided is an introduction to the egg masses and reproduction of some of our opisthobranchs.

CEPHALASPIDEANS AND ACTEONACEANS

Included here are highly variable taxa—some have conspicuous external shells, others have reduced internal shells, and others have no shell at all. Species with conspicuous external shells are also treated in the section on Shelled Gastropods.

NOTASPIDEA

A limpetlike shell is found in one common notaspidean in central California and Oregon (*Tylodina fungina*), but this species differs from limpets in having the single gill restricted to the right half of the body. Most other notaspideans do not have external shells.

ANASPIDEA

These include the well-known herbivorous "sea hares," *Aplysia,* famous for medical and neurological research and for being the world's largest snails. Common locally is a small green-striped anaspidean, *Phyllaplysia,* that lives on eelgrass blades.

NUDIBRANCHIA AND SACOGLOSSA

Included here are the "sea slugs" familiar to most tide pool visitors. Nudibranchs are largely (with some exceptions) predators on sea anemones and colonial and encrusting animals, whereas sacoglossans are largely herbivores.

A great deal of knowledge regarding the nudibranchs in particular has accumulated in the past three decades. See David Behrens and Alicia Hermosillo (2005, *Eastern Pacific Nudibranchs,* Sea Challengers Publishers) for a field guide to nudibranchs and other opisthobranchs. For information on opisthobranch natural history, see the chapter by Robert Beeman and Gary Williams (1980) in *Intertidal Invertebrates of California* (edited by R. H. Morris, D. P. Abbott, and E. C. Haderlis, Stanford University Press). Predator/prey interactions of nudibranchs with other organisms is covered by Gary McDonald and James Nybakken (1978, Veliger 21: 110–118) and Jeffrey Goddard (1998, Opisthobranch Newsletter 24: 11–14). It is best to observe nudibranchs while alive because color and shape are profoundly altered or lost in preservation. High-quality color photography of living material is the best way to record color and body shape. See the section on Methods of Preservation at the beginning of this manual for recommendations about how material should be handled for later study. Specimens for molecular analysis should be preserved in ethanol.

SLUG LOOK-ALIKES

ONCHIDIACEANS

The geographically widespread family Onchidiidae is represented in California and Oregon by a single species, *Onchidella borealis* Dall, 1871 (plate 386). It ranges from Alaska to at least as far south as San Luis Obispo County. The taxonomic affinities of onchidiaceans have long been disputed—regarded as opisthobranchs by some authors but here considered as marine pulmonates. *Onchidella* has a lung in the posterior part of the body for aerial respiration; gills are absent. Our species is commonly encountered in the upper- to middle-rocky intertidal of northern California, in mostly exposed situations. It is known to feed on diatoms that are epiphytic on the alga *Odonthalia floccosa* and perhaps also on the surface cells of the alga. Body color is variable, from uniform red-brown to yellow with red-brown blotches. The foot, oral lappets, and eye stalks are white; the body margin bears 20–24 papillae. Most individuals are <10 mm in length. See Young et al. 1986, Biol. Bull. 171: 391–404 (defensive secretions).

LAMELLARIANS

Members of the velutinid genera *Marsenina, Marseniopsis,* and *Lamellaria* are superficially opisthobranchlike and may remind one of some dorids. They are usually found on (and eating) encrusting ascidians, which they often mimic remarkably in color. These snails have flattened spiral shells that are reduced and completely or partly hidden and an anterior siphonal opening. They are treated in the previous section on shelled gastropods.

CHLAMYDOCONCHA

At first glance this "naked clam" looks like a small white sea slug. It is a galeommatid with a rudimentary, internal shell. The name *Chlamys* means "cloak." *Chlamydoconcha orcutti* is keyed out in the bivalve section.

REFERENCES

Bieler, R. 1992. Gastropod phylogeny and systematics. Annual Review of Ecology and Systematics 23: 311–338.

Haszprunar, G. 1988. On the origin and evolution of major gastropod groups, with special reference to the Streptoneura. Journal of Molluscan Studies 54: 367–441.

Mikkelsen, P. M., 1996. The evolutionary relationships of Cephalaspidea *s.l.* (Gastropoda: Opisthobranchia): a phylogenetic analysis. Malacologia 37: 375–442.

Valdés, Á., and Y. Camacho-García. 2004. "Cephalaspidean" heterobranchs (Gastropoda) from the Pacific Coast of Costa Rica. Proceedings of the California Academy of Sciences 55 (26): 459–497.

Reproduction and Egg Masses of Benthic Opisthobranchs

JEFFREY H. R. GODDARD

(Plates 387)

Most opisthobranchs are simultaneous hermaphrodites that copulate reciprocally or unilaterally using the penis and common genital opening on the right side of the body. A few nudibranchs and sacoglossans copulate by hypodermic injection through the body wall. Copulation may last from seconds to hours and in some species (e.g., *Hermissenda crassicornis*) is preceded by a sequence of distinct precopulatory behaviors. Mating chains of unilaterally copulating individuals are frequently formed among the Aplysiidae and also in some notaspideans and cephalaspideans. In most opisthobranchs, however, copulation is between two individuals only. Spermatozoa obtained from mating partners may be stored for days or weeks, and with few exceptions, fertilization does not occur until ova and sperm are mixed during egg-mass deposition.

Benthic opisthobranchs lay gelatinous egg masses in the form of strings, ribbons, and sacs (plate 387), often on or near their prey. Variations on these three basic forms are numerous, and a few are species specific. For example, the cephalaspidean *Navanax inermis* lays its eggs strings in loose gelatinous skeins, while *Phyllaplysia taylori* forms compact rectangular packets on the leaves of the eelgrass *Zostera marina*. The infaunal cephalaspidean *Rictaxis punctocaelatus* and the aeolid nudibranch *Cumanotus* sp. lay distinctive helical coils (plate 387B, right) anchored in soft sediments or attached to the hydroid *Pinauay crocea,* respectively.

Other distinctive nudibranch egg masses include the tightly coiled, inward slanting, white ribbons of *Cadlina marginata*, the small tight coils laid by *Rostanga pulchra* on red-orange sponges, and the salmon-colored, tangled strings laid by *Flabellina iodinea* on the hydroid *Eudendrium ramosum*. *Crimora coneja*, *Conualevia alba*, *Limacia cockerelli*, and *Okenia rosacea* all lay spirally-wound, pink or white ribbons flat against the substratum, rather than on edge like most dorids. In the absence of spawning adults, however, the egg masses of many opisthobranchs are difficult to identify to species without knowledge of the specific habitat and prey requirements of the adults.

The jelly matrix and mucous layers surrounding opisthobranch eggs are transparent and colorless, and the egg masses are the color of the yolk in the eggs. As a result, egg-mass colors fade with time as this yolk is transformed into tissues and organs during embryonic development. Egg masses can also be tainted by the color of surface-fouling microorganisms. Some nudibranch egg masses are chemically defended from fish and crustacean predators (Pawlik et al. 1988), and those of the dorid nudibranch *Doris montereyensis* are preferentially deposited in shaded microhabitats, providing protection from the lethal effects of solar radiation and surface-fouling microalgae (Biermann et al. 1992).

Within an egg mass, eggs are encapsulated singly or in groups, depending on the species. Nurse eggs are unknown. Development time from fertilization to hatching varies with temperature, mode of development, and egg size, but for most species in the region takes between 10 and 20 days at typical temperatures of 11°C to 14°C (Goddard 1984; Strathmann 1987). Owing to the transparency of the egg coverings, embryonic development is easily observed by examining small pieces of egg masses under a compound microscope.

Most northeast Pacific opisthobranchs hatch as free-swimming planktotrophic veliger larvae with an operculum, a clear unsculptured shell, and a small foot lacking a propodium (Goddard 2001, 2004). These larvae feed and grow in the plankton for weeks or months and become competent to settle and metamorphose only after they have acquired eyespots, a propodium, and sufficient tissue mass and lipid reserves to fuel the transformation into a functional benthic juvenile. Settlement and metamorphosis in many species is triggered by chemical cues emanating from the prey or habitat of the adults.

Doto amyra and *Phidiana hiltoni* develop from larger eggs into lecithotrophic veligers that do not have to feed in the plankton and are competent to metamorphose within a few days of hatching (Goddard 2004). Four more species (*Acteocina inculta*, *Runcina macfarlandi*, *Phyllaplysia taylori* and *Dendrodoris behrensi*) lack a larval stage entirely and hatch as crawl-away juveniles (Goddard 2004, 2005). Two estuarine species, the sacoglossan *Alderia willowi*, which occurs from Bodega Bay south, and the introduced aeolid *Tenellia adspersa*, each produce two sizes of eggs and exhibit variable developmental mode, a rare reproductive polymorphism known as poecilogony (Eyster 1979; Krug 1998; Goddard 2004; Ellingson and Krug 2006; Krug et al. 2007). Additionally, both the introduced cephalaspidean *Haminoea japonica* and the lecithotrophic morphs of *Alderia willowi* exhibit a dispersal dimorphism in which some embryos metamorphose prior to hatching (Gibson and Chia 1989; Krug et al. 2007).

Most of the above native species with non-feeding modes of development occur either (1) largely south of Point Conception, where surface waters are warmer, lower in nutrients and less productive than those to the north, or (2) are known from habitats, like estuaries, that are small and sparsely distributed on the Pacific coast of North America (Goddard 2004).

REFERENCES

Behrens, D. W. and A. Hermosillo. 2005. Eastern Pacific nudibranchs. Sea Challengers: Monterey, California. 137 pp.

Biermann, C. H., G.O. Schinner, and R.R. Strathmann. 1992. Influence of solar radiation, microalgal fouling, and current on deposition site and survival of embryos of a dorid nudibranch gastropod. Marine Ecology Progress Series 86: 205–215.

Ellingson, R. A. and P. J. Krug. 2006. Evolution of poecilogony from planktotrophy: cryptic speciation, phylogeography, and larval development in the gastropod genus *Alderia*. Evolution 60: 2293–2310.

Eyster, L. S. 1979. Reproduction and developmental variability in the opisthobranch *Tenellia pallida*. Marine Biology 51: 133–140.

Gibson, G., and F.-S. Chia. 1989. Developmental variability (pelagic and benthic) in *Haminoea callidegenita* (Opisthobranchia: Cephalaspidea) is influenced by egg mass jelly. Biological Bulletin 176: 103–110.

Goddard, J. H. R. 1984. The opisthobranchs of Cape Arago, Oregon, with notes on their biology and a summary of benthic opisthobranchs known from Oregon. Veliger 27: 143–163.

Goddard, J. H. R. 2001. Mollusca: Gastropoda. pp. 86–128. In An identification guide to the larval marine invertebrates of the Pacific Northwest. A. L. Shanks, ed. Corvallis, OR: Oregon State University Press, 314 pp.

Goddard, J. H. R. 2004. Developmental mode in opisthobranch molluscs from the northeast Pacific Ocean: feeding in a sea of plenty. Canadian Journal of Zoology 82: 1954–1968.

Goddard, J. H. R. 2005. Ametamorphic direct development in *Dendrodoris behrensi* (Nudibranchia: Dendrodorididae) with a review of development mode in the family. Proceedings of the California Academy of Sciences 56: 201–211.

Hadfield, M. G., and M. Switzer-Dunlap. 1984. Opisthobranchs, pp. 209–350, In The Mollusca, vol. 7, Reproduction. A. S. Tompa, N. H. Verdonk, and J. A. M. van den Biggelar, eds. New York: Academic Press, 486 pp.

Hurst, A. 1967. The egg masses and veligers of thirty northeast Pacific opisthobranchs. Veliger 9: 255–288.

Hyman, L. H. 1967. The invertebrates, Vol. VI, Mollusca I. New York: McGraw-Hill, 792 pp.

Krug, P. J. 1998. Poecilogony in an estuarine opisthobranch: planktotrophy, lecithotrophy, and mixed clutches in a population of the ascoglossan *Alderia modesta*. Marine Biology 132: 483–494.

Krug, P. J., R. A. Ellingson, R. Burton and A. Valdés. 2007. A new poecilogonous species of sea slug (Opisthobranchia: Sacoglossa) from California: comparison with the planktotrophic congener *Alderia modesta* (Lovén, 1844). Journal of Molluscan Studies 73.

Pawlik, J. R., M. R. Kernan, T. F. Molinski, M. K. Harper, and D. J. Faulkner. 1988. Defensive chemicals of the Spanish dancer nudibranch *Hexabranchus sanguineus* and its egg ribbons: macrolides derived from a sponge diet. Journal of Experimental Marine Biology and Ecology 119: 99–109.

Rasmussen, E. 1944. Faunistic and biological notes on marine invertebrates I. The eggs and larvae of *Brachystomia rissoides* (Hanl.), *Eulimella nitidissima* (Mont.), *Retusa truncatula* (Brug.) and *Embletonia pallida* (Alder & Hancock), (Gastropoda marina). Videnskabelige Meddelelser fra Danske naturhistorisk Forening i København 107: 207–233.

Strathmann, M. F. 1987. Reproduction and development of marine invertebrates of the northern Pacific coast. Seattle: University of Washington Press, 670 pp.

Thompson, T. E. 1976. Biology of opisthobranch molluscs, Vol. I. Ray Society, London, 207 pp.

Todd, C. D., W. J. Lambert, and J. Davies. 2001. Some perspectives on the biology and ecology of nudibranch molluscs: generalizations and variations on the theme that prove the rule. Bolletino Malacologico 37: 105–120.

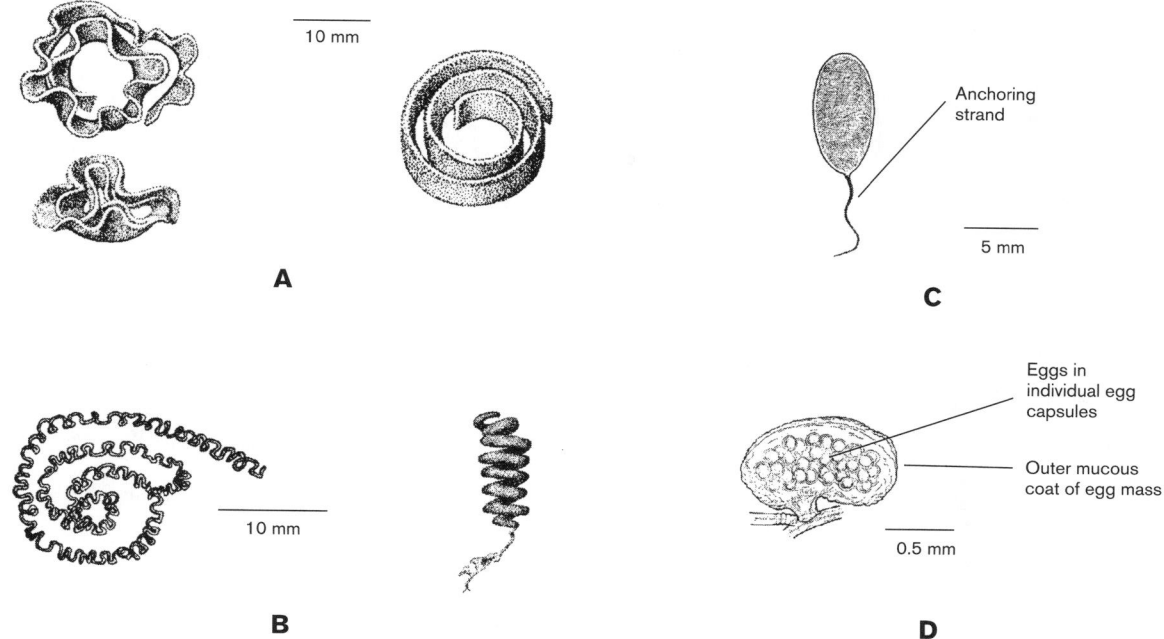

PLATE 387 Types of egg masses of benthic opisthobranchs (based on Hurst 1967): A, ribbons, common in notaspideans and dorid nudibranchs, also laid by some cephalaspideans and some dendronotacean nudibranchs; B, strings, common in anaspideans, sacoglossans, and aeolid, arminacean, and dendronotacean nudibranchs; C, ovoid or globular sacs anchored in soft sediments by a thin strand of mucous, common in cephalaspideans; D, small rounded sections of ribbons or cords, common small aeolids and small dendronotaceans (A, B, from Thompson 1976; C, from Hyman 1967; D, from Rasmussen 1944).

KEY TO MAJOR OPISTHOBRANCH CLADES

TERRENCE M. GOSLINER AND GARY C. WILLIAMS

(Plates 388–390)

1. Shell present, wholly or at least partially visible externally, spirally coiled . . . . (in part) Cephalaspidea and Acteonacea (Key A)
— Shell internal or absent, or if external, cap-shaped as in a limpet 2
2. Dorsal surface with projections, which may be either fin-gerlike or a circlet of gills around the anus 3
— Dorsal surface without projections, although lateral flaps (parapodia) of the foot may project dorsally 4
3. Rhinophores rolled . . . . (in part) Sacoglossa (except *Stiliger*) (next section)
— Rhinophores not rolled, but solid with various elaborations . . . . (in part) Nudibranchia (next section)
4. Head shield present; rhinophores absent . . . . (in part) Cephalaspidea and Acteonacea (Key A)
— Head shield absent; rhinophores present 5
5. Gill enclosed or partially hidden in a slit on the dorsal surface . . . . Anaspidea (Key B)
— Gill not enclosed in a slit on the dorsal surface 6
6. Gills, if present, located posteroventrally on both sides of the body between dorsum and foot margin . . . . (in part) Nudibranchia (next section)
— Gill absent or located exclusively on right side of the body 7
7. Gill on right side of animal, under overlapping dorsum between dorsum and foot; if external shell is present, it is cap-shaped, resembling a limpet (plate 390D) . . . . "Notaspidea" (Key C)
— Gill absent, rhinophores rolled . . . . (in part) Sacoglossa (next section)

KEY A: CEPHALASPIDEA AND ACTEONACEA

1. Shell absent or internal, not externally visible 2
— Shell external and visible, may be thin and transparent or thick and calcified 9
2. Shell absent; body color orange to rust-brown; animal minute (<5 mm); possesses four gizzard plates (plate 389F) . . . . *Runcina macfarlandi*
— Shell internal; body tan, cream white, to dark brown or black, may be mottled or striped; animal usually larger than 5 mm. 3
3. Body uniformly white to pale yellow; three gizzard plates present 4
— Body variously colored or mottled usually with spots and/or stripes; gizzard plates absent 5
4. Gizzard plates with prominent central bar (plate 389G, 389H) . . . . *Philine auriformis*
— Gizzard plates with paired circular pores (plate 389I) . . . . *Philine* spp.
5. Posterior end of foot rounded, without prominent projections; parapodia expanded, winglike and used for swimming; color red (plate 388F) . . . . *Gastropteron pacificum*
— Posterior end of foot with prominent projections; parapodia not winglike or used for swimming 6

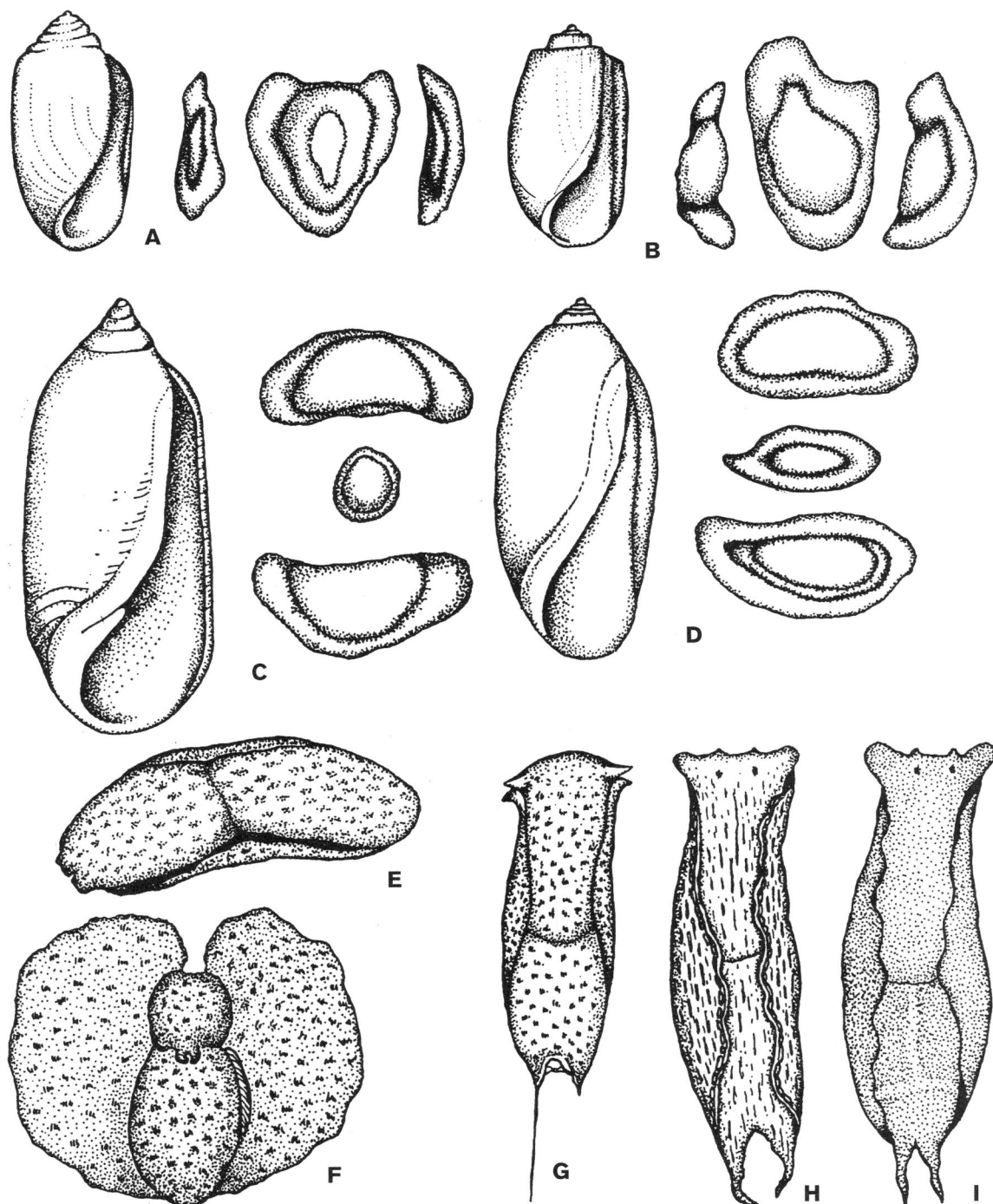

PLATE 388 A, *Acteocina inculta* (shell 3.8 mm long; gizzard plates 1.0 mm maximum length); B, *Acteocina harpa* (shell 3.9 mm long; gizzard plates 1.9 mm maximum length); C, *Acteocina culcitella* (shell 21 mm long; gizzard plates 1.3 mm maximum length); D, *Acteocina cerealis* (shell 6 mm long; gizzard plates 2.5 mm maximum length); E, *Melanochlamys diomedea* (body length 15 mm); F, *Gastropteron pacificum* (body length 20 mm); by G. C. Williams; G, *Aglaja ocelligera* (body length 20 mm); H, *Navanax inermis* (body length 120 mm); I, *Navanax polyalphos* (body length 50 mm) (A, D–I, by G. C. Williams; B, C, shell figure modified from Keen and Pearson 1952).

6. Posterior projection of foot short and truncate (plate 388E)
    . . . . *Melanochlamys diomedea*
—   Posterior projection of foot elongate and acute. 7
7. Anterior margin of head with four shallow lobes; eyes conspicuous dorsally. 8
—   Anterior margin of head rounded with two laterally pro-

jecting lobes; with an elongate filamentous appendage on the left posterior foot extension; eyes not conspicuous (plate 388G) . . . . *Aglaja ocelligera*
8. Parapodial margin yellow to orange with sub-marginal electric blue spots (plate 388H) . . . . *Navanax inermis*
—   Body color dark brown to black with uniformly distributed

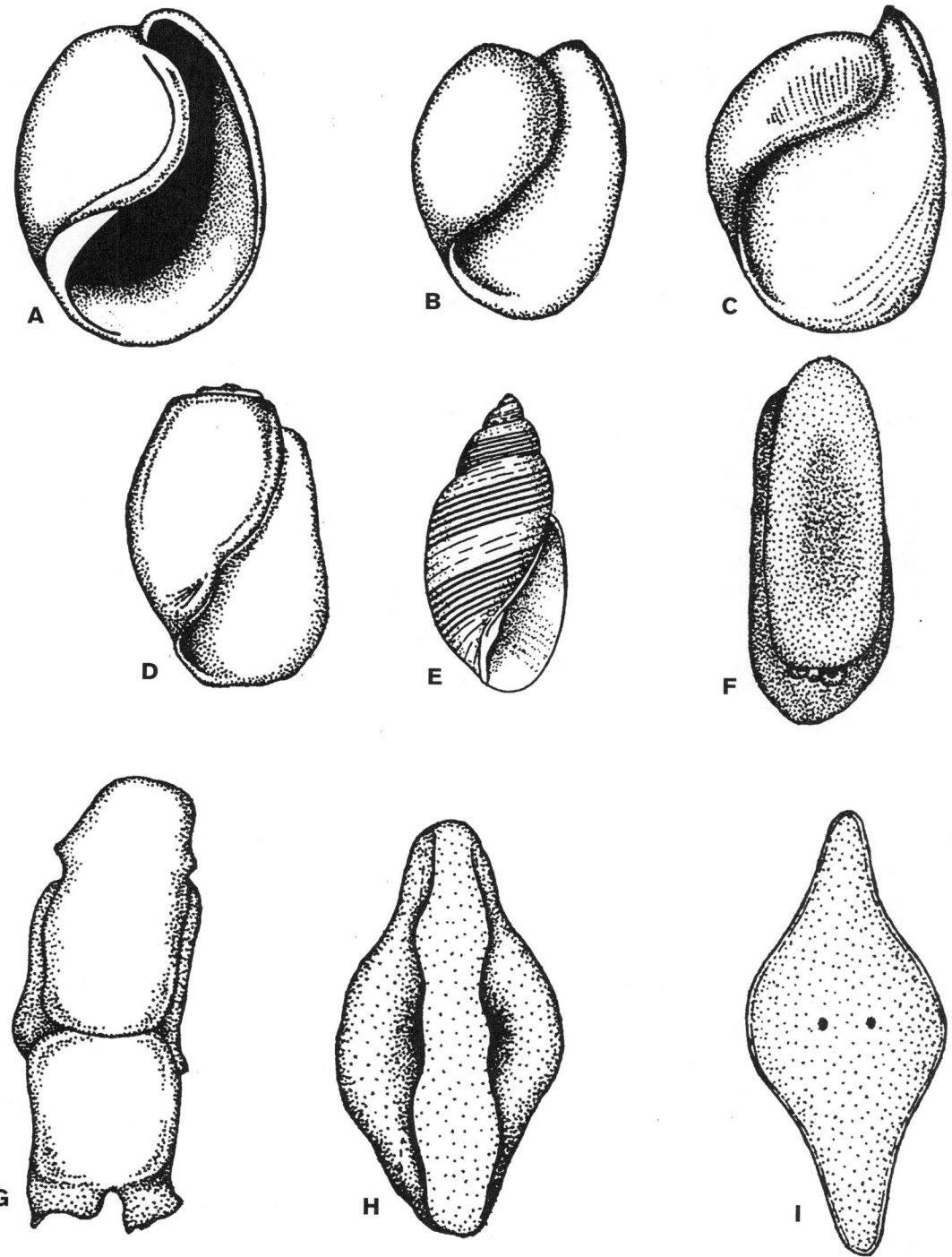

PLATE 389 A, *Bulla gouldiana* (shell 40 mm long); B, *Haminoea vesicula* (shell 15 mm long); C, *Haminoea virescens* (shell 17 mm long); D, *Diaphana californica* (shell 3 mm long); E, *Rictaxis punctocaelatus* (shell 15 mm long); F, *Runcina macfarlandi* (body length 3 mm); G, *Philine auriformis* (body length 40 mm); H, *Philine auriformis*, gizzard plate (5 mm long); I, *Philine* sp., gizzard plate (8 mm long) (A, modified from Keen and Pearson 1952; E, E. Reid; rest by G. C. Williams).

small white spots (plate 388I) . . . . . . *Navanax polyalphos*
9. Shell thickly calcified with alternating black and white banding (plate 389E). . . . . *Rictaxis puntocaelatus*
— Shell white to pale yellow or variously mottled 10
10. Shell globose with no portion of the spire visible externally; body color brown to green, variously mottled 11

— Shell cylindrical with spire visible externally, uniformly white to pale yellow; body color uniformly white 14
11. Shell reddish brown to brownish gray with dark and light mottling (plate 389A) . . . . *Bulla gouldiana*
— Shell clear or milky white without mottling, but mottled pigment of underlying the mantle is visible 12

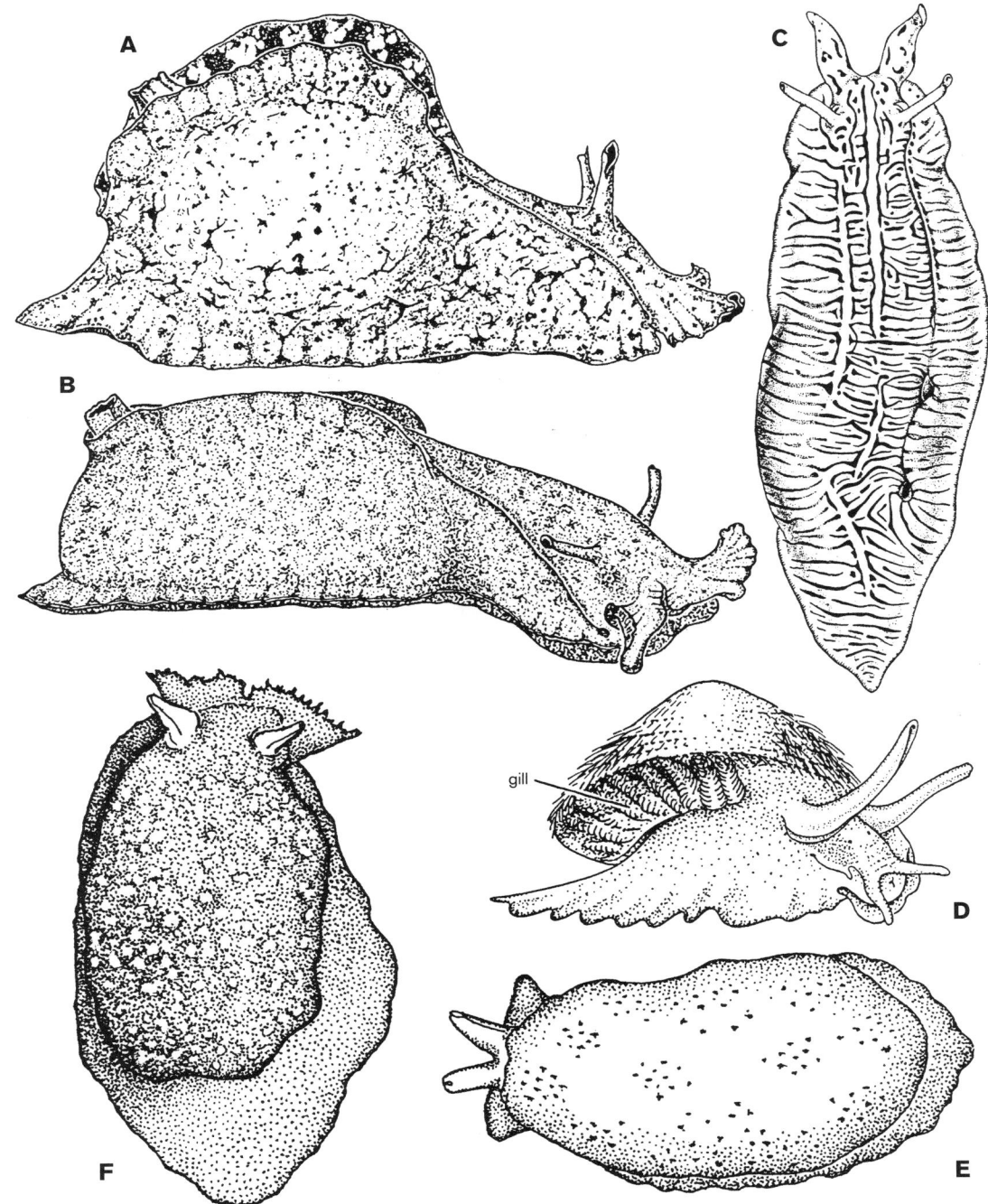

PLATE 390 A, *Aplysia californica* (body length 300 mm); B, *Aplysia vaccaria* (body length 240 mm); C, *Phyllaplysia taylori* (body length 45 mm); D, *Tylodina fungina* (body length 45 mm); E, *Berthella californica* (body length 37 mm); F, *Pleurobranchaea californica* (body length 90 mm) (A–C, by R. D. Beeman; D, redrawn by Emily Reid after MacFarland 1966; E, F, by G. C. Williams).

12. Posterior end of head shield deeply notched (see plate 373I in Shelled Gastropoda) . . . . *Haminoea japonica*
— Posterior end of head shield slightly notched or entire 13
13. Shell aperture ≤half shell length, shell usually 15 mm–24 mm long, thin, translucent white (plate 389B) . . . . *Haminoea vesicula*
— Shell aperture >half shell length, shell usually <18 mm long, somewhat thickened, transparent to milky white (plate 389C) . . . . *Haminoea virescens*

14. Head shield deeply bilobed anteriorly; shell translucent; rocky habitats (plate 389D) . . . . *Diaphana californica*
— Head shield rounded to shallowly lobed; shell opaque; muddy or sandy habitats 15
15. Shell white or pale yellow with brownish spiral lines on surface; unpaired gizzard plates smaller than paired ones (plate 388C, 388D) 16
— Shell uniformly whitish or brownish; unpaired gizzard plates larger than or equal to paired ones (plate 388A, 388B) 17

16. Shell ≤11 mm; spire poorly developed (plate 388D).....
.... *Acteocina cerealis*
— Shell up to 22 mm; spire strongly developed (plate 388C)
.... *Acteocina culcitella*
17. Shell with strongly carinate keeled shoulder (plate 388B)
.... *Acteocina harpa*
— Shell with rounded shoulder (plate 388A)............
.... *Acteocina inculta*

LIST OF SPECIES

ACTEONIDAE

Rictaxis puntocaelatus (Carpenter, 1864) (=*Acteon punctocae-latus*). Fairly common on mud- and sand flats; a predator on cirratulid polychaetes.

CYLICHNIDAE

Acteocina cerealis (Gould, 1853) (=*Cylichnella cerealis*). Relatively uncommon but has been found on intertidal flats at Bodega Harbor.

Acteocina culcitella (Gould, 1853) (=*Cylichnella culcitella, Acteocina rolleri*). Found in sandy habitats around Monterey Bay; most common in the shallow subtidal from 5 m to 10 m depth.

Acteocina harpa (Dall, 1871) (=*Retusa harpa, Cylichnella harpa*). Relatively uncommon in central California.

Acteocina inculta (Gould, 1856) (=*Cylichnella inculta*). A common intertidal species on mudflats, especially from Monterey south.

AGLAJIDAE

Aglaja ocelligera (Bergh, 1894). Seasonally common in muddy habitats both intertidally and subtidally.

Melanochlamys diomedea (Bergh, 1894) (=*Aglaja nana* Steinberg & Jones, 1960). Commonly found on mudflats.

Navanax inermis Cooper, 1862 (=*Chelidonura inermis*). A voracious predator of other opisthobranchs such as *Bulla gouldiana*; on mudflats and intertidal pools from Bodega Harbor to Baja California.

Navanax polyalphos (Gosliner and Williams, 1972) (=*Chelidonura polyalphos*). On mudflats from the Channel Islands to Mexico; feeds on *Haminoea vesicula*. See Gosliner and Williams 1972, Veliger 14: 424–436.

BULLIDAE

Bulla gouldiana Pilsbry, 1893. On mudflats; the primary food of *Navanax inermis*.

DIAPHANIDAE

Diaphana californica Dall, 1919. Seasonally common on algal blades (e.g., *Mazzaella*) in low rocky intertidal pools.

GASTROPTERIDAE

Gastropteron pacificum Bergh, 1894. Occasionally encountered on intertidal mudflats, but far more common subtidally.

See Mills 1994, pp. 313–319, in: Wilson, Stricker, and Shinn 1994. Reproduction and Development of Marine Invertebrates, Johns Hopkins Univ. Press, Baltimore (seasonal swimming and population dispersal).

HAMINOEIDAE

*Haminoea japonica (Pilsbry, 1895) (=*H. callidegenita* Gibson and Chia, 1989). An introduced Asian species mistakenly redescribed from Puget Sound as new; first found in San Francisco in the 1990s on dock floats in marinas. Shell figured in Shelled Gastropod section. See Gosliner and Behrens 2006, Proceedings of the California Academy of Sciences 57(37): 1003–1007.

Haminoea vesicula (Gould, 1855). On boat landings or mudflats of bays.

Haminoea virescens (Sowerby, 1833). In higher tide pools of the rocky intertidal.

PHILINIDAE

Philine auriformis (Suter, 1913). Introduced to San Francisco Bay from New Zealand presumably during the summers of 1992 and 1993 in freighters' ballast water; found in Bodega Harbor in 1994; now widespread from Oregon to Baja California. See Gosliner 1995, Marine Biology 122: 249–255 (introduction into San Francisco Bay).

Philine spp. Three additional introduced species inhabit estuaries of the San Francisco Bay Region. Other native species of the genus are recorded from California; see Behrens 2004, Pacific Coast Nudibranchs, Supplement II, Proceedings of the California Academy of Sciences 55(2): 11–54.

RUNCINIDAE

Runcina macfarlandi Gosliner, 1991. Found in the high intertidal among the algae *Endocladia* and *Cladophora*, as well as the tubes of the polychaete *Phragmatopoma*; Oregon to Monterey Bay.

KEY B: ANASPIDEA

1. Body dorsoventrally flattened; length under 10 cm; parapodia reduced; body color green with black and white striping (plate 390C) .... *Phyllaplysia taylori*
— Body laterally compressed, often exceeding 10 cm in length; parapodia highly developed; body color not green, unstriped 2
2. Parapodia joined posteriorly: body color uniform dark brown or black (plate 390B)............ *Aplysia vaccaria*
— Parapodia not obviously joined posteriorly; body color mottled tan or brown (plate 390A) .... *Aplysia californica*

LIST OF SPECIES

APLYSIIDAE

Aplysia californica Cooper, 1863. Common in southern California and present in northern and central California (including San Francisco Bay) especially during El Niño events.

* = Not in key.

Aplysia vaccaria Winkler, 1955. The world's largest opisthobranchs: animals may be up to 1 m in length and 14 km in weight have been recorded; Monterey Bay to the Gulf of California.

NOTARACHIDAE

Phyllaplysia taylori Dall, 1900 (=*Phyllaplysia zostericola* Mc-Cauley, 1960). Common on the eelgrass *Zostera* in the waters and mudflats of bays.

KEY C: "NOTASPIDEA" (UMBRACULACEA AND PLEUROBRANCHACEA)

1. With limpetlike external shell often covered with bristles, body yellowish in color (plate 390D). . . . . *Tylodina fungina*
— External shell absent 2
2. Bases of rhinophores widely separated (plate 390F) . . . . .
. . . . *Pleurobranchaea californica*
— Bases of rhinophores very close together 3
3. Body color uniformly bright orange. . . . .
. . . . *Berthellina ilisima*
— Body color white to pale yellow 4
4. Color cream to white, covered with small white dots; dorsum outlined by fine white line (plate 390E) . . . .
. . . . *Berthella californica*
— Color very pale yellow, finely punctate with darker yellow; dorsum without conspicuous white outline . . . .
. . . . *Berthella strongi*

LIST OF SPECIES

PLEUROBRANCHIDAE

Berthella californica (Dall, 1900) (=*Pleurobranchus californicus*). The most common notaspidean on the central California coast.

Berthella strongi MacFarland, 1966. Occasionally found in rocky intertidal pools from San Mateo County to Santa Cruz Island.

Berthellina ilisima Marcus and Marcus, 1967 (=*Berthellina citrina* and *B. engleli* of previous authors). *Berthellina citrina* was thought to be a single circumtropical species but actually represents a series of distinct species globally; may be encountered as far north as central California during El Niño events.

Pleurobranchaea californica MacFarland, 1966. From the shallow subtidal to at least 300 meters; feeds on various invertebrates and fish.

TYLODINIDAE

Tylodina fungina Gabb, 1865. Found commonly on the yellow sponge *Verongia thiona* (note: sponge not keyed in this manual), from San Luis Obispo County south to Ecuador.

REFERENCES

Beeman, R. D. 1968. The Order Anaspidea. Veliger 3 (Supplement): 87–102.
Beeman, R. D., and G. C. Williams. 1980. Opisthobranchia and Pulmonata: The Sea Slugs and Allies. In Intertidal Invertebrates of California. Morris, R. H., D. P. Abbott, and E. C. Haderlie, eds. Stanford, CA: Stanford University Press, 690 pp.
Behrens, D. W., and A. Hermosillo. 2005. Eastern Pacific nudibranchs. Monterey, California: Sea Challengers, 137 pp.
Gosliner, T. M. 1996. The Opisthobranchia. In Taxonomic Atlas of the Benthic Fauna of the Santa Maria Basin and western Santa Barbara Channel. Vol. 9, The Mollusca Part 2—The Gastropoda. Scott, P. H., J. A. Blake, and A. Lissner, eds. Santa Barbara, CA: Santa Barbara Museum of Natural History: 159–213.
Keen, A. M., and J. C. Pearson. 1952. Illustrated key to West North American gastropod genera. Stanford: Stanford University Press, 39 pp.
MacFarland, F. M. 1966. Studies of Opisthobranchiate Mollusks of the Pacific Coast of North America. Memoirs of the California Academy of Sciences 6, 546 pp.

Sacoglossa and Nudibranchia

GARY R. McDONALD

(Plates 391–392)

1. Rhinophores in the form of longitudinally rolled plates (plate 391A, 391S) (except in *Stiliger*) or very reduced, or absent; oral tentacles absent; usually found on algae . . . . . .
. . . . Order Sacoglossa 2
— Rhinophores not in form of longitudinally rolled plates; oral tentacles usually present 8
2. With two parapodia carried folded together in a vertical position over back; ground color rich green to yellowish-tan, with small spots of yellow, red, and blue; 10 mm (plate 391S) . . . . *Elysia hedgpethi*
— Lacking parapodia; with dorsal cerata 3
3. Cerata few (usually <10), occurring only on posterior one-half of dorsum; rhinophores quite short; body mottled greenish-brown, 8 mm . . . . *Olea hansineensis*
— Cerata usually more than 10, not confined to posterior one-half of dorsum 4
4. Rhinophores quite reduced; anus borne on a long tube originating on median line among posterior cerata and resembling cerata; ground color greenish to yellowish-tan with small black spots on body; 8 mm . . . .
. . . . *Alderia* spp.
— Rhinophores prominent; anus not borne on a long tube
. . . . 5
5. Body, cerata, and proximal one-half of rhinophores with a branching system of olive to dark green lines; tail long and obvious; cerata rather long; ground color pale yellow; 8 mm . . . . *Placida dendritica*
— Branching system of lines not continuing up onto rhinophores 6
6. With a triangular patch of dark brown spots extending from base of rhinophores forward to the midline; neck with a band of brown forming a collar; cerata cores chocolate brown; 5 mm . . . . *Hermaea vancouverensis*
— Lacking triangular brown patch and brown cerata cores
. . . . 7
7. Body covered with irregular patches of greenish to brownish-black; head uniformly pigmented with dark green to brownish-black; ground color yellowish white; 22 mm . . . . *Aplysiopsis enteromorphae*
— With a Y-shaped mahogany line running posteriorly from bases of rhinophores to midline on top of head; body with irregular lines of mahogany; a bright pink spot below and behind eyes; ground color pale yellow; 10 mm
. . . . *Hermaea oliviae*
8. Ground color translucent grayish white; body with reddish brown spots and irregular lines; two parallel, reddish brown lines running from rhinophores posteriorly; rhinophores simple; 3 mm . . . .
. . . . (Order Sacoglossa) *Stiliger fuscovittatus*
— Not as above . . . . Order Nudibranchia 9

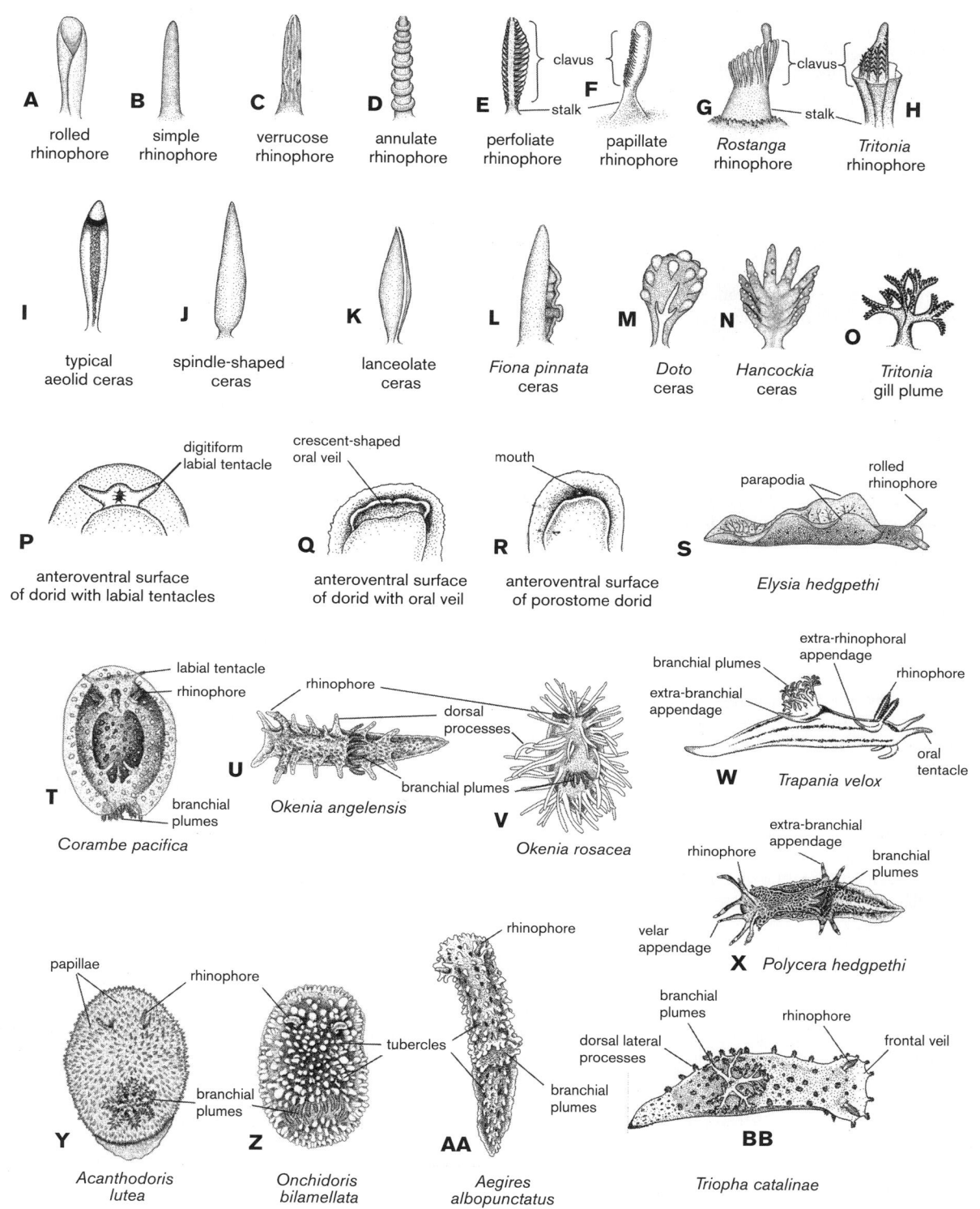

A rolled rhinophore

B simple rhinophore

C verrucose rhinophore

D annulate rhinophore

E perfoliate rhinophore — clavus, stalk

F papillate rhinophore

G *Rostanga* rhinophore — clavus

H *Tritonia* rhinophore — stalk

I typical aeolid ceras

J spindle-shaped ceras

K lanceolate ceras

L *Fiona pinnata* ceras

M *Doto* ceras

N *Hancockia* ceras

O *Tritonia* gill plume

P anteroventral surface of dorid with labial tentacles — digitiform labial tentacle

Q anteroventral surface of dorid with oral veil — crescent-shaped oral veil

R anteroventral surface of porostome dorid — mouth

S *Elysia hedgpethi* — parapodia, rolled rhinophore

T *Corambe pacifica* — labial tentacle, rhinophore, branchial plumes

U *Okenia angelensis* — rhinophore, dorsal processes, branchial plumes

V *Okenia rosacea*

W *Trapania velox* — branchial plumes, extra-branchial appendage, extra-rhinophoral appendage, rhinophore, oral tentacle

X *Polycera hedgpethi* — rhinophore, extra-branchial appendage, branchial plumes, velar appendage

Y *Acanthodoris lutea* — papillae, rhinophore, branchial plumes

Z *Onchidoris bilamellata* — tubercles, branchial plumes

AA *Aegires albopunctatus* — rhinophore, branchial plumes

BB *Triopha catalinae* — branchial plumes, rhinophore, frontal veil, dorsal lateral processes

PLATE 391 A–E, G–P, S, W, by Milos Radakovich; F, Q, R, T–V, X–AA, by Cleo Vilett; BB, by Emily Reid, redrawn from MacFarland 1966.

9. Gills (branchial plumes) located posteriorly on dorsum (plates 391U–391BB, 392A–392C) or in three groups directed posteriorly 10
— Gills (branchial plumes) not located posteriorly on dorsum; respiratory surface otherwise. 73
10. With extra-branchial appendages (plate 391W–391X), or large, obvious tubercles on dorsum, lateral to branchial plumes. 11
— Without extra-branchial appendages or large tubercles lateral to branchial plumes 19
11. With extra-rhinophoral appendages (plate 391W); three branchial plumes 12
— Without extra-rhinophoral appendages; branchial plumes usually more than three. 14

12. Each rhinophore shaft with one extra-rhinophoral appendage; ground color whitish; longitudinal brownish lines running most of the length of the body; distal tips of rhinophores, oral tentacles, tips of branchial plumes, extra-branchial, and extra-rhinophoral appendages orange-yellow; 10 mm (plate 391W) . . . . *Trapania velox*
— Each rhinophore shaft with two extra-rhinophoral appendages 13
13. With one extra-branchial appendage on each side; ground color translucent tan to whitish; head, tail, and sides of body with irregular patches of dark reddish-brown; 5 mm . . . . *Ancula lentiginosa*
— With four or more extra-branchial appendages on each side; ground color translucent yellowish-white; three longitudinal yellow lines running most of the length of the body; rhinophores and extra-branchial appendages with korange-yellow band near tip; 8 mm . . . . . *Ancula gibbosa*
14. Body covered with numerous tubercles or papillae 15
— Body not covered with numerous tubercles or papillae 17
15. Ground color black with oval, orange spots; papillae elongate; six branchial plumes, black 12 mm . . . . *Polycera alabe*
— Ground color translucent grayish-white to cream 16
16. With three to six extra-branchial appendages (plate 391W, 391X) on each side; branchial plumes three to five; with numerous small, brown to black dots and a lesser number of yellow dots over entire body (except ventral surface of foot); 10 mm . . . . *Palio zosterae*
— With one extra-branchial appendage (plate 391W, 391X) on each side; branchial plumes six to seven; dorsum with irregular brown streaks and blotches, and with yellow-white and dark brown blotches; 8 mm . . . . *Polycerella glandulosa*
17. Branchial plumes five to six, blackish with yellow lines; overall body color translucent grayish-white; velar appendages (plate 391X) and extra-branchial appendages with black on proximal one-third and a yellow ring immediately distal to black pigment; 15 mm . . . . *Polycera tricolor*
— Branchial plumes seven to 11 18
18. With oblong orange spots occurring between longitudinal black lines running the length of the body (black lines may be very sparse or very abundant); branchial plumes eight to 11; 12 mm . . . . *Polycera atra*
— With closely set, small black dots on body; rhinophores, corners of foot, velar and extra-branchial appendages (plate 391W, 391X) with yellowish-orange pigment; caudal crest and upper edge of foot with streaks of yellow-orange; branchial plumes seven to nine; 12 mm (plate 391X) . . . . *Polycera hedgpethi*
19. Rhinophores simple (plate 391B); ground color opaque white 20
— Rhinophores not simple; ground color variable 21
20. Dorsum with large cylindrical tubercles; usually with many irregularly scattered, dark brown to black spots; branchial plumes three; rhinophore sheaths with five to six high, rounded tubercles; 12 mm (plate 391AA) . . . . *Aegires albopunctatus*
— Dorsum nearly smooth, tubercles minute; lacking any brown or black pigment; branchial plumes about eight; rhinophores long and tapered; lateral edge of dorsum with opaque white glands; 12 mm . . . . *Conualevia alba*

21. With dorsal processes other than tubercles or papillae on dorsum or sides of body 22
— Dorsum without dorsal processes, but with tubercles (plates 391Z, 392C) or papillae (plate 391Y) or entirely smooth (plate 392A, 392B) 29
22. Dorsal processes simple 23
— Dorsal processes bifurcated, branched, or warty (tuberculate) distally 26
23. Body color entirely rose pink; with many long dorsal processes covering entire dorsum; branchial plumes seven to 14 and usually darker pink than dorsal processes; 15 mm (plate 391V) . . . . *Okenia rosacea*
— Body color not pink 24
24. Dorsal processes club-shaped and tipped with orange-red; branchial plumes five; ground color white to pale yellow; white or orange tubercles medially on dorsum; rhinophores tipped with orange-red; 15 mm . . . . *Limacia cockerelli*
— Dorsal processes not club-shaped, not tipped with orange-red 25
25. Dorsum with a single median process; branchial plumes eight to 11; body flattened dorsoventrally; ground color whitish with minute brownish spots; five marginal processes on each side; 8 mm . . . . *Okenia plana*
— Dorsum with five to seven median processes; branchial plumes five to seven; body limaciform (sluglike); ground color translucent white with small yellow flecks and greenish-gray patches; with numerous rather long papillae; 5 mm (plate 391U) . . . . *Okenia angelensis*
26. Dorsum with tubercles or processes medially, in addition to dorso-lateral (plate 391BB) and frontal marginal process; ground color whitish 27
— Dorsum smooth medially; with dorso-lateral and frontal marginal processes; ground color yellow or orange 28
27. Dorsal processes and frontal marginal processes with a subapical orange ring, often tipped with black; branchial plumes three, translucent grayish white; 20 mm . . . . *Crimora coneja*
— Dorsal processes and frontal marginal processes orange-tipped; branchial plumes five, translucent grayish, tipped with orange; 40 mm (plate 391BB) . . . . *Triopha catalinae*
28. Branchial plumes whitish with reddish tips; ground color pale yellow to yellowish-brown; dorsal processes tipped with reddish-orange; powder blue spots on dorsum; 40 mm . . . . *Triopha occidentalis*
— Branchial plumes yellowish with orangish tips; ground color orange to orange-brown, usually quite dark; dorsal processes usually tipped with dark orange; dorsum with numerous small blue dots in larger specimens; 15 mm . . . . *Triopha maculata*
29. Branchial plumes in two semicircles (∩∩) or in three groups directed posteriorly 30
— Branchial plumes in a single circlet or single arch 31
30. Branchial plumes 16–32, in form of two semicircles united anteriorly (∩∩); ground color brownish-white, usually with two to three irregular, longitudinal, brownish stripes on dorsum; many large, conspicuous tubercles on dorsum; 15 mm (plate 391Z) . . . . *Onchidoris bilamellata*
— Branchial plumes three, directed posteriorly from beneath three thick, bluntly triangular lobes located posteriorly on dorsum; ground color raw umber with many small dark brown to black spots on dorsum; dorsum with many

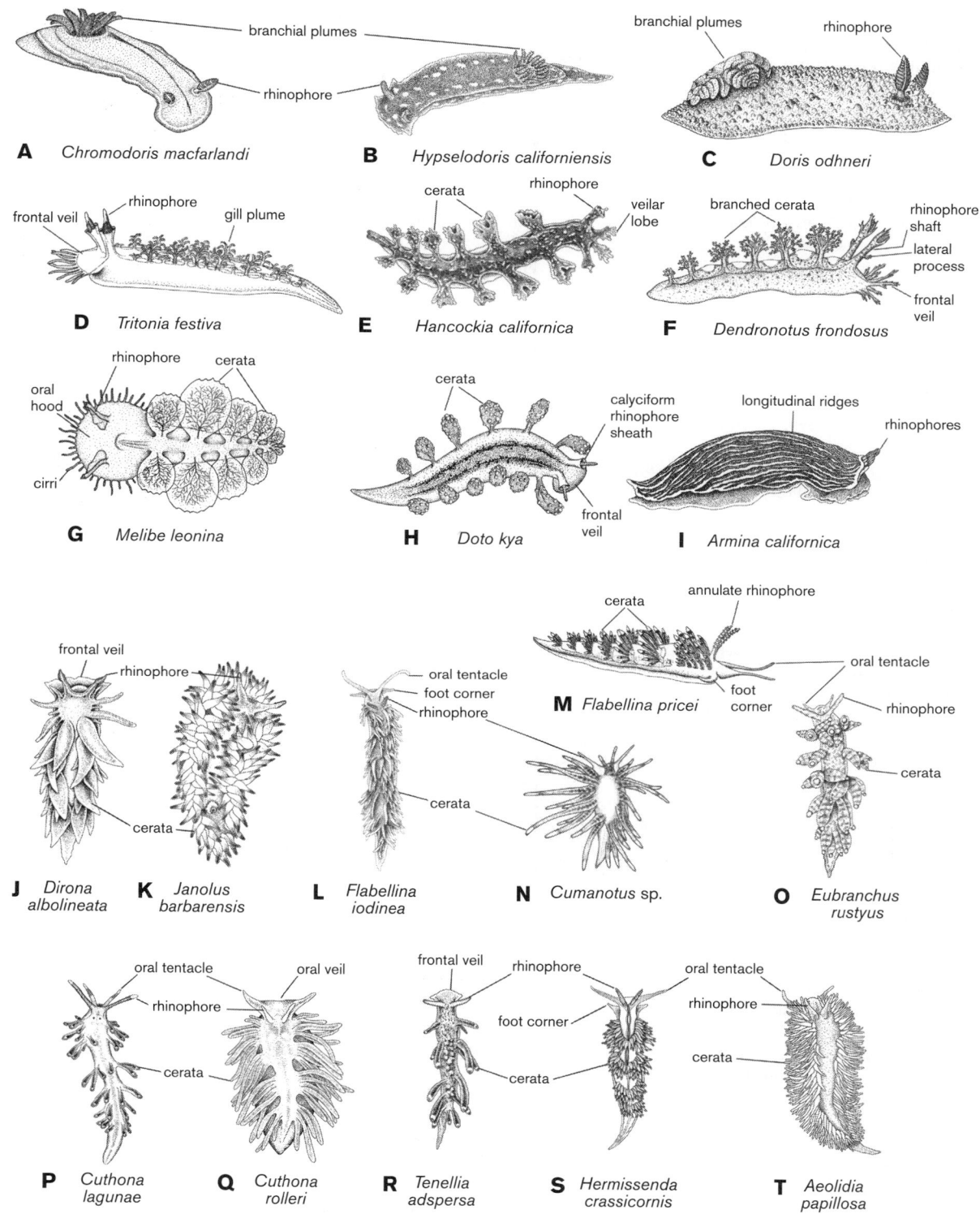

A *Chromodoris macfarlandi*

branchial plumes
rhinophore

B *Hypselodoris californiensis*

branchial plumes
rhinophore

C *Doris odhneri*

frontal veil
rhinophore
gill plume

D *Tritonia festiva*

cerata
rhinophore
veilar lobe

E *Hancockia californica*

branched cerata
rhinophore shaft
lateral process
frontal veil

F *Dendronotus frondosus*

rhinophore
cerata
oral hood
cirri

G *Melibe leonina*

cerata
calyciform rhinophore sheath
frontal veil

H *Doto kya*

longitudinal ridges
rhinophores

I *Armina californica*

frontal veil
rhinophore

J *Dirona albolineata*

cerata

K *Janolus barbarensis*

oral tentacle
foot corner
rhinophore
cerata

L *Flabellina iodinea*

cerata
annulate rhinophore
foot corner

M *Flabellina pricei*

cerata

N *Cumanotus* sp.

oral tentacle
rhinophore
cerata

O *Eubranchus rustyus*

oral tentacle
rhinophore
cerata

P *Cuthona lagunae*

oral veil
cerata

Q *Cuthona rolleri*

frontal veil
rhinophore
foot corner
cerata

R *Tenellia adspersa*

rhinophore
cerata

S *Hermissenda crassicornis*

oral tentacle
rhinophore
cerata

T *Aeolidia papillosa*

PLATE 392 A, C, D, F–H, M, by Milos Radakovich; figures B, E, I–L, N–T, by Cleo Vilett.

hispid papillae; dorsum with a prominent, irregular ridge extending along midline; 12 mm . . . . *Atagema alba*

31. Labial tentacles absent or minute and rudimentary (plate 391R). 32

— Labial tentacles present in form of tentacles or veil (plate 391P, 391Q). 37

32. With iridescent blue band around edge of dorsum; ground color creamy white to yellow-orange; branchial plumes four to five, bipinnate; labial tentacles small and grooved laterally; 50 mm . . . . *Doriopsilla spaldingi*

— Without iridescent blue band around edge of dorsum 33

33. Clavus (plate 391E) of rhinophores and gills deep purple; ground color opaque white to creamy tan; with white band around margin of dorsum; 50 mm............. .... *Dendrodoris azineae*
— Clavus of rhinophores and gills not deep purple..... 34
34. Ground color white to grayish-white 35
— Ground color yellow to orange or chestnut brown..... 36
35. Ground color white; dorsum with numerous brown to black blotches and flecks; mantle margin undulating; body rather long and narrow; 20 mm.......... *Dendrodoris behrensi*
— Ground color grayish-white; dorsum with numerous minute, brownish flecks which are more concentrated medially; about four to five opaque white blotches dorso-laterally on dorsum; 8 mm .... *Dendrodoris nigromaculata*
36. Branchial plumes white to pale yellowish; tubercles on dorsum with a small, apical, white dot; ground color yellow to reddish-brown; 25 mm..................... .... *Doriopsilla albopunctata*
— Branchial plumes orangish; white dots on dorsum and on tubercles; ground color yellow-orange; 25 mm........ .... *Doriopsilla gemela*
37. Body color red, or if orange, then dorsum velvety (with minute papillae) and branchial plumes eight or more and same color as body 38
— Body color not red; if orange, then dorsum with large, distinct tubercles or papillae 41
38. Clavus (plate 391G) of rhinophores with a blunt, cylindrical process above the lamellae, nearly one-quarter the length of the entire rhinophore; frequently with minute brownish-black dots on dorsum; 10 mm .... *Rostanga pulchra*
— Rhinophores perfoliate (plate 391E), lacking blunt, cylindrical process above lamellae.................. 39
39. With small, black flecks on dorsum; usually with distinct black dots on midline of dorsum; branchial plumes eight to 10, unipinnate 40
— Lacking black spots; dorsum deep dark red to pink, rather velvety smooth; body greatly flattened dorso-ventrally; branchial plumes seven, bi- and tripinnate; 30 mm .... *Platydoris macfarlandi*
40. With one or two black spots on midline of dorsum, one just posterior of rhinophores and the other just anterior of the branchial plumes (rarely one or both may be lacking); ground color yellow-orange to dark crimson, occasionally yellow-green; 20 mm .... *Aldisa sanguinea*
— With six to 10 small black spots on midline of dorsum; ground color lemon yellow to cadmium orange; 20 mm .... *Aldisa cooperi*
41. Ground color blue or deep violet; dorsum smooth, lacking any obvious tubercles or papillae 42
— Ground color not blue or deep violet; dorsum with tubercles or papillae or smooth.................... 44
42. Ground color deep violet; dorsum with three bright yellow, longitudinal lines; edge of dorsum with a submarginal white band; 25 mm (plate 392A) .... *Chromodoris macfarlandi*
— Ground color blue 43
43. Dorsum with two longitudinal series of about seven yellow to golden spots; anterior edge of dorsum and sides of foot also with yellow spots; ventral surface of posterior edge of dorsum with a series of about 8 small, white dots; 40 mm (plate 392B) .... *Hypselodoris californiensis*

— Dorsum with two bright yellow, longitudinal stripes, ending anteriorly at the rhinophores, and a single bright yellow stripe anterior of rhinophores; a median, light blue line extending from between the rhinophores to the branchial plumes; margins of dorsum with narrow, white band; 15 mm .... *Mexichromis porterae*
44. Labial tentacles large and auriculate (ear-shaped), reaching nearly to edge of dorsum; dorsum thickly set with long papillae (plate 391Y) 45
— Labial tentacles small and digitiform (plate 391P) or triangular, extending less than halfway to edge of dorsum or in form of small crescent-shaped veil (plate 391Q); dorsum with or without papillae.................. 50
45. Ground color orange; dorsum with numerous flecks of yellow; branchial plumes nine, whitish; usually with a pungent odor of cedar or sandalwood; 20 mm (plate 391Y) .... *Acanthodoris lutea*
— Body color not orange 46
46. Ground color brownish, flecked with irregular blotches of black; branchial plumes seven, tipped with lemon yellow, usually with a pungent odor of cedar; 15 mm .... *Acanthodoris brunnea*
— Ground color white to gray-white................ 47
47. Dorsal papillae tipped with black 48
— Dorsal papillae tipped with yellow 49
48. Edge of dorsum with an outer yellow and inner black line; branchial plumes five; rhinophores and branchial plumes tipped with reddish-brown to black; 15 mm .... *Acanthodoris rhodoceras*
— Edge of dorsum without an outer yellow and inner black line; branchial plumes five to nine; 15 mm .... *Acanthodoris pilosa*
49. Rhinophores and branchial plumes tipped with brown or maroon; branchial plumes five to nine; 15 mm .... *Acanthodoris nanaimoensis*
— Rhinophores and branchial plumes tipped with yellow; branchial plumes five; edge of dorsum with yellow band; 15 mm .... *Acanthodoris hudsoni*
50. Dorsum with obvious black or brown pigment, not including rhinophores or branchial plumes.......... 51
— Dorsum lacking obvious black or brown pigment 62
51. Dorsum with a lateral series of small, brown to black dots with yellow centers; ground color salmon to yellowish-pink; branchial plumes 12; 15 mm...... *Cadlina sparsa*
— Dorsum lacking black dots with yellow centers 52
52. Rhinophores tipped with maroon to brownish-black to black 53
— Rhinophores not tipped with maroon to brownish-black to black 54
53. Branchial plumes eight, bi- and tripinnate, tipped with dark purple to black; clavus of rhinophores dark purple to black with 15–17 lamellae; ground color cream to yellow, with scattered brown to black, leopardlike spots; dorsum covered with small, tall, bluntly rounded papillae; 50 mm .... *Jorunna pardus*
— Branchial plumes 12–14, unipinnate, translucent yellowish-white, with small, reddish-maroon spots near base of each plume; clavus (plate 391E) of rhinophores brownish-maroon with eight to 12 lamellae; ground color yellowish with numerous reddish-brown flecks of various sizes; dorsum covered with numerous tubercles which are larger and more numerous near margin; 15 mm .... *Hallaxa chani*

54. Branchial plumes eight to 10, tripinnate, sprinkled with minute brownish flecks, whitish distally; blackish pigment in form of many very small dots, with major concentration of pigment forming a diffuse blotch just anterior of branchial plumes; ground color yellow to yellowish-brown; tubercles small; 30 mm . . . . *Geitodoris heathi*

— Branchial plumes <8; blackish pigment on dorsum not in a single concentration forming a diffuse blotch. . . 55

55. Dorsum with small tubercles giving dorsum a velvety appearance; branchial plumes six, tripinnate 56

— Dorsum with large, conspicuous tubercles; branchial plumes various (plates 391Z, 392C). 57

56. Ground color whitish-yellow to very pale brown; dorsum with brown to almost black rings, or occasionally blotches of various sizes; 30 mm (up to 150 mm) . . . .
. . . . *Diaulula sandiegensis*

— Ground color pale cream to bright yellow to golden-orange; dorsum, underside of mantle and sides of body with irregular brown blotches; 50 mm . . . .
. . . . *Peltodoris mullineri*

57. Dorsum with numerous large, irregular tubercles that bear smaller tubercles; ground color light yellowish-tan; dorsum with irregularly round, dark tan to brown blotches between the large tubercles; ventral surface of foot and mantle margin with flecks of brown; branchial plumes seven; 25 mm. . . . . *Doris tanya*

— Dorsum with numerous large rather regular tubercles which do not bear smaller tubercles 58

58. Dorsum with two brown to blackish blotches, formed of smaller spots, on midline, ground color whitish to yellow or orange; dorsum with many pointed papillae; branchial plumes six; 25 mm . . . . . *Thordisa bimaculata*

— Dorsum lacking two brown to blackish spots on midline
. . . . 59

59. Dorsum covered with numerous villous papillae and with a few larger, evenly scattered irregular papillae; ground color dusky yellow with patches of white and dark brown on dorsum; a subtle cream line connecting the rhinophore pits; branchial plumes six, tripinnate, cream with brown flecks; 30 mm . . . . *Taringa aivica*

— Dorsum with numerous larger, low, rounded tubercles (plate 392C); ground color translucent white to light yellow to orange-yellow 60

60. Blackish pigment in blotches on both dorsum and tubercles; branchial plumes 7, yellowish, tri- and quadripinnate; ground color light yellow to orange-yellow; 50 mm (up to 150 mm) . . . . *Doris montereyensis*

— Blackish blotches on dorsum only, not on tubercles; branchial plumes six 61

61. With many, small, blackish blotches on dorsum; ground color light yellow to yellow-orange; branchial plumes whitish, tri- and quadripinnate; rhinophores with 18–24 lamellae; 50 mm (up to 200 mm) . . . . . . *Peltodoris nobilis*

— With a few, large, blackish blotches on dorsum; ground color translucent white to pale yellow; branchial plumes slightly darker than body, quadripinnate; 60 mm . . . . .
. . . . *Peltodoris lentiginosa*

62. Clavus (plate 391E) of rhinophores brown to black. . . . .
. . . . 63

— Clavus of rhinophores white, yellow, orange, scarlet, or brownish-orange. 64

63. Branchial plumes brown to black; ground color whitish; dorsal and ventral surfaces of foot and sides of body with minute black spots; 15 mm . . . . . *Cadlina limbaughorum*

— Branchial plumes white to yellow; ground color whitish to yellow; lateral edges of dorsum with six to 10 yellow dots on each side, anterior-most dot usually posterior of rhinophores; 15 mm . . . . *Cadlina flavomaculata*

64. Dorsum with yellow line around margin 65

— Dorsum lacking yellow line around margin 66

65. Rhinophores white to light yellow; dorsum with yellow-tipped tubercles; ground color whitish to pale yellow; branchial plumes six, tipped with yellow; 35 mm . . . . . .
. . . . *Cadlina luteomarginata*

— Rhinophores scarlet; dorsum lacking yellow-tipped tubercles; ground color white; branchial plumes seven to 16, scarlet; dorsum with scarlet spots surrounded by yellow ring; 20 mm. . . . . *Chromodoris galexorum*

66. Tail extending posteriorly beyond dorsum and bearing a slight, mid-dorsal ridge; dorsum with many long, slender, tapered papillae giving animal a fuzzy appearance; branchial plumes four to nine, unipinnate; ground color white to yellowish-white; dorsum with many opaque white flecks; 12 mm. . . . . *Diaphorodoris lirulatocauda*

— Tail normally not extending posteriorly beyond dorsum and lacking a mid-dorsal ridge. 67

67. Ground color red-orange, dorsum encrusted with gold flecks; branchial plumes six, white to orange proximally and brownish distally; dorsum covered with numerous inflated papillae of various sizes; rhinophore stalk white to yellow-orange, clavus (plate 391E) orange to brown with white flecks; 50 mm . . . . *Thordisa rubescens*

— Ground color white to yellow or orange. 68

68. Dorsum with numerous, elongate or bulbous tubercles
. . . . 69

— Tubercles, if present, low and rounded, or hispid, not elongate or bulbous; ground color white to yellow. 71

69. With spicules projecting in a radiating pattern from the apex of the tubercles on dorsum; body elongate, wider anteriorly; ground color white; usually with five tubercles surrounding each rhinophore; 15 mm . . . . . *Adalaria* sp.

— Without spicules projecting in a radiating pattern from tubercles, body oval, neither elongate nor wider anteriorly 70

70. Branchial plumes darker than body; digestive gland brown to reddish-brown; ground color pale yellow to orange, semitranslucent; body outline oval, not tapered posteriorly; margin of rhinophore sheath usually with three tubercles; 5 mm . . . . *Adalaria jannae*

— Branchial plumes lighter than body; digestive gland bright red, occasionally brown; ground color white to pale yellow; body outline oval, tapered posteriorly; margin of rhinophore sheath usually with two tubercles; 5 mm
. . . . *Onchidoris muricata*

71. Dorsum with several small yellow dots along lateral edges, anterior-most dots extending in front of rhinophores; ground color pale yellow; branchial plumes 10–12, yellowish-white; rhinophores with 10–12 lamellae; 25 mm . . . . *Cadlina modesta*

— Dorsum lacking yellow dots along lateral edges 72

72. Branchial plumes seven, white, tri- and quadripinnate; rhinophores with 20–24 lamellae; ground color pure white with white tubercles of various sizes on dorsum (rarely dusky yellow with dusky yellow tubercles); 60 mm (up to 200 mm) (plate 392C) . . . . *Doris odhneri*

— Branchial plumes seven to nine, white, bipinnate; rhinophores with 10–14 lamellae; ground color yellow with numerous small white dots on dorsum; with small,

hispid papillae on dorsum; 20 mm . . . .
. . . . *Baptodoris mimetica*

73. Without cerata 74
— With cerata (plate 391I–391N), or branched branchial plumes (plate 391O) arranged laterally on dorsum . . . 77

74. Dorsum with undulating, longitudinal ridges (plate 392I) of white on a pinkish-brown background; gills located laterally below dorsum edge on either side; edge of dorsum with an antero-median notch through which the rhinophores project; clavus of rhinophores with many longitudinal grooves; 25 mm (plate 392I). . . . .
. . . . *Armina californica*
— Dorsum without undulating longitudinal ridges . . . . . .
. . . . 75

75. Gills located laterally, between the dorsum and the foot; dorsum with tubercles; rhinophores project through dorsum, posterior of anterior edge of dorsum; ground color white to grayish-white; dorsum with numerous low tubercles; 20 mm . . . . *Phyllidiopsis blanca*
— Gills located postero-ventrally; dorsum lacking tubercles
. . . . 76

76. Dorsum with posterior notch through which gills may be seen; ground color pale gray, with small, yellowish-brown blotches and reticulate lines; rhinophore shaft with platelike expansion; 5 mm (plate 391T) . . . . . *Corambe pacifica*
— Dorsum without posterior notch; rhinophores smooth and tapering; ground color pale gray, with small, yellowish-brown blotches and reticulate lines; 5 mm . . . . .
. . . . *Corambe steinbergae*

77. Cerata or branchial plumes branched 78
— Cerata not branched. 89

78. Cerata palmately branched (plate 391N), each with four to 16 digitiform projections; cerata in four to seven pairs; head with a broad, palmate velar lobe on either side, each with six to 10 or more unequal digitiform processes; ground color reddish-brown in mature individuals, younger individuals translucent greenish-brown; 15 mm (plate 392E) . . . . *Hancockia californica*
— Cerata not palmately branched 79

79. Processes on frontal veil simple (plate 392D) or lacking
. . . . 80
— Processes on frontal veil branched (plate 392F) 84

80. Dorsum distinctly tuberculate, each tubercle tipped with white; ground color deep orange-yellow to yellow-brown; body margins with an irregular series of low, white branchial plumes; foot edged with a narrow white band; 120 mm (up to 300 mm) . . . . *Tochuina tetraquetra*
— Dorsum smooth, not distinctly tuberculate 81

81. Ground color rose pink; frontal veil with 10–30 low, white processes; body margins with about 20–30 branchial plumes (plate 391O); margins of foot, margins of dorsum, and the edge of rhinophore sheath bear a narrow, white line; 150 mm . . . . *Tritonia diomedea*
— Ground color not rose pink; body margin with <20 branchial plumes on each side. 82

82. Dorsum with fine reticulate, opaque white lines; ground color white to yellowish-white; frontal veil with seven to 12 digitiform processes; body margins with 11–15 branchial plumes on each side; foot edged with a narrow band of opaque white; 25 mm (plate 392D) . . . .
. . . . *Tritonia festiva*
— Dorsum lacking reticulate lines, but with broad white longitudinal band medially or white patches; foot not edged with a narrow band of opaque white 83

83. Dorsum with broad white longitudinal band medially; ground color translucent white; frontal veil with two to seven digitiform processes; body margins with seven to 13 branchial plumes on each side; 10 mm . . . .
. . . . *Tritonia pickensi*
— Dorsum with distinctive white patches (often as a matched pair on either side of midline); ground color pale, dirty, orange-brown; frontal veil with four to seven digitiform processes; body margins with about nine branchial plumes on each side; 10 mm . . . . *Tritonia myrakeenae*

84. Rhinophore shaft with a lateral process (plate 392F) . . . . .
. . . . 85
— Rhinophore shaft lacking a lateral process 88

85. Posterior face of rhinophore shaft with vertical row of three to six small, slightly branched processes; lateral process arising near base of rhinophore shaft; usually four pairs of branched velar processes; dorsal edge of broad foot usually with an opaque white line; ground color may be white, gray, brownish, orange-red, or maroon; four to seven pairs of cerata that may be tipped with orange, yellow, or purple; 60 mm (up to 300 mm). . . . .
. . . . *Dendronotus iris*
— Posterior face of rhinophore shaft lacking vertical row of small, slightly branched processes; margin of rhinophore sheath usually with five long processes. 86

86. With a narrow, median, opaque white line extending posteriorly from between the last cerata to tip of tail 87
— Without a white line dorso-medially on tail; frontal veil usually with two to four pairs of branched processes; often with yellow flecks on body; cerata in three to nine pairs; ground color variable, translucent grayish-white to greenish or even brownish; 15 mm (plate 392F) . . . . . .
. . . . *Dendronotus frondosus*

87. Frontal veil with two pairs of branched processes; general ground color translucent grayish-white; four to eight pairs of cerata; usually tipped with orange-red; 15 mm . . . . . .
. . . . *Dendronotus albus*
— Frontal veil with three pairs of branched processes; ground color translucent grayish-white or lilac; four to five pairs of cerata; 20 mm . . . . . *Dendronotus diversicolor*

88. With solid, white line on edge of broad foot; ground color pale pink to red-brown to translucent yellow-white; four pairs of cerata; red-brown spots and tiny opaque white spots scattered over body; low tubercles on dorsum; frontal veil with six, branched processes; margin of rhinophore sheath with four (three to seven) unbranched processes which are about as high as clavus; 30 mm. . . . .
. . . . *Dendronotus albopunctatus*
— Lacking white line on edge of foot; ground color extremely variable, may be yellow, brown, orange, greenish, or white; about three to six pairs of cerata; dorsum with four (sometimes diffuse) longitudinal, light brown lines running from posterior of rhinophore shaft to tip of tail; frontal veil with four to six stout, branched processes; margin of rhinophore sheath with five to seven short, blunt processes which are shorter than clavus; 15 mm . . . . *Dendronotus subramosus*

89. With large oral hood (plate 392G), having two rows of cirri on margin, outer series long, inner series shorter; about five to six pairs of petallike cerata; ground color yellowish-brown to greenish-brown; rhinophore shaft with a thin, triangular, saillike expansion on inner margin; 35 mm (up to 300 mm) (plate 392G) . . . .
. . . . *Melibe leonina*

—   Lacking large oral hood with cirri; lacking saillike expansion on inner margin of rhinophore shaft 90

90.   Rhinophores perfoliate (plate 391E), annulate (plates 391D, 392M), or papillate (plate 391F), not smooth (plate 391B) nor verrucose (plate 391C) 91

—   Rhinophores smooth (simple) or verrucose, not perfoliate, annulate, nor papillate 107

91.   Rhinophores papillate (plate 391F), with papillae on posterior surface, not perfoliate or annulate; a red-orange stripe middorsally on head; cerata opaque white at base, cores brick red; dorsal and lateral surfaces of body with opaque white; 8 mm . . . . *Noumeaella rubrofasciata*

—   Rhinophores perfoliate (plate 391E) or annulate (plate 391D), not papillate 92

92.   Cerata extending well in front of rhinophores and same color as rhinophores 93

—   Cerata not extending well in front of rhinophores . . . . . .
. . . . 98

93.   Cerata spindle-shaped (plate 391J); lacking frontal veil; oral tentacles present; with a pale orange to red crest between rhinophores 94

—   Cerata lanceolate (plate 391K); with a distinct frontal veil; oral tentacles absent 96

94.   Cerata highly papillate, with brown and white flecks; with irregular brown blotches on body . . . .
. . . . *Janolus annulatus*

—   Cerata not highly papillate, lacking brown and white flecks 95

95.   Cerata with a subapical gold band, blue apex, and branched cores; without irregular, dorso-lateral, red line; with minute, blue flecks on dorsum; 20 mm (plate 392K)
. . . . *Janolus barbarensis*

—   Cerata with subapical yellow band, white apex, and unbranched cores; with irregular, dorso-medial, red line; lacking minute, blue flecks on dorsum; 20 mm . . . .
. . . . *Janolus fuscus*

96.   Ground color orange to reddish-orange; cerata same color as body, with opaque white line on medial edge of each ceras, from near base to tip; with opaque white spots on dorsum; 50 mm . . . . *Dirona pellucida*

—   Ground color not orange to reddish-orange 97

97.   Cerata usually with irregular bumps; ground color light brown to greenish-gray with fine yellowish-white, olive-green, and pink dots; with a pale red spot on the outer side of each ceras; 20 mm . . . . *Dirona picta*

—   Cerata smooth, the lateral margin of each with an opaque white line; anterior edge of frontal veil and median crest of tail with opaque white line; ground color translucent grayish-white; 40 mm (up to 180 mm) (plate 392J) . . . . *Dirona albolineata*

98.   Rhinophores arising from a single median stalk, Y-shaped; cerata pinkish-red on proximal half, followed distally by a wide band of opaque white, then a narrow band of yellow-orange below translucent tip; ground color white to pinkish; 15 mm . . . . *Babakina festiva*

—   Rhinophores not on a single, common shaft 99

99.   With an opalescent blue line along edge of foot; double, opalescent blue median line running the length of the body and enclosing a bright orange-yellow elongate spot between and posterior to rhinophores and another in cardiac region; cerata with a subapical band of orange-yellow, cores extremely variable in color; 30 mm (plate 392S) . . . . *Hermissenda crassicornis*

—   Lacking opalescent blue lines. 100

100.   Rhinophores annulate (plate 391D) 101

—   Rhinophores perfoliate (plate 391E) 104

101.   Oral tentacles grayish-white, encrusted with opaque white distally 102

—   Oral tentacles brilliant vermilion to scarlet-red or yellow to deep cadmium orange distally. 103

102.   Rhinophores encrusted with white; cerata opaque white distally, lacking subterminal band of rich brown, cores deep brown; opaque white line dorso-medially from last group of cerata to tip of tail; opaque white line proximally on oral tentacles, changing to opaque white dots distally; head long and thick; 20 mm . . . . *Flabellina trophina*

—   Rhinophores pale yellowish-green distally; cerata with subterminal band of rich brown, cores deep olive- to brownish-green; ground color grayish-white; 20 mm (plate 392M) . . . . *Flabellina pricei*

103.   With an opaque white line dorso-medially and similar line dorso-laterally; tips of cerata whitish, cores usually reddish to orange, occasionally brownish; 20 mm . . . . .
. . . . *Flabellina trilineata*

—   Lacking opaque white lines on dorsum; ground color light pinkish; with an irregular band of scarlet-orange to vermilion on either side of head and a band of similar color below tip of each ceras; 20 mm . . . .
. . . . *Austraeolis stearnsi*

104.   Ground color deep purple; rhinophores perfoliate lamellae deep maroon; cerata violet proximally, grading to flaming scarlet or orange distally; 30 mm (plate 392L) . . . . . .
. . . . *Flabellina iodinea*

—   Ground color not deep purple. 105

105.   With a narrow orange-red line between and on the proximal half of the oral tentacles; cerata with pink tinge throughout, cores dark brown to black; 30 mm . . . .
. . . . *Phidiana hiltoni*

—   Lacking narrow, orange-red line between oral tentacles
. . . . 106

106.   Cerata orange-red with white tips; ground color grayish-white; dorsum along midline free of cerata; head usually with a pale orange spot; 20 mm . . . . . . *Aeolidiella oliviae*

—   Cerata cores greenish-brown, cerata with pink tinge, tipped with white; ground color dull orange; midline of dorsum with irregular patches of opaque white; oral tentacles very stout; 20 mm . . . . *Aeolidiella chromosoma*

107.   Oral tentacles present and digitiform (plate 392L–392Q, 392S, 392T). 108

—   Oral tentacles rudimentary or absent (plate 392N) or in the form of a frontal veil (plate 392H, 392R) that lacks distinct oral tentacles on either side 110

108.   With obvious oral veil connecting oral tentacles (plate 392Q); ground color white; body flattened dorso-ventrally; cerata flattened antero-posteriorly, with very fine brown flecks scattered over surface, tips frosted with white, cores salmon pink; 10 mm (plate 392Q) . . . .
. . . . *Cuthona rolleri*

—   Lacking obvious oral veil between oral tentacles . . . . . .
. . . . 109

109.   Each ceras with thin, saillike expansion on posterior edge from base upward for at least three-quarters of length of ceras (plate 391L); ground color translucent gray to brownish; 20 mm . . . . *Fiona pinnata*

—   Cerata lacking posterior, saillike expansion 115

110.   Rhinophore sheath calyciform (funnellike) (plate 392H); cerata large and bulbous with a small, plumelike gill on inner surface (plate 391M). 111

— Rhinophore sheath absent; cerata without plumelike gills
.... 114

111. With brown to black pigment on body or cerata..... 112
— Lacking brown to black pigment on body or cerata; cerata cores orange to pinkish, cerata rather bulbous; ground color pale yellowish-white; tubercles on cerata elongate; anal papilla high; 10 mm............... *Doto amyra*

112. Cerata with black rings at base of tubercles; ground color whitish to light brown; with brown to black pigment on head, back, and sides 113
— Cerata lacking black rings at base of tubercles; upper border of rhinophores sheaths somewhat scalloped; ground color white to yellowish; cerata long with yellowish to brown cores; 10 mm (plate 392H)........... *Doto kya*

113. Distal tubercles on cerata usually with an apical blackish spot; upper border of rhinophore sheaths somewhat scalloped; 10 mm .... *Doto lancei*
— Distal tubercles on cerata lacking apical blackish spot; upper border of rhinophore sheaths smooth; 10 mm .....
.... *Doto columbiana*

114. Cerata usually in groups of one to two, longest cerata shorter than body length; ground color white to yellowish-white; a few black specks on dorsum; 3 mm (plate 392R)
.... *Tenellia adspersa*
— Cerata very long, longest about equal to body length, some groups with more than two cerata; ground color grayish-green to brownish; cerata cores brownish-yellow to bluish-green, with network of similar colored lines on dorsum between groups; 6 mm (plate 392N) ....
.... *Cumanotus* sp.

115. Dorsum with a distinct (sometimes discontinuous) medial opaque white line 116
Dorsum without a distinct medial opaque white line
.... 118

116. Usually with a greenish-blue patch, anterior and to the left of the cardiac region; cerata with few, very minute, opaque white dots; cerata cores reddish-brown; ground color white with pinkish tinge; 20 mm.............
.... *Flabellina cooperi*
— Lacking a greenish-blue patch in cardiac region; cerata with opaque white dots 117

117. With longitudinal, opaque white line on dorsal surface of oral tentacles; cerata tipped with white, cores yellow-orange to reddish-brown, occasionally green; 25 mm ...
.... *Sakuraeolis enosimensis*
— Without longitudinal, opaque white line on dorsal surface of oral tentacles; with opaque white line extending from base of either oral tentacle to between rhinophores and continuing to tip of tail as an irregular (often discontinuous) line; cerata with numerous white dots, tips translucent grayish-white, cores pinkish-brown distally and greenish-brown proximally; 25 mm....... *Flabellina* sp.

118. Rhinophores with various amounts of orange to red pigment 119
— Rhinophores lacking orange to red pigment....... 124

119. Dorsum with distinct medial opaque white blotches; foot corners (plate 392L, 392M, 392S) distinct; vermilion on oral tentacles, rhinophores, head, and dorsum; cerata tipped with white, and narrow white band below tip, cores brownish; 30 mm......... *Anteaeolidiella indica*
— Dorsum lacking distinct medial opaque white blotches; foot corners indistinct 120

120. Tail with a median, opaque white line; oral tentacles with white pigment on at least distal half............. 121

— Tail lacking median, opaque white line; white pigment on oral tentacles lacking or confined to distal third..... 122

121. Rhinophores white distally; light orange pigment covering front of head, extending one-quarter of the way up the rhinophores and on bases of oral tentacles; cerata with small, opaque white spots and with cores brownish to dark green; 8 mm.............. *Cuthona flavovulta*
— Rhinophores orange-red for nearly entire length; large, orange-red spot between oral tentacles and anterior of rhinophores; cerata with orange-red tips and with blackish cores; 10 mm (plate 392P) .... *Cuthona lagunae*

122. Rhinophores with a subapical orange band and opaque white on distal third; oral tentacles with opaque white on dorsal surface of distal third and usually with a subapical orange band; cerata with opaque white at tip, cores yellow, orange, pink, red-brown, burgundy, or brownish-green; 10 mm .... *Catriona rickettsi*
— Rhinophores lacking subapical orange band....... 123

123. Cerata with small brown flecks on surface, cores orange-brown; oral tentacles translucent white to orange; 10 mm
.... *Cuthona phoenix*
— Cerata recumbent with broad, opaque white band extending from base to tip of each ceras, covering distal third, cores brownish; oral tentacles orange-red with white tips; 8 mm .... *Catriona columbiana*

124. Cerata irregularly inflated, warty, or rugose in appearance (plate 392O) 125
— Cerata not irregularly inflated, not warty 128

125. Anterior foot corners tentaculiform; rhinophores frequently with two to three brown rings; oral tentacles frequently with subapical brown ring; ground color yellowish-white; numerous distinct brown dots on dorsum extending up onto cerata; each ceras with a subapical yellow ring; 20 mm .... *Eubranchus misakiensis*
— Anterior foot corners not produced into tentaculiform processes 126

126. With dark olive-green, longitudinal stripes on dorsum connecting the cerata groups; cerata with dark green dots, cores cream colored; rhinophores and oral tentacles speckled with olive-green proximally, tipped with white, with a subapical band of dark olive-green; body color tan; 6 mm...................... *Eubranchus steinbecki*
— Lacking dark olive-green, longitudinal stripes on dorsum connecting the cerata groups................... 127

127. With a fairly broad, wavy band of deep olive-green dorsomedially on the dorsum; cerata with distinct, subterminal band of reddish-brown to olivaceous-green, cores deep olivaceous-green; rhinophores frosted with small, opaque white dots distally and with a subterminal band of red-brown to olivaceous-green; body color pale, translucent yellowish-green, with numerous light brown spots; 8 mm .... *Eubranchus rupium*
— Lacking a band of deep olive-green dorso-medially on dorsum; cerata with diffuse, subapical band of brownish to light gray or greenish, cores yellowish-green to brown; rhinophores and oral tentacles with subapical band of brownish to light gray or greenish; body color grayish-white to light cream or yellowish to light brownish-yellow, with numerous irregular specks of brownish to light gray or greenish; sides of body with small yellow dots; 8 mm (plate 392O)................. *Eubranchus rustyus*

128. With two median, light orange, oval spots, one spot anterior and one posterior of base of rhinophores; opaque white dots on distal third of rhinophores and oral tentacles;

an irregular series of white blotches medially on dorsum; cerata with three dark brownish-black bands, core ochre; 10 mm . . . . *Emarcusia morroensis*

— Lacking median, light orange spots anterior and posterior to rhinophores. 129

129. With band of purple midway on oral tentacles and rhinophores; cerata cores yellowish at base, olive-green to brown in middle, and yellow at tip; 8 mm. . . . .
. . . . *Cuthona abronia*

— Lacking band of purple midway on rhinophores and oral tentacles 130

130. Cerata cores red or dark wine-red to purple 131

— Cerata cores not red or dark wine-red to purple 132

131. Cerata with broken, opaque white line on outer surface, cores dark wine-red to purple, tips white; oral tentacles with broad, opaque white line on dorsal surface; rhinophores with opaque white flecks which are densest posteriorly and apically; body color translucent white to pale peach; 20 mm . . . . *Cuthona punicea*

— Cerata without broken, opaque white line on outer surface, cores red, tips white; oral tentacles and rhinophores opaque white on distal two-thirds; body color translucent white with opaque white dorsally; 10 mm . . . .
. . . . *Flabellina bertschi*

132. Cerata frosted with opaque white on distal two-thirds; dorsum encrusted with opaque white; cerata cores pale green to raw umber; distal third of rhinophores encrusted with white; 8 mm. . . . . *Cuthona albocrusta*

— Cerata not frosted with opaque white on distal two-thirds 133

133. With yellow bands near tips and bases of cerata, or with yellow on frontal margin between tentacles. 134

— Lacking yellow bands on cerata or between oral tentacles 136

134. With yellow on frontal margin between oral tentacles; cerata with orange flecks, cores green with some brown; rhinophores and oral tentacles with white dots on distal half; 8 mm . . . . *Cuthona virens*

— Lacking yellow on frontal margin 135

135. Medial third of rhinophores and oral tentacles with dark brown band, distal third encrusted with opaque white, proximal two-thirds with white speckling; cerata usually with white speckles on proximal two-thirds, cores kelly-green at base, tan to orange to salmon above, with white tips; with irregular patches of white and dark brown on dorsum and white spots on head; 10 mm . . . . *Cuthona hamanni*

— Lacking brown band on medial third of rhinophores and oral tentacles; cerata with yellow band near tip and another near base, cores yellow-brown to dark brown; body surface with white dots; 8 mm . . . . *Cuthona fulgens*

136. With opaque white, subapical band on each ceras; cerata core olive-green to brown; dorsum and cerata with opaque white dots; head may be light orange; 8 mm. . . . .
. . . . *Cuthona perca*

— Lacking band of opaque white below tip of each ceras 137

137. Cerata slightly flattened and decumbent, not round in cross-section 138

— Cerata not flattened, but round in cross-section, cores dark chocolate brown to reddish-brown, to pink. 140

138. With a triangular to crescentic patch of encrusting white to cream in the area between the rhinophores and oral tentacles and in cardiac region; body color dirty white, dull gray, mauve, rose, or pinkish; 25 mm (plate 392T) . . . . .

. . . . *Aeolidia papillosa*

— Lacking a patch of white to cream between rhinophores and oral tentacles or in cardiac region 139

139. Cerata encrusted with brown to reddish-brown; 10 mm
. . . . *Cerberilla mosslandica*

— Cerata pinkish-brown with white tips; 20 mm. . . . .
. . . . *Cerberilla* cf. *pungoarena*

140. Rhinophores and oral tentacles with opaque white distally; cerata cores deep reddish-brown to dark chocolate brown; ground color translucent white; 10 mm. . . . .
. . . . *Cuthona cocoachroma*

— Lacking opaque white on rhinophores and oral tentacles; ground color translucent cream to brownish-pink; cerata cores reddish-brown, pink, or burnt umber; 15 mm. . . . .
. . . . *Cuthona divae*

LIST OF SPECIES

Most species are widespread (Alaska or Canada to southern California or Mexico or further south), unless otherwise noted. The size listed for each species is the length of an average individual; your specimens may be smaller or larger.

SACOGLOSSA

PLACOBRANCHIDAE

Elysia hedgpethi Marcus, 1961 (=*Elysia bedeckta* MacFarland, 1966). Color photos in Behrens and Hermosillo 2005: 44, fig. 56; Beeman and Williams 1980, pl. 98, fig. 14.12; color illustration of *E. bedeckta* in MacFarland, 1966, pl. 4, figs. 1–2. Usually found on *Bryopsis corticulans* or on *Codium fragile*.

LIMAPONTIIDAE

Alderia spp. *A. modesta* (Loven, 1844), 4.5 mm, Alaska to San Francisco Bay (and Atlantic Ocean); *Alderia willowi* Krug, Ellingson, Burton and Valdes, 2007, 2mm, Bodega Harbor to Mexico; both on alga *Vaucheria* at edges of *Salicornia* marshes. In *A. modesta* (color photo, Behrens and Hermosillo 2005: 48, fig. 68), dorsum is smooth and squared off near the cephalic lobes, dark pigment is speckled over dorsal surface and over cerata on top of a yellow background; in *A. willowi*, dorsum tapers toward cephalic lobes and is raised into a hump, down which runs a stripe of yellow background color; dark pigment evenly covers most of dorsal surface and cerata. See Krug et al. (2007, J. Molluscan Studies 73).

Olea hansineensis Agersborg, 1923. Color photo in Behrens and Hermosillo 2005: 49, fig. 71. Usually found in *Zostera* beds where it feeds on eggs of opisthobranchs.

Placida dendritica (Alder and Hancock, 1843) (=*Hermaea ornata*, MacFarland, 1966). Color photo in Behrens and Hermosillo 2005: 47, fig. 65; color illustration of *H. ornata* in MacFarland 1966, pl. 4, fig. 3. Usually found on *Bryopsis corticulans* or *Codium fragile*. Also on Atlantic coast; Japan; Mediterranean.

Stiliger fuscovittatus Lance, 1962. Color photos in Beeman and Williams 1980, pl. 98, fig. 14.11; Lance 1962, Veliger 4: 155–159. Usually found on *Polysiphonia* spp. and other red algae.

HERMAEIDAE

Aplysiopsis enteromorphae (Cockerell and Eliot, 1905) (=*Aplysiopsis smithi* Marcus, 1961, =*Hermaeina smithi*). Color photo in Behrens and Hermosillo 2005: 48, fig. 69. Usually on *Chaeto-morpha* spp., *Rhizoclonium,* or *Cladophora* spp.

Hermaea oliviae (MacFarland, 1966) (=*Aplysiopsis oliviae,* =*Hermaeina oliviae*). Color photo in Behrens and Hermosillo 2005: 48, fig. 70; color illustration of *H. oliviae* in MacFarland 1966, pl. 4, figs. 4–5. Feeds on red algae *Griffithsia pacifica* and *Polysiphonia hendryi.* Very rare.

Hermaea vancouverensis (O'Donoghue, 1924). In bays, feeding on the diatom *Isthmia nervosa* growing on *Zostera, Laminaria,* and other algae. Alaska to central California.

NUDIBRANCHIA

CORAMBIDAE

Corambe pacifica MacFarland and O'Donoghue, 1929. Color photos in Behrens and Hermosillo 2005: 50, fig. 72; Beeman and Williams 1980, pl. 104, fig. 14.35; McDonald and Nybakken, 1980: 41, fig. 1; color illustration in MacFarland 1966, pl. 22, fig. 6. Usually on its prey, the bryozoan *Membranipora* growing on *Macrocystis,* closely resembling the bryozoan.

Corambe steinbergae (Lance, 1962) (=*Corambella bolini* MacFarland, 1966; =*Doridella steinbergae*). Color photos in Behrens and Hermosillo 2005: 50, fig. 73; and of *Doridella steinbergae* in McDonald and Nybakken 1980: 41, fig. 2; color illustration of *Corambella bolini* in MacFarland 1966, pl. 22, figs. 9–10. Valdés, and Bouchet, 1998 evaluated several characters for the species of the Corambidae and determined that only three genera are valid, making *Doridella* a junior subjective synonym of *Corambe.* Usually on its prey, the bryozoan *Membranipora* growing on *Macrocystis,* closely resembling the bryozoan, see McBeth (1968 Veliger 11: 145–146) for a discussion of feeding.

*Loy thompsoni (Millen and Nybakken, 1991) (=*Corambe thompsoni*). Color photo in Behrens and Hermosillo 2005: 50, fig. 74. Ground color white, with dense, chocolate-brown spots on dorsum; rhinophores smooth and conical; 5 mm. Subtidal, Alaska to central California.

GONIODORIDIDAE

Ancula gibbosa (Risso, 1818) (=*Ancula pacifica* MacFarland, 1905). Color photos in Behrens 2004: 29, pl. 3 fig. D–E; Behrens and Hermosillo 2005: 51, fig. 75; and of *Ancula pacifica* in Beeman and Williams 1980, pl. 104, fig. 14.36; McDonald and Nybakken 1980: 41, fig. 6; and color illustration in Thompson and Brown 1984, pl. 12a–c; color illustration of *A. pacifica* in MacFarland 1966, pl. 21, fig. 1. Thompson and Brown 1984: 47 state that specimens of *A. pacifica* from California have extra pigment patches on the dorsum but lack other morphological differences from *A. gibbosa* and therefore consider the two species synonymous. Feeds on the kamptozoan *Barentsia ramosa.* Also on the Atlantic coast and in Europe.

Ancula lentiginosa Farmer in Farmer and Sloan, 1964. Color photos in Behrens and Hermosillo 2005: 51, fig. 76; McDonald and Nybakken 1980: 41, fig. 5. Northern California to Mexico; Galapagos Is., rare.

Okenia angelensis Lance, 1966. Color photos in Behrens and Hermosillo 2005: 51, fig. 77; McDonald and Nybakken 1980: 41, fig. 3. Gosliner and Bertsch (2004, Proc. Calif. Acad. Sci. (4) 55: 414–430) give an extended description of this species. Usually on pilings in bays. San Francisco Bay to Mexico; Chile.

Okenia plana Baba, 1960. Color photos in Behrens and Hermosillo 2005: 53, fig. 81; Gosliner 2004: 145, fig. 20B; McDonald

and Nybakken 1980: 41, fig. 4. Feeds on the bryozoans *Membranipora membranacea, Cryptosula pallasiana,* and *Membranipora tuberculata.* San Francisco Bay and Southern California; Australia; Hong Kong; Japan; New Zealand.

Okenia rosacea (MacFarland, 1905) (=*Hopkinsia rosacea*). Color photos in Behrens and Hermosillo 2005: 53, fig. 82; and of *Hopkinsia rosacea* in Beeman and Williams 1980, pl. 104, fig. 14.37; McDonald and Nybakken 1980: 41, fig. 8; color illustration of *H. rosacea* in MacFarland 1966, pl. 21, figs. 2–3. Gosliner 2004: 144, 158 discusses synonymy of *Hopkinsia* with *Okenia.* See McBeth 1971 for feeding. McBeth (1972) shows that the pink pigment hopkinsiaxanthin is derived from its bryozoan prey, *Eurystomella bilabiata.* Oregon to Mexico; rare north of Cape Mendocino.

*Okenia sp. Behrens, 2004. Color photo of preserved specimen in Behrens 2004: 27, pl. 1 fig. I. Found on worm tubes and *Molgula* sp. Subtidal in southern California.

Trapania velox (Cockerell, 1901). Color photos in Behrens and Hermosillo 2005: 54, fig. 86; McDonald and Nybakken 1980: 41, fig. 7; color illustration in MacFarland 1966, pl. 20, figs. 1–2. Central California to Mexico, rare.

ONCHIDORIDIDAE

Acanthodoris brunnea MacFarland, 1905. Color photos in Behrens and Hermosillo 2005: 55, fig. 87; McDonald and Nybakken 1980: 43, fig. 9; color illustration in MacFarland 1966, pl. 20, fig. 5. Has pungent cedar odor; rare.

Acanthodoris hudsoni MacFarland, 1905. Color photos in Behrens and Hermosillo 2005: 55, fig. 88; McDonald and Nybakken 1980: 43, fig. 10. Has mild cedar odor; rare.

Acanthodoris lutea MacFarland, 1925. Color photos in Behrens and Hermosillo 2005: 55, fig. 89; Beeman and Williams 1980, pl. 104, fig. 38; McDonald and Nybakken 1980: 43, fig. 11. Feeds on the bryozoan *Alcyonidium* sp.; has pungent cedar odor.

Acanthodoris nanaimoensis O'Donoghue, 1921. Color photos in Behrens and Hermosillo 2005: 56, fig. 90; Beeman and Williams 1980, pl. 104, fig. 14.39; McDonald and Nybakken 1980: 43, fig. 12. Feeds on the bryozoan *Alcyonidium* sp. Alaska to central California.

Acanthodoris pilosa (Abildgaard in Müller, 1789). Color photos in Behrens and Hermosillo 2005: 56, fig. 91; McDonald and Nybakken 1980: 43, fig.13. A highly variable species, the specimens found in our area may belong to an undescribed species. Feeds on bryozoans such as *Alcyonidium* spp. and *Flustrellidra* sp. Circumboreal.

Acanthodoris rhodoceras Cockerell in Cockerell and Eliot, 1905. Color photos in Behrens and Hermosillo 2005: 57, fig. 93; Beeman and Williams 1980, pl. 105, fig. 14.40; McDonald and Nybakken 1980: 43, figs. 14–15. Feeds on the bryozoan *Alcyonidium* sp. Oregon to Mexico.

Adalaria jannae Millen, 1987. As *Onchidoris muricata* in third edition. Color photos in Behrens and Hermosillo 2005: 57, fig. 94; and of *Onchidoris* sp. in McDonald and Nybakken 1980: 45, fig. 17; color illustration of Onchidorididae sp. B in Just and Edmunds 1985, pl. 34. Millen (1987, Can. J. Zool. 65: 2696–2702) discusses the differences between this species and *Onchidoris muricata.* Feeds on the bryozoans *Lyrula hippocrepis, Membranipora,* and *Reginella* sp. Alaska to central California.

Adalaria sp. Color photo in Behrens and Hermosillo 2005: 58, fig. 96. Feeds on the bryozoan *Hincksina minuscula.* Alaska to Oregon.

* = Not in key.

Diaphorodoris lirulatocauda Millen, 1985. [=*Onchidoris hystricina* (Bergh, 1878) partim.]. Color photos in Behrens and Hermosillo 2005: 58, fig. 98; and of *Onchidoris hystricina* in McDonald and Nybakken 1980: 45, fig. 18. Millen 1985: 89 distinguished this species from *Onchidoris hystricina*. Feeds on the bryozoans *Nolella stipitata* and *Tubulipora* sp.

Onchidoris bilamellata (Linnaeus, 1767). Color photos in Behrens and Hermosillo 2005: 59, fig. 99; Beeman and Williams 1980, pl. 105 fig. 14.41; McDonald and Nybakken 1980: 43, fig. 16; Picton and Morrow 1994: 63. See Todd (1979, J. Exper. Mar. Biol. Ecol. 41: 213–255) for ecology. Usually found with the barnacles *Balanus* spp., on which it feeds. Also on Atlantic coast and in Europe.

Onchidoris muricata (Müller, 1776) (=*Onchidoris hystricina*). Color photos in Behrens and Hermosillo 2005: 59, fig. 100; Picton and Morrow 1994: 65; and color illustration in Thompson and Brown pl. 14g. See Millen (1985) for systematics and ecology of this species on the Pacific coast. Feeds on numerous species of bryozoans, including *Alcyonidium* spp., *Callopora* spp., *Cryptosula pallasiana*, *Electra pilosa*, *Membranipora* spp., and *Microporella* spp. Alaska to central California and also in Europe.

TRIOPHIDAE

Crimora coneja Marcus, 1961. Color photos in Behrens and Hermosillo 2005: 60, fig. 103; Beeman and Williams 1980, pl. 106, fig. 14.45; McDonald and Nybakken 1980: 45, fig. 23. See Goddard (1987) for biology of this species. Feeds on the bryozoan *Hincksina minuscula*. Oregon to southern California; rare.

Limacia cockerelli (MacFarland, 1905) (=*Laila cockerelli*). Color photos in Behrens and Hermosillo 2005: 60, fig. 104; of *Laila cockerelli* in Beeman and Williams 1980, pl. 107, fig. 14.49; McDonald and Nybakken 1980: 47, figs. 25–26; color illustration of *Laila cockerelli* in MacFarland 1966, pl. 20, fig. 4. Specimens from the southern part of the range tend to have a longitudinal row of orange-tipped tubercles medially on the dorsum, while specimens from the northern part of the range tend to have smaller, white tubercles more randomly distributed on the dorsum. Ortea et al. (1989, Revista de Biología de la Universidad de Oviedo 7: 99–107) discuss synonymy of *Laila* with *Limacia*. Feeds on the bryozoan *Hincksina velata*.

Triopha catalinae (Cooper, 1863)[=*Triopha carpenteri* (Stearns, 1873)]. Color photos in Behrens and Hermosillo 2005: 65, fig. 116; Beeman and Williams 1980, pl. 105, fig. 14.43; McDonald and Nybakken 1980: 45, fig. 16; color illustration of *T. carpenteri* in MacFarland 1966, pl. 19, figs. 3–4. See Nybakken and Eastman (1977, Veliger 19: 279–289) for feeding preferences. Feeds on numerous species of bryozoans, including: *Caulibugula ciliata*, *Cellaria mandibulata*, *Crisia* spp., *Dendrobeania* spp., *Filicrisia franciscana*, and *Scrupocellaria* spp.

Triopha maculata MacFarland, 1905. Color photos in Behrens and Hermosillo 2005: 66, fig. 117; Beeman and Williams 1980, pl. 105, fig. 144.44a; McDonald and Nybakken 1980: 45, figs. 21–22; color illustration in MacFarland 1966, pl. 19, fig. 5. See Nybakken and Eastman (1977, Veliger 19: 279–289) for feeding preferences. Feeds on numerous species of bryozoans, including *Caulibugula ciliata*, *Cauloramphus californiensis*, *Crisia occidentalis*, *Membranipora membranacea*, and *Scrupocellaria* cf. *S. californica*. Oregon to Mexico.

Triopha occidentalis (Fewkes, 1889) (=*Triopha grandis* MacFarland, 1905). Color photo in McDonald and Nybakken 1980:

45, fig. 19; and of *Triopha maculata* Beeman and Williams 1980, pl. 106, fig. 14.44b; color illustration of *T. grandis* in MacFarland 1966, pl. 19, figs. 1–2. Some authors consider *T. grandis* a synonym of *T. maculata*. McDonald (1983: 218) discusses reasons for maintaining both species and for synonymy of *T. grandis* with *Cabrilla occidentalis* Fewkes, 1889. Usually found on *Macrocystis* among bryozoans which it probably eats.

AEGIRIDAE

Aegires albopunctatus MacFarland, 1905. Color photos in Behrens and Hermosillo 2005: 59, fig. 101; Beeman and Williams 1980, pl. 106, fig. 14.46; McDonald and Nybakken 1980: 45, fig. 24; color illustration in MacFarland 1966, pl. 18, figs. 5–6. Bertsch (1980, Veliger 22: 222–224) discusses feeding on the sponge *Leucilla nuttingi*; also feeds on the sponge *Leucetta losangelensis*.

POLYCERATIDAE

Palio zosterae (O'Donoghue, 1924) (=*Polycera zosterae*). Color photos of *Polycera zosterae* in McDonald and Nybakken 1980: 47, fig. 30; and of *Palio dubia* in Behrens and Hermosillo 2005: 61, fig. 106. Thompson and Brown 1984: 73 state that *P. zosterae* is very similar to *P. dubia*. Feeds on the bryozoans *Bowerbankia gracilis* and *Membranipora membranacea*. See Rivest (1974, Biol. Bull. 167: 534–554) (copulation). Alaska to Northern California; Europe.

Polycera alabe Collier and Farmer, 1964. Color photos in Behrens and Hermosillo 2005: 62, fig. 107; Collier and Farmer 1964: 387, pl. 4. Southern California to Mexico; Chile, rare.

Polycera atra MacFarland, 1905. Color photos in Behrens and Hermosillo 2005: 63, fig. 108; Beeman and Williams 1980, pl. 106, fig. 14.47; McDonald and Nybakken 1980: 47, fig. 27; color illustration in MacFarland 1966, pl. 18, figs. 1–3. Feeds on the bryozoans: *Bugula* sp. and *Membranipora membranacea*. Oregon to Mexico.

Polycera hedgpethi Marcus, 1964. Color photos in Behrens and Hermosillo 2005: 63, fig. 110; Beeman and Williams 1980, pl. 106, fig. 14.48; McDonald and Nybakken 1980: 47, fig. 28. Feeds on *Bugula* spp. Northern California to Panama; Australia; New Zealand; Japan; South Africa; Spain.

Polycera tricolor Robilliard, 1971. Color photos in Behrens and Hermosillo 2005: 64, fig. 111; McDonald and Nybakken 1980: 47, fig. 29; mostly subtidal.

Polycerella glandulosa Behrens and Gosliner, 1988. Color photo in Behrens and Hermosillo 2005: 64, fig. 112. See Behrens and Gosliner (1988, Veliger 30: 319–324) for natural history and morphology. Found on the bryozoans *Bugula* sp. and *Zoobotryon* sp. Central California to Costa Rica.

ACTINOCYCLIDAE

Hallaxa chani Gosliner and Williams, 1975. Color photos in Behrens and Hermosillo 2005: 66, fig. 118; Beeman and Williams 1980, pl. 101, fig. 14.25; McDonald and Nybakken 1980: 39, fig. 49. See Gosliner and Williams (1975) for natural history and morphology. Feeds on the sponge *Halisarca* sp. Alaska to central California.

CHROMODORIDIDAE

Cadlina flavomaculata MacFarland, 1905. Color photos in Behrens and Hermosillo 2005: 78, fig. 153; Beeman and

* = Not in key.

Williams 1980, pl. 99, fig. 14.18; McDonald and Nybakken 1980: 47, fig. 31; color illustration in MacFarland, 1966, pl. 23, fig. 1. Feeds on *Aplysilla glacialis* and other sponges.

Cadlina limbaughorum Lance, 1962 (=*Cadlina limbaughi*). Color photos in Behrens and Hermosillo 2005: 78, fig. 154; Beeman and Williams 1980, pl. 99, fig. 14.17; McDonald and Nybakken 1980: 47, fig. 32; Lance 1962, Veliger 4: 155–9, pl. 38. Behrens (1982, Veliger 24: 291) discusses the spelling of name. Feeds on the sponges *Axinella* sp., *Dysidea amblia*, and *Leiosella idia*. Southern California to Mexico.

Cadlina luteomarginata MacFarland, 1905 (=*Cadlina marginata* MacFarland, 1905). Color photos in Behrens and Hermosillo 2005: 79, fig. 156; Beeman and Williams 1980, pl. 99, fig. 14.19; and of *Cadlina marginata* in McDonald and Nybakken 1980: 49, fig. 33; color illustration in MacFarland 1966, pl. 23, figs. 2–3. See Thompson et al. (1982, Tetrahedron 38: 1865–1873) and Penney (2004, J. Moll. Stud. 70: 399–400) for chemical defenses of this species. Feeds on numerous species of sponges, including *Aplysilla glacialis*, *Dysidea amblia*, *Halichondria panicea*, and *Myxilla incrustans*.

Cadlina modesta MacFarland, 1966. Color photos in Behrens and Hermosillo 2005: 79, fig. 157; Beeman and Williams 1980, pl. 100, fig. 14.20; McDonald and Nybakken 1980: 49, fig. 34. Feeds on the sponges *Aplysilla glacialis* and *Halisarca* sp.

Cadlina sparsa (Odhner, 1921). Color photos in Behrens and Hermosillo 2005: 80, fig. 158; Beeman and Williams 1980, pl. 100, fig. 14.21; McDonald and Nybakken 1980: 49, fig. 35. Feeds on the sponge *Aplysilla glacialis*. Northern to southern California; Chile, rare.

Chromodoris galexorum Bertsch, 1978. Color photo in Behrens and Hermosillo 2005: 80, fig. 160. Southern California to Mexico.

Chromodoris macfarlandi Cockerell, 1901 (=*Glossodoris macfarlandi*). Color photos in Behrens and Hermosillo 2005: 81, fig. 161; Beeman and Williams 1980, pl. 100, fig. 14.22; McDonald and Nybakken 1980: 49, fig. 36; color illustration of *Glossodoris macfarlandi* in MacFarland 1966, pl. 22, figs. 1–3. Feeds on sponges *Aplysilla polyraphis* and *Aplysilla glacialis* (see Goddard 2000 Opisthobranch Newsletter 26: 9–10). Central California to Mexico.

Hypselodoris californiensis (Bergh, 1879) (=*Chromodoris californiensis*). Color photos in Behrens and Hermosillo 2005: 84, fig. 170; Beeman and Williams 1980, pl. 101, fig. 14.24; McDonald and Nybakken 1980: 49, fig. 38; color illustration of *C. californiensis* in MacFarland 1966, pl. 24, figs. 1–3. Feeds on sponges *Dysidea amblia*, *Haliclona* sp., and *Stelletta estrella*. Central California to Mexico.

Mexichromis porterae (Cockerell, 1901) (=*Chromodoris porterae*). Color photo in Behrens and Hermosillo 2005: 85, fig. 174; and of *Chromodoris porterae* in Beeman and Williams 1980, pl. 100, fig. 14.23; McDonald and Nybakken 1980: 49, fig. 37; color illustration of *C. porterae* in MacFarland 1966, pl. 24, figs. 4–5. Bertsch (1977) created the genus *Mexichromis* for chromodorids whose radular teeth are acuspid, with pectinate denticulation. May feed on the sponge *Dysidea amblia*. Central California to Mexico.

DORIDIDAE

Aldisa cooperi Robilliard and Baba, 1972. Color photos in Behrens and Hermosillo 2005: 67, fig. 120; Millen and Gosliner 1985: 197, fig. 1C. See Millen and Gosliner (1985: 208) for ecology and morphology. Feeds on sponges *Anthoarcuata graciae*, and *Hymedesmia* sp.. Alaska to Northern California; Japan.

Aldisa sanguinea (Cooper, 1863). Color photos in Behrens and Hermosillo 2005: 67, fig. 121; Beeman and Williams 1980, pl. 102, fig. 14.27; McDonald and Nybakken 1980: 51, figs. 41–42; color illustration in MacFarland 1966, pl. 25, fig. 8. Roller (1969, Veliger 11: 280–281) describes a color variation in this species. See Millen and Gosliner (1985: 212) for ecology and morphology. Feeds on the sponge *Clathria brepha*. Oregon to Mexico.

Conualevia alba Collier and Farmer, 1964. Color photos in Behrens and Hermosillo 2005: 77, fig. 151; Beeman and Williams 1980, pl. 99, fig. 14.16; Collier and Farmer 1964: 387, pl. 3; McDonald and Nybakken 1980: 49, fig. 40. Central California to Galapagos Islands.

Doris montereyensis Cooper, 1863 (=*Archidoris montereyensis*). Color photos in Behrens and Hermosillo 2005: 71, fig. 132; of *Archidoris montereyensis* in Beeman and Williams 1980, pl. 102, fig. 14.30; McDonald and Nybakken 1980: 53, fig. 50; color illustration of *A. montereyensis* in MacFarland 1966, pl. 27, fig. 8. Valdes (2002) discusses the synonymy of *Archidoris* with *Doris*. Feeds on sponges *Halichondria panicea*, *Haliclona* sp., *Mycale* spp., *Suberites ficus*, *Terpios* sp., and *Hamacantha hyaloderma*. See Cook (1962, Veliger 4: 194–197, feeding); Knowlton and Highsmith 2005, J. Exp. Mar. Biol. Ecol. 327: 36–46 (benefits of feeding on *Halichondria* with zoochlorellae).

Doris odhneri (MacFarland, 1966) (=*Archidoris odhneri*; =*Austrodoris odhneri*). Color photos in Behrens and Hermosillo 2005: 71, fig. 133; of *Archidoris odhneri* in Behrens 1991: 65, fig. 111; Beeman and Williams 1980, pl. 102, fig. 14.29; McDonald and Nybakken 1980: 53, fig. 51; color illustration in MacFarland 1966, pl. 26, fig. 1. Feeds on sponges *Halichondria panicea*, *Haliclona permollis*, *Hymeniacidon ungodon*, *Lissodendoryx firma*, *Mycale* spp., *Myxilla incrustans*, *Stylissa stipitata*, *Syringella amphispicula*, and *Hamacantha hyaloderma*. Found mostly subtidally in California.

Doris tanya Marcus, 1971 (=*Sclerodoris tanya*). Color photos in Behrens and Hermosillo 2005: 72, fig. 134; McDonald and Nybakken 1980: 53, fig. 43. Southern California to Mexico.

DISCODORIDIDAE

Atagema alba (O'Donoghue, 1927) (=*Atagema quadrimaculata* Collier, 1963; =*Petelodoris spongicola* MacFarland, 1966). Color photos in Behrens and Hermosillo 2005: 68, fig. 123; McDonald and Nybakken 1980: 53, fig. 52; and of *Atagema quadrimaculata* in Beeman and Williams 1980 pl. 102, fig. 14.28; color illustration of *P. spongicola* in MacFarland 1966, pl. 27, figs. 1–2. McDonald (1983) and Bertsch and Gosliner (1986) discuss synonymy of *A. quadrimaculata* and *Glossodoridiformia alba* O'Donoghue. Central California to Mexico, rare.

Baptodoris mimetica Gosliner, 1991. Color photos in Behrens and Hermosillo 2005: 69, fig. 125; Gosliner 1991: 273, fig. 1B. Superficially resembles *Dendrodoris albopunctata* from which it is easily distinguished by presence of labial tentacles. Central California to Mexico, rare.

Diaulula sandiegensis (Cooper, 1863) (=*Doris* s.l. sp. MacFarland, 1966). Color photos in Behrens and Hermosillo 2005: 70, fig. 128; Beeman and Williams 1980, pl. 103, fig. 14.32a–b; and of *Discodoris sandiegensis* in McDonald and Nybakken 1980: 51, figs. 46–47; color illustration of *D. sandiegensis* in MacFarland 1966, pl. 27, fig. 6; and *Doris* s.l. sp. in MacFarland 1966, pl. 25, figs. 1–2. Behrens and Valdes (2001) discuss synonymy of *Doris* s.l. sp. and *Diaulula sandiegensis*. See Elvin (1976, Veliger 19: 194–198) for feeding on the sponge *Haliclona permollis*. Also feeds on sponges *Axinella* sp., *Biemma rhadia*, *Halichondria panicea, Lissodendoryx firma, Mycale* spp., *Neoesperiopsis rigida, Sigmadocia* sp., *Terpios* sp., *Toxadocia* sp., and *Xestospongia trindanea.*

Geitodoris heathi (MacFarland, 1905) (=*Discodoris heathi*). Color photos in Behrens and Hermosillo 2005: 73, fig. 137; and of *Discodoris heathi* in Beeman and Williams 1980, pl. 103, fig. 14.33; McDonald and Nybakken 1980: 50, fig. 45; color illustration of *D. heathi* in MacFarland 1966, pl. 27, fig. 7. Miller (1996, J. Moll. Stud. 62: 433–442) and Schrödl (2000, Veliger 43: 197–209) discuss placement in *Geitodoris* based on the shape of the marginal radular teeth. Feeds on sponges *Adocia gellindra, Halichondria panicea, Haliclona permollis, Lissodendoryx firma, Mycale* spp., *Myxilla incrustans,* and *Terpios* sp.

Jorunna pardus Behrens and Henderson, 1981. Color photo in Behrens and Hermosillo 2005: 73, fig. 138. California to Mexico.

Peltodoris lentiginosa (Millen, 1982) (=*Anisodoris lentiginosa*). Color photo in Behrens and Hermosillo 2005: 74, fig. 142. Valdes (2002) discusses the relationship of *Anisodoris* and *Peltodoris.* Millen (1982, Can. J. Zool. 60: 2694–2705) discusses the ecology and morphology of this species. Feeds on sponges *Mycale adhaerens, Mycale psila,* and *Tedania gurjanovae.* Alaska to Oregon.

Peltodoris mullineri Millen and Bertsch, 2000. Color photos in Behrens and Hermosillo 2005: 75, fig. 143; Millen and Bertsch 2000: 356, fig. 1B. See Millen and Bertsch (2000) for the natural history and morphology of this species. Southern California to Columbia.

Peltodoris nobilis (MacFarland, 1905) (=*Anisodoris nobilis*). Color photos in Behrens and Hermosillo 2005: 75, fig. 144; of *Anisodoris nobilis* in Behrens 1991: 66, fig. 113; Beeman and Williams 1980, pl. 103, fig. 14.31a–b; McDonald and Nybakken 1980: 53, fig. 49; color illustration of *A. nobilis* in MacFarland 1966, pl. 28, figs. 1, 3. See Kitting (1981, Biol. Bull. 161: 126–140) for feeding in this species. Feeds on sponges *Antho lithophoenix, Hymedesmia arndti, Clathria originalis, Biemma rhadia, Halichondria panicea, Haliclona permollis, Acanthancora cyanocrypta, Lissodendoryx firma, Mycale* spp., *Myxilla* spp., *Clathria pennata, Mycale psila, Prianos* sp., *Suberites ficus, Tedania gurjanovae, Tetilla arb,* and *Hamacantha hyaloderma.* Common.

Platydoris macfarlandi Hanna, 1951. Color photo in Behrens and Hermosillo 2005: 75, fig. 145. Behrens and Henderson (1983, Veliger 25: 365–369) redescribed this rare species; Bertsch et al. (2000) discussed rarity; occasionally found on *Suberites ficus.* Central California to Mexico, subtidal.

Rostanga pulchra MacFarland, 1905. Color photos in Behrens and Hermosillo 2005: 76, fig. 146; Beeman and Williams 1980, pl. 101, fig. 14.26; McDonald and Nybakken 1980: 51, figs. 43–44; color illustration in MacFarland 1966, pl. 5, fig. 7. See Cook (1962, Veliger 4: 194–197) for feeding in this species; most often found on red sponges, which it closely matches in color. Feeds on sponges *Acarnus erithacus, Antho lithophoenix, Clathria originalis, Clathria pennata,* and *Antho karykina.* Muniaín and Valdés (2000, Proc. Calif. Acad. Sci. 52: 1–10) discuss disjunct range of this species.

Taringa aivica Marcus and Marcus, 1967 (=*Taringa aivica timia*). Color photos in Behrens and Hermosillo 2005: 76, fig. 147; Behrens 2004: 30, pl. 4 fig. A; Hermosillo-Gonzalez 2003: 25 fig. 7. Occasionally found on *Hamacantha hyaloderma.* Southern California to Mexico.

Thordisa bimaculata Lance, 1966. Color photos in Behrens and Hermosillo 2005: 76, fig. 148; McDonald and Nybakken 1980: 51, fig. 48. Central California to Mexico.

Thordisa rubescens Behrens and Henderson, 1981. Color photos in Behrens and Hermosillo 2005: 77, fig. 149; Valdes 2002:

547, fig 4F. Usually found on or near the red sponges *Axinella* and *Lissodendoryx.* Southern California to Mexico.

PHYLLIDIIDAE

Phyllidiopsis blanca Gosliner and Behrens, 1988. Color photo in Behrens and Hermosillo 2005: 89, fig. 186. Subtidal; southern California to Mexico.

DENDRODORIDIDAE

Dendrodoris azineae Behrens and Valdés, 2004. Color photos in Behrens and Hermosillo 2005: 86, fig. 177; Behrens and Valdés 2004: 409, fig. 1. Subtidal, Central to Southern California; rare.

Dendrodoris behrensi Millen & Bertsch, 2005. Color photos in Behrens and Hermosillo 2005: 86, fig. 178; Millen and Bertsch 2005: 191, figs. 1 C–D; and of *Dendrodoris* species B in McDonald and Nybakken 1980: 55, fig. 58. Central California to Mexico.

Dendrodoris nigromaculata (Cockerell in Cockerell and Eliot, 1905). Color photos in McDonald and Nybakken 1980: 53, fig. 56; and of *Doriopsilla nigromaculata* in Behrens and Hermosillo 2005: 88, fig. 184. Southern California to Costa Rica, rare.

Doriopsilla albopunctata (Cooper, 1863) (=*Dendrodoris albopunctata;* =*Doriopsis fulva* MacFarland, 1905). Color photos in Behrens and Hermosillo 2005: 87, fig. 181; Beeman and Williams 1980, pl. 104, fig. 14.34; and of *Dendrodoris albopunctata* in McDonald and Nybakken 1980: 53, fig. 55; color illustration of *Dendrodoris fulva* in MacFarland 1966, pl. 28, fig. 2, and *Dendrodoris albopunctata* pl. 28, fig. 4. Gosliner, Schaffer, and Millen (1999) discuss synonymy of *D. fulva.* Feeds on sponges *Acarnus erithacus, Cliona californiana, Ficulina suberea,* and *Suberites ficus.* Northern California to Mexico, very common.

Doriopsilla gemela Gosliner, Schaefer, and Millen, 1999 (=*Dendrodoris* sp. A McDonald and Nybakken, 1980; McDonald, 1983; =*Dendrodoris* sp. 1 Behrens, 1991). Color photo in Behrens and Hermosillo 2005: 88, fig. 183; and of *Dendrodoris* sp. A in McDonald and Nybakken 1980: 55, fig. 57. Feeds on sponge *Pseudosuberites pseudos.* Central California to Mexico.

Doriopsilla spaldingi Valdés and Behrens, 1998. Color photos in Behrens 2004: 28, pl. 2 fig. C.; Behrens and Hermosillo 2005: 89, fig. 185. Southern California; subtidal. See Valdés and Behrens (1998, Proc. Calif. Acad. Sci. 50: 307–314).

TRITONIIDAE

Tochuina tetraquetra (Pallas, 1788). Color photos in Behrens and Hermosillo 2005: 90, fig. 189; McDonald and Nybakken 1980: 55, fig. 59. Wicksten and DeMartini (1973, Veliger 15: 195) report feeding on the alcyonacean *Alcyonium* sp. (as *Gersemia rubiformis*); also feeds on the sea pen *Ptilosarcus gurneyi* and the gorgonian *Lophogorgia chilensis.* Mainly subtidal.

Tritonia diomedea Bergh, 1894 (=*Duvaucelia gilberti* MacFarland, 1966; =*Tritonia exsulans* Bergh, 1894). Color photos in Behrens and Hermosillo 2005: 91, fig. 190; McDonald and Nybakken 1980: 55, fig. 60; color illustration of *Duvaucelia exsulans* in MacFarland 1966, pl. 39, fig. 7. Feeds on the sea pens *Ptilosarcus gurneyi, Stylatula elongata,* and *Virgularia* sp. Subtidal.

Tritonia festiva (Stearns, 1873) (=*Duvaucelia festiva*). Color photos in Behrens and Hermosillo 2005: 91, fig. 191; Beeman and Williams 1980, pl. 107, fig. 14.50; McDonald and Nybakken 1980: 55, fig. 61; color illustration of *D. festiva* in MacFarland

1966, pl. 39, figs. 1–2. Gomez (1973, Veliger 16: 163–165) found this species feeding on the gorgonian *Lophogorgia chilensis* in the subtidal and in so doing creating a settlement area for the barnacle *Balanus galeatus*. See Jaeckle (1984) for discussion of feeding. Feeds intertidally on alcyonaceans *Clavularia* sp. (southern California only), *Cryptophyton goddardi, Discophyton rudyi,* and *Gersemia rubiformis;* also feeds subtidally on sea pens.

Tritonia myrakeenae Bertsch and Mozqueira, 1986. Color photo in Behrens and Hermosillo 2005: 91, fig. 192. Southern California to Costa Rica.

Tritonia pickensi Marcus and Marcus, 1967. Color photos in Behrens 2004: 28, pl. 2 fig. D; Behrens and Hermosillo 2005: 92, fig. 193; Bertsch and Gosliner 1984 (Shells and Sea Life 16: 138–139, figs. 1–2); Hermosillo-Gonzalez 2003: 25, figs. 9–11. Hermosillo-Gonzalez (2003) reports this species on the gorgonian *Lophogorgia* sp. Feeds on the alcyonacean *Psammogorgia arbuscula*. Southern California to Costa Rica.

HANCOCKIIDAE

Hancockia californica MacFarland, 1923. Color photos in Behrens and Hermosillo 2005: 96, fig. 205; McDonald and Nybakken 1980: 55, figs. 62–63; Hermosillo-Gonzalez 2003: 25, fig. 1; color illustration in MacFarland 1966, pl. 38, fig. 9. Usually on *Macrocystis*, and occasionally found in numbers in the grooves of the kelp fronds where it resembles the hydroids (upon which it probably feeds) growing on the fronds. Northern California to Costa Rica.

DENDRONOTIDAE

Dendronotus albopunctatus Robilliard, 1972. Color photo in Behrens and Hermosillo 2005: 93, fig. 196. See Robilliard (1972, Can. J. Zool. 50: 421–432) for ecology and morphology. Feeds on hydroid *Thuiaria argentea*. Washington to Oregon.

Dendronotus albus MacFarland, 1966. Color photos in Behrens and Hermosillo 2005: 93, fig. 197; Beeman and Williams 1980, pl. 107, fig. 14.52; McDonald and Nybakken 1980: 55, fig. 64; Robilliard 1970, pl. 64, fig. 34; color illustration in MacFarland 1966, pl. 40, fig. 1. See Robilliard (1970) for morphology, ecology, and taxonomy. Feeds on hydroids *Abietinaria* sp., *Thuiaria argentea,* and *Tubularia* sp.

Dendronotus diversicolor Robilliard, 1970. Color photos in Behrens and Hermosillo 2005: 94, fig. 199; McDonald and Nybakken 1980: 57, fig. 65; Robilliard 1970, pl. 64, figs. 35–36. See Robilliard (1970) for morphology, ecology, and taxonomy. Feeds on hydroids *Abietinaria amphora* and *Abietinaria traski*.

Dendronotus frondosus (Ascanius, 1774) (=*Dendronotus venustus* MacFarland, 1966). Color photos in Behrens and Hermosillo 2005: 94, fig. 200; McDonald and Nybakken 1980: 57, figs. 66–67; Robilliard 1970, pl. 63, fig. 29; and color illustration in Thompson and Brown, pl. 5c; color illustration of *D. venustus* in MacFarland 1966, pl. 40, figs. 2. See Robilliard (1970) for morphology, ecology, and taxonomy. Feeds on a wide variety of hydroids, including *Abietinaria* spp., *Aglaophenia struthionides, Bougainvillia glorietta, Coryne* spp., *Dynamena pumila, Hydractinia echinata, Obelia* spp., *Sarsia eximia, Sertularia* spp., *Thuiaria argentea,* and *Tubularia* spp. and on the tunicate *Botryllus schlosseri*. Cosmopolitan in temperate waters of northern hemisphere.

Dendronotus iris Cooper, 1863. Color photos in Behrens and Hermosillo 2005: 94, fig. 201; Beeman and Williams 1980, pl. 107, fig. 14.53; McDonald and Nybakken 1980: 57, figs. 68–69;

Robilliard 1970, pl. 63, fig. 30. See Robilliard (1970) for morphology, ecology, and taxonomy; Wobber (1970, Veliger 12: 383–387) describes feeding behavior. Feeds on burrowing anemone *Pachycerianthus fimbriatus;* usually subtidal.

Dendronotus subramosus MacFarland, 1966. Color photos in Behrens and Hermosillo 2005: 95, fig. 203; Beeman and Williams 1980, pl. 107, fig. 14.51; McDonald and Nybakken 1980: 57, figs. 70–71; Robilliard 1970, pl. 64, fig. 33; color illustration in MacFarland 1966, pl. 40, figs. 3. See Robilliard (1970) for morphology, ecology, and taxonomy. Feeds on hydroids *Aglaophenia struthionides, Obelia* sp., and *Pinauay marina*.

*Dendronotus sp. Color photo in Behrens and Hermosillo 2005: 95, fig. 204. A large red species with white spots on the body and white line on cerata; 150 mm.

TETHYIDAE

Melibe leonina (Gould, 1852). Color photos in Behrens and Hermosillo 2005: 99, fig. 215; Beeman and Williams 1980, pl. 108, fig. 14.54; McDonald and Nybakken 1980: 57, fig. 72; color illustration of *M. leonina* in MacFarland 1966, pl. 41, figs. 1–3. Usually on *Macrocystis* where it uses its oral hood to capture amphipods, copepods, isopods, megalops larvae, ostracods, veliger larvae, and zoea larvae. See Ajeska and Nybakken 1976, Veliger 19: 19–26 (biology); Mills 1994, pp. 313–319 in: Wilson, Stricker, and Shinn 1994, Reproduction and Development of Marine Invertebrates, Johns Hopkins Univ. Press (seasonal swimming and population dispersal); Caldwell and Donovan 2003, Veliger 46: 355–361 (energetics of swimming and crawling).

DOTIDAE

Doto amyra Marcus, 1961 (=*Doto varians* MacFarland, 1966, in part). Color photos in Behrens and Hermosillo 2005: 96, fig. 206; Beeman and Williams 1980, pl. 108, fig. 14.55; McDonald and Nybakken 1980: 59, fig. 73; color illustration of *D. varians* in MacFarland 1966, pl. 42, figs. 1, 3. Feeds on hydroids *Abietinaria* sp., *Aglaophenia* sp., *Bougainvillia* sp., *Obelia* sp., and *Sertularia furcata*. Goddard (1996) indicates that specimens currently identified as *D. amyra* maybe two or even four separate species.

Doto columbiana O'Donoghue, 1921. Color photos in Behrens and Hermosillo 2005: 96, fig. 207; McDonald and Nybakken 1980: 59, fig. 74. Feeds on hydroid *Obelia* sp. Canada to northern California.

Doto kya Marcus, 1961 (=*Doto varians* MacFarland, 1966, in part). Color photos in Behrens and Hermosillo 2005: 97, fig. 208, fig. 151; McDonald and Nybakken 1980: 59, fig. 75; color illustration of *D. varians* in MacFarland 1966, pl. 42, figs. 2, 4. Feeds on hydroids *Abietinaria* sp., *Aglaophenia struthionides, Eudendrium californicum, Obelia dichotoma, Plumularia* sp., *Sarsia* sp., and *Sertularella* sp.

Doto lancei Marcus and Marcus, 1967. Color photo in Behrens and Hermosillo 2005: 97, fig. 209. Feeds on hydroid *Aglaophenia* sp. Southern California to Costa Rica.

ARMINIDAE

Armina californica (Cooper, 1863). Color photos in Behrens and Hermosillo 2005: 101, fig. 219; Beeman and Williams

* = Not in key.

1980, pl. 108, fig. 14.56; McDonald and Nybakken 1980: 59, fig. 76; color illustration in MacFarland 1966, pl. 38, figs. 1–2. Bertsch (1968, Veliger 10: 440–441) describes this species feeding on sea pansy *Renilla koellikeri*; also feeds on the sea pen *Ptilosarcus gurneyi*. Usually subtidal.

Armina cordellensis Gosliner and Behrens, 1996. Body color gray; dorsum granular with irregularly spaced, low, rounded tubercles; 60 mm. Subtidal, Northern California.

DIRONIDAE

Dirona albolineata MacFarland in Cockerell and Eliot, 1905. Color photos in Behrens and Hermosillo 2005: 101, fig. 221; Beeman and Williams 1980, pl. 108, fig. 14.57; McDonald and Nybakken 1980: 59, fig. 79; color illustration in MacFarland 1966, pl. 56, figs. 1–2. Robilliard (1971, Pac. Sci. 25: 429–435) describes predation by *D. albolineata* on the prosobranchs *Lacuna carinata* and *Margarites* spp.; also feeds on the bryozoans *Bugula* sp. and *Dendrobeania lichenoides*.

Dirona pellucida Volodchenko, 1941. (=*Dirona aurantia* Hurst, 1966; see Martynov, 1997). Color photos in Behrens and Hermosillo 2005: 102, fig. 222; of *Dirona aurantia* in Hurst 1966, pl. 2, fig. 1–4. Feeds on bryozoan *Bugula pacifica* but Hurst (1966, Veliger 9: 9–15) notes wide food range.

Dirona picta MacFarland in Cockerell and Eliot, 1905. Color photos in Behrens and Hermosillo 2005: 102, fig. 223; Beeman and Williams 1980, pl. 109, fig. 14.58; McDonald and Nybakken 1980: 59, fig. 80; color illustration in MacFarland 1966, pl. 56, fig. 5. Feeds on hydroid *Aglaophenia* sp., and on bryozoans *Celleporella hyalina* and *Scrupocellaria californica*. Oregon to Mexico; also in Japan.

ZEPHYRINIDAE

Janolus annulatus Camacho-Garcia and Gosliner 2007. Color photo of *Janolus* sp. 1 in Behrens & Hermosillo 2005: 103, fig. 226. Southern California.

Janolus barbarensis (Cooper, 1863) (=*Antiopella barbarensis*; =*Antiopella aureocincta* MacFarland, 1966). Color photo in Behrens and Hermosillo 2005:102, fig. 224; and *Antiopella barbarensis* in McDonald and Nybakken 1980: 59, fig. 77; color illustration of *A. aureocinta* [sic] in MacFarland 1966, pl. 57, fig. 1. Feeds on the bryozoan *Bugula californica* and the hydroid *Corymorpha palma*. Northern California to Mexico.

Janolus fuscus O'Donoghue, 1924. Color photo in Behrens and Hermosillo 2005: 103, fig. 225. Gosliner (1982) clearly differentiates this species from *J. barbarensis*. Feeds on the bryozoans *Bugula pacifica* and *Tricellaria* spp. Alaska to central California; Japan.

FLABELLINIDAE

Flabellina bertschi Gosliner and Kuzirian, 1990. Color photos in Behrens 2004: 29, pl. 3 fig. B; Gosliner and Kuzirian 1990: 3, fig. 1B. Southern California to Panama.

Flabellina cooperi (Cockerell, 1901) (=*Coryphella cooperi*). Color photo in Behrens and Hermosillo, 2005: 105, fig. 230; and of *Coryphella cooperi* in McDonald and Nybakken 1980: 61, fig. 81. Feeds on hydroid *Pinauay crocea*. Central California to Mexico; rare.

Flabellina iodinea (Cooper, 1862) (=*Flabellinopsis iodinea*). Color photos in Behrens and Hermosillo 2005: 106, fig. 232; Beeman and Williams 1980, pl. 109, fig. 14.61; and *Coryphella*

iodinea in McDonald and Nybakken 1980: 61, fig. 82; color illustration of *Flabellinopsis iodinea* in MacFarland 1966, pl. 58, fig. 1. Feeds on tunicate *Diplosoma pizoni* and hydroid *Eudendrium ramosum*. Canada to Mexico, (rare north of Marin County); Galapagos Islands.

Flabellina intermedia (Bergh, 1900). Cerata cores pink, tips white; oral tentacles sprinkled with white distally; 40 mm. Subtidal, Oregon; rare.

Flabellina pricei (MacFarland, 1966) (=*Coryphella pricei*). Color photo in Behrens and Hermosillo 2005: 1–7, fig. 236; and *Coryphella pricei* in McDonald and Nybakken 1980: 61, fig. 83; color illustration of *C. pricei* in MacFarland 1966, pl. 58, fig. 6. Feeds on the hydroid *Halecium* sp. Alaska to southern California; rare.

Flabellina trilineata (O'Donoghue, 1921) (=*Coryphella trilineata*; =*Coryphella fisheri* MacFarland, 1966). Color photo in Behrens and Hermosillo 2005: 108, fig. 238; and *Coryphella trilineata* in Beeman and Williams 1980, pl. 109, fig. 14.60; McDonald and Nybakken 1980: 61, fig. 84; color illustration of *C. fisheri* in MacFarland 1966, pl. 58, figs. 3, 5. Feeds on the hydroids *Eudendrium californicum*, *Garveia annulata*, and *Tubularia* spp.

Flabellina trophina (Bergh, 1894) (=*Coryphella fusca* O'Donoghue, 1921 [see Roginskaya, 1990]; =*Flabellina fusca*). Color photos in Behrens 2004: 30, pl. 4 fig. C; Behrens and Hermosillo 2005: 108, fig. 239. Feeds on the hydroids: *Bougainvillia glorietta*, *Obelia commissuralis*, and *Syncoryne eximia*; and on other aeolids and polychaete worms. See Roginskaya (1990, pp. 47–57, 151–152, 156. In: A. P. Kuznetsov, ed., Feeding and bioenergetics of marine bottom invertebrates. Academy of Sciences of the U.S.S.R. Institute of Oceanology, Moscow, 162 pp. [In Russian]). Alaska to Oregon.

Flabellina verrucosa (Sars, 1829) [=*Flabellina longicaudata* (O'Donoghue, 1922)]. Color photo in Behrens and Hermosillo 2005: 109, fig. 241; and of *Coryphella verrucosa* in Picton and Morrow 1994: 97; color illustration in Just and Edmunds 1985, pls. 43–44; and in Thompson and Brown, pl. 28d. Opaque white line on dorsal surface of oral tentacles and from pericardium to tail; cerata white tipped, cores light brown, red-brown, maroon, or crimson; 20 mm. Feeds on a wide variety of hydroids; Alaska to Washington.

Flabellina sp. Color photo of *Coryphella* sp. A in McDonald and Nybakken 1980: 61, fig. 85. Usually found on the hydroid *Pinauay crocea*. Central California; rare.

EUBRANCHIDAE

Cumanotus sp. Thompson and Brown, 1984 [=*Cumanotus beaumonti* (Eliot, 1906) in part; =*Cumanotus fernaldi* Thompson and Brown, 1984 in part]. Color photos of *Cumanotus* sp. 1 in Behrens and Hermosillo 2005: 110, fig. 245; and *Cumanotus beaumonti* in McDonald and Nybakken 1980: 61, fig. 86. Usually found on the hydroid *Pinauay crocea* on which it feeds and whose polyps it resembles, and where it lays distinctive "cork screw–shaped" egg masses. Oregon to southern California.

Eubranchus misakiensis Baba, 1960. Color photos in Behrens and Hermosillo 2005: 111, fig. 248; McDonald and Nybakken 1980: 61, fig. 87. Behrens (1971, Veliger 14: 214–215) reported *E. misakiensis* in San Francisco Bay. Baba (1971, Venus 30: 63–66) reviews the anatomy. Feeds on hydroids *Obelia geniculata* and *Obelia plana*. San Francisco Bay; Japan; Hong Kong.

* = Not in key.

Eubranchus rupium (Moller, 1842) [=*Eubranchus olivaceus* (O'Donoghue, 1922); see Martynov, 1998]. Color photos in Behrens and Hermosillo 2005: 111, fig. 249; of *Eubranchus olivaceus* in McDonald and Nybakken 1980: 61, fig. 88; color illustration in Just and Edmunds 1985, pl. 51. Feeds on hydroids *Laomedea flexuosa*, *Obelia* spp., and *Plumularia* sp. Also in Greenland, Denmark, Norway.

Eubranchus rustyus (Marcus, 1961). (=*Eubranchus occidentalis* MacFarland, 1966; =*Capellinia rustya*). Color photos in Behrens and Hermosillo 2005: 112, fig. 250; McDonald and Nybakken 1980: 63, fig. 89; color illustration in Just and Edmunds 1985, pl. 54A; and of *E. occidentalis* in MacFarland 1966, pl. 62, fig. 7. Hirano, Goddard, and Hirano (1987, Zool. Sci. (Tokyo) 4: 1116) state that when found together on *Plumularia*, *E. rustyus* eats the hydranth while *E. misakiensis* eats the hydrorhiza. Feeds on the hydroids *Obelia* sp. and *Plumularia lagenifera*.

*Eubranchus sanjuanensis Roller, 1972. Color photo in Behrens and Hermosillo 2005: 112, fig. 251; color illustration in Just and Edmunds 1985, pl. 52A. Cerata cores red to reddish-tan, cerata with opaque white band apically; usually with opaque white on dorsum, anterior of pericardium; 5 mm. Feeds on the hydroid *Sertularella tricuspidata*. Alaska to Washington.

Eubranchus steinbecki Behrens, 1987. Color photo in Behrens and Hermosillo 2005: 112, fig. 252. Southern California to Mexico.

PSEUDOVERMIDAE

*Pseudovermis sp. An interstitial species reported by J. Nybakken and R. Higgins from coarse intertidal sand in Monterey Bay.

TERGIPEDIDAE

Catriona columbiana (O'Donoghue, 1922) [=*Cuthona columbiana*; =*Catriona alpha* (Baba and Hamatani, 1963); =*Cratena spadix* MacFarland, 1966]. Gosliner and Griffiths (1981) discuss the differences between *Catriona* and *Cuthona*. Color photos in Behrens and Hermosillo 2005: 124, fig. 286; and of *Cuthona columbiana* in McDonald and Nybakken 1980: 63, fig. 93; color illustration of *Cratena spadix* in MacFarland 1966, pl. 60, fig. 4. Williams and Gosliner (1979) and McDonald (1983) discuss the synonymy of *Amphorina columbiana* O'Donoghue and *Cuthona alpha* Baba and Hamatani, 1963. Feeds on hydroids *Obelia* sp., *Sarsia eximia*, and *Pinauay crocea*.

Catriona rickettsi Behrens, 1984. Color photo in Behrens and Hermosillo 2005: 124, fig. 287. Found on the hydroid *Pinauay crocea*. Oregon to Mexico.

Cuthona abronia (MacFarland, 1966) (=*Cratena abronia*; =*Trinchesia abronia*). Color photos in Behrens and Hermosillo 2005: 125, fig. 288; McDonald and Nybakken 1980: 63, fig. 91; color illustration of *Cratena abronia* in MacFarland 1966, pl. 59, fig. 3.

Cuthona albocrusta (MacFarland, 1966) (=*Cratena albocrusta*; =*Trinchesia albocrusta*). Color photos in Behrens and Hermosillo 2005: 125, fig. 289; McDonald and Nybakken 1980: 63, fig. 92; color illustration of *Cratena albocrusta* in MacFarland 1966, pl. 61, figs. 1–2. Found on pilings and floating docks in bays and harbors, as well as rocky intertidal; feeds on the hydroid *Pinauay crocea*.

Cuthona cocoachroma Williams and Gosliner, 1979. Color photos in Behrens and Hermosillo 2005: 125, fig. 290; Williams

and Gosliner 1979: 205, fig. 1A. Feeds on the hydroid *Thuiaria* sp. Washington to northern California.

Cuthona divae (Marcus, 1961) (=*Cuthona rosea* MacFarland, 1966; =*Precuthona divae*). Color photos in Behrens and Hermosillo 2005: 126, fig. 292; Beeman and Williams 1980, pl. 110, fig. 14.63; and *Precuthona divae* in McDonald and Nybakken 1980: 63, fig. 90; color illustration of *C. rosea* in MacFarland 1966, pl. 59, fig. 1. Brown (1980: 229) considers *Precuthona* a junior synonym of *Cuthona*. Feeds on the hydroids *Corymorpha* sp. and *Hydractinia*.

Cuthona flavovulta (MacFarland, 1966) (=*Cratena flavovulta*; =*Trinchesia flavovulta*). Color photos in Behrens and Hermosillo 2005: 126, fig. 293; and in McDonald and Nybakken 1980: 63, fig. 94; color illustration of *Cratena flavovulta* in MacFarland 1966, pl. 60, fig. 2. Occasionally found at base of *Laminaria* spp., among hydroids on which it probably feeds. Washington to central California.

Cuthona fulgens (MacFarland, 1966) (=*Cratena fulgens*; =*Trinchesia fulgens*). Color photos in Behrens and Hermosillo 2005: 127, fig. 294; McDonald and Nybakken 1980: 63, fig. 95; color illustration of *Cratena fulgens* in MacFarland 1966, pl. 60, fig. 3. Feeds on the hydroid *Sarsia* sp. Washington to central California.

Cuthona hamanni Behrens, 1987. Color photo in Behrens and Hermosillo 2005: 127, fig. 295. Southern California.

Cuthona lagunae (O'Donoghue, 1926) (=*Cratena rutila* MacFarland, 1966; =*Trinchesia lagunae*). Color photos in Behrens and Hermosillo 2005: 127, fig. 296; Beeman and Williams 1980, pl. 109, fig. 14.62; McDonald and Nybakken 1980: 63, fig. 96; color illustration of *C. rutila* in MacFarland 1966, pl. 60, fig. 1. Occasionally found at base of *Laminaria* spp., amongst hydroids on which it probably feeds. Oregon to Mexico.

Cuthona perca (Marcus, 1958) (=*Trinchesia* sp. in 3rd edition). Color photos in Behrens and Hermosillo 2005: 128, fig. 299; and *Cuthona* sp. A in McDonald and Nybakken 1980: 65, fig. 98. Feeds on the anemones *Aiptasia* sp. and *Diadumene lineata*. On the West Coast, found in Lake Merritt, in Oakland, San Francisco Bay; Florida; Jamaica; Brazil; Barbados; Hawaii; New Zealand.

Cuthona phoenix Gosliner, 1981. Color photo in Behrens and Hermosillo 2005: 129, fig. 300. Feeds on the hydroid *Gonothyraea serialis*. Central California to Costa Rica.

Cuthona punicea Millen, 1986. Color photo in Behrens and Hermosillo 2005: 129, fig. 301. Feeds on an undescribed purple hydroid. Subtidal; Canada.

Cuthona rolleri Behrens and Gosliner, 1988. Color photo in Behrens and Hermosillo 2005: 130, fig. 303. Found on mudflats in bays. Central California.

Cuthona virens (MacFarland, 1966) (=*Cratena virens*; =*Trinchesia virens*). Color photos in Behrens and Hermosillo 2005: 130, fig. 304; McDonald and Nybakken 1980: 65, fig. 97; color illustration of *Cratena virens* in MacFarland 1966, pl. 61, fig. 5. Northern to southern California; rare.

*Cuthona sp. With eight to nine rows of cerata with brown cores and white tips; 8 mm. Found on the hydroid *Tubularia* sp. on docks. Southern California.

Tenellia adspersa (Nordmann, 1844) (=*Tenellia pallida* Alder and Hancock, 1855). Color photos in Behrens and Hermosillo 2005: 133, fig. 313; McDonald and Nybakken 1980: 65, fig. 99; and color illustration in Thompson and Brown, pl. 31g. Widely distributed, usually in bays and harbors, feeds on a wide variety of hydroids, including: *Obelia* spp. and *Pinauay crocea*. Oregon to Southern California; Europe.

* = Not in key.

FIONIDAE

Fiona pinnata (Eschscholtz, 1831). Color photos in Behrens and Hermosillo 2005: 133, fig. 314; Beeman and Williams 1980, pl. 110, fig. 14.64; McDonald and Nybakken 1980: 65, fig. 100; and color illustration in Thompson and Brown, pl. 36c. See Holleman (1972, Veliger 15: 142–146) for feeding and reproduction of this species. Usually found on floating objects. Feeds on the barnacles *Lepas* spp., and the hydrozoans *Porpita porpita* and *Velella velella*. Cosmopolitan in Northern hemisphere.

PHIDIANIDAE

Austraeolis stearnsi (Cockerell, 1901) (=*Facelina stearnsi*). Color photo in Behrens and Hermosillo 2005: 119, fig. 272; and *Phidiana stearnsi* in McDonald and Nybakken 1980: 67, fig. 105. Chan and Gosliner 2006, Rec. West. Australian Mus. Suppl. 69: 111–117 (assignment to *Austraeolis*). Central California to Mexico, rare.

Babakina festiva (Roller, 1972) (=*Babaina festiva*). Color photos in Behrens and Hermosillo 2005: 118, fig. 268; McDonald and Nybakken 1980: 67, fig. 106. Northern California to Mexico; Bahamas; Japan; New Zealand; Spain, rare.

Emarcusia morroensis Roller, 1972. Color photos in Behrens and Hermosillo 2005: 119, fig. 271; McDonald and Nybakken 1980: 65, fig. 104. Most often found on pilings and floating docks in bays and harbors. Washington to southern California; rare.

Hermissenda crassicornis (Eschscholtz, 1831). Color photos in Behrens and Hermosillo 2005: 121, fig. 278; Beeman and Williams 1980, pl. 110, fig. 14.66; McDonald and Nybakken 1980: 65, fig. 102; color illustration in MacFarland 1966, pl. 55, fig. 1. See Avila et al. (1998 Mar. Fresh. Behav. Physiol. 31: 1–19) (feeding behavior). Feeds on a wide variety of organisms, including the hydroids *Abietinaria* spp., *Aglaophenia struthionides*, *Campanularia ritteri*, *Eudendrium* spp., *Halecium corrugatum*, *Obelia* spp., *Sertularella conica*, *Sertularia furcata*, *Stauridiosarsia japonica*, and *Tubularia* spp., the anemones *Anthopleura elegantissima*, *Epiactis prolifera*, *Diadumene lineata*, and *Metridium senile*; cannibalistic in captivity. Very common.

Noumeaella rubrofasciata Gosliner, 1991. Color photos in Behrens and Hermosillo 2005: 122, fig. 281; Gosliner 1991: 273, fig. 1C. Southern California to Costa Rica.

Phidiana hiltoni (O'Donoghue, 1927) (=*Phidiana nigra* MacFarland, 1966; =*Phidiana pugnax* Lance, 1962). Color photos in Behrens and Hermosillo 2005: 123, fig. 282; McDonald and Nybakken 1980: 65, fig. 103; and *Phidiana pugnax* in Beeman and Williams 1980, pl. 110, fig. 14.63; Lance 1962, pl. 38; color illustration of *P. nigra* in MacFarland 1966, pl. 62, fig. 1. McDonald (1983: 202) discusses the synonymy of *Facelina hiltoni* O'Donoghue and *P. pugnax*. Lance (1962: 159) mentions this species attacking other aeolids. Feeds on the hydroid *Hydractinia* sp., and preys on other aeolids when in captivity. Northern California to Mexico.

Sakuraeolis enosimensis (Baba, 1930). Color photo in Behrens and Hermosillo 2005: 123, fig. 284. Behrens (1982 Veliger 24: 359–363) reports this species in San Francisco Bay. Feeds on the hydroid *Tubularia* sp. San Francisco Bay; Japan; Hong Kong.

AEOLIDIIDAE

Aeolidia papillosa (Linnaeus, 1761) Color photos in Behrens and Hermosillo 2005: 114, fig. 256; Beeman and Williams 1980, pl. 111, fig. 14.67; McDonald and Nybakken 1980: 67,

fig. 107; and color illustration in Thompson and Brown, pl. 41a. See Edmunds (1983, J. Moll. Stud. 49: 80–81); Edmunds, et al. (1974, J. Mar. Biol. Assoc. U.K. 54: 939–947) and (1976, ibid., 56: 65–83) (all, predation); Hall, Todd, and Gordon (1982, J. Anim. Ecol. 51: 907–921; 1984, J. Exp. Mar. Biol. Ecol. 82: 11–33) (predation); Harris (1976, J. Moll. Stud. 42: 301) (predation); Seavy and Muller-Parker (2002, Invert. Biol. 121: 115–125) (feeding on *Anthopleura*); and Waters (1973, Veliger 15: 174–192) (feeding); Hall and Todd (1986, J. Moll. Stud. 52: 193–205) (growth and reproduction); Howe and Harris (1978, J. Chem. Ecol. 4: 551–561) (anemone response to feeding); Longley and Longley (1984, Can. J. Zool. 62: 8–14) (mating behavior). Mauch and Elliot (1997, Veliger 40: 148–151) report that the mucus of *A. papillosa* protects it from the nematocysts of *Anthopleura elegantissima*. The color of the species varies with the species of anemones it has eaten. Widely distributed, feeds on a wide variety of sea anemones. Cosmopolitan in Northern Hemisphere.

Aeolidiella chromosoma (Cockerell in Cockerell and Eliot, 1905) (=*Spurilla chromosoma*). Color photo in Behrens and Hermosillo 2005: 114, fig. 258; and *Spurilla chromosoma* in Beeman and Williams 1980, pl. 111, fig. 14.68; McDonald and Nybakken 1980: 67, fig. 111. Feeds on the anemone *Metridium senile*. Central California to Costa Rica; also Japan.

Aeolidiella oliviae MacFarland, 1966 (=*Spurilla oliviae*). Color photos in Behrens and Hermosillo 2005: 115, fig. 259; and *Spurilla oliviae* in Beeman and Williams 1980, pl. 111, fig. 14.69; McDonald and Nybakken 1980: 67, fig. 110; color illustration in MacFarland 1966, pl. 62, fig. 4. Williams (1971, Veliger 14: 215–216) reports a specimen with light salmon pink cerata. Feeds on the anemones *Corynactis californica*, *Epiactis prolifera*, and *Metridium senile*. Northern California to Mexico.

Anteaeolidiella indica (Bergh, 1888) (=*Aeolidiella indica*; =*Aeolidiella takanosimensis* Baba, 1930). Color photo Behrens and Hermosillo 2005: 115, fig. 260; of *Aeolidiella takanosimensis* in McDonald and Nybakken 1980: 67, fig. 108. Miller (2001) established the genus *Anteaeolidiella*. Usually in bays and harbors, feeds on the anemones *Anthothoe stimpsonii* and *Sagartia troglodytes*. Southern California to Mexico; Australia; Florida; Hawaii; Galapagos Islands; Japan; Italy; New Zealand; South Africa.

Cerberilla mosslandica McDonald and Nybakken, 1975. Color photos in Behrens and Hermosillo 2005: 116, fig. 262; McDonald and Nybakken 1980: 67, fig. 109. Feeds on burrowing sea anemones. Subtidal, Oregon to southern California.

Cerberilla cf. *pungoarena* Collier and Farmer, 1964. Color photos in Behrens 2004: 30, pl. 4 fig. F; Collier and Farmer 1964: 390, pl. 6. Subtidal, southern California to Mexico.

REFERENCES

INTERNET WEB SITES

Bibiliographia nudibranchia. http://repositories.cdlib.org/ims/Bibliographia_Nudibranchia/.

A list of the worldwide food habits of nudibranchs. http://www.theveliger.org/nudibranch_food.html.

The opisthobranch site. http://www.seaslug.com/.

Sea slug forum. http://www.seaslugforum.net/.

The slug site. http://slugsite.tierranet.com/.

PUBLISHED WORKS

Beeman, R. D., and G. C. Williams. 1980. Chapter 14. Opisthobranchia and Pulmonata: the sea slugs and allies, pp. 308–354, pls. 95–111. In: Intertidal invertebrates of California, ix + 690 pp., 200 pls. R. H. Morris, D. P. Abbott, and E. C. Haderlie, eds. Stanford University Press.

Behrens, D. W. 1984. Notes on the tergipedid nudibranchs of the northeastern Pacific, with a description of a new species. Veliger 27: 65–71.

Behrens, D. W. 1987. Two new aeolid nudibranchs from southern California. Veliger 30: 82–89.

Behrens, D. W. 2004. Pacific coast nudibranchs, supplement II new species to the Pacific coast and new information on the oldies. Proceedings of the California Academy of Sciences 55: 11–54.

Behrens, D. W. 2005. Nudibranch behavior, 176 pp. New World Publications, Inc., Jacksonville, Florida.

Behrens, D. W., and T. M. Gosliner. 1988a. A new species of tergipedid nudibranch from Morro Bay, California. Veliger 31: 262–266.

Behrens, D. W., and R. Henderson. 1981. Two new cryptobranch dorid nudibranchs from California. Veliger 24: 120–128.

Behrens, D. W., and A. Hermosillo. 2005. Eastern Pacific nudibranchs, a guide to the opisthobranchs from Alaska to Central America. vi + 137 pp., 314 photos. Monterey, CA: Sea Challengers.

Behrens, D. W., and Á. Valdés. 2001. The identity of Doris (s.l.) species MacFarland, 1966 (Mollusca, Nudibranchia, Discodorididae): A persistent mystery from California solved. Proceedings of the California Academy of Sciences 52: 183–193.

Behrens, D. W., and Á. Valdés. 2004. A new species of Dendrodoris (Mollusca: Nudibranchia: Dendrodorididae) from the Pacific Coast of North America. Proceedings of the California Academy of Sciences 55: 408–413.

Bertsch, H. 1977. The Chromodoridinae nudibranchs from the Pacific coast of America.—Part I. Investigative methods and supra-specific taxonomy. Veliger 20: 107–118.

Bertsch, H., O. Angulo Campillo, and J. L. Arreola. 2000. New distributional records of opisthobranchs from the Punta Eugenia region of the Baja California peninsula: a report based on 1997–1998 CONABIO-sponsored expeditions. Festivus (San Diego Shell Club) 32: 99–104.

Bertsch, H., and T. M. Gosliner. 1986. Anatomy, distribution, synonymy, and systematic relationships of Atagema alba (O'Donoghue, 1927) (Nudibranchia: Doridacea). Veliger 29: 123–128.

Bertsch, H., and A. Mozqueira Osuna. 1986. A new species of Tritonia (Nudibranchia) from southern California and Baja California. Nautilus 100: 46–49.

Brown, G. H. 1980. The British species of the aeolidacean family Tergipedidae (Gastropoda: Opisthobranchia) with a discussion of the genera. Zoological Journal of the Linnean Society 69: 225–255.

Camacho-García, Y. E., and T. M. Gosliner. 2007. New species of zephyrinid genus Janolus Bergh, 1884 (Opisthobranchia:Nudibranchia) from North America and Costa Rica. Revista de Biología Tropical 54(4): 1295–1305.

Camacho-García, Y. E., T. M. Gosliner, and Á. Valdés. 2005. Guía de campo de las babosas marinas del Pacifica esto tropical/Field guide to the sea slugs of the tropical eastern Pacific. California Academy of Sciences, 129 pp.

Collier, C. L., and W. M. Farmer. 1964. Additions to the nudibranch fauna of the east Pacific and the Gulf of California. Transactions of the San Diego Society for Natural History 13: 377–396, pls. 1–6.

Engle, J. M., and D. V. Richards. 2001. New and unusual marine invertebrates discovered at the California Channel Islands during the 1997–1998 El Niño. Bulletin Southern California Academy of Sciences 100: 186–198.

Goddard, J. H. R. 1984. The opisthobranchs of Cape Arago, Oregon, with notes on their biology and a summary of benthic opisthobranchs known from Oregon. Veliger 27: 143–163.

Goddard, J. H. R. 1987. Observations on the opisthobranch mollusks of Punta Gorda, California, with notes on the distribution and biology of Crimora coneja. Veliger 29: 267–273.

Goddard, J. H. R. 1990. Additional opisthobranch mollusks from Oregon, with a review of deep-water records and observations on the fauna of the south coast. Veliger 33: 230–237.

Goddard, J. H. R. 1996. Lecithotrophic development in Doto amyra (Nudibranchia: Dendronotacea), with a review of developmental mode in the genus. Veliger 39: 43–54.

Goddard, J. H. R. 1997. Range extensions of eight northeastern Pacific nudibranchs. Opisthobranch Newsletter 23: 13.

Goddard, J. H. R. 1998. A summary of the prey of nudibranch molluscs from Cape Arago, Oregon. Opisthobranch Newsletter 24: 11–14.

Goddard, J. H. R., and N. R. Foster. 2002. Range extensions of sacoglossan and nudibranch mollusks (Gastropoda: Opisthobranchia) to Alaska. Veliger 45: 331–336.

Goddard J. H. R., T. A. Wayne, and K. R. Wayne. 1997. Opisthobranch mollusks and the pulmonate limpet Trimusculus reticulatus (Sowerby, 1835) from the outer Washington coast. Veliger 40: 292–297.

Gosliner, T. M. 1981. A new species of tergipedid nudibranch from the coast of California. Journal of Molluscan Studies 47: 200–205.

Gosliner, T. M. 1982. The genus Janolus (Nudibranchia: Arminacea) from the Pacific coast of North America, with a reinstatement of Janolus fuscus O'Donoghue, 1924. Veliger 24: 219–226.

Gosliner, T. M. 1991. Four new species and a new genus of opisthobranch gastropods from the Pacific coast of North America. Veliger 34: 272–290.

Gosliner, T. M. 1996. 2. The Opisthobranchia, pp. 159–213. In Taxonomic atlas of the benthic fauna of the Santa Maria Basin and western Santa Barbara Channel. Vol. 9—the Mollusca Part 2—the Gastropoda.: P. Scott, J. A. Blake, and A. Lissner, eds. Santa Barbara Museum of Natural History.

Gosliner, T. M. 2004. Phylogenetic systematics of Okenia, Sakishimaia, Hopkinsiella, and Hopkinsia (Nudibranchia: Goniodorididae) with descriptions of new species from the Tropical Indo-Pacific. Proceedings of the California Academy of Sciences 55: 125–161, 29 figs.

Gosliner, T. M., and D. W. Behrens. 1988. A review of the generic divisions within the Phyllidiidae with the description of a new species of Phyllidiopsis (Nudibranchia, Phyllidiidae) from the Pacific coast of North America. Veliger 30: 305–314.

Gosliner, T. M., and R. J. Griffiths. 1981. Description and revision of some South African aeolidacean Nudibranchia (Mollusca, Gastropoda). Annals of the South African Museum 84: 105–150.

Gosliner, T. M., and A. M. Kuzirian. 1990. Two new species of Flabellinidae (Opisthobranchia: Aeolidacea) from Baja California. Proceedings of the California Academy of Sciences 47: 1–15.

Gosliner, T. M., M. C. Schaefer, and S. V. Millen. 1999. A new species of Doriopsilla (Nudibranchia: Dendrodorididae) from the Pacific coast of North America, including a comparison with Doriopsilla albopunctata (Cooper, 1863). Veliger 42: 201–210.

Hermosillo-Gonzalez, A. 2003. New distributional records of opisthobranch mollusks for Bahia de Banderas, Mexico (tropical eastern Pacific). Festivus (San Diego Shell Club) 35: 21–28.

Jaeckle, W. B. 1984. The opisthobranch mollusks of Humboldt County, California. Veliger 26: 207–213.

MacFarland, F. M. 1966. Studies of opisthobranchiate mollusks of the Pacific coast of North America. Memoirs of the California Academy of Sciences 6: 1–546, pls. 1–72.

Marcus, Er. 1961. Opisthobranch mollusks from California. Veliger 3(suppl.): 1–85, pls. 1–10.

Martynov, A. V. 1994. Materials for the revision of nudibranchiate molluscs of the family Corambidae (Gastropoda, Opisthobranchia) taxonomy. Communication 1. Zoologicheskii Zhurnal 73: 3–15. [In Russian].

Martynov, A. V. 1997. Opisthobranchian gastropods at the Commander Islands with remarks on their fauna of the Russian far eastern seas, pp. 230–241. In Benthic flora and fauna of the shelf zone of the Commander Islands. A. V. Rzhavsky, ed. Kamchatka Institute of the Ecology and Environment, Russian Academy of Sciences, Far Eastern Branch. [In Russian].

Martynov, A. V. 1998. Opisthobranch molluscs (Gastropoda, Opisthobranchia) of the family Eubranchidae: taxonomy and two new species from the Sea of Japan. Zoologichesky Zhurnal 77: 763–777. [In Russian].

McBeth, J. W. 1971. Studies on the food of nudibranchs. Veliger 14(2): 158–161.

McBeth, J. W. 1972. Carotenoids from nudibranchs—II. The partial characterization of hopkinsiaxanthin. Comparative Biochemistry and Physiology 41(1B): 69–77.

McDonald, G. R. 1983. A review of the nudibranchs of the California coast. Malacologia 24: 114–276.

McDonald, G. R. 1986. Bibliographia Nudibranchia. Special Publication #8, Institute of Marine Sciences, University of California, Santa Cruz, California, pp. 1–332.

McDonald, G. R., and J. W. Nybakken. 1978. Additional notes on the food of some California nudibranchs with a summary of known food habits of California species. Veliger 21: 110–118.

McDonald, G. R., and J. W. Nybakken. 1980. Guide to the nudibranchs of California, including most species found from Alaska to Oregon, 72 pp., 112 photos. American Malacologists.

McDonald, G. R., and J. W. Nybakken. 1991. A preliminary report of a world-wide review of the food of nudibranchs. Journal of Molluscan Studies 57(4, supplement): 61–63.

McDonald, G. R., and J. W. Nybakken. 1997. A worldwide review of the food of nudibranch mollusks. I. Introduction and the suborder Arminacea. Veliger 40: 157–159.

McDonald, G. R., and J. W. Nybakken. 1999. A worldwide review of the food of nudibranch mollusks. II. The suborder Dendronotacea. Veliger 42: 62–66.

Millen, S. V. 1985. The nudibranch genera *Onchidoris* and *Diaphorodoris* (Mollusca, Opisthobranchia) in the northeastern Pacific. Veliger 28: 80–93.

Millen, S. V., and H. Bertsch. 2000. Three new species of dorid nudibranchs from southern California, USA, and the Baja California peninsula, Mexico. Veliger 43: 354–366.

Millen, S. V., and H. Bertsch. 2005. Two new species of Porostome nudibranchs (family Dendrodorididae) from the coasts of California (USA) and Baja California (Mexico). Proceedings of the California Academy of Sciences 56: 189–199.

Millen, S. V., and T. M. Gosliner. 1985. Four new species of dorid nudibranchs belonging to the genus *Aldisa* (Mollusca, Opisthobranchia), with a revision of the genus. Zoological Journal of the Linnean Society 84: 195–233.

Miller, M. C. 2001. Aeolid nudibranchs (Gastropoda: Opisthobranchia) of the family Aeolidiidae from New Zealand waters. Journal of Natural History 35: 629–662.

Nybakken, J. W., and G. R. McDonald. 1981. Feeding mechanisms of west American nudibranchs feeding on Bryozoa, Cnidaria, and Ascidiacea, with special respect to the radula. Malacologia 20: 439–450.

Ortea, J. A., A. Quero, G. Rodríguez, and Á. Valdés. 1989. Estudio de *Limacia clavigera* (Müller, 1776) (Mollusca: Nudibranchia), con nota sobre su distribución geográfica y la validez del género *Laila* MacFarland, 1905. Revista de Biología de la Universidad de Oviedo 7: 99–107.

Picton, B. E., and C. Morrow. 1994. A field guide to the nudibranchs of the British Isles, 143 pp. Immel Publishing.

Robilliard, G. A. 1970. The systematics and some aspects of the ecology of the genus *Dendronotus*. Veliger 12: 433–479, pls. 63–64.

Thompson, T. E. 1976. Nudibranchs, 96 pp., 30 pls. T. F. H. Publ., New Jersey.

Thompson, T. E. 1976. Biology of opisthobranch molluscs, vol. 1, 207 pp., 21 pls. Ray Society, no. 151.

Thompson, T. E., and G. H. Brown. 1984. Biology of opisthobranch molluscs, vol. 2, 229 pp., 41 pls. Ray Society, no. 156.

Trowbridge, C. D. 2002. Northeastern Pacific sacoglossan opisthobranchs: natural history review, bibliography, and prospectus. Veliger 45: 1–24.

Valdés, Á. 2002. A phylogenetic analysis and systematic revision of the cryptobranch dorids (Mollusca, Nudibranchia, Anthobranchia). Zoological Journal of the Linnean Society 136: 535–636.

Valdés, Á., and P. Bouchet. 1998. A blind abyssal Corambidae (Mollusca, Nudibranchia) from the Norwegian Sea, with a reevaluation of the systematics of the family. Sarsia 83: 15–20.

Vallès, Y., Á. Valdés, and J. A. Ortea. 2000. On the phanerobranch dorids of Angola (Mollusca, Nudibranchia): a crossroads of temperate and tropical species. Zoosystema 22: 15–31.

Williams, G. C., and T. M. Gosliner. 1979. Two new species of nudibranchiate molluscs from the west coast of North America, with a revision of the family Cuthonidae. Zoological Journal of the Linnean Society 67: 203–223.

Bivalvia

EUGENE V. COAN AND PAUL VALENTICH-SCOTT

(Plates 393–434)

The Bivalvia, which includes clams, cockles, scallops, oysters, mussels, piddocks, and shipworms, are fundamentally bilaterally symmetrical mollusks in which the mantle encloses the head, foot, and visceral mass and secretes a shell in the form of two lateral valves, hinged dorsally. With the retreat of the body from direct contact with the substratum, a unique mode of feeding using the ctenidia has developed.

Although the exact mode of feeding of ancestral bivalves is not known, it is likely that suspension feeding by palps and ctenidia was developed with the enclosure of the body by the mantle and shell. Modern representatives of the order Nuculoida retain several primitive morphological characters and are suspension feeders. However, such forms also possess specialized, elongate, palp appendages that can be extended from within the shell to sweep up detritus with cilia and convey it to the mouth. The radula and other structures of the head were lost as this indirect mode of feeding developed. The protobranch ctenidia of the primitive bivalve, presumably similar to those of *Ennucula* or of *Acila,* consist of a central axis bearing, on either side, a series of flattened, ciliated filaments.

The evolution of lamellibranch ctenidia, which characterize the vast majority of modern bivalves, was accomplished by the elongation of the filaments, their folding back on themselves so each ctenidium commonly resembles a tall, narrow W in cross-section, and the binding of adjacent filaments into extensive lamellae. Diagrams of the various types of fusions and shapes of the resulting demibranchs are available in invertebrate zoology textbooks. Complex feeding and rejection tracts of cilia on the ctenidia transport food and rejecta (pseudofeces) respectively, and the palps further sort the food before passing it into the mouth (plate 393A). Thus equipped, bivalves have become nearly complete introverts, using their ciliated ctenidia both for respiration and filtering food from the water. To keep in touch with their environments, they have developed sensory tentacles on mantle edges and at siphonal apertures and may even possess distinctive eyes along the edges of the mantle, as in *Crassadoma* and other scallops.

Ctenidial food collecting, successful as it is, has imposed certain limitations; no bivalve can lead a terrestrial existence, and enclosure of the body within the mantle precludes a really active life. The ability to collect food from the water has ensured a steady source of nutrition, and this has made possible the retreat of many bivalves into protected crevices or burrows.

Bivalves furnish splendid examples of evolutionary diversification and adaptive radiation. Easily recognizable adaptive modifications create features of taxonomic importance. On one hand, this is of great value to the student because closely related families, superfamilies, and orders generally show uniformity in way of life. On the other hand, similar ways of life have also produced parallelisms in structure and adaptation, like we see among distantly related genera that attach themselves to hard substrata (e.g., *Chama, Pododesmus,* and *Crassadoma*). Certain structures, such as the hinge, often permit the recognition of affinities despite outward dissimilarity. Thus, the ligament and cardinal teeth in *Tivela, Mactrotoma,* and *Tresus* readily demonstrate the taxonomic affinity of the latter two genera, whereas in outward form and way of life, *Mactrotoma* and *Tivela* are most similar.

Modern classification of bivalves is based on a wide spectrum of characters, the most important of which are: (1) the structure of the ctenidia, including their relationship to the palps and the types of cilia on them; (2) the mode of life, such as burrowing, boring, attaching with a byssus, cementing to a substratum, or free-living; (3) the morphology of the shell, particularly the hinge teeth and the ligament and the relative sizes and degree of gape of the two valves; (4) the surface sculpture of the valves; (5) the size and position of the adductor muscles that pull the shells closed, which create distinctive scars on the insides of the valves; (6) the degree of fusion of the mantle edges and the presence and nature of siphons, reflected by readily visible scars on the insides of the valves; (7) the microstructure and mineralogy of the shell; (8) the morphology of the stomach; (9) the form of the foot and the presence on it of attachment threads forming a byssus, and (10) information now available through biochemical and genetic methods.

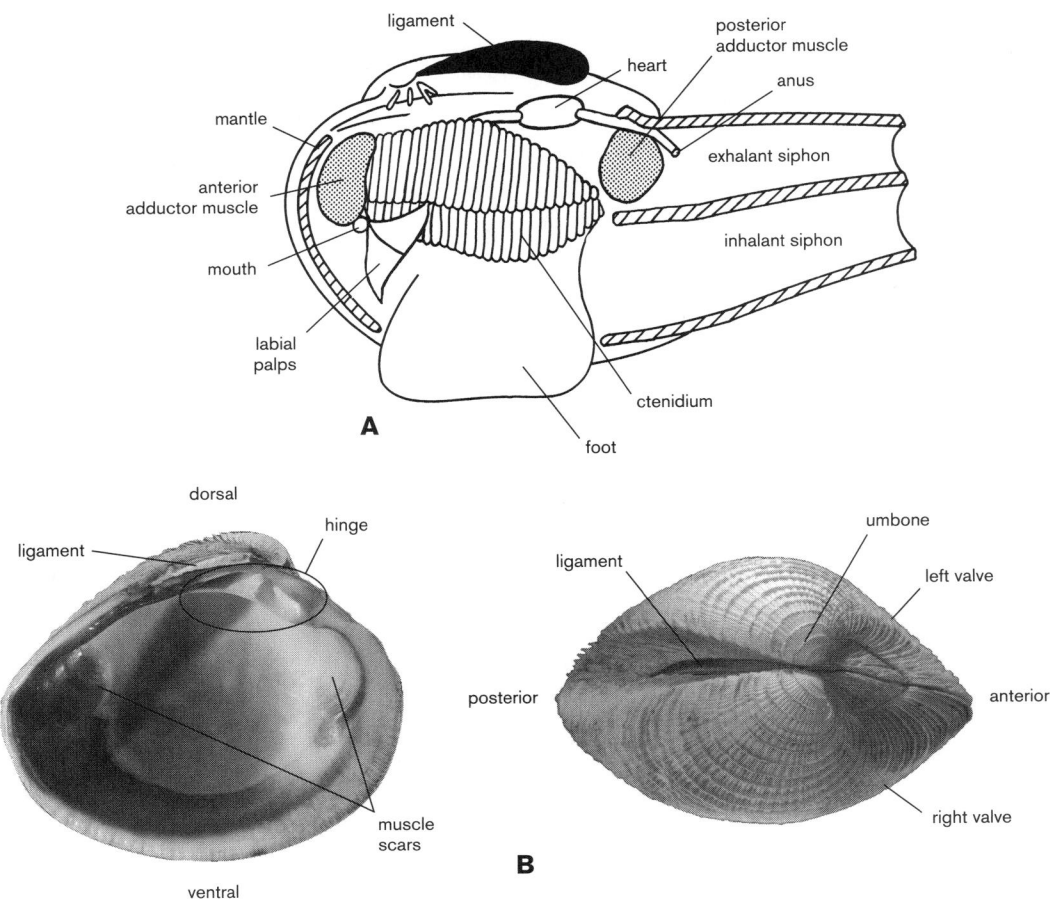

labels in figure A:
ligament
heart
posterior adductor muscle
anus
mantle
exhalant siphon
anterior adductor muscle
mouth
inhalant siphon
labial palps
ctenidium
foot

A

labels in figure B:
dorsal
hinge
ligament
umbone
left valve
ligament
posterior
anterior
muscle scars
right valve
ventral

B

PLATE 393 A, diagrammatic anatomy of a heterodont bivalve; B, diagnostic characters that aid in the identification of bivalve mollusks.

A long and abundant bivalve fossil record has also enabled systematists to establish relationships and rankings that would be difficult to discern using only living forms. The number of characters now employed in bivalve classification, and the degree of parallel evolution in the expression of these features, has led workers to adopt names for orders not based on any one set of characters.

Bivalve Diversity and Classification

PROTOBRANCHIA

A compact, probably natural group composed of two orders.

SOLEMYOIDA

The genera *Solemya* and *Acharax* have primitive, protobranch ctenidia, which are the main organs of feeding, and, as in the Nuculoida, a flattened foot. Solemyoida are subtidal, with *Solemya valvulus* and *S. reidi* occurring offshore in waters of moderate depth in central California.

NUCULOIDA

These have protobranch ctenidia, a primitive taxodont hinge with a row of similar teeth, and they feed in large part by the palps. They are also subtidal. Valves of *Nuculana taphria* and *Yoldia cooperi* are occasionally washed ashore, and *Acila castrensis* and *Ennucula tenuis* may be dredged in water of moderate depth.

PTERIOMORPHA

Most systematists agree that this subclass is a natural group, a conclusion based both on fossil evidence and on the overall similarity of living representatives. Most members are epifaunal, attached to surfaces by a byssus or by cementation, and, as a result, the foot is reduced or entirely absent. The mantle margins are less fused than in the subclass Heterodonta.

ARCOIDA

This order has filter-feeding, filibranch ctenidia in which the elongate filaments are reflected, so each gill appears as a tall, narrow W in cross-section; adjacent filaments are united by patches of interlocking cilia to form lamellae. This is chiefly but not exclusively a tropical order, and most have a taxodont hinge. Of the Arcidae, only *Acar bailyi* has occasionally been found in the intertidal zone of central California, and of the Glycymerididae, shells of *Glycymeris septentrionalis* have been found washed ashore. The Philobryidae, represented locally by *Philobrya setosa*, which resembles a small mytilid, is also placed here.

MYTILOIDA

Members of this order have either filibranch or eulamellibranch ctenidia, the latter with actual bridges of tissue between adjacent filaments. The adductor muscles are unequal in size (heteromyarian). The order includes the Mytilidae, the well-known mussels, which lack conspicuous hinge teeth. Most are found epifaunally attached to rocks or pilings by a byssus, although some are infaunal in soft substrata and others bore into shale.

PTERIOIDA

This group includes many taxa that are mostly tropical, with only the Ostreidae (true oysters) found in central California.

LIMOIDA

This order contains the family Limidae, two species of which occur offshore in water of moderate depth in central California.

PECTINOIDA

Members of this order include two families in the central California intertidal zone, Pectinidae (scallops) and Anomiidae (rock jingles or rock oysters). Only the much-enlarged posterior adductor muscle is present, a condition termed "monomyarian." Some attach to the substratum, whereas others are free living.

HETERODONTA

This group includes most of the familiar clams. Ctenidia are eulamellibranchiate; the mantle margins are well fused, and elongate siphons are present in most. Distinctive patterns of hinge teeth and ligament characterize the different families.

VENEROIDA

Most heterodonts are members of this order, in which hinge teeth are generally well developed. Many veneroids are shallow to deep burrowers. Local families include Lucinidae, Thyasiridae, Ungulinidae, Carditidae, Chamidae, Galeommatidae, Lasaeidae, Corbiculidae, Pisidiidae, Cardiidae, Neoleptonidae, Veneridae, Petricolidae, Tellinidae, Donacidae, Psammobiidae, Semelidae, Solecurtidae, Solenidae, Pharidae, Mactridae, and Hiatellidae.

MYOIDA

In this group, burrowing and boring are characteristic ways of life; most have long siphons, and the hinge has few teeth. Local families include Myidae, Corbulidae, Pholadidae, and Teredinidae (shipworms).

PHOLADOMYOIDA

Members of this group have siphons, and many burrow into the substratum; the shells are generally thin, and many are nacreous within; hinge teeth are inconspicuous or absent. There is only one order, the Pholadamyoida. In the central California intertidal zone, there are a few members of this order, which also includes the offshore, carnivorous Septibranchia. The only local intertidal families are the Lyonsiidae, Thraciidae,

and Laternulidae. Taxa occasionally dredged in water of moderate depth include *Pandora punctata*, *P. bilirata*, and *P. filosa* of the Pandoridae, *Thracia trapezoides* of the Thraciidae, *Periploma discus* of the Periplomatidae, *Cardiomya pectinata* of the Cuspidariidae, and *Trigonulina novemcostatus* of the Verticordiidae.

Ecology and Habitats

Most local bivalves are free-living infaunal burrowers or nestlers, or epifaunal and attach to the substratum by cementation or a byssus. A number of species often occupy empty pholad holes; the external shape in these and other nestlers may vary considerably, and such situs forms have sometimes been given separate names of no taxonomic significance. Only a few of our local species are commensal; for example, *Cryptomya californica* lives in association with *Urechis* and burrowing anomuran shrimps, tapping their burrows with its short siphons; *Mytilimeria nuttallii* lives embedded in compound ascidian tests; *Neaeromya rugifera* occurs byssally attached beneath the abdomen of *Upogebia*; and *Mysella pedroana* is found on the legs and gills of the large sand crab *Blepharipoda occidentalis*.

Other galeommatoideans, such as *Lasaea* and *Kellia*, are nestlers among the byssal threads of mussels, in crevices, even in marine-laboratory seawater systems. Best represented in our intertidal zone fauna are the mytilids (with about 16 species), the venerids (about 12 species), the tellinids (about 10 species), and the pholads (about 10 species).

It may be noted that some of the most abundant bivalves in our bays and lagoons were introduced from other provinces; these include *Mytilus galloprovincialis*, *Geukensia demissa*, *Gemma gemma*, *Petricolaria pholadiformis*, *Mya arenaria*, *Lyrodus pedicellatus*, and *Teredo navalis* from the Atlantic, and *Musculista senhousia*, *Venerupis philippinarum*, *Nuttallia obscurata*, *Theora lubrica*, *Corbula amurensis*, and *Laternula marilina* from Asia. Others, such as the quahog *Mercenaria mercenaria*, are not established locally, but specimens may be encountered. The Asian oysters *Crassostrea sikamea* and *Crassostrea gigas*, and the eastern oyster, *Crassostrea virginica*, although raised commercially in Tomales Bay, Drake's Estero, and other areas, do not usually reproduce here; rare free-living individuals are occasionally reported. However, *Crassostrea gigas*, the most important commercially raised oyster, reproduces in southern California. Further introductions may be expected.

Morphology and Identification

Plate 393B illustrates most of the basic terminology used in the keys to the bivalves. The first-formed part of the shell is the beak. The highest or most prominent point of each valve, at or near the beak, is called the umbo (pleural, umbones). The outer surface of the valves may be covered with a fibrous or horny layer, the periostracum. Beneath this is the calcareous shell, which may be variously sculptured with radial and/or commarginal ridges. The valves are joined dorsally at the hinge, where there is a horny, elastic ligament. The ligament may be partly or entirely internal; if within, it is called a resilium, housed in a resilifer, and if a projecting calcareous shelf is built up for it, this structure is called a chondrophore. In most bivalves, interlocking teeth strengthen the hinge. In bivalves with a few strong teeth, those radiating directly from the beaks are cardinals, whereas those lying posterior or anterior to the beaks are called laterals. In the boring pholads, there may be a calcareous projection in each valve below the hinge called a myophore for the attachment of muscles.

The inner surface of the valves bears the scars of muscle attachments, the most prominent being those of the adductors. The pallial line marks the attachment of the mantle. Posteriorly, this line may be indented to form a pallial sinus, marking the position of muscles for the siphons, if present.

There is no single technique for relaxing all species of bivalves. We have had varying results relaxing bivalves by adding menthol crystals, magnesium chloride (7% in seawater), or dilute ethyl alcohol (approximately 3%) to native seawater until the specimens are not responsive to touch. Members of the family Veneridae can be especially difficult to relax, and we have had to resort to a combination of the above methodologies to properly relax them.

When a specimen is nonresponsive to touch it can then be fixed in 5% buffered formalin (a handful of sodium borate [Borax] per gallon sample—not scientific but the reality of fieldwork). Formalin should be buffered to prevent decalcification, especially of small shells that can rapidly dissolve. Most bivalve specimens should not be allowed to remain in formalin for more than 48 hours, or severe damage can occur to the shell. As usual, use care with formalin; this carcinogen should only be used outdoors or in well-ventilated conditions.

After fixation, bivalves should be rinsed and soaked in distilled water to remove any formalin residue. Finally, they should be place in 80% ethyl alcohol for long-term preservation. If DNA studies are anticipated, formalin should not used. Instead, the specimens should be frozen or placed directly into 95% alcohol, although this makes the specimens poorer candidates for histological examination.

Perhaps the most difficult process in examining a bivalve is to safely open tightly closed (and frequently fragile) valves to observe details of the hinge, pallial line, and muscle scars. Depending on the final deposition of the specimens, one of the following two techniques should provide acceptable results. With either method one must be sure to examine the outside characters (e.g., ligament, lunule, escutcheon, etc.) of the specimen first, as these may be unobservable once the specimen is opened.

If there is no need to examine the soft parts of the bivalve, the specimen can be placed in a dilute solution (approximately 30%–50%) of household bleach in distilled water. Usually after 15 minutes for small shells, longer for larger specimens, all soft tissue will dissolve, allowing easy access to the inside of the shell. This method destroys all soft tissue and the periostracum.

To preserve the soft tissue and periostracum of the specimen, one can open a bivalve shell by using a single-edge razor blade. Carefully place the specimen, ventral side up, in a stiff adhesive (e.g., poster mount) or clay. Then, while carefully holding the specimen with forceps, gently slice the blade into the opening between the valves. With small bivalves, one must accomplish this task while viewing the specimen through a dissecting microscope. While this technique takes practice, it is possible to open specimens as small as 1 mm in length with little or no damage. This method is usually not effective with species that have a heavily crenulated ventral margin.

External observations should include shell sculpture, ligament (if external), beak size and orientation, lunule and escutcheon (if present), and periostracum. Internal shell observations should include type and number of teeth, type of resilifer (if present), pallial line and sinus (if present), and adductor muscle scars (plate 393B).

In smaller specimens, it is often difficult to observe the pallial line and muscle scars. Dyeing the shell in crystal violet allows easier examination of these features. Place the specimen in a solution of crystal violet (no definite concentration, just mix a small amount—maybe 3 g—in 30 cc–50 cc of water and dispense from a dropper bottle) for approximately 15 minutes, or until the pallial line becomes more visible. Crystal violet stain can be removed from the shell by soaking it in ethyl alcohol.

The following keys attempt to achieve a balance between demonstrating adaptive, phylogenetically related groups and providing somewhat artificial keys for easy identification. The keys will be most useful for fresh, mature, unworn specimens and are designed for intertidal species occurring between the Oregon coast and Point Conception. The keys contain both dichotomous and multichotomous choices.

In addition to the general references on Bivalvia, the general references on Mollusca include valuable sources of information on the ecology, biology, physiology, and systematics of bivalves. See especially Coan et al. (2000).

Key to Intertidal Zone Bivalve Families

1. Shell with file-like denticulations anteriorly (plate 394A), and internally with a myophore (except in *Netastoma*) (plate 394B); boring into heavy mud, clay, shale, wood, or shell 2
— Shell various, not as above 3
2. Boring into wood; pallets at siphon tips (plate 394C); anterior end of shell indented with an angular notch (plate 394D) (shipworms) . . . . Teredinidae
— Boring into a variety of substrata; no pallets on siphon tips; anterior end pointed or evenly curved, not notched (plate 394A, 394B) . . . . Pholadidae
3. Dorsal margin produced into triangular "ears" at least in young; sculpture of radial ribs (plate 394E) . . . . Pectinidae
— Dorsal margin without ears 4
4. Shell firmly cemented to substratum (plate 394F); valves irregular and/or distinctly different from one another 5
— Shell not cemented to substratum; valves more or less regular, similar 7
5. Adductor muscles coalesced (plate 394G), resulting in one large, sometimes complex, muscle scar near center of shell 6
— Adductor muscles at opposite ends of shell, not coalesced (plate 394F) . . . . Chamidae
6. Adductor muscle scar complex, with central area showing two to three superimposed secondary scars; one valve with a hole ventral to beaks in most (plate 394H); attached to hard substrata . . . . Anomiidae
— Adductor muscle scar simple (plate 395A); in bays, often attached to rocks and other shells by left valve . . . . . Ostreidae
7. Without a chondrophore or projecting, interlocking teeth on hinge (irregular denticles or a nonprojecting resilifer may be present) 8
— Hinge plate with a projecting chondrophore and/or true teeth 11
8. Shell minute, triangular, with a single central adductor muscle scar (plate 395B) . . . . Philobryidae
— Shell with two adductor muscle scars 9
9. Adductor muscles and their scars not equal in size (plate 395C); anterior muscle smaller, located at or near beaks; shell

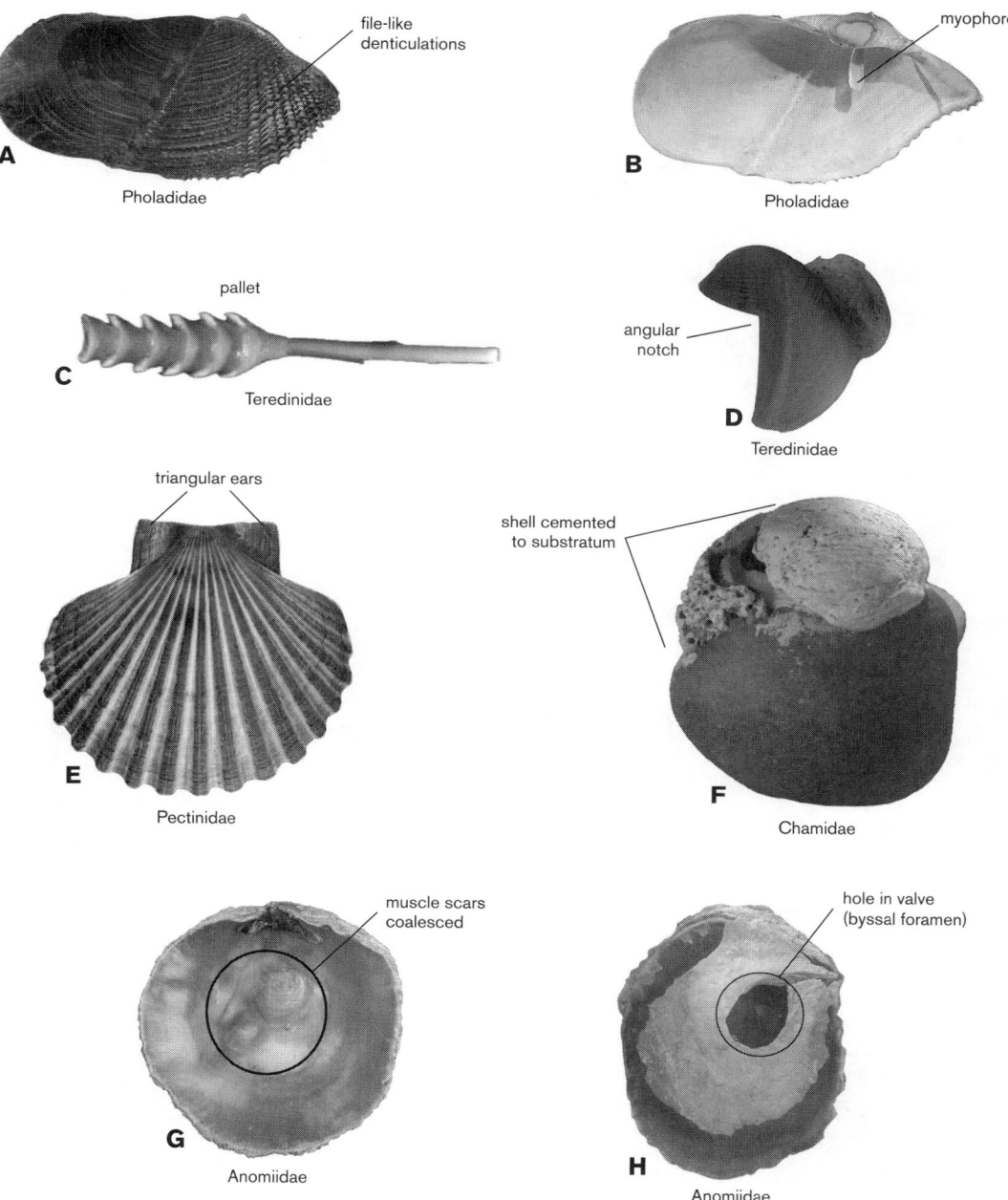

PLATE 394 A, file-like denticulations of Pholadidae; B, projecting myophore of Pholadidae; C, pallet attached to siphons of Teredinidae; D, notch in shell of Teredinidae; E, triangular ears of Pectinidae; F, Chamidae shell cemented to rock; G, coalesced muscle scars of Anomiidae; H, byssal foramen of some Anomiidae.

brown or black, cylindrical or tapering anteriorly . . . .
. . . . Mytilidae
— Adductor muscles and their scars approximately equal in size (but not necessarily in shape) 10
10. Ligament external; shell porcelaneous within; pallial line broken into patches (plate 395D) in some (others not in patches) . . . . Hiatellidae (in part)
— Ligament both external and internal in a resilifer; shell porcelaneous; pallial line never in patches (plate 395E) . . . .
. . . . Thraciidae
— Ligament only internal; shell nacreous within (plate 395F); pallial line not in patches . . . . Lyonsiidae

11. Hinge with a row of many similar appearing teeth (taxodont) (plate 395G) . . . . Arcidae
— With only a few (heterodont) teeth and/or a projecting chondrophore (plates 395H, 396D, 396H) 12
12. Hinge with an internal ligament in a distinct resilifer or chondrophore (an external ligament may also be present) (plates 395H, 396A) 13
— Hinge with ligament entirely external, on dorsal surface (or slightly sunken into dorsal margin of hinge plate) (plate 397F) 20
13. Hinge with a spoon- or peg-shaped chondrophore in left valve only (plate 395H) 14

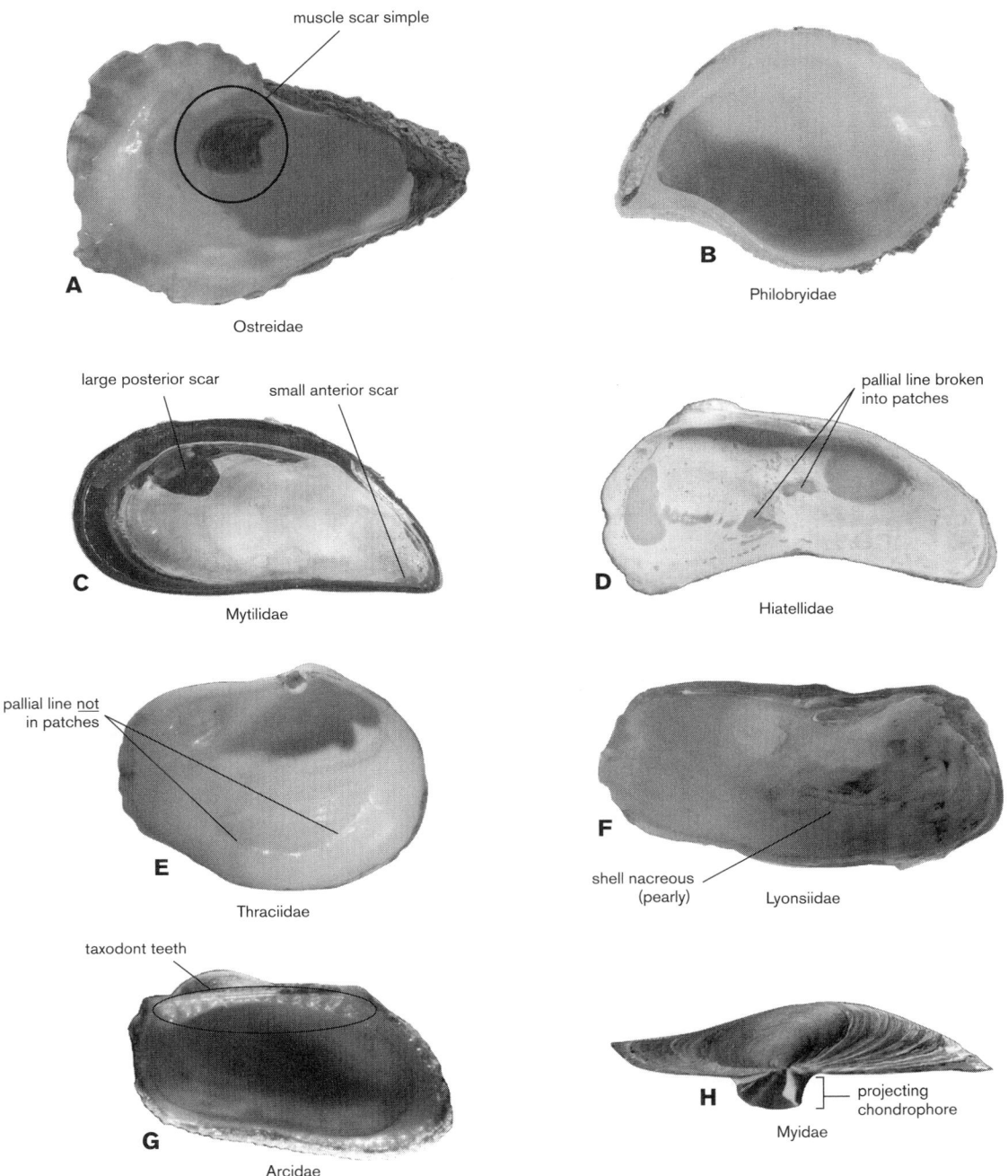

PLATE 395 A, simple muscle scar of Ostreidae; B, triangular shape of Philobryidae; C, unequal adductor scars of Mytilidae; D, patchy pallial line of some Hiatellidae; E, unbroken pallial line of Thraciidae; F, nacreous interior of Lyonsiidae; G, taxodont teeth of Arcidae; H, projecting chondrophore of Myidae.

— Ligament chiefly internal; projecting chondrophore, if present, in both valves 15
14. Chondrophore strongly projecting (plate 395H). . . . . Myidae
— Chondrophore weakly or not projecting (plate 396A). . . . . Corbulidae
15. Adult shells mostly small (<25 mm); pallial line without a sinus. 16
— Adult shells mostly large (>25 mm); pallial line with a distinct sinus (plate 396G) 18
16. Hinge with two to three cardinal teeth and at least one lateral

tooth in each valve (plate 396B) . . . . Neoleptonidae
— Hinge simple, with only one to two hinge teeth 17
17. Animal larger than shell (plate 396C) . . . . Galeommatidae
— Shell able to cover entire animal . . . . Lasaeidae
18. Shell nacreous within (plate 396E) . . . . Laternulidae
— Shell porcelaneous within (plate 396G). 19
19. Anterior cardinal tooth in left valve inverted V-shaped (plate 396F) . . . . Mactridae
— Cardinal teeth not inverted V-shaped (plate 396G) . . . . Semelidae

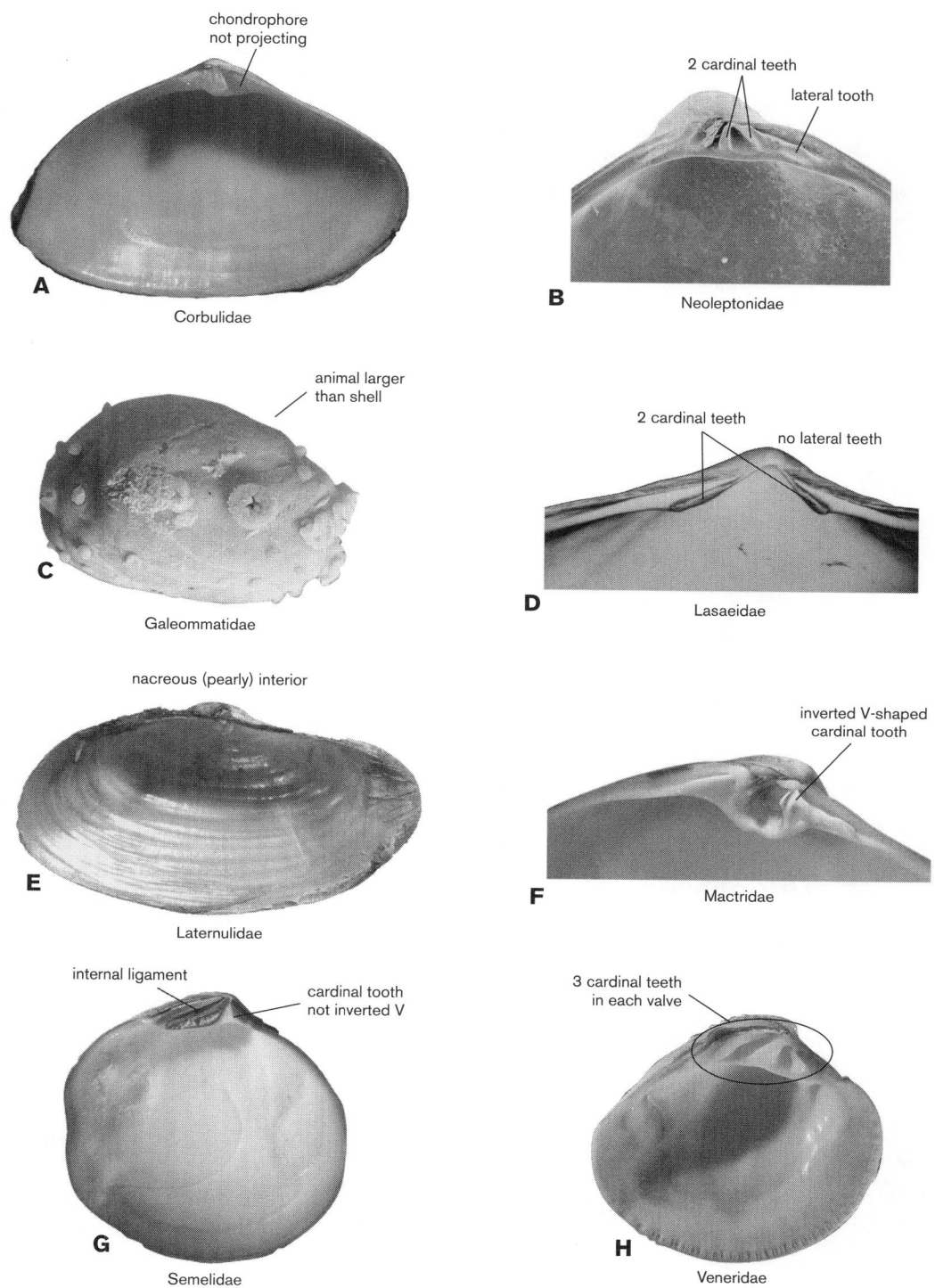

PLATE 396 A, nonprojecting chondrophore of Corbulidae; B, two cardinal teeth and one lateral tooth of Neoleptonidae; C, shell completely internal, Galeommatidae; D, two cardinal teeth of Lasaeidae; E, nacreous interior of Laternulidae; F, inverted V-shaped cardinal tooth of Mactridae; G, porcelaneous interior and simple cardinal tooth of Semelidae; H, three cardinal teeth of Veneridae.

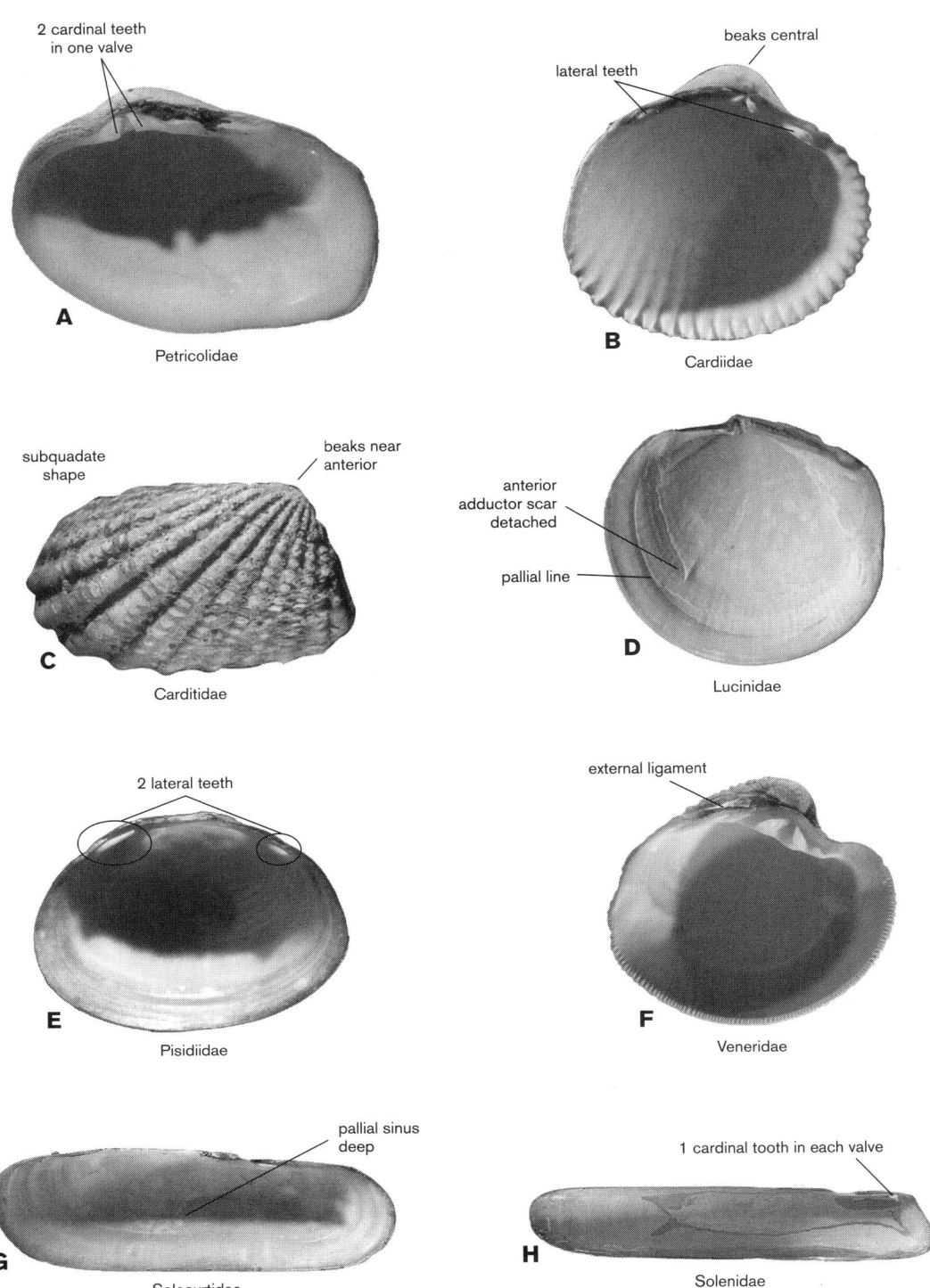

A
Petricolidae

2 cardinal teeth
in one valve

B
Cardiidae

beaks central

lateral teeth

C
Carditidae

subquadate
shape

beaks near
anterior

D
Lucinidae

anterior
adductor scar
detached

pallial line

E
Pisidiidae

2 lateral teeth

F
Veneridae

external ligament

G
Solecurtidae

pallial sinus
deep

H
Solenidae

1 cardinal tooth in each valve

PLATE 397 A, two cardinal teeth of Petricolidae; B, lateral teeth of Cardiidae; C, anteriorly placed beaks of Carditidae; D, detached adductor muscle scar of Lucinidae; E, two lateral teeth of Pisidiidae; F, external ligament of Veneridae; G, deep pallial sinus of Solecurtidae; H, single cardinal tooth of Solenidae.

— Shell subquadrate, not particularly inflated; beaks near anterior end (plate 397C) . . . . Carditidae
24. Anterior adductor muscle scar narrower than posterior, its lower end detached from pallial line and bent inward (plate 397D) . . . . Lucinidae
— Adductor muscle scars approximately equal in shape . . . . .
. . . . 25

25. Adult small to minute, <15 mm in length 26
— Adult >15 mm in length 27
26. Shell thin; ligament partly sunken below dorsal surface; hinge with one to two cardinal and two lateral teeth in each valve (plate 397E) . . . . Pisidiidae
— Shell relatively thick and heavy; ligament completely external; hinge with three cardinal teeth in each valve and,

at most, one lateral tooth in each valve (plate 397F). . . . .
. . . . Veneridae (in part)
27 Shell evenly cylindrical, length about 2.5 times height (plate
    397G, 397H) 28
—   Shell otherwise 30
28. Beaks at or near anterior end; pallial sinus relatively shal-
    low; some with a prominent internal radial rib 29
—   Beaks nearly central; pallial sinus relatively deep; never with
    prominent radial strengthening rib (plate 397G) . . . .
    . . . . Solecurtidae
29. One cardinal tooth in each valve (plate 397H) . . . .
    . . . . Solenidae
—   One valve with two cardinal teeth, the other with four
    (plate 398A) . . . . Pharidae
30. Fewer than three cardinal teeth in each valve 31
—   Three cardinal teeth in one or both valves (plate 398C)
    32
31. Hinge with very elongate, serrate lateral teeth (plate 398B)
    . . . . Corbiculidae
—   Lateral teeth, if present, not very elongate or serrate 33
32. Three cardinal teeth in each valve (plate 398C) . . . .
    . . . . Veneridae (in part)
—   Three cardinal teeth in one valve, two in the other (plate
    398D) . . . . Petricolidae (in part)
33. Pallial sinus deep (plate 398E) . . . .
    . . . . Psammobiidae, Tellinidae
—   Pallial sinus shallow or absent 34
34. Shell oval; pallial line narrow, simple (plate 398F) . . . . . .
    . . . . Ungulinidae
—   Shell elongate; pallial line thick, patchy in some (plate
    398G) . . . . Hiatellidae

Species Keys and Lists

SOLEMYIDAE

Solemya reidi Bernard, 1980 (plate 399A). Offshore on soft
bottoms with high organic content and low oxygen level.
Coan et al. 2000: 65, 67.

Solemya valvulus Carpenter, 1864 (plate 399B). Offshore in
fine sediments as far north as Monterey Bay in warm-water
years. Coan et al. 2000: 66–67.

NUCULIDAE

Acila castrensis (Hinds, 1843) (plate 399C). Offshore in fine
sediments. Coan et al. 2000: 75–76.

Ennucula tenuis (Montagu, 1808) (plate 399D). Offshore in
fine sediments. Coan et al. 2000: 76–77.

NUCULANIDAE

Nuculana taphria (Dall, 1896) (plate 399E). Offshore in fine
sediments, occasionally washed up on beaches. Coan et al.
2000: 89–90.

YOLDIIDAE

Yoldia cooperii Gabb, 1865 (plate 399F). Offshore in fine sediments, occasionally washed up on beaches. Coan et al. 2000:
112–113.

ARCIDAE

The southern Californian *Acar bailyi* (Bartsch, 1931) (plate
400A), has been found in the intertidal zone as far north as
Cayucos, San Luis Obispo County. Coan et al. 2000: 130–131;
Rost 1955, Alan Hancock Pacific Exped. 20: 190–191.

GLYCYMERIDIDAE

Glycymeris septentrionalis (Middendorff, 1849) (plate 400B).
Occasionally washed ashore on Californian beaches. Coan et al.
2000: 143–144.

PHILOBRYIDAE

Philobrya setosa (Carpenter, 1864) (plate 400C). Resembles a
small mussel (height to about 5 mm), lives attached to rocks or
algae by a byssus in the low intertidal zone. Coan et al. 2000: 151.

MYTILIDAE

KEY TO MYTILIDAE

1. Adult shell minute (5 mm or less), stubby, inflated, with a
    forwardly directed, anteroventral protuberance; ventral mar-
    gin sinuous posteriorly (plate 401A) . . "*Musculus*" *pygmaeus*
—   Adult shell not minute 2
2. Beaks terminal, at anterior end of shell (plates 401B, 402A)
    3
—   Beaks near anterior end, but not terminal (plate 402C) 6
3. Anterior end bridged by a shelly septum internally; shell
    with prominent radial ribs; black externally, purplish in-
    ternally (plates 401B, 402A) . . . . *Septifer bifurcatus*
—   Anterior end without internal septum . . . . *Mytilus* 4
4. Shell generally with irregular radial ribs especially on pos-
    terior end (plate 401C) . . . . *Mytilus californianus*
—   Shell generally smooth, without strong radial ribs 5
5. Members of this group can not be reliably diagnosed by
    shell characters (see text) (plate 401D–401F) . . . .
    . . . . . *Mytilus trossulus, Mytilus galloprovincialis, Mytilus edulis*
6. Shell cylindrical, with dorsal and ventral margins more or
    less parallel (plates 402B, 403A–403D) 7
—   Shell not cylindrical 10
7. Posterodorsal slope with rough, chalky encrustations; bor-
    ing in rock (plates 402B, 403A) . . . . . . *Lithophaga plumula*
—   Posterodorsal slope with a thick mat or wrinkles, often with
    mud and debris, not chalky encrustations . . . . *Adula* 8
8. Periostracum with irregular striae; boring in soft shale
    (plate 403B) . . . . *Adula gruneri*
—   Smooth or with a few radiating striations anteriorly 9
9. Shell elongate, generally tapering posteriorly, posterior end
    not wider than anterior end; generally boring in soft shale
    (plate 403C) . . . . *Adula californiensis*
—   Shell stout, generally distinctly wider posteriorly; free-
    living (plate 403D) . . . . *Adula diegensis*
10. Shell with prominent radial ribbing; dark brown to black-
    ish (plate 403E) . . . . *Geukensia demissa*
—   Shell without ribs 11
11. Shell smooth, thin, without hairs; often with wavy brown
    bands on greenish background (plate 403F) . . . .

* = Not in key.

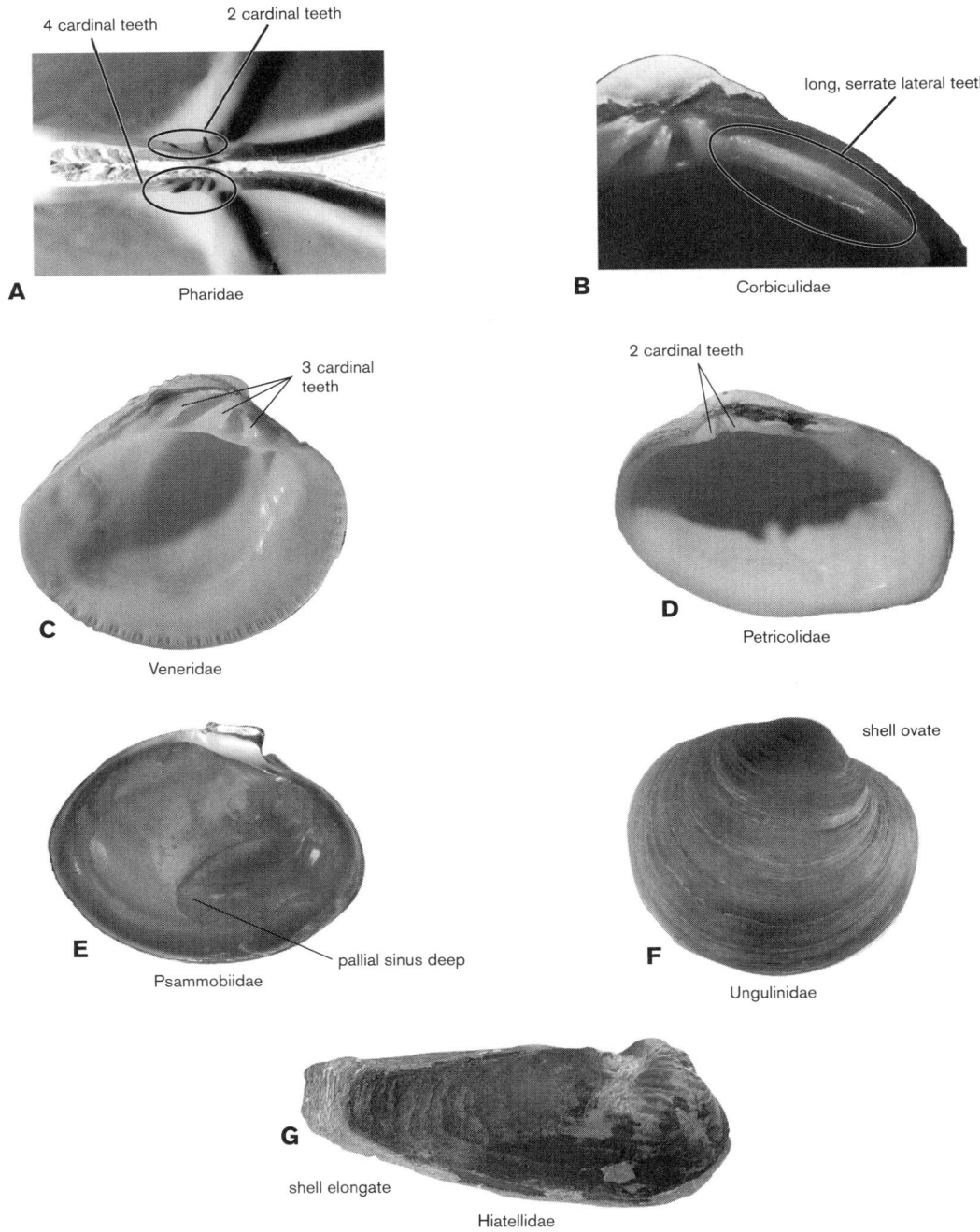

A
4 cardinal teeth
2 cardinal teeth
Pharidae

B
long, serrate lateral teeth
Corbiculidae

C
3 cardinal teeth
Veneridae

D
2 cardinal teeth
Petricolidae

E
pallial sinus deep
Psammobiidae

F
shell ovate
Ungulinidae

G
shell elongate
Hiatellidae

PLATE 398 A, four cardinal teeth of Pharidae; B, long, serrate lateral teeth of Corbiculidae; C, three cardinal teeth of Veneridae; D, two cardinal teeth of Petricolidae; E, deep pallial sinus of Psammobiidae; F, oval shell of Ungulinidae; G, elongate shell of Hiatellidae.

. . . . *Musculista senhousia*
— Shell with periostracal hairs (plate 402D–402E) . . . .
. . . . *Modiolus* 12
12. Periostracum with serrate hairs (plates 402D, 404A) . . . . .
. . . . *Modiolus capax*
— Periostracal hairs smooth (plate 402E) 13
13. Shell elongate; umbones set well back from anterior margin; adult to 120 mm or more (plate 404B) . . . . . *Modiolus rectus*
— Shell rhomboidal; beaks not projecting beyond anterior end; adult to 140 mm (plates 402C, 404C) . . . .
. . . . *Modiolus modiolus*

— Shell short, stout, inflated, umbones protruding beyond anterior margin; adult to about 40 mm (plate 404D) . . . . . .
. . . . *Modiolus carpenteri*
— Shell short, stout; anteroventral margin projecting beyond beaks; adult to 70 mm (plates 402E, 402F, 404E) . . . .
. . . . *Modiolus sacculifer*

LIST OF SPECIES

Soot-Ryen 1955, Allan Hancock Pac. Expeds. 20: 175 pp.

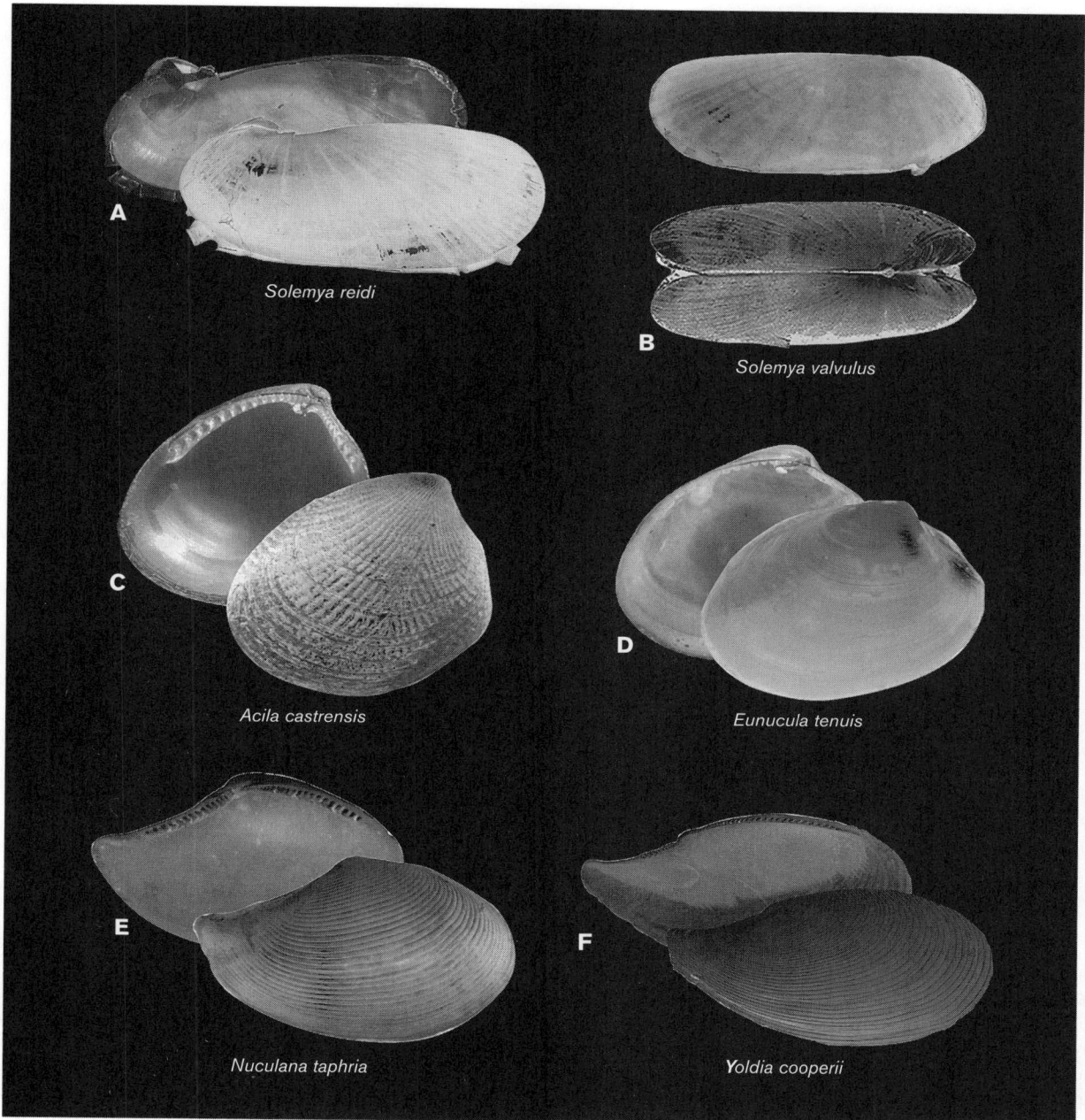

PLATE 399 A, *Solemya reidi*, length 35 mm; B, *Solemya valvulus*, length 11 mm; C, *Acila castrensis*, length 8 mm; D, *Ennucula tenuis*, length 4 mm; E, *Nuculana taphria*, length 5 mm; F, *Yoldia cooperii*, length 39 mm.

Adula californiensis (Philippi, 1847) (plate 403C). Boring mechanically in soft shale, mudstone, but occasionally free-living; intertidal to sublittoral zones. Coan et al. 2000: 176–178; Lough and Gonor 1971, Mar. Biol. 8: 118–125 (embryology, developmental rate); Lough and Gonor 1973, Mar. Biol. 22: 241–250 (larvae); Soot-Ryen 1955: 90–91; Yonge 1955, Quart. J. Micr. Sci. 96: 383–410 (boring).

Adula diegensis (Dall, 1911). Free-living on mud flats, pilings, with *Mytilus*; distinguished from *A. californiensis* by its more flaring dorsal margin and sparse periostracal mat on the posterior slope. Coan et al. 2000: 178; Soot-Ryen 1955: 91.

Adula gruneri (Philippi, 1851) (=*A. falcata* [Gould, 1851]). Boring mechanically in soft shale or clay. Coan et al. 2000: 159,

178–179; Fankboner 1971, Biol. Bull. 140: 28–45 (feeding, ciliary currents); Soot-Ryen 1955: 89–90; Yonge 1955, Quart. J. Micr. Sci. 96: 383–410 (boring).

Geukensia demissa (Dillwyn, 1817) (formerly placed in *Volsella, Modiolus, Arcuatula,* and *Ischadium*). An introduced western Atlantic mussel now abundant in mud in sloughs, bays, in cracks on pilings. There is a substantial literature on the biology of this species, chiefly in the western Atlantic; for a bibliography: Coan et al. 2000: 187–188.

Crenella decussata (Montagu, 1808) (plate 405A). A small, rounded species in soft sediments just offshore. Coan et al. 2000: 163, 166.

* = Not in key.

PLATE 400 A, *Acar bailyi,* length 7 mm; B, *Glycymeris septentrionalis,* length 17.5 mm; C, *Philobrya setosa,* length 6 mm (left pair), length 5 mm (right specimen with brood).

*Gregariella coarctata (Carpenter, 1857) (plate 405B). A trape-zoidal species that nestles in offshore rocks. Coan et al. 2000: 163, 166–167.

Lithophaga plumula (Hanley, 1843) (=*L. subula* [Reeve, 1857] and *L. plumula kelseyi* [Hertlein and Strong, 1946]). Boring in calcareous shale, shells; inner mantle fold secretes acid mucus. Coan et al. 2000: 180; Hodgkin 1962, Veliger 4: 123–129 (boring); Kleemann 1990, Senckenbergiana Maritima 21: 101–154 (boring); Soot-Ryen 1955: 96–97; Yonge 1955 (boring, above, under *Adula*).

Modiolus capax (Conrad, 1837). Intertidal, on rocks, pilings; southern; central California records require confirmation. Coan et al. 2000: 183–185; Orduña, Rojas, and Farfán 1991, Veliger 34: 302–308 (development); Soot-Ryen 1955: 60–62.

Modiolus carpenteri Soot-Ryen, 1963. Occasional in low in-tertidal zone, largely sublittoral, and commonly washed ashore; among rocks, shells, gravel. Coan et al. 2000: 184–185; Soot-Ryen 1955: 62.

*Modiolus neglectus Soot-Ryen, 1955 (plate 404F). Offshore in soft substrata. Coan et al. 2000: 184, 186; Soot-Ryen 1955: 64–65.

Modiolus modiolus (Linnaeus, 1758). This northern species has been found in the intertidal zone as far south as Monterey. In boreal waters of both the Atlantic and Pacific; there is a sub-stantial literature on the biology of this species, cited in Coan et al. 2000: 185–186.

Modiolus rectus (Conrad, 1837) (=*M. flabellatus* [Gould, 1850]). Largely sublittoral, rare in low intertidal zone; in mud, anterior end embedded in soft substrata. Coan et al. 2000: 165, 184, 186–187; Soot-Ryen 1955: 63–64.

Modiolus sacculifer (Berry, 1953). Generally offshore in hold-fasts, shell not elongate, umbones set back from produced an-terior end, not overhanging anterior margin as in *M. carpenteri.* Coan et al. 2000: 184, 186; Soot-Ryen 1955: 65–66.

"Musculus" pygmaeus Glynn, 1964. High intertidal, attached to blades or holdfast of alga *Endocladia muricata* from Marin County to San Luis Obispo County. Coan et al. 2000: 169–170; Glynn 1964, Veliger 7: 121–128.

Musculista senhousia (Benson, 1842) (formerly placed in *Modiolus* and *Musculus; M. senhousei* is an invalid emendation). Abundant in mud, forming extensive mats; in fouling on pil-ings, among algae; introduced from Japan. Coan et al. 2000: 163, 167 (including references to many other papers); Morton 1974, Pac. Sci. 28: 19–33 (functional morphology, etc.); Soot-Ryen 1955: 74–75.

Mytilus californianus Conrad, 1837. California mussel; abun-dant in exposed intertidal rocks; also in bays, on pilings. Coan et al. 2000: 156–158 (with references to more than 60 papers); Coe and Fox 1942, J. Exp. Zool. 90: 1–30; Fox and Coe 1943, J. Exp. Zool. 93: 205–249; Coe and Fox 1944, Biol. Bull. 87:

* = Not in key.

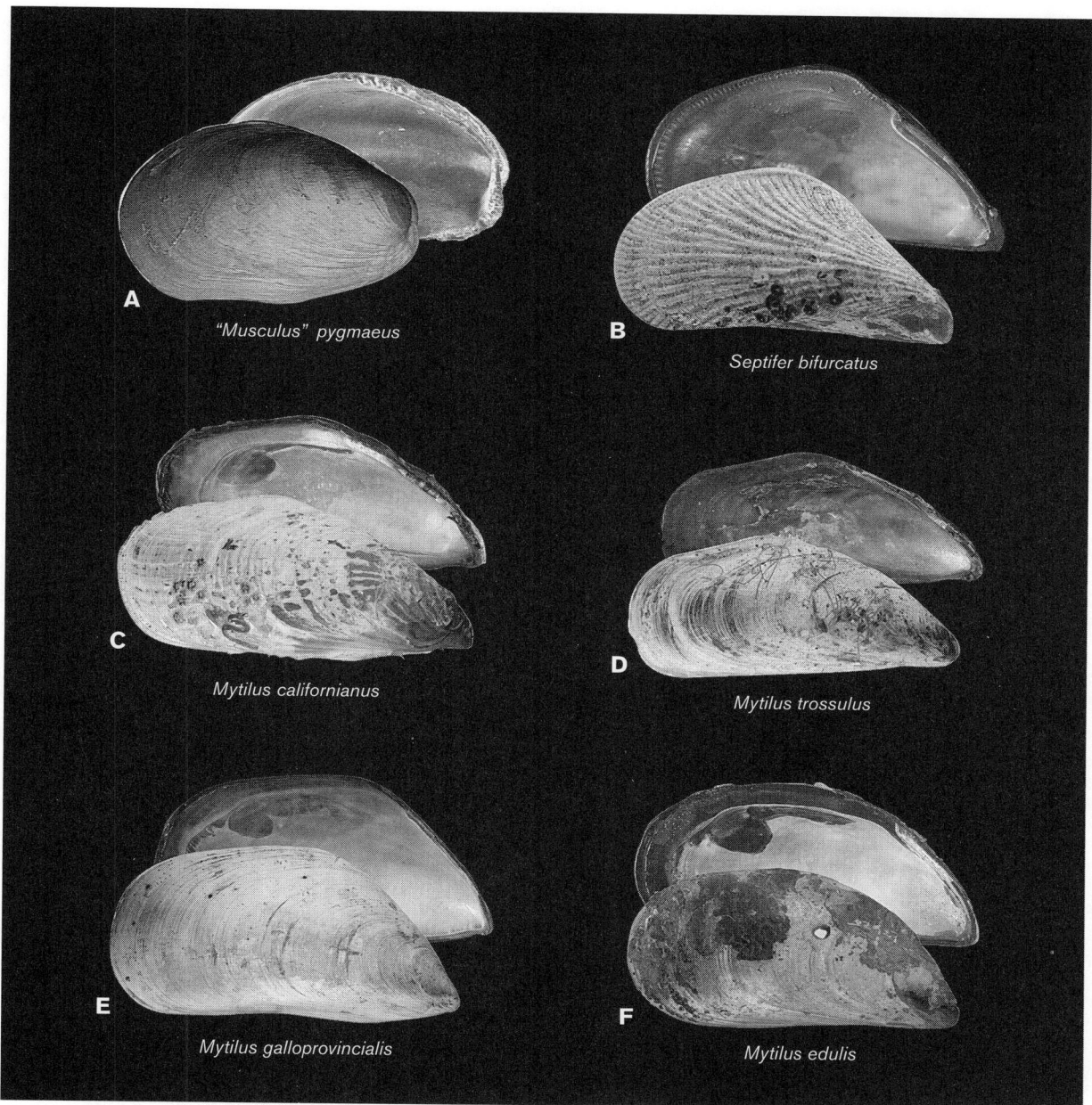

PLATE 401 A, *"Musculus" pygmaeus*, length 3.2 mm; B, *Septifer bifurcatus*, length 27 mm; C, *Mytilus californianus*, length 97 mm; D, *Mytilus trossulus*, length 46 mm; E, *Mytilus galloprovincialis*, length 87 mm; F, *Mytilus edulis*, length 64 mm.

59–72; Coe 1948, J. Mar. Res. 7: 586–601; aspects of biology; papers by Harger, and Harger and Landenberger, in Veliger 11–14 (1968–1971) on ecology, biology of *M. "edulis"* and *M. californianus*.

Mytilus trossulus Gould, 1851; *M. galloprovincialis* Lamarck, 1891; *M. edulis* Linnaeus, 1758. Genetic studies have shown that the native smooth mussel in the North Pacific is *M. trossulus*, which occurs from the Arctic south to central California, where it now forms a hybrid zone with the introduced European *M. galloprovincialis*; the latter occurs from central California to Baja California; wild specimens of *M. galloprovincialis* have also been found in southern British Columbia, but establishment there is uncertain. The native North

Atlantic *M. edulis* is subject to experimental mariculture in British Columbia and may spread and hybridize as well. Members of this complex are widespread; abundant on wharf pilings, floats, docks, rocks; occasional on outer coast. There is a vast literature on all aspects of the biology of this group, with many papers cited in Coan et al. 2000: 157–160. Other important reviews of this group are in Gosling 1992, The genus Mytilus: ecology, physiology, genetics and culture, Amsterdam (Elsevier), 589 pp.; Bayne 1976, Marine mussels: their ecology and physiology, Cambridge, U.K. (Cambridge Univ.), 506 pp. See also Kafanov 1999, Bull. Inst. Malac., Tokyo 3: 103–108; Buianovskii 2000, Ruthenica 10: 43–48; Martel et al. 1999, Invert. Biol. 118: 149–164 (distinguishing early juveniles).

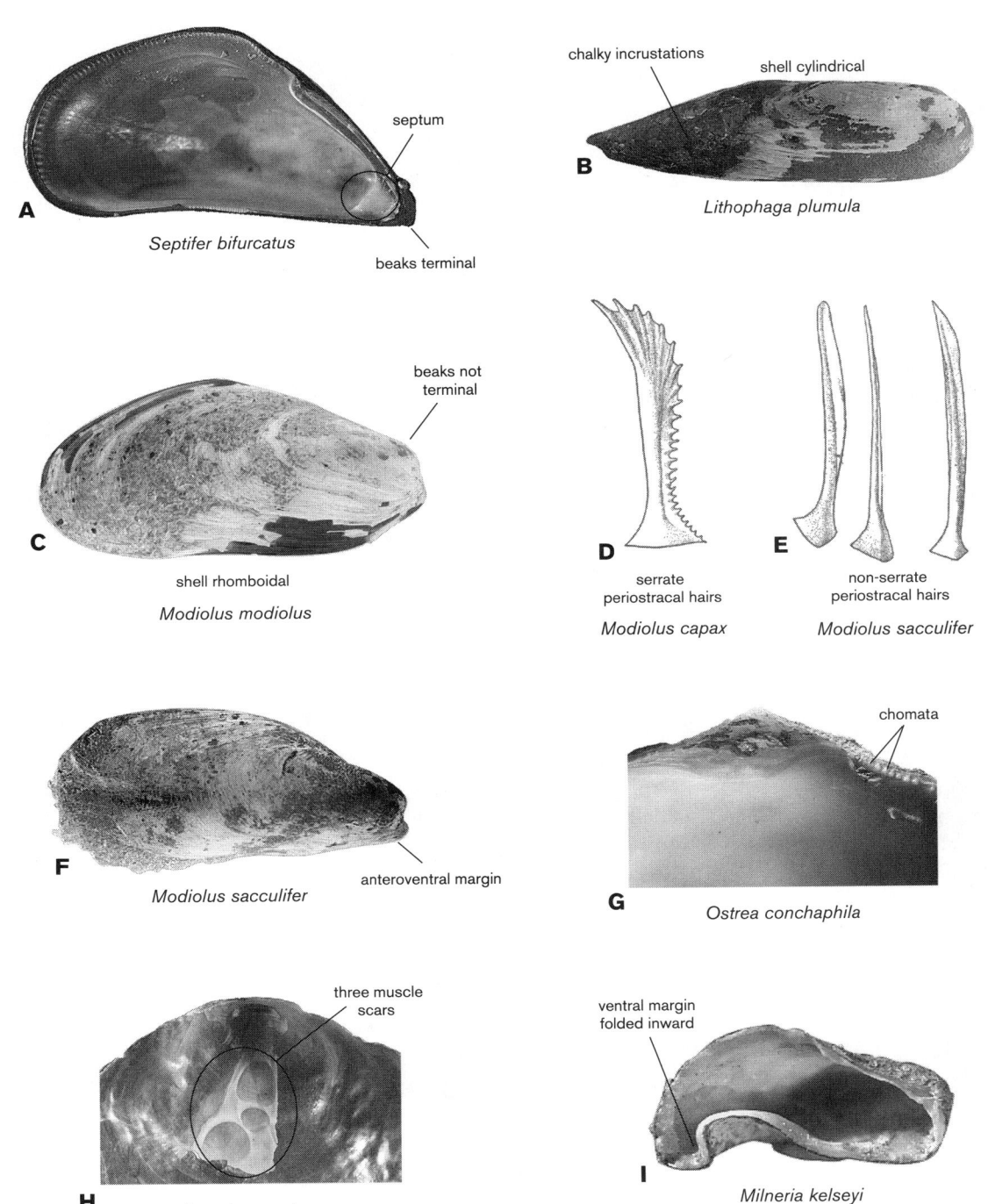

PLATE 402 A, *Septifer bifurcatus*, length 27 mm; B, *Lithophaga plumula*, length 51 mm; C, *Modiolus modiolus*, length 66 mm; D, detail of periostracal hair of *Modiolus capax*; E, detail of periostracal hair of *Modiolus sacculifer*; F, *Modiolus sacculifer*, length 74 mm; G, detail of chomata on dorsal margin of *Ostrea conchaphila*; H, three muscle scars of *Anomia peruviana*; I, infolded ventral margin of *Milneria kelseyi*, length 5 mm (D, E, from Soot-Ryen 1955).

Septifer bifurcatus (Conrad, 1837). Low intertidal, under rocks. Coan et al. 2000: 188, 190; Soot-Ryen 1955: 41–42; Yonge and Campbell 1968, Trans. Roy. Soc. Edinburgh 68: 34 (functional morphology).

*Solamen columbianum (Dall, 1897) (plate 405C). A rounded, inflated species characteristic of offshore soft sediments; larger than *Crenella decussata*, with finer sculpture. Coan et al. 2000: 171–172.

LIMIDAE

*Limaria hemphilli (Hertlein and Strong, 1946) (plate 405D). In shallow water offshore in warm-water years. Coan et al. 2000: 205–206.

*Limatula saturna Bernard, 1978 (plate 405E). In shallow offshore waters. Coan et al. 2000: 206–208.

* = Not in key.

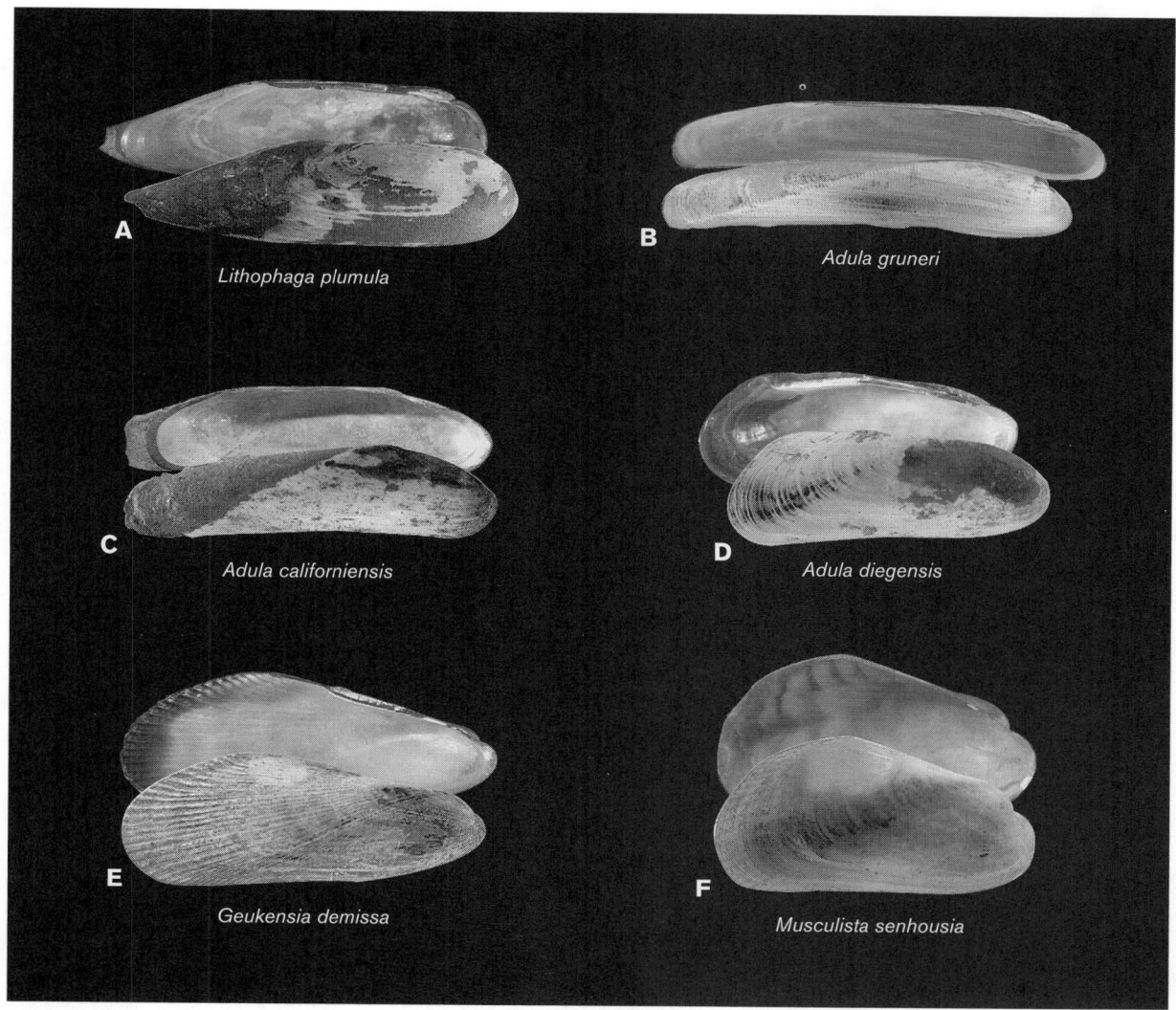

PLATE 403 A, *Lithophaga plumula*, length 51 mm; B, *Adula gruneri*, length 64 mm; C, *Adula californiensis*, length 38 mm; D, *Adula diegensis*, length 24 mm; E, *Geukensia demissa*, length 61 mm; F, *Musculista senhousia*, length 20 mm.

OSTREIDAE

KEY TO OSTREIDAE

(With assistance from Patrick Baker, Andrew Cohen, and James Carlton)

1. Hinge region with chomata in most specimens (plate 402G)..2
— Hinge region without chomata 3
2. Outline of shell usually circular or broadly oval; sculpture of right (upper) valve usually with conspicuous, thin, frilly, concentric lamellae; ventral margin of shell (opposite hinge) may be undulate, but not plicate; maximum shell height 170 mm (plate 406C)..............*Ostrea edulis*
— Shell outline variable; sculpture of right valve variable, with obscure concentric lamellae, frequently abraded; ventral margin of shell may be smooth or distinctly plicate (shell edges interlock in wavy pattern); maximum shell height 80 mm (plate 406A)..........*Ostrea conchaphila*
3. Maximum shell height 60 mm; left (attached) valve deeply cupped, with three or more distinct radial ridges; right valve relatively smooth; ventral shell margin usually plicate (plate 406E)..................*Crassostrea sikamea*
— Maximum shell height exceeding 60 mm; left valve may be deeply cupped when compared to right valve; right (upper) valve often with spine-like features or fluting, well-raised above the shell; ventral shell margin often plicate (plate 406B)......................*Crassostrea gigas*
— Maximum shell height exceeding 60 mm; left valve usually not deeply cupped when compared to right valve.. 4
4. Adductor muscle scar usually solid deep violet or dark brown (plate 406F)...............*Crassostrea virginica*
— Adductor muscle scar frequently white, or with blotchy dark color, usually not a solid color (plate 406D) ....
...............................*Crassostrea ariakensis*

LIST OF SPECIES

Yonge 1960, Oysters, London: Collins, 209 pp.; Stenzel 1971, Treatise on Invertebrate Paleontology, N (Mollusca) (6)3: 953–1224. Introduced species do not generally reproduce on the central California coast.

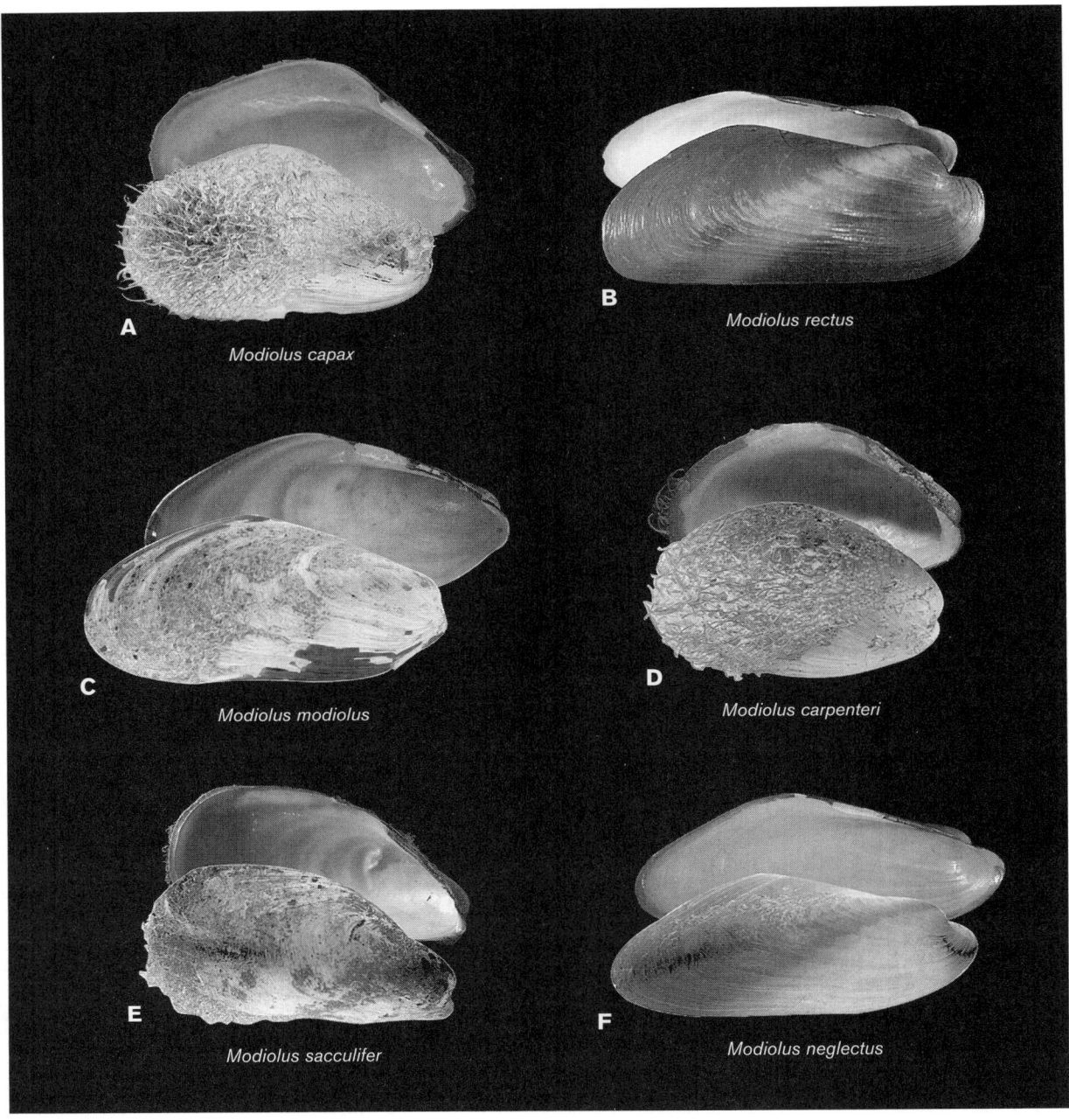

PLATE 404 A, *Modiolus capax*, length 65 mm; B, *Modiolus rectus*, length 67 mm; C, *Modiolus modiolus*, length 66 mm; D, *Modiolus carpenteri*, length 18 mm; E, *Modiolus sacculifer*, length 74 mm; F, *Modiolus neglectus*, length 55 mm.

Ostrea conchaphila Carpenter, 1857 (=*Ostrea lurida* Carpenter, 1864; in the subgenus *Ostreola*, treated by some as a genus). Olympia or native oyster; common in mud, on rocks, pilings, in bays, often in clumps. Locally abundant in Humboldt Bay and Tomales Bay; still common in parts of San Francisco Bay but rare on open coastline. Adults normally 2 cm–6 cm in shell height. Coan et al. 2000: 214–215; Hopkins 1936, Ecology 17: 551–566; 1937, Bull. Bur. Fish. 48: 439–503 (reproduction, larval development); Barrett, below; Baker 1996, J. Shellfish Res. 14: 501–518 (ecology, fishery); Couch and Hassler 1989, U.S. Fish Wildl. Serv. Biol. Rep. 82 (11.124, 8 pp., review of biology).

Ostrea edulis Linnaeus, 1758. European flat oyster, native to western Europe; plantings have been made in Tomales Bay and Drake's Estero. Adults 6 cm–10 cm in shell height, rarely to 17 cm. Seldom reproduces or occurs in the wild in this range. Coan et al. 2000: 215–216; Leonard 1969, Veliger 11: 382–390.

Crassostrea ariakensis (Fujita, 1913). Suminoe oyster; native to eastern Asia and occasionally cultured in bays and estuaries north of San Francisco Bay. Adults 10 cm–15 cm in shell height, rarely to 20 cm. Reproductive status in this range unknown. Also mistakenly known as *C. rivularis*. Langdon and Robinson 1996, Aquaculture 144: 321–338; Torigoe 1981, J.

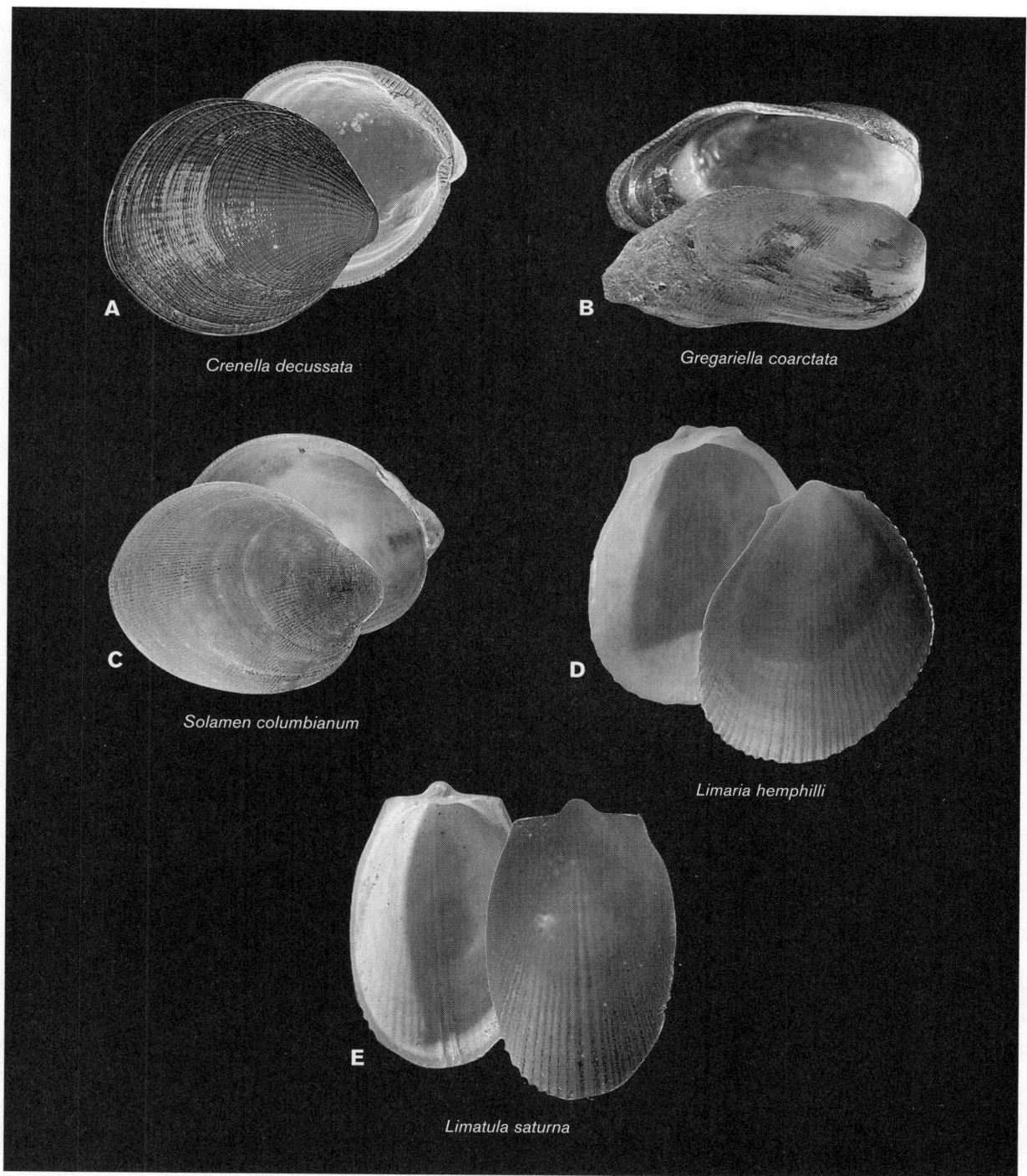

PLATE 405 A, *Crenella decussata*, length 3.8 mm; B, *Gregariella coarctata*, length 12 mm; C, *Solamen columbianum*, holotype, length 12 mm; D, *Limaria hemphilli*, height 28 mm; E, *Limatula saturna*, height 8 mm.

Sci. Hiroshima Univ. Ser. B, Div. 1, 29: 291–419; Zhou and Allen 2003, J. Shellfish Res. 22: 1–20 (literature review).

Crassostrea gigas (Thunberg, 1795). Japanese or Pacific oyster, native to Asia; introduced; large oyster farms in Drake's Estero and Tomales Bay in central California; shells are common. Adults 8 cm–20 cm in shell height, rarely to 40 cm. Frequently reproduces; scattered individuals common in estuaries with oyster culture, and rare on open coastline. Coan et al. 2000: 215, 217–218; Barrett 1963, Calif. Fish Game Fish Bull. 123:

103 pp.; Quayle 1988, Pacific oyster culture in British Columbia. Can. Bull. Fish. Aquatic Sci. 218, 241 pp.; Berg 1969, Veliger 12: 27–36; 1971, Calif. Fish Game 57: 69–75 (causes of mortality); see also Torigoe, above.

Crassostrea sikamea (Amemiya, 1928). Kumamoto oyster; native to Japan. Cultured in bays and estuaries north of San Francisco Bay. Adults 4 cm–8 cm in shell height. Reproduction in California unknown. Morphology frequently overlaps strongly with small *C. gigas*, of which it was formerly considered a subspecies

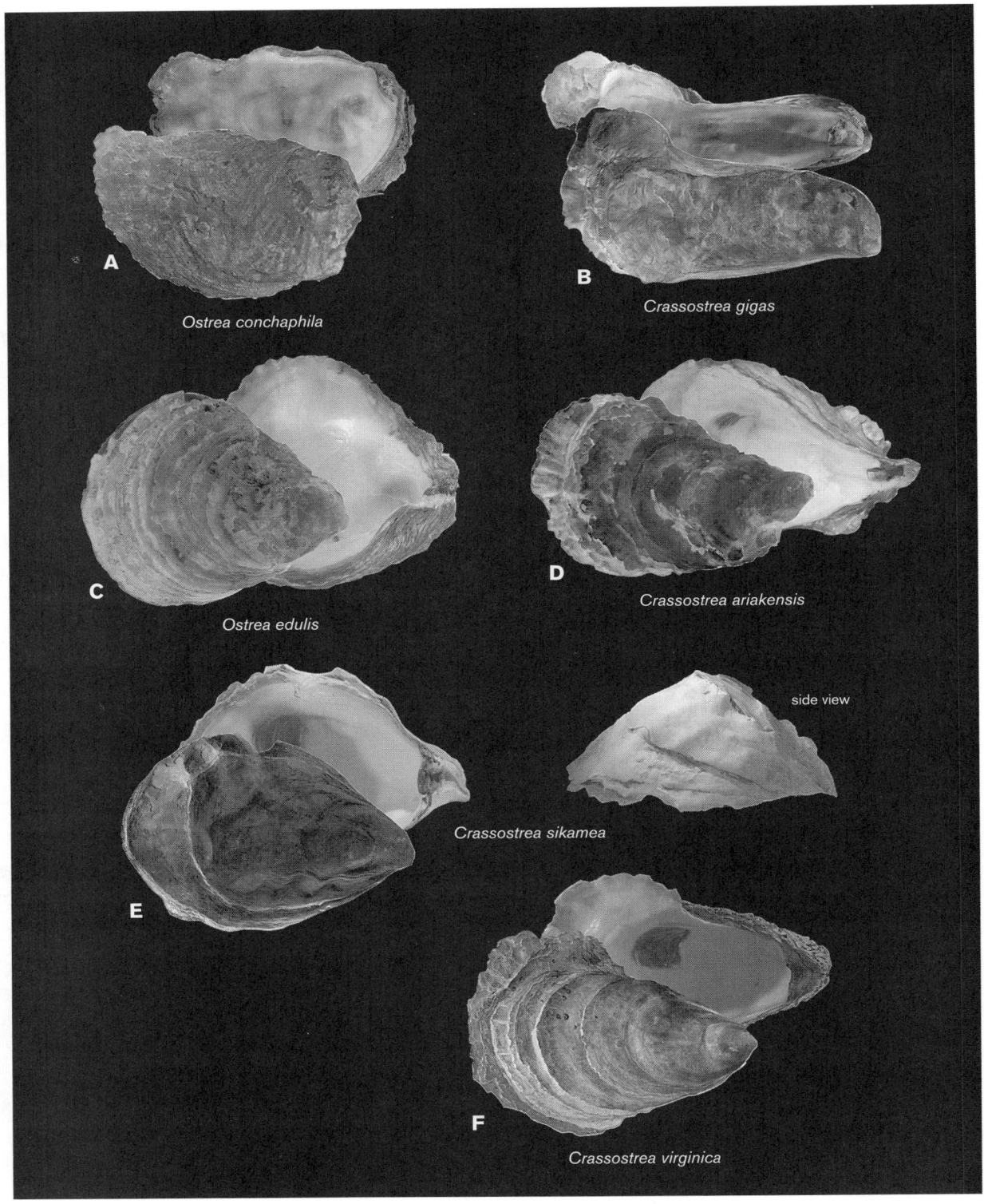

PLATE 406 A, *Ostrea conchaphila*, height 50 mm; B, *Crassostrea gigas*, height 105 mm; C, *Ostrea edulis*, height 67 mm; D, *Crassostrea ariak-ensis*, height 170 mm; E, *Crassostrea sikamea*, height 58 mm, width of sideview 39 mm; F, *Crassostrea virginica*, height 96 mm.

PLATE 407 A, *Crassadoma gigantea,* height (adult) 53 mm, height (juvenile) 35 mm; B, *Leptopecten latiauratus,* height 17 mm; C, *Chlamys hastata,* height 56 mm.

or variety. Coan et al. 2000: 218–219; Banks et al. 1994, Mar. Biol. 121: 127–135; Hedgecock et al. 1993, J. Shellfish Res. 12: 215–221.

Crassostrea virginica (Gmelin, 1791). The eastern oyster, native to Atlantic and Gulf of Mexico coasts, occasionally cultured in bays and estuaries from San Francisco Bay north. Adults 7 cm–15 cm in shell height, rarely to 20 cm. Reproduction in this range rare. A comprehensive treatment with references to hundreds of papers is in Kennedy et al. 1996, The eastern oyster, Crassostrea virginica, College Park, Maryland, Maryland Sea Grant Book, 734 pp.; Coan et al. 2000: 218–219.

PECTINIDAE

KEY TO PECTINIDAE

— Radial ribs imbricated (with shingle-like spines); height greater than length in small specimens; hinge purple; adult attached to rocks or pilings (plate 407A)......*Crassadoma gigantea*

— Radial ribs not imbricated; length greater than height; hinge not purple; adults free-living (plate 407B)........
.............................*Leptopecten latiauratus*

Chlamys rubida

Euvola diegensis

Patinopecten caurinus

PLATE 408 A, *Chlamys rubida*, height 60 mm; B, *Euvola diegensis*, height 90 mm; C, *Patinopecten caurinus*, height 117 mm.

LIST OF SPECIES

Grau 1959, Pectinidae of the eastern Pacific, Allan Hancock Pacific Expeditions 23, 308 pp.

*Chlamys hastata (G. B. Sowerby II, 1842) (plate 407C). The spiny scallop is most common offshore pectinid in central California. Coan et al. 2000: 232–234; Grau 1959: 85–91.

*Chlamys rubida (Hinds, 1845) (plate 408A). Less common, the red scallop is occasionally dredged offshore. Coan et al. 2000: 232, 234; Grau 1959: 76–81.

Crassadoma gigantea (Gray, 1838) (=*Hinnites giganteus*). The giant rock-scallop is free-living when young, attaching as an adult to rocks and pilings, where growth becomes irregular. Coan et al. 2000: 236, 238; Grau 1959: 134–137; Yonge 1951, Univ. Calif. Publ. Zool. 55: 409–420.

*Euvola diegensis (Dall, 1898) (plate 408B). The San Diego scallop is occasionally found in shallow water as far north as Bodega Bay. Coan et al. 2000: 225–226; Grau 1959: 143–145.

Leptopecten latiauratus (Conrad, 1837). Coan et al. 2000: 236–237; Grau 1959: 107–113; McPeak and Glantz 1982, Festivus 14: 63–69 (ecology); Morton 1994,Veliger 37: 5–22 (functional morphology).

*Patinopecten caurinus (Gould, 1850) (plate 408C). The weathervane scallop occurs offshore. In Alaska it has supported a modest fishery. Coan et al. 2000: 240–241; Grau 1959: 221–222.

* = Not in key.

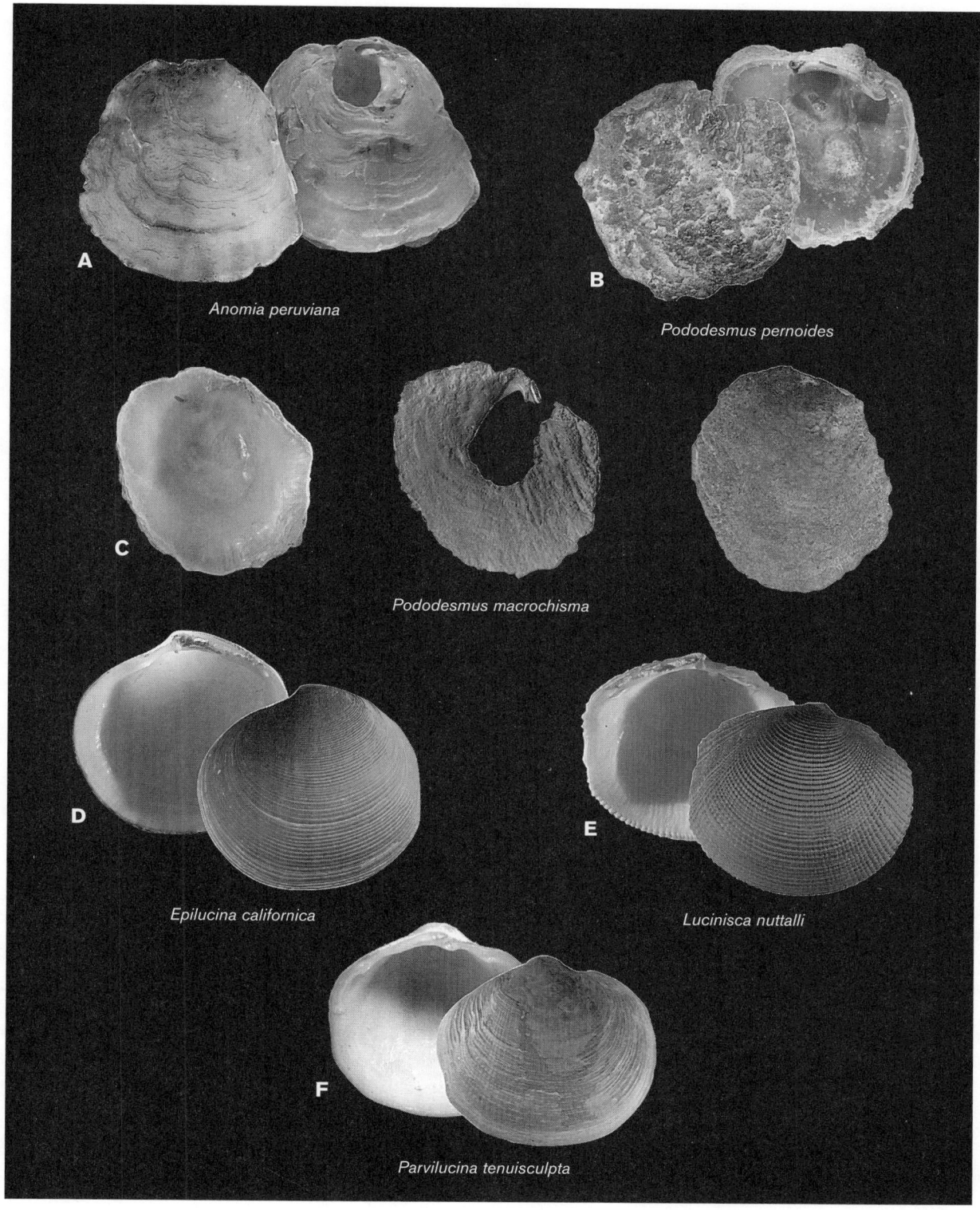

A

Anomia peruviana

B

Pododesmus pernoides

C

Pododesmus macrochisma

D

Epilucina californica

E

Lucinisca nuttalli

F

Parvilucina tenuisculpta

PLATE 409 A, *Anomia peruviana,* height 20 mm; B, *Pododesmus pernoides,* height 36 mm; C, *Pododesmus macrochisma,* height 91 mm; D, *Epilucina californica,* length 35 mm; E, *Lucinisca nuttalli,* length 25 mm; F, *Parvilucina tenuisculpta,* length 9 mm.

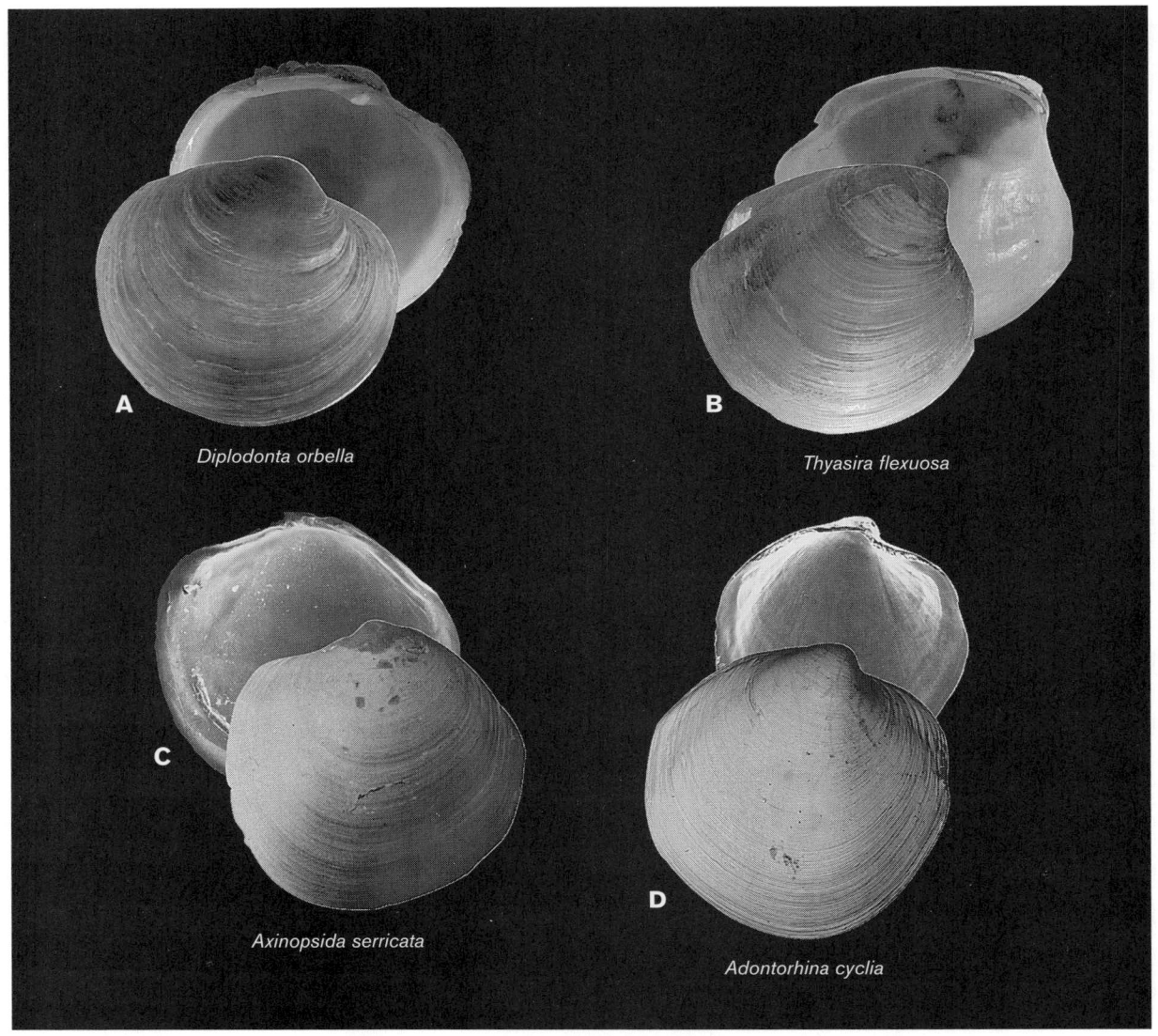

PLATE 410 A, *Diplodonta orbella*, length 27 mm; B, *Thyasira flexuosa*, length 17 mm; C, *Axinopsida serricata*, length 4 mm; D, *Adontorhina cyclia*, length 2 mm.

ANOMIIDAE

KEY TO ANOMIIDAE

1. Two smaller muscle scars inside central area of valves (plate 394G) . . . . *Pododesmus* 2
— Three scars inside central area (plates 402H, 409A). . . . . . .
. . . . *Anomia peruviana*
2. Hole in right valve for attachment of calcified byssus medium-size to large; shell greenish (plate 409C). . . . . . .
. . . . *Pododesmus macrochisma*
— Hole in right valve small; shell chocolate in color (plate 409B) . . . . *Pododesmus pernoides*

LIST OF SPECIES

Pododesmus macrochisma (Deshayes, 1839) (=*P. cepio* [Gray, 1850]; *P. macroschisma*, invalid emendation). Conclusive proof has yet to be offered that Alaskan material, described as *P. macrochisma*, represents a different species than *P. cepio*, described from California. This species is green because of algae within the shell and is common on rocks, abalone shells, and in dead shells. Coan et al. 2000: 256–257; Yonge 1977, Philos. Trans. Roy. Soc. London (B) 276: 459–476 (functional morphology); Leonard 1969, Veliger 11: 382–390 (reproduction); Anderson 2000, Festivus 32: 139–142 (algae in shell).

Pododesmus pernoides (Gray, 1853). Much less common, it is generally attached to the shells of other mollusks. Coan et al. 2000: 255–257.

Anomia peruviana Orbigny, 1846. Reported as far north as Monterey in association with El Niño years. Coan et al. 2000: 255, 257.

LUCINIDAE

Taylor and Glover 2000, pp. 207–225, in: Harper et al., eds., The evolutionary biology of the Bivalvia. Geol. Soc. London, Spec.

Publ. 177; Taylor and Glover 2006, Zool. J. Linn. Soc. 148: 421–438 (functional anatomy, evolution).

Epilucina californica (Conrad, 1837) (plate 409D). In the intertidal zone nestling in gravel and crevices on rocky coast. Coan et al. 2000: 263–264.

Lucinisca nuttalli (Conrad, 1837) (plate 409E). A southern Californian species recorded as far north as Monterey in El Niño years. Coan et al. 2000: 263–265.

Parvilucina tenuisculpta (Carpenter, 1864) (plate 409F). Very common in soft bottoms immediately offshore. Coan et al. 2000: 264–266; Hickman 1994, Veliger 37: 53–55.

UNGULINIDAE

Represented in local rocky intertidal zone by *Diplodonta orbella* (Gould, 1851) (plate 410A), which lives in holes in rocks and forming a nest of agglutinated detritus and sand under rocks. Coan et al. 2000: 270–271; Haas 1943, Zool. Ser. Field Mus. Natl. Hist. 29: 9–12; Hertz et al. 1982, Festivus 14: 76–87.

THYASIRIDAE

Although not present in the intertidal zone of central California, the following members of this family are relatively common on soft bottoms immediately offshore.

Thyasira flexuosa (Montagu, 1803) (plate 410B). There is a considerable literature about the biology of this species, listed in Coan et al. 2000: 275–276.

Axinopsida serricata (Carpenter, 1864) (plate 410C). Coan et al. 2000: 278–280.

Adontorhina cyclia Berry, 1947 (plate 410D). Coan et al. 2000: 279–280.

CARDITIDAE

KEY TO CARDITIDAE

1. Ventral margin of shell folded inward (in female); radial ribs absent anteroventrally (plate 402I) 2
— Ventral margin evenly arched, not bent inward; strong radial ribs throughout (plate 411C) . . . . *Glans carpenteri*
2. With a strong keel, 10 strong ribs, and a narrow, short lunule (plates 402I, 411A) . . . . *Milneria kelseyi*
— Without a keel, four strong ribs, and a long, broad lunule (plate 411B) . . . . *Milneria minima*

LIST OF SPECIES

Yonge 1969, Proc. Malac. Soc. London 38: 493–527 (functional morphology); Coan 1977, Veliger 19: 375–386 (systematic review).

Cyclocardia ventricosa (Gould, 1850) (plate 411D). This is the most common carditid found offshore on soft bottoms. Coan et al. 2000: 302–304.

Glans carpenteri (Lamy, 1922). Attains only about 10 mm in length and attaches by a byssus to the undersurfaces of rocks on open coast; broods young. Coan et al. 2000: 305–306; Yonge 1969: 505–509.

Milneria minima (Dall, 1871). Reaches about 8 mm and also attaches with a byssus to rocks from the low intertidal zone to 50 m. Ventral margin is folded inward in the female to form a brood pouch. Coan 1974, Veliger 17: 183–184; Coan et al. 2000: 306–307.

Milneria kelseyi Dall, 1916. Reaches about 18 mm, this southern species occasionally gets as far north as Monterey Bay. Also with brood pouch in female. Coan 1974: 183–184; Coan et al. 2000: 306–307; Yonge 1969: 517.

CHAMIDAE

KEY TO CHAMIDAE

1. Attached by left valve; growth counter-clockwise, markedly foliose (plate 411E) . . . . *Chama arcana*
— Attached by right valve; growth clockwise, valves scaly, but not markedly foliose (plate 411F) . . . .
. . . . *Pseudochama exogyra*

LIST OF SPECIES

Yonge 1967, Phil. Trans. Roy. Soc. London (B) 252: 49–105 (functional morphology); Bernard 1976, Contrib. Sci., Los Angeles Co. Mus. 278: 43 pp. (systematic review).

Chama arcana Bernard, 1976. Long misidentified as *C. pellucida* Broderip, 1835, a South American species. Cemented to protected surfaces of mid-intertidal rocks; on pilings. Yonge 1967; Bernard 1976: 14–15; Coan et al. 2000: 309–311; Hamada and Matsukuma 1995, Kyushu Univ., Dept. Earth and Planetary Sci., Sci. Repts. 19: 93–102 (shell structure).

Pseudochama exogyra (Conrad, 1837). Cemented to algae-covered rocks on open coast from mid- to low intertidal zone. Yonge 1967; Bernard 1976: 26–27; Coan et al. 2000: 310–311.

Pseudochama granti Strong, 1934 (plate 411G). An offshore species that has been confused with *P. exogyra*. Coan et al. 2000: 301–311.

GALEOMMATOIDEA

KEY TO GALEOMMATOIDEA

1. Hinge without teeth; shell entirely internal (plates 396C, 412A, 413A) . . . . *Chlamydoconcha orcutti*
— Cardinal teeth developed in at least one valve; animal fitting within shell . . . . Lasaeidae 2
2. Beaks near posterior end 3
— Beaks subcentral 5
3. Left valve without teeth; right valve with a long anterior cardinal tooth and a short posterior cardinal; teeth not serrate (plate 413B) . . . . *Mysella pedroana*
— Left valve without teeth; right valve with subequal cardinal teeth; teeth not serrate . . . . *Rochefortia* 4
— Both valves with a long posterior cardinal and a short, ventrally directed anterior cardinal; teeth serrate (plate 413C)
. . . . *Pristes oblongus*
4. Shell thin, subovate (plate 413D) . . . . *Rochefortia coani*
— Shell thick, subtrigonal to subquadrate (plate 413E) . . . . .
. . . . *Rochefortia tumida*
5. Hinge with cardinal teeth only; no lateral teeth; shell quadrangular, yellow; on abdomen of *Upogebia* (plate 414A) . . . . .
. . . . *Neaeromya rugifera*
— Hinge with lateral teeth or lamellae 6

* = Not in key.

Milneria kelseyi

Milneria minima

Glans carpenteri

Cyclocardia ventricosa

Chama arcana

Pseudochama exogyra

Pseudochama granti

PLATE 411 A, *Milneria kelseyi*, length 5 mm; B, *Milneria minima*, length 4 mm; C, *Glans carpenteri*, length 9 mm; D, *Cyclocardia ventricosa*, length 15 mm; E, *Chama arcana*, height 57 mm; F, *Pseudochama exogyra*, height 56 mm; G, *Pseudochama granti*, height 17 mm.

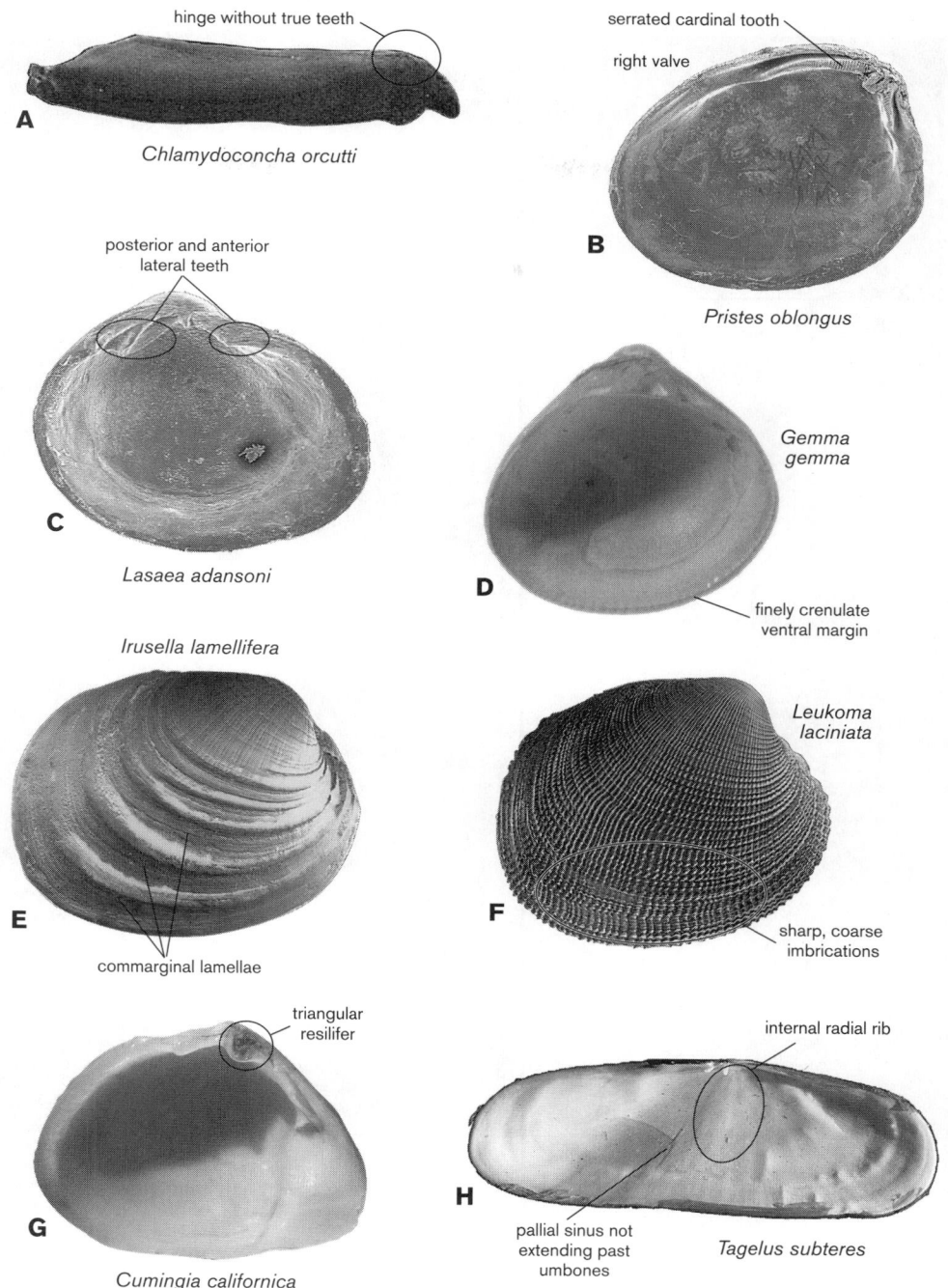

PLATE 412 A, *Chlamydoconcha orcutti*, length 7 mm; B, *Pristes oblongus*, length 4.1 mm; C, *Lasaea adansoni*, length 3.2 mm; D, *Gemma gemma*, length 2.5 mm; E, *Irusella lamellifera*, length 28 mm; F, *Leukoma laciniata*, length 69 mm; G, *Cumingia californica*, length 28 mm; H, *Tagelus subteres*, length 34 mm.

6. Hinge with posterior lateral teeth, no anterior laterals; shell nearly round to elliptical; with fine, commarginal, growth lines and thin, smooth, yellowish periostracum; length to 25 mm (plate 414C) . . . . *Kellia suborbicularis*

— Hinge with posterior and anterior lateral teeth; shell reddish; periostracum wavy, wrinkled; length to 3 mm (plates 412C, 414D) . . . . *Lasaea adansoni*

— Hinge with only thin lateral lamellae on each side of beaks (plate 414E) . . . . *Rhamphidonta retifera*

LIST OF SPECIES

Popham 1940, J. Mar. Biol. Assoc. U.K. 24: 549–587 (functional morphology of British species).

GALEOMMATIDAE

Chlamydoconcha orcutti Dall, 1884. Appearing more like a nudibranch than a clam, this species is rarely observed. Rocky

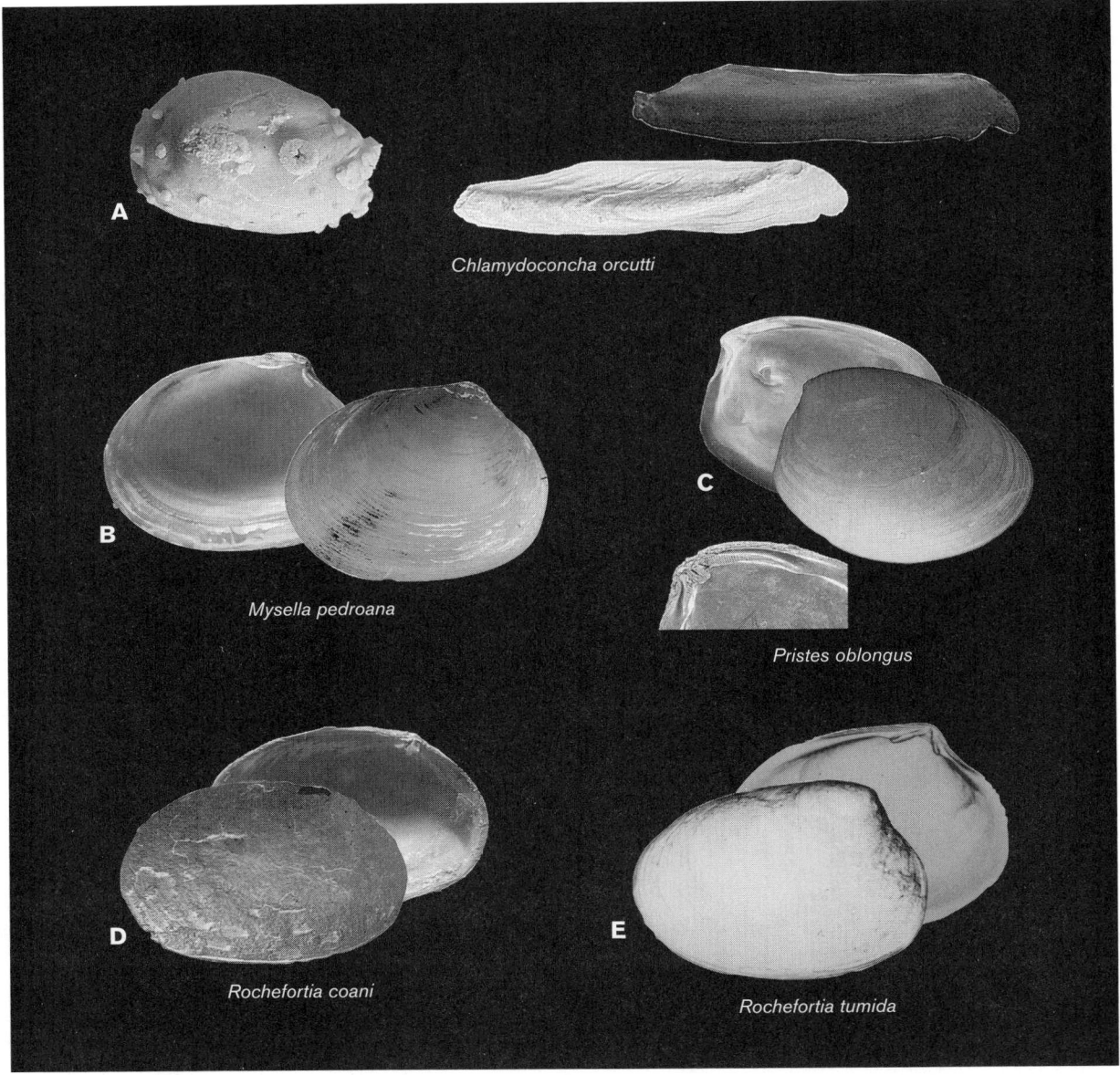

PLATE 413 A, *Chlamydoconcha orcutti,* (left) mantle completely covering shells, length 8 mm, (right) internal shells, length 7 mm; B, *Mysella pedroana,* length 6 mm; C, *Pristes oblongus,* length 4.1 mm; D, *Rochefortia coani,* holotype, length 5.1 mm; E, *Rochefortia tumida,* length 2.8 mm.

low intertidal and subtidal. Coan et al. 2000: 313, 315; Carlton 1979, Veliger 21: 375–378.

LASAEIDAE

Lasaea adansoni (Gmelin, 1791) (=*L. rubra* [Montagu, 1803], *L. cistula* Keen, 1938). In algal holdfasts and among byssal threads of mussels, abundant, on open and semiprotected rocky coast. These strange little bivalves are now known to represent hermaphroditic, polyploid clones ranging from 3°N to 6°N, which brood their young. Coan et al. 2000: 317, 320–321, with additional references to the now substantial literature; Ballentine and Morton 1956, J. Mar. Biol. Assoc. U.K. 35: 241–274 (filtering, feeding, digestion); Ó Foighil and Smith 1996, Evolution 49: 140–150 (genetics).

Kellia suborbicularis (Montagu, 1803) (=*K. laperousii* [Deshayes, 1839]). There is evidence that Pacific material may represent a separate species. Often abundant, nestling on pilings and rocks and in mussel beds, empty barnacles, bottles, and pholad holes; a common fouler in seawater systems of marine laboratories. This species broods its young in the suprabranchial chamber, with dwarf males occurring in the infrabranchial chamber. Coan et al. 2000: 321, 323; Yonge 1952, Univ. Calif. Publ. in Zool. 55: 451–454 (functional morphology); Howard 1953, Wasmann J. Biol. 11: 233–240 (larvae); Kamenev 2004, Malacologia 46: 57–71 (taxonomy); Oldfield 1961, Proc. Malac. Soc. London 34: 255–295 (functional morphology); Oldfield 1964, Proc. Malac. Soc. London 36: 79–120 (reproduction, development).

Mysella pedroana Dall, 1899. In the gill chambers or externally attached to the sand crab *Blepharipoda occidentalis.* Coan

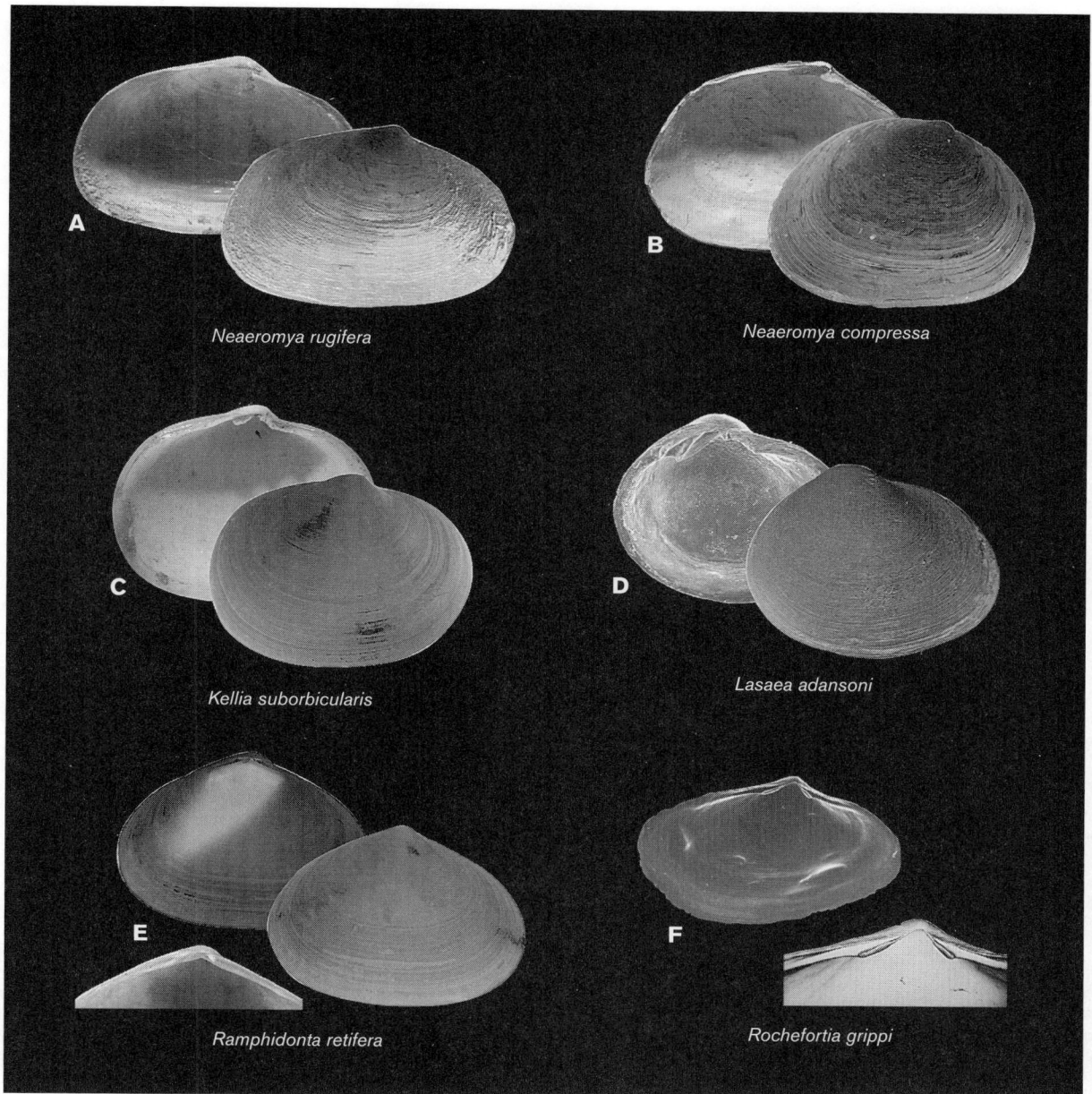

PLATE 414 A, *Neaeromya rugifera,* length 9.5 mm; B, *Neaeromya compressa,* length 13 mm; C, *Kellia suborbicularis,* length 26 mm; D, *Lasaea adansoni,* length 3.2 mm; E, *Ramphidonta retifera,* length 10 mm, detail of hinge (holotype); F, *Rochefortia grippi,* length 3.8 mm, detail of hinge.

et al. 2000: 321, 323–324; Boyko and Mikkelsen 2002, Zoologischer Anzeiger 241: 149–160 (anatomy, biology, commensal relationship with crab).

*Neaeromya compressa (Dall, 1899) (plate 414B). An offshore species, possibly commensal with the sea urchin *Brisaster latifrons.* Coan et al. 2000: 324, 326.

Neaeromya rugifera (Carpenter, 1864) (=*Pseudopythina rugifera*). Intertidal, attached by byssus to abdomen of *Upogebia* and in the setae of the polychaete *Aphrodita.* Coan et al. 2000: 325–326; Narchi 1969, Veliger 12: 43–52 (morphology); Ó Foighil 1985, Veliger 27: 245–252 (reproduction).

Pristes oblongus Carpenter, 1864. Commensal with chiton *Stenoplax heathiana,* Monterey south. Coan et al. 2000: 326–327.

Rhamphidonta retifera (Dall, 1899). Possibly associated with enteropneusts. Coan et al. 2000: 326–327; Bernard 1975, J. Conchyl. 112: 105–115 (morphology).

Rochefortia coani Valentich-Scott 1998. Uncommon, mostly offshore. Coan et al. 2000: 328–329.

*Rochefortia grippi Dall, 1912 (plate 414F). Just offshore. Coan et al. 2000: 328–329.

Rochefortia tumida (Carpenter, 1864). In bays and offshore. Coan et al. 2000: 329–330; Maurer 1967, Veliger 9: 305–309, 10: 72–76; 1969, Veliger 11: 243–249 (filtering, feeding, ecology); Ó Foighil 1985, Biol. Bull. 169: 602–614 (reproduction); Ó Foighil 1985, Zoomorph. 105: 125–135 (reproduction).

* = Not in key.

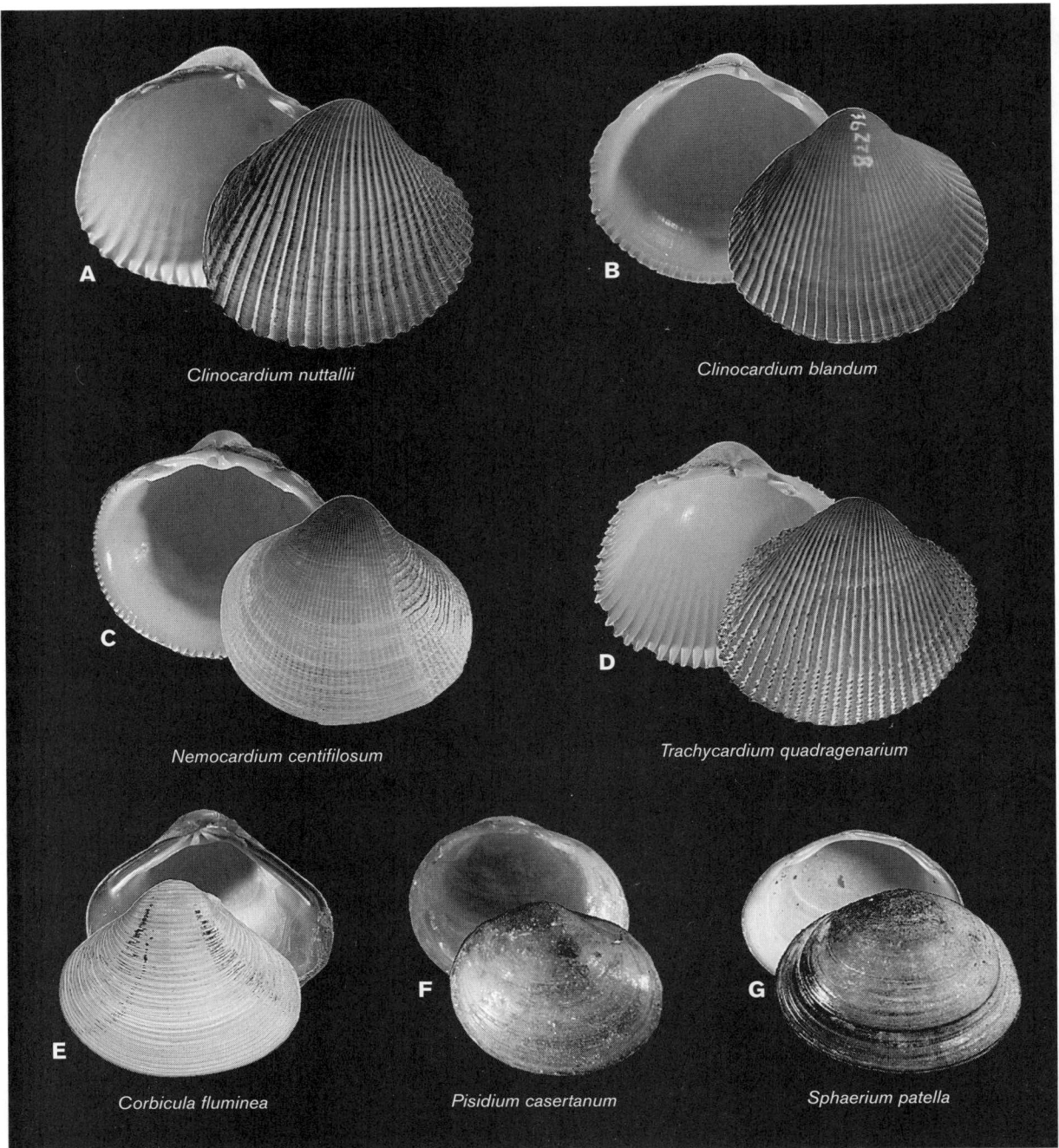

PLATE 415 A, *Clinocardium nuttallii*, length 56 mm; B, *Clinocardium blandum*, length 24 mm; C, *Nemocardium centifilosum*, length 13 mm; D, *Trachycardium quadragenarium*, length 72 mm; E, *Corbicula fluminea*, length 37 mm; F, *Pisidium casertanum*, length 4.5 mm; G, *Sphaerium patella*, length 7 mm.

CARDIIDAE

Schneider 1995, Zool. Scripta 24: 321–346 (evolution).

Clinocardium nuttallii (Conrad, 1837) (plate 415A). Mid-intertidal to offshore in sandy areas of bays. Coan et al. 2000: 351–354; Cooke 1975, Phycologia 14: 35–39 (endozoic green algae); Evans 1972, Science 176: 416–417 (growth); Fraser 1931, Trans. Roy. Soc. Can. Sec. 5, 25: 59–72 (ecology); Gallucci and Gallucci 1982, Mar. Ecol. Prog. Ser. 7: 138–145 (reproduction, ecology); Kafanov 1998, Bull. Mizunami Fossil Mus. 25: 30–35 (taxonomy, distribution); Schneider 1994, Veliger 37: 36–42 (anatomy); Weymouth and Thompson 1931, Bull. Bur. Fish. 46: 633–641 (age, growth).

Clinocardium blandum (Gould, 1850) (plate 415B). A more northerly species that is found just offshore as far south as Sonoma County. Coan et al. 2000: 353–354.

Nemocardium centifilosum (Carpenter, 1864) (plate 415C). Offshore in fairly shallow water. Coan et al. 2000: 358, 361.

* = Not in key.

Trachycardium quadragenarium (Conrad, 1837) (plate 415D). More characteristic of southern California, but has been recorded at Monterey in association with El Niño years. Coan et al. 2000: 358, 362.

PISIDIIDAE

The small, freshwater fingernail clams are occasionally encountered in deltas or at river mouths. *Pisidium casertanum* (Poli, 1791) (plate 415F), with an exhalant siphon and anterior end much longer than posterior, is common in springs, ponds, and slow creeks, whereas *Sphaerium patella* (Gould, 1850) (plate 415G), with the posterior end longer than the anterior and with exhalant and inhalant siphons, is common in such streams as Salmon Creek (Sonoma County). Herrington 1962, Univ. Mich. Mus. Zool. Misc. Pub. 118 (systematic review). Burch 1972, U.S. Environmental Protection Agency, Biota of Freshwater Ecosystems, Identification Manual 3, 33 pp.

CORBICULIDAE

The introduced Asian *Corbicula fluminea* (Müller, 1774) (plates 398B, 415E) has a thick, trigonal shell with low commarginal ridges covered by a heavy black-brown periostracum. Sometimes encountered on bay and ocean beaches as discarded fish bait, it is abundant in freshwater canals and irrigation channels, and large aggregations have locally clogged canal systems and water pipes. Coan et al. 2000: 349; Britton 1986, Proceedings of the Second International *Corbicula* Symposium, American Malacological Bulletin, Special Ed. 2: 239 pp.; Britton and Morton 1982, A dissection guide, field and laboratory manual for the introduced bivalve *Corbicula fluminea*, Malacological Review, Suppl. 3: 82 pp.

NEOLEPTONIDAE

KEY TO NEOLEPTONIDAE

1. Shell with strong commarginal ribs (plate 416A) . . . . *Bernardina bakeri*
— Shell with commarginal striae only 2
2. With an anterior lateral tooth in the left valve (plate 416B) . . . . *Neolepton salmoneum*
— Without such a tooth (plate 416C) . . . . *Neolepton subtrigoneum*

LIST OF SPECIES

Coan 1984, Veliger 27: 227–237 (systematic account). Formerly known on this coast as the Bernardinidae, this family may represent neotenous derivatives of the Veneridae. All species are very small-size nestlers in rubble from the intertidal zone to just offshore.

Bernardina bakeri Dall, 1910. Coan et al. 2000: 326–327; Coan 1984: 345–346.

Neolepton salmoneum (Carpenter, 1864). Coan et al. 2000: 346–347; Coan 1984: 231–233.

Neolepton subtrigonum (Carpenter, 1857). Coan et al. 2000: 346–347; Coan 1984: 231.

VENERIDAE

KEY TO VENERIDAE

1. Adult shell small, <10 mm in length. 2
— Adult shell >10 mm in length 4
2. Shell no longer than high, triangular; inner ventral margin finely crenulate; pallial sinus directed sharply upward; maximum length 5 mm (plates 412D, 416D) . . . . *Gemma gemma*
— Shell elongate to oval; inner ventral margin with obscure, oblique striae only; pallial sinus directed anteriorly . . . . *Nutricola* 3
3. Shell uniform white to cream; siphons fused nearly to tips; anterior lateral tooth relatively weak (plate 416E). . . . . *Nutricola tantilla*
— Shell light to deep purple; siphons with prominent cleft; anterior lateral tooth strong (plate 416F) . . . . . . *Nutricola confusa*
4. Shell smooth, with a shiny, adherent periostracum; large, trigonal (plate 417A) . . . . *Tivela stultorum*
— Shell sculptured; without a shiny periostracum 5
5. Sculpture commarginal, some with very fine radial striae 6
— Sculpture radial and commarginal, the commarginal sculpture not predominating and occasionally obsolete 10
6. Shell short; pronounced, widely spaced, commarginal lamellae (see note in species list); nestling among rocks or in borer holes (plates 412E, 417B) . . . . *Irusella lamellifera*
— Shell elongate or ovate, without prominent commarginal lamellae 7
7. Hinge without anterior lateral teeth; valves not gaping. 8
— Hinge with short anterior lateral tooth, close to cardinals; valves with narrow gape posteriorly. 9
8. Shell trigonal; valves convex, somewhat inflated; commarginal sculpture fine; beaks prominent; lunule deep (plate 417C) . . . . *Mercenaria mercenaria*
— Shell elongate-ovate; valves flattened; sculpture of thin, sharp, widely spaced commarginal ribs and very fine radial striae; beaks not prominent; lunule shallow (plate 417D) . . . . *Callithaca tenerrima*
9. Commarginal ribs heavy, conspicuous, well spaced, frequently stronger posteriorly; shell elongate, thinner than *S. gigantea*; lunule absent, escutcheon narrow (plate 417E) . . . . *Saxidomus nuttalli*
— Commarginal ribs thin, low, and closely spaced, giving shell a relatively smooth appearance; shell ovate, more rounded than *S. nuttalli*; lunule and escutcheon absent (plate 417F) . . . . *Saxidomus giganteus*
10. Inner ventral margin smooth; ligament prominent, elevated above dorsal margin; siphons separate at tips (plate 418A) . . . . *Venerupis philippinarum*
— Inner ventral margin crenulate; ligament sunken, not markedly elevated above dorsal margin; siphons fused for entire length 11
11. Radiating ribs numerous, fine; commarginal ridges faint to lacking in some specimens (see note in species list) (plate 418B). . . . . *Leukoma staminea*
— Radiating ribs and commarginal ribs both predominant, forming a sharp, coarse imbrications (plate 418C) . . . . *Leukoma laciniata*

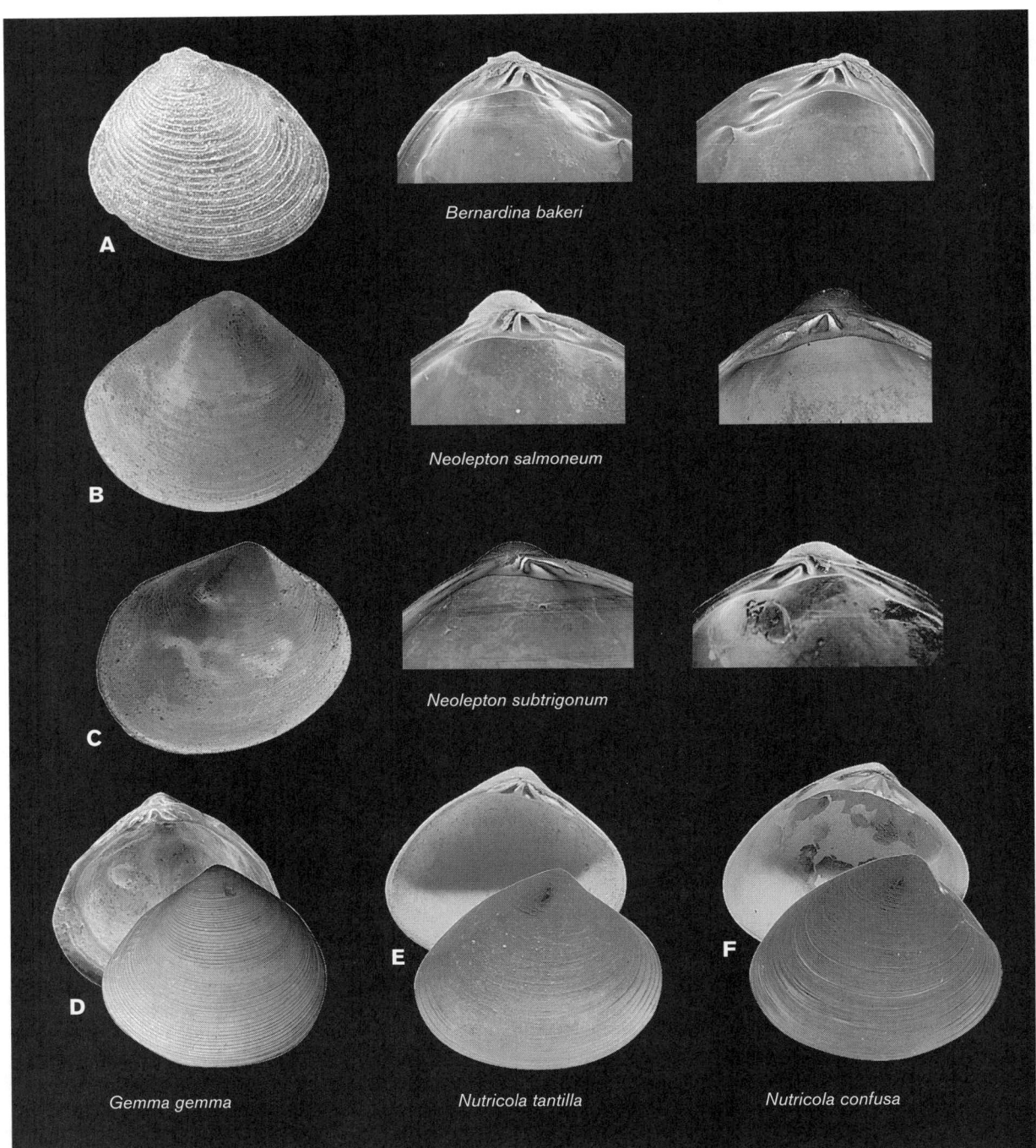

PLATE 416 A, *Bernardina bakeri,* length 2 mm; B, *Neolepton salmoneum,* length 2.5 mm; C, *Neolepton subtrigonum,* length 3 mm; D, *Gemma gemma,* length 2.5 mm; E, *Nutricola tantilla,* length 6 mm; F, *Nutricola confusa,* length 7 mm.

LIST OF SPECIES

Callithaca tenerrima (Carpenter, 1857) (=*Protothaca tenerrima*). Uncommon, in semiprotected areas of bays and offshore in sandy mud. Coan et al. 2000: 375–377.

Compsomyax subdiaphana (Carpenter, 1864) (plate 418D). Shell with weak commarginal striae; on soft offshore substrata. Coan et al. 2000: 375, 377.

Gemma gemma (Totten, 1834). Gem clam; introduced from Atlantic; common to abundant in mud of bays. Coan et al. 2000: 378, 380, with references to many additional biological papers; Sellmer 1967, Malacologia 5: 137–223 (functional morphology, ecology, life history); Narchi 1972, Bull. Mar. Sci. 21: 866–885 (anatomy).

Globivenus fordii (Yates, 1890) (plate 418E). A southern species occasionally found just offshore in rubble areas, particularly in warm-water years. Coan et al. 2000: 370–371.

* = Not in key.

PLATE 417 A, *Tivela stultorum*, length 93 mm; B, *Irusella lamellifera*, length 28 mm; C, *Mercenaria mercenaria*, length 115 mm; D, *Callithaca tenerrima*, length 90 mm; E, *Saxidomus nuttalli*, length 113 mm; F, *Saxidomus giganteus*, length 76 mm.

Humilaria kennerleyi (Reeve, 1863) (plate 418F). A northern species found offshore on soft bottoms. Coan et al. 2000: 370–372.

Irusella lamellifera (Conrad, 1837) (=*Irus lamellifer*). Nestling among rocks and in pholad burrows along outer coast; on pilings, in fouling. In bays. See note under *Leukoma staminea*. Coan et al. 2000: 370, 372.

* = Not in key.

Leukoma laciniata (Carpenter, 1864) (=*Protothaca laciniata*). Less common than *L. staminea*, in sandy mud of bays. Coan et al. 2000: 374–375.

[†]*Leukoma staminea* (Conrad, 1837) (= *Protothaca staminea*; = *Protothaca restorationensis* [Frizzell, 1930], once thought to be a hybrid). The Pacific littleneck is common both in sand of bays and nestling among rocks (often in empty pholad holes) on outer coast. Coan et al. 2000: 374–376, with references to many

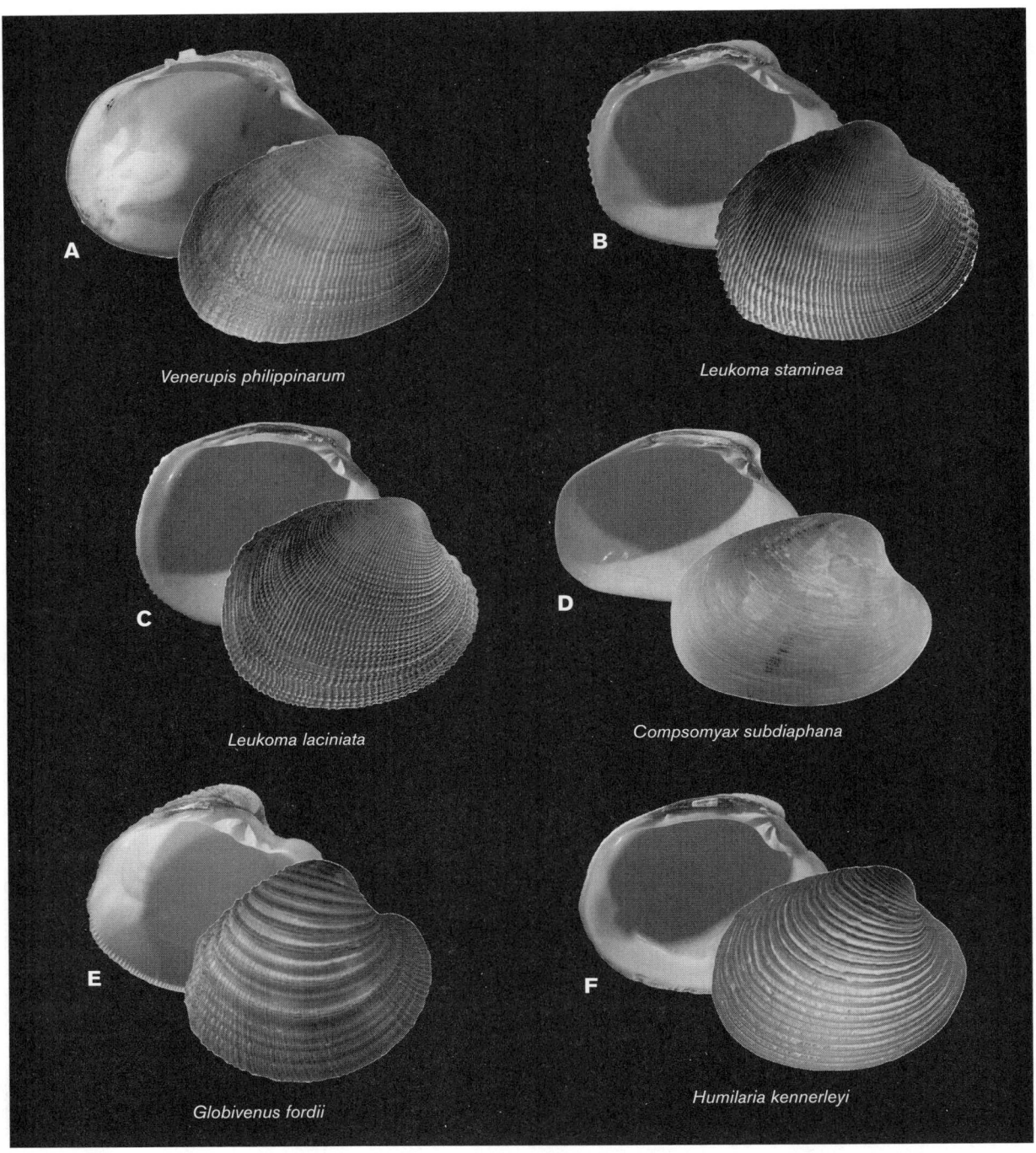

PLATE 418 A, *Venerupis philippinarum,* length 41 mm; B, *Leukoma staminea,* length 60 mm; C, *Leukoma laciniata,* length 69 mm; D, *Compsomyax subdiaphana,* length 49 mm; E, *Globivenus fordii,* length 46 mm; F, *Humilaria kennerleyi,* length 73 mm.

other biological papers; Fraser and Smith 1928, Trans. Roy. Soc. Canada (3) 22: 249–269 (ecology); Harrington 1987, Veliger 30: 148–158 (growth); Peterson 1982, Ecol. Mono. 52: 437–475 (ecology); Peterson 1983, J. Exper. Mar. Biol. Ecol. 68: 145–158 (ecology); Peterson and Ambrose 1985, Lethaia 18: 257–260 (growth); Schmidt and Warme 1969, Veliger 12: 193–199 (population characteristics); Boulding and Labarbera 1986 Biol. Bull. 171: 538–547 (crab predation by fatigue damage). A nestling form of this species may have raised, commarginal lamellae like

Irusella but can be distinguished by its more prominent radial sculpture.

†Our recent review of many of the species that have been placed in the New World genus *Protothaca,* most of which are tropical, demonstrates that the several subgenera proposed for this genus are not helpful, being based on minor sculptural difference. Moreover, one of those subgenera, *Leukoma* Römer, 1857, is older than *Protothaca* Dall, 1902, and is therefore the genus to be used for the whole group. Further, we have concluded that *Callithaca* Dall, 1902, is sufficiently different to be regarded as a separable genus.

Mercenaria mercenaria (Linnaeus, 1758). The quahog; introduced from Atlantic, reproducing populations present in southern California and in southern British Columbia; in mud in bays; not common. Coan et al. 2000: 373–375; Kraeuter and Castagna 2001, Biology of the hard clam. New York (Elsevier), 732 pp., both with many references.

Nutricola confusa (Gray, 1982) (=*Transennella confusa*). In muddy sand of bays. Coan et al. 2000: 381–382; Gray 1982, Malac. Rev. 15: 107–117 (description, anatomy).

Nutricola lordi (Baird, 1862) and *N. ovalis* (Dall, 1902) (both formerly in *Psephidia*). Just offshore on soft bottoms. Coan et al. 2000: 382–383, plate 79.

Nutricola tantilla (Gould, 1853) (=*Transennella tantilla*). Common in sand or sandy mud in semiprotected situations in bays as well as offshore. Coan et al. 2000: 382, 384, with references to additional biological papers; Asson-Batres 1986, Veliger 30: 257–266 (reproduction, growth); Hansen 1953, Vidensk. Medd. Dansk. Naturh. Foren. 115: 313–324 (brood protection, sex ratio); Kabat 1985, J. Exp. Mar. Biol. Ecol. 91: 271–279 (brooding); Narchi 1972, Bull. Mar. Sci. 21: 866–885 (anatomy); Pamatmat 1969, Amer. Zool. 9: 419–426 (respiration).

Saxidomus gigantea (Deshayes, 1839). Uncommon in central California, a more northern species, in same habitat as *S. nuttalli*. Coan et al. 2000: 3384–3385, with many references to biological literature; Fraser and Smith 1928, Trans. Roy. Soc. Canada (3) 22: 271–277 (ecology).

Saxidomus nuttalli Conrad, 1837. California butter clam; common in bays and lagoons in mud or sand; also on outer coast in sand among rocks. Coan et al. 2000: 385–386; Peterson 1977, Mar. Biol. 43: 343–359 (ecology).

Tivela stultorum (Mawe, 1823). Pismo clam; on exposed sandy beaches from Half Moon Bay south. Coan et al. 2000: 379–380; Weymouth 1923, Calif. Fish Game Fish Bull. 7: 120 pp. (life history and growth); Herrington 1930, ibid. 18: 67 pp. (growth, populations); Coe and Fitch 1950, J. Mar. Res. 9: 188–210 (growth, reproduction); Fitch 1950, Calif. Fish Game 36: 285–312.

Venerupis philippinarum (A. Adams and Reeve, 1850) (=*V. semidecussata* and *V. japonica*; variously placed in *Paphia, Tapes, Protothaca,* and *Ruditapes,* the latter now used as a subgenus). Japanese littleneck, introduced from Asia; common in mud of bays. Coan et al. 2000: 385, 387–388, with references to the considerable biological literature.

PETRICOLIDAE

KEY TO PETRICOLIDAE

1. Shell with moderate to heavy radial sculpture 2
— Shell smooth, without heavy radial sculpture 3
2. With moderate radial sculpture, strongest anteriorly; nestling in crevices (plate 419A) . . . . *Petricola californiensis*
— With heavy radial sculpture, strongest on anterior end; shell elongate; burrowing in mud (plate 419B) . . . . *Petricolaria pholadiformis*
3. Periostracum adherent, shiny or with satin sheen (plate 419C). . . . . *Cooperella subdiaphana*
— Periostracum eroded, shell dull; with very fine radial striae, may be eroded; shell shape extremely variable; nestling in rock crevices (plate 419D) . . . . *Petricola carditoides*

* = Not in key.

LIST OF SPECIES

Coan 1997, Veliger 40: 298–340 (systematic account).

Cooperella subdiaphana (Carpenter, 1864). A thin, fragile species more common offshore, but occasionally found on low tidal mudflats. Often found living in mucus-lined "mud-balls." Coan et al. 2000: 391, 393–394; Morton 1995, Veliger 38: 162–170 (functional morphology); Hertz 2004, Festivus 36: 60–62 (habitat).

Petricola carditoides (Conrad, 1837). Common in rocky intertidal, nestling in rock crevices and in pholad holes, which it can enlarge by limited boring. Coan 1997: 302–307; Coan et al. 2000: 390–392; Yonge 1958, Proc. Malac. Soc. London 33: 25–31 (functional morphology).

Petricola californiensis Pilsbry and Lowe, 1832. A southern species that occasionally reaches central California in warmwater years. Coan 1997: 311–312; Coan et al. 2000: 391–392.

Petricolaria pholadiformis (Lamarck, 1818). Burrowing in mud; introduced from Atlantic coast in San Francisco Bay, although living populations have not been confirmed there in recent decades; remains established in Willapa Bay, WA. Coan 1997: 391, 394; Coan et al. 2000: 372, 375, both with references to a number of biological papers.

TELLINOIDEA

KEY TO TELLINOIDEA

1. Part of ligament internal, seated in a resilifer, either below beaks or in a posteriorly directed furrow (plate 420A). . . . . Semelidae 2
— Ligament entirely external 5
2. Shell smooth, shiny; in mud of bays (plate 419E). . . . . *Theora lubrica*
— Shell with conspicuous sculpture 3
3. Resilifer triangular, projecting below beaks; conspicuous commarginal sculpture; shell white internally (plates 412G, 419F) . . . . *Cumingia californica*
— Resilifer elongate, a posteriorly directed furrow . . . . *Semele* 4
4. Pallial sinus extending well past beaks; shell with heavy, frequently irregular, commarginal undulations; posterior end truncate (plate 420A). . . . . *Semele decisa*
— Pallial sinus not extending past beaks; shell with fine to medium commarginal sculpture and growth checks; posterior end not truncate, usually rounded (plate 420B). . . . . *Semele rupicola*
5. Shell elongate, cylindrical (plate 420C). . . . Solecurtidae 6
— Shell ovate, not cylindrical 7
6. Shell with weak internal radial strengthening rib extending ventrally from beaks; light purple to white in color; pallial sinus never extending to umbones (plates 412H, 420C) . . . . *Tagelus subteres*
— Shell white, without radial strengthening rib; pallial sinus never extending to umbones (plate 420D). . . . . *Tagelus californianus*
— Shell white, without radial strengthening rib; pallial sinus usually extending to or past umbones (plate 420E) . . . . *Tagelus affinis*
7. External ligament seated on a particularly conspicuous nymph (buttress for the external ligament); conspicuously colored with rays or purple color; without lateral teeth

PLATE 419 A, *Petricola californiensis*, length 32 mm; B, *Petricolaria pholadiformis*, length 45 mm; C, *Cooperella subdiaphana*, length 13 mm; D, *Petricola carditoides*, length 32 mm; E, *Theora lubrica*, length 9 mm; F, *Cumingia californica*, length 28 mm.

(plate 398E). . . . . Psammobiidae 8
— Nymph not conspicuous; some with reddish tinge (not rays); some with lateral teeth 10
8. Shell subquadrate, heavy, white internally; periostracum dull; with pink to purple radial rays externally (plate 420F) . . . . *Gari californica*
— Shell ovate, thin, purple internally; periostracum, shiny brown . . . . *Nuttallia* 9
9. Posterior end tapered; left valve more inflated than right; white to light purple internally (plate 421A). . . . .
. . . . *Nuttallia nuttallii*

— Posterior end rounded; equivalve; color always deep purple internally (plate 421B). . . . . *Nuttallia obscurata*
10. Shell with fine radial sculpture (plate 421C) . . . .
. . . . *Donax gouldii*
— Shell with commarginal ribs or striae only . . . .
. . . . Tellinidae 11
11. Hinge with lateral teeth, especially evident in right valve
. . . . *Tellina* 12
— Hinge without lateral teeth . . . . *Macoma* 14
12. Shell with regular, strong commarginal sculpture; to 60 mm in length (plate 421D) . . . . *Tellina bodegensis*

PLATE 420 A, *Semele decisa*, length 92 mm; B, *Semele rupicola*, length 24 mm; C, *Tagelus subteres*, length 38 mm; D, *Tagelus californianus*, length 114 mm; E, *Tagelus affinis*, length 41 mm; F, *Gari californica*, length 112 mm.

— Shell smooth externally or weak commarginal sculpture or growth lines; to 20 mm in length 13

13. Shell ovate-elongate, thin; color white; periostracum inconspicuous; with internal strengthening rib anteriorly (plate 421E) . . . . *Tellina modesta*

— Shell ovate-trigonal, heavy; color pink to white; often with olive periostracum; without internal strengthening rib (plate 421F) . . . . *Tellina nuculoides*

14. With a tendency to produce a posterordorsal flange from posterior end to ligament (plate 422A) 15

— Without a posterior dorsal flange. 16

15. Posterior end truncate (plates 422A, 423A) . . . .
. . . . *Macoma secta*

— Posterior end produced, pointed; posteroventral margin often flexed (plate 423B) . . . . *Macoma indentata*

16. Anterior ventral edge of pallial sinus detached for at least one-quarter distance to posterior adductor muscle scar and more or less paralleling pallial line (true of both valves); length <30 mm (plate 422B) 17

— Anterior ventral end of pallial sinus not detached from pallial line for a substantial distance, although it may overlap slightly near point of juncture (usually only in one valve); length to 110 mm (plate 422D) 18

17. Pointed to produced posteriorly (plates 422B, 423C) . . . .
. . . . *Macoma yoldiformis*

— Broadly rounded posteriorly (plates 422C, 423D) . . . .

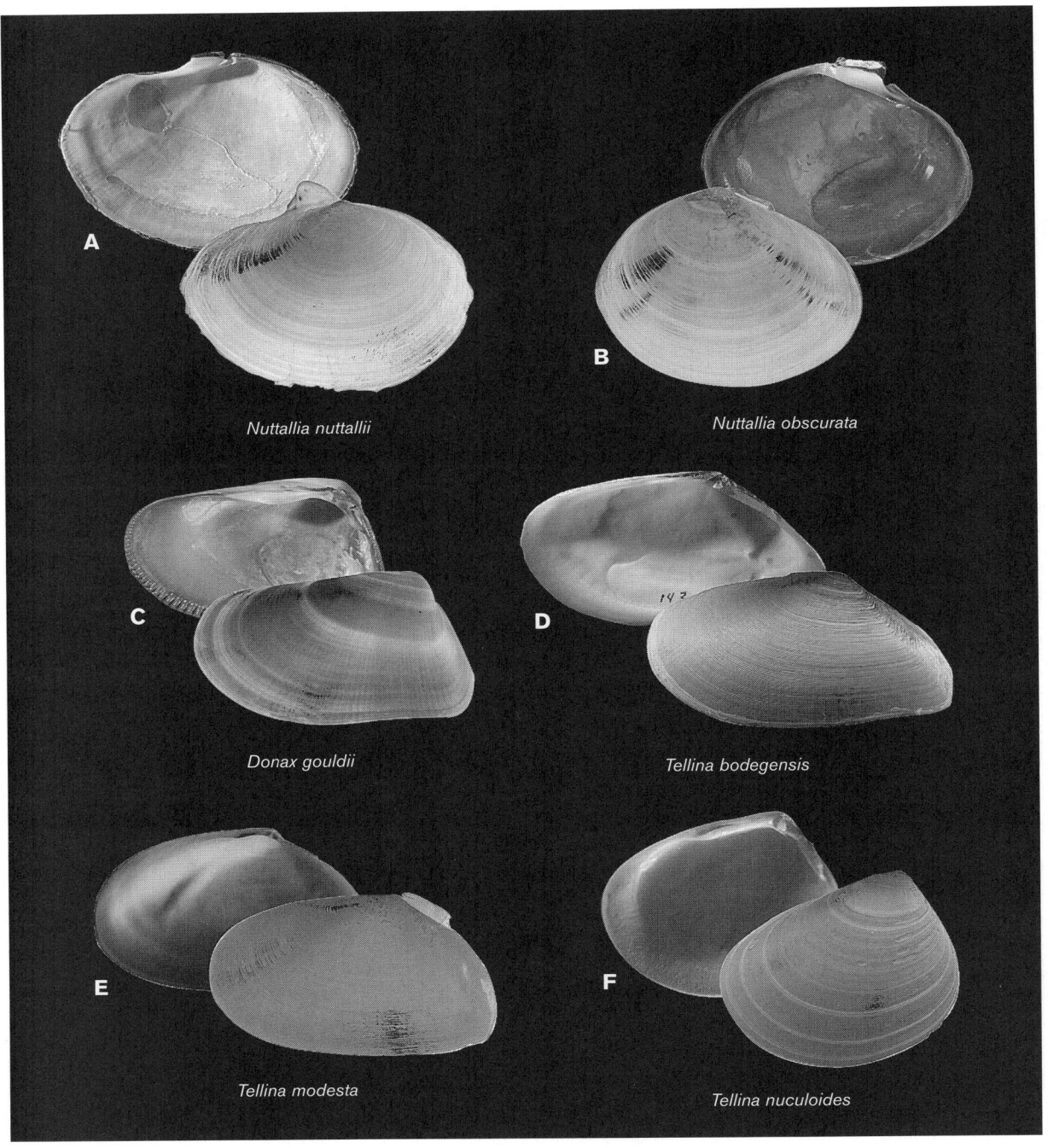

PLATE 421 A, *Nuttallia nuttallii*, length 84 mm; B, *Nuttallia obscurata*, length 44 mm; C, *Donax gouldii*, length 18 mm; D, *Tellina bodegensis*, 59 mm; E, *Tellina modesta*, length 14 mm; F, *Tellina nuculoides*, length 13 mm.

. . . . *Macoma acolasta*

18. Pallial sinus terminating only one-quarter of way to anterior adductor muscle scar; often with pinkish tinge (plates 422D, 423E) . . . . *Macoma balthica/M. petalum*

   Note: (see species list)

— Pallial sinus reaching to or almost to anterior adductor muscle scar in one or both valves; whitish 19

19. Pallial sinus in left valve nearly always reaching anterior adductor muscle scar and fusing with it; shell bent to right

posteriorly (plates 422E, 423F) . . . . *Macoma nasuta*

— Pallial sinus not quite reaching ventral end of anterior adductor muscle scar; equivalve, not bent posteriorly (plates 422F, 424A) . . . . *Macoma inquinata*

LIST OF SPECIES

Yonge 1949, Phil. Trans. Roy. Soc. London (B) 234: 29–76; Pohlo 1969, Proc. Malac. Soc. London 38: 361–364 (both functional morphology).

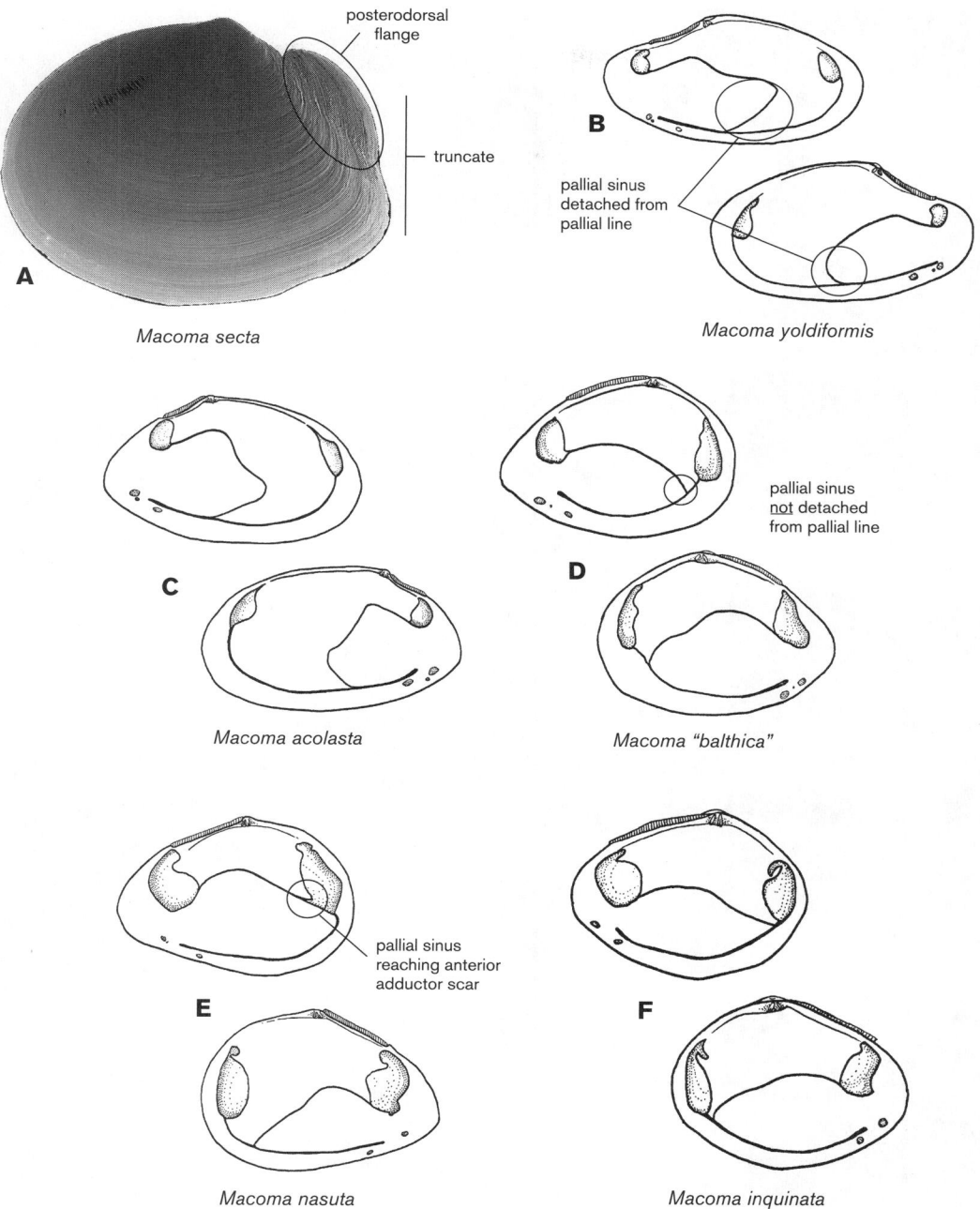

posterodorsal flange

truncate

A

Macoma secta

B

pallial sinus detached from pallial line

Macoma yoldiformis

C

Macoma acolasta

D

pallial sinus <u>not</u> detached from pallial line

Macoma "balthica"

E

pallial sinus reaching anterior adductor scar

Macoma nasuta

F

Macoma inquinata

PLATE 422 A, *Macoma secta,* length 51 mm; B–F, diagrammatic sketches of adductor muscle scars, pallial line, and pallial sinus of member of the genus *Macoma.*

DONACIDAE

Coan 1983, Veliger 25: 273–298 (systematic account).

    Donax gouldii Dall, 1921. Coan 1983: 290–291; Coan et al. 2000: 423, with references to additional biological papers; Pohlo 1967, Veliger 9: 330–337 (functional morphology). Normally not north of Pismo Beach, but sporadically to Santa Cruz in warm-water periods.

PSAMMOBIIDAE

Coan 1973, Veliger 16: 40–57; Coan 2000, Malacologia 42: 1–29 (systematic accounts).

    Gari californica (Conrad, 1837). Among rubble, low intertidal zone. Coan 1973: 42–43; Coan 2000, Malacologia 42: 3–7; Coan et al. 2000: 426–427.

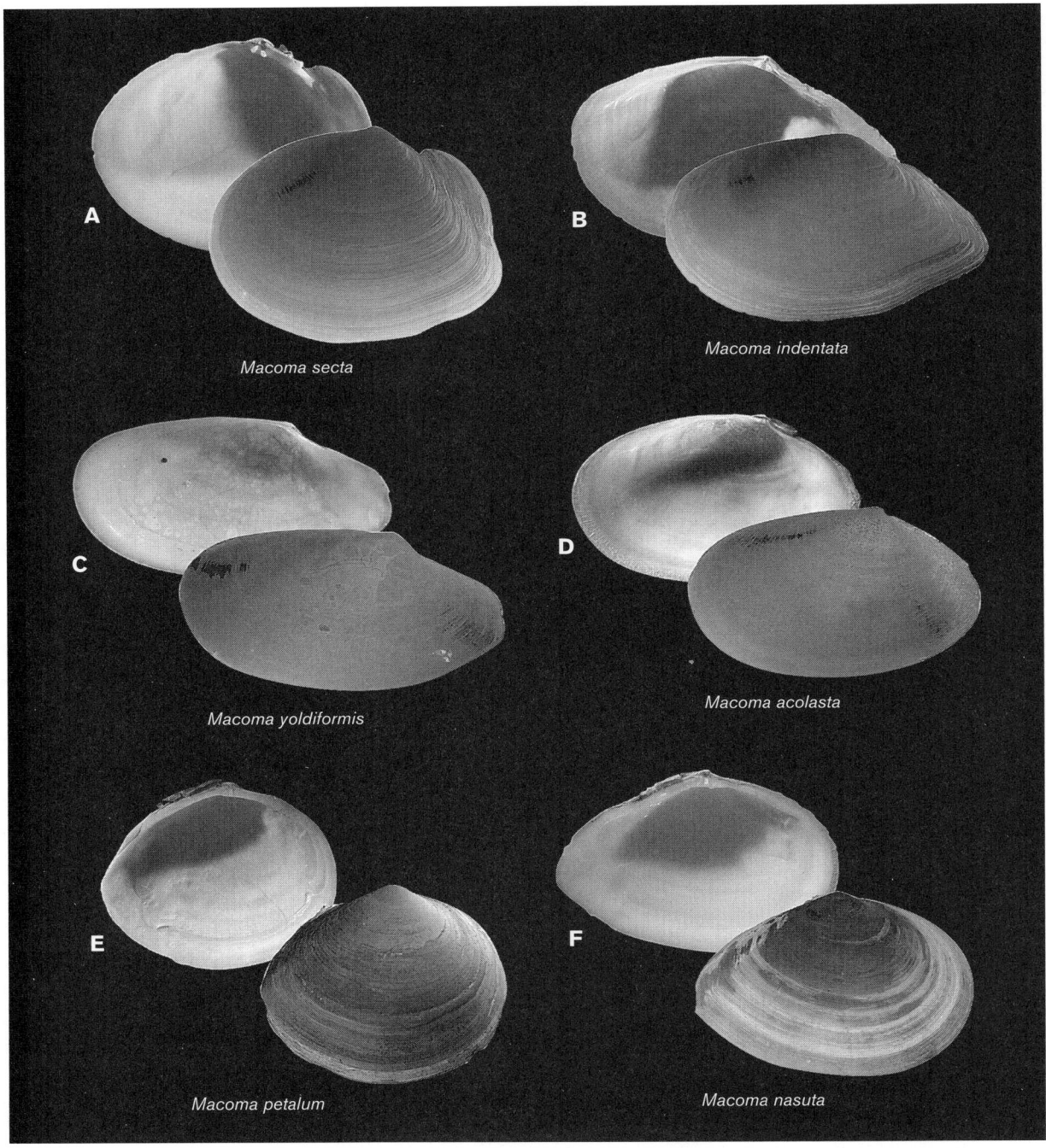

PLATE 423 A, *Macoma secta*, length 51 mm; B, *Macoma indentata*, length 43 mm; C, *Macoma yoldiformis*, length 14 mm; D, *Macoma acolasta*, length 11 mm; E, *Macoma petalum,* length 23 mm; F, *Macoma nasuta*, length 37 mm.

Nuttallia nuttallii (Conrad, 1837) (=*Sanguinolaria nuttallii*). In sand and sandy mud of low intertidal, protected bays. Reported as far north as Humboldt Bay (Yoshimoto 2004, Festivus 36: 128–129), and Coos Bay (J. Carlton, personal communication, 2004). Coan 1973: 48–49; Coan et al. 2000: 427, 429; Pohlo 1972, Veliger 14: 298–301 (feeding, associated morphology).

Nuttallia obscurata (Reeve, 1857). Introduced from Asia by ballast water. Coan et al. 2000: 427, 429.

SEMELIDAE

Coan 1973, Veliger 15: 314–329; Coan 1988, Veliger 31: 1–42 (systematic reviews).

Cumingia californica Conrad, 1837. Common nestler in rocky intertidal zone. Coan 1973: 323–325; Coan et al. 2000: 436–437.

Semele decisa (Conrad, 1837). A southern California species that occurs as far north as Point Arguello, nestling in rocky situations. Coan 1988: 6–7; Coan et al. 2000: 432–433.

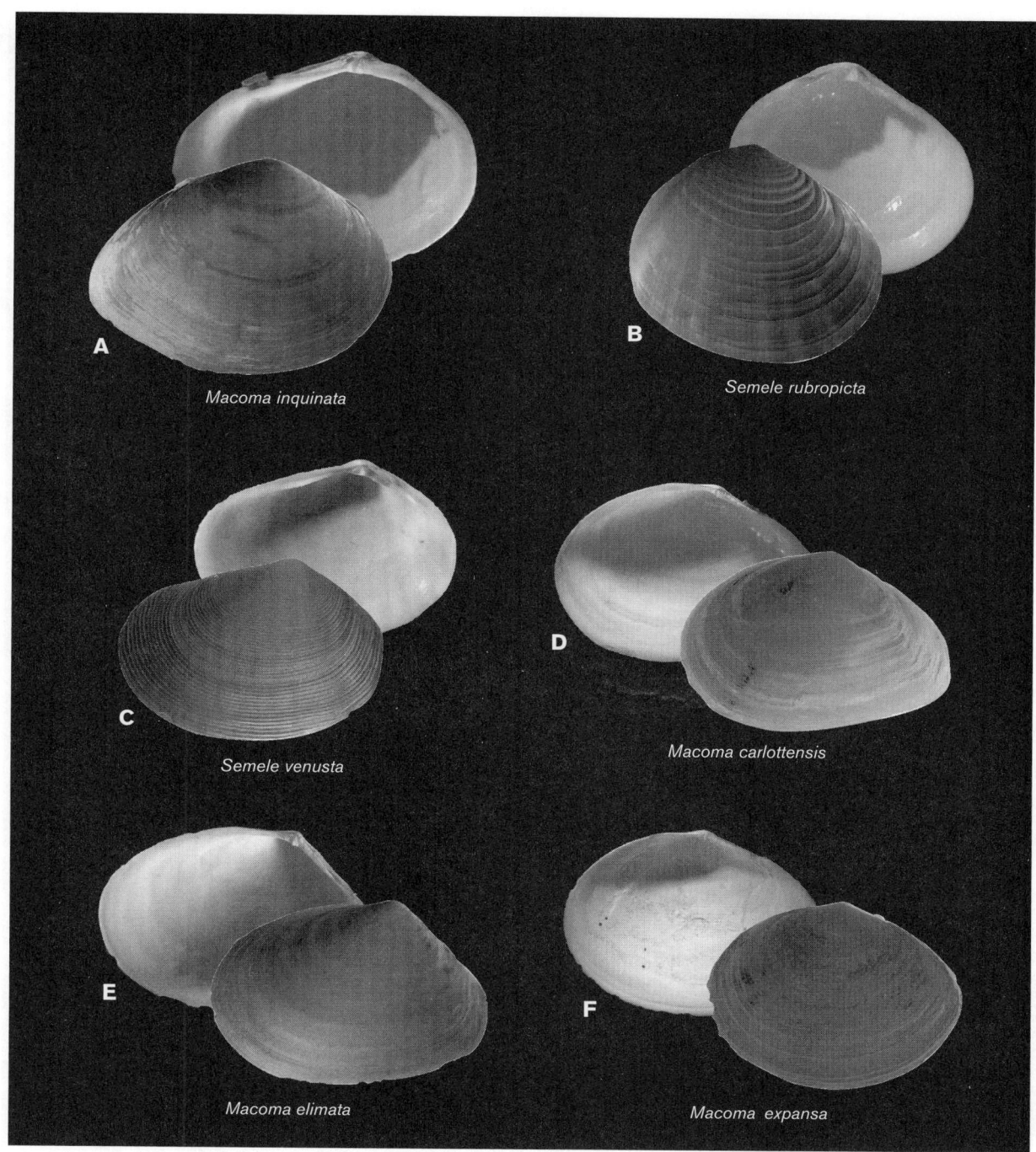

PLATE 424 A, *Macoma inquinata,* length 53 mm; B, *Semele rubropicta,* length 33 mm; C, *Semele venusta,* length 21 mm; D, *Macoma carlottensis,* length 18 mm; E, *Macoma elimata,* length 15 mm; F, *Macoma expansa,* length 43 mm.

Semele rubropicta Dall, 1871 (plate 424B). Uncommon, offshore species; valves occasionally wash ashore; characterized by conspicuous radial and commarginal sculpture and red radial rays. Coan 1988: 12–13; Coan et al. 2000: 432–434.

Semele venusta (Reeve, 1853) (plate 424C). An offshore species in mud bottoms. Coan 1988: 21–23; Coan et al. 2000: 433–434.

* = Not in key.

Semele rupicola Dall, 1915. Nestler in rocky intertidal; uncommon. Coan 1988: 23–24; Coan et al. 2000: 433–435.

Theora lubrica Gould, 1861. Introduced from Asia, this species occurs in several Californian bays, including San Francisco Bay. Coan et al. 2000: 4436, 438.

SOLECURTIDAE

Coan 1973, Veliger 16: 40–57 (systematic account).

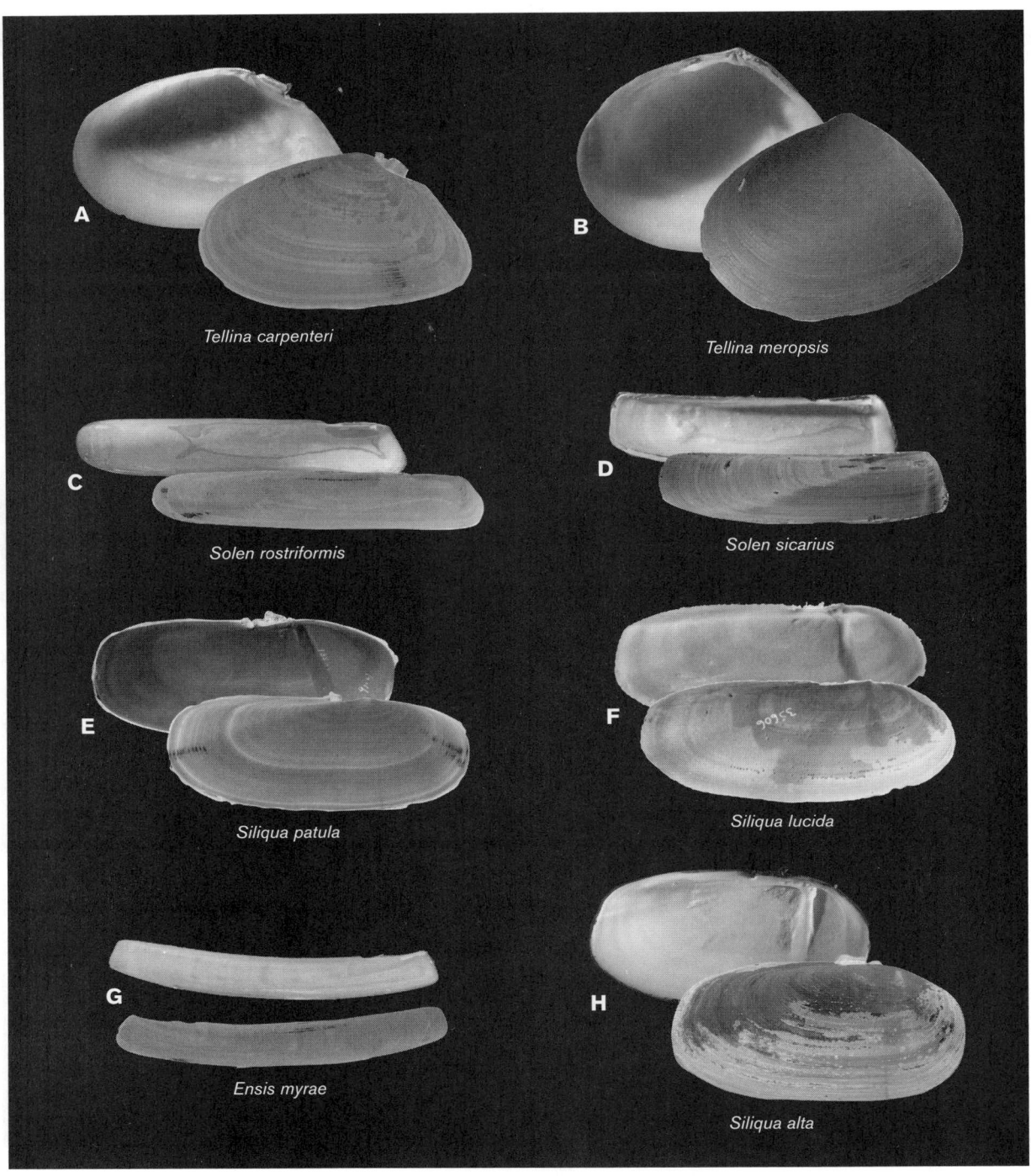

PLATE 425 A, *Tellina carpenteri*, length 10 mm; B, *Tellina meropsis*, length 16 mm, C, *Solen rostriformis*, length 57 mm; D, *Solen sicarius*, length 78 mm; E, *Siliqua patula*, length 125 mm; F, *Siliqua lucida*, length 52 mm; G, *Ensis myrae*, length 56 mm; H, *Siliqua alta*, length 120 mm.

Tagelus affinis (C. B. Adams, 1852). Low intertidal zone in sand and mud of protected bays; as far north as Morro Bay. Coan 1973: 50–51; Coan et al. 2000: 441–442.

Tagelus californianus (Conrad, 1837). Low intertidal zone in sand and mud of protected bays. Coan 1973: 51–52; Coan et al. 2000: 441–442; Pohlo 1966, Veliger 8: 225 (feeding).

Tagelus subteres (Conrad, 1837). Low intertidal zone in sand and mud of protected bays. Southern, occasionally north to Morro Bay. Coan 1973: 52–53; Coan et al. 2000: 442.

TELLINIDAE

Coan, 1971, Veliger 14 Suppl.: 63 pp. (systematic review); Dunnill and Ellis 1969, Veliger 12: 207–291 (*Macoma* ecology); Reid and Reid 1969, Can. J. Zool. 47: 64–57 (*Macoma* feeding).

Macoma acolasta Dall, 1921 . Protected bays in low intertidal zone in sand; uncommon. Coan 1971: 34–35; Coan et al. 2000: 414, 416.

Macoma balthica (Linnaeus, 1758) (=*Macoma inconspicua* [Broderip and G. B. Sowerby I, 1829]). The common brackish water macoma from at least Coos Bay, Oregon, north through the Arctic. See Vainola 2003 (above); Meehan et al. 1985, Mar. Biol. 102: 235–241 (genetics of San Francisco Bay and Coos Bay populations); Coan 1971: 44–46; Coan et al. 2000: 417–419, with references to the extensive biological literature. See also *Macoma petalum.*

Macoma carlottensis Whiteaves, 1880 (plate 424D). Offshore on soft bottoms. Coan 1971: 35–37; Coan et al. 2000: 414, 416.

Macoma elimata Dunnill and Coan, 1968 (plate 424E). Offshore on soft bottoms. Coan 1971: 22–23; Coan et al. 2000: 405, 410.

Macoma expansa Carpenter, 1864 (plate 424F). Of the same group as *M. secta* and *M. indentata,* but thinner, more inflated, with an only slightly developed posterior dorsal flange: rare, in sand offshore, in exposed areas; valves occasionally washed ashore. Coan 1971: 40–41; Coan et al. 2000: 415–416.

Macoma indentata Carpenter, 1864. In silt to sand of bays, uncommon. Coan 1971: 39–40; Coan et al. 2000: 415, 418.

Macoma inquinata (Deshayes, 1855). Common, in silt and mud in protected areas, most common in bays, but also below surf zone offshore. Coan 1971: 42–43; Coan et al. 2000: 418–419; Levinton 1991, Mar. Biol. 110: 375–383 (feeding).

Macoma nasuta (Conrad, 1837). Bent-nosed clam; common, in mud and muddy sand in protected areas, most common in bays at mid-tide, also below surf zone offshore. Coan 1971: 41–42; Coan et al. 2000: 418, 420; Gallucci and Hylleberg 1976, Veliger 19: 59–67 (growth); Hylleberg and Gallucci 1975, Mar. Biol. 32: 167–178 (feeding); Levinton 1991, Mar. Biol. 110: 375–383 (feeding); Rae 1978, Biol. Bull. 155: 207–291 (reproduction); Rae 1979, Veliger 21: 384–299 (population study).

Macoma petalum (Valenciennes, 1821) (=*Macoma balthica* of San Francisco Bay and other central California localities). Common in mud in upper intertidal of bays, especially in brackish water. The common small pink and white macoma of San Francisco Bay is the North American Atlantic coast species *M. petalum,* now recognized as genetically distinct from the boreal and European *M. balthica* (Vainola 2003, Mar. Biol. 143: 935–946). It was introduced with oysters.

Its range along the Pacific coast has not been determined. Morphological distinctions between *M. petalum* and *M. balthica* have not been worked out; they are keyed together here, and a specimen of *M. petalum* from Chesapeake Bay is shown in the plate. See Vassallo 1969, Veliger 11: 223–234; 1971, 13: 279–284 (ecology in San Francisco Bay).

Macoma secta (Conrad, 1837). Sand clam; common, intertidal in sand in semiprotected areas of bays and offshore of sandy beaches. Coan 1971: 37–39; Coan et al. 2000: 417–418; Levinton 1991, Mar. Biol. 110: 375–383 (feeding); Rae 1978, Biol. Bull. 155: 207–291 (reproduction); Rae 1979, Veliger 21: 384–299 (population study).

Macoma yoldiformis Carpenter, 1864. In silt to sand, in protected areas in low intertidal of bays; rare. Coan 1971: 33–34; Coan et al. 2000: 414, 416.

Tellina bodegensis Hinds, 1845. Low intertidal in sand of exposed beaches; in bays. Coan 1971: 10–11; Coan et al. 2000: 404–405.

Tellina carpenteri. Dall, 1900 (plate 425A). Similar to *T. modesta,* but light pink in color; below low tide on various bottoms. Coan 1971: 15–16; Coan et al. 2000: 400–401.

Tellina meropsis Dall, 1900 (plate 425B). A southern form that is sometimes transported northward in warm-water years;

on soft bottoms just offshore. Coan 1971: 14–15; Coan et al. 2000: 401–402.

Tellina modesta (Carpenter, 1864). In sand to silty sand of bays, to well offshore. Coan 1971: 16–17; Coan et al. 2000: 401–402; Maurer 1967a–c, Veliger 9: 305–309, 376–381, 10: 72–76, 1969, Veliger 11: 243–249 (biology).

Tellina nuculoides (Reeve, 1854). On various bottoms in protected areas of bays. Coan 1971: 12–13; Coan et al. 2000: 401, 403; Maurer 1967a–c, Veliger 9: 305–309, 376–381, 10: 72–76, 1969, Veliger 11: 243–249 (aspects of biology).

SOLENIDAE

KEY TO SOLENIDAE

1. Shell thin, tapered; periostracum light olive to light brown (plate 425C) . . . . *Solen rostiformis*
— Shell thick, blunt posteriorly; periostracum dark brown (plate 425D) . . . . *Solen sicarius*

LIST OF SPECIES

Cosel 1990, Veliger 35: 366–380 (taxonomy).

Solen rostiformis Dunker, 1862 (=*Solen rosaceus* Carpenter, 1864). In mud in bays as far north as Morro Bay. Coan et al. 2000: 444–445; Cosel 1990, Veliger 35: 366–380 (taxonomy; with *S. rostiformis* and *S. rosaceus* regarded as separate species); Pohlo 1963, Veliger 6: 98–104 (functional morphology).

Solen sicarius Gould, 1850. Protected areas of bays in mud or muddy sand; forming permanent burrows in which it moves freely up and down. Coan et al. 2000: 445.

PHARIDAE

KEY TO PHARIDAE

1. Internal radial rib sloping anteriorly, relatively wide; posterior end tapered in specimens less than 55 mm; large, to 190 mm in length (plate 425E) . . . . *Siliqua patula*
— Internal radial rib vertical, narrow; posterior end truncate; not more than 55 mm in length (plate 425F) . . . .
. . . . *Siliqua lucida*

LIST OF SPECIES

Cosel 1990, An introduction to the razor shells (Bivalvia: Solenacea), pp. 283–311, in B. Morton, ed., The Bivalvia, University of Hong Kong Press (systematics).

Ensis myrae Berry, 1953 (plate 425G). A long, thin, offshore species. Coan et al. 2000: 446–447, 449.

Siliqua alta (Broderip and G. B. Sowerby I, 1829) (=*S. sloati* Hertlein, 1961) (plate 425H). A northern species occurring offshore in central California, where it is smaller than in Alaska. Coan et al. 2000: 448–449.

Siliqua lucida (Conrad, 1837). Protected sandy areas of bays. Coan et al. 2000: 448–449; Hertlein 1961, Bull. So. Calif. Acad. Sci. 60: 12–18 (comparisons). Possibly only young of *S. patula;* more research needed.

Siliqua patula (Dixon, 1789). Razor clam; semiprotected, clean-sand beaches. Coan et al. 2000: 448–449, with references

* = Not in key.

to many additional biological and fisheries papers; Bourne and Quayle 1970, Fish. Res. Bd. Can. Tech. Report 232: 42 pp. (breeding, growth); Hertlein 1961, as above (comparisons); Lassuy and Simons 1989, U.S. Fish and Wildlife Serv. Biol. Rept. 82: 16 pp. (general review); LeClair and Phelps 1994, J. Shellfish Res. 13: 207–216 (genetics); Pohlo 1963, Veliger 6: 98–104 (morphology, burrowing); Taylor 1959, J. Conseil 25: 93–101 (growth); Weymouth and McMillin 1930, Bull. U.S. Bur. Fish. 46: 543–567 (growth); Weymouth et al. 1925, ibid. 41: 201–236 (growth); Weymouth et al. 1931, J. Exp. Biol. 8: 228–249 (growth); Yonge 1952, Univ. Calif. Publ. Zool. 55: 421–438 (aspects of morphology, reproduction, growth, biology, ecology).

MACTRIDAE

KEY TO MACTRIDAE

1. Shell broadly gaping posteriorly . . . . *Tresus* 2
— Shell narrowly gaping or closed posteriorly 3
2. Shell ovate-elongate; posterior end markedly longer than anterior (plate 426A) . . . . *Tresus nuttallii*
— Shell ovate to rhomboidal; posterior end not markedly longer than anterior (plate 426B) . . . . *Tresus capax*
3. Chondrophore projecting ventrally (plate 427A) . . . .
 . . . . *Mactromeris* 4
— Chondrophore not projecting past hinge line, or projecting only slightly past it (plate 427B) 5
4. Anterior end shorter than posterior, broad; anterodorsal margin straight (plate 426C) . . . . *Mactromeris catilliformis*
— Anterior end longer than or equal to posterior; anterodorsal margin concave (plates 426D, 427A) . . . .
 . . . . *Mactromeris hemphillii*
5. Beaks with commarginal undulations (plate 426E) . . . . . .
 . . . . *Mactrotoma californica*
— Beaks without undulations . . . . *Simomactra* 6
6. Shell longer anteriorly; pallial sinus of moderate depth; anterior lateral tooth long in left valve, aligned with anterior cardinal tooth (plate 426F) . . . . *Simomactra falcata*
— Shell subequilateral; pallial sinus shallow, narrow; anterior lateral tooth short in left valve, not aligned with cardinal (plates 426G, 427B) . . . . *Simomactra planulata*

LIST OF SPECIES

Mactromeris catilliformis (Conrad, 1867) (=*Spisula catilliformis*). On sand beaches on open coast, often washing ashore after large storms. Coan et al. 2000: 454–455.

Mactromeris hemphillii (Dall, 1894) (=*Mactra hemphillii*). Occurs in the same situations, only as far north as Cayucos. Coan et al. 2000: 454–455.

Mactrotoma californica (Conrad, 1837) (=*Mactra californica*). In sandy mud of bays. A southern species found north of Point Conception in warm-water years. Coan et al. 2000: 457.

Simomactra falcata (Gould, 1850) (=*Spisula falcata*). Uncommon in sandy areas, chiefly offshore. Coan et al. 2000: 458–459.

Simomactra planulata (Conrad, 1837) (=*Mactra planulata*). Occurs in similar situations. Coan et al. 2000: 458–459.

Tresus allomyax Coan and Valentich-Scott, 2000 (plate 426H). Offshore, northern California; occasionally washes ashore. Coan et al. 2000: 460–463.

* = Not in key.

Tresus capax (Gould, 1850) (=*Schizothaerus capax*). The fat gaper; a more northern species and uncommon in central California. Coan et al. 2000: 461, 463, with references to many additional biological papers; Bourne and Smith 1972, Proc. Natl. Shellfisheries Assn. 62: 35–37 (larvae); Bourne and Smith 1972, Proc. Natl. Shellfisheries Assn. 62: 38–46 (reproduction, growth); Breed-Willecke and Hancock 1980, Proc. Natl. Shellfisheries Assn. 70:1–13 (reproduction, growth); Campbell and Bourne 2001, J. Shellfish Res. 19: 933–942 (populations, growth); Machell and DeMartini 1971, Calif. Fish Game 57: 274–282 (reproduction); Pearce 1965, Veliger 7: 166–170 (ecology); Smith and Davis 1965, J. Exp. Biol. 43: 171–180 (physiology); Swan and Finucane 1952, Nautilus 66: 19–26 (distinctions between *T. capax* and *T. nuttallii*).

Tresus nuttallii (Conrad, 1837) (=*Schizothaerus nuttallii*). The Pacific gaper, in sand in bays. Coan et al. 2000: 461, 463–464; Campbell et al. 1990, J. Shellfish Res. 9: 273–278 (growth); Campbell and Bourne 2001, J. Shellfish Res. 19: 933–942 (populations, growth); Clark et al. 1975, Calif. Fish Game 61: 215–227 (life history); Pearce 1965, as above (ecology); Pohlo 1964, Malacologia 1: 321–330 (ontogeny, ecology); Illg 1949, Proc. U.S. Nat. Mus. 99: 391–428 (parasitic copepods); Smith and Davis 1965, as above (physiology); Stout 1970, Veliger 13: 67–70 (epizoics on siphonal plates). The two intertidal species of *Tresus* are further distinguished by the presence of a "visceral skirt" (a prolongation of the inner palp lamellae, which forms a curtain-like structure hanging from the dorsal extremities of and covering much of posterior of the visceral mass) in *T. capax* and its absence in *T. nuttallii* (see Pearce 1965, above).

MYIDAE

KEY TO MYIDAE

1. Shell heavy, with wavy commarginal sculpture; anterior round, posterior truncate, gaping; periostracum thick at posterior end; boring into rock, hard clay (plate 428A) . . . . . .
 . . . . *Platyodon cancellatus*
— Shell thin; burrowing in mud or sand 2
2. Pallial sinus deep; length to 120 mm or more (plates 395H, 428B) . . . . *Mya arenaria*
— Pallial sinus shallow, inconspicuous; length to 30 mm (plate 428C) . . . . *Cryptomya californica*

LIST OF SPECIES

Cryptomya californica (Conrad, 1837). In sand or mud in bays; using its very short siphons, *Cryptomya* "taps" the burrows of other invertebrates, particularly of *Urechis* and *Upogebia*; also on open coast in gravel, among rocks; Coan et al. 2000: 474–475; Lawry 1987, Veliger 30: 46–54 (various aspects of biology); Yonge 1951, Univ. Calif. Publ. Zool. 55: 395–400 (functional morphology).

Mya arenaria Linnaeus, 1758. Soft-shelled or long-necked clam; in mud and sand of bays, burrowing to 30 cm deep; introduced from Atlantic coast. Coan et al. 2000: 470, 472, with many references to the substantial biological and fisheries literature; Bernard 1979, Venus 38: 185–204 (taxonomy); Checa and Cadée 1995, J. Molluscan Studies 63: 157–171 (burrowing); MacNeil 1965, U.S. Geol. Surv. Prof. Paper 483G: 33–35 (taxonomy, evolution).

Platyodon cancellatus (Conrad, 1837). A rock borer, common in shale, also in sandstone and hard clay. Coan et al. 2000:

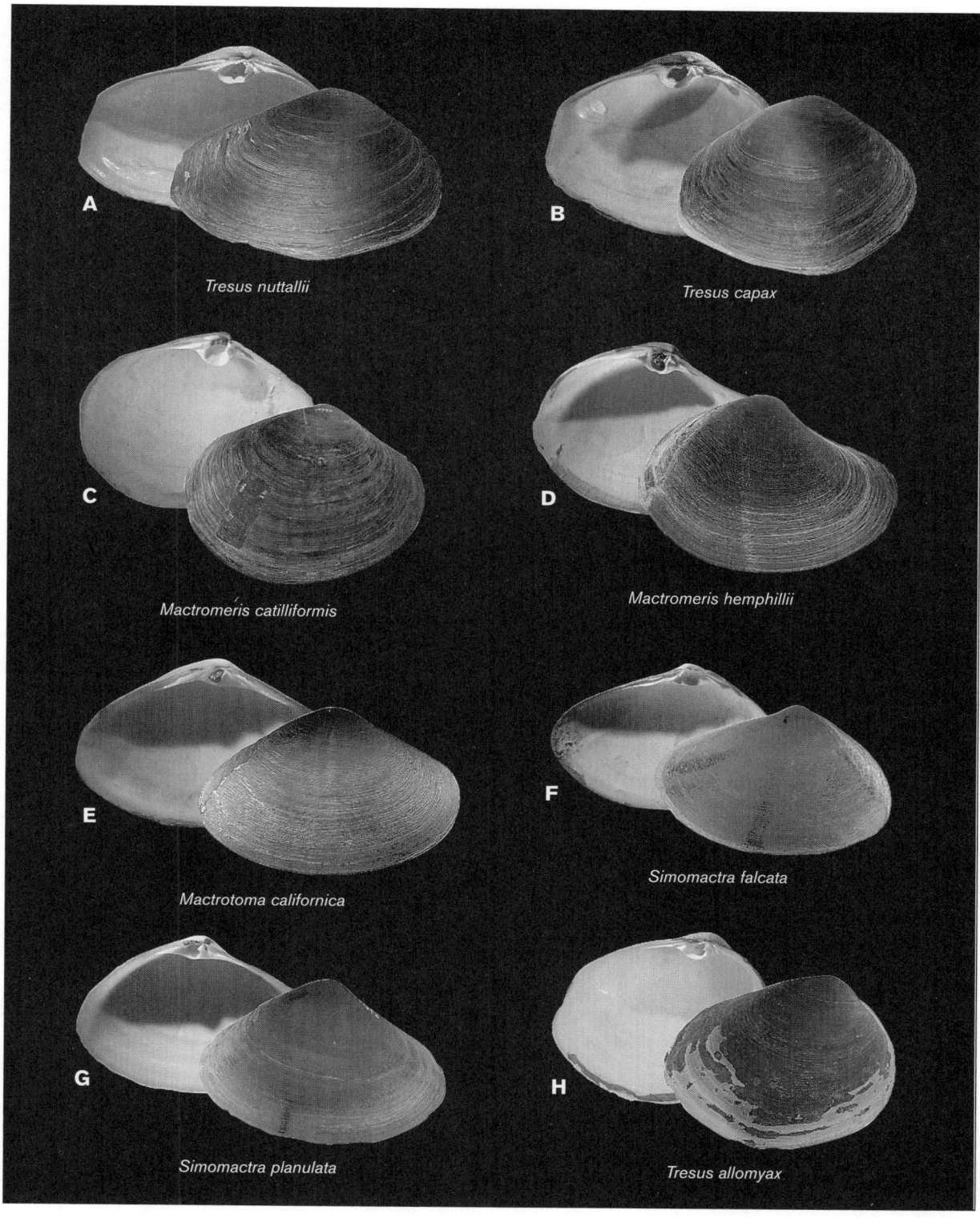

A — *Tresus nuttallii*
B — *Tresus capax*
C — *Mactromeris catilliformis*
D — *Mactromeris hemphillii*
E — *Mactrotoma californica*
F — *Simomactra falcata*
G — *Simomactra planulata*
H — *Tresus allomyax*

PLATE 426 A, *Tresus nuttallii*, length 142 mm; B, *Tresus capax*, length 120 mm; C, *Mactromeris catilliformis*, length 140 mm; D, *Mactromeris hemphillii*, length 87 mm; E, *Mactrotoma californica*, length 32 mm; F, *Simomactra falcata*, length 32 mm; G, *Simomactra planulata*, length 38 mm; H, *Tresus allomyax*, holotype, length 142 mm.

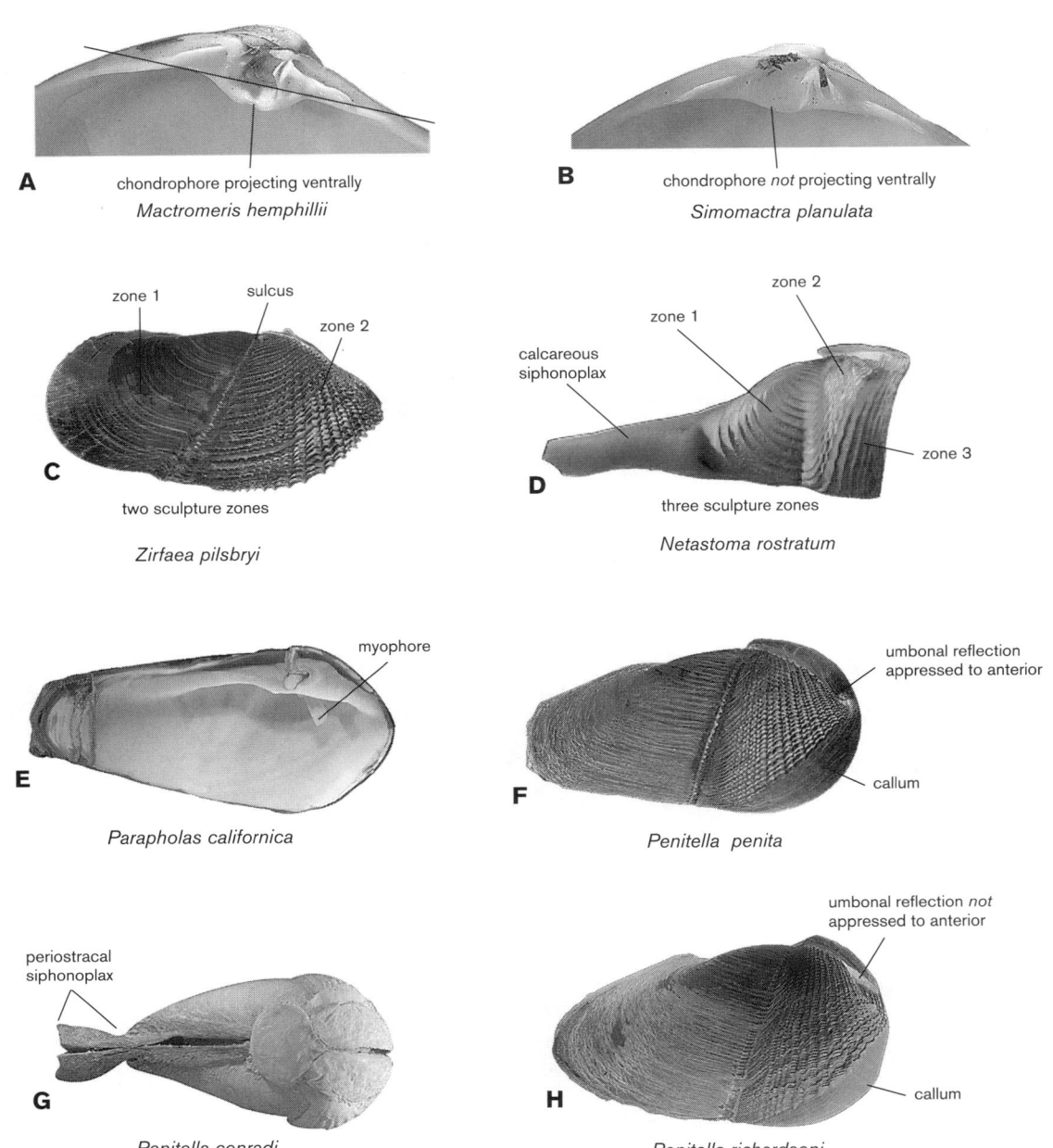

PLATE 427 A, ventrally projecting chondrophore of *Mactromeris hemphillii*; B, *Simomactra planulata* without vertically projecting chondrophore; C, sculpture zones of *Zirfaea pilsbryi*, length 92 mm; D, sculpture zones and siphonoplax of *Netastoma rostratum*, length 18 mm; E, projecting myophore of *Parapholas californica*, length 83 mm; F, appressed umbonal reflection of *Penitella penita*, length 50 mm; G, periostracal siphonoplax of *Penitella conradi*, length 20 mm; H, free umbonal reflection of *Penitella richardsoni*, length 46 mm.

472–473; Yonge 1951, Univ. Calif. Publ. Zool. 55: 401–418 (functional morphology).

*Sphenia luticola (Valenciennes, 1846) (plate 428D). An off-shore species nestling in rock cavities; occasionally washed ashore in rocks entwined in kelp holdfasts. Coan 1999, Nautilus 113: 112–115; Coan et al. 2000: 475–476.

CORBULIDAE

Coan 2002, Malacologia 44: 47–105 (systematic treatment).

*Corbula luteola Carpenter, 1864 (plate 428E). A southern Californian species recorded in shallow water in Monterey Bay in warm-water years. Coan et al. 2000: 479–480; Coan 2002: 84–87.

Corbula amurensis Schrenck, 1861 (=*Potamocorbula amurensis*) (plate 428F). An Asian species introduced into San Francisco Bay, where it has occurred in vast numbers. Coan et al. 2000: 479–480; Coan 2002: 71–72; Carlton et al. 1990, Mar. Ecol. Prog. Ser. 66: 81–94 (introduction); Nichols et al. 1990, Mar. Ecol. Prog. Ser. 66: 95–101 (ecology).

* = Not in key.

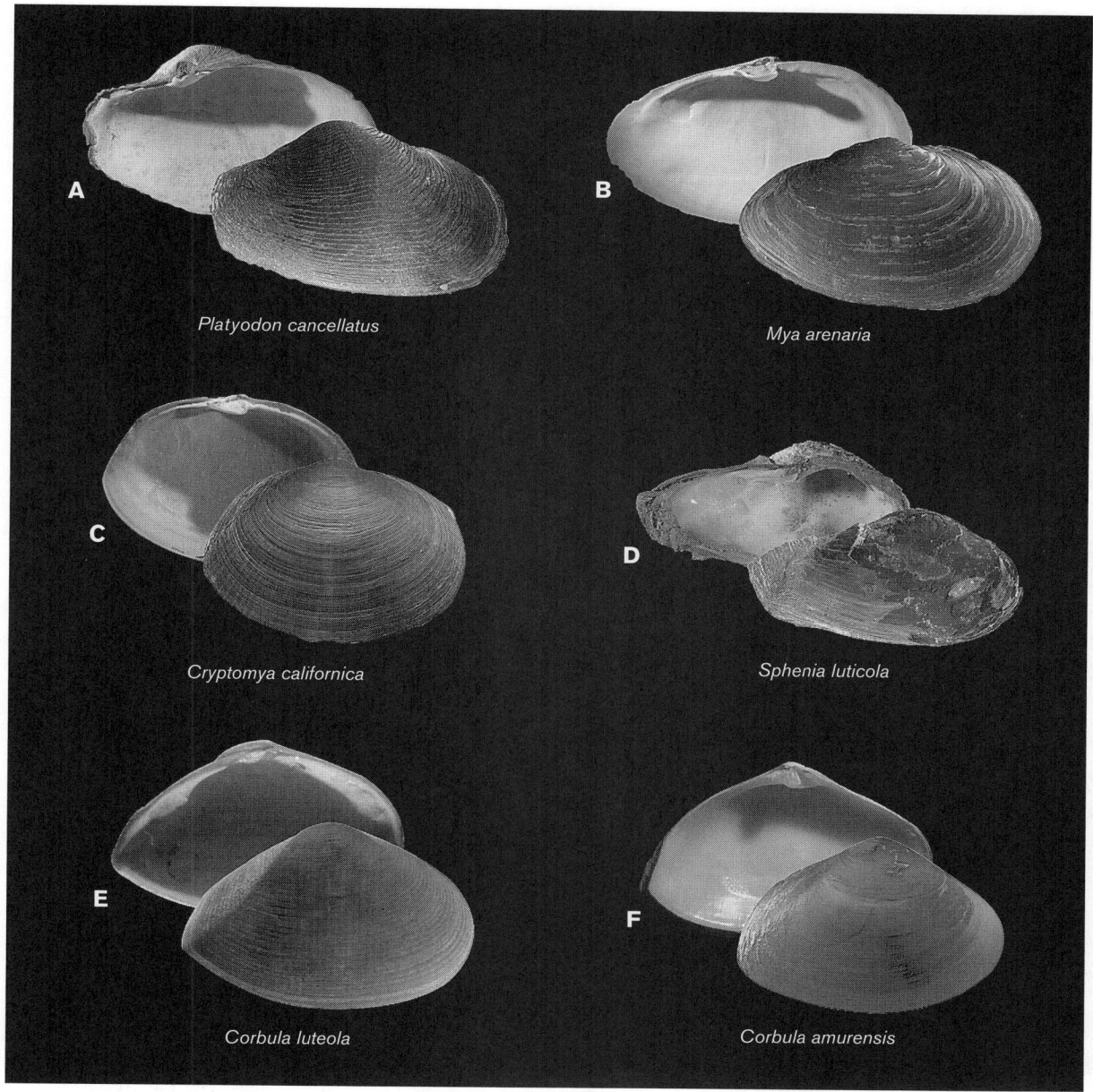

PLATE 428 A, *Platyodon cancellatus*, length 52 mm; B, *Mya arenaria*, length 78 mm; C, *Cryptomya californica*, length 27 mm; D, *Sphenia luti-cola*, length 15 mm; E, *Corbula luteola*, length 8 mm; F, *Corbula amurensis*, length 17 mm.

HIATELLIDAE

KEY TO HIATELLIDAE

1. Shell length to 200 mm, quadrate; pallial line entire, broad; posterior end broadly gaping; shell not distorted (plate 429A) . . . . *Panopea abrupta*
— Shell length to 78 mm, extremely variable shape, often distorted by boring or nestling habit pallial line faint, broken into irregular patches (plates 395D, 429B) . . . .
. . . . *Hiatella arctica*

LIST OF SPECIES

Yonge 1971, Malacologia 11: 1–44 (functional morphology).

Hiatella arctica (Linnaeus, 1767). (=*Saxicava arctica*; other species names in occasional use include *H. pholadis* [Linnaeus, 1771] and *H. solida* [G. B. Sowerby I, 1834]). Attaches with a byssus but can also bore; in bays, on pilings, in fouling clumps, open coast in algal holdfasts, abandoned pholad holes, and *Mytilus* beds. This highly variable species is a dominant byssally attached bivalve of arctic and boreal regions. There may be more than one species in this genus: different spawning times, eggs of two colors (red and pinkish cream), two post-larval forms (oval and triangular), juvenile shell with or without spinose external radial ridges, animal with or without a byssus, position of the adductor muscles and their scars relative to the dorsal margin, and siphon tips of two colors (red and white). Unfortunately, these traits have not been correlated with each other or with type material. Full taxonomic understanding of

A
B *Hiatella arctica*
Panopea abrupta
C *Saxicavella pacifica*
D *Zirfaea pilsbryi*
E *Barnea subtruncata*
F *Netastoma rostratum*

PLATE 429 A, *Panopea abrupta,* length 110 mm; B, *Hiatella arctica,* length 36 mm; C, *Saxicavella pacifica,* length 5.5 mm; D, *Zirfaea pilsbryi,* length 92 mm; E, *Barnea subtruncata,* length 59 mm; F, *Netastoma rostratum,* length 18 mm.

this genus is a problem that cries out for attention with modern methodologies grounded in sound nomenclature and with type specimens. One might expect that the situation will prove to be genetically complex as well, in part because of human introductions through shipping. Coan et al. 2000: 485–487; Beu 1971, New Zealand J. Geol. Geophysics 14: 64–6 (taxonomy); Narchi 1973, Mar. Biol. 332–337 (functional morphology of *H. "solida"*); Russell-Hunter 1949, Proc. Roy. Soc. Edinburgh (B) 43: 271–289 (morphology, biology of British material); Yonge 1971, Malacologia 11: 1–44 (functional morphology).

Panopea abrupta (Conrad, 1849). (*Panope* is an invalid emendation; =*P. generosa* Gould, 1850). The geoduck (several variant spellings, but usually pronounced gooey-duck, possibly from the Chinook *gweduc*); a very deep burrower in soft bottoms from low intertidal to offshore; siphons may be several feet long. Coan et al. 2000: 486, 490, with many additional references to the bi-

ological and fisheries literature; Goodwin 1977, Veliger 20: 155–158 (ecology); Illg 1949, Proc. U.S. Nat. Mus. 99: 391–428 (parasitic copepods); Shaul and Goodwin 1982, Can. J. Fish. Aquatic Sci. 39: 632–636 (growth); Yonge 1971, as above (functional morphology).

*Saxicavella pacifica Dall, 1916 (plate 429C). A small species offshore on soft bottoms. Coan et al. 2000: 491.

PHOLADIDAE

KEY TO PHOLADIDAE

1. Burrowing into sand or mud; adult shell without a callum
   2

* = Not in key.

— Boring into clay, shale, or shell; adult shell with a callum in form of a band or complete anterior covering (plate 427F) 3
2. Sculpture with two distinct zones, divided by a sulcus (plates 427C, 429D) . . . . *Zirfaea pilsbryi*
— Sculpture without two distinct zones or a dividing sulcus (plate 429E) . . . . *Barnea subtruncata*
3. Shell without myophores; callum present only as an anterior band (sculptured with high, thin flutes); adult with a long, tapering, calcareous siphonoplax; sculpture with three distinct zones; shell often irregular in shape (plate 429F) . . . . *Netastoma rostratum*
— Myophore present; callum present in adult as an anterior covering; siphonoplax never wholly calcareous (plate 427E) 4
4. Sculpture divided into three distinct zones; posterior end of shell with overlapping chitinous plates (plates 427E, 430A) . . . . *Parapholas californica*
— Sculpture divided into two distinct zones; posterior end without overlapping plates 5
5. Umbonal reflection appressed to anterior end (plate 427F) 6
— Umbonal reflection not appressed to anterior end (plate 427H) 7
6. Siphonoplax absent; callum complete (when present); siphons smooth (plate 430B) . . . . *Penitella turnerae*
— Siphonoplax heavy, laminate; callum (when present) with gap; siphons smooth (plate 430C) . . . . *Penitella fitchi*
— Siphonoplax heavy, flexible, chitinous flaps not lined with calcareous granules; siphons smooth (plates 427F, 430D) . . . . *Penitella penita*
— Siphonoplax heavy, not flexible, composed of a chitinous outer layer lined with coarse calcareous granules; callum complete (when present); siphons smooth (plates 427G, 430E) . . . . *Penitella conradi*
7. Callum in adult with gape; shell ovate; umbonal reflection only slightly appressed dorsally; siphons with conspicuous, orange, chitinous patches and warty tips that cannot be retracted into shell (plate 430F) . . . . *Chaceia ovoidea*
— Callum in adult complete (when present); shell ovate-elongate; umbonal reflection free anteriorly, appressed dorsally; siphonoplax absent; siphons pustulose (plates 427H, 430G) . . . . *Penitella richardsoni*

LIST OF SPECIES

Turner 1954–1955, Johnsonia 3: 1–160; Kennedy 1974, Mem. San Diego Soc. Nat. Hist. 8: 128 pp. (monographic treatments).

Barnea subtruncata (G. B. Sowerby I, 1834). Burrowing in mud or clay of well-protected bays. Coan et al. 2000: 497, 500; Turner 1954: 31–34.

Chaceia ovoidea (Gould, 1851). Boring into shale. Coan et al. 2000: 500, 503; Turner 1955: 66–70.

Netastoma rostratum (Valenciennes, 1846). Boring in shale at low tide and occurring offshore. Coan et al. 2000: 500, 502; Turner 1955: 143–145; Turner 1962, Mus. Comp. Zool., Occ. Papers Mollusks 2: 289–308.

Parapholas californica (Conrad, 1837). Boring into a variety of substrata from clay to rock; siphonoplax absent, instead, a thick tube ("chimney") of finely cemented particles is formed by the siphons as a result of boring activity. Coan et al. 2000: 500, 503; Turner 1955: 124–128.

Penitella conradi Valenciennes, 1846. Often found boring into such shells as *Mytilus* or *Haliotis* (forming "blister pearls" inside abalone shells); occasionally in shale or soft rock. Coan et al. 2000: 504–505; Hansen 1970, Veliger 13: 90–94 (habitat); Smith 1969, Amer. Zool. 9: 869–880 (functional morphology); Turner 1955: 75–80.

Penitella fitchi Turner, 1955. Found in rocks from the low intertidal zone to 25 m, rare. Coan et al. 2000: 504–505; Turner 1955: 71–74.

Penitella penita (Conrad, 1837). Boring into a variety of substrata from stiff clay to sandstone and cement; common. Coan et al. 2000: 505–506; Evans 1967, Veliger 10: 148–149 (ecology); Evans 1968, Ecology 49: 619–628 (growth rate); Evans 1968, Palaeogeo., Palaeoclim., Palaeoecol. 4: 271–278 (boring); Evans 1968, Proc. Malac. Soc. London 38: 111–119 (morphology); Evans 1968, Ecol. 49: 156–159 (ecology); Evans 1970, Can. J. Zool. 48: 625–627 (reproduction); Evans and LeMessurier 1972, Can. J. Zool. 50: 1251–1258 (functional morphology); Turner 1955: 80–85.

Penitella richardsoni Kennedy 1989 (=*P. gabbi*, of authors). Found along with *P. penita*, but much less common. Coan et al. 2000: 505–506; Kennedy 1989, Veliger 32: 313–319.

Penitella turnerae Evans and Fisher, 1966. In rocks, not common. Coan et al. 2000: 505, 507; Evans and Fisher 1966, Veliger 8: 222–224.

Zirfaea pilsbryi Lowe, 1931. In bays, burrowing into mud and clay. Coan et al. 2000: 500–501; MacGinitie 1935, Amer. Midl. Nat. 16: 731–735 (burrowing); Turner 1954: 58–62.

TEREDINIDAE

KEY TO TEREDINIDAE

1. Pallets with elongate blade, composed of distinct, cone-shaped segments (plates 394C 431A) . . . . . *Bankia setacea*
— Pallets not segmented 2
2. Distal half of pallet blade with dark colored periostracal cap overlapping calcareous base (plates 431B, 432A) . . . . *Lyrodus pedicellatus*
— Pallet almost entirely calcareous, periostracum covers less than half of blade (plates 431C, 432B) . . . . . *Teredo navalis*

LIST OF SPECIES

All local species of shipworms occur in wood, such as wharf pilings, in bays and harbors. Because of their economic importance, there is a substantial literature. Key references include: Turner 1966, A Survey and Illustrated Catalogue of the Teredinidae (Mollusca: Bivalvia), Mus. Comp. Zool., Harvard, 265 pp.; C. L. Hill and C. A. Kofoid, eds. 1927, Marine Borers and their Relation to Marine Construction on the Pacific Coast, Final Report, San Francisco Bay Marine Piling Comm., San Francisco, 357 pp.

Bankia setacea (Tyron, 1863). Possesses external fertilization and planktonic larval stage. Coan et al. 2000: 515; Haderlie and Mellor 1973, Veliger 15: 265–286 (aspects of biology); Turner 1966: 38, 121, 248–249.

Lyrodus pedicellatus (Quatrefages, 1849). Introduced; young are retained to late veliger stage. Coan et al. 2000: 511, 513; Eckelbarger and Reish 1972, Bull. So. Calif. Acad. Sci. 71: 48–50 (self-fertilization); Turner 1966: 70–71, 116, 132–141.

Teredo navalis Linnaeus, 1758. Introduced; young are retained until veliger stage. Coan et al. 2000: 511–512; Miller et al. 1922–1924, Univ. Calif. Publ. Zool. 22 and 26: 41–80 (biology,

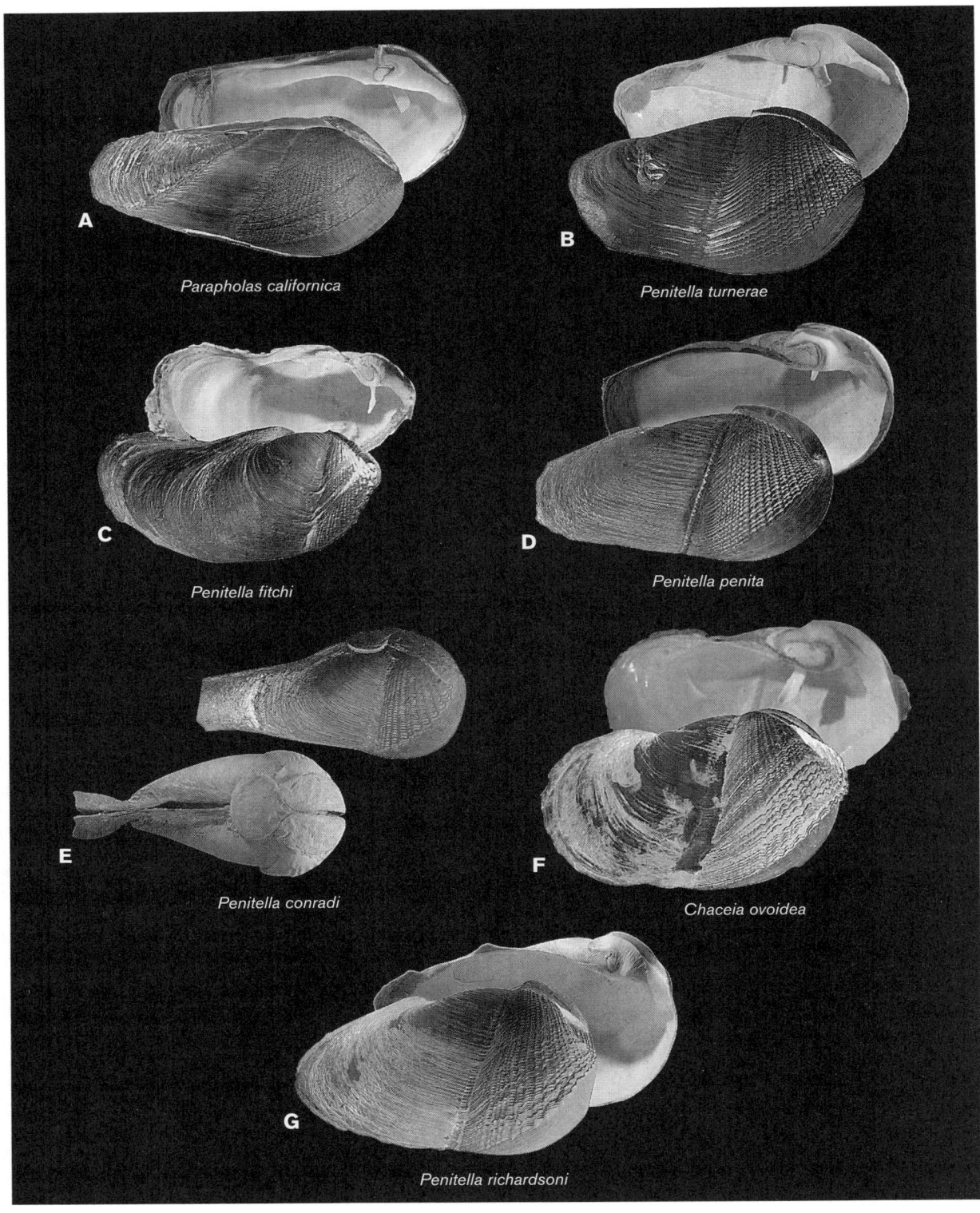

PLATE 430 A, *Parapholas californica*, length 83 mm; B, *Penitella turnerae*, length 60 mm; C, *Penitella fitchi*, length 26 mm; D, *Penitella penita*, length 50 mm; E, *Penitella conradi*, length 20 mm; F, *Chaceia ovoidea*, length 94 mm; G, *Penitella richardsoni*, length 46 mm.

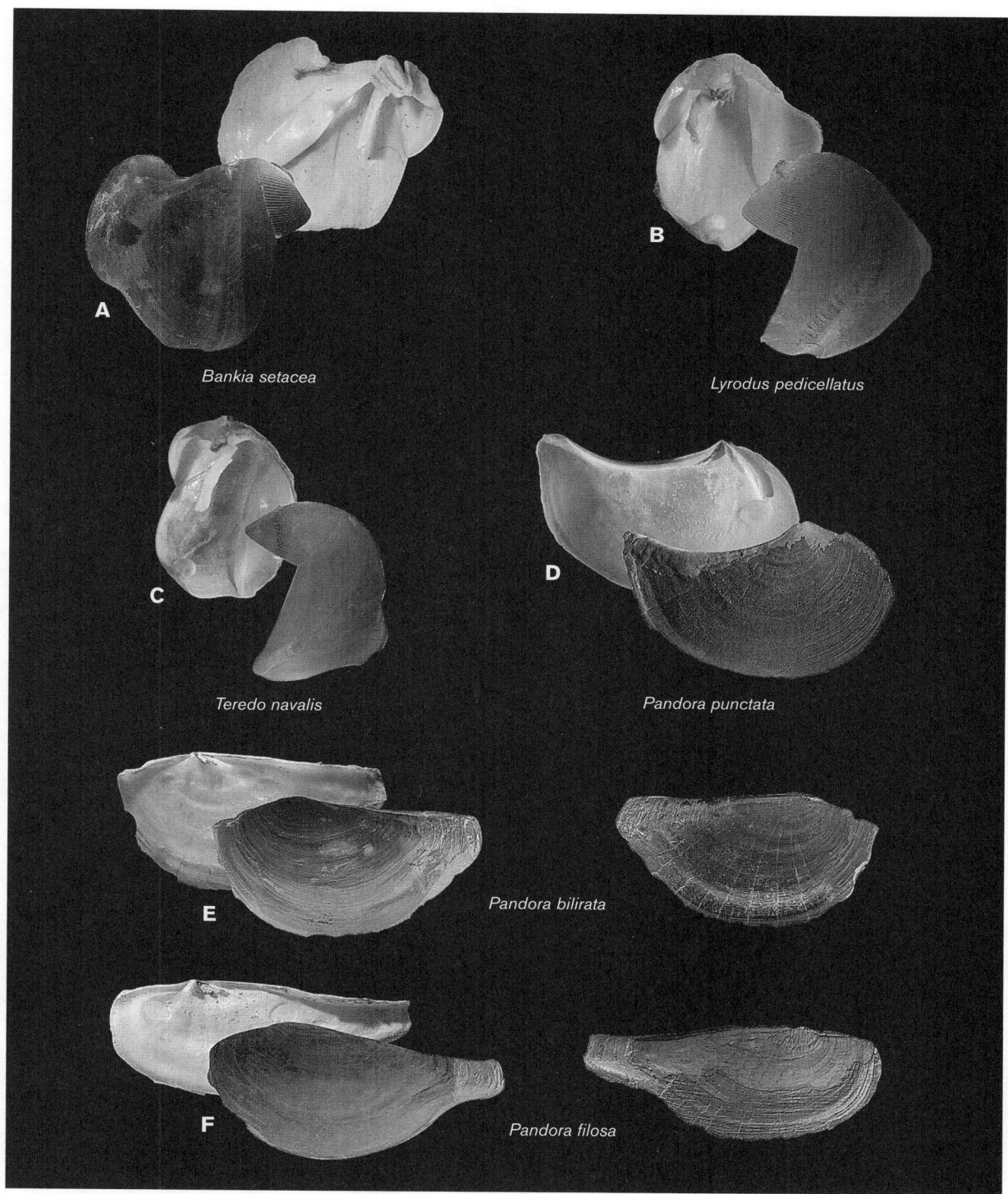

PLATE 431 A, *Bankia setacea*, length 11 mm; B, *Lyrodus pedicellatus*, length 4 mm; C, *Teredo navalis*, length 6 mm; D, *Pandora punctata*, length 42 mm; E, *Pandora bilirata*, length 12 mm; F, *Pandora filosa*, length 22 mm.

morphology, boring mechanism); Turner 1966: 112–113, 158–161.

PANDORIIDAE

This family occurs only offshore on soft sediments. The following three species may be dredged in fairly shallow water.

Pandora punctata Conrad, 1837 (plate 431D). Occasionally washed up on beaches. Coan et al. 2000: 520, 522.

Pandora bilirata Conrad, 1855 (plate 431E). Coan et al. 2000: 520, 522.

Pandora filosa (Carpenter, 1864) (plate 431F). Coan et al. 2000: 521–522.

* = Not in key.

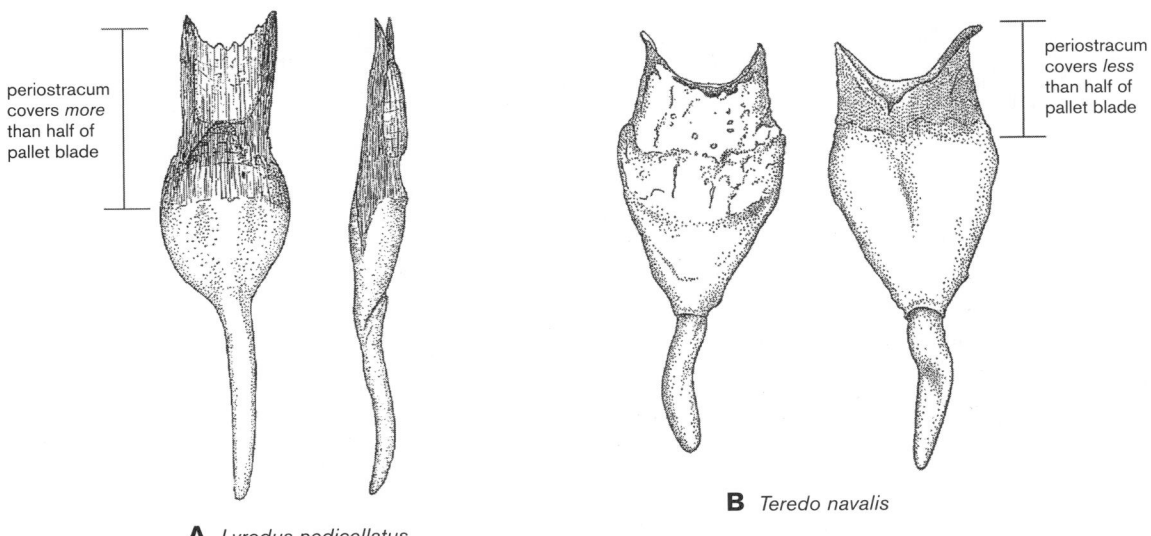

periostracum covers *more* than half of pallet blade

periostracum covers *less* than half of pallet blade

A *Lyrodus pedicellatus*

B *Teredo navalis*

PLATE 432 A, pallets of *Lyrodus pedicellatus*; B, pallets of *Teredo navalis* (both from Turner 1966). For pallets of *Bankia*, see plate 394C.

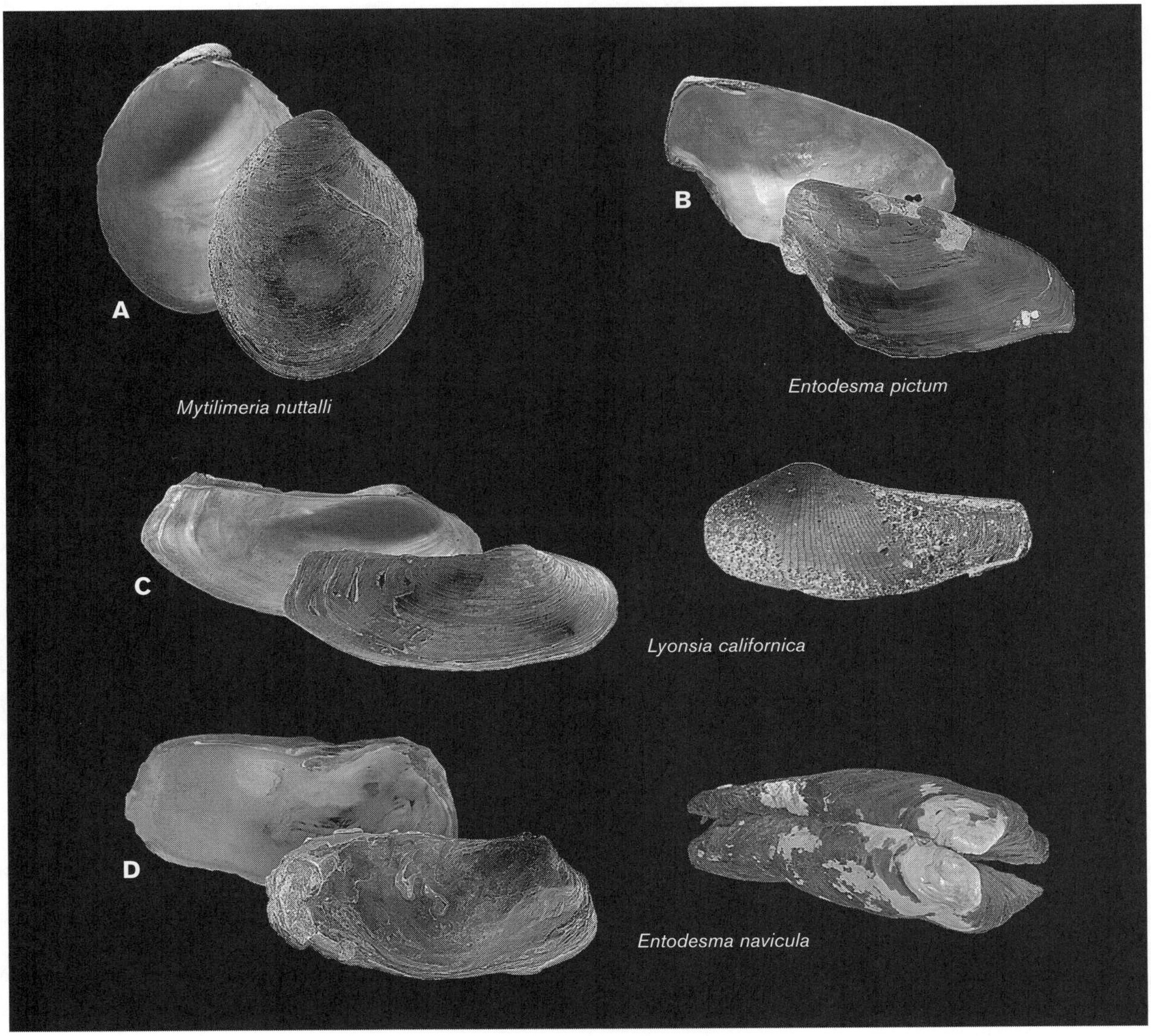

A

Mytilimeria nuttalli

B

Entodesma pictum

C

Lyonsia californica

D

Entodesma navicula

PLATE 433 A, *Mytilimeria nuttalli*, length 21 mm; B, *Entodesma pictum*, length 25 mm; C, *Lyonsia californica*, length (left pair) 32 mm, length (right valve with sand grains) 18 mm; D, *Entodesma navicula*, length (left pair) 112 mm, length (dorsal view) 80 mm.

PLATE 434 A, *Thracia curta*, length 32 mm; B, *Thracia trapezoides*, length 48 mm; C, *Laternula marilina*, length 5.5 mm; D, *Periploma discus*, length 42 mm; E, *Cardiomya pectinata*, length 11 mm; F, *Trigonulina novemcostatus*, length 5 mm.

LYONSIIDAE

KEY TO LYONSIIDAE

1. Outline circular or ovate; beaks twisted; living in compound ascidians (plate 433A) . . . . *Mytilimeria nuttallii*
— Posterior end produced; beaks not twisted; free-living or byssally attached 2
2. Periostracum thin, striate; shell thin, elongate, pearly; in muddy substrata (plate 433C) . . . . *Lyonsia californica*
— Shell irregular in shape; nestling or attached to hard sub-

strata . . . . *Entodesma* 3
3. Shell thick; anterior end one-third of length; periostracum heavy (plate 433D) . . . . *Entodesma navicula*
— Shell thin; anterior end much less than one-third of length; periostracum thin (plate 433B) . . . . *Entodesma pictum*

LIST OF SPECIES

Entodesma pictum (G. B. Sowerby I, 1834 (=*E. inflatum* [Conrad, 1837]). Small, lighter in color and more regular in shape than *E. navicula*; reported in sponges and compound ascidians. Coan et al. 2000: 526, 528; Morton 1987, J. Moll. Stud. 52: 139–151 (mantle).

Entodesma navicula (A. Adams and Reeve, 1850) (=*E. saxicola* Baird, 1863). Byssally attached under rocks in low intertidal, in fouling on wharf pilings. Coan et al. 2000: 526–527; Morgan and Allen 1976, Malacologia 15: 233–240 (functional morphology); Morton 1987, J. Moll. Stud. 52: 139–151 (mantle); Yonge 1952, Univ. Calif. Publ. Zool. 55: 439–450 (functional morphology); Yonge 1976, J. Moll. Stud. 42: 395–408 (ligament).

Lyonsia californica Conrad, 1837. In muddy substrata in protected areas of bays. Coan et al. 2000: 525–527; Narchi 1968, Veliger 10: 305–313 (functional morphology); Maurer 1967, Veliger 10: 72–76 (aspects of biology).

Mytilimeria nuttallii Conrad, 1837. In compound ascidians *Eudistoma* and *Distaplia* in rocky intertidal; commonly washed ashore (e.g., at Horseshoe Cove, Bodega Head) embedded in ascidians. Coan et al. 2000: 526, 528; Prezant and Carriker 1983,Veliger 24: 326–328 (lithodesma); Yonge 1952, Univ. Calif. Publ. Zool. 55: 439–450 (functional morphology).

THRACIIDAE

Thracia curta Conrad, 1837 (plate 434A). Nestling in rock crevices from the low intertidal zone to offshore. Coan et al. 2000: 533–534; Coan 1990, Veliger 33: 333–335.

*Thracia trapezoides Conrad, 1849 (plate 434B). An offshore species on soft substrata. Coan et al. 2000: 523–533; Coan 1990, Veliger 33: 30–31.

LATERNULIDAE

Laternula marilina (Reeve, 1860) (plate 434C). Introduced from Asia in ballast water; in mud in Humboldt Bay. Coan et al. 2000: 539; Miller et al. 1999, Veliger 42: 282–284.

PERIPLOMATIDAE

*Periploma discus Stearns, 1890 (plate 434D). Offshore on soft sediments. Coan et al. 2000: 542, 544.

CUSPIDARIIDAE

*Cardiomya pectinata (Carpenter, 1864) (plate 434E). Offshore on soft sediments. Coan et al. 2000: 555–556.

VERTICORDIIDAE

*Trigonulina novemcostatus (A. Adams and Reeve, 1850) (=T. *pacifica* Jung, 1996) (plate 434F). Offshore in mud. Coan et al. 2000: 563–564.

References

See also general molluscan references.

Adamkewicz, S. L., M. G. Harasewych, J. Blake, D. Saudek, and C. J. Bult. 1997. A molecular phylogeny of the bivalve mollusks. Molecular Biology and Evolution 14: 619–629.

Allen, J. A. 1985. The Recent Bivalvia: their form and evolution. pp. 337–403. In The Mollusca, Vol. 10, Evolution. E. R. Trueman and M. R. Clarke, eds. Orlando, FL: Academic Press.

Bayne, B. L. 1987. Genetic aspects of physiological adaptation in bivalve molluscs, pp. 169–189. In Evolutionary physiological ecology. P. Calow, ed. Cambridge, England: Cambridge University, 239 pp.

Berry, W. B. N., and R. M. Barker. 1975. Growth increments in fossil and modern bivalves, pp. 9–25, In Growth rhythms and the history of the Earth's rotation. G. D. Rosenberg, and S. K. Runcorn, eds. London: Wiley.

Bieler, R., ed. 2006. Bivalvia—a look at the branches. Zoological Journal of the Linnean Society 148: 119–552.

Bonnot, P. 1935. The California oyster industry. California Fish and Game 21: 65–80.

Boss, K. J. 1982. Mollusca, pp. 946–1166, Vol. 1, and pp. 1092–1096, Vol. 2. In Synopsis and classification of living organisms. S. P. Parker, ed. New York: McGraw-Hill.

Carter, J. G., ed. 1990. Skeletal biomineralization: patterns, processes and evolutionary trends. New York: Van Nostrand Reinhold, 2 volumes.

Carter, J. G., D. C. Campbell, and M. Campbell. 2000. Cladistic perspectives on early bivalve evolution, pp. 47–79. In The evolutionary biology of the Bivalvia. E. M. Harper, J. D. Taylor, and J. A. Crame, eds. Geological Society of London, Special Publication 166, 494 pp.

Coan, E. V., P. Valentich-Scott, and F. R. Bernard. 2000. Bivalve seashells of western North America. Santa Barbara, CA: Santa Barbara Museum of Natural History, 764 pp.

Cope, J. C. W. 1995. The early evolution of the Bivalvia, pp. 361–370. In Origin and evolutionary radiation of the Mollusca. J. D. Taylor, ed. Oxford: Oxford University, 392 pp.

Cox, L. R., and 24 other authors. 1969. Part N [Bivalvia], Mollusca 6, vols. 1 and 2, 952 pp. In Treatise on invertebrate paleontology. R. C. Moore, ed. Lawrence, KS: Geological Society of America and University of Kansas.

Crosby, N. D., and R. G. B. Reid. 1971. Relationships between food, phylogeny, and cellulose digestion in the Bivalvia. Canadian Journal of Zoology 49: 617–622.

Dame, R. E. 1993. Bivalve filter feeders in estuarine and coastal ecosystem processes. Berlin: Springer-Verlag, 579 pp.

Eagar, R. M. 1978. Shape and function of the shell: a comparison of some living and fossil bivalve molluscs. Biological Reviews 53: 169–210.

Fisher, W. S., ed. 1988. Disease processes in marine bivalve molluscs. American Fisheries Society, Special Publication 18, 315 pp.

Gosling, E. M. 1998. Bivalve molluscs. Biology, ecology and culture. Oxford: Fishing News Books, 200 pp.

Harper, E. M., J. D. Taylor, and J. A. Crame. 2000. Evolutionary biology of the Bivalvia. Geological Society of London Special Publication 177, 494 pp.

Healy, J. M. 1996. Molluscan sperm ultrastructure: correlation with taxonomic units within the Gastropoda, Cephalopoda, and Bivalvia, pp. 99–113. In Origin and evolutionary radiation of the Mollusca. J. D. Taylor, ed. Oxford: Oxford University.

Howard, D. W., and C. S. Smith. 1983. Histological techniques for marine bivalve mollusks. National Marine Fisheries Service Technical Memorandum NMFS-F/NEC 25, 97 pp.

Inaba, Akihiko, and Yoko Otani. 2003. Oysters of the world. Part 1. General remarks. Nishinomiya Shell Museum, Bulletin 2: 59 + 21 pp., 2 fold-out charts.

Inaba, Akihiko, Kenji Torigoe, and Yoko Otani. 2004. Oysters of the world. Part 2. Systematic descriptions of Recent oysters. Bulletin of the Nishinomiya Shell Museum3: 63 + 10 pp., 13 pls.

Johnston, P. A., and J. W. Haggart. 1998. Bivalves: an eon of evolution. Calgary, Alberta, Canada: University of Alberta Press, 461 pp.

Jørgensen, C. B. 1990. Bivalve filter feeding: hydrodynamics, bioenergetics, physiology, and ecology. Fredensborg, Denmark: Olsen and Olsen, 140 pp.

Kas'ianov, V. L., G. A. Kriuchkova, V. A. Kulikova, and L. A. Medvedeva, ed. by D. L. Pawson. 1998. Larvae of marine bivalves and echinoderms. Washington, D.C. (Smithsonian Institution Libraries), 288 pp.

Loosanoff, V. L., H. C. Davis, and P. E. Chanley. 1966. Dimensions and shapes of larvae of some marine bivalve mollusks. Malacologia 4: 351–435.

Lutz, R. A. 1985. Identification of bivalve larvae and postlarvae: a review of recent advances. American Malacological Bulletin, Special Edition 1: 59–78.

Lutz, R. A., et al. 1982. Preliminary observations on the usefulness of hinge structures for identification of bivalve larvae. Journal of Shellfish Research 2: 65–70.

Lutz, R. A., and D. C. Rhoads. 1980. Growth patterns within the molluscan shell; an overview, pp. 203–254. In Skeletal growth of aquatic

* = Not in key.

organisms; biological records of environmental change. D. C. Rhoads and R. A. Lutz, eds. New York and London: Plenum.

Manzi, J. J., and M. Castagna. 1989. Clam mariculture in North America. Amsterdam: Elsevier, 461 pp.

Morris, N. J. 1979. On the origin of the Bivalvia, pp. 381–413. In The origin of major invertebrate groups. M. R. House, ed. London: Academic Press.

Morse, M. P., and J. D. Zardus. 1997. Bivalvia, pp. 7–118. In Microscopic anatomy of invertebrates 6A(II). Frederick W. Harrison and Alan J. Kohn, eds. New York: Wiley-Liss.

Morton, B. 1983. Feeding and digestion in Bivalvia, pp. 65–147. In The Mollusca, Vol. 5 (Physiology, Part 2). A. S. M. Saleuddin and K. M. Wilbur, eds. New York: Academic Press.

Morton, B., ed. 1990. Proceedings of a memorial symposium in honour of Sir Charles Maurice Yonge (1899–1986), Edinburgh, 1986. Hong Kong (Hong Kong University). 355 pp.

Morton, B., 1995. The evolutionary history of the Bivalvia, pp. 337–359. In Origin and evolutionary radiation of the Mollusca. J. D. Taylor, ed. Oxford: Oxford University, 392 pp.

Morton, B., R. S. Prezant, and B. Wilson. 1998. Class Bivalvia, pp. 195–429. In Mollusca: the southern synthesis. Fauna of Australia, Vol. 5. P. L. Beesley, G. J. B. Ross, and A. Wells, eds. Melbourne, Australia: CSIRO Publishing.

Newell, N. D., and D. W. Boyd. 1978a. Iteration of ligament structures in pteriomorphian bivalves. American Museum of Natural History, Novitates 2875: 1–11.

Newell, N. D., and D. W. Boyd. 1978b. A paleontologist's view of bivalve phylogeny. Philosophical Transactions of the Royal Society of London (B) 284: 203–215, pl. 1.

Owen, G. 1974. Feeding and digestion in the Bivalvia. Advances in Comparative Physiology and Biochemistry 5: 1–35.

Owen, G. 1978. Classification and the bivalve gill. Royal Society of London, Philosophical Transactions (B) 284(1001): 377–385, pls. 1, 2.

Pojeta, J., Jr. 1978. The origin and early taxonomic diversification of pelecypods. Philosophical Transactions of the Royal Society of London (B) 284: 225–246, pls. 1–15.

Pojeta, John, Jr., and B. Runnegar. 1985. The early evolution of diasome molluscs, pp. 295–336. In The Mollusca, Vol. 10, Evolution. E. R. Trueman and M. R. Clarke, eds. Orlando, Florida: Academic Press.

Popham, J. D. 1979. Comparative spermatozoon morphology and bivalve phylogeny. Malacological Review 12: 1–20.

Prieur, D. 1987. A review of the relationships between bivalve molluscs and bacteria in the marine environment. Symbiosis 4: 37–50.

Purchon, R. D. 1968a. Feeding methods and evolution in the Bivalvia, pp. 101–145. In The biology of the Mollusca. Oxford: Pergamon.

Purchon, R. D. 1968b. Adaptive radiation in the polysyringian bivalves, pp. 147–206. In The biology of the Mollusca. R. D. Purchon. Oxford: Pergamon.

Purchon, R. D. 1971. Digestion in filter feeding bivalves—a new concept. Proceedings of the Malacological Society of London 39: 253–262.

Purchon, R. D. 1987a. The stomach in the Bivalvia. Philosophical Transactions of the Royal Society of London (B) 316: 183–276.

Purchon, R. D. 1987b. Classification and evolution of the Bivalvia: an analytical study. Philosophical Transactions of the Royal Society of London (B) 316: 277–302.

Purchon, R. D. 1990. Stomach structure, classification and evolution of the Bivalvia, pp. 73–82. In Proceedings of a memorial symposium in honour of Sir Charles Maurice Yonge (1899–1986). B. Morton, ed., Hong Kong: Hong Kong University.

Purchon, R. D., and D. Brown. 1970. Phylogenetic interrelationships among families of bivalve molluscs. Malacologia 9: 163–171.

Purchon, R. D., with an appendix by G. Clarke. 1978. An analytical approach to a classification of the Bivalvia. Philosophical Transactions of the Royal Society of London (B) 284: 425–436.

Rosenberg, G. D. 1980. An ontogenetic approach to the environmental significance of bivalve shell chemistry, pp. 133–168. In Skeletal growth of aquatic organisms. D. C. Rhoads and R. A. Lutz, eds. New York and London: Plenum.

Sastry, A. N. 1979. Pelecypoda (excluding Ostreidae), pp. 113–292. In Reproduction of marine invertebrates, Vol. 5. Molluscs: Pelecypoda and lesser classes. A. C. Giese and J. S. Pearse, eds. New York: Academic Press.

Seed, R. 1980. Shell growth and form in the bivalve shell, pp. 23–67. In Skeletal growth of aquatic organisms. D. C. Rhoads and R. A. Lutz, eds. New York and London: Plenum.

Shorbatov, G. L., and Y. I. Starobogatov, eds. 1990. Metody izucheniia dvustovorchatykh molliuskov. [Methods for the study of bivalve molluscs]. Trudy Zoologicheskii Institut 219: 208 pp.

Sindermann, C. J. 1990. Principal diseases of marine fish and shellfish. Vol. 2. Diseases of marine shellfish. Orlando, Florida: Academic Press, 516 pp.

Skelton, P. W., and M. J. Benton. 1993. Mollusca: Rostroconchia, Scaphopoda and Bivalvia, pp. 237–263. In The fossil record 2. M. J. Benton, ed. London: Chapman and Hall.

Stanley, S. M. 1970. Relation of shell form to life habits of the Bivalvia (Mollusca). Geological Society of America, Memoir 125, 296 pp.

Stanley, S. M. 1972. Functional morphology and evolution of byssally attached bivalve mollusks. Journal of Paleontology 46: 165–212.

Stanley, S. M. 1975. Adaptive themes in the evolution of the Bivalvia (Mollusca). Annual Review of Earth and Planetary Sciences 3: 361–385.

Stanley, S. M. 1977. Trends, rates, and patterns of evolution in the Bivalvia, pp. 209–250. In Patterns of evolution as illustrated by the fossil record. A. Hallam, ed. Amsterdam: Elsevier.

Stanley, S. M. 1981. Infaunal survival: alternative functions of shell ornamentation in the Bivalvia (Mollusca). Paleobiology 7: 384–393.

Stanley, S. M., and X. Yang. 1987. Approximate evolutionary stasis for bivalve morphology over millions of years: a multivariate, multilineage study. Paleobiology 13: 113–139.

Stehli, F. G., A. L. McAlester, and C. E. Helsley. 1967. Taxonomic diversity of Recent bivalves and some implications for geology. Bulletin of the Geological Society of America 78: 455–466.

Steiner, G., and M. Müller. 1996. What can 18S rDNA do for bivalve phylogeny. Journal of Molecular Evolution 43: 58–70.

Steiner, G., and S. Hammer. 2000. Molecular phylogeny of the Bivalvia inferred from 18S rDNA sequences with particular reference to the Pteriomorphia, pp. 11–29. In The evolutionary biology of the Bivalvia. E. M. Harper, J. D. Taylor, and J. A. Crame, eds. Geological Society of London, Special Publication 166.

Valentich-Scott, P. H. 1998. Class Bivalvia, pp. 97–173. In Taxonomic atlas of the benthic fauna of the Santa Maria Basin and the western Santa Barbara Channel. Volume 8. The Mollusca Part 1. The Aplacophora, Polyplacophora, Scaphopoda, Bivalvia, and Cephalopoda. P. H. Valentich-Scott and J. A. Blake. Santa Barbara, CA: Santa Barbara Museum of Natural History.

Vokes, H. E. 1980. Genera of the Bivalvia: a systematic and bibliographic catalogue (revised and updated). Ithaca, New York: Paleontological Research Institution, 307 pp.

Vokes, H. E. 1990. Genera of the Bivalvia: a systematic and bibliographic catalogue—addenda and errata. Tulane Studies in Geology and Paleontology 23: 97–120.

Waller, T. R. 1990. The evolution of ligament systems in the Bivalvia, pp. 49–71. In The Bivalvia— Proceedings of a memorial symposium in honour of Sir Charles Maurice Yonge, Edinburgh, 1986. B. Morton, ed. Hong Kong: Hong Kong University.

Waller, T. R. 1998. Origin of the molluscan class Bivalvia and a phylogeny of major groups, pp. 1–45. In Bivalves: an eon of evolution. Johnston, P. A. and J. W. Haggart, eds. Calgary, Alberta: University of Alberta Press.

Walne, P. R. 1991. Culture of bivalve molluscs: 50 years' experience at Conwy, 2nd ed. Oxford (Blackwell; Fishing News Books), 190 pp.

Watters, G. T. 1993. Some aspects of the functional morphology of the shell of infaunal bivalves (Mollusca). Malacologia 35: 315–342.

Yonge, C. M. 1936. The evolution of the swimming habit in the Lamellibranchia. Mémoires de la Musée Royal d'Histoire Naturelle de Belgique (2)3: 77–100.

Yonge, C. M. 1953. The monomyarian condition in the Lamellibranchia. Transactions of the Royal Society of Edinburgh 62: 443–478.

Yonge, C. M. 1957. Mantle fusion in the Lamellibranchia. Pubblicazionni di Stazione Zoologica, Naples 29: 151–171.

Yonge, C. M. 1962. On the primitive significance of the byssus in the Bivalvia and its effects in evolution. Journal of the Marine Biological Association of the United Kingdom 42: 113–125.

Yonge, C. M. 1978. Significance of the ligament in the classification of the Bivalvia. Proceedings of the Royal Society of London (B) 202(1147): 231–248.

Yonge, C. M. 1979. Cementation in bivalves, pp. 83–106. In Pathways in malacology. Utrecht: Bohn, Scheltema and Holkema, and The Hague: Junk.

Yonge, C. M. 1982. Mantle margins with a revision of siphonal types in Bivalvia. Journal of Molluscan Studies 48: 102–103.

Yonge, C. M. 1983. Symmetries and the role of the mantle margins in the bivalve Mollusca. Malacological Review 16: 1–10.

Phoronida

(Plate 435)

RUSSEL ZIMMER

Phoronids are slender marine animals, each with a long, unsegmented trunk that has a horseshoe-shaped ring of tentacles, the lophophore, at its distal end. The tentacles encircle the mouth but not the closely adjacent anus. The trunk contains the U-shaped gut, the gonads, and paired metanephridia that serve also as gonoducts. The lophophore and trunk have separate coelomic cavities, and there is a small preoral flap called the epistome, which also has coelomic spaces and is often considered a separate body region. The well-developed closed vascular system with nucleated red blood cells sends blind capillaries into each of the tentacles. The lophophore serves in feeding and respiration and, in most species, has seasonally developed accessory reproductive glands. Their tubiculous nature, lack of segmentation, and presence of a vascularized lophophore collectively distinguish phoronids from all other animals.

On the bases of morphological and developmental criteria, phoronids are usually recognized as a distinct phylum and have been commonly allied with the phyla Bryozoa and Brachiopoda as the Tentaculata or, more commonly, Lophophorata. However, the relationships of these three phyla to each other and other phyla have been subject to multiple interpretations and remain controversial. A useful review of many of the proposed relationships is in Emig (1997). Molecular studies place phoronids as members of the subkingdom Protostomia (rather than Deuterostomia) in the superphylum Lophotrochozoa (basically, the lophophorates plus phyla with trochophore larvae; see for example Halanych et al. 1995).

Although usually considered a phylum, phoronids have, on the basis of morphological and developmental information, been proposed as one of three classes in the Phylum Lophophorata (Emig 1997). On the basis of molecular studies, Cohen (2000) concluded that phoronids and brachiopods shared a common ancestor (are monophyletic) and that phoronids should be recognized as the subphylum Phoroniformea of the phylum Brachiopoda or, potentially, even as the class Phoronata of the inarticulate brachiopod subphylum Linguliformea.

Phoronids are here considered to constitute a small phylum. As only two genera (*Phoronis* and *Phoronopsis*) are recognized, orders, classes, families, and so forth, are not used.

I am indebted to Dr. Daniel Geiger for permission to use his photograph of *Phoronopsis californica* and for his assistance in preparing the plate.

Keys to Common Phoronids

The positive identification of phoronids to species usually requires the study of microscopically thin sections to obtain information on internal anatomy. Taxonomic characters include "muscle formulae" (the number of longitudinal muscles within a cross-section of the trunk and their distribution among the right and left oral and right and left anal coelomic compartments); structure of longitudinal muscles; whether metanephridia have one or two coelomic funnels; whether there are 0, 1, or 2 longitudinal nerve bundles; whether the species is gonochoristic or hermaphroditic; and the nature of accessory reproductive glands. Greatest weight is assigned to muscle formulae, but this varies within an individual, between individuals of a given population, and between different populations. As a consequence, taxonomists often have to make an arbitrary decision as to how much variation is acceptable for a given species, leading to different interpretations of how many valid species should be recognized. Fortunately, most West Coast phoronids can be identified with considerable confidence using external features alone and only these features are used in the key.

Six species of phoronids have been first described from the West Coast: *Phoronis pacifica*, *Phoronis striata*, *Phoronis vancouverensis*, *Phoronopsis californica*, *Phoronopsis harmeri*, and *Phoronopsis viridis*. Of these, the first two were inadequately described and are considered probable synonyms of *Phoronopsis harmeri*. *Phoronopsis viridis*, a green form familiar in several central California bays has also been united with *Phoronopsis harmeri* and that synonymy is used here. *Phoronis vancouverensis* has been submerged with the Japanese *Phoronis ijimai* Oka, but that synonymy is questionable; see the diagnosis for *Phoronis vancouverensis* below.

Phoronis architecta, originally described from the East Coast, and three species described from Europe, *Phoronis pallida*, *Phoronis psammophila*, and *Phoronis ovalis*, have also been reported from the West Coast. Of these, *Phoronis architecta* has been made a synonym of *Phoronis psammophila* (e.g., Emig

1982), but it is clear that both species should be recognized because the former is free spawning and the latter is a brooder; in addition, distinctive larval forms are known for each.

Considering the above, we here recognize from the West Coast five species of the genus *Phoronis*—*architecta, ovalis, pallida, psammophila,* and *vancouverensis*—and two species of *Phoronopsis*—*californica* and *harmeri.* Of these, all but *Phoronopsis californica* (originally described from mudflats at Balboa Bay) have recently been found intertidally, although their main populations may be subtidal. All but *Phoronis ovalis* have planktotrophic, long-lived actinotroch larvae that are species-specific. *Phoronis ovalis* has a slug-shaped, lecithotrophic larva that is only briefly demersal. At least two additional actinotrochs, for which no plausible adult can be assigned, are found in California. One of these "orphans" is the second most common actinotroch in southern California waters, and the other is uncommon but has been found sporadically from Coos Bay, Oregon, to near Los Angeles, California. Several local actinotrochs, including the two orphans, are described in Johnson and Zimmer (2001).

Most species of phoronids—even *Phoronis ovalis* with a briefly demersal larvae—are now known to have quite broad geographic distributions; only local distributions are provided below.

The reader should be cautioned that the tube constructed by a given phoronid may be considerably longer than the phoronid itself and that the animal's contractility makes length a subjective criterion except in broad terms.

Key to Species

1. Adults with a distally directed epidermal fold or "collar" at base of lophophore (this may be inconspicuous); lophophore may be white (or pink or flesh-colored), green, orange, or peach-colored . . . . *Phoronopsis* 2
— Adults without epidermal collar at base of lophophore; lophophore usually white (or pink or flesh-colored) . . . . .
. . . . *Phoronis* 3
2. Adults with orange or peach-colored lophophore, in complex spiral with four or more turns on each side (plate 435D) . . . . *Phoronopsis californica*
— Adults with white (or pink or flesh-colored) or green lophophore (plate 435E) . . . . *Phoronopsis harmeri*
3. With membranous tubes intertwined to form a compact mass attached to rocks, shells, or floats (but sometimes separately attached to various substrates or burrowing into calcareous materials); may have brood masses during spring and summer (plate 435B) . . . .
. . . . *Phoronis vancouverensis* (or *Phoronis ijimai*)
— Tubes otherwise, usually with attached sand grains and arranged vertically in substrate 4
4. Adults longer than 2 cm is length (plate 435A) . . . .
. . . . *Phoronis architecta* (or *Phoronis psammophila* or *Phoronis muelleri*)
— Adults smaller than 2 cm in length 5
5. Adults about 0.5 cm in length, boring in calcareous substrates, including limestone and shells of mollusks and barnacles . . . . *Phoronis ovalis*
— Adults about 1.5 cm in length, commensal in burrow walls of *Upogebia pugettensis* (plate 435C) . . . . . . *Phoronis pallida*

List of Species

Phoronis architecta Andrews, 1890 (with information on *Phoronis psammophila*). Locally abundant subtidally, but also found intertidally in southern California. Specimens have been collected as far north as Washington, but records are wanting for most areas north of Point Conception. Typically with straight, sand-encrusted, vertically positioned tubes, but these can be attached to undersides of rocks and shells. This species was made a synonym of *Phoronis psammophila* (see Emig 1982). The two forms may be indistinguishable if not alive and reproductive, but *P. architecta* is free-spawning whereas *P. psammophila* broods its early embryos. The larva of *P. architecta* is easily distinguished from that of *P. psammophila* and is by far the most abundant actinotroch collected in the southern California Bight. An additional difference is that *P. psammophila,* as do most phoronids, proliferates new tentacles from a single site between the mouth and anus, whereas *P. architecta,* like *P. muelleri,* also produces tentacles on the opposite (adanal) side of the mouth. Finally, architecta has only white pigmentation on the tentacles of the lophophore and at its base, whereas psammophila, at least from Europe, may have conspicuous red, green, and yellow pigmentation, although such pigmentation would be evident only in living specimens. Nonreproductive adults of the two forms may be inseparable, but this does not justify their synonymy.

Phoronis ijimai Oka, 1897. (See *Phoronis vancouverensis.*)
Phoronis ovalis Wright, 1856. This, the smallest phoronid, is commonly about 0.5 cm long and has a lophophore in the shape of an indented circle. It has not yet been reported from California, but it is easily overlooked because it forms burrows beneath the surface of calcareous substrates, including mollusk and barnacle shells and limestone.

Phoronis pallida Schneider, 1862. This small (about 1.5 cm in length) phoronid has been collected subtidally from a number of worldwide locations. Thompson (1972) discovered it as a commensal in the mucus-consolidated walls of the mud shrimp *Upogebia pugettensis.* The phoronid can be collected from the deeper parts of intertidal *Upogebia* burrows (but is not found with the often sympatric mud shrimp *Neotrypaea californiensis*) and has been collected as far north as the San Juan Islands. In addition to the association with mud shrimp, three conspicuous annular constrictions of the trunk are unique to this species and provide positive identification. Santagata (2004) successfully reared the larvae through metamorphosis and has studied their behavior and metamorphosis.

Phoronis psammophila Cori, 1889 (see *Phoronis architecta*). *Phoronis psammophila* has occasionally been collected subtidally (Los Angeles and Long Beach Harbors) and intertidally (Tijuana River Estuary). Reports that suggest it is found widely along the West Coast should be accepted with caution, as records may involve *Phoronis architecta* not *Phoronis psammophila.* Because *Phoronis architecta* is common in southern California and *Phoronis psammophila* is apparently quite rare, care should be taken in accepting the identification of local forms as the latter.

Phoronis vancouverensis Pixell, 1912. This species is found in shallow water along the length of our West Coast, occasionally as abundant large clumps attached to shells or rocky outcroppings (e.g., Garrison Bay, WA), but more commonly as isolated clumps in similar environments or attached to floats in marinas. Specimens burrowing in or attached to the surface of limestone outcroppings also occur and may represent incipient clumps. Although this species can regenerate two complete worms if bisected through the muscular region, the aggregates are probably the result of gregarious settlement rather than asexual proliferation. *Phoronis vancouverensis* has been submerged with the Japanese *Phoronis ijimai* Oka, but that synonymy can

PLATE 435 A, Adult of *Phoronis architecta* removed from tube; B, adult of *Phoronis pallida* (removed from burrow wall of *Upogebia*); C, adult of *Phoronis vancouverensis*—there are paired white masses consisting of embryos of different ages within the lophophoral concavity of this hermaphroditic form. Several eggs can be seen in the trunk coelom preparatory to their spawning; D, adult of *Phoronopsis californica* in situ; E1, male of *Phoronopsis harmeri* seen from the anal side—the paired white areas on the distal part of the trunk are the nephridia filled with sperm, portions of the spermatophoral glands can be seen within the lophophore, compare with E2; E2, male of *Phoronopsis harmeri*, looking into the lophophoral concavity at the base of which are paired spermatophoral glands—the white mass in the spermatophoral gland on the left is a spherical accumulation of sperm being molded into a spermatophore (D, courtesy of Daniel Geiger).

be questioned. Ikeda (1901), who studied the actinotrochs of Japan, believed one of the four types that he collected belonged to *Phoronis ijimai*; that form, and the other three, were all significantly different from the larva of *Phoronis vancouverensis* (Johnson and Zimmer 2001). It is possible that Ikeda failed to collect the proper larva of *Phoronis ijimai* in his careful study, but it is plausible that there are two species with adults that are similar but larvae that can be readily distinguished.

Phoronopsis californica Hilton, 1930. This is the largest local phoronid and has a complex lophophore with four to seven turns on each side. The lophophore, with many hundreds of tentacles, is usually a clear orange, but may be peach- to flesh-colored. The tentacles are typically flecked with white, and the papilla containing the anus is almost solidly white. The animals measure up to about 8 cm in length, but the tubes may be much longer and vary greatly depending on the substratum. Animals with tubes a half meter in length and studded with basalt cubes several millimeters in dimension occur at San Clemente Island, whereas specimens from Santa Catalina Island have tubes that are rarely much longer than the animal and are coated with fine sediment.

Phoronopsis harmeri Pixell, 1912 (=*Phoronopsis viridis* Hilton, 1930). The large phoronids with green lophophores found in massive beds in California (e.g., Morro and Tomales Bays and Bodega Harbor) have long been identified as *Phoronopsis viridis* and distinguished from a less common "white" form (*Phoronopsis harmeri*) originally described from Vancouver Island, Canada, and now known from Oregon and Washington. However, the two cannot be distinguished using most taxonomic criteria (color is proving to be a variable character). Marsden reported that her specimens from Washington and British Columbia were in fact green, but "white" forms are common there. Whether "green" or "white," the tentacles have numerous white flecks. Although better known, *Phoronopsis viridis* is the junior name to *Phoronopsis harmeri*. The tubes are typically straight with sand encrustation, but gravel may be attached and the tubes may then be meandering. Tubes average about 10 cm long with the lower end shaped as a closed cone. See Everett 1991, J. Exp. Mar. Biol. Ecol. 150: 223–247 (ecology in Bodega Harbor).

References

Ayala, F. J., J. W. Valentine, L. G. Barr, and G. S. Zumwalt. 1975. Genetic variability in a temperate intertidal phoronid, *Phoronopsis viridis*. Biochem. Genet. 11: 413–427.

Bergen, M., S. B. Weisberg, R. W. Smith, D. B. Cadien, A. Dalkey, D. E. Montagne, J. K. Stull, R. G. Velarde, and J. A. Ranasinghe. 2001. Relationship between depth, sediment, latitude, and the structure of benthic infaunal assemblages on the mainland shelf of southern California. Marine Biology 138: 637–647.

Cohen, B. L. 2000. Monophyly of brachiopods and phoronids: reconciliation of molecular evidence with Linnaean classification (the sub-phylum Phoroniformea nov.). Proceeding of Royal Society, London, Series B, 267: 225–231.

Emig, C. C. 1982. The biology of Phoronida. Advances in Marine Biology 19: 1–89.

Emig, C. C. 1997. Les Lophophorates consitituent-ils un embranchement? Bulletin Soc. Zool. Fr. 122: 279–288.

Halanych, K. M., J. D. Bacheller, A. M. A. Aguinaldo, S. M. Liva, D. M. Hillis, and J. A. Lake. 1995. Evidence from 18S ribosomal DNA that lophophorates are protostome animals. Science 267: 1641–1643.

Ikeda, I. 1901. Observations on the development, structure and metamorphosis of Actinotrocha. J. Coll. Sci., Imperial University, Tokyo 13: 507–591.

Johnson, A. S. 1988. Hydrodynamic study of the functional morphology of the suspension feeder *Phoronopsis* (Phoronida). Marine Biology 100: 117–126.

Johnson, K., and R. Zimmer. 2001. Phylum Phoronida. In *Atlas of Marine Invertebrate Larvae*. C. Young and M. Rice, eds. Chapter 22, pp. 477–487. London: Academic Press.

Marsden, J. R. 1957. Regeneration in *Phoronis vancouverensis*. Journal of Morphology 101: 307–323.

Marsden, J. C. R. 1959. Phoronidea from the Pacific coast of North America. Canadian Journal of Zoology 37: 89–111.

Nielsen, C. 1998. Morphological approaches to phylogeny. American Zoologist 38: 942–952.

Nielsen, C., N. Scharf, and D. Eibye-Jacobsen. 1996. Cladistic analyses of the animal kingdom. Biological Journal of the Linnean Society 57: 385–410.

Peterson, K. J., and D. J. Eernisse. 2001. Animal phylogeny and the ancestry of bilaterians: Inferences from morphology and 18S rDNA gene sequences. Evolution and Development 3: 1–35.

Rattenbury, J. C. 1953. Reproduction in *Phoronopsis viridis*: The annual cycle in the gonads, maturation and fertilization of the ovum. Biological Bulletin 104: 182–196.

Rattenbury, J. C. 1954. The embryology of *Phoronopsis viridis*. Journal of Morphology 95: 289–349.

Ronan, T. E. 1978. Food resources and the influence of spatial patterning on feeding in the phoronid *Phoronopsis viridis*. Biological Bulletin 154: 472–484.

Santagata, S. 2004. A waterborne behavioral cue for the actinotroch larva of *Phoronis pallida* (Phoronida) produced by *Upogebia pugettensis* (Decapoda: Thalassinidea). Biol. Bull. 207: 103–115.

Silén, L. 1954. Developmental biology of the Phoronidea of the Gullmar Fjord area of the West coast of Sweden. Acta Zool. 35: 21–57.

Strathmann, R. 1973. Function of lateral cilia in suspension feeding of lophophorates (Brachiopoda, Phoronida, Ectoprocta). Marine Biology 23: 129–136.

Thompson, R. K. 1972. Functional morphology of the hindgut gland of *Upogebia pugettensis* (Crustacea, Thalassinidea) and its role in burrow construction. PhD dissertation, Zoology, University of California Berkeley, Berkeley, CA, 202 pp.

Zimmer, R. L. 1964. Reproductive biology and development of Phoronida. PhD dissertation, Zoology, University of Washington, Seattle, WA, 416 pp.

Zimmer, R. L. 1991. Phoronida. In Reproduction of marine invertebrates Vol. VI. Echinoderms and lophophorates. A. C. Giese, J. S. Pearse, and V. B. Pearse, eds. pp. 1–45, 7 plates. Pacific Grove, CA: Boxwood Press.

Zimmer, R. L. 1997. Phoronids, brachiopods, and bryozoans, the lophophorates. In Embryology—constructing the organism. S. F. Gilbert and A. M. Raunio, eds. pp. 279–305. Sunderland, MA: Sinauer.

Zimmer, R. L. 1998. Phoronida. In Encyclopedia of reproduction. E. Knobil and J. D. Neill, eds. Vol. 3, pp. 770–778. San Diego, CA: Academic Press.

Brachiopoda

(Plates 436 and 437)

F. G. HOCHBERG

Brachiopods are a phylum of entirely marine and sessile organisms. They have a two-valve shell that at first glance may be mistaken for a clam or bivalved mollusk. The valves are dorsal and ventral in orientation, not lateral as in bivalves. The shell is attached to the substrate by a stalk (pedicle or peduncle). There are two classes: **ARTICULATA**, whose valves are hinged by interlocking teeth and sockets, and whose pedicel emerges through a hole (foramen) in the ventral valve; and **INARTICULATA**, whose valves are not hinged but are held together by muscles. The stalk (pedicel) in the inarticulates is often long and muscular and used to retract the shell into a burrow in sandy or muddy substrates, or may be reduced or absent.

Key to Brachiopoda

1. Shell flattened dorsal-ventrally flattened (depth less than one-half width) 2
— Shell not flattened, deep or inflated (depth greater than one-half width) 3
2. Shell large (to 35 mm) elongate, much longer than wide; valves thin and smooth, anterior valve margin (commissure) straight; pedicle very long, fleshy; color brown-white; infaunal, live in soft substrates (plate 436) . . . . Glottidia albida
— Shell very small (to 5 mm), dorsal-ventrally flattened, round in outline; valves thin, translucent with small punctae; anterior valve margin straight; pedicle very short, foramen large; color light brown; lives on or attached to hard substrates . . . . Platidia hornii
3. Shell thin, translucent; valves smooth; shape globular to elongate, longer than wide; anterior valve margin straight; foramen very small; color typically yellow-brown; length to 53 mm . . . . Laqueus californianus
— Shell thick, not translucent; valves typically with distinct ribs 4
4. Shell usually longer than wide; ovate to triangular; anterior valve margin straight or very slightly sulcate; valves with numerous radiating ribs; foramen large; color light yellow-white; length to 25 mm . . . . Terebratulina unguicula
— Shell usually wider than long; subquadrate; anterior valve margin with pronounced undulation (sulcate); valves thick, typically deeply ribbed; foramen large; color red; length to 45 mm (plate 437) . . . . . . Terebratalia transversa

List of Species

All species occur from Alaska or British Columbia to southern or Baja California, except for *Glottidia*.

Inarticulata

LINGULIDAE

Glottidia albida (Hinds, 1844). Tomales Bay to Todos Santos Estero, Baja California; 0 m–150 m; often common in soft sediments in the low intertidal. (See Jones and Barnard, 1963: distribution, abundance in southern California).

Articulata

CANCELLOTHYRIDIDAE

Terebratulina unguicula (Carpenter, 1864). 10 m–850 m; may be locally abundant in intertidal in areas of broken rocks and coralline algae in northern part of range.

DALLINIDAE

Terebratalia transversa (Sowerby, 1846). 0 m–1800 m; often attached to the underside of rocks at very low tides in the northern part of range.

Glottidia albida

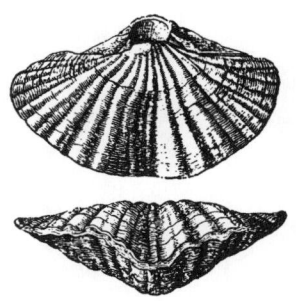

Terebratalia transversa

PLATE 437 *Terebratalia transversa* (Keep 1904).

PLATE 436 *Glottidia albida* (original, Emily Reid).

LAQUEIDAE

Laqueus californianus (Koch, 1848) (=*Terebratula californiana, Waldheimia californica*). 0 m–1500 m; abundant intertidally in northern part of range.

PLATIDIIDAE

Platidia hornii (Gabb, 1861) (=*Morrisia hornii*). 0 m–500 m; often attached to rocks or shells in low intertidal in northern part of range.

References

Bernard, F. R. 1972. The living Brachiopoda of British Columbia. Syesis 5: 73–82.

Foster, M. W. 1982. Brachiopoda, pp. 773–780. In Synopsis and classification of living organisms. S. P. Parker, ed., Vol. 2. New York: McGraw-Hill.

Hertlein, L. G., and U. S. Grant. 1944. The Cenozoic Brachiopoda of Western North America. Publ. Univ. Calif., Math. Phys. Sci. 3: 1–236.

Hochberg, F. G. 1996. The Brachiopoda, pp. 1–47. In Taxonomic atlas of the benthic fauna of the Santa Maria Basin and the Western Santa Barbara Channel. Vol. 14. Miscellaneous Taxa. J. A. Blake, P. H. Scott, and A. Lissner, eds. Santa Barbara: Santa Barbara Museum of Natural History.

Jones, G. F., and J. L. Barnard. 1963. The distribution and abundance of the inarticulate brachiopod *Glottidia albida* (Hinds) on the mainland shelf of southern California. Pac. Nat. 4: 27–52.

Keep, J., and J. L. Baily. 1935. Lamp shells and allied forms. pp, 18–24. In West coast shells. Stanford: Stanford University Press.

Mattox, N. T. 1955. Observations on the brachiopod communities near Santa Catalina Island, pp. 73–86. In Essays in the natural sciences in honor of Captain Allan Hancock, on the occasion of his birthday, July 26, 1955. Los Angeles: University of Southern California Press.

Paine, R. T. 1969. Growth and size distribution of the brachiopod *Terebratalia transversa* Sowerby. Pac. Sci. 23: 337–343.

Rudwick, M. J. S. 1970. Living and Fossil Brachiopods. London: Hutchinson University Library, 199 pp.

Smith, A. G., and M. Gordon. 1948. The marine mollusks and brachiopods of Monterey Bay, California, and vicinity. Proc. Calif. Acad. Sci. (4) 26: 147–245.

Williams, A., et al. 1965. Brachiopoda. In Treatise on invertebrate paleontology. R.C. Moore, ed., Part H. Vol. 1, 522 pp.; Vol. 2, 405 pp.

Bryozoa

(Plates 438–452)

DOROTHY F. SOULE, JOHN D. SOULE, PENNY A. MORRIS,
AND HENRY W. CHANEY

Bryozoans are minute, mostly colonial marine organisms, familiar to casual observers in temperate climates as intertidal organisms that form part of the fouling community found on boat hulls and pilings, or comprise part of the "bioturf," feltlike growths of many small colonial organisms. Thus the colonies observed were named moss (*bryo*-) animals (-*zoans*) by early observers. There are probably some 6,000 described living species, many as yet undescribed species, and far more fossil species.

Bryozoan colonies may be erect, slender or bushy, flaccid or loosely attached, or may form calcareous encrustations on kelp, rock, wood, concrete, and other substrates such as rocks, molluscan shells, crustacean carapaces, other bryozoan colonies, and even on sea snake skins. They are prolific on the older, shady, protected surfaces of coral reefs in the tropics, but not where live coral polyps are active. Most bryozoans do not adapt well to sunny exposures; sandy, silty bottoms; or areas of high currents or turbulence, although there are exceptions. Bryozoans occur worldwide, from intertidal to abyssal depths and from tropical to polar environments, but are most abundant in the littoral and neritic zones down to about 200 m.

The phylum Bryozoa, *sensu lato,* also known in older British literature as Polyzoa, was divided in the nineteenth century into the classes Ectoprocta and Entoprocta (Nitsche 1869). Although both groups have a U-shaped gut with a circlet of tentacles surrounding the mouth, the anus lies outside the tentacles on a retractable lophophore in the Ectoprocta, while in the Entoprocta the anus lies within a circlet of tentacles on a nonretractable calyx that is not a true lophophore. Hyman (1951, 1959) elevated both classes to phylum status. The standard terminology adopted by researchers of the International Bryozoology Association employs the term "Bryozoa," instead of "Ectoprocta," and this usage is now widespread.

Some bryozoan species cast ova directly into the sea, but many retain ova, which undergo initial cleavage within the zooid and have various brooding strategies. Some brood ova in the body cavity, some have membranous sacs suspended from the soft-bodied zooids, and most have calcified brood chambers, modified zooids called ovicells or ooecia. The Cyclostomatida have simple or complex calcified zooids called gonozooids in which eggs may undergo total cleavage many times to form a large number of embryos (polyembryony).

Larvae are of several types. The *cyphonautes* larvae, which occur in a few species, are bivalved, triangulate forms, some with a complete gut that makes them capable of feeding and remaining in the plankton for a number of days. The *cyphonautes* has a ciliary corona, an apical tuft, and a pyriform organ that is involved in testing the substrate prior to settlement, plus an eversible adhesive sac for attachment. Other larval forms include nonfeeding coronate larvae that resemble a flattened hat or cap with a mid-body ciliary girdle, and a nonfeeding, ball-shaped larva that is almost completely covered with cilia (Zimmer and Woollacott 1977). Some larvae are incapable of planktonic existence, settling close to or on the parent colony. None of these forms can be readily classified as the trochophore or tornaria larvae typical of Spiralia.

Most bryozoan larvae undergo a complete metamorphosis, when a single or twinned ancestrula is formed that differs from adult zooids, although some species show a direct development, with the ancestrula resembling a small adult. Asexual budding then produces a colony, which includes a zone of astogeny and a zone of growing zooids that may differ in minor respects from the adult zooids comprising the mature portion of typically extensive colonies.

Classification

Classification of the Bryozoa has become quite complicated, with construction of infraorders, supraorders, and superfamilies that may represent evolutionary relationships but do not assist in identification of species, as there is no general consensus on keys to superfamilies and families. We use the categories employed in the literature most available for identifications. Researchers have traditionally relied on external morphologies for identification, which has proven to be reliable—confirmed, for example, by protein electrophoresis (Jackson and Cheetham 1990). Newer mitochondrial ribosomal DNA research (e.g., Dick et al. 2000) has tended to confirm some higher-level relationships but not others. Different techniques may be required to pursue this promising line of inquiry, but still may not soon assist in separating closely related species.

The Class Phylactolaemata is composed of freshwater species, mostly with a horseshoe-shaped lophophore. rDNA work suggests that the phylactolaemates may be more closely allied with phoronids than with bryozoans (Wood and Lore 2003). Species in the marine classes Stenolaemata and Gymnolaemata have circular lophophores bearing the mouth and tentacles. The Stenolaemata include many fossil species dating from the Ordovician (McKinney and Jackson 1989), but currently living species are found only in the Order Cyclostomatida. All cyclostomatids consist of calcareous tubular individuals with terminal apertures, although colonies may be erect, slender with few individuals per branch, or fused into thick bundles, or they may be encrusting, forming circular, lobate, or irregularly branched growths. Polyembryony occurs only in cyclostomatids. There has been scant systematic research on the eastern Pacific cyclostomatids since the work of Osburn, 1953.

The Class Gymnolaemata is divided into the Orders Ctenostomatida and Cheilostomatida. Individual zooids of Ctenostomatida are soft bodied, fleshy, and squat with contiguous zooids (suborder Carnosina), or stolonate or erect (suborder Stoloniferina). Individuals are tubular with terminal apertures that are closed by either purse-string sphincter muscles or a folding flap musculature. Southern California Ctenostomatida were discussed in Soule (1953) and Soule and Soule (1969, 1976a), but little research has been done more recently on the California groups and is badly needed. Earlier studies relied mostly on the British and European literature, and the perspective was based on similarities of species rather than differences in the eastern Pacific species (Soule, Soule, and Morris 2001).

By far the most numerous species are in the Order Cheilostomatida, which dates from the Jurassic and probably were derived from ctenostome ancestors (e.g., Todd 2000). Most living cheilostomes are encrusting, boxlike, with membranous or calcareous frontal (ventral), basal (dorsal), and vertical (lateral and transverse) walls, although a number of shallow-water species are erect and branching, with nearly tubular or terete zooids. The apertures through which the lophophore bearing the tentacles and mouth are extruded are subterminal on the frontal surface and are usually closed by an extension of the frontal membrane modified to become the operculum.

The Cheilostomatida have traditionally been divided into Suborders Anascina (=Anasca), lacking an ascus, or hydrostatic chamber, and Ascophorina (=Ascophora) that possess an ascus, a hydrostatic chamber that aids in extrusion of the lophophore. However, anascines and those ascophorines having a cryptocystidial frontal wall have a hypostegal coelom between the frontal membrane and the cryptocyst, an extension of the internal body coelom, instead of an ascus. Muscles attached to the frontal membrane and the body wall assist in eversion and retraction of the lophophore.

Ascophorines developed fully calcified frontal walls. Frontal wall calcification developed in several ways, indicating that the Ascophorina are polyphyletic (e.g., Gordon and Voigt 1995), and the suborders are being revised. Suborder Cribrimorphina (=Cribrimorpha, Cribriomorphina) was added to encompass species having a frontal shield formed from pinnate hollow spines that fuse in an arch over the frontal membrane, but others place the cribrimorphs in the suborder Ascophorina with a separate infraorder and a superfamily. Other modes of formation include frontal walls derived from fusion of spines, or the folding of the lateral wall gymnocyst over the frontal membrane, development of the cryptocystal shelf beneath the frontal membrane into a complete frontal wall, and calcification of the frontal wall itself, forming a solid gymnocyst that required development of the ascus beneath to extrude the lophophore. These variously formed frontal wall structures allowed the cheilostomes to achieve much greater species diversity.

Adult Morphology

Bryozoan literature often contains terminology that is unduly complex. A glossary is included herein for reference. The colony (ZOARIUM) is a functional group of tubular or boxlike individuals budded from a single, sexually produced larva that has settled and metamorphosed into an ANCESTRULA. Individuals are known as ZOOIDS (=ZOIDS), but the exoskeleton alone is technically a ZOOECIUM (PL. ZOOECIA), such as remains after drying or bleaching for scanning microscopy, although the terms have been used interchangeably.

Within the zooid, the soft tissues of the body wall lining the exoskeleton form the CYSTID. The digestive tract, reproductive organs, musculature, and the lophophore (consisting of the ciliated tentacles and mouth) form the POLYPIDE. In cheilostomes, the FUNICULUS is a mesenchymal cord of tissue connecting the polypide to adjacent zooids by means of club-shaped cells that pass through various types of pores in vertical walls, forming a combined nerve pathway and nutritional transport system. The "brain" is a small ganglion at the base of the tentacles. There is no excretory system, but diffusion occurs through the ciliated tentacles and epithelium, and in some species BROWN BODIES composed of wastes are sequestered in the body cavity. When a polypide degenerates, the brown bodies may be released, or if a new polypide with digestive tract is regenerated, the brown body may be incorporated and expelled through the anus. Polypides are generally similar throughout the group, varying in minor ways such as the presence/absence or the complexity of a gizzard or difference in number of tentacles, so identification has traditionally relied on external skeletal criteria (see plate 438). Tentacle count is not a reliable criterion for identification because research indicates that it can be influenced by the environment or by food.

The frontal wall of soft-bodied ctenostomatids does not differ greatly from lateral and so-called basal walls in stolonate forms. In anascines, the FRONTAL MEMBRANE covers the OPESIA, which may be a widely open oval or partly or mostly closed by CRYPTOCYST, which grows from the vertical walls as a shelf beneath the membrane, leaving one large or several small openings through which muscles pass to the membrane to extrude the lophophore. In a few anascine species, the opesial rim is formed of GYMNOCYST, which grows as a narrow fold from the vertical walls but does not cover the opesial space. Hollow SPINES may extend from the margins of the zooids over the frontal membrane. In cribrimorphs, pinnate spines have evolved into structures fused over the frontal membrane to form a FRONTAL SHIELD.

Ascophorines develop the calcified frontal wall or shield in various ways as the CRYPTOCYST comes to form the entire frontal wall, or in UMBONULID forms, the body wall forms a fold from the proximal and sometimes lateral body walls over the frontal wall, or the frontal wall itself becomes calcified, forming a GYMNOCYSTIDIAL frontal shield (see Boardman et al. 1983; Hayward and Ryland 1998). The ASCUS (HYDROSTATIC CHAMBER) is produced beneath the calcified gymnocystal frontal wall by various methods (e.g., Gordon 1993; Gordon and Voigt 1995).

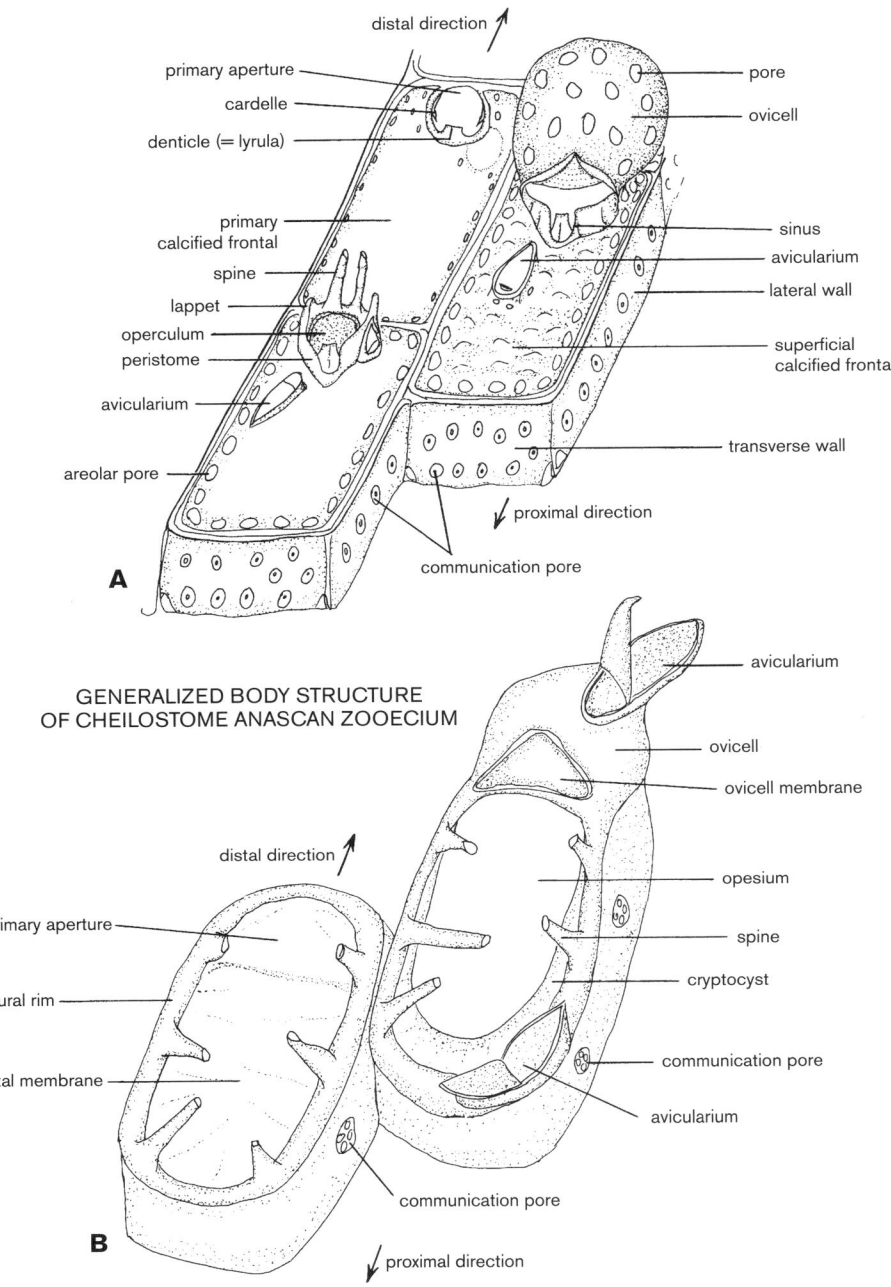

GENERALIZED BODY STRUCTURE
OF CHEILOSTOME ASCOPHORAN ZOOECIUM

distal direction

primary aperture
cardelle
denticle (= lyrula)

pore
ovicell

primary
calcified frontal

spine
lappet
operculum
peristome

sinus
avicularium
lateral wall

superficial
calcified frontal

avicularium

areolar pore

transverse wall

proximal direction

communication pore

A

GENERALIZED BODY STRUCTURE
OF CHEILOSTOME ANASCAN ZOOECIUM

avicularium
ovicell
ovicell membrane

distal direction

opesium

primary aperture

mural rim

spine
cryptocyst

frontal membrane

communication pore
avicularium

communication pore

B

proximal direction

PLATE 438 A, B, generalized cheilostome structure.

Colonies may include **KENOZOOIDS**, which are membrane-covered spaces on the surface that lack a polypide, and **HETEROZOOIDS** such as the pincerlike **AVICULARIUM**, most without a polypide, in which the operculum is modified as an upper mandible, rounded, triangular, or setose. Some avicularia are capable of capturing small predators or clearing the apertural area. We have observed cooperative action between avicularia on adjacent zooids in capturing, tearing up, and feeding on small prey by *Scrupocellaria*. **AVICULARIA** may be small or large, originating on the frontal wall or lateral walls of zooids, sessile, or on a stalk on individual zooids (**ADVENTITIOUS AVICU-** **LARIA**), or they may arise between zooids as **INTERZOOECIAL (VICARIOUS) AVICULARIA**, which sometimes migrate onto adjacent frontal walls or onto ovicells. **OVICELLS** (=**OOECIA**) are heterozooids, zooids modified to brood larvae, and may be mostly internal (**ENDOZOOECIAL**) or be larger globular structures lying externally at the distal end of zooids (**HYPERSTOMIAL**). The ovicell bud forms as a fold or double layer forming an inner layer (**ENTOOECIUM,** =**ENDOOECIUM**) that may have pores or striations, and the **ECTOOECIUM**, which may cover the entooecium totally or form only a rim around it. Differences in the ovicell are important in identification, and young colonies that may not yet have developed ovicells are often difficult to assign to species.

Colony Form

Early bryozoan identification was based on colony form, largely because optics were not adequate to observe the details of individual zooecia. However, it was found that colony form was not accurate because some species may encrust extensively in a turbulent or otherwise marginal environment, but rise in branching, foliaceous, or frondose forms under other conditions. The growth forms in some cases even appear to overlap in major categories. Nevertheless, the following grow patterns may be observed.

Recumbent or Encrusting Forms

SOFT-BODIED SPECIES

STOLONATE OR STOLONIFORM

Forms with creeping branches extending over the substrate.

Certain ctenostomes have individual zooids that lack polypides and are modified as stolons (e.g., *Bowerbankia*). A few cheilostomes are stoloniform, but the extensions are formed by the basal portion of the zooid.

Penetrating species form stolonate colonies by burrowing into shells and, rarely, other invertebrates (Cheilostomatida: *Penetrantia*; Ctenostomatida: *Terebripora, Immergentia* (D. F. and J. D. Soule 1969).

FLESHY (CARNOSE)

Species that form flattened, gelatinous colonial masses, sometimes of considerable size (Ctenostomatida: *Alcyonidium, Flustrellidra*).

ENCRUSTING, LIGHTLY OR HEAVILY CALCIFIED SPECIES

FLEXIFORM

Species that form a lightly calcified, flat crust over soft, flexible substrates like algal blades (e.g., Cheilostomatida, Anascina: *Membranipora, Dendrobeania*).

RIGID

Heavily calcified species that form unilaminar or multilaminar crusts over solid, inflexible substrates, such as shell or rock. Some species form thick, knobby crusts over originally flexible surfaces such as algae or twigs, creating solid masses (most ascophorine cheilostomes and some anascine species are of the rigid type).

**TUBULAR, HEAVILY CALCIFIED SPECIES
WITH TERMINAL APERTURES**

RECUMBENT

Colonies of recumbent, tubular bundles with terminal apertures; although single or multiple and fused, they may curve upward from their encrusting bases (e.g., Cyclostomatida:

Tubulipora, Diaperoforma, formerly placed in *Diaperooecia* in the eastern Pacific).

DISCOIDAL

Colonies with circular or irregularly lobate discoidal bases, having tubules extending upward singly or in bundles (Cyclostomatida: *Disporella*, includes California species formerly placed in *Lichenopora*).

Erect, Branching, Foliaceous, or Fenestrate Forms

BRANCHING

LIGHTLY CALCIFIED

Some species of cheilostomatid anascines with various modes of branching, with or without joints, attached to substrate with rootlets (radicles); most are not tolerant of turbulent waters. Easily confused with some hydroids (Cheilostomatida, Anascina: e.g., *Bugula, Caberia, Scrupocellaria*).

MORE HEAVILY CALCIFIED

Small cyclostomes forming erect, branching colonies that mesh into bioturf mats in shallow water (Cyclostomatida: e.g., *Crisia, Bicrisia, Crisulipora*). Some deeper-water species form larger thickened, branching colonies.

FOLIACEOUS (FRUSTRAFORM OR FRONDOSE)

Species sometimes forming leaflike colonies from encrusting bases that resemble small cabbage plants or lichens (Cheilostomatida, Anascina: e.g., *Thalamoporella*; Cheilostomatida, Ascophorina; e.g., some *Hippodiplosia, Watersipora* in warm waters).

FENESTRATE

Erect species with a "chicken-wire" appearance (=reteporid), forming large colonies in deeper waters (Cheilostomatida, Ascophorina, e.g., *Phidolophora*).

Distribution

Bryozoans apparently do not conform to traditional distribution patterns and depths, being generally far more tolerant of temperature and depth than most invertebrate species. Northeastern Pacific bryozoans do not conform to a typical Oregonian Region, but some species ranges extend farther north to Alaska or British Columbia, and south through the Channel Islands to San Diego, Baja California, or the Gulf of Panama and beyond, perhaps to the Galapagos Islands and Colombia. Also, very few species are limited to intertidal depths, although some are limited from intertidal to about a 30 m depth, while others extend from a few meters to 100 m or more, and out to the continental shelf.

Many older distribution records have been questioned in recent years because use of SEM has enabled researchers to distinguish differences in structure that were either not visible by light microscope or were considered to be within a wider range of variation than considered acceptable.

When Osburn (1950, 1952, 1953) assembled all available records for the eastern Pacific and described many new species, some distributions extended from Alaska or British Columbia to the Gulf of California, Panama, Colombia or the Galapagos Islands. Many of these species have not been reexamined by SEM, nor have new collections and studies been made, but in instances where they have been restudied, species and generic level differences have been found.

Soule, Soule, and Chaney (1995) described 41 new species and six new genera. D. F. Soule, J. D. Soule and H. W. Chaney (1995) described 41 new species and six new genera, based on collections from the Santa Maria Basin and western Santa Barbara Channel, reexamination of similar species from Hancock Foundation Pacific collections and comparisons with the Natural History Museum, London, type collections. Soule, Soule and Morris (2001) discussed the changing concepts in species diversity in the northeastern Pacific, emphasizing the revisions required of old identifications based on the British and European literature. Soule and Soule (2002) reexamined *Parasmittina* from the eastern Pacific, resulting in six new species and one from the western Caribbean. Soule, Chaney, and Morris (2003, 2004) reexamined Microporellidae from the eastern Pacific, resulting in a new genus, two new subgenera, and eight new species. The new species had been considered part of existing species, some of which are now known not to occur in the eastern Pacific. Others are included herein because the species from which they were split have a shallow-water distribution component, and the newly described species will surely be found in future sampling.

Because bryozoans settle on small rocks, crab carapaces, molluscan shells, algae, and other substrates, they may be carried from intertidal depths into deeper water or vice versa. Some of these species would probably not be collected on panels suspended from rafts or docks because of unfavorable light exposure, turbulence, sedimentation, or other adverse environmental conditions.

There have been many opportunities in the past two centuries for introduction of species on ships' hulls or in ballast for those limited species capable of colonizing and surviving the varied temperature and salinity regimes encountered. Natural conditions may also effect distribution because there are large fluctuations in coastal eastern Pacific water temperatures and rainfall runoff. At times, cold Aleutian and cool California Current waters that normally flow southward along the coast are deflected seaward, also eliminating upwelling. Coastal waters can warm as far north as British Columbia during ENSO events, and temperatures as high as 25°C can occur in the Southern California Bight. Such episodes offer opportunities for recurrent establishment of warmer-water species in more northern areas, and survival can occur in sheltered embayments or near power-plant cooling water entrainments. The same factors could also reduce the incidence of supposedly introduced species forming established populations, so that greater care is needed in making identifications without examining type species and in using SEM to illustrate species collected.

When using bryozoans to calculate species diversity, it is essential to consider the new species identified based on morphological differences observed by higher SEM magnifications and comparisons with supposed cosmopolitan distribution. To continue using out of date information gives the impression that these species are so ubiquitous it is unnecessary to consider them in environmental impact analysis and a unique fauna may be eliminated inadvertently.

D. F. and J. D. Soule (1979, 1985) first mentioned this regarding differences in certain bryozoan species in the Galapagos Islands, and Soule et al. (1995) discussed the potential impacts on bryozoan speciation along the northeastern Pacific coast. Consequently we have included in this chapter some of the newer species which may be found outside their initially recorded ranges.

The work of Hayward (1985) and Hayward and Ryland (1985, 1998, 2002) have helped to delineate differences in Atlantic/Mediterranean species that call into question many of the synonymies previously listed in the northeastern Pacific. We have examined many types at the Natural History, London, and elsewhere to determine differences, but detailed discussions are not possible herein.

Preservation and Microscopy

Bryozoans are not easily recognized by non-specialists, some species resembling coralline algae, others colonial tunicates, sponges or hydroids. In the field, a hand lens will show patterns of multiple zooecia, each with its individual aperture, and some distinctive genera may be identified. Magnifications of 75x to 100x in a dissecting microscope are needed for species identifications, and SEM may be necessary for definitive separation of similar species. It is strongly recommended that encrusting colonies not be scraped from substrates, if at all possible, because it is unlikely that sufficient intact zooecia will be obtained by scraping.

Bryozoans may be preserved in the field with 10% reagent grade formalin in seawater, or fresh water (distilled water) may be used as a diluent if calcium carbonate is added as a buffer; 70% alcohol is much less effective as the initial preservative. Formalin-preserved specimens must be transferred to 70% alcohol for storage, or the calcareous exoskeleton, which contains many of the key characters for identification, may be damaged. Well-calcified colonies may be air dried and retain their skeletal characters. Some specimens may be cleaned with an ultrasonic cleaner for a few minutes to remove debris. Treatment with sodium hypochlorite (bleach) is necessary to remove external membranes and internal soft parts for microscope or SEM examination.

Dried specimens should be permanently mounted on glass slides with water-soluble casein (white) glue to the avoid acidity of other cements and consequent decalcification. When removal from glass slides for SEM risks fragile specimens, some SEM instrument specimen chambers will accommodate slides that have been wrapped in aluminum foil, except for the specimen, and then coated with gold, carbon, gold-paladium, or platinum before mounting with carbon on stubs, thus protecting original labels. It is more difficult to remove slides from stubs affixed with carbon tape than with liquid carbon. SEM is usually done at 10 Kv–15 Kv, unless charging is a problem, in which case lower Kv should be used. Newly developing genetic studies will require different procedures and more complex methodologies now being addressed in research papers; e.g., Dick 2000, Hughes et al. 2004.

Glossary

ADNATE Recumbent, resting on substrate, not entirely cemented to it.

ANCESTRULA First zooid of colony formed by settlement of sexually produced larva.

APERTURE Opening through which tentacles and lophophore are extruded from body to feed; in Cheilostomatida, closed by operculum; also termed "orifice."

AREOLA (PL., AREOLAE) Tissue-filled pore(s) occurring in one or more rows around the margin of zooid frontal wall, in Cheilostomatida, Ascophorina.

AREOLAR PORE Used as equivalent to areola; or used only as opening at base of areola connecting to interior visceral coelom.

ASCOPORE Opening to ascus (compensation sac) in some cheilostomatid ascophorines; opening in some genera in the calcified frontal wall; in other genera opening inside proximal lip of aperture.

ASCUS Membranous sac (asc) in some cheilostomatid ascophorines beneath frontal wall opened by ascopore; sac performs hydrostatic function to extrude or retract lophophore containing mouth, anus, and tentacles.

ASTOGENY Changes in zooid structure in colony in zone of development from ancestrula to mature zooids by asexual budding.

AUTOZOOECIUM (PL., AUTOZOOECIA) Skeletal structure of autozooid; calcified structure remaining when tissues are dissolved away for light microscopy or SEM.

AUTOZOOID Normal feeding individual zooid with complete digestive tract.

AVICULARIUM (PL., AVICULARIA) Highly modified zooid (heterozooid) in Cheilostomatida, with chitinous mandible in place of operculum, usually with reduced nonfeeding "polypide"; functions in defense, capture of prey or in cleaning of colony surface.

BASAL WALL The dorsal surface, attached side of encrusting cheilostomes; aboral side in groups having subterminal apertures.

BIOTURF Feltlike mat formed by small erect bryozoan species, sometimes mixed with recumbent species in intertidal waters.

BISERIAL Two rows of zooids growing side by side, attached at inner and transverse lateral walls.

CARDELLES (=CARDELLA) Paired lateral projections in aperture on which operculum hinges, in Cheilostomatida, Ascophorina. See "condyles."

CAUDAL Tapering at one end like a tail.

CHITIN Amorphous cuticular substance; glycosaminoglycans (mucopolysaccharides) linked covalently with a protein core; a long-chain polymer of N-acetylglucose-amine units.

CHITINOUS Chitinlike; noncalcareous, hardened surfaces.

COMMUNICATION PORES Openings in vertical walls between zooids filled with tissue for passage of nutrients, nerve fibers (funiculus); each pore may be single, rimmed, or in a plate containing many smaller pores. See "dietella," "septulum," "rosette plate," "pore chamber."

CONDYLES Paired lateral hinge teeth within aperture for pivoting of operculum when aperture opens or closes.

CONNATE Single or double row of tubules fused together; in some cyclostomatids.

CONTIGUOUS Adjacent; zooids fused together.

COSTA (PL., COSTAE) Modified hollow spines (ribs) fused variously above frontal membrane in cribrimorph cheilostomatids to form calcified frontal shield.

COSTATE Ribbed in appearance.

CRYPTOCYST Horizontal calcified frontal rim under frontal membrane around opesia, or more extensive wall shield developed by calcification inward from the epithelium lining the vertical walls, in some Cheilostomatida.

CYSTID Boxlike or vaselike body wall, including calcified exoskeleton if present, cellular lining and musculature, containing the "polypide" (viscera).

DENTAL LEDGE In some ascophorines, visible primary calcified layer in proximal apertural rim; may form condyles, median denticle, or some sinus margins, or may show only as rim on which operculum rests.

DENTICLE Median tooth ("lyrula") in proximal aperture in some ascophorines; or paired teeth ("condyles, cardelles") on which operculum hinges; or small spinules located variously.

DIETELLA (PL., DIETELLAE) Minute chamber in vertical wall containing one or more pores though which funicular tissue passes between zooids.

DISTAL Toward the oral end of zooid; away from point of origin of zooid.

ECTOOECIUM Outermost calcified layer of cheilostomatid ovicell.

ENDOOECIAL OVICELL Ovicell immersed, internal, intruding beneath next distal zooid, in some cheilostomatids.

ENTOOECIUM Inner layer of ovicell wall; "endooecium" of some authors.

FASCICLE Bundle of tubules fused together; in some cyclostomatids.

FOLIACEOUS Foliate; leafy colonies, some resembling small cabbages.

FRONTAL Ventral surface, oral (apertural) side in cheilostomes, membranous or calcified frontal wall.

FRONTAL SHIELD Calcified ventral wall of the oral (apertural) side of some cheilostomatids; formed by fusion of spines ("spinocyst") over frontal membrane, or by extension of lateral wall ("cryptocyst") to lie beneath part or all of frontal membrane, or by internal calcification of body wall fold (extension of "gymnocyst") that extends over frontal membrane or by calcification of the frontal membrane itself.

GONOZOOECIUM Exoskeleton of individual modified as a brood chamber, in some cyclostomatids. See "gonozooid."

GONOZOOID Individual zooid modified as brood chamber in some cyclostomatids.

GYMNOCYST Calcified frontal membrane; in some cheilostomatid ascophorine genera.

HETEROZOOID Zooid specialized in function, may be lacking a polypide. See "avicularium," "vibraculum," "ovicell," "kenozooid".

HYPERSTOMIAL OVICELL (OOECIUM) Ovicell that develops distal to aperture, usually resting on or partially embedded in next distal zooid.

INTERCOSTAL PORES Pores (pseudopores) between ribs (costae) in Cribrimorphina. See also "intracostal pores," "lacunae".

INTRACOSTAL PORES Pores (pseudopores) that occur on the surface of costae in some Cribrimorphina.

KENOZOOID Specialized zooid lacking a polypide. Usually applied to spaces between autozooids when colonies adapt to substrate contours, or modified as anchoring devices (rhizoids, radicles, rootlets), sometimes applied to include avicularia, vibracula.

LABELLUM Liplike extension; downward extension of entooecium in some ascophorines.

LACUNA (PL., LACUNAE) Intercostal pore or space between costae; or suboral space between proximal apertural rim and first pair of costae, in some Cribrimorphina.

LOPHOPHORE Mound bearing ring of hollow, ciliated tentacles surrounding mouth, extruded from the cystid while autozooid is feeding, otherwise retracted inside body.

LUCIDUM (PL., LUCIDA) Lightly calcified area bordered by heavier calcification; the unfused space or slit between the "ectooecium" and "entooecium" in some species of cheilostomatids.

LUMEN (PL., LUMINA) Interior channel of a tubule or cavity in a structure.

LYRULA Calcified median proximal tooth or "denticle" in the proximal rim of apertures in some cheilostomatid ascophorines, especially in the family Smittinidae.

MULTISERIAL Multiple zooids growing in serial rows side by side attached at transverse and lateral walls.

MURAL RIM The medial margin of the calcified body wall, sometimes bearing spines, in cheilostomatid anascines.

NANOTUBULE Small zooid tubule with reduced polypide, attached to or separate from autozooid tubule in some cyclostomatids.

OCCLUSOR LAMINAE Calcified shelves beneath cryptocyst to which muscles that close the operculum are attached, in chaperiid cheilostomatid anascines.

OLOCYST Primary calcified layer of frontal wall (obsolete term still in literature).

OOECIUM (PL., OOECIA) Calcified structure for brooding of ova, larvae in cheilostomatids. See also "ovicell."

OOECIOSTOME Raised tubular extension or opening of gonozooid, in Cyclostomatida.

OPERCULUM Chitinized extension of frontal wall forming a usually separate flap closing aperture by contraction of occlusor muscles, in cheilostomatids, counterpart formed otherwise in a few cyclostomatids.

OPESIA (PL., OPESIAE) Uncalcified opening beneath frontal membrane, medial to cryptocyst, or to gymnocyst in absence of cryptocyst, in cheilostomatid anascines.

OPESIULE Small openings in cryptocyst through which muscles to the frontal wall pass to interior vertical wall or basal wall in species with the opesia reduced to the apertural area, in some anascines.

ORIFICE: Opening of mouth; or opening of "aperture" for extrusion of lophophore, of some authors.

OVICELL Heterozooid for brooding of ova or larvae resulting from sexual reproduction, budded from between zooids, sometimes with secondary contributions from adjacent frontal walls, in cheilostomatids (="brood chamber," "ooecium").

PERICYST Calcified frontal shield formed of fused spines above ectocyst, frontal membrane, in cribrimorphines; (obsolete term, still in literature).

PERISTOME Secondarily calcified collar or rim surrounding primary aperture (orifice), in ascophorine cheilostomatids.

PLEUROCYST Secondary calcified layer of frontal wall, in ascophorines (obsolete term still in literature).

POLYPIDE The lophophore, viscera and musculature (the soft body parts) in autozooids.

PORE CHAMBER Chamber in vertical wall containing pores though which funicular tissues connect contiguous zooids. See also "pore plate, dietella."

PORE PLATE Circular plate in vertical wall that contains one or more pores. See also "dietella, pore chamber."

PRIMARY APERTURE Initial opening for extrusion of tentacles in young zooid. See also "secondary aperture."

PROXIMAL Direction of origin of zooid or ancestrula, opposite end from aperture; toward central portion of zooid.

PSEUDOPORE Tissue-filled pore that penetrates all or some of frontal wall skeletal layer.

RADICLE Rootlet attaching colonies to substrate in some species.

ROSETTE PLATE Multiporous subcircular plate in vertical walls of Cheilostomatida for passage of funicular tissues between contiguous zooids.

SCUTUM Spine sometimes modified as paddle or multipronged projection, curved above frontal membrane, in some cheilostomatid anascines.

SECONDARY APERTURE Opening for extrusion of tentacles formed by growth of surrounding body wall until "primary aperture" may be hidden, usually differing in shape.

SINUS Slit, U- or V-shaped notch or curve on proximal margin of primary aperture; also used for notch in secondary aperture, in some cheilostomatid ascophorines.

SPIRAMEN (PL., SPIRAMINA) Median pore in proximal wall of peristome; also used for opening remaining when sides of peristomal collar merge, leaving a small opening; in some cheilostomatid ascophorines.

TERETE Tapered at both ends; sweet potato–shaped.

TREMOCYST Calcified frontal wall bearing more or less uniform array of pores (pseudopores), in cheilostomatid ascophorines; (obsolete term); see also "pericyst, pleurocyst".

UMBO (PL., UMBONES) Raised, calcified mound or knob on frontal wall, often proximal to aperture, or atop ovicell, in some cheilostomatid ascophorines.

UMBONULOID Mode of formation of frontal wall from an epifrontal fold of gymnocyst from proximal end of zooid that becomes calcified on interior side above frontal membrane; in some cheilostomatid ascophorines.

UNISERIAL Zooids attached together at transverse walls to form row of colony but not attached at lateral walls. See also "biserial, multiserial".

UVULATE PROCESS, UVULA Narrow structure extending proximally, in bryozoans, from distal margin of ascopore, in Microporellidae; entooecial projection extending over aperture, in some Phidolophoridae.

VIBRACULUM (PL., VIBRACULA) Avicularium mandible modified into a long, whiplike process capable of sweeping over colony, in some cheilostomatids.

ZOARIUM (PL., ZOARIA) Bryozoan colony budded asexually from single, sexually produced ancestrula (obsolete term).

ZOID ("ZO-ID") Individual autozooid in colony, based on Reichert's, 1870, "Bryozoid"; American and some European pronunciation. See "zooid".

ZOOECIUM (PL., ZOOECIA) Exoskeleton of single autozooid; sometimes used by some authors as equal to "zooid".

ZOOECIULE A miniature zooid having an aperture but usually lacking some autozooid parts.

ZOOID Individual autozooid in colony, based on Huxley's (1851) "animallike thing"; "zoo-id" preferred pronunciation of British authors, commonly accepted spelling regardless of pronunciation.

ACKNOWLEDGMENTS

Senior authors John D. Soule and Dorothy Soule passed away in 2001 and 2005; they were mentors to the coauthors and a great many other bryozoologists around the world. We thank Paul D. Taylor and Mary Spencer Jones for assistance with examination and SEM of selected type specimens. We also thank Alicia Thompson, who provided valuable assistance with SEM, and Richard Hayduk, who assisted with computer graphics and plate production. Susan Soule Harrison provided the line drawings in the previous edition of this manual, and those used herein. Thanks also to James Carlton for recent updated informaton on estuarine species.

Key to the Orders and Suborders of Bryozoa

1. Colonies soft bodied, gelatinous, and contiguous; or separate, linked by stolonate kenozooids; apertures terminal, closed by "purse-string" sphincter muscles or liplike flaps; no avicularia; ovicells, if present, transitory uncalcified sacs . . . . Key A: Order Ctenostomatida (=Ctenostomata)
— Colonies extremely varied; zooid external surfaces usually encased partly or entirely by lightly to heavily calcified exoskeleton (zooecium), often with pores, spines or other structures; with or without external calcified ovicells 2
2. Zooecia heavily calcified whitish tubules with terminal apertures, rarely with spines or opercula, no avicularia; tubules often fused partly or totally together; colonies branching, arborescent, or recumbent, encrusting, discoid, or lobate; gonozooids are single zooids or communal chambers serving multiple zooids, may contain multiple developing embryos . . . . Key B: Order Cyclostomatida (=Cyclostomata, =Tubuliporata)
— Zooecia rarely lightly calcified tubules, otherwise boxlike, apertures partly or fully subterminal at distal end of frontal (ventral) wall, aperture closed by operculum; often with spines, avicularia, or vibracula; calcified ovicells, if present, internal or external; colonies arborescent, foliaceous, flaccid, recumbent or encrusting, often varying with environment . . . . Order Cheilostomatida (=Cheilostomata) 3
3. A few tubular species with calcium imbedded in flexible body wall; most others with frontal wall uncalcified, membranous, or with a calcified cryptocyst beneath frontal membrane; no ascus beneath frontal membrane . . . . Key C: Suborder Anascina (=Anasca)
— Frontal wall a shield of fused costae or otherwise completely calcified, with an ascus beneath frontal wall, opening to exterior by ascopore in frontal wall or to interior below proximal lip of aperture 4
4. Frontal wall a shield formed by fusion of pinnate costae above frontal membrane . . . . Key D: Suborder Cribrimorphina (=Cribrimorpha, =Cribriomorpha)
— Originally described as cheilostomes having an ascus beneath a calcified frontal wall or shield; ascus and calcification now known to be formed variously, leading to further division of group into various new infraorders, superfamilies . . . . Key E: Suborder Ascophorina (=Ascophora)

Key A: Order Ctenostomatida

Key to genera of species described briefly herein.

1. Colonies gelatinous, fleshy, or leathery; zooids squat, contiguous, matlike, encrusting, covered with silt; apertures round, oval, or slitlike. 2
— Colonies soft bodied; autozooids tubular, not contiguous, separate at distal ends, connected by kenozooid stolons, erect or recumbent; apertures mostly liplike 3
2. Interzooecial kenozooids forming thicket of branching chitinous spines; aperture slitlike. . . . . Flustrellidra
— Zooids smooth or papillate, no spines; aperture puckered, closed by sphincter . . . . Alcyonidium
3. Zooids tubular, arising directly from stolon. 4
— Zooids attached to stolon by thin stalk (pedicel, =peduncle) 5
4. Zooids slim with stolonate basal area, secondary zooids may arise from primary zooid at base or distally. . . . .
. . . . Victorella

— Zooids without expanded basal area, arising in clusters from stolons, zooids often with spurlike caudal extension . . . . Bowerbankia
5. Zooid pedicels attached to stolons by jointed bases; distal (oral) end of zooids squared or irregularly rounded . . . . Triticella
— Zooid pedicels attached at bases in tangled clusters without jointed bases; distal end of zooids bilobate, liplike . . . . Farella

Key B: Order Cyclostomatida

Key to genera of species described briefly herein.

1. Colonies encrusting, or discoid, or recumbent (adnate), uniserial or multiserial, or spreading, fan shaped 2
— Colonies erect or partially erect 6
2. Colonies recumbent, not discoid or fan shaped; uniserial except at gonozooid or branches, or irregularly biserial, sometimes with raised branches; gonozooid large, with large pores, erect ooeciostome 3
— Colonies discoid or fan shaped 4
3. Zooid tubules small, with small pores, short raised ends sometimes curved outward . . . . Stomatopora
— Zooid tubules large, thick, with many coarse pores . . . . Proboscina
4. Colonies discoid, zooid tubules erect in radiate, connate rows or clusters of rows; brood chambers beneath central area or between rows; new colonies budding laterally or upward; now includes living species formerly placed in fossil genus Lichenopora . . . . Disporella
— Colonies fan shaped, sometimes lobate with zooid tubules raised, erect, single, or in fascicles 5
5. Zooid tubules not immersed, erect, single, or in fascicles; no small erect nanotubules; gonozooid a single large inflated tubule with flared ooeciostome . . . . Tubulipora
— Zooid tubules erect, separate but immersed for most of length beneath common crustlike surface; smaller separate or connate nanozooid tubules; gonozooid spreading between tubules, small, short ooeciostome . . . . . . Diplosolen
6. Colonies various, branching, with joints between groups of zooid tubules (internodes) 7
— Colonies large, stiff, without joints, tubules fasciculate, eastern Pacific species formerly placed in Diaperooecia. . . . .
. . . . Diaperoforma
7. Colonies erect, feathery, attached by rootlets, one zooid per beginning internode (between joints), two to three per sterile internode, more with gonozooid; gonozooid an expanded zooid with terminal ooeciostome dorsal; long, whiplike spines attached beside aperture on outer margin . . . . Bicrisia
— Colonies, gonozooid otherwise; no whiplike spines beside aperture 8
8. Colonies frail; one to three zooids in sterile internodes, three to five in internodes with gonozooid; joints black in eastern Pacific species; gonozooid lying against next distal autozooid, short ooeciostome toward frontal . . . . Filicrisia
— Colonies otherwise; internodes three to 40 autozooids; joints yellowish to brown, gonozooid otherwise. 9
9. Zooids in alternating series; joints pale at first, becoming brown with age, not attached to substrate by radicles; gonozooid adnate to multiple autozooid tubules, expanded dorsoventrally. . . . . Crisia

— Zooids randomly arranged, partly immersed, long internodes, yellowish joints; attached to substrate by jointed radicles; gonozooid variable, expanded laterally . . . . *Crisulipora*

Key C: Order Cheilostomatida, Suborder Anascina

Key to the genera of species described briefly herein.

1. Colonies fuzzy masses of uniserial zooids; zooids erect annulated or striated stalks, some with spinules, tubercles, polypide can be withdrawn into stoloniform base; opesia beneath frontal membrane near distal end, aperture subterminal but may extend over distal end; ovicell a temporary membranous sac hung outside apertural area to brood ova; no frontal budding; no spines no avicularia or vibracula. . . . . *Aetea*
— Colonies otherwise; opesia occupying some or most of frontal wall area; ovicells, if present, calcified, not a membranous external sac; spines, avicularia, or vibracula present or absent 2
2. Colonies encrusting, firmly attached to substrate, zooids usually contiguous, sometimes rising in bilaminar folds, frills or multilaminar stems 3
— Colonies loosely attached by tubular kenozooids, or erect, branching, attached by rootlets (radicles) 15
3. Zooids mostly rectangular, cryptocyst and/or gymnocyst narrow, opesia large, covered by uncalcified frontal membrane, a knob or spine at each distal corner; ancestrula twinned . . . . *Membranipora*
— Zooids less rectangular, opesia more ovoid, cryptocyst and/or gymnocyst more developed; ancestrula single. 4
4. Calcified ovicells absent, or not known to occur; opesia extensive 5
— Calcified ovicells present, internal (endozooecial) or external (hyperstomial) located distal to aperture 7
5. Cryptocyst somewhat more developed than gymnocyst; thin spines arising between cryptocyst and gymnocyst; triangular pair of kenozooids in some species at distal corners . . . . *Conopeum*
— Gymnocyst more extensive than cryptocyst, cryptocyst narrow, crenulate. 6
6. Gymnocyst wider proximally, with single proximal conical base topped by chitinous spine; no avicularia . . . . *Electra*
— Gymnocyst more uniform, many large spines between gymnocyst and cryptocyst; some species with thick spines leaning over opesia, spines distal to aperture; avicularia erect, stalked . . . . *Cauloramphus*
7. Ovicells small, mostly endozooecial but sometimes visible as shallow hood. 8
— Ovicells larger, hyperstomial 9
8. Cryptocyst narrow or wider, granular, gymnocyst narrow, some species with thin spines leaning over opesia; large interzooecial avicularia; ovicell a small hood with rounded distal end internal, sometimes projecting into next distal zooid. . . . . *Hincksina*
— Cryptocyst wide, narrowing at distal end, descending to large opesia; some with tiny spinules; ovicell inconspicuous, shallow with distal marginal ridge, granular proximal face, slitlike opening . . . . *Alderina*
9. Cryptocyst, gymnocyst narrow, opesia large 10
— Cryptocyst wider, or forming frontal wall with opesia limited to apertural area, paired opesiules for passage of muscles controlling frontal membrane. 13

10. Cryptocyst not rolled, spines between cryptocyst and gymnocyst around opesia or near aperture; large avicularium interzooecial between transverse walls moving onto ovicell; smaller interzooecial avicularia between lateral walls in some species; ovicell imperforate, but slit between ectooecium, entooecium . . . . *Tegella*
— Cryptocyst heavily rolled, surrounded or flanked by spines or not; interzooecial avicularia present; ovicells perforate or not. 11
11. Thick spines surrounding opesia; small marginal avicularia; ovicell wide, granular with slit between granular ectooecium, thinner entooecium . . . . *Callopora*
— Spines not surrounding opesia; ovicell porous or not. 12
12. Thick spines around distal half of opesia only, medial pair developed into branches; large, lanceolate interzooecial avicularia; ovicell a shallow imperforate hood . . . . ?*Copidozoum*
— No spines around rolled cryptocyst; interzooecial avicularia long, slender, often beside ovicell; ovicell quadrate, with many pores. . . . . *Copidozoum*
13. Cryptocyst saucerlike, granular, opesia large; paired calcified *occlusor laminae* below cryptocyst; a median distal triangular avicularium. . . . . *Chaperiopsis*
— Cryptocyst perforate, forming frontal wall, opesia limited to apertural area; slitlike or oval lateral opesiules for muscle passage flanking distal portion of cryptocyst at lateral walls. 14
14. Colonies encrusting, not becoming raised; lateral opesiules small, slitlike; frontal wall pores denticulate, no body cavity spicules; interzooecial avicularium distal to aperture directed laterally; ovicell perforate, sometimes an umbo on top . . . . *Micropora*
— Colonies encrusting but sometimes rising into jointed branches or bilaminar leaves; opesiules large; cryptocyst perforate, pores not denticulate; body cavity with minute spicules shaped like compasses (dividers) and/or calipers (tongs); ovicell imperforate, like two halves fused, sometimes keeled. . . . . *Thalamoporella*
15. Colonies ramifying, loosely attached by tubular rhizoids, forming recumbent fronds; frontal membrane extensive; few or many spines extending over membrane; no avicularia; ovicells large, globular . . . . *Dendrobeania*
— Colonies erect, sometimes branching, attached by nontubular radicles 16
16. Colonies with erect, club-shaped internodes, brown or black joints; cryptocyst extensive; no spines; apertures arched distally and proximally with strong proximal condyles, sometimes distal condyles; interzooecial avicularia large; ovicells endozooecial, opening by separate pore distal to aperture . . . . *Cellaria*
— Colonies bushy, turbinate or fan-shaped; not composed of club-shaped internodes; gymnocyst extensive, cryptocyst much reduced or absent; avicularia present, variously sized, or absent; ovicells not endozooecial 17
17. Colonies stalked, jointed with fanlike or palmate tips . . . . *Caulibugula*
— Colonies erect, biserial or multiserial; not palmate; slender, or tufted, bushy, or turbinate; zooids in internodes 18
18. Colonies biserial or multiserial; no scutum; avicularia, if present, pedunculate, resembling birds' heads; ovicells may be mere curved distal corners, shallow hoods, or imperforate sculptured globes, poorly or heavily calcified . . . . . . *Bugula*
— Colonies multiserial; scutum present; avicularia not pe-

dunculate; ovicells flattened or rounded with entooecial frontal area uncalcified or thinly calcified 19

19. Colonies stiff, branching; zooids alternating, with barbed dorsal vibracula curving over ventral face in tangle; long radicles originating in vibracular chambers, grouped in cords extending down to anchor colonies; ovicells flattened, imperforate with uncalcified frontal entooecium . . . .
. . . . *Caberia*

— Colonies various; radicles, if present, originating in dorsal, dorsolateral, or on ventral (frontal) wall above avicularium chambers, radicles not grouped in dorsal cords; ovicells otherwise 20

20. Colonies erect, tufted, or flaccid branches; zooids tapered, wider at distal ends, opesia in distal half of zooid, thick spines around opesia, one modified to bend over opesia as a blunt scutum; radicles absent or originating on frontal wall above avicularia; ovicells large, with scattered pores or imperforate, striate . . . . *Tricellaria*

— Colonies tufted, stiff; zooids alternating, biserial; ovoid with large opesia, scutum a single spine or branched, extending over opesia, other spines distal to aperture; small triangular to giant lateral avicularia with curved, acute mandibles; dorsal vibracula with radicle attached at base; ovicells imperforate, flattened or globular with frontal entooecium calcified or not . . . . *Scrupocellaria*

Key D: Order Cheilostomatida, Suborder Cribrimorphina

Key to the genera and subgenera of species described briefly herein.

1. Zooids small, <600 μm in length; distal apertural rim curved, five distal apertural spines in eastern Pacific species, proximal rim straight, both rims originating from distal gymnocyst; modified costae below proximal rim leaving small lacuna below aperture; ovicells with or without pores . . . . *Puellina* (subgenus *Cribrilaria*)

— Zooids large, >600 μm in length; distal apertural rim formed from distal gymnocyst, no apertural spines; proximal rim formed by fusion of modified frontal costae no suboral lacunae; ovicell with large costal pores 2

2. No spines or avicularia known; frontal wall a shield with five to seven large flat costae, gymnocyst extensive outside costae, one large pyriform pore near center of each costa, one to two intercostal pores; ovicell formed by pair of modified costae, with one large intracostal pore on each side of median suture . . . . *Reginella* (subgenus *Figularia*)

— Frontal wall shield with smaller costae, gymnocyst reduced, one to two smaller costal pores near ends of costae; four or more intercostal pores; avicularia present or absent
. . . . 3

3. Interzooecial avicularium almost as large as autozooid, with spatulate mandible; four to six pairs of costae with large intracostal pore at outer end, three to four dumbbell-shaped intercostal pores; first pair of costae extending distally at outer ends of proximal apertural rim; ovicell a pair of modified costae, one pair of large intracostal pores, one pair of smaller intercostal pores, indistinct median suture
. . . . *Reginella* (subgenus *Jullienula*)

— No avicularia; aperture bell-shaped, proximal rim formed of modified pair of costae turned distally at outer ends, raised at sides of aperture; five to eight pairs of frontal wall costae, transverse near aperture, radiating proximally from center of shield, one to four intracostal pores, many inter-

costal pores; ovicells large with few or many pores, often obscured . . . . *Reginella* (subgenus *Reginella*)

Key E: Order Cheilostomatida, Suborder Ascophorina

Key to the genera of species described briefly herein.

1. An external ascopore in calcified frontal wall 2

— Ascopore opening inside proximal aperture, not in external calcified frontal wall 4

2. Ascopore not separated from aperture by frontal wall pores; D-shaped aperture arched distally, no spines, straight proximally; frontal wall with pores, smooth walled or denticulate or cribrate; distal apertural spines; avicularia usually present; *Microporella*, spp. with imperforate ovicell not found in eastern Pacific; species with perforate ovicell now placed in . . . . *Microporelloides*

— Ascopore separated from proximal apertural rim by large frontal pores; avicularia very rare; ovicells present or absent
. . . . 3

3. Zooids with distinct gymnocystal rim around lateral and proximal margins; row of frontal pores interior to gymnocyst; five to seven long distal apertural spines; one to two rows of pores between proximal apertural margin and denticulate ascopore, a few scattered frontal pores with denticles at surface; no avicularia; ovicells not known in California species . . . . *Fenestrulina*

— Zooids with little or no gymnocyst showing except around ovicell; frontal wall becoming completely covered with pores except for proximal imperforate area between aperture and umbo, or raised ascopore in some species; frontal pores with few or many spicules at surface; avicularia rare; ovicell imperforate with large areolae . . . . . . *Fenestruloides*

4. Aperture bell shaped, without oral spines, frontal pores around distal and lateral apertural margins; suboral sulcus, sometimes with small avicularium; frontal wall reticulate with large pores; no external ovicells . . . . *Cryptosula*

— Aperture otherwise; frontal wall various; avicularia present or absent; ovicells present 5

5. Three kinds of zooids: large autozooids, smaller female zooids with large ovicells, smaller male zooids 6

— Not with three kinds of zooids 7

6. Frontal walls imperforate with transverse striations, large flared fenestrae on margins for frontal budding, smaller marginal areolae; aperture with U- or V-shaped sinus, small condyles ovicells large with large pores, or imperforate; no avicularia (formerly included in *Hippothoa*) . . . .
. . . . *Celleporella*

— Frontal walls with numerous large pores; aperture with wide, shallow proximal sinus flanked by strong condyles; small interzooecial avicularia with spatulate mandible directed distally; ovicell large with many small pores . . . . .
. . . . *Trypostega*

7. Aperture arched distally, with or without raised peristome, proximally curved or a wide sinus with large pore or groove within peristome or not, some bearing small rounded avicularium; ovicells imperforate, ribbed or nodular . . . .
. . . . *Hippoporina*

— Aperture, ovicells mostly otherwise 8

8. Zooids large, subhexagonal, frontal walls imperforate, no areolae; aperture large, subtriangular with large notch at lateral ends of proximal rim for opercular muscles; ovicell a small, shallow subtriangular hood with one large pore . . . .
. . . . *Eurystomella*

— Frontal wall, zooid size and shape, aperture otherwise.....
....9
9. Primary distal apertural rim not beaded 10
— Primary distal apertural rim beaded.................29
10. Primary proximal apertural rim with median denticle (lyrula) or wide dental ledge with median notch; avicularia present..17
— Aperture without lyrula or dental ledge with median notch; with or without avicularia 11
11. Aperture elliptical, wider than high, or D-shaped with straight or slightly curved proximal lip, no avicularia.....
....12
— Aperture otherwise; avicularia present or absent......13
12. Aperture rounded distally, straight or a shallow curve proximally, condyles present or not; frontal wall with pores, reticulate with larger areolae; ovicell with single large pore, sometimes other smaller irregular pores, larger areolae
......................................'Dakaria'
— Proximal aperture a wide curve or widely V-shaped sinus, very large condyles; ovicell with frontal entooecium a subtriangular array of pores surrounded laterally and distally with imperforate ectooecium ....Neodakaria
13. Primary aperture rounded without a distinct sinus....26
— Primary aperture various, not D-shaped; with distinct proximal sinus 14

    Note also the bright-red, often foliaceous Watersipora, an abundant fouling organism; see species list.

14. Frontal wall almost or completely filled with pores 15
— Frontal wall imperforate or with few small pores, or developing from reticulate to imperforate; with marginal areolar pores....................................24
15. Aperture with V- or U-shaped proximal sinus, large or small condyles, not pyriform; lateral avicularia single, paired or absent....................................16
— Aperture pyriform, a median suboral avicularium as part of sinus dental ledge or proximal on frontal wall, sometimes on umbo; some species with other paired lateral avicularia
....Schizomavella
16. Aperture with widely U-shaped sinus; ovicells of most species with many pores, some ridged; others with central imperforate area or perforate but immersed in next distal zooid may belong in other genera ....Schizoporella
— Aperture with narrowly U-shaped sinus, operculum stem fitted into sinus; no spines, no avicularia; ovicell entirely imperforate ....Arthropoma
17. Frontal wall mostly or completely perforate 18
— Frontal wall mostly imperforate with areolae, or reticulate in development, becoming imperforate with areolae 21
18. Ovicells raised, central entooecium with many pores, rimmed distally or surrounded by imperforate ectooecium
....19
— Ovicell composed of one to three flaps, with one or few pores, becoming immersed in next distal zooid 20
19. Aperture with dental ledge containing U-shaped sinus, wide flat condyles above dental ledge flanking sinus, median suboral avicularium on pedestal; ovicell raised, with large irregular central pores, imperforate ectooecium rim
....Schizosmittina
— Primary aperture with median lyrula, secondary aperture a raised peristome, suboral avicularium recumbent on base of lyrula or proximal to it; other avicularia present or absent; ovicell with large or small pores.........Smittina
20. Ovicell with one large distal flap of frontal wall having one large pore, meeting rolled lateral margins of peristome; frontal wall pores without granules or spinules inside rims; small oval avicularium on small lyrula inside peristome
....Dengordonia
— Ovicell formed of one distal, two lateral frontal wall flaps from adjacent zooids, with one or a few small pores; frontal wall pores with granules or spinules inside rims, lyrula flanked by condyles, avicularium on lyrula base or proximal to it ....Raymondcia
21. Frontal wall imperforate in center but with two to three rows of marginal pores merging, forming tubules between primary and secondary frontal wall layers moving pores up frontal wall, sometimes other small scattered pores; peristome rounded, thickened, or erect with spines, spine bases fused into raised collar, no proximal sinus; a suboral umbo, or a mucro extending down inside collar forming column leading to deep lyrula; no avicularia; ovicell on distal peristome or recumbent, imperforate or with few scattered pores.............................Haywardipora
— Frontal wall imperforate with areolae, no frontal tubules extending upward; primary aperture with lyrula or dental ledge, no mucro with column extended downward to lyrula; avicularia present.......................22
22. Ovicell an imperforate hood......................28
— Ovicell with entooecial pores, bordered distally by ectooecium or not 23
23. Primary aperture with proximal sinus, lyrula, paired lateral condyles; secondary aperture forming raised peristome or not; median suboral avicularium within sinus on lyrula or proximal to base peristome ....Smittoidea
— Avicularia never median suboral, paired or single, lateral oral, or lateral frontal, other avicularia present or absent, sometimes large, interzooecial; ovicell with porous entooecium, rimmed by imperforate ectooecium or not........
.............................Parasmittina
24. Frontal wall imperforate except a few central pores, larger areolae; aperture with curved sinus; ovicell immersed, with single median pore ....Stomachetosella
— Frontal wall mostly imperforate with few or many areolae, reticulate or not during development; aperture with V-shaped or curved sinus; ovicell with lunate frontal entooecium with or without costae, ectooecium imperforate on distal, lateral margins 25
25. Colony coarse, heaped; frontal wall reticulate in development or frontal budding, becoming imperforate with areolae; ovicell raised, entooecium forming radiate costae bordered by imperforate ectooecium; raised paired lateral suboral avicularia, some species with median suboral avicularium, large oval interzooecial avicularia .....Celleporina
— Colony vitreous, encrusting, imperforate with few areolae; primary aperture with wide sinus, strong condyles; ovicell with imperforate horseshoe-shaped entooecium ringed by tiny marginal slits, imperforate ectooecium; avicularia small, acute, on frontal wall, or larger, interzooecial.....
.............................Buffonellaria
26. Frontal wall mostly imperforate with one or more rows of areolae; suboral, acute avicularium, short or long, median or skewed to side, but originating at proximal apertural pore; ovicell with many pores, rimmed by imperforate ectooecium......................Pleurocodonellina
— Frontal wall mostly reticulate or perforate; avicularia, ovicells various...................................27
27. Colonies encrusting, becoming erect, sturdy; zooids flask shaped; frontal walls reticulate or perforate; primary aper-

ture round or oval, secondary aperture a solid tubular peristome or tall peristome like fused spines, some species with added spines or projections; avicularia small, paired, lateral oral, raised on peristomal rim, sometimes a suboral shelf or lip; ovicells suspended from distal peristome, with reticulate crescent of entooecium, imperforate rim of ectooecium; eastern Pacific species formerly placed in *Lagenipora* which has imperforate frontal walls . . . . *Lagenicella*

— Colony encrusting, fragile; zooids elongate, rectangular; frontal wall porous; aperture with strong condyles, low peristome; avicularium not on raised peristome; avicularia small, median suboral, acute, or replaced by giant spatulate avicularium; ovicell not suspended from peristome, mostly porous with thin rim of imperforate ectooecium. . . . . *Codonellina*

28. Colony heaped, coarse, may form irregularly erect cylinders; primary aperture symmetrical, dental ledge with condyles on top almost meeting to form median notch in California species, spines present or not; secondary aperture asymmetrical due to suboral avicularium, may form umbo; giant brown spatulate interzooecial avicularia present; ovicell a shallow imperforate hood. . . . . . .
. . . . *Celleporaria*

— Primary aperture with wide lyrula, with median avicularium suboral or on lyrula; distal spine bases in some species, secondary aperture symmetrical; colonies thin, flat; no interzooecial avicularia; ovicells, if present, with large imperforate hood in California species . . . . *Porella*

29. Colony erect, fenestrate, all zooids opening on ventral side, frontal wall with a few small pores, few areolae; dorsal side with kenozooid sutural lines; primary aperture symmetrical with small median sinus, sometimes with distal spines; secondary aperture symmetrical with high peristome having proximal notch or closed spiramen; scattered hooked acute frontal wall avicularia; large hooked avicularia on dorsal side at base of fenestrae; ovicell with imperforate entooecial area, may form median labellum extending downward, distal ectooecium imperforate . . . . *Phidolopora*

— Colony encrusting, zooids irregular, heaped; frontal wall imperforate with large areolae; primary aperture symmetrical, sometimes with two distal spines, wide proximal sinus flanked by strong condyles; small, acute, transverse suboral avicularium not median, making secondary aperture asymmetrical, apertures becoming immersed; many acute frontal avicularia; ovicell with circular imperforate entooecium, almost surrounded by wide imperforate ectooecium, narrowing at aperture . . . . *Rhynchozoon*

List of Species

Gymnolaemata

CTENOSTOMATIDA

The taxonomy of northeastern Pacific Ctenostomatida is in great need of revision since the only extensive work was by J. D. Soule in 1953. New studies on freshly collected material require whole mounts and histological sections, while electrophoresis or molecular genetics would serve to determine whether the organisms are introduced Atlantic/Mediterranean species or are distinct Pacific species, as has seemed the case in the species that have been reexamined. The species names given in the earlier eastern Pacific literature were based prima-

rily on British and European literature, where similarities rather than differences were emphasized and optics were limited. Hayward 1985 and Hayward and Ryland 1985, 1998, 1999, have clarified the northern Atlantic and many Mediterranean species, assisting greatly in comparisons. Ryland (personal communication 2005) has added the following comments on the *Alcyonidium* of the region.

Alcyonidium. Ryland and Porter (2006) have demonstrated that the reproductive mode may represent a potentially valuable tool in distinguishing between species of *Alcyonidium*. Most Pacific coast species appear to resemble *A. mytili*, a European species, as they are oviparous and develop planktonically. During some periods of the year the colonies never develop conspicuous embryos and individual members possess a minute, ciliated funnel that is located between the most dorsal tentacles. Larviparous species, such as the European species *A. gelatinosum* and *A. polyoum*, can possess clusters of developing oocytes and embryos. This latter mode of development appears to be uncommon in Pacific coast species and it is unlikely that any of the European species occur on the west coast of North America naturally.

California and Oregon representatives are placed in *A.* cf. *parasiticum* and probably represent at least five species that incrust rocks, shells, kelp holdfasts, various algal species and crab carapaces (Ryland, personal communication 2006). One species is found on the gastropod *Ilyanassa* in San Francisco Bay. The species probably include one larviparous species with pale peach embryo clusters (European *A.* cf. *gelatinosum*) and at least four others that appear to have planktotrophic development (European *A.* cf. *mytili*). Common traits shared by all of the species are the hexagonal, brown to gray gelatinous zooids that are generally flat, at least near the growing margins (Plate 439B), but sometimes upright. Orifice is central and located towards the middle of the colony. Peristomes circling the orifice may or may not be raised. The colony surface can vary from smooth to mammillate to raised in lobes.

Publications since 1996 have greatly improved understanding of the European species *Alcyonidium gelatinosum* (Linnaeus 1761), *A. mytili* Dalyell 1847, and *A. polyoum* (Hassall 1841) (summarized by Ryland and Porter 2006). In particular, they have demonstrated the importance of reproductive mode in the biology of *Alcyonidium*. Most Pacific coast species seem to resemble *A. mytili* in being oviparous, with planktotrophic development. At some (but not all) times of the year, they will have a minute ciliated funnel (intertentacular organ) between the dorsalmost tentacles, and the colony will never contain conspicuous developing embryos. Larviparous species will, at certain times of the year, contain clusters of developing oocytes and embryos, as in *A. gelatinosum* and *A. polyoum*. This mode of development seems much the less common in Pacific coast *Alcyonidium*. It is most unlikely that any of the European species occur here naturally, despite the use of their names in the past.

In addition to the distinctive *A.* cf. *parasiticum*, at least five encrusting species, and probably more, are present in California and Oregon; all await description (J. S. Ryland, personal communication, 2006). They include one larviparous species with clusters of pale peach embryos (cf. European *A. gelatinosum*), and four others that probably or certainly have planktotrophic development (cf. European *A. mytili*). All of these have roughly hexagonal, brown to gray gelatinous zooids; generally flat (at least near the growing margins: plate 439B) but sometimes upright, with the orifice central, toward the middle of the colony. The colony surface may be essentially level, or mammillate or raised in lobes; and peristomes around the

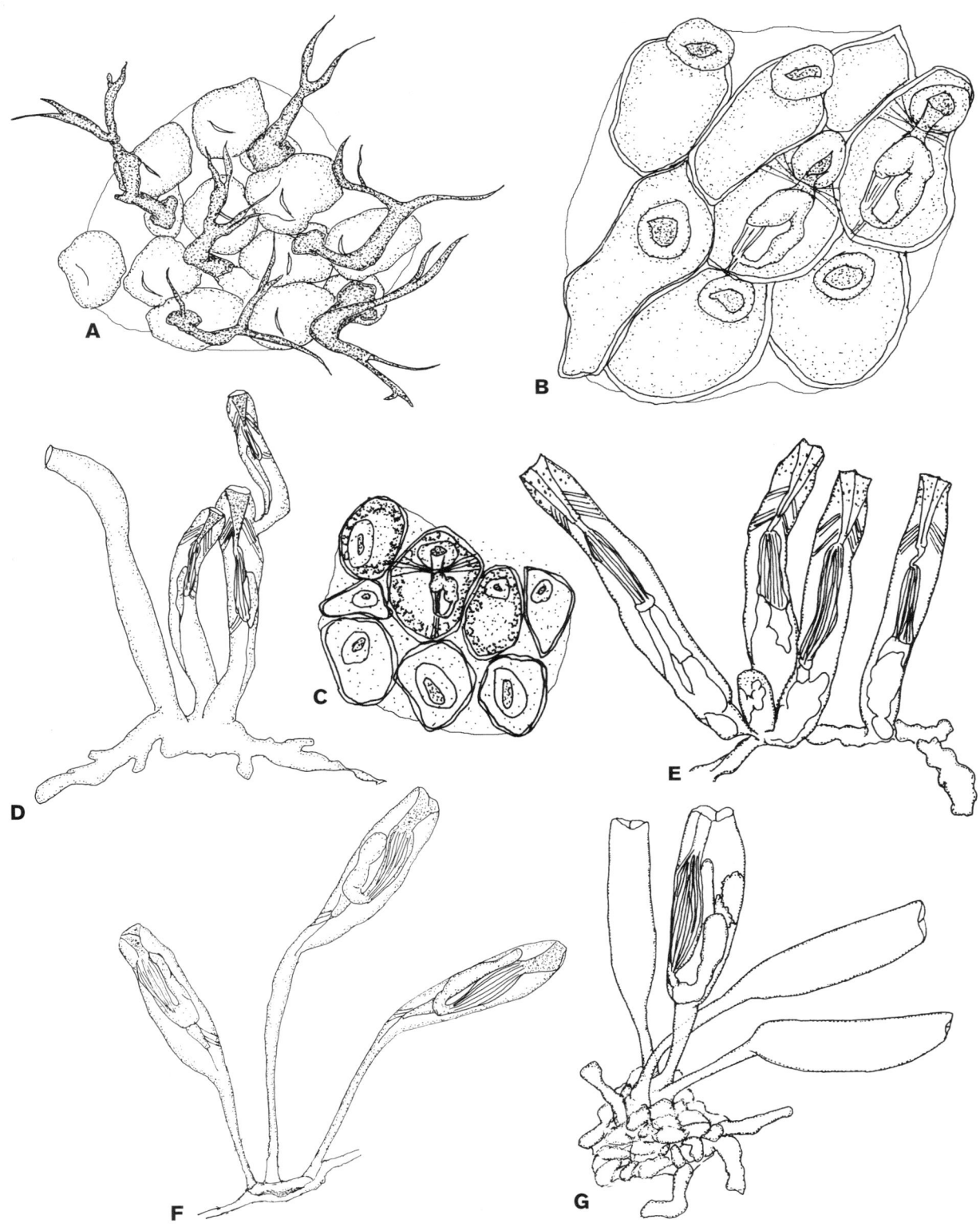

PLATE 439 Ctenostomatida. Carnosina. A, *Flustrellidra spinifera*; B, *Alcyonidium* cf. *gelatinosum*; C, *Alcyonidium* cf. *parasiticum*; Stoloniferina: D, *Victorella* ?*pavida*; E, *Bowerbankia* ?*gracilis*; F, *Triticella* ?*elongata*; G, *Farella* ?*elongata*.

orifice may or may not be raised. Northeast Pacific species occur on rock, shells, kelp holdfasts, other algae, and crab carapaces from the intertidal to about 80 m. A species of *Alcyonidium* is common at times on the shells of the mudsnail *Ilyanassa* in San Francisco Bay.

Alcyonidium cf. *parasiticum* (Fleming, 1828). See Soule 1953. Plate 439C. Eastern Pacific form probably an undescribed species; colonies flat with gelatinous, papillate surface, covered with sand, silt; zooids small with papillate border; British *A. parasiticum* zooids have many short frontal filaments; thickly encrusts erect hydroids, bryozoans; northern Pacific to central California; Tomales Bay colonies on shell; shallow water to 10 m.

*Anguinella palmata van Beneden, 1845. Reported by Cohen and Carlton (1995) on floating docks in San Francisco Bay.

Bowerbankia ?*gracilis* (Leidy, 1855). Plate 439E. See Soule 1953, Soule et al. 1980. May be cosmopolitan species or complex of undescribed species in eastern Pacific; see Hayward 1985 for characters of Atlantic species; northeast Pacific specimens form brown tangled masses of tubular zooids with puckered terminal apertures on stolonate stems; Puget Sound, Monterey Bay to Gulf of California; abundant fouling organism on rocks, harbor pilings; intertidal.

Farella elongata (van Beneden, 1845) (?=*F. tegeticula* C. H. and E. O'Donoghue, 1923). Plate 439G. See Soule 1953. Eastern Pacific form(s) may be one or more undescribed species; stolonate colony with tubular zooids having pedunculate base, bilabiate terminal aperture; arising in tangled clusters within internodes; a cool-water species from Britain, Adriatic Sea; in Pacific, from Coos County, Oregon, and Tomales Bay; intertidal.

Flustrella corniculata (Smitt, 1871): Soule 1953. Cold-water European species; see *Flustrellidra spinifera*.

Flustrellidra spinifera (C. H. and E. O'Donoghue, 1923). Plate 439A. Colonies cylindrical to foliaceous or flattened, tan to dark brown; large chitinous spines with one to six irregular prongs arising from kenozoids between autozooids; zooids ovate, aperture slitlike, sometimes on raised papillae. Examination of O'Donoghues' types show northeast Pacific specimens have larger zooids and stouter, less branched spines than *F. corniculata*; see Cook 1964; Alaska to Morro Bay; on algal stipes, especially at bases of *Laminaria sinclairii*; intertidal to about 70 m.

Triticella ?*elongata* (Osburn, 1912). Plate 439F. See Soule 1953, Soule et al. 1980. Tubular zooids with long pedicels, jointed bases arising from side branches off main stolons; may be the Atlantic species ranging from Massachusetts to North Carolina; collected from gill chambers of pea crab *Scleroplax granulata*; at Elkhorn Slough on various crabs; perhaps introduced with attempts to rear eastern oysters, or may be undescribed species; intertidal to unknown depth of crab substrate.

Victorella ?*pavida* Saville Kent, 1870. Plate 439D. See Hayward 1985. Colonies form chains or clumps of tubular zooids; new zooids bud from bases of autozooids or from peristomal area; originally described from London docks; reported from northern Atlantic, Mediterranean, Black and Baltic Seas, India, and Japan, if all are same species. Common in some areas of San Francisco Bay, including Lake Merritt (Oakland); also reported from Salton Sea where it was supposedly introduced on small boat hulls transported from the western Atlantic; brackish waters in marinas, harbors; on wood, stones, shells, barnacles, water plants, hydroids, other bryozoans; intertidal.

*Zoobotryon cf. *verticilliatum* (Delle Chiaje, 1828). A problematic species complex. Forms are abundant in southern California harbors, where massive colonies (resembling large clumps of transparent spaghetti) may reach several meters in size. Occasionally occurs in San Francisco Bay and would likely be found in other northern locations during periods when coastal waters are warmer.

CYCLOSTOMATIDA

Research on eastern Pacific Cyclostomatida species has not been extensively performed since the work of Osburn (1953). Soule et al. (1995) updated some information, illustrating 11 species with SEM, three of which were new. Hayward and Ryland (1985) detailed the northeast Atlantic fauna, making better comparisons of European and Pacific fauna. Much more needs to be done to recharacterize the eastern Pacific fauna.

Bicrisia edwardsiana (d'Orbigny, 1839): Osburn 1953. Sterile internodes one to two, zooids, fertile internodes three to five zooids. See *Bicrisia robertsonae*.

Bicrisia robertsonae Soule, Soule and Chaney, 1995. Plate 440D. Colonies erect, feathery, small, jointed, branched; zooids one to two per internode, widened distally, new zooid growing from beside aperture on either side, or a curved, jointed spine or whip outside at that site; gonozooid a simple expanded individual with raised ooeciostome, terminal ooeciostome dorsal on top. See Soule et al. 1995 for discussion; Alaska to San Diego, ? Peru; intertidal in bioturf, on rock, red algae and *Laminaria* holdfasts; to >120 m.

Crisia maxima Robertson, 1910: Soule et al. 1995. Plate 440F. Colonies erect, coarse, shrub-like; zooids in double rows, fused, with short, raised peristomes, 12–20 zooids in internode but sometimes up to 40, internode joints brown with age; gonozooid large; lying against multiple zooids, ooeciostome short, straight, forward of adjacent autozooid peristome; British Columbia to Coronados Islands, Baja California, and ? Galapagos; intertidal to 126 m.

Crisia occidentalis Trask, 1857: Soule et al. 1995. Plate 440G. Colonies erect, delicate, five to 12 zooids to internode, internode joints whitish to yellow; biserial zooids fused almost to tip, tips pointed on outer ends; gonozooid an inflated individual with short ooeciostome beside autozooid; Puget Sound and south; intertidal to 74 m.

Crisia serrulata Osburn, 1953 (=*Crisia serrata* Gabb and Horn, 1862, preoccupied by d'Orbigny 1853). See Soule et al. 1995. Colony erect, bushy, stiff, with yellowish joints; zooids biserial, alternating, immersed to tips, apertures turned outward with tip extended giving serrate appearance; gonozooid large with flat distal end, ooeciostome a short adnate tube below adjacent zooid aperture; Pleistocene, southern California; Recent, British Columbia to southern California, ? Gulf of California, ? Galapagos Islands; intertidal to 135 m.

Crisulipora occidentalis Robertson, 1910. See Soule et al. 1980, 1995. Large, stiff, tangled masses to 30 mm high, attached to substrate by jointed radicles; three to five zooids near base, 40 or more in longer internodes, yellowish joints; zooid tubes long, not connate, with circular apertures; gonozooid a simple inflation on internode surface, or larger, between tubes, ooeciostomes on same colony may be straight tube fused to adjacent autozooid with round or hooded aperture, or short,

* = Not in key.

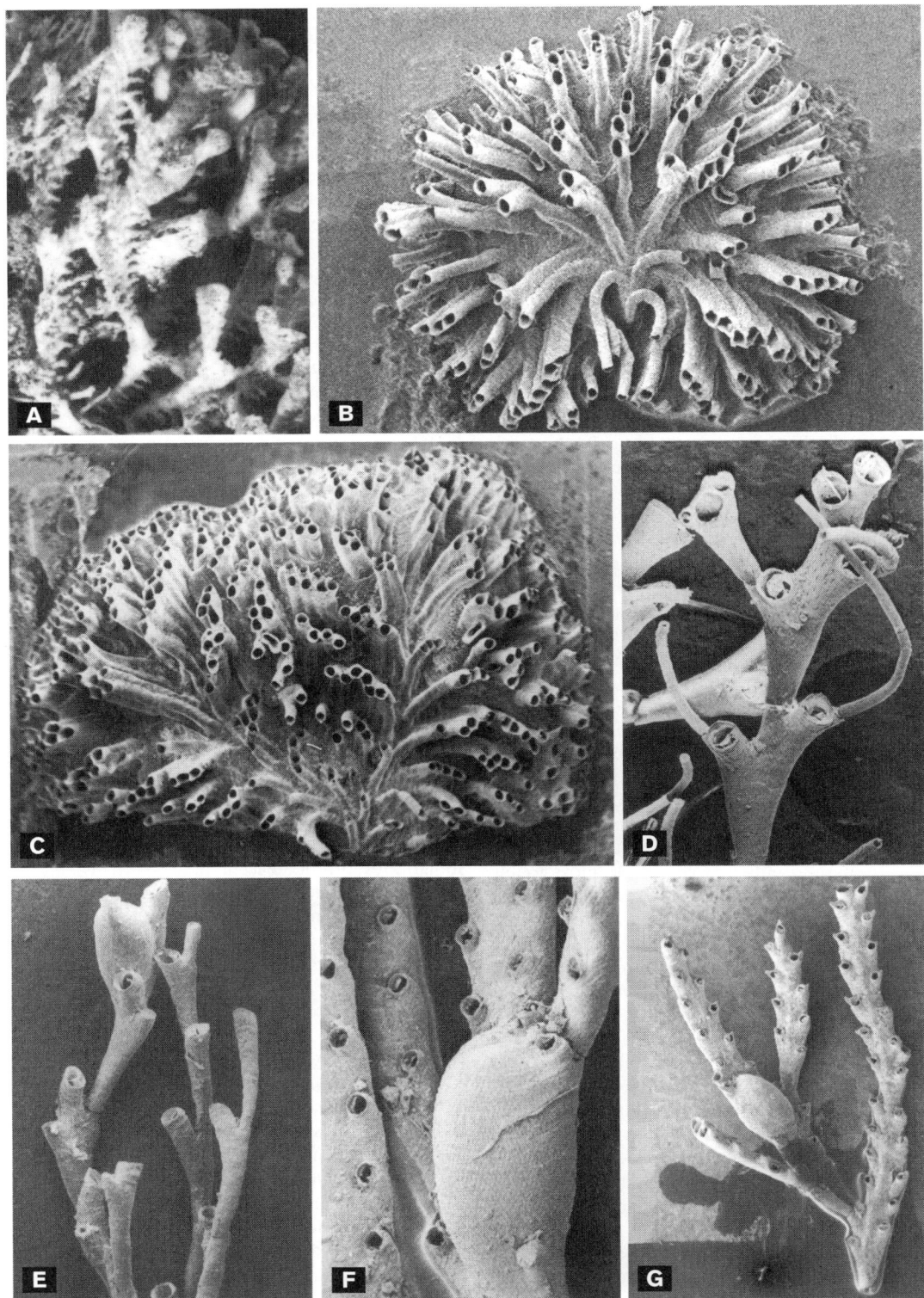

PLATE 440 Cyclostomatida. A, *Diaperoforma californica*, tangled branches; B, *Tubulipora pacifica*, ooeciostomes short, ovals between tall zooids; C, *Tubulipora tuba*, ooeciostome short, oval, mid-right; D, *Bicrisia robertsonae*; E, *Filicrisia franciscana*, gonozooid at top left; F, *Crisia maxima*, gonozooid on right; G, *Crisia occidentalis*, gonozooid mid-left.

freestanding tube with flared opening; Point Conception, possibly central California, Galapagos Islands; low tide to 86 m.

Diaperoecia californica (d'Orbigny, 1852). See *Diaperoforma californica*.

Diaperoforma californica (d'Orbigny, 1852) (=*Diaperoecia californica, Idmonea californica*). Plate 440A. See Osburn 1953, Soule et al. 1995. Large, branching, subtidal colonies forming thickets on tube molluscans, worm colonies, or heavy balls on submerged lines; zooids are long tubes fused into bundles (fascicles) of four to eight tubules, apertures slightly raised; interior of tubes with numerous minute hooks; dorsal surface with striations, no perforations; gonozooid at bifurcating branches, surrounding autozooid tubules, ooeciostome short with flared rim; British Columbia to the Coronada Islands, Mexico; described from Nicaragua and "Vermillion Sea" (Gulf of California), but not recorded there since 1852; shallow water to >200 m.

Diplosolen harmelini Soule et al. 1995; =*Diplosolen obelium*: Johnston 1838: Osburn 1953; fan-shaped colonies with straight, stout autozooid tubules embedded in porous surface crust, nanozooids are small accessory tubules scattered among autozooids, sometimes connate with autozooids below, not above, colony surface. *D. obelium*, a north Atlantic species, has smaller nanozooid tubules connate with each autozooid above colony surface; recorded Arctic Alaska to Santa Cruz Island; shallow water to 160 m.

Disporella spp. Genus tentatively includes living '*Lichenopora*' spp. of Osburn 1953 and others; taxonomy of *Disporella* and '*Lichenopora*' worldwide needs revision; discoid colonies with erect connate or nonconnate tubules radiating from center; on algal blades, hydroids, shell, stone; intertidal to 200 m.

Filicrisia franciscana (Robertson, 1910). Plate 440E. See Soule et al. 1980, 1995. Sparse delicate branches, white with black joints; zooids mostly uniserial, long slender tubules with small pores, round, terminal apertures, widened gonozooid a single zooid, flattened on top, tubular ooeciostome extended from frontal margin toward autozooid dorsal to it, or erect; Alaska to ? Baja California, common in San Francisco Bay, Monterey Bay; undersides of rocks, bioturf; intertidal to 100 m.

Filicrisia cf. *geniculata* (Milne Edwards, 1838). See Osburn 1953. Delicate, whitish branches with large black joints; gonozooid slender, tapered, connate with short ooeciostome tube originating dorsally, straight or bending forward; undersides of rocks; British Colombia to San Pedro, CA; common in Monterey Bay; intertidal, shallow water. *F. geniculata* (Hayward and Ryland 1985) is a north Atlantic/Mediterranean species with pale joints; gonozooid adnate, club shaped.

Lichenopora spp. *Lichenopora* is a Cretaceous-Miocene genus with inverted conical colonies (Gordon and Taylor 1997). Pacific colonies formerly placed in the genus are flat discs or with a few layers of similar size; now tentatively placed in *Disporella*; prominent on algal blades, shell, stone, hydroids, coral in tropics; intertidal to deep water.

Proboscina cf. *major* (Johnston, 1847). Hayward and Ryland (1985) placed British species *P. major* in new genus *Annectocyma,* but view records from outside temperate northeast Atlantic with caution; northeast Pacific colonies encrusting, or raising erect growths; autozooid tubules large, coarsely perforate, irregularly extending from base; gonozooid at distal ends of lobes or at branchings, with many fine pores, simple or ex-

panded around tubules, ooeciostome short; Puget Sound in shallow water, Monterey Bay, Channel Islands to ? Galapagos Islands; more common in deeper water.

Stomatopora cf. *granulata* (Milne-Edwards, 1838). Hayward and Ryland 1985 placed *S. granulata* of Hincks 1880, but not of Milne-Edwards, 1836, in *Entalophoroecia deflexa*. The Milne-Edwards species is supposedly uniserial, with very short tubules, but this may represent only youngest colonies; additional growth may produce multiserial, sometimes erect branches with tubules curving outward all around branch. Northeast Pacific specimens differ, need redescription; colony encrusting; zooids smaller than European species, uniserial except at branches and where gonozooid grows around multiple tubules, autozooids curving upward to round apertures, walls finely perforate; gonozooid walls with larger pores, ooeciostome shorter. Shallow water to 100 m.

Tubulipora aliciae Soule, Soule and Chaney, 1995 (=*T. pulchra* of Robertson, 1910; ? C. H. and E. O'Donoghue 1923; Osburn 1953). Small fan-shaped colonies recumbent, loosely attached to algae by small projections; raised and recumbent separate autozooid tubules anchored by serrate, discoid base of kenozooids; gonozooid a striated, enlarged tubule or expanded between tubules, ooeciostome short, flaring, may be compressed against adjacent tubule; central California coast, to Clarion Island off Mexico, ? Galapagos Islands, ? British Columbia; intertidal to >60 m.

Tubulipora pacifica Robertson, 1910. Plate 440B. See Osburn 1953; Soule et al. 1995. Small, fan-shaped to circular or lobate colonies with initial tubules separate, others connate to tips in radiate rows, with scattered small pores, smooth discoid base; often on algae; gonozooid a single chamber with one to four lobes, large pores, ooeciostome short, oval, flaring; British Columbia; Monterey Bay and south; shallow water to about 100 m.

Tubulipora pulchra MacGillivray, 1885: Osburn 1953. Australian species, not found in eastern Pacific. See *Tubulipora aliciae*.

Tubulipora tuba (Gabb and Horn, 1862). Plate 440C. See Soule et al. 1995. Large colonies of bundles of thick tubules, gray, white, or purplish; tubules tall, raised, radiating from base in connate series, with small pores; gonozooid large, inflated, with large pores, ooeciostome tall, ovate, flaring, compressed by adjacent tubules; encrusting algae, rock, shell; British Columbia to Baja California, Pleistocene to Recent; intertidal zone to 235 m.

T. tuba var. *fasciculifera* may be younger stage of *T. tuba*.

CHEILOSTOMATIDA

ANASCINA

Aetea anguina (Linnaeus, 1758). See *Aetea pseudoanguina*. British species with zooids tubular, 600 μm–800 μm long; see Hayward and Ryland 1998.

Aetea ligulata Busk, 1852. See *Aetea paraligulata*. Busk's species collected by Darwin from tip of South America, described as having tubules 450 μm long, no annulations, with terminal apertures.

Aetea paraligulata Soule, Soule and Chaney, 1995 (=*Aetea ligulata* of Osburn, 1951, in part). Plate 441B. Zooids erect annulated tubules about 700 μm long, rising from stoloniform, terete bases; zooid distal half with opesia beneath membranous frontal, aperture subterminal extending to distal tip; see

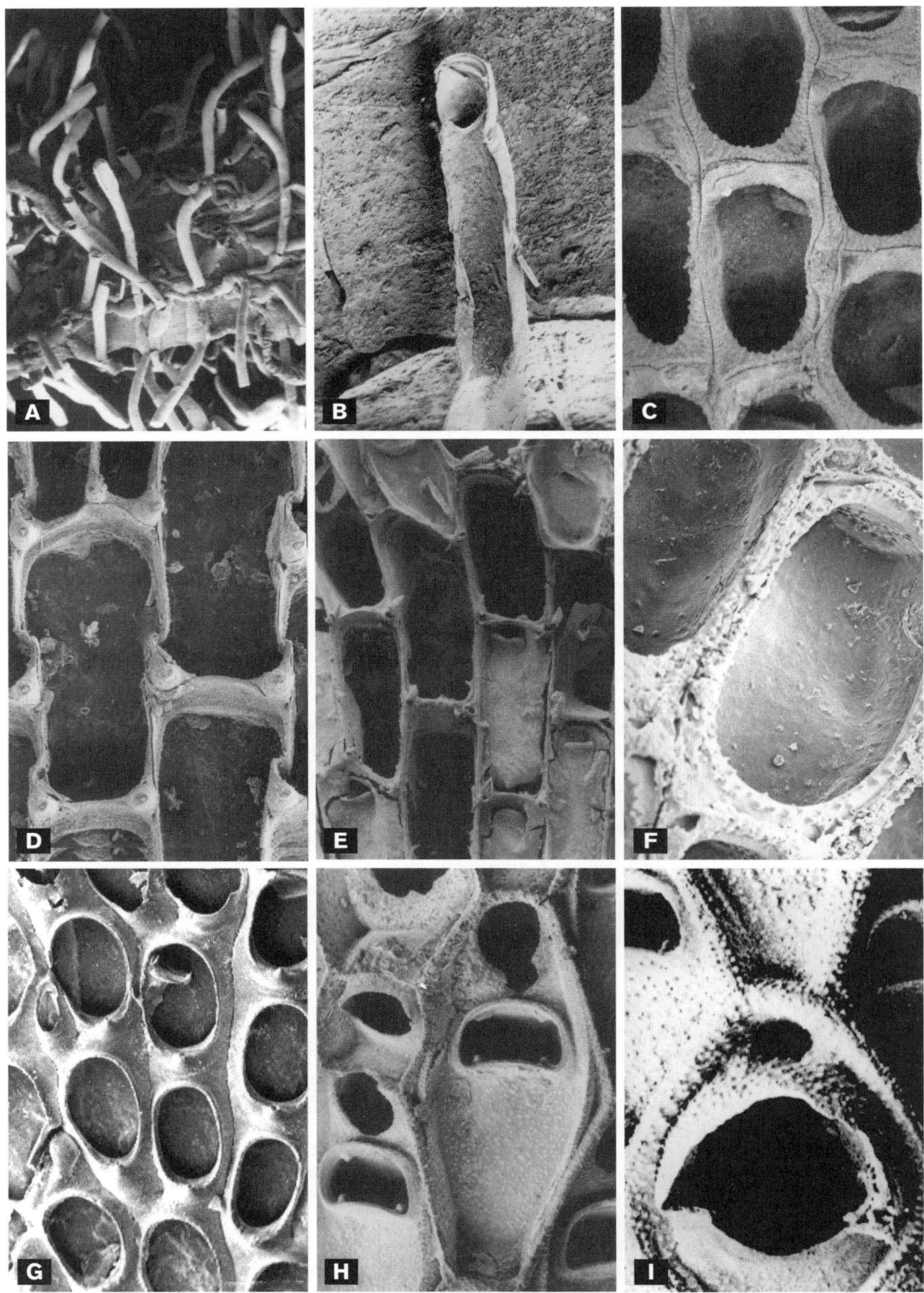

PLATE 441 Cheilostomatida. Anascina. A, *Aetea pseudoanguina*, tangled zooids with spoon-shaped distal ends; B, *Aetea paraligulata*, long opesia, subterminal aperture; C, *Membranipora fusca*, with distal knobs; D, *Membranipora membranacea* (France), note mid-lateral wall bends; E, *Membranipora serrilamella*; F, *Conopeum* cf. *reticulum*, many small pores in distal wall, upper right; G, *Electra venturaensis*, with proximal spine, small oval kenozooids on left; H, *Cellaria diffusa*, D-shaped apertures with distal ovicell openings; I, *Cellaria mandibulata*, large avicularium center, zooid apertures above.

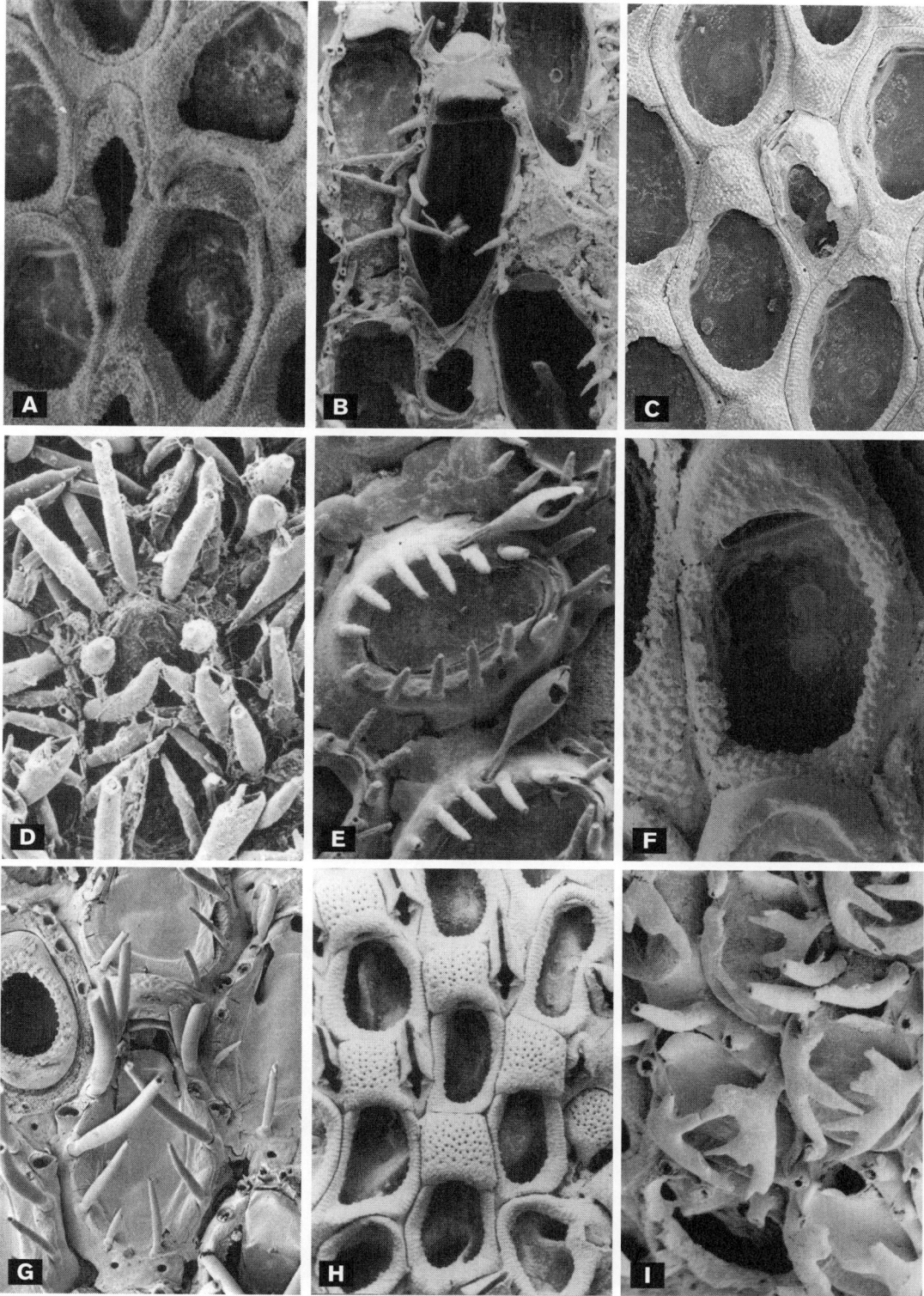

PLATE 442 Anascina, continued. A, *Hincksina alba*, ovicell shallow hood lower right, avicularium middle; B, *Hincksina pallida*, acute avicularium mid-bottom; C, *Hincksina velata*, ovicell center, avicularium right; D, *Cauloramphus californiensis*, single zooid with stout spines, tall avicularia with open mandibles; E, *Cauloramphus echinus*, shorter spines, tall avicularia; F, *Alderina ?brevispina*, zooid with shallow hood distally, tiny spine scars on opesial rim, central zooid; G, *Callopora corniculifera*; H, *Copidozoum adamantum*, diamond-shaped avicularia, lanceolate mandibles; I, *Copidozoum protectum*, interdigitating spines arching over frontal membrane of opesia.

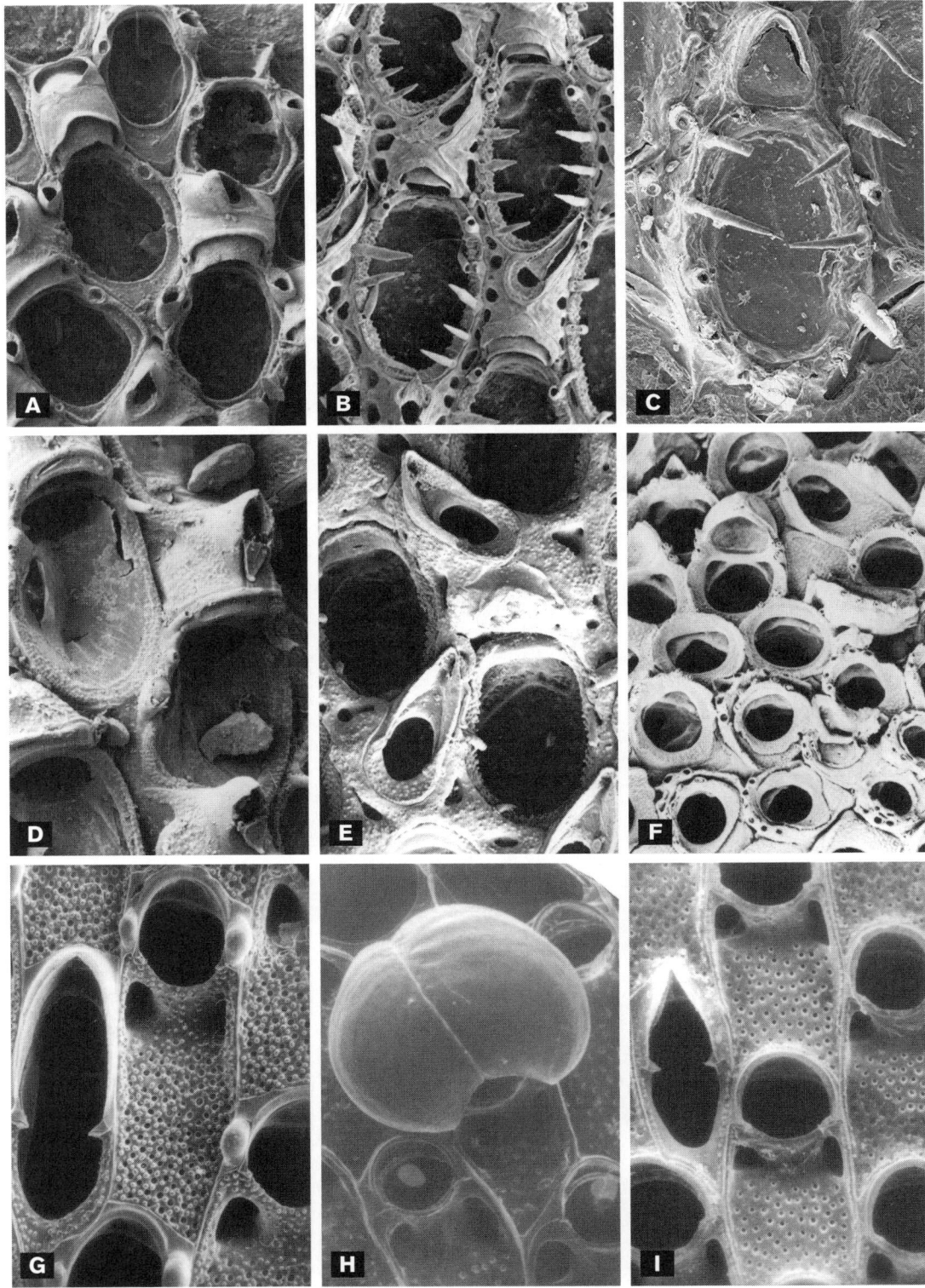

PLATE 443 Anascina, continued. A, *Tegella armifera*, ovicell with slit between layers, interzooecial avicularium migrates to ovicell top; B, *Tegella circumclathrata*, loose connections between zooids, many spines, acute avicularium atop ovicell; C, *Tegella horrida*, avicularium with ovicell directed distally; D, *Tegella laruensis*; E, *Tegella robertsonae*; F, *Chaperiopsis patula*, with paired distal shelves below opesia; G, *Thalamoporella californica*, paired opesiules below aperture (opesia) large, bluntly acute avicularium to left; H, *Thalamoporella californica* ovicell typical of genus; I, *Thalamoporella gothica*, smaller avicularium shaped like gothic arch at tip.

Soule et al. 1995 for discussion; central, southern California; shallow water to 100 m.

Aetea pseudoanguina Soule, Soule and Chaney, 1995 (=*Aetea anguina* of Osburn 1951, in part). Plate 441A. Whitish, tangled colonies of stoloniform bases with erect striate tubules, zooids 1,200 μm–1,500 μm long with tiny spinules or tubercles; distal area spoon shaped with oval opesia beneath frontal membrane, aperture subterminal, 20%–25% of length of opesia; on algae, rock, other bryozoans; California, ? Baja California; shallow waters to 100 m.

Alderina ?brevispina (C. H. O'Donoghue and E. O'Donoghue, 1926). Plate 442F. See Soule et al. 1995. Small encrusting colonies on rock, shell, other hard substrates; zooids small with widely open opesia, granular cryptocyst wider proximally, gymnocyst almost absent; sometimes tiny spinous processes around opesia; ovicell shallow, inconspicuous, granular proximal entooecial front, distal ectooecial rim, slitlike aperture; tiny spinules not seen in the O'Donoghues' specimens. British Columbia to ? Santa Barbara Channel; shallow waters to 90 m.

Bugula californica Robertson, 1905. Plate 445F. See Soule et al. 1980, 1995. Distinctive spiral colony growth, often confused with tufted growth of other species; zooids elongate with large opesia, two large blunt outer distal spines, one thinner inner distal spine; numerous "birds-heads" avicularia; ovicells large, numerous, sculpted with thin entooecial area almost surrounded by thicker, wider ectooecial area; British Columbia to southern California, ? Gulf of California, ? Galapagos Islands; fouling community, shallow water to about 70 m.

Bugula cf. *mollis* Harmer, 1926: Osburn, 1950. Not Harmer's tropical western Pacific species with three long distal spines jointed at bases. Northeast Pacific material with short points on distally truncate zooids, no spines; small avicularia medially, ovicell a small distal flap; needs redescription; San Francisco, ? Panama, ?Galapagos Islands; shallow to >40 m.

Bugula longirostrata Robertson, 1905. Plate 445G. See Soule et al. 1995. Colony erect with slender branches, zooids elongate; distinctive, long, slender, terete avicularia; ovicells like shallow bowls; British Columbia to Mexico, and the Galapagos Islands; intertidal on pilings to 230 m.

Bugula neritina Linnaeus, 1758. Plate 445D. See Soule et al. 1980, 1995. Colonies bushy, reddish to purple, biserial branching, zooids with large frontal membrane, outer distal wall forming acute tip, no spines; ovicell large, globose at inner distal corner of zooid; obverse proximal end of zooid like forked tail or Y-shaped at branches; no avicularia; in fouling community worldwide in warm temperate to tropical waters, but is most likely a complex of species as discussed by Davidson et al. (1999, Biol. Bull. 196: 273–280). Common on harbor rocks, pilings, ship's hulls; Monterey Bay to Gulf of California, Panama, Galapagos Islands; intertidal to about 100 m.

Bugula pacifica Robertson, 1905. Plate 445E. See Soule et al. 1995. Colonies delicate, erect, in pinkish tufts; zooids biserial, elongate, gymnocyst curved over frontal membrane with two spines on outer distal tip, one on inner tip; avicularia large, pedunculate birds' heads; ovicell a shallow incomplete hood; Bering Sea to Channel Islands; intertidal to 123 m.

Bugula pugeti Robertson, 1905. See Soule et al. 1995. Erect, branching, multiserial colonies, four to seven rows wide; zooids on margins with two outer blunt spines, one inner; avic-

ularia birds' heads; ovicells unknown, but distal corners of central zooids curve over until almost touching, possibly to shelter ovum; Alaska to Channel Islands; mostly northern; in fouling community; low tide to 117 m.

Bugula stolonifera Ryland, 1960. An introduced Atlantic species common in San Francisco Bay fouling communities as reported by Okamura (1984, J. Exp. Mar. Biol. Ecol. 83: 179–193) and discussed by Cohen and Carlton (1995).

Caberia ellisii (Fleming, 1818). Plate 444D, 444E. See Soule et al. 1995. Colony erect, fan shaped, branches two to four zooids wide, tangled; zooids with long, barbed vibracula, radicles grouped in rigid cords on dorsal side; no scutum; ovicell a shallow imperforate hood with incompletely calcified frontal entooecium; arctic Atlantic to English Channel and Cape Cod; reported from Alaska, British Columbia, Puget Sound, Channel Islands, and Baja California; no significant differences found between Pacific, Atlantic specimens; shoreline to >450 m.

Callopora circumclathrata (Hincks, 1881): Osburn 1950. See *Tegella circumclathrata*.

Callopora corniculifera (Hincks, 1882). Plate 442G. See Soule et al. 1995. Zooids ovate with large opesia, rolled cryptocyst bristling with three to four pair of thinner marginal spines proximally, two to three pair of stout marginal spines distally; avicularia small, on lateral wall outside spines distal to midzooid; ovicells wide, shallow, coarsely granular, with slit-like opening between ectooecium, entooecium; encrusting rock, shell; British Columbia to Santa Catalina Island; intertidal to 126 m.

Caulibugula californica (Robertson, 1905). Plate 445H. See Soule et al. 1995. Erect, jointed, stalked colonies with palmate tufted tips, zooids resemble those of *Bugula* spp.; two spines on outer distal margin; ovicell pedunculate on inner margin; intertidal in British Columbia, to 231 m off La Jolla.

Caulibugula ciliata (Robertson, 1905). Plate 445I. See Soule et al. 1995. Tiny, delicate, erect fanlike tufts with five or more long, incurved spines on outer margin; pedunculate ovicell on inner margin of distal end; common on red algae; British Columbia to Channel Islands; depths from lower intertidal to over 100 m.

Cauloramphus californiensis Soule, Soule and Chaney, 1995 (=*C. spiniferum* of Osburn, 1950, in part). Plate 442D. Brown to tan, unilaminar colony; oval opesia with lateral walls bearing 13–15 stout spines covering frontal membrane; one to two long, stalked avicularia flanking operculum, similar to spines in size; ovicells not known; encrusts shell, rock, kelp holdfasts; Monterey Bay to La Jolla; common in Channel Islands, perhaps from Alaska to Galapagos Islands; shallow water to >100 m.

Cauloramphus echinus (Hincks, 1882). Plate 442E. See Soule et al. 1995. Unilaminar colonies, ovate zooids loosely connected by small tubules; large opesia, four erect distal spines, seven to eight pair of short marginal spines extended over opesia but not touching; paired pedunculate avicularia flanking aperture; ovicells not known; British Columbia to Santa Barbara Channel; intertidal to >100 m.

Cauloramphus spiniferum (Johnston, 1832): Osburn 1950, in part. A north Atlantic species not found on Pacific coast. See *Cauloramphus californiensis*.

Cellaria diffusa Robertson, 1905. Plate 441H. See Soule et al. 1995. Erect, club-shaped internodes, brown or black joints; branching colony; zooids bluntly diamond shaped

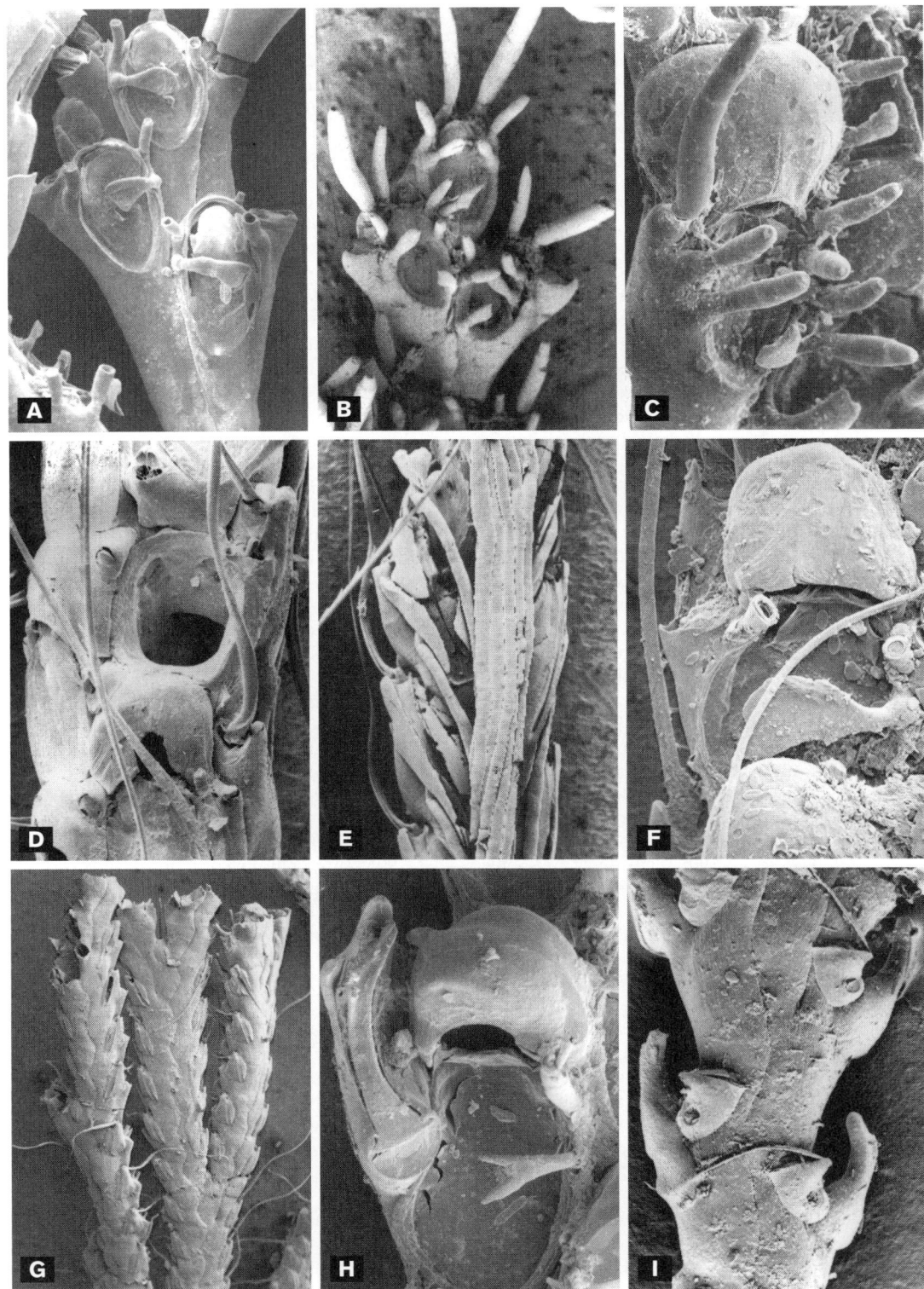

PLATE 444 Anascina, continued. A, *Tricellaria circumternata*, with paddle-shaped central scutum; B, *Tricellaria occidentalis*, tall spines, lateral avicularia directed outward; C, *Tricellaria occidentalis*, obverse with pores, three pair long spines, scutum at bottom; D, *Caberia ellisi*, ovicell below, lateral vibracula; E, *Caberia ellisi*, obverse, radicles forming cords down back; F, *Scrupocellaria diegensis*, with large scutum; G, *Scrupocellaria diegensis*, obverse view; H, *Scrupocellaria varians*, with giant avicularium, forked scutum; I, *Scrupocellaria varians*, obverse view with vibracula.

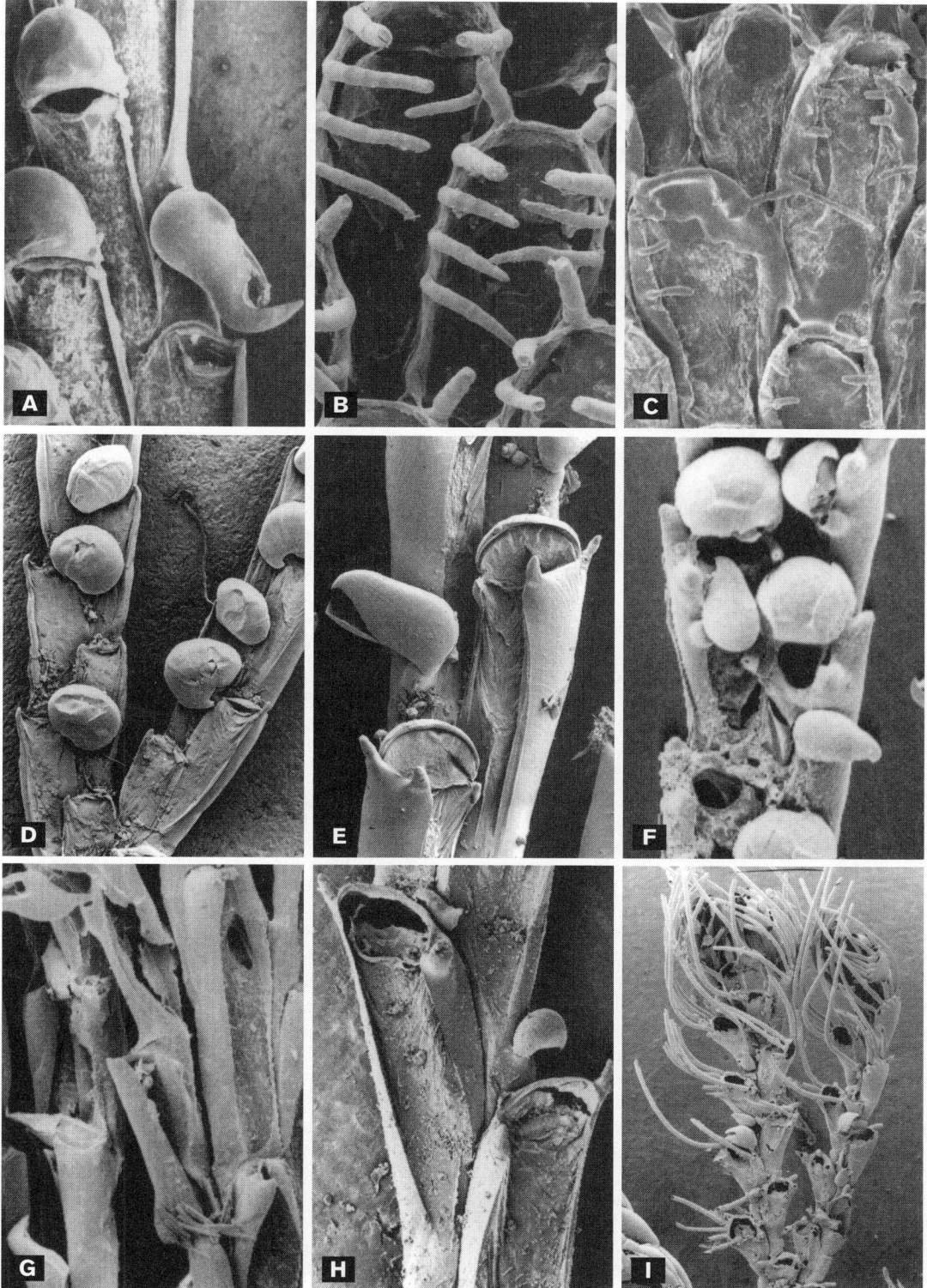

PLATE 445 Anascina, continued. A, *Dendrobeania curvirostris*, zooids with large opesiae, hooked lateral avicularium; B, *Dendrobeania laxa*, stout spines over opesiae; C, *Dendrobeania lichenoides*, with thin, short spines; D, *Bugula neritina*, with ovicells, no spines, no avicularia; E, *Bugula pacifica*, shallow ovicell hoods, stout avicularium on pedicell; F, *Bugula californica*, with well-calcified ovicells, avicularia on pedicles; G, *Bugula longirostrata*, with long, thin zooids; H, *Caulibugula californica*, with slanted apertures, small marginal avicularia; I, *Caulibugula ciliata*, with long curved spines, small erect avicularia.

with large, depressed cryptocyst; opesia consists only of aperture, arched proximally and distally, with strong blunt condyles on proximal lip, smaller distal denticular rests for operculum; avicularium interzooecial, between transverse walls, almost square, with elliptical rostrum, rounded mandible; ovicell internal but opening by large pore distal to aperture, pore becoming eroded, larger than avicularium opesia; Pleistocene of Santa Monica; Recent, Puget Sound to southern California, ? Galapagos Islands; intertidal to >200 m.

Cellaria mandibulata Hincks, 1882. Plate 441I. See Soule et al. 1995. Erect, branching internodes, brown or black joints; zooids diamond shaped, aperture rounded distally and proximally with two large condyles on proximal lip; giant avicularia with large brown semicircular mandible (looks like an open mouth); ovicells internal with large pore distal to aperture; on hard substrates; British Columbia to Baja California; intertidal to >140 m.

Chapperia patula (Hincks, 1881). =*Chapperia*: Willey 1905; Osburn 1950, a misspelling of *Chaperia*, preoccupied; see *Chaperiopsis patula*.

Chaperiopsis patula (Hincks, 1881). Plate 443F. See Soule et al. 1995. Encrusting, reddish purple; opesia a wide oval, cryptocyst raised like saucer rim, *occlusor laminae* shelves distally; four to six large spines around distal half; ovicell a large imperforate hood, sometimes sculpted, slit between ectooecium, entooecium open or fused; British Columbia to Baja California; shallow water to 90 m.

Colletosia radiata (Moll, 1803). See Cribrimorphina: *Puellina (Cribrilaria) californiensis*.

Conopeum osburni Soule, Soule and Chaney, 1995 (=*Electra crustulenta* var. undescribed of Osburn, 1950). Encrusting; linear, loosely connected oval zooids bordered with three to six pair of thin spines, two tiny spines at distal corners, no proximal spine; no ovicells known; Oregon to Gulf of California; intertidal in north, to about 73 m in south.

Conopeum cf. *reticulum* (Linnaeus, 1767). Plate 441F. See Soule et al. 1995. May be one or more, probably undescribed, encrusting species; zooids with narrow cryptocyst and gymnocyst, cryptocyst crenulate, sometimes with small spinules extending over frontal membrane, opesia large, oval; small triangular kenozooids at distal corners, sometimes absent; internal distal transverse wall with large multiporous communication plate or sometimes smaller single pore plates flanked by vertical buttresses, in some colonies causing confusion with British species; no ovicells known: brackish water of San Francisco Bay, Berkeley Yacht Harbor, often found with ctenostome *Victorella* cf. *pavida* and entoproct (kamptozoan) *Barentsia* sp; coastal embayments south to La Jolla, CA; intertidal, shallow water. San Francisco Bay is also home to the abundant *Conopeum tenuissimum* (Canu, 1928) which was introduced with Atlantic oysters in the 19th century.

Copidozoum adamantum Soule, Soule and Chaney, 1995 (=*C. tenuirostre* of Osburn 1950, includes *C. planum* of Osburn 1950; not *M. tenuirostris* Hincks 1880). Plate 442H. Encrusting, oval zooids, separated, with raised crenulated cryptocyst, diamond-shaped interzooecial avicularia with lanceolate mandibles; squared, porous ovicells; British Columbia to Channel Islands, ? Galapagos Islands and ? Peru; intertidal to 129 m.

Copidozoum planum (Hincks, 1880): Osburn, 1950. See *C. adamantum*. *C. planum* is an Australian species, not found in eastern Pacific.

?*Copidozoum protectum* (Hincks, 1882). Plate 442I. See Osburn, 1950: Soule et al. 1995. Generic placement questioned because *C. protectum* has imperforate ovicells and large acute interzooecial avicularia as in *Copidozoum*, but has thick branching spines on gymnocyst in distal half of zooid arching over large opesia as in *Chaperiopsis*, although there are no *occlusor laminae*; British Columbia to Gulf of California; encrusting hard substrates; shallow water to 50 m or more.

Copidozoum tenuirostre (Hincks, 1880). Osburn, 1950. See *Copidozoum adamantum*.

Dendrobeania curvirostrata (Robertson, 1905). Plate 445A. See Soule et al., 1995. Erect or recumbent strap-like colonies four to eight zooids wide, attached by rootlets; zooids elongate, one very short inner spine, one outer distal spine, one to two tiny spines flanking operculum; beaked avicularium very large, at base of outer zooids; ovicells large, imperforate but sculpted due to uncalcified areas in entooecium; British Columbia to southern California; shallower water in north, deeper in south, to >100 m.

Dendrobeania laxa (Robertson, 1905). Plate 445B. See Soule et al. 1995. Loosely attached by rootlets; recumbent colonies on hard substrates, zooids separated by small gaps; two distal spines, four pair medium large lateral spines arching over long, slim opesia; no avicularia; large, globose, striate ovicells; British Columbia to southern California; intertidal to 100 m.

Dendrobeania lichenoides (Robertson, 1900). Plate 445C. See Soule et al. 1995. Greenish, loosely attached recumbent fronds; opesia large, ovoid; two tiny distal spines, two to three pair small lateral spines present or absent; ovicell globular, imperforate with striations; often associated with green colonial ascidian *Perophora*; British Columbia to southern California; common in low, shaded intertidal to 100 m.

Electra crustulenta arctica (Borg, 1931). See Osburn 1950. A north Atlantic brackish water species; var. *arctica* described from Spitsbergen. See *Electra venturaensis*.

Electra venturaensis Banta and Crosby, 1994 (=*E. crustulenta* of Osburn, 1950; =*E. crustulenta arctica* of Osburn1950, in part: see Soule et al. 1995). Plate 441G. Single proximal spine typical of genus; colonies encrusting, large opesia, gymnocyst wide with transverse wrinkles as in *Celleporella hyalina*, cryptocyst narrow; no known ovicells; oval dwarf zooids on San Francisco specimens, small rectangular kenozooids on Alaskan material (may be separate species); ? Alaska; central, southern California; encrusting rocks, shells, algae, intertidal to >30 m.

Figularia hilli Osburn, 1950. See Cribrimorphina: *Reginella (Figularia) hilli*.

Hincksina alba (C. H. O'Donoghue and E. O'Donoghue, 1923). Plate 442A. See Soule et al.1995. Encrusting species with large open opesia, granular cryptocyst, no spines; large, winged interzooecial avicularium between lateral walls of adjacent zooids; ovicell visible as small, shallow, granular hood; encrusting rock, molluscan and brachiopod shells; British Columbia to Channel Islands; shallow water to 100 m.

Hincksina pallida (Hincks, 1884). Plate 442B. See Soule et al. 1995. Encrusting, loosely attached to hard substrates; large opesia, with six pair small lateral spines outside narrow cryptocyst between lateral walls; interzooecial avicularium between transverse zooid walls; ovicell small, with shallow brim, projecting into body cavity of next distal zooid; British Columbia to central California, shallow water to 90 m.

Hincksina velata (Hincks, 1882). Plate 442C. See Osburn 1950. Encrusting; opesia large, no spines, small, triangular to curved interzooecial avicularia between lateral walls directed distolaterally; ovicell endozooecial but visible as small subtriangular distal knob with separate operculum; encrusting shells; British Columbia to ? Colombia; common at Monterey, Pebble Beach; shallow water to >120 m.

Lyrula hippocrepis (Hincks, 1882). See *Reginella* (*Lyrula*) *hippocrepis*.

Membranipora fusca Osburn, 1950. Plate 441C. Frontal wall at first a thick, clear membrane, developing heavy dark border, becoming yellowish-brown to black; two distal knobs in older zooids (may belong in genus *Conopeum*); no spines, no avicularia, no ovicells; encrusting shells, stones; northern, central California; intertidal to about 12 m.

Membranipora membranacea (Linnaeus, 1767). Plate 441D. Common European species, oblong to dumbbell-shaped zooids, small knobs at distal corners may form taller tower cells, may aid colony in water current flow; no ovicells; heavy encrustations on giant kelp in northeast Atlantic; transitory if introduced in eastern Pacific, similar to *M. serrilamella*; see Soule et al. 1995; intertidal to depth of kelp fronds. See Schwaninger, 1999, Mar. Biol. 135: 411–423 (genetics, population structure).

Membranipora serrilamella Osburn 1950 (?=*M. perfragilis* of Osburn, 1950). Plate 441E. See Soule et al.1995. Zooids thin walled, elongate with pair of short, acute, curved spines distally beside base of operculum; a few tiny spinules directed centrally from narrow cryptocyst; may include *M. villosa* Hincks, 1880, with small chitinous spinules induced on frontal membrane in early spring, summer by juvenile molluscan predation (Harvell, 1984); common on algae; British Columbia to southern California; shallow water to 30 m, perhaps to bottom of photic zone.

Membranipora tuberculata (Bosc, 1802) (?=*Jellyella tuberculata* Taylor and Monks, 1997). See Soule et al. 1980, 1995. Encrusting oblong zooids with thick, crenulate, widely oval cryptocyst, with two proximal tubercles; in Atlantic specimens, walls thin, tubercles small, spinules small; on floating *Sargassum*; tropical specimens have heavier walls, much larger tubercles and spinules, if all the same species; central California during El Niño periods; encrusting kelp, other algae; may be reintroduced on algae on ships' hulls, surviving in warm bays, harbors temporarily; intertidal to 50 m.

Membranipora villosa Hincks, 1880. See *Membranipora serrilamella*.

Micropora coriacea (Johnston, 1847). See Osburn 1950. See *Micropora santacruzana*. *M. coriacea* from Britain has no avicularia, opesiules smaller than in *M. santacruzana*.

Micropora santacruzana Soule, Soule and Chaney, 1995 (=*M. coriacea* of Osburn 1950 in part; ? of Hincks1882; ? of Robertson1908; ? of C. H. and E. O'Donoghue 1923, 1925). Encrusting, frontal wall with complete perforate cryptocyst, small to medium opesiules at lateral walls proximal to straight aperture, proximal rim flanked by tiny round knobs; small transverse bulbous avicularium with acute mandible between transverse walls; ovicell perforate with apical umbo; ? British Columbia to Channel Islands; shallow water to 168 m.

Scrupocellaria cf. *californica* Trask, 1857. See Soule et al. 1995. Colony erect, biserial, jointed; opesia about half zooid length; ovicell imperforate, striated; originally described *S. californica* has seven to nine zooids per internode, one spine per zooid, scutum a small spine or paddle shaped, on proximal lobe, small vibraculum or none; currently identified "*S. californica*," needs new name; has three to four zooids per internode, sometimes up to nine per internode, opesia a large oval, two to three outer spines, one inner spine; scutum a single prong or oar shaped;

reported from British Columbia, common in central California, Baja California; shallow water to >120 m.

Scrupocellaria diegensis Robertson, 1905. Plate 444F, 444G. See Osburn, 1951. Colony coarse, up to 12 zooids per internode; opesia large, more than half of zooid length, two outer, three inner spines; scutum paddle shaped; small raised proximal frontal avicularia, large lateral avicularia; ovicells large with small pores, striations; British Columbia, Gulf of California to ? Colombia, on floats and pilings; shallow water to 20 m–40 m.

Scrupocellaria varians Hincks, 1882. Plate 444H, 444I. See Soule et al. 1995. Colony erect, bushy, opesia covering most of zooid frontal area, forked scutum; giant curved lateral avicularium on outer margin, rostrum troughlike, small frontal avicularium proximal to opesia; dorsal vibraculum small with short bristle, short radicle; ovicell ectooecium bordering imperforate central entooecium; British Columbia, California, ? Gulf of California; shallow water to about 60 m.

Scrupocellaria spp. common under rocks, floats, bushy, whitish, greenish or brownish, bristly colonies or feltlike bioturf.

Tegella armifera (Hincks, 1880). Plate 443A. Soule et al., 1995; Osburn, 1950 (in part). Colony encrusting, oval zooids with grooves between;1pair tiny spines flanking aperture, one large spine or one proximally directed small avicularium on each side of opesia; ovicell with lucidum slit between ectooecium, entooecium, one pair of sunken pores at ends of slit; large interzooecial avicularium overgrowing ovicell from between transverse walls; Arctic seas, North Atlantic, Alaska, Oregon, California; intertidal in northern waters to >100 m in southern California if all are the same species.

Tegella cassidata (C. H. and E. O'Donoghue, 1923) (=*T. armifera* of Osburn, 1950, in part; see also *T. laruensis* Soule, Soule and Chaney 1995). Zooids ovate, cryptocyst with rolled granular rim; one pair of bulbous acute avicularia flanking aperture raised on conical chambers on gymnocyst; one large spine interior of, proximal to, base of avicularium; ovicell with wider lucidum, strong ectooecial rim, no lateral pores; small or large raised, transverse interzooecial avicularium, curved distolaterally; encrusting rock, shell; British Columbia, to ? southern California; distribution, depths uncertain due to confusion with other species.

Tegella circumclathrata (Hincks, 1881) (=*Callopora circumclathrata* of Osburn, 1950). Plate 443B. See Soule et al. 1995. Genus *Tegella* now limited to species with large interzooecial avicularium between transverse walls, avicularium may migrate to top of ovicell; zooids ovate, loosely connected by tubules; opesia large with crenulate rim, large marginal pores, pair of spines flanking aperture, three to four pairs curving over frontal membrane; avicularium with long acute mandible directed proximally in absence of ovicell, directed distolaterally atop ovicell if present; ovicell with gap (lucidum) between ectooecium, entooecium, ovicell not closed by operculum; encrusting rock, shell; ? British Columbia to ? Baja California; shallow water to >100 m.

Tegella horrida (Hincks, 1880) (=*Callopora horrida*). Plate 443C. Colonies encrusting; differs from *T. circumclathrata* in having contiguous zooids without loose, large, tubular connections; with two to three pairs distal spines, two to three pairs lateral spines; British Columbia, California, ? Galapagos Islands; shallower water in north, to >60 m in Galapagos Islands.

Tegella laruensis Soule, Soule and Chaney, 1995 (? =some earlier citations of *T. armifera* or *T. cassidata*). Plate 443D. Zooids with ovate opesia, narrower at operculum; one large and one small spine on one side of aperture, one on other; a raised acute avicularium on lateral gymnocyst at mid-opesia, a larger acute interzooecial avicularium between transverse walls raised on pedicel, directed proximally on zooids without ovicell; ovicell raised, ectooecium and entooecium separated by rimmed slitlike lucidum, interzooecial avicularium lying atop ovicell directed distally; central, southern California, encrusting, especially arenaceous worm tubes, shallow to deep waters.

Tegella robertsonae C. H. and E. O'Donoghue, 1926. Plate 443E. Species has one to three small spines extending over opesia, no avicularia on lateral wall gymnocyst; very large interzooecial avicularia, acute with raised tip, moving atop ovicells, directed distolaterally; ovicell with slit between ectooecium, entooecium; encrusting shell, sponge, larger algae, tunicates; Alaska, British Columbia, Dillon Beach, Monterey Bay, Channel Islands; intertidal to >45 m.

Thalamoporella californica (Levinsen, 1909). Plate 443G, 443H. See Soule et al. 1980; Soule, Soule, and Chaney 1999. Colonies encrusting, sometimes rising in small, jointed, club-shaped branches; zooids large, vase shaped, paired large opesiules in perforate cryptocystal frontal wall; interzooecial avicularium arched distally to subacute tip, about same length as zooid; large bilobate ovicells common; internal body cavities with tiny calcified, curved caliper-shaped (ice tong) spicules; Pleistocene, southern California; Recent, Monterey Bay and south; on kelp, other algal substrates; intertidal to 15 m.

Thalamoporella gothica (Busk, 1856). Plate 443I. Soule et al. 1999. Colonies encrusting, sometimes rising in bilaminar, foliate lobes; zooids large, similar to *T. californica* with two opesiules; interzooecial avicularium smaller, with more blunt distal tip; ovicells rare, large; internal spicules both calipers and compasses; on rock, shell; shallow, warmer California waters especially during El Niño years, more common in Gulf of California.

Tricellaria circumternata Soule, Soule and Chaney, 1995 (=*Tricellaria ternata* of Osburn 1950, in part). Plate 444A. Colonies erect in tufts or flaccid branches; three to seven small zooids per internode with light-colored joints; two to three spines on outer margin, one to two on inner margin; scutum a single spine or paddle shaped; a few zooids with a small frontal avicularium, some with large lateral avicularia; ovicells large, imperforate, striate; attached to hydroids, arenaceous worm tubes, erect bryozoan colonies, seaweeds; British Columbia, Dillon Beach and Big Sur, central California; low tide to 50 m.

Tricellaria occidentalis (Trask, 1857). Plate 444B, 444C. See Soule et al. 1995. Erect, bushy colonies, internodes usually three zooids, sometimes up to five to six; opesia about half of frontal wall, three smaller spines on inner margin, three larger spines on outer margin, longest spines most distal; scutum a single or bifurcate prong in northern material, multipronged in southern; larger lateral avicularium on outer side, mandible triangular, hooked; ovicell with large pores, sometimes a median suture on brim; common in fouling on boat hulls, other hard substrates, algae; ? British Columbia, California, ? Baja California; intertidal to 40 m.

Tricellaria ternata (Solander 1786): Osburn 1950. See *Tricellaria circumternata*.

CRIBRIMORPHINA

Colletosia radiata form *innominata* (Couch 1844): Osburn 1950. See *Puellina* (*Cribrilaria?*) *perplexa*.

Puellina (*Cribrilaria*) *californiensis* Soule, Soule and Chaney, 1995 (=*Puellina setosa* of Osburn, 1950, in part; *Colletosia radiata* form *innominata*: Osburn 1950, in part; *Colletosia radiata* of J. Soule 1959). Plate 446F. Zooids small, shiny, translucent, a small suboral lacuna below proximal apertural rim, five distal spines; frontal shield composed of 13–16 radiate costae with knobs at periphery with marginal buttressed portion slanting upward, five intercostal pores on each side above first costae below lacuna, five to seven on each side between rows of larger costae; gymnocyst wider in young zooids, becoming narrow; acute interzooecial avicularia; ovicells imperforate; encrusting rock, shell; warmer California waters, Channel Islands, Gulf of California; shallow waters to >180 m.

Puellina (*Cribrilaria?*) *perplexa* Soule, Soule and Chaney, 1995 (=*Colletosia radiata* form *innominata* of Osburn 1950, in part; not *P.* (*C.*) *innominata* from Britain with fewer costae; see Bishop and Househam, 1987). Plate 446G. Colonies encrusting; zooids small, oval, five distal spines; medium to large suboral lacuna flanked by two smaller pores, three intercostal pores on each side between costal rows one and two, five between other costae, frontal shield composed of 15–17 costae, gymnocyst hardly visible; interzooecial avicularia acute, directed distolaterally; ovicell with pores sometimes obscured by advanced calcification, sometimes umbonate; ? British Columbia to southern California, ? Galapagos Islands, ? Peru; shallow water to 200 m.

Reginella (*Figularia*) *hilli* (Osburn, 1950). Plate 446E. See Soule et al. 1995. Colony encrusting; zooids large, one fused center proximal costa with two to three pairs of costae; single intracostal pores large, pyriform; aperture wider than high, ovoid; ovicell formed from modified costae, with pair of very large pores, central suture; no spines or avicularia. Monterey Bay, Channel Islands; Baja California; intertidal to >130 m.

Reginella (*Jullienula*) *hippocrepis* (Hincks, 1882) (=*Lyrula hippocrepis* of Osburn, 1950; =*Cribrilina hippocrepis*). Plate 446C, 446D. See Soule et al. 1995. Colony encrusting, sometimes raised, bilaminar; zooids large, frontal shield raised, four to six pairs of costae, separated by three to four irregular, dumbbell-shaped intercostal pores, one large intracostal pore at outer end of each costa; large interzooecial avicularia with spatulate mandible directed distally; ovicells small, indistinct, formed of modified costae with two intracostal pores, one median pore; Alaska to southern California; shallow water to 160 m.

Reginella (*Reginella*) *furcata* (Hincks, 1882). See also *Reginella nitida* of Osburn 1950; Soule et al. 1995. Plate 446A. Colonies encrusting; six to eight pairs of costae radiating from central suture, four to six intercostal pores, each costa with two small infundibular intracostal pores; aperture bell shaped, flanked by single or weakly bifurcate spines; ovicells porous, sometimes with faint median keel, blending into next distal frontal wall; British Columbia to San Benito Islands off Baja California; shallow waters to >160 m.

Reginella (*Reginella*) *nitida* (Osburn, 1950). Plate 446B. See Soule et al. 1995. Very similar to *R. furcata*. Colonies encrusting; five to eight pairs of flattened costae, five to seven intercostal pores per rib, three to four smaller infundibular intracostal pores atop each costa; aperture ovate, with or without small lateral projections; ovicell with scattered pores, median suture raised into keel; reported from Puget Sound to San Benito Islands off Baja California, shallow water to >160 m.

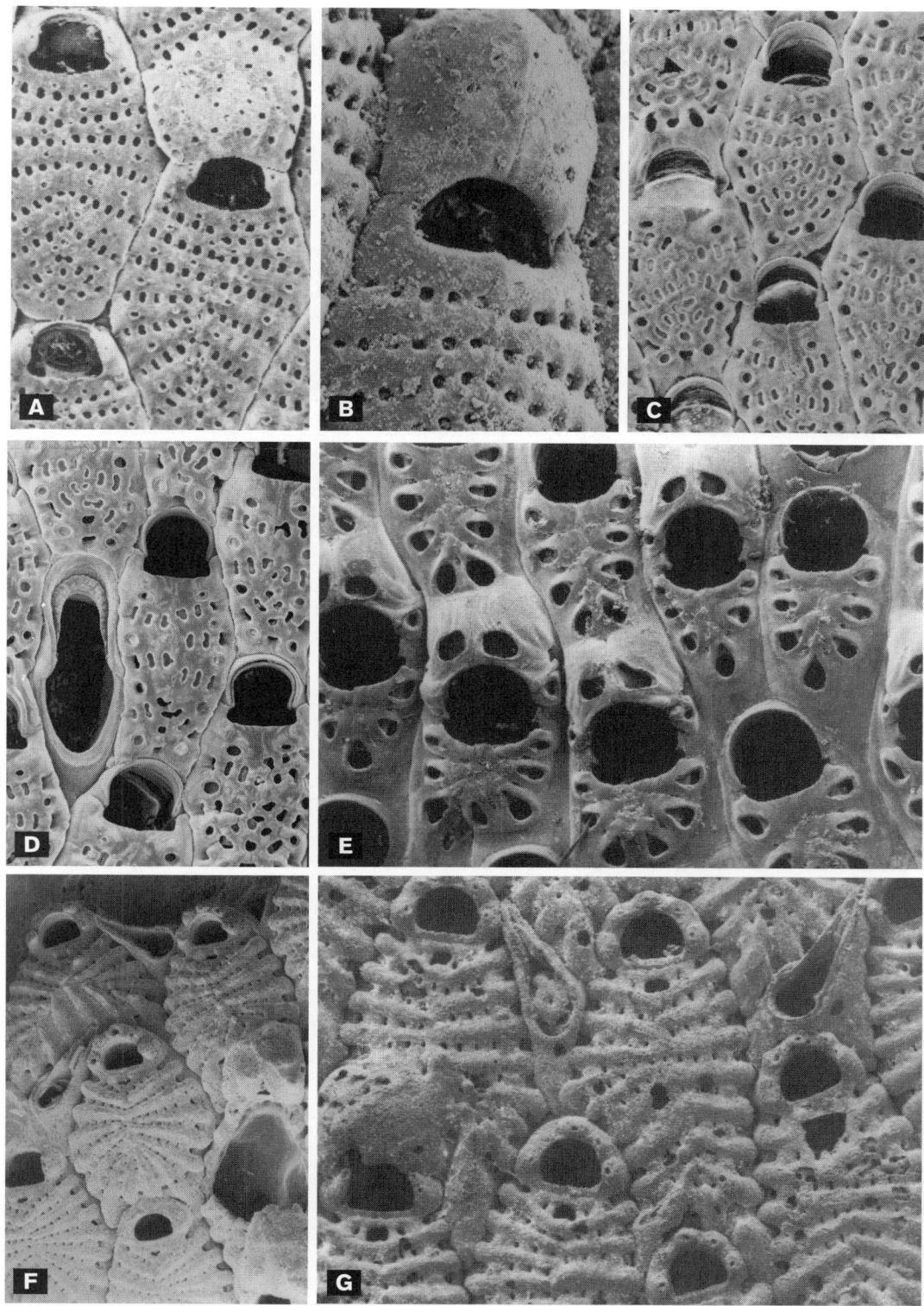

PLATE 446 Cribrimorphina. A, *Reginella (Reginella) furcata*, bell-shaped aperture, ovicell with pores on right; B, *Reginella (Reginella) nitada*, aperture more ovate with scattered pores on ovicell; C, *Reginella (Jullienula) hippocrepis*, ovicell on left with two large pores, small median pore; D, *Reginella (Jullienula) hippocrepis*, giant avicularium on left; E, *Reginella (Figularia) hilli*, ovicells with two large pores, median suture; F, *Puellina (Cribrilaria) californiensis*, zooids with median suboral lacuna, imperforate ovicell on right, interzooecial avicularium at top; G, *Puellina (Cribrilaria?) perplexa*, perforate ovicell bottom left, interzooecial avicularia hypercalcified.

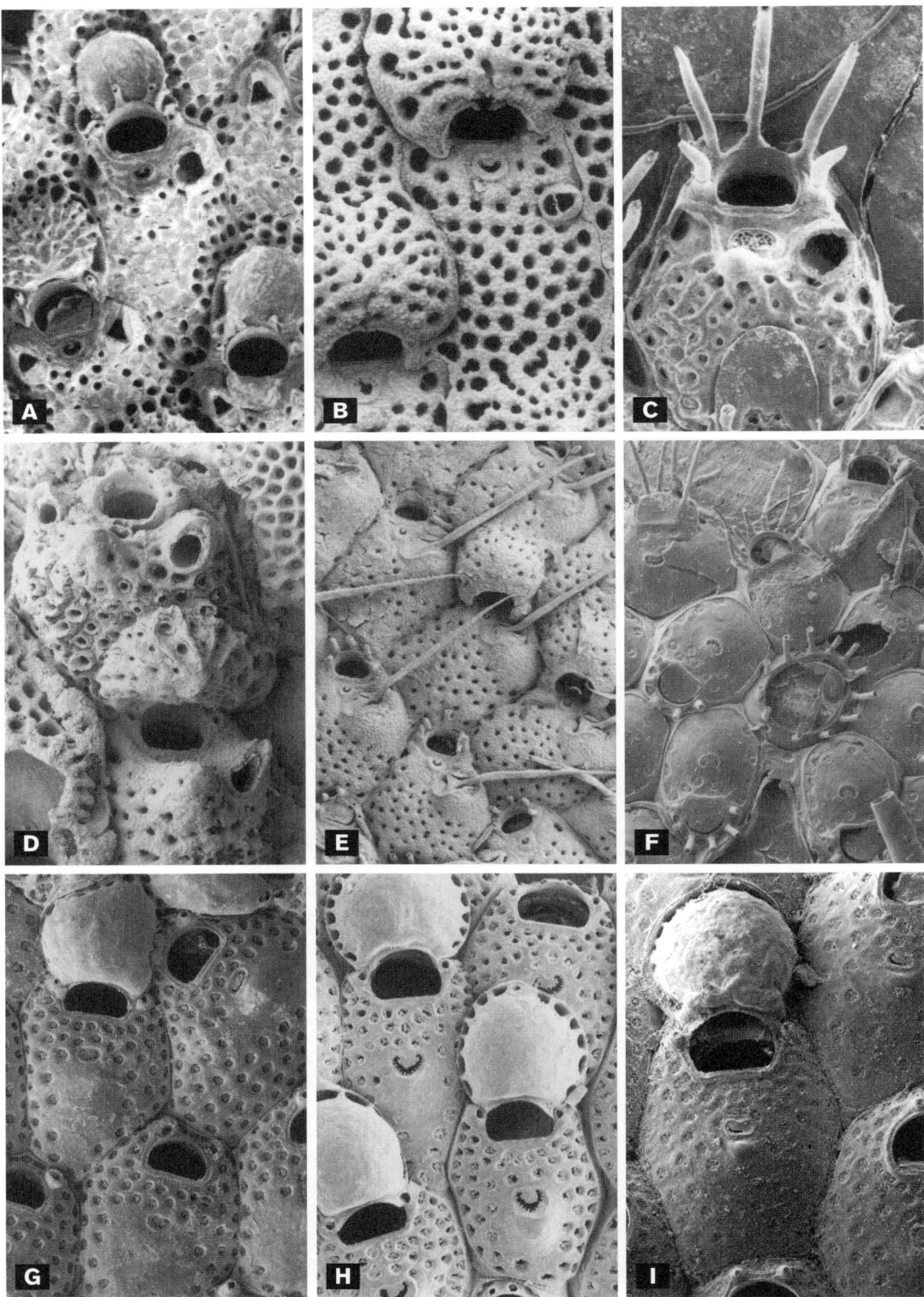

PLATE 447 Ascophorina. A, *Microporelloides californica*, suboral ascopore with median labellum, acute lateral oral avicularia; B, *Microporelloides catalinensis*, with reticulate frontal wall, ovicell pores; C, *Microporelloides cribrosa*, with suboral cribrate ascopore; D, *Microporelloides infundibulipora*, with sunken pores, ridged ovicells, suboral umbo proximal to ascopore; E, *Microporelloides vibraculifera*, avicularia with long, setose mandibles; F, *Fenestrulina farnsworthi*, developing colony with ancestrula (metamorphosed larva) surrounded by spines, frontal wall with marginal pores, central ascopore; G, *Fenestruloides blaggae*, with imperforate ovicell, frontal wall with pores, ascopore, proximal umbo; H, *Fenestruloides miramara*, pair of spines on distal rim of aperture or flanking aperture of ovicells, ascopore on umbo, frontal wall pores except for umbo, imperforate ovicells; I, *Fenestruloides umbonata*, frontal wall with pores, small umbo proximal to small ascopore, ovicell rugose.

Arthropoma cecilii (Audouin, 1826). Plate 449D. See Osburn 1952. Colony encrusting, white; zooids irregularly hexagonal; frontal wall with many small pores, aperture a semicircle with narrow U-shaped proximal sinus, operculum stem fits in sinus; small suboral umbo in some zooids; ovicell large, irregular, imperforate, leaning over aperture; reported in circumtropical, warm temperate waters; ? British Columbia, California, to ? Costa Rica; intertidal to 80 m.

Buffonellaria vitrea (Osburn, 1952) (=*Stephanosella vitrea*). Plate 451E. See Soule et al. 1995. Colonies porcellanous, encrusting; zooids small, only distinct at growing margins, set off by a few areolar pores; aperture with V-shaped sinus; paired or single suboral frontal avicularia with rounded rostral opesia, narrowed to acute tip, directed laterally, sometimes a larger interzooecial avicularium resting on frontal but set off by areolae; ovicell globose, with large horseshoe-shaped frontal entooecium area bordered by slitlike pores, ectooecium hood thick, imperforate; on shell, rock, worm tubes; Puget Sound and south; shallow water to 170 m.

Buffonelaria bolini is a similar species from deep water, with larger zooids, a wider sinus; off Pt. Sur and Santa Rosa Island.

Cellepora costazii Audouin, 1826 (=*Costazia costazii*) See Osburn 1952. Not found in northeastern Pacific. See *Celleporina robertsoniae, C. souleae.*

Celleporaria brunnea (Hincks, 1884) (=*Holoporella brunnea*). Plate 451A. See Osburn, 1952, Soule et al. 1980, 1995. Forms thick brown or gray encrustations and clumps; colony very irregular, with large interzooecial avicularia having dark brown mandibles, tips curved upward distally; primary aperture symmetrical, rounded distally, almost straight proximally with shelflike condyles almost meeting in center to leave small notch; secondary aperture asymmetrical due to small oval suboral avicularium directed laterally; ovicell a very shallow hood becoming immersed; encrusting algae, stone, shell, Oregon to Ecuador; abundant in shallow water; intertidal to 50 m.

Celleporella hyalina (Linnaeus, 1767) (=*Hippothoa hyalina*). Plate 448C. Genus *Hippothoa* now limited to species with uniserial colonies. Colonies multiserial with three kinds of zooids: small tapered autozooids with aperture having wide, shallow sinus; female zooids without sinus, sometimes small suboral umbo, large ovicells with large pores; and tiny dwarf male zooids; frontal walls smooth with transverse striations, large, flaring areolae on margins where frontal budding occurs; smaller areolae also present; *C. hyalina* becomes multilaminar during reproduction. A northern Atlantic species; *C. hyalina sensu lato* in the eastern Pacific is a complex of species; reported from Miocene of Alaska, Pliocene, Pleistocene of California; Recent, reported from Alaska, San Francisco to Channel Islands; Galapagos Islands; encrusting rock, shell, algae, hermit crabs; from intertidal to >130 m. Similar species *C. cornuta* has a suboral umbo, a U-shaped sinus flanked by notches on autozooids, an imperforate ovicell on female zooids, found from Bodega Bay to Baja California. *C. santacruzana* colonies remain unilaminar, the latter has a narrow V-shaped sinus, imperforate ovicells. See Morris (1976, 1979, 1980).

Celleporina robertsoniae (Canu and Bassler, 1923) (=*Cellepora costazi* Robertson 1908, in part; *Costazia robertsoniae* of Osburn 1952, part; *Celleporina ventricosa* of Morris 1979, in part; see Soule et al. 1995). Plate 451B. Colony encrusting; initial zooids recumbent, frontally budded zooids erect, frontal wall with scattered small pores, areolae; primary aperture rounded with deep, V-shaped proximal sinus, flanked by paired lateral, oral,

rounded avicularia on pedestals forming secondary aperture; other small avicularia proximal or distal to aperture; larger shoe-shaped interzooecial avicularia; ovicells raised, globose, ectooecium imperforate, striated, with triangular entooecial frontal area having 12–16 radiating costae; on erect stems of algae, hydroids, seagrasses, bryozoans, wood, shell; Alaska, common at Dillon Beach to Tanner Bank off California-Mexican border; shallow waters to 100 m. Original *C. costazi* has imperforate frontal wall with small areolae, 16–20 costae on ovicell entooecial face.

Celleporina souleae Morris, 1979 (=*Costazia robertsoniae* of Osburn 1952, in part; see Soule et al. 1995). Plate 451C, 451D. Encrusting, multilaminar from frontal budding; frontal wall with many small pores, becoming reticulate; aperture with shallow, V-shaped sinus flanked by small paired avicularia raised on pedestals sometimes turned toward each other, rare interzooecial avicularia with tan to brown mandibles smaller than in *C. robertsoniae*; ovicell with large crescentic frontal entooecial area with five to eight costae, ectooecium a shallow imperforate hood, slit between layers; on erect stems; Bodega Head to southern California, Pleistocene of Oregon, California; shallow water to 60 m.

Coleopora gigantea (Canu and Bassler, 1923). See Osburn 1952. Encrusting, distinctive yellowish nodules, among largest cheilostome zooids, up to 2 mm in length; perforate frontal wall, high, flaring peristome; ovicell with an arcuate thin entooecial area, roughened, imperforate ectooecial outer layer; Recent, Monterey Bay, Channel Islands, Baja California; Pleistocene of Santa Monica; not common; shallow water to >200 m.

Costazia costazi (Audouin 1826): Osburn 1952. See *Celleporina souleae* Morris 1979.

Costazia robertsonae Canu and Bassler 1923. Plate 451B. See Osburn 1952. See *Celleporina robertsonae.*

Cryptosula pallasiana (Moll, 1803). See Soule et al., 1995. Colonies encrusting, thin; apertures bell shaped, frontal wall with reticulate pores extending distal to aperture; sometimes a small, round, suboral avicularium; no spines or ovicells; yellow, pink, orange embryos visible, brooded in body cavity; Alaska to Mexico, Chile; Atlantic from Nova Scotia to Florida, Norway to Red Sea, if all are same species; intertidal to 60 m.

'Dakaria' dawsoni (Hincks, 1883). See Soule et al. 1995. *Dakaria* type belongs in the genus *Watersipora*. The species is different from *Watersipora* and *Neodakaria*. Its generic status needs to be re-assigned. Colonies encrusting; zooids with distinct lateral walls, transverse walls extending down to middle of aperture, aperture ovate or D-shaped; frontal walls with reticulate pores, no spines or avicularia; ovicell immersed with single rounded opening or with several irregular pores in ectooecium, entooecium may be visible with pores in openings; British Columbia to Channel Islands; shallow water to >130 m.

Dengordonia uniporosa Soule, et al. 1995; =*Smittina bella*; Osburn, 1952, part. Plate 450A. Encrusting, zooids large; frontal wall coarsely reticulate, lateral walls indistinct; aperture higher than wide, pyriform margins becoming raised, with median proximal denticle, pronounced lateral condyles, median suboral avicularium; ovicell imbedded, a flap with a single median pore extended from next distal frontal, meeting raised lateral oral margins; British Columbia to southern California; intertidal to 50 m.

Eurystomella bilabiata (Hincks 1884). Encrusting; reddish, pink or red-brown; zooids large, hexagonal but rounded distally, aperture resembling profile of derby hat, curved proximally with notches at ends for passage of opercular muscles;

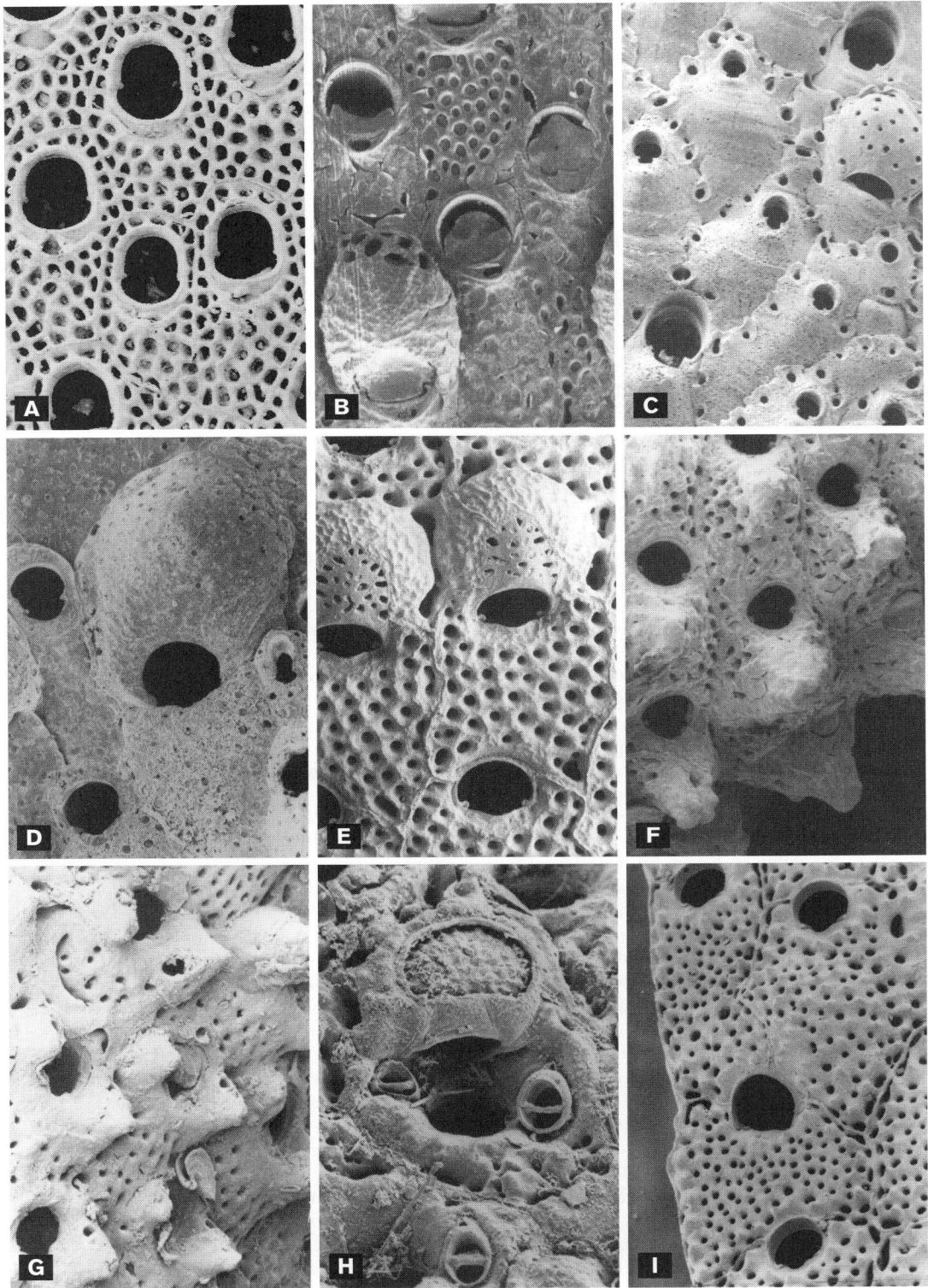

PLATE 448 Ascophorina, continued. A, *Cryptosula pallasiana*, with bell-shaped apertures, median suboral sulcus bearing avicularium in some; B, *Hippoporina insculpta*, with large frontal pores, ribbed ovicell lower left; C, *Celleporella hyalina*, with large porous ovicell on right, dwarf male proximal, transverse frontal striations; D, *Trypostega claviculata*, dwarf zooid left of ovicell, tiny avicularium to right; E, *Neodakaria islandica*, ovicells with central pores; F, *Neodakaria umbonata*; G, ?*Schizoporella cornuta*; H, *S. cornuta* ovicell; I, *Schizoporella inarmata*.

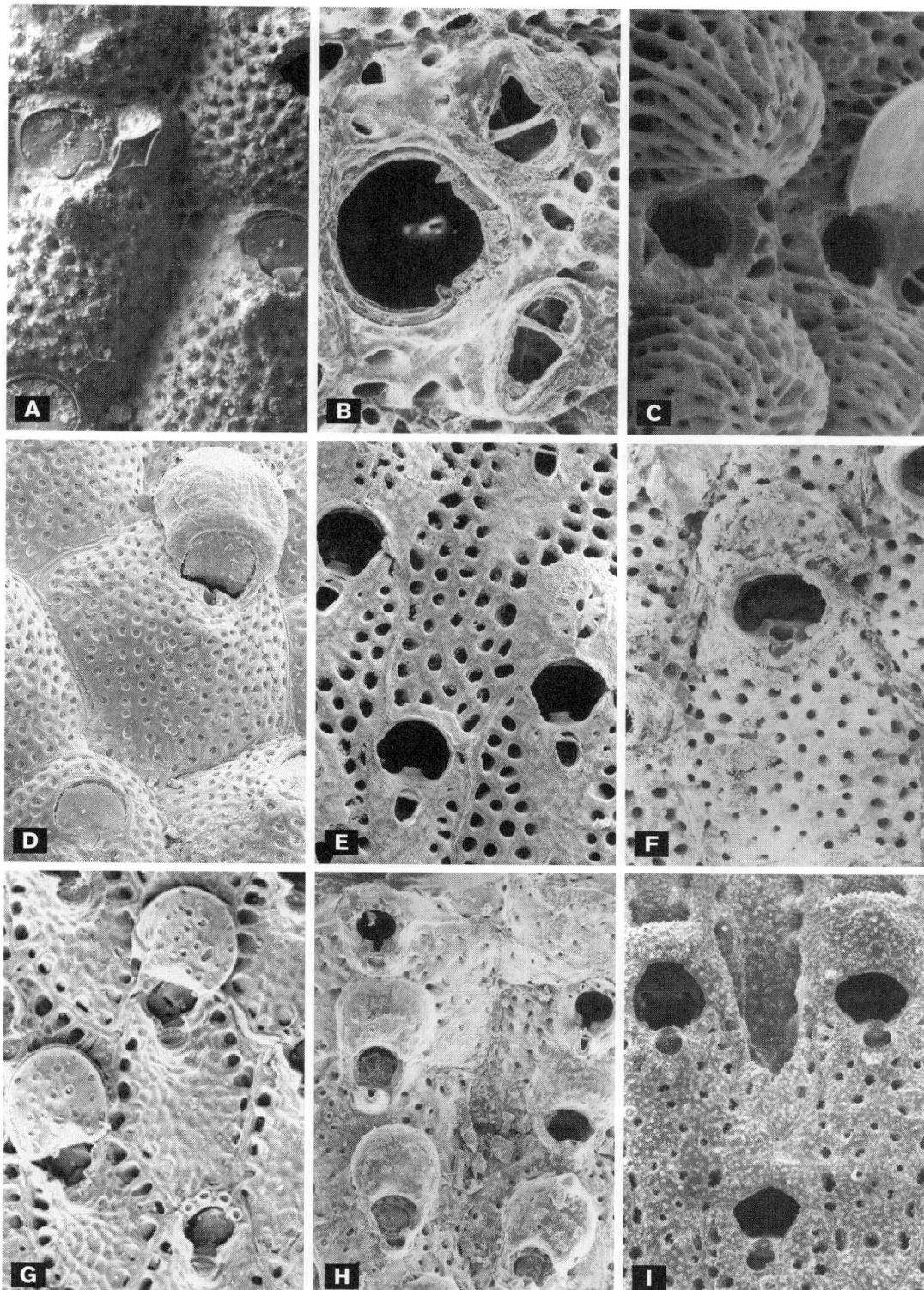

PLATE 449 Ascophorina, continued. A, *Schizoporella unicornis* (lectotype, Britain), with very acute lateral oral avicularium, suboral umbo; B, *Schizoporella pseudoerrata*, rotated 90 degrees left, with sharp oral condyles, paired lateral suboral avicularia; C, *Schizoporella pseudoerrata*, ovicell typical of genus; D, *Arthropoma cecilii*, with proximal apertural notch, operculum fill snotch; E, *Smittina ovirotula*, median lyrula, suboral avicularium, ovicell with wagon-wheel pore pattern; F, *Smittina veleroa*, with median lyrula bearing avicularium, ovicell with many pores; G, *Smittoidea prolifica*, ovicells with many pores, four spine bases on zooid without ovicell; H, *Schizosmittina pedicellata*, with suboral avicularium on pedistal; I, *Raymondcia osburni*, with suboral avicularium, upper ovicells developed from distal, lateral flaps, bottom ovicel submerged.

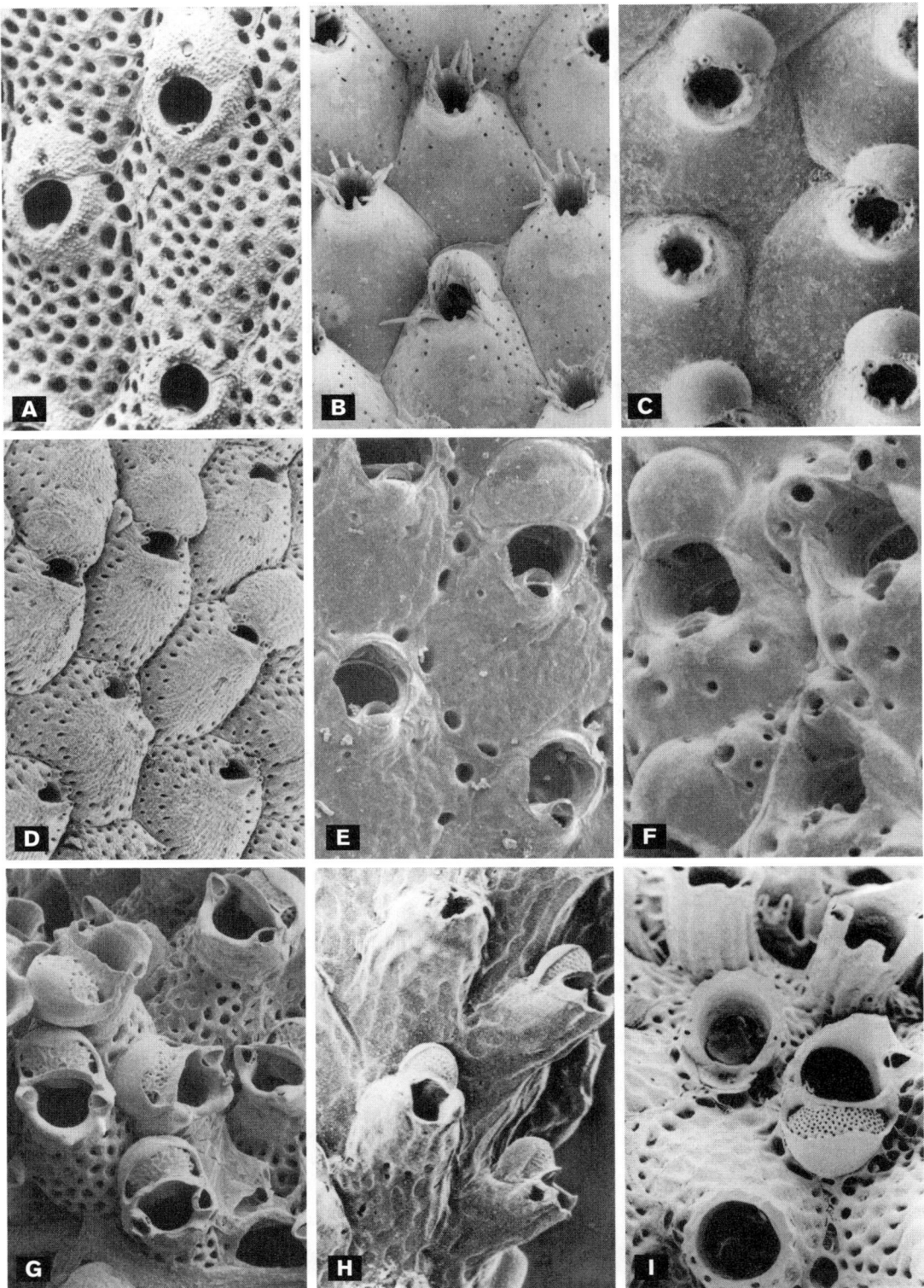

PLATE 450 Ascophorina, continued. A, *Dengordonia uniporosa*, ovicell distal flap with single pore; median lyrula bearing avicularium; B, *Haywardipora major*, slender spines around aperture, median suboral mucro, ovicell small, imperforate, central frontal wall without pores; C, *Haywardipora orbicula*, raised subterminal aperture with spines, median mucro, frontal wall with pores, ovicell small, imperforate; D, *Haywardipora rugosa*, with suboral umbo, mid-frontal rugose with marginal pores; E, *Porella columbiana*, median lyrula bearing raised avicularium, frontal imperforate with few large areolae, ovicell small, imperforate; F, *Porella porifera*, median suboral avicularium on umbo set off by pores, frontal wall imperforate with few areolae, imperforate ovicell; G, *Lagenicella neosocialis*, raised tubular apertures flanked by avicularia, reticulate frontal wall, ovicells with perforate frontal area; H, *Lagenicella punctulata*, tall peristomes with fused spines flanked by two small avicularia, frontal wall reticulate, ovicell suspended on distal peristome rim, with perforate central area; I, *Lagenicella spinulosa*, tall peristomes with irregular points topped with spines, avicularia on tall posts, ovicells suspended from distal peristome rim, with central area perforate.

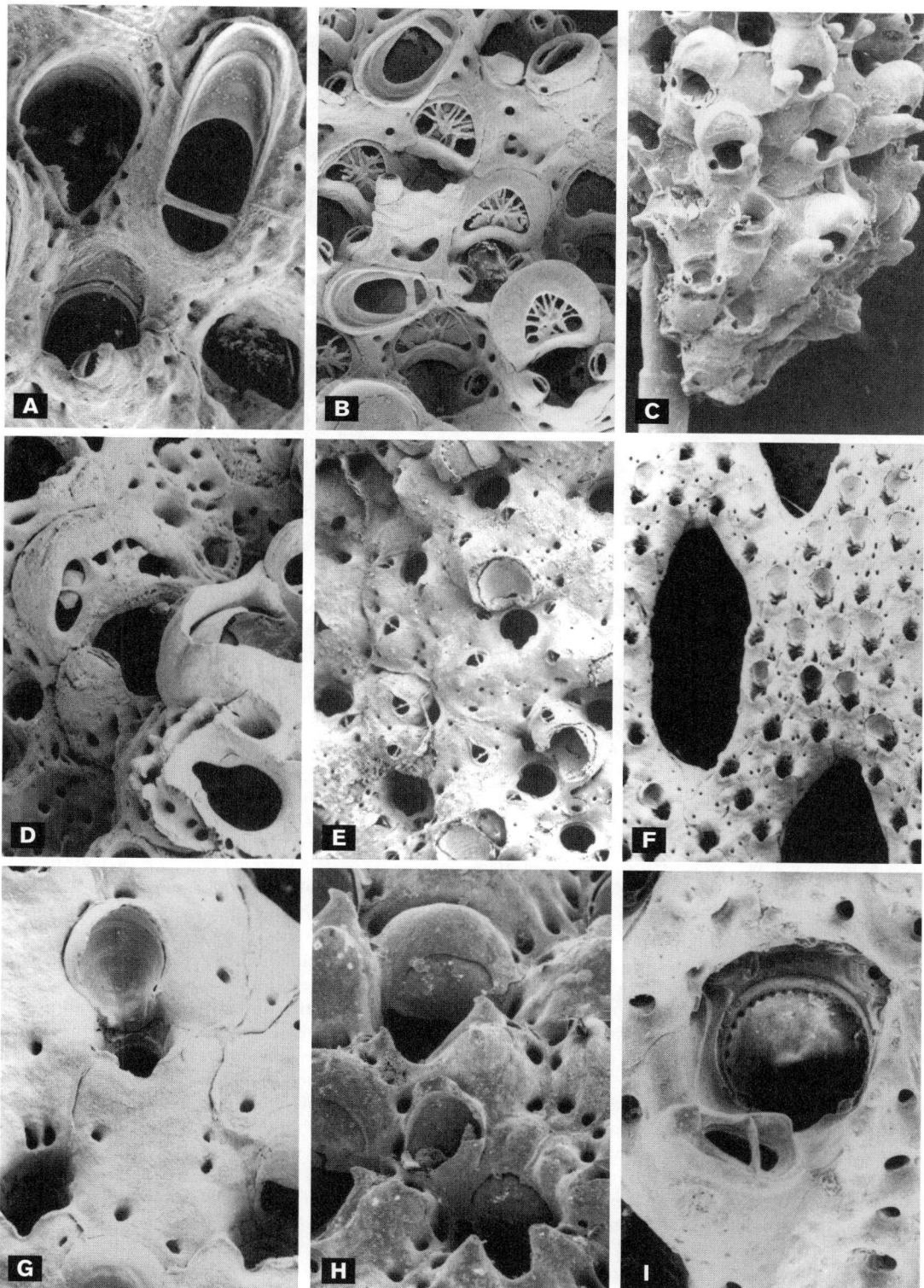

PLATE 451 Ascophorina, continued. A, *Celleporaria brunnea*, with serrate avicularium in aperture, giant spatulate interzooe-cial avicularium on right; B, *Celleporina robertsoniae*, with paired lateral oral avicularia on posts, shoe-shaped interzooecial avicularia, ovicells raised with radiate ribs in central area; C, *Celleporina souleae*, colony; D, *Celleporina souleae*, primary aperture lower right with deep sinus, ovicell with radiate ribs; E, *Buffonellaria vitrea*, with developing and immersed ovicells with imperforate central area set off by slits; many interzooecial avicularia; F, *Phidolopora pacifica*, fenestrate colony; G, *Phidolopora pacifica*, developing ovicell, sunken aperture with beaded distal rim, frontal wall imperforate with few areolae; H, *Rhynchozoon rostratum*, ovicell with central labellum, apertures hidden by avicularium on umbo; I, *Rhynchozoon rostratum*, primary aperture with beaded rim, asymmetrical suboral avicularium.

operculum dark brown; no avicularia or spines; ovicell a sub-triangular shallow hood with large median pore; reddish color is a caratinoid, *hopkinsiaxanthin,* transferred intact to the predator nudibranch *Okenia rosacea;* Alaska to Tenacatita, Mexico; intertidal to >200 m.

Fenestrulina farnsworthi Soule, Soule and Chaney, 1995 (*=Fenestrulina malusi* of Osburn, 1952, in part). Plate 447F. Colony encrusting; zooids small, separated by gymnocystal walls, frontal wall with crescentic ascopore containing branched denticles; aperture curved distally with five to seven spines on young zooids; scattered pores on frontal wall and between ascopore and aperture with denticles that form grid over pores; no avicularia, no ovicell; ancestrula with 12 marginal spines; Pleistocene, southern California; Recent, Channel Islands, La Jolla; ? British Columbia; 15 m–20 m, but may be on intertidal drift algae or shell.

Fenestrulina malusii (Audouin, 1826). Species probably restricted to eastern Atlantic, Mediterranean, not an eastern Pacific species; =*F. malusi* of Osburn, 1952, in part: see Soule et al. 1995. See *Fenestrulina farnsworthi, Fenestruloides blaggae, F. eopacifica, F. miramara, F. morrisae, F. unibonata.*

Fenestruloides blaggae Soule, Soule and Chaney, 1995 (*=Fenestrulina malusi* of Osburn, 1952, in part). Plate 447G. Colonies encrusting, porcellanous; small zooids irregular in size, shape; young zooids with single row of marginal pores, frontal wall with oval ascopore with median uvulate process, both with serrate rims, in mature zooids two rows of pores between aperture and ascopore; pores increasing in number to fill frontal wall except proximal to umbo below ascopore; pores containing denticles; extra umbones on proximal frontal wall; ovicell imperforate except for marginal areolar pores; two spines flanking aperture of ovicell; Carmel Bay on settling plates, perhaps more widely distributed as part of Osburn's *F. malusi;* shallow water.

Fenestruloides eopacifica Soule, Soule and Chaney, 1995 (*=Fenestrulina malusi* of Osburn 1952, in part). Colonies encrusting; zooids small; young zooids with single row of marginal pores; frontal pores filled with denticles, one to three rows between aperture and ascopore, older zooids with pores over entire frontal wall; ascopore raised, no other umbo, one to three spines distal to aperture; ovicell imperforate except for marginal areolar pores, a rim separating ovicell from next zooid; ancestrula resembles true *Fenestrulina,* with single row of marginal and suboral pores; on kelp off San Onofre, probably occurs in other shallow waters as part of "*F. malusi*" of Osburn.

Fenestruloides miramara Soule, Soule and Chaney, 1995 (*=Fenestrulina malusi* of Osburn, 1952, in part). Plate 447H. Colonies encrusting; zooids small, aperture semicircular with two to three distal spines, two lateral spines; ancestrula and young zooids with one row of marginal pores, pores covering frontal except proximal to umbonate ascopore, ascopore with small uvulate process, denticulate rims, pores denticulate; ovicells large, imperforate except for large marginal pores, aperture flanked by two spines; on drift algae off Montecito (Santa Barbara), beach, probably elsewhere as part of "*F. malusi*"; shallow water.

Fenestruloides morrisae Soule, Soule and Chaney, 1995 (*=F. malusi* of Osburn, 1952: 387, in part). Colonies encrusting; zooids larger than most species of genus, irregularly hexagonal, without gymnocystal rim; pores with denticles covering frontal wall, ascopore wide, opening between uvulate process and rim narrow, denticulate; ovicell flattened, imperforate except for one to two closed pores, but with marginal pores

bound by gymnocystal rim; the only *Fenestrulina* or *Fenestruloides* with a rare avicularium, acute, at proximal margin of zooid; on shell, stone; Channel Islands, California coast, Revillagigedo Islands off Baja California, Gulf of California, perhaps north elsewhere as "*F. malusi*"; on shell, stone; shallow water to 100 m.

Fenestruloides umbonata (C. H. O'Donoghue and E. O'Donoghue, 1926) (*=Fenestrulina malusi* var. *umbonata* of Osburn 1952). Plate 447I. See Soule et al. 1995. Colonies encrusting; zooids hexagonal with little or no gymnocyst rim on margins; frontal wall with numerous denticulate pores, ascopore wide, almost closed by uvulate process, both rims denticulate, a small umbo proximal to ascopore but separate from it; ovicell large, rugose, with ribbed front margin like two spines folded across it, flanked by two spines, marginal pores small, rimmed by gymnocyst; hard substrates; British Columbia to Channel Islands; shallow water to >100 m.

Haywardipora major (Hincks, 1884) (*=Mucronella major* of Osburn 1952, in part). Plate 450B. See Soule et al. 1995. Colonies encrusting, zooecia large, up to 1 mm, zooids hexagonal, raised, aperture subterminal with high peristome, spine bases imbedded in peristome, a mucro on median proximal lip extending down as ridge to wide, anvil shaped lyrula in primary aperture; frontal wall with two to four irregular rows of tubular pores; no avicularia; ovicell small, imperforate suspended on distal peristome, becoming recumbent; on rock, shell; British Columbia to southern California, ? Galapagos Islands, identifications confused; shallow water.

Haywardipora orbicula Soule, Soule and Chaney, 1995 (*=Mucronella major* of Osburn, 1952 in part). Plate 450C. Colonies encrusting; zooids large, hexagonal with inflated frontal wall, subterminal peristome tall with 10 long spines forming collar, large pores over entire frontal wall including distal to peristome; hemicylindrical ridge extending down to primary aperture, an inverted triangular median denticle at base of ridge; ovicells small, with imperforate ectooecium, distal to spines, appearing to dangle from peristome; perhaps from Vancouver Island to Baja California; shallow water to >200 m.

Haywardipora rugosa Soule, Soule and Chaney, 1995 (*=Mucronella ventricosa* of Osburn, 1952, in part). Plate 450D. Colonies encrusting; zooids ovoid to hexagonal, raised distally, frontal wall rugose, two to three rows of marginal pores with denticles projecting inward, tubules extending beneath secondary calcification up frontal wall; aperture with four large spines, sometimes two small spines flanking proximal rim, two blunt lateral condyles, a low truncate median denticle with suboral umbo; ovicells large, imperforate, recumbent on next zooid, aperture with four stout spines; on hard substrates, other bryozoans; ? Arctic; Puget Sound, Oregon, Channel Islands; shallow water to ? >400 m.

Haywardipora rylandi Soule, Soule and Chaney, 1995 (*=Mucronella major:* Osburn 1952, in part). Colonies encrusting, zooids large, irregularly hexagonal, frontal wall raised, with many small pores extending upward from areolae; peristome raised, with eight spines, a sunken median, anvil-shaped denticle; a pointed external suboral mucro or umbo; ovicells small, dimpled, imperforate or with a few scattered marginal pores; Point Barrow, Alaska, possibly to Baja California; distributions uncertain due to confused identifications; shallow water in north to deeper water in south.

Hippoporina insculpta (Hincks, 1882) (*=Hippodiplosia insculpta* of Osburn, 1952, in part). Plate 448B. See Soule et al. 1995. Colonies encrusting; young colonies flat, sometimes raised in fan shapes or yellowish to orange frills; zooids flattened, reticulate with coarse pores; aperture with low, curved distal rim ending at small upturned condyles, proximal rim a widely curved sinus, large median suboral pore sometimes bearing a tiny avicularium; ovicells with radiate ribs, set off by areolar pores; on rock, shell, algae, bryozoans, hydroids; Alaska to central California, common in shallow water, to >200 m.

Hippoporina mexicana Soule, Soule and Chaney, 1995 (=*Hippodiplosia insculpta* of Osburn 1952, in part). Colony encrusting; zooid frontal walls arched, reticulate with large pores; aperture rounded, higher than wide, peristome raised, encloses a suboral pore, sometimes bearing a tiny thin blunt, proximally directed avicularium, down-curved condyles hidden by peristome; ovicells nodular, raised; forming large aperture with peristome; on shell, stone; Channel Islands to Cocos Island off Costa Rica, but if recognized range may overlap that of *H. insculpta* in central California; shallow water to deep.

Hippomonavella longirostrata (Hincks 1882): Osburn. See *Pleurocodonellina longirostrata*.

Hippothoa spp. See *Celleporella hyalina*.

Lagenicella neosocialis Dick and Ross, 1988. See Soule et al., 1995. Not *Lagenipora socialis* (Hincks, 1877); not *Lagenipora socialis* of Osburn 1952, nor of C. H. and E. O'Donoghue 1923 or J. D. Soule 1961. Plate 450G. Colonies small, encrusting; zooids flask shaped, frontal wall perforate, reticulate, primary aperture round with spine scars, peristome becoming tall, thick; proximal lip thick, curved, flanked by pair of tiny, acute avicularia raised above peristome; ovicell with distal imperforate ectooecial hood with arcuate, perforate frontal entooecial plate with pores above tubular peristome, aperture flanked by pedicellate avicularia; on algae, rock, shell, bryozoans; intertidal to >120 m.

Lagenicella ?punctulata (Gabb and Horn, 1862) (=*Lagenipora punctulata*). Plate 450H. See Soule et al. 1995. Colonies erect, branching, cylindrical; zooids flask shaped, frontal wall coarsely punctate, reticulate; primary aperture circular, becoming obscured by tall imperforate peristome resembling fused spines; with paired low spinules flanking proximal rim, pair of tiny, raised, acute avicularia proximal to spinules or replacing them; ovicell appears hung from side of peristome, ectooecial hood distally imperforate, arcuate frontal entooecium perforate, becoming occluded; no type known for Pleistocene fossil from Santa Barbara, so identity is questionable; Recent, ? British Columbia, California, Baja California, Gulf of California,? Galapagos Islands; shallow waters to 200 m.

Lagenicella spinulosa (Hincks 1884) (=*Lagenipora spinulosa*). Plate 450I. See Soule et al. 1995. Colonies small, encrusting; zooids small, flask shaped, frontal walls reticulate, covered by thick cuticle, peristomes raised, resembling partly fused spines with irregular tips; paired small acute avicularia raised above spines; primary aperture round with three to four spine scars; ovicells with imperforate ectooecial hood distally, arcuate perforate entooecial plate frontally, appearing suspended from side of peristome, paired pedicellate avicularia flank aperture, with large median suboral umbo or shelf; British Columbia, southern California Pleistocene, Recent; Gulf of California; ? Galapagos Islands; intertidal to >120 m.

Lagenipora spp. See *Lagenicella*. *Lagenipora* spp. have imperforate frontal walls and ovicells; apparently not present in eastern Pacific. *Lagenicella* spp. have perforate frontal walls and ovicells; tall peristomes imperforate, resembling fused spines.

Microporelloides (*Cribriporella*) *californica* (Busk, 1856). See Soule et al., 2003; =Osburn 1952, in part; see also *M. infundibulipora* Soule, Soule and Chaney, 1995. Plate 447A.

Colonies encrusting; zooids ovoid, separated by grooves, frontal wall inflated, with numerous reticulate pores containing wheellike sieve plates; aperture curved distally with four spines or spine scars, straight proximally with small condyles part of a dental ledge, a large ascopore with slim uvulate process, both denticulate, separated from proximal lip and from secondary frontal wall calcification; flanked by pair of raised acute avicularia directed distolaterally; ovicells globular, immersed, perforate to reticulate; ? British Columbia; central, southern California; intertidal to 150 m.

Microporelloides (*Microporelloides*) *catalinensis* (Soule, Soule and Chaney, 1995) (=*M. ciliata* of Osburn, 1952, in part. See Soule et al. 2003). Plate 447B. Colonies encrusting, yellowish; zooids ovate to quadrate, frontal wall granular with large pores sometimes merging into slits, no pore spicules or plates, circular ascopore proximal to aperture with small uvulate process, both finely denticulate, small umbo proximal to ascopore; aperture rimmed, wider than high, curved distally, straight proximally with small lateral condyles; sometimes one acute avicularium on lateral frontal wall directed outward; ovicell with pores radiating from proximal center of hood, distal area immersed in next distal zooid; on shell, algae; Channel Islands, probably recorded elsewhere in California as "*M. californica*"; shallow water to ? 27 m.

Microporella ciliata (Pallas 1766). See Osburn 1952; Atlantic species with imperforate ovicells; see Hayward and Ryland 1998, not found in eastern Pacific waters; see *Microporelloides catalinensis*, *M. planata*.

Microporelloides (*Cribriporella*) *cribrosa* (Osburn, 1952) (=*Microporella californica* Robertson 1908, in part). Plate 447C. See Soule et al. 2003. Encrusting; zooids small, frontal wall raised with many small pores having stellate or cribrate plates at bottom; aperture arched distally, with five to six spines, almost straight proximally, condyles present or worn away; ascopore uvulate process denticles grow to meet marginal denticles forming sieve plate; an umbo proximal to ascopore; avicularia single or paired flanking ascopore, directed distolaterally, acute with setose mandible; ovicell a perforate, ribbed hood with rib on margin, a median umbo on top, aperture flanked by two spines; on rock, shell, algae, pilings; Mussel Point, central California, to Gulf of California; shallow water (pilings) to >120 m.

Microporelloides (*Cribriporella*) *infundibulipora* (Soule, Soule and Chaney, 1995) (=*Microporella californica* of Osburn 1952, in part). Plate 447D. See Soule et al. 2003. Colonies encrusting; zooids ovoid with deep separating grooves, frontal wall reticulate having large infundibuliform pores with sunken pore plates like spoked wheel, aperture curved distally with five spine scars, straight proximally with tiny condyles, small ascopore proximal to lip with small uvulate process, small denticles on pore rim, umbo sometimes quite large, reticulate; frontal wall proximal to umbo very reticulate; ascopore flanked by paired, bluntly acute avicularia directed distolaterally; encrusting both sides of shells; Pleistocene, Recent of southern California, probably central California; reports of "*M. californica*" range from British Columbia to Gulf of California, Baja California, Galapagos Islands; *M. infundibulipora,* described from Channel Island; shallow water to >150 m.

Microporelloides (*Cribriporella*) *planata* (Soule, Soule and Chaney, 1995). See Soule et al. 2003. Colony encrusting; flat, thin, fragile, yellowish; zooids varied in size, aperture curved distally, four fragile spines, almost straight proximally; frontal wall with small pores containing reticulate plates, areolae larger; suboral ascopore small, denticulate, with rim, median uvulate process, denticulate; avicularia paired, proximolateral to ascopore, directed distolaterally, with acute, setose mandibles; ovicells perforate, ribbed except on central imperforate area, hood margin curved away from frontal wall at lateral wall; on shell, stone; found in collections labeled *M. ciliata* by Osburn, reported as cosmopolitan; from off Channel Islands; specimens identified in collections from nearshore to >180 m.

Microporelloides (Cribriporella) setiformis (C. H. and E. O'Donoghue 1923:). See Soule et al. 2003. Colonies encrusting; zooids with frontal wall inflated, numerous small frontal pores with deep set cribrate plates, aperture horseshoe-shaped distally, straight proximally, ascopore tiny, with round flat collar, tiny uvulate process, no denticles on either; avicularia paired, small, flanking ascopore, triangular bases, complete hinge bar, short setose mandibles directed distally; ovicell prominent, many small pores, becoming ribbed, pores closed, immersed; shallow water to more than 100 m.

Microporelloides (Microporelloides) umboniformis (Soule, Soule and Chaney, 1995) (=*Microporella umbonata* of Osburn 1952, in part). Colony encrusting; zooids oval, frontal wall inflated with many tiny pores; aperture rounded with five to six distal spine scars, less curved proximal margin, a small ascopore with slender uvulate process, both denticulate; umbo proximal; one or two setose avicularia proximolateral to umbo directed laterally, or no avicularia; ovicell small, rounded with umbo, small pores; on worm tubes; southern California, ? northern California, ? Baja California; shallow water to >100 m. *M. umbonata* (Hincks) has large pores, three umbones; a Puget Sound species, it may occur in northern, central California.

Microporelloides (Microporelloides) vibraculifera (Hincks, 1883) (=*Microporella vibraculifera* of Osburn 1952). Plate 447E. See Soule et al. 1995. Colony encrusting, zooids irregularly hexagonal, with numerous frontal pores; aperture semicircular with five to seven hollow spines; median suboral ascopore small with small median uvulate process, both finely denticulate; avicularium a bulbous chamber proximolateral to ascopore with very long, setose mandible sweeping over adjacent zooids; ovicell raised, with many small pores; British Columbia to Baja California; tolerant of sediment; 5 m to >125 m.

Mucronella major (Hincks, 1884). See *Haywardipora major*.

Mucronella ventricosa (Hassall, 1842). Osburn, 1952. See *Haywardipora rugosa*; not *M. ventricosa*: Hincks 1880; ? C. H. and E. O'Donoghue; ? Kluge 1975.

Neodakaria islandica Soule, Soule and Chaney, 1995 (=*Dakaria ordinata* of Osburn 1952, in part). Plate 448E. Encrusting; zooids quadrate, frontal wall reticulate with large pores, aperture semicircular distally, pushing into next distal zooid, large condyles, proximal lip a shallow curve or straight, not V-shaped; no spines, no avicularia; ovicells with arched subtriangular entooecium with large pores, surrounded by granular ectooecial hood; on rock, shell, algae; Channel Islands, ? northern California to Baja California; shallow water to 50 m.

Neodakaria ordinata (C. H. and E. O'Donoghue, 1923) Soule et al. 1995 (=*Dakaria ordinata* of Osburn, 1952, in part). Encrusting; zooids quadrate to hexagonal, frontal wall reticulate with large pores becoming closed by calcification, aperture arched distally with large condyles, proximal lip a wide, V-shaped sinus; ovicell with frontal triangular porous area, rimmed by rugose ectooecium without pores; on rock, shell, algae; British Columbia, ? to Dillon Beach; intertidal to 70 m.

Neodakaria umbonata Soule, Soule and Chaney, 1995 (=*Dakaria ordinata* of Osburn 1952, in part). Plate 448F. Encrusting pebbles on sandy bottoms; zooids rectangular with large frontal wall pores, large marginal areolae, aperture curved distally, a wide shallow curve proximally with strong condyles, a large frontal suboral umbo; ovicells large with large frontal area, large pores, little encroaching rim distally; recognized from Monterey Bay to San Pedro Bay, and Channel Islands; intertidal on pebbles to more than 50 m.

Parasmittina collifera (Robertson 1908). See Osburn 1952. Colonies encrusting, becoming nodular, heaped, multilaminar; frontal wall rugose, imperforate with pillars (colli), marginal areolae; aperture rounded with two distal spines on young, with secondary peristomal collar, truncate median denticle, lateral condyles; avicularia small to large triangular, erect, directed distally, very large in older colonies, small to large ovate frontal avicularia directed variously; ovicells with few large irregular pores, ectooecium overgrowing pores with pillars in older zooids; British Columbia to Coranados Islands of Mexico; common intertidally off Oregon, central California, deeper water off Mexico.

Parasmittina trispinosa (Johnston): Osburn, 1952. Atlantic-Boreal species, *P. trispinosa* does not occur in eastern Pacific (see *P. regularis*, D. F. Soule and J. D. Soule, 2002, from off Dillon Beach, *P. aviculifera*, D. F. Soule and J. D. Soule, 2002, from off Monterey). *P. californica* (Robertson, 1908) (see Soule et al. 2002) is a southern California–Baja California species with large serrate avicularia may range north only to Point Dume; on gravel, sponge, shell; shallow water to 100 m.

Parasmittina tubulata Osburn, 1952. Encrusting; zooids large, raised, frontal wall finely granular, with large, irregular marginal areolae; aperture with anvil-shaped median denticle, lateral condyles in primary aperture, secondary aperture tubular with sinus, hiding denticle; a large or small spatulate avicularium directed proximolaterally, small acute frontal avicularia directed proximally, small acute avicularia near areolae directed variously; ovicells large, entooecium with many pores, peristome a complete tube in front of ovicell, ovicell becoming immersed with encroaching ectooecium leaving only a small crescent of pores; northern California to Scammons Lagoon, Baja California; shallow water to 150 m.

Phidolopora pacifica (Robertson, 1908). Plates 451F, 451G. See Soule et al. 1995. Colony erect, fenestrate, forming large meshwork from rounded base attached to hard substrate by kenozooids; fenestrae ovoid to diamond shaped; zooids all opening on ventral surface, dorsal surface showing outlines of kenozooids with acute avicularium at base of fenestrae; primary aperture rounded with beaded rim distally, small sinus proximally, sometimes flanked by one to two elongate spines, aperture becoming sunken with notch on proximal border; occasional large raised avicularium with hooked rostrum directed proximally, recumbent on frontal wall; ovicell ectooecium imperforate with concentric lines, a frontal uvulate extension sometime hanging down over aperture; British Columbia to Galapagos Islands; shallow water to >200 m.

Pleurocodonellina longirostrata (Hincks, 1882) (=*Hippomonavella longirostrata* of Osburn, 1952, in part). See Soule 1961, Soule et al. 1995. Encrusting; zooids elongate with one row of areolar pores, a few other scattered frontal pores; aperture curved distally ending at large down-curved condyles sometimes with two small distal spines, proximal lip a shallow curve connected by sulcus to avicularium either median or skewed laterally, avicularium with V-shaped hinge bar; ovicells with medium sized pores, surrounded by secondary frontal wall of next distal zooid; shallow water to ? 200 m; may be confused with *P. californica* Soule, Soule and Chaney, 1995, which has three to five tiny distal spines, frontal wall with three rows of marginal areolae, leaving only central area imperforate, long or short acute avicularium median or skewed, connected to proximal aperture; ovicell with large pores raised above next zooid; known only from waters 90 m–150 m off southern California.

Porella columbiana C. H. and E. O'Donoghue, 1923. Plate 450E. See Soule et al. 1995. Colonies encrusting, thin, shiny, yellowish white; zooids with frontal wall raised, ventricose, imperforate, a few large areolar pores becoming sunken, buttressed; primary aperture rounded distally with four spines, strong lateral condyles, wide median lyrula, becoming hidden by raised lateral, distal lappets, forming peristome, a median, raised, bluntly acute suboral avicularium with two to three small pores at base of avicularium chamber; ovicell imperforate, separated from next distal zooid, with a brim above ovicell opening; British Columbia to southern California, ? Galapagos Islands; intertidal on floating kelp to 110 m.

Porella major Hincks, 1884 (?=*Porella acutirostris* of Osburn, 1952, Kluge, 1975; Dick and Ross 1988). See Soule et al. 1995. Colony encrusting, light brown; zooids regular, smooth, lateral walls raised with single row of six to 10 areolae; primary aperture rounded distally, with shallow rim, no spines, tiny lateral condyles, a short, wide lyrula; a median suboral avicularium proximal to lyrula, within peristome, originating from one to two lateral areolae, forming raised chamber, with few small pores at base; avicularium bluntly acute, directed distally; ovicell imperforate; on shell; Alaska to southern California; shallow water to >100 m.

Porella porifera (Hincks, 1884) (=*P. porifera* of Osburn, 1952, in part; see also *Porella taylori*). See Soule et al. 1995. Plate 450F. Encrusting; zooids ovate, quadrate, frontal wall smooth, imperforate, with three to four lateral marginal areolae, four frontal pores at base of median avicularium; aperture rounded distally with four thin spines or spine bases, lateral condyles small, proximal rim with small truncate lyrula, aperture becoming sunken, surrounded by peristome; large median avicularium originating from two lateral frontal pores; large and small ovoid interzooecial avicularia set off by areolae on mounds between zooids, sometimes absent; ovicells small, imperforate, set off by brim, becoming immersed; on rock, shell; British Columbia to southern California, ? west coast of Baja Calfornia; intertidal to 250 m.

Porella taylori Soule, Soule and Chaney, 1995. See also *P. porifera*. Encrusting; zooids subhexagonal, frontal wall imperforate, granular; primary aperture rounded to quadrate, becoming sunken within peristome, tiny median lyrula, no visible lateral condyles; median suboral avicularium mostly outside peristome, ovoid, with acute tip directed proximally, chamber raised, with five to six pores around base; tiny avicularia flanking distal corners of aperture directed distolaterally, originating at lateral walls (interzooecial); ovicells imperforate, raised; on shell; only known from Santa Barbara Channel, 69 m–74 m; perhaps unrecognized along California coast.

Raymondcia osburni Soule, Soule and Chaney, 1995 (=*Smittina landsborovi* [sic] of Osburn 1952 in part). Plate 449I. Encrusting; zooids ovoid to rectangular, primary aperture wider than high, becoming pyriform, distal rim formed by division in transverse wall, lateral, proximal rims bordered by frontal wall; truncate median lyrula, large paired condyles depressed; frontal wall beaded with large pores showing large coarse granules inside, median suboral oval avicularium directed proximally, mostly outside primary rim; ovicells composed of merging segments from adjacent zooid frontal wall, with suture lines, sometimes a small pore, becoming immersed to appear as solid frontal wall

surrounded by areolae; on shell, rock; *S. landsborovii* Johnston, 1847, is an eastern Atlantic boreal species mistakenly listed by Osburn from Alaska to Galapagos Islands; shallow water to >100 m. *R. osburni*, described from the Channel Islands, may range from Alaska to California, ? Galapagos Islands.

Raymondcia macginitei Soule, Soule and Chaney, 1995 (=*Smittina bella* of Osburn 1952, in part). Encrusting; aperture pyriform, wider distally, composed of flaps from adjacent frontal, distal walls; large median avicularium originating proximal to lip, moving into aperture to rest on lyrula; frontal wall granular with large pores having irregular spicules inside; ovicell level with frontal wall, formed by frontal wall flaps from sides, leaving slit, then merging at sutures with distal flap; on shell, rock; cool water–arctic species, may occur in Washington, Oregon, California; shallow water.

Rhynchozoon rostratum (Busk, 1856). See Soule, Soule and Chaney, 1995: includes *R. tumulosum* (Hincks, 1882). Plate 451H, 451I. Encrusting; zooids irregular, heaped except at growing margins, frontal wall smooth, inflated, with 12–14 marginal areolae, sometimes with ridges between; primary aperture round, beaded, with strong condyles, young with two distal spines, proximal border a sinus, becoming obscured by secondary peristome, a transverse, acute avicularium set to side of aperture forming a secondary sinus, becoming bulbous giving colony honeycomb appearance, other large acute frontal avicularia randomly placed; ovicell imperforate, becoming immersed, with distinct circular frontal entooecium area mostly surrounded by ectooecium hood; on rock, shell, other hard substrates; Alaska to South America if all are the same species; shallow water to 200 m.

Schizomavella acuta Soule, Soule and Chaney, 1995. Not a variety of *S. auriculata* as indicated by Osburn 1952. Encrusting; zooids quadrate with distinct lateral, transverse wall, frontal wall rugose with small pores, sometimes nodular; aperture curved distally, wider than high, shallow sinus proximally, large, cogged condyles; a median, elongate, acute, suboral avicularium directed proximally outside apertural rim, sometimes skewed; ovicells reticulate with pores over entire surface, becoming immersed in next distal frontal wall; on rock, shell; Channel Islands south to Baja, Gulf of California; common; 33 m to more 100 m.

Schizomavella auriculata (Hassall, 1842). See Soule et al. 1995; not Robertson 1908; not Osburn, 1952. See *S. robertsonae*. Original British *S. auriculata* has pyriform aperture with narrow sinus, suboral avicularium within aperture, small, rounded or elongate directed proximally; ovicell immersed, showing only a crescent of large, irregular entooecial pores, ectooecium immersed in rugose porous frontal of next distal zooid.

Schizomavella robertsonae Soule, Soule and Chaney, 1995 (=*S. auriculata* of Osburn, 1952, in part). Encrusting; zooids irregular, with rugose, porous frontal walls, raised, riblike in center supporting round, proximally directed avicularium; aperture wider than high, with wide, sometimes four spine scars; proximal shallow sinus, blunt condyles; ovicell nodular, small pores becoming hidden, set off by areolae from distal zooid; on shell; Channel Islands to Coronados Islands, 182 m, but "*S. auriculata*" of Osburn recorded from Oregon to Baja California, Gulf of California, shallow water to deep.

Schizomavella triavicularia Soule, Soule and Chaney, 1995 (=*S. auriculata* of Osburn, 1950, in part). Encrusting; zooids quadrate, lateral walls distinct, frontal wall with numerous pores; aperture round, sometimes with three small spine scars, U-shaped sinus proximally within peristome, incised at lower corners; three small rounded avicularia outside peristome, one suboral directed proximally, two flanking aperture directed proximally; ovicell with crescentic perforate ectooecium, with imperforate ectooecium surrounding distally and laterally, bounded by areolae; on rock, shell; Channel Islands to Galapagos Islands; >60 m, but may range, as *S. auriculata*: Osburn, from Oregon to Baja California, Gulf of California, shallow water to deep.

Schizosmittina pedicellata Soule, Soule and Chaney, 1995 (=*Schizomavella auriculata*: of Osburn, 1952, in part). Plate 449H. Encrusting; zooids irregular, rugose with large frontal wall pores except on pedestal bearing median suboral avicularium originating outside aperture, directed proximally, aperture rounded with five spine scars distally, a deep, U-shaped sinus proximally, condyles shelf-like; ovicell entooecium with large irregular pores, imperforate ectooecium distally, raised above distal frontal wall, set off by areolae; on rock, shell; recently described from the Channel Islands, but may range from Oregon to Gulf of California; shallow water to deep; new species from Channel Islands; >100 m.

?*Schizoporella cornuta* (Gabb and Horn, 1862). See Soule et al. 1955: ?=*Buffonellaria*. Plate 448G, 448H. Encrusting; zooids raised with frontal pores becoming sunken, occluded; primary aperture rounded distally, with strong paired condyles, wide V- or U-shaped sinus, becoming immersed; avicularia absent or one to two acute flanking aperture directed distolaterally, becoming erect (cornuate); ovicell not typically porous schizoporellid, globose, imperforate with central granular entooecium surrounded by thick ectooecial rim and bar across frontal area; on rock, shell, other hard substrates; reported from Pleistocene of Santa Barbara; shallow waters to 200 m.

?*Schizoporella inarmata* (Hincks, 1884). Plate 448I. See Soule et al. 1995. Encrusting; zooids irregular, flattened with large immersed frontal pores, distinct lateral walls; aperture higher than wide, with strong, burred, down-curved condyles, a narrower V-shaped sinus; avicularia absent; ovicell aperture without sinus, ovicell immersed, almost indistinguishable from adjacent porous frontal wall except for sutural lines above transverse walls; on rock, shell; British Columbia to Costa Rica; 3 m–4 m to >120 m.

Schizoporella pseudoerrata Soule, Soule and Chaney, 1995. Plate 449B, 449C. Encrusting; zooids regularly oriented in first layer, becoming heaped with frontal budding; frontal wall irregularly porous, margins indistinct; aperture wider than high, with thin, sharp condyles, wide, sinus within proximal ledge; avicularia one, two-paired, or absent proximolateral to aperture, raised, acute, directed distolaterally, mandibles not setose; ovicells perforate, sometimes with ribs, becoming immersed; on shell, rock, ships' hulls, pilings; Elkhorn Slough; other distribution uncertain, confused with other species; intertidal, shallow water, sometimes on shells in deeper water.

Schizoporella japonica Ortmann, 1890 (=*Schizoporella unicornis* of northeastern Pacific authors; see Dick et al., 2005). See also *Schizoporella pseudoerrata*. Plate 449A. Encrusting, spreading in circular patches, often multilaminar; zooids regular, with distinct grooves at lateral walls, frontal wall with many small and large pores, aperture wider than high or rounded with very shallow proximal sinus, condyles small, rounded; avicularia absent, single or paired, acute but not setose, proximolateral to aperture directed distolaterally; ovicells raised, with pores and strongly ribbed; Japanese species introduced with oyster culture (Powell 1970, J. Fish. Res. Bd. Canada 27: 1847–1853; Ross and McCain 1976 Northwest. Sci. 50: 160–171), Plate 449A is of *S. unicornis*, similar in general morphology.

Smittina landsborovii (Johnston 1847). See Soule et al. 1995 See also *Smittina veleroa, Raymondcia osburni*.

Smittina bella (Busk 1860). See Osburn 1952. See *Raymondcia macginitei, Dengordonia uniporosa*.

Smittina ovirotula Soule, Soule and Chaney, 1995 (=?*Smittina spathulifera* of Osburn 1952). Plate 449E. Encrusting; reddish brown; zooids elongate, reticulate with large pores except in central suboral area around avicularium; aperture rounded with strong median truncate denticle, lateral condyles hooked downward, a suboral avicularium formed outside peristome directed proximally; ovicell with central pores like wagon wheel in entooecium, surrounded by imperforate ectooecium; *S. spathulifera* (Hincks), originally described from British Columbia has a larger shoe-shaped avicularium lying partly within peristome on lyrula; on hard substrates; *S. ovirotula*, Channel Islands to Baja and Gulf of California, depths to 150 m.

Smittina veleroa Soule, Soule and Chaney, 1995 (=*Smittina landsborovii* of Robertson 1908, Osburn 1952; Soule and Duff 1957, Soule, 1961). Plate 449F. Encrusting; zooids with numerous frontal pores, becoming heavily calcified, aperture rounded, with large median denticle bearing small median, bluntly acute avicularium directed proximally, tiny condyles, peristome becoming raised, sometimes enclosing avicularium; avicularia in *S. landsborovi*, described from Ireland, are small, round, outside tall thin peristome, or large, transverse, shoe-shaped suboral avicularium, sometimes both kinds, larger one sometimes further down frontal wall; ovicell shallow, porous, becoming indistinct; on shell, rock; Oregon to ? Galapagos Islands but confused with other species; low tide to deeper waters.

Smittoidea prolifica Osburn, 1952. Plate 449G. Colonies encrusting, small white patches; zooids irregularly hexagonal, small, distinct margins, frontal wall imperforate, smooth or granular, single row of marginal areolar pores; aperture rounded distally with large condyles, truncate median denticle, peristomal collar low distally with two to four spine scars; small, rounded, median suboral avicularium, no other avicularia; large ovicells on most zooids, with many pores turning colony pink or yellow when ova are present; on stone, shell, stems; San Francisco Bay, Channel Islands to Baja California; intertidal to >100 m.

Stephanosella biaperta (Michelin, 1845): Osburn 1952. Not that species, described from Miocene of France. See *Buffonellaria vitrea*.

Stomachetosella condylata Soule, Soule and Chaney, 1995. Not *S. sinuosa* (Busk, 1860, an Arctic-boreal species; ?=*S. sinuosa* (Hincks, 1884); =*S. sinuosa* of Osburn, 1952, in part; ?=*S. sienna* Dick and Ross, 1988). Encrusting, rose-colored to purple patches; frontal wall raised, rugose, with marginal areolae plus a few frontal pores; aperture almost circular, with transverse walls meeting proximal to distal curve, proximal lip almost straight with U-shaped sinus, wide shelf-like condyles not present in *S. sinuosa*; no spines, no avicularia; ovicells immersed, set off only by areolae, a single central pore, becoming occluded; on shell, rock; Alaska to Channel Islands; shallow water to >125 m.

Trypostega claviculata (Hincks, 1884). Plate 448D. Encrusting; three kinds of zooids: quadrate to hexagonal autozooids, larger female zooids with very large ovicells and small zooids beside ovicells, possibly males; frontal walls with large pores over all surface including distal to aperture; aperture rounded with rocker-shaped proximal rim ending at sharp condyles directed proximally; small interzooecial avicularia (zooeciules of some

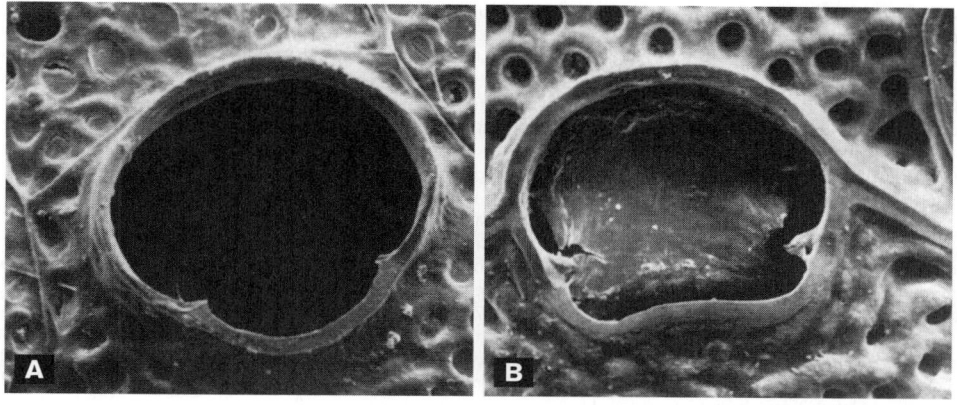

PLATE 452 Ascophorina, continued. A, *Watersipora subtorquata*, from St. Thomas, West Indies; B, *Watersipora arcuata* (A, B, from Soule and Soule 1975).

authors) with single row of frontal pores, spatulate mandible directed distally; ovicells large, raised, with many pores; on shell, rock; south to Morro Bay; shallow water to 180 m.

*Watersipora subtorquata (d'Orbigny, 1852). Plate 452A. An often abundant orange-red to black bryozoan, occurring both as crusts and (in quiet water) forming very large foliaceous masses (60 cm and more in length, and 30 cm and more in height) on floats and pilings in estuaries and harbors along the California and Oregon coasts. Handling fresh colonies of this species will stain the hands orange. Opercula black or dark brown. *Watersipora arcuata* Banta, 1969 (plate 452B) occurs in southern California, and the two species may co-occur. They are distinguished by the shape of the lower border of the aperture: in *W. subtorquata,* the lower border of the aperture is curved outward, whereas in *W. arcuata* the lower border of the aperture is curved inward. The species-level taxonomy of *Watersipora* remains to be worked out, and a number of additional names are in use, including *W. cucullata* and *W. subovoidea.* Note also the analysis of invasion patterns of *W. subtorquata* and *W. arcuata* by Mackie et al. (2006, Mar. Biol. 149: 285–295).

References

For more extensive references see Soule et al. 1995.

Bishop, J. D. D., and B. C. Househam. 1987. *Puellina* (Bryozoa; Cheilostomata; Cribrilinidae) from British and adjacent waters. Bulletin of the British Museum (Natural History), Zoology 53: 1–63.

Boardman R. S., A. H. Cheetham, D. B. Blake, J. Utgaard, O. L. Karklins, P. L. Cook, P. A. Sandberg, G. Lutaud, and T. S. Wood. 1983. Treatise on Invertebrate Zoology. Part G. Revised, Vol. 1. Geological Society of America, Boulder, Colorado, and University of Kansas, Lawrence, Kansas. 625 pp.

Cohen, A. N., and J. T. Carlton. 1995. Biological Study. Nonindigenous Aquatic Species in a United States Estuary: A Case Study of the Biological Invasions of the San Francisco Bay and Delta. A Report for the United States Fish and Wildlife Service, Washington, D.C., and The National Sea Grant College Program, Connecticut Sea Grant, NTIS Report Number PB96–166525, 246 pp.

Cook, P. L. 1964. Notes on the Flustrellidae (Polyzoa, Ctenostomata). Annals and Magazine of Natural History (13) 7: 278–300.

Dick, M. H., J. R. Freeland, L. P. Williams, and M. Coggeshall–Burr. 2000. Use of 16S mitochondrial ribosomal DNA sequences to investigate sister-group relationships among gymnolaemate bryozoans, pp. 197–210. In Proceedings of the Eleventh International Bryozoology Association Conference. A. Herrera Cubilla and J. B. C. Jackson, eds. Smithsonian Tropical Research Institute, 448 pp.

Dick, M. H., and J. R. P. Ross. 1988. Intertidal Bryozoa (Cheilostomata) of the Kodiak vicinity, Alaska. Occasional Paper 23, 133 pp. Western Washington University, Center for Pacific Northwest Studies, Bellingham, Washington.

Dick, M. H., A. V. Grischenko, S. F. Mawatari. 2005. Intertidal Bryozoa (Cheilostomata) of Ketchikan, Alaska. Journal of Natural History 39: 3687–3784.

Gordon, D. P. 1993. Bryozoan frontal shields: Studies on umbonulomorphs and impacts on classification. Zoologica Scripta 22: 203–221.

Gordon, D. P., and P. D. Taylor. 1997. The Cretaceous-Miocene genus *Lichenopora* (Bryozoa), with a description of a new species from New Zealand. Bulletin of the Natural History Museum, London 53: 71–78.

Gordon, D. P., and E. Voigt. 1995. The kenozooidal origin of the ascophorine hypostegal coelom and associated frontal shield, pp. 89–107. In Proceedings of the Tenth International Bryozoology Association Conference, Bryozoans in Time and Space. D. P. Gordon, A. M. Smith, and J. A. Grant-Mackie, eds. NIWA, Wellington, New Zealand, 442 pp.

Hayward, P. J. 1985. Ctenostome Bryozoans. Synopses of the British Fauna (new series), No. 33, 169 pp. The Linnean Society of London.

Hayward, P. J., and J. S. Ryland. 1985. Cyclostome Bryozoans. Synopses of the British Fauna (new series), No. 34, 147 pp. The Linnean Society of London.

Hayward, P. J., and J. S. Ryland. 1998. Cheilostomatous Bryozoa. Part I. Aetoidea–Cribrilinoidea. No. 10 (2nd ed.), 366 pp. The Linnean Society of London.

Hayward, P. J., and J. S. Ryland. 2002. Cheilostomatous Bryozoa. Part 2. Hippothoidea-Celleporoidea. No. 14 (Second Edition), 416 pp., The Linnean Society of London.

Hughes, R. S., A. Gomez, P. J. Wright, D. Lunt, G. R. Carvahlo, J. M. Cancino, and H. I. Moyano G. 2004. Phylogeography and sibling speciation in *Celleporella hyalina*. Boletin de la Sociedad de Biologia de Concepcion 74: 70.

Hyman, L. H. 1951.The Invertebrates: Acanthocephala, Aschelminthes and Entoprocta.Vol. 3, 572 pp. New York: McGraw-Hill Book Co.

Hyman, L. H. 1959. The Invertebrates: Smaller Coelomate Groups. Vol. 5. 784 pp. New York: McGraw-Hill Book Co.

Jackson, J. B. C. and A. H. Cheetham. 1990. Evolutionary significance of morphospecies: A test with cheilostome Bryozoa. Science 248: 521–636.

Kluge, G. A. 1962. Bryozoa of the Northern Seas No. 76: 1–711. Moscow: Zoological Institute, Academy of Sciences (in Russian; 1975, English translation).

McKinney, F. K., and J. B. C. Jackson. 1989. Bryozoan Evolution, 238 pp. Boston: Unwin Hyman.

Morris, P. A. 1976. Middle Pliocene temperature implications based on the Bryozoa *Hippothoa* (Cheilostomata-Ascophora). Journal of Paleontology 50: 1143–1149.

Morris, P. A. 1979. Pacific coast *Celleporina* Gray (1848): fossil and Recent, pp. 467–490. In Advances in Bryozoology. G. P. Larwood and M. B. Abbott, eds. The Systematics Association Spec. Vol. 13: 639 pp. London: Academic Press.

Morris, P. A. 1980. The bryozoan family Hippothoidae (Cheilostomata–Ascophora) with emphasis on the genus *Hippothoa*. Monograph

* = Not in key.

series of the Allan Hancock Foundation No. 10: 1–115. Allan Hancock Foundation and Institute for Marine and Coastal Studies. Los Angeles: University of Southern California.

Nitsche, H. 1869. Beiträge zur Kenntniss der Bryozoen. Zeitschrift für wissenschaftliche Zoologie 20: 1–36.

O'Donoghue, C. H., and E. O'Donoghue.1923. A preliminary list of Bryozoa (Polyzoa) from the Vancouver Island region. Contributions to Canadian Biology, new series (10): 145–201.

O'Donoghue, C. H., and E. O'Donoghue. 1925. List of Bryozoa from the vicinity of Puget Sound. Publications of the Puget Sound Biological Station 5: 91–108.

O'Donoghue, C. H., and E. O'Donoghue. 1926. A second list of Bryozoa (Polyzoa) from the Vancouver Island region. Contributions to Canadian Biology and Fisheries, new series, 3: 49–131.

Osburn, R.C. 1950. Bryozoa of the Pacific Coast of America. Part 1, Cheilostomata—Anasca. Allan Hancock Pacific Expeditions 14(1): 1–270. Los Angeles: University of Southern California Press.

Osburn, R. C. 1952. Bryozoa of the Pacific Coast of America. Part 2, Cheilostomata-Ascophora. Allan Hancock Pacific Expeditions 14(2): 271–612. Los Angeles: University of Southern California Press.

Osburn, R. C. 1953. Bryozoa of the Pacific Coast of America. Part. 3, Cyclostomata, Ctenostomata, Entoprocta, and Addenda. Allan Hancock Pacific Expeditions 14(3): 613–843. Los Angeles: University of Southern California Press.

Pinter, P. 1969. Bryozoan-algal associations in southern California waters. Bulletin of the Southern California Academy of Sciences 68: 199–218.

Robertson, A. 1903. Embryology and embryonic fission in the genus Crisia. University of California Publications in Zoology 1: 115–1256. (based on species from Land's End, San Francisco).

Robertson, A. 1905. Non-incrusting chilostomatous Bryozoa of the west coast of North America. University of California Publications in Zoology 2: 235–322.

Robertson, A. 1908. The incrusting chilostomatous Bryozoa from the west coast of North America. University of California Publications in Zoology 4: 253–344.

Robertson, A. 1910. The cyclostomatous Bryozoa of the west coast of North America. University of California Publications in Zoology 6: 225–284.

Ryland, J. S., and J. S. Porter. 2006. The identification, distribution and biology of encrusting species of Alcyonidium (Bryozoa: Ctenostomatida) around the coasts of Ireland. Biology and Environment: Proceedings of the Royal Irish Academy 106: 19–33.

Soule, D. F., H. W. Chaney, and P. A. Morris. 2003. New taxa of Microporellidae from the northeastern Pacific Ocean. Irene McCulloch Foundation Monograph Series, No. 6: 1–38. Los Angeles: Hancock Institute for Marine Studies, University of Southern California.

Soule, D. F., H. W. Chaney, and P. A. Morris. 2004. Additional new species of Microporelloides from Southern California and American Samoa. Irene McCulloch Foundation Monograph Series, No. 6A: 1–15. Los Angeles: Hancock Institute for Marine Studies, University of Southern California.

Soule, D. F., and J. D. Soule. 1975. Species groups in Watersiporidae, pp. 299–309. In Proceedings of the Third International Bryozoology Conference. Bryozoa 1974, Lyon, France: Documents Sciences de Lyon, Université Claude Bernard, H.S. 3(1), 690 pp.

Soule, D. F., and J. D. Soule. 2002. The eastern Pacific Parasmittina trispinosa complex (Bryozoa, Cheilostomatida): New and previously described species. Irene McCulloch Foundation Monograph series No. 5, 1–40. Los Angeles: Hancock Institute for Marine Studies, University of Southern California.

Soule, D. F., J. D. Soule, and H. W. Chaney. 1992. The genus Thalamoporella worldwide (Bryozoa, Anasca). Morphology, evolution and speciation. Irene McCulloch Foundation Monograph Series, No. 1: 193. Los Angeles: Los Angeles: Hancock Institute for Marine Studies, University of Southern California.

Soule, D. F., J. D. Soule, and H. W. Chaney. 1995. The Bryozoa. In Taxonomic Atlas of the Benthic Fauna of the Santa Maria Basin and Western Santa Barbara Channel, Vol. 13: 1–344. Santa Barbara, CA: Santa Barbara Museum of Natural History.

Soule, D. F., J. D. Soule, and H.W. Chaney. 1999. New species of Thalamoporella (Bryozoa) with acute of subacute avicularium mandibles and review of known species worldwide. Irene McCulloch Foundation Monograph Series, No. 4: 1–57. Los Angeles: Hancock Institute for Marine Studies, University of Southern California.

Soule, D. F., J. D. Soule, and P. A. Morris. 2001. Changing concepts in species diversity in the northeastern Pacific, pp. 299–306. In Bryozoan Studies, Proceedings of the Twelfth International Bryozoology Association Conference, Trinity College, Dublin, Ireland. Lisse: Swets and Zeitlinger.

Soule, D. F., J. D. Soule, and P. A. Pinter. 1975. Phylum Ectoprocta (Bryozoa), pp. 579–608. In Light's manual: intertidal invertebrates of the central California Coast. 3rd ed. R. I. Smith and J. T. Carlton, eds., 716 pp. University of California Press.

Soule, J. D. 1953. Order Ectoprocta, Suborder Ctenostomata. In Bryozoa of the Pacific Coast of America, Part 3. R.C. Osburn, ed. Allan Hancock Pacific Expeditions, 14(3): 726–755. Los Angeles: University of Southern California Press.

Soule, J. D. 1957. Anascan Cheilostomata (Bryozoa) of the Gulf of California. Results of the Puritan-American Museum of Natural History Expedition to Western Mexico, No. 6. American Museum of Natural History Novitates No. 1969: 1–54.

Soule, J. D. 1961. Ascophoran Cheilostomata (Bryozoa) of the Gulf of California. Results of the Puritan-American Museum of Natural History Expeditions to western Mexico, No. 13. American Museum of Natural History Novitates. No. 2053: 1–66.

Soule, J. D. 1963. Cyclostomata, Ctenostomata (Ectoprocta) and Entoprocta of the Gulf of California. Results of the Puritan-American Museum of Natural History Expeditions to western Mexico, No. 18. American Museum of Natural History Novitates No. 2144: 1–34.

Soule, J. D., and M. M. Duff. 1957. Fossil Bryozoa from the Pleistocene of southern California. Proceedings of the California Academy of Sciences (4) 29: 87–146.

Soule, J. D., and D. F. Soule. 1969. Systematics and biogeography of burrowing bryozoans. American Zoologist 9: 791–802.

Soule, J. D., and D. F. Soule. 1976. Spathipora. Its anatomy and phylogenetic affinities, pp. 247–253 in Proceedings of the Third International Bryozoology Conference. Bryozoa 1974, Lyon, France: Documents Sciences de Lyon, Université Claude Bernard, H.S. 3(1), 690 pp.

Soule, J. D., D. F. Soule, and D. P. Abbott. 1980. Bryozoa and Entoprocta: The Moss Animals, pp. 91–107. In Intertidal invertebrates of California. R. H. Morris, D. P. Abbott, and E. C. Haderlie. Stanford University Press.

Soule, J. D., D. F. Soule, and H. W. Chaney. 1998. Two new tropical Pacific species of Cribralaria (Bryozoa: Cribrilinidae) and a review of known species. Irene McCulloch Foundation Monograph Series. No. 3: 1–24. Los Angeles: Hancock Institute for Marine Studies, University of Southern California.

Taylor, P. D. 2000. Cyclostome systematics: Phylogeny, suborders and the problem of skeletal organization, pp. 87–103. In Proceedings of the Eleventh International Bryozoology Association Conference, Smithsonian Tropical Research Institute.

Todd, J. A. 2000. The central role of ctenostomes in bryozoan phylogeny. pp. 104–135. In Proceedings of the Eleventh International Bryozoology Association Conference, Smithsonian Tropical Research Institute.

Taylor, P. D., and N. Monks. 1997. A new cheilostome bryozoan genus pseudoplanktonic on molluscs and algae. Invertebrate Biology 116: 39–51.

Wood, T. S., and M. Lore. 2003. The higher phylogeny of phylactolemate bryozoans inferred from 18S ribosomal DNA sequences. Boletin de la Sociedad de Biologia de Concepción 74: 126.

Woollacott, R. M., and W. J. North. 1971. Bryozoans of California and northern Mexico kelp beds. Nova Hedwigia 32: 455–475.

Zimmer, R. L., and R. M. Woollacott. 1977. Structure and classification of gymnolaemate larvae, pp 57–89 in R. M. Woollacott and R. L. Zimmer, eds., Biology of Bryozoans, New York: Academic Press.

Chaetognatha

(Plates 453 and 454)

ERIK V. THUESEN

More than 35 species of chaetognaths (arrow worms) are found in the eastern north Pacific Ocean. Eight species are treated here, including the six most likely to turn up in short surface tows made nearshore with small plankton nets from small boats or piers.

The two shallow-living benthic chaetognaths known from southern California are also included. Although the West Coast of the United States has no known tide-pool chaetognaths, small benthic chaetognaths are collected in tide pools in Japan and other locations. Their unique mating dance has been studied in detail (Goto and Yoshida 1985). It is possible that tide pools of the West Coast harbor benthic chaetognaths, and the two species of benthic chaetognaths are included here in the hopes that more records of these interesting animals will be forthcoming.

Temperature and salinity affect morphological characteristics of chaetognaths during development, and the identification of chaetognath species is more difficult as a result of various intraspecific ecotypes. Most studies of chaetognaths are undertaken on preserved specimens, but the fragile external structures of chaetognaths preserve poorly. Living mature specimens are the easiest to identify, but these are often absent in samples. Extent of ovaries, shape and position of seminal vesicles, relative length of the tail section to total body length, and the shape and position of the fins and extent of fin rays are all important characteristics used to identify chaetognaths.

The use of vital stains (e.g., aniline blue or rose bengal) and various lighting angles under the microscope highlight these characteristics and make chaetognaths easier to identify. The shape of eye pigmentation and the number of hooks and teeth have often been used as identification characteristics, but these often overlap between species and can change with maturity. The general body aspect of juvenile specimens often matches the adult, even though important distinguishing features have not yet developed fully. The degree of transparency is another identifying characteristic, but this is often dependent on the condition of the specimen. Chaetognaths quickly lose transparency if damaged during collection. The key is best used with mature living specimens that have been collected carefully to preserve their fragile body structures.

The taxonomic outline of Bieri (1991) is followed throughout this section, and a complete list of chaetognaths is available online at http://academic.evergreen.edu/t/thuesene/chaetognaths/chaetognaths.htm. Descriptions and illustrations of other Pacific chaetognaths can be found in Alvariño (1967), Yamaji (1980), and Chihara and Murano (1997). Reviews of specific aspects of the biology and ecology of chaetognaths are found in Bone et al. (1991). The general anatomy of chaetognaths is presented in plate 453A.

Key to Some Common Nearshore Chaetognaths

1. Heavy transverse musculature along the trunk segment gives opaque appearance, small (<10 mm), one pair of lateral fins, benthic or epibenthic, often attached to algae or other substrates. . . . . Spadellidae 2
— No transverse musculature along the trunk segment, transparent, size can exceed 20 mm, two pairs of lateral fins, planktonic. . . . . Sagittidae 3

2. Without adhesive organs, body length to 6.5 mm (plate 453B) . . . . Spadella bradshawi
— With adhesive organs, body length to 3.8 mm (plate 453C) . . . . Paraspadella pimukatharos

3. Inner edges of grasping spines with small serrations (400x), seminal vesicles with anterior nipple, matures at 14 mm–17 mm (plate 453D) . . . . Serratosagitta bierii
— Grasping spines without serrations, no anterior nipple on seminal vesicles. 4

4. Fins with complete rays 5
— Fin rays very sparse or only filling outer edges 6

5. Two rows of eggs in each ovary extend as far as or further than the middle of the anterior fins, collarette extensive, seminal vesicles with terminal knobs when fully mature, body length to 26 mm (plate 453E) . . . . . Sagitta bipunctata
— Ovaries do not reach the anterior fins 7

6. Ovaries may extend past the leading edge of the posterior fins; sperm usually forms a heavy "V" pattern in the tail section, whereby the anterior and central portion of the tail section is devoid of sperm, seminal vesicles are simple bulbs, total body length when mature is >15 mm (plate

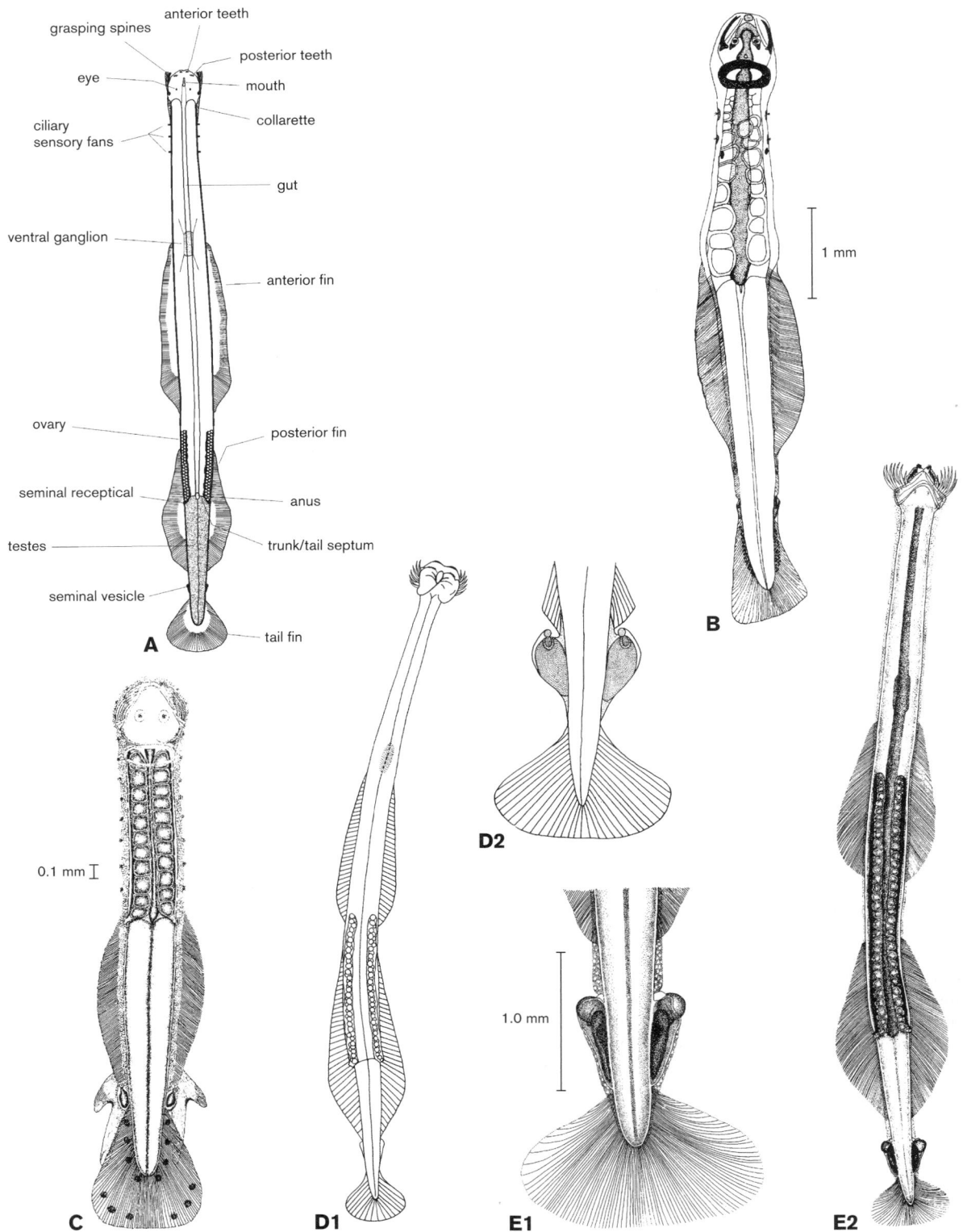

PLATE 453 A, General anatomy of a sagittid chaetognath; B, *Spadella bradshawi*; C, *Paraspadella pimukatharos*; D1, D2, *Serratosagitta bierii*; E1, E2, *Sagitta bipunctata* (A, Thuesen; B, from Bieri, revised by Thuesen; C–E, from Alvariño, with permission of the Biological Society of Washington, University of Hawaii Press, and Scripps Institution of Oceanography, respectively).

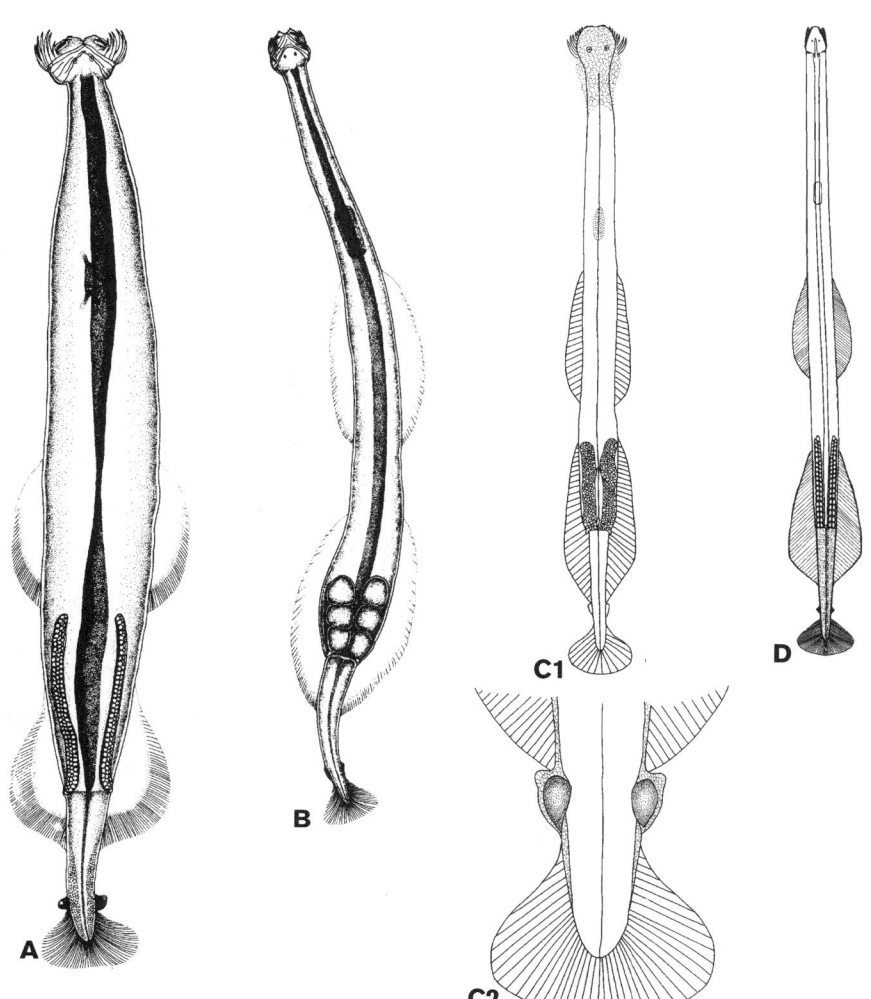

PLATE 454 A, *Flaccisagitta enflata*; B, *Mesosagitta minima*; C1, C2, *Parasagitta euneritica*; D, *Parasagitta elegans* (A, B, from Alvariño with permission of the Scripps Institution of Oceanography; C, from Alvariño with permission of the University of Hawaii Press; D, Thuesen).

454A). . . . . *Flaccisagitta enflata*
— Ovaries shorter than the length of the posterior fins with a few large ova, highly transparent with buoyancy sac mesenteries often visible, seminal vesicles are very simple, total mature body length <10 mm (plate 454B) . . . .
. . . . *Mesosagitta minima*
7. Anterior fins, seminal vesicles, and posterior fins may all be touching in alignment, collarette-type tissue extends the length of the body, mature body length of 8 mm–16 mm (plate 454C). *Parasagitta euneritica*
— Posterior fins not touching seminal vesicles, mature body length of 24 mm–48 mm (plate 454D) . . . .
. . . . *Parasagitta elegans*

List of Species

Sagittoidea

SPADELLIDAE

Spadella bradshawi Bieri, 1974. Point Loma south in 25 m– 100 m, may occur shallower further north; prefers course sand; also on silt. Chestnut-colored markings on body and brilliant green eyes visible in living specimens (see Bieri 1974, Bieri et al. 1987).

Paraspadella pimukatharos (Alvariño, 1987). Reaches densities over 3,500 individuals m^{-2} at Catalina Island (~10 m depth); most abundant on sediment with coralline algae fragments. Coloration pattern unknown. Possible indicator of El Niño (see Alvariño 1987).

SAGITTIDAE

Flaccisagitta enflata (Grassi, 1881).
Mesosagitta minima (Grassi, 1881).
Parasagitta elegans (Verrill, 1873). Boreal species with southern limit ~40°N, indicative of northern intrusions of cold water south of this point (see Terazaki and Miller 1986). Obvious ammonia-filled buoyancy sacs in fresh-caught specimens (see Bone et al. 1991).

Parasagitta euneritica (Alvariño, 1961). Dominant chaetognath in the California Current (see Bieri 1959, as *Sagitta friderici,* and Alvariño 1966) and nearshore waters, including Bodega Bay (see Renshaw 1962), Monterey Bay (see Bigelow and Leslie 1930, as *Sagitta bipunctata*), Santa Barbara Channel (see Thuesen and Childress 1993), Anaheim Bay (see Felts 1973), Newport Bay (see Kinoshita 1981) and San Diego (see Michael 1911, as *Sagitta bipunctata*). Possibly synonymous with *S. friderici.*

Sagitta bipunctata Quoy and Gaimard, 1827. Off California; this species has been described as *Sagitta californica* (see Michael 1913, Bieri 1959).

Serratosagitta bierii (Alvariño, 1961). Common coastal species off California (see Bieri 1959, as *Sagitta* sp., and Alvariño 1966), usually further offshore than *P. euneritica*. Serrations appear as bright sheen on inner side of hooks under a dissecting microscope but seen clearly under a compound microscope.

References

Alvariño, A. 1961. Two new chaetognaths from the Pacific. Pac. Sci. 15: 67–77.

Alvariño, A. 1966. Zoogeografía de California: Quetognatos. Rev. Soc. Mex. Hist. Nat. 27: 199–243.

Alvariño, A. 1967. The Chaetognatha of the NAGA Expedition (1959–1961) in the South China Sea and the Gulf of Thailand. I. Systematics. Naga Reports. 4: 1–197.

Alvariño, A. 1987. *Spadella pimukatharos*, a new benthic chaetognath from Santa Catalina Island, California. Proc. Biol. Soc. Wash. 100: 125–133.

Bieri, R. 1959. The distribution of planktonic Chaetognatha in the Pacific and their relationship to the water masses. Limnol. Oceanogr. 4: 1–28.

Bieri, R. 1974. A new species of *Spadella* (Chaetognatha) from California. Pub. Seto Mar. Biol. Lab. 21: 281–286.

Bieri, R. 1991. Systematics of the Chaetognatha. In The Biology of Chaetognaths. Q. Bone, H. Kapp, and A. C. Pierrot-Bults, eds. Oxford: Oxford University Press, pp. 122–136.

Bieri, R., M. Terazaki, E. V. Thuesen, and T. Nemoto. 1987. The colour pattern of *Spadella angulata* (Chaetognatha: Spadellidae) with a note on its northern range extension. Bull. Plankton Soc. Japan 34: 83–84.

Bigelow, H. B., and M. Leslie. 1930. Reconnaissance of the waters and plankton of Monterey Bay, July, 1928. Bull. Mus. Comp. Zool. Harvard 70: 427–581.

Bone, Q., H. Kapp, and A. C. Pierrot-Bults, eds. 1991. The Biology of Chaetognaths. Oxford: Oxford University Press, 173 pp.

Chihara, M., and M. Murano, eds. 1997. An illustrated guide to marine plankton in Japan. Tokai University Press, Tokyo, 1574 pp.

Felts, R. W. 1973. Seasonal distribution and abundance of the chaetognaths from Anaheim Bay, California, and the adjacent waters. MA thesis, California State University, Long Beach, 118 pp.

Goto, T., and M. Yoshida. 1985. The mating sequence of the benthic arrowworm *Spadella schizoptera*. Biol. Bull. 169: 328–333.

Kinoshita, P. B. 1981. Population structure, vertical migration and feeding of the chaetognath *Sagitta euneritica* in Newport Bay, California. MA thesis, California State University, Fullerton, 72 pp.

Michael, E. L. 1911. Classification and vertical distribution of the Chaetognatha of the San Diego region. Univ. Calif. Publ. Zool. 8: 20–186.

Michael, E. L. 1913. *Sagitta californica*, n. sp., from the San Diego region. Univ. Calif. Publ. Zool. 11: 89–126.

Renshaw, R. W. 1962. The Chaetognaths of the Dillon Beach area and their possible use as indicators of water movements. MA thesis, University of the Pacific, Stockton, California, 71 pp.

Terazaki, M., and C. B. Miller. 1986. Life history and vertical distribution of pelagic chaetognaths at Ocean Station P in the subarctic Pacific. Deep Sea Res. 33: 323–337.

Thuesen, E. V., and J. J. Childress. 1993. Enzymatic activities and metabolic rates of pelagic chaetognaths: Lack of depth-related declines. Limnol. Oceanogr. 38: 935–948.

Yamaji, I. 1980. Illustrations of the Marine Plankton of Japan. Osaka, Japan: Hoikusha Publishing Company, 369 pp.

Hemichordata

(Plate 455)

KEITH H. WOODWICK AND CHRISTOPHER B. CAMERON

The Enteropneusta (acorn worms) are soft-bodied worms found intertidally most commonly in sand and mud. They also occur in subtidal and deeper waters. The first enteropneust was identified as a sea cucumber, but later findings of gill slits and structures similar to a notochord and a dorsal hollow nerve cord led to investigation of chordate relationships. Most workers consider enteropneusts to be a class of Hemichordata (Hyman 1959, Ruppert 1997, Cameron et al. 2000, Cameron 2005), but Nielsen (1998) supported his earlier elevation of Enteropneusta to phylum status. He placed them close to chordates and separated them from echinoderms and pterobranchs.

Enteropneusts vary in size in extremes from 2.5 cm to 2 m, but most specimens are 10 cm–40 cm in length. They have three body divisions: proboscis, collar, and trunk (plate 455A). The trunk may have as many as four distinct regions: branchiogenital, esophageal, hepatic, and intestinal. The straight gut begins with an anterioventral mouth at the interphase of the muscular proboscis and the cufflike collar; it ends in a terminal anus. The branchiogenital region has a few to hundreds of pairs of small gill pores or eternally visible gill slits. Gonads appear as surface bumps, ridges, or genital wings (plate 455E, 455F5) or lappets (plate 455D4). Some genera have external sacculations (caeca) in the hepatic region. They are finger- or ear-shaped.

In the field enteropneusts can be found in sandy mudflats by utilizing surface clues and shoveling up clods of substratum. Worms may leave fecal strings or coils at the openings of their burrows. Some extend their proboscis and collar to the surface to deposit feed utilizing mucus and ciliary action (Barrington 1965). Several species collect food from the water by filter-feeding using the gill slits in the pharynx (Cameron 2002). At this time body colors of red, orange, to light yellow may reveal their presence; however, many forms are drab in color (Hyman 1959). More complete specimens may retain the hepatic region with its brown and green pigment. Clods of substratum can be hand-processed, but enteropneusts are fragile and it is difficult to collect complete specimens. The proboscis, collar, and anterior trunk are least fragile and are the body parts most often taken in the field, found in museum collections, and utilized here in the key. Specimens in the substratum are covered with mucus and sediment and contain ciliary collected sediment (food) in their digestive tract. Fixation in this condition increases the difficulty of appropriate processing of specimens for identification and morphological and other studies.

When possible, live specimens should be sorted from the sediment. Sometimes breaking the sediment into large pieces will reveal the burrow and the organism, and further careful breaking will free the specimen. Live specimens should then be placed in large containers of clean seawater to permit evacuation of the gut. Several changes of seawater may be needed and surface sediment may be released with their movement or removed with a camels hair brush or forceps.

Clean specimens should be fixed by holding them at the posterior end with a pair of forceps and dipping them in a tall jar of fixative (e.g., Bouin's fluid). Gravity will assist in straightening the specimen as fixative is added to the posterior end and allowed to flow down into the jar. When fixation is complete, release the specimen into the jar and store on its side for several days. Transfer specimens to 50% alcohol and then wash several times in 70% alcohol (not water) to clear the picric acid (Galigher and Kozloff 1971). Specimens can then be stored or run through an alcohol, toluene, paraffin series for embedding. Prepared paraffin blocks are then serially sectioned at 10 μm–15 μm and slide sections stained with Harris's Hematoxylin and Eosin Y. The anatomy is reconstructed from microscopic study of the serial cross and sagittal sections. Most published descriptions have included labeled drawings of critical diagnostic features and areas (Woodwick 1996, pp. 252–253). In addition to the external features emphasized here, the nature of and presence and absence of internal structures are important in placing specimens in family, genus, and species (Spengel 1893, Horst 1939, Dawydoff 1948, Hyman 1959, Benito and Pardos 1997).

Classification of Enteropneusta

Characteristics important to the key and field observations are listed first.

Harrimaniidae

Proboscis short, conical, or elongate; trunk has four enlarged genital lappets or lacks genital lappets or wings; proboscis lacks

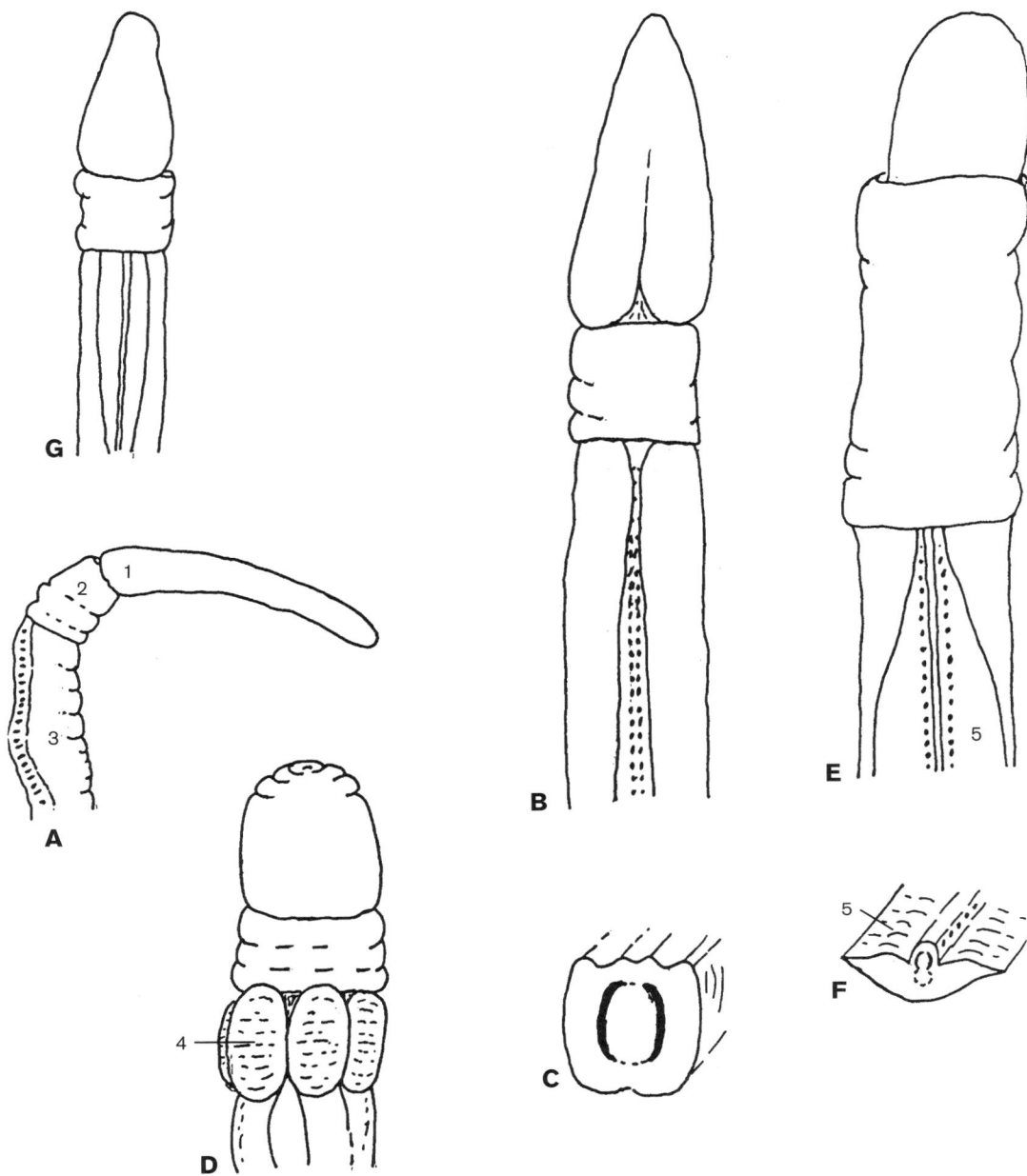

PLATE 455 A, *Saccoglossus*; B, *Schizocardium*; C, cross-section of *Schizocardium*; D, *Stereobalanus*; E, *Balanoglossus*; F, cross-section of *Balanoglossus*; G, *Glossobalanus* (1, proboscis; 2, collar; 3, trunk; 4, genital lappets; 5, genital wing).

cauliflower organ and vermiform process; peripharyngeal cavities and neurocord nerve roots of the collar not present; trunk lacks parabranchial ridges, synapticules, and hepatic caeca; development direct, eggs large (150 μm–400 μm).

GENERA: *HARRIMANIA, PROTOGLOSSUS, SACCOGLOSSUS, STEREOBALANUS, XENOPLEURA*

SACCOGLOSSUS

Proboscis elongate, cylindroid, one proboscis pore; collar broad as long; genital pores medio-dorsal; gonads simple; longitudinal proboscis muscles in concentric rings.

STEREOBALANUS

Proboscis short, conical; collar shorter than broad, usually with two proboscis pores; paired dorsal and ventral genital lappets; lappets partially cover dorso-ventral gill slits that open directly to the exterior; lappets not in esophageal region; longitudinal muscles of proboscis form radial pattern.

Spengeliidae

Proboscis short, ovate; collar shorter than broad; hepatic caeca present in some genera; proboscis muscle includes thick layer of circular muscle encompassing homogenously arranged longitudinal fibers; stomochord with a vermiform process; pericardium and glomerulus with paired anterior diverticula more or less developed; chondroid tissue well-developed; eggs small, development indirect, tornaria larva.

GENERA: *GLANDICEPS, SCHIZOCARDIUM, SPENGELIA, WILLEYIA*

SCHIZOCARDIUM

Hepatic caeca present and finger-shaped; esophageal pores present, both single and paired; long vermiform process; gill slits almost equaling the pharynx in depth so the ventral non-pharyngeal part of the pharynx is reduced to a mere groove, long gill bars with synapticles and a narrow hypobranchial area; gonads lateral only.

Ptychoderidae

Proboscis short, conical; collar as long as broad with nerve-roots; trunk external regionation pronounced; ventral part of pharynx large and sometimes more or less separated from branchial part by parabranchial ridges; genital wings extend into esophageal region; hepatic caeca ear-shaped; cauliflower organ may be present; proboscis muscle radial pattern of bands; gill bars short, curved, with synapticules; parabranchial ridge present; longitudinal ciliated intestinal grooves; eggs small, development indirect, tornaria larvae (Hadfield 1975).

GENERA: *BALANOGLOSSUS, GLOSSOBALANUS, PTYCHODERA*

BALANOGLOSSUS

Proboscis may be very small, partially enclosed by collar; genital wings long, well-developed; genital openings near branchial pores, branchial pores small; lack cauliflower organ; hepatic caeca not regularly arranged posteriorly; intestinal grooves paired.

GLOSSOBALANUS

Proboscis as long as wide and mainly free of collar; genital wings not well-developed; genital openings on margin of genital wings; lack cauliflower organ; hepatic caeca in two regularly arranged rows; intestinal groove on left only.

ACKNOWLEDGMENTS

This section is dedicated to Professor Theodore H. Bullock, who kept alive the hope for further study of West Coast enteropneusts.

Key to Enteropneusta

1. Proboscis elongate (plate 455A) . . . . *Saccoglossus* 2
— Proboscis not elongate 3
2. Proboscis orange, collar darker orange; 60 or less pairs of gill pores; one pair of esophageal pores; eggs 145 μm–155 μm

. . . . *Saccoglossus pusillus*
— Proboscis creamy white; collar brick red to reddish orange; 60 or more pairs of gill pores; four to six pairs of esophageal pores; eggs ca. 250 μm . . . . . . *Saccoglossus bromophenolosus*
3. Enlarged genital lappets or genital wings absent (plate 455B, 455C) . . . . *Schizocardium*
— Enlarged genital lappets or genital wings present 4
4. Four enlarged genital lappets extend above and below branchial openings; lappets not present in esophageal region (plate 455D) . . . . *Stereobalanus*
— Ventro-lateral wings only, extend below branchial openings; wings present in esophageal region 5
5. Genital wings well-developed; genital openings near branchial pores (plate 455E, 455F) . . . . *Balanoglossus*
— Genital wings not well-developed; genital openings on margin of genital wings (plate 455G) . . . . *Glossobalanus*

List of Species

For subtidal and deeper water forms see Woodwick, 1955, Allan Hancock Pac. Exped. 19:166–167; Woodwick 1996; Holland et al. 2005, Nature 434: 374–376.

Harrimaniidae

Saccoglossus sp. Newport, Half Moon Bay; see Bullock, 1944, J. Comp. Neur. 80: 355–367.

Saccoglossus pusillus (Ritter, 1902) Duxbury Reef (Bolinas Bay) at base of seaweeds; southern and Baja California (e.g., San Pedro, Newport, San Diego in sand or mud, Ensenada), north to British Columbia; see Ritter and Davis 1904, Univ. Calif. Publ. Zool. 1: 171–210 and Davis 1908, Univ. Calif. Publ. Zool. 4: 187–226 (reproduction and development); Evans 1919, Pomona Coll. J. Ent. and Zool. 11: 28–33 (general morphology); Horst 1930, Vidensk. Medd. Dansk Nat. Foren. 87: 135–200 (includes original description of species by Ritter, pp. 154–156); Bullock 1944, and Bullock 1945, Quart. J. Microsc. Sci. 86: 55–111 (nervous system); Smith et al. 2003, Can. J. Zool. 81:131–141. (molecular biogeography of Pacific Northwest saccoglossids).

Saccoglossus bromophenolosus King, Giray, and Kornfield, 1994. Washington (Willapa Bay sandy mudflats), Oregon; see King, Giray, and Kornfield 1994, Proc. Biol. Soc. Wash. 107: 383–390 (original description and biochemical systematics); Giray and King 1996, Proc. Biol. Soc. Wash. 109: 430–445 (table of 14 species of *Saccoglossus*); Smith et al. 2003, Can. J. Zool. 81:131–141. (molecular biogeography of Pacific NW saccoglossids).

Stereobalanus sp. Newport, San Diego; Spengel, 1893 (described only known species); Reinhard 1942, J. Wash. Acad. Sci. 32: 309–311 (described complete, live specimen).

Spengeliidae

Schizocardium sp. Morro Bay; see Bridges and Woodwick, 1994, Acta Zool. 75: 371–378. (hepatic caeca); Spengel 1893 (described the two known species).

Ptychoderidae

Balanoglossus sp. Newport, Laguna, Mission Bay, also Puget Sound and Baja California; see Bullock 1944 and 1945.

Glossobalanus sp. La Jolla; see Bullock 1944 and 1945; also found at Moss Beach, Shelter Cove and Puget Sound.

Glossobalanus berkeleyi (Willey, 1931). See Willey 1931, Trans Roy. Soc. Can. 25: 19–28 (original description; collected at Nanaimo, British Columbia, but more abundant at Penrose Point, Puget Sound).

References

Barrington, E. J. W. 1965. The Biology of Hemichordata and Protochordata. Freeman, San Francisco, 176 pp.

Benito, J., and F. Pardos. 1977. Hemichordata. In F. W. Harrison and E. E. Ruppert, eds. Microscopic Anatomy of Invertebrates. 15. New York: Wiley-Liss, pp. 15–101.

Cameron, C. B. 2005. A phylogeny of the hemichordates based on morphological characters. Canadian Journal of Zoology 83: 196–215.

Cameron, C. B. 2002 Particle retention and flow in the pharynx of the enteropneust worm *Harrimania planktophilus*: the filter feeding pharynx may have evolved prior to the chordates. Biological Bulletin 202:192–200.

Cameron, C. B., B. J. Swalla, and J. R. Garey. 2000. Evolution of the chordate body plan: New insights from phylogenetic analysis of deuterostome phyla. Proceedings of the National Academy of Sciences 97: 4469–4474.

Dawydoff, C. 1948. Stomochordes. In P. Grassé, ed. Traité de Zoologie. Masson, Paris, pp. 367–532.

Galigher, A. E., and E. N. Kozloff. 1971. Essentials of practical microtechnique. 2nd ed., Lea and Febiger, Philadelphia, 531 pp.

Hadfield, M. G. 1975. Hemichordata. In Reproduction of marine invertebrates. 2. A. C. Giese and J. S. Pearse, eds. London: Academic Press, pp. 185–240.

Horst, C. J. van der. 1939. Hemichordata. In Bronn, Klassen und Ordnungen des Tierreichs. 4, 737 pp.

Hyman, L. H. 1959. Phylum Hemichordata. In The invertebrates 5. New York: McGraw-Hill, pp. 72–207.

Nielsen, C. 1998. Origin and evolution of animal life cycles. Biological Reviews 73: 125–155.

Ruppert, E. E. 1997. Introduction: Microscopic Anatomy of the Notochord, Heterochrony, and Chordate Evolution. In Microscopic anatomy of invertebrates. 15. F. W. Harrison and E. E. Ruppert, eds. New York: Wiley-Liss, pp. 1–13.

Spengel, J. W. 1893. Die Enteropneusten des Golfes von Neapel. Fauna und Flora des Golfes von Neapel. 18, 757 pp.

Woodwick, K. H. 1996. Phylum Hemichordata, Class Enteropneusta. In Taxonomic atlas of the benthic fauna of the Santa Maria Basin and Western Santa Barbara Channel. 14. J. A. Blake, P. H. Scott, and A. Lissner, eds. Santa Barbara, CA: Santa Barbara Museum of Natural History, pp. 251–259.

Echinodermata

(Plates 456–475)

Introduction

JOHN S. PEARSE AND RICH MOOI

The Echinodermata (*echino* = spiny; *derm* = skin) is a phylum of deuterostome macroinvertebrates sharing morphological features not found in any other phylum. The phylum includes the **CRINOIDEA** (sea lilies and feather stars), **ASTEROIDEA** (sea stars or starfishes), **OPHIUROIDEA** (brittle stars), **ECHINOIDEA** (sea urchins, heart urchins, and sand dollars), and **HOLOTHUROIDEA** (sea cucumbers) in addition to a less familiar assemblage of fossil groups. A sixth, somewhat controversial group, the **CONCENTRICYCLOIDEA** (sea daisies), with three species known from sunken wood in the deep sea, may be highly derived asteroids.

Echinoderms are entirely marine and extremely intolerant of fresh water. Most echinoderms have pelagic larvae with bilateral symmetry, but following a drastic metamorphosis, during which the left side comes to predominate in the adult, they develop into bizarre forms with unusual symmetries. The adults typically have pentaradial (five-sided radial) symmetry with a water-vascular ring that encircles the esophagus. This ring canal gives off five radial canals along which are arranged tube feet (podia) used for locomotion, respiration, and feeding. Within the body wall are the ossicles (spicules, plates, spines) of the internal skeleton (stereom), composed of closely aligned, magnesium-rich, fenestrated crystals of calcium carbonate. In most echinoderms, five series of paired ambulacral columns of ossicles form an axial skeleton that supports the radial canals; other ossicles fill in between these columns to form the extraxial skeleton. The growth and organization of the skeletal ossicles in large part determines the form of the adults: feathery, star-shaped, or globular or flattened discs. In contrast, sea cucumbers lack spines and are generally soft-bodied, wormlike animals with only tiny ossicles in the body wall and a calcareous ring around the esophagus.

Echinoderm adults range in lifestyle from pelagic holothurians to infaunal burrowers and inhabit nearly every marine environment, from sandy beaches and coral reefs to the greatest depths of the sea. The phylum has left a relatively complete and detailed fossil record of more than 13,000 species stretching back 500 million years to the Cambrian.

Crinoids are not found in shallow waters off the California and Oregon coasts, and therefore we treat them here only briefly. In asteroids and ophiuroids, the axial skeleton forms five or more arms (rays). These distinctive, starlike animals are mainly active predators, scavengers, or detritivores of the benthos, and together form the most species-rich group of echinoderms considered here. Echinoids lack arms and have a rigid test of tightly fitted plates and movable spines. About 14 species occur along the shores of California and Oregon, but only three—two sea urchins and one sand dollar—are commonly found in the intertidal and shallow subtidal, and these can be important herbivores and planktivores. Holothuroids are also represented by about a dozen species along Oregon and California shores, but the few that are commonly found in the intertidal and shallow subtidal are usually inconspicuous.

References

Binyon, J. 1972. Physiology of Echinoderms. Oxford: Pergamon Press, 264 pp.

David, B., and R. Mooi. 1996. Embryology supports a new theory of skeletal homologies for the phylum Echinodermata. Comptes Rendus de l'Academie des Sciences, Paris, 319: 577–584.

Giese, A. C., J. S. Pearse, and V. B. Pearse, eds. 1991. Reproduction of Marine Invertebrates, Volume VI, Echinoderms and Lophophorates. Pacific Grove: Boxwood Press, 808 pp.

Harrold, C., and J. S. Pearse 1987. The ecological role of echinoderms in kelp forests. In Echinoderm studies 2. M. Jangoux and J. M. Lawrence, eds., Balkema, Rotterdam, 137–233 pp.

Jangoux, M., and J. M. Lawrence, eds. 1982. Echinoderm Nutrition. Balkema, Rotterdam, 654 pp.

Lawrence, J. 1987. A Functional Biology of Echinoderms. Johns Hopkins University Press, Baltimore, 340 pp.

Mooi, R. 2000. Not all written in stone: Interdisciplinary syntheses in echinoderm paleontology. Canadian Journal of Zoology, 79: 1209–1231.

Mooi, R., and B. David. 1997. Skeletal homologies of echinoderms. The Paleontological Society Papers, 3:305–335.

Moore, R. C., ed. 1966. Treatise on Invertebrate Paleontology. Part U. Echinodermata 3. Geological Society of America, New York, 695 pp.

Moore, R. C., ed. 1967. Treatise on Invertebrate Paleontology. Part S. Echinodermata 1. Geological Society of America, New York, 650 pp.

Moore, R. C., and C. Teichert, eds. 1978. Treatise on Invertebrate Paleontology. Part T. Echinodermata 2. Lawrence: University of Kansas Press, 1027 pp.

Readers can also consult the proceedings volumes of the International Echinoderm Conferences, and the following websites:

The California Academy of Sciences Echinoderm Webpage: http://www.calacademy.org/research/izg/echinoderm/

Crinoidea

JOHN S. PEARSE AND CHARLES G. MESSING

The class Crinoidea includes the most exquisite members of the Echinodermata. Moreover, they have the longest fossil record, dating back almost 500 million years, and once were dominant forms in shallow seas. They differ from other echinoderms by having (1) the oral surface with both mouth and anus facing away from the substrate, (2) five flexible, usually branched, featherlike rays that extend into the water and gather food using tube feet, and (3) a cuplike body (calyx) that contains most of the viscera. Most of the fossil forms were sessile, stalked forms known as sea lilies; only a small fraction remain today, all in deep water. Nearly all living crinoids shed their stalks at an early age and cling to the bottom using small, hooklike appendages (cirri). These are called feather stars or comatulids. Some swim short distances by undulating their rays.

There are approximately 540 living species of feather stars and about 95 sea lilies. Though found in all oceans, feather stars are especially abundant and diverse in the shallow tropical western Pacific and Indian Oceans, as well as in the deep sea.

Only one is common along the coast of the eastern Pacific, the feather star *Florometra serratissima* (A. H. Clark, 1907). A second species, *F. asperrima* (A. H. Clark, 1907) is probably a synonym. It is often abundant along the outer continental shelf and slope to about 1,500 m. Although occurring in <30 m depths in a few places in British Columbia, often attached to kelp, it lives in deeper water off California and Oregon. Several other species, including sea lilies, have been observed and collected much deeper (500 m–3,000 m) in the eastern Pacific, but these are very poorly known.

References

Clark, A. H., and A. M. Clark. 1967. A monograph on the existing crinoids, 1(5). Bulletin of the U.S. National Museum 82, 860 pp.

McEdward, L. R., S. F. Carson, and F.-S. Chia. 1988. Energetic content of eggs, larvae, and juveniles of *Florometra serratissima* and the implications for the evolution of crinoid life histories. International Journal Invertebrate Reproduction and Development 13: 9–22.

Meyer, D. L., and D. B. Macurda, Jr. 1977. Adaptive radiation of the comatulid crinoids. Paleobiology 3: 74–82.

Mladenov, P. V. 1987. Phylum Echinodermata, Class Crinoidea. In: Reproduction and Development of Marine Invertebrates of the Northern Pacific Coast, M. F. Strathmann, ed. Seattle: University of Washington Press, pp. 597–606.

Rasmussen, H. W., and H. Sieverts-Doreck. 1978. Articulata Classification. In Treatise on invertebrate paleontology. R. C. Moore and C. Teichert, eds. Part T, Echinodermata 2(3), Geological Society of America, Boulder, CO, pp.T813–T928.

Roux, M., C. G. Messing, and N. Améziane. 2002. Artificial keys to the genera of living stalked crinoids (Echinodermata). Bulletin of Marine Science 70(3): 799–830.

Websites

Echinoidea

JOHN S. PEARSE AND RICH MOOI

(Plates 456–462)

Echinoids are rounded to flattened echinoderms encased in an interlocking endoskeleton of calcareous plates covered with spines. Tube feet extend through ambulacral plates, arranged in five paired columns from the mouth on the oral surface to the opposite, aboral or apical pole. On each side of each ambulacral column is a column of interambulacral plates; these dominate the body wall, and together the ambulacral and interambulacral plates form a rigid "test" upon which are movable spines and pedicellariae.

Most echinoids have an intricate jaw apparatus, Aristotle's lantern, with five teeth that aid in the ingestion of food. The opposite, aboral side of the animal carries the apical ring of small plates, including the genital plates with gonopores and the madreporite; in sea urchins, these plates surround the periproct with the anus. The anus is shifted to one side in sand dollars and heart urchins, forming a "posterior" end; in most heart urchins, the mouth is shifted to the opposite "anterior" end of the animal. Heart urchins also do not have a jaw apparatus. Most echinoids are omnivorous grazers and scavengers, scraping algae and encrusting animals off hard surfaces, or ingesting detritus and debris caught from the water or other organic material in soft substrates.

The nearly 1,000 species of echinoids were traditionally divided into two groups: rounded "regular" echinoids that generally occur on hard surfaces (sea urchins) and flattened or heart-shaped "irregular" echinoids that occur on or within soft substrates (sand dollars, cake urchins, heart urchins). Only the latter group is today considered a natural taxonomic assemblage. The class is currently divided into three subclasses: **PERISCHOECHINOIDEA** (an ancient stock with no living representatives), **CIDAROIDEA** (pencil-spine urchins occurring mainly in the tropics, deep sea, and Antarctic, plate 456), and **EU-ECHINOIDEA** (including all the other living echinoids divided among 14 orders, plates 457–462).

The shores of central California and Oregon are relatively poor in species of echinoids, with only three species of sea urchins (*Strongylocentrotus purpuratus*, *S. franciscanus*, and *S. droebachiensis*, the last found as far south as southern Oregon) and one species of sand dollar (*Dendraster excentricus*) commonly found. However, additional species occur in southern California that could occasionally be found along these shores, especially as general global warming continues. In addition, other species found subtidally may be cast ashore, either live or as empty tests, from time to time. All these species are included in the following key and annotated list.

Key to Echinoidea

1. Body flattened, diameter several times greater than thickness 2
— Body not flattened, nearly spherical, rounded, or heart-shaped 4
2. Six holes or marginal slots (lunules) penetrate flattened test (plate 459E, 459F) . . . . *Encope micropora*
— Test without lunules 3

PLATE 456 Cidaroida, Diadematoida, Arbacioida: A, *Eucidaris thouarsii*; B, *Centrostephanus coronatus*; C, *Arbacia stellata*; D, *Arbacia stellata*, denuded test.

3. Gray-lavender, brown, or purple color; apical system off-center and in posterior half of test; anterior ambulacral petaloid ("petal") conspicuously larger than other four (plate 459A, 459B). . . . . *Dendraster excentricus*
— Pale yellow or tan; apical system nearly central; ambulacral petaloids all nearly same size (plate 459C, 459D) . . . .
. . . . *Dendraster terminalis*
4. Body strongly bilaterally symmetric, heart-shaped 5
— Body nearly rounded; not asymmetrical and heart-shaped
. . . . 10
5. Test elongate with conspicuous long spines articulating on deeply sunken large tubercles (plate 462B) . . . .
. . . . *Lovenia cordiformis*
— Test not elongate; covered with spines of nearly the same length. 6
6. Petaloids deeply sunken, band of minute spines encircling petaloids (peripetalous fasciole), on either side of the periproct (anal fasciole) and forming a ring below the periproct (subanal fasciole) 7
— Petaloids not deeply sunken, fascioles not like above 8
7. Width of posterior petaloid <11% length of the test (plate 461A) . . . . *Brisaster latifrons*

— Width of posterior petaloid >11% length of the test (plate 461B) . . . . *Brisaster townsendi*
8. Test outline nearly round, not indented anteriorly or posteriorly, both peripetalous and subanal fascioles (plate 462A)
. . . . *Brissopsis pacifica*
— Test indented anteriorly or posteriorly, subanal fasciole only 9
9. Deep anterior indentation (plate 460) . . . .
. . . . *Spatangus californicus*
— No anterior indentation, slight posterior indentation (plate 462C). . . . . *Nacospatangus laevis*
10. Primary spines not covered by living epidermis, eroded, usually encrusted with coralline algae, sponges, bryozoans, and other organisms (plate 456A) . . . .
. . . . *Eucidaris thouarsii*
— Primary spines covered with epidermis, not encrusted with other organisms. 11
11. Primary spines long, black, hollow, covered with forward-pointing imbricating spinelets, and easily broken (plate 456B). . . . . *Centrostephanus coronatus*
— Primary spines otherwise, if black, not hollow or easily broken 12

PLATE 457 Echinoida. A, *Strongylocentrotus purpuratus*, living specimen with tube feet fully extended (photo courtesy of Ralph Buchsbaum); B, *Strongylocentrotus franciscanus*, living specimen lifted from tide pool (photo courtesy of Ralph Buchsbaum); C, *Strongylocentrotus fragilis*; D, *Strongylocentrotus fragilis*, denuded test; E, *Lytechinus pictus*; F, *Lytechinus pictus*, denuded test.

12. Periproct with four to five large plates around anus; apical portion of ambulacral plates conspicuously devoid of spines (plate 456C, 456D) . . . . *Arbacia stellata*
— Periproct with many small plates; test uniformly covered with spines 13

13. Color of test and spines yellow or pale tan, often with blotches of grey or pale purple; ambulacral plates with single primary spine and no secondary spines (plate 457E, 457F) . . . . *Lytechinus pictus*
— Color of test and spines green, pink, red, purple, or nearly

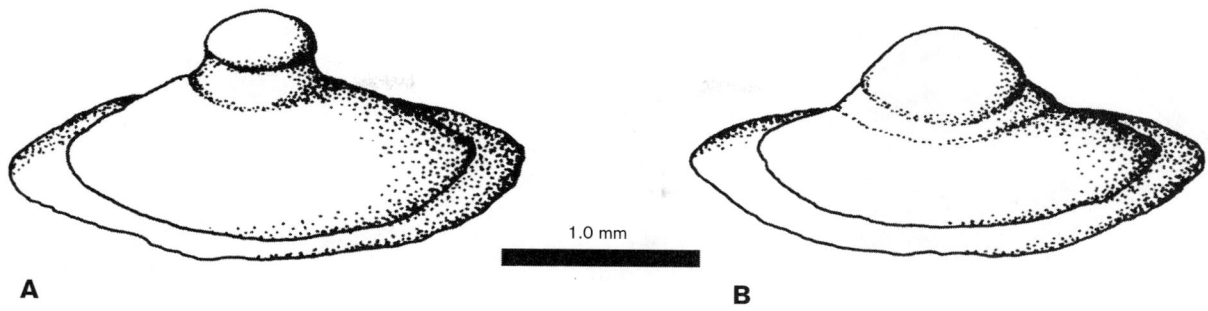

A   1.0 mm   **B**

PLATE 458 Tubercle morphology of *Strongylocentrotus*: A, oblique view of denuded primary spine tubercle of *S. droebachiensis*; B, same view of tubercle from *S. purpuratus*.

black; ambulacral plates with single primary and several secondary spines 14

14. Color of test and spines pale orange-pink; spines and test very fragile and easily broken (plate 457C, 457D) . . . . *Strongylocentrotus fragilis*
— Color of test and spines not pale orange-pink; spines and test not easily broken 15
15. Primary spines nearly as long or longer that half test diameter, and colored orange, red, maroon, or nearly black; peristome never green, orange in the smallest animals (plate 457B) . . . . *Strongylocentrotus franciscanus*
— Primary spines much shorter, often stubby looking and colored green or purple; peristome often green, especially in juveniles 16
16. Primary spines purple . . . . *Strongylocentrotus purpuratus* (in part)
— Primary spines green 17
17. Primary spine tubercles with distinct neck undercutting mammelon (area articulating with spine) (plate 458A) . . . . *Strongylocentrotus droebachiensis*
— Primary spine tubercles rounded, without distinct neck below mammelon (plates 457A, 458B) . . . . *Strongylocentrotus purpuratus* (in part)

List of Species

CIDAROIDA

CIDARIDAE

Eucidaris thouarsii (Valenciennes, 1846). Slate-pencil urchin; a southern species common in rocky low intertidal to 140 m in the Gulf of California south to the Galapagos Islands. Rarely found in southern California; fossils known from central California; see Glynn et al. 1979, Science 203: 47–49 (feeding on corals).

DIADEMATOIDA

DIADEMATIDAE

Centrostephanus coronatus (Verrill, 1867). Long-spined, black sea urchin; a southern species occasionally common from southern California to the Galapagos Islands. Mainly rocky subtidal to 110 m; see Kennedy and Pearse 1975, J. Exp. Mar. Biol. Ecol. 17: 323–331 (lunar reproductive cycle), Vance 1979, Ecology 60: 537–546 (grazing effects).

ARBACIOIDA

ARBACIIDAE

Arbacia stellata Gmelin, 1872. Sharp-spined, black sea urchin; a southern species common in rocky low intertidal to 90 m in the Gulf of California south to Peru. Rarely found in southern California.

ECHINOIDA

TOXOPNEUSTIDAE

Lytechinus pictus (Verrill, 1867). White sea urchin; common in embayments and rocky-sandy areas of southern California and Mexico, shallow subtidal to 300 m; occasionally found subtidally below 20 m in Monterey Bay. *L. anamesus* H. L. Clark, 1912, is a synonym for the ecoform that has longer spines and is generally found in deeper, open-coast water; see Dean et al. 1984, Mar. Biol. 78: 301–313 (grazing effects), Zigler and Lessios 2004, Evolution 58: 1225–1241 (speciation).

STRONGYLOCENTROTIDAE

Strongylocentrotus droebachiensis (O. F. Müller, 1776). Green sea urchin; an Arctic and circum-boreal species extending into north Atlantic and north Pacific, south to Cape Blanco in southern Oregon; low rocky intertidal to 300 m. Although there are no confirmed records of this species from California, it is easily confused with juveniles of *S. purpuratus,* which are also green; see Scheibling and Hatcher in Lawrence, 2007, pp. 353–392 (ecology, mainly in North Atlantic).

Strongylocentrotus fragilis Jackson, 1912. Pink sea urchin; an abundant deep-water species on rocky and fine sand bottoms, often occurring in large numbers from 30 to more than 1,200 m. The monotypic genus *Allocentrotus* Mortensen is abandoned because it leaves *Strongylocentrotus* paraphyletic (Biermann et al. 2003).

Strongylocentrotus franciscanus (A. Agassiz, 1863). Red sea urchin; large, over 100 mm in diameter, with long spines. Reported presence in Japan almost certainly misidentification of sister species, *S. nudus.* Uncommon in crevices and pools of low intertidal; mainly subtidal in or near kelp forests, but extending to 125 m. Main species of commercial fishing industry in northeast Pacific; see Rogers-Bennett in Lawrence 2007, pp. 393–425 (natural history and commercial harvest), Rogers-Bennett et al. 2003, Fish. Bull. 101: 614–626 (growth and longevity).

Strongylocentrotus purpuratus (Stimpson, 1857). Purple sea urchin; generally smaller echinoid (usually 50 mm diameter or

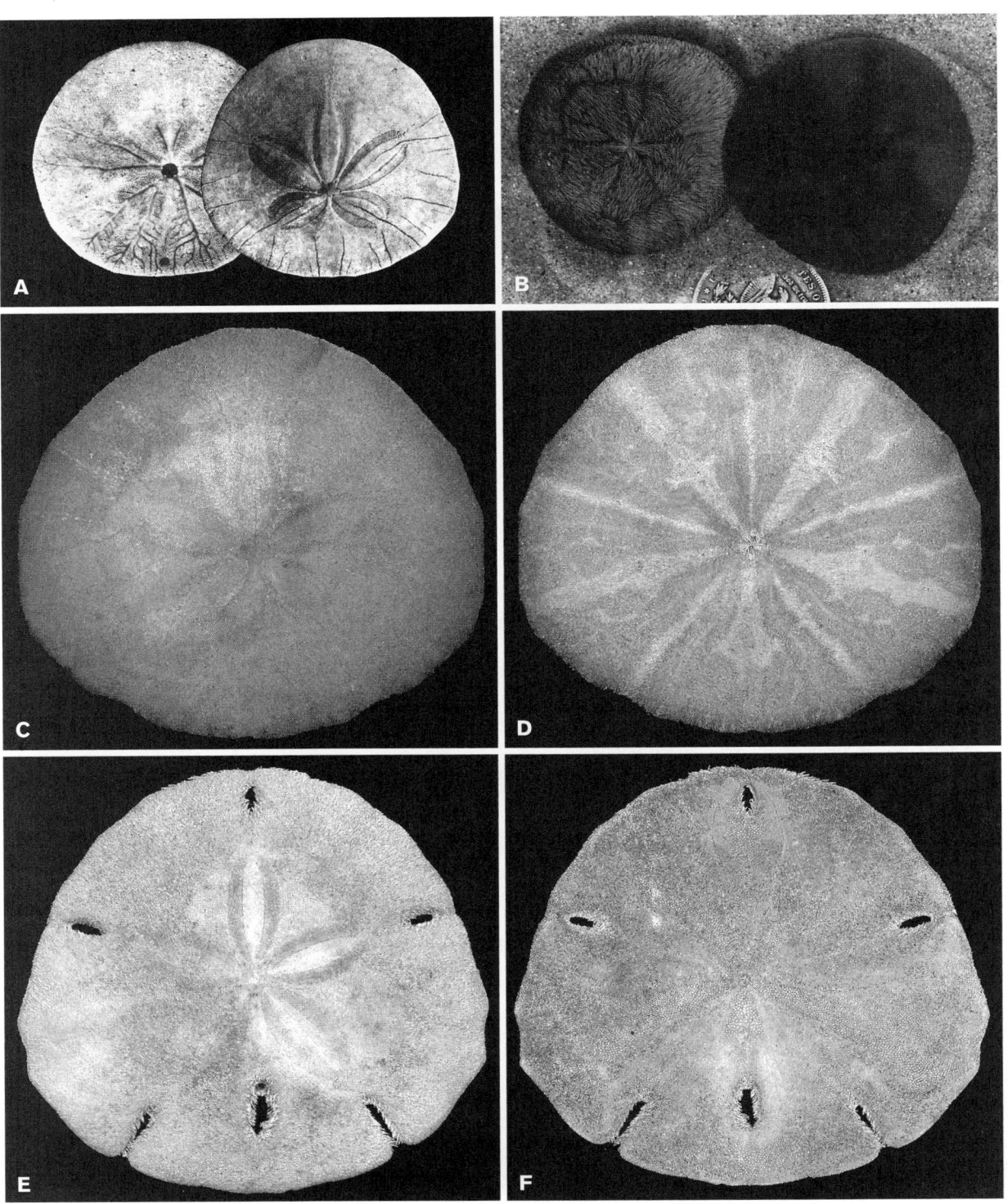

PLATE 459 Clypeasteroida: A, *Dendraster excentricus*, lower (oral) and upper (aboral) surfaces of dead, denuded tests as they would appear washed up on the beach (anterior end is towards top of page (photo courtesy of Ralph Buchsbaum); B, *Dendraster excentricus*, living specimens from a subtidal population (photo courtesy of Ralph Buchsbaum); C, *Dendraster terminalis*, aboral surface; D, *Dendraster terminalis*, oral surface; E, *Encope micropora*, aboral surface; F, *Encope micropora*, oral surface.

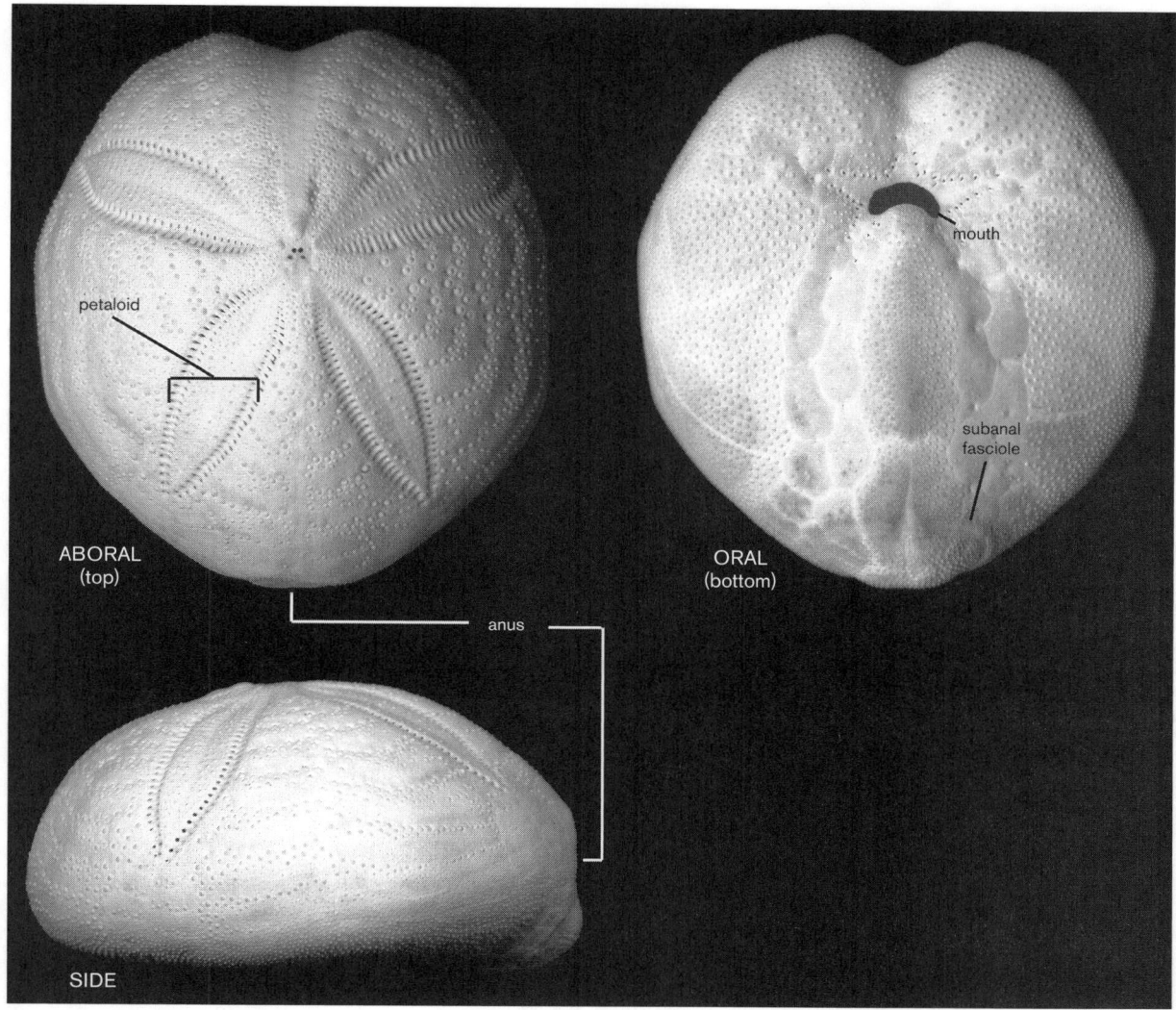

PLATE 460 Spatangoida: A, General features of a heart urchin as shown by a denuded test of *Spatangus californicus*.

less) than *S. franciscanus*, with short spines. Often common in crevices, pools, and mussel beds in mid- to low-intertidal, extending subtidally to 90 m. Action of teeth and spines can erode hollows in soft rock, within which animals nestle. Sometimes dense subtidal populations graze areas of most macroalgae, forming "barrens." A model organism for genome sequencing (see http://www.ncbi.nlm.nih.gov/genome/ guide/sea-urchin); see Rogers-Bennett in Lawrence 2007, pp. 393–425 (natural history), Behrens and Lafferty 2004, Mar. Ecol. Prog. Ser. 279: 129–139 (diseases); Pearse 2006, Science 314: 940–941 (ecology).

> Note: Species of *Strongylocentrotus* readily hybridize in lab, and hybrids are likely to occur in field. Lab-reared hybrids of *S. franciscanus* × *S. purpuratus* have varied appearances—some look like either parent or only a little different from either; they would be difficult to distinguish in the field (J. S. Pearse and M. E. Steele, pers. obs.).

CLYPEASTEROIDA

DENDRASTERIDAE

Dendraster excentricus (Eschscholtz, 1831). Common sand dollar; locally abundant low intertidal-shallow subtidal in sand or sandy-mud of inlets and embayments, and subtidally along open coast to 90 m; dead tests frequently encountered on beaches. In higher currents, embeds anterior end into sand to stand "upright"; see Cameron and Rumrill 1982, Mar. Biol. 71: 197–202 (recruitment), Morin et al. 1985, Mar. Ecol. Prog. Ser. 27: 163–185 (subtidal aggregations), Mooi 1997 (systematics, distribution, and overview).

Dendraster terminalis (Grant & Hertlein, 1938). Southern sand dollar; subtidal species found in sandy areas of southern California and the West Coast of Baja California, 6 m–55 m depth. Reaches sexual maturity at smaller sizes (<20 mm test length) than any other *Dendraster,* leading to its being mistaken for miniaturized forms in a completely different suborder (Mooi 1997).

MELLITIDAE

Encope micropora L. Agassiz, 1841. Keyhole sand dollar; southern species found in intertidal and subtidal to 30 m, sandy areas from Baja California to Peru. Rarely encountered in southernmost California; see Ebert and Dexter 1975, Mar. Biol. 32: 397–407 and Dexter 1977, Bull. Mar. Sci. 27: 5445–551 (natural history of related species).

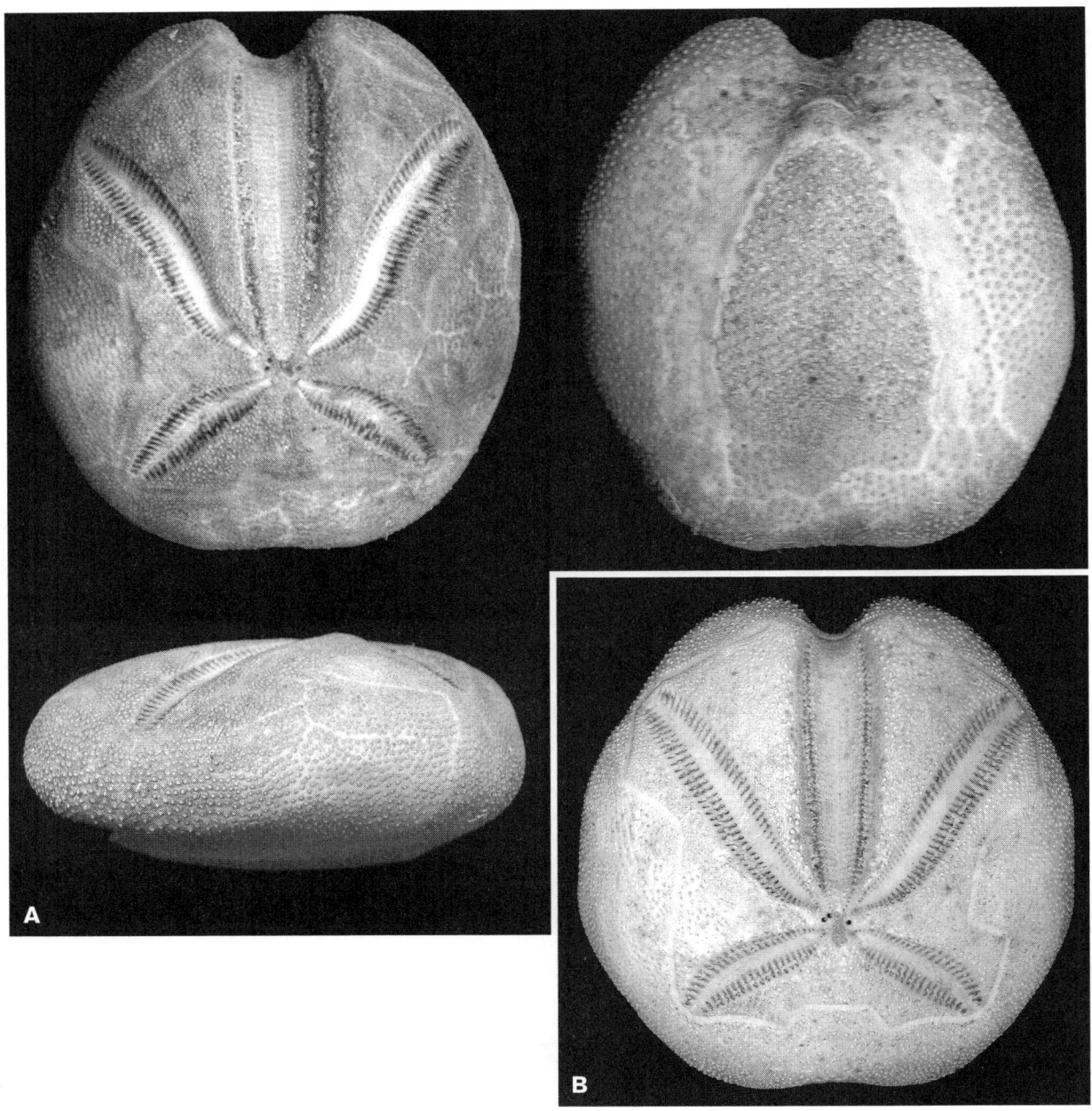

PLATE 461 Spatangoida: A, *Brisaster latifrons*, aboral, oral, and side views; B, *Brisaster townsendi*, aboral view.

SPATANGOIDA

SCHIZASTERIDAE

Brisaster latifrons (A. Agassiz, 1898). Common heart urchin; subtidal species burrowing in sandy-mud bottoms, 35 m–1,800 m, see Hood and Mooi 1998 (systematics, evolution).

Brisaster townsendi (A. Agassiz, 1898). Very similar to *B. latifrons,* especially young individuals, several measurements from several specimens usually required to distinguish these species; 35 m–1,900 m; see Hood and Mooi 1998 (systematics, evolution).

BRISSIDAE

Brissopsis pacifica (A. Agassiz, 1898). Southern heart urchin; subtidal species burrowing in sandy-mud bottoms from southern California to the Galapagos Islands, 9 m–75 m.

LOVENIIDAE

Lovenia cordiformis (A. Agassiz, 1872). Sea porcupine; southern species extending from southern California to the Galapagos; burrows just below the surface of sand from the low intertidal to 140 m.

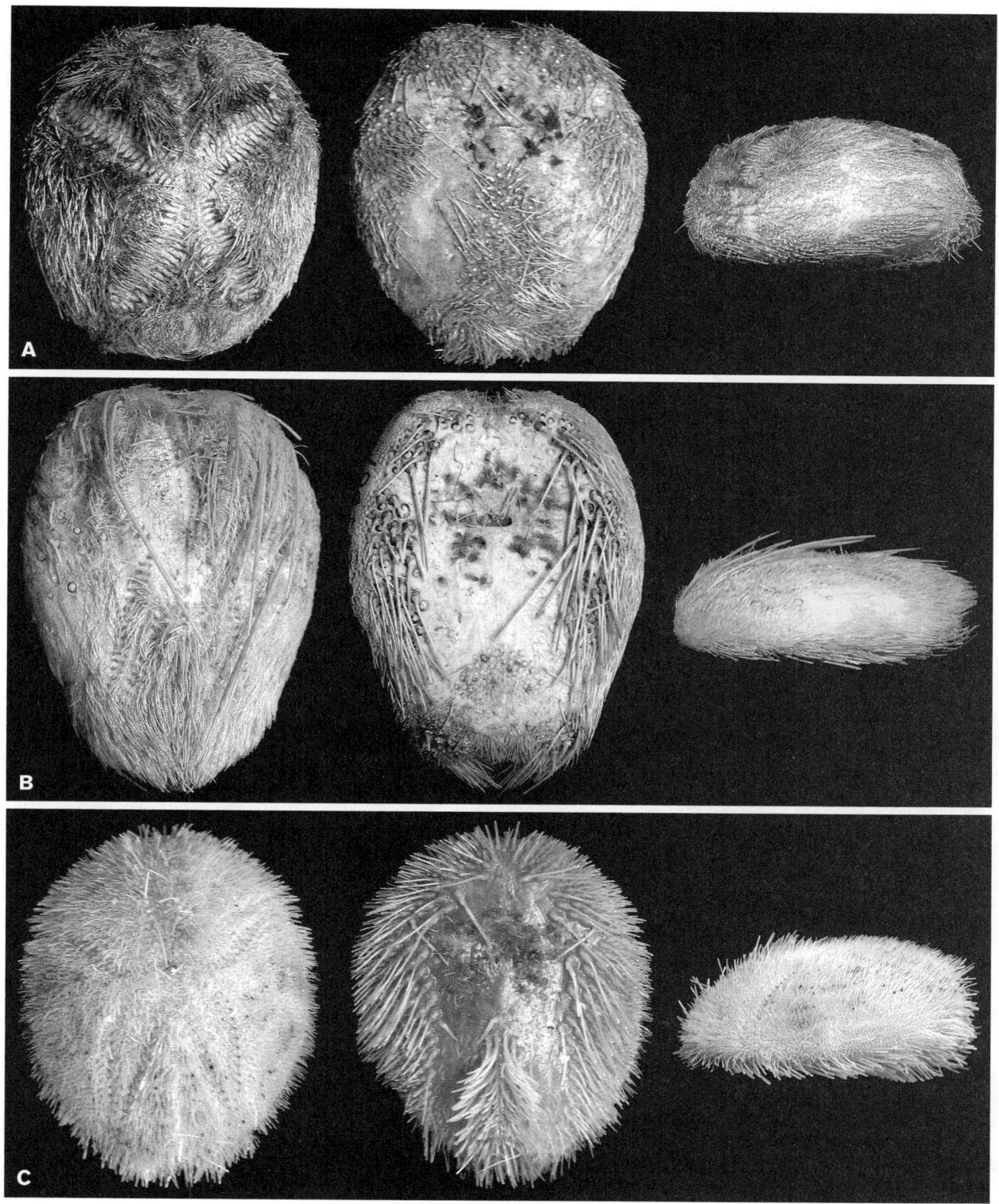

PLATE 462 Spatangoida: A, *Brissopsis pacifica*, aboral, oral, and side views; B, *Lovenia cordiformis*, aboral, oral, and side views; C, *Nacospatangus laevis*, aboral, oral, and side views.

Nacospatangus laevis (H. L. Clark, 1917). Sea mouse; southern California to Gulf of California, 5 m–410 m. The nomenclature for this species is greatly confused. Although listed in some works as *Gonimaretia laevis* or *Pseudomaretia laevis,* studies suggest that *Nacospatangus, Gonimaretia* (under the name *Goniomaretia*) and *Pseudomaretia* should all be submerged under the oldest name *Nacospatangus.* Without formal revision, we assume that *Nacospatangus depressus* Clark, 1917 is a junior synonym of *N. laevis.*

Spatangus californicus H. L. Clark 1917. Santa Barbara County to Gulf of Mexico, 5 m–300 m.

References

Biermann, C. H., B. D. Kessing, and S. R. Palumbi. 2003. Phylogeny and development of marine model species: strongylocentrotid sea urchins. Evolution and Development 5: 360–371.

Durham, J. W., C. D. Wagner, and D. P. Abbott. 1980. Echinoidea: The sea urchins, pp. 160–176, In R. H. Morris, D. P. Abbott, and E. C. Haderlie, Intertidal Invertebrates of California, Stanford University Press.

Grant, U. S. IV, and L. G. Hertlein. 1938. The west American Cenozoic Echinoidea. Publ. UCLA Math. Phys. Sci. 2: 1–225.

Hood, S., and R. Mooi. 1998. Taxonomy and phylogenetics of extant *Brisaster* (Echinoidea: Spatangoida), pp. 681–686. R. Mooi and M. Telford, eds. Echinoderms: San Francisco, A. A. Balkema: Rotterdam.

Jensen, M. 1974. The Strongylocentrotidae (Echinoidea), a morphologic and systematic study. Sarsia 57: 113–148.

Lawrence, J. M., ed. 2007. Edible sea urchins: biology and ecology. 2nd Edition. Ansterdam: Elsevier, 529 pp.

Lee, Y.-H. 2003. Molecular phylogenies and divergence times of sea urchin species of Strongylocentrotidae, Echinoida. Mol. Biol. Evol. 20: 1211–1221.

McCauley, J. E., and A. G. Carey, Jr. 1967. Echinoidea of Oregon. J. Fish. Res. Bd. Canada 24: 1385–1401.

Mooi, R. 1997. Sand dollars of the genus *Dendraster* (Echinoidea: Clypeasteroida): phylogenetic systematics, heterochrony, and distribution of living species. Bulletin of Marine Science, 61: 343–357.

Mortensen, T. 1928–1951. A Monograph of the Echinoidea. 5 vols. Copenhagen: C. A. Reitzel.

Pearse, J. S., and R. A. Cameron. 1991. Echinodermata: Echinoidea, pp. 513–662, In Reproduction of marine invertebrates. Vol. VI, Echinoderms and Lophophorates. A. C. Giese, J. S. Pearse, and V. B. Pearse, eds. Pacific Grove, CA: Boxwood Press.

Asteroidea

CHRISTOPHER MAH

(Plates 463–466)

Asteroids, also known as starfishes or sea stars, are familiar inhabitants of intertidal and subtidal California and Oregon. The asteroid fauna of the northeast Pacific is among the most diverse in the world, including many endemic species, such as the well-known sunflower star *Pycnopodia helianthoides.* Two of the largest sea stars in the world, *Pycnopodia helianthoides* and *Pisaster brevispinus,* are found in our area, with adults reaching radii of 35 cm–40 cm. Clark (1962) summarized general biology and diversity. Hyman (1955) remains a dependable account of functional morphology. Blake (1989) reviewed functional morphology, classification, and phylogeny. Feder (1980) provided excellent coverage of most shallow-water California and Oregon sea stars.

Because of the important role sea stars play in marine ecosystems, there is an extensive literature concerning in-

tertidal and subtidal interactions of many species, such as the ochre star *Pisaster ochraceus.* Interspecific interactions between asteroids and other invertebrates were reported by Mauzey et al. (1968) and Birkeland (1974); Harrold and Pearse (1987) reviewed the role of asteroids and other echinoderms in kelp forest ecosystems. Wobber (1975) documented interspecific behavioral interaction between certain California sea stars. Jangoux (1982) and Sloan (1980) provided comprehensive reviews of asteroid diet and feeding behavior. As a group, sea stars feed opportunistically (Jangoux 1982). Some species are predators on specific prey, such as *Solaster dawsoni,* which feeds upon other asteroids (Van Veldhuizen and Oakes 1981); others, such as *Patiria miniata,* are omnivores or detritivores.

Since the previous edition of this book, there have been several comprehensive summaries of northwest Pacific Asteroidea. Feder (1980) reviewed most California shallow-water species and provided brief summaries of each species' biology, ecology, and natural history. Lambert (2000) provided a key and short accounts of biology and ecology for asteroids of British Columbia, and Kozloff (1987) provided a key to the shallow-water Asteroidea of Washington and Oregon. Brusca and Brusca (1978) listed asteroid species from northern California. Hopkins and Crozier (1966) provided a key to and natural history of southern California asteroids. Among the earliest taxonomic monographs of Pacific coast asteroids is that of Verrill (1914). Fisher (1911, 1928, 1930) remains the authoritative monograph for asteroids from this region.

Asteroid identification is based primarily on endoskeletal characteristics. The elements of the endoskeleton, known as ossicles or plates, form regular series and complex arrangements that are often decorated with spines, granules, or other armament or accessories. In most cases, observation with a dissection microscope or hand lens is all that is required for recognizing characteristics. Analysis of more subtle characters (e.g., valves of pedicellariae, plate form) may require minor preparation with sodium hypochlorite (common household bleach) for proper identification.

Preservation in ethanol is best for observation of soft characters such as tube feet; however, most skeletal characters are best displayed in dry specimens. For many, the color of the living animals is the most useful diagnostic tool, but unfortunately color quickly fades in preserved specimens. Some species, such as *Solaster stimpsoni* and *Solaster dawsoni,* are immediately recognizable when alive but are more difficult to distinguish when preserved.

Terminology and Glossary

The mouth-bearing or "bottom" side is referred to as the actinal, or oral, surface and the nonmouth-bearing, "top" side is the abactinal or aboral surface. Terms such as "dorsal" and "ventral" are discouraged in adult echinoderms owing to the incorrect suggestion of homology with the larvae or other bilateral animals. Actinal and abactinal also refer to structures located on those surfaces. Thus, granules on the mouth-bearing side are referred to as actinal, or oral, granules. "R" and "r" refer to standard descriptive measurements used in reference to the body: R, the major radius, from the center of the disk to the arm tip; r, the minor radius from the disk center to the interradial edge (plate 463A).

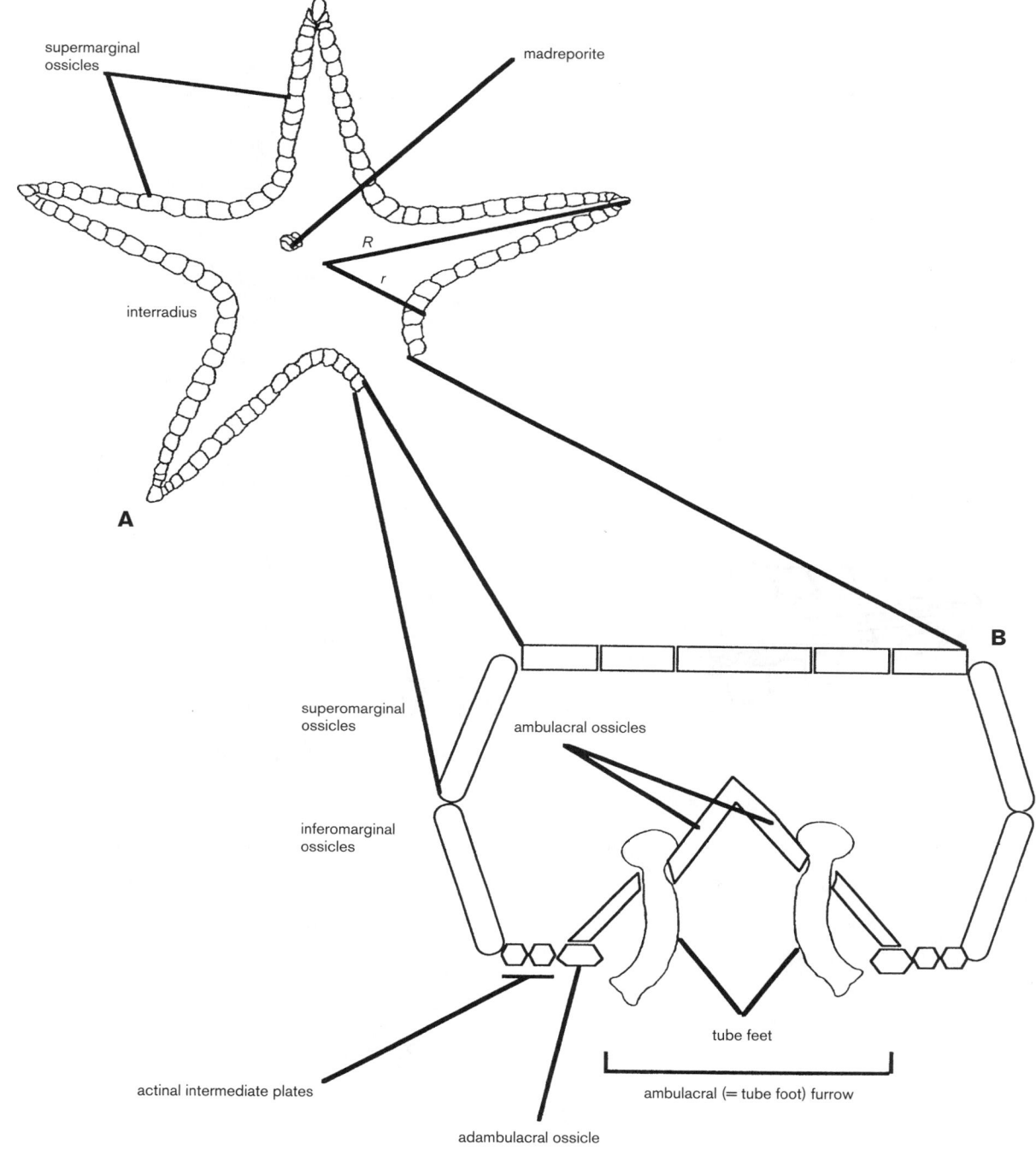

supermarginal
ossicles

madreporite

interradius

R

r

A

superomarginal
ossicles

ambulacral ossicles

inferomarginal
ossicles

B

tube feet

actinal intermediate plates

adambulacral ossicle

ambulacral (= tube foot) furrow

PLATE 463 A, B, abactinal (aboral) surface of generalized sea star.

Glossary

Terminology follows Lambert (2000) and Clark and Downey (1992). Useful glossaries of terms may also be found in Spencer and Wright (1966).

AMBULACRUM Midradial groove on lower surface of arm containing radial elements of water vascular system. Also referred to as a tube-foot groove (plate 464).

ACTINAL INTERMEDIATE OSSICLES Area of actinal surface between marginals and adambulacrals (plate 464).

AMBULACRAL OSSICLES One of a paired series of plates within the ambulacrum, forming an arched channel for the radial water vessel, and between which the tube feet are extended (plate 463B).

ADORAL CARINA One or more pairs of contiguous adambulacral behind each pair of mouth angle ossicles (plate 466C).

ADAMBULACRAL OSSICLES One row on each side of the ambulacrum between and articulating with two successive ambulacral plates, defining the edge of the furrow (plates 463B, 464).

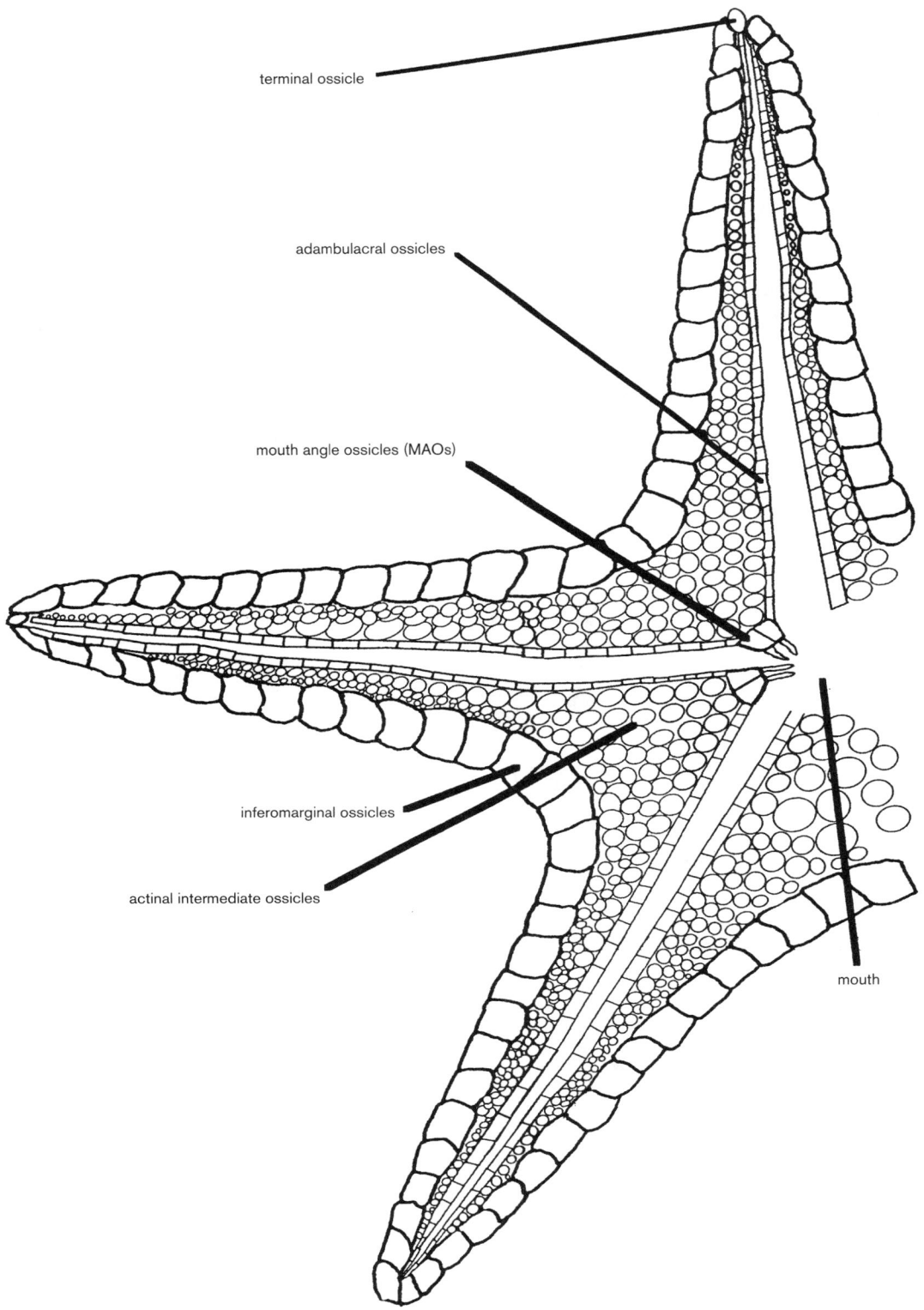

terminal ossicle

adambulacral ossicles

mouth angle ossicles (MAOs)

inferomarginal ossicles

actinal intermediate ossicles

mouth

PLATE 464 Actinal (oral) surface of generalized sea star.

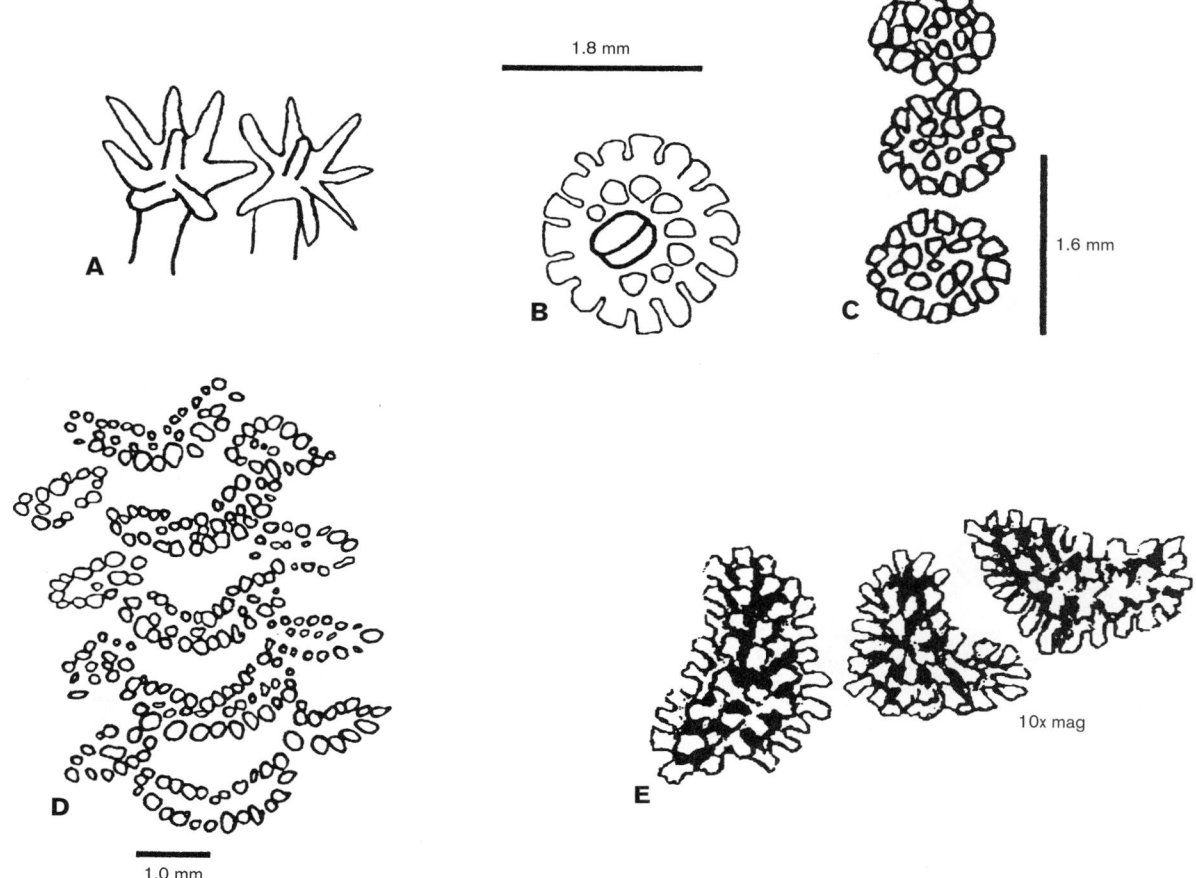

1.8 mm

1.6 mm

1.0 mm

10x mag

PLATE 465 A, abactinal view of paxillae of *Astropecten verrilli*; B, C, sessile pedicellariae on a tabulate ossicle (B) and tabulate ossicles (C) of *Mediaster aequalis*; D, Crescentic abactinal ossicles of *Patiria miniata*; E, abactinal plates of *Henricia leviuscula* (original; E, from Verrill, 1914, plate 87, 1a).

ADAMBULACRAL SPINES Spines that sit upon adambulacral ossicles that project into the ambulacrum. These arrays are comprised of several spines in comb or forklike arrangements.

ARM (=RAY) Radial part of sea star distal to disk (plate 463A).

COMET Term referring to either a newly regenerated or post-fissiparous sea star in which all but one partially or fully formed arm is attached to the disk, giving a cometlike appearance. This term is used particularly in reference to ophidiasterids, such as *Linckia columbiae*.

DISK Central part (body) of sea star from which arms project distally (plate 463A).

FURCATE PEDICELLARIAE A type of stalked pedicellariae specific to *Pisaster* (plate 466B).

INFEROMARGINAL OSSICLES (=INFRAMARGINAL PLATES) The lower of the two marginal ossicle series. Defines the ambitus of the actinal surface (plates 463B, 464).

MADREPORITE (=SIEVE PLATE) A specialized, perforated, interradial plate on the abactinal surface of the disk, forming a sievelike opening for the water vascular system (plate 463A).

MARGINAL PLATES: The two horizontal series of plates (ossicles), inferomarginal and superomarginal, usually defining the ambitus. Often larger and more regularly aligned than the other series of plates extending to the terminal plate (plates 463A, 463B, 464).

MOUTH ANGLE OSSICLE One of the pair of plates opposite the mouth at the apex of each lower interradius.

PAPULAE Respiratory finger or glovelike pockets of the coelom that protrude through the body wall.

PAXILLA(E) A columnar plate with the base usually expanded and the top crowned with a cluster of spinelets or granules. Found in sand or mud-dwelling sea stars such as *Luidia* or *Astropecten*. They are believed to inhibit clogging of papulae and act as protection from predators (plate 465A).

PEDICELLARIA(E) Small pincerlike organ on body surface, variously modified in shape and number of component valves. Plates 465B, 466A, and 466B show different types of pedicellariae. A review of pedicellarial form and function can be found in Jangoux and Lambert (1988). See also sessile, furcate, and stalked pedicellariae.

PSEUDOPAXILLAE Superficially resemble paxillae, but the top crowned with spines which are extensions of the plate rather than articulated with the paxillae. Household bleach applied to the latter will remove the tissue articulating the spines to the paxillae. Found in *Solaster*.

SESSILE OR BIVALVE PEDICELLARIAE Pedicellariae of two tonglike valves, sitting in a pit nearly flush with the asteroid surface or on tabulae, or paxillae. Found on tabulae and actinal surface of *Mediaster* (plate 465B).

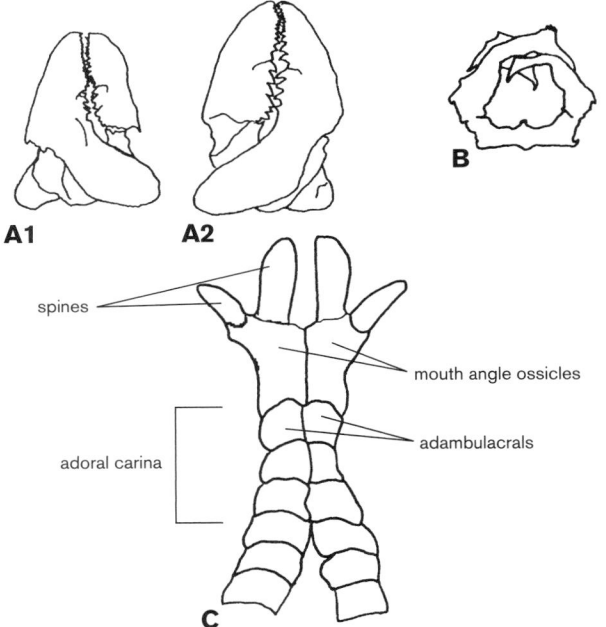

PLATE 466 A, crossed pedicellariae of *Leptasterias aequalis*; B, furcate pedicellariae of *Pisaster ochraceus*; C, adoral carina of *Orthasterias koehleri* (A, from Fisher, 1930, plate 46, fig. 3a,b; B, from Fisher, 1930, plate 73, fig. 8a; C, from Fisher, 1928, plate 65, fig. 9).

STALKED (=CROSSED OR FORCIPULATE) PEDICELLARIAE Pedicellariae composed of two wrench-shaped valves that articulate at the base (plate 466A). Stalked pedicellariae are sheathed by epithelium and attached to the body surface with a stalk and frequently occur in wreaths or batteries around spines. Forciform pedicellariae are a form of stalked pedicellariae (plate 466B).

SUPEROMARGINAL PLATES (=SUPRAMARGINAL PLATES) The upper of the two marginal plate series. These plates define the ambitus of the abactinal/aboral surface (plates 463A, 464).

SUPRADORSAL MEMBRANE Membrane found only in *Pteraster tesselatus* and other pterasterids. A second membrane suspended by pseudopaxillae over the abactinal surface, forming a tentlike cavity between the two surfaces.

TABULAE Columnar structures, similar to paxillae. Crowned with prismatic or blunt spines rather than narrow stellate ones (plate 465C).

TUBE FEET Two to four rows of tube feet occupy the tube foot groove along the radii of the underside, or actinal surface, of the sea star (plate 463B). Tube feet may be suckered or pointed. The latter are found only in the sand- or mud-inhabiting paxillosidan asteroids such as *Luidia* or *Astropecten*.

Key to Asteroidea

Species in the key and species list are those encountered in the intertidal to 2 m. Some subtidal species are included in the species list but not keyed out. Excellent color photographs are found in Gotshall (1994).

1. Eight to 24 or more arms 2
—   Five to six arms (as a result of injury and regeneration, individuals with other than the normal number of arms also

occur) 4
2. Four rows of tube feet; with stalked or crossed pedicellariae (plate 466A) arranged in prominent rosettes around spines; pseudopaxillae absent; disk and arms, soft, fleshy, and flexible; typically 20–24 arms; color in life pink, purplish, or brown, less often red, orange, or yellow. . . . . *Pycnopodia helianthoides*
—   Two rows of tube feet; pedicellariae absent; pseudopaxillae present; disk and arms firm, not fleshy, and rigid; typically eight to 13 . . . . *Solaster* 3
3. Typically 10 arms (exceptionally nine or 11); arms long and slender; color in life red/orange with dark blue, purplish-gray strip radiating from disk along arms. . . . . *Solaster stimpsoni*
—   Typically 12–13 arms (exceptionally eight to 11); arms short and stocky; color in life solid brown to gray-yellow and occasionally red or orange . . . . *Solaster dawsoni*
4. Tube feet suckered; paxillae absent on surface 5
—   Tube feet pointed, lacking a flaring disk; abactinal surface of body covered with paxillae (plate 465A); large marginal plates bearing short, stout spines . . . . *Astropecten armatus*
5. Four rows of tube feet; stalked (=crossed or forcipulate) pedicellariae (plate 466A) never sessile; pedicellariae in wreaths around spines; adoral carina present in adults (plate 466C). 6
—   Two rows of tube feet; pedicellariae sessile or absent, never stalked (=crossed or forcipulate); adoral carina absent 12
6. Typically with six rays; in California adults not greater than R = 4.0 cm (Washington, Oregon, and northerly species of *Leptasterias* can be larger); R = 3r to 5r; with brooding embryos . . . . *Leptasterias* 7
—   Typically five rays; adults easily with R > 4.0 cm; R = 5r to 10r; no brooding embryos or eggs 8
7. Abactinal skeletal ossicles in distinct radial rows, midradial row conspicuous (denuding of abactinal spines may be necessary for observing the midradial ossicle series); arms stouter, thicker at base. . . . . *Leptasterias aequalis*
—   Abactinal skeletal ossicles not in distinct radial rows, midradial row not conspicuous, arms more slender . . . . *Leptasterias pusilla*
8. Abactinal spines blunt or low in irregular series; adambulacral spines with pedicellarial clusters; arms thick and relatively short, R = 5.0–6.0r; relatively wider at base; larger central disk than above. 9
—   Prominent sharp abactinal spines in regular series; adambulacral spines without attached pedicellariae; arms slender and deciduous; R = 6.0–10.0r; base of arms relatively slender; with small disk 11
9. Two to three adambulacral spines with clusters of pedicellariae; pedicellariae not furcate; abactinal spines neither blunt nor clublike, never in regular series; irregular reticulations over abactinal surface; arms more slender, R = 4.5–8.0r; color in life extremely variable . . . . *Evasterias troschelii*
—   Single adambulacral spines; furcate pedicellariae (plate 466B); arms more stout, R = 2–5r; blunt, clublike abactinal spines sometimes forming regular series . . . . *Pisaster* 10
10. Midradial row of spines typically distinct, straight; spines on lateral and abactinal surfaces of arms not forming reticulated pattern (except rarely); color in life consistently whitish pink to deep pink sometimes mottled with gray-green or maroon purple . . . . *Pisaster brevispinus*

— Spines on lateral and abactinal surfaces of arms very short and blunt, forming extensive irregular, reticulated pattern, or separate convex, curved groups at arm tips; distinct midradial row of spines not typically present, or at least not straight; color in life uniform orange, yellow, brown to reddish purple . . . . *Pisaster ochraceus*

— Abactinal (aboral) spines typically blunt, club-shaped, relatively large, surrounded at base by rings of flesh (blue in life), with ring of pedicellariae outside fleshy ring; spines dense but typically not arranged in distinct radial or concentric rows; color in life yellow to gray with blue rings around bases of spines/pedicellariae . . . . *Pisaster giganteus*

11. Inferomarginal plates with two spines, each of which possesses clusters of crossed pedicellariae (plate 466A); color in life bright red/brick red-orange, often with yellow to white banding or mottling on arms. . . . . *Orthasterias koehleri*

— Inferomarginal plates with two spines, only one of which possesses a cluster of crossed pedicellariae; color in life blue-based spines and bright orange tips; surface is mottled gray-green to brown with distinct dark bands across the arms . . . . *Astrometis sertulifera*

12. Slender arms with small disk 13

— Body with large disk and broad, triangular arms. 14

13. Surface covered with coarse hemispherical granules; two madreporites (exceptionally one, three, or four); variable number of blunt, cylindrical rays; fissiparous-occasionally found in "comet" stage; brooding absent; adambulacral spines, rounded-granular; color in life red with yellow to orange spots. . . . . *Linckia columbiae*

— Surface covered with reticulate network covered by fine spination (plate 465E); single madreporite; typically five slender arms; nonfissiparous; brooding present or absent; adambulacral spines needlelike; color in life bright red, light orange to dark brown with mottled tan-cream (see species description for explanation of quotation marks) . . . . *Henricia* spp.

14. Abactinal skeleton composed of crescentic-shaped plates forming a chain-maillike mesh (plate 465D); papulae emerge through mesh; abactinal surface with minute bifurcate or trifurcate spines; no distinctive odor; color in life highly variable, red-orange, blue, purple to cream or purple with light or dark mottling and blotches . . . . *Patiria miniata*

— Surface almost completely smooth to the touch, covered with a velvety skin; papulae present in widespread patches over surface; no pedicellariae; pungent garlic-sulfurous odor; color in life red to orange with gray-red imbrications . . . . *Dermasterias imbricata*

— Surface covered with tabulate plates (plate 465C); with prominent marginals, covered by large prismatic granules, forming border around perimeter of body; sessile pedicellariae present on tabulae (plate 465B); color in life bright red-orange . . . . *Mediaster aequalis*

List of Species

PAXILLOSIDA

ASTROPECTINIDAE

Astropecten armatus Gray, 1840 (=*A. brasiliensis armatus*). Southern California species extending south from San Pedro. Shallow water (0–1 m) to subtidal depths. Frequently mistaken for *A. verrilli* but possesses short, stout spines on superomarginal plates. An infaunal predator on sand to mud bottoms. Diet includes sea pansies, sand dollars, mud snails, and jack-knife clams. See Hopkins and Crozier (1966) and Segal 1988, Bull. So. Cal. Acad. Sci. 87: 35–38. Morin et al. (1985) treats interaction with other sand community inhabitants and ectoparasitism by the eulimid gastropod *Polygireulima rutila*.

*Astropecten verrilli deLoriol, 1899 (=*A. californicus* Fisher). Gotshall (1994) treats this species as a synonym of *A. armatus*, but the two are retained as distinct here. Uncommon, poorly known; primarily subtidal. An infaunal predator occupying sandy bottoms; see Morin et al. (1985).

LUIDIIDAE

*Luidia foliolata Grube, 1866. Uncommon; occurs on sandy bottoms at subtidal depths. Abactinal surface with subquadrate paxillae and long, straplike arms (R = 7.1–8.7r). A burrowing predator of infaunal invertebrates, mostly bivalves, but also ophiuroids, echinoids, holothurians, scaphopods, bivalves, polychaetes, and crustaceans (Sloan and Robinson 1982, Jangoux 1982). The brittle star *Ophiura lütkeni* displays a strong escape response from this species (Lambert 2000). Pettibone (1953) reports two commensal polynoids, *Arctonöe pulchra* and *A. vittata*.

VALVATIDA

ASTERINIDAE

Patiria miniata (Brandt, 1835) (=*Asterina miniata*; see Clark 1983, Bull. Br. Mus. Nat. Hist. (Zool). 45; 359–380). This species was reassigned to *Patiria* by O'Loughlin and Waters 2004, Mem. Mus. Victoria 61: 1–40. "Bat stars" or "sea bats." Common from low intertidal to subtidal depths. Bat stars are omnivores/predators, consuming echinoids, algae, sponges, bryozoans, tunicates and bryozoans (see Harrold and Pearse 1987). The worm *Ophiodromus pugettensis* is a commensal (Pettibone 1953). See Wobber (1975) (ecological interaction with asteroids); Schroeter et al. 1983, Oecologia 56: 141–147 (ecological interaction with the echinoid *Lytechinus anamesus*); Rumrill 1989, Mar. Ecol. Prog. Ser. 56: 37–47 (growth, reproduction and population biology); Leonard 1994, J. Exp. Mar. Biol. Ecol. 179: 81–98 (effect on recruitment of kelp *Macrocystis*).

GONIASTERIDAE

Mediaster aequalis Stimpson, 1857. Common in the intertidal to subtidal depths. Sloan and Robinson (1983) report *M. aequalis* as a microphagous feeder with a very broad diet. Diet includes encrusting sponges, bryozoans, and sea pens (Mauzey et al. 1968, Birkeland 1974); also scavenges dead animals (Hopkins and Crozier 1966). See Birkeland, Chia, and Strathmann 1971, Biol. Bull. 141: 99–108 (development and growth).

OPHIDIASTERIDAE

Linckia columbiae Gray, 1840. Common intertidally to subtidally in southern California; absent north of San Pedro. Possesses developed regenerative abilities, but seldom are individuals observed with all five rays perfectly intact, and frequently found as

* = Not in key.

"comets." Fisher (1911) provides extensive morphological evaluation. Food and feeding behavior are unknown, but possibly a particulate suspension feeder (Anderson 1960, Biol. Bull. 119: 371–398); other species of *Linckia* are substrate film feeders and microherbivores (Jangoux 1982). Digestive morphology is described by Anderson 1962, Amer. Zool., 2: 387. See McAlary 1993, pp. 233–248, Third CA Islands Symposium (population structure).

PORANIIDAE

Dermasterias imbricata (Grube, 1857). The leather star. Common, intertidal to 91 m (Maluf 1988). Ricketts et al. (1985: 285) note that *D. imbricata* is most numerous in the northern half of its range and that specimens ". . . found in Puget Sound are gigantic (diameter up to 25 cm) by comparison with those in the tide pools of the protected outer coast." Leather star diets vary with region; feeds on sponges, hydroids, bryozoans, colonial tunicates, algae, pennatulids, holothurians, echinoids, and even other asteroids (Jangoux 1982, Harrold and Pearse 1987). Elicits a dramatic escape response from actinarian sea anemones; see Dalby et al. 1988, Can. J. Zool. 66: 2484–2491; Elliott et al. 1989, Biol. Bull. 176: 73–78 (*Stomphia*) and Elliott et al. 1985, Can. J. Zool. 63: 1921–1929 (*Urticina*). Potentially mutualistic relationship with the polynoid worm *Arctonoe vittata* (Wagner et al. 1979, J. Exp. Mar. Biol. Ecol. 39: 205–210).

Poraniopsis inflata (Fisher, 1906). Primarily subtidal, seldom encountered. White to orange; pointed spines with a fleshy surface texture. Probably feeds on sponges. See Anderson and Shimek 1993, Zool. Biol. 12: 499–503 (diet and feeding habits in aquaria). *Poraniopsis jordani* Gotshall, 1994, is a *nomen nudum* and should be removed from use.

VELATIDA

PTERASTERIDAE

Pteraster tessellatus Ives, 1888. Subtidal depths, seldom encountered. Stellate (R = 1.6–1.6r) along most of its range with the more pentagonal subspecies (R = 1.44r). Very thick, fleshy species with unique supradorsal membrane. Color a cream/tan with dark mottling. *Pteraster tessellatus arcuatus* Fisher, 1928, in Monterey Bay is a probable synonym. Utilizes mucus as a defense mechanism against predators (see Nance and Braithwaite 1979, J. Exp. Mar. Biol. Ecol. 40: 259–266; 1981, J. Exp. Mar. Biol. Ecol. 50: 21–31). McEdward 1992, Biol. Bull. 182: 177–187; McEdward 1995, Biol. J. Linn. Soc. 54: 299–327 (evolution of pelagic direct development); Harley et al. 2006, Biol. Bull. 211: 248–262 (color polymorphism).

SOLASTERIDAE

Solaster dawsoni Verrill, 1880. Uncommon. Largely subtidal, rarely very low intertidal; primarily northern. Feeds on asteroids, preferentially on its congener *Solaster stimpsoni* (Van Velduizen and Oakes 1981). Feder (1980) reports that *S. dawsoni* also feeds on holothurians and, in the laboratory, elicits escape responses from the nudibranch *Tritonia*. See Lambert (2000) for commensal polynoids *Arctonöe fragilis* and *A. vittata*.

Solaster stimpsoni Verrill, 1880. Uncommon. Predominantly subtidal, but in the low intertidal of northern California (Hum-

* = Not in key.

boldt County) (Brusca and Brusca 1978). See Mauzey et al. (1968) (general diet); Van Veldhuizen and Oakes (1981) (escape response from *S. dawsoni*); two commensal scale worms, *Arctonöe pulchra* and *A. vittata* (Pettibone, 1953).

SPINULOSIDA

ECHINASTERIDAE

Henricia spp. In both shallow and deep waters; common intertidally. This genus forms a puzzling complex on the Pacific northwest coast, and the systematics of species along the California and Oregon coasts require revision. Fisher (1911) treated *Henricia* from California as *H. leviuscula* (Stimpson, 1857), but also noted that this name may not be applicable to California populations. He also (Fisher 1911) described six distinct "varieties" of *Henricia leviuscula* from California. Some of these morphotypes may be separate species (D. Eernisse and M. Strathmann, pers. comm. 2005). Diet includes encrusting invertebrates (Mauzey et al. 1968), particulate suspension feeding (Anderson 1960; Rasmussen 1965) and detritus (Hopkins and Crozier 1966). Morphology of the digestive system is reported by Anderson (1960, Biol. Bull. 119: 371–398). See Lambert (2000) and Feder (1980) for reproductive biology. Brooding in Fisher's variety "F" is described by Chia (1966, as *H. leviuscula*).

FORCIPULATIDA

ASTERIIDAE

Astrometis sertulifera (Xantus, 1860). Rare in southern California. Brusca (1980) reports that in the Gulf of California smaller specimens often occur in the low intertidal zone, apparently migrating offshore as they mature. Larger specimens are always found subtidally. The smaller individuals are considerably less flexible than their larger, older kin. This species is considered a "voracious predator" of gastropods, pelecypods, and barnacles (Brusca 1980). Further prey items include chitons, the ophiuroid *Ophiothrix spiculata* and occasionally sea urchins (Feder 1980). Small crabs and even fishes as large as the sea star itself can be held by pedicellariae for feeding (Jennings 1907, U.C. Publ. Zool. 4: 43–185, but misidentified there as *Asterias forreri* (=*Stylasterias forreri*)).

Evasterias troschelii (Stimpson, 1862). Intertidal to 70 m; not south of Monterey. Frequently mistaken for *Pisaster*. Fisher (1930) describes variation in *E. troschelii* and considered it as among the most variable of sea stars. Lambert (2000) noted that it displaces *Pisaster ochraceus* in sheltered inlets. Mauzey et al. (1968) report that the diet varies with the relative abundance of prey, which includes bivalves, limpets, snails, brachiopods (*Terebratalia*), barnacles (*Balanus glandula*), and tunicates. See Christensen 1957, Limnol. Oceanogr. 2: 180–197 (feeding on bivalves); Young 1984, Echinodermata, Proc. 5th IEC, 577–583 (feeding on ascidians). The polynoid *Arctonoe fragilis* is a commensal, occurring in the ambulacral groove and on the body surface (Feder 1980). See also Patterson et al. 1978, J. Exp. Mar. Biol. Ecol. 33: 51–56.

Leptasterias spp. A species complex exists on the northwest Pacific coast. See Foltz 2001, Mar. Biol. 139: 475–483 and references therein. Papers pertaining to biology of *Leptasterias* (mostly *L. hexactis*) include Chia 1966, Biol. Bull. 130: 304–315; Chia 1968, Acta Zool. 49: 321–364; Menge 1975, Mar Biol. 31:

87–100, Menge and Menge 1974, Ecol. Monogr. 44: 596–600 (competition with *P. ochraceus*)

Leptasterias aequalis (Stimpson, 1862). In the previous edition as *L. hexactis*; see Foltz et al., 1996, Can. J. Zool. 74: 1275–1283. Common, south to San Simeon, CA. MacGinite and MacGinitie 1968, Nat. Hist. Marine Animals, 2nd ed., notes eggs brooding in February and March. Pearse and Beauchamp 1986, Intl. J. Inv. Reprod. Develop. 9: 289–297 report photoperiod relative to feeding and reproduction.

Leptasterias pusilla (Fisher, 1930). The validity of this species is unclear. See Foltz 2001 (cited above). A predator on small snails and limpets (Feder 1980). See also Smith 1971, Ph.D., Biological Sciences, Stanford University, 229 pp.

Orthasterias koehleri (de Loriol, 1897). Rainbow star. An uncommon, chiefly subtidal species (to 250 m), but occasionally found in the low intertidal (Ricketts et al. 1985). Diet extremely varied but predominantly composed of bivalves, as well as limpets, barnacles, chitons, snails, tunicates, and brachiopods (*Terebratalia*). Mauzey et al. (1968) suggest an ontogenetic diet shift. Lambert (2000) reports breeding from June to August. The polynoid polychaete *Arctonöe fragilis* is a commensal (Lambert, 2000).

Pisaster brevispinus (Stimpson, 1857). Pink sea star. Common, intertidal to 102 m. Sandy to muddy bottoms; a highly specialized predator on infaunal bivalves and gastropods and able to extend the central tube feet into the substrate to capture prey for a distance roughly equal to the radius of the arm (Van Veldhuizen and Phillips 1978, Mar. Biol. 48: 89–97). Diet also consists of sand dollars *Dendraster excentricus* (Farmanfarmaian et al. 1958) and the olive snail *Callianax biplicata* (Feder, 1980). Both of these species display escape responses to *P. brevispinus*. On hard substrate, this species also preys upon barnacles (*Balanus* spp.), mussels (*Mytilus* spp.), and tube-dwelling annelids (Feder 1980). See Wobber (1975) (agonistic bouts) and Farmanfarmaian et al. (1958) (reproductive cycles).

Pisaster giganteus (Stimpson, 1857). Intertidal to 374 m depth. Only exceptionally found north of San Francisco; although recorded as far north as Vancouver Island by Fisher (1930), this has not been confirmed in recent years (P. Lambert, pers. comm., 1999). Fisher (1930) recognized a southern subspecies *Pisaster gigantean capitatus*, which possesses more clublike abactinal spines. *P. giganteus* is a predator with a broad-based diet consisting primarily of gastropods, bivalves, barnacles and detritus (Harrold and Pearse 1987). See also Landenberger 1968, Ecology 49: 1062–1075 (feeding), and 1969, Physiol. Zool. 42: 220–230 (distributional ecology); Stubbs 1998, pp. 293–297, in Echinoderms: Proc. 9th, Intern. Echinoderm Conf. (diurnal behavior patterns); Farmanfarmaian et al.1958; Pearse et al. 1988 (reproductive biology).

Pisaster ochraceus (Brandt, 1835). Ochre star. Among the most studied sea stars on the Pacific coast. Feder (1980) provides an older review. For morphological variation see Fisher (1930). See Menge 1975, Ecology 31: 87–100; Menge and Menge 1974, Ecology 44: 189–209 (competition with *Leptasterias*); Palumbi and Fred 1988, Ecology 69: 1624–1627 (agonistic interactions); Miller 1986, Veliger 28: 394–396 (interaction with gastropods); Feder 1970, Ophelia 8: 161–185 (feeding and ecological interaction with the mussel, *Mytilus*); Patton et al. 1991, Bull. Mar. Sci. 48: 623–634 (ecology, interaction with *Corynactis californica*); Paine 1976, Ecology 57: 858–873 (role of *P. ochraceus* as a "keystone species" affecting distributional patterns); Abed and Crawford 1986, J. Morphology 188: 239–250 (development and growth). Harley et al. 2006, Biol. Bull. 211: 248–262 (color polymorphism).

Sclerasterias heteropaes Fisher, 1924. A rarely encountered, poorly known species. Subtidal from Monterey to southern California. Co-occurs with *Astrometis sertulifera*. Relatively large with R = 10 cm. Oral surface is pale with lack of color on spines. Found in deeper kelp beds among rocks and holdfasts (Hopkins and Crozier 1966).

Stylasterias forreri (de Loriol, 1884). Subtidal. Can approach considerable arm radius (R = 20 cm–25 cm). Catches motile prey such as fish with its pedicellariae. See Robilliard 1971, Syesis 4: 191–195; Chia and Amerongen 1975, 53: 745–755. Reproductive biology in Pearse et al. (1988).

PYCNOPODIIDAE

Pycnopodia helianthoides (Brandt, 1835). The sunflower star. Low intertidal to 435 m. Below 100 m depth, this species is frequently confused with *Rathbunaster californicus* Fisher. In Baja California/Mexico *P. helianthoides* is confused with *Coronaster marchennus* Ziesenhenne. Prey of *P. helianthoides* includes gastropods, but also chitons, pelecypods, barnacles, and echinoids. This species often elicits alarm responses in other invertebrates (e.g., Weightman and Arsenault 2002, Can. J. Zool. 80: 185–190. Feder (1980) provides a summary of the biology; see also Shivji et al. 1983, Pac. Sci. 37: 133–140 (feeding and distribution); Sloan and Robinson (1983); Harrold and Pearse (1987, prey in Alaska, Canada, and central California); McClintock 1989, Comp. Biochem. Phys. A, Comp. Phys. 93: 695–698 (growth); Wobber 1975 (agonistic bouts with other sea stars), Lawrence 1991, Mar. Behav. Phys. 19: 39–44 (alarm response to other *P. helianthoides*); Mladenov et al. 1989, Biol. Bull. 176: 169–175; Wilkie et al. 1995, pp. 137–146 in Proc. 4th European Echinoderms Colloquium. (autotomy).

ACKNOWLEDGMENTS

Thanks to John Pearse, Gil Van Dykhuizen, Rich Mooi, James Watanabe, Steve Lonhart, Jon Flowers, Dave Foltz , and Megumi Strathmann for assistance, and Patricia Liu, CAS-Education, who provided line drawings for plates 465A and 466.

References

Anderson, R. C., and R. L. Shimek. 1993. A note on the feeding habits of some uncommon sea stars. Zoo Biology 12: 499–503.

Birkeland, C. 1974. Interactions between a sea pen and seven of its predators. Ecological Monographs 44: 211–232.

Blake, D. B. 1989. Asteroidea: Functional morphology, classification and phylogeny. Echinoderm Studies 3: 179–223.

Brusca, G. C., and R. C. Brusca. 1978. A Naturalist's Seashore guide: common marine life of the northern California coast and adjacent shores. Mad River Press, Eureka, CA, 205 pp.

Brusca, R. C. 1980. Common intertidal invertebrates of the Gulf of California. 2nd ed. University of Arizona Press, 513 pp.

Chia, F. S. 1966. The development of two brooding sea stars, *Henricia leviuscula* and *Leptasterias hexactis*. American Zoologist 5: 331–332 (abstract).

Clark, A. M. 1962. Starfishes and their relations. London, British Museum (Nat. Hist.) 119 pp.

Clark, A. M., and Downey, M. E. 1992. Starfishes of the Atlantic. Chapman and Hall, London, 794 pp.

Farmanfarmaian, A., A. C. Giese, R. A. Boolootian, and J. Bennett. 1958. Annual reproductive cycles in four species of west coast starfishes. Journal of Experimental Zoology 138: 355–367.

Feder, H. M. 1980. Asteroidea: The Sea Stars, pp. 117–135. In Intertidal invertebrates of California. R. H. Morris, D. P. Abbott, and E. C. Haderlie, eds., 690 pp. Stanford University Press.

* = Not in key.

Fisher, W. K. 1911. Asteroidea of the North Pacific and adjacent waters. Part 1. Phanerozonia and Spinulosa. Bulletin of the United States National Museum 76, 419 pp.

Fisher, W. K. 1928. Asteroidea of the North Pacific and Adjacent Waters, Pt. 2: Forcipulata (Part). Bulletin of the United States National Museum 76, 245 pp.

Fisher, W. K. 1930. Asteroidea of the North Pacific and Adjacent Waters, Pt. 3: Forcipulata (Concluded). Bulletin of the United States National Museum 76, 356 pp.

Gotshall, D. W. 1994. Guide to Marine Invertebrates: Alaska to Baja California. Sea Challengers Press, Monterey, CA, 105 pp.

Harrold, C., and J. S. Pearse. 1987. The ecological role of echinoderms in kelp forests. Echinoderm Studies 2: 137–233.

Hopkins, T. S., and G. F. Crozier. 1966. Observations on the asteroid echinoderm fauna occurring in the shallow water of southern California (intertidal to 60 m). Bulletin of the Southern California Academy of Sciences 65: 129–145.

Hyman, L. H. 1955. The Invertebrates: Vol. IV. Echinodermata. McGraw-Hill, 1–763.

Jangoux, M. 1982. Food and feeding mechanisms: Asteroidea, pp. 117–159. In Echinoderm nutrition. M. Jangoux and J. M. Lawrence, eds. Rotterdam: A. A. Balkema.

Jangoux, M., and A. Lambert. 1988. Comparative anatomy and classification of asteroid pedicellariae, pp. 719–723. In Echinoderm biology. R. D. Burke et al., eds. Rotterdam: Balkema.

Kozloff, E. N. 1987. Marine Invertebrates of the Pacific Northwest. University of Washington Press, 511 pp.

Lambert, P. 2000. The Sea Stars of British Columbia. British Columbia Provincial Museum Handbook 39. 186 pp.

Maluf, L. Y. 1988. Composition and distribution of the Central Eastern Pacific Echinoderms. Natural History Museum of Los Angeles County Technical Report No. 2. 1–242.

Mauzey, K. P., C. Birkeland, and P. K. Dayton. 1968. Feeding behavior of asteroids and escape responses of their prey in the Puget Sound region. Ecology 149: 603–619.

Morin, J. G., J. E. Kastendiek, A. Harrington, and N. Davis. 1985. Organization and patterns of interactions in a subtidal sand community on an exposed coast. Marine Ecology Progress Series 27: 163–185.

Pearse, J. S., D. J. McClary, M. A. Sewell, W. C. Austin, A. Perez-Ruzafa, and M. Byrne. 1988. Simultaneous spawning of six species of echinoderms in Barkley Sound, British Columbia. Invertebrate Reproduction and Development 14: 279–288.

Pettibone, M. H. 1953. Some scale-bearing polychaetes of Puget Sound and adjacent waters. University of Washington Press, Seattle, 89 pp.

Rasmussen, B. 1965. On taxonomy and biology of the North Atlantic species of the asteroid genus Henricia Gray. Medd. Danm. Fisk.-Havunders 4: 157–213.

Ricketts, E. F., J. Calvin, J. W. Hedgpeth, and D. W. Phillips. 1985. Between Pacific Tides. 5th ed. Stanford University Press, 652 pp.

Sloan, N. A. 1980. Aspects of the feeding biology of asteroids. Ocean. Mar. Biol. Ann. Rev. 18: 57–124.

Sloan, N. A., and S. M. C. Robinson. 1983. Winter feeding by asteroids on a subtidal sandbed in British Columbia. Ophelia 22: 125–140.

Spencer, W. K., and C. W. Wright. 1966. Asterozoans, Part U: Echinodermata In R. C. Moore, ed. Treatise on Invertebrate Paleontology 3(1): U4–U107. Lawrence: University of Kansas Press.

Van Veldhuizen, H. D., and V. J. Oakes. 1981. Behavioral responses of seven species of asteroids to the asteroid predator, Solaster dawsoni. Oecologia. 48: 214–220.

Verrill, A. E. 1914. Monograph of the shallow-water starfishes of the North Pacific coast from the Arctic Ocean to California. United States National Museum, Harriman Alaska series 14, 408 pp.

Wobber, D. R. 1975. Agonism in asteroids. Biological Bulletin 148: 483–496.

Ophiuroidea

GORDON HENDLER

(Plates 467–470)

As a general rule, ophiuroids are more numerous and diverse, more agile, and frequently more colorful and attractive than other echinoderms. In central California and Oregon, intertidal species represent a minor but resilient subset of much more populous, wide-ranging, subtidal ophiuroid populations. This is also the case worldwide, as the number of ophiuroid species is typically higher in comparatively stable deeper-water environments than in shallow water. Intertidal species are a relatively small component of total ophiuroid biodiversity. Of more than 2,000 species of ophiuroids that have been described, only one well-studied species, the subtropical and tropical *Ophiocoma scolopendrina* (Lamarck, 1816), is clearly adapted to the intertidal zone.

Although the first account of eastern Pacific ophiuroids was published more than 150 years ago, they have not been extensively studied, and their representation in museum collections and coverage in the scientific literature is limited. A better appreciation of their natural history will emerge with further investigation, and future research will reveal previously unreported and undescribed species along the Pacific coast. The present section provides updates and revisions since the previous edition in 1975. For example, the ophiuroids referred to as *Amphiodia* sp. and *Ophionereis eurybrachiplax* H. L. Clark, 1911, in past editions are now *Amphiodia akosmos* Hendler and Bundrick, 2001, and *Ophionereis diabloensis* Hendler, 2002. *Ophiopholis aculeata var. kennerlyi* is accorded species-level status as *Ophiopholis kennerlyi* (Lyman, 1860). In addition, *Ophionereis eurybrachiplax* and *Amphiodia urtica* (Lyman, 1860) are expunged from the list of the intertidal fauna from central California and Oregon.

Ophiuroids are identified in different languages by vernacular names, which bear little relation to their classification or phylogeny. In English, ophiuroids with unbranched arms are called brittle stars, and sometimes serpent stars or snake stars. The attribution "brittle" refers to the ability of ophiuroids to voluntarily cast off (autotomize) their arms when adversely stimulated. However, the propensity to do so varies widely among different species. During autotomy, which is under nervous control, the mutable connective tissue linking the arm joints abruptly deteriorates, and the arm immediately disarticulates at the weakened junctures. Pieces of the arm that separate from the disk can remain active for many hours. The stump of arm remaining attached to the disk forms a new growing tip and may eventually regenerate to its original length. Moreover, some ophiuroids, chiefly in the family Amphiuridae, can autotomize a portion of the disk as well as the arms. Afterwards, they regenerate the body wall, gonads and digestive tract from remnants of the body.

The English term for ophiuroids with branching arms is basket stars, a name based on the tendency of individuals pulled from the water to coil and compact their arborescent arms into a massive knot, creating the appearance of a woven basket. Individuals in the two families that have branching arms, Gorgonocephalidae and Euryalidae, begin life with five simple arms. During development, the arms bifurcate, and the sister branches may continue to subdivide dichotomously, the pattern of branching nodes depending on the species. The rings of microscopic hooks that encircle the distal arm tendrils of gorgonocephalids are used to secure the zooplankters on which they feed, and to hold the animals fast to rocky reefs. *Gorgonocephalus eucnemis* (Müller and Troschel, 1842) is the only local basket star species. Off the California coast, it inhabits deep water, but it thrives in greater numbers in the frigid shallow water of the north Pacific.

Many ophiuroid species are cryptic, particularly those in shallow waters. They occupy shelter afforded by inanimate objects and sessile biota, often tucked in inaccessible and virtually undetectable crannies. As they increase in age and body size, many ophiuroids, including California and Oregon intertidal species, migrate from one microhabitat to another. Most large intertidal ophiuroids live in bedrock, beneath boulders and cobbles. The smallest species, and juveniles of larger species, occur among fronds, holdfasts, and rhizomes of plants, and in protective clumps of bryozoans, hydroids, worm tubes, echinoid spines, and so on. A search among fouling communities on pilings, buoys, and marina floats will also often produce smaller species. Burrowing species remain loosely or deeply covered in sediment.

Although ophiuroids are almost exclusively marine, a few species tolerate brackish environments, and to varying degrees, intertidal species can tolerate freshwater runoff and rain, sedimentation, pollution, and desiccation. They generally avoid anoxic conditions, and for that reason ophiuroids flourish where water circulation is relatively unobstructed, such as beneath rocks that rest loosely on coarse sediment. Species inhabiting sponges depend on water currents created by the host, and those that burrow in fine sediment create currents that irrigate their burrows.

Many echinoderms are so incredibly slow-moving that their behavior is clearly perceptible only in a time-lapse recording. However, some ophiuroids defy that rule, and their intricately jointed and muscled arms execute lithe, rapid feeding and escape responses. It is commonly presumed that, excepting basket stars and related species, ophiuroids are capable of moving their arms only in a horizontal plane. However, numerous species can conform their arms to the irregular dimensions of crevices, lock their arms rigidly when disturbed, nimbly flex their arms during locomotion, and deftly coil their slender, distal arm tips. Brittle stars can gallop by sweeping their arms in rowing strokes, or crawl by combining dragging and pushing efforts. Long-armed species often employ a combination of tube-foot and muscular whole-arm movements. Small and juvenile ophiuroids, and the adults of some species, advance solely by flexing their tube feet, much as a centipede uses its legs. Further, certain species respond to sudden suspension in the water column by reflexively folding the arms above the disk, to minimize drag and maximize the rate of sinking. Although none of the ophiuroids discussed in the manual can swim, there are specialized, highly active species with that capability.

Ophiuroids have been mischaracterized as strict deposit feeders with a monotonous diet and a limited repertoire of feeding behaviors. However, all ophiuroids are to some degree selective feeders, and various species are specialized to pursue and capture other animals, collect particles of sediment and debris, and scavenge plant and animal remains, offal, and feces. Macrophagous ophiuroids ensnare edible items in coils of their arms, which like an elephant's trunk convey food to the mouth. Microphagous species, whether suspension or deposit feeders, entangle particles in the mucus on their arm spines and tube feet. The tube feet compact particles into a bolus, which is passed down the arm and transferred to the mouth. Ophiuroids can open their mouths to a surprising extent, and their flexible disks can be distended to a considerable degree, permitting large masses of particulate material or bulky prey to be accommodated within the gut. After material is processed in the stomach, it is expelled from the mouth and may then be conveyed out along the arm by the tube feet. In addition to ingesting solid food, ophiuroids can extract dissolved nutrients from seawater and obtain soluble organic material from symbiotic bacteria living within their tissues.

The sensory abilities of ophiuroids are exceedingly acute. Sensitive chemoreceptors enable them to detect and rapidly respond to minute quantities of waterborne chemicals liberated by food. Touch receptors enable them to deftly manipulate minute particles, and trigger escape reactions in response to potential predators and physical disturbance. It is unlikely that ophiuroids can locate prey visually, but species in the photic zone typically rely on photoreception to find shelter. Their preference for shadow and shelter is related to the avoidance of predation by fish, crustaceans, asteroids, and other benthic predators. Different species show contrasting responses to illumination. Epifaunal ophiuroids may "freeze" in response to shadow, and cryptic ophiuroids preferentially move away from bright light. Photoreceptors appear to be situated within the skeleton, which in some cases is modified and acts in concert with chromatophores to concentrate and direct light.

The details of reproductive biology have been elucidated for relatively few ophiuroid species. Although it is presumed that most ophiuroids have planktonic feeding larvae, few such species have been reared through metamorphosis. Referred to as an ophiopluteus, the planktotrophic larva is bilaterally symmetrical, with up to eight arms supported by a glassy calcite skeleton, and bears a single, sinuous band of cilia used for locomotion and feeding. In some cases, the ophiopluteus resorbs the larval arms prior to metamorphosis and transforms into an irregularly ellipsoidal, secondary larva with transverse rings of cilia positioned between the developing appendages of the ophiuroid rudiment. Species with planktonic, nonfeeding larvae have reduced, yolky larvae resembling the armed ophiopluteus or the secondary larva. The planktonic larvae develop from eggs that are released in the water and fertilized externally. Brooding species hold the eggs and embryos in internal bursae. Brooders are usually hermaphroditic, and at least one species is capable of self-fertilization, whereas broadcast-spawning ophiuroids generally have separate sexes that are superficially indistinguishable. Nonfeeding larval development can take as few as three days, but feeding larvae may require several months to complete metamorphosis.

Ophiuroid life spans have not been documented with precision, but depending on the species, individuals may mature, reproduce, and die within one or two years, or live several decades and longer. Longevity is particularly difficult to pinpoint among the small number of species that reproduce asexually while larvae, or as adults in a process termed fissiparity. The fissiparous species generally have six arms and split across the disc to produce two genetically identical brittle stars that have part of a disk and three arms. The animals regenerate the missing components of arms and disk and thereafter are capable of further fissiparity and of perpetuating a prolific, long-lived clone.

Morphology and Terminology

The critical structures for the identification of ophiuroids are the superficial skeletal elements, but to appreciate ophiuroid biology it is necessary to "lift the hood" and study their internal structures. Further, the present account is tailored to the identification of a small number of species and barely suggests the profound anatomical adaptations that have evolved in the ophiuroid body plan. The literature cited below may be consulted for more extensive, detailed information on morphology. Plate 467 should be referred to for illustrations of anatomical structures and also bolded terms, and their abbreviations in parentheses in this section.

Despite the common misconception that they have an "exoskeleton," the skeleton of ophiuroids is internal, consisting of

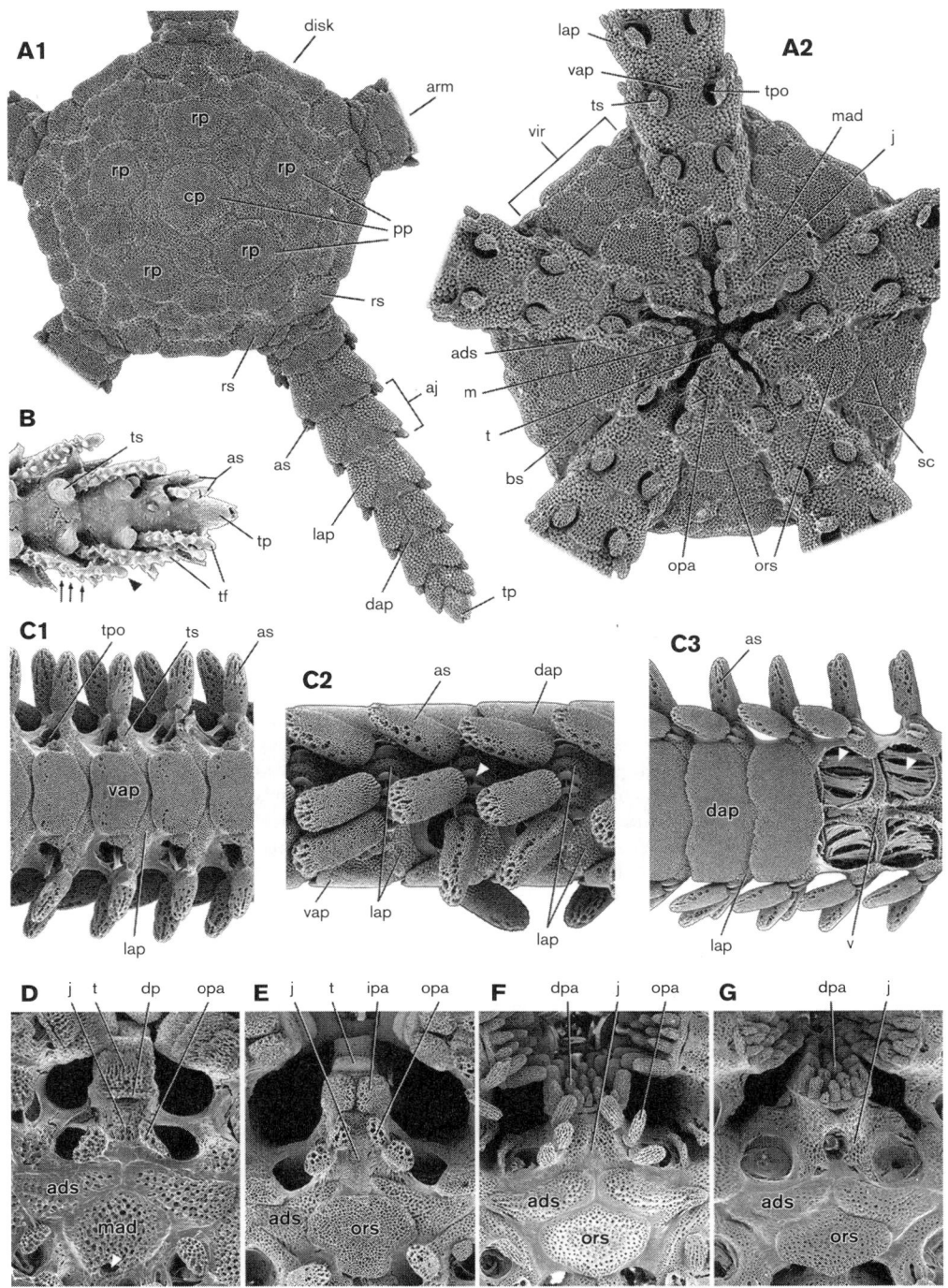

PLATE 467 Ophiuroid anatomy illustrated with scanning electron micrographs, integument is removed from structures to reveal the skeleton: A1, A2, juvenile *Ophioplocus esmarki* labeled to show structures of the dorsal surface (A1), ventral surface (A2) of the disk and arms; B, distal tip of the arm of *Ophiactis simplex* with intact soft tissue structures, showing integument covering the skeleton and the extended tube feet; small arrows, tube foot papillae; arrowhead, terminal bulb of tube foot; a pore through which the terminal tube foot can be extended is visible at the tip of the terminal plate; C1–C3, sections of the arm of *Amphiodia occidentalis*, with tube feet removed, showing structures of the ventral surface (C1), lateral surface (C2), dorsal surface (C3), with several dorsal arm plates removed to reveal the vertebrae within the arm joints; C2, arrowhead arm-spine articular surface on the lateral arm plate; C3, arrowheads in intervertebral muscle tissue; D–G, details of the oral structures of *Ophiactis simplex* (arrowhead, hydropore of madreporite) (D); *Amphiodia occidentalis* (E); *Ophiopteris papillosa* (F); *Ophiothrix spiculata* (G), labeled to show the arrangement of oral papillae, dental papillae, and infradental papillae in several types of oral armament. Abbreviations: ads, adoral shield; aj, arm joint; as, arm spine; bs, bursal slit; cp, central plate; dap, dorsal arm plate; dp, dental plate; dpa, dental papilla; ipa, infradental papilla; j, jaw; lap, lateral arm plate; m, mouth; mad, madreporite; opa, oral papilla; ors, oral shield; pp, primary plate; rp, radial plate; rs, radial shield; sc, scale; t, tooth; tf, tube foot; tp, terminal plate; tpo, tentacle pore; ts, tentacle scale; vap, ventral arm plate; v, vertebra; vir, ventral interradius.

OSSICLES that are each composed of a single calcite crystal enveloped in a thin layer of integument. The names used for specific ossicles are inconsistent, even in English, and descriptive terms like plate, shield, scale, spine, and papilla are somewhat arbitrary. The nomenclature adopted here is widely but not universally in use (below, some common alternative expressions are presented parenthetically). Also, be aware that the key features of an ophiuroid, including its overall appearance, color, and the numbers and shapes of specific ossicles, change markedly as an individual grows.

The ophiuroid body consists of a central hub, the **DISK**, from which radiate five or more **ARMS** (plate 467A). The **PROXIMAL** direction is toward the center of the disk, and the **DISTAL** direction is toward the tips of the arms. The arms are composed of numerous **ARM JOINTS (aj)**, also called segments, that are generated at the distal tip of the arm. Those nearest the disk are the oldest and largest, as they increase in girth as the animal grows. On the **VENTRAL** (=oral) surface of the animal, which usually faces the substrate, are a star-shaped **MOUTH (m)** at the center of the disk and pairs of tube feet that emerge along the length of the arm (plate 467A2, 467B). The opposite side of the animal is referred to as **DORSAL** (=aboral).

The joints are composed of a protective ossicle on each principal surface, including a **DORSAL ARM PLATE (dap)**, a **VENTRAL ARM PLATE (vap)**, and a pair of **LATERAL ARM PLATES (lap)**, which surround a **VERTEBRA (v)** that superficially resembles one of the bones of a spinal column (plate 467C). The vertebrae each have articulating protuberances and attachment surfaces for the intervertebral muscle, ligament, and connective tissue that link the arm joints. The dorsal arm plate may be a large solitary plate (plate 467C3), or consist of a dominant plate associated with one or more minute **ACCESSORY DORSAL ARM PLATES** at its edges (plates 468D1, 469C1), or it may be composed of a mosaic array of associated plates (plate 468C1). The lateral arm plates typically bear one or more moveable **ARM SPINES (as)** that are used for protection, locomotion, and feeding (plate 467C).

The first ventral arm plate is positioned at the edge of the mouth, and the most distal arm ossicle is the cylindrical **TERMINAL PLATE (tp)** (plate 467A1, 467B), through which projects a terminal tube foot. In addition, there are two **TUBE FEET (tf)** (=podia, tentacles) per arm joint, and each protrudes through a **TENTACLE PORE (tpo)** situated between a lateral arm plate and the ventral arm plate on the ventral surface of each arm joint (plate 467A2, 467C1). Protecting the retracted tube foot are one or more small ossicles, **TENTACLE SCALES (ts)**, which arise from the lateral or ventral arm plate and overlap the tentacle pore. The tube feet are appendages of a **WATER-VASCULAR SYSTEM**, an organ system found exclusively in echinoderms and composed of flexible, muscular, fluid-filled tubes. It consists of a circular canal within the disk, and a radial canal in each arm from which the tube feet branch. They are used as sensory and respiratory structures and may function in feeding and locomotion. Particularly in suspension feeding species, tube feet are equipped with microscopic **PAPILLAE** rich in sensory structures and mucus secreting glands (plate 467B). Those of deposit feeders have a smooth shaft lacking papillae, and a **TERMINAL BULB** with secretory glands (plate 467B).

The disk develops in the ophiuroid larva, beginning as a series of six ossicles that consist of a **CENTRAL PLATE (cp)** and five concentrically arranged **RADIAL PLATES (rp)**, which are referred to collectively as **PRIMARY PLATES (pp)** (plate 467A1). Depending on the species, they may comprise most of the dorsal surface of the adult disk, be barely discernible, or be entirely resorbed in the adult. Generally, the primary plates are separated or entirely supplanted by the numerous intercalary ossicles called **SCALES (sc)**, which may be large and robust, or small and delicate. And often the disk scales and plates are overlain by an armament of microscopic ossicles, such as slender sharp **SPINES** (plate 469A1) or blunt **PAPILLAE**, short **STUMPS**, spherical **GRANULES** (plate 468A1, 468B1), or large **TUBERCLES**. In some cases, the scales themselves take the form of spines, tubercles, or nearly invisible granules that are resorbed and incorporated in the connective tissue of the body wall. **RADIAL SHIELDS (rs)** are frequently the most conspicuous ossicles on the dorsal disk surface (plate 467A1). They consist of a pair of mirror-image ossicles at the base of each arm, and together with an inconspicuous series of internal ossicles, they fasten the disk and arms together.

On the ventral surface of the disk are triangular **JAWS (j)** that extend inward from between the bases of the arms and work to open and close the centrally located mouth (plate 467A2). Each jaw is composed of modified vertebrae (**ORAL PLATES**) and carries two pairs of tube feet (**ORAL TENTACLES**) that project into the mouth, manipulate incoming food, and disperse outgoing waste. Most of the digestive tract is a blind, pouch-shaped stomach that fills the disk and opens at the mouth. Ophiuroids lack an anus. The most basal arm joints are connected to the jaws and disk, and the wedge of disk between pairs of arms is referred to as a **VENTRAL INTERRADIUS (vir)** (pl. interradii). The outer edges of the interradius are creased alongside the base of the arms, and in each crease an orifice, the **BURSAL SLIT (bs)**, opens into a **BURSA** (=respiratory bursa, genital bursa). The edge of the slit may bear rounded granules or minute, blunt spines (**GENITAL PAPILLAE**). The bursa itself is a thin pouch, an inpocketing of the disk integument, which is inserted within the disk between the stomach and the body wall. The ciliated walls of the bursa (pl. bursae) circulate seawater for respiratory exchange, and the gonoducts discharge gametes near the bursal slit. In viviparous species, embryos develop within the bursa, and the eggs and embryos absorb nutrients from the surrounding tissues.

A vertical column of **TEETH (t)** extends dorsally into the mouth, along the proximal edge of each jaw, attached to an inconspicuous **DENTAL PLATE (dp)** (plate 467D). In most ophiuroids, a series of small scale- or spine-shaped ossicles, the **ORAL PAPILLAE (opa)**, are attached to the edge of each jaw, nearly flush with the ventral surface of the disk (plate 467D–467F). Instead of or in addition to oral papillae, some species have structures at the apex of the jaw, ventral to the teeth. They are called **INFRADENTAL PAPILLAE (ipa)** if they are a pair of block-like ossicles (plate 467E) and are called **DENTAL PAPILLAE (dpa)** (=tooth papillae) if they consist of a close-set group of blunt ossicles (plate 467F, 467G). A trio of conspicuous ossicles lies between the jaw and the ventral interradius, composed of an **ORAL SHIELD (ors)** flanked on each side by an **ADORAL SHIELD (ads)** (plate 467D–467G), the latter ossicle usually connecting the jaw and arm. One of the oral shields serves as a **MADREPORITE (mad)** and may often be distinguished by its relatively large size and sometimes by the presence of one or more **HYDROPORES** that open into the water vascular system (plate 467D).

Classification

Several major classification schemes and phylogenies have been proposed for the living Ophiuroidea, of which more than 2,000 species have been described. According to a dominant

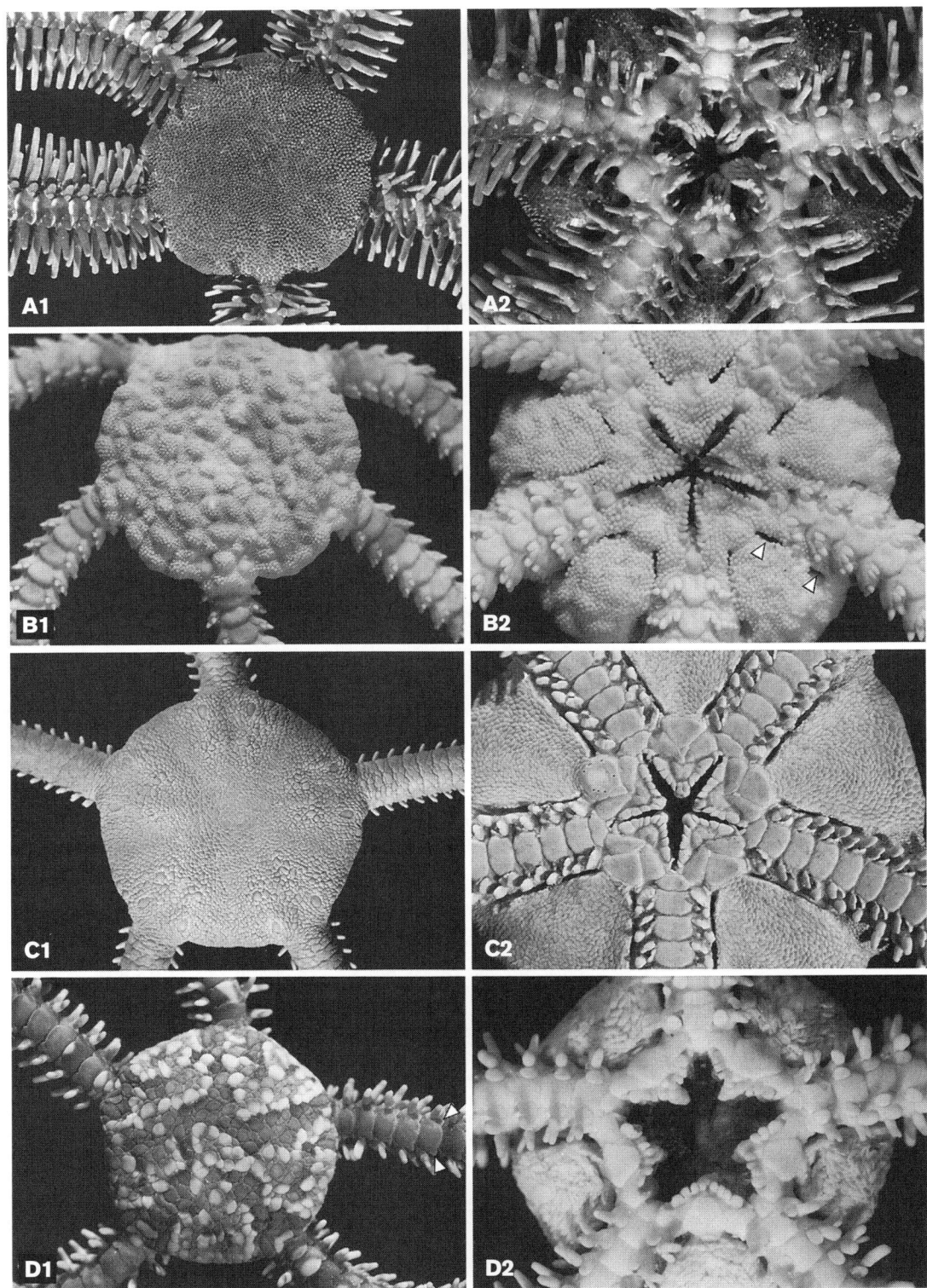

PLATE 468 Disk and basal portion of the arm of: A1, *Ophiopteris papillosa*, dorsal; A2, ventral; B1, *Ophioncus granulosus*, dorsal; B2, ventral, arrowheads point to divided bursal slit; C1, *Ophioplocus esmarki*, dorsal; C2, ventral; D1, *Ophionereis diabloensis*, dorsal; D2, ventral; in D1, a pair of accessory dorsal arm plates (arrowheads) is distal to the lateral arm plate.

nineteenth century school of thought, a distinction was drawn between Euryalae (basket stars and related ophiuroids with unbranched arms) and Ophiurae (the other brittle stars) based on the nature of the vertebral articulation, arm spine position, integument histology, and the reduction and loss of principal skeletal elements. Beginning in the twentieth century, insights into the significance of internal structures such as genital, oral, and dental plates allowed for the discrimination of the order Ophiurida, consisting of three subgroups, from the order Phrynophiurida, which combined the thick-skinned family Ophiomyxidae and the "Euryalae." In addition, there is evidence, open to question, that one "living fossil" species represents an ancient sister-group to all other extant ophiuroids.

That controversy aside, the most inclusive, contemporary, cladistic classification divides the class Ophiuroidea into the order Euryalida, equivalent to the classical Euryalae, and the order Ophiurida. The prime division within the latter group segregates Ophiomyxidae from five other major clades (infraorders) that do not readily lend themselves to recognition by their general appearance. Even in this scheme, relationships among the clades of Ophiurida are not clearly resolved by the available morphological and molecular data, possibly because they all evolved relatively contemporaneously during the early Mesozoic. The approximately 20 ophiuroid families have been somewhat more taxonomically stable than the higher taxa, and that will probably remain the case until the genera are critically revised.

Collection, Preservation, and Examination

Any method devised to extract invertebrates from benthic substrates can yield ophiuroids. Because they are rarely found in clear sight, a thorough survey of ophiuroids requires disruptive or destructive techniques, which should be performed in a manner that minimizes environmental damage. Looking between and below loose-set rocks and sifting the sediment beneath them can be productive. A chisel-tipped geologist's pick can provide useful leverage, and patches of soft sediment may be shoveled and sieved to find burrowing forms. Small animals, which are frequently the most unusual and interesting, are extracted by washing rocks overgrown with plants and animals, clumps of sessile invertebrates, or plants in a container of seawater; carefully examining and discarding the cleaned substrate; and sorting the remnants under low magnification. Ophiuroids may also be found by dissecting sponges and kelp holdfasts. Individuals can be prevented from autotomizing if they are handled carefully and not lifted by their arm tips.

In the field, ophiuroids are best maintained in containers of clean seawater with some pebbles or a bit of sediment, and preferably kept in the dark. The containers can be transported in an insulated box, to which a sealed bag of ice may be added. Examining living animals is indispensable for an understanding of the organism, but microscopic observations are most easily made on anesthetized or preserved individuals.

Ophiuroids can be rapidly relaxed by transferring them to a magnesium chloride solution (74 g per liter of fresh water), epsom salts solution (200 g per liter of fresh water), or by placing them in a minimal amount of seawater and gradually adding small amounts of epsom salts. Afterward, they may be revived by placing them in fresh seawater or preserved for further study. That is accomplished by straightening the arms of a relaxed specimen, laying it flat in a tray of alcohol for several minutes to "harden," and storing it in a container of al-

cohol. Because natural color deteriorates after preservation, it is advantageous to photograph the living animal or add notes on color to the sample label along with date and locality information.

Preservation and storage in 70%–95% ethyl alcohol is generally preferred, as it permits the study of soft tissue and skeleton and DNA extraction. Preserved specimens may be partially or completely dried to facilitate identification and examination of the skeleton. Bear in mind, however, that ophiuroids directly removed from seawater and dried eventually disintegrate because of the hygroscopic salts remaining in their tissues. Taxonomic specimens are vulnerable to acidic fixatives that dissolve the skeleton, and for that reason storage in formalin must be avoided. Cotton should not be used as a cushioning material because the fibers invariably damage delicate structures. However any specimen, no matter how badly damaged, is potentially informative and useful for identification.

Preservation for electron microscopy or histology requires toxic fixative solutions that must be handled using special procedures discussed elsewhere in the literature.

Adult individuals of the intertidal species of California and Oregon can be distinguished using the key, photographs, and comments provided. However, keys to ophiuroids are virtually useless for the identification of small and juvenile individuals because the shape and number of crucial characters drastically change during ontogenesis (e.g., compare plates 467A and 468C). Thus, accurate identification may require additional information gleaned from the primary literature and museum specimens, or the advice of an expert.

In the descriptions of ophiuroid species, standard measurements of body size include disk diameter (dd) measured from the outer edge at the radial shields to the distal periphery of the opposite interradius, and arm length (AL) from the distal tip of the arm to the edge of the disk. Dimensions provided herein are representative of large individuals. These data and other information summarized in the following list of species are based on the author's personal observations and the literature cited. Note that the pigmentation described is that of living animals, unless otherwise specified. The term **STRIPE** refers to a pigment pattern running the length of the arm, and **BANDS** are patterns that run across the arm. References to numbers of arm spines and tentacle scales refer to the greatest number found on one lateral arm plate near the base of an arm, and counts of oral papillae refer to the number on one side of the jaw. The specimens photographed are from the collections of the Natural History Museum of Los Angeles County.

Key to Ophiuroidea

1. Number of arms five 2
— Number of arms six; fissiparous (plate 469D) . . . .
. . . . *Ophiactis simplex*
2. Dorsal arm plate is composed of a mosaic of ossicles (plate 468C) . . . . *Ophioplocus esmarki*
— Dorsal arm plate is single; accessory dorsal arm plates are present 3
— Dorsal arm plate is single; accessory dorsal arm plates are absent 4
3. One accessory dorsal arm plate between each lateral arm plate and the dorsal arm plate (plate 468D) . . . .
. . . . *Ophionereis diabloensis*
— Numerous accessory dorsal arm plates surround the dorsal arm plate (plate 469C) . . . . *Ophiopholis kennerlyi*

4. Disc scales bear spherical granules (plate 468A1, 468B1) 5
— Disc scales bear serrated or cylindrical spines, or tall cylindrical granules (plate 469A1, 469B1); jaws bear dental papillae but lack oral papillae (plate 467G) 6
— Disc covered by naked scales (plate 470A1); jaws bear paired infradental papillae and oral papillae (plate 467E) 7
5. Dorsal surface of disk smooth; two bursal slits in each interradius (plate 468A) .... *Ophiopteris papillosa*
— Dorsal surface of disk bumpy; four bursal slits in each interradius (plate 468B) .... *Ophioncus granulosus*
6. Arm spines with jagged edges (plate 469A) .... *Ophiothrix spiculata*
— Arm spines smooth (plate 469B) .... *Ophiothrix rudis*
7. Most distal oral papilla considerably enlarged, middle oral papilla diminutive (plate 470A2, 470B2) 8
— Most distal oral papilla and middle oral papillae similar in size and shape (plates 467E, 470C2, 470D2) 9
8. Arm spines with bulbous base, short and abruptly tapering; primary plates absent (plate 470A) .... *Amphipholis squamata*
— Arm spines gradually tapering; longest spines have slender shaft and may have a bulbous tip; primary plates usually present (plate 470B) .... *Amphipholis pugetana*
9. Moderately large species, with two tentacle scales (plates 467C, 467E, 470D) .... *Amphiodia occidentalis*
— Small species, with single elongated tentacle scale (plate 470C) .... *Amphiodia akosmos*

List of Species

AMPHIURIDAE

Amphiodia akosmos Hendler and Bundrick, 2001. Previous editions of this manual treated this small species (dd = 4.2, AL = 24 mm) as an unnamed "ovoviviparous species" or a "variety of *A. occidentalis*." White to pale beige, with some disk scales and arm plates a contrasting gray or brown. Distinguished from similar juveniles of *Amphiodia occidentalis,* together with which it can occur, by the single, relatively elongate tentacle scale, prominent wedge-shaped scales separating the radial shields, and the absence of primary plates. Individuals suspended in the water swiftly retract their arms into a tightly coiled mass atop the disk. Known only from the intertidal at Monterey Peninsula, Santa Cruz County, and Farallon Island. Lives under rocks that rest on coarse sand. It is the only *Amphiodia* species known to bear live young. See Hendler and Bundrick 2001.

Amphiodia occidentalis (Lyman, 1860) (=*Diamphiodia occidentalis*). Large, with long, slender arms and small disk (dd = 12, AL = 220 mm). Dorsal surface tan or yellow, densely variegated with gray, brown, red, and cream; ventral surface of arms dark gray. Distinguished from *A. akosmos* by two tentacle scales, and dorsal arm plates two and a half to three times wider than long. Adults burrow in sediment beneath rocks and seagrass, often in protected tide pools. Extend their arms to deposit feed and possibly to gather suspended material. Suspended individuals behave similarly to *A. akosmos.* Broadcast yolky eggs that undergo rapid development within a benthic fertilization envelope. Reports of the species from deep water and south of central California are suspect. See May

1924, Austin and Hadfield 1980, Rumrill and Pearse 1984, Hendler and Bundrick 2001, Emlet 2006.

Amphiodia urtica (Lyman, 1860) (=*Amphiodia barbarae*). Moderately small, with long slender arms (dd = 4.0, AL = 50 mm). Disk red-brown with dark primary plates, radial shields gray with pale distal tips, arms red gray with red medial stripe. Allied with other California species in the subgenus *Amphispina,* which have hyaline, barbed disk scales and proximally curving arm spines beneath the disk. Reported from the intertidal and at scuba-accessible depths, but reliable central California and Oregon records are strictly for the subtidal where populations may attain densities up to several thousand per square meter. Its larva is a planktotrophic ophiopluteus. The parasitic nauplius stage of a thespesiopsyllid copepod can be found within its stomach. The species is monitored as an indicator of water quality. See H. L. Clark, 1911, Nielsen 1932, Barnard and Ziesenhenne 1961, Bergen 1995, Hendler 1996, Maurer and Nguyen 1996, Ho et al. 2003.

Amphipholis pugetana (Lyman, 1860). Small (dd = 5.8, AL = 32 mm). Disk red-brown, red-gray, or gray, radial shields brown or gray-brown with white mark at distal tip; arms paler than disk, blotched with red-brown, green, gray, pale brown, or cream. Large individuals distinguished from adult *Amphipholis squamata* by arm spines longer than arm joint; middle arm spine longest, up to one and half times length of joint, with slender shaft between the base and the thickened or bulbous tip. A putative difference in relative arm length that has been suggested to distinguish between individuals of *A. pugetana* and *A. squamata* is unreliable. The two species have often been misidentified, and small individuals may be impossible to identify with certainty. *A. pugetana* rarely occurs intertidally, despite reports to the contrary; nearly all reliable records are for depths from 15 m–600 m. The large number and small size of orange eggs found in the ovaries of ripe individuals indicate that the larva is planktonic. See H. L. Clark 1911, Nielsen 1932, Strathmann and Rumrill 1987, Hendler et al. 1995, Hendler 1996.

Amphipholis squamata (Delle Chiaje, 1828) (=*Axiognathus squamatus*). Diminutive, rarely reaching dd = 4.1, AL = 21 mm. Widely characterized as worldwide in distribution, but molecular evidence indicates that it may consist of several cryptic taxa. However, individuals from wide-ranging localities are luminescent, viviparous, and hermaphroditic. Some populations are capable of self-fertilization as well as outcrossing. California individuals white, gray, or tan in color, with pale patch at distal tip of radial shield, often with several thin black bands on arms. Adults distinguishable from *Amphipholis pugetana* by arm spines bulbous at base, abruptly tapering, and no longer than an arm joint. Symbionts of *A. squamata, sensu lato,* include an orthonectid, a turbellarian, a polychaete, external copepods, internal copepod parasites that may castrate the host, and bacteria that play a role in host nutrition. Although references to its biology are too numerous to list in full, see Austin and Hadfield 1980; Emson and Wilkie 1982; Radner 1982; Rumrill and Pearse 1984; A. M. Clark, 1987, Bull. Zool. Nom. 44: 246–247 (nomenclature); Strathmann and Rumrill 1989; Hendler et al. 1995; Østergaard and Emson 1997; Poulin et al. 1999; Sponer and Roy 2002; Fauville et al. 2003.

OPHIACTIDAE

Ophiactis simplex (Le Conte, 1851). Fissiparous, usually with three to six regenerating arms. Usually very small, but some

PLATE 469 Disk and basal portion of the arm of: A1, *Ophiothrix spiculata*, dorsal; A2, ventral; B1, *Ophiothrix rudis*, dorsal; B2, ventral; C1, *Ophiopholis kennerlyi*, dorsal; C2, ventral; arrowheads indicate accessory dorsal arm plates surrounding the dorsal arm plate; D1, *Ophiactis simplex*, dorsal; D2, ventral.

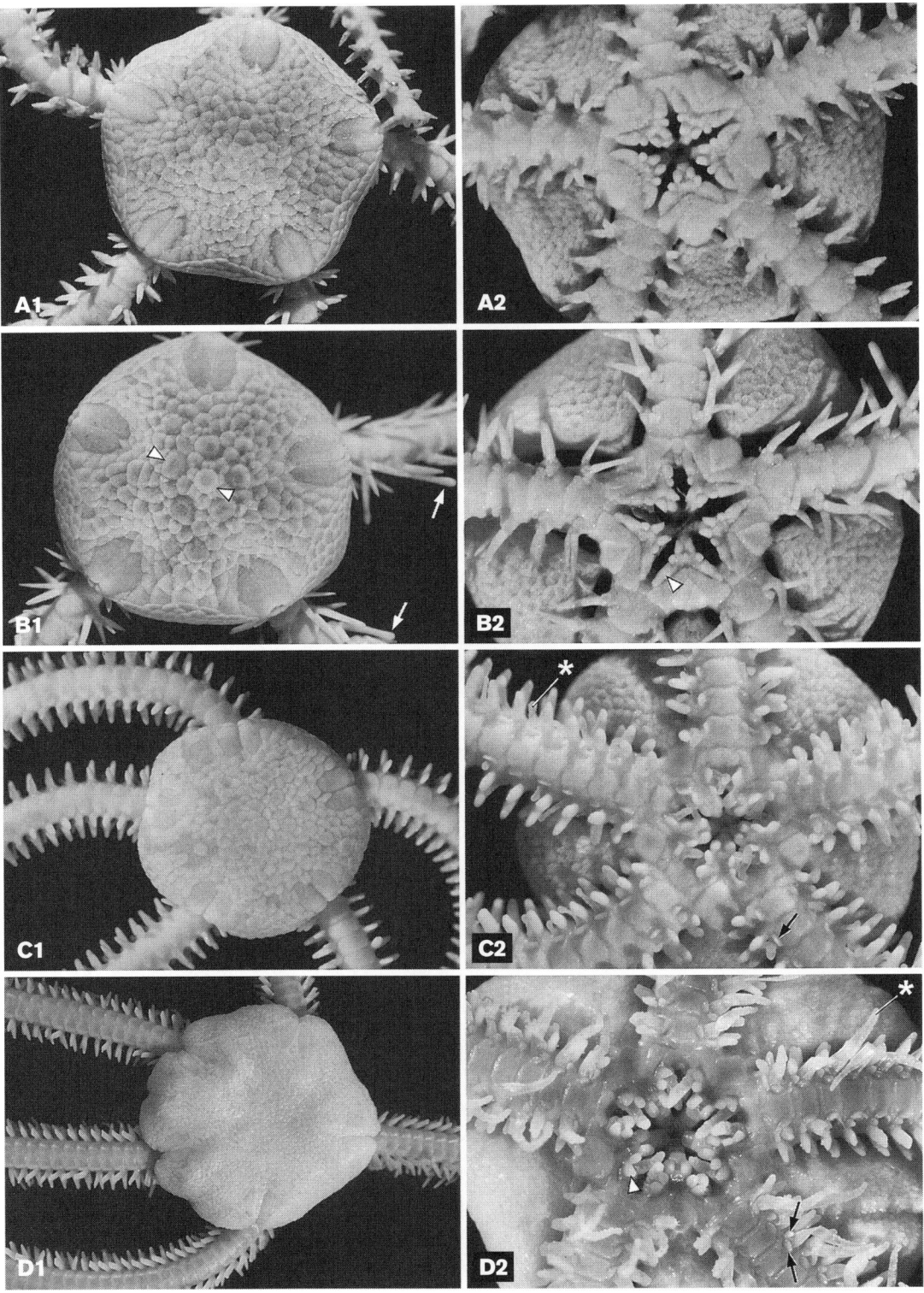

PLATE 470 Disk and basal portion of the arm of: A1, *Amphipholis squamata*, dorsal; A2, ventral; B1, *Amphipholis pugetana*, dorsal; B2, ventral; C1, *Amphiodia akosmos*, dorsal; C2, ventral; D1, *Amphiodia occidentalis*, dorsal; D2, ventral. Primary plates (arrowheads point to the central and a radial plate) and modified, bulbous tipped arm spines (arrow) are indicated in B1; compare the most distal oral papillae (arrowheads), which are considerably enlarged in B2, but not so in D2; compare the single, elongated tentacle scale in C2 and the small, paired tentacle scales in D2 (small arrows). Tube feet (asterisks) remain visible in the specimens preserved in alcohol (C2, D2).

individuals reach moderate size (dd = 7.1, AL = 25 mm) and may have five arms. Variegated with patches of brown, green-brown, and tan pigmentation; arms distinctly banded. Tube feet of living specimens red, presumably from hemoglobin-containing coelomocytes. First discovered in Panama, it has been considered identical to *Ophiactis orstedii* Lütken, 1856, and *Ophiactis arenosa* Lütken, 1856, which were originally found in Costa Rica and Nicaragua. *Ophiactis simplex* was long known from as far north as the Channel Islands, California, and was unexpectedly recorded in central California (Lonhart and Tupen 2001). Cause of the apparent range extension is unresolved (Schiel et al. 2004). Moreover, it appears to occur in Texas and may be identical to *Ophiactis rubropoda* Singletary, 1973, first described from Florida. Distinguished from *Ophiactis savignyi* (Müller and Troschel 1842), a related tropicopolitan species found in southern California, by the single oral papilla and relatively small radial shields that are completely separated by small scales. Can be exceedingly abundant in sponges, algae, and around serpulid tubes. Capable of absorbing dissolved organic material from seawater. See Lyman 1865, Nielsen 1932, Stephens and Virkar 1966, Christensen and Christensen 2003, Christensen and Dean 2003, and Christensen 2004.

Ophiopholis kennerlyi Lyman, 1860 (=*Ophiopholis aculeata* var. *kennerlyi*). First described based on a specimen from Puget Sound collected by Caleb Kennerly, naturalist and surgeon of the United States Northwest Boundary Survey, who died during the voyage. Afterward, taxonomically merged with *Ophiopholis aculeata* (Linnaeus, 1767) and assigned an infrasubspecific name, which is today taxonomically unavailable. Individuals from California are here tentatively accorded specific rank as *O. kennerlyi* to distinguish them from other eastern Pacific species of *Ophiopholis*. Uncommon in California and appreciably smaller (dd = 8.5, AL = 38 mm) than conspecifics from Alaska. Disk granule-covered, with few dorsal naked plates and lacking dorsal spines; six to 10 relatively large, angular, accessory dorsal arm plates per joint; arm spines usually shorter than arm joint, broad at the base, blunt. Coloration highly variable and attractive; disk and arms often red to pink, with contrasting mottling, marbling, and bands of black, brown, cream, and green hues. Slow-moving, it lives under rocks intertidally or in varied substrates subtidally. Spawns in summer, producing an ophiopluteus larva. The numerous publications on *O. aculeata, sensu lato* are not cited herein, as the biology of *O. kennerlyi* must differ to some degree from *O. aculeata,* but see H. L. Clark 1911; May 1924; Nielsen 1932; LaBarbera 1978, 1982; Austin and Hadfield 1980; Hart 1982; Strathmann and Rumrill 1987; and Balser 1998.

Ophiopholis bakeri McClendon, 1909. The previous edition of this manual suggested intertidal occurrence, but it is reliably reported only at depths from 18 m–902 m. Individuals are similar in size to *O. kennerlyi* (dd = 10.5, AL = 67 mm). Distinguished by the disk covered with pointed granules and short, slender spines with several terminal spikes; accessory dorsal arm plates separated from one another, small, and often with one or two sharp spikes. Red or orange-brown to pink with some gray, cream, or brown markings; arms banded but disk not boldly patterned with sharply contrasting colors. Typically associated with holdfasts and sessile animals on outcrops of hard substrate. See McClendon 1909, H. L. Clark 1911, and Hendler 1996.

OPHIOCOMIDAE

Ophiopteris papillosa (Lyman, 1875). Reaches moderate size (dd = 19, AL = 70 mm) in the subtidal; intertidal individuals uncommon and usually small. Distinguished by densely granulated disk, prominent dental and oral papillae, arm spine closest to dorsal arm surface small and scalelike, lower arm spines long, dorso-ventrally compressed, smooth. Coloration black to pale brown or red, with darker bands on the arms; sometimes variegated with cream-colored splotches or bands; distal ends of arms sometimes yellow. Very active, readily autotomizes, and shows rapid escape behavior in response to predators, which include sea stars, fish, crabs, and lobster. Individuals suspended in water assume a teardrop shape, with the arms extended vertically above the disk and pressed together, which facilitates rapid descent. Carnivorous, using arm spines to tear apart material held in the mouth; also able to feed on suspended material. Adhesive tube feet enable it to climb glass surfaces. Spawns in winter, releasing minute eggs of a size and number characteristic of species with planktotrophic ophiopluteus larvae. See May 1924, Austin and Hadfield 1980, Wallerstein 1982, Rumrill and Pearse 1984, Yee et al. 1987, and Pomory 2001.

OPHIODERMATIDAE

Ophioncus granulosus Ives, 1889. Small species (dd = 7.1, AL = 17 mm), rarely encountered intertidally. Distinguished by bursal slits divided into two parts; disk distinctly lumpy, granule covered. Resembles the juveniles of large *Ophioderma* species that occur south of central California. Light gray, cream, or pale yellow-brown; disk sometimes with several black or brown marks, arms often with several brown to pale yellow-brown bands. Crawls using the rowing motions of laterally positioned arms. Individuals suspended in midwater quickly fold the arms in a concentric pattern atop the disk, and sink like a pebble.

OPHIOLEPIDIDAE

Ophioplocus esmarki Lyman, 1874. Reaches moderate size (dd = 22, AL = 60 mm), although individuals are smaller and less numerous in the intertidal than the subtidal. Distinguished by the dorsal arm plates composed of small plates arranged in a mosaic pattern. General coloration gray or brown-pink, red-brown, or brown, finely stippled with a contrasting darker hue; arm tips often pale. Slow-moving, relatively unresponsive, and loath to autotomize. Adhesive tube feet are used to climb smooth surfaces, and to transport and ingest small prey and food particles. Reproductive when approximately 10 years old. Females produce yolky eggs, and large individuals brood more than 1,500 embryos. Restricted dispersal of the juveniles results in genetic differences among populations. See Austin and Hadfield 1980, Gaarde and McClenaghan 1982, Rumrill and Pearse 1984, and Medeiros-Bergen and Ebert 1995.

OPHIONEREIDIDAE

Ophionereis diabloensis Hendler, 2002. A considerably smaller species (dd = 6, AL = 17 mm) than *Ophionereis eurybrachiplax*, with which it has been confused. Differs from *O. eurybrachiplax* by its conspicuous, rounded disk scales, absence of genital papillae, absence of a patch of dark brown pigment distal to oral shield. In addition, distinguished from a much more similar, southerly species *Ophionereis amphilogus* (Ziesenhenne, 1940) by its coarser disk scales, more robust arms, and oral shield with truncate proximal tip. Disk predominantly brown and green-brown with dark brown and cream-colored patches; arms green-brown with dark brown bands and cream patches. Occurs in low intertidal algal turf on rock substrate. Known solely from Diablo Cove, San Luis Obispo County, although in the first half of the twentieth century, it was collected at the Monterey Peninsula. Broods crawl-away juveniles. See Hendler 2002.

Ophionereis eurybrachiplax H. L. Clark, 1911. Moderately large (dd = 25, AL = 215 mm). Contrary to information in previous editions of this manual, it is strictly subtidal, in depths 53 m–45 m. Disk brown with black and cream mottling, or rusty brown with white spots; radial shields·orange-brown with white distal edge; dorsal surface of arms orange-brown with mottled bands of cream, brown, and mottled gray, often with a pale discontinuous medial stripe; ventral side of arms sometimes a salmon color.

OPHIOTRICHIDAE

Ophiothrix rudis Lyman, 1874. Moderately small (dd = 10, AL = 50 mm). Distinguished by large, naked radial shields and smooth, somewhat dorso-ventrally compressed arm spines; disk armament, depending on the individual, consisting of smooth-sided cylindrical granules, stumps, or spines. Disk green, tinged reddish or bluish; radial shields sometimes with red or orange spots; arms green or yellow with narrow dusky and broad red bands. In alcohol, specimens become blue, and thin dark bands are apparent across and distal to each dorsal arm plate. Lives under rocks, in coarse sand, and associated with seagrass rhizomes. May occur together with *Ophiothrix spiculata*. Females with ripe eggs in summer. See McClendon 1909, Nielsen 1932, Austin and Hadfield 1980.

Ophiothrix spiculata Le Conte, 1851. Moderately small (dd = 15, AL = 85 mm). Distinguished by serrate or conspicuously prickly spines on the disk and arms. Pigmentation and color patterns extraordinarily variable; frequently green, green-brown, or green-yellow, and more or less extensively marked with white, gray, orange, red, garnet, maroon, brown, pink, or gray. Alcohol preserved specimens turn blue, as do many ophiotrichid species. The species occupies a wide range of substrates, and in current-swept habitats it can occur ". . . in almost unbelievable numbers. . . The bottom in deeper water may be covered to depths of an inch or more by millions of these active animals" (Limbaugh 1955). Prone to autotomize. Individuals extend several arms rigidly from crevices to capture microscopic particles using mucus-coated arm spines and tube feet, or gracefully flex the arm tips to secure larger particles. Broadcast immense numbers of minute eggs in the spring, which develop into planktotrophic ophiopluteus larvae. See: May 1924, Nielsen 1932, and Austin and Hadfield 1980.

ACKNOWLEDGMENTS

Specimens, personal observations, opportunities for field work, and comments on the manuscript were offered by Shane Anderson, William Austin, Michael Behrens, Karin Boos, Carla Bundrick, Ana Christensen, Paula Cisternas, Cathy Colloff, Dimitri Deheyn, Richard Emlet, John Engle, Constance Gramlich, Cathy Groves, Florence Nishida, Steven Rumrill, Freya Sommer, Richard Strathmann, Anne Richmond, Jeffery Tupen, Waldo Wakefield, and James Watanabe. Giar-Ann Kung, William Ormerod, and Michelle Schwengel helped prepare the plates. Curatorial staff at the California Academy of Sciences, Museum of Comparative Zoology, National Museum of Natural History, and Santa Barbara Museum of Natural History provided access to specimens and data. I am grateful to all who provided assistance and am indebted to Capt. F. C. Ziesenhenne, who established the collection of eastern Pacific ophiuroids, now in the Natural History Museum of Los Angeles County, on which much of this contribution is based.

References

Austin, W. C., and M. G. Hadfield. 1980. Ophiuroidea: The brittle stars. In Intertidal invertebrates of California. R. H. Morris, D. P. Abbott, and E.C. Haderlie, eds., pp. 146–159, figs. 10.1–10.12. Stanford, CA: Stanford University Press.

Balser, E. J. 1998. Cloning by ophiuroid echinoderm larvae. Biol. Bull. 194: 187–193.

Barnard, J. L., and F. C. Ziesenhenne. 1961. Ophiuroid communities of southern Californian coastal bottoms. Pacific Nat. 2: 131–152.

Bergen, M. 1995. Distribution of brittlestar *Amphiodia* (*Amphispina*) spp. in the southern California Bight in 1956 to 1959. Bull. Southern Calif. Acad. Sci. 94: 190–203.

Boolootian, R. A., and D. Leighton. 1966. A key to the species of Ophiuroidea (brittle stars) of the Santa Monica Bay and adjacent areas. Contr. Sci. Los Angeles County Museum 93: 1–20.

Byrne, M. 1994. Ophiuroidea. In Microscopic anatomy of invertebrates, Vol. 14. Echinodermata. F. W. Harrison and F.-S. Chia, eds., pp. 247–343. New York: Wiley-Liss.

Christensen, A. B., and E. F. Christensen. 2003. Molecular comparison of a Texas population of ophiactid brittle star with *Ophiactis simplex* and *Ophiactis rubropoda*. In Echinoderms. München. T. Heinzeller, and J. H. Nebelsick, eds., p. 574. Balkema, Leiden.

Christensen, A. B., and D. K. Dean. 2003. Population structure in a fissiparous ophiactid brittlestar possessing hemoglobin. http://sicb.org/meetings/2003/schedule/abstractdetails.php3?id=245.

Christensen, A. B. 2004. A new distribution record and notes on the biology of the brittle star *Ophiactis simplex* (Echinodermata: Ophiuroidea) in Texas. Texas J. Sci. 56: 175–179.

Clark, H. L. 1911. North Pacific ophiurans in the collection of the United States National Museum. Bull. U.S. Natl. Mus. 75: 1–302.

Clark, H. L. 1915. Catalog of Recent ophiurans: Based on the collection of the Museum of Comparative Zoölogy. Mem. Mus. Comp. Zoöl. Harvard 62: 265–338, pls. 1–8.

Emlet, R. B. 2006. Direct development of the brittle star *Amphiodia occidentalis* (Ophiuroidea, Amphiuridae) from the northeastern Pacific Ocean. Invertebrate Biol. 125: 154–171.

Emson, R. H., and I. C. Wilkie. 1982. The arm-coiling response of *Amphipholis squamata* (Delle Chiaje). In Echinoderms: Proceedings of the International Conference, Tampa Bay. J. M. Lawrence, ed., pp. 11–18. Rotterdam: Balkema.

Fauville, G., S. Dupont, and J. Mallefet. 2003. Comparison of mechanically and chemically induced luminescence in the brittle star *Amphipholis squamata*. In Echinoderm research 2001. J.-P. Feral and B. David, eds., pp. 189–192. Balkema: Lisse.

Fell, H. B. 1960. Synoptic keys to the genera of Ophiuroidea. Zool. Publs. Victoria Univ. Wellington 26: 1–44.

Gaarde, W. A., and L. R. McClenaghan Jr. 1982. Genetic variability, dispersal, and differentiation of two species of ophiuroids from southern California. Southwestern Nat. 27: 255–262.

Hart, M. W. 1982. Particle captures and the method of suspension feeding by echinoderm larvae. Biol. Bull. 180: 12–27.

Hendler, G. 1991. Echinodermata: Ophiuroidea. In Reproduction of marine invertebrates. Vol. VI. Echinoderms and Lophophorates. A. C. Giese, J. S. Pearse, and V. B. Pearse, eds., pp. 355–511. Pacific Grove, CA: Boxwood Press.

Hendler, G. 1996. Class Ophiuroidea. In Taxonomic atlas of the benthic fauna of the Santa Maria Basin and Western Santa Barbara Channel. Vol. 14–Miscellaneous taxa. J. A. Blake, P. H. Scott, and A. Lissner, eds., pp. 111–179. Santa Barbara, CA: Santa Barbara Museum of Natural History.

Hendler, G. 2002. Account of *Ophionereis diabloensis*, a new species of brittle star, and of *O. amphilogus*, with information on their brooding reproduction and distribution (Echinodermata: Ophiuroidea: Ophionereididae). Proc. Biol. Soc. Wash. 115: 57–74.

Hendler, G. 2004. An echinoderm's eye view of photoreception and vision. In Echinoderms. München, T. Heinzeller, and J. H. Nebelsick, eds. pp. 339–349. Balkema: Leiden.

Hendler, G., and C. J. Bundrick. 2001. A new brooding brittle star from California (Echinodermata: Ophiuroidea: Amphiuridae). Contr. Sci., Los Angeles County Natural History Museum 486: 1–11.

Hendler, G., J. E. Miller, D. L. Pawson, and P. M. Kier. 1995. Sea stars, sea urchins, and allies. Echinoderms of Florida and the Caribbean. Smithsonian Institution Press, Washington, 390 pp.

Ho, J.-S., M. Dojiri, G. Hendler, and G. B. Deets. 2003. A new species of Copepoda (Thaumatopsyllidae) symbiotic with a brittle star from California, U.S.A., and designation of a new order Thaumatopsylloida. J. Crust. Biol. 23: 582–594.

Kyte, M. A. 1969. A synopsis and key to the Recent Ophiuroidea of Washington State and southern British Columbia. J. Fish. Res. Board Can. 26: 1727–1741.

LaBarbera, M. 1978. Particle capture by a Pacific brittle star: Experimental test of the aerosol suspension feeding model. Science. 201: 1147–1148.

LaBarbera. 1982. Metabolic rates of suspension feeding crinoids and ophiuroids (Echinodermata) in a unidirectional laminar flow. Comp. Biochem. Physiol. 71A: 303–307.

Limbaugh, C. 1955. Fish life in the kelp beds and the effects of kelp harvesting. Univ. of California, Inst. Mar. Sci. IMR Ref. 55–9:1–158, 20 figs.

Lonhart, S. I., and J. W. Tupen. 2001. New range records of 12 marine invertebrates: The role of El Niño and other mechanisms in southern and central California. Bull. Southern California Acad. Sci. 100: 238–248.

Lyman, T. 1865. Ophiuridae and Astrophytidae. Ill. Cat. Mus. Comp. Zoöl. Harv. 1: 1–200, pls. 1–2.

Maurer, D., and H. Nguyen. 1996. The brittlestar *Amphiodia urtica*: a candidate bioindicator? P.S.Z.N.I: Mar. Ecol. 17: 617–636.

May, R. M. 1924. The ophiurans of Monterey Bay. Proc. Calif. Acad. Sci.. Ser. 4, 13: 261–303.

McClendon, J. F. 1909. The ophiurans of the San Diego Region. Univ. Calif. Publs. Zool. 6: 33–64.

Medeiros-Bergen, D. E., and T. A. Ebert. 1995. Growth, fecundity, and mortality rates of two intertidal brittlestars (Echinodermata: Ophiuroidea) with contrasting modes of development. J. Exp. Mar. Biol. Ecol. 189: 47–64.

Nielsen, E. 1932. Ophiurans from the Gulf of Panama, California, and the Strait of Georgia. Vidensk. Medd. fra Dansk naturh. Foren. 91: 241–346.

Østergaard, P., and R. Emson. 1997. Interactions between the life histories of a parasitic copepod, *Parachordeumium amphiurae*, and its brittle-star host, *Amphipholis squamata*. J. Crust. Biol. 17: 621–631.

Pomory, C. M. 2001. An escape response behaviour in the brittle star *Ophiopteris papillosa* (Echinodermata: Ophiuroidea). Mar. Fresh. Behav. Physiol. 34: 171–180.

Poulin, E. J.-P. Féral, M. Florensa, L. Cornudella, and V. Alva. 1999. Selfing and outcrossing in the brood protecting ophiuroid *Amphipholis squamata*. In Echinoderm Research 1998. Proceedings of the Fifth European Conference on Echinoderms, Milan, Italy, 7–12 September 1998. M.D. Candia Carnevali and F. Bonasoro. eds., pp. 147–150. Rotterdam: Balkema.

Radner, D. N. 1982. Orthonectid parasitism: effects on the ophiuroid, *Amphipholis squamata*. In Echinoderms: Proceedings of the International Conference, Tampa Bay. J. M. Lawrence, ed., pp. 395–401. Rotterdam: Balkema.

Rumrill, S. S., and J. S. Pearse. 1984. Contrasting reproductive periodicities among north-eastern Pacific ophiuroids. In Echinodermata. B. F. Keegan and B. D. S. O'Connor, eds., pp. 633–638. Balkema, Rotterdam.

Schiel, D. R., J. R. Steinbeck, and M. S. Foster. 2004. Ten years of induced ocean warming causes comprehensive changes in marine benthic communities. Ecology 85: 1833–1839.

Smith, A. B., G. L. J. Pateron, and B. LaFay. 1995. Ophiuroid phylogeny and higher taxonomy: morphological, molecular and palaeontological perspectives. Zool. J. Linn. Soc. 114: 213–243.

Sponer, R., and M. S. Roy. 2002. Phylogeographic analysis of the brooding brittle star *Amphipholis squamata* (Echinodermata) along the coast of New Zealand reveals high cryptic genetic variation and cryptic dispersal potential. Evolution 56: 1954–1967.

Stephens, G. C., and R. A. Virkar. 1966. Uptake of organic material by aquatic invertebrates. IV. The influence of salinity on the uptake of amino acids by the brittle star, *Ophiactis arenosa*. Biol. Bull. 131: 172–185.

Strathmann, M. F., and S. S. Rumrill. 1987. Phylum Echinodermata, Class Ophiuroidea. In: Reproduction and development of marine invertebrates of the northern Pacific coast, M. F. Strathmann, ed., pp. 556–573. Seattle: University of Washington Press, Seattle, Washington.

Wallerstein, M. C. 1982. An examination of the roles of predation and

competition in determining the distributions of the ophiuroids *Ophiothrix spiculata* Le Conte, 1851 and *Ophiopterus* [sic] *papillosa* (Lyman, 1875) in a shallow marine community. Ph.D. Dissertation. University of Southern California, 142 pp.

Warner, G. 1982. Food and feeding mechanisms: Ophiuroidea. In Echinoderm nutrition. M. Jangoux and J. M. Lawrence, eds., pp. 161–181. Rotterdam: Balkema.

Yee, A., J. Burkhardt, and W. F. Gilly. 1987. Mobilization of a coordinated escape response by giant axons in the ophiuroid, *Ophiopteris papillosa*. J. Exp. Biol. 128: 287–305.

Holothuroidea

PHILIP LAMBERT

(Plates 471–475)

Sea cucumbers have a soft body wall containing circular and longitudinal muscles and a vestigial skeleton made up of isolated calcite particles, called ossicles, and a calcareous ring around the oesophagus. Typically, a sea cucumber is an elongate cylinder lying on its side with the mouth at one end and the anus at the other. It has five longitudinal rows of tube feet, which are part of a closed hydraulic network called the water vascular system. One or two circles of feeding tentacles surround the mouth.

Tube feet, or podia, usually consist of a cylindrical shaft with a sucker at the tip. The three ventral rows tend to be more robust than the two on the dorsal side, but they can be quite variable. *Leptosynapta albicans,* for example, has no tube feet, and on the dorsal side of *Parastichopus californicus,* the tube feet are modified into pointed warts, or papillae.

The feeding tentacles, being part of the water vascular system, are extended and retracted by hydraulic pressure. Sea cucumbers have four types of tentacles (plate 471). Treelike, dendritic tentacles (plate 471A) pick up small suspended particles on a coating of mucus; the tentacle is then placed in the mouth and "licked" clean. Detritus feeders, like *Parastichopus californicus* have moplike, peltate tentacles (plate 471C), which when pressed onto the substratum pick up particles and transfer them to the mouth. The digestive system processes the organic matter, and the bits of shell and sand particles pass through the gut. Sea cucumbers that burrow push sediment into the mouth with fingerlike, digitate (plate 471E) or pinnate (plate 471B) tentacles.

Species with dendritic tentacles have a thin-walled "neck" region just behind the tentacles called the introvert. Five internal retractor muscles pull the introvert and tentacles into the body cavity to prevent predatory fish from nipping them.

Plate 471D shows the internal anatomy of a generalised *Cucumaria.* To dissect a sea cucumber, determine the dorsal side (usually slightly darker or with tube feet less well-developed) and make a lengthwise incision to the left of the midline with sharp-pointed scissors. This will leave the dorsal mesentery intact. The digestive tract in this genus is two or three times the length of the body. The anterior part of the intestine hangs from the mid-dorsal body wall by a transparent sheet of tissue, or mesentery. Other mesenteries connect the next part of the intestine to the left side and finally to the ventral body wall. The position of these mesenteries varies in different groups of sea cucumbers and this arrangement is an important character in classification.

Five bands of longitudinal muscle, and in taxa with an introvert the tentacle retractor muscles, are plainly visible. The rest of the body wall consists of a layer of circular muscles, connective tissue, and skin. The contraction of these circular and longitudinal muscle layers produces a wormlike or peristaltic action. Imagine a long thin, water-filled balloon. When you squeeze

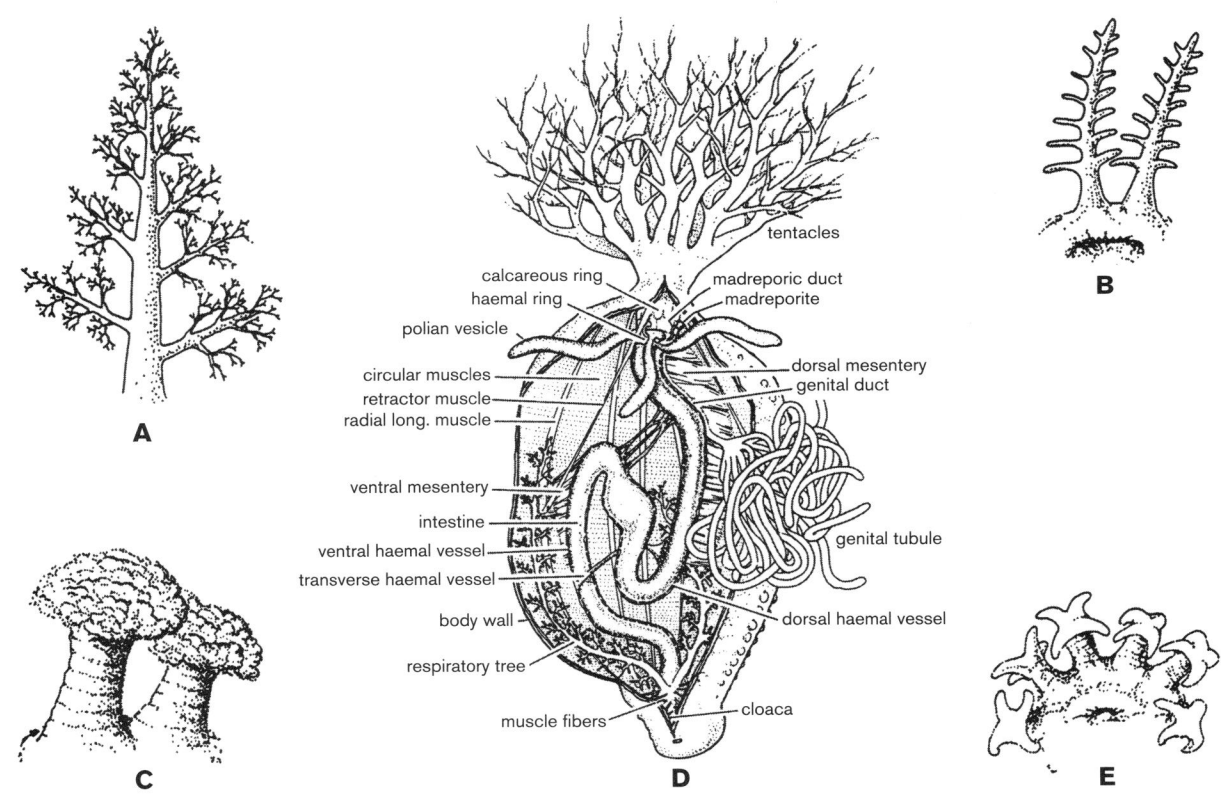

PLATE 471 A, dendritic tentacles; B, pinnate tentacles; C, peltate tentacles; D, internal anatomy of a generalized cucumarid sea cucumber; E, digitate tentacles.

one end the other extends forward. Some species use these waves of contraction to move along the ocean floor or through the mud.

Respiratory trees are the "lungs" of a sea cucumber. These hollow, branched organs lie inside the body cavity on either side of the posterior intestine. The base of the tree connects to a muscular cavity, or cloaca. Circular muscles, or sphincters, close each end of the cloaca. A sea cucumber "breathes" by expanding the cloaca to draw oxygenated water in through the anus. The posterior sphincter then closes, then cloacal muscles contract to force water up into the respiratory trees. Oxygen is transferred across the thin membrane into the fluids of the body cavity. When the oxygen is depleted, the main body wall contracts to squeeze water out of the trees.

The ring canal circles the oesophagus just behind the mouth and may have saclike extensions called polian vesicles. They are thought to function as expansion chambers for the system. One or more stone canals, calcified tubes with a perforated swelling at the end, called the madreporite, also attach to the ring canal. The function of stone canals is not clear but it may have something to do with maintaining the fluid levels in the water-vascular system. The number and position of stone canals is another useful taxonomic character.

Five radial canals emerge forward of the ring canal sending branches to the tentacles, then curve back to run the length of the body between the longitudinal and the circular muscle layers. Short branches go to the tube feet and a small sac, or ampulla, marks the internal end of each tube foot. When contracted, these ampullae force fluid into the tube foot, causing it to extend. In some species like *Parasti-*

chopus californicus, the base of each feeding tentacle also has an ampulla.

The calcareous ring is one of the few obvious, internal hard parts of a sea cucumber. It is composed of a series of plates, usually 10, joined side by side like a collar around the oesophagus. The tentacle retractor muscles attach to this structure. The plates vary in shape in different species, some plates having long tails and others having anterior projections. The shape of the ring is important in the classification of sea cucumbers. For example, those that have long posterior tails on the ring are placed in the same family. Each plate may be a solid piece or in some species, a mosaic of smaller segments. Being one of the few hard structures in a sea cucumber, the calcareous ring is often the only part that fossilises, thus providing a way of relating extinct and living forms.

The rest of the sea cucumber skeleton is represented by isolated pieces of calcite, embedded in the outer layers of skin (plate 472). These microscopic ossicles are complex and varied in shape and along with the calcareous ring are important in verifying the identity of a species. A description or illustration of them is an important part of a scientific account.

The reproductive organs of a sea cucumber consist of one or two tufts of tubules in the forepart of the body cavity. They combine into a single duct leading to an external gonopore near the tentacles. Most sea cucumbers have separate sexes, but the sex is difficult to determine by examining external features.

Sea cucumbers usually spawn annually, either by broadcasting eggs or sperm into the surrounding water, or by brooding the fertilized eggs. An environmental cue, such as a certain temperature, consecutive sunny days, or a plankton bloom,

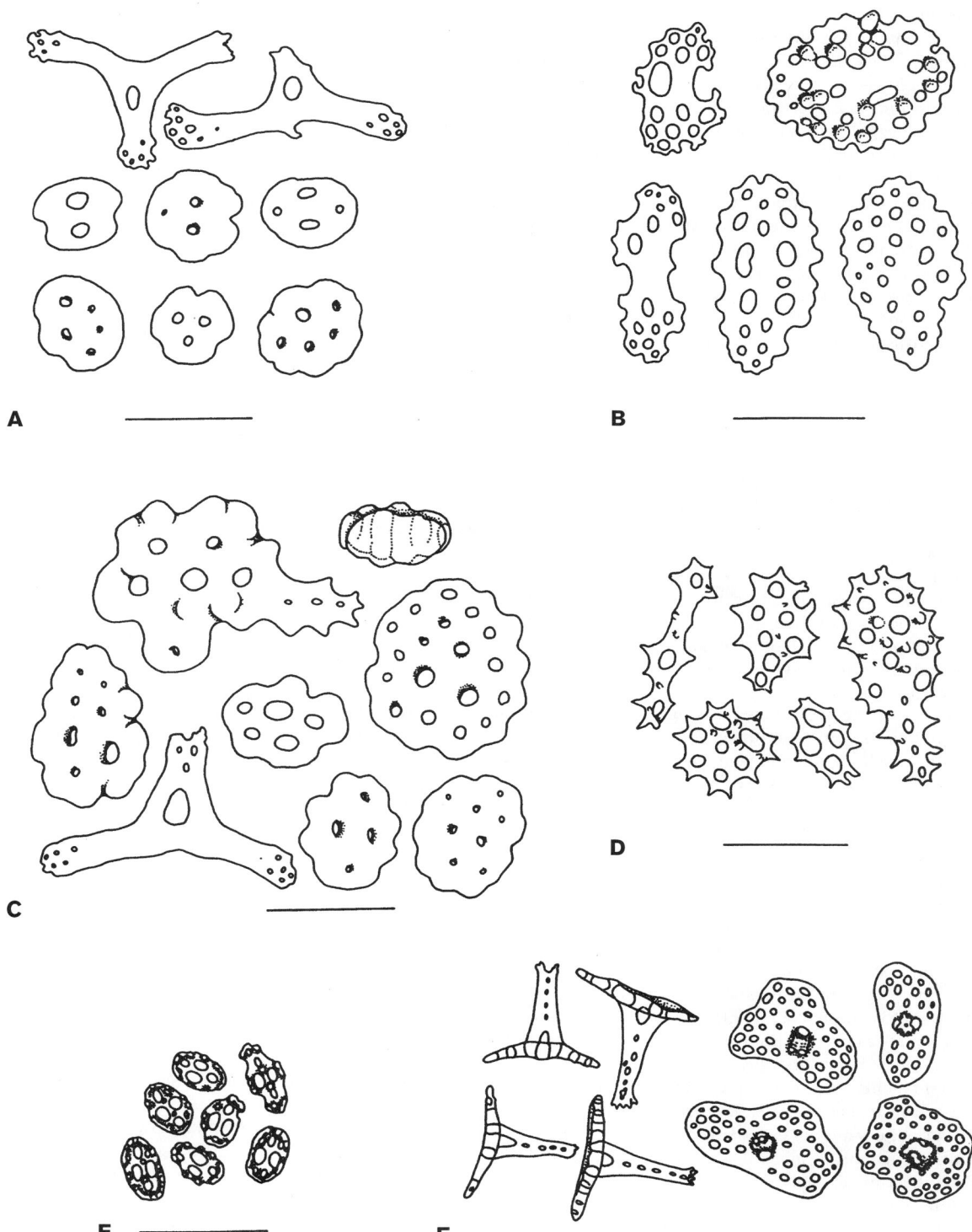

PLATE 472 Scale line = 100 μm: Skin ossicles of A, *Pseudocnus curatus*; B, *Cucumaria pseudocurata*; C, *Pseudocnus lubricus*; D, *Cucumaria piperata*; E, *Pentamera montereyensis*; F, *Pentamera charlottae*.

may cause many individuals to spawn simultaneously, thus increasing the chance for successful fertilisation. A female *Cucumaria miniata* releases clusters of green eggs as a buoyant pellet. Males release sperm nearby and fertilization takes place by chance in midwater. The pellet of eggs eventually breaks apart and each egg develops into a swimming larva. The larva develops and grows for days or weeks in the plankton and then settles on rocky areas inhabited by the adults.

A few species retain their eggs, rather than releasing them. As the eggs emerge, the female collects and holds them underneath her body. Sperm shed by nearby males fertilize the eggs. The larvae develop for a few months until the juveniles are large enough to crawl away. *Cucumaria pseudocurata, Pseudocnus lubricus,* and *Pseudocnus curatus* all reproduce this way during the winter. *Leptosynapta* broods its young internally.

Preparation of Ossicle Slides

If, after using the key, you are still unsure of the identification, use the following procedure to isolate skin ossicles to compare with the more detailed descriptions in Bergen (1996) or Lambert (1997).

It is a relatively simple procedure to isolate ossicles, but you will need access to a microscope and, if you want to prepare a permanent slide, some mounting medium. You need some household bleach, a glass slide, a coverslip, and a small dropper or pipette. Simply cut a small piece of skin (about 2 mm square) from the mid-dorsal side between the rows of tube feet. On the slide, cover the skin sample with a drop of bleach. The flesh will dissolve away in about five minutes, leaving the ossicles behind. Placing a fine-tipped pipette at the edge of the drop, carefully draw off the bleach. An alternative method is to let the bleach and ossicles dry on the slide. The ossicles will then stick to the slide and the bleach crystals can be redissolved with water.

For a good, permanent preparation, it is crucial to wash the ossicles with water at least three times to remove bleach crystals. This part is tricky because it is easy to suck up the ossicles while removing the water. Swirl the slide quickly in tight circles to cause the ossicles to collect in the center, making it easier to draw the water off. Using the alternate method of drying first, the sample should be flooded with water several times until no bleach crystals are left behind when the slide is dried. If you only want to confirm your identification with a temporary wet mount, place a coverslip on the water-covered ossicles and view them under the microscope.

For a permanent slide, remove the last wash water and let it air dry or, if available, use a slide warmer to speed up the process. Once the water has evaporated, place two small drops of mounting medium (Canada Balsam or any synthetic substitute) on the sample, apply a coverslip, and place on a horizontal surface to set.

The following key was modified from the one in the third edition by James Rutherford. With the exception of plate 473D by Richard Hermann and plate 475A by Shane Anderson, all other drawings and photographs were produced by Royal BC Museum staff—James Cosgrove (plate 473C), Gerald Luxton (plate 471), and all others by the author.

Key to Holothuroidea

1. Tube feet absent 2
— Tube feet present, scattered or in rows. 3

2. Wormlike, delicate, white or semitransparent; five muscle bands visible through the skin in live specimens (plate 473A) . . . . *Leptosynapta albicans*
— Stout body tapering to a wormlike tail; color in life purplish gray to dark brown with an opaque, smooth, wrinkled skin (plate 473B) . . . . *Paracaudina chilensis*
3. Obvious tube feet only on ventral surface; ventrum easily distinguished from dorsum by form and color 4
— Tube feet on all sides of body; ventral surface not sharply defined from dorsal. 7
4. Large cylindrical species (up to 50 cm) with tube feet in rows on the ventral side and fleshy papillae or warts on the dorsum; short feeding tentacles like mops; crawls on the sea bottom. 5
— Small species (<12 cm) with three rows of tube feet on a soft flat sole; overlapping plates on dorsum, obvious or partially hidden. 6
5. Body with about 40 prominent fleshy papillae and many smaller ones on dorsal surface; color in life mottled reddish brown; common subtidally (plate 473C) . . . .
. . . . *Parastichopus californicus*
— Body with many small black-tipped papillae on dorsal surface; brownish in color (plate 473D) . . . .
. . . . *Parastichopus parvimensis*
6. Domed dorsal surface of coarse overlapping plates; about four transverse rows of plates between the mouth and anus; body color reddish-orange, tentacles red with white tips; body up to 7 cm long (plate 473E). . . . . . *Psolus chitonoides*
— Dorsum scattered uniformly with small tube feet; a few plates embedded in skin around mouth; color usually bright red with a pinkish sole; rarely exceeds 2 cm in length; few large eggs are brooded in pits on the dorsal surface (plate 473F) . . . . *Lissothuria nutriens*

Note: Also *Psolidium bidiscum* (plate 475C) found subtidally but uncommon.

7. Body color in life beige, yellowish, or whitish, with or without spots. 8
— Body color black, orange, or red. 12
8. Tube feet in five rows; rigid and not retractable, giving a bristly appearance; body fairly stiff; 10 short whitish tentacles around the mouth, two ventral tentacles much smaller (plate 474A) . . . . *Eupentacta quinquesemita*

Note: Two other uncommon species, up to 6 cm long, may key out here, *Pentamera montereyensis* (plates 472E) and *Pentamera charlottae* (plates 472F). Identification requires analysis of ossicles. See species list.

— Tube feet in rows or scattered on the dorsum; retractable; body not stiff. 9
9. Body color whitish to yellowish, occasionally tan or faint pink 10
— Body color whitish to yellowish with brown spots on body or tentacles 11
10. Small species (<5 cm) with tube feet like dimples scattered on dorsum. Body yellowish white to light brown, plain or with fine brown specks on dorsum; often subtidal in large aggregations on rock; brood young; skin ossicles: knobby buttons with four to eight holes (plates 472C, 474B) . . . .
. . . . *Pseudocnus lubricus*
— Larger species (up to 26 cm) with tube feet in rows; body pale whitish orange or tan; 10 thin white tentacles of equal size; body usually hidden beneath rocks (plate 474C)
. . . . *Cucumaria pallida*

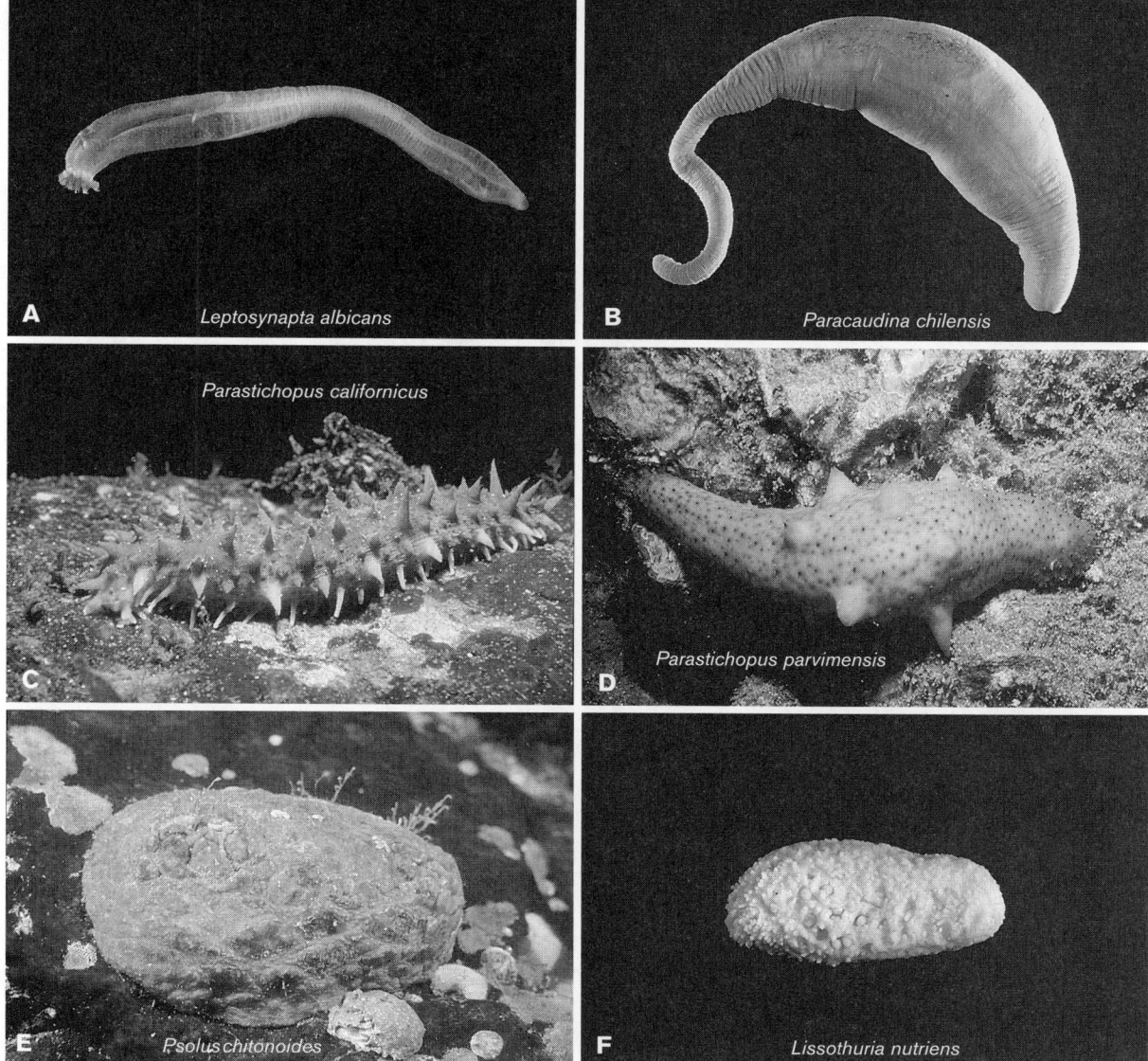

PLATE 473 A, *Leptosynapta albicans* (preserved, from San Francisco Bay); B, *Paracaudina chilensis* (preserved, from Queen Charlotte Islands); C, *Parastichopus californicus* (from Saanich Inlet); D, *Parastichopus parvimensis* (from Bird Rock, Catalina); E, *Psolus chitonoides* (from Victoria); F, *Lissothuria nutriens* (preserved) (D, photo by Richard Herrmann).

11. Small species (<5 cm) with tube feet reduced to scattered dimples on dorsum. Body yellowish white to light brown, plain or with fine brown specks on dorsum; occur subtidally in aggregations on rock; brood young; skin ossicles: knobby buttons with four to eight holes (plates 472C, 474B) . . . . *Pseudocnus lubricus*
— Medium-size whitish species with black or brown spots all over the body, on the 10 equal tentacles, and around the mouth; body usually under rocks or buried in shell gravel; skin ossicles flat, circular perforated plates with serrated edges, some have a narrow handlelike extension at one end (plates 472D, 474D) . . . . *Cucumaria piperata*
12. Body color dark brown to black; small species to 5 cm 13
— Body color orange, orange-brown, or red 14

13. Small (to 2 cm) black or gray species with five rows of tube feet; 10 tentacles with two smaller ventral ones; usually found among mussels at the mid-tide level; often in large numbers on rock surfaces; skin ossicles: oval perforated plates (plates 472B, 474E) . . . . *Cucumaria pseudocurata*
— Small (to 3 cm) black species; tube feet like dimples scattered on dorsum, in rows on ventral side; low intertidal among coralline algae; skin ossicles: small round buttonlike plates with two to six holes (plates 472A, 474F) . . . . *Pseudocnus curatus*
14. Small species (up to 4 cm); dorsum red or reddish orange; usually white ventrally; tube feet scattered (plate 475A1, 475A2) . . . . *Pachythyone rubra*
— Large species (up to 25 cm) with robust bushy tentacles; body and tentacles orange to reddish brown; tube feet in

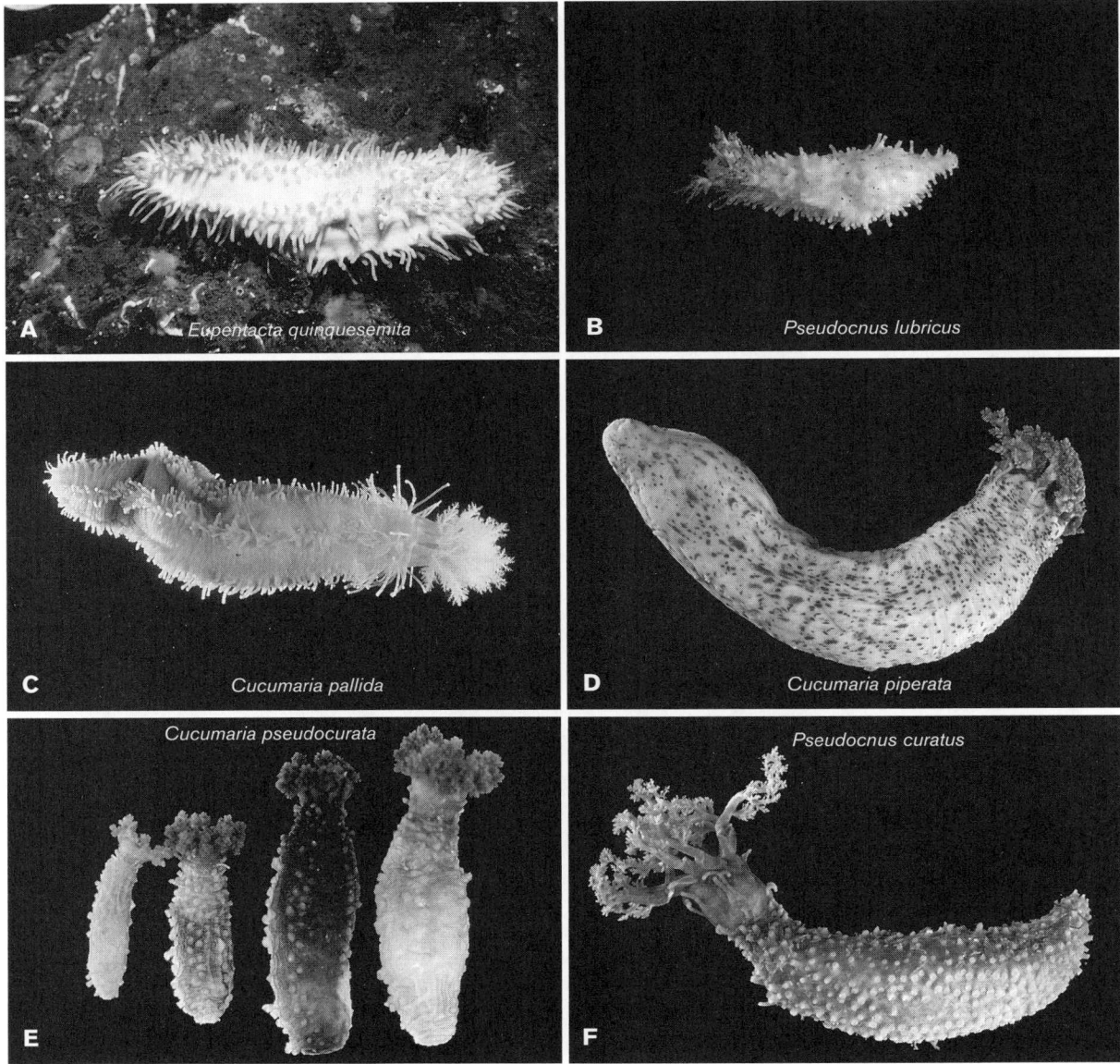

PLATE 474 A, *Eupentacta quinquesemita* (from Victoria); B, *Pseudocnus lubricus* (from Sooke); C, *Cucumaria pallida* (from Saanich Inlet); D, *Cucumaria piperata* (preserved, from Quatsino Sound); E, *Cucumaria pseudocurata* (preserved); F, *Pseudocnus curatus* (preserved, from Esquimalt).

rows; usually found under rocks in the shallow subtidal (plate 475B) . . . . *Cucumaria miniata*

Note: *Cucumaria salma* may key out here (see species list); not yet known north of Point Conception.

List of Species

ASPIDOCHIROTIDA

STICHOPODIDAE

Parastichopus californicus (Stimpson, 1857). Common from Alaska to central California, but as far south as Cedros Island, Baja California. Twenty moplike feeding tentacles.

Parastichopus parvimensis (Clark, 1913). Common from southern California to Baja California but uncommon in central California. Twenty moplike feeding tentacles.

DENDROCHIROTIDA

PSOLIDAE

Psolus chitonoides Clark, 1901. Common subtidally on rock, from the Aleutian Islands to Baja California. Ten equal tentacles. Chemicals (saponins) in tentacles reduce predation by fish. In northern part of range also in low intertidal and at times found washed ashore.

Lissothuria nutriens (Clark, 1901). Formerly *Thyonepsolus*. Low intertidal and shallow subtidal to 20 m. Monterey Bay to

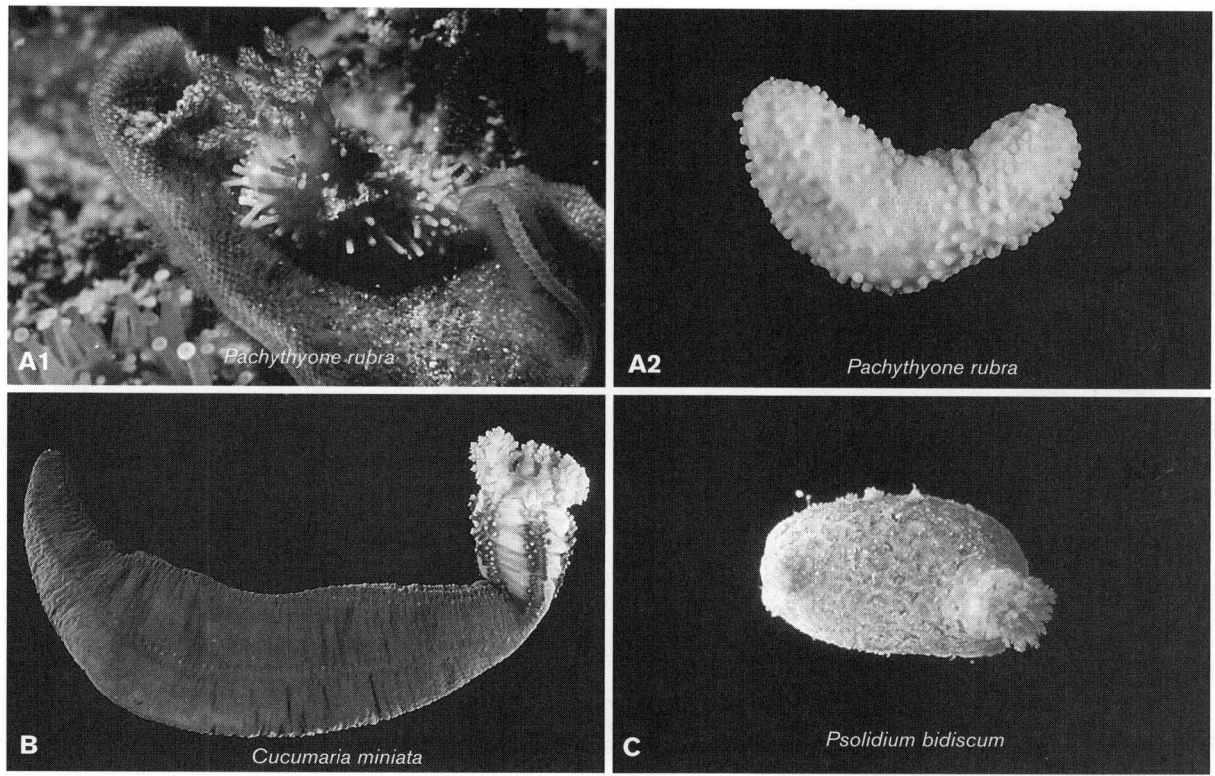

PLATE 475 A1, *Pachythyone rubra* (from Dutch Harbor); A1, *Pachythyone rubra* (preserved, from Monterey Bay); B, *Cucumaria miniata* (preserved, from Sooke); C, *Psolidium bidiscum* (from Moresby Island, Prevost Passage) (A1, photo by Shane Anderson).

southern California. On rocks and in sandy deposits among holdfasts. Broods eggs in pits on the dorsal surface. Eight large and two small tentacles.

Psolidium bidiscum Lambert, 1996. Uncommon. Small (up to 3 cm), pale pink with dorsal plates perforated by tiny tube feet; flat sole; subtidal on stones and pieces of shell. Previously identified as *Psolidium bullatum,* which only occurs in the Aleutian Islands. Southern Alaska to central California. Collected near Monterey at 12 m. Eight large and two small tentacles.

CUCUMARIIDAE

CUCUMARIINAE

Cucumaria miniata (Brandt, 1835). Common, bright orange; in low intertidal or shallow subtidal rocky areas from the Aleutian Islands to San Benito Island, Baja California. Ten robust, equal tentacles.

Cucumaria pallida Kirkendale and Lambert, 1995. Subtidal; from Southeast Alaska to Santa Rosa Island, California. Ten equal tentacles. Tentacles might be confused with *Eupentacta quinquesemita,* but the latter has eight large tentacles and two short.

Cucumaria piperata (Stimpson, 1864). Can be confused with small speckled *Pseudocnus lubricus* (formerly *Cucumaria fisheri astigmatus* Wells). *C. piperata* is larger and the body is usually buried under rocks or in the substratum. Ten equal tentacles.

Cucumaria pseudocurata Deichmann, 1938. Small brooding species in mussel beds or in large aggregations on intertidal

rocks. Often confused with *Pseudocnus curatus* (formerly *Cucumaria curata* Cowles). Check arrangement of tube feet and form of the skin ossicles (plate 472B). Eight large and two small tentacles.

Cucumaria salma Yingst, 1972 Common subtidally in the Channel Islands, but not known north of Point Conception. Lives in rock crevices. Body salmon color; 10 equal tentacles, black with white mottling. See Bergen (1996).

Pseudocnus curatus (Cowles, 1907). In previous edition as *Cucumaria curata*; ossicles on plate 146, fig. 12 in the previous edition are incorrect. Skin ossicles (plate 472A). Northern British Columbia to central California. Ten equal tentacles. See Lambert (1998b).

Pseudocnus lubricus (H. L. Clark, 1901). In previous edition as *Cucumaria lubrica*. Usually subtidal from southeast Alaska to Cortez Bank, southern California. Commonly in aggregations on the exposed surfaces of rock in areas of tidal currents or wave surge. Skin ossicles (plate 472C). Ten equal tentacles.

PHYLLOPHORIDAE

Pentamera charlottae Deichmann, 1938. Type locality: tidepools in Pacific Grove. Bergen (1996) records it largely from subtidal areas in southern California as *Havelockia charlottae.* While the type locality is a tide pool in Pacific Grove, California, it is not known whether this species is common at Monterey. The author has identified this species from two subtidal sites at Monterey. Superficially like *Eupentacta* with nonre-

* = Not in key.

tractable tube feet in crowded rows. Skin ossicles: tables (0.090–0.110 μm diam.) with tall spires (plate 472F). See Lambert (1998b). Eight large and two small tentacles.

Pentamera montereyensis Deichmann, 1938. Similar in appearance to *P. charlottae* and *E. quinquesemita*; skin ossicles: tiny oval plates (0.025–0.055 mm diam.) with no central spire but with bumps around the edges (plate 472E). Eight large and two small tentacles. Bergen (1996) lists it from Long Beach, California and Santa Rosa Island as *Thyone montereyensis*. The type locality is the intertidal in Pacific Grove. The author has collected this species from two locations on the west coast of Vancouver Island and two in the Queen Charlotte Islands.

SCLERODACTYLIDAE

Eupentacta quinquesemita (Selenka, 1867). Common; white; intertidal and shallow rocky areas from Sitka, Alaska to Baja California. Tube feet and body stiff due to heavy ossicle content. Skin ossicles large, porous, egg-shaped bodies and a few much smaller hollow baskets. Eight large and two small tentacles.

Pachythyone rubra (Clark, 1901). Intertidal to 548 m from Monterey to Los Angeles. Often very abundant in rocky habitats. Broods eggs internally. Eight large and two small tentacles.

MOLPADIIDA

CAUDINIDAE

Paracaudina chilensis (Müller, 1850). Usually burrows in clean sand with its mouth down and long wormlike tail at the surface. Produces a characteristic mound of sand. Wideranging species all along the Pacific coast of North and South America but not common in California. Fifteen short digitate tentacles.

> * = Not in key.

APODIDA

SYNAPTIDAE

Leptosynapta albicans (Selenka, 1867). In soft sediment among roots of eelgrass and in sand under rocks on the outer coast. Five species of *Leptosynapta* are described from the Pacific coast: *albicans, clarki, lens, roxtona* and *transgressor*. *L. lens* Heding, 1928 is a synonym of *L. albicans* (Selenka, 1867) and *L. roxtona* Heding, 1928 is a synonym of *L. clarki* Heding, 1928 (Sewell, Thandar, and Chia 1995). *L. clarki* is a northern species from the Queen Charlotte Islands to central California and *L. albicans* occupies a southern range to Baja California. The northern extent of *L. albicans* is not clear. Bergen (1996) was not able to clearly separate the two species. Until the identification can be clarified, all leptosynaptids in central California should be identified as *L. albicans*. All species have between nine and 14 short pinnate tentacles.

References

Bergen, M. 1996. 9. Class Holothuroidea Including keys and descriptions to all continental shelf species from California. In Taxonomic atlas of the benthic fauna of the Santa Maria Basin and the Western Santa Barbara Channel. Vol. 14 Miscellaneous Taxa. J. A. Blake, P. H. Scott, and A. Lissner, eds. pp. 195–250. Santa Barbara: Santa Barbara Museum of Natural History.

Lambert, P. 1997. Sea cucumbers of British Columbia, Puget Sound, and southeast Alaska. Vancouver: Royal BC Museum and University of BC Press.

Lambert, P. 1998a. *Pentamera rigida* and *Pentamera pediparva*, two new species of sea cucumber from the west coast of North America (Echinodermata: Holothuroidea). Proceedings of the Biological Society of Washington 111: 535–550.

Lambert, P. 1998b. A taxonomic review of five northeastern Pacific sea cucumbers (Holothuroidea). In Echinoderms: San Francisco. R. Mooi and M. Telford, eds., pp. 473–477. Rotterdam: A. A. Balkema.

Sewell, M. A., A. S. Thandar, and F.-S. Chia. 1995. A redescription of *Leptosynapta clarki* Heding (Echinodermata: Holothuroidea) from the northeast Pacific, with notes on changes in spicule form and size with age. Canadian Journal of Zoology 73: 469–85.

Chordata

(Plates 476–483)

Chordates are characterized by possessing, at least in embryonic stages, three distinctive features: (1) a notochord, (2) pharyngeal clefts or pouches ("gill slits"), and (3) a tubular dorsal nerve cord. In addition, all chordates possess either an endostyle or its evolutionary derivative, a thyroid gland, and all manufacture the hormone thyroxin. The best known subphylum of chordates, the Vertebrata, is not included in this manual. The remaining chordates, subphyla Tunicata (Urochordata) and Cephalochordata, are invertebrates.

Cephalochordates, typified by the well-known "amphioxus," are represented on the central California coast by *Branchiostoma californiense* Andrews, 1893, which is taken subtidally by trawl or grab sampler in Monterey and Tomales Bays and has occasionally been encountered in sand in the very low intertidal (Elkhorn Slough).

Tunicates are divided into the classes Appendicularia (Larvacea), Thaliacea, and Ascidiacea. Ascidians are abundant subtidally, in the lowest levels of our rocky intertidal zone, and on marina floats in our harbors.

Ascidiacea

DONALD P. ABBOTT, CHARLES C. LAMBERT, GRETCHEN LAMBERT, AND A. TODD NEWBERRY

(Plates 476–483; 479–483 compiled by Deniz Haydar)

Ascidians are sessile as adults and usually found on the sides and undersurfaces of rocks low in the intertidal zone, where they occur with hydroids, sponges, and bryozoans. In addition, they often dominate artificial surfaces such as floating docks and boat hulls in harbors and bays.

Simple (or "solitary") ascidians are ovoid, elongate, or somewhat irregular in shape. Most species are attached directly to the substratum by one side or by the base, but some are borne on a conspicuous stalk. The body is encased in a protective outer tunic, or test, provided with a pair of apertures, which are often borne on tubular extensions called siphons. When the animal is undisturbed, a current of water generated by cilia in the branchial sac enters through the oral aperture (mouth), bringing food (plankton) that is trapped on a mucous sheet inside the branchial sac. Water leaves by the atrial aperture, carrying off waste products and sometimes spawned gametes.

Simple ascidians reproduce only by sexual means, each of the numerous, small eggs developing into a swimming tadpole larva which settles, metamorphoses, and grows into a single adult. Spawning and development are usually pelagic, but in a few species, the eggs are retained and tadpoles brooded in the atrial cavity surrounding the pharynx (e.g., *Styela truncata*). Adults are usually solitary but may occur in clusters because of gregarious settlement of the larvae. Direct development of anural larvae occurs in a few molgulids.

Colonial ascidians, although basically similar to simple ascidians in structure, produce only a few large eggs, and these are retained until hatching. They also reproduce asexually by budding. As a result, several to many individuals are usually found connected together in a colony or clone. The form of the colony varies greatly. In "social" ascidians, the individuals, while joined at the base by stolons, are largely distinct from one another. In "compound" ascidians, the small individuals or zooids are embedded in a continuous mass of common tunic that frequently has a characteristic growth form of its own. The separation of ascidians into simple, social, and compound types is convenient but has no taxonomic significance. Some colonial species form colonies that are intermediate between the social and the compound types, and some simple ascidians are most closely related to colonial species.

In some compound ascidians, the zooids are scattered more or less randomly throughout the common test, and both apertures of each zooid open independently to the outside. In others, the zooids are arranged in recognizable clusters or "systems." Where this occurs, the atrial apertures may open independently at the colony surface or the atrial apertures of all zooids may empty into common cloacal cavities or pits in the test, which in turn open to the outside through common cloacal apertures. The distinction between the three types of colonial organization is important in identification. The key uses external features as far as is possible, but ascidians are so variable in size, shape, and color that it is often necessary to refer to the internal anatomy to confirm identifications. Internal features are best seen in preserved individuals.

Ascidians taken in the field can be placed directly into small plastic bags of seawater with a few crystals of menthol; they will often be narcotized by the time one returns to the lab. If not, they should be allowed to relax in a bowl of seawater. When apertures are expanded, add menthol crystals or several drops of a saturated solution of menthol in 95% ethanol to the surface and cover the bowl tightly. Complete narcotization takes 20 minutes to several hours, during which time, add a few more drops of menthol solution to the bowl each hour. After the animals no longer close their apertures when probed, they should be rinsed quickly in fresh seawater to remove extra menthol crystals and then fixed for several hours or more in 10% seawater formalin buffered with sodium borate. Such treatment permits easy observation of features that are difficult or impossible to see in contracted, living or preserved specimens.

Color notes should be made prior to fixation. It is becoming more and more important to identify and compare species not only on their gross morphology but also their DNA; in these cases small bits of colony or gonadal tissue of solitaries should be preserved directly into 95% ethanol, preferably without exposure to menthol or other relaxants, and stored in a freezer. Didemnids and *Cystodytes,* which contain calcium carbonate spicules, should have a subsample preserved directly into 95% ethanol to preserve the spicules, which will dissolve in formalin even if it is buffered.

For internal anatomy, simple ascidians are conveniently dissected in the following manner (plate 476A). Insert one point of a pair of scissors into the oral siphon and cut downward along, or very slightly to the right of, the median line (for most simple ascidians the median sagittal plane is defined roughly by the positions of the apertures: a plane passing posteriorly through the centers of the two apertures will divide the body bilaterally). Continue the cut around the base of the body, cutting through the tissues of the tunic, body wall (mantle), and pharynx, and then below the gut loop until the two attached halves can be spread apart like an opened book. Remove the body gently from the tunic, especially at the siphon tips, pin the opened animal down in a wax-bottomed petri dish or pan, and cover it with water. The inner surface of the pharynx thus exposed shows many features of taxonomic importance (plate 476B, 476D). In most simple ascidians, the stomach and intestinal loop lie on the body's left side; gonads may lie in the gut loop or in the mantle on both sides of the pharynx (plate 476C). To expose these structures, cut the numerous fine tissue strands that attach the pharynx (branchial sac) to the mantle and other organs on each side and fold it back. In *Ciona,* the gut loop and gonads lie posterior to the pharynx and can be seen without further dissection.

Colonial ascidians with large zooids may be dissected like simple ascidians. For those with small zooids completely buried in a massive common test, slice the colony parallel to the long axes of the zooids, observe how the zooids lie in the tunic, then remove several for study. Or probe carefully through the tunic surface and free a few individual zooids. Select well-expanded individuals in which the rows of pharyngeal stigmata are clearly visible. In most cases, zooids can be removed intact by grasping them with fine forceps by their anterior ends (oral siphons) and gently pulling them out of the tunic. You may only be able to obtain fragments from several individuals; however examination of these will allow you to reconstruct the intact zooid. Place them in a small dish of seawater and observe under a dissecting microscope with illumination from the side as with a fiberlight. Staining with a few drops of methylene blue or 1% toluidine blue often helps delineate anatomical details. Body organization and anatomy are shown in plates 476–478.

Colonies containing sexually mature individuals are much more easily identified than juvenile colonies or old, degener-

ating ones. Sexually ripe zooids possess well-developed gonads. Colonial ascidians brood the embryos after fertilization—in the zooidal atrial chamber in many species, but in the tunic matrix in didemnids and some botryllids—and are released only when ready to swim. In most species gonadal and larval morphological details are necessary for species identification.

More information on life history, reproduction, physiology etc. for most of the species listed below can be found in Abbott and Newberry (1980), a valuable resource for anyone interested in the biology of Pacific coast ascidians.

Key to Ascidiacea

There are some groups of animals for which keys can be made that really work in a considerable number of instances, but the ascidians are not among them.

W. G. Van Name, 1945

1. Colonial ascidians: zooids connected by stolons or thin sheets of tunic or embedded in thick sheets or lobes of common tunic. 2
— Solitary ascidians: individuals may be in clusters but are separated within their own tunics 3
2. "Social" colonial ascidians: individual zooids appear to be more or less discrete but are joined basally by stolons or sheets of tunic 24
— "Compound" colonial ascidians: zooids completely embedded together in common tunic; colony may form flat sheets, stalked or unstalked lobes, mounds, or slabs, according to species 30
3. Tunic transparent (may have isolated pigment flecks) or translucent. 4
— Tunic opaque; may be thick and leathery 13
4. Body is little low dome 1 cm across and 3 mm high when relaxed, somewhat flattened with simple slitlike apertures when disturbed; thin translucent tunic encloses tan-rose body; broods dozens of larvae in posterior atrial cavity; no branchial folds; only four internal longitudinal vessels per side in branchial sac; always present in aggregations. . . . .
. . . . *Dendrodoa abbotti*
— Larger than 1 cm, not a small dome; branchial folds present or absent, but many more than four longitudinal vessels per side in branchial sac. 5
5. Both oral and atrial apertures on a flattened terminal disk made up of numerous distinct plates, six of which surround each opening; body usually attached to substrate by basal region; 2 cm–5 cm long . . . .
. . . . translucent specimen of *Chelyosoma productum*
— Oral and atrial apertures not on a flattened terminal disk
. . . . 6
6. Body elongate or about as long as it is wide; spiral stigmata
. . . . 7
— Body decidedly longer than it is wide; straight stigmata
. . . . 9
7. Roughly rectilinear in outline, laterally compressed, 1 cm–5 cm long, 0.5 cm–3 cm wide; siphons short and both terminal; tunic and most of body clear, colorless, strikingly transparent; white pigment spots in body visible externally; oral tentacles simple; gut and intestine on right side of body; single gonad . . . . *Corella willmeriana*
— Gut and intestine on left side of body; bean-shaped "kidney" in right body wall often visible through tunic; oral tentacles branched; a gonad on each side of the body 8

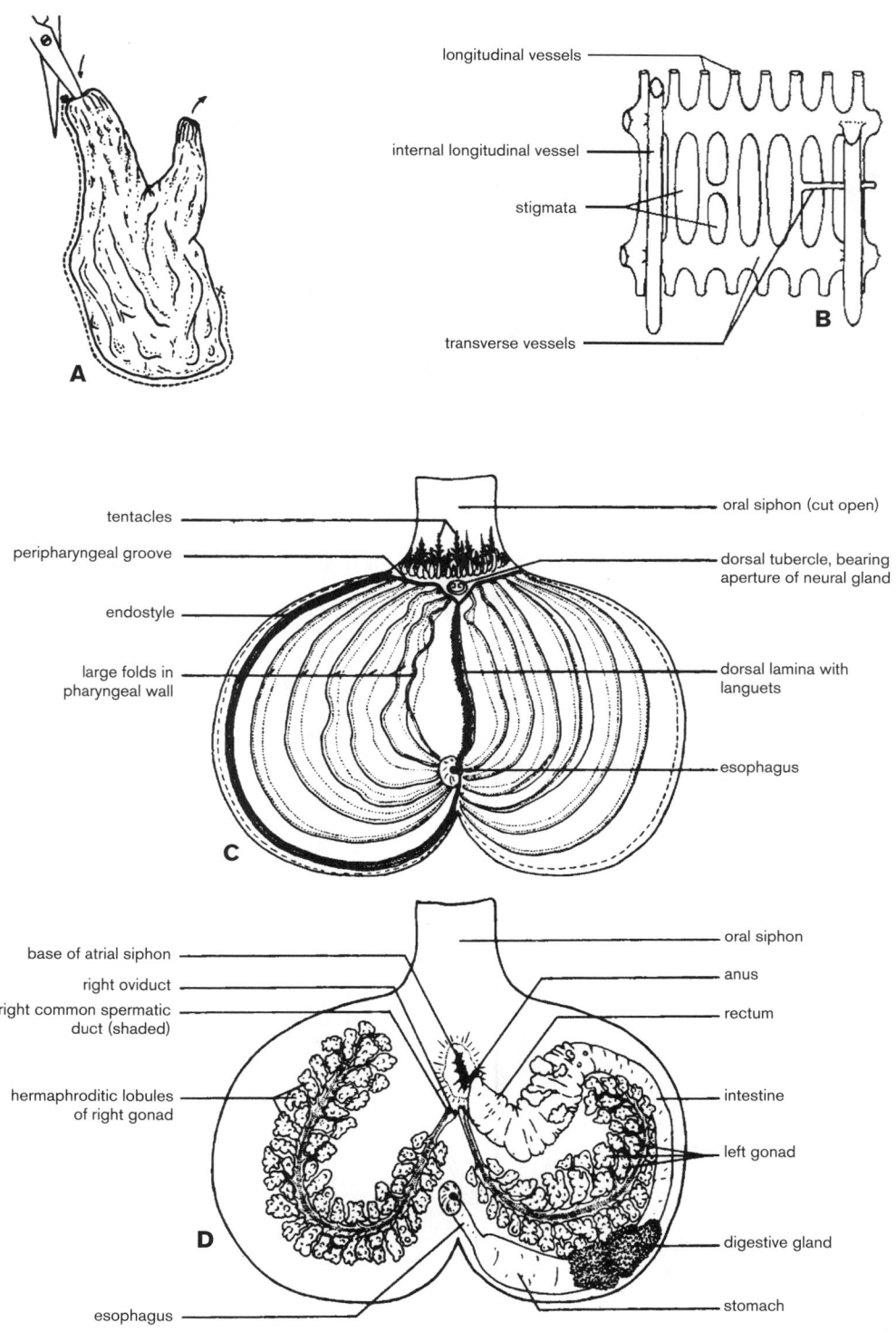

longitudinal vessels

internal longitudinal vessel

stigmata

transverse vessels

B

A

tentacles

peripharyngeal groove

endostyle

large folds in
pharyngeal wall

oral siphon (cut open)

dorsal tubercle, bearing
aperture of neural gland

dorsal lamina with
languets

esophagus

C

base of atrial siphon

right oviduct

right common spermatic
duct (shaded)

hermaphroditic lobules
of right gonad

oral siphon

anus

rectum

intestine

left gonad

digestive gland

stomach

esophagus

D

PLATE 476 Method of dissection and anatomy of taxonomic importance in a representative simple ascidian (*Pyura haustor*): A, method of dissection: cut along dotted line; B, a small area (one "mesh") of a pharyngeal wall, much enlarged (diagrammatic); C, specimen cut along midventral line and spread open to show the inner surfaces of oral siphon and pharynx; D, same view as in C, but with pharynx removed, tentacles and dorsal tubercle not shown.

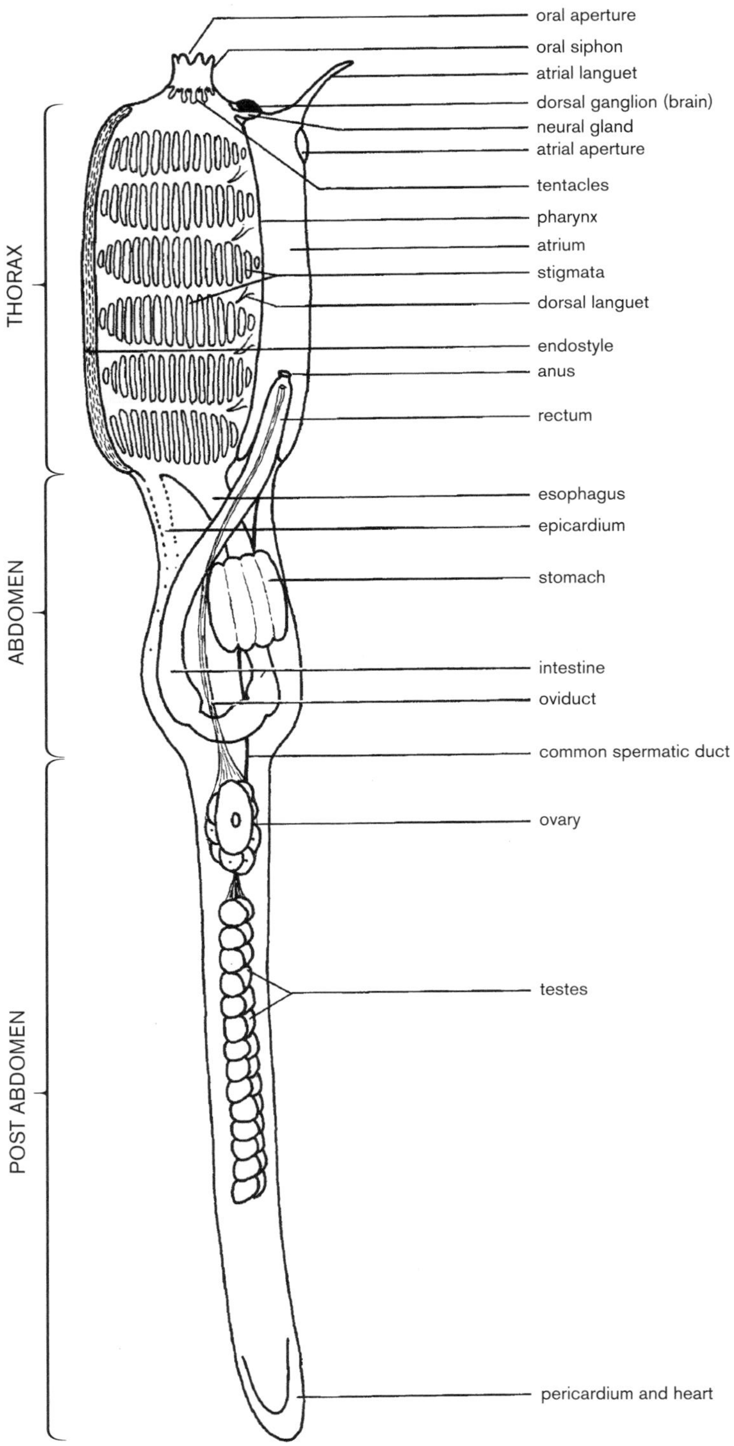

oral aperture
oral siphon
atrial languet
dorsal ganglion (brain)
neural gland
atrial aperture
tentacles
pharynx
atrium
stigmata
dorsal languet
endostyle
anus
rectum
esophagus
epicardium
stomach
intestine
oviduct
common spermatic duct
ovary
testes
pericardium and heart

THORAX
ABDOMEN
POST ABDOMEN

PLATE 477 Diagram of generalized compound ascidian (*Aplidium*) zooid with a postabdomen.

8. Pharynx with six branchial folds per side; free spawner (plate 483E) . . . . *Molgula manhattensis*
— Pharynx with seven branchial folds per side; broods larvae (plate 483F) . . . . *Molgula pugetiensis*
9. Body cylindrical, 4 cm–15 cm long, with restricted basal attachment to substrate 10
— Body irregular, 2 cm–6 cm long; broadly attached by left side to substrate 12

10. Conspicuous warty protrusions scattered profusely over thick white tunic, including on siphons (plate 481E) . . . . *Ascidia paratropa*
— Tunic more or less smooth, thin, lacking warts. 11
11. Siphons close together at distal end of body; translucent tough jellylike tunic without pigment flecks; red spot at end of sperm duct visible through tunic or through open atrial aperture (plate 481A) . . . . *Ciona intestinalis*

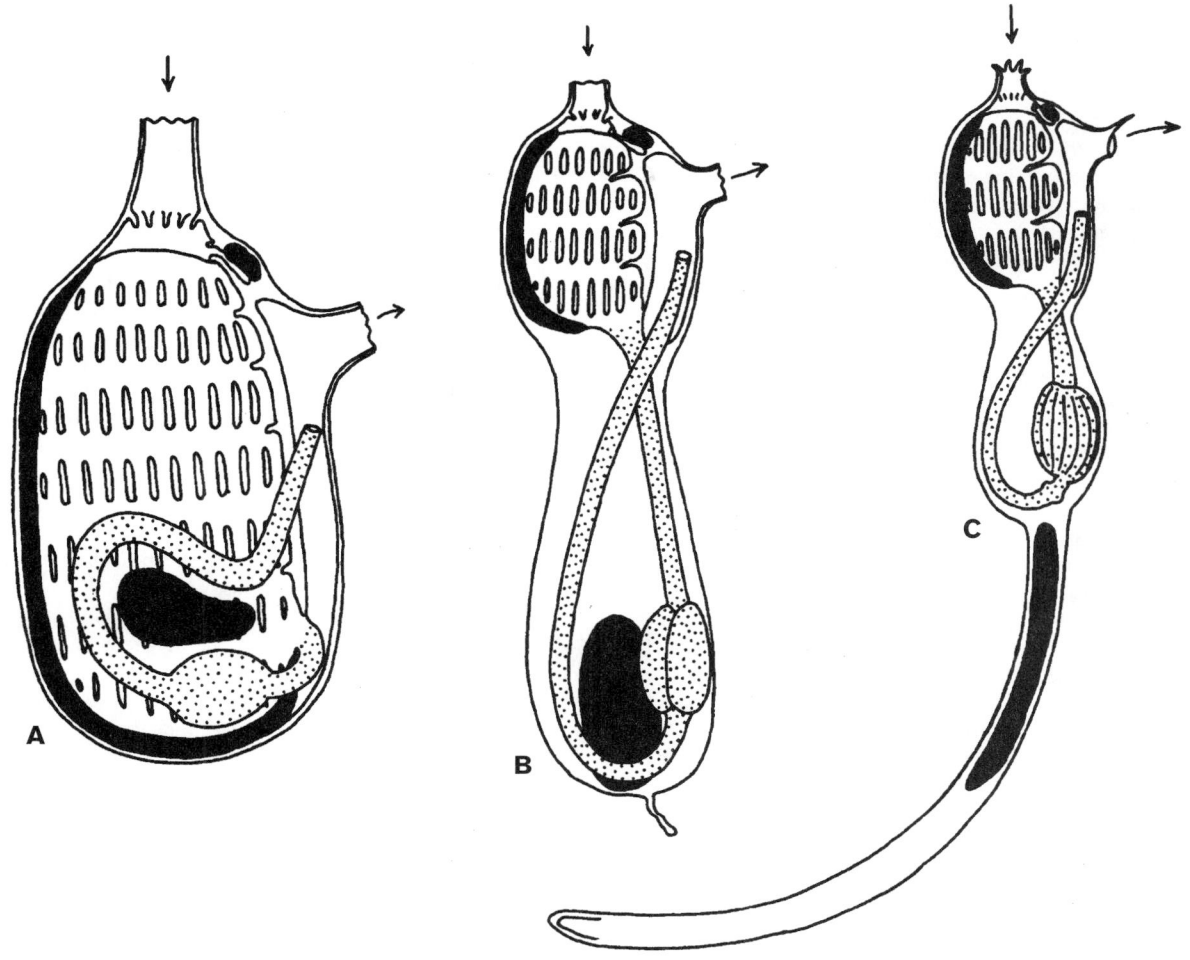

PLATE 478 Diagrams showing common variations in body form and arrangement of gut and gonads: gut stippled; brain, endostyle, and gonads solid black; A, intestinal loop lying beside and at left of the pharynx, body not divided into regions, gonads on one or both sides, attached to gut or to mantle—this condition is common in simple ascidians and is found in numerous colonial species; B, C, zooids of compound ascidians showing the subdivision into more or less distinct regions; B, a generalized clavelinid or polycitorid zooid (e.g., *Cystodytes*) showing division of body into two regions, thorax and abdomen; C, a generalized polyclinid zooid (e.g., *Aplidium*) showing division of body into three regions, thorax, abdomen, and postabdomen.

— Siphons separated at distal end of body; clear yellowish-green tunic with numerous white or yellow pigment flecks; no red spot at end of sperm duct. . . . . *Ciona savignyi*

12. Both siphons usually within anterior third of body; tunic thin, easily torn, covered with short papillae (may be only around siphons in some individuals); about 30 large oral tentacles . . . . *Ascidia zara*

— Atrial siphon in middle third of body; tunic thick, tough, smooth, semi-opaque; 150–200 oral tentacles (plate 481F) . . . . *Ascidia ceratodes*

13. Tunic covered with flexible, often branched or clustered spines or large bristly hairs. 14

— Tunic smooth or wrinkled, not covered with bristles or spines; hairs if present are fine. 15

14. Globular or heart-shaped body 1 cm–3 cm across, borne on distinct stalk 1 cm–4 cm long; tunic bristles simple or with many tiny secondary bristles irregularly distributed on them; tunic orange, reddish, orange-brown; apertures red (plate 483A) . . . . *Boltenia villosa*

— Body unstalked, stout, 2 cm–5 cm across; thickly covered with long thorny tunic spines, the largest spines topped with circle of prominent spinelets; spines mostly obscure the dark brown tunic; apertures red or orange . . . . *Halocynthia igaboja*

15. Body elongate, borne on a distinct stalk; tunic pale tan to dark brown 16

— Body elongate, globular, or flattened to substrate, unstalked, attached by side or base; color various. 17

16. Slender body tapers into thin stalk much longer than body; entire length 10 cm–25 cm; tunic without tubercles but with coarse longitudinal surface pleats; siphons terminal, oral curves away from atrial; one to three (usually two) tubular ovaries per side in atrial wall with numerous tiny, ricelike testes alongside each (plate 482H). . . . . *Styela montereyensis*

— Stalk shorter than or equal in length to body; body elongate but stout; entire length 5 cm–15 cm; tunic with distinct tubercles, at least near siphons, and with fine longitudinal pleats; siphons terminal, both of them straight or nearly so; tubular ovaries in atrial wall, four to nine on right, two to five on left; numerous tiny testes coalesce into irregular little lobes alongside ovaries (plate 482D) . . . . *Styela clava*

17. Both oral and atrial apertures situated on a flattened terminal disk consisting of several distinct plates (see key item 5); usually attached to substrate by basal region; 2 cm–5 cm long (plate 481D) . . . .
. . . . Opaque specimen of *Chelyosoma productum*
— Oral and atrial apertures not situated on a flattened terminal disk 18

18. Body low, dome-shaped or flattened, broadly attached to substrate (but review key item 4); 1 cm–3 cm wide, 2 cm–4 cm long; tunic thin but tough, smooth and shiny, red or rose, rarely white; five to 12 tubular hermaphroditic gonads in atrial wall on each side (usually more on right than on left) . . . . *Cnemidocarpa finmarkiensis*
— Body cylindrical or globular 19

19. Globular body <1.5 cm across; tough tunic covered with fine hairs and tendrils and usually coated with sand; irregular tentaclelike processes around siphon openings; spiral stigmata highly subdivided and thus mostly straight or slightly curved, not spiral; bean-shaped "kidney" in right atrial wall; broods larvae in atrium (plate 483G) . . . .
. . . . *Molgula verrucifera*
— Body hemispherical to elongate; thick or thin, tough tunic without tendrils, sand-free but may be fouled by debris or epibionts; "kidney" absent but lobulated "liver" attached to stomach 20

20. Elongate, 5 cm–10 cm from siphon to siphon; attached by one side; siphons arise from opposite ends of body; tunic clean and whitish to pale brown (plate 483D) . . . .
. . . . *Pyura mirabilis*
— Elongate or hemispherical but siphons never at opposite ends of body 21

21. As long as or longer than wide, 2 cm–4 (rarely to 8) cm long; thick, leathery, red-brown tunic deeply and irregularly creased, often covered by debris and epibionts leaving only red apertures exposed; in smaller specimens red siphons nearly as long as body; one long gonad (double row of hermaphroditic sacs) per side in atrial wall, left gonad inside curve of gut-loop; branchial sac with six folds per side (plate 483C). . . . . *Pyura haustor*
— Hemispherical or elongate; siphons short, stubby; tunic clean; multiple gonads in each atrial wall; branchial sac with four folds per side. 22

22. Ovoid, to 5 cm–7 cm from base to siphons; siphons close together; clean, thick tunic has many rounded areas that give it a lumpy look; white to tan with black stripes on siphons; four to seven tubular gonads in right atrial wall, usually two in left (plate 482C) . . . . *Styela plicata*
— Tunic brown, tan, ochre, or suffused with red; smooth or wrinkled but not lumpy; usually two tubular gonads per side. 23

23. Elongate to cylindrical, 2 cm–6 cm long, 0.5 cm–2 cm wide; body attached basally, short siphons close together at distal end; thin, tough brown tunic finely ridged longitudinally and wrinkled transversely in contracted animals; three to six longitudinal vessels between middle branchial folds; does not brood larvae (plate 482F) . . . . *Styela gibbsii*
— Hemispherical, globular, or stubby, 1 cm–3 cm long; thin, tough, yellowish to red-brown tunic smooth in relaxed animals but often deeply creased in contracted ones; one

longitudinal vessel between each branchial fold; broods larvae (plate 482G) . . . . *Styela truncata*

24. "Social" ascidians with elongate zooids, forming a densely packed bouquetlike colony 25
— "Social" ascidians with roughly globular zooids, often forming a densely packed encrusting colony or mass 27

25. Tunic transparent, sand-free except at base of colony where zooids interconnect; zooids 2 cm–4 cm long, 5 mm–10 mm wide; each zooid has two parallel fluorescent pink or orange lines along endostyle (mid-ventral line of thorax), visible through tunic; 16–20 rows of stigmata (plate 479A) . . . . *Clavelina huntsmani*
— Tunic usually heavily sand-encrusted except distally, but some colonies are fairly sand-free; zooids in individual cylindrical tunics, interconnected basally, densely packed into carpetlike masses; zooids gray or orange, thorax visible through a cleaned tunic; seven to 13 rows of stigmata 26

26. Zooids 1 cm–2 cm long, 1 mm–2 mm across in sandy cylinders; zooids' orange or yellow pharynxes make bright flecks against a sandy background; seven rows of stigmata; stomach smooth-walled; on sandy rocks . . . .
. . . . *Pycnoclavella stanleyi*
— Zooids 3 cm–5 cm long, 2 mm–4 mm across in individual sand-encrusted tubes; pharynx pale gray or colorless; 12–13 rows of stigmata; stomach surface has six axial ridges (plate 480B) . . . . *Euherdmania claviformis*

27. Transparent pale green tunic; pale yellow-green globular zooids 2 mm–4 mm across in clusters that resemble tiny grapes connected by threadlike stolons; four rows of stigmata; ovary and testis within curve of gut-loop; broods larvae until hatching. 28
— Opaque bright red tunic and zooids (rarely tinged green); zooids globular to low cylindrical, 4 mm–6 mm; young colonies show dendritic encrusting patterns, older ones are dense masses of zooids 29

28. Stolons with star-shaped flattened yellow terminal buds (plate 481C) . . . . *Perophora japonica*
— Stolons without such terminal buds (plate 481B) . . . .
. . . . *Perophora annectens*

29. Zooids connected only basally by stolons or thin sheet of tunic; about 12 rows of stigmata; multiple gonads in atrial wall; broods larvae (plate 482B) . . . . . *Metandrocarpa taylori*
— Zooids embedded almost to their siphons in tough common tunic, giving colony a "compound" appearance; in other respects colony and zooids like *M. taylori* . . . .
. . . . *Metandrocarpa dura*

30. Colony forms encrusting sheet or thick slab, or mounds or hanging fingerlike lobes widely attached to the substrate; colony may be lumpy but lacks distinctly stalked masses 31
— Colony distinctly stalked: one to many club- or mushroom-shaped, paddlelike lobes, or stalked globular or mounded masses 46

31. Body of zooids in one part, colony variously colored; compact, bean-shaped zooids 2 mm–3 mm long, organized into conspicuous systems; abundant blind sacs (ampullae) of tunic's blood-vessel system prominent at colony's growing edges 32
— Body of zooids in two or three parts; tunic lacks vascular ampullae 34

32. Colony 3 mm–4 mm thick, 2 cm–10 cm across; orange, yellow, red, gray-black; star-shaped or round systems 0.5 cm–1 cm across, usually containing fewer than 20 zooids;

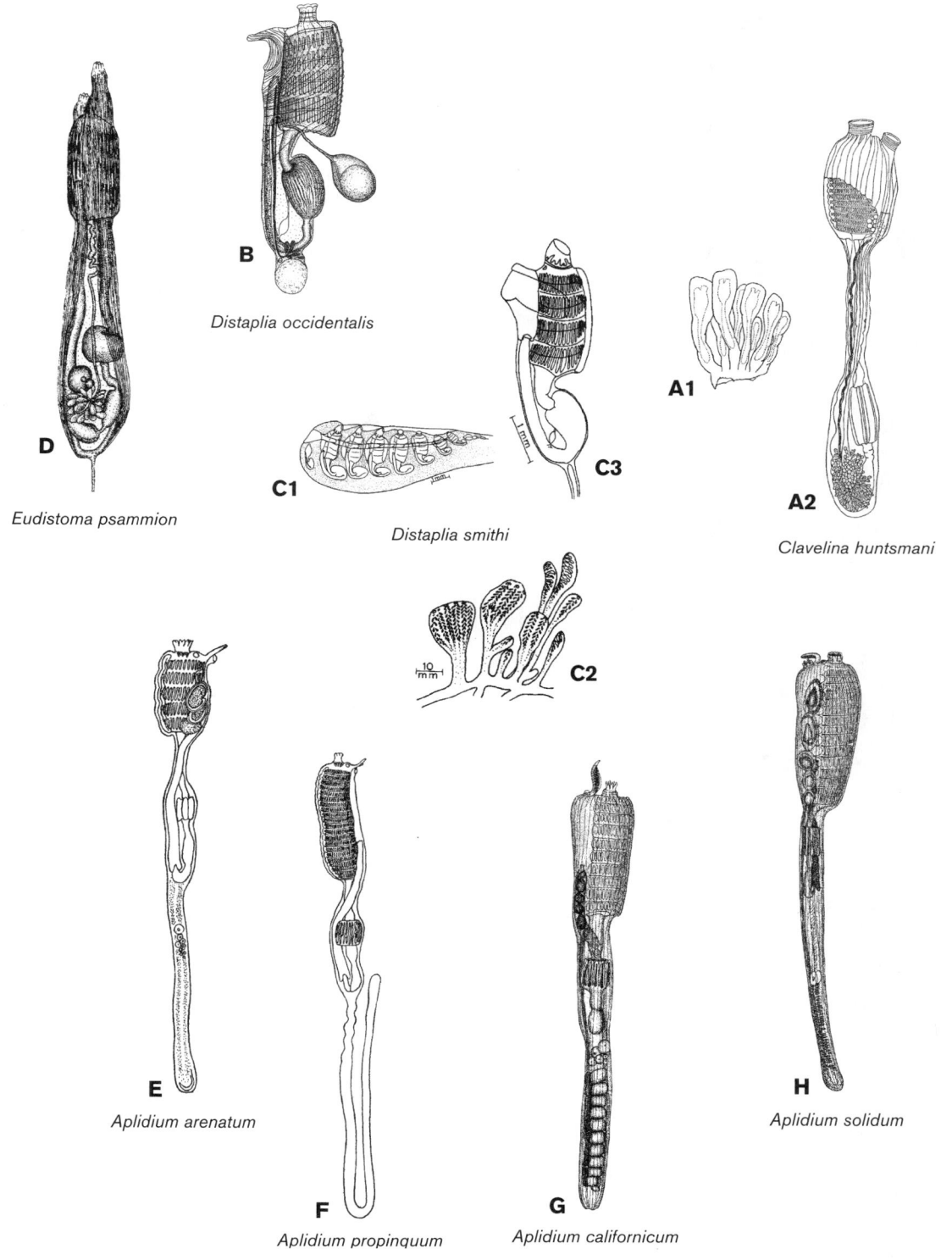

B

Distaplia occidentalis

D

Eudistoma psammion

C1

C3

Distaplia smithi

C2

A1

A2

Clavelina huntsmani

E

Aplidium arenatum

F

Aplidium propinquum

G

Aplidium californicum

H

Aplidium solidum

PLATE 479 Species as indicated: A, Clavelinidae; B, C, Holozoidae; D, Polycitoridae; E-H, Polyclinidae (A, E, F, from Van Name 1945; B, D, G, H, from Ritter and Forsyth 1917; C, from Abbott and Trason 1968).

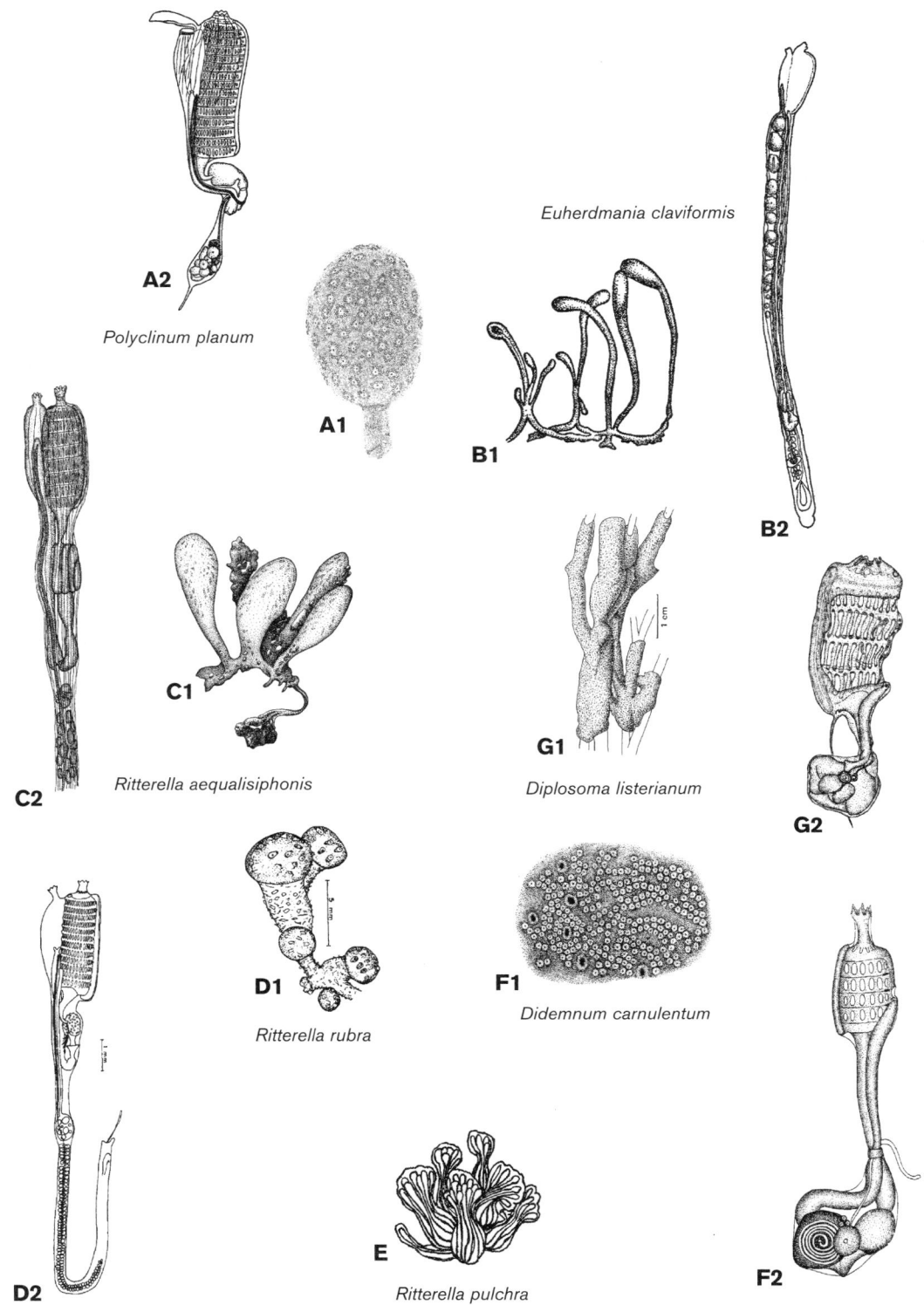

A2

Polyclinum planum

A1

Euherdmania claviformis

B1

B2

C1

Ritterella aequalisiphonis

C2

G1

Diplosoma listerianum

G2

D1

Ritterella rubra

F1

Didemnum carnulentum

E

Ritterella pulchra

D2

F2

PLATE 480 Species as indicated: A, B, Polyclinidae; C, D, E, Ritterellidae; F, G, Didemnidae (A, B, E, from Van Name 1945; C1, F, from Ritter and Forsyth 1917; C2, from Tokioka 1967; D, from Abbott and Trason 1968; G, from Hayward and Ryland 1990.

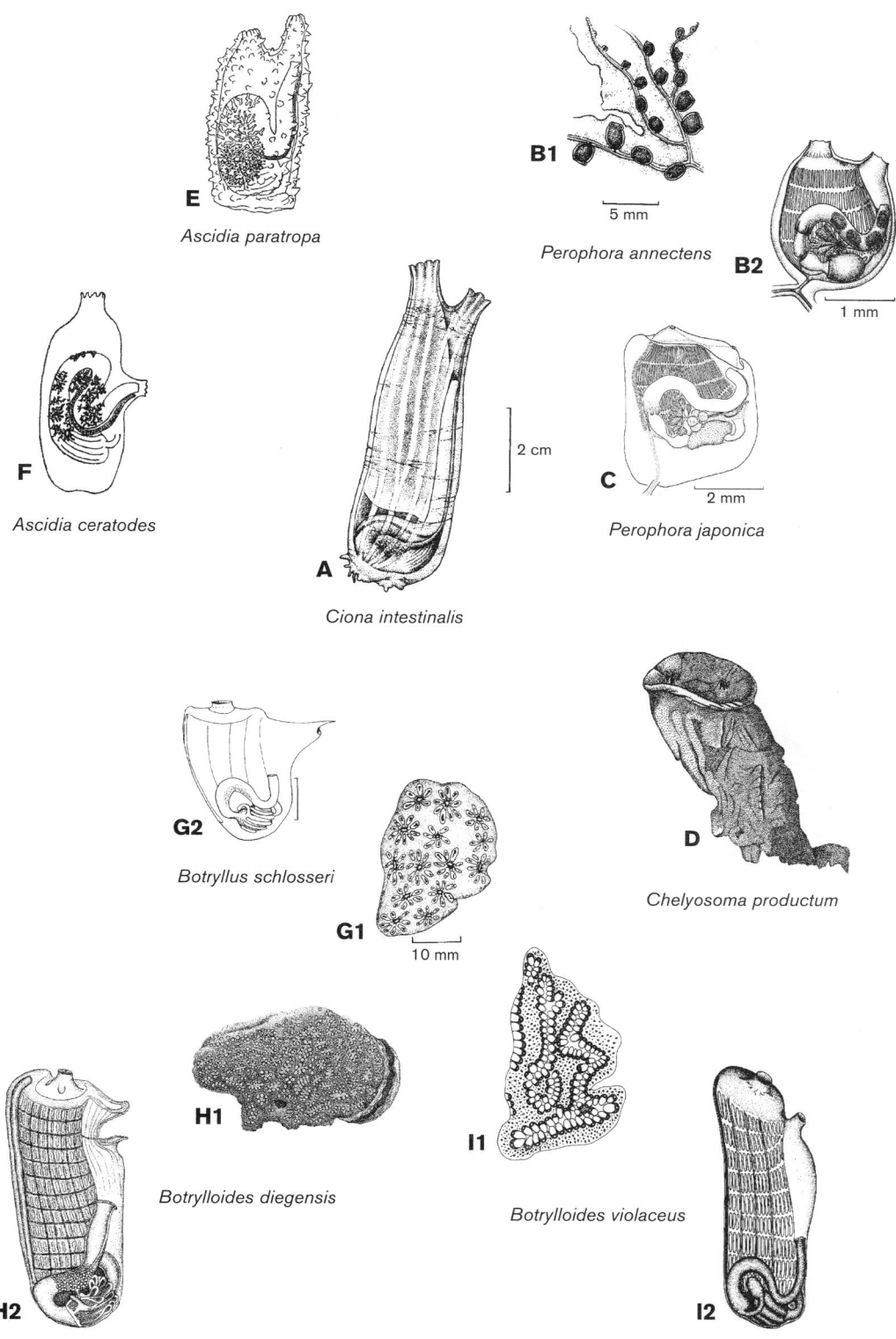

E
Ascidia paratropa

B1
5 mm

Perophora annectens

B2
1 mm

F
Ascidia ceratodes

A
2 cm
Ciona intestinalis

C
2 mm
Perophora japonica

G2
Botryllus schlosseri

G1
10 mm

D
Chelyosoma productum

H1
Botrylloides diegensis

H2

I1
Botrylloides violaceus

I2

PLATE 481 Species as indicated: A, Cionidae; B, C, Perophoridae; D, Corellidae; E, F, Ascidiidae; G–I, Styelidae (A, B, from Abbott 1997; C, from Monniot and Monniot 1985; D, E, F, from Van Name 1945; G, from Millar 1970; H, from Ritter and Forsyth 1917; I, from Tokioka 1967).

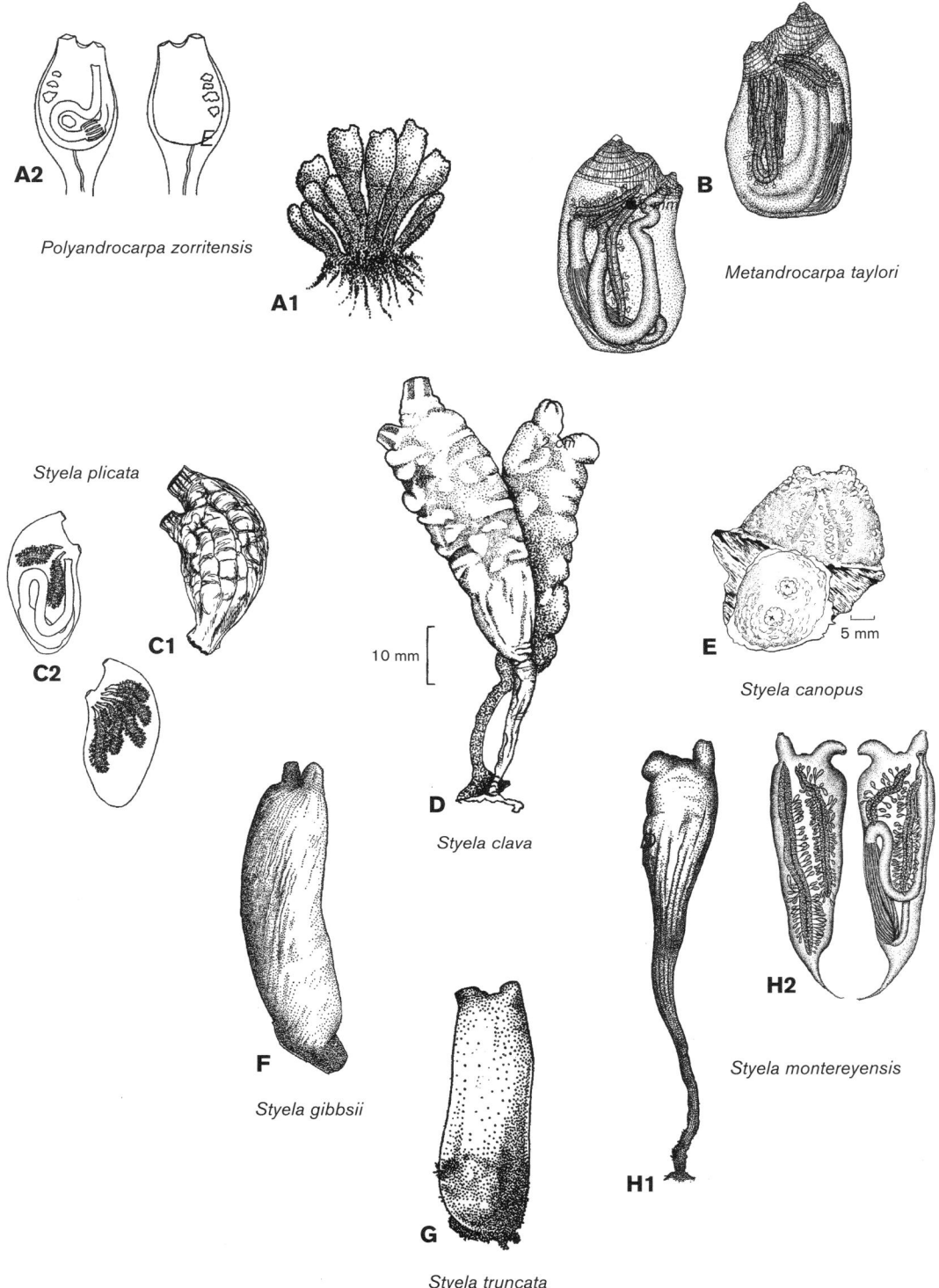

A2

Polyandrocarpa zorritensis

A1

B

Metandrocarpa taylori

Styela plicata

C2

C1

D

10 mm

Styela clava

E

5 mm

Styela canopus

F

Styela gibbsii

H1

H2

Styela montereyensis

G

Styela truncata

PLATE 482 Species as indicated: Styelidae (A, C2, F, G, H1 from Van Name 1945; B, H2, from Ritter and Forsyth 1917; C1, from Allen 1976; D, E, from Millar 1970).

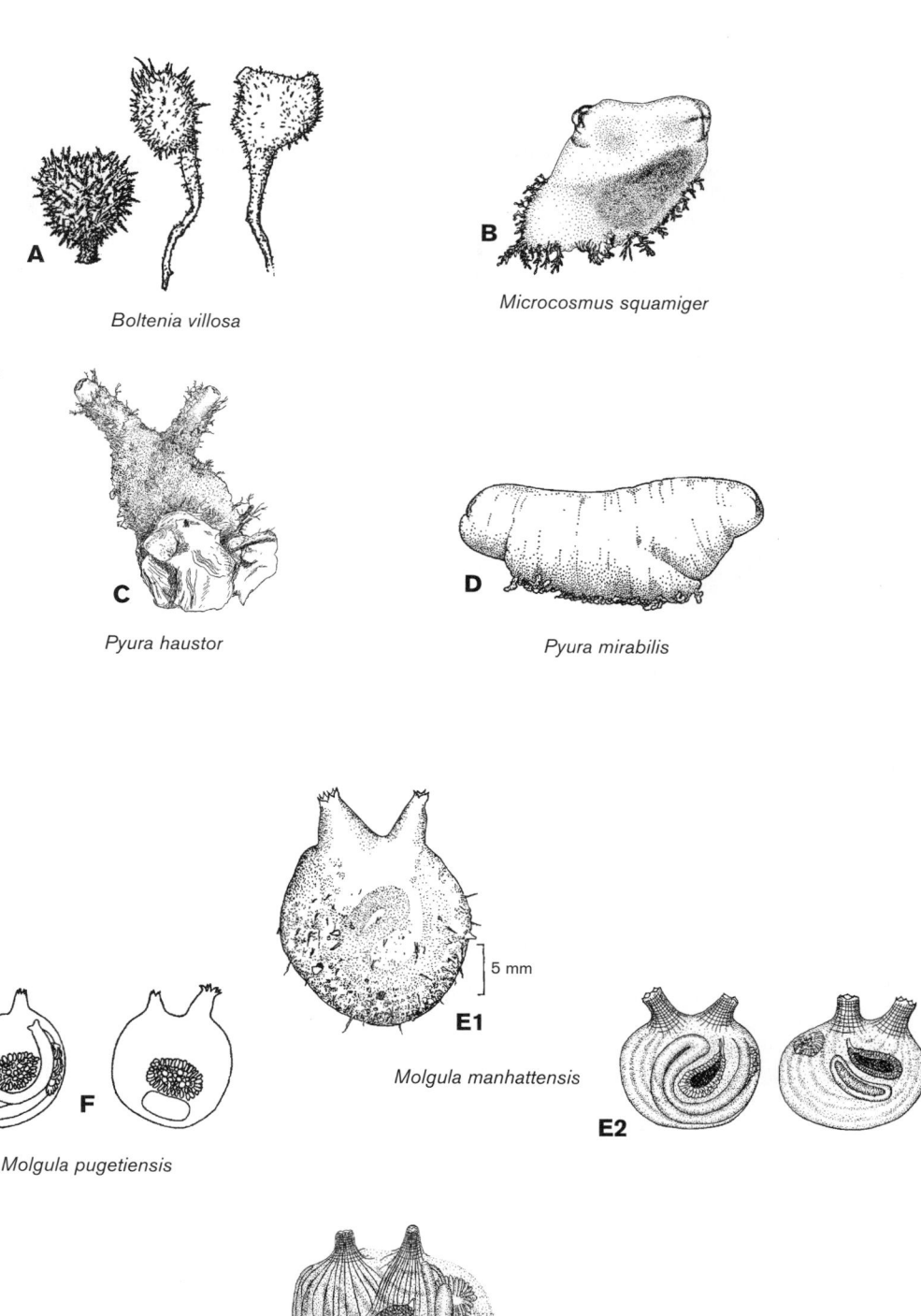

Boltenia villosa

Microcosmus squamiger

Pyura haustor

Pyura mirabilis

5 mm

Molgula manhattensis

Molgula pugetiensis

Molgula verrucifera

PLATE 483 Species as indicated: A–D, Pyuridae; E–G, Molgulidae (A, D, F, from Van Name 1945; B, from Kott 1985; C, from Johnson and Snook 1927; E1, from Millar 1970; E2, from Hayward and Ryland 1990; G, from Ritter and Forsyth 1917).

zooid has gaping atrial aperture beneath canopylike "languet"; nine to 10 rows of stigmata; tadpoles with eight lateral ampullae on trunk, incubated inside adult (plate 481G) . . . . *Botryllus schlosseri*

— Systems are mostly long, sinuous, sometimes branched double-rows of zooids, systems 0.5 cm–1 cm wide; often many more than 20 zooids per system 33

33. Zooids always of two colors, some combination of white, tan, orange, yellow, purple, or brown with a white or bright ring of pigment around the oral aperture contrasting with the darker overall colony color; eight oral tentacles, 10–12 "complete" rows of stigmata; small tadpole with eight ampullae, incubated inside adult (plate 481H) . . . .
. . . . *Botrylloides diegensis*

— Colony all one color, usually some shade of orange, yellow, pink, tan, purple, or brown; 16 oral tentacles, 10–11 rows of stigmata of which the "incomplete" second row stops well short of the mid-dorsal line; very large tadpoles (>1 mm body diameter) with up to 34 lateral ampullae, incubated in individual brood sacs in tunic (plate 481I) . . . .
. . . . *Botrylloides violaceus*

34. Body of zooids in two parts: thorax and abdomen; three or four rows of stigmata 35

— Body of zooids in three parts: thorax, abdomen, and postabdomen; more than four rows of stigmata 45

35. Tunic with tiny (20–50 μm) stellate calcareous spicules
. . . . 36

— Tunic with flat disc-shaped spicules around the zooids' abdomens, or lacking spicules 39

36. White or pale gray, slightly rough-surfaced, occasionally wrinkled sheets, swollen above zooids, 2 mm–10 mm thick, 5 cm–15 cm across, with widely scattered, large common excurrent openings; rays of microscopic stellate spicules in tunic longer than one fourth the total spicule diameter; tiny, pale zooid with three rows of stigmata and funnel-shaped atrial siphon opening downward . . . . *Trididemnum opacum*

— Zooids with four rows of stigmata; atrial aperture is a gaping hole 37

37. Testis with two lobes, covered by coiled sperm duct . . . . .
. . . . *Didemnum albidum*

— Testis with single lobe, covered by coiled sperm duct . . . . .
. . . . 38

38. White or pale pink, smooth-surfaced encrusting sheets 1 mm–4 mm thick, 1 cm to a few centimeters across, many small common excurrent openings; microscopic spicules in tunic have rather stubby sharp-tipped rays, less than one fourth the total spicule diameter; larvae unknown (plate 480F). . . . . *Didemnum carnulentum*

— Similar to *D. carnulentum* but colonies often huge, lumpy, usually >4 mm thick, and usually with long, hanging, fingerlike lobes especially on marina floats. Pink to cream or tan colored tunic when living (color disappears in alcohol) with tiny (20–40 μm) white spherical calcareous spicules with many short pointed rays, not abundant and mostly confined to surface layer of colony, more abundant around the zooids with dark meandering narrow lines in the tunic between groups of zooids where the spicules are absent; tunic flaccid, lobes break off easily; larvae with six pairs of lateral ampullae, incubated in spicule-free inner part of tunic . . . . *Didemnum* sp.

39. Tunic with layer of 0.5 mm–1 mm diameter overlapping discoid calcareous spicules around each zooid's abdomen (colony must be torn open to see spicules); colony opaque white, pale pink to lavender, gray, or orange; sand-free

"cakes" 1 cm–3 cm thick, 5 cm–25 cm across, often lumpy or convoluted; tunic has densely packed bladder cells and usually produces copious mucus in response to touching or injury. . . . . *Cystodytes lobatus*

— Tunic lacking calcareous spicules but may have pigment granules 40

40. Zooids with four rows of stigmata; gelatinous sheets 1 mm–3 mm thick, many centimeters across; soft, easily torn, transparent, colorless, gray, olive, tan, often with white pigment flecks (do not confuse with stellate calcareous spicules); tiny (1 mm–2 mm) pale zooids have stout thorax and abdomen connected by narrow esophagus; atrial apertures empty into extensive chambers inside tunic; living colony's occasional common excurrent apertures marked on surface by delicate little conical swellings (plate 480G) . . . . *Diplosoma listerianum*

— Zooids with three rows of stigmata 41

41. Soft and yielding lobes or cushions 1 cm–2 cm thick, to 10 cm across; tunic translucent, soapy white to pale pink, densely speckled with bright red zooids (color disappears in alcohol); zooids less than 10 mm long. . . . .
. . . . *Eudistoma molle*

— Colonies not as above; zooids not bright red 42

42. Small encrusting sheets or low rounded mounds 1 cm–3 cm across; tunic transparent and colorless to gray; tiny zooids (3 mm) pale yellow or gray . . . . *Eudistoma diaphanes*

— Colonies form thick sheets or slabs 43

43. Colony sand-free, purple or lavender, occasionally gray; may be massive with undulating lobes; many small purple granules near the tunic surface and lining the zooids' three rows of stigmata. . . . . *Eudistoma purpuropunctatum*

— Colony coated or impregnated with sand at least in part; may have several to many broad-based projecting lumps, lobes, or knobs 44

44. Deep brown, maroon to red, or somber purple-gray opaque slabs 1 cm–2 cm thick; tunic very firm, resilient, usually sand-impregnated but tough and leathery even if fairly sand-free; zooid 5 mm–10 mm long, with long atrial siphon; colonies especially prominent in heavily surf-swept habitats (plate 479D) . . . . *Eudistoma psammion*

— Sheet to 1 cm thick, often with lumps or pendant lobes 3 cm–5 cm long, especially if colony grows on overhanging substrate; sheet and basal parts of colony may be sand-coated but lobes are usually sand-free; colony usually pale yellow or tan-ochre due to abundance of yellow zooids about 10 mm long in almost transparent tunic . . . .
. . . . *Eudistoma ritteri*

45. Smooth, glossy-surfaced sheets 1 cm–3 cm thick, may be many cm across; tunic soft, translucent white, pale yellow to tan, gray to almost colorless; zooids tan, yellow, or dull ochre, 5 mm–10 mm long, in clear oval or elongate systems; usually 10 to 12 rows of stigmata; stomach surface with 15–20 axial ridges; zooid's atrial aperture smooth-rimmed, with large languet (plate 479G) . . . . *Aplidium californicum*

— Usually sand-free fleshy lobes or slabs to 3 cm thick, many centimeters across; tunic traits and colors like those of *A. californicum* though generally bulkier, more opaque and often pitted; zooids often have bright ochre or red thoraxes; most zooids 10 mm–15 mm long but some much longer; 12–16 (usually 13–15) rows of stigmata; stomach surface with about eight axial ridges; atrial aperture's rim five-toothed, atrial languet small (plate 479H) . . . . *Aplidium solidum*

46. Colonies sand-free; zooids have two body regions, four rows of stigmata 47

— Colonies may be sand-free or sand-encrusted; zooids have three body regions, five or more rows of stigmata 48

47. Multiple club-shaped or mushroomlike masses 1 cm–3 cm high, 0.5 cm–3 cm across at top, or broad, flat, stalked or V-shaped mounds to 10 cm across; pale orange to deep purple-red, gray, yellow, cream; zooids 3 mm long in distinct little round systems around common excurrent pores; peripheral zooids usually visible in side view of lobe above stalk (plate 479B) . . . . *Distaplia occidentalis*

— Clusters of paddle-shaped blades to 3 cm–4 cm long, 1 cm–2 cm wide; zooids 5 mm long arranged in parallel linear systems each with its common excurrent aperture at distal edge of paddle; zooids' oral apertures visible on one surface of paddle, bodies visible on the other; colonies cream-yellow to brown (plate 479C) . . . . . *Distaplia smithi*

48. Zooids organized into systems with common excurrent apertures 49

— Both apertures of each zooid open at colony surface . . . . . .
. . . . 52

49. Sand-encrusted and -impregnated, multiple club-shaped lobes 1 cm–3 cm long with narrow stalks and swollen, distally rounded heads; usually deep red-gray; only a few zooids per lobe; zooids 10 mm long, with five rows of stigmata; stomach surface has five axial ridges (plate 479E) . . . .
. . . . *Aplidium arenatum*

— Lobes sand-free or externally sand-encrusted, but tunic is not sand-impregnated; zooids with at least 14 rows of stigmata 50

50. Large colony a single, largely sand-free, roughly discoid lobe up to 8 cm across, attached edge-on to substrate by short stalk, small colonies more globular but still pedunculate; tunic firm but not hard; tan to brown-orange; zooid 5 mm long including short post-abdomen; 14–18 rows of stigmata (plate 480A) . . . . *Polyclinum planum*

— Colony lightly or heavily sand-encrusted 51

51. Many club-shaped lobes to 5 cm long but mostly shorter, usually with flattened distal ends; colonies dull red beneath sand coating; mature zooids 35 mm long, of which 25 mm is exceptionally long, thin post-abdomen; 17–21 rows of stigmata; stomach surface with about 20 axial ridges (plate 479F) . . . . *Aplidium propinquum*

— Many lightly sand-coated, club- or pear-shaped lobes to 8 cm long, rounded distally, stalks tapering gradually to narrow bases; opaque tunic dirty to bright orange, zooids orange; many zooids per lobe, to 40 mm long including prominent post-abdomen, lying parallel to one another but diverging as they approach colony surface; 14–16 rows of stigmata; stomach surface smooth except for one deep axial seam . . . . *Synoicum parfustis*

52. Densely massed, heavily sand-encrusted, slender, somewhat capitate lobes 1 cm–2 cm long, 5 mm–10 mm across top; lobes gray to red, zooids red-brown to orange; zooids 10 mm–15 mm long, with seven to 10 rows of stigmata; stomach surface with about six broad axial folds (plate 480C) . . . . *Ritterella aequalisiphonis*

— Colony only lightly coated with sand or sand-free 53

53. Several rounded lobes often piled into masses up to 3 cm high; deep or bright ruby red; zooids 5 mm–15 mm long including long, thin post-abdomen; 10–13 rows of stigmata; stomach surface has blister-like tubercles (plate 480D) . . . . *Ritterella rubra*

— Somewhat capitate lobes 2 cm–4 cm high; tunic clean and transparent; brown to bright orange zooids visible through sides of lobes; zooids 20 mm long including moderately long

post-abdomen; eight to 10 rows of stigmata; stomach with six to eight wide ridges (plate 480E) . . . . . *Ritterella pulchra*

List of Species

APLOUSOBRANCHIA (ALL COLONIAL)

CLAVELINIDAE

Clavelina huntsmani Van Name, 1931. The "light-bulb tunicate." Common to abundant on somewhat exposed rocky shores; very favorable for class study.

Pycnoclavella stanleyi Berrill and Abbott, 1949. Common on sandy rocks and among roots of surfgrass *Phyllospadix*; see Trason 1963, Univ. Calif. Publ. Zool. 65: 283–326 (life cycle).

HOLOZOIDAE

Distaplia occidentalis Bancroft, 1899. Common on intertidal rocks and on marina floats.

Distaplia smithi Abbott and Trason, 1968. On low intertidal rocky shore in high-flow areas. Named in honor of the late Professor Ralph I. Smith, the editor of this manual in 1954 and in 1975. See Abbott and Trason 1968, Bull. So. Calif. Acad. Sci. 67: 143–154.

POLYCITORIDAE

*Cystodytes lobatus (Ritter, 1900). Usually subtidal, occasionally washed ashore in debris. Preyed on by the snail *Lamellaria diegoensis* (see Lambert 1980, Veliger 22: 340–344). The amphipod *Polycheria osborni* and the clam *Mytilimeria nuttallii* are embedded in this and other colonial ascidians. There may be an additional *Cystodytes* sp., thin colony, orange.

Eudistoma diaphanes Ritter and Forsyth, 1917. This and other *Eudistoma* spp. were earlier placed in the genus *Archidistoma*. Rocky shores, common.

Eudistoma molle (Ritter, 1900). Common in beds of surfgrass *Phyllospadix*.

Eudistoma psammion Ritter and Forsyth, 1917. Common under low intertidal rocks in surf-swept outer coast; the fragile clam *Mytilimeria nuttallii* lives buried in the tunic of this and other ascidians.

Eudistoma purpuropunctatum Lambert, 1989. Common in Washington, rare subtidally in northern California. See Lambert 1989, Can. J. Zool. 67: 2700–2703.

Eudistoma ritteri Van Name, 1945. Common on rocks near surfgrass *Phyllospadix*, and on channel and cave walls; see Levine 1962, J. Morph. 111: 105–138 (morphology, development, reproduction) The amphipod *Polycheria osborni* is found in distinctive depressions in the tunic in this and other colonial tunicates, including *Aplidium* and *Ritterella*.

POLYCLINIDAE

Aplidium arenatum (Van Name, 1945). This and our other *Aplidium* spp. were earlier placed in the genus *Amaroucium*. Rocky shores.

Aplidium californicum (Ritter and Forsyth, 1917). Very common on intertidal rocks all along coast.

*Species not included in the key are found south of Point Conception, but their ranges may expand into our region.

Aplidium propinquum (Van Name, 1945). Common, low intertidal, rocks and pilings.

Aplidium solidum (Ritter and Forsyth, 1917). Fairly common on intertidal rocks and pilings; Abbott and Newberry (1980) note that the seaslug *Hermissenda* "often eats the upper parts of the zooids" of this species in Monterey Harbor.

*Aplidium sp. Small rounded colonies with clear tunic and red zooids with only eight to nine rows of stigmata. Possibly introduced, often common on marina floats.

Euherdmania claviformis (Ritter, 1903). Common in surfgrass *Phyllospadix* beds; see Trason 1957, J. Morph. 100: 509–545 (larval structure and development).

Polyclinum planum (Ritter and Forsyth, 1917). Common, rocky low intertidal; excellent for class study. See Holyoak 1997, Biol. Bull. 192: 87–97 (growth in Monterey Bay).

Synoicum parfustis (Ritter and Forsyth, 1917). Associated with rhizomes of surfgrass *Phyllospadix*.

*Synoicum pellucidum (Ritter and Forsyth, 1917). Uncommon; rocky shores at La Jolla and Horseshoe Cove, Bodega Head (the latter collected by Ralph Smith, and identified by Donald Abbott). The pedunculate colonies are small, transparent, and sand-free. Zooids have 11 rows of stigmata.

RITTERELLIDAE

Ritterella aequalisiphonis (Ritter and Forsyth, 1917) (=*Sigillinaria aequalisiphonis*). Rocky shores.

Ritterella pulchra (Ritter, 1901) (=*Sigillinaria pulchra*). Rocky shores; also on cave walls and other deeply shaded areas.

Ritterella rubra Abbott and Trason, 1968. Rocky shores of Monterey Bay. See Abbott and Trason, Bull. So. Calif. Acad. Sci. 67: 143–154; Abbott and Newberry, 1980.

DIDEMNIDAE

Didemnum albidum (Verrill, 1871). Circumpolar; common in northern Europe and New England, subtidal.

Didemnum carnulentum Ritter and Forsyth, 1917. Low intertidal; abundant subtidally; on a wide variety of substrates, such as rocks, pilings, mollusk shells, and algae.

Didemnum sp. Introduced, but provenance uncertain. Often abundant on marina floats, and may cover many square centimeters of subtidal hard surface; at scattered locations along much of West Coast from northern California to British Columbia, appearing in California in the 1990s. See http://woodshole. er.usgs.gov/project-pages/stellwagen/didemnum/index.htm.

Diplosoma listerianum (Milne-Edwards, 1841) (=*Diplosoma macdonaldi* Herdman, 1886). Abundant fouling species on floats and pilings.

Trididemnum opacum (Ritter, 1907) (=*Trididemnum dellavallei* Ritter and Forsyth, 1917). Common low intertidal to subtidal species, including in caves and crevices sheltered from sunlight (Abbott and Newberry 1980). The snail *Marsenina stearnsii* is camouflaged on this ascidian, which it eats (Ghiselin 1964, Veliger 6: 123–124). See Lambert 2003, Zoosystema 25: 665–679 (comparison of *Trididemnum* species of the northeast Pacific).

PHLEBOBRANCHIA (ALL SOLITARY EXCEPT *PEROPHORA*)

CIONIDAE

Ciona intestinalis (Linnaeus, 1767). An introduced North Atlantic fouling species, common in harbors on floats and pilings;

in marine and brackish water. See Abbott and Newberry 1980 (review of biology and reproduction); Hoshino and Nishikawa 1985, Publ. Seto Mar. Biol. Lab 30: 61–79 (taxonomy).

Ciona savignyi Herdman, 1882. Introduced from Japan, and now abundant on floats and pilings in harbors and marinas. See Hoshino and Nishikawa 1985, above; Lambert 2003, Zoosystema 25: 665–679 (occurrence on Pacific coast).

PEROPHORIDAE

Perophora annectens Ritter, 1893. Common on rocks and algae, forming pale green clusters in lower midtidal zone, sometimes common on marina floats. Also superb for class study to observe heart action and feeding.

Perophora japonica Oka, 1927. A Japanese species that appeared in Humboldt Bay in 2003 on marina floats. See Nishikawa et al. 2000, J. Mar. Biol. Assoc. U.K. 80: 949–950 (introduction to England); Lambert 2005, J. Mar. Biol. Assoc. UK 85: 1–2 (introduction to California).

CORELLIDAE

Chelyosoma productum Stimpson, 1864. Rocky shores, pilings, and floats.

Corella willmeriana Herdman, 1898. Subtidal California to Alaska; see Lambert, Lambert and Abbott 1981, Can. J. Zool. 59: 1493–1504 (taxonomy).

ASCIDIIDAE

Ascidia ceratodes (Huntsman, 1912) (=*Ascidia californica* Ritter and Forsyth, 1917). Common intertidally on pilings and protected floats, and subtidally.

Ascidia paratropa (Huntsman, 1912). Subtidal California to Alaska.

Ascidia zara Oka, 1935. Introduced from Japan, and largely on marina floats.

STOLIDOBRANCHIA

STYELIDAE (COLONIAL AND SOLITARY)

*Botryllus tuberatus Ritter and Forsyth, 1917. An uncommon southern California species in low intertidal and subtidal habitats. The zooids, with only four rows of stigmata, occur in small star-shaped systems that somewhat resemble *B. schlosseri*.

Botrylloides diegensis Ritter and Forsyth, 1917. A common fouling organism.

Botrylloides violaceus Oka, 1927. An abundant fouling organism on floats and pilings; introduced from Japan and also abundant in New England; see Lambert 2003, Zoosystema 25: 665–679 (occurrence on Pacific coast).

*Botrylloides perspicuum Herdman, 1886. A fouling species formerly known from Indonesia and now established in southern California. Easily distinguished by the long zooids with 18 or more rows of stigmata. Colonies are thick and reddish brown.

Botryllus schlosseri (Pallas, 1766). An introduced European species. See Boyd et al. 1986, Biol. Bull. 170: 91–109 (growth and

---

*Species not included in the key are found south of Point Conception, but their ranges may expand into our region.

sexual maturation); Boyd et al. 1990, Biol. Bull. 178: 239–250 (genetic match of Monterey Bay and Woods Hole populations); Chadwick-Furman and Weissman 2003, Biol. Bull. 205; 133–143 (allorecognition and life-history traits in Monterey Bay).

Cnemidocarpa finmarkiensis (Kiaer, 1893). An arctic-boreal species uncommon intertidally in California.

Dendrodoa abbotti Newberry, 1984. Rocky intertidal and subtidal from northern California to southern British Columbia. Aggregative habit resembles budding but individuals are easily separated. See Newberry 1984, Proc. Calif. Acad. Sci. 43: 239–248.

Metandrocarpa dura (Ritter, 1896) (=*Metandrocarpa dermatina* Huntsman, 1912). Usually on algae; generally subtidal but occasionally washed up on beaches.

Metandrocarpa taylori Huntsman, 1912 (=*Metandrocarpa michaelseni* Ritter and Forsyth, 1917). Moderately common on low intertidal rocks. See Abbott 1953, Univ. Calif. Publ. Zool. 61: 1–78 (asexual reproduction); 1955, J. Morph. 97: 569–594 (larvae); Newberry 1965, Mem. Acad. Roy. Belg. Sci. (2) 16: 1–57 (morphology, budding); Haven 1971, Biol. Bull. 140: 400–415 (reproduction).

*Polyandrocarpa zorritensis (Van Name, 1931). A recently introduced species of uncertain origin although described from South America; common at times in marina fouling communities of southern California harbors (plate 482A).

*Styela canopus Savigny, 1816. (=*Styela partita* (Stimpson, 1852). An introduced Asian species found in fouling communities in southern California (plate 482E).

Styela clava Herdman, 1881 (=*Styela barnharti* of Van Name, 1945, not of Ritter and Forsyth 1917). An introduced Japanese fouling species, often in robust and dense colonies in protected bays; see Abbott and Johnson 1972 (review of *Styela*); Abbott and Newberry 1980 (biology, ecology); Sims 1984, below (osmoregulation).

Styela gibbsii (Stimpson, 1864). Rare intertidally; commoner subtidally California to Alaska.

Styela montereyensis (Dall, 1872). Common, very large individuals exceeding 25 cm (and reaching a length of about 23 cm in three years; outer coast and open bays; also found on pilings and piers (as in Monterey Harbor) and marinas in more open marine situations. See Young and Braithwaite, 1980, Biol. Bull. 159: 428–440 (orientation and current-induced flow); Sims, 1984, J. Exp. Mar. Biol. Ecol. 82: 117–129 (osmoregulation).

Styela plicata (Lesueur, 1823) (=*Styela barnharti* Ritter and Forsyth, 1917). An introduced Asian fouling species, occurring in massive numbers in some southern California marinas. See Sims 1984, above (osmoregulation).

Styela truncata Ritter, 1901. Common intertidally in California; also common on floats and pilings.

*Symplegma reptans (Oka, 1927). Recently introduced from Japan; in fouling communities in southern California. Individual flattened zooids 2 mm–4 mm arranged randomly within a thin, translucent tunic; both siphons open independently on the tunic surface. Only one hermaphroditic gonad per side.

PYURIDAE (ALL SOLITARY)

Boltenia villosa (Stimpson, 1864). Rocky lower intertidal and subtidal.

Halocynthia igaboja Oka, 1906 (=*Halocynthia hilgendorfi igaboja*). Rare intertidally in California, commoner subtidally to Alaska.

*Microcosmus squamiger Hartmeyer & Michaelsen, 1928. An introduced species of uncertain warm-water origins; in marina and harbor fouling communities in southern California. See Lambert and Lambert 1998, 2003 (plate 483B).

Pyura haustor (Stimpson, 1864). Low intertidal to subtidal depths on rocks, pilings, floats, in mussel beds, and kelp holdfasts.

Pyura mirabilis (von Drasche, 1884). Rare; rocky shores or in sand attached to shells and gravel.

MOLGULIDAE (ALL SOLITARY)

*Molgula ficus (MacDonald, 1859). Introduced from western Pacific, possibly Australia. Common in southern California harbors, first recorded in San Francisco Bay in 2005. See Lambert 2007, Cah. Biol. Mar. 48: 95–102. Tough, opaque tunic, up to 4 cm in diameter, 7 rows of stigmata per side, delicate fingerlike tunic projections around both apertures. Does not brood larvae.

Molgula manhattensis (DeKay, 1843). Introduced from Atlantic; often in huge numbers in brackish waters and protected bays. The fascinating funguslike heterotrophic protist *Nephromyces* is an endosymbiont in the renal sac of this and other molgulids. See Saffo 1982, Biol. Bull. 162: 95–104; Saffo and Davis 1982, Biol. Bull. 162: 105–112; Saffo 1990, Mar. Biol. 107: 291–296.

Molgula pugetiensis Herdman, 1898. Little-known species, usually subtidal in sand or gravel but has been recorded from marina floats in Humboldt Bay. Northern California to southern British Columbia.

Molgula verrucifera Ritter and Forsyth, 1917. Rocky shores and subtidal; Abbott and Newberry (1980) note that it is often attached to *Euherdmania claviformis*.

References

Abbott, D. P., and A. T. Newberry. 1980. Urochordata: the tunicates, pp. 177–226. In Intertidal invertebrates of California. Morris, R. H., D. P. Abbott, and E. C. Haderlie, eds. Stanford Univ. Press, 690 pp.

Abbott, D. P., and J. V. Johnson. 1972. The ascidians *Styela barnharti, S. plicata, S. clava,* and *S. montereyensis* in Californian waters. Bull. So. Calif. Acad. Sci. 71: 95–105.

Berrill, N. J. 1950. The Tunicata. With an Account of the British Species. London: Ray Society, 354 pp.

Fay, R. C., and J. V. Johnson. 1971. Observations on the distribution and ecology of the littoral ascidians of the mainland coast of southern California. Bull. So. Calif. Acad. Sci. 70: 114–124.

Fay, R. C., and J. A. Vallee. 1979. A survey of the littoral and sublittoral ascidians of southern California, including the Channel Islands. Bull. So. Calif. Acad. Sci. 78: 122–135.

Huntsman, A. G. 1912. Holosomatous ascidians from the coast of western Canada. Contrib. Canad. Biol. 1906–1910: 103–185.

Lambert, C. C., G. Lambert, and E. N. Kozloff. 1987. Chapter 23: Phylum Urochordata, Class Ascidiacea, pp. 467–476. In Marine invertebrates of the Pacific Northwest. E. N. Kozloff, ed. Seattle: Univ. of Wash. Press, 511 pp.

Lambert, C. C., and G. Lambert. 1998. Non-indigenous ascidians in southern California harbors and marinas. Mar. Biol. 130: 675–688.

Lambert, C. C., and G. Lambert. 2003. Persistence and differential distribution of nonindigenous ascidians in harbors of the Southern California Bight. Mar. Ecol. Prog. Ser. 259: 145–161.

Lambert, G. 2005. Ecology and natural history of the protochordates. Can. J. Zool. 83: 43–50.

MacGinitie, G. E. 1939. The method of feeding in tunicates. Biol. Bull. 77: 443–447.

Millar, R. H. 1971. The biology of ascidians. Adv. Mar. Biol. 9: 1–100.

Monniot, C., and F. Monniot. 1991. Coral Reef Ascidians of New Caledonia. Paris: Orstom, 247 pp. *Contains much interesting general information about ascidian taxonomy and biology.*

*Species not included in the key are found south of Point Conception, but their ranges may expand into our region.

Reverberi, G. 1971. Ascidians. In G. Reverberi, ed., Experimental Embryology of Marine and Fresh-Water Invertebrates, pp. 507–550. New York: American Elsevier, 587 pp.

Ritter, W. E. 1913. The simple ascidians from the northeastern Pacific in the collection of the United States National Museum. Proc. U.S. Nat. Mus. 45: 427–505.

Ritter, W. E., and R. A. Forsyth. 1917. Ascidians from the littoral zone of southern California. Univ. Calif. Publ. Zool. 16: 439–512.

Satoh, N. 1994. Developmental Biology of Ascidians. Cambridge Univ. Press, 234 pp.

Sawada, H., H. Yokosawa, and C. C. Lambert. 2001. The Biology of Ascidians. Springer-Verlag, Tokyo, 470 pp.

Tokioka, T. 1963. The outline of Japanese ascidian fauna as compared with that of the Pacific coasts of North America. Publ. Seto. Mar. Biol. Lab. 11: 131–156.

Van Name, W. G. 1945. The North and South American ascidians. Bull. Amer. Mus. Nat. Hist. 84: 1–476.

Appendicularia (Larvacea) and Thaliacea

LAURENCE P. MADIN

Although ascidian tunicates are commonly found on rocks, pilings, piers and other hard substrates largely in the subtidal zone, their planktonic relatives, the pelagic salps, doliolids, pyrosomes, and appendicularians, are rarely seen in the intertidal. These pelagic tunicates are usually offshore species, distributed widely in the slope water and open ocean. They may occasionally come into shallow water or even be washed up on beaches when offshore populations are pushed in by currents or wind.

Pelagic tunicates comprise two classes of the Tunicata—the small, tailed Appendicularia (Larvacea), and the larger gelatinous Thaliacea, including salps, doliolids and pyrosomes. All are free-swimming filter-feeders that use filters to trap phytoplankton and other small cells from a feeding current generated by ciliary or muscular force. Appendicularians reproduce sexually, and almost all species are simultaneous hermaphrodites. They secrete elaborate mucous-filtering structures called houses, which trap small particulate matter and contribute importantly to the cycling of organic matter in the ocean.

Thaliaceans reproduce in both sexual and asexual modes, the latter producing colonylike assemblages of individual zooids. Pyrosomes occur mainly as cylindrical colonies of hundreds or thousands of individuals, attaining an overall size of 1 m or more. Doliolids produce specialized zooids for feeding, dispersal, reproduction, and locomotion, while salps alternate between sexual and asexual stages. Chains of the asexually produced aggregate salps can be several meters long, containing hundreds of individuals.

About a dozen species of salps occur fairly commonly in offshore waters of the Pacific coast, along with two species of doliolids and probably one or two pyrosomes (Lavaniegos and Ohman 2003).

The common doliolids are *Dolioletta gegenbauri* and *Doliolum denticulatum*; pyrosomes are often difficult to identify, but any found nearshore are likely to belong to the genus *Pyrosoma*.

Commonly occurring salps include *Cyclosalpa affinis*, *C. bakeri*, *Iasis zonaria*, *Pegea confoederata*, *P. socia*, *Ritteriella picteti*, *Salpa aspera*, *S. fusiformis* and *S. maxima*, *Thalia democratica*, and *Thetys vagina*. The last species occasionally washes ashore on Oregon beaches, for example. A recent illustrated key to these taxa is Godeaux (1998).

Surveys along the California coast found 27 species of appendicularians (Fenaux and Dallot 1980), of which three species, *Oikopleura longicauda*, *Fritillaria venusta*, and *F. tenella*, constituted 72% of the total numbers. A concise key to appendicularians is Fenaux (1998). Because of their small size and fragility, appendicularians are unlikely to be recognizable in the intertidal zone, unless large swarm populations should be washed ashore.

References

Fenaux, R. 1998. The classification of Appendicularia, pp. 295–306. In The biology of pelagic tunicates. A. Bone, ed. Oxford: Oxford Univ. Press.

Fenaux, R., and S. Dallot. 1980. Répartition des Appendiculaires au large des côtes de Californie. J. Plankton Res. 2: 145–167.

Godeaux, J. 1998. The relationships and systematics of the Thaliacea, with keys for identification, pp. 273–294. In The biology of pelagic tunicates. A. Bone, ed. Oxford: Oxford Univ. Press, pp. 273–294.

Lavaniegos, B. E., and M. D. Ohman. 2003. Long-term changes in pelagic tunicates of the California Current. Deep-Sea Res. II, 50: 2473–2498.

INDEX

NOTE: Generic names in parentheses are for cross-referencing purposes and are not necessarily subgenera.

montereyensis, 440, 443*pl*
sp., 440
Austraeolis stearnsi, 795, 805
Austrodoris. See Doris
Austrognathia sp., 276, 278*pl*
Autolytinae, 345, 346*pl*, 349–50
Autolytus, 344
varius, 349, 350
Axinella sp., 800, 801
Axinopsida erricata, 828*pl*, 829
Axiognathus. See Amphipholis squamata
Axiothella rubrocincta, 26, 69, 332, 391, 392*pl*, 393
Axocielita originalis. See Thalysias originalis
Axonolaiminae, 261
Axonolaimus, 258*pl*

Babaina festiva, 805
Babakina festiva, 795, 805
Baginnidae, 68
Baikaloperla, 670
Bairdiidae, 435, 443*pl*
Bairdiocopina, 423, 435
Balanidae, 483
Balanoglossus, 910, 910*pl*, 911
Balanoidea, 476, 482–83
Balanomorpha, 476–77, 481, 482
Balanophyllia elegans, 14, 15, 177, 183
Balanus, 476, 477*pl*, 512
amphitrite, 483 (*See also Amphibalanus amphritite*)
aquila, 483 (*See also Menesiniella aquila*)
cariosus, 483 (*See also Semibalanus cariosus*)
crenatus, 480*pl*, 481, 483
eburneus, 483
flos, 483
galeatus, 482
glandula, 12, 478, 480*pl*, 481–83, 684, 928
hesperius laevidomus, 483 (*See also Hesperibalanus hesperius*)
improvisus, 483 (*See also Amphibalanus improvisus*)
nefrens, 482 (*See also Armatobalanus nefrens*)
nubilus, 478*pl*, 479*pl*, 481, 483
pacificus, 483
tintinnabulum californicus, 483 (*See also Megabalanus californicus*)
trigonus, 480*pl*, 481, 483
spp., 15, 476, 799, 929
Balcis
rutila, 739
thersites, 739
Bamboo worms. *See* Maldanidae
Bankia setacea, 78, 853, 855*pl*
Baptodoris mimetica, 794, 800
Barbonematinae, 248
Barentsia, 284*pl*, 286
benedeni, 30, 285*pl*, 286, 287
conferta, 285*pl*, 286, 287
discreta, 285*pl*, 286, 287
hildegardae, 285*pl*, 286, 287
parva, 285*pl*, 286, 287
ramosa, 285*pl*, 286, 287
sp., 888
Barentsiidae, 286, 287
Bariaka pamelae, 473
Barleeia, 544
acuta, 729, 730*pl*
dalli, 729
haliotiphila, 729, 730*pl*
marmorea, 729
oldroydi, 729, 730*pl*
Barleeidae, 718, 729
Barnacles. *See* Cirripedia

Barnea
pacifica, 853
subtruncata, 852*pl*, 853
Bartschella laminata, 749*pl*, 750
Baseodiscus, 222
punnetti, 224*pl*, 228, 230
Basiliochiton heathii, 711
Basket stars. *See* Ophiuroidea
Basommatophora, 715, 751
Bassia bassensis, 154*pl*, 165
Batea lobata, 546*pl*, 602*pl*, 603, 604
Bateidae, 602
Bathycopea, 512, 517
Bathocyroidae, 198
Bathycyroe fosteri, 193*pl*, 194, 198
Bathydrilus litoreus, 302
Bathyechiniscus tetronyx, 296
See also Styraconyx sargassi
Batillaria
attramentaria, 25, 29, 717, 727, 728*pl*
cumingi, 727
zonalis, 727
Batillariidae, 718, 727, 728*pl*
Batillipedidae, 293, 296, 297
Batillipes
gilmartini, 294*pl*, 296, 297
mirus, 294*pl*, 296, 297
orientalis, 294*pl*, 296, 297
tridentatus, 294*pl*, 296, 297
Bdelloidea, 282
Beach hoppers. *See* Talitridae
Beetles. *See* Coleoptera
Belbolla, 244
Bemlos concavus, 553*pl*, 561*pl*, 562
Bernardina bakeri, 835, 836*pl*
Beroe, 191
adyssicola, 194, 196*pl*, 199
cucumis, 194, 196*pl*, 199
forskalii, 194, 196*pl*, 199
gracilis, 194, 196*pl*, 199
mitrata, 194, 196*pl*, 199
Beroida, 191, 199
Berthella
californica, 786*pl*, 788
strongi, 788
Berthellina
citrina, 788
ilisima, 788
Betaeus, 635, 636
gracilis, 637, 638*pl*, 650
harfordi, 637, 638*pl*, 650
harrimani, 637, 638*pl*, 650
longidactylus, 637, 638*pl*, 650
macginitieae, 637, 638*pl*, 650
setosus, 637, 638*pl*, 650
Bicrisia, 869, 873
edwardsiana, 879
robertsonae, 879, 880*pl*
Biemna rhadia, 800
Bimeria franciscana. See Garveia franciscana
Bispira sp., 400, 401*pl*, 402, 472
Bittium
attenuatum, 727
eschrichtii, 727
eschrichtii montereyense, 727
interfossa, 727
latifilosum, 727
purpureum, 727
See also under Stylidium
Bivalvia, 34, 694, 807–58
preservation of, 39, 41, 809
symbiotic ciliates of, 77
Blackfordia virginica, 130*pl*, 131, 139*pl*, 148, 160
Blackfordiidae, 160
Bledius, 689

fenyesi, 690
monstratus, 690
Blepharipoda occidentalis, 275, 648, 654, 809, 832
Boccardia
berkeleyorum, 371, 373*pl*, 375
columbiana, 371, 373*pl*, 375
proboscidea, 371, 375
tricuspa, 371, 373*pl*, 375
uncata, 375
Boccardiella
hamata, 371, 373*pl*, 375
ligerica, 371, 373*pl*, 375
truncata, 371, 375
Bolbolaimus sp., 240*pl*
Bolinopsidae, 198
Bolinopsis
cydippid, 193*pl*
infundibulum, 193*pl*, 194, 198
Bolivina
acuminata, 50*pl*, 58, 59*pl*, 65
acutula, 49*pl*, 55, 58, 59*pl*, 65
advena, 65
compacta, 58, 59*pl*, 65
pseudoplicata, 54*pl*, 58, 59*pl*, 65
striatula, 65
subadvena, 65
subexcavata, 58, 59*pl*, 65
vaughani, 58, 59*pl*, 65
Boltenia villosa, 79, 472, 953, 959*pl*, 963
Bolivinidae, 65
Bombyliidae, 682, 684
Bomolochidae, 471
Bomolochus
constrictus, 471
cuneatus, 471
longicaudatus, 471
paucus, 471
pectinatus, 471
Bonellia viridis, 291
Boonea
bisuturalis, 750
lucca, 749*pl*, 750
oregonensis, 749*pl*, 750
Bopyrella macginitei, 535
Bopyridae, 508, 512, 535–36
Bopyriscus calmani, 512, 513*pl*, 535
Botrylloides
diegensis, 957*pl*, 960, 962
perspicuum, 962
violaceus, 957*pl*, 960, 962
Botryllophilus abbotti, 471
Botryllophilidae, 467, 471
Botryllus
schlosseri, 802, 957*pl*, 960, 962–63
tuberatus, 962
Bougainvillia
glorietta, 802, 803
muscus, 125*pl*, 128, 143, 145*pl*, 151
ramosa, 151
spp., 143, 145*pl*, 151, 155
Boveria, 69, 71
subcylindrica, 78
teredinidi, 71*pl*, 78
sp., 77
Bowerbankia, 286, 869, 873
gracilis, 799, 878*pl*, 879
Bowmaniella banneri, 492
Brachionus, 281*pl*, 282
Brachiopoda, 864–65
preservation of, 40
symbionts of, 78
Brachycera, 681
Brachystomia angularis, 749*pl*, 750
Brachyura, 76, 632, 633*pl*, 636, 637*pl*, 640*pl*, 642*pl*, 651–54

Cnidaria, 21, 33, 35, 118
    symbionts of, 73*pl*, 74–75, 75*pl*, 467
    See also Anthozoa; Hydrozoa;
        Hydromedusa; Siphonophora
Cnidopus ritteri, see Epiactis ritteri
Coboldus hedgpethi, 555*pl*, 586, 587*pl*
Cochliophilus
    depressus, 77
    minor, 71*pl*, 77
Cockles. *See* Cardiidae; Veneridae
Codium fragile, 797
Codonellina, 877
Coelenterata. *See* Cnidaria
Coelogynopora
    brachystyla, 212
    paracnida, 212
    tenuiformis, 212
Coelopa (Neocoelopa) vanduzeei, 685
Coelopidae, 683, 685
Coeloplanidae, 198
Colanthura bruscai, 506, 507*pl*, 531
Coleopa (Neocoleopa) vanduzeei, 685
Coleopora gigantea, 893
Coleoptera, 412*t*, 671, 686–91
Coleosphaeridae, 114
Colidotea, 525
    findleyi, 525, 526*pl*, 538
    rostrata, 525, 526*pl*, 538
Collastoma pacifica, 212
Collembola, 23, 412*t*, 671, 672–80
Colletosia
    innominata, 890
    radiata, 888, 890
Colligocineta, 72
    furax, 72*pl*, 75
Collisella
    asmi, 760
    austrodigitalis, 760
    digitalis, 482, 759–60
    instabilis, 4, 760
    limatula, 759
    ochracea, 760
    paradigitalis, 760
    pelta, 759
    scabra, 760
    triangularis, 760
    See also under Lottia
Colobomatus embiotocae, 472
Coloboneura. See Chersodromia
Columbaora cyclocoxa, 553*pl*, 561–62, 561*pl*
Columbellidae, 718, 743
Columbiaemysis ignota, 492, 494*pl*
Comb jellies. *See* Ctenophora
Comesoma, 258*pl*
Comesomatinae, 261
Commensals. *See* Parasites and commensals
Commensipleustes commensalis, 598, 599*pl*,
    601
Compsomyax subdiaphana, 836, 838*pl*
Concentricycloidea, 913
Conchoderma auritum, 629
Conchopthirus caryoclada, 71*pl*, 78
Conchopus borealis, 684
Conchostraca, 414
Conica, 160
Conidae, 718, 748
Coninckiinae, 261
Conocyema decca, 201
Conocyemidae, 200
Conopea galeata, 479*pl*, 481–82
Conopeum, 874
    reticulum, 882*pl*, 888
    osburni, 888
    tenuissimum, 30, 888
Conophoralia, 276, 278*pl*
Conualevia alba, 790, 800

Conus californicus, 747*pl*, 748
Cooperella subdiaphana, 839, 840*pl*
Copepoda, 412*t*, 446–75
    commensal and parasitic, 464–74
    free-living, 446–47
Copidognathus
    curtus, 667*pl*, 668*pl*
    pseudosetosus, 666, 668*pl*
    sp., 666
Copidoplana tripyla, 210, 213
Copidozoum, 874
    adamantum, 883*pl*, 888
    planum, 888
    protectum, 883*pl*, 888
    tenuirostre, 888
Corallanidae, 513, 514, 536
Corallimorpharia, 177, 183
Corallina, 562
    chilensis, 53
    gracilis, 442
Corambe
    pacifica, 789*pl*, 794, 798
    steinbergae, 794, 798
    thompsoni, 798
Corambella bolini, 798
Corambidae, 798
Corbicula fluminea, 834*pl*, 835
Corbiculidae, 835
Corbula
    amurensis, 29, 809, 850, 851*pl*
    luteola, 850, 851*pl*
Corbulidae, 812, 813*pl*, 815, 816*pl*, 850
Cordylophora
    caspia, 125*pl*, 128, 155
    lacustris, 155
Corella willmeriana, 742, 950, 962
Corellidae, 957*pl*, 962
Corisella
    decolor, 680
    inscripta, 680
Corixidae, 670, 671, 680
Corniger hilgendorfi, 664
    See also Ammothea hilgendorfi
Cornuspira involvens, 64
    See also Cyclogyra involvens
Corolla spectabilis, 774*pl*, 775
Coronaster marchennus, 929
Corophiidae, 551*t*, 553*pl*, 554, 570–73
Corophium, 30
    alienense, 553*pl*, 571, 571*pl*, 573
    dentalium, 573
    heteroceratum, 571, 571*pl*, 573
    See also Americorophium; Crassicorophium;
        Laticorophium; Monocorophium
Corycaeus anglicus, 447, 451*pl*
Corymorpha
    palma, 126*pl*, 128, 157
    sp., 126*pl*, 157, 804
Corymorphidae, 157
Corynactis californica, 14, 177, 178*pl*, 183,
    805, 929
Coryne
    cliffordi, 157
    eximia, 127*pl*, 128, 141, 142*pl*, 157
    japonica, 124, 127*pl*, 141, 142*pl*, 157
    spp., 127*pl*, 128, 141, 142*pl*, 157, 158, 802
Corynidae, 157–58
Coryphella
    cooperi, 803
    fisheri, 803
    fusca, 803
    iodinea, 803
    pricei, 803
    trilineata, 803
    verrucosa, 803
    See also under Flabellina

Cossura
    bansei, 384, 384*pl*
    candida, 384, 384*pl*
    pygodactylata, 383, 384, 384*pl*
    rostrata, 383, 384, 384*pl*
Cossuridae, 323, 383–84, 384*pl*
Costazia
    costazii, 893
    robertsoniae, 893
Cothurnia
    fecunda, 74*pl*
    limnoriae, 74*pl*, 76
    sp., 45,75
Cotylea, 213–14
Coullana canadensis, 453*pl*, 458
Coxicerberus abbotti, 22, 521, 524*pl*, 538
Crabs. *See* Brachyura
Crangon, 16, 634, 635, 636
    alaskensis, 636, 640*pl*, 650
    franciscorum, 535, 634*pl*, 636, 637*pl*, 640*pl*,
        650
    handi, 636, 640*pl*, 650
    nigricauda, 636, 640*pl*, 650
    nigromaculata, 636, 640*pl*, 650
Crangonidae, 650
Crangonyctidae, 551*t*, 552, 556, 557*pl*,
    610–11
Crangonyx
    floridanus, 610, 611*pl*
    pseudogracilis, 557*pl*, 610–11, 611*pl*
Craniella arb, 110, 112*pl*, 113
Craspedacusta
    sowerbii, 124, 136*pl*, 144*pl*, 148, 164
    sowerbyi, 164
Crassicorophium
    baconi, 553*pl*, 571*pl*, 572*pl*, 573
    bonellii, 573
Crassadoma, 807
    gigantea, 12, 14, 825, 825*pl*, 826
Crassostrea
    ariakensis, 822, 824*pl*
    gigas, 29, 467, 472, 809, 821, 823, 824*pl*
    rivularis, 822
    sikamea, 809, 821, 823, 824*pl*
    virginica, 29, 432, 809, 821, 824*pl*, 825
Cratena
    abronia, 804
    albocrusta, 804
    flavovulta, 804
    fulgens, 804
    rutila, 804
    spadix, 804
    virens, 804
    See also under Cuthona
Crayfish. *See* Astacura
Crebricoma carinata, 77
Crenella decussata, 817, 820, 823*pl*
Crenopharynx, 249*pl*
Crepidula, 26, 716
    adunca, 733
    convexa, 29, 733, 734*pl*
    fornicata, 29, 733, 734*pl*
    glauca, 733
    nivea, 733
    nummaria, 733, 734*pl*
    perforans, 733, 734*pl*
    plana, 29, 717, 733, 734*pl*
Crepipatella
    dorsata, 733
    lingulata, 733, 734*pl*
Creseis virgula, 773*pl*, 775
Cribrilina hippocrepis, 890
Cribrillaria, 875
Cribrimorpha, 867
Cribriomorphina, 867, 871, 873, 875, 890,
    891*pl*

Mycale, 800, 801
    (Aegograpila) adhaerens, 102, 105pl, 115
    (Carmia) richardsoni, 102, 105pl, 115
    (Mycale) hispida, 102, 105pl, 115
    (Mycale) macginitiei, 88, 102, 105pl, 115
    (Paresperella) psila, 88, 102, 103pl, 115, 801
Mycalidae, 115
Mycodinium, 73
    pilisuctor, 73pl, 76
Myicola ostreae, 472
Myicolidae, 472,
Myidae, 812, 812pl, 848
Myodocopa, 422pl, 428, 431
    tabular keys to, 424–26t
Myodocopida, 428, 431, 433–36pl
    tabular keys to, 424–26t
Myoida, 809
Myoisophagos sanguineus, 231
Myosoma spinosa, 285pl, 287
Myosotella myosotis, 11, 31, 77, 717, 751, 752pl
Myrianida
    convoluta, 346pl, 349, 350
    inermis, 346pl, 349, 350
Myriapoda, 412t
Myriochele spp., 323, 391
Myriowenia californiensis, 391, 392pl
Mysella pedroana, 654, 809, 829, 832, 832pl
Mysida, 489
Mysidacea, 489–95
Mysis relicta, 489
Mystacocarida, 23, 412t, 413
Mytilicola
    orientalis, 30, 467, 468pl, 472
    ostrea, 30
Mytilidae, 811, 812pl, 815–20
Mytilimeria nuttallii, 26, 809, 856pl, 857, 858, 961
Mytiloida, 809
Mytilopsis sallei, 31
Mytilus, 35, 161, 742, 760
    californianus, 12, 17, 26, 32, 84, 472, 482, 652, 664, 759, 815, 818, 819pl
    edulis, 32, 815, 819, 819pl
    galloprovincialis, 32, 809, 815, 819, 819pl
    trossulus, 32, 77, 205, 815, 819, 819pl
    spp., 10, 467, 472, 929
Myxicola
    aesthetica, 400, 402
    infundibulum, 400, 401pl, 402, 472
Myxilla (Myxilla)
    agennes, 107, 108pl, 114
    incrustans, 107, 109pl, 114, 800–01
    parasitica, 114

Nacospatangus
    depressus, 922
    laevis, 915, 921pl, 922
Naidinae, 300
Naineris dendritica, 367, 368pl
Nais communis, 302
Najna
    consiliorum, 573, 574
    kitamati, 573, 574
Najnidae, 551t, 552, 553pl, 554, 573–74, 574pl
Namakosiramia californiensis, 471
Namanereis pontica, 11, 352
Nanaloricidae, 272
Nanaloricus mysticus, 270pl, 272
Nannastacidae, 500
Nannomesochra arupinensis, 458
Nannopus palustris, 460
Nannosquilla anomala, 631
Nannotheres moorei, 413
Nanomia bijuga, 150, 152pl, 153pl, 164

Naobranchia occidentalis, 473
Narcomedusae, 118, 120, 139pl, 149pl, 166
Nassariidae, 718, 743–45
Nassarius, 743
    cooperi, 745
    fossatus, 745
    fraterculus, 745
    mendicus, 745
    obsoletus, 745 (See also Ilyanassa obsoleta)
    perpinguis, 745
    rhinites, 745
    See also under Caesia, Hima
Naticidae, 718, 736–38
Nautilus [Mollusca], 697
Nautilus [Protista]
    auriculus, 68 (See also Cancris auricula)
    becarii, 67 (See also Ammonia becarii)
    crispum, 66 (See also Elphidium crispum)
    inflata, 64 (See also Trochammina inflata)
    lobatulus, 68 (See also Cibicides lobatulus)
    repandus, 67 (See also Poroeponides cibrorepandus)
Navanax
    inermis, 471, 781, 784, 784pl, 787
    polyalphos, 784pl, 785, 787
Neaeromya
    compressa, 833, 833pl
    rugifera, 655, 809, 829, 833, 833pl
Neanthes
    brandti, 352, 354pl
    diversicolor, 352
    lighti, 352
    limnicola, 5, 352, 354pl
    succinea, 352, 354pl
    virens, 352
Nebalia, 280, 485
    daytoni, 486, 487pl, 488
    gerkenae, 486, 487pl, 488
    hessleri, 486, 487pl, 488
    kensleyi, 488
    pugettensis, 76, 486, 488
    sp., 488
Nebaliella, 486
Nebaliopsis, 486, 487pl, 488, 489
Nectobranchia indivisa, 473
Nectonema, 274
    agile, 274, 274pl
Nematocera, 681
Nematoda, 21, 234–64
    preservation of, 38
Nematomorpha, 274
    preservation of, 38
Nematostella vectensis, 174, 179pl, 180, 183
Nemertea, 21, 24, 221–33
    Heteronemertea, 221, 222, 230–31
    Hoplonemertea, 221, 222, 231–33
    preservation of, 41, 221
Nemertes, see Nemertopsis
Nemertopsis
    bullocki, 232
    gracilis, 207pl, 228, 232
Nemocardium centifilosum, 834, 834pl
Neoamphitrite robusta, 397, 399
Neochthebius, 688
    vandykei, 690
Neocoleopa. See Coleopa
Neoconorbina
    opercularis, 53, 54pl, 61, 66
    terquemi, 54pl, 61, 66
Neodakaria, 876, 893
    islandica, 894pl, 900
    ordinata, 900
    umbonata, 894pl, 900
Neodexiospira pseudocorrugata, 405, 406
Neoesperiopsis rigida, 801
Neogastropoda, 715, 739

Neoleprea californica, 398, 398pl, 399
Neolepton
    salmoneum, 835, 836pl
    subtrigonum, 835, 836pl
Neoleptonidae, 812, 813pl, 835
Neomachilis, 672
    halophila, 11, 672
    sp., 672
Neomenioids, 695
Neomolgus littoralis, 666
Neomunna. See Munna
Neomysis
    awatschensis, 492
    franciscorum, 495
    intermedia, 492
    japonica, 492
    kadiakensis, 492, 493pl
    macropsis, 492
    mercedis, 467, 469pl, 474, 489, 492, 493pl, 495
    rayii, 492, 493pl, 495
    sculpta, 492
    tonion, 495
    spp., 491pl, 493pl
Neonesidea sp., 435, 443pl
Neopachylopus, 687, 689
    sulcifrons, 690
Neopetrosia vanilla, 87, 94, 96pl, 115
Neosabellaria cementarium, 393
Neotachidius triangularis, 463
Neotonchinae, 251
Neotonchus, 254pl
Neotrypaea, 5, 633, 635, 648, 650
    biffari, 648, 655
    californiensis, 5f, 220, 332, 471, 634pl, 648, 649pl, 652–3, 655, 861
    gigas, 471, 648, 649pl, 652–3, 655
Neoturris
    breviconis, 146, 147pl, 155
    fontata, 146, 147pl, 155
    pelagica, 146, 147pl, 155
    pileata, 147pl, 155
    spp., 146, 147pl, 155
Neoutelga inermis, 212
Nephromyces, 963
Nephtyidae, 317, 358, 359pl
Nephtys
    caeca, 319pl, 358
    caecoides, 358, 359pl
    californiensis, 358, 359pl
    cornuta, 358, 359pl
    cornuta franciscana, 358
    ferruginea, 358
    parva, 358
Nereididae, 312pl, 320, 351–53
Nereiphylla castanea, 338pl, 339, 340
Nereis, 309, 310pl, 311
    eakini, 351, 352, 353pl
    grubei, 351, 352, 354pl
    latescens, 351, 352, 354pl
    mediator, 352
    natans, 352
    pelagica, 352
    pelagica neonigripes, 351, 354pl
    procera, 351, 352, 353pl
    vexillosa, 13, 351, 352–53, 353pl
    sp., 309–11, 310pl
Nereocystis, 492, 517
Nerilla sp., 323pl
Nerillidae, 319, 323pl, 326
Nerinides
    acuta, 376
    tridentata, 376
    See also under Scolelepis

Indexer: Live Oaks Indexing

Composition: Aptara, Inc.

Text: 8.25/12 Stone Serif

Display: Akzidenz Grotesk, Stone Serif

Printer and Binder: Transcontinental Gagné